The
ENCYCLOPEDIA
of
SEDIMENTOLOGY

ENCYCLOPEDIA OF EARTH SCIENCES SERIES

Volume
- I THE ENCYCLOPEDIA OF OCEANOGRAPHY/*Rhodes W. Fairbridge*
- II THE ENCYCLOPEDIA OF ATMOSPHERIC SCIENCES AND ASTROGEOLOGY/*Rhodes W. Fairbridge*
- III THE ENCYCLOPEDIA OF GEOMORPHOLOGY/*Rhodes W. Fairbridge*
- IVA THE ENCYCLOPEDIA OF GEOCHEMISTRY AND ENVIRONMENTAL SCIENCES/*Rhodes W. Fairbridge*
- VI THE ENCYCLOPEDIA OF SEDIMENTOLOGY/*Rhodes W. Fairbridge and Joanne Bourgeois*
- VIII THE ENCYCLOPEDIA OF WORLD REGIONAL GEOLOGY, PART 1: Western Hemisphere (Including Antarctica and Australia)/*Rhodes W. Fairbridge*

ENCYCLOPEDIA OF EARTH SCIENCES, VOLUME VI

The
ENCYCLOPEDIA
of
SEDIMENTOLOGY

EDITED BY

Rhodes W. Fairbridge
Columbia University

Joanne Bourgeois
University of Wisconsin

Dowden, Hutchinson & Ross, Inc.
Stroudsburg Pennsylvania

Copyright © 1978 by **Dowden, Hutchinson & Ross, Inc.**
Library of Congress Catalog Card Number: 78-18259
ISBN: 0-87933-152-6

All rights reserved. No part of this book may be reproduced or transmitted in any form or by any means—graphic, electronic, or mechanical, including photocopying, recording, taping, or information storage and retrieval systems—without written permission of the publisher.

82 81 80 79 78 5 4 3 2 1
Manufactured in the United States of America

Library of Congress Cataloging in Publication Data

Main entry under title:
The Encyclopedia of sedimentology.

(Encyclopedia of earth sciences series ; v. 6)
Bibliography: p.
Includes index.
 1. Rocks, Sedimentary—Dictionaries. 2. Sediments (Geology)—Dictionaries. I. Fairbridge, Rhodes Whitmore 1914- II. Bourgeois, Joanne. III. Series.
QE471.E49   551.3'03   78-18259
ISBN 0-87933-152-6

Distributed world wide by Academic Press, a subsidiary of Harcourt Brace Jovanovich, Publishers.

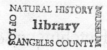

PREFACE

The *Encyclopedia of Sedimentology* is a comprehensive, alphabetical treatment of the discipline of sedimentology. It is intended to be a reference book for sedimentologists, geologists, and others who come in contact with sediments. In the broadest sense, this group includes most of the world's population because over 75 percent of the earth's surface is covered with sediments and sedimentary rocks. The book should be particularly useful, however, to petroleum and coal geologists, soil scientists, hydrologists, archaeologists, and other professionals in related fields.

Sedimentology is a broad and growing discipline (see *Sedimentology; Sedimentology—Yesterday, Today and Tomorrow*) that has seen the publication of several new texts and innumerable more specialized volumes in the last decade. We could not hope in this encyclopedia to cover the field in depth; but in breadth—from *Microbiology in Sedimentology* to *Radioactivity in Sediments* to *Martian Sedimentation*—we intend to provide readers with basic information and to direct them to more specialized literature.

Some attempt has been made to define terms and to adhere to definitions in this volume, but an encyclopedia is *not* a dictionary. It is a compendium of knowledge, and in some instances there are differences of opinion and in choice of terminology among the people who practice sedimentology. Nor is an encyclopedia a symposium volume; although we have tried to present an up-to-date book, controversial issues or new and untested ideas are not emphasized in an encyclopedia. The reader should be aware, however, that each entry may reflect the individual outlook of an author, from a broad, international spectrum of contributors. Partly for this reason, there are numerous entries that overlap in their coverage. The extensive cross-referencing should allow the reader to note differences in opinion or in approach. We hope that this diverse coverage will enhance the usefulness of the volume.

Literature

It is customary in the encyclopedia series to provide a list of basic references that would complement the volume. Below is a list of only some of the most recent fundamental sedimentology texts and older reference books. Books marked with an asterisk (*) contain comprehensive reference lists.

*Bathurst, R. G. C., 1975. *Carbonate Sediments and Their Diagenesis*. Amsterdam: Elsevier, 658p.

Berner, R. A., 1971. *Principles of Chemical Sedimentology*. New York: McGraw-Hill, 240p.

*Blatt, H., Middleton, G., and Murray, R., 1972. *Origin of Sedimentary Rocks*. Englewood Cliffs, N.J.: Prentice-Hall, 634p.

*Boswell, P. G. H., 1933. *On the Mineralogy of Sedimentary Rocks*. London: Murby, 393p.

Carozzi, A., 1960. *Microscopic Sedimentary Petrography*. New York: Wiley, 485p.

Carver, R. W., ed., 1971. *Procedures in Sedimentary Petrology*. New York: Wiley, 458p.

*Englehardt, W. V., 1977. *The Origin of Sediments and Sedimentary Rocks*. Stuttgart: E Schweizerbart'sche; New York: Halsted, 359p.

*Friedman, G. M., and Sanders, J. E., 1978. *Principles of Sedimentology*. New York: Wiley, 700p.

*Füchtbauer, H., 1974. *Sediments and Sedimentary Rocks 1*. Stuttgart: E. Schweizerbart'sche; New York: Halsted, 464p.

Garrels, R. M., and Mackenzie, F. T., 1971. *Evolution of Sedimentary Rocks*. New York: Norton, 397p.

Hatch, F. H., Rastall, R. H., and Greensmith, J. T., 1971. *Petrology of Sedimentary Rocks*, 5th ed. New York: Hafner, 502p.

Krumbein, W. C., and Pettijohn, F. J., 1938. *Manual of Sedimentary Petrography*. New York: Plenum, 549p.

Krumbein, W. C., and Sloss, L. L., 1963. *Stratigraphy and Sedimentation*, 2nd ed. San Francisco: Freeman, 660p.

Kukal, Z., 1970. *Geology of Recent Sediments*. Prague: Czech. Acad. Sci., 490p.

*Milner, H. B., 1962. *Sedimentary Petrography*, 2 volumes. New York: Macmillan, 643p and 715p.

*Pettijohn, F. J., 1975. *Sedimentary Rocks*, 3rd ed. New York: Harper & Row, 628 p.

*Pettijohn, F. J., Potter, P. E., and Siever, R., 1972. *Sand and Sandstone*. New York: Springer, 618p.

PREFACE

*Reineck, H.-E., and Singh, I. B., 1973. *Depositional Sedimentary Environments.* New York: Springer, 439p.
Selley, R. C., 1976. *An Introduction to Sedimentology.* London: Academic Press, 408p.
*Shrock, R. R., 1948. *Sequence in Layered Rocks.* New York: McGraw-Hill, 507p.
*Strakhov, N. M., 1967, 1969, 1970. *Principles of Lithogenesis,* 3 vols. (translated from 1962 Russian edition) New York: Consultants Bureau; Edinburgh: Oliver & Boyd, 245p, 609p, and 577p.
Twenhofel, W. H., 1950. *Principles of Sedimentation.* New York: McGraw-Hill, 673p.
*Wilson, J. L., 1975. *Carbonate Facies in Geologic History.* New York: Springer, 417p.

Three major journals deal exclusively with sedimentology:

Journal of Sedimentary Petrology (1931), sedimentological journal of the Society of Economic Paleontologists and Mineralogists, Tulsa, Oklahoma.
Sedimentary Geology (1967), international journal of pure and applied sedimentology, published by Elsevier, Amsterdam.
Sedimentology (1962), journal of the International Association of Sedimentologists, published by Blackwell, Oxford.

Other journals that emphasize sedimentology include:

Bulletin of the American Association of Petroleum Geologists
Bulletin of Canadian Petroleum Geology
Clays and Clay Minerals
Deep-Sea Research
Estuarine and Coastal Marine Science
Geochimica et Cosmochimica Acta
Geologie en Mijnbouw
Lethaia
Limnology and Oceanography
Lithology and Mineral Resources
Lithos
Marine Geology
Maritime Sediments
Palaeogeography, Palaeoclimatology, Palaeoecology
Senckenbergiana Maritima

The American Association of Petroleum Geologists publishes a series of *Memoirs,* and the Society of Economic Paleontologists and Mineralogists, *Special Publications;* both societies produce reprint series, short-course notes, and various other publications. A comprehensive series of volumes entitled *Developments in Sedimentology* is published by Elsevier. The *Initial Reports of the Deep Sea Drilling Project* contain a wealth of information. A particularly valuable reference is *GEO Abstracts E: Sedimentology,* which attempts to abstract all papers in sedimentology, in six issues a year.

Organization and Style

Major entries, for example *Limestones,* appear alphabetically. If the reader does not find an entry where expected, it is best to consult the Index. Most articles are extensively cross-referenced, both in the body of the text and at the end of the reference list for each entry. The abbreviation q.v. (*quod vide*), for example "flow regimes (q.v.)," is used to indicate that an entry by that title appears in the volume. Cross-references to other volumes in the Encyclopedia of Earth Science Series are also included. The following is a list of published and planned volumes:

- I: Oceanography
- II: Atmospheric Sciences and Astrogeology
- III: Geomorphology
- IVA: Geochemistry and Environmental Sciences
- IVB: Mineralogy
- V: Petrology
- VII: Paleontology
- VIII: World Regional Geology, Part 1: Western Hemisphere
- VIII: World Regional Geology, Part 2: Europe and Asia
- VIII: World Regional Geology, Part 3: Africa and Middle East
- IX: Stratigraphy
- X: Structural Geology
- XI: Pedology
- XII: Soil Science
- XIII: Applied Geology
- XIV: Petroleum Geology
- XV: Beaches and Coastal Environments
- XVI: Natural Resources and Energy Conservation
- XVII: Snow, Ice and Glaciology
- XVIII: Geohydrology and Water Supply
- XIX: World Ore Deposits
- XX: Ore Genesis and Metallogeny
- XXI: Mining and Mineral Resources
- XXII: Volcanoes and Volcanology
- XXIII: Geophysics
- XXIV: History of Geology

Each entry is followed by a list of references.

There may be references included that are not cited in the text, but that may lead the reader to additional, more detailed information.

Abbreviations in the volume include:

i.e. (*id est*) = that is
e.g. (*exempli gratia*) = for example
≈ = approximately
μ (mu) = micrometers (S.I. abbreviation has now been established as μm)
B.P. = Before Present (1950)
m.y. = million years
‰ = parts per thousand
ppm = parts per million
< = less than
> = greater than

Standard abbreviations are used for measures of length, area, and volume. An attempt has been made to use the metric system whenever possible. Where definitions are involved, metric equivalents are usually given in parentheses. Some tables, figures, and maps, however, retain the English system of measurement; the following conversion tables may, therefore, be useful:

Metric to English Units—Equivalents of Length
1μ = 0.001 mm = 0.00004 in
1 mm = 0.1 cm = 0.03937 in
1000 mm = 100 cm = 1 m = 39.37 in
  = 3.2808 ft
1 in = 2.54 cm
12 in = 1 ft = 0.3048 m
1 cm = 0.39370 in
1 km = 0.62137 mi
1 fathom = 1.8288 m
1 nautical mile = 1.85325 km
1 statute mile = 1.60935 km = 5280 ft

Square Measures
1 acre = 43560 ft^2 = 0.0015625 mi^2
1 yd^2 = 0.836127 m^2
1 mm^2 = 0.00155 in^2
1 cm^2 = 0.155 in^2 = 0.0011 ft^2
1 m^2 = 10.764 ft^2
1 km^2 = 0.3861 mi^2
1 in^2 = 6.452 cm^2
1 ft^2 = 0.09290 m^2 = 929 cm^2
1 mi^2 = 2.59 km^2

Cubic Measures
1 gal (UK) = 4.5461 liters = 1.201 gal (US)
1 liter = 0.22 gal (UK) = 0.264 gal (US)
1 gal (US) = 3.7854 liters = 0.83 gal (UK)
1 in^3 = 16.387 cm^3
1 ft^3 = 0.0283 m^3
1 mi^3 = 4.1681 km^3
1 mm^3 = 0.000061 in^3
1 cm^3 = 0.061 in^3
1 m^3 = 35.315 ft^3
1 km^3 = 0.24 mi^3

Acknowledgments

No endeavor of this magnitude would be possible without the support of innumerable people, not the least of whom are the nearly 200 contributors; many of them have been with the Encyclopedia of Earth Science project since its inception and have patiently revised their manuscripts time and again. In some instances, an author could not be found for a desired entry, and it was expedient for the editors to write the necessary article—in these cases we are particularly appreciative of the experts upon whom we relied heavily in abstracting their knowledgeable writing.

Over half of the contributors reviewed other manuscripts in the volume, and many agreed to write second or third (or more) entries. Among them the following deserve special mention: Arnold Bouma, Wolfgang Berger, Benno Brenninkmeyer, Robert Carver, John Conolly, Gerald Friedman, Peter Gascoyne, David Jablonski, David Kinsman, David Krinsley, Curt Olsen, Richard Orme, John Sanders, Daniel Stanley, Diana Thurston, and Karl Wolf. Many thanks are due to Gerard Middleton and Donn Gorsline, whose untiring assistance and moral support were invaluable to the completion of this volume. Charlotte Schreiber began the editorial work on this volume and continued to help in later stages; many of her contributions were incorporated into larger articles, and she deserves extra credit.

We note with sadness the deaths of the following contributors—they will surely be missed: Otto Braitsch, Walter Häntzschel, Henry B. Milner, and W. Sugden.

The generous people who reviewed manuscripts are too numerous to list, but those who made especially significant contributions, and others who in some way contributed (e.g., photos) include: P. L. Abbott, Z. S. Altschuler, E. Angino, B. W. Beebe, J. Bruun, W. B. Bull, C. S. Chen, R. Davidson-Arnott, G. R. Davies, E. T. Degens, S. Dzulinski, R. L. Folk, J. Gentilli, C. Ghenea, B. Greenwood, G. A. Gross, W. K. Hamblin, R. S. Harmon, B. C. Heezen, C. D. Holmes, J. Hudson, A. Kamel, C. A. Kaye, R. Kosanke, C. Lalou, D. A. McManus, G. K. Merrill, R. Miller, D. G. Moore, T. A. Mutch, C. H. Nelson, R. Q. Oaks, Jr., W. A. Pryor, M. L. Rhodes, A. R. Rice, D. A. Ross, S. O. Schlanger, R. A. Schweickert, B. M. Shaub, E. B. Shykind, L. L. Sloss, J. B. Southard, D. J. Spearing, D. L. Stevenson, T. H. Van

Andel, R. G. Walker, E. G. Williams, and H. L. Windom.

Joel J. Lloyd, former Director of Sciences Information of the American Geological Institute, must be thanked for lending us the original cards prepared for the sedimentological entries of the 1972 A. G. I. *Glossary of Geology*. The Society of Economic Paleontologists and Mineralogists, ably managed by Ruth Tener (now retired), deserves special acknowledgment for their permission to reprint numerous figures from their journal (*Journal of Sedimentary Petrology*) and other publications. Numerous other societies, particularly the American Association of Petroleum Geologists, and publishers have also been generous with their permissions; each figure caption acknowledges its source.

Janet Stroup and Tamar Gordon provided editorial assistance in early stages; Pat Manley and Susan Lundstedt drafed many figures. Milt Pappas worked long hours checking references and obtaining figures and permissions; he retyped, sometimes twice, nearly every manuscript in the volume. The copy editor, Mary Lewis, was painstakingly careful. The publishers, particularly Chuck Hutchinson, are thanked for their infinite patience and good will. The production manager, Shirley End, has done an incredible job overseeing the completion of this volume.

The junior editor would like to make the following acknowledgments. John Sanders and Rhodes Fairbridge provided encouragement from the inception of my task. The interest, support, and forebearance of Curt Olsen were invaluable. The volume was completed while I was a Van Hise Fellow at the University of Wisconsin, and I would like to thank my fellow graduate students there for their geniality. My special thanks to C. W. Byers, David Larson, Jane Porter, Lloyd Pray, and Karen Schwartz for their generous support during the final stages of editing.

JOANNE BOURGEOIS
RHODES W. FAIRBRIDGE

CONTRIBUTORS

TORBJORN ALEXANDERSSON, Section for Sedimentology, Paleontological Institute, Box 558, University of Uppsala, S-751 22 Uppsala 1, Sweden. *Carbonate Sediments—Bacterial Precipitation.*

J. R. L. ALLEN, Dept. of Geology, University of Reading, Reading RG6 2AB, England. *Deltaic Sediments; Fluvial Sedimentation; Fluvial Sediments; Ripple Marks.*

PAUL AVERITT, U. S. Geological Survey, Federal Center, Denver, Colorado 80225. *Coal.*

IMRE V. BAUMGAERTNER, Dept. of Geology-Geography, C. W. Post College, Long Island University, P. O. Greenvale, New York 11548. *Bagnold Effect; Flame Structures; Primary Sedimentary Structures; Settling Velocity.*

WOLFGANG H. BERGER, Geological Research Division A-105, University of California, San Diego, La Jolla, California 92093. *Lysocline; Paleobathymetric Analysis; Pelagic Sedimentation, Pelagic Sediments.*

LEOPOLD BERTHOIS, Ecole Nationale d'Agriculture, 35-Rennes, Ille-et-Vilaine, France. *Estuarine Sedimentation.*

EDWIN W. BIEDERMAN, JR., 501G Keller Building, Pennsylvania State University, University Park, Pennsylvania 16820. *Crude-Oil Composition and Migration.*

PIERRE E. BISCAYE, Lamont-Doherty Geological Observatory of Columbia University, Palisades, New York 10964. *Clays, Deep-Sea.*

ARNOLD H. BOUMA, U. S. Geological Survey, 345 Middlefield Road, Menlo Park, California 94025. *Bouma Sequence; Scour-and-Fill; Sedimentological Methods; Slump Bedding.*

JOANNE BOURGEOIS, Dept. of Geology and Geophysics, University of Wisconsin, Madison, Wisconsin 53706. *Alluvial-Fan Sediments; Beachrock; Conglomerates; Flocculation; Flood Deposits; Fluidized and Liquefied Sediment Flows; Graded Bedding; Grain Flows; Gravity Flows; Ironstone; Saltation; Submarine (Bathyal) Slope Sedimentation; Turbidity-Current Sedimentation; Volcanism—Submarine Products.*

FREDERICK A. BOWLES, Sea Floor Division, Naval Ocean Research and Development Activity, Bay St. Louis, Mississippi 39529. *Clay as a Sediment.*

DONALD W. BOYD, Dept. of Geology, University of Wyoming, Laramie, Wyoming 82070. *Coral-Reef Sedimentology.*

OTTO BRAITSCH, deceased. *Evaporites—Physicochemical Conditions of Origin; Marine Evaporites—Diagenesis and Metamorphism.*

BENNO M. BRENNINKMEYER, S. J., Dept. of Geology and Geophysics, Boston College, Chestnut Hill, Massachusetts 02167. *Heavy Minerals; Littoral Sedimentation.*

LOUIS I. BRIGGS, Dept. of Geology and Mineralogy, University of Michigan, Ann Arbor, Michigan 48104. *Evaporite Facies; Hydraulic Equivalence.*

R. GRANVILLE BROMLEY, Univ. Mineralogisk-Geol., Mineralogisk Museum, Østervolgade 7, Copenhagen K, Denmark. *Hardground Diagenesis.*

MARGARETHA BRONGERSMA-SANDERS, Geologisch en Mineralogisch Inst. der Rijksuniversiteit, Leiden, The Netherlands. *Sapropel.*

PAUL BROQUET, Dépt. de Géologie, Place LeClerc, 25 Besançon, France. *Olistostrome, Olistolite.*

WAYNE E. BROWNELL, College of Ceramics, Alfred University, Alfred, New York 14802. *Clay.*

HUGH BUCHANAN, Dept. of Geology and Geography, West Virginia University, Morgantown, West Virginia 26506. *Bahama Banks Sedimentology.*

A. L. BURLINGAME, Space Science Lab, Uni-

CONTRIBUTORS

versity of California, Berkeley, California 94720. *Petroleum—Origin and Evolution.*

ANDRÉ CAILLEUX (A. de Cayeux), 9 Ave de Trémouille, 94100 St. Maur, Val de Marne, France. *Niveo-Eolian Deposits.*

MAURICE A. CARRIGY, Oil Sands Technology and Research Authority, 9915 108th Street, Edmonton, Alberta, Canada. *Tar Sands.*

ROBERT E. CARVER, Dept. of Geology, University of Georgia, Athens, Georgia 30601. *Abrasion pH; Redeposition, Resedimentation; Sedimentary Petrology; Udden Scale; Wentworth Scale.*

JOHN A. CATT, Dept. of Pedology, Rothamsted Experimental Station, Harpendon, Hertsfordshire, England. *Glacigene Sediments.*

ROGER H. CHARLIER, University of Northeastern Illinois, 5500 N. St. Louis Avenue, Chicago, Illinois 60625. *Tangue.*

G. V. CHILINGARIAN, Petroleum Engineering Dept., University of Southern California, Los Angeles, California 90007. *Compaction in Sediments.*

PHILIP W. CHOQUETTE, Marathon Oil Company, P. O. Box 269, Littleton, Colorado 80121. *Oolite.*

H. EDWARD CLIFTON, U. S. Geological Survey, 345 Middlefield Road, Menlo Park, California 94025. *Lateral and Vertical Accretion.*

PATRICK J. COLEMAN, Dept. of Geology, University of Western Australia, Nedlands, W.A. 6009, Australia. *Tsunami Sedimentation.*

DONALD J. COLQUHOUN, Dept. of Geology, University of South Carolina, Columbia, South Carolina 29208. *Paralic Sedimentary Facies.*

JOHN R. CONOLLY, 31 Noroton Avenue, Darien, Connecticut 06820. *Contourites; Glacial Marine Sediments; Hadal Sedimentation; Maturity; Top and Bottom Criteria.*

JOHN C. CROWELL, Dept. of Geology, University of California, Santa Barbara, California 93106. *Pebbly Mudstones.*

ERICH DIMROTH, Sciences de la Terre, University of Quebec, Chicoutimi, P.Q., Canada. *Cherty Iron Formation (BIF).*

JACK DONAHUE, Dept. of Earth and Planetary Sciences, University of Pittsburgh, Pittsburgh, Pennsylvania 15213. *Pisolite.*

LARRY J. DOYLE, Dept. of Marine Science, University of South Florida, 830 First St. South, St. Petersburg, Florida 33701. *Abyssal Sedimentation; Mud Belt, Mud Zone, Mud Line; Neritic Sedimentary Facies.*

ALEKSIS DREIMANIS, Dept. of Geology, University of Western Ontario, London, Ontario, Canada. *Till and Tillite.*

GILBERT DUNOYER DE SEGONZAC, Faculté des Sciences de Strasbourg, Inst. de Géologie, 1 rue Blessig, 67 Strasbourg, France. *Clay—Diagenesis.*

THOMAS E. EASTLER, Dept. of Geology, University of Maine, Farmington, Maine 04938. *Casts and Molds; Raindrop Imprint.*

GORDON P. EATON, Hawaiian Volcano Observatory, Hawaii National Park, Hawaii 96718. *Volcanic Ash Deposits.*

A. J. EHLMANN, Dept. of Geology, Texas Christian University, Fort Worth, Texas 76129. *Glauconite.*

PAUL ENOS, Dept. of Geological Sciences, State University of New York, Binghamton, New York 13901. *Dolomite, Dolomitization.*

GEORGE E. ERICKSEN, U. S. Geological Survey, Federal Center, Reston, Virginia 22092. *Salar, Salar Structures.*

J. A. FAGERSTROM, Dept. of Geology, University of Nebraska, Lincoln, Nebraska 68508. *Clay-Pebble Conglomerate and Breccia.*

RHODES W. FAIRBRIDGE, Dept. of Geological Sciences, Columbia University, New York, New York 10027. *Black Sea Sediments; Breccias, Sedimentary; Dedolomitization; Delta Sedimentation; Diagenesis; Duricrust; Eolianite; Ferricrete; Kara-Bogaz Gulf Evaporite Sedimentology; Lamina, Laminaset, Laminite; Lithology, Lithofacies, Lithotope, Lithification, Lithogenesis; Midden; Reef Complex; Sabkha*

CONTRIBUTORS

Sedimentology; Submarine (Bathyal) Slope Sedimentation.

KURT FIEGE, Geologe, 34 Gottingen-Geismar, Charlottenburgerstr. 19–C719, Germany B.R.D. *Cyclic Sedimentation.*

MICHAEL E. FIELD, U. S. Geological Survey, 345 Middlefield Road, Menlo Park, California 94025. *Continental-Rise Sediments; Outer Continental-Shelf Sediments.*

CHARLES W. FINKL, JNR., P. O. Box 2473, Ft. Lauderdale, Florida 33303. *Duricrust.*

ROBERT J. FINLEY, Bureau of Economic Geology, University of Texas at Austin, Austin, Texas 78712. *Tidal-Current Deposits.*

RICHARD V. FISHER, Dept. of Geological Sciences, University of California, Santa Barbara, California 93106. *Base-Surge Deposits; Volcaniclastic Sediments and Rocks.*

RICHARD S. FISKE, Dept. Mineral Sciences, NHB-119, Smithsonian Institution, Washington, D. C. 20560. *Pyroclastic Sediment.*

DAVID W. FOLGER, U. S. Geological Survey, Woods Hole, Massachusetts 02543. *Eolian Dust in Marine Sediments; Extraterrestrial Material in Sediments.*

LAWRENCE A. FRAKES, Dept. of Earth Science, Monash University, Clayton, Victoria 3168, Australia. *Diamictite.*

PAUL C. FRANKS, Dept. of Geology, University of Akron, Akron, Ohio 44304. *Concretions; Geodes.*

ROBERT W. FREY, Dept. of Geology, University of Georgia, Athens, Georgia 30601. *Bioturbation.*

GERALD M. FRIEDMAN, Dept. of Geology, Rensselaer Polytechnic Institute, Troy, New York 12812. *Carbonate Sediments–Lithification; Grain-Size Parameters–Environmental Interpretation; Sedimentology–Organizations and Associations; Staining Techniques.*

HANS FÜCHTBAUER, Geol. Inst. Der Ruhr-Universitat, Universitatstr. 150, 4630 Bochum-Querenberg, Postfach 2148, Germany B.R.D. *Clastic Sediments–Lithification and Diagenesis.*

PETER GASCOYNE, Soil Survey of England and Wales, Rothamsted Experimental Station, Harpendon, Hertfordshire, England. *Gravity-Slide Deposits; Mass-Wasting Deposits; Mudflow, Debris-Flow Deposits.*

BILLY P. GLASS, Dept. of Geology, University of Delaware, Newark, Delaware 19711. *Extraterrestrial Material in Sediments; Sedimentation Rates, Deep-Sea.*

K. W. GLENNIE, Shell U. K. Explor. & Prod. Ltd, Shell Centre, London SE1 7NA, England. *Desert Sedimentary Environments; Eolian Sands.*

MARTIN GOLDHABER, U. S. Geological Survey, Uranium and Thorium Research Branch, Federal Center, Denver, Colorado 80225. *Euxinic Facies.*

ROLAND GOLDRING, Dept. of Geology, University of Reading, Whiteknights, Reading RG6 2AB, England. *Sedimentation–Paleoecologic Aspects.*

DONN S. GORSLINE, Dept. of Geological Sciences, University of Southern California, Los Angeles, California 90007. *Marine Sediments; Turbidity-Current Sedimentation.*

JOHN C. GRIFFITHS, 239 Deike Building, Pennsylvania State University, University Park, Pennsylvania 16802. *Sediment Parameters.*

DOUGLAS E. HAMMOND, Dept. of Geological Sciences, University of Southern California, Los Angeles, California 90007. *Gases in Sediments.*

J. M. HANCOCK, Dept. of Geology, King's College, University of London, Strand WC2R 2LS, London, England. *Chalk.*

WALTER HÄNTZCHEL, deceased. *Bioturbation.*

W. B. HARLAND, Dept. of Geology, University of Cambridge, Sedgwick Museum, Downing Street, Cambridge CB2 3EQ, England. *Fulgurites.*

RICHARD L. HAY, Dept. of Geology and Geophysics, University of California, Berkeley, California 94720. *Volcanic Ash–Diagenesis.*

JOHN B. HAYES, Marathon Oil Company, Box

269, Littleton, Colorado 80160. *Concretions; Geodes.*

MILES O. HAYES, Coastal Research Division, Dept. of Geology, University of South Carolina, Columbia, South Carolina 29208. *Tidal-Current Deposits.*

ALBERT C. HINE, III, University of North Carolina at Chapel Hill, Institute of Marine Science, P. O. Drawer 809, Morehead City, North Carolina 28557. *Tidal-Current Deposits.*

GORDON W. HODGSON, Director, Environmental Science Centre, University of Calgary, Alberta T2N 1N4, Canada. *Hydrocarbons in Sediments.*

SHOJI HORIE, Institute of Paleolimnology and Paleoenvironment on Lake Biwa, Kyoto University, Takashima-chō, Shiga-ken, 520-11, Japan. *Lacustrine Sedimentation.*

KENNETH J. HSÜ, Geologisches Inst., Eth., Sonegstrasse 5, Zurich, Switzerland. *Geosynclinal Sedimentation.*

RICHARD V. HUGHES, Rte. 5, Box 866. Golden, Colorado 80401. *Oil Shale.*

DEREK W. HUMPHRIES, Dept. of Geology, University of Sheffield, Mappin Street, Sheffield, Yorkshire, England. *Clastic Sediments and Rocks; Continental Sedimentation; Eolian Sedimentation; Paludal Sediments; Provenance.*

JOHN M. HUNT, Woods Hole Oceanographic Institute, Woods Hole, Massachusetts 02543. *Organic Sediments.*

TAKASHI ICHIYE, Dept. of Oceanography, Texas A & M University, College Station, Texas 77843. *Bernoulli's Theorem.*

DAVID JABLONSKI, Dept. of Geology, Yale University, New Haven, Connecticut 06520. *Grapestone.*

TOGWELL A. JACKSON, Freshwater Institute, 501 University Crescent, Winnepeg 19, Manitoba, Canada. *Humic Matter in Sediments; Kerogen.*

MARIAN B. JACOBS, 7 Robin Road, Mahwah, New Jersey 07430. *Abyssal Sedimentary Environments; Nepheloid Sediments and Nephelometry; Red Clay.*

DONALD LEE JOHNSON, Dept. of Geography, University of Illinois, Urbana, Illinois 61801. *Eolianite.*

ALAN V. JOPLING, Dept. of Geography, University of Toronto, Ontario, Toronto MS5 1A1, Canada. *Vackset Bedding; Bedding Genesis.*

CLIFTON F. JORDAN, Research and Development Dept., Continental Oil Company, Ponca City, Oklahoma 74601. *Tropical Lagoonal Sedimentation.*

JOHN KALDI, Dept. of Earth and Environmental Sciences, Queens College, CUNY, Flushing, New York 11367. *Beach Sands.*

MIRIAM KASTNER, Scripps Institute of Oceanography, La Jolla, California 92093. *Silica in Sediments.*

JOHN F. KENNEDY, Director, Institute of Hydraulic Research, University of Iowa, Iowa City, Iowa 52242. *Bed Forms in Alluvial Channels.*

DENNIS V. KENT, Lamont-Doherty Geological Observatory of Columbia University, Palisades, New York 10964. *Anisotropy in Sediments; Remanent Magnetism in Sediments.*

DAVID J. J. KINSMAN, Freshwater Biological Association, Windermere Laboratory, The Ferry House, Ambleside, Cumbria LA22 OLP, England. *Evaporites—Physicochemical Conditions of Origin; Marine Evaporites—Diagenesis and Metamorphism; Persian Gulf Sedimentology; Sabkha Sedimentology.*

GEORGE DE VRIES KLEIN, Dept. of Geology, University of Illinois, Urbana, Illinois 61801. *Sands and Sandstones.*

R. JOHN KNIGHT, Div. of Sedimentology, EG-13 MNH, Smithsonian Institution, Washington, D. C. 20560. *Bay of Fundy Sedimentology.*

CHARLES R. KOLB, 3314 Highland Drive, Vicksburg, Mississippi 39180. *Prodelta Sediments.*

YEHOSHUA KOLODNY, Dept. of Geology,

Hebrew University, Jerusalem, Israel. *Porcelanite.*

LOUIS S. KORNICKER, Smithsonian Institution, Washington, D. C. 20560. *Coral-Reef Sedimentology.*

DAVID KRINSLEY, Dept. of Geology, Arizona State University, Tempe, Arizona 85281. *Beach Sands; Eolian Sands; Sand Surface Textures.*

HANS G. KUGLER, Naturhistorisches Museum Basel, Augustinergasse 2, CH-4051 Basel, Switzerland. *Volcanism, Sedimentary.*

NARESH KUMAR, Atlantic Richfield Company, P. O. Box 2819, Dallas, Texas 75221. *Barrier Island Facies; Storm Deposits; Tidal-Inlet and Tidal-Delta Facies.*

E. R. LANDIS, U. S. Geological Survey, Branch of Organic Fuels, Federal Center, Denver, Colorado 80225. *Coal.*

RALPH L. LANGENHEIM, JR., Dept. of Geology, University of Illinois, Urbana, Illinois 61801. *Resin and Amber.*

D. E. LAWRENCE, Geological Survey of Canada, 601 Booth Street, Ottowa, Ontario K1A OE8, Canada. *Rapid Sediment Analyzer.*

LOUIS LEINE, Shell International Petr. MIJ., B. V., Oostduinlaan 75, The Hague, The Netherlands. *Boxwork; Rauhwacke.*

GILLIAN C. LEWARNE-SHEEHAN, 3 Sheelin Road, Caherdavin Park, Caherdavin, Limerick, Ireland. *Spongolite.*

JOHN F. LINDSAY, The Lunar Science Institute, 3303 NASA Road 1, Houston, Texas 77058. *Lunar Sedimentation.*

BRIAN W. LOGAN, Dept. of Geology, University of Western Australia, Nedlands, W. A. 6009, Australia. *Shark Bay Sedimentology.*

AUGUSTIN LOMBARD, 52, Chemin Naville, 1211 Conches, Genève, Switzerland. *Sedimentology.*

DONALD R. LOWE, Dept. of Geology, Louisiana State University, Baton Rouge, Louisiana 70803. *Water-Escape Structures.*

KAZIMIERZ ŁYDKA, Warsaw University, Warsaw, Poland. *Marl.*

PAUL J. F. MACAR, Inst. de Géologie, 7 Place du XX, Université de Liège, Liège, Belgium. *Penecontemporaneous Deformation of Sediments.*

EARLE F. McBRIDE, Dept. of Geology, University of Texas, Austin, Texas 78712. *Novaculite; Parting Lineation.*

ALISTAIR W. McCRONE, 15 Robert Court West, Arcata, California 95521. *Mudstone and Claystone; Sedimentation.*

VINCENT E. McKELVEY, U.S. Geological Survey, 12201 Sunrise Valley Drive, Reston, Virginia 22092. *Phosphate in Sediments.*

BERNARD L. MAMET, Dépt. de Géologie, Université de Montréal, P. O. Box 6128, Montréal, P.Q., Canada. *Limestone Fabrics.*

FRANK T. MANHEIM, Woods Hole Oceanographic Institute, Woods Hole, Massachusetts 02543. *Interstitial Water in Sediments.*

ROBERT MARSCHALKO, Geol. Inst. SAV, Dúbravská Cesta, Bratislava 88625, Czechoslovakia. *Clastic Dikes.*

I. PETER MARTINI, Dept. of Land Resource Science, University of Guelph, Guelph, Ontario, Canada. *Fabric, Sedimentary.*

FLORENTIN MAURRASSE, Dept. of Physical Sciences, Florida International University, Tamiami Trail, Miami, Florida 33144. *Diatomite; Fucoid; Ignimbrites; Pseudomorphs; Secondary Sedimentary Structures; Siliceous Ooze.*

GERARD V. MIDDLETON, Dept. of Geology, McMaster University, Hamilton, Ontario L8S 4M1, Canada. *Facies; Flow Regimes; Sedimentology–History.*

GEORGES MILLOT, Inst. de Géologie, 1 Rue Blessig, 67 Strasbourg, France. *Clay–Genesis.*

HENRY B. MILNER, deceased. *Sedimentary Petrography.*

A. H. G. MITCHELL, c/o UNDP, P. O. Box

650, Rangoon, Burma. *Sedimentary Facies and Plate Tectonics.*

NILS-AXEL MÖRNER, Geologiska Institutionen, Kungstensgatan 45, Box 6801, 113 86 Stockholm, Sweden. *Gyttja, Dy; Varves and Varved Clays.*

A. S. NAIDU, Inst. of Marine Science, University of Alaska, Fairbanks, Alaska 99701. *Continental-Shelf Sediments in High Latitudes.*

VLADIMÍR NÁPRSTEK, Katedra Geológie, Přírodovědecké Fakulty, Karlovy University, 128 43 Praha 2 - Albertov 6, Czechoslovakia. *Black Sands.*

JAMES T. NEAL, Civil Engineering Research Division, Air Force Weapons Laboratory, Kirtland AFB, New Mexico 87117. *Mud Cracks (Contraction Polygons); Syneresis.*

H. D. NEEDHAM, Centre Oceanographique de Bretagne, B. P. 337, 29.273 Brest, France. *Color of Sediments.*

CARL F. NORDIN, JR., WRD USGS, Box 25046, MS 413, DFC, Denver, Colorado 80225. *Fluvial Sediment Transport.*

IRWIN D. NOVAK, University of Maine, 150 Science, Portland, Maine 04103. *Beach Gravels.*

REGINALD T. O'BRIEN, Leighton Mining N. L., BP House, 1 Albert Road, Melbourne, Victoria 3004, Australia. *Tuff.*

HAKUYU OKADA, Lab. Marine Geology and Sedimentology, Geoscience Institute, Faculty of Science, Shizuoka University, Shizuoka 422, Japan. *Inverse Grading; Quartzose Sandstone; Subgraywacke.*

CLIFFORD D. OLLIER, Research School of Pacific Studies, National University, P. O. Box 4, Canberra, Australia. *Weathering in Sediments.*

CURTIS R. OLSEN, Lamont-Doherty Geological Observatory of Columbia University, Palisades, New York 10964. *Cementation; Estuarine Sediments; Sedimentation Rates.*

CARL H. OPPENHEIMER, Marine Sciences Institute, University of Texas, Port Aransas, Texas 78373. *Microbiology in Sedimentology.*

G. RICHARD ORME, Dept. of Geology and Mineralogy, University of Queensland, St. Lucia, Queensland 4067, Australia. *Authigenesis; Diagenesis; Diagenetic Fabrics.*

THOMAS E. PICKETT, Delaware Geological Survey, University of Delaware, Newark, Delaware 19711. *Lagoonal Sedimentation.*

GEORGE PLAFKER, U. S. Geological Survey, Branch of Alaskan Geology, 345 Middlefield Road, Menlo Park, California 94025. *Avalanche Deposits.*

PAUL EDWIN POTTER, H. N. Fisk Laboratory of Sedimentology, University of Cincinnati, Cincinnati, Ohio 45221. *Sedimentology—Yesterday, Today, and Tomorrow.*

J. E. PRENTICE, Dept. of Geology, King's College, Strand, London WC2R 2LS, England. *Sedimentation—Tectonic Controls.*

HAROLD G. READING, Dept. of Geology, University of Oxford, Oxford OX1 3PR, England. *Sedimentary Facies and Plate Tectonics; Turbidites.*

C. C. REEVES, JR., Dept. of Geosciences, Texas Tech College, P. O. Box 4109, Lubbock, Texas 79409. *Caliche, Calcrete.*

ROBIN REID, Dept. of Geology, Queen's University, Belfast, Northern Ireland. *Sponges in Sediments.*

HANS-ERICH REINECK, Forschunsinstitut, Senckenberg, Schleusenstr. 39A, 2940 Wilhelmshaven, Germany B.R.D. *Tidal-Flat Geology.*

H. H. RIEKE, School of Mines, West Virginia University, Morgantown, West Virginia 26506. *Compaction in Sediments.*

L. J. ROSEN, Du Pont Company, Engineering Test Center, Newark, Delaware 19711. *Sedimentation Rates, Deep-Sea.*

JOHN N. ROSHOLT, U. S. Geological Survey, Federal Center, Denver, Colorado 80225. *Radioactivity in Sediments.*

JOHN E. SANDERS, Dept. of Geology, Barnard College, Columbia University, New York, New York 10027. *Current Marks; Graywacke;*

Intraformational Disturbances; Sedimentary Structures; Storm Deposits.

B. CHARLOTTE SCHREIBER, Dept. of Earth and Environmental Sciences, Queens College, CUNY, Flushing, New York 11367. *Accretion, Accretion Topography; Agglomerate, Agglutinate; Aggradation; Alling Scale; Allochem; Arenite; Arkose, Subarkose; Atterberg Scale; Birdseye Limestone; Bitumen, Bituminous Sediments; Blanket Sand; Calcarenite, Calcilutite, Calcirudite, Calcisiltite; Coquina, Criquina; Elutriation; Fissility; Flocculation; Impact Law; Intraclast; Kurtosis; Micrite; Phi Scale; Reduction, Reduction Number; Roundness and Sphericity; Sand Waves; Skewness; Sparite; Sternberg's Law; Stokes' Law; Traction; Tripoli, Tripolite; Zebra Dolomite.*

JOHANNES H. SCHROEDER, Institute of Oceanography, P. O. Box 24, Port Sudan, Sudan. *Algal Reef Sedimentology; Cementation, Submarine.*

MAURICE SCHWARTZ, Dept. of Geology, Western Washington University, Bellingham, Washington 98225. *Deposition; Dispersal; Transportation.*

ADOLF SEILACHER, Dept. Geologie Paläontolcgie, University of Tübingen, Tübingen, Germany B.R.D. *Biostratinomy.*

RICHARD C. SELLEY, Dept. of Geology, Royal School of Mines, Imperial College of Science and Technology, Prince Consort Road, London SW7, England. *Sedimentary Environments.*

B. W. SELLWOOD, Dept. of Geology, University of Reading, Whiteknights, Reading, England. *Biogenic Sedimentary Structures.*

SUPRIYA SENGUPTA, Geological Studies Unit, Indian Statistical Institute, Calcutta 700 035, India. *Channel Sands; Granulometric Analysis; Paleocurrent Analysis; Sorting.*

ROBERT P. SHARP, Division of Geology and Planetary Science, California Institute of Technology, Pasadena, California 91125. *Martian Sedimentation.*

RUSSELL G. SHEPHERD, 3917 Xavier Street, Denver, Colorado 80212. *Experimental Sedimentology.*

ROY J. SHLEMON, P. O. Box 3066, Newport Beach, California 92663. *Silt.*

H. JAMES SIMPSON, Lamont-Doherty Geological Observatory of Columbia University, Palisades, New York 10964. *Estuarine Sediments.*

INDRA BIR SINGH, Geology Department, Lucknow University, Lucknow (U.P.), India. *Scour Marks; Tool Marks.*

LESLIE SIRKIN, Dept. of Earth Sciences, Adelphi University, Garden City, New York 11530. *Peat-Bog Deposits.*

ROGER M. SLATT, Dept. of Geology, Arizona State University, Tempe, Arizona 85281. *Glacial Gravels; Glacial Sands.*

ALEC J. SMITH, Dept. of Geology, Bedford College, Regents Park, London NW1 4NS, England. *Sedimentary Rocks.*

NORMAN D. SMITH, Dept. of Geology, University of Illinois, Chicago Circle, Chicago, Illinois 60680. *Braided-Stream Deposits; River Gravels.*

JAMES E. SORAUF, Dept. of Geology, State University of New York, Binghamton, New York 13901. *Pillow Structures.*

DANIEL JEAN STANLEY, Div. of Sedimentology, E-109 MNH, Smithsonian Institution, Washington, D. C. 20560. *Submarine Fan (Cone) Sedimentation.*

ROBERT J. STANTON, JR., Dept. of Geology, Texas A & M University, College Station, Texas 77843. *Solution Breccias.*

C. G. STEPHENS, 2 Glenside Avenue, Myrtle Bank, South Australia 5064, Australia. *Silcrete.*

GEORGE E. STOERTZ, U. S. Geological Survey, Federal Center, Reston, Virginia, 22092. *Salar, Salar Structures.*

W. SUGDEN, deceased. *Faceted Pebbles and Boulders.*

DONALD J. P. SWIFT, NOAA-AOML, 15 Rickenbacker Causeway, Virginia Key, Florida

33149. *Continental-Shelf Sedimentation; Continental-Shelf Sediments.*

ADA SWINEFORD, Dept. of Geology, Western Washington University, Bellingham, Washington 98225. *Loess; Spring Deposits.*

TARO TAKAHASHI, Dept. of Earth and Environmental Sciences, Queens College, CUNY, Flushing, New York 11367. *Sand Surface Textures.*

WILLIAM F. TANNER, Dept. of Geology, Florida State University, Tallahassee, Florida 32306. *Cross-Bedding; Grain-Size Studies; Reynolds and Froude Numbers; Sediment Transport–Initiation and Energetics.*

MARLIES TEICHMÜLLER, Geologisches Landesamt Nordrhein Westfalen, De-Greiff-Str. 195, 4150 Krefeld, Germany B.R.D. *Coal–Diagenesis and Metamorphism.*

ROLF TEICHMÜLLER, Geologisches Landesamt Nordrhein Westfalen, De-Greiff-Str. 195, 4150 Krefeld, Germany B.R.D. *Coal–Diagenesis and Metamorphism.*

CHARLES P. THORNTON, Dept. of Geosciences, 201 Deike Building, Pennsylvania State University, University Park, Pennsylvania 16802. *Paragenesis of Sedimentary Rocks.*

DIANA R. THURSTON, 16 Crossfields, Bromley Cross, Bolton, Lancashire, England. *Chert and Flint; Pressure Solution and Related Phenomena.*

L. M. J. U. VAN STRAATEN, Geologisch Instituut, Rijksuniversiteit, Melkweg-1, Groningen, The Netherlands. *Salt-Marsh Sedimentology; Wadden Sea Sedimentology.*

GLENN S. VISHER, Dept. of Earth Sciences, 1133 N. Lewis, University of Tulsa, Tulsa, Oklahoma 74110. *Grain-Size Frequency Studies.*

STEPHEN VONDER HAAR, Dept. of Geology and Geophysics, University of California, Berkeley, California 94720. *Gypsum in Sediments.*

THEODORE R. WALKER, Dept. of Geology, University of Colorado, Boulder, Colorado 80302. *Silicification.*

CHARLES E. WEAVER, School of Geophysical Science, Georgia Institute of Technology, Atlanta, Georgia 30332. *Argillaceous Rocks; Bentonite; Clay Sedimentation Facies.*

MALCOLM P. WEISS, Dept. of Geology, Northern Illinois University, DeKalb, Illinois 60115. *Calcarenite, Calcilutite, Calcirudite, Calcisiltite; Corrosion Surface.*

J. H. McD. WHITAKER, Dept. of Geology, The University, Leicester LE1 7RH, England. *Submarine Canyons and Fan Valleys, Ancient.*

WILLIAM B. WHITE, Materials Research Lab, Pennsylvania State University, University Park, Pennsylvania 16802. *Cave Pearls; Speleal Sediments; Travertine.*

A. M. WINKELMOLEN, Geologisch Instituut, Rijksuniversiteit, Melkweg-1, Groningen, The Netherlands. *River Sands.*

KARL H. WOLF, Directorate General of Mineral Resources, Kingdom of Saudi Arabia, and Watts, Griffith and McOuat Ltd, Consulting Geologists, P. O. Box 345 and 5219, Jeddah, Saudi Arabia. *Carbonate Sediments–Diagenesis; Limestones; Sedimentology–Models; Stromatolites; Sulfides in Sediments.*

M. GORDON WOLMAN, Dept. of Geography, Johns Hopkins University, Baltimore, Maryland 21218. *Alluvium.*

FREDERICK F. WRIGHT, NOAA, Box 537, Douglas, Alaska 99824. *Nuée Ardente; Relict Sediments; Talus.*

P. C. WSZOLEK, Dept. of Agronomy, Cornell University, Ithaca, New York 14853. *Petroleum–Origin and Evolution.*

DAN H. YAALON, Dept. of Geology, Hebrew University, Jerusalem, Israel. *Nodules in Sediments; Shale.*

WARREN E. YASSO, Dept. of Science Education, Teachers College, Columbia University, New York, New York 10027. *Salcrete; Tracer Techniques in Sediment Transport.*

V. P. ZENKOVICH, Moscow, 109240, Kotelnicheskaya nab., 1/15, kv. 196. U.S.S.R. *Attrition.*

The
ENCYCLOPEDIA
of
SEDIMENTOLOGY

A

ABRASION pH

Abrasion pH is defined as the pH of a slurry produced by grinding a mineral in a small quantity of water (Stevens and Carron, 1948). To determine abrasion pH, soft minerals that do not absorb large quantities of water are abraded on a wet streak plate, and hard minerals, or minerals that absorb water, are ground in an agate mortar with a few drops of water. Indicator papers or solutions are used to find the pH of the resultant slurry. Stevens and Carron investigated the abrasion pH of 280 minerals and suggested that abrasion pH, ranging from 1 to 12, is a valuable tool for field identification of minerals.

Abrasion pH is dependent on hydrolysis of the mineral tested. Minerals that can be considered to be salts of strong acids and strong bases give nearly neutral (7) abrasion pH values; minerals consisting of relatively passive cations and highly active anions (iron sulfate minerals) give low abrasion pH values; and carbonates give very high values. The effect is strongly dependent on the quantity of water added to the crushed mineral and on solute concentration of the water added.

According to Stumm and Morgan (1970) the phenomenon of abrasion pH is explained in general by electrical double-layer theory and specifically by the Gouy-Chapman theory. Because the broken surfaces of a silicate mineral are negatively charged and, when in contact with water, hydrated, the hydrogen ion concentration at the surface of the mineral may be quite different than the concentration in the bulk of the surrounding water. The difference between pH at the surface and in the bulk of the surrounding fluid can be estimated from the Gouy-Chapman theory.

ROBERT E. CARVER

References

Chapman, D. L., 1913. A contribution to the theory of electrocapillarity, *Phil. Mag.*, ser. 6, **25**, 475-481.
Gouy, G., 1910. Sur la constitution de la change électique à la surface d'un électrolyte," *J. Phys.* ser. 4, **9**, 457-458.
Keller, W. D., 1957. *The Principles of Chemical Weathering.* Columbia, Mo.: Lucas Bros., 111p.
*Stevens, R. E., and Carron, M. K., 1948. Simple field test for distinguishing minerals by abrasion pH, *Am. Mineral.*, **33**, 31-49.
Stumm, W., and Morgan, J. J., 1970. *Aquatic Chemistry.* New York: Wiley, 583p.

*Additional bibliographic references may be found in this work.

Cross-references: *Authigenesis; Diagenesis; Silicification; Sulfides in Sediments; Weathering in Sediments.* Vol. IV A: *Abrasion pH.*

ABYSSAL SEDIMENTARY ENVIRONMENTS

Abyssal pertains to the marine environment of the deep sea having depths >1000 m. The outstanding features of the abyssal province are the broad spatial expanses and the great ranges of depth. The boundaries between the overlying bathyal region and the underlying hadal realm are somewhat gradational and indistinct. Bruun (1957) describes the boundary between the bathyal and abyssal zones as coinciding with the 4°C isotherm, which is also a faunal boundary. In the Atlantic Ocean the 4°C isotherm occurs at 2000 m, while in the Indian and Pacific Oceans it occurs between 1000 and 1500 m. The lower limit of the abyssal region is probably about 6000 m, which means that a vertical range of 4000 m exists for the abyssal environment. The area between 2000 and 6000 m covers 76% of all the area of the oceans, or more than half the surface of the globe. In a recent endeavor to describe the abyssal environment, Menzies et al. (1973) assert that the bathyal/abyssal boundary cannot be established at a constant depth (the 2000 m isobath) or at a constant temperature (the 4°C isotherm). They hypothesize that the zone oscillates, depending on water masses and bottom currents. They suggest a bentho-abyssal scheme of zonation, based upon faunal composition.

Although there is some controversy regarding the boundaries of the abyssal zone, agreement exists concerning its physical characteristics. There is no solar light, so that no photosynthesis is possible. Bruun (1957) observes that the food supply of the abyssal zone is not plentiful but that actual depth is of less importance than the proximity of productivity zones at the surface. The fauna are carnivores and detritus feeders. The hydrostatic pressure is tremendous, rising with depth at the rate of 1 atm/10 m and resulting in the wide range 200-600 atm. Small variation in salinity has been observed in the abyssal zone; an average figure is 34.8‰, with a variation of 0.2‰. The temperature of the water is uniformly low, from $4°C$ to $-1°C$. There is ample circulation in the open ocean to supply well aerated water from deep vertical movements in the high latitudes.

The bottom topography of the province consists of abyssal plains, submarine channels and fans, and abyssal hills. Abyssal plains occur at the base of a continental rise, where turbidity flows transport clastic material in suspension, erode and redeposit bottom sediments, and produce the smooth stratification and remarkably flat surfaces (slopes < 1:1000) characteristic of abyssal plains.

The assortment of sediment types that have been observed in the abyssal province suggest that a description of abyssal sedimentation should emphasize the factor concerning the extreme depth of water, not the processes of deposition. The nature of abyssal sedimentation is influenced by the foregoing environmental factors. The greater part of the abyssal floor is covered with a soft bottom of deep-sea red clay, while biogenous ooze is found on the rises. Suspended terrigenous sediments, carried by wind, water, or ice, settle with extreme slowness, so that all matter soluble in sea water has been removed, leaving only ferric oxide and aluminum silicate, the main constituents of red clay. Similarly, as skeletal parts of planktonic tests settle, the most fragile parts are dissolved, leaving the larger, more resistant ones. Calcareous oozes of foraminifera and coccolithophores occur at depths < 4000 m, and pelagic red clays are found at the greater depths, the boundary between them determined by the position of the calcite compensation level (see *Pelagic Sedimentation*). Siliceous oozes of radiolarian and diatomaceous skeletal parts are prevalent beneath the high productivity regions of surface waters of high latitudes and the Equatorial Pacific. Sediment cores raised from abyssal plains give evidence of the action of turbidity currents in sedimentation processes of the abyssal plains. Shallow-water foraminifera, graded quartz sand, gray clays, and silts have been observed interbedded with deep-sea red clay (Heezen and Laughton, 1963). Sediments of cosmic and volcanic origin, including ash and pumice, and coarse rock detritus, rafted by ice and sometimes by the roots of plants, may settle to the abyssal floor; and secondary deposits, including authigenic zeolites and manganese nodules, are also found there. In summary, the extreme depth of water and the spatial expanse of the abyssal marine environment account for the variety of sediment types that reach the deep abyss and accumulate there.

MARIAN B. JACOBS

References

Bruun, A. F., 1957. Deep sea and abyssal depths, in J. Hedgpeth, ed., *Treatise on Marine Ecology and Paleoecology. Geol. Soc. Am. Mem. 67*, vol. 1, 641-672.

Hamilton, E. L., 1973. Marine geology of the Aleutian Abyssal Plain, *Marine Geol.*, **14**, 295-325.

Heezen, B. C., and Laughton, A. S., 1963. Abyssal plains, in M. N. Hill, ed., *The Sea*, vol. 3. New York: Wiley, 312-364.

Hessler, R. R., and Jumars, P. A., 1974. Abyssal community analysis from replicate box cores in the central North Pacific, *Deep-Sea Research*, **21**, 185-209.

Malone, T. C.; Garside, C.; Anderson, R.; and Roels, O. A., 1973. The possible occurrence of photosynthetic microorganisms in deep-sea sediments of the North Atlantic, *J. Phycology*, 9, 482-488.

Menzies, R. J.; George, R. Y.; and Rowe, G. T., 1973. *Abyssal Environment and Ecology of the World Oceans*. New York: Wiley, 488p.

Tucholke, B. E., 1974. The history of sedimentation and abyssal circulation on the Greater Antilles Outer Ridge, *Woods Hole Oceangr. Inst., Tech. Rept. 74-1*, 318p.

Cross-references: *Hadal Sedimentation; Pelagic Sedimentation; Red Clay; Siliceous Ooze; Submarine (Bathyal) Slope Sedimentation; Turbidity-Current Sedimentation.* Vol. I: *Abyssal Hills; Abyssal Plains; Abyssal Sediment (Thickness); Abyssal Zone.*

ABYSSAL SEDIMENTATION

Abyssal sedimentation occurs in the deep ocean basins, generally at depths > 2000 m. Included within its scope are the standard parameters of sedimentation: provenance, transportation, deposition, and diagenesis.

Abyssal sedimentation can be subdivided in a variety of ways, none of which is wholly satisfactory or universally accepted. A simple division based upon deep-sea sedimentary processes, into pelagic sedimentation, authigenesis, and sedimentation involving near-bottom and gravity-driven density currents, is used here. Pelagic sedimentation (q.v.), a loosely but widely used term, occurs slowly by settling of particles through the water column. Authigenesis (q.v.)

is the formation of sedimentary deposits either on or beneath the ocean bottom by chemical or biochemical action. The amount and types of sediments which form the deposits of the world's oceans are determined by depth, distance from continental provenance areas, and by bathymetry of deep-sea floor.

Pelagic Sedimentation

Biogenic. A number of planktonic marine animals and plants secrete tests, or skeletons, of calcium carbonate and silica which, when the animals die, sink and form the widespread deep ocean deposits called oozes. Foraminifera, coccolithophorids, and pteropods are the principal components of carbonate oozes; radiolarians and diatoms are the major constituents of the siliceous oozes. These light- (white, cream, and straw) colored deposits are found isolated from large inputs of continentally derived sediments in those portions of the ocean basin that are 2-5 km in depth. They accumulate at rates of 1-5 cm/1000 yr if not subjected to erosion by currents. Siliceous oozes predominate in the areas of high productivity such as the Subarctic Convergence and the Equatorial Divergence.

The relationship between biological productivity and rate of solution of skeletal parts, especially those of calcium carbonate, is a major determining factor in the formation of abyssal sedimentary deposits. Calcite and aragonite are soluble under the low temperatures and high pressures that occur in the deep oceans. High partial pressure of CO_2, which results from oxidation of organic matter in bottom sediments, and rapid removal of the solution further enhance dissolution of calcium carbonate minerals (Arrhenius, 1963). These processes are most effective at depths greater than about 4000 m; hence carbonate content is much reduced in deposits beyond that depth. Soon after death of the animals or plants, some of the siliceous tests are dissolved as the skeletons sink into deep water. Many of the rest are dissolved in the interstitial waters of the sediment after deposition.

Detrital. Where terrigenous components are not masked by biogenic oozes, they form a major part of the deposits called red clays, which are found in depths >5000 m. Red clays are actually brown in color and are mostly composed of particles less than 30μ in diameter. River-borne clays, wind-blown dust, volcanic debris, residue from dissolved plankton tests, and micrometeorites may all make up these deposits, which form at the extremely low rate of 0.1-1 mm/1000 yr, if not subject to redistribution by bottom currents.

Rafted detritus is an important contributor to some deep-sea deposits. Many of the sediments found on the deep-sea floor in high latitudes were transported to depositional sites by floating ice. Transportation by kelp and in the guts of animals may be important in the formation of lower latitude deep-sea deposits.

Closer to the continents, on the continental rise in the uppermost portion of the abyssal zone, green to black terrigenous muds are found. These are composed of sands, silts, and clays which have been swept off the continental shelves or reworked from the continental slopes. Organic material may make up 10-15% of these deposits, which form rather quickly compared with other abyssal facies (see *Euxinic Facies*).

Authigenic. Closely associated with pelagic sedimentation and included within it by many is authigenic mineral formation. Manganese nodules and zeolites are the two most common examples. About 10% of the Pacific sea floor is covered by manganese nodules which accrete through incompletely understood chemical actions in onion-like layers at rates of about $1 \text{ cm}/10^5-10^6$ yr.

Currents

Turbidity Currents. Turbidity currents, thought to be initiated mostly in submarine canyons by catastrophic failure, are gravity-driven density flows, the submarine equivalent of highly fluid landslides. Failure may be caused by instability of the slope, or by earthquakes. Once initiated, a turbid turbulent mass is formed, flows downslope, is channelized at first in the canyon and on the submarine fan, and then spreads out on the deep-sea floor. Although turbidity currents have not been directly observed in the ocean, flows have been deduced to travel thousands of miles at high velocities (Heezen and Ewing, 1952). Turbidity-current deposits, or turbidites (q.v.), are major components of the abyssal plains of the world's oceans. Other deep-sea deposits dominate only where isolated by bathymetry from turbidity currents or by distance from turbidity current source areas. Turbidites are distinguished by presence of coarse grain sizes, graded bedding (q.v.), sedimentary structures, and a sequence of intervals within the deposit (Bouma, 1962).

Near-Bottom Currents. Near-bottom currents are agents of reworking and deposition in the abyssal zone. Some of the material which they redistribute is fine-grained sediment suspended by passage of turbidity currents. Some near-bottom currents which, because of the Coriolis effect, follow the contours of the deep ocean basins form deposits called contourites (q.v.). Found along the edges of deep ocean basins, contourites are deposited by relatively slow-moving currents with much lower sediment concentrations than turbidity currents. Contourites may be distinguished from turbidites by texture and by their distinctive suite of sedimentary structures (Hollister and Heezen, 1972).

Methods of Study

Piston cores have long been used to study sedimentary processes on the deep-sea floors. Up to 6 m long, these cores give direct evidence of diagenetic changes and authigenesis within the upper portion of the sedimentary column. Many inferences about other parameters of sedimentation may also be made from them. With the advent of deep-sea drilling carried out by the *Glomar Challenger,* sedimentologists have been able to obtain much longer cores, which allow study of the entire sedimentary column of the ocean basins. Hence, deductions about abyssal sedimentation may be carried back in geologic time, in some cases as far as the Mesozoic Era. In addition to cores, continuous-reflection seismic profiling and refraction studies have added to the knowledge of abyssal sedimentation. Within the past few years, direct methods of studying abyssal sedimentary processes as they occur have been developed. Submarine photography, direct observation from submersibles capable of penetrating the deepest parts of the ocean basins, and reliable current meters have all been important. Large-volume in-situ water samplers, continuous pumping and centrifugation, and instruments which indicate the quantity of particulate matter present in the water column by scattering of light have proved valuable in collection of data about sedimentary processes as they occur.

LARRY J. DOYLE

References

Arrhenius, G., 1963. Pelagic sediments, in M. N. Hill, ed., *The Sea,* vol. 3. New York: Wiley, 655-727.
Bouma, A. H., 1962. *Sedimentology of Some Flysch Deposits—a Graphic Approach to Facies Interpretation.* Amsterdam: Elsevier, 168p.
Cleary, W. J., and Conolly, J. R., 1974. Petrology and origin of deep-sea sands: Hatteras Abyssal Plain, *Marine Geol.,* 17, 263-279.
Hamilton, E. L., 1973. Marine geology of the Aleutian - Abyssal Plain, *Marine Geol.,* 14, 295-325.
Heezen, B. C., and Ewing, M., 1952. Turbidity currents and submarine slumps, and the 1929 Grand Banks earthquake, *Am. J. Sci.,* 250, 849-873.
Hollister, C. D., and Heezen, B. C., 1972. Geologic effects of ocean bottom currents: Western North Atlantic, in A. L. Gordon, ed., *Studies in Physical Oceanography.* New York: Gordon and Breach, 37-66.
Hollister, C. D.; Johnson, D. A.; and Lonsdale, P. F., 1974. Current controlled abyssal sedimentation, *J. Geol.,* 82, 275-300.
Lonsdale, P., and Malfait, B., 1974. Abyssal dunes of foraminiferal sand on the Carnegie Ridge, *Geol. Soc. Am. Bull.,* 85, 1697-1712.
Rupke, N. A., 1975. Deposition of fine-grained sediments in the abyssal environment of the Algero-Balearic Basin, western Mediterranean Sea, *Sedimentology,* 22, 95-110.

Cross-references: *Abyssal Sedimentary Environments; Authigenesis; Contourites; Pelagic Sedimentation; Turbidity-Current Sedimentation.*

ACCRETION, ACCRETION TOPOGRAPHY

Accretion is a process by which inorganic bodies grow larger by the addition of fresh particles to their outside surfaces, as do concretions (q.v.), ooids, and pisoliths (see *Oolite* and *Pisolite*). This term may also be applied to a topographic feature built by the accumulation of sediment, as in the case of a meandering stream, the process sometimes being called *lateral accretion.* Such sediment is usually carried by the stream as bed load and is deposited on the inner side of a meander. Related deposits which accumulate from suspended load materials have been termed *vertical accretion deposits* (Happ et al., 1940).

The term *accretion topography* was apparently first used in a legal study and statement by Sellards et al. (1923) concerning the disposition of accreted lands along the Red River (Oklahoma-Texas).

Accretion may also be applied to the extension of a beach; Fisk (1959) employed the term *accretion ridge* to describe an ancient beach ridge inland of a modern beach deposit, showing that the coastline had grown seaward.

B. CHARLOTTE SCHREIBER

References

Fisk, H. N., 1959. Padre Island and the Laguna Madre Flats, coastal south Texas, *2nd Coastal Geogr. Conf., Baton Rouge, La., Proc.,* 103-151.
Happ, S. C.; Rittenhouse, G.; and Dobson, G. C., 1940. Some principles of accelerated stream and valley sedimentation, *U.S. Dept. Agr. Tech. Bull. 695,* 22-31.
Sellards, E. H.; Tharp, B. C.; and Hill, R. T., 1923. Investigations on the Red River made in connection with the Oklahoma-Texas boundary suit, *Univ. Texas Bull.,* 2327, 9-174.
Thornbury, W. D., 1954. *Principles of Geomorphology.* New York: Wiley, 618p.

Cross-references: *Aggradation; Concretions; Lateral and Vertical Accretion; Oolite; Pisolite.*

AEOLIAN—See EOLIAN

AGGLOMERATE, AGGLUTINATE

Lyell first employed the term *agglomerate* to designate accumulations of angular fragments of rock which are thrown up by volcanic eruptions and are showered around the center of eruption. If carried farther by running water, thus often becoming worn and rounded, the fragments may be incorporated in a conglomerate (q.v.).

The term *agglomerate* is also applied to a contemporaneous, loose, pyroclastic accumulation which contains a predominance of angular or subangular fragments >32 mm in diameter. The term is more often used to describe a consolidated deposit composed primarily of bombs and fragments in a tuff (q.v.) matrix. The fragments may include not only volcanic material but also blocks of country rock from the walls of the pipe or magma chamber.

Agglutinate is a lithologic term applied to a pyroclastic accumulation made up of originally plastic, volcanic ejectamenta, primarily of volcanic bombs and driblets, which are cemented only by the thin glassy skin of the fragments at their point of contact. It may be distinguished from an agglomerate by the presence of this glassy cement, by the occurrence of fragments of spalled-off scoria in the interstices between the bombs, and by the general absence of a tuff matrix.

Agglutinated refers to a kind of foraminiferal test; *agglutinating value* is a coal production term; *agglutination* is a synonym of *cementation* (q.v.).

B. CHARLOTTE SCHREIBER

References

Blyth, F. G. H., 1940. The nomenclature of pyroclastic deposits, *Bull. Vulcan.*, ser. 2, **6**, 145-156.
Carozzi, A. V., 1960. *Microscopic Sedimentary Petrography.* New York: Wiley, 485p.
Fisher, R. V., 1958. Definition of volcanic breccia, *Bull. Geol. Soc. Am.*, **69**, 1071-1073.
Lyell, C., 1832. *Principles of Geology*, vol. 2. London: John Murray, 330p.

Cross-references: *Pyroclastic Sediment; Tuff; Volcaniclastic Sediments and Rocks.*

AGGRADATION

Aggradation, a term first used by Salisbury (1893) and brought into general use by Davis (1909), usually refers to the fluvial build-up of a land surface (see *Accretion*) on which the gradient of a stream has remained constant for an appreciable period. This accumulation represents deposition in the stream channel of sediment which is in excess of the stream's carrying power; the grade of the stream is thereby maintained.

A valley is said to be *aggraded* when the deposit of alluvium over the valley floor becomes thicker than the depth of the stream channel; this "valley fill" may eventually become hundreds of meters thick. Such valleys tend to become broad and rather flat and are called *aggraded valley plains.* The channels of aggraded streams often become clogged, and they are forced to change their courses or become braided streams (see *Braided-Stream Deposits*).

Aggradation is caused by any increase in sediment load or by a decrease in stream flow, either of which results in so-called "overloading." This condition may be found in streams affected by such processes as stream capture; climatic desiccation; loss of vegetation; increase in underground water losses (seepage); and, in coastal areas, submergence.

B. CHARLOTTE SCHREIBER

References

Cotton, C. A., 1942. *Geomorphology.* Christchurch: Whitcombe & Tombs, 505p.
Davis, W. M., 1909. *Geographical Essays.* D. W. Johnson, ed. Boston: Ginn, 777p.
Emmett, W. W., 1974. Channel aggradation in western United States as indicated by observations at Vigil Network sites, *Z. Geomorph. Supplbd.*, **21**, 52-62.
Salisbury, R. D., 1893. Surface geology, *Geol. Survey N. Jersey Rept., 1892-1893*, 35-246.
Smith, D. G., 1974. Aggradation of the Alexandra-North Saskatchewan River, Banff Park, Alberta, in M. Morisawa, ed., *Fluvial Geomorpholoy.* Binghamton: State Univ. of New York, 201-219.

Cross-references: *Accretion; Alluvium; Fluvial Sedimentation.* Vol. III: *Terraces, Fluvial.*

ALGAL-REEF SEDIMENTOLOGY

Algal reefs are defined as rigid, potentially wave-resistant, in-situ topographic structures formed primarily by algae which secrete calcareous skeletons. They are distinguished from two kinds of nonskeletal algal deposits sometimes also called reefs: (1) *stromatolites* (q.v.), structures of sediment trapped by filamentous algae and subsequently cemented, and (2) *algal tufa* deposits of calcium carbonate precipitated in a meshwork of filamentous algae largely by inorganic processes. Further, in their in-situ position, algal reefs differ from mobile nodular structures composed of algal skeletons, which are called rhodolites, rhodoids, or rhodoliths (Adey and Macintyre, 1973), and from accumulations of these nodules, known as algal banks.

Morphology

Algal reefs assume three shapes.

Ridges. Algal ridges are narrow, elongate structures of m to several tens of m width and several hundred m to several km length found on the seaward rim of atolls and platforms. The designation "Lithothamnion ridge" has been abandoned because of its misleading taxonomic implication (Adey and Macintyre, 1973). A system of growth spurs and intermittent grooves may extend seaward from the ridge (Tracey et al., 1948[1]). Micromorphological features include channels, blow-holes, and terrace systems (Kuenen, 1950[2]) as well as open and roofed depressions. Examples include those of the Pacific atolls such as Bikini (Tracey et al., 1948[1]) and Raroia (Newell, 1956[2]); more recently such ridges have been reported from the Atlantic Ocean and the Caribbean Sea, e.g., off Brazil (Laborel, 1969), Panama (Glynn, 1971[2]), the Virgin Islands (Adey and Macintyre, 1973), and Colombia (Geister, pers. comm., 1973).

Vase or Cup Shape. These structures, referred to as microatolls, boilers, breakers, bosses, or cup reefs, are subcylindrical aggregates of up to 30 m in diameter and attain heights of 5-12 m. Micromorphologically they are characterized by raised rims, central depressions, overhanging lateral faces, and internal caves and tunnels of up to meter dimensions. Examples have been observed on Bikini atoll (Tracey et al., 1948[1]), off Yucatan (Boyd et al., 1963[1]), and on the Bermuda Platform (Agassiz, 1895[1], and many others; Ginsburg and Schroeder, 1973). Tracey et al. (1948[1]) reported from Bikini atoll that these structures may coalesce to form large roof-and-pillar structures.

Visor or Lip Shape. Known as corniche, trottoir, algal bench, or bioconstructional lip, these elongate structures extend tens to hundreds of m and protrude horizontally for decimeters or meters from cliffs, benches, or terraces. The outer rim may be elevated to form a depression behind, which may be subdivided into bowl- or dish-shaped pools by means of small ridges. Examples found include those from the Mediterranean and the Marianas (Termier and Termier, 1963), and from the Bermuda Islands (Oertel, 1970).

[1] Reference in Ginsburg and Schroeder, 1973.
[2] Reference in Adey and Macintyre, 1973.

Distribution

Many algal reefs reported have not been studied in section, and it is not known if the algae really built the structure by themselves or just veneered preexisting substrate. Algal reefs as defined above have been reported from Pacific and Atlantic tropical shallow waters as well as from the Mediterranean. Numerous reports indicate that coralline algae occur at greater depths and up to high latitudes (Adey and Macintyre, 1973), but there is no evidence of living in-situ topographic structures in such locations. Ridges similar to algal reefs have been found at various shelf edges in depths of 40-180 m and interpreted to be relics of shallow-water algal ridges of lower sea-level stands (Milliman, 1974).

Algae Involved in Forming Algal Reefs

Taxonomy and Ecology. The reefs are constructed by crustose and ramose algae of the phylum *Rhodophyta*, more specifically of the family *Corallinaceae*. The genera common in tropical waters include *Prolithon, Lithophyllum,* and *Neogoniolithon;* the latter two are the main components in the Mediterranean. Contrary to widespread reports *Lithothamnium* is not a major contributor (Adey and Macintyre, 1973).

The reefs usually extend from the subtidal into the intertidal zone, thus they receive a maximum of light and turbulence, which satisfies the two basic requirements of calcareous algae (Johnson, 1961). The tops of the reefs are awash, if not dry (Glynn, 1971[2]), for a portion of the tidal cycle. The algae either use preexisting hard substrate or produce it by firmly binding loose material present.

Composition. The skeletons of the calcareous algae involved in the formation of algal reefs are composed of high magnesium calcite containing up to 33, commonly 10-20, mole% $MgCO_3$. The magnesium content is positively correlated to mean water temperature and may reflect seasonal temperature variations. Additional magnesium may be present in another phase, possibly brucite (see references in Milliman, 1974). Analyses on trace element contents are presented by Milliman (1974).

Growth Rates. Radiocarbon dates from Bermuda cup reefs (Ginsburg and Schroeder, 1973, and unpub.) provide a basis for computing the rate of accretion (= thickening) of algal crusts. The rate ranges from 1.1-0.05 mm/yr. These figures are averages over at least 700 years; actual rates over short terms may be higher or lower. Observed local and temporal variations of growth rates reflect variations of environmental factors, presumably mainly of turbulence and light intensity.

Sedimentological Processes in Algal Reefs

Sedimentological processes are studied best where the interior of the reef is accessible; this, however, is rarely the case. The following summary is largely based on the report on Bermuda cup reefs (Ginsburg and Schroeder, 1973) where artificial outcrops and large samples were available; this report was confirmed by comparative studies of cores from the algal reef of Eniwetok, which were first described by Ladd and Schlanger (1960[1]).

Construction. By definition the framework of algal reefs is constructed primarily of skeletons of coralline algae (Fig. 1). Secondarily, several encrusting or attached organisms participate in frame building, among them the hydrozoan *Millepora*, the vermetid gastropods *Vermetus* and *Dendropora,* the foraminifer *Homotrema rubrum*, the barnacle *Cataphragmus imbricatus,* and ahermatypic corals such as *Astrangia solitaria.* Hermatypic corals are lacking or a minor component; inasmuch as they occur, crustose and platy growth forms predominate.

The resulting framework has a high porosity; pores vary widely in origin, shape, and size; particularly characteristic are sheet pores between successive algal crusts. The pores in turn may provide substrate for further attached organisms, which then contribute to the frame, or shelter for vagile organisms, for example crustaceans.

Sedimentation. Alternatively the pores may serve as receptacles for internal sediment. Sedimentary particles are derived primarily from the skeletons of reef-dwelling organisms, invertebrates as well as green and red articulate calcareous algae. Other sources include the reef structure, from which portions are broken off by wave action, and the seawater surging through the maze of internal cavities, from which skeletons of pelagic organisms, such as coccoliths and diatoms, are strained.

The sediment is characterized by its variety of components and grain sizes. Horizontal layering and geopetal structures are common, and bioturbation may occur.

Cementation. Precipitation of calcium carbonate in various intra- and interskeletal and particle pores is another void-filling, thus porosity-reducing, process. The resulting submarine cements are composed of high magnesium calcite and aragonite; micrite and needles, respectively, are their most common habits, but several other habits have been recognized (Schroeder, 1972[1]). Cementation is significantly controlled by the microenvironment; hence the variety in nature and size of skeletal and particle pores accounts for the considerable heterogeneity in cements observed. Submarine cementation (q.v.) begins within mm below the growing surface; within cm or decimeters the algal reef rock may be fully lithified, ringing under the hammer, and breaking across particles.

Destruction. The above constructive processes are countered by the destruction by organisms, also known as bioerosion. Particularly, boring organisms such as mollusks (e.g., *Lithophaga*), sponges (e.g., *Cliona*), and endolithic green or blue-green algae (e.g., *Ostreobium*) are continuously removing portions of the reef frame, thus weakening the structure and at the same time providing new pores for organisms to live in, for sedimentation, or for cementation. Other organisms such as echinoids, crustaceans, fish, and turtles destroy the frame by rasping the surface, leaving grooves on it and converting part of the structure into sediment.

Spatial Relationships and Sequences of Processes. Hardly any portion of the reef is formed by one process only and never affected again; almost any thin section of algal reef rock will reveal the intimate spatial and temporal interrelationship between various reef-forming processes (Fig. 2). While an organism lives in one pore, sediment is deposited in the next and cement precipitated in the third pore; while sediment is deposited on the bottom, organisms may encrust the top of a pore. Inasmuch as there is space left, any process may follow earlier ones, and where no space is left it may be provided by borers, which again and again may continue sequences seemingly terminated earlier. Considerable diversity in reef formation arises from the fact that many organisms can substitute for each other, either in construction

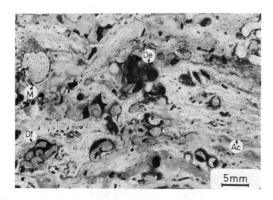

FIGURE 1. Fabric of recent algal reef rock from Bermuda cup reefs in section. Crinkled laminae are sections of algal crusts (Ac). Dark shells are *Dendropora irregulare,* a vermetid gastropod; these may be empty (De) or filled with sediment (Df). Boring mollusks (M) frequently destroy the original fabric; their holes are filled with sediment, which in turn is cemented.

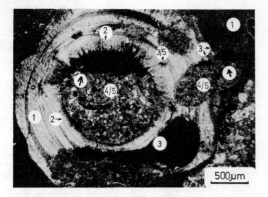

FIGURE 2. Sequence of sedimentological processes in recent algal reef rock from Bermuda cup reef (thin section; polarized light). (1) Construction of framework by coralline red alga and vermetid gastropod (2) Cementation by aragonite needle cement (3) Destruction by boring sponge (4) Sedimentation (note geopetal structures indicated by arrows) (5) Cementation by Mg-calcite micrite.

or destruction. Additional complexity results from the potential of some organisms to assume roles in various processes simultaneously, for example in boring and cementation. The final product of all these processes, the reef rock, reflects the dynamic interplay of construction, sedimentation, cementation, and destruction in its fabric (Schroeder and Zankl, 1974).

Fossil Algal Reefs

Detailed sedimentological studies of fossil algal reefs are virtually lacking. There are many reports of algae in reefs throughout the stratigraphic column, but the roles of various algae in fossil reefs are not well understood (Wray, 1977). Frame-building algae include blue-green and green algae such as *Renalcis* and *Sphaerocodium* in the Devonian, ancestral coralline algae such as *Archeolithophyllum* in the Pennsylvania and Permian, and crustose coralline red algae since Cretaceous (Wray, 1977). Wray's (1977) book provides more information and discussion on fossil algae in reefs.

JOHANNES H. SCHROEDER

References

Adey, W. H., and Macintyre, I. G., 1973. Crustose coralline algae: a re-evaluation in the geological sciences, *Geol. Soc. Am. Bull.*, 84, 883-904.

Ginsburg, R. N., and Schroeder, J. H., 1973. Growth and submarine fossilization of algal cup reefs, Bermuda, *Sedimentology*, 20, 575-614.

Johnson, J. H., 1961. *Limestone-building Algae and Algal Limestones*. Golden, Colorado: Colorado School of Mines, 297p.

Laborel, J., 1969. *Les Peuplements des madréporaires des Çotes Tropicales du Brésil*. Ann. Univ. Abidjan, ser. E, II, Fasc. 3, 261p.

Milliman, J. D., 1974. *Marine Carbonates*. Berlin: Springer, 375p.

Oertel, G. F., 1970. Preliminary investigation of intertidal bioconstructional features along the south shore of Bermuda, in R. N. Ginsburg and S. M. Stanley, eds., *Reports of Research 1969 Seminar on Organism-Sediment Interrelationships*. Bermuda Biol. Sta. Spec. Publ. 6, 99-107.

Schroeder, J. H., and Zankl, H., 1974. Dynamic reef formation: a sedimentological concept based on studies of recent Bermuda and Bahama reefs, in A. M. Cameron et al., eds., *Proc. 2nd Internat. Coral Reef Symposium*, vol. 2, Brisbane, Australia: Great Barrier Reef Comm., 413-428.

Termier, H., and Termier, G., 1963. *Erosion and Sedimentation*. London: Nostrand, 433p.

Wray, J. L., 1977. *Calcareous Algae*. New York: Elsevier, 185p.

Cross-references: *Bahama Banks Sedimentology; Coral-Reef Sedimentology; Limestones; Reef Complex; Stromatolites; Tropical Lagoonal Sedimentation.* Vol. III: *Algal Reefs; Atolls.* Vol. VII: *Algae; Stromatolites.*

ALLING SCALE

This scale was devised by Alling (1943) as an essentially two-dimensional approach for measuring mineral grain sizes in the study of

TABLE 1. The Size Scale Proposed by Alling[a]

Major Divisions	Minor Divisions	Size Range (mm)		Aggregate Term
Boulder	Coarse Medium Fine Very fine	560 320 180 100	−1,000 − 560 − 320 − 180	Boulderstone
Cobbles	Coarse Medium Fine Very fine	56 32 18 10	− 100 − 56 − 32 − 18	Cobblestone
Gravel	Coarse Medium Fine Very fine	5.6 3.2 1.8 1.0	− 10.0 − 5.6 − 3.2 − 1.8	Gravelstone
Sand	Coarse Medium Fine Very fine	0.56 0.32 0.18 0.10	− 1.0 − 0.56 − 0.32 − 0.18	Sandstone
Silt	Coarse Medium Fine Very fine	0.056 0.032 0.018 0.010	− 0.10 − 0.056 − 0.032 − 0.018	Siltstone
Clay	Coarse Medium Fine Very fine	0.0056 0.0032 0.0018 0.001	− 0.01 − 0.0056 − 0.0032 − 0.0018	Claystone
Colloid	Coarse Medium Fine Very fine	0.00056− 0.00032− 0.00018− 0.0001 −	0.001 0.00056 0.00032 0.00018	Colloidstone

[a]Alling, 1943.

thin sections and polished surfaces. The subdivisions of size were based on intervals of the fourth root of 10, in the metric scale, along the lines first proposed by Hopkins (1899). These sizes and their associated "aggregate terms" are shown in Table 1.

B. CHARLOTTE SCHREIBER

References

Alling, H. I., 1943. A metric grade scale for sedimentary rocks, *J. Geol.,* **51,** 259-269.
Hopkins, C. G., 1899. A plea for a scientific basis for the division of soil particles in mechanical analysis, *U.S. Dept. Chem. Bull.,* **56,** 64-66.

Cross-reference: *Grain-Size Studies.*

ALLOCHEM

The term *allochem,* or *allochemical constituent,* was proposed by Folk (1959) to describe "all materials that have formed by chemical or biochemical precipitation *within* the basin of deposition, but which are organized into discrete aggregated bodies and for the most part have suffered some transportation ('allo' is from the Greek meaning 'out of the ordinary,' in the sense that these are not just simple, unmodified chemical precipitates, but have a higher order or organization). Allochems are by far the dominant constituent of limestones, and only four types of allochems are of importance: intraclasts, oolites, fossils, and pellets."

Folk further indicated the interrelationship of the major components of sedimentary rocks in graphic form (Fig. 1). He divided allochemical rocks on the basis of their matrix materials (Fig. 2 on page 10).

Other considerations that enter into the nomenclature of allochems include (1) dominant fossil type in the biosparite and biomicrite rocks, and (2) the grain size of the allochem component. Dominant fossil type is specified, e.g., "brachiopod biosparite." The grain size is divided into three groups: *calcirudite* (>1 mm), *calcarenite* (0.062-1 mm), and *calcilutite* (<0.062 mm), of which the last seems to be the least common.

Under Folk's classification it is assumed that allochems, except for certain sedimentary fossils, are transported constituents. Although this is true for the majority of carbonate rocks, some limestones may contain pseudo-allochems,

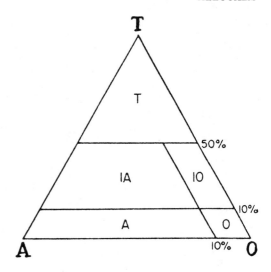

FIGURE 1. Main divisions of sedimentary rocks, based on relative proportions of terrigenous (T), allochemical (A), and orthochemical (O) constituents (from Folk, 1959). The five divisions correspond to: T, terrigenous rocks (sandstones, mudrocks, etc.); IA, impure allochemical rocks (sandy oolitic limestones, silty pellet limestones, etc.); IO, impure orthochemical rocks (clayey microcrystalline limestones, silty primary dolomites); A, allochemical rocks (intraclastic, oolitic, biogenic, pelodial limestones, etc); O, orthochemical rocks (microcrystalline limestones, primary dolomite, halite, chert, etc.).

which are objects that simulate the appearance of allochems but which have formed in place by processes of diagenesis. Furthermore, there are some instances in which it is logical to infer that oolites have grown in situ while remaining stationary, completely embedded in carbonate mud. Other intraclast-like objects are produced when homogeneous rock recrystallizes in patches to sparry calcite, and the isolated, unaltered remnants mimic intraclasts.

It may be important to recognize and describe such unusual features, but it does not seem necessary to expand and complicate classification by the erection of new pigeonholes for each of these lithologic oddities.

B. CHARLOTTE SCHREIBER

References

Folk, R. L., 1959. Classification of limestones, *Bull. Am. Assoc. Petrol. Geologists,* **43,** 1-38.
Folk, R. L., 1968. *Petrology of Sedimentary Rocks.* Austin: Univ. of Texas, Hemphill's, 170p.

Cross-references: *Calcarenite, Calcilutite; Limestone Fabrics; Limestones.*

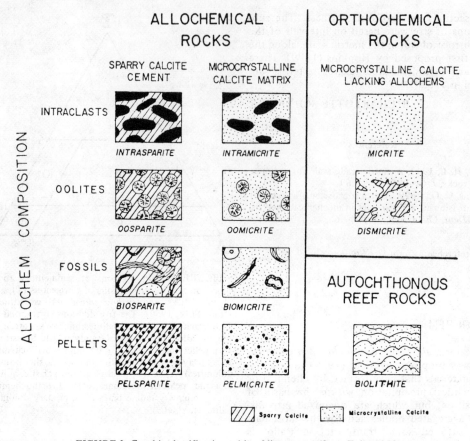

FIGURE 2. Graphic classification table of limestones (from Folk, 1959).

ALLUVIAL-FAN SEDIMENTS

An alluvial fan is a low, cone-shaped, gently sloping (usually <5°) mass of loose sediment deposited by a stream where it issues from a narrow mountain valley onto a broad plain or valley, i.e., where the constriction of a stream abruptly ceases; change in stream gradient is rarely significant. Alluvial fans are associated with regions of rapid uplift, especially vertical fault movement, and are most common in arid to semiarid climates (see *Desert Sedimentary Environments*), although they do occur in humid regions (see Bull, 1972).

An alluvial fan is steepest near the mouth of the valley, sloping gently and concavely outward. Modern fans range in area from 1 to 900 km^2; radii of several hundred meters to a few kilometers are common, ranging up to over 50 km (Spearing, 1974). The thicknesses of modern fans are usually <1 km, but fans in the geologic record may be as thick as 10 km, as a result of progressive subsidence and infilling. Fans composed primarily of mud are generally larger and thicker than fans composed of sand and gravel; fans in humid climates are usually flatter (Spearing, 1974).

Deposits

Alluvial-fan sediments are generally coarse, ranging from boulder to clay size. Sorting and roundness are variable, and grain size generally decreases away from the fan apex. Armored mud balls—spherical masses of clayey pebbles studded with pebbles—are common in some fan deposits (Bell, 1940). The two depositional processes most active on alluvial fans are sheetfloods and debris flows.

Debris-Flow Deposits. A debris flow is a high-density, viscous flow, usually consisting of coarse material in a muddy matrix. Debris flows usually occur when there is intense intermittent rainfall; lack of vegetation and steep slopes encourage these flows. Debris-flow deposits are poorly sorted, often containing boulders weighing >1 ton, are sheet-like with lobate extensions, and have abrupt contacts (see *Mudflow Deposits*).

Sheetflood Deposits. Sheets of sand or gravel are deposited on a fan by a network of rapidly shifting, sediment-charged braided streams (see *Braided Stream Deposits*). The deposits are usually well sorted, contain little clay, and may be cross-bedded, plane-bedded, or massive.

Deposits that backfill channels temporarily entrenched into a fan (*stream-channel deposits, cut-and-fill, channel fill;* Bull, 1972) are generally coarser grained and less well sorted than sheet-flood deposits. In addition, bedding may not be as well defined. If the source area supplies little sand or finer-sized debris, water from a flood discharge may quickly soak into the coarse debris, promoting the deposition of lobes of gravel, or *sieve deposits* (Hooke, 1967). Sieve deposits are composed primarily of angular blocks and are well sorted; they are much less common than other alluvial-fan deposits but are among the most distinctive (Bull, 1972).

Diagnostic Criteria. Bull (1972) listed ten diagnostic features that distinguish alluvial fans as products of a distinctive terrestrial depositional environment: (1) oxidation; organic material rarely preserved; (2) thick sequences of mudflow and braided-stream deposits; (3) numerous sheets with length:width ratios of roughly 5 to 20; (4) decrease in proportion of debris-flow deposits downfan; (5) decrease in particle size downfan, but fanhead trenches may deposit coarse debris on the middle or lower parts of the fan; (6) cut-and-fill structures near the fan apex; (7) wide variations in grain size, sorting, and thickness among beds; (8) distinctive patterns in logarithmic plots of the coarsest 1-percentile particle phase and the median particle size (Bull, 1962); (9) intertonguing relations with alluvial or lacustrine deposits; and (10) depositional structures reflecting a radial flow direction from the apex.

Ancient Deposits. Ancient fan deposits have been recognized from many geologic periods (Bull, 1972). Lithified deposits from the upper part of the alluvial fan were termed *fanglomerates* by Lawson (1913). Examples include the Late Precambrian Keweenawan red beds of the Canadian Shield and the Torridonian fan deposits of Scotland (see Bull, 1972); the Devonian fanglomerates of Norway (Nilsen, 1969); the Pennsylvanian Fountain Formation of Colorado (Howard, 1966); the Permo-Triassic New Red Sandstone of Scotland (Steel, 1974); the Triassic Newark Group of the Appalachians (Krynine, 1950; Klein, 1962), and Triassic conglomerates in Wales (Bluck, 1965); and the Neogene intermontane basin deposits of southern California (Crowell, 1954; 1974).

JOANNE BOURGEOIS

References

Bell, H. S., 1940. Armored mud balls–their origin, properties and role in sedimentation, *J. Geol.,* 48, 1-31.

Bluck, B. J., 1965. The sedimentary history of some Triassic conglomerates in the Vale of Glamorgan, South Wales, *Sedimentology,* 4, 225-245.

Bull, W. B., 1962. Relation of textural (CM) patterns to depositional environment of alluvial-fan deposits, *J. Sed. Petrology,* 32, 211-216.

Bull, W. B., 1972. Recognition of alluvial-fan deposits in the stratigraphic record, *Soc. Econ. Paleont. Mineral. Spec. Publ. 19,* 290-303.

Crowell, J. C., 1954. Geology of the Ridge Basin Area, *Bull. Calif. Div. Mines,* 170, map sheet 7.

Crowell, J. C., 1974. Sedimentation along the San Andreas Fault, California, *Soc. Econ. Paleont. Mineral. Spec. Publ. 19,* 292-303.

Denny, C. S., 1967. Fans and pediments, *Am. J. Sci.,* 265, 81-105.

Hooke, R. LeB., 1967. Processes on arid-region alluvial fans, *J. Geol.,* 75, 438-460.

Howard, J. D., 1966. Patterns of sediment dispersal in the Fountain Formation of Colorado, *Mtn. Geol.,* 3(4), 147-153.

Klein, G. deV., 1962. Triassic sedimentation, Maritime Provinces, Canada, *Geol. Soc. Am. Bull.,* 73, 1127-1146.

Krynine, P. D., 1950. Petrology, stratigraphy, and origin of Triassic sedimentary rocks of Connecticut, *Bull. Conn. Geol. Nat. Hist. Surv.;* 73, 247p.

Lawson, A. C., 1913. The petrographic designation of alluvial fan formations, *Bull., Univ. Calif. Dept. Geol.,* 7, 328-334.

Nilsen, T. H., 1969. Old Red sedimentation in the Buelandet-Vaerlandet Devonian district, western Norway, *Sed. Geol.,* 3, 35-57.

Spearing, D. R., 1974. Alluvial fan deposits, *Summary Sheets of Sedimentary Deposits,* Sheet 1, Boulder, Colo.: Geol. Soc. America.

Steel, R. J., 1974. New Red Sandstone floodplain and piedmont sedimentation in the Hebridean Province, Scotland, *J. Sed. Petrology,* 44, 336-357.

Cross-references: *Arkose; Braided-Stream Deposits; Continental Sedimentation; Desert Sedimentary Environments; Flood Deposits; Mudflow, Debris-Flow Deposits; Sedimentation-Tectonic Controls; Submarine Fan (Cone) Sedimentation.*

ALLUVIUM

A natural, mature river channel wanders or meanders within a broad valley of its own making. As it does so, the river deposits sediments both within the migrating channel and over the valley flats beyond the banks of the channel itself. These sedimentary deposits laid down by the river constitute the alluvium. By definition, *alluvium* is a subaerial deposit of a river.

The process of river migration and the associated depositional sequence can be seen in an idealized plan view and cross section of an alluvial (river) valley (Fig. 1). At any bend, erosion is concentrated on the outside or concave bank, and deposition occurs along the convex bank. Rates of both processes tend to be episodic and commonly out of phase. The convex depositional feature is called a *point bar;* because of episodic accumulation, point bars develop as discrete ridges (accretion ridges). As shown by the arrows in Plan A of Fig. 1, the locus of erosion causes the river to migrate both downstream and laterally across its valley, adding to the point bar deposition as it does so. In addition, inasmuch as the river channel is rarely large enough to contain flood flows, sediment-laden waters overtopping the channel banks deposit sediment both on the point bars and on the adjacent flood-plain surface.

A classic alluvial landscape as found in a major river valley includes several distinctive depositional environments (Fig. 1, Profile B): channel and point-bar deposits dominated by relatively uniform or better sorted sands, natural levees of coarser sand deposited at the channel margins when channel banks are just overtopped, backswamp sediments of finer, poorly sorted silts and clays deposited on the flood-plain surface between valley walls and natural levees, and clays and silts deposited in slack water in the upstream portions of abandoned cut-off meander channels (Fisk, 1947). The spatial distribution of these local subfacies or types of deposits may be highly variable, and repeated movement of the river channel creates complex stratigraphic relationships when seen in cross-section. Similarly, at a particular point, during the waning stage of a flood, sediments deposited from the flow become progressively finer. Thus the physiography of the valley bottom and the associated deposits are subject both to the regime of flow of the river and to the characteristics of the available sediment. As combinations of these are highly variable in nature, departures from the classic conditions are common, and a wide range of alluvial sequences is found in nature (Leopold et al., 1964).

Because larger particles are usually transported nearer the channel bed, a vertical section of alluvium from river bed to flood-plain surface will often be characterized by a decrease in particle size from coarser, cross-bedded sands at the base to finer, laminated sands and silts near the surface. This simple sequence may be interrupted by aberrant strata or lenses of coarser particles deposited by rapid flow during floods or by flow in a rising or aggrading channel (Jahns, 1947). Remnants of such channels and channel fills of coarse sediments may be distinctive in exposed sections of alluvium. The geologic record in many places displays alluvial deposits hundreds and even thousands of m thick, which have resulted from aggradation or building up of river deposits. Such a process is clearly in evidence in Holocene and Quaternary deposits along the lower courses of the Yellow River in China, and along the Ganges-Brahmaputra and Indus systems of the Indian subcontinent.

From the standpoint of the geologist or stratigrapher, alluvium is characterized by several distinctive sedimentary properties. It is customarily described as a poorly sorted sediment, ranging from clay to boulders in mean diameter. Texture, composition, and particle size are all highly variable. Sorting coefficients for river sands often exceed 1.3 and for river gravel 4.0. Similarly, standard deviations of frequency distributions of fluvial sediments based upon weight % may equal 15–20% of the mean diameter. Not infrequently river sands are log-normally distributed, but size distributions alone are not diagnostic inasmuch as river deposits may be exceedingly uniform where available source materials are well sorted. On the other hand, glacial stream deposits may contain large, ice-rafted blocks. Many alluvial gravels are bimodal, with a paucity of size grades 4–8 mm. Particles transported by rivers customarily

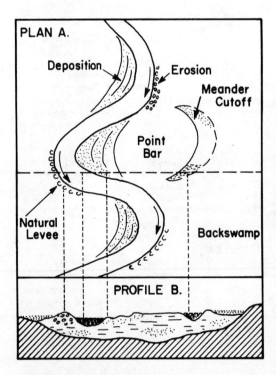

FIGURE 1. Plan of a meandering river and profile showing alluvial sediments.

become rounded and made more spherical in the process; thus alluvium is likely to include rounded particles.

Primary sedimentary structures (q.v.) are common, particularly in sandy alluvium. These may include festoon and planar cross-bedding (q.v.), imbrication (see *Conglomerates*), and large-scale cut-and-fill channels. Festoon forms appear to occur most frequently, and their origin has been attributed to the movement of dunes along the beds of river channels (see *Bed Forms in Alluvial Channels*).

A number of fluvial or river deposits have distinctive physiographic or other connotations. *Placers*, for example, are a distinctive deposit of alluvium containing valuable minerals or rocks. When a river cuts into a prior deposit of alluvium, the abandoned flood plain or erosional surface constitutes a *terrace*. The term terrace is sometimes erroneously used to denote both the physiographic surface and the underlying alluvium. *Deltas* also consist largely of river sediments (see *Deltaic Sediments*). However, although channel sands, levees, and backswamp or fine overbank deposits may be found in deltas, particularly in those built by distinctive distributaries, a delta is primarily a subaqueous and not a subaerial deposit.

From the definition and examples it is evident that alluvium (synonymous with alluvial deposit) is a sedimentary facies. Although in Europe at one time Alluvium connoted the Holocene Epoch as distinct from *Diluvium*, the old term for Pleistocene, such usage is obsolete. Alluvium has no time connotation.

M. GORDON WOLMAN

References

Fisk, H. N., 1947. Fine-grained alluvial deposits and their effects on Mississippi River activity, *U.S. Waterways Exp. Sta.*, 80p.

Jahns, R. H., 1947. Geologic features of the Connecticut valley, Mass., as related to recent floods, *U.S. Geol. Surv. Water Supply Pap. 996*, 158p.

Leopold, L. B.; Wolman, M. G.; and Miller, J. P., 1964. *Fluvial Processes in Geomorphology*. San Francisco: Freeman, 522p.

Schumm, S. A., ed., 1972. *River Morphology*. Stroudsburg. Pa.: Dowden, Hutchinson & Ross, 421p.

Spearing, D. R., 1974. Alluvial valley deposits, *Summary Sheets of Sedimentary Deposits*, Sheet 2. Boulder, Colo.: Geol. Soc. Am.

Cross-references: *Braided-Stream Deposits; Fluvial Sediments; River Gravels; River Sands*. Vol. III: *Alluvial Fan, Cone; Alluvium*. Vol. XI: *Alluvial Soils*. Vol. XIII: *Engineering Geology in Alluvial Plains*.

ANISOTROPY IN SEDIMENTS

The grains of most clastic sediments possess some degree of preferred orientation brought about by the gravitational and hydrodynamic forces acting during deposition. Determination of sedimentary fabric therefore can allow inference about conditions of sedimentation. For example, the measurement of the orientation of elongated grains can be used to estimate the direction of currents during deposition because the long axes of the grains usually tend to align in the direction of flow (see *Fabric, Sedimentary*). Several other methods exist for finding paleocurrent directions, but these usually depend on the presence in the rock of some macroscopic structure such as cross-bedding or sole marks (see *Paleocurrent Analysis*).

Magnetic Susceptibility Anisotropy

Magnetic susceptibility anisotropy analysis provides an indirect but more rapid method of obtaining grain orientation information. Several common magnetic mineral grains, as of magnetite, often are present in small proportions in sedimentary rock and have shape-dependent susceptibilities, i.e., they are easier to magnetize along their long axes than in other directions. An assemblage of such grains having a preferred orientation will give rise to a resultant susceptibility anisotropy that is a measure of the rock's magnetic fabric. In favorable circumstances, the magnetic fabric is closely related to the fabric of the nonmagnetic grains forming the bulk of the sediment.

A number of techniques have been developed for the measurement of susceptibility anisotropy; a description of the widely used torque meter method is given by King and Rees (1962). The usual method of measurement and calculation estimates a susceptibility in the form of a second rank symmetric tensor K which relates the magnetization J_i induced in a specimen by a field H_j: $J_i = K_{ij}H_j$, $K_{ij} = K_{ji}$. Geometrically, this describes the attitude and shape of a triaxial ellipsoid having maximum (K_{max}), intermediate (K_{int}), and minimum (K_{min}) principal susceptibility axes forming an orthogonal system. The magnetic foliation plane contains the K_{max} and K_{int} axes while the magnetic lineation is specified by the K_{max} axis within that plane.

Primary Magnetic Fabric. Susceptibility anisotropy measurements made on a variety of laboratory and recently deposited natural sediments have established the characteristics of a primary magnetic fabric that is attributed to normal depositional processes (see references in

Hamilton and Rees, 1970). The susceptibility ellipsoid is invariably more oblate than prolate, and there is commonly a well-developed magnetic foliation in or near the bedding plane caused by the effect of gravity in bringing the long axes of magnetic grains as near as possible to the horizontal at the time of deposition. A magnetic lineation may be present within the foliation plane due to some alignment of long axes of magnetic grains by water currents, but it may also be due to other causes, e.g., the effect of depositional slope (Rees, 1966). These observations under controlled or known conditions of deposition provide control for the interpretation of naturally occurring magnetic fabrics observed elsewhere.

Application to Paleocurrent Directions. A number of studies have been made to establish the applicability of the susceptibility anisotropy method in natural situations by comparing its results with those obtained by use of standard paleocurrent methods. In many cases good agreement has been found between the orientation of the K_{max} susceptibility axes and current or sediment transport direction deduced from (1) optical determination of grain long axes in Scripps beach sand (Rees, 1965); (2) azimuth of sole markings in Silurian siltstones from Wales (Hamilton, 1963); (3) channel morphology and cross-bedding in a deep-sea fan off California (Rees et al., 1968); and (4) a comprehensive set of paleocurrent determinations over a wide area of exposure of Tertiary continental gravels, sands, and silts in California (Galehouse, 1968). It has occasionally been found, however, that maximum susceptibility and paleocurrent directions are not simply related. For example, an apparently discordant result was obtained from a turbidite horizon within Eocene strata exposed in Wheeler Canyon, California, in which the magnetic lineations were nearly perpendicular to the current direction (Rees, 1965). Optical grain-orientation measurements on thin sections cut parallel to the bedding plane showed two alignments of long axes, one parallel to the sole marks and one at right angles to them. The susceptibility anisotropy method cannot resolve more than a single lineation with the magnetic foliation plane; when two are present, as appears to be the case cited above, only the stronger is distinguished.

Anomalous Magnetic Fabrics

Additional mechanisms have been recognized which may produce discordant magnetic fabrics. In very fine-grained sediments, the magnetic grain orientation may be affected significantly by the geomagnetic field present during deposition (Rees, 1961), although the effect of the magnetic field should decrease in importance with increasing particle size. No evidence of magnetic field control has been observed in the fabric of laboratory-deposited sand. On the other hand, mechanical deformation of sediment or sedimentary rock can modify an original primary fabric to produce a fabric that is often distinctive for the type of deformation. The effect of tectonic deformation on magnetic fabric may be systematic and related closely to other evidence of strain in the rock (e.g., Graham, 1967). Another type of deformation involves the activity of plants and burrowing organisms in the sediment, which will tend to randomize any preferred grain orientation and eventually destroy any recognizable primary fabric (Rees et al., 1968). Visible evidence of rock deformation may be difficult to detect, and criteria have been developed to help recognize the existence of anomalous fabrics (Crimes and Oldershaw, 1967).

DENNIS V. KENT

References

Crimes, T. P., and Oldershaw, M. A., 1967. Paleocurrent determinations by magnetic fabric measurements on the Cambrian rocks of St. Tudeval's Peninsula, North Wales, *Geol. J.,* **5**, 217-232.

Galehouse, J. S., 1968. Anisotropy of magnetic susceptibility as a paleocurrent indicator: A test of the method, *Geol. Soc. Am. Bull.,* **79**, 387-390.

Graham, J. W., 1967. Significance of magnetic anisotropy in Appalachian sedimentary rocks, in J. S. Steinhart, and T. J. Smith, eds., *The Earth Beneath the Continents.* Am. Geophys. Union Monogr. 10, 627-648.

Hamilton, N., 1963. Susceptibility anisotropy measurements on some Silurian siltstones, *Nature,* **197**, 170-171.

Hamilton, N., and Rees, A. I., 1970. The use of magnetic fabric in paleocurrent estimation, in S. K. Runcorn, ed., *Paleogeophysics.* New York: Academic Press, 445-464.

King, R. F., and Rees, A. I., 1962. The measurement of the anisotropy of magnetic susceptibility of rocks by the torque method, *J. Geophys. Research,* **67**, 1565-1572.

Rees, A. I., 1961. The effect of water currents on the magnetic remanence and anisotropy of susceptibility of some sediments, *Geophys. J.,* **5**, 235-251.

Rees, A. I., 1965. The use of anisotropy of magnetic susceptibility in the estimation of sedimentary fabric, *Sedimentology,* **4**, 257-271.

Rees, A. I., 1966. The effect of depositional slopes on the anisotropy of magnetic susceptibility of laboratory deposited sands, *J. Geol.,* **74**, 856-867.

Rees, A. I.; von Rad, U.; and Shepard, F. P., 1968. Magnetic fabric of sediments from the La Jolla submarine canyon and fan, California, *Marine Geol.,* **6**, 145-178.

Cross-references: *Fabric, Sedimentary; Paleocurrent Analysis; Sediment Parameters; Sedimentary Structures.*

ARENITE

Arenites, or arenaceous rocks, and corresponding sediments constitute a major group of clastic deposits, the average grain size of which ranges from 1/16 to 2 mm, i.e., they have a sandy or medium texture (see *Sands and Sandstones*). The mineral constituents may show a great range in bulk composition from associations of many different mineral species to dominantly monomineralic accumulations. Limestones composed of sand-sized particles are termed calcarenites (q.v.).

B. CHARLOTTE SCHREIBER

References

Crook, K. A. W., 1960. Classification of arenites, *Am. J. Sci.,* 258, 419-428.

Pettijohn, F. J.; Potter, P. E.; and Siever, R., 1973. *Sand and Sandstone.* New York: Springer-Verlag, 618p.

Cross-references: *Arkose; Calcarenite, . . . ; Clastic Sediments and Rocks; Grain-Size Studies; Graywacke; Sands and Sandstones.*

ARGILLACEOUS ROCKS

Argillaceous rocks are those rocks containing a high proportion of clay minerals. They include slates, shales, claystones, and mudstones. *Mudstones* are nonfissile rocks containing approximately equal amounts of clay and silt (largely quartz); *claystones* contain >2/3 clay. *Shales* are fissile (thin cleavage) rocks usually containing >50% clay minerals. *Slates* are low-grade metamorphic, clay-rich rocks. Shales are by far the most abundant.

Abundance

Calculations based on the proportions of average shale, sandstone, and limestone required to make the average igneous rock (from which the sedimentary rocks were derived) indicate shales should comprise from 70-83% of all sedimentary rocks. Stratigraphic measurements indicate that shales in fact comprise in the vicinity of 40-50% of all sedimentary rocks. The discrepancy is believed to represent the large amounts of clay in the deep sea. Although these calculations are at best estimates, there is little doubt that argillaceous rocks comprise more than half the total volume of the earth's sediment, deposited primarily in marine and deltaic environments.

Composition

Chemical. Most of the major chemical variations in shale (Table 1) are due to the nonclay minerals. The variations in SiO_2 content are largely due to variations in quartz content. Much of the CaO and some (relatively minor amount) of the MgO variation is due to the content of carbonate minerals. Any CaO in excess of 2-3% is usually present in nonclay minerals. Al_2O_3, Fe_2O_3, FeO, Na_2O, K_2O are largely present in the clay minerals, and much of the variation is due to changes in the clay suite. Feldspars, iron oxides, pyrite, gibbsite, and others contribute to the chemical variations; but, in most shales, their contribution is relatively minor. High Al_2O_3 values usually indicate the presence of above-average amounts of kaolinite or aluminum hydroxides.

Shales have a wider range of variation in composition than do sandstones and limestones, largely because many more minerals may occur as a major component in shales. The chemical composition of shales is controlled by the environmental conditions in the weathering regime and at the site of deposition. Postdepositional alterations affect the mineralogy of shales but probably do not materially change the bulk chemistry.

Table 1 contains a series of average analyses of shales from various environments. The SiO_2 content of moderately weathered continental shales is higher than that of marine shale. X-ray analysis indicates shales from the former environment usually have more detrital quartz than those from the latter, a result of the limited amount of mechanical separation that take place in most continental environments. Most of the elements, other than silica, reflect the intensity and thoroughness of weathering more so than any specific type of environment.

Mineral. Using recent mineralogical information, and average chemical composition data, Yaalon (1961) calculated the mineralogical composition of a variety of shales and clays (see *Shale,* Table 1). The major change from earlier calculations was a major decrease in feldspar and an approximate doubling of the clay content.

Shaw and Weaver (1965), using quantitative x-ray methods, determined directly the mineral composition of 400 samples from 60 formations in North America. Only shale samples in which silt and sand could not be obviously detected by the unaided eye were used. The average clay mineral content of 61% was similar to that

TABLE 1. Average Chemical Composition of Shales from Various Environments (percent)

	1	2	3	4	5	6	7	8	9	10
SiO_2	63.06	53.32	58.32	51.72	63.8	57.8	57.0	59.19	54.48	57.05
TiO_2	0.69	1.21	0.91	0.82	0.87	0.77	0.75	0.79	0.98	1.27
Al_2O_3	13.53	29.79	16.60	15.44	17.6	18.2	18.8	15.82	15.94	17.22
Fe_2O_3	5.25	2.46	6.13	6.72	5.4	6.9	5.4	7.34	8.66	5.07
FeO									0.84	2.30
MnO	–	0.06	0.06	0.07	–	–	–	–	–	0.12
MgO	1.57	0.54	2.22	3.31	1.27	1.80	1.82	3.30	3.31	2.17
CaO	3.16	0.93	3.76	8.18	0.17	1.53	3.01	3.07	1.96	2.04
Na_2O	1.98	0.56	0.67	0.75	0.55	0.64	0.79	2.05	2.05	1.05
K_2O	3.97	1.24	3.07	3.71	2.94	2.87	3.24	3.93	2.85	2.25
SO_3	0.31	0.22	0.43	0.50	–	–	–	–	–	–
CO_2				5.37						
	7.08	10.01	7.97							
I. L.				5.55	–	–	–	–	7.04	8.99
	100.60	100.34	100.14	102.14	92.60	90.51	90.81	95.49	98.11	99.53

I. L. = Ignition loss
1. Ronov and Khlebnikova (1957), Av. comp. of continental clays of the cold and temperate belt of the Russian Platform.
2. *Ibid.*, Av. comp. of continental clays of the tropical belt.
3. *Ibid.*, Av. comp. of marine clays.
4. *Ibid.*, Av. comp. of marine clays and pools and lakes in arid regions.
5. Murray (1954), Av. comp. of Pennsylvanian nonmarine shales, 8 samples.
6. *Ibid.*, Av. comp. of Pennsylvanian brackish-water shales, 8 samples.
7. *Ibid.*, Av. comp. of Pennsylvanian marine shales, 10 samples.
8. Hougen, et al. (1925), Av. comp. of Quaternary Norwegian glacial clays, 96 samples.
9. Clark (1924), Av. comp. of 51 red deep-sea clays, calcium carbonate removed.
10. *Ibid.*, Av. Comp. of 52 blue and green deep-sea muds, calcium carbonate removed.

calculated by Yaalon. The quartz content (31%) was significantly higher. The average feldspar (4.5%), carbonate (3.6%), and other mineral (<2.0%) values are about half of Yaalon's values.

The clay-mineral suites in shales and clays are quite variable. Any of the major clay-mineral types, with the apparent exception of chlorite, can be the only clay mineral present. Nearly monomineralic clay suites are formed where the source material and weathering conditions are relatively uniform. A source area composed largely of volcanic material and situated in a mild weathering environment will produce a clay suite composed largely of smectite. A feldspar-rich, granitic source rock under intense weathering conditions will afford a kaolin-rich clay suite (some moderately weathered muscovite usually persists). Where weathering conditions are less severe, illite-dominated clay suites are produced from the feldspars.

Ancient Argillaceous Rocks. The most common clay minerals in the Precambrian and Paleozoic are illite and chlorite. Perhaps as many as 50% of pre-Pennsylvanian sediments contain illite or mixed-layer illite-smectite (of minor importance) as the only clay mineral. Most of the other sediments contain illite and chlorite. Chlorite may be present in amounts as high as 70%, but the average value is between 10 and 20%.

Smectite and mixed-layer illite-smectite (with smectite layers the more abundant) dominate the post-Lower Cretaceous clay suite, though the dominance is not as great as that of illite in the Paleozoic. Both illite and kaolinite occur as the dominant clay mineral in some formations, with the latter being more common.

The post-Lower Mississippian and pre-Upper Cretaceous sediments contain complex and highly variable clay suites. Clay suites commonly contain three or more clay minerals. Illite, smectite, and kaolinite all occur as the dominant clay in many formations. Mixed suites (with illite dominant) are characteristic of the Pennsylvanian, and smectite is abundant in the Permian. Illite is the most common clay mineral in sediments of this age interval, but considerably less so than in the earlier Paleozoic.

Other Properties

Color. Most argillaceous rocks are colored various shades of gray and gray-green. This is due to the presence of fine-grained organic material (black) and illite and chlorite (green).

The blackness of shales increases with increasing organic content.

The green color of shales may also be due to the presence of ferrous (reduced) iron in the clay minerals. With oxidation, the proportion of ferric (oxidized) iron in the clay increases to the extent that colors are various shades of purple and red.

Texture. Argillaceous rocks, by definition, have more than 50% clay-sized minerals. Most of these are clay minerals, but quartz and feldspar also commonly occur in clay sizes. The nonclay minerals in shales are dominantly in the silt and very fine sand-sized range. Sand-sized grains tend to be common in residual clays, claystones, and mudstones. Clay pebbles and clay grains are present in many argillaceous sediments.

Structure. Argillaceous rocks have a wide variety of structures. The most significant structure is fissility (q.v.). Fissile shales have well-oriented clay minerals and/or organic material and little or no chemical cement. These sediments also occur with burrows, slump structures, ripple marks, cross-bedding, and the other structural features commonly observed in the coarser grained sediments.

CHARLES E. WEAVER

References

Clark, R. W., 1924. Data of geochemistry, *U.S. Geol. Surv. Bull.,* 770, 839p.

Englund, J.-O., and Jorgensen, P., 1973. A chemical classification system for argillaceous sediments and factors affecting their composition, *Geol. Fören. Förh.,* 95, 87-98.

Hougen, H.; Klüver, E.; and Lökke, D. A., 1925. Undersöklser over norske lerer (V). Kjemiske analyser urfört for statens rästoffkomité, *Statens Rästoffkomite Publikation,* 22, 3-21.

Murray, H. H., 1954. Genesis of clay minerals in some Pennsylvanian shales of Indiana and Illinois, in A. Swineford and N. V. Plummer, eds., *Clays and Clay Minerals: Natl. Research Council Publ. 327,* 47-67.

Ronov, A. B., and Khlebnikova, Z. V., 1957. Chemical composition of the main genetic clay types, *Geochemistry,* 1957, 527-552.

Shaw, D. B., and Weaver, C. E., 1965. The mineralogical composition of shales, *J. Sed. Petrology,* 35, 213-222.

Weaver, C. E., 1959. The clay petrology of sediments, *Natl. Research Council Publ. 566,* 154-187.

Wickman, F. E., 1954. The "total" amount of sediments and the composition of the average igneous rock, *Geochim. Cosmochim. Acta,* 5, 97-100.

Yaalon, D. H., 1961. Mineral composition of the average shale, *Clay Minerals Bull.,* 5, 31-36.

Cross-references: *Clay as a Sediment; Clay Sedimentation Facies; Color of Sediments; Shale.*

ARKOSE, SUBARKOSE

The term *arkose* was introduced by Brongniart (1826) to describe a granular rock composed of clastic feldspar and quartz. Today it is generally defined as a light-colored sandstone containing at least 25% feldspar. Arkoses are commonly poorly cemented (silica or calcite) and contain various matrix minerals (usually <15%). These sediments are usually derived from the disintegration of granite and granitic rocks and may accumulate in place ("grus") or be transported (Twenhofel, 1937). Arkose is thus generally composed of angular, coarse sand-sized quartz and feldspar (chiefly microcline), with up to 10-15% of other materials such as clay, mica, and rock fragments.

Subarkose is a term independently suggested by both Folk (1954) and Pettijohn (1954) for sandstones of 5-25% feldspar content, in place of Pettijohn's term "feldspathic sandstone" (1949). In this same book, Pettijohn (1949) had redefined arkose as a sandstone having 25% or more labile constituents, of which feldspar is not less than half. Arkose has more recently been variously redefined (McBride, 1963; Folk, 1968).

Many sandstones that are highly feldspathic are not properly termed either arkose or subarkose, but fall under the term *graywacke* (q.v.) or "lithic arenite" because they contain large quantities of matrix materials such as clay or of lithic fragments.

Arkose has long been recognized as a paleoclimatic indicator of arid environments, either hot or cold, in view of the ease with which feldspars are reduced to kaolinite under humid tropical, acid weathering conditions. It was demonstrated by Krynine (1941), however, that rapid tectonic uplift, erosion, and deposition could generate arkosic sediments even in the humid tropics; and it has been found that many arkoses originate in such environments as rift valleys. Today, therefore, one recognizes either "tectonic arkoses" or "climatic arkoses."

B. CHARLOTTE SCHREIBER

References

Brongniart, A., 1826. L'Arkose caractères minéralogiques et histoire géognostique de cette roche, Paris: *Ann. Sci. Naturelles,* 8, 113-163.

Folk, R. L., 1954. The distinction between grain size and mineral composition in sedimentary rock nomenclature, *J. Geol.,* 62, 344-359.

Folk, R. L., 1968. *Petrology of Sedimentary Rocks.* Austin, Texas: Hemphill's, 170p.

Krynine, P. D., 1941. Paleogeographic and tectonic significance of sedimentary quartzites (abstr.), *Geol. Soc. Am. Bull.,* 52, 1915-1916.

McBride, E. F., 1963. A classification of common sandstones, *J. Sed. Petrology,* 33, 664-669.
Pettijohn, F. J., 1949. *Sedimentary Rocks.* New York: Harper & Row, 526p.
Pettijohn, F. J., 1954. Classification of sandstones, *J. Geol.,* 62, 360-365.
Shrock, R. R., 1948. A classification of sedimentary rocks, *J. Geol.,* 56, 118-129.
Twenhofel, W. H., 1937. Terminology of the fine-grained mechanical sediments, *Natl. Research Council Ann. Rept. 1936-1937,* Appendix 1, Rept. Comm. Sedimentation, 81-106.

Cross-references: *Clastic Sediments and Rocks; Graywacke; Sands and Sandstones.*

ASH—See VOLCANIC ASH DEPOSITS

ATMOSPHERIC DUST—See EOLIAN DUST IN MARINE SEDIMENTS

ATTERBERG SCALE

This geometric and decimal grade scale (Table 1), first devised by Atterberg (1905), is based on the settling rates of sediment in a water column. He noted that in dealing with sizes of less than 0.002 mm, *Brownian movement* (q.v., Vol. II) is of great significance; but that above 0.003 mm this molecular motion has little effect on settling rates. Of considerable importance was his standardization of the water temperatures, pH, and purity in his experiments. Atterberg also made percentage composition studies of various typical sediments such as dune sands, moraine tills, and riverbank and lake deposits to illustrate the use of his new, standardized size scale.

The scale is based on the unit value 2 mm, and each grade is based on a fixed ratio of 10. Subdivisions are the geometric means of the grade limits. The scale has been widely used in Europe and was adopted by the International Commission on Soil Science (but not by the U.S. Bureau of Soils) (AGI, 1972).

B. CHARLOTTE SCHREIBER

References

AGI, 1972. *Glossary of Geology,* Margaret Gary, Robert McAfee, Jr., and Carol L. Wolf, eds., Washington, D.C.: American Geological Institute.
Atterberg, A., 1905. Die rationelle Klassifikation der Sande und Kiese, *Chem. Zeitung,* 29, 195-198.

Cross-references: *Grain-Size Studies.* Vol. XIII: *Atterberg Limits and Indices.*

ATTRITION

Attrition is defined as the loss of weight and decrease in diameter of sediment particles in motion and is a composite phenomenon resulting from grinding, rubbing, and impact, primarily affecting coarser materials. If particles of unequal size are considered, the largest appear to undergo the greatest amount of loss, the loss becoming almost negligible for sand grains of < 0.3 mm in diameter.

Numerous laboratory attrition experiments have been undertaken, beginning with that of Daubrée (1879). Among the more recent experiments, should be mentioned, those by Marshall (1927), in which he used a special cylinder rotating in an inclined plane. He found that size reduction was accomplished by three separate processes: (1) abrasion, (2) impact, and (3) grinding. *Abrasion* was restricted in meaning to the effect of rubbing one pebble or particle against the other; *impact* is the effect of definite blows of relatively large fragments on others; and *grinding* is the crushing of small grains by pebbles of somewhat large size. He showed that the largest particles (pebbles) that he used had the greatest percentage of loss, i.e., 300-400 times that of sand-sized debris.

Kuenen (1964) confirmed Marshall's observations and, in addition, noted that the rate of attrition appears to diminish very little with rounding, but that it is very strongly dependent upon the relative resistance of different rock types. The losses for a quartzite are about one third those of a dense limestone and three times as much as for chert. He also showed that if a beach is composed of various sizes of identical rock pebbles, those falling in the median size range are the most reduced.

TABLE 1. Nomenclature and Size Boundaries for the Classification of Clastic Sediments with Original German and Swedish Terms.[a]

		Size Boundaries
Block (Blöcke)	Klippblock Stenblock Blocksten	above 2 meters 20-6 decimeters 6-2 decimeters
Klapper (Geröll)	Grofklapper (Grobgeröll) Singel (Schotter)	20-6 cm 6-2 cm
Grus (Kies)	Mal (Grobkies) Gryske (Kleinkies)	20-6 mm 6-2 mm
Sand	Grand (Grand) Dyne	2.0-0.6 mm 0.6-0.2 mm
Mo	Fimma Mjäla	0.2-0.06 mm 0.06-0.02 mm
Lättler (Lehm)	Vesa Mjuna	0.02-0.006 mm 0.006-0.002 mm
Ler (Ton)		0.002 mm or less

[a] See Atterberg, 1905.

A related type of attrition, combining abrasion and impact, is *sand blasting,* the movement of fine particles against or past a larger, relatively stationary fragment or outcrop of bedrock. Sand blasting is achieved in any fluid medium, e.g., in a high-velocity water jet or the strong winds of a desert sandstorm. Salminen (1935) experimented on this effect by placing granite and gneiss fragments at the Finnish shore of the Baltic Sea and collecting and weighing them after a given time of exposure. The results, expressed in terms of wind-hours, showed that the specimens had a maximum weight loss of 5.07% during an exposure of 1200 hr.

Zhdanov (1958) randomly placed 2500 specially marked pebbles into the intergroin space at the Caucasus-Black Sea shore, and observations were continued for nearly five years. Collection and weighing of specimens was repeated nineteen times, during which time nearly 80% of the material had disappeared (leaving only 500 pebbles). The conclusions, after appropriate calculations, were as follows: the average weight loss in the surf zone is about 5%/yr. It differed, depending on rock types, from 1.6% (basalt) to 8% (marble). The values thus obtained may be expressed as a function of wave energy, making it possible to use the calculations for practical purposes. For instance, it has been determined that the loss of material per 1 km beach length is 900-1600 m^3/yr, from this cause alone.

Lithologic comparisons of natural sediment distributions represent another method of attrition evaluation. Landon (1930) first obtained qualitative results for shingle beaches. Marshall (1929) investigated a New Zealand beach near the Mohaka River, and noted that the grain diameter of the prevailing size fraction diminished after an apparent drift of 56 km, from 12.7 mm to 0.6 mm although it is entirely possible that this observed reduction in size is the result of the sorting action of longshore currents, rather than attrition. The latter interpretation was made in a study of sediment distribution along the New Jersey shore (eastern US) by Shepard and Cohee (1936).

When several different methods of study are applied to an investigation, such as the three approaches just mentioned (laboratory, in situ but contrived, actual observation) the final results present a much more convincing picture than when only one method is employed. Zarva's (1959) lithologic studies of about 100 km of Black Sea shore beaches showed results that corresponded with Zhdanov's experiments with pebbles. This confirmation made the two works far more meaningful, where the one-method approach left many unanswered questions.

Durability Index

Wentworth (1931) used the term *durability index* to indicate the relative resistance to abrasion exhibited by a sedimentary particle in the course of transportation. The abrasion resistance of various minerals has been examined experimentally, and these studies indicate that hardness, cleavage, and tenacity seem to be the most significant factors; elasticity may also be significant. Among those minerals examined, quartz ranks as the most durable; in most cases, tourmaline ranks next. Pettijohn (1957) devised a very general formula for the durability index: $DI = (x/a)^n$, where x = Moh's hardness; a = the hardness of quartz (7); and n is a constant. Erratic DI numbers are frequent and are probably a reflection of size-reduction processes other than abrasion, of the nonuniform nature of the hardness scale, and of other physical properties of minerals, such as elasticity.

V. P. ZENKOVICH

References

Cozzens, S. B., 1931. Rates of wear of common minerals, *Wash. Univ. Studies Sci. Technol.,* 5, 71-80.

Daubrée, A., 1879. *Etudes Synthétique de Géologie Expérimentale.* Paris: Dunod, 828p.

Drake, L. D., 1972. Mechanisms of clast attrition in basal till, *Geol. Soc. Am. Bull.,* 83, 2159-2165.

Friese, F. W., 1931. Untersuchung von Mineralen auf Abnutzbarkeit bei Verfrachtung im Wasser, *Mineralog. Petrog. Mitt.,* 41, 1-7.

Gregory, K. J., and Cullingford, R. A., 1974. Lateral variations in pebble shape in northeast Yorkshire, *Sed. Geol.,* 12, 237-248.

Kuenen, P. H., 1964. Experimental abrasion: surf action, *Sedimentology,* 3, 29-43.

Landon, R. E., 1930. An analysis of beach pebble abrasion and transportation, *J. Geol.,* 38, 437-446.

Marshall, P., 1927. The wearing of beach gravels, *Trans. Proc. New Zealand Inst.,* 58, 507-532.

Marshall, P., 1929. Beach gravels and sands, *Trans. Proc. New Zealand Inst.,* 60, 324-365.

Pettijohn, F. J., 1957. *Sedimentary Rocks.* New York: Harper & Row, 718p.

Salminen, A., 1935. On the weathering of rocks and the composition of clays, *Suom. Tiedeakat., Toimit.,* ser. A, 44(6), 149p.

Schumm, S. A., and Stevens, M. A., 1973. Abrasion in place: a mechanism for rounding and size reduction of coarse sediments in rivers, *Geology,* 1, 37-40.

Shepard, F. P., and Cohee, G. V., 1936. Continental shelf sediments off the mid-Atlantic States, *Geol. Soc. Am. Bull.,* 47, 441-458.

Wentworth, C. K., 1931. Pebble wear on the Jarvis Island beach, *Wash. Univ. Studies Sci. Technol.,* 5, 11-37.

Zarva, A. V., 1959. The shingle beaches investigation at the Caucasus Black Sea shore between Tuapse and Adler, *Trans. Oceanog. Comm. Acad. USSR,* 1 (in Russian).

Zhdanov, A. M., 1958. Pebble attrition under surf action, *Oceanog. Comm. Bull. Acad. Sci. USSR,* 1 (in Russian).

Cross-references: *Beach Gravels; Beach Sands; Littoral Processes; Maturity; River Gravels; Sternberg's Law; Till and Tillite.*

AUTHIGENESIS

The term *authigenesis* is derived from the Greek *authigenes,* meaning "formed on the spot," and was originally employed by Kalkowsky (1880) to describe the origin of any newly formed, or secondary, mineral. *Allothigene* (commonly shorted to allogene) is the contrasting term. Authigenesis is currently applied specifically to sediments and sedimentary rocks, and, like so many other terms in this field, frequently has been liberally interpreted and randomly redefined since its introduction, owing to disparity of opinion as to the scope of interrelated terms applied to sediment genesis.

There have been attempts to restrict authigenesis in time and with regard to the form of the new minerals. Thus Baturin (1937, see Fairbridge, 1967) equated authigenic mineralization with early diagenesis (*halmyrolysis*), and he described subsequent mineralization as epigenic; this usage is customary in the USSR, as pointed out by Fairbridge (1967); however, halmyrolysis refers to reactions between sediment and sea water, and is effective for only a part of the sedimentational (syndiagenetic) phase. Tester and Atwater (1934) reserved the term for discrete crystallographic units, as distinct from rock-forming components. Thus, feldspars formed in situ would qualify for the description "authigenic," but locally generated cements and overgrowths of similar composition and optical orientation to the host grains would not. Yet limitations of this kind are not implicit in the term, nor are they equated with the present consensus.

Nebulous and conflicting views regarding the definition of authigenesis are a consequence of such attempts to restrict the application of the term. Too rigid a definition is impractical because the limits of authigenesis are obscure, the processes involved merging into the provinces of metasomatism and metamorphism. Authigenesis is most valuable as a collective term encompassing those processes, operating throughout diagenetic evolution (Fairbridge, 1967), which result in the formation of new minerals, regardless of form, provided they are generated more or less within the sedimentary body. Some authors (e.g., Pettijohn, 1957) consider authigenesis to be quite distinct from diagenetic (low temperature) metasomatism. Others (e.g., Fairbridge, 1967) regard the formation of new minerals through metasomatism to be an aspect of authigenesis. Clearly, individual experience and opinion govern the interpretation and usage of these terms; nevertheless, it is widely held that authigenesis refers to minerals formed more or less in place, whereas *metasomatism* refers to widespread mineral replacement brought about by mineralizing fluids from sources external to the sedimentary body. Authigenic minerals, therefore, bear the stamp of sedimentary accumulation and physicochemical parameters prevailing at the time of their formation, whereas metasomatic minerals may not show such a close relationship. Widespread metasomatism would be more likely to occur during the later stages of diagenetic evolution when the deposit would pass through a variety of geological situations and be subject to the influences of a wide range of percolating fluids and physicochemical parameters.

Principles of Authigenesis

Although data concerning many aspects of authigenesis are lacking, the general principles involved are recognized. Fundamental to an appreciation of authigenesis is the realization that most solid and fluid phases in the sedimentary environment are intrinsically unstable, and reaction between ions of the fluid phase and the sedimentary particles may result in the formation of a new mineral in response to changing physicochemical factors affecting the sedimentary accumulation.

Controlling Parameters and Processes. The particular stage set for authigenesis and, to some extent, the course subsequently taken, is determined by the physical state and mineralogical constitution of the original sediment, and the physicochemical characteristics of the depositional environment. These in turn are related to provenance, climate, and biological factors. The parameters of major importance are hydrogen ion potential (pH), oxidation-reduction potential—or *redox* potential—(Eh), ionic absorption phenomena, and the ionic potential of components such as Na^+, Mg^{2+}, Fe^{2+}, Mn^{2+}, Al^{3+}, Fe^{3+}, Si^{4+}, P^{5+}, etc. Where CO_2 content of the fluid phase largely controls pH, as is the case in early diagenetic environments, cation concentration is important (Packham and Crook, 1960). Other fundamental controls on authigenesis are pressure, temperature, and time.

The pH is influenced by temperature and the availability of CO_2 (released through organic oxidation), and is therefore very susceptible to organic control. Removal of oxygen and consequent lowering of Eh is accomplished chiefly by oxygen-consuming organisms. Aerobic bacteria may bring this about rapidly once free circulation ceases owing to entrapment of the fluid phase by sediment. Baas Becking et al. (1960) investigated the Eh-pH characteristics of biological systems and natural aqueous environments. Through adsorption, ions of elements such as copper, lead, zinc, mercury, etc. may be removed from natural waters and taken up by colloidal particles such as iron hydroxides. Various elements may be concentrated through the physiological processes of organisms. The important reactions involved in authigenesis may be listed as: (1) oxidation–reduction; (2) hydration–dehydration; (3) hydrolysis–dehydrolysis; (4) ionic adsorption; (5) cation or base exchange; and (6) carbonatization. For a concise account the reader is referred to Fairbridge (1967).

Phases of Authigenesis. The sediment complex can be expected to undergo a sequence of physicochemical changes which promote the formation of distinctive suites of authigenic minerals. As the sediment is buried, it is less and less influenced by the depositing medium, and changes occur which ultimately create totally new physicochemical conditions.

The complex of fluid and solid phases, with its attendant microorganisms, which constitutes the freshly accumulated sediment in aqueous basins, is physicochemically unstable. Under normal marine conditions of well-ventilated basins, the water at the sediment-water interface and in the interstices of the upper layer of sediment contains free oxygen which is gradually absorbed by organisms; the reduction of various hydroxides and sulfates follows. Thus, the oxidation-reduction potential is lowered, anaerobic bacteria replaced aerobic bacteria, and the pH rises after an initial drop.

Strakhov et al. (1954) consider the thickness of the layer characterized by an oxidizing or neutral environment to average 10–15 cm, reaching 40 cm, and lasting for periods of a few days to thousands of years. This early phase of diagenesis is characterized by the formation of glauconite, phosphorite, iron-manganese deposits such as nodules and crusts, and some zeolites, all commonly found in present-day seas and sometimes forming extensive deposits. The next phase of authigenesis takes place in the underlying zone where reduction of sulfates and oxides occurs, e.g., ferric oxide is reduced to ferrous oxide.

In basins with restricted circulation such as the Black Sea, however, where the bottom waters are foul (see *Black Sea Sediments*), reducing conditions exist at and above the sea bed. Consequently the sediment is initially deposited in a reducing environment.

Accompanying these changes from an oxidizing to a reducing regime, certain solid substances such as silica, calcium carbonate, magnesium carbonate, etc. dissolve, and the pore water may thereby achieve saturation or supersaturation with respect to these components. Decomposition of organic matter proceeds, the products contributing to both fluid and solid phases of the sediment complex. Cation exchange occurs between solid and fluid phases; clay minerals release cations to the pore water and adsorb others from it. There may be a complete change in the nature of the fluid phase; and under the influence of this new geochemical regime there is a trend toward enrichment in alkalies, phosphorus, organic matter, iron, silica, etc. Hydrogen sulfide, ammonia, carbon dioxide, and hydrogen increase as the oxygen content decreases. This does not lead to a static situation, however, for an exchange takes place between the sediment complex and the overlying water mass; various gases, together with phosphorous, iron, silica, calcium carbonate, manganese, etc. pass from the sediment complex, while oxygen, and calcium and magnesium ions may be absorbed by it. By such means, certain ions may accumulate in the sediment in excess of their original concentration, e.g., S and Mg^{2+}, and may ultimately give rise to economically viable mineral deposits.

It is unlikely that conditions will remain uniform over geographically extensive regions. Furthermore, with every phase of diagenesis (q.v.), variations in Eh and pH and ion concentrations may cause solution and redistribution of authigenic deposits; e.g., the formation of nodules and concretions (the "diagenetic differentiation" of Pettijohn, 1957), a more advanced stage of diagenesis, is characterized by minimal bacteriological activity.

Compaction and cementation are concomitant with the geochemical changes outlined above. The former will tend to facilitate pore fluid migration towards the surface; the latter will tend to inhibit it, resulting in the entrapment of connate water. Dehydration and recrystallization commonly accompany compaction. Although these processes may operate soon after deposition they become more effective at lower levels; Strakhov et al. (1954) maintain that interstitial water continues to be squeezed out to a depth of approximately 300 m. This compaction-maturation phase is characterized by hypogene authigenesis and is described as the phase of anadiagenesis by Fairbridge (1967). To give some order of magnitude, it is said to extend from 1–10,000 m in depth and to cover a period of $10^3 - 10^8$ years. The end of this phase could be marked by general lithification, with no further circulation of fluids, but this condition is scarcely ever achieved before uplift occurs and the final phase, epidiagenesis or epigenesis, begins.

If the course of diagenetic evolution follows what is considered to be a normal sequence in a mobile belt, the sediment complex, now cemented and compacted, will be deeply buried and may become involved in orogenic buckling, thereby reaching the threshold of metamorphism. Alternatively, epeirogenic uplift may bring the rocks within the sphere of descending meteoric water, the epigene or supergene zone (see Perel'man, 1967). During this "emergent pre-erosion phase," therefore, (epi-) diagenetic processes are influenced by subaerial factors. Ground water, normally saturated with oxygen and carbon dioxide, deeply penetrates permeable strata, inducing a further change in the geochemical environment, and is effective in the downward transport of ions in regions of ore deposits. With only local exceptions, oxidation prevails and the pH falls (in sulfide ore deposits, pH becomes extremely low). Oxidation of iron compounds and other substances ensues, and solution of various salts heralds a new phase of cementation, often accompanied by mineralization along joints and faults. This phase of diagenesis, termed epidiagenesis by Fairbridge (1967), is an outstanding feature of permeable strata, and is much less apparent in very fine-grained, homogeneous deposits sealed from subaerial influences except by jointing and interrelated permeable horizons. Controversy reigns regarding the point at which diagenesis ends and weathering begins, but it seems logical to restrict the latter to surface processes of rock decay and soil formation.

Authigenic Environments and Diagenetic Facies

The most significant result of the changes wrought in the physicochemical nature of the sediment complex is the formation of authigenic minerals such as siderite, sulfides of iron, lead, zinc, etc., which register phases of equilibrium with prevailing physicochemical parameters. Conditions are usually transient, changing with time and situation, and authigenesis is, therefore, a dynamic multiphase process which may evolve a complex paragenesis (q.v.), there being a correlation between authigenic minerals and diagenetic processes. After formation, during early diagenesis, authigenic minerals may survive changes in physicochemical conditions that could be inimical to their initial development, and a progression of changes may thereby be recorded by individual mineral indicators or a series of minerals, e.g., iron hydroxides → glauconite → iron chlorites → siderite → pyrite, which indicates a gradual change from oxidizing to reducing conditions.

The Eh and pH characteristics of principal sedimentary and diagenetic environments have been examined by Krumbein and Garrels (1952), see Fig. 1; and a study of the same parameters enabled Teodorovich (see Chilingar, 1955) to recognize thirteen early diagenetic facies in subaqueous sediments. The mineralogical, pH, and Eh characteristics of the six principal geochemical environments are outlined in Larsen and Chilingar (1967). The Eh of these environments ranges from oxidizing (ferric oxides and hydroxides zone), through weakly oxidizing (glauconite zone and oxykertschenite zone), neutral (leptochlorite and kertschenite zone), weakly reducing (siderite and vivianite zone), reducing (carbonates, and iron sulfides zone), to strongly reducing (sulfide zone proper). The pH ranges given are: acid (5.5–2.1), slightly acid (6.6–5.5), neutral (7.2–6.6), weakly alkaline (8.0–7.2), alkaline (9.0–8.0), and strongly alkaline (pH >9.0). Packham and Crook (1960) extended this facies concept by the examination of authigenic minerals in certain depth sequences. They suggested that higher ranking diagenetic facies characterized by heulandite-analcite, laumontite, prehnite-pumpellyite, and albite-epidote (in order of increasing rank) may be recognizable in certain rocks subject to an alkaline regime.

Evidence for Some Authigenic Effects

Among the criteria which testify to the authigenic nature of minerals in sedimentary rocks are the following:

1. Euhedra, and signs that crystals have pushed aside the ground mass through force of crystallization.
2. Zones of inclusions arranged parallel to idiomorphic faces of euhedral crystals; certain configurations testify to successive growth stages.
3. In contrast to (1), delicately spired boundaries are also indicative of growth in situ.
4. Cross-cutting relationships between crystal boundaries and host rock fabrics (see *Diagenetic Fabrics*).
5. Authigenic minerals that form under somewhat similar physicochemical conditions may show alternating relationships.
6. Crystal aggregates such as spherulites and radiated clusters are often characteristic of authigenic development.
7. The partial or complete replacement of one mineral by another; the presence of pseudomorphs.
8. Certain forms of secondary rims.
9. The sharp reducing nature of a mineral in normal marine deposits.

Significance of the Products of Authigenesis

Some authigenic minerals such as glauconite are good environmental indicators and give in-

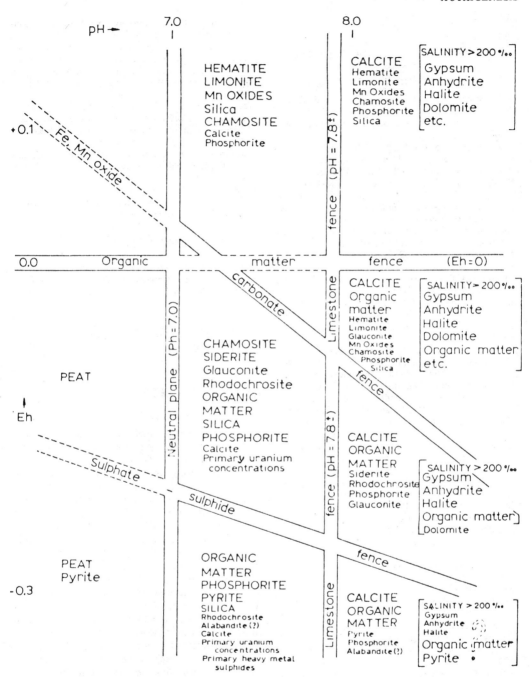

FIGURE 1. "Fence diagram" illustrating principal environments of sedimentation and diagenesis according to Eh and pH (from Fairbridge, 1967, after Krumbein and Garrels, 1952).

formation regarding the physicochemical nature of the sea bed. Other minerals provide clues to the subsequent trend of diagenesis, and are of great academic importance.

The concentration of rare elements during early diagenesis may form economically viable deposits. The properties of road metal and building stones may be governed by an authigenic cementing agent. Further, Elliston (1963) suggests that diagenetic events involving authigenesis are significant in the formation of ore bodies.

G. RICHARD ORME

References

Baas Becking, L. G. M.; Kaplan, I. R.; and Moore, D. M., 1960. Limits of the natural environment in terms of pH and oxidation-reduction potentials, *J. Geol., 68*, 243-284.

Buyce, M. R., and Friedman, G. M., 1975. Significance of authigenic K-feldspar in Cambrian-Ordovician carbonate rocks of the Proto-Atlantic shelf in North America, *J. Sed. Petrology, 45*, 808-821.

Chilingar, G. V., 1955. Review of Soviet literature on petroleum source-rocks, *Bull. Am. Assoc. Petrol. Geologists, 39*, 764-768.

Cronan, D. S., 1974. Authigenic minerals in deep-sea sediments, in E. D. Goldberg, ed., *The Sea*, vol. 5. New York: Wiley, 491-525.

Davis, A. F., and McKenzie, J. A., 1975. Experimental authigenesis of phyllosilicates from feldspathic sands, *Sedimentology, 22*, 147-155.

Elliston, J., 1963. The diagenesis of mobile sediments in syntaphral tectonics and diagenesis, a symposium, Hobart: Geology Department, University of Tasmania, Q1-Q11.

*Fairbridge, R. W., 1967. Phases of diagenesis and authigenesis, in Larsen and Chilingar (1967), 19-89.

Funk, H., 1975. The origin of authigenic quartz in the Helvetic Siliceous Limestone (Helvetischer Kiselkalk), Switzerland, *Sedimentology, 22*, 299-306.

Garrels, R. M., and MacKenzie, F. T., 1971. *Evolution of Sedimentary Rocks*. New York: Norton, 397p.

Hardjosoesastro, R., 1971. Note on chamosite in sediments on the Surinam Shelf, *Geol. Mijnbouw, 50*, 29-33.

Kalkowsky, E., 1880. Uber die Erforschung der archaeischen Formationen, *Neues Jahrb. Min. Geol. Palaeont., Monatsh., 1*, 1-29.

Krumbein, W. C., and Garrels, R. M. 1952. Origin and classification of chemical sediments in terms of pH and oxidation-reduction potentials, *J. Geol., 60*, 1-33.

Larsen, G., and Chilingar, G. V., 1967. Introduction, *Diagenesis in Sediments*. Amsterdam: Elsevier, 1-17.

Lawrence, J. R., and Kastner, M., 1975. O^{18}/O^{16} of feldspars in carbonate rocks, *Geochim. Cosmochim. Acta, 39*, 97-102.

Middleton, G. V., 1972. Albite of secondary origin in Charny Sandstones, Quebec, *J. Sed. Petrology, 42*, 341-349. See also discussion by J. Lajoie, *43*, 575-576.

Packham, G. H., and Crook, K. A. W., 1960. The principle of diagenetic facies and some of its implications, *J. Geol., 68*, 392-407.

Perel'man, A. I., 1967. *Geochemistry of Epigenesis*. New York: Plenum, 266p. (Russian ed. published by Nedra Press, Moscow, 1965).

Pettijohn, F. J., 1957. *Sedimentary Rocks*. New York: Harper & Row, 718p.

Sartori, R., 1974. Modern deep-sea magnesian calcite in the central Tyrrhenian Sea, *J. Sed. Petrology, 44*, 955-965.

Stewart, R. J., 1974. Zeolite facies metamorphism of sandstone in the western Olympic Peninsula, Washington, *Geol. Soc. Am. Bull., 85*, 1137-1142.

Strakhov, N. M.; Brodskaya, N. G.; Khyazeva, L. M.; Raszhivina, A. N.; Rateev, M. A.; Sapozhnikov, D. G.; and Shiskova, E. S., 1954. *Formation of Sediments in Recent Basins*. Moscow: Izd. Acad. Nauk SSSR, 791p.

*Teodorovich, G. I., 1961. *Authigenic Minerals in Sedimentary Rocks*. New York: Consultants Bureau, 120p. (Russian text published by the USSR Academy of Sciences Press, Moscow, 1958).

Tester, A. C., and Atwater, G. I., 1934. The occurrence of authigenic feldspars in sediments, *J. Sed. Petrology, 4*, 23-31.

Woodward, H. H., 1972. Syngenetic sanidine beds from Middle Ordovician Saint Peter Sandstone, Wisconsin, *J. Geol., 80*, 323-332.

*Contains numerous references.

Cross-references: *Clay—Genesis; Diagenesis*. Vol. IVA: *Authigenesis of Minerals—Marine; Authigenesis of Minerals—Nonmarine*. Vol. IVB: *Authigenic Minerals*.

AUTHIGENESIS OF CLAYS—See CLAY—GENESIS

AVALANCHE DEPOSITS

Avalanche deposits are large masses of snow, ice, earth, rock, or mixtures of these materials that are emplaced by high-speed motion under the force of gravity. These deposits may be classified, depending upon their predominant constituent material as *snow-* and *ice-avalanche deposits, debris-avalanche deposits* and *rock-avalanche deposits*. Parallel terms may be applied to the dynamic processes that produce the deposits, e.g., *rock avalanche,* depending on the parent material involved. Although they may be wet, the movement of avalanches does not necessarily depend upon the presence of water as is the case for mudflows and debris flows. The term *debris avalanche* has been applied to mass movements of incoherent earth materials in which flowage is due chiefly to loss of shear strength upon water saturation (see Vol. III: *Avalanche;* Varnes, 1958), but as used herein the term *debris-avalanche deposit* carries no implication regarding water content. The following discussion is concerned only with avalanche deposits of earth materials and primarily with the deposits of very large catastrophic avalanches. The deposits of ice and snow avalanches, which are ephemeral and probably unimportant in the stratigraphic record, are not considered further. The deposits of most smaller avalanches, which commonly form irregular half-cones of blocky detritus at the base of the source slope, are discussed elsewhere (see *Mass-Wasting Deposits*).

Occurrence

Catastrophic avalanches of earth materials are generally thought to involve lubrication by air and/or fluidization to reduce sliding and internal friction. They appear to require a combination of (1) large vertical drop in the source area in order to attain high velocities, and (2) some minimum mass of material that is probably on the order of several million cubic meters. As a result of these requirements, the deposits of avalanches tend to be large, and they are virtually restricted in occurrence to valley floors and the bases of steep escarpments in rugged, mountainous terrain. The locations and some of the distinctive characteristics of better studied historic and prehistoric terrestrial avalanches, as well as one possible lunar avalanche, are given in Table 1. Comparable deposits undoubtedly occur in all the major mountain ranges of the world.

Reports of avalanche deposits in the pre-Quaternary stratigraphic record are rare, possibly because the coarse material may be confused with tectonic breccias, glacial moraines, or the deposits of debris flows. Emplacement by avalanching has been suggested for megabreccias interbedded with steeply dipping Miocene playa and alluvial deposits near Ray, Arizona (Krieger, 1973). The breccias are conformable bed-like or lens-like masses with a maximum strike length of 4 km and a thickness of as much as 185 m. Preservation of stratal continuity within the breccias and the lack of disturbance of the underlying sands and clays suggest to Krieger that the breccias may have been emplaced as air-cushioned sheets that were derived from highlands to the west.

Some historic catastrophic avalanches, most notably at Goldau, Switzerland, were initiated by slope failures without any obvious triggering mechanism. Others have been triggered by earthquakes, undercutting of slopes by quarrying operations, or possibly volcanic steam explosions. Several of the historic avalanches that occurred in inhabited areas have had disastrous consequences to life and property. The destructive Huascarán avalanche of May 30, 1970, involved an estimated 50-100 million m^3 of rapidly moving rock, ice, snow, mud, and soil. The avalanche devastated an area of about 22.5 km^2, and its two main lobes of debris covered a densely populated area of about 8 km^2 to an estimated average depth of 5 m. Boulders and angular blocks of rock up to 7000 metric tons in size comprise an estimated 10-50% of the total volume of the debris lobes. Locally, boulders weighing several tons were hurled as much as 1 km beyond its margins. The rock consists almost entirely of fresh biotite-rich quartz diorite of the type that makes up most of the sheer W flank of Huascarán Mountain. The interstitial material is a gravelly mud derived by fragmentation of granitic rock and by admixture of glacial rock flour and soil which was incorporated into the avalanche along its path. At its distal end, the Huascarán avalanche deposit gave rise to a debris flow that swept down the Río Santa 160 km to the sea, causing additional death and destruction along its path.

Recognition

Catastrophic avalanche deposits are distinguishable from other types of landslide or debris-flow deposits primarily by geomorphic criteria suggestive of high-velocity movement facilitated by fluidization or by entrained air. To a lesser extent, they are characterized by lithologic criteria that indicate a large localized source of shattered rock or debris and, in some cases, nonturbulent transport of material. Geomorphic features characteristic of avalanche deposits include: (1) occurrence as sheets that are very thin relative to their areal extent; (2) lobate form varying from tongue-like in confined valleys to fan-like, with marginal fingers, where not confined; (3) runout far beyond the base of source slopes; (4) relatively low relief of the lobes from head to toe; (5) hummocky surfaces; (6) arcuate ridges, furrows, and soil streaks aligned in the apparent movement direction; (7) local pressure ridges where flowage was impeded by barriers; (8) evidence of movement up or over ridges or slopes; (9) lateral ridges; (10) surficial debris cones; and (11) volumes measurable in millions of m^3. Important lithologic characteristics common in avalanche deposits include: (1) angularity and poor sorting of clasts within the deposit including numerous exceptionally large blocks; (2) a limited amount, or absence of, abrasion of constituent clasts; (3) similarity of rock type to source area except for material picked up en route; (4) local internal imbricate structure, especially near distal margins; (5) a three-dimensional, "jig-saw puzzle" brecciation resulting from pervasive shattering of large blocks that remain undispersed during movement of the avalanche; (6) high porosities; (7) a general absence of size-sorting from the proximal to the distal end of the deposits, and (8) relict stratigraphy.

Many features of avalanche deposits may also be applicable to the deposits of various other origins, especially more slowly moving wet debris flows and mudflows. In the absence of

TABLE 1. Characteristics of Some Avalanche Deposits

Name, Location, and Reference	Date or Age	Principal Lithology of Deposit	Volume of Deposit (Million m³)	Area of Deposit (km²)	Maximum Vertical Drop (km)	Maximum Horizontal Distance Moved (km)	$\frac{V^a}{H}$	Maximum Run-up[b] (m)	Minimum Velocity[c] (km/hr)	Remarks
Historic Avalanches										
Goldau, Switzerland (Heim, 1932)	1806	carbonate	39	1.3	≈0.6	1.8	0.33	—	—	Destroyed four villages; killed 457 persons
Elm, Switzerland (Heim, 1932)	1881	slate	11	≈0.6	0.5+	1.6	0.26	103	160	Destroyed villages; killed 115 persons; triggered by quarry operations
Frank, Alberta (in Shreve, 1968, and Mudge, 1965)	1903	limestone	37	2.7	0.9	3.2	0.28	120	180	Destroyed part of Frank; killed 70 persons
Madison Canyon, Montana (in Mudge, 1965)	1959	schist, gneiss, dolomite	28	0.5	0.4	0.5	0.81	131	160	Killed 26 persons; triggered by earthquake
Little Tahoma Peak, Washington (Crandell and Fahnestock, 1965)	1963	andesite	11		1.9	6.9	0.27	91	155	Deposited in part on a glacier; possibly triggered by steam explosion on flank of volcano
Sherman Glacier, Alaska (Shreve, 1966; Plafker, 1968)	1964	graywacke with minor argillite	26	8.5	1.1	5.3	0.21	135	185	Deposited almost entirely on a glacier; triggered by major earthquake
Huascaran, Peru (Plafker et al., 1971)	1970	granite	50–100	22.5	3.7	16.0	0.22	140	280–335	Buried Yungay and smaller communities with loss of over 18,000 lives; triggered by major earthquake
Prehistoric Avalanches										
Blackhawk, California (Shreve, 1968)	Pleistocene?	limestone	283	14	1.1	9.7	0.13	60	120	
Flims, Switzerland (Heim, 1932)	Pleistocene	carbonate	12,500	41.4	1.3	17.4	0.07	—	—	
Saidmarreh, Iran (Watson and Wright, 1970)	—	carbonate	20,840	165	1.0	15.6	0.10	450	335	Largest known avalanche deposit
Sawtooth Ridge, Montana (Mudge, 1965)	Late Pleistocene	carbonate	≈649	11.6	0.1–0.4	4.3	0.09	—	—	
Apollo 17, Moon (Howard, 1973)		soil derived from impactites[d]	≈200	21	1.9	9	0.2	—	—	Possibly triggered by ejecta impact

[a] Ratio of net vertical drop (from top of source scar to toe of avalanche deposit) to horizontal distance traveled.
[b] Vertical height of ridges or slopes overridden by avalanche.
[c] From eyewitness accounts or calculations based on vertical height of avalanche run-up.
[d] Keith Howard, personal commun., 1973.

eyewitnesses, recognition criteria may be especially ambiguous in certain wet-avalanche deposits that contain high percentages of fine-grained material. Some wet-avalanche deposits, such as the Huascarán deposits, may even grade into debris-flow or mudflow deposits at their distal margins, but, unlike the avalanches, these flows are confined to drainage channels. The deposits of small avalanches differ from the larger ones in that they are localized at the base of the source slope, the surface is at or near the angle of repose of the constituent fragmental material, they vary greatly in thickness, and there may be a considerable degree of gravity separation, with smaller fragments remaining near the apex of the deposit and the larger ones accumulating around its outer edge.

Mechanism

The characteristic features of most catastrophic avalanches are explained by one, or a combination, of two mechanisms: reduction of sliding friction by entrapment of air beneath the moving debris as suggested by Shreve (1966, 1968) and/or fluidization by entrained air (Kent, 1966), mud (Heim, 1933), or dust (Hsü, 1975). The apparent occurrence of an avalanche deposit on the moon that is in some respects comparable to terrestrial avalanches, suggests to Howard (1973) that avalanching may also occur in environments where air buoyancy cannot be a factor in reducing friction.

Air lubrication and fluidization can explain many of the details of geomorphology, particle-size distribution, and internal structure in avalanche deposits that cannot be accounted for otherwise. They are the only known mechanisms proposed that account for the low ratio of vertical fall to horizontal runout (ranging from as little as 0.07 to 0.33) and the high observed or calculated velocities (from 120–335 km/hr) of the terrestrial avalanches listed in Table 1. The observation that this ratio shows a general inverse relationship to avalanche size (Howard, 1973) suggests that large avalanches are more efficient in reducing internal and/or sliding friction than are small avalanches.

GEORGE PLAFKER

References

Crandell, D. R., and Fahnestock, R. K., 1965. Rockfalls and avalanches from Little Tahoma Peak on Mount Rainier, Washington: *U.S. Geol. Surv. Bull. 1221-A*, 30p.

Heim, A. 1932. Bergsturz und Menschenleben, *Naturf. Gesell. Zurich Vierteljahrsschr.*, 77(20), 218p.

Howard, K., 1973. Avalanche mode of motion: implications from lunar examples, *Science*, 180, 1052-1055.

Hsü, K. J., 1975. Catastrophic debris streams (Sturzstroms) generated by rockfalls, *Geol. Soc. Am. Bull.*, 86, 129-140.

Kent, P. E., 1966. The transport mechanism in catastrophic rock falls, *J. Geol.*, 74, 79-83.

Krieger, M. H., 1973. Megabreccias (large landslide blocks) interbedded in Miocene playa and alluvial deposits, south of Ray, Arizona, *Geol. Soc. Am. Abstr.*, 5(7), 699-700.

Mudge, M. R., 1965. Rockfall-avalanche and rockslide avalanche deposits at Sawtooth Ridge, Montana, *Geol. Soc. Am. Bull.*, 76, 1003-1014.

Plafker, G., 1968. Source areas of the Shattered Peak and Pyramid Peak landslides at Sherman Glacier, in The great Alaska earthquake of 1964–Hydrology, Pt. A, *Natl. Acad. Sci. Publ.*, 1603, 374-382.

Plafker, G.; Ericksen, G. E.; and Fernandez Concha, J., 1971. Geological aspects of the May 31, 1970 Perú earthquake, *Bull. Seismol. Soc. Am.*, 61(3), 543-578.

Shreve, R. L., 1966. Sherman landslide, Alaska, *Science*, 154(3757), 1639-1643.

Shreve, R. L., 1968. The Blackhawk Landslide, *Geol. Soc. Am. Spec. Pap. 108*, 47p.

Varnes, D. J., 1958. Landslide types and processes, *Highway Research Brd. Spec. Rept. 29*, 20-47.

Watson, R. A., and Wright, H. E., Jr., 1970. The Saidmarreh Landslide, Iran, *Geol. Soc. Am. Spec. Pap. 123*, 115-139.

Cross-references: *Breccias; Diamictite; Gravity-Slide Deposits; Mass-Wasting Deposits; Mudflow, Debris-Flow Deposits; Till and Tillite.* Vol. III: *Avalanche; Landslides; Snow Avalanche.*

B

BACKSET BEDDING

The term *backset bedding* was coined by Davis (1890) to describe the upslope deposition at the head of glaciofluvial outwash plains, eskers, and kame deltas. According to his theory of origin, backsets form where glacial streams, emerging from underneath the glacier ice, are constrained to rise to the level of the outwash surface. A backward growth of the outwash deposit caused by glacier melting would result in upslope deposition with bedding dipping in the *upstream* direction (Fig. 1).

The term *backset* is now used in a much broader context to refer to inclined bedding of primary origin that is formed by current action in various sedimentary environments. Davis' criterion that the bedding dip *into* the current is still a valid one. On the basis of present-day usage, backset bedding includes the bedding developed on the back (stoss) sides of ripples, dunes, and other sand waves of subaqueous origin, as well as the bedding deposited on the windward slopes of transverse sand dunes of eolian origin. Also included in this category is the bedding formed by the settling of suspended sediment (as in a lake) onto a depositional surface inclined in a general upstream direction; here the configuration of the bedding is controlled in part by the morphology of the depositional surface.

Backset bedding can form in the lower regime of flow, where the bed forms include ripples and dunes, or in the upper regime of flow, where the bed forms include standing waves and antidunes. Occurrences in the upper regime of flow have evoked by far the most interest. Twenhofel (1939), in discussing the bedding formed by Gilbert's antidune mode of sediment transport, stated that deposition takes place on the upstream side of the antidunes. He also stated (p. 522) that "such lamination as exists is inclined up-current," but he qualified his observation by noting that there is little chance of preservation. Indeed, reported field occurrences of the antidune type of backset bedding are few and far between, and, in some cases, of questionable origin.

Although the preservation of backset bedding appears to be relatively uncommon in the geological record, it may be more common than Twenhofel's statement implies. Within the last decade, for example, backset bedding of (inferred) standing wave and antidune origin has been described from fluvial deposits (Hand et al., 1969), and from turbidite beds (Skipper, 1971). Recently, backset bedding has been reported in phreatic pyroclastic flows of very high flow energy (Schmincke et al., 1973). Here the backset bedding is apparently associated with antidune and chute-and-pool structures (see *Base-Surge Deposits*).

Backset bedding dips at a relatively low angle (commonly $< 15°$) in the upcurrent direction, and caution must be exercised in the interpretation of its directional significance; it can easily be mistaken for "normal" cross-bedding, which dips in the general downcurrent direction. The association of other cogenetic sedimentary structures may provide the confirmatory clue to paleoflow direction.

The flume researches of Middleton (1965), Jopling and Richardson (1966), and others demonstrate that backset bedding can form both in antidune flow and in chute-and-pool flow. Characteristically, the backset bedding is

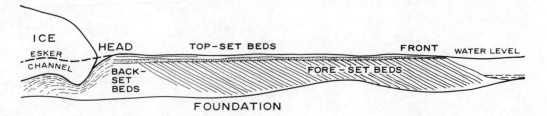

FIGURE 1. Backset beds forming at the head of a glaciofluvial sand plain (redrawn after Davis, 1890).

FIGURE 2. Backset beds developed in a flume experiment (from Jopling and Richardson, 1966). The material is a gravelly sand and flow is from right to left.

intercalated with horizontal, undulose, or lenticular bedding, or with bedding gently inclined in the downcurrent direction. An example of backset bedding developed in a laboratory flume is shown in Fig. 2.

ALAN V. JOPLING

References

Davis, W. M., 1890. Structure and origin of glacial sand plains, *Geol. Soc. Am. Bull.,* **1,** 195-202.

Hand, B. M.; Wessel, J. M.; and Hayes, M. O., 1969. Antidunes in the Mount Toby Conglomerate (Triassic), Massachusetts, *J. Sed. Petrology,* **39,** 1310-1316.

Harms, J. C., and Fahnstock, R. F., 1965. Stratification, bed forms, and flow phenomena (with an example from the Rio Grande), in G. V. Middleton, ed., *Primary Sedimentary Structures and Their Hydrodynamic Interpretation, Soc. Econ. Paleont. Mineral. Spec. Publ. 12,* 84-115.

Jopling, A. V., and Richardson, E. V., 1966. Backset bedding developed in shooting flow in laboratory experiments, *J. Sed. Petrology,* **36,** 821-825.

Middleton, G. V., 1965. Antidune cross-bedding in a large flume, *J. Sed. Petrology,* **35,** 922-927.

Schmincke, H.-U.; Fisher, R. V.; and Waters, A. C., 1973. Antidune and chute and pool structures in the base surge deposits of the Laacher See area, Germany, *Sedimentology,* **20,** 553-574.

Skipper, K., 1971. Antidune cross-stratification in a turbidite sequence, Cloridorme Formation, Gaspé, Quebec, *Sedimentology,* **17,** 51-68.

Twenhofel, W. H., 1939. *Principles of Sedimentation.* New York: McGraw-Hill, 610 p.

Cross-references: *Base-Surge Deposits; Bedding Genesis; Bed Forms in Alluvial Channels; Cross-Bedding; Flow Regimes; Fluvial Sediment Transport; Sedimentary Structures.*

BAGNOLD EFFECT

The term *Bagnold effect* (Bagnold dispersive stress) was proposed by Sanders (1963) for the collective collision among individual grains during sediment transport by traction. The name honors R. A. Bagnold, who first demonstrated the existence of such collisions and has also quantified the effects of these collisions on the current flow.

The Bagnold effect occurs only where the individual grains of sand are free to collide with one another (i.e., they are cohesionless) while being driven forward by the shearing effect of the current. The subsequent grain-to-grain collisions result in the transfer of momentum and inertia, thus giving rise to the physical sediment-transport process of *traction*. In traction the saltating individual sediment particles are not influenced by the turbulent eddies of the current, as is the case in suspension, but follow straight lines paralleling the main current flow.

Bagnold not only explained for the first time the unique attributes of traction but also indicated the importance of shear forces at the liquid/sediment interface. Rippled sediment surface has been interpreted by geologists since Bucher (1919) to represent a surface of equilibrium of lesser frictional resistance to current flow. Bagnold, however, has proved that ripple marks produce a surface of greater frictional resistance to the current. The spontaneous construction of ripples from cohesionless sediment grains results when shearing stress from the current is applied more rapidly to the topmost layer of sediment than resistance to this shearing can be supplied by the bottom through grain-to-grain collisions. The resulting ripples create greater resistance to shearing by forming a rougher sediment surface. Ripple marks are thus equilibrium forms, but instead of providing less frictional resistance to the liquid (Bucher, 1919; and others), they form a surface of greater friction to the current.

IMRE V. BAUMGAERTNER

References

Bagnold, R. A., 1937. The transport of sand by wind. *Geogr. J.,* **89,** 409-438.

Bagnold, R. A., 1956. The flow of cohesionless grains in fluids, *Roy. Soc. London Phil. Trans.,* ser. B, **249,** 235-297.

Bagnold, R. A., 1966a. An approach to the sediment transport problem from general physics, *U.S. Geol. Surv. Prof. Pap. 411-1,* 11-137.

Bagnold, R. A., 1966b. The shearing and dilation of dry sand, *Proc. Roy. Soc. London,* ser. A, **295,** 219-232.

Bagnold, R. A., 1968. Deposition in the process of hydraulic transport, *Sedimentology,* **10,** 45-56.

Bucher, W. H., 1919. On ripples and related sedimentary surface forms and their paleogeographic interpretation, *Am. J. Sci.*, n.s., 47, 149-210, 241-269.

Sanders, J. E., 1963. Concepts of fluid mechanics provided by primary sedimentary structures, *J. Sed. Petrology*, 33, 173-179.

Cross-references: *Eolian Sedimentation; Grain Flows; Inverse Grading; Saltation.*

BAHAMA BANKS SEDIMENTOLOGY

The Bahama Banks have long been recognized as containing within their complex sedimentary environments the modern analogues of many of the kinds of carbonate rocks found in the geologic record. The scale and variability of the sedimentary environments, as well as the relative accessibility of the Bahama Banks to European and North American geologists, have attracted countless large- and small-scale investigations and, in addition, field expeditions to train geologists in the recognition of ancient carbonate facies. Thus, a great deal of the published and unpublished interpretative stratigraphy of ancient carbonate terranes is influenced by conceptual models of carbonate sedimentation based upon data from the Bahama Banks.

Setting

The geography and hydrography of the Bahama Banks have been described in several geological reports; Bathurst (1971) contains a recent summary. The Bahama Banks are located on the continental shelf of North America, E and SE of the coast of Florida. The Banks are shallowly submerged platforms separated from the mainland of the United States and from Cuba by deep (>500 fathoms) channels or straits, a fact of great significance to the sedimentology of the Banks. The deeps that surround the Banks are evidently sinks for terrigenous detritus. Very little silicate material is found in Bahamian sediments, and what there is may be from wind-blown material and the breakdown of the skeletons of siliceous sponges (Newell and Rigby, 1957). Thus, the isolation of the Banks from sources of terrigenous detritus gives rise to a very pure form of carbonate deposition, one undiluted by extrabasinal materials.

The climate of the Bahamas is mild and humid. Temperatures range from 19-32°C (Smith 1940); and rainfall is unevenly distributed throughout the year, most of it falling from May to December, the total varying from about 100-150 cm/yr. (Purdy, 1963). Tropical storms (called *hurricanes* in this region) are common. The relatively moist climate of the Bahamas, together with a generally open circulation of marine waters, evidently precludes severe hypersalinity of marine waters or the widespread formation of evaporites. Exceptions are the elevated salinities (to 46‰ according to Smith) that occur to the west of Andros Island, the dolomite or "proto-dolomite" found in a few areas on the W coast of Andros Island, and the commercial salt deposits on the island of Inagua.

The physiography of the Bahama Banks also strongly affects sedimentation. The Banks are steeply margined; in many places the 100-fathom and 1000-fathom contours are only about a km apart. The upper surface of the platforms are regionally quite flat—more than 50% of the submarine area of the platform is < 3 fathoms deep, and most of the islands are of low relief. Local, small changes in relief, however, are important with respect to sedimentary processes. The slope break between the steep margins of the platform and their flattish upper surfaces has been termed the *outer platform.* (Many workers have contributed to the special physiographic terminology used to describe the Bahamas, see Newell and Rigby, 1957.) This feature is a submarine terrace, or, at places, a series of terraces, floored by Pleistocene bedrock or by a thin sediment cover. Toward the interiors of the Banks, the outer platform is bordered by a *barrier rim,* which in places may be a reef, an island (or cay), or a shoal of unconsolidated (commonly oolitic) sediment. Although the relief of most of the barrier rim is not great (on the order of a meter, except for some of the islands, which may rise as high as 125 cm), it constitutes a significant barrier to the completely free circulation of oceanic waters onto the Banks. Behind the barrier rim, and occupying by far the greatest areal extent of any of the physiographic regions of the Banks, is the *shelf lagoon* with water depths mostly <3 fathoms.

The Holocene sediments of the Bahamas constitute as extremely thin (±6-7 m—Ball, 1967; Bathurst, 1971; Buchanan, 1970, etc.) layer of carbonate sediments atop an immense pile of carbonate rocks, of which close to 5700 m has been penetrated by drilling. The carbonate rocks are as old as early Cretaceous (see Spencer, 1967, and Khudoley, 1967, cited in Goodell and Garman, 1969). The constitution of the rocks below the Lower Cretaceous carbonates is unknown, but geophysical data (Khudoley, 1967) indicate the total sedimentary thickness may be 10.5 kilometers. Goodell and Garman (1969) have interpreted the rocks taken from an exploratory well drilled on Andros Island as having formed from shallow-water sediments essentially the same as those now forming on the shelf lagoon.

Sedimentary Facies

The significance of the sediments of the Bahama Banks to the interpretation of ancient carbonate rocks was recognized by Sorby at least as early as 1879. In this century, other workers have added knowledge of these sediments, to the point that a highly predictive model for the interpretation of ancient carbonate rocks has been constructed.

The early work of Vaughan, Agassiz, Drew, Field, Thorp, and Black (see references in Bathurst, 1971) made known the general distribution of sediment types, and to some extent, the origins of many of the kinds of carbonate grains. Later work by Illing, Newell and his coworkers (including Imbrie and Purdy), Cloud, Bathurst, Taft, and Ginsburg and his coworkers (see the bibliographies of Purdy, 1963 or Bathurst, 1971) has provided detailed descriptions of sedimentary distributions and a firm theoretical basis for understanding the origin of grain types. Imbrie and Buchanan (1965) have described internal sedimentary structures of several of the sedimentary facies.

Figure 1 shows the distribution of sedimentary facies on the Great Bahama Banks (see also *Tropical Lagoonal Sedimentation*). Most sediments on the Banks are mixtures of several kinds of grains. The facies mapped reflect recurring combinations of various kinds of grains as well as the most abundant or "dominant" grain type.

The *coralgal facies* is a skeletal sand which floors much of the outer platform. It is typified by an abundance of coral and coralline algal fragments as well as other skeletal grains. A source of much of this sediment is the sessile benthos which dominates the biota of the outer platform. The reef facies constitutes an extensive coral reef tract on the windward side of Andros Island.

The *oolite* (q.v.) and *oolitic facies* form sand shoals on the barrier rim. The two facies grade into one another, the oolitic facies having a greater admixture of grains other than ooids than the oolite facies, which occurs in the shallowest, most turbulent parts of oolite shoals. Internally, bodies of oolite sand are complexly cross-bedded (Imbrie and Buchanan, 1965).

The *grapestone* (q.v.) *facies* occupies large areas of the shelf lagoon. The distinctive grain type in this facies is the grapestone grain (so named by Illing, 1954) which is a sand-sized aggregate grain in which smaller grains are cemented together, apparently by lithified lime mud (Purdy, 1963). Cores taken in level bottom

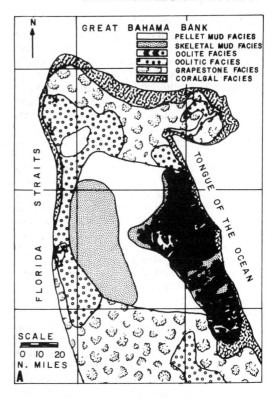

FIGURE 1. Sedimentary facies of the Great Bahama Bank (from Imbrie and Purdy, 1962).

areas of the grapestone facies usually show the effects of bioturbation; grapestone shoals are often cross-bedded, probably resulting from major storms (Imbrie and Buchanan, 1965).

Both mud facies, which occupy the center of the shelf lagoon, contain abundant amounts of lime mud in the form of aragonite needles only a few microns long. The *pellet-mud facies* contains abundant fecal pellets, which are absent or much less abundant in samples from the *skeletal-mud facies,* where either pellet-producing organisms are not present or else the pellets are not preserved (Purdy, 1963).

The origin of the sand-sized skeletal grains, or the organically produced grains, such as fecal pellets, is fairly obvious and undebatable. However, much discussion has been given to the origin of the other grains which so typify Bahamian sediments—ooids, grapestone, and lime mud. Bathurst (1971) has summarized and commented on the many theories of origin of these grains, and there is not space here to discuss all the arguments. Much of the difficulty lies in determining the role, if any, played by organisms, especially algae and bacteria in (1) forming oolitic laminae (see *Oolite*); (2) cementing grains to form grapestone (q.v.) and (3) producing aragonite needles (see *Limestones*).

Mineralogically, Bahamian sediments consist mainly of aragonite and high-magnesian calcite with some low-magnesian calcite. Dolomite occurs in supratidal areas of the W coast of Andros Island, where it is associated with blue-green algal mats. As mentioned above, Bahamian sediments are remarkably free of noncarbonate minerals.

The role of organisms in producing and modifying sediments is extremely important in Bahama Banks sedimentology. A close association of foraminiferal facies with sedimentary facies was illustrated by Streeter (1963). Clearly the foraminifers respond to some of the same physical influences as the sediments (of course, the foraminifers, on death, become part of the sediment). The complex interplay of organisms, sediments, and physical parameters is discussed at length by Newell et al. (1959). Imbrie and Buchanan (1965) illustrate the effects of burrowing and root growth on sedimentary structures.

Purdy (1963) has suggested that the major physical factors affecting the distribution of sedimentary facies on the Great Bahama Bank are (1) current strength (which decreases toward the center of the Bank), (2) salinity (which increases toward the center of the Bank), and (3) the geometry of the karst surface over which the sea rose at the end of the latest glacial episode (notice the evident effect of Andros Island, a bedrock high, in providing a wind-shadow for the accumulation of lime mud in its lee).

Ball (1967) and Buchanan (1970) have added the dimension of depth to the areal distribution of sedimentary facies. Buchanan's cross section (Fig. 2) shows, in a much smaller area, a facies distribution much like that of Purdy's map. The pellet-mud facies does not exist at the sediment surface in the southern Berry Islands, but is preserved beneath the grapestone facies in a swale in the underlying Pleistocene karst surface.

Summary

The Bahama Banks provide a useful model for carbonate sedimentary processes and facies distribution. It must be kept in mind, however, that the Bahamas are a humid, subtropical area surrounded by deep, normal-marine water masses. That conditions in the vast epeiric seas of the geologic past were different is almost a certainty. Conditions, for example, in the present-day arid and restricted Persian Gulf (q.v.) are obviously different. So, caution is the rule in applying the Bahamian (like any other) modern model of sedimentation. That the application can be successful when done properly is evident from work such as that of Roehl (1967).

HUGH BUCHANAN

References

Ball, M. M., 1967. Carbonate sand bodies of Florida and the Bahamas, *J. Sed. Petrology*, 37, 556-591.

Bathurst, R. G. C., 1971. *Carbonate Sediments and Their Diagenesis*. Amsterdam: Elsevier, 620p.

Buchanan, H., 1970. Environmental stratigraphy of Holocene carbonate sediments near Frazers Hog Cay, British West Indies, Ph.D. thesis, Columbia University, New York, 229p.

Goodell, H. G., and Garman, R. K., 1969. Carbonate geochemistry of Superior Deep Test Well, Andros Island, Bahamas, *Bull. Am. Assoc. Petrol. Geologists*, 53, 513-536.

Husseini, S. I., and Matthews, R. K., 1972. Distribution of high-magnesium calcite in lime muds of the Great Bahama Bank: diagenetic implications, *J. Sed. Petrology*, 42, 179-182.

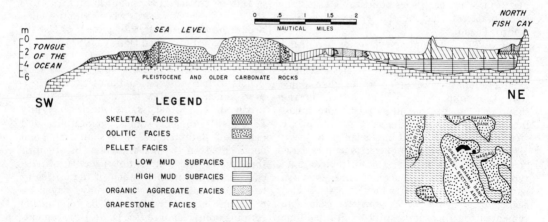

FIGURE 2. Sedimentary facies, southern Berry Islands area, Great Bahama Bank (from Buchanan, 1970).

Illing, L. V., 1954. Bahaman calcareous sands, *Bull. Am. Assoc. Petrol. Geologists,* 38, 1-95.
Imbrie, J., and Buchanan, H., 1965. Sedimentary structures in modern carbonate sands of the Bahamas, *Soc. Econ. Paleont. Mineral. Spec. Publ. 12,* 149-172.
Imbrie, J., and Purdy, E. G., 1962. Classification of modern Bahamian carbonate sediments, in W. E. Ham, ed., *Classification of Carbonate Rocks.* Tulsa, Okla: Am. Assoc. Petrol. Geologists, 253-272.
Kornicker, L. S., 1963. The Bahama Banks: a "living" fossil-environment, *J. Geol. Educ.,* 11(1), 17-25.
Lynts, G. W., Judd, J. B., and Stehman, C. F., 1973. Late Pleistocene history of Tongue of the Ocean, Bahamas, *Geol. Soc. Am. Bull.,* 84, 2665-2684.
Margolis, S., and Rex, R. W., 1971. Endolithic algae and micrite envelope formation in Bahamian oolites as revealed by scanning electron misroscopy, *Geol. Soc. Am. Bull.,* 82, 843-852.
Newell, N. D., and Rigby, J. K., 1957. Geological studies on the Great Bahama Bank, *Soc. Econ. Paleont. Mineral. Spec. Publ. 5,* 15-72.
Newell, N. D.; Imbrie, J.; Purdy, E. G.; and Thurber, D. L., 1959. Organism communities and bottom facies, Great Bahama Bank, *Bull. Am. Mus. Nat. Hist.,* 117, 181-228.
Purdy, E. G., 1963. Recent calcium carbonate facies of the Great Bahama Bank, *J. Geol.,* 71, 334-355, 472-497.
Roehl, P. O., 1967. Stony Mountain (Ordovician) and Interlake (Silurian) facies analogs of Recent low-energy marine and subaerial carbonates, Bahamas, *Bull. Am. Assoc. Petrol. Geologists,* 51, 1979-2032.
Shattuck, G. B., and Miller, B. L., 1905. Physiography and geology of the Bahama Islands, in G. B. Shattuck, ed., *The Bahama Islands.* New York: Macmillan, 1-20.
Smith, C. L., 1940. The Great Bahama Bank, 1. General hydrographic and chemical factors, 2. Calcium carbonate precipitation, *J. Marine Research,* 3, 1-31; 147-189.
Streeter, S. S., 1963. Forminifera in the sediments of the northwestern Great Bahama Bank, Ph.D. thesis, Columbia University, New York, 228p.

Cross-references: *Algal-Reef Sedimentology; Carbonate Sediments—Bacterial Precipitation; Carbonate Sediments—Lithification; Coral-Reef Sedimentology; Grapestone; Limestone; Oolites; Tropical Lagoonal Sedimentation.*

BALL-AND-PILLOW STRUCTURES— See PILLOW STRUCTURES

BANDED IRON FORMATION—See CHERTY IRON FORMATIONS

BARRIER ISLAND FACIES

Barrier islands are elongate depositional features occurring in most of the coastal regions of the world. Johnson (1919) used the term *offshore bar* for these features; but because they remain emerged most of the time and because in some cases they may be up to 100 km long, 20 km wide, and 50 m high, *barrier island* seems a more appropriate term for features that trend roughly parallel to the coastline but are separated from the mainland by a relatively narrow body of water or marsh.

Barrier islands are typical of both submergent and stable coasts (Hoyt, 1967; Schwartz, 1971). According to Leontyev (1965), 10% of the world's shorelines contain barrier/lagoon complexes. If a similar situation has prevailed during times of general submergence in the geologic past, barrier islands must be represented in a significant percentage of *strand-line deposits* in the geologic record. Their recognition is therefore of great paleogeographic significance.

Sediments in barrier environments are deposited in relatively shallow, and often agitated, waters; therefore, barriers are mainly composed of mature sediments which are highly porous and permeable. Flanking lagoonal and offshore sediments are characterized by finer-grained, organic-rich deposits which act not only as natural traps but also as potential source beds for petroleum (Davies et al., 1971). Hence, recognition of ancient barriers is also economically very important—potentially, barriers may be among the most productive reservoir sand bodies.

Even though some of the highly productive "shoestring sands" of Oklahoma and Kansas were compared with the modern barriers as early as 1934 by Bass (1934, 1939), it was not until the 1950s that criteria were developed for the recognition of barrier environments (Shepard and Moore, 1955). The earlier workers interpreted linear sand bodies as barriers chiefly on the basis of trends and overall geometry; some of the more recent workers have also followed the same approach (Rittenhouse, 1961). According to concepts based on geometry, barrier sands should have a flat bottom and a convex-up upper surface and should trend parallel to the former coastline. Chiefly as a result of the work carried out by Shepard and others in the Gulf Coast, it was recognized that barrier islands represent a complex of various subenvironments. Hence the identification of a particular unit as an ancient barrier island depends not only on its trend and overall geometry, but also on the vertical sequence of sedimentary structures, sediment textures, and the relation of this suspected barrier unit to the units above and below.

Various subenvironments of barriers and the criteria for recognizing them in both modern and ancient settings are described here; some criteria for recognizing ancient barrier coastlines are described by Dickinson et al (1972) and Spearing (1974). Also discussed here are some theoretical models for expected sequences in the case of marine transgression or regression along a coastline of barrier/lagoon complexes. Their recognition can thus greatly aid any reconstruction of the history of transgressions and regressions. The origin of barriers (see Hoyt, 1967; Schwartz, 1971, 1973) will not be discussed here.

Major Barrier Environments

Four major environments (Fig. 1) can be recognized within the barrier island: *backbarrier, dunes and beach, inlet,* and *barrier "toe"* (Kumar, 1973). The dunes and beach, and "toe" environments would also be present on shorelines lacking barrier/lagoon complexes, and only inlets and backbarrier environments are unique to barrier coastlines. These four major divisions do not cover all the related environments; many subenvironments can also be recognized within modern barriers (Shepard, 1960; Hayes and Scott, 1964). Three out of the above four major environments produce characteristic vertical sequences, and these sequences have a good chance of preservation in the geologic record. The geometry of these sequences as they are preserved on the shelf during a transgression or regression reflects the history of barriers through geologic time. The fourth major environment, the dunes and beach environment, has very little chance of preservation, especially during a transgression. Ironically, this environment constitutes the most "visible" part of the barrier today and has been studied in the most detail.

Study of barrier sediments in Holland has been carried out by Van Straaten (1965). Usually, however, Gulf Coast barriers are cited as a "type" model although, in fact, they represent a rather special geologic setting where the coastline is relatively stable and the supply of sediment abundant; the coastline is therefore prograding. In such a setting, a "depositional regression" (Curray, 1964) would take place, and the overall sequence would be, from top to bottom: nonmarine, lagoon, barrier, nearshore, and offshore (Fig. 2). In this situation the dune belt and beach facies may also be preserved.

In contrast to the geologic setting of the Gulf Coast, which cannot be considered as very typical of shorelines all over the world, the

FIGURE 1. Aerial view of Fire Island barrier, Long Island, New York showing various subenvironments of barrier-lagoon complex on the south shore of Long Island. The "toe" of the barrier lies seaward of the beach in water depths between 5 and 21 m. (View looking E, Oct. 21, 1972; the bridge in the picture is 1.2 km long. (Photo: Bruce Caplan; courtesy of John Sanders.)

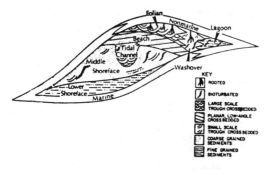

FIGURE 2. Expected sequence during a depositional regression on a coastline with barrier-lagoon complex (after Davies et al., 1971).

extensive barrier systems of the US Atlantic coast, at least N of the Carolinas, occur in a geologic setting where the coast is submerging and the sediment supply to the shelf from rivers is minimal, a situation more usual along world coastlines. This situation represents a setting favorable for transgression, and two possibilities exist for the barriers in this case (Fig. 3). First, the barriers may migrate more or less continuously landward as the sea level rises (see *Continental-Shelf Sedimentation*). Second, the barriers may remain in place as the sea rises to the level of the top of the dunes; then the surf zone may "jump" landward to establish a new shoreline, and thus the old barrier may "drown" in place (Sanders and Kumar, 1975). In either case, the chances of preserving the dune facies are minimal.

During a transgression, the geometry and setting of inlet facies would reflect the barrier behavior during the submergence. Because the inlets migrate laterally (i.e., parallel to the coast) at a relatively fast rate (up to several km per century), a major part of the length of barriers may contain inlet-filling sediments beneath the dune-belt facies (Kumar, 1973; Hoyt and Henry, 1967). Also, because the inlets are generally deeper than the lagoon on the landward side of the barrier and the shelf immediately seaward of the barrier, any sediment deposited in the inlet environment has a higher probability of escaping erosion during a submergence. In this way, if the barriers migrate continuously landward, the inlet facies would form a continuous blanket as a result of a combination of the following two processes: (1) lateral migration of inlets parallel to shore, and (2) continuous landward retreat of barriers perpendicular to the shore. The sequence under the shelf after a continuous retreat of barriers would be, from top to bottom: offshore; nearshore; beach; inlet-filling sediments; and, possibly, nonmarine. When the barriers "drown" in place, the inlet-filling sediments are represented by long, linear bodies of sand marking each successive position of the barrier (Fig. 4). The sequence in the backbarrier environment of the drowned barrier would be, from top to bottom: offshore, or nearshore, lagoon, and backbarrier. The facies underlying the backbarrier facies would reflect the history of the barrier before the "in-place drowning" took place.

Recognition of Barrier Facies in Modern Sediments

Backbarrier facies (Shepard's barrier flats or marshes), as well as dune and beach facies and

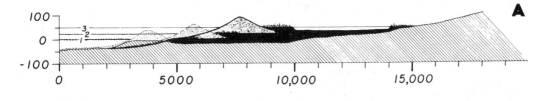

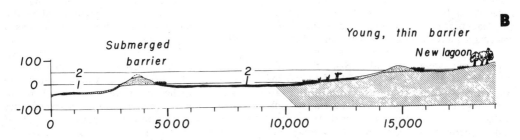

FIGURE 3. Johnson's concept of barrier retreat during a rise of sea level (levels 1 through 3). Width of lagoon remains constant. Horizontal and vertical scales give relative dimensions only. V.E. = 10. (Both drawings by courtesy of John Sanders.)

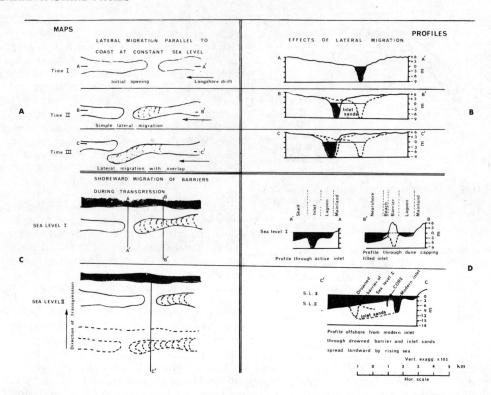

FIGURE 4. A graphic summary of the effects of submergence on a shelf with barrier-lagoon system (from Kumar, 1973). As the inlets migrate parallel to the shore (plan views A), inlet-filling sands form the "roots" of those parts of the barrier through which the inlet has migrated (sections B). As the sea level rises from one level to another (plan view C), the barriers may migrate landward continuously with the sea level. In this situation, a blanket of inlet-filling sands would underlie that part of the shelf through which the sea has transgressed (section D). However, if the barriers "drown" in place, long lenses of inlet-filling sands would underlie the shelf at the former locations of barriers (plan view C and section D).

criteria for recognizing them will be discussed here. For description of inlet-filling sediments and barrier "toe" sediments, see *Tidal-Inlet and Tidal-Delta Facies* and *Storm Deposits*, respectively.

Barrier-flat sediments contain a high percentage of silt and clay. Calcareous aggregates are common (Shepard, 1960). Caldwell (1971) has described the following subenvironments in a tidal delta: channel, mussel beds, active sand lobes, tidal flats, and marsh. Washover fans in the backbarrier environment have been described by Pierce (1970). Backbarrier sediments commonly contain peat beds (Kumar, 1973) and brown-over-gray sands (Sanders et al., 1970; reference in Kumar, 1973). The brown-over-gray sands consist of a top layer, 20–50 cm thick, of coarse to medium-grained, well-sorted brown sand containing shell fragments, and a bottom layer which consists of gray, fine, silty sand with shells. A sharp color and size boundary separates the brown and gray sands.

Dune sediments are quite similar to beach sediments in texture (see *Beach Sands* and *Eolian Sands*). However, the silt content is higher in dune sands than in beach sands; shells are generally rarer than in beach, and the plant material is more common. Dunes should be most easily recognized by their sedimentary structures (see *Eolian Sedimentation*). Dickinson et al. (1972) have discussed the problems of recognizing dune sediments on the basis of sedimentary textures, structures, and mineralogy.

Beach environment has been defined as the zone between outermost breakers and swash limit, the uppermost limit of wave action at high tide. Clifton et al. (1971) discuss subenvironments of the beach zone. Shideler (1973) has proposed a model, according to which the sands under the foreshore (beachface), berm, and dunes have unique textural characteristics acquired as a response to processes in each of these zones as the sand is transported across the barriers from beachface to berm zone and from berm zone to dunes. Bedding in the beach zone has gentle to moderate dips both landward and seaward (Thompson, 1937), and is generally even but variable in thickness; banding is common and may be imparted by laminae of heavy minerals or layers of coarse shell fragments (Dickinson et al., 1972).

Ancient Barrier Facies

Dickinson et al. (1972) mention several examples of ancient barrier islands preserved in the geologic record. According to Davies et al. (1971), however, the differentiation in the ancient rocks classified as "barrier" or "bar" is rarely undertaken. They have discussed three examples: Galveston Barrier Island, Texas; Lower Cretaceous of Montana; and Lower Jurassic of England. In each of these examples the following sequence was seen from bottom upwards: (1) irregular interlamination of siltstone and claystone at the base, representing lower shoreface; (2) burrowed and generally structureless sandstone, representing middle shoreface; (3) low-angle and micro-trough cross-laminated sandstone, representing upper shoreface beach; and finally, (4) structureless and rooted sandstone, representing eolian environment.

The examples cited by Davies et al. (1971) and Miller (1962) are from regressive sequences, and in these cases preservation of eolian environments is consistent with the theoretical model discussed earlier. Swift (1968) has attempted to explain the transgressive sequence in the Cretaceous of North Carolina through continuous landward retreat of barriers, and Fischer (1961) has invoked a similar process to explain the cyclothems of Pennsylvanian to early Permian age in SE Kansas. Interpretations based on "in-place drowning" of barriers are not available from the geologic literature, but the writer believes that the sequence in the Cretaceous Mesaverde Group in NW Colorado (Masters, 1967) and in the Cliff House and Menefee formations (Cretaceous) in the San Juan Basin (Hollenshead and Pritchard, 1961) can be explained only by invoking the in-place drowning mechanism. A major problem in predicting the consequences of transgression on a coastline with barrier-lagoon systems has been that until recently no reliable criteria had existed for recognizing inlet-filling sediments (Kumar, 1973). Kumar and Sanders (1970) have suggested that once inlet-filling sediments become "household words" for geologists, many *blanket sands* and *shoestring sands* in basal parts of transgressive sequences would be recognized as *inlet-filling sands*.

The recognition of ancient barrier islands can only be carried out by recognizing all the individual environments within a barrier-island complex. The reconstruction of paleogeography and history of transgression or regression can be carried out only by recognizing all the facies and through an imaginative use of Walther's Law.

NARESH KUMAR

References

Bass, N. W., 1934. Origin of Bartlesville Shoestring sands, Greenwood and Butler Counties, Kansas, *Bull. Am. Assoc. Petrol. Geologists,* 18, 1313-1345.

Bass, N. W., 1939. Verden Sandstone of Oklahoma, an exposed shoestring sand of Permian age, *Bull. Am. Assoc. Petrol. Geologists,* 23, 559-581.

Caldwell, D. M., 1971. A sedimentological study of an active part of a modern tidal delta, Moriches Inlet, L. I., New York, Master's thesis, Columbia Univ., New York, N.Y., 70p.

Clifton, H. E., Hunter, R. E., and Phillips, R. L., 1971. Depositional structures and processes in the non-barred high energy nearshore, *J. Sed. Petrology,* 41, 651-670.

Curray, J. R., 1964. Transgressions and regressions, in R. L. Miller, ed., *Papers in Marine Geology.* New York: Macmillan, 175-203.

Davies, D. K., Ethridge, F. G., and Berg, R. R., 1971. Recognition of barrier environments, *Bull. Am. Assoc. Petrol. Geologists,* 55, 550-565.

Dickinson, K. A., Berryhill, H. L., and Holmes, C. W., 1972. Criteria for recognizing ancient barrier coastlines, *Soc. Econ. Paleont. Mineral. Spec. Publ. 16,* 192-214.

Fischer, A. G., 1961. Stratigraphic record of transgressive seas in light of sedimentation on Atlantic coast of New Jersey, *Bull. Am. Assoc. Petrol. Geologists,* 45, 1656-1661.

Hayes, M. O., and Scott, A. J., 1964. Environmental complexes south Texas coast, *Gulf. Coast Assoc. Geol. Soc. Trans.,* 14, 237-240.

Hollenshead, C. T., and Pritchard, R. L., 1961. Geometry of producing Mesaverde sandstones, San Juan Basin, in Peterson and Osmond (1961) 98-118.

Hoyt, J. H., 1967. Barrier island formation, *Geol. Soc. Am. Bull.,* 78, 1125-1136.

Hoyt, J. H., and Henry, V. J., 1967. Influence of island migration on barrier-island sedimentation, *Geol. Soc. Am. Bull.,* 78, 77-86.

Johnson, D. W., 1919. *Shore Processes and Shoreline Development.* New York: Wiley, 584p.

Kumar, N., 1973. Modern and ancient barrier sediments: new interpretations based on stratal sequence in inlet-filling sands and on recognition of nearshore storm deposits, *Ann. N.Y. Acad. Sci.,* 220, 245-340.

Kumar, N., and Sanders, J. E., 1970. Are basal transgressive sands chiefly inlet-filling sands? *Maritime Sediments,* 6, 12-14.

Leontyev, O. K., 1965. Flandrian transgression and the genesis of barrier bars, in H. E. Wright, Jr., ed., Quaternary geology and climate, Proc. 7th INQUA Congr., 16, *Natl. Acad. Sci. Publ. 1701,* 146-149.

Masters, C. D., 1967. Use of sedimentary structures in determination of depositional environments, Mesaverde Formation, Williams Fork Mountains, Colorado, *Bull. Am. Assoc. Petrol. Geologists,* 51, 2033-2043.

Miller, D. N., Jr., 1962. Patterns of barrier bar sedimentation and its similarity to Lower Cretaceous Fall

River stratigraphy, *Wyo. Geol. Assoc. Guidebook, 17th Field Conf.,* 232–347.
Peterson, J. A., and Osmond, J. C., eds., 1961. *Geometry of Sandstone Bodies.* Tulsa, Okla.: Am. Assoc. Petrol. Geologists, 240p.
Pierce, J. W., 1970. Tidal inlets and washover fans, *J. Geol.,* 78, 230–234.
Rittenhouse, G., 1961. Problems and principles of sandstone-body classification, in Peterson and Osmond (1961), 98–118.
Sanders, J. E., and Kumar, N, 1975. Evidence of shoreface retreat and in-place "drowning" during Holocene submergence of barriers, shelf off Fire Island, New York, *Geol. Soc. Am. Bull.,* 86, 65–76.
Schwartz, M. L., 1971. The multiple causality of barrier islands, *J. Geol.,* 79, 91–94.
Schwartz, M. L. ed., 1973. *Barrier Islands.* Stroudsburg, Pa.: Dowden, Hutchinson & Ross, 451p.
Shepard, F. P., 1960. Gulf Coast barriers, in F. P Shepard, F. B. Phleger and T. H. Van Andel, eds., *Recent Sediments, Northwest Gulf of Mexico.* Tulsa, Okla.: Am. Assoc. Petrol. Geologists, 197–220.
Shepard, F. P., and Moore, D. G., 1955. Central Texas coast sedimentation: characteristics of sedimentary environment, recent history, and diagenesis, *Bull. Am. Assoc. Petrol. Geologists,* 39, 1463–1593.
Shideler, G. L., 1973. Evaluation of a conceptual model for the transverse sediment transport system of a coastal barrier chain, Middle Atlantic Bight, *J. Sed. Petrology,* 43, 748–764.
Spearing, D. R., 1974. Barrier island deposits, *Summary Sheets of Sedimentary Deposits,* Sheet 5, Boulder, Colo.: Geol. Soc. Am.
Swift, D. J. P., 1968. Shoreface erosion and transgressive stratigraphy, *J. Geol.,* 76, 444–456.
Thompson, W. O., 1937. Original structures of beaches, bars and dunes, *Geol. Soc. Am. Bull.,* 48, 723–752.
Van Straaten, L. M. J. U., 1965. Coastal barrier deposits in south and north Holland in particular in the areas around Scheveningen and Ijmuiden, *Meded. Geol. Stichting,* 17, 41–75.

Cross-references: *Beach Sands; Eolian Sands; Heavy Minerals; Littoral Sedimentation; Sands and Sandstones; Storm Deposits; Tidal-Inlet and Tidal-Delta Facies.* Vol. III: *Barriers–Beaches and Islands; Cuspate Foreland or Spit.*

BASE-SURGE DEPOSITS

The Base Surge

The base surge is a ring-shaped basal cloud that moves outward as a turbulent density current from the base of a vertical explosion column. Base surges are produced naturally from volcanic eruptions and artificially by thermonuclear and chemical explosions and probably can be produced by supersonic impact explosions. According to the nature of the explosion and available material, the base surge may carry only water droplets and gas, as during the shallow, underwater nuclear explosion produced at Bikini atoll in 1946, or they may carry fragmental volcanic ejecta.

First mention of the base surge in geological literature was by Richards (1959). He compared the behavior of the eruption column at Bárcena Volcano, Isla San Benedicto to the base surge developed by atomic explosion at Bikini. Subsequently, the concept of the base surge was mentioned as a possible mechanism of origin for ash flows (Fisher, 1966). Observations of horizontally moving clouds from the 1965 phreato-magmatic eruptions of Taal Volcano, Philippines led to further development of the base-surge concept and application to phreato-magmatic volcanic eruptions (Moore, 1967). Subsequent descriptions of base-surge deposits are found in studies by Crowe, Fisher, Heiken, Lörenz, Nakamura, Schmincke, and Waters (see references in Crowe and Fisher, 1973, and Schmincke et al., 1973).

Mechanisms

The development and radial dispersion of volcanic base surges occur in essentially three stages, although continuous or pulsating eruptions may cause considerable overlapping of the stages: (1) White steam with few, if any, solid particles surges outward immediately from the base of the emerging eruption column as water vapor concentrated along its periphery escapes and condenses. (2) Black plumes of solid particles shoot radially on ballistic trajectories from the walls of the rising eruption column as it is torn apart by internal steam bursts. (3) The turbulent mixture of solid particles, water vapor or steam, and air tumbles en masse to the ground (bulk subsidence) and flows outward behind and becomes incorporated within the steam surge (Waters and Fisher, 1971). The base surges observed at Taal traveled at hurricane velocities, with external flow characteristics similar to a *nuée ardente* (q.v.).

Some *nuées ardentes* are developed as turbulent clouds and/or glowing avalanches moving outward from the base of a volcanic eruption column (Hay, 1959) and could, strictly speaking, be called base surges; but the name is reserved here for the relatively cool, steam-generated volcanic base surges first recognized at Taal.

At Taal, base surges sandblasted trees and other objects at distances up to 6 km; the absence of charred wood showed that the base surges were little, if any, hotter than steam. Ash, lapilli, and blocks up to 1 m in diameter were carried. Significantly, the base-surge deposits developed surface dunes, and their migration produced internal low-angle cross-laminae with stoss sides usually steeper than the lee sides.

Evidence of abundant water in the initial deposition of base-surge deposits consists of cohesive fine-grained ash laminae plastered onto vertical surfaces, and bedding-plane sags where airborne blocks fell into wet cohesive ash, bending underlying beds downward when still in a plastic condition. Similar features are found in the ejecta rims of maar volcanoes. Maar volcanoes are characterized by shallow bowl- or dish-like craters indicative of shallow explosions; nearly all have been formed in an environment where abundant water was available either on or below the surface.

Deposits

Base-surge deposits include ash, lapilli, and block-sized fragments composed of quickly chilled bits of glass mixed in various proportions with fragments of the underlying bedrock, or shattered debris from the volcano itself. As described from the rim sequences of maar volcanoes, they are characterized by one or more of the following structural features: (1) *Low-angle cross-bedding;* current directions from the cross-bedding show that flows move radially outward from the crater; in some instances, as at Ubehebe Craters, the flows have moved uphill. Dune-like structures with wavelengths of over 20 m and wave heights of up to 2 m, down to those with a wavelength of 47 cm and wave height of 2.4 cm have been measured (Schmincke et al., 1973; Crowe and Fisher, 1973). Stoss-side laminae are commonly preserved, and lee-side laminae usually have dip angles less than the angle of repose. Stoss-side laminae commonly dip at steeper angles than lee-side laminae within the same structure. Figures 1 and 2 show some of these features. Spacing and morphologic features are similar to antidunes deposited in the high flow regime of alluvial channels. Some of the dune-like features described from Laacher See, Germany (Schmincke et al., 1973) appear to have been deposited in the chute-and-pool phase of the high flow regime. (2) *Thin, laterally continuous ash or tuff beds associated with cross-bedded tuffs;* many of these beds may be of airfall origin, but some of the plane-parallel beds have very subtle internal cross-bedding and are transitional with cross-bedded bedding sets, which suggests that they were emplaced by flow (Fisher and Waters, 1970; Crowe and Fisher, 1973). (3) *Ash or tuff laminae plastered onto steeply inclined or vertical surfaces;* Waters and Fisher (1971) illustrate an instance where the tuff is plastered beneath an overhanging cornice. (4) *Rounded or irregularly shaped cavities or vesicles within the ash or tuff,* which develop by steam expansion after deposition. (5) *Thinning and thickening of individual tuff beds over low-relief topographic highs and lows within short distances without evidence of reworking.* (6) *Small U-shaped cut-and-fill channels* (rare). (7) *Flame structures* (rare).

Many beds associated with base-surge deposits contain accretionary lapilli or cored lapilli. The accretionary lapilli are pellets, mostly 2–10 mm in diameter, composed of concentric shells of volcanic ash. Moore and Peck (1962) state that in normal airfall tuff, accretionary lapilli are formed by accretion of ash around a moist nucleus in airborne eruption clouds and thus are usually indicative of airfall deposits rather than base-surge deposits. Concentrically structured accretionary lapilli have been found, however, in cross-bedded base-surge deposits at Salt Lake Crater, Hawaii (Fisher and Waters, 1970) and thus are not unique to airfall tuffs. More common are cored lapilli that consist of a solid fragment coated with a rind of ash. The rock fragment that forms the core may be

FIGURE 1. Sequence of cross-bedded and plane parallel tuff and lapilli tuff beds in rim of maar volcano near Macdoel, California. Base surges moved from left to right. Hammer (in circle) for scale.

FIGURE 2. Close view of dune-like structure in rim sequence of maar volcano near Macdoel, California. Base surge moved from left to right. Abundant thin laminae show successive building up of the composite structure from an initial plane parallel surface. Crest migrates downcurrent. Lee-side laminae dip no more than 10°, far less than the angle of repose. Laminae are well developed, and bed forms are preserved at each level within the composite structure because of high sedimentation rate and high sediment cohesion. The structure was probably constructed during the passage of one base surge flow.

sharply angular. The coating of ash is usually a single layer up to 1-2 cm thick, but some show two or more coatings.

RICHARD V. FISHER

References

Crowe, B. M., and Fisher, R. V., 1973. Sedimentary structures in base-surge deposits with special reference to cross-bedding, Ubehebe Craters, Death Valley, California, *Geol. Soc. Am. Bull.*, **84**, 663-682.

Fisher, R. V., 1966. Mechanism of deposition from pyroclastic flows, *Am. J. Sci.*, **264**, 350-363.

Fisher, R. V., 1977. Erosion by base-surge density currents: U-shaped channels, *Geol. Soc. Am. Bull.*, **88**, 1287-1297.

Fisher, R. V., and Waters, A. C., 1970. Base surge bed forms in maar volcanoes, *Am. J. Sci.*, **268**, 157-180.

Hay, R. L., 1959. Formation of the crystal-rich glowing avalanche deposits of St. Vincent, B.W.I., *J. Geol.*, **67**, 540-562.

Lörenz, V., 1974. Vesiculated tuffs and associated features, *Sedimentology*, **21**, 273-291.

Mattson, P. H., and Alvarez, W., 1973. Base surge deposits in Pleistocene volcanic ash near Rome, *Bull. Volcan.*, **37**, 553-572.

Moore, J. G., 1967. Base surge in recent volcanic eruptions, *Bull. Volcan.* **30**, 337-363.

Moore, J. G., and Peck, D. L., 1962. Accretionary lapilli in volcanic rocks of the western continental United States, *J. Geol.*, **70**, 182-193.

Richards, A. F., 1959. Geology of the Islas Revillagigedo, Mexico, 1. Birth and development of Volcan Barcena, Isla San Benedicto, *Bull. Volcan.* **22**, 73-123.

Schmincke, H. -U.; Fisher, R. V.; and Waters, A. C., 1973. Antidune and chute and pool structures in the base surge deposits of the Laacher See area, Germany, *Sedimentology*, **20**, 553-574.

Waters, A. C., and Fisher, R. V., 1971. Base surges and their deposits: Capelinhos and Taal Volcanoes, *J. Geophys. Research*, **76**, 5596-5614.

Cross-references: *Backset Bedding; Flow Regimes; Inverse Grading; Nuée Ardente; Pyroclastic Sediment; Tuff; Volcanic Ash Deposits; Volcaniclastic Sediments and Rocks.*

BASINAL SEDIMENTATION
—*See* VOL. IX

BATHYAL ZONE—*See* SUBMARINE (BATHYAL) SLOPE SEDIMENTATION

BAUXITE—*See* Vol. XI: BAUXITE, BAUXITIZATION

BAY OF FUNDY SEDIMENTOLOGY

The Bay of Fundy, known to sedimentologists for its large tides and tidal-current deposits (q.v.), is an elongate body of water separating New Brunswick from Nova Scotia. Toward the northern and eastern end of the bay, it splits into Chignecto Bay and Minas Basin. The Bay of Fundy system is approximately 144 km in length and up to 100 km in width at its mouth to the SE. The average depth of the bay is about 75 m (Fig. 1).

Tides

The mean tidal range increases from about 3.5 m at the mouth of the bay to 11.7 m at Burncoat Head on the southeastern shore of Minas Basin. The mean spring range at Burncoat Head is 12.3 m at lunar apogee and 15.4 m at lunar perigee (Dawson, 1917). The tides are semidiurnal and of the anomalistic type. The diurnal equality is relatively small (maximum of about 0.8 m). Geostrophic forces produce a counterclockwise motion in the bay which causes the tidal range along the south shore to be about 0.45 m greater than that along the N shore (Miller, 1966). During flood tide, the water surface is banked up along the S shore, but during ebb tide, the water tends to pile up along the N shore.

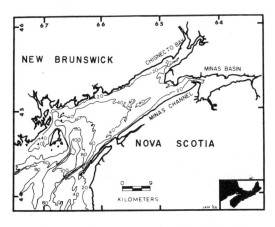

FIGURE 1. Location map of the Bay of Fundy showing bathymetry in m (from Pelletier and McMullen, 1972).

The large tides in the Bay of Fundy have often been thought to be the result of resonant amplification of the lunar semidiurnal component of the tide within the bay, but several recent studies have found this concept to be incorrect. It has been proposed that they are caused by either a two-stage amplification (Harleman, 1966) (a primary amplification due to shoaling on the contenental shelf, and a secondary amplification of the tide within the bay itself), or by resonance (Garrett, 1972) if the Bay of Fundy system is enlarged to include adjacent parts of the continental shelf.

Processes

The tides generate currents with speeds as high as 5.5 m/sec in the Minas Passage (Cameron, 1961), but generally velocities do not exceed 2 m/sec and decrease to 0.5 to 1.5 m/sec over intertidal areas. As a result of the geostrophic effects in the bay, stronger currents are recorded on the S shore than on the N (Harleman, 1966). A counterclockwise, residual current system exists with velocities up to 0.1 m/sec. Tidal currents are the most important agent of deposition and erosion in the Bay of Fundy.

The prevailing winds in the bay are from the W, so wave activity is most effective from this direction onto the bay's generally exposed coasts. Drifting ice during the winter and early spring months transports sediment within the bay. The sediment is picked up by accretionary freezing of the turbid bay waters on cake ice, and by rafting of sediments when cake ice grounds in intertidal and shallow areas.

Sediments

The St. John River contributes about 70% (mean yearly discharge rate 936.7 m^3/sec) of fresh water entering the Bay of Fundy system. For the most part, however, the rivers are seasonal and are not presently contributing large volumes of water or sediment to the bay (Miller, 1966). Shoreline erosion is probably the most important source of detrital suspended sediments, particularly in Chignecto Bay and Minas Basin (Atlantic Tidal Power Programming Board, 1969).

The average concentration of suspended sediment in the bay is about 6.6 mg/liter (Miller, 1966). The concentration increases from the mouth to the head of the bay, from the Nova Scotia shore to the New Brunswick shore, and generally with depth. The composition of the suspended sediment is largely silt and organic debris (Miller, 1966).

Most of the sediments in the Bay of Fundy have been derived from the erosion of Pleistocene, Triassic, and Carboniferous deposits in

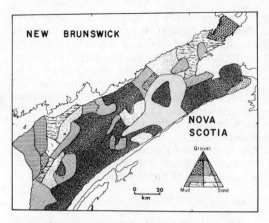

FIGURE 2. Textural facies in the Bay of Fundy (from Pelletier and McMullen, 1972).

and around the bay. The Bay of Fundy has incised itself, through repeated Pleistocene glaciations, into red continental sandstones and mudstones, and tholeiitic basalts of a Triassic half-graben. Most of the Bay of Fundy coasts (75%) are flanked by bedrock, which consists primarily of an intertidal wave-cut bench overlain by less than 1 m of poorly sorted sediments. This thin veneer of sediments shows similar textural and compositional characteristics to the underlying bedrock. The intertidal sediments are commonly backed by bedrock sea cliffs, up to 30 m in height, which are presently being eroded at rates up to 2 m/yr (ATPPB, 1969). Topographic lows in the bedrock shoreline are occupied by glacially deposited materials, of which outwash gravels are an important constituent (Swift et al., 1973). The cliffline may be cut locally by deeply incised valleys 60 m below present sea level that have been infilled with fluvioglacial deposits during the Wisconsin ice retreat, and later have been partly scoured by tidal currents.

In bays and other areas protected from wave action and strong tidal currents, extensive tidal mudflats occur backed by tidal marshes. The mudflats display many of the topographic, textural, structural and faunal features and zonations described from other areas, e.g., the Dutch Wadden Sea (q.v.). The tidal marshes are flooded only during high spring tides.

Almost 60% of the Bay of Fundy is floored with medium to coarse pebble gravels in a thin lag. Swift et al. (1973) have interpreted these as relict sediments from the Pleistocene. The gravel occurs mainly at depths greater than

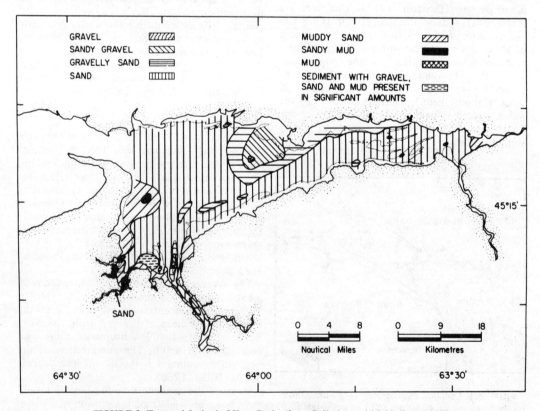

FIGURE 3. Textural facies in Minas Basin (from Pelletier and McMullen, 1972).

40 m over most of the central and southeastern parts of the bay (Fig. 2). The mineralogical composition of the gravel generally reflects the underlying bedrock lithology, with less than 20% far-traveled materials except where the sample contains friable red Triassic mudstone, in which case, the percentages are reversed unless there is a sea-floor outcrop nearby.

Sand occupies about 22% of the bay floor (almost 80% of the Minas Basin floor) (Swift et al., 1973; Pelletier and McMullen, 1972). The median grain size of the sand is medium to very coarse except in areas of transition to mud deposits, where it is fine to very fine. The main mass of sand near the head of the bay (Figs. 2 and 3) is believed to be derived from Pleistocene fluvioglacial outwash. The sand body in the eastern part of the Minas Basin has been molded into longitudinal sand bars covered by extensive bedform fields. The upper parts of some of these sand bars are exposed during low-tide periods. The pattern of sand movement on individual sand bars suggests a circulation of sand around the bar, similar to that suggested for subtidal bars in the North Sea by Houbolt (1968; see Swift and McMullen, 1968; Klein, 1970; and Dalrymple et al., 1975). Continuous seismic records indicate that the sand body in the eastern end of the Minas Basin reaches thicknesses up to 25 m.

Muddy sediments comprise about 20% of the Bay of Fundy floor and occur along the NW side of the bay from Grand Manan Island to the entrance of Chignecto Bay. The mud deposits are localized by weaker tidal currents on the NE side of the bay and are redistributed by the counterclockwise residual tidal current which flows to the SW along the north shore.

R. JOHN KNIGHT

References

Atlantic Tidal Power Programming Board, 1969. *Report to Atlantic Tidal Power Programming Board on Tidal Power Development in the Bay of Fundy, Halifax, Nova Scotia.* 181p. (circulation restricted).

Cameron, H. L., 1961. Interpretation of high altitude, small scale photography, *Canadian Surveyor,* **15,** 567-573.

Dalrymple, R. W.; Knight, R. J.; and Middleton, G. V., 1975. Intertidal sand bars in Cobequid Bay (Bay of Fundy. Ottawa: Dept. Naval Services, 34p. Vol. 2. New York: Academic Press, 293-307.

Dawson, K. R., 1917. Tides at the head of the Bay of Fundy. Ottawa: Dept. Naval Services, 34p.

Garrett, C., 1972. Tidal resonance in the Bay of Fundy and Gulf of Maine, *Nature,* **238,** 441-443.

Harleman, D. R. L., 1966. Real estuaries, in A. T. Ippen, ed., *Estuary and Coastline Hydrodynamics.* New York: McGraw-Hill, 522-545.

Klein, G. de V., 1970. Depositional and dispersal dynamics of intertidal sand bars, *J. Sed. Petrology,* **40,** 1095-1127.

Middleton, G. V., 1972. Brief guide to intertidal sediments, Minas Basin, Nova Scotia, *Maritime Sediments,* **8,** 114-122.

Miller, J. A., 1966. The suspended sediment system in the Bay of Fundy, M.Sc. thesis, Univ. Dalhousie, Halifax, N.S., 100p.

Pelletier, B. R., and McMullen, R. M., 1972. Sedimentation patterns in the Bay of Fundy and Minas Basin, in T. J. Gray and O. K. Gashus, eds., *Tidal Power.* New York: Plenum, 153-187.

Swift, D. J. P., and McMullen, R. M., 1968. Preliminary report on intertidal sand bodies in the Minas Basin, Bay of Fundy, *Can. J. Earth Sci.,* **5,** 175-183.

Swift, D. J. P.; Pelletier, B. R.; Lyall, A. K.; and Miller, J. A., 1973. Quaternary sedimentation in the Bay of Fundy, in P. J. Hood, ed., Earth Science Symposium on Offshore Eastern Canada, *Geol. Soc. Canada Pap.,* **71**(23), 113-151.

Cross-references: *Continental-Shelf Sedimentation; Estuarine Sedimentation; Relict Sediments; Tidal-Current Deposits; Wadden Sea Sedimentology.*

BEACH GRAVELS

Gravel is an unconsolidated, natural accumulation of rounded rock fragments resulting from erosion and consisting predominantly of particles larger than sand (AGI, 1972). This general definition may be equally applied to beach and fluvial gravels, the two being distinguished by their place of deposition and, according to some authors, by certain dimensionless shape indices and/or size parameters. Early studies of beach gravels were done by Wentworth, Lewis, Landon, and Marshall, while more recent work has been published by Cailleux, Carr, Kidson, Krumbein, Kuenen, King, Bluck, and Kirk (see references in Humbert, 1968). A summary of Russian literature can be found in Zenkovich (1967).

Sources. Beach gravels occur where coastal formations yield debris of suitable size: fragments eroded from stratified sedimentary rocks, conglomerates or gravel deposits forming the back-beach area, or from intricately fissured igneous outcrops. Coasts bordering regions which are, or have been, subject to glaciation or periglaciation have relatively abundant supplies of rock fragments derived from the erosion of glacial drift, gravelly moraines, and drumlins (Bird, 1969).

Size. Most US sedimentologists use the Udden-Wentworth-Krumbein scale to describe the size of beach gravels (see *Grain-Size Studies*). British workers (King, 1959) use the term *shingle* when referring to beach gravels, but this term is often strictly applied to beach *pebbles* and *cobbles* of the same size. Russian workers also use shingle in the general sense of material larger than sand size, but divide gravel and shingle (chips) into separate grain size classes (Zenkovich, 1967).

Shape. Various ways of describing particle shape have been used in order to provide a useful environmental discriminator between fluvial and beach gravels; two useful parameters are sphericity (q.v.) and roundness (q.v.).

In his evaluation of several sphericity indices Humbert (1968) found that the maximum projection sphericity, ψ_p, of Sneed and Folk (1958) was the most satisfactory in beach gravel studies. Dobkins and Folk (1970) suggest that a ψ_p value of 0.65-0.66 can serve as a boundary between fluvial and beach gravels of the same lithology within the 16-265 mm size range. Beach gravels would generally have ψ_p values 0.65-0.66. Studies of pebbles on Chesil Beach, England, by Carr (1971) and Carr et al. (1970) suggest the ψ_p boundary between fluvial and beach gravels may be closer to 0.7. A study by Stratten (1974) has confirmed that beach pebbles are statistically flatter and less spherical than river gravels.

Dobkins and Folk (1970) review the methods used for determining the roundness of beach gravels, which generally increases as the energy input on a beach increases, and also increases along the beach in the direction of longshore drift. Thus, rounding of particles on beaches results mainly from abrasion of particles during transport. It must be remembered that there is a reduction in size as well (see *Attrition*). There are some instances where particle transport results in more angular rather than more rounded beach gravels, particularly in larger sizes, where splitting, spalling, and crushing are important (Bluck, 1969).

Sorting. By sorting (q.v.) processes, gravels displaying a similarity of size or shape are naturally separated from associated but dissimilar particles by the action of waves. Considerable attention has been given to the problem of whether size or shape is the dominant factor in the sorting of gravels along and across a beach. Size-sorting plays the major role on some beaches because wave energy may be high enough to move all shapes (Carr, 1971). On other beaches, shape-sorting is dominant. Whether by size or by shape, beach gravels are usually well sorted.

IRWIN D. NOVAK

References

AGI, 1972. *Glossary of Geology*, M. Gary, R. McAfee, Jr., and C. L. Wolf, eds. Washington, D.C.: Am. Geol. Inst., 805p.

Bird, E. C. F., 1969. *Coasts.* Cambridge: MIT Press, 246p.

Bluck, B. J., 1969. Partical rounding in beach gravels, *Geol. Mag.,* **106,** 1-14.

Carr, A. P., 1971. Experiments on longshore transport and sorting of pebbles: Chesil Beach, England, *J. Sed. Petrology,* **41,** 1084-1104.

Carr, A. P.; Gleason, R.; and King, A., 1970. Significance of pebble size and shape in sorting by waves, *Sed. Geol.,* **4,** 89-101.

Dobkins, J. E., and Folk, R. L., 1970. Shape development on Tahiti-Nui, *J. Sed. Petrology,* **40,** 1167-1203.

Humbert, F. L., 1968. Selection and wear of pebbles on gravel beaches, *Univ. Groningen Geol. Inst. Publ. 190,* 144p.

McManus, J., 1973. Graded beach gravels as paleowind indicators, *J. Sed. Petrology,* **43,** 844-847.

Orford, J. D., 1975. Discrimination of particle zonation on a pebble beach, *Sedimentology,* **22,** 441-463.

Sneed, E. D., and Folk, R. L., 1958. Pebbles in the lower Colorado River, Texas, a study in particle morphogenesis, *J. Geol.,* **66,** 114-150.

Stratten, T., 1974. Notes on the applicability of shape parameters to differentiate between beach and river deposits in southern Africa, *Trans. Geol. Soc. S. Africa,* **77,** 59-64.

Zenkovich, V. P., 1967. *Processes of Coastal Development.* New York: Wiley, 1295p.

Cross-references: *Attrition; Conglomerates; Littoral Sedimentation; River Gravels; Roundness and Sphericity.*

BEACHROCK

Beachrock is a friable to well-cemented sedimentary rock lithified in the intertidal and sea-spray zones. Because it may be found on tidal flats and in tidal channels as well as on beaches, the term *beachrock* is not entirely appropriate (Bricker, 1971). The lithified material may extend up creeks into fresh-water zones and should not here be called beachrock. Beachrock has been reported from tropical and subtropical areas throughout the world (see Matthewson, 1972; Bricker, 1971).

Description

Clasts in beachrock range from sand to gravel size (Scoffin, 1970); no fine-grained beachrocks are known. The clasts range in composition from volcaniclastics to quartz, to carbonate grains of many varieties. Beachrocks containing human artifacts as young as 10 to 20 yr old attest to the rapid process of cementation in the intertidal zone. Beachrock may form scattered nodules, isolated layers, or whole thicknesses of beach sediment; the beds dip gently seaward, conforming with typical beach lamination.

Beachrock is cemented first by simple pore filling; the cement may occur in three main morphologies involving two minerals (Bricker, 1971): (1) micritic coatings, often of irregular thickness and lumpy or pelleted, composed of either aragonite or Mg calcite; (2) fibrous or bladed crusts (10–200 μ thick) aligned perpendicular to the encrusted grain surface, commonly composed of aragonite, but also often Mg calcite; and (3) equant crusts of Mg calcite (rare). In a review of beachrocks of the Mediterranean, Alexandersson (1972) found Mg calcite was being directly precipitated from sea water; aragonite, which is the major cement of Caribbean beachrocks, was only a minor constituent. Calcite (low-Mg) cement has been found in older beachrock, but Stoddart and Cann (1965) suggest that the calcite is diagenetic; no primary low-Mg calcite has been found as beachrock cement. Alexandersson (1972) reports dolomitization of Mg-calcite beachrock cement.

Genesis

Beachrock forms very rapidly; Captain Robert Moresby (1835) reported that Indian Ocean natives harvested beachrock annually, and within a year there was a new lithified crop. Apparently, all beachrock forms in the tropics or subtropics but has no connection with the amount of rainfall. J. D. Dana's theory that beachrock forms by evaporation of sea water at times of low tide explains most of the occurrences today (Bricker, 1971). The sea water may range up to twice normal salinity; the interstitial water in the beach sediment ranges from 1.2–4 times normal salinity. Fresh water apparently inhibits cementation; it has been found that beachrock forms in zones of mixing of meteoric brackish water and sea water (Moore, 1973), but beachrock has also been reported in areas where there is no possibility of fresh water influx. It is suspected, but not proven, that microorganisms play a part in beachrock cementation (Moore, 1973; Davies and Kinsey, 1973); Alexandersson (1972) reviews this question and concludes, in his study, that the Mg calcite is inorganically precipitated.

Fossil (relict) beachrock has only recently been reported (Siesser, 1974), radiocarbon dated at about 26,000 B.P.; in addition Mistardis (1963) and others have noted beachrock used in ancient tombs, some over 3500 years old. Beachrock may be difficult to distinguish from material cemented subaerially, e.g., eolianite (q.v.), or in a submarine environment (Allen et al., 1969; Macintyre et al., 1975).

JOANNE BOURGEOIS

References

*Alexandersson, T., 1972. Mediterranean beachrock cementation: marine precipitation of Mg-calcite, in D. J. Stanley, ed., *The Mediterranean Sea*. Stroudsburg, Pa.: Dowden, Hutchinson & Ross, 203–223.

Allen, R. C.; Eliezer, G.; Friedman, G. M. and Sanders, J. E., 1969. Aragonite-cemented sandstone from outer continental shelf off Delaware Bay: Submarine lithification mechanism yields product resembling beachrock, *J. Sed. Petrology*, 39, 136–149.

Bricker, O. P., ed., 1971. *Carbonate Cements*. Baltimore: Johns Hopkins Univ. Press, 376p., see Part 1: Beachrock and intertidal cement, Introduction, 1–3, 9 short entries, 4–46.

Davies, P. J., and Kinsey, D. W., 1973. Organic and inorganic factors in recent beach rock formation, Heron Island, Great Barrier Reef, *J. Sed. Petrology*, 43, 59–81.

Macintyre, I. G., Blackwelder, B. W., Land, L. S., and Stuckenrath, R., 1975. North Carolina shelf-edge sandstone: Age, environment of origin, and relationship to pre-existing sea levels, *Geol. Soc. Am. Bull.*, 86, 1073–1078.

*Matthewson, C. C., 1972. A note on the relationship between beachrock and the environment, *Compass*, 49(4), 119–124.

Mistardis, G. G., 1963. On the beachrock of Southern Greece, *Geol. Soc. Greece Bull.*, 5, 1–19 (in Greek with English summary, 15–19).

Moore, C. H., 1973. Intertidal carbonate cementation, Grand Cayman, West Indies, *J. Sed. Petrology*, 43, 591–602.

Moresby, R., 1835. Extracts from Commander Moresby's report on the northern atolls of the Maldives, *Geog. J.*, 5, 398–403.

Russell, R. J., 1962. Origin of beachrock, *Z. Geomorph.*, N.F., 6, 1–16.

Scoffin, T. P., 1970. A conglomeratic beachrock in Bimini, Bahamas, *J. Sed. Petrology*, 40, 756–758.

Siesser, W. G., 1974. Relict and recent beachrock from Southern Africa, *Geol. Soc. Am. Bull.*, 85, 1849–1854.

*Stoddart, D. R., and Cann, J. R., 1965. Nature and origin of beachrock, *J. Sed. Petrology*, 35, 243–247.

West, I. M., 1973. Carbonate cementation of some Pleistocene temperate marine sediments, *Sedimentology*, 20, 229–249.

*Contains numerous references.

Cross-references: *Carbonate Sediments–Bacterial Precipitation; Carbonate Sediments–Diagenesis; Carbonate Sediments–Lithification; Cementation; Cementation, Submarine; Diagenesis; Dolomitization.* Vol. III: *Beachrock.*

BEACH SANDS

The majority of the world's beaches are made up of sand-sized sediments (1/16 to 2.0 mm). This material is derived from one of three sources: (1) from inland, via streams, glaciers, and, to a lesser degree, wind; (2) from offshore, limited by the zone of influence of wave action in water less than 18 m deep (Trask, 1955); and (3) from along the shore, including material eroded from nearby headlands and reworked material originally derived from one of the first two sources and transported by longshore currents. Beach sands of terrigenous origin generally contain quartz as the most abundant mineral (95%), accompanied by feldspars, micas, and varying proportions of heavy minerals (1–3%). Slight mineralogical variations in sands supplied by erosion of adjacent headlands or transported by local streams may reflect different provenances. Beach sands supplied from offshore to modern low-latitude beaches may have high proportions of calcareous debris (Pettijohn et al., 1972).

Particle size along a shore is a function of the processes of accretion and removal of sand. Most mean sizes of beach sand lie between 0 and 4ϕ. Generally, grain size is greater where wave energy is higher, and hence there are seasonal influences on this parameter. The grain size tends to vary normal to the shore: finer sand on the upper foreshore, coarser on the lower foreshore and berm crests (King, 1972). The lower foreshore receives more energy than the upper foreshore, which is at the limit of wave action. Coarser sand on the berms may be explained by: (1) the rapid percolation of swash so that backwash is unable to carry coarse particles, or (2) the winnowing of fines from this lag deposit by the wind.

The *sorting* of beach sands involves the readjustment of varying grain sizes and specific gravities on a slope in accord with certain dynamic parameters such as wave height, wave length, still-water depth, and particle terminal-fall velocity (Bowen and Inman, 1966). Major size differences in shore sands can be attributed to sorting (i.e., fines "outrun" coarse material). Folk and Robles (1964) have suggested that σ_I = 0.3 to 0.6ϕ is the limiting value for sorting of beach sands (where σ_I is the Inclusive Standard Deviation). King (1972) notes that the best sorting in beach sediments occurs between the mean sizes of 3.25ϕ and 2.25ϕ, below which sorting deteriorates. *Skewness* of beach sands has been found to be negative by many authors (Duane, 1964; Hails and Hoyt, 1969; Friedman, 1962), probably indicative of the winnowing action of wind and waves, which remove sediment to dunes and shelf respectively. *Kurtosis* is found to be near normal for beach sediments. Kolmer (1973) has found that grain-size frequency distribution (q.v.) of beach sands exhibits a double saltation population attributable to the swash-backwash action of waves (Fig. 1).

Beach sands tend to be extremely well

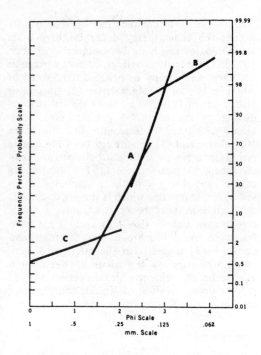

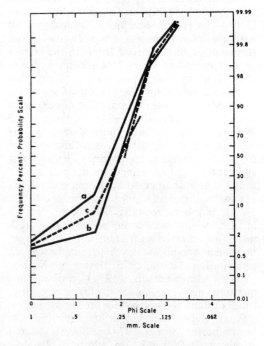

FIGURE 1. *Above:* A model of beach grain-size distribution of the U.S. Gulf and Atlantic coast. A—Saltation population (note *double saltation population*); B—Suspension population; C—Rolling population. *Below:* Composite grain-size distribution of swash and backwash samples taken from a wave tank, a—Swash distribution; b—Backwash distribution; c—Composite distribution of swash and backwash. (After Kolmer, 1973.)

rounded, a result of the tremendous length of time that sediments spend in the high energy surf zone (King, 1972), where minute projections are gradually chipped away from the sand grain which experience repeated back-and-forth motion. In this process, argillaceous and other soft rock fragments are removed, leaving shore sands remarkably homogeneous (Pettijohn et al., 1972). Ridgway and Scotton (1973) found that it is well-sorted, well-rounded beach sand that produces squeaks or whistles when walked upon.

Relief on grain surfaces, or *surface texture*, is another parameter of some importance in beach sands. Most sands from medium- to high-energy beaches contain somewhat irregular, V-shaped depressions on rounded corners, ranging from 0.5 to several or more μ in diameter (Fig. 2). Additionally, irregular depressions from 1.0 to as much as 25μ long appear in conjunction with the V's. These depressions are frequently found with V's forming one limb of the corresponding depression. Both of the above features are characteristic of modern beaches from all over the world, but are occasionally also found in other subaqueous environments, such as river systems, where energy input is sufficiently great (Krinsley and Doornkamp, 1973). Beach sands may have a complex history illustrated by several surface-texture features (see *Sand Surface Textures*).

DAVID KRINSLEY
JOHN KALDI

References

Bowen, A. J., and Inman, D. L., 1966. Budget of littoral sands of Point Arguello, California, *U.S. Army Coastal Engin. Research Center Tech. Memo. 19,* 41p.

Duane, D. B., 1964. Significance of skewness in recent sediments, West Pamlico Sound, N.C., *J. Sed. Petrology,* **34,** 864-874.

Folk, R. L., and Robles, R., 1964. Carbonate sands of Isla Perez, Alacran reef complex, Yucatan, *J. Geol.,* **72,** 255-292.

Friedman, G. M., 1962. On sorting, sorting coefficients, and the lognormality of the grain size distribution of sandstone, *J. Geol.,* **70,** 737-753.

Hails, J. R., and Hoyt, J. H., 1969. Significance and limitations of statistical parameters for distinguishing ancient and modern sedimentary environments of the lower Georgia coastal plain, *J. Sed. Petrology,* **39,** 559-580.

King, C. A. M., 1972. *Beaches and Coasts.* New York: St. Martin's, 570p.

Kolmer, J. R., 1973. A wave tank analysis of the beach foreshore grain size distribution, *J. Sed. Petrology,* **43,** 200-204.

Krinsley, D. H., and Doornkamp, J. C., 1973. *Atlas of Quartz Sand Surface Textures.* Cambridge: Cambridge Univ. Press, 91p.

Pettijohn, F. J.; Potter, P. E.; and Siever, R., 1972. *Sand and Sandstone.* Berlin: Springer-Verlag, 618p.

Ridgway, K., and Scotton, J. B., 1973. Whistling sand beaches in the British Isles, *Sedimentology,* **20,** 263-279.

Setlow, L. W., and Karpovich, R. P., 1972. "Glacial" microtextures on quartz and heavy mineral sand grains from the littoral environment, *J. Sed. Petrology,* **42,** 864-875.

Slatt, R. M., 1973. Frosted beach-sand grains on the Newfoundland continental shelf, *Geol. Soc. Am. Bull.,* **84,** 1807-1812.

Stapor, F. W., and Tanner, W. F., 1975. Hydrodynamic implications of beach, beach ridge and dune grain size studies, *J. Sed. Petrology,* **45,** 926-931.

Trask, P. D., 1955. Movement of sand around the Southern California promontories, *Beach Erosion Brd. Tech. Memo. 76,* 60p.

Cross-references: *Barrier Island Facies; Beach Gravels; Eolian Sands; Grain-Size Frequency Studies; Glacial Sands; Heavy Minerals; Littoral Sedimentation; River Sands; Sand Surface Textures.* Vol. III: *Singing Sands.*

BEACH SEDIMENTS—See LITTORAL SEDIMENTATION

BEDDING GENESIS

The term *bedding* is synonymous with stratification (Latin: *stratum,* a layer). By tradition

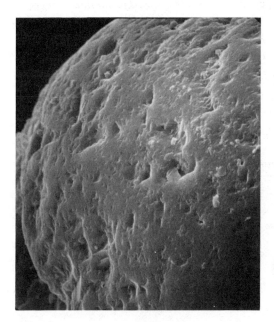

FIGURE 2. Beach grain (relict) from shelf sediments off Argentina, showing typical V-shaped patterns (2000X).

and long-established usage, it is used in a general context to describe the layered arrangement within sedimentary rocks. In precise stratigraphic terminology, however, the term *bed* refers to the smallest rock-stratigraphic unit recognized in classification (AGI, 1972). Ideally, a bed may be of any thickness, but in practice the term is applied to relatively homogeneous or gradational material ranging in thickness from 1 cm to a meter or more. In the widely accepted terminology of McKee and Weir (1953), a layer or stratum having a thickness of 1 cm or more would be defined arbitrarily as a *bed*, and a stratum having a thickness less than 1 cm as a *lamina* (see Table 1). More recently, however, Campbell (1967) has argued cogently in favor of dropping these arbitrary thickness limits; according to his definition, a bed has the attributes of a time-stratigraphic unit of limited areal extent and short time span. It should be noted that the terms *bedding, stratification,* and *lamination* refer to *attributes* of a sediment and should not be used to denote actual layers (beds, strata, laminae) within a deposit.

A bed (lamina) may be delimited from the rock material above and below it by boundaries that commonly, but not invariably, take the form of bedding planes. The presence of bedding planes is generally indicative of changes in the conditions of deposition. Such changes may be manifest by color differences and partings between beds; by variations in particle texture, hardness, cementation, mineralogy, lithology, and organic content; or by changes in thickness, shape, and internal structure. Bedding varieties include tabular, lenticular, trough, wedge-like, rippled, irregular, and contorted (see Andrée, 1915; Payne, 1942; Shrock, 1948, for a discussion of early literature; see Campbell, 1967, for a review of terminology).

Although bedding attracted the attention of early naturalists such as Nicolaus Steno (1638-1687; see Mather and Mason, 1939, p. 33-39), a detailed study of its origin was not attempted until comparatively recently. Some of the stimulus for modern research on bedding genesis stems from detailed flume studies on sediment transport (see *Fluvial Sediment Transport*), some from the current interest in the geochemistry and biochemistry of modern sediments, and some from the recognition of the key role played by sedimentary structures in studies of environmental reconstruction and discrimination.

Bedding may be the product of either physical (mechanical), chemical, or biologic processes, or a combination of these processes. In conformity with the concept of the sedimentation unit, it is reasonable to assume that each bed or lamina forms under essentially constant conditions of deposition. Bedding genesis may be conceived as a process-response to changes in the overall environment controls: weather changes—short-term, seasonal, and long-term; extreme events such as floods and storms; changes in current strength and direction; interaction of meteoric and marine waters; tidal action, and changes in wave intensity and sea level; changes in the biogenic and geochemical environments of deposition. Depending on the type of sedimentation, the time required for the deposition of a lamina or bed probably ranges from a second or two for coarse clastics to hundreds (perhaps thousands) of years for pelagic sediments (see *Sedimentation Rates*).

Origin Related Primarily to Physical Processes

Most of the bedding in this category develops as a process-response to the local fluid/dynamic environment of deposition. Recent advances in

TABLE 1. Comparison of Quantitative Terms Used in Describing Layered Rocks

Terms to Describe Stratification	Terms to Describe Cross-Stratification	Thickness	Terms to Describe Splitting Property
Beds	*Cross-beds*		
Very thick-bedded	Very thickly cross-bedded	> 120 cm	Massive
Thick-bedded	Thickly cross-bedded	60-120 cm	Blocky
Thin-bedded	Thinly cross-bedded	5-60 cm	Slabby
Very thin-bedded	Very thinly cross-bedded	1-5 cm	Flaggy
Laminae	*Cross-laminae*		
Laminated	Cross-laminated	0.2-1 cm	Shaly
			Flaggy
Thinly laminated	Thinly cross-laminated	< 2 mm	Papery

After McKee and Weir, 1953.

the study of sediment transport mechanics and experimental sedimentology (q.v.) have shed much light on the mechanics of bedding formation in clastic sediments; in particular, the elucidation of the *flow regime concept* (see *Flow Regimes*) has greatly facilitated interpretation (Middleton, 1965). Nevertheless, the study of clastic bedding genesis is still in its infancy.

Current Action. The three modes of sediment accumulation promoted by subaqueous current action are: (1) deposition from suspension, (2) deposition from bed load, saltation, and surface creep, and (3) sliding or avalanching of grains down a slip face or foreset slope. Clastic detritus, such as clay and silt, that settles from suspension in a low-energy environment is generally well laminated, and in some cases may be graded. Varves (q.v.), for example, are graded sediments of glaciolacustrine origin that consist of alternating layers of silt and clay.

Bedding formed in cohesionless materials is often composite in origin, i.e., the three cited modes of transport and accumulation are pertinent to genesis. Such bedding is produced when bed forms move along a fluid/sediment interface. The sequence of bed forms generated along a sandy bed by increasing stream power (velocity of flow × bed shear stress) includes: ripples and dunes (lower flow regime); washed-out dunes (transition); and plane bed, standing waves, antidunes, and chutes and pools (upper regime). In addition, there are other higher rank members in the bed-form hierarchy, including bars, ridges, and large sand waves (Allen, 1968).

The movement of ripples along a fluid-sediment interface characteristically deposits parallel (horizontal), wavy, or lenticular laminae, as well as sets of trough cross-laminae. The movement of dunes produces sets of tabular or trough cross-beds, and, in the case of low-amplitude sand waves a sequence of parallel to slightly undulatory beds. An aggrading plane bed in the upper regime deposits plane-parallel beds, whereas the movement of antidune and chute-and-pool bed forms deposits a mixture of parallel and inclined beds together with backset beds (see *Backset Bedding*). In the transition region, the bedding may include plane, wavy, lenticular, and cross-bedding. For clastics composed of coarse sand and larger sizes, a flat bed may persist just above the threshold for particle movement (lower regime), thereby favoring the deposition of parallel beds (Southard, 1975). In fine clastics where "fallout" from suspension dominates over bed-load traction, a type of bedding is formed known as ripple-drift cross-lamination (Fig. 1).

Cross-beds are produced by the avalanching of grains down the lee (foreset) slopes of ripples, dunes, bars, and small deltas (Fig. 2). Some fluvioglacial deltas with topset, foreset, and bottomset components are many m thick and display foresets dipping at $30°$ or more. Here the thickness of the foresets is controlled directly by the depth of water, whereas for ripple and dune cross-bedding the thickness of the cross-bedded set is related directly to the size of the bed forms and only indirectly to water depth. In general, there seems to be direct correlation between grain size and thickness of the cross-bedded set (see *Cross-Bedding*).

There are a number of *hydrodynamic sorting*

FIGURE 1. Ripple-drift cross-lamination of the type A variety (from Jopling and Walker, 1968). The cross-lamination consists of climbing sets of lee-side laminae, with no preservation of laminae on the stoss sides. The upper part of the photograph shows sinusoidal lamination. The current flowed from left to right; scale in inches.

BEDDING GENESIS

FIGURE 2. A laboratory delta with cross-laminae (from Jopling, 1964; copyrighted by American Geophysical Union). Some of the particles carried in suspension over the front of the delta settle on the foreset slope to form a cross-lamina of relatively fine material. This cross-lamina may be preserved if buried under a large slip of relatively coarse material (mostly bed load) sliding down from the top of the slope; depth of water about 15 cm.

processes responsible for differentiating a water-laid deposit into laminae or beds (Jopling, 1964; Moss, 1972). The most obvious explanation hinges on the concept of flow competency, i.e., the relationship between current strength and particle size. Fluctuations in current strength will selectively remove or concentrate particles of certain size grades, and each successive layer bears the imprint of this selectivity. The relationship between flow intensity and the nature of the entrained and depositing sediment can also be studied within the context of sediment transport mechanics, using, e.g., the transport equations of Einstein (1972).

Random segregation of grain sizes along the fluid/sediment contact that is causally related to the structure of fluid turbulence is also a factor in bedding genesis. Systematic segregation related to secondary or transverse currents in the flow may lead to parting lineation (q.v.). Sorting may also be caused by small-scale pulses in transport and by dispersive pressure generated by grain collisions (see *Bagnold Effect*). Textural differences related to grain shape, packing, and orientation may suffice to delineate bedding; for example, the entity of a bed may stem from the distinctive current imbrication of its gravel clasts. A current expanding over a foreset slope effectively disperses and sorts sediment. Experimental studies demonstrate that fine grains of the suspended load may settle on a foreset slope and be overridden intermittently by coarser material (mostly bed load) sliding down from the top of the slope (Fig. 2), resulting in cross-lamination. Temporal and spatial variability in the grain-size composition of the material accumulating at the top of the slope is also important because successive slips or avalanches (and therefore cross-laminae) will then differ slightly in composition. Many of these sorting processes also obtain in an aerodynamic environment (see *Eolian Sedimentation*).

Cross-bedding is also prominently displayed in point-bar deposits developed on the inside bends of river meanders. Such deposits commonly have a tripartite structure and a fining-upward sequence, i.e., a basal lag gravel, overlain by trough cross-beds and then small-scale ripple cross-laminae (Fig. 3). The trough cross-beds which comprise the bulk of the deposit, are formed by the movement of dunes, whereas the ripple cross-laminae are formed by the movement of ripples on or near the crest of the bar. The point bar grows laterally as the river migrates, and the boundaries between the sets of beds are therefore time transgressive.

Wave and Tidal Cycles. In general, the complex interaction between wind, waves, and tidal cycles, coupled with variable sediment input from coastal streams, means that sediment is shifted back and forth with continuous scour-and-fill, concomitant with the formation and destruction of bedding. Many changes may accompany storms.

Bedding in the nearshore zone can be explained within the general context of wave mechanics theory. When the oscillatory (orbital) water motion induced by waves impinging on the bottom reaches a threshold value, loose particles are moved back and forth on the bed, a process referred to as wave-surge transport (see Ingle, 1966; Pettijohn et al., 1972 for general discussion). The intensity and direction of this movement depends on (1) bottom slope, (2) grain size (settling velocity), (3) the average hydrodynamic drag force (a measure of wave intensity), (4) frictional grain resistance,

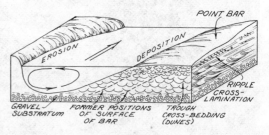

FIGURE 3. Vertical cross-section of a river bend showing the (simplified) internal structure of a point bar deposited on the inner side of the bend. As the river migrates because of erosion on the outer bend, the bar grows by lateral accretion. Successive positions of the surface of the bar are shown. The main flow is into the paper, and the direction of secondary flow is shown by the arrow.

and (5) the gravity component of the particle's weight acting along the bottom slope. Depending on the balance between these forces, a given particle may move offshore or onshore, or it may oscillate back and forth about a mean position (null-line concept).

Under appreciably steady-state conditions, a dynamic equilibrium is reached between wave energy, bottom slope, particle size, sediment supply, and beach exposure. Changes in wave energy (height, period, angle of attack on the shore) will be reflected in the beach and nearshore zones by changes in bedding characteristics (sorting, angle of dip; discordant, wedging, and truncating contacts between sedimentation units of the nearshore zone). Clifton (1969) has described the bed-flow segregation of heavy minerals in twofold sedimentation units produced by swash and backwash on the beach face (see also *Littoral Sedimentation*).

The shifting back and forth of offshore bars in response to the depositional dynamics leads to a complex internal structure for the bar/trough complex (Fig. 4). The internal structure may include ripple bedding caused by wave oscillation and cross-bedding produced by megaripple migration. Reineck (1963, cited in Reineck and Singh, 1973) has described the morphology and internal structure of ebb and flood sand waves on the floor of the North Sea. Box-core samples taken from these sand waves showed bimodal cross-bedding deposited by the reversing tidal current (see also *Tidal-Current Deposits*).

Flaser, wavy, lenticular, and irregular types of lamination and bedding are characteristically produced on intertidal mud flats when rippled sand is shifted back and forth by tides over a muddy sea floor (see Reineck and Singh, 1973; see also *Wadden Sea Sedimentation*). The "flaser" are irregular laminae and pockets of mud deposited over the rippled sand during periods of slack water in the tidal cycle (Fig. 5). Flaser bedding can also form in association with wave-oscillation ripples.

Transgression and regression related to sea-level changes, crustal movement, and sediment supply may cause contrasting bedding types deposited in different coastal and shallow marine subenvironments to be sharply juxtaposed in the stratigraphic record, often in a time-transgressive context. This sort of relationship commonly holds for deltaic and barrier-island complexes where onlap and offlap of depositional subenvironments (e.g., delta plain, distributary, delta front, prodelta, barrier beach, lagoonal, tidal flat) reflects in a very sensitive way the vagaries of change (see *Deltaic Sedimentation; Barrier Island Facies*).

Turbidity-Current Action. Turbidity currents

FIGURE 4. Cross-bedding from a box-core sample of an offshore bar, Kouchibouguac Bay, New Brunswick. The bar is situated in a low-energy coastal environment in a water depth of 3 m. The trough cross-bedding near the top of the core was formed by the movement of lunate megaripples; the plane bedding near the bottom formed by strong current action in the upper regime of flow; width of core approximately 30 cm. (Photo: courtesy of B. Greenwood and R. Davidson-Arnott.)

are density underflows consisting of turbidly suspended sediment. Reputedly, they provide the mechanism whereby relatively coarse detritus is moved downslope into the deeper water of ocean basins. As the turbidity current slows down in the lower reaches of its descent, the load of detritus is deposited as a *turbidite* (q.v.) bed with a distinctive sequence of sedimentary structures. In the ideal Bouma sequence (q.v.), of which there are several varieties, there is a characteristic size gradation from sand at the base to mud at the top of the bed (see *Graded Bedding*).

Davies and Ludlam (1973) have described graded clastic carbonate beds of turbidite origin that are interbedded with the basal Muskeg laminites (Devonian) of Alberta (Fig. 6). This sequence formed around the flanks of carbon-

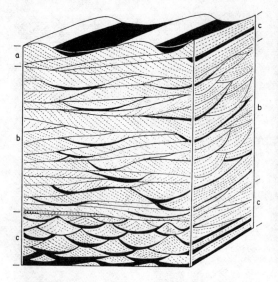

FIGURE 5. Types of flaser bedding deposited by (a) straight-crested current ripples; (b) current ripples with curved crests; (c) wave ripples (from Reineck and Wunderlich, 1968).

ate reefs in a partly enclosed, shallow marine environment.

Wind Action. Eolian beds are deposited by the movement of dunes in desert and coastal areas, and in proglacial areas associated with wasting ice sheets. Sediment movement takes place by suspension, saltation, and surface creep (see *Eolian Sedimentation; Desert Sedimentary Environments*). Large, lenticular to wedge-shaped sets of cross-beds (Fig. 7) are common in eolian sands and sandstones, the morphology of the sets depending on the size and type of sand dune (transverse, barchan, parabolic, or longitudinal). The beds may range from backsets to high-angle foresets. Horizontal and low-angle accretion beds formed by surface creep are common on the crests and windward slopes, and around the flanks, of modern dunes (Glennie, 1970).

Horizontal and gently inclined laminae are characteristic of eolian sheet-sand deposits of the interdune areas. Here the sediment movement takes place by saltation and surface creep, and the optimum conditions for laminae formation appear to be high wind velocity, well-sorted sand, and rapid deposition. Adhesion ripples may form when wind-blown sand adheres to damp, salt-encrusted surfaces in sabkha environments, giving rise to irregular, wavy bedding.

Other. Crude stratification may develop in many different types of deposits, e.g., in talus deposits in response to temporal variations in the texture and composition of the depositing material. Mudflows, debris flows, tills, and flow

FIGURE 6. Section of basal Muskeg laminite with interbedded turbidites, Devonian of northern Alberta (from Davies and Ludlam, 1973). The dolomitic turbidite beds (T) show a fining upward sequence with organic detritus and platy clasts (C). The detritus was derived from the flanks of carbonate buildups. The laminite is interpreted as a chemical (possibly biochemical) precipitate of carbonate in an enclosed basin of hypersaline water chemistry.

tills sometimes display a crude stratification related to size segregation during transport. The origin is as yet unclear.

Some beds also originate as lag deposits or residues. River gravel may form an "armor" on a streambed, the finer material having been selectively washed away. Similarly, wind action may winnow the fines from a desert mudflow consisting of mud, sand, and pebbles, leaving a lag layer of coarse material on the desert floor. The finer material may eventually settle out from suspension to form a poorly stratified loess deposit.

"Massive" bedding may owe its origin to rapid deposition by a sediment-laden current, to high viscosity of the entraining fluid (mudflows), to

FIGURE 7. Large-scale set of eolian cross-bedding in the Navajo (Jurassic) Sandstone. (Photo: courtesy of W. K. Hamblin.)

subaqueous slumping, or to bioturbation (q.v.) by burrowing organisms. Some thick, massive beds of marine sandstone have been tentatively explained by invoking Bagnold's (1966) concept of "grain flow" (q.v.). The technique of X-radiography has demonstrated unequivocally that some "massive" beds do indeed have a stratified structure (see Campbell, 1967, for reference to Hamblin's work).

The distinctiveness of some clastic beds may be enhanced by specific sedimentary structures: concretionary or deformation structures, mudballs, shale chips, mudcracks, root casts, salt crystals, ventifacts, animal trails, etc. Pseudobedding has been recorded from climbing ripple sequences rich in mica flakes; here the concentration of mica flakes on the foreset slopes of superimposed ripples gives a false impression of bedding (Spurr, 1894).

Schmincke et al. (1973) have recently described upper-regime stratification formed in high-energy, phreatic, pyroclastic deposits in Germany. These deposits originated when a volcanic eruption forcefully ejected a sediment-laden cloud of pyroclastic material (see *Base-Surge Deposits*). Most pyroclastic (ash) beds are deposited from suspension by simple "fallout" either under water or on the land surface.

Contorted bedding may commonly be the result of hydroplastic deformation under load, fluid drag in streamflow, collapse caused by melting ice, failure due to seismic vibration, and reduction in shear strength caused by high pore-water pressure (see *Penecontemporaneous Deformation of Sediments*).

Origin Related Primarily to Chemical Processes

Under this heading is included bedding associated with the deposition of many chemically precipitated limestones, dolomites, evaporites, cherts, ironstones, phosphates, and zeolites. The geochemical environment of deposition is controlled by hydrogen ion concentration (pH, connoting acidity or alkalinity); oxidation-reduction potential (Eh); temperature; pressure; ionic exchange; and, of course, by the availability of water and the concentration of the chemical ingredients (see Blatt et al., 1972, for general discussion).

The precipitation of aragonite ($CaCO_3$) and calcite (also $CaCO_3$ and commonly magnesium rich) is often cited as a classic example of chemical deposition. Deep ocean currents rising toward the surface and carrying the calcium ion

(Ca^{++}) in solution trigger the deposition of calcium carbonate because the increase of temperature, decrease in pressure, and increase in fluid turbulence results in the loss of carbon dioxide from the HCO_3^- ion. The solubility of calcium carbonate is decreased as a result of this loss, and the pH of the ocean water increases. Precipitation then takes place, as in the environs of the Bahama Banks (q.v.). Temporal and spatial variations in the conditions of deposition are reflected in the stratified nature of the deposited sediment; adjacent beds may be visually distinct because of slight differences in texture, sorting, bioclastic content, color, porosity, bioturbation, and degree of diagenetic alteration.

Evaporite deposits, with their beds of contrasting chemical composition, nicely illustrate the principles of chemical sedimentation. Important precipitates include sodium chloride (halite), calcium sulphate (anhydrite and gypsum), calcium carbonate, and, less commonly, potassium and magnesium salts. The laminites of Fig. 6 illustrate carbonate deposition in an evaporitic succession. Evaporite deposits generally form when free circulation is impeded between a shallow arm of the sea, e.g., a lagoon, and its parent water body. The enclosed basin acts as a giant evaporating dish in which the saline constituents are progressively concentrated and eventually precipitated, the order depending on relative solubility and concentration (see *Evaporites—Physicochemical Conditions of Origin*).

In desert areas, salt crusts formed by evaporation in temporary lakes may be covered by clay to form crude beds (Glennie, 1970); and in saline alkaline lakes, the best-documented example of which is Lake Natron, Tanzania, the lake-bottom sediments may consist of alternating layers of clay and zeolites (Hay, 1966). Zeolite-rich beds are also found in volcaniclastic sequences, as in the Olduvai Gorge deposits, Tanzania, where six zeolitic facies have been recognized from an ancient alkaline lake. During its life history, Olduvai Lake must have been rich in sodium, potassium, and carbonate ions derived in part from the surrounding volcanogenic sediments (Hay, 1966).

Horizontal zonation related to pedogenic processes may impart a crudely layered structure to soils (see Vol. XI) and paleosols that is akin to bedding; here chemical weathering plays an important role in the transformation.

Origin Related Primarily to Organic Processes

Locally, a depositional surface may be carpeted with organisms, as, for example, when brachiopods and gastropods proliferated on the sea floor, leaving an extensive layer of organic remains. Such remains may be preserved as complete skeletons, or they may be abraded, fractured and comminuted into organic debris that is subsequently washed and sorted into beds by current and wave action (see *Biostratinomy*). Where the skeletons of shellfish are preserved intact on shallow banks on the ocean floor, the resulting deposit is often referred to as a coquina (q.v.) bed.

Changes in the environmental conditions of deposition (temperature, depth, salinity, Eh, turbidity, wave energy, etc.) may make the bottom uninhabitable for a given assemblage of organisms, leading to the influx of another assemblage, or even to successive assemblages. Superposition of these organic remains imparts a crude bedded structure to the deposit, as revealed in some biostromes and bioherms. Lamination is rare because of bioturbation (q.v.).

Many ancient carbonate rocks are finely laminated and may consist of thinly interbedded dolomite and limestone (Fig. 8). The dolomitic beds (laminae) are now thought to have been derived from algal mat layers, and the limestone from detrital, calcite-rich layers. The magnesium ions are complexed organically in the algal mat layers, resulting in secondary dolomite enrichment (Gebelein and Hoffman, 1973; includes reference to Aitken's work on cryptalgal laminites and stromatolites—q.v.)

Calcareous ooze is a widespread deposit in the deep-sea basins. Although admixed with some terrigenous and cosmic material, the ooze consists primarily of the skeletons of marine plankton that have settled from suspension, accumulating at rates as high as 1-4 cm/1000 yr. The laminated structure displayed in some cores taken from the ocean floor attests to variations in the texture and organic composition of the settling detritus, with modifications imposed by the diagenetic environment. Siliceous oozes and organically formed cherts (q.v.) also have well-defined bedding. Bedded cherts found in eugeosynclinal sequences commonly contain radiolarian tests, sponge spicules, and diatom frustules. Such deposits form in quiet water of low Eh value, where clastic sedimentation is minimal (see *Pelagic Sedimentation*).

Coal beds (seams) attest the cyclothemic accumulation of vegetal debris (including pollen and spores) in ancient swamps associated with unstable shelves and intracratonic basins (see *Coal*). Glauconitic beds on the sea floor appear to be partly biogenic; here the role of organisms is to provide a reducing microenvironment (see *Glauconite*). Organisms, especially bacteria, play a role in the deposition of some sedimentary iron deposits; for example chamosite, a primary iron silicate, accumulates in fecal pellets and in other organic remains (see *Cherty Iron Formations*).

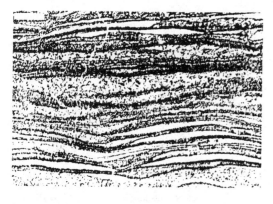

FIGURE 8. Lamination in Proterozoic cryptalgalaminite (from Gebelein and Hoffman, 1973). Thin, dark dolomite layers (probably of algal-mat origin) alternate with coarse, clastic calcite layers; calcite layers show evidence of ripple forms; the scale is given by the 4-mm bar.

Origin Related to a Combination of Processes

Many and perhaps most sedimentary deposits are composite in origin. Deep-sea sediments nicely illustrate this point because they generally consist of a mixture of terrestrially derived clays (partly windborne), submarine volcanic materials, skeletal remains of calcareous and siliceous marine plankton, extraterrestrial cosmic dust (spherules of nickel-iron), and inorganic precipitates, including trace elements.

In conclusion, it is the temporal and spatial variations in the supply of the various components that result in bed formation. These variations, in turn, may be related to the rates of sediment supply, availability of material, local scour-and-fill, minor diastemic episodes, transgression and regression, interaction of meteoric and sea water, hydrodynamic sorting, diagenetic change, etc. Potter and Blakely (1968) have studied bedding sequences from a probability point of view, and have shown that dependent Markov processes are useful for predicting lithologic transitions.

ALAN V. JOPLING

References

AGI, 1972. *Glossary of Geology*. M. Gary, R. MacAfee, Jr., and C. L. Wolf, eds. Washington, D.C.: Am. Geol. Inst., 805p.

Allen, J. R. L., 1968. *Current Ripples*. Amsterdam: North-Holland, 433p.

Andrée, K., 1915. Wesen, Ursachen, und Arten der Schichtung, *Geol. Rundsch.*, 6, 351-397.

Bagnold, R. A., 1966. An approach to the sediment transport problem from general physics, *U.S. Geol. Surv. Prof. Pap. 422-I*, 37p.

Blatt, H.; Middleton, G. V.; and Murray, R., 1972. *Origin of Sedimentary Rocks*. Englewood Cliffs, N.J.: Prentice-Hall, 634p.

Campbell, C. V., 1967. Lamina, laminaset, bed and bedset, *Sedimentology*, 8, 7-27.

Clifton, H. E., 1969. Beach lamination: nature and origin, *Marine Geol.*, 7, 553-559.

Davies, G. R., and Ludlam, S. D., 1973. Origin of laminated and graded sediments, Middle Devonian of Western Canada, *Geol. Soc. Am. Bull.*, 84, 3527-3546.

Einstein, H. A., 1972. The bed-load function for sediment transportation in open channel flow, with Appendix C: Bed load transport as a probability problem, in W. H. Shen, ed., *Sedimentation* (Einstein Vol.). Fort Collins: Colo. St. Univ. Water Research Publ., 1-78, C1-C105.

Gebelein, C. D., and Hoffman, P., 1973. Algal origin of dolomitic laminations in stromatolitic limestone, *J. Sed. Petrology*, 43, 603-613.

Glennie, K. W., 1970. *Desert Sedimentary Environments*. Amsterdam: Elsevier, 222p.

Hay, R. L., 1966. Zeolites and zeolite reactions in sedimentary rocks, *Geol. Soc. Am. Spec. Pap.* 85, 130p.

Ingle, J. C., Jr., 1966. *The Movement of Beach Sand*. Amsterdam: Elsevier, 221p.

Jopling, A. V., 1964. Laboratory study of sorting processes related to flow separation, *J. Geophys. Research*, 69, 3403-3418.

Jopling, A. V., and Walker, R. G., 1968. Morphology and origin of ripple-drift cross-lamination, with examples from the Pleistocene of Massachusetts, *J. Sed. Petrology*, 38, 971-984.

Mather, K. F., and Mason, S. L., 1939. *A Source Book in Geology*. New York: McGraw-Hill, 702p.

Keary, R., and Keegan, B. F., 1975. Stratification by in-fauna debris: A structure, a mechanism and a comment, *J. Sed. Petrology*, 45, 128-131.

McKee, E. D., and Weir, G. W., 1953. Terminology for stratification and cross-stratification in sedimentary rocks, *Geol. Soc. Am. Bull.*, 64, 381-390.

Middleton, G. V., ed., 1965. Primary Sedimentary Structures and their Hydrodynamic Interpretation, *Soc. Econ. Paleont. Mineral. Spec. Publ. 12*, 265p.

Moss, A. J., 1972. Bed-load sediments, *Sedimentology*, 18, 159-219.

Payne, T. G., 1942. Stratigraphic analysis and environmental reconstruction, *Bull. Am. Assoc. Petrol. Geologists*, 26, 1697-1770.

Pettijohn, F. J.; Potter, P. E.; and Siever, R., 1972. *Sand and Sandstone*. New York: Springer-Verlag, 618p.

Potter, P. E., and Blakely, R. F., 1968. Random processes and lithologic transitions, *J. Geol.*, 76, 154-170.

Reineck, H.-E., 1974. Vergleich dünner Sandlagen verschiedener Ablagerungsbereiche, *Geol. Rundsch.*, 63(3), 1087-1101.

Reineck, H.-E., and Wunderlich, F., 1968. Classification and origin of flaser and lenticular bedding, *Sedimentology*, 11, 99-104.

Reineck, H.-E., and Singh, I. B., 1973. *Depositional Sedimentary Environments*. New York: Springer-Verlag, 439p.

Schmincke, H.-U.; Fisher, R. V.; and Waters, A. C.; 1973. Antidune and chute and pool structures in base surge deposits of the Laacher Sea area, Germany, *Sedimentology,* **20,** 553-574.

Shrock, R. R., 1948. *Sequence in Layered Rocks.* New York: McGraw-Hill, 507p.

Southard, J. B., 1975. Bed configurations, *Soc. Econ. Paleont. Mineral. Short Course,* 2 (April 5, 1975), 5-44.

Spurr, J. E., 1894. False bedding in stratified drift deposits, *Am. Geol.,* **13,** 43-47.

Cross-references: *Bed Forms in Alluvial Channels; Cross-Bedding; Cyclic Sedimentation; Deposition; Graded Bedding; Intraformational Disturbances; Lamina, Laminaset, Laminite; Lateral and Vertical Accretion; Primary Sedimentary Structures; Sedimentary Structures; Sedimentation; Sedimentation Rates; Slump Bedding; Storm Deposits;* see numerous cross-references in the text. Vol. IX: *Stratification.*

BED FORMS IN ALLUVIAL CHANNELS

The most striking feature of the interface between an erodible, granular bed and a fluid flowing over it is that the moving fluid interacts with the bed and deforms it into a somewhat regular pattern of two- or three-dimensional troughs and crests. The characteristics (shape, wavelength, height, velocity of movement, etc.) of these bed forms depend on the depth and velocity of flow and the properties of the fluid and sediment. Although there are numerous variations in the details of the shape and behavior of different bed forms, there appear to be relatively few truly distinct types. The following nomenclature for alluvial channel bed forms was adopted by the Task Force on Bed Forms in Alluvial Channels of the Sedimentation Committee, Hydraulics Division, American Society of Civil Engineers (1966):

1. *Bed Configuration:* Any array of bed forms, or absence thereof, generated on the bed of an alluvial channel by the flow.
2. *Flat Bed:* A bed surface devoid of bed forms.
3. *Bed Form:* Any deviation from a flat bed that is readily detectable by eye or higher than the largest sediment size present in the parent bed material, generated on the bed of an alluvial channel by the flow.
4. *Ripples:* Small bed forms with wavelengths less than about 1 ft (30 cm) and heights less than about 0.1 ft (3 cm; Fig. 1).

   Observations—Ripples occur at velocities slightly higher than that required for incipient motion, but at lower velocities than flat bed or antidunes. Portions of upstream slopes of dunes and bars are occupied by ripples under some flow conditions. In plan view, a ripple configuration can vary from an irregular array of three-dimensional peaks and pockets to a very regular array of continuous parallel crests and troughs transverse to the direction of flow. In longitudinal section, ripple profiles vary from approximately triangular, with long, gentle upstream slopes and steep downstream slopes, to symmetrical, nearly sinusoidal shapes. Ripples move downstream with velocities that are small compared to the mean velocity of the generating flow. They are observed to occur only rarely in sediment coarser than about 0.6 mm.

5. *Bars:* Bed forms having lengths of the same order as the channel width or greater, and heights comparable to the mean depth of the generating flow.

   Observations—Several different types of bars are observed, as follows: *Point bars* occur near convex banks of channel bends. Their shape may vary with changing flow conditions, but they do not move relative to the bends. *Alternating bars* tend to be distributed periodically along a channel, with alternate bars near opposite channel banks. Their lateral extent is significantly less than the channel width. Alternating bars move slowly downstream. *Transverse bars* occupy nearly the full channel width. They occur both as isolated and as periodic forms along a channel, and move slowly downstream. *Tributary bars* occur immediately downstream from points of lateral inflow into a channel. In longitudinal section, bars are approximately triangular, with very long, gentle, upstream slopes and short downstream slopes that are approximately the same as the angle of repose of the bed material. Bars generated by high flows frequently appear as small islands during low flows. Portions of the upstream slopes of bars are often covered with ripples or dunes.

6. *Dunes:* Bed forms smaller than bars but larger than ripples that are out of phase with any water surface gravity waves that accompany them (Fig. 2).

   Observations—Dunes generally occur at higher velocities and sediment transport rates than ripples, but smaller velocities and transport rates than antidunes; ripples do occur on the upstream slopes of dunes at the lower velocities in the dune regime, however. The lengths of dune crests are usually of the same order as the wavelength. In longitudinal profile, dunes are approximately triangular with fairly gentle upstream slopes and downstream slopes that are approximately equal to the angle of repose of the bed material. Dunes move slowly downstream. The large lee eddies that occur in the dune troughs often cause surface boils of intense turbulence and high sediment concentration to exist above and slightly downstream from dune crests.

7. *Transition:* A bed configuration consisting of a heterogeneous array of bed forms, primarily low-amplitude ripples or dunes and flat areas (Fig. 3).

   Observations—The transition bed is a configuration generated by flow conditions intermediate to those producing dunes and flat bed.

8. *Antidunes:* Bed forms that occur in trains and that are in phase with and strongly interact with gravity water-surface waves (Fig. 4).

   Observations—Antidunes can move upstream or downstream or can remain stationary, depending on the properties of the flow, fluid, and sediment. The free-surface waves have larger amplitudes than the antidunes. At higher velocities and Froude numbers, the surface waves usually grow until they become unstable and break in the upstream direction. The agitation accompanying breaking obliterates the antidunes, and the process of antidune initiation

FIGURE 1. Side view of ripples in a laboratory flume. (ASCE, 1966). The flow is from right to left. Aluminum powder was added to the water to make the streamlines visible. Note that the free surface appears to be unaffected by the ripples.

FIGURE 2. Side view of dunes in a laboratory flume. The flow is from right to left (ASCE, 1966). Aluminum powder was added to the water to make the streamlines visible. Note the surface wave that is out of phase with the bed configuration, and the region of separation in the lee of the dune.

and growth is then repeated. At lower velocities, the antidunes will diminish in amplitude without the surface waves ever breaking. In longitudinal section, antidune profiles vary with flow and sediment properties from approximately triangular to sinusoidal. The sharp-crested, triangular-shaped antidunes have been observed only in laboratory flumes, however. The crest lengths of antidunes are usually of the same order as the wavelength.

9. *Chutes and Pools:* A sediment bed configuration occurring at relatively large slopes and sediment discharges, consisting of large mounds of sediment, which form chutes on which the flow is supercritical, connected by pools, in which the flow may be supercritical or subcritical.

Observations—A hydraulic jump often forms at the downstream end of each chute where it joins a pool, or at the downstream end of the pools where the flow approaches the next chute. The chutes move slowly upstream. Chutes and pools are infrequently observed in the field. The principal types of bed forms are depicted schematically in Fig. 1 in *Flow Regimes,* adapted from Simons and Richardson (1962).

FIGURE 3. A transition configuration in a laboratory flume (ASCE, 1966). The flow was toward the viewer.

Most of the terms defined above have one to several synonyms in common use in the literature and among those working in this subject. The following summary lists several of the more widely used synonyms.

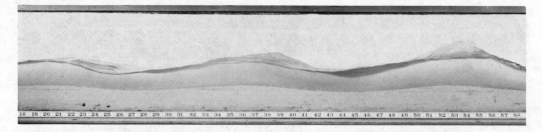

FIGURE 4. Side view of antidunes and stationary waves in a laboratory flume (ASCE, 1966). The flow is from right to left. The antidunes move upstream. Note the wave at the right is at incipient breaking.

Bed Configuration
  Bed geometry
  Forms of bed roughness
  Bed form
  Bed regime
  Bed phase
  Bed irregularities
  Sand waves
  Bed material forms
  Bed shape
Flat Bed
  Smooth bed
  Plane bed
Bed Form
  Bed irregularity
  Bed wave
  Bed feature
  Dune
  Ripple
  Sand bar
  Gravel bar
  Sand wave
Bars
  Sand waves
  Banks
  Sand banks
  Deltas
Ripples
  Dunes
  Sand waves
  Ripple marks
  Current ripples
Dunes
  Ripples
  Sand waves
  Sand bars
Transition
  Sand waves
  Washed-out dunes
Antidunes
  Standing waves
  Antiripples
  Sand waves (also applied to water waves)
Chutes and Pools
  Violent antidunes

When examining bed forms in natural alluvial channels, several factors should be considered. First, the bed configuration observed may not be the equilibrium configuration for the flow occurring at the time of the observation. This point is especially significant when the flow is changing rapidly or during low flows, when the sediment transport rate is small (or possibly zero) and the rate of adjustment of the bed configuration to the flow is very slow. In these cases, the bed configuration may be characteristic of earlier flows. Second, the depth of flow and the bed configuration usually vary laterally across a natural channel, and in extreme cases the whole spectrum of bed configurations will exist simultaneously at the same section, various bed features occurring at different locations across the section. Hence the local bed configuration should be viewed as related to the local depth and velocity of flow rather than to the average flow properties of the whole section.

Most alluvial channel bed forms are inherently three-dimensional, with comparable characteristic lengths normal and parallel to the mean flow direction. In laboratory flumes, particularly those with narrow channels, one frequently observes only a "slice" of the bed configuration and hence may receive an erroneous impression of the true geometric character of the bed. This consideration is worthy of note when examining bed configurations generated in channels that are significantly narrower than the characteristic length of the bed forms (see *Experimental Sedimentology*).

The mechanism of the instability at the fluid/bed interface that causes the various bed forms to occur, and the related stability that makes the flat bed possible for certain flow conditions, is by no means well understood at the present time. Several theories have been proposed to explain this intriguing instability (see references), but none of them is entirely satisfactory. These theories have been of two general types. One type of analysis attempts to explain the formation of the bed forms as a result of an instability between two different continuous media, similar to either the well-known Helmholtz instability which occurs at the interface of two different moving fluids with different densities, or to a shear instability resulting from an inability of the bed material to support without deformation the shear stress applied by the overlying fluid and sediment. The analyses of Liu (1957) and Bagnold (1956) are in this category. The other approach to the problem consists of an examination of the kinematics and/or dynamics of particle motion over a bed that is only slightly deformed. These analyses attempt to ascertain what factors might cause the sediment to be preferentially scoured from the troughs and deposited on the crests to make the initial, small bed undulation grow. The analyses of Anderson (1953), Exner (1925), Kennedy (1963, 1969), and Matsunashi (1959) utilize this approach. The only results of these analyses that have been found to be in fair agreement with observation are those given by Kennedy (1961, 1963, 1969) for the wavelength of antidunes and the maximum height (at breaking) of the accompanying water waves:

$$L = \frac{2\pi V^2}{g} \quad \text{and} \quad \frac{2a}{L} = 0.142$$

where L = wavelength of the antidunes, V = mean velocity of flow, g = gravitational constant, and $2a$ = water-wave height (trough to crest) when the waves break. These formulas, which are identical with the celerity-wavelength and maximum amplitude-wavelength relations for gravity waves in a deep fluid, have been found to give fairly reliable results for flow in laboratory flumes (Kennedy, 1961) and in natural alluvial channels (Nordin, 1964).

Nordin (1971) initiated a new departure from the second line of inquiry described above by undertaking to analyze the statistical properties, as described by the spectra of bed profiles, of ripples and dunes, and to infer therefrom the interaction between the flow and bed that is responsible for the instability (see *Fluvial Sediment Transport*). Hino (1968) succeeded in deriving from dimensional considerations the "minus-three-power" law that has been observed to characterize ripple and dune spectra at the higher spatial frequencies. Jain and Kennedy (1974) analyzed the evolution of the spectra of ripples and dunes as they form in an initially flat bed. They concluded that the equilibrium spectrum is achieved when the rate of variance production at high frequencies just equals the rate of dissipation at low frequencies. From an analysis of the kinematical interaction on the different frequency components they succeeded in obtaining a formal derivation of the "minus-three-power" law.

JOHN F. KENNEDY

References

Allen, J. R. L., 1973. Phase differences between bed configuration and flow in natural environments, and their geological relevance, *Sedimentology*, 20, 323-329.

Allen, J. R. L., and Collinson, J. D., 1974. The superimposition and classification of dunes formed by unidirectional aqueous flows, *Sed. Geol.*, 12, 169-178.

American Society of Civil Engineers, Task Force on Bed Forms in Alluvial Channels, 1966. Nomenclature for bed forms in alluvial channels, *Proc. ASCE, J. Hyd. Div.*, 92 (HY3), 51-64.

Anderson, A. G., 1953. The characteristics of sediment waves formed by flow in open channels. *Proc. 3rd Mid-West. Conf. Fluid Mech.*, Univ. of Minnesota, March 1953, 379-395.

Bagnold, R. A., 1956. The flow of cohesionless grains in fluid, *Proc. Roy. Soc. London*, 249, 235-297.

Exner, R., 1925. Uber die wechselwirkung zwischen wasser und geschiebe in flussen, *Sitz. Akad. Wiss.*, Wien, *Math.-Nat. Kl., Abt. IIa*, 134, 166-204.

Hino, M., 1968. Equilibrium-range spectra of sand waves formed by flowing water, *J. Fluid Mech.*, 34, 565-573.

Jain, S. C., and Kennedy, J. F., 1974. The spectral evolution of sedimentary bed forms, *J. Fluid Mech.*, 63, 301-314.

Keller, E. A., and Melhorn, W. N., 1974. Bedforms and fluvial processes in alluvial channels: selected observations, in M. Morisawa, ed., *Fluvial Geomorphology*. Binghamton: State Univ. of New York, 253-283.

Kennedy, J. F., 1961. Stationary waves and antidunes in alluvial channels, Cal. Inst. Tech.: *W.M. Keck Lab. Hydraulics Water Research Rept. KH-R-2*.

Kennedy, J. F., 1963. The mechanics of dunes and antidunes in erodible-bed channels, *J. Fluid Mech.*, 16, 521-544.

Kennedy, J. F., 1969. The formation of sediment ripples, dunes and antidunes, *Ann. Rev. Fluid Mech.*, 1, 147-168.

Liu, H. K., 1957. Mechanics of sediment-ripple formations, *Proc. ASCE, J. Hyd. Div.*, 83(HY2), Pap. 1197.

Matsunashi, J., 1959. On the instability of movable bed in open channel flow, *Mem. Faculty Engin., Kobe Univ.*, 6.

Nordin, C. F., Jr., 1964. Aspects of flow resistance and sediment transport, Rio Grande near Bernalillo, New Mexico, *U.S. Geol. Surv. Water Supply Pap. 1498H*, 41p.

Nordin, C. F., 1971. Statistical properties of dune profiles, *U.S. Geol. Surv. Prof. Pap. 562F*, 41p.

Simons, D. B., and Richardson, E. V., 1961. Forms of bed roughness in alluvial channels, *Am. Soc. Civ. Engin. Proc.*, 87(HY3), 87-105.

Simons, D. B., and Richardson, E. V., 1962. The effect of bed roughness on depth-discharge relations in alluvial channels, *U.S. Geol. Surv. Water Supply Pap. 1498E*, 26p.

Vanoni, V. A., 1974. Factors determining bed forms of alluvial streams, *J. Hyd. Div., ASCE*, 100(HY3), *Proc. Pap. 10396*, 363-377.

Cross-references: *Backset Bedding; Bedding Genesis; Cross-Bedding; Experimental Sedimentology; Flow Regimes; Fluvial Sediment Transport; Parting Lineation; Ripple Marks.*

BENTONITE

Bentonite is defined as rock composed of a crystalline, clay-like material formed by the devitrification and the accompanying chemical alteration of a glassy igneous material, usually a tuff or volcanic ash (Ross and Shannon, 1926). The chemical composition of the clays and the resistant minerals of bentonites indicate that the volcanic ash and tuff from which they were formed are most commonly felsic (high silica content) in composition.

Bentonite beds are most common in Cenozoic and Mesozoic rocks but have been found in all the Paleozoic systems except the Silurian. In Cretaceous and younger rocks, bentonites are usually composed of montmorillonitic clays (smectites), and the terms have been considered synonymous; most of the Paleozoic bentonites, however, are composed of mixed-layer illite-smectite with the illite layers usually predominating. Examples have been found of chlorite, illite, glauconite, attapulgite, and kaolinite formed from the alteration of volcanic glass. In the present day, southern Pacific Ocean volcanic glass is altering to a smectite and zeolites. The original volcanic glass is easily altered and has little or no influence on the type of clay that will be formed from it (see *Volcanic Ash–Diagenesis*).

Bentonites occurring as thin beds in limestone and as relatively thick beds in shales may be easily recognized. Individual bentonite beds (0.3–1.5 m thick) can be traced for hundreds of kilometers and are accurate time planes in the stratigraphic section; zircon, biotite, and feldspar can give reliable isotopic age dates and provide good control points in the section (Weaver, 1963). Unless volcanic glass is present, however, bentonite zones can go unrecognized in shale and sandstone sections.

Bentonite beds may cover areas >2.5 million km^2, but they are probably volumetrically unimportant compared to the large amount of disseminated and reworked bentonitic material that is mixed with other sediments. Even the preserved record is impressive, however. The volume of Ordovician bentonite in the central and eastern US is on the order of 800 km^3; the volume of Cretaceous bentonite in the northern Great Plains is estimated to be 14–20,000 km^3.

CHARLES E. WEAVER

References

Grim, R. E., 1968. *Clay Mineralogy*. New York: McGraw-Hill, 596p.
Ross, C. S., 1955. Provenience of pyroclastic materials, *Geol. Soc. Am. Bull.*, 66, 427–434.
Ross, C. S., and Hendricks, S. B., 1945. Mineralogy of the montmorillonite group, *U.S. Geol. Surv. Prof. Pap. 205B*, 23–79.
Ross, C. S., and Shannon, E. V., 1926. The minerals of bentonite and related clays and their physical properties, *J. Am. Ceram. Soc.*, 9, 77–96.
Schultz, L. G., 1969. Lithium and potassium absorption, dehydroxylation temperature, and structural water content of aluminous smectites, *Clays Clay Minerals.*, 17, 115–150.
Weaver, C. E., 1953. Mineralogy and petrology of some Ordovician K-bentonites and related limestones, *Geol. Soc. Am. Bull.*, 64, 921–943.
Weaver, C. E., 1963. Interpretive value of heavy minerals from bentonites, *J. Sed. Petrology*, 33, 343–349.

Cross-references: *Clay as a Sediment; Tuff; Volcanic Ash Deposits; Volcanic Ash–Diagenesis*. Vol. IVB: *Bentonite*.

BERNOULLI'S THEOREM

This useful statement of conservation of energy was established by Daniel Bernoulli in 1738, for the particular case of incompressible, steady fluid flow. The more general form may be written as

$$\tfrac{1}{2}\rho v^2 + p + \rho\phi + \mu = \text{constant}$$

where ρ is the density, $\tfrac{1}{2}\rho v^2$ represents the kinetic energy, the pressure p represents the mechanical work done, $\rho\phi$ represents the potential energy, and μ represents the internal energy (roughly the total of thermal energy and work done due to compressibility). The sum of these terms is the total energy, which remains constant along a streamline but which may vary from streamline to streamline. The internal energy need not be considered for isentropic flow.

In the context above, this relationship is severely restricted as it applies only to frictionless fluids; few, if any, are found in nature. The appropriate generalization is to include friction effects. An engineering artifice that provides the theorem with its universality is to employ an average velocity for v. This effectively lumps individual streamlines together, so that only the cross-sectional area of the flow need be considered. Dividing through by g, this form of the equation, which is used in hydraulics, is,

$$\frac{1}{2}\frac{v_1^2}{g} + \frac{\Delta p_1}{\rho g} + y_1 = \frac{1}{2}\frac{v_2^2}{g} + \frac{\Delta p_2}{\rho g} + y_2 + h_l$$

where internal energy is neglected; the pressure term is replaced with deviation from the reference pressure; the subscripts indicate separate upstream and downstream stations; y_1 and y_2 are distances from some reference datum (e.g., sea level); and h_l, the *head loss*, is a function of v^2 which includes a constant of proportionality commonly called the *friction factor* and usually determined experimentally. Each term in this equation has a dimension of length. The kinetic energy term is called the *velocity head; pressure head* and *elevation head* apply to the other terms; and the total energy is called the *total head, E*. The sum $\Delta p/\rho g + y$ is called the *potential head* (the term *head* came from the manometer column, which indicates pressure in units of length).

For open channel flow (a river as opposed to a pipe), the potential head is taken to be the depth of the flow y plus the height h_0 of the bottom of the channel about some datum. Limiting concern to only one station, we drop the h_l term, and then $H = E - h_0$ represents the elevation of the head above the bottom and is called the *specific energy*. The volume of flow or rate of discharge per unit width of the channel at this one station is $q = vy$. Hence, $H = q^2/2gy^2 + y$. For a given q, the right hand side becomes minimum for a certain value of y called the critical value y_c. The value of H corresponding to $y = y_c$ is the minimum possible specific energy. If $H < H_c$, there is no positive y which satisfies this relationship and thus no flow is possible. If $H > H_c$, there are two values of y satisfying this relationship. For $H = H_c$, the value of y satisfying this equals y_c. This can be confirmed by plotting the right side against y, and by intersecting the curve with a straight line of height H. The critical value y_c is obtained from $\partial H/\partial y = 0$, i.e., $y_c = (2/3) H_c = (q^2/g)^{1/3}$. For a given q, and $H > H_c$, the current velocity v is smaller for $y > y_c$ (deep channel) than for the $y < y_c$ (shallow channel). For the former case, the flow is said to be "tranquil," for the latter, "shooting" or "rapid" (analogous to supersonic flow).

Therefore, for any specific energy above the critical value there are two flow regimes with different speeds (rapid and tranquil flows). This concept can be applied to determine change of speed and surface elevation for one-dimensional flow like a river with variable depth, where the total head is constant but specific energy changes with depth. The flow regimes exert a considerable influence on types of sand waves. When the flow is tranquil, ripples and dunes can be present. When the flow is rapid, antidunes can occur (see *Bed Forms in Alluvial Channels*).

TAKASHI ICHIYE

References

Batchelor, C. K., 1967. *An Introduction to Fluid Mechanics.* Cambridge: Cambridge Univ. Press, 156-164.

Raudkivi, A. J., 1967. *Loose Boundary Hydraulics.* New York: Pergamon, 178-198.

Rouse, H., 1961. *Fluid Mechanics for Hydraulic Engineers.* New York: Dover, 281-290.

Rouse, H., and Ince, S., 1963. *History of Hydraulics.* New York: Dover, 95-100, 104-105.

Yalin, M. S., 1973. *Mechanics of Sediment Transport.* Elmsford, N. Y.: Pergamon, 208-215.

Cross-references: *Bed Forms in Alluvial Channels; Flow Regimes; Fluvial Sediment Transport; Reynolds and Froude Numbers.*

BIOGENIC SEDIMENTARY STRUCTURES

The disturbance of sediment by living agents is termed bioturbation (q.v.), and fossilized tracks, trails, burrows, borings, and fecal pellets are known as *trace fossils* (Frey, 1975). The digestion of sediment by organisms may constitute the first phase of diagenesis of the sediment. At one extreme, the intensity of bioturbation may result in the complete obliteration of original sedimentary structures (see *Bioturbation*). Other sediments may retain some relict lamination between discrete burrow (or rootlet) systems (Fig. 1). Well-laminated sediments either accumulate too rapidly, as with cross-bedding, or are deposited in environments unfavorable to the activities of burrowers, e.g., black shales. On the other hand, Richter (1936) proved that the black Devonian Hünsruck Shales of West Germany were *not* euxinic because they contained numerous trails.

Phytogenic Structures

Rootlet structures are usually restricted to terrestrial and very shallow water deposits, notably in gannisters beneath coal seams (see *Coal*). The presence of black, carbonaceous tube linings generally signifies a botanic (phytogenic) rather than a zoologic origin.

Zoogenic Structures

Biogenic structures (trace fossils) produced by animals may be present in sedimentary environments ranging from terrestrial to abyssal-marine. In the geological record, however, they are more common in marine and paralic sediments from the late Precambrian onward. In many bioturbated sediments, distinct burrow-styles may be recognized (Fig. 2); and, in recent years, significant advances have been made in relating assemblages of burrow styles to particular sedimentary environments (Fig. 2; Seilacher, 1967a; Sellwood, 1972). As our knowledge of modern ichnocoenoces (burrow assemblages) increases we shall be able to make more refined interpretations of fossil ichnocoenoces. If biogenic structures are present, they are nearly always in situ and may thus be very useful environmental indicators.

A particular burrow style may bear a "generic" name (Fig. 2). These trace-fossil (or ichnofossil) "genera" represent the styles employed by many, often unrelated, animals; and similar burrow forms simply reflect the comparable modes of life of their constructors. Many of the "generic" names applied to burrow forms date back to the days when they were believed to be the impressions of fossil sea-weeds (see *Fucoid*), and thus a large and complicated nomenclature exists (Häntzschel, 1975; see Fig. 2):

Chondrites, which are believed to be dwelling/feeding burrows (producers unknown), are plant-like ramifying burrow systems; the tunnels are circular in cross section, of constant diam-

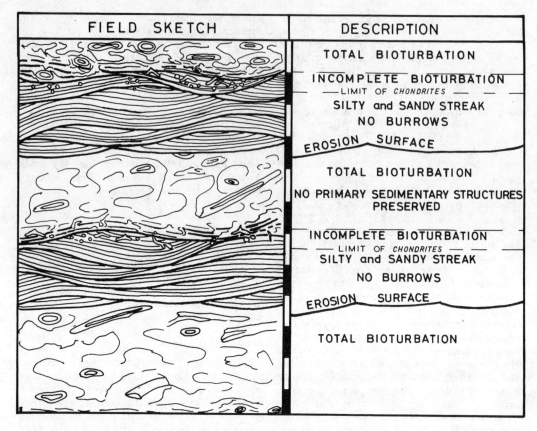

FIGURE 1. Successive storm-deposited sand units from the Pliensbachian of Dorset. Each unit was originally entirely current bedded; but, following its deposition, burrowers penetrated from the top. Scale divisions 5 cm.

eter, with a smooth burrow wall. Branching is lateral, never equal, with many orders of branching. Burrow diameters range from 1-6 mm.

Diplocraterion are U-shaped burrows oriented normal to bedding, with sediment reworked between the arms of the U, believed to be a dwelling/feeding burrow of a part-suspension feeding organism.

Rhizocorallium are U-shaped burrows oriented parallel or oblique to bedding. The sediment between the arms is reworked due to lengthening of the burrow. The walls of the tube are often covered with reticulate ridges; septa may also bear ridges. Systems may be up to 30 cm long and are probably arthropod feeding/dwelling burrows.

Thalassinoides, which are undoubtedly produced by decapod crustaceans, are largely horizontal burrow systems tending to follow bedding planes. The burrows are 2-20 cm in diameter and typically show Y-shaped branching patterns which may join to form polygons. Filling may consist of cylindrical fecal pellets and may show concave/convex laminae. Vertical elements occur and are common in hardground (see *Hardground Diagenesis*) forms.

Pelecypodichnus are resting traces and burrows of bivalves. *Mya* burrows consist of reworked sediment in myoid burrows due to migration of animals either up or down in response to sedimentation or erosion.

Zoophycos is a complex burrow with swirling form produced by the progression of an obliquely inclined burrow in a circular path with one entrance remaining fixed; it is constructed by an unknown (annelid?) group usually in deep water.

Skolithos are vertical, tubular burrows with structureless, sandy fillings and thin, clayey burrow lining. Believed to be produced by suspension-feeding annelids in neritic conditions, they may occur in great numbers in sandy sediments.

Organisms

The most important burrowing organisms are the annelids, arthropods and mollusks. Banks

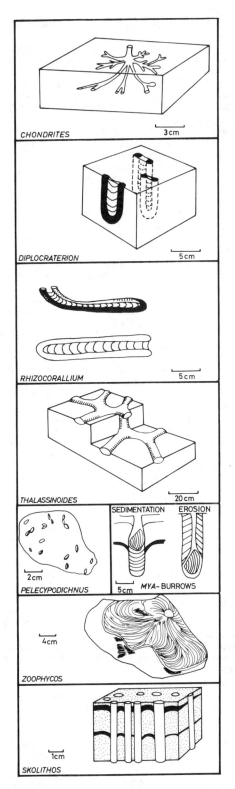

FIGURE 2. Some common trace fossils (see text).

BIOGENIC SEDIMENTARY STRUCTURES

(1970) suggested that the increasing diversity of biogenic structures in the late Precambrian and early Cambrian predominantly reflected the evolutionary development of these three groups.

Although burrow structures are well represented in Phanerozoic sediments, the burrowers themselves are seldom fossilized. Burrowers with the greatest fossilization potential are the mollusks; infaunal bivalves and gastropods may frequently be found in their burrowing positions (Fig. 2). Burrowing bivalves may ascend in response to sediment accretion, and their burrows may then give some indication of sedimentation rates (Fig. 2; Goldring, 1964). Annelids are never preserved in bioturbated sediments, where they probably abounded. Burrowing arthropods (particularly crustaceans) generally have thin carapaces, and only the more strongly calcified chaelipeds are commonly preserved. Anthropods dig with their appendages, and their burrows frequently bear characteristic scratch marks. Rare examples of fossil burrows containing their crustacean constructors were cited by Waage (1968), Shinn (1968) and Sellwood (1971).

We interpret the modes of life of ancient, and mostly unknown, burrowers by comparing fossil burrows with similar modern ones. Our knowledge of modern burrow assemblages has been increased by Reineck et al. (1967, 1968), Shinn (1968), Weimer and Hoyt (1964), Hertweck (1973), and many others. Simple, open burrow systems generally serve as the dwelling burrows of suspension feeders, while deposit feeders produce elaborate trails and follow complex mining programs to exploit large volumes of sediment (Fig. 2; Seilacher 1967a, b).

Sedimentologic Interpretation

Environmental interpretations may be tested by comparing the trace fossil/body fossil/sedimentary data from the same bed, and, in numerous published cases, Seilacher's predictions have been confirmed. Biogenic structures have also been used to estimate relative rates of sedimentation. Seilacher (1962) showed that graded sandstones from the Spanish Flysch (Tertiary) had been deposited instantaneously. He recognized scoured burrows in the underlying shales, which had suffered erosion during the emplacement of sandstone units; and he noted the escape tubes of animals that ascended through the sand units from the shales below; and, finally, he noticed postdepositional burrows which penetrated the sandstone units from above during the phases of clay sedimentation. Goldring (1962) attributed ascending and descending structures in *Diplocraterion* (Fig. 2) to a response by the burrower to depositional and erosive events respectively at the sediment/water interface. In general, phases of slow sedimentation are marked by intensely bioturbated beds. Intense bioturbation in fine-grained substrates may promote thixotropic conditions within the top few centimeters of the sea bed, which then remains uncolonized by epifaunal benthos (Rhoads, 1970).

Hard substrates are recognized by the presence of encrusting epifaunas and boring infaunas. The complex histories of hardening in such beds can often be unraveled from studies on the burrows and boring present. From his studies on *Thalassinoides* burrows (Fig. 2) in British Chalk, Bromley (1970) recognized prolific prelithification bioturbation, synlithification burrow restriction followed by bed surface hardening, and encrustation. His observations gave an early confirmation of the process of submarine lithification in carbonates during nondepositional phases (see *Hardground Diagenesis*). Exhumed concretions may also provide hard surfaces which are bored by cirripedes, mollusks, algae, and worms. As in the case of hardgrounds, shells and other hard material are penetrated by the borers. Sometimes even spar-filled septarian cracks are perforated. Some concretions were shown to have hardened, suffered exhumation, boring and reburial in little more than an ammonite zone (Hallam, 1969). Thus the borings provide evidence of the time required for concretion formation.

B. W. SELLWOOD

References

Banks, N. L., 1970. Trace fossils from the late Precambrian and Lower Cambrian of Finnmark, Norway, in Crimes and Harper (1970), 19-34.

Bromley, R. G., 1970. Some observations on burrows of Thalassinidean Crustacea in chalk hardgrounds, *Quart. J. Geol. Soc. London*, 123, 157-177.

Crimes, T. P., 1973. From limestones to distal turbidites: a facies and trace fossil analysis in the Zumaya Flysch (Paleocene-Eocene), North Spain, *Sedimentology*, 20, 105-132.

Crimes, T. P., and Harper, J. C., eds., 1970. *Trace Fossils*. Liverpool: Seel House Press, 547p.

Frey, R. W., 1973. Concepts in the study of biogenic sedimentary structures, *J. Sed. Petrology*, 43, 6-19.

Frey, R. W., ed., 1975. *The Study of Trace Fossils*. New York: Springer-Verlag, 562p.

Goldring, R., 1962. The trace fossils of the Baggy Beds (Upper Devonian) of North Devon, England, *Paleont. Z.*, 36, 232-251.

Goldring, R. 1964. Trace fossils and the sedimentary surface in shallow marine environments, in L. M. J. U. Van Straaten, ed., *Deltaic and Shallow Marine Deposits*. Amsterdam: Elsevier, 136-143.

Hallam, A., 1969. A pyritized limestone hardground in the Lower Lias of Dorset, England, *Sedimentology*, 12, 231-240.

Häntzschel, W., 1975. Trace fossils and problematica, in C. Teichert, ed., *Treatise on Invertebrate Paleontology*, W1-W269.

Hertweck, G., 1973. Der Golf von Gaeta (Tyrrenisches Meer). VI. Lebensspuren einiger Bodenbewohner und Ichnofaciesbereiche, *Senckenberg. Marit.*, 5, 179-197.

Lessertisseur, J., 1955. Trace fossiles d'activité animale et leur signification paleobiologique, *Soc. Géol. Fr. Mem. 74*, 7-150.

Pryor, W. A., 1975. Biogenic sedimentation and alteration of argillaceous sediments in shallow marine environments, *Geol. Soc. Am. Bull.*, 86, 1244-1254.

Reineck, H.-E.; Dörjes, J.; Gadow, S.; and Hertweck, G., 1968. Sedimentologie Faunenzonierung und Faziesabfolge vor der Ostküste der Inneren Deutschen Bucht, *Senckenberg. Leth.*, 49, 261-309.

Reineck, H.-E.; Gutmann, W. F.; and Hertweck, G., 1967. Das Schlickgebiet sudlich Helgoland als Beispiel rezenter Schelfablagerungen, *Senckenberg. Leth.*, 48, 219-275.

Rhoads, D. C., 1970. Mass properties, stability, and ecology of marine muds related to burrow activity, in Crimes and Harper (1970), 391-406.

Richter, R., 1936. Marken und Spuren in Hunsrück-Schiefer, II; Schichtung und Grund-Leben, *Senckenbergiana*, 33, 1-12.

Roth, P. H.; Mullin, M. M.; and Berger, W. H., 1975. Coccolith sedimentation by fecal pellets: laboratory experiments and field observations, *Geol. Soc. Am. Bull.*, 86, 1079-1084.

Seilacher, A., 1962. Paleontological studies on turbidite sedimentation and erosion, *J. Geol.*, 70, 227-234.

Seilacher, A., 1964. Biogenic sedimentary structures, in J. Imbrie and N. Newell, eds., *Approaches to Paleoecology*. New York: Wiley, 296-316.

Seilacher, A., 1967a. The bathymetry of trace fossils, *Marine Geol.*, 5, 413-428.

Seilacher, A., 1967b. Fossil behavior, *Sci. Am.*, 217, 72-80.

Sellwood, B. W., 1970. The relation of trace fossils to small scale sedimentary cycles in the British Lias, in Crimes and Harper (1970), 489-504.

Sellwood, B. W., 1971. A Thalassinoides burrow containing the crustacean *Gryhaea udressieri* (Meyer) from the Bathonian of Oxfordshire, *Palaeontology*, 14, 589-591.

Sellwood, B. W., 1972. Regional environmental changes across a Lower Jurassic stage boundary in Britain, *Palaeontology*, 15, 125-157.

Shinn, E. A., 1968. Burrowing in recent lime sediments of Florida and the Bahamas, *J. Paleont.*, 42, 879-894.

Waage, K. M., 1968. The type Fox Hills Formation, Cretaceous (Maestrichtian), South Dakota, L. The stratigraphy and paleoenvironments, *Bull. Peabody Mus. Nat. Hist.*, 27, 175p.

Weimer, R. J., and Hoyt, J. H., 1964. Burrows of *Callianassa major* Say, geologic indicators of shallow neritic environments, *J. Paleont.*, 38, 761-767.

Cross-references: *Bedding Genesis; Biostratinomy; Bioturbation; Fucoid; Tidal-Flat Geology.* Vol. VII: *Bioturbation: Coprolite; Trace Fossils.*

BIORHEXISTASY—See VOL. XI

BIOSTRATINOMY

Biostratinomy (Weigelt, 1919) is the study of the sedimentational history of organic remains. It is part of the broader field of *taphonomy* (Efremov, 1940), the postmortem history of fossils, which also includes *necrolysis,* the biological breakdown of nonmineralized parts before burial, and *fossilization,* the history of diagenetic alteration of a fossil after burial (Müller, 1951). Thus biostratinomy emphasizes the physical behavior of fossils as sedimentary particles. Biostratinomy was developed mainly by paleontologists interested in removing the "taphonomic overprint" (Lawrence, 1968) before beginning their paleobiological studies. The significance of these studies to the depositional histories of sediments has often been overlooked (Seilacher, 1973).

The first major biostratinomic study was a study of the paleobiological significance of modern vertebrate corpses done by Weigelt (1927), a student of Johannes Walther. A recent book which summarizes the early studies, most of them carried out in Germany, is Schäfer (1962, 1972), who built on the pioneer work of Rudolph Richter.

Sedimentary Transport

Because the original attitudes, shapes, and relative frequencies of many fossils are known, their present orientation, fragmentation, and sorting may serve as a clue to the kind, direction, and intensity of their postmortem sedimentary transport.

Orientation. Few fossils are perfect spheres; most show various degrees of asymmetry that enable us to describe, measure, and compare their orientations much more precisely than the orientation of sand grains. So rarely is fossil alignment random that "chaotic fossil orientation" has only recently been recognized as a possible case indicative of slumping or complete bioturbation (Toots, 1965). In most cases, fossils are aligned at least with their long axes parallel to the bedding plane; it is therefore usually sufficient to describe their orientation within this plane (Seilacher, 1959).

Perhaps the most familiar orientation phenomenon is the preferred *convex-up* position of bowl-shaped fossils exposed to tractional

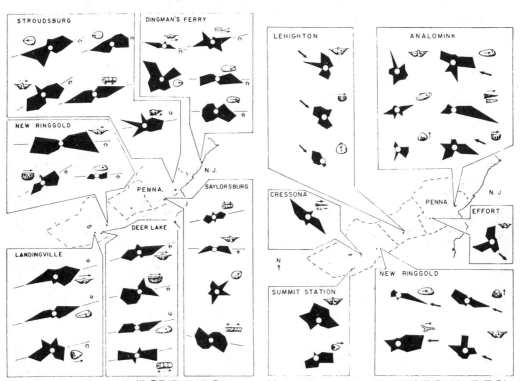

FIGURE 1. Transition from dominant wave to current orientation marks the deepening below wave base of the Appalachian trough during the Devonian (after Nagle, in Seilacher, 1973).

currents; if *convex-down* positions prevail, lack of currents or vertical sedimentation is indicated. Preferred *azimuth orientations* of elongate fossils or their flexible appendages may serve in paleocurrent analysis (q.v.; DeWindt, 1972). If enough measurements are taken to produce rose diagrams, the shape of these diagrams may distinguish *current* orientation and *wave* orientation (Fig. 1; Nagle, 1967), a useful means of paleobathymetric analysis (q.v.). More detailed studies of individual fossils and group relationships can produce a wealth of information (Seilacher, 1973, Brenner, 1976).

Fragmentation. Various sedimentological processes should be reflected in particular kinds of breakage and abrasion (Fig. 2). Rolled and tumbled fossil fragments may also create distinctive marks (see *Current Marks*).

Sorting. Sorting is well known to distort the original size distribution in shell assemblages. It may be caused by differences in durability of fossil parts, but also by morphological differences leading to selective transport. Echinoderms and vertebrates are particularly well suited for studying sorting because they produce a regular number of diverse skeletal elements (Fig. 3). Even the right/left symmetry of perfectly equal valves such as *Donax,* however, may be sorted by the right/left symmetry in the hydrodynamic system of the surf (Fig. 4; Lever et al., 1964).

Internal Sedimentation

Many fossils have acted not only as sedimentary particles but also as sediment recipients before they were buried. Depending on

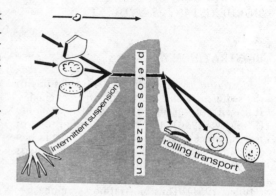

FIGURE 3. Crinoidal sedimentation (from Seilacher, 1973). Fresh echinoderm ossicles are so light that they are carried away mainly in suspension, suffering practically no abrasion and very little sorting, except by size. After prefossilization has mineralized the pore spaces in the calcite skeleton, however, bottom transport may separate and abrade the ossicles according to shape and rollability.

their shapes, apertures, and orientations, as well as the hydrodynamic situation, biogenic sediment traps may provide important information for the sedimentologist. For example, they often retain material that did not come to rest on the level bottom because it either rolled along or remained in suspension (Fig. 5). If internal sedimentation is incomplete, the remaining voids may be used as geopetal criteria. Broadhurst and Simpson (1967) have shown that they may also help to distinguish paleoslope from secondary tilting. Sediment fill may also

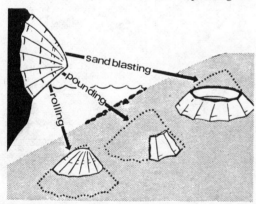

FIGURE 2. Fragmentation (from Seilacher, 1973). Abrasion and breakage may produce diagnostic fragments in different environments. If the shell remains fairly stable (muddy bottoms or dune sand), abrasion attacks only the top side. Breakage on a rocky shore will largely follow radial and concentric shell structures, while rolling and shifting will affect mainly the free edge.

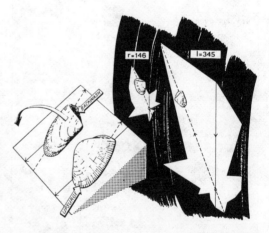

FIGURE 4. In recent beaches the swash orients right and left valves of equivalve pelecypods, such as *Donax,* to mirror-image rose diagrams, but the backwash tends to flip or rotate them both in opposite directions. The left-right sorting (unequal numbers) becomes more efficient if the shell asymmetry is combined with a left-right asymmetry of the hydrodynamic system. (from Seilacher, 1973.)

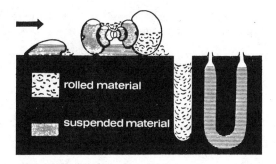

FIGURE 5. Biogenic sediment traps (from Seilacher, 1973). Depending on the hydrodynamic situation and on the width of the openings, biogenic cavities remain empty, or become gradually filled with material that is often coarser or finer than the surrounding sediment. In flat-lying ammonite shells, internal sedimentation may differ between body and gas chambers and between upper and lower umboes.

give a clue to the particular hydrodynamic situation in which it was deposited, particularly in the case of complex cavity systems such as in ammonitic shells (Seilacher, 1967, 1973).

Prefossilization

In neritic sediments, fossils rarely remain undisturbed at the place of their first burial. Most become reworked, many after enough time in the sediment to have undergone early diagenetic alterations. They reenter sedimentation unchanged in shape, but different in durability and specific weight and accordingly different in behavior.

Many *prefossilized* remains are more friable and disintegrate very rapidly, but others, such as phosphatic vertebrate remains, may become heavier and more durable than they were before, so that reworked coprolites, bone fragments, and teeth are concentrated and sorted, together with coarse sand grains, to form typical *bone beds* (Reif, 1971). The spongy but calcitic stereome of echinoderms is in a way similar to bone so that prefossilization would drastically change the behavior of echinoderm particles (Fig. 5; Ruhrmann, 1971). Prefossilization may also increase the durability of mollusk shells, but in this case by cementing the enclosed sediment fill earlier than the surrounding mud. Prefossilized steinkerns often survive subsequent resedimentation so perfectly that their reworked nature can only be demonstrated by the tilt of internal fill structures (Seilacher, 1971). It may well be that studies of this nature will better illustrate the amount and the duration of intraformational recycling, particularly in shallow marine sediments.

ADOLF SEILACHER

Note: This entry is largely abstracted from Seilacher (1973), courtesy Johns Hopkins University Press.

References

Brenner, K., 1976. Ammoniten-Gehäuse als Anzeiger von Paleoströmungen, *Neues. Jahrb. Geol. Palaeont. Abhandl.*, **151**, 101-118.

Broadhurst, F. M., and Simpson, I. M., 1967. Sedimentary infillings of fossils and cavities in limestone at Treak Cliff, Derbyshire, *Geol. Mag.*, **104**, 443-448.

Carter, R. W. G., 1974. Feeding sea birds as a factor in lamellibranch valve sorting patterns, *J. Sed. Petrology*, **44**, 689-692.

DeWindt, J. T., 1972. Vertebrate fossils as paleocurrent indicators in the Upper Silurian of the central Appalachians, *Compass*, **49**(4), 125-137.

Efremov, I. A., 1940. Taphonomy, a new branch of paleontology, *Acad. Nauk. SSSR Biul., Biol. Ser.*, **3**, 405-413 (in Russian). Translation in *Pan-Am. Geol.*, **74**(2), 181-193.

Geyer, O. F., 1973. *Grundzuege der Stratigraphie und Facies-Kunde*, vol. 1. Stuttgart: E. Schweizerbart, 279p.

Kranz, P. M., 1974. The anastrophic burial of bivalves and its paleoecological significance, *J. Geol.*, **82**, 237-265.

Lawrence, D. R., 1968. Taphonomy and information losses in fossil communities, *Geol. Soc. Am. Bull.*, **79**, 1315-1330.

Lever, J.; Bosch, M. van den; Cook, H.; Dijk, T. van; Thiadens, A. J. H., Thijssen, S. J.; and Thijssen, R., 1964. Quantitative beach research. III. An experiment with artificial valves of *Donax vittatus*, *Neth. J. Sea Research*, **2**, 458-492.

Müller, A. H., 1951. Grundlagen der Biostratinomie, *Abhandl. Deut. Akad. Wiss. Berlin, Kl. Math. Allgem.*, 1950(3), 147p.

Munthe, K., and McLeod, S. A., 1975. Collections of taphonomic information from fossil and recent vertebrate specimens with a selected bibliography, *PaleoBios* (Contrib. Univ. Calif. Mus. Paleont., Berkeley), **19**, 12p.

Nagle, J. S., 1967. Wave and current orientation of shells, *J. Sed. Petrology*, **37**, 1124-1138.

Reif, W., 1971. Bonebeds an der Muschelkalk-Keuper-Grenze in Ostwürttemberg, *Neues Jahrb. Geol. Palaeont. Abhandl.*, **139**, 369-404.

Ruhrmann, G., 1971. Riff-ferne Sedimentation unterdevonischer Krinoidenkalke im Kantabrischen Gebirge (Spanien), *Neues Jahrb. Geol. Palaeont., Monatsh.*, 1971, 231-248.

Schäfer, W., 1962. *Actuo-Paläontologie nach Studien in der Nordsee*. Frankfort: Kramer, 666p. Trans. in 1972: *Ecology and Paleoecology of Marine Environments*. Chicago: Univ. Chicago Press, 568p.

Seilacher, A., 1959. Fossilien als Strömungsanzeiger, *Aus der Heimat. Öhringen*, **67**, 170-77.

Seilacher, A., 1967. Sedimentationsprozesse im Ammonitengehäusen, *Akad. Wiss. Literatur, Abhandl. Math.-Naturwiss. Kl.*, 1967, 191-203.

Seilacher, A., 1971. Preservational history of Ceratite shells, *Palaeontology*, **14**, 16-21.

Seilacher, A., 1973. Biostratinomy: The sedimentology of biologically standardized particles, in Robert N. Ginsburg, ed., *Evolving Concepts in Sedimentology*. Baltimore: Johns Hopkins Univ. Press., 159-177.

Toots, H., 1965. Random orientation of fossils and its significance, *Contr. Geol. Univ. Wyo.*, 4, 59-62.

Weigelt, J., 1919. Geologie und Nordseefauna, *Der Steinbruch*, 14, 228-231, 244-246.

Weigelt, J., 1927. Wirbeltierleichen in Gegenwart und geologischer Vergangenheit, *Natur und Museum*, 57, 97-106.

Cross-references: *Biogenic Sedimentary Structures; Bioturbation; Current Marks; Paleobathymetric Analysis; Paleocurrent Analysis; Sorting.* Vol VII: *Taphonomy.*

BIOTURBATION

Rudolf Richter (1952) proposed the term *bioturbation* to comprise all kinds of displacements within unconsolidated sediments and soils produced by the activity of organisms. The sedimentologic results of such activities as the search for food or for resting places, or a flight movement, are termed *bioturbation structures* (see Frey, 1973, Table 2). These may be divided into *zooturbation structures* (i.e., caused by burrowing animals) and *phytoturbation structures* (i.e., caused by the growth of root systems of plants; see Sarjeant, 1975). Both forms of bioturbation may be found in subaquatic (marine, fluvial, limnic) as well as in intertidal and subaerial environments. Such deformations of the primary structures of sediments may begin immediately after their deposition (Moore and Scruton, 1957), and may continue up to the time of final consolidation of the rock (Bromley, 1975). Somewhat similar phenomena may be developed under cold climatic conditions and are called "cryoturbation structures." Not all activities of organisms in sediments result in bioturbation structures, however; some organisms produce special layering, or "biostratification structures" (Frey, 1973, Fig. 1; Simpson, 1975).

The most important bioturbation structures, other than those caused by plant roots, are those made by bottom-living animals. These disclose a so-called *burrow texture* (in German, *Wühlgefüge;* in French, *structure de bioturbation*) or *fossitexture*-a term proposed by Richter (1952). In general, surficial as well as intrastratal tracks and trails are included in this "class" of bioturbation structures (Frey, 1973, Tables 2-3, Appendix I).

The following main forms of bioturbation structures may be distinguished:

1. The original internal structure of the sediment is altered by the organism so that the laminae are bowed up or down, or the stratification may have been partially or totally obliterated (biodeformational structures); in the latter case, the sediment either looks *mottled* or is completely homogeneous—Schäfer's (1956) *fossitextura deformativa* (in German, *Verformungswühlgefüge;* in French *structure de biodéformation*).

2. The primary structure of the sediment is mainly preserved; but it is penetrated by burrows, the walls of which are typically consolidated (a *burrow lining*). The walls are agglutinated with mucus or packed with sediment, or are comprised by *dwelling tubes*, and they border distinctly against the surrounding sediment—Schäfer's *fossitextura figurativa* (in German, *Gestaltungswühlgefüge*). This kind of biogenic sedimentary structure (q.v.) is preserved in "full relief" (in German, *Vollformen* or *Vollrelief;* in French, *plein relief*).

3. *Semireliefs*—the "demireliefs" of earlier literature—(in German, *Halbformen* or *Halbrelief*) are produced by animals along sediment/water, sand/clay, or similar interfaces, or along lamination within the same stratum. They may be preserved as epireliefs on the top surface of a layer (grooves or ridges) or as hyporeliefs at the lower surfaces of beds (Frey, 1973, Figs. 4, 6 and Table 5). Surface tracks and trails are observed very frequently on recent unconsolidated sediments, but they are very transient structures; only a small fraction of them have some chance of being preserved as trace fossils.

The "mottled structure" or *fossitextura deformativa* is extremely frequent in occurrence and may be produced by vagile benthic animals of all phyla, the individuals including endobionts and epibionts. The reworking of sediments by organisms is thus a common phenomenon, important both sedimentologically and geochemically (Hanor and Marshall, 1971).

If a bottom-living organism tries to escape the danger of being buried by great quantities of quickly deposited sediment, it will crawl upward in order to reach the surface—or at least a higher level within the sediment layers. If erosion occurs, the organism may dig deeper. In either case, layers of stratified sediment may be bowed upward or downward. The resulting disruptions of physical sediment texture are called *escape structures*. As shown by experiments in aquaria utilizing artificial and colored strata, this structural deformation is produced by a specific kind of movement made while the animal creeps up or down (Elders, 1975). The parapodia of worms, and their peristaltis, give rise to loops directed downward. If pelecypods burrow through the sediment (whether downward or upward), the layers are virtually always displaced downward (Fig. 1).

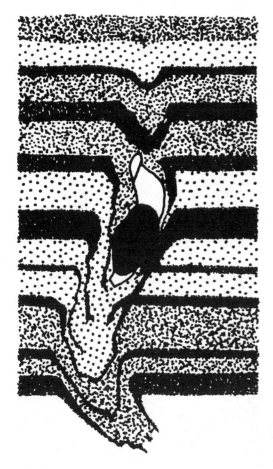

FIGURE 1. The recent pelecypod *Mya arenaria* creeping upward and penetrating stratified sediment; strata are bent downward. Example of *fossitextura deformativa* (from Schäfer, 1962).

FIGURE 2. Resting trace of the recent crustacean *Carcinides* burrowing in laminated sand (from Schäfer, 1962). The structure is now covered by sandy layers deposited after biogenic disturbance of the original lamination.

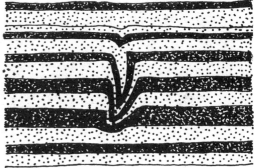

FIGURE 3. Trace of a single walking dactylus of the recent decapod *Eriocheir*, poked into laminated sediment (from Schäfer, 1962). The layers are bent down but their connection is not broken; the structure is now covered by later sedimentation.

Original stratification may also be deformed by the digging activity of an animal for purposes of repose or defense—"resting traces" (Fig. 2). Disturbances of this sort may be caused by various crustaceans (*Corystes, Cumacea*), and similar deformation may result from the digging activity of pelecypods after being uncovered suddenly by erosion of the host sediment. Crustaceans having sharp appendages (phalanges) walking over stratified sediment may pierce and deform it (Fig. 3). Such structures, if preserved, are potential "cleavage reliefs" (Frey, 1973, Table 5, Appendix I).

Layered sediments may lose their stratification largely or entirely by the activity of burrowing inhabitants (Fig. 4), and the process operates on both large and small scales (Fig. 5). Considering all sediments everywhere, complete homogenization probably occurs only rarely and in most cases only in isolated places; but in certain environmental settings, the process is widespread and volumetrically very important (Howard, 1975). A lack of physical stratification in homogeneous sediments that contain traces of former benthic animals should always be regarded as evidence of secondary modification; indeed, even in "massive" beds apparently lacking any kind of internal structure, one should suspect bioturbation and its ultimate influence upon sediment textures and fabrics. Special techniques may be necessary to disclose certain kinds of bioturbate textures (Farrow, 1975).

The *fossitextura figurativa* may be produced by organisms of all phyla, and may be observed

BIOTURBATION

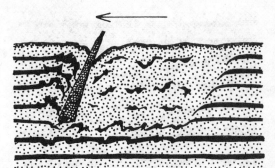

FIGURE 4. The recent polychaete *Pectinaria koreni*, with its "trumpet tube," creeping through stratified sandy sediment (direction of movement indicated by arrow) destroys the stratification within its feeding swath (from Schäfer, 1962).

rhythmic evacuation of the animals' intestine (e.g., the trace fossils *Keckia, Saportia, Taenidium*, etc.; *Stopf-Tunnel* of the German literature). Other burrows are filled by animals that pass sediment around their bodies and pack it behind themselves (Frey and Howard, 1972; see Fig. 6).

In ancient sediments, bioturbation structures are usually visible only when rock sections are cut transverse to the bedding plane and then polished. Some of these structures are made visible by erosion or by weathering, especially in dolomitized limestones ("mottled limestone"). In recent marine sediments, particularly mud, these structures are observable with x-radiography or (as with plastic) after artifical cementation (see Farrow, 1975). In some cases, however,

 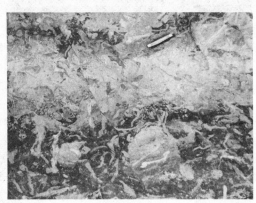

FIGURE 5. *Left:* Small-scale bioturbation by amphipods, Florida Pleistocene sediments; the otherwise sharp laminae have become fuzzy or blurred in appearance. The term *cryptobioturbation* has been proposed for this process by Howard and Frey (1975). *Right:* Intermediate- and large-scale bioturbate textures in Kansas Cretaceous chalk. (from Frey, 1971.)

in ancient sediments of almost all geologic ages and occurring through virtually all facies. Generally, this texture is represented by burrows originally made as tunnels, without much alteration or deformation of the surrounding sediment. Later on, these burrows are filled with stratified (e.g., *meniscus fill*) or unstratified sediment. Their walls are somewhat consolidated by mucus (e.g., the polychaete worms), or may be agglutinated. Some burrows are lined either by the fixing of in-situ sediment particles or pelletal aggregates to the wall, or by pressing detrital sediment fallen into the burrow against its walls. Concentric layers are clearly visible in a transverse section of such a tunnel (Schäfer, 1962).

The burrows of some mud-eating animals are filled with their own excrement (*coprogenic fossitexture*). Burrows of this kind are characterized by crescentic segments, caused by the

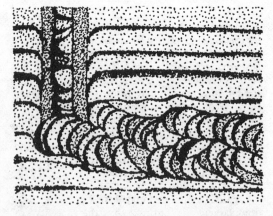

FIGURE 6. Bioturbate textures made by the recent echinoid *Echinocardium*. At left, a vertical burrow filled with disturbed sediment, the bordering strata bent downward; below, crescentic structures produced by back-filling the nearly horizontal burrow. Example of *fossitextura figurativa* (from Schäfer, 1962).

as in tidal flats, they are also made visible by current erosion.

All in all, bioturbation is a significant, ubiquitous sedimentological process that has left an imposing stratigraphic record.

WALTER HÄNTZSCHEL
ROBERT W. FREY

Note: The main text of this manuscript was written by Walter Häntzschel before his death. I am honored to review and complete the final version, and I dedicate it to his memory (*R. W. Frey*).

References

Bromley, R. G., 1975. Trace fossils at omission surfaces, in Frey (1975), 399-428.

Byers, C. W., 1974. Shale fissility: relation to bioturbation, *Sedimentology*, 21, 479-484.

Davis, R. B., 1974. Stratigraphic effects of turbificids in profundal lake sediments, *Limnol. Oceanogr.*, 19, 466-488.

Elders, C. A., 1975. Experimental approaches in neoichnology in Frey (1975), 513-536.

Farrow, G. E., 1975. Techniques for the study of fossil and recent traces, in Frey (1975), 537-554.

Frey, R. W., 1971. Ichnology—the study of fossil and recent lebensspuren, in B. F. Perkins, ed., *Trace Fossils*. Louisiana St. Univ., School Geosci., Misc. Publ. 71-1, 91-125.

Frey, R. W., 1973. Concepts in the study of biogenic sedimentary structures, *J. Sed. Petrology*, 43, 6-19.

Frey, R. W., ed., 1975. *The Study of Trace Fossils*. New York: Springer-Verlag, 562p.

Frey, R. W., and Howard, J. D., 1972. Georgia coastal region, Sapelo Island, U.S.A.: sedimentology and biology. VI. Radiographic study of sedimentary structures made by beach and offshore animals in aquaria, *Senckenberg. Marit.*, 4, 169-182.

Hanor, J. S., and Marshall, N. F., 1971. Mixing of sediment by organisms, in B. F. Perkins, ed., *Trace Fossils*. Louisiana St. Univ., School Geosci., *Misc. Publ.*, 71-1, 127-135.

Howard, J. D., 1975. The sedimentological significance of trace fossils, in Frey (1975), 131-146.

Howard, J. D., and Frey, R. W., 1975. Estuaries of the Georgia coast, U.S.A.: sedimentology and biology. II. Regional animal-sediment characteristics of Georgia estuaries, *Senckenberg. Marit.*, 7, 33-103.

Moore, D. G., and Scruton, P. C., 1957. Minor internal structures of some recent unconsolidated sediments, *Bull. Am. Assoc. Petrol. Geologists*, 41, 2723-2751.

Reineck, H. -E., 1967. Parameter von Schichtung und Bioturbation, *Geol. Rundschau*, 56, 420-438.

Richter, R., 1952. Fluidal-Textur in Sediment-Gesteinen und über Sedifluktion überhaupt, *Notizbl. Hess. Landesamt Bodenforschg.*, 6(3), 67-81.

Sarjeant, W. A. S., 1975. Plant trace fossils, in Frey (1975), 163-179.

Schäfer, W., 1956. Wirkungen der Benthos-Organismen auf den jungen Schichtverband, *Senckenberg. Leth.*, 37, 183-263.

Schäfer, W., 1962. *Aktuo-Paläontologie nach Studien in der Nordsee.* Frankfurt: W. Kramer, 666p. Translation and update, 1972: *Ecology and Palaeoecology of Marine Environments.* Oliver & Boyd (Edinburgh) and Univ. Chicago Press, 568p.

Simpson, S., 1975. Classification of trace fossils, in Frey (1975), 39-54.

Stanley, D. J., 1971. Bioturbation and sediment failure in some submarine canyons, *Vie et Milieu: Troisième Symp. Eur. Biol. Mar., Suppl.*, 22, 541-555.

Cross-references: *Biogenic Sedimentary Structures; Biostratinomy; Fucoid; Sedimentary Structures; Tidal-Flat Geology.* Vol. VII: *Trace Fossils.*

BIRDSEYE LIMESTONE

Birdseye limestone is a type of fine-grained limestone or dolomite in which spots or tubes of sparry calcite appear at irregular intervals (*dismicrite* of Folk, 1959). Hall (1847) first used "Birdseye Limestone" as a formation name (equivalent to the Louisville Formation of the Middle Ordovician). This formation contained worm burrows, polygonal mudcracks, current ripples and crossbeds; and the fossils were rather poorly preserved and fragmented. From this association it was an early thought that this formation was deposited in shallow water and that the "eyes" were either coralites or the remains of shallow-water reeds.

Generally, "calcite eyes" (*fenestrae*) are common to pelsparites, and may have resulted from one or more of the following: (1) precipitation of sparry calcite in animal burrows, or in worm tubes; (2) soft-sediment slumping or mud cracking; (3) precipitation of sparry calcite in tubules resulting from escaping gas bubbles; (4) reworking and rapid deposition of soft sediment containing semicoherent clouds of calcareous mud and spar; (5) recrystallization of calcareous (or dolomitic) mud in patches; and (6) "arrested" dolomitization (Bissell and Chilingar, 1967).

A *fenestra* has been defined as a primary or penecontemporaneous gap in a rock framework, larger than grain-supported interstices (Tebbutt et al., 1965). Many fenestral fabrics are believed to be indicative of intertidal or supratidal environments and to be created as gas pockets or shrinkage cavities.

Birdseye coal involves a different phenomenon. The term is used for an anthracite with numerous small fractures, the convex surfaces developing eye-like forms (Moore, 1940).

B. CHARLOTTE SCHREIBER

References

Bissell, H. J., and Chilingar, G. V., 1967. Classification of sedimentary carbonate rocks, in G. V. Chilingar, H. J. Bissell, and R. W. Fairbridge, eds., *Carbonate Rocks.* Amsterdam: Elsevier, 87-168.

Boyd, D. W., 1975. Fenestral fabric in Permian carbonates of the Bighorn Basin, Wyoming, *Wyo. Geol. Assoc. Guidebook,* **27,** 101–106.

Folk, R. L., 1959. Classification of limestones, *Bull. Am. Assoc. Petrol. Geologists,* **43,** 1–38.

Hall, J., 1847. Paleontology of New York, *Natural History of New York,* **1,** 37–46.

Moore, E. S., 1940. *Coal.* New York: Wiley, 473p.

Shinn, E. A., 1968. Practical significance of birdseye structures in carbonate rocks, *J. Sed. Petrology,* 38, 215–223.

Tebbutt, G. E.; Conley, C. D.; and Boyd, D. W., 1965. Lithogenesis of a distinctive carbonate rock fabric, *Contr. to Geology, Geol. Record,* **4,** 1–13.

Cross-references: *Bioturbation; Limestone Fabrics; Limestones.*

BITUMEN, BITUMINOUS SEDIMENTS

Originally applied to mineral pitch or asphalt, the term *bitumen* is now used as a general name for many forms of solid and semisolid hydrocarbons. Bituminous substances can be divided into two groups: those that are soluble and those that are insoluble, in carbon disulfide. On this basis, the major types of bituminous compounds, both those appearing in nature and those produced synthetically, are indicated in Fig. 1. In the study of soils, the term *humic* is commonly the equivalent of bituminous (see *Humic Matter in Sediments*).

Bituminous sediments include: "bituminous coal," a soft coal, high in carbonaceous material, having 15–50% volatile matter; "bituminous lignite," a descriptive term used somewhat indiscriminately in the coal literature to apply to a certain kind of lignite; "bituminous sandstone," a sandstone containing bituminous material; and "bituminous shale," a shale containing hydrocarbons (bituminous material), which may yield gas or oil or distillation. Bituminous shale includes both black shale (see *Euxinic Facies*) and oil shale (q.v.); the organic constituents of bituminous shale are termed *sapropel* (q.v.). *Kerogen* (q.v.) is the petroleum distillate of oil shale.

B. CHARLOTTE SCHREIBER

References

Abraham, H., 1960/62. *Asphalts and Allied Substances.* New York: Van Nostrand, 4 vol.

Degens, E. T., 1965. *Geochemistry of Sediments.* Englewood Cliffs, N.J.: Prentice-Hall, 342p.

Cross-references: *Coal; Euxinic Facies; Humic Matter in Sediments; Kerogen; Oil Shale; Organic Sediments; Petroleum—Origin and Evolution; Sapropel; Shale.*

BLACK SANDS

Black sands are an unconsolidated, clastic sediment containing a large portion of dark heavy minerals (see *Heavy Minerals*). These minerals resist chemical and mechanical weathering and are preserved in weathered rocks and especially in marine placers (Figs. 1 and 2). The term *black sand* is sometimes used as a synonym for *concentrate* or the Russian term *shlikh*. At present, these terms are replaced by the more common term, *heavy minerals*.

SOLUBILITY IN CARBON DISULFIDE

```
                    Soluble                                    Insoluble
                    Bitumens                                   Pyrobitumens
         Liquid              Solid
                         Fusible    Difficulty          Low              High
                                    Fusible             Oxygen           Oxygen

Petroleum   Mineral    Asphalt     Asphaltites      Asphaltic        Non-Asphaltic
            Wax                                     Pyrobitumens     Pyrobitumens

All Crudes  Ozocerite  Bermudez    Gilsonite        Wurtzilite       Peat
Oil Seeps   Montan Wax Pitch       Grahamite        Elaterite        Lignite
                       Tabbyite    Glance Pitch     Albertite        Coal
                                                    Impsonite        Kerogen
```

FIGURE 1. Terminology and classification of naturally occurring bituminous substances (from Degens, 1965).

Origin

Black-sand deposits originate in all places where weathered rocks (igneous, metamorphic, or sedimentary) exposed to erosion contain a suitable mineral assemblage, which undergoes selective sorting in water and/or air by various means. This is a long-lasting exodynamic process linked with the morphological development of the area studied. Regions covered with glaciers, for example, do not offer favorable conditions for accumulation of heavy minerals (although the heavy-mineral content in glacial and glaciofluvial sediments has not yet been treated in detail), nor do areas where littoral sediments are rimmed with coral reefs (King, 1972; Zenkovich, 1962).

Occurrence

The most important accumulations of light heavy minerals are found in foreshore and backshore (berm) beach sediments as layers from 1 cm to many cm thick (Fig. 3). Increased concentration of heavy minerals has been found in the crests of submarine bars, in terminal parts of spits, and also in dunes, whose silt fraction is richest in heavy minerals. The highest concentrations (up to 100%) originate especially in sites where earlier accumulations of beach sands are exposed to intensive erosion (Figs. 1 and 4). If the coastline retreats, the sediments, enriched in heavy minerals, remain in situ.

Fluvial sediments are never as enriched in heavy minerals as marine. Their common content is 10–40%; maximum concentrations (up to 75%) may occur at the base of the alluvium. The concentration of small streams depends more upon the mineralogical composition of surrounding rocks, whereas in large rivers, whose

FIGURE 1. Cross-section of Ras Luale Island, northern part (after Cílek, 1971).

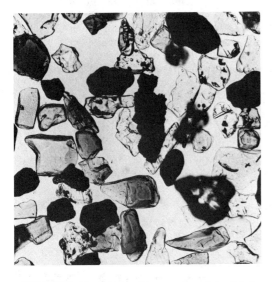

FIGURE 2. Assemblage of heavy minerals in vicinity of Dar-es-Salaam in Tanzania. The average content of heavy minerals equals 45.8 weight percent (16.6% ilmenite, 1.3% rutile, 19.6% garnet, 1.6% zircon, 6.0% kyanite and 0.7% magnetite). The average grain size is about 0.2 mm. (Microphoto: Náprstek.)

Black sands most often occur in littoral deposits, from which they have been mined industrially, less often in eolian, alluvial, and glacial deposits. Thus, genetic and quantitative relations among heavy minerals have to be considered with respect to their site of occurrence. The study of black sands is important in the determination of source areas during the search for mineral deposits and for conclusions of paleogeographic and paleogeologic character (see *Provenance*).

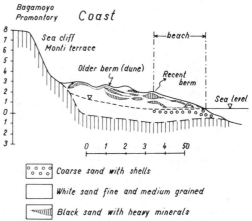

FIGURE 3. Generalized section of Kaole deposit in Tanzania; scales in m (after Cílek, 1971).

FIGURE 4. Older beach sands exposed to recent erosion. Black laminae are enriched in heavy minerals. (Photo: Cílek.)

alluvium represents a kind of independent geological unit, the concentration of heavy minerals depends rather on hydrodynamic regime.

Differences in the density of heavy minerals are more pronounced in water than in air, and when sufficient data are available, eolian and water-laid sediments can be distinguished with relative precision using the grain-size distribution of light and heavy fractions.

The sorting of heavy minerals makes it possible to detect the direction of transport, but complicates the determination of provenance, from quantitative proportions of individual heavy minerals. For example, amphibole becomes enriched with respect to garnet in the direction of transport. The Nubian sandstone of Late Cretaceous age (Said, 1962) overlying the peneplaned basement complex corresponds by its granulometric parameters to beach sediments and contains black bands of heavy minerals. Other sandstones in the vicinity of the basement complex are more enriched in zircon and anatase (see Fig. 5).

Heavy mineral composition may also vary with grain size. The coarse-grained sediments of the Rhône delta are rich in augite, and the fine-grained sediments abound in epidote (Andel, 1955), although both materials had the same source. Study showed that the size fraction 0.15 mm contained 98% of all augite in the sediment, but only 37% of all epidote.

Numbers of basic problems still remain to be solved both from qualitative and quantitative points of view: mechanical and chemical resistance and the role of grain size and density in different types of transport must be studied. The investigation of heavy minerals is very time-consuming, and it is sometimes convenient to carry out a detailed study of one mineral only.

Economic Importance

Monazite has been mined as a source of thorium since 1885, when the mixture of Th and Ce nitrates was first used for the production of gas lamps. Germans mined and transported Brazilian black sands for this purpose; in 1911 the Travancoore deposit in India assumed the leading position in mining output. In the US, the exploitation of heavy minerals began during World War I; the effective manner of exploitation has enabled the utilization of sands with a heavy mineral content of less than 2% (ilmenite, rutile, zircon, and monazite).

Australia is the chief world producer of black sands, possessing also the most extensive deposits—about 640 km on the E coast of Queensland and New South Wales. Rutile is the major product there; ilmenite is the primary product on the W coast. Ilmenite, zircon, rutile, and monazite are also mined in New Zealand. Australia's annual output (1972) is about 700,000 tons of ilmenite, about 360,000 tons of rutile, and 36,400 tons of zircon.

Heavy minerals are mined from marine sediments in a number of African states. Sierra Leone has probably the largest deposit of rutile in the world (estimated reserves about 12 million tons). Important producers also are

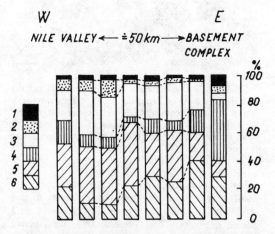

FIGURE 5. Average distribution of nonopaque heavy minerals in Nubian sandstone from the Central Eastern Desert of Egypt (Barramyia area), illustrating the changes in mineralogy of sandstones along the Edfu-Mersa Alam Desert Road (after El-Hinnawi et al., 1973). 1, others; 2, garnet; 3, tourmaline; 4, anatase; 5, rutile; 6, zircon.

Senegal, Malagasy Republic, South African Republic, Mozambique, Tanzania, and Egypt.

Asian deposits occur in Sri Lanka, Taiwan, Japan, Thailand, and in Malaysia, where heavy minerals are a by-product of the production of tin. In South America, deposits are known from Argentina and Uruguay. In Europe, marine placers are known from the coast of Tyrrhenian Sea (Italy), from the coast of the Black Sea (Bulgaria and Turkey), from the coasts of Spain, Norway, both the German states, Poland, and the USSR; there are reports of titanium mineral deposits in Greenland and from the bottom of the North Sea.

In the future, more black sands will be mined from alluvium, presently unfeasible because of the low heavy-mineral concentration. There is an increasing demand for minerals containing rare and trace elements, and black sands represent an important source of the following elements: Ti, Ce, Fe, Zr, Th, Cr, Sn, Y, Nb, Ta, W, Sc, Hf, Ag, Au, Pt, and U.

VLADIMÍR NÁPRSTEK

References

Andel, T. van, 1955. Sediments of Rhône delta. II. Sources and deposition of heavy minerals, *Verhandl. v. konikl. Nederl. Geol. Mijnb. Gen. Geol.*, ser. D., 15, 515–556.

Cílek, V., 1971. Geomorphological development of seashore of Indian Ocean in vicinity of Dar-es-Salaam, *Sborník čsl. spol. zeměpisné*, 76(4), 237–255.

El-Hinnawi, E. E.; Kabesh, M. L.; and Zahran, I., 1973. Mineralogy and chemistry of Nubian sandstones from the central eastern desert of Egypt, *N. Jb. Mineral. Abhand.*, 118(3), 211–234.

Emery, K. O., and Noakes, L. C., 1968. Economic placer deposits of the continental shelf, *ECAFE Tech. Bull.*, 1, 95–111.

King, C. A. M., 1972. *Beaches and Coasts.* New York: St. Martin's, 570p.

Náprstek, V., 1970. Black sand from the Bibijagua beach, Isla de Pinos, Cuba, *Acta Universitatis Carolinae-Geologica*, 2, 137–141.

Said, R., 1962. *Geology of Egypt.* Amsterdam: Elsevier, 337p.

Stapor, F. W., Jr., 1973. Heavy mineral concentrating processes and density/shape/size equilibria in the marine and coastal dune sand of the Apalachicola, Florida, Region, *J. Sed. Petrology*, 43, 396–407.

Zenkovich, V. P., 1962. Mineral resources of the beaches and marine nearshore aquatories, *U.N. Conf. on the Use of Sci. and Techn. Knowledge for Underdeveloped Countries.*

Zimmerle, W., 1973. Fossil heavy mineral concentrations, *Geol. Rundsch.*, 62(2), 536–548.

Cross-references: *Beach Sands; Heavy Minerals; Littoral Sedimentation; Provenance; River Sands; Sedimentary Petrography.*

BLACK SEA SEDIMENTS

The Black Sea is of peculiar interest to sedimentologists (Degens and Ross, 1974) as the type area of the *euxinic facies* (q.v.), a term based on the Greek name for this almost landlocked water body. The Black Sea is a basin of some 423,000 km^2, with a very shallow appendage (Sea of Azov) covering 38,000 km^2, together having a volume of 537,000 km^3. A larger part of the Black Sea is over a km deep, and it has a sill depth of 27.5 m in the Bosporus, connecting it via the Dardanelles to the Mediterranean. Its connection to the Sea of Azov, the Kerch Strait, is only 5 m deep (see Fig. 1).

The surface water of the Black Sea has a relatively low salinity (16–18‰ in the center, much fresher in many places near the shore); there is, however, an inflow at depth through the Bosporus of denser Mediterranean water (36‰ or more). As a result, there is limited mixing with the deep water. In the bottom sediment anaerobic bacteria decompose proteins and break down the sulfate ions to liberate H$_2$S, which is present in all the deep water to within 150–200 m of the surface.

The present fauna is of three sorts: Mediterranean immigrants of euryhaline potential; relics of the original Pliocene fauna; and fluvial, freshwater species that have adapted. The total number of species is quite impoverished (20%) with respect to the Mediterranean. Benthonic fauna becomes scarce below 50 m, and there are none at all below 150 m. The Sea of Azov has very few species (1.5% of Mediterranean), but has a vast biomass of essentially three species of mollusks.

The peculiar character of Black Sea sedimentation has been recognized for almost a century and has become well known through early work by Arkhangel'skii and Strakhov. Deep-sea cores up to 4 m long were obtained as early as 1927, when it was discovered that in the seismic belt near the shelf margin south of Crimea there was very extensive submarine slumping. In places on the shelf, all sediments back to the Miocene are missing, while at the foot of the slope there is a jumble of all the Pliocene/Quaternary formations (Fairbridge, 1947). Cores up to 30 m long have since been obtained, and slump areas are now known all around the Caucasus and along the Turkish shore.

The deep floor of the SW Black Sea today has a layer of over 50% carbonate mud, a coccolith ooze (the hydrology shows a high carbonate, low sulfate character, reflecting the carbonate-rich river flow; Fig. 2). In the SE depression, the carbonates are interbedded with gray turbidite silts with sand lenses particularly from the slump areas off SE Crimea. Marginally,

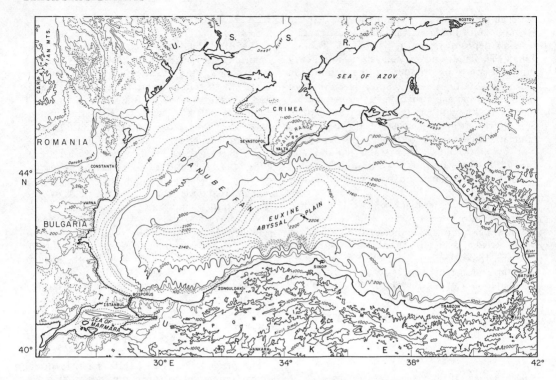

FIGURE 1. Bathymetric chart of the Black Sea (from Ross et al., 1974). Note change in contour interval at +200 m and -2000 m.

these sediments pass into gray clays. On the outer shelf there is a *Phaseolina* mud (the mussel *Modiolus phaseolina*). In the broad shelf area in the northwest, extending to south of the Danube delta, there are extensive gravels and coquinas, evidently relics of late Pleistocene low sea levels. From about 60-150 m there is a successive zonation of *Mytilus, Phyllophora* and *Modiolus phaseolina*. The silty clays are black when wet but dry out to a gray color. The deeper muds today, oddly enough, are not black at all, but in the carbonate (marly) areas there is a subsurface layer of truly black sapropelic mud. This is rubbery when wet, but dries hard and brittle, and carries up to 35% organic matter and 6-12% $CaCO_3$. Geochemically, they resemble oil shale (q.v.) in its early stages of formation (Ross, Neprochnov, et al., 1975). Black muds are also found today in the nearly landlocked lagoons ("limans").

History

The geological history of the Black Sea can be traced back in considerable detail to the middle Tertiary. It has evolved through an extraordinary variety of sedimentary environments. In Miocene times it was still part of a seaway that extended from north of the Alps through the Black Sea region to the Caspian and beyond; there was no direct connection with the eastern Mediterranean. In the Middle Miocene this seaway was open marine; in the Upper Miocene (Sarmatian) it became brackish, and in the Maeotian the Carpathian Orogeny cut the basin off completely from the west. There was a connection to the Aegean at this time, but, as today, the central basin was euxinic. In the latest Miocene (Pontian) the Black Sea became part of the freshwater lake that extended over the present basin, southernmost Ukraine, and most of Crimea to the Caspian depression. At the same time in the Mediterranean, the Messinian stage, there was a contrasting general desiccation, with evaporites apparently marking the total closure at Gibraltar. Evaporite phases are indicated by thick, dolomite-rich sequences which resemble those found in lakes (Ross, Neprochnov, et al., 1975).

The Quaternary history of the basin has been even more varied. With terraces around 95-105 m, the oldest stage (*Chaudian*) was a time marked by exclusively Caspian-type brackish faunas and appropriate sediments. During the Calabrian and Sicilian, high-level stages of the Mediterranean overflowed into the Black Sea, introducing Mediterranean species (*Old Euxi-*

and freshening of the water. Isotopic studies (Degens, 1971) show that the cold phases are marked by lithified bands in the sedimentary record, presumably due to chemical precipitation of carbonates.

In the Holocene, the Flandrian transgression (c. 9000 B.P.) caused a large waterfall at the head of the Bosporus, followed by a rapid rise of water level (compared by some writers to the Noachian Flood). Breccia from such an event was cored on Leg 42B of the DSDP (Ross, Neprochnov, et al., 1975). The *Old Black Sea* stage was succeeded by the *New Black Sea* stage of mid-Holocene time, when there have been minor oscillations in sea level and salinity. With the small drop of sea level during the last few thousand years there has been a rise in the anoxic, H_2S level (Deuser, 1970).

RHODES W. FAIRBRIDGE

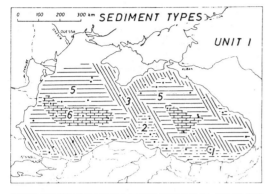

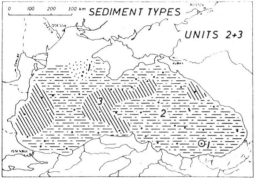

FIGURE 2. Black Sea sediment types (from Müller and Stoffers, in Degens and Ross, 1974, 200-248, fig. 20). Unit 1 represents 0-3000 B.P.; Units 2 and 3, 3000-7000 B.P. and >7000 B.P. respectively (see Ross et al., 1970). Sediment types are (1) silt (plus sand); (2) calcareous clay silt; (3) calcareous silt clay; (4) highly calcareous clay silt; (5) highly calcareous silt clay and silty clay; (6) marly clay and silt clay.

nian stage) and leaving two high-level terrace systems at 60-65 and 42-44 m (Hey, 1971). In the later Quaternary, there were progressively lower levels, with extremely low levels during glacial phases. The salinity was high and temperatures warm during interglacials; during glacials there was an alternation between milder times when the water was practically fresh and times when temperatures were very low and the hinterland was arid, leading to hypersaline conditions. During deglacial phases, meltwaters from the Scandinavian, Alpine, and Caucasian ice-fields lowered the salinity.

The Würm glacial stage is marked by the *New Euxinian*, with the level of the Black Sea very low and the waters mostly brackish, alternating with hypersaline phases. Pollen analyses of southern European sites indicate the widespread existence of an *Artemesia* steppe in the period 25,000-15,000 B.P. Milder phases or *interstades* during the Würm stage were marked by partial melting of continental ice, increased rainfall,

References

Andrussov, N., 1897. Ueber die Northwendigkeit der Tiefseeuntersuchungen im Schwarzen Meere, *Istv. Kais. Russ. Geog. Gesell. St. Petersburg,* **26**, 171-185.

Arkhangel'skii, A. D., and Strakhov, N. M., 1932. The geologic history of the Black Sea, *Byull. Moskov. Obshch. Ispyt. Prir., Otdel Geol.,* **40**, 1-104.

Baturin, G. N., 1975. *Iran: in Sovremennom Morskom Osadkoobrazovannie.* Moscow: Atomizdat, 152p.

Bogdanova, A. K., 1963. The distribution of Mediterranean waters in the Black Sea, *Deep-Sea Research,* **10**, 665-672.

Brewer, P. G., and Murray, J. W., 1973. Carbon, nitrogen and phosphorus in the Black Sea, *Deep-Sea Research,* **20**, 803-818.

Caspers, H., 1957. Black Sea and the Sea of Azov, in J. W. Hedgpeth, ed., *Treatise on Marine Ecology, Geol. Soc. Am. Mem.,* **67**(1), 803-889.

Degens, E. T., 1971. Sedimentological history of the Black Sea over the last 25,000 years, in A. S. Campbell, *Geology and History of Turkey.* Tripoli: Petrol. Expl. Soc., 407-429.

Degens, E. T., and Ross, D. A., 1974. The Black Sea—geology, chemistry, and biology, *Am. Assoc. Petrol. Geologists, Mem.* **20**, 633p.

Deuser, W. G., 1970. Carbon-13 in Black Sea waters and implications for the origin of hydrogen sulfide, *Science,* **168**, 1575-1577.

Fairbridge, R. W., 1947. Coarse sediments on the edge of the continental shelf, *Am. J. Sci.,* **245**, 146-153.

Hey, R. W., 1971. Quaternary shorelines of the Mediterranean and Black seas, *Quaternaria,* **15**, 273-284.

Laking, P. N., ed., 1974. *The Black Sea–Its Geology, Chemistry, Biology–A Bibliography.* Mass.: Woods Hole Oceanographic Inst., 368p.

Muratov, V. M.; Ostrovsky, A. B.; and Fridenberg, E. O. 1974. Quaternary stratigraphy and paleogeography on the Black Sea coast of western Caucasus, *Boreas,* **3**, 49-60.

Murray, J., 1900. On the deposits of the Black Sea, *Scot. Geogr. Mag.,* **16**, 673-702.

Ross, D. A.; Degens, E. T.; and MacIlvaine, J., 1970. Black Sea: Recent sedimentary history, *Science,* 170, 163-165.

Ross, D. A.; Uchupi, E.; Prada, K. E.; and MacIlvaine, J. C., 1974. Bathymetry and microtopography of Black Sea, *Am. Assoc. Petrol. Geologists, Mem. 20,* 1-10.

Ross, D.; Neprochnov, Y.; and 11 others, 1975. *Glomar Challenger* drills the Black Sea, *Geotimes,* 20(10), 18-20.

Smirnow, L. P., 1958. The Black Sea Basin; its position in the Alpine structure and its richly organic Quaternary sediments, in L. G. Weeks, ed., *Habitat of Oil.* Tulsa: Am. Assoc. Petrol. Geologists, 982-994.

Strakhov, N. M., 1969. *Principles of Lithogenesis,* vol. 2. Edinburgh: Oliver & Boyd, 609p.

Cross-references: *Euxinic Facies; Oil Shale; Sulfides in Sediments;* Vol. I: *Black Sea; Mediterreanean Sea;* Vol. III: *Caspian Sea.*

BLACK SHALES—*See* EUXINIC FACIES

BLANKET SAND

A blanket sand, or *sheet sand,* is a uniformly thin, widespread body of sand or sandstone. These sand sheets usually occur in the stratigraphy of broad and stable shelf areas (cratons). A typical example is the Ordovician St. Peter Sandstone, which originally extended in an unbroken sheet, 30-50 m thick, from Colorado to Indiana (Dapples, 1955; Fig. 1).

Most blanket sands are believed to originate in an area undergoing slow submergence, in such a way that strandline deposits are progressively developed as a nearly continuous succession of parallel beaches which are then buried by the transgressive offshore (shelf) sediments. Similar but regressive shoreline deposits are not as readily preserved, due to erosion, but regressive tongues are well known within transgressive deposits.

Blanket sands are often quartz arenites (orthoquartzites), presumably because the source is either an intensely weathered coastal plain deposit or an older deposit of sands. These sands are then subjected to intense reworking by wind, waves, and/or currents.

B. CHARLOTTE SCHREIBER

References

Dapples, E. C., 1955. General lithofacies relationship of St. Peter Sandstone and Simpson group. *Bull. Am. Assoc. Petrol. Geologists,* 39, 444-467.

Goldring, R., and Bridges, P., 1973. Sublittoral sheet sandstones, *J. Sed. Petrology,* 43, 736-747.

Potter, P. E., and Pryor, W. A., 1961. Dispersal centers of Paleozoic and later clastics of the Upper Mississippi valley and adjacent areas, *Geol. Soc. Am. Bull.,* 72, 1195-1250.

Spearing, D. R., 1975. Stratification and sequence in prograding shoreline deposits, *Soc. Econ. Paleont. Minerol. Short Course,* 2 (April 1975), 81-132.

Cross-references: *Barrier Island Facies; Continental-Shelf Sedimentation; Sands and Sandstones.*

BOUMA SEQUENCE

The genetic term *turbidite* (q.v.) represents the deposit of a turbidity current (see also *Turbidity-Current Sedimentation;* pr. Vol. IX: *Flysch and Flysch Facies*). A turbidite formation is in part characterized by its general appearance of regular bedding, which usually consists of alternating arenaceous and lutaceous beds; the sequence may be found frequently in clastic limestones, also in volcaniclastics, and even in evaporite sediments (Mutti and Ricci Lucchi, 1972).

Signorini (see Bouma, 1972) was probably the first to note characteristic sedimentary structures in the turbidite unit, and Kuenen and Migliorini (1950) used graded bedding as a tool to identify these sediments as deposits laid down by turbidity currents. Bouma (1962) indicated

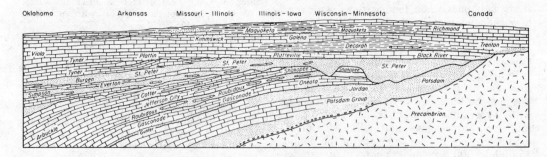

FIGURE 1. Generalized, reconstructed N-S cross section of the early Paleozoic of the Mississippi Valley, showing several transgressive blanket sands, including the Potsdam and St. Peter sandstones (Potter and Pryor, 1961).

that there was a well-defined sequence of sedimentary structures, now known as the *Bouma sequence,* within each arenite-lutite bed couple. Other models have been described by Hubert (1966) from Paleozoic flysch-type beds in Scotland and from Mesozoic deposits in Switzerland, and by Van der Lingen (1969) from Tertiary turbidites in New Zealand.

Several investigators have related the sedimentary-structure division to flow regimes (q.v.) using data from experimental studies (e.g., Simons et al., 1965; Walker, 1965). The Bouma sequence can be explained as the result of a waning turbidity current passing from an upper flow regime to lower flow regimes (Fig. 1). It is not as simple as the figure may indicate, however (Walton, 1967). The maturity of the turbidity current and the size distribution are only a few of the variables that may influence the deposits. A good summary of flume experiments with an extensive bibliography is presented by Middleton and Hampton (1973).

The Sequence

The Bouma sequence describes a complete turbidite unit and is characterized by a fixed vertical succession of five intervals (later renamed divisions by Walker, 1965), each identified by one sedimentary structure (Fig. 1). According to this concept, an arenaceous bed together with its overlying lutaceous bed is considered as one layer, the result of one turbidity current. The five successive divisions (Bouma, 1962, 1972) from bottom (a) to top (e) are:

Graded or Massive Division (a). No primary sedimentary structures other than graded bedding have been observed in this division, except possibly antidune-like structures (Hubert, 1966). The grading may be reversed in the lower few centimeters of a very thick layer but rapidly changes to positive graded bedding (q.v.). This grading may be rather irregular (Fig. 1) or totally absent (massive bedding). The latter case may occur in well-sorted material, in which it is almost impossible to develop graded bedding, or in sediments with a wide variation in grain size, often including coarse-grained pockets (Walton, 1967) which possibly imply deposition from immature currents (Walker, 1965). Inclusions of marl, shale, or large plant fragments are also common. Dish structures may be observed in this division.

Lower Division of Parallel Lamination (b). Parallel laminae are characteristic of this division (b). The transition from (a) to (b) normally is gradual and in many cases consists of a zone of rather indistinct thin bedding and/or thick laminae (called "poorly laminated interval" by Van der Lingen, 1969). Graded bedding usually continues into (b) but is sometimes masked completely by the lamination. These vague transitions make it very difficult to measure accurate thicknesses in the field.

Division of Current Ripple Lamination (c). Several sets of unidirectional foreset bedding characterize division (c); in a few locations complete current ripples have been observed. The foreset bedding may be deformed to various degrees, often having a convolute appearance or

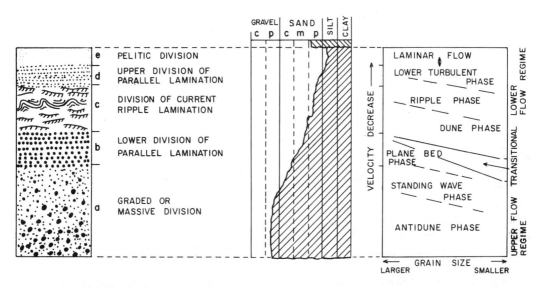

FIGURE 1. Schematic presentation of the Bouma sequence with its characteristic divisions (modified after Simons et al., 1965; Bouma, 1972). The lithology-texture curve is based on actual data. On the right, a schematic flow regime chart is given revealing the possible relationship to the sequence.

a consolidation-influenced flow pattern. If convolute lamination is present in the sequence, its occurrence is restricted to division (c). Climbing ripples have also been observed, and clayey laminae may occur between ripple sets. The transition between divisions (b) and (c) may be either distinct or gradational. In the former case, it represents the boundary between the arenaceous and the (overlying) lutaceous beds.

Upper Division of Parallel Lamination (d). Division (d) is normally part of the lutaceous bed and may contain varying amounts of fine sand and silt-sized material. Sand content and the degree of weathering influence whether the characteristic fine, parallel laminae are visible, poorly discernible, or not visible at all. The transition from division (c) to (d) may correspond to the break in grain size between arenite and lutite, but in most cases it is more or less gradational.

Pelitic Division (e). The uppermost division consists of pelitic or lutaceous material in which no primary sedimentary structures are visible. Bioturbation is rather common, increasing upward in intensity. The contact with division (d) is often transitional and difficult to locate.

This sequence may be covered by a hemipelagic deposit, often noted by a color difference and a deeper water fauna than the turbidite. If the hemipelagic deposit is thin, it may easily be destroyed by burrowing animals or by erosion. It should be noted that the contact between the arenaceous and the lutaceous layers may be found anywhere between the top of division (b) and the lower part of (d) and may range from very distinct to nonexistent.

Occurrence

The complete Bouma sequence, as discussed above, can only be observed in a small percentage of all turbidites encountered in a series. The lower one, two, or three divisions may be absent, and occasionally one or more top divisions may be missing, presumably eroded by a succeeding turbidity current. In a few instances, both top and bottom divisions may be missing. Bouma (1962), however, stressed the observation that within the set of divisions present no gaps exist. It should be realized that thin divisions and effects of weathering and of cementation may make any division difficult to recognize. Walker (1965) described sequences from North Devon in which he found middle divisions to be absent (e.g., ae, ace, abe, etc.).

Cases where the basal divisions of a sequence do not exist can be explained in two ways. First, if a turbidity current reaches an upper flow-

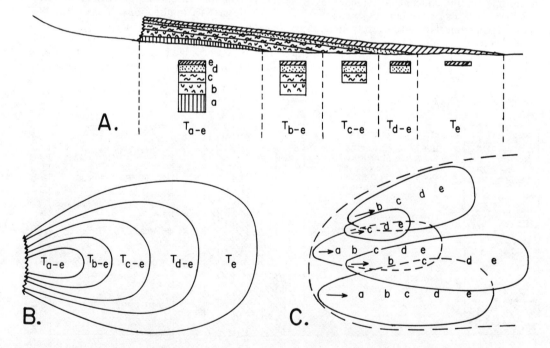

FIGURE 2. Hypothetical shapes of turbidites (modified after Bouma, 1962, 1972). No scale factors. A: Decrease in thickness and sequence completeness in down-current direction. B: Depositional cone in plan with decrease of completeness of the sequence. C: Filling of a basin by a number of turbidites; the letters indicate the division at the bottom of its deposit.

regime velocity capable of producing an antidune bedform, graded beds will be deposited. If the current never reaches this phase, however, sedimentary structures characteristic of the obtained velocity will be formed. Secondly, in vertical and in horizontal directions, the current velocity decreases and enters successively lower phases of the flow regime, resulting in varying sedimentary structures. Consequently, one will observe a decrease in the completeness of the sequence in the downcurrent direction and away from the current axis (Fig. 2A and B). The deposit normally has an elongate shape, which is influenced by local microbathymetry (Fig. 2B), and it will form a local high that influences any subsequent turbidity current (Fig. 2C). The decrease in the completeness of the Bouma sequence is accompanied by a decrease in layer thickness (Fig. 2A) and in maximum grain size. Although the sand:clay ratio usually decreases in the same direction, it is not apparent that thickness ratios of the divisions follow a rigid pattern. In areas close to the source, often called proximal areas, division (a) may comprise over 90% of the thickness of a several m thick layer, divisions (d) and (e) being very thin.

Flutes and grooves are often visible on bedding planes when (a) or (b) forms the bottom division, but become scarce when (c) forms the bottom. Loadcasted scour-and-fill structures are quite common in (a); normal loadcasts can be observed in (b) and (c) and sometimes in (d) when it is lowermost.

ARNOLD H. BOUMA

References

Bouma, A. H., 1962. *Sedimentology of Some Flysch Deposits: A Graphic Approach to Facies Interpretation.* Amsterdam: Elsevier, 168p.

Bouma, A. H., 1972. Recent and ancient turbidites and contourites, *Gulf Coast Assoc. Geol. Soc. Trans.,* **22,** 205-221.

Hubert, J. F., 1966. Modification of the model for internal structures in graded beds to include a dune division, *Nature,* **211**(5049), 614-615.

Kuenen, P. H., and Migliorini, C. I., 1950. Turbidity currents as a cause of graded bedding, *J. Geol.,* **58,** 91-127.

Middleton, G. V., and Hampton, M. A., 1973. Sediment gravity flows: mechanics of flow and deposition, in G. V. Middleton and A. H. Bouma, eds., *Turbidites and Deep-water Sedimentation* (Short course lecture notes). Anaheim: SEPM Pacific Section, 1-38.

Mutti, E., and Ricci Lucchi, F., 1972. Le torbiditi dell' Appennino settentrionale: introduzione all' analisi di facies, *Soc. Geol. Ital. Mem.,* **11,** 161-199.

Simons, D. B.; Richardson, E. V.; and Nordin, C. F., Jr., 1965. Sedimentary structures generated by flow in alluvial channels, *Soc. Econ. Paleont. Mineral. Spec. Publ. 12,* 34-52.

Van der Lingen, G. J., 1969. The turbidite problem, *New Zeal. J. Geol. Geophys.,* **12,** 7-50.

Walker, R. G., 1965. The origin and significance of the internal sedimentary structures of turbidites, *Yorkshire Geol. Soc. Proc.,* **35,** 1-32.

Walton, E. K., 1967. The sequence of internal structures in turbidites, *Scot. J. Geol.,* **3,** 306-317.

Cross-references: *Flow Regimes; Graded Bedding; Intraformational Disturbances; Penecontemporaneous Deformation of Sediments; Scour-and-Fill; Scour Marks; Turbidites;* Vol. IX: *Flysch and Flysch Facies.*

BOXWORK

Boxwork is a cellular structure containing roughly angular or box-like cavities separated by resistant plates or septa that intersect one another at various angles. Larger and smaller cavities occur side by side without any regularity. The larger cells usually measure 1-2 cm across, although limonitic boxworks of gossans generally show very small meshes. Boxwork structure originates as a result of weathering and oxidation processes acting upon parent rocks intersected by a network of resistant veins. These veins originally develop along fractures, cleavage and bedding planes, and shrinkage cracks. When the parent material is dissolved from the space between the tabular veins, the protruding network of veins—boxwork—remains. The plates of this honeycomb-like structure are usually more or less straight but may show curves and windings. In some cases there is a marked tendency for a majority of the plates to lie parallel, being intersected at irregular intervals by other plates, which cut the former ones at any angle. Residues of leached or altered parent rocks sometimes fill the cavities of the boxwork or occur as crusts upon the cell walls. Various kinds of boxwork may be distinguished by their mineralogical composition, such as boxworks of quartz, limonitic jasper, limonite, calcite, dolomite, siderite, smithsonite, malachite, azurite, etc.

Occurrence

The mode of occurrence of boxwork structures varies widely. Boxworks essentially consisting of limonitic jasper are found in leached outcrops of oxidized sulfide ore. These limonitic boxworks may gradually merge into more spongy structures which differ from boxwork in possessing more rounded, irregular cells with thicker walls. The pattern of boxwork and spongy structures is governed by the pattern of the cleavages, fractures, and crystal boundaries in the original, unoxidized sulfidic mass;

boxwork structures in gossans may thus be diagnostic of the character of the underlying sulfide ore. The veinlets of supergene limonitic jasper, formed during incipient oxidation of the ore, may be accompanied by quartz veins of hypogene origin. The low solubility of limonitic jasper and quartz in acidic solutions explains their preservation during further oxidation and leaching. The cells of such boxworks are usually partly filled with indigenous limonite derived from the original sulfides or other minerals which formerly occupied the cavities. Where oxidizing sulfide deposits are underlain by limestones, boxwork siderite and smithsonite may form below the ore. Boxwork structures are also found in the zone of oxidation of pitchblende ore.

Typical boxworks may also occur as a result of selective weathering of fractured dolostones and dolomitic marls that are intensely seamed by resistant calcite veins. Boxworks resulting from veined dolostones are common in Permian and Triassic carbonate complexes in Europe, where they are known as *Rauhwackes* (q.v.) or *Zellendolomits* (Fig. 1). Cavities in these rauhwackes often still contain pulverulent dolomite or dolomitic sand formed by partial weathering of the dolomitic parent rock.

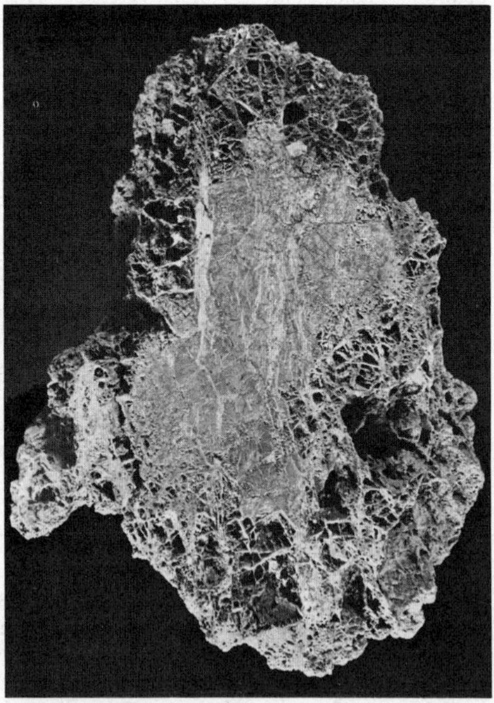

Boxwork is also a term used in speleology in order to denote a protruding network of calcite veins found on dolomitic parts of cavern walls. Resistant fillings of desiccation cracks in septarian nodules may stand out prominently through selective weathering, thus forming a boxwork pattern. Similar structures, not related to nodules, have been named *Melikaria* (Greek for honeycomb).

LOUIS LEINE

References

Boswell, P. F., and Blanchard, R., 1929. Cellular structure in limonite, *Econ. Geol.*, **24**, 791-796.

Bretz, J. H., 1942. Vadose and phreatic features of limestone caverns, *J. Geol.*, **50**, 675-811.

Leine, L., 1968. *Rauhwackes in the Betic Cordilleras, Spain—Nomenclature, Description and Genesis of Weathered Carbonate Breccias of Tectonic Origin.* Rotterdam: Princo N.V., 112p.

Trischka, C.; Rove, O. N.; and Barringer, D. M., 1929. Boxwork siderite, *Econ. Geol.*, **24**, 677-686.

Cross-references: *Diagenesis; Diagenetic Fabrics; Dolomite; Limestone; Rauhwacke; Speleal Sediments; Weathering in Sediments.*

BRAIDED-STREAM DEPOSITS

Braided streams are characterized by divisions and rejoinings of flow around islands or bars composed of alluvial sediment. The underlying causes for the occurrence of braided streams are incompletely understood, but they tend to be favored by high regional slopes, easily eroded banks, abundant supplies of coarse sediment, and highly variable discharges (see Fahnestock, 1963, for review of ideas concerning braided patterns). Arid and semiarid regions, alluvial fans, mountainous areas, and glacial outwash plains are typical, but not exclusive, sites for braided streams (Miall, 1977).

Braided streams are highly transient environments. Shifting bars and channels coupled with normal discharge fluctuations cause wide variations in local flow conditions, resulting in varied assemblages of grain sizes, textures, and sedimentary structures. Deposition is primarily coarse bed load in bars and channels; suspended fines (silt and clay) may accumulate locally in depressions, cut-off channels, bar surfaces, and overbank areas. Rarely abundant, such fine-grained deposits are usually thin and lenticular. Most braided streams in high-slope regions and on glacial outwash plains (sandar) deposit gravel and coarse sand; others, including some very large rivers such as the Brahmaputra

FIGURE 1. Boxwork structure as shown by a monomict rauhwacke, the central part of which is composed of nonweathered parent rock. Length of specimen 13 cm.

(Coleman, 1969) and Yellow (Chien, 1961), may deposit medium and fine sand.

Alluvial islands, or braid bars, are local accumulations of bed material formed during high flows and exposed at lower water stages. Such exposed bars may be relatively unaltered from their original depositional forms (unit bars), but others are erosional remnants of earlier bar and/or channel deposits. Thus, the internal characteristics of braid bars may vary considerably depending upon the nature of the original deposit and the extent of modification prior to and during exposure. Active unit bars assume various geometric forms depending on the character of flow, sediment size, and channel shape. A precise bar classification system does not yet exist, and there is considerable overlap in usage of existing terminology. Transverse, longitudinal, linguoid, diagonal, spool, point, sheet, and lateral bars are some of the more common terms applied to both active and exposed bars. There have been some recent attempts to relate sedimentary sequences to bar types (Smith, 1974; Hein, 1974; Walker, 1975).

Braided-stream gravels are mostly deposited as sheet-like layers on bar surfaces and in shallow channels, yielding a crude horizontal stratification in cross section. Lateral growth of bars with riffle or slip-face margins produces some low- to high-angle cross-stratification as well. Ripples and especially dunes are abundant in sandy braided streams; such deposits are rich in large- and small-scale trough cross-strata. Sand-bed braided streams also typically contain an assortment of wide, flat-topped bars with slip-face margins which yield tabular bodies of planar cross-stratification. Common features in both gravel and sand-bed braided-stream deposits include irregular bedding surfaces and variable bed thicknesses resulting from uneven bar-channel topography, scour-and-fill structures, abrupt small-scale variations in grain size and texture including lateral and vertical grading within single beds, and locally abundant mud intraclasts.

In vertical sections, most braided-stream deposits appear to be a random array of varying grain sizes, textures, structures, and bed geometries. Costello and Walker (1972) describe upward-coarsening sequences in Pleistocene braided-stream deposits thought to be due to gradual infilling of abandoned channels. Occasional upward-fining sequences might be expected in deposits of some large bars formed by lateral accretion. Regional down-slope trends in braided-stream sediments include decreases in grain size (gravel to sand) and horizontal stratification, and increased proportions of planar cross-strata, bed surface regularity, and vegetation (Smith, 1970; Rust, 1972).

Ancient Deposits. Few ancient (pre-Quaternary) braided-stream deposits have been studied in detail. Some examples include the Silurian Shawangunk Conglomerate and related formations (Smith, 1970); the Precambrian Gulneselv Formation, Norway (Banks, 1973); the Pennsylvanian Sharon Conglomerate, Ohio (Mrakovich and Coogan, 1974); the (?)Permo-Triassic Stornoway Formation, Scotland (Steel and Wilson, 1975); and the Devonian Battery Point Formation (Cant, 1973).

NORMAN D. SMITH

References

Banks, N. L., 1973. Falling-stage features of a Precambrian braided stream, *Sed. Geol.,* 10, 147-154.

Bluck, B. J., 1974. Structure and directional properties of some valley sandur deposits in southern Iceland, *Sedimentology,* 21, 533-544.

Boothroyd, J. C., 1972. Coarse-grained sedimentation on a braided outwash fan, northeast Gulf of Alaska, *Univ. Mass. Amherst Coastal Research Cen. Tech. rept.,* 127p.

Cant, D. J., 1973. Devonian braided stream deposits in the Battery Point Formation, Gaspé Est, Quebec, *Maritime Sed.,* 9, 13-20.

Chien, N., 1961. The braided stream of the Lower Yellow River, *Scientia Sinica* (Peking), 10, 734-754.

Church, M. A., 1972. Baffin Island sandurs: a study of Arctic fluvial processes, *Geol. Surv. Canada Bull.,* 216, 208p.

Coleman, J. M., 1969. Brahmaputra River: channel processes and sedimentation, *Sed. Geol.,* 3, 129-239.

Costello, W. R., and Walker, R. G., 1972. Pleistocene sedimentology, Credit River, southern Ontario: a new component of the braided river model, *J. Sed. Petrology,* 42, 389-400.

Eynon, G., and Walker, R. G., 1974. Facies relationships in Pleistocene outwash gravels, southern Ontario: a model for bar growth in braided rivers, *Sedimentology,* 21, 43-70.

Fahnestock, R. K., 1963. Morphology and hydrology of a glacial stream, White River, Mount Rainier, Washington, *U.S. Geol. Surv. Prof. Pap. 422A,* 1-70.

Hein, F. J., 1974. Gravel transport and stratification origins, Kicking Horse River, British Columbia, M.Sc. Thesis, McMaster University, 135p.

Miall, A. D., 1977. A review of the braided-river depositional environment, *Earth-Science Reviews,* 13, 1-62.

Mrakovich, J. V., and Coogan, A. H., 1974. Depositional environment of the Sharon Conglomerate member of the Pottsville Formation in northeastern Ohio, *J. Sed. Petrology,* 44(4), 1186-1199.

Rust, B. R., 1972. Structure and process in a braided river, *Sedimentology,* 18, 211-245.

Smith, N. D., 1970. The braided stream depositional environment: comparison of the Platte River with some Silurian clastic rocks, north-central Appalachians, *Geol. Soc. Am. Bull.,* 81, 2993-3014.

Smith, N. D., 1974. Sedimentology and bar formation in the upper Kicking Horse River, a braided outwash stream, *J. Geol.,* 81, 205-223.

Steel, R. J., and Wilson, A. C., 1975. Sedimentation and tectonism (? Permo-Triassic) on the margin of the North Minch Basin, Lewis, *J. Geol. Soc.,* **131,** 183-202.
Walker, R. G., 1975. Conglomerate: sedimentary structures and facies models, *Soc. Econ. Paleont. Mineral., Short Course,* **2** (Dallas, 1975), 133-161.
Williams, P. F., and Rust, B. R., 1969. The sedimentology of a braided river, *J. Sed. Petrology,* **39,** 649-679.

Cross-references: *Alluvial-Fan Sediments; Alluvium; Conglomerates; Fluvial Sediments; River Gravels.* Vol. III: *Alluvial Fan, Cone; Braided Streams; Rivers: Meandering and Braiding.*

BRECCIAS, SEDIMENTARY

A breccia is a cemented, clastic, rudaceous rock made up of angular or subangular fragments (commonly defined as consisting of > 80% angular particles over 2 mm in diameter). There are three broad categories: one is exogenetic—*sedimentary breccia;* and two are endogenetic—*pyroclastic* (see *Pyroclastic Sediment*), *volcanic,* or *igneous breccias*—e.g., *explosion breccias, intrusion breccias,* or *agglomerates* (see *Agglomerate*)—and *cataclastic, diastrophic,* or *orogenic breccias*—*fault, gash,* or *crush breccias.* Sedimentary breccia is not to be confused with *conglomerate* (see *Conglomerates*) which is a rudite with rounded clasts. Breccia clasts are often referred to as *sharpstone,* in contrast to a conglomerate's *roundstone.* Shrock (1948) found "breccia" too loose a term and preferred "sharpstone conglomerate," as opposed to "roundstone conglomerate," but the suggestion has not found much favor.

The word *breccia* originated in Italy, where it has long been used for the crushed rubble used in walls, and for *breccia marble,* the well-known ornamental stone (a metamorphic crush breccia). In several European countries, the term also embraces conglomerates.

The clasts in a breccia may be (1) monomictic and/or autoclastic, i.e., fractured in situ so that the bits without the cement could fit back together; or (2) polymictic. *Autoclastic breccias* are usually intraformational and synsedimentary, i.e., they are formed during sedimentation or early diagenesis. The clasts are penecontemporaneous mudstones, limestones (or dolomites), sandstones, and cherts. *Polymictic breccias,* in contrast, usually comprise exotic clasts of pre-existing sedimentary or igneous rocks.

Cements of both types of breccias may be carbonate, silica, claystone, or, more rarely, iron oxides, or gypsum, or other evaporite minerals. The cements may form at any stage of diagenesis (q.v.), and their study may be very rewarding in any analysis of postdepositional history. The cements may be designated thus: *quartzose breccia, arkose breccia, graywacke breccia, lime breccia* (calcirudite), etc.

Penecontemporaneous Breccias

In penecontemporaneous breccias, the clast material, the fragmentation, and the cements—at least in part—are all closely related in time. *Autoclastic breccia* (or *autobreccia*) produced by submarine slumping and landsliding is the best known of this category. When beds of different lithology are subjected to tension, the fine-grained member generally fractures and is slightly pulled apart, and the adjacent, water-saturated silts or sands flow in to form the cement.

The volume loss of dolomitization (see *Dolomite*) may cause autoclastic brecciation; both clasts and cement are usually dolomitized, but with differing fabrics and trace elements. Another important type of autoclastic brecciation occurs during dehydration of chert gels. The outer layers become desiccated first and fracture, only to become sealed by extrusion of the still-plastic gel within Gignoux and Avnimelech (1937) called this phenomenon *éclatement.*

Subaerially, during the formation of a calcrete duricrust (see *Duricrust*) or caliche (q.v.), there is a phase when a calcareous (or gypseous) paste develops near the lower boundary of vadose water infiltration; this phenomenon is limited to semiarid climates (less than about 20 cm precipitation). The paste gradually desiccates and crystallizes to calcite or gypsum, developing pull-apart clasts. These clasts are subsequently recemented by downward percolating waters of a similar or somewhat disparate composition, e.g., a red sand or illuvial clay. Other forms of subaerial autoclastic breccias common in carbonate terrains and also in certain evaporite formations are the various *solution* or *collapse breccias,* also known as *karst breccias.* A particular horizon (carbonate, gypsum, halite, etc.) is dissolved, the roof caves in, and the resulting clasts are recemented. The cement is commonly of the same composition as the clasts because the process occurs within the zone of seasonal ground-water saturation and undersaturation. A spectacular example is the Mackinac Breccia (Middle Devonian) of northern Michigan, where Devonian limestone forms the clasts, undermined by the solution of the underlying Silurian salt beds (Landes, 1945).

A rather special autoclastic form is the *bone breccia,* which is normally of reptile bones in the Mesozoic and mammalian bones in the Cenozoic. The bones may have been contemporaneously crushed by predators, or postdepositionally crushed by loading. Bone beds (apart from cave types) are sometimes important in stratigraphy, marking formation boundaries; in this case they are often phosphatized or associated with glauconite (q.v.), indicating a phase of shallowing water, increased bottom currents, winnowing of fine particles, etc. (see *Biostratinomy*).

Closely related to autoclastic breccia is *desiccation breccia* (also called *edgewise conglomerate* or *clay-pebble conglomerate*, q.v.). These breccias are generated from a mud-cracked surface, where the desiccated layer curls up, and the flakes are broken up and distributed by the next flood or high tide. The clasts may be angular or rounded, and the cement is penecontemporaneous, perhaps only a day or two younger than the clasts.

Transitional between the autoclastic and the polymictic types are two breccias related to submarine talus or landslides. Usually associated with slump structures (see *Slump Bedding*), *intraformational breccia* contains a mixture of clasts of the subjacent formations and is clearly connected with submarine landslides, sea-floor displacements that have transported clastic debris downslope. It may be considered as intermediate between a simple autoclastic slump and a totally fluidized turbidity flow. A special type of submarine talus breccia is the *reef breccia*—a typical forereef facies of a bioherm, which may be coral, algal, or mixed. In the normal upward growth of bioherms, the slopes tend to become oversteepened; and, under wave turbulence, offreef slides are common. In this way, a complex breccia of shallow-water reef organisms and sediment becomes embedded in more bathyal facies, often fine-grained pelagic muds. In Paleozoic shallow-shelf seas, the reef knolls were often quite small, but, being wave resistant, established centers of fragmentation. Widespread *shoal breccias* are recognized in these limestones.

Polymictic Breccias

In this second major category, the clasts are in no way related either to their cements or to formations into which they are introduced. Emplacement is in almost every case by gravitational sliding.

Exotic breccias (see also *Olistostromes*) develop from submarine landslides where active faulting (normal, thrust, nappe) on the sea floor forces basement rocks to the surface, and the angular fragments are then transported downslope by gravity and incorporated into marine deposits. Breccias of this sort are particularly well known in the Alps and comparable orogenic belts; a very extensive one in the Jurassic of the Western Alps has given its name to the "Breccia Nappe" (see Hendry, 1972). Another, in the Hercynian orogenic belt of the Ardennes in Belgium, is known as "La Grande Brèche" (Kaisin, 1942). A celebrated North American example is the Cambrian Cow Head Breccia of Newfoundland (Schuchert and Dunbar, 1934). At colliding plate margins, as might be expected, there is every transition between a simple sedimentary breccia, to *wildflysch* to *sedimentary klippes*, to a tectonic *melange* (see pr. Vol. X: *Mélange*). Not all exotic breccias are plate boundary indicators, although they usually bear witness to large structural anomalies that seem to have been active at the time of emplacement. A classic example is the late Jurassic breccia situated near the Great Glen Fault in Scotland (Bailey and Weir, 1933).

Subaerially, talus, landslides, and mudflows may produce breccias. *Talus breccia* (or *scree breccia* in England) is usually cemented by ground-water solutions. Harrington (1946) proposed the (rarely used) term *cenuglomerate* for mudflow breccia. Subaerial mudflows may produce extensive breccias, especially in semiarid regions, in mountains, and in volcanic terrain. In semiarid regions, such as the Basin and Range Province, these deposits are so striking that they have been called *bajada breccias* (Norton, 1917), but the term *fanglomerate* has precedence and has been most widely adopted (see *Alluvial-Fan Deposits*).

Giant examples of *fanglomerates*, or *megabreccias* (Longwell, 1951), are produced by landslides in favorable structural settings, e.g., southern Nevada and Arizona, where clasts up to 1000 m in length are recorded; tectonics may or may not be involved. The expression *chaos*, proposed earlier by Noble (1941) for a megabreccia associated with thrusting in the Death Valley region of California, is not a desirable term inasmuch as chaos and chaotic are common descriptive words in the language, and their use in a technical sense could be ambiguous.

Talus breccia resulting from solifluction under periglacial climatic conditions is extremely widespread, a product of the periglacial climates which extended from glacier fronts to about $30°N$ and S latitudes during the Pleistocene glaciations. These *solifluction breccias* (or *colluvial breccias*) are usually cemented with calcite, and noncarbonate Pleistocene solifluction rubble is usually still uncemented today. The cemented solifluction breccias are particularly striking along the limestone coasts of the Mediterranean, where they can often be seen disappearing below present sea level, indicating an origin during the last glaciation. A special name for solifluction breccias in France is *grèzes litées* (Guillien, 1951).

The term *moraine breccia* is sometimes used for a cemented glacigene deposit, characteristically a lateral moraine in a region of carbonate-rich ground water. An unusual glacigene rock, at least one not widely identified in this sense, is a breccia generated by ice-plucking along shorelines, then transportation by floe ice (or river ice), and redeposition of still-angular clasts in a setting perhaps quite remote from its source. This special form of *tillite* (see *Till and Tillite*) is a product of floating ice (*glaciel* in French) rather than glacier ice (see *Glacial Marine Sediments*).

Residual Breccia. Weathering of breccias can lead to curious results, either by differential solution of a calcitic cement from a chert or dolomite, for example, or by removal of the clasts, leaving a honeycomb or alveolar structure (*boxwork*). Examples of the first case may result in a skeletal soil, an insoluble rubble, or chert breccia fragments, from which the original cement has been completely leached. These residual breccias are known from the Flint Hills of Kansas, in the Boone Chert (Mississippian) of the Ozark foothills, and in the Lower Ordovician Knox Dolomite of the southern Appalachians (see Dunbar and Rodgers, 1957). In the last case the "lag" deposits of residual chert were developed by weathering before the Middle Ordovician, so that there is a *basal breccia* in the Blackford and Lenoir limestones (Bridge, 1955). Similar basal breccias occur at the base of the Pennsylvanian shales overlying the Boone Chert.

In modern deserts, thermal cracking combined with sheet wash and deflation of fine-grained debris often leaves a brecciated surface (a reg or a gibber plain) that is likely to be incorporated into the next phase of accumulation. Quaternary alluvial breccias (usually with calcite or silica cements) are thus quite common in semiarid regions.

RHODES W. FAIRBRIDGE

References

Bailey, E. B., and Weir, J., 1933. Submarine faulting in Kimmeridgian times: East Sutherland, *Roy. Soc. Edinburgh Trans.*, 57(2), 429-467.

Bonatti, E.; Emiliani, C.; Ferrara, G.; Honnorez, J.; and Rydell, H., 1974. Ultramafic-carbonate breccias from the equatorial Mid-Atlantic Ridge, *Marine Geol.*, 16, 83-102.

Bridge, J., 1955. Disconformity between Lower and Middle Ordovician series at Douglas Lake, Tennessee, *Geol. Soc. Am. Bull.*, 66, 725-730.

Dunbar, C. O., and Rodgers, J., 1957. *Principles of Stratigraphy*. New York: Wiley, 356p.

Gignoux, M., and Avnimelech, M., 1937. Genèse de roches sédimentaires bréchoïdes: intrusion et éclatement, *Soc. Géol. France Bull.*, 7, 27-33.

Guillien, Y., 1951. Les grèze litées de Charente, *Rev. Géogr. Pyrenées Sud-Ouest*, 22, 154-162.

Harrington, H. J. 1946. Las corrientes de barro ("mud-flows") de "El Volcan," quebrada de Humahuaca, Juyuy, *Soc. Geol. Arg. Rev.*, 1(2), 149-165.

Hendry, H. E., 1972. Breccias deposited by mass flow in the Breccia Nappe of the French pre-Alps, *Sedimentology*, 18, 277-292.

Kaisin, F., 1942. Age géologique et 'milieu générateur' de la Grande Brèche, *Soc. Belge. Géol. Bull.*, 51, 84-92.

Landes, K. K., 1945. Mackinac breccia, *Mich. Geol. Biol. Surv. Publ. 44*, 123-153.

Longwell, C. R., 1951. Megabreccia developed downslope from large faults, *Am. J. Sci.*, 249, 343-355.

Noble, L. F., 1941. Structural features of the Virgin Spring area, Death Valley, California, *Geol. Soc. Am. Bull.*, 52, 941-999.

Norton, W. H., 1917. A classification of breccias, *J. Geol.*, 25, 160-194.

Schuchert, C., and Dunbar, C. O., 1934. Stratigraphy of western Newfoundland, *Geol. Soc. Am. Mem. 1*, 123p.

Shrock, R. R., 1948. A classification of sedimentary rocks, *J. Geol.*, 56, 118-129.

Cross-references: *Agglomerate; Agglutinate; Alluvial-Fan Deposits; Avalanche Deposits; Clastic Sediments and Rocks; Clay-Pebble Conglomerate and Breccia; Conglomerates; Diamictite; Glacigene Sediments; Gravity-Slide Deposits; Intraformational Disturbances; Mudflow and Debris-Flow Deposits; Olistostrome, Olistolite; Pebbly Mudstones; Penecontemporaneous Deformation of Sediments; Rauhwacke; Storm Deposits; Syneresis; Talus; Till and Tillite; Volcaniclastic Sediments and Rocks.*

C

CALCARENITE, CALCILUTITE, CALCIRUDITE, CALCISILTITE

Calcarenite is a term used first ("calcarenyte") by Grabau (1904) to describe mechanically deposited carbonate rocks of sand-grain size (Fig. 1) that are composed of 50% or more of carbonate detritus. The carbonate particles consist commonly of foraminiferal and molluscan shell fragments, both intact and broken; debris of calcareous algae; echinoderms; crinoids; as well as pebbles, lumps, and granules of calcilutite, and oolites. In some cases, the detritus is biofragmental (resulting from biologic destruction), some is strictly clastic (e.g., broken by hydraulic action), and some is chemical in origin; but the distinction between these various materials does not alter the basic classification. There may even be a mixture within a single grain, as in the case of a particle of fossil debris

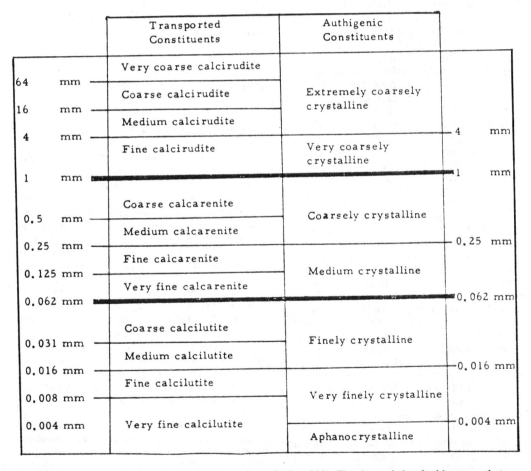

FIGURE 1. Grain-size scale for carbonate rocks (from Folk, 1959). The size scale is a double one so that one can distinguish between physically transported particles and authigenic constituents such as cement or products of recrystallization and replacement (e.g., coarse calcirudites may be cemented with very finely crystalline dolomite). The size scale for transported constituents uses the terms of Grabau. For dolomites of allochemical (see *Allochem*) origin, terms such as "dolarenite" would be used.

coated with one or more layers of precipitated carbonate (pseudo-oolith). Calcarenitic sediments are invariably transported by wind or water, sorted, and deposited; and as a consequence, display the normal range of sedimentary structures commonly observed in sandy facies. When a calcarenite is largely made up of grains of one type, the primary constituent name is used as a modifying term (e.g., "oolitic calcarenite") to make the name as descriptive as possible. Where a specific genesis is indicated by the gross structure or texture, such as in paleodune deposits, the appropriate name is *eolian calcarenite* (also called *eolianite*, q.v.). Most *beachrocks* (q.v.) are ancient offshore bars, spits, or related littoral deposits.

The term "calcilutyte" also was coined by Grabau (1904, 1913) as part of his wholly genetic classification of sedimentary rocks. It referred to rocks consolidated from lime mud of silt and/or clay-sized grains. This usage has persisted, with spelling (*calcilutite*) and textural modifications. With improvement of size analyses, silt and clay could be distinguished, and the use of the term *calcisiltite* (see below) relegated use of *calcilutite* to rocks of subsilt grain size—equivalent terms being microcrystalline or microgranular calcite, or limestone, and lithographic limestone. Unfortunately, Grabau's textural suffixes carry the connotation of milling of existing grains to ever-finer grades, and when microgranular calcite was found to be a primary biogenic phenomenon in many or all cases, the word "calcilutite" lost currency. The modern and prevailing term for limestones composed of particles of $< 5\mu$ diameter is *micrite* (q.v.; Folk, 1959). Coarser microgranular limestone (15 to 20μ grains) is still appropriately called calcilutite (Bathurst, 1971).

The term *calcirudite*, also introduced (as "calcirudyte") by Grabau (1913), designates conglomerate or breccia composed of mechanically deposited carbonate rocks in which the grain size is larger than sand. It must be composed of 50% or more of angular to rounded fragments >2 mm, having a matrix and/or a cementing material of variable composition.

The term *calcisiltite* (Kay, 1951) represents an extention of Grabau's (1913) terminology for rocks composed of clastic sediments of a specific grain-size range. It applies to a limestone composed of calcareous fragmental sediment of silt size (1/16 mm to 1/256 mm). Grabau's classification was limited to rudite, arenite, and lutite, but Kay felt that grain-size classification for rocks should follow that of the sediments themselves.

B. CHARLOTTE SCHREIBER
MALCOLM P. WEISS

References

Bathurst, R. G. C., 1971. *Carbonate Sediments and Their Diagenesis*. Amsterdam: Elsevier, 620p.

Folk, R. L., 1959. Practical petrographic classification of limestones, *Bull. Am. Assoc. Petrol. Geologists*, 43, 1-38.

Folk, R. L., 1962. Spectral subdivision of limestone types, in W. E. Ham, ed., Classification of carbonate rocks, *Am. Assoc. Petrol. Geologists Mem. 1*, 62-84.

Folk, R. L., 1968. *Petrology of Sedimentary Rocks*. Austin, Texas: Hemphill's, 170p.

Grabau, A. W., 1904. On the classification of sedimentary rocks, *Am. Geol.*, 33, 228-247.

Grabau, A. W., 1913. *Principles of Stratigraphy*. New York: Seiler, 1185p.

Kay, M., 1951. North American Geosynclines, *Geol. Soc. Am. Mem. 48*, 143p.

Cross-references: *Limestones; Micrite; Oolite; Pisolite; Sand and Sandstone.*

CALCAREOUS SEDIMENTARY ROCKS—See LIMESTONES

CALCITIZATION—See DEDOLOMITIZATION

CALICHE, CALCRETE

A type of calcareous *duricrust* (q.v.), a product of pedogenesis in semiarid regions, *caliche* is a term derived from the Spanish of the American Southwest. It first appeared in print in 1719 (Corominas, 1954). Basically, the word comes from the Latin *calix* (lime) and likewise the Spanish term *calizo* (lime-rich rocks). Unfortunately the term has also been applied to various other duricrusts: siliceous, aluminous, ferruginous, and saline deposits (Bretz and Horberg, 1949). In Chile and Peru the term *caliche* also refers to impure nitrate of soda or gravel with sodium nitrate; associated sulfates, iodates, phosphates; and banks of clay, sand, and gravel. Well-known synonyms for the calcareous crusts are *nari* (Middle East), *kunkur* (India), *croute calcaire* (France), and *kalkkruste* (Germany). Internationally, it is synonymous with *calcrete*, Lamplugh's term (see *Duricrust*), which is ambiguous.

The first scientific description of a caliche horizon was by Darwin (1846), who used the term *tosca* to describe the regional caliche profile of the Pampean Formation of South America. In 1902, Blake used the term caliche to describe the petrocalcic zones in southern Arizona. Considerable confusion has resulted from the fact that the term caliche may refer to different types of deposits in different countries. Likewise, petrocalcic zones may be designated by several different terms within one country. Goudie (1971) proposed Lamplugh's term *calcrete* as "the most suitable international term for materials formed by the cementation and/or alteration of pre-existing soil or rock by (dominantly) calcium carbonate." It seems unlikely, however, in view of its priority and entrenchment, that the term *caliche* will be superseded in the US, where it is ordinarily restricted to friable, indurated, light-colored, pisolitic (Fig. 1), cylindroidal, nodular, and concretionary layers (paleosols) of secondary calcium and/or magnesium carbonate, always closely associated with colloidal and fragmental quartz (see also *Pisolite*). Classifications based on physical properties have been proposed by Price (1933), Durand (1949; 1963), and Gile (1961); Reeves (1970), Netterberg (1971) and Goudie (1973) also classify on the basis of physical and chemical properties.

Caliche forms in single or multiple layers (Reeves and Suggs, 1964) in the Cca horizon of aggrading pedocal soils, usually in semiarid areas where evaporation exceeds precipitation (Fig. 2). The underlying bedrock may or may not consist of carbonate rocks. Although thin, incipient caliche forms in many humid areas (Ontario, Canada; Idaho; Channel Islands, California), thick, massive, regional caliches, which often exist over tens of thousands of km^2, characteristically occur bordering the downwind sides of the world's major desert areas (Sahara, Arabian, Great American, Chihuahuan, Australian, South African, Asian, and South American). Thicknesses range from several cm to >50 m in South Africa (Goudie, 1971), and decrease in the downwind direction

FIGURE 2. The Ogallala (Pliocene) caliche in Terry County, Texas, an example of old caliche. The upper zone is laminated and consists of pisolitic and brecciated caliche. The lower zone (the massive zone just above person's head) consists of caliche impregnated with opal. The lower zone is transitional with underlying sands.

from the carbonate source areas. Recently, analogies have been drawn between modern caliche and ancient concretionary carbonates (cornstone) in Britain (Steel, 1974).

Bedding varies from indistinct massive to thin lamination, often with involved contortions known as *pseudoanticlines* or *tepee structures*. Caliche which forms in lacustrine clays characteristically exhibits tepee structures, probably due to the bentonitic content (Reeves, 1976).

Caliche consists primarily of $CaCO_3$ and silica, accessories being principally alumina, iron oxides, and some salts. Goudie (1972) calculated the world-wide mean value of $CaCO_3$ in caliche to be 79.28% and silica 12.30%; the $CaCO_3$ may range from 61.02 to 97.00%. Reeves and Suggs (1964) and Brown (1956) found gypsum ranging from 2-30% in the caliche of W Texas. Chemistry is a function of age of the caliche (Reeves, 1970), silica gradually replacing the calcium carbonate. Thus, in young caliches, silica is predominantly in the form of sand grains, but in old caliches the silica is an amorphous form such as opal (Reeves, 1976).

Caliche of W Texas and S Africa is commercially used for road metal, and the W Texas caliche is also used in the manufacture of Portland cement and in sugar refining.

C. C. REEVES, Jr.

FIGURE 1. Pisolitic and brecciated caliche from top of Ogallala Formation (Pliocene), Lynn County, Texas.

References

Blake, W. P., 1902. The caliche of southern Arizona: an examination of deposition by vadose circulation, *Am. Inst. Min. Met. Eng. Trans.*, 31, 220-226.

Bretz, J. H., and Horberg, C. L., 1949. Caliche in southeastern New Mexico, *J. Geol.,* 57, 491-511.

Brown, C. N., 1956. The origin of caliche on the northeastern Llano Estacado, Texas, *J. Geol.,* 64, 1-15.

Corominas, E., 1954. *Diccionario Criticio Etimologico de la Lengua Castellana.* Madrid: Editorial Greuos. 933p.

Darwin, C., 1846. *Observations on South America.* London: Smith, Elder & Co., 278p.

Durand, J. H., 1949. Essai de nomenclature des croutes, *Soc. Sci. Natur. Tunisie, Tunis, Bull.,* 3-4, 141-142.

Durand, J. H., 1963. Les croutes calcaires et gypseuses en Algérie: formation et age, *Soc. Géol. France Bull.,* ser. 7, 5, 959-968.

Esteban, M., 1976. Vadose pisolite and caliche, *Bull. Am. Assoc. Petrol. Geologists,* 60, 2048-2057.

Gile, L. H., 1961. A classification of Ca horizons in soils of a desert region, Dona Ana County, New Mexico, *Soil Sci. Soc. Am. Proc.* 25, 52-61.

Goudie, A. S., 1971. A regional bibliography of calcrete, in W. G. McGinnies, B. J. Goldman, and P. Paylore, eds., *Food, Fiber and the Arid Lands.* Tucson: Univ. Arizona Press, 421-437.

Goudie, A. S., 1972. The chemistry of world calcrete deposits, *J. Geol.,* 80, 449-463.

Goudie, A. S., 1973. *Duricrusts in Tropical and Subtropical Landscapes.* Oxford: Clarendon, 174p.

Nagtegaal, P. J. C., 1969. Microtextures in recent and fossil caliche, *Leidse Geol. Meded.,* 42, 131-142.

Netterberg, F., 1971. Calcrete in road construction, *Nat. Inst. Road Research Bull.,* 10, 73p.

Price, W. A., 1933. Reynosa problem of South Texas, and origin of caliche, *Bull. Am. Assoc. Petrol. Geologists,* 17, 488-522.

Reeves, C. C., Jr., 1970. Origin, classification and geologic history of caliche on the southern High Plains, Texas and eastern New Mexico, *J. Geol.,* 78, 353-362.

Reeves, C. C., Jr., 1976. *Caliche, Origin, Classification, Morphology and Uses.* Lubbock, TX: Estacado Books, 233p.

Reeves, C. C., Jr., and Suggs, J. D., 1964. Caliche of central and southern Llano Estacado, Texas, *J. Sed. Petrology,* 34, 669-672.

Steel, R. J., 1974. Cornstone (fossil caliche)—Its origin, stratigraphic, and sedimentological importance in the New Red Sandstone, Western Scotland, *J. Geol.,* 82, 351-369.

Walls, R. A.; Harris, W. B.; and Nunan, W. E., 1975. Calcareous crust (caliche) profiles and early subaerial exposure of Carboniferous carbonates, northeastern Kentucky, *Sedimentology,* 22, 417-440.

Cross-references: *Duricrust; Oolite; Pisolite; Weathering in Sediments.* Vol. XII: *Crusts, Soil; Soil Pans.*

CARBONATE FABRICS—See LIMESTONE FABRICS

CARBONATE SEDIMENTS—BACTERIAL PRECIPITATION

The question of bacterial precipitation is one aspect of a very complex issue, namely, the genesis of carbonate precipitates in sea water. (This article considers marine conditions only, although bacteria may also play a part in the formation of various nonmarine carbonates, e.g., calcareous tufas, travertines, and speleothems.) Marine carbonate precipitates occur both in the deep sea and in shallow waters, and it is obvious that the mechanisms leading to their formation may be many and varied. For example Mg-calcite cement occurs under conditions as different as those in: (1) beachrock (q.v.) ledges along shorelines in the Mediterranean Sea and the Persian Gulf; (2) cemented reefs at depths of 1-100 m in most tropical waters; and (3) in lithified *Globigerina* oozes at depths of 1000-3000 m in the Mediterranean Sea, the Red Sea, and the Pacific. Evidently, the opportunities for chemical and biochemical variations in these environments of precipitation must be great. Similarly, there is no single, universal mechanism of precipitation responsible for precipitates as varied as: (1) aragonite beachrock cement; (2) aragonite muds; and (3) aragonite oolites.

Physicochemical Versus Biochemical Precipitation

The debate concerned with microbial aspects of precipitation deals mainly with the shallow marine environment, approximately corresponding to the photic zone, where organic production and decomposition are ubiquitous processes. Under these conditions, the issue is whether the carbonate supersaturation in warm, shallow sea water in itself is sufficient to cause precipitation, or whether a triggering biological activity is needed. The two alternatives can be summarized as follows.

The Physicochemical Model. Sea water is supersaturated with calcium carbonate, and this supersaturation, being a metastable state, leads to slow precipitation of carbonate minerals under the appropriate conditions. The processes are, however, too slow to be observed in the laboratory.

The Biochemical Model. The carbonate supersaturation found in normal sea water is not in itself sufficient for nucleation and growth of a carbonate precipitate. Crystal growth is triggered, directly or indirectly, by the activities of organisms, and carbonate precipitates thus should be regarded as biogenic. Although of paramount interest in this article, bacteria are not the only organisms supposed to be active as biochemical triggers; photosynthesizing algae and various burrowing and boring organisms may also take part in the processes (see *Algal-Reef Sedimentology; Carbonate Sediments—Lithification; Stromatolites*).

Aspects of Bacterial Precipitation

The idea that bacteria are responsible for a considerable part of marine calcium carbonate precipitation has been proposed and opposed time and again for perhaps a hundred years, and it is still a controversial idea. The controversy originates in the obvious discrepancy between experimental data and field observations. Numerous experimental studies have yielded bacteriogenic calcium salts, but natural occurrences of marine carbonate precipitates are not readily explained in the light of these studies. In this review, the experimental work is therefore treated in one section, followed by another section on field observations.

Experimental Precipitation

Experiments. A long suite of experimental studies, ranging back about a hundred years, has shown beyond doubt that bacterial decomposition of organic substances in sea-water media may lead to precipitation of calcium carbonate and various other salts. The basic experiment is quite simple: organic compounds, bacteria, and sea water are brought together in a suitable vessel and within days or weeks a precipitate can be observed. Murray and Irvine, with a natural approach to the problem, performed this experiment as early as in 1889 using 750 ml urine in 3 liters of North Sea sea water; slightly more sophisticated procedures have been repeatedly employed up to the present day.

Minerals. The minerals commonly observed in microbial cultures are: aragonite, Mg-calcite, monohydrocalcite ($CaCO_3 \cdot H_2O$), and struvite ($NH_4MgPO_4 \cdot 6H_2O$). Of these, only aragonite and Mg-calcite are found as authigenic minerals in marine sediments. The crystals formed by bacteria are, as a rule, in the size range $1-10\mu$, although no definite size limits are known to exist. Aragonite occurs in the form of needles, whereas Mg-calcite usually grows as poorly defined rhombohedrons. As a rule, the crystals grow attached to a solid surface, e.g., a pre-existing grain or the walls of the experimental vessel, but they may also grow downward from nuclei floating on the liquid surface. Bacterial precipitation may lead to interparticle cementation of sediment grains (Fig. 1).

Specific Calcium Bacteria. A pioneer in the experimental field was Harold Drew (1881-1913); his few papers, published 1911-1914, had great impact, even to the extent that the term *drewite*, now obsolete, was used for

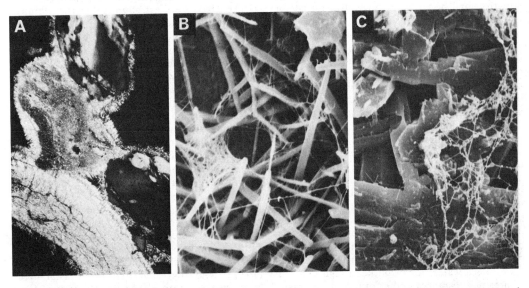

FIGURE 1. Aragonite in microbial culture. 30 g calcareous sand from Discovery Bay, Jamaica, was immersed in 1 litre natural sea water in a lightly stoppered flask, thus providing both organic substances and bacteria. After 3 months, aragonite was found (1) as a coat on the topmost sediment grains, and (2) as flaky aggregates floating at the water surface. (A) Sediment grains cemented together by bacteriogenic aragonite. The coat is about 15μ thick. Four grains are visible in the picture, the background is black. Thin section, polarized light. Width of field is 280μ. (B) Bacteriogenic aragonite needles on grain surface. Glutaraldehyde fixation, critical-point drying in liquid carbon dioxide. SEM, width of field 6μ. (C) Aragonite from floating flaky aggregate, same preparation as in B. SEM, width of field 10μ.

shallow-marine lime muds of inferred bacterial origin. According to Drew and his followers, calcium carbonate precipitation was specific to certain denitrifying calcium bacteria, e.g., *Bacterium calcis* (Drew, 1912), renamed *Pseudomonas c.* by Kellerman and Smith (1914), and *Pseudomonas calcipraecipitans* and *P. calciphila* (Molisch, 1925). The formation of extracellular calcium salts from sea water was assumed to be a normal and characteristic effect of the metabolism in these organisms. In this respect, the calcium bacteria were supposed to be different from the majority of bacteria in the sea.

The concept that carbonate precipitation is limited to a few species of specific calcium bacteria was virtually erased by Lipman (1924). He isolated 20 species of bacteria from South Pacific sea water and found (1) that all of these organisms could precipitate calcium carbonate in calcium-lactate growth medium, a medium previously used by Drew, and (2) that nevertheless none of these bacteria, not even an inferred *Pseudomonas calcis*, caused precipitation in natural filtered sea water. Based on his experiments, Lipman concluded that a majority of marine bacteria (and some soil bacteria) have the power of precipitating calcium carbonate, given growth media including organic calcium salts and/or soluble organic substances, but Lipman did not believe that the appropriate growth conditions were normally found in natural marine sediments.

Bacteria as Microenvironmental Regulators. Since the days of Drew, the scope and emphasis of the debate has changed. Originally, the bacterial precipitation hypothesis was thought to account for the vast accumulations of lime mud on the Bahama Banks (q.v.) and in S Florida, and, by analogy, for aragonite muds in general. The environment studied was the open shallow-marine environment as a whole, and bacteria were regarded as major carbonate-sedimentological contributors. Figuratively speaking, carbonate-precipitating bacteria have since then been pushed back to more obscure niches in the sedimentological ecosystem, such as mangrove swamps, restricted bays, algal mats, buried carcasses, and intragranular cavities in the sediments. Apparently, bacterial precipitation requires considerable changes in the chemical state of sea water, for example, a high content of organic substances in the form of dead organisms, increased carbon dioxide partial pressure, and/or anaerobic conditions, and thus can proceed in certain microenvironments only.

Berner (1968) studied experimentally the microbial breakdown of fish and clams under conditions simulating shallow burial in sediments. The experiments did not yield crystalline calcium carbonate, but precipitates of calcium soaps were formed after periods of 65-205 days. Calcium soaps may alter to calcium carbonate during continued diagenesis, and decaying dead animals thus may become fossilized within calcium carbonate concretions in the sediments. Experimental data are not yet available dealing with the possible bacterial influence on carbonate cementation in various other sedimentary microenvironments, e.g., beachrock, stromatolites, algal-coral reefs, burrows, and endolithic borings.

Literature. Biological aspects of bacterial carbonate precipitation are briefly treated in ZoBell (1946, 100-103). Short reviews are available also in Cloud (1962, 99-101), Malone and Towe (1970), Pautard (1970), and Krumbein (1972). Most of the recent work, however, is field oriented. Various mechanisms postulated for bacterial calcium carbonate precipitation in the sea are summarized in Cloud (1962).

Field Observations

Bacteria in the Sea. According to colony counts, sea water in the biologically active littoral zone commonly contains 10^2-10^3 bacteria per ml water; the extremes range from <1 to 10^8 bacteria per ml. Bacterial populations normally found in bottom sediments range from <10 to 10^8 bacteria per g wet sediment (ZoBell, 1963). They are most abundant in the topmost sedimentary layers and the number falls off rapidly within the first m beneath the sediment surface. The deposit of lime mud west of Andros Island, Great Bahama Bank, contains 10^6-10^{10} bacteria per g wet sediment, unusually large bacterial populations (Sisler, in Cloud, 1962). For intertidal sediment from a mangrove swamp at Andros Island, the count was 10^6 bacteria per g with as many as 10^4 of these being sulfate reducers. To quote Sisler, however, "proof is lacking that the bacteria are in any large degree responsible for the precipitation observed."

The Great Bahama Bank—A Case Study. The bacterial hypothesis formulated by Drew (1911-1914) was widely accepted, and according to his theory the white lime muds of the Bahama Banks (q.v.) were formed by bacterial precipitation taking place in the muds on the sea floor. The model was slightly modified by Bavendamm (1931), who suggested that the coastal mangrove swamps, rather than the deposits of lime mud, were the sites of bacterial activity and that the resulting aragonite needles were washed out from the swamps by tidal currents. Black (1933) came to similar conclusions, and in addition pointed out that concentration by evaporation on the shallow banks would cause some physicochemical precipitation. The first systematic investigation of the carbonate chemistry of the banks was made by Smith (1940). His data were directly opposed to the bacterial hypothesis, and he stressed that although some bacterial precipitation may be going on in the mangrove swamps, it "is believed to be very small compared with that produced by chemical precipitation over the hundreds of square miles of the Great Bahama Bank." The physicochemical model, arrived at by Smith, received further support from the extensive studies of Cloud (1962), and Broecker and Takahashi (1966). In addition, investigators have pointed out that a considerable part of the lime mud is not a precipitate but consists of comminuted and disintegrated skeletal carbonate, mostly from the green calcareous alga *Halimeda*. The estimated proportion ranges from about 20 wt % (Cloud, 1962) to virtually all the sedimentary aragonite needles (Lowenstam and Epstein, 1957). It is now evident that the bacterial hypothesis does not explain the genesis of the very sediment which was a starting point for the theory, namely, the lime muds of the Great Bahama Banks.

Shallow-Marine Cementation. Since the middle 1960s, great interest has been given to various processes of carbonate cementation in the shallow marine environment. The cementing minerals are aragonite and Mg-calcite with about 15 mole % magnesium carbonate in solid solution in the calcite (Fig. 2). These cements have been found under a variety of conditions, such as: littoral beachrock (q.v.) and stromatolites (q.v.), sublittoral sands, algal crusts, coral reefs, and intragranular cavities in sediment particles (reviews in Bathurst, 1971; Milliman, 1974; references in Alexandersson, 1974).

In particular, the intragranular cementation in borings and other hollows of sediment grains seems to take place in close connection with normal sea water. The grains are commonly < 1 mm in diameter, and the distances between the ambient sea water and the loci of crystal growth are measured in tens or hundreds of micrometers. The precipitation is not related to the microbial breakdown of the tissues of the original cavity dwellers, e.g., the mollusk tissue in protoconchs, the foraminiferal cytoplasm in foraminifer tests, or the boring algae and fungi in microscopic endolithic borings. The crystals grow from the clean walls of the cavities, and growth may proceed until the voids are completely occluded. The occurrence of intragranular cementation is closely related to the state of carbonate saturation in the sea water—ubiquitous and abundant cementation in highly supersaturated regions like the West Indies, but absence of cementation in undersaturated or barely saturated areas, e.g., Skagerrak Bay in the North Sea. Should a bacterial influence

FIGURE 2. Shallow-marine aragonite and Mg-calcite cements. (A) Aragonite beachrock cement, Barbados. A grain surface is visible in the lower left part of the picture. Needles radiate out from the grain surfaces, the crossing needles to the right grow on a grain outside the picture. Sample rinsed in distilled water and air dried. SEM, width of field 45μ. (B) Mg-calcite beachrock cement with 15 mole percent $MgCO_3$ in solid solution. Rhodes, Greece. Grain surface in the lower left part of the picture. Growth in rosette-like aggregates is characteristic of Mg-calcite cement; individual crystals are usually $4-6\mu$ in size. Same preparation as in A. SEM, width of field 22μ. (C) Fringe of Mg-calcite cement (arrow), followed by aragonite needle cement. Fractured foraminifer chamber, water depth 3 m, Cuba. Same preparation as in A. SEM, width of field 21μ.

exist in these processes, it is not as simple and direct a relationship as the decaying of organisms and the concomitant precipitation of aragonite or Mg-calcite at the sites of decay.

Literature. Sedimentological and geochemical aspects of marine carbonate genesis are reviewed in Bathurst (1971) and Milliman (1974). Several papers on carbonate cementation in particular are published in Füchtbauer (1969) and Bricker (1971). The works of Cloud (1962) and Broecker and Takahashi (1966) are comprehensive studies of the carbonate sedimentation on the Great Bahama Bank. Microbial aspects are covered by Sisler, in Cloud (1962).

Summary

The question of bacterial precipitation of calcium carbonate in the sea is a complex and controversial problem. In the laboratory as well as in the field, bacterial activity is but one of several possible ways to cause precipitation. Precipitation in laboratory cultures is readily achieved, but field observations give little support to the bacterial hypothesis. The lime muds of the Great Bahama Bank, a starting point for the theory, are now supposed to be partly detrital, partly the product of physicochemical precipitation. An evaluation of the geochemical/sedimentological significance of bacterial carbonate precipitation in the sea cannot be made on the basis of available data.

TORBJÖRN ALEXANDERSSON

References

Alexandersson, E. T., 1974. Carbonate cementation in coralline algal nodules in the Skagerrak, North Sea: Biochemical precipitation in undersaturated waters, *J. Sed. Petrology*, 44, 7–26.

Bathurst, R. G. C., 1971. *Carbonate Sediments and Their Diagenesis.* Amsterdam: Elsevier, 620p.

Bavendamm, W., 1931. Die Frage der bakteriologischen Kalkfällung in der tropischen See, *Berichte Deut. Botan. Ges.,* 49, 282–287.

Berner, R. A., 1968. Calcium carbonate concretions formed by the decomposition of organic matter, *Science,* 159, 195–197.

Black, M., 1933. The precipitation of calcium carbonate on the Great Bahama Bank, *Geol. Mag.,* 70, 455–466.

Bricker, O. P., ed., 1971. *Carbonate Cements.* Baltimore: Johns Hopkins Univ. Press, 376p.

Broecker, W. S., and Takahashi, T., 1966. Carbonate precipitation on the Bahama Banks, *J. Geophys. Research,* 71, 1575–1602.

Cloud, P. E., Jr., 1962. Environment of calcium carbonate deposition west of Andros Island, Bahamas, *U.S. Geol. Surv. Prof. Pap. 350,* 138p.

Drew, G. H., 1912. Report of investigations on marine bacteria carried out at Andros Island, Bahamas, British West Indies, in May 1912, *Carnegie Inst., Washington, Yearbook,* 11, 136–144.

Füchtbauer, H., ed., 1969. Lithification of carbonate sediments, I, *Sedimentology* (Special Issue), 12, 5–159.

Kellerman, K. F., and Smith, N. R., 1914. Bacterial precipitation of calcium carbonate, *Washington Acad. Sci. J.,* 4, 400–402.

Krumbein, W. E., 1972. Rôle des microorganismes dans la genèse, la diagenèse et la dégradation des roches en place, *Rev. Ecol. Biol. Sol,* 9, 283–319.

Lipman, C. B., 1924. A critical and experimental study of Drew's bacterial hypothesis on $CaCO_3$ precipitation in the sea, *Pap. Dept. Marine Biol. Carnegie Inst. Washington,* 19, 181–191.

Lowenstam, H. A., and Epstein, S., 1957. On the origin of sedimentary aragonite needles of the Great Bahama Bank, *J. Geol.,* 65, 364–375.

Malone, P. G., and Towe, K. M., 1970. Microbial carbonate and phosphate precipitates from sea water cultures, *Marine Geol.,* 9, 301–309.

Milliman, J. D., 1974. *Marine Carbonates.* Berlin: Springer-Verlag, 375p.

Molisch, H., 1925. Uber Kalkbakterien und andere kalkfällende Pilze, *Centralbl. Bakteriol., Abt. 2,* 65, 130–139.

Murray, J., and Irvine, R., 1889. On coral reefs and other carbonate of lime formations in modern seas, *Roy. Soc. Edinburgh Proc.,* 17, 79–109.

Pautard, F. G. E., 1970. Calcification in unicellular organisms, in H. Schraer, ed., *Biological Calcification.* Amsterdam: North-Holland, 105–201.

Smith, C. L., 1940. The Great Bahama Bank. II. Calcium carbonate precipitation, *J. Marine Research,* 3, 171–189.

ZoBell, C. E., 1946. *Marine Microbiology.* Waltham, Mass.: Chronica Botanica Company, 240p.

ZoBell, C. E., 1963. Domain of the marine microbiologist, in C. H. Oppenheimer, ed., *Symposium on Marine Microbiology.* Springfield, Ill.: C. C. Thomas, 3–24.

Cross-references: *Algal-Reef Sedimentology; Authigenesis; Bahama Banks Sedimentology; Beachrock; Carbonate Sediments–Lithification; Cementation; Cementation, Submarine; Microbiology in Sedimentology; Stromatolites.*

CARBONATE SEDIMENTS–DIAGENESIS

Diagenesis of carbonate sediments and equivalent rocks comprises more than thirty different processes that are controlled by local and regional physical, chemical, and biological factors, all of which alter the original composition and texture of the sediments, sometimes completely obliterating them. The investigation of *diagenesis* (q.v.) has become a complex, specialized field, the results of which are significant not only in theoretical petrology, but also in general geology applied to petroleum search (i.e., gas and oil) and to metalliferous (e.g., lead-zinc-barite) and nometalliferous (e.g., barite-fluorite) exploration.

The scope of diagenesis depends not only on boundaries set by the definition, but also on the multitude of factors and processes involved. Diagenesis of carbonate sediments is influenced by the following: (1) geographic factors—e.g., climate, type of terrestrial weathering; (2) geotectonism—e.g., platform vs. miogeosynclinal vs. eugeosynclinal environments; (3) geomorphologic situation—e.g., basinal vs. lagoonal vs. deltaic milieux; (4) geochemical factors in a regional sense—e.g., surface water chemistry; abyssal stagnation; (5) rate of erosion and sediment accumulation; (6) initial composition of the sediments—e.g., aragonite vs. high-Mg calcite vs. low-Mg calcite, amount of organic matter, amount and type of clastic admixtures like clay minerals, isotope and trace element content; (7) grain size and degree of sorting; (8) presence of cavity systems to permit influx of material and fluids; (9) physicochemical conditions above and within the sediments—e.g., pH, Eh, partial pressure of gases, CO_2 content; and (10) previous diagenetic history, if any, of the sediments. It should be noted, these variables are complexly interrelated; and also, that the numerous large-scale, regional environmental parameters in one way or another control the more locally operative factors, and these in turn influence the micromilieux.

Environments of Carbonate Diagenesis

Probably less information is available on the relationships between the sedimentary milieux and secondary changes of the sediments than on any other problem in the field of carbonate diagenesis. The difficulties are enormous because innumerable factors play a role simultaneously, whereas our investigations are restricted to a few of them. As studies of regional diagenetic variations become more plentiful, comparative data can be obtained. Differences between marine and nonmarine carbonates, for example, show that sedimentary environments produce distinctive variations of carbonate rocks and their trace elements. Though challenging, eventually it should be possible to list the various physical, chemical, and biological processes and parameters characteristic of each of these niches, which, in turn, lead to possible variations in diagenesis.

Borchert (1965) offered a geochemical explanation for the distribution of iron-ore facies, which may be considered with limestone diagenesis inasmuch as the same chemical variables are operative in carbonate mineral precipitation, and iron and carbonate facies are frequently associated with each other (Fig. 1). In the investigation of regional diagenesis in older sediments, one has to consider changes in sea level, so that one limestone unit may actually have undergone progressive diagenesis in stages, corresponding to regressive-transgressive oscillations. Purser (in Füchtbauer, 1969), illustrated changes with time (and sea level) of Holocene diagenetic lithification, using radiocarbon dating.

Blatt et al. (1972) have presented a simple

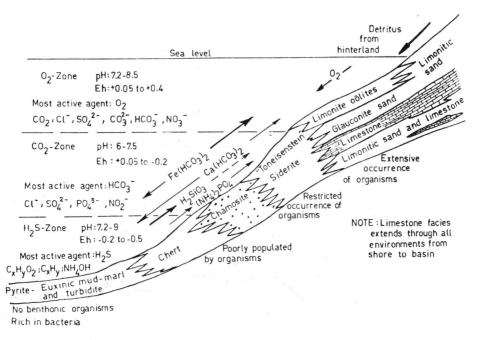

FIGURE 1. Diagrammatic presentation of physicochemical zones related to diagenesis of carbonate sediments (after Borchert, 1965).

diagenetic model, using a three-fold subdivision of environments of secondary changes: (a) subsea or just below the sea floor, (b) subaerial exposure in the vadose or shallow ground-water zone, and (c) the deep subsurface (see Fig. 2). Independently, Folk (1973) presented a conceptual model depicting the influence of environments on diagenesis (Fig. 3), supported by illustrating the relationship between the mineralogy and texture (= morphology) of carbonate cements. Details of carbonate diagenesis in the vadose and/or phreatic zones may be found in Dunham (in Friedman, 1969) and Purdy (1968; Fig. 4). The data confirm Folk (Fig. 3), pointing out the important diagenetic control exerted by the composition of the subsurface solutions as

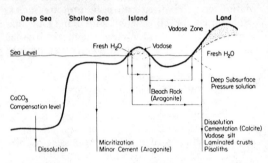

FIGURE 2. Environments of calcium carbonate diagenesis (Blatt et al., © 1972; reprinted by permission of Prentice-Hall, Inc.).

reflected by the type of calcium carbonate precipitated. Certain variations and similarities of diagenesis in various sedimentary environments are summarized by Füchtbauer and Müller (1975).

Chemical Composition Related to Diagenesis

The trace-element composition of sedimentary carbonates is a function of (1) element availability, i.e., composition of the solution; (2) sedimentary environments in general—fresh water, brackish, marine, or supersaline water; (3) physicochemical milieu—e.g., pH, Eh, temperature, and pressure; (4) biological activity; and others (see Bathurst, 1971; Wolf et al., 1967; Chilingar et al., 1967; Friedman, 1969). On the other hand, it should again be noted that the trace elements within a calcium carbonate-precipitating solution control (1) the morphology, and (2) the composition of the precipitate.

Morphology. The investigation by Usdowski (1963) demonstrated that the origin of amorphous calcium carbonate, vaterite, calcite, and aragonite minerals, and their form (as spherulite, radial-fibrous and concentric oolites, and prismatic and granular crystals), are, in part, dependent on the Mg:Ca ratio and the salinity. The experiment indicated a sequence of precipitation of finely granular or crystalline vaterite, calcite spherulites with no radial-fibrous structure, aragonite spherulites with radial-fibrous struc-

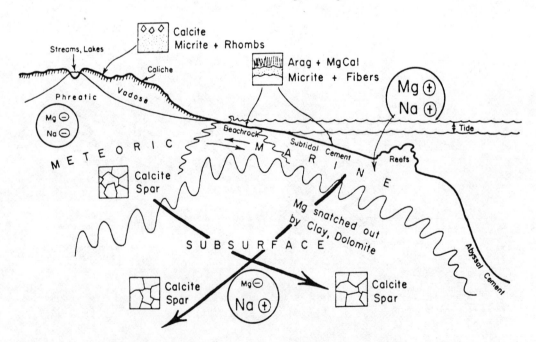

FIGURE 3. Schematic diagram showing the relationship between the mineralogy and morphology of carbonate cements and their diagenetic environments (Folk, 1973). For details on vadose zone, see Fig. 4.

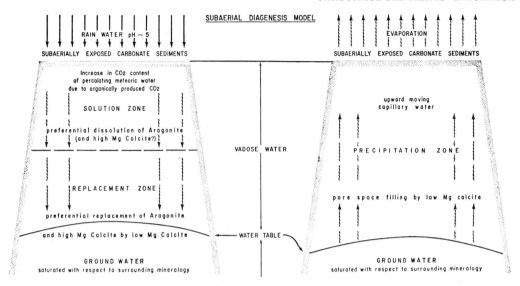

FIGURE 4. Subaerial diagenesis on an oceanic island as a consequence of downward-percolating meteoric water and upward-moving capillary water. Preferential dissolution of high-Mg calcite has not been documented, although dissolution effects are in evidence; hence the question mark with respect to high-Mg calcite within the solution zone. Below the water table, ground water seems to be saturated with respect to the predominant surrounding mineralogy, which in the case of recent carbonate sediments is aragonite; however it is not known with certainty whether or not subaerial diagenetic effects are restricted or largely limited to the area above the water table (Purdy, 1968).

ture, and finally amorphous calcium carbonate gel that accompanied a progressive increase of Mg:Ca ratio from various solutions ranging in salinity from 165‰ down to 36‰. His experiments suggest that the Mg:Ca ratio determines the $CaCO_3$ polymorph, and that the salinity controls the habit or structural form of the carbonate precipitate.

The influence of temperature on the Mg content of calcareous shells is shown in Fig. 5. Temperature also controls the oxygen isotope compositions of calcium carbonate (Bathurst, 1971). The temperature-magnesium trend of a specific class or phylum of organisms may be characteristic, but it should be emphasized that a reexamination of this two-variable relationship showed that other factors (e.g., physiological influences, such as rate of calcification) are influential in causing variations (Weber, 1973). The original Mg content of calcium carbonate controls the subsequent diagenetic history of limestones and dolomites, e.g., rate of inversion to calcite and release of Mg for dolomitization (q.v.).

Porosity in Carbonate Rocks

The origin, preservation, and blocking of porosity in carbonate sediments is a direct result of diagenesis and has long been given close attention by sedimentary petrologists. Both porosity and permeability are of practical importance in controlling the movement of subsurface fluids, e.g., petroleum, and in the deposition of ore deposits. The control of the movement of solutions in turn controls subsequent diagenetic processes. This complex interrelationship usually proceeds until the rock becomes totally impermeable, although subsequent reinstitution of

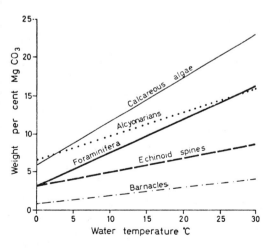

FIGURE 5. Relation between temperature and wt.% $MgCO_3$ in $CaCO_3$ skeletons of several taxa (Bathurst, 1971).

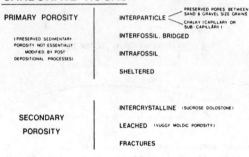

FIGURE 6. Porosity types in carbonate sediments (formed during deposition) and carbonate rocks. More than one porosity can develop in the same rock; "primary" and "secondary" porosity is an arbitrary classification used for practical purposes. (Courtesy of Gulf Oil of Canada Ltd.)

permeability and porosity may be facilited by tectonic deformation and epigenesis. Types of porosities are outlined in Fig. 6.

The genetic aspects of porosity development in grain-supported sediments and mud-supported carbonates and the equivalent rocks are summarized in Figs. 7 and 8 (see also *Limestones*).

Morphologic and Genetic Diagenetic Calcium Carbonate

All calcium carbonate of both physicochemical and biochemical origin is crystalline and can occur in any of the morphologies listed in Table 1. Both the size and morphology are taken into account (left side of Table 1), but it has been found that a number of processes can give rise to similar or identical products so that a number of prefixes have been suggested that can be used when the origin of the calcium carbonate is known (right side of Table 1). In this way, the crystal-size and morphological terminology can be employed in describing any material under investigation, and the genetic prefixes can be added as soon as enough data have been accumulated to permit genetic interpretation.

Cementation and Lithification. *Lithification* is the process that changes unconsolidated sedi-

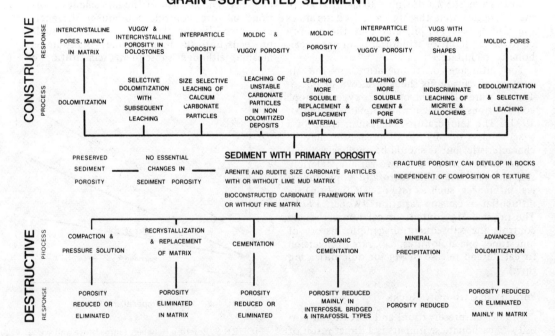

FIGURE 7. Porosity development through diagenesis of grain-supported sediment. Constructive—increasing porosity. Destructive—decreasing porosity. (Courtesy of Gulf Oil of Canada, Ltd.)

TABLE 1. Descriptive and Genetic Nomenclature for Micrite-Sparite Range of Aragonite and Calcite

Descriptive				Genetic descriptive	Origin
indicating crystal size	approx. size	indicating crystal morphology and size	approx. proportions		
Sparite	> 0.02 mm	Granular sparite	Equidimensional	Orthosparite	Open-space precipitation, i.e., void fillings
		Drusy sparite (size and morphology change distally)	Elongate	Pseudosparite	Recrystallization or grain growth
		Fibrous sparite	Elongate		
Microsparite	0.005–0.02 mm	Granular microsparite	Equidimensional	Orthomicrosparite	Open-space precipitation
		Drusy microsparite	Elongate	Pseudomicrosparite	Recrystallization or grain growth
		Fibrous microsparite	Elongate		
Micrite (often called calcilutite, ooze, lime-mud) Cryptocrystalline	< 0.005 mm	Too small to observe visually morphologic differences except by use of electron-microscope		Pseudomicrite	Degradation recrystallization (= grain or crystal diminution)
				Orthomicrite	"Genuine primary" micrite
				Allomicrite	Allochthonous micrite
				Automicrite	Autochthonous micrite

(Genetic descriptive column grouping: Open-space versus recrystallization and grain growth / Detrital versus in situ)

From Wolf, 1965b.

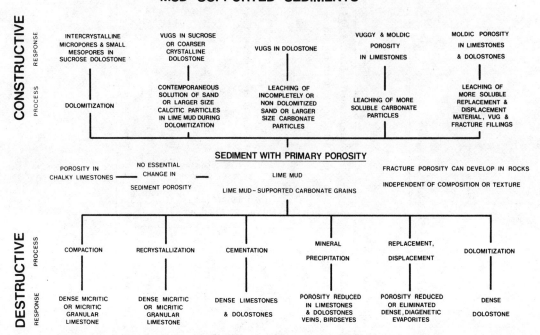

FIGURE 8. Porosity development through diagenesis of mud-supported sediments. Constructive—increasing porosity. Destructive—decreasing porosity. (Courtesy of Gulf Oil of Canada, Ltd.)

ments into weakly to strongly consolidated rocks. Lithification of carbonate sediments (q.v.) may occur through recrystallization, replacement (e.g., dolomitization), crystallographic welding of limy particles (especially clay-sized ones), and by desiccation. One of the major processes is *cementation* (q.v.), the infilling of open spaces by the chemical precipitation of calcium carbonate or other minerals. For details on the geochemical aspect of precipitation, source problem of the chemicals involved, textural criteria to recognize openspace cements from recrystallization products, cement fabrics as environmental indicators, etc., the reader should consult, for example, Folk (1965), Bathurst (1971), Füchtbauer (1969), and Chilingar et al. (1967). A summary of six major cement types is given in Fig. 9, representing observation made on deposits that originated in shallow-marine, tidal, deep-sea, and subaerial environments (see also *Limestone Fabrics*).

Inversion and Recrystallization. The especially widespread occurrence of these two processes in carbonate-rich sediments is the result of the great susceptibility of aragonite, high- and low-Mg calcite, vaterite, and dolomite minerals to secondary changes. The results of numerous studies are summarized by Folk (1965, 1973), Bathurst (1971), and Chilingar et al. (1967), among others.

Neomorphism, a collective term for inversion and recrystallization, is a destructive process unless pseudomorphism is involved during inversion. In particular, *recrystallization*, with either an increase or a decrease in crystal size, leads to the obliteration and often complete elimination of the original textures and fabrics. Neomorphism can be selective (= preferential) in that it may affect only aragonite and/or high-Mg calcite in a sediment composed of a mixture of calcium carbonate minerals, or it may affect only the fine-grained material in a calcarenite with a micrite matrix. In the latter instance, the recrystallized matrix would be difficult to distinguish from a chemically precipitated sparry calcite cement unless great care is taken in petrologic work, and certain discriminatory textural criteria are employed. The investigation of neomorphism also is important in geochemical interpretations. The secondary alterations in trace elements, and in carbon and oxygen isotope compositions depend directly on the rate and degree of neomorphism. X-ray diffraction techniques may be used to

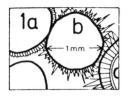
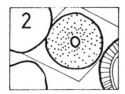

SHALLOW MARINE

1a fibrous (Mg-)calcite cement
1b " aragonite "
1c " homoaxial "

homoaxial overgrowth
of ?Mg-calcite
on crinoid fragments

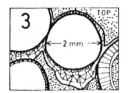
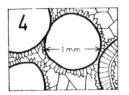

TIDAL

microstalactitic cement
(Mg-calcite, aragonite)
with matrix

SHALLOW

LATE DIAGENETICAL CEMENT
fibrous cem., by calcitization
of 1b.-Drusy (Fe-)calcite cement

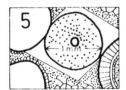
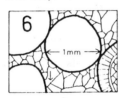

MARINE

1a, 2 with cryptocrystalline
Mg-calcite cement and
LATE DIAGENETICAL
drusy (Fe-)calcite cement

SUBAERIAL | DEEP SEA

calcite | Mg-calcite
        | drusy cement
(arrow: enfacial junction)

FIGURE 9. Cement types; early diagenetic if not indicated (Füchbauer, 1969). Numbers 4 and 5 are shallow marine.

determine the degree of recrystallization, provided that one knows the mineralogy and crystal morphology of the original material. Some of the frequently observed changes during neomorphism are shown in Table 2, and Table 3 summarizes the *degrading* and *aggrading* processes. The former mechanism results in a general decrease in crystal size, whereas the latter results in an increase.

The macroenvironmental factors controlling the mineralogy and morphology of carbonate precipitates have already been discussed briefly. Folk (1973) illustrates the relationships between carbonate morphology, Mg-ion, and rate of crystallization. Certain combinations of these factors are known to occur in specific sedimentary environments. Care should be exercised, however,

CARBONATE SEDIMENTS—DIAGENESIS

in realizing that any combination could be independent of a particular sedimentary milieu, and more research is required to be more specific. Purdy (1968) provided information on both subsea and subaerial recrystallization.

Destructive Diagenesis. A number of destructive processes operative on particles and sediments, as well as on their lithified equivalents, can alter and obliterate primary textures and structures and may remove material. Inversion and recrystallization are destructive in many instances, but there are additional destructive mechanisms such as compaction, disintegration, dissolution and corrosion, and abrasion (Neumann, 1965; Bathurst, 1971; Purdy, 1968), decementation, and the boring-drilling activity of organisms (Bathurst, 1971).

Neumann (1965) discussed the effects of selective solution, which apparently produces important changes in the mineralogic character of fine-grained carbonate sediments. With a progressive selective solution of the more soluble material, there is an increase in mineral stability with decreasing grain size of the carbonate. The variations in the degree of selective dissolution superimposed upon the supply of organic and/or inorganic carbonate particles is reflected in the texture and mineralogy of the fine fraction of a deposit, because it is the most sensitive to effects of solution; various environments represent the range of the degree of early diagenetic modification by selective solution. The Bahama Banks (q.v.) show the effects of precipitation of carbonate particles rather than selective solution. Florida Bay shows effects of neither solution nor precipitation. In Harrington Sound (Neumann, 1965), solutional equilibrium is maintained in the high-Mg calcite phases with a degree of dissolution sufficient to overcome the effects of supply.

Bermuda and the Yucatan Shelf are areas where the supply of high-Mg calcite is great enough to insure its presence in fine deposits, but dissolution effects are distinct enough to overtake supply and lead to a consistent decrease in both aragonite and high-Mg calcite content with grain size; dissolution modifies both Mg calcite and aragonite content in fine fractions of carbonate sediments. In the Red Sea, under low rates of sediment accumulation and constant supply of relatively cool undersaturated water, *all* the fine-grained calcite remains. The most extreme case of dissolution takes place in deep oceanic environments, where the temperature, water circulation, and CO_2 pressure are such that all $CaCO_3$ is removed by solution (see *Pelagic Sediments*).

Compaction. Limestones normally show little evidence of compaction (Bathurst, 1971). Apparently, compaction in carbonate sediments

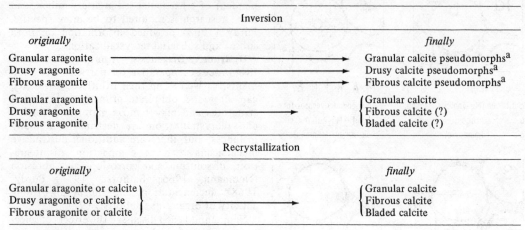

TABLE 2. Possible Grain or Crystal Morphology Changes During Inversion and Recrystallization

From Wolf, 1965b.
[a] These types of pseudomorphs are usually referred to as paramorphs because no change in composition occurs during inversion (except for possible changes in trace element and isotope contents).

TABLE 3. Degrading and Aggrading Processes of Neomorphism

From Folk, 1965.

can take place only prior to cementation (q.v.), and there is abundant evidence that carbonates are lithified rapidly, even while still under water (see *Cementation, Submarine*). Fischer and Garrison (1967) point out that this lithification may take place in slightly compact or even uncompacted sediments. Fruth et al. (1966) showed experimentally that the degree of compaction of unlithified carbonate sediments varies considerably at pressures up to 30 bars; the compaction of well-sorted oolite grains was negligible; grapestone, with its mixture of friable and hard grains and aggregates, was moderately compacted; and fine-grained lime mud and porous skeletal debris were compacted more than 30%. After initial compaction, under higher pressures, compaction rates were similar; these data do not, however, necessarily illustrate the effect of millions of years on compaction (Stevenson, 1973). Even at higher pressures (120 bars), carbonate sediments maintain approximately 35% porosity (Fruth et al., 1966). Fruth et al. discuss the effects of compaction on carbonate fabrics (see *Diagenetic Fabrics; Pressure Solution Phenomena*).

Other Aspects of Carbonate Diagenesis

Internal Sedimentation. Internal sedimentation is the process by which particles accumulate *within* a sediment or rock framework by the action of subsurface fluids. The particles may have come from outside (i.e., above), or from within the framework as a result of inter-erosion, and the sediment has the tendency to settle to the bottom of the open spaces, forming *geopetal structures*. In some instances these grains are not detrital in origin but have originated as chemically precipitated particles prior to transportation and settling within the sediment. In combination with chemically formed cement encrusting the walls of the openings as well as the earlier deposited sediment itself, complex "open-space" structures can form, sometimes exhibiting cyclic precipitation and sediment accumulation. Internal sedimentation has been studied for decades (see *Biostratinomy*), but renewed interest has led to detailed information useful in general petrology of diagenesis as well as in paleoenvironmental reconstructions (e.g., Wolf, in Larsen and Chilingar, 1967; Dunham, in Friedman, 1969). Figure 10 presents a number of these open-space structures with both mechanical and chemical internal sediments and chemically precipitated cement linings. For a possible environmental application, see Wolf (1965a, b).

Noncalcareous Replacements. Limy material may easily be replaced by noncarbonate minerals and mineraloids, in particular, silica in the form of chert (q.v.), and also iron, phosphate, and manganese, among others, without requiring high temperature or pressure. Because of this susceptibility, carbonate rocks are known to make excellent host rocks for ore bodies formed from metal-bearing solutions.

In particular, during low-temperature diagenetic replacements, the process is often atom-by-atom, or molecule-by-molecule, so that the external original textures are preserved, although

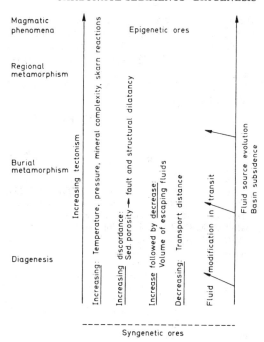

FIGURE 11. Primary ore precipitation in limestones (from Beales and Onasick, 1970).

important exceptions exist. The most widespread examples are silicified ("petrified") fossils, which can be easily removed from the limestone country rock by dissolving the unreplaced carbonate minerals in dilute hydrochloric acid, leaving the silicified fossils free for laboratory investigations.

Economic Importance. Carbonate rocks are host or reservoir rocks for fluid hydrocarbons (gas and oil) as well as for certain varieties of ore deposits. Depending on the statistical data and methods used, it is estimated that 50-70% of the world's hydrocarbons are located within limestones and dolomites. Two conspicuous examples of oil-producing carbonate complexes cover thousands of km^2 in western Canada and in the Middle East.

The Mississippi Valley type of Pb-Zn-Ba-F deposits are probably the best-known ores within carbonate host rocks, but only relatively recently have sedimentologists become involved in studying the possible relationships between paleoenvironments and diagenesis on one hand, and ore genesis on the other (see, e.g., Jackson and Beales, 1967; and Wolf, 1976, for summary). Figure 11 illustrates that the whole geologic history of a sedimentary basin, with its various processes and stages of evolution, must be considered in unraveling the genesis of an ore body within sediments. Metal-bearing fluids derived

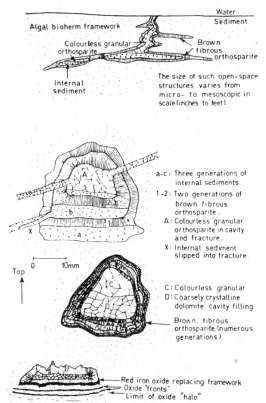

FIGURE 10. Open-space structures in Devonian algal bioherms, Nubrigyn Formation, N.S.W. (Chilingar et al., 1967, after Wolf).

from the deep-water basinal clays, which released metallic ions into compaction solutions, migrated up-dip into shallow-water porous and permeable carbonate facies where precipitation of the ore minerals (e.g., galena and sphalerite) plus gangue minerals (e.g., barite, fluorite, and calcite) took place.

KARL H. WOLF

References

Balcon, J., ed., 1973. Sédimentation et diagenèse des carbonates, *Bull. Centre Rech. Pau-SNPA*, 7, 97-289.

Bathurst, R. G. C., 1971. *Carbonate Sediments and their Diagenesis*. Amsterdam: Elsevier, 620p.

Bathurst, R. G. C., 1974. Marine diagenesis of shallow water calcium carbonate sediments, *Ann. Rev. Earth Planetary Sci.*, 2, 257-274.

Beales, F. W., and Onasick, E. P., 1970. Stratigraphic habit of Mississippi Valley type ore bodies, *Trans. Inst. Mining Metal.*, 79, B145-B154.

Blatt, H.; Middleton, G. V.; and Murray, R. C., 1972. *Origin of Sedimentary Rocks*. Englewood Cliffs, N.J.: Prentice Hall, 634p.

Borchert, M., 1965. Formation of marine sedimentary iron ores, in J. P. Riley, and A. Skirrow, eds. *Chemical Oceanography*, vol. 2. New York: Academic Press, 159-204.

Chilingar, G. V.; Bissel, H. J.; and Wolf, K. H., 1967. Diagenesis of carbonate rocks, in Larsen and Chilingar, 1967, 176-322.

Fischer, A. G., and Garrison, R. E., 1967. Carbonate lithification on the sea floor, *J. Geol.*, 75, 488-496.

Folk, R. L., 1965. Some aspects of recrystallization in ancient limestones, *Soc. Econ. Paleont. Mineral. Spec. Pub. 13*, 14-48.

Folk, R. L., 1973. Carbonate petrography in the post-Sorbian age, in R. N. Ginsburg, ed., *Evolving Concepts in Sedimentology*. Baltimore: Johns Hopkins, 118-158.

Friedman, G. M., ed., 1969. Depositional environments in carbonate rocks—a symposium, *Soc. Econ. Paleont. Mineral. Spec. Publ. 14*, 209p.

Friedman, G. M., 1975. The making and unmaking of limestones or the downs and ups of porosity, *J. Sed. Petrology*, 45, 379-398.

Fruth, L. S., Jr.; Orme, G. R.; and Donath, F. A., 1966. Experimental compaction effects in carbonate sediments, *J. Sed. Petrology*, 36, 747-754.

Füchtbauer, H., ed., 1969. Lithification of carbonate sediments, Pts. 1 and 2, *Sedimentology*, 12, 7-159, 163-322.

Füchtbauer, H., and Müller, G., 1975. *Sediments and Sedimentary Rocks*, Vol. 2, Pt. 1 of Sedimentary Petrology series, Englehardt, W. V., ed. Stuttgart: E. Schweizesbart'sche Verlagsbuchhandlung; New York: Halsted, 464p. (German edition, 1970, 726p.)

Gulf Oil Canada, Ltd., 1971. *Diagenesis and Porosity in Carbonate Rocks*. Calgary, Alberta (pamphlet for exhibit).

Jackson, S. A., and Beales, F. W., 1967. An aspect of sedimentary basin evolution: the concentration of Mississippi Valley-type ores during late stages of diagenesis, *Bull. Canadian Petroleum Geol.*, 15, 383-433.

Larsen, G., and Chilingar, G. V., eds., 1967. *Diagenesis in Sediments*. Amsterdam: Elsevier, 551p.

Lippmann, F., 1973. *Sedimentary Carbonate Minerals*. Berlin: Springer-Verlag, 228p.

Milliman, J. D., 1974. *Marine Carbonates*. New York: Springer-Verlag, 375p.

Neumann, A. C., 1965. Processes of recent carbonate sedimentation in Harrington Sound, Bermuda, *Bull. Marine Sci.*, 15, 987-1035.

Pittman, E. D., 1974. Porosity and permeability changes during diagenesis of Pleistocene corals, Barbados, West Indies, *Geol. Soc. Am. Bull.*, 85, 1811-1820.

Pray, L. C., and Murray, R. C., eds., 1965. Dolomitization and limestone diagenesis—a symposium, *Soc. Econ. Paleont. Mineral. Spec. Publ. 13*, 180p.

Purdy, E. G., 1968. Carbonate diagenesis: an environmental survey, *Geol. Romana*, 7, 183-228.

Purser, B. H., ed., 1973. *The Persian Gulf: Holocene Carbonate Sedimentation and Diagenesis in a Shallow Epicontinental Sea*. New York: Springer, 471p.

Stevenson, D. L., 1973. The effect of buried Niagaran reefs on overlying strata in southwestern Illinois, *Ill. Geol. Surv. Environ. Geol. Notes*, 482, 22p.

Usdowski, M. E., 1963. Der Rogenstein des norddentschen Unteren Buntsandsteins, ein Kalkoolith des marinen Faziesbereichs, *Fortschr. Geol. Rheinland Westfalen*, 10, 337-342.

Weber, J. N., 1973. Temperature dependence of Mg in echinoid and asteroid skeletal calcite: a reinterpretation of its significance, *J. Geol.*, 81, 543-556.

Wolf, K. H., 1965a. Littoral environment indicated by open-space structures in algal reefs, *Palaeogeogr., Palaeoclim., Palaeoecol.*, 1, 183-223.

Wolf, K. H., 1965b. Petrogenesis and paleoenvironment of Devonian algal limestones of New South Wales, *Sedimentology*, 4, 113-178.

Wolf, K. H., 1976. Ore genesis influenced by compaction, in G. V. Chilingar, and K. H. Wolf, eds. *Compaction of Coarse-Grained Sediments*, Vol. II. Amsterdam: Elsevier, 1976, 475-675.

Wolf, K. H.; Chilingar, G. V.; and Beales, F. W., 1967. Elemental composition of carbonate skeletons, minerals, and sediments, in G. V. Chilingar, H. J. Bissell, and R. W. Fairbridge, eds., *Carbonate Rocks, B*. Amsterdam: Elsevier, 1967, 23-150.

Cross-references: *Authigenesis; Bahama Bank Sedimentology; Beachrock; Birdseye Limestone; Caliche; Carbonate Sediments—Bacterial Precipitation; Carbonate Sediments—Lithification; Cementation; Chert and Flint; Coral-Reef Sedimentology; Corrosion Surface; Dedolomitization; Diagenesis; Diagenetic Fabrics; Dolomite, Dolomitization; Evaporite Facies; Interstitial Water in Sediments; Limestone Fabrics; Limestones; Micrite; Pelagic Sedimentation; Pressure Solution and Related Phenomena; Rauhwacke; Sabkha Sedimentology; Weathering in Sediments.* Vol. IVA: *Calcium Carbonate: Geochemistry.* Vol. IVB: *Aragonite; Calcite; Dolomite.*

CARBONATE SEDIMENTS— LITHIFICATION

The question of how carbonate sediments become lithified is one of the most important in the study of geology. When reconstructing the depositional environment of ancient rocks, we may find that the process of lithification has modified the depositional texture of the original sediment, even partially obliterated it; or it may have superimposed a new pattern which may make interpretation of the depositional pattern difficult or even impossible. From the economic point of view, lithification is a stage at which porosity may be partially or entirely obliterated and thus may determine whether the resultant limestone (q.v.) becomes a petroleum reservoir or not.

The sometimes drastic changes which carbonate sediments undergo during lithification are manifest in textural characteristics and mineralogical composition. For example, modern-day shallow-water carbonate sediments are composed of aragonite and high-Mg calcite, with subordinate amounts of low-Mg calcite; high-Mg calcite has 4% Mg carbonate in solid solution and may contain up to 30% Mg carbonate (Chave, 1954). In ancient limestones, however, high-Mg calcite and aragonite are rare. Stehli and Hower (1961) and Friedman (1964) have shown that lithified carbonates as young as Pleistocene which were deposited as high-Mg calcite and aragonite are composed now of low-Mg calcite. The pore spaces between the constituent particles have been filled by a *drusy calcite mosaic,* i.e., the size of the crystals that make up the pore fillings increases away from the walls of the particles (Bathurst, 1958). This mosaic is diagnostic of the infilling of space, whether between particles or as an infilling of molds.

In a comparative study of Holocene and Pleistocene carbonates (Friedman, 1964), it was shown that at an early stage in the lithification process, the aragonite skeletons of mollusks and of *Halimeda* are dissolved out to form *moldic porosity.* The outline of the skeleton is usually preserved by calcite cement or, more commonly, by a thin *micritic envelope,* a dark, finely crystalline rim which bounds the skeleton (Bathurst, 1964). Intermediate stages can be observed in which aragonite has been partially removed from the skeleton, creating moldic porosity. At a more advanced stage, drusy calcite partially fills the molds. Complete infilling of molds by a drusy calcite mosaic is very common. Bathurst (1964) only deduced the existence of empty micritic envelopes because in ancient limestones he saw micritic envelopes enclosing shell-shaped spaces now filled with drusy mosaic. In Pleistocene limestones it is possible to trace the entire sequence of mold formation and elimination.

In contrast to the solution-deposition of aragonitic grains, high-Mg calcite grains are not affected by dissolution. Magnesium is preferentially removed from high-Mg calcite to yield low-Mg calcite without a textural change of the grains affected. Thus, the original texture of the high-Mg calcite grains remains completely intact, and it is impossible to distinguish the fragment of an organism composed of high-Mg calcite and the same organism after magnesium has been removed to produce low-Mg calcite.

Conditions Necessary for Lithification

Source of Cement. The source of calcium carbonate to provide for the infilling of molds as well as the intergranular cement is the dissolved aragonite. Lithification of carbonates, therefore, is the result of the dissolution of aragonite grains and the reprecipitation in pore spaces of calcium carbonate in the form of calcite. High-Mg calcite grains are not a source of potential cement. The magnesium that is removed from these grains is carried away in solution.

Environment. Fresh-water conditions are conducive for dissolution of aragonite and the preferential removal of magnesium from high-Mg calcite. If aragonite and high-Mg calcite grains continue to be exposed to marine water, they usually retain their original mineralogical and textural characteristics and resist lithification (see, however, *Cementation, Submarine*). Thus, Pleistocene carbonate sediments that have remained in a marine environment cannot usually be distinguished, mineralogically or texturally, from Holocene sediments, whereas carbonates that have been exposed subaerially are well lithified and have gone through the complete cycle of textural and mineralogical changes.

pH and Salinity. Experimental evidence (Friedman, 1964) indicates that both an acid pH and a low salinity promote the mineralogical and, by inference, the textural changes that lead to lithification. In nature, the meteoric water with which carbonate sediments come in contact when exposed subaerially is usually acid and of low salinity; thus, subaerial exposure promotes lithification of carbonates. If waters of acid pH and low salinity are available in the subsurface, lithification would also occur there. The range of pH of formation waters in the subsurface is from 5 to 7 (after allowance is made for CO_2 partial pressure), and the salinity can be as low as fresh water; it appears reasonable, therefore,

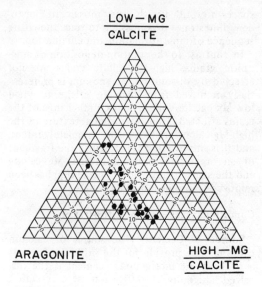

FIGURE 1. Mineralogy of Holocene skeletal carbonate sands from Bermuda.

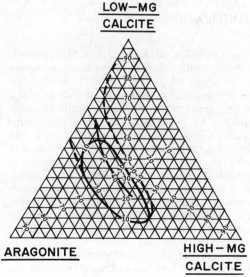

—— SKELETAL SAND (RECENT)
– – – SKELETAL SAND (PLEISTOCENE)

FIGURE 3. Mineralogy of Recent (Holocene) and Pleistocene skeletal sands from Bermuda. The arrow indicates the direction of diagenetic alteration of those carbonate sands exposed to fresh-water leaching.

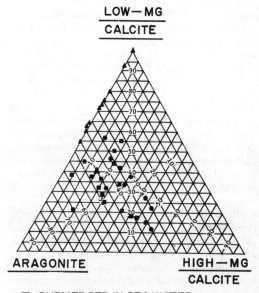

■ SUBMERGED IN SEA WATER
● FROM MARINE SPRAY ZONE
▲ BEYOND REACH OF MARINE SPRAY

FIGURE 2. Mineralogy of Pleistocene skeletal carbonate sands from Bermuda. Note that Pleistocene carbonate sands exposed to marine water contain high-Mg calcite, whereas those unexposed to sea water are, with one exception, devoid of Mg (see text).

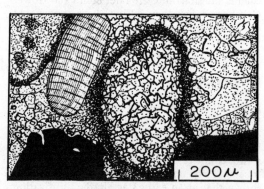

FIGURE 4. Thin-section sketch of Pleistocene limestone from St. George's Island, Bermuda (see text).

to assume that carbonate sediments can be lithified in the subsurface. More extended discussions on lithification of carbonate sediments are provided by Bathurst (1971), Friedman et al. (1974), and Friedman (1975).

Model: Bermuda Holocene and Pleistocene Carbonates. The mineralogy of Holocene unconsolidated skeletal sands is shown in Fig. 1. These skeletal sands are composed of aragonite and high-Mg calcite but have been partially contaminated by Pleistocene detrital debris, which contributes to the low-Mg calcite fraction. Fig. 2 shows the mineralogy of Pleistocene skeletal sands. These Pleistocene samples were obtained from both marine and subaerial postdepositional environments. Pleistocene samples from the marine environment include carbonates below sea level and in the intertidal marine spray zones. Subaerial samples were collected from Pleistocene carbonates beyond the reach of marine spray. With one exception, all of these subaerial carbonates are devoid of high-Mg calcite, whereas samples taken from carbonates in contact with marine water contain variable amounts of high-Mg calcite. Both types of skeletal sands, exposed or not exposed to marine waters, are made up of the same faunal assemblage, both qualitatively and quantitatively; therefore, their initial mineralogical composition was essentially identical.

The mineralogical determinations of the Holocene (Recent) and Pleistocene skeletal sands are compared in Fig. 3. The Pleistocene carbonates in the field overlapped by Holocene carbonates have been exposed to marine water, and the mineralogy and magnesium content have remained essentially unchanged. The arrow in Fig. 3 indicates the direction of diagenetic alteration of those carbonates exposed to freshwater leaching. High-Mg calcite is converted to low-Mg calcite, and aragonite undergoes a slower change to low-Mg calcite via a solution-deposition stage. Fig. 4, a thin-section sketch of a Pleistocene skeletal sand from Bermuda, illustrates these changes. In the center of this figure is the outline of a fossil fragment that was once composed of aragonite. It is now rimmed by a micritic envelope and filled by a drusy calcite mosaic. The skeletal fragment with the rectangular pattern to its left is a coralline algal fragment made of low-Mg calcite. This fragment, which was originally made up of high-Mg calcite, has retained its original texture even though magnesium has been removed. The original interparticle porosity has been occluded by drusy calcite mosaic.

GERALD M. FRIEDMAN

References

Bathurst, R. G. C., 1958. Diagenetic fabrics in some Dinantian limestones, *Liverpool and Manchester Geol. J.*, 2, 11-36.
Bathurst, R. G. C., 1964. The replacement of aragonite by calcite in the molluscan shell, in J. Imbrie and N. D. Newell, eds., *Approaches to Paleoecology*. New York: Wiley, 357-376.
Bathurst, R. G. C., 1971. *Carbonate Sediments and their Diagenesis.* New York: Elsevier, 620p.
Benson, L. V., 1974. Transformation of a polyphase sedimentary assemblage into a single phase rock: a chemical approach, *J. Sed. Petrology*, 44, 123-135.
Chave, K. E., 1954. Aspects of the biogeochemistry of magnesium, *J. Geol.*, 62, pt. 1, Calcareous marine organisms, 266-283; pt. 2, Calcareous sediments and rocks, 587-813.
Friedman, G. M., 1964. Early diagenesis and lithification of carbonate sediments, *J. Sed. Petrology*, 34, 777-813.
Friedman, G. M., 1975. The making and unmaking of limestones or the downs and ups of porosity, *J. Sed. Petrology*, 45, 379-398.
Friedman, G. M.; Amiel, A. J.; and Schneidermann, N., 1974. Submarine cementation in reefs; example from the Red Sea, *J. Sed. Petrology*, 44, 816-825.
Lindholm, R. C., 1974. Fabric and chemistry of pore filling calcite in Septarian veins; models for limestone cementation, *J. Sed. Petrology*, 44, 428-440.
Stehli, F. G., and Hower, J., 1961. Mineralogy and early diagenesis of carbonate sediments, *J. Sed. Petrology*, 31, 358-371.

Cross-references: *Authigenesis; Beachrock; Carbonate Sediments-Bacterial Precipitation; Carbonate Sediments-Diagenesis; Cementation; Cementation, Submarine; Chalk; Dedolomitization; Diagenetic Fabrics; Dolomite, Dolomitization; Interstitial Water in Sediments; Limestone Fabrics; Limestones.* Vol. IVA: *Calcium Carbonate; Geochemistry.*

CASTS AND MOLDS

Cast and *mold* are paleontological as well as sedimentological terms. Unfortunately, both terms are quite often misused in sedimentology without regard to, and in the opposite sense of, their well-defined paleontological definitions. Both terms, in the strictest sense, refer to the relationship between a particular structure under consideration and some original structure (either a fossil organism or a primary sedimentary structure). A *cast,* then, is a copy of the original, and a *mold* is the counterpart of a cast, i.e., a structure which when filled with or surrounded by a given material will produce a copy of the original in that material.

Paleontologists refer to internal and external casts and molds when discussing the replacement of fossil organisms. All such reference is formal, and both terms are well defined. Many sedimentologists, on the other hand, when discussing primary sedimentary structures (q.v.), refer to such features as, e.g., "flute casts." Such reference is informal, not well defined, and should be avoided. For instance, a *flute cast* (Crowell, 1955) refers generally to the positive primary structures found on the bottoms of some arenaceous beds that overlie lutaceous material (see *Turbidite*). The positive feature is in fact the *counterpart,* or *mold,* of the original flute (which was a scour mark, q.v., on the surface of the lutaceous material) and not *a copy of the original flute,* which would be called a *cast.* Hence the proper term for the counterpart of a flute is *flute mold* (Fig. 1). By pressing a flute mold into plaster of paris, the resultant depression would become a true flute cast. The same can be said for *groove cast* (Shrock, 1948, p. 162; Fig. 2).

FIGURE 1. Massive counterparts of flutes (flute molds) exposed on vertically dipping volcanic arenite beds of the Change Islands Formation near Red Rock Cove, Change Islands, Notre Dame Bay, Newfoundland; hammer for scale (left center).

FIGURE 2. Transverse scour molds (counterparts of scour marks) and bounce molds exposed on the Sansom graywacke, Ninepin Arm, New World Island, Notre Dame Bay, Newfoundland.

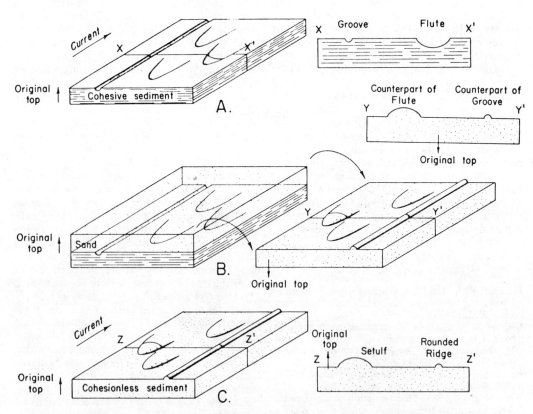

FIGURE 3. Sketches comparing origin of counterparts of flutes and grooves with setulfs (positive relief features resembling flute molds) (Friedman and Sanders, 1974). A: Current sculpts flutes and grooves into cohesive substrate. B: Sand is deposited on cohesive substrate; turned-over slab of sandstone shows counterparts of flutes and grooves. C: Setulfs and linear rounded ridge fashioned by current in cohesionless sediment. Sections YY′ and ZZ′ are identical and in absence of other data on internal structures could be confused.

Load casts (Kuenen, 1953) or *flow casts* (Shrock, 1948, p. 156), on the other hand, are also positive features on the base of arenitic beds which, in this case, originally were deposited on top of thixotropic lutaceous material (see *Pillow Structures*). But, unlike flute casts, load casts are actually the original sedimentary feature, and the counterparts of these structures would be a depression in the lutite, which, if hollow, would be a true mold.

Many sedimentologists are aware of the problem in semantics (see Dzulinski and Sanders, 1962, pp. 63-64); incorrect usage is widespread in the literature. It is important to clarify this usage because positive-relief bed forms resembling the counterparts of negative-relief bed forms have been discovered (Friedman and Sanders, 1974; Fig. 3), and scour and tool marks (q.v.) and their counterparts (sole marks) are used extensively as top and bottom criteria (q.v.).

THOMAS E. EASTLER

References

Crowell, J. C., 1955. Directional current structures from the Pre-Alpine Flysch, Switzerland, *Geol. Soc. Am. Bull.*, **66**, 1351-1384.

Dzulynski, S., and Sanders, J. E., 1962. Current marks on firm mud bottoms, *Conn. Acad. Arts Sci., Trans.*, **42**, 57-96.

Friedman, G. M., and Sanders, J. E., 1974. Positive-relief bedforms on modern tidal flat that resemble molds of flutes and grooves; implications for geopetal criteria and for origin and classification of bedforms, *J. Sed. Petrology*, **44**, 181-189.

Kuenen, P. H., 1953. Significant features of graded bedding, *Bull. Am. Assoc. Petrol. Geologists*, **37**, 1044-1066.

Shrock, R. R., 1948. *Sequence in Layered Rocks*. New York: McGraw-Hill, 507p.

Cross-references: *Current Marks; Primary Sedimentary Structures; Scour Marks; Tool Marks.* Vol. VII: *Fossils and Fossilization.*

CAVE PEARLS

Cave pearls are unattached, roughly equant concretions found in caves and mines. They are also known as pisolites and occasionally (erroneously) as oolites. Cave pearls vary in size from a few mm to several cm. The essential structure consists of a nucleus, usually of some foreign material, and a sequence of concentric growth bands, usually very thin compared to the diameter of the pearl. They are composed of calcite, although intermediate layers of aragonite may occur.

Cave pearls are found in deep, cone-shaped pockets beneath active drip points or as strewn fields of pearls in wide shallow basins, sometimes covering the entire floor of the cave. Usually, large numbers of pearls are found together, the diameter varying considerably within any pearl population. The deposition of pearls is associated with the deposition of flowstone. The pearl pockets may begin as a drip hole in the clay and silt sediments of the cave floor, achieve a thin calcite lining, and become filled with unattached pearls. In later stages, the pearls are cemented together and covered with dripstone, which may eventually evolve into a stalagmite. The shallower basins may be completely filled with flowstone at later stages.

The nucleus of the pearl may be almost anything. Sand grains, bits of rock fragments, and even bat bones have been observed. In some pearls, the nucleus is a tiny bit of material at the center. Other pearls consist of thin multi-layered coatings over a large fragment of foreign material. Pearls formed in an abandoned commercial cave were observed to consist of smooth calcite coatings on chips of the gravel used to pave the walkway.

Cave pearls appear in a variety of shapes and perfection of surface finish. They can be divided into three broad categories: spheres with smooth or polished surfaces; polyhedral or irregular shapes with smooth outer surfaces; or roughly spherical concretions with spongy, irregular outer surfaces. The highly spherical forms (Fig. 1) are usually found in shallow basins. Rounded polyhedral forms typical of a random packing arrangement obtain when the pearls are packed into deep pockets; irregular shapes may be disc-like or simply reflect the shape of the nucleus. The degree of polish varies, but the outer surface is usually smooth and is typically white. Smooth-surfaced pearls show an intricate growth banding, often with very thin layers, and the constituent minerals occur as very tiny crystals with random orientation. In contrast, the rough-surfaced, highly porous concretions have a radial structure, with individual crystals crossing the growth layers, and tend to be yellow or brown from included clay and silt.

There has been considerable disagreement concerning the amount of agitation necessary for the formation of pearls. Cave pearls in pockets occur where dripping waters are available, and pearls in wide, shallow basins are also stirred by dripping water. Certainly, in some densely packed cave-pearl nests, completely free rotation would not be possible. Spherical forms require some rolling to maintain their shape; simple agitation may produce polyhedral or irregular shapes. The polished surface is apparently obtained by the pearls' buffing against each other. Rough or porous surfaces appear

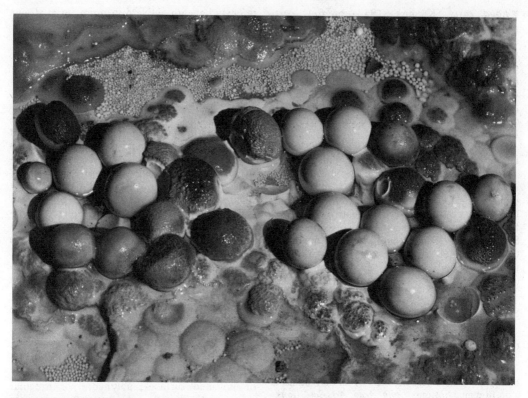

FIGURE 1. Spherical cave pearls with two size populations from Castlegard Cave, Columbia Ice Field, British Columbia. (Photo: courtesy of Russell S. Harmon.)

not to have been buffed, and it may be that the growth of an unattached concretion as such does not require any agitation at all. If a special agitation mechanism is not required, one must turn to the solution chemistry as a means of explaining why pearls appear in profusion in some locations where calcite is being deposited and not at all in others, a problem that has not been investigated in detail.

WILLIAM B. WHITE

References

Baker, G., and Frostick, A. C., 1947. Pisoliths and ooliths from some Australian caves and mines, *J. Sed. Petrology,* 17, 39-67.

Baker, G., and Frostick, A. C., 1951. Pisoliths and ooliths and calcareous growths in limestone caves at Port Campbell, Victoria, Australia, *J. Sed. Petrology,* 21, 85-104.

Gradzinski, R., and Radomski, A., 1967. Pizolity z Jaskin Kubanskich [Pisoliths from Cuban caves], *Ann. Soc. Geol. Poland,* 37, 243-265.

Hahne, C.; Kirchmayer, M.; and Otteman, J., 1968. "Höhlenperlen" (Cave Pearls), besonders aus Bergwerken des Ruhrgebietes, *N. Jahrb. Geol. Paläont. Abh.,* 130, 1-46.

Mackin, J. H., and Coombs, H. A., 1945. An occurrence of "cave pearls" in a mine in Idaho, *J. Geol.,* 53, 58-65.

Otteman, J., and Kirchmayer, M., 1967. Uber Höhlenperlen und die Mikroanalyse von Ooiden mit der Elektronensonde, *Naturwissenschaften,* 14, 360-365.

Cross-references: *Oolite; Pisolite; Speleal Sediments.*

CAVE SEDIMENTS—See SPELEAL SEDIMENTS

CEMENTATION

Cementation, the precipitation of mineral matter within the open pore spaces of sediments, is one of the major diagenetic processes that converts loose sediments into lithified rock (see *Diagenesis*). Inorganic precipitation from supersaturated pore waters, precipitation resulting from the activities of organisms, and precipitation after pressure solution are some of the processes through which cementation may occur. The most common cementing materials are low-Mg calcite, high-Mg calcite, dolomite, aragonite, siderite, quartz, chalcedony, opal, hematite, limonite, pyrite, barite, gypsum and anhydrite. The mineralogy of the cement precipitated depends on several factors: (1) the chemical composition of the pore waters; (2) nucleation phenomena; (3) crystallization kinetics; (4) temperature; (5) pressure; (6) Eh/pH relationships; and (7) the effects of macroorganism and bacterial processes. If the mineralogy of the cementing material and the detrital grain are the same, the resulting secondary overgrowth is frequently in optical continuity with the original grain, probably because the minimum surface energy for a solid-solid mineral boundary is obtained between crystals of the same composition in parallel crystallographic orientation (Blatt et al., 1972). Cementation may occur rapidly after deposition or during later stages of burial, compaction, and diagenesis.

Carbonate Cements

Low-Mg calcite, high-Mg calcite, dolomite, aragonite, and siderite are the most common carbonate cements and may occur as micrite (q.v.); as fibers; and as coarser, euhedral to anhedral, subequant crystals (Folk, 1974). High-Mg calcite and aragonite frequently occur as fibrous carbonate cements, whereas dolomite, siderite, and low-Mg calcite more commonly occur as coarser, subequant crystals (spar).

In older rocks, low-Mg calcite and dolomite are the most common carbonate cements; aragonite and high-Mg calcite become increasingly abundant in more recent rocks (see *Carbonate Sediments—Lithification*). The factors that promote the precipitation of aragonite and high-Mg calcite under conditions of low-Mg calcite stability are still uncertain Daniels (1961) suggested that trace amounts of Sr^{++} may lead to the precipitation of aragonite because Sr^{++} can be accommodated best in the aragonite lattice. Kitano and Hood (1965) have shown that different types of organic material in solution may be an important influence on the crystal structure of precipitating carbonates. Bischoff and Fyfe (1968) have proposed that specific ions such as Mg^{++} and SO_4^{--} in the pore water may inhibit the growth of low-Mg calcite and permit the precipitation of aragonite. Folk (1974) proposed that the presence of magnesium in the precipitating solution selectively poisons the sideward growth of low-Mg calcite and, consequently, $CaCO_3$ prefers to crystallize as fibrous aragonite or as micritic high-Mg calcite in marine (Mg-rich) environments. In meteoric waters, where concentrations of Mg^{++} are low, coarse, subequant grains of low-Mg calcite form when precipitation is relatively slow; micritic, low-Mg calcite cements form when precipitation is very rapid (Folk, 1974).

Dolomite cement is common in ancient rocks but quite rare in recent rocks as mentioned above (MacKenzie and Bricker, 1971). Coarse, subequant grains of dolomite cement are generally considered to be a result of replacement during late-stage diagenetic processes (see *Dolomite, Dolomitization*). Mg^{++} does not inhibit the growth of large dolomite crystals because Mg^{++} is incorporated into specific layers in the dolomite crystal structure (Folk, 1974).

The presence of siderite ($FeCO_3$) cement requires low oxidation potentials (Eh) and reducing conditions generally associated with an abundance of organic material. Siderite may form as a replacement of calcite if the concentration of calcium ions in solution is less than 150 times that of ferrous ions (Blatt et al., 1972). Siderite is rarely seen in outcrops because of its instability in the presence of oxygen; consequently, many of the rocks containing hematite and limonite cements may have once been cemented by siderite which has since been hydrated and oxidized.

Inorganic Precipitation of Carbonate Cement. The inorganic precipitation of carbonate from supersaturated pore waters is a function of temperature, pressure, pH, P_{CO_2}, and ionic concentrations. The pore waters of marine sediments are approximately saturated with respect to calcite (Blatt et al., 1972), that is, the ion activity product $aCa^{++}aCO_3^{--}$ in the pore water is approximately equal to the thermodynamic solubility constant, K_{sp}. The inorganic precipitation of $CaCO_3$ in the pore waters can occur when the ion activity product is greater than (or equal to) the solubility constant: $aCa^{++}aCO_3^{--} \geqslant K_{sp}$. Consequently, either decreasing the solubility constant or increasing the ion activity product will increase the probability of $CaCO_3$ precipitation.

The solubility constant (K_{sp}) is dependent on temperature and pressure (Broecker and Oversby, 1971) and in the case of $CaCO_3$ may be decreased by increasing the temperature or reducing the pressure. Fig. 1 illustrates the decrease in the solubility constant with increasing temperature for aragonite, calcite, and dolomite. It is thus apparent that increasing temperatures, associated with burial, decrease the solubility of carbonate minerals and promote the precipitation of carbonate cement.

The ion activity product may be raised by increasing the concentration of Ca^{++} and/or CO_3^{--}. The concentration of Ca^{++} increases with evaporation or the diagenetic weathering of calcic minerals. The concentration of CO_3^{--} increases with increasing pH of the pore water,

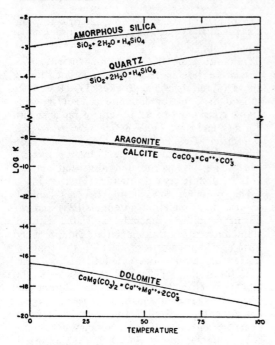

FIGURE 1. Equilibrium (solubility) constants for amorphous silica, quartz, aragonite, calcite, and dolomite as a function of temperature (from MacKenzie and Bricker, 1971).

and Fig. 2 illustrates the relationship of pH and $CaCO_3$ solubility. Several mechanisms have been proposed to increase the pH of the pore water and thereby precipitate $CaCO_3$: (1) loss of CO_2 by degassing or photosynthetic extraction; (2) the uptake of H^+ and the release of K^+, Na^+, Ca^{++}, and Mg^{++} into the pore water during diagenetic weathering of minerals; and (3) the more rapid production of bases by bacterial

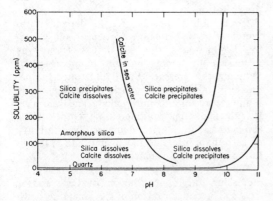

FIGURE 2. The solubility of calcite, amorphous silica, and quartz as a function of pH (from Blatt et al., ©1972; reprinted by permission of Prentice-Hall, Inc.).

processes, such as sulfate reduction and ammonia formation (Berner, 1971).

Constant flushing of the sediments by supersaturated water is necessary for appreciable quantities of cement to form, and the amount of carbonate precipitated will be almost negligible if the pore waters are static. Thus the cementation of carbonate sediments in the deep-marine environment is extremely slow because there is no mechanism to cause advection of pore waters through the sediment. On the other hand, in the intertidal zone, which is subject to subaerial exposure, evaporation, and a constant flushing of sea water through the sediments, lithification of carbonate sediments to beachrock (q.v.) occurs so rapidly that human artifacts, 10–20 years old, are sometimes included (see also *Cementation, Submarine*).

Biogenic Precipitation of Carbonate Cement. There has been some debate whether carbonate supersaturation is itself sufficient to cause nucleation and precipitation of carbonate cements, or whether precipitation is triggered by the activities, directly or indirectly, of organisms. Photosynthetic extractions of CO_2 and bacterial processes may increase the pH of the pore waters and thereby indirectly cause the precipitation of $CaCO_3$ cement (Berner, 1971). Also the precipitation of $CaCO_3$ cement within or near borings suggests that the metabolic activity of boring microorganisms may be responsible for the precipitation of the cement.

Silica Cements

Siliceous cements may occur as subequant blocks, secondary overgrowths, or as replacements of some preexisting material (Blatt et al., 1972). The most common and the only thermodynamically stable siliceous cement is macrocrystalline quartz; under certain conditions, however, silica may be deposited as opal or chalcedony.

Opal (hydrous, amorphous silica) occurs as a cement in sandstones which contain abundant volcanic glass and detritus. The devitrification of the glass at shallow depths and the release of excess silica favors the nucleation of opal (Blatt et al., 1972). Opal, however, is thermodynamically unstable with respect to quartz and is more soluble at higher temperatures (Fig. 1). Consequently, the temperature rise with burial results in the replacement of amorphous opal by microcrystalline quartz or chalcedony. Chalcedony often occurs as a replacement cement in materials such as wood, bone, and carbonates.

Inorganic Precipitation of Silica Cement. Silica exists in pore waters as silicic acid, H_4SiO_4. The first dissociation constant of H_4SiO_4 is $10^{-9.9}$; consequently, the solubility of silica remains fairly constant at pH values less than 9, as illustrated in Fig. 2. The equilibrium solubility of amorphous silica at 25°C is approximately 120 ppm, whereas the solubility of quartz is only 10 ppm (Krauskopf, 1967). The concentration of silica in ground waters is generally 10–60 ppm, which is well below the solubility of amorphous silica, but supersaturated with respect to quartz. Consequently, quartz precipitates from the interstitial waters as overgrowths on detrital grains (Siever, 1959). The precipitation of quartz, however, even from supersaturated solutions, is extremely slow, and this is possibly the reason for the limited extent of silica cementation in recent sediments.

As illustrated in Fig. 1 and 2, the solubility of silica is temperature and pH dependent, but the dependence is reversed when compared to carbonates. The solubility of both quartz and opal increases with increasing temperature, whereas the solubility of $CaCO_3$ and dolomite decreases with temperature. This inverse relationship may account for the occurrence of carbonate replacing quartz cements at higher temperatures associated with deep burial (Siever, 1959). Also $CaCO_3$ is very soluble at pH values less than 7.5, whereas the solubility of silica is unaffected at pH values below 9, which may explain the replacement of carbonate by siliceous cements in acidic environments.

Other Cementing Processes

Pressure Solution. Upon deep burial, sediment is compacted and its porosity decreases, interfering with the migration of pore waters. Here the process of pressure solution (q.v.) becomes more important as a mechanism for cementation. Waldschmidt (1941) proposed that sand grains may dissolve at their points of contact due to high pressures resulting from burial and compaction. The material released (silica or carbonate) migrates and precipitates in the pore spaces between the grains where pressures are less extreme. Waldschmidt cited the interpenetration of sand grains as evidence for solution at the point of contact and precipitation in the voids. Deelman (1975), however, has suggested that the interpenetration may be a result of compression only, not associated with pressure solution.

Heald (1956) noted a positive correlation between clay films at grain contacts and quartz interpenetration. Weyl (1959) and Thomson (1959) have provided mechanisms for the involvement of the clay film in enhancing pressure solution. Weyl (1959) proposed that the numerous films of interlayer and absorbed water in the clay structure act to increase the rate of diffusion of the pressure-dissolved material from the site of solution. Thomson (1959) has shown that clay films are frequently composed of illite and suggested that during diagenetic alteration K^+ ions are leached from the illite structure and replaced by H^+. The loss of K^+ from the clay structure and the formation of K_2CO_3 increases the pH of the pore solution. As illustrated in Fig. 2, the solubility of silica increases abruptly at pH >9, resulting in the solution of silica in regions of highest pH, i.e., near the clay-quartz grain contact, and precipitation in regions of lower pH, away from the clay zone.

Whalley (1974) has proposed that an increase in pH may also be responsible for the quartz cementation of a lateral moraine in the Feegletscher area of Switzerland. This increase in pH is a result of the chemical weathering of feldspars, muscovite, and other minerals in the moraine. Glacially crushed quartz may thus be dissolved and precipitated in areas of lower pH, providing cohesive strength to the moraine.

Iron Oxide and Sulfide Cements. During chemical weathering, iron in minerals such as hornblende, biotite, chlorite and magnetite may be oxidized, or dissolved as ferrous ion, Fe^{++}. In reducing and slightly acidic solutions, this ion may be transported considerable distances and precipitated in sediments as pyrite where the concentration of sulfide is unusually high, as siderite where the concentration of carbonate ion is high, or as iron oxides and hydroxides in oxidizing environments. Walker (1967) maintains that the iron oxide pigment and cement of many shales and sandstones is not derived from migrating solutions but is produced entirely within the sediments themselves, accomplished by the transfer of iron to the exterior of decomposing minerals during diagenesis.

Sulfate Cements. Gypsum ($CaSO_4 \cdot 2H_2O$) and anhydrite ($CaSO_4$) occur as cements in environments where there is extensive evaporation. Intense evaporation may form pore water brines supersaturated with respect to gypsum (q.v.). The precipitation of gypsum can occur very rapidly from supersaturated brines, and continued migration of the pore waters (dense brines downward and lighter saline waters upward) may result in large quantities of a gypsum cement, and even evaporite formations (Blatt et al., 1972). The formation of anhydrite is favored by higher salinities, temperatures, and pressures generally associated with burial. Cementation by evaporation is also important in the formation of salcrete (q.v.), silcrete (q.v.), and other duricrusts (q.v.).

CURTIS R. OLSEN

References

Berner, R. A., 1971. Bacterial processes affecting the precipitation of calcium carbonate sediments, in O. P. Bricker, ed., *Carbonate Cements*. Baltimore: Johns Hopkins Univ. Press, 247-251.

Bischoff, J. L., and Fyfe, W. S., 1968. Catalysis, inhibition, and the calcite-aragonite problem I. The aragonite-calcite transformation, *Am. J. Sci.*, 266, 65-79.

Blatt, H.; Middleton, G.; and Murray, R., 1972. *Origin of Sedimentary Rocks*. Englewood Cliffs, N.J.: Prentice-Hall, 323-373, 501-530.

Broecker, W. S., and Oversby, V. M., 1971. *Chemical Equilibria in the Earth*. New York: McGraw-Hill, 318p.

Daniels, R. 1961. Kinetics and thermoluminescence in geochemistry, *Geochim. Cosmochim. Acta*, 22, 65.

Deelman, J. C., 1975. Pressure solution or indentation? *Geology*, 3, 23-24.

Folk, R. L., 1974. The natural history of crystalline calcium carbonate: effect of magnesium content and salinity, *J. Sed. Petrology*, 44, 40-53.

Heald, M. T., 1956. Cementation of Simpson and St. Peter sandstones in parts of Oklahoma, Arkansas, and Missouri, *J. Geol.*, 64, 16-30.

Kitano, Y., and Hood, D. W., 1965. The influence of organic material on the polymorphic crystallization of calcium carbonate, *Geochim. Cosmochim. Acta*, 29, 29-41.

Krauskopf, K. B., 1967. *Introduction to Geochemistry*. New York: McGraw-Hill, 721p.

MacKenzie, F. T., and Bricker, O. P., 1971. Cementation of sediments by carbonate minerals, in O. P. Bricker, ed., *Carbonate Cements*. Baltimore: Johns Hopkins Univ. Press, 239-246.

Siever, R., 1959. Petrology and geochemistry of silica cementation in some Pennsylvanian sandstones, *Soc. Econ. Paleont. Mineral. Spec. Publ.*, 7, 55-79.

Thomson, A., 1959. Pressure solution and porosity, *Soc. Econ. Paleont. Mineral. Spec. Publ.*, 7, 92-110.

Waldschmidt, W. A., 1941. Cementing materials in sandstones and their probable influence on migration and accumulation of oil and gas, *Bull. Am. Assoc. Petrol. Geologists*, 25, 1839-1879.

Walker, T. R., 1967. Formation of red beds in ancient and modern deserts, *Geol. Soc. Am. Bull.*, 78, 353-368.

Whalley, W. B., 1974. A possible mechanism for the formation of interparticle quartz cementation in recently deposited sediments, *Trans. N.Y. Acad. Sci.*, ser. 11, 36, 108-123.

Weyl, P. K., 1959. Pressure solution and the force of crystallization—a phenomenological theory, *J. Geol. Research*, 64, 2001-2025.

Cross-references: *Beachrock; Carbonate Sediments—Bacterial Precipitation; Carbonate Sediments—Diagenesis; Carbonate Sediments—Lithification; Cementation, Submarine; Clastic Sediments—Lithification and Diagenesis; Diagenesis; Dolomite, Dolomitization; Duricrust; Gypsum in Sediments; Pressure Solution and Related Phenomena; Silica in Sediments; Silicification.*

CEMENTATION, SUBMARINE

Cementation (q.v.) is an important process in diagenesis (q.v.), specifically in lithification. It is distinguished from biogenic precipitation of skeletal carbonates and from all transformations of preexisting material, such as neomorphism and recrystallization of skeletal material or void fillings (see *Carbonate Sediments: Diagenesis*). *Submarine cementation* occurs in submarine environments, at or a little below the sea floor, distinguished from cementation in meteoric, littoral, and subsurface diagenetic environments.

Deep Sea

Cemented crusts are known from areas of slow or nondeposition on the sea floor or on seamounts, mostly in 200-1000 m depths (see *Hardground Diagenesis*). They have been found off Barbados on the Blake Plateau, on Atlantis Seamount, Gerda Guyot, and Great Meteor Seamount. Milliman (1974) has compiled the data on these and other occurrences and also discussed deep-sea cementation in volcanic areas and semi-enclosed basins such as the Mediterranean and Red seas.

Shallow Water

Nodules and crusts have also been reported from shelf areas of slow or nondeposition, among others from the Bahamas, the Persian Gulf, and the Mediterranean (Milliman, 1974). Reef structures contain pores of many kinds and sizes suitable for cementation, which has been observed in the reefs of the Bermudas (Ginsburg and Schroeder, 1973), off Jamaica (Land and Goreau, 1970), in the Red Sea (Friedman et al., 1974), off Yucatan, and on the Great Barrier Reef. Shallow-water submarine cementation is reviewed by Bathurst (1974).

Composition

Virtually all submarine cements reported to date are composed of calcium carbonate in the form of aragonite or high-Mg calcite; these magnesian calcites frequently contain 10-20 mol % $MgCO_3$. The calcite cement is commonly micritic, the aragonite needly; several other habits have been observed, however (Schroeder, 1972; Milliman, 1974; Alexandersson, 1974), and causes of variations in composition and habit are not yet understood.

Genesis

Ion Transport. To precipitate one volume of $CaCO_3$, the ions of 3000-100,000 volumes of sea water must be brought to the site of precipitation (Bathurst, 1974). In reefs, this is probably accomplished by continuous flushing due to surf, tidal and/or other currents. In shelf and deep-sea crusts diffusion is a more likely process (Bathurst, 1971).

Mechanisms of Precipitation. Supersaturation with respect to $CaCO_3$ in tropical shallow seas is certainly conducive to cementation, but it alone is not a sufficient cause; otherwise cements should be widespread in such waters. In addition, cementation in cold and/or deep waters, which are undersaturated with respect to $CaCO_3$, calls for other mechanisms. Organic influences such as coatings by organic matter, metabolism, activities or catalytic effects of organisms, decay of organic matter, or bacterial action may be effective in cementation (see *Carbonate Sediments—Bacterial Precipitation*).

Factors Influencing Cementation. Substrate properties such as morphology, composition, and fabric, as well as organic coating and microorganisms present, may determine nucleation, composition, and habit of the cement crystals; the rate of sea-water circulation or diffusion through the pore system determines growth rates. The determining factors vary from one pore to the next and even within given pores; correspondingly, cements may vary on the same scale. Conditions within the microenvironment may change with time; cement crystals, as a result, may show discontinuous growth and zoning (Shroeder, 1973), or several generations of cement may succeed in one pore. Up to three submarine cement generations have been observed (Shroeder, 1973, 1974).

Recognition of Fossil Submarine Cements

Composition and fabric of submarine cements have been traced into the Pleistocene (Schroeder, 1973, 1974). These characteristics are subject to diagenetic alterations, however, and the alteration products are not known or not sufficiently distinct; presently criteria of circumstantial evidence are used to recognize fossil contemporaneous submarine cements. In crusts these include single or multiple borings by organisms such as mollusks, alternations of cementation and sedimentation (Shinn, 1969), and several features of the top surface, as well as facies relationships (see *Corrosion Surface*). In reefs, alternations of growth, sedimentation, boring, and cementation are useful (see references in Schroeder, 1973).

JOHANNES H. SCHROEDER

References

Alexandersson, T., 1974. Carbonate cementation in coralline algal nodules in the Skagerrak, North Sea: biochemical precipitation in undersaturated waters, *J. Sed. Petrology*, **44**, 7-26.

Bathurst, R. G. C., 1971. *Carbonate Sediments and Their Diagenesis.* Amsterdam: Elsevier, 620p.
Bathurst, R. G. C., 1974. Marine diagenesis of shallow water calcium carbonate sediments, *Ann. Rev. Earth Planet. Sciences*, **2**, 257-274.
Friedman, G. M.; Amiel, A. J.; and Schneidermann, N., 1974. Submarine cementation in reefs: example from the Red Sea, *J. Sed. Petrology*, **44**, 816-825.
Ginsburg, R. N., and Schroeder, J. H., 1973. Growth and submarine fossilization of algal cup reefs, Bermuda, *Sedimentology*, **20**, 575-614.
Land, L. S., and Goreau, T. F., 1970. Submarine lithification of Jamaican reefs, *J. Sed. Petrology*, **40**, 457-461.
Milliman, J. D., 1974. *Marine Carbonates.* Berlin: Springer, 375p.
Schroeder, J. H., 1972. Fabrics and sequences of submarine carbonate cements in Holocene Bermuda cup reefs, *Geol. Rundsch.*, **61**, 708-730.
Schroeder, J. H., 1973. Submarine and vadose cements in Pleistocene Bermuda reef rock, *Sed. Geol.*, **10**, 179-204.
Schroeder, J. H., 1974. Carbonate cements in Recent reefs of the Bermudas and Bahamas—keys to the past? (Summary), *Ann. Soc. Géol. Belgique*, **97**, 153-158.
Shinn, E. A., 1969. Submarine lithification of Holocene carbonate sediments in the Persian Gulf, *Sedimentology*, **12**, 109-144.

Cross-references: *Beachrock; Carbonate Sediments—Bacterial Precipitation, Diagenesis,* and *Lithification; Cementation; Chalk; Coral-Reef Sedimentology; Corrosion Surface; Diagenesis; Hardground Diagenesis.*

CHALK

Chalk is a porous, fine-grained, friable white limestone which gives its name to the Cretaceous system (Latin *creta* = chalk) but which is usually limited to the upper half of the system. The word "chalk" generally only indicates a rock type; formerly, however, and still occasionally today, it is also used to mean "Upper Cretaceous" or even "Cretaceous," as in the French *La Craie*, the German *Die Kreide*, and the Russian *Mel.*

Considered as a rock type, no definition properly embraces every variation to be found in chalk and yet excludes rocks not generally regarded as chalk. The simplest definition is "a coccolith-rich limestone," but such rocks are also found in nonchalky sequences, e.g., in the Kimmeridgian of Dorset, England; on the other hand, the Turonian of Devon includes chalk that is largely made up of fragments of the bivalve *Inoceramus* (Beer Freestone). For many people, chalk's friability or apparent softness is its definitive characteristic, but chalks of Ireland, Yorkshire, and the Teutoburger Wald in Germany are hard and well cemented; and there are early Tertiary limestones from seamounts off the Iberian coast which are porous, uncemented, and chalky in appearance but which contain relatively few coccoliths. Another characteristic, often of great industrial value, is chalk's purity, often >97% $CaCO_3$; but limestones of similar purity are known in many other lithologies, e.g., the Orgon Limestone (Barremian) of SE France.

Study History

The very fine particles that make up so much of the chalk were generally called *crystalloids* in the 19th century. As early as 1861, H. C. Sorby recognized their organic origin, equating them with the coccoliths of Atlantic oozes. Later, J. Murray showed that these little calcite crystals were derived from planktonic, flagellate algae which were named coccospheres or *coccolithophores*.

Near the turn of the century, two major works on chalk were published (Cayeux, 1897; Jukes-Browne and Hill, 1903–04), both of which contain many valuable observations. Hill thought that all the smaller calcite particles that he could resolve in his microscope (there were many he could not) were coccoliths, but the zoologists Murray and Blackman disagreed. Neither Cayeux nor Hill followed the late 19th-century fashion of equating the chalk with modern globigerina ooze. Cayeux thought that the maximum depth of deposition was not more than 300 m, but Hill considered the depth range to have been as much as 800–1400 m.

Early in the 20th century, descriptions were published of calcium carbonate muds deposited in only a few meters of water in Florida and the West Indies, either as simple chemical precipitates or by the action of ammonifying bacteria. The sediments seemed sufficiently analogous to the chalk for a consensus to develop that the chalk also was an inorganic precipitate. This idea was demolished by Black (1953), who showed by mechanical analysis that the carbonate mud of the West Indies was bimodal, whereas typical chalk was far more complex, contained much more shell material; the smaller particles of the chalk were coccolith scales and plates, and all the material was organic. These conclusions have been confirmed by electron microscopy (e.g., Hancock and Kennedy, 1967).

Distribution

Although chalk is best known, and occupies the greatest timespan without interruption, in NW Europe (Upper Cretaceous), it ranges further in both time and space. The Cretaceous chalk extends eastward into Russia (Bushinsky, 1954), is found in western Australia at Gingin, and covers large areas in central North America.

Lithology

Chemistry. Few analyses have been published since 1904, but an average of five samples from the Senonian of England (Upper Chalk) gives $CaCO_3$ 98.4%, $MgCO_3$ 0.64%, Al_2O_3 0.48%, SiO_2 0.34%, and Fe_2O_3 0.09% (Wolfe, 1968). In the Lower Cenomanian chalk (Chalk Marl) of southern England and northern France a higher clay content reduces the $CaCO_3$ to 70%.

Mineralogy. The 98% $CaCO_3$ is in the form of low-Mg calcite, and the overwhelming majority originated as calcite. The only source of aragonite in the original sediment seems to have been from macrofossils. Eighty percent or more of the noncarbonate portion is clay minerals and clay-sized quartz. In the Cenomanian chalk of England there are two main antipathetic assemblages: (1) smectite, illite, and quartz; and (2) illite, kaolinite, chlorite, and vermiculite. In the Turonian and Senonian chalks, there are roughly equal proportions of smectite and illite. Although the sand content is normally <0.05%, over twenty detrital minerals have been recorded. Authigenic minerals include low-temperature tridymite and cristobalite, orthoclase, sodic plagioclase, limonite, pyrite, marcasite, glauconite, collophane, apatite, siderite, barite, and manganese oxides.

At the base of chalk successions, and sometimes on erosional breaks in the succession, there is commonly a concentration of phosphatic nodules, often accompanied by phosphatic internal molds of fossils (see *Hardground Diagenesis*). Concentrations of granular phosphate are rarer and much more local.

Dolomitic chalk is similarly very local in distribution and is known only from the upper Turonian and Senonian of the northern half of the Paris Basin, and the Upper Senonian of Mullaghcarton in Ireland.

Organic Components. Normally, not less than 60% of the rock is of coccolith material, consisting of tablet-shaped plates (or elements) of low-Mg calcite derived from planktonic algae of the class Haptophyceae. In the original plants most of the plates, often overlapping one another, were grouped into annular shields (or scales), and such complete shields can still be readily found in the chalk (Fig. 1). The majority, however, occur as isolated plates 0.2–3µ across. Stereoscan photographs of the Fort Hays Chalk (Lower Senonian) of Kansas (by Hancock and McTurk, in Frey, 1972) show many complete coccolith shields, unlike the European chalks where they are largely fragmented. Many chalks have not yet been examined with an electron microscope.

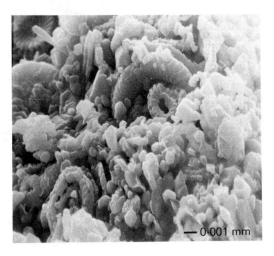

FIGURE 1. Photomicrograph of soft, white chalk, composed primarily of coccolith plates, from: Mucronata zone (Upper Campanian), Clarendon, Wiltshire, England.

The remainder of the rock, the particles still largely <100μ across, consists of foraminifera (5-10%), fragments of *Inoceramus,* and, to a lesser degree, other bivalves, polyzoa, ostracodes, and fragments of other macrofossils.

Diagenesis. Chalk is an unusual limestone in that most of it is uncemented, although it should be remembered that the original sediment contained no aragonite other than as scattered macroshells (see *Carbonate Sediments—Lithification*). It is strange that pressure solution (q.v.) has seldom occurred, even under loads of >1000 m of sediment. Recent study has shown that chalk is able to offer resistance to pressure solution because its pore fluid may have a Mg:Ca ratio similar to that of sea water. A pore fluid of this kind is able to prevent pressure solution of calcite up to a nonhydrostatic pressure of 250-1000 atm (Neugebauer, 1973). The chalk that is cemented has probably reached that state in several different ways. Within some chalk, there are pebbles of chalk which have served as anchorages to animals such as oysters, and must therefore have been hardened contemporaneously. There are also many examples of hardgrounds (see *Hardground Diagenesis*), which are sharply topped beds, typically about 1/3 m thick, in which the chalk becomes harder upwards, but which may contain burrows infilled with soft sediment, also indicating contemporaneous cementation. In contrast, the chalk of Ireland is hard throughout and probably suffered considerable compaction before pressure-solution cementation set in (Wolfe, 1968), possibly associated with a region of greater heat flow (P. A. Scholle, 1974).

The most conspicuous diagenetic feature in chalk is flint: nodules and sheets of fine black chert (q.v.). There are several generations of flint, and although none is contemporaneous, the flint nodules show so many close connections with the carbonate sedimentation that it is probable they are penecontemporaneous. Their courses mark the bedding even where the chalk has slumped. Some fill burrows, and others have distinct forms that are difficult to explain: the *paramoudras* of Ireland and Norfolk in England are barrel-shaped flints averaging 1 m in height, with a flint-free vertical core.

Conditions of Deposition

The nearest present-day analogue of chalk is the carbonate sediment in the shelf lagoon off British Honduras, which contains only 7-20% coccoliths, however; these samples were from water 27-43 m deep (Scholle and Kling, 1972). On the evidence of its sponge fauna, chalk seems to have been deposited at a much greater depth (100-300 m), but compared with the deep-water globigerina ooze, chalk is richer in molluscan fragments and relatively poor in planktonic foraminifera. The coccolith-bearing algae must have been about nine times as abundant during the Upper Cretaceous as they are at the present day. Deposition was not continuous, however; the hardgrounds are often capped by submarine erosional surfaces.

The purity of chalk is largely accounted for by a scarcity of land to provide detritus, but in NW Europe this factor may have been accentuated by a nonseasonal arid climate on the land (Hancock, 1976). The distribution of fossil animals and plants and the oxygen isotope ratios in Cretaceous chalk all indicate that the seas were appreciably warmer than they are today.

J. M. HANCOCK

References

Black, M., 1953. The constitution of the Chalk, *Geol. Soc. Lond. Proc.,* 1499, lxxxi-lxxxvi.

Bushinsky, G. I., 1954. Lithology of the Cretaceous sediments of the Dneprovsk-Donetz basin, *Akad. Nauk. SSSR, Trudy Inst. Geol. Nauk.,* **156** (in Russian).

Cayeux, L., 1897. Contribution à l'étude micrographique des terrains sédimentaires. 2 Craie du Bassin de Paris, *Soc. Geol. Nord Mem. 4,* 207-589.

Frey, R. W., 1972. Paleoecology and depositional environment of Fort Hays Limestone Member, Niobrara Chalk (Upper Cretaceous), West-Central Kansas, *Univ. Kansas Paleont. Contr. 58,* 72p.

Håkansson, E., Bromley, R., and Perch-Nielsen, K., 1974. Maastrichtian chalk of north-west Europe—a pelagic shelf sediment, *Internat. Assoc. Sedimentologists, Spec. Pub. 1,* 211-233.

Hancock, J. M., 1976. The petrology of the Chalk, *Proc. Geol. Assoc.,* **86,** 499-535.

Hancock, J. M., and Kennedy, W. J., 1967. Photographs of hard and soft chalks taken with a scanning electron microscope, *Geol. Soc. Lond. Proc.,* **1643,** 249-252.

Hancock, J. M., and Scholle, P. A., 1975. Chalk of the North Sea, in A. W. Woodland, ed., *Petroleum and the Continental Shelf of Northwest Europe. I. Geology.* London: Applied Science, 413-427.

Huxley, T. H., 1868. On a piece of chalk, in T. H. Huxley, *Collected Essays,* **8** (1909). New York: Appleton, 1-36.

Juignet, P., and Kennedy, W. J., 1974. Structures sédimentaires et mode d'accumulation de la craie du Turonien supérieur et du Sénonien du Pays de Caux, *Bull. Bur. Rech. Géol. Min., Sec. IV,* 1974, 19-47.

Jukes-Browne, A. J., and Hill, W., 1903-04. The Cretaceous rocks of Britain, *Mem. Geol. Serv. G. B.,* **2** and **3,** 568p and 566p.

Mapstone, N. B., 1975. Diagenetic history of a North Sea chalk, *Sedimentology,* **22,** 601-614.

Neugebauer, J., 1973. The diagenetic problem of chalk—the role of pressure solution and pore fluid, *N. Jahrb. Geol. Paläont. Abh.,* **143**(2), 223-245.

Neugebauer, J., 1974. Some aspects of cementation in chalk, *Internat. Assoc. Sedimentologists, Spec. Publ.* **1,** 149-176.

Noël, D., 1970. *Coccolithes Crétacés (La Craie campanienne du Bassin de Paris).* Paris: C.N.R.S., 130p.

Reid, R. E. H., 1973. The Chalk Sea, *Irish Naturalists' J.,* **17,** 357-375.

Scholle, P. A., 1974. Diagenesis of Upper Cretaceous chalks from England, Northern Ireland and the North Sea, *Internat. Assoc. Sedimentologists, Spec. Pub. 1,* 177-210.

Scholle, P. A., and Kling, S. A., 1972. Southern British Honduras: lagoonal coccolith ooze, *J. Sed. Petrology,* **42,** 195-204.

Wolfe, M. J., 1968. Lithification of a carbonate mud; Senonian Chalk in Northern Ireland, *Sed. Geol.,* **2,** 263-290.

Cross-references: *Carbonate Sediments—Diagenesis; Chert and Flint; Hardground Diagenesis; Limestones; Pelagic Sedimentation....* Vol. VII: *Coccoliths and Rhabdoliths;* Vol. IX: *Cretaceous.*

CHANNEL SANDS

The sands that accumulate in stream channels are coarse, often silty, immature, and poorly sorted. They occur as thin elongate bodies, sinuous in plan and prismatic to lenticular in cross section, and are one of a group of similarly shaped bodies known as *shoestring sands.* Individual channel sands may coalesce to form complex sheet deposits which are often only partly preserved. Strong erosional scours at the bottom characterize many channel sands. The deposits are invariably coarse at the bottom and gradually fine upward until they merge with fine overbank deposits at the top (Fig. 1). Coarse channel sands sometimes show intertonguing relationships with the flat, clayey, and silty backswamp deposits that lie alongside the channel. Fauna are rarely preserved in channel sands.

Cross-beds of *trough* (lenticular) and *planar* (tabular) varieties occur in profusion in these sands. The cross-strata trough axes parallel the trends of the elongate sand bodies in which they occur. The directions of cross-bedding foresets consistently point downstream, so that the dispersion in the cross-bedding azimuth is small; this relationship does not hold for braided-stream deposits (q.v.). Ripple marks (q.v.), parting lineation (q.v.), and various deformational structures are also common in the channel sands.

Accounts of channel sands of both ancient and modern origin are frequent in geological literature. A general review of the published literature on this subject will be found in Pettijohn et al. (1972, p. 464) and Potter (1967).

SUPRIYA SENGUPTA

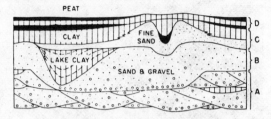

A. Braided river, with probable channel bottom deposits
B. Large meandering river
C. Small meandering river, strongly branched
D. Swamp and peat formation

FIGURE 1. Idealized sequence and structure of alluvial channel deposits in the Netherlands (Potter, 1967, after Zagwijn).

References

Peterson, J. A., and Osmond, J. C., eds. 1961. *Geometry of Sandstone Bodies.* Tulsa, Okla.: Am. Assoc. Petrol. Geologists, 240p.

Pettijohn, F. J.; Potter, P. E.; and Siever, R., 1972. *Sand and Sandstone.* Berlin: Springer-Verlag, 618p.

Potter, P. E., 1967. Sand bodies and sedimentary environments: a review, *Bull. Am. Assoc. Petrol. Geologists,* **51,** 337-365.

Cross-references: *Alluvium; Barrier Island Facies; Fluvial Sediments; River Sands; Sand and Sandstones.* Vol. IX: *Channel Sands.*

CHERT AND FLINT

Chert is a fine-grained sedimentary rock composed of chemically precipitated silica together with minor impurities. Silica is present in one or more forms, ranging from amorphous (opal) to progressively more highly crystalline forms (cristobalite, chalcedony, crypto- and microcrystalline quartz); the hardness is slightly less than 7. Crystal size varies from 1μ to 20μ or more. Cherts occur as nodules and as bedded deposits; *flint* is a variety of nodular chert.

Occurrence

Cherts occur as (1) *nodules* in carbonate sequences, commonly formed in miogeosynclinal environments, e.g., flint nodules in chalk (q.v.) of southern England; and (2) *bedded deposits* with mudstones or ironstones (see *Cherty Iron Formations*), frequently associated with volcanics in eugeosynclinal sequences, e.g., the Franciscan and Monterey cherts of the Cordilleran eugeosyncline and the Alpine radiolarian cherts of the Tethyan eugeosyncline. The areal extent and volume of bedded and nodular chert is large; it is probably the most abundant chemically precipitated rock type after carbonate.

Mineralogy

Opal is isotropic, $n = 1.40-1.47$, water content 3-10%. It may be truly amorphous (e.g., biogenous opal) or poorly crystalline, either disordered cristobalite or a cristobalite-tridymite mixture. Biogenous opal yields cristobalite on heating, whereas diagenetic cristobalite has a mixed-layer tridymitic structure (Greenwood, 1973).
Chalcedony is an optically fibrous variety of quartz and may appear brownish in thin-section. X-ray diffraction analysis yields normal quartz, but the mean n of 1.54 and the birefringence are lower and used to be attributed to admixed opal. Fluid-filled pores (0.1μ) occur in cherts; these *cryptopores* are present in abundance in chalcedony and may account for the observed anomalous optical and physical properties (Folk and Weaver, 1952). The pore fluid is saline (Smith, 1960). Smith described three types of chalcedony, all varieties of crypto- and microcrystalline quartz: (1) *Granulo*-chalcedony has equal grains of $<30\mu$. (2) *Crypto*-chalcedony, which is isotropic, can be distinguished from opal by its n of 1.535. (3) *Fibro*-chalcedony occurs as fibers displaying a random, parallel, divergent, radiating, spherulitic, or aggregate arrangement. *Lutecite* is a variety of chalcedony found in evaporite sequences, the fibers of which are length slow (Folk and Pittman, 1971).
Cryptocrystalline quartz is isotropic, resulting from the presence of submicroscopic pores which cause aggregate polarization. Because it has a similar x-ray diffraction trace it may be a variety of chalcedony (Lancelot, 1973).
Microcrystalline quartz appears as a mosaic of more-or-less equant crystals, commonly 1-5μ in diameter with a maximum of 20μ. The crystals exhibit curved faces due to interference during growth.
Opal becomes increasingly rare in older sediments, and quartz becomes more common, e.g., Tertiary cherts are opaline, Paleozoic cherts are quartzose. The mineralogical nature of the chert is also influenced by the composition of the sediment in which it develops; e.g., porcelanitic chert made of disordered cristobalite is found in clay-rich sediments, and quartzose nodules are restricted to carbonate environments (Lancelot, 1973).

Texture

Chert is largely composed of fibrous chalcedonic quartz and microcrystalline quartz (Folk and Weaver, 1952). Euhedral crystals are absent because of mutual interference during growth. Textures include metamorphic (granoblastic and porphyroblastic), igneous (seriate and merocrystalline), and sedimentary-hydrothermal textures and structures (spherulitic, botryoidal, and color-banded). The smallest crystal size reported is 0.2μ; thermodynamically, quartz crystals in chert should have a minimum stable size of a few tenths of a μ (Blatt et al., 1972).
Electron microscopy of fracture surfaces reveal two types of surface morphology: *novaculite* (q.v.), consisting of polyhedral crystals, and *spongy*, resulting from numerous cavities, corresponding to the microcrystalline and chalcedonic quartz respectively; gradations may occur. Surface morphology is determined by the spacing between sites of crystal nucleation (Folk and Weaver, 1952).
Kneller (1968) compared the morphology of 22 varieties of chert with their solubility and potential alkali reactivity. A change in surface area and crystallinity occurs with grain growth; the latter influences the physical and chemical behavior of the chert and is a function of time, depth of burial, tectonics, and diagenesis.

Diagenesis

Mizutani (1970) experimentally studied the conversion of amorphous silica to quartz. The transformation involves two first-order steps, from amorphous silica to low cristobalite, and from low cristobalite to low quartz. Rate of transformation is dependent on temperature. All diagenetic facies represent intermediate stages on the way to equilibrium; the cristobalite stage is referable to the early stage of diagenesis. Studies of deep-sea samples have confirmed these results and earlier hypotheses for the diagenetic sequence: solution of biogenous silica within the sediment, then precipitation of poorly ordered cristobalite, and gradual inversion to microcrystalline quartz (Greenwood, 1973). Thus cherts go through an evolutionary development during which there is a progressive loss of combined water, a transformation of crystal structure from opaline cristobalite through chalcedonic quartz to quartz, and a progressive coarsening of crystal size (Mizutani, 1970).

Simonds (1972) found that the rate of grain growth is controlled by temperature, pressure, initial grain size, and the composition of the fluids; recrystallization is promoted by basic pH solutions. Silica in a calcite matrix may crystallize directly to quartz (Greenwood, 1973), presumably because of these high-pH solutions, whereas cherts formed in clays and mudstones all contain cristobalite because the ratio of silica to foreign cations is much lower (Lancelot, 1973). The cristobalite-tridymite structure is more difficult to convert to quartz than the cristobalite structure. Simonds (1972) showed experimentally that a few preexisting grains grow and consume adjacent material until the entire sample consists of equant polygonal grains, which then continue to grow, with larger grains consuming smaller grains, e.g., chlorite-grade metamorphism yields polygons 25μ across. Many thin-bedded quartzites in metamorphic terranes evolved from chert, as did many of the so-called quartzites occurring in the Archean terranes of the world.

Varieties of Chert

Several varieties of chert can be distinguished based on their occurrence, texture, and impurities; only the more important ones are mentioned here.

Flint is a term applied to dense, irregular nodules of chert occurring typically in the Cretaceous chalk deposits of southern England and western Europe. Fracture surfaces are smooth and conchoidal, with a waxy luster. Flint is translucent in thin pieces, but opaque in mass and is composed of cryptocrystalline and chalcedonic quartz. The difference between flint and chert is only a matter of grain size, porosity, and dehydration features (Buurman and Van Der Plas, 1971). Because flints form sharp cutting edges they were used by early humans for tools. The color varies from gray to black, probably due to organic matter (for further information see Shepherd, 1972).

Novaculite (q.v.) is a dense, hard, even-textured, light-colored chert. It owes its whitish color partly to its purity and partly to extra-crystalline water. *Porcelanite* (q.v.) resembles porcelain, has a dull luster resulting from calcareous and argillaceous impurities, and is composed of disordered cristobalite. *Lydite* is a gray-black radiolarian chert with a dull, velvety luster, such as occur in the Culm Measures of Germany and SW England.

Fossiliferous cherts are bedded cherts that contain siliceous skeletal remains (radiolarians, diatoms, and sponge spicules). They have been compared with siliceous oozes; oozes may age to earths, and with further diagenesis to radiolarites, diatomites, etc., and ultimately to chert.

Jasper, usually red, also green, yellow, or brown, is chert containing iron minerals. *Taconite,* cherty (banded) iron formation (q.v.) of the Lake Superior region, contains 25–40% iron. *Itabarite,* found in Brazil, is metamorphosed banded iron formation composed of specular hematite and megascopic quartz. *Beekite,* also called *welded chert,* has a distinctive form and color, consisting of coral pink to red-brown spherules (4 mm; King and Merriam, 1969).

Sinter is silica precipitated from siliceous springs, and around geysers (geyserite), forming white, porcelaneous, low-density rocks, with a botryoidal or irregular surface (see *Spring Deposits*).

Relationship Between Impurities, Color, and Texture. The colors and textures exhibited by cherts are determined by contained impurities, which can best be determined by chemical analysis. Because the impurities have distinctive chemical compositions their mineralogical interpretation is simplified; nevertheless, analyses must be made in conjunction with petrographic work. The nonsilica components: carbonates (calcite, dolomite, siderite), clay, and iron minerals (hematite, pyrite, ferrous silicates) contribute Ca, Mg, CO_2, Al, Ti, Na, K, and Fe to the cherts (see Table 1).

In general, white cherts are pure silica (Table 1, B); gray to blackish cherts contain either organic material or finely disseminated pyrite; red cherts contain hematite granules or flakes (Table 1, A); green cherts contain ferrous silicates (Table 1, C); whitish to buff cherts contain calcite, dolomite, or siderite; more purplish bands may result from an increased manganese content.

Diatomaceous and radiolarian cherts are

TABLE 1. Composition of Various Types of Chert

	A	B	C	D	E	F
SiO_2	90.02	95.45	91.92	99.82	70.78	95.11
Al_2O_3	3.48	2.14	3.19	0.11	0.45	0.14
Fe_2O_3	2.84	N.D.	1.27	–	0.02	0.40
FeO	0.96	0.86	1.34	0.07	0.30	0.44
MgO	1.33	0.57	1.08	–	1.88	0.71
CaO	0.24	0.16	0.36	–	12.90	1.11
Na_2O	0.16	0.09	0.05	–	0.05	<0.01
K_2O	0.50	0.33	0.49	0.03	0.06	0.01
TiO_2	0.19	0.15	0.17	–	0.03	0.03
MnO	0.24	0.21	0.09	–	0.02	0.01

A–C: Mesozoic radiolarian cherts from a Steinmann association (Thurston, 1972).
    A. Red and pink chert
    B. Palest (whitish) cherts
    C. Green chert
D–F: From Blatt et al. (1972) after Cressman (1962).
    D. Novaculite from Arkansas
    E. Chert nodule, Delaware Limestone, Devonian, Ohio (also contains 12% CO_2)
    F. Chert layer in dolomite, Bisher Formation of Foerste, Silurian, Ohio

relatively enriched in clay minerals; nodular cherts are higher in calcite (Table 1, E). The latter usually contain <1% total iron, whereas bedded cherts containing >17% have been recorded by the author for radiolarian cherts. In general, red cherts contain more Fe_2O_3 than FeO and green cherts more FeO than Fe_2O_3. Because many ferrous silicates also contain Fe_2O_3, the division between Fe_2O_3 "red" and FeO "green" is not distinct. Al_2O_3 is often higher in the greener cherts. As mentioned before, the mineralogical nature of chert, which is determined by the initial composition of the sediment, determines the textures observed.

Origin of Chert

Theories on the origin of chert are concerned with the source of the silica and its mode of precipitation. The chemical behavior of silica in sedimentary environments imposes strict limits on such hypotheses (see *Silica in Sediments; Silicification*).

Nodular Cherts. Nodular cherts are almost entirely a product of diagenesis. The formation of nodules in some cases and beds in others is determined by the volume of silica available for silicification. Suggested sources for the silica are dissolution of skeletons deposited with the carbonate or migration of silica-rich waters from adjacent strata. The actual sites of silicification are related to sedimentary structures or processes, e.g., changes in grain size or zones of biological activity through a change of P_{CO_2} and pH. Once crystalline silica has formed, a solution gradient develops which aids in the crystallization of more silica and its removal from the surrounding limestone by diffusion (Buurman and Van Der Plas, 1971). Such a concentration gradient could explain concentric features within nodules, and incompletely silicified outer rims.

Buurman and Van Der Plas (1971) suggested that an intermediate calcium silicate mineral is important in the transformation of calcite to silica, enabling particle-by-particle replacement. Studies of chert nodules from deep-sea Tertiary chalk have shown the presence of cristobalite microspherulites but no calcium silicate intermediary (Weaver and Wise, 1973). These spherules are thought to represent the initial stage of silicification of carbonate rock in the deep-sea environment (Wise and Kelts, 1972). The cristobalite is authigenic and is derived from the dissolution of siliceous microfossils and possibly some volcanic glass. Weaver and Wise (1972) observed no gel or gel phase during the conversion of biogenous opal to cristobalite. In contrast Lancelot (1973) and Greenwood (1973) suggest that quartz can precipitate directly in a calcite matrix. Lancelot finds cristobalite spherules only in the outer part of the nodule where impurities have concentrated as the nodule was accreted. The source of the silica is biogenic; several meters above the nodules, radiolarian skeletons are well preserved, and at one meter above dissolution has begun. Because the interstitial waters contain 10–50 ppm silica, they are undersaturated with respect to amorphous silica but supersaturated with respect to crystalline silica; thus the skeletons dissolve, and cristobalite and quartz can precipitate out (Lancelot, 1973).

Bedded Cherts. Two sources have been suggested for the silica now comprising bedded cherts, either volcanic or nonvolcanic. The mode of precipitation of the silica has been attributed to either biogenic processes, or chemogenic (inorganic), for which several mechanisms are proposed. The precipitation of inorganic silica requires a tremendous increase in the silica concentration in sea water from about 3 ppm to over 120 ppm. The direct precipitation of quartz from sea water has not been observed in the marine environment (Krauskopf, 1959). To achieve the high concentrations necessary for amorphous silica precipitation requires special conditions, and many theories invoke the role of volcanism in achieving this build-up, e.g., Davis (1918), and Bailey et al. (1964). Berner (1971) predicted the most favorable environment for possible volcanic-inorganic chert formation: a deep, restricted basin, with submarine volcanism, where the water is stratified, resulting in depletion of O_2 in the lower layers. Such conditions would lead to a build-up in the Si concentration in the lower layers, resulting in the chemical precipitation of colloidal silica with sulfides. He cites the association of black shales with cherts and volcanics as evidence that this process might have occurred in nature. Although Berner (1971) suggests that the Gulf of California fulfills most of the requirements, he cites other studies which have shown that all the silica is precipitated by diatoms and, furthermore, that sufficient silica to account for all diatom sedimentation is provided by inflow from the open ocean; the nearby volcanic source is unnecessary.

Indeed, recent studies of the ocean floor have not revealed the presence of any masses of silica gel on the sea floor. Lisitsyn (1969) found that the entire silica contribution to the sediment in the Sea of Okhotsk, was accomplished by diatom accumulation and not volcanism. All the silica in suspension in the oceans is organogenic even in close proximity to active volcanoes (Lisitsyn, 1967). At the present day, siliceous organisms are the most active agents in keeping the silica concentration in the oceans very low; such conditions have obtained since the appearance of those organisms around 1000 to 600 m.y. ago. Organisms occur in greatest abundance where upwelling, nutrient-rich water reaches the surface, and also downstream from active volcanoes.

Bramlette (1946) attributed the formation of the Monterey cherts, which are composed of diatomite (amorphous silica) overlying porcelanite (fine-grained cristobalite) overlying chert (quartz), to the diagenetic rearrangement of biogenous silica. Ernst and Calvert (1969) confirmed these conclusions experimentally; they produced chert from porcelanite in pure water; at 20°C the reaction takes 180 m.y. and at 50°C, 5 m.y.

Heath and Moberly (1972) suggest that Eocene chert from the Pacific forms in two stages: (1) biogenous opal is dissolved and reprecipitated as finely crystalline cristobalite, and (2) the cristobalite inverts to quartz. The second-stage inversion may be a solid-solid zero-order reaction of the type described by Ernst and Calvert (1969). The second stage excludes the possibility of a solution phase except in a thin film at the inversion fronts. Growth of quartz masses is accompanied by severe degradation and destruction of textural features within the cristobalite (such as the tests of siliceous organisms; Heath and Moberly, 1972).

Similar results have been obtained by Thurston (1972) in a study of radiolarian cherts from a classic Steinmann association. Studies by optical and scanning electron microscopy revealed that all gradations in preservation of radiolarians were present, both between beds and vertically and laterally within a single bed. The effects of diagenesis (solution, reprecipitation, and recrystallization) were least in the dark red cherts, where hematitic granules and clay material inhibited the movement of pore waters, and greatest in the palest cherts, which represent the purest accumulation of radiolarian skeletons (see Table 1 A–C). For a review of the history of organic siliceous sediments in oceans, see Ramsay (1973).

It is of interest that Eocene opaline claystones of the Atlantic and Gulf Coastal Plain deposits, long considered volcanic in origin, have recently been shown to be highly altered diatomite (Weaver and Wise, 1974). The only places where inorganic precipitation of silica is occurring at the present day are in alkaline lakes and certain restricted lagoonal areas, but not on the deep, open-ocean floor. It is possible that mechanisms similar to those suggested by Harder (1965) and Bischoff (1969) may have played a part in the precipitation of some of the silica now contained in bedded cherts, but their role is thought to be minor. The depth of formation is variable, from >3 km to shallow depths. The rhythmic bedding has been attributed to turbiditic deposition (Nisbet and Price, 1974).

At the present day, the main process for the precipitation of silica is biogenic. Such conditions must have obtained since the appearance of siliceous organisms; prior to that time it would seem likely that the cherts of the Precambrian represent chemical precipitations of silica (see *Cherty Iron Formation*).

DIANA R. THURSTON

References

Bailey, E. G.; Irwin, W. P.; and Jones, D. L., 1964. Franciscan and related rocks, and their significance in the geology of western California, *Calif. Div. Mines and Geol. Bull.,* 183, 177p.

Berner, R. A., 1971. *Principles of Chemical Sedimentology.* New York: McGraw-Hill, 240p.

Bischoff, J. L., 1969. Red Sea geothermal brine deposits; their mineralogy, chemistry and genesis, in E. T. Degens and D. A. Ross, eds., *Hot Brines and Recent Heavy Metal Deposits in the Red Sea.* New York: Springer, 368-401.

Blatt, H.; Middleton, G.; and Murray, R., 1972. *Origin of Sedimentary Rocks.* Englewood Cliffs, N.J.: Prentice-Hall, 634p.

Bramlette, M. N., 1946. The Monterey Formation of California and the origin of its siliceous rocks. *U.S. Geol. Surv. Prof. Pap.* 212, 1-55.

Buurman, P., and Van Der Plas, L., 1971. The genesis of Belgian and Dutch flints and cherts, *Geol. Mijnbouw,* 50, 9-28.

Davis, E. F., 1918. The radiolarian cherts of the Franciscan Group, *Univ. Calif. Publ. Geol.,* 11, 285-432.

Ernst, W. G., and Calvert, S. E., 1969. An experimental study of the recrystallization of porcelanite and its bearing on the origin of some bedded cherts, *Am. J. Sci., Schairer Vol.,* 267A, 114-133.

Folk, R. L., and Pittman, E. D., 1971. Length-slow chalcedony: A new testament for vanished evaporites, *J. Sed. Petrology,* 41, 1045-1058.

Folk, R. L., and Weaver, C. E., 1952. A study of the texture and composition of chert, *Am. J. Sci.,* 250, 498-510.

Greenwood, R., 1973. Cristobalite: its relationship to chert formation in selected examples from the Deep Sea Drilling Project, *J. Sed. Petrology,* 43, 700-708.

Harder, H., 1965. Experiments zur "Ausfällung der Kieselsaure," *Geochim. Cosmochim. Acta,* 29, 429-442.

Heath, G. R., and Moberly, R., Jr., 1972. Cherts from the western Pacific, Leg 7, Deep Sea Drilling Project, *Init. Repts. Deep Sea Drilling Project,* 7, 991-1007.

King, R. J., and Merriam, D. F., 1969. Origin of the "welded chert" Morrison Formation (Jurassic), Colorado, *Geol. Soc. Am. Bull.,* 80, 1141-1148.

Knauth, L. P., and Epstein, S., 1975. Hydrogen and oxygen isotope ratios in silica from the JOIDES Deep Sea Drilling Project, *Earth Planetary Sci. Lett.,* 25, 1-10.

Kneller, W. A., 1968. A study of chert aggregate reactivity based on observations of chert morphologies using electron optical techniques, in Proc. 17th Ann. Highway Geol. Symp., Iowa State Highway Comm.; Ames, Iowa: *Iowa State Univ., Dept. Earth Sci. Publ. 1,* 73-83.

Krauskopf, K. B., 1959. The geochemistry of silica in sedimentary environments, *Soc. Econ. Paleont. Mineral. Spec. Pub.* 7, 4-19.

Lancelot, Y., 1973. Chert and silica diagenesis in sediments from the central Pacific, Leg 17, Deep Sea Drilling Project, *Init. Repts. Deep Sea Drilling Project,* 17, 577-405.

Lisitsyn, A. P., 1967. Basic relationships in distribution of modern siliceous sediments and their connection with climatic zonation, *Internat. Geol. Rev.,* 9, 631-652, 842-865, 980-1004, 1114-1130.

Lisitsyn, A. P., 1969. Recent sedimentation in the Bering Sea, in P. L. Bezrukov, ed., *Izdatel'stvo Nauka, Moska.* Israel Program for Scientific Translations, Jerusalem, 614p.

Mizutani, S., 1970. Silica minerals in the early stage of diagenesis, *Sedimentology,* 15, 419-436.

Namy, J. N., 1974. Early diagenetic chert in the Marble Falls Group (Pennsylvanian) of central Texas, *J. Sed. Petrology,* 44, 1262-1268.

Nisbet, E. G., and Price, I., 1974. Siliceous turbidites: bedded cherts as redeposited ocean ridge-derived sediments, *Internat. Assoc. Sedimentologists, Spec. Pub. 1,* 351-366.

O'Neil, J. R., and Hay, R. L., 1973. $^{18}O/^{16}O$ ratios in cherts associated with the saline lake deposits of East Africa, *Earth Planetary Sci. Lett.,* 19, 257-266.

Rad, U. von, and Rösch, H., 1974. Petrography and diagenesis of deep-sea cherts from the central Atlantic, *Internat. Assoc. Sedimentologists, Spec. Pub. 1,* 327-349.

Ramsay, A. T. S., 1973. A history of organic siliceous sediments in oceans, in Organisms and Continents Through Time, *Palaeont. Assoc. Lond., Spec. Pap. 12,* 199-234.

Shepherd, W., 1972. *Flint: Its Origin, Properties, and Uses.* London: Faber and Faber, 255p.

Sheppard, R. A., and Gude, A. J., 3d., 1974. Chert derived from magadiite in a lacustrine deposit near Rome, Malheur County, Oregon, *J. Research U.S. Geol. Surv.,* 2(5), 625-630.

Simonds, C. H., 1972. Kinetics of chert recrystallization, *Geol. Soc. Am. Abs.,* 4, 666-667.

Smith, W. E., 1960. The siliceous constituents of chert, *Geol. Mijnbouw,* 22, 1-5.

Thurston, D. R., 1972. Studies on bedded cherts, *Contr. Mineralogy Petrology,* 36, 329-334.

Weaver, F. W., and Wise, S. W., 1972. Ultramorphology of deep sea cristobalite chert, *Nature Phys. Sci.,* 237, 56-57.

Weaver, F. W., and Wise, S. W., 1973. Ultramorphology of carbonate and silicate phases associated with deep-sea chert, *Bull. Am. Assoc. Petrol. Geologists,* 57, 811.

Weaver, F. W., and Wise, S. W., 1974. Opaline sediments of the southeastern Coastal Plain and Horizon A: biogenic origin, *Science,* 184, 899-901. See also discussion by T. G. Gibson and K. M. Towe, and reply, 1975, 188, 1221-1222.

Wise, S. W., and Kelts, K. R., 1972. Inferred diagenetic

history of weakly silicified deep sea chalk, *Bull. Am. Assoc. Petrol. Geologists*, **56**, 1906.
Wise, S. W., Jr., and Weaver, F. M., 1974. Chertification of oceanic sediments, *Internat. Assoc. Sedimentologists, Spec. Pub. 1*, 301-326.

Cross-references: *Chalk; Cherty Iron Formation; Corrosion Surface; Diagenesis; Diatomite; Novaculite; Pelagic Sediments; Porcelanite; Silica in Sediments; Silicification; Spring Deposits; Volcanic Ash–Diagenesis.* Vol. IVB: *Chalcedony; Opal.* Vol. VII: *Radiolaria.*

CHERTY IRON FORMATION (BIF)

Lithology

Cherty iron formation is chert with > 15% Fe (Gross, 1965; James 1966). A characteristic alternation of layers, not uncommonly lenticular or nodular, that are rich in quartz and iron minerals has given rise to the name *banded iron formation* (*BIF*). *Cherty ironstone* and *banded ironstone* are synonyms used in Europe and South Africa (see also *Ironstone*). Transitions of BIF to chert and clastic rocks (sandstone, siltstone, shale) are common. BIF is characteristic of older Precambrian terranes (> 1800 m.y.; Goldich, in James and Sims, 1973), but some large BIF are latest Precambrian and Paleozoic (see Dimroth, 1976, for discussion of ages).

FeO, Fe_2O_3, and SiO_2 make up >90% of BIF; contents of Al_2O_3 (<2%), MgO, CaO, and P_2O_5 (<0.1%) are characteristically low (Lepp and Goldich, 1964). BIF is composed of cherty quartz (quartz, chalcedony, quartzine), iron oxides (magnetite, hematite), iron silicates (greenalite, minnesotaite, stilpnomelane, 14-Å chamosite, riebeckite), and carbonates (siderite, ankerite, dolomite, calcite, manganoan carbonates). Pyrite generally is subordinate. Goethite, martite, maghemite, and certain varieties of hematite are products of surficial alteration. Metamorphosed BIF contains a very large number of silicate minerals (Klein, in James and Sims, 1973).

James (1954) defined iron-formation facies based on the predominant iron mineral: hematite, magnetite, silicate, carbonate, and sulfide facies. He, Goodwin (see references in James and Sims, 1973), and others demonstrated lateral transitions between these facies. In addition Gross (1965) distinguished two types of BIF on the basis of sedimentary textures and structures: Algoma-type BIF has thin, parallel laminae, whereas Lake Superior-type BIF is thin- to thick-bedded, characteristically has oolitic or intraclastic (= granular) textures, and commonly has sedimentary structures indicating shallow-water environments (dune crossbedding, channels, etc.).

Depositional Environment

Cayeux (1911) pointed out that the sedimentary textures of BIF and Minette-type iron ores (and thus also of limestones) are very similar. This similarity permitted interpretation of depositional processes and depositional environment of BIF by comparison with Holocene limestones (Chauvel and Dimroth, 1974).

Analysis of the crystallization textures of BIF (Mukhopadhyay and Chanda, 1972; and others) has shown that all minerals now present in BIF formed during diagenesis, and that complex redox reactions play a considerable role in diagenesis of BIF.

Cayeux (1911) saw in BIF silicified Minette-type iron ore, i.e., limestone replaced by Fe and SiO_2; this hypothesis has not been widely accepted. Most authors (see Van Hise and Leith, 1911; Gruner, 1922; James, 1954) believe cherty iron formations are siliceous-ferriferous precipitates from an aqueous environment. James (1954, 1966) and Gross (1965) suggested that the mineral facies of BIF are depositional and reflect oxidation potential and acidity of the precipitating aqueous medium. This concept has served as basis for many paleogeographic interpretations (see Goodwin, in James and Sims, 1973). The mineral facies present, however, are diagenetic at least in certain cases, and then are determined by pH and Eh during diagenesis (Dimroth, 1976).

Lepp and Goldich (1964) proposed deposition of BIF under reducing atmospheric conditions because large amounts of iron ore are not easily transported in the presence of O_2. General validity cannot be claimed for this hypothesis, however, because a number of very large BIFs have been deposited in oxidizing sedimentary environments (Dimroth, 1976). Cloud suggested biochemical precipitation of iron, and LaBerge of silica; Eugster and Chou proposed a magadiite-type precursor for chert in BIF (see references in James and Sims, 1973).

ERICH DIMROTH

References

Brandt, R. T.; Gross, G. A.; Gruss, H.; Semenenko, N. P.; and Dorr, J. V. N., II, 1972. Problems of nomenclature for banded ferruginous-cherty sedimentary rocks and their metamorphic equivalents, *Econ. Geol.*, **67**, 682-684.
Cayeux, L., 1911. Comparaisons entre les minéraux de fer huroniens des Etats-Unis et les minéraux de fer oolithique de France, *Compt. Rend. Acad. Sci.*, **153**, 1188-1190.
Chauvel, J.-J., and Dimroth, E., 1974. Facies types and depositional environment of the Sokoman

Iron Formation, central Labrador Trough, Quebec, Canada, *J. Sed. Petrology,* **44,** 299-327.
Cloud, P., 1973. Paleoecological significance of the banded iron-formation, *Econ. Geol.,* **68,** 1135-1143.
Dimroth, E., 1976. Aspects of sedimentary petrology of cherty iron-formation, in K. H. Wolf, ed., *Handbook of Strata-Bound and Stratiform Ore Deposits,* vol. 7. Amsterdam: Elsevier, 203-254.
Drever, J. I., 1974. Geochemical model for the origin of Precambrian banded iron formations, *Geol. Soc. Am. Bull.,* **85,** 1099-1106.
Floran, R. J., and Papike, J. J., 1975. Petrology of low-grade rocks of the Gunflint Iron-Formation, Ontario-Minnesota, *Geol. Soc. Am. Bull.,* **86,** 1169-1190.
Gross, G. A., 1965. Geology of iron deposits in Canada. Vol. 1: general geology and evaluation of iron deposits, *Geol. Surv. Canada Econ. Geol. Rept. 22(1),* 1-181.
Gruner, I. W., 1922. Organic matter and the origin of the Biwabik Iron-bearing Formation, *Econ. Geol.,* **17,** 415.
James, H. L., 1954. Sedimentary facies of iron formation, *Econ Geol.,* **49,** 235-293.
James, H. L., 1966. Chemistry of iron-rich sedimentary rocks, *U.S. Geol. Surv. Prof. Pap. 440-W,* 1-61.
James, H. L., and Sims, P. K. 1973. Precambrian iron formations of the world, *Econ. Geol.,* **68,** 913-1179.
Lepp, H., ed., 1975. *Geochemistry of Iron.* Stroudsburg, Pa.: Dowden, Hutchinson & Ross, 464p.
Lepp, H., and Goldich, S. S., 1964. The origin of Precambrian iron-formation, *Econ. Geol.,* **59,** 1025-1060.
Mukhopadhyay, A., and Chanda, S. K., 1972. Silica diagenesis in the banded hematitic jasper and bedded chert associated with the Iron Ore Group of Jamda-Koira Valley, Orissa, India, *Sed. Geol.,* **8,** 113-135.
Pichamuthu, C. S., 1974. On the banded iron formations of Precambrian age in India, *J. Geol. Soc. India,* **15,** 1-30.
Van Hise, C. R., and Leith, C. K., 1911. The geology of the Lake Superior region, *U.S. Geol. Surv. Mon. 52,* 641p.
Young, G. M., ed., 1973. Huronian stratigraphy and sedimentation, *Geol. Assoc. Can. Spec. Pap. 12,* 271p.

Cross-references: *Chert and Flint; Ironstone; Silica in Sediments.* Vol. IVA: *Iron; Economic Deposits.* Vol. IVB: *Hematite; Jasper.*

CLASTIC DIKES

Clastic dikes (and sills) are tabular, intrusive bodies resembling igneous dikes and sills in appearance and field relations. Unlike molten igneous intrusions, however, clastic intrusions result from injection of mobilized sediments into fracture zones (Smyers and Peterson, 1971). They are particularly well preserved in clay-rich rocks, forming uninterrupted walls or wall-like ramparts that may be followed by means of surface geological mapping or aerial photography. Pavlow (1896) called clastic dikes neptunic, meaning "filled from above" as opposed to the plutonic, magmatic dikes; but his term does not include all modes of origin of clastic dikes. Gareckij (1956) classified clastic dikes as *injection clastic dikes,* if formed by injection of sedimentary material from below (Fig. 1). and *neptunian,* if formed by injection

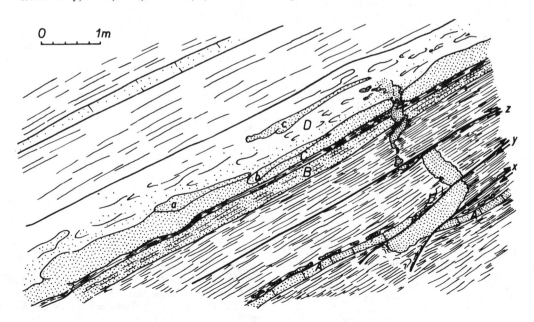

FIGURE 1. A sand injection has penetrated from the bottom into the mobile slump and dispersed; the original structure has been preserved in the laminated B bed (after Dzulynski and Radomski, 1956).

from above (Fig. 2). The rich geological literature on the genesis of clastic dikes evokes the supposition that they are found only in clastic sedimentary formations; yet they are not excluded from volcanic formations, nor from crystalline or granite massifs.

Clastic dikes are described by their composition, e.g., sandstone, conglomerate, claystone, and limestone dikes. The majority consist of fine- to medium-grained sandstones, well cemented by calcite. The sandstones are usually structureless, although recent research by Laubscher (1961) and Peterson (1968) pointed out parallel laminae and laminae resembling crossbeds in sandstone dikes. They supposed that the intruding sands were abruptly turbulent or experienced laminar flow. The source beds from which the intrusions arise must be loosely packed and particularly susceptible to spontaneous liquefaction; when pressured, they flow into fissures (see *Water Escape Structures*).

Clastic dikes arise in all stages of sedimentary basin formation and sensitively record the dynamic evolution of the basin. Detailed analysis of clastic dikes may contribute not only to the investigation of hydroplastic deformation but also to seismic and paleotectonic studies (Marschalko, 1965). Dionne and Shilts (1974) review the occurrences of clastic dikes in consolidated and unconsolidated sediments.

ROBERT MARSCHALKO

References

Diller, J. S., 1890. Sandstone dikes, *Geol. Soc. Am. Bull.*, **1**, 411-442.

Dionne, J. -C., and Shilts, W. W., 1974. A Pleistocene clastic dike, Upper Chaudière Valley, Québec, *Can. J. Earth Sci.*, **11**, 1594-1605.

Dzulynski, S., and Radomski, A., 1956. Clastic dikes in the Carpathian flysch, *Soc. Geol. Pologne, Ann.*, **26**, 255-264.

Gareckij, R. G., 1956. Clastic dikes (in Russian), *Izvest. Akad. Nauk SSSR, Ser. Geol.*, **3**, 81-103.

Laubscher, H. P., 1961. Die Mobilisierung klastischer Massen. I. Die Sandsteingänge in der San Antonio Formation (Senon) des Rio Querecual, Ostvenezuela, 11. Teil: Die Mobilisierung Klastischer Massen und ihre geologische Dokumentation, *Eclogae Geol. Helv.*, **54**(2), 283-334.

Lewis, D. W., 1973. Polyphase limestone dikes in the Oamoru region, *J. Sed. Petrology*, **43**, 1031-1045.

Marschalko, R., 1965. Clastic dikes and their relations to synsedimentary movements (Flysch of Central Carpathians), *Geol. Práce, Zprávy*, **36**, 139-148.

Pavlow, A. P., 1896. On dikes of Oligocene sandstone in the Neocomian clays of the district of Alatyr in Russia, *Geol. Mag.*, **3**, 49-53.

Peterson, G. L., 1968. Flow structures in sandstone dikes, *Sed. Geol.*, **2**, 177-190.

Smyers, N. B., and Peterson, G. L., 1971. Sandstone dikes and sills in the Moreno Shale, Panoche Hills, California, *Geol. Soc. Am. Bull.*, **82**, 3201-3208.

Truswell, J. F., 1972. Sandstone sheets and related intrusions from Coffee Bay, Transkei, South Africa, *J. Sed. Petrology*, **42**, 578-583.

Verzilin, N. N., 1963. Melovye otlozenija severa ferganskoj vpadiny i icgneftenosnost, *Trudy Leningrad. Obshch. Estest.*, 70(2), 219.

Cross-references: *Intraformational Disturbances; Penecontemporaneous Deformation of Sediments; Sedimentary Structures; Water Escape Structures.*

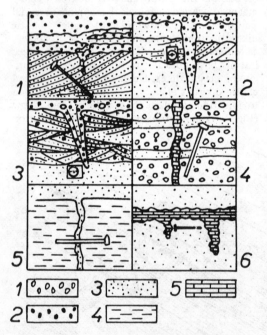

FIGURE 2. Sandy conglomerate and limestone dikes of wedge form. In majority of cases, the material composing the dikes originates in the immediately overlying bed (after Verzilin, 1963). The dikes are perpendicular or slightly inclined with respect to the surface beds. Some dikes are still in the initial stage (1). The limestone dikes in variegated claystones (6) are connected with the slumps by means of transition. (1) Conglomerates; (2) fine-grained conglomerates; (3) sandstones and siltstones; (4) many-colored claystones; (5) limestones.

CLASTIC SEDIMENTS AND ROCKS

Clastic sediments and rocks, are, by definition, essentially composed of fragments of preexisting materials that have been eroded from their source, transported from that source in particulate form, and ultimately deposited by mechanical agents. A *clast* is an individual particulate constituent of a detrital sediment or sedimentary rock, produced by the disintegration of a larger mass. Thus, any material, irrespective of its chemical composition, may contribute detritus to a clastic sediment in appropriate circumstances. Clastic sediments are formed principally as a result of the operation of physical (e.g., hydrodynamic) laws, in contrast with nonclastic sediments whose formation is controlled by chemical or physicochemical laws. It must be borne in mind, however, that the conversion to a rock of a sediment deposited by mechanical agencies is usually the result of the operation of chemical processes (see *Clastic Sediments—Lithification and Diagenesis*).

The properties of a terrigenous sedimentary rock are texture, structure, and mineralogical composition. These properties constitute evidence of the sedimentary processes and depositional environments under which a sedimentary rock has been formed. It is very important, therefore, to analyze them properly so that the history of a clastic sediment or rock can be deduced.

Texture

Classification by Particle Size. The classification of clastic sediments must, as a result of definition, be based initially on a physical or mechanical property of the detritus and only secondarily on its chemical composition. Thus, the classification of clastic sediments is based on the size of the component particles. Although the size of an irregularly shaped rock or mineral fragment can be defined and measured in many ways, it is common practice to define the size in terms of the diameter of the particle. A sedimentary fragment is generally of irregular shape and can therefore have an almost infinite number of diameters; it is therefore tacitly assumed that the particles are spherical or ellipsoidal. The sphere has a unique diameter, and the ellipsoid can be defined in terms of three unique "diameters," of which the intermediate one is generally adopted as defining the size of the particle. Although such measures of size are only approximations to the true size, they can be readily determined by sieving or direct microscopic measurement. For very small particles, it is more common to determine size on the basis of rate of settling of the particles in a fluid.

Range of Particle Size. The size of sedimentary particles shows a complete gradation from large boulders, several m across, weighing several thousand tons, to submicroscopic particles, $<1\mu$ in diameter. Thus, any classification on the basis of size must set arbitrary limits to the subdivisions of the scale of sizes. The scales proposed by several authors and the descriptive terms given to the various particle sizes are shown in Fig. 1. The characteristics of such "grade scales" have been discussed by Krumbein and Pettijohn (1938). Krumbein (1934) applied a logarithmic transformation equation to the Wentworth grade scale (q.v.) and obtained a phi (Φ) scale (q.v.) in which integers for the class limits increase with decreasing grain size. This scale has found almost universal favor for the quantitative description of clastic sediments.

Common to many grade scales are the terms *gravel*, *silt*, and *clay*, whose common usage has tended to imply a compositional as well as a size meaning. Thus Grabau (1904, 1913) proposed the terms *rudyte*, *arenyte*, and *lutyte* for particle sizes of gravel, sand, and clay, and added prefixes to further describe the rock.

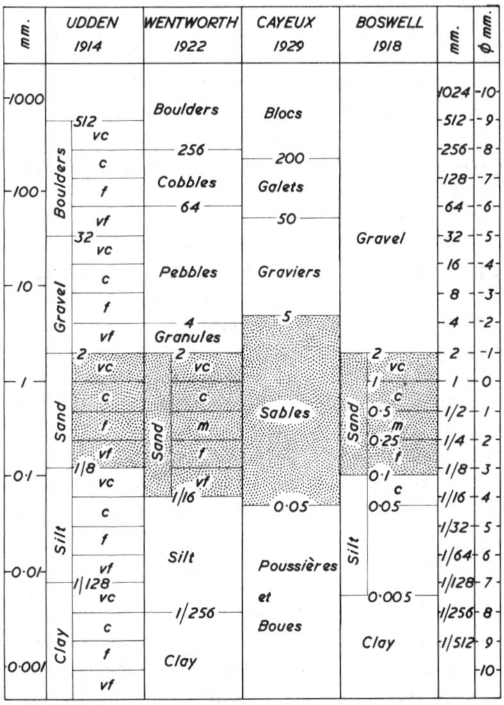

FIGURE 1. The classification of clastic sediments on the basis of particle size, according to Udden (1914), Wentworth (1922), Cayeux (1929), and P. G. H. Boswell (1918). The Udden scale is a geometric scale based on a factor of 2 in which each class is identified. The Wentworth scale is similar, but several classes are grouped together to form a single grade. Cayeux's scale is nonregular, but appears to be based on a multiple of 10 (i.e., a logarithmic scale). Boswell's scale is a mixture of regular and nonregular scales. Logarithmic, geometric, and phi scales are given at the margins. The diagram illustrates the variety of meaning given to the size terms (e.g., sand, shown stippled) and the need for specification of the scale used and for standardization.

127

Strictly, the terms were applied only to consolidated rocks, but current usage (spelled *rudite, arenite, lutite*) also applies them to unconsolidated sediments. Some authors have used the terms psephite, psammite, and pelite as equivalents, but, following Tyrrell (1921), it is modern practice to use these terms only for the metamorphic equivalents.

Rocks having particles of gravel, sand, or clay size are spoken of as rudaceous, arenaceous, or lutaceous rocks. (Many authors prefer "argillaceous" to "lutaceous".) Of the many prefixes proposed by Grabau, few have survived, though calc- or calci-, e.g., calcarenite, for rocks of calcium carbonate composition, are firmly entrenched in the literature.

Particle Shape. The component particles of clastic sediments and rocks have two other important physical attributes—*shape* (or sphericity) and the *roundness* of the particle (see *Roundness and Sphericity*). Both sphericity and roundness can be expressed in quantitative terms, although it must be admitted that the methods of measuring them are subject to a number of criticisms. More commonly, sediments are described and classified on the basis of qualitative inspection. For example, a distinction is drawn within the rudaceous rocks between conglomerates (q.v.) and breccias (q.v.) according to the presence or absence of evidence of abrasion of the larger fragments.

Sorting, Fabric, and Packing. The attributes of individual particles must not be allowed to obscure the fact that sediments and rocks are actually aggregates of particles; therefore, variations in size of particles within a sediment, the position of those particles relative to each other and the orientations of those particles are attributes of the rock as a whole and can lead to an understanding of the nature and behavior of sediments. These three attributes are referred to as *sorting* (q.v.), *packing*, and *fabric* (see *Fabric, Sedimentary*).

Porosity and Permeability. Both packing and fabric are closely linked with the porosity and permeability of sedimentary rocks (see *Fabric, Sedimentary*). Packing is principally concerned with the dimensions of the pore spaces in a sediment (i.e., the porosity). Interconnection between pore spaces, provided they are not too small, is essential if the rock is to be permeable. The fabric has a marked effect on permeability because the orientation of grains in a particular direction enables easier fluid movement in that direction compared with other directions. Recent development of the scanning electron microscope with its high depth of field has greatly improved the study of pore geometry of sandstone and other clastic sedimentary rocks (Weinbrandt and Fatt, 1969; Pittman and Duschatko, 1970; and Kieke and Hartmann, 1973).

Structures

The structures of clastic sediments and rocks depend more upon relations between sedimentary aggregates than on the grain-to-grain relation that controls texture; therefore, most sedimentary structures are larger features and can be studied only on outcrops and in cores. The usefulness of sedimentary structures in interpreting ancient clastic deposits has recently been greatly enhanced by studies made on modern sediments and by laboratory experiments.

The study of sedimentary structures (q.v.) has generated a very large literature. Only the most recent papers and earlier classic studies which summarize most of the earlier literature are cited. The reader is referred to such compilations as that by Shrock (1948) on the utility of structures in identifying the proper stratigraphic sequence, and to those of Khabakov (1962), Pettijohn and Potter (1964), Gubler et al. (1966), and Conybeare and Crook (1968), which deal with all the types of structures. There are four broad types of sedimentary structures found in clastic (and sometimes other) sedimentary rocks (Pettijohn et al., 1972, p. 103): (1) current structures, (2) deformed structures, (3) biogenic structures, and (4) chemical structures.

Mineralogical and Chemical Composition

The mineralogical and chemical composition of clastic rocks varies between extreme simplicity (as in many quartz arenites) to considerable complexity (as in graywackes or wackes). The principal factors controlling the composition of a sediment are the composition of the source (or sources) of the detritus, the rigor of erosion and transportation, the nature of the depositional environment, and postdepositional or diagenetic changes.

Quartz is the commonest mineral in coarser grained clastic sediments, because of its high chemical stability and resistance to abrasion. Similarly, clay minerals are most abundant in fine-grained clastics. Feldspar is less resistant to mechanical abrasion than quartz and is much less stable chemically. Feldspar can accumulate in considerable quantity provided it is transported quickly and buried rapidly (see *Arkose*). Rock (lithic) fragments are common in coarse sediments, but tiny particles of very fine-grained rocks may persist in many sediments. Many other minerals may occur in clastic sediments; there is usually a small proportion (<0.02%) of these accessory minerals even in very pure sands (Humphries, 1961). In specialized circumstances, the proportion of an accessory mineral might rise to predominant proportions, as in monazite sands, glauconite sands, and many placer deposits (see *Black Sands*).

The chemical composition of clastic rocks is closely related to grain size, especially in sand-sized and smaller material. In these sediments, silica is usually the predominant component, either as free quartz or in clay minerals. In sands, the silica content may be close to 100% but more often is of the order of 70–80%; but in fine clay, it may fall to around 40% (see *Shale*). Aluminum is usually the second most important component, averaging about 10% in many fine sands and reaching almost 20% in fine clays. Other oxides, such as TiO_2, MgO, CaO, and iron oxides, and alkalies rarely exceed 7%. For a further discussion of the chemical composition of clastic sediments see, for example, Pettijohn (1963, 1975).

Classification of Clastic Sedimentary Rocks

The classification of clastic sediments on the basis of mineralogical composition is based on the end-member concept. Any clastic sedimentary rock is composed of two, three, four, or more major components, of which one or more may be of "primary" (clastic) origin and the remainder are of "secondary" (chemical) origin. The composition of a rock can then be expressed in terms of the proportion of these end members present. Although two- and four-component systems have been used, the three-component system is most common. Such systems are often expressed in the form of ternary or triangular diagrams. It is also common practice to use ternary diagrams for the description of unconsolidated sediments, taking for the three components gravel, sand, and clay, or sand, silt, and clay. Subdivisions of such a triangular diagram can be used as a basis for erecting a nomenclature for the sediments described by the system. An example each of two, three, and four component systems is shown in Figs. 2, 3, 4. It should be noted that a four-component system is represented by a tetrahedral model and cannot be adequately represented in two dimensions.

A single, unified classification of clastic sediments based on the end-member concept does not appear to exist, and it is common practice to make a threefold division (based on grain size) into the three groups: rudites, arenites, and argillites (lutites). Classification of the rudites is generally based on the roundness and shape of the pebbles and on the diversity or uniformity of rock types represented among the pebbles. The mode of origin of rudites is

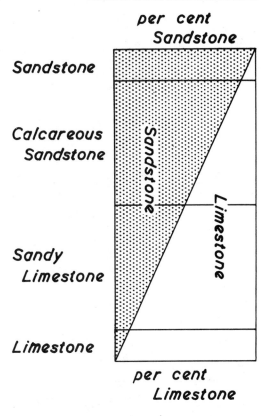

FIGURE 2. Two-component (two end-member) classification of sandstone-limestone mixtures. The sandstone component is the detrital phase and the limestone is regarded as the matrix.

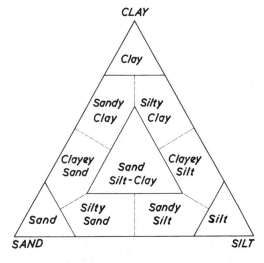

FIGURE 3. Three-component (three end-member) classification of sand-silt-clay mixtures (after Shepard, 1954).

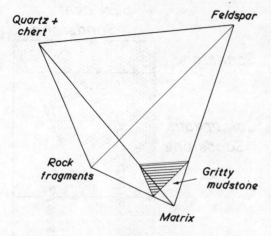

FIGURE 4. Four-component (four end-member) classification of clastic sediments based on a tetrahedron. The faces may be subdivided in the manner of Fig. 5, but ideally the tetrahedron should be subdivided into solid segments, one of which is shown here at the matrix apex.

also significant in their classification (see *Conglomerates* and *Breccias*).

The classification of the arenite group has led to considerable discussion, but no universally acceptable scheme (or terminology) has evolved (see *Sands and Sandstones*). The end members are often composite, e.g., quartz and chert, micas and chlorite, feldspar and kaolin. The reasoning is that quartz and chert are stable components; mica and chlorite are taken to include the clay minerals and form the matrix; and feldspar and its alteration product kaolin comprise a moderate proportion of many arenites. Cementing materials such as carbonates or iron compounds, provided they are present in only small proportion, are usually ignored in such classifications. The application of a classification based on these end members gives rise to three main groups of arenites, viz., quartz arenites, graywackes or wackes, and arkoses, which may themselves be further subdivided (Fig. 5). If detrital carbonate fragments form a large part of the rock, the classification may involve a four-member system and a tetrahedral model. For the classification of carbonate rocks containing a large proportion of detrital carbonate, the reader is referred to Folk (1959), Krumbein and Sloss (1963), or Okada (1971).

The classification of argillites on a mineralogical basis is virtually nonexistent. Commonly, this group is subdivided using clay as one end member, with carbonate, silica and organic matter as the other (Pettijohn, 1957). Picard (1971) suggested that the classification of fine-grained sedimentary rocks and sediments should be based on combination of several criteria such as texture, mineral composition, chemical composition, fissility, color, tectonic association, and degree of metamorphism.

A classification of water-laid clastic sediments (the basis of mode of deposition) was proposed by Natland (1967). Natland stated that there are four principal sedimentary processes at work in both shallow and deep water—traction flow, turbidity current, gravity flow, and slow accumulation on the deep-sea floor. The sediments produced by these processes are designated accordingly as tractionites, turbidites, gravitites, and hemipelagites.

The classification of clastic sediment may lead to its neat pigeonholing, but to build up a complete picture of the history of a rock, classification is not enough. It is necessary to consider the processes that led to the initial breakdown of the parent rock, the nature of the transporting agent, the conditions of the depositional environment, the processes of diagenesis that converted the newly formed sediment into a rock, and possibly the deformation of the rocks. The final character of the rock will depend largely on the length of time and the vigor with which various agencies acted, thus affecting the *maturity* (q.v.) of the rock (Folk, 1951). It is also important to consider the nature of the associated rocks, because associations are indicative of different sedimentary and tectonic environments.

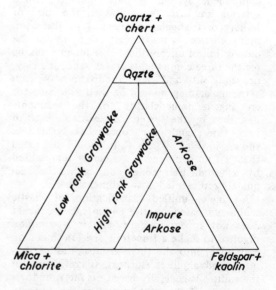

FIGURE 5. Triangular classification of sandstones (after Krynine, 1948). Oqzte = orthoquartzite.

Sedimentary Environments

Clastic sedimentary environments may be classified by any of several criteria, namely, (1) transporting agents, (2) water depth, and (3) landforms, depending upon the features to be emphasized. The ancient sedimentary environments, like the recent sedimentary environments, can generally be divided into subenvironments. For example, a linear shoreline is often composed of a complex of barrier islands, lagoons, and tidal flats lying between alluvial and shelf major environments (see *Sedimentary Environments*). The following genetic classification of depositional environments where clastic sediments may be deposited is based mainly on the general geographic setting and the processes that are dominant: (1) *Nonmarine* (continental)—fluvial (river), lacustrine (lake), desert, glacial. (2) *Transitional* (shorelines)—beaches and barriers, tidal flat, lagoon, estuary, delta (marine). (3) *Marine* —continental shelf (neritic zone), carbonate shelf and reef, continental slope (bathyal zone), deep-sea basin (abyssal zone).

Total Volume and Relative Abundance of Clastic Sediments

Various attempts have been made to assess the total amount of sediment in the known lithosphere. By volume, sedimentary rocks comprise <5%, although they cover some 75% of the present land surface. By far the most abundant are the shales (or argillites), which are estimated to make up perhaps 80% of the total of known sediments. Sandstones rank second with some 15%, while (usually nonclastic) limestones occupy less than 5% of the total volume.

In the examination of a specific geologic section, the ratio of the thickness of clastic sediments to the total nonclastic thickness is termed the *clastic ratio* (Krumbein, 1948). Change in the clastic ratio with time (vertically at any specific site) will reflect the variation in regional tectonics, just as it does from site to site within a depositional basin. Although it is not invariably true, the higher clastic ratios observed along the margins of an ancient basin have been taken to indicate proximity to the strand line. Interestingly, this interpretation, which holds true in ancient sediments, is at variance with modern marine sediments, possibly a result of the changing shoreline caused by Pleistocene glaciation (Shepard, 1959).

DEREK W. HUMPHRIES

References

Cayeux, L., 1929. *Les Roches Sédimentaires de France: Roches Silicieuses.* Paris: Impr. Nationale, 774p.

Conybeare, C. E. B., and Crook, K. A. W., 1968. Manual of sedimentary structures, *Austral. Bur. Min. Research, Geol. Geophys. Bull. 102,* 327p.

Fisher, W. L., and Brown, L. F., Jr., 1972. *Clastic Depositional Systems—a Genetic Approach to Facies Analysis, Annotated Outline and Bibliography.* Austin: Bur. Econ. Geol., Univ. Texas, 211p.

Folk, R. L., 1951. Stages of external maturity in sedimentary rocks, *J. Sed. Petrology,* 21, 127-130.

Folk, R. L., 1959. Practical petrographic classification of limestones, *Bull. Am. Assoc. Petrol. Geologists,* 43, 1-38.

Grabau, A. W., 1904. On the classification of sedimentary rocks, *Am. Geologist,* 33, 228-247.

Grabau, A. W., 1913. *Principles of Stratigraphy.* New York: A. G. Seiler, 1185p.

Gubler, Y.; Bugnicourt, D.; Faber, J.; Kübler, B.; and Nyssen, R., 1966. *Essai de Nomenclature et Caracterisation des Principales Structures Sédimentaires.* Paris: Editions Technip, 291p.

Humphries, D. W., 1961. The Upper Cretaceous white sandstone of Loch Aline, Argyll, Scotland, *Proc. Yorksh. Geol. Soc.,* 33, 47-76.

Khabakov, A. V., ed., 1962. *Atlas Tekstur i Struktur Osadochyhk Gornykh Porod* (an Atlas of Textures and Structures of Sedimentary Rocks). Moscow: Vsegei, 578p.

Kieke, E. M., and Hartmann, D. J., 1973. Scanning electron microscope application to formation evaluation, *Trans. Gulf Coast Assoc. Geol. Soc.,* 10, 60-67.

Krumbein, W. C., 1934. Size frequency distributions of sediments, *J. Sed. Petrology,* 4, 65-77.

Krumbein, W. C., 1948. Lithofacies, maps and regional sedimentation—stratigraphic analysis, *Bull. Am. Assoc. Petrol. Geologists,* 32, 1909-1923.

Krumbein, W. C., and Pettijohn, F. J., 1938. *Manual of Sedimentary Petrography.* New York: Appleton-Century-Crofts, 549p.

Krumbein, W. C., and Sloss, L. L., 1963. *Stratigraphy and Sedimentation.* San Francisco: Freeman, 660p.

Krynine, P. D., 1948. The megascopic study and field classification of sedimentary rocks, *J. Geol.,* 56, 130-165.

Lützner, H.; Falk, F.; Ellenberg, J.; and Grumbt, E., 1974. Tabellarische Documentation klastischer Sedimente, *Akad. Wiss. Veröff. Zentral. Phys. Erde,* 20, 153p.

Natland, M. L., 1967. New classification of water-laid clastic sediments (abstr.), *Bull. Am. Assoc. Petrol. Geologists,* 51, 476.

Okada, H., 1971. Classification of sandstone: analysis and proposal, *J. Geol.,* 79, 509-525.

Pettijohn, F. J., 1957. *Sedimentary Rocks,* 2nd ed. New York: Harper & Row, 718p.

Pettijohn, F. J., 1963. Chemical composition of sandstones, excluding carbonate and volcanic sands, *U.S. Geol. Surv. Prof. Pap. 440-S,* 21p.

Pettijohn, F. J., 1975. *Sedimentary Rocks,* 3rd ed. New York: Harper & Row, 628p.

Pettijohn, F. J., and Potter, P. E., 1964. *Atlas and*

Glossary of Primary Sedimentary Structures. New York: Springer-Verlag, 370p.

Pettijohn, F. J.; Potter, P. E.; and Siever, R., 1972. *Sand and Sandstone.* New York: Springer-Verlag, 618p.

Picard, M. D., 1971. Classification of fine-grained sedimentary rocks, *J. Sed. Petrology,* **41,** 179-195.

Pittman, E. D., and Duschatko, R. W., 1970. Use of pore casts and scanning electron microscope to study pore geometry, *J. Sed. Petrology,* **40,** 1153-1157.

Shepard, F. P., 1954. Nomenclature based on sand-silt-clay ratios, *J. Sed. Petrology,* **24,** 151-158.

Shepard, F. P., 1959. *Submarine Geology.* New York: Harper & Row, 557p.

Shrock, R. R., 1948. *Sequence in Layered Rocks.* New York: McGraw-Hill, 507p.

Tyrrell, G. W., 1921. Some points in petrographic nomenclature, *Geol. Mag.,* **58,** 501-502.

Udden, J. A., 1914. Mechanical composition of clastic sediments, *Geol. Soc. Am. Bull.,* **25,** 655-744.

Weinbrandt, R. M., and Fatt, I., 1969. A scanning electron microscope study of the pore structure of sandstone, *J. Petrol. Technol.,* **21,** 543-548.

Wentworth, C. K., 1922. A scale of grade and class terms for clastic sediments, *J. Geol.,* **30,** 377-392.

Cross-references: *Alluvial-Fan Sediments; Alluvium; Argillaceous Rocks; Arkose, Subarkose; Bedding Genesis; Breccias, Sedimentary; Clastic Sediments—Lithification and Diagenesis; Clay as a Sediment; Conglomerates; Continental Sedimentation; Deltaic Sediments; Desert Sedimentary Environments; Diamictite; Eolian Sedimentation; Fabric, Sedimentary; Fluvial Sediments; Glacigene Sediments; Grain-Size Studies; Graywacke; Lacustrine Sedimentation; Littoral Sedimentation; Mass-Wasting Deposits; Pelagic Sedimentation, Pelagic Sediments; Provenance; Pyroclastic Sediment; Roundness and Sphericity; Sands and Sandstones; Sedimentary Petrography; Sedimentary Petrology; Sedimentary Rocks; Sedimentary Structures; Sedimentation-Tectonic Controls; Sediment Parameters; Shale; Sorting; Volcaniclastic Sediments and Rocks.*

CLASTIC SEDIMENTS—LITHIFICATION AND DIAGENESIS

Although there are numerous varieties of clastic rocks, an adequate review of their lithification and diagenesis is possible by considering only sandstones and shales. Most shales can be classified as siltstones; with respect to diagenesis and grain size, silt ($2-63\mu$) lies between sand and clay. Diagenesis of conglomerates does not differ markedly from that of sandstone.

Diagenesis (q.v.) implies equilibration with increasing pressure (by mechanical and chemical compaction) and temperature, as well as with changing interstitial fluids (by cementation and replacement) during subsidence and burial of sediments. Compaction (q.v.) and cementation (q.v.) are both involved in lithification, which occurs during early subsidence, whereas replacement is more common at depth, where the rock is already lithified.

Sandstone

Compaction. Compaction, the reduction of the thickness of beds, results in a decrease in porosity (Fig. 1). It is subdivided into mechanical compaction due to the rearrangement of particles and chemical compaction due mainly to pressure solution (q.v.) and reprecipitation

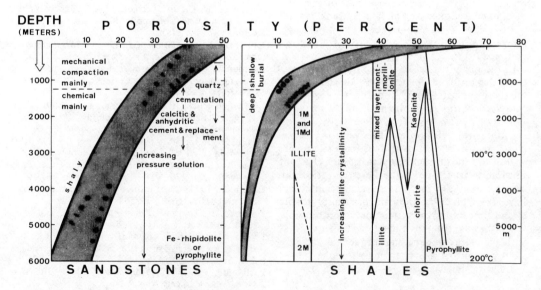

FIGURE 1. Porosity decrease of sandstones and shales with increasing maximum depth of burial, covering the whole range of diagenesis. For details, see text.

in the remaining voids. Mechanical compaction is more or less restricted to shallow burial depth and is more effective in shales than in sandstones (Fig. 1). Pressure solution, the main agent of chemical compaction, affects in particular quartz and feldspar grains, especially where they are in contact with detrital mica or clay seams (both of which are hardly affected; Fig. 2a), and is therefore more obvious in "dirty" sandstones and siltstones than in clean quartzose sandstones and claystones.

Mechanical compaction is more effective in coarse- than in fine-grained sands, because of the generally higher grain roundness and the higher pressure per grain contact in coarse sands. Chemical compaction and cementation are more effective in fine-grained sands, because their higher specific surface area provides nuclei for the main cement, i.e., silica. Accordingly the porosity decrease in coarse sands is rapid soon after burial and slow at deep burial (providing relatively high porosity in deep-seated coarse sandstones), whereas for fine-grained sands the opposite is true.

Chemical compaction and cementation are time-consuming processes. For this reason, the porosity of older sandstones is lower than that of younger sandstones at the same depth. The effective depth is the maximum depth of burial ever reached in the history of the deposit; later uplift does not increase the porosity (Füchtbauer and Müller, 1974).

Cementation. Cementation, i.e., chemical precipitation in voids, is more obvious in coarse clastics (conglomerates, sandstones, coarse siltstones) than in shales. As in limestones, the fabric of cement in coarse clastics depends upon the degree of supersaturation—the fibrous cement representing higher supersaturation than the isometric cement. This applies especially to silica (Figs. 2b, 2c): At high supersaturation, soluble types of silica including opal and chalcedony precipitate, forming tiny crusts and fibrous mosaics (Fig. 2b), whereas at lower supersaturation, the less soluble quartz occurs, forming homoaxial overgrowths on the detrital quartz grains (Fig. 2c). Supersaturation of silica is a function of elevated pH, and of availability of silica or soluble silicates. Other possible cements include carbonates, sulfates (e.g., anhydrite, barite, celestite), halite, zeolites, and clay minerals. The time sequence of cement precipitation can be investigated by different methods:

1. The *minus-cement porosity* (Rosenfeld, 1949), i.e., the porosity that would occur if the cement were dissolved, indicates the porosity of the sand prior to cementation. In Fig. 3a, the minus-cement porosity is 33%, indicating an early cementation; whereas the lower minus-cement porosity of 16% in Fig. 3b points to a cementation that began at greater burial depth.
2. The *contact strength* (Füchtbauer and Müller, 1974) indicates the prevailing type of contacts between detrital grains using the formula $(a + 2b)/(a + b)$,

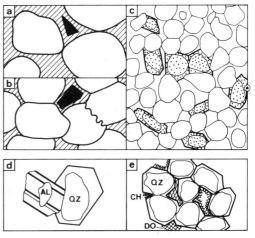

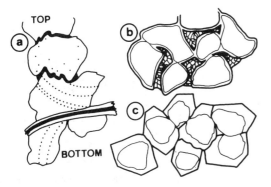

FIGURE 2. (a) Pressure solution of quartz grains bordering mica or clay seams. Rows of bubbles in the grains indicate that they are detrital, not authigenic. (b) Rapid cementation of quartz grains near the land surface by (1) opal seams and subsequently by (2) chalcedony filling the rest of the voids of a Tertiary sand. (c) Slow homoaxial cementation of quartz grains in the subsurface by quartz—the most frequent type of cement.

FIGURE 3. (a) and (b) demonstrate different minus-cement porosities (a: 33%, b: 16%, see text), as well as different contact strengths (a: only point contacts, b: only narrow contacts including planar, concave-convex, and sutured grain contacts). Hatched areas = cement; black = voids. (c) Sketch of a typical thin-section of a Triassic Buntsandstein sandstone of northern Germany. Dots are feldspars (primary albite and K feldspar) with albite overgrowths. White are quartz grains with quartz overgrowths, the contacts of which are not shown in the figure. No porosity is present. (d) and (e) are qualitative criteria of cement sequences. AL = Albite, QZ = Quartz, CH = Chlorite, DO = Dolomite (see text).

with a = number of point contacts and b = number of narrow contacts (Figs. 3a, 3b). In Fig. 3c, the contact strength of grains separated by albite cement (dotted) is 1.2 [10 point contacts, 3 narrow contacts: $(10 + 2 \times 3)/(10 + 3) = 1.2$], whereas the contact strength of grains separated by quartz cement (white) is 1.6 (13 point contacts, 19 narrow contacts). The contacts become closer with increasing burial depth, before the main cementation occurs; thus albite cementation must have occurred earlier than quartz cementation in this example.

3. Qualitative fabric criteria of the succession of cements include the relationship shown in Fig. 3d, in which evidently the albite overgrowth occurred prior to the quartz overgrowth, and the relationship shown in Fig. 3e: If different cement minerals occur in a pore space, the earlier cement (in Fig. 3e quartz) is found adjacent to the pore walls, the later cements (in Fig. 3e: chlorite and dolomite) in the pore center.

The source of quartz cement during early diagenesis is amorphous silica, e.g., diatom or radiolarian skeletons or glass shards; whereas during late diagenesis, quartz dissolved by pressure solution in adjacent siltstones, or even in the sandstones, is the main source of quartz overgrowths. Feldspar cements and clay mineral cements may be derived from the dissolution of unstable clay minerals or zeolites, possibly also of feldspars, but this hypothesis is still open to question. Carbonate cements, which generally are richer in iron when formed later, may originate from dissolved skeletons. Anhydrite and halite cement occur in the neighborhood of evaporitic rocks. In all cases, the compaction water which percolates upward through the rocks plays a major role as a transport mechanism of ions.

Replacements. Replacements are *pseudomorphs* if they have the shape of the replaced mineral. *Alterations* use part of the crystals of the original mineral (e.g., K-feldspar → albite, illite → muscovite). Most replacements occur at high burial depth—e.g., the replacement of quartz by calcite under conditions of alkaline pH—although some replacements—e.g., kaolinite forming from feldspar at low pH—are early diagenetic or even weathering products (see *Weathering in Sediments*).

Diagenetic Sequences. With increasing depth, sequences of diagenetic events occur. These sequences depend on the primary constituents of the sediments as well as on the interstitial fluids and their alterations (such as increasing salinity and pH) during increasing burial depth. In many cases, the following sequence of cements and replacements has been observed.

1. silicate minerals including clay minerals and feldspars;
2. quartz, frequently simultaneous with (1) and continuing for a long time, so that, in quartz sandstones, quartz overgrowths can be used to date events such as oil accumulations (Fig. 4); and
3. carbonate and sulfate minerals (e.g., calcite, ankerite, barite, celestite).

Transition to Metamorphism. The "very low stage of metamorphism" (Winkler, 1970) begins at about 200°C, corresponding to a burial depth of about 6000 m, under normal geothermal conditions. It is characterized by the zeolite mineral laumontite, by pyrophyllite (in rocks poor in K, Mg, Fe), by coalification rank corresponding to a maximum reflectance of vitrinite of 4–6%, and by an illite peak width at half height (fraction 2–6μ) of 1.1–1.5 compared with quartz (Weber, 1972), whereas the chlorite mineral "Fe-rhipidolite" is found at shallower depths (see Fig. 1). A general decrease of sedimentary fabric, including spine-like intergrowths (Kossovskaya and Shutov, 1970) and strong porosity reduction, is characteristic of the transitional stage (Füchtbauer and Müller, 1974).

Shales

Compaction. The bladed shape and the negative charge of clay mineral particles cause them to coagulate to form a loose, random arrangement, in which cardhouse and/or honeycomb structures prevail (Fig. 5). The distances between the particles may vary considerably; the larger

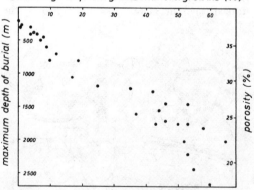

FIGURE 4. Increase of the intensity of quartz overgrowth with increasing maximum burial depth (Füchtbauer and Müller, 1974, Fig. 3-50, with permission of Schweizerbart-Verlag, Stuttgart). The percentage used as abscissa is derived from the formula $100(0.5b + c)/(a + b + c)$, with a = number of quartz grains without overgrowths, b = number of quartz grains with overgrowths covering less than half of the surface, c = number of quartz grains with overgrowths covering more than half of the grain surface. Counts are made in dry smear slides.

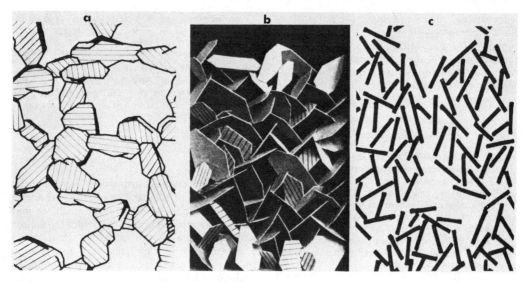

FIGURE 5. Fabric types occurring in clay mineral coagulations (from Füchtbauer-Müller (1974), Fig. 4-40, with permission of Schweizerbart-Verlag, Stuttgart). (a) honeycomb structure occurring in illite, (b) cardhouse structure occurring in kaolinite, (c) aggregate structure.

pores help in dewatering of the clay sediments (Engelhardt and Gaida, 1963). Upon loading, these open structures collapse to form a more or less parallel arrangement of the blades. Most of this dewatering occurs within the first 500 m of burial (Fig. 1). During compaction, most of the water incorporated in the sediments during deposition is replaced by interstitial water from below; pore fluids are generally not expelled into the depositional basin. For these reasons, a relatively small difference is found between the interstitial fluids of the uppermost sediments and the sea water: Si, Ca, and K are enriched; Mg, HCO_3, and SO_4 are diminished in the sediment (Siever et al., 1965).

Replacements and Diagenetic Sequences. Diagenetic sequences that are mainly caused by alterations and replacements can be investigated by (1) analyzing samples from various depths of a presumably homogeneous sequence (Fig. 6); (2) comparing the same sequence in different bore holes, at different depths; or (3) comparing the clastic minerals in concretions with the surrounding shales (Füchtbauer and Goldschmidt, 1963). The following diagenetic sequence is frequently found: (a) formation and/or crystallinity increase of kaolinite, then (b) formation of chlorite or pyrophyllite (Frey, 1970) on the account of kaolinite. Another sequence is: (a) montmorillonite (smectite), (b) mixed layer montmorillonite-illite, (c) illite (Fig. 1).

Illite crystallinity increases with increasing depth. At the same time, the K_2O content increases. According to Moort (1972), the average K_2O content of pre-Silurian shales is about 4%, of younger shales 2%. Much of the K_2O may be provided diagenetically by potassium feldspars in graywackes, which are partially replaced by sodium feldspars in older rocks (Pettijohn et al., 1972; Middleton, 1972). Moreover, the raw material of pre-Silurian shales may have contained more K than the raw material of younger shales, a result of the ab-

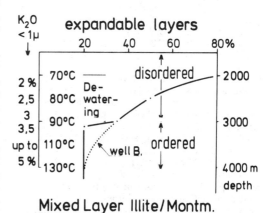

FIGURE 6. Alterations of the mixed-layer composition reflected by the percentage of expanded layers, the ordering degree, and the K_2O content in the fraction $<1\mu$ (containing the bulk of the mixed layers), with increasing depth in bore hole E; Tertiary of the Gulf Coast (after Perry and Hower, 1970).

sence of land plants to remove the potash from clays during weathering (Middleton, pers. comm.). In rocks that are poor in cations, kaolinite can be altered to dickite, which in turn may transform into pyrophyllite at the beginning of the lowest stage of metamorphism.

HANS FÜCHTBAUER

References

Engelhardt, W. v., 1973. *Die Bildung von Sedimenten und Sedimentgesteinen,* Sediment-Petrologie, Teil III. Stuttgart: Schweizerbart, 378p. Now in English.
Engelhardt, W. v., and Gaida, K. H., 1963. Concentration changes of pore solutions during the compaction of clay sediments, *J. Sed. Petrology,* 33, 919-930.
Frey, M., 1970. The step from diagenesis to metamorphism in pelitic rocks during Alpine orogenesis, *Sedimentology,* 15, 261-279.
Füchtbauer, H., 1974a. Some problems of diagenesis in sandstones, *Bull. Centre. Recherches, Pau,* 8, 391-403.
Füchtbauer, H., 1974b. Zur Diagenese fluviatiler Sandsteine, *Geol. Rundschau,* 63, 904-925.
Füchtbauer, H., and Goldschmidt, H., 1963. Beobachtungen zur Tonmineral-Diagenese, *Internat. Clay Conf., Stockholm* (Pergamon Press), 1, 99-111.
Füchtbauer, H., and Kossovskaya, A. G., eds., 1970. Lithification of clastic sediments, *Sedimentology,* special issue, 15(1-4) 440p.
Füchtbauer, H., and Müller, G., 1974. *Sediments and Sedimentary Rocks.* Sedimentary Petrology, p. II (in 2 parts). Stuttgart: Schweizerbart; New York: Halsted, part 1, 464p; part 2, in press.
Kossovskaya, A. G., and Shutov, V. D., 1970. Main aspects of the epigenesis problem, *Sedimentology,* 15, 11-40.
Larsen, G., and Chilingar, G. V., eds., 1967. *Diagenesis in Sediments.* Amsterdam: Elsevier, 551p.
Merino, E., 1975. Diagenesis in Tertiary sandstones from Kettleman North Dome, California. I. Diagenetic mineralogy, *J. Sed. Petrology,* 45, 320-336.
Middleton, G. V., 1972. Albite of secondary origin in Charny sandstones, Quebec, *J. Sed. Petrology,* 42, 341-349.
Moort, J. C. van, 1972. The K_2O, CaO, MgO, and CO_2 contents of shales and related rocks and their implications for sedimentary evolution since the Proterozoic, *24th Intern. Geol. Congr.,* sec. 10, 427-439.
Perry, E., and Hower, J., 1970. Burial diagenesis in Gulf Coast pelitic sediments, *Clays Clay Minerals,* 18, 165-177.
Pettijohn, F. J.; Potter, P. E.; and Siever, R., 1972. *Sand and Sandstone.* Berlin: Springer, 618p.
Rosenfeld, M. A., 1949. Some aspects of porosity and cementation, *Producers Monthly,* 13, 39-42.
Siever, R.; Beck, K.; and Berner, R., 1965. Composition of interstitial waters of modern sediments, *J. Geol.,* 73, 39-73.
Weber, K., 1972. Notes on determination of illite crystallinity, *Neues Jahrb. Mineral. Monatsh.,* 1972, 267-376.
Winkler, H. G. F., 1970. Abolition of metamorphic facies, introduction of the four divisions of metamorphic stage, and of a classification based on isograds in common rocks, *Neues Jahrb. Mineral. Monatsh.,* 1970, 189-248.

Cross-references: *Cementation; Clastic Sediments and Rocks; Clay—Diagenesis; Compaction in Sediments; Diagenesis; Pressure Solution and Related Phenomena; Shale; Silica in Sediments; Weathering of Sediments.*

CLAY

A sedimentary deposit is called *clay* if the clay mineral content is greater than 30%. Otherwise, the deposit may be called a sand, silt, or marl (q.v.). Clay mineral content provides the deposit with a unique plastic behavior when damp. The term *clay* or *clay-sized* may also connote the smallest size fraction ($<4\mu$) in grain-size analysis of sediments.

Plasticity of clays is the ability of the wet material to be shaped and to have the strength to hold its shape after the deforming pressure is removed. Clay minerals provide this property to the overall clay deposit because of certain fundamental characteristics of the clay particles. Their flake-like shape allows them to be oriented in such a way as to strengthen the plastic mass in the direction of applied stresses. The high surface energy of the clay particles is responsible for the rigidity and thickness of the water structure adsorbed on the particles. In sedimentary deposits, the clay mineral grains are seldom greater than 2μ in their longest dimension. The behavior of particles in the clay-size range is dominated by their surface forces.

The plastic behavior of clay deposits is unique in several ways. In addition to plasticity as defined above, clays exhibit a tendency to retain large volumes of water, increased strength on drying, and a substantial drying shrinkage. All of these properties tend to make sedimentary clays useful, as humans have found from the beginning of civilization.

Clay Minerals

The clay minerals that are so essential to a clay deposit are crystals built up in layers, hence the flake-like shape of most of the clay minerals. The clay minerals belong to the disilicate (phyllosilicate) classification, along with the micas, talc, and pyrophyllite, whose structures are all dominated by a two-dimensional, silicon-oxygen framework. The *disilicates* are characterized by the silicon-oxygen ratio, $(Si_2O_5)^{-2}$. In the clay mineral layers, this silica sheet is combined with either an alumina sheet, resembling gibbsite, $Al(OH)_3$, or a magnesia sheet similar to brucite, $Mg(OH)_2$. In *kaolinite*, $Al_2(Si_2O_5)(OH)_4$, one silica sheet and one gibbsite sheet are combined to form a layer, a crystal being composed of many layers. Kaolinite is the purest aluminosilicate of all clay minerals.

Smectite (montmorillonite) has two silica sheets on either side of a gibbsite-like or a brucite-like sheet. The most distinguishing feature of the structure of this clay mineral is a partial substitution of Al^{+3} ions and S^{+4} ions in the silica sheets. The charge balance is made up by variable and exchangeable interlayer cations. The most common, montmorillonite, also has some Mg^{+2} ions substituting for Al^{+3} ions in the gibbsite-like sheet. A typical structural formula for montmorillonite is $(Al,Mg)_2 (Al_{0.7}Si_{3.3}O_{10})(OH)_2 X^+_{0.7+}$. This mineral is characterized by its expansion perpendicular to the layers due to interlayer water adsorption and its ultra-fine-grained particles. *Illite* is a clay mineral similar in crystal structure to the smectites, but, like micas, the layers are tied together by K^+ ions. Expansion perpendicular to the layers is absent, and this mineral is more like micas in all respects. A typical structural formula for illite is $(Al_{1.5}Fe_{0.3}Mg_{0.2}) (Al_{0.6}Si_{3.4}O_{10})(OH)_2 K_{0.6}$. Illite always contains iron and fires to brick red colors in ceramic applications.

Deposits

Sedimentary clay deposits are formed by the transportation by water or wind of various mineral fragments which are deposited as the carrying force subsides. The nature and volume of material deposited often varies with time depending on weather conditions, giving the deposits a varved or layered appearance. Such deposits are found on the floors of the seas and oceans, where they are classified as *marine* (see *Clays, Deep-Sea; Red Clay*). Airborne deposits are *eolian* clays.

As might be expected, all kinds and quantities of mineral substances can be found in sedimentary clay deposits. One or more of the clay minerals may be deposited together with other disilicate minerals, such as muscovite, biotite, chlorite, pyrophyllite, and vermiculite. Quartz is the predominant nonplastic mineral in clay deposits, but fragments of many other minerals —e.g., feldspars, rutile, anatase, zircon, hornblende, epidote, and tourmaline—are also common in lesser quantities. Carbonaceous matter from decayed animal and vegetable life is often closely associated with minerals in clay deposits. Calcite and dolomite are also found intimately mixed with the clay minerals, especially in marine deposits. Their origin is generally directly or indirectly related to organic precipitation.

In an existing deposit, it is possible for alterations to occur in the mineral fabric to change or create a clay deposit. In these cases, the acting diagenetic forces are leaching solutions; dissolution of slightly soluble ingredients; and chemical reactions, such as oxidation-reduction, decomposition, and precipitation. Pressure may also be a factor in alteration. Some clay minerals are formed or changed in the deposit by these processes; other minerals—such as chlorite, pyrite, microcline, albite, quartz, goethite, limonite, hematite, gypsum, and barite—are created (see *Clay–Diagenesis*).

Sedimentary clay deposits are abundant all over the world, and a few deposits are commercially valuable because of their degree of purity or unique assemblage of minerals. *Sedimentary kaolins* are examples of high-purity deposits, some of them containing only kaolinite and quartz. These white clays can be beneficiated by the removal of the quartz; a fine-grained, plastic kaolin results. Many sedimentary kaolin deposits are rendered practically useless by the presence of nontronite, goethite, limonite, or hematite, all of which are iron minerals. *Ball clays* are important deposits because they contain a large amount of fine-grained kaolinite. They may contain small amounts of other clay minerals, and the principal nonplastic constituent is silt-sized quartz. *Fireclays* are kaolinitic, less plastic, and more impure than the clays already mentioned. In these clays, the kaolinite is usually associated with illite, quartz, pyrite, and carbonaceous material. A special type of fireclay is so compacted into a dense, rock-like mass that it is called *flint clay*. Flint clays do not slake or soften in water.

A very unusual but important clay deposit is *bentonite* (q.v.), essentially composed of smectite with only traces of coarse-grained, nonplastic materials. Upon prolonged exposure to potassium-bearing solutions, these clays change their character to a clay closely resembling illite. Such deposits are called metabentonites, and they have little commercial use. Probably the most abundant, *illitic clays* are the least pure aluminosilicate clays, almost always containing quartz and a wide variety of other mineral fragments, including micas. Many deposits also contain carbonaceous matter, which maintains a chemically reducing environment. These deposits are blue to black in color. Examples of red illitic deposits are common where oxidation conditions prevailed. The mineralogy of these different-appearing illitic clays is remarkably similar. *Shales* (q.v.) are a kind of sedimentary clay deposit which has been consolidated into layers by the pressure of overlying formations.

Types of Clays

Ball Clay. A ball clay is a fine-grained sedimentary deposit composed essentially of kaolinite (50–90%). The other major minerals are quartz, micas, and illite, or hydrous mica. As with all sedimentary deposits, ball clays may contain small amounts of rutile, anatase, montmorillonite, carbonaceous matter, and other fine-grained mineral fragments. The total iron oxide content usually does not exceed 2.5%. The small particle size (80% <1µ) and high clay mineral content of these deposits cause them to be highly plastic and to exhibit great drying shrinkage. The deposits are usually lenticular in shape and lacustrine in origin; some deposits are fluvial, however. In ceramics, ball clays are used to increase plasticity of white-firing bodies. They are also used in molding sands and as fillers.

Blue clay. Blue clay is a sedimentary deposit in which the clay mineral content may vary from 30–80%. Its characteristic color is derived from the presence of carbonaceous matter, which causes chemically reducing conditions and allows the ferrous iron minerals such as FeS and FeO to persist. The deposits may be lacustrine, estuarine, deltaic, or marine in origin. The clay minerals are usually illite, but kaolinite and smectite can also be found in some deposits. Quartz grains are the usual major nonplastic mineral, and sand to silt-sized fragments of other minerals are often present. Some blue clays are weathered and redeposited shales, but this derivation does not alter the above description.

Brick Earth. A sedimentary clay deposit suitable for making red bricks or pottery is sometimes called *brick earth* or *brick clay*. Such deposits may be fluvial, estuarine, lacustrine, marine, or eolian. In order to be suitable, the clay mineral content must be 30–50%. The remainder of the deposit should be quartz sand and silt—the silt fraction usually containing small amounts of feldspar, iron oxide, rutile, anatase, zircon, and other stable mineral fragments. Sulfates and carbonates of the alkaline earth metals are usually considered undesirable. The ultimate suitability comes only when firing tests show the proper color, porosity, shrinkage, and strength.

Fat Clay. If a sedimentary clay contains a large fraction (50–80%) of clay minerals, and the amount of silt-sized nonplastic mineral fragments is small, it is described as a *fat clay*. This type of deposit is highly plastic, dense, and heavy when damp. It is impervious to water because of low porosity and the large amount of water adsorbed by the clay minerals. Fat clays generally require the addition of a nonplastic, fine material to control shrinkage in order for them to be useful in plastic-forming operations.

Lean Clay. If a sedimentary clay deposit contains a small fraction (20–30%) of clay minerals together with sand or silt-sized nonplastic mineral fragments, it is described as a *lean clay*. This type of deposit drains well when wetted, and the porosity of the body is higher than usual for clay deposits. It is sometimes referred to as a *silty clay*.

Flint Clay. A flint clay is a massive, dense, hard, nonslakable, nonplastic clay which has flint-like characteristics (see *Chert and Flint*) and is highly refractory when fired. It is composed of well-crystallized kaolinite in very small particle sizes. Sometimes there is an excess of silica and aluminum present in the form of either diaspore (AlOOH) and/or boehmite (AlOOH). These aluminum-rich minerals often serve to cement the kaolinite particles together, although the precise reasons for the distinctive physical properties of flint clays have not been established. The origin of these clays has been variously ascribed to leaching action associated with the formation of sinkholes, to direct precipitation from colloidal solutions, and to selective leaching of a weathered basalt flow in the presence of organic matter. These clays are used in the making of furnace linings, where they convert into mullite ($3Al_2O_3 \cdot SiO_2$).

Underclay. An underclay is an argillaceous, gray, nonbedded layer (substrate) found directly under a coal bed. It may vary in thickness (a few cm to 10 m) and in horizontal extent, both of which are apparently unrelated to the dimensions of the overlying coal. Occasionally, underclays form layers unrelated to coals over part or all of their traceable extent. Underclays that contain no alkaline substances may be able to withstand high temperatures (alkalines being fluxing agents) and are then termed *fireclays*. Underclays are composed of kaolinite, illite, chlorite, and mixed-layer clays; some underclays are calcareous, in which case they are primarily composed of illite. The upper few cm, directly below the coal, usually contain some organic matter, such as stumps, roots, and small plants, and may be leached. Often the underclays of a region are related to the coal beds in a recognizably cyclic manner; this relationship was first termed a *cyclothem* by Wanless and Weller (see *Cyclic Sedimentation*). The study of underclays has been spurred by their varied economic uses.

WAYNE E. BROWNELL

References

Boswell, P. G. H., 1961. *Muddy Sediments: Some Geotechnical Studies for Geologists, Engineers, and Soil Scientists.* Cambridge: W. Heffer & Sons, 140p.

Grim, R. E., 1968. *Clay Mineralogy.* New York: McGraw-Hill, 596p.

Keller, W. D.; Wescott, J. F.; and Bledsoe, A. O., 1954. The origin of Missouri fireclays, *Proc. 2nd Natl. Conf., Clays and Clay Minerals, 1953,* 7-46.

Millot, G., 1970. *Geology of Clays.* New York: Springer-Verlag, 429p. Originally in French: *Géologie des Argiles.* Paris: Masson, 1964.

Odom, I. E., and Parham, W. E., 1968. Petrology of Pennsylvanian underclays in Illinois and their application to some mineral industries, *Ill. St. Geol. Surv. Circ. 429,* 36 p.

Van Schoick, E., 1963. *Ceramic Glossary.* Columbus: American Ceramic Society, 31p.

Weaver, C. E., and Pollard, L. D., 1973. *The Chemistry of Clay Minerals.* Amsterdam: Elsevier, 224p.

Cross-references: *Argillaceous Rocks; Clay as a Sediment; Clay Sedimentation Facies; Shale.* Vol. IVA: *Clay Minerals-Base Exchange.* Vol. IVB: *Clays and Clay Minerals.* Vol. XI: *Clay Minerals, Distribution in Soils of the World.*

CLAY AS A SEDIMENT

The word *clay* (q.v.) carries with it a double meaning: it is a mineralogic term which refers to a particular group of minerals; and it is a size classification which refers to the fine-grained fraction of a sediment or soil, including all types of organic and inorganic products that have nothing to do with clay minerals. The clay-sized fraction of most soils and sediments consists largely of clay minerals, however, and this discussion will deal only with the clay mineral component.

Clay minerals are essentially hydrous aluminum silicates but contain appreciable magnesium, potassium, calcium, sodium, and iron, as well as other less important ions. Despite their varied chemical composition, the clay minerals fall largely into a few major groups: kaolinite, smectite (montmorillonite), illite (hydrous mica), and chlorite. Other clay minerals exist but are volumetrically unimportant. In addition, a great range of mixed-layer clay minerals occur, i.e., interstratifications involving usually two or more different clay minerals. Grim (1968) provides a more complete discussion on clay minerals (see Vol. IVB: *Clay Minerals*).

Clay minerals make up an enormous part of the sedimentary column. It has been estimated (Weaver, 1958) that 50% of the mineral content of sedimentary rocks are clay minerals, and approximately 95% of all sedimentary rocks contain some clay minerals. Shales (q.v.), which consist largely of clay minerals, comprise 50-80% of the entire stratigraphic section. Clay minerals are formed in, or transported into a wide variety of different environments such as soils, lakes, estuaries, and the deep sea.

Soils

Weathering, Clay Mineral Formation. Most soils contain clay minerals as major constituents. Strictly speaking, the clay minerals occur in soils not as a sediment but as products of rock weathering at the earth's surface, where they are formed by the chemical breakdown of other silicate minerals. Chemical weathering of the silicates is primarily a hydrolysis reaction, although other processes, e.g., oxidation, hydration, and ion exchange, are involved as well. Clay minerals may form in a gel phase (in which they are crystallized), in ionic solution (from which they are precipitated), or by the direct restructuring of the parent silicate lattice into that of a clay mineral (see *Weathering in Sediments*).

Pedogenesis, or soil formation, is essentially a late stage in the chemical weathering process, of which the end or near-end products are the clay minerals. Hence, both the type and the amount of clay minerals formed are largely a function of the major soil-forming factors: climate, parent material, drainage (surface and subsurface), vegetation, and time. Each of these factors has a significant, but rarely independent, effect on both soil and clay mineral formation. Climate and time are generally regarded as the most significant of the soil-forming factors (see Vol. XI, *Encyclopedia of Pedology*).

The effect of climate on soils is evident in the distribution of mature "zonal" soils which comprise the Great Soil Groups (Fig. 1). The different soil types reflect broad geographic variations in both rainfall and temperature. As a general rule, 60 cm of rainfall per year represents the division between the humid-zonal soils or *pedalfers*—characterized by concentrations of aluminum (Al) and iron (Fe) in parts of the soil profile—and arid-zonal soils or *pedocals*—characterized by concentrations of calcium (Ca) salts. Variations within the pedalfers result chiefly from differences in temperature, whereas, within the pedocals, the variations are the result of differences in rainfall (Barshad, 1959).

A broad correlation between climate and soil clay mineralogy exists as well. Wet, hot (tropical) climates produce soils that are rich in 1:1 clay minerals (kaolinite) and iron and

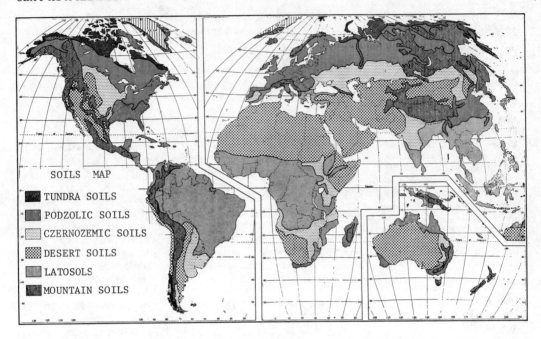

FIGURE 1. World distribution of great soil groups (from Winters and Simonson, 1951). *Note:* important areas of organic and saline soils, intrazonal soils, and alluvial soils are omitted.

aluminum sesquioxides such as gibbsite, bauxite, and hematite. Moist, warm to cool (temperate) climates, on the other hand, favor the development of structurally and chemically more complex 2:1 (illite, smectite) and 2:2 (chlorite, vermiculite) clay minerals (Papadakis, 1969).

In general, chemical weathering and clay mineral formation is enhanced by increasing moisture content of a soil. The weathering process, however, is most effective if rainfall is evenly distributed annually, thereby resulting in the continuous and efficient removal of soluble soil constituents. Increasing soil temperature, although usually conducive to weathering, may retard it in the case of very high temperature. The movement of water in the weathering zone may actually be upward (*illuviation,* as opposed to *eluviation,* or downward movement), causing the soluble constituents to be deposited as salts at or near the surface. Crusts consisting of $CaSO_4$ (gypsum), and Ca and Mg carbonates (caliche, q.v.) are common to soils of arid regions.

The effects of climate are frequently offset, however, by the predominance of some local variation in one or more of the other soil-forming factors. Under such circumstances "intrazonal" soils develop. The clay minerals formed, although locally stable, may be quite unstable under the prevailing regional climatic conditions. Therefore, the use of climate in describing environments of clay mineral formation is not altogether satisfactory except in the broadest sense. More definitive parameters such as soil pH, Eh, and concentration of reacting ions should also be used to describe environments of weathering.

A major consideration in the formation of clay minerals is the availability of magnesium (Mg), potassium (K), sodium (Na), calcium (Ca), and iron (Fe). The presence of these cations in the weathering environment, particularly Mg, is critical to the formation of smectite; the illite structure requires an abundance of K; and the absence of all these cations is necessary for kaolinite formation. Composition of the parent rock establishes the initial presence or absence of any particular cation; but, during weathering, all the cations may be lost through leaching of the parent material.

Compositionally, kaolinites are characterized by a higher aluminum (Al) to silica (Si) ratio than are the other clay minerals, and this usually indicates the removal of Si during kaolinization. The process of Si removal during weathering is facilitated by the presence of K and Na, which tend to keep Si in solution (below a pH of 8, Si is significantly more soluble than Al). Ca and Mg, on the other hand, tend to flocculate Si and ensure its retention in the weathering environment. In terms of parent materials, then, acid (granitic) rocks rich in K and Na can be more readily altered directly to kaolinite than can basic (basaltic) rocks rich in Ca and

Mg. In the latter case, kaolinization usually involves an "intermediate" stage of smectite, illite, or chlorite development.

Whether or not cations remain in a weathering environment depends primarily on how freely the soluble constituents are removed by percolating waters. Leaching is most effective where surface and subsurface drainage is good, usually areas of gentle slopes and high permeability of the parent material. Little water penetrates the parent material on steep slopes. In flat, low-lying areas, drainage tends to be very poor so that leaching is weak; concentrations of Mg, Ca, etc., build up, promoting the formation of clay minerals other than kaolinite. In water-saturated, low-lying areas, breakdown of the parent material is inhibited because of extremely sluggish leaching, so that overall clay mineral development may be low.

Leaching of a soil is most severe, and weathering most accelerated, when the percolating waters are highly acidic. The acidity may be due to mineral acids (from oxidation of sulfides such as pyrite in the presence of water to form sulfuric acid), carbonic acid (reaction of CO_2 with water), or organic acids. The importance of vegetation as a soil-forming factor is thus obvious. Strongly acid soils are especially prominent in the temperate, well-vegetated areas of the world where the accumulation of organic matter is promoted by slow decay (oxidation).

Inheritance. Many clay minerals in soils are not primary, i.e., they are not newly formed from nonclay-mineral parent silicates. Soils developing from sedimentary or metasedimentary rocks commonly inherit previously formed clay minerals directly from these parent materials (Van Houten, 1953). Once formed, clay minerals resist modification despite later changes of environment; thus, clay minerals tend to be indicative only of their initial environment of formation, which may have occurred long ago, as well as far away, from where they are presently found. Clay mineral modifications do occur, however, particularly if a new environment differs greatly from the initial one (see *Clay–Genesis*). For example, Van Houten (1953) suggests that smectite (montmorillonite) inherited from the Cenozoic sediments of the Alabama coastal plain has been converted to kaolinite through intense leaching.

A few broad generalizations can be made concerning inheritance and parent material. For example, the fact that illite is the overwhelmingly dominant clay mineral in most shales (q.v.) implies that soils developing on shales (and slates) will inherit abundant illite. Chlorite also occurs abundantly in both shales and slates, whereas smectite has not been reported in slates (Grim, 1968). Presumably the smectite is completely destroyed or its structure altered to that of another clay mineral (illite) during metamorphism. Soils formed on limestone or dolomite seem to inherit little kaolinite, but relatively abundant illite (Booth and Osborne, 1971). Although there is evidence that calcite and kaolinite have developed together in some caliche profiles (Aristarain, 1971), there is some suggestion that the presence of calcium in soils tends to block the development of kaolinite. Even in calc-alkalic granite massifs, Millot (1970) notes the absence of hydrothermal deposits of kaolin, yet he does find such deposits in alkaline granites. Kaolinite found in carbonate rocks can usually be considered as detrital. In most instances, both kaolinite and smectite are commonly thought to be the products of weathering, whereas most illite and chlorite are inherited from parent rocks (Van Houten, 1953; Grim, 1968).

Inherited clay minerals are particularly prominent in soils formed in arid and cold regions. The predominance of physical rather than chemical weathering in these environments leads to the formation of immature soils, i.e., soils in which there is little development of new clay minerals and the nature of the parent material is still retained.

Distribution. Kaolinite occurs in regions of abundant rainfall, good drainage, and acid soils. Under these conditions, Ca, Mg, Fe, Na, and K are removed freely from the weathering environment. Hence, kaolinite is frequently the dominant clay mineral of podsol and podsolic soils, which typically develop in the temperate, damp, forested areas of the world. Podsolization also occurs under grass cover (prairie soils); but different organic products result, which appear to act with less intensity than those produced by forest vegetation (Grim, 1968). Consequently, kaolinite is present but usually not predominant in prairie soils.

In carbonate-rich soils the kaolinite content is usually low. Such soils resist acidification because of the neutralization of organic acids by Ca and Mg (Papadakis, 1969) to the extent that kaolinization does not take place until the Ca is removed. Even in regions of intense chemical weathering, Millot (1970) notes that kaolinite development may be delayed while Ca is present.

Tropical soils, although neutral to slightly alkaline because of rapid oxidation of organic matter (inhibiting its accumulation), are very rich in kaolinite. Intense chemical weathering is brought about by a combination of high temperatures and abundant rainfall. The occurrence of kaolinite in these soils suggests that an acid environment is not an absolute prerequisite to kaolinite formation. From equilibrium considerations Keller (1970) notes that acidity of weathering solutions need not be particularly high to form kaolinite provided the ratio of $[K^+]$ to $[H^+]$ is appropriately low.

Under extreme leaching conditions, complete removal of cations and desilicification reduce parent materials to Fe and Al sesquioxide end products. Such soils are called laterites or bauxites depending on whether Fe or Al predominates. The *red soils* of the southern and southeastern United States (Fig. 1), although not true laterites, are lateritic in nature and contain mineral assemblages still rich in kaolinite and sesquioxides. To the north, these lateritic soils grade into kaolinite podsols.

In direct contrast to kaolinite formation, smectite is favored by low rainfall, poor drainage, and alkaline conditions. Weak leaching allows Ca, Mg, etc., to remain in the environment. Predictably, smectite is a predominant clay mineral of many arid-region soils, of which the best known and agriculturally most important are the *chernozem* or *black earth soils* found in the Ukraine and the central plains of North America (Fig. 1). Warmer temperatures result in quicker oxidation of organic matter and the higher latitude chernozems and *chestnut soils* (slightly more arid than chernozems) give way to the *red-brown earths* in the lower latitudes and to more pronounced development of kaolinite.

The occurrence of smectite in soils of hot, humid regions, however, is not uncommon. In fact, tropical smectite has been reported forming near laterite, under virtually the same weathering conditions except for differences in drainage (See Mohr and Baren, 1954), the laterite developing at the better-drained site. The predominance of smectite in the Coastal Plain Province of the S and SE United States reflects to some extent the lower relief and poorer drainage of this area than in the adjacent Piedmont Province, where kaolinite prevails. Much of this smectite, however, may be inherited from the Coastal Plain deposits (Van Houten, 1953).

The environmental aspects of illite formation are very similar to those of smectite with the exception of the need for a K-rich environment. Hence, illite is also typical of soils in arid regions. Because illite occurs so abundantly in sedimentary rocks, however, it is commonly inherited by the soils forming on these rocks, even in humid climates; Johnson (1970) found inheritance of illite to be of major importance in 348 Pennsylvania soil profiles. Quarternary glacial deposits provide a rich source of illite to soils developing on these deposits.

In the Arctic, cold temperatures greatly inhibit rock and plant decay, and permafrost results in boggy, poorly drained soils. Consequently, *tundra soils* are highly organic and low in clay mineral content. Illite is often the predominant clay mineral of these soils.

According to Degens (1965), the requirements for chlorite development are an alkaline environment and Mg and Fe concentrations far greater than those necessary for smectite formation, possibly accounting for the fact that although chlorite occurs widely in soils it is seldom the dominant clay mineral. The susceptibility of chlorite to chemical attack restricts its formation in soils of warm humid regions, and probably in most acid soils. The distribution of chlorite in soils is, therefore, theoretically reciprocal to that of kaolinite.

Rivers, Lakes, Swamps

Literature dealing with fluvial clay minerals shows clearly that rivers and streams are characterized by diverse clay mineral types which closely reflect the nature of both the source materials and the climate in their watershed. Thus, excepting the possibility of some leaching or minor cation exchange, major transformation of clay minerals during river transport does not occur, i.e., the clay minerals as evolved during pedogenesis remain structurally intact.

River clay-mineral assemblages may vary considerably along the course of a river as more tributaries (eroding different parent materials) enter the river or, perhaps less significantly, as the river passes over rock types having distinctively different clay-mineral types or abundances. Weaver (1967) took samples from the Arkansas River at the same place over a 2-year period and correlated the variations with the flood pattern of the river and the distribution of precipitation in the drainage basin.

Rivers and streams (and their deposits) can be expected to contain varied mixtures of clay minerals, but ones that are always detrital in origin. Ordinarily, there is little reason for the clay minerals in a lake to differ from those transported by its streams or rivers provided the lake waters remain fresh, so that lacustrine clays also reflect source materials to the extent that lake sediments are likely to be a fairly homogeneous blend of all the clay minerals found in a drainage area. The detrital, rather than diagenetic, nature of nearly all lacustrine clays is emphasized by the fact that there is apparently not a single association of clay minerals that can be considered typical of lakes (Picard and High, 1972). Millot (1952) correlated lacustrine clay minerals with environment of deposition and concluded that *aggressive lakes,* either because of acid waters or movement of water through the sediments, are predominantly kaolinitic as a result of leaching of cations. Illite and smectite, on the other hand, are more typical of *nonaggressive lakes* where water movement through the sediment is slight and/or the water is alkaline. It should be noted that alkaline lakes are to some extent characteristic of arid regions, and a predominance of illite or smectite in these lakes may simply reflect the formation of these clay minerals under arid weathering conditions. In a like manner, aggressive lakes may be more characteristic of regions of acid weathering where kaolinite formation takes place.

According to Millot (1970), when the salt content of lake waters increases greatly (by evaporation), conditions for clay mineral transformations and neoformation, or authigenesis, are created (see *Clay–Genesis*). Investigations of saline lakes and playas of arid regions, however, show that smectite and illite derived from the surrounding desert soils are the dominant clay minerals (Parry and Reeves, 1968). Except for some authigenic formation of sepiolite and palygorskite, which requires a high magnesium environment, the high cation concentrations of saline lakes do not appear to promote significant diagenesis in the sediments (Droste, 1961). Both illite and smectite are formed in cation-rich environments to begin with and would not be expected to undergo major modification upon passing into a highly saline lake. Quaide (1958) found no evidence of diagenetic changes in clay minerals taken from salt concentration ponds after exposure to hypersaline conditions for 27 years.

The mineralogy of swamp deposits, such as *underclays,* can be quite varied because of the variety of rocks from which they originate (Millot, 1970). Generally, swamps tend to be very acid (aggressive) because of accumulating organic matter, producing a strong leaching environment, and progressively transforming deposited clay minerals to kaolinite. The kaolinite may also be inherited, particularly if the swamp is located in an area with a warm, humid climate. Illite seems to be the only other clay mineral occurring abundantly in underclays (Grim, 1968).

Flint clays occur frequently with underclays but differ in that they consist almost entirely of kaolinite. Patterson and Hosterman (1962) have suggested that flint clays represent the end stage in the progressive leaching of swamp sediments. An alternative explanation is that the flint clays represent detrital deposits (Loughnan, 1970).

Coasts: Bays, Lagoons, Estuaries

It is generally agreed that clay mineral modifications are very likely to occur in areas where the clay minerals first pass from fresh-water to salt-water conditions. Diagenetic clay minerals have been suggested for a number of such *transitional environments.* Transformation of smectite to illite and chlorite has been observed in some Texas bays and lagoons, and in the sediments in the area of the Mississippi River delta; there is also evidence that diagenetic illite and chlorite are forming in estuaries and bays along the eastern United States. The interpretation of these changes is difficult, and the evidence for them is not unequivocal. In general, the evidence takes the form of a progressive increase of one clay mineral in a downstream direction (toward higher salinities) at the apparent expense of another clay mineral. Usually, the new phase lacks some of the attributes of a well-crystallized phase (of the same mineral). For example, the "chloritic" material forming in some east-coast estuaries does not show the thermal stability of "true" chlorite (see references in Grim, 1968).

Because well-crystallized clay minerals usually do not change upon entering the marine environment, diagenetic clay minerals are thought to form from weakly crystalline, degraded constituents, that is, clay minerals in which the interlayer cations have been largely removed by leaching during pedogenesis, and perhaps during stream transport. Degraded illites exhibit some of the characteristics of smectite (14–15 Å basal reflections, expandable structure), but upon exposure to sea water they quickly readsorb K or Mg and are reconstituted back to normal illites and chlorites. If the strict definition that diagenesis includes only alterations that structurally modify the basic lattice (Weaver, 1958) is accepted, the observed modifications are not diagenetic. Weaver argues that the basic lattice of degraded illite is not seriously changed, and that the alterations are nothing more than exchange reactions which are termed diagenesis "because different names are attached to the clay before and after exchange." Upon contact with sea water, all clay minerals undergo some cation exchange without any apparent change. It may be necessary, however, to consider exchange reactions such as potassium fixation as part of the diagenetic process when they greatly alter the clay's (001) spacings, one of the major criteria in clay mineral identification.

A number of investigators believe that the clay mineral suites of individual coastal environments reflect the source materials and climate in the drainage area rather than diagenetic trends. Neiheisel and Weaver (1967) and Windom et al. (1971) show that the clay minerals in rivers (and their estuaries) draining the southeastern Coastal Plain (U.S.) are largely smectitic, whereas those draining the more northern Piedmont Province are primarily kaolinitic. They also conclude that much of the illite in estuarine sediments is carried in from the continental shelf by flood tides, accounting in part, perhaps entirely, for clay mineral distribution patterns that have been offered as proof of diagenesis (see *Clay–Diagenesis*).

To some extent, the observed "diagenetic" trends noted above may involve clay mineral authigenesis, or neoformation (see *Clay–Genesis*). The suspended loads of streams usually contain some amorphous alumina, silica, and aluminosilicates which provide suitable materials for the formation of clay minerals. Mackenzie and Garrels (1966) have proposed that these stream-derived constituents react with the cations in sea water to form new clay minerals. Poorly crystalline phases develop first, later separating into illite, chlorite, and smectite, either as discrete particles or in mixed-layer associations. Although experimental and field evidence suggest that clay minerals can develop in this manner, such development is not absolutely proven, primarily because no positive criterion exists for distinguishing between authigenic and allogenic (detrital) clay minerals. If authigenic clay minerals do exist, their contribution to estuarine sediments would be slight, because only about 1–10% of the suspended load of rivers is amorphous material, some of which is utilized by organisms (Kennedy, 1965).

Differential settling and transport also affect the distribution of clay minerals and may explain variations attributed to diagenesis. Laboratory evidence shows that clay minerals flocculate in saline water and settle differentially according to their mineralogy, in decreasing order: kaolinite and chlorite, illite, and smectite (see *Flocculation*). One would expect, for example, decreasing amounts of kaolinite to be deposited as one proceeds away from the source. Several investigators have used this concept to help explain clay mineral variations in their study areas (e.g., Biscaye, 1965; Johnson, 1973).

Although there are few, if any, reliable diagenetic changes recorded in modern nearshore sediments which would convincingly exclude alternative explanations, it is nearly indisputable that there is diagenesis of clay minerals in ancient sediments. As they become buried, the clay minerals experience changes which occur chiefly in response to increasing temperature and pressure (see *Clay–Diagenesis*).

Deep Sea

There are numerous published studies dealing with the clay mineralogy of deep-sea sediments (see references in *Clays, Deep-Sea*). The combined results of these works strongly support the continental origin of most marine clay minerals. Occurrences of marine authigenic clay mineral formation have been described notably in the case of smectite. These occurrences do not alter the general global distribution pattern of most clay minerals (Figs. 2, 3, 4, 5), which reflect very clearly the relationship between climate and terrigenous sedimentation. The trends shown by Griffin et al. (1968) are generally in good agreement with those defined by other investigators. Recent work in the Indian Ocean has improved on the distribution patterns for the area north of $30°S$. One of the more generalized observations that can be made about the clay minerals in ocean sediments is their tendency to be distributed latitudinally, in response to the latitudinal character of the global climatic regimes. This distribution is modified to some extent by the action of latitude-crossing oceanic surface currents, bottom currents, and contour and turbidity currents. An obvious latitudinal zonation is still apparent, however, particularly in the case of kaolinite and chlorite.

The appearance of kaolinite in tropical weathering profiles is well known, and the distribution of kaolinite in deep-sea sediments is correspondingly "equatorial" in character (Fig. 2). Frequently associated with the kaolinite is gibbsite, a typical end product of lateritic weathering. The decreasing kaolinite concentrations away from the equator reflect the gradual change toward cooler, less humid climates, and hence, a diminishing of the chemical weathering intensity on land. Exceptions exist, however, as in the Beaufort Sea, where relatively high concentrations of kaolinite (7–29%) have been linked to kaolinite-bearing Mesozoic sediments in Alaska (Naidu et al., 1971). The kaolinite content of the Indian and Pacific ocean is less than in the Atlantic because both oceans lack the expansive source area of lateritic soils which characterize the equatorial Atlantic, and the Pacific is not fed by any large, tropical rivers. In the Pacific the most notable concentrations of kaolinite are found off northern and eastern Australia, the source being lateritic and podsolic soils.

Chlorite is rarely the dominant clay mineral in deep-sea sediments, reflecting the fact that

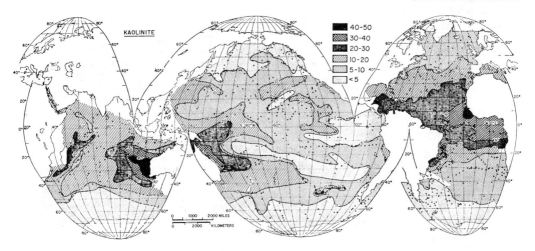

FIGURE 2. Kaolinite concentrations in the <2μ size fraction of sediments in the world ocean (from Griffin et al., 1968).

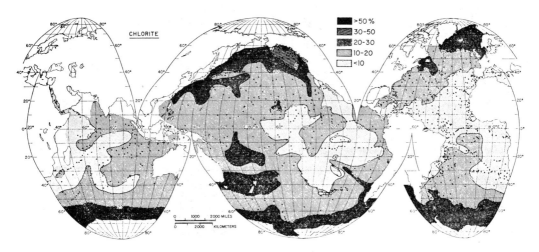

FIGURE 3. Chlorite concentrations in the <2μ size fraction of sediments in the world ocean (from Griffin et al., 1968).

it is seldom dominant in continental source areas. The distribution of chlorite is striking because of its pronounced "bipolar" character (Rateyev et al., 1968). Chlorite is susceptible to acid weathering in warm, moist climates, and hence, chlorite is consistently most abundant in the glacial marine sediments of the polar regions, where it is inherited from the metamorphic and sedimentary parent rocks of the Arctic and Antarctic. Chlorite is also found, however, in the very weak chemical-weathering environment of arid regions (Fig. 1), for example off the SE coast of Australia. Some chloritic material is also present near oceanic ridges and volcanic islands, as an alteration product (Siever and Kastner, 1967; Copeland et al., 1971). The majority of the chlorite in marine sediments, however, seems clearly to be detrital, as evidenced by the close association of chlorite abundance with continental sources.

The distribution of illite may also be described as "bipolar," although the latitudinal character of its distribution is not very striking, probably because illite is both abundant and widespread on the continents. In the Atlantic the latitudinal distribution of illite is apparently a reflection of the dilution effect of large amounts of kaolinite introduced from equatorial Africa and South America. Even so, it is noteworthy that illite still predominates throughout most of the equatorial Atlantic sediments. The dominance of illite in North Pacific sediments is a result

CLAY AS A SEDIMENT

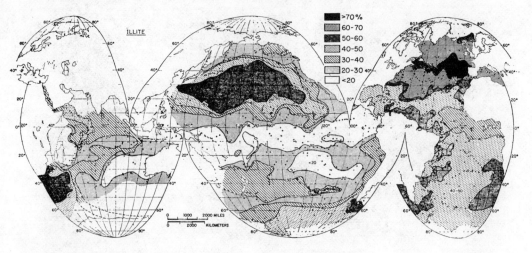

FIGURE 4. Illite concentrations in the $<2\mu$ size fraction of sediments in the world ocean (from Griffin et al., 1968).

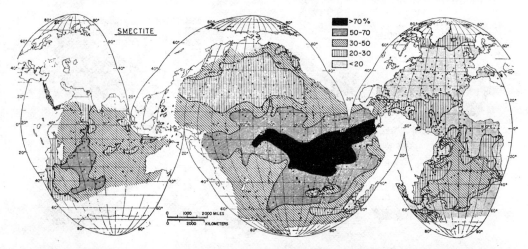

FIGURE 5. Smectite (montmorillonite) concentrations in the $<2\mu$ size fraction of sediments in the world ocean (from Griffin et al., 1968).

of eolian transport of the clay from arid regions on the Asiatic mainland.

Whereas the average illite concentration is higher N of the equator than S, the reverse is true for smectite, emphasizing the greater importance of detrital clay minerals in the northern hemisphere compared with authigenic clay minerals in the southern hemisphere. Sediments of the South Pacific are characterized by an overwhelming abundance of smectite, coinciding roughly with basaltic volcanic provinces, where it forms as a direct and rapid alteration product of basic volcanic glass (see *Volcanic Ash–Diagenesis*). Although the N/S pattern of illite/smectite distribution is much the same for the Atlantic, no strong case can be made for large-scale authigenesis of smectite in the South Atlantic, where volcanism is less widespread and there is not a striking correlation between the volcanism and smectite abundance (Biscaye, 1965). In the Caribbean, however, increased crystallinity of the marine smectite suggests its in-situ formation from volcanic material because intense chemical weathering on land would produce poorly crystalline smectite (Biscaye, 1965; see also Griffin and Goldberg, 1969).

The influence of continental detrital sources in the Indian Ocean, particularly the northern part, can readily be seen in its clay mineral distribution patterns (see *Clays, Deep-Sea*). Turbidity currents and wind are here recognized as key transporting agents in the dispersal of the clays. High concentrations of smectite are derived from the weathering of the poorly drained Deccan Plateau basalts. The Ganges and Indus rivers contribute large quantities of illite and chlorite; the lateritic soils surrounding the Bay of Bengal (Fig. 1) contribute relatively small amounts of kaolinite, principally because the major rivers drain either illite-rich soils, or, in the case of peninsular India, smectite-rich soils. Abundant kaolinite is, however, transported by winds from the ancient lateritic soils of western Australia. Coincident with the kaolinite maximum off western Australia is a smectite minimum. The remainder of the Indian Ocean, except in the glacial marine sediments off Antarctica, is characterized by abundant smectite, thought to be the product of alteration of submarine volcanics.

Pre-Quaternary Deep-Sea Clay Minerals. Information on the distribution of clay minerals in pre-Quaternary marine sediments has recently increased considerably as a result of the Deep Sea Drilling Project, and the amount of information stands to increase even more in the future as more detailed analyses are conducted. Although the effects of changes in climate, source area, transport, volcanism, etc. on clay mineral distributions in Pre-Quaternary deep-sea sediments are still not well known, a few important generalizations have already been noted. For example, some clay minerals appear to form in deep-sea sediments after deposition, e.g., *sepiolite* and/or *palygorskite* (attapulgite), which have been recovered at several drilling sites. Both of these minerals apparently characterize certain saline or hypersaline environments and have been reported in lacustrine and lagoonal sediments (Millot, 1970; Singer et al., 1972). They have only recently been encountered in the deep sea (Hathaway and Sachs, 1965), where rather special conditions, such as an unusually high concentration of magnesium ion, are necessary in order to satisfy the requirements of their formation. The association of sepiolite and palygorskite with deep-sea volcanic and tectonic features suggests the possibility that hydrothermal solutions emanating from the sea floor may be the source of the magnesium (Bonatti and Joensuu, 1968; Bowles et al., 1971; Ugolini, 1974). If so, the occurrence of sepiolite-palygorskite in ancient deep-sea sediments may serve as an indicator of past hydrothermal activity on the sea floor. Also notable is the abundant occurrence of volcanogenic smectite and the development of highly zeolitic clays (phillipsite and clinoptilolite) in the Tertiary sediments of nearly all the ocean basins, suggesting increased importance of continental and/or marine volcanic sources of supply to the deep sea (see *Clays, Deep-Sea*).

The equatorial Pacific provides an unusual opportunity to study the clays of the past because of the relatively frequent occurrence of pre-Pleistocene sediments at or near the sediment/water interface. Cores from this area date back to middle Eocene, and smectite dominates the mineralogy of sediments older than 2 m. y. In the northern equatorial Pacific these abundances were diluted during the Quaternary by large influxes of illite resulting from the increased intensity of continental glaciation (Jacobs and Hays, 1972).

Quaternary Clay-Mineral Variations. It is probably safe to assume that the distribution patterns of clay minerals in the surface sediments of the deep sea are representative for the Holocene as a whole because the conditions (climate, source, transport, volcanism) shaping these patterns have remained essentially constant during this time.

Investigations by Duncan et al. (1970) in the NE Pacific and by Zimmerman (1972) in the North Atlantic indicated lower concentrations of illite in Holocene sediments than are found in late Pleistocene sediments. In the equatorial Atlantic, Biscaye (1964) observed changes in the kaolinite:chlorite ratio associated with the Pleistocene/Holocene boundary, but he found no significant variations in clay mineral ratios in a core taken near the area studied by Zimmerman. A lack of variation has also been observed by others, and investigations of relative abundances of clay minerals in the equatorial Atlantic by Goldberg et al. (1963), Murray (1970), and Bowles (1970), conflict with some of the findings of Biscaye (1964). It is apparent that more detailed work on the clay sediments of the Quaternary is necessary.

FREDERICK A. BOWLES

References

Aristarain, L. F., 1971. Clay minerals in caliche deposits of eastern New Mexico, *J. Geol.,* 79, 75–90.

Barshad, I., 1959. Factors affecting clay formation, *Clays Clay Minerals,* 6, 110–132.

Biscaye, P. E., 1964. Mineralogy and sedimentation of the deep sea sediment fine fraction in the Atlantic Ocean and adjacent seas and oceans, Thesis, Yale University.

Biscaye, P. E., 1965. Mineralogy and sedimentation of recent deep-sea clay in the Atlantic Ocean and adjacent seas and oceans, *Geol. Soc. Am. Bull.,* 76, 803–832.

Bonatti, E., and Joensuu, O., 1968. Palygorskite from Atlantic deep sea sediments, *Am. Mineralogist,* 53, 975–983.

Booth, J. S., and Osborne, R. H., 1971. The American Upper Ordovician standard. XV. Clay mineralogy of insoluble residues from Cincinnatian limestones, Hamilton County, Ohio, *J. Sed. Petrology,* 41, 840–842.

Bowles, F. A., 1970. Mineralogy of sediments from

the equatorial Atlantic Ocean, *Am. Geophys. Union Trans.*, **51**(4), 328.

Bowles, F. A.; Angino, E. E.; Hosterman, J. W.; and Galle, O. K., 1971. Precipitation of deep-sea palygorskite and sepiolite, *Earth Planetary Sci. Lett.*, **11**, 324-332.

Copeland, R. A.; Frey, F. A.; and Wones, D. R., 1971. Origin of clay minerals in a Mid-Atlantic Ridge sediment, *Earth Planetary Sci. Lett.*, **10**, 186-192.

Degens, E. T., 1965. *Geochemistry of Sediments.* Englewood Cliffs, N.J.: Prentice-Hall, 342p.

Droste, J. B., 1961. Clay minerals in sediments of Owens, China, Searles, Panamint, Bristol, Cadiz and Danby Lake Basins, California, *Geol. Soc. Am. Bull.*, **72**, 1713-1722.

Duncan, J. R.; Kulm, L. D.; and Griggs, G. B., 1970. Clay mineral composition of Late Pleistocene and Holocene sediments of Cascadian Basin, Northeastern Pacific Ocean, *J. Geol.*, **78**, 213-221.

Goldberg, E. D.; Koide, M.; Griffin, J. J.; and Peterson, M. N. A., 1963. A geochronological and sedimentary profile across the North Atlantic Ocean, in H. Craig et al., eds., *Isotopic and Cosmic Chemistry.* Amsterdam: North Holland, 211-232.

Griffin, G. M., 1962. Regional clay mineral facies—products of weathering intensity and current distribution in the northeastern Gulf of Mexico, *Geol. Soc. Am. Bull.*, **73**, 737-767.

Griffin, J. J., and Goldberg, E. D., 1969. Recent sediments of Caribbean Sea, *Am. Assoc. Petrol. Geologists Mem. 11*, 258-268.

Griffin, J. J.; Windom, H.; and Goldberg, E. D., 1968. The distribution of clay minerals in the world ocean, *Deep-Sea Research,* **15**, 433-459.

Grim, R. E., 1968. *Clay Mineralogy.* New York: McGraw-Hill, 596p.

Hathaway, J. D., and Sachs, P. L., 1965. Sepiolite and clinoptilolite from the Mid-Atlantic Ridge, *Am. Mineralogist,* **50**, 852-867.

Jacobs, M. B., and Hays, J. D., 1972. Paleo-climatic events indicated by mineralogical changes in deepsea sediments, *J. Sed. Petrology,* **42**, 889-898.

Johnson, C. M., 1973. Occurrence and alteration of clay minerals in the Caribbean Sea, *Texas A&M Univ. Tech. Rept. 73-4-T.*

Johnson, L. J., 1970. Clay minerals in Pennsylvania soils—relation to lithology of the parent rock and other factors, *Clays Clay Minerals,* **18**, 247-260.

Keller, W. D., 1970. Environmental aspects of clay minerals, *J. Sed. Petrology,* **40**, 788-854.

Kennedy, V. C., 1965. Mineralogy and cation-exchange capacity of sediments from selected streams, *U.S. Geol. Surv. Prof. Pap. 433-D,* 28p.

Loughnan, F. C., 1970. Fine clay in the coal-barren Triassic of the Sydney Basin, Australia, *J. Sed. Petrology,* **40**, 822-828.

Mackenzie, F. T., and Garrels, R. M., 1966. Chemical mass balance between rivers and oceans, *Am. J. Sci.*, **264**, 507-525.

Millot, G., 1952. The principal sedimentary facies and their characteristic clays, *Clay Minerals Bull.*, **1**, 235-237.

Millot, G., 1970. *Geology of Clays.* New York: Springer-Verlag, 429p. (Translated from French 1964 edition.)

Mohr, E. C. J., and Baren, J. van, 1954. *Tropical Soils.* New York: Interscience, 498p.

Murray, H. H., 1970. The clay mineralogy of marine sediments in the North Atlantic at 20°N latitude, *Earth Planetary Sci. Lett.*, **10**, 39-43.

Naidu, A. S.; Burrell, D. C.; and Hood, D. W., 1971. Clay mineral composition and geologic significance of some Beaufort Sea sediments, *J. Sed. Petrology,* **41**, 691-694.

Neiheisel, J., and Weaver, C. E., 1967. Transport and deposition of clay minerals, southeastern United States, *J. Sed. Petrology* **37**, 1084-1116.

Papadakis, J., 1969. *Soils of the World.* Amsterdam: Elsevier, 208p.

Parry, W. T., and Reeves, C. C., Jr., 1968. Clay mineralogy of pluvial lake sediments, southern high plains, Texas, *J. Sed. Petrology,* **38**, 516-529.

Patterson, S. H., and Hosterman, J. W., 1962. Geology and refractory clay deposits of the Haldeman and Wrigley Quadrangle, Kentucky, *U.S. Geol. Surv. Bull. 1122-F,* 113p.

Picard, M. D., and High, L. R., Jr., 1972. Criteria for recognizing lacustrine rocks, *Soc. Econ. Paleont. Mineral. Spec. Publ. 16,* 108-145.

Quaide, W., 1958. Clay minerals from salt concentration ponds, *Am. J. Sci.,* **256**, 431-437.

Rateyev, M. A.; Kheirov, M. B.; and Shantar, A. A., 1968. Alteration of volcanic ash as a function of the physicochemical environment of diagenesis as illustrated by marine Miocene sediments of Western Kamchatka, *Acad. Sci. U.S.S.R., Dokl, Earth Sci. Sect.,* **176**, 176-179.

Siever, R., and Kastner, M., 1967. Mineralogy and petrology of some Mid-Atlantic Ridge sediments, *J. Marine Research,* **25**, 263-278.

Singer, A.; Gal, M.; and Banin, A., 1972. Clay minerals in recent sediments of Lake Kinneret (Tiberias), Israel, *Sed. Geol.,* **8**, 289-308.

Ugolini, F. C., 1974. Hydrothermal origin of the clays from the upper slopes of Mauna Kea, Hawaii, *Clays Clay Minerals,* **22**, 189-194.

Van Houten, F. B., 1953. Clay minerals in sedimentary rocks and derived soils, *Am. J. Sci.,* **251**, 61-82.

Weaver, C. E., 1958. A discussion of the origin of clay minerals in sedimentary rocks, *Nat. Acad. Sci., Publ. 566,* 159-173.

Weaver, C. E., 1967. Variability of a river clay suite, *J. Sed. Petrology,* **37**, 971-974.

Windom, H. L.; Neal, W. J.; and Beck, K. C., 1971. Mineralogy of sediments in three Georgia estuaries, *J. Sed. Petrology,* **41**, 497-504.

Winters, E., and Simonson, R. W., 1951. The subsoil, *Advanc. Agron.,* **3**, 2-92.

Zimmerman, H. B., 1972. Sediments of the New England continental rise, *Geol. Soc. Am. Bull.,* **83**, 3709-3724.

Cross-references: *Argillaceous Rocks; Clay; Clay–Diagenesis; Clay–Genesis; Clays, Deep-Sea; Clay Sedimentation Facies; Deltaic Sediments; Desert Sedimentary Environments; Estuarine Sedimentation; Fluvial Sediments; Lacustrine Sedimentation; Mudstone and Claystone; Oil Shale; Pelagic Sedimentation, Pelagic Sediments; Shale; Varves and Varved Clays; Weathering in Sediments.* See also Vol. XI, *Encyclopedia of Pedology.*

CLAY–DIAGENESIS

Sediments consisting of clays undergo obvious diagenetic changes that are expressed not only by physical characteristics (compaction, induration, fissility), but also by important mineralogical transformations. These can be studied equally well in sediments that are not clays, but contain a clay fraction. Consequently, the study of clay diagenesis concerns almost all sediments and sedimentary rocks.

The concept of diagenesis (q.v.; Dunoyer de Segonzac, 1968) is related to the transformation sediments undergo after deposition on basin floors. It is essential to define the field of diagenetic transformations because the term *diagenesis* has been used incorrectly by a school of clay mineralogists in order to name all transformations of clay minerals in sedimentary environments before their deposition (see references in Dunoyer de Segonzac, 1970). In addition, the various stages of diagenesis have been given different names by several authors, and in some cases, different stages have been given the same name.

Early Diagenesis

Early diagenesis (initial or preburial stage; *syngenesis; syndiagenesis; diagenesis* of Russian authors) concerns the first stages of burial, i.e., the top few cm or m of sediment. Ten m may represent from 10,000 yr in a zone of detrital sedimentation (deltas, alluvial fans) to 10 m.y. in a zone of deep-sea brown clays. It is apparent, therefore, that early diagenetic phenomena may present very different aspects.

In the case of terrigenous or hemipelagic sediments, early diagenetic transformations of clay minerals are very weak, if they occur at all, because clay minerals inherited from land are generally stable in the marine environment. Aggradation of illite-smectite mixed layers to illite by adsorption of K^+ has been described in many estuaries (see references in Millot, 1970), but this transformation also may be interpreted as an effect of differential settling. Illite recrystallization in marine and interstitial environments (Grim, 1968) is rarely observable and always weak. It is, moreover, masked by variations in crystallinity caused by the continental history in relation to climatic changes (Chamley, 1971). Chlorite crystallinity appears to increase in some cases of deep-sea volcanogenic sedimentation (Carroll, 1969). Smectites have shown the possibility of cationic exchange: Fe^{++} is lost to form pyrite, whereas Mg^{++} is adsorbed from interstitial solution (Drever, 1971).

In the case of oceanic sedimentation far from land, clay minerals appear to originate principally from the submarine weathering of volcanic material and generally to consist of ferriferous smectites (nontronites) which are concentrated beneath the carbonate compensation depth in so-called deep-sea brown clays together with the zeolites, phillipsite (Peterson and Griffin, 1964), and clinoptilolite (Venkatarathnam and Biscaye, 1973). Magnesian fibrous clays may also be formed on active volcanic ridges by submarine weathering (Bonatti and Joensuu, 1968).

Thus, detrital clays that have reached the floor of marine basins undergo practically no early diagenetic change, and their mineralogical composition reflects the great climatic weathering zones of the earth (see *Clay as a Sediment*). Recrystallization is very weak, and neoformation is rare, being related mainly to volcanic activity.

Late Diagenesis

Late diagenesis (burial stage; *anadiagenesis; epigenesis* of Russian authors) concerns the evolution of unconsolidated or consolidated sediments in the burial conditions of a sedimentary basin away from any abnormal geothermal gradient or any influence of regional metamorphism. Two categories of processes can be distinguished.

Compaction of Argillaceous Sediments (Rieke and Chilingarian, 1974). Recent argillaceous muds generally have a water content of 50–100%. Burial diagenesis begins with the expulsion of a part of this pore water from between the particles (Skempton, 1970) and the microstructural rearrangement of particles (Meade, 1968). This mechanical process causes changes in the chemical composition of interstitial fluids due to the ion filtration of charged clay membranes (Siever et al., 1965; Degens and Chilingar, 1967). Bivalent cations are more strongly adsorbed on clay surfaces than are monovalent cations; consequently, the latter are enriched in the solutions. The expulsion of the interlayer water of smectites requires not only a load charge but also a rise in temperature (Burst, 1969). The migration of diagenetic fluids in argillaceous formations is of great importance in the exploration of petroleum and minerals (Rumeau and Sourisse, 1972).

Transformation of Clay Minerals (Dunoyer de Segonzac, 1970). Clay minerals are sensitive to rises in temperature and to their chemical environment, i.e., the composition of the interstitial solutions. Variations of these two parameters in sedimentary formations during their diagenetic history induces mineralogical changes in pore spaces (kaolinization, illitization, chloritization, etc.). Many examples are discussed by Millot (1970). Rises in temperature and pressure, however, cause irreversible changes (Perry and Hower, 1970; Heling, 1974), which can be summarized as follows:

- *Kandites* (kaolinite group) can withstand high temperature, provided that the environment is acidic, in which case they recrystallize to form dickite, or occasionally, nacrite. In the general case of basic environments, illitization and chloritization occur.
- *Smectites* (montmorillonite group) are hydrated minerals. Rises in temperature and pressure during burial expel water from the interlayers, and the concentrated interstitial solutions provide cations which replace the molecules of water between the layers. This is an irreversible process which produces 14Å minerals (chlorites) or 10Å minerals (illites), generally passing via mixed-layer structures. The lack of smectite is usual in formations that have undergone a marked burial.
- *Mixed-layer clays* are intermediate stages that occur during degradation by weathering and during aggradation by deep diagenesis. This aggradation is the result of an incorporation of certain cations from interstitial solutions and of a rearrangement within the lattice. There are two major pathways (Fig. 1): a potassium and sodium pathway, which produces the illites, followed by the micas, possibly via a regular mixed layering of the allevardite-rectorite type; and a magnesium pathway, which produces the chlorites, possibly via a regular mixed layering of the corrensite type.

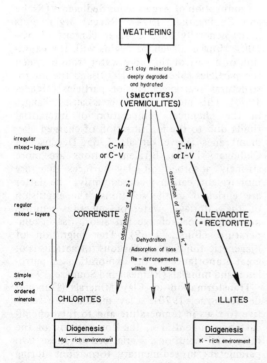

FIGURE 1. The transformation of degraded 2:1 clay minerals during diagenesis.

- *Micaceous clay minerals or illites* form a heterogeneous group in sediments that have been weakly diagenetized, and particles of diverse origin are found. They become more regular during burial and provide a valuable index of the intensity of recrystallization in deep diagenesis (see below).
- *Chlorites* are the least well-known clay minerals in diagenesis. Detrital particles can be aggraded to chlorite via the mixed-layer stage of corrensite (Fig. 1). A massive growth of chlorite is observed in deep diagenesis. Illite and chlorite are the dominant, if not the only, clay minerals present in formations that have undergone advanced diagenesis.

Advanced Diagenesis and Regressive Diagenesis

Anchimetamorphism. So-called "illite crystallinity" ("sharpness ratio" of Weaver, "half-height width" of Kübler, see references in Dunoyer de Segonzac, 1970), may define the limits of the transitional zone of *anchimetamorphism* (*metagenesis* of Russian authors). At this point, clays have lost their plasticity, and a fracture cleavage begins to develop. Sheet silicates found here are illite and chlorite; dickite, pyrophyllite, and regular mixed-layer clays sometimes occur. Corrensite becomes unstable at the top of the anchizone (Kübler, 1973), whereas allevardite persists up to the epizone (Dunoyer de Sergonzac and Heddebaut, 1971). Illites become micas (layer charge > 0.7), mostly phengites, of the 2 M polymorphic type. They accept Na^+ in the interlayer space without the occurrence of a paragonite phase. Paragonite begins to grow at the top of epizone. Temperature and pressure conditions of the anchimetamorphic zone (200–300°C, 1–2 kb) can be approached in the laboratory in the case of the formation of zeolites or by the reflectance of the carbonaceous particles (Kisch, 1969). These conditions of advanced diagenesis can be realized in two very different types of geological situation: (1) deep burial ("burial metamorphism") in geothermal fields (Muffler and White, 1969) where minerals of the zeolitic facies occur widely; and (2) peri-orogenic regions at the external fringes of a regional metamorphism ("anchimetamorphism"; Artru et al., 1969).

Regressive Diagenesis (Catamorphism). Compaction and recrystallization of clay minerals during burial are irreversible. Consequently, the uplift of sedimentary formations induces very few modifications in clays, except in the case of soils, where clay minerals may undergo dissolution, transformation, and neoformation. This retrograde diagenesis before outcropping is thus of little importance for argillaceous rocks. On the other hand, the uplift of igneous and metamorphic rocks will be the seat of clay minerals genesis—illite in feldspars and chlorite in mafic minerals (Lelong and Millot, 1966). This aspect of the silicate cycle in the earth's crust is often neglected, although quantitatively of primary importance. Moreover, aggradation of mica by progressive diagenesis and degradation of mica by regressive diagenesis show an interesting symmetry in the geochemical processes.

Clay Diagenesis in the Geochemical Cycle

Clays are the stable phase of aluminosilicates at the earth's surface; they result from the interaction of the lithosphere and the hydrosphere, and are formed during regressive diagenesis and weathering. Because of their sheet structure, small size, water content, and ion-exchange capacity, the clay minerals are very sensitive to the geochemical and thermodynamic environment. They are thus marked by the environments through which they pass and are therefore valuable markers of diagenetic events.

The crystal chemistry of clays shows a diagenetic evolution characterized by the following features: dehydration of the interlayer spaces; adsorption of K and Mg ions from the interlayer solution; redistribution of ions within the lattices; ordering of the lattices; and growth of crystals. These conclusions are supported by field observations (Dunoyer de Segonzac, 1970) and also by experimental data (Velde, 1972). The dioctahedral 2:1 clay minerals furnish the most demonstrative examples (Fig. 2): diagenetic reactions consist of an increase in the tetrahedral charge (Al) and in layer charge which permits the fixation of alkaline ions (K and Na). During their diagenesis, clays permit a partial reincorporation of Na, K, Mg, and Fe, from solutions back into the aluminosilicate mineral phase, before feldspars and biotites begin to grow.

GILBERT DUNOYER DE SEGONZAC

References

Artru, P.; Dunoyer de Segonzac, G.; Combaz, A.; and Giraud, A., 1969. Variations d'origine sédimentaire et évolution diagénétique des caractères palynologiques et géochimiques des Terres Noires jurassiques en direction de l'arc alpin (France, sud-est), *Bull. Centre Rech. SNPA, Pau,* **3,** 357-376.

Bonatti, E., and Joensuu, O., 1968. Palygorskite from Atlantic deep sea sediments, *Am. Mineralogist,* **53,** 975-983.

Burst, J. F., 1969. Diagenesis of Gulf Coast clayey sediments and its possible relation to petroleum migration, *Bull. Am. Assoc. Petrol. Geologists,* **53,** 73-93.

Carroll, D., 1969. Chlorite in central-north Pacific Ocean sediments, *Internat. Clay Conf., Tokyo, Proc.,* **1,** 335-338.

Chamley, H., 1971. Recherches sur la sédimentation argileuse en Méditerranée, *Sci. Géol. Strasbourg,* **35,** 209p.

Degens, E. J., and Chilingar, G. V., 1967. Diagenesis of subsurface waters, in G. Larsen and G. V. Chilingar, eds., *Diagenesis in Sediments.* Amsterdam: Elsevier, 477-502.

Drever, J. I., 1971. Early diagenesis of clay minerals, Rio Ameca Basin, Mexico, *J. Sed. Petrology,* **41,** 982-994.

Dunoyer de Segonzac, G., 1968. The birth and development of the concept of diagenesis, *Earth-Sci. Rev.,* **4,** 153-201.

Dunoyer de Segonzac, G., 1970. The transformation of clay minerals during diagenesis and low grade metamorphism: a review, *Sedimentology,* **15,** 281-346.

Dunoyer de Segonzac, G., and Heddebaut, C., 1971. Paléozoïque anchimétamorphique à illite, chlorite, pyrophyllite, allevardite et paragonite dans les Pyrénées basques, *Bull. Serv. Carte Géol. Alsace-Lorraine, Lorr. Strasbourg,* **24,** 277-290.

Grim, R. E., 1968. *Clay Mineralogy.* New York: McGraw-Hill, 596p.

Heling, D., 1974. Diagenetic alteration of smectite in argillaceous sediments of the Rhinegraben (SW-Germany), *Sedimentology,* **21,** 463-472.

Kisch, H. J., 1969. Coal-rank and burial-metamorphic mineral facies, in P. A. Schenk and I. Havenaar, eds., *Advances in Organic Geochemistry.* Oxford: Pergamon, 407-425.

Kübler, B., 1973. La corrensite, indicateur possible de milieux de sédimentation et du degré de transformation d'un sédiment, *Bull. Centre Rech. SNPA, Pau,* **7,** 543-556.

Lelong, F., and Millot, G., 1966. Sur l'origin des minéraux micacés des altérations latéritiques. Diagenèse régressive. Minéraux en transit, *Bull. Serv. Carte Géol. Alsace-Lorraine, Strasbourg,* **19,** 271-287.

Meade, R. H., 1968. Compaction of sediments underlying areas of land subsidence in central California, *U.S. Geol. Surv. Prof. Pap. 497-D,* 39p.

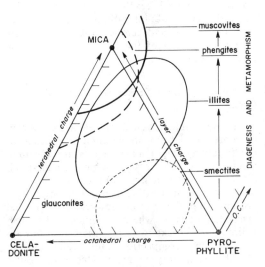

FIGURE 2. Diagenetic and metamorphic evolution of 2:1 clay minerals as shown by lattice parameters. Tetrahedral charge + octahedral charge = layer charge.

Millot, G., 1970. *Geology of Clays*. New York: Springer; Paris: Masson, 429p.

Muffler, L. J. P., and White, D. E., 1969. Active metamorphism of Upper Cenozoic sediments in the Salton Sea geothermal field and the Salton trough, southeastern California, *Geol. Soc. Am. Bull.,* **80**, 157–182.

Perry, E., and Hower, J., 1970. Burial diagenesis in Gulf Coast pelitic sediments, *Clays Clay Minerals,* **18**, 165–177.

Peterson, M. N. A., and Griffin, J. J., 1964. Volcanism and clay minerals in the south-eastern Pacific, *J. Marine Research* **22**, 13–21.

Rieke, H. H., and Chilingarian, G. V., 1974. *Compaction of Argillaceous Sediments*. Amsterdam: Elsevier, 424p.

Rumeau, J. L., and Sourisse, C., 1972. Compaction, diagenèse et migration dans les sédiments argileux, *Bull. Centre Rech. SNPA, Pau,* **6**, 313–345.

Siever, R., Beck, K. C., and Berner, R. A., 1965. Composition of interstitial waters of modern sediments, *J. Geology,* **73**, 39–73.

Skempton, A. W., 1970. The consolidation of clays by gravitational compaction, *J. Geol. Soc. London Quart.,* **125**, 373–412.

Velde, B., 1972. Phase equilibria for dioctahedral expandable phases in sediments and sedimentary rocks, *Internat. Clay Conf. Proc.,* **1**, 285–300.

Venkatarathnam, K., and Biscaye, P. E., 1973. Deep-sea zeolites: variations in space and time in the sediments of the Indian Ocean, *Marine Geol.,* **15**, M11–M17.

Cross-references: *Argillaceous Rocks; Authigenesis; Clastic Sediments–Lithification and Diagenesis; Clay; Clay–Genesis; Clays, Deep-Sea; Compaction in Sediments; Diagenesis; Shale.*

CLAY–GENESIS

Silicate rocks comprise 90% of the earth's crust, and are generally composed of quartz, feldspars, micas, amphiboles, pyroxenes, and various accessory minerals. When exposed to the atmosphere, these minerals are subject to attack by water, oxygen, and carbon dioxide. Water penetrates the pores, cleavages, and micro-openings in the minerals and dissolves the more soluble constituents. The reconstitution of the residue with water, oxygen, CO_2, and some of the dissolved ions, forms new silicate minerals (clays) which are in equilibrium at atmospheric conditions on the surface of the earth. Clays are principally composed of clay minerals which are very small flakes or crystals usually $< 2\mu$ in size.

Origins of Clay

The processes of neoformation, inheritance, and transformation are responsible for the clay minerals in soils, marls, argillites, and shales. *Neoformation,* or *authigenesis* (q.v.), is the in-situ crystallization of clay minerals from ions present in the environment. *Inheritance* is the detrital accumulation of previously formed clay minerals without any modification. *Transformation* is the alteration of previously formed clay minerals due to geochemical changes in the environment. There are two types of transformation: (1) degradation, the removal of ions from the clay structure, and (2) aggradation, the addition of ions to the clay structure.

All three processes are governed by two environmental types: (1) a *leached* environment and (2) a *confined* environment. The leached environment destroys the primary silicate minerals, degrades the layer silicates, and transforms clay minerals. The confined environment is responsible for the aggradation of degraded clay minerals and for the formation of new clay minerals from concentrated and confined solutions.

Thermodynamic Diagrams. Thermodynamic diagrams permit investigators to predict the succession of mineral formation as the pH and the ionic concentration of the solution change. Fig. 1 consists of two thermodynamic diagrams which show the stability fields of (1) gibbsite, kaolinite, Na-montmorillonite (smectite) and albite; and (2) gibbsite, kaolinite, muscovite (illite), and K-feldspar. Arrow W represents the path of reactions in the course of weathering as the solution gradually becomes depleted in $[H_4SiO_4]$, $[Na^+]$, $[K^+]$, $[Ca^{++}]$, and $[Mg^{++}]$. In contrast, arrow D represents the reaction path in the diagenetic zone if the solution becomes more and more enriched in silica and in the various cations. From these diagrams, the possible chemical reactions involved in the successive mineral formations can be determined—pressure and temperature act only to change the speed of the reactions and to displace the equilibria.

Clay Minerals in the Crustal Geochemical Cycle

The geochemical cycle of the earth's crust can be represented by a simple circle consisting of the following stages: (1) the *weathering zone,* where the earth's crust is subject to physical and chemical weathering; (2) the *sedimentation zone,* where the weathered products are transported, deposited, and reworked; (3) the *diagenetic zone,* where sediments become sedimentary rocks by processes of burial, dehydration, compaction, and cementation; (4) the *metamorphic zone,* where the sedimentary rocks are heated, compressed and partially or completely melted into metamorphic and granitic crystalline rocks; and

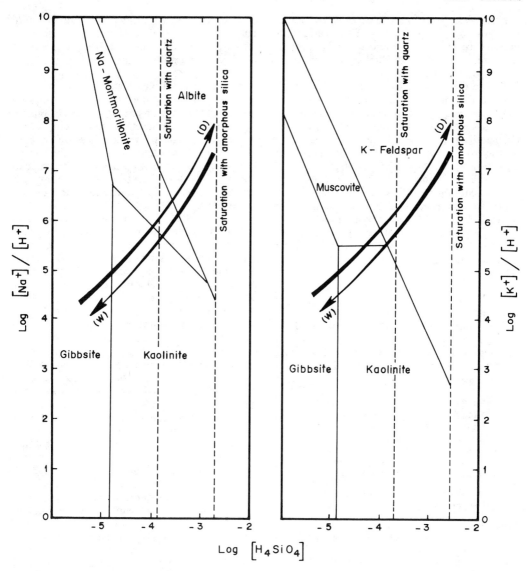

FIGURE 1. Stability diagrams of some minerals as a function of pH and of concentrations in silica and alkaline ions (after Garrels and Christ, 1965; Tardy, 1969; Millot et al., 1974). The arrows W show the succession of minerals during weathering: feldspar, mica, clay minerals, gibbsite. The arrows D indicate the opposite successions of diagenesis.

(5) the *retrograde metamorphic and diagenetic zones*, where the crystalline rocks are brought back to the surface (the weathering zone) by volcanic and tectonic processes.

The genesis of clay minerals takes place to some extent in the retrograde diagenetic zone and to a much greater extent in the weathering zone. These clay minerals are transported, deposited, and reworked in the sedimentation zone and are transformed and diversified in the diagenetic zone. Ultimately they are destroyed in the metamorphic zone, leading to the formation of micas and feldspars.

Retrograde Diagenetic Zone. During tectonic ascent, minerals that were in equilibrium at high temperatures and pressures are no longer in equilibrium; and, as the diagenetic zone is traversed, primary silicate minerals give rise to clay minerals by retrograde diagenesis. Feldspars and other aluminum silicates give rise to sericites, damourites, pinites, and illites by neoformation. Also, ferromagnesian minerals (amphiboles, pyroxenes, biotite) give rise to chlorites by degradation and neoformation. Inheritance is of little importance because no clay minerals occur in the crystalline rocks deep in the earth's crust.

Weathering Zone. The weathering zone is the most conducive place for the genesis of clay minerals. Weathered materials and soils may inherit clay minerals originating from retrograde diagenesis and from the weathering of clay-rich parent rocks, such as shales and argillites, which have been brought back to the surface without entering the deep zone of metamorphism. Consequently, inheritance plays an important role in the accumulation of clay minerals in the weathering zone.

Also, in the weathering zone, solutions penetrate the pores, cleavages, and microcracks of the primary rock-forming materials. The reaction:

primary mineral ⇋ solution ⇋ secondary mineral

proceeds to the right, and a sequence of secondary minerals develop by neoformation from nonlayered minerals and by transformation (through degradation) from layered silicates. Neoformation from nonlayered silicates may progress as follows:

Plagioclase → (sericite) → vermiculite
 → montmorillonite → Al-smectite
 → kaolinite → gibbsite
K-feldspar → montmorillonite → kaolinite
 → gibbsite
Glass → gels (allophanes) → montmorillonite
 → halloysite → kaolinite → gibbsite

Transformation by degradation of layered silicates includes the following process:

Biotite → (chlorite) → vermiculite
 → montmorillonite → Al-smectite
 → kaolinite → gibbsite
Chlorite → mixed layer (chlorite-vermiculite)
 → vermiculite → mixed layer (vermiculite-montmorillonite) → montmorillonite → gibbsite
Illite → mixed layer (illite-vermiculite)
 → vermiculite → mixed layer (vermiculite-montmorillonite) → montmorillonite
 → kaolinite → gibbsite

On a larger scale, the neoformation of clay minerals results as solutions migrate downward through sediments and soil profiles in weathering environments. This migration takes place through the macropores in sedimentary material. Clay minerals form in the following order: *gibbsite* from alumina (gels and complexes) in an extremely silica-desaturated medium (<1 ppm SiO_2); *montmorillonite* and *Al-Mg-Fe smectites* in a slightly alkaline medium, by combination of Si, Fe, and Mg with autochthonous Al; *attapulgite* and *sepiolite*, magnesian clay minerals, in the calcareous crusts of semiarid regions; and *sodium silicates* (analcime, magadiite, mordenite, etc.) in arid lowlands.

At the termination of weathering, in a constant environment, the clay minerals forming in the micropores of primary minerals and the clay minerals originating in the macropores of weathered sedimentary materials are in equilibrium with a single solution impregnating the whole system. A single secondary clay mineral develops, e.g., montmorillonite in vertisols, kaolinite in lateritic soils, and gibbsite in bauxites.

The genesis of clays in the weathering zone is extremely dependent on climate. In glacial regions, the clays are inherited because chemical weathering is limited, and physical weathering only provides unaltered, fine-grained particles (illites and chlorites). The mild weathering of temperate regions gives rise to mixed-layer clays, vermiculite, and montmorillonite. In semiarid and subtropical zones, ferriferous smectites prevail in the lowlands. In equatorial and humid tropical regions, the landscape is covered by a kaolinitic and sometimes a gibbsitic mantle. All these clay minerals, formed by weathering, may be transported and deposited elsewhere.

Sedimentation Zone. In the sedimentation zone, clay minerals are principally inherited, sometimes transformed, and rarely neoformed. Sedimentary basins receive material in two forms: (1) detrital mineral particles and (2) ions in solution. When the clay accumulation is entirely due to detrital, mechanical, or allochthonous processes, and when the sedimentary environment does not impose any change on the clay minerals, the clay sediments are inherited. It is generally agreed that the chief mechanism in the sedimentation of clay minerals is *inheritance*. Numerous eolian, glacial, fluvial, lacustrine, and marine deposits are detrital, i.e., the clay minerals come from elsewhere. The clay minerals of many geological sequences, such as present-day marine muds, molasse and flysch, piedmont sediments, iron- and coal-bearing facies, and red beds are inherited. The knowledge of the clay minerals of these sequences permit paleogeographic and paleoclimatic reconstructions (see *Clay as a Sediment*).

Many clay minerals are sensitive to changes in the geochemical environment and are, consequently, subject to *transformations* by degradation and aggradation. During degradation, the interlayer cations are leached, and the composition of the clay mineral's lattice is modified. If these degraded minerals are carried to an environment more saturated in cations than that from which they originated, they will undergo aggradation, reorganize their structure, and even grow. Excellent examples of aggradation have been described in the Triassic clays of France and Morocco (Millot, 1964). Under the influence of a cation-rich environment, there is a progressive transition from degraded illites to mixed-layer clays (illite-chlorite), to corrensites, and then to well-crystallized, trioctahedral magnesian chlorites. These transformations are rarely spectacular and often tenuous; nevertheless, they have been observed in numerous cases, and several indirect proofs that they occur have been pointed out: (1) boron is much more abundant in the structure of illites that have formed in saline environments, due to transformations; (2) Rb-Sr age determinations in clay minerals require transformations which allow rubidium to enter the clay structure; and (3) the balance of silica in marine sedimentation indicates that siliceous fauna and flora are not sufficient to utilize all the silica brought by rivers; consequently, silica must be fixed by clay minerals by transformation.

In-situ crystallization from solution (*neoformation*) is the third mechanism for the genesis of sedimentary clay minerals. This mechanism does not occur frequently and is even rare, but it has an important geochemical significance in that the mineral exactly characterizes the environment in which it originates.

Transformed and neoformed clays occur with chemically precipitated carbonates, silicates, phosphates, and sulfates in great quantities. The suite of clay minerals accompanying these precipitates is montmorillonite, attapulgite, sepiolite, and stevensite. The first two clays are inherited and transformed; the last two are neoformed, and experimental synthesis (under natural conditions) of these minerals has been performed. Other clays that can form by neoformation in specific geochemical environments are glauconite (q.v.), which may also be a product of transformation of inherited clay minerals, and montmorillonite, which may form from the dissolution products of deep-sea volcanic ash (see *Volcanic Ash–Diagenesis*). If the environment is enriched in silica but depleted in the various cations, the genesis of clay minerals may be retarded by the direct precipitation of silica in the form of opal, cristobalite, or microcrystalline quartz.

Topography may also influence the genesis of clay minerals in the sedimentary environment. If the land has a rugged relief, the clay material released from the rocks and soil is delivered directly to the sedimentary basin, and inheritance prevails. If the land has an eroded relief and is affected by chemical weathering, clay particles are still delivered to the basin, but these minerals tend to react with interstitial, cation-rich solutions, which originate from chemical weathering, and transformations by aggradation occur. If the land has a very flat relief and is affected by intense evaporation, the detrital supply is negligible, and neoformed clay minerals prevail.

Diagenetic Zone. The diagenetic zone is the zone of the earth's crust located between sedimentation and metamorphism. Temperatures do not exceed 200°C and pressures remain lower than 3 kb. In the diagenetic zone, clay minerals are inherited from sedimentation and are buried, dehydrated, and transformed. The succession of these transformations can be seen in Fig. 1, by following arrow D. Kaolinite recrystallizes into common clay minerals, and smectites and mixed-layer clays transform by aggradation into illite in a potassium-rich environment and into chlorite in a magnesium-rich environment (see Fig. 1 in *Clay–Diagenesis*). Illites (sericites) and chlorites are reconstructed during diagenesis (as they were constructed during retrograde diagenesis) by mechanisms of aggradation, which act in the reverse direction of weathering.

Finally, in the metamorphic zone the clay minerals are destroyed, and new minerals (micas and feldspars) are formed in equilibrium with the higher temperatures and pressures of the metamorphic environment.

GEORGES MILLOT

References

Bocquier, G., 1971. Genèse et evolution de deux toposéquences de sols tropicaux du Tchad. Interprétation biogéodynamique, *Mém. O.R.S.T.O.M.*, 62, 1973, 350p.

Garrels, R. M., and Christ, C. L., 1965. *Solutions, Minerals, and Equilibria.* New York: Harper & Row, 450p.

Glazovskaya, M. A., 1968. Geochemical landscapes and types of geochemical soil sequences, *IXth Intern. Congr. Soil Sci. Adelaide,* 4, 303-312.

Helgeson, H. C., 1969. Thermodynamics of hydrothermal systems at elevated temperatures and pressures, *Am. J. Sci.,* 267, 729-804.

Isphording, W. C., 1973. Discussions of the occurrence and origin of sedimentary palygorskite-sepiolite deposits, *Clays and Clay Minerals,* 21, 391-402.

Jackson, M. L., 1959. Frequency distribution of clay minerals in major great soil groups as related to the factors of soil formation, *Proc. 6th Natl. Conf. Clays and Clay Minerals, 1957.* New York: Pergamon 133-143.

Jackson, M. L., 1965. Clay transformation in soil genesis during the Quaternary, *Soil Sci.,* 99, 15-22.

Jackson, M. L., 1968. Weathering of primary and secondary minerals in soils, *IXth Intern. Congr. Soil Sci. Adelaide*, **4**, 281–292.

Keller, W. D., 1964. Processes of origin and alteration of clay minerals, in C. I. Rich and G. W. Kunze, eds., *Soil Clay Mineralogy: A Symposium*. Chapel Hill: Univ. of North Carolina Press, 3–76.

Loughnan, F. C., 1969. *Chemical Weathering of the Silicate Minerals*. New York: Elsevier, 154p.

Lucas, J., 1962. La transformation des minéraux argileux dans la sédimentation. Etudes sur les argiles du Trias, *Serv. Carte Géol. Alsace-Lorraine, Mém. 23*, 202p. Trans available, U.S. Dept. Commerce, Clearinghouse Fed.-Sci. Tech. Info., Springfield, Va.

Millot, G., 1964. *Géologie des Argiles:* Paris: Masson, 499p. Translated, 1970: *Geology of Clays*. New York: Springer-Verlag; London: Chapman and Hall, 429p.

Millot, G.; Dunoyer de Segonzac, G.; Lucas, J; Paquet, H.; Tardy, Y.; and Trauth, N., 1978. The chemistry of formation of natural clays, Chapter XI of *Chemistry of Clays and Clay Minerals*. London: Clay Minerals Group, British Mineralogical Society, in press.

Paquet, H., 1969. Evolution des minéraux argileux dans les altérations et les sols des climats méditerranéens et tropicaux à saisons contrastées, *Serv. Carte Géol., Alsace-Lorraine, Mém, 30*, 1970, 210p.

Pedro, G., 1968. Distribution des principaux types d'altération chimique à la surface du globe. Présentation d'une esquisse géographique, *Rev. Géogr. phys. Géol. Dynam.*, **10**, 457–470.

Tardy, Y., 1969. Géochimie des altérations. Etude des arènes et des eaux de quelques massifs cristalins d'Europe et d'Afrique, *Serv. Carte Géol. Alsace-Lorraine, Mém.* 31, 199p.

Tardy, Y.; Bocquier, G.; Paquet, H.; and Millot, G., 1973. Formation of clay from granite and its distribution in relation to climate and topography, *Geoderma*, **10**, 271–284.

Weaver, C. E., 1959. The clay petrology of sediments, *Proc. 6th Natl. Conf. Clays and Clay Minerals, 1957*. New York: Pergamon, 154–187.

Cross-references: *Authigenesis; Clay; Clay as a Sediment; Clay–Diagenesis; Clays, Deep-Sea; Diagenesis; Shale; Weathering of Sediments.* Vol. XI: *Clay Formation, Genesis.*

CLAY-PEBBLE CONGLOMERATE AND BRECCIA

Clay-pebble (or flat-pebble) conglomerates and breccias are relatively uncommon rocks characterized by numerous rudaceous fragments composed of well-indurated, clay-sized particles (Fig. 1). The lithology of both the rudaceous fragments and the surrounding matrix is extremely variable. The fragments are generally tabular, of heterogeneous sizes, and may be either randomly or uniformly arranged with respect to bedding. The matrix

FIGURE 1. Flat-pebble breccia (polished section). Bertie-Akron Dol. (U. Sil.), Hagersville, Ont. Diameter of coin 18 mm.

may resemble the clay-sized particles of the rudaceous fragments or be strikingly dissimilar; the fragments and matrix may be composed of carbonate or terrigenous clastic materials.

Origin

The rudaceous fragments most commonly originate as mud crusts—the loosely attached, upper laminae of partially dried and cracked sediment (see *Mud Cracks*). The planar dimensions of the crusts are determined by such factors as (1) amount, pH, and salinity of the interstitial water; (2) chemical and physical properties of the mud; (3) rate and period of dehydration; and (4) organisms on or in the mud. Mud-crust thickness depends upon the depth to significant changes in texture or internal sedimentary structures (Fig. 2).

The induration of mud crusts during dehydration is a result of natural cohesion of the clays, "setting" of included lime, or binding by included algal filaments. Thoroughly dried mud crusts maintain their identity as tabular constituents in clay-pebble conglomerates and breccias. In cases where the original cracks surrounding the crusts are filled with sediment and the crusts are incorporated, essentially in situ, into later sediments, the resulting rocks are called *intraformational* or *desiccation conglomerates* and *breccias* (Pettijohn, 1957; Hyde, 1908).

FIGURE 2. Slightly curled mud crust, a potential tabular constituent of clay-pebble conglomerate or breccia.

Rewetting does not necessarily destroy well-indurated mud crusts. Softened crusts with abundant binding algal filaments may even float and undergo significant transportation (Fagerstrom, 1967); wind may also cause movement of curled crusts (Young, 1935). Transportation by either water or wind may result in burial of the crusts in environments quite alien to the one in which they originated. If the arrangement of the mud crusts at the time of burial is imbricate, the resulting rock is called an *edgewise conglomerate* or *breccia* (Fig. 3).

Geologic Significance

Clay-pebble conglomerates and breccias associated with mud cracks (q.v.) and other shrinkage structures represent essentially in-situ accumulations of mud crusts and are excellent evidence of deposition in such subaerial and ephemeral aquatic environments as ancient flood plains, lakes, and deltas (Barrell, 1906). They are also generally associated with unconformities and are particularly abundant in limestones, dolomites, and red beds containing such primary sedimentary structures as ripple marks, raindrop impressions, and rill marks.

Conversely, edgewise conglomerates composed of transported mud crusts may be of subaqueous origin, and are associated with slump structures, contorted bedding, oolites, stromatolites, or stream-channel deposits; they may occur in either conformable or unconformable rock sequences.

J. A. FAGERSTROM

References

Barrell, J., 1906. Relative geological importance of continental, littoral, and marine sedimentation. Part III, Mud cracks as a criterion of continental sedimentation, *J. Geol.*, 14, 524-568.

Fagerstrom, J. A., 1967. Development, flotation, and transportation of mud crusts—neglected factors in sedimentology, *J. Sed. Petrology*, 37, 73-79.

Hyde, J. E., 1908. Desiccation conglomerates in the coal-measures limestone of Ohio, *Am. J. Sci.*, 25, 400-408.

Pettijohn, F. J., 1957. *Sedimentary Rocks.* New York: Harper & Row, 718p.

Pettijohn, F. J., and Potter, P. E., 1964. *Atlas and Glossary of Primary Sedimentary Structures.* New York: Springer-Verlag, 370p.

Young, R. B., 1935. A comparison of certain stromatolitic rocks in the Dolomite Series of South Africa with marine algal sediments in the Bahamas, *Geol. Soc. S. Africa, Trans.*, 37, 153-162.

Cross-references: *Breccias, Sedimentary; Conglomerates; Mud Cracks; Syneresis.* Vol. X: *Pseudo-stretched Pebble Conglomerate.*

CLAYS, DEEP-SEA

The term *deep-sea clays* carries ambiguities as does much of the nomenclature of deep-sea sedimentation. If the user's viewpoint is granulometric, the term can include all the sediments less than 1/256 mm (4μ) in diameter that are deposited in the deep ocean, irrespective of composition or origin. This usage overlaps several other terms commonly found in the literature which themselves are ambiguous because of inadequate or variable definition and usage—terms such as red clay (q.v.), brown clay, ooze, lutite, or pelagic sediment (see *Pelagic Sedimentation*). These terms carry a range of descriptive, granulometric and/or generic implications and have become almost useless. A more useful definition for deep-sea clays is a mineralogic one in which "clays" refers primarily to a group of sheet silicates called "clay minerals," which also happen to

FIGURE 3. Edgewise breccia overlain by stromatolitic dolomite. Bertie-Akron Dol. (U. Sil.), Hagersville, Ont. Length of scale 6 inches (15.2 cm).

be fine-grained, thereby excluding from discussion fine-grained biogenic material such as calcium carbonate and opaline silica, while including some associated nonsheet aluminosilicates.

Because the clay minerals are generally very fine grained (almost entirely smaller than 20μ but predominantly smaller than 2μ) they have been principally studied using x-ray diffraction techniques. These studies began in the 1930s and were continued on a limited scale through the 1950s. It was not until the 1960s, however, that x-ray studies were attempted on sufficient scale to permit a view of the broad outlines of clay mineral distribution in the world oceans. The work of this era included studies of the Atlantic by Yeroschev-Shak (1962), and Biscaye (1965); of the Pacific Ocean by Gorbunova (1963), Griffin and Goldberg (1963), and Heath (1969); and of the Indian Ocean by Goldberg and Griffin (1970) and Venkatarathnam and Biscaye (1973). Other studies of smaller areas such as the Mediterranean and Caribbean seas and the Gulf of Mexico have also been published (see references in *Clay as a Sediment*). During this period, several workers summarized knowledge of clay mineral distributions for the world oceans, notably Rateev et al. (1969) and Griffin et al. (1968), whose figures showing global scale clay mineral distributions appear in *Clay as a Sediment*.

In a complex multiphase system like that of marine sediments, it is extremely difficult to identify individual clay mineral species. Most workers have characterized the sediments with respect to the relative abundance of four major clay mineral groups or families—the kaolinite, chlorite, smectite (or montmorillonite), and illite groups—each of which has different origins and therefore carries different information about the processes controlling its distribution in the deep sea.

Kaolinite

The kaolinite group, or kaolinite, is formed in soils of continental areas undergoing intense chemical weathering, a process that leaves the soil relatively enriched in alumina and is characteristic of tropical climates. Its occurrence in deep-sea sediments, therefore, represents a continental detrital influence of primarily low-latitude origins. In the Equatorial Atlantic Ocean a kaolinite-rich zone extends from South America to Africa, with highest concentrations immediately adjacent to the continents. Kaolinite abundance decreases toward higher latitudes in the North and South Atlantic. A similar pattern is seen in the distribution of the mineral gibbsite, which is an aluminum hydroxide and represents a further step in the tropical aluminum-enrichment weathering processes. It is not strictly a clay mineral, but its parallel distribution to that of kaolinite confirms the continental detrital origin of those sediments enriched in it. In the Pacific and Indian oceans, the origin of kaolinite in the continental tropics is evidenced by its abundance in sediments adjacent to Australia and Madagascar. The less extensive distribution of kaolinite in these oceans is consistent with the lower ratio of adjacent tropical land mass to oceanic area. The distribution of this mineral of continental origin provides perhaps the most unequivocal index of the influence of detrital contribution to deep-sea sedimentation, and its origin in a distinctive continental weathering regime makes it potentially useful as an indicator of paleoclimate.

Chlorite

By contrast, the clay mineral chlorite is most abundant in deep-sea sediments at high latitudes. Chlorite-group minerals occur as common mineral phases in both low-grade metamorphic and old (pre-Mesozoic) sedimentary rocks as well as, more rarely, being formed in modern soils at temperate to high latitudes. They are extremely sensitive to chemical weathering and are, therefore, more common in sediments derived from continental regions where physical, as opposed to chemical, weathering processes are dominant. Bands of sediment relatively enriched in chlorite, therefore, rim the North Atlantic and Pacific and the Antarctic continent, with diminishing abundance equatorward. Because the distributions of chlorite and kaolinite are controlled by opposing climate- (latitude-) related processes, the ratio of abundances of the two minerals provides a latitude-sensitive parameter and has been used to deduce oceanic transport mechanisms such as deep boundary currents which tend to displace sediment from high to low latitude.

Smectite

The distribution of smectite (or montmorillonite) in global deep-sea sediments provides a less clear-cut indication of its provenance because of several possible origins. Smectites occur in several types of continental soils ranging from poorly drained tropical regions to desert areas, thus precluding a clear, climate-related signal in smectites of detrital origin. In addition, smectites have long been known as common submarine and subaerial alteration products of volcanic material. Indications of each of these origins can be seen in different areas. The clearest such indication comes from the very high concentration of smectite in sediments of the South Pacific. The parallel occurrence of the zeolite phillipsite (or, in some sediments, clinoptilolite), also common alteration products of volcanic material, confirms the volcanogenic origin of much of the sediment in this ocean, which has the highest ratio of oceanic area to adjacent continental runoff. Large areas in the western Indian Ocean, centering principally on the Mid-Indian Ridge, and possibly in the central South Atlantic, centering on the Mid-Atlantic Ridge, are also influenced by volcanogenic smectite, although these sediments do not contain the abundant phillipsite of the South Pacific. Other areas such as the western Bay of Bengal and the North Atlantic contain significant smectite probably of continental detrital origin, as indicated by known adjacent soil types and by the distribution of smectite crystallinity (Biscaye, 1965). Thus, to a first approximation, high smectite concentrations (and associated zeolites) may be considered indicative of volcanogenic influence, or the lack of detrital influence, on deep-sea sedimentation. If other ocean basins were protected from detrital sediment sources as is the South Pacific, they probably would also be characterized by low nonbiogenic rates of sedimentation and high concentrations of smectite and zeolite.

Illite

The fourth clay mineral group—illite—is the most abundant and consists of a family of fine-grained micaceous minerals. The minerals operationally defined as illite in the literature have a broader range of origins and occur throughout a wider range of weathering regimes than the other clay minerals, factors which contribute to illite being overall the most abundant clay mineral in deep-sea sediments. Because of its widespread continental occurrence, it is more difficult to discern specific sources of illite in deep-sea sediment and to deduce transport mechanisms from the distributions than it is for a clay such as kaolinite. The Ganges, Amazon, and other major rivers draining mountainous regions, however, are discernable as point sources of illite-rich sediments to the deep sea. The most striking illite-rich zone is the broad band across the North Pacific which coincides with a band of high quartz concentration and represents wind-borne transport of continental dust from arid regions (see *Eolian Dust in Marine Sediments*). The same feature is not seen in the North Atlantic because continental detritus from other sources masks this eolian effect.

Several other fine-grained mineral particles may be associated with deep-sea clays, for example, quartz, amphibole, and pyrophyllite. As with the clay minerals, the occurrence and distributions of these minerals have been used in some areas to extract information from deep-sea cores concerning genesis, provenance, and continental, atmospheric and oceanic mechanisms of transport.

PIERRE E. BISCAYE

References

Biscaye, P. E., 1965. Mineralogy and sedimentation of recent deep-sea clay in the Atlantic Ocean and adjacent seas and oceans, *Geol. Soc. Am. Bull.,* 76, 803–832.

Goldberg, E. D., and Griffin, J. J., 1970. The sediments of the northern Indian Ocean, *Deep-Sea Research,* 17, 513–537.

Gorbunova, Z. N., 1963. Clay minerals in the sediments of the Pacific Ocean (in Russian), *Litologia y Poleznye Iskopamye,* 1, 28–42.

Griffin, J. J., and Goldberg, E. D., 1963. Clay mineral distributions in the Pacific Ocean, in M. N. Hill, ed., *The Sea,* vol. III. New York: Interscience, 728–741.

Griffin, J. J.; Windom, H.; and Goldberg, E. D., 1968. The distribution of clay minerals in the world ocean, *Deep-Sea Research,* 15, 433–459.

Heath, G. R., 1969. Mineralogy of Cenozoic deep-sea sediments from the Equatorial Pacific Ocean, *Geol. Soc. Am. Bull.,* 80, 1997–2018.

Rateev, M. A.; Gorbunova, Z. N.; Lisitsyn, A. P.; and Nosov, G. L., 1969. The distribution of clay minerals in the ocean, *Sedimentology,* 13, 21–43.

Venkatarathnam, K., and Biscaye, P. E., 1973. Clay mineralogy and sedimentation in the eastern Indian Ocean, *Deep-Sea Research,* 20, 727–738.

Yeroschev-Shak, V. A., 1962. The clay minerals in the Atlantic Ocean sediments, *Okeanologiia, Akad. Nauk. SSSR,* 2, 98–105.

Cross-references: *Clay; Clay as a Sediment; Clay–Genesis; Eolian Dust in Marine Sediments; Marine Sediments; Pelagic Sedimentation, Pelagic Sediments; Red Clay; Volcanic Ash–Diagenesis.*

CLAY SEDIMENTATION FACIES

It is likely that more than 50% of the minerals in sedimentary rocks are clay minerals. The bulk of these occur in shales, which include approximately half of the rocks in the geologic section. Shale (q.v.) is extremely variable in composition. The most common components in addition to the clay minerals are silt- and sand-sized quartz (silty and sandy shales), chert and/or opal (siliceous shales), carbonate minerals (calcareous shales) and organic material (carbonaceous shales). There appears to be a complete gradation between pure, clay-sized shale and siltstone, sandstone, chert, limestone, and coal. If the rock contains >50% clay-sized material, the appropriate modifier is used, e.g., calcareous shale. If the clay-sized material is <50% of the rock, the terms shaley or argillaceous precede the name characterizing the dominant lithologic type, e.g., shaley or argillaceous siltstone.

The chemical composition of shales is closely related to their mineralogy; however, trace elements—in clay and nonclay minerals—and organic material vary widely and commonly reflect the depositional environment. The minerals themselves, and the texture and structure of the shales are closely related to the depositional environment. The general composition and texture of shales is in part determined by tectonic activity. The tectonic activity affects the composition of source area, controls the relief, and in part determines the amount of chemical weathering in the source area. As most clays are originally derived from the weathering of other minerals and volcanic material, climate and weathering conditions in general play a major role in determining the ultimate composition of shales. For example, in the humid tropics, the relatively unstable clay mineral montmorillonite rather than kaolinite may form in areas of poor drainage; but, eventually, it will be converted to the acid, stable kaolinite (see *Clay as a Sediment*).

Association of Shales with Other Sediments

Because most of the minerals of sandstones and shales are detrital in origin, there is a correlation between the composition of sandstones and associated shales. This similarity is more apparent in the coarser silty and sandy shales than in the pure shales, i.e., the silty shales associated with arkosic sandstones can contain as much feldspar as the sandstone.

Arkose. The formation of arkose (q.v.) depends on the composition of the source rock, assuming it is sufficiently complex to contain the necessary elements, the character of the leaching solution, and the rate and duration of leaching. The latter factors are affected by topography, permeability, temperature, and amount and distribution of rainfall. In an arctic or desert environment, arkose is minimally altered. In a temperate environment with limited rainfall, the feldspar will be weathered to form a moderate amount of illite and montmorillonite. In the humic tropics, the feldspar will quickly alter to kaolinite, and a relatively large amount of clay material will be produced. Under the latter conditions, if erosion and deposition are not sufficiently rapid, most of the feldspar will be destroyed, the end products being kaolinitic soils and orthoquartzitic sandstones.

The arkosic facies is characteristically representative of rapid erosion and deposition predominantly in a continental environment. Alluvial fan, channel flood plain, and swamp environments are the major sites of deposition. Shales comprise a small (10-20%) portion of this facies, and many of these shales are silty. Even the relatively pure clay-sized shales contain over 40% of nonclay detrital material. Kaolinite, illite, montmorillonite, and mixed-layer illite-montmorillonite all can occur as the dominant clay in these shales. By the nature of the environments of deposition, the shale beds are irregular and not very extensive. The shales are commonly red in color but some of the swamp shales are black.

Graywacke. The formation of graywacke (q.v.) is favored by moderate tectonic activity and/or exposure of micaceous metamorphic rocks and shales. Under temperate and arctic climatic conditions, weathering is moderate, and relatively thin soil zones are formed. If there is any appreciable relief, a large part of the clay-sized material carried by the streams may be directly derived from the relatively unweathered outcrops. This type of source area provides the maximum amount of clay material for the basin of deposition. The mobile flanks of geosynclines often exhibit these characteristics and produce graywacke sandstones.

The graywacke (and subgraywacke) suite of sediments has been called the argillaceous facies. Shales and silty shales commonly comprise 60-70% of a graywacke sequence. In addition the sandstone portion contains 20-40% of clay matrix and 10-20% argillaceous rock fragments. Thus, a typical sequence will contain 80-85% shale-like material, or approximately 50% clay minerals. Illite and chlorite clay minerals are dominant in the source area and are deposited rapidly in the sedimentary trough with little modification. In the graywacke suite, illite is 2-4 times as abundant as chlorite. The clay content of the nonsilty shales averages 60-65%, and the quartz/feldspar ratio is approximately 8. The clay and accessory sand-sized material is deposited in alluvial fans, flood plains, and deltas, and in relatively deep marine environments by turbidity currents.

Most of the chlorite in sediments occurs in this type of shale. It is most abundant in the source rocks (micaceous metamorphic) associated with graywacke deposition. Because chlorite is easily weathered, it is seldom preserved in soils and must be obtained from relatively unweathered outcrop material.

Graywackes which are derived from the erosion of a continental land mass (epicontinental graywacke) have a much more variable clay suite. Montmorillonite, mixed-layer illite-montmorillonite, and kaolinite occur in abundance, along with illite and chlorite.

As a result of the green color of chlorite and illite and the general reducing nature of the depositional environments, graywacke shales are usually various shades of green and gray. Individual shale beds are irregular and of limited extent. Larger shale units (formations, facies) are commonly lens shaped due to their deposition in deltaic environments and in actively sinking basins and geosynclines. Although shales occur throughout the stratigraphic section, they are most abundant in geosynclinal sequences—on both an absolute and a relative basis.

Orthoquartzite-Carbonate Suite. The orthoquartzite-carbonate association is indicative of structural stability of the site of deposition. Shales commonly comprise 10-20% of this facies and occur as relatively thin units. These shales have a relatively high clay content. They are composed almost entirely of illite with minor amounts of chlorite, mixed-layer illite-montmorillonite and chlorite-montmorillonite. Feldspar is relatively rare, and the shales commonly contain only a small amount of carbonate.

Under the conditions of crustal stability and low relief, chemical deposition is dominant and relatively little terrigenous material is supplied to the basins of deposition. Kaolinite, which is stable in an acid environment, is considered to be the clay mineral characteristic of the end product of weathering and should be the clay associated with the orthoquartzite-carbonate sequence. When relief and permeability are low, however, it is difficult to develop kaolinite. Where weathering is sufficiently severe, many of the minerals will be completely destroyed and silica and alumina carried to the sea, where illite, the clay mineral most stable in a marine environment, will be formed. In many shallow lakes and seas and extensive tidal flats, evaporitic conditions may prevail, and magnesium-rich minerals will be formed—chlorite, attapulgite, sepiolite, mixed-layer chlorite-montmorillonite.

It appears that the authigenic formation of clay minerals under marine conditions is at a maxiumum under these periods of crustal stability. Also, under these conditions, extremely slow accumulation of carbonaceous shales takes place. Similar black shales are found throughout the geologic record; the extensive, thin, blanket-type, marine black shales, however, are usually associated with the orthoquartzite-carbonate facies. These shales are seldom more than about 100 m thick but may cover hundreds of thousands of km^2. They contain 5-15% organic carbon, abundant pyrite (10%), scarce faunas, and a number of trace elements (V, U, Ni, Cu) in unusual concentrations. The terrigenous quartz content is usually <25%. Illite is the dominant (>80%) and frequently the only clay mineral present. Mixed-layer illite-montmorillonite and chlorite are common accessory clays. Kaolinite is usually rare, though in some instances dominant.

Whether these shales were deposited in shallow or deep water has not been established. They indicate slow deposition in a large basin of stagnant water in which reducing conditions predominated. The surrounding land area must have had little relief and been heavily vegetated. The absence of kaolinite in these shales could indicate a cool and/or dry climate, but the abundant vegetation indicates the climate was reasonably warm and humid. It is likely that rainfall was sufficiently high so that the soils were largely water logged, leaching was restricted, and there was relatively little bacterial action to form carbon dioxide and increase the acidity of the soil waters.

Siliceous shales may be associated with black shales and cherts in the orthoquartzite-carbonate facies, but they are also found in the graywacke facies. The siliceous character is due to authigenic, fine-grained silica in the form of opal or quartz (chert). The silica content may vary from a trace up to predominance. In the latter case the rock is called an argillaceous chert. Many of these shales have a relatively high organic content, and illite is by far the most abundant clay mineral. The silica in some shales is biochemical in origin.

Calcareous, ferriferous, and phosphatic shales are also usually associated with this general type of sedimentary facies, as are most extensive, blanket-type shale formations.

Depositional Environments

Because of their extremely small size, clay minerals are normally deposited in environments where water currents are slow or are effective for only short periods of time. Environments of this type include lakes, flood plains, tidal flats, lagoons and bays, and moderate to deep seas. Shales deposited in the marine environment are considerably more abundant than those in other environments. The charged surface of the clay minerals plays a more important role than their small size in controlling where most of the clay will be deposited. When a clay carried in suspension in fresh river water comes into contact with the cation-rich saline waters of the sea, a large portion of the clay minerals flocculate to form aggregates sufficiently large to be deposited. The general slackening of the current at the fresh-water/salt-water interface also accelerates deposition. Thus, the deltaic environment is one of the major sites of deposition of clay material, and clay deposition rates in deltaic areas can be extremely high (see *Sedimentation Rates*). Over 70% of the Mississippi delta is composed of silt and clay, clay being predominant.

Much of the mud deposited at the delta front and at river mouths without deltas may be transported down the continental slope in sediment-laden turbidity currents. Many of the voluminous geosynclinal graywacke shales and sands, whose fauna and clay minerals indicate mixing of shallow- and deep-water material, are believed to have originated in this manner.

Marine shales are common in the geologic record but usually occur as relatively thin, blanket-type units of minor volumetric importance (see the foregoing section on orthoquartzite-carbonate facies). As with the black shales, many of these shales were deposited in shallow, extensive, epicontinental seas during periods when large areas of stable interior were flooded. There are many large, wedge-shaped bodies of shale in ancient deposits, often fringed above and below by beach and lagoonal deposits, representing a cycle of transgression and regression of the sea. Although these wedges contain some open marine shales, much of the shale appears to have been deposited in delta and prodelta environments.

Lagoonal and tidal flat shales are common but not volumetrically important. In addition to being of limited extent they commonly occur as thin (5 cm to 3 m) units horizontally interbedded with thin sandstones and siltstones.

Continental flood-plain and lacustrine shale deposits are usually small in extent but occur in many parts of the stratigraphic section. They commonly occur as discontinuous and irregular beds and lenses mixed with channel sandstones, flood-plain siltstones, and marsh and swamp coal beds. Not only are these shales generally irregular in shape and distribution, but their composition and texture show wide variations, much more so than those deposited in the deltaic and marine environments. Geosynclinal marine shales may have a nearly uniform (<5% variation) clay suite (illite and chlorite) through 1500 m of vertical section. The thin, epicontinental shales may have a uniform clay suite (>90% illite) over areas greater than two million km^3. Continental shales can show a complete change in the composition of the clay suite at intervals of a few cm or less.

The shales deposited in these various environments have distinctive sedimentary structures, fabric, mineralogy, organic content and composition, flora and fauna, etc. For example, continental shales are more apt to be red or brown, marine shales gray to green. The latter usually have better-developed fabric and fissility than the former. Detrital quartz and feldspar are more abundant and carbonate minerals less abundant in continental shales. Continental shales are likely to contain bituminous (saprophytic) material, marine shales carbonaceous (humic). The chlorine content is higher in the pore water of marine shales than in continental shales, and marine shales commonly have higher amounts of such trace elements as B, Li, F, Sr, and U, among others.

Clay Suites and Depositional Environments

In nearly all instances, different depositional environments have different clay mineral suites. These variations may be less than 2 or 3%, or as much as 100%. Although subtle environmental differences can be detected by changes in the clay suite and adjacent environments can be readily identified, similar environments commonly do not have the same clay mineral suite. A clay mineral distribution pattern that may characterize two different environments in one area may be completely reversed in similar environments of another area.

Most of the clay minerals deposited in the major environments of deposition are detrital in origin. Thus, source material and climatic conditions determine the clay suite at the site of deposition. The distribution of clay minerals in the recent muds of the Atlantic Ocean are closely related to the climatic pattern (see *Clay as a Sediment; Clays, Deep-Sea*).

Kaolinite. Kaolinite is formed under acid conditions and is most abundant in continental sediments, usually systematically decreasing in marine environments with increasing distance from shore. Kaolinite in any abundance must come from a continental source. The seaward decrease is caused by the relatively early flocculation of kaolinite and/or its dilution by clay which has been transported by marine currents. In the Cretaceous Lewis Formation of Wyoming, kaolinite has a minimum value of 2-3% in the central marine portion of the basin. The kaolinite content gradually increases to 20% in littoral shales and attains a maximum value of 30% in the continental marsh and flood-plain deposits. These values are quite similar over an area of at least 26,000 km² and through a vertical section of as much as 300 m. When detritus from a different deltaic system enters the basin, however, the significance of the absolute kaolinite percentages changes, and new criteria must be established.

The kaolinite content of black shales is an excellent environmental indicator. Marine black shales are composed largely of illite and rarely contain any kaolinite. Lagoonal and brackish-water shales usually contain appreciable amounts of kaolinite in addition to illite, and in continental black shales kaolinite is usually dominant and may be the only clay present.

Montmorillonite. The environmental significance of montmorillonite is more difficult to interpret than that of kaolinite. Montmorillonite can form from volcanic glass and various minerals in fresh-water environments where there is a reasonable supply of magnesium and limited drainage, and thus be characteristic of continental environments. It may be eroded from these environments or from ancient shales and then deposited in deltaic and marine environments as detrital montmorillonite (Mississippi Delta and Gulf of Mexico), or may form by the alteration of volcanic glass on the sea floor, as in much of the South Pacific. When montmorillonite is detrital in origin there is generally less contrast between environments than when it is authigenic in origin. In many situations the kaolinite:montmorillonite ratio is an extremely sensitive environmental indicator, particularly when the montmorillonite is formed in the marine environment.

These three types of occurrences may also be observed in ancient sediments. In the Cretaceous Lewis Formation of the Washakie Basin, montmorillonite is restricted to the deeper marine facies. In the Lewis Formation bentonitic beds (altered volcanic ash falls) are abundant in deep marine shales and are largely absent from the shallow marine and other environments. Apparently, only in the deeper marine environment was deposition sufficiently slow that the process of alteration could proceed, and sufficient magnesium could be obtained from the sea water. In some continental environments, this same ash may be altered to kaolinite.

The Mississippian Springer Formation of Oklahoma and parts of the Miocene of the Gulf Coast are largely deltaic in origin, and the shales are composed largely of detrital montmorillonite, which is also the dominant clay in the marine Porters Creek Formation (Paleocene) of the Gulf Coast and the Rierdon (Jurassic) of Canada. Continental montmorillonite is abundant in lacustrine Eocene Green River deposits and the Lower Pennsylvanian Fountain Shales of Colorado.

Smectites (montmorillonites) are an extremely complex group of clay minerals. Frequently differences in the type of interlayer (adsorbed) cation and variations in the amount and nature of the lattice charge will afford excellent environmental distinctions. Fresh-water smectites have calcium as the dominant interlayer cation, hold two layers of water, and have a lattice approximately 15Å thick. In the marine environment, much of the calcium is replaced by magnesium and sodium; the sodium smectites usually contain only one layer of water and adsorb enough sodium to reduce lattice thickness to somewhere between 12.5 and 14Å.

Many smectites (beidellites), particularly those that have formed by the leaching of potassium from micas and illites, have enough lattice charge so that when they reach a marine environment, potassium is readsorbed, some (or all) of the layers collapse, and a mixed-layer illite-montmorillonite clay is formed. Though the ratio of illite to montmorillonite layers is a function of the inherited charge, the distribution of mixed-layer clays in some ancient sediments is environmentally controlled. In some shales the ratio of illite to montmorillonite layers in the mixed-layer clay shows a gradual, systematic increase from the deeper marine to the shallower marine environment. The explanation may be that the more highly charged material usually has a larger grain size and thus settles out first.

Magnesium-rich Clay Minerals. Magnesium hydroxide in the form of brucite will precipitate between montmorillonite layers and form a mixed-layer chlorite-montmorillonite (or vermiculite) or even a chlorite if sufficient magnesium is available. Mixed-layer chlorite-montmorillonite is commonly associated with evaporitic and limestone environments where magnesium is abundant. It is common in tidal flat limestones (Ordovician Arbuckle, Knox, and Ellenburger limestones) where temporary evaporitic conditions must have occurred between tides.

Under more advanced evaporitic conditions, sufficient magnesium is available so that the magnesium-rich clay minerals attapulgite and sepiolite may form. These clay minerals apparently form from montmorillonite and volcanic material; whether they form from other clays has not been shown. Under evaporitic conditions where calcium and sodium are dominant and magnesium is rare, the detrital clays do not appear to be altered. For example, illite and kaolinite are abundant in halite deposits.

Under normal marine conditions, most of a volcanic glass will alter to montmorillonite, but a portion, usually small, will be converted to chlorite. Thus in the deep marine Lewis Shales of Wyoming, chlorite increases directly with the montmorillonite content. In the Ordovician K-bentonites of the eastern United States, chlorite is commonly present and increases from the center of the beds outward, suggesting that the formation of chlorite is related to the availability of sea water. In many marine sandstones, volcanic glass and montmorillonite are altered to a poorly crystallized chlorite. This has happened after burial and is presumably due to the flushing of large amounts of sea water from the surrounding shales through the permeable sandstones.

Illite. Illite is by far the most abundant clay mineral (perhaps as much as 60-70% of all the clays). It is stable in, and can apparently form in, marine environments; it also forms, however, from the alteration of K feldspars under mild weathering conditions where the potassium is not leached too rapidly. Much of the illite has been formed from montmorillonite and mixed-layer illite-montmorillonite by minor chemical changes caused by deep burial (3000 m) and low-grade metamorphism. Holocene illite is more abundant in soils and marine muds of the temperate and arctic zones. Muscovite and illite are quite resistant to weathering, second in stability to kaolinite under conditions of continental weathering. As with montmorillonite, there are several possible origins for illite.

The only appreciable development of marine authigenic illite, outside of sandstones and glauconites, appears to be in the thin-blanket shales of the orthoquartzite-carbonate facies (see above). The illites in these shales commonly contain very few expanded layers, and the fine fraction is apt to be poorly crystallized. The great predominance of illite (90-100%) in the clay suite of these shales (other clays being chlorite or mixed-layer clays) generally indicates a marine origin.

Under conditions of rapid sedimentation, characteristic of the geosynclinal and arkosic facies, most of the illite is detrital in origin, as is the bulk of chlorite, largely derived from shales and metamorphic rocks. Many Paleozoic shales and carbonates contain nothing but illite and chlorite, but the variation (with environment) of the ratio of these two detrital clays is discernable. In the Pennsylvanian shales of the Midland Basin the deltaic shales are composed of 70% illite and 30% chlorite. As the outer edge of the delta is approached the illite:chlorite ratio gradually increases. There is a relatively abrupt change in passing from the deltaic into the marine shales, in which the chlorite content gradually increases shoreward. The illite:chlorite ratio in this situation is extremely sensitive to environment.

Illite is commonly a dominant clay in limestone. In some cases the more clastic (and presumably shallower water) limestone facies can be distinguished by an increase in chlorite (Jurassic: Smackover) in other cases by an increase in kaolinite (Pennsylvanian: Toronto) relative to illite.

Interpretation. There are many examples indicating close relation between clay-mineral changes and environmental changes, and in many instances, subtle changes within environments can be more readily detected and traced with clay minerals than with the more conventional techniques. In evaporitic, soil, and marsh environments, and under some marine conditions (when volcanic glass is present), authigenic clay minerals may form and provide strongly diagnostic clay suites (see *Clay–Genesis*). The great bulk of shales and clay minerals are deposited in brackish water, deltaic, and marine environments; here authigenic and diagenetic changes are minor, and physical, chemical and biological segregation of the clay minerals provides the variations that characterize the various environments. Because the energy necessary to accomplish partial sorting of the detrital clay minerals is considerably less than that required to form authigenic minerals from solution or alter preexisting clay minerals, the detrital clays are usually much more sensitive environmental indicators than the chemically modified clays. To modify an environment in which montmorillonite is stable to one in which chlorite can form may necessitate the evaporation of large amounts of water. To physically sort chlorite from a montmorillonite-chlorite mixture may only require mild current action over a short period of time.

CHARLES E. WEAVER

References

Keller, W. D., 1970. Environmental aspects of clay minerals, *J. Sed. Petrology,* **40**, 788-859.

Muller, G., 1967. Diagenesis in argillaceous sediments, in G. Larson and G. V. Chilingar, eds., *Diagenesis in Sediments.* New York: Elsevier, 127-177.

Weaver, C. E., 1959. The clay petrology of sediments, *Proc. 6th Natl. Conf. Clays and Clay Minerals.* New York: Pergamon Press, 154-187.

Weaver, C. E., 1967. The significance of clay minerals in sediments, in B. Nagy, and U. Colombo, eds., *Fundamental Aspects of Petroleum Geochemistry.* New York: Elsevier, 37-76.

Cross-references: *Argillaceous Rocks; Authigenesis; Clay; Clay as a Sediment; Clay–Genesis; Clays,*

Deep-Sea; Graywacke; Oil Shale; Organic Sediments; Sedimentary Environments; Sedimentation–Tectonic Controls; Shale; Weathering in Sediments.

CLAYSTONE–*See* MUDSTONE AND CLAYSTONE

COAL

Coal is a compact, stratified organic rock composed largely of metamorphosed plant remains mixed with a variable but subordinate amount of inorganic material. Oil shales (q.v.) and carbonaceous shales are excluded by the above definition because they contain larger amounts of inorganic material, and peat is excluded because it is not metamorphosed and is generally only slightly compacted. Peat, however, cannot be excluded from a discussion of the origin of coal because it is the initial phase of the peat-coal-graphite metamorphic series.

As with other strata, coal deposits vary considerably in physical characteristics, from beds 1 m or a few cm thick covering thousands of square km to pod-like bodies, measuring several m or less and nearly equidimensional. Coal deposits are subject to the same deforming forces—folding, faulting, and igneous intrusion—that affect the shape, attitude, and chemical composition of other stratified rocks such as shale and sandstone.

As pointed out by Hendricks (1945, p. 1), "no two coals are absolutely identical in nature, composition, or origin." It has long been agreed, however, that a sequence of events beginning with the accumulation of plant material, followed by burial and compaction, and ending with the application of varying amounts of heat and pressure explains the origin of coals. Within this sequence of events, enough variation can occur to account for the various types and compositions of coals.

Depositional Environment

The coal-forming plant material may have accumulated where it grew or may have been transported to the depositional site by water or wind. Most large deposits of coal are believed to have had their origin in fresh- or brackish-water swamps or in lakes, and appear to have been derived from plant materials that grew at or near the place of accumulation.

The accumulation of peat requires a humid climate to support a rich growth of vegetation and a high water table to permit prolonged accumulation of plant material in a reducing environment. Most of the large peat deposits of Pennsylvanian age were formed near sea level—some in estuaries or coastal lagoons; others in ponds on large deltas or many coalescing deltas; others in low-lying, broad coastal marshes. Deltas, estuaries, coastal lagoons, and related features characteristically form in areas of gentle downwarping along the margins of a slowly eroding landmass. Intermittent gentle downwarping or eustatic rises in sea level permit periodic transgressions of the sea during which topographically low places are flooded. Some thick coal beds of very wide areal extent must have required a very large and wide coastal plain; a prolonged optimum rate of plant growth and accumulation; a slow, continuous rate of subsidence; and an equally slow encroachment of the sea over centuries.

The transgressive sea would ultimately cover the peat-forming swamp, while the eroding landmass continued to supply sand, silt, and mud to the sea; this material would settle in layers over the submerged peat swamp. In time, depending in length on the rate of sedimentation, the depth of the transgressive sea, and the rate of subsidence, this sedimentary material builds up new deltas, lagoons, and coastal plains conducive to the development of new, younger, peat-forming swamps.

This sequence of deposition may be repeated many times, but individual sequences might be prolonged, shortened, or terminated at any time by relatively minor movements of land relative to sea level. In the very delicate balance between sedimentation, subsidence, and uplift of the land, the sea may also regress from time to time. Peat swamps obviously form during the regressive phases of the cycle, but the peat may thus be subject to oxidation and be less commonly preserved. Cyclic repetitions of the conditions allowing the formation of coal are documented in many of the world's coal fields, but rarely as strikingly as in a sequence of more than a thousand meters of sedimentary rock in West Virginia that contain 117 coal beds of sufficient geologic and economic interest to have been described and named.

In the southwestern part of the US, the San Juan Basin of New Mexico and Colorado contains large amounts of coal (Shomaker et al., 1971) in strata that have been the subject of many classic studies (e.g., Sears et al., 1941; Pike, 1947). These studies describe sedimentary sequences deposited in environments that changed locally in response to large lateral shifts in ancient shorelines. Fig. 1 illustrates the vertical and lateral relationships of marine and nonmarine rock units during a generally regressive depositional phase (Fassettt and Hinds, 1971). In general, when a shoreline is stable or moving very slowly, the organic material ac-

COAL

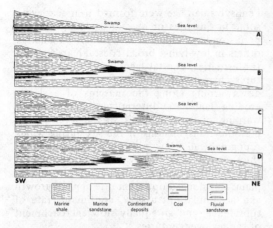

FIGURE 1. Diagrammatic cross sections showing the relations of the continental, beach, and marine deposits of Pictured Cliffs after (A) shoreline regression, (B) shoreline stability, (C) shoreline transgression, and (D) shoreline regression (after Fassett and Hinds, 1971). V.E. ≈ 60.

cumulating in the paludal (swamp) environment shoreward of the beach environment is thicker than organic material accumulating during more rapid lateral shoreline shifts. Also, the resulting coal beds generally are of limited extent normal to the shoreline as compared to their greater linear extent parallel to the shoreline.

During the accumulation phase of the depositional process, differences in the original plant material and associated inorganic material, both among beds and internally within beds, account for many of the differences among the types of coals ultimately formed. During the burial and compaction phases, biochemical processes such as rotting, digestion, and maceration alter the original plant material and produce further differences. At this stage, when the plant material is compacted but not lithified, it is classified as peat. Application of small to moderate amounts of heat and pressure to the compacted and biochemically altered plant material results in further changes leading to characterization of the material as coal. Increased heat and/or pressure causes progressive increase in the amount of carbon in the coal relative to other organic constituents and ultimately results in complete carbonization of all the original plant material. The foregoing is, of course, an extreme simplification of the complex process called coalification (see *Coal–Diagenesis and Metamorphism*).

As shown in Fig. 2, the coals of the world range in age from Silurian to Quaternary. Although coals of Paleozoic and Mesozoic age are of prime importance in most of the coal-producing parts of the world, younger coals of Tertiary and Quaternary age are of great quantitative and economic significance in some areas. In general, older coals are likely to be of higher rank than younger coals because they have been buried more deeply and subjected for a longer period of time to greater dynamic and thermal metamorphism. Deformation and igneous intrusion may, however, locally raise the rank of a coal considerably above that of coals of equal or greater age elsewhere.

Classification

Tomkeieff (1954) classifies coals in three series: humic, humic-sapropelic, and sapropelic. The three series are distinguished by differences in the original plant materials. The *humic series* includes the common (bright) banded coals having bright luster, and the compact dull-luster splint coals. The *humic-sapropelic series* includes the nonbanded coals—the cannel (spore-rich) and boghead (algal-rich) coals. The *sapropelic series* is composed of hydrogen-rich components, largely pollen, waxes, and resins. The sapropelic series and the humic-sapropelic series are quantitatively insignificant when compared to the humic series, which contains most of the coals of the world, particularly those of economic importance.

Coals in each of the subdivisions of the series

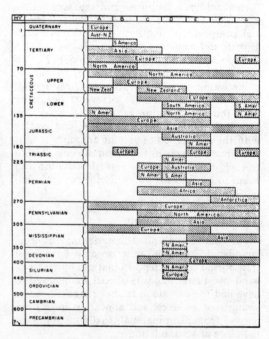

FIGURE 2. Geologic age of various ranks of coal throughout the world (after Simon and Hopkins, 1973; © Soc. Mining Eng., AIME). Ranks: A, lignite; B, subbituminous; C–E, bituminous, F, low-volatile bituminous and semianthracite; G, anthracite.

classification have a wide range in properties. Many other classifications have been used (Rose, 1945), but subdivision by *rank*, which is dependent on the degree of metamorphism, and by *grade*, which is dependent on the amount and types of deleterious inorganic material in the coal, are the commonly used systems, in some form or another, throughout the world.

The categorization of coals by degree of metamorphism is known as rank classification. Most rank classification sytems are based on the following four changes that occur with increased metamorphism. These changes, as determined by laboratory tests, are: (1) decrease in the amount of volatile matter yielded during destructive distillation; (2) increase in heat value until the volatile matter content has decreased to between 14 and 22%, after which the heat value decreases also; (3) increase in carbon content; and (4) decrease in hydrogen content (see *Coal-Diagenesis and Metamorphism*).

The standard coal analysis consists of two parts, the proximate analysis and the ultimate analysis, though commonly only the less expensive proximate analysis is made. During the *proximate analysis,* which is a destructive distillation process, the volatile matter, ash, moisture (either inherent or total), and fixed carbon contents are determined. The heat value, reported in Btu/lb or cal/g, and the sulfur content are also generally determined with the proximate analysis. In the *ultimate analysis,* the carbon, hydrogen, oxygen, nitrogen, sulfur, and ash contents of the coal are determined. Obviously, standard methods must be used if analytical results are to be compared.

Coal classification according to grade or quality is generally based on the amount of ash and sulfur present, although the amount of other deleterious constituents, such as phosphorus or carbonates, is sometimes also used. Grade classifications are commonly used to supplement rank classification, particularly in determining the commercial usability of a coal or its desirability from an environmental standpoint.

E. R. LANDIS
PAUL AVERITT

References

American Society for Testing and Materials, 1966. Tentative specifications for classification of coals by rank (A.S.T.M. Designation: D388-66), *A.S.T.M. Standards,* Part 19, *Gaseous Fuels,* coal and coke, 73-78.

Averitt, P., and Lopez, L., 1972. Bibliography and index of U.S. Geological Survey publications relating to coal, 1882-1970, *U.S. Geol. Surv. Bull., 1377,* 173p.

Bennett, A. J. R., 1963. Origin and formation of coal seams. A literature survey, *Australia Sci. Indust. Res. Org., Div. Coal Res. Misc. Rept. 239,* 13p.

Chandra, D., and Chatterjee, K. K., 1971. On the origin of ball coals, *J. Sed. Petrology, 41,* 770-779.

Dapples, E. C., and Hopkins, M. E., eds., 1969. Environments of coal deposition, *Geol Soc. Am. Spec. Publ. 114,* 204p.

Fassett, J. E., and Hinds, J. S., 1971. Geology and fuel resources of the Fruitland Formation and Kirtland Shale of the San Juan Basin, New Mexico and Colorado, *U.S. Geol. Surv. Prof. Pap. 676,* 76p.

Francis, W., 1961. *Coal, Its Formation and Composition.* London: Edward Arnold, 806p.

Hendricks, T. A., 1945. The origin of coal, *in* Natl. Research Council Committee, *Chemistry of Coal Utilization,* vol. 1. New York: John Wiley, 1-24.

Krevelen, D. W. van, 1961. *Coal.* Amsterdam: Elsevier, 514p.

Murchison, D., and Westoll, T. S., eds., 1968. *Coal and Coal-Bearing Strata.* New York: American Elsevier, 418p.

Parks, B. C., 1963. Origin, petrography, and classification of coal, in Natl. Research Council Committee, *Chemistry of Coal Utilization* suppl. vol. New York: John Wiley, 1-34.

Pike, W. S., Jr., 1947. Intertonguing marine and nonmarine Upper Cretaceous deposits of New Mexico, Arizona, and southwestern Colorado, *Geol. Soc. Am. Mem. 24,* 103p.

Rose, H. W., 1945. Classification of coal, in Natl. Research Council Committee, *Chemistry of Coal Utilization,* vol. 1. New York: John Wiley, 25-85.

Schopf, J. M., 1948. Variable coalification-the processes involved in coal formation, *Econ. Geol., 43,* 207-225.

Schopf, J. M., 1956. A definition of coal, *Econ. Geol., 51,* 521-527.

Sears, J. D.; Hunt, C. B.; and Hendricks, T. A., 1941. Transgressive and regressive Cretaceous deposits in southern San Juan Basin, New Mexico, *U.S. Geol. Surv. Prof. Pap. 193-F,* 101-121.

Simon, J. A., and Hopkins, M. E., 1973. Geology of coal, in S. M. Cassidy, ed., *Elements of Practical Coal Mining.* Soc. Min. Engrs., AIME, 11-39.

Shomaker, J. W.; Beaumont, E. C.; and Kottlowski, F. E., eds., 1971. Strippable low-sulfur coal resources of the San Juan Basin in New Mexico and Colorado, *N. Mexico Bur. Mines Min. Res. Mem. 25,* 189p.

Tomkeieff, S. I., 1954. Coals and bitumens and related fossil carbonaceous substances; nomenclature and classification. London: Pergamon, 122p.

Cross-references: *Coal-Diagenesis and Metamorphism; Cyclic Sedimentation; Lagoonal Sedimentation; Oil Shale; Organic Sediments; Peat-Bog Deposits.* Vol. IX: *Carboniferous.* Vol. XIII: *Coal Gasification and Liquefaction; Coal Geology; Coal Petrology.*

COAL-DIAGENESIS AND METAMORPHISM

The process of transformation of coal, starting from the peat stage, passing the stage of lignites and bituminous (hard) coals, up to the stage of anthracites, is called *coalification,* or *carbonification.* Only the first part of this development—from peat to the boundary between soft brown coal and hard brown coal (see Fig. 1)—can be called *diagenetic.* The later transformations are so severe that they must be considered *metamorphic,* although this metamorphism takes place in temperature-pressure-time conditions under which the *minerogenic* rocks are still in the stage of diagenesis (see below). Because the diagenesis of coals (i.e., the early coalification) is supported by the activity of microorganisms, the first part of the coalification process is often referred to as *biochemical coalification,* whereas the following metamorphism of coals is called *geochemical coalification.* The international term *rank* is used to designate the degree of coalification (ICCP, 1963, 1971, 1975).

The chemical and physical alterations during the coalification process are complicated, and, depending on the different ranges, also very heterogeneous. In most cases, the following chemical properties are used to determine the rank of coal: contents of carbon, oxygen, hydrogen, volatile matter. The moisture content, the calorific value, and the reflectivity of vitrinite also serve as rank parameters. The importance of the various measures of rank varies with the rank range (Fig. 1). The classification of coals on the basis of their rank differs in various countries. In the United States, the ASTM classification is used (Fig. 1).

Peat Stage

The parent matter of coals is plant remains, which usually were deposited in forest swamps (e.g., *Taxodium* swamps as in southern Georgia) or in reed moors (e.g., sawgrass swamps as in southern Florida). Because, in topographically low areas, the ground water reaches the soil zone, or even stands above it, the dead vegetal material of the swamp is quickly introduced into a reducing environment, its total decomposition is prevented, and it is transformed to peat. At the peat surface, where slightly oxidizing conditions exist, fungi and aerobic bacteria are active in peat formation. They attack mainly the more easily hydrolyzable plant constituents such as cellulose, hemicellulose, sugars, pentosanes, pectines, and proteins. These substances are decomposed wholly or partly, whereby metabolic products of the microorganisms are left. Resistant substances such as lignin, tannins, and, above all, the "bituminous" plant remains (resins, waxes, fats, oils, sterins, sporopollenin, cutine, and suberine) are enriched. Most bituminous matter does not change much before the stage of low-rank bituminous coals, but lignins and tannins are attacked by microbiological and chemical processes in the later peat stage.

Typical products of peat formation are the *humic acids,* i.e., colloidal, dark-colored substances of high molecular weight, which are soluble in alkaline solutions. Their molecules consist of an aromatic core which is surrounded by bridges and chains of nonaromatic character.

A great part of the original plant structures disappears during peat formation. The parenchymatic tissues, which are rich in cellulose, are easily attacked, whereas the lignified wood tissues are preserved relatively well, especially if they contain resins and tannins. The shapes and structures of "bituminous" plant matter, i.e., phyterals such as spores, pollen, cuticles, and suberinized tissues are preserved best. Later, in the coal stage, these constituents represent the maceral group (*macerals* are the microscopically visible units of coals) of *liptinites* (or *exinites*), whereas the humic matter of peats form the maceral group of *huminites* (brown coal stage), and *vitrinites* (hard coal stage). Constituents that are relatively rich in carbon and are chemically less reactive (like charcoal from forest fires), are called *inertinites.*

During peat formation, the carbon content of the dry, ash-free substance increases, mainly as a result of the decomposition of cellulose and hemicellulose and the enrichment of lignin and humic matter. Already in the *peatigenic layer* near the surface, the carbon content may grow from 45% up to 60%. At greater depth, it increases only very slowly.

A typical property of peats is their high moisture content, which is 90% near the surface. Part of the water fills the numerous large pores; the greater part is physically adsorbed by the humic substances. As the peat bed is subsiding and being covered by younger sediments, the density increases, and the pore volume and the moisture content decrease (ca. 1% H_2O/10 m depth).

The transition from peat to brown coal is a gradual one. Usually the boundary between peat and brown coal is placed at a moisture content of 75% (inherent bed moisture). The presence of cellulose also serves as a characteristic for the distinction of peat from brown coal: peat still contains free cellulose; whereas, in brown coals, cellulose is preserved only where it was protected by lignin or cutin.

Braunkohle Stage

With increasing thickness of overburden, peat is converted to *brown coal* (at a depth of 200–400 m), *lignite,* and *subbituminous coal* (ASTM

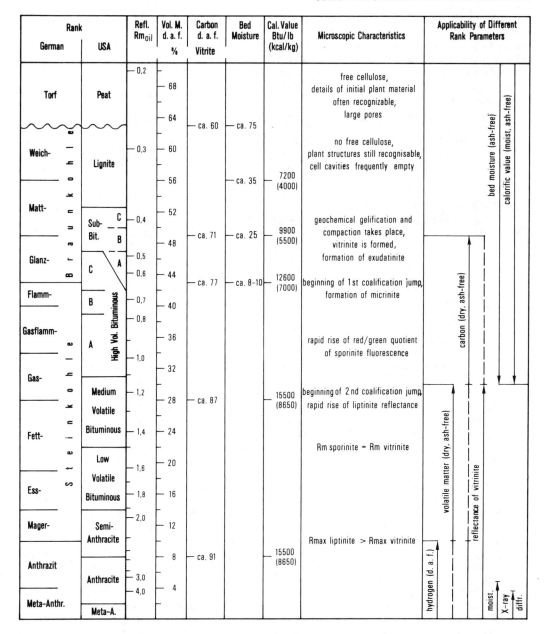

FIGURE 1. The main stages of coalification distinguished by chemical and microscopical characteristics, and the applicability of various measures of rank (after Patteisky and M. Teichmüller, 1960; *International Handbook of Coal Petrography,* 1963, 1971, 1975; and M. Teichmüller, 1974).

classification). Lignites are more consolidated than brown coals. The distinction between lignites and subbituminous coals is based on calorific value.

In Europe, the following rank types of *braunkohle* are distinguished by % inherent bed moisture (Fig. 1): soft brown coal, and hard brown coal—which is subdivided into dull brown coal and bright brown coal. This classification originally was based on the petrographic habit. Soft brown coals are light to dark brown, dull, and earthy. In most cases, one still can recognize with the naked eye some well-preserved plant remains, mainly wood. Under the microscope, the various components of soft brown coal are loosely packed and only slightly compressed, and reflectance of incident light is very weak. In some plant tissues, small quantities of cellulose and lignin are preserved.

The hard brown coals are darker, more consolidated, and more or less well stratified. A dull, hard brown coal usually is dark brown in color, and does not yet show the typical brightness of bituminous coals, whereas a bright, hard brown coal cannot be distinguished from a bituminous coal by its petrographic (megascopic and microscopic) properties.

The transition from dull to bright hard brown coal is the most striking petrographic alteration during the whole coalification process. The process causing this alteration seems to be mainly colloidal (M. and R. Teichmüller, 1954). The humic substances are homogenized and strongly compacted, passing through a soft, pulpy stage. This process is called *geochemical gelification*. During this process, the different humified plant remains (e.g., stems, leaves, roots) and the humic detritus are transformed into *vitrinite* which is, petrographically, a nearly homogeneous component (maceral). Bands of vitrinite appear as the bright coal layers of bituminous coal seams. Dull coal layers consist mainly of protobituminous components (liptinites or exinites), and/or of primarily strongly decomposed constituents rich in carbon (inertinites). Mineral matter also causes a dull appearance of coal.

During and after geochemical gelification, the newly formed vitrinitic coal-gel may be compressed as a result of overburden pressure. The smallest colloidal elements, the micelles, are arranged parallel to the bedding plane, causing an optical tension anisotropy which first is visible under the microscope in thin sections (later also in reflected light).

In the braunkohle stage, the average carbon content increases and the volatile matter decreases. The chemical alterations depend mainly on the splitting off of certain oxygen-rich marginal molecular groups ($-OCH_3$, $-COOH$, $=C=O$, and $-OH$) from the humin molecules, which themselves consist of a relatively unreactive aromatic core and of reactive functional groups of nonaromatic character. The remaining lignin and the last remnants of cellulose are converted into humic substances. The humic acids lose their acid character, and become insoluble in alkali. According to the *International Handbook of Coal Petrography* (1963, 1971) and Patteisky and Teichmüller (1960), the boundary between brown coal and bituminous hard coal is represented by the following values: 8–10% moisture, 77% carbon (dry, ash-free), 16% oxygen (dry, ash-free), 7000 kcal/kg (=12,600 Btu/lb) calorific value (moist, ash-free).

Bituminous (Hard) Coal and Anthracite Stages

Further subsidence of the coal into regions of higher temperature brings about the formation of bituminous coals and finally anthracite. The transitions between the different rank stages of these coals are gradual ones. Coalification "breaks" occur only, (1) at the boundary subbituminous/bituminous coal (ASTM System): (42% volatile matter, 77% carbon, 0.6% reflectance Rm_{oil}); (2) in the stage of medium-volatile bituminous coal (29% vol. m., 87% C, 1.2% Rm_{oil}); (3) at the boundary semianthracite/anthracite (8% vol. m., 91% C, 2.5% Rm_{oil}); and (4) at the boundary anthracite/metaanthracite (4% vol. m., 93.5% C, 4% Rm_{oil}). At these "breaks" certain chemical reactions are emphasized. In the stage of subbituminous and high-volatile bituminous coal (first *coalification jump*), a bituminization of the nonaromatic components of liptinites and vitrinites takes place. This "bituminization range" corresponds to the generation of petroleum in oil source rocks (M. Teichmüller, 1974). In the stage of coking coals, the newly formed bituminous matter is cracked into progressively smaller molecules, finally into gaseous hydrocarbons (second *coalification jump*).

Generally speaking, during the coalification of bituminous coals, the carbon content increases and the oxygen decreases. The hydrogen content changes relatively little until the anthracite stage is reached. In a certain rank range (medium-volatile bituminous coal to semianthracite) a strong decrease of volatile matter is characteristic. Correspondingly, the concentration of aromatic carbon increases. The more or less laminated aromatic clusters of the humic molecules grow and align with the bedding plane. These changes result in an increase of optical anisotropy and in x-ray diagrams resembling those of graphite (bituminous coals give diagrams with wide, diffuse graphite bands, whereas diagrams of anthracites and especially of metaanthracites show a pronounced graphitization).

In the bituminous coal stage, microporosity and the internal surface of the bituminous coals (which in the early stages reaches 180 m²/g coal) decrease, and the density increases up to a rank stage characterized by a carbon content of 89% C and a volatile-matter content of 20%.

Many physical properties of the coal begin to change in this stage, which is the rank stage in which the intermolecular spaces have become particularly small because the bulky, nonaromatic bridges and chains of the humic molecules have been destroyed to a great extent. The greatest part of the relatively heavy oxygen has been given off, whereas the relatively light hydrogen has been concentrated because, during coalification, carbon dioxide is released first; only when the stage of medium-volatile bituminous coal is reached does hydrogen-rich methane become the predominant coalification gas. In the stage of bituminous coals, about 100 m^3 CO_2 and about twice as much CH_4 (per kg final coal) are produced. Most methane is liberated during anthracitization. Part of the methane remains adsorbed to the coal and is first released during mining, sometimes causing fire damps. The greater part migrates upward and—under favorable conditions—may collect in reservoirs to form dry natural-gas deposits.

The micropetrographic changes during coalification of bituminous coals and anthracites are mainly characterized by an increasing similarity of the various macerals, which originally differ markedly in their microscopic appearance. The reflectivity of vitrinite (humic substance) increases so progressively that it is possible to determine the rank of coal by reflectance measurements of vitrinites alone. This method is used also for finely disseminated, coaly particles in minerogenic sediments to determine the degree of rock diagenesis.

The increase of reflectivity is caused by an increase of refractive index and of absorption index which, for their part, depend on the increasing aromaticity and condensation of the humic molecules of vitrinite. The liptinite macerals (sporinite, cutinite, resinite, alginite, etc.) change strongly during the first and second coalification jump (see above). As a result of the second coalification jump, which occurs in the stage of medium-volatile bituminous coal (29–20% volatile matter), the reflectivity of liptinites grows so quickly—therewith approaching the reflectance power of vitrinites—that finally it is impossible to distinguish, e.g., the maceral sporinite from vitrinite. This stage is reached in high-rank bituminous coals with about 20% volatile matter. In high-rank anthracite, only very strongly reflecting inertinites (fusinites) can be distinguished from vitrinite.

The final stage of coalification is reached with meta-anthracites, which have properties very similar to those of graphite (very high reflectivity parallel to stratification, strong anisotropy, especially strong pleochroism, graphite-like x-rays, and electron-diffraction diagrams.

Causes of Carbonification

Microbiological processes play a decisive role in the peat stage only. They influence primarily the facies of coal that depends upon the original plant matter as well as upon the environment of peat formation (climate, water supply, redox potential, pH value).

Experiments of artificial coalification and many geological observations have shown that it is the temperature of the earth's crust which plays the decisive role for the chemical alterations during coalification. Usually the rank of a coal depends upon the maximum temperature to which the coal was exposed during its geological history, i.e., upon the maximum depth to which it has subsided into the crust. In a sequence of coal-bearing rocks, therefore, the older seams commonly are higher in rank than younger ones (C. Hilt's rule), and the increase of rank with increasing depth is stronger in areas with high geothermal gradients.

Static pressure will retard the various chemical reactions involved in coalification. The reason why intensely folded coal basins often bear high-rank coals is not the folding pressure, but rather the high temperature before the folding, for the orogenesis of coal-bearing foredeeps commonly is preceded by a particularly deep subsidence of the beds. On the other hand, pressure causes changes in certain physical properties of coal. The porosity, and therewith the moisture content, is reduced considerably by the overburden pressure, mainly in the low-rank stages (peat to high-volatile bituminous coal). The optic anisotropy of vitrinites—an effect of the orientation of aromatic clusters parallel to the bedding plane—is caused by the overburden pressure also. The graphitization of meta-anthracites and the formation of graphite proper is promoted by shearing stress.

Time has an influence on the degree of coalification insofar as the duration of heat exposure is important, provided that the coalification temperature was high enough. The more severe the heating, the greater is the influence of time. If the duration of heating is known, it is possible to estimate from the rank of coal the temperature to which it was exposed (Karweil, 1956). The temperatures necessary for the formation of bituminous coals commonly are no higher than 100–150°C. Thus the rank of coal may serve as a relatively sensitive geological thermometer in a low temperature range. This has a great practical importance in prospecting for hydrocarbons because the generation and alteration of petroleum and natural gas also depend upon the heat exposure of the source rocks. Vassoyevich et al. (1970) demonstrate

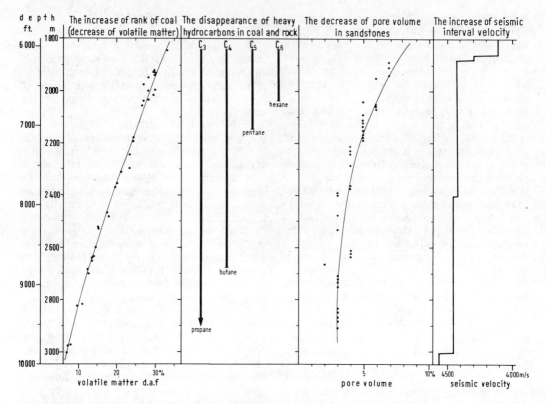

FIGURE 2. The increase of coal rank and of bituminous matter with degree of diagenesis of associated rocks (after M. and R. Teichmüller, 1968): volatile matter of coal, heavy hydrocarbons, pore volume in sandstones at 1800–3000 m depth in the borehole Münsterland 1.

the relationship between temperature, pressure, time, oil and gas generation and degree of coalification.

Coalification in Relation to Diagenesis and Metamorphism of Rocks

Coalification corresponds to diagenesis in minerogenic rocks (see Fig. 2). Early coalification, characterized by a loss of pore volume and of moisture, parallels the decrease of porosity in sandstones and claystones (see *Compaction in Sediments*). Disregarding the transformation of smectite into mixed-layered minerals, which corresponds with the stage of subbituminous coals, more severe mineral transformations in clayey and sandy rocks usually begin after the coal has already reached the stage of anthracite. This is because coal, as an organic substance, is much more sensitive to temperature rise than minerals. Exact correlations between rank of coal and mineral transformation are not possible because, in contrast to coalification, the conversion of minerals depends not only upon the diagenetic factors temperature, pressure, and time, but also, to a great extent, upon the cation supply. Nevertheless, rough correlations can be seen. For instance, the formation of pyrophyllite from kaolinite may be correlated with the late anthracite stage. Likewise, the crystallinity of illites increases rapidly when coal is anthracitized. The transformation of meta-anthracite to graphite corresponds to the greenschist facies of rock metamorphism. A survey of our knowledge concerning the relations between coalification and rock diagenesis and metamorphism has been given recently by Kisch (1974).

MARLIES TEICHMÜLLER
ROLF TEICHMÜLLER

References

Dulhunty, J. A., 1954. Geological factors in the metamorphic development of coal, *Fuel,* 33, 145-152.

Epstein, A. G.; Epstein, J. B.; and Harris, L. D., 1975. Conodont color alteration—an index to organic metamorphism, *U.S. Geol. Surv. Open-file Rept. 379,* 59p.

International Committee for Coal Petrology, 1963, 1971, 1975. *International Handbook of Coal Petrography.* 2nd ed. and 1st and 2nd suppl. to the 2nd ed. Paris: Centre National de la Recherche Scientifique, 337p.

Karweil, J., 1956. Die Metamorphose der Kohlen vom Standpunkt der physikalischen Chemie, *Z. deutsch geol. Ges.,* **107,** 132-139.

Kisch, H. J., 1974. Anthracite and meta-anthracite coal ranks associated with "anchimetamorphism" and "very-low-stage" metamorphism, *Koninkl. Nederl. Akad. Wetensch. Proc.,* ser. B., **77,** 81-118.

Patteisky, K., and Teichmüller, M., 1960. Inkohlungs-Verlauf, Inkohlungs-Masstäbe und Klassifikation der Kohlen auf Grund von Vitrit-Analysen, *Brennstoff-Chemie,* **41,** 79-84, 97-104, 133-137.

Stach, E.; Mackowsky, M. T.; Teichmüller, M.; Taylor, G. H.; Chandra, D.; and Teichmüller, R., 1975. *Textbook of Coal Petrology.* Stuttgart: Gebruder Borntrager, 428p.

Stopes, M. C., 1919. On the four visible ingredients in banded bituminous coal. Studies in the composition of coal, *Roy. Soc. London Proc.,* ser. B, **90,** 470-487.

Teichmüller, M., 1962. Die Genese der Kohle, *4th Congr. Avanc. Etud. Strat. Géol. Carbonifère, C. R.,* **3,** 699-722.

Teichmüller, M., 1974. Entstehung and Veränderung bituminöser Substanzen in Kohlen in Beziehung zur Entstehung und Umwandlung des Erdöls, *Fortschr. Geol. Rheinl. u. Westf.,* **24,** 65-112.

Teichmüller, M., 1974. Generation of petroleum-like substances in coal seams as seen under the microscope, *Adv. Org. Geochem.,* **1973,** Technip Paris, 379-407.

Teichmüller, M., and Teichmüller, R., 1954. Die stoffliche und strukturelle Metamorphose der Kohle, *Geol. Rundsch.,* **42,** 265-296.

Teichmüller, M., and Teichmüller, R., 1968. Geological aspects of coal metamorphism, in D. Murchison and T. S. Westoll, eds., *Coal and Coal-Bearing Strata.* New York: Elsevier, 233-267.

Vassoyevich, N. B.; Korchagina, Ju. I; Lopatin, N. W.; and Chernyshev, V. V., 1970. Principal phase of oil formation, *Internat. Geol. Rev.,* **12,** 1276-1296.

Cross-references: *Bitumen, Bituminous Sediments; Coal; Compaction in Sediments; Diagenesis; Humic Matter in Sediments; Lagoonal Sedimentation; Organic Sediments; Peat-Bog Deposits.* Vol. XIII: *Coal Petrology.*

COLOR OF SEDIMENTS

Various substances absorb and transfer electromagnetic radiation of different frequencies. Transferred energy, when selectively distributed within the visible range, is perceptible to the eye as color. When uniformly distributed, it is seen as *white* (colorless) or as some shade of neutral *gray,* darkening toward *black* as the level of visible radiation approaches zero. Sediment colors all include an achromatic component; they may relate quite sensitively to physical and chemical properties and, therefore, may be useful guides to small variations within a sedimentary sequence, to provenance, depositional environments, changing facies, and weathering and diagenetic effects on sediments (McBride, 1974).

Causes of Color

The minerals quartz, feldspar, calcite, and dolomite are commonly more or less white and may form sedimentary deposits of corresponding appearance, e.g., some sands, sandstones, deep-sea oozes, and limestones. Most pure clay structures are also inherently colorless (or pale green) and occasionally so occur under special conditions. Other sedimentary minerals, such as black ilmenite or dark green glauconite, are relatively rare in terms of total abundances but are found concentrated in local accumulations.

It is the exception rather than the rule, however, for sediments to show the unmodified colors of their major components. These colors may be partially or completely obscured by the effect of a relatively small content of colorful or dark material, chiefly iron compounds and organic matter. Susceptibility to exotic coloring increases rapidly with decreasing grain size. Hence the mass color of a sediment, and its significance, is also importantly associated with textural composition.

Iron Compounds. Mineral colors are ubiquitously, although not exclusively, related to various electronic processes involving the presence of one or more transition metals, especially some of those belonging to the first series (Sc, Ti, V, Cr, Mn, Fe, Co, Ni, Cu, Zn). The particular absorbtion spectrum depends partly on the particular anions or dipolar groups that surround the metallic ions; and it is for this reason that color tends to be a constant property only of those compounds that have transition metals as essential elements.

Iron is by far the most abundant of the transition metals; this circumstance, and the attribute of strong color which is characteristic of some of its compounds in sediments, combine to make iron also the most important. The relatively intense colors (greens, yellows, reds) are partly due to charge-transfer absorption at high energies (short wavelengths), a process facilitated by the ability of iron to exist in more than one oxidation state. The colors tend to darken with increasing iron content, and some compounds may appear black.

Iron-bearing minerals, released during the weathering of crystalline rocks, may survive transportation and, in the form of dark, dispersed, detrital grains or their more stable derivatives, substantially influence the appearance of sediments and soils, e.g., black sands (q.v.). Iron transferred from mineral structures to an aqueous phase may form ionic complexes and be precipitated when corresponding solubility products are exceeded, or remain temporarily mobile as a metastable colloid, or be physically or chemically absorbed by clay particles.

Silicate, sulfide, and carbonate ions may all lead to the deposition of iron. Regardless of its immediate parentage, however, the metal is found in surface sediments quite predominantly as an oxyhydrate or an oxide, most conspicuously and most abundantly in the ferric state; sulfides may predominate at depth. Ferrous compounds may persist, or form by in-situ reduction, under conditions of negative redox potential and/or high hydrogen-ion concentration.

Amorphous ferric hydrate is generally brown and tends to age, e.g., to brown-yellow goethite or lepidocrocite. Both the amorphous and crystalline hydrates may age and dehydrate to brown-red, anhydrous Fe_2O_3 (preeminently hematite and also, rarely, its less stable dimorph, maghemite). These pigments may be dispersed throughout the sediment's matrix or cement as interstitial aggregates, threads, or small crystals; or they may form rims, even very thin skins, over individual particles. Iron in the ferrous state may cause greenish hues; the strong greens of some sediments indicate the presence of glauconite or, possibly, chlorite.

A relationship of sediment color to the oxidation state of iron is illustrated by observations that some red and purple shales have a higher Fe_2O_3:FeO ratio than do similar greenish sediments containing approximately the same amount of total iron. Other examples are found in modern oceanic sediments: rapidly flocculated and buried river detritus may remain oxidized and retain a yellow, red, or brown color, whereas the greens of many slope and shelf sediments are consistent with (although not necessarily caused by) reducing conditions. On the other hand, the brown "red clays" (q.v.) of the pelagic environment represent deposition slow enough for oxidizing bottom waters to more than balance the effects of decaying organic matter. Many of the paler brown and gray hemipelagic muds of intermediate depths probably develop under variously intermediate oxidation potentials.

Organic Matter. In the majority of fine-grained sediments, and with allowances for the influence of iron compounds, gray to black (and, more rarely, dark brown) shades can be partially or quite predominantly attributed to the strong light-absorptive properties of carbonaceous compounds derived from the breakdown of plant tissues. Indeed, since the extensive terrestrial adaptation of plants in the late Paleozoic, organic matter is likely to have been incorporated in a wide variety of nondesert, nonmarine sediments and some marine shales. Preservation, however, implies limited winnowing and also limited bacterial action under, for example, conditions of relatively low temperatures and/or poor drainage or circulation.

Easily decomposable organic detritus is generally not as abundantly supplied to oceanic as it is to terrestrial sediments, and it is usually further drastically reduced by the activity of benthic mud-feeders. Thus, few marine deposits are dark gray or black. In particular, however, hydrotroilite ($FeS \cdot nH_2O$) may form in some modern marine muds as a bacterial product of included organic compounds under conditions of early burial, and so induce black or gray colors that are rapidly lost on exposure to the atmosphere. Exceptionally, in stagnant and semi-stagnant environments, oxygen depletion facilitates the accumulation both of abundant organic matter and of reduced iron compounds; sediments found under these conditions are characteristically black to green (see *Euxinic Facies*).

Pigments such as the green porphyrins and yellow carotinoids have been extracted from marine muds without apparent effects, and it seems unlikely that organic matter itself is more than a trivial contributor of chromatic color to soils and sediments.

Texture. The sedimentary processes leading to the accumulation of coarse-grained sediments are normally inimical to the concentration and retention of organic detritus. Carbonaceous matter is, therefore, most widely responsible for darkening soils and other sediments with a high proportion of fine silt and clay. Graywackes may contain abundant mud, as well as rock fragments (including dark mud chips), and in this sense are exceptional among sandstones. Furthermore, the absolute quality of dark or coloring pigment is not as important as its distribution. Coarse-grained sediments, therefore, are more likely to approach the colors of constituent particles. A few % pigment may be more than enough to color a sandstone red; and it has been observed that in some marine muds as small a fraction as 1% pigment (apparently as particles of colloidal dimensions) is responsible for imparting, e.g., a distinct brownish or grayish color to more or less white silt and clay.

It is a fair generalization that the finer the grain size, the more striking, more variable, and hence more useful, color becomes as a property, and the more troublesome analytical work becomes with respect to individual phases. Indeed, the color of some sediments may be the only readily observed symptom that reveals differences in sedimentary history. On the other hand, similarly colored sediments may reflect very different depositional environments, as in red beds (Van Houten, 1973).

Other Considerations. The color of sediments is the product of a combination of the presence of chromatic pigment (e.g., red hematite) and of a neutral background of black, gray, or white (e.g, dark organic matter, white carbonate). Only iron compounds and organic matter have been considered here, and it has been convenient to discuss them, and texture, separately. Self-evidently, the mass appearance of a sediment should be viewed as the overall product of all possible determinants. Manganese, for example, may be of considerable importance locally; it is widely associated with iron, although the dark brown and black results of its oxidation are less conspicuous. The desert varnish of rock surfaces represents a skin of manganese oxide.

Description of Color

Observation. Outcrop colors differing from those of fresh, broken surfaces are shown by numerous sediments, particularly rocks that were deposited under subaqueous conditions and subsequently uplifted and exposed. The weathered zone is normally thinner and has more abrupt contacts in dense, fine-grained rocks than in rocks with a high porosity.

On exposure, ferrous compounds oxidize; bluish-gray sideritic shales, for example, develop a brownish surface, and iron sulfide alters to sulfate and then to ferric hydrate. Liesegang-type banding appears to be the result of inward diffusion of oxidation. Carbonaceous matter may be bleached and many, but, conspicuously, not all dark shales may be paler in outcrop; some dark limestones are also superficially bleached. Cores of modern marine sediments become paler when desiccated. This is an easily reversible phenomenon, but, more seriously, drying may be accompanied by color changes due to oxidation and bacterial activity (Moberly and Klein, 1976).

The colors of exposed sediments should be recorded from a fresh, dry surface showing a typical texture. The colors of modern marine sediments and other subaqueous deposits should be described immediately upon obtaining a sample. Whether or not this practice is followed, the condition of the surface being described should be noted.

Furthermore, viewing conditions are important and may be critical, espcially when members of a suite of sediments are examined under varying light sources. The problem may be further aggravated if any color matches are metameric. Hence, if color observations of sediments are not made under natural daylight, it should be so recorded. Ordinary incandescent light has a very high predominance of relative energy in the red-yellow range and represents the most unsatisfactory of normally available artificial illuminants.

Designation of Colors. Color can be specified by making spectrophotometric measurements of percentage reflection at each of small intervals of visible wavelength. Standard (CIE) practice, recommended in 1931 by the International Commission on Illumination, is to calculate from these numbers the so-called tristimulus values X, Y, and Z, which respectively relate to the red, green, and blue-violet primary responses in the Young-Helmholz theory of color vision. From the tristimulus values, it is customary to compute the chromaticity coordinates, x and y, and to plot these on a chromaticity diagram from which can be read the intuitively meaningful dominant wavelength (analogous to the concept of *hue*) of the color, and its purity (analogous to *chroma*). These terms, together with the reflectance Y (analogous to *value*), specify a color with an accuracy potentially equivalent to the separation of about one hundred thousand different colors. The human eye can distinguish between about ten million surface colors in daylight; color memory, however, is poor and color vocabulary worse.

As Grawe observed in 1927, routine quantitative measurements in geological studies would resolve many difficulties but, at least at present, they are impracticable. Even in the laboratory, there are limitations which relate principally to the problem of surface texture effects and to the relatively small variations of color that may be of interest. Accordingly, material standards, i.e., color charts, remain the most useful method for designating sediment colors.

Werner organized an arrangement of colors and color names for minerals which was published under Jameson's guidance in 1814. Subsequently various other systems were worked out. In particular Ridgway's *Color Standards and Color Nomenclature* (1912) and Maerz and Paul's *Dictionary of Color* (1940) were widely used by geologists until the *Rock-Color Chart* (Goddard et al., 1948) was developed in the United States by the National Research Council and similar *Soil Color Charts* were produced by the Munsell Color Company. These rock and soil color charts enable colors to be designated in terms of the Munsell system. This system, although purely psychological, can be related to the basic CIE system and is fundamental to the comprehensive ISCC-NBS *Method of Designating Colors and a Dictionary of Color Names* (Kelley and Judd, 1955), and to the companion *ISCC-NBS Centroid Color Charts*.

H. D. NEEDHAM

References

De Ford, R. K., 1944. Rock colors, *Bull. Am. Assoc. Petrol. Geologists,* 28, 128-137.

Goddard, E. N.; Trask, P. D.; De Ford, R. K.; Rove, O. N.; Singewald, J. T., Jr.; and Overbeck, R. M., 1948. *Rock-Color Chart.* Boulder, Colo.: Geol. Soc. Am., 16p.

Grawe, O. R., 1927. Quantitative determination of rock color, *Science,* 66, 61-62.

ISCC-NBS Centroid Color Charts, Standard Sample no. 2106, Office of Standard Reference Materials, National Bureau of Standards, Washington, D.C.

Kelly, K. L., and Judd, D. B., 1955. The ISCC-NBS method of designating colors and a dictionary of color names, NBS Circular 553.

Kornerup, A., and Wanscher, J. H., 1967. *Methuen Handbook of Color.* London: Methuen, 243p.

McBride, E. G., 1974. Significance of color in red, green, purple, olive, brown, and gray beds of Difunta Group, northeastern Mexico, *J. Sed. Petrology,* 44, 760-773.

Maerz, A., and Paul, M. R., 1940. *A Dictionary of Color.* New York: McGraw-Hill.

Moberly, R., and Klein, G. de V., 1976. Ephemeral color in deep-sea cores, *J. Sed. Petrology,* 46, 216-225.

Van Houten, F. B., 1973. Origin of red beds: a review—1961-1972, *Ann. Rev. Earth Planetary Sci.,* 1, 29-61.

Cross-references: *Black Sands; Cherty Iron Formation; Diagenesis; Euxinic Facies; Glauconite; Ironstone; Red Clay; Shale; Weathering of Sediments.* Vol. IVB: *Color of Minerals.* Vol. XII: *Color, Coloration of Soils.* Vol XIII: *Rock-Color Chart.*

COMPACTION IN SEDIMENTS

Gravitational compaction of sediments under the influence of their own weight has long been a recognized geologic phenomenon. In the seventeenth century, Steno attributed variations in the attitude of sedimentary formations to compaction (see Hedberg, 1936). Sorby (1908) applied quantitative methods to the study of compaction and presented original data on the porosity (void volume/bulk volume) of natural sediments, discerned an inverse relationship between porosity and geologic age, and recognized that the compaction of sediments is due to the applied load, which causes the expulsion of interstitial fluids and reduction in porosity.

The mechanics of compaction have been approached from two divergent viewpoints, i.e., the one of the geologist-sedimentologist and that of a soils engineer. Most of the knowledge of properties affecting sediment compaction has come from the field of soil mechanics. In soil mechanics terminology, the geologic term *compaction* is called *consolidation*. Inasmuch as these two terms are used by both disciplines, but with different connotations, it is necessary to clarify the terms as they are used in each field. *Compaction* in soil mechancis expresses the concept of a density increase in the soil (sediment) by hydraulic or mechanical means such as tamping, loading, and vibrating. The term *compaction* in a geological sense refers to a decrease in the bulk volume (void volume + solids volume) of the sediment due to an overburden. *Consolidation* in civil engineering terminology refers to a decrease in the bulk volume by the expulsion of the pore fluid (gas or liquid) upon the application of a load (force per unit area) on the soil or sediment. In the field of geology, the term *consolidation* is commonly understood to imply not only the process of compaction but also the cementation and induration of the sediment (Allen and Chilingarian, 1975).

In addition to the load exerted by overlying sediments, the compaction of clastic sediments in depositional basins is influenced by other factors, such as (1) the chemical and mineralogical composition of sediments, (2) the chemical composition of interstitial fluids, and (3) granulometric composition. One's understanding of the process of compaction, therefore, involves the comprehension of how all the above factors interact with each other and with increasing overburden loads to inhibit or to enhance the removal of fluids and the reduction of pore volume in sediments.

Role of Stresses in Compaction

Only recently have sedimentologists begun to pay attention to the magnitude and direction of stresses in thick sedimentary sequences (see Rieke and Chilingarian, 1974). Knowledge of the vertical and lateral stress patterns in a depositional basin is important inasmuch as it will aid in the understanding of the expulsion and migration of pore fluids during basin subsidence.

Gravitational compaction is defined as the expulsion of pore fluids and the pore volume decrease in a sedimentary rock column as a result of normal and shear compressional stresses initiated by the overburden load. *Differential compaction* refers to the gravitational compaction of sediments over and around a positive buried geomorphological feature such as a hill or a reef (Stevenson, 1973; O'Connor and Gretener, 1974).

Forces Involved in Gravitational Compaction. Compaction of a sedimentary sequence having certain particle characteristics is governed by the overburden load placed upon it. This load is normally a result of the continual deposition of new sediments. In nature, the overburden load is divided between the sediment matrix (solids) and the interstitial fluid in the pores, so that the total vertical stress at any point consists of the sum of two components: the pore-fluid stress and the intergranular stress. Terzaghi (see Terzaghi and Peck, 1968) used the term *effective pressure* to designate the difference between the total overburden pressure and the pore pressure:

$$\sigma = \sigma_e + \sigma_p \quad \text{or} \quad p_t = p_e + p_p$$

where σ or p_t is the total stress on the sedimentary system due to an overburden-geostatic load, σ_e or p_e is the effective stress (intergranular stress), and σ_p or p_p is the pore-fluid stress (neutral stress) within the void space.

If the vertical permeability of a sediment column allows the pore fluid to move out in response to the overburden load, then the pressure distribution in the pore fluids is the same as that of a continuous column of ground water extending to the water-table surface. The pressure in the fluid can be expressed as $p = \gamma z$, where p is the hydrostatic pressure in kg/cm^2 (or lb/ft^2); γ is the specific weight of the fluid in kg/cm^3 (or lb/ft^3); and z is the vertical distance (depth) in cm (or ft). The terms *pressure* and *stress* are used interchangeably in geologic literature.

Commonly, the overburden weight creates the major stress, which acts in a vertical direction. The directions of the minor and intermediate stresses are assumed to be perpendicular to the axis of the major stress. Accumulation of additional sediments upon older ones will cause a gradual increase in the major stress throughout the sediment column. Thus, matrix pressure is redistributed by the grains squeezing closer together so that they will bear more of the load. In thick shale sequences having low permeability, compaction is a slow process, the intergranular stress remains practically the same; and, consequently, the additional load must be supported by the fluid, which creates abnormally high pore-fluid pressures.

A useful method of recording what portion of the load is borne by the pore fluid is to express the pressures in terms of a ratio, λ, which is equal to $\lambda = \sigma_p/\sigma$, where σ_p is the pore stress (pressure) and σ is the total overburden stress (pressure).

Forces Involved in Tectonic Compaction. In regions where tectonic forces are distorting sedimentary rocks by folding and faulting, the greatest stress can occur in the horizontal direction and have a magnitude of 2 to 3 times the overburden stress. The overburden load would be the least principal stress in this model. Horizontal compacting stress, σ_h, can be expressed by the equation

$$\sigma_h = \frac{\mu}{1-\mu} \sigma_z \quad (1)$$

where μ is the Poisson's ratio and σ_z is the vertical stress. Poisson's ratio is equal to

$$\mu = \frac{\Delta B/B}{\Delta L/L} \quad (2)$$

where ΔB is the lateral strain (deformation) and B is the original lateral dimension; ΔL is the longitudinal strain; and L is the original vertical dimension. Equations 1 and 2 assume an elastic state in the sedimentary rock mass.

Normally, the average Poisson's ratio for rocks is regarded to be equal to 0.25. A ratio of 0.5 indicates plastic behavior in the rock mass, whereas a ratio of 1.0 would refer to liquid rock mass, as the stresses then would obey hydrostatic pressure concepts. Assuming that μ normally ranges from 0.18 to 0.27 for consolidated sedimentary rocks, the horizontal compressive stress would range from 0.22 to 0.37 psi/ft of depth. (A pressure gradient of 0.433 lb/in^2/ft corresponds to a specific weight of 1 g/cm^3 or 1 g/cm^2/cm.) Soft shale and unconsolidated sands in the northern Gulf of Mexico depositional basin have horizontal stresses in excess of 0.37 psi/ft depth and are considered to be in a plastic rather than elastic state of stress.

Compressibilities of Sand and Clayey Sediments

Compressibility. Compressibility, c, can be defined as the rate of change of volume, ∂V, with respect to the applied stress, σ, per unit volume, V. That is, $c = -(1/V)(\partial V/\partial \sigma)$. Several different usages of the term compressibility appear in the literature, depending on the method of its determination. Bulk compressibility and pore compressibility are most commonly used in sediment studies.

Laboratory Investigations. Fatt (1958a,b) studied the relationship between compressibility and rock composition. He reported that unconsolidated sediments, which are poorly sorted and contain clay, have higher compressibilities than do consolidated and well-sorted sands. Fatt (1958b) found that his measured bulk compressibilities in sandstones are a function of rock composition for a given grain shape and sorting. If sandstones are divided into two groups (one with well-sorted, well-rounded grains and the other with poorly sorted, angular grains), then for each group the compressibility is a linear function of the amount of intergranular material.

Van der Knaap (1959) noted that pore compressibility increases with decreasing porosity. It has been suggested by some investigators that between certain minimum and maximum pressures, the relation between pore compressibility and logarithm of pressure can be approximated by a straight line. A straight-line relationship has been found to exist between the log of the bulk compressibility and the log of the "effective" pressure, which in this case was equal to the direct applied axial load, because pore pressure was atmospheric (Van der Knaap and Van der Vlis, 1967). Bulk compressibilities of unconsolidated clays and sands decreased with increasing overburden pressure. From their studies, Van der Knaap and Van der Vlis concluded that clay and sand layers compact almost to the same extent, the main difference being that the low permeability to water of the clay prevents instantaneous compaction, and time effects become important.

Compressibilities of unconsolidated sands and clays are of the order of 10^{-3} to 10^{-5} psi^{-1} in the 100-10,000 psi range (1 kg/cm^2 = 14.223 psi and 1 psi^{-1} = 14.223 cm^2/kg). Chilingarian et al. (1973a) found that bulk compressibilities of unconsolidated sands range from 7.4×10^{-4} to 3×10^{-5} psi^{-1} at effective pressure range of 0-3000 psi, whereas the pore volume compressibilities range from 1×10^{-3} to 1×10^{-4} in the same pressure range using hydrostatic compaction apparatus. These values are greater by about 55-100% than those obtained using uniaxial compaction equipment. Various possible loading conditions are presented in Fig. 1.

The bulk compressibilities obtained on compacting montmorillonite clay saturated in sea water using the hydrostatic compaction apparatus (5×10^{-4} to 2.9×10^{-5} psi^{-1}) were found to be about 300-500% higher than those obtained on using uniaxial loading (1.85×10^{-4} to 5.4×10^{-6} psi^{-1}) in the applied pressure range of 400-20,000 psig (pounds per square inch, gage).

Compressibilities of unconsolidated sands appear to be very close to those of clays. Unconsolidated sands are as compressible as clays, or even more compressible. Compressibility values obtained in a hydrostatic compaction apparatus are usually about twice as high as those determined on using uniaxial compaction equipment.

Compressibilities of consolidated sandstones, shales, and carbonates are lower and range from 10^{-5} to 10^{-7} psi^{-1} in the 500-15,000 psi pressure range.

Effect of Compaction on the Chemistry of Interstitial Fluids in Sediments

Enormous amounts of water are squeezed out of continental and marine sediments during compaction and lithification. In geosynclinal basins, the overburden pressures on sediments may reach magnitudes as high as 3000 kg/cm^2, or even higher. The chemistry of expelled solutions largely depends on the types of clay minerals present. Most of the dissolved salts present in the interstitial fluids, which are trapped during sedimentation, are squeezed out during the initial stages of compaction. Laboratory results (Kryukov, 1971, see Rieke and Chilingarian, 1974) showed that mineralization of expelled solutions progressively de-

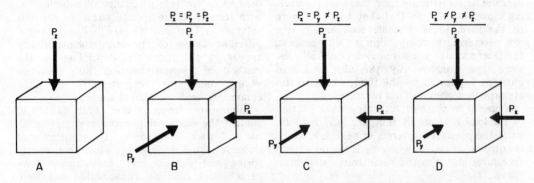

FIGURE 1. Compaction loading classification. A: Uniaxial loading (force is perpendicular to one face of the cube, while the four faces perpendicular to this face are kept stationary); referred to as *biaxial loading* by some investigators who reserve the term *uniaxial* to cases where there is no lateral constraint and lateral strain does occur. B: Hydrostatic loading (three principal stresses are equal, $p_x = p_y = p_z$). C: Triaxial loading (two out of three principal stresses are equal, $p_x = p_y \neq p_z$); referred to as biaxial loading by some investigators. D: Polyaxial loading (all three principal stresses have different magnitude, $p_x \neq p_y \neq p_z$); referred to as triaxial loading by some investigators.

creases with increasing overburden pressure. These experimental results led researchers to conclude that the concentrations of interstitial solutions in shales should be lower than those in associated sandstones (see Chilingar et al., 1969). A corollary of this premise suggests that solutions squeezed out at the beginning of compaction should have higher concentration than the interstitial solutions initially present in argillaceous sediments.

Laboratory Investigations. As shown previously, because of the intimacy of the relations between clay-sized mineral grains and water, a reduction of pore volume in sediments under increasing loads can best be considered in terms of the removal of the pore fluids. The factors that are known to influence the water content of argillaceous sediments under applied loads are the type of clay minerals, particle size, adsorbed cations, temperature, pH, Eh, and type of interstitial electrolyte solutions. The general effects of all but pH, Eh, and temperature are shown in Fig. 2. With the exception of particle size, the influence of these factors is deduced mainly from laboratory compaction experiments, i.e., squeezing of monomineralic clays mixed with simple electrolytes.

Rieke et al. (1964) determined the percentage increase in the resistivity of expelled solutions from a marine mud with increasing overburden pressure. The mud was obtained from the Santa Cruz Basin, off the coast of southern California. Their results indicate that the mineralization of squeezed-out solutions decreases with pressure.

Kazintsev (1968, see Rieke and Chilingarian, 1974) observed in laboratory experiments a gradual decrease in chlorine concentration on squeezing samples of Maykop Clay (Eastern Pre-Caucasus) having an initial moisture content of 20 and 25%. The final moisture content after compaction constituted 8.83 and 10.88%, respectively. He also determined the effect of temperature (heating to $80°C$) on concentration of Cl^- and Na^+ decreases with increasing pressure, whereas temperature does not seem to have any appreciable effect. The Mg^{++} ion concentration increases (about 1.5 times) with increasing pressure, but the absolute values are lower at high temperatures than at low temperatures. The concentration of K^+ decreases with pressure. Concentrations of K^+, Li^+, and I^- are higher in solutions expelled at higher temperatures, whereas that of SO_4^- is lower.

Krasintseva and Komarova (1968, see Rieke and Chilingarian, 1974) studied the variation in chemistry of solutions expelled from unlithified marine muds from the Black Sea. At room temperature, the chlorine concentration definitely decreases with increasing pressure, whereas concentration of some components go through a maximum at pressure of 500–1000 kg/cm^2. The Br^- and B^{+3} contents increase with increasing compaction pressure.

Some investigators, however, disagree with the above-described findings. For example, the study by Manheim (1966), who used pressures ranging from 41–844 kg/cm^2, indicated that pressure does not appreciably affect the composition of extracted waters. Shishkina (1968) investigated interstitial solutions in marine muds from the Atlantic and the Pacific Oceans and from the Black Sea and did not observe any appreciable changes in chemistry of squeezed-out solutions up to a pressure of 1260 kg/cm^2 in some samples and up to a pressure of 3000 kg/cm^2 in others. There was some increase in Ca^{++} concentration at a pressure range of 675–1080 kg/cm^2 and then a decrease at higher pressures. Shishkina (1968) stated that at very low pressures at which 80–85% of interstitial water is squeezed out, there are no changes in concentration. Obviously, the chemistry of the remaining 20–15% interstitial fluid is also of great interest and should be determined in most studies.

Chilingarian et al. (1973b) showed that the concentrations of the solutions squeezed out at the initial stages of compaction (up to 35 kg/cm^2) are slightly higher than concentration of the interstitial solution initially present in montmorillonite clay saturated in sea water. Concentrations of expelled solutions seem to go through a maximum (peak), or at least remain constant, before starting to decrease with increasing overburden pressure.

As discussed before, according to many investigators, the salinity of squeezed-out solutions progressively decreases with increasing overburden pressure. Consequently, the salinity of interstitial solutions in shales is possibly less than that of waters in associated sandstones, because practically all of the interstitial fluids were expelled in many of the laboratory experiments. It has been also observed that, during production of crude oil from sandstones surrounded by thick shale sequences, the salinity of produced water gradually decreases with time, possibly owing to the influx of fresher water from the associated shales.

The mineralization of solutions moving upward through a thick shale sequence as a result of compaction probably will progressively increase in salinity. It should be remembered, however, that if water from a sandstone bed moves through a shale layer into another sandstone bed, the water in the latter bed may be less mineralized because of filtration through a charged-net membrane. The effect of compaction on the chemistry of the fluids con-

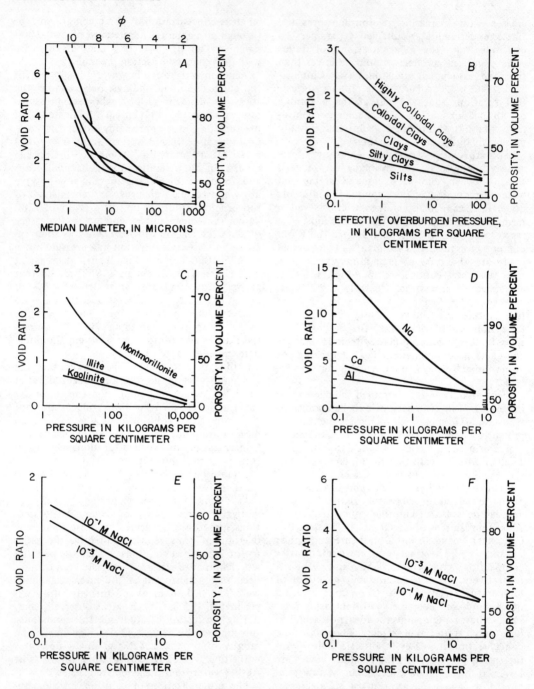

FIGURE 2. Influence of various factors on the relationship between void ratio and pressure in clayey materials (after Meade, 1968, Fig. 1). A: Relationship between void ratio and median particle diameter at overburden pressures less than 1 kg/cm². B: Generalized influence of particle size. C: Influence of clay-mineral species. D: Influence of cations adsorbed by montmorillonite. E: Influence of NaCl concentrations in unfractionated illite, about 60% of which was coarser than 2μ in size. F: Influence of NaCl concentration in illite finer than 0.2μ.

tained in sediments of a depositional basin is very complex. Further research in this area must be carried out in order to achieve a complete understanding of the mechanisms involved (see *Interstitial Water in Sediments*).

Subsidence

The best examples of the role of fluid pressure and load transfer in compaction have been inadvertently created by man's use of his natural resources, primarily ground water and petroleum. Extraction of these fluids has caused widespread, and sometimes destructive, subsidence of the ground surface in many areas of the world (see Allen et al., 1971).

Since 1925, the problems of surface subsidence have plagued the oil industry. Owing to reservoir compaction, as the crude oil is removed from the pore space, a subsidence of as much as 9 m has occurred over the Wilmington Oil Field, Los Angeles County, California. Geologists have also reported that an area of about 9000 km^2 in the San Joaquin Valley, California, was subsiding to some extent as a result of ground-water extraction.

As stated by Van der Knaap and Van der Vlis (1967), one can assume that a certain volume of subsidence at the surface of an oil field is caused by an equal volume of reservoir compaction. Subsidence at the surface is spread over a larger area, however, and consequently the subsidence at a given point in the field may not be equal in magnitude to the formation compaction directly below this point. Estimates of subsidence caused by compaction can be made if the thicknesses of the clay and shale layers, net thicknesses of the producing horizons, rate of pore-pressure decline, and the compressibility-pressure relationships are known.

H. H. RIEKE
G. V. CHILINGARIAN

References

Allen, D. R., and Chilingarian, G. V., 1975. Mechanics of sand compaction, in G. V. Chilingarian, and K. H. Wolf, eds., *Compaction of Coarse-Grained Sediments*, vol. I. Amsterdam: Elsevier, 43–77.

Allen, D. R.; Chilingarian, G. V.; Mayuga, M. N.; and Sawabini, C. T., 1971. Studio e previsione della subsidenza, in A. Mondadori, ed., *Enciclopedia della Scienza e della Tecnica*. Annuario della EST, Milano, Italy, 281–292.

Berner, R. A., 1975. Diagenetic models of dissolved species in the interstitial waters of compacting sediments, *Am. J. Sci*, 275, 88–96.

Borradaile, G., 1973. Curves for the determination of compaction using deformed cross-bedding, *J. Sed. Petrology*, 43, 1160.

Bull, W. B., 1973. Geologic factors affecting compaction of deposits in a land-subsidence area, *Geol. Soc. Am. Bull.* 84, 3783–3802.

Carpenter, C. B., and Spencer, G. B., 1940. Measurements of compressibility of consolidated oil-bearing sandstones, *U.S. Bur. Mines, Rept. Inv. 3540*, 20p.

Chilingar, G. V.; Rieke, H. H., III; and Sawabini, C. T., 1969. Compressibilities of clays and some means of predicting and preventing subsidence, in *Symp. on Land Subsidence: Internat. Assoc. Sci. Hydrology and UNESCO*, Tokyo, 89, 377–393.

Chilingarian, G. V.; Sawabini, C. T.; and Rieke, H. H., III, 1973a. Comparison between compressibilities of sands and clays, *J. Sed. Petrology*, 43, 529–536.

Chilingarian, G. V.; Sawabini, C. T.; and Rieke, H. H., III, 1973b. Effect of compaction on chemistry of solutions expelled from montmorillonite clay saturated in sea water, *Sedimentology*, 20, 391–398.

Fatt, I., 1958a. Pore structure in sandstones by compressible sphere pack models, *Bull. Am. Assoc. Petrol. Geologists*, 42, 1914–1923.

Fatt, I., 1958b. Compressibility of sandstones at low to moderate pressures, *Bull. Am. Assoc. Petrol. Geologists*, 42, 1924–1957.

Hedberg, H. D., 1936. Gravitational compaction of clays and shales, *Am. J. Sci.*, 5th ser., 31, 241–287.

Knutson, C. F., and Bohor, F. B., 1963. Reservoir rock behavior under moderate confining pressure, in Fairhurst, C. ed., *Rock Mechanics*. New York: Pergamon, 627–658.

Manheim, F. T., 1966. A hydraulic squeezer for obtaining interstitial water from consolidated and unconsolidated sediments, *U.S. Geol. Surv. Prof. Pap. 550-C*, 256–261.

Meade, R. H., 1968. Compaction of sediments underlying areas of land subsidence in central California, *U.S. Geol. Surv. Prof. Pap. 497-B*, 1–23.

O'Connor, M. H., and Gretener, P. E., 1974. Quantitative modelling of the processes of differential compaction, *Bull. Canadian Petrol Geol.*, 22, 241–268.

Perrier, R., and Quiblier, J., 1974. Thickness changes in sedimentary layers during compaction history; methods for quantitative evolution, *Bull. Am. Assoc. Petrol. Geologists*, 58, 507–520.

Rieke, H. H., III; Chilingar, G. V.; and Robertson, J. O., Jr., 1964. High-pressure (up to 500,000 psi) compaction studies on various clays, *22nd Intern. Geol. Cong.*, New Delhi, India, 15, 22–38.

Rieke, H. H., and Chilingarian, G. V., 1974. *Compaction of Argillaceous Sediments*. Amsterdam: Elsevier, 424p.

Shishkina, O. V., 1968. Methods of investigating marine and ocean mud waters, in G. V. Bogomolov et al., eds., *Pore Solutions and Methods of Their Study (A Symposium)*. Minsk, USSR: Izd. Nauka i Tekhnika, 167–176.

Smith, J. E., 1971. The dynamics of shale compaction and evolution of pore fluid pressures, *J. Int. Assoc. Math. Geol.*, 3, 239–263.

Sorby, H. C., 1908. On the application of quantitative methods to the study of the structure and history of rocks, *J. Geol. Soc. Lond.*, 64, 171–232.

Stevenson, D. L., 1973. The effects of buried Niagaran reefs on overlying strata in Southwestern Illinois, *Ill. St. Geol. Surv., Circ. 482*, 22p.

Terzaghi, R. D., 1940. Compaction of lime mud as a

cause of secondary structures, *J. Sed. Petrology*, **10**, 78-90.

Terzaghi, K., and Peck, R. P., 1968. *Soil Mechanics in Engineering Practice*. New York: Wiley, 729p.

Van der Knaap, W., 1959. Non-linear behavior of clastic porous media, *Soc. Petrol. Engineers, A.I.M.E., Trans.*, **216**, 179-186.

Van der Knaap, W., and Van der Vlis, A. C., 1967. On the cause of subsidence in oil-producing areas, *7th World Petrol. Cong.*, Mexico City, Mexico, **3**, 85-95. Amsterdam: Elsevier.

Cross-references: *Carbonate Sediments–Diagenesis; Clastic Sediments–Lithification and Diagenesis; Clay–Diagenesis; Diagenesis; Interstitial Water in Sediments; Penecontemporaneous Deformation of Sediments.*

CONCRETIONS

Concretions are discrete segregations of mineral matter found in sedimentary rocks, particularly in shales and sandstones. Concretions have a concentric structure, or show other evidence of growth *outward* from a nucleus. Most concretions are spheroidal, ellipsoidal, or discoidal, although irregular shapes may arise if two or more concretions unite during growth, or if the nucleus was a rather large and irregular object. Concretions range in size from a few millimeters to a few meters in long dimension. They have wide geographic and stratigraphic distribution; they have even been recognized in highly metamorphosed sedimentary rocks of Precambrian age. Pettijohn (1975), Hayes (1964), and Hallam (1967), among others, discuss concretions and related phenomena, and give numerous references to specific examples.

Mineralogy

Concretions are composed of common sedimentary minerals, but such minerals are only minor constituents of the particular host rock. Concretions of calcite, iron oxides and hydroxides, silica, pyrite, marcasite, barite, siderite, gypsum, and dahlite are known (Pettijohn, 1975). Concretions may also incorporate dominant minerals of the host rock, e.g., calcite concretions in shale may contain silt-sized quartz and clay minerals, and barite concretions in sandstone may contain sand grains. By virtue of better cementation, lower porosity, and different mineral composition, concretions tend to resist weathering better than the host rock, explaining the discreteness of concretions (see Fig. 1).

Genesis

Many concretions form early in the history of the host rock and are termed syngenetic or

FIGURE 1. Calcite concretion in mudstone of Warsaw Formation (Mississippian) of SE Iowa. (Photo: J. B. Hayes.)

early diagenetic. The minerals of such concretions may have been deposited as dispersed minor phases within the host sediments, and subsequent solution of the minor constituents, migration in solution, and precipitation around a nucleus produced a segregation or concretion. Others may have formed by direct precipitation of ions from interstitial solutions during diagenesis. Either way, the molal free energy of a minor phase is less if it is segregated rather than dispersed; hence, concretionary growth is spontaneous. In many concretions, the nucleus was a dead organism, such as a fish, cephalopod, gastropod, or plant fragment. The decay of the organism created local chemical conditions favorable to the precipitation of dispersed phases, especially carbonates (Weeks, 1953; Berner, 1968). Studies of carbon isotopes in calcite and siderite concretions show that the carbon dioxide entrained in them as carbonate ion was derived by bacterial decay of organic matter (Fritz et al., 1971; Hodgson, 1966). The mineral composition of early diagenetic concretions reflects not only the chemistry of the environment established in the host rock during dewatering and consolidation of the sediments, but it may also reflect the chemistry and salinity of the environments in which the sediments initially were deposited (Curtis, 1967; Franks, 1969).

Some concretions, particularly those in sandstone, are thought to have formed long after consolidation and uplift of the host rock, and are termed epigenetic. The concretion mineral migrates in ground-water solutions and precipitates at a nucleus as cement between sand grains. Swineford (1947) and Franks (1969), however, offered evidence that concretionary masses of calcite in some sandstone, like concretions in associated clay rocks, formed during early diagenesis. Some concretions reveal nothing about the cause of nucleation or time of formation.

JOHN B. HAYES
PAUL C. FRANKS

References

Berner, R. A., 1968. Calcium carbonate concretions formed by the decomposition of organic matter, *Science,* **159,** 195-197.
Curtis, C. D., 1967. Diagenetic iron minerals in British Carboniferous sediments, *Geochim. Cosmochim. Acta,* **31,** 2109-2123.
Franks, P. C., 1969. Nature, origin, and significance of cone-in-cone structures in the Kiowa Formation (Early Cretaceous), north-central Kansas, *J. Sed. Petrology,* **39,** 1438-1454.
Fritz, P.; Binda, P. L.; Folinsbee, F. E., and Krouse, H. R., 1971. Isotopic composition of diagenetic siderites from Cretaceous sediments in western Canada, *J. Sed. Petrology,* **41,** 282-288.
Hallam, A., 1967. Siderite- and calcite-bearing concretionary nodules in the Lias of Yorkshire, *Geol. Mag.,* **104,** 222-227.
Hayes, J. B., 1964. Geodes and concretions from the Mississippian Warsaw Formation, Keokuk region, Iowa, Illinois, Missouri, *J. Sed. Petrology,* **34,** 123-133.
Hodgson, G. A., 1966. Carbon and oxygen isotope ratios in diagenetic carbonates from marine sediments, *Geochim. Cosmochim. Acta,* **30,** 1223-1233.
Pettijohn, F. J., 1975. *Sedimentary Rocks,* 3rd ed. New York: Harper & Row, 628p.
Swineford, A., 1947. Cemented sandstones of the Dakota and Kiowa Formations in Kansas, *Kansas Geol. Surv. Bull.,* **70**(4), 53-104.
Weeks, L. G., 1953. Environment and mode of origin and facies relationships of carbonate concretions in shales, *J. Sed. Petrology,* **23,** 162-173.

Cross-references: *Carbonate Sediments–Diagenesis; Cementation; Chalk; Clastic Sediments–Lithification and Diagenesis; Diagenesis; Geodes; Loess; Nodules in Sediments.*

CONGLOMERATES

A conglomerate is a rudaceous sedimentary rock consisting of cemented, rounded particles of pebble or gravel size (>2 mm). A conglomerate is distinguished from a breccia (see *Breccias, Sedimentary*) in that the particles of the latter are angular instead of rounded. Conglomerate (containing >25% clasts) and conglomeratic sandstone (<25% clasts) are the lithified equivalents of gravel and gravelly sand, respectively.

Since the time of Grabau (1913), texture and composition have been most commonly used in erecting classifications of conglomerates (Fig. 1). *Oligomictic* (monogenetic) and *polymictic* (polygenetic) types are recognized according to the uniformity or variability of the composition and source of the pebbles (Shvetzov, 1934). Conglomerates may be further subdivided by percent matrix and labile constituents (Fig. 1). The matrix of most conglomerates is sand sized, but may consist of silt- or even clay-sized particles so that many diamictites (q.v.) fall under the classification of conglomerates.

Descriptive Parameters

Walker (1975a) has discussed the sedimentary features that contribute to a descriptive framework for conglomerates (Fig. 2): sorting and size distribution; fabric, stratification, and grading.

Sorting and Size Distribution. Whether the clasts are closely or loosely packed and whether clast size relative to matrix particle size is small or large is usually a measure of the fluidity of the transporting medium. There appears to be a spectrum between conglomerates that are either (1) bimodal (clasts, matrix), well sorted, and clast supported; or (2) polymodal, poorly sorted, and matrix supported (Walker, 1975a). Thus Pettijohn (1957; see Fig. 1) distinguished between *orthoconglomerates* (<15% matrix), which he believed were laid down by normal aqueous currents, and more poorly sorted *paraconglomerates* (conglomeratic mudstone, diamictite; >15% matrix), usually deposited

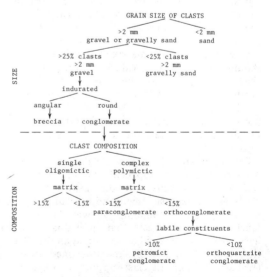

FIGURE 1. A classification of conglomerates by texture and composition (courtesy E. G. Williams, written communication). Pettijohn (1975) would place orthoquartzite conglomerate under oligomict, <10% labile constituents.

CONGLOMERATES

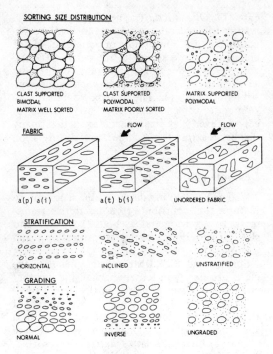

FIGURE 2. Descriptive features for conglomerate (from Walker, 1975b). Under fabric, the coding a(p)a(i) means a-axis parallel to flow and imbricated and a(t)b(i) means a-axis transverse to flow with b-axis imbricated.

by more viscous media such as subaqueous slides, mudflows, and ice. Selection of a sorting value to separate environmentally significant types, however, is relatively arbitrary. When examining a conglomerate, two other important questions about matrix-clast relationships should be answered (Walker, 1975a): (1) If clast supported, was the finer material deposited with the clasts, or did it filter in later? (2) If matrix supported, were the clasts transported with the matrix or were they transported by different mechanisms, e.g., as dropstones into muds?

Fabric. Fabric (q.v.) of conglomerates has been studied often, especially in tills (see *Till and Tillite*), because the dimensions of the clasts are more easily seen and measured than in finer-grained sediments. A well-developed fabric strongly suggests that individual clasts have moved freely and have assumed an orientation that is imposed by the flow mechanism (Walker, 1975a). In clast-supported conglomerates composed of pebbles with long (a-) axes >2 cm, well-developed fabrics are of two kinds (see Fig. 2). The a(t)b(i) fabric is characteristically the result of clasts rolling on the bed; the less common a(p)a(i) fabric has been found in tills and clast-supported, deep-water, turbidite-like conglomerates (Walker, 1975a,b). Walker suggests that the a(p)a(i) fabric, associated with stratification and graded bedding, may be a fairly reliable indicator of transport processes that maintain cobbles and boulders in dispersion (see *Grain Flows*) above the bed.

Stratification. Stratification, although well developed in most orthoconglomerates, is poorly developed or absent in many paraconglomerates, partly as a result of the absence of currents that fluctuate in velocity and direction during sedimentation. An exception, laminated pebbly mudstones are plane-bedded argillites, i.e., quiet-basin (especially lake) deposits into which ice-rafted pebbles have been dropped (Pettijohn, 1975). Mudflow deposits (q.v.) may also display stratification, resulting from successive influxes of mudflow. Horizontal stratification in conglomerates may be shown by changes in clast size or composition, sorting, or fabric; if the layers grade into each other, the depositional process may be fluctuating or pulsating, but if the layers are sharply bounded, different processes may be responsible for each layer (Walker, 1975a). Walker shows that alternating pebbly and sandy layers require only a very slight current fluctuation. Cross-stratification may be produced by migration of bed forms, by lateral accretion on point bars and braided bars, or by the filling of depressions. If clasts occur in stringers along individual bedding planes, the probability is high that they represent a *lag gravel* and that the matrix is secondary.

Grading. Normal graded bedding (q.v.) indicates deposition of progressively finer material from one waning current. At the bases of some conglomerate beds, there is a relatively thin zone of inverse graded bedding (q.v.), usually capped by normal grading (Walker, 1975a,b). Both inverse and normal grading imply that the current was able to sort material and that the clasts were free to move relative to one another according to size (Walker, 1975a).

Composition. Composition varies in conglomerates because of variations in composition of source rocks, distance of transport, degree of reworking, and rates of erosion and weathering. These variables are often related to tectonism and the stage of the geomorphic cycle (Krynine, 1948; Folk, 1954). Composition is therefore of significance in determining provenance and maturity of conglomerates, although texture must also be considered. Pettijohn (1957, 1975), for example, subdivides orthoconglomerates into orthoquartzite (oligomict, <10% labile constituents) and petromict conglomerate (>10% labile constituents).

Facies

There are relatively few modern environments of conglomerate deposition, the major sites being (1) alluvial fans, (2) braided streams, (3) shorelines, (4) submarine fans, and (5) glacial environments. Common associations of the descriptive features discussed above have only recently begun to be determined. Walker (1975b) suggests the relationships shown in Fig. 3. The various environments of conglomerate deposition are described in numerous entries (see cross-references).

Ancient Conglomerates

In the stratigraphic column, conglomerates have been described most often from rock sequences of fluvial (flood-plain, braided-stream, alluvial-fan) origin, as in the Silurian Shawangunk and Pennsylvanian Pottsville formations of the Appalachians, the Devonian Old Red Sandstone of the British Isles, the Triassic Newark conglomerates of eastern North America, and the Tertiary Molasse of the Alps (for references, see *Alluvial-Fan Sediments; Braided-Stream Deposits; Fluvial Sediments*). Individual conglomerate bodies are often elongate in shape as a result of their occurrence within channels, although the form of the larger unit in which they occur may be tabular, or wedge-shaped, as in the case of polymictic conglomerates resulting from the rapid erosion of land masses of high relief. The sands associated with such conglomerates are marked by their very poor rounding. This type is a common associate of sediments infilling intermontane basins, and notable examples include the early Precambrian conglomerates of the Lake Superior region and the Molasse of the Alps. Exceptionally, conglomerates composed almost entirely of limestone may occur, recording unusual circumstances of rapid erosion and burial, usually along a fault scarp, as illustrated by the limestone conglomerates of the Triassic of Maryland and Tennessee.

The marginal marine environment is represented by the Cretaceous Gosau Conglomerate of the eastern Alps and parts of the Cambrian Baraboo Quartzite in north central US and the Silurian Red Mountain Formation of Alabama. Thin sheets of oligomictic conglomerate of wide extent occur at the base of sandstone layers, marking periods of widespread transgression of an advancing sea. Though commonly referred to as *basal conglomerates*, they are strictly of marginal origin and are comparable to the well-sorted gravels of many modern beaches. Such a conglomerate is commonly found at or near the base of the Cambrian, where it is usually relatively thin, exceptionally several meters thick. Erosion of limestone cliffs by sea water (presumably) saturated with calcium carbonate may produce limestone conglomerate, e.g., the Liassic conglomerate of southwest England and South Wales, which is made up of large boulders of the underlying Carboniferous limestone.

Ancient examples of clast-supported, resedimented (deep-water) conglomerates, reviewed by Walker (1975a,b) include Archean conglomerates of Minnitake Lake, Ontario, the Cambro-Ordovician Cap Enragé Formation, and the Cretaceous Wheeler Gorge conglomerate and Pliocene Pico Formation, both in California.

Excluding tillite (see *Till and Tillite*), most paraconglomerates have been produced by subaerial and subaqueous gravity movement, including slides, slumps, plastic mass flows, and viscous fluid flows. Subaerial processes such as mudflows (see *Mudflow Deposits*) produce paraconglomerates on a small scale and are often associated with alluvial fans. Ancient examples of subaqueous origin, some of which are still contested, include the widespread Eocambrian diamictites, the mid-late Paleozoic Squantum "tillite" of Massachusetts, the Pennsylvania Haymond Formation in Texas (McBride, 1966), and the Jura-Cretaceous pebbly mudstones of western North America (see references in *Diamictite*).

JOANNE BOURGEOIS

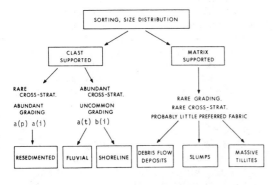

FIGURE 3. Common associations of features in conglomerate (from Walker, 1975b). The diagram is not intended to show all possible types but rather those that are abundant. The features are illustrated in Fig. 2.

References

Bearce, D. H., 1973. Origin of conglomerates in Silurian Red Mountain Formation of central Alabama; their paleogeographic and tectonic significance, *Bull. Am. Assoc. Petrol. Geologists*, 57, 688-701.

Brock, E. J., 1974. Coarse sediment morphometry: a comparative study, *J. Sed. Petrology*, 44, 663-672.

Folk, R. L., 1954. The distinction between grain size and mineral composition in sedimentary-rock nomenclature, *J. Geol.*, **62**, 344–359.
Grabau, A. W., 1913. *Principles of Stratigraphy.* New York: A. G. Seiler, 1185p.
Krynine, P. D., 1948. The megascopic study and field classification of sedimentary rocks, *J. Geol.*, **56**, 130–165.
McBride, E. F., 1966. Sedimentary petrology and history of the Haymond Formation (Pennsylvanian), Marathon Basin, Texas, *Univ. Texas Bur. Econ. Geol., Rept. Inv. 57,* 101p.
Pettijohn, F. J., 1957. *Sedimentary Rocks*, 2nd ed. New York: Harper & Row, 718p.
Pettijohn, F. J., 1975. *Sedimentary Rocks*, 3rd ed. New York: Harper & Row, 628p.
Schvetzov, M. S., 1934. *Petrography of Sedimentary Rocks* (in Russian). Moscow, 375p.
Walker, R. G., 1975a. Conglomerate: sedimentary structures and facies models, *Soc. Econ. Paleont. Mineral. Short Course,* **2,** (April, 1975), 133–161.
Walker, R. G., 1975b. Generalized facies models for resedimented conglomerates of turbidite association, *Geol. Soc. Am. Bull.,* **86,** 737–748.

Cross-references: *Agglomerate, Agglutinate; Alluvial-Fan Sediments; Arkose, Subarkose; Avalanche Deposits; Beach Gravels; Braided-Stream Deposits; Breccias, Sedimentary; Clay-Pebble Conglomerate and Breccia; Diamictite; Fabric, Sedimentary; Flood Deposits; Fluvial Sediments; Glacial Gravels; Glacigene Sediments; Graded Bedding; Gravity Flows; Gravity-Slide Deposits; Intraformational Disturbances; Mass-Wasting Deposits; Mudflow and Debris-Flow Deposits; Olistostrome, Olistolite; Pebbly Mudstones; Pyroclastic Sediment; River Gravels; Sedimentation–Tectonic Controls; Storm Deposits; Submarine Fan (Cone) Sedimentation; Till and Tillite; Volcaniclastic Sediments and Rocks.*

CONTINENTAL-RISE SEDIMENTS

The continental rises bordering the world's oceans are a major and often final depositional site for sediments eroded from the continental land masses and transported across the continental margin toward the abyssal sea floor. Characterized by a seaward gradient of 1:100 to 1:700, a depth range of 1500–5600 m, and a width range of about 100 to several hundred km (Heezen et al., 1959; Guilcher, 1963), rises comprise thick sedimentary sequences of geosynclinal proportions that have accumulated through steady build-up coincident with gentle subsidence. Distinctive suites of sediments and structures occur which record the relative dominance of various depositional regimes (Fig. 1).

Depositional Regimes of the Continental Rise

Recognition of significant downslope transport of shelf-edge and continental-slope sediments to the rise in high-velocity, sediment-

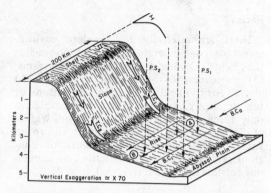

FIGURE 1. Schematic diagram of depositional regimes of the lower continental rise (after Stanley, 1969). Areas I, II, and III represent major source regions which supply sediment to the rise by turbidity currents ($TC_{1,2}$), slumps (not shown), and fine detritus which is deposited along with open-ocean pelagic sediments ($PS_{1,2}$). Bottom currents (BC) redistribute material brought to the rise by hemipelagic and turbidity-current sedimentation. Note that sediments at site A would show a mixture of materials transported by TC_1, PS_1, and BC_0. At site B, the sediments are derived from TC_2, PS_2, and bottom currents, which at this point are a composite of TC_1, PS_1, and BS_0, hence cource of site-B sediments consists of detritus from all three regions (I, II, and III). The relative proportion of each source depends on many variables such as continental-margin width and relief, bottom-current competence, etc.

charged flows (see *Turbidites*) led many to conclude that turbidity currents are the major agent of rise construction (Menard, 1964, p. 226; Dietz, 1965). This view has persisted until recently, when the collection of various other kinds of evidence indicated that other processes, such as alongslope bottom currents, may be equally important or even dominant (see *Contourites*). Added to the detritus accumulated by turbidity and bottom currents is a steady rain of inorganic and biogenic sediments settling from the overlying surface waters (see *Pelagic Sediments*). The major avenues of sediment transport may then be generalized as *downslope, alongslope,* and *vertical settling.* The nature of the deposits resulting from sediment travel along each of these routes is often, but not always, distinctive.

Downslope transport of sediments to the rise occurs in a series of separate, gravity-driven events such as slumps, sand falls, debris flows, turbidity currents, and other similar mechanisms. The initiating force is usually the result of some phenomenon that induces a stress on shelf-edge deposits such as: sudden increase in shelf sediment build-up (coastal floods); seismic disturbance (earthquakes); or direct agitation of the bottom by waves and currents (storms, internal waves, or tide-related phenomena, etc.). Most downslope transport occurs at relatively high velocity (up to 200 cm/sec). Therefore the competence exists to carry whatever material is present at the point of origin and to erode additional material en route.

Transport of sediment along the rise by deep, low-velocity, bottom currents flowing at nearly constant velocities (around 20 cm/sec) is a continuous, uninterrupted process. These currents, such as the Western Boundary Undercurrent in the North Atlantic, are driven by the flow of dense bottom water away from the poles and held against the base of the continents by the *Coriolis effect;* hence the currents are parallel to regional contours, and the deposits are called *contourites* (q.v.). Sediments transported by these relatively weak currents are restricted to the silt, clay, and occasionally fine-sand sizes. Such bottom currents are secondary conduits which transport sediment initially either injected by downslope movement or settled from overlying surface waters.

Biogenic and detrital sediments slowly settling and accumulating on the surface of the rise are termed *hemipelagic,* because they are a mixture of continentally derived (terrigenous) detritus and *pelagic* sediments that comprise the accumulations of the deep ocean basins. The diluting (terrigenous) material is both organic debris and nonorganic weathering products, usually in the fine-silt and clay size, delivered to surface waters over the rise via wind and rivers or winnowed from the shelf and slope. The pelagic sediments (see *Pelagic Sedimentation, Pelagic Sediments*) accumulating throughout the ocean basin are made up of calcareous microfossils (coccoliths, forams, tetrapods) and siliceous microfossils (diatoms, radiolarians) as well as atmospheric dust (extraterrestrial, volcanic, and continental) and terrigenous clays. Included in the pelagic sediments are the secondary or authigenic sediments formed at and just below the water/sea-floor interface.

Through these three main avenues of transport, continental-rise sediments are accumulating at a rate of about 4–10 cm/1000 yr. During the Pleistocene, the rate was increased by at least a factor of 2.

Factors Influencing the Character of Rise Sediments

All aspects of continental-rise sediments, including dominant texture and mineralogy, color, rate of sedimentation, and major depositional processes depend on a hierarchy of variables ranging from water depth and temperature to climate and petrology of source regions hundreds of kilometers inland. The major factor influencing overall character of rise deposits is tectonic stability of the continental margin. Continental margins can be broadly classed as tectonically stable (i.e., those on a trailing edge of a crustal plate) and tectonically active (i.e., those on a leading edge of a crustal plate). Tectonically stable margins tend to be broader and lower in relief than tectonically active margins; hence rate of sedimentation is lower, sediments are finer and more mature, and the importance of alongslope and vertical sedimentation increases. Active coastlines often have narrow shelves indented by steep submarine canyons which intercept littoral and fluvial sediments and channel them to adjacent basins, trenches, or rises, thus increasing the relative importance of downslope sedimentation.

Properties of the overlying water mass are also important in determining lithology: strong surface currents introduce fine sediments from distant sources; cool waters stimulate diatom productivity, whereas in warm water radiolarians predominate; and in deep water, below the carbonate compensation level, tests of calcitic and aragonitic microorganisms are selectively removed. Small variations in source rocks are homogenized in rise sediments; major regional rock types and climate, however, strongly influence grain size and mineralogy of particles and clays. In summary, the character of rise sediments depends on source material and weathering products available; distance, time, and mechanism of transport; and the nature of contribution from the water mass itself.

Lithology and Structure of Rise Deposits

Rise sediments range in size from clay to fine sand, with a modal size in the medium silt range. Coarse sands and gravels occur locally but usually are restricted to submarine fans (see *Submarine Fan Sedimentation*). Turbidity currents deposit their load in a sheet deposit which is commonly <50 cm thick and normally graded from fine sand at the base upward to fine silt and clay. Repetitive flows truncate the upper units, resulting in multistoried sand-silt units with abrupt truncation of grading trends.

Deposits laid down by gentle bottom currents are often difficult to decipher because of their secondary nature. Their load consists of planktonic organisms, fine terrestrial detritus, and outer-shelf sands. Variations in load and current velocity may result in normal and reverse grading, horizontal and cross-stratification, and heavy mineral laminae. Fine sand and coarse

TABLE 1. General Sedimentary Structures of Continental Rise Sediment

Massive sand beds thicker than 50 cm	Rare
Cross beds of heavy mineral placers	Very common
Cross beds of clay concentrations	None
Lamination of heavy mineral placers in sand or silt bed	Very common
Lamination of clay concentrations in sand or silt bed	Rare
Alternating thin beds of laminae of silt and clay	Very common
Graded silts and sands	Common

Modified from Hollister and Heezen, 1972, p. 51.

TABLE 2. Comparison of Turbidite vs. Contourite Sands and Silts

		Turbidite	Contourite
	Size sorting	moderate to poorly sorted > 1.50 (Folk)	well to very well sorted < 0.75 (Folk)
	Bed thickness	usually 10–100 cm	usually < 5 cm
Primary sedimentary structure	Grading	normal grading ubiquitous, bottom contacts sharp, upper contacts poorly defined	normal and reverse grading, bottom and top contacts sharp
	Cross lamination	common, accentuated by concentrations of lutite	common, accentuated by concentrations of heavy minerals
	Horizontal lamination	common in upper portion only, accentuated by concentration of lutite	common throughout, accentuated by concentrations of heavy minerals or foraminifera shells
	Massive bedding	common, particularly in lower portion	absent
Principal constituents of sand and silt beds	Grain fabric	little or no preferred grain orientation in massive graded portions	preferred grain orientation parallel to the bedding plane is ubiquitous throughout bed
	Matrix (< 2μ)	10–20%	0–5%
	Microfossils	common and well preserved, sorted by size throughout bed	rare and usually worn or broken, often size sorted in placers
	Plant and skeletal remains	common and well preserved, sorted by size throughout bed	rare and usually worn or broken
	Classification (Pettijohn)	graywacke and subgraywacke	subgraywacke, arkose, and orthoquartzite

Modified from Hollister and Heezen, 1972, p. 61.

silt remnants from recent turbidity flows are found distributed randomly throughout thick sequences of typical pelagic sediments (Field and Pilkey, 1971).

Hemipelagic sediments accumulate in a random fashion, leading to a homogeneous structure. Grain sizes range from clay to medium silt for terrestrial materials. Mixture of sources for hemipelagic sediments often yields a strong bimodality in sediment size: terrigenous materials are usually clay and fine silt, whereas the biogenic fraction is medium silt to fine sand (principally forams). In areas where hemipelagic sediments are the dominant deposit, the rate of sedimentation is low, and bottom deposits are often bioturbated.

Hollister and Heezen (1972) have summarized aspects of continental-rise sediment properties, and their data are shown in Tables 1 and 2. Although the bulk of rise sediments is derived from the adjacent continental landmass, the sediments are usually more clayey, feldspathic, and micaceous, and less quartzose, than are immediately adjacent shelf sediments, and are comparable in many ways to ancient flysch sediments (Stanley, 1970; Horn et al., 1972).

MICHAEL E. FIELD

References

Dietz, R. S., 1965. Collapsing continental rises: an actualistic concept of geosynclines and mountain building, *J. Geol.*, **73**, 901-906.

Field, M. E., and Pilkey, O. H., 1971. Deposition of deep-sea sands: Comparison of two areas of the Carolina continental rise, *J. Sed. Petrology*, **41**, 526-536.

Fritz, S. J., and Pilkey, O. H., 1975. Distinguishing bottom and turbidity currents on the continental rise, *J. Sed. Petrology*, **45**, 57-62.

Guilcher, A., 1963. Continental shelf and slope (continental margin), in M. N. Hill, ed., *The Sea*, vol. III. New York: Interscience, 281-311.

Heezen, B. C.; Tharp, M.; and Ewing, M., 1959. The floors of the oceans. I. The North Atlantic, *Geol. Soc. Am. Spec. Pap. 65*, 122p.

Hollister, C. D., and Heezen, B. C., 1972. Geologic effects of ocean bottom currents, western North Atlantic, in A. L. Gordon, ed., *Studies in Physical Oceanography*, Vol. 2. New York: Gordon and Breach, 37-66.

Horn, D. R.; Ewing, J. I.; and Ewing, M., 1972. Graded bed sequences emplaced by turbidity currents north of 10°N in the Pacific, Atlantic and Mediterranean, *Sedimentology*, **18**, 247-275.

Klasik, J. A., and Pilkey, O. H., 1975. Processes of sedimentation on the Atlantic continental rise off the southeastern U.S., *Marine Geol.*, **19**, 69-90.

Menard, H. W., 1964. *Marine Geology of the Pacific*. New York: McGraw-Hill, 271p.

Stanley, D. J., 1969. Sedimentation in slope and base of slope environments, in D. J. Stanley, ed., *The New Concepts of Continental Margin Sedimentation*. Washington: American Geological Institute, DJS-8-1-25.

Stanley, D. J., 1970. Flyschoid sedimentation of the outer Atlantic margin off northeastern North America, *Geol. Assoc. Can. Spec. Pap. 7*, 179-210.

Cross-references: *Continental-Shelf Sedimentation; Contourites; Pelagic Sedimentation, Pelagic Sediments; Submarine (Bathyal) Slope Sedimentation; Submarine Fan (Cone) Sedimentation; Turbidity-Current Sedimentation*. Vol. I: *Continental Rise*.

CONTINENTAL SEDIMENTATION

Continental sediments are those sediments deposited in nonmarine environments. Although generally above sea level, the surface of deposition may be below sea level, as in the Dead Sea.

The principal environments of continental sedimentation may be listed as follows:

- Desert (see *Desert Sedimentary Environments*)
- Glacial (see *Glacigene Sediments*)
- Fluvial (see *Fluvial Sediments*)
- Lacustrine (see *Lacustrine Sedimentation*)
- Paludal (see *Paludal Sediments*)
- Speleal (see *Speleal Sediments*)

An additional environment, though generally regarded as a region of erosion rather than deposition, is *high mountain,* a source of talus (q.v.), which, upon consolidation, produces a type of breccia (q.v.); regions of high relief may also be classified by process as *gravity-dominated* environments (see *Mass-Wasting Deposits*). Another special, partially sedimentary, environment is the *volcanic terrane* (see *Volcaniclastic Sediments and Rocks*).

The sediments formed in continental environments may be *terrestrial* or *aqueous* depending whether water has played a subordinate or major role in their formation. This water is not necessarily fresh, but may be brackish or saline and may or may not contain sodium chloride, the principal salt present in marine waters.

The Geologic Record. Presumably, continental sediments have been deposited on all land masses during the earth's history; their rarity in the geologic column must imply that their preservation was largely a matter of chance, although it may also be that many continental deposits have not been recognized. Even slight uplift will increase the activity of erosive agents and will tend to bring about the destruction of the superficial sediments. Thus, such continental sediments as have been preserved must represent periods of continuous downwarping or the excess of sedimentation over erosion. Continental sediments in the geologic column are best known, therefore, in fault-block depressions or rift valleys and in isolated intracratonic basins.

Soils. One of the principal products of the continental environment is soil (see Vol. XI, *Encyclopedia of Pedology*), although soils are rarely preserved in situ. There can be no doubt that their characteristics have a marked effect on the sediments in which, by erosion, transportation, and deposition, they are ultimately deposited (see *Clay as a Sediment*).

DEREK W. HUMPHRIES

References*

Beuf, S.; Biju-Duval, B.; Charpal, O. de; Rognon, P.; Gabriel, G.; and Bennacef, A., 1971. *Les Grès du Paléozoique Inférieur au Sahara; Sédimentation et Discontinuités, Evolution Structurale d'un Craton.* Paris: Editions Technip, 464p.

Geologische Rundschau, **63**(3), 1974. Contains numerous papers from the annual meeting of Geologische Vereinigung, on Continental Sedimentation.

Klein, G. de V., ed., 1968. Late Paleozoic and Early Mesozoic sedimentation, northeastern North America, *Geol. Soc. Am. Spec. Pap. 106*, 309p.

Termier, G., and Termier, H., 1963. *Erosion and Sedimentation*. London: Van Nostrand, 433p.

*See also references under various cross-referenced entries.

Cross-references: *Alluvial-Fan Sediments; Arkose, Subarkose; Avalanche Deposits; Braided-Stream Deposits; Breccias, Sedimentary; Caliche, Calcrete; Clastic Sediments and Rocks; Clay Sedimentation Facies; Coal; Conglomerates; Desert Sedimentary Environments; Diamictite; Eolian Sedimentation; Faceted Pebbles and Boulders; Flood Deposits; Fluvial Sediments; Glacigene Sediments; Gravity-Slide Deposits; Lacustrine Sedimentation; Loess; Marl; Mass-Wasting Deposits; Mudflow and Debris-Flow Deposits; Niveo-Eolian Deposits; Oil Shale; Organic Sediments; Paludal Sediments; Peat Bog Deposits; Pyroclastic Sediment; Salar, Salar Structures; Sedimentary Environments; Sedimentation—Tectonic Controls; Speleal Sediments; Spring Deposits; Till and Tillite; Tuff; Varves and Varved Clays; Volcanic Ash Deposits; Volcaniclastic Sediments and Rocks; Weathering in Sediments.*

CONTINENTAL-SHELF SEDIMENTATION

Continental shelves comprise a major subsystem in the world's dominant sediment transport system, i.e., that system whereby continents are weathered, the soil transported to the shore by rivers, across shelves and over the shelf break by shallow marine currents, and down slopes on to the abyssal plains by turbidity currents and other processes.

Shelf Hydraulics

The Shelf Velocity Field. An analysis of shelf-sediment transport requires an understanding of the shelf velocity field. The pattern of shelf flow is complex in time and space (Fig. 1). Shelves such as the North and Celtic seas of Europe experience tidal ranges generally in excess of 3 m as a consequence of shelf geometry, which amplifies the tidal wave by a shoaling process or causes the shelf tide to co-oscillate with the oceanic tide. On such *tide-dominated shelves* mid-tide surface velocities may exceed 200 cm/sec over broad areas; and rates of sediment transport, both as bed and suspended load, are high. On *weather-dominated shelves*, where tides are weaker,

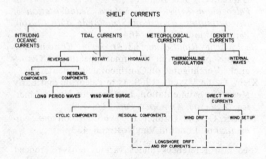

FIGURE 1. Components of the shelf velocity field.

storm currents drive sedimentation. During fair weather, a weak thermohaline and wind-driven circulation distributes suspended fine sediment; but significant transport of sand seaward of the surf is accomplished mainly by high-velocity pulses of water movement associated with storms.

Transport is most efficient during the winter months, when thermohaline stratification has broken down, and the downward transfer of momentum from surface wind-drift currents is unimpeded. Such wind-drift currents dominate the inner shelf; in deeper water, where the effects of friction are proportionately less, a coast-parallel geostrophic flow results in response to the pile-up of water against the coast by winds. Oscillatory wave surge plays an important but poorly understood secondary role in shelf-sediment transport; its peak velocities affect bottom roughness by generating ripples and aid entrainment of sediment into the flow field.

Numerical Modeling of Shelf-Sediment Transport. Numerical modeling of shelf-sediment transport may eventually be feasible, but both data and techniques are inadequate at present. Models of the tidal flow field may be applied directly to sediment-transport calculations (McCave, 1971). Models of the storm velocity field (Jelesnianski, 1967) may be modified to describe resulting shear stress on the bottom. Regional wave-propagation models are available, although these also must be modified, to determine the effect of wave energy input into the sea floor (Cherry, 1966).

Tidal models are directly applicable to assessment of mean annual sediment transport because plots of surface tidal velocities against time at a given point on the shelf are (relatively) simple polyharmonic curves. In order to determine mean annual transport on wave-dominated inner shelves, or storm-dominated shelves, however, the problem of climatic weighting of the aperiodic velocity field must be faced. Wave climatology has been explored by a number of workers (e.g., see Harris, in Swift et al., 1972) as a consequence of the importance of such studies to coastal erosion and structural stability, but no attempts have been made to apply these studies to sediment transport. The hydraulic climate of the shelf water column as a function of wind-driven currents has not been considered.

Numerical modeling of shelf-sediment transport requires estimates of the instantaneous or time-integrated coupling of the velocity field to the sea floor. Bottom shear stresses may be estimated from a point velocity determination by the quadratic stress law, or if a vertical array of three meters are used, by means of the

Von Karman-Prandtl velocity profile equation (Sternberg, in Swift et al., 1972). With greater vertical resolution of the velocity field, the Landweber method may be used (Ludwick, 1973).

A variety of threshold criteria for sediment transport are available in the literature. All are empirical relationships determined by experimentation in shallow flumes, and their utility in the marine environment is uncertain (Sternberg, in Swift et al., 1972). Volume and mass transport rates are calculated from such threshold criteria by application of transport functions such as those of Bagnold (1963), which relate the transport rate to the power expended by the fluid moving over the boundary. Such calculations are probably simplistic in that they do not take into account the possible synergistic effects of oscillatory wave surge on sediment transport or other complexities of the shelf flow field.

Smith and Hopkins (in Swift et al., 1972) have performed such calculations on current meter data from the Washington shelf. They estimate that a typical winter storm with current speeds of 60 cm/sec 3 m off the bottom transports on the order of 6 m^3/hr per meter transverse to the transport direction. A single one-month current-meter record from the Virginia shelf indicates that during this particular November, the threshold velocity of the finest sediment present was exceeded 17% of the time, and that 5.2 m^3 was transported per meter transverse to transport direction during that time (Fig. 2). Equivalent estimates are not available from tide-dominated shelves, although a semiquantitative approach has been considered by McCave (1971).

Models of the shelf velocity field modified by means of such equations to permit estimates of bottom shear stress and volume transport rates, should lead to finite difference models of shelf-sediment transport. Bottom-sediment grain-size data, and a time history of shear stress at a grid of bottom stations should permit assessment of sediment transport by means of a continuity equation of the form

$$\nabla \cdot (\vec{v}\bar{c}) = -\frac{\rho}{d}\frac{dh}{dt}$$

where v is velocity, $\bar{c}$ is sediment concentration averaged over depth, t is time, h is height above a reference level, ρ is density, d is water depth, and ∇ is an operator indicating net difference. The equation states that the difference between sediment input and sediment output at a grid point is a function of the change in bottom elevation with time.

Such a finite-difference model must be verified by means of direct observation. Radioisotope

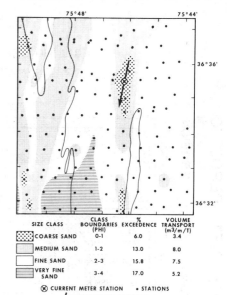

FIGURE 2. Calculations of volume sediment transport through the sand ridges of the Virginia Inner Shelf, based on record of bottom-mounted geodyne current meter, recorded during November 1972.

tracers appear to be a promising technique for verification (Duane, 1970; see *Tracer Techniques*).

Patterns of Shelf Sedimentation

The Secular Controls. Numerical process modeling is a potentially powerful tool for assessing and predicting shelf-sediment transport. Such assessment becomes meaningful, however, only in the context of a broader conceptual model that considers the secular or long-term variables of shelf-sediment transport. Students of shelf sedimentation (e.g., Curray, 1964) tend to isolate the following as major variables: rate of sediment input to the shelf, S; sea-level rise relative to land, R; rate of input of energy available to transport sediment, E; and character of sediment (grain size and mineral composition), G (see Fig. 3). These variables may be related to each other in quasi-quantitative fashion, $(S/E)_G - R = K$, to indicate that for a given sediment character (for instance, grain size) the ratio between the rate of sediment input, S, and the energy, E, available to disperse it must be balanced by the relative rate of sea level rise, R; otherwise the shoreline will advance or retreat.

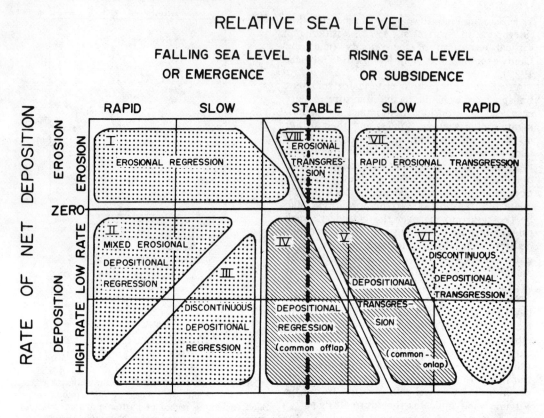

FIGURE 3. Varieties of shelf sedimentary regimes, as determined by rate of sediment input and sense of relative sea-level change (from Curray, 1964; Copyright © 1964, Macmillan Publishing Co., Inc.).

Shoreface Translation. The shoreline is the permeable boundary across which sediment is introduced into the shelf transport system, and its movements (as governed by the variables described above) and dynamic geometry thus determine the nature of the sedimentary regime.

One of the oldest concepts of shelf sedimentation, often presented but rarely evaluated, is the concept of the equilibrium profile (Swift, 1970). This hypothesis states that the profile of the shelf surface approximates an exponential curve, concave up, with the steepest part being the shoreface. This profile is seen as a substrate response to the hydraulic climate such that the slope at each point is just adequate to bypass sediment of the available grain size and concentration; the curve as a whole is seen as the best adjusted profile to absorb the energy of shoaling waves of the prevailing height and length.

The simplistic model is most nearly valid on the inner shelf of low, unconsolidated coasts, where the break in slope between the inner shelf and the wave-agitated shoreface generally occurs at 15 m or less. In fact, a true wave-graded profile is rarely attained even here, and the actual profile is instead a resultant response to the constructional, wave-dominated processes of fair weather on one hand, and strong, wind-driven contour currents which sweep the shoreface during storms on the other. On tide-dominated shelves, tidal currents are also important in shaping the lower shoreface.

The model is useful, however, in that the translation of the shoreface profile with the relative rise and fall of sea level determines the character of sediment supplied to the shelf (Fig. 4). In a relatively rapid transgression, as experienced by most shelves during the post-Pleistocene rise of sea level, river mouths are drowned and become traps for shelf-derived as well as river-derived sediment. As a consequence, the shoreface profile translates landward by means of shoreface erosion (Swift et al., in Swift et al., 1972). During storms, onshore winds generate coast-parallel shoreface currents with an onshore surface component and a seaward bottom component. These currents strip off the fair-weather veneer of fine sand deposited from rip currents and erode the underlying

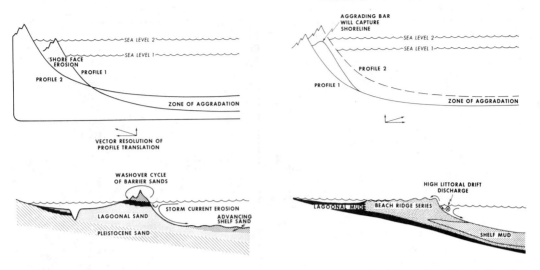

FIGURE 4. *Left*: Landward translation of the shoreface profile during a sea-level rise, and resulting stratigraphy. *Right*: Seaward translation of shoreface profile during a sea-level rise, and resulting stratigraphy.

barrier base and the pre-Holocene substrate on which it rests (Fig. 4). The debris of shoreface erosion is left behind on the beveled shelf surface as an intermittent sheet of relatively clean and well-sorted sand. Most grains in this sheet have probably rested on a beach at least once, but they now comprise an entirely new deposit which is properly termed a marine sand. Beach deposits are replaced in the stratigraphic section by a disconformity.

Autochthonous Shelf Sedimentation. Shelf deposits thus derived from shoreface erosion during a transgression are the product of *autochthonous shelf sedimentation*. The stratigraphy of autochthonous shelves is threefold in nature, consisting of a pre-Holocene substrate, the Holocene back-barrier sequence, and the Holocene nearshore deposits.

The nearshore sand sheet, where well developed, tends to have a complex structure (Fig. 5). First-order morphologic and stratigraphic elements consist of shelf-transverse

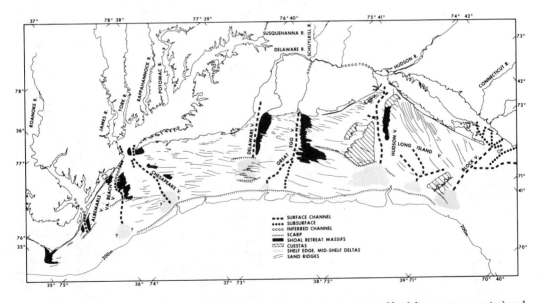

FIGURE 5. First-order morphologic elements (shelf valleys, shoal retreat massifs, deltas, scarps, cuestas) and second-order morphologic elements (sand ridges) of the Middle Atlantic Bight, and autochthonous, storm-dominated shelf.

thickenings of the sand sheet, which may have surface expression as shelf-transverse ridges and valleys. These are the retreat paths of zones of deposition and erosion associated with the transgressing shoreline (Swift et al., in Swift et al., 1972). Zones of littoral drift convergence at estuary mouths and at cuspate forelands result in shelf-transverse sand ridges or *shoal-retreat massifs*. Estuarine shoal-retreat massifs may be paired with troughs that are the retreat paths of flood channels built by the reversing tides of estuary mouths. Such paired morphologic elements comprise *shelf valley complexes*. On high, indurated coasts, however, such relict components of the topographic pattern of constructional nearshore marine origin are less conspicuous than relict components of subaerial erosional origin such as cuestas and drowned fluvial or glacial valleys.

Few shelves are sufficiently quiescent to preserve first-order, shelf-transverse, morphologic and stratigraphic elements without modification. In the Middle Atlantic Bight, a well-studied, storm-dominated shelf, dominantly southerly storm currents have molded the surficial sand sheet into a subdued, pervasive *ridge and swale topography* (see Swift et al., 1972). Ridges are up to 10 m high, tens of kms long, and converge southward with the coastline. Side slopes are usually <1°. Shoreface-connected ridges are initiated at the leading edge of the sand sheet by storm currents and storm-wave trains, and are detached and isolated during shoreface retreat. Shoal-retreat massifs constrict and accelerate the storm flow field; they are remolded into massif-transverse ridges and swales. Zones of flow expansion and deceleration in the lee of shoal-retreat massifs and other highs become sinks for fine sand, which may be carried for long distances during periods of peak velocity. The precise relationship of the ridge topography of the Middle Atlantic Bight to the structure of the storm flow field is not clearly understood, but the apparent mobility of the sediment suggests that such a relationship does exist, and that the ridges are large-scale substrate responses to hydraulic process, analogous in some respects to longitudinal desert dunes.

Remobilization of the surficial sand sheet of high-energy, tide-dominated shelves is yet more striking (Fig. 6). European studies of the tide-swept shelf of Western Europe (e.g., Kenyon and Stride, 1970) reveal that over large areas, a thin surficial sand sheet is in transit over its basal gravel. *Sand streams*, or paths of net sand transport, may be traced for great distances. Such paths tend to start at "bed-load partings," or zones of divergence

FIGURE 6. Pattern of sediment transport on the western European shelf, an autochthonous tide-dominated shelf (from Kenyon and Stride, 1970).

of the residual component of tidal currents and of divergence of the sense of bed-load transport.

A sequence of sediment facies tends to occur along such transport paths, down the velocity gradient of mid-tide surface currents, from the erosional source area, to the depositional zone. Zones of scour and bedrock outcrop tend to occur where mid-tide surface velocities exceed 3 knots. Ribbons of sand in transit over the basal Holocene gravel occur where mid-tide surface velocities exceed 2 knots. Extensive beds of sand, and, locally, fields of sand waves occur at the depositional ends of sand streams (sand transport paths), where mid-tide surface velocities range from 1–2 knots. Sand-wave field development is locally enhanced by elongation of the tidal ellipse, so that the destructive effects of mid-tide cross flow is minimized. Farther down the velocity gradient, beyond the main depositional center, there may be patchy growth of the sand sheet. Patches may be current transverse ("starved sand waves") or current parallel, but without the elongation ratio (40:1) that Kenyon and Stride (1970) have set for true sand ribbons.

Allochthonous Shelf Sedimentation. A very different sedimentary regime prevails on continental shelves when the rate of sediment input is sufficiently high relative to the rise of sea level and to the ability of the hydraulic climate to disperse it, such that the sense of shoreface translation is that of very slow transgression, or depositional regression (Fig. 4). A broad zone of subaerial (fluvial or eolian) and transitional (lagoonal, estuarine, or deltaic) environments develops, and the regional gradient is reduced. If the sense of shoreline movement is still negative (transgressional), estuaries are no longer drowned river valleys with dendritic patterns (e.g., Chesapeake Bay), but are instead simple, trumpet-shaped, constructional forms, whose dimensions are in equilibrium with tidal discharge (i.e., any cross-sectional area therein is a logarithmic function of the tidal prism landward of that point).

Such estuaries are no longer efficient sediment traps and tend to bypass some size fraction of the sediment load delivered to them by rivers. "Stratigraphic" bypassing of sediment also occurs as sediment delivered to and deposited in estuaries and lagoons is released to the shelf by shoreface erosion and by tidal scour at estuary mouths. The inner and outer estuary environments serve as a seaward series of filters, retaining decreasingly coarser fractions and bypassing increasingly finer fractions of the original sediment load, so that material ultimately released to the continental shelf by either mode of bypassing consists primarily of fine sand, silt, and clay. This material travels in suspension at relatively low threshold velocities and is readily transported to all portions of the shelf. This regime of shelf aggradation by fine, far-traveled sediment bypassed through the shoreface is termed *allochthonous shelf sedimentation,* as distinct from the autochthonous sedimentation characteristic of rapid or sediment-starved transgressions in which the shelf cannibalizes its own substrate. Should the rate of sediment input increase to the point that the sense of shoreface translation is reversed and the shoreline progrades, then deltaic deposits build out over inner marine deposits. Deltas, however, are equally efficient filters and fractionators of sediment; fine sediment continues to be bypassed to the shelf, and the allochthonous regime is maintained (Fig. 7).

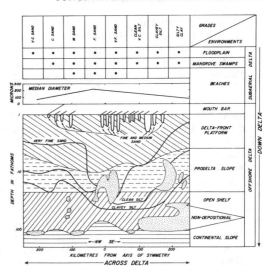

FIGURE 7. Textural gradients of the Niger shelf, an allochthonous shelf (from Allen, 1965).

The input of sediment to allochthonous shelves may be highly episodic as a consequence of river flooding. Exceptionally heavy flooding may deposit enormous tonnages of sediment beneath the turbidity plumes of river mouths in a matter of weeks, to be redistributed over the shelf during ensuing years (Drake et al., in Swift et al., 1972). Reentrainment may occur with the aid of the oscillatory surge of storm waves, as in the case of the coarser sediment of autochthonous shelves; unlike the latter material, fine allochthonous sediment is retained in suspension and is transported for long distances by thermohaline shelf circulation, and wind-drift currents and tidal residuals too weak to affect coarser material (Fig. 8). Advective

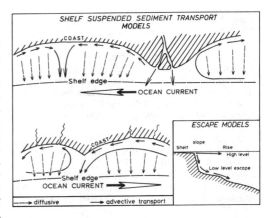

FIGURE 8. Modes of suspended sediment transport on shelves (from McCave, in Swift et al., 1972).

transport of suspended fine sediment is dominantly coast-parallel, but the more nearly continuous nature of transport enhances cross-shelf diffusion. Advective cross-shelf transport occurs at coastal current convergences. As indicated by McCave in Swift et al., 1972), the localization of fine sediment deposition depends not on the absolute hydraulic activity of the shelf floor, but rather on the relationship of sediment concentration in the water column and hydraulic activity.

The fine sediments of allochthonous shelf surfaces have lower effective angles of repose and do not as a rule sustain the large-scale bedforms of autochthonous shelves. The greater effective mobility of fine allochthonous sediment results in gentle textural gradients that are maintained over many kilometers. Such gradients reflect decreasing values of maximum orbital velocity attendant on water deepening. They are also the consequence of progressive sorting, whereby sediment at rest during intermittent transport over an aggrading surface will tend to leave behind its coarsest fraction if reentrained by the currents of a storm weaker than the previous one.

DONALD J. P. SWIFT

References

Allen, J. R. L., 1965. Late Quaternary Niger Delta, and adjacent areas: Sedimentary environments and lithofacies, *Bull. Am. Assoc. Petrol. Geologists,* **49,** 547-600.

Bagnold, R. A., 1963. Mechanics of marine sedimentation, in M. N. Hill, ed., *The Sea,* vol. 3. New York: Wiley Interscience, 507-525.

Cherry, J. A., 1966. Sand movement along equilibrium beaches north of San Francisco. *J. Sed. Petrology,* **36,** 341-357.

Curray, J. R., 1964. Transgressions and regressions, in R. L. Miller, ed., *Papers in Marine Geology.* New York: MacMillan, 175-203.

Curray, J. R.; Emmel, F. J.; and Crampton, P. J. S., 1969. Holocene history of a strand plain, lagoonal coast, Nayarit, Mexico, in Lagunas costeras, un simposio, *Mem. Simp. Intern. Lagunos Costeras, UNAM-UNESCO,* 1967, Mexico, 63-100.

Duane, D. B., 1970. Tracing sand movement in the littoral zone: progress in the radioisotope sand tracer (RIST) study July 1968-February 1969, *Coastal Eng. Research Cen., Misc. Pap., April 1970,* 52p.

Jelesnianski, C. P., 1967. Numerical computations of storm surges with bottom stress, *Monthly Weather Rev.,* **95,** 740-750.

Kenyon, N. H., and Stride, A. H., 1970. The tide swept continental shelf sediments between the Shetland Isles and France, *Sedimentology,* **15,** 154-173.

Ludwick, J. C., 1973. Tidal currents and zig-zag sand shoals in a wide estuary entrance, *Old Dominion Univ. Inst. Oceanog., Tech. Rept., Sept. 7,* 23p.

McCave, I. N., 1971. Sand waves in the North Sea off the coast of Holland, *Marine Geol.,* **10,** 199-277.

Smith, D. B., 1974. Sedimentation of upper Artesia (Guadalupian) cyclic shelf deposits of northern Guadalupe Mountains, New Mexico, *Bull. Am. Assoc. Petrol. Geologists,* **58,** 1699-1730.

Stanley, D. J., and Swift, D. J. P., 1976. *Marine Sediment Transport and Environmental Management.* New York: Wiley, 602p.

Swift, D. J. P., 1970. Quaternary shelves and the return to grade, *Marine Geol.,* **8,** 5-30.

Swift, D. J. P.; Duane, D. B.; and Pilkey, O. H., 1972. *Shelf Sediment Transport: Process and Pattern.* Stroudsburg, Pa.: Dowden, Hutchinson & Ross, 656p.

Werner, F., and Newton, R. S., 1975. The pattern of large-scale bed forms in the Langeland Belt (Baltic Sea), *Marine Geology,* **19,** 29-59.

Cross-references: *Continental-Rise Sediments; Continental-Shelf Sediments; Deltaic Sediments; Estuarine Sedimentation; Flow Regimes; Lagoonal Sedimentation; Littoral Sedimentation; Neritic Sedimentary Facies; Outer Continental-Shelf Sediments; Paralic Sedimentary Facies; Relict Sediments; Storm Deposits; Tidal-Current Deposits.* Vol. I: *Continental Shelf, Classification; Grand Banks.*

CONTINENTAL-SHELF SEDIMENTS

Stratigraphy

On shelves undergoing autochthonous sedimentation (see *Continental-Shelf Sedimentation*) the stratigraphy is threefold in nature (see Fig. 1). A pre-Holocene substrate (C to P_1) may be overlain by Holocene back-barrier deposits (H_6 to H_4) and Holocene nearshore sands (H_2 to H_3). The pre-Holocene substrate is variable. On low coasts (Fig. 1A, B), it may consist of fine-grained regressive shelf and lagoonal deposits which date from the retreat of the mid-Wisconsin highstand of the sea. On high coasts (Fig. 1E), the indurated substrate may have considerable relief and may be the dominant sea-floor element, with other facies occurring mainly as lenses in low areas. A Holocene sequence is variable in thickness and complexity. It may start with lagoonal or estuarine deposits and may pass upward into fine-grained, barrier-base sands (Fig. 1B) which have escaped destruction by the process of erosional shoreface retreat. These deposits rest disconformably on the old subaerial erosional surface. This back-barrier sequence, if present, has been beveled by shoreface retreat, resulting in a major disconformity within the Holocene sequence. It is overlain by the shelf surficial sand sheet, starting with a sporadic basal gravel of terrigenous clasts (if available) or shell hash, and grading upward

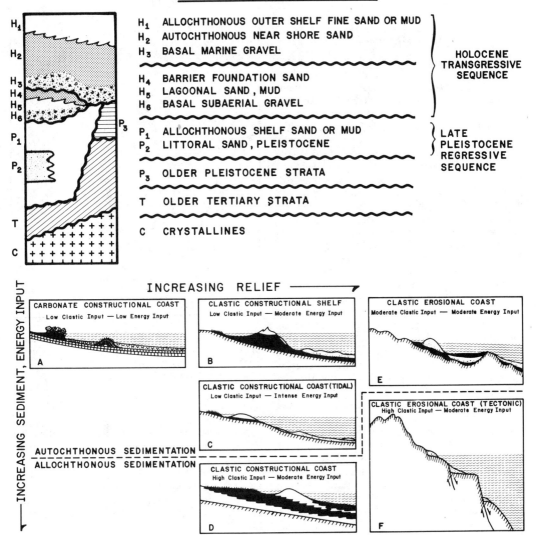

FIGURE 1. *Above*: Generalized stratigraphic column on the continental shelf. *Below*: Continuum of shelf depositional schemes in terms of variables of shelf sedimentation and relief.

into a well-sorted sand. On tide-dominated shelves (Fig. 1C), the nearshore sand may be swept into large-scale bed forms, leaving the basal gravel exposed over broad areas. On low, carbonate coasts (Fig. 1A), a basal coquina may be overlain by calcite mud. In many cases, shoreface erosion cuts through barrier-base, lagoonal, or estuarine deposits. the intra-Holocene and basal Holocene disconformities merge, and the surficial sand sheet rests directly and disconformably on the pre-Holocene substrate (Fig. 1C).

The Holocene nearshore sequence is transitional upward with Holocene outer-shelf deposits, which are, by definition, poorly developed on autochthonous shelves, (see *Outer Continental-Shelf Sediments*). The stratigraphy of allochthonous shelves is similar to that of autochthonous shelves, except that in many cases, facies H_4 to H_6 (back-barrier deposits) and in all cases facies H_1 (outer-shelf deposits) are better developed than facies H_2 to H_3 (nearshore deposits). In fact, many modern shelves (e.g., northern Gulf of Mexico) are in a state of transition between these two sedimentary regimes or have recently shifted to the allochthonous regime—a consequence of the stabilization of the shoreline after the late Holocene reduction in the rate of sea-level rise about 4000 years ago.

Petrography and Classification

There is nothing unique about shelf sediments observed out of context, and operational or descriptive classification must be by means of the textural and compositional parameters employed for other sediments. As in the case of other sediments, the size-frequency distribution is composed of log-normal subpopulations that reflect the building of the bed (see *Grain-Size Frequency Studies*). "Clogs" of coarse, traction sediment occur in a framework population of medium size. A third population is interstitial to the other two. The relative percentages of the three populations are a function of the intensity of the depositional regime (Moss, 1972). Mean diameter of a sample reflects that of the source area as modified by transport and site history. It varies with the local hydraulic climate, and also with the location of the sample in the regional dispersal pattern. Such patterns may be broad and simple on outer allochthonous shelves, but many are localized and complex near shore, or on autochthonous shelves. Shelf highs tend to be mantled by coarser lags, surrounded by aprons of out-winnowed finer sediment. Constructional forms such as current-parallel sand ridges usually comprise sand circulation cells, whose textural patterns tend to reflect the radial dispersive component of sediment transport, rather than the concentric advective component. Grain size tends to vary rhythmically over fields of transverse sand waves or current-parallel sand ribbons, in sympathy with the hydraulic microclimate.

As noted by Emery (1968), the composition and genesis of shelf sediments vary with latitude (Fig. 2). Polar sediments are compositionally immature, being enriched in such labile (easily weathered) constituents as feldspar and rock fragments. Such sediments become increasingly quartzose toward the equator. On the eastern margins of ocean basins, shelf-edge sediments may be enriched with such authigenic chemical precipitates as phosphorite and silica. Toward low latitudes, eastern margin shelves may be deserts during low stands and may then accumulate eolian sediments. Biogenic precipitation of calcium carbonate dominates equatorial shelves.

Such source considerations, and process and environmental considerations, led Emery (1968) to propose a genetic classification of shelf sediments: *authigenic* (glauconite or phosphorite), *organic* (foraminifera, shells), *residual* (weathered from underlying rocks), *relict* (rem-

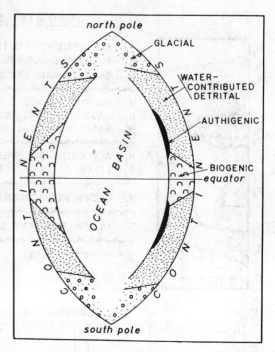

FIGURE 2. Idealized distribution of the chief classes of sediment on continental shelves (from Emery, 1968). Assumes asymmetric, clockwise water circulation, with upwelling on E coast.

nant from a different earlier environment such as a now-submerged beach or dune), and *detrital*, which includes material supplied by rivers, coastal erosion, and glacial or eolian activity.

The implications of this classification should be carefully considered. First, it is a genetic, rather than a descriptive, or operational, classification; it can thus be applied only after inferences are drawn from sediment attributes. As such, it is a paradigm, or conceptual model, for shelf-sediment genesis. It modifies an earlier view which held that shelf sediments are size graded as a result of a seaward (depth) decreasing gradient of energy input into the sea floor; or in other words, that all shelf sediments other than organic ones are detrital.

Recent advances in shelf studies suggest that a further modification is in order. The problem lies in the definition of relict and residual sediments on one hand, versus detrital on the other. These are in part environment-determined categories. The concept of environment is itself a high-order abstraction; the geographic locus over which a given dynamic system (another abstraction) or subsystem of sediment distribution prevails.

Relict shelf sediments have exchanged, in situ, one environment and its dynamic system for another during the Holocene transgression. The term *relict* however, tends to distort our thinking by overemphasizing the effects of the first environment. Relict sediments are acknowledged to have been "reworked"; the reworking, however, is as systematic and complex as the original "working." Such reworking may be referred to as autochthonous sedimentation (deposition of sediments whose origin within the present sedimentary cycle is local). The reworking deserves equal or greater emphasis than the earlier cycle; the attributes of shelf sediments may reflect both environments, but those of the later tend to obliterate those of the earlier. If sediment attributes reflecting both cycles are present, then the sediments are *palimpsest* (similar to a medieval scroll, partly erased, then rewritten; Swift et al., 1971). The term *relict* is perhaps better applied to a given sediment attribute such as the surface textures of a shelf sand sample dominated by high-velocity impact scars acquired in the surf. The sample's size-frequency distribution, however, will almost certainly reflect reconstitution of the granular fabric in a deeper environment, after the erosional retreat of the shoreface. Thus the sample is a palimpsest sand because attributes of both a former and the present environment are detected; because the sediment is of local origin, it is the product of autochthonous sedimentation.

DONALD J. P. SWIFT

References

Alexander, A. E., 1934. A petrographic and petrologic study of some continental shelf sediments, *J. Sed. Petrology,* 4, 12-22.

Curray, J. R., 1965. Late Quaternary history, continental shelves of the United States, in H. E. Wright, Jr. and D. G. Frey, eds., *The Quaternary of the United States.* Princeton: Princeton Univ. Press, 723-736.

Dott, R. H., Jr. and Dalziel, I. W. D., 1972. Age and correlations of the Precambrian Baraboo Quartzite, *J. Geol.,* 80, 552-568.

Emery, K. O., 1968. Relict sediments on continental shelves of world, *Bull. Am. Assoc. Petrol. Geologists,* 52, 445-464.

Field, M. E., and Duane, D. B., 1974. Geomorphology and sediments of the inner continental shelf, Cape Canaveral, Florida, *U.S. Army Coastal Eng. Research Center, Tech. Memo 42,* 87p.

Frank, W. M., and Friedman, G. M., 1973. Continental-shelf sediments off New Jersey, *J. Sed. Petrology,* 43, 224-237.

Goldring, R., and Bridges, P., 1973. Sublittoral sheet sandstones, *J. Sed. Petrology,* 43, 736-747.

Hite, R. J., 1972. Shelf carbonate sedimentation controlled by salinity in the Paradox basin, south east Utah, *Mtn. Geol.,* 9, 329-344.

Hommeril, P.; Larsonneur, C.; and Pinot, J-P., 1972. Les sédiments du précontinent armoricain, *Soc. Géol. Fr. Bull.,* 7, 237-247.

McKinney, T. F., and Friedman, G. M., 1970. Continental shelf sediments of Long Island, New York, *J. Sed. Petrology,* 40, 213-248.

McManus, D. A., 1975. Modern versus relict sediment on the continental shelf, *Geol. Soc. Am. Bull.,* 86, 1154-1160.

Moss, A. J., 1972. Bed-load sediments, *Sedimentology,* 18, 159-220.

Owens, J. P., and Sohl, N. F., 1969. Shelf and deltaic paleoenvironments in Cretaceous-Tertiary formations of the New Jersey coastal plain, in S. Subitsky, ed., *Geology of Selected Areas in New Jersey and Eastern Pennsylvania.* New Brunswick: Rutgers University Press, 235-278.

Schlee, J., 1973. Atlantic continental shelf and slope of the United States sediment texture of the northeastern part, *U.S. Geol. Surv. Prof. Pap. 529-L,* 64p.

Schopf, T. J. M., 1968. Atlantic continental shelf and slope of the United States: nineteenth century exploration, *U.S. Geol. Surv. Prof. Pap. 529-F,* 19p.

Shepard, F. P., 1932. Sediments of the continental shelves, *Geol. Soc. Amer. Bull.,* 43, 1017-1040.

Slatt, R. M., 1974. Formation of palimpsest sediments, Conception Bay, southeastern Newfoundland, *Geol. Soc. Amer. Bull.,* 85, 821-826.

Spearing, D. R., 1975. Shallow marine sands, *Soc. Econ. Paleont. Mineral. Short Course,* 2 (April 5, 1975), 103-132.

Stanley, D. J., ed., 1969. *The NEW Concepts of Continental Margin Sedimentation.* Washington, D.C.: Amer. Geol. Inst.

Stetson, H. C., 1938. The sediments of the continental shelf off the eastern coast of the United States, *Mass. Inst. Tech. Woods Hole Oceanog. Inst. Pap. Phys. Oceanog. Meteorol.,* 5(4), 49p.

Swift, D. J. P.; Stanley, D. J.; and Curray, J. R., 1971. Relict sediments on continental shelves: a reconsideration, *J. Geol.,* 79, 322-346.

Trumbull, J. V. A., 1972. Atlantic continental shelf and slope of the United States sand-size fraction of bottom sediments, New Jersey to Nova Scotia, *U.S. Geol. Surv. Prof. Pap. 529-K,* 45p.

Wantland, K. F., and Pusey, W. K., III, eds., 1975. Belize (British Honduras) Shelf—carbonate sediments, clastic sediments, and ecology, *Am. Assoc. Petrol. Geologists, Stud. Geol.,* 2, 599p.

Wilson, J. L., 1974. Characteristics of carbonate-platform margins, *Bull. Am. Assoc. Petrol. Geologists,* 58, 810-824.

Cross-references: *Authigenesis; Barrier Island Facies; Blanket Sands; Continental-Shelf Sedimentation; Continental-Shelf Sediments in High Latitudes; Glauconite; Littoral Sedimentation; Neritic Sedimentary Facies; Outer Continental-Shelf Sediments; Relict Sediments.*

CONTINENTAL-SHELF SEDIMENTS IN HIGH LATITUDES

The lithological and chemical composition of high-latitude continental-shelf sediments is largely determined by the extent of ice cover, terrigenous source, shelf topography, biological productivity, and Late Cenozoic paleogeography of each area. Wide variations in size distributions of surficial bottom sediments of both polar regions are presumably related to differences in depositional settings.

Arctic-shelf sediments (Herman, 1974) are predominantly poorly sorted, terrigenous muds with varying admixtures of sand and gravel containing scarce foraminifera—chiefly arenaceous when present. Coarser and better-sorted sediments are observed only locally in some inner shelves where wave sorting and current deposition (Creager, 1963), as well as ice-rafting of gravels, prevail, and in places where residual and relict deposits are exposed (McManus et al., 1969; Barnes, 1970). Contrary to common belief (see Creager and Sternberg, 1972, for several references), gravel transport by contemporary ice-rafting over the northern polar shelf is generally unimportant; it may have been significant in the Beaufort Sea, however, during the Woronzofian Transgression (Naidu and Mowatt, 1973). The shelves skirting the Canadian Archipelago and Greenland are the only likely areas in the Arctic for extensive transport of gravels by icebergs. Prevalence of mud on Arctic shelves is attributed partly to ice-rafting, and also to sedimentation under low-energy conditions, entrapment of coarse terrigenous detritus by low-gradient coastal rivers, and the derivation of fine-grained sediments from the permafrost hinterland. Postdepositional sediment reworking by bioturbation, iceberg plough, and ice gouging is widespread (Belderson and Wilson, 1973; Reimnitz et al., 1972). "Strudel scour" (due to freshwater drainage) characterizes nearshore areas (Reimnitz et al., 1974). Ice-rafting of boulders, as well as iceberg gouging and shore erosion, is observed today as far south as the St. Lawrence (Dionne, 1972).

Latitudinal lithofacies changes are observed on the Antarctic shelf (Lisitsyn, 1972). The inner shelf, extending seaward from Antarctica to the Antarctic Divergence, is blanketed by "iceberg sediments," which are poorly sorted terrigenous (chiefly morainic) muddy sands with large admixtures of gravels. In this zone, icebergs calved from glaciers are the principal transporting agent; sediment dispersal by sea ice is insignificant. There is no influence of continental drainage on shelf sedimentation because of the absence of river runoff on Antarctica. On the middle and outer shelves, the influence of icebergs is limited, and biological productivity very high; the sediments are essentially biogenous oozes, and consist of siliceous diatoms and calcareous globigerina on the middle and outer shelves respectively. Lateral migrations of the boundaries of these principal lithological zones possibly reflect climatic changes (Lisitzin, 1972; Hays, 1967).

Several heavy-mineral suites are delineated in the shelves of the Antarctic (Edwards and Goodell, 1969) and the Arctic (Lisitzin, 1972). Distribution of these suites is influenced by mineralogy of the hinterland rocks as well as the dispersing agents, e.g., currents and ice-rafting. Generally, total heavy minerals in the sands are relatively low, and their distribution in the various sand and subfractions is not determined by the hydraulic equivalence (q.v.), concept (Naidu and Sharma, 1972; Naidu and Mowatt, 1975).

Clay minerals of the Arctic shelves are detrital and consist predominantly of illite (about 60%) with small amounts of chlorite, kaolinite, and smectite. Dispersal patterns of these minerals in the Kara-Barents, E Siberian, Chukchi, and Beaufort seas are correlative to source and current movements. On the Antarctic shelf, detrital chlorite and illite predominate in the widely distributed terrigenous clays, whereas smectite is enriched only locally in submarine volcanogenic clays. Of paleoclimatic significance is kaolinite, a typical low-latitude clay mineral, found locally on some Arctic shelves. Chemical studies of polar-shelf sediments indicate low levels of several major and minor elements compared with similar marine sediments elsewhere (Angino, 1966; Barnes, 1970). Organic carbon and carbonate are also relatively scarce (Naidu and Mowatt, 1975).

Several speculations have been made about the Cenozoic paleogeography of the N and S polar shelves. During the Late Pleistocene glaciation and lowered sea level, extensive portions of the Kara-Barents and Norwegian seas were covered by glaciers (Dibner, 1970; Holtedahl and Bjerkli, 1975), whereas the East Siberian, Laptev and Chukchi seas, were exposed to subaerial deltation. Presence of several Late Pleistocene and Holocene sea-level stillstands have been recognized in the Laptev and Chukchi seas; dates of these stillstands correlate well with those given by Fairbridge (1961) on a worldwide basis. Several authors (in Hopkins, 1967) have shown that the Chukchi Sea shelf up to the Wisconsin age served as a subaerial land bridge connecting Siberia and Alaska. Subbottom reflection profiles substantiate this conclusion. The exposure of relict sediments on the Arctic shelves is apparently much less extensive than predicted by Emery (1968). Knowledge on the Cenozoic paleogeography of the Antarctic shelf is limited and controversial (see Kemp, 1972, for references), although the JOIDES Antarctic Shelf program is shedding some light on the subject (Hayes and Frakes, 1975). The great depth of the Antarctic shelf (400-600 m av., 800 m max.) is believed by some to have resulted from crustal downwarp caused by continental ice-loading during glacial maxima.

More intensive study of the polar shelves is called for because of the extensive presence there of productive and prospective strata for oil and gas, of ferromanganese concretions (Goodell et al., 1971), and of information they presumably hold on the origin of oceans.

A. S. NAIDU

References

Angino, E. E., 1966. Geochemistry of Antarctic pelagic sediments, *Geochim. Cosmochim. Acta*, 30, 939-961.

Barnes, P., 1970. Preliminary results of geologic studies in the eastern central Chukchi Sea, *U.S. Coast Guard Oceanogr. Rept.*, 50, 87-110.

Belderson, R. H., and Wilson, J. B., 1973. Iceberg plough marks in the vicinity of the Norwegian Trough, *Norsk Geol. Tidsskr.*, 53, 322-328.

Creager, J. S., 1963. Sedimentation in a high energy embayed, continental shelf environment, *J. Sed. Petrology*, 33, 815-830.

Creager, J. S., and Sternberg, R. W., 1972. Some specific problems in understanding bottom sediment distribution and dispersal on the continental shelf, in D. J. P. Swift et al., eds., *Shelf Sediment Transport: Process and Pattern*. Stroudsburg, Pa.: Dowden, Hutchinson & Ross, 347-362.

Dibner, V. D., 1970. "Ancient clays" and the relief of the Barents-Kara shelf as direct proof of its sheet glaciation during the Pleistocene, in M. I. Belov, ed., *Problems of Polar Geography*. Washington: Nat. Sci. Foundation, 124-128.

Dionne, J. C., 1972. Caractéristiques des blocs de rives de l'estuaire du Saint-Laurent, *Rev. Géogr. Montréal*, 26, 125-152.

Edwards, D. E., and Goodell, H. G., 1969. The detrital mineralogy of ocean floor surface sediments adjacent to the Antarctic Peninsula, Antarctica, *Marine Geol.*, 7, 207-234.

Emery, K. O., 1968. Relict sediments on continental shelves of the world, *Bull. Am. Assoc. Petrol. Geol.*, 52, 445-464.

Fairbridge, R. W., 1961. Eustatic changes in sea level, in L. H. Ahrens, et al., eds., *Physics and Chemistry of the Earth*, vol. 4. London: Pergamon, 99-185.

Goodell, H. G.; Meylan, A.; and Grant, B., 1971. Ferromanganese deposits of the South Pacific Ocean, Drake Passage, and Scotia Sea, *Am. Geophys. U., Antarctic Research Ser.*, 15, 27-92.

Gordon, A. L., 1971. Oceanography of Antarctic waters, *Am. Geophys. U., Antarctic Research Ser.*, 15, 169-203.

Hayes, D. E.; Frakes, L. A.; *et al.,* 1975. *Initial Reports of the Deep Sea Drilling Project, Volume 28*. Washington: U.S. Govt. Printing Office, 1017p.

Hays, J. D., 1967. Quaternary sediments of the Antarctic Ocean, in M. Sears, ed., *Progress in Oceanography*, vol. 4. Oxford: Pergamon, 117-131.

Herman, Y., ed., 1974. *Marine Geology and Oceanography of the Arctic Seas*. New York: Springer, 397p.

Holtedahl, H., and Bjerkli, K., 1975. Pleistocene and recent sediments of the Norwegian continental shelf (62°N-71°N) and the Norwegian Channel area, *Norges Geol. Undersökelse*, 316, 241-252.

Hopkins, D. M., ed., 1967. *The Bering Land Bridge*. Stanford, Calif.: Stanford Univ. Press, 495p.

Kemp, E. M., 1972. Reworked palynomorphs from the West Ice Shelf area, East Antarctica, and their possible geological and paleoclimatological significance, *Marine Geol.*, 13, 145-157.

Klenova, M. V., 1960. *Geologiya Barentsova morya* (Geology of the Barents Sea). Moscow: Akad. Nauk SSSR, 365p.

Lisitsyn, A. P., 1972. Sedimentation in the world ocean, *Soc. Econ. Paleont. Mineral. Spec. Publ. 17*, 218p.

McManus, D. A.; Kelley, J. C.; and Creager, J. S., 1969. Continental shelf sedimentation in an Arctic environment, *Geol. Soc. Am. Bull.*, 80, 1961-1983.

Naidu, A. S., and Mowatt, T. C., 1975. Depositional environments and sediment characteristics of the Colville and adjacent deltas, northern Arctic Alaska, in M. L. Broussard, ed., *Deltas: Models for Exploration*. Houston: Houston Geol. Soc., 283-309.

Naidu, A. S., and Sharma, G. D., 1972. Texture, mineralogy and chemistry of Arctic Ocean sediments, *Inst. Marine Sci., Univ. Alaska Fairbanks, Rept. R72-12*, 30p.

Reed, J. C., and Satev, J. E., eds. 1974. *The Coast and Shelf of the Beaufort Sea*. Arlington, Va.: Arctic Inst. of N. America, 750p.

Reimnitz, E.; Barnes, P.; Forgatsch, T.; and Rodeick, C., 1972. Influence of grounding ice on the Arctic shelf of Alaska, *Marine Geol.*, 13, 323-334.

Reimnetz, E.; Rodeick, C. A.; and Wolf, S. C., 1974. Strudel scour: a unique Arctic marine geologic phenomenon, *J. Sed. Petrology*, 44, 409-420.

Wright, P. L., 1974. Recent sediments of the southwestern Barents Sea, *Marine Geol.*, 16, 51-81.

Cross-references: *Continental Shelf Sedimentation; Continental Shelf Sediments; Glacial Marine Sediments; Glacigene Sediments.* Vol. I: *Arctic Ocean; Southern Ocean.* Vol. III: *Arctic Beaches.*

CONTOURITES

Contourites are sedimentary rocks deposited by contour currents. Contour currents, first described by Heezen et al. (1966), are bottom currents resulting from thermohaline circulation on a rotating earth. They flow parallel to bathymetric contours and occur most commonly on the continental rise of the modern deep ocean.

In a recent review of continental-rise sediments, Hollister and Heezen (1972) list the properties of contourites and compare them with turbidites (see Table 1 in *Continental-Rise Sediments*). The best descriptions of contourites are from piston cores that have sampled sediment layers for depths of 1–10 m below regions of the sea floor where contour currents are known to be moving the surface sediments. These studies include Heezen et al. (1966) and Schneider et al. (1967)—Blake Bahama Outer Ridge; Heezen and Schneider (1966)—the Bermuda Rise; and Rona (1969)—the lower continental-rise hills off Cape Hatteras. All these examples cite cores that contain mostly lutite (silty clays), but there are commonly many laminae of silt and even fine-grained sand within the lutite. Hence the western North Atlantic has provided the best examples of contour currents and of the sediments that immediately underlie them, which are presumably contourites.

From these core descriptions, contourites appear to be mainly fine-grained sediment (silty clay or clayey silt) containing many distinct laminae. These laminae may be caused by just subtle color changes; for example, the upper 5–10 m of sediment of the lower continental rise of the western North Atlantic are characterized by rose-colored mud laminae which apparently have their source in the red-colored Permo-Carboniferous rocks of Nova Scotia and Newfoundland (Needham et al., 1969). These red sediments were eroded during the Pleistocene glaciations and redistributed from the Gulf of St. Lawrence outward across the continental shelf and down the slope, then redistributed by contour currents south almost 2400 km along the continental rise to the Blake Bahama Outer Ridge (Emery and Uchupi, 1972). As described by Hollister and Heezen (1972), contourites are also characterized by very well-sorted, cross-laminated sand layers, generally <5 cm thick. These laminae are often accentuated by concentrations of heavy minerals. The late Tertiary and Quaternary sediment laid down by contour currents is generally appreciably thicker than time equivalents of the adjacent deep-sea floor, indicating that contourite deposition is, in comparison, much faster.

Contourites commonly have a conspicuous character on shallow seismic reflection or echo-sounding records. Fox et al. (1968) describe abyssal "anti-dunes" or mud dunes as one of the characteristic large-scale sediment forms of contourites. The "dunes" are actually abyssal hills on the sea floor, 3–7 km wide and 50–100 m high, with gradients of 1:120 to 1:180. They are characterized by offlap deposition on one flank and by erosion on the other flank. Hyperbolic echos seen on echo-sounding records parallel bathymetric contours along the continental rise of the western North Atlantic and are presumed to be caused by areas underlain by contourites (Heezen et al., 1966). Similar small regions have been mapped by Damuth (1975) on the continental rise north of Brazil.

Detailed descriptions of cores from many parts of the sea floor dominated by contour currents are lacking, and it can only be inferred that these sediments exist and have characteristic properties. Areas of the deep-sea floor where bottom currents, hyperbolic echo, or abyssal mud dunes occur, or where thick sediments have accumulated away from any obvious source upslope, are probably contourites. In the South Atlantic the flow of Antarctic bottom water north over the Argentine Basin may have helped build the lower continental rise to thicknesses exceeding 3000 m (Ewing et al., 1973). Similar accumulations of acoustically transparent sediment occur throughout the world's oceans.

Ancient Contourites

The paucity of accurate descriptions of modern contourites makes it difficult to aptly describe ancient counterparts (Bouma, 1972a); Wezel (1969) was apparently the first to do so. Bouma (1972b) described what he believed to be ancient contourites from the Lower Niesenflysch near Adelboden, Switzerland. This sequence consists of shale/shale-sandstone alternations similar to the sediments described by Hubert (1964) from the continental rise of the western North Atlantic.

Many ancient thick silty shales may have been deposited by contour currents on the continental slope or rise. It is often difficult, however, to distinguish thick prodelta sequences (see *Prodelta Sediments*) from deeper slope muds. Much more detailed studies of provenance and paleoenvironment are needed to establish the existence of a "fossil" continental-slope rise, and hence of a region of fossil contourite deposition.

JOHN R. CONOLLY

References

Bouma, A. H., 1972a. Recent and ancient turbidites and contourites, *Trans. Gulf Coast Assoc. Geol. Soc.*, 22, 205-221.

Bouma, A. H., 1972b. Fossil contourites in lower Niesenflysch, Switzerland, *J. Sed. Petrology*, 42, 917-921.

Damuth, J., 1975. Echo character of the western equatorial Atlantic floor and its relationship to the dispersal and distribution of terrigenous sediments, *Marine Geol.*, 18, 17-45.

Emery, K. O., and Uchupi, E., 1972. Western North Atlantic Ocean, *Am. Assoc. Petrol. Geol. Mem. 17*, 532p.

Ewing, M.; Carpenter, G.; Windisch, C.; and Ewing,

J., 1973. Sediment distribution in the oceans: the Atlantic, *Geol. Soc. Am. Bull.,* **84,** 71-88.

Fox, P. J., Heezen, B. C., and Harian, A. M., 1968. Abyssal anti-dunes, *Nature,* 220, 470-472.

Heezen, B. C.; Hollister, C. D.; and Ruddiman, W. F.; 1966. Shaping of the continental rise by deep geostrophic contour currents, *Science,* **152,** 502-508.

Heezen, B. C., and Schneider, E. D., 1966. Sediment transport by the Antarctic Bottom Current on the Bermuda Rise, *Nature,* **211,** 611-612.

Hollister, C. D., and Heezen, B. C., 1972. Geologic effects of ocean bottom currents, western North Atlantic, in A. L. Gordon, ed., *Studies in Physical Oceanography,* vol. 2. New York: Gordon and Breach, 37-66.

Hubert, J. F., 1964. Textural evidence for deposition of many western North Atlantic deep-sea sands by ocean bottom currents rather than turbidity currents, *J. Geol.,* 72, 575-785.

Needham, H. D.; Habib, D.; and Heezen, B. C. 1969. Upper Carboniferous palynomorphs as a tracer of red sediment dispersal patterns in the northwest Atlantic, *J. Geol.,* 77, 113-120.

Rona, P. A., 1969. Linear lower continental rise hills off Cape Hatteras, *J. Sed. Petrology,* 39, 1132-1141.

Schneider, E. D.; Fox, P. J.; Hollister, C. D.; Needham, H. D.; and Heezen, B. C., 1967. Further evidence of contour currents in the western North Atlantic, *Earth Planetary Sci. Lett.,* **2,** 351-359.

Wezel, F. C., 1969. Prossimalità, distalità e analisi die bacini dei flysch: un punto di vista attualistico, *Mem. Soc. Natur. in Napoli, suppl. al Boll.,* 78, 11p.

Cross-references: *Abyssal Sedimentation; Continental Rise Sediments; Pelagic Sedimentation, Pelagic Sediments; Submarine (Bathyal) Slope Sedimentation; Submarine Fan (Cone) Sedimentation; Turbidites; Turbidity-Current Sedimentation.* Vol. I: *Dynamics of Ocean Currents; Ocean Bottom Currents.* Vol. III: *Submarine Geomorphology.*

CONVOLUTE LAMINATION—See PENECONTEMPORANEOUS DEFORMATION OF SEDIMENTS

COQUINA, CRIQUINA

Coquina (Kindle, 1923) is a carbonate rock consisting wholly or largely of mechanically sorted, weakly to moderately cemented fossil debris, in which the interstices are not necessarily filled with material. This term is commonly applied to cemented shell debris, most of which has experienced abrasion and transport before reaching the depositional site, thus forming a true detrital rock.

When the shell detritus is size sorted and primarily of coarse sand size or less, the term *mesocoquina* may be applied; *microcoquina* usually implies finer grain sizes, as in chalk. The limits between these classes overlap, and most often microcoquina is used to cover the entire size range. Occasionally, a microcoquina is made up of just one type of debris. The rock that forms when a coquina becomes indurated is termed a *coquinite.* Most of the discrete particulate fossil material is >2.0 mm in size, and the shells may still be intact. They are entirely cemented by a largely crystalline calcite matrix, which is probably a chemical precipitate. A distinction must be made between coquina and coquinite on the one hand, and a *coquinoid* on the other. The latter is an autochthonous (in situ) deposit of shelly material, usually having a fine-grained matrix, which may build up, under certain conditions, to *biostromes.*

A coquina made up of crinoidal debris is termed a *criquina. Crinoidal limestone (criquinite)* is the indurated equivalent of criquina and is firmly cemented, compact, and matrix bound. If crinoidal fragments make up 10-50% of the bulk of the rock, it may properly be termed an *encrinal limestone,* whereas if there is >50% crinoidal material it is an *encrinite.*

The crinoid fragments are usually coarse grained, well worn, and sorted into layers that differ in the grain size of their fragments (Cayeux, 1935). Each calyx plate or stem fragment of a crinoid is a single calcite crystal, and when deposited in a sediment, secondary calcite grows in optical continuity with the original crystal, producing large and apparently structureless grains with rounded outlines. Careful investigation, however, will reveal remnants of the original organic structure within the crystals when the central canal contains a concentration of iron oxide and other impurities (Carozzi, 1960). In some varieties of criquinite, the matrix may be almost entirely absent, and the organic fragments have been tightly welded together by pressure and differential solution along their boundaries. In other cases, rhombs of calcite encroach on the crinoid boundaries or even develop within them.

B. CHARLOTTE SCHREIBER

References

Bissell, H. J., and Chilingar, G. V., 1967. Classification of sedimentary carbonate rocks, in H. J. Bissell, G. V. Chilingar, and R. W. Fairbridge, eds., *Carbonate Rocks.* Amsterdam: Elsevier, 154.

Carozzi, A. V., 1960. *Microscopic Sedimentary Petrography.* New York: John Wiley, 485p.

Carozzi, A. V., and Soderman, J. G. W., 1962. Petrography of Mississippian (Borden) crinoidal limestones at Stobo, Indiana, *J. Sed. Petrology,* 32, 397-414.
Cayeux. L., 1935. *Les Roches Sédimentaires de France—Roches Carbonatées.* Paris: Masson, 447p.
Kindle, E. M., 1923. Nomenclature and genetic relations of certain calcareous rocks, *Pan-Am. Geol.,* 39, 365-372.

Cross-references: *Biostratinomy; Limestone Fabrics; Limestones.*

CORAL-REEF SEDIMENTOLOGY

Types of Deposits

In general, contemporary sedimentation in coral-reef areas (for ancient reefs, see *Reef Complex*) results in two major types of accumulations: rigid frames and loose sediment.

Rigid Frames. A reef frame is the three-dimensional network of calcareous skeletons, in growth position, which would essentially maintain its form if interstitial sediment were removed. The most important type of frame is associated with reefs that have attained sufficient wave resistance and topographic relief to significantly influence sedimentation in their vicinity. In areas exposed to open-ocean surf, only the most rigid frames can be constructed, built by a succession of overgrowths involving the spread of new coral colonies and calcareous algal crusts over the skeletons of their predecessors. The resulting irregular pavement, dotted with living frame builders, forms the reef crest and upper slope. The rock-like mass beneath the surface bears innumerable cavities and is highly susceptible to modification by recrystallization, dissolution, and precipitation (Land and Goreau, 1970).

Loose Sediment. Particulate calcium carbonate is quantitatively more important than reef frame in most reef areas. Typically, this loose calcium carbonate consists of mud, sand, and, to a lesser extent, gravel, produced by organisms on the reef and living in adjacent environments (Swinchatt, 1965; Stockman et al., 1967). The sedimentary particles are whole shells, fragments and segments of skeletons, and fragments of reef rock. Inorganically precipitated sediment is rarely important (Milliman, 1967).

Size and sorting of particles are influenced both by the extent of water agitation and by source of the skeletal debris. Influence of wave action, tidal currents, and wind-driven currents varies greatly from one part of a reef complex to another. Furthermore, some skeletal material is more susceptible to abrasion than others in the same size range, and some is more susceptible to transport. In this sense, porous plates of the green alga *Halimeda* have lower bulk specific gravity than, e.g., pelecypod shell fragments of the same size.

Where reefs border the coast of a major land area, carbonate sediment from the reef tract is mixed with terrigenous clastics (Maxwell, 1973).

Organisms as Agents of Sedimentation

Frame Builders. Many genera of corals and a few species of carbonate-secreting hydrozoans are included among the frame builders. Some species produce colonies with diameters of only a few cm, whereas others include branching or massive colonies more than 1 m in diameter and height (Goreau and Goreau, 1973). Some species can only colonize reef-rock surfaces, whereas others initiate colony growth on loose sand as well as on hard substrate. In some western Atlantic reefs (see, e.g., Kornicker and Boyd, 1962), presence of large pinnacles of massive corals on the sediment slope seaward of a cavernous reef front suggests that the ability of the massive forms to build large structures on loose sediment may be the key to seaward advance of the reef front. After the massive colonies create a foundation, other species requiring a solid substrate can colonize and contribute to the reef frame.

Cementing Agents. Many reef communities throughout the oceans include organisms whose encrusting mode of growth contributes greatly to reef enlargement by cementing loose debris into the reef frame. Red algae are especially important in adding calcium carbonate laminae to surfaces of broken coral and shells, thereby welding adjacent objects at points of contact (see also *Algal-Reef Sedimentology*). All gradations can be observed between recently dead coral branches and those that have been encrusted to a point where only gross aspects of the original form can be recognized.

Sediment Contributors. Many species of the reef complex play an important part in sedimentation through contribution of skeletal parts to the loose sediment. Whole shells in the sediment of a reef tract include tests of foraminifera, valves of pelecypods, and shells of gastropods. Much more sediment is fragmental material resulting from mechanical breakup of coral, mollusk shells, and algal plates (Ginsburg et al., 1963). In many reef areas, a surprising amount of the sand consists of segments of the bushy green alga *Halimeda*. Commonly several cm high, a living *Halimeda* consists of many aragonitic plates held together by nonmineralized organic material. After death of the organism, the plates are contributed to the surrounding bottom as discrete sedimentary particles.

Sediment Binders. Various species influence sedimentation by trapping and stabilizing loose sediment. Many animals pelletize sediment as it passes through their digestive tracts. Plant roots and holdfasts bind sediment in which they grow, and the presence of plants above the substrate disrupts currents that would otherwise winnow sediment.

Frame Destroyers. Many organisms erode reef frames by their activities. Coral heads are penetrated by boring worms, pelecypods, and sponges (Goreau and Hartman, 1963). Boring filamentous algae exploit exposed carbonate surfaces and are known to be important agents of skeletal destruction. An abundance of chisel-like marks on algal-encrusted reef rock testifies to the activity of many rock-scraping fish. These fish scrape the reef-rock surface while obtaining algae and erode the rock in the process. They ingest a considerable amount of rock, as shown by the contents of their digestive tracts. Cloud (1959) estimated that excrement of such fish adds 400–600 tons/km^2 of calcareous detritus annually to sediments of the principal lagoon area of Saipan.

Sediment Movers. Organisms not only transport or rework sediment, they may also change the character of the sediment by passing it through their digestive tracts. Cloud (1959) estimated that holothurians in the lagoon area at Saipan probably pass several times as much sediment during a year as the annual addition of new shoal sediment. Holothurian gut fluids have a low pH, so that long-term chemical effects on the lagoonal sediments would include reduction in size of individual grains and partial elimination of the finest sediment fraction.

Burrowing crustaceans move significant quantities of buried sediment to the surface (Shinn, 1968). Many lagoon sand bottoms are dotted with their conical mounds several centimeters high and up to 30–40 cm in diameter.

Sedimentary Environments

Coral reefs typically exhibit a variety of sedimentary environments, including reef crests, slopes between reef crest and deeper water, lagoons, and islands (Logan, 1969). These major environments may be subdivided from the standpoint of either sedimentary processes or character of deposits (Bonet, 1967). Excellent summaries of the varied sedimentary environments in several modern reef tracts have been provided by Bathurst (1971).

Reef Crests. The seaward edge of a major reef standing in the path of prevailing winds and the resulting waves is the site of vigorous growth of encrusting calcareous red algae. In fact, most Pacific reefs of this type have a narrow rim at their windward edges created by the pavement-building algae. This algal ridge rises slightly above the reef flat to its lee, and is exposed at low tide. Although calcareous red algae are important encrusting agents in this same zone in Atlantic reefs, the algal ridge as a topographic feature is relatively rare (Boyd et al., 1963).

The reef front seaward of the surf line pitches downward and is commonly indented by a system of grooves normal to the reef front. The gravel- and sand-floored grooves are several tens of m long, as much as 8 m deep, and several m wide. The intergroove areas are spurs of reef rock supporting living frame builders. Investigators of various reef tracts have reached different conclusions concerning the relative importance of erosion and differential reef growth in the origin of groove and spur systems (Shinn, 1963).

Relatively little loose sediment is found in the surf zone except as pockets of sand or gravel trapped in irregularities of the reef-rock surface. Behind their seaward margins, upper surfaces of major reefs commonly form relatively smooth flats at about low-tide level or <1 m below low tide. Such flats are as much as 100 m wide. Emery et al. (1954) noted at Bikini that rich coral growth is present across the flat from sea to lagoon in some areas where islands are not present. Many reef flats have sparse or no growth of frame builders, however, especially on the inner half of the flat. Such areas are commonly covered by loose sediment, from sand to boulder size, consisting mainly of skeletal material and debris broken from the reef margin. Displaced blocks of reef rock as much as 9 m long have been observed on reef flats at Bikini. The blocks are not generally found on the side of the atoll facing the prevailing winds, however, because the reef is adjusted to steady surf conditions, and major storm waves seem to have little additional effect. On other parts of the atoll margin, major storms can cause greater havoc as the waves pound overhanging lips and projections constructed by frame builders during long intervals of relatively weak wave action.

Where circular reef growth encloses a lagoon, the reef surface on the leeward side of the atoll is typically covered by more water and is more irregular than on the windward side. Coarse coral sand accumulates between clumps of living coral colonies. The inner edge of a major reef enclosing a lagoon is the site of either accumulation of sand and gravel or of frame-building activity. At Bikini, extensive sand flats or gravel deposits line some parts of the lagoonal side of the windward reef, whereas other parts of the lagoon border have thriving coral development.

Outer Slopes. Debris resulting from wave action along the reef edge is worked down the outer slope of a major reef to contribute to a growing apron of sand, gravel, and boulders. At Bikini, one of many deep-ocean atolls of the Pacific, the outer slopes down to 360 m average about 45°, becoming gentler below that depth. Dredging indicates that the slopes are partly covered down to 150 m and probably farther, with large blocks such as heavy chunks of coral.

The large blocks are embedded in fine sediment. Sampling by Emery et al. (1954) off the Bikini reef margin produced coarse coral fragments and *Halimeda* plates near the surface. At greater depths, the samples were fine sand and silt. Coral sand was found down to 545 m, and turned up sporadically in sampling down to 2545 m. On the outer slopes beyond the reefs at Guam, Emery (1962) found *Halimeda* plates to be abundant in samples from depths as great as 354 m, even though the plant lives principally in water shallower than 60 m. Sampling at Guam indicates that reef debris is generally restricted to depths <900 m.

In contrast to the abrupt change in depth adjacent to deep-ocean atolls such as Bikini, a shallow-sea environment lies seaward of the barrier reef at Saipan. This shallow bottom is characterized by banks of lime sand and interspersed reef patches. The sediments were reported by Cloud (1959) to be better sorted and locally finer grained than those of the lagoon on the landward side of the barrier reef.

Lagoons. Reef-bordered lagoons are floored mainly by carbonate sand and mud derived both from skeletal material of lagoon-dwelling organisms and from carbonate debris swept into the lagoon by currents passing over the peripheral reef. (For details, see *Tropical Lagoonal Sedimentation*.)

Islands. Small sand islands are formed on some reef flats and in some shallow lagoons by piling up of loose sediment by currents, waves, and wind (Folk and Cotera, 1971). At Bikini, some of the islands have a platform-like foundation of lithified boulder conglomerate. It seems likely that ridges of rubble heaped on the reef surface by great storms serve as nuclei for many islands by creating eddies which lead to additional deposition of sand.

Wave action continually influences the character of the sediment on the beaches (Folk and Robles, 1964), and wind piles the sand into dunes on some islands. Long, low ridges of gravel composed of coral fragments and mollusk shells have been formed by storm waves along some island shorelines.

DONALD W. BOYD
LOUIS S. KORNICKER

References

Bathurst, R. G. C., 1971. *Carbonate Sediments and Their Diagenesis.* Amsterdam: Elsevier, 620p.

Bonet, F., 1967. Biogeologia subsuperficial del Arrecife Alacranes, Yucatan, Universidad Nacional Autonoma de Mexico, *Inst. Geol. Bol.,* 80, 192p.

Boyd, D. W.; Kornicker, L. S.; and Rezak, R., 1963. Coralline algal microatolls near Cozumel Island, Mexico, *Univ. of Wyoming, Contrib. to Geology,* 2, 105-108.

Chave, K. E.; Smith, S. V.; and Roy, K. J., 1972. Carbonate production by coral reefs, *Marine Geol.,* 12, 123-140.

Clausad, M.; Gravier, N.; Picard, J.; Pichon, M.; Roman, M.-L.; Thomassin, B.; Vasseur, P.; Vivien, M.; and Weydart, P., 1971. Coral reef morphology in the vicinity of Tulear (Madagascar): Contribution to a coral reef terminology, *Tethys,* Suppl. 2, 72p.

Cloud, P. E., Jr., 1959. Geology of Saipan, Mariana Islands, pt. 4, submarine topography and shoal water ecology, *U.S. Geol. Surv. Prof. Pap. 280-K,* 361-445.

Emery, K. O., 1962. Marine geology of Guam, *U.S. Geol. Surv. Prof. Pap. 403-B,* 73p.

Emery, K. O.; Tracey, J. I., Jr.; and Ladd, H. S., 1954. Geology of Bikini and nearby atolls, *U.S. Geol. Surv. Prof. Pap. 260-A,* 254p.

Folk, R. L., and Cotera, A. S., 1971. Carbonate sand cays of Alacran Reef, Yucatan, Mexico: sediments, *Atoll Res. Bull.,* 137, 1-16.

Folk, R. L., and Robles, R., 1964. Carbonate sands of Isla Perez, Alacran reef complex, Yucatan, *J. Geol.,* 72, 255-292.

Ginsburg, R. N.; Lloyd, R. M.; Stockman, K. W.; and McCallum, J. S., 1963. Shallow-water carbonate sediments, in M. N. Hill, ed., *The Sea,* Vol. 3. New York: Interscience, 554-582.

Goreau, T. F., and Goreau, N. I., 1973. The ecology of Jamaican coral reefs. II. Geomorphology, zonation, and sedimentary phases, *Bull. Marine Sci.,* 23, 399-464.

Goreau, T. F., and Hartman, W. D., 1963. Boring sponges as controlling factors in the formation and maintenance of coral reefs, *Publ. Am. Assoc. Advan. Sci.,* 75, 25-54.

Hoffmeister, J. E., 1974. *Land from the Sea: The Geologic Story of South Florida.* Coral Gables, Fla.: Univ. Miami Press, 143p.

Kornicker, L. S., and Boyd, D. W., 1962. Shallow-water geology and environments of Alacran reef complex, Campeche Bank, Mexico, *Bull. Am. Assoc. Petrol. Geologists,* 46, 640-673.

Land, L. S., and Goreau, T. F., 1970. Submarine lithification of Jamaican reefs, *J. Sed. Petrology,* 40, 457-462.

Laporte, L., ed., 1974. Reefs in time and space: selected examples from the Recent and ancient, *Soc. Econ. Paleont. Mineral. Spec. Publ. 18,* 256p.

Logan, B. W., 1969. Carbonate sediments and reefs, Yucatan shelf, Mexico, Part 2. Coral reefs and banks, *Am. Assoc. Petrol. Geologists Mem.,* **11,** 129-198.

Maxwell, W. G. H., 1973. Sediments of the Great Barrier Reef province, in O. A. Jones and R. Endean, eds., *Biology and Geology of Coral Reefs.* New York: Academic Press, 299-345.

Milliman, J. D., 1967. Carbonate sedimentation on Hogsty Reef, a Bahamian atoll, *J. Sed. Petrology,* 37, 658-676.

Shinn, E. A., 1963. Spur and groove formation on the Florida reef tract, *J. Sed. Petrology,* **33,** 291-303.

Shinn, E. A., 1968. Burrowing in recent lime sediments of Florida and the Bahamas, *J. Paleont.,* **42,** 879-894.

Stockman, K. W.; Ginsburg, R. N.; and Shinn, E. A., 1967. The production of lime mud by algae in south Florida, *J. Sed. Petrology,* 37, 633-648.

Swinchatt, J. P., 1965. Significance of constituent composition, texture, and skeletal breakdown in some Recent carbonate sediments, *J. Sed. Petrology,* 35, 71-90.

Zankl, H., and Schroeder, J. H., 1972. Ecology, sedimentology, and diagenesis of Recent and fossil reefs, *Geol. Rundsch.,* **61**(2), 480-483.

Cross-references: *Algal-Reef Sedimentology; Bahama Banks Sedimentology; Carbonate Sediments—Lithification; Cementation, Submarine; Limestones; Reef Complex; Tropical Lagoonal Sedimentation.* Vol. I: *Coral Sea.* Vol. III: *Coral Reef.*

CORING TECHNIQUES—See Vol. I: SEDIMENT CORING

CORROSION SURFACE

A corrosion surface is a bedding surface in carbonate rocks that shows evidence of lithification, followed by partial solution, of the underlying bed prior to deposition of the overlying bed. Corrosion surfaces are inherently diastemic or disconformable surfaces; they are often called corrosion zones or corrosion conglomerates (Weiss, 1954), and are also referred to as *discontinuity surfaces* (Jaanusson, 1961).

Reports by numerous authors make clear that an individual surface may be smooth in one place and rough elsewhere. Only the hackly, pitted surfaces are properly ascribable to corrosion; the smooth or planed surfaces must result largely from abrasion. Jaanusson (1961) and Lindström (1963) include the whole range of carbonate surface types under discontinuity surface, restricting the terms *diastem* or *disconformity* to noncarbonate rocks. The usage of Weiss (1954; 1958) and other American workers has been equally misleading by including surfaces with obvious abrasion effects (Prokopovich, 1955).

The singular features produced by chemical corrosion in carbonates seem worthy of the special name corrosion surface. It refers to those diastemic surfaces within carbonate rocks that are partly or wholly the result of chemical effects on the underlying bed. Where locally abraded, such a surface may be called a diastem or disconformity. Conkin and Conkin (1973) have introduced *paracontinuity,* a synonym of discontinuity surface, and include corrosion surfaces as examples.

A single characteristic shared by all the disconformable surfaces that have been called paracontinuities, or corrosion or discontinuity surfaces, is that the carbonate beds on which the surfaces are developed were consolidated and had some strength before the deposition of superjacent beds. This quality is expressed in the term *hartgründe,* or *hardgrounds,* sometimes given to etched or bored surfaces in Mesozoic chalks and marls of Europe (see *Hardground Diagenesis; Cementation, Submarine*).

Characteristics

A corrosion surface may be irregular, hackly, bored, or locally even smooth or undulant. Roughness may result from corrosion or boring of the subjacent bed. Although Jaanusson (1961) states that borings are typical of abraded rather than corroded surfaces, some modern bored beds are not abraded (Shinn, 1969). The corroded surfaces of limestones in the Minnesota Ordovician and in the Straits of Florida are not bored. The irregularity of the surface may be such that pouches or bore holes in the subjacent bed are filled by younger rock, and the beds are locked together as in a dove-tailed joint. The rocks below and above a corrosion surface may be similar or different. The uppermost part of the underlying rock may be mineralized, most commonly with pyrite, goethite or limonite, glauconite, or phosphatic minerals; if bituminous, the limestone may be bleached, but to a depth independent of relief of its surface (Jaanusson, 1961). The corrosion surface may be encrusted by sessile benthonic animals, particularly in the chalks of Europe. The basal part of the younger bed may contain corroded particles of the subjacent rock; may be conspicuously glauconitic; typically contains much collophane as accretionary grains, flakes, and organic fragments; and may contain conspicuous amounts of allochthonous detritus (Fig. 1). Figure 2 shows a corrosion surface from the Middle Ordovician of Minnesota. A modern instance of a subaerial corroded surface is illustrated by Ginsburg (1953), and the pitted and bored submarine surfaces described by Shinn (1969) are perhaps even more characteristic of the class.

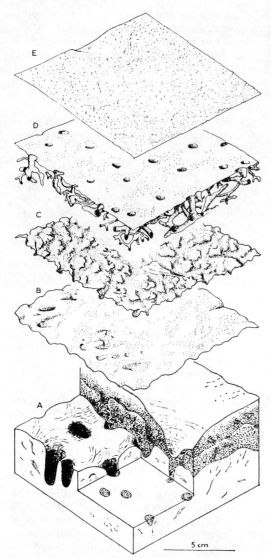

FIGURE 1. Slightly schematized morphology of discontinuity-surfaces, lower Ordovician limestones in Sweden (from Lindström, 1963). A is typical for the lowermost Ordovician limestone beds. Glauconite is shown by black spots, phosphorite with vertical shading. C contains no glauconite crust. The surface of E is greenish (in an otherwise red limestone) and contains as ridges the exposed fillings of branching tubules. The surface is also studded with slightly protruding shell fragments.

FIGURE 2. Corrosion surfaces lie between A and B, and B and C. The hydroclastic biogenic limestones are darkened by pyrite in the few mm of rock beneath each corrosion surface. Abundant collophane pellets and flakes occur in the lower part of bed B. The lower edge of the specimen is the locus of bottoms of pits in rock A. Whitish areas are clean calcite. McGregor Member of Platteville Formation, Minn. (Photo: by author.)

the later Mesozoic and earliest Tertiary from the Alps to England and Denmark. It may be that they are not more widely recognized partly because of a failure to consider them worthy of note. Corrosion surfaces are known, but not so named, from certain other places, e.g., the Cobbleskill-Onondaga contact near Phelps, New York, and the Devonian-Mississippian contact in Illinois and Missouri.

The "nonsequential beds" in the Paleozoic carbonates of the Williston Basin (Cumming et al., 1959) extend over great areas. They lie disconformably on surfaces known to have abrasional features, and the beds themselves have current-formed features. The surfaces are considered to have been eroded during regional emergence. Corrosion surfaces may occur locally in connection with the "nonsequential beds," but the two are not the same.

It is probable that accelerated study of the marine realm will yield ever more examples (Bramlette et al., 1959) or ones still being developed (Shinn, 1969).

Origin

The characteristics of corrosion surfaces have led most authors to believe that they were formed beneath the sea, even before modern examples of submarine lithification of carbonates were known. Examples of intertidal cementation and corrosion are known (Fairbridge, 1948; Ginsburg, 1953; Revelle and Emery, 1957), and Weiss (1958) has postulated the formation of corrosion surfaces within broad intertidal areas. Lindström (1963) describes primary folds that persisted in limestone through one or more episodes of corrosion, and considers this proof of submarine origin; because such folds were not modified mechanically, he believes that the corrosion must have occurred in a

Occurrence

Corrosion surfaces and associated mechanically eroded diastems are known to be common in only a few places: the Ordovician and Silurian of Baltoscandia, the Middle Ordovician of North America, the Devonian of Russia, and tranquil environment at a "safe" depth. Rose (1970) studied Cretaceous surfaces in Texas, applied criteria for subaerial or submarine origin, and concluded they were the latter. In a model for the precipitation of early diagenetic chert, Namy (1974) concluded that a Pennsylvanian corrosion surface was produced subaerially by pore waters undersaturated in $CaCO_3$ and saturated in SiO_2.

On-going submarine cementation (q.v.) of carbonates has become axiomatic as a result of dives in submersibles in Jamaica, British Honduras, and Florida. Another example, cited by Milliman (1966), foreshadowed the very recent discoveries. Subsequent boring, etching, and burial will produce a corrosion surface.

MALCOLM P. WEISS

References

Bramlette, M. N.; Faughn, J. L.; and Hurley, R. J.; 1959. Anomalous sediment deposition on the flank of Eniwetok Atoll, *Geol. Soc. Am. Bull.,* 70, 1549-1551.
Conkin, J. E., and Conkin, B. M., 1973., The paracontinuity and the determination of the Devonian-Mississippian boundary in the Lower Mississippian type area of North America, *Univ. Louisville Stud. Paleont. Strat.,* 1, 36p.
Cumming, A. D.; Fuller, J. G. C. M.; and Porter, J. W., 1959. Separation of strata: Paleozoic limestones of the Williston Basin, *Am. Jour. Sci.,* 257, 722-733.
Fairbridge, R. W., 1948. Notes on the geomorphology of the Pelsart Group of the Houtman's Abrolhos Islands, *J. Roy. Soc. W. Austral.,* 33, 1-43.
Ginsburg, R. N., 1953. Beachrock in south Florida, *J. Sed. Petrology,* 23, 85-92.
Jaanusson, V., 1961. Discontinuity surfaces in limestones, *Geol. Inst. Univ. Uppsala Bull.,* 40, 221-241.
Lindström, M., 1963. Sedimentary folds and the development of limestone in an Early Ordovician sea, *Sedimentology,* 2, 243-276.
Milliman, J. D., 1966. Submarine lithification of carbonate sediments, *Science,* 153, 995-997.
Namy, J. N., 1974. Early diagenetic chert in the Marble Falls Group (Pennsylvanian) of central Texas, *J. Sed. Petrology,* 44, 1262-1268.
Neumann, A. C., and Ball, M. M., 1970. Submersible observations in the Straits of Florida: geology and bottom currents, *Geol. Soc. Am. Bull.,* 81, 2861-2874.
Prokopovich, N., 1955. The nature of corrosion zones in the Middle Ordovician of Minnesota, *J. Sed. Petrology,* 25, 207-215.
Purser, B. H., 1969. Syn-sedimentary marine lithification of Middle Jurassic limestones in the Paris Basin, *Sedimentology,* 12, 193-204.
Revelle, R., and Emery, K. O., 1957. Chemical erosion of beach rock and exposed reef rock, *U.S. Geol. Surv. Prof. Pap. 260-T,* 699-709.
Rose, P. R., 1970. Stratigraphic interpretation of submarine versus subaerial discontinuity surfaces: an example from the Cretaceous of Texas, *Geol. Soc. Am. Bull.,* 81, 2787-2798.
Shinn, E. A., 1969. Submarine lithification of Holocene carbonate sediments, *Sedimentology,* 12, 109-144.
Weiss, M. P., 1954. Corrosion zones in carbonate rocks, *Ohio J. Sci.,* 54, 289-293.
Weiss, M. P., 1958. Corrosion zones: A modified hypothesis of their origin, *J. Sed. Petrology,* 28, 486-489.

Cross-references: *Beachrock; Carbonate Sediments–Diagenesis; Cementation, Submarine; Hardground Diagenesis.* Vol. IX: *Corrosion Surface.*

COSMOGENOUS SEDIMENTS—See EXTRATERRESTRIAL MATERIAL IN SEDIMENTS

CROSS-BEDDING

Cross-bedding refers to an internal arrangement of subsets of sedimentary strata inclined to the general stratification. The term was restricted by McKee and Weir (1953) to stratification characterized by foreset beds > 1 cm thick; smaller cross-strata would be termed *cross-laminae*. The term *cross-stratification* is recommended to describe all thicknesses of inclined subsets, but the term *cross-bedding* has commonly been used without restriction (see also *Bedding Genesis*).

Two or more cross-strata make up a *set*. Two or more sets, taken together, compose a *coset*. Cross-stratification may be planar or curved and may be inclined at various angles to the general stratification. McKee and Weir (1953) classified cross-strata by the character of their lower boundary and by their geometry (Table 1). Allen (1963) revised McKee and Weir's classification and extended it into the third dimension (Fig. 1). Jacob (1973) has proposed a descriptive classification of cross-stratification developed from the previous two classifications.

Origin

Most cross-stratification is the product of migrating bed forms produced by water or wind flowing over a cohesionless sediment surface (see *Bedding Genesis*). The smallest scale cross-strata (cross-laminae) are produced by migrating ripples (see *Ripple Marks*). Recent studies have shown that curved (festoon, trough) cross-stratification is produced by migrating dunes (in the bed-form sense; see *Bed Forms in Alluvial Channels*); and planar (tabular) cross-stratification, having a constant dip, is produced by migrating sand waves (Fig. 2). Dunes are produced by slightly higher flow velocities than are sand waves. Low-angle cross-stratification,

TABLE 1. Summary of Cross-Stratification Classification of McKee and Weir[a]

Basic Criterion of Sets of Cross-Strata	Subordinate Criteria of Cross-Strata				
Character of Lower Boundary Surface	Shape of Sets	Attitude of Axis of Sets	Symmetry of Sets	Arching	Dip (°)	Length (ft)
Nonerosional surfaces (simple cross-stratification)	lenticular	plunging	symmetric	concave	high angle (>20)	small scale (<1)
Planar surfaces of erosion (planar cross-stratification)	tabular	nonplunging	asymmetric	straight		medium scale (1–20)
Curved surfaces of erosion (trough cross-stratification)	wedge shaped			convex	low angle (<20)	large scale (>20)

[a]From McKee and Weir, 1953.

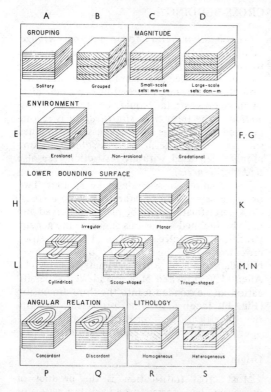

FIGURE 1. Block diagrams illustrating descriptive terms applicable to cross-stratified units (from Allen, 1963).

typically with dips of 5–8° or less, is common on beaches and is not related to migrating bed forms, but is rather the product of accreting beach surfaces (e.g., foreshore, berm, see *Littoral Sedimentation*); other cross-stratification may be produced by similar processes. Large-scale cross-stratification may be produced by the progradation of the foreset slope of a delta (q.v.).

Cross-stratification dip angles as high as 45° are known to occur in loose, modern sediments, exceeding by a considerable margin the angle of repose obtained in laboratory tests on clean, dry sand. Several explanations are possible. Backflow eddies may form on the lee side of a bed form, counteracting the tangential component of gravity. Angular grains tend to stack at a steeper angle than rounded or subrounded grains. Under water, quartz is lighter than in air, and therefore may pile up at a higher angle. Wind-blown sands may be cemented almost imperceptibly by tiny amounts of salt or other material, which will be leached out or masked as lithification proceeds. In addition, cross-strata may be oversteepened by penecontemporaneous deformation of sediments (q.v.).

Analysis

Cross-stratification is common in sandstones, conglomerates, and some detrital limestones—any clastic sediment of silt size or above, is easy to observe in the field, and provides more information than merely current direction. Hence, this feature is commonly the core around which vector studies are made. It is not, however, widely suited to subsurface studies, inasmuch as electric logs designed to record cross-strata have been run in relatively few deep holes. Subsurface vector studies are typically based on size, roundness, and maturity changes from well to well.

Many sources for paleocurrent analysis (q.v.) are available in sedimentry rocks; the commonest, and most available, is cross-stratification. At a given exposure, it is standard practice to determine bearing (direction of cross-bed dip), amount of dip, geometry of the individual laminae (straight, curved, or doubly curved; changes in rate of curvature), nature of the top

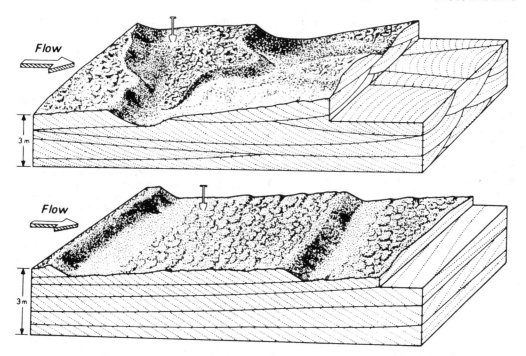

FIGURE 2. *Top*: Large-scale, trough cross-stratification produced by migrating dunes. *Bottom*: Tabular cross-stratification formed by migrating sand waves. (After Harms, 1975.)

and bottom contacts for a given set of crossbeds, thicknesses of both laminae and sets, and uniformity of direction from one set to another. All angular measurements (such as direction and amount of dip) must be corrected for tectonic or other tilting that may have occurred since the materials were deposited. Where the probable error in attempting to reestablish the original plane of deposition exceeds the original dip angle, no meaningful conclusions as to direction or inclination can be drawn. Dott (1973) has evaluated the validity of paleocurrent analysis using trough cross-stratification. Cross-bedding, like other oriented features and particles, exhibits a great deal of variability. This variability may be reduced to useful measures by statistical means of various kinds, including modal and vector analyses.

Modal Analysis. A modal analysis of directional data can be made by dividing the compass into an arbitrary number of classes, e.g., 10 ($36°$ each). Quadrants produce results too coarse to be valuable, whereas as many as 18 classes require a much larger number of field observations. A plot or *rose diagram* can be made by entering a dot in the proper class, for each measurement made at a given locality. Classes which contain more than the average number of dots indicate preferred directions, whereas deficient classes may suggest the general position of the source. A *net transport* picture can be obtained by noting which classes individually exceed their opposites. The class having the largest number of entries is the mode. There may be some question about the significance of a mode. For instance, if the measurements recorded in eight classes were 10, 10, 10, 11, 10, 10, 10, 9, a mode is obviously present, but it is undoubtedly meaningless. In order to separate significant modes from other modes, the standard deviation can be computed. Significant modes can be identified by taking all those modes outside one standard deviation, or, if a sharper discrimination is desired, two standard deviations can be used.

Vector Analysis. Individual measurements can be plotted, with pencil and paper, as unit vectors (arrows having arbitrary unit length). If these arrows are arranged as required for ordinary vector addition, the vector resultant, drawn from the origin of the first arrow to the terminus of the last arrow, shows the preferred direction. The vector resultant is a true vector, having variable length, which depends on the scatter, or spread, of the data. For large numbers of measurements, trigonometric computation can be substituted for graphic procedures.

Vector resultants have the advantage that they are sharply defined, compared with modes, which are inherently uncertain by some few tens of degrees. If high precision appears useful, vector methods should be employed. Vector resultants have the disadvantage of obscuring important information in polymodal data, unless the investigator is able to make a rigorous separation of modes prior to plotting unit vectors. In instances where some doubt is entertained, both modes and vectors should be derived.

Vector resultants are also better suited to the mapping of statistical scatter over relatively large field areas. It should be noted that R_v, the vector resultant, has a length directly proportional to the degree of uniformity of the original measurements (and therefore inversely proportional to the scatter). This length can be normalized by dividing it into n, the number of unit vectors used, to produce n/R_v, an important measure of uniformity which is theoretically invariant through changes in sample size. This measure might be called a "consistency ratio." A set of isopleths, drawn normal to plotted vector resultants, and spaced according to the various consistency ratios, is a first approximation to a map of the energy gradient down which the sands were being transported. If the cross-bedding was produced by downhill movement, then a regional slope has been pictured. If eolian cross-bedding was studied, then a map of wind consistency has been produced. Marine cross-bedding develops under such diverse conditions that it can be interpreted in this fashion only if the investigator is able to make a satisfactory separation and identification of the several modes.

An inverse form, R_v/n, may appear at first sight to be more logical; but it has the disadvantage that it must be fitted by isopleths that are crowded where the gradient is gentle, and widely spaced where the gradient is steep, a situation directly contrary to standard practice with isopleth maps or any of the gradients known to geologists, physicists, or other scientists.

It is important to remember that a regional average, no matter how obtained, may mask *much or all* of the information that is geologically pertinent to the problem at hand. Consider two exposures, at each of which a suitable number of measurements can be made. The vector resultant at one is E, at the other, W; the two consistency ratios are identical. Any combination of these two results yields a null vector, or the conclusion that no transport has taken place. Or consider two exposures, one of which produces the direction E, the other, S. If these are adjacent reaches of a single stream, then the mean direction (SE) may represent the regional slope. But if one is a channel result, the other a littoral product, any average we compute is less meaningful than the two arrows plotted side by side.

WILLIAM F. TANNER

References

Allen, J. R. L., 1963. The classification of cross-stratified units with notes on their origin, *Sedimentology,* 2, 93-114.

Allen, J. R. L., 1973. Features of cross-stratified units due to random and other changes in the forms, *Sedimentology,* 20, 189-202.

Boersma, J. R.; van de Meene, E. A.; and Tjalsma, R. C., 1968. Intricate cross-stratification due to interaction of a mega-ripple with its lee-side system of backflow ripples, *Sedimentology,* 11, 147-162.

Dott, R. H., Jr., 1973. Paleocurrent analysis of trough cross-stratification, *J. Sed. Petrology,* 43, 779-783.

Elliott, D., 1965. The quantitative mapping of directional minor structures, *J. Geol.,* 73, 865-880.

Harms, J. C., 1975. Stratification produced by migrating bed forms, *Soc. Econ. Paleont. Mineral. Short Course* 2, (April 5, 1975), 45-62.

Hunter, R. E., 1973. Pseudo-crosslamination formed by climbing adhesion ripples, *J. Sed. Petrology,* 43, 1125-1127.

Jacob, A. F., 1973. Descriptive classification of cross-stratification, *Geology,* 1, 103-105; comments by M. T. Roberts (105-106) and D. R. Spearing (106).

McKee, E. D., and Weir, G. W., 1953. Terminology for stratification and cross-stratification in sedimentary rocks, *Geol. Soc. Am. Bull.,* 64, 381-390.

Mardia, K. V., 1972. *Statistics of Directional Data.* New York: Academic Press, 352p.

Shrock, R. R., 1948. *Sequence in Layered Rocks.* New York: McGraw Hill, 507p.

Steinmetz, R., 1972. Sedimentation of an Arkansas River sand bar in Oklahoma: A cautionary note on dipmeter interpretation, *Shale Shaker,* 23(2), 32-38.

Tanner, W. F., 1959. The importance of modes in cross-bedding data, *J. Sed. Petrology,* 29, 221-226.

Underwood, J. R., and Lambert, W., 1974. Centroclinal cross strata, a distinctive sedimentary structure, *J. Sed. Petrology,* 44, 1111-1113.

Cross-references: *Bedding Genesis; Bed Forms in Alluvial Channels; Braided-Stream Deposits; Delta Sedimentation; Fabric, Sedimentary; Flow Regimes; Fluvial Sedimentation; Fluvial Sediment Transport; Paleocurrent Analysis; Provenance; Ripple Marks; Sedimentary Structures.*

CRUDE-OIL COMPOSITION AND MIGRATION

Crude oil contains a number of clues concerning its history and the path it took to the reservoir. The objective of this review is to begin with simple measurements which reflect crude-oil composition, and to show how these change with depth of burial, age, and in some cases, environment of deposition. The pioneer work in this area was by D. C. Barton for crude oils of the Gulf Coast; since that date, numerous others have provided insight and data on various facets of the subject (see references in Cordell, 1972). Some of the more recent articles include those of Cordell (1972; 1973), Johns and Shimoyama (1972), Nixon (1973), Bailey et al. (1973), and Ikan et al. (1975).

Influence of Age and Depth of Burial

A brief quantitative review of what happens to crude oils with increasing age and depth of burial will aid in understanding oil migration. The most common measurements obtained from crude oils are API gravities and specific gravities. The American Society for Testing Materials defines both types of gravity measurements as follows:

Specific Gravity — The specific gravity of a petroleum oil and of mixtures of petroleum products with other substances is the ratio of the weight of a given volume of the material at a temperature of 60°F (15.56°C) to the weight of an equal volume of distilled water at the same temperature, both weights being corrected for the buoyancy of air.

API Gravity — The API gravity scale is an arbitrary one related to the specific gravity of a petroleum oil in accordance with the formula

$$\text{degrees API} = \frac{141.5}{\text{sp. gr. } 60°F} - 131.5$$

Looking at the problem in terms of API gravities, it is instructive to consider the following classes of crude oils (Biederman, 1965):

1. Mesozoic and Cenozoic oils produced from a depth of 2000 feet (600 m) or less (young shallow).
2. Mesozoic and Cenozoic oils produced from 10,000 feet (3000 m) or below (young deep).
3. Paleozoic oils produced from 2000 feet (600 m) or less (old shallow).
4. Paleozoic oils produced from 10,000 feet (3000 m) or below (old deep).

Distributions of API gravity data that fall into these classes for three different areas (Fig. 1) were obtained from the *International Oil and Gas Development Yearbook, 1963.* Distributions of West Texas crude oils illustrate differences between the shallow and deep Paleozoic oils (Fig. 1A). A statistical comparison (Student's t, Dixon and Massey, 1957) indicates that the probability of the two mean values ($\bar{X}$) being drawn from the same population is less than one in a thousand. It can be argued that the mean ages for the two distributions are appreciably different; hence, a look at US Gulf Coast young oils is in order. In this case, although the average age for both deep and shallow distributions is Miocene to Eocene, there is an even

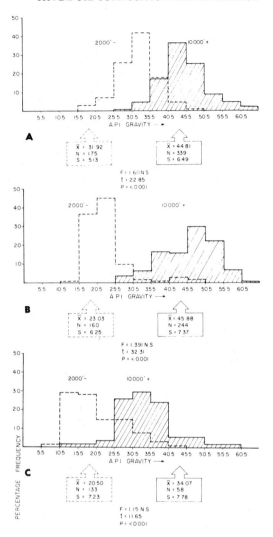

FIGURE 1. Distribution of API gravities for (A) Texas crudes of Permian age and older; (B) crude oils Cretaceous age and younger from the Texas Gulf Coast and the SW Texas Corpus Christi district; and (C) California crude oils of Cretaceous age and younger. $\bar{X}$ = means; N = number of observations; S = standard deviation of the means; F = variance ratio; P = probability.

broader spread between the means (Fig. 1B). As has been suggested by several authors, environment contributes to this difference, in that the shallower crude oils are often found in brackish to nonmarine sediments, whereas the deeper oils are more often from marine strata. A comparison of the old shallow crudes (Fig. 1A) average 10 API degrees higher than young shallow crudes on the Gulf Coast (Fig. 1B), indicating that age is significant.

In a similar study of Tertiary oils from California, where marine and nonmarine environments are about equally divided between the two classes, effects of both age and environment should be minimized (Fig. 1C). The difference in mean values for the depth grouping is still statistically significant, and it is most unlikely that they were from the same population.

Both age and depth of burial influence the API gravity of oil. Environment appears to have an additional influence which is superimposed upon the first two. The fact that age is significant means that some chemical changes occur very slowly over long periods of time. Depth of burial provides heat and pressure, both of which can greatly accelerate the maturing process.

Other factors have recently been shown to have significant influences upon crude oil composition—viz, bacterial degradation, water washing, and inorganic oxidation (Bailey et al., 1973). Relatively shallow reservoirs exposed to meteoric waters moving in from an outcrop along an aquifer may experience any of these three influences in varying degrees. Bacteria metabolize the normal paraffins and, to a somewhat lesser degree, the lighter naphthenes and aromatics. Crude oils attacked by bacteria have reduced API gravities and increased sulfur contents. *Water washing* is the term applied to the removal of the more soluble hydrocarbons from crude oils by the flow of waters undersaturated in these hydrocarbons. The lighter hydrocarbons are in general more soluble; hence, this process also tends to leave oils that have lower API gravities. Inorganic oxidation involves exposing the oil to substantial amounts of molecular oxygen. The overall magnitude of the changes brought about by these processes is given by Bailey et al. (1973) who indicate that the API gravities of oils from the Mississippian Mission Canyon carbonates of Saskatchewan were changed from 40° to 15° API and sulfur contents rose from 0.5 to 3.0%.

Classification of Oils for Compositional Comparisons

Changes that occur in crude oils with increasing age and depth of burial can be illustrated with the help of statistical inference. Martin et al. (1963) have published analyses of the hydrocarbons (boiling through 111°C) for a number of crude oils, and although their samples do not fall into the age and depth classes used above, nearly all of them can be classified in a similar way (Table 1).

A statistical comparison has been made, based on this study, wherein the API gravities for the various age and depth classes were compared (Table 2A). A similar statistical test of total volume (in %) of hydrocarbons boiling through

TABLE 1. Classification of Crude Oils Analyzed by Martin et al.[a]

Class	Field	API Gravity	Depth (Feet)	Age	Total Boiling Through 111° C Vol. %
Old Shallow Crude Oils	Alida	38.1	3,707	Miss.	19.21
	Hendricks	33.1	2,887	Perm.	14.07
	Kawkawlin	35.0	2,789	Dev.	10.47
	Lee Harrison	25.3	4,888	Perm.	14.30
	Ponca City	42.0	3,904	Ord.	17.90
	Red Water	35.3	3,182	Dev.	17.54
	Average	34.7	3,726		15.58
Young Shallow Crude Oils	South Houston	24.0	4,888	Mio.	3.54
	Wafra	18.3	6,693	Eoc.	4.58
	Wilmington	19.3	4,003	Mio.	4.44
	Uinta Basin	30.6	4,883	Eoc.	4.50
	Average	23.1	5,117		4.27
Old Deep Crude Oils	Beaver Lodge	40.7	8,398	Miss.	18.43
	Eola	36.6	10,006	Ord.	15.69
	Eola (Deep)	44.6	10,597	Ord.	22.83
	North Smyer	43.3	10,006	Penn.	27.33
	Teas	40.0	8,399	Ord.	23.46
	Average	41.0	9,481		21.55
Young Deep Crude Oils	Darius (Iran)	29.0	11,713	Cret.	9.18
	Swanson River	31.3	9,514	Tert.	12.11
	Average	30.2	10,614		10.65

[a]From Martin et al., 1963.

TABLE 2. Statistical Comparison of % Propane Volume (A) and Volume % of Hydrocarbons Boiling Through 111° C (8), in Crude Oils Described in Table 1[a]

Category			Probabilities That Samples Were Drawn from Same Population
A			
1. Old Shallow Oils	vs.	Young Shallow Oils	P = 0.01
2. Old Deep Oils	vs.	Young Deep Oils	P = 0.047
3. Old Deep Oils	vs.	Old Shallow Oils	P = 0.026
4. Young Shallow Oils	vs.	Young Deep Oils	P = 0.133
5. Young Shallow Oils	vs.	Old Deep Oils	P = 0.008
6. Old Shallow Oils	vs.	Young Deep Oils	P = 0.143
B			
1. Old Shallow Oils	vs.	Young Shallow Oils	P = 0.005
2. Old Deep Oils	vs.	Young Deep Oils	P = 0.047
3. Old Deep Oils	vs.	Old Shallow Oils	P = 0.026
4. Young Shallow Oils	vs.	Young Deep Oils	P = 0.067
5. Young Shallow Oils	vs.	Old Deep Oils	P = 0.008
6. Old Shallow Oils	vs.	Young Deep Oils	P = 0.071

[a] From Martin et al., 1963.

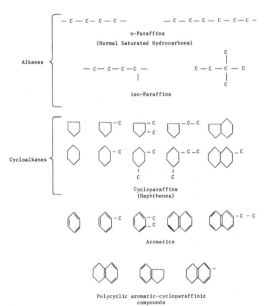

FIGURE 2. Important types of hydrocarbons in crude oils (modified after Welte, 1965).

General Composition. The distribution of some of the important structural hydrocarbon types (Fig. 2) among the various crude oils representative of the four previously established general categories is significant (Fig. 3). From these specific examples it appears that the young shallow crudes are characterized by more cycloparaffins, aromatics, and heavier polycyclic compounds—which is indeed the case. The time of development is short so that the thermal breakup of the soluble organic material and kerogen is closer to the beginning stage, and relatively few low-molecular-weight hydrocarbons have been formed. Splitting off of organic acid groups (decarboxylation) and other functional groups is only partially completed.

Young, deep crude oils exhibit a large and significant change toward enrichment in paraffins, which suggests that thermal cracking is the most important factor in maturation. It should be pointed out that the Darius crude is of definite marine origin, and the paraffinicity is increased due to environmental effects. Old shallow oils show an evolutional stage which, when compared to young shallow oils, also emphasizes the formation of lighter paraffins. The old deep crudes show an even greater change to paraffins of lower molecular weight. Ultimately, with increasing temperatures at greater depths, the major final product is carbon dioxide. The depth at which carbon dioxide becomes the major component depends upon the geothermal gradient.

Individual Components. By averaging the data for the various categories in Table 1 and

111°C was also run (Table 2B). It appears that enough evidence is present to justify the use of the above classes, despite the weakness of the probabilities involving comparisons of young deep crudes. Their lack of significance at the 0.05 level is likely to be due to the low sample number in that one category.

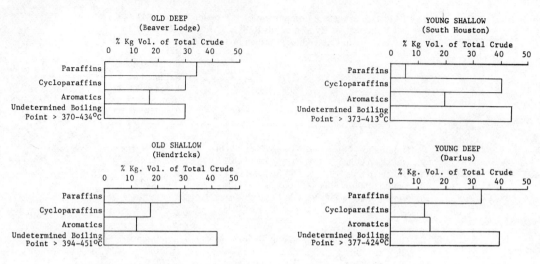

FIGURE 3. Examples of general crude-oil composition (data from Martin et al., 1963).

ranking the components in order of abundance, the gradual evolution of crude oil is established (Fig. 4). The next step involves ranking all the components in the order of abundance obtained for the old deeply buried crude oils. The assumption involved in this sequence is that old deeply buried oils have evolved to the highest degree and represent an end point. Shifts in volume % can be followed in terms of this ultimate ordering.

The most obvious aspect of this analysis is that there is a progressive increase in the volume % for practically all the components with increasing age and depth of burial. Secondly, certain components are enriched much more than others; e.g., methylcyclohexane is one of the dominant components in young shallow oils, whereas n-heptane becomes dominant in deeply buried young oils. Taking the dominant component for each class, the evolutionary series appears in Fig. 5.

The cyclic nature of young shallow crudes is changed to one of straight-chain predominance with depth of burial. Similarly, old crude oils from shallow depths show a preference for straight-chain paraffin hydrocarbons of the n-hexane and n-heptane type. The combined effect of age and depth of burial is exhibited by the old deeply buried oils in which n-pentane and n-butane dominate. In short, crude oils evolve toward normal paraffins with fewer carbon atoms per molecule.

One feature of this study that bears particular mention is the difference in the volume of propane between old shallow crudes and old deep crudes. The relatively small amount of propane in old shallow crudes suggests that this component has been removed. How this removal took place opens an area for additional experiment.

Petroleum Migration

Petroleum migration is generally divided into two types: *primary* and *secondary*. Primary migration includes the movement of oil from the source rock into the reservoir formation; secondary migration refers to movement of petroleum from one place to another within the reservoir rock and into traps where further movement is stopped.

Primary Migration. Considerable effort has been devoted during recent years to determining the timing of maximum primary migration. Albrecht and Ourisson (1969) located a thick sedimentary sequence in Douala, Cameroun (West Africa), which was deposited over a relatively short time interval. The depositional environment is reported to be similar throughout, and the total organic carbon content is roughly comparable over the section. This particular sedimentary sequence allowed Albrecht and Ourisson to study the effect of increasing depth upon organic matter without the compounding influence of age differences, changing environments of deposition, and differing amounts of organic matter. Results of the analysis of the total bitumen and total alkane hydrocarbons (paraffins) indicated that at a depth of around 4920–8200 ft (1500–2400 m), where the bitumen content rises only moderately, the alkane hydrocarbon concentration increased sharply (Fig. 6). At a depth of about 8200 ft (2400 m), the alkane concentration dropped sharply along with the drop in bitumen content. The authors

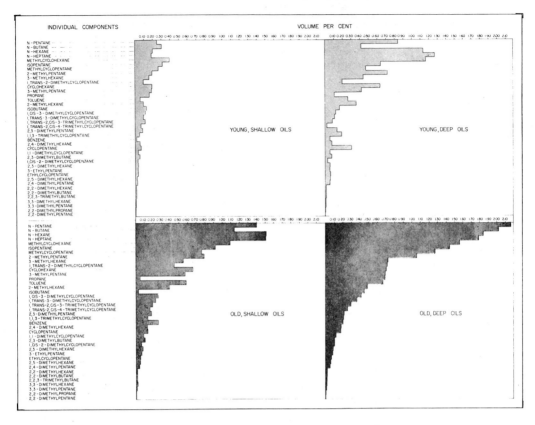

FIGURE 4. Average amounts of individual components boiling through 111°C for crude oils classified by age and depth of burial. Lack of lightest ends is striking in case of young, shallow oils.

concluded that the apparent losses were a function of either (1) removal because of primary migration or (2) cracking to form lighter hydrocarbons not measured in their analysis. It seems likely that both processes were occurring at about the same time and that the lighter hydrocarbons migrated more readily, leaving the heavier, more polar materials behind.

Another study, by Tissot et al. (1971), indicated that the Jurassic "Schistes carbons" formation of the Paris Basin shows increased hydrocarbon generation beginning at depths of around 4600-4900 ft (1400-1500 m) at temperatures of 60-65°C. This formation also exhibited little variation in either mineralogy or depositional environment. The maximum value was reached in the 5500-6600 ft (1650-2000 m) range and decreased thereafter.

If the above data are combined with the observations and theories of the clay mineralogists, a picture emerges whereby the water released from clay minerals due to the effects of heat and pressure becomes an important mechanism for moving oil from the organic-rich environment into the reservoir rock. Returning to the Cameroun section, the clay mineralogy work of Dunoyer de Segonzac (1964) shows that the zone where the change from a predominantly montmorillonite (smectite) section to a predominantly illite section occurs at a depth that corresponds with the depth of maximum hydrocarbon generation as determined by Albrecht and Ourisson (1969).

Similar work carried out in the US by a number of investigators approached the same general conclusions. Specifically, Powers (1967) concluded that in the Texas Gulf Coast sediments, the release of interlayer water as montmorillonite (smectite) changes to illite and furnishes the medium for the primary migration of oil. Burst (1969) suggested that the water-loss stage is a function of the geothermal gradient so that where the gradient is high, migration begins at around 4,000 ft (1200 m) and where it is low, primary migration can extend down to 16,000 ft (4800 m).

Although petroleum migration is not restricted to the above mechanism, a number of lines of

1 (Young Shallow)
Methylcyclohexane

2 (Young Deep)
N-Heptane
C—C—C—C—C—C—C

3 (Old Shallow)
N-Hexane
C—C—C—C—C—C

4 (Old Deep)
N-Pentane
C—C—C—C—C

FIGURE 5. Evolution of dominant individual components from young, shallow oil to old, deep oil.

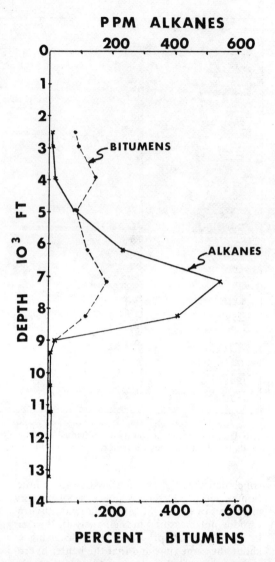

FIGURE 6. Changes in bitumen and alkane hydrocarbon content with increasing depth, Turonian to Paleocene section, Douala, Cameroun (after Albrecht and Ourisson, 1969).

evidence appear to favor it as one of the more plausible ones. One area that has received little attention with respect to both primary and secondary migration is the role played by microfractures. Careful examination of a number of shales has indicated that it is rare to find a fragment of shale as much as 15 cm square that does not have a microfracture in it. Examination of the history and productivity of the Florence-Canyon City Field in Fremont County, Colorado, shows that the fractured shales of the Pierre Formation have been producing oil with little or no water for over 100 years. The suggestion from this example and others like it is that oil can move out of a consolidated source shale and into a reservoir by means of fractures, whether they are flushed with water or not. Furthermore, oil may be stored for considerable time in such a vast fracture network and released when earth movements provide a method of escape.

Secondary Migration. Once the oil has moved into a reservoir rock, there occur movements of various kinds, collectively called secondary migration. Silverman (1965) has emphasized what he calls separation migration where the original oil has been buried to the point where the temperature and pressure require that it exist as a single phase. Pressure reduction by faulting or erosion converts the deposit into two phases, one of which is more mobile than the other. Migration of the more mobile phase to a new trap occurs, and the whole process may repeat itself again. Large differences occurring in oils that apparently came from the same source can be explained by this mechanism.

Both the evolution of crude oils toward normal paraffins with fewer carbon atoms and Silverman's migration-separation mechanism should produce changes in viscosity which, in turn, should have a direct bearing upon ease of migration. Viscosity is as good a measure of hydrocarbon mobility as is readily available. The most viscous oils (presumably those that migrate with greatest difficulty) are those from <2000 ft (600 m) in depth, of Mesozoic and Cenozoic age (Table 3). If the mean viscosities are ranked for each category of depth and age, it becomes clear that the deep Paleozoic crudes have the lowest viscosity, deep Mesozoic and Cenozoic are the next most mobile, and shallow Paleozoic crudes are third with respect to ease of migration. Compaction and decreases in permeability due to cementation work against this viscosity change, but the general concept is valid.

It thus becomes clear that both the proponents of long-distance migration and proponents of short-distance migration can be right. Those favoring long-distance migration are more likely to be right if they are working with deep Paleozoic sediments. Those in favor of short-distance migration are likely to be correct for shallow Cenozoic sediments. The fact that oil changes in its ability to migrate with time and depth of burial indicates that migration of oil probably takes place in several steps at different times and is by no means confined to an early compaction stage. Such movements are in addition to those caused by faulting, folding, and tilting.

Sulfur Content. Nickel and vanadium content was measured for a large number of crude oil samples from Canada (Hodgson and Baker, 1959). These values were found to parallel the variations in sulfur, resins, and asphaltene content. Hodgson and Baker concluded that the passage of crude oil through a porous solid may result in the retention of high molecular weight and polar material, implying a retention by the porous solids of the asphaltenes and resins, and the vanadium, nickel, and sulfur compounds in crude oils. Thus, the farther a crude oil has migrated, the less sulfur it should contain. The problem in using this idea lies in (1) finding a base line against which a comparison can be made and (2) determining whether or not the crude has suffered bacterial alteration and sulfur enrichment. However, the concept has been applied to the sulfur data for oil from 63 wells in the Cardium sand in the Pembina field, Alberta, Canada (Fig. 7); and the fact that a regular pattern was observable for Pembina, as well as a number of other fields, suggests that the concept is correct. The distance between the extreme values in the Pembina field is roughly 15 km, suggesting that the crude oil containing 0.232% sulfur upon entry to the field lost 0.118% in traversing the distance to the other side. If the idea of increased migration ability of old oils is generally correct, there should be less sulfur with increasing depth; and, in general, this is true (Fig. 8). As soon as age is considered,

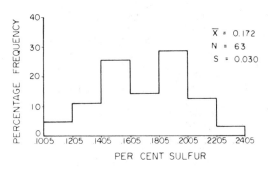

FIGURE 7. Distribution of percent sulfur in Pembina crude-oil data (from Hodgson and Baker, 1959). The distance between the samples at opposite ends of the distribution is about 15 km.

TABLE 3. Average Sabolt Viscosity Based on Data from 62 Different Crude Oils (based on 100° F)

Age	Mean Viscosity of Oils from	
	2,000 ft and Shallower[a] (sec)	10,000 ft and Deeper[b] (sec)
Paleozoic	61.87	36.75
Mesozoic and Cenozoic	1175.07	54.88

After Martin et al. 1963.
[a]The probability that the viscosities are from the same population is <0.05.
[b]The probability that the viscosities are from the same population is <0.002.

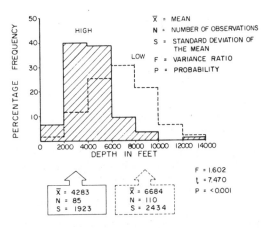

FIGURE 8. Distribution of high- (1.00%) and low- (0.15%) sulfur crude oils according to depth—USA.

it becomes obvious that the oils from West Texas Permian carbonates are high in sulfur content (Jones and Smith, 1965). A background sulfur content, therefore, has to be obtained for each problem. Smith (1952) provided a good starting place by recording the overall values for various producing regions; however, beyond this point the answers are much harder to obtain.

As may readily be seen, the subject of petroleum migration provides one of the most varied and fruitful fields for future scientific research. It utilizes every possible branch of study in order to find a coherent relationship between the origin, content, and occurrence of oil.

EDWIN W. BIEDERMAN, JR.

References

Albrecht, P., and Ourisson, G., 1969. Diagénèse des hydrocarbures saturés dans une série sédimentaire épaissé (Douala, Cameroun), *Geochim. Cosmochim. Acta,* 33(1), 138-142.

Bailey, N. J. L.; Krouse, H. R.; Evans, C. R.; and Rogers, M. A., 1973. Alteration of crude oil by waters and bacteria—evidence from geochemical and isotope studies, *Bull. Am. Assoc. Petrol. Geologists,* 57, 1276-1290.

Biederman, E. W., Jr., 1965. Crude oil composition— a clue to migration, *World Oil,* 161, 78-82.

Burst, J. F., 1969. Diagenesis of Gulf Coast clayey sediments and its possible relation to petroleum migration, *Bull. Am. Assoc. Petrol. Geologists,* 53, 73-93.

Cordell, R. J., 1972. Depths of oil origin and primary migration: a review and critique, *Bull. Am. Assoc. Petrol. Geologists,* 56, 2029-2067. See also discussion by Philippi, G. T. (1974), 58, 149-150.

Cordell, R. J., 1973. Colloidal soap as proposed primary migration medium for hydrocarbons, *Bull. Am. Assoc. Petrol. Geologists,* 57, 1618-1643. See also discussion by Saxby, J. D., and Shibokaoka, M., (1975), 59, 721-723.

Dixon, W. J., and Massey, F. J., Jr., 1957. *Introduction to Statistical Analysis.* New York: McGraw-Hill, 488p.

Dunoyer de Segonzac, G., 1964. Les argiles du crétacé superieur dans le bassin de Douala (Cameroun): Problèmes de diagénèse, *Alsace-Lorraine Serv. Carte Geol. Bull.,* 17, 287-310.

Hodgson, G. W., and Baker, B. L., 1959. Geochemical aspects of petroleum migration in Pembina, Redwater, Joffre, and Lloydminster oil fields of Alberta and Saskatchewan, Canada, *Bull. Am. Assoc. Petrol. Geologists,* 43, 311-328.

Ikan, R.; Aizenshtat, Z.; Baedecker, M. J.; and Kaplan, I. R., 1975. Thermal alteration experiments on organic matter in recent marine sediments, *Geochim. Cosmochim. Acta,* 39, 173-203.

Johns, W. D., and Shimoyama, A., 1972. Clay minerals and petroleum-forming reactions during burial and diagenesis, *Bull. Am. Assoc. Petrol. Geologists,* 56, 2160-2166.

Jones, T. S., and Smith, H. M., 1965. Relationships of oil composition and stratigraphy in the Permian Basin of West Texas and New Mexico, *Am. Assoc. Petroleum Geologists Mem. 4,* 101-224.

Martin, R. L.; Winters, J. C.; and Williams, J. A., 1963. Composition of crude oil by gas chromatography: geological significance of hydrocarbon distribution, *6th World Petrol. Cong.,* Frankfurt, Germany, sec. V, paper 13.

Nixon, R. P., 1973. Oil source beds in Cretaceous Mowry shale of northwestern interior United States, *Bull. Am. Assoc. Petrol. Geologists,* 57, 136-161.

Powers, M. G., 1967. Fluid-release mechanisms in compacting marine mudrocks and their importance in oil exploration, *Bull. Am. Assoc. Petrol. Geologists,* 51, 1240-1254.

Silverman, S. R., 1965. Migration and segregation of oil and gas fluids in subsurface environments, *Am. Assoc. Petrol. Geologists Mem. 4,* 53-65.

Smith, H. M., 1952. Composition of United States crude oils, *Indust. Eng. Chem.,* 44, 2577-2585.

Tissot, B.; Califte-Debyser, Y.; Deroo, G.; and Oudin, J. L., 1971. Origin and evolution of hydrocarbons in early Toarcian shales, Paris basin, France, *Bull. Am. Assoc. Petrol. Geologists,* 55, 2177-2193.

Welte, D. H., 1965. Relation between petroleum and source rock, *Bull. Am. Assoc. Petrol. Geologists,* 49, 2246-2268.

Cross-references: *Clay—Diagenesis; Gases in Sediments; Hydrocarbons; Kerogen; Organic Sediments; Petroleum—Origin and Evolution; Shale.*

CUMULATIVE FREQUENCY—See GRAIN-SIZE FREQUENCY STUDIES

CURRENT MARKS

Although first defined as "irregular structures produced by erosion" (Twenhofel, 1939), *current marks* have been generally used to describe all features made on a cohesive sediment- or a bedrock substrate by a current. The current may or may not have been carrying a sediment load; included under the term *current* are liquefied coarse-particle flows. Current marks include such things as potholes eroded in bedrock, but exclude rhythmic bed forms created by currents in cohesionless sediment. Examples of these excluded features are sand waves, dunes, ripple marks, and related bed forms (see *Bed Forms in Alluvial Channels*).

Most current marks known from the geologic record have been buried by sand-sized or coarser sediments, usually but not always deposited by the current that made the marks. After burying the current marks, the sand is cemented into a discrete bed and thus preserves the shapes of the buried features as counterparts on its bottom surface. Because this preservation by burial and sand cementation is so common, it is important to make careful distinction among (1) the original current marks and other features formed on a cohesive substrate (i.e., the original, or *mark*); (2) the counterpart preserved on the base of the overlying sandstone (a negative, or *mold*); and (3) a replica of the original such as can be made by pressing clay against the bottom of the covering sandstone bed (a *cast;* see *Casts and Molds*).

Description

Current marks include all varieties of scour marks (q.v.; Fig. 1), including flutes, current crescents, transverse scour marks (Allen, 1971) and scour in the lee of obstacles; flow-aligned ridges; meandering ridges (Fig. 2); shear wrinkles; and various *tool marks* (q.v.; Fig. 3), including certain grooves and *impact marks,* such as prod marks, brush marks, bounce marks, skip marks, and roll marks (Dzulynski and Sanders, 1962). The shapes of individual marks may be linear, spatulate, circular, or crescentic. These marks may cover large areas of the substrate uniformly, or they may be arranged in clusters which form random or linear groups. The long axes of individual linear current marks and of linear groups of current marks typically are parallel to threads of fluid flow in the boundary layer of the current. Some marks and groups of marks, however, are oriented in diagonal positions or even are perpendicular to the main flow direction of the current.

Origin

Most information on the origin of current marks has come from geologic analysis of features preserved on the bottoms of sandstone beds (Shrock, 1948). Such information has led to the generally accepted conclusions concerning orientation of current marks with respect to the active current, but has contributed little to the understanding of the mechanisms responsible for creating these marks. A few very suggestive experiments by Dzulynski and Walton (1963), Dzulynski (1965), and Allen (1969) have created many of the kinds of features seen in the geologic record. Flow-aligned ridges were created experimentally by passing thin mixtures of water and plaster of Paris over a clay substrate. These ridges were interpreted by Dzulynski as being products of tiny, cylindrical *flow tubes* in the boundary layer within which the threads of current traveled in spiral paths. Such flow tubes are thought to be comparable to the large-scale Langmuir cells described from the flow of air over the ocean. The long axes of these tiny flow tubes are parallel to the direction of the main current.

Allen's experiments created numerous current marks, which varied with the intensity of the current. As current intensity was increased, the progression of current marks was longitudinal rectilinear grooves, longitudinal meandering grooves, flutes, transverse erosional marks, and shear wrinkles.

Geologic Significance

Current marks have been employed as top and bottom criteria (q.v.), in paleocurrent analysis (q.v.), in paleogeographic reconstructions, and as indicators of flow conditions within the current.

The use of current marks in geopetal (or "way up") studies has depended on the proposition that the marks were made in a cohesive substrate and buried and preserved by deposition and cementation of a covering layer of sand. Until recently, the counterparts of flutes and grooves, two of the most common kinds of current marks, were reliably regarded as diagnostic indicators of the bases of covering sandstone beds, nothing comparable having been seen on the tops of sandstone beds. Friedman and Sanders (1974), however, have described

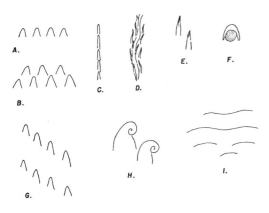

FIGURE 1. Schematic plan view summary of various kinds of scour marks (Dzulynski and Sanders, 1962). Current flowed from top to bottom in all examples. (A) Flutes aligned side-by-side perpendicular to current direction. (B) Flutes arranged in alternate positions. (C) Flutes aligned end-to-end in a line parallel to current flow direction. (D) Curved, rill-like pattern of flutes. (E) Scour marks formed downcurrent from small obstacle in bottom. (F) Current crescent formed upcurrent from large obstruction on bottom. (G) Flutes arranged in rows diagonal to current flow direction. (H) Flutes with spirally twisted upcurrent ends. (I) Transverse scour marks which resemble transverse current ripple marks formed in sand.

FIGURE 2. Counterpart of anastomosing, subparallel, meandering rill marks; current from right to left. Base of sandstone slab, Carpathians (Dzulynski and Sanders, 1962). For a remarkably similar result produced experimentally see Fig. 6 in Allen, 1969.

positive relief bedforms resembling counterparts of flutes (*setulfs*) and of grooves on the tops of cohesionless sediment. Accordingly, the setulfs and ridges will be preserved in the geologic record on the top surfaces of arenaceous beds (see *Casts and Molds*).

Numerous paleocurrent studies have been based on field measurement of the orientations of current marks preserved on the bottoms of sandstone beds. The chief attempt of many of these studies has been to determine one single direction, the presumed ancient slope of the sea floor. In seeking this one direction, the data from many beds have been averaged, generally without regard to stratigraphic position. More recently, the emphasis has been on locating measurements within a detailed stratigraphic framework and trying to assess the significance of all the readings (Enos, 1969).

The paleogeographic emphasis on current marks has been concerned with their use for interpreting environments of deposition. Because current marks are so common in the suite of strata known as flysch (Dzulynski and Walton, 1965), many workers have tended to equate these marks with flysch. Current marks can be expected to occur wherever sand can be transported over cohesive substrates. These conditions are met in certain deep marine basins where turbidity currents and other bottom-following sediment flows are active; on river floodplains (Stanley, 1968); on deltas (Coleman and Gagliano, 1965); on the floors of desert basins; and in large lakes (Belt, 1968). Current marks alone, therefore, have not proved to be as diagnostic for environmental indicators as some have thought.

The use of current marks as indicators of the conditions within the flow that deposited the covering sand is only in its infancy, but even so, has already given indications that ideas about the randomness of turbulence, for example, require reexamination. Large exposures of individual bedding surfaces are not common, but wherever they are found, they should be studied carefully to see if they exhibit large-scale patterns of current marks that can be employed to reconstruct the flow lines within the ancient current. Experiments have shown that the current itself does not need to be carrying a load of sediment in order to create current marks (Allen, 1966).

JOHN E. SANDERS

References

Allen, J. R. L., 1966. Flow visualization using plaster of Paris, *J. Sed. Petrology*, 36, 806-811.

Allen, J. R. L., 1969. Erosional current marks of weakly cohesive mud beds, *J. Sed. Petrology*, 39, 607-623.

Allen, J. R. L., 1971. Transverse erosional marks of mud and rock; their physical basis and geological significance, *Sed. Geol.*, 5, 165-385.

Belt, E. S., 1968. Carboniferous continental sedimentation, Atlantic Provinces, Canada, *Geol. Soc. Am. Spec. Pap. 106*, 127-176.

Coleman, J. M., and Gagliano, S. M., 1965. Sedi-

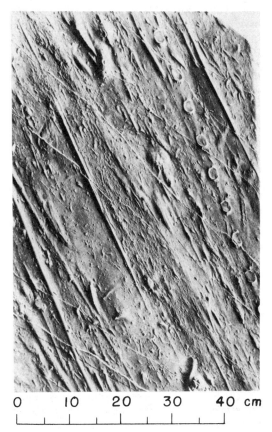

FIGURE 3. Counterpart of circular skip marks made by fish vertebra (top right), grooves, bounce marks (center and upper right), and roll marks of uncertain origin (center and lower left). Scour marks are absent. Current is from lower right to upper left (based on skip marks and bounce marks). Base of sandstone slab, Carpathians. (Dzulynski and Sanders, 1962.)

mentary structures: Mississippi River, deltaic plains, *Soc. Econ. Paleont. Mineral. Spec. Publ. 12*, 133-148.

Dzulynski, S., 1965. New data on experimental production of sedimentary structures, *J. Sed. Petrology*, 35, 196-212.

Dzulynski, S., and Sanders, J. E., 1962. Current marks on firm mud bottoms, *Conn. Acad. Arts Sci. Trans.*, 42, 57-96.

Dzulynski, S., and Walton, E. K., 1963. Experimental production of sole markings, *Edinburgh Geol. Soc. Trans.*, 19, 279-303.

Dzulynski, S., and Walton, E. K., 1965. *Sedimentary Features of Flysch and Greywackes.* Amsterdam: Elsevier, 300p.

Enos., P., 1969. Anatomy of a flysch, *J. Sed. Petrology*, 39, 680-723.

Friedman, G. M., and Sanders, J. E., 1974. Positive relief bedforms on modern tidal flats that resemble molds of flutes and grooves: implications for geo-

petal criteria and for origin and classification of bedforms, *J. Sed. Petrology*, 44, 181-189.

Shrock, R. R., 1948. *Sequence in Layered Rocks.* New York: McGraw-Hill, 507p.

Stanley, D. J., 1968. Graded bedding—sole marking—graywacke assemblage and related sedimentary structures in some Carboniferous flood deposits, Eastern Massachusetts, *Geol. Soc. Am. Spec. Pap. 106*, 211-239.

Twenhofel, W. H., 1939. *Principles of Sedimentation.* New York: McGraw-Hill, 610 p.

Cross-references: *Bed Forms in Alluvial Channels; Bouma Sequence; Casts and Molds; Paleocurrent Analysis; Scour Marks; Sedimentary Structures; Tool Marks; Top and Bottom Criteria; Turbidites; Turbidity-Current Sedimentation.*

CUT-AND-FILL—See SCOUR-AND-FILL

CYCLIC SEDIMENTATION

Cyclical phenomena in sedimentation are described in various terms: sedimentary cycles (cycles of sedimentation or simply cycles), rhythms, cyclothems, cyclical sequences (or simply sequences). The term *sequence* is useless because any succession of sediments, systematically deposited or not, may be called a sequence; the term *cyclical sequence* is simply a synonym for the term *cycle* or *rhythm* and should be waived in deference to the older, established terms. The term *cyclothem*, introduced in 1930 by Weller (see Wanless and Weller, 1932), has been used in a general sense but should be restricted to denoting the special cycles of the Carboniferous as developed in central North America and other similar sequences (Weller, 1958).

Rhythm and *cycle* may be considered synonymous, but the word *rhythm* is often used to describe the alternation of two different layers, e.g., dolomite and salt, or clay and sandstone, or clay and limestone. Although the term is not restricted in the literature, it is here recommended that *sedimentary cycle* (= cycle) denote a repeated sequence of at least three different units of sediment.

Each part of a cycle is called a *lithophase*. Each lithophase may be composed of several lithologies or fabrics, e.g., a rhythmic sequence of limestone and clay, or of sandstone with varying fabrics. Cycles may also be divided into *biophases*, which designate faunal and floral changes in the sediment. There is no set rule dictating where to place lithophase boundaries, but in all cases it is undesirable to divide a sequence into an excess of phases. It has been much debated where the boundaries of the cycles should be drawn. Let us assume, for example, that a cycle consists of the five phases *a, b, c, d,* and *e*. The sand content increases from *a* to *c* and decreases from *c* to *d;* and phase *e* consists of a coal seam, which is overlain by phase *a* of the next cycle. Phase *d* contains plant material and peat with rootlets, whereas phase *a* yields marine or nonmarine fossils. In this case the most important change of facies is between *e* and *a,* although significant changes occur between *b* and *c,* and between *d* and *e.*

In the cycle discussed above, the phase with the smallest grain size is the *low point,* and the phase with the largest grain size or with the highest sand content is the *high point.* The segment from the low point to the high point is the *ascending (progressive) branch (aufsteigender Ast, séquence négative, coarsening upward sequence),* from the high point to the next low point is the *descending (recessive) branch (absteigender Ast, séquence positive, fining upward sequence).* In a *symmetrical cycle,* the branches are equal or quasi-equal, either by equal ascendance and descendance of the grain size or by equal development of the phases of the cycle; otherwise the cycle is termed *asymmetrical.* If the phases of the descending part of the cycle are not developed, e.g., if the phase with the largest grain size or of the highest lime content is at the top of the cycle, it is termed a *Dachbank cycle (roofband cycle, cycle à banc majeur en tête).* If the phases of the ascending part of the cycle are missing, it is called a *Sohlbank cycle (bottomband cycle, cycle à banc majeur au pied);* at the base of a Sohlbank cycle an erosional disconformity is not uncommonly developed.

The nature of a cycle need not be consistent throughout a sedimentary basin, but may change from profile to profile. The ascending or descending branches may become reduced, or the number of lithophases present may change. Indeed, in nearly all sedimentary basins the good development of cycles is restricted to the area between the nearshore, where only the coarse-grained phases may be deposited, and the deepest region, where only clays may accumulate. Cyclic sedimentation may occur in deep basins, however, where the content of fine sand and silt varies, and the strata with the smallest sand content contain carbonate concretions, as in the Middle Lias and Lower Cretaceous of NW Germany. Although a sedimentary cycle within a geographical area is often incomplete, the significance of the missing parts may aid in paleoenvironmental studies. Many terms have been created to designate the varied development (completeness) of cycles, discussed at length by Duff et al. (1967). Recently, mathematical methods have also been introduced to describe the temporal and spatial variations of cycles (Merriam, 1972; Schwarzacher, 1975).

The thicknesses of individual phases and of the complete cycle are also important in the description and analysis of sedimentary cycles. First, it is necessary to distinguish between *minor* and *major* cycles *(Kleinzyklen* and *Grosszyklen),* not to be confused with micro-, macro-, and megacycles (see below). Minor cycles are in the range of mm to cm, rarely reaching meter proportions. Major cycles are predominantly in the range of meters to tens of meters, occasionally to hundreds of meters. The distinction between the two, however, is not dependent on the difference in thickness—indeed they may overlap in this respect—but rather on the differences in lithology and genesis. Minor cycles are related to short-term variations, e.g., season, climate, circulation; major cycles are related to major geologic events, e.g., orogenies, transgressions and regressions.

Minor Cycles

Varves. Perhaps the best-known minor sedimentary cycle is the *varve,* an annual rhythm, of which probably the most familiar examples are *glacial varves,* layers of alternating fine sand and clay deposited in a glacial lake (see *Varves and Varved Clays*).

Another simple type of minor cycle, annual in nature, is the alternation of quartz or calcite grains with dark bituminous material, one cycle usually < 1 mm thick, sometimes only measuring in microns. The boundary between the two phases can be sharp (2-phase rhythm) or transitional (3-phase cycle). This type of minor cycle is widespread in marine as well as fresh-water basins. Recent examples have been found, for example, in the Santa Barbara Basin off California, in the Zürich See (Switzerland), and in the Black Sea (q.v.). There are also many examples in ancient sediments, notably in oil shales (q.v.), e.g., the Eocene Green River Formation in the western US (see also *Sapropel*). The dark shales of the Upper Liassic Posidonien-Schiefer in western Europe also exhibit this alternation, as do the minute laminae in coal layers of varying rank (see *Coal*). It has been claimed that these alternations are seasonal in origin, especially because in several sequences each eleventh (or multiple of 11) layer has a noticeable anomaly, which has been attributed to the 11-year sunspot cycle.

Graded Beds. Some marine, and less often nonmarine, Sohlbank cycles in the range of cm to about 1 m in thickness are characterized by diminishing grain size, e.g., from sandstone (graywacke) to claystone, or from clastic limestone to claystone. The term *graded bedding* (q.v.) has been applied to this fining-upward cycle. It should be noted, however, that Bailey (1930) first introduced the term graded bedding to describe only the coarse member of this sequence; but the entire cycle does exhibit a diminishing grain size. Originally, this type of minor cycle was related to climatic fluctuations, ranging from one year (then comparable with varves) to 21,000 years! A tectonic origin was also suggested but was found to be unlikely. Kuenen and Migliorini (1950) proposed that these cycles were the products of turbidity currents and should, therefore, be named *turbidites* (q.v.). Turbidites are most often found in flysch facies, but they may also occur in shallow, epicontinental seas and even in fresh-water lakes. Thus turbidity currents and turbid suspensions may account for many graded-bed cycles (see *Graded Bedding*); these beds need not be cyclic, however.

Limestones. Cyclic limestone units with clastic or other components are worth noting, although only a few examples are known. In the Upper Visean and Lower Namurian of the Hainaut in SW Belgium are limestone and chert (lydite) beds roughly 5–20 cm thick, which are divided by partings (lamination) with diminishing thicknesses from bottom to top and/or with increasing clay content toward the top. The limestones may have different types of stratification; for instance, from bottom to top: parallel, wavy, flaser, and flame lamination; the top part is often massive (Fiege et al., 1970). Another type of calcareous minor cycle is found in the lower part of the Middle Triassic (Wellenkalk) in western Germany. Here phases of predominantly crystalline or dense platy or wavy limestones are intercalated with limestone layers averaging 4 cm in thickness, composed of a middle layer of crystalline limestone and smaller layers of dense limestone above and below. Numerous other examples of this type of cycle have been noted (personal observation; also see Schüller, 1967).

Evaporites. Minor cycles in evaporite sequences are very well developed. In the pelite phase of the major evaporite cycle (see below), minor cycles consist of minute, alternating layers of pelitic and carbonaceous sediments. The same is the case in the major carbonate phase, where the minor pelitic layers become much thinner; in the upper part of the carbonate phase, minor anhydrite layers are introduced. In the major anhydrite phase, alternation of carbonate and anhydrite makes up the minor cycle, the carbonate layers gradually becoming thinner until halite is instead produced, and so on (see *Evaporite Facies*). There seems to be agreement that these cycles are of climatic origin; it has been claimed that they are yearly, showing the 11-year solar period, but this opinion is not completely accepted.

Other Cycles. The *banded iron ores* also belong in the category of the minor cycles, the lower unit consisting of minute quartz grains or chert, the upper unit of iron oxides, iron carbonates, or iron silicates, with an average total thickness of 1 cm. They are almost all confined to the Precambrian and commonly are hundreds of meters thick. Their origin is still a matter of discussion (see *Cherty Iron Formation*).

Some of the above-mentioned cycles cannot be categorized with certainty as minor cycles or major cycles by their thicknesses and/or their fabrics. For example, the Wellenkalk facies consists, apart from the intercalated crystalline layers, of dense, grey, calcareous flagstone with wavy bedding-plane partings with or without very thin clay films. In medium thicknesses of 30 cm, clayey or marly intercalations occur with thicknesses of several mm, rarely of a few cm. Each such portion may be differentiated, e.g., a layer beginning with more marly limestones with more planar surfaces and followed by clearly wavy flagstone richer in carbonates may be again overlain by more platy and finely crystalline flagstone. The whole is covered by the above-mentioned crystalline limestones. The upper part of the flagstone is often cross-bedded, contorted, and convoluted. In such a case, the top crystalline layer is not developed. The following portion is again flat-bedded, indicating that a small diastem is present. The complete succession of the units suggests cycles of a higher order, in spite of the thinness. Limestone units with thicknesses of ±6 m of the Carboniferous Yoredale in England often have a sandy and argillaceous basal part; the middle part is purer, whereas the upper part again becomes increasingly argillaceous, sometimes grading into the overlying calcareous shales. Other limestones of the Yoredale are composed from bottom to top of a coral faunal phase to "normal" limestones with coral-brachiopod fauna or brachiopod fauna to algal fauna with chert beds (see references in Duff et al., 1967). Similar biocycles have been detected in the Lower Carboniferous of Belgium.

Major Cycles

In describing major cycles, it is essential to establish the thicknesses of the cycles, the various isochronous facies types, as well as the hierarchy of cycles of different thicknesses but of the same facies. A cycle of greater thickness may be made up of cycles of lesser thickness, these units again divided by cycles of still smaller thickness. These orders have been named variously: *cycles, megacycles, magnacycles.* For the sake of simplicity, it is best to speak only of cycles of first, second, and third order, etc., the first order being the smallest one. It is not unusual within a cyclic sequence for a cycle of a higher order to have the same thickness as a cycle of the lower order elsewhere in the sequence. Whether a cycle of higher order with the abnormally small thickness belongs to the higher or lower category can be judged by comparing the younger and the older cycles of the same profile and by comparing neighboring profiles of the same age.

Cycles of a certain order often are twice the thickness of the normal cycle; individual phases may also be abnormal in development or thickness. The comparison of this cycle with nearby contemporaneous cycles may show that the abnormally thick cycle is the equivalent of two cycles elsewhere, the result of reduction of one phase or several phases in the apparent single cycle.

Red Beds. If cyclic sedimentation is influenced by disturbing factors, irregular and badly developed cycles may be produced which may be difficult to recognize. This situation is more often the case under continental and coastal conditions than in marine basins, and cycles in continental basins have only recently been recognized. Such is the case of red-bed cycles of terrestrial, mixed terrestrial, and marine or brackish basins. These cycles are often of the fining-upward type (Allen, 1970) and consist of three phases: (1) mostly thick sandstone units, with subordinate clay components; (2) an alternation of sandy and clayey strata and mixed clay and sand; and (3) claystones, often with a brackish or lacustrine fauna, intercalated carbonate concretions, and thin sandstone layers. This cycle may be 1,2,3 (Sohlbank), or 3,2,1 (Dachbank) and may be hundreds of meters thick, but cycles of a minor order (5–20 m) are commonly observed, e.g., in the Buntsandstein of Germany (Wohlberg, 1961).

Shale and Graywacke. Cyclic shale and graywacke have been deposited since early Precambrian times, in nonmarine and especially in marine basins. Symmetrical and asymmetrical types are found, but the symmetrical and coarsening-upward types are more frequent than in red beds. The phases are granulometrically similar to the red-bed phases. Carbonate (Ca, Mg, Fe) concretions can occur in all phases, but large ones (> 1 m across) are developed only in phase 1. The clays may be normal or organic-rich (bituminous). Various faunal assemblages occur, often beginning with marine faunas of mixed types (i.e., benthic inhabitants with different requirements of oxygen, nektic forms which are the only fossils embedded in black shales) overlain by brackish faunas, and then nonmarine (lacustrine) faunas. The transition from marine to lacustrine elements can occur within a few decimeters of sediment, indicating that very soon after a marine incursion the sea retreated and a nonmarine regime ruled the rest of the cycle, so that often only 10% of a cycle is of marine origin. In such cases, the lithophases above the *a* phase very rarely yield fossils, because nonmarine organisms are seldom preserved as fossils. There can be exceptions, and even sediments of the *c* phase sometimes yield fossils. Plant remains and rootlets can be found in every phase, increasing upward.

The cycles vary in thickness up to >1000 m. The variation of an individual cycle can be >200 m—essential knowledge if one is comparing isochronous cycles or elaborating different orders. Indeed, several orders can be distinguished in deposits within these cycles. A good example is the Middle Siegenian of the Lower Devonian in Germany. Here deltaic sediments ranging from shales to graywackes, with very different fabrics, are arranged in cycles of 0.7, 4, 23, 73, 185, and 300 m. The last two are somewhat problematical. In contrast to other cycles, here the thicknesses do not overlap. In the Lower Namurian in NW Germany, the strata are made up of shale → shale + graywackes → graywacke cycles of the Dachbank type including minor shaly-sandy cycles. The major cycles can be arranged in three orders of (median values) 4, 10, and 50 m; the last one may be divisible into two equal parts (Fiege, 1937).

As stated previously, the well-developed cycles are restricted to the intermediate parts of the sedimentary basin. The cycles in question commonly are incomplete toward the edge of the basin, and the *a* phase, *b* phase, and, if existing, *d* phase may be suppressed, so that only the *c* phase—a sandstone unit—remains, sometimes with graded bedding. It is not necessary that the other phases disappear, but they may become more like the sand unit. The thickness of sands (beach facies) near the basin edge is often much higher than in the area of typical cyclic sedimentation, as noted long ago by Brinkmann (1929). The sandstone units are mostly divided by thin, clayey partings.

Well-developed minor cycles grade into

clayey sediments toward the center of the basin, and major cycles show a similar pattern. In the Lower Cretaceous of NW Germany, in the portions with the smallest sand content, carbonate concretions occur. The concretions in these layers can differ layer by layer in thickness and shape, or by the median between concretionary layers. The layers have a median thickness of 0.5 m, and are thus intermediate between minor and major cycles. The thicknesses of individual layers can systematically diminish from bottom to top or vice versa, and the thickness of the concretions within single layers may also diminish or enlarge respectively. These cycles have a median thickness of 2 m. Furthermore, there can exist an alternation of several Sohlbank and Dachbank cycles (as described above) of the smaller order, producing cycles of a third order, with a median thickness of 6 m, which can be followed for > 100 km distance. Rhythmic alternation of concretionary units can be seen every 10–15 m.

This type of cycle has also been divided on the basis of variations in foraminiferal species (Bettenstaedt, 1960). Brinkmann (1929) divided the Callovian Oxford Clay in SE England into cycles of only about 1 m thickness, but still belonging to this type of major cycle, as follows: greenish shale → bituminous brown shales → greenish shales → comminuted shells (coarsening upward) → pavement of lamellibranch and ammonite shells. In the nearshore region of the same basin, the cycles consist of: clay → fine sand → medium-grained sand → sandy iron oolite → iron oolite.

Cyclothems. The above discussed cycles of the clay → sand type are often completed by an *e* phase consisting of coal and equivalent layers. The coal seam may lie on the *c* phase if the *d* phase is missing. Usually, but not always, an underclay in the form of fireclay or a ganister is developed, and rootlets are often found in the whole *d* phase, as well as in the *c* phase. All other parameters, including thickness, are the same as in the previously described cycle. In a coal-bearing sequence, the 10-m cycles often contain an *e* phase. The 50-m cycles consist of a lower, predominantly clayey marine phase; the middle phase is mostly nonmarine sands, possibly divided by a thin transgressive marine unit, and coal seams are badly developed; the upper phase has mostly underdeveloped coal phases in nearly every 10 m unit. It is easy to see that normally the *a* → *e* cycles lay above the type without the *e* phase, but this transition is not sharp. On the contrary, if a number of isochronous cycles are compared, it is seen that the coal-forming conditions occur at one place earlier than at another place. These anachronous conditions also hold for greater stratigraphic differences; for instance, in Devonshire the Carboniferous succession without coal goes on up to the Lower Westphalian, whereas in the area of the Yoredale facies the coal seams appear as early as the lower third of the Visean. This type of cycle is found in areas with coal-bearing strata (see, e.g., Ramsbottom, 1969; Calver, 1969; and many references in Duff et al., 1967).

Major fining-upward cycles are developed predominantly in intermontane basins and alluvial plains, but in such situations may be masked by noncyclic factors. Coarse clastic material (sandstone, graywackes, arkoses, etc.), often conglomeratic, grades into a mixture of sands and clays (interbedding of sandstones and shales or similar features), then to clay, underclay, and finally, coal. This type of coal can be seen in the Innersudetic Basin (Poland and Czechoslovakia), in Karaganda Basin (Asiatic USSR), and elsewhere. In the North America Carboniferous, this type has been named the Piedmont, or Alluvial, type of cycle (= cyclothem). The Piedmont type usually begins with an unconformity beneath a conglomerate; fossils apart from plant remains are missing, and coals are seldom seen. The Alluvial type often begins with black shales lacking fossils; an ordinary cyclical sequence follows, up to sandstones, beneath which an unconformity can occur (see Wanless, 1969).

It is surprising that in very many instances, thicknesses of these cycles average ±10 m. In many regions, these 10-m cycles can be subdivided; for example, in Belgium and the Ruhr District the 10-m cycles of the Namurian and Westphalian are often split into 4-m cycles of the first order; third-order cycles of 50-m thickness also occur. A fourth order, with thicknesses up to several hundred meters, has a rhythmic character. The thicknesses of the cycles of the Upper Silesian Basin have been carefully computed and analyzed by Zeman (1964); there are cycles from 4–60 m, some being more frequent than others, producing a zigzag pattern. The thicknesses are greater in stable regions of the basin than in mobile areas, and also greater in times of epeirogenic unrest.

Coal-bearing cycles may also incorporate one or several limestone units, either by only a rhythmic interbedding, i.e., clay → limestone → clay, or by a cyclical pattern, i.e., clay → marl → limestone → marl → clay. This symmetrical cycle can be reduced to either a Dachbank or a Sohlbank cycle. The papers of Weller were the start of an extensive exploration of the Carboniferous cycles in North

America, which Weller called *cyclothems*. These cycles are described comprehensively by Branson (1962) and Wanless (1969). The typical succession of the cycles under discussion is in the Eastern Interior Coal Basin (Weller, 1931): black fissile shale → calcareous shale → limestone → shale with thin limestone layers in the lower part and ironstone units → sandstone (erosional base) → sandy and micaceous shale → freshwater limestone → underclay → coal. The phases beneath the sandstone (ascending branch) are marine to brackish, the part beginning with the sandstone (descending branch) are nonmarine. These cycles were deposited in a paralic environment; of course, there are variations due to areal as well as to stratigraphical conditions. To the east, the nonmarine environments prevail; for instance, in the Northern Appalachian Basin the cycles of the Monongahela Group consist of the following phases: brackish water limestone + shales → sandstone → freshwater limestone → underclay → coal. On the other hand, the marine influence dominates to the west; the thicknesses of the shales are diminished in favor of the limestone. In other profiles, sandstones and coals can be totally absent.

The thicknesses of these cycles vary from only 2 m to > 20 m; the smaller cycles generally belong to the Alluvial type, the thicker ones to the Piedmont (delta). In Kansas, five cycles make up one megacyclothem, whose middle cycle consists of marine limestones divided by only thin shaly layers. The cycles beneath and above have increasingly more shale, nonmarine sandstone, and coal.

Cycles of the type under discussion are well developed in parts of Europe and elsewhere, for example the cycles of the Yoredale facies in England and Scotland. The Yoredale facies is situated between nearshore shale-sandstone-coal facies and the shaly facies of the open sea. The limestones die out in both directions, the sandstones and coal toward the open sea. The shales and the coals also disappear toward the shore, so that only a coarse-grained sandstone remains. Corresponding facies occur in the central part of England and in the continental NW European Coal Belt (Ramsbottom, 1969); here limestones are not developed apart from locally and stratigraphically limited occurrences in southern Belgium and concretions in several marine horizons. Carboniferous cycles also occur in the Bug River district and in the Donetz Basin (see Duff et al., 1967).

Clastic-Calcareous Cycles. In the Tertiary Molasse of the northern Alps, cycles are developed consisting of conglomerates (*nagelfluh*) → sandstones → marls with several modifications. Thicknesses are on the average 10-20 m (Kraus, 1923). Similar but thicker cycles are developed in the Late Precambrian of the Urals: conglomerate → sandstone → arkose → shale → limestone, representing a geosynclinal facies (Bubnoff, 1951). In the oil-producing Devonian of the Ural-Volga area, the following cycles are developed: fine sandy shales → argillite → fine sands in sandy shales with pyrite → fine sandy shales with siderite concretions → thin-shelled fossils, among them *Lingula* → clayey limestones → (sometimes) limestones + dolomites with shallow-water fauna (Teodorovich, 1952). In the Devonian of the Tschat Kal Mts (Asia), cycles of the type limestone/dolomite → clastic carbonate sediments → red sandstones have been observed; the lowermost phase has an enrichment of Pb-Mn-ores (Lur'e, 1958).

Tertiary cycles in the Gulf of Mexico are a good example of clastic cycles with minimal calcareous material. Five cycles were deposited during the Eocene, being of exceptional duration and thickness (600-3900 m). Each cycle begins with a marine incursion, producing thin marls, limestones, glauconite sands and black shales. These sediments were deposited only in a comparatively small area; up- and downdip are found normal shales. The beginning of the regression of the sea is marked by silty and sandy shales, followed by nonmarine or brackish facies of interbedded sand and clay. The cycle is then repeated; the paleogeography is defined by foraminifera (Lowman, 1949). The Eocene of the English-Parisian-Belgian Basin also exhibits a five-fold alternation of marine and nonmarine sediments, generally with only a few intercalations of marly or limy sediments, with exception of the Parisian part of the basin (Stamp, 1921). The cyclical pattern of the entire basinal sequence is rather indistinct, understandable because this basin is tectonically heterogeneous, whereas the Gulf geosyncline is a homogeneous unit.

Clay-Marl-Lime Cycle. This cycle is one of the most widespread. The Lower Permian of Kansas can be divided into symmetrical cycles of ±15 m thickness. The shallowing cycle is: limestone with fusulinids and then with brachiopods and corals → shaly limestones and massive mudstones with mixed fauna → clayey shales, mudstones to bedded limestones with mollusks → gray shales and silts with *Lingula* → red shales, deposited above sea level. As sea level rises, the reverse sequence is deposited (Elias, 1937). The existence of megacycles is still a matter of dispute.

Cycles of the clay-marl-lime type occur in nearly every stratigraphic system. The epicontinental Silurian of the Isle of Gotland in the Baltic Sea is divided into three cycles (> 200 m maximum thickness), each beginning with marls, above which are different types of limestones (Hadding, 1959). Symmetrical and asymmetrical cycles, mostly a few tens of meters thick, occur in Devonian rocks above and on the flanks of swells within the Variscan Geosyncline of middle Europe; whereas in the troughs, clays of a much greater thickness have been deposited. In the Mississippian, shales, marls, and limestones often show a cyclical pattern in both North America and Europe.

Clay-marl-lime cycles, mostly of the Dachbank type, occur in the Lower and Middle Jurassic basin of NW Europe. Here, this type of cycle is known in several stratigraphical units. The uppermost limestones are often capped by hardgrounds (q.v.), usually overlain by thin layers of fossils, some of which are phosphatized, and partly rounded rock fragments from the underlying layers and attached organisms such as oysters, serpulids, and bryozoans. The relative thickness of the phases of the cycles vary, in general, from a few meters to >10 m; over rising topographic highs that may diminish to only a few decimeters. These cycles have been described by many authors, enumerated in Duff et al. (1967, p. 163ff.). They occur either near coastlines or upon and on the flanks of swells (as in the Variscan Geosyncline); whereas in the deeper parts of the basin, clays are deposited, which indeed can be also cyclically arranged.

If clay-marl-lime cycles indicate a diminishing of the depth of water, then sequences of marl → limestone or only limestone with cyclically varying fabrics, or limestone → dolomite must have been deposited in very shallow water. In all these cyclical sequences, various fossil communities play an important role. Sequences of this sort occur in many systems. In the lower Ordovician of Maryland cycles ±1.25 m thick with a maximum of eight phases of dolomite and limestone have been described by Sarin (1962). In the Mississippian and Pennsylvanian of North America, especially in the Rocky Mountains, cycles of this sort have been described by several authors (see Wanless, 1969). The limestones differ by their faunal content, degree of fragmentation, sorting of shell detritus, and depth and agitation of the water. In the European Lower Carboniferous in the facies of the *Kohlenkalk*, cycles of this kind have long been studied (see Duff et al., 1967, p. 158).

Also, the Upper Triassic so-called *Dachsteinkalk* of the eastern Alps has drawn the attention of several geologists. Zankl (1967) has described the conditions of sedimentation, repeating the cyclical sequences as given by A. G. Fischer: erosional discontinuity with a red to gray, sometimes green, clayey carbonate residual sediment → 10-50 cm dolomitic rhythmites → up to 5 m massive calcareous beds with megalodonts and other shallow-water organisms. In the Upper Jurassic of Swabia in SW Germany, clay (only occasionally) → marl → limestone cycles were described by Beurlen (1927).

The calcareous cycles discussed above are sometimes combined with glauconite (q.v.), especially in the Cretaceous. Much attention has been paid to the cyclical sedimentation of the Swiss Helvetian Cretaceous since 1910 (Buxtorf). The Upper Jurassic and Lower Cretaceous can be divided from the Kimmeridgian through the Albian into eleven cycles or rhythms of the Dachbank type (rarely with a descending branch) with thicknesses of 3 to >400 m, each with an increasing lime content, from shales or marls to thickly layered limestone beds and increasing size of the quartz and glauconite grains; several of the cycles end with a small layer of condensed sedimentation with many fossils, often phosphatized and highly glauconitic. Near the southern border of the Upper Cretaceous of Westphalia, four transgressions occurred, with glauconitic marls at the beginning of each. The quantity of the glauconite decreases in the course of the expanding sea, whereas the lime content increases, so that limestones with a small glauconite content result.

Ironstones. Much more widespread stratigraphically and areally are cycles with sedimentary ironstones (see *Ironstone*)—oolitic and conglomeratic. These sequences are mostly deposited near shores, and when the basins are of a complicated structure, the cycles often change patterns, or certain strata may even become acyclic. The cyclical nature of such deposits, therefore, has in some cases only recently been discerned. The following types can be distinguished: (1) Sohlbank type: conglomeratic iron ores → oolitic iron ores (partly sandy, partly limy or mixed) → shales; an example is the Lower Cretaceous iron ore of Salzgittern in Lower Saxony (Kolbe, 1962). (2) Dachbank type, with sandstones and limestones: carbonaceous shales with bone beds and rock fragments → marls → sandstones → iron ores; examples are the iron ores of the Upper Liassic and Lower Doggerian in Lorraine, NE France, and the cycles in the Liassic of Yorkshire, England. Similar also are the cycles in the Fig Tree Series of the Swaziland System in South Africa: coarse graywackes → shales with iron ores → siliceous shales → parallel-bedded graywackes (Gribnitz, 1964 and pers. comm.) and the cycles in the Precambrian Jatulian of the Baltic Shield: conglomeratic quartzites →

dolomites with *Carelozoon* → shales and sandstones with iron ores → shaly marls with shungite (a peculiar anthracite). (3) Cycles with predominantly limestone: coarser-grained clastics → finer-grained clastics → iron ooids in sandy marls and shales → dense and oolitic limestones; this facies occurs in the Corallian oolite in the eastern part of Westphalia. The cycles are limited to only a part of the whole iron facies of the area (Klüpfel, 1926; Thienhaus, 1957).

Evaporites. If a normal open saline basin with clay-marl-lime sedimentation becomes isolated, and if evaporation then exceeds the influx of water, such a basin becomes hypersaline. Subsequently, after the limestone or dolomite deposition, anhydrite, rock salt, and finally potash salts follow, according to solubility (see *Evaporites—Physicochemical Conditions of Origin*). Sometimes also a descending branch is developed whose phases, naturally, are the reverse of the phases of the ascending branch, and are usually thinner than the ascending branch and often incomplete. The whole sequence is not always deposited. In many basins, only halites or even only anhydrite precipitation is reached. It is understandable that the area of supersaturation will decrease as concentration of the brine increases. Thus, the potash salts will occupy a much smaller area than the halite, the halite less than the anhydrite, and so forth. From this follows that if potash salts are being deposited in the center of the basin, then halite will lie in a surrounding ring, and still nearer to the coast we find anhydrite, and then dolomite or calcium carbonate, and last perhaps clastics. In other words, the sequence of the phases of a cycle is not only vertically realized, but also horizontally—this is also true in cycles of other sedimentological types. The transition from one phase to the next is not sudden, but transitional (see *Evaporite Facies; Kara-Bogaz Gulf Evaporite Sedimentology; Persian Gulf Sedimentology*).

Causes of Cyclicity

The causes of cyclic sedimentation have been much debated, but no agreement has yet been reached. We will exclude here the "minor cycles," discussed already under that heading.

The thicknesses of the major cycles and their lateral extent are different; it is obvious, therefore, that to determine the controlling factor or factors of the cyclicity one must at first establish the median thickness and the extent of a group of cycles belonging together. One will discover that the smaller cycles have a smaller extent than the cycles of greater thickness. A special argument is not necessary to show, for example, that a cycle of the 4-m type and of limited extent (e.g., a few km) is caused by a limited efficaceous and external reason. Such may be found in a fluviatile or deltaic environment, as in the crevassing of levees and the overflow of a marshland or in the wandering of rivers. Some of the causes worked out by several authors, e.g, "plant control," "compaction control," and so on may effect cyclic sedimentation, but only locally. The different possible factors controlling cyclic sedimentation are enumerated and discussed by Robertson (1948), Weller (1956), and Duff et al. (1967, p. 34).

For cycles of higher order and of an extent on the order of 100 km or even more, one has to search for adequate factors. One controlling factor is the epeirogenic movement of the continents, both up and down. Sometimes it is said that one could not imagine great parts of a continent, e.g., the stable platform of North America (Kansas to Pennsylvania) during the Pennsylvanian, moving down and up, but one knowledgeable in epeirogenic tectonics knows that it is a possibility.

Another possible controlling factor inducing cycles of a higher order is eustatic change in sea level. This effect can be induced by melting and freezing of polar ice, but ice ages are limited events in earth history. In many cases, a tectonic factor must be added to effect these changes (Fairbridge, 1961). But we must ask, Are the tectonic movements causing the changes of sea level of such short duration and equality as the cycles are? The duration of the 10-m cycles of the Carboniferous is estimated at 50,000–150,000 yr. From a tectonic standpoint, these figures are rather short measures of time. Factors causing cyclic sedimentation have been efficaceous since early Precambrian times. Are we allowed to assume that the primary factor inducing the primary cycles has changed from time to time, even though our evidence indicates that the tectonic events are principally the same as far back as we can see them?

Many studies have been carried out in the field of cyclic sedimentation, but much more must still be done, and then the light on this field of sedimentology will be brighter.

KURT FIEGE

References

Allen, J. R. L., 1970. Studies in fluviatile sedimentation: A comparison of fining-upward cyclothems with special reference to coarse-member composition and interpretation, *J. Sed. Petrology*, **40**, 298–323.

Arbenz, P., 1919. Probleme der Sedimentation und ihre Beziehungen zur Gebirgsbildung in den Alpen,

Vierteljahresschriften nat. forsch. Ges. in Zurich, 64, 246-275.
Bailey, E. B., 1930. New light on sedimentation and tectonics, *Geol. Mag.,* 67, 77-92.
Bettenstaedt, F., 1960. Die stratigraphische Bedeutung phylogenetischer Reihen in der Mikropaläontologies, *Geol. Rundsch.,* 49, 51-69.
Beurlen, K., 1927. Stratigraphische Untersuchungen in Weissen Jura Schwabens, *Neues Jb. Min., Geol., Pal.,* 56(B), 78-124, 161-229.
Bouroz, A., 1960. La sédimentation des séries houillères dans leur contexte paléogéographique, *C. R. 4th Congr. Avan. Etudes Stratigr. Géol. Carbonifère,* Heerlen, 1958, 1, 65-78.
Branson, C. C., ed., 1962. *Pennsylvanian System in the United States.* Tulsa, Okla.: Am. Assoc. Petrol. Geologists, 505p.
Briggs, G., ed., 1974. Carboniferous of the southeastern United States, *Geol. Soc. Am. Sp. Pap. 148,* 361p.
Brinkmann, R., 1929. Statistisch-biostratigraphische Untersuchungen an mitteljurassischen Ammoniten über Artbegriff und Stammesentwicklung, *Abh. Ges. Wiss. Göttingen, Math. Physik Kl.,* 13, 1-249.
Bubnoff, S. von, 1951. Neue geologische Forschungen im Ural in ihrer grundsätzlichen Bedeutung, *Abh. Deutsch. Akad. Wiss. Berlin, Kl. Math. allgem. Nat. Wiss.,* 3, 18p.
Buxtorf, A., 1910. Erläuterungen zur geologischen Karte des Bürgenstocks, *Geol. Karte Schw., Erläuterungen,* Spezialkarte 27a, 48p.
Cailleux, A., 1969. Les remaniements cycliques de l'écorce terrestre, *Sci. Prog. Découverte,* 3416, 479-480.
Calver, M. A., 1969. Westphalian of Britain, *C. R. 6th Congr. Intern. Stratigr. Géol. Carbonifère,* 1, 233-254.
Duff, P. McL. D.; Hallam, A.; and Walton, E. K., 1967. *Cyclic Sedimentation.* New York: Elsevier, 280p.
Elias, M. K., 1937. Depth of deposition of the Big Blue (late Paleozoic) sediments in Kansas, *Geol. Soc. Am. Bull.,* 48, 403-432.
Fairbridge, R. W., 1961. Eustatic changes of sea level, in L. H. Ahrens, ed., *Physics and Chemistry of the Earth,* vol. 4. Oxford: Pergamon, 99-185.
Fiege, K., 1937. Untersuchungen über zyklische Sedimentation geosynklinaler und epikontinentaler Räume, *Abh. Preus. Geol. Landesamtes,* 177, 218p.
Fiege, K.; Scheere, J.; and van Tassel, P., 1970. Die stratinomische und petrologische Entwicklung des oberen Vise und untersten Namur im Kanal-Einschnitt des Mont des Grosseilliers bei Blaton, Hainaut (Hennegua) Belgien, *C. R. 6th Intern. Stratigr. Geol. Carbonifère,* 2, 755-770.
Füchtbauer, H., 1974. *Sediments and Sedimentary Rocks, I.* New York: Wiley, 464p.
Grebe, W. H., 1957. Zur Gliederung des oberen Kohlenkalkes (Visé-Stufe) bei Aachen, *Mitt. Geol. Staatsinst. Hamburg,* 26, 45-54.
Gribnitz, K. H., 1964. Notes on the Barberton Goldfield, in S. H. Haughton, ed., *Geology of Some Ore Deposits of Southern Africa,* vol. 2, Johannesburg: Geol. Soc. S. Africa, 77-90.
Hadding, A., 1959. Silurian algal limestones of Gotland, *Lunds Univ., Årsskr. N. F., Avd. 2,* 56(7), 26p.

Klüpfel, W., 1926. Beziehungen zwischen Tektonik, Sedimentation und Paläogeographie in der Weser-Erzformation des Ober-Oxford, *Zeitschr. Deutsch. Geol. Ges.,* 78, 178-192.
Kolbe, H., 1962. Die Eisenerzkolke im Neokom-Eisenerzgebiet Salzgitter, *Mitt. Geol. Staatsinst., Hamburg,* 31, 276-308.
Kraus, E., 1923. Sedimentationsrhythmus im Molassetrog des bayerischen Allgäu, *Abh. Danzig. naturf. Ges.,* 1, 1-25.
Kuenen, P. H., and Migliorini, C. I., 1950. Turbidity currents as a cause of graded bedding, *J. Geol.,* 58, 91-127.
Lowman, S. W., 1949. Sedimentary facies in Gulf Coast, *Bull. Am. Assoc. Petrol. Geologists,* 33, 1939-1997.
Lur'e, A. M., 1958. The association of the increased concentration of lead and manganese with cyclic sediment-accumulation in the Devonian deposits of the Chatkal Range (the Gava-Kassan Interfluve), *Akad. Nauk. SSSR Doklady,* 123, 145-147.
Mamet, B., 1972. Quelques aspects de l'analyse séquentielle, *Extrait du Mém. du B. R. G. M.,* 77, 663-677.
Merriam, D. F., ed., 1972. *Mathematical Models in Sedimentary Processes, an International Symposium.* New York: Plenum, 271p.
Ramsbottom, W. H. C., 1969. The Namurian of Britain, *C. R. 6th Congr. Intern. Stratigr. Géol. Carbonifère,* 1, 219-232.
Richter-Bernburg, G., ed., 1972. Geology of saline deposits, *UNESCO Earth Sci. Sess.,* 7, 316p.
Robertson, T., 1948. Rhythm in sedimentation and its interpretation with particular reference to the Carboniferous sequence, *Edinb. Geol. Soc. Trans.,* 14, 141-175.
Sarin, D. D., 1962. Cyclic sedimentation of primary dolomite and limestones, *J. Sed. Petrology,* 32, 451-471.
Schüller, M., 1967. Petrographie und Feinstratigraphie des unteren Muschelkalkes in Südiedersachsen und Nordhessen, *Sed. Geol.,* 1, 353-401.
Schwarzacher, W., 1975. *Sedimentation Models and Quantitative Stratigraphy.* Amsterdam: Elsevier, 396p.
Stamp, L. D., 1921. On cycles of sedimentation in the Eocene strata of the Anglo-Franco-Belgian Basin, *Geol. Mag.,* 58, 108-114, 146-157, 194-200.
Talbot, M. R., 1973. Major sedimentary cycles in the Corallian Beds (Oxfordian) of southern England, *Palaeogeogr., Palaeoclimat., Palaeoecol.,* 14, 293-317.
Teodorovich, G. I., 1952. Rhythm in prospective petroleum source rocks (with examples from the Ural-Volga region), *Akad. Nauk SSSR Doklady,* 86, 1025-1028.
Thienhaus, R., 1957. Zur Paläogeographie der Korallenoolitherze des Wesergebirges, *Zeitschr. Deutsch. Geol. Ges.,* 109, 49-62.
Trendall, A. F., 1972. Revolution in earth history, *J. Geol. Soc. Australia,* 19, 287-311.
Wanless, H. R., 1969. Marine and non-marine facies of the Upper Carboniferous of North America, *C. R. 6th Congr. Intern. Stratigr. Géol. Carbonifère,* 1, 293-336.
Wanless, H. R., and Weller, J. M., 1932. Correlation

and extent of Pennsylvanian cyclothems, *Geol. Soc. Am. Bull.,* **43,** 1003-1016.

Weller, J. M., 1931. The conception of cyclical sedimentation during the Pennsylvanian period, *Ill. St. Geol. Surv. Bull.,* **60,** 163-177.

Weller, J. M., 1956. Argument for diastrophic control of Late Paleozoic cyclothems, *Bull. Am. Assoc. Petroleum Geologists,* **40,** 17-50.

Weller, J. M., 1958. Cyclothems and larger sedimentary cycles of the Pennsylvanian, *J. Geol.,* **66,** 195-207.

Wolburg, J., 1961. Sedimentationzyklen und Stratigraphie des Buntsandsteins in NW-Deutschland, *Geotekt. Forsch.,* **14,** 7-74.

Zankl, H., 1967. Die Karbonatsedimente der Obertrias in den nördlichen Kalkalpen, *Geol. Rundschau,* **56,** 128-139.

Zeman, J., 1964. Die zyklische Sedimentation in einem Kohlenbecken als Spiegelbild seiner tektonischer Entwicklung, *C. R. 5th Congr. Intern. Stratigr. Géol. Carbonifère,* **2,** 873-884.

Cross-references: *Bouma Sequence; Cherty Iron Formation (BIF); Coal; Delta Sedimentation; Evaporite Facies; Fluvial Sediments; Graded Bedding; Ironstone; Lacustrine Sedimentation; Limestone; Oil Shale; Sedimentation—Tectonic Controls; Turbidites; Varves and Varved Clays.* Vol. I: *Sunspots and Sedimentation.* Vol. IVA: *Cycles, Geochemical.* Vol. IX: *Cyclic Sedimentation and Plate Tectonics; Cyclic Sedimentation: Markov Chain Analysis; Cyclothems.*

D

DEBRIS FLOW—See
MUDFLOW, DEBRIS-FLOW DEPOSITS

DEDOLOMITIZATION

Dolomitization (see *Dolomite, Dolomitization*) is essentially a phenomenon of diagenesis which proceeds, according to thermodynamic theory, as follows:

$$2CaCO_3 + Mg^{++} \to CaMg(CO_3)_2 + Ca^{++} \quad (1)$$

The reaction is sluggish in any system generally encountered at the earth's surface; but with a more or less continuous motion of the requisite dissolved component (Mg^{++}), such as is found in many ground-water systems, and unlimited time, the reaction will proceed to the right, provided that the activity ratio $\alpha Mg^{++}/\alpha Ca^{++}$ is >1 (Hsü, 1963, 1966; Hanshaw et al., 1971).

By the same thermodynamic argument, Abbott (1974) claims that reaction (1) is reversible when the activity ratio $\alpha Mg^{++}/\alpha Ca^{++}$ is <1, a condition that also exists in many ground waters. In this event, Mg^{++} is progressively leached out; and dolomite is replaced by calcite, a process commonly termed *dedolomitization*.

In both cases, there are three essential physical criteria: there must be an appropriate Mg/Ca activity ratio, a very long period of time, and a reasonable porosity and permeability in the formation to permit a free flow of ground water. The porosity may be primary, as in many coral reefs and related biohermal facies, and in fine-grained carbonates interbedded with poorly cemented sandstones or siltstones. It may be achieved or amplified by various secondary processes. For example, intrastratal solution may cause the development of collapse breccias, or brecciation by tectonic stresses may take place. In addition, during dolomitization within a fairly stable framework, there will normally be an increase of porosity by some 10–13%, a fact that has long attracted the notice of oil geologists.

The term *dedolomitization* was proposed by von Morlot (1847), who believed that groundwater leaching of preexisting evaporite deposits which contained gypsum ($CaSO_4 \cdot 2H_2O$) or anhydrite ($CaSO_4$) would be enriched with respect to calcium and sulfate, thus

$$CaMg(CO_3)_2 + CaSO_4 \cdot 2H_2O \text{ (sol.)}$$
$$\to 2CaCO_3 + MgSO_4 \text{ (sol.)} + 2H_2O \quad (2)$$

thereby achieving *calcitization* of a former dolomite. This reaction is in fact merely an example of reaction (1) reversed, but interesting inasmuch as it was written well over a century earlier.

It should be borne in mind that some evaporites, particularly those of "sulfate lakes" (typical of continental interiors, cf. *Karabogaz Gulf*), are rich in Mg^{++}, accumulating the various $MgSO_4$ series minerals. It has been pointed out (Fairbridge, 1957) that there is an alternative to von Morlot's model, thus

$$CaMg(CO_3)_2 + MgSO_4 \to CaSO_4$$
$$+ 2MgCO_3 \quad (3)$$

In this case, gypsum or anhydrite replaces dolomite, again conforming to the thermodynamics of Eq. (1).

It is evident from the above that there are varied physical conditions of porosity, and varied sources of the dedolomitizing fluids. In addition, there are various types of dolomite. Lippmann (1973, p. 166) has pointed out that no solution experiments appear to have been carried out on disordered, magnesium-deficient dolomites, and it might be reasonable to assume that they could be more soluble than ordinary dolomite, in the same way that magnesian calcites are more soluble than regular, ordered calcite. Indeed, Katz (1968) has established that, within the limits of his sampling, dedolomitization only seems to affect the calcian dolomites. Lippmann therefore suggests that the process of dedolomitization is actually "the preferential dissolution of calcian dolomites accompanied by the reprecipitation of calcium carbonate." The process would be analogous to the behavior of magnesian calcites, shown by the experiments of Jansen and Kitano (1963).

Field Observations

In the field, dedolomitization takes a number of forms. First, there is simple weathering along the crystal faces that produces a loose "dolomite sand." No replacement is involved here. Evidently the CO_2 pressure and the presence of weak acids generated in soil development, notably carbonic acid, would greatly assist this reaction (Murray, 1930; Sander, 1936).

Numbers of examples of calcitization associated with dedolomitization were studied and their fabrics described by Teodorovich (1945); and Tatarski (1949) analyzed their distribution (see review by Chilingar, 1956). In the Alps, Sander (1936) observed that calcitization of dolomites has been more important than the dolomitization of calcites; here anadiagenesis was favored and hydrothermal solutions were widespread. Hydrothermal dolomitization is probably rather unusual, however, and limited to special circumstances, e.g., a case described from the Dinarides (Ilich, 1974). The potential role of hydrothermal waters in dedolomitization seems to be equally restricted (e.g. to plate margins).

The simplest form of calcitic dedolomitization is well illustrated by an example from the Edwards Group (Cretaceous: Albian) in the Balcones Fault Zone aquifer of Texas. Abbott (1974) has shown that a large volume of high Ca:Mg-ratio ground water has passed through the artesian aquifer in the Edwards Group, causing the calcitization of many of the dolomite beds. These beds, formed on the San Marcos platform, were mainly tidal flat facies (Persian Gulf type) with thin-bedded dolomites alternating with evaporites. Most of the latter have been removed by solution, leading to the development of collapse breccias. The aquifer apparently originated in Miocene time, and the ground water commonly has a Ca:Mg ratio of 6:12. Thus the three criteria for dedolomitization are met: porosity and permeability, time, and high Ca:Mg ratio.

With reference to reaction (3), where $CaSO_4$ (anhydrite) is the replacing mineral, examples are found at every stage. This phenomenon was first reported by Stewart (1949) in a Permian dolomite overlain by Triassic evaporites; in this section, dolomite crystals were observed with fine-grained anhydrite "eating" into them. Calcite may equally well be found replacing anhydrite (Shearman and Fuller, 1969).

In the French Alps, there is an even more complex lithogenesis in Triassic rocks known as *cornieule*, represented in formations of from a few meters up to about 300 m in thickness. These are limestone rocks with a veined, brecciated, or honeycombed appearance that today are interpreted as dedolomitized. In the German Alps, these brecciated dolomites are called *Rauhwacke* (q.v.) or *Zellendolomit* (Brückner, 1941). As clearly demonstrated by Warrak (1974), they are restricted to an association with evaporites. Under tectonic stress or hydraulic fracturing (Masson, 1972) brecciation may occur in both gypsum and dolomites. Gypsum then may be replaced by calcite and subsequently dolomitized in an anadiagenetic phase. Finally, in an epidiagenetic phase, calcium sulfate solutions lead to dedolomitization (for an analysis of diagenetic phases, see *Diagenesis*).

Terminology

Smit and Swett (1969) complain that the term *dedolomitization* is ambiguous, and for this reaction, the term *calcitization* is appropriate. They submit that a process of this sort ought to be identified in terms of the replacing mineral and not of the mineral replaced. It should be recognized, however, that calcitization may apply to many processes; calcite can replace many other minerals. Furthermore, it is clear from the above discussion that dedolomitization embraces many processes, only one of which is calcitization.

The expression *dedolomite* as a rock term should be avoided (see comment by M. H. Battey, in Warrak, 1974) because interpretive terms should be kept to an adjectival role, thus "dedolomitized limestone." Dedolomitized limestone should be mapped as "limestone" because today it *is* a limestone.

RHODES W. FAIRBRIDGE

References

Abbott, P. L., 1974. Calcitization of Edwards Group dolomites in the Balcones Fault Zone aquifer, south-central Texas, *Geology*, 2(7), 359-362.

Brückner, W., 1941. Über die Entstehung der Rauhwacken und Zellendolomite, *Eclogae Geol. Helv.*, 34, 117-134.

Chilingar, G. V., 1956. Dedolomitization: A review, *Bull. Am. Assoc. Petrol. Geologists*, 40, 762-764.

Fairbridge, R. W., 1957. The dolomite question, in R. J. Le Blanc and J. G. Breeding, eds., Regional aspects of carbonate deposition, *Soc. Econ. Paleont. Mineral. Spec. Publ. 5*, 125-178.

Hanshaw, B. B.; Back, W.; and Deike, R. G., 1971. A geochemical hypothesis for dolomitization by ground water, *Econ. Geol.*, 66, 710-724.

Hsü, K. J., 1963. Solubility of dolomite and composition of Florida ground waters, *J. Hydrology*, 1, 288-310.

Hsü, K. J., 1966. Origin of dolomite in sedimentary sequences: A critical analysis, *Mineralium Deposita*, 2, 133-138.

Ilich, M., 1974. Hydrothermal-sedimentary dolomite: The missing link? *Bull. Am. Assoc. Petrol. Geologists*, 58, 1331-1347.

Jansen, J. F., and Kitano, Y., 1963. The resistance of

recent marine carbonate sediments to solution, *J. Oceanogr. Soc. Japan* 18(42)–(52), 208–219.
Katz, A., 1968. Calcian dolomites and dedolomitization, *Nature*, 217, 439–440.
Lippmann, F., 1973. *Sedimentary Carbonate Minerals.* New York: Springer-Verlag, 228p.
Masson, H., 1972. Sur l'origine de la Cornieule par fracturation hydraulique, *Eclogae Geol. Helv.*, 65, 27–41.
Morlot, A. von, 1847. Ueber die Dolomit und seine künstliche Darstellung aus Kalkstein, *Haidinger, Natl. Abhandl.*, 1, 305.
Murray, A. N., 1930. Limestone oil reservoirs of the northeastern United States and of Ontario, Canada, *Econ. Geol.*, 25, 452–469.
Sander, B., 1936. Beiträge zur Kenntniss der Anagerungsegefüge: (Rhythmische Kalke und Dolomite der Trias), *Mineral. Petrogr. Mitt.*, 48, 27–139. English Trans. 1951: Contribution to the study of depositional fabrics (Rhythmically deposited Triassic limestones and dolomites), *Amer. Assoc. Petrol. Geologists, Tulsa, Spec. Publ.*, 207p.
Shearman, D. J., and Fuller, J. G. C. M., 1969. Phenomena associated with calcitization of anhydrite rocks, Winnepegosis Formation, Middle Devonian of Saskatchewan, Canada, *Proc. Geol. Soc. Lond.*, 1658, 235–239.
Smit, D. E., and Swett, K., 1969. Devaluation of "dedolomitization," *J. Sed. Petrology*, 39, 379–380.
Stewart, F. H., 1949. The petrology of the evaporites of the Eskdale no. 2 boring, east Yorkshire, pt. 1, *Mineralog. Mag.*, 28, 621–675.
Tatarski, V. B., 1949. Distribution of rocks in which dolomite is replaced by calcite, *Doklady Acad. Sci. U.S.S.R., Moscow*, 69, 849–851.
Teodorovich, G. E., 1945. Principal types of chemogenous $CaCO_3$ in carbonate sedimentary rocks, *Doklady Akad. Nauk S.S.S.R.*, 49, 281–283.
Warrak, M., 1974. The petrography and origin of dedolomitized, veined or brecciated carbonate rock, the "cornieules" of the Frejus region, French Alps, *J. Geol. Soc. London*, 130(3), 229–247.

Cross-references: *Dolomite, Dolomitization; Duricrust; ˉRauhwacke; Solution Breccias.* Vol. IVA; *Calcium Carbonate: Geochemistry.*

DEEP-SEA SEDIMENTATION—
See PELAGIC SEDIMENTS, PELAGIC SEDIMENTATION

DELTAIC SEDIMENTS

Our concept of deltas stems from Herodotus' observation nearly 2500 years ago that the land area reclaimed from the sea by deposition of alluvium is deltoid in shape. A delta is a relatively localized and generally protrusive mass of stream-borne and partly stream-deposited sediment, in some measure visible as land but partly concealed beneath water, and laid down close to the mouth of a stream which is usually bifurcated into distributaries.

Deltas rank as complex and important constructional land forms (Guilcher, 1958), and ample evidence exists that deltaic sediments contribute substantially to the geological record. Today, deltas are numerous, divergent in form, and variable in size and situation. The largest, found at the mouths of great rivers that empty into seas or oceans, are hundreds of km across. The smallest occur where small streams debouch into lakes or larger but more sluggish rivers, and measure no more than a few tens or hundreds of m in width. Many deltas comprise smaller subdeltas. Deltas are encountered in almost all latitudes and climates, from the tropics to the arctic tundra (Shirley and Ragsdale, 1966).

Controlling Factors

An active delta records the regression of a water body; sediment is introduced at a point on the periphery of a water body faster than it can be removed from that vicinity by coastal agencies (waves, tides, circulatory currents). At every moment, a proportion of the space occupied by the water becomes replaced by new deltaic deposits, i.e., the delta progrades. Studies in the Quaternary reveal that the development and history of deltas is strongly influenced by fluctuations in the relative level of the land and the water body (see *Delta Sedimentation*). A relative rise will increase the volume of the water body, but the delta continues to prograde provided the rate of local replacement of the water body by deltaic sediment exceeds the rate at which the water body increases in volume. If sediment is added less rapidly than the volume increases, the delta is inevitably transgressed.

Bernard (1965) related the form of a delta, and in particular the nature of its coastline, to the ratio of the sediment supply to the energy of the water body as expressed by coastal processes. Listed in order of decreasing supply and increasing energy, the better known forms are (Fig. 1): the Gilbert-type (Baha California), birdfoot (Mississippi), lobate (Lena), cuspate (Nile), arcuate (Niger), and estuarine (Ganges).

The sediment size supplied to a delta must also influence its form, for grain size determines the rate of net sediment accumulation at any given energy level of a water body, other things being equal. Coarse loads accompany high net rates, fine loads, low net rates. In addition, the energy level controls the dispersal of sediment away from distributary mouths by determining the competence and capacity of the nearshore currents. At a given

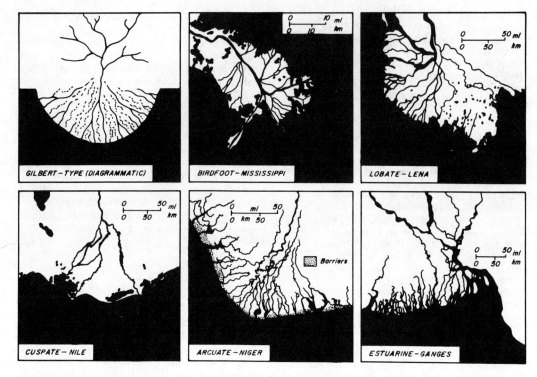

FIGURE 1. Types of deltas.

energy level, coarse sediments will be distributed more slowly and in smaller quantities than will fine sediments. An additional complicating factor, emphasized by Bates (1953), is the density contrast between the inflowing sediment-laden water and the water body. Thus, the relationship Bernard advanced is clearly not an oversimplification, but on the other hand is certainly affected by numerous other factors.

A Model Delta

A delta is a high-ranking sedimentary environment composed of numerous environments of subordinate rank. All deltas include subaqueous and subaerial realms of sedimentation. Stream action determines the character of the proximal part of the subaerial delta: the landscape is alluvial, showing channels, levees, and flood basins, often with swamps or marshes (see *Alluvium*). The delta's distal part is controlled by coastal processes: barrier beaches are not uncommon, and with these may go back-barrier lagoons, bays, swamps, tidal flats, or variable-salinity lakes, depending on other circumstances. When there is no well-defined coastal barrier, as in deltas of estuarine type, the distal part of the subaerial delta is a complex of islands, mudflats, tidal guts, and estuaries with river inflow. Through this complex, the delta gradually becomes submarine. It is in their subaqueous realms, affected by waves and, commonly, by tidal and circulatory currents, that deltas are most uniform, because the offshore environments generally occupy larger areas and are more stable than the environments of the subaerial delta. The depositional surface slopes gently away from the delta coast, and features of substantial relief (mouth bars, offshore bars, longitudinal shoals) occur only in the inshore shallows where conditions are most disturbed.

The facies geometry of deltas is closely related to delta type and is shown best by the distribution of their major sand bodies. These bodies have either a radial or concentric relationship to the head of the delta where the distributaries begin to bifurcate. Radially arranged sand bodies are produced in the stream channels that cross the subaerial delta. These bodies represent meander belts, for example in cuspate and arcuate deltas, or prograding mouth bars capped by levees and with channel fills, as in deltas of birdfoot type. Concentrically arranged sand bodies occur only where barrier beaches are conspicuous. Deltas of birdfoot and estuarine types usually lack significant barriers and, therefore, generally show only radial sand bodies. Lobate, cuspate, and arcuate deltas show concentric as well as radial elements.

It is impossible to represent the whole of deltaic sedimentation in a single model because of the great diversity and inherent complexity of deltas. The model shown in Fig. 2 is for cuspate-arcuate deltas, which occupy a middle position in Bernard's scheme. Many principles of deltaic sedimentation can be illustrated by reference to this model, which consists of three submodels: the subaerial delta divided into delta floodplain (proximal) and delta barrier (distal), and the subaqueous delta.

In the floodplain, the distributaries fashion meander belts beneath which point-bar gravels and sands with embedded, fine-grained channel fills are surmounted by fine-grained levee and crevasse-splay deposits. Cross-bedding (mainly large scale), plane lamination, and scour-and-fill typify the bar deposits; and drifted plant remains are common. The levee and splay deposits are characterized by small-scale cross-bedding, erosional surfaces, evidence of exposure, and the rapid interbedding of silts and clays with the finer grades of sand. Many of the plant remains found are autochthonous. The "sand" bodies represented by the meander belts are linear but arranged radially. Between them are thick, linear-radial masses of floodbasin silts and clays with swamp and marsh beds.

The delta barrier is dominated by sand ridges which define earlier positions of a prograded coast (see *Barrier Island Facies*). The ridges represent storm berms and offshore bars driven onto the beaches; they are capped by eolian dune sands blown up by onshore winds. The barrier sands are finer grained and better sorted than the point-bar deposits of the floodplain (see *Fluvial Sediments*). Landward of the barriers are shallow, low-energy lakes and lagoons of variable salinity, fringed locally by marshes and tidal mudflats, but elsewhere by low, wave-cut cliffs in the barrier sands. The back-barrier deposits are predominantly silts and clays with a fauna restricted in number of species. Thin lenses and sheets of sand occur only near the clifflines. Wave ripples, bioturbation, and desiccation features are common. As the delta progrades, the barrier and back-barrier deposits suffer partial removal by floodplain processes.

The environments of the subaqueous delta decrease in energy level from the strand out into deeper water. Wave and tidal currents are strongest on the beach itself, the shoreface, and the distributary mouth bars. The deposits here are well-sorted, relatively fine-grained sands with wave and current ripples, cross-bedding on large and small scales, plane lamination, scour-and-fill structures, and few organisms. The surface waters are highly turbid because

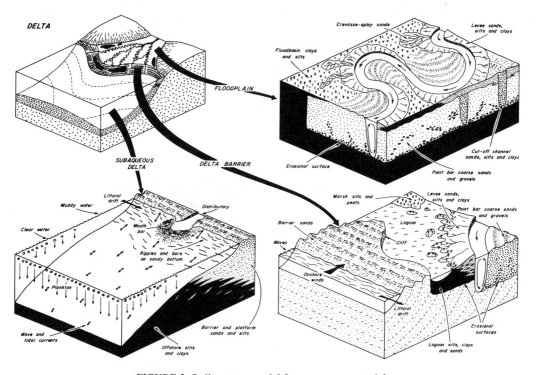

FIGURE 2. Sedimentary model for arcuate-cuspate deltas.

DELTAIC SEDIMENTS

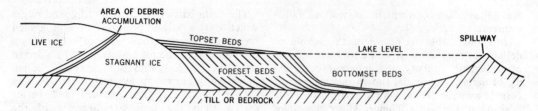

FIGURE 3. Diagrammatic profile through an ice-front/proglacial lake situation, showing development of typical "Gilbert delta sequence." The bottomset beds commonly interfinger with varved lacustrine facies. (Diagram: courtesy of Carl Koteff, 1974.)

the silt and clay are kept in suspension. With increase in depth and distance from shore, the energy level of the environments gradually declines, and the relatively coarse sands of the nearshore grade into finer sands and coarse but clean silts. Wave ripples and plane laminae are the characteristic structures; and benthonic faunas, limited in number of species, appear beneath the somewhat muddy waters. In still deeper water, the currents are insufficiently strong to prevent the intermittent deposition of silty clays and clayey silts, which become interbedded with the coarser detritus spread from the shoreface. Well off shore, where the water is clear and bottom currents weak, silty clay and clayey silt only are laid down. Bioturbation is common, and the sediments include planktonic faunas rained down from above and an abundant and diverse benthos.

The three components of the deltaic model discussed above depend on delta progradation and regression of the water body. If transgression were to commence, the subaerial delta would become covered erosively by a thin but coarse, sheet-like deposit of littoral origin, which would grade up into progressively finer-grained sediments, indicative of steadily deepening water. Peats may also be deposited.

Examples of Deltas

Gilbert-type deltas of Pleistocene Lake Bonneville. The type of delta first described from Lake Bonneville, Utah, and now associated with G. K. Gilbert's name, is the only one for which the terms *topset, foreset,* and *bottomset* have any physical significance (Gilbert, 1885, 1890; Fig. 3). Essentially, these deltas are alluvial fans prograded into a deep, low-energy water body. The subaerial parts, where the topsets form, are semicircular in plan, steep in slope (5-10 m/km), and not more than several km in radius. They arise by the repeated crevassing of streams, which deposit coarse gravels and sands. The subaqueous parts slope even more steeply, (10-25°), and the deposits here are sands and gravels laid down partly by swift, dense currents and partly by gravity-controlled avalanches. Commonly, the height of these delta foresets is many m (Fig. 4). Down slope, the foreset layers become less steep and grade into bottomset fine sands and silts, which settle onto the lake floor. Similar deltas have been studied experimentally.

Mississippi Delta. The active and inactive Holocene deltas of the Mississippi River dominate the Louisiana coast (Kolb and Van Lopik, 1958; see Fig. 1). The Balize, or birdfoot, delta is the present locus of active deposition; it comprises a deep-water delta and at least six shoal-water subdeltas (Fisk et al., 1954); Scruton, 1955; Shepard, 1956; Coleman and Gagliano, 1964). The deep-water birdfoot delta comprises four very nearly straight distributaries which empty onto steep submarine slopes in deep but low-energy waters near the shelf edge. These distributaries, flanked by levees grading out into marshes, are terminated by large mouth bars, with deep-lying toes, that have prograded rapidly. Beneath the distributaries lie extensive bar-finger sands up to 90 m thick and 7 km across. The bar fingers are flanked by sandy silts, clayey silts, and silty clays formed in the shallow, open bays that lie between the distributaries (see *Prodelta Sediments*). A sheet-like mass of marine clay de-

FIGURE 4. Foreset beds of a proglacial delta, Pennsylvania. (Photo: R. W. Fairbridge.)

posited in deep offshore waters lies beneath the bar-finger and bay sediments.

The subdeltas are also of birdfoot type but formed in water less than about 6 m deep. They are underlain by thin (3-12 m), extensive lenses of sediment which accumulated rapidly from prograding channels originating as crevasses in the natural levees of the major channel. The oldest of the subdeltas formed prior to 1500 A.D.; the other four date from 1839, 1860, 1874, and 1891. Closure of the parent crevasse results in rapid transgression as subsidence takes place. A vertical section through a typical subdelta shows a marine clay at the base succeeded by prodelta clays and then channel, levee and marsh sands, silts, and peats. The transgressive deposits include thin beach sands and oyster reefs.

Niger Delta. The Niger delta lies on the tropical W coast of Africa, is arcuate in form, and almost perfect in symmetry (Allen, 1965, 1970; see Fig. 1). The sediment influx is small, and the Gulf of Guinea, into which the delta is being built, is subject to moderate tides and powerful waves. A long chain of barrier islands, segmented by tidal passes which are closed off by mouth bars, defines the coastline. Powerful, wave-generated littoral currents sweep the coast in directions away from the tip of the delta. Landward of the barriers is a broad tidal swamp thickly populated by mangrove trees and fed by a net-like system of channels. Behind this again lies a forested delta floodplain crossed by meandering distributaries. The subaqueous part of the delta is broad and gently sloping. The shoreface descends to a wide delta-front platform, surmounted by the mouth bars of the tidal passes, with an outer break of slope in depths of 10-20 m. Beyond lie the steeper slopes of the prodelta and open-shelf environments.

The Niger delta is a sandy one, in contrast to that of the Mississippi. Its deposits become finer grained from the floodplain to the open shelf and, to a lesser degree, from the axis of symmetry of the delta toward the extremities of the delta flanks. In the floodplain are found channel sands, levee sands and silts, and floodbasin silty clays with swamp debris. The tidal swamps are characterized by organic-rich clayey silts with only thin beds of sand. Well-sorted fine sands with plane lamination typify the barrier sediments. Generally the mouth bar and platform deposits show finer-grained, clean sands and silts. Clean sand and silt interbedded with silty clay and clayey silt dominate the prodelta slope. The typical deposits of the open shelf are clayey silts and silty clays devoid of laminae but with a rich planktonic fauna.

These deposits have prograded across a thin sheet of sand, parts of which remain unburied today, laid down on the continental shelf at the advancing strand of the late Pleistocene sea as it rose from its Weichselian lowstand. Subsequently, the Niger gave rise to only one delta, whereas the Mississippi created several, none of which was active for longer than about 1000 yr.

J. R. L. ALLEN

References

Allen, J. R. L., 1965. Late Quaternary Niger delta and adjacent areas; sedimentary environments and lithofacies, *Bull. Am. Assoc. Petrol. Geologists,* **49,** 547-600.

Allen, J. R. L., 1970. Sediments of the modern Niger delta; A summary and review, *Soc. Econ. Paleont. Mineral. Spec. Publ. 15,* 138-151.

Bates, C. C., 1953. Rational theory of delta formation, *Bull. Am. Assoc. Petrol. Geologists,* **37,** 2119-2162.

Bernard, H. A., 1965. A resume of river delta types (abstr.), *Bull. Am. Assoc. Petrol. Geologists,* **49,** 334-335.

Broussard, M. L., ed., 1975. *Deltas—Models for Exploration.* Texas: Houston Geol. Soc., 555p.

Coleman, J. M., and Gagliano, S. M., 1964. Cyclic sedimentation in the Mississippi River deltaic plain, *Trans. Gulf Coast Assoc. Geol. Soc.,* **14,** 67-80.

Fisk, H. N., 1961. Bar-finger sands of Mississippi delta, in J. A. Peterson, and J. C. Osmond, eds., *Geometry of Sandstone Bodies.* Tulsa, Okla.: Am. Assoc. Petrol. Geologists, 29-52.

Fisk, H. N.; McFarlan, E.; Kolb, C. R.; and Wilbert, L. J., 1954. Sedimentary framework of the modern Mississippi delta, *J. Sed. Petrology,* **24,** 76-99.

Gilbert, G. K., 1885. The topographic features of lake shores, *U.S. Geol. Surv., 5th Ann. Rept.,* 69-123.

Gilbert, G. K., 1890. Lake Bonneville, *U.S. Geol. Surv. Monogr. 1,* 438p.

Guilcher, A., 1958. *Coastal and Submarine Morphology.* London: Methuen, 274p.

Kolb, C. R. and Van Lopik, J. R., 1958. *Geology of the Mississippi River Deltaic Plain Southeastern Louisiana.* Vicksburg: Mississippi River Commission, 120p.

Koteff, C., 1974. The morphologic sequences concept and deglaciation of southern New England, in D. R. Coates, ed., *Glacial Geomorphology,* Binghamton: SUNY, Publ. Geomorph., 121-144.

LeBlanc, R. J., ed., 1976. Modern deltas, *Am. Assoc. Petrol. Geologists Reprint Ser.* **18,** 205p.

LeBlanc, R. J., ed., 1976. Ancient deltas, *Am. Assoc. Petrol. Geol. Reprint Ser.* **19,** 226p.

Morgan, J. P., and Shaver, R. H., eds., 1970. Deltaic sedimentation: modern and ancient, *Soc. Econ. Paleontol. Mineral., Spec. Publ. 15,* 312p.

Oomkens, E., 1967. Depositional sequences and sand distribution in a deltaic complex, *Geol. Mijnbouw,* **46,** 265-278.

Scruton, P. C., 1955. Sediments of the eastern Mississippi delta, *Soc. Econ. Paleont. Mineral. Spec. Publ. 3,* 21-50.

Scruton, P. C., 1960. Delta building and the deltaic sequence, in F. P. Shepard et al., eds., *Recent Sediments, Northwest Gulf of Mexico.* Tulsa, Okla.: Am. Assoc. Petrol. Geologists, 82–102.

Shepard, F. P., 1956. Marginal sediments of Mississippi delta, *Bull. Am. Assoc. Petrol. Geologists,* **40,** 2537–2623.

Shirley, M. L., and Ragsdale, J. A., 1966. *Deltas in Their Geologic Framework.* Houston, Texas: Houston Geol. Soc., 251p.

Sutton, R. G., and Ramsayer, G. R., 1975. Association of lithologies and sedimentary structures in marine deltaic paleoenvironments, *J. Sed. Petrology,* **45,** 799–807.

Cross-references: *Alluvium; Barrier Island Facies; Delta Sedimentation; Estuarine Sedimentation; Fluvial Sediments.*

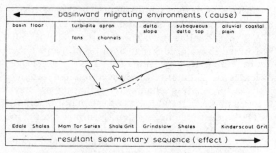

FIGURE 1. Cross-section illustrating the relationship between facies and environments in the Carboniferous rocks of the Central Pennine Trough (Selley, 1970). In Rich's (1951) terminology, the first two environments (shallow to deep) would constitute the *undathem,* the second two the *clinothem,* and the deepest the *fondothem.*

DELTA SEDIMENTATION

Delta sedimentation may take place in either lacustrine or marine basins, being characterized by the more or less abrupt deceleration or termination of linear fluvial flow upon reaching a deeper and less turbulent body of water; even in marine delta environments most of the sediment is deposited under fluvial or paludal (swamp) conditions or under varying degrees of brackishness. A vast literature on deltas is available, and a useful bibliography has been compiled by Smith and Broussard (1971).

Geometry and Dynamics

The third dimension of deltaic accumulation could not be appreciated by early investigators, and it was not until Gilbert (1890) studied the late Pleistocene/early Holocene lacustrine deltas of Glacial Lake Bonneville that the modern concept was developed. He recognized that the delta had a prograding history, the stream mouth advancing over its own deposits into a standing body of water, developing in this way *bottomset, foreset,* and *topset* structures. The same structural relationship was recognized by Rich (1951) when he characterized the megafacies of the continental margin in general into *undathem* (neritic), *clinothem* (slopes), and *fondothem* (bottom) *facies* (See Fig. 1). The corresponding structural terms were *undaform, clinoform,* and *fondoform.* These terms, however, have achieved very limited acceptance, although the concept was well founded. The Gilbert model can be easily reproduced in tank experiments and is repeatedly illustrated in the sand and gravel deposits that mark the retreat stages of the late Pleistocene ice sheets in North America, Europe, and elsewhere. These fluvioglacial (glaciolacustrine) deltas were generally small, single cycle, and of classic simplicity.

The complex marine deltas of the world's great rivers are more difficult to study. Visible only in two dimensions, they require drilling and geophysical surveys to disclose their varied geometry. They are all polycyclic, inasmuch as the delta locus and elevation have migrated with the Quaternary oscillations of sea level. The delta environment is subject to further modification due to other eustatic and local sea-level changes, the latter due to compaction or crustal movements; regional shifts (e.g., glacial/interglacial) in wind-wave regimes and ocean currents; variations in littoral vegetation; regional changes in tidal characteristics; and variations in fluvial discharge and sediment load, which are dependent on climate and vegetation variations in the drainage basin. Of all these variables, the two overriding ones are sea-level oscillations and climate-related variations in discharge and load. In the Quaternary, both of these factors have been more or less contemporaneous; they are world-wide but affect certain latitudinal belts more than others.

Regional factors such as tidal range, exposure to prevailing winds, local climate, and vegetation vary on a smaller scale and influence the detailed morphology of deltas; Table 1 summarizes these aspects. Differences in the plan shape of the delta are related mainly to wave energy, currents, and tides. *Lunate* forms are found on high-energy coasts; *cuspate* forms indicate unidirectional energy; *lobate* and *birdfoot* deltas are low-energy forms. Models must relate rate of sediment discharge to slope (and width) of shelf versus combined wave-current-tide energy. The theory has been discussed by Bates (1953), Berthois (1960)

TABLE 1. Factors Influencing Deltaic Sedimentation

RIVER REGIME (Variations influence sediment load and transport capacity)	Flood stage	Sediment load	Quantity of suspended load and bed load (that is, stream capacity) increases during flood
		Particle size	Particle size of suspended load and bed load (that is, stream competence) increases during flood
	Low river stage	Sediment load	Stream capacity diminishes during low river stage
		Particle size	Stream competence diminishes during low river stage
COASTAL PROCESSES	Wave Energy		High wave energy with resulting turbulence and currents erode, rework, and winnow deltaic sediments
	Tidal range		High tidal range distributes wave energy across an extended littoral zone and creates tidal currents
	Current strength		Strong littoral currents, generated by waves and tides, transport sediment alongshore, offshore, and inshore
STRUCTURAL BEHAVIOR (With respect to sea level datum)	Stable area		Rigid basement precludes delta subsidence and forces deltaic plain to build upward as it progrades
	Subsiding area		Subsidence through structural downwarping coupled with sediment compaction allows delta to construct overlapping sedimentary lobes as it progrades
	Elevating area		Uplift of land (or lowering of sea level) causes river distributaries to cut downward and rework their sedimentary deposits
CLIMATIC FACTORS	Wet area	Hot or warm	High temperature and humidity yield dense vegetative cover, which aids in trapping sediment transported by fluvial or tidal currents
		Cool or cold	Seasonal character of vegetative growth is less effective in sediment trapping; cool winter temperature allows seasonal accumulation of plant debris to form delta plain peats
	Dry area	Hot or warm	Sparse vegetative cover plays minor role in sediment trapping and allows significant aeolian processes in deltaic plain
		Cool or cold	Sparse vegetative cover plays minor role in sediment trappings; winter ice interrupts fluvial processes; seasonal thaws and aeolian processes influence sediment transportation and deposition

From Morgan, 1970.

and Keulegan (1966). For example, the birdfoot and cuspate deltas of the Java Sea reflect optimal growth conditions; the Bodri progrades at up to 200 m/yr, and the Bengawan Solo progrades at up to 100 m/yr (see Vol. I: *Java Sea*). The discharge is very high, the shelf is wide and shallow, and the oceanographic energy factors are minimal. Prevalence of mangrove on these tropical coasts assists rapid "fixing" of the new sediment. The Ganges-Brahmaputra complex, in contrast and despite high wave-tide energy, has an immense discharge and progrades at 10–40 m/yr onto the sloping Bengal Fan, loosing sediments into the Ganges Canyon. The digitate delta is open to the full effects of the Indian monsoon combined with large tidal range (Morgan, 1970).

Modern Delta History

Sedimentational histories of modern deltas, apart from local dynamic factors, are governed by two basic universal eustatic trends: the "Flandrian" rise, with minor oscillations, (16,000–6,000 B.P.) from about −130 m to +3 m (Fairbridge, 1961) and the late Holocene stable phase (6000 B.P. to present) with minor oscillations, whose amplitude is controversial, but appears to be in the range 1–3 m above and below present MSL (the arguments pro and con are summarized by Maxwell, 1973).

The first trend resulted in a displacement of delta loci and facies from the outer shelf margin to the inner shelf, and a backing up and filling of the main discharge channel. As a result, a modern delta facing a wide shelf, like that of the Rhine (Van Straaten, 1960) or the Po (Nelson, 1970) is so far from its late Pleistocene counterpart that its recent history was only initiated about 8000 B.P. Even in a delta discharging nearly directly into a deep basin, like the Mississippi or the Rhone (Oomkens, 1970), one finds little trace of glacial-stage deposition because sediments were funneled through a eustatically lowered channel and a submarine canyon to be dispersed onto a deep-sea fan, sometimes known as a *submarine delta* (see *Submarine Fan (Cone) Sedimentation*). Thus the first stage of any Holocene delta is marked by an alternation between dominant *trangressive facies* and short-lived *regressive* (prograding) *facies*. Marine clays may be deposited on intertidal peat or sands during a transgression, whereas regressions favor broad peat growth behind prograding sand barriers.

About 6000 B.P., the rapid Flandrian eustatic rise stopped, and there ensued a stable interval with small cyclic (about 500-1500 yr) oscillations which have been related to paleoclimate patterns. Statistical analyses of C^{14} dates of recent littoral sediments (peats, shells, archeological debris) show very distinctive histograms with peaks corresponding to transgressive (deepening) phases and troughs to regressive phases (Geyh, 1971). Regression on a coastal deltaic marsh is often indicated by oxidation of peat and the spread of human habitation sites. Progressive delta subsidence may obscure the records of minor eustatic oscillations; gaps in the sequence and numerous cut-and-fill structures may record the change of loci of sub-deltas from oscillation to oscillation.

Ancient Delta Sedimentation

Similar patterns may be expected in ancient delta records. One must look for thick clastic sections (the Mississippi has accumulated 2000 m in 2 m.y.) which may contain numerous coarsening-upward sequences passing from marine shales up to fluvial clastics, and possibly coal. These cycles need not be related to glacio-eustasy, although this cause can be properly invoked for Permo-Carboniferous (Wanless and Shepard, 1936) and late Ordovician deltas. Sea-floor spreading and pole shifts (causing rapid geoidal sea-level adjustments) may also cause episodic changes on a scale of 10^3-10^5 yr. In the Permo-Carboniferous deposits, each cyclic unit is termed a *cyclothem* (see Moore, 1959), and this term has been applied to similar sequences (see *Cyclic Sedimentation*). These deposits also include *shoestring sands*—linear sand bodies such as filled channels, which are essentially radial in plan—and barrier beach deposits, which are found in peripheral belts.

Ancient deltas are to be found in rocks of all ages, but they are perhaps most prevalent from the Permo-Carboniferous. It can scarcely be purely coincidental that this was the last great globally effective glaciation before the Quaternary. The most widespread coal deposits in the Northern Hemisphere are Carboniferous (both Mississippian and Pennsylvanian), but in the Southern Hemisphere they are mainly Permian and correspond to the major interglacial megacycles.

The late Paleozoic cycles of the central US have been very closely studied, partly in connection with coal geology (Wanless et al., 1969, 1970; Ferm, 1970; Wermund and Jenkins, 1970). Paleogeographically these Paleozoic deltas differ very notably from modern ones, due to the existence of wide epicontinental seas at the time. Deltas in these broad, shallow seas were often 1000 km or more from the abyssal oceanic depressions; thus, cyclic deltaic events are traceable over immense horizontal distances, but vertical sequences are limited.

A structurally more mobile deltaic setting is found in the Carboniferous of NW Europe. Although the late Paleozoic (Hercynian) orogenic revolution was essentially restricted to the eastern and southern (Appalachian/Oua-

TABLE 2. Deltaic Facies of the Yoredale Series, Which Was Built Out into Open Marine Carbonate Shelf Facies

Facies	Environment	
Coals and rootlet beds	Swamp	
Cross-bedded sands	Alluvial distributary channels	
Rippled sands	Sub-aqueous delta platform	Deltaic
Laminated silts	Prodelta	
Limestones	Marine shelf	

From Selley, 1970.

chita) sector in the US, the axial belt of the same orogeny crossed Ireland-southern England-France-Germany so that their deltaic sequences are often found debouching into active geosynclinal environments, and are thus more closely related to flysch and molasse associations (see *Geosynclinal Sedimentation*). Nevertheless cyclothems are present, e.g., in Britain (see Selley, 1970, p. 78). Typical facies in the Yoredale Series of northern England are interpreted in Table 2 (see also Fig. 2). The cycles in the more tectonically active region of the SW of England have been discussed by De Raaf et al. (1964). Comparable studies may be found in the literature from other European areas (see Gignoux, 1955).

Perhaps the most celebrated of all deltaic sequences is the Catskill Group of the Upper Devonian, first studied in detail by Barrell (1912-1914). Actually it was not one but a series of deltas (see, e.g., Pepper et al., 1954) that developed as a result of the Acadian (mid-Devonian) orogeny; these deposits are typical *molasse facies*. A useful study of the Sonyea Group, which outcrops in New York State, discloses typical open-shelf sediments, prodelta silts and clays, delta-front sands, distributary mouth bars, channel deposits, and estuarine and marsh sediments. Each has its distinctive fossils and paleoecologic setting (Sutton et al., 1970).

Epicontinental cyclic sedimentation combined with deltaic accumulation during another major ice-age period, the Ordovician, is widely recognized in central North America, and delta forms along the Appalachian geosynclinal margin have been discerned by Horowitz (1966).

Economic Aspects

Although not all coals are deltaic, it is the cyclic/deltaic sequence that is the hallmark of the world's major coal deposits. Delta and prodelta facies, with their rapid accumulation and entrapment of fine organic particles, are also ideal sites for petroleum hydrocarbon distillation. The porous shoestring sands are excellent potential reservoirs, and other isolated coarse-grained bodies in deltaic sediments often serve as oil traps (Visher et al., 1971).

RHODES W. FAIRBRIDGE

References

Barrell, J., 1912. Criteria for recognition of ancient delta deposits, *Geol. Soc. Am. Bull.*, **23**, 377-446.

Barrell, J., 1913-1914. The Upper Devonian delta of the Appalachian geosyncline, *Am. J. Sci.*, **36**, 429-472; **37**, 225-253.

Bates, C. C., 1953. Rational theory of delta formation, *Bull. Am. Assoc. Petrol. Geologists*, **37**, 2119-2162.

Berthois, L., 1960. Etude dynamique de la sédimentation dans la Loire, *Cahiers Océanographiques*, **12**, 631-657.

Coleman, J. M., and Wright, L. D., 1975. Modern river deltas; variability of processes and sand bodies, in M. L. Broussard, ed., *Deltas: Models for Exploration*. Houston: Houston Geol. Soc., 99-149.

Delfaud, J., 1974. La sédimentation deltaique ancienne–examples nord-Sahariens, *Bull. Centre Rech. de Pau*, **8**, 241-262.

De Raaf, J. F. M.; Reading, H. G.; and Walker, R. G., 1964. Cyclic sedimentation in the Lower Westphalian of north Devon, *Sedimentology*, **4**, 1-52.

Fairbridge, R. W., 1961. Eustatic changes of sea level, *Physics and Chemistry of the Earth*, vol. 4, New York: Pergamon Press, 99-185.

Ferm, J. C., 1970. Allegheny deltaic deposits, in Morgan and Shaver, 1970, 246-255.

Geyh, M. A., 1971. Versuch einer chronologischen Gliederung des marinen Holozäns an der Nordseekuste, *Z. Deutsch. Geol. Ges.*, **116**, 351-360.

Gignoux, M., 1955. *Stratigraphic Geology*. San Francisco: Freeman, 682p.

Gilbert, G. K. 1890. Lake Bonneville, *U.S. Geol. Survey Mon. 1*, 438p.

Horowitz, D. H., 1966. Evidence for deltaic origin of an Upper Ordovician sequence in the Central Appalachians, in Shirley and Ragsdale, 1966, 159-169.

Keulegan, G. H., 1966. The mechanism of an arrested saline wedge, in A. T. Ippen, ed., *Estuary and*

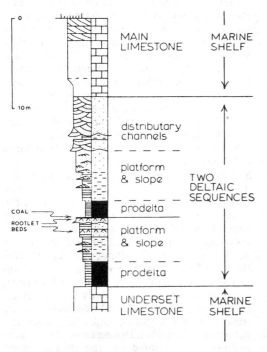

FIGURE 2. Measured section of the Yoredale Series, England (Selley, 1970). Note how the overall upward-coarsening clastic sequence between the two limestones contains two subsequences, only the lower one of which is capped by a coal.

Coastline Hydrodynamics. New York: McGraw-Hill, 546-574.
Maxwell, W. G. H., 1973. Geomorphology of eastern Queensland in relation to the Great Barrier Reef, in O. A. Jones, and R. Endean, eds., *Biology and Geology of Coral Reefs.* New York: Academic Press, 233-272.
Moore, D., 1959. Role of deltas in the formation of some British Lower Carboniferous cyclothems, *J. Geol.,* 67, 522-539.
Moore, D., 1966. Deltaic sedimentation, *Earth Sci. Rev.,* 1, 87-104.
Morgan, J. P., 1970. Depositional processes and products in the deltaic environment, in Morgan and Shaver, 1970, 31-47.
Morgan, J. P., and Shaver, R. H., eds., 1970. Deltaic sedimentation, *Soc. Econ. Paleont. Mineral. Spec. Publ. 15,* 312p.
Nelson, B. W., 1970. Hydrography, sediment dispersal and recent historical development of the Po River delta, Italy, in Morgan and Shaver, 1970, 152-184.
Oomkens, E., 1970. Depositional sequences and sand distribution in the postglacial Rhone delta complex, in Morgan and Shaver, 1970, 198-212.
Pepper, J. F.; DeWitt, W., Jr.; and Demarest, D. F., 1954. Geology of the Bedford Shale and Berea Sandstone in the Appalachian Basin, *U.S. Geol. Surv. Prof. Pap. 259,* 111p.
Rich, J. L., 1951. Three critical environments of deposition, and criteria for recognition of rocks deposited in each of them, *Geol. Soc. Am. Bull,* 62, 1-19.
Selley, R. C., 1970. *Ancient Sedimentary Environments.* Ithaca, N.Y.: Cornell Univ. Press, 237p.
Shirley, M. L., and Ragsdale, J. A., eds., 1966. *Deltas in Their Geologic Framework.* Houston: Houston Geol. Soc., 251p.
Smith, A. E. Jr., and Broussard, M. L., eds., 1971. *Deltas of the World: Modern and Ancient—Bibliography,* Houston: Houston Geol. Soc., 42p.
Sutton, R. G.; Bowen, Z. P.; and McAlester, A. L., 1970. Marine shelf environments of the Upper Devonian Sonyea Group of New York, *Geol. Soc. Am. Bull.,* 81, 2975-2992.
Van Straaten, L. M. J. U., 1960. Some recent advances in deltaic sedimentation, *Liverp. Manch. Geol. J.,* 2, 411-443.
Van Straaten, L. M. J. U., ed., 1967. *Deltaic and Shallow Marine Deposits.* Amsterdam: Elsevier, 464p.
Visher, G. S.; Sandro Saitta, B.; and Phares, R. S., 1971. Pennsylvanian delta patterns and petroleum occurrences in eastern Oklahoma, *Bull. Am. Assoc. Petrol. Geologists,* 55, 1206-1230.
Wanless, H. R., and Shepard, F. P., 1936. Sea level and climatic changes related to late Paleozoic cycles, *Geol. Soc. Am. Bull.,* 47, 1177-1206.
Wanless, H. R.; Barroffio, J. R.; and Trescott, P. C., 1969. Conditions of deposition of Pennsylvanian coal beds, *Geol. Soc. Am. Spec. Pap. 114,* 105-142.
Wanless, H. R., and nine others, 1970. Late Paleozoic deltas in the central and eastern United States, in Morgan and Shaver, 1970, 215-245.
Wermund, E. G., and Jenkins, W. A., Jr., 1970. Recognition of deltas by fitting trend surfaces to Upper Pennsylvanian sandstones in north-central Texas, in Morgan and Shaver, 1970, 256-269.

Cross-references: *Channel Sands; Coal; Cyclic Sedimentation; Deltaic Sediments; Estuarine Sedimentation; Flocculation; Fluvial Sedimentation; Scour-and-Fill.* Vol. III: *Delta Dynamics; Deltaic Evolution; Glacial Lakes.*

DEPOSITION

Deposition is the laying down of material as transportation of it ceases (see also *Sedimentation*). The nature of the deposit formed is dependent upon many variables, among which are the character of the material transported, the effects of transportation upon the material, the conditions causing deposition, and the environment of deposition. A review is best accomplished in terms of depositional environments, the three main categories of classification being terrestrial (nonmarine), paralic (marginal marine), and marine.

Terrestrial Environments

Deposits made on land are small, scattered, and temporary compared with deposits made in the sea; and because terrestrial conditions are more varied, the deposits are more irregularly constituted.

Fluvial Deposition. Fluvial deposits (see *Fluvial Sediments*) consist of material deposited on land by running water. As the velocity or volume of the water decreases, particles held in turbulent suspension begin to settle out depending on their size, shape, and density. A reduction in velocity follows a lessening in gradient or change to a wide shallow course; the volume may decrease through evaporation or seepage. Fluvial deposits are usually well sorted and their particles well rounded (see *River Sands*).

Where streams pour forth from high lands onto lower slopes, alluvial fans (see *Alluvial-Fan Sediments*) and alluvial plains form as the sudden decrease in velocity causes deposition of the transported load (see *Braided-Stream Deposits*). Occasional heavy rains may produce mudflows which flow out over the fans. The alluvial fans are composed of well-sorted, small-sized particles bedded parallel to the sloping surface, whereas the mudflow deposits (q.v.) consist of an unsorted mass ranging from viscous silt to large, erratic boulders. On more gently graded terrain, gravel and sand are generally deposited in the stream channel as shoals, banks, bars, spits, and cut-and-fill structures. During flood stages, most sediment is deposited on the river's natural levees, grading into silt and clay away from the main course, on the flood plain.

Lacustrine Deposition. Lake deposits are in many ways small-scale replicas of those found in the oceans (see *Lacustrine Sedimentation*).

Sand and gravel form beaches, bars, and spits when waves are present; and the fine debris is moved to deeper regions. Deltas develop at the mouths of feeder streams, where deposition of the transported load takes place with the sudden reduction in velocity encountered by running water upon entering the stillness of a pond or lake. Silts may settle out in the calm of the lake, followed by colloidal clay if the surface freezes over and thus further damps out any disturbances in the water. In arid climates, or elsewhere if salinities are high enough, salts in solution may be deposited as a result of concentration through evaporation. Swamps, bogs, and marshes are special cases of poorly drained ponds and lakes gradually filled in by little sediment and much vegetation.

Glacial Deposition. Material deposited directly by glacial ice in the form of drumlins, moraines, or till plains is not sorted or stratified (see *Glacigene Sediments; Till and Tillite*). As glaciers melt, the sedimentary material formerly transported by the flowing ice is deposited in situ or is carried by meltwater, which may deposit well-sorted and stratified deposits of the finer sediments. The angularity, flattened surfaces, and striae associated with transport by ice may be seen on individual fragments in till, although fluvioglacial processes may impart some degree of rounding to the material thus transported.

Eolian Deposition. Eolian deposits are formed from small wind-blown particles such as sand, dust, or volcanic ash (see *Eolian Sedimentation*). Where sand is available, the wind may move the grains about by traction, i.e., surface creep and/or saltation (q.v.). Depending upon the sand supply and wind regime, various dune forms may be developed. Crescentic barchans (scarce sand and moderate winds) and transverse dunes (more sand and stronger winds) are asymmetrical in shape, their steep lee sides facing in the direction of transport. An abundant sand supply and high wind velocity develop longitudinal or seif dunes. Eolian sands (q.v.) are well sorted and rounded, many with surficial frosting. Complex, steep cross-bedding is developed by the changing wind and shifting sand.

Silt-sized dust, carried aloft by turbulent air currents, may be deposited on the lee side of deserts and glacial outwash plains to form *loess* (q.v.). Similarly, volcanic ash may be carried in the atmosphere beyond the immediate vicinity of the erupting cone, to be deposited as *tuff* (q.v.; see also *Volcanic Ash Deposits*). Airborne sediments also play an important role in marine deposition (see *Eolian Dust in Marine Sediments*).

Other Terrestrial Deposition. The unconsolidated rubble produced under the influence of gravity by earthflow, slumping, and landslides, may be preserved as breccias or conglomerates (see *Gravity-Slide Deposits*). More commonly, however, these deposits provide material for reworking by other agents of transportation and deposition. In arid regions, *talus cones,* formed by accumulating rockfall, may retain their conical cross section, circular plan, and crude sorting, when incorporated into larger depositional features.

Moving ground water, carrying material in solution, may concentrate it in the lower soil zones (*illuviation*) or carry it upward and form a hard surface layer (*duricrust*, q.v.). Similarly in the limestone caverns of karst topography, dripping carbonate-rich solutions form stalactites and stalagmites (see *Speleal Sediments*). Other solutions may deposit minerals as veins in cracks or crevices. Geodes (q.v.) are formed in much the same way through cavity filling.

Paralic Environment

Deposition in the coastal zone is characterized by both terrestrial and marine features. The sediments accumulated are to be found near the shoreline, and slightly above or below mean sea level.

Deltaic Deposition. As rivers enter gulfs or seas their velocity is arrested, and deposition of their transported sediments takes place. These deltaic sediments (q.v.) in turn are intertongued with the sediments of other contemporary nearshore deposition, the whole building a delta complex out into the water. Ideally, the delta consists of fine- to medium-grained sediments deposited in bottomset, foreset, and topset beds, which grade tangentially from one into the other. New distributary channels are developed as the old ones become aggraded, tending to form a fan-like plan. In fact, however, the typical cross section and plan are considerably modified by the various constructive and destructive forces of a deltaic environment.

Coastal Lagoon and Marsh Deposition. Coastal lagoons are shallow bodies of water separated from the open sea by bars, spits, or reefs (see *Lagoonal Sedimentation; Tropical Lagoonal Sedimentation*). Tidal currents transport fine to sandy, wave-sorted sediments in through tidal inlets; and local streams contribute material from the mainland. As these lagoons are filled, they become coastal marshes. Subsequent deposits in the upper layers then consist mainly of mud and organic matter that settles out from the tidal water as it moves slowly between the stems of growing marsh plants (see *Salt-Marsh Sedimentology; Tidal-Flat Geology*).

Beach Deposition. Beach material can be supplied by headland erosion, transverse drift from offshore sources, stream transport, and in certain regions the fragmentation of shells and coral. The two prime forces that transport and deposit beach material are turbulent wave action and shore drift, the latter being the combined effect of beach- and alongshore drift. Wave turbulence in the breaker zone throws particles of all sizes up on the beach in the swash, and only some of them are carried back by the backwash. It is here that considerable rounding and sorting take place (see *Barrier Island Facies; Littoral Sedimentation*).

Marine Environment

The marine environment constitutes 70% of the earth's surface and includes the deepest basins; it is, therefore, the final depository for most transported sediment. Several zones and types of environments are recognized as unique in depositional characteristics (see *Marine Sediments*).

Neritic Zone. The relatively shallow portion of the ocean extending from the low-tide mark to a depth of 100 fathoms is designated as the neritic zone. Wave action and stream discharge provide sediment, which ideally is deposited in decreasing sizes as it moves into deeper water. Wave action is not usually effective beyond depths of 5-6 fathoms, but tidal or turbidity currents may transport sand to deeper zones (see *Neritic Sedimentary Facies*). Other coarse-grained deposits may be relict sediments (q.v.).

The neritic zone is a region of mixing of river and oceanic waters, leading to chemical precipitation and colloidal flocculation. In addition, calcareous deposits composed of skeletal, coral, or algal fragments of local origin accumulate in this zone. Under certain conditions, oolites are formed on wave- or current-swept shoal bottoms. Outstanding, too, are the organic reefs constructed by colonial corals, calcareous algae, and other carbonate or silica organisms.

Evaporitic Basins. In warm, arid regions restricted basins receiving an influx of saline water are the site of evaporite deposition (see *Evaporite Facies*). Intense concentration of the saline solution through evaporation results in the deposition of dolomite, calcite, gypsum, anhydrite, halite, and potassium-magnesium-bromide salts. A complex alternation and geometry of the various beds follows from the ever-changing chemical and physical parameters within each basin.

Bathyal Zone. The ocean floor from depths of 100 to 1000 fathoms is called the bathyal zone, usually including the continental slope and rise. Deposition is effected by slow drift from the neritic zone, settling from suspension, turbidity currents, or slumping. Though the latter may take place in other zones, its occurrence is favored by the steeper bathyal slopes, with highly contorted folds being subsequently deposited intact on top of younger sediments. In general, bathyal deposits bridge the gap between those of the neritic and abyssal zones. (see *Continental-Rise Sediments; Submarine (Bathyal) Slope Sedimentation*).

Anaerobic or Euxinic Environment. Restricted basins that have poor water circulation may be the site of black, carbonaceous deposits, produced by the settling of sediment and organic matter into the reducing environment common to stagnant basins (see *Euxinic Facies*). The presence of anaerobic bacteria in the oxygen-scarce regime leads to the partial decay of organic matter and the production of hydrogen sulfide characteristic of typical black shale facies.

Abyssal Zone. The ocean basins, troughs, and trenches below 1000 fathoms constitute the abyssal zone (see *Abyssal Sedimentary Environments*). Much of the bottom is covered by thick deposits of siliceous or calcareous ooze, the result of the continual raining down of the microscopic tests of dead planktonic organisms (see *Pelagic Sedimentation*). Considerable amounts of silt and sand, transported from higher levels by turbidity currents, are deposited as partially graded and stratified submarine fans (q.v.) or cones. Ponding of sediments within the basins tends to obscure all structural relief, forming an almost flat surface called an *abyssal plain*.

In addition to the foregoing deposits, dust of terrestrial, volcanic, or cosmic origin may fall into the seas and settle to the bottom; accumulations over long periods of time may be considerable. Hydrogenous or authigenic sediments, formed by precipitation of minerals out of solution, constitute a second category. To a lesser extent, particles ranging from sand grains to pebbles and boulders are rafted over the oceans by floating ice or plants, or carried as gastroliths in the digestive tracts of various animals. These erratics may ultimately be deposited in sediments of entirely different character within the marine environment.

MAURICE SCHWARTZ

References*

Blatt, H.; Middleton, G.; and Murray, R., 1972. *Origin of Sedimentary Rocks*. Englewood Cliffs, N.J.: Prentice-Hall, 634p.

Krumbein, W. C., and Sloss, L. L., 1963. *Stratigraphy and Sedimentation*. San Francisco: Freeman, 660p.

Pettijohn, F. J., 1975. *Sedimentary Rocks*. New York: Harper & Row, 628p.

Shen, H. W., ed., 1972. *Sedimentation*. Colo. St. Univ.: H. W. Shen, 814p.

Termier, H., and Termier, G., 1963. *Erosion and Sedimentation*. London: Van Nostrand, 433p.

Twenhofel, W. H., 1932. *Treatise on Sedimentation.* New York: Dover, 926p.

*See also references under cross-referenced entries.

Cross-references: *Dispersal; Flocculation; Fluvial Sediment Transport; Sedimentary Environments; Sedimentation; Sedimentology; Sediment Transport—Initiation and Energetics; Settling; Transportation;* see also numerous cross-references in text.

DESERT SEDIMENTARY ENVIRONMENTS

A classical desert is an almost barren tract of land, within or bordering the tropics, over which rainfall is too limited and spasmodic to support vegetation adequately. The upper limit for rainfall is about 25 cm/yr, and because of the high temperature and general lack of humidity, the potential rate of evaporation far exceeds precipitation. Tropical deserts cover about 20% of the world's present land surface and tend to be concentrated in the regions of prevailing trade winds which flow roughly between the latitudes 10° and 30° N and S of the equator (Fig. 1). The reasons for their occurrence seem to be largely meteorological. Desert conditions also occur in other latitudes, such as in the rainshadow of mountain ranges or in areas far from the sea. Only about 20% of the desert surface is covered by sand dunes. The remainder comprises the deposits of ephemeral streams and large areas of outcrop, which are subjected to processes of desert weathering and erosion (Ollier, 1969; Glennie, 1970).

Desert Erosion. Areas of outcrop are continually subjected to weathering processes. The relative humidity in some desert areas may reach 100% before dawn, and heavy dews cover the rock surfaces, leading to chemical corrosion, especially of carbonate rocks (Clark et al., 1974). Moisture, in very small quantities, reaches the surface from the water table, and its evaporation close to the surface results in the growth of

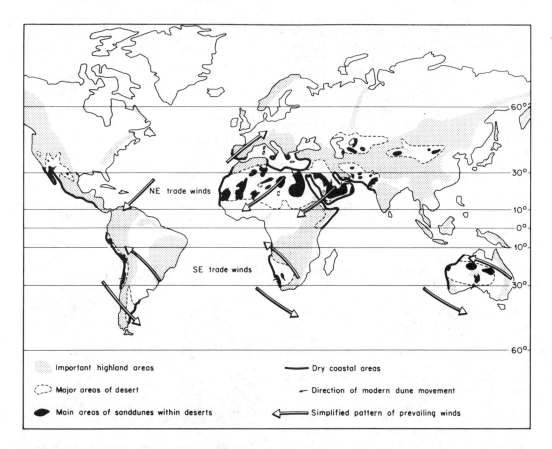

FIGURE 1. Quaternary deserts of the world. Tropical deserts are confined largely to the belt of prevailing winds between 10° and 30° north and south of the equator. This simplified pattern is further modified by the large land masses, which heat up rapidly in summer and cool rapidly in winter. Other deserts, in more temperate latitudes, are in the rainshadow of high mountain ranges, and sometimes far from the sea.

gypsum and halite crystals which exert an expansion force on the surrounding host rock, causing it to split. Rapid diurnal changes in temperature, especially if accompanied by a rainstorm, cause differential expansion and contraction stresses between the surface and underlying rock, resulting in spalling of the rock surface and even splitting of boulders. Tucker (1974) cites a possible Triassic example of this phenomenon. Siliceous rocks and fine-grained homogenous limestones tend to form angular boulders, whereas more argillaceous boulders, granites, and dolerites become rounded by exfoliation. Softer rocks can be abraded by wind-driven sand and silt.

The wind can force dry silt- and sand-sized products of weathering into motion. With removal of the finer weathering products by the wind (*deflation*), the larger, more resistant pebbles and boulders form a lag deposit that may be abraded into *ventifacts*.

Water in Deserts

Fluvial Sediments. Soil is thin or absent on the barren highlands, and little rainwater soaks into the ground. When rainfall ceases, the flow of surface water soon stops, and ephemeral streams, along which sediment was being transported, dry out. These dry watercourses are known as *wadis* in Arabia and North Africa. After heavy rainstorms, flowing water may extend the length of the wadi to reach the sea or the center of a basin of inland drainage. Overloading occurs at the nickpoint between hill and plain, and braiding follows with the formation of alluvial fans (see *Alluvial-Fan Sediments*). Over the plains, however, waters of a flash flood may overflow from the channel and cover the surrounding plains. During the Central Australian floods of 1967, for example, all except the crestal parts of a system of dunes were covered by water (Williams, 1970). The bed forms of the resulting deposits of sand and clay were ascribed to both the upper and lower flow regimes (see *Flood Deposits*).

When there is a high sediment:water ratio, viscous muds capable of transporting large boulders may fill the channel with mudflow conglomerates (see *Mudflow Deposits*). A stream may also become overloaded as water soaks into the stream bed before reaching the lowest point in the channel, which becomes clogged with sediment.

Desiccation. After surface flow has ceased, the water percolating through the stream bed is commonly saturated with respect to calcium carbonate, and evaporation results in initial cementation of the sediment. The surface sands and silts dry out rapidly, however, and generally are not cemented; they can be readily carried away by the wind. Clay lags in stagnant pools may dry, curl, and crack. Thin and fragile clay flakes are removed by the wind, but thicker and heavier flakes may be preserved in place, at least temporarily, by a covering of wind-blown sand (Fig. 2).

Desert Lakes and Inland Sabkhas. Temporary lakes may form in the center of a basin of inland drainage or where eolian sands block a stream channel. As the water evaporates, salts are concentrated; and the sands, silts, and clays become encrusted with halite and gypsum crystals (Glennie, 1970). Such saline areas are known in Arabia and North Africa as *sabkhas* (see *Sabkha Sedimentology*), in the Western Hemisphere as *playas* or *salars* (q.v.). Not all inland sabkhas are supplied by surface water. In many stream channels, water continues to percolate through the sediment long after surface flow has ceased and is brought to the surface of the sabkha by the pressure of the head of water or by capillary action, and then evaporated. Other inland sabkhas appear to be supplied solely by ground water. Two sedimentary structures characteristic of inland sabkhas are sand dikes and adhesion ripples (see *Salar, Salar Structures*). The former seem to be confined to inland sabkhas and desert fluvial environments, but the latter may also occur in nondesert areas.

Eolian Sediments

Uncemented fluvial sediments form an important source for windblown sand. Wind is capable of transporting very fine (100μ) particles in suspension. Consequently, much of the clay and silt originally deposited by stream action is removed by the wind and redeposited beyond the desert as *loess* (q.v.). Coarser sand

FIGURE 2. Preservation of curled clay flakes by wind-blown sand about two months after the wadi channel had flowing water in it; Wadi Amayri, Oman, Arabia (from Glennie, 1970).

grains (100–2000μ) travel by saltation and cause still larger grains to move over the desert by surface creep (see *Eolian Sedimentation*).

Eolian sands (q.v.) are deposited as well-sorted, laminated sediments (A in Fig. 3) when the velocity of the driving wind becomes insufficient to transport them farther. A given wind can drive sand over a hard, immobile surface faster than over a surface of loose sand, so that sand tends to accumulate in areas that are already sand covered.

The axes of eolian ripples are transverse to the wind, and the coarsest grains are concentrated at the crest. Ripples tend to flatten out at higher wind strengths and greater rates of deposition (Bagnold, 1941); these factors possibly result in extensive areas of horizontally laminated sheet sands. Transverse dunes, or barchans, are produced where the supply of sand is limited, are formed by winds of moderate velocity, and are composed mostly of beds deposited as avalanche slopes at the 34° angle of repose for dry sand; they are unstable in strong winds. The axes of longitudinal (seif) dunes are parallel to the dominant sand-transporting wind; their bedding dips away from their axes with a down-wind component, but avalanche slopes are rare (see *Eolian Sedimentation*).

Coastal Desert Sediments

Many deserts border coastlines, but because in most deserts there are no rivers with a continuously flowing supply of water, coastlines that border deserts generally have no fluvial deltas. Exceptions include the Colorado, Indus, and Nile rivers. Subsiding coastal areas are subject to marine transgressions; and, in the clear tropical seas that border them, organisms produce large quantities of calcium carbonate which is available for transport by tidal and longshore currents. Also, strong tidal flow into and out of lagoons formed behind submarine bars causes the formation of oolite deltas (Fig. 4).

Extensive low-lying coastal areas are subjected to flooding by abnormally high tides and seepage from adjacent saline waters, producing a coastal sabkha environment (see *Sabkha Sedimentology*). These coastal sabkhas are characterized by crusts of evaporites produced by

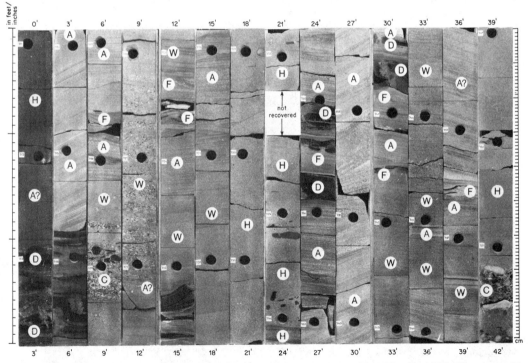

FIGURE 3. Sequence of Permian Rotliegend cores from the North Sea (from Glennie, 1972). They comprise alternations of conglomerates (C; including claystone pebbles and flakes); curled and broken clay layers (black) that are probably still in their site of deposition; sandstone dikes (D); fluvial sand (F); poorly sorted sands that are possibly former eolian sands homogenized during water transport (H); and undisturbed well-sorted subhorizontal to inclined eolian sand laminae (A). Other sands (W) could have been deposited after either eolian or water transport. Numbered holes are of plugs cut in cores for porosity/permeability determinations.

FIGURE 4. Tidal creek (t) with outward-facing submarine delta (sd), coastal sabkha (cs) with black algal mats in the intertidal zone behind the protection of sand spit (s) north of the creek, and probable former coastal sabkha complex of prograding coastline now covered by dunes (d) south of the creek. South of Ra's Ghanadah, Abu Dhabi, Arabia (from Glennie, 1970).

evaporation of saline ground waters; by a belt of algal mat (Fig. 4) that binds the underlying sediment and which, when exposed to desiccation, hardens, curls, and cracks; and by adhesion ripples, which may form over the damp sabkha surface. Accumulation of these sediments can result in a rapidly prograding coastline (Evans et al., 1964). The sediments of a coastal sabkha are commonly rich in skeletal (high Mg) carbonate which, after burial, is subject to dolomitization (q.v.). The products of deflation of a coastal sabkha are rich in foraminifera, which become constituents of nearby dunes (Glennie, 1970).

Ancient Desert Sediments

Ancient desert sediments are characterized, as are many continental deposits, by a reddish coloration, caused for the most part by a post-depositional coating of ferric oxides (goethite, hematite) on sand grains (Walker, 1967). Other colors may occur locally.

Ancient desert sediments may be identified from an association of any of the following criteria (see also Glennie, 1970, ch. 8), which are grouped into two categories.

Water-laid sediments are characterized by sedimentary features similar to those of water-laid sediments of nondesert continental environments having a seasonal rainfall, i.e., sedimentary structures of both the upper and lower flow regimes, but modified by one or more of the following:

1. Commonly calcite cemented; locally cemented by gypsum or anhydrite.
2. Conglomerates may be common and, locally, several cycles of deposition may lack a sand-sized fraction at the top of the cycle (deflation of sand and silt).
3. Presence of mudflow conglomerates (Bluck, 1967).
4. Sharp upward decrease in grain size (especially from sand to clay), indicating rapid fall in water velocity.
5. Common presence of clay pebbles and curled clay flakes.
6. Presence of mud cracks and sandstone dikes.

Wind-laid sediments may exhibit any of the following characteristics.

1. Sequences of sandstones that may vary in thickness from a few centimeters to several hundred meters and whose laminae dip at angles from horizontal to 34° (less if compacted)—dips either of constant or multiple orientation.
2. Laminae commonly planar with only sparse poorly developed ripples (Glennie, 1972).
3. Individual laminae well sorted, especially in finer grain sizes; sharp size differences between larger grains of adjacent laminae are common.
4. Large sand grains tend to be well rounded.
5. Grain size commonly ranges from silt (60μ) to coarse sand (2000μ) with bulk about $125-300\mu$; and, apart from authigenic clay, silt and clay generally well below 5% of sediment.
6. Clay drapes very rare and usually accompanied by evidence that they were water-laid.
7. Adhesion ripples with associated increase in clay or silt content and common presence of gypsum or anhydrite cement (Glennie, 1972).
8. Mica generally absent.
9. Under scanning electron microscope, quartz grains exhibit pattern of meandering ridges and upturned fracture plates; and under optical microscope, quartz grains appear frosted (see *Eolian Sands*).

From criteria such as those listed above, ancient desert sediments of various ages have been identified in many parts of the world (Bigarella, 1972). The sediments must have accumulated in areas of subsidence and in many cases were preserved beneath the wave base of transgressing seas. Desert sediments such as the Nubian type sandstones of N Africa and Arabia (McKee, 1962; Busson, 1967; Klitzsch, 1968) recur interbedded with marine sequences from the Cambrian to the Cretaceous. In NW Europe, deposition of Permian desert sediments (Fig. 3) was succeeded by salts of the Zechstein Sea, followed again by continental desert sediments during the Triassic. Analysis of the bedding attitudes of Permian dune sands indicates that the Permian winds blew over NW Europe from northeast to southwest (Glennie, 1972) and over North America in the same direction (Poole, 1964). These paleowind directions show that the Permian deserts of these two areas probably existed within the trade-wind belt of the Northern Hemisphere (Fig. 1). Desert conditions persisted in NW Europe until the end of the Triassic, as seen, for example, from local development of dune sands (Thompson, 1969) and more widespread continental evaporites. Dune sands were still being deposited during the Jurassic in parts of the western US (Poole, 1964); dune sands of Triassic/Jurassic age also occur in South America (Bigarella and Salamuni, 1961).

The deposits of coastal sabkhas of widely differing age have also been recognized in different parts of the world (see *Sabkha Sedimentology*). They are characterized by sedimentary cycles comprising from bottom (shallow marine) to top (supratidal): algal boundstones and grainstones; dolomites with birdseye structures overlain by dolomitized algal mat with an upward increase in anhydrite cement; nodular or chicken-mesh anhydrite in a dolomite matrix. Unconformities (deflation surfaces) are common at the top of each cycle.

K. W. GLENNIE

References

Bagnold, R. A., 1941. *Physics of Blown Sand and Desert Dunes.* London: Methuen, 265p.

Bigarella, J. J., 1972. Eolian environments: their characteristics, recognition, and importance, *Soc. Econ. Paleont. Mineral. Spec. Publ. 16*, 12-62.

Bigarella, J. J., and Salamuni, R., 1961. Early Mesozoic wind patterns as suggested by dune bedding in the Botucatú Sandstone of Brazil and Uruguay, *Geol. Soc. Am. Bull.,* 72, 1089-1106.

Bluck, B. J., 1967. Deposition of some Upper Old Red Sandstone conglomerates in the Clyde area: A study in the significance of bedding, *Scot. J. Geol.,* 3, 139-167.

Busson, G., 1967. Mesozoic of southern Tunisia, in L. Martin, ed., *Guidebook to the Geology and History of Tunisia.* Tripoli: Petrol. Expl. Soc. Libya, 9th Ann. Field Conf., 131-151.

Clark, D. M.; Mitchell, C. W.; and Varley, J. A., 1974. Geomorphic evolution of sediment filled solution hollows in some arid regions, *Z. Geomorph. Supplbd.,* 20, 130-139.

Cooke, R. U., and Warren, A., 1973. *Geomorphology in Deserts.* Berkeley: Univ. Calif. Press, 374p.

Evans, G.; Kendall, C. G.St.C.; and Skipwith, P., 1964. Origin of the coastal flats, the sabkha, of the Trucial Coast, Persian Gulf, *Nature,* 202, 759-761.

Glennie, K. W., 1970. *Desert Sedimentary Environments.* Amsterdam: Elsevier, 222p.

Glennie, K. W., 1972. Permian Rotliegendes of northwest Europe interpreted in light of modern desert sedimentation studies, *Bull. Am. Assoc. Petrol. Geologists,* 56, 1048-1071.

Klitzsch, E., 1968. Outline of the geology of Libya, in F. T. Barr, ed., *Geology and Archeology of Northern Cyrenaica, Libya.* Tripoli: Petrol. Expl. Soc. Libya, 10th Ann. Field Conf., 71-78.

McKee, E., 1962. Origin of the Nubian and similar sandstones, *Geol. Rundsch.,* 52, 551-587.

Ollier, C. D., 1969. *Weathering.* Edinburgh: Oliver and Boyd, 304p.

Picard, M. D., and High, L. R., Jr., 1973. *Sedimentary Structures of Ephemeral Streams.* Amsterdam: Elsevier, 233p.

Poole, F. G., 1964. Palaeowinds in the western United States, in A. E. M. Nairn, ed., *Problems in Palaeoclimatology.* New York: Interscience, 394-405.

Thompson, D. B., 1969. Dome-shaped aeolian dunes in the Frodsham member of the so called "Keuper" Sandstone Formation (Scythian to Anisian?, Triassic) at Frodsham, Cheshire (England), *Sed. Geol.,* 3, 263-289.

Tucker, M. E., 1974. Exfoliated pebbles and sheeting in the Triassic, *Nature,* 252, 375-376.

Walker, T. R., 1967. Formation of red beds in modern and ancient deserts, *Geol. Soc. Am. Bull.,* 78, 353-368.

Walther, J., 1924. *Das Gesetz der Wüstenbildung in Gegenwart und Vorzeit,* 4th ed. Leipzig: Verlag von Quelle und Mayer, 436p.

Williams, G. E., 1970. The Central Australian stream floods of February-March 1967, *J. Hydrol.,* 11, 185-200.

Cross-references: *Alluvial-Fan Sediments; Alluvium; Braided-Stream Deposits; Eolianite; Eolian Sands; Eolian Sedimentation; Evaporite Facies; Faceted Pebbles and Boulders; Flood Deposits; Flow Regimes; Fluvial Sediments; Sabkha Sedimentology; Salar, Salar Structures; Synersis; Weathering of Sediments.* Vol. III: *Alluvial Fan, Cone; Deserts and Desert Landforms; Playa; Sand Dunes; Wind Action.* Vol. XI: *Desert Soils.*

DESICCATION—See MUD CRACKS; SYNERESIS

DEWATERING—See WATER ESCAPE STRUCTURES

DIAGENESIS

The term *diagenesis* was proposed by Von Gümbel in 1868 but was not generally accepted until recognized by Walther (1894) in his *Lithogenesis der Gegenwart*, which must be regarded as the first systematic textbook treating actualistic sedimentology, lithification, and diagenesis. Walther defined diagenesis as "all those physical and chemical changes which a rock (i.e., a sediment) undergoes after its deposition, without the introduction of rock pressure or igneous heat."

In present usage, the boundary between diagenesis and metamorphism is not well defined. Metamorphism is the process by which consolidated rocks are altered in composition, texture, or internal structure by pressure, heat, and the introduction of new chemical substances, but *not* including changes brought about by burial and loading. Many igneous petrologists regard diagenesis as incipient metamorphism. Van Hise (1904) subdivided metamorphism into *anamorphism*, the building of complex minerals under elevated pressure and heat, and *katamorphism*, the near-surface alteration, including weathering, that reduces complex minerals to simple ones. The latter may be called "delithification," but this is simplistic because many new minerals, some more complex than before, are formed during this near-surface evolution.

Widely disparate definitions of diagenesis have been adopted over the years, partly as the result of a lack of linguistic communication. Difficulty also stems from vagueness about the initiation of metamorphic processes "at elevated pressures and temperatures"—*How* elevated? Coring and experiment have shown that no metamorphism may take place at considerable depth (7-10 km). Yet metamorphic petrologists apply the term *zeolite facies* to altered ash deposits which begin changing even before they are buried (Fyfe et al., 1958; see *Volcanic Ash–Diagenesis*). There is a difference of opinion as to whether the rock fabric or mineral composition should be used as a guide to the onset of metamorphism. Pseudo-metamorphic fabrics are often produced by purely diagenetic processes in evaporites, and very little stress is required to bring about fabric modifications in certain limestones; indeed the friction of a cutting wheel has been known to produce strain effects in calcite crystals.

Diagenetic Phases

In order to classify the range and boundary limits of diagenesis, Fairbridge (1967) proposed the concept of *diagenetic phases*, with three categories (Fig. 1).

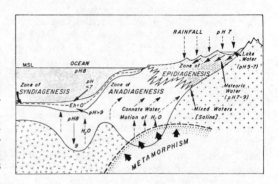

FIGURE 1. Idealized profile through a continental margin, showing the sites of contemporary marine sedimentation and the three phases of diagenesis (from Fairbridge, 1967). Note the (1) diffusion potential during syndiagenesis; (2) upward liquid motion in anadiagenesis; and (3) downward motion in epidiagenesis.

1. Syndiagenesis. Bissel (1959) introduced the term *syndiagenesis* for the phase of syndepositional changes during early burial (order of 0-100 m) marked by close interactions with the sedimentary environment, and either oxidizing or reducing conditions, often leading to early lithification and syngenetic authigenesis.

2. Anadiagenesis. Anadiagenetic changes take place during deep burial (on the order of 1000-10,000 m). These changes are varied and include compaction, secular dehydration, and hypogene authigenesis; they occur under the *upward* and lateral migration of originally connate waters and other fluids (essentially brines and petroleum, at temperatures that may reach 100-200°C).

3. Epidiagenesis. Following the reintroduction of surface waters (of meteoric origin moving *downward*), the diagenetic changes in a rock result in oxidation, weathering, and what Van Hise (1904) classified as *katamorphic alteration*, together with distinctive authigenic mineralization.

Processes and Changes. Numbers of characteristic diagenetic processes (e.g., dolomitization, q.v., and dedolomitization, q.v.) can take place at any of the three phases, under appropriate thermodynamic controls. Stress is not placed so much on temperature or pressure, but rather on fluid chemistry and direction of circulation, and on the reactions between these fluids and the initial mineral grains. Phase I is marked by all extremes of surface and near-surface environments within the Eh/pH catenary limits established by Baas Becking et al. (1960); phase 2 is marked generally by a high pH and an Eh of about 0; phase 3 is influenced by rain water and ground water usually with an initially low pH and strongly positive Eh.

The considerable differences that may exist between the state of a modern sediment and its ancient lithified counterpart are attributed to processes called diagenetic. The changes wrought by such processes may be limited to the binding together of the sediment particles by a cement, or there may be drastic reorientations of original fabrics induced by compaction and recrystallization, *sensu lato,* which may be accompanied by the in-situ formation of new minerals. Unlike metamorphic studies, the study of diagenetic changes has in the past been severely neglected. In recent years, however, studies of various aspects of diagenesis have received a new impetus, due in part to the appreciation of its economic significance. Diagenesis exercises a fundamental control on the location and migration of oil, natural gas, and water, and is a prime factor also in the emplacement of certain types of ores, all of which depend upon the distribution of pores and fissures in rocks.

Diagenesis is clearly initiated while the sediment is still very much under the influence of the depositional environment. Solution and etching of detrital grains and the formation of aphanitic rims to clastic carbonate grains is an early occurrence, but biochemical and chemical conditions may be such as to cause the decay of the original materials and the in-situ formation of new minerals (*authigenesis*, q.v.). The early changes that take place at the boundary of the lithosphere and marine hydrosphere, i.e., "submarine weathering," were defined as *halmyrolysis* by Hummel (1922). In so doing, he restricted the term *diagenesis* to postburial changes. Nevertheless, most geologists take the precise moment a sediment grain touches its accumulation site as the beginning of diagenesis. It is more difficult to obtain agreement as to where diagenesis ends. Many current definitions exclude postlithification changes, which are often described as epigenetic, and would constitute the *metharmosis* of Kessler (see Fairbridge, 1967, for references). Lithification is one of the most conspicuous results of diagenesis, but it is only a link in the chain of events that eventually convert a sediment into a rock. The speed with which lithification takes place and the conditions under which it occurs, even in sediments of the same type, are extremely variable factors which are still poorly understood.

It has been quite usual (particularly in the USSR) to exclude from diagenesis those processes which bring about the delithification of a rock, but many weathering processes parallel those of halmyrolysis, and many new and complex minerals are formed at this stage. Sometimes it is impossible to separate katamorphism from diagenesis in its limited sense; in a striking example, Moore (1964) conclusively showed that the Erda Tonstein was the result of soil-forming processes, and he presented evidence in favor of an early diagenetic, biochemical origin for the authigenic kaolinite macrocrystals which occur in the tonstein.

Diagenetic Environment

The two components in the aqueous diagenetic environment are sediment particles and interstitial fluids; the influential factors are hydrogen-ion potential (pH), oxidation-reduction potential (Eh), ion adsorption phenomena, temperature, and pressure. Minor diagenetic changes of a purely chemical nature may occur in shell material for instance, without the accompaniment of recrystallization, with only minor temperature changes as the important control. The detection and significance of such changes with regard to paleoenvironmental interpretations are discussed by Curtis and Krinsley (1965). Temperature and pressure generally exert greater influence, however, in later, deep-seated diagenesis. In the early stages, organisms are of fundamental importance because their metabolic processes and decay may control the pH-Eh characteristics of the environment. The chief organisms responsible are algae and bacteria; fungi and related organisms are also important in particular environments (see *Microbiology in Sedimentology*). Baas Becking et al. (1960), concluded that the pH-Eh limits of biological systems almost coincide with those of the naturally occurring aqueous environments.

Conditions at the sediment/water interface vary from usually alkaline and oxidizing under an open sea, to acid and reducing beneath a body of water with restricted circulation; the conditions that exist at the surface of a sedimentary accumulation, however, may be quite different from those that prevail at shallow depths below the sediment/water interface. Changes in the pH-Eh values may extend gradually and progressively downward from the interface or may be rapid and irregular due chiefly to biological activities (Taylor, 1964). As the depth of burial increases, extremes of pH and Eh cease to be maintained, and the pH-Eh factors progressively become less important. Little is known about the milieu or reactions of later diagenesis, but speculations can be made from the composition of fluids encountered in deep bore holes.

Diagenetic Phenomena

Following deposition, a sediment may be disturbed by wave or current action, organisms, or earth tremors; the purely physical rearrangement of the sediment that results—e.g., slump structures, bioturbation, etc.—would be diagenetic according to most current definitions. It is, however, customary to exclude such features from diagenesis in spite of the influence these *sedimentary structures* (q.v.) have on the migration of fluids throughout the history of the rock. Phenomena accredited to diagenesis are manifold, and a few may be noted briefly here.

Activities of Organisms. Highly disturbed sediments, both ancient and modern, are often the legacies of burrowing organisms (see *Bioturbation*); bedding planes and other primary structures may be completely destroyed. In an aqueous environment, bottom-dwelling creatures such as worms, echinoderms, crustaceans, mollusks, and holothurians may produce aggregated lumps of fine material which has passed through their intestinal tracts (fecal pellets). Comminution of sediment is also effected by boring algae and other organisms that are active in breaking down skeletal material, particularly in marine carbonate deposits (Swinchatt, 1965; see also *Diagenetic Fabrics*).

Compaction. As a sediment is buried, the weight of the overburden increases sediment packing density, pore space is reduced, and the connate water is expelled, finding its way to higher levels (see *Compaction in Sediments; Water-Escape Structures*). The degree of compaction achieved depends upon depth of burial, nature of the sediment, and degree of cementation. In general, the rate of compaction is most rapid shortly after deposition and decreases with time. Porosity thus decreases with age, with a few exceptions, e.g., due to dolomitization (q.v.).

Cementation. The most common cementing materials are calcite, dolomite, siderite, iron oxides, and silica. These are usually precipitated in the pores and interstices in a sediment, thus binding the grains together and reducing porosity. The cementing material may be of the same composition as the sediment or may be quite different; various cement fabrics are described in *Diagenetic Fabrics*. Cementation may occur early and rapidly, or may be considerably delayed. The conditions necessary to bring about cementation, and also loss of cementation, are clearly related to circulating fluids, pressure, and other factors still not fully understood.

Intrastratal Solution. During the course of the postdepositional history of a rock, modifications of porosity and mineral composition may be brought about by intrastratal solution. Material from the sediment may be dissolved and the porosity thereby increased. On the other hand, simultaneous precipitation may take place so that the grains of sediment become firmly cemented together, resulting in a net decrease in porosity. Certain limestones frequently show solution channels that have been subsequently filled with drusy calcite. Shell fabrics may be partly or completely dissolved and replaced by drusy mosaics; molds of shells are common in dolomites (q.v.).

The effects of corrosion may be evident on some mineral grains; various unstable minerals—e.g., olivine, pyroxene—will be eliminated, particularly from older rocks. Thus, there may be an apparent decrease in complexity of heavy-mineral assemblages with increasing geological age. In chemically homogeneous rocks such as limestones or quartzitic sandstones, intrastratal solution may be concentrated along well-marked planes which develop normal to pressure, therefore often coinciding with bedding planes (see *Diagenetic Fabrics; Pressure Solution and Related Phenomena*).

Recrystallization and Polymorphic Transformation. In considering ancient limestones, Folk (1965) distinguished between polymorphic transformations, e.g., inversion of aragonite to calcite in skeletal material, and recrystallization with no change of mineral species, which may occur where there are differences in grain size, morphology, or orientation. He introduced the term *neomorphism* to include both processes, which is particularly useful in cases where it is uncertain whether in-situ changes involve polymorphic transformation and/or recrystallization.

It requires very little stress to cause an evaporite such as rock salt, gypsum, or anhydrite to flow from an area of high pressure to a region of lower pressure, and important new minerals are thus formed by recrystallization. Limestones are also prone to reorganization under suitable environmental conditions, and the new fabrics commonly obscure the original fabric partly or completely. According to the conditions prevailing, a net increase or decrease in grain size will result. Change in monomineralic fabrics may take place in the solid state by ionic transfer between crystal lattices. Redistribution of fabrics can also be accomplished by intrastratal solution, according to grain size, solubility, temperature, and pressure.

Authigenesis and Metasomatism. A mineral developed in place in a sedimentary rock at any time during its diagenetic evolution is described as authigenic (see *Authigenesis*). If large-scale replacement of the host rock by a mineral of different chemical compositon occurs through the introduction of material from outside, i.e., the new minerals are not generated merely as a result of internal adjustment within the sedimentary body, *metasomatism* has taken place. Many dolomites (q.v.) are believed to be complete replacements of limestones. Metasomatic replacement of limestones by silica as chert (q.v.) or "quartz rock" is well known (Orme, 1974). Many siderites and phosphate rocks are generated in the same manner. Textures and fabrics may provide clues to the process of replacement.

Diagenetic Evolution, Grade, Facies

A freshly accumulated sediment is not an equilibrium mixture in a state of complete adjustment with the physicochemical factors of the depositional environment; consequently, thermodynamic changes are effected in response to the prevailing conditions of temperature, pressure, and ionic concentrations. The gradual change in these variables promotes a succession of adjustment processes which control the diagenetic evolution of the rock. The state reached in this evolution is reflected by the resultant fabric, textures, and mineralogy of the rock, and is described as the *diagenetic grade*. It is, therefore, logical to consider that the farther a sediment is divorced from its depositional environment by burial, the more advanced will be the diagenetic grade.

A succession of rocks showing a sequence of diagenetic mineral assemblages with depth permits the definition of a system of diagenetic facies analogous to the well-established metamorphic facies. In early diagenesis, pH and Eh are the most important environmental parameters, becoming less significant with increase of temperature and pressure. Teodorovich (1954) recognized thirteen facies by the presence of diagnostic authigenic mineral assemblages, defined by pH and Eh limitations. These subdivisions of low-grade diagenetic facies are called *parafacies* by Packham and Crook (1960).

The hydrated alkali aluminous silicates are very susceptible to temperature and pressure changes, and are therefore considered to be a suitable guide to the progress of diagenesis. Work in this direction was carried out by Coombs et al. (1959), Packham and Crook (1960), Kossovskaya and Shutov (1965, 1970), and Hay (1966); also see articles in Mumpton (1977). Two diagenetic depth sequences in sediments of andesitic affinities have been studied in detail (Coombs, 1954, from the South Island of New Zealand, and Crook, 1959, in New South Wales, Australia; see references in Packham and Crook, 1960). From the data obtained, Packham and Crook suggest that under alkaline conditions facies characterized by heulandite-analcite, laumontite, prehnite-pumpellyite, and albite-epidote, may be recognizable in rocks of suitable composition. However, as indicated earlier, the point at which metamorphism begins is debatable.

Dapples (1959, 1962), who was particularly concerned with the development of diagenetic grade in sandstones, recognized three stages in the diagenetic evolution of the rock: namely, initial or depositional, intermediate or early burial, and late burial or premetamorphic. The concept of diagenetic evolution is discussed at length by Fairbridge (1967), who also recognizes the three distinctive phases already discussed. A sediment need not necessarily pass through all three phases of diagenetic evolution, for the sequence may be interrupted, reactivated, or disrupted by eustatic or tectonic movements.

Investigations of modern sedimentary environments, although yielding much useful data, give information only on the early stages of diagenesis. Careful petrological studies, if carried out in conjunction with adequate field investigations of the geological setting, often provide clues to the diagenetic evolution of a rock, and the fluids encountered in deep bore holes may give an indication of the reactions occurring in later stages of diagenesis (see, e.g., Perry and Hower, 1972). Study of interstitial waters (q.v.) in DSDP cores is providing clues to the nature of marine diagenesis (Gieskes, 1975; Price, 1976). Our knowledge of the conditions and processes that function throughout the greater part of diagenetic evolution, however, is very inadequate. This deficiency may in part be remedied by systematic experimentation (see, e.g., Fruth et al., 1966; Hiltabrand et al., 1973).

G. RICHARD ORME
RHODES W. FAIRBRIDGE

References

Baas Becking, L. G. M.; Kaplan, I. R.; and Moore, D. M., 1960. Limits of the natural environment in terms of pH and oxidation-reduction potentials, *J. Geol.,* 68, 243-284.

Bissell, H. J., 1959. Silica in sediments of the Upper Paleozoic of the Cordilleran area, *Soc. Econ. Paleont. Mineral. Spec. Publ. 7,* 150-185.

Coombs, D. S.; Ellis, A. J.; Fyfe, W. S.; and Taylor, A. M., 1959. The zeolite facies; with comments on the interpretation of hydrothermal syntheses, *Geochim. Cosmochim. Acta,* 17, 53-107.

Curtis, C. D., and Krinsley, D., 1965. The detection of minor diagenetic alteration in shell material, *Geochim. Cosmochim. Acta,* 29, 71-84.

Dapples, E. C., 1959. The behavior of silica in diagenesis, *Soc. Econ. Paleont. Mineral. Spec. Publ. 7,* 36-54.

Dapples, E. C., 1962. Stages of diagenesis in the development of sandstones, *Geol. Soc. Am. Bull.,* 73, 913-934.

Fairbridge, R. W., 1967. Phases of diagenesis and authigenesis, in G. Larsen and G. V. Chilingar, eds., *Diagenesis in Sediments.* Amsterdam: Elsevier, 19-89.

Folk, R. L., 1965. Some aspects of recrystallization in ancient limestones, *Soc. Econ. Paleont. Mineral. Spec. Publ. 13,* 14-48.

Fruth, L. S.; Orme, G. R.; and Donath, F. A., 1966. Experimental compaction effects in carbonate sediments, *J. Sed. Petrology,* 36, 747-754.

Fyfe, W. S.; Turner, F. J.; and Verhoogen, J., 1958. Metamorphic reactions and metamorphic facies, *Geol. Soc. Am. Mem. 73*, 259p.

Fyfe, W. S.; Turner, F. J.; and Verhoogen, J., 1974. Chemical history of the oceans deduced from post-depositional changes in sedimentary rocks, *Soc. Econ. Paleont. Mineral. Spec. Publ. 20*, 193-204.

Gieskes, J. M., 1975. Chemistry of interstitial waters of marine sediments, *Ann. Rev. Earth Planet. Sci.*, 3, 433-453.

Hay, R. L. 1966. Zeolites and zeolite reactions in sedimentary rocks, *Geol. Soc. Spec. Pap. 85*, 130p.

Hiltabrand, R. R.; Ferrell, R. E., Jr.; and Billings, G. K., 1973. Experimental diagenesis of Gulf Coast argillaceous sediment, *Bull. Am. Assoc. Petrol. Geologists*, 57, 338-348.

Hummel, K., 1922. Die Entstehung eisenriecher Gesteine durch Halmyrolse, *Geol. Rundsch.*, 31, 40-81.

Kossovskaya, A. G., and Shutov, V. D., 1965. Facies of regional epi- and metagenesis, *Isvestia A.N. SSSR, ser. Geol.* 1963, 7, 3-18; in English: *Internat. Geol. Rev.*, 7, 1157-1167.

Kossovskaya, A. G., and Shutov, V. D., 1970. Main aspects of the epigenesis problem, *Sedimentology*, 15, 11-40.

Kübler, B., 1973. La corrensite: indicateur possible de milieux de sédimentation et du degrée de transformation d'un sédiment, *Bull. Centre Réch. de Pau*, 7, 543-556.

Larsen, G., and Chilingar, G. V., eds., 1967. *Diagenesis in Sediments.* Amsterdam: Elsevier, 551p.

Moore, L. R., 1964. The microbiology, mineralogy and genesis of a tonstein, *Proc. Yorkshire Geol. Soc.*, 34, 235-292.

Mumpton, F. A., ed., 1977. Mineralogy and geology of natural zeolites, *Mineral. Soc. Am. Short Course Notes*, 4, 233p.

Orme, G. R., 1974. Silica in the Visean Limestones of Derbyshire, England, *Proc. Yorkshire Geol. Soc.*, 40, 63-104.

Packham, G. H., and Crook, K. A. W., 1960. The principle of diagenetic facies and some of its implications, *J. Geol.*, 68, 392-407.

Perry, E. A., and Hower, J., 1972. Late-stage dehydration in deeply buried pelitic sediments, *Bull. Am. Assoc. Petrol. Geologists*, 56, 2013-2021.

Price, N. B., 1976. Chemical diagenesis in sediments, *Chem. Oceanog.*, 6, 1-58.

Swinchatt, J. P., 1965. Significance of constituent composition, texture and skeletal breakdown in some recent carbonate sediments, *J. Sed. Petrology*, 35, 71-96.

Taylor, J. H., 1964. Some aspects of diagenesis, *Advanc. Sci.*, 20, 417-436.

Teodorovich, G. I., 1954. Towards questions of studying oil-producing formations (source-rocks), *Byull. Mosk. Obshch. Ispytatelei Prirody, Otdel. Geol.*, 29, 59-66.

Van Hise, C. R., 1904. A treatise on metamorphism, *U.S. Geol. Surv. Monogr. 47*, 1286p.

Von Gümbel, C. W., 1868. Geognostische Beschreibung des ostabayerischen grenzgebirges., Vol. 2 of Geognostische beschreibung des Köigreichs. Bayern. Kassel: Fischer, 700p.

Von Gümbel, C. W., 1888. *Grundzüge der Geologie.* Kassel: Fischer, 1144p.

Walther, J., 1894. *Einleitung in die Geologie als Historische Wissenschaft. 3: Lithogenesis der Gegenwart.* Jena: Fischer, 1036p.

Cross-references: *Authigenesis; Beachrock; Bentonite; Bioturbation; Caliche, Calcrete; Carbonate Sediments—Diagenesis; Carbonate Sediments—Lithification; Cementation; Chert and Flint; Clastic Sediments—Lithification and Diagenesis; Clay—Diagenesis; Coal—Diagenesis and Metamorphism; Color of Sediments; Compaction in Sediments; Concretions; Corrosion Surface; Crude-Oil Composition and Migration; Dedolomitization; Diagenetic Fabrics; Dolomite, Dolomitization; Duricrust; Gases in Sediments; Geodes; Hardground Diagenesis; Interstitial Waters in Sediments; Marine Evaporites—Diagenesis and Metamorphism; Microbiology in Sedimentology; Nodules in Sediments; Paragenesis of Sedimentary Rocks; Petroleum—Origin and Evolution; Pressure Solution and Related Phenomena; Pseudomorphs; Rauhwacke; Secondary Sedimentary Structures; Silicification; Solution Breccias; Syneresis; Volcanic Ash—Diagenesis; Weathering in Sediments.* Vol. IVB: *Diagenetic Changes in Minerals.*

DIAGENETIC FABRICS

The terms *diagenetic fabric* and *diagenetic texture* are often confused. In sedimentary petrology "fabric" (q.v.) is a function of the shapes and arrangement of rock components whether crystalline or otherwise. "Texture" has a wider meaning; it refers to crystallinity, granularity and fabric (compare definitions in the AGI *Glossary of Geology*, 1972; also, Holmes, 1928). Thus, an arenite, for example, may have the same fabric as a rudite, i.e., particles in both may show similar spatial relationships produced by the same depositional or diagenetic influences; it is also possible for similar fabrics to occur in a limestone and a sandstone. *Diagenetic fabric* depends upon the shapes and arrangements of components produced during diagenesis (q.v.). Similar diagenetic fabrics can be produced in chemically and mineralogically dissimilar rocks; diagenetic fabric, therefore is more closely related to process than to composition. Diagenetic fabric largely governs the strength and porosity of a sedimentary rock, and is, therefore, a vital concern in the quality and strength of building stones and aggregates, as well as being a control of the location and movement of fluids and gases through rocks. Many fabric studies, of carbonate rocks in particular, are now in progress, largely because of their economic importance (see *Carbonate Sediments—Diagenesis*). It is likely that current ideas concerning crystallization and recrystallization fabrics are oversimplified, and the value of criteria for the identification and interpretation of specific diagenetic fabrics is under constant review.

Types of Diagenetic Fabrics

Modifications of depositional and post-depositional environments may cause reorganization within a sedimentary accumulation. The extent of such changes will depend upon the intrinsic chemical and physical stability of the sedimentary particles and the relative activity of the pore fluids, and may be limited to partial solution of clastic grains or the filling of pore spaces by cements deposited from solution. Under conditions of increased pressure and temperature, more acute reorganization resulting in widespread recrystallization and/or replacement of partly or completely lithified sediments might occur. Thus, each phase in the diagenetic evolution of a rock may be represented by characteristic fabrics which record the events that have taken place. Diagenetic fabrics, therefore, may be broadly classified according to the processes believed to be responsible for their formation.

Fabrics Resulting from the Activities of Organisms. Microorganisms are very active during the early stages of diagenesis (see *Microbiology in Sedimentology*). The work of boring algae in breaking down skeletal remains (Fig. 1-1) and other types of clastic grains is testified by the numerous fine borings which are common in Holocene carbonate sediments and many ancient limestones. One product of this process is a micrite envelope (Bathurst, 1966; for reference see Bathurst, 1971) believed to originate by the filling of algal borings with microscopic carbonate grains. Fungi and similar organisms are also significant in this disintegration process and leave a network of hyphae which invest and penetrate many types of carbonate grains. Boring algae, fungi, and other microorganisms are important in the micritization of carbonate particles (see Alexandersson, 1972). The importance of fungi in the genesis of certain types of deposits is well documented by Moore (1964).

Compaction. Skeletal elements that enclose cavities—e.g., bivalve shells, brachiopod spines—if not quickly supported by cement, may collapse under the superincumbent load of sediment, resulting in a fabric of crushed and/or dislocated skeletal fragments which may be preserved, more or less in place, by subsequent cementation (Fig. 1-2; see *Compaction in Sediments*). Grain interpenetration and spalling of oolite borders may result (see Fruth et al., 1966). Differential compaction often results in the reorientation of grains in fine-grained sediments. Thus, laminae in fine-grained sediment may be diverted around nodules, lenses of coarser material, or large fossils.

Pressure Solution. Commonly in limestones and dolomites, and sometimes in quartzites and sandstones, i.e., in rocks that are homogeneous in composition, stylolites may be present (Figs. 1-3, 1-4, 1-5; see *Pressure Solution and Related Phenomena*). The fabrics of shells, oolites, and other clastic grains, depositional fabrics, and cement fabrics are very often transected by stylolitic seams (Fig. 1-4). It is, therefore, generally believed that stylolites are pressure-solution phenomena (q.v., for discussion). Pressure solution need not necessarily be restricted to a limited number of seams, as in the case of stylolites, but may operate between grains in three dimensions so as to produce general grain suturing as in many quartzites; some limestones also show this effect. Short intergrain microstylolites are sometimes produced by normal pressure solution in sandstones and limestones (Fig. 1-3; see Bathurst, 1971).

Cementation and Cavity Filling. The outward growth of crystalline material from free surfaces is accomplished by precipitation in pore spaces and cavities, resulting in the formation of characteristic fabrics (Sander, 1951). By this means, multigranular particles in a sediment may acquire fibrous fringes of small crystals oriented with long axes and optic axes normal to the multigranular surface (see *Cementation*). The initial stages in this process can often be clearly seen in recent marine carbonates (Fig. 1-6). Sedimentary quartzites often have such a fibrous cement of chalcedony, and this fabric may be adopted by other minerals, e.g., a lithic (volcanic) sandstone cemented by fibrous chlorite (Fig. 1-8). Such fabrics may be composite (Fig. 1-7).

The cavities in skeletal remains, e.g., ostracods, brachiopods, and corals, are sometimes completely filled by fibrous calcite (Fig. 1-13); but in most limestones, the infilling fabric consists of two elements, the initial peripheral prismatic crystals, and the large, equant grains in the central areas which mark the final stages of filling. The equant crystals in such a mosaic are characterized by plane intergranular boundaries and generally increase in size toward the center. Mosaics have been observed, however, in which the equant grains are of uniform size and are differentiated from the zone of peripheral crystals by an abrupt size increase which suggests a pause in the infilling process. The application of stains to distinguish between iron-bearing and iron-free calcites (see Evamy and Shearman, 1965) may help to establish phases of cementation. A further criterion

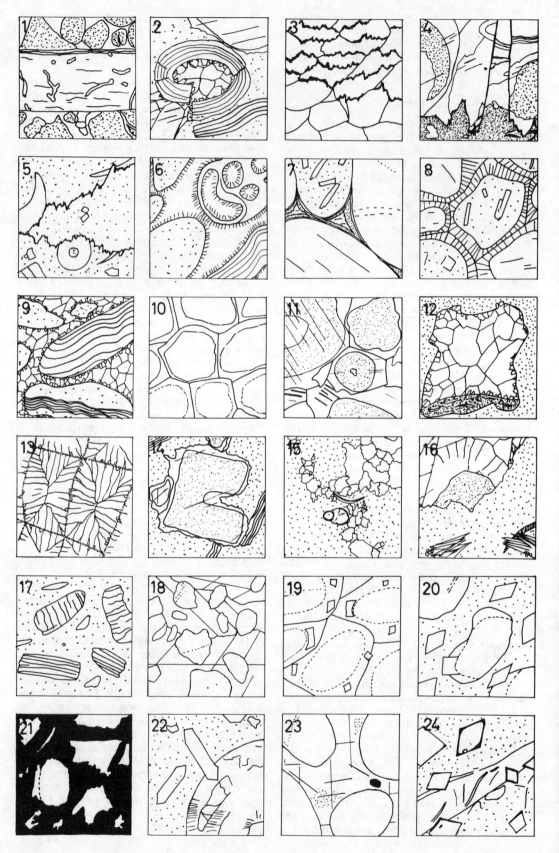

FIGURE 1. *Some Common Diagenetic Fabrics:* (1) Algal borings in a mollusk shell in artificially consolidated grapestone from the Bahamas. (2) Brachiopod spine in transverse section, crushed due to compaction, and later cemented by calcite; the brachiopod spine is associated with crinoid fragments in a calcarenite.
Pressure-Solution Fabrics: (3) Microstylolitic junctions developed between quartz grains as a result of normal pressure solution; dark areas at sutured contacts are clay (after Thomson, 1959). (4) Microstylolite in limestone roughly separating calcarenite from calcisiltite (below); the microstylolite cuts across shell fabrics and the syntaxial cement rims of crinoid fragments. (5) Stylolites in limestone truncating brachiopod shells and crinoid plates.
Cementation and cavity-filling fabrics: (6) An early phase in the cementation process; calcarenite partly cemented by fibrous crystals of aragonite which have grown outward from the boundaries of the clastic grains. (7) The pores in a lithic (volcanic) sandstone filled with a composite drusy fabric composed of a micaceous mineral, chlorite (stippled) with minute serrations directed inward, and feldspar (in the center); the micaceous mineral may be a reconstituted clay film on the clastic grains (Arrowsmith Sandstone, after Glover, 1963). (8) Drusy cement fabric of fibrous chlorite in a lithic (volcanic) sandstone—the simple micro-drusy texture of Glover, 1963. (9) Brachiopod shell fragments and intraclasts in a calcarenite cemented by calcite in the form of a granular cement fabric. (10) Quartz sandstone in which the detrital quartz grains have acquired syntaxial cement rims of quartz; the quartz overgrowths meet along plane or curved boundaries and completely occupy the original pore spaces. (11) Crinoid calcarenite cemented by clear calcite developed as syntaxial cement rims around individual crinoid fragments; note that the rims meet along plane or slightly curved boundaries and that the cleavages in the rims are continuous with those of the original crinoid grains. (12) Para-axial drusy mosaic—a cavity in fine-grained limestone is floored by a layer of internal sediment (microcrystalline calcite containing skeletal debris) and is occupied by clear drusy calcite in the form of a para-axial mosaic; note the plane intergranular boundaries in the mosaic, and the increase in the grain size of the mosaic away from the wall of the cavity. (13) Fibrous calcite occurring as a void filling between coral septa; fibers of calcite have grown from the coral septa towards the center of the area which they now occupy.
Neomorphic Fabrics: (14) Syntaxial replacement rim (neomorphic overgrowth)—a rim of calcite developed syntaxially with a crinoid fragment has replaced the original fine mosaic of the limestone into which it has advanced; note the irregular outer boundary of the rim. (15) Mosaic of coarse neomorphic spar—a mosaic of clear calcite crystals which have grown at the expense of the original fine mosaic of the limestone; note the irregular crystal boundaries, the uneven size distribution, and their cross-cutting relationship with the shell fragments. (16) Neomorphic fibrous calcite—fibrous calcite developed at the expense of the original fine mosaic, and cross-cutting shell fragments in the limestone; a reduction in grain size is apparent in the cement fabrics occupying the upper part of the figure, and here the syntaxial cement rim of a crinoid fragment (stippled) has become fibrous. (17) Vermicular kaolinite crystals in claystone; "fragility of the crystals is proof of *in situ* formation"—Glover (1963) distinguishes (17) and (18) as reorganization textures. (18) Sand grains cemented by large grains of calcite—the calcite of the matrix has recrystallized to form large crystals which enclose the sand grains (Fontainebleau Sandstone, after Glover, 1963); this fabric can be seen in some limestones when the granular cement fabric has recrystallized to form rafts of calcite which enclose the original clastic grains in an incipient poikilitic manner. (19) Dolomitic sandstone showing the "enclosure texture" of Glover (1963)—dolomite is completely enclosed by quartz and some grains are molded (upper center), therefore the dolomite must have formed before the enclosing quartz; note how a molded dolomite has retreated marginally (left center) due to slight solution during silicification.
Replacement Fabrics: (20) Dolomitic sandy marl—dolomite rhombs are partly surrounded by quartz overgrowths; one of the dolomite grains is molded onto a clastic quartz grain which suggests that dolomite developed before secondary quartz (indentation texture; Glover, 1963). (21) Ferruginous quartz sandstone—the quartz grains have been corroded by the ferruginous matrix, and evidence of an earlier diagenetic phase is provided by the overgrowth visible on one quartz grain; corrosion of sand grains is a common feature of calcite cemented sandstones (after Glover, 1963). (22) Euhedral quartz crystals with calcite inclusions replacing the fine calcite mosaic of a limestone, and cross-cutting the fabric of a partly recrystallized foraminifer, Visean Limestone, Derbyshire, England. (23) "Quartz sandstone cemented by sparry barite; the original pyritic and argillaceous matrix is represented by patches of argillaceous impurity, and isolated pyrite grains" (after Glover, 1963). (24) Dolomite rhombs with iron-rich inclusions concentrated along their borders, replacing the original fine calcite mosaic and cross-cutting a large shell fragment, Ordovician limestones, Lake Champlain, US.

for the identification of this type of mosaic was proposed by Bathurst (1964), who described the enfacial junction, in which one of the three angles formed where three intercrystalline boundaries meet in two dimensions is 180°. This is not a feature of mosaics resulting from aggrading neomorphism. If the resultant drusy fabric occupies pore spaces between clastic particles in a sediment, it is called *granular cement* (Bathurst 1958; reference in Bathurst, 1971; Fig. 1-9), but the same drusy fabric in any other cavity is a paraaxial mosaic (Fig. 1-12; Bathurst 1964). The less common drusy mosaic in which the grains are fibrous rather than equant has been called a radiaxial mosaic by Bathurst (1959). Limestones and dolomites commonly show drusy fabrics, which also occur in some cherts (Orme, 1974).

Chemical precipitation of material onto crystal nuclei of the same or similar composition will produce a different kind of cement fabric. This process operates in sediments consisting of grains which are themselves unit crystals; such grains acquire rims of cementing material which is deposited syntaxially, and nonporous fabric results (Figs. 1-10, 1-11). The process is known by a variety of names, e.g., cementation by enlargement, secondary enlargement; the most useful term is *rim cementation* (Bathurst 1958; reference in Bathurst, 1971). It is characteristic of many quartzose sandstones and is a common feature of many calcilutites, calcisiltites, and crinoid calcarenites. Cementation commonly is a polyphase process, and fabric elements attributed to successive phases of cementation may respond to staining techniques (q.v.; Davies, 1972).

Neomorphism. The term *recrystallization* as employed by Folk (1965) is restricted to changes of fabric without changes of mineralogy, as distinct from *inversion,* which involves replacement of a mineral by its polymorph, e.g., aragonite by calcite in a shell. The two processes are collectively termed *neomorphism* (see *Diagenesis*). Extensive and extreme fabric reorganizations often occur during and after compaction and lithification, and the controlling factors usually favor replacement of finer crystal mosaics by coarser crystal mosaics in situ (*aggrading neomorphism*); the reverse is termed *degrading neomorphism.* Such extensive reorganization is conspicuous in some carbonate deposits and evaporites, which are relatively unstable and readily react to changes in the physicochemical parameters of the environment.

The mosaic produced by aggrading neomorphism in limestone was previously known as "coarse grain growth mosaic" (Fig. 1-15; see Bathurst, 1971), which, unlike cement fabrics, shows an irregular variation in size and irregular grain boundaries. Areas of neomorphic spar are sometimes delimited by a well-defined boundary and may cut various textural elements of the limestone, but in other cases it grades into the original fine mosaic; it usually has a patchy distribution.

In poorly washed limestones, clastic grains that are unit crystals, e.g., crinoid plates, are often the loci of recrystallization. This is manifested by the conversion of the original fine mosaic in the vicinity of the crinoid plate to a rim of clear crystalline calcite in optical continuity with it (Fig. 1-14). This rim, which has been called a *syntaxial replacement rim* (Orme and Brown, 1963), is a syntaxial overgrowth of neomorphic spar which has a highly irregular outer boundary. It has clearly advanced at the expense of the fine mosaic and may be seen to truncate the fabrics of skeletal fragments that lie in its path.

Fibrous calcite occurs in some drusy fabrics, but there is evidence that a similar fabric is produced by recrystallization of the fine mosaics of certain limestone (Orme and Brown, 1963; Fig. 1-16); it may also result from the breakdown of larger crystal units—e.g., cement fabrics, crinoid fragments—and it is probably most commonly found in tectonically disturbed beds.

Mica is sometimes produced by the diagenetic reorganization of clay minerals in fine-grained sediment; the development of authigenic kaolinite crystals has also been recorded. Such features usually result in a general grain size increase. Under favorable circumstances, calcite may recrystallize by the seeding of a new mosaic along strain or intergranular boundaries; thus large crystal units may be converted to a granular mosaic. Fibrous calcite developed in cement fabrics, crinoid plates, etc., also breaks up the original crystal units into finer mosaics. Neomorphic processes and products are discussed by Folk (1965), and Bathurst (1971).

Replacement. The replacement of one mineral by another is very common in sedimentary rocks. Occasionally the form of the replaced mineral will be retained by the new mineral, and a partial or complete pseudomorph will result, e.g., gypsum after anhydrite, calcite after gypsum, lutecite after gypsum (see, e.g., West, 1964). Various calcite fossils in limestones are sometimes pseudomorphed by chalcedony. A form of replacement involving simple inversion from aragonite to calcite is common in some skeletal material, and this process is included under the general heading of neomorphism. Replacements are most often discordant, however, and exhibit cross-cutting relationships with the pre-existing fabrics (Fig. 1-20; Orme, 1974). Inclusions of replaced material may survive in the new fabric (Fig. 1-23), or a complete matrix replacement may be expected (Fig. 1-21). The matrices of some rocks, whether consisting of fine detrital material or of crystalline cement, may also suffer replacement.

Methods of Study. In addition to the standard use of the petrographic microscope for the examination of thin sections and peels, electron microscope, electron probe, and x-ray techniques are becoming of increasing importance. Much useful information can be obtained by the application of differential staining methods (see *Staining Techniques*), but the instrument most vital to diagenetic fabric studies is the universal stage (Glover, 1964).

Relationship of Diagenetic Fabrics to Facies and Tectonics

Often in a monomineralic rock the distinction between diagenetic fabrics is not clear, and in the intrinsically unstable carbonates and evaporites the superimposition of one fabric upon another often makes genetic interpretation difficult and complex. Attempts to decipher the course of events recorded by the fabrics must take account of the geological environment. There is, to some extent, a relationship between diagenetic fabrics and facies; for instance, the various cement fabrics are more likely to be found in a well-sorted sandstone or a well-washed calcarenite than in a poorly sorted, sand-sized sediment or a fine-grained accumulation. Coarse mosaics resulting from coalescive neomorphism (Folk, 1965) are most likely to occur in calcilutites or in the fine-grained matrices of poorly washed calcarenites. The nature and complexity of the interstitial fluids in a heterogeneous sediment determine the probability, rate, and course of recrystallization or replacement. A relationship between the type of diagenetic fabric displayed by a fossil shell and the enclosing sediment is postulated by Wilson (1967).

The incidence of aggrading neomorphism often increases toward faults and fold axes, which is not surprising because it is here that the most radical changes of the late diagenetic environmental parameters would occur. Although many of these fabrics have been called diagenetic because there is no evidence of metamorphic forces having been involved, study of the relationships among diagenetic fabrics, facies, and tectonic features is in its infancy. There are, however, prospects that data derived from experimental projects aligned with systematic field investigations will shed light on some of these problems.

Ideas presented over the past few years in the quest for an understanding of diagenetic fabrics have advanced a general appreciation of the complexities involved and have provided an important stimulus to present thought and discussion. It is becoming increasingly clear, however, that the use of optical and morphological criteria alone for the recognition and interpretation of diagenetic fabrics and textures is of limited value and can be very misleading. Anomalies have been pointed out by several researchers; for example Swett (1965) presented evidence for a recrystallization origin of the "drusy" fabric in some ancient oolites, and a replacement origin for some "drusy" quartz in certain cherts is evident (Orme, 1974). Attempts to decipher diagenetic fabrics divorced from geochemical and wider petrological considerations are likely to be abortive and to lead to erroneous conclusions. An excellent discussion of these problems in relation to ancient limestones is given by Folk (1965).

G. RICHARD ORME

References

Alexandersson, T., 1972. Micritization of carbonate particles: Processes of precipitation and dissolution in modern shallow marine sediments, *Bull. Geol. Inst. Univ. Uppsala, N. S.,* **3**, 201-236.

Bathurst, R. G. C., 1959. The cavernous structure of some Mississippian Stromatactis reefs in Lancashire, England, *J. Geol.,* **67**, 506-521.

Bathurst, R. G. C., 1964. The replacement of aragonite by calcite in the molluscan shell wall, in J. Imbrie, and N. Newell, eds., *Approaches to Paleoecology.* New York: Wiley, 357-376.

Bathurst, R. G. C., 1971. *Carbonate Sediments and Their Diagenesis.* Amsterdam: Elsevier, 619p.

Davies, P. J., 1972. Trace element distribution in reef and subreef rocks of Jurassic age in Britain and Switzerland, *J. Sed. Petrology,* **42**, 183-194.

Evamy, B. D., and Shearman, D. J., 1965. The development of overgrowths from echinoderm fragments, *Sedimentology,* **5**, 211-233.

Folk, R. L., 1965. Some aspects of recrystallization in ancient limestones, *Soc. Econ. Paleont. Mineral. Spec. Publ.,* **13**, 14-48.

Fruth, L. S., Jr.; Orme, G. R.; and Donath, F. A., 1966. Experimental compaction effects in carbonate sediments, *J. Sed. Petrology,* **36**, 747-754.

Glover, J. E., 1963. Studies in the diagenesis of some Western Australian sedimentary rocks, *J. Roy. Soc. W. Austral.,* **46**, 33-56.

Glover, J. E., 1964. The universal stage in studies of diagenetic textures, *J. Sed. Petrology,* **34**, 851-854.

Holmes, A., 1928. *The Nomenclature of Petrology,*

with References to Selected Literature. London: Murby, 284p.

Moore, L. R., 1964. The microbiology, mineralogy and genesis of a tonstein, *Proc. Yorkshire Geol. Soc.*, 34, 235-292.

Orme, G. R., 1974. Silica in the Visean Limestones of Derbyshire, England, *Proc. Yorkshire Geol. Soc.*, 40, 63-104.

Orme, G. R., and Brown, W. W. M., 1963. Diagenetic fabrics in the Avonian limestones of Derbyshire and North Wales, *Proc. Yorkshire Geol. Soc.*, 34, 51-66.

Paterson, M. S., and Weiss, L. E., 1961. Symmetry concepts in the structural analysis of deformed rocks, *Geol. Soc. Am. Bull.*, 72, 841-882.

Sander, B., 1951. *Contributions to the Study of Depositional Fabrics. Rhythmically Deposited Triassic Limestones and Dolomites.* Tulsa, Okla.: Am. Assoc. Petrol. Geologists, 207p.

Swett, K., 1965. Dolomitization, silicification and calcitization patterns in Cambro-Ordovician oolites from northwest Scotland, *J. Sed. Petrology*, 35, 928-938.

Thomson, A., 1959. Pressure solution and porosity, *Soc. Econ. Paleont. Mineral. Spec. Pap. 7*, 92-110.

West, I. M., 1964. Evaporite diagenesis in the lower Purbeck beds of Dorset, *Proc. Yorkshire Geol. Soc.*, 34, 315-330.

Wilson, R. C. L., 1967. Diagenetic carbonate fabric variations in Jurassic limestones, *Proc. Geol. Assoc.*, 78, 535-554.

Cross-references: *Carbonate Sediments—Diagenesis; Carbonate Sediments—Lithification; Cementation; Clastic Sediments—Lithification and Diagenesis; Diagenesis; Dolomite, Dolomitization; Fabric, Sedimentary; Microbiology in Sedimentology; Pressure Solution and Related Phenomena; Silica in Sediments; Staining Techniques.*

DIAMICTITE

Terrigenous sedimentary rocks ranging in particle size from clay to boulder dimensions are termed *diamictites*, and their unlithified equivalents, *diamictons* (from the Greek, *diamignymi*—to mingle thoroughly; Flint et al., 1960b). "Diamict" has been used as an even more general term to include both consolidated and unconsolidated types (Harland et al., 1966). Diamictites vary from poorly sorted cobble conglomerate through boulder clay to well-sorted mudstone or shale containing isolated clasts larger than 2 mm; the strictly descriptive definition relates to range of particle size and not to relative abundance of any or all size classes. A practical method of further classifying diamictites has been introduced by Folk (1954), in which the percentage of gravel present and the sand:mud ratio determine the rock name. Under this approach, a rock containing, for example, 5-30% pebbles and having a sand:mud ratio between 1:1 and 9:1 would be termed a *pebbly muddy sandstone*.

Diamictite includes both glacial rocks, *tillites* (see *Till and Tillites*), and nonglacial rocks, *tilloids*, and therefore is somewhat comparable to Pettijohn's (1957) *paraconglomerate*, except that the former may apply to rocks in which clasts are in contact as well as to those in which they are not. The need for such a broad term as diamictite has been justified, first, by the textural variability of the poorly sorted rocks, even within a single stratigraphic unit, and second, by the difficulty inherent in deciphering their mode of origin. For example, diamictite was the only suitable term for Late Paleozoic rocks of varied and problematic origin in the Falkland Islands (Frakes and Crowell, 1967). The term originally proposed for diamictite by Flint et al. (1960a) was *symmictite;* after it was pointed out that this name was used earlier for an homogenized eruptive breccia (including a mixture of country rock plus intrusive rock), however, the name was altered.

Nonsorted and poorly sorted terrigenous sediments that contain a wide range of particle size have been observed in many localities and may originate from many processes other than glacial flow. These include landslides, earthflows, mudflows, solifluction, flowtill activity, subaqueous slumping and sliding, and subaqueous deposition and deformation by floating ice (see Table 1 pages 264-265). Many names have been applied to these deposits in the literature: Pettijohn's "conglomerate mudstone" and "paraconglomerate"; Ackermann's (1951) "Geröllton," equivalent to "tilloid" (nonglacial conglomeratic mudstones); "conglomeratic sandy mudstone"; "mixtite"; and "pebbly mudstone" (q.v.). Flint et al. (1960a, b) pointed out that many of these terms, unfortunately, imply genetic origins where none is known, or are too specific or cumbersome in description for convenient usage. Furthermore, many of the "tillites" described in the past should well be termed diamictites because they might not survive careful genetic scrutiny.

LAWRENCE A. FRAKES

References

Ackermann, E., 1951. Geröllton, *Geol. Rundsch.*, 39, 237-239.

Flint, R. F., 1971. *Glacial and Quaternary Geology.* New York: Wiley, 892p.

Flint, R. F.; Sanders, J. E.; and Rodgers, J., 1960a. Symmictite, a name for nonsorted terrigenous sedimentary rocks that contain a wide range of particle sizes, *Geol. Soc. Am. Bull.*, 71, 507-510.

Flint, R. F.; Sanders, J. E.; and Rodgers, J., 1960b. Diamictite, a substitute term for symmictite, *Geol. Soc. Am. Bull.,* 71, 1809.

Folk, R. D., 1954. The distinction between grain size and mineral composition in sedimentary rock nomenclature, *J. Geol.,* 62, 344-359.

Frakes, L. A., and Crowell, J. C., 1967. Facies and paleogeography of the Late Paleozoic Lafonian Diamictite, Falkland Islands, *Geol. Soc. Am. Bull.,* 78, 37-58.

Harland, W. B.; Herod, K. N.; and Krinsley, D. H., 1966. The definition and identification of tills and tillites, *Earth Sci. Rev.,* 2, 225-256.

Pettijohn, F. J., 1957. *Sedimentary Rocks.* New York: Harper & Row, 718p.

Washburn, A. J.; Sanders, J. E.; and Flint, R. F., 1963. A convenient nomenclature for poorly sorted sediments, *J. Sed. Petrology,* 33, 478-480.

Cross-references: *Avalanche Deposits; Breccias, Sedimentary; Conglomerates; Gravity-Slide Deposits; Mudflow and Debris-Flow Deposits; Pebbly Mudstones; Slump Bedding; Till and Tillite.*

DIATOMITE

Origin

Diatomite, or diatomaceous earth, is one of the biogenic siliceous rocks long known as *tripoli,* or *tripolite,* after the city of that name in N Africa. The rock is characteristically friable and very light in weight (specific gravity 0.4-0.85), porous (70-90%) and light gray to yellowish gray to nearly white in color. This same kind of rock has also been known under various names in different parts of Europe: *Kieselguhr* in Germany, and *randanite* in France, from the Oligocene fresh-water lake of the locality of Randane in the Puy-de-Dome.

In 1836, Christian Fisher, owner of a porcelain factory near Karlsbad, Germany, was the first to unravel the composition of a Kieselguhr, from Franzensbad in Bohemia. Ehrenberg, who presented Fisher's discovery to the Berlin Academy of Sciences, erroneously classified the siliceous remains among the *Infusoria;* hence diatomite was also incorrectly called "infusorial earth."

As the name suggests, a diatomite consists primarily of siliceous (opaline silica) skeletons of diatoms—microscopic, unicellular algae in which the cell wall is impregnated with silica. The term *diatom* (diatoma), given by de Candolle in 1805, is from the Greek "diatomos" (= cut through). The cell wall, or frustule, is indeed formed of two halves, like a box and its lid, generally 30-90μ in diameter, but with extreme sizes from 1μ to 2000μ.

Distribution

Known diatomites have been formed in fresh, brackish, and salt-water environments, the latter essentially from the continental margin provinces, in contrast to present extensive diatomaceous deposits of the deep-sea provinces (see *Pelagic Sediments*). Diatomites are very abundant in the geologic record; but although the oldest diatom genus, *Pyxidicula* is found in Jurassic rock, diatomites are known only from Cenozoic deposits. Marine and brackish-water diatomites occur throughout the world. Freshwater diatomites are equally widespread and quite important in California (Pit River Valley), France, and northern Africa. Certain diatomites in N Africa (Morocco, Algeria, Tunisia) are associated with Eocene phosphatic rocks. Most of the frustules there are epigenized by calcium phosphate. In most diatomites, however, the frustules are well preserved and retain their opaline composition.

Significance

There is often close correlation between diatomaceous deposits and evidence of extensive volcanic activities. This is well illustrated in the well-known Monterey Formation in California and in some diatomites in the Central Massif of France. A volcanic source of silica often seems to be an important factor in the genesis of diatomites in shallow-marine and fresh-water environments. In present deep-sea environments, however, recent oceanographic data indicate that some diatom oozes may depend on volcanic activities, but the limiting factors appear to be nitrogen and phosphate complexes, hence their distribution in areas of upwelling such as Antarctica, the Equatorial Pacific and Indian Oceans, off the Congo River along Africa, and the NW Pacific Ocean.

Diatomites have long been used for various purposes, and their use as abrasives goes back to antiquity. They are also extensively used for filtering in modern industry, and for insulation of safes. In 1867, Nobel first discovered that nitroglycerin (glyceryl trinitrate) could be made safer by absorption in a porous material such as Kieselguhr, which is chemically inert. It is the wealth resulting from this discovery, i.e., dynamite, which is now used to fund the Nobel Prize. Diatomites are sometimes used in building construction, and one such classic use is the marvelous 30-m diameter dome of the Church of Saint Sophia in Istanbul, Turkey.

FLORENTIN MAURRASSE

(References on page 266.)

TABLE 1. Characteristics of Tills and Tillites Compared with Those of Other Diamictons and Diamictites*

Sediment	Clasts			
	Composition	Provenance	Shape	Surface Markings
Subaerial				
Till Drift deformed and injected by ice-pressing	any composition possible—clast size related to crushing strength of rock types	mainly local; minor fraction exotic	some with facets separated by rounded edges; a few "flat-irons"	striations and polish on many
Sliderock Rock-avalanche debris Debris-flow sediments Solifluction sediments	any composition possible	local	predominantly angular	striations exceptionally on soft-rock clasts
Mudflow sediments, including "flow till" Flash-flood alluvium Fan alluvium	any composition possible	confined to drainage basin	predominantly worn	
Subaqueous				
Slump debris ("Fluxoturbidites," etc.)	any composition possible	inherited	inherited	inherited
Glacial-marine drift, in part modified by slump and sliding	any composition possible	resembles that of related till	some with facets separated by rounded edges	some with striations
Subaerial or Subaqueous				
Volcanic breccia	volcanic	local	angular; not faceted	
Fault breccia	any composition possible	local	angular; slickensided clasts faceted	slickensides may resemble glacial striations and/or polish, but may include internal as well as external ones
Collapse breccia	any composition possible; commonly carbonate rocks	local	angular; some may have solution surfaces	solution markings

*After Flint, 1971

Stratification	Primary Fossil Content	Color	Thickness, Extent, and Shape of Body	Stratigraphic Relationships
none, apart from thin lenses and transported bodies of stratified sediment—may possess distinctive fabric	none, apart from rare, broken, transported individuals	inherited, mostly light colored	broad blankets, tongues; may be discontinuous; thickness rarely exceeds 100 m	underlying floor commonly polished, with striations, grooves, and crescentic marks
			tongues, blankets; may be thick	on, or near bases of, slopes
			commonly, tongues; relatively thin	
graded layers in some volcanic mudflow sediments			tongues and fan forms; may be thin	
distorted or completely destroyed	if present, may include mixture from several depth zones	mainly dark colored	local masses; thickness variable up to 500 m	may be interstratified with graded layers and fine-grained marine or lacustrine sediments
nonstratified to stratified, with or without graded layers and/or distortion	may be present, broken or unbroken	color of matrix governed mainly by depth conditions of normal marine sediment. color of glacial fraction inherited	broad blankets; thickness may exceed 300 m	commonly, associated with till in landward direction
may be crudely stratified; more commonly not stratified	charred logs in some volcanic-mudflow sediments		local, irregular bodies and extensive layers	local bodies are related to volcanic vents
			body narrow; thicknesses rarely greater than 15 m	may be associated with slickensided surfaces resembling glaciated floors
			local, irregular bodies, mostly thin	may overlie solution remnants of soluble rocks

References (DIATOMITE)

Benda, L., 1963. Die Diatomeen aus dem Eozan Norddeutschlands, *Paläonf. Zeitschrift,* 39(3-4), 165-187.
Bramlette, M. N., 1946. The Monterey Formation of California and the origin of its siliceous rocks, *U.S. Geol. Surv. Prof. Pap. 212,* 57p.
Hanna, G. D., and Grant, W. M., 1929. Brackish-water Pliocene diatoms from the Etchegoin Formation of Central California, *J. Paleont.,* 3, 87-101.
Ichikawa, W., 1950. Geological studies on the diatomite in Japan, *J. Geogr.,* 58, 675-676.
Ichikawa, W., 1955. On fossil diatoms from the Onikoube basin, Miyagi-prefecture, collected by Dr. N. Katayama, *Kanazawa Univ. Sci. Rept.,* 4(1), 151-175.
Moore, B. N., ed., 1937. Non-metallic mineral resources of eastern Oregon, *U.S. Geol. Surv. Bull.* 875, 180p.
Moret, L., 1964. *Manuel de Paléontologie Végétale.* Paris: Masson, 244p.
Taliaferro, N. L., 1933. The relation of volcanism to diatomaceous and associated siliceous sediments, *Univ. Calif. Publ. Geol. Sci.,* 23(1), 1-56.
Van Landingham, S. L., 1966. Origin and development of Dry Lake diatoms in the Great Basin region of Nevada (Abs.), *J. Phycol.,* 2, 8.
Varley, E. R., 1940. Diatomite in Brazil, *Bull. Imp. Inst.,* 38, 240-242.

Cross-references: *Chert and Flint; Pelagic Sedimentation, Pelagic Sediments; Phosphates in Sediments; Silica in Sediments.* Vol. VII: *Diatomacea.* Vol. XIII: *Diatomaceous Earth and Diatomite Technology.*

DISH STRUCTURES—See WATER-ESCAPE STRUCTURES

DISMICRITE—See BIRDSEYE LIMESTONE

DISPERSAL

The term *dispersal* combines the closely related processes of transportation (q.v.) and deposition (q.v.). From the moment that weathered material is separated from the parent mass until its final resting place is reached, the forces of transportation and the depositional regime act to determine the characteristics of the accumulated product. The region over which these actions take place, in relation to the origin or provenance (q.v.), is known as the dispersal shadow. An area of sediment derived from a common provenance is called a *sedimentary petrologic province*.

Transportation. Products of weathering, in the form of particles or solution, may be transported by one or more agents such as water, atmosphere, ice, gravity and temperature change, and organisms. Transportation may be effected directly from the source to the ultimate site of deposition in a brief span of time, or in stages covering an enormously long geologic interval.

Deposition. The nature of a deposit is dependent on the character of the material transported, the effects of transportation on the material, the conditions causing deposition, and the environment of deposition. The three main classes of depositional environments are: terrestrial, paralic, and marine.

MAURICE SCHWARTZ

Cross-references: *Deposition; Provenance; Transportation.*

DOLOMITE, DOLOMITIZATION

Dolomite is used in two senses in geology; first, for a mineral, a carbonate double salt, $CaMg(CO_3)_2$; and also for a sedimentary rock composed predominantly of the mineral dolomite. *Dolostone,* introduced by Shrock (1948) as a rock name specifically to avoid the ambiguity which sometimes arises from the double usage, has found some acceptance. Dolomite is named after D. Dolomieu, a French engineer who studied the rock.

Mineral

The mineral properties of dolomite are similar to calcite and other less common carbonate minerals (Lippmann, 1973). Dolomite may be distinguished from calcite by its relatively slow reaction in dilute acid, by staining techniques, and by its commonly euhedral rhombic form in thin section, as well as by x-ray diffraction techniques. Useful stains for distinguishing dolomite, which may be used in rock slabs or thin section, are titan yellow, alizarin red S, and potassium ferricyanide (see *Staining Techniques*).

The crystal structure of dolomite is, essentially, regularly alternating layers of $CaCO_3$ (with calcite structure) and $MgCO_3$. Random distribution of Mg and Ca atoms, produced by marine organisms, results in distorted calcite structure with $MgCO_3$ substitution limited to less than about 20 mole%. $CaCO_3$ substitution up to about 10 mole% in the dolomite lattice is common in sedimentary and low-temperature experimental dolomites. Such substitution results in altered lattice dimensions, which may be used to estimate the degree of calcite substitution, and, commonly, in the absence of certain superstructure or ordering reflections from x-ray diffractograms. These carbonates were termed *protodolomite* by Graf and Goldsmith (1956). Iron substitutes for magnesium in the dolomite lattice to form an apparently complete solid-solution series with $CaFe(CO_3)_2$. Varieties with more than 20 mole% Fe are generally called *ankerite*. Small amounts of iron produce the characteristic brown or tan weathering color of many sedimentary dolomites. Manganese also substitutes for Fe in dolomite and ankerite; it may form an analogous solid-solution series.

Sedimentary Rock

Dolomite is a common sedimentary rock, and its relative abundance increases systematically with age at least back through the Phanerozoic. Estimates of the abundance of carbonate rocks (limestone and dolomite) range from 5-15% of all sedimentary rocks; the proportion of this amount that is dolomite varies from nearly zero in very young rocks to about 60% in the oldest carbonates (Fig. 1). Although many carbonate rocks are mixtures of the minerals calcite and dolomite (as well as detrital quartz and clay minerals), there exists a striking tendency toward the single-mineral end members (Fig. 2).

Origin of Dolomite

The origin of dolomite is a long-standing problem in geology. Recent studies have shown convincingly that all dolomite is not of the same origin, but it is still difficult to establish the origin of dolomite in many cases and to answer some general questions about its origin. Some of the current theories may be tested by reviewing various dolomite occurrences.

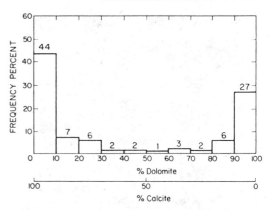

FIGURE 2. Percent of dolomite and calcite computed from 1148 analyses of carbonate rocks of North America plotted by frequency of occurrence (after Steidtmann, from Blatt et al., © 1972, by permission of Prentice-Hall, Inc.).

Textures. Dolomite often replaces calcite, as demonstrated, e.g., by dolomite rhombs which cross the boundaries of $CaCO_3$ fossils or other grains in limestone (Fig. 3a). Outlines of former grains may be preserved even in rocks composed entirely of dolomite (Fig. 3b). In partially dolomitic rocks, fossils (even specific kinds of fossils), coarse-grained lenses, fine-grained matrix, mud fragments, fossil burrows, or other inhomogeneities may be selectively composed of dolomite *or* preserved as calcite in contrast to the surrounding rock. This selectivity is commonly interpreted to reflect control on dolomitization by slight differences in permeability, in crystal size, or in impurities in the original mineral. Evidence of this type has led to the generalization that many, if not most, dolomites are formed by the replacement of $CaCO_3$, as concluded long ago by Steidtmann (Friedman and Sanders, 1967, cite early basic papers; Blatt et al., 1972, review the evidence for dolomitization of limestone). On the other hand, many dolomites lack clear evidence of replacement, possibly because either dolomitization destroyed the outlines of preexisting grains (partial destruction can often be observed), or the original sediment was homogeneous lime mud, or the dolomite was not formed by replacement.

Distribution Patterns. The large-scale distribution of dolomite follows two or possibly three very generalized patterns: (1) some dolomite occurs as distinct beds or formations interlayered with other sedimentary rocks, including limestone. The contacts with adjacent beds may be sharp or gradational and a dolomite bed may grade laterally on a scale of

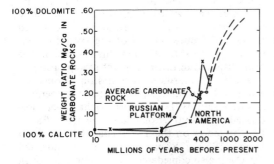

FIGURE 1. Weight ratios of Mg to Ca in carbonate rocks from North America and the Russian Platform as a function of age (log scale) (after Garrels and Mackenzie, 1971).

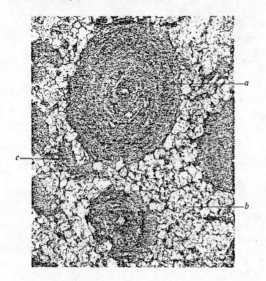 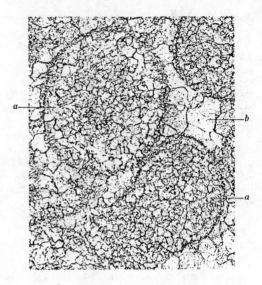

FIGURE 3. Replacement dolomite (from Cayeux, 1970, copyright © 1970 Hafner Publishing Company). *Left:* Partial dolomitization of ooids (a) in matrix of dolomite crystals, (b), which probably replaced fine-grained calcium carbonate matrix (c). Large ooid (a) is 0.7 mm across. *Right:* Complete dolomitization of ooid grains (a), with outlines preserved, cemented by coarser-grained dolomite (b). Large ooid is 0.7 mm across.

kilometers into another rock type, typically limestone or evaporite (gypsum and anhydrite). (2) Other dolomites cut across bedding so that their lateral contacts with limestone or other rocks are sharp or gradational on a scale of centimeters or meters. The distribution of dolomites of this type is often demonstrably related to fractures or joints in the rocks that can control the movement of fluids. (3) Dolomite may also occur as local pods or regional masses beneath unconformities in the continental interiors. There is general agreement that the second and third pattern types are of replacement origin because of their relationships to adjacent limestones. Origin of many of the first type is controversial; arguments are reminiscent of those among economic geologists on the hydrothermal vs. sedimentary origin of stratabound ore deposits.

Precipitation of Dolomite. Dolomite is not known to be precipitating from sea water of normal salinity (35‰) anywhere in the world at present, either directly or as a replacement of $CaCO_3$. This fact, coupled with the greater abundance of dolomite in older rocks, has been used to argue that the composition of sea water has changed during the past 600 m.y. It could be argued, however, that older dolomites are of replacement origin and that the odds of being in an environment favorable to replacement increase with time available.

Possible modern environments of direct precipitation of dolomite, with some degree of ordering, are the Coorong Lagoon in Australia and Deep Springs Lake, California. Dolomite crystals from the sediment at each locality are as young as 300 yr. The Coorong is a series of shallow, ephemeral lakes now isolated from the ocean but containing essentially sea water modified by annual cycles of runoff and evaporation and by carbonate precipitation (von der Borch, 1965). Among a series of carbonate minerals formed in the various lakes, partially ordered dolomite is found in lakes with high $Mg^{++}:Ca^{++}$ ratios, high pH, and very slow sedimentation rates. The possibility of dolomite formation by alteration of $CaCO_3$ sediment cannot be conclusively ruled out. Deep Springs Lake is a shallow playa whose spring-fed water bears little relation to sea water. Dolomite formation has been explained as extremely slow direct precipitation of a calcium-rich precursor which is modified to dolomite by diffusion of Mg^{++} inward and Ca^{++} outward through the outer 100 Å of the crystal (Peterson et al., 1966), although this conclusion has been disputed in favor of direct precipitation.

Direct submarine precipitation from sea water with normal Mg:Ca ratios but perhaps elevated salinity (<2× normal) has been postulated for 3000 yr-old dolomite from Baffin Bay, a lagoon on the Texas coast (Behrens and Land, 1972). Replacement of precipitated $CaCO_3$ is also considered a possibility, although no evidence of replacement was observed. Direct precipitation has been postulated for many older dolomites but in most cases can be neither proved or disproved.

Solution Chemistry. Direct precipitation of dolomite at sedimentary temperatures has not been accomplished experimentally. Ordered dolomite has been produced by replacement of Mg-rich calcium carbonate at temperatures above several hundred degrees C, and *protodolomite* or Mg-calcites have been produced by direct precipitation or replacement at lower temperatures. In no case did the experimental conditions very closely approach conditions in sea water or in other natural settings. Typical starting materials were solutions of $CaCl_2$, $MgCl_2$, and $NaHCO_3$; CO_2 pressures were elevated in many cases; or Mg:Ca ratios were very high (Zenger, 1972a, reviews experimental work).

The difficulty, if not impossibility, of precipitating dolomite has precluded accurate determination of the equilibrium constant for its formation or its solubility product, fundamental values which can be considered reliable only when equilibrium is approached from both solution and precipitation. Using various estimates, Hsü (1967) concluded that for the reaction $Ca + Mg + 2CO_3 \rightleftharpoons CaMg(CO_3)_2$, the best value for the equilibrium constant

$$K_{dol} = \alpha_{Ca^{++}} \cdot \alpha_{Mg^{++}} \cdot \alpha^2_{CO_3^=}$$

is about 10^{-17}. Older estimates range as widely as 3×10^{-17} to 4.7×10^{-20}.

Applying this value to normal surface sea water, for which relevant ionic activities have recently been estimated (see Berner, 1971), indicates great supersaturation with respect to dolomite. In sea water, K_{dol} is about 10^{-15}. Dolomite also appears to be stable in sea water relative to calcite and to aragonite, the form of $CaCO_3$ most commonly precipitated from sea water. The reason that dolomite does not form readily, if at all, from normal sea water may be that formation of the regular interlayering of $CaCO_3$ and $MgCO_3$ in the dolomite lattice is a very slow process.

Stable Isotopes. Studies of the stable isotopes of carbon and oxygen, which have been very useful in understanding origin and alteration of other carbonates, have been severely handicapped by the inability to synthesize dolomite under earth-surface conditions. The concentration of O^{18} (δO^{18})—usually reported as the ratio $O^{18}:O^{16}$ in the sample compared to a standard—is controlled in carbonates by the $O^{18}:O^{16}$ and the temperature of the water in which the reaction occurs. δO^{18} values measured in a carbonate mineral for which the temperature dependence is known can aid in interpretation of its history if a reasonable guess can be made as to the δO^{18} value of the water or the temperature of reaction. In dolomite, however, the temperature dependence can be determined with confidence only at the higher temperatures at which dolomite can be synthesized. Extrapolation of these values to surface conditions leads to predicted δO^{18} in dolomite of +8 relative to calcite precipitated under the same conditions (temperature dependence in calcite is well known). Isotopic data on natural dolomites apparently formed at low temperatures and on coexisting calcites suggest that the difference between dolomite and calcite formed at the same temperatures may be much less than 8, if a difference indeed exists, a controversial point. Some workers, confident that a difference does exist, interpret similar values in coexisting dolomite and calcite to mean that the dolomite formed by alteration of the calcite, inheriting the calcite isotopic values with little or no alteration.

Quaternary Occurrences of Dolomite. Until the 1960s, the problems already mentioned were compounded by the fact that virtually no locations were known where dolomite was actively forming. The situation has since changed dramatically with the discovery of a number of occurrences of young dolomite in various parts of the world (Table 1). Generalizations about the occurrences are possible only if a number of exceptions are permitted. (Exceptions are noted below in parentheses by numbers keyed to Table 1.)

1. Dolomite replaces a preexisting $CaCO_3$, i.e., it is penecontemporaneous dolomitization rather than dolomite precipitation (1?,8?,9?,12,13?,14?).
2. The dolomites are only partially ordered (1,12) and generally have more than 50 mole % $CaCO_3$. They are something between protodolomite (unordered) and ordered stoichiometric dolomite.
3. The dolomitizing fluid is sea water (9–14,16?, 17?).
4. The dolomitizing waters are of elevated salinity, greater than normal sea water (11,12,16,17; 8 is slight).
5. Mg:Ca ratios are elevated above normal sea water where sea water is the dolomitizing fluid (8?).
6. Dolomitization occurs in environments with intermittent subaerial exposure, resulting from tides or seasonal evaporation (8,12?,15?16–18), a primary cause of elevated salinities.
7. Slow sedimentation characterizes environments of young dolomitization. Quantitative estimates are generally lacking, particularly as to contrasting rates between dolomitized sediment and adjacent undolomitized sediment.

These generalizations have led to considerable stress being placed on the importance of elevated Mg:Ca ratios (see, for example, Folk and Land, 1975), of sea water of elevated salinites, and of supratidal environments in penecontemporaneous dolomitization. The fact that exceptions to each generalization do exist strongly suggests that there is more than one way to

TABLE 1. Some Occurrences of Quaternary Dolomite

Location	Environment	Mechanism	Age (yr B.P., min.)	CaCO$_3$ in Dolomite (%)	pH	Salinity (‰)	Mg^{++}/Ca^{++} in water	Reference
1. Coorong Lagoon, S. Australia	ephemeral lake, altered sea water	direct precipitation? replacement? of Mg calcite	300 ± 250	54	8.3–10.2	30–85	to 15	von der Borch, 1965
2. Bonaire, Neth. Antilles	supratidal margin of saline pond	replacement of aragonite	1480 ± 140	54–56	—	60–240	to 47	Deffeyes et al., 1965
3. Qatar, Persian Gulf	supratidal flat (sabkha)	replacement of aragonite	2450 ± 130	53–55	6.1–6.6	to 270	5–20	Illing et al., 1965
4. Trucial Coast, Persian Gulf	supratidal flat (sabkha)	replacement of Ca carbonate	contemporary?	—	6.0–6.4	to 295	to 35	Butler, 1969
5. Andros Island, Bahamas	supratidal flat	replacement of aragonite	<160	55–62	—	30–83	5.5–7.5	Shinn et al., 1965
6. Sugarloaf Key, Florida	supratidal flat	replacement of aragonite	300	60 / 56–70	7.4–8.3	to 200 / 45–100	to 40 / 3.7–8.5	Shinn, 1968; Atwood and Bubb, 1970
7. Fuerteventura, Canary Is.	shore zone terrace, supratidal	replacement of Mg calcite in Pleistocene calcarenite	Holocene?	56	—	140–350??	"high"	Müller and Tietz, 1966
8. Baffin Bay, Texas	shallow marine lagoon	precipitation? replacement?	2,310 ± 60	53–56	—	45–85	5.3?	Behrens and Land, 1972
9. Deep Springs Lake, Calif.	playa lake	precipitation with Ca-rich precursor	290 ± 50	55	9.5–10	hypersaline	—	Petersen et al., 1963
10. Salt Lake, Turkey	playa lake	replacement of Ca carbonate	6,000 ± 40 and Pleistocene	57	—	hypersaline	85–150	Müller and Irion, 1969
11. Carlsbad Caverns, N. Mex.	cave	replacement of aragonite	contemporary?	54–57	6.7–7.5	low	to 14	Thrailkill, 1968
12. Coast Ranges, Calif.	perennial streams	direct precipitation	contemporary? (2000)	50–60	8.2–11.8	<2	to 31	Barnes and O'Neill, 1971
13. Salt Flat, Texas	playa lake	direct precipitation? replacement of aragonite?	20,300 ± 825	—	—	—	—	Friedman, 1966
14. Great Salt Lake Desert, Utah	former playa lake	direct precipitation? replacement	11,150 ± 250	50	—	hypersaline?	—	Graf et al., 1961
15. Bonaire, Neth. Antilles	marine foreslope, reflux from lagoon?	replacement	Plio-Pleistocene	—	—	brine?	—	Deffeyes et al., 1965
16. Falmouth Fm., N. Jamaica	reef, in meteoric water	replacement and precipitation in voids	30,000?	56	—	low	—	Land, 1973b
17. Hope Gate Fm., N. Jamaica	reef, in meteoric water	replacement and precipitation in solution voids	Yarmouth? >300,000	58	—	low?	—	Land, 1973a
18. Pacific Atolls	lagoon-reef-slope limestones, sea water with periodic exposure	replacement of calcite, precipitation in voids	Plio-Pleistocene	54–62	—	—	—	Schlanger, 1963

form dolomite, an impression reinforced by the number of models that have been proposed to explain various examples of young dolomite and ancient dolomitic rocks.

Conditions for Dolomite Formation. If dolomite is to form by direct precipitation, which has yet to be demonstrated, then (1) the solution must be saturated with respect to dolomite; (2) ions must be supplied at an adequate rate to produce the observed volume of dolomite in the time available; and (3) time appears necessary to allow ordering of the dolomite lattice, which might also be expressed as an activation energy because dolomite forms rapidly in the range of 400°C, but very slowly, if at all, at low temperatures.

For dolomite to replace $CaCO_3$, which seems geologically the important case, added conditions are necessary: (4) The solution must be undersaturated with respect to the $CaCO_3$ mineral replaced. The simultaneous constraints of conditions (1) and (4) have led to the practice of formulating dolomitizing environments in terms of the Ca:Mg ratio. Condition (2) is modified in that it is only necessary to supply Mg^{++} in the requisite volume; Ca^{++} and $CO_3^=$ are already present in $CaCO_3$. Because Ca^{++} is replaced, an added requirement is: (5) Ca^{++} must be removed from the system to maintain undersaturation relative to $CaCO_3$. Conditions (2) and (5) are functions of fluid movement, in many cases through a porous sediment, which requires consideration of sediment permeability and a driving mechanism.

Each of the above five conditions can be defined as necessary, but it is not certain that these are sufficient conditions. Certainly many other conditions have been suggested, e.g., elevated salinity, high pH, increased pressure, depth of burial, elevated temperature, presence of algal mats, reducing environment, Mg-calcite precursor, etc.

Models for Dolomite Formation

Direct precipitation from sea water continues to be suggested for some ancient dolomites where evidence of other processes is lacking, but present surface sea water is highly supersaturated for dolomite and is nowhere known to be precipitating dolomite directly or dolomitizing sediment on the open sea floor, except possibly in Baffin Bay (no. 8, Table 1) and the African continental slope (Siesser, 1972). Skepticism is warranted. The time needed to achieve the ordering of dolomite is commonly cited to explain the lack of dolomite in open-marine sediments; the slow sedimentation rates of the deep sea and the physicochemical conditions near the lysocline (q.v.) should, therefore, favor dolomitization. However, no significant case of dolomitization of deep-sea surface sediments has been reported. Dolomite has been found by JOIDES drilling in deep-sea sediments as young as Oligocene. It is highly unlikely that these sediments have ever been exposed to waters with salinity appreciably different from normal sea water, and the origin of such dolomites has not been adequately explained.

Many young dolomites are found in the *supratidal zone* (Table 1, nos. 1–7), and many dolomites in the rock record are associated with sedimentary structures and facies patterns suggestive of deposition on tidal flats (see *Sabkha Sedimentology*). The young dolomites result from penecontemporaneous dolomitization of calcareous sediments, generally by marine pore waters of at least slightly elevated salinities and elevated Mg:Ca ratios. The Caribbean dolomites (Table 1, nos. 2,5,6) are cemented crusts forming near the high-tide zone, but the associated pore-water salinity varies. Persian Gulf dolomites (nos. 3,4) are unconsolidated sediments underlying sabkhas. Salinities are elevated by evaporation of pore water and the Mg:Ca ratio by precipitation of gypsum (nos. 2,3,4) or $CaCO_3$ (nos. 5,6). These conditions endure for periods of one to several months, between influxes of new sediment and sea water with a fresh supply of Mg. The dolomitized sediment contains metastable $CaCO_3$ phases (aragonite and Mg calcite), which appear to be preferentially replaced. It is not at all clear, however, that aragonite and Mg calcite should not also be highly supersaturated in the pore-water solutions. Similar conditions exist in the Coorong Lagoon (no. 1), which is not supratidal in the usual sense. Supratidal penecontemporaneous dolomitization appears likely for the numerous ancient dolomites associated with sedimentary structures and facies suggestive of tidal-flat deposition (Gebelein and Hoffman, 1973; Zenger, 1972b). Dolomitization in young tidal flats is less extreme than in the ancient analogs (Fritz and Katz, 1972), suggesting later diagenetic dolomitization. Greater contrasts exist in scale; ancient analogues are much more extensive, and although such contrasts may be explained by the greater time intervals of dolomitization, it is also possible that the difference is more fundamental.

Many ancient dolomites are associated with extensive deposits of anhydrite ($CaSO_4$) and gypsum ($CaSO_4 \cdot 2H_2O$). This association led to the model of dolomitization by *"seepage refluxion"* (Adams and Rhodes, 1960). Evaporation of sea water to the concentration for gypsum precipitation (> 140‰) requires some restriction in circulation, such as a geographic barrier. Continuous precipitation of gypsum without precipitation of halite (NaCl) or bitter salts (K and Mg salts) requires removal of the brine before its concentration reaches 350‰. Such dense brines might escape from an evaporating shelf by sinking down into the underlying sediments, displacing lighter pore waters, and seeping back to the deep ocean. Calcium carbonate sediments could be dolomitized by the brine, which would have a high Mg:Ca ratio through gypsum precipitation and would provide a continuous flux of Mg^{++} as well as a water flow to remove Ca^{++} liberated by dolomitization. This mechanism appears capable of extensive dolomitization on a regional scale. It also appears compatible with the geologic setting of such extensively dolomitized rocks as the Permian shelf carbonates in W Texas, where it was originally deduced, and the Edwards Formation in central Texas (Fisher and Rodda, 1969).

Modern examples of this model have been difficult to identify. Very small-scale refluxion with slight evaporative concentration has been suggested in the coastal zone of Fuerteventura, Canary Islands (Table 1, no. 6), and on a larger scale from gypsum pans on Bonaire, Netherlands Antilles (Deffeyes et al., 1965), although study has shown that the reflux in Bonaire may be by rapid seasonal outflow through conduits rather than continuous seepage through sediment pores. Bonaire is widely cited as a modern example of dolomitization by seepage refluxion, but the only documented modern dolomite from Bonaire is from crusts in the supratidal zone of the gypsum pan; the postulated reflux dolomite is from Pleistocene rocks inferred to have underlain Pleistocene gypsum pans now removed by erosion.

Models for *dolomitization by ground water* have not been widely accepted, in part because the quantity of Mg^{++} that can be supplied at normal ground-water circulation rates is too small to effect large-scale dolomitization. Recently, however, it has been suggested that significant dolomitization may occur in the zone of mixing of fresh water and sea water, which terminates most fresh water aquifers near the shore line. Circulation is dynamic; sea water, entrained by the outflowing freshwater wedge, provides a constant supply of Mg^{++}. Theoretical studies show that another condition for dolomitization may be satisfied in that mixing of sea water and fresh water in some proportions can produce a mixture that is undersaturated with respect to calcite and supersaturated with respect to dolomite, even though both phases may have been at saturation in each of the original waters (Fig. 4)—possible

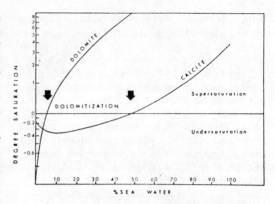

FIGURE 4. Relative saturation of dolomite and calcite in mixtures of fresh water and sea water (from Badiozamani, 1973). Mixtures containing 5-50% sea water (arrows) should be supersaturated with respect to dolomite and undersaturated with calcite, a condition for dolomitization.

because of nonlinear variations in the activity coefficients of Mg^{++} and Ca^{++}. Badiozamani (1973) combined these conditions in a model for the extensive Ordovician dolomites in Wisconsin; essentially the same model has been applied to Pleistocene dolomites in Jamaica (Table 1, nos. 16, 17).

Heating of ground water at depths to produce convective circulation has been suggested as another means of providing sufficient flux of Mg^{++} for dolomitization by ground water (Lovering, 1969). Dolomitization by waters at elevated temperatures (several hundred °C) is indicated by oxygen isotope values (low δO^{18}) in extensive dolomites associated with some ore deposits. Lovering's work illustrates that increased temperature, in addition to providing the activation energy apparently required for formation of ordered dolomite, shifts the $Mg^{++}:Ca^{++}$ ratio at equilibrium with dolomite and calcite to progressively lower values (Fig. 5). This combination may also account for dolomites associated with igneous intrusions encountered in the Deep Sea Drilling Project.

Dolomite occurrences in *continental saline lakes* (Table 1, nos. 9,10,13,14) elucidate some factors in the formation of dolomite, but the individuality of water chemistry in land-locked lakes prevents their use as general models for dolomite formation. A number of dolomites in the geologic record appear to be adequately explained by *detrital reworking* of preexisting dolomites.

The variety of models, each of which seems satisfactory for some dolomites, illustrates why no single answer has been forthcoming to the "dolomite problem" over the years. Despite

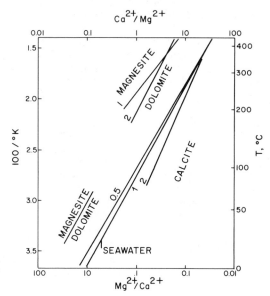

FIGURE 5. Ionic ratios Ca^{++} and Mg^{++} in chloride solutions precipitating or replacing calcite, dolomite, and magnesite at varying temperatures (after Lovering, 1969). Apparent stability fields are determined by coexistence, precipitation, or replacement by the mineral indicated. Inclined lines indicate molarity of $(Ca^{++} + Mg^{++})$. All data are between 18-35°C (sea water), or 290-420°C (experimental solutions). Normal sea water Mg^{++}/Ca^{++} is plotted for reference; solution chemistry suggests this should be in the dolomite field.

the mineralogical complexity of dolomite, it apparently may form in a variety of geologic settings if the basic conditions are satisfied.

PAUL ENOS

References

Adams, J. E., and Rhodes, M. L., 1960. Dolomitization by seepage refluxion, *Bull. Am. Assoc. Petrol. Geologists,* 44, 1912-1920.

Atwood, D. K., and Bubb, J. N., 1970. Distribution of dolomite in a tidal flat environment, Sugarloaf Key, Florida, *J. Geol.,* 78, 499-505.

Badiozamani, K., 1973. The dorag dolomitization model—application to the Middle Ordovician of Wisconsin, *J. Sed. Petrology,* 43, 965-984.

Barnes, I., and O'Neil, J. R., 1971. Calcium-magnesium carbonate solid solutions from Holocene conglomerate cements and travertines in the Coast Ranges of California, *Geochim. Cosmochim. Acta,* 35, 699-711.

Behrens, E. W., and Land, L. S., 1972. Subtidal Holocene dolomite, Baffin Bay, Texas, *J. Sed. Petrology,* 42, 155-161.

Berner, R. A., 1971. *Principles of Chemical Sedimentology.* New York: McGraw-Hill, 41-73, 148-157.

Blatt, H.; Middleton, G.; and Murray, R., 1972. *Origin of Sedimentary Rocks.* Englewood Cliffs, N.J.: Prentice-Hall, 477-500.

Borch, C. von der, 1965. The distribution and preliminary geochemistry of modern carbonate sediments of the Coorong area, South Australia, *Geochim. Cosmochim. Acta,* 29, 781-799.

Butler, G. P., 1969. Modern evaporite deposition and geochemistry of coexisting brines, the sabkha, Trucial Coast, Arabian Gulf, *J. Sed. Petrology,* 39, 70-90.

Cayeux, L., 1970. *Carbonate Rocks.* Darien, Conn.: Hafner, 506p. Original French edition, 1935.

Chilingar, G. V.; Bissell, H. J.; and Fairbridge, R. W.; eds., 1967. *Carbonate Rocks,* Amsterdam: Elsevier, A, 471p., B, 413p.

Deffeyes, K. S.; Lucia, F. J.; and Weyl, P. K., 1965. Dolomitization of Recent and Plio-Pleistocene sediments by marine evaporite waters on Bonaire, Netherlands Antilles, *Soc. Econ. Paleont. Mineral. Spec. Publ. 13,* 71-88.

Fisher, W. L., and Rodda, P. U., 1969. Edwards Formation (Lower Cretaceous), Texas: Dolomitization in a carbonate platform system, *Bull. Am. Assoc. Petrol. Geologists,* 53, 55-72.

Folk, R. L., and Land, L. S., 1975. Mg/Ca ratio and salinity: Two controls over crystallization of dolomite, *Bull. Am. Assoc. Petrol. Geologists,* 59, 60-68.

Friedman, G. M., 1966. Occurrence and origin of Quaternary dolomite of Salt Flat, West Texas, *J. Sed. Petrology,* 36, 263-267.

Friedman, G. M., and Sanders, J. E., 1967. Origin and occurrences of dolostones, in Chilingar et al., 1967, pt. A, 267-348.

Fritz, P., and Katz, A., 1972. The sodium distribution of dolomite crystals, *Chem. Geol.,* 10, 237-244.

Garrels, R. M., and MacKenzie, F. T., 1971. *Evolution of Sedimentary Rocks.* New York: Norton, 397p.

Gebelein, C. D., and Hoffman, P., 1973. Algal origin of dolomite laminations in stromatolitic limestone, *J. Sed. Petrology,* 43, 603-613.

Graf, D. L., and Goldsmith, J. R., 1956. Some hydrothermal syntheses of dolomite and protodolomite, *J. Geol.,* 64, 173-186.

Graf, D. L.; Eardley, A. J.; and Shimp, N. F., 1961. A preliminary report on magnesium carbonate formation in Lake Bonneville, *J. Geol.,* 69, 219-223.

Hsü, K. J., 1967. Chemistry of dolomite formation, in Chilingar et al., 1967, pt. B, 169-192.

Ilich, M., 1974. Hydrothermal-sedimentary dolomite: The missing link? *Bull. Am. Assoc. Petrol. Geologists,* 58, 1331-1347.

Illing, L. V.; Wells, A. J.; and Taylor, J. C. M., 1965. Penecontemporary dolomite in the Persian Gulf, *Soc. Econ. Paleont. Mineral. Spec. Publ. 13,* 89-111.

Land, L. S., 1973a. Contemporaneous dolomitization of Middle Pleistocene reefs by meteoric water, north Jamaica, *Bull. Marine Sci.,* 23, 64-92.

Land, L. S., 1973b. Holocene meteoric dolomitization of Pleistocene limestones, North Jamaica, *Sedimentology,* 20, 411-424.

Lippmann, F., 1973. *Sedimentary Carbonate Minerals.* Berlin: Springer-Verlag, 228p.

Lovering, T. S., 1969. The origin of hydrothermal and low temperature dolomite, *Econ. Geol.,* 64, 743-754.

Müller, G., and Irion, G., 1969. Subaerial cementation and subsequent dolomitization of lacustrine carbonate muds and sands from Paleo-Tuz (Salt Lake), Turkey, *Sedimentology*, **12**, 193-204.
Müller, G., and Tietz, G., 1966. Recent dolomitization of Quaternary biocalcarenites from Fuerteventura (Canary Islands), *Contr. Mineral. Petrology*, **13**, 89-96.
Peterson, M. N. A.; Bien, G. S.; and Berner, R. A., 1963. Radiocarbon studies of Recent dolomite from Deep Spring Lake, California, *J. Geophys. Research*, **68**, 6493-6505.
Peterson, M. N. A.; von der Borch, C. C.; and Bien, G. S., 1966. Growth of dolomite crystals, *Am. J. Sci.*, **264**, 257-272.
Schlanger, S. O., 1963. Subsurface geology of Eniwetok Atoll, *U.S. Geol. Surv. Prof. Pap. 260-BB*, 991-1038.
Shinn, E. A., 1968. Selective dolomitization of Recent sedimentary structures, *J. Sed. Petrology*, **38**, 612-616.
Shinn, E. A.; Ginsburg, R. N.; and Lloyd, R. M., 1965. Recent supratidal dolomite from Andros Island, Bahamas, *Soc. Econ. Paleont. Mineral. Spec. Publ. 13*, 112-123.
Shrock, R. R., 1948. A classification of sedimentary rocks, *J. Geol.*, **56**, 118-129.
Siesser, W. G., 1972. Dolostone from the South African continental slope, *J. Sed. Petrology*, **42**, 694-699.
Thrailkill, J., 1968. Dolomite cave deposits from Carlsbad Caverns, *J. Sed. Petrology*, **38**, 141-145.
Zenger, D. H., 1972a. Dolomitization and uniformitarianism, *J. Geol. Ed.*, **29**, 107-124.
Zenger, D. H., 1972b. Significance of supratidal dolomitization in the geologic record, *Geol. Soc. Am. Bull.*, **83**, 1-12.

Cross-references: *Beachrock; Carbonate Sediments–Diagenesis; Carbonate Sediments–Lithification; Cementation; Dedolomitization; Diagenesis; Interstitial Water in Sediments; Limestones; Persian Gulf Sedimentology; Sabkha Sedimentology.* Vol. IVA: *Calcium Carbonate: Geochemistry.*

DURICRUST

An indurated soil or *paleosol* (Woolnough, 1927; Goudie, 1973), duricrust is characteristic of subtropical to tropical regions that have two very contrasting seasons, wet and dry, and those regions that passed through comparable cycles during the Pleistocene. It has three characteristic compositional types (and several minor ones): *calcrete* (or *caliche*, q.v.), *ferricrete* (q.v.), and *silcrete* (q.v.), respectively cemented by carbonate, iron, and silica (Lamplugh, 1902). Silicified clay or shale crusts are known as *porcelanite*. An aluminous crust is called *bauxite*. Less frequent cements include gypsum (gypcrete) and halite (salcrete, q.v.), but these are unstable under moist conditions.

Duricrusts of the silcrete-ferricrete type are produced homotaxially over broad continental areas, especially those that have reached an advanced stage of planation. Intensive (deep) chemical weathering has furnished the iron or silica; the parent soil is frequently lateritic. Calcretes, on the other hand, are distinctive of littoral regions and desertic downwind semi-arid belts where the carbonate sand or dust has been brought in by wind. Duricrusts may have great longevity (10^7 to 10^8 yr) at the land surface and serve as evidence of ancient paleoclimates (Dury and Knox, 1971).

The process of duricrust formation is comparable to *induration*, a general term for diagenetic hardening especially where the invading solutions are epigene (epidiagenetic). Both upward and downward movement of the solutions can be deduced. Many authors believe that a two-phase sequential process is involved.

CHARLES W. FINKL, JNR.
RHODES W. FAIRBRIDGE

References

Dury, G. H. and Knox, J. C., 1971. Duricrusts and deep-weathering profiles in southwestern Wisconsin, *Science*, **174**, 291-292.
Fairbridge, R. W., 1948. The geology and geomorphology of Point Peron, Western Australia, *J. Roy. Soc. W. Austral.*, **34**, 35-72.
Finkl, C. W., Jr., and Churchward, H. M., 1973. The etched land surfaces of southwestern Australia, *J. Geol. Soc. Austral.*, **20**, 295-307.
Goudie, A., 1973. *Duricrusts in Tropical and Subtropical Landscapes.* Oxford: Clarendon Press, 174p.
Lamplugh, G. W., 1902. Calcrete, *Geol. Mag.*, Dec. 4, **9**, 575.
Stephens, C. G., 1971. Laterite and silcrete in Australia: A study of the genetic relationships of laterite and silcrete and their companion materials, and their collective significance in the formation of weathered mantle, soils, relief and drainage of the Australian continent, *Geoderma*, **5**, 5-52.
Woolnough, W. C., 1927. Presidential address. Part I. The chemical criteria of peneplanation; Part II. The duricrust of Australia, *J. Proc. Roy. Soc. New S. Wales*, **61**, 1-53.

Cross-references: *Caliche, Calcrete; Carbonate Sediments–Diagenesis; Cementation; Clastic Sediments–Lithification and Diagenesis; Ferricrete; Pisolite; Porcelanite; Salcrete; Silcrete; Silicification; Weathering in Sediments;* Vol. III: *Bauxite; Duricrust; Induration; Laterization; Paleosols;* Vol. XII: *Duricrust; Soil Pans.*

DUST (ATMOSPHERIC) IN SEDIMENTS–See EOLIAN DUST IN MARINE SEDIMENTS

E

EARTH-FLOW DEPOSITS—See MUDFLOW DEPOSITS

ELUTRIATION

Elutriation is the purification of a particulate material by the washing and straining (or decanting) of the lighter suspended portion of the fluid. Specific usage in sedimentology and also in industrial parlance is applicable to particle-size classification by moving fluids, usually air but sometimes water. Water elutriation is most efficient in the silt range (Follmer and Beavers, 1973). The basic premise is that at certain velocities a fluid is capable of supporting only particles smaller than a certain size, i.e., the fluid velocity is equal to the terminal velocity for particles of a given size, as calculated from *Stokes Law* (q.v.). Sizing by elutriation, therefore, is the opposite of the process of sedimentation (see also *Settling Velocity*).

To elutriate is to remove all the particles that are finer than a given size from a bed of particles by passing a fluid over or through the powder or sediment (*winnowing*). The particles thus collected are usually removed from the fluid by filtration and weighed. If the separation is carried out in several steps, using successively greater fluid velocities, the original sample is divided into a number of particle-size classes (Fig. 1). The size-distribution data obtained in this manner are therefore cumulative and are plotted on the basis of weight or frequency for samples larger than a certain size (settling results are plotted on the basis of frequency or weight for samples smaller than a given size).

The major problem with elutriation methods is that the fluid velocity is not constant across the ducts so that the assumption of particle velocity being equal to fluid velocity is usually invalid. A second and less serious objection is that the size fractions do not have sharp size distributions so that estimated sizes have to be assigned. Despite these problems, however, Follmer and Beavers (1973), using the apparatus of Beavers and Jones (1966), attained reproducibility of ±1%, recovering 94–99% of the original sample.

B. CHARLOTTE SCHREIBER

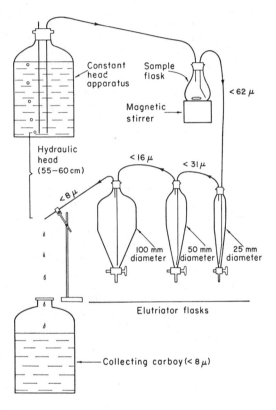

FIGURE 1. Schematic diagram of elutriator apparatus (from Follmer and Beavers, 1973).

References

Beavers, A. H., and Jones, R. L., 1966. Elutriator for fractionating silt, *Soil Sci. Soc. Am. Proc.*, 30, 126–128.

Dallavalle, J. M., 1948. *Micromeritics.* New York: Putman, 555p.

Drani, R. R., and Callis, C. F., 1963. *Particle Size: Measurement, Interpretation and Application.* New York: Wiley, 165p.

Follmer, L. R., and Beavers, A. H., 1973. An elutriator method for particle-size analysis with quantitative silt fractionation, *J. Sed. Petrology*, 43, 544–549.

Cross-references: *Grain-Size Studies; Settling Velocity; Stokes Law.*

EOLIAN DUST IN MARINE SEDIMENTS

Nineteenth century interest in atmospheric transport of dust has recently been revived mainly because it accounts for a significant fraction of deep-sea sediments and because it is responsible for widespread distribution of some pollutants. In the deep sea, where deposition and erosion rates are commonly slow, the eolian contribution can be assessed more easily than on continents. Delineation of sinks for natural airborne material provides an estimate of areas where some pollutants can also be expected to accumulate. Many observations of the distribution of volcanic detritus by winds have provided a good basis for predicting direction and distance that deflated material might also be atmospherically transported over land and sea (Fig. 1; Shaw et al., 1974).

Early Studies

Much information now available on eolian sediments was compiled during the last half of the 19th century (see Free, 1911). Many studies of dust falls on the continent were initiated after Darwin (1846) noted their importance to sediment accumulation on the floor of the E Atlantic Ocean. He gave some dust that he had collected from HMS *Beagle's* mainmast to the great naturalist, C. G. Ehrenberg, for microscopic analysis. Ehrenberg's (1847) study of this and several other dust samples collected at sea and in S Europe resulted in the most complete single compilation of data concerning airborne biogenic detritus ever assembled. On the basis of erroneous geologic and biologic assumptions, however, Ehrenberg concluded that the source of the dust was South America and not Africa. This view was championed by the foremost oceanographer of the time, M. F. Maury (1856); and, hence, transoceanic sediment transport by wind was widely accepted before the turn of the century. Hellmann (1878) later proved that the source of the dust was northern Africa and demonstrated that large dust falls observed at sea are mainly limited to the eastern Atlantic. Subsequently Hellmann, with Meinardus (1902), traced the great dust fall of 1901 for more than 3000 km from its probable source area in southern Algeria to England and southern Scandinavia, and even calculated a sedimentation rate for African dust in central Europe.

From such European studies and from his own work on airborne particles in the US, Udden (1898) concluded that deflated mineral grains $31-62\mu$ in diameter could remain airborne for about 300 km, those $15-31\mu$ for 1600 km, and material $<15\mu$ around the globe. Though controversy still surrounds estimates of distances that deflated material may be transported in the troposphere, Udden's early estimates for the coarser components are certainly too conservative.

Distribution

Pacific Ocean. Rex and Goldberg (1958) launched the most recent phase of investigation with a study suggesting that much of the quartz in North Pacific pelagic sediments is derived from Asia and transported mainly by

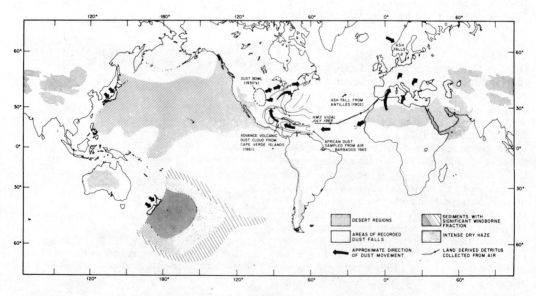

FIGURE 1. Some principal desert areas from which dust is deflated and transported by winds to adjacent lands and oceans.

TABLE 1. Composition, Source and Transporting Wind System of Atmospheric Dusts for Various Ocean Basins

Ocean Basin	Wind System	Source Area	Composition of Dust					Composition of Marine Sed.				
			<2μ Clay Mineralogy[a]				Wt. % Quartz	<2μ Clay Mineralogy[a]				Wt. % Quartz
			S	I	C	K		S	I	C	K	
Equatorial N Atlantic	NE Trades	N Africa	25	34	12	29	9	18	44	12	26	14
North Pacific	N Westerlies	Asian Desert	17	48	21	7	11	35	20	18	8	10
South Pacific	S Westerlies	Australia	0	50	23	27	—	29	34	28	9	4
Indian Ocean												
Northern	monsoons	N India Deserts	6	66	19	9	6					
Southwestern	monsoons		8	58	24	10	6	41	30	15	13	9
Bay of Bengal	monsoons		14	48	27	11	13					
Average								41	33	12	17	8

Courtesy of H. L. Windom.
[a]Relative percent of smectite (montmorillonite), illite, chlorite, and kaolinite in the <2μ size fraction.

winds. Arrhenius (1966) also demonstrated that the distribution of mica in South Pacific sediments outlines the area where much dust has been transported by wind from desert regions of Australia (Fig. 1). Numerous subsequent studies of deep-sea sediments and of dust collected in the air and from snowfields (see Windom, 1969) show that up to 75% of the detrital fraction of some Pacific pelagic sediment is transported by wind. Abundant constituents include illite, mica, chlorite, kaolinite, quartz, and feldspars. A smectite component in the clays, however, is attributed to in-situ degradation of volcanic debris (Griffin et al., 1968).

Sedimentation rates of this terrigenous material have been estimated to be 0.01 to 0.2 g/cm^2/10^3 years. Measurements in the in-situ dust load carried by Pacific winds range from 0.2–15 μg/m^3 (Goldberg, 1971). Thus, in broad areas of the Pacific inaccessible to turbidity currents, much, and in some areas most, of the land-derived erosional debris being added to the ocean floor is windborne.

Atlantic Ocean. Though dust falls, volcanic dust movement, and haze frequency were well documented in the early 1900s, trans-Atlantic eolian dust transport was not firmly established until the mid 1960s (Delany et al., 1967; Folger and Heezen, 1968). Previous studies of equatorial Atlantic deep-sea sediment components suggested that some of the material in pelagic areas east of the Mid-Atlantic Ridge in the equatorial region was wind transported. Subsequently, dust collected on islands, ships, and aircraft (summarized by Goldberg, 1971; Carlson and Prospero, 1972) have shown that highest dust loadings (10–100 μg/m^3) occur in the NE trades; the northern westerlies carry about 0.01–20 μg/m^3, the SE trades, about 0.1–1.0 μg/m^3. Most material collected is below medium silt, but a small percentage of coarse silt is present more than 300 km from source areas in Saharan Africa. Predominant minerals in equatorial dust include quartz, feldspars, mica, kaolinite, amphibole, and calcite. There is a decreasing concentration gradient going westward from the African coast. Wind-transported biogenic constituents include opal phytoliths, fresh-water diatoms, and fungus spores, which together make up about 1% (by count) of the 10μ dust fraction (Fig. 2). Similar variations in concentrations of biogenic constituents in both air and surface water suggest that the trade winds are the main transporting agent for material derived from equatorial Africa, rather than the Canary or North Equatorial currents (Folger, 1970). The daily advance of N African dust clouds has now been traced westward on satellite photos from Africa to the Caribbean (Carlson

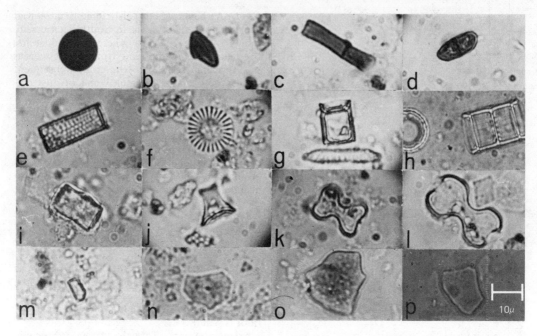

FIGURE 2. Typical airborne detritus derived from land and transported by wind over the North Atlantic. a: Spherule (probably industrial in origin). b–d: Fungus spores and hypha. e–h: Freshwater diatoms. i–l: Phytoliths (opaline bodies initially precipitated in the epidermal cells of common land grasses). m–p: Mineral grains which commonly comprise the bulk of most airborne suspensions.

and Prospero, 1972). An average accumulation rate for Saharan eolian components of pelagic sediments in the N Equatorial Atlantic has been estimated at about 0.4 g/cm^2/10^3 years (Prospero and Carlson, 1972), amounting to 25–37 million tons annually.

Both terrigenous and biogenous detritus are more abundant in deep-sea sediments of the Equatorial Atlantic from glacial times, probably due primarily to intensified wind circulation (Bowles, 1973) and to increased equatorial aridity during high-latitude glacial advances.

Indian Ocean. Data from the Indian Ocean are sparse, but the few measurements available indicate that atmospheric dust loads in northern areas are about 6–83 μg/m^3 and that the source of the material is mainly the Rajputana Desert (Aston et al., 1973); despite such heavy loadings, oceanic sedimentation is dominated by detritus carried by rivers (Goldberg, 1971). High concentrations of kaolinite in the eastern Indian Ocean are most probably dust deflated from Miocene laterites in western Australia (Griffin et al., 1968). In the Bay of Bengal, however, winds from Asia introduce a high illite concentration, although nearshore sediments are dominated by smectite weathered from the Deccan Traps.

DAVID W. FOLGER

References

Arrhenius, G., 1966. Sedimentary record of long period phenomena, in P. M. Hurley, ed., *Advances in Earth Science.* Cambridge, Mass.: M.I.T. Press, 155-174.

Aston, S. R.; Chester, R.; Johnson, L. R.; and Padgham, R. C., 1973. Eolian dust from the lower atmosphere of the eastern Atlantic and Indian Oceans, China Sea and Sea of Japan, *Marine Geol.,* 14, 15-28.

Bowles, F. A., 1973. Climatic significance of quartz content in sediments of the eastern equatorial Atlantic, *Trans. Am. Geophys. Union,* 54, 328.

Carlson, T. N., and Prospero, J. M., 1972. The large scale movement of Saharan air outbreaks over the equatorial North Atlantic, *J. Appl. Meterol.,* 11, 283-297.

Darwin, C., 1846. An account of the fine dust which often falls on vessels in the Atlantic Ocean, *Quart. J. Geol. Soc. London,* 2, 26-30.

Delany, A. C.; Delany, A. C.; Parkin, D. W.; Griffin, J. J.; Goldberg, E. D.; and Reiman, B. E. F., 1967. Airborne dust collected at Barbados, *Geochim. Cosmochim. Acta,* 31, 885-909.

Ehrenberg, C. G., 1847. Passatstaub und Blutregen, *Abhandl. K. Akad. Wiss.* (Phys.), 269-460.

Folger, D. W., 1970. Wind transport of land-derived mineral, biogenic, and industrial matter over the North Atlantic, *Deep-Sea Research,* 17, 337-352.

Folger, D. W., and Heezen, B. C., 1968. Trans-Atlantic sediment transport by wind (abstr.), *Geol. Soc. Am. Spec. Pap. 115,* 68.

Free, E. E., 1911. The movement of soil material by the wind, *U.S. Dept. Agric. Bur. Soil Bull. 68,* 272p.

Goldberg, E. D., 1971. Atmospheric dust, the sedimentary cycle and man, *Geophysics,* **1,** 117-132.

Griffin, J. J.; Windom, H. L.; and Goldberg, E. D., 1968. The distribution of clay minerals in the world ocean, *Deep-Sea Research,* **15,** 433-459.

Hellmann, G., 1878. Uber die auf dem Atlantischen Ocean in der Hohe der Capverdischen Inseln häufig vorkommenden Staubfälle, *Mber. K. Preuss. Akad. Wiss. Berl.,* 364-403.

Hellmann, G., and Meinardus, W., 1902. Hauptergebisse einer Untersuchung über den grossen staubfälle vom 1-12 Marz, 1901 in Nordafrika, sud und mitteleuropa, *Meteorol. Z.,* **19,** 180-184.

Maury, M. F., 1856. *The Physical Geography of the Sea.* New York: Harper & Row, 348p.

Prospero, J. M., and Carlson, T. N., 1972. Vertical and areal distribution of Saharan dust over the western equatorial North Atlantic Ocean, *J. Geophys. Research,* **77,** 5255-5265.

Rex, R. W., and Goldberg, E. D., 1958. Quartz contents of pelagic sediments of the Pacific Ocean, *Tellus,* **10,** 153-159.

Shaw, D. M.; Watkins, N. D.; and Huang, T. C., 1974. Atmospherically transported volcanic glass in deep-sea sediments: theoretical considerations, *J. Geophys. Research,* **79,** 3087-3094.

Udden, J. A., 1898. The mechanical composition of wind deposits, *Publ. Augustana Libr.,* **1,** 69p.

Windom, H. L., 1969. Atmospheric dust records in permanent snow fields: Implications to marine sedimentation, *Geol. Soc. Am. Bull.,* **80,** 761-782.

Windom, H. L., 1975. Eolian contributions to marine sediments, *J. Sed. Petrology,* **45,** 520-529.

Cross-references: *Clays, Deep-Sea; Desert Sedimentary Environments; Eolian Sedimentation; Extraterrestrial Material in Sediments; Pelagic Sedimentation, Pelagic Sediments.* Vol. I: *Cosmic Dust.* Vol. II: *Atmospheric Circulation-Global; Cosmic Dust.*

EOLIANITE

The term *eolianite* was coined by Sayles (1931) to include "all consolidated sedimentary rocks which have been deposited by the wind." In his study of Bermudan stratigraphy, Sayles called the calcite-cemented, calcareous dune rocks "calcareous eolianite," clearly implying that other noncalcareous dune rocks should be identified as eolianites; in the older terminology of Grabau the rock would be termed an *eolian calcarenite.* The term *eolianite* as used by Sayles has not gained wide acceptance and has generally been limited to calcite-cemented coastal dunes of Quaternary age; carbonate dunes are rare in nonlittoral settings. For convenience, coastal eolianite may be arbitrarily divided into two subtypes, *quartzose eolianite* ($<50\%$ $CaCO_3$) and *carbonate eolianite* ($>50\%$ $CaCO_3$). Petrographically, the coastal eolianites generally consist of skeletal, algal, chemical (oolitic), volcanic, or quartzose grains, but invariably have a calcitic cement.

Unlithified coastal eolian sands of primarily carbonate composition become cemented so quickly that the process is almost syndiagenetic. Older Quaternary eolianites are found to be progressively more cemented with age. Beneath the protection of a calcrete crust many youthful eolianites are hardly lithified, and the sand may become readily remobilized; except for the crust, the rock has a very high porosity and permeability.

Quaternary Coastal Eolianite

Pleistocene and Holocene dune rocks occur along many littoral, insular, and shelf areas of the world, from 55°N to 45°S; they are exceptionally common in the high tropics and subtropics in semiarid and arid coastal regions but are almost missing in the equatorial belt (15°N to 15°S). These accumulations are believed to be genetically related to oscillating Quaternary seas.

Coastal eolianite has a wide variety of synonyms throughout the world, some of the more common of which are *kurkar* (Lebanon), *ramleh* (Israel), *grès dunaire* (French Mediterranean, North Africa, and Madagascar), *miliolite* (Persian Gulf, India, Arabia), *cay rock* (Bahamas and Antilles), *dune rock* (South Africa and Australia), *coastal limestone, consolidated limestone,* and *consolidated dunes* (Australia), *eolian calcarenite* (US and Australia), and a number of others (Fairbridge and Teichert, 1953; Bauer, 1961; see Vol. III, *The Encyclopedia of Geomorphology*). The question of whether the "miliolite" of Arabia is of eolian or marine origin has generated some discussion. Commonly, it is of eolian origin, as shown by its sorting, the presence of rhizoconcretions, large-scale trough cross-bedding directed onshore, and lack of coarse skeletal debris or evidence of other marine activity such as coral growths and burrowing.

Coastal eolianites frequently form littoral cordons which act as barriers to drainage systems—resulting in fresh-water lakes, mangrove swamps, and peat deposits. In places, there are several offshore belts recently drowned by the Flandrian transgression, now forming the foundation of reefs, some of which (e.g., off NW Australia, Bahamas, Bermuda) are partly overgrown by coral or algae. Coastal eolianites are distinguished from continental, noncarbonate dunes by their genesis, location, and internal characteristics. The constituent grains of coastal eolianites are primarily products of the sea, although they may be transported inland and mixed with terrigenous source material.

It has been variously postulated that the sands that form coastal eolianites were originally transported inland from Quaternary shorelines during (1) regressional phases, (2) transgressional phases, (3) both regressional and transgressional phases, (4) stillstands, or (5) some combination of these time ranges relative to sea-level position. In any case, eolianite has been recognized by almost all investigators who have studied it as a Quaternary phenomenon genetically related in some manner to the relatively rapid glacioeustatically fluctuating sea levels and attendant climatic fluctuations. Coastal eolianites are not at all well known in pre-Quaternary sequences (Johnson, 1968), and it may be that eolianite is exclusively an ice-age phenomenon.

Extensive deposits of Quaternary coastal eolianites occur along tens of thousands of km of the world's coastlines; they range inland usually not more than 3 to 5 km, although they have been reported inland as far as 160 km along the Trucial Coast on the Arabian Peninsula (Kendall, pers. comm., Johnson, 1967). They attain a thickness of some hundreds of m, as in W and S Australia, S Africa, the Middle East, and Madagascar, although as a rule they are generally in units 10–50 m thick. Extensive bodies occur also in the southern Mediterranean and Atlantic Morocco and some western Atlantic islands, e.g., Bermuda and the Bahamas. Smaller occurrences are reported in the Canary Islands, Madeira, the Azores, Cape Verde Islands, Mauritius, Hawaii, and on the islands off E Yucatan (Ward, 1975). In some areas in South Africa and parts of Australia, the rock may actually be much more abundant below sea level than above. There is an exceptional lacustrine coastal eolianite formed of calcareous, cemented oolite grains on the shores of the Great Salt Lake, Utah (Jones, 1953).

Paleogeography and Paleoclimatology

Coastal eolianites are important as paleogeographic and paleoclimatic indicators (Fairbridge, 1956). Apparently, the rock is restricted to tropical and subtropical coastlines affected by glacio-eustasy. The greatest accumulations are found along coasts that lie within the present semiarid belts where continental shelves are widest and where strong and constant onshore winds prevailed during the times of receding and low-level Pleistocene seas. It was first noted in Corsica that the eolianites began to accumulate at the end of an interglacial interval.

In contrast to such Quaternary glacial features as tills and glacial pavements, which are geographically intimately associated with upper-middle and high-latitude continental glaciers, coastal eolianites are lower-middle and low-latitude, indirect glacial indicators. Thus, if found in ancient rocks, not only would they serve as indicators of paleoclimate and paleocoasts, but they could also indicate paleolatitudes.

If the size and thickness of till sheets may be used as rough indices to duration and intensity of glaciations, then Paleozoic and Quaternary sea-level fluctuations were probably comparable. Thus, if coastal eolianite formation is genetically a function of glacio-eustasy, the probability of its development during Paleozoic times should have been equally great. On the basis of paleomagnetic studies and on wind directions inferred from statistical analyses of continental dune cross-strata, Opdyke and Runcorn (1960) postulated that during Carboniferous and Permian times both the British Isles and parts of the western US lay within the trade wind belt, at about latitudes 10–15°N, some 25–30° closer to the equator than today. Some coastal eolianites might well be found near the former shorelines; thus far, they have not been specifically recognized, although Permian eolian deposits are widespread (see *Desert Sedimentary Environments*).

The prerequisites for coastal eolianite development would be (1) an abundant littoral sand supply with sufficient carbonate content to allow for rapid cementation; (2) strong and persistent onshore winds of long duration coinciding with a falling sea level; and (3) a semiarid or strongly seasonal climate with alternating wet and dry periods. It would be curious if coastal eolianites were really restricted in time solely to the Quaternary, for then the principle of uniformitarianism would in a sense be violated.

The alternative explanation for the absence of recognized analogues of pre-Quaternary coastal eolianite is that they were in fact developed, but (1) were subsequently eroded away leaving no record, (2) have never been discovered, (3) have been found but misinterpreted because of confusion of marine and eolian features or gross alteration, e.g., by silicification or other diagenetic processes. The erosion theory (1) seems unlikely in view of the massive bulk of eolianites which have accumulated over such a wide latitudinal and longitudinal expanse during the Quaternary. Coastal eolianite is generally highly porous, however, and if subjected to leaching, the carbonate fraction could be removed, leaving, e.g., quartz sand, which could be redistributed by wave action during the next rise of sea level or could be blown inland to form continental dune sands; such quartz sand bodies are very common in Paleozoic rocks and conceivably could in part represent the insoluble residues of ancient coastal eolianites. It is certainly possible that (2) pre-Quaternary examples are not exposed at the surface today. In addition, it is very probable that Paleozoic coastal eolianites

FIGURE 1. Numerous soil bands in eolianite, San Clemente Island, California (figures at left for scale).

have either gone unrecognized due to alteration subsequent to deposition or have been misinterpreted as being noneolian in origin (Johnson, 1968).

Recognition of Coastal Eolianites

At least six characteristics are fairly distinctive and apparently worldwide in expression wherever coastal eolianite occurs: (1) eolian structures associated with intercalated terra rossa paleosols (Fig. 1), often with caliche hardpans; (2) beachrock—the only known occurrence of beachrock on the W coast of the US lies beneath a coastal eolianite on one of the California Channel Islands (Fig. 2; Johnson, 1969); (3) the presence in the paleosols of *rhizoconcretions* (Kindle, 1925) or calcified vegetation, mainly roots (Fig. 3; Johnson, 1967); (4) the frequent presence of solution pipes ("organ pipes"), often filled with soil derived from superjacent paleosols; (5) varying, high-angle, convex-upward cross-strata (normally ranging up to 32° but exceptionally up to 41°; Fig. 4); and (6) the

FIGURE 2. Eolianite (upper half) overlying tectonically raised beachrock, San Miguel Island, California.

presence in eolianite paleosols of terrestrial fossils including pollen, land snails, and even mammoths (Johnson, 1967, 1969, 1971, 1972).

Eolianite Karst Development

The rapid cementation of freshly accumulated calcareous eolian sediments is widely noted and has been closely studied by Ward (1973); a calcrete duricrust or caliche 10-50 cm thick is eventually developed. The carbonate cement is furnished by rainwater leaching of the youngest sands, and reprecipitation often begins within the top 10 cm (Fig. 5). The role of rapid desiccation under the hot tropical sun seems to be an important factor. No sooner does a crust start to form but heavier rainfall leads to the opening of vertical pipes or drainage channels

FIGURE 3. Rhizoconcretions in eolianite, San Miguel Island, California. *Left:* Exhumed rhizoconcretions and caliche. *Right:* Pencil-sized rhizoconcretions exposed by wind erosion.

FIGURE 4. Mediterranean quartzose eolianite on an ancient (Tyrrhenian) barrier now separated from the sea by a modern barrier and (filled) lagoon. (Photo: E. Gazda.)

FIGURE 5. Eolianite showing lithification by downward-precipitating caliche from overlying (now eroded) soil, San Miguel Is., California. Note change in dip of cross-strata.

through it, particularly in the swales. Studies of the cementation of the Yucatan (Quintana Roo) Pleistocene eolianites suggest that they developed under more arid conditions than those of today (glacial maxima), in view of the preservation of metastable Mg-rich calcite (Ward, 1973). In contrast, the Holocene eolianites developing under less arid conditions are less well cemented and rapidly lose their metastable carbonates.

Typical karst features include lapies, sink holes, "organ pipes," collapsed dolinas, caves, and associated speleothems. Bermuda's topography was described by Bretz (1960) as a "Pleistocene karst," but the landscape is very little lowered by karst erosion, although the rock itself is very greatly modified (Land et al., 1967). Most of the world's coastal eolianites show this karstification, giving the formations a remarkably similar aspect.

RHODES W. FAIRBRIDGE
DONALD LEE JOHNSON

References

Amiel, A. J., 1975. Progressive pedogenesis of eolianite sandstone, *J. Sed. Petrology*, 45, 513-519.

Bauer, F. H., 1961. Chronic problems of terrace study in Southern Australia, *Z. Geomorph.*, 3, 57-72.

Bretz, J. H., 1960. Bermuda: A partially drowned late mature Pleistocene karst, *Geol. Soc. Am. Bull.*, 71, 1729-1754.

Fairbridge, R. W., 1956. Eolian calcarenite as a paleoclimatic indicator (abstr.), *Geol. Soc. Am. Bull.*, 67, 1813.

Fairbridge, R. W., and Teichert, C., 1953. Soil horizons and marine bands in the coastal limestones of Western Australia, *J. Roy. Soc. N. S. Wales*, 86, 68-87.

Jones, D. J., 1953. Gypsum-oolite dunes, Great Salt Lake Desert, Utah, *Bull. Am. Assoc. Petrol. Geologists*, 37, 2530-2538.

Johnson, D. L., 1967. Caliche on the Channel Islands, *Mineral Info. Service*, 20(12), 151-158.

Johnson, D. L., 1968. Quaternary coastal eolianites: do they have ancient analogues? (abstr.), *Program with Abstracts*. Mexico City: Geol. Soc. Am. Annual Meeting, 150-151.

Johnson, D. L., 1969. Beach rock (water-tablerock) on San Miguel island, in D. W. Weaver, ed., *Geology of the Northern Channel Islands*. Am. Assoc. Petrol. Geologists–Soc. Econ. Paleont. Mineral., Pac. Sec., Publ., 105-108.

Johnson, D. L., 1971. Pleistocene land snails on the California Channel Islands, *Nautilus*, 85(1), 32-35.

Johnson, D. L., 1972. *Landscape Evolution on San Miguel Island, California*. Ann Arbor: Xerox University Microfilms, Cat. No. 73-11902, 482p.

Kindle, E. M., 1925. A note on rhizoconcretions, *J. Geol.*, 33, 744-746.

Land, L. S.; MacKenzie, F. T.; and Gould, S. J., 1967. Pleistocene history of Bermuda, *Geol. Soc. Am. Bull.*, 78, 933-1006.

Opdyke, N. D., and Runcorn, S. K., 1960. Wind direction in the western United States in the late Paleozoic, *Geol. Soc. Am. Bull.*, 71, 959-971.

Sayles, R. W., 1931. Bermuda during the Ice Age, *Am. Acad. Arts Sci. Proc.*, 66, 382-467.

Ward, W. C., 1973. Influence of climate on early diagenesis of carbonate eolianites, *Geology*, 1, 171-174.

Ward, W. C., 1975. Petrology and diagenesis of carbonate eolianites of northeastern Yucatan Peninsula, Mexico, *Am. Assoc. Petrol. Geologists, Stud. Geol.*, 2, 500-571.

Cross-references: *Barrier Island Facies; Beachrock; Calcarenite; Caliche, Calcrete; Cementation; Desert*

Sedimentary Environments; Duricrust; Eolian Sands; Eolian Sedimentation; Littoral Sedimentation; Paralic Sedimentary Facies.

EOLIAN SANDS

Eolian sands (Fig. 1) are commonly well laminated (horizontal, or inclined at angles of up to 34°), with each lamina comprising moderately to very-well-sorted grains ranging in size from silt (60μ) to medium sand (1000μ; Glennie, 1970) and locally, particularly near the base of an intraformational sequence, to very coarse sand (2000μ or over).

At the base of a dune sequence there are sharp differences in grain size between individual laminae, with the finer-grained laminae being very well sorted and the laminae containing coarser grains being only moderately so because of an admixture of finer sand and silt. Upward, the overall grain size diminishes, and individual laminae of differing grain size tend to become sets of laminae of more uniform grain size (Glennie, 1972). Vossmerbaumer (1974) found that Quaternary eolian sands from Germany were coarser-grained, less well-sorted, and more skewed than desert sands from Algeria and Iran. Folk (1968) cited a bimodality of sand grains as evidence for eolian action.

The smaller eolian grains are subangular, the coarser ones subrounded to well rounded. When seen under the optical microscope, the majority of eolian sand grains have a typical frosted appearance, apparently caused by light diffrac-

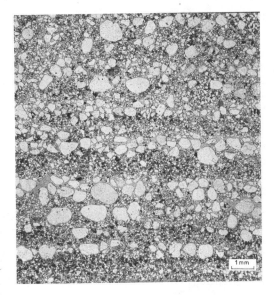

FIGURE 1. Photomicrograph of Permian "Yellow Sands" illustrating eolian laminae, grain-size differences between laminae, the poor sorting of the coarser-grained eolian laminae, and grain roundness. Durham, England (from Glennie, 1970, fig. 126). Plane polarized light.

tion associated with small irregularities on the surface of the grain. Small angular grains and parts of some larger ones have conchoidal fracture surfaces which are not frosted. These conchoidal surfaces are apparently the result of spalling of smaller flakes as the result of im-

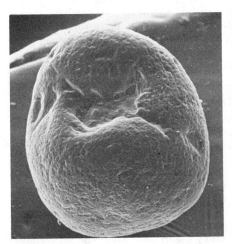 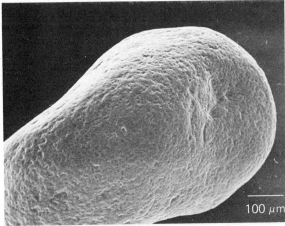

FIGURE 2. *Left:* Irregular depressions on rounded grain. Sabha Sand Sea, south-central Libya. Modern dune sand. Very well-rounded, large eolian grain with several irregular depressions in the middle, probably due to mechanical chipping during particularly powerful sand storms. *Right:* Elongate grain, rounded corners, upturned plates. Sabha Sand Sea, south-central Libya. Modern dune sand. Corners are beautifully rounded, but the shape is elongate, probably inherited. Upturned plates cover the grain; one shallow depression can be seen, but it too is covered with upturned plates. (From Krinsley and Doornkamp, 1973.)

pact during transport or the effects of insolation.

Examination of eolian sand grains with a scanning electron microscope (see Margolis and Krinsley, 1971; Fig. 2) reveals a pattern of meandering ridges and upturned fracture plates of quartz which possibly dip in a direction consistent with crystallographic planes; it has been suggested that these ridges represent quartz cleavage (Krinsley and Smalley, 1973) resulting from mechanical abrasion. The close placing of these planes is possibly responsible for the frosted appearance of many desert sands. The plates seem to be subdued by chemical solution and precipitation during periods of reduced wind in the desert. In the coastal dune sands of more temperate climates, the upturned plates are larger than those of desert sands and are less affected by solution and precipitation. In periglacial dune sands, these features are associated with others of typical glacial origin (Krinsley and Cavallero, 1970).

K. W. GLENNIE
DAVID KRINSLEY

References

Folk, R. L. 1968. Bimodal supermature sandstones. Product of the desert floor, XXIII *Intern. Geol. Congr. Proc.*, 8, 9–32.

Glennie, K. W., 1970. *Desert Sedimentary Environments.* Amsterdam: Elsevier, 222p.

Glennie, K. W., 1972. Permian Rotliegendes of Northwest Europe interpreted in light of modern desert sedimentation studies, *Bull. Am. Assoc. Petrol. Geologists*, 56, 1048–1071.

Krinsley, D., and Cavallero, L., 1970. Scanning electron microscope examination of periglacial eolian sands from Long Island, New York, *J. Sed. Petrology*, 40, 1345–1350.

Krinsley, D., and Doornkamp, J., 1973. *Atlas of Quartz Sand Surface Textures.* Cambridge: Cambridge Univ. Press, 91p.

Krinsley, D., and Smalley, I., 1973. Shape and nature of small sedimentary quartz particles, *Science*, 180, 1277–1279.

Margolis, S., and Krinsley, D., 1971. Submicroscopic frosting on eolian and subaqueous quartz sand grains, *Geol. Soc. Am. Bull.*, 82, 3395–3406.

Vossmerbaumer, H., 1974. Grain-size data of some aeolian sands: inland dunes in Franconia (Southern Germany), Algeria and Iran, a comparison, *Geol. Fören. Förh.*, 96, 261–274.

Cross-references: *Beach Sands; Desert Sedimentary Environments; Eolian Sedimentation; Glacial Sands; River Sands.*

EOLIAN SEDIMENTATION

The wind is probably the most persistent and universal agent of transportation of loose debris on the earth's surface. It is necessary that loose particles should be abundant, that they are not protected by vegetation, and that they are not bound together by water. Such conditions are found in a variety of environments, mainly deserts, beaches, and man-made barrens. The major factor favoring movement of particles by wind is aridity, and wherever the mean annual rainfall is <300 mm per year, eolian erosion and transportation is likely to be significant (see *Desert Sedimentary Environments*).

Although eolian conditions are often associated with high temperatures, they are also associated with low-temperature deserts. For example, the vast blankets of loess (q.v.) across Europe, Asia, and North America were undoubtedly the result of wind transport of finely divided dust particles from the outwash and moraines left by the retreat of Pleistocene glaciers. Extensive sand dunes, derived from the outwash plains of the Scandinavian and North American ice sheets, are today preserved by vegetation (Black, 1951; Seppälä, 1972). Calcareous coastal dunes were generated on the continental shelves during eustatic withdrawal, and they are now cemented to form eolianites (q.v.). Where the ice-age winds were offshore, as along the western edge of the Sahara, the dunes advanced across the continental shelf to be transported to great ocean depths by turbidity currents as a new type of "eolian-marine" deposit (Sarnthein and Walger, 1974).

It is notable that the thickness of eolian accumulations in modern deserts commonly measures <100 m, rarely >300 m (McKee and Tibbitts, 1964); whereas ancient sediments of similar origin, in the Triassic system, for example, may be 1000 m or more thick. Modern deserts appear to be progressively expanding, so that it may be that their eolian sediments represent only the beginnings of accumulations.

Source of Debris

The formation of sand-sized and smaller particles necessarily involves the disintegration of preexisting rocks. The bulk of desert sand is composed of quartz grains that are ultimately derived from acid igneous and metamorphic rocks. Insolation is the most commonly invoked agent of rock disintegration in desert environments (Blackwelder, 1933), but because of its exponentially decreasing effectiveness with decreasing grain size (Rice, 1976), it tends to produce a stony desert (*hamada*). The significance of chemical processes has been grossly underestimated; granites are particularly susceptible to chemical weathering. Exfoliation of granites due to insolation undoubtedly occurs, but apparently on a limited scale, where the rock-fragment size often reaches a mean of about 5-10 cm. Even in the driest deserts, granites are found to be deeply weathered and only superficially stripped of loose material. It is highly likely, however, that much of the sand has been derived from regions of higher rainfall or that it is residual debris formed at times when the region was less arid, for example in the Sahara, where large areas are covered with unconsolidated sediments of undoubtedly pluvial, fluvial, or lacustrine origin.

Eolian Transportation

The transportation of dust and sand by the wind is a familiar phenomenon both within and outside desert regions (see Vol. III: *Eolian Tranport*). The short-distance transportation of pebbles and small boulders by the wind is also possible in certain circumstances (W. E. Sharp, 1960). The high-atmospheric transport of dust has also been studied (see *Eolian Dust in Marine Sediments*).

The classical investigations into the physics of sand movements by wind action were carried out by Bagnold (1941), who distinguished three aspects of the subject: (1) the physics of grain movement; (2) small-scale effects occurring during removal, transportation, and deposition of sand grains; and (3) large-scale effects resulting in the construction of sand dunes. More recent work has added detail to Bagnold's studies, but none has seriously challenged his conclusions.

The Physics of Grain Movement. It is common practice (e.g., Scheidegger, 1961) to suppose that there are two fundamentally different modes of transport, depending on whether a particle is held in suspension for long periods of time by the turbulence of the wind alone or whether the particle returns to the ground at short intervals of time. Particles held in suspension for a long time are commonly called *dust*, the others *sand*. The two modes of transport are termed *suspension* and *saltation* (q.v.) respectively and are completely analogous to the movement of silt and sand grains in moving water (see *Fluvial Sediment Transport*). Although the two modes are fundamentally different, there can be no rigid boundary between them because a grain may at a certain velocity move by saltation but at a higher velocity may become suspended and so carried for considerable distances.

A third mode of transport is the movement of grains over the surface by *surface creep* without losing contact with the ground. Bagnold observed that grains moving this way do so not as a result of wind pressure, but rather by the impact of saltating grains. This third mode of transport is not, therefore, analogous to the "rolling" of pebbles on a stream bed. Bagnold states that surface creep accounts for 20-25% of the movement of sand; the rest moves by saltation.

In order to start sand grains moving, the drag exerted by the wind must reach a critical value in order that the resistance of the grain due to gravity may be overcome. The fundamental equation relating velocity and grain diameter is

$$V_{cr} = A \sqrt{[(\delta - \rho)/\rho] \cdot gd}$$

where V_{cr} is the critical drag velocity at which grains will just start to move, δ is the density of the sand (2.65 g/cm^3 for quartz), ρ is the density of air, g the gravity acceleration, and d the diameter of the grains. The quantity A is a constant approximately equal to 0.1. For grains of 0.08 mm diameter, Bagnold (1941) gives the critical (or threshold) velocity as approximately 15 cm/sec, and for a fine dune sand of 0.25 mm diameter as about 20 cm/sec.

Once grain movement has started, some particles can be kept in motion by winds below the critical velocity because a moving grain will transfer its energy on impact with another grain, causing the latter to rise from the ground; this grain in turn will gain momentum as it falls back to strike yet another grain. Thus the energy required from the wind will be less by the amount of energy acquired by the grain falling under gravity once movement (and saltation) have commenced. At least in theory, grains could thus continue to move by saltation for a short time if the wind speed were instantaneously reduced to zero.

Williams (1964) has described experimental studies of some characteristics of the eolian saltation load. Wind strength, the initial surface size distribution, and particle shape each exerted a certain influence on the average particle size trapped at a given height above a sand surface, whereas sorting and skewness were affected to a significant extent only by changes in the initial surface size-frequency distribution. The saltation load moved closer and closer to the ground as the grain shape departed from the true spherical form. Scheidegger (1961) has drawn attention to the evidence that sand particles blown by the wind may become electrically charged and suggests that charges on sand grains on dunes could lead to a layering of the sand.

In constrast to the movement of sand by saltation, dust particles are transported in suspension. Provided particles are light enough, the turbulence of the atmosphere will carry them to great heights (10-20 km) above the ground into the region of geostrophic flow, and they will then be carried over very large distances (see *Eolian Dust in Marine Sediments*).

Deposition

Small-Scale Effects of Transport and Deposition. The deposition of sand being transported by the wind may occur either as a result of a general decrease in the strength of the wind or in the lee of some obstruction in the path of the wind. The principal small-scale effect of sand deposition is the formation of wind ripples. R.P. Sharp (1963) distinguishes two types of ripples: (1) *sand ripples,* with grains of median size about 0.30–0.35 mm (this term was also used by Bagnold, 1941), and (2) *granule ripples,* consisting of lag concentrates of grains mostly >1 mm diameter, and generally in the 2–4 mm size range. Both Bagnold and Sharp believe that ripples are formed by the irregular forward movement of sand by surface creep. Bagnold, however, infers that there is a considerable exchange of grains between the surface creep and the saltation curtain, whereas Sharp believes the exchange to be negligible. As a consequence, Bagnold believes that the ripple wavelength is directly related to the characteristic path length of the saltation grains and that the wind velocity could be determined by direct measurement of the wavelength (see *Saltation*). Sharp, however, contends that the size of creeping grains as well as the wind velocity control the wavelength and also the height of the ripples. Clearly, further experimental and field study is needed to resolve this question.

Large-Scale Effects. *Dunes.* Dunes tend to be semipermanent features, whereas ripples are usually ephemeral. Three principal types of dunes are recognized: the *barchan,* a horse-shoe or horn-shaped dune, the *transverse* or ripple dune, and the *seif* or *longitudinal* dune. Barchan dunes may occur singly or in groups in which the dunes appear to be arranged en echelon. The horns of the barchan point downwind, and there is a slip face (30–35°) on the inside of the horseshoe. If the dunes become asymmetrical, they may link up to form a seif-type dune, or they may join laterally to form a transverse dune. Cooper (1958) believes that the barchan dune is a transverse sand ridge laterally compressed and claims that the setting up of the barchan dune as a distinct (or fundamental) unit of dune morphology is not justified.

The barchan dune is highly characteristic of many desert areas where there is a limited supply of sand resting on a hard substrate. In areas where there is a large supply of sand, true barchans do not form, but imperfact barchans may form on slopes. Instead, there is a variety of irregular ridges, which may in places be broken through ("blow-outs"), or there may be a mass of apparently irregular parabolic dunes. These *parabolic* dunes may be a fourth category, associated with a line source, and are often characteristic along shorelines.

The origin of dunes has been discussed at length by Bagnold (1941). In the consideration of dunes, it is necessary to distinguish between "gentle" winds which will move the finer particles of sand and "strong" winds which will move the coarse sands as well. The "gentle" winds tend to lead to the flattening of dunes, whereas it is the strong winds which tend to build them up. It is also important to note that barchan and transverse dunes are formed by unidirectional winds, whereas seif dunes are the result of strong, bidirectional winds. This difference leads to a difference in internal structure in that barchan and transverse dunes show only one set of steeply dipping cross-stratification, whereas seif dunes show two sets (McKee, 1957; McKee and Tibbitts, 1964).

The manner in which a dune is initiated may best be observed along a shoreline. Sand is trapped in hummocks by beach grasses. Enhanced wind velocities between the hummocks build out tongues of parabolic shape, which, in a humid region, quickly die out (with loss of sea-breeze velocity). On desert coasts, however, they merge gradually into a sand sheet on which transverse dunes develop. Farther inland, with increasing and steady high wind velocities, the seifs take over. Barchans are limited to bare, flat areas. Longitudinal dunes may also originate as *sand shadows* in the lee of an escarpment, or even a boulder or a small plant. In suitable circumstances, such a sand shadow may grow to form a dune of considerable magnitude. The seif dunes of the Sahara and Arabia commonly run for 100 km or more without interruption and are not associated with any particular cliffs or escarpments. They seem to pass abruptly from regions of transverse dunes as blow-outs, related to critical higher wind velocities. There are routinely broad areas of parallel seif dunes about 1–2 km apart. At present, there seems to be no precise observational data to explain their origin, although, once formed, it is apparent that bidirectional, strong winds (varying through about 30°) will stabilize and preserve them.

Morton (1961) has investigated the formation of a dune in a wind tunnel and has shown that saltation is largely responsible for the build-up of the dune. He has also shown that there is a cyclic or pulsating movement of the sand due to the drag of the saltation load on the wind velocity. This movement results in a lowering of the intensity of saltation, and hence there is less saltation drag; as the wind velocity increases again, more sand is picked up and the cycle is repeated.

One aspect of sand deposition on dunes that has been largely neglected concerns the manner of packing and the porosity of the resulting sediment. Kolbuszewski (1953) showed that the relationship between wind velocity and sand size is a complex one, but that in general terms, it appears that low velocities produce dense packing and high velocities loose packing. Undoubtedly, roundness and sphericity of the grains are likely to be significant factors.

The Lee Eddy. Early writers on the subject of wind ripples and sand dunes (Cornish, 1897, 1914; Darwin, 1884) believed that a fixed eddy situated in the lee of the ripple or dune played a significant role in its development. They believed that grains falling to the foot of the lee slope were held back or even lifted up the slope by this eddy. This view has been widely held despite Bagnold's (1941) strong denial that such an eddy existed. Cooper (1958) also dissented from this hypothesis, as did R. P. Sharp (1963), who reported that never had he seen "a grain of sand moving in a direction or in a manner that would suggest the influence of lee-side eddy." Nevertheless, smudge-pot experiments show some eddy development, especially on seif dunes.

Loess. The sedimentation of dust particles carried by the wind can lead to widespread areas of loess (q.v.) and its related soils. Little is known of the sedimentation mechanism with respect to dust particles. Gravity alone appears to be insufficient because even moderate turbulence can keep dust particles suspended for long periods. Accretion of the particles seems essential in order that they shall achieve sufficient velocity to reach the ground. If the particles become moisturized in the atmosphere, it is possible that random collisions may produce sufficient collisions to allow accretion to occur. Nevertheless, the process of sedimentation and accumulation of such particles must be relatively slow.

The movement of clay pellets ("parna" in Australia) to form clay dunes has been described by Price (1963) and Bowler (1973), but here the pellets are generated by salt efflorescence or the mechanical break-up of desiccation curls. Their origin throws little light on the problem of loess formation. Clay dunes formed on the lee side of playas or sabkhas have their convexity in the opposite sense to barchans; slope angles rarely exceed 15°.

Eolian Deposits

Dunes. Spearing (1974) listed the following internal structures as characteristic of dunes: (1) medium- to large-scale cross-bed sets facing downwind, with foreset dips at angle of repose; (2) series of boundary surfaces, separating individual cross-sets, which are horizontal or dip downwind at low angles; (3) progressively flatter cross-sets toward the top of dunes, with more truncation and thinning upward; (4) progressively larger foresets downwind; (5) contorted bedding; and (6) rare ripple laminae. Spearing also summarized characteristic structures of individual dune types. Hunter (1977) described the basic types of stratification found in small eolian dunes.

Ancient Deposits. There are several eolian or partly eolian sand deposits in the western US (Spearing, 1974), including the Jurassic Navajo (Kiersch, 1950) and Entrada formations, the Permian DeChelly, Lyons, and Coconino (Reiche, 1938) formations, and the Triassic Wingate Formation. Shotten (1937) identified as eolian the Lower Bunter Sandstone of England, and Bigarella and Salamuni (1961) the Botucatu Sandstone in Brazil.

DEREK W. HUMPHRIES

References

Bagnold, R. A., 1941. *The Physics of Blown Sand and Desert Dunes.* London: Methuen, 265p.
Becker, R. D.; Bigarella, J. J.; and Suarte, G. M., 1969. Coastal dune structures from Parana (Brazil), *Marine Geol.,* 7, 5-55.
Bigarella, J. J., 1972. Eolian environments–Their characteristics, recognition, and importance, *Soc. Econ. Paleont. Mineral. Spec. Publ. 16,* 12-62.
Bigarella, J. J. and Salamuni, I., 1961. Early Mesozoic wind patterns as suggested by dune bedding in the Botucatu Sandstone of Brazil and Uruguay, *Geol. Soc. Am. Bull.,* 72, 1089-1105.
Black, R. F., 1951. Eolian deposits of Alaska, *Arctic,* 4, 89-111.
Blackwelder, E., 1933. The insolation hypothesis of rock weathering, *Am. J. Sci.,* ser. 5, 26, 97-113.
Bowler, J. M., 1973. Clay dunes: Their occurrence, formation and environmental significance, *Earth-Sci. Rev.,* 9, 315-338.
Cooper, W. S., 1958. Coastal sand dunes of Oregon and Washington, *Geol. Soc. Am. Mem. 72,* 169p.
Cornish, V., 1897. On the formation of sand dunes, *Geogr. J.,* 9, 278-309.
Cornish, V., 1914. *Waves of Sand and Snow.* London: Unwin, 383p.
Darwin, G. H., 1884. On the formation of ripple marks in sand, *Proc. Roy. Soc. London,* 36, 18-43.
Higgins, G. M.; Baig, S.; and Brinkman, R., 1974. The sand of Thal: wind regimes and sand ridge formations, *Z. Geomorph.,* 18, 272-290.
Hunter, R. E., 1977. Basic types of stratification in small eolian dunes, *Sedimentology,* 24, 361-387.
Kiersch, G. A., 1950. Small-scale structures and other features of Navajo Sandstone, northern part of San Rafael swell, Utah, *Bull. Am. Assoc. Petrol. Geologists,* 34, 923-942.
Kolbuszewski, J., 1953. Note on factors governing the porosity of wind deposited sands, *Geological Mag.,* 90, 48-56.

McKee, E. D., 1957. Primary structures in some recent sediments, *Bull. Am. Assoc. Petrol. Geologists,* **41,** 1704-1747.

McKee, E. D., 1966. Structures of dunes at White Sands National Monument, New Mexico (and a comparison with structures of dunes from other selected areas), *Sedimentology,* **7,** 1-69.

McKee, E. D. and Tibbitts, G. C., Jr., 1964. Primary structures of a seif dune and associated deposits in Libya, *J. Sed. Petrology,* **34,** 5-17.

Morton, J. A., 1961. Some problems involved in the deposition of an aerodynamic sand dune in the laboratory, *Proc. Midland Soil Mech. Fdn. Eng. Soc.,* **4,** 177-185.

Norris, R. M., 1966. Barchan dunes of Imperial Valley, California, *J. Geol.,* **74,** 292-306.

Opdyke, N. D., and Runcorn, S. K., 1960. Wind direction in the western United States in the late Paleozoic, *Geol. Soc. Am. Bull.,* **71,** 959-972.

Poole, F. G., 1962. Wind direction in late Paleozoic to middle Mesozoic time on the Colorado Plateau, *U.S. Geol. Survey Prof. Pap. 450-D,* 147-151.

Price, W. A., 1963. Physicochemical and environmental factors in clay dune genesis, *J. Sed. Petrology,* **33,** 766-778.

Reiche, P., 1938. An analysis of cross-lamination in the Coconino Sandstone, *J. Geol.,* **46,** 905-932.

Rice, A. R., 1976. Insolation warmed over, *Geology,* **4,** 61-62.

Sarnthein, M., and Walger, E., 1974. Der aölische Sandström aus der W-Sahara sur Atlantikküste, *Geol. Rundsch.,* **63,** 1065-1087.

Scheidegger, A. E., 1961. *Theoretical Geomorphology.* Berlin: Springer, 333p.

Seppälä, M., 1972. Location morphology and orientation of inland dunes in northern Sweden, *Geogr. Annaler,* **54,** 85-104.

Sharp, R. P., 1963. Wind ripples, *J. Geol.,* **71,** 617-636.

Sharp, W. E., 1960. The movement of playa scrapers by the wind, *J. Geol.,* **68,** 567-572.

Shotten, F. W., 1937. The Lower Bunter Sandstones of North Worcestershire and East Shropshire, *Geological Mag.,* **74,** 534-553.

Spearing, D. R., 1974. Eolian sand deposits, *Summary Sheets of Sedimentary Deposits,* Chart 3. Boulder, Colo.: Geol. Soc. America.

Stokes, W. L., 1968. Multiple parallel truncation bedding planes—a feature of wind-deposited sandstone formations, *J. Sed. Petrology,* **38,** 510-515.

Williams, G., 1964. Some aspects of the eolian saltation load, *Sedimentology,* **3,** 257-287.

Cross-references: *Bagnold Effect; Bedding Genesis; Desert Sedimentary Environments; Eolian Dust in Marine Sediments; Eolian Sands; Fluvial Sediment Transport; Saltation.*

ESTUARINE SEDIMENTATION

An *estuary* (from the Latin *aestus,* meaning tide) is that part of a river subject to oceanic influences. The salinity of the river is augmented; and the currents are modified, both in speed and in direction, being reversed for a period during flood tide. The periodic rise and fall of the tides changes the character of the river both dynamically and chemically. The result is in every case a particular distribution of sedimentary facies (see *Estuarine Sediments*).

As Guilcher (1954) points out, when a river ends on a coastal plain, tidal flats (q.v.) and estuaries are often found together. Most modern tidal marshes and estuaries are continental zones submerged by the Flandrian transgression, sometimes coinciding with isostatic movements, as in Flanders. Bedrock morphology and ancient sediments up to 30 m thick govern the present morphology of the estuaries. The longitudinal profiles of estuaries are extremely irregular, frequently with a shallow bottom at the mouth followed by a series of deeply depressed channels upstream (55-59 m in the bed of the Escault; Guilcher, 1954). The tidal currents (ebb and flood) deepen the channels, which sometimes have very peculiar courses (Fig. 1).

Estuarine Dynamics

Estuarine sedimentation depends completely upon the dynamics of the estuary, which in turn depend upon the equilibrium of river and tidal flow. The various states of equilibrium were discovered after many measurements were taken in the Loire River (France) and

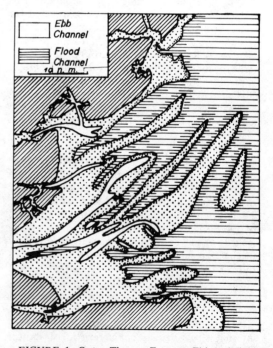

FIGURE 1. Outer Thames Estuary—Ebb and flood channels. Scale in nautical miles (after Robinson, 1963).

reconfirmed by measurements taken in numerous estuaries around the globe. The principle of this balance is the following: The flowing water develops inertia, resulting in a collision of river and ocean water in the estuary. The intensity depends upon the river's flow. During periods of low water level, the intensity of collision will be smaller than during high water. The ocean water pressure upon the estuary varies considerably depending upon time and location. It depends first of all upon the amplitude of the tides, which can vary in large proportions. For example, high tide can reach about 22 m at certain points of the northern coast of Brittany, about 6 m in the Loire estuary, and <3 m in the mouth of the Mahury River (French Guiana). In given regions, the oceanic pressure against the estuary varies with the tidal coefficient and can reduce the abovementioned maximum amounts by up to 50%. Moreover, secondary factors such as direction and intensity of wind, atmospheric pressure, and temperature also influence this balance, but to a lesser degree.

Hydrology. This meeting in an estuary of fresh river water and saline marine water that is denser at the same temperature creates very particular hydrologic conditions despite the slight difference in density (about 0.025 g/cm^3 at 15°C). This condition is what led Pritchard (1955) to define an *estuary* as a body of water open to the ocean on one side and open to a river on the other, where the fresh water is diluted by sea water.

Numerous studies, notably Pritchard (1955, 1958, 1967), Ketchum (1951), Stommel and Farmer (1952), and Hansen and Rattray (1966), have enabled us to connect different aspects of estuarine circulation to two basic parameters (Allen, 1972): (1) volume brought by the tide and (2) fluvial flow and morphology of the estuary. Generally, the more substantial the river discharge, the weaker the mixing, and, therefore, the greater the salinity gradient. In effect, a sizeable fluvial flow blocks the entrance of the sea water into the estuary, as is the case in the Amazon River.

In his classification, Pritchard (1955) at first considered an estuary without the tide, Ω, whose waters are very stratified; by progressively decreasing the fluvial flow Q_{fl}, he diminished the ratio Q_{fl}/Ω to end in a "homogeneous estuary" without a vertical salinity gradient. In this classification, the status of an estuary therefore depends on *density gradients*, which are produced by the process of *advection*, in which the currents and displacement of the fresh- and salt-water masses operate, or *diffusion*, which is brought about by whirlpool or turbulent movements.

Accordingly, at a given point, by using the cartesian coordinates which are parallel and perpendicular to the axis of an estuary, one will have, according to Pritchard

$$\frac{\partial s}{\partial t} = -V_x \frac{\partial s}{\partial x} - V_y \frac{\partial s}{\partial y} - V_z \frac{\partial s}{\partial z} \left(K_x \frac{\partial s}{\partial x} \right) + \frac{\partial}{\partial y} \left(K_y \frac{\partial s}{\partial y} \right) + \frac{\partial}{\partial z} \left(K_z \frac{\partial s}{\partial t} \right)$$

where s is salinity; V_x, V_y, V_z are components of speed all along each axis; and K is a diffusion constant. The first terms represent the advective processes, the last three terms represent the diffusive processes. If an estuary is in equilibrium for a sufficiently long time, at a given point, one will have $\partial s/\partial t = 0$. According to Pritchard, each type of estuarine circulation is defined by the predominance of one or more terms of the general equation (see Fig. 2).

Salt-Wedge Estuary Without Tidal Influence. Where the tide is very weak, as is the case with the Mississippi, the river water flows out above the denser, saline marine water, which forms a firm salt wedge at the bottom (Fig. 2), and the stratification of the water is complete. The subjacent fresh water is in direct contact with the salt water at the bottom. The only important processes are those of longitudinal and vertical advection. The equation of continuity is of the form

$$0 = -V_x \frac{\partial s}{\partial x} - V_z \frac{\partial s}{\partial z}$$

Salt-Wedge Estuary with Tidal Influence. Introducing a tide into the preceding model will have the effect of increasing the vertical exchanges; consequently, the salinity of the fluvial waters will increase in the downstream direction, and the salinity of the lower-level waters will decrease in the upstream direction (Fig. 2). Although there can be a reversal of the direction of the currents at the peak of the tidal flow, the middle or "residual" flow is directed in the upstream direction at the bottom, and in the downstream direction on the surface. The equation of continuity will cover, therefore, a term of vertical diffusion:

$$0 = -V_x \frac{\partial s}{\partial x} - V_y \frac{\partial s}{\partial y} + \frac{\partial}{\partial y} \left(K_z \frac{\partial s}{\partial y} \right)$$

Partially Mixed Estuary. If the fluvial flow is deficient compared to the volume of water introduced by the tide, there will be an important decrease in the ratio Q_{fl}/Ω; the result of diffusion and the mixing due to tides will be such that the vertical gradient of salinity will be very reduced. According to Pritchard (1955) the average speed of the flow is di-

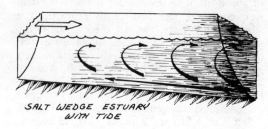
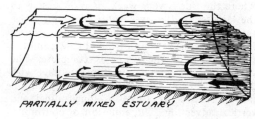
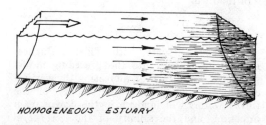

FIGURE 2. Types of estuarine circulation according to Pritchard (after Allen, 1972).

rected downstream. In estuaries that are sufficiently large, such as the Gironde in France, one observes (Fig. 2) the formation of a saltier tidal channel on the right ascending the estuary, and a relatively unsalty ebb-tide channel on the left side (in the Northern Hemisphere; Allen, 1972). The equation of continuity will therefore be

$$0 = V_x \frac{\partial s}{\partial x} - V_z \frac{\partial s}{\partial z} + \frac{\partial}{\partial z}\left(K_z \frac{\partial s}{\partial z}\right)$$

Homogeneous Estuary. Here the ratio Q_{fl}/Ω is extremely deficient, and no vertical gradient of salinity exists. The movement of salt upstream only results from the diffusive process. Bowden (1967), and even Pritchard (1967), doubts the existence in nature of this theoretical estuary. There always exists, unquestionably, a very deficient but positive gradient of salinity, even in enormous bays, such as Galway Bay (Ireland) or the harbor of Brest. In this last bay, Q_{fl}/Ω is extremely reduced. In spite of this, the journey of fluvial waters may be followed along the length of the bay (Berthois and Auffert, 1968, 1969).

Bowden (1967) reverted to this classification, but used the ratio Ω/Q_{fl}. The estuary will be moderately stratified if $10 > \Omega/Q_{fl} > 1$. The estuary will be of the well-stratified type or a salt-wedge one if $\Omega/Q_{fl} < 1$. By applying this ratio to the mouth of the Gironde (France), Allen (1972) put forth $\Omega = 1.25 \times 10^9$ m^3 in 6.25 hr for an average tide, and he produced the graph in Fig. 3. From this graph, as a function of river flow one has a vertical homogeneity for a very deficient flow; then, due to the augmentation of this flow, one passes on to a flow of two layers with vertical mixing; then, for the high flows, a flow of two layers with entrainment.

Sedimentology

Taking a more strictly sedimentological viewpoint, the balance is schematically represented in Fig. 4, which is based on measurements taken in the Loire River. At the moment of the high tide (highest level of oceanic water in the

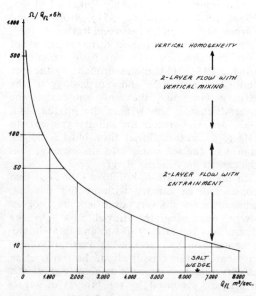

FIGURE 3. Graph of the $\Omega/Q_{fl} \times 6h$ ratio as a function of fluvial discharge at the mouth of the Gironde (after Allen, 1972).

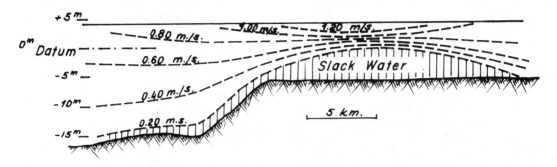

FIGURE 4. Loire Estuary—an example of fluvio-oceanic equilibrium maintained for >20 km for a period of 4 hr after high tide.

estuary), all the water in the estuary is motionless. Then the outward flow of the water starts as a faint current on the surface toward the mouth of the estuary. Commencing as a thin layer only, it increases progressively in thickness as well as in velocity, reaching and even surpassing 1 m/sec. The water flow becomes stratified. The salinity of the upper layer can vary from $2-4^o/_{oo}$, but at 2 m depth it surpasses $13-15^o/_{oo}$, which is maintained without appreciable differences all the way down to the bottom.

Thus a deep, motionless, or only very slowly flowing lens of water (velocity <20 cm/sec) forms in the Loire. This motionless lens of water reaches a length of 9–11 km. Its thickness diminishes progressively during the ebb tide until it is carried away by the surface flow, the velocity of which continues to increase until mid-ebb.

During periods of high water, the equilibrium of the deep waters can exist for the duration of the ebb and even through the beginning of the flood, a period of >7 hr. Evidently, the location and duration of such an equilibrium in any estuary depends on the fresh-water flow and on the amplitude of the oceanic tides. In the Loire estuary, where the flow may be 4000 m³/sec or more, the zones of equilibrium between the river and oceanic tides is next to the estuary's mouth. For a flow of 750–1000 m³/sec, the zone is about 20 km upstream; at the lowest water level and a flow of <200 m³/sec, it is about 30 km from the mouth.

Clearly, whenever there is a lens of deep water which is quasi-motionless for periods of up to 8–10 hr (as in the Mahury River), there is a characteristic sedimentation in the estuary. Only the particularly fine sediment floating in the surface waters above the motionless lens can reach the oceanic waters. Coarser particles carried by the middle zone of the channel flow settle because they have insufficient velocity. Finally, coarse sediments carried near the bottom by rolling and saltation must settle on the edge of the motionless lens.

These conditions of sedimentation have been verified by many observations in numerous estuaries, especially in the following rivers: Rance, Ic, Trieux, and Loc'h rivers (France); Konkouré River (Republic of Guinea); Betsiboka River (Malagasy Republic); Kangerdlugssuaq Fiord (Greenland); Mahury, Maroni, and Cayenne rivers (French Guiana).

This dynamic new concept of sedimentation in estuaries is implicitly contained in the theory of deltas, expressed by G. K. Gilbert in 1885. The formation of a delta depends almost entirely on the following law: The capacity and competence of a stream vary with the velocity. Thus, if the velocity diminishes quickly at the mouth, deposition will occur on the bottom of the channel. This deposition forms an obstacle to the current and promotes continuous deposition upstream to some more steeply inclined part of the channel. The stream constantly maintains a balance between its slope and the work it has to do.

Berthois (1958) has shown that the formation of deltas is based on the same phenomenon as takes place in estuaries, i.e., the basic idea of the river-oceanic equilibrium which causes the slowing or the complete standstill of a deep-water lens that regulates the sedimentation. Thus, given a constant difference between the highest and lowest levels of the tides, the zone of equilibrium will be located in the mouth proper if the fresh-water flow is weak. Such is the case with those estuaries that contain submerged deltas. If, on the contrary, there is a considerable flow, the equilibrium will be beyond the mouth and will result in a higher sedimentation output. Hence a delta will emerge and will build itself up (see *Delta Sedimentation*).

Classification of Estuaries and Deltas

The combined concept of sedimentation in deltas and in estuaries suggests a classification of all types of river mouths (Berthois, 1958).

Upstream Equilibrium Zones. Strong tides and weak river flow promote the development of river mouths with upstream equilibrium zones. There are many examples, in all parts of the world, e.g., in all the rias along the Armorican peninsula (Brittany), where tidal amplitudes vary from 6-12 m. The sea penetrates considerably upstream into the ria channel; there, the zone of river-oceanic equilibrium occurs and extensive marshes develop. In the subpolar regions, where the fiords are blocked by large glaciers, there is an equilibrium of the same type because the flow of the subglacial streams is very weak, even in the great fiords of Greenland. This type of sedimentation can also be observed in currents of middle or low capacity in tropical zones during periods of very low water, as in the Konkouré (Republic of Guinea).

Zone of Equilibrium Near the Coast But Somewhat Upstream. In the estuary that is a delta entirely submerged at high tide, the flow is often subject to considerable fluctuations and can carry notable amounts of sediment at times of high water. At these times, the river-oceanic balance zone is located near the mouth, and the sediments settle there; during low-water periods, the equilibrium moves 20 km upstream and even farther. The sediments settled in the estuary are more or less strongly remodified by the sea, which deepens the flood channels, whereas the river deepens the ebb channels; the major channels are relatively stable and are separated by vast sand banks which may emerge at low tide. Most estuaries in the temperate latitudes are of this type.

Zone of Equilibrium in the Coastal Region. The water currents associated with development of equilibrium zones in the coastal region of a river-mouth have a variable flow at low water and may be assimilated with a river of the first type, but they always have one period of considerable high water during the year. At that moment, the thrust of the flowing waters is significant, and the river-oceanic equilibrium zone will establish itself in the vicinity of the mouth. In numerous cases at low flow, a sandy ridge will develop, encroaching slightly upon the realm of the sea and under whose shelter fine sediments can settle. Examples of this type of sedimentation are the rivers of French Guiana, in particular the Approuaqgue, Mahury, and Cayenne, and the Konkouré River, Republic of Guinea. An estuary of this type may temporarily transform into a delta and, therefore, be intermediate between an estuary proper and a permanent delta.

Zone of Equilibrium Outside the Coastal Zone. The river-oceanic equilibrium zone of some river mouths is in the sea, far from the coast. To develop a delta there must be a considerable load of sandy sediments sufficiently coarse to resist oceanic erosion and littoral currents. Evidently, when a river empties into a tideless sea, the flow required to overcome the oceanic thrust and form the equilibrium zone in the widest part of the estuary need not be as strong as in the case of strong tides; thus, the majority of deltas are found in seas without appreciable tides, or in lakes (see *Delta Sedimentation*).

LEOPOLD BERTHOIS

References

Allen, G. P., 1972. Etude des processus sédimentaires dans l'estuaire de la Gironde, Thèse Fac. Sci. Bordeaux, 310p.

Allen, G. P., and Klingebiel, A., 1974. La sédimentation estuarienne: Exemple de la Gironde, *Bull. Centre Rech. Paris,* 8, 263-293.

Berthois, L., 1958. La formation des estuaires et des deltas, *Comptes Rendus Acad. Sci.,* 247(13), 947-950.

Berthois, L., and Auffert, G., 1968, 1969. Contribution à l'étude des conditions de sédimentation dans la rade de Brest, Ch. 1-4, *Cahiers Océanographiques,* 1968, 10, 893-920; 1969, 5, 469-485; 7, 701-726; 10, 981-1010.

Biggs, R. B., 1970. Sources and distribution of suspended sediment in northern Chesapeake Bay, *Marine Geol.,* 9, 187-201.

Bouquet de la Grye, 1866. Rapport sur le régime de Loire, *Recherches hydrographiques sur le régime des côtes,* Cahier 3, 1864-1866, 69p.

Bowden, K. F., 1967. Circulation and diffusion, in Lauff (1967), 25-63.

Brajnikov, B., 1947. Le problème de l'origine des alluvions dans l'estuaire de la Seine, *Bull. Inst. Oceanogr.,* 47, 1-28.

Christmas, J. Y., 1973. *Cooperative Gulf of Mexico Estuarine Inventory and Study, Mississippi,* pt. 3; *Sedimentology.* Ocean Springs, Miss.: Gulf Coast Research Lab., 433p.

Dyer, K. R., 1973. *Estuaries: A Physical Introduction.* New York: Wiley, 140p.

Francis-Boeuf, C., 1947. La sédimentation dans les estuaires, in *La géologie des terrains récents dans l'ouest de L'Europe.* Brussels: Soc. Belg. Geol., 174-185.

Gallenne, B., 1974. Study of fine material in suspension in the estuary of the Loire and its dynamic grading, *Estuarine and Coastal Marine Sci.,* 2, 261-272.

Gilbert, G. K., 1885. The topographic features of lake shores, *U.S. Geol. Survey Ann. Rept. 5,* 69-123.

Guilcher, A., 1954. *Morphologie Littorale et Sous Marine.* Paris: Presses Univ. France, 216p.

Hansen, D. V., and Rattray, M., 1966. New dimensions in estuary classification, *Limnol. Oceanogr.,* **11,** 319-326.

Ippen, A. T., ed., 1966. *Estuary and Coastline Hydrodynamics.* New York, McGraw-Hill, 744p.

Lauff, G. H., ed., 1967. *Estuaries.* Washington, D.C.: Am. Assoc. Adv. Sci., Publ. 83, 757p.

Ketchum, B. H., 1951. The exchanges of fresh and salt waters in tidal estuaries, *J. Marine Research,* **10,** 18-38.

McManus, J., 1972. Estuarine development and sediment distribution with particular reference to the Tay, *Proc. Roy. Soc. Edinburgh,* ser. B, **71,** 97-113.

Nelson, B. W., ed., 1973. Environmental framework of coastal plain estuaries, *Geol. Soc. Am. Mem. 133,* 619p.

Postma, H., 1954. Hydrography of the Dutch Wadden Sea, *Arch. néerland. Zool.,* **10,** 406-511.

Pritchard, D. W., 1955. Estuarine circulation patterns, *Proc. Am. Soc. Eng.,* **81,** 1-11.

Pritchard, D. W., 1958. The equations of mass continuity and salt continuity in estuaries, *J. Marine Research,* **17,** 412-423.

Pritchard, D. W., 1967. Observations of circulation in coastal plain estuaries, in Lauff (1967), 37-44.

Schubel, J. R., ed., 1971. *The Estuarine Environment: Estuaries and Estuarine Sedimentation.* Washington, D.C.: Am. Geol. Inst., 324p.

Schubel, J. R., 1972. The physical and chemical conditions of the Chesapeake Bay, *J. Washington Acad. Sci.,* **62,** 56-87.

Spencer, R. S., 1956. Studies in Australian estuarine hydrology the Swan River, *Austral. J. Marine Freshwater Research,* **7,** 193-253.

Stommel, H., and Farmer, H. G., 1952. Abrupt change in width in two-layer open channel flow, *J. Marine Research,* **11,** 205-214.

Sundermann, J., and Vollmers, H., 1973. Gezeitenbedingte Zukulationssysteme in Meeresbuchten und Flussmundungen, *Die Küste,* **24,** 72-82.

Tricart, J., 1960, Les types de lits fluviaux, *Inf. Geogr.,* **24**(5), 210-214.

Tucker, M. E., 1973. The sedimentary environments of tropical African estuaries: Freetown Peninsula, Sierra Leone, *Geol. Mijnbouw,* **52,** 203-215.

Voorthuisen, J. H. van, ed., 1960. *The Ems-Estuary.* K. Nederl. Geol.-Mijnb. Gen., Verh. Geol. Ser., 300p.

Cross-references: *Delta Sedimentation; Estuarine Sediments; Fluvial Sedimentation; Paralic Sedimentary Facies; Tidal-Current Deposits; Tidal-Flat Geology; Wadden Sea Sedimentology.*

ESTUARINE SEDIMENTS

The character and distribution of estuarine sediments are influenced by numerous physical, chemical, and biological processes. These include alternating tidal currents (Schubel, 1968; Visher and Howard, 1974); estuary morphology and its associated local hydrologic characteristics; storm events (Schubel, 1972); flocculation (Edzwald and O'Melia, 1975); bioagglomeration (Meade, 1972) and bioturbation; bacterial processes (Oppenheimer, 1960); multiple sediment sources (Guilcher, 1967); human activities (Roberts and Pierce, 1974); and a nontidal, two-layer estuarine circulation pattern (see *Estuarine Sedimentation*). Nontidal estuarine flow in a partially mixed estuary is characterized by a lower salinity surface layer with a net seaward flow and a denser, more saline bottom layer having a net landward flow. The landward transport of sedimentary material in the bottom layer, under either the influence of tidal processes (Van Straaten and Kuenen, 1957) or nontidal estuarine flow (Meade, 1969), causes estuaries to act as sediment traps and to be sites of rapid deposition (see *Sedimentation Rates*). Substantial changes in sedimentation rates and accumulation patterns have occurred during the last few decades in a number of estuaries as a result of human activities, such as dredging channels, constructing piers and barriers to wave action, diverting fresh water into or from the system, and dumping wastes (Simmons and Hermann, 1972).

Sources

The sedimentary material deposited in estuaries may be riverborne or derived from marine or marginal sources. Much of the riverborne sediment is trapped by estuarine circulation; and deposition is often pronounced in areas where tidal or estuarine flow is hindered or dissipated, such as in coves, around piers and natural points, and near the upstream end of the salinity intrusion. Bed-load material tends to accumulate in this latter zone, where the net downstream flow in the bottom water approaches zero (Postma, 1967). Suspended matter is carried downstream in the upper layer but gradually settles into the more saline bottom layer. In the bottom layer it is transported upstream and some accumulates near the tip of the salt intrusion, frequently forming a turbidity maximum (Schubel, 1968). The extent of salt penetration varies—diurnally due to currents and seasonally with river flow. Consequently, the turbidity maximum and zone of pronounced deposition moves along the stream axis, smoothing differences in sedimentary properties (Postma, 1967).

Tidal processes, littoral drift along beaches, and the net landward flow of bottom waters may also transport and deposit sediment from marine sources in estuary mouths. Kulm and Byrne (1967) have reported that marine sand is the principal shoaling material near the mouth of Yaquina Bay and that marine sand is transported 6 miles (10 km) upstream in the estuary. The movement of sediment into the mouths of estuaries from nearby beaches or adjacent continental shelves has also been reported for the Rhine (Terwindt et al., 1963), Rio de la Plata (Ottman and Urien, 1965), Seine (Guilcher, 1967) and the Dutch Wadden Sea (see *Wadden Sea Sedimentology*). In general, it appears that marine sources of sediment are most important for estuaries with low water and sediment discharge.

Estuaries also receive large quantities of sediment from marginal sources, such as bank erosion, landfill, and human and industrial wastes. Waste discharge in the Hudson Estuary from the New York–New Jersey metropolitan area may amount to ≈ 0.3 million tons per year (Gross, 1974) and may account for $\approx 10\%$ of the shoaling material in the lower estuary (Panuzio, 1965). In addition, Roberts and Pierce (1974) have shown that erosion from areas undergoing urbanization greatly increases the rate of sediment input to estuaries.

Texture and Structures

Estuaries usually contain both fine-grained and coarse-grained deposits. In general, the coarsest sediment accumulates in areas where the strongest currents flow, e.g., in channels and tidal inlets, near the estuary mouth, or along the shoreline where wave energy is high. Finer sediments tend to accumulate where energy levels are low, e.g., in coves and intertidal flats, along subtidal banks, and near the landward end of the salinity intrusion. Rapid changes in sediment texture can occur with distance and with depth in the sediment.

Estuaries may, in general, be divided into three depositional environments on the basis of sedimentary textures and structures, water depth, and estuary morphology: (1) channel bottom, (2) subtidal bank adjacent to the channel, and (3) marginal zones that include broad, shallow reaches; coves; and intertidal flats.

Channel-bottom deposits are frequently coarse grained (Fig. 1), containing abundant shells, pebbles, wood, and anthropogenic detritus. Clifton et al. (1976) have shown that channel-bottom deposits in Willapa Bay show the following general change away from the estuary mouth: somewhat bioturbated, well-sorted sand, rippled and cross-bedded by tidal currents → bioturbated muddy sand with abundant shells → bioturbated mud with shells, pebbles, and wood → mud, pebbles, and coarse sand near the fresh-water mouth of the river.

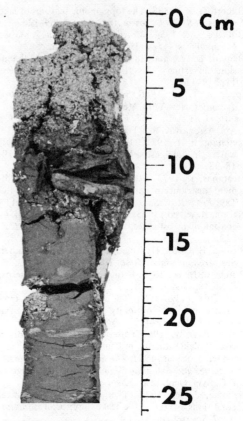

FIGURE 1. The top 25 cm of a core taken in the Hudson Estuary channel, ≈ 30 km upstream from the southern tip of Manhattan, illustrating the coarse sediment on the surface. Photo taken ≈ 7 days after core extrusion to emphasize sedimentary structures; parting, due to shrinkage, generally occurs along fine sand layers.

Subtidal bank deposits are frequently laminated; but nonlaminated zones, 1–10 cm thick, separate the zones of lamination (Fig. 2). Clifton et al. (1976) have also found that the subtidal bank deposits of Willapa Bay change systematically away from its mouth: cross-bedded sand → mixed sand and mud → interbedded fine sand or silt and mud → interbedded silt, clay, and plant debris.

Marginal coves and intertidal flat deposits may be either sandy or muddy, depending on the intensity of wave activity (Clifton et al., 1976) and the source of sediments. The sedimentary deposits of coves are frequently extensively bioturbated, whereas intertidal mud-flat deposits are generally laminated (see *Tidal-Flat Geology*).

The bioturbation of fine-grained estuarine

FIGURE 2. Photo illustrating a laminated zone (48–53 cm) and a bioturbated zone (53–58 cm) in a core taken on the subtidal bank adjacent to the channel core pictured in Fig. 1. Photo taken ≈ 7 days after core extrusion; parting generally occurs along fine-sand layers.

sediments may result from the release or entrapment of biochemically formed gases (see *Gases in Sediments*) as well as from mechanical mixing of sediment by organisms. The entrapment of CH_4 bubbles in the sediment may result in the formation of secondary gas fissures which obliterate the primary sedimentary structures to varying degrees.

Composition

Estuarine sediments consist of mineral detritus, biogenic debris, anthropogenic waste products, and authigenic components; inorganic detritus is generally predominant. Folger (1972) has provided an excellent summary of the bottom-sediment mineralogy for estuaries in the US. He reports that the minerals of the sand fraction consist mostly of quartz and feldspar and that the highest concentrations of heavy minerals are found along the Pacific coast, reflecting the proximity to mafic source rocks. The relative abundance of specific clay minerals reflects the composition of the source region (fluvial, marginal, or marine), the type of weathering, and the extent of diagenetic alteration (see *Clay Sedimentation Facies*).

Frequently, distinct changes in mineral composition can be observed downstream in an estuary. These changes may reflect different sources of sediment (Kulm and Byrne, 1967; Windom et al., 1971), diagenetic alteration, or differential settling due to differences in the rate of flocculation (q.v.) for different clays (Edzwald and O'Melia, 1975).

Biogenic debris, principally calcareous shells and organic matter, is frequently found in estuarine sediments and may be locally abundant, occurring as shell banks, channel-lag shell layers, or peat layers. Calcareous debris, however, is only dominant in estuaries where terrigenous inorganic constituents are lacking, such as in Florida Bay where carbonates make up as much as 90% of the sediment (Folger, 1972). Concentrations of organic matter (consisting of soft-bodied plants and animals and their waste and decay products) tend to be highest in deposits of fine-grained silts or clays. In most unpolluted estuaries, the concentration of organic carbon is rarely >5% and is generally <1% in sands (Folger, 1972). In areas affected by sewage waste disposal, however, organic carbon concentrations may be as high as 15% (Folger, 1972). In addition to higher organic carbon concentrations, higher trace-metal concentrations and other anthropogenic wastes, such as bricks, coal, sewage, fly ash, and metalliferous slags are frequently observed in the near-surface sediments of estuaries in populated areas.

Authigenic components of estuarine sediments form in situ, either as primary precipitates or from diagenetic reactions. Although authigenic components generally contribute a small fraction of the sediment, they may play an important role in governing the concentration of trace metals, nutrients, and other elements during sediment, water, and interstitial water interactions.

CURTIS R. OLSEN
H. JAMES SIMPSON

References

Clifton, H. E.; Phillips, R. L.; and Scheihing, J. E., 1976. Modern and ancient estuarine-fill facies,

Willapa Bay, Washington, (abstr.), *Bull. Am. Assoc. Petrol. Geologists,* **60,** 657-658.

Edzwald, J. K., and O'Melia, C. R., 1975. Clay distributions in recent estuarine sediments, *Clays Clay Minerals,* **23,** 39-44.

Folger, D. W., 1972. Characteristics of estuarine sediments of the United States, *U.S. Geol. Surv. Prof. Paper 742,* 94p.

Gross, M. G., 1974. Sediment and waste deposition in New York Harbor, *Ann. N.Y. Acad. Sci.,* **250,** 112-128.

Guilcher, A., 1967. Origin of sediments in estuaries, in Lauff (1967), 149-157.

Kulm, L. D., and Byrne, J. V., 1967. Sediments of Yaquina Bay, Oregon, in Lauff (1967), 226-238.

Lauff, G. H., ed., 1967. *Estuaries.* Washington, D.C.: Am. Assoc. Adv. Sci., Publ. 83, 757p.

Meade, R. H., 1969. Landward transport of bottom sediments in estuaries of the Atlantic coastal plain, *J. Sed. Petrology,* **39,** 222-234.

Meade, R. H., 1972. Transport and deposition of sediments in estuaries, *Geol. Soc. Am. Mem. 133,* 91-120.

Oppenheimer, C. H., 1960. Bacterial activity in sediments of shallow marine bays, *Geochim. Cosmochim. Acta,* **19,** 244-260.

Ottman, F., and Urien, C. M., 1965. Observaciones preliminares sobre la distribucion de los sedimentos en la zona externa del Rio de la Plata, *Acad. Brasilera Cienc. Anais.,* **37,** supp., 283-288.

Panuzio, F. L., 1965. Lower Hudson River siltation, *Proc. Fed. Interagency Sed. Conf., 1963, Misc. Publ. 970,* 512-550.

Postma, H., 1967. Sediment transport and sedimentation in the estuarine environment, in Lauff (1967), 158-179.

Roberts, W. P., and Pierce, J. W., 1974. Sediment yield in the Patuxent River (Md.) undergoing urbanization, 1968-1969, *Sed. Geol.,* **12,** 179-197.

Schubel, J. R., 1968. Turbidity maximum of the northern Chesapeake Bay, *Science,* **161,** 1013-1015.

Schubel, J. R., 1972. Distribution and transportation of suspended sediment in Upper Chesapeake Bay, *Geol. Soc. Am. Mem. 133,* 151-167.

Simmons, H. B., and Hermann, F. A., Jr., 1972. Effects of man-made works on the hydraulic, salinity and shoaling regimens of estuaries, *Geol. Soc. Am. Mem. 133,* 555-570.

Terwindt, J. H.; DeJong, J. J. D.; and Van der Wilk, E., 1963. Sediment movement and sediment properties in the tidal area of the Lower Rhine (Rotterdam Waterway), *Koninkl. Nederlands Geol. Mijnb. Genoot. Verh., Geol. Series,* **21,** 243-258.

Van Straaten, L. M. J. U., and Kuenen, P. H., 1957. Accumulation of fine-grained sediments in the Dutch Wadden Sea, *Geol. Mijnbouw,* **19,** 329-354.

Visher, G. S., and Howard, J. D., 1974. Dynamic relationship between hydraulics and sedimentation in the Altamaha estuary, *J. Sed. Petrology,* **44,** 502-521.

Windom, H. L.; Neal, W. J.; and Beck, K. C., 1971. Mineralogy of sediments in three Georgia estuaries, *J. Sed. Petrology,* **41,** 497-504.

Cross-references: *Estuarine Sedimentation; Flocculation; Fluvial Sedimentation; Gases in Sediments; Paralic Sedimentary Facies; Tidal-Current Deposits; Tidal-Flat Geology; Wadden Sea Sedimentology.*

EUXINIC FACIES

The term *euxinic* is an adjective referring to marine and brackish environments in which the depositional waters are strongly depleted in dissolved oxygen and in which hydrogen sulfide may be present (i.e., stagnant conditions). The euxinic facies comprises sedimentary rocks and recent sediments deposited under such conditions. As discussed by Pettijohn (1975) and Krumbein and Sloss (1963), such sediments may have the following characteristics: (1) high organic-matter content; (2) dark color; (3) fine grain size (limited to the silt and clay size classes); (4) unusual trace-element assemblages; (5) impoverished faunas; and (6) small-scale lamination. By criteria (3), rocks of this facies would be classified as shale (or mudstones). A descriptive term, *black shale,* is frequently used synonymously with euxinic shale (Pettijohn, 1975), although not all black shales need form under euxinic conditions. Other terms such as "humic shale," "sapropelite," "graptolitic shale," etc. are also commonly applied to particular shales. Much of the early work on black shales is reviewed by Conant and Swanson (1961).

Stagnant Marine Conditions

In order to discuss further the characteristics of the sediments of the euxinic facies it is necessary to have an understanding of the physical and chemical processes leading to stagnation.

Origin of Stagnation. Below the photosynthetic composition depth there is a net consumption of oxygen in marine waters as a consequence of the respiratory and biochemical oxidation of organic matter [see Eq. (1) below]. Normally, processes of advection (flow) and eddy diffusion are sufficiently vigorous in the open ocean to resupply oxygenated water to depth from surface waters in contact with the atmosphere. There are environments, however, in which circulation is sufficiently restricted and/or consumption rates of oxygen sufficiently rapid so that oxygen is completely removed. The chemistry and oceanography of such systems are described in detail by Richards (1965). Restrictions on circulation occur in enclosed or semienclosed water masses such as inland seas, silled basins, and fjords, where movement of the deeper water is limited horizontally by the geomorphology. Vertical circulation may be inhibited by stability of the water column arising from a vertical density gradient, due to either salinity or temperature stratification. High oxygen-consumption rates are known to occur in areas of upwelling of deeper nutrient-rich waters to the photic zone. Such areas occur on the western edges of continents. Carried to an extreme, euxinic conditions could develop in shelf or open ocean waters via this mechanism.

Chemistry of Anoxic (Oxygen-Free) Waters. The principle biochemical reactions in stagnant water bodies may be conveniently described by invoking the concept of the "average" composition of marine organic matter, with a carbon:nitrogen:phosphorus ratio of 106:16:1. The initial reaction is aerobic respiration resulting in oxygen removal:

$$(CH_2O)_{106}(NH_3)_{16}(H_3PO_4) + 138O_2$$
$$= 106CO_2 + 16HNO_3 + H_3PO_4 + 122H_2O \quad (1)$$

Oxygen removal is followed sequentially by nitrate (and nitrite) reduction [Eq. (2)] and sulfate reduction [Eq. (3)], in which both nitrate and sulfate act as terminal electron acceptors in the bacterially catalyzed oxidation of organic matter.

$$(CH_2O)_{106}(NH_3)_{16}H_3PO_4 + 94.4HNO_3$$
$$= 106CO_2 + 55.2N_2 + 177.2H_2O + H_3PO_4 \quad (2)$$

$$(CH_2O)_{106}(NH_3)_{16}H_3PO_4 + 53SO_4^{--}$$
$$= 106HCO_3^- + 53H_2S + 16NH_3 + H_3PO_4 \quad (3)$$

As a consequence of these reactions, ammonia and phosphate can build up to concentrations more than an order of magnitude greater than sea-water values. Dissolved sulfide is not present in oxygenated water and is extremely toxic to aerobic organisms.

Modern Anoxic Systems. Both "permanently" and intermittently anoxic systems are known. Some fjords in Norway (e.g., Dramsfjord, an arm of the Oslo-fjord) become anoxic but are flushed, on the time scale of a few years, with oxygenated water. The same is true of Saanich Inlet, British Columbia, which is a typical fjord north of Victoria on Vancouver Island. Euxinic conditions can develop in quite shallow water (Conant and Swanson, 1961), as in the Baltic and the upper basin of the Pettaquamscutt River (Rhode Island). The type example of a basin in which anoxic conditions have been maintained for several thousand years is the Black Sea (see *Black Sea Sediments*). The term *euxinic*, in fact; is derived from the Latin name for the Black Sea, *Pontus Euxinus*. Surface waters of the Black Sea are of a lower salinity than deeper waters owing to a large influx of fresh water, which leads to a relatively stable water column. Oxygen disappears at depths between 150 m in the central parts to 250 m or more along the coasts. Below the depth of zero oxygen, dissolved sulfide builds up, reaching a maximum concentration of about 0.3 millimolar. The Cariaco Trench off the coast of Venezuela and Lake Nitanat off the southern coast of Vancouver Island, British Columbia are additional examples of anoxic waters exhibiting relatively long-term stagnation.

Areas of oxygen depletion resulting from upwelling of nutrient-rich waters occur in the E Tropical Pacific Ocean (where nitrate reduction is known to be proceeding) and off the W coast of Africa in the area of Walvis Bay (where free sulfide may periodically occur in the water column).

Characteristics of Euxinic Sediments

High Organic-Matter Content. The deposition of high percentages of organic matter in sediments is clearly favored by high productivity in the surface waters. Also favorable are low wave and current energy, a relatively rapid rate of deposition, and restricted bottom-water circulation (Berner, 1972). Low wave and current energy allow organic matter (with a low specific gravity) to settle out. Rapid burial isolates organic matter from prolonged exposure to oxygenated water and therefore affords protection from oxidation by aerobic organisms. Such protection is, of course, not required to allow abundant organic matter to accumulate in sediments overlain by anoxic (low-oxygen) waters, in which case a higher percentage of organic matter may be preserved. This condition in part explains the correlation of organic-matter content with restricted bottom-water circulation and may also account for the high organic-carbon contents of euxinic sediments. The organic-carbon contents of some modern euxinic sediments are presented in Table 1. These values are quite high when compared to average shale (q.v.), but similar to the "average" black shale as determined by Vine and Tourtelot (1970; Table 2). The presence of extremely high organic-carbon content is by itself not evidence for deposition in an euxinic environment. For example, in the deep basins of the southern Gulf of California, sediments being deposited from oxygenated waters (≈ 1.5 ml/liter O_2) are known to contain up to 5% organic carbon, probably due to high productivity in surface waters.

Dark Color. Shales deposited under euxinic conditions are commonly dark in color, ranging from dark gray through grayish black, or even black. Darkening of rocks results from the presence of organic matter, but the darkening is not closely tied to the % organic carbon. This lack of a clear relationship is due to differences

TABLE 1. Modern Euxinic Environments

Location	Organic Carbon (% dry wt)	Approximate Water Depth (m)	Maximum H_2S Concentration in Water Column (mmol)	Reference
Walvis Bay, continental shelf of SW Africa	2-26.5	80-500	(present)	Calvert and Price, 1971
Upper basin of the Pettaquamscutt River (Rhode Island)	8-16	20	2.0	Orr and Gaines, 1974
Kiel Bay; Baltic Sea	1-6	25	a	Hartmann and Nielsen, 1969
Framvaren Fjord, Norway	1-15	180	0.9	Piper, 1971
Lake Nitinat, Vancouver Island, BC	—	200	0.3	Richards, 1965
Saanich Inlet, BC	1-5	240	0.02	Richards, 1965; Presley et al., 1972
Black Sea	1-5	2000	0.3	Richards, 1965; Strakhov, 1967

[a]Present below 20 m.

in the nature of the organic matter itself. Some oil shales (q.v.), for example, are brown. The sediments presently being deposited in the central Black Sea are gray, although with diagenesis and lithification they will presumably darken as a consequence of polymerization of organic matter into a more condensed form by a process similar to that in which brown peat changes to black lignite. Dispersed manganese oxides can also act as a black pigmenting agent, but their effect has yet to be established. The black color of shales is not to be confused with the black color of some recent muds, the latter due to the presence of iron "monosulfides" (e.g., the mineral mackinawite, $FeS_{0.9}$), which subsequently are quantitatively transformed to pyrite, FeS_2. Pyrite is not a black pigmenting agent.

Trace Elements. Some black shales are at least locally enriched in trace elements (Swanson, 1961; Vine and Tourtelot, 1970), although values of most minor elements are similar to those of average shale (Table 2). The greater content of molybdenum and copper in the average black shale may reflect the association of these elements with organic matter. The uranium content of black shales has been studied extensively. The euxinic Chattanooga shale of the SE US and the phosphatic black shales of Pennsylvanian age in Kansas and Oklahoma were found by Swanson (1961) to be the most extensive uraniferous shales in the US, with uranium contents in the range 0.005 and 0.010%, compared to 0.0001 to 0.0004% for more typical marine shales. Again, organic matter is directly or indirectly responsible for concentrating the uranium.

Impoverished Faunas. Literature on organism-oxygen relationships in recent marine environments is reviewed by Rhoads and Morse (1971). The range of oxygen content of 0.1– 1 ml/liter dissolved oxygen (termed the *dysaerobic range*) is restrictive to many, but not all, benthic invertebrates. Fig. 1 shows the species abundances of major invertebrate groups in the Black Sea off the Rumanian coast over the *peri-azoic zone*. This zone is transitional between shallow depths rich in benthic life and deeper, sulfide-bearing, azoic depths. Diversity in aschelminths (nematods), benthic crustaceans, and polychaetes decreases below 130 m where dissolved oxygen falls below 1.0 ml/liter. Well-calcified taxa such as benthic mollusks are restricted to oxygen levels >1.0 ml/liter. Similar relationships are observed in other low-oxygen environments such as parts of the Gulf of California and some basins off southern California. Thus, an impoverished fauna is indicative of low oxygen but not necessarily anoxic or sulfide-bearing conditions, with the most tolerant species persisting in all except the most extreme conditions.

Lamination. Only a few examples of fine-grained, recent marine sediments are known to contain laminae with a thickness on the order of a few millimeters (Reineck and Singh, 1973). It is well established that under most conditions, small-scale sedimentary structures

TABLE 2. Estimates of Element Abundance in Average Shale and Average Black Shale (%)

Element	Black Shale (Average) a	Shale (Average) b	c
Al	7.0	7.8	8.0
Fe	2.	4.3	4.72
Mg	.7	1.9	1.5
Ca	1.5	5.2	2.21
Na	.7	.65	.96
K	2.	2.8	2.66
C_{org}	3.2	.65	–
C_{min}	.33	1.62	–
Ti	.2	.44	.46
Mn	.015	.67	.085
Ag	< .0001	.00009	.000007
B	.005	.012	.01
Ba	.03	.08	.058
Be	.0001	.0007	.0003
Co	.001	.0012	.0019
Cr	.01	.016	.009
Cu	.007	.0038	.0045
Ga	.002	.004	.0019
La	.003	.004	.0092
Mo	.001	.000074	.00026
Ni	.005	.0021	.0068
Pb	.002	.002	.002
Sc	.001	.001	.0013
Sr	.02	.0299	.03
V	.015	.013	.013
Y	.003	.0033	.0026
Zn	< .03	.008	.0095
Zr	.007	.02	.016

a: Vine and Tourtelot, 1970.
b: Green, 1959, cited by Vine and Tourtelot.
c: Turekian and Wedepohl, 1961, cited by Vine and Tourtelot.

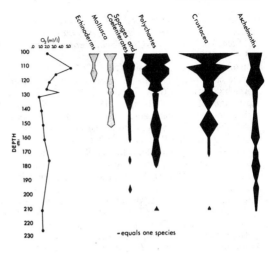

FIGURE 1. Species abundance of major benthic invertebrate groups in the Black Sea related to depth and dissolved oxygen (from Rhoads and Morse, 1971). Heavily calcified taxa represented by a dotted pattern.

are effectively obliterated by burrowing activities of benthic organisms, and it is therefore not surprising that all modern, fine-grained, laminated sediments occur in sulfide-bearing water bodies (e.g., the Black Sea and Saanich Inlet, British Columbia) or under dysaerobic conditions (e.g., the oxygen minimum zone of the Gulf of California and Santa Barbara basin off the coast of Southern California). The presence of such laminae retained in ancient sediments thus gives information of conditions at the sediment/water interface at the time of deposition.

Fine Grain Size. The fine grain size of euxinic sediments is consistent with the "starved basin" model of Adams et al. (1951). Such a basin would receive a minimum of clastic material by reason of distance from the source or by intervening troughs or basins which trap incoming sediment. This concept has been questioned in the case of the Chattanooga shale by Conant and Swanson (1961), who argue that these extensive euxinic sediments were formed in shallow water from sediment derived from an extensive peneplain drained by small and sluggish streams. Under such conditions, only fine sediment would have reached the Chattanooga Sea.

Ancient Euxinic Sediments

Pettijohn (1975) includes the following as representing euxinic sedimentation: (1) the black pyritic and carbonaceous slates of the Precambrian of Iron County, Michigan; (2) the dark graptolitic shales of the Ordovician of New York; (3) the black shales and limestones of the Upper Devonian and Lower Carboniferous of North America and Europe; (4) the Kupferschiefer of the western European Permian; (5) the black cherts, phosphorites, and black shales of the Phosphoria formation (Permian) of the north Rocky Mountain area of the US.

MARTIN GOLDHABER

References

Adams, J. E.; Frenzel, H. N.; Rhodes, M. L.; and Johnson, D. P., 1951. Starved Pennsylvanian Midlandbasin, *Bull. Am. Assoc. Petrol. Geologists*, 35, 2600-2606.

Berner, R. A., 1972. Sulfate reduction, pyrite formation, and the oceanic sulfur budget, in D. Dyrssen and D. Jagner, eds., *The Changing Chemistry of the Oceans*, vol. 20 of *Nobel Symp.* Stockholm: Almquist and Wiksell, 347-361.

Calvert, S. E., and Price, N. B., 1971. Recent sedi-

ments of the South West African Shelf, in *The geology of the east Atlantic continental margin, Africa. Gt. Brit. Inst. Geol. Sci., Rept. 70/16,* vol. 4, 171-185.
Conant, L. C., and Swanson, V. E., 1961. Chattanooga shale and related rocks of central Tennessee and nearby areas, *U.S. Geol. Surv. Prof. Pap. 357,* 91p.
Hartmann, M., and Nielsen, H., 1969. δ^{34}S—Werte in rezenten Meeressedimenten und ihre Deutung am Beispiel eineger Sedimentoprofile aus der Westlichen Ostsee, *Geol. Rundsch.,* 58, 621-655.
Krumbein, W. C., and Sloss, L. L., 1963. *Stratigraphy and Sedimentation.* San Francisco: Freeman, 660p.
Orr, W. L., and Gaines, A. G., Jr., 1974. Observations of rate of sulfate reduction and organic matter oxidation in the bottom waters of an estuarine basin: The upper basin of the Pettagnamscutt river (Rhode Island), *Proc. 6th Internat. Cong. Organic Geochem.* Paris: Editions Technip, 791-812.
Pettijohn, F. J., 1975. *Sedimentary Rocks.* New York: Harper & Row, 628p.
Piper, D. Z., 1971. The distribution of Co, Cr, Cu, Fe, Mn, Ni and Zn in Framvaren, a Norwegian anoxic fjord, *Geochim. Cosmochim. Acta,* 35, 531-550.
Presley, B. J.; Kolodny, Y.; Nissenbaum, A.; and Kaplan, I. R., 1972. Early diagenesis in a reducing fjord, Saanich Inlet, British Columbia: II, Trace element distribution in interstitial water and sediment, *Geochim. Cosmochim. Acta,* 36, 1073-1090.
Reineck, H. E., and Singh, I. B., 1973. *Depositional Sedimentary Environments.* New York: Springer-Verlag, 439p.
Rhoads, D. C., and Morse, J. W., 1971. Evolutionary and ecologic significance of oxygen-deficient marine basins, *Lethaia,* 4, 413-428.
Richards, F. A., 1965. Anoxic basins and fjords, in J. P. Riley and G. Skirrow, eds., *Chemical Oceanography,* vol. I. New York: Academic Press, 611-645.
Strakhov, N. M., 1967. *Principles of Lithogenesis,* vol. 2. New York: Consultants Bureau, 609p.
Swanson, V. E., 1961. Geology and geochemistry of uranium in marine black shales; A review, *U.S. Geol. Surv. Prof. Pap. 356-C,* 112p.
Tourtelot, E. B., 1970. Selected annotated bibliography of minor-element content of marine black shales and related sedimentary rocks, 1930-1965, *U.S. Geol. Surv. Bull.,* 1293, 118p.
Vine, J. D., and Tourtelot, E. B., 1970. Geochemistry of black shale deposits: A summary report, *Econ. Geol.,* 65, 253-272.

Cross-references: *Black Sea Sediments; Humic Matter in Sediments; Kerogen; Oil Shale; Organic Sediments; Shale.* Vol. IX: *Black Shales.*

EVAPORITE DIAGENESIS—See MARINE EVAPORITES—DIAGENESIS AND METAMORPHISM

EVAPORITE FACIES

The evaporite sedimentary rocks (evaporites) are salts deposited by chemical precipitation from natural brines during evaporation. The most common evaporites are limestone (see *Limestones*), dolomite (q.v.), gypsum (see *Gypsum in Sediments*), anhydrite, rock salt; the marine potash salts (polyhalite, sylvite, kieserite, and carnallite); and the terrestrial sulfates (glauberite, thenardite, and mirabilite) and alkali carbonates (shortite and trona) of the desert playas.

The process of evaporite formation by evaporation of brine correlates well with the origin of the major evaporite deposits by the concentration of ocean water in arid climates (see *Evaporites—Physicochemical Conditions of Origin*). The oceans are too large to change greatly in salt concentration and to become in themselves salt basins; thus, most evaporite basins lie along the borders of the seas; as the Black Sea and the Caspian Sea of today. Evaporite basins may be relatively small *lagoons* or *salinas,* or they may be major *salt seas* which cover tens or hundreds of thousands of km^2. The Silurian Salina salt basin in NE United States and Canada is estimated to have covered about 400,000 km^2, and the Devonian Prairie salt basin of western Canada was probably somewhat larger, comparable in size to the present Caspian Sea.

Evaporite minerals and rocks are deposited today in marginal salt pans; salinas; lagoons and supratidal flats; cut-off arms of the seas; and, to a lesser extent, in saline lakes. The principal evaporite rocks in the majority of the marine basins are limestone, gypsum, and halite. Occasionally residual brines precipitate to give polyhalite, carnallite, sylvite, and other complex potassium, magnesium; and calcium chlorides and sulfates. The inland salts are characterized by sodium and calcium sulfates and carbonates. Evaporite facies and periodic vertical repetition characterize the evaporite deposits. The theoretical depositional pattern developed from considerations of the chemistry of evaporation of sea water and salinity patterns in salt basins closely approximates the distribution of evaporite mineral deposition in ancient salt seas.

Environments of Evaporite Deposition

Marginal salt pans characteristically are flooded with oceanic brine during high tides and storms. The Rann of Cutch in W India was long considered to be the classic example, but Platt (1962) showed that the Rann is a normal alluvial plain where salt is brought to the surface by capillary action. Better examples of salt pans are found along the Red Sea coast of Ethiopia and scattered along the Nile Delta. In general, no thick accumulations of salts occur in salt pans, owing to the frequent flooding by normal marine water which tends to dilute and carry off the residual brines.

Salinas receive marine brines by percolation through a permeable ridge or barrier of sand or similar material. The salt lake of Larnaca, Cyprus, is cited as a typical example (Grabau, 1920). It has an area of 5 km^2 and receives brines across a 1.5 km^2 barrier of permeable sands and gravels; salt is deposited only during the summers. Lake Assal, French Territory of Afars and Issas, is some 15 km from the Gulf of Aden and lies 160 m below sea level; water percolating inland from the gulf emerges as a series of salt springs along the eastern margin of the lake (Borchert and Muir, 1964).

Lagoons and relict seas are important sites of thick salt deposition; the Kara-Bogaz Gulf is the most significant example (see *Kara-Bogaz Gulf Evaporite Sedimentology*). Lagoons precipitating salts also exist along the coast of the Gulf of Mexico, the Persian Gulf, and Red Sea. Associated with these lagoons are supratidal flats, recently recognized as an important environment of evaporite deposition (see *Sabkha Sedimentology*). Relict seas include the Caspian Sea, the Salton Sea, and the Dead Sea.

Lacustrine salts are deposited largely in desert regions with internal drainage. The Great Salt Lake in Utah, the Lop Nor of the W China desert, and the numerous playa lakes in desert regions of the world are examples. The salt in many of these lakes is recycled from older deposits exposed in the drainage areas around their rims (see *Desert Sedimentary Environments;* also *Salar, Salar Structures*).

The Bocana de Verrila, Peru is a modern example of an estuary where gypsum and halite have been deposited, and modern polyhalite is recorded from the Laguna Ojo de Liebre, Baja California. Sea spray is another source for evaporite deposition.

Fossils in Salt Deposits. Most salt basins are noted for their paucity of living plants or animals. Tasch (in Bersticker et al., 1963, p. 96–103) summarizes the major occurrences of fossils in evaporite deposits. He lists plant debris, including spores and pollen blown into the salt basins, blue-green algae, crustaceans, reptiles, birds, amphibians, mammals, mollusks, insects, and fish. Of particular interest is the finding of viable bacteria of Permian age from fluids trapped in salt crystals, which are residues of the ancient salt seas. Many such organisms may have been washed or blown into the evaporite basin during storms or high tides only to die in the highly concentrated brine. Some modern forms live in highly saline waters, as in the Great Salt Lake and the Dead Sea. Most characteristic are bacteria (no fewer than nine distinct varieties are known from the Great Salt Lake), fungi, algae, and the euryhaline crustacea (brine shrimp); most of these organisms are unlikely to leave a fossil record. To most organisms, the environment is truly *terra hospitibus ferox*.

Evaporite Facies

The Kara-Bogaz Gulf on the eastern shore of the Caspian Sea was cited by C. Ochsenius in support of his bar theory for the origin of evaporites (see *Kara-Bogaz Gulf Evaporite Sedimentology*). Lötze (1957) described it as a flat, shallow lagoon with a lateral salinity distribution that increased from inlet to distal edge. This salinity distribution is characteristic of both large and small evaporite basins. It readily follows that there will be a lateral distribution of salt precipitation, i.e., evaporite facies, that parallels the pattern of salinities.

Evaporation of the brine, producing steadily increasing salinities, will produce the sequence of mineral precipitation: calcite → calcite + gypsum → gypsum → gypsum + halite → halite → halite + potash salts. This sequence was established experimentally over 100 years ago by the chemist Usiglio (1849; Table 1). The distribution of the evaporite facies being precipitated at any time in the basin will exhibit a lateral pattern much like the brine densities in the Gulf of Kara-Bogaz. Several important features

TABLE 1. Properties of the Brine and Weight of Minerals Precipitated from the Evaporation of Sea Water

Density	Volume	Salinity[a]	Minerals Precipitated (mg/ml evaporated)		
			CaCO$_3$	CaSO$_4$	NaCl
1.0258	1.000	37.5	–	–	–
1.050	0.533	68.5	0.137	–	–
1.084	0.316	111.9	tr.[b]	–	–
1.1037	0.245	141.0	tr.	–	–
1.1264	0.190	157.1	0.963	10.20	–
1.1604	0.1445	222.5	–	15.80	–
1.1732	0.131	240.5	–	13.62	–
1.2015	0.112	274.0	–	8.42	–
1.2138	0.095	291.0	–	2.99	192
1.2212	0.064	305.2	–	4.76	312
1.2363	0.039	325.7	–	2.80	316
1.2570	0.030	339.5	–	1.64	298
1.2778	0.023	357.5	–	–	315
1.3069	0.016	399.0	–	–	207

[a]Salinity is calculated as the weight proportion of the dissolved solids per kg of brine having the density shown in the left column.
[b]tr. = trace.

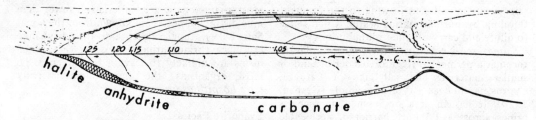

FIGURE 1. Cross section of an evaporite basin showing brine circulation (arrows), lines of equal brine density in cross section and on surface of the basinal brine, and distribution of evaporite mineral deposition produced when the brine density distribution is as shown (from Briggs, 1957).

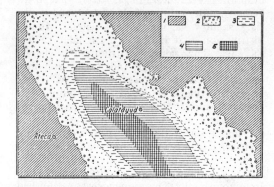

FIGURE 2. Facies relationships in the Miocene Basin of Calatayud, Spain (after Lötze, 1957). 1, Paleozoic basement; 2, conglomerate and sand; 3, chalk; 4, gypsum; 5, gypsum with salt.

evolve from this consideration: (1) The sequence of mineral deposition, calcite → gypsum → halite → potash salts, duplicates the evaporite facies pattern from an inlet and edges of a basin toward the center. Lines of equal % of salt in the mineral deposits depict this map pattern in the halite facies. (2) The evaporite facies pattern is arcuate about an inlet, the concave side toward the inlet. (3) The evaporites are thickest in the center of the basin where halite precipitates (salt facies) and thinnest around the edges where calcite precipitates (limestone facies). The evaporite basin, its current, salinities, and mineral deposition pattern are illustrated in Fig. 1.

Temporal changes in the position of precipitation of a particular mineral would be produced by a change to more extreme evaporation or to refreshening of the basinal brine. Such changes, periodic with time, produce layering and varving which characterizes almost all evaporite deposits.

Ancient Evaporite Facies. Many ancient evaporite deposits exhibit these characteristic lateral facies. The Miocene Basin of Calatayud, Spain (Fig. 2) is an excellent example.

The Late Silurian evaporites of the Michigan Basin and adjoining Appalachian Basin illustrate both facies variability and the relationship between evaporite facies and inlets connecting to the open ocean (Heckel and O'Brien, 1975). The Michigan evaporite basin developed as a result of the growth of an extensive Niagaran reef platform which ringed the edge of the basin. Within this fringe of reefs, the evaporites were deposited (Fig. 3).

The pattern in Fig. 3 illustrates the concentration of halite toward the center of the basin and the inlet across the platform reef on the NE and SE sides. The southern inlet coincides with a low "valley" through the reef joining southern Michigan with open ocean in NW Ohio and in Indiana (Fig. 4). A similar inlet can be seen in E central Ohio (Fig. 3) which fed ocean brine into the evaporite basin in northern Ohio. The open ocean lay farther to the south.

LOUIS I. BRIGGS

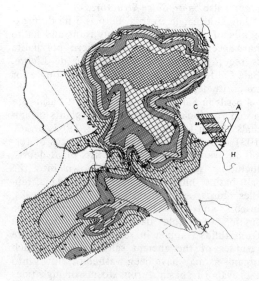

FIGURE 3. Evaporite facies, Michigan Basin, Upper Silurian Salina Formation. A-Anhydrite; C-Carbonate; H-Halite.

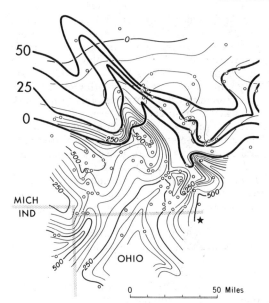

FIGURE 4. Valley across the Niagaran reef platform, an inlet to the Salina evaporite basin.

References

Bersticker, A. C.; Hoekstra, K. E.; and Hall, J. F., eds., 1963. *Sympsium on Salt*. Cleveland: Northern Ohio Geol. Soc., 661p.

Borchert, H., 1959. *Ozeane Salzlagerstätten*. Berlin: Gebrüder Borntraeger, 227p.

Borchert, H., and Muir, R. O., 1964. *Salt Deposits*. London: Van Nostrand, 338p.

Bosellini, A., and Hardie, L. A., 1973. Depositional theme of a marginal marine evaporite, *Sedimentology*, 20, 5-27.

Briggs, L. I., 1957. Quantitative aspects of evaporite deposition, *Mich. Acad. Sci. Arts Lett.*, 42, 115-123.

Busson, G., 1974. Sur les évaporites marines: sites actuels ou récent de dépôts d'évaporites et leur transition dan les séries du passé, *Rév. Géog. Phys. Géol. Dynam.*, 16, 189-208.

Drooger, C. W., ed., 1973. *Messinian Events in the Mediterranean*. Amsterdam: North-Holland, 272p.

Grabau, A. W., 1920. *Geology of the Non-Metallic Mineral Deposits Other than Silicates*. vol. 1: *Principles of Salt Deposits*. New York: McGraw-Hill, 435p.

Hardie, C. A., and Eugster, H. P., 1971. The depositional environment of marine evaporites: A case of shallow clastic accumulation, *Sedimentology*, 16, 187-220.

Heckel, P. H., and O'Brien, G. D., 1975. *Silurian reefs of Great Lakes region of North America*. Am. Assoc. Petrol. Geologists, Reprint Series 14, 205p.

Kinsman, D. J. J., 1974. Evaporite deposits of continental margins, in A. H. Coogan, ed., *Fourth Symposium on Salt*, vol. 2. Cleveland: Northern Ohio Geol. Soc., 255-259.

Kirkland, D. W., and Evans, R., eds., 1973. *Marine Evaporites*. Stroudsburg, Pa.: Dowden, Hutchinson & Ross, 426p.

Krumbein, W. C., 1951. Occurrence and lithologic associations of evaporites in the United States, *J. Sed. Petrology*, 21, 63-81.

Landes, K. K., 1960. The geology of salt deposits, in D. W. Kauffman, ed., *Sodium Chloride*. Am. Chem. Soc. Monogr. Ser. 145, 28-69.

Lötze, F., 1957. *Steinsalz und Kalisalz*, vol. 1. Berlin: Gebrüder Borntraeger, 465p.

Mesolella, K. J.; Robinson, J. D.; McCormick, L. M.; and Ormiston, A. R., 1974. Cyclic deposition of Silurian carbonates and evaporites in Michigan basin, *Bull. Am. Assoc. Petrol. Geologists*, 58, 34-62.

Platt, L. B., 1962. The Rann of Cutch, *J. Sed. Petrology*, 32, 92-98.

Richter-Bernburg, G., ed., 1972. *Geology of saline deposits*. Internat. Union Geol. Sci., Earth Sci. Ser. 7, 316p.

Stewart, F. H., 1963. Marine evaporites, *U.S. Geol. Surv., Prof. Pap. 440-Y*, 53p.

Usiglio, M. J., 1849. Etudes sur la composition de l'eau de la Mediterranée et sur l'exploitation des sels qu'elle contient, *Ann. Chim. Phys.*, 27(3), 172-191.

Cross-references: *Black Sea Sediments; Cyclic Sedimentation; Desert Sedimentary Environments; Dolomite, Dolomitization; Evaporites–Physicochemical Conditions of Origin; Gypsum in Sediments; Kara-Bogaz Gulf Evaporite Sedimentology; Limestones; Persian Gulf Sedimentology; Sabkha Sedimentology; Salar, Salar Structures; Salt-Marsh Sedimentology; Tropical Lagoonal Sedimentation*. Vol. IVA: *Evaporite Processes; Halides*. Vol. IVB: *Anhydrite; Evaporite Minerals*.

EVAPORITES–PHYSICOCHEMICAL CONDITIONS OF ORIGIN

The primary concern here is with the evaporites originating by precipitation from marine brines. The mixed types of marine and continental origin, which are often characterized by sodium sulfate minerals such as thenardite and glauberite, and the essentially nonmarine evaporites dominated by sodium sulfates or nitrates, sodium carbonates, or borate minerals are considered to be beyond the scope of this article (see Vol. IVB: *Evaporite Minerals* for remarks and references).

Significant Physicochemical Parameters

The most important physicochemical parameters in the formation of evaporites are temperature and composition of the parent brine. In thermodynamic considerations, however, instead of brine composition, other intensive parameters such as the activities of the respective components (H_2O, Na^+, etc.) are much more significant. Unfortunately, at the present time many of these parameters are not known with sufficient precision. These uncertainties are due to the inapplicability of existing electrolytic theories to saturated solutions of highly soluble salts. Therefore, in this article the classic approach of Van't Hoff (Eugster, 1973), using experimentally determined solubilities, is followed. The pressure effect is neglected because of its minor influence on the equilibria in the pressure range encountered.

Additional physicochemical factors are the hydrogen ion activity (pH) and the oxidation potential (Eh). Normally they are not reflected in the mineral associations of the main components but are indicated in the accessory minerals. Evaporites are very sensitive, however, to subsequent changes of temperature and to changing solution compositions. Thus, post-depositional or diagenetic history is found to markedly influence evaporite deposits in many ways. These secondary features often dominate ancient evaporites and render the analysis of their primary history difficult or ambiguous. Therefore, the conditions of diagenesis and metamorphism need to be investigated and understood as intensively as the physicochemical factors responsible for primary mineralogies and textures (see *Marine Evaporites–Diagenesis and Metamorphism*).

Measurable Quantities in Relation to Origin. Deductions may be based on observations of the chemical compositions, mineral assemblages, trace elements, accessory minerals, and textural and structural characteristics of the rocks. Even when all this information is available, quantitative interpretations are difficult and unique answers rarely obtained. A very useful approach is to compare theoretical models of evaporite sequences with observed sequences. These models must be calculated from the stability relationships of the evaporite minerals for different solutions and for different sets of conditions. In most cases, mineral solubilities are taken as measures of stability. It should be noted that it is the quantitative aspect of the models that is required to derive the compositional variables of interest, in particular in the discussion of deviations of the composition of the parent brine from ordinary sea water. In some cases, a more direct determination of physicochemical parameters such as temperature and composition of the parent brine is possible by means of partition equilibria of trace elements. For a more general picture, however, the qualitative sequence of mineral assemblages derived from the stability diagrams, without calculation, is sufficient.

Earlier Work and Literature. The first theoretical models were provided by Van't Hoff and his associates (see Eugster, 1973). They calculated the primary sequence of evaporite minerals that should be precipitated at 25° and at 83°C. Jänecke, in addition, considered subsequent thermometamorphic transformations. D'ans and Kühn and, in more detail, Baar, as well as Stöcke formulated the very important concept of solution metamorphism. Borchert (1965) discussed this earlier work and calculated models for intermediate temperatures. Kurnakov and his school have examined the effect of metastable equilibria on the mineral sequence. All these previous models were reviewed and enlarged upon by Braitsch (1962, 1971), paying particular attention to coincidence between the predicted mineral sequences and those actually found in natural evaporite deposits.

For references to the extensive evaporite literature the texts of Borchert (1965) and Braitsch (1962, 1971) and the valuable summary of Stewart (1963) should be consulted. The compilation of Kirkland and Evans (1973) is also a valuable reference source.

Differentiation Trends in Marine Brines

Primary Evaporation. Geochemical and petrographic evidence has revealed two types of marine evaporites with intermediate or transitional members known in only a few instances: (1) *normal marine*, with magnesium sulfates and complex salts such as polyhalite, kainite, langbeinite, but without primary sylvite; and (2) *modified marine* without these minerals but containing primary sylvite. The first type is well developed in the classical Stassfurt area of the German Zechstein, the second type by the Oligocene evaporites of the Upper Rhine Valley.

The chemical differentiation of sea water (metamorphization in the sense of Kurnakov) forming the two types is thought to take place primarily during the earlier stages of evaporite formation by: (a) a variety of reactions whereby Mg^{++} is removed from the brine and Ca^{++} added in its place; the commonest of these reactions is probably the dolomitization (q.v.) of earlier carbonate rocks:

$$2CaCO_3 + Mg^{++} \rightarrow CaMg(CO_3)_2 + Ca^{++}$$
(calcite) (brine)   (dolomite)   (brine)

or, (b) contamination of the marine brines by influx of calcium-rich continental waters. Another possible explanation of the process of brine differentiation is presented by Borchert (1965).

In both (a) and (b), the added calcium ions are usually precipitated as calcium sulfate minerals. The magnesium is fixed mainly in carbonate minerals and to a smaller extent in authigenic chlorite, talc, and other impurities. After the precipitation of the greater part of the sulfate ions, the process of modification by means of magnesium metasomatism may continue, the excess calcium ions giving rise to calcium chloride minerals on extreme evaporation. On water of the Dead Sea represents such a strongly modified brine with excess calcium chloride.

The two types (normal and modified) of marine evaporites are clearly recognized only in the late stages of the evaporation process because the diagnostic minerals have high solubilities and, therefore, occur rather late in the mineral sequences. Additional direct criteria for distinguishing them in the earlier stages might be found in the composition of primary brine inclusions within early-formed minerals.

Second-Cycle Marine Brines. Several authors strongly deny, for no very justifiable reason, the second-cycle nature of some large salt deposits (e.g., Borchert, 1965). Some salt sequences, however, when compared with theoretical models, cannot be classified as normal marine or as modified marine. In the Carlsbad deposit (New Mexico), for instance, the abnormal primary sequence and the abnormally low bromine contents (S. Adams, pers. comm.) point, at least in part, to the secondary nature of the brine. The geological details regarding the older evaporitic source rocks and the nature of the leaching solutions (probably fresh sea water) at present remain obscure.

It seems probable that in the last stages of evaporite formation, leaching and reprecipitation processes are often widespread. The Upper Rhine Valley evaporites provide an example. Possible large-scale reworking should be considered in Saskatchewan, in the Ural foredeep region, and in several other examples.

Conditions of Calcium Sulfate Deposition

Many experimental and thermochemical studies on calcium sulfate deposition have been conducted since the classic work of Van't Hoff, unfortunately with conflicting results. Hardie (1967), however, was able to prove the essential correctness of Van't Hoff's results by reversibly locating the equilibrium curve (Fig. 1). The solubility measurements of other workers (with the exception of Zen, 1965), approaching from the undersaturated side alone, are unreliable.

At constant temperature and pressure, the equilibrium constant of

$$CaSO_4 \cdot 2H_2O \rightleftharpoons CaSO_4 + 2H_2O$$
$$\text{(gypsum)} \quad \text{(anhydrite)} \quad \text{(water)}$$

depends only on the activity of water ($K = a_w^2$), which is a function of the concentration of dis-

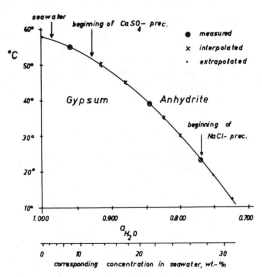

FIGURE 1. Gypsum-anhydrite equilibrium temperature at different activities of water (modified from Hardie, 1967).

solved solids (ionic strength of the solution). The equilibrium temperature decreases from 58°C in pure calcium sulfate solutions to 23°C at an a_w corresponding to the beginning of halite precipitation. These temperatures are respectively 16° and 5°C higher than previously assumed.

As Fig. 1 suggests, in many cases of natural evaporation, gypsum will be the stable phase of $CaSO_4$. Only at high salinities or at comparatively high temperatures will anhydrite be stable. Even in these cases, however, anhydrite does not form immediately from the solution—a result of serious kinetic barriers—instead, metastable gypsum nearly always precipitates. Nevertheless, Hardie's experiments and the observations by Kinsman (1966) of anhydrite and gypsum in Holocene sediments of the Persian Gulf demonstrate at least the possibility of contemporaneous conversion of gypsum to anhydrite without seeding, and probably the primary precipitation of anhydrite under certain conditions.

Precipitation of $CaSO_4$ from a sea-water brine starts at a concentration factor of 3.5 (about 120⁰/₀₀ salinity). Some ancient, thick anhydrite deposits are diagenetic transformation products from gypsum, established by the observation of anhydrite pseudomorphs after swallow-tail twins of gypsum (see Borchert, 1965; see also *Gypsum in Sediments*). The time interval between gypsum precipitation and conversion to anhydrite remains unknown, but according to experimental data the dehydration reaction proceeds fairly rapidly.

Only 3 m of anhydrite (or 4.8 m of gypsum) are deposited from the static evaporation of 8500 m of sea water. Such depths of evaporating seas are surely unreasonable. Thick evaporite deposits, therefore, must result from the repeated supply of new brines into the basin of deposition.

Stability Relations of Saline Minerals

After precipitation of $CaSO_4$, sea water may be approximated by the Na, K, Mg, Cl, SO_4, H_2O system with five independent components, inasmuch as the calcium activity is too low to affect the equilibria of calcium-free phases (see subsection on polyhalite). Under natural conditions, only the phase relations in NaCl-saturated solutions are important, and so we are left with four compositional variables. In evaporite-brine systems depleted in $MgSO_4$ (SO_4^{--} absent or rare) we need to consider only three independent variables. In a more advanced treatise, complex formation and ion association should be considered, as has been done for mineral equilibria in normal sea water. At the high concentrations of salt-precipitating brines, however, no electrolyte theory applies, so that experimental data must be used empirically to determine the mineral stability.

The stability relations of saline minerals are shown in a concentration-temperature diagram in Fig. 2 and in an isothermal section in Fig. 3—based on half a century of experimental work initiated by Van't Hoff (see D'Ans, 1933; Braitsch, 1962, 1971). The increasing $MgCl_2$ content in the brine used as abscissa in Fig. 2 roughly corresponds with the decreasing activity of water. Only calcium-free phases and neutral solutions are considered. These are saturated with halite and the indicated phase (or phases). Fig. 2a shows a confining quaternary system without potassium. It explains many of the mineral associations in the sulfate subgroup of nonmarine evaporites (in this case glauberite is normally the Ca phase), whereas this system in marine evaporites is realized only in metamorphic associations and mainly on the $MgCl_2$-rich side. A second confining quaternary system without sulfate, which is responsible for the important subgroup of modified marine evaporites, is discussed below. The third confining quaternary system without magnesium is of minor importance (e.g., in some fumarolic efflorescences). Figs. 2b and 2c show sections through the quinary system. Both figures delimit the stability area of double salts of magnesium plus potassium, or of sodium plus potassium: Fig. 2b on the potash-poor side, and Fig. 2c on the potash-rich side of the system.

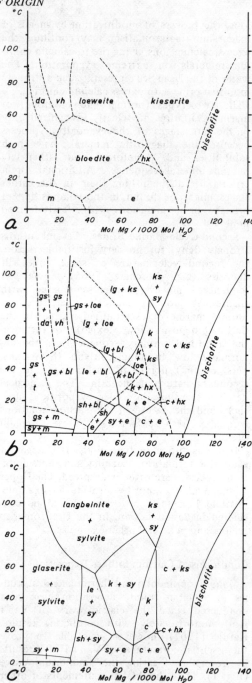

FIGURE 2. Stability relations of evaporite minerals, all at saturation with respect to NaCl (halite): (a) System Na^+, Mg^{++}, Cl^-, SO_4^{--}, H_2O (potash absent); (b) System Na^+, K^+, Mg^{++}, Cl^-, SO_4^{--}, H_2O (potash-poor side of the system); (c) System Na^+, K^+, Mg^{++}, Cl^-, SO_4^{--}, H_2O (potash-rich side of the system), Symbols: bi = bischofite; bl = bloedite; c = carnallite; da = dansite; e = epsomite; gs = glaserite; hx = hexahydrite; k = kainite; ks = kieserite; lg = langbeinite; le = leonite; loe = loeweite; m = mirabilite; sh = schoenite (shönite); sy = sylvite; t = thenardite; vh = vanthoffite.

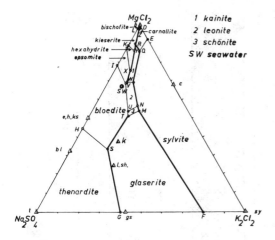

FIGURE 3. 25°C-isotherm of the system Na$^+$, K$^+$, Mg^{++}, Cl$^-$, SO$_4^{--}$, H$_2$O in NaCl (halite) saturated solutions at 1 atm (modified from Braitsch, 1962). The fields of the diagram show the composition of solutions from which any solid phase crystallizes (in association with halite). The compositions of solid phases are indicated with small triangles. In most cases these lie outside the corresponding mineral stability field (because of incongruent solubility). Heavy lines are crystallization paths with precipitation of solid phases from both adjoining fields. Thin field boundaries are transgression lines being transgressed during crystallization. Invariant points (at constant pressure) are indicated with capital letters; most of them are reaction points. b = bischofite; bl = bloedite; c = carnallite; e = epsomite; gs = glaserite; h = hexahydrite; k = kainite; ks = kieserite; l = langbeinite; sh = shoenite (shönite); sy = sylvite; t = thenardite.

To derive the crystallization path on evaporation, isothermal projections of MgCl$_2$, Na$_2$SO$_4$, and K$_2$Cl$_2$ are more convenient (Fig. 3). Data for other temperatures are reproduced in Borchert (1965) and Stewart (1963) and discussed in Braitsch (1962, 1971). Inside the diagram, only point Z is a true eutectic point; all other isothermally invariant points are reaction points.

From these data, quantitative models of the resulting mineral assemblages may be derived by means of graphical or numerical methods. A few examples will be presented.

Calculated Models of Primary Evaporite Sequences

Modified Marine Evaporites. In this case only halite, sylvite, carnallite, and bischofite are possible phases. Fig. 4 shows an isothermal section of the relevant system with the concentrations of KCl on the abscissa and of MgCl$_2$ on the ordinate. The NaCl-saturation concentrations are projected onto the same plane as

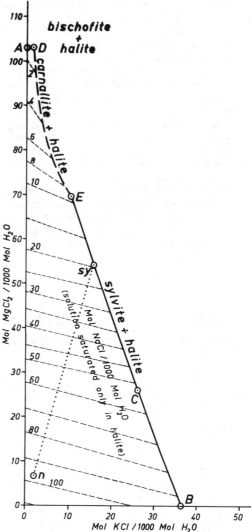

FIGURE 4. 20°C-isothermal section of the system Na$^+$, K$^+$, Mg^{++}, Cl$^-$, H$_2$O, saturated in halite (modified from Braitsch, 1962 and D'Ans, 1933). $n-sy$ = crystallization path of MgSO$_4$-free sea water during halite deposition.

dashed *isohalines* (Mol NaCl/1000 Mol H$_2$O). In solutions of this composition, sylvite may react with the brine and form carnallite:

KCl + MgCl$_2$ + 12H$_2$O $\rightleftharpoons$ KMgCl$_3$(H$_2$O)$_6$
(sylvite) (brine)   (carnallite)
                          + 6H$_2$O
                          (brine)

Actually, in SO$_4^{--}$-free brines, the reaction mechanism of sylvite is of minor importance because the sylvite is mantled by thin reaction rims of carnallite, so that the sylvite for the most part is preserved. Fig. 5 shows evaporite sequences calculated for different temperatures, assuming that no reaction takes place at point R in Fig. 3 (this might occur if the sylvite crystals settled to the crystal-brine interface). Thicknesses are those that would be developed if the potash phase of precipitation was preceded by 100 m of halite. As already stated, such thicknesses require a repeated, continuous, or discontinuous supply of additional brines. Furthermore, the bischofite layer is of only theoretical interest, inasmuch as in nature the evaporation process only exceptionally reaches this stage. As will be seen, the proportion of halite to sylvite and to carnallite depends slightly on temperature. This proportion is independent of the initial ratio of sodium to potassium, as long as the system does not contain additional components.

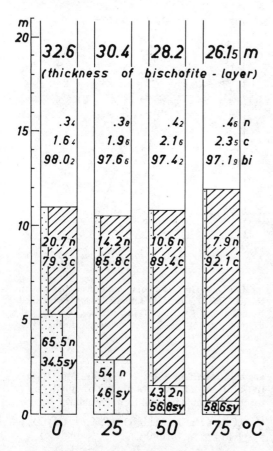

FIGURE 5. Theoretical sequence of evaporite minerals precipitated from $MgSO_4$-free sea water at different temperatures. Thicknesses (meter) normalized to a 100-m thick basal halite layer, below the potash seam. n = halite; sy = sylvite; c = carnallite; bi = bischofite (wt.%).

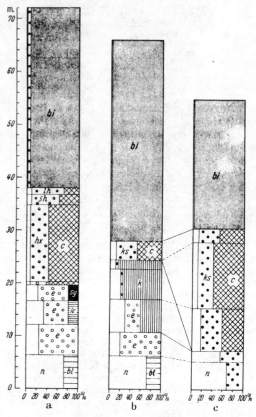

FIGURE 6. Theoretical sequence of potash salts precipitated by progressive evaporation of a normal marine brine at 25°C (from Braitsch, 1962, 1971). Thicknesses are normalized to a 100-m thick basal halite layer below the sequences shown. (a) metastable equilibria; (b) stable equilibria, without reaction at the transition points (c) stable equilibria, with complete reaction at the transition points. n = halite; $5h$ = pentahydrite; lh = leonhardtite (otherwise symbols as in Fig. 2).

Normal Marine Evaporites. Starting from normal sea water, represented by point SW in Fig. 3, or for that matter any other solution in the system, the crystallization path will terminate at point Z, saturated in bischofite, carnallite, kieserite, and halite. But, on the crystallization path, different reaction points have to be passed and different reaction sequences are possible. In addition, metastable phases may be developed.

Fig. 6 illustrates three theoretical models of evaporite mineral sequences formed under different reaction conditions, but at the same temperature (25°C). The thicknesses are related to a halite rock thickness of 100 m below the sequence indicated. This mineral sequence and thickness would result from complete evaporation of 8500 m of sea water (supplied by influx, from the sea or from a larger reservoir). Fig. 6b is similar to Fig. 5 and corresponds with static evaporation without reaction at the possible reaction points. This is the model exclusively considered by Van't Hoff and by Jänecke. The essential feature is a massive kainite layer. A major part of the potassium is fixed in this mineral. The kainite layer is to be expected in the temperature interval between $11°C$ and $72°C$. Fig. 6a shows the sequence, if only metastable phases are considered. This model is favored mainly by Russian scientists, particularly Kurnakov's group. The main features are the absence of kainite and the occurrence of primary sylvite and metastable magnesium sulfate hydrates. The third model (Fig. 6c), originally considered by Braitsch, is the reaction model with reaction of the primary phases at the different reaction points. In this model, the only phases to survive are those that are stable in the final solution Z (halite, carnallite, kieserite, bischofite). The potassium is fixed only in carnallite. The carnallite layer, therefore, is considerably enlarged, whereas the bischofite layer is reduced.

The most important step in the reaction model is elimination of the kainite originally formed (or, in the case of metastable parageneses, elimination of the association epsomite + sylvite). Kainite will react with the evaporating brine, and carnallite + kieserite (+ halite) will precipitate. The solution remains unchanged in composition but is reduced in volume. In a similar way, reactions may be calculated for other reaction points in any of the three theoretical models presented. Another important reaction is the penecontemporaneous reaction of earlier stable (or sometimes metastable) gypsum to anhydrite in later, more saline, brines. Some authors, however, prefer to classify this as a diagenetic transformation (see *Marine Evaporites—Diagenesis and Metamorphism*).

Comparison of these models with natural sequences leads to the following conclusions: (1) Metastable equilibria dominate the crystallization processes in many modern saline lakes. In the case of potassium-poor continental salt lakes, however, stable equilibrium assemblages are developed. (2) Evaporites of the past contain only the more stable mineral associations, which are at least partly the product of diagenetic alteration. (3) The elimination of primary phases may start penecontemporaneously as a variant on the reaction model.

At different temperatures, different sequences may be precipitated, as illustrated qualitatively in the phase diagrams of Fig. 2 and quantitatively in Fig. 5. Other quantitative models were calculated by Borchert (1965) and Braitsch (1962, 1971). For example, the weight ratio of kieserite to carnallite in the carnallite layer decreases from 0.8 at $25°C$ to 0.1 at $55°C$.

The Significance of Polyhalite. In more saline marine brines, after the initial period of halite precipitation, polyhalite becomes the stable calcium-bearing mineral. Under conditions of progressive evaporation of sea water, however, the calcium ion activity, after calcium sulfate precipitation, is too low to produce significant amounts of polyhalite. A different situation develops, however, when renewed influx of sea water or relatively dilute marine brines occurs, as these waters still contain appreciable amounts of calcium. These brines will remain stratified above the denser, more saline bottom brine. Upon evaporation, gypsum will precipitate from the surface brine; but as they settle out, the gypsum crystals should react with the lower brines because they are normally unstable in solutions of rather low water activity. Textural observations indicate anhydrite to be the first alteration product, replaced in turn by polyhalite:

$2CaSO_4 + 2K^+ + Mg^{++} + 2SO_4^- + 2H_2O$
(anhydrite)   (from the solution)
$\rightleftharpoons K_2MgCa_2(SO_4)_4 \cdot 2(H_2O)$
(polyhalite)

Saturation with respect to polyhalite is achieved well before saturation with the other solid phases of the quinary system (Braitsch, 1962, 1971); therefore some K^+ and Mg^{++} will be removed from solution rather early in the crystallization process. The influence on Mg^{++} and SO_4^- ion activities is not very important. The activity of K^+, however, may significantly decrease and, in the later stages of evaporation, therefore, the potential kainite layer may be completely absent. Only in the rather infrequent deposits where thick polyhalite beds occur as at Carlsbad, New Mexico, has this mechanism been particularly effective. It is certainly not sufficient as an explanation for the absence of primary kainite in the Stassfurt area, where the amount of polyhalite present is rather small.

Unfortunately the stability field of polyhalite and its boundary against the glauberite stability field in equilibrium with other phases of the quinary system is not yet known with sufficient precision, and therefore, quantitative models including polyhalite are not yet available. A qualitative discussion may be found in Braitsch (1962, 1971).

Influence of Temperature Variations. The temperature coefficient of sylvite and carnallite solubility is positive and rather strong; we can conclude, therefore, that these two minerals may, in many cases, have been formed by means of cooling of the saturated brines. Calculation of the effect of a 1°C cooling of a 10-m deep brine body initially at 20°C indicates that 6.3 mm sylvite + halite or 20 mm carnallite + halite will be precipitated. The sylvite rock will contain about 78 wt.% KCl, and the carnallite rock will contain about 98 wt.% carnallite. Because the cooling effect is reversible, an amount of water must be evaporated in the subsequent warming period to maintain the solutions at saturation with respect to the solid phases precipitated by cooling, if net crystal accumulation is to occur. The importance of temperature differences has been stressed by Jänecke and by Borchert (1965), and is discussed by Braitsch (1962, 1971).

Behavior of Accessory and Trace Components

The rarer components of sea water may (1) be isomorphously substituted for the main elements (e.g., bromine, rubidium, strontium); (2) form accessory minerals (e.g., borates, magnesite, celestite, hematite, pyrite); (3) be adsorbed on clay and other mineral surfaces; or (4) be enriched in the final brine. Most elements occur in several associations.

Information on hydrogen ion activities and on the redox potentials prevailing during the mineral formation may be gained from (2). Little is known of (3), and evidence of (4) is usually lost as final brines are ultimately purged from the system and may leave little or no memory behind. If a minor element is isomorphously coprecipitated within the lattice of an abundant mineral, however, several aspects of the depositional basin history may be elucidated. As an example, the chemical evaluation of bromine will be considered.

Behavior of Bromine. Bromine almost exclusively substitutes for chlorine in the haloids. At concentrations of up to 1 wt.% Br in the brine, the bromine content in the haloid is proportional to the bromine content of the brine. The experimentally determined partition coefficient b = wt.% Br (crystals) ÷ wt.% Br (solution) is always less than unity; and, in nature, partition equilibrium between precipitated solid phases and coexisting solutions is achieved only at the surface of the solid phases. Therefore, the bromine content of the brine during evaporation increases. In addition, of course, the ratio $Br^-:Cl^-$ progressively increases in the brine. The bromine distribution within the evaporite mineral and brine sequence at

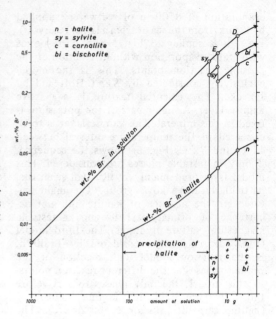

FIGURE 7. Bromine partition between solution and solid phases at different brine concentration stages in $MgSO_4$-depleted sea-water system (25°C, 1 atm).

25°C for the $MgSO_4$-free system under progressive continuous evaporation is shown in Fig. 7. In comparison with halite, the bromine contents of sylvite, carnallite, and bischofite, crystallizing from the same solution at the same moment (i.e., equilibrium crystallization) are higher by factors of 10, 7, and 9 respectively. Lower ratios point to secondary processes (diagenetic or metamorphic recrystallization). Higher ratios point to particular disequilibrium conditions of primary crystallization, for instance brine stratification and crystallization of the associated haloids from different levels of the brine.

Because of the large positive temperature coefficients of sylvite, carnallite, and bischofite solubility (which are accompanied by a slight increase in the bromine partition coefficients with rising temperature), points sy, E, and D in Fig. 7 shift to the left at lower temperatures and to the right at higher temperatures; whereas the gradient of bromine increase remains nearly constant. The bromine contents of the haloids crystallizing from these univariant points are therefore potentially suitable for geothermometry. Using this approach, a detailed study has been completed of the Upper Rhine Valley potash deposit. Brine temperature during sylvite precipitation ranged between 10°C and 50°C, increasing upward through the sequence (Braitsch and Hermann, 1965).

OTTO BRAITSCH*
DAVID J. J. KINSMAN

*Deceased

References

Arthurton, R. S., 1973. Experimentally produced halite compared with Triassic layered halite-rock from Cheshire, England, *Sedimentology,* 20, 145–160.
Borchert, H., 1965. Principles of oceanic salt deposition and metamorphism, in J. P. Riley, and G. Skirrow, eds., *Chemical Oceanography,* vol. 2. London: Academic Press, 205–276.
Borchert, H., and Muir, R. O., 1964. *Salt Deposits.* London: Van Nostrand, 338p.
Braitsch, O., 1962. *Entstehung und Stoffbestand der Salzlagerstätten.* Berlin: Springer, 232p.
Braitsch, O., 1971. *Salt Deposits, Their Origin and Composition.* Berlin: Springer, 297p.
Braitsch, O., and Hermann, A. G., 1965. Konzetrations, Dichte und Temperaturverteilung in der unteroligozanen Salzlagune des Oberrheins, *Geol. Rundschau,* 54, 344–356.
Braune, K.; Heimann, K.-O.; and Fabricius, F., 1972. Zum Problem der evaporitführenden neogenen Sedimente in mediterranen Raum, *Z. Deutsch. Geol. Ges.,* 123, 555–558.
D'Ans, J., 1933. *Die Lösungsgleichwichte der Systeme der Salze ozeanischer Salzablagerungen.* Berlin: Verlagsgesellschaft für Ackerbau, 254p.
Dean, W. E., 1975. Shallow-water versus deep-water evaporites: discussion, *Bull. Am. Assoc. Petrol. Geologists,* 59, 534–542.
Eugster, H. P., 1973. Experimental geochemistry and the sedimentary environment: Van't Hoff's study of marine evaporites, in R. N. Ginsburg, ed., *Evolving Concepts in Sedimentology.* Baltimore: Johns Hopkins Univ. Press, 38–65.
Hardie, A. A., 1967. The gypsum-anhydrite equilibrium at one atmosphere pressure, *Am. Mineralogist,* 52, 171–200.
Holland, H. D., 1974. Marine evaporites and the composition of sea water during the Phanerozoic, *Soc. Econ. Paleont. Mineral. Spec. Pap.* 20, 187–192.
Kinsman, D. J. J., 1966. Gypsum and anhydrite of Recent age, Trucial Coast, Persian Gulf, in J. L. Rau, ed., *Proc. 2nd Salt Symp.,* vol. 1. Cleveland: Northern Ohio Geol. Soc., 302–326.
Kirkland, D. W., and Evans, R., eds., 1973. *Marine Evaporites.* Dowden, Hutchinson & Ross, 426p.
Ross, D. A., and 11 others, 1973. Red Sea drillings, *Science,* 179, 377–380.
Sozansky, V. I., 1973. Origin of salt deposits in deep-water basins of Atlantic Ocean, *Bull. Am. Assoc. Petrol. Geologists,* 57, 589–595.
Stewart, F. H., 1963. Marine evaporites (data of geochemistry 6th ed.), *U.S. Geol. Surv. Prof. Pap. 440-Y,* 52p.
Stuart, W. D., 1973. Wind-driven model for evaporite deposition in a layered sea, *Geol. Soc. Am. Bull.,* 84, 2691–2704.
Zen, E-an, 1965. Solubility measurements in the system $CaSO_4$-$NaCl$-H_2O at 35°, 50° and 70°C and one atmosphere pressure, *J. Petrology,* 6, 124–164.

Cross-references: *Carbonate Sediments–Diagenesis; Dolomite, Dolomitization; Evaporite Facies; Experimental Sedimentology; Gypsum in Sediments; Kara-Bogaz Gulf Evaporite Sedimentology; Marine Evaporites–Diagenesis and Metamorphism; Sabkha Sedimentology; Tropical Lagoonal Sedimentation.* Vol. IVA: *Evaporite Processes; Halides.*

EXPERIMENTAL SEDIMENTOLOGY

To experiment is to conduct trials under controlled conditions in order to prove or illustrate the validity of previously determined assumptions or hypotheses. Experiments in sedimentology may be conducted in the field as well as in the laboratory. They may be physical as well as chemical, and they may be designed to test hypotheses concerning sedimentary processes or sediment properties. Sedimentological experiments may be entirely qualitative or, as is usual, they may be designed on a numerical measurement basis. Some advantages of experimental data are listed in Table 1.

Experimentation connotes modeling. A model is a supposedly representative structured simulation of a natural prototype. Models may be used to test theory or to simulate known natural processes (discussed in detail by King, 1966). Most physical, chemical, field, or laboratory experiments can be considered models. If the model and prototype have equal ratios in the fundamental dimensions of mass, length, and time, then dynamic, geometric, or kinematic similitude, respectively, is obtained. If dimensional similitude is not obtained, the model is an analog. Analog models generally have distortions in material or time properties. Statistical mathematical models are used to quantitatively determine experimental precision, accuracy, correlations, and thus, validity. They help to establish the model–prototype similitude.

Experiments have been conducted in virtually all aspects of sedimentology. Although field studies of modern and ancient sediments provide the basis for their description, comparison, and classification, the detailed mechanics of many sedimentary processes can be only presumed in the field. For example, the aggradation in rivers and on deltas during floods, the deposition on hurricane washover fans, and the diagenetic compaction of muds or altera-

TABLE 1. A Comparison of Observational and Experimental Data in Geology

Observational Data	Experimental Data
Numerous variables operate simultaneously in a complexly interlocked manner	Selected variables are "permitted to operate" one or several at a time; interactions can generally be controlled
Data tend to be "noisy" because of local geographic and temporal variations in the operative processes	Measurement error and "unexplained variability" can be kept very small; experimental results are independent of geographic location or time of experiment
Variables are not controlled by man	Experiment is performed as completely as possible under man's control
Observations can be made only where natural phenomena occur or can be reached by probes	Essentially all of a given phenomenon can be studied within the range of experimental conditions
Heavy reliance is placed on sample observations, sometimes greatly limited by lack of access to points of observations	Measurements ("samples") can be taken at will, essentially continuously if desired
Many observations are not yet quantifiable, thus full numerical analysis is limited	Experiments can be deliberately confined to measurable processes and responses

From Krumbein, 1962.

tion of sandstone cements are extremely difficult processes to document as they occur. Thus, from field studies only inferences can be drawn concerning the dynamic factors responsible for the resulting deposits. Through scale or analog models the dynamic properties of the sediment and its transporting medium, plus other associated variables, can be isolated and studied.

Statistically validated correlations derived from experiments provide for better understanding of natural process-response relationships. The results of experiments in sedimentology have numerous applications, but they must be applied with conservatism and caution. The foundation of experimentation is control, and when we seek to control the variabilities of nature we distort the processes and products of natural change.

Historical Basis

Theophrastus reported the earliest known sedimentological experiments, which consisted of decantation methods which ultimately produced concentrated cinnabar (Krumbein, 1932). Frizi (1762) conducted probably the first experiment on particle abrasion by placing a number of rock fragments in a box and shaking it for a long period of time. He noted little reduction in weight, and hypothesized that rock fragments could not be reduced to a fine powder by transport abrasion within the length of transport ordinarily found in rivers. Sorby (1862) conducted probably the first experiment in carbonate sedimentology, as he successfully changed aragonite to calcite by heating it in a boiler. Many early experiments were discussed by Daubrée (1879) and Adams (1918).

During the last centennium, the numbers and diverse types of sedimentological experiments have increased markedly. Van't Hoff's experimental geochemical studies of marine evaporites, conducted between 1893 and 1908, effectively laid the foundations for the quantitative studies of chemical thermodynamics and kinematics so important in carbonate sedimentology today (reviewed by Eugster in Ginsburg, 1973). Although most early experiments with detrital sediments were qualitative, Gilbert's (1914) flume experiments were remarkably advanced quantitatively. Gilbert identified the relationships between many of the complex variables which control sediment transport and bed forms. Useful quantitative relationships between grain size and ripple size, however, were obtained earlier by F. A. Forel and M. C. de Candolle (1883, reported in Evans and Ingram, 1943).

Essentially all of the sediment analysis and experimental sedimentary process techniques used today were originated and developed through experimentation. The methods and equipment for grain-size analysis, sedimentary-structure simulation, microscopic analysis, x-ray analysis, heavy-mineral separation, and geochemical analysis are examples (reviewed in Müller, 1967). Experiments of the processes of deltas, rivers, and beaches conducted by engineers in the last 40 years have contributed much to our knowledge of these sedimentological environments, and to experimental methods in general (for examples, see reports of the US Waterways Experiment Station, Vicksburg, Miss.). The need for quantification and detailed analysis of the many complex variables in sedimentary processes has resulted in modern experimental methods using electronic analog models, huge scale-model flume and wave tanks, and the most complex experimental chemical apparatus.

Experiments in Field and Laboratory

Field Experiments. Natural sedimentary environments may be used to test preconceived hypotheses of natural processes or properties. Field experiments thus formulated employ samples of natural populations upon which some type of control has been exerted. For example, experimental results will be obtained by marking a group of pebbles and recording the distribution of sizes and the distance of transport of individual pebbles over a specified river distance. Such an experiment and its results form a model for transport processes in the specified river distance, and that model may apply to other rivers.

Field experiments can only be conducted in natural situations where sufficient control can be maintained. Because little control can be exerted on most natural processes, the control for field experiments is usually obtained by segregating and recording morphological and spatial properties as they are altered by uncontrollable processes. For example, Carr (1971) studied rates of movement and size segregations of pebbles by longshore transport in the surf zone at Chesil Beach, England. Driscoll (1967) compared abrasion of shell fragments on beaches with that in the sublittoral zone. Such studies are of value because they provide data on processes that cannot usually be simulated completely in the laboratory. They provide much of the observational data that is the foundation for subsequent hypotheses and laboratory experiments. Furthermore, field experiments provide numerical data which may demonstrate statistical trends, provided adequate qualitative explanation is given for experimental error and natural variability.

Laboratory Experiments. Control and measurement of the rates and products of sedimentary processes can be effectively accomplished in the laboratory. Variables can be segregated, and the effect of altering some while others are unaltered can be determined. The most directly applicable experimental results are derived from scale models, which are constructed with dimensional similitude between the model and a natural prototype. Such similitude is reflected in dimensionless parameters which provide specific ratios of important physical properties. Examples of dimensionless parameters are ratios of inertial to gravitational forces (Froude parameter) and of inertial to viscous forces (Reynolds parameter), common parameters in fluid flow problems (see *Reynolds and Froude Numbers*). The experiments of flow regime and bed forms conducted by Simons and Richardson (*in* Middleton, 1965) in a flume 46 m long by 2.5 m wide may be considered scale models because the hydraulics and sediments were of the same order of dimensional magnitude as those of many natural rivers. Excellent discussions of dimensional analysis and similitude have been given by Hubbert (1937) and Strahler (1958).

There is usually some degree of distortion in most experiments, especially when the material in the model is, of necessity, different than that in the prototype. For example, the kaolin model glaciers of Lewis and Miller (1955) were analog models. Almost all sedimentological models are analog models because they are usually considerably distorted in at least the time dimension, and commonly in mass and length dimensions as well. Indeed, distortions can occasionally be considered valuable because it is practically impossible to model time effectively; yet it is essential to model changes through time in order to extrapolate back in geologic history.

Physical and Chemical Experiments

In sedimentology, the distinction between physical and chemical processes is not precise—witness the importance of physical chemistry in geochemistry—there are, however, some prominent distinctions which justify their separate treatment on an experimental basis. Thus, experiments of transport and deposition of detrital particles are experimentally studied most effectively with physical models, whereas experiments of carbonate and evaporite sedimentation and solution are essentially chemical in nature. Experiments of the pre- or post-depositional changes of buried sediments involve both physical and chemical changes, as in sediment compaction, porosity, cementation, permeability, and weathering.

Physical Experiments. The transport and deposition of sediments by water and air have been experimentally studied intensively and with some success. Experiments of particle entrainment by flowing water have produced methods for calculating sediment load amounts and the conditions under which entrainment occurs (see Blatt et al., 1972). The abrasion of particles by various transporting agents was studied experimentally by Kuenen (1965). Experimental studies of turbidity currents (reviewed by Walker in Ginsburg, 1973) have provided evidence for the interpretation of many sandstone beds in flysch sequences. Experimental studies of the origin of laminae, bedding, and primary sedimentary structures (see Brush in Middleton, 1965) have made it possible to better infer paleoenvironments from ancient sedimentary rocks. Experimental studies of glacial deposits and wind dynamics have increased our understanding of these processes. Many types of physical experiments are referenced in Charlesworth and Passero (1973) and Blatt et al. (1972).

Chemical Experiments. Experimental studies of evaporite and carbonate deposition, diagenesis, and chemical weathering are based on the principles of chemical thermodynamics and kinetics. Experiments in chemical sedimentology are generally concerned with determinations of the reaction rates and reaction mechanisms of gases, liquids, and solids. Chemical experiments have been successful in determining quantitatively the chemical equilibria between elements and compounds in natural environments. They have helped to explain the occurrence and distribution of evaporites and carbonates, and the processes of weathering of different source rocks in different environments. High-temperature geochemical studies have reproduced processes and products which occur naturally at lower temperatures. Low-temperature experiments have encountered problems of model-prototype kinematic distortions, for example in the experimental production of dolomite, but the experimental results at least provide insight into the direction and type of process that occurs. Chemical experiments are reviewed by Billings et al. (1972), Berner (1971), Degens (1965), Garrels and Christ (1965), and Billings (1973).

Experimental Design

Probably the most important factor in the success of sedimentological experiments is their design. The objectives, materials, available equipment, procedures, similitude, and techniques of data collection and data analysis are all extremely important considerations in design. The objectives should be attainable and

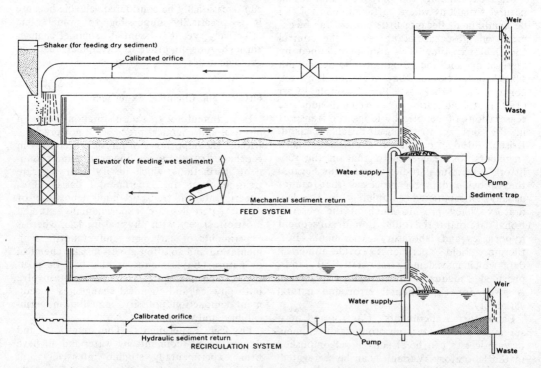

FIGURE 1. The feed system (top) of flume operation involves the controlled rate of supply of sediment to the water, which continually recirculates. Methods for both the wet and dry feeding of sediment are shown. In the sediment recirculation system of flume operation (bottom), both sediment and water are continually transferred through the pump and channel. For most sedimentological research the nonrecirculating (feed) system will usually give the most flexibility and control. (From Rathbun et al., 1969, p. D2).

clearly specified. The materials and equipment should be chosen so that the model has the desired similitude.

The materials used in most sediment transport and deposition experiments are simply sand and water. Flumes, which are rigid-boundary, open, and artificial water-conveyance channels, may either recirculate water and sediment or recirculate only water, with the sediment fed by some type of mechanical device (Fig. 1). Flume design and operation are discussed by Williams (1970; 1971). Flumes are especially useful in modeling fluvial sedimentological processes (see *Bed Forms in Alluvial Channels; Fluvial Sediment Transport*). Delta and beach processes have also been modeled in flumes, but wave tanks have better geometric similitude for such investigations (see Allen, 1970, for references). Wind tunnels, such as that used by Bagnold (1941), can be effectively used to model eolian processes. Scoffin (1968) developed an underwater flume to study the in-situ erosion of carbonate sands. Circular flumes have been used to study wave and turbidity current phenomena (Kuenen, 1966). The many and varied types of apparatus used for geochemical studies are discussed in Vol. IVA, *Geochemistry and Environmental Sciences*.

Inexpensive equipment can be used to model many sedimentological processes. McKee (1947) modeled the development of fossil animal tracks with lizards that ran over sand in a covered trough with an attractive light source at one end. Folk (1971) modeled seif dune forms with a roller and grease on paper. Menard (1950) modeled the transportation of sediment particles by floating bubbles. Numerous other examples of simple yet illustrative experiments are referenced in Charlesworth and Passero (1973).

Entirely qualitative experiments may provide insight into processes that have been insoluble in the field. For example, turbidity currents have never been directly observed in the deep ocean, but their existence and operation can be observed by stirring up the mud at the edge of a puddle and watching the denser mud mixture flow toward the middle under the less dense clear water. This example illustrates one of the simplest experimental approaches to complex sedimentological problems; it is often surprising how simple approaches can be the most rewarding. As Kuenen (1958) aptly stated, experiments in sedimentology can be used to observe and measure known processes, to test simplified mental pictures, and simply to stimulate the imagination.

RUSSELL G. SHEPHERD

References

Adams, F. D., 1918. Experiment in geology, *Geol. Soc. Am. Bull., 29,* 167-186.

Allen, J. R. L., 1970. *Physical Processes of Sedimentation.* London: Allen and Unwin, 248p.

Bagnold, R. A., 1941. *The Physics of Blown Sand and Desert Dunes.* London: Methuen, 265p.

Berner, R. A., 1971. *Principles of Chemical Sedimentology.* New York: McGraw-Hill, 240p.

Billings, G. K., 1973. Chemical geology: An annotated bibliography, *Counc. Educ. Geol. Sci. Prog. Publ. 11,* 40p.

Billings, G. K.; Garrels, R. M.; and Lewis, J. E., eds., 1972. Papers on low temperature geochemistry, *J. Geol. Educ., 20,* 217-272.

Blatt, H.; Middleton, G. V.; and Murray, R., 1972. *Origin of Sedimentary Rocks.* Englewood Cliffs, N.J.: Prentice-Hall, 634p.

Brady, L. L., and Jobson, H. E., 1973. An experimental study of heavy-mineral segregation under alluvial-flow conditions, *U.S. Geol. Surv. Prof. Pap. 562-K,* 38p.

Carr, A. P., 1971. Experiments on longshore transport and sorting of pebbles, Chesil Beach, England, *J. Sed. Petrology, 41,* 1084-1104.

Charlesworth, L. J., Jr., and Passero, R. N., 1973. Physical modeling in the geological sciences: An annotated bibliography, *Counc. Educ. Geol. Sci. Prog. Publ. 16,* 85p.

Daubrée, A., 1879. *Etudes Synthétiques de Géologie Experimental.* Paris: Dunod, 828p.

Degens, E. T., 1965. *Geochemistry of Sediments.* Englewood Cliffs, N.J.: Prentice-Hall, 342p.

Driscoll, E. G., 1967. Experimental field study of shell abrasion, *J. Sed. Petrology, 37,* 1117-1123.

Evans, O. F., and Ingram, R. L., 1943. An experimental study of the influence of grain size on the size of oscillation ripple marks, *J. Sed. Petrology, 13,* 117-120.

Folk, R. L., 1971. Genesis of longitudinal and oghurd dunes elucidated by rolling upon grease, *Geol. Soc. Am. Bull., 82,* 3461-3468.

Frizi, P., 1762. *Treatise on Rivers and Torrents with the Method of Regulating their Course and Channels.* Milan.

Garrels, R. M., and Christ, C. L., 1965. *Solutions, Minerals, and Equilibria.* New York: Harper & Row, 450p.

Gilbert, G. K., 1914. The transportation of debris by running water, *U.S. Geol. Survey Prof. Paper 86,* 263p.

Ginsburg, R. N., ed., 1973. *Evolving Concepts in Sedimentology.* Baltimore: Johns Hopkins Univ. Press, 191p.

Gross, M. G., 1972. Waste discharges as sedimentological experiments, *Geol. Soc. Am. Mem. 132,* 623-630.

Hubbert, M. K., 1937. Theory of scale models as applied to the study of geologic structures, *Geol. Soc. Am. Bull., 48,* 1459-1520.

King, C. A. M., 1966. *Techniques in Geomorphology.* New York: St. Martin's, 342p.

Krumbein, W. C., 1932. A history of the principles and methods of mechanical analysis, *J. Sed. Petrology, 2,* 89-142.

Krumbein, W. C., 1962. The computer in geology, *Science, 136,* 1987-1092.

Kuenen, P. H., 1958. Experiments in geology, *Trans. Geol. Soc. Glasgow,* **23,** 1-28.

Kuenen, P. H., 1965. Value of experiments in geology, *Geol. Mijnbouw,* **44,** 22-36.

Kuenen, P. H., 1966. Experimental turbidite lamination in a circular flume, *J. Geol.,* **74,** 523-545.

Lewis, W. V., and Miller, M. M., 1955. Kaolin model glaciers, *J. Glaciology,* **2,** 533-538.

McKee, E. D., 1947. Experiments on the development of tracks in fine cross-bedded sandstone, *J. Sed. Petrology,* **17,** 23-28.

Menard, H. W., 1950. Transportation of sediment by bubbles, *J. Sed. Petrology,* **20,** 98-106.

Middleton, G. V., ed., 1965. Primary sedimentary structures and their hydrodynamic interpretation, *Soc. Econ. Paleont. Mineral. Spec. Publ. 12,* 265p.

Müller, G., 1967. *Methods in Sedimentary Petrology.* New York: Hafner, 283p.

Rathbun, R. E.; Guy, H. P.; and Richardson, E. V., 1969. Response of a laboratory alluvial channel to changes of hydraulic and sediment-transport variables, *U.S. Geol. Surv. Prof. Pap. 562-D,* 31p.

Scoffin, T. P., 1968. An underwater flume, *J. Sed. Petrology,* **38,** 244-246.

Sorby, H. C., 1862. On the cause of difference in the stage of preservation of different kinds of fossil shells, *Brit. Assoc. Advance. Sci. Rept.,* **32**(2), 95-96.

Strahler, A. N., 1958. Dimensional analysis applied to fluvially eroded landforms, *Geol. Soc. Am. Bull.,* **69,** 279-300.

Williams, G. P., 1970. Flume width and water depth effects in sediment transport experiments, *U.S. Geol. Surv. Prof. Pap. 562-H,* 37p.

Williams, G. P., 1971. Aids in designing laboratory flumes, *U.S. Geol. Surv. Open-file Rept.,* 294p.

Cross-references: *Bed Forms in Alluvial Channels; Evaporites—Physicochemical Conditions of Origin; Flow Regimes; Fluvial Sediment Transport; Reynolds and Froude Numbers; Sedimentology—Models.*

EXTRATERRESTRIAL MATERIAL IN SEDIMENTS

The Swedish scientist, Nordenskjold (1874), reported that material of probable cosmic origin was present in samples of snow and ice collected in the Arctic. At about the same time, Murray (1876) described microscopic, black, magnetic spherules (up to 0.2 mm in diameter), which he had extracted from deep-sea sediments recovered during the *Challenger* Expedition, 1872-1876. Murray called them *cosmic spherules,* and suggested in the *Challenger* reports that they were formed as molten droplets ablated from the surface of meteorites traversing the earth's atmosphere. Since that time, particles recovered from sediment and thought to be of extraterrestrial origin have been studied extensively. They have been referred to variously as cosmic spherules, cosmic dust, caudites, meteoritic dust and micrometeorites (Schmidt, 1965).

Such microscopic particles have been recovered from various types of sedimentary deposits ranging in age from Lower Paleozoic to recent. These include deep-sea sediments, manganese nodules (Jedwab, 1970; Finkelman, 1970), unconsolidated sediments on land (Funiciello and Fulchignoni, 1969; Marvin and Einaudi, 1967), consolidated rocks such as evaporites, carbonates (Mutch, 1964; Lougheed, 1966; Ivanov and Florenskiy, 1970) and glacial ice (Schmidt, 1965). In general, particles identified as having a cosmic origin can be divided into two groups: (1) magnetic spherules, and (2) glassy spherules.

Magnetic Spherules. Murray (1876) was able to separate 20-30 magnetic spherules from a quart of red clay. Later investigators, using more sophisticated electromagnetic separators, were able to recover several hundred to several thousand from a kilogram of pelagic sediment.

Black *magnetic spherules* found in sediments range in diameter from a few microns up to about 800μ, but most are <200μ in diameter. Like meteorites, they can be divided into stony spherules and iron spherules. *Stony spherules* have a density of approximately 3 and are composed of fine-grained, magnesium-rich olivine similar to chondrules (Hunter and Parkin, 1960). *Iron spherules* have an average density of approximately 5-6 and are composed of magnetite with or without a metallic nucleus of iron-nickel-cobalt (Fig. 1). The composition of the metallic nucleus is similar

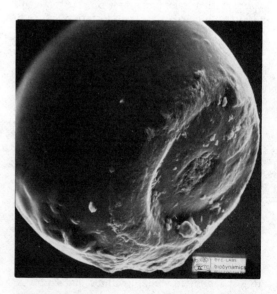

FIGURE 1. Scanning electron photomicrograph (800X) of a cosmic spherule with a metallic nucleus. (Photo: courtesy of R. B. Finkelman.)

to that of iron meteorites (Smales et al., 1958; Castaing and Fredriksson, 1958).

Pettersson and Fredriksson (1958) found evidence that the frequency of spherule deposition in deep-sea deposits during the last few thousand years has been higher than in the past. Utech (1971), however, maintains that cosmic-spherule influx has been constant over long geological periods. A study of particles collected from the atmosphere at Barbados revealed no spherules of extraterrestrial origin (Parkin and Tilles, 1968), possibly indicating that spherules collected from deep-sea sediments may have accumulated primarily as ablation products of meteorites arriving at random intervals.

Magnetic spherules similar to those recovered from sedimentary deposits can be produced by volcanic and industrial processes (Fredriksson and Martin, 1963; Brownlow et al., 1965). Industrial contaminants can be avoided by studying samples deposited prior to industrialization. The following criteria have been used by one or more investigators to distinguish between extraterrestrial and terrestrial iron spherules.

Titanium. Volcanic spherules are generally titaniferous magnetite with Ti content as high as 30%. Meteoritic magnetites generally do not contain Ti. Thus a low Ti content indicates that a particle is extraterrestrial in origin (El Goresy, 1967).

Manganese. A Mn content >0.01% is not consistent with an iron meteorite origin (Millard and Finkelman, 1970).

Nickel and Cobalt. Ni and Co in concentrations similar to those in meteorites is a good indication of extraterrestrial origin, but does not preclude an industrial origin (Millard and Finkelman, 1970). The absence of Ni, however, does not exclude an extraterrestrial origin because during oxidation of Ni-Fe at high temperature (as during atmospheric ablation) the Ni tends to concentrate in the metallic phase while the oxide phase is depleted in Ni (Marvin and Einaudi, 1967).

Ni-Fe Cores. Presence of Ni-Fe cores with meteoritic composition (Fredriksson and Martin, 1963) is considered to be indicative of extraterrestrial origin.

Wüstite. The presence of wüstite in a particle indicates that it was formed by ablation in the earth's atmosphere (Millard and Finkelman, 1970). Wütite is virtually unknown in terrestrial rocks but can be produced by industrial processes (Marvin and Einaudi, 1967).

After much debate over the criteria used to distinguish extraterrestrial material, the only particles generally accepted as having a cosmic origin are the magnetic spherules with Ni-Fe cores (Fig. 1).

Estimates of the influx rate of extraterrestrial material to the earth's surface based on spherule concentrations in sedimentary deposits vary by many orders of magnitude (<10 to 500,000 tons/yr; Schmidt, 1965). Most estimates, however, are around 10^4-10^5 tons/yr (Schmidt, 1965; Ivanov and Florenskiy, 1970). The presence and relative abundance of some stable and radioactive isotopes distinguish meteorites from terrestrial rocks. Some of these differences are due to corpuscular radiation in space or complex events during the meteorite's history. Detection of these elements or isotopes in deep-sea sediments and glacial ice has, therefore, been used to estimate the influx of extraterrestrial material to the earth's surface. It has been suggested, for example, that the high nickel content of deep-sea sediments, as compared to the average terrestrial rock, is due partly to extraterrestrial material (Pettersson and Rotschi, 1950; Opik, 1950). Laevastu and Mellis (1955) questioned this assumption, however, because they were not able to find a correlation between the concentration of magnetic spherules and the nickel content in a deep-sea core. Measurements of Al^{26}, Cl^{36}, rare-gas, and noble-metal (Ir and Os) content of deep-sea sediments suggest that the influx rate of extraterrestrial material is about 10^5-10^6 tons/yr (Parkin and Tilles, 1968; Barker and Anders, 1968).

Glassy Spherules. Microscopic glassy spherules (Fig. 2) have also been recovered from sedimentary rocks, deep-sea sediments, and glacial ice (Lougheed, 1966; Schmidt, 1964). Many (<1 mm in diameter) have been recovered from deep-sea sediments in the Indian Ocean, Philippine Sea, Equatorial Atlantic Ocean, and the Caribbean Sea. Based on their geographical distribution, age of deposition, fission-track age, and chemistry they have been identified as *microtektites*. Those from the Philippine Sea and Indian Ocean belong to the Australasian tektite strewnfield, and those from the eastern Atlantic Ocean and Caribbean Sea belong to the Ivory Coast and North American strewnfields (Glass, 1968; 1972; Glass et al. 1973). Some glassy particles similar to microtektites have been found in Antarctic ice (Schmidt, 1964; Shima, 1966), but their relationship to the ones recovered from deep-sea sediments is unclear.

The Australasian, Ivory Coast, and North American microtektites are recognized in layers that are approximately 0.7, 1.0, and 35 m.y. old, respectively. The events that produced the Australasian and Ivory Coast microtektites may have occurred simultaneously with reversals of the earth's magnetic field at approximately 0.7 and 1.0 m.y. ago. The total mass of microtektites in the Australasian strewnfield is approximately 10^{14} g (Glass and Heezen, 1967). The geographical extent of the Ivory Coast and

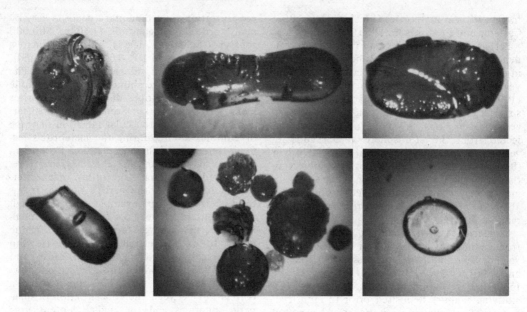

FIGURE 2. Photographs of Ivory Coast microtektites. *Top* (left to right): sphere with pitted and grooved surface, diameter 280μ; dumbbell with elongated grooves and pits, length 1.24 mm; flattened, oval-shaped microtektite with flow structures and caplike protrusion, length 862μ. *Bottom* (left to right): fragment of dumbbell—note large bubble cavity, length 350μ; group microtektites, largest diameter 400μ; disc-shaped microtektite; maximum diameter 180μ.

North American strewnfields has not yet been delineated.

Although the origin of tektites (and microtektites) is still being debated, most investigators now believe that they were produced by impact on the earth or moon. Recent investigation of lunar samples, however, has not lent much support to the lunar-origin hypothesis (see *Lunar Sedimentation*). Although microtektites themselves are not necessarily extraterrestrial in origin, they are most likely produced by the impact of an extraterrestrial body with the earth.

BILLY P. GLASS
DAVID W. FOLGER

References

Barker, J. L., Jr., and Anders, E., 1968. Accretion rate of cosmic matter from iridium and osmium contents of deep-sea sediments, *Geochim. Cosmochim. Acta, 32,* 627-645.

Brownlow, A. E.; Hunter, W.; and Parkin, D. W.; 1965. Cosmic dust collections at various latitudes, *Geophys. J., 9,* 337-388.

Castaing, R., and Fredriksson, K., 1958. Analyses of cosmic spherules with an x-ray microanalyser, *Geochim. Cosmochim. Acta, 14,* 114-117.

El Goresy, A., 1967. Electron microprobe analysis and microscopic study of polished surfaces of magnetic spherules and grains collected from the Greenland Ice, *Smithsonian Astrophys. Observ. Spec. Rept. 251,* 29p.

Finkelman, R. B., 1970. Magnetic particles extracted from manganese nodules: Suggested origin from stony and iron meteorites, *Science, 167,* 982-984.

Fredriksson, K., and Martin, L. R., 1963. The origin of black spherules found in Pacific Islands, deep-sea sediments and Antarctic ice, *Geochim. Cosmochim. Acta, 27,* 245-248.

Funiciello, R., and Fulchignoni, M., 1969. First remarks on the abundance and structure of cosmic spherules in central Italy sediments, *Geol. Romana, 8,* 117-128.

Glass, B., 1968. Glassy objects (microtektites?) from deep-sea sediments near the Ivory Coast, *Science, 161,* 891-893.

Glass, B., 1972. Australasian microtektites in deep-sea sediments, *Antarctic Research Ser., 19,* 335-348.

Glass, B. P.; Baker, R. N.; Storzer, D.; and Wagner, G. A., 1973. North American microtektites from the Caribbean Sea and their fission track age, *Earth Planetary Sci. Lett., 19,* 184-192.

Glass, B. P., and Heezen, B. C., 1967. Tektites and geomagnetic reversals, *Nature, 214,* 372.

Hunter, W., and Parkin, D. W., 1960. Cosmic dust in recent deep-sea sediments, *Roy. Soc. London Proc., 255A,* 382-397.

Ivanov, A. V., and Florenskiy, K. P., 1970. The rate of fall of cosmic dust on the Earth, *Geokhimiya, 11,* 1365-1372.

Jedwab, J., 1970. Les spherules cosmiques dans les nodules de manganèse, *Geochim. Cosmochim. Acta, 34,* 447-457.

Laevastu, T., and Mellis, O., 1955. Extraterrestrial material in deep-sea deposits, *Trans. Am. Geophys. Union, 36,* 385-389.

Lougheed, M. S., 1966. A classification of extraterrestrial spherules found in sedimentary rocks and till, *Ohio J. Sci.,* **66**(3), 274–283.

Marvin, U. B., and Einaudi, M. T., 1967. Black, magnetic spherules from Pleistocene and Recent beach sands, *Geochim. Cosmochim. Acta,* **31**, 1871–1884.

Millard, H. T., Jr., and Finkelman, R. B., 1970. Chemical and mineralogical compositions of cosmic and terrestrial spherules from marine sediment, *J. Geophys. Research,* **75**, 2125–2134.

Murray, J., 1876. On the distribution of volcanic debris over the floor of the ocean, *Proc. Roy. Soc. Edinburgh,* **9**, 247–261.

Mutch, T., 1964. Extraterrestrial particles in Paleozoic salts, *Ann. N.Y. Acad. Sci.,* **119**, 166–185.

Nordenskjold, N. A. E., 1874. On the cosmic dust which falls on the surface of the earth with the atmospheric precipitation, *Philos. Mag.,* ser. 4, **48**, 546p.

Opik, E. J., 1950. Astronomy at the bottom of the sea, *Irish Astronom. J.,* **1**, 145–158.

Parkin, D. W., and Tilles, D., 1968. Influx measurements of extraterrestrial material, *Science,* **159**, 935–946.

Pettersson, H., and Fredriksson, K., 1958. Magnetic spherules in deep-sea deposits, *Pacific Sci.,* **12**, 71–81.

Pettersson, H., and Rotschi, H., 1950. Nickel content in deep-sea deposits, *Nature,* **166**, 308.

Schmidt, R. A., 1964. Microscopic extraterrestrial particles from the Antarctic Peninsula, *Ann. N.Y. Acad. Sci.,* **119**, 186–204.

Schmidt, R. A., 1965. A survey of data on microscopic extraterrestrial particles, *NASA Techn. Note,* NASA TN D-2719, 132p.

Shima, M., 1966. Glassy spherules (microtektites?) found in ice at Scott Base, Antarctica, *J. Geophys. Research,* **71**, 3595–3596.

Smales, A. A.; Mapper, D.; and Wood, A. J., 1958. Radioactivation analysis of cosmic and other magnetic spherules, *Geochim. Cosmochim. Acta,* **13**, 123–126.

Utech, K., 1971. On the constancy of cosmic spherule influx during the Quaternary, *Meteoritics,* **6**, 237–239.

Cross-references: *Eolian Dust in Marine Sediments; Lunar Sedimentation; Martian Sedimentation; Pelagic Sedimentation, Pelagic Sediments; Sedimentation Rates.* Vol. I: *Cosmic Dust.* Vol. II *Australites: Meteorites; Tektites.*

EXTRATERRESTRIAL SEDIMENTATION—See LUNAR SEDIMENTATION; MARTIAN SEDIMENTATION

FABRIC, SEDIMENTARY

In sedimentary geology, *fabric* refers to spacial arrangement (packing) and orientation of rock components (for magnetic fabric, see *Anisotropy in Sediments*). Fabric is a term vexed by several different meanings; when used it must be defined with respect to *concept* and *methods of measurement*. In addition, the terms *fabric, texture,* and *structure* are frequently confused. To clarify these concepts, consider that a sediment is a layered deposit made up of particles. The particle population is uniquely characterized when its composition and texture, i.e., size, shape, orientation, and packing of particles have been measured. For the determination of size and shape, each particle can be considered a discrete entity, and no external reference coordinate system is necessary. Such a coordinate system is required to measure grain orientation and packing, i.e., what here is called *fabric*.

Variations in texture or composition may produce regular or irregular domains, layers, and similar features referred to as *sedimentary structures* (q.v.). In this context, *fabric* is part of *texture*, which is part of *structure*. Other geologists hold different views; for example, the German school considers *fabric* to include both *texture* and *structure* (Sander, 1961; Müller, 1967). Occasionally, the term fabric is used as a synonym of *orientation*. This does not reflect an intention to modify the basic concept; rather, only orientation has been measured in certain sediments, such as tills. The term *appositional fabric* is a more proper synonym of "primary orientation."

Grain Orientation

Orientation refers to the geometric relation, expressed in degree azimuth, between definable linear or planar properties of particles and a reference coordinate system. Operational constraints may require that, rather then measuring true orientation in three-dimensional space, it be estimated from measurements of apparent orientations on two-dimensional surfaces such as faces of outcrops, cores, and thin sections. The optical orientation of selected crystallographic axes of anisotropic minerals and the dimensional orientation of elongated particles are the two types most commonly measured in sedimentary geology.

Optical orientation is measured in thin sections under the petrographic microscope. Rapid determinations of the position of the crystallographic axes are used to identify minerals and to determine the relative optical orientation between detrital grains and overgrowths and other types of cements. The true orientation, in space, of selected crystallographic axes of specific minerals, commonly the C-axis of calcite, dolomite, quartz, and others, can be measured with the help of a three-axis universal stage. The results are plotted in structural diagrams (Vistelius, 1966), and they are used in estimating primary dimensional orientation or patterns of recrystallization in relation to tectonic stresses.

Dimensional orientation is measured in the field and in laboratories on oriented samples with the major objectives of analyzing processes active in the depositional environment, reconstructing paleocurrents and paleoslopes (see *Paleocurrent Analysis*), and recognizing old strandlines. The expectation is, and it has been found often, that the longest dimensions of elongated particles tend to assume a preferred position parallel or perpendicular to the direction of movement (orientation) and are slightly inclined (imbricated) upflow with respect to the surface upon which they come to rest (i.e., the "deposition surface," not necessarily horizontal). Different kinds and degrees of orientation and imbrication result from the various distributions of stresses exerted on the particles by the transporting medium, by the type of movement (rolling, sliding, or a combination), and by the type of readjustment of position that the object is undergoing on the deposited layer before or during burial (Parkash and Middleton, 1970). The theory of grain orientation is discussed by Rusnak (1957).

In experimental conditions and in older sediments and rocks, preferred particle orientations have been detected, often when considering fossils, minerals, or other materials of submicroscopic to boulder size and environments from small pores in rocks to glacial environments, to fluvial and shallow and deep marine environments (Potter and Pettijohn, 1977). Preferred orientations of quartz sand grains have been found to be parallel to directions of recent and ancient flows determined by other indicators such as cross-beds (see *Cross-Bedding*), ripple marks (q.v.), and current marks (q.v.).

Methods of Measurement. The many methods devised for measurement of dimensional grain orientation fall into two broad categories: (1) bulk or aggregate measurements correlated to the crystallographic and/or dimensional orientation of the particles; and (2) the treatment of particles as discrete entities. Aggregate properties of quartz-rich samples such as dielectric anisotropy or acoustic anisotropy are analyzed on the premise that the crystallographic C-axes parallel the longest dimensions of the grains (Sippel, 1971). Other bulk analyses consider visual properties of sediments in thin sections or special photographs to detect preferred disposition of the grains by recording the relative amount of darker matrix in traverses of different directions (Bonham and Spotts, 1971).

The particulate methods of measurement are more commonly used both in the field and in laboratories. True orientation and imbrication are recorded for coarser particles in sediments and are plotted in structural diagrams. Apparent orientations and imbrications are measured on cut surfaces (e.g., outcrops, core slabs, thin sections) of consolidated rocks and impregnated sediments. For microscopic studies of sands, two or three thin sections are cut: one parallel to the depositional surface to measure "apparent orientation" and one or two perpendicular to the depositional surface to measure "apparent imbrication" in directions parallel and perpendicular to the average trend of the apparent orientation. Except for elongated and flat particles, the apparent measurements are reasonable estimates of the true three-dimensional orientation (Sippel, 1971).

Statistical Analysis. Standardization of measurements and presentation of data should be achieved so that information can be collated and general models can be built. Visual presentation and statistical manipulations can be varied to take advantage of the information contained in the data set. Techniques for handling both two-dimensional and three-dimensional data are available (Martini, 1965; Andrews and Shimizu, 1968; Jizba, 1971; Reyment, 1971). In each method, the orientations of particles are considered as vectors with unit dimensions. Vector means and measures of dispersion are calculated, and the measured distributions are compared with either uniform circular or random theoretical distributions. If statistically significant differences exist, then the sample is considered to have a preferred grain orientation. For grain imbrication to be a reliable indicator of flow direction, its vector mean must be significantly different from zero, tested by a simple t test (Martini, 1971). Only preferred orientations and imbrications can be used as paleocurrent indicators; the analysis of modes of the distributions, however, may provide useful information about the processes active in the environments of deposition (Tanner, 1959).

Packing

Packing is the spacing or density pattern of mineral grains in an aggregate. Packing of clastic sediments and sedimentary rocks is highly correlated to porosity. Theoretical and experimental work has emphasized the analysis of primary porosity and its changes during diagenesis, and bulk measurements of porosity have been used to estimate packing. However, the interactions among textural variables and the modifications that occur during and after lithification defy any accurate prediction, except in the most simple condition of packing of spheres (Graton and Fraser, 1935; Rittenhouse, 1971; Beard and Weyl, 1973). In this ideal case porosity varies from a maximum of 47.6% for a cubic packing of spheres to 26% for a rhombohedral one.

Although time consuming, direct measurements of properties related to packing are necessary to understand the relationship among the fundamental textural variables, the processes active during compaction (q.v.) and cementation (q.v.), and the modification of bulk properties such as porosity, permeability, and density of aggregates. In sedimentary geology, packing is measured occasionally in gravelly outcrops and commonly in thin sections cut perpedicular to the depositional surfaces of sandy units. Two theoretical analogue models to the real aggregate can be considered. One is based on a random distribution of particles in the aggregate, and a tentative methodology for actual measurements has been devised (Smalley, 1964). The second and most frequently used analogue model is the regular packing of equal-radius spheres (Graton and Fraser, 1935). A simple operational methodology has been devised, and some knowledge exists on the statistical relationships among the measured variables of packing and their relations to other textural variables (Kahn, 1956; Martini, 1972). A measure of the mutual spacial relationship among particles that constitute the framework of sedimentary samples is obtained from thin sections by measuring packing density (P_d) and packing proximity (P_p) along traverses of known length (t in mm). Packing density relates to an aggregate property that is a measure of the amount of space occupied by framework grains. Its mathematical expression is

$$P_d = \left[\left(m \sum_{i=1}^{n} q_i\right) \div t\right] \times 100$$

$$\left(0 < \sum_{i=1}^{n} q_i \leq t\right)$$

where m is the correction to reduce micrometric values to millimeters; and q_i is the micrometric value of the intercept grain size (I_s), i.e., the length of the grain cut by the cross hair along the prefixed line of the traverse. This can be used as a measure of grain size for comparative purposes.

Packing proximity is related to the presence or absence of contacts between grains. It is expressed as

$$P_p = \frac{q}{n} \times 100 \quad (0 \leq q \leq n)$$

where q is the number of grain-to-grain contacts and n is the total number of grain contacts.

During this type of analysis, it is customary and advisable to obtain an estimate of the amount of winnowing of fines, of modifications of primary packing by interpenetration between grains, of porosity, and of amount of cement by recording the types of contacts of the grains (Taylor, 1950). Along the traverse, a decision is made as to whether the edge of each grain has a tangential, long, concave-convex, or sutured contact with another grain, or if it is in contact with void, matrix, or cement.

I. PETER MARTINI

References

Andrews, J. T., and Shimizu, K., 1968. Three-dimensional vector analysis for till fabric: Discussion and Fortran program, *Geogr. Bull.*, 8, 151-165.

Beard, D. C., and Weyl, P. K., 1973. Influence of texture on porosity and permeability of unconsolidated sand, *Bull. Am. Assoc. Petrol. Geologists*, 57, 349-369.

Bonham, L. C., and Spotts, J. H., 1971. Measurements of grain orientation, in Carver, 1971, 285-312.

Carver, R. E., ed., 1971. *Procedures in Sedimentary Petrology.* New York: Wiley, 653p.

Gillott, J. E., 1969. Study of the fabric of fine-grained sediments with scanning electron microscope, *J. Sed. Petrology*, 39, 90-105.

Graton, L. C., and Fraser, H. J., 1935. Systematic packing of spheres, with particular relation to porosity and permeability, *J. Geol.*, 43, 785-909.

Jizba, Z. V., 1971. Mathematical analysis of grain orientations, in Carver, 1971, 313-333.

Kahn, J. S., 1956. The analysis and distribution of packing in sand-size sediments, *J. Geol.*, 64, 385-395.

Krumbein, W. C., 1939. Preferred orientation of pebbles in sedimentary deposits, *J. Geol.*, 47, 673-706.

Martini, I. P., 1965. Fortran IV Programs (IBM 7040 Computer) for Grain Orientation and Directional Sedimentary Structures, *Tech. Memo* 65-2 (unpubl.). McMaster University, Hamilton, Ontario, 10p.

Martini, I. P., 1971. A test of validity of quartz grain orientation as a paleocurrent and paleoenvironmental indicator, *J. Sed. Petrology*, 41, 60-68.

Martini, I. P., 1972. Studies of microfabrics: An analysis of packing in the Grimsby Sandstone (Silurian), Ontario and New York State, *Proc. 24th Int. Geol. Cong.*, Sec. 6, 415-423.

Müller, G., 1967. *Methods in Sedimentary Petrology.* New York: Hafner, 283p.

Parkash, B., and Middleton, G. V., 1970. Downcurrent textural change in Ordovician turbidite greywackes, *Sedimentology*, 14, 259-293.

Potter, P. E., and Pettijohn, F. J., 1977. *Paleocurrents and Basin Analysis*, 2nd ed. New York: Springer, 460p.

Reyment, R. A., 1971. *Introduction to Quantitative Paleoecology.* New York: Elsevier, 226p.

Rittenhouse, G., 1971. Pore-space reduction by solution and cementation, *Bull. Am. Assoc. Petrol. Geologists*, 55, 80-91.

Rusnak, G. A., 1957. The orientation of sand grains under conditions of unidirectional fluid flow, *J. Geol.*, 65, 384-409.

Sander, B., 1951. *Contributions to the Study of Depositional Fabrics; Rhythmically Deposited Triassic Limestones and Dolomites.* Tulsa, Okla.: Am. Assoc. Petrol. Geologists, 120p.

Sippel, R. F., 1971. Quartz grain orientation-1. (the photometric method), *J. Sed. Petrology*, 41, 38-59.

Smalley, I. J., 1964. Representation of packing in a clastic sediment, *Am. J. Sci.*, 262, 242-248.

Tanner, W. F., 1959. The importance of modes in cross-bedding data, *J. Sed. Petrology*, 29, 221-226.

Taylor, J. M., 1950. Pore-space reduction in sandstones, *Bull. Am. Assoc. Petrol. Geologists*, 34, 701-716.

Vistelius, A. B., 1966. *Structural Diagrams.* Oxford: Pergamon, 178p.

Winkelmolen, A. M.; Van Der Knaap, W.; and Eijpe, R., 1968. An optical method of measuring grain orientation in sediments, *Sedimentology*, 11, 183-196.

Cross-references: *Anisotropy in Sediments; Cementation; Compaction in Sediments; Conglomerates; Diagenesis; Diagenetic Fabrics; Fluvial Sediment Transport; Limestone Fabrics; Paleocurrent Analysis; Sedimentary Structures; Sediment Parameters; Till and Tillite.* Vol. XII: *Soil Fabric.*

FACETED PEBBLES AND BOULDERS

Ventifacts

Most faceted pebbles and boulders, commonly known as *ventifacts,* owe their shape and surface features partly to abrasion by wind-driven sand. They may be composed of any material capable of forming a pebble or boulder, and are ordinarily shaped in the first instance by weathering and abrasion processes that are noneolian. These stones, of any shape from ellipsoidal to angular, then become modified by eolian abrasion. The modification of the shape, and of the surface smoothness and texture of the stone, varies from slight to profound according to the effectiveness of sand blasting and the hardness of the stone (see references). Ventifacts are found associated with many desert sediments of all ages, but they are best known in Pleistocene periglacials, where they are widespread (Cailleux, 1942; King, 1936; Minard, 1966).

Facet Formation

Bryan (1931) discussed early contention as to the manner of facet development, explaining some now-abandoned terms such as *dreikanter* and *zwikanter*. Sharp (1964) and others have shown that in natural eolian environments sand abrasion increases from the ground level upward to heights of 25 cm or more. Thus, when a boulder of that order of size becomes subject to sand abrasion, the windward side is cut back, receding most rapidly at the top. Recent study has suggested that suspended dust particles accomplish much of the abrasion (Whitney and Dietrich, 1973). The shape of the face developed depends on the original shape and the distribution of cutting (Schoewe, 1932), but if the face is ultimately cut away to a low angle it will become flattened, developing quite a sharp angle with the leeward, uneroded portion of the boulder. In deposits such as lag gravels, pebbles may, if abrasion is severe, be cut off almost to a pavement. Most often, however, pebbles or boulders will be undermined and turned in some way, and so will have new faces developed, perhaps several in succession.

Facets subjected to eolian abrasion may be simply smoothed and polished, as is commonly the case with small, hard, fine-grained pebbles, or may become deeply pitted or grooved, as in the case of a conglomerate of heterogeneous material (Minard, 1966). Some materials, especially limestones, tend to give rise to fluting reminiscent of *rillensteine*, and Whitney and Brewer (1968) have suggested that such features may be produced by the motion of turbulent air with vacuum effect as much or more than by sand impact.

Other Faceted Rocks

Engeln (1930) studied the shapes of glacial cobbles and concluded that the ideal glacial cobbles was a faceted flat-iron shape, i.e., roughly triangular with a longer base. Wentworth (1936) verified Engeln's work, and it is now generally accepted that *glacial clasts* are often faceted. It is likely that some ventifacts were originally faceted during glacial transport; a glacially faceted clast may be distinguished from a ventifact by a highly polished or deeply striated surface.

Pebbles abraded on beaches, particularly sandy beaches, may in some cases assume a subtriangular shape, caused by the action of sand-laden swash and backwash (Lenk-Chevitch, 1959). These pebbles are not as distinctively faceted as clasts of glacial or eolian origin. Only the facets of a ventifact tend to meet along a sharp ridge.

W. SUGDEN

References

Bryan, K., 1931. Wind worn stones or ventifacts; a discussion and bibliography, *Rept. Comm. Sedimentation 1929-1930, Natl. Research Council Circ. 98*, 29-50.

Cailleux, A., 1942. Les actions éoliennes périglaciaires en Europe, *Mem. Soc. Geol. France*, 46, 176p.

Engeln, O. D. von, 1930. Type form of faceted and striated glacial pebbles, *Am. J. Sci.*, ser. 5, 19, 9-16.

King, L. C., 1936. Wind faceted stones from Marlborough, New Zealand, *J. Geol.*, 44, 201-213.

Lenk-Chevitch, P., 1959. Beach and stream pebbles, *J. Geol.*, 67, 103-108.

Minard, J. P., 1966. Sandblasted blocks on a hill in the coastal plain of New Jersey, *U.S. Geol. Surv. Prof. Pap. 550-B*, 87-90.

Schoewe, W. H., 1932. Experiments on the formation of wind-faceted pebbles, *Am. J. Sci.*, 24, 111-134.

Sharp, R. P., 1964. Wind driven sand in Coachella Valley, California, *Geol. Soc. Am. Bull.*, 75, 785-803.

Sugden, W., 1964. Origin of some faceted pebbles in some recent desert sediments of southern Iraq, *Sedimentology*, 3, 65-74.

Wentworth, C. K., 1936. An analysis of the shape of glacial cobbles, *J. Sed. Petrology*, 6, 85-96.

Whitney, M. I., and Brewer, H. B., 1968. Discoveries in aerodynamic erosion in wind tunnel experiments, *Mich. Acad. Sci., Arts, Lett., Pap. 53*, 91-104.

Whitney, M. I., and Dietrich, R. V., 1973. Ventifact sculpture by windblown dust, *Geol. Soc. Am. Bull.*, 84, 2561-2582.

Cross-references: *Attrition; Beach Gravels; Desert Sedimentary Environments; Eolian Sedimentation; Glacial Gravels; Littoral Sedimentation; River Gravels; Till and Tillite*. Vol. III: *Ventifacts*.

FACIES

The modern concept of facies was introduced into geology by Amand Gressly (1814-1865). Gressly was a student at the University of Strasbourg, when, under the direction of Jules Thurmann, he began studies of the Jurassic rocks in the northwest part of Switzerland. In his first major publication (1838), he explains how he was led to develop the idea of facies:

In the regions that I have studied, perhaps more than anywhere else, very variable modifications, both petrographic and paleontologic, everywhere interrupt the universal uniformity that, up until now, has been maintained for different stratigraphic units ("*terrains*") in different countries. They even reappear successively in many stratigraphic units and astonish the geologist who wishes to study the nature of our Jurassic ranges. . . .

There are two principal points that always characterize the group of modifications that I call *facies* or *aspects of stratigraphic units:* one is that *a similar petrographic aspect of any unit necessarily implies, wherever it is found, the same paleontological assemblage; the other, that a similar paleontological assemblage rigorously excludes the genera and species of fossils frequent in other facies.* (Gressly, 1838, pp. 10–11: Gressly's own emphasis.)

Gressly recognized the intimate relationship of lithological and paleontological aspects of facies and wisely did not attempt to separate the two. He identified two major facies, a "coralline facies" and a facies consisting of "rocks of muddy nature." And he distinguished a number of subfacies, including transported deposits, which were derived from the coralline facies but differed from it because the fossils were fragmental, not in position of growth, and consequently did not constitute "any characteristic zoological community." Gressly interpreted his facies in environmental terms: the coralline facies was littoral or shallow water, and the muddy facies had two types, one characteristic of littoral, quiet-water environments and the other formed in deeper water, pelagic, or subpelagic environments.

Later writers have attempted to separate the lithological from the paleontological aspects of facies: thus, *lithofacies* and *biofacies.* As the presence or absence of fossils (including trace fossils and fragmental fossil material) constitutes an essential part of the lithology of many sedimentary rocks, however, the concepts are intermixed.

The term *sedimentary facies* was specifically defined by Moore (1949, p. 32) as "any areally restricted part of a designated stratigraphic unit which exhibits characters significantly different from those of other parts of the unit." It seems fairly clear that this is *not* the sense in which the term facies was used by Gressly or other European pioneers such as Mojsisovics and Walther (see Teichert, 1958). Moore distinguished sedimentary facies from lithofacies, which he defined as "the rock record of any sedimentary environment, including both physical and organic characters." This definition is closer to the early European tradition, though it introduces an element of abstraction that seems superfluous.

Modern stratigraphers and sedimentologists have tended to employ "facies" in two ways, corresponding roughly to the distinction made by Moore. The less common usage is the more restricted sense named "sedimentary facies" by Moore: a formation (or other stratigraphic unit) is broken down into a number of areally restricted subunits called facies (and generally designated "facies A," "facies B," etc. "of the X Formation"). The more common usage is exemplified by De Raaf et al. (1965) who subdivided a group of three formations into a cyclical repetition of a number of facies distinguished by "lithological, structural and organic aspects detectable in the field." The facies may be given informal designations ("facies A," etc.) or brief descriptive designations (e.g., "laminated siltstone facies"); and it is understood that they are units that will ultimately be given an environmental interpretation; but the facies definition is itself quite objective and based on the total field aspect of the rocks themselves. This usage of the term *facies* seems entirely in accordance with Gressly's original intentions and more straightforward than suggested by Moore's definition of lithofacies.

The term *facies* is also very frequently used in a very imprecise way, to refer to the overall characteristics of rocks formed in a particular environment, e.g., "fluvial facies," or setting, e.g., "geosynclinal facies." Such usages are useful, even though no exact definition is implied (see AGI *Glossary,* 1972).

Facies Analysis

The importance of facies in sedimentary geology is that the definition and analysis of facies is the basis for an environmental interpretation of stratigraphic units. Furthermore, the interrelationships of sedimentary environments, and therefore of facies, are not chaotic or random, because they are subject to controls imposed by geological setting, tectonics, and climate. Thus, the distribution of facies is subject to a number of regularities; and the key to the interpretation of facies is to combine observations made on their spatial relations and internal characteristics (lithology and sedimentary structures) with comparative information obtained from other well-studied stratigraphic units, and particularly from studies of modern sedimentary environments. The value of such studies was clearly seen by Walther (1893–1894), who gave them the name *comparative lithology.* The term is still used in Russia, but has generally been replaced in the west by the terms *facies analysis, facies models,* or *comparative sedimentology.* Walther also enunciated one of the rules frequently used in facies analysis: Only those facies can be superimposed (without a break) which were deposited in adjacent sedimentary environments (see Middleton, 1973, for a fuller discussion of Walther's original statement). To the geologist studying ancient sediments, the vertical transitions between facies are generally much more readily observed than the lateral transitions: Walther's "Law" is therefore a key concept in applying observations made on modern sediments to the environmental interpretation of sedimentary rocks.

Melguen (1974) presents an example of statistical facies analysis, and Walker (1975) discusses facies modeling. Romanovskiy (1975) reviews the problem of substantive content in facies analysis.

GERARD V. MIDDLETON

References

Braunstein, J., ed., 1974. Facies and the reconstruction of environments, *Am. Assoc. Petrol. Geologists Reprint Ser.,* **10,** 224p.

De Raaf, J. F. M.; Reading, H. G.; and Walker, R. G., 1965. Cyclic sedimentation in the Lower Westphalian of north Devon, *Sedimentology,* **4,** 1-52.

Gressly, A., 1838. Observations géologiques sur le Jura Soleurois, *Nouv. Mém. Soc. Helv. Sci. Natur.* **2,** 349p.

Melguen, M., 1974. Facies analysis by "correspondence analysis": Numerous advantages of this new statistical technique, *Marine Geol.,* **17,** 165-182.

Middleton, G. V., 1973. Johannes Walther's law of the correlation of facies, *Geol. Soc. Am. Bull.,* **84,** 979-988.

Moore, R. C., 1949. Meaning of facies, *Geol. Soc. Am. Mem. 39,* 1-34.

Romanovskiy, S. I., 1975. Problem of substantive content of facies analysis, *Internat. Geol. Rev.,* **17,** 78-82.

Teichert, C., 1958. Concept of facies, *Bull. Am. Assoc. Petrol. Geologists,* **42,** 2718-2744.

Walther, J., 1893-1894. *Einleitung in die Geologie als historische Wissenschaft.* Jena: Von Gustav Fischer, 3 Vols., 1055p.

Walker, R. G., 1975. From sedimentary structures to facies models: Example from fluvial environments, *Soc. Econ. Paleont. Mineral. Short Course,* **2** (April 5, 1975), 63-80.

Weller, J. M., 1960. *Stratigraphic Principles and Practice.* New York: Harper & Row, 725p.

Cross-references: *Cyclic Sedimentation; Sedimentary Environments.*

FECAL PELLETS—See BIOGENIC SEDIMENTARY STRUCTURES

FERRICRETE

Proposed by Lamplugh (1902), along with *silcrete* (q.v.) and *calcrete* (see *Caliche*), ferricrete is a paleosol cemented by iron oxides. A comparable aluminum-rich formation is *bauxite.* A term very similar to ferricrete, according to Tieje (1921), is *ferrite* (*Brauneisen* in German). Ferricrete is a cemented, iron-rich sediment or conglomerate, the particles of which do not interlock, i.e., a particulate ironstone, which is usually associated with an alluvial weathering profile and developed under Mediterranean or subtropical seasonal contrasts. The cementing iron compounds are reddish-brown, amorphous chemical-weathering products, often referred to as *sesquioxides.* In Europe, the ferruginous paleosols are sometimes referred to as *ferreto* and the process *ferritization* (Baulig, 1956).

When the iron oxide is concentrated in a soil B horizon, the A horizon is often stripped off by eolian transport. If a drier climate ensues, there is an upward capillary motion leading to iron (and silica) concentration near the surface, as a crust, often referred to as *ferruginous duricrust, ferricrust, ironstone cap, cuirasse de fer,* or *iron hardpan.*

It should be noted that ferricrete and ferrite are not to be confused with *ferrilite (ferrilith, ferrilyte),* a term proposed by Grabau (1924) for an iron-rich sedimentary rock. *Ferralite (ferrallite)* is a term widely used by the French in West Africa, where the silica:sesquioxide ratio is < 2; it is generated from basic igneous rocks in humid tropical conditions by mildly acidic to neutral solutions that leach silica and bases, concentrating Fe, Al, and possibly Mn and Ti.

RHODES W. FAIRBRIDGE

References

Baulig, H., 1956. *Vocabulaire franco-anglo-allemand de géomorphologie.* Paris: Soc. Ed. Belles Lettres, 229p.

Grabau, A. W., 1924. *Principles of Stratigraphy* (2nd ed.). New York: A. G. Seiler, 1185p.

Lamplugh, G. W., 1902. Calcrete, *Geol. Mag.* (Dec. 4), **9,** 575.

Tieje, A. J., 1921. Suggestions as to the description and naming of sedimentary rocks, *J. Geol.,* **29,** 650-666.

Cross-references: *Duricrust; Ironstone.* Vol. XI: *Ferralitic Soils.*

FISSILITY

The property of splitting easily along closely spaced parallel planes is termed *fissility.* This property is in part an effect of parallel orientation of the micaceous constituents of a sediment at the time of deposition. Fissility varies directly with the organic content and inversely with the content of calcareous and siliceous matter and with the extent of bioturbation (Byers, 1974). Another factor is the secondary growth of oriented clay-mineral grains in response to external pressure.

Several authors (Alling, 1945; Ingram, 1953; McKee and Weir, 1953) have attempted to establish relative fissility scales ranging from massive (nonfissile) to fissile, flaggy, or papery.

B. CHARLOTTE SCHREIBER

References

Alling, H. L., 1945. Use of microlithologies as illustrated by some New York sedimentary rocks, *Geol. Soc. Am. Bull.*, **56**, 737-756.
Byers, C. W., 1974. Shale fissility: Relation to bioturbation, *Sedimentology*, **21**, 479-484.
Ingram, R. L., 1953. Fissility of mudrocks, *Geol. Soc. Am. Bull.*, **64**, 869-878.
McKee, E. D., and Weir, G. W., 1953. Terminology for stratification and cross-stratification in sedimentary rocks, *Geol. Soc. Am. Bull.*, **64**, 381-390.
Rubey, W. W., 1931. Lithologic studies of fine-grained Upper Cretaceous sedimentary rock of the Black Hills region, *U.S. Geol. Surv. Prof. Pap. 165-A*, 54p.

Cross-references: *Argillaceous Rocks; Bedding Genesis; Sedimentary Structures; Shale.*

FLAME STRUCTURES

Flame structures, a term originated by Walton (1956), are syndepositional deformation structures, formed at sand/mud interfaces, which resemble "streaked-out" ripples or flame wisps in cross section (Figs. 1 and 2). They are most commonly found in turbidite (q.v.) sequences.

The origin of flame structures is related to (1) differential loading along the sand/mud interface and (2) a turbulent current passing above the bottom, resulting in deposition of

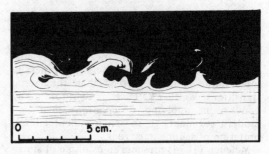

FIGURE 1. H. C. Sorby's specimen of green slate from Langdale (from Sanders, 1960, by permission, Cambridge University Press). Volcanic sand (black) overlies laminated volcanic mud. Along the contact the mud has been skimmed off and streaked out into attenuated "ripples." Note extreme deformation of "ripple" at left, which has been bent forward onto the next "ripple" to its right and further deformed. Also note spray-like wisps on downcurrent side of some "ripples."

sand around ripple-like wisps of mud drawn upward by the current-carrying sand (Fig. 3). The syndepositional origin of some flame structures is established by the existence of smaller wisps on larger ones (Fig. 2); cross-laminated sand on the downcurrent side of some wisps; and graded sand surrounding flame structures, which often interrupt the sequence of grading.

J. E. Sanders (pers. comm.) has proposed a third mechanism which may produce flame structures. He suggests that they may be formed normal to the direction of the main current by small, turbulent, rotational eddies similar to Langmuir cells or Langmuir circulation observed at boundary layers in the atmosphere and in the ocean (Fig. 2).

IMRE V. BAUMGAERTNER

FIGURE 2. Photographs of specimen of mud wisps covered by deformed cross-laminated sand, viewed parallel to (long photograph) and perpendicular to (short photograph) current. Although interpreted by Walker (1963) as "load casts," these wisps are inferred by Sanders (pers. comm.) to have originated syndepositionally. The view perpendicular to the current suggests that the wisps were produced by Langmuir circulation. (Photo: courtesy of Roger Walker.)

FIGURE 3. Sketch showing pointed mud wisps (black) and lee-side cross-laminae differentially deformed during deposition by effects of current drag during sediment fallout from overlying turbulent suspension (Sanders, 1965). Three sets of cross-laminae marked I, II, III, are shown in the lee-side fillings. Set I was deposited and slightly deformed prior to deposition of Set II. Set II was deformed and Set I further deformed prior to deposition of Set III, which is undeformed. Undeformed layers will be deposited on top of depression fillings. Fallout from current is indicated by the small dots and arrows. Current from left to right.

References

Dzulinski, S., and Simpson, F., 1966. Experiments on interfacial current markings, *Estr. Geol. Romana*, 5, 197-214.

Kelling, G., and Walton, E. K., 1957. Load-cast structures: their relationship to upper-surface structures and their mode of formation, *Geological Mag.*, 94, 481-490.

Sanders, J. E., 1960. Origin of convoluted laminae, *Geological Mag.*, 97, 409-421.

Sanders, J. E., 1965. Primary sedimentary structures formed by turbidity currents and related resedimentation mechanisms, *Soc. Econ. Paleont. Mineral. Spec. Publ. 12*, 192-219.

Sanders, J. E., 1969. Syndepositional deformational structures at sand/mud interfaces (abstract), *Geol. Soc. Am. Abstr., 1969*, pt. 1, 52-53.

Walker, R. G., 1963. Distinctive types of ripple-drift cross-lamination, *Sedimentology*, 2, 173-188.

Walton, E. K., 1956. Limitations of graded bedding and alternative criteria of upward sequence in the rocks of the Southern Uplands, *Trans. Edinburgh Geol. Soc.*, 15, 262-271.

Cross-references: *Bouma Sequence; Current Marks; Penecontemporaneous Deformation of Sediments; Pillow Structures; Ripple Marks; Sedimentary Structures; Turbidites; Water-Escape Structures.*

FLOCCULATION

Flocculation is the process by which the dispersed particles in a colloidal system (a sol) coagulate, so that a visible precipitate is readily formed. This phenomenon, also variously termed precipitation or coagulation, was first observed by Graham (1861). The two phases involved in a colloidal system (the medium and the particles) are together termed a *disperse system*, the particulate portion being the *disperse phase* and the medium in which the particles are distributed is termed the *disperse medium*.

The range of particulate grain size of colloids is usually given as from 0.2μ, which is the lower limit of microscopic visibility, to $5 \times 10^{-3}\mu$, which is the upper size limit of some molecular particles. The particles of a colloidal system have a relatively large surface area, around which (when in suspension) they apparently carry an electrical charge—the so-called "double layer." Because of this surface charge, colloidal particles are mutually repulsive. A colloidal particle, together with its double layer, is sometimes termed a *micelle*, an equivalent in a colloidal solution of the molecule in a true solution.

Colloidal solutions, with a liquid dispersant, can be divided roughly into two categories, *lyophobic* (Gr: liquid hating) and *lyophilic* (Gr: liquid loving). If the medium is water, they are specifically termed *hydrophobic* and *hydrophilic*, respectively. Generally speaking, most systems belong to one category or the other; there are materials, however, which share some qualities of both systems, e.g., hydroxides. Lyophobic sols require the addition of only a small quantity of an electrolyte to flocculate, whereas small quantities have little effect on sols. Moreover, upon evaporation or cooling, lyophobic systems yield solids that cannot be converted by reversing the physical change, while in lyophilic systems, flocculation is reversible, and the solids thus obtained are true *gels*. In general, hydrophobic sols include those of metals, sulfur, sulfides, and silver halides; hydrophilic sols include gums, starches, proteins, soaps, and most clays.

Traces of electrolites, termed *dispersants* (peptizers) are essential to the stability of sols in water. The addition of somewhat larger amounts, however, may also cause flocculation. In comparing the precipitating effects of various ions it is essential that the procedures be carried out under exactly comparable conditions, as the results may vary greatly. If, for example, an electrolytic component is added slowly, or all at once, the result may be totally different. The coagulative effect of additives upon colloids apparently depends upon both the ionic sign and valence; approximately following the *Schulze-Hardy Rule* (which is the analogue of Perrin's Rule for electroosmosis). The flocculative power of electrolytes usually increases very rapidly with increase in valence, so that a tetravalent ion is from 600-1000 times as effective as a univalent ion. The flocculation of different types of sols does not always appear to follow the simplest pattern because materials are rarely pure in nature. The mixing of sols of opposite signs can cause mutual coagulation;

the addition of a small amount of a hydrophilic sol to one which is hydrophobic may cause the latter to become more sensitive to precipitation by electrolytes. If the two have the same charge, however, or if the hyrophilic sol is of opposite sign and is added in large amounts, it serves to desensitize the hydrophobic sol to the effect of electrolytes.

Sedimentology

Flocculation, as commonly used in sedimentology, is the process by which minute, suspended particles are brought together into small lumps, clusters, or granules, as in the deposition or settling out of clay flocs from salt water. Flocculation occurs most often in areas where fresh water, carrying suspended clays, meets salt water, as in estuarine, deltaic, and lagoonal environments.

Whitehouse et al. (1960) concluded that clay minerals do not settle as single unit phases, grains, or solid aggregates, but rather as *coacervates,* thermodynamically reversible assemblies of solid clay particles or strands within a solid-rich liquid unit phase (Fig. 1). Coacervation takes place simultaneously with and after flocculation.

In a statistical study of settling speed and grain-size distribution, Kranck (1975) found that flocculated suspensions have stable size distributions whose modal size is dependent on the modal size of the deflocculated single grains. Because flocs settle out faster as large grains, the grain-size distribution of muddy sediments (deflocculated) will be asymmetrical.

Using previous studies as a base, Kranck (1973) proposed the following mechanism for flocculation of suspended sediment with high inorganic content: In salinities of $>3‰$, fine particles in a suspension are unstable and flocculate readily on contact, brought about by turbulence- and gravity-caused collisions. Smaller particles flocculate more readily, and as the flocs grow, larger particles adhere to them; as flocculation progresses, the number of collisions decreases. Eventually the flocs reach their settling velocity, flocculation ceases, and settling occurs. Kranck (1975) also found that the ratio of floc mode to grain size mode decreased with increase in particle size, suggesting that as flocs become larger they also become more unstable, and the stage where all particles have the same speed is approached but not reached. She also suggested that the smaller grains contain more organic material, and hence flocs consisting of smaller grains are less dense, and therefore have larger diameters.

Whitehouse et al. (1960) found that the flocculation of clay minerals occurs differentially at the marine/fresh water interface, e.g., kaolinite flocculates more rapidly at low salinities than does montmorillonite (smectite); montmorillonite flocculates at an increased rate at increased salinities. Edzwald and O'Melia (1975) modified Whitehouse et al.'s laboratory conditions (quiescent) to determine the effect of flocculation in a dynamic (estuarine) environment. They found that the distribution of clays in the Pamlico River Estuary can be explained by the relative stability of clays, where kaolinite, which aggregates rapidly, is found relatively upstream of illite. Pryor and Vanwie (1971) used Whitehouse et al.'s experimental results and other studies to produce a flocculation model for the Eocene "Sawdust Sand," an unconsolidated sediment composed largely of aggregate grains of two compositional types. They suggested that the units rich in kaolinite-montmorillonite aggregates were produced by flocculation in a marine-to-brackish water environment and that the kaolinite aggregates could have formed in a fresh to brackish water environment. Floccules have not been widely recognized in ancient sediments, but Pryor and Vanwie

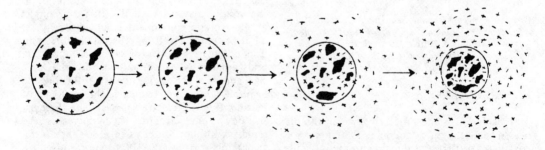

FIGURE 1. Idealized representation of the effective settling entity (coacervate) of a clay mineral in saline water; increase in effective density, by ionic dehydration and coulombic force-induced deocclusion of water, without change in solid mass content. Solid black areas, clay with residual negative charge; plus and minus signs, ionic constituents in water; circles, settling coacervates (liquid-solid phases); arrows, direction of increasing ionic concentration of sea water. (From Whitehouse et al., 1960.)

(1971) suggest that this may be oversight rather than poor preservation.

B. CHARLOTTE SCHREIBER
JOANNE BOURGEOIS

References

Edzwald, J. K., and O'Melia, C. R., 1975. Clay distributions in recent estuarine sediments, *Clays Clay Minerals*, 23, 39-44.
Eitel, W., 1954. *The Physical Chemistry of the Silicate.* Chicago: Univ. Chicago Press, 1592p.
Graham, T., 1861. Liquid diffusion applied to analysis, *Roy. Soc. London Phil. Trans.*, ser. A, 1861, 183-224.
Gripenberg, S., 1934. A study of the sediments of the North Baltic and adjoining seas, *Fennia*, 60(3), 1-231.
Kranck, K., 1973. Flocculation of suspended sediment in the sea, *Nature*, 246, 348-350.
Kranck, K., 1975. Sediment deposition from flocculated suspensions, *Sedimentology*, 22, 111-124.
Pryor, W. A., 1975. Biogenic sedimentation and alteration of argillaceous sediments in shallow marine environments, *Geol. Soc. Am. Bull.*, 86, 1244-1254.
Pryor, W. A., and Vanwie, W. A., 1971. The "Sawdust Sand"—an Eocene sediment of floccule origin, *J. Sed. Petrology*, 41, 763-769.
Van Olphen, H., 1966. *An Introduction to Clay Colloid Chemistry.* New York: Interscience, 301p.
Whitehouse, U. G.; Jeffrey, L. M.; and Debrecht, J. D., 1960. Differential settling tendencies of clay minerals in saline waters, *Natl. Conf. Clays Clay Minerals*, 7, 1-76.

Cross-references: *Clay as a Sediment; Estuarine Sedimentation, Shale.* Vol. IVA: *Electrolytes: Flocculation of Colloids.*

FLOOD DEPOSITS

Early mention of flood deposits was tied to diluvial theories, which explain geological phenomena by a deluge or flood far larger than any within human experience. In the 17th century, René Descartes suggested a great deluge had taken place during the formation of the earth, and Nicolaus Steno asserted that fossiliferous strata were deposited by the Noachian deluge. In the 18th century, Buffon (G. L. Leclerc, Comte de Buffon) asserted that the Noachian deluge had been geologically inconsequential; A. G. Werner attributed only unconsolidated strata to a deluge. In the 19th century, Georges Cuvier and William Buckland supported what is now known as the *Diluvial Theory:* that the poorly sorted, unconsolidated sediments of northern Europe, now known to be glacial deposits, were deposited by a flood, or floods. The term *diluvium* is still used in continental Europe to denote Pleistocene glacial deposits and probably should be avoided in the description of true flood deposits.

Although the Diluvial Theory has been discredited, large and often catastrophic floods are indeed associated with glacial melt and with glacial lake bursts. In the late 19th century, postglacial flood deposits were recognized by J. D. Dana in Connecticut and by T. F. Jamieson in Scotland. The best-known and, as first suggested by Bretz in 1923 (see Bretz et al., 1956), most catastrophic flood caused by a glacial lake burst was the great Spokane Flood. Baker (1973) studied the paleohydrology and sedimentology of the Spokane Flood deposits, which include huge bars and giant current ripples (mean chords 20-125 m, heights 0.5-7.0 m), and slack-water deposits, including possible turbidity-current deposits in tributary valleys. Modern glacial melt (annual) and glacial lake-burst (*jökullhlaup*) deposits are studied on outwash plains (*sandar*). Church and Gilbert (1975) review the hydrology and sedimentology of such events.

Proglacial streams are usually braided, and the studies of outwash plains and valley trains are part of the study of braided-stream deposits (q.v.). Braided streams are also common on alluvial fans (see *Alluvial-Fan Sediments*), and flood deposits on alluvial fans have been described by Chawner (1935), Allen (1965) and Bull (1972). These deposits include armored mudballs; scour-and-fill structures and large-scale festoon cross-strata; pebble pavements; and sheetflood deposits of sand and gravel, often exhibiting plane lamination produced in the upper flow regime, with erosive lower contacts.

Picard and High (1973) depict many unusual structures produced during flash floods of ephemeral streams. McKee et al. (1967) described widespread plane lamination (upper flow regime) in "flash" sheetflood deposits of Bijou Creek, Colorado. On the other hand, flash floods restricted to channels usually produce deposits exhibiting structures from the lower flow regime (Williams, 1970). Krumbein (1942) studied flash-flood deposits in a very restricted channel; he related the deposits to the morphology of the channel and to changes in velocity caused by changes in gradient and channel dimensions, and log and boulder jams. Not mentioned by Krumbein were "high eddy deposits" (Pardee, 1942), produced by eddying currents from a flooded main channel into tributary valleys. Baker (1973) described these deposits in detail.

Flood deposits associated with meandering streams, i.e., flood-plain deposits, have been described in detail by Allen (1965). Flood-plain deposits constitute the upper section of the fining-upward sequence of fluvial sediments (q.v.) and include natural levee deposits, crevasse-splay deposits, flood-plain (backswamp, stilling basin) deposits, cut-off meander fills, and swale-fill deposits (see *Fluvial Sedimentation*). Stanley (1968) described an ancient cyclic fluvial sequence, made up predominantly of levee and crevasse-splay deposits, which in many ways resembled turbidites (q.v.). Kesel et al. (1974) described sediments produced by the 1973 Mississippi River flood.

In conclusion, the study of floods and their deposits is important because their catastrophic nature may result in preferential preservation in the stratigraphic record. The enormity of the event, however, dictates that most studies must be conducted after the fact and are not directly confirmable by experiment. Some phenomena that might be expected in flood deposits include large-scale bed forms, climbing ripples with complete preservation of cross-laminae, scour-and-fill and basal erosional features, and sedimentary structures associated with the upper flow regime. Floods are intermittent and commonly periodic; their deposits, therefore, are often cyclic in nature, with abundant evidence of subaerial exposure.

JOANNE BOURGEOIS

References

Allen, J. R. L., 1965. A review of the origin and characteristics of recent alluvial sediments, *Sedimentology*, 5, 89-191.
Baker, V. R., 1973. Paleohydrology and sedimentology of Lake Missoula flooding in eastern Washington, *Geol. Soc. Am. Spec. Pap. 144*, 79p.
Bretz, J. H.; Smith, H. T. U.; and Neff, G. E., 1956. Channeled scabland of Washington: New data and interpretations, *Geol. Soc. Am. Bull.*, 67, 957-1049.
Bull, W. B., 1972. Recognition of alluvial-fan deposits in the stratigraphic record, *Soc. Econ. Paleont. Mineral. Spec. Publ. 16*, 63-83.
Chawner, W. F., 1935. Alluvial fan flooding: The Montrose, California flood of 1934, *Geogr. Rev.*, 25, 255-263.
Church, M., and Gilbert, R., 1975. Proglacial fluvial and lacustrine environments, *Soc. Econ. Paleont. Mineral. Spec. Publ. 23*, 22-100.
Kesel, R. H.; Dunne, K. C.; McDonald, R. C.; Allison, K. R.; and Spicer, B. E., 1974. Lateral erosion and overbank deposition on the Mississippi River in Louisiana caused by 1973 flooding, *Geology*, 2, 461-464.
Krumbein, W. C., 1942. Flood deposits of Arroyo Seco, Los Angeles County, California, *Geol. Soc. Am. Bull.*, 53, 1355-1402.
Malde, H. E., 1968. The catastrophic late Pleistocene Bonneville flood in the Snake River plain, Idaho, *U.S. Geol. Survey Prof. Pap. 596*, 52p.
McKee, E. D.; Crosby, E. J.; and Berryhill, H. L., Jr., 1967. Flood deposits, Bijou Creek, Colorado, June, 1965, *J. Sed. Petrology*, 37, 829-851.
Pardee, J. T., 1942. Unusual currents in glacial Lake Missoula, *Geol. Soc. Am. Bull.*, 53, 1569-1600.
Picard, M. D., and High, L. R., Jr., 1973. *Sedimentary Structures of Ephemeral Streams*. Amsterdam: Elsevier, 224p.
Scott, K. M., and Gravlee, G. C., Jr., 1968. Flood surge on the Rubicon River, California–hydrology, hydraulics and boulder transport, *U.S. Geol. Survey Prof. Pap. 422-M*, 40p.
Stanley, D. J., 1968. Graded bedding–sole marking–graywacke assemblage and related sedimentary structures in some Carboniferous flood deposits, Eastern Massachusetts, *Geol. Soc. Am. Spec. Pap. 106*, 211-239.
Williams, G. E., 1971. Flood deposits of the sand-bed ephemeral streams of central Australia, *Sedimentology*, 17, 1-40.

Cross-references: *Alluvial-Fan Sediments; Braided-Stream Deposits; Desert Sedimentary Environments; Flow Regimes; Fluvial Sedimentation; Fluvial Sediments; River Gravels; River Sands; Scour-and-Fill; Storm Deposits.*

FLOW REGIMES

In hydraulics, the term *flow regime* is commonly used in two different ways: in a broad sense, it refers to many different types of flow; in a restricted sense, it refers specifically to flows in alluvial channels, i.e., channels having a bed of sand or gravel that can be moved under "normal" discharges. In the broad sense, the term is used to distinguish two or more flows that are distinguished from each other by the general nature of the flow phenomena, or by the different hydraulic equations valid for each range of conditions. It is implied, therefore, that there is some critical condition where a rapid transition takes place from one regime to another. Examples include the transition from laminar to turbulent flow (at some critical Reynolds Number) and the transition from subcritical to supercritical flows (at a Froude Number equal to one). By combining both of these criteria, Robertson and Rouse (1941) distinguished four regimes of open-channel flow. Another pair of regimes of interest to the sedimentologist are the two types of grain flow—viscous and inertial—distinguished by Bagnold (1956).

More commonly, the term *flow regime* is used by sedimentologists in a restricted sense, as applied to flow in alluvial channels by Simons and Richardson (1961). Sorby (1908) was one of the first to record the sequence of different bed configurations and flow conditions as the velocity of flow increased over a sand bed: he noted "two important limits . . . one just sufficient to wash (the sand) along, and the other to wash it away." Gilbert (1914) made the first thorough investigations of sediment movement in flumes and recognized the following sequence, as the slope and velocity were increased: (1) plane bed with no movement; (2) ripples (called "dunes" by Gilbert who explicitly made no distinction according to scale); (3) bed "without waves and approximately plane"; and (4) antidunes. Gilbert's investigation remained the single most complete experimental investigation of sediment transport in sand-bed channels until the studies begun in 1956 by the U.S. Geological Survey at Fort Collins, Colorado, under the direction of Daryl B. Simons.

Experimental Studies

Simons and coworkers made use of a large flume, 8 feet (2.44 m) wide and 150 feet (45.72 m) long, and they observed that two distinctly different types of "ripples" formed: the smaller were called ripples, and the larger [with a wave length greater than 2 ft (0.61 m) and a height greater than 0.2 ft (6.1 cm)] were called dunes. The sequence in which the bed forms appeared as the slope and velocity in the flume was increased is shown in Fig. 1. A complete summary of the experimental data has been given by Guy et al. (1966); and reviews of the experimental results are given in Middleton (1965) by Simons et al. and Harms and Fahnstock, and in Shen (1971) by Simons and Richardson. An experimental study by Williams (1967) has extended the results to very coarse sands. For the classic papers and a review of more recent studies see the reprint volume edited by Middleton (1977).

Simons and coworkers distinguished two basic flow regimes: A lower regime in which bed forms are out of phase with surface (water) waves, and an upper regime in which bed forms are in phase with surface waves. The upper regime also includes the plane bed condition. The terms "upper" and "lower" regime were suggested by the analogy between flow over alluvial bed forms and flow over channel obstructions for supercritical and subcritical flows. It has been established experimentally, however, that the transition between the upper and lower regime for alluvial flows does not take place at a Froude Number of one, but generally at some lower value. For large channels, this value may be as low as 0.5 (Simons and Richardson, in Shen, 1971).

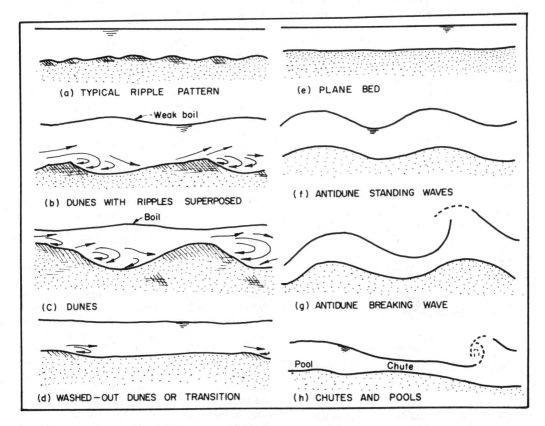

FIGURE 1. Forms of bed roughness in alluvial channels: those on the left define the lower flow regime, those on the right define the upper flow regime (from Simons et al., in Middleton, 1965).

The problem of defining the hydraulic criteria for different bed forms, and therefore for the flow regimes that are defined on the presence or absence of the bed forms, is still not entirely solved. Simons and co-workers proposed the use of stream power (average velocity, U, times the bottom shear-stress, τ_0) and fall diameter (the diameter of the quartz sphere with the same fall velocity as the bed material) as the basic hydraulic criteria. A modification of their diagram, prepared by Allen (1970) to include the data of Williams (1967), is shown in Fig. 2. Southard (1971) has shown that the different bed forms produced experimentally plot without overlap on depth-velocity-size diagrams. This suggests that these three variables are the most important ones controlling bed forms and flow regimes.

It has been suggested by Boothroyd and Hubbard (1971) from field observations, and by Costello and Southard (1974) from laboratory experiments that a third, lower, flow-regime bed form, called "sand waves" or "bars" should be distinguished from dunes. Sand waves or bars are lower and longer than dunes, have more regular crestlines than most dunes but more irregular spacing. Fig. 3 suggests a comparison between laboratory and field observations for medium sands. None of the extrapolations of laboratory studies to larger scales has yet been adequately tested with field data (but see Simons and Richardson, in Shen, 1971, for discussion of some river data).

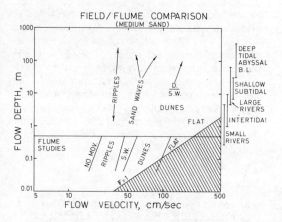

FIGURE 3. Graph of flow depth vs. flow velocity for sediment of medium sand size, showing sequence of bed configurations in flume experiments (based mainly on the work of Simons and Richardson) and in a general way, in various natural environments (courtesy of John Southard, unpublished).

Interpretation

The main use of the flow-regime concept by sedimentologists has been to interpret sequences of sedimentary structures observed in vertical sections through sedimentary deposits. A sequence from plane lamination up through large-scale cross-bedding to small-scale (ripple) cross-lamination, suggests a sequence of bed forms from upper-regime plane bed through lower-regime dunes and ripples (see, e.g., *Bouma Sequence*). Such a sequence is predicted by the experimental studies (e.g., Fig. 2). Harms and Fahnstock (in Middleton, 1965) have emphasized that the geologist will rarely be able to reconstruct all of the paleohydraulic conditions—e.g., the water temperature is generally unknown, yet it may determine whether the bed is plane or covered by dunes. An interpretation in terms of flow regimes, however, may be possible where a full hydraulic interpretation is not.

The major difficulties with such an interpretation stem from: (1) the fact that flow regimes have been established experimentally for "equilibrium flow" conditions—in nature, rapidly changing flow conditions and locally high, unsteady rates of scour and sedimentation may lead to the development of bed forms under conditions much different from those generally studied in flumes (Allen, 1973); (2) the flow-regime concept applies only to regular bed forms, of small to medium scale, formed by unidirectional flows—in nature, cross-bedding may be formed or modified by bed forms different from or larger than those studied in flumes, or by waves or reversing flows (e.g., tides); (3) the

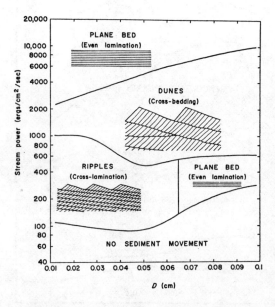

FIGURE 2. Fields of stability of the main bed forms, based on stream power and sediment fall diameter (from Allen, 1970, based on experimental data of Guy et al., 1966, and Williams, 1967).

uncertainty of interpretation of certain sedimentary structures—e.g., it is unlikely that all plane lamination is formed under upper-regime, plane-bed, flow conditions; and (4) types of flows other than flows of water in open channels, e.g., density currents, may not adhere to flow-regime criteria without substantial modification (Hand, 1974).

GERARD V. MIDDLETON

References

Allen, J. R. L., 1970. *Physical Processes of Sedimentation.* London: George Allen and Unwin, Ltd., 248p.
Allen, J. R. L., 1973. Phase differences between bed configuration and flow in natural environments, and their geological relevance, *Sedimentology,* 20, 323-329.
Bagnold, R. A., 1956. The flow of cohesionless grains in fluids, *Roy. Soc. London, Phil. Trans.,* ser. A, 249, 235-297.
Boothroyd, J. C., and Hubbard, D. K., 1971. Genesis of estuarine bedforms and crossbedding, *Geol. Soc. Am. Abstr. Programs,* 3(7), 509-510.
Costello, W. R., and Southard, J. B., 1974. Development of sand bed configurations in coarse sands, *Am. Assoc. Petrol. Geologists Ann. Meeting Abstr.,* 1, 20-21.
Gilbert, G. K., 1914. The transportation of debris by running water, *U.S. Geol. Surv. Prof. Pap. 86,* 263p.
Guy, H. P.; Simons, D. B.; and Richardson, E. V., 1966. Summary of alluvial channel data from flume experiments, 1956-61, *U.S. Geol. Surv. Prof. Pap. 462-I,* 96p.
Hand, B. M., 1974. Supercritical flow in density currents, *J. Sed. Petrology,* 44, 637-648.
Middleton, G. V., ed., 1965. Primary sedimentary structures and their hydrodynamic interpretation, *Soc. Econ. Paleont. Mineral. Spec. Publ. 12,* 265p.
Middleton, G. V., ed., 1977. Sedimentary processes: hydraulic interpretation of primary sedimentary structures, *Soc. Econ. Paleont. Mineral. Reprint Ser.,* 3, 285p.
Robertson, J. M., and Rouse, H. 1941. On the four regimes of open-channel flow, *Civ. Eng.,* 11(March), 169-171.
Shen, H. W., ed., 1971. *River Mechanics.* Box 606, Fort Collins, Colo.: H. W. Shen, 2 vols, 1236p.
Simons, D. B., and Richardson, E. V., 1961. Forms of bed roughness in alluvial channels, *Am. Soc. Civ. Eng., Proc.,* 87(HY3), 87-105.
Sorby, H. C., 1908. On the application of quantitative methods to the study of the structure and history of rocks, *Geol. Soc. [London] Quart. J.,* 64, 171-232; Discussion, 232-233.
Southard, J. B., 1971. Representation of bed configurations in depth-velocity-size diagrams, *J. Sed. Petrology,* 41, 903-915.
Southard, J. B., 1975. Bed configuration, *Soc. Econ. Paleont. Mineral. Short Course,* 2 (April, 1975), 5-44.
Williams, G. P., 1967. Flume experiments on the transportation of a coarse sand, *U.S. Geol. Surv. Prof. Pap. 562-B,* 31p.

Cross-references: *Bed Forms in Alluvial Channels; Bernoulli's Theorem; Bouma Sequence; Fluvial Sediment Transport; Grain Flows; Reynolds and Froude Numbers; Settling Velocity; Stokes' Law.*

FLUIDIZED AND LIQUEFIED SEDIMENT FLOWS

Fluidization occurs when the upward drag exerted by moving pore fluids exceeds the effective weight of the grains in a body of sediment, lifting the grains against the force of gravity. When this upward movement exceeds the minimum fluidization velocity (Fig. 1), the bed expands rapidly, porosity increases, and the sediment becomes fluid-supported rather than grain supported, i.e., the sediment is fluidized (Lowe, 1975). Fluidization has been shown or *suggested* to be effective in the horizontal mass flow of granular detritus, including nuées ardentes (q.v.), avalanches (see *Avalanche Deposits*), and sediment gravity flows (Middleton and Hampton, 1973). Middleton and Hamptom used the term *fluidized sediment flows* to include both the process of fluidization and also liquefaction. Lowe (1976), however, showed that only liquefaction, not fluidization, is significant in sediment gravity flows; Middleton and Southard (1977) replaced the term fluidized sediment flow with *liquefied sediment flow*.

Liquefaction occurs when metastable, loosely packed grain frameworks break down, the grains becoming temporarily suspended in the pore fluid and settling rapidly through the fluid until a grain-supported structure is reestablished (Lowe, 1975). Andreson and Bjerrum (1967) described several examples of sediment gravity flows caused by liquefaction, and Van der Knaap and Eijpe (1969) suggested that liquefaction might be the cause of some turbidity currents. Lowe (1976) described both liquefaction and fluidization processes and provided an extensive bibliography.

Sands subject to liquefaction are loosely packed or have textural properties that are favorable to breakdown of the fabric; densely packed sands resist liquefaction because it is necessary to expand the fabric (increase the porosity) in order to shear the sand (Middleton and Hampton, 1973). The liquefaction and fluidization of natural sands usually accompanies the collapse of loosely packed, crossbedded units (Lowe, 1976). This collapse is commonly initiated by water which is forced

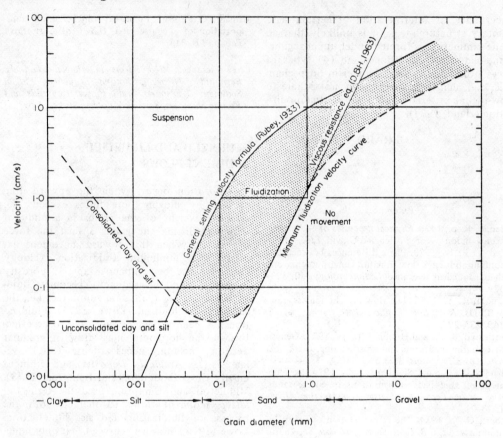

FIGURE 1. Generalized minimum fluidization velocity curve and range of fluidization of natural sediments (stippled) as a function of sediment grain size (from Lowe, 1975).

into the units when underlying beds, especially clay-rich sediments, consolidate. The consolidation of subjacent units is often triggered by the rapid deposition of the sand itself, although disturbances such as earthquakes are probably influential in some instances.

Once the fabric is destroyed, the grains no longer form a rigid framework but must be supported at least in part by the pore fluids; consequently pore fluid pressures rise much above normal hydrostatic pressures. So long as grains are supported by the pore fluid, the sand has little strength and behaves as a fluid with a viscosity on the order of 1000 times the viscosity of water. Thus, a liquefied body of sand can flow rapidly down relatively gentle slopes (3–10°; Middleton, 1969, 1970).

Fluidization is probably not by itself a mechanism for sediment transport because excess pore pressures are rapidly dissipated by loss of pore fluids, resulting in deposition. Many sedimentary structures are produced by the escape of this fluid (see *Water-Escape Structures*). It is likely that fluidization is combined with other mechanisms, such as turbulence (see *Turbidity-Current Sedimentation*) or dispersive pressure (see *Grain Flows*).

JOANNE BOURGEOIS

References

Andresen, A., and Bjerrum, L., 1967. Slides in subaqueous slopes in loose sand and silt, in A. F. Richards, ed., *Marine Geotechnique*. Urbana: Univ. Illinois Press, 221-239.

Castro, G., 1969. Liquefaction of sands, *Harvard Soil Mech. Ser.*, 81, 112p.

Lowe, D. R., 1975. Water escape structures in coarse-grained sediments. *Sedimentology*, 22, 157-204.

Lowe, D. R., 1976. Subaqueous liquefied and fluidized sediment flows and their deposits, *Sedimentology*, 23, 285-308.

Middleton, G. V., 1969. Turbidity currents and grain flows and other mass movements down slopes, in D. J. Stanley, ed., *The NEW Concepts of Continental Margin Sedimentation*. Washington, D.C.: Am. Geol. Inst. GM-A-1–GM-B-14.

Middleton, G. V., 1970. Experimental studies related

to problems of flysch sedimentation, *Geol. Assoc. Canada Spec. Pap.* **7**, 253-272.

Middleton, G. V., and Hampton, M. A., 1973. Sediment gravity flows: Mechanics of flow and deposition, *Soc. Econ. Paleont. Mineral. Pacific Section, Short Course* (May, 1973), 1-38.

Middleton, G. V., and Southard, J. B., 1977. Mechanics of sediment movement. *Soc. Econ. Paleont. Mineral. Short Course 3*, 246p.

Van der Knaap, W., and Eijpe, R., 1968. Some experiments on the genesis of turbidity currents, *Sedimentology*, **11**, 115-124.

Cross-references: *Grain Flows; Gravity Flows; Turbidity-Current Sedimentation; Water-Escape Structures.*

FLUVIAL SEDIMENTATION

Fluvial sedimentation covers the processes of transport and deposition associated with overland flows of sediment-charged water, and the process resultants known as fluvial sediments or alluvium (q.v.; Fig. 1). Overland flows, representing runoff derived from precipitation, vary from shallow, extensive sheetfloods highly charged with debris to deep streams, low in sediment load, occupying narrow and definite courses. Although participating in the general denudation of land areas, overland flows may deposit sediments locally, as in intermontane basins and interior lowlands, and also extend them as deltas and coastal plains into water bodies. Overland flows are the principal agents carrying weathering products off the land.

Processes

Runoff acts upon very poorly sorted regolith produced by the mechanical and chemical weathering of bedrock. Entrainment of rock waste by overland flow is followed by turbulent downstream transport, during which particles become size-sorted according to definite physical laws (Sundborg, 1956), rounded by abrasion, and perhaps altered further chemically.

Transport. In streams low in sediment, generally more or less meandering, gravel- and sand-sized debris is transported mainly as bed load, the particles erratically sliding or rolling over the bed (Sundborg, 1956). This bed-load transport is manifested by migrating asymmetrical ripples and dunes (Simons et al., 1961), producing cross-stratified sediments (Allen, 1963; see *Bed Forms in Alluvial Channels*). Stationary or moving antidunes and plane-bed transport surfaces may also be important (Simons et al., 1961), producing plane- and wavy-bedded sediments. Wave processes are minor, only affecting bed loads near channel-edge shores. Alluvial islands and bars arise from bed-load transport in braided streams (Leopold et al., 1964; see *Braided-Stream Deposits*).

Silt- and clay-sized debris is carried as a suspension load (Sundborg, 1956). Particles are distributed over the entire flow cross-section, for they are small enough to be buoyed up by vertical velocity components present in the turbulent flow structure.

Grains of intermediate coarseness saltate over the bed and form the saltation load (Sundborg, 1956), being neither wholly in suspension nor wholly in continuous contact with the bed. The suspension load travels faster than the bed load; hence the finer particles travel faster than the coarser ones, and alluvial sediments generally become finer downstream through differential transport. Similarly, in a turbulent flow, fine particles are more readily dispersed upward against gravity than are coarse ones, and the total stream load fines upward across the line of flow at a point.

Sheetfloods and streamfloods on alluvial fans and in the channels of ephemeral streams in arid-semiarid regions accomplish little

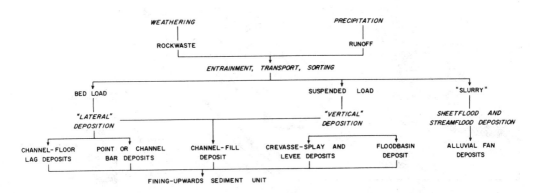

FIGURE 1. Flow diagram illustrating materials, processes, and process resultants involved in fluvial sedimentation. Processes are shown in italic.

sorting (see *Alluvial-Fan Sediments*). Generally, bed and suspension loads cannot be distinguished, the flows resembling slurries forced along by the weight of contained particles.

Accretion. Grossly, accretionary processes in streams of low sediment concentration are either "lateral" or "vertical" (Fig. 1). In meandering streams, lateral deposition depends on the secondary flow (Leopold et al., 1964) within their curved channels (Fig. 2). High-velocity flow lines impinge on the outer meander bank and erode it; the bottom secondary flow is toward the inner bank, where stream velocity and bed shear are less, building the bank sideways and upward to form a point bar through the intermediate agency of bed forms such as ripples and dunes. Because of the secondary flow, sediment deposited deep on the bar is coarser than in the shallows. Thus, point-bar sequences fine upward, even though only bed-load materials are generally present. At the base of the bar is concentrated the coarsest fraction of the stream load, moved only when currents are exceptionally strong. This fraction forms a lag deposit winnowed of fines. Channel cutoff generally accompanies meander growth and lateral deposition, creating arcuate lakes, the familiar oxbows (Fisk, 1944).

Lateral deposition differs in braided streams. Coarse debris carried over the shallow tops of alluvial islands is often deposited as foresets on their flanks, the islands being delta-like channel bars (Leopold et al., 1964).

Vertical deposition on interfluve areas following floods leads to construction of levees, crevasse-splays, and flood basins (Fisk, 1944, 1947; Figs. 1, 2; see *Flood Deposits*). In most streams, such overbank flows occur on an average of once every one or two years (Leopold et al., 1964). Most overbank flows contain suspended fines only; vertical deposition, therefore, produces essentially horizontal layers of fine sediment that become thinner and finer away from active channels (Fisk, 1947). Vertical deposits are coarsest in the levees and crevasse-splays—the latter built out from crevasse channels breaching levees—which immediately border active streams. Only the finest suspended sediment reaches the more distant flood basins.

The sheetfloods deposit differently (Blissenbach, 1954). Streamfloods tend to aggrade already existing channels, whereas sheetfloods—and streamfloods building alluvial fans—produce deposits laterally more extensive. Levee-like structures only occasionally border fan channels and invariably consist of coarse materials.

Environments and Sediments

Table 1 classifies modern alluvial environments and sediments (see also Fig. 2) other than those of alluvial fans.

Channel or substratum deposits are those relatively coarse sediments of bed-load origin laid down as point bars, alluvial islands, and lag pavements on channel floors, thus forming the lower layers in the typical flood-plain sequence (Fig. 2). Point- and channel-bar deposits are mainly well-sorted sands and/or gravels with drifted plant remains and mud pebbles derived by erosion, e.g., bank cutting, of fine-grained overbank sediments (Fisk, 1944; Leopold et al., 1964). Trough cross-stratification (small and large scale) and horizontal bedding dominate many sandy point-bar deposits (Harms et al., 1963). Channel-bar gravels of braided streams are commonly cross-stratified, but may also fill channel forms, where large-scale festoon structure may be produced (Doeglas, 1962).

Overbank or top-stratum deposits are fine-

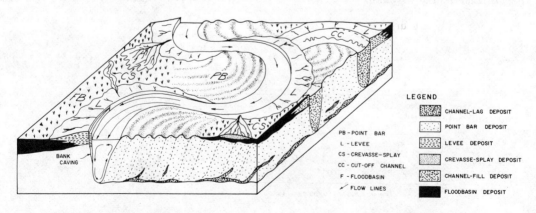

FIGURE 2. Diagrammatic representation of sedimentation in the flood plain of a strongly meandering stream (vertical scale much exaggerated).

TABLE 1. Classification of Fluvial Sedimentary Environments and Sediments

Environment	Deposit	Origin Reflected in Typical Stratigraphic Position
Channel floor	channel-lag deposit	channel or sub-stratum deposit
Point bar	point-bar deposit	
Alluvial island	channel-bar deposit	
Point-bar swale or in abandoned braided-stream channel	swale-fill deposit	overbank or top-stratum deposit
Levee	levee deposit	
Crevasse-splay	crevasse-splay deposit	
Flood basin	flood-basin deposit	
Within abandoned or decaying channel	channel-fill deposit	transitional deposit

After Allen, 1965.

grained sediments (very fine sand, silt, clay) of suspension-load origin accumulated from overbank flows (Fisk, 1947). They typically overlie channel deposits and form the uppermost layers of flood plains (Fig. 2). The ribbonlike levees generally comprise thin, rapidly alternating, very fine sand and clayey silt layers (Fisk, 1944, 1947). Autochthonous plant remains and desiccation features are often present. Swale-fill deposits are similar but plug arcuate hollows on point-bar tops (Fisk, 1944, 1947). Crevasse-splay deposits are normally tongue-like and usually rather sandy because the crevasse channels tap loads carried low in the stream's vertical profile. Thick, unbedded silty clays and clayey silts dominate flood-basin deposits (Fisk, 1944, 1947). In these, in addition to plant roots, fresh-water shells representing ephemeral ponds may be found. In drier regions, calcretes, ferricretes, alkali concentrations, and sun cracks abound.

Arcuate channel-fill deposits (Fig. 2) represent the transitional category, occurring through a considerable thickness in flood plains (Fisk, 1944, 1947). Suspended-load fines are dominant but interstratified with bed-load sediments. Plant roots and shells are common, and sometimes fish remains and vertebrates are found.

Gross geometry of such flood-plain deposits seems dependent on channel sinuosity (Allen, 1965). Markedly sinuous streams, flowing in meander belts, create linear bodies of coarse channel sediment in lateral and vertical contact with thick masses of overbank origin. Braided or weakly meandering streams, combing more freely across their flood plains, produce sheets of channel sediment associated with little or no overbank material.

Alluvial-fan deposits (see *Alluvial-Fan Sediments*) are mainly of gravel grade, consist of poorly sorted and rounded labile debris, and are poorly stratified to unstratified.

J. R. L. ALLEN

References

Allen, J. R. L., 1963. Asymmetrical ripple marks and the origin of water-laid cosets of cross-strata, *Liverpool Manchester Geol. J.,* 3, 187-236.

Allen, J. R. L., 1965. A review of the origin and characteristics of recent alluvial sediments, *Sedimentology,* 5, 89-191.

Allen, J. R. L., 1970. Studies in fluviatile sedimentation: A comparison of fining-upwards cyclothems, with special reference to coarse-member composition and interpretation, *J. Sed. Petrology,* 40, 298-323.

Blissenbach, E., 1954. Geology of alluvial fans in semi-arid regions, *Geol. Soc. Am. Bull.,* 65, 175-190.

Bridge, J. S., 1975. Computer simulation of sedimentation in meandering streams, *Sedimentology,* 22, 3-44.

Doeglas, D. J., 1962. The structure of sedimentary deposits of braided rivers, *Sedimentology,* 1, 167-190.

Fisk, H. N., 1944. *Geological Investigation of the Alluvial Valley of the Lower Mississippi River.* Vicksburg: Miss. River Comm., 78p.

Fisk, H. N., 1947. *Fine Grained Alluvial Deposits and Their Effect on Mississippi River Activities.* Vicksburg: Miss. Riv. Comm., 82p.

Harms, J. C.; McKenzie, D. B.; and McCubbin, D. G., 1963. Stratification in modern sands of the Red River, Louisiana, *J. Geol.*, **71**, 566–580.
Hjulström, F., 1935. Studies of the morphological activity of rivers as illustrated by the River Fyris, *Bull. Geol. Inst. Uppsala*, **25**, 221–527.
Leopold, L. B.; Wolman, M. G.; and Miller, J. P., 1964. *Fluvial Processes in Geomorphology*. San Francisco: Freeman, 522p.
Morisawa, M., ed., 1974. *Fluvial Geomorphology*. Binghamton: State Univ. of New York, 314p.
Simons, D. B.; Richardson, E. V.; and Albertson, M. L., 1961. Flume studies using medium sand, *U.S. Geol. Surv. Water-Supply Pap. 1498-A*, 76p.
Sundborg, Å., 1956. The River Klarälven: A study of fluvial processes, *Geogr. Ann.*, **38**, 127–316.
Temple, P. H., and Sundborg, Å., 1972. The Rufiji River, Tanzania: Hydrology and sediment transport, *Geogr. Ann.*, **54A**, 345–368.
Visher, G. S., 1965. Fluvial processes as interpreted from ancient and recent fluvial deposits, *Soc. Econ. Paleont. Mineral. Spec. Publ. 12*, 116–132.

Cross-references: *Alluvial-Fan Sediments; Alluvium; Bedding Genesis; Bed Forms in Alluvial Channels; Braided-Stream Deposits; Delta Sedimentation; Deposition; Flood Deposits; Flow Regimes; Fluvial Sediment Transport; Lateral and Vertical Accretion; Sedimentation; Transportation.* Vol. III: *Fluvioglacial Processes; Rivers; Rivers: Meandering and Braiding.*

FLUVIAL SEDIMENTS

Many sedimentary formations of past ages are fluvial in origin and have properties that distinguish them from strata formed in other ways. Geographically, fluvial deposits are commonly found at the margins of sedimentary basins, between areas underlain by marine strata and the upland regions where the sediment originated. On a large scale, most fluvial strata are complex in facies geometry, with mappable units of no great lateral extent. The coarser-grained units, lying nearly at right angles to the depositional strike, commonly are elongate and underlain by concave-upward erosional surfaces representing infilled stream valleys or channels. The finer-grained units are less markedly elongate and probably record nonchannel environments distant from active streams. Fining-upward cycles (conglomerate–sandstone–siltstone) up to tens of meters thick are especially characteristic of sections wholly fluvial in origin (Allen, 1965). On a small scale, the beds of fluvial formations generally have restricted lateral extents; evidence of erosion and redeposition abounds.

Fluvial strata commonly display an extremely wide range of lithologies, and it is not uncommon for claystones, siltstones, sandstones, exotic and intraformational conglomerates, lignites, calcretes, and biogenic limestones (fresh water) all to be found in one succession. Typically, the sandstones, in addition to the conglomerates, are mineralogically immature and replete with labile components, indicating transport histories short in distance and duration. In addition, the sandstones show structures indicating deposition from strong, turbulent currents. Large-scale cross-bedding (q.v.) is perhaps the commonest, but plane lamination, parting lineation (q.v.), and small-scale cross-bedding with current ripples are seldom absent. Proofs of exposure (mud cracks, rain-drop imprints, tracks and trails, roots and rootlets) and of fluctuating water levels abound in the finer-grained deposits. Paleontologically, post-Silurian fluvial sediments are characterized by the abundance of plant remains, both drifted and autochthonous, and the presence of fresh water or terrestrial bivalves, gastropods, crustaceans, and vertebrates.

Occurrence

Rocks of fluvial origin are ubiquitous through the height and breadth of the geological column, and only a few of the most important occurrences are mentioned here.

Those of Paleozoic age are best known from Europe and North America. In Devonian times, the Caledonian Orogeny produced mountains from which detritus was shed onto broad coastal plains lying to the east in S Britain and NW Europe, and to the W in the Appalachian region. The red or green conglomerates, sandstones, and siltstones of the Old Red Sandstone in Europe (Allen, 1964) and the Catskill sequence in the Appalachians (Barrell, 1913–1914) are the deposits of these coastal plains. In both these regions, fining-upward cycles, comparable in many respects to the deposits of modern flood plains, occur through hundreds of meters of strata.

Fluvial deposits of a somewhat different nature accumulated over wide areas of North America and Europe during Permo-Carboniferous times (Bluck and Kelling, 1962; Potter, 1963; Jablokov et al., 1963; see also *Cyclic Sedimentation*). These "coal measure" facies, in which the stream-laid sediments are closely associated with shallow-marine beds, contain elongate sandstone bodies, in the form of pods, ribbons, dendroids, and belts. They often cut down erosively into sheet-like, finer sandstone, marine limestones, and shales, beneath which lie persistent coal seams which seem to represent widespread coastal swamps. A number of the sandstones may have been formed in delta distributaries, for they bifurcate downcurrent and cover large areas.

Thick successions of fluvial origin occur in the Tertiary Molasse of the Alpine foredeep in Switzerland and southern France (Crouzel, 1957; Bersier, 1958). They include fining-upward cycles through long sections, and also important conglomerates, such as the Nagelfluh, which record braided streams on alluvial fans at the margin of the rising mountains.

J. R. L. ALLEN

References

Allen, J. R. L., 1964. Studies in fluviatile sedimentation: Six cyclothems from the Lower Old Red Sandstone, Anglo-Welsh Basin, *Sedimentology*, 3, 163-198.

Allen, J. R. L., 1965. Fining-upwards cycles in alluvial successions, *Liverpool Manchester Geol. J.*, 4, 229-246.

Barrell, J., 1913-1914. The Upper Devonian delta of the Appalachian geosyncline, *Am. J. Sci.*, ser. 4, 36, 429-472; 37, 87-109, 225-253.

Bersier, A., 1958. Séquences détritiques et divagations fluviales, *Eclogae geol. Helvetiae*, 51, 854-893.

Bluck, B. J., and Kelling, G., 1962. Channels from the Upper Carboniferous coal measures of South Wales, *Sedimentology*, 2, 29-53.

Crouzel, F., 1957. Le Miocène continental du Bassin D'Aquitaine, *Bull. Serv. Carte Géol. de la France*, 248, 264p.

Jablokov, V. S.; Botvinkina, L. N.; and Feofilova, A. P., 1963. Sedimentation in the Carboniferous and the significance of alluvial deposits, *C. R. 4ième Congrès pour l'avancement des études de stratigraphie Carbonifère, Heerlen*, 2, 293-299.

Jonker, R. K., 1972. Fluvial sediments of Cretaceous age along the southern border of the Cantalabrian Mountains, Spain, *Leides Geol. Meded.*, 48, 276-380.

Potter, P. E., 1963. Late Paleozoic sandstones of the Illinois Basin, *Ill. St. Geol. Surv., Rept. Inv. 217*, 92p.

Schumm, S. A., 1972. Fluvial paleochannels, *Soc. Econ. Paleont. Mineral. Spec. Publ. 16*, 98-107.

Spoljaric, N., 1971. Recognition of some primary sedimentary structures of fluvial sediments in the subsurface, *Geol. Soc. Am. Bull.*, 82, 1051-1054.

Walker, R. G., 1975. From sedimentary structures to facies models: Example from fluvial environments, *Soc. Econ. Paleont. Mineral. Short Course*, 2(April, 1975), 63-80.

Cross-references: *Alluvial-Fan Deposits; Braided-Stream Deposits; Cross-Bedding; Cyclic Sedimentation; Deltaic Sediments; Flood Deposits; Fluvial Sedimentation; River Gravels; River Sands; Sand and Sandstones; Sedimentary Structures.* Vol. III: *Flood Plains; Terraces, Fluvial.* Vol. IVB: *Placer Deposits.*

FLUVIAL SEDIMENT TRANSPORT

The waters flowing in natural channels carry both dissolved minerals and discrete particles of minerals and rock fragments that are the products of weathering of the earth's crust. The solid particles moved by a stream are called *fluvial sediments*, or sometimes the *sediment load* of a stream, and the processes by which they are transported are termed *fluvial sediment transport* (Graf, 1971; Yalin, 1972; Middleton and Southard, 1977).

The sediment load of a stream may be classified, according to its mode of transport, as bed load or suspended load. The *bed load*, sometimes called *contact load*, consists of the coarser particles, which travel along the bed by *traction* (q.v.) and/or *saltation* (q.v.), essentially in continuous contact with the bed, under the influence of shearing stress or drag imposed by the flowing water. The *suspended load* consists of those particles suspended in the flow by turbulence or by colloidal suspension. One may also classify the sediment load either as bed-material load or as fine-material load. The *bed-material load* is composed of particles that are of the same range of sizes as the material comprising the stream bed. The *fine-material load*, sometimes called *wash load*, consists of particles finer than material found in appreciable quantities in the stream bed.

Table 1 is a summary of the general classification of fluvial sediments. The size break between fine material and bed material, and between suspended load and contact load is only approximate; the actual value for any stream will depend on the characteristics of the sediment and the flow (Table 2). Usually, the sediment load of a stream is given in terms of a transport rate, which is the weight of sediment passing a given cross section of a stream per unit time. The terms *transport rate* and *sediment discharge* are used interchangeably. The sediment concentration is defined as the ratio of the dry weight of sediment discharge to the total weight of discharge of the water-sediment mixture.

All natural streams may be considered agents of transportation, and the rate at which they transport fluvial sediments can be limited either by the rate at which sediment is made available to the stream or by the ability of the flow to move the particles. An extremely important class of streams includes those whose beds and banks are composed of material that has been or can be transported by the flow. The channels of those streams are called *alluvial channels*, and it is with the transport

TABLE 1. General Classifications of Fluvial Sediments

	Size (mm)	Mode of Transport	Type of Load
Clay	<0.004	suspended load	fine material
Silt	0.004–0.062		
Fine sand	0.062–0.25	suspended load or contact load	bed material
Medium sand	0.25–0.5		
Coarse sand	0.5–2.0	contact load	
Gravel	2.0–64		
Cobbles	64–250		
Boulders	>250		

processes in such channels that the following discussion is primarily concerned.

Transport Relations in Alluvial Channels

Channels in Equilibrium. Many alluvial channels attain a state of approximate equilibrium such that over a period of time, there is no net aggradation or degradation in a given reach of the stream. Under these conditions, the sediment load introduced at the upstream end of a reach must equal the sediment load transported through and out of the reach. Because the stream can adjust its slope and its cross section, it can be assumed that for the prevailing flow, the channel slope and the cross sectional area (and, hence, the mean flow depth and mean velocity) are adjusted to transport only the sediment load introduced into the reach. If more sediment is delivered to a reach than the stream is capable of transporting, the sediment will be deposited on the bed, and if the stream is capable of transporting more sediment than is delivered to it, it will scour its bed, adjusting its slope and cross-sectional area until equilibrium is reestablished. Thus, for streams in equilibrium, it is possible to develop approximate relations between the sediment transport rate and the characteristics of the flow—the velocity, depth, and slope.

Streams are usually capable of transporting more fine sediment than is delivered to the channel by sheet and rill flow over the drainage surfaces of the watershed. These fine sediments are not deposited at the streambed, but are simply washed through the reach with the flow to be deposited in backwater or deadwater areas, such as floodplain or overbank flows during floods, or in the still waters of reservoirs, lakes, and ocean depths. The transport rate of fine material is not functionally related to the flow. Thus, any functional relations between flow parameters and sediment discharge apply only to the bed-material load.

In reality, unique relations between bed-material transport rates and simple flow characteristics seldom exist because flow and transport processes in natural channels are much more complex than described by the simple considerations of equilibrium given above. Stream banks usually are stabilized by deposits of clay or by vegetation, restricting a stream's ability to adjust its cross-sectional area. Bedrock outcrops or artificial controls may limit the channel slope. The stream-bed sediment is a complex mixture that can be described only statistically, and flows and transport rates usually vary in time and space. In addition, the ability of a stream to transport sediment depends not only on mean velocity, depth, and slope, but also on such factors as the water temperature, the velocity distribution, the characteristics of turbulence of the flow, and the variations of velocity and depth in a cross section. Nonetheless, for many practical problems, it is essential to be able to estimate at least crudely the equilibrium transport rates of bed material, because it is these transport rates which determine the stability of navigation and conveyance channels in alluvium and the aggradation or degradation downstream of reservoirs, diversions, and channel-control structures.

Bed-Load Equations. Historically, sediment transport studies dealt only with bed-load, and the equations proposed by Du Boys, Chang, O'Brien, Meyer-Peter, Schoklitsch and other early workers (see references in Rouse, 1950,

TABLE 2. Typical Transport Characteristics of Selected Rivers

River Station	Flow	Discharge (cfs)	Mean Depth (ft)	Mean Velocity (ft/sec)	Slope (ft/ft)	Median Diameter of Bed Material (mm)	Suspended Sediment Concentration (ppm)	Percent Finer Than 0.062 mm
Rio Puerco near Bernardo, New Mexico, US	Low	160	0.98	3.07	0.0010	0.22	112,000	96.8
	Medium	1,430	2.48	6.48	.0015	.24	240,000	64.9
	High	6,620	6.97	7.54	.001	—	72,000	84.0
Rio Grande near Bernalillo, New Mexico, US	Low	81.2	0.58	0.130	.00086	.29	1,000	94.0
	Medium	1,000	1.50	2.51	.00086	.29	2,600	51.2
	High	10,100	4.80	7.71	.00080	.29	3,040	31.8
Colorado River at Lee's Ferry, Arizona, US	Low	3,560	4.30	2.47	.0013[a]	.20	1,010	75.0
	Medium	10,000	7.73	3.55	.0013[a]	.39	3,000	73.0
	High	62,800	18.1	8.95	.0013[a]	.38	6,280	53.0
Mississippi River at St. Louis, Missouri, US	Low	51,000	16.1	2.12	.000081	.24	11[b]	—
	Medium	172,000	27.4	3.90	.000063	.19	90[b]	—
	High	763,000	56.7	7.53	.00014	.70	181[b]	—
Amazon River at Obidos, Brazil[c]	Low	2,560,000	134	2.58	.0000017	.22	60–110	—
	Medium	5,810,000	152	5.10	.0000056	.22	60–110	—
	High	8,400,000	158	6.45	.0000088	.22	235	78.0

[a]Approximate channel slope.
[b]Concentration of sand only (material coarser than 0.062 mm).
[c]Data provided by F. C. Ames, R. E. Oltman, and R. H. Meade, U.S.G.S. Slopes were computed.

and Scheidegger, 1961) are of the general functional forms

$$q_B = f(\tau) \quad (1)$$

or

$$q_B = f(\tau - \tau_c) \quad (2)$$

in which q_B is the transport rate of bed load per unit width of channel and τ is the shear stress or tractive force at the bed. For steady uniform flow, τ equals the product (unit weight of the fluid) × (mean depth) × (channel slope). The critical tractive force, τ_c, is the shear stress required to initiate movement of the bed material. The functional forms of Eq. (1) and (2) and values of τ_c for a particular grain size have been determined experimentally. Because a relation exists between mean velocity and shear stress, Eqs. (1) and (2) can easily be expressed in terms of mean velocity.

Suspended Sediment Discharge. The discharge of suspended sediment per unit width of channel, q_s, between the distances a and D from the bed is given by

$$q_s = \int_a^D \gamma C_y V_y dy \quad (3)$$

wherein V_y and C_y are the fluid velocity and sediment concentration at distance y from the bed and γ is the unit weight of the water-sediment mixture. Eq. (3) is precise, but can be integrated only if γ, V_y, and C_y can be expressed as functions of y.

No methods exist for determining the absolute concentration, C_y, but assuming suspension to be analogous to a diffusion process, an approximate relation for the relative concentration distribution C_y/C_a is obtained that permits integrating Eq. (3) for any assumed velocity distribution if the reference concentration C_a at the distance a from the bed is known (see Rouse, 1950, p. 799).

Theory and Applications. Notable contributions to studies of fluvial transport processes have been made by Hjülstrom, Sundborg, Leopold and Maddock, Brooks, Simons and Richardson, Blench, Vanoni and many others (see references in Middleton, 1965; Leopold et al., 1964; and Scheidegger, 1961). Studies of theoretical aspects of sediment transport are too numerous to detail, but reference should be made to the work of Velikanov, Bagnold, Einstein, Vanoni, Laursen and Colby, (see, e.g., references in Shen, 1972).

Of special interest are Bagnold's (1956) suspended-load equation and the gravitational theory of Velikanov (1936), which treat suspension from energy considerations rather than as a diffusion process. Velikanov's and Bagnold's equations are similar, although developed quite independently and from different assumptions, and both are somewhat controversial. Although the relations are useful in demonstrating the nature of the forces involved in transporting sediment, they are not especially useful as practical formulas for computing rates of sediment movement.

Colby (1964), comparing the bed-load equations of Einstein, Meyer-Peter and Muller, and Bagnold, noted that the three equations have the same general functional form—a dimensionless form of Eq. (1) and (2), but that the Meyer-Peter and Muller and Einstein equations are likely to be more applicable because both contain empirically defined corrections to account for the effects of form roughness due to dunes, bars, and other irregularities at the bed.

Einstein's bed-load function permits calculation of both the bed load and the suspended bed material, and consists of estimating the bed-load discharge from a dimensionless shear-transport relation and then integrating Eq. (3) using a logarithmic velocity distribution and equating the reference concentration C_a to the concentration of the bed layer, assumed to be two grain diameters thick. The method, although laborious and somewhat inexact, is widely used, and it served as a basis for developments by Colby and his co-workers (see references in Colby, 1964) to compute bed-material discharges from the basic data of sediment samples; bed-material samples; and mean width, depth, and velocity of the flow. This method, known generally as the modfied Einstein method, consists basically of extrapolating the sampled suspended load through an unsampled zone of the flow down to the bed, to include the bed load, using approximately the equations given by Einstein. The method, although criticized as highly empirical, has been found to yield quite reliable results.

On the basis of many determinations of bed-material transport by this method, Colby (1964) defined the approximate graphical relation of the transport rate of bed material as a function of velocity, depth, and median diameter of bed material (Fig. 1). Auxiliary graphs are given to correct for effects of water temperature or high concentrations of fine sediment. These graphs provide a rapid and reasonably accurate estimate of bed-material transport rates for sand-bed streams.

Daily and periodic sediment samples are collected on many rivers of the world, and some of the records extend back for several decades. From these records, long-term sediment yields can be determined and probable future yields can be estimated using flow-duration curves and

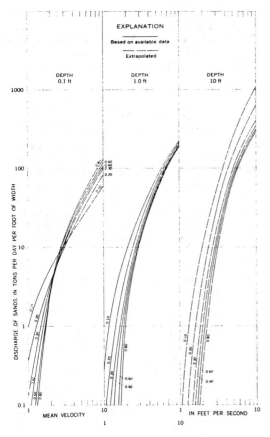

FIGURE 1. Relationship of discharge of sands to mean velocity for six median sizes of bed sands, three depths of flow, and a water temperature of 60°F (15.6° C) (from Colby, 1964). Numbers on graph (e.g., 0.60) are median size of bed sand in mm. 1 ft/sec = 1.1 km/hr. 1 ft = 0.3 m.

relations between water discharge and sediment discharge, or other techniques. A general review of sampling programs, sediment records, and various applications of sediment data is given by Colby (1963).

CARL F. NORDIN, JR.*

*Approved for publication by the Director, U.S. Geological Survey.

References

Ackers, P., and White, W. R., 1973. Sediment transport: New approach and analysis, *J. Hyd. Div. ASCE*, 99(HY11), 2041-2060.
Bagnold, R. A., 1956. The flow of cohesionless grains in fluids, *Roy. Soc. London, Phil. Trans.*, 249, 235-297.
Colby, B. R., 1963. Fluvial sediments—a summary of source, transportation, deposition and measurement of sediment discharge, *U.S. Geol. Surv. Bull.*, 1181-A, 47p.
Colby, B. R., 1964. Discharge of sands and mean-velocity relationships in sand-bed streams, *U.S. Geol. Surv. Prof. Pap. 462-A*, 47p.
Graf, W. H., 1971. *Hydraulics of Sediment Transport*. New York: McGraw-Hill, 513p.
Leopold, L. B.; Wolman, M. G.; and Miller, J. P., 1964. *Fluvial Processes in Geomorphology*. San Francisco: Freeman, 522p.
Middleton, G. V., ed., 1965. Primary sedimentary structures and their hydrodynamic interpretation, *Soc. Econ. Paleont. Mineral. Spec. Publ. 12*, 265p.
Middleton, G. V., and Southard, J. B., 1977. Mechanics of sediment movement. *Soc. Econ. Paleont. Mineral. Short Course 3*, 246p.
Nanson, G. C., 1974. Bedload and suspended-load transport in a small steep mountain stream, *Am. J. Sci.*, 274, 471-486.
Rouse, H., ed., 1950. *Engineering Hydraulics*. New York: Wiley, 1039p.
Scheidegger, A. E., 1961. *Theoretical Geomorphology*. Berlin: Springer-Verlag, 327p.
Shen, H. W., ed., 1972. *Sedimentation: Symposium to honor Professor H. A. Einstein*. Colo. State Univ.: Dept. Civil Eng., 814p.
Velikanov, M. A., 1936. Formation of sand ripples on the stream bottom, *Internat. Assoc. Sci. Hydrology, Comm. de Potomologie*, 3, 13.
Yalin, M. S., 1972. *Mechanics of Sediment Transport*. New York: Pergamon, 290p.

Cross-references: *Bed Forms in Alluvial Channels; Cross-Bedding; Flow Regimes; Fluvial Sedimentation; Ripple Marks; Sedimentation; Sediment Transport—Initiation and Energetics; Transportation*. Vol. III: *Sediment Transport: Fluvial and Marine*. Vol. XIII: *Hydraulic Movement of Sand; Stream Gauging*.

FRESHWATER LIMESTONE—See LIMESTONE; SPELEAL SEDIMENTS; SPRING DEPOSITS

FUCOID

Marks in sedimentary rocks assumed to be the result of organic activity are called *trace fossils, ichnofossils,* or *Lebensspuren* (see *Biogenic Sedimentary Structures*). Ichnologists have been able to describe certain stratigraphically characteristic trace fossils to the genus, and even the species level. Many individually diagnostic ichnofossils, however, show indefinite markings that cannot be ascribed to known ichnogenera. Thus, the terms *fucoid, fucoidal,* and even a genus *Fucoides* were used in the past to designate the doubtful ichnofossil groups that show a dendritic pattern resembling the marks that the common marine alga *Fucus* (Greek:

phukos; Latin: fucus = seaweed) might leave if buried under favorable conditions. Many of these problematic marks have been subsequently interpreted as being nonbiologic, and, rather, mechanical in origin (Nathorst, 1886; James, 1894). For instance, Clarke (1918) interpreted *Fucoides graphica* from the Devonian Portage beds of New York as counterparts of ice-crystal impressions, and other fucoidal markings are most likely the fillings of incomplete mud cracks. Although the term *fucoid* persists in modern literature, it is usually informal and does not necessarily carry a genetic connotation as in early usage. Use of the term is discouraged.

FLORENTIN MAURRASSE

References

Clarke, J. M., 1918. Strand and undertow markings of Upper Devonian time, *N.Y. State Mus. Bull.*, **196**, 199-238.

Fuchs, T., 1895. Studien über Fucoiden un Hieroglyphen, *Akad. Wiss. Wien, math-naturwiss. kl. Denkschr.*, **62**, 369-448.

James, J. F., 1894. Studies in problematic organisms, 2: The genus *Fucoides, Cincinnati Soc. Nat. Hist. J.*, **16**, 19-39.

Nathorst, A. G., 1886. Nouvelles observations sur des traces d'animaux et autres phénomènes d'origine mécanique décrises comme "algues fossiles," *K. Svenska Vetenskapsa Kad. Handl.*, **21**(14), 1-58.

Cross-references: *Biogenic Sedimentary Structures; Bioturbation; Mud Cracks; Sedimentary Structures.* Vol. VII: *Trace Fossils.*

FULGURITES

The term *fulgurite* (Latin: *fulguritus,* "struck by lightning"; German: *Blitzrohr*), was first used by Withering (1790) for the tubes he found in sand that had been fused by a lightning discharge. Such tubes had been known for at least 80 years previously without being understood. They have since been extensively studied, being found in contemporary deserts and sand dunes throughout the world. They are sometimes known as *sand fulgurites,* to distinguish them from *rock fulgurites,* which are common on exposed rock surfaces struck by lightning.

Sand fulgurites are commonly long tubes, occasionally branching downward, from 1-3 cm in diameter. They have been consolidated by fusion of sand which is otherwise loose, allowing them to be readily distinguished from modern, unlithified sands. Such tubes have been traced downward as deep as 20 m, and their lower termination may appear as a bulbous irregular expansion of the tube. The tubes show an irregular, longitudinally striated surface like a bit of dried root, and they are often mistaken for fossil roots or stems (see Fig. 3 in *Eolianite*). The striation or corrugation apparently results from the sudden collapse of the tube, which is originally inflated by the hot gases at the time of fusion.

Under the microscope, the sand grains show degrees of melting to a completely isotropic glass, commonly with radially elongated bubbles (Fig. 1). Where the sand grains have only melted on their surfaces, they may be fused together or connected by minute filaments of glass resulting from sudden movement apart of grains in contact. The glass itself has been given the name *lechatelierite* (Lacroix, 1915), but it is not a distinct mineral species, its composition depending on that of the sand fused, as does a tektite. They are typically 90-99.5% silica (Frondel, 1962).

The distribution of contemporary sand fulgurites is clearly related to relatively dry, loose, quartz sand typical of deserts or well-drained sands. There would seem to be an optimum moisture content, for on the one hand if the sand is too wet the lightning discharge is conducted through a larger mass; and, as the water is converted to steam, the material is likely to explode and no structure preserved. On the other hand, the sand may be too dry to give adequate conductivity. In the Sahara, fulgurites are commonest in the (moister) hollows; and in temperate latitudes they are found in well-drained, raised sand.

Thousands of clearly recent occurrences of sand fulgurites have been described; the deposits in which they were formed however, have been as old as Cretaceous. Ancient or paleofulgurite has been identified in the New Red Sandstone desert dune deposits of Permian age in Arran, Scotland (Harland and Hacker, 1966). The sandstone is fully consolidated, and the tubes are very delicate, only made evident by

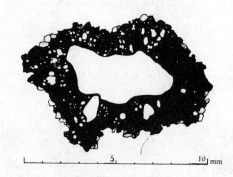

FIGURE 1. Enlargement of cross section of fulgurite showing vesicular texture (Harland and Hacker, 1966). The glass is colored black and sand grains are stippled.

FIGURE 2. Axial section of a fulgurite, showing the surface relief due to differential weathering, and the angle between the approximately vertical fulgurite and the dipping dune bedding (from Harland and Hacker, 1966). The intensity of stippling indicates the degree of red staining; × 2/3; natural horizontal scale.

differential weathering (Fig. 2). A penecontemporaneous pebble of such a paleofulgurite was also found.

The general *magnetic properties* of fulgurites have long been known from the effect on compass bearings near some mountain peaks. Paleomagnetic studies take into consideration the hazard that surface samples might have been remagnetized by recent storms. Such anomalous results can be recognized, for the orientation and relative intensity of the magnetic field around a fulgurite is distinctive, the latter being greater by two or three orders of magnitude. The magnetic field probably results from the electromagnetic effect rather than from its thermal consequence, as is clear from its wider distribution and the ease of its removal by an AC current, leaving the magnetic field characteristic of the rest of the rock.

W. B. HARLAND

References

Frondel, C., 1962. *The System of Mineralogy of J. D. and E. S. Dana,* 7th ed., Vol. 3, *Silica Minerals.* New York: Wiley, 334p.

Glover, J. E., 1974. Sand fulgurites from Western Australia, *J. Roy. Soc. W. Austral.,* **57,** 97-104.

Harland, W. B., and Hacker, J. L. F., 1966. Fossil lightning strikes 250 million years ago, *Advanc. Sci.,* **22,** 663-671.

Lacroix, A., 1915. La silice fondue considérée comme minéral (Lechatelierite, *Bull. Soc. Fr. Minéral.,* **38,** 182-186.

Lacroix, A., 1942. Les fulgurites du Sahara, *Bull. Serv. Des Mines,* Gouvern. A.O.F., **6,** 28-35.

Van Tassel, R., 1955. Fulgurites in situ a Zonhohoven (Limbourg belge), *Bull. Instn. Roy. Sci. Nat. Belg.* (5), **31**(9), 1-24.

Withering, W., 1790. An account of some extraordinary effects of lightning, *Phil. Trans.,* ser. B., **80,** 293-295.

Cross-references: *Remanent Magnetism in Sediments; Sand and Sandstones.* Vol. II: *Lightning.*

G

GASES IN SEDIMENTS

Types and Phases

Gaseous compounds found in sediments may be present in dissolved form, as a gas phase, or as a solid phase (in hydrate form) depending on abundance and on temperature and pressure. Usually, methane is the only gas produced in sediments in sufficient quantities to initiate formation of either a vapor or a hydrate phase. If such a phase is formed, however, other gases will partition among all phases present. For information on the physical chemistry of gas solubility, see Berner (1971) or a physical chemistry text.

Chemical and isotopic analysis of gaseous compounds is usually accomplished by gas chromatography or mass spectrometry. To extract gases from sediments for analysis, two techniques have commonly been used. One involves squeezing sediments and purging the expressed interstitial water with a carrier gas, and the other involves preparing a slurry of sediment and water and purging the slurry with a carrier gas. With both techniques, care must be taken to avoid contamination of the sample with atmospheric gases and also to avoid the loss of gas from bubbles which can form as pressure is reduced from in situ to one atmosphere.

Gaseous compounds observed in sediments may be grouped in two classes: (1) those reactive with water (CO_2, H_2S, NH_3); and (2) those nonreactive with water (N_2, O_2, CH_4, C_2H_6, He, Ne, Ar, Kr, Xe, Rn). The distribution and isotopic composition of these gases reflects the combined impact of diagenetic reactions and mass transport processes. Diffusion through interstitial waters (see *Interstitial Water in Sediment*) is primarily responsible for the migration of the first group. Migration of bubbles (if they are present) can also affect the second group.

Reactions Involving Gases

A simplified diagram (Fig. 1) illustrates the behavior of many gases in sediments. A sequence of zones can be defined in sediments on the basis of the biologically mediated reactions that can occur (see *Microbiology in Sedimentology*). The order of these zones corresponds to the free energy change that can be obtained by combining reduced organic matter with various oxygen sources (Claypool and Kaplan, 1974; Vol IVA: *pH-Eh Relations*). The thickness of each of these zones is extremely variable, depending on the environment, but is largely a function of the amount and type of organic material present. In the uppermost zone, dissolved oxygen is utilized, producing CO_2. Below this is a zone in which anoxic processes dominate as SO_4^{--} is utilized and CO_2, NH_3, and H_2S are produced. The majority of H_2S produced is probably precipitated, a fate that may be shared by part of the NH_3 and CO_2. The remaining NH_3 and CO_2 will diffuse along gradients in the interstitial water (as NH_4^+ and HCO_3^- at the pH normally found in sediments). Below this zone lies the region of methane production. The dominant mechanism of methane production in recent sediments (at low temperatures) is probably CO_2 reduction by bacteria. This methane can diffuse upward where it may be oxidized by methane-utilizing bacteria in one of the overlying zones. If sufficient methane accumulates, a gas phase can be formed. Portions of this methane-rich gas phase can escape from sediments by sedimentary flatulence, carrying with it significant fractions of gases of the second group. This process is very common in shallow-water sediments; but although some evidence supports the presence of gas phases in deep-sea sediments, flatulence from such environments has not been documented. In cold, deep environments, a solid hydrate phase may form rather than a gas phase.

A second mechanism for methane production can become important as temperature increases. Above $\approx 50°C$, thermal processes, like those associated with the maturation of hydrocarbons into petroleum (see *Petroleum—Origin and Evolution*), can become significant on time scales of 10^6 yr. The reaction rate increases exponentially, having an activation energy of ≈ 40 kcal/mole. Claypool and Kaplan (1974) have shown that methane associated with petroleum can be distinguished from bio-

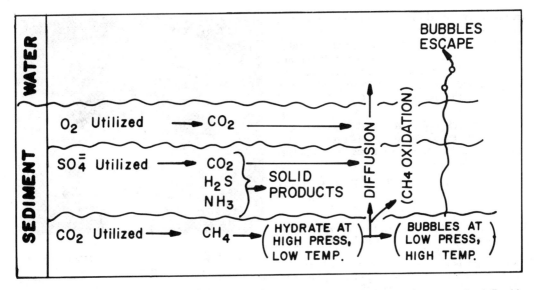

FIGURE 1. Schematic drawing of the behavior of gases in sediments. A sequence of zones can be defined in sediment on the basis of microbiology (see text). Methane is produced below the sulfate reduction zone and can diffuse into overlying zones, where it may be oxidized slowly by sulfate reducers, or more rapidly in the oxygenated zone. If sufficient methane accumulates, a gas phase (under warm, shallow waters) or a hydrate phase (under cold, deep waters) may form, with other gases partitioned among existing phases.

genic methane by two characteristics: (1) It is isotopically heavier than biogenic methane ($\delta C^{13} \approx -40$ to $-50‰$ vs. $\delta C^{13} \approx -60$ to $-80‰$ for biogenic methane). (2) The CH_4:C_2H_6 ratio for gases associated with petroleum is about 10–100, and biogenic gases have ratios of 10^3-10^4.

A second process, in addition to bubble migration, can generate gradients in noble-gas distribution. He^4, Ar^{40}, and Rn^{222} are all produced by radioactive decay. If the production rate of these gases is known, their distribution may be modeled to determine the rate of diffusion in sediments.

Gases and Sedimentary Structures

Gas hydrates, or *clathrates,* are solid compounds in which a cage-like structure of "host" water molecules is stabilized by "guest" molecules of gases at certain sites in a crystal lattice. High pressures and low temperatures favor their formation (Fig. 2), and sufficient methane must be present to exceed in-situ solubility (≈ 220 m mol/liter at 440 atm and 25°C). These compounds are believed to be a storage reservoir for large amounts of natural gas under permafrost areas.

In deep-sea sediments, it has been proposed that clathrates may serve as a mechanism for trapping large quantities of gas. Methane bubbles can migrate upward from a deep source,

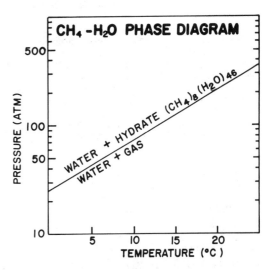

FIGURE 2. If sufficient methane accumulates in interstitial waters, either a hydrate or gas phase will form. Pressures and temperatures to the upper left of the phase boundary are required to form a hydrate; conditions to the lower right will produce a gas. Addition of 35% NaCl will shift the boundary about 0.5°C to the left. (After Deaton and Frost, 1946.)

until the geothermal gradient intersects the hydrate phase boundary, at which point the hydrate becomes stable. The abrupt transition from an upper layer of hydrate-present, bubble-free sediments to a lower layer of bubble-present, hydrate-free sediments would create a seismic reflector and acoustic turbidity which parallel topography and cut across other sedimentary structures. This argument has been used by Lancelot and Ewing (1972) to explain such a reflector on the Blake-Bahama Outer Ridge.

The escape from sediment of gases such as methane or gases associated with the origin and evolution of petroleum (q.v.) may produce mud volcanoes and related structures (see *Volcanism, Sedimentary*).

DOUGLAS E. HAMMOND

References

Berner, R. A., 1971. *Principles of Chemical Sedimentology*. New York: McGraw-Hill, 240p.

Claypool, G. E., and Kaplan, I. R., 1974. The origin and distribution of methane in marine sediments, in I. R. Kaplan, ed., *Natural Gases in Marine Sediments*. New York: Plenum, 99-139.

Deaton, W. M., and Frost, E. M., Jr., 1946. Gas hydrates and their relation to the operation of natural-gas pipe lines, *U.S. Bur. Mines Mono.*, 8, 101p.

Goldhaber, M., and Kaplan, I. R., 1974. The sulfur cycle, in E. D. Goldberg, ed., *The Sea*, vol. 5. New York: Wiley, 569-655.

Kaplan, I. R., ed., 1974. *Natural Gases in Marine Sediments*. New York: Plenum, 324p.

Lancelot, Y., and Ewing, J. I., 1972. Correlation of natural gas zonation and carbonate diagenesis in Tertiary sediments from the north-west Altantic, *Init. Repts. Deep Sea Drilling Project*, 11, 791-799.

Vilks, G.; Rashid, M. A.; and van der Linden, W. J. M., 1974. Methane in recent sediments of the Labrador shelf, *Canad. J. Earth Sci.*, 11, 1427-1434.

Whelan, T., 1974. Methane, carbon dioxide and dissolved sulfate from interstitial water of coastal marsh sediments, *Estuarine Coastal Marine Sci.*, 2, 407-416.

Cross-references: *Diagenesis; Interstitial Water in Sediments; Microbiology in Sedimentology; Organic Sediments; Petroleum—Origin and Evolution; Volcanism, Sedimentary.*

GEOCHEMISTRY OF SEDIMENTS, ANCIENT AND MODERN
—*See* Vol. IVA

GEODES

Geodes are discrete bodies of mineral matter, commonly spherical, hollow, and lined inside with crystals of various minerals (Fig. 1). Geodes necessarily originate from cavities in rock. Deposition of minerals proceeds *inward* from the cavity wall, producing a characteris-

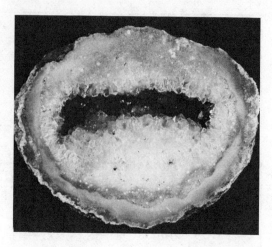

FIGURE 1. Geode from shale of Warsaw Formation of SE Iowa. Outer rim of specimen is siliceous, whereas the cavity is nearly filled with quartz crystals; polished surface.

tic drusy structure, so that the youngest mineral generation is near the center and may fill the void completely. Thus, geodes differ from concretions (q.v.), which grow *outward* by accretion about an original nucleus. In a geode, the mineral layers lining the cavity must be more resistant to weathering and erosion than the host rock, so that upon weathering, a discrete, hollow mineral body is released. Not every crystal-lined cavity, or *drusy vug*, in rock is a geode. Van Tuyl (1925), Hayes (1964), and Chowns and Elkins (1974) discuss geodes, and include several references to important examples.

Mineralogy

Most geodes have siliceous outer shells; the physical and chemical durability of silica is responsible for the discreteness and preservation of such geodes. Geodes with outer shells of calcite or iron oxides and hydroxides are less common. A wide variety of minerals occur as euhedral crystals in geode cavities. In the well-known geodes from the Mississippian Warsaw Formation near Keokuk, Iowa (Hayes, 1964), at least fifteen distinct minerals have been identified: aragonite, barite, calcite, chalcopyrite, dolomite, goethite, gypsum, kaolinite, malachite, marcasite, millerite, pyrite, quartz, smithsonite, and sphalerite. Because a cavity in buried rock represents a place of low free energy, minerals that are dispersed throughout the host rock or in solution in pore fluids can be expected to precipitate as euhedra in geode cavities.

FIGURE 2. Enhydrites, or water-stones, from Victoria, Australia. (a) Broken specimen showing hollow interior; (b) Another specimen showing boxwork interior. (Photos: courtesy of Victoria Dept. of Mines.)

Genesis

Any cavity in rock is potentially the site of geode formation. In a particular geode occurrence, the key question is the origin of the cavity. Geodes have formed in vesicles of lava flows, such as those of the Columbia Plateau. The "thunder eggs" and agates of that region are related phenomena. Miarolitic cavities in plutonic igneous rocks, the original voids or body cavities of fossils, and a host of solution cavities in sedimentary rocks have given rise to geodes. Although Pettijohn (1957) cited evidence of expansion during the formation of geodes, cavities in which the famous Warsaw geodes formed resulted from solution of calcareous concretions in the less soluble mudstones of the Warsaw Formation (Hayes, 1964). In a study of all geode occurrences related to the Keokuk geodes, Chowns and Elkins (1974) concluded that the geodes are pseudomorphs after early diagenetic anhydrite; this *silicification* (q.v.) involves both replacement and void filling. They reviewed geode occurrences of the United States, found a strong correlation with sulfate evaporites (interpreted as sabkha deposits), and suggested that geodes be redefined as sedimentary nodules with drusy internal structure formed as pseudomorphs after earlier concretions, especially calcium sulfate concretions.

Enhydrites

Popularly known as "water-stones," *enhydrites* (*enhydros*) are chalcedony geodes containing water. The water is gradually lost by diffusion, and the interior is then gradually filled with quartz. The best-known examples are found in Australia, at Beechworth, Victoria (Fig. 2). Other occurrences have been reported from North Carolina and Idaho (Lindgren, 1900), South Africa and New Zealand (Dunn, 1913). Specimens vary in shape and form and are about 1–10 cm in size, with walls from 0.1–5.0 mm thick. The chalcedony is laminated along flat planes parallel to the bounding surfaces (Fig. 3). It is fibrous, parallel to the surface near the contact, but radial inside. There are often layers of clay between the laminae and in the interior, and in some enhydrites there is boxwork. Enhydrites are evidently epigenetic; they are not pseudomorphs, but are cavity fillings between the pre-existing faces of pegmatitic feldspar, calcite, and other minerals now removed by deep weathering.

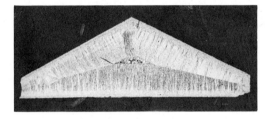

FIGURE 3. Section through a chalcedony-filled enhydrite, showing parallel lamination against its walls with radial system within. (Photo: courtesy of Victoria Dept. of Mines.)

JOHN B. HAYES
PAUL C. FRANKS

References

Boutakoff, N., and Whitehead, S., 1952. Enhydros or water-stones, *Mining and Geol. J.*, 4(5), 14–18.

Chowns, T. M., and Elkins, J. E., 1974. The origin of quartz geodes and cauliflower cherts through the silicification of anhydrite nodules, *J. Sed. Petrology*, 44, 885–903.

Dunn, E. J., 1913. The Woolshed Valley, Beechworth, *Bull. Geol. Surv. Vict.*, 25, 1–20.

Hayes, J. B., 1964. Geodes and concretions from the Mississippian Warsaw Formation, Keokuk region,

Iowa, Illinois, Missouri, *J. Sed. Petrology,* **34,** 123–133.
Lindgren, W., 1900. The gold and silver veins of Silver City, De Lamar, and other mining districts of Idaho, *20th Ann. Rep., U.S. Geol. Surv.,* pt. 3, 75–256.
Pettijohn, F. J., 1957. *Sedimentary Rocks,* 2nd ed. New York: Harper & Row, 718p.
Van Tuyl, F. M., 1925. The stratigraphy of the Mississippian formations of Iowa, *Iowa Geol. Surv. Ann. Rept.,* **30,** 33–349.

Cross-references: *Concretions; Diagenesis; Nodules in Sediments; Sabkha Sedimentology; Silicification.*

GEOPETALITY—See TOP AND BOTTOM CRITERIA

GEOSYNCLINAL SEDIMENTATION

The word *geosyncline,* strictly speaking, means a large downwarped surface. The concept was introduced by James Hall (1859) as part of a working hypothesis for the origin of mountains; and although the role of the geosyncline in orogenesis has been largely rejected (Dickinson, 1970), the innumerable observations that have been made using geosynclinal models are significant in many disciplines, especially in sedimentology (Dott and Shaver, 1974). Geosynclines are (almost always) centers of deposition.

History

To James Hall, the downwarping was related to the accumulation of an unusually thick sedimentary sequence, so that the line of maximum depression coincided with the line of maximum accumulation. Such a sequence in the Appalachians includes formations that contain abundant littoral faunas, which led Hall to conclude that geosynclinal sediments accumulated in shallow waters in environments similar to their coeval deposition on a tectonically stable shelf.

The European geologists studying the Alps took issue with Hall. Suess (1875) first advocated the idea that geosynclinal sediments represent a pelagic facies. He cited the Triassic of the eastern Alps, however, now known to be a tidal-flat complex. He also erred in interpreting Mesozoic benthonic faunas of the Alps as relic Paleozoic forms which had survived in oceanic depths. Neumayr (1875), however, was more successful with an actualistic approach to paleobathymetric analysis (q.v.). He correctly compared the Mesozoic radiolarian cherts of the Alps to Holocene radiolarian oozes of the equatorial Pacific and Indian oceans. Neumayr's belief that the Tertiary radiolarian chert of Barbados had been uplifted from a deep ocean bottom could be considered proven as a result of JOIDES drilling (Bader et al., 1970).

Despite an attempted conciliation by Walther (1897) the American-European schism persisted, exemplified by Haug's classic study at the turn of the century. In mapping the ammonitic black shales of the *dauphinois* Alps, Haug (1900) was impressed with the facies contrast between the shales and the shelf carbonate sequence in the Jura and Provence (SE France), suggesting two entirely different environments of deposition. Haug considered the geosynclinal sediments the bathyal (80 or 100 to 900 m) equivalent of neritic stable-shelf deposits. His postulate was based upon: (1) the absence of unconformities within a geosynclinal sequence; (2) the presence of pelagic fossils in geosynclinal sediments; and (3) the monotonous, shaly geosynclinal succession which suggests deposition at bathyal depth, in contrast to the (neritic) shelf sediments which show great vertical and lateral facies variations in response to transgressions and regressions. He cited graptolitic shale, *Posidonomia* shale, *Aptychus* shale, and other formations containing pelagic fossils as examples of typical geosynclinal deposits.

Haug's scheme was supported by Jones (1938), who contrasted the geosynclinal graptolite facies with the stable-shelf shelly facies in Britain's Paleozoic. Other British geologists characterized geosynclinal sediments by their graywacke lithology (Tyrrell, 1933) or by their graded bedding (Bailey, 1936).

If nature had cooperated so that sediments *thicker* than their coeval sequences were always deposited in *deeper* waters, Haug's new connotation would not lead to difficulty. Exceptions, however, such as the thin Middle to Upper Devonian Chattanooga Shale (eastern US), which was deposited in much deeper waters than its coeval, thick deltaic Catskill and prodeltaic Chemung Formations of the Appalachian geosyncline led to further difficulty. One had either to adhere to Hall's original definition or to select bathymetry as a new criterion. The Americans stood by Hall, the Europeans followed Haug, resulting in the so-called "American" and "European" concepts of geosynclines (see Trümpy, 1960; Aubouin, 1965). To conform with the European concept, a "lepto-geosyncline" had to be fabricated to designate deep-sea deposits that are thinner than their shelf equivalents.

The Geosynclinal Sequence

Despite the lack of consensus on the definition of geosyncline, there is one common denominator underlying all schools of thought: *geosyncline* implies mobility, and this mobility is represented by an unusually high subsidence rate, giving rise to either unusually thick sedimentary sequences or unusually deep-water deposits. The original concept proposed by Hall and developed by Dana (1873) was designated to emphasize the existence of such mobile zones within the earth's crust. Haug (1900), Schuchert (1923), and Bucher (1933) all portrayed geosynclines as long linear belts, mediterranean, or marginal to continents. Under different geographical settings, this mobility has found manifold expressions, resulting in the wide varieties of sedimentary prisms. Classic works by Stille (1940) and Kay (1951) attempted to characterize those various sequences, leading to a complicated nomenclature which emphasized the spatial variation of the geosynclinal mobility. The question of whether variations in tempo and mood of this mobility existed and whether any pattern or rhythm in the geosynclinal sedimentary sequence could be detected was answered in the affirmative by Bertrand (1897) through his recognition of the flysch-molasse succession.

Bertrand (1897) innovated the concept of recurrent facies, established the principle of geosynclinal cycles, and presented evidence for the ensialic origin of geosynclines. He recognized four "facies" or *formations de montagne* which constitute a geosynclinal cycle (see Fig. 1):

1. *gneiss*, which constitutes the basement of a primary geosyncline;
2. *flysch schisteux* (= *schistes lustrés*), which filled up the central zone of a primary geosyncline;
3. *flysch grossier* (= *flysch*), which filled up the border of the geosyncline after the elevation of the central axis; and
4. *grès rouges* (= *molasse*), which was developed at the foot of the newly elevated mountain chain.

The subsequent development on this theme mirrored the trend of sedimentological research in the twentieth century. In the first two decades, when embryotectonics (Argand, 1916) was new and fashionable, the tendency was to relate sedimentation to geosynclinal growth. Arbenz (1919) recognized an "epeirogenic facies" of cyclothemic deposits in the Helvetic Alps, followed by an "orogenic facies" of flysch and molasse. Those facies are not defined on the basis of their sedimentary environments but are manifestations of the mood and tempo of tectonic mobility.

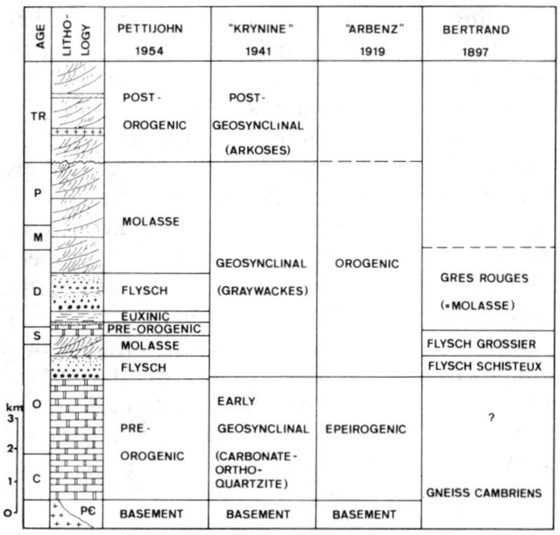

FIGURE 1. Evolution of the concept of geosynclinal cycles: The Appalachian sedimentary sequence may serve as an illustration. Bertrand (1897) referred to this section; although his interpretation, based upon the existing literature, necessarily contained some grave errors, he did grasp the essence of geosynclinal development. Arbenz (1919) did not actually discuss the Appalachians, but he probably would have come to this conclusion had he visited North America. Similarly, the column styled after Krynine is a portrayal of his views (Hsü, 1973). The last column is from Pettijohn (1957).

The terms *flysch* and *molasse* and the concept of orogenic facies were introduced to North America in 1931 by Waterschoot van der Gracht, who described flysch as "sediments deposited previous to the major paroxysm" and molasse as sediments deposited "during or immediately after the major diastrophism." The preoccupation of geologists with orogenic chronology resulted in the emphasis of temporal relationships between flysch, molasse, and orogeny. Adding *pre-, syn-,* and *post-* to the term *orogenic facies* was predestined to cause confusion when the "one big bang" concept of orogeny was disproven.

The growth of sedimentary petrography during the 1920s led Jones (1938) to attempt a petrographic characterization of geosynclines, and graywacke was singled out as its essential deposit. This idea was supported by Krynine (1941, 1942), Pettijohn (1943), and Krumbein et al. (1949). Krynine divided geosynclinal stages into three cycles (see Fig. 1): "(1) Peneplanation (or nearly geosynclinal) stage: cyclic deposition on fluctuating flat surface after much weathering, characterized by *first-cycle quartzites*. . . . (2) Geosynclinal stage: basinal deposition interrupted by local vertical buckling.... Typical sediment: *graywackes*. ... (3) Post-geosynclinal stage: uplift ... after folding and magmatic intrusion of geosyncline. Typical sediments: *arkoses*." Krynine found his model in the Appalachians. Yet the situation in the Alps was quite different: there the "geosynclinal" flysch sediments are arkosic, having been derived from an old granitic landmass, whereas the "postorogenic" molasse detritus are largely rock fragments that came from the cover-nappes topping the nearby elevated mountains. Although mineralogical composition is certainly an important attribute of a rock, it is now widely recognized that the mineralogy depends not only upon tectonic mobility but also upon too many other complex factors to be an infallible indicator of geosynclinal development.

The significance of sedimentary structures, specifically cross-bedding and graded beds, to tectonics and sedimentary facies was noted by Bailey in 1930. Brinkmann (1933), Cloos (1938), and Migliorini (1943) also noted the paleogeographic implications of those structures. In the US, Pettijohn (1943) related Bailey's "graded facies" and Jones' "graywacke suite" to the Alpine "flysch-type" sediments. Through his work on the Archean rocks of the Lake Superior region, he further noted a kinship between flysch and eugeosynclinal sedimentation.

A flood of flysch research followed Kuenen and Migliorini's (1950) milestone paper (see Walker, 1973) on turbidity currents. In their studies of the Appalachians, Pettijohn (1957) and his students spearheaded an entirely new approach to basin analysis (see Potter and Pettijohn, 1963). Furthermore, their carefully compiled data permitted a new appraisal of the concept of the geosynclinal cycle. Pettijohn (1957) emphasized that the main influence of tectonism "is the effect on the total sediment supply and the rate of subsidence." Following Krynine, he recognized a pregeosynclinal (or early geosynclinal) stage of orthoquartzite-carbonate sedimentation. Euxinic facies (q.v.) or flysch facies followed, as tectonic mobility accelerated. The sedimentary record is, then, "essentially a record of basin filling," culminating in molasse deposition (see Fig. 1). This analytical approach enabled him to perceive a polycyclic geosynclinal development in the Appalachians, which could be related to the multiple orogenic events there.

The traditional approach culminated in the theory of Aubouin (1965), who attempted to represent geosynclines by concrete time-space models describing regular sequences of events. He particularly emphasized the presence of eu- and miogeosynclinal couplets in the tectonic development of ancient mountains. During the last decade, however, there has been a tendency to interpret geosynclines in the framework of plate tectonic theory (e.g., Dickinson, 1971; Dewey and Bird, 1971). In the view of plate-tectonicians, geosynclines characterize tectonic settings of sedimentary sequences and their associated rocks existing at the present continental or plate margins: the Pacific type of geosyncline represents the setting at converging plate margins, the Atlantic type describes stable continental margins within plates, and the Mediterranean type denotes intercontinental realms of sedimentation. Reviews of the classical theory and new interpretations are among the papers included in a symposium on geosynclinal sedimentation (Dott and Shaver, 1974).

KENNETH J. HSÜ

References

Arbenz, P., 1919. Probleme der Sedimentation und ihre Beziehungen zur Gebirgsbildung in den Alpen, *Naturf. Gesell. Zurich, Vierteljahresschrift, Jahrg.* 64, 246-275.

Argand, E., 1916. Sur l'arc des Alpes Occidentales, *Eclogae Geol. Helv.*, 14, 145-191.

Aubouin, J., 1965. *Geosynclines*. Amsterdam: Elsevier, 335p.

Bader, R. G., et al., 1970. *Initial Reports of the Deep Sea Drilling Project*, 4. Washington, D.C.: U.S. Govt. Printing Office, 753p.

Bailey, E. B., 1930. New light on sedimentation and tectonics, *Geol. Mag.*, 67, 77-92.

Bailey, E. B., 1936. Sedimentation in relation to tectonics, *Geol. Soc. Am. Bull.*, 47, 1713-1726.

Bertrand, M., 1897. Structure des Alpes françaises et récurrence de certain faciès sédimentaires, *VI Int. Geol. Congr., Zurich, 1894, Compt Rend.*, 163-177.

Brinkmann, R., 1933. Ueber Kreuzschichtung im deutschen Buntsandsteinbecken, *Nachr. Wiss. Göttingen, Math.-physik. kl.*, 1933, 1-2.

Bucher, W. H., 1933. *The Deformation of the Earth's Crust*. Princeton: Princeton Univ. Press, 518p.

Cloos, H., 1938. Primiäre Richtungen in Sedimenten der rheinischen Geosynklinale, *Geol. Rundschau*, 29, 357-367.

Dana, J. D., 1873. On some results of the earth's contraction from cooling, including a discussion of the origin of mountains and the nature of the earth's interior, *Am. J. Sci.*, ser. 3, 5, 423-443; 6, 6-14, 104-155, 161-172.

Dewey, J. F., and Bird, J., 1971. Geosynclines in terms of plate-tectonics, *Tectonophysics*, 40, 625-635.

Dickinson, W. R., 1970. Second Penrose Conference: the new global tectonics, *Geotimes*, 15(4), 18-22.

Dickinson, W. R., 1971. Plate-tectonic models of geosynclines, *Earth Planetary Sci. Lett.*, 10, 165-174.

Dott, R. H., Jr., and Shaver, R. H., eds., 1974. Modern and ancient geosynclinal sedimentation, *Soc. Econ. Paleont. Mineral. Spec. Publ. 19*, 380p.

Fisk, H. N., 1952. Sedimentation and orogeny with particular reference to the Gulf Coast geosyncline, *Geol. Soc. Am. Bull.,* **63** (abstr.), 1328.

Glaessner, M. F., and Teichert, C., 1947. Geosynclines: A fundamental concept in geology, *Am. J. Sci.,* **245**, 465-482, 571-591.

Hall, J., 1859. Paleontology, *N.Y. Geol. Surv.,* **3**(1), 66-96.

Haug, E., 1900. Les geosynclinaux et les aires continentales, *Soc. Geol. France Bull.,* ser. 3, **28**, 617-711.

Hsü, K. J., 1973. The odyssey of geosyncline, in R. N. Ginsburg, ed., *Evolving Concepts in Sedimentology.* Baltimore: Johns Hopkins, 66-92.

Jones, O. T., 1938. On the evolution of a geosyncline, *Quart. J. Geol. Soc. London,* **94**, lx-cx.

Kay, M., 1951. North American geosynclines, *Geol. Soc. Am. Mem.* **48**, 143p.

Knopf, A., 1948. The geosynclinal theory, *Geol. Soc. Am. Bull.,* **59**, 649-670.

Krumbein, W. C.; Sloss, L. L.; and Dapples, E. C., 1949. Sedimentary tectonics and sedimentary environments, *Bull. Am. Assoc. Petrol. Geologists,* **33**, 1859-1891.

Krynine, P. D., 1941. Differentiation of sediments during the life history of a landmass, *Geol. Soc. Am. Bull.,* **52** (abstr.), 1915.

Krynine, P. D., 1942. Differential sedimentation and its products during one complete geosynclinal cycle, *First Pan-Am. Cong. Mining Eng. Geol.,* **2**, pt. 1, 537-561.

Kuenen, P. H., and Migliorini, C., 1950. Turbidity currents as a cause of graded bedding, *J. Geol.,* **58**, 91-127.

Migliorini, C., 1943. Sul modo di formazione dei complessi tipo macigno, *Boll. Soc. Geol. Ital.,* **62**, xlviii.

Neumayr, M., 1875. *Erdgeschichte,* 1. Leipzig: Bibliographisches Inst., 364p.

Pettijohn, F. J., 1943. Archean sedimentation, *Geol. Soc. Am. Bull.,* **54**, 925-972.

Pettijohn, F. J., 1957. *Sedimentary Rocks.* New York: Harper & Row, 718p.

Potter, P. E., and Pettijohn, F. J., 1963. *Paleocurrents and Basin Analysis.* Berlin: Springer, 296p.

Schuchert, C., 1923. Sites and natures of the North-American geosynclines, *Geol. Soc. Am. Bull.,* **34**, 151-260.

Stille, H., 1940. *Einführung in den Bau Amerikas.* Berlin: Borntraeger, 717p.

Seuss, E., 1875. *Die Entstehung der Alpen.* Wien: W. Braunmüller, 168p.

Trümpy, R., 1960. Paleotectonic evolution of the central and western Alps, *Geol. Soc. Am. Bull.,* **71**, 843-908.

Tyrrell, G. W., 1933. Greenstones and graywackes, *Reun. Internat. pour l'Etude du Précamb. et des Vieilles Chaines de Montagnes en Finlande, Comp. Rend.,* **1931**, 24-26.

Waterschoot van der Gracht, W. A. J. M. van, 1931. Permo-Carboniferous orogeny in the south-central United States, *Bull. Am. Assoc. Petrol. Geologists,* **15**, 991-1057.

Walker, R. G., 1973. Mopping up the turbidite mess, in R. N. Ginsburg, ed., *Evolving Concepts in Sedimentology,* Baltimore: Johns Hopkins, 1-37.

Walther, J., 1897. Ueber Lebensweise fossiler Meeresthiere, *Z. Deutsch, Geol. Ges.,* **49**, 209-273.

Cross-references: *Arkose, Subarkose; Graded Bedding; Graywacke; Paleocurrent Analysis; Turbidites.*

GLACIAL GRAVELS

Glacial gravel is an accumulation of rock and mineral clasts coarser than 2 mm in diameter that have been derived by glacial erosion of underlying rock, or by rockfalls onto a glacier. Glacial quarrying produces most gravel, particularly from hard rocks with closely spaced joints. Abrasion and frost wedging may also produce gravel.

Gravel-bearing glacial deposits on land, in lakes, or in the sea include: till deposited directly by a glacier; outwash and valley trains deposited by glacial streams; avalanches, debris flows, and stream sediments deposited upon or adjacent to a glacier; lag gravels or boulder pavements derived by postglacial sheet erosion of a till surface; free boulders deposited individually or within matrix that has since been removed; deposits from beneath floating or grounded glaciers and ice shelves, and rafted upon icebergs (Flint, 1971).

Grain Size and Composition. It is known that there is a wide range of sizes of clasts comprising glacial gravel. Till gravels are more poorly sorted than individual beds of outwash gravel (Fig. 1) because of their different modes of formation. Any lithology, or large minerals, may comprise glacial gravel. Gravels commonly are composed of a variety of lithologies. Gravel in till is usually locally derived, so clast composition varies with composition of underlying or nearby bedrock. Clasts of a different composition than that of underlying bedrock are called *erratics.*

Gravel in basal till is contained within a matrix of finer-sized particles (Dreimanis, 1969), but a matrix is sparse or absent in outwash gravel. In basal till, the proportion of gravel exceeds that of matrix near the source (Dreimanis, 1969; Slatt, 1971). With increasing distance of transport, the proportion of matrix increases due to abrasion and crushing of clasts (Dreimanis, 1969). Beyond a transport distance of 25-50 km, most gravel clasts are reduced to finer sizes (Dreimanis, 1969; Drake, 1972), but erratics have been found up to 1200 km from their source (Flint, 1971).

Fabric. Elongate clasts in basal till and outwash are often aligned with their long axis par-

GLACIAL GRAVELS

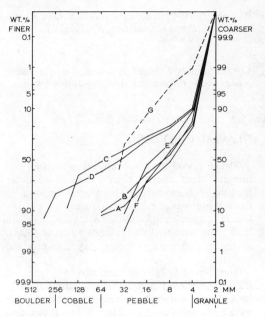

FIGURE 1. Grain-size distribution of gravel fraction of glacial till (solid lines) and outwash (dashed line) composed of clasts of dolomite (A), shale (B), phyllitic graywacke (C), quartz-mica schist (D), igneous, metamorphic, and sedimentary rocks (E), granitic and metamorphic rocks (F), and gneiss (G). Curves A and B recalculated from Dreimanis (1969) so as to exclude particles finer than 2 mm in diameter. Curves C, D, and G are unpublished analyses, and E and F are recalculated from Slatt (1971).

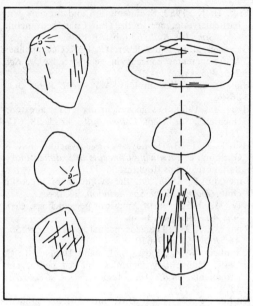

FIGURE 2. Outlines of flatiron-shaped gravel exhibiting striae, facets, and radiating fractures (modified from Pettijohn, 1957, and Flint, 1971). Clast on the right is viewed in each of three dimensions.

FIGURE 3. Progressive abrasion of limestone pebbles during glacial transport (courtesy Chauncy D. Holmes; also in Flint, 1971). Samples collected from basal till (1) 5 km, (2) 13 km, and (3) 19 km downstream from the outcrop of limestone bedrock.

allel to the direction of transport, and inclined upstream. A secondary transverse orientation is often preserved in till (Dreimanis, 1969).

Shape. "Flatiron-" (pentagonal-) shaped clasts (Fig. 2) are generally considered diagnostic of a glacial origin (see *Faceted Pebbles and Boulders*). Spheres are the most stable shape of transported gravel clasts in till, followed by discs and rods, then blades (Drake, 1972). Near their source, gravel clasts in till are angular or subangular (Slatt, 1971), but roundness increases with distance of transport due to repeated crushing and abrasion (Fig. 3; Drake, 1972). Outwash gravel clasts are more rounded than those in till, and roundness increases downstream. Rounded outwash gravel may be present in till if the outwash has been overridden by an advancing glacier (Slatt, 1971).

Surface Marks. Subparallel to random striae, facets and radiating fractures (Fig. 2) on gravel clasts are generally considered diagnostic of a glacial origin (Flint, 1971). Chatter marks and crescentic gouges may also occur. These surface marks are most common on softer rocks. About 10% of gravel clasts in till contain striae (Flint, 1971). Diagnostic glacial marks are not generally present on outwash gravel.

ROGER M. SLATT

References

Drake, L. D., 1972. Mechanisms of clast attrition in basal till, *Geol. Soc. Am. Bull.*, 83, 2159-2166.
Dreimanis, A., 1969. Selection of genetically significant parameters for investigations of tills, *Zesz.*

Nauk. Univ. A. Mickiewicza, Geografia, Poznan, Poland, 8, 15-29.

Flint, R. F., 1971. *Glacial and Quaternary Geology.* New York: Wiley, 892p.

Pettijohn, F. J., 1957. *Sedimentary Rocks.* New York: Harper & Row, 718p.

Slatt, R. M., 1971. Texture of ice-cored deposits from ten Alaskan valley glaciers, *J. Sed. Petrology,* **41**, 828-834.

Cross-references: *Beach Gravels; Faceted Pebbles and Boulders; Glacial Sands; Glacigene Sediments; River Gravels; Till and Tillite.*

GLACIAL MARINE SEDIMENTS

Glacial marine (or glaciomarine) sediment is composed mainly of material that has been transported from land or shallow water by floating ice and later dropped to the bottom when the ice melted. The ice-rafted debris varies in size from boulders to clay and may be mixed in varying proportions with almost any other type of fairly slowly accumulating marine sediment; or it may make up the bulk of sea-floor sediment, forming a till.

Charles Darwin (1842) was among the first scientists to realize the widespread abundance of sediments deposited from the melting of floating ice masses. From observations he made in eastern Tierra del Fuego, Darwin concluded, "it appears that masses of floating ice by which fragments of rock are conveyed are produced in two ways, and under circumstances considerably different although often acting together, namely, by the breaking off of icebergs from glaciers descending into the sea, and by the actual freezing of the surface of the sea or its tributary streams."

Since Darwin, discoveries of glacial marine sediments on the sea floor have been made by Phillipi (1912) around Antarctica, Kindle (1924) and Bramlette and Bradley (1940) in the Arctic and North Atlantic, and Menard (1953) in the NE Pacific. Carey and Ahmad (1960) suggest that there are two radically different types of floating glacier that produce two different types of glacial marine sedimentation, according to whether meltwater is present or absent at the base of the glacier. They define several distinctive environments (Fig. 1) of glacial marine sedimentation in relation to a glacier that extends to the sea, as follows: (1) *terrestrial;* where the base of the glacier is above sea level; (2) *grounded shelf;* where the base of the glacier is below sea level, but not floating; (3) *floating shelf;* where the glacier is floating; (4) *inner iceberg zone;* from the ice-barrier to the limit of winter pack ice; and (5) *outer iceberg zone;* beyond the limit of winter pack ice, but within the limit of icebergs. A wet-base glacier may grade into a dry-base glacier farther inland, but it generally rests on a pile of sediment that is continually being deposited under the ice.

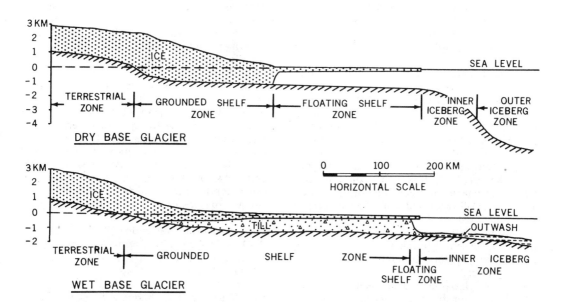

FIGURE 1. Sections through typical dry-base and wet-base glaciers in Antarctica (after Cary and Ahmad, 1960). The dry-base glacier is a section through the Pencksokka glacier of Dronning Maud Land and the wet-base glacier a profile of the Ross Barrier.

Modern Sediments

The lower, sediment-rich layers of the wet-base glacier generally drop their entrained sediment quickly; by the time the ice reaches the ice barrier where icebergs are calved, little sediment may be left. Hence, at the front of wet-base glaciers, the iceberg-zone sediments are characterized by silt, sand, and clay deposited by foreset till flows and containing very few dropped erratics.

The dry-base glacier carries most of its sediment to the buoyancy line, and erratics and clay are frozen into the floating ice shelf, which is generally many times wider than the buoyant shelf zone of the wet-base glaciers. The sediments formed beneath this ice shelf usually consist of muds with abundant dropped erratics (*dropstones*). Care should be taken in the interpretation of infrequent dropstones, which may be biologically transported.

Detritus is also frequently trapped during the freezing of river or sea ice onto river banks, beaches, and arid rocky coasts, or it can be windblown onto floating ice. The nature of the detritus varies considerably. Commonly it is very poorly sorted, containing fragments from boulder to clay size, many of which may be fresh and angular, particularly if they have been produced by freezing onto fresh bedrock. However, the occurrence of sand and pebbles that have been rounded before finally being entrapped in ice which moved out into the sea may be just as common. On tidal shores, the freezing process is vastly accelerated by the diurnal rise and fall of the sea, whereby the "ice foot" forming at low tide tears up the bedrock as it buoyantly rises with the next tide. The same tidal action often transports ice-foot material seaward as ice floes, and the process is repeated in later cycles.

In general, the best criterion that can be used for recognizing glacial marine sediments is the occurrence of a poorly sorted mixture of erratic fragments or fresh angular dropstones in a marine sediment. For instance, the occurrence of large blocks of granitic rocks in a deep-sea ooze deposited far from land could be considered as unequivocal evidence of ice rafting, although small sizes and amounts could possibly be rafted by floating kelp, or could be carried in the stomachs of large marine animals. An ancient equivalent may be a radiolarian chert or foraminiferal marl containing erratic material.

When the glacial marine sediment is a poorly sorted and nonstratified deposit, it may be mistaken for terrestrial tills or mudslide deposits or submarine debris flows. In these instances, it will be necessary to check for any evidence of a marine environment in the former cases, or the proximity of a glaciated land mass within the source area in the latter (see *Diamictite; Pebbly Mudstones*).

The widespread distribution of glacial marine sediments has been revealed by recent investigations of the ocean floors using coring and photographic techniques, indicating that they were deposited over almost one third of the ocean floor during the Pleistocene, reaching from polar latitudes to $20°N$ and $30°S$ in the respective hemispheres. The intensity of ice rafting was greatest in the intermediate high latitudes ($60-45°$), and the composition of the debris generally reflects the composition of the rocks of the closest glaciated continent. For instance, the glacial marine sands and gravels covering the sea floor off Labrador consist mainly of fragments of gneiss and schist from the adjacent Canadian Shield, whereas the rafted rocks off Iceland are commonly volcanic.

In the North Pacific, ice-rafted debris is most abundant in the Gulf of Alaska, where it forms a glacial marine remnant in the surficial sediment and is known to outcrop on Middleton Island. Farther west, S of the Aleutian Island Arc, the amount of ice-rafted debris deposited during the Pleistocene makes up only minor proportions of the Pleistocene diatom oozes and gray muds (Kent et al., 1971). Ice-rafted sand and gravel also occur in abundance in the Bering Sea near Kamchatka and are known to cover most of the floor of the Arctic Ocean.

One of the largest areas of glacial marine sediment now occurs surrounding Antarctica. This sediment was deposited throughout the Quaternary and probably earlier (Blank and Margolis, 1975), and is still being deposited today. A wide zone of almost pure glacial marine sediment occurs outward from the continental shelf to the limit of the winter pack ice. North of this limit of the Polar Front, the glacial marine sediments are generally dispersed within diatom ooze. North of this zone of diatom ooze, to about $40-50°S$, the pelagic sediments are generally calcareous ooze at depths $<4,100-4,700$ m and abyssal clays in greater depths; these sediments contain minor amounts of ice-rafted debris. Northward displacement of the ice-debris limit occurred with every glaciation crescendo (Watkins et al., 1974), and large banks of erratics were dumped, for example, off South Africa (about lat. $32°S$; Needham, 1962).

Ice-rafted detritus occurs in abundance throughout the diatom oozes S of the Polar Front (Conolly and Ewing, 1965), indicating continuity of glaciation in Antarctica. A few cores have reached a depth below which no ice-rafting material is found. Similar zones of ice-rafted debris occur in deep-sea cores in the North Atlantic (Bramlette and Bradley, 1940), and boulders are found as far south as the Canaries on the E side, and the Blake Plateau on the W.

Many glaciomarine sands are mechanically weathered; they are, therefore, commonly rich in fresh feldspars, carbonates, and other lithic clasts that otherwise tend to break down under chemical weathering. In favorable sites, such as fjords and bays adjacent to forested slopes, organic matter encourages bacterial activity in the interstitial (connate) waters, and rapid hydrodiagenesis is facilitated (Sharma, 1970).

Ancient Glacial Marine Sediments

There is evidence that there were several major glacial periods prior to the Pleistocene; several in the Precambrian, one in the Ordovician, and one in the Permo-Carboniferous. It is strange that there are very few well-documented occurrences of ancient glacial marine sediments associated with these glaciations. Late Precambrian glacial marine sediments in Scandinavia (Finnmark) lie on a striated rock floor and contain boulders that may have been ice rafted, but very few other Precambrian glacial sediments have been recognized as having a specifically glacial marine origin. There is probable ice-foot erosion and glaciomarine transport of some of the Saharan Ordovician tillites (Fairbridge, pers. comm.), although glacial grooving of the continental shelf sediments is also widespread (see *Continental-Shelf Sediments in High Latitudes*).

Du Toit (1921) sketched the distribution of the Permo-Carboniferous glaciation in the ancient continent of "Gondwanaland," showing the probable occurrence of Carboniferous glacial marine sediments in South America, South Africa, Antarctica, and Australia. Du Toit concluded that the boulder shale and boulder mudstone of the Cape Province in South Africa were glacial marine sediments deposited during the retreat of the Carboniferous ice sheet. Near Capetown, there are also glaciomarine deposits of Ordovician age. More recently, Permo-Carboniferous glacial marine sediments have been described from several localities in Australia (Carey and Ahmad, 1960; Whetten, 1965). Reconstructions of the distribution of Permo-Carboniferous glacial sediments have been made (published in parts since 1969 in *Geol. Soc. Am. Bull.*) in an intensive study by J. C. Crowell and L. A. Frakes.

JOHN R. CONOLLY

References

Blank, R. G., and Margolis, S. V., 1975. Pliocene climatic and glacial history of Antarctica as revealed by southeast Indian Ocean deep-sea cores, *Geol. Soc. Am. Bull.,* 86, 1058-1066.

Bramlette, M. N., and Bradley, W. H., 1940. Geology and biology of North Atlantic deep-sea cores between Newfoundland and Iceland, *U.S. Geol. Surv. Prof. Pap. 196-A,* 56p.

Carey, L. W., and Ahmad, N., 1960. Glacial marine sedimentation, in G. O. Raosch, ed., *Geology of the Arctic,* vol. 2. Toronto: Univ. Toronto Press, 865-894.

Conolly, J. R., and Ewing, M., 1965. Ice-rafted detritus as a climatic indicator in Antarctic deep-sea cores, *Science,* 150, 1822-1824.

Darwin, C., 1842. Remarks on the glaciers of Tierra del Fuego and on the transportal of boulders, *Trans. Geol. Soc. London,* 6, 415-431.

Du Toit, A. L., 1921. The Carboniferous glaciation of South Africa, *Geol. Soc. S. Africa,* 24, 188-227.

Kent, D.; Opdyke, N. D.; and Ewing, M., 1971. Climatic change in the North Pacific using ice-rafted detritus as a climatic indicator, *Geol. Soc. Am. Bull.,* 82, 2741-2754.

Kindle, E. M., 1924. Observations on ice-borne sediments by the Canadian and other Arctic expeditions, *Am. J. Sci.,* 7, 251-286.

Menard, J. W., 1953. Pleistocene and Recent sediments from the northeastern Pacific Ocean, *Geol. Soc. Am. Bull.,* 64, 1279-1294.

Needham, H. D., 1962. Ice-rafted rocks from the Atlantic Ocean off the coast of the Cape of Good Hope, *Deep-Sea Research,* 9, 475-486.

Phillipi, E., 1912. Die Grundproblem der deutschen Sudpolar-Expedition, in E. von Drygalski, *Deutsches Sudpolar-Expedition, 1901-1903,* 2, 431-434.

Sharma, G. D., 1970. Sediment-seawater interaction in glaciomarine sediments of Southern Alaska, *Geol. Soc. Am. Bull.,* 81, 1097-1106.

Watkins, N. D.; Keany, J.; Ledbetter, M. T.; and Huang, T. -C., 1974. Antarctic glacial history from analyses of ice-rafted deposits in marine sediments: New model and initial tests, *Science,* 186, 533-536.

Whetten, J. T., 1965. Carboniferous glacial rocks from the Werrie Basin, New South Wales, Australia, *Geol. Soc. Am. Bull.,* 76, 43-56.

Cross-references: *Continental-Shelf Sediments in High Latitudes; Diamictite; Glacigene Sediments; Marine Sediments; Pelagic Sedimentation, Pelagic Sediments; Till and Tillite;* Vol. I: *Icebergs.*

GLACIAL SANDS

Glacial sand is an accumulation of mineral and lithic grains ranging from 0.063-2 mm in particle diameter that have been produced mainly by crushing and abrasion of larger clasts glacially eroded from underlying rock. Glacial sand may either be of local derivation or be transported a long distance from its source (Gravenor, 1951; Shepps, 1953).

GLACIAL SANDS

Sand-bearing glacial deposits on land and in the sea include till deposited directly by a glacier; outwash and kames deposited by glacial streams; lacustrine deposits; eolian sands; sediments formed by mass downslope movement; deposits from beneath floating or grounded glaciers and ice shelves, and rafted upon icebergs (Flint, 1971). The sand content of these deposits varies from subordinate amounts to 100% (Fig. 1) as a result of their different modes of formation.

Grain Size and Composition

The sand fraction of till consists of a continuous range of particle sizes, but reworking and transport by water or wind sort the sand into narrower size ranges (Fig. 1; Slatt, 1971). A wide variety of mineral and/or lithic grains may comprise glacial sand, depending upon the variability of lithologies being eroded. Chemical weathering is minimal in the glacial environment, so glacial sand is often characterized by a high feldspar:quartz ratio and an abundance of

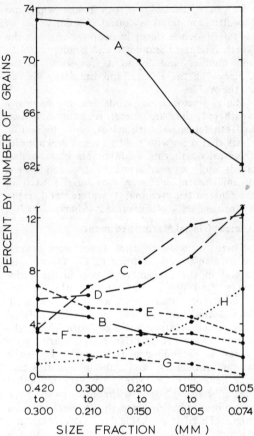

FIGURE 2. Average composition by size fraction of 17 sands of glacial origin, Illinois (modified from Hunter, 1967, Fig. 2). (A) Quartz. (B) Polycrystalline quartz. (C) Potash feldspar. (D) Plagioclase. (E) Feldspathic rock fragments. (F) Nonfeldspathic rock fragments. (G) Chert. (H) Heavy minerals.

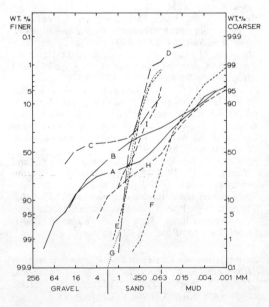

FIGURE 1. Grain-size distribution of sand-bearing glacial deposits. (A) Pleistocene ice-sheet till, Ontario (Dreimanis and Vagners, 1972). (B) Recent valley glacier till, Alaska (Slatt, 1971). (C) Recent glacial outwash, Alaska (Slatt and Hoskin, 1968). (D) Pleistocene outwash sand, Iowa (Udden, 1914). (E) Recent intertidal sand of glacial origin, Alaska (reference in F). (F) Recent lake shoreline deposit of glacial origin, Alaska (Slatt and Hoskin, 1968). (G) Pleistocene dune sand, Illinois (Hunter, 1966). (H) Sediment on a floating iceberg, Antarctica (Warnke and Richter, 1970). (I) Ice-rafted glacial marine sediment, Alaskan Abyssal Plain (Slatt and Piper, 1974).

chemically unstable heavy minerals such as amphiboles and pyroxenes.

Concentrations of different minerals and lithic grains comprising glacial sand vary with grain size (Fig. 2). Lithic grains comprise coarse sand. With decreasing grain size, lithic grains and quartz decrease, and feldspars and heavy minerals increase (Fig. 2; Hunter, 1967). In till, these minerals are reduced during transport to specific sizes (within the sand or finer-size ranges), called *terminal grades*, which depend upon the original sizes of the mineral grains in the source rock and their resistance to mechanical breakdown (Dreimanis and Vagners, 1972). Proportions of sand and coarser- or finer-sized particles are commonly used to correlate or differentiate tills (Fig. 3A), as are proportions of light-mineral varieties, lithic grains, and heavy-mineral varieties (Fig. 3B;

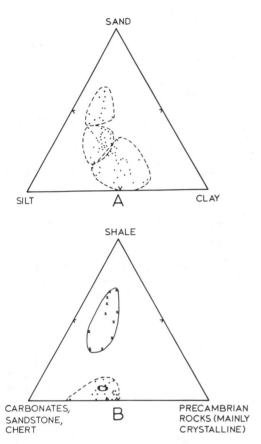

FIGURE 3. Texture and composition as a means of correlating and differentiating glacial till. (A) Texture of the finer than 2 mm size fraction of 76 till samples from three tills in NE Ohio (Shepps, 1953, Fig. 2). (B) Composition of lithic grains in the 0.5–1.0-mm size fraction of 34 samples from two tills in Ontario (modified from Dreimanis and Reavely, 1953).

Arneman and Wright, 1959), and sometimes heavy-mineral concentrations. Light- and heavy-mineral varieties in till are also used to determine provenance and ice-flow direction.

Fabric and Grain-Surface Textures

Ostry and Deane (1963) suggest the fabric of elongate sand grains in till is similar to that of gravel. There is a primary orientation of the long axis of grains parallel to the direction of transport and there may be a secondary transverse orientation.

The following microtextures on the surface of quartz sand grains observed under a transmission electron microscope are considered to be diagnostic of a glacial origin: (1) conchoidal breakage patterns of varying size; (2) high surface relief; (3) semiparallel steps; (4) arc-shaped steps; (5) parallel striations of varying length; (6) imbricated breakage blocks; (7) irregular small-scale indentations; (8) prismatic patterns (Krinsley and Donahue, 1968); and (9) chattermark trails on garnet (Folk, 1975). These features are more readily observed under a scanning electron microscope (Fig. 4; Walley and Krinsley, 1974), and scanning electron microscopy has become a particularly useful means of identifying glacial sand, particularly ice-rafted, deep-sea sand (e.g., Margolis and Kennett, 1971). Diagnostic microtextures have been recognized on sand grains deposited in different environments, so identification of superimposed sets of microtextures on grains has led to interpreting sequential events in the depositional history of glacial sand (Krinsley and Cavallero, 1970).

ROGER M. SLATT

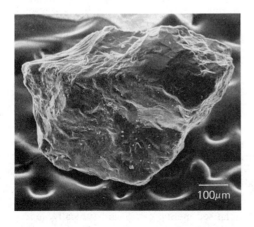 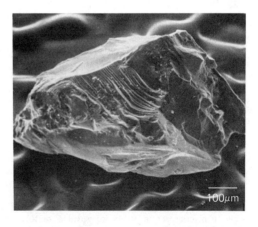

FIGURE 4. Sand grains from delta in fjord at margin of modern glacier, Norway, showing conchoidal breaking patterns, high surface relief, semi-parallel steps, arc-shaped steps, and irregular small-scale indentations. (Photos: D. H. Krinsley.)

References

Arneman, H. F., and Wright, H. E., 1959. Petrography of some Minnesota tills, *J. Sed. Petrology,* **29,** 540–554.

Dreimanis, A., and Reavely, G. H., 1953. Differentiation of the lower and the upper till along the north shore of Lake Erie, *J. Sed. Petrology,* **23,** 238–259.

Dreimanis, A., and Vagners, U. J., 1972. The effect of lithology upon texture of tills, in E. Yatsu and A. Falconer, eds., *Research Methods in Pleistocene Geomorphology.* 2nd Guelph Symposium on Geomorphology, 1971, Dept. Geography, Univ. Guelph Geographical Publ. 2, 66–82.

Flint, R. F., 1971. *Glacial and Quaternary Geology.* New York: Wiley, 892p.

Folk, R. L., 1975. Glacial deposits identified by chattermark trails in detrital garnets, *Geology,* **3,** 473–474.

Gravenor, C. P., 1951. Bedrock source of tills in southwestern Ontario, *Am. J. Sci.,* **249,** 66–71.

Hunter, R. E., 1966. Sand and gravel resources of Tazewell County, Illinois, *Ill. St. Geol. Surv. Circ. 399,* 22p.

Hunter, R. E., 1967. The petrography of some Illinois Pleistocene and Recent sands, *Sed. Geol.,* **1,** 57–75.

Krinsley, D. H., and Cavallero, L., 1970. Scanning electron microscopic examination of perigalcial eolian sands from Long Island, New York, *J. Sed. Petrology,* **40,** 1345–1350.

Krinsley, D. H., and Donahue, J., 1968. Environmental interpretation of sand grain surface textures by electron microscopy, *Geol. Soc. Am. Bull.,* **79,** 743–748.

Margolis, S. V., and Kennett, J. P., 1971. Cenozoic paleoglacial history of Antarctica recorded in subantarctic deep-sea cores, *Am. J. Sci.,* **271,** 1–36.

Ostry, R. C., and Deane, R. E., 1963. Microfabric analyses of till, *Geol. Soc. Am. Bull.,* **74,** 165–168.

Rehmer, J. A., and Hepburn, J. C., 1974. Quartz sand surface textural evidence for a glacial origin of the Squantum "Tillite," Boston Basin, Massachusetts, *Geology,* **2,** 413–415.

Shepps, V. C., 1953. Correlation of the tills of northeastern Ohio by size analysis, *J. Sed. Petrology,* **23,** 34–48.

Slatt, R. M., 1971. Texture of ice-cored deposits from ten Alaskan valley glaciers, *J. Sed. Petrology,* **41,** 828–834.

Slatt, R. M., and Hoskin, C. M., 1968. Water and sediment in the Norris Glacier outwash area, upper Taku Inlet, southeastern Alaska, *J. Sed. Petrology,* **38,** 434–456.

Slatt, R. M., and Piper, D. J. W., 1974. Sand-silt petrology and sediment dispersal in the Gulf of Alaska, *J. Sed. Petrology,* **44,** 1061–1071.

Udden, J. A., 1914. Mechanical composition of clastic sediments, *Geol. Soc. Am. Bull.,* **25,** 655–744.

Warnke, D. A., and Richter, J., 1970. Sedimentary petrography of till from a floating iceberg in Arthur Harbor, Antarctic Peninsula, *Rev. Geogr. Phys. Géol. Dyn.,* **12,** 441–448.

Whalley, W. B., and Krinsley, D. H., 1974. A scanning electron microscope study of surface textures of quartz grains from glacial environments, *Sedimentology,* **21,** 87–106.

Cross-references: *Beach Sands; Eolian Sands; Glacial Gravels; Glacigene Sediments; River Sands; Sands and Sandstones; Sand Surface Textures; Till and Tillite.*

GLACIGENE SEDIMENTS

Glacigene, glacigenic, or glaciogene sediments include all deposits composed largely or entirely of material eroded by ice from older rocks and their weathered mantles, and subsequently transported mainly by a glacier. Sediments composed of material carried long distances from a glacier by wind or water and deposited in periglacial or temperate regions are excluded. The debris carried by a glacier (Table 1) may be deposited either directly from the ice (as *glacial sediment*), or indirectly by meltwaters on, beneath, or within the glacier or near its margin (as *glaciofluvial, glaciolacustrine,* or *glacial marine* sediments). Although glacigene sediments include all these types, the word "glacigene" is rarely used, and many writers use "glacial" to include glaciofluvial and sometimes glaciolacustrine deposits as well as the sediments deposited directly from ice. Flint (1971) and Embleton and King (1968) described the origin and morphology of many glacigene sediments and the characteristic landforms associated with them.

Most pre-Quaternary glacigene sediments are indurated rocks, but those of Quaternary age are usually soft, unconsolidated deposits with little or no cementation. Indigenous fossils do not occur in sediments deposited directly from ice and are rare in most other glacigene sediments. Fossils derived from other deposits are sometimes abundant, and a few pre-Pleistocene species are known only from erratic blocks in glacigene sediments.

The glacial environment inhibits chemical weathering during erosion, transportation, and deposition, but promotes the physical disintegration of rock material into particles with a very wide size range. Consequently, glacigene sediments often contain unusually large quantities of chemically unaltered minerals, which readily decompose in temperate conditions and release plant nutrients. This factor accounts partly for the agricultural importance of many soils developed on Pleistocene glacigene sediments in temperate parts of North America and Europe.

Sediments Deposited Directly from Ice (Glacial Sediments)

Glacial sediments, or *till,* are typically deposited on a land surface, but some may be subsequently drowned by rising water level in a lake

TABLE 1. Classification of Glacial Deposits

Dominant Materials		Conjectured Situation		Shape or Morphology
Lodgment till subglacial, basal, nonbedded, compact	*Till* till dominated, poorly sorted	*Ground moraine* under ice and off retreating edge	I	*Till plain* or rolling hills washboard moraine *minor moraines
		Streamlined sliding melting base	II	*Drumlin* grooved till crag-tail
Ablation till superglacial, loose lens, bedded, contorted		*End moraine* at standing or advancing ice edge	III	*Lobate/looped moraine* push/thrust boulder belt lateral/interlobate moraine *kame moraine
		Disintegration stagnant, decaying marginal area, buried ice masses	V	*Controlled/uncontrolled disintegration* dead ice knobs/rings disintegration ridge inverted lake
Glaciofluvial coarse cobble to silt, channeled or cross-bedded	*Wash* well sorted	*Ice contact* dipping, deformed, irregular beds in ice pit, channel, or tunnel	IV	*Esker* *crevasse filling chain (of kames)
				Kame *field/kame and kettle kame moraine moulin kame kame terrace/plain
Glaciolacustrine fines: fine sand to colloid clay, laminated		*Proglacial* at grade, uniform beds extending away from ice	V	*Outwash plain/fan* valley train kettled/pitted outwash *collapsed outwash
				Lacustrine/marine delta strand/raised beach glacial *varve*

From Goldthwait, 1975.
*Other forms often attributed to stagnant ice disintegration.

or on the sea floor. Glaciers deposit their load either while still active or by melting in situ when movement has largely ceased. Dry-base glaciers deposit very little material subglacially; and some melting, however minor and localized, is always involved in subglacial or supraglacial release of the englacial detritus. To this extent, most glacial sediments are, in a strict sense, waterlaid, but there is often too little water to abrade, sort, or even disaggregate the sediment particles. Lithified tills are called tillites (see *Till and Tillite*).

Subglacial Deposits. The commonest sediment deposited directly from ice, and possibly the most widespread of all glacigene deposits, is *lodgment till* (= ground moraine), which is plastered irregularly on the subglacial surface by moving ice. It is unsorted and may have any particle-size distribution; *boulder clay* is a common variety, in which far-traveled rock fragments (erratics) of gravel, cobble, or boulder size are set in a firm, clay-rich matrix (see *Pebbly Mudstones*). Most lodgment tills are structureless and uniformly colored; some have a crude horizontal fissility caused by compression or shearing, and some are color-banded as a result of incomplete mixing of the detritus. Both large (cobbles, boulders) and medium (sand, silt) particles are often preferentially oriented with their long axes parallel to the direction of ice movement; a common subsidiary preferred orientation at 90° to the ice movement results from shearing or contortion either in the parent ice or during subsequent readvances. The shape of larger particles is variable (Holmes, 1941), and usually only softer rock fragments (e.g., limestone) are striated. The surface of lodgment till is often ridged parallel or transverse to the direction of ice movement, or may form drumlins.

Carruthers (1953) described finely laminated clays and silts or fine sands in northern England, which he suggested were formed subglacially by the melting of stagnant ice containing banded dirt horizons separated by layers of cleaner ice. He distinguished these "shear clays" from lacustrine deposits by their sharply divided and often extremely thin (<0.01 mm) laminae, the rare occurrence of graded bedding, the lack of organic matter, and the presence of strongly contorted layers bounded above and below by flat-bedded laminae. Their flame-like minor structures were attributed to the suspension of partly released clay laminae into shallow subglacial streams depositing sand.

Supraglacial Deposits. On or near the surface of a glacier where englacial detritus is released by ablation (melting and sublimation of surface ice), supraglacial sediments accumulate. On valley glaciers, broken rock material from adjacent land surfaces may locally contribute to this accumulation. Final melting lowers the supraglacial deposits onto the subglacial, thus forming a two-tiered sequence. With modern temperate glaciers (e.g., Alps, Norway), the upper unit comprises a thin *ablation till*, which is less compact, more permeable, and often slightly coarser than the relatively thick, subjacent lodgment till.

In arctic glaciers (e.g., Spitsbergen, Greenland), there is less subglacial melting and correspondingly less lodgment till, but a greater development of supraglacial deposits, especially near the ice margin (Boulton, 1972). Here, the amounts of meltwater are often sufficient to sort the debris, the finer fractions being removed by small sinuous streams, or to cause local mass movement of unsorted debris as *flowtill*. The supraglacial cover decreases the rate of ablation, and further slow melting of the buried ice produces *melt-out till*, which cannot flow and retains most of its englacial structure. Collapse structures such as faults and flexures may develop in the complex of sorted and unsorted supraglacial deposits as the buried ice melts. Differential ablation, caused by variable thicknesses of supraglacial material, produces an irregular, hummocky topography, with mounds, enclosed hollows (kettle holes), ridges, and valleys. Unless the subsurface drainage is good, however, the downslope flow of waterlogged debris can reduce such features to a complex of low mounds or an almost flat plain.

Glaciofluvial Sediments

Streams of glacial meltwater flow within, beneath, and on the ice, or away from the ice front. Englacial, subglacial, and supraglacial streams occur mainly in large, almost stagnant ice sheets; in contrast, the ice of valley glaciers is active even near the snout, and most of the meltwater is carried in proglacial streams. The term *fluvioglacial* is commonly used to describe proglacial stream deposits, which are composed mainly of sand and coarser particles, as the silt and clay are removed in suspension. Those sediments deposited in close contact with the ice are often poorly sorted and commonly may contain intercalated rafts of flow till and armored mudballs. Sediment deposited farther from the ice is better sorted. The conventional classification of glaciofluvial deposits is based mainly on the geomorphological features they produce.

Eskers. Sometimes known as *osar* (Swedish: åsar, pl. of ås), eskers are elongate ridges of glaciofluvial sand, gravel, and coarser material deposited usually in englacial or subglacial meltwater channels. An unusual form, the *beaded esker* consists of many separate elongate mounds, which probably contain the sediment released at the ice front by retreat during successive summers. Most eskers are continuous, however, and range in size from 100 m to several km long, 50–500 m wide and 5–50 m high. The sediments are often current-bedded, with the foreset beds dipping either parallel to the axis of the ridge or outward from the center line of the esker; stratification is sometimes disturbed by the melting of ice cores. Eskers usually extend approximately parallel to the direction of the latest ice movement, which may be locally uphill, the flow of water maintained by hydrostatic pressure under the weight of ice.

Kames. The irregular mounds of glaciofluvial deposits usually occupying relatively large areas extending roughly parallel to the ice front are termed *kames;* occasionally these show no clear relationship to other glacial features. The deposits range from well to poorly sorted and often show ice-collapse structures. Many originate in a supraglacial position and often have a hummocky surface; others accumulate as proglacial deltas in meltwater temporarily ponded by stagnant ice and produce a much more even surface. *Kame terraces* are composed of glaciofluvial sediment deposited between the side of a valley glacier and the enclosing valley wall.

They are not continuous down the valley and usually have a gently sloping surface, which ends on the side away from the valley wall in a steep ice-contact slope.

Valley Trains. Deposits of streams carrying meltwater away from the snout of a valley glacier and outwash deposits occupying valleys in continental glacier areas are called valley trains. The detritus is derived either directly from the glacier or by erosion of till and other sediments previously deposited by the glacier. Near the snout, the deposits are poorly sorted and often gravelly. At greater distances from the ice, the deposits are finer and well sorted; these deposits often bury earlier glacial deposits and may build up an almost flat valley floor.

Outwash Plains. Outwash plains (= sandar) are composed of the sediment deposited by meltwater streams flowing from a broad ice front in a lowland area. The streams are braided, and migrate laterally, depositing mainly sand over a wide flat area (see *Braided-Stream Deposits*).

Glaciolacustrine Sediments

At the present day, lakes of glacial origin outnumber all other lake types put together, but the term *glacial lake* includes many different lake forms (Hutchinson, 1957). Most of these are subaerial, but a few lakes are sub- or englacial. There are three types of glaciolacustrine sediment: delta, strandline, and lake-floor deposits.

Deltaic Deposits. Deltas form where a meltwater stream or the glacier itself enters a glacial lake. The sediments are texturally more variable than deltaic deposits (see *Deltaic Sediments*) of nonglacial origin because the fast-flowing steams may carry cobbles or boulders into the lake, and masses of till or large erratic blocks may be deposited from floating ice. The usual sequence of topset, foreset, and bottomset beds is sometimes disturbed by deposition against steep ice-contact slopes and slumping after ice collapse.

Strandline Deposits. Deposits on the shores of glacial lakes are usually not very extensive and often occupy narrow shoreline terraces cut by wave action. Washed gravel, cobbles, and boulders accumulate where waves erode older glacigene deposits (e.g., till); these coarse beach deposits do not occur, however, where the lake is too small to produce waves of adequate erosive power, where the water does not remain at the same level for long enough, or where the rocks exposed at the lake margin are too resistant. A few shore terraces are thinly veneered with silt or clay; but most are covered with sand, which may be built into spits and bars or blown inland to form dunes. Floating pack ice driven onshore by strong winds may push the beach deposits into low ridges known as lake ramparts.

Lake-Floor Deposits. Sediments deposited on the lake floor are primarily fine material which has been carried in suspension away from the lake margins, but some sand or coarser sediment may be introduced by floating ice or turbidity currents. Many of these deposits show the effect of strong rhythmic sedimentation, which often depends on annual climatic fluctuations. Seasonally alternating lake-floor sediments of this type are commonly called *varves* (q.v.); not all strongly laminated glacigene sediments are varves, however, because a few lake-floor sediments show nonannual rhythmic sedimentation, and alternations of sand and clay are possibly deposited subglacially.

Glacial Marine Sediments

Glacial marine sediments (q.v.) are deposited where meltwater streams carry glacially transported material into the sea or where a glacier flows off the land surface into the sea. The coarse sediment is deposited mainly near the shore, often as a delta, though some is carried long distances in icebergs and eventually deposited in deep water. Clay and silt are spread more widely, but varves do not form, mainly because the clay is flocculated and therefore deposited sooner than in fresh water. Glacial marine silty clays are poorly stratified and often light-colored; they are the main deposit of some seas bordering large ice caps, but elsewhere they are commonly interbedded with marine clays of nonglacial origin. Much of the coarse, nearshore, glacial marine detritus is incorporated in beach deposits by strong wave action.

JOHN A. CATT

References

Boulton, G. S., 1972. Modern Arctic glaciers as depositional models for former ice sheets, *J. Geol. Soc. London,* 128, 361-393.

Carruthers, R. G., 1953. *Glacial Drifts and the Undermelt Theory.* Newcastle-upon-Tyne: H. Hill, 42p.

Charlesworth, J. K., 1957. *The Quaternary Era,* vol. 1: *Glaciology and Glacial Geology.* London: E. Arnold, 591p.

Embleton, C., and King, C. A. M., 1968. *Glacial and Periglacial Geomorphology.* London: E. Arnold, 608p.

Flint, R. F., 1971. *Glacial and Quaternary Geology.* New York: Wiley, 892p.

Goldthwait, R. P., ed., 1975. *Glacial Deposits.* Stroudsburg, Pa.: Dowden, Hutchinson & Ross, 464p.

Holmes, C. D., 1941. Till fabric, *Geol. Soc. Am. Bull.* 52, 1299-1354.

Hutchinson, G. E., 1957. The origin of lake basins, in

A Treatise on Limnology, vol. 1, *Geography, Physics and Chemistry.* New York: Wiley, 1-163.
Jago, J. B., 1974. The terminology and stratigraphic nomenclature of proven and possible glaciogenic sediments, *J. Geol. Soc. Austral.,* **21,** 471-474. See also discussion by M. Schwarzbach and reply, **22,** 255-256.
Jopling, A. V., and McDonald, B. C., eds., 1975. Glaciofluvial and glaciolacustrine sedimentation, *Soc. Econ. Paleont. Mineral. Spec. Publ. 23,* 325p.
Lliboutry, L., 1958. *Traité de glaciologie,* 2 vols. Paris: Masson, 1040p.
Price, R. J., 1973. *Glacial and Fluvioglacial Landforms.* Edinburgh: Oliver & Boyd, 242p.

Cross-references: *Braided-Stream Deposits; Delta Sedimentation; Diamictite; Fabric, Sedimentary; Faceted Pebbles and Boulders; Glacial Gravels; Glacial Marine Sediments; Glacial Sands; Pebbly Mudstones; Sedimentary Environments; Till and Tillite; Varves and Varved Clays.*

GLAUCONITE

The term *glauconite* was first used in the 19th century for dark green to greenish-black, generally smooth grains, usually of ovaloid and/or lobate shape, found in marine sedimentary rocks and in modern marine sediments; this morphologic usage of the term is still popular today. Mineralogic research now has shown conclusively (Burst, 1958) that the grains may be, and usually are, mixtures of minerals rather than a single mineral species; hence, in the strict sense, they are rocks. A constituent common to these grains, however, is a nonexpandable, dioctahedral member of the illite clay mineral family, unfortunately also called glauconite. Currently, therefore, nomenclatural confusion exists in which "glauconite" is used as both a morphologic term and a mineral term. Herein, "glauconite grain" will be used as the morphological term and "glauconite" will be used to indicate the specific clay mineral. Millot (see Bjerkli and Ostmo-Saeter, 1973) has suggested the term *glauconie* for the heterogeneous grains. McRae (1972) gives an excellent detailed review of glauconite, including an exhaustive bibliography on glauconite in both the morphologic and mineralogic sense.

Mineralogy and Morphology

Glauconite grains are extremely variable in mineralogy, even when the common embedded grains of calcite, quartz, feldspar, and other nonclay grains are not considered. The clay minerals found in glauconitic grains include glauconite, smectite, kaolinite, chamosite, chlorite, vermiculite, and mixed layering of many of these clay minerals. The clay minerals with mica structure appear to be among the most abundant constituents, suggesting that smectite-like clays (common in modern glauconite grains) gradually convert to illite-like clays (common in ancient glauconitic grains).

The mineral glauconite has a somewhat variable composition, resulting mainly from two factors: a large amount of ionic substitution and the difficulty of purifying mixtures of clay minerals. A sample of Franconia glauconite, often used as a reference material in glauconite studies, has a calculated formula: $K_{0.76}(Na, Ca)_{0.13}(Fe^{+3}_{0.49}Al_{0.40}Mg_{0.40})(Si_{3.42}Al_{0.58})O_{10}(OH)_2$; ionic substitution is indicated by grouping the elements within the parentheses. This represents a statistical ratio of elements in an extremely large number of clay particles, many of which probably vary individually in composition. The chemical variation in glauconite grains is summarized by Weaver and Pollard (1973).

Glauconite grains have many different external morphologies but are usually ovaloid, often multilobed, sand-sized particles (Fig. 1). The surfaces of the grains commonly have cracks which may be filled with sediment of diverse types. The morphologic classification of glauconite grains (Triplehorn, 1966) reflects a great diversity of forms, including ovoidal, tabular, discoidal, mammillary, lobate, capsule-shaped, composite pellets, vermicular grains, casts, and molds. The internal morphology of glauconite grains is also diverse, the most common form being aggregates of randomly oriented flakes. Some grains appear to have patches of oriented flakes, particularly in peripheral rinds. These rinds may represent replacement of fibrous calcite or aragonite of shells that formerly enclosed the glauconitic grains or may represent recrystallization of original random flakes.

Origin and Occurrence

Occurrences in Recent Sediments. In modern sediments, glauconite grains are most commonly found in calcareous oozes (Fig. 1). There is a wide range of fauna associated with the grains, but foraminifera commonly are abundant (Murray and Renard, 1891; Ehlmann et al., 1963). The glauconite grains in these oozes generally cover a wide range in color, grading from creamy white to greenish black. Most of the darker grains are not associated with shells; but lighter-colored grains often are found as internal molds of foraminifera, best observed after solution of the sediment in dilute HCl. A single cast may be several different shades of green. In these oozes and in some other sediments, glauconite grains are also found which have morphologies that strongly suggest they are fecal pellets (Takahashi, 1939; Porrenga, 1966).

FIGURE 1. *Left:* The magnetic fraction of a foraminiferal ooze from off the SE coast of the United States. Some glauconite grains are free of shells and other grains are partly or completely enclosed. *Right:* The magnetic fraction of the same sample after solution of the foraminiferal shells.

In Monterey Bay, California, glauconite grains are found in association with generally arkosic sands and silts being derived from adjacent granitic terrane (Galliher, 1939). The glauconite grains in this sediment appear to grade imperceptibly into biotite flakes, suggesting a genetic relationship. Laterally, however, these arkosic sediments grade into shell sands, a common matrix for glauconite grains. Although glauconite grains typically are associated with organic debris it appears that where the glauconite grains are less abundant, organic debris may be relatively rare and detrital grains more abundant.

Origin. The origin of glauconite grains remains a matter of dispute despite the many occurrences that have been described. Environmental boundary conditions are wide (Cloud, 1955), but all appear to be marine; depths of occurrence range from about a meter up to 500 m (Murray and Renard, 1891). Chemically, the environment usually appears to be oxidizing. This fact has caused speculation that a reducing environment localized within foraminifera tests, or within digestive tracts of animals, accounts for the ferrous iron present in glauconite grains. Other specific variables, such as pH and salinity, appear to be poorly defined.

More generalized environmental conditions apparently favorable for the formation of glauconite grains are (1) open marine waters; (2) nearby continental shores; and (3) the presence of biological activity. No occurrence of glauconite grains (molds, fecal pellets, and/or lobate grains) has been described from nonmarine waters. The mineral glauconite, however, has been found in various nonmarine environments, where it replaces various minerals or even occurs as bedded deposits (Keller, 1958). The general correlation between proximity to shore and glauconite-grain occurrence was noted by Murray and Renard (1891), who found foraminiferal oozes far from continental masses to be barren of glauconite grains.

The case for the necessity of biologic activity for the formation of glauconite grains rests on the common association of the grains with tests of microorganisms and their common fecal-pellet morphology. It must be inferred that biological activity, if not necessary, certainly is conducive to glauconite grain formation. Even when organic activity is closely related, however, as when glauconite grains are found as foraminiferal casts, the nature of origin is unclear. The material within the foraminifera test either sifted in from associated mud or was precipitated directly from solution in sea water. It seems more probable that the material sifted in, because of the mineral heterogeneity of glauconite grains, particularly the lighter-colored casts.

Considering the possibility of alteration of detrital grains to glauconite grains, Galliher (1939) described grains that range from unquestionable biotite to unquestionable glauconite grains. The final morphology of the grains, however, suggests that most of the grains in that area had been passed through the digestive tracts of the fauna. Hence, the necessity of biologic activity for the alteration of biotite to glauconite grains is not precluded, and the occurrence described by Galliher can not be used as proof of strictly inorganic formation of glauconite grains. Detailed observations by Hein et al. (1974) do not support Galliher's study.

Occurrence in the Stratigraphic Column. Assuming the origin of glauconite grains has not varied throughout geologic time, occurrences in sedimentary rocks should reflect modern environments of formation, possibly modified by significant diagenetic changes. Being relatively soft, glauconite grains are easily distorted, and are even squeezed into calcite and quartz-grain interstices. Also it is not uncommon to observe glauconite replacing carbonate shell fragments and even whole microfossils on a volume-to-volume basis. Less commonly, replacement of detrital grains such as feldspar is observed.

Glauconite grains occur widely in sedimentary rocks of virtually all ages. Not unusually, they are associated with phosphatic pellets and pyrite above unconformities (Debrabant and Paquet, 1975), a relationship which appears to have resulted from the formation of these pellets in environments of low depositional rates. Glauconite grains may commonly have been redeposited locally.

Scientifically, glauconite grains are syngenetic and reflect the time of sedimentary formation rather than that of a preexisting rock supplying the weathered particles. Incorrect dates, however, may be common because of included impurities and chemical gains and/or losses commonly found in these grains.

Uses

Glauconite grains do not have great commercial value, although they have been used for soil conditioning and fertilizing. Their chemical heterogeneity and high bulk probably exclude extensive future use as fertilizers except in local areas of highly leached soils where their ion-exchange properties might be beneficial.

A. J. EHLMANN

References

Bjerkli, K., and Ostmo-Saeter, J. S., 1973. Formation of glauconie in foraminiferal shells on the continental shelf off Norway, *Marine Geol.*, 14, 169-178.

Burst, J. F., 1958. Mineral heterogeneity in "glauconite" pellets, *Am. Mineral.*, 43, 481-497.

Cloud, P. E., Jr., 1955. Physical limits of glauconite formation, *Bull. Am. Assoc. Petrol. Geologists*, 39, 484-492.

Debrabant, P., and Paquet, J., 1975. L'association glauconites-phosphates-carbonates (Albien de la Sierra de Espuña, Espagne Méridionale), *Chem. Geol.*, 15, 61-75.

Ehlmann, A. J.; Julings, N. C.; and Glover, E. D., 1963. Stages of glauconite formation in modern foraminiferal sediments, *J. Sed. Petrology*, 33, 87-96.

Galliher, E. W., 1939. Biotite-glauconite transformation and associated minerals, in Trask, 1939, 513-515.

Giresse, P., and Odin, G. S., 1973. Nature minéralogique et origine des glauconies du plateau continental du Gabon et du Congo, *Sedimentology*, 20, 457-488.

Hein, J. R.; Allwardt, A. O.; and Griggs, G. B., 1974. The occurrence of glauconite in Monterey Bay, California, diversity, origins, and sedimentary environmental significance, *J. Sed. Petrology*, 44, 562-571.

Keller, W. D., 1958. Glauconitic mica in the Morrison Formation in Colorado, *Clays Clay Minerals*, 5, 120-128.

McRae, S. G., 1972. Glauconite, *Earth Sci. Rev.*, 8, 397-440.

Murray, J., and Renard, A. F., 1891. Report on deep-sea deposits based on specimens collected during the voyage of H.M.S. *Challenger* in the years 1872-1876. London: H.M.S.O., 525p.

Owens, J. P., and Sohl, N. F., 1973. Glauconites from New Jersey-Maryland Coastal Plain: their K-Ar ages and application in stratigraphic studies, *Geol. Soc. Am. Bull.*, 84, 2811-2838.

Porrenga, D. H., 1966. Clay minerals in recent sediments of the Niger delta, *Clays Clay Minerals*, 14, 221-233.

Takahashi, J., 1939. Synopsis of glauconitization, in Trask, 1939, 503-512.

Trask, P. D., ed., 1939. *Recent Marine Sediments*. Tulsa, Okla.: Am. Assoc. Petrol. Geologists, 736p.

Triplehorn, D. M., 1966. Morphology, internal structure and origin of glauconite pellets, *Sedimentology*, 6, 247-266.

Weaver, C. E., and Pollard, L. D., 1973. *The Chemistry of Clay Minerals.* New York: Elsevier, 213p.

Cross-references: *Authigenesis; Clay; Continental-Shelf Sediments; Diagenesis; Organic Sediments; Phosphates in Sediments.* Vol. IVB: *Glauconite.*

GRADED BEDDING

A *graded bed* (Bailey, 1930) is a clastic sedimentary unit grading from coarsest material at the base to finest at the top, and varying in thickness from a few mm to 1 m or more. Thick graded beds may be correlated for many km (Hesse, 1974). In general, the thicker the unit the coarser the material at the base of the bed (Potter and Scheidegger, 1966). Grading in the opposite sense (coarsest at the top) is called reverse or *inverse grading* (q.v.) and is not uncommon. Thus, caution must accompany the use of graded bedding as a top-and-bottom criterion (q.v.). Graded beds commonly occur in sequences of several units, quite often with erosive (scoured) contacts at the base of each unit. Because graded beds are deposited by currents carrying a suspended load of sediment, a current carrying less than its capacity will scour the underlying sediment, producing *scour marks* (q.v.). Graded beds are most commonly found in turbidites (q.v.), and it is here also that scour marks and their accompanying molds (*sole marks*) are abundant. Subunits of varying structure may also typify a graded unit (see *Bouma Sequence*).

As a current slows down, it loses its carrying capability; first, the heaviest (usually the coarsest) particles fall, and then finer and finer material settles out, thus forming a graded bed. Fluctuations in the current will result in irregularities in the graded bed, but the overall structure reveals the upward decrease in grain size. As the current wanes, the distal areas of the deposit will consist only of finer material; consequently, gradation occurs not only vertically but also laterally in the downcurrent direction. This final distribution only partly reflects the dynamics of the turbidity current as it flows, however, because at this time the faster part of the current, which carries the coarser material, travels at the head of the turbidity flow (see *Turbidity-Current Sedimentation;* Kuenen, 1953). Graded beds in marine environments may be capped by hemipelagic or pelagic sediments (q.v.), or they may be separated by other types of deposits, e.g., contourites (q.v.) or, in the case of lake deposits, by varves (q.v.).

Pettijohn (1957) suggested that graded bedding produced by turbidity currents may be distinguished from graded beds produced by waning, shallow-water currents because much fine-grained material is incorporated in the coarse bed of a turbidite (sorting improving upwards); in waning currents (e.g., ebb tides), however, the sorting should be better throughout because the fines are excluded from the bed load. Brush (1965) found this not always to be the case. The hydrodynamic conditions within a turbidity current which give rise to graded bedding are still not completely understood.

Most of the commonly observed variations in graded bedding (Fig. 1) were described and discussed by Kuenen (1953). Birkenmajer (1959) later devised an exhaustive and detailed classification. Graded bedding may form in a variety of ways (Klein, 1965), including the activity of burrowing organisms (Rhoads and Stanley, 1965) and the effects of hurricanes (Hayes,

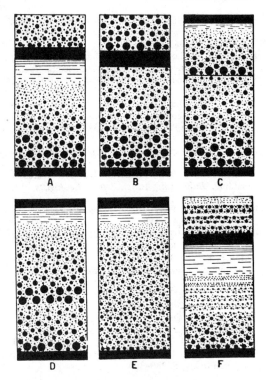

FIGURE 1. Diagram showing various types of graded bedding (from Kuenen, 1953). Deep-water lutites are in black. A, ideal type; B, fine top is missing below lutite; C, fine top is missing below the following graded bedding (from Kuenen, 1953). Deep-water lower part not graded; F, graded bed is laminated.

1967; see *Storm Deposits*). Reineck (1972) noted graded beds <1 mm thick in tidal-flat deposits. A form of grading may also be produced by seasonal or daily variations in sedimentation (see *Varves and Varved Clays*). The most common origin however, i.e., turbidity-current sedimentation (q.v.), has been confirmed by studies of deep-sea cores (e.g., Horn et al., 1972). Graded beds are common throughout the stratigraphic column, from the early Precambrian to the present (see Pettijohn, 1975).

JOANNE BOURGEOIS

References

Bailey, E. B., 1930. New light on sedimentation and tectonics, *Geol. Mag.,* 67, 77-92.

Birkenmajer, K., 1959. Classification of bedding in flysch and similar graded deposits, *Stud. Geol. Polonica,* 3, 1-133 (Polish: English Summary).

Brush, L. M., 1965. Analysis of graded bedding (abstr.), *Geol. Soc. Am. Spec. Pap. 82,* 21.

Conybeare, C. E. B., and Crook, K. A. W., 1968. Manual of sedimentary structures, *Australia Bur. Mineral Resources, Geol. Geophys. Bull.,* 102, 327p.

Hayes, M. O., 1967. Hurricanes as geological agents, south Texas coast, *Bull. Am. Assoc. Petrol. Geologists,* 51, 937-942.

Hesse, R., 1974. Long-distance continuity of turbidites: possible evidence for an early Cretaceous trench-abyssal plain in the East Alps, *Geol. Soc. Am. Bull.,* 85, 859-870.

Horn, D. R.; Ewing, J.; and Ewing, M., 1972. Graded-bed sequences emplaced by turbidity currents north of 20°N in the Pacific, Atlantic, and Mediterranean, *Sedimentology,* 18, 247-275.

Klein, G. de Vries, 1965. Diverse origin of graded bedding (abstr.), *Geol. Soc. Am. Spec. Pap. 82,* 109.

Kuenen, P. H., 1953. Significant features of graded bedding, *Bull. Am. Assoc. Petrol. Geologists,* 37, 1044-1066.

Pettijohn, F. J., 1957. *Sedimentary Rocks,* 2nd ed. New York: Harper, 718p.

Pettijohn, F. J., 1975. *Sedimentary Rocks,* 3d ed. New York: Harper & Row, 628p.

Potter, P. E., and Scheidegger, A. E., 1966. Bed thickness and grain size: graded beds, *Sedimentology,* 7, 233-240.

Reineck, H. E., 1972. Tidal flats, *Soc. Econ. Paleont. Mineral. Spec. Publ. 16,* 146-159.

Rhoads, D. C., and Stanley, D. J., 1965. Biogenic graded bedding, *J. Sed. Petrology,* 35, 956-963.

Scheidegger, A. E., and Potter, P. E., 1965. Textural studies of graded bedding: Observation and theory, *Sedimentology,* 5, 289-304.

Cross-references: *Bouma Sequence; Inverse Grading; Storm Deposits; Tidal-Flat Geology; Turbidites; Turbidity-Current Sedimentation; Varves and Varved Clays.*

GRAIN FLOWS

The concept of grain flow came about as a result of the experimental work of R. A. Bagnold (1954, 1956). He claimed to have measured an upward supportive stress acting on cohesionless grains within flowing sediments, resulting from grain-to-grain interaction rather than fluid turbulence. This *dispersive pressure* is proportional to the shear stress transmitted between grains, and it counteracts the tendency for grains to settle out of the flow (see *Bagnold Effect*). Bagnold applied his analysis to the case of *grain flow,* or sand avalanching, where concentrated dispersions of cohesionless sediment move downslope in response to the pull of gravity, and grains remain in a dispersed state essentially by bouncing off one another. Middleton and Hampton (1973) suggest that grain flow is one of four types of sediment *gravity flow* (see *Gravity Flows*).

Lowe (1976) restricted the term *grain flow* to sediment gravity flows in which not only do the grains remain dispersed by grain interaction, but also the fluid interstitial to the grains is the same as the ambient fluid above. Flows in which a dense interstitial fluid, a current, or escaping pore fluid aids in maintaining the dispersion are termed by him *modified grain flows.* Lowe pointed out the differences in dynamics and characteristics between Bagnold's (experimental) fully confined, gravity-free dispersions and natural flows, which are neither fully confined nor gravity free.

Shepard and Dill (1966) reported numerous occurrences of small-scale grain flows in the upper reaches of submarine canyons. Sanders (1965) speculated that grain flows may be a companion to some turbidity currents and used his model to interpret sedimentary structures of some turbidites. Lowe (1976) believed that Sanders' model is unlikely in the absence of additional modifying influences. Inferred large-scale grain flows preserved in ancient rocks were described by Stauffer (1967). His careful description of the morphology, texture, and sedimentary structures has served as the basic model for recognition of grain-flow deposits (e.g., see Klein et al., 1972). Caution must be used, however, in applying Stauffer's model, because the transport mechanism (grain flow) is highly inferential, and there is a serious lack of data on the features of unequivocal grain-flow deposits (Middleton and Hampton, 1973). For example, Link (1975) reinterpreted the sequence studied by Stauffer as proximal turbidites and possibly fluidized sediment-flow deposits, with grain-flow mechanisms playing only a minor role.

Mechanics of Flow*

Bagnold's experiment consisted of shearing cohesionless sediments in the annular space between two concentric drums. He rotated the outer drum and measured shear and dispersive stresses produced on the inner drum. Bagnold found that grain flow can occur in a viscous regime, where fluid viscosity is important in producing dispersive pressure, or in an inertial regime, where viscosity is insignificant and grain inertia dominates. In both regimes, dispersive pressure is produced by momentum change associated with grain interaction when grains approach each other or actually bounce off one another.

The relationship between shear stress (T) and dispersive pressure (P), expressed as $T/P = \tan \alpha$, has distinct constant values for the inertial and the viscous regimes. The angle α is interpreted as a "dynamic angle of internal friction" and was determined by Bagnold (1954) to be 37° for viscous grain flow, and 18° for inertial

---
*Based on Middleton and Hampton, 1973; see also Lowe, 1976.

grain flow. Bagnold later (1973) recognized that his work on perfect spheres did not apply to natural grains, where the correct value for both viscous and inertial flows would be ~30°. The dynamic angle of internal friction can be used to calculate the slope angle necessary for sustained movement of grain flows. For the case of a subaqueous grain flow with plain water as the interstitial fluid, the minimum slope angle necessary for movement is equal to the angle of internal friction, i.e., 30°. Some grain flows may move on lesser slopes, however. If the density of the interstitial fluid is greater than that of plain water, the effective weight of the solid grains is reduced by buoyancy, reducing the magnitude of dispersive pressure necessary to support the grains. The added weight of a high-density fluid also increases the downslope driving force.

The role of turbulence in grain flows in unclear. Bagnold (1954, 1956) concluded from his experiments that high solid-grain concentrations suppress turbulence in interstitial fluids, and thus that the role of turbulence in grain flow is negligible. Stauffer (1967) and others have described grain flow as laminar shear of cohesionless sediment, implying the absence of turbulence. At high energy conditions, however, flow probably becomes turbulent, the turbulence helping to support the solid grains. Thus, a continuum may exist between grain flows and turbidity currents, and some flows may be combinations of the two types (see *Gravity Flows; Turbidity-Current Sedimentation*).

Deposition. Deposition is fundamentally by mass emplacement, which involves a sudden "freezing" of the flow resulting in simultaneous deposition of a layer several grains thick. "Freezing" occurs when the driving gravity stress becomes less than the yield strength of the sediment, either by reduction of the slope angle or by consolidation of the flow.

Textures and Structures

Stauffer (1967) described supposed grain-flow beds in Tertiary rocks of California as being thick and sharply bounded; internally, the grain-flow beds described by Stauffer typically are massive and ungraded, but diffuse parallel lamination and dish structure occur in some beds. Stauffer also reported the presence of numerous large lutite and siliceous clasts at any level within a bed, commonly floating in a sandy or muddy matrix. Lowe (1976), however, concluded that grain flows cannot account for thick, featureless sand beds, but that associated sandy conglomerate units may have been deposited by density-modified grain flows.

Inverse grading (q.v.) should be a feature of coarse-grained grain-flow deposits. Bagnold (1954, 1968) concluded that dispersive pressure acts to size-sort grains, pushing larger grains to the top of the flow (the area of least shear-strain rate), and smaller grains move to the bottom. Middleton (1970) disagreed with this proposed mechanism, suggesting instead that inverse grading is the product of a kinetic sieve mechanism whereby small grains fall downward between large grains during flow, displacing the large grains upward. Inverse grading can be observed as sand cascades down the slip face of a dune, the large grains moving to the top and periphery of the flow (Middleton and Hampton, 1973). Fisher and Mattinson (1968) reported inverse grading in conglomerates in California and suggested a high-density, highly fluid emplacement mechanism, perhaps grain flow.

Rees (1968) showed that grain flow in sands (and probably also in coarser sediments) produces grain orientation parallel to the flow and imbricated upflow. This fabric, however, is similar to the grain orientation produced in sands by deposition from suspension, with minor traction (Middleton and Hampton, 1973).

Positive identification of grain-flow deposits is still problematical because all features proposed to be characteristic of grain flows have also been produced by other transport processes. Another point to consider is that in the absence of excess pore pressures, relatively steep slopes are required for true grain flow (Middleton and Hampton, 1973; Lowe, 1976). Slopes of this magnitude are rare in the ocean, suggesting that most grain-flow deposits are localized features. Finally, grain flows are only part of the spectrum of sediment gravity flows and are probably often combined with liquefied sediment flows and turbidity currents (Middleton and Hampton, 1973; Link, 1975; see *Gravity Flows*). Because the deposits of grain flows and of liquefied (fluidized) flows are difficult to separate, Stanley (1975) suggested the term *sandflow* to include both (or either) types of flow.

JOANNE BOURGEOIS

References

Bagnold, R. A., 1954. Experiments on a gravity-free dispersion of large solid spheres in a Newtonian fluid under shear, *Proc. Roy. Soc. London,* ser. A., 225, 49-63.

Bagnold, R. A., 1956. The flow of cohesionless grains in fluids, *Roy. Soc. London, Philos. Trans.,* ser. A, 249, 235-297.

Bagnold, R. A., 1968. Deposition in the process of hydraulic transport, *Sedimentology,* 10, 45-56.

Bagnold, R. A., 1973. The nature of saltation and of "bed load" transport in water, *Proc. Roy. Soc. London,* ser. A, 332, 473-504.

Fisher, R. V., and Mattinson, J. M., 1968. Wheeler Gorge turbidite-conglomerate series, California: Inverse grading, *J. Sed. Petrology,* 38, 1013-1023.

Klein, G. deV.; deMelo, U.; and Favera, J. C. D., 1972. Subaqueous gravity processes on the front of Cretaceous deltas, Reconcavo basin, Brazil, *Geol. Soc. Am. Bull.,* 83, 1469-1492.

Link, M. H., 1975. Matilija Sandstone: A transition from deep-water turbidite to shallow-marine deposition in the Eocene of California, *J. Sed. Petrology,* 45, 63-78.

Lowe, D. R., 1976. Grain flow and grain flow deposits, *J. Sed. Petrology,* 46, 188-199.

Middleton, G. V., 1969. Turbidity currents and grain flows and other mass movements down slopes, in D. J. Stanley, ed., The New Concepts of Continental Margin Sedimentation, *Am. Geol. Inst. Short Course Notes,* GM-A-1-GM-B-14.

Middleton, G. V., 1970. Experimental studies related to problems of flysch sedimentation, in J. Lajoie, ed., Flysch sedimentology in North America, *Geol. Assoc. Canada Spec. Pap. 7,* 253-272.

Middleton, G. V., and Hampton, M. A., 1973. Sediment gravity flows: Mechanics of flow and deposition, *Soc. Econ. Paleont. Pac. Sec. Short Course* (May 12, 1973), 1-38.

Rees, A. I., 1968. The production of preferred orientation in a concentrated dispersion of elongated and flattened grains, *J. Geol.,* 76, 457-465.

Sanders, J. E., 1965. Primary sedimentary structures formed by turbidity currents and related resedimentation mechanisms, *Soc. Econ. Paleont. Mineral. Spec. Publ. 12,* 192-219.

Shepard, F. P., and Dill, R. F., 1966. *Submarine Canyons and Other Sea Valleys.* Chicago: Rand McNally, 381p.

Stanley, D. J., 1975. Submarine canyon and slope sedimentation (Gres d'Annot) in the French Maritime Alps, *Guidebook, 9th International Congress of Sedimentology, Nice, France,* 129p.

Stauffer, P. H., 1967. Grain-flow deposits and their implications, Santa Ynes mountains, California, *J. Sed. Petrology,* 37, 487-508.

Cross-references: *Bagnold Effect; Fluidized Sediment Flows; Gravity Flows; Gravity-Slide Deposits; Inverse Grading; Mudflow and Debris-Flow Deposits; Turbidity-Current Sedimentation.* Vol. I: *Sand Flows and Sand Falls.*

GRAIN-SIZE FREQUENCY STUDIES

The analysis of the textural aspects of clastic sediments is a particularly valuable tool to the sedimentologist in the reconstruction of sedimentary processes. The grain-size distribution of a sediment is sensitive to very minor changes in water depth, shear stress, turbulence, and flow velocity. These factors cause the modification of the typically log-normal source distribution (Fig. 1). This source distribution is variable in grain size, sorting, and character of truncation of both the coarse and fine tails (see *Grain-Size*

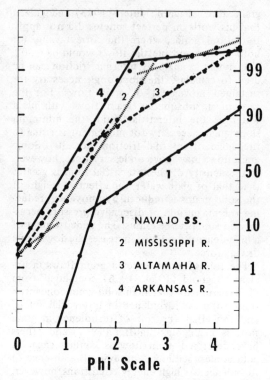

FIGURE 1. The form of the grain-size distribution is a modification of size distributions formed by weathering and crushing. The form reflects transportational history which causes truncation and segregation into bedload and suspension populations.

Studies). These aspects are the product of the transportational and depositional history of the clastic sediment.

The essentially log-normal character of the source distribution is an important consideration in the interpretation of depositional mechanics. Of particular importance in the analysis of size distributions is the modification that occurs at both ends of the distribution. For this reason, simple bar graphs, cumulative curves, and statistics describing normal curves are not sensitive enough to provide sufficient data to determine depositional processes. The log-normal plot of grain-size data provides a mechanism to illustrate the changes in the tails of the distribution (Fig. 2). In addition, small changes in the amounts and sorting of segments of the grain-size distribution are illustrated and can be readily compared to simple log-normal distributions and to other distributions (Fig. 3).

Analysis

Interpretation of depositional processes depends upon the analysis of truncation points

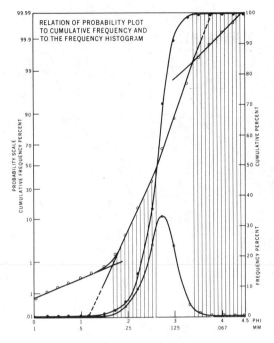

FIGURE 2. Comparison of methods of plotting grain-size distribution curves. Variability is most pronounced in the tails of the distribution, and the probability scale emphasizes changes in these regions.

and slopes of curve segments. These aspects are well illustrated on log-probability plots, and textural responses reflect the mechanics of deposition. A heterogeneous source distribution can be modified in a few days to produce characteristic truncation points and segment slopes (Fig. 4). Characteristic patterns are developed that are directly comparable to those found in modern environments. Similar processes produce nearly identical textural responses.

The process analysis of textural patterns is illustrated by studies of the beach environment (Kolmer, 1973). Laboratory studies indicate that fundamental differences occur in the mechanisms of transport and grain selection at the depositional interface by swash and backwash (Fig. 5). The combination of laminae resulting from these two events produces a characteristic curve shape with a very slight but identifiable change in sorting in the center of the distribution. This inflection point is a product of mixing of two different populations resulting from slightly different physical conditions of transport and deposition.

Analysis of grain-size distributions requires the interpretation of mechanisms of sedimentation. Of particular importance is the extraction of a well-sorted population from the bed load. This central portion of the curve is a product of grain selection from the moving bed load and reflects rates of sediment transport and deposition. The slope or sorting of this curve segment can be shown to vary with water depth, velocity, and bed shear (Fig. 6). The range of sizes selected also can be related to bed shear, turbulence, water depth, and current velocity. This aspect is reflected by the position of truncation points bounding the central curve segment (Fig. 7). Bed shear appears to have the dominant influence on positioning the coarser truncation point, and turbulence on positioning the finer truncation point.

The combination of position of truncation points, and the slope of sorting of the central curve segment provides the principal components for the interpretation of the genesis of a grain-size distribution. Other aspects, such as limits placed on the source distribution and mixing or recombination of separated populations, can be interpreted by adding or subtracting segments of the grain-size distribution, obtained by mathematically mixing end-member populations (Klovan, 1966; Visher, 1974). The analysis of shapes of log-probability curves does allow the reconstruction of the processes that were important in their development.

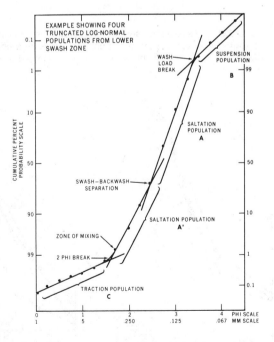

FIGURE 3. Subtle variations in slopes, and positions of truncation points are exemplified by the log-probability cumulative frequency plots. These are easily compared to other distributions to determine similarities and contrasts. Portions of each distribution may be compared to modes of deposition.

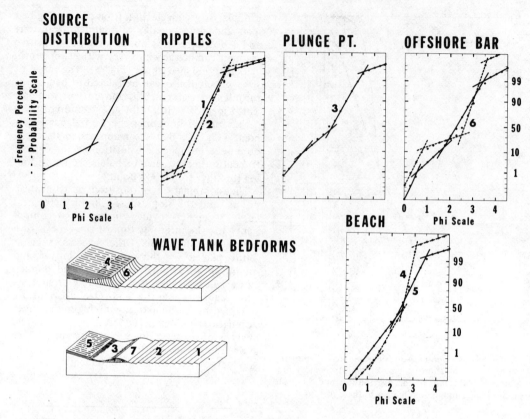

FIGURE 4. Changes in slopes and positions of truncation points reflect depositional processes. Curves 1–7 illustrate the effect of process within a few days on a heterogeneous size distribution. These distributions produced in a wave tank are comparable to those sampled from modern beaches.

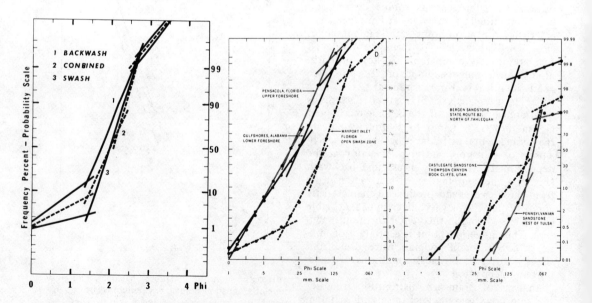

FIGURE 5. Subtle changes in depth, bed shear, and sediment concentration produce differing slopes and positions of truncation points. Interlamination of grain-size distributions formed under differing physical conditions may produce characteristic log-probability cumulative frequency distributions.

GRAIN-SIZE FREQUENCY STUDIES

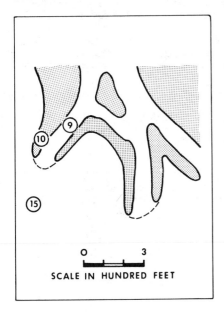
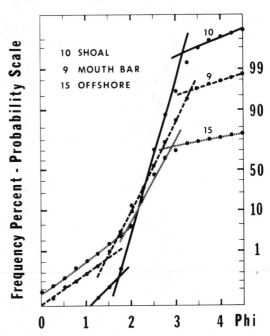

FIGURE 6. Systematic changes in size frequency of truncation of grain-size distribution may be directly related to water depth, bed shear, and sediment concentration.

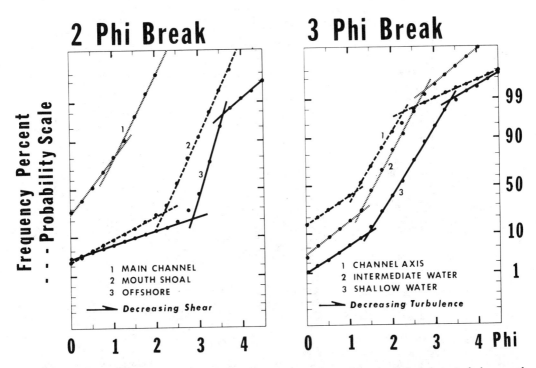

FIGURE 7. Systematic changes in the grain size at truncation points reflects depth, bed shear, turbulence, and sediment concentration. Positions of truncation points may be used for reconstructing depositional processes.

Applications to Environmental Analysis

The process aspect of textural analysis is a developing field in the study of sedimentary deposits. New insights are being developed, and the application to interpretation of the genesis of grain-size distribution is rapidly progressing. The shapes of grain-size curves from known depositional sites can easily be used as patterns for comparison to grain-size distributions from ancient sedimentary units. This technique does not always provide a unique interpretation of the sedimentary process or depositional environment, but it does typically limit the range of possibilities and is an important supporting method for the determination of the origin of ancient stratigraphic units. This method has been described and characteristic patterns presented for most depositional conditions (Sindowski, 1958; Visher, 1969).

A combined approach using C/M plots (Passega, 1967), statistical measures (see next entry), factor analysis (Klovan, 1966), and discriminant function analysis can be important in determining depositional environments (Moiola and Spencer, 1973; Reed et al., 1975). In conjunction with process analysis of grain-size distributions and empirical correlation of curve shapes, a powerful tool in the interpretation of the depositional mechanics of clastic sediments can be obtained. Textural analysis is one of the principal methods of interpretating the history of sedimentation in ancient stratigraphic units.

GLENN S. VISHER

References

Blatt, H.; Middleton, G. V.; and Murray, R. C., 1972. *Origin of Sedimentary Rocks.* Englewood Cliffs, N.J.: Prentice-Hall, 634p.

Klovan, J. E., 1966. The use of factor analysis in determining depositional environments from grain-size distributions, *J. Sed. Petrology,* 36, 115-125.

Kolmer, J. R., 1973. A wave tank analysis of the beach foreshore grain size distribution, *J. Sed. Petrology,* 43, 200-204.

Moiola, R. J., and Spencer, A. B., 1973. Sedimentary structures and grain size distribution, Mustang Beach Island, Texas, *Trans. Gulf Coast Assoc. Geol. Soc., 23d Annual Meeting,* Houston, Texas, 324-332.

Passega, R.; Rizzini, A.; and Borghetti, G., 1967. Transport of sediments by waves, Adriatic coastal shelf, Italy, *Bull. Am. Assoc. Petrol. Geologists,* 51, 1304-1319.

Reed, W. E.; LeFever, R. E.; and Moir, G. J., 1975. Depositional environment interpretation from settling-velocity (Psi) distributions, *Geol. Soc. Am. Bull.,* 86, 1321-1328.

Sindowski, K. H., 1958. Die synoptische Methode des Korkurven—Vergleichesch zur Aussenzur Fossiler Sedimentationsraume, *Geol. Jahrb.,* 73, 235-275.

Visher, G. S., 1969. Grain size distributions and depositional processes, *J. Sed. Petrology,* 39, 1074-1106.

Visher, G. S., and Howard, J. D., 1974. Dynamic relationship between hydraulics and sedimentation in the Altamaha estuary, *J. Sed. Petrology,* 44, 502-521.

Cross-references: *Beach Sands; Eolian Sands; Glacial Sands; Grain-Size Parameters—Environmental Interpretation; Grain-Size Studies; River Sands; Sands and Sandstones.*

GRAIN-SIZE PARAMETERS— ENVIRONMENTAL INTERPRETATION

Textural parameters, such as grain size, sorting (q.v.), and skewness (q.v.), which are based on the shape of size-frequency curves, are environment sensitive. They reflect the mode of transportation and the energy conditions of the transporting medium. The literature on this subject has been reviewed by Folk (1966) and Friedman (1961, 1967; see also *Grain Size-Frequency Studies*).

Data

In Fig. 1, the skewness of the size-frequency distribution for dune and beach sand is plotted against the mean grain size for these samples, using the phi (ϕ) scale. The phi scale (q.v.) is a commonly employed scale in studies of sedimentary rocks (Krumbein, 1934). Between the dune and beach sands a separation is found for the two depositional environments, the dune sands being for the most part positively skewed and the beach sands generally negatively skewed. Polymodal beach sands or beach sands that have not yet come to equilibrium with their environment (such as some of the beach sands of the Louisiana coast, see Friedman, 1967) may be positively skewed (Fig. 1).

In Fig. 2, skewness has been plotted against

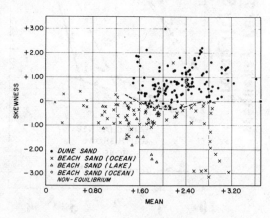

FIGURE 1. Plot of mean and skewness, using phi (ϕ) scale, for beach and dune sands (from Friedman, 1961).

GRAIN-SIZE PARAMETERS—ENVIRONMENTAL INTERPRETATION

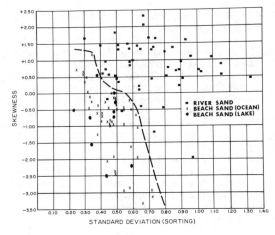

FIGURE 2. Plot of skewness and sorting (standard deviation), using phi (ϕ) scale, for beach and river sands (Friedman, 1961).

sorting for medium to fine and very fine-grained beach and river sands. This figure indicates that beach sands are for the most part better sorted than river sands and tend to be negatively skewed, whereas river sands tend to be positively skewed. On the basis of this plot, it is possible to distinguish beach from river sands, but it should be noted that exceptions exist. Coarse-grained beach and river sands can be either positively or negatively skewed; for these sands there is no simple pattern; and, therefore, it is not possible to distinguish between coarse-grained sands of different environments on the basis of standard deviation and skewness. Other textural parameters exist, however, which permit differentiation of coarse-grained beach and river sands (Friedman, 1967).

Sorting of sands and depositional environment are correlatable, as shown in Table 1 (Friedman, 1962).

TABLE 1. Genetic Sorting Classification Based on Standard Deviation

Sorting Interval	Sorting Designation	Environment of Deposition
Medium to fine- and very-fine-grained sands (mean >1.0-2.0ϕ):		
<0.35	very well sorted	most coastal, barrier-bar, and lake-dune sands, many beach sands; many marine sands above wave base; many lagoonal sands.
0.35-0.50	well sorted	most beach sands; many or most marine sands above wave base; many lagoonal sands; many inland dune sands; some river sands.
0.50-0.80	moderately well sorted	most river sands; many beach sands (0.80 is approx. upper limit for beach sands); many lagoonal sands from restricted lagoons; most continental shelf sands below wave base; most inland dune sands (0.80 is approx. upper limit except for some stable dunes).
0.80-1.40	moderately sorted	many river sands (1.40 is approx. upper limit for river sands); some lagoonal sands from restricted lagoons; some continental shelf sands below wave base; many glaciofluvial sands.
1.40-2.00	poorly sorted	many glaciofluvial sands.
2.00-2.60	very poorly sorted	many glaciofluvial sands.
>2.60	extremely poorly sorted	some glaciofluvial sands.
Coarse-grained sands (mean $<1.00\phi$):		
0.50-0.80	moderately well sorted	many beach sands.
0.80-1.40	moderately sorted	most river sands; many or most beach sands; most continental shelf sands.
1.40-2.00	poorly sorted	some river sands; some continental shelf sands; many glaciofluvial sands.
2.00-2.60	very poorly sorted	many glaciofluvial sands.
>2.60	extremely poorly sorted	some glaciofluvial sands.

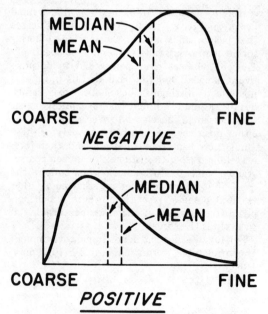

FIGURE 3. Curves showing (top) negatively skewed and (bottom) positively skewed distributions. Phi value increases and grain size decreases to the right.

Explanation of Empirical Data. The waves that deposit sand on the beach have a greater competency than the wind transporting sands onto dunes. This difference in competency between water and wind explains why many beach sands are coarser than dune sands. Fig. 1 puts the upper limit for mean grain size of dune sands at 1.50 phi, whereas beach sands commonly are considerably coarser (i.e., have smaller phi values). In a high-energy, nearshore environment such as that of a beach, the sand is more effectively sorted than in a fluvial environment (Fig. 2). Thus, beach sands tend to be better sorted than river sands. On a marine beach, sand is exposed to incoming waves and outgoing wash, which remove fine-grained particles. The frequency distribution curve of a winnowed sand lacks the "tail" at the fine-grained end of the curve and the distribution is negatively skewed (Fig. 3), if the phi system is employed in plotting the data. By contrast, the fine-grained particles in dune and river sands are trapped between the coarser grains and are thereby incorporated in the sand, leading to a positive skew of the size-frequency curve.

GERALD M. FRIEDMAN

References

Cronan, D. S., 1972. Skewness and kurtosis in polymodal sediments from the Irish Sea, *J. Sed. Petrology,* 42, 102-106.

Folk, R. L., 1966. A review of grain-size parameters, *Sedimentology,* 6, 73-93.

Friedman, G. M., 1961. Distinction between dune, beach and river sands from their textural characteristics, *J. Sed. Petrology,* 31, 514-529.

Friedman, G. M., 1962. On sorting, sorting coefficients, and the log normality of the grain-size distribution of sandstones, *J. Geol.,* 70, 737-753.

Friedman, G. M., 1967. Dynamic processes and statistical parameters compared for size frequency distribution of beach and river sands, *J. Sed. Petrology,* 37, 327-354.

Hails, J. R.; Seward-Thompson, B.; and Cummings, L., 1973. An appraisal of the significance of sieve intervals in grain size analysis for environmental interpretation, *J. Sed. Petrology,* 43, 889-893.

Krumbein, W. C., 1934. Size frequency of sediments, *J. Sed. Petrology,* 4, 65-77.

Cross-references: *Beach Sands; Grain-Size Frequency Studies; Grain-Size Studies; Kurtosis; Phi Scale; River Sands; Skewness; Sorting.*

GRAIN-SIZE STUDIES

Mineral particles, carried by various transporting agencies such as wind or running water, are conveniently described in terms of their sizes. Thus we have a range of common sizes, from boulders (the largest), down to sand, silt, and clay (clay, q.v., also has a mineralogical meaning). A *boulder* must be at least 256 mm in diameter. A clay particle, on the other hand, should be no larger than 1/256 mm (about 0.004 mm) in diameter. This large range in sizes suggests that an ordinary arithmetic size scale will not be equally useful at all points in the range. In addition, the techniques used for study are greatly varied (see *Granulometric Analysis*).

The finest material, clay, is not easily separable and is excluded from many size studies for this reason. Silt is easier to handle than clay, but nevertheless requires techniques that may not be identical with those used for sand. Sand is the most easily studied grain size, using sieving techniques. Gravel is best measured in the field, inasmuch as quantities great enough for detailed study are too bulky to bring into the laboratory, and special measuring techniques must, therefore, be employed. Nevertheless, a complete treatment of grain size requires a study of gravel, sand, silt, and clay, even though some of these sizes are measured with methods not appropriate for other sizes. Whatever the technique used, the data obtained represent an approximation of grain diameter. Highly accurate studies are generally impossible, partly because very few grains are spherical (see *Roundness and Sphericity*), and partly because of cost.

Size Classification

An early size classification was produced in the US by J. A. Udden, in 1898, using powers of 2, both positive and negative, to mark the limiting sizes of classes, e.g., 2^{-1} (1/2) mm, 2^1 mm (see *Udden Scale*). In 1905, Atterberg proposed a decimal system (i.e., 2 mm, 0.2 mm, 0.02 mm, 0.002 mm, etc.; see *Atterberg Scale*). These boundaries were based on physical behavior, specifically on the differences between water capacities near 2 mm diameter, "wetness" near 0.2 mm, visibility near 0.02 mm, and Brownian movement near 0.002 mm. This scale has been widely used in Europe. In 1899, C. G. Hopkins had also suggested a decimal system, starting at 1 mm. Both the Atterberg and Hopkins scales provide classes that are too large, however, and must, therefore, be subdivided in some way. In 1943, H. L. Alling divided the classes of the Hopkins scale on the basis of the fourth root of 10, thereby producing four times as many classes (see *Alling Scale*). The scale adopted by C. K. Wentworth in 1922, a slightly modified version of Udden's proposal, is still in wide use (see *Wentworth Scale*). W. C. Krumbein applied a transformation, $\phi = -\log_2 x$, where x is the grain diameter, to the Wentworth scale. This transformation has proven extremely useful and popular in sediment size studies. Because the *phi scale* (q.v.) does not furnish sufficiently small subdivisions for certain analyses, "half-phi" and "quarter-phi" schemes have been adopted. Quarter-phi divisions permit enough precision for most research purposes. Page (1955) published a detailed phi-to-millimeter conversion table, accurate to four digits throughout most of its range, and designed to be entered from either kind of notation. A comparison of various studies, their interrelation and absolute values, may be seen in Table 1.

The Rubey scale, following a comment by G. W. Robinson, was based on settling velocities rather than on diameters. Eight divisions were established, from very fine sand to fine clay, with settling velocities varying from $>3840\mu$/sec to $<0.9375\mu$/sec. The ratio between the successive size fractions is a constant 1:4 (see *Settling Velocity*). Shea (1973) has proposed a new grade scale based on 10.

For several years geologists, soil scientists, engineers, and others concerned with grain-size measurements have been attempting to adopt a standard size scale. An initial report of this work (Tanner, 1969) showed that the geologically acceptable boundary between silt and clay, at about 0.004 mm, would have to be replaced by the more widely used nongeological boundary at 0.002 mm (a small change having no real effect as far as interpretation of data is concerned), and that the sand/silt-boundary question was not resolved. More recently (1973) the soil scientists have initiated a move to have the sand/silt boundary fixed at 1/16 mm (0.0625 mm), which has been used by American geologists for decades. The gravel/sand boundary remains at 2 mm.

Practical Methods of Measurement

The most commonly used method of obtaining actual grain measurements is by sieving. For this purpose, standard sieves, in full-phi, half-phi, or quarter-phi sets, are obtainable. Nested screens are generally used on automatic shaking equipment, such as the Ro-Tap machine. Inasmuch as the machine capacity for a stack of sieves is limited, it is obvious that measurement in quarter-phi intervals may require two or three times as many machine runs as measurement in full-phi units. Hence the operator may have to weigh machine time against precision desired, before undertaking an analysis. At least 30 min of shaking time, per stack of sieves, is necessary if reproducible results are desired. For coarse sand to silt sizes, sonic sifters, which are less time consuming, may be used.

Settling in water or air, either with or without screens, is also used for size analysis, especially in the silt and clay sizes. In making settling-time measurements (Gibbs, 1972), one assumes that all of the particles have the same density and approximately the same shape and roundness, and further that local large grains or clusters do not set up any serious turbulence. This last assumption is certainly not correct for sand-sized particles, inasmuch as individual grains of medium or fine sand, when dropped in a column of still water, generally develop pronounced von Karman vortex trails behind (e.g., above) them, and, therefore, distance between grains within the settling tube becomes an important variable. On the other hand, sieve analysis is based on the more-or-less tacit assumption that all grains of one size are essentially equant in shape and have equal opportunities to pass through any specified screen.

Actual measurement of separate grains (e.g., grain counts), although tedious and time consuming, is occasionally carried out under the microscope. This technique is particularly useful where well-lithified sediments are to be studied. Sieving, settling, counting, and other techniques, are *not* freely interchangeable. The results of one procedure cannot be combined with the results of other methods, except in rare cases (see *Granulometric Analysis*).

Treatment of Data

Histograms. In view of the large numbers of grains present on or near the surface of the

TABLE 1. Comparison of Grain Size Scales (after Truesdell and Varnes, 1950)

This chart compares grain-size grade scales. Grain size (in mm) runs horizontally from 1.0 mm (left) to 0.0001 mm (right); the several classification systems are shown as stacked horizontal bands. The size-class labels within the cells are printed as rotated (vertical) text.

Size-class headers (top band):

Coarse | Medium | Fine | Very fine | Coarse | Medium | Fine | Very fine | Coarse | Medium | Fine | Very fine | Coarse | Medium | Fine | Very fine

mm scale: —1.0— ... —.10— ... —.01— ... —.001—

Consolidated equivalents: **SANDSTONE | SILTSTONE | CLAYSTONE | COLLOIDSTONE**

Band 1:
—1.00(No.18).0394— Coarse sand —.50(No.35).0197— Medium sand —.20—.0079— Fine sand —.10—.0039— Coarse Mo —.05—.0020— Fine Mo —.02—.00079— Coarse silt —.006—.0024— Fine silt —.002—.00079— Coarse clay —.0006-.000024— Fine clay —.00002-.0000079— Ultra clay

Band 2:
—1.00—.0394— Coarse sand —.50—.0197— Medium sand —.20—.0079— Fine sand —.10—.0039— Very fine sand —.05—.0020— Coarse silt —.02—.00079— Medium silt —.01—.00039— Fine silt —.005—.00020— Coarse clay —.002—.000079— Medium clay —.001—.000039— Fine clay —.0005-.000020— Colloids

Band 3:
Coarse sand | Medium sand | Fine sand | Very fine sand | Silt | Clay

Band 4:
Coarse sand | Medium sand | Fine sand | Very fine sand | Coarse silt | Medium silt | Fine silt | Very fine silt | Coarse clay | Medium clay | Fine clay

Band 5 (German):
Grobsand (Coarse sand) —.20—.0079— Feinsand (Fine sand) —.02—.00079— Schluff (Silt) —.002—.000079— Kolloidale Teilchen oder Rohton (Colloidal particles or raw clay)

Band 6:
—1.00(No.18).0394— Coarse sand —.50(No.35).0197— Medium sand —.25(No.60).0098— Fine sand —.10—.0039— Very fine sand —.05—.0020— Silt —.002—.000079— Clay

Band 7:
—1.00—(No.18)—.0394— Coarse sand —.50—(No.35)—.0197— Medium sand —.25—(No.60)—.0098— Fine sand —.125—(No.120)—.0049— Very fine sand —.0625— Coarse silt —.0313—.0013— Medium silt —.0156—.00062— Fine silt —.0078—.00031— Very fine silt —.0039—.00015— Coarse clay size —.00195—.000077— Medium clay size —.00098—.000038— Fine clay size —.00049—.000019— Very fine clay size —.00024—.0000096—

Band 8:
Medium sand —.297(No.50)—.0117— Fine sand —.074(No.200).0029— Clay (plastic) to Silt (nonplastic)

Band 9:
Coarse sand —.42—(No.40)—.065— Fine sand —.074-(No.200)(.0029)— Silt: If liquid limit is 27 or less and plasticity index (based on -40 fraction) is less than 6, or Clay: If liquid limit is over 27 and plasticity index (based on -40 fraction) is 6 or greater.

Band 10:
Coarse sand —.25—(No.60)—.0098— Fine sand —.05—.0020— Silt —.005—.00020— Clay —1/1024—.000038— A.S.T.M. / A.A.S.H.O. Colloids

Phi (φ) scale:
0 | +1 (1/2) | +2 (1/4) | +3 (1/8) | +4 (1/16) | +5 (1/32) | +6 (1/64) | +7 (1/128) | +8 (1/256) | +9 (1/512) | +10 (1/1024) | +11 (1/2048) | +12 (1/4096)

Millimetre ruler: 1.0 ... .10 ... .01 ... .001 ... .0001

Sieve / particle-diameter data table (bottom):

U.S. sieve no.	U.S. sieve no.	Opening (mm)	Diameter (mm)
16	18	1.00	.0394
20	25	0.840	.0331
24	30	0.710	.0280
28	35	0.590	.0232
32	40	0.500	.0197
35	45	0.420	.0165
42	50	0.350	.0138
48	60	0.297	.0117
60	70	0.250	.0098
65	80	0.210	.0083
80	100	0.177	.0070
100	120	0.149	.0059
115	140	0.125	.0049
150	170	0.105	.0041
170	200	0.088	.0035
200	230	0.074	.0029
250	270	0.062	.0024
270	325	0.053	.0021
325	400	0.044	.0015
400	400	0.037	.0005

earth, size data must be handled statistically. Historically, graphic *models* were adopted early; these bar graphs and *histograms* were drawn so that each bar represented one class, and the bar heights (actually the areas) indicated the relative proportions. For ease of work, it was common practice to limit the classes by geometrically spaced boundaries (Fig. 1). Histograms had the advantage that a large number of grains (perhaps 10^7 or 10^8) could be represented, after sieving, by a few bars. With such a large sample, one felt confident that the original grain population (i.e., part of a dune, beach, sand bar, or rock) had been sampled adequately. The bar pattern was easy to read: tall bars indicated large amounts of material, short bars relatively little. The disadvantage lay in the fact that only intuitive interpretations could be made from the bar graph. If systematic methods of treatment could be devised, the results could be presented in a more rigorous and meaningful manner; various statistical methods have made this possible.

Probability Plots. Routine laboratory practice is to dry and weigh the study sample prior to separation. For sieving, a sample of 15–50 g is adequate; for settling, only about 2 g. After separation, subsamples are weighed and stored in suitable vials, envelopes, or plastic boxes. The weights are converted to percentages, and summed from one end of the scale to the other, generally from the coarser sizes down to the finer sizes. The cumulative weight percent, from coarse to fine, is then calculated. If the cumulative percent does not add up to exactly 100% because of various common discrepancies, it should be recalculated (to 99.99% for convenience in plotting), distributing the error proportionately throughout the column. The corrected cumulated percents can then be entered on probability paper as dots (points), which can be connected with straight line segments; other methods of plotting are not recommended. The *median* value can be read directly off of this plot, at the point where the sediment size curve intersects the 50% line (see *Grain Size-Frequency Studies*).

Ideally, a sample is expected to plot on probability paper as a straight line. Most samples do not. This result has been interpreted in at least three different ways: (1) the Gaussian assumption, underlying the construction of probability paper, is not applicable to sediment sizes; (2) serious errors have been made in sampling, weighing, sieving, computing, or plotting; or (3) the departures from a straight line are sources of valuable information. Most analysts have been reluctant to accept the first, and inclined to prefer the third. If the basic Gaussian assumption is accepted, and it is thought that errors are minimal, departure from a straight line can be interpreted in terms of mixing of two or more sediment types, or subtraction of a statistically meaningful fraction of a single sediment type. Each of these possibilities can be investigated by either statistical or field methods (see *Grain-Size Frequency Studies*).

Statistical Parameters. For general comparison of numerous samples, moment measures are useful. This sytem includes the *mean, standard deviation, skewness* (q.v.), and *kurtosis* (q.v.); higher moments do not appear to be sufficiently stable to be reliable (Hoel, 1958). These four measures can be interpreted as, respectively, the average size, the scatter of various sizes about the average, the asymmetry about the average, and the peakedness of the curve (Fig. 2). Of the four, the first two have

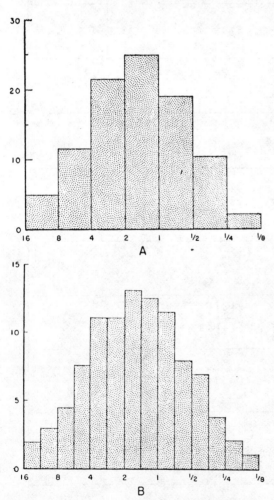

FIGURE 1. Histograms of size distributions in a clastic sediment (from Pettijohn, 1957). A, based on Udden size grades; B, based on half Udden size grades.

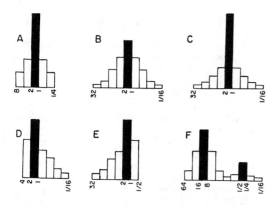

FIGURE 2. Types of size frequency distributions (from Pettijohn, 1957). A, B, and C have similar median size; A and B differ in the sorting; C, though similar in median and sorting to B, differs in its kurtosis; D and E are markedly skewed and differ in direction of skewing; F is a bimodal distribution; all others are unimodal.

been used much more extensively than the last two; they are also easier to interpret. Various graphic approximations have been suggested, but they are less reliable (Folk, 1966).

For simple, unimodal distributions (straight lines on probability paper), the median = 50% value = mean = mode. The standard deviation can be approximated by the graphic device $\sigma_I = 0.5(\phi_{84} - \phi_{16})$; where the subscripts indicate phi sizes to be obtained from the probability plot by reading the intersection of the curve with the 84% and 16% lines (see *Grain-Size Frequency Studies*). A more generalized graphic method produces a deviation measure, $D_k = (1/k)(P_a - P_b)$, in which P_a and P_b are selected percentiles, and k is the number of standard deviations between those two percentiles and hence is dependent on a and b. A small value for D_k (whether a true standard deviation, or σ_I, or some other approximation) indicates a very narrow range of sizes in the sample, whereas a large number indicates a wide range. Typical values are 0.4 for narrow range, 2.5 for wide range.

An older parameter which served about the same purpose is the *sorting coefficient*, So, obtained from the size plot as follows: $So = (Q_3/Q_1)^{0.5}$. In this equation, Q represents quartile sizes, taken from the size curve at the 25% and 75% lines, with the stipulation that Q_3 is the larger (e.g., coarser) quartile. The two basic measures of scatter are related (for simple cases) by means of the equation $D_k = \log_2 So^{1.48}$. Early use of the sorting coefficient led to expressions such as "well sorted," and "poorly sorted." A poorly sorted sediment might have $So = 2.5$ ($D_k \approx 2$); a well-sorted sediment, $So = 1.2$ ($D_k \approx 0.4$). Dune, beach, and some river sands have D_k values below 0.5 (see *Sorting*).

At one time it was thought that mean and standard deviation ($\approx$ sorting) might provide sufficient information to permit identification of depositional environment for any given sediment. A widely used procedure was to plot two such parameters on coordinate paper, and then to circle, as distinct "fields" of dots, the plots made by river sediments, dune sands, beach sands, and so forth. In a number of local areas, plots of a few hundred samples did fall into distinct and recognizable "fields." The addition of more data, from other areas, invariably has blurred the boundaries between the designated "fields." That is, one must know the environment of deposition (so as to exclude troublesome areas) before an interpretation can be made (see *Grain-Size Parameters—Environmental Interpretation*).

More recently, three- (or four-) dimensional plots, using all or most of the first four moment measures, have been tried, with modest success. In some areas, the available reservoir of quartz sand is so uniform that estimates of the mean size are almost useless. In other studies, comparing sediments from different areas, the grain sizes may differ so much that this one measure tends to overwhelm all of the others. A variety of additional procedures has been undertaken, in an effort to remedy the failings of simple multidimensional plots of grain-size moment data or approximations.

Much of this work was summarized at a grain-size symposium held in November 1973 (see *Geol. Soc. Am. Abstracts with Programs*, 5(7), xlvi and li). Papers presented at this symposium revealed three trends: (1) toward using raw classified weight-percent data obtained by techniques such as sieving or settling, rather than moment measures, as the basis for discriminant analysis; (2) toward using some form of discriminant analysis coupled with canonical plotting, which involves computer generation of four new axes in optimum orientation, each axis representing projections of the original four; and (3) toward making some kind of a hydrodynamic interpretation of segments and internal shapes of individual curves.

The symposium also revealed a clear dichotomy in goals: some wanted paleogeographic information (environment of deposition), whereas others proposed to describe hydrodynamics. The former was generally attempted by placing each sample on a discriminant or canonical plot having boundaries between different environments (such as beach, dune, river). The latter was done in two different and mutually exclusive ways: (1) by interpreting each segment of a non-Gaussian size-distribution curve as if it existed independently of adjacent segments, and therefore represented a unique hydrodynamic size fraction, and (2) by examining the character of the entire curve, in terms of such statistical processes as mixing, filtering, and truncation, which are taken to represent certain hydrodynamic processes rather than certain hydrodynamic size fractions.

Other reports represent these various points of view in more detail but perhaps with less data and sophistication. The paleogeographic approach has been used by Hails (1967), Friedman (1967), Landim and Frakes (1968), Buller and McManus (1972), Glaester and Nelson (1974), and many others; it is successful in some instances but commonly cannot be extended from a given study to a totally different one. The curve-segment approach to hydrodynamic interpretation has been used by Visher (1969), Allen et al. (1972), Kolmer (1973), and others. The use of curve character as a clue, or clues to hydrodynamic process, was spelled out by Tanner (1960, 1964), and has been applied to field problems subsequently (Stapor, 1973a; Stapor and Tanner, 1975). The curve-segment technique is attractive at first glance, but assumes a consistency of hydrodynamic fractionation of sizes, under widely varied energy conditions, which is not warranted. The curve-character approach is the most difficult method of interpretation and provides neither precise paleogeographic information nor hydrodynamic labels for the individual curve segments. The sizes of various heavy-mineral varieties also yield information of a hydrodynamic character (Stapor, 1973b; May, 1973).

WILLIAM F. TANNER

References

Allen, G. P.; Castaing, P.; and Klingebiel, A., 1972. Distinction of elementary sand populations in the Gironde Estuary (France) by R-mode factor analysis of grain-size data, *Sedimentology*, 19, 21-35.

Buller, A. T., and McManus, J., 1972. Simple metric sedimentary statistics used to recognize different environments, *Sedimentology*, 18, 1-21.

Folk, R. L., 1966. A review of grain-size parameters, *Sedimentology*, 6, 73-93.

Friedman, G. M., 1967. Dynamic processes and statistical parameters compared for size frequency distributions of beach and river sands, *J. Sed. Petrology*, 37, 327-354.

Gibbs, R. J., 1972. The accuracy of particle-size analyses utilizing settling tubes, *J. Sed. Petrology*, 42, 141-145.

Glaester, R. P., and Nelson, H. W., 1974. Grain-size distribution, an aid in facies identification, *Bull. Canadian Petrol. Geol.*, 22, 203-240.

Hails, J. R., 1967. Significance of statistical parameters for distinguishing sedimentary environments in New South Wales, Australia, *J. Sed. Petrology*, 37, 1059-1069.

Hoel, P. G., 1958. *Introduction to Mathematical Statistics*. New York: Wiley, 331p.

Irani, R. R., and Callis, C. F., 1963. *Particle Size: Measurement, Interpretation and Application*. New York: Wiley, 165p.

Kolmer, J. R., 1973. A wave tank analysis of the beach foreshore grain size distribution, *J. Sed. Petrology*, 43, 200-204.

Landim, P. M. B., and Frakes, L. A., 1968. Distinction between tills and other diamictons based on textural characteristics, *J. Sed. Petrology*, 38, 1213-1223.

May, J. P., 1973. Selective transport of heavy minerals by shoaling waves, *Sedimentology*, 20, 203-211.

Page, H. G., 1955. Phi-millimeter conversion table, *J. Sed. Petrology*, 25, 285-292.

Pettijohn, F. J., 1957. *Sedimentary Rocks*, 2nd ed. New York: Harper & Row, 718p.

Rubey, W. W., 1933. Settling velocities of gravel, sand, and silt, *Am. J. Sci.*, 225, 325-338.

Shea, J. H., 1973. Proposal for a particle-size grade scale based on 10, *Geology*, 1, 3-8.

Stapor, F. W., 1973a. Coastal sand budgets and Holocene beach ridge plain development, Northwest Florida. Ph.D. dissertation, Florida State Univ., Tallahassee.

Stapor, F. W., 1973b. Heavy mineral concentrating processes and density/shape/size equilibria in the marine and coastal dune sands of the Apalachicola, Florida, region, *J. Sed. Petrology*, 43, 396-407.

Stapor, F. W., and Tanner, W. F., 1975. Hydrodynamic implications of beach, beach ridge and dune grain size studies, *J. Sed. Petrology*, 45, 926-931.

Tanner, W. F., 1960. Filtering in geological sampling, *Am. Statist.*, 14(5), 12.

Tanner, W. F., 1964. Modification of sediment size distributions, *J. Sed. Petrology*, 34, 156-164.

Tanner, W. F., 1969. The particle size scale, *J. Sed. Petrology*, 39, 809-812.

Truesdell, P. E., and Varnes, D. J., 1950. *Chart Correlating Various Grain-Size Definitions of Sedimentary Material*. Washington, D.C.: U.S. Geol. Survey.

Vandenburghe, N., 1975. An evaluation of CM patterns for grain size studies of fine grained sediments, *Sedimentology*, 22, 615-622.

Visher, G. S., 1969. Grain size distributions and depositional processes, *J. Sed. Petrology*, 39, 1074-1106.

Cross-references: *Alling Scale; Atterberg Scale; Biostratinomy; Clay; Elutriation; Grain-Size Frequency Studies; Grain-Size Parameters—Environmental Interpretation; Granulometric Analysis; Hydraulic Equivalence; Kurtosis; Phi Scale; Rapid Sediment Analyzer; Roundness and Sphericity; Sands and Sandstones; Sedimentary Petrography; Sedimentary Petrology; Sediment Parameters; Settling Velocity; Skewness; Sorting; Udden Scale; Wentworth Scale.*

GRANULOMETRIC ANALYSIS

The techniques commonly used for grain size (granulometric) analysis are: (1) sieving and pipetting, (2) settling, and (3) microscopic analysis. Particles coarser than sand can be measured directly, whereas electron microscopes might be needed for grains of clay size (Fig. 1).

In the *sieving method,* the material is shaken through a series of screens, and the particles retained on each screen are weighed to obtain the "weight-frequency distribution" of the sample (see *Grain Size Studies*). The method is easy to use (hence popular), but manufacturing defects in sieves are common, and the results of sieving are affected by particle shapes. Square apertures create additional difficulties. The materials finer than sand size are analyzed by *settling method* using a simple pipette or a sedimentation balance. The settling velocities of the grains are converted to sizes with the help of Stokes' Law (q.v.). Such composite analyses often show abrupt "breaks" in size distribution because the principles of separation by sieving and sedimentation are basically different. To avoid composite analysis, some investigators prefer to use decantation or elutriation (q.v.) methods (see Krumbein and Pettijohn, 1938). Statistical techniques for analyzing the weight-frequency data are not fully solved; hence it is customary to compute the different measures by the conventional statistical methods, treating the weight frequencies as number frequencies.

In a *settling tube,* the material is allowed to settle through a long column of water, and the amount of materials settled are automatically recorded against time with the help of a balance or a pressure-sensing device. Most modern equipment yields *continuous* records of settling times (see Sengupta and Veenstra, 1968, for a review of the common settling tubes). Whereas Stokes' Law is not valid for sizes greater than 60μ, the settling times are converted to grain sizes with the help of experimentally determined curves for single, spherical quartz grains. Corrections for grain shape are necessary because shape has a great influence on settling velocity. Grain density also influences the settling velocity.

Size measurement under the microscope can be achieved either directly by an eyepiece micrometer or by a microprojector. Loose grains can be measured directly. When measured in a thin section (as in the case of an indurated rock), a correction has to be applied to remove the effect of random sectioning (Krumbein and Pettijohn, 1938). Microscopic measurements are tedious, but the size-frequency distributions obtained through grain counts are amenable to treatment by standard statistical formulae. The problem of transformation of weight to number frequency has been handled both empirically and mathematically (Friedman, 1958, 1962; Sahu, 1964, 1965, 1966). Many attempts have been made to correlate grain-size distributions of sediments to the environments of sedimentation (see *Grain-Size Frequency Studies;* Pettijohn et al., 1972).

SUPRIYA SENGUPTA

References

Friedman, G. M., 1958. Determination of sieve-size distribution from thin-section data for sedimentary petrological studies, *J. Geol.,* 66, 394-416.

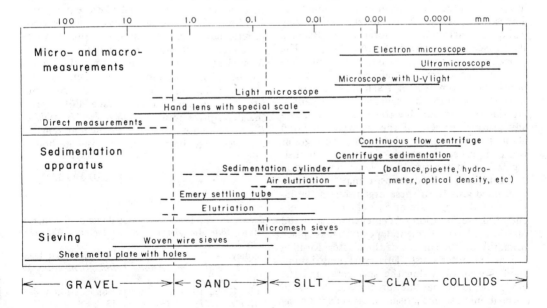

FIGURE 1. Range of grain size and methods of size analyses (from Pettijohn, 1975, modified from Mueller, 1967).

Friedman, G. M., 1962. Comparison of moment measures for sieving and thin-section data in sedimentary petrological studies, *J. Sed. Petrology,* **32,** 15-25.

Krumbein, W. C., and Pettijohn, F. J., 1938. *Manual of Sedimentary Petrography.* New York: Appleton-Century-Crofts, 549p.

Pettijohn, F. J., 1975. *Sedimentary Rocks,* 3rd ed. New York: Harper & Row, 628p.

Pettijohn, F. J.; Potter, P. E.; and Siever, R., 1972. *Sand and Sandstone.* New York: Springer-Verlag, 86-88.

Sahu, B. K., 1964. Transformation of weight frequency and number frequency data in size distribution studies of clastic sediments, *J. Sed. Petrology,* **34,** 768-773.

Sahu, B. K., 1965. Transformation of weight- and number-frequencies for phi-normal size distributions, *J. Sed. Petrology,* **35,** 973-975.

Sahu, B. K., 1966. Thin-section analysis of sandstones on weight-frequency basis, *Sedimentology,* **7,** 255-259.

Sengupta, S., and Veenstra, H. J., 1968. On sieving and settling techniques for sand analysis, *Sedimentology,* **11,** 83-98.

Cross-references: *Elutriation; Grain-Size Frequency Studies; Grain-Size Parameters—Environmental Interpretation; Grain-Size Studies; Sorting.*

GRAPESTONE

The term *grapestone* was proposed by Illing (1954) for clusters of rounded, calcareous pellets or other sand-sized grains cemented together by incipient aragonite precipitation. The protruding rounded grains give the aggregate a lumpy outer surface resembling a bunch of grapes. Grapestone forms in shallow marine carbonate environments, notably the Bahama Banks (see *Bahama Banks Sedimentology*). The processes of grapestone formation are not selective, and small grains of all types, skeletal or nonskeletal, complete or fragmentary, may be incorporated into the aggregates. Mechanical erosion and abrasion due to current action usually prevents growth of the aggregates beyond about 1 mm in diameter. Initial binding of grains is accomplished by growth of algae and encrusting forams in the substrate; later cementation and infilling of the composite may be due to continued growth of these organisms in the interior of the aggregate and to chemical or biochemical precipitation of cement (Winland and Matthews, 1974). Cement deposition takes place primarily on the surface of the cluster, forming a resistant outer crust. This crust may inhibit fluid movement within the aggregate, and the aragonite here may be precipitated in well-formed, microscopic, prismatic crystals radiating outward from the grain surfaces. With prolonged cementation, the component grain boundaries become obscured until, finally, the whole aggregate is composed of a uniform matrix, with no trace of its original composite structure.

When grapestone is subjected to oolitic accretion (see *Oolite*), each protruding grain is coated with a shiny calcareous layer, and the spaces between them are partly filled in, forming what Illing (1954) calls *botryoidal lumps*. He notes that the "very irregular, somewhat nodose white grains" described by Eardley (1938) from the oolite sands of the Great Salt Lake appear to be similar to the Bahaman botryoidal lumps.

Winland and Matthews (1974) proposed that the recrystallized ooids making up grapestone were formed as the sea transgressed onto the Bahaman Platform at the end of the last glacial lowstand, i.e., that grapestone is produced through a sequence of depositional conditions rather than a single set of conditions. They listed a supply of firm grains, uneven water turbulence, high water-circulation rates, and very low sedimentation rates as favorable for the formation of grapestone.

The term *bahamite* was proposed by Beales (1958) to denote ancient shallow-water carbonate deposits closely resembling those described by Illing (1954) on the Bahama Banks. These deposits are very pure, generally fine-grained, massively bedded deposits composed of concretionary and, commonly, composite grains. Bahamite is a descriptive term, and is not meant to imply formation under genetic conditions identical with those prevailing on the present-day Bahama Banks. Folk (1959), however, objected to the term because he felt that it does imply that the aggregates formed like the grapestone of Illing (1954) and hence has a very restricted genetic meaning. According to Folk, it would be almost impossible to prove that bahamite formed by this exact mechanism rather than as a result of ordinary modes of erosion, especially after abrasion had smoothed off the characteristic "grapestone" outer surface. Folk prefers the descriptive term *intraclast* (see *Intraclastic Rocks*) because it covers the whole range of particles regardless of the precise method of origin.

DAVID JABLONSKI

References

Beales, F. W., 1958. Ancient sediments of Bahaman type, *Bull. Am. Assoc. Petrol. Geologists,* **42,** 1845-1880.

Eardley, A. J., 1938. Sediments of Great Salt Lake, Utah, *Bull. Am. Assoc. Petrol. Geologists,* **22,** 1305-1411.

Folk, R. L., 1959. Classification of limestones, *Bull. Am. Assoc. Petrol. Geologists,* **43,** 1-38.

Illing, L. V., 1954. Bahama calcareous sands, *Bull. Am. Assoc. Petrol. Geologists,* **38,** 1-95.

Winland, D. H., and Matthews, R. K., 1974. Origin and significance of grapestone, Bahama Islands, *J. Sed. Petrology,* **44,** 921-927.

Cross-references: *Bahama Banks Sedimentology; Carbonate Sediments—Bacterial Precipitation; Carbonate Sediments—Lithification; Cementation; Limestones; Oolite.*

GRAVITY FLOWS

Sediment gravity flows (*mass flows, sediment flows*) are flows consisting of sediment or a sediment-fluid mixture moving downslope under the action of gravity (Middleton and Hampton, 1973). This term is more or less synonymous with *density current,* as used by Benjamin (1968). Gravity flows are distinguished from *gravity slides* (see *Gravity-Slide Deposits*) or *slumps* (see *Slump Bedding*) on the basis of degree of internal deformation, which is extensive in flows, slight in slides, intermediate in slumps. Flows may be either subaerial or subaqueous; submarine gravity flows are most abundant in the geologic record (Middleton and Hampton, 1973).

Middleton and Hampton (1973) distinguished four types of gravity flows by the nature of the sediment support mechanism (Fig. 1). *Turbidity currents* (see *Turbidity-Current Sedimentation*) are flows in which the sediment is supported mainly by the upward component of fluid turbulence. In *liquefied sediment flows* (Middleton and Southard, 1977; originally called fluidized flows by Middleton and Hampton; see *Fluidized and Liquefied Sediment Flows*), sediment is supported by the upward movement of fluid escaping from between the grains as the grains settle out by gravity. In *grain flows* (q.v.), the sediment is supported by grain-to-grain interactions. *Debris flows* (see *Mudflow, Debris-Flow Deposits*) are flows in which the larger grains are supported by a matrix of interstitial fluid and fine sediment which has a finite yield strength.

It should be expected that in real sediment flows, more than one of these mechanisms will be important; in addition, mechanisms such as saltation and traction may operate in some types of sediment gravity flow (Middleton and Hampton, 1973; see *Fluvial Sediment Transport*). The probable sequence of events leading to formation of sediment gravity-flow deposits is shown in Fig. 2.

JOANNE BOURGEOIS

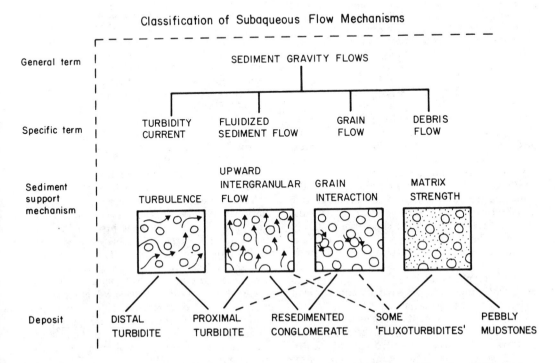

FIGURE 1. Classification of subaqueous sediment gravity flows (after Middleton and Hampton, 1973; courtesy G. V. Middleton). The term fluidized flow has been replaced by liquefied flow (Middleton and Southard, 1977).

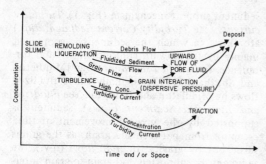

FIGURE 2. Hypothetical evolution of a single gravity flow, either in time or in space (after Middleton and Hampton, 1973; courtesy G. V. Middleton). The term fluidized flow has been replaced by liquefied flow (Middleton and Southard, 1977).

References

Benjamin, T. B., 1968. Gravity currents and related phenomena, *J. Fluid Mech.*, **31**, 209-248.

Middleton, G. V., and Hampton, M. A., 1973. Sediment gravity flows: Mechanics of flow and deposition, *Soc. Econ. Paleont. Mineral Pac. Sec., Short Course* (May 1973), 1-38.

Middleton, G. V., and Southard, J. B., 1977. Mechanics of sediment movement, *Soc. Econ. Paleont. Mineral. Short Course 3*, 246p.

Cross-references: *Avalanche Deposits; Bouma Sequence; Fluidized Sediment Flows; Fluvial Sediment Transport; Grain Flows; Gravity Flows; Gravity-Slide Deposits; Mass-Wasting Deposits; Mudflow and Debris-Flow Deposits; Slump Bedding; Traction; Turbidites; Turbidity-Current Sedimentation.*

GRAVITY-SLIDE DEPOSITS

Gravity-slide deposits are formed by the downslope movement of a mass of rock or rock debris, principally under the influence of gravity, along a clearly defined plane of rupture (slide plane or shear plane); slides may be subaqueous or subaerial. The mass retains contact with the slope throughout the movement, although this may be through a film of water or air. The discrete surfaces on which the slide occurs sharply define the moving mass. Deposits resulting from distributed shear, either between grains or along many irregularly distributed planes, are excluded. Falling, bouncing, and rolling are also excluded because there is irregular contact with a clearly defined slide surface.

Many mass movements of plastic or quasi-plastic material, e.g., slump earthflows, appear to flow; but movement still occurs along discrete shear surfaces. Thus, some so-called *mudflows* (see *Mudflow Deposits*) appear to flow, but they are more properly termed *slides* because they advance along clear basal-shear surfaces (Hutchinson and Bhandari, 1971). Virtually all sediment and rock types can slide, but the following appear to be particularly prone: clays, especially over- or underconsolidated types; silts; marls; fissile rocks, such as slates and schists (especially those with a large mica content); sedimentary sequences which involve an alternation of permeable and impermeable strata; and regolith, which has often inherited an unstable state of equilibrium and may lie on a future slide plane in the subjacent bedrock.

Slide movements may be promoted in many ways, but two fundamental divisions can be made: (1) those slides resulting from the overcoming of the shear strength (resistance to shear), as a result of, e.g., tectonic shock; and (2) slides resulting from the reduction of shear strength, e.g., by removal of an overburden leaving an overconsolidated material which will progressively lose its shear strength. Subaqueous slides are often caused by underconsolidation, when the lithification process (expulsion of water) has failed to keep pace with sediment accumulation. These materials have a very small shear strength. Many marine clays have very large natural-water contents and are very sensitive; their natural water content exceeds the *liquid limit,* i.e., the moisture content at which the properties of a remolded clay change from those of a solid to those of a viscous fluid. Subaerial *slump earthflows* normally develop in such material. In this type of failure, retrogressive rotational slips in a sensitive clay are removed from the foot by viscous flow along clear boundary-shear surfaces. Remolding of the clay usually occurs during the slide movement. Yatsu (1967) believes the presence of a large content of strongly swelling clay minerals, such as smectite, make this a distinctive landslide type. Karrow (1972) illustrates the typical morphological features of a slump earthflow.

Flow slides result from the collapse of a metastable structure in fine sands or silts, occurring both subaqueously and subaerially. The exact relationship between subaqueous slides and mass flows is still somewhat problematic, and many of the suggested debris-flow deposits formed subaqeuously may in fact be types of flow slides (see *Mudflow Deposits*). Subaerial flow slides produce deposits lacking the clay bridges between silt particles which give dry strength to the original loess and silt deposits, or in the case of sands, lead to a repacking of the grains at a higher density than the original. Recent subaerial slides can show a distinctive surface morphology (Fig. 1).

In ancient sediments, the presence of large, wedge-shaped blocks of allochthonous origin

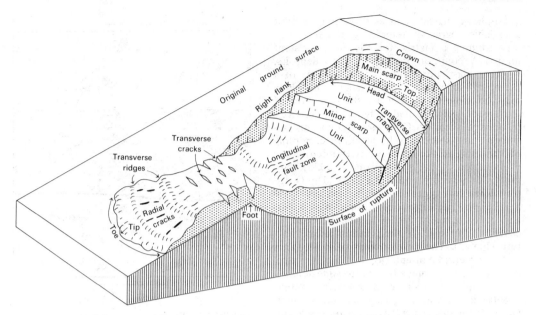

FIGURE 1. Features of an idealized rotational slide (after Varnes, 1958).

in marine sediments, the covering of younger sediments by older ones, and areas of apparent nondeposition can all suggest a slide origin. Folding and contortion of strata can indicate the direction of the slide movement, as can slickensides at the contact of a slide block and the underlying stratum. Boudinage and pull-apart structures may also occur through sliding of a rock mass or sedimentary layer. Other features present can include load structures, which may or may not be oriented; piling up of beds; and topographic evidence of submarine gullies and valleys, normally on delta fronts or on the continental slope. In cohesionless or slightly cohesive sediments, however, few of these are evident. Davies (1975) describes slide planes in Permo-Carboniferous cherty limestones and shales along fiord cliffs in northern Canada (viewed by helicopter). The planes are almost a km long and cut out about 100 m of section. Along the strike, the contact relationship changes from an angular unconformity where the slide runs oblique to the underlying bedding, to a disconformity where the slide becomes parallel with bedding. There is little or no deformation of the beds below the contact, and the sediments filling the trough thicken into the trough. In the field, however, with less outcrop continuity, such a discontinuity would be detectable only by carefully tracing bed truncation along strike.

Rotational slides (slips) are subaerial slides with a variable, but generally small, movement from the seat of the slide; they can be single, multiple, or successive, and normally occur within clays or shales. Becoming increasingly translational in character, rotational slides may progress through successively rotational to an undulation type (producing a series of contour ridges or irregular hummocks) on clay slopes of decreasing gradient (Chandler, 1970). The caprock on the clay involved in a rotational slide forms a series of cross-slope scarp features, increasing in dip the farther the blocks have moved from the main scarp.

Translational slides move either as discrete units, or as essentially cohesionless material, both involving movements of a relatively thin layer parallel to the ground surface. The first group comprises *rock slides* (*block slides*) and *slab slides*, both of which may be multiple failures. *Rock slides* involve the sliding of a lithified unit along a bedding, cleavage, or joint plane; *slab slides* involve material which, though uncemented, still slides as a single unit. *Mudflows* advancing by basal shearing, i.e., *mudslides*, may also be included in this group. The second group comprises masses known as *detritus* or *debris slides*, involving much more deformation, especially the ones acting as free-running sand; this type generally occurs on steeper slopes than do mudslides.

Internal Characteristics

In relatively rigid materials, original stratification and bedding characteristics will be largely retained, whereas slides involving plastic or quasi-plastic masses such as clays may exhibit a variable degree of remolding, the remolded clay forming a matrix for the more resistant lumps. A common feature in slide deposits, whether plastic or rigid, is the presence of a peripheral slide plane. In clay slides, this plane may be striated parallel to the direction of movement, slickensided, and polished. Any secondary shear surfaces are usually discontinuous, and there may be localized crumpling or rupturing of the laminae. After microscopic examination of the shear plane in a clay sediment, Morgenstern and Tchalenko (1967) distinguished two types of secondary shears, some parallel to the direction of slide movement, and others, usually Riedel shears, arranged *en echelon* at an angle to the principal shear direction (Fig. 2). The development of both types and of a preferred orientation in the clay matrix depends on the composition of the slide deposit; for example, increasing organic matter content results in increasing structural randomness. Other distinctive structures in sheared clays, including kink bands and compressional textures (Fig. 3) have been recognized by Tchalenko (1968). Shear zones can be a different color from the main body of the rock, due to the percolating reducing water. Thus, in the shear zone of a slide investigated by Early and Skempton (1972), the clay was a light gray, contrasting with the adjacent yellow-brown clay with orange mottling.

There are often no distinctive features in slide deposits apart from the presence of shear planes, but negative evidence may also suggest a slide origin, e.g., the absence of features depending on the relatively free movement of particles in a fluid, such as size sorting, particle rounding, or graded bedding.

PETER GASCOYNE

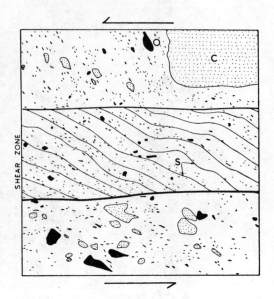

FIGURE 3. Example of compressional texture (from Tchalenko, 1968). C = clay pellets; O = organic matter; S = particle orientation in the shear zone.

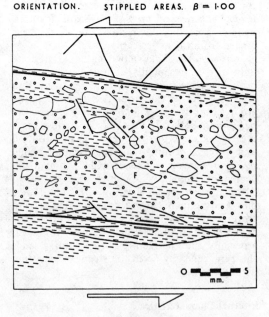

FIGURE 2. Scale drawing from a photomicrograph of a shear zone (from Morgenstern and Tchalenko, 1967).

References

Chandler, R. J., 1970. The degradation of the Lias Clay slopes in an area of the East Midlands, *Quart. J. Eng. Geol. [London]*, 2, 161-181.

Chandler, R. J., 1972. Periglacial mudslides in Vestspitsbergen and their bearing on the origin of fossil "solifluction" shears in low angled clay slopes, *Quart. J. Eng. Geol. [London]*, 5, 223-241.

Cox, D. P., and Pratt, W. P., 1973. Submarine chertargillite slide-breccia of Paleozoic age in the southern Klamath Mountains, California, *Geol. Soc. Am. Bull.*, 84, 1423-1438.

Davies, G. R., 1975. Penecontemporaneous slide and fill megastructures in Upper Paleozoic slope-facies carbonates, Arctic Canada (abstr.), *Am. Assoc. Petrol. Geologists–Soc. Econ. Paleont. Mineral., Annual Meeting Abstr.*, 2, 17-18.

Early, K. B., and Skempton, A. W., 1972. Investigations of the landslide at Waltons Wood, Staffordshire, *Quart. J. Eng. Geol. [London]*, 5, 19-41.

Hutchinson, J. N., and Bhandari, R. K., 1971. Un-

drained loading, a fundamental mechanism of mudflows and other mass movements, *Geotechnique,* 21, 353-358.
Karrow, P. F., 1972. Earthflows in the Grondines and Trois Rivieres areas, Quebec, *Canadian J. Earth Sci.,* 9, 561-573.
Morgenstern, N. R., and Tchalenko, J. S., 1967. Microstructural observations on shear zones from slips in natural clays, *Proc. Geotech. Conf., Oslo,* 1, 147-152.
Tchalenko, J. S., 1968. The evolution of kink-bands and compressional textures in sheared clays, *Tectonophysics,* 6, 159-174.
Terzaghi, K., 1956. Varieties of submarine slope failures, *Proc. Texas Conf. Soil Mech. Foundation Engin.,* 8, 41p. See also 1957, *Teknisk Ukeblad,* 43-44, 1-16.
Varnes, D. J., 1958. Landslide types and processes, *Hwy. Research Bd., Spec. Rept.* 29, 20-47.
Yatsu, E., 1967. Some problems on mass movements, *Geogr. Ann.,* ser. A, 49, 396-401.

Cross-references: *Delta Sedimentation; Gravity Flows; Mass-Wasting Deposits; Mudflow, Debris-Flow Deposits; Olistostrome, Olistolite; Penecontemporaneous Deformation of Sediments; Slump Bedding; Submarine Canyons and Fan Valleys, Ancient; Turbidites; Turbidity-Current Sedimentation.* Vol. III: *Earthflow; Landslides; Mass Movement; Mass Wasting; Submarine Geology.*

GRAYWACKE

The venerable term *graywacke* probably ranks with "flysch" and with "geosyncline" as a part of the geological vocabulary that means something special to every geologist, yet has persistently defied all proposals at either precise definition or complete abandonment. Precedent, factual descriptive data, theories of classification, and tectonic philosophy have all left their imprint on the voluminous literature on graywacke. About the only points of general agreement seem to be that the term refers to a special kind of sandstone (q.v.) and that it entered geology from the everyday usage of eighteenth-century miners in the Harz Mountains of Germany [according to Dott, 1964, p. 626, one of the first published descriptions of graywacke (*grauwacke*) was by Lasius, 1789]. The miners doubtless had become very well acquainted with those dark-colored sedimentary rocks which required so much effort to break but which could be hammered into smooth-sided blocks.

The initial geologic formulation of what a graywacke should be was made by Abraham Gottlob Werner. According to Werner's model of Earth history, the oldest rocks, granites, were chemical precipitates from the Universal Ocean. As the water level from this primordial sea dropped, high areas on its bottom became islands, around whose sloping sides were deposited, with steep initial dip, the sediments which became the next group of rocks, including the notorious *grauwacke.* This second group, the *Flotzgebirge,* were interpreted by Werner as partly chemical, partly mechanical sediments. Taken literally, and expressed in modern terms, we would have to state that Werner was talking about a granite-cemented arkose! (This follows from the fact that only preexisting rocks were granites. The *grauwackes,* being partly mechanical and partly chemical, would have consisted of quartz, feldspar, and granitic rock fragments—the only things available—cemented by more of what the Universal Ocean was precipitating, namely granite).

It is clear that Werner's fanciful ideas do not in any way relate to the Harz or to any other graywackes. But Werner's approach, defining rocks in terms of his private vision of reality, blazed a trail which, in their own ways, numerous other authors have followed. This is evident from the fact that much more has been written about what a graywacke *ought to be* than what graywackes actually are.

After surviving many attempts to expunge it from usage, graywacke made a comeback in the US. In 1926, in an influential book, Twenhofel and collaborators, following Fay (1920), decided that a graywacke is the mafic equivalent of an arkose, resulting from the disintegration of granular mafic rocks (Twenhofel, 1926). This definition seems to have influenced the definition adopted by the Committee on Sedimentation of the National Research Council (V. T. Allen, Chairman, Report for 1936; see Boswell, 1960), viz, that graywacke is "a sandstone composed of 33 or more percent of easily destroyed minerals and rock fragments derived by rapid disintegration of basic igneous rocks, slates, and dark colored rocks. It may or may not be intensely indurated or metamorphosed." In this definition, the principal stress is laid on provenance. Graywacke is presumed to contrast with arkose (q.v.) in being derived from mafic, rather than felsic, igneous rocks; also included was the concept that the particles are derived from parent rocks which ordinarily do not survive chemical weathering. The "matrix," to be much emphasized later, was not even mentioned.

Krynine (1948) was the first to note the importance of micaceous materials in defining graywacke. According to Krynine, a graywacke is a fine-grained, gray micaceous grit containing quartz, variable proportions of feldspar, and metamorphic rock fragments. Krynine inferred that the paste of chlorite and other "hydromicas" had come from the breakup and dispersal within the interstices of the sediment of micaceous particles from the sand-sized and larger framework particles. From this point of view, two ideas branched: (1) the matrix was inferred to be a primary factor and to indicate poor sorting; and (2) the provenance of the framework particles was held to be a terrain composed chiefly of metamorphic rocks.

The first idea branching from Krynine's viewpoint on the mica content was emphasized by Pettijohn (1949, 1957), who defined graywacke as a sandstone containing >15% detrital matrix; but he included the idea that the rock must also contain >25% of unstable materials such as rock fragments and feldspar.

The second idea branching from Krynine became the basis of Folk's (1954) definition that a graywacke is a sandstone containing >25% of sand-sized or coarser debris from a metamorphic source terrain; that is, graywacke contains detrital micas, chlorite, metaquartzite, phyllites, schists, and so forth.

Without going into all the other classifications, it should be mentioned that the alleged associations of graywackes with geosynclines (see *Geosynclinal Sedimentation*) soon was transmuted to the idea that graywackes are hallmarks of eugeosynclines, hence should contain volcanic debris. Perhaps the most appropriate definition of graywacke was spoken in jest by Adolph Knopf (personal communication, 1951), and prompted by studies of thin sections of rocks brought to Yale by students working on the Canadian Shield. After seeing the great variety of rocks they were naming graywacke, Knopf decided that a graywacke on the Canadian Shield is "any dark rock that is not positively known to be igneous."

The other recent developments in graywacke defining have centered on (1) the method of deposition—graywacke has been equated to "turbidites" (Packham, 1954) or to "flysch arenites" (Crook, 1973, 1974); and (2) the significance of the matrix—the polygenetic origin of the matrix was pointed out by Cummins (1962) and has been demonstrated experimentally (Hawkins and Whetten, 1969; Whetten and Hawkins, 1970). Matrix of graywackes also has been discussed by Audley-Charles (1967); Beall and Fischer (1969); Brenchley (1969); Dickinson (1970); Kuenen and Sengupta (1970); Dott, (1964); Kuenen, (1966a,b); Whetten, (1972); and Pettijohn (1975).

Perhaps a significant point in defining graywacke is the chemical composition. Because of the presence of chlorite and the general absence of microcline and orthoclase, the analyses of many graywackes show an abundance of MgO and a $K_2O:Na_2O$ ratio <1 (Middleton, 1960). This chemical composition has been ascribed to the kind of source rock, potassium-enriching regional metamorphism, and "incomplete weathering." The closest in composition to Na_2O-rich graywackes is the volcanic rock spilite, which contains abundant albite; most Na_2O-rich graywackes, however, do not contain much volcanic detritus. In an example studied from Australia, the graywackes displayed the usual enrichment of Na_2O, whereas the interbedded mudstones and shales did not (Reed, 1957). If the matrix of the graywackes is detrital, as was formerly believed by so many geologists, then the chemical composition of both graywackes and interbedded mudstones should be the same. If, however, diagenetic reactions have been important factors in creating the graywacke matrix, then this difference can be readily explained. Originally porous sands, future graywackes, when deeply buried and subjected to appropriate geothermal temperatures and migrating solutions, could experience reactions that might not take place in associated mudstones. Original potassium feldspar might be dissolved. The conversion of smectite to illite and chlorite would also take place (but in both graywackes and mudstones; Burst, 1969). It remains to be seen whether deep-burial diagenesis would affect only deep-water marine sands (the only true graywackes, according to Packham, Crook, and others).

In summary, the much-used term *graywacke* is presently being applied to a great variety of rocks, some of which match the general characteristics of the rocks in the Harz Mountains of Germany (petrographic aspects discussed by Helmbold, 1952; Helmbold and van Houten, 1958; Huckenholz, 1963; Mattiat, 1960; summarized in Pettijohn, 1975; sedimentological features described by Kuenen and Sanders, 1956), and some of which do not (the volcanic graywackes, e.g., as described by Rogers, 1966, Chappell, 1968, Dickinson, 1971; and the serpentinite graywackes described by Zimmerle, 1968). Whether graywacke will ever be precisely defined to the satisfaction of all remains doubtful. Meanwhile, scientific communication will be advanced if users of the term define what they mean by it.

JOHN E. SANDERS

References

Audley-Charles, M. G., 1967. Graywackes with a primary matrix from the Viqueque Formation (Upper Miocene-Pliocene), Timor, *J. Sed. Petrology*, 37, 5-11. See also comment by R. A. Rahmani (1968), 38, 271-273.

Beall, A. O., Jr., and Fischer, A. G., 1969. Sedimentology, *Initial Reports Deep Sea Drilling Project*, 1, 521-529.

Boswell, P. G. H., 1960. The term graywacke, *J. Sed. Petrology*, 30, 154-157.

Brenchley, P. J., 1969. Origin of matrix in Ordovician

graywackes, Berwyn Hills, North Wales, *J. Sed. Petrology,* **39**, 1297-1301.

Burst, J. F., 1969. Diagenesis of Gulf coast clayey sediments and its possible relation to petroleum migration, *Bull. Am. Assoc. Petrol. Geologists,* **53**, 73-93.

Chappell, B. W., 1968. Volcanic graywackes from the Upper Devonian Baldwin Formation, Tamworth-Barraba district, New South Wales, *J. Geol. Soc. Australia,* **15**, 87-102.

Crook, K. A. W., 1973. Graywackes, *Encyclopaedia Britannica,* 15th ed., vol. 8, 295-299.

Crook, K. A. W., 1974. Lithogenesis and tectonics: the significance of compositional variation in flysch arenites (graywackes), *Soc. Econ. Paleont. Mineral. Spec. Publ. 19,* 304-310.

Cummins, W. A., 1962. The greywacke problem, *Liverpool and Manchester Geol. J.,* **3**, 51-72.

Dickinson, W. R., 1970. Interpreting detrital modes of graywacke and arkose, *J. Sed. Petrology,* **40**, 695-707.

Dickinson, W. R., 1971. Detrital modes of New Zealand graywackes, *Sed. Geol.,* **5**, 635-638.

Dott, R. H., Jr., 1964. Wacke, graywacke, and matrix — what approach to immature sandstone classification? *J. Sed. Petrology,* **34**, 625-632.

Fay, A. H., 1920. *A Glossary of the Mining and Mineral Industry.* Washington: Govt. Printing Office, 754p. (*U.S. Bur. Mines Bull.,* 95).

Fischer, G., 1933. Die Petrographie der Grauwacken, *Preuss. Geol. Landesanst., Jahrb.,* **54**, 320-343.

Folk, R. L., 1954. The distinction between grain size and mineral composition in sedimentary rock nomenclature, *J. Geol.,* **62**, 344-359.

Hawkins, J. W., and Whetten, J. T., 1969. Graywacke matrix minerals: Hydrothermal reactions with Columbia River sediments, *Science,* **166**, 868-870.

Helmbold, R., 1952. Beitrag zur Petrographie der Tanner Grauwacken, *Beitrage Min. Petrog.,* **3**, 253-288.

Helmbold, R., and van Houten, F. B., 1958. Contribution to the petrography of the Tanner graywacke, *Geol. Soc. Am. Bull.,* **69**, 301-314.

Huckenholz, H. G., 1963. Mineral composition and texture in graywackes from the Harz Mountains (Germany) and in arkoses from the Auvergne (France), *J. Sed. Petrology,* **33**, 914-918.

Krynine, P. D., 1948. The megascopic study and field classification of sedimentary rocks, *J. Geol.,* **56**, 130-165.

Kuenen, P. H., 1966a. Geosynclinal sedimentation, *Geol. Rundschau,* **56**, 1-19.

Kuenen, P. H., 1966b. Matrix of turbidites: Experimental approach, *Sedimentology,* **7**, 267-297.

Kuenen, P. H., and Sanders, J. E., 1956. Sedimentation phenomena in Kulm and Flözleeres graywackes, Sauerland and Oberharz, Germany, *Am. J. Sci.,* **254**, 649-671.

Kuenen, P. H., and Sengupta, S., 1970. Experimental marine suspension currents, competency and capacity, *Geol. Mijnbouw,* **49**, 89-118.

Lasius, G., 1789. *Beobachtungen über die Harzgebirge* (2 vol.). Hannover: Helwingeschen Hofbuchhandlung, 559p. (Grauewacke, p. 132-152, according to Dott, 1964).

Mattiat, B., 1960. Beitrag zur Petrographie der Oberharzer Kulmgrauwacke, *Beitr. Min. Petrog.,* **7**, 242-280.

McBride, E. F., 1962. The term graywacke, *J. Sed. Petrology,* **32**, 614.

Middleton, G. V., 1960. Chemical composition of sandstones, *Geol. Soc. Am. Bull.,* **71**, 1011-1026.

Okada, H., 1971. Classification of sandstone: analysis and proposal, *J. Geol.,* **79**, 509-525.

Packham, G. H., 1954. Sedimentary structures as an important factor in the classification of sandstones, *Am. J. Sci.,* **252**, 466-476.

Pettijohn, F. J., 1949. *Sedimentary Rocks.* New York: Harper & Row, 526p.

Pettijohn, F. J., 1957. *Sedimentary Rocks,* 2nd ed. New York: Harper & Row, 718p.

Pettijohn, F. J., 1963. Chemical composition of sandstones—excluding carbonate and volcanic sands, in M. Fleischer, ed., Data of geochemistry, 6th ed., *U.S. Geol. Surv. Prof. Pap. 440-S,* 21p.

Pettijohn, F. J., 1975. *Sedimentary Rocks,* 3rd ed. New York: Harper & Row, 628p.

Reed, J. J., 1957. Petrology of the lower Mesozoic rocks of the Wellington District, *N. Zeal. Geol. Surv. Bull.* (n.s.), **57**, 60p.

Rogers, J. J. W., 1966. Geochemical significance of the source rocks of some graywackes from western Oregon and Washington, *Texas J. Sci.,* **18**, 5-20.

Twenhofel, W. H., and collaborators, 1926. *Treatise on Sedimentation.* Baltimore: Williams & Wilkins, 661p.

Walker, R. G., and Pettijohn, F. J., 1971. Archaean sedimentation: Analysis of the Minnitaki Basin, northwestern Ontario, Canada, *Geol. Soc. Am. Bull.,* **82**, 2099-2130.

Whetten, J. T., 1972. Matrix-rich Pleistocene sediments from western Washington: Incipient graywacke-type sedimentary rocks? *Geol. Soc. Am. Mem. 132,* 573-584.

Whetten, J. T., and Hawkins, J. W., 1970. Diagenetic origin of graywacke matrix, *Sedimentology,* **15**, 347-361. (See also discussion by J. P. B. Lovell, 1972, **19**, 141-143; and reply, 144-146).

Zimmerle, W., 1968. Serpentinite graywackes from the North Coast basin, Colombia, and their geotectonic significance, *N. Jahrb. Min., Abhandl.,* **109**, 156-182.

Cross-references: *Arkose, Subarkose; Clastic Sediments and Rocks; Geosynclinal Sedimentation; Sands and Sandstones; Subgraywacke; Turbidites.*

GYPSUM IN SEDIMENTS

A variety of sedimentary environments contain gypsum ($CaSO_4 \cdot 2H_2O$) as crystals, crystal aggregates, and cleavage fragments which assume a range of habits, textures, and sedimentary structures. Because gypsum has relatively fast reaction rates, insight is permitted into formative and diagenetic processes not so readily discernible in other minerals. In addition, gypsum that originates from sea water yields important information as to the constancy and universality of the chemicals found in the ocean throughout geologic time, such as the isotopic ratios of sulfur and sulfate. The monumental work of Strakhov (1962) synthesizes worldwide investigations of continental and marine evaporite deposits, and Blatt et al. (1972) provide a concise textbook summary of gypsum as a sediment. Major papers on marine evaporites have now been reprinted in Kirkland and Evans (1973).

Environments of Formation

Most gypsum now being deposited in sediment is diagenetic and is found within the upper phreatic zone and vadose zone of desert playas and continental sabkhas or on supratidal flats and coastal sabkhas, such as along the Trucial Coast (see *Persian Gulf Sedimentology*). The host sediment in these environments may be dominantly carbonates, siliciclastics, or mixed evaporites. Gypsum may form up to 3 m below the groundwater table in these host sediments, whose fauna and sedimentary structures often indicate a depositional environment incompatible with the gypsum precipitates. Brines derived from sea water are the probable source for most gypsum, but continental-water brines or mixed marine and continental brines are not uncommon. Evaporation rate, water chemistry, and hydrologic constraints are key parameters influencing gypsum, whether it is precipitated out of (1) interstitial water within a host sediment (subaerially), or (2) a standing body of water (subaqueously). Small to medium quantities of gypsum occur in hypersaline lakes, hot springs, marsh salt pans, beachrock, ephemeral shoreline crusts, and in salty spray blown from the ocean and playa ponds.

Habits and Textures

Gypsum crystals in sediment display habits that range from acicular through prismatic, and tabular to lensoid or discoid. Each habit reflects the particular chemical, biological, and physical properties of the environments in which it exists; but definitive, quantitative correlation of properties to a given habit has not been well documented especially in the case of transitional forms. Individual crystals and aggregates may be clear or clouded with particles. Sand grains tend to be incorporated into the gypsum crystal, whereas clays are usually excluded. Repeated phases of dissolution and reprecipitation are common, with internal, concentric growth surfaces often accentuated by incorporated grains. Gypsum also replaces carbonate and evaporite minerals and organic materials such as plant roots.

In a hot arid climate, gypsum can dehydrate to bassanite ($CaSO_4 \cdot \frac{1}{2}H_2O$), but seasonal rainfall will cause rehydration, forming a very fine-grained gypsum flour. This process is exhibited in some mounds of gypsum crystals within the Pleistocene Lisan Marls of the Jordan Valley, where the original gypsum, and gypsum flour and bassanite coexist.

The acicular habit grows in a nearly vertical orientation from the floor of brine pans. Often, the crystals are free of inclusions and grow to a few cm in length, probably reflecting rapid crystal growth resulting from a rapid rate of brine concentration.

Intermixed prismatic and tabular habits, both usually twinned, are interlayered with the halite rock that forms in depressions at Saline Ometepec along the northwestern Gulf of California (Shearman, 1970).

Lensoid crystals are the most abundant and occur in coastal and continental sabkhas throughout the world (Fig. 1). They are convex in cross section and diamond shaped to circular in plan, with flattening in a plane approximately perpendicular to the crystal c axis. The long diameter of individual crystals ranges from < 0.05 mm to 20 mm and may reach 240 mm. These crystals precipitate interstitially within the host sediment, which may be sand, silt, or clay layers, or previously formed layers of acicular or lensoid gypsum. Growth seemingly is concentrated along

FIGURE 1. Lensoid gypsum crystals, aggregated into blebs, that grow by displacement of surrounding clays and/or previous gypsum sands. 16.5 cm depth, Laguna Mormona, Baja California.

high permeability interfaces. As the crystals increase in size, they tend either to mechanically displace the host grains owing to the force of crystallization, or to enclose the grains within their crystal structure. Algal filaments also may be incorporated into an individual gypsum crystal. There is an apparent gradation from pore-filling gypsum cement to lensoid microcrystalline gypsum of the displacement type.

In addition to isolated crystals scattered throughout host sediment or algal mats, gypsum may form aggregates of interpenetrating crystals, rosettes (intergrown masses of discoid crystals) being a common type. Some rosettes reach 50 cm in diameter. Bladed gypsum aggregates range in size from 0.5 mm to 100 mm and are composed of interlocked clusters and twins of elongate euhedra. Bladed forms grow by displacement; examples include deposits on the shore of Lake Lucero near White Sands, New Mexico, and the tidal flats of Laguna Madre, Texas (Murray, 1964). Intergrown tabular to lensoid crystal mounds that range from 10-50 cm in diameter exist in the southern Laguna Mormona ponds of the Pacific coast of Baja California, Mexico. These mounds extend upward from the bottom of brine-filled channels, with the crystals growing freely into the surrounding brine rather than displacing sediment.

Sedimentary Structures

When gypsum precipitates from a standing body of water, it forms beds or laminae (Lucia, 1972) that may show wind ripples and, less often, current ripples (Fig. 2). Sediment and organic material carried by storm currents or wind is frequently interlayered with the gypsum in a varve-like sequence. An extensive literature, however, has shown that care must be taken in assigning a temporal scale to these "varved" units.

Diagenetic precipitation of gypsum out of interstitial hypersaline water within preexisting sediment results in a variety of structures. Lensoid crystals aggregate into blebs similar to anhydrite nodules but less cohesive because of the greater size of the gypsum crystals. Coalescence of such blebs along a surface produces lenticular pseudo-bedding or lamination that may persist laterally for tens of m. Locally, lensoid displacement crystals may form a mushy, moist layer from one to tens of cm in thickness. These layers may exhibit inverse and normal graded bedding or may consist of crystals having a uniform or mixed range of diameters.

Irregular, wavy lamination and cm-size diapir-like structures in gypsum sabkha deposits are common, and are attributed to adhesion ripple or adhesion wart formation that results when the wind blows grains across a moist surface (see Reineck and Singh, 1973). Similar eolian action sorts and abrades gypsum crystals while either depositing them as sheet layers on the sabkha surfaces or piling them into dunes (see *Sabkha Sedimentology*). Massive gypsum rock composed of lensoid crystals that enclose the host sediment form hard layers of up to 0.5 m in thickness on continental sabkhas and playas. Similar layers are not known from coastal evaporite deposits. Blebs of clear lensoid gypsum also fill voids in algal mats and stromatolites (q.v.). If these blebs increase in total volume by either increasing individual crystal diameter or increasing the number of crystals, they may disrupt the algal mat structures or render them unrecognizable.

Crusts that consist of clastic sediment, gypsum, and other evaporite minerals form an array of structures including "tepees" and edgewise conglomerates. More ephemeral, porous, and friable "puffy earth" mounds contain gypsum cement, crystals, and aggregates. Both of these structure types are present in marine and continental evaporites.

STEPHEN VONDER HAAR

References

Bebout, D. G., and Maiklem, W. R., 1973. Ancient anhydrite facies and environments, Middle Devonian Elk Point Basin, Alberta, *Bull. Canadian Petrol. Geol.,* **21,** 287-343.

Blatt, H.; Middleton, G.; and Murray, R., 1972. *Origin of Sedimentary Rocks.* New York: Prentice-Hall, 634p.

Fontes, J.-C., 1968. Le gypse du bassin de Paris historique et données récents, *Bur. Rech. Géol. Minières* **58,** 359-386.

Kinsman, D. J. J., 1974. Calcium sulphate minerals of evaporite deposits: Their primary mineralogy, in

FIGURE 2. Laminar gypsum with current ripples, Eraclea-Minoa (Capo Bianco), Sicily. (Photo: B. C. Schreiber).

A. H. Coogan, ed., *Fourth Symposium on Salt*, vol. 2. Cleveland: Northern Ohio Geol. Soc., 343–348.

Kirkland, D., and Evans, R., eds., 1973. *Marine Evaporites–Origin, Diagenesis and Geochemistry.* Stroudsburg, Pa.: Dowden, Hutchinson & Ross, 426p.

Lucia, F. J., 1972. Recognition of evaporite-carbonate shoreline sedimentation, *Soc. Econ. Paleont. Mineral. Spec. Publ. 16*, 160–191.

Masson, P. H., 1955. An occurrence of gypsum in southwest Texas, *J. Sed. Petrology*, 25, 72–77.

Mossop, G. D., and Shearman, D. J., 1973. Origins of secondary gypsum rocks, *Inst. Min. Metal. Trans.*, 82, B147-B154.

Murray, R. C., 1964. Origin and diagenesis of gypsum and anhydrite, *J. Sed. Petrology*, 34, 512–523.

Reineck, H. E., and Singh, I. B., 1973. *Depositional Sedimentary Environments*. New York: Springer-Verlag, 439p.

Schreiber, B. C., and Kinsman, D. J. J., 1975. New observations on the Pleistocene evaporites of Montallegro, Sicily and a modern analog, *J. Sed. Petrology*, 45, 469–479.

Shearman, D. J., 1970. Recent halite rock, Baja California, Mexico, *Inst. Mining Met. Trans.*, 79, 155–162.

Strakhov, N. M., 1962. *Principles of Lithogenesis*, vol. 3. *The Arid Zone*. New York: Plenum, 577p.

Tanner, W. F., 1972. Large gypsum mounds in the Todilto Formation, New Mexico, *Mt. Geol.*, 9(1), 55–58.

Cross-references: *Desert Sedimentary Environments; Evaporite Facies; Evaporites–Physicochemical Conditions of Origin; Marine Evaporites–Diagenesis and Metamorphism; Sabkha Sedimentology; Salar, Salar Structures; Stromatolites; Tropical Lagoonal Sedimentation.* Vol. IVB: *Gypsum.* Vol. XIII: *Gypsum and Anhydrite.*

GYTTJA, DY

Gyttja is a Scandinavian word for organic detrital matter. Boiled in alkaline solution, gyttja does not give the solution a dark brown color as do dy, humus, and pitchy soil. Coarse detrital gyttja contains many macro- and microfossils and fragments of fossils; fine detrital gyttja is formed of organic fragments usually too small to be identified. Algal gyttja is made up of cyanophytes and chlorophytes. Gyttja is found mixed with clay and silt, with dy, and with carbonate (lime gyttja, marl and shell gyttja). Mixed with mineralogic sediment (clay and silt), the following terminology is applied depending upon the percent of gyttja: 3–6%, *gyttja clay;* 6–12%, *clay gyttja;* >12%, *gyttja. Lime gyttja* is gyttja with a moderately high percentage of $CaCO_3$, usually rich in shells of mollusks and gastropods. It is the usual limnic sediment in carbonate rich areas (see *Marl*).

Pitchy soil, or *dopplerite*, is a coal-black, "fatty" soil, found as thin layers or impregnation in the upper part of another layer (clay, sand, peat, etc.). Microscopically, pitchy soil consists of black, formless lumps and aggregates in a varying matrix. Pitchy soil consists of humus residues from a decomposed vegetation layer or chemical precipitation of humus in the upper part of a buried layer. Boiled in alkaline solution, pitchy soil (like dy) gives the solution a dark brown color. Pitchy soil is formed when an overgrown land surface is buried, e.g., by transgression or by artificial burials; pitchy soil below a shore deposit or a marine layer may be used as evidence for a transgression.

Dy is a Swedish word for the precipitate of humic acids in waterlaid sediments. Dy is black and occurs as separate layers (usually thin) or mixed with gyttja (e.g., dyey gyttja). Boiled in alkaline solution, dy gives the solution a distinct dark brown color. Dy is a common lake deposit in areas where the water has a high content of humic acids (e.g., water from bogs and swamps).

NILS-AXEL MÖRNER

References

Ishiwatari, R., 1970. Structural characteristics of humic substances in recent lake sediments, in G. D. Hobson, and G. C. Speers, eds., *Advances in Organic Geochemistry*. Oxford: Pergamon, 285–311.

Kononova, M. M., 1966. *Soil Organic Matter*. Oxford: Pergamon, 544p.

Kubiena, W. L., 1953. *The Soils of Europe*. London: Murby, 318p.

Schnitzer, M., and Khan, S. U., 1972. *Humic Substances in the Environment*. New York: Marcel Dekker, 237p.

Cross-references: *Bitumen, Bituminous Sediments; Humic Matter in Sediments; Kerogen; Lacustrine Sedimentation; Marl; Organic Sediments.*

H

HADAL SEDIMENTATION

The term *hadal* refers to that part of the benthonic zone occurring in oceanic depressions beneath the general level of the deep sea bottom. Such areas are almost exclusviely found in deep-sea trenches. The trenches are all related to recognized island arcs or lie adjacent to active continental margins and form one of the most distinctive topographic features related to subduction zones. Deep-sea trenches are long and narrow depressions of the ocean floor with relatively steep sides; in this summary, only some of these major trenches will be discussed (Fig. 1).

An understanding of the deposition and nature of sediments in trenches has long been considered desirable, as they may possibly be the modern equivalents of certain ancient sediments supposed to have been deposited in deep, subsiding troughs or geosynclines marginal to the cordilleran belts of ancient continents (see *Geosynclinal Sedimentation*).

The maximum depth of trenches ranges from 6660 m for the Middle American Trench to 10,865 m for the Marianas Trench (Fisher and Hess, 1963). The trenches characteristically have a V-shaped profile, and the oceanward flank generally slopes 2-6° and is shallower than the continental flank, which slopes 5-15°. The lower slopes of the trenches, particularly on the continental side, are generally much steeper, with slopes as much as 25-35°. These lower slopes are characterized by outcrops of

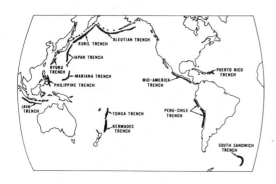

FIGURE 1. The major deep-sea trenches of the world.

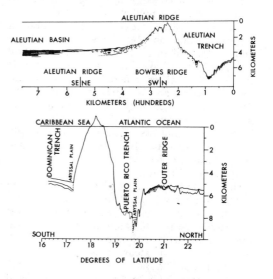

FIGURE 2. Seismic reflection profiles across the Aleutian Trench (redrawn from Ewing et al., 1965) and the Puerto Rico Trench (redrawn from Ewing and Ewing, 1962).

volcanic rocks or Tertiary sediments, and contain many steps or benches attributed to faulting (Fig. 2). Trench floors commonly contain 0.5-5 km-wide abyssal plains with up to 1.5 km of sedimentary fill; parts of the floor may contain little or no sediment, however, suggesting that the trenches are youthful geological features, or that little sediment is reaching the trench.

Sediment Distribution in Some Major Trenches

Seismic reflection profiling is the most accurate method available to measure the distribution and thickness of the sediment in trenches (Fig. 2); this technique has only been actively used across deep-sea trenches in the last decade. Scholl and Marlow (1974) summarize the sedimentary sequences filling most trenches, using a combination of seismic profiling with data from deep-sea drilling and also from estimates made from topographic profiles. They found that the trenches of the eastern Pacific bordering North and South America typically contain a two-layer sequence. The upper layer is wedge shaped and consists mainly of terrigenous turbidites; the lower layer is primarily hemipelagic sediments (Fig. 3). On the western and southwestern sides of the Pacific, however, the trenches are either essentially empty or partially filled (100–400 m) with mainly pelagic or hemipelagic sequences.

The Aleutian Trench, which bridges these two major provinces, is predictably characterized by the eastern style of *thick* sedimentary fill in its eastern portion and by the western style of *thin* pelagic fill in its western portion (Scholl and Marlowe, 1974). There are some notable exceptions to these generalizations, however; e.g., the pelagic and hemipelagic deposits in the Kamchatka Trench may be >2000 m thick. Also, the northern part of the Peru-Chile Trench is very young and has <200 m of sediment fill. The 2700 km-long Tonga-Kermedec Trench also appears to be extremely youthful, with <100 m of sediment fill.

The "rapid" filling of modern trenches in the Pacific probably occurs primarily in those trenches that have received a significant amount of turbidity-current sediment during the Pleistocene. According to Scholl and Marlow (1974), this turbidite fill occurred as a direct result of the middle to late Pleistocene glaciation, which caused increased rates of erosion and sedimentation.

Two distinctive trench systems, the Puerto Rico Trench and the South Sandwich Trench, border islands on the western side of the Atlantic Ocean. The floor of the South Sandwich Trench is intermittently flat, and the trench plain attains a width of 6 km in some places, probably indicating up to 500 m of sedimentary fill. One core recovered from the trench floor contains graded turbidites of volcanic ash and sand. Using seismic reflection profiles, Ewing et al. (1966) have shown that there are three sedimentary units in the Puerto Rico Trench: (1) a relatively thin (100–300 m), transparent layer covering most of the north wall and extending beneath the turbidites at the bottom of the trench; (2) a layer of highly stratified sediments, resting conformably on the upper part of the south wall, up to 500 m thick; (3) up to 1.7 km of level-bedded turbidites beneath the abyssal plain of the floor of the trench. The four major reflectors that occur in the turbidites beneath the abyssal plain suggest correlation of increased turbidity-current action with Pleistocene glaciations (Ewing et al., 1966). It is interesting to note that these sediments have suffered only very minor amounts of deformation.

Using piston cores, Conolly and Ewing (1967) demonstrated that it is possible to correlate graded beds with distinct color and textural properties over large distances of the floor of the Puerto Rico Trench. There is a general decrease in grain size and in the thickness of the basal sandy section in each graded bed down from a high-level abyssal plain to the lower trench floor. The sand is a heterogeneous mixture of calcareous and terrigenous material mainly derived from the Puerto Rico–Virgin Islands Shelf, but also consists of fragments of Pleistocene globigerina and pteropod oozes,

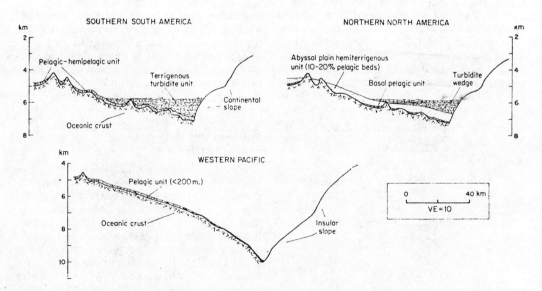

FIGURE 3. Major depositional units in typical North America, South America (lat 274°S), and western Pacific Trenches (from Scholl and Marlow, 1974).

Miocene muds, and serpentine and limestones that crop out on the southern flank of the trench (Fig. 4). In at least two instances, the turbidity currents were powerful enough to deposit 20-50 cm of fine and light sand 100 km away from and 10 m above the point where they entered the lower trench floor. Apart from a few scattered tests of planktonic organisms that occur on the top of the uppermost burrowed portions of the graded beds, there is no pelagic sediment in the cores, indicating the efficiency of the turbidity-current process, and perhaps suggesting that all the sediment in the trench floor could have been deposited during the Pleistocene.

The small amount of available data suggests that the hadal sediments in trenches are thin and consist in their upper parts of graded beds of clays, silts, and sands deposited by turbidity currents that originate either on the flanks or ends of the trench or both. Most data also indicate that the principal trenches have been formed by subduction and superficially modified by block faulting and that the thin sedimentary fill is relatively undisturbed, suggesting a relatively youthful origin for the trench and an even more recent origin for the sedimentary fill.

Presumably, many hadal sequences are eventually uplifted or plastered onto the continental side of the trench due to processes associated with subduction. These sequences, and the continental slope sequences deposited adjacent to them, may be preserved in some island arcs. Such a sequence deposited during the Cretaceous is found along the Alaska Peninsula—Bering Sea Shelf, and the rocks in the sequence have been well described from the Shumagin and Sanak Islands by Moore (1973).

JOHN R. CONOLLY

References

Conolly, J. R., and Ewing, M., 1967. Sedimentation in the Puerto Rico Trench, *J. Sed. Petrology,* 37, 44-59.

Ewing, J., and Ewing, M., 1962. Reflection profiling in and around the Puerto Rico Trench, *J. Geophys. Research,* 67, 4729-4739.

Ewing, M.; Lonardi, A. G.; and Ewing, J., 1966. The sediments and topography of the Puerto Rico Trench and Outer Ridge, *Trans. 4th Caribbean Geological Conference, Port-of-Spain, Trinidad,* 325-334.

Ewing, M.; Ludwig, W. J.; and Ewing, J., 1965. Oceanic structural history of the Bering Sea, *J. Geophys. Research,* 70, 4593-4600.

Fisher, R. L., and Hess, H. H., 1963. Trenches, in M. N. Hill, ed., *The Sea,* vol. 3. New York: Interscience, 411-436.

Moore, J. C., 1973. Cretaceous continental margin sedimentation, southwestern Alaska, *Geol. Soc. Am. Bull.,* 84, 595-614.

Piper, D. J. W., 1972. Trench sedimentation in the Lower Paleozoic of the Southern Uplands, *Scot. J. Geol.,* 8, 289-291.

Piper, D. J. W.; Von Huene, R.; and Duncan, J. R., 1973. Late Quaternary sedimentation in the active Eastern Aleutian Trench, *Geology,* 1, 19-22.

Scholl, D. W., and Marlow, M. S., 1974. Sedimentary sequence in modern Pacific trenches and the deformed circum-Pacific Eugeosyncline, *Soc. Econ. Paleont. Mineral. Spec. Publ. 19,* 193-211.

Von Huene, R., 1974. Modern trench sediments, in C. A. Burk, and C. L. Drake, eds. *The Geology of Continental Margins.* New York: Springer, 207-211.

Cross-references: *Abyssal Sedimentation; Geosynclinal Sedimentation; Pelagic Sedimentation, Pelagic Sediments; Submarine (Bathyal) Slope Sedimentation; Turbidites.* Vol. I: *Hadal Zone; Trenches and Related Deep Sea Troughs.*

FIGURE 4. Sand from 230-cm in core from the Puerto Rico Trench. Consists of a mixture of calcareous and terrigenous detritus derived from shallow and deep-water environments. The large round calcareous fragment in the center of the field of view has a maximum diameter of 1 mm.

HARDGROUND DIAGENESIS

In oceanography, a "hardground" was originally any sea floor which failed to return a mud sample on the sounding lead. In geology, it came to indicate any discontinuity surface that was evidently lithified before the next beds were laid down. Presently, the term *hardground* is applied in geology only to the discontinuity surface (*omission surface*) at which early diagenetic cementation has produced a lithified sea floor. Thus the classic hardground is the product of submarine preburial diagenesis alone. In practice, however, it is often impossible to distinguish purely submarine hardgrounds from horizons that were cemented in the vadose environment (e.g., by beachrock processes) and subsequently submerged and exposed as hard sea floor. This restriction of the term to horizons of synsedimentary (syndiagenetic) lithification was first established by Voigt (1959), who summarized previous literature, and this usage has been followed by the majority of later workers (see references in Bromley, 1975).

Occurrence

A hardground so defined is characteristically intraformational and generally marks a relatively minor hiatus in the depositional record. The term *diastem* is appropriate for the time break represented by this sort of submarine nondeposition (Heckel, 1972). Commonly these "omission surfaces" that terminate units in a cyclic sequence are hardened, producing cyclic repetition of hardgrounds. In lateral extent, individual hardgrounds may cover thousands of km^2, usually with noticeable diachronism. The majority of observed hardgrounds, however, are of only local or very local significance, dying out laterally within a few hundred meters.

Hardgrounds are almost entirely restricted to carbonate sequences and are frequent in the stratigraphic record. Detailed studies have been made of only a few cases, including particularly the hardgrounds in the NW European Cretaceous chalk (see Bromley, 1975), the Lower Cretaceous of central Texas (Rose, 1970), the condensed pelagic limestones of the Tethyan Jurassic (Jenkyns, 1971), the nearshore Jurassic limestones in France and SW England (Purser, 1969; references in Bromley, 1975), the Triassic Muschelkalk of Germany and Poland (e.g., Kaźmierczak and Pszczółkowski, 1969), and the Ordovician of Sweden (Lindström, 1963).

Contemporary or Holocene hardgrounds in modern seas have been described from the Persian Gulf, Red Sea, Mediterranean Sea and elsewhere (see *Cementation, Submarine*). The majority lie under a few tens of m of water, but some have been found in deeper water, to depths of over 3000 m.

Description

Hardgrounds may be identified in ancient settings by several characteristics:

1. The discontinuity surface may exhibit an overhanging sculpture which could not have supported itself unlithified.
2. The surface is often strewn with angular intraclasts.
3. An encrusting hard-substrate epifauna is frequently present on the discontinuity surface.
4. True organic borings (as opposed to burrows) are commonly present.
5. Mineral crusts on the surface are mutually intergrown with the encrusting fauna or cut by borings of organisms.
6. The petrographic fabric and burrow systems reveal zero compaction.

In most cases, several cm of sediment are cemented below the discontinuity surface, reaching some 30 cm in some chalk hardgrounds. The lithified rock characteristically contains networks of uncompacted burrows, protecting uncompacted fills, which were "fossilized" by the rapid hardening of the sediment (Bromley, 1975).

Hardground surfaces show a great variety of sculptural topography (Bromley, 1975). In its simplest form the surface represents the "normal" topography of the advancing depositional interface. In most cases, however, it has been modified by (1) prelithification scour (to produce a flat surface); (2) scour during the period of lithification (to produce highly irregular, convoluted sculpture and lag intraclasts); or (3) postlithification erosion or corrosion. If the postlithification process is one of bioerosion, an extremely complex sculpture may result. Physical erosion subsequent to lithification may produce a flat surface; dissolution of the lithified sea floor by sea water produces a highly intricate sculpture varying with the chemical heterogeneity of the rock (see *Corrosion Surface*).

Finally, it is particularly characteristic of many chalk hardgrounds that a rich fauna of aragonitic mollusks occurs in the lithified rock, preserved as voids and as reworked internal molds. This fauna is virtually absent from the soft chalk or late-cemented limestone above and below the hardground and owes its presence to rapid lithification of the hardground matrix, which predated the dissolution of at least some of the aragonite. The special constitution of the fauna probably reflects not only special conditions of preservation but also the change in depositional environment that brought about the genesis of the hardground.

Diagenetic Processes

Prelithification Diagenesis. Prelithification diagenesis essentially involves initial dewatering, bioturbation, and loss of unstable organic matter. In some cases, this phase has been short and simple, as when the hardground represents a nonerosive omission surface, and the topmost sediment was lithified rapidly after deposition. On the other hand, where more deeply buried sediment has been reexposed at the sea floor by scour, this sediment will already have gone through the earliest phases of diagenesis, and its reexposure will submit it to a recycle. For example, in chalk hardgrounds, preservation of aragonitic faunas will only be expected in nonerosive hardgrounds, because aragonite is lost at and very close beneath the depositional interface in chalks.

Cementation. Most hardgrounds are lithified in an irregular nodular manner in the early stages, probably as a result of interference by burrowing animals (Bathurst, 1971; Bromley, 1975). If burial by renewed deposition does not arrest the process at this stage, lithification continues until a tight, splintery limestone is produced, commonly containing a ramifying tubular cavity system representing active burrows during the lithification process. The lithification process appears to be quite rapid. The cement usually only partly fills the pores of the sediment, sufficient to rigidify the fabric. Many hardgrounds, therefore, retain a high porosity owing to lack of later compaction. Various hardgrounds have revealed different kinds of cement (see *Cementation, Submarine*). Recrystallization and the addition of late diagenetic cement, however, have hampered the study of ancient hardground cements.

Postlithification, Preburial Diagenesis. While the hardground is exposed as a rocky sea floor it is vulnerable to the attack of boring organisms, physical abrasion, and/or solution. Bodies of loose sediment migrating over the bottom may temporarily cover the hardground and locally arrest the retrogressive processes. Thin covers of new sediment may also become cemented onto the hardground surface. The newly hardened sediment may then be attacked by erosive or corrosive processes; repeated alternation of these phases can produce highly complicated fabrics (Bromley, 1975; see *Corrosion Surface*). In addition, the hardened sea floor may be mineralized. Glauconite and phosphate often replace carbonate in the Cretaceous chalk hardgrounds; pyrite is partly replacive and partly void filling. Direct precipitates of phosphorite and manganese oxide produce crusts on some hardgrounds.

Late Diagenesis. The rigidity of the fabric deters later compaction from overburden and, therefore, supports a relatively high porosity which may become filled with late cement. In a sequence in which silicification has produced chert nodule horizons, the hardground escapes such replacement, whereas the uncompacted sediment in the burrow systems within the hardground offer favorable sites for chert formation. Kennedy and Juignet (1974) have described hardgrounds in chalk which have preferentially suffered dolomitization and have since been dedolomitized, both apparently late processes.

Environment of Formation

Before it was recognized that widespread cementation of sea bottoms is occurring today, hardgrounds were generally interpreted as emergent horizons such as beachrock (q.v.) or caliche (q.v.), despite recognition that in many cases the geological setting militated strongly against subaerial exposure. Opinions have changed, however, since the detailed reports of Holocene submarine hardgrounds, and it is now recognized that the presence of hardgrounds in a rock sequence does not in any way imply emergence at these horizons. On the contrary, geological setting, paleobiology, and petrography of many ancient hardgrounds suggest that they, like their modern counterparts, formed in a wide variety of water depths. Some hardgrounds appear to have formed in very shallow situations so that there is a possibility that beachrock processes in a littoral environment have contributed to the cement. In such cases it is sometimes difficult to demonstrate emergence; structures produced by drying out would be easily destroyed by a new cycle of diagenesis initiated upon resubmergence. Purser (1969) and Schroeder (1973) have, however, offered two diagnostic fabrics of cement which can only form in the vadose environment: stalactitic cement and concave surface or meniscus cement.

It is far from clear what changes in the environment bring about the formation of hardgrounds. Depositional hiatus alone is not sufficient; uncemented omission surfaces are abundant in carbonate sequences. On the other hand, cement may be precipitated without the intervention of periods of nondeposition, either as intergranular cement producing nodules, or exclusively as intragranular cement. For the formation of true hardgrounds, however, an interruption in deposition is necessary, with or without scour or erosion. In most cases, the nature of the fauna and mineralization suggest relative shallowing of water or other oceanographic change. Paleogeographic setting indicates that hardgrounds are usually regressive phenomena, typically formed on seamounts or submerged highs in the pelagic environment or on positive shelf areas in carbonate depositional basins.

R. GRANVILLE BROMLEY

References

Bathurst, R. G. C., 1971. *Carbonate Sediments and Their Diagenesis.* Amsterdam: Elsevier, 620p.

Bromley, R. G., 1975. Trace fossils at omission surfaces, in R. W., Frey, ed., *The Study of Trace Fossils.* New York: Springer, 399-428.

Heckel, P. H., 1972. Recognition of ancient shallow

marine environments, *Soc. Econ. Paleont. Mineral. Spec. Publ. 16*, 226-286.

Jenkyns, H. C., 1971. The genesis of condensed sequences in the Tethyan Jurassic, *Lethaia*, 4, 327-352.

Kaźmierczak, J., and Pszczółkowski, A., 1969. Burrows of Enteropneusta in Muschelkalk (Middle Triassic) of the Holy Cross Mountains, Poland, *Acta Palaeont. Polonica*, 14, 299-324.

Kennedy, W. J., and Juignet, P., 1974. Carbonate banks and slump beds in the Upper Cretaceous (Upper Turonian-Santonian) of Haute Normandie, France, *Sedimentology*, 21, 1-42.

Kennedy, W. J., and Garrison, R. E., 1975. Morphology and genesis of nodular chalks and hardgrounds in the Upper Cretaceous of southern England, *Sedimentology*, 22, 311-386.

Lindström, M., 1963. Sedimentary folds and the development of limestone in an early Ordovician sea, *Sedimentology*, 2, 243-292.

Palmer, T. J., and Fürsich, F. T., 1974. The ecology of a Middle Jurassic hardground and crevice fauna, *Palaeontology*, 17, 507-524.

Purser, B. H., 1969. Synsedimentary marine lithification of Middle Jurassic limestones in the Paris Basin, *Sedimentology*, 12, 205-230.

Rose, P. R., 1970. Stratigraphic interpretation of submarine versus subaerial discontinuity surfaces: An example from the Cretaceous of Texas, *Geol. Soc. Am. Bull.*, 81, 2787-2798.

Schroeder, J. H., 1973. Submarine and vadose cements in Pleistocene Bermuda reef rock, *Sed. Geol.*, 10, 179-204.

Voigt, E., 1959. Die ökologische Bedeutung der Hartgründe ("Hardgrounds") in der oberen Kreide, *Paläontol. Z.*, 33, 129-147.

Cross-references: *Beachrock; Biogenic Sedimentary Structures; Carbonate Sediments—Diagenesis; Cementation; Cementation, Submarine; Chalk; Chert and Flint; Corrosion Surface; Diagenesis; Duricrust; Pelagic Sedimentation, Pelagic Sediments; Silicification.*

HEAVY MINERALS

Heavy minerals are minerals with a specific gravity greater than quartz (s.g. = 2.65) or feldspar (s.g. = 2.54-2.76); a density of 2.8 is generally the accepted lower limit. Heavy mineral concentrations are composed primarily of ilmenite, magnetite, garnet, zircon, hornblende, rutile, monazite, or olivine. The most frequently occurring heavy minerals are listed in Table 1. There is less variety in many deposits because of differential transport and mechanical instability (Table 1). Heavy minerals can be divided into three groups according to their density and economic importance (Emery and Noakes, 1968):

- Heavy heavy minerals—gold, tin, platinum: density 6.8-21.0 g/cm³

TABLE 1. List of Heavy Minerals, Occurrence (a), Specific Gravity (S. G.), Habit, and Stability (b)

Mineral	a	S.G.	Habit	b
Anatase	C	3.9	equant	S
Andalusite	L	3.2	prismatic	M
Apatite	R	3.2	prismatic	U
Augite	C	3.2- 3.4	prismatic	M
Barite	R	4.5	tabular	S
Biotite	C	2.8- 3.2	platy	M
Brookite	C	3.9- 4.1	tabular	S
Cassiterite	L	6.8- 7.1	prismatic	S
Chlorite	C	2.6- 3.0	platy	SP
Chromite	R	4.3- 4.6	equant	S
Columbite	VR	5.3- 7.3	prismatic	S
Cordierite	L	2.6- 2.7	prismatic	U
Corundum	L	4.0- 4.1	equant	S
Diamond	VR	3.5	equant	S
Epidote	C	3.3- 3.5	equant	M
Fluorite	L	3.2	equant	S
Garnet	C	3.5- 4.3	equant	S
Glauconite	C	2.2- 2.9	platy	U
Glaucophane	R	3.0- 3.2	prismatic	U
Gold	VR	15.0-19.3	irregular	S
Hematite	A	5.3	platy	SP
Hornblende	C	3.0- 3.3	prismatic	M
Hypersthene	L	3.2- 3.5	prismatic	M
Ilmenite	C	4.5- 5.0	prismatic	M
Kyanite	L	3.5- 3.7	platy	S
Leucoxene	C	3.5- 4.5	equant	S
Limonite	A	3.6- 4.0	coating	SP
Marcasite	L	4.8- 4.9	equant	M
Magnetite	C	5.2- 6.5	equant	S
Monazite	L	4.9- 5.3	equant	S
Muscovite	C	2.8- 3.1	platy	S
Olivine	L	3.3- 4.4	irregular	U
Pyrite	L	5.0- 5.1	equant	U
Pyrolusite	R	4.7- 4.8	irregular	M
Pyrrhotite	L	4.6- 4.7	irregular	U
Rutile	C	4.2- 4.3	acicular	S
Siderite	L or A	3.8- 3.9	equant	SP
Sillimanite	L	3.2	acicular	S
Spinel	R	3.5- 4.0	equant	S
Staurolite	L	3.6- 3.7	prismatic	S
Titanite	R	3.4- 3.6	prismatic	S
Topaz	L	3.4- 3.6	platy	S
Tourmaline	L	3.0- 3.3	prismatic	S
Wolframite	L	7.0- 7.5	prismatic	S
Xenotime	VR	4.5- 4.6	acicular	S
Zircon	C	4.2- 4.9	acicular	S

C = common   S = stable
L = local   M = moderately stable
R = rare   U = unstable
VR = very rare   SP = stable as a secondary product
A = alteration product

- Light heavy minerals—ilmenite, rutile, zircon, monazite: density 4.2-5.3 g/cm³
- Gems—mainly diamonds: density 2.9-4.1 g/cm³

The first group is found mainly in fluvial sediments; the second group in beach sediments, dune sands, and shelf placers; the third group in stream sediments, and, exceptionally, in shelf sediments. All these occurrences are known from ancient as well as modern sediments.

The nearshore environment provides the greatest opportunity for wave action to sort and concentrate the different heavy minerals; a similar environment existed on the exposed Pleistocene continental shelf, where there are now relict beach, lagoon, sound, or coastal stream deposits (Emery and Noakes, 1968).

The most important determinant for heavy mineral concentration is availability. They must be moved to the nearshore zone by rivers, inlet flushing, rip currents, storm-induced undertow, or coastal erosion, and from the continental shelf.

Behavior

The mechanisms for the concentration of heavy minerals can be illustrated by their behavior in the littoral environment (see also *Black Sands; Littoral Sedimentation*). Heavy mineral deposits on beaches are commonly found at or near the surface of the backshore, close to the foot of the dunes or wave-cut cliffs, and, less commonly or in diminished thickness, on the foreshore (Fig. 1).

Minerals coming to the beach are separated as they are moved alongshore as a result of their differing hydraulic behavior. Settling velocity (q.v.) is dependent on the density of the particle. When lifted above the bottom by waves, a 100μ garnet grain will fall at least 0.35 cm/sec faster than the same size quartz grain, so that the lighter mineral can be transported farther by the alongshore current. Only if the garnet is approximately 70% of the size of quartz (and the same shape) will they be deposited together (see *Hydraulic Equivalence*). This difference in size is sufficient to segregate grains of different densities. Also, if the grains are of the same size originally, the heavy minerals lag behind during alongshore drifting because they are suspended and set in motion less easily than are the light-density minerals (Rittenhouse, 1943). The sorting action of the surf zone can be seen in spits where "super heavy" minerals (s.g., >7) are found nearest the source: the concentration of the usual suite of heavy minerals follows; and, finally, the light grains are found around the end of the spit.

When sand grains are deposited on a beach, the heavy minerals are concentrated by the swash and backwash. Rolling of sediment under swash impact favors the removal of larger, lighter grains, leaving a lag deposit of the heavy minerals. Once entrained, grains will be deposited according to their hydraulic equivalence. A denser grain is smaller than a light mineral with the same settling velocity, however, and thus is shielded from water flow, its entrainment blocked by the larger grains (Hand, 1967). This shielding is enhanced if the heavy mineral

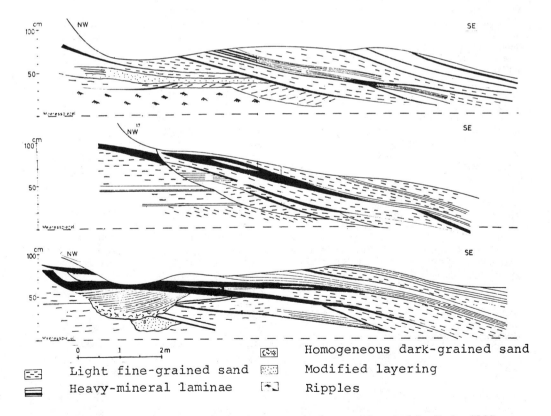

FIGURE 1. Various types of heavy mineral concentration in coastal deposits (after Cordes, 1966).

is elongated or platy. After deposition by the swash, the minerals are reentrained by the backwash, where dense clouds of sediment 1 or 2 cm thick are sorted by inertial effects (Inman et al., 1966) or shear sorting (Bagnold, 1954). Here the coarser grains of any particular density move toward the top of the flow and the finer grains toward the bottom (see *Grain Flow*). For grains of equal size the denser ones will work their way down relative to the lighter grains in this moving layer. Thus, each backwash results in a sedimentation unit with small, heavy minerals concentrated along the bottom and coarser, lighter grains on top (Clifton, 1969). Larger, more erosive waves enhance the thickness of these units.

BENNO M. BRENNINKMEYER, S. J.

References

Bagnold, R. A., 1954. Experiments on a gravity-free dispersion of large solid spheres in a Newtonian fluid under shear, *Proc. Royal Soc.*, ser. A., 225, 49-63.

Clifton, H. E., 1969. Beach lamination—nature and origin, *Marine Geol.*, 7, 553-559.

Cordes, E., 1966. Aufbau und Bildungbedingunen der Schwermineralseifen bei Skagen (Danemark), *Meyniana*, 16, 1-35.

Emery, K. O., and Noakes, L. C., 1968. Economic placer deposits of the continental shelf, *ECAFE Tech. Bull.*, 1, 95-111.

Everts, C. H., 1972. Exploration for high energy placer deposits: Part 1—field and flume tests, North Carolina Project, *Univ. Wisconsin Sea Grant Tech. Rept.*, WIS-SG-72-210, 192p.

Gadow, S., 1972. Georgia Coastal Region II, provenance and distribution of heavy minerals, *Senckenb. Marit.*, 4, 15-45.

Hand, B. M., 1967. Differentiation of beach and dune sands, using settling velocities of light and heavy minerals, *J. Sed. Petrology*, 37, 514-520.

Inman, D. L.; Ewing, G. C.; and Corliss, J. B., 1966. Coastal sand dunes of Guerrero Negro, Baja California, Mexico, *Geol. Soc. Am. Bull.*, 77, 787-802.

May, J. P., 1973. Selective transport of heavy minerals by shoaling waves, *Sedimentology*, 20, 203-212.

Rittenhouse, G., 1943. Transportation and deposition of heavy minerals, *Geol. Soc. Am. Bull.*, 54, 1725-1780.

Rizzini, A., 1974. Holocene sedimentary cycle and heavy-mineral distribution Romagna-Marche coastal plain, Italy, *Sed. Geol.*, 11, 17-37.

Cross-references: *Beach Sands; Black Sands; Continental-Shelf Sediments; Hydraulic Equivalence; Littoral Sedimentation; Provenance; Sedimentary Petrography; Settling Velocity.*

HUMIC MATTER IN SEDIMENTS

When the dead cells and extracellular products of plants, algae, and other organisms undergo microbial decomposition, they are partially transformed into an assortment of complex, heterogeneous, brown or yellow organic acids of rather high molecular weight which are collectively termed *humic substances*. Humic matter is formed from a variety of biochemical precursors (carbohydrates, proteins, lipids, lignin and other phenols, pigments, etc.) under various environmental conditions. Being relatively stable, it tends to accumulate, even under moderately oxidizing conditions. Humic substances comprise the bulk of the organic matter in soil, natural waters, and recent sediments (Bordovskiy, 1965; Schnitzer and Khan, 1972).

Humic matter has immense geochemical, ecological, economic, and possibly paleobiological importance. This material represents a major "leak" in the cycles of carbon and nitrogen. It is the principal precursor of coal (q.v.), and of kerogen (q.v.), which in turn is considered the main source of petroleum (q.v.). Humic matter is a strong chelating agent, and it participates in cation exchange reactions; as a result, humic matter plays a significant role in plant nutrition, chemical weathering, the migration of metals, and the concentration of metals in sediments (Schnitzer, 1969; Szalay, 1964; Rashid, 1971). Humic substances may stimulate (or sometimes may depress) the activity of plants and phytoplankton (Prakash, 1971). In soil, the formation of clay-humic complexes (Greenland, 1971) creates physical conditions favorable for the growth of plants. Moreover, sorbed humic matter tends to impede the flocculation of clay minerals entering the sea and may therefore be a factor in sedimentation (see *Flocculation*). The binding of environmental poisons such as heavy metals and pesticides by humic matter, and the possible detoxifying effect of such interactions, deserve further study. The possible value of humic matter in paleobiological research has been demonstrated by the work of Jackson (1975b).

The most effective extractant for humic matter is NaOH solution. On acidification of the extract, the "humic acid" fraction precipitates, the "fulvic acid" fraction remaining in solution. These fractions are mixtures which can be further separated by techniques such as gel filtration and fractionation with organic solvents. There is commonly a nonextractable fraction called "humin." A number of relatively mild reagents such as $Na_4P_2O_7$, NaF, Na_2CO_3/$NaHCO_3$, EDTA, and organic solvents have been used as substitutes for NaOH; physical dispersion by ultrasonic vibration is also used. The ability of metal-binding agents to solubilize humic matter suggests that in its native state humic matter consists largely of a network of molecules bonded to each other and to mineral particles by metal "bridges." Analytical methods used to study the molecular properties of humic matter include ultraviolet, visible, and infrared spectrophotometry, wet-chemical determination of functional groups, determination of breakdown products following partial degradation, methods of ultimate analysis, methods for determination of molecular weight, x-ray diffraction, and electrometric titrations. Various methods for extraction, fractionation, and analysis of humic matter are summarized by Schnitzer and Khan (1972) and Felbeck (1971).

Humic substances consist chiefly of carbon and oxygen, with lesser amounts of nitrogen and hydrogen. In general, a molecule of humic matter consists of a more or less condensed aromatic "core" (Haworth, 1971) with various side-chains and functional groups, notably COOH, phenolic OH, C=O, alcoholic OH, OCH_3, and aliphatic C—H chains; biochemical compounds such as proteins, carbohydrates, and fatty acids are commonly complexed with the molecule. The composition and molecular size and structure of humic substances vary with the biological source material, environment, and stage of humification. Although the complexity of humic substances has impeded the exact determination of molecular structures, quantitative comparisons between different samples of humic matter can be made by the methods listed above. Aquatic or sedimentary humic matter derived from algae can be distinguished from soil humic matter derived from higher plants by the lower degree of condensation of its aromatic rings, its high nitrogen and sulfur content, its high $^{13}C:^{12}C$ ratio, its tendency to be more highly aliphatic, and by the presence of constituents that fluoresce in ultraviolet light. Marine humic matter is largely of marine algal origin, but appreciable amounts of humic matter eroded from soil are transported to coastal waters by rivers (Bordovskiy, 1965). Similarly, lacustrine humic matter includes both autochthonous and allochthonous components, the relative amount of each depending on local conditions (Otsuki and Hanya, 1967).

TOGWELL A. JACKSON

References

Bordovskiy, O. K., 1965. Accumulation and transformation of organic substances in marine sediments, *Marine Geol.,* 3, 1-114.

Felbeck, G. T., Jr., 1971. Chemical and biological characterization of humic matter, in A. D. McLaren, and J. J. Skujins, eds., *Soil Biochemistry,* vol. 2. New York: Marcel Dekker, 36-59.

Greenland, D. J., 1971. Interactions between humic and fulvic acids and clays, *Soil Sci.,* 111, 34-41.

Haworth, R. D., 1971. The chemical nature of humic acid, *Soil Sci.,* 111, 71-79.

Ishiwatari, R., 1970. Structural characteristics of humic substances in recent lake sediments, in G. D. Hobson and G. C. Speers, eds. *Advances in Organic Geochemistry.* Oxford: Pergamon, 285-311.

Jackson, T. A., 1975a. Humic matter in natural waters and sediments, *Soil Sci.,* 119, 56-64.

Jackson, T. A., 1975b. "Humic" matter in the bitumen of pre-Phanerozoic and Phanerozoic sediments and its paleobiological significance, *Am. J. Sci.,* 275, 906-953, 276, 560.

Kononova, M. M., 1966. *Soil Organic Matter.* Oxford: Pergamon, 544p.

Manskaya, S. M., and Drozdova, T. V., 1968. *Geochemistry of Organic Substances.* Oxford: Pergamon, 347p.

Nissenbaum, A., and Kaplan, I. R., 1972. Chemical and isotopic evidence for the in situ origin of marine humic substances, *Limnol. Oceanogr.,* 17, 570-582.

Otsuki, A., and Hanya, T., 1967. Some precursors of humic acid in Recent lake sediment suggested by infra-red spectra, *Geochim. Cosmochim. Acta,* 31, 1505-1515.

Prakash, A., 1971. Terrigenous organic matter and coastal phytoplankton fertility, in J. D. Costlow, ed., *Fertility of the Sea,* vol. 2. London: Gordon and Breach, 351-368.

Rashid, M. A., 1971. Role of humic acids of marine origin and their different molecular weight fractions in complexing di- and trivalent metals, *Soil Sci.,* 111, 298-305.

Schnitzer, M., 1969. Reactions between fulvic acid, a soil humic compound and inorganic soil constituents, *Soil Sci. Soc. Am. Proc.,* 33, 75-81.

Schnitzer, M., and Khan, S. U., 1972. *Humic Substances in the Environment.* New York: Marcel Dekker, 327p.

Stevenson, F. J., and Goh, K. M., 1971. Infrared spectra of humid acids and related substances, *Geochim. Cosmochim. Acta,* 35, 471-483.

Szalay, A., 1964. Cation exchange properties of humic acids and their importance in the geochemical enrichment of UO_2^{++} and other cations, *Geochim. Cosmochim. Acta,* 28, 1605-1614.

Cross-references: *Bitumen, Bituminous Sediments; Clay; Coal; Crude-Oil Composition and Migration; Flocculation; Gyttja, Dy; Hydrocarbons in Sediments; Kerogen; Lacustrine Sedimentation; Microbiology in Sedimentology; Organic Sediments; Petroleum—Origin and Evolution; Weathering in Sediments.* Vol. XI: *Humification; Humus and Humic Acids.*

HYDRAULIC EQUIVALENCE

Sedimentary materials have hydraulic equivalency when they exhibit the same hydraulic properties in a stream (water or air). Hydraulic equivalency is controlled by a combination of properties which makes a sedimentary particle indistinguishable to the fluid, i.e., the mineral grains respond alike to the hydraulic properties of the moving water or air. Particles are said to be hydraulically equivalent when their terminal fall velocities are equal (see *Settling Velocity*). The fall velocity of the particle in a fluid can be measured by timing its fall over a measured course after the particle has accelerated to a constant velocity. In practice hydraulic engineers and geologists use the term *fall diameter* in reference to the size of particles relative to fall velocities. *Fall diameter* is the diameter of a quartz sphere having the same fall velocity as the particle. Hydraulically equivalent particles have equal fall velocities and equal fall diameters, although the particles may be different minerals having varying properties of specific gravity, shape, and surface texture.

When the falling particle has reached its fall velocity, the force of particle weight equals the resistive fluid force due to hydraulic drag. Small particles, e.g., quartz grains 50μ or smaller, are retarded entirely by the viscous forces which produce fluid molecular viscosity. Stokes' Law (q.v.) describes this force as

$$F_v = C_1 d^2 w$$

where C_1 is a coefficient, d is particle diameter, and w is fall velocity. Larger particles are resisted also by inertial forces, defined as $F_i = C_2 d^2 w^2$. For sand-sized particles the ratio of inertial to viscous forces ranges from 1–100; thus, the inertial forces predominate for sand in water and air.

At terminal fall velocity the weight is counterbalanced by the resistive forces, $F_i + F_v$. Particle size and specific gravity (γ) define the weight; particle diameter (d) controls the cross-sectional area; and particle shape (S) affects the drag. Consequently, the fall velocity is a function primarily of these sediment properties in addition to viscosity (ν) of the fluid. Stated in general mathematical form, $w = f(d, \gamma, \nu, S)$. Hydraulic equivalency exists where $w_1 = w_2 = w_i$, where i is the subscript for the general particle.

LOUIS I. BRIGGS

References

Allen, J. R. L., 1970. *Physical Processes of Sedimentation*. London: George Allen, 248p.

Inman, D. L., 1949. Sorting of sediments in the light of fluid mechanics, *J. Sed. Petrology*, 19, 51–70.

Lowright, R. H., 1973. Environmental determination using hydraulic equivalence studies, *J. Sed. Petrology*, 43, 1143–1147.

White, J. R., and Williams, E. G., 1967. The nature of the fluvial process as defined by settling velocities of heavy and light minerals, *J. Sed. Petrology*, 37, 530–539.

Cross-references: *Fluvial Sediment Transport; Grain-Size Studies; Heavy Minerals; Settling Velocity; Sorting; Stokes' Law; Transportation.*

HYDROCARBONS IN SEDIMENTS

Hydrocarbons are compounds composed solely of hydrogen and carbon. They occur as gases, liquids, and solids, and are widely distributed in the lithosphere, hydrosphere, and atmosphere, as well as specifically in the biosphere (see Vol. IVA: *Hydrocarbons*).

Occurrences

Hydrocarbons occur commonly up to a few ppm in soils. The distribution of structural types is wide, including normal, branched, cyclic, and aromatic hydrocarbons. The normal alkanes commonly exhibit odd-carbon-number preference. Branched aliphatic hydrocarbons usually include members of the isoprenoid series, i.e., compounds including one or more units with the branched isoprene skeleton of five carbon atoms. Polycyclic aromatic hydrocarbons are probably present, but evidently at levels <0.01 ppm.

Hydrocarbons in modern sediments are more fully known. Odd-carbon preference in normal alkanes is common. Aliphatic hydrocarbons occur in the range of 10–1000 ppm, and aromatics are frequently equally abundant. Isoprenoid hydrocarbons are common, with most falling in the C_{15} (farnesane) to C_{20} (phytane) range. Polycyclic aromatics are present in small amounts in modern fresh-water and marine sediments.

Hydrocarbons of ancient sedimentary rocks are very well known. The level of concentration is generally the same or greater than that for modern sediments (Bray and Evans, 1961). Alkanes of all structures are found, including isoprenoids, simple aromatics, and polycyclic aromatics. Of the isoprenoids, pristane (C_{19}) is usually the most abundant, followed by C_{16} and phytane (C_{20}), for overall abundances of up to about 10 ppm of the rock (Hodgson et al., 1968). Isoprenoids are roughly 5–50% as abundant as the *n*-alkanes. Normal alkanes in sedimentary rocks commonly show odd-carbon preference, usually most strongly in the C_{20}–C_{30} range. Alkanes occur in rocks of all ages, even older than three billion years.

Petroleum. Hydrocarbons in petroleum (q.v.) accumulations are special occurrences of hydrocarbons in the lithosphere and have been studied extensively, as reviewed comprehensively by Smith (1966) and Hunt (1968). Normal alkanes range from one carbon atom (methane) to >40, with chains of as many as 74 and 78 carbons reported. High-melting waxes containing about 25 carbon atoms are largely normal alkanes, but branched and cyclic structures are probably also included. Nearly 100 branched alkanes in the C_4–C_{13} range have been identified in crude oil (q.v.); in the higher ranges, principal attention has focused on particular series of compounds, mainly the isoprenoids, which are most abundant in the C_{15}–C_{20} range. In general, the branched alkanes with the fewest and simplest branches are most prevalent. Unsaturated hydrocarbons are uncommon in petroleum, but C_6, C_7, and C_8 olefins have been identified.

Cyclic alkanes are ring compounds, usually comprising five- or six-membered rings. All cycloalkanes are saturated, and all except the parent members have alkyl side chains. Thirty-one monocyclic pentanes and forty-two monocyclic hexanes have been identified in petroleum. Two- and three-ring cycloalkanes have been identified; and recently several pentacyclic triterpanes have been isolated (Hills et al., 1968).

Aromatic hydrocarbons are common in petroleum. The simplest are benzene and its alkylated homologs, which are generally less prevalent than either the corresponding alkanes or cycloalkanes. Toluene is almost always more abundant than benzene, generally by a factor of 3 to 6. Polycyclic aromatics with 2 to 5 fused rings have been found in petroleum. These occur commonly in alkylated homologous series. As molecular weight increases, more and more compounds are found that are mixtures of alkanes, cycloalkanes, and aromatics. Included are the indanes, tetralins, fluorenes, and cyclopentaphenanthrenes. Asphaltene and resin portions of the heavy gasoil fraction of petroleum consist primarily of graphite-like clusters of condensed aromatic rings joined together by alkyl groups. Naphthalenic ring systems are also present but are not fused to the aromatic clusters (Erdman, 1965). Hydrocarbon waxes are very common in petroleum and range widely in content from 0 to 33% (Hedberg, 1968).

Origin of Hydrocarbons

Primordial Origin. Hydrocarbons existing in crustal rocks of the earth may have come in part from the primordial matter from which the earth accumulated, through continuing degassing of the planet. No reliable criteria exist for the identification of primordial hydrocarbons, and no clear-cut examples of accumulations of such hydrocarbons are known.

Biogenic Origin. Hydrocarbons are synthesized by living organisms. Normal alkanes in the range from C_{25} to C_{35} appear to be widely distributed in the plant kingdom, occurring in roots, stems, leaves, flowers, fruits, and seeds of a variety of plants (Oro et al., 1965). Aromatic hydrocarbons are derived from biogenetic sources, e.g., from pine-wood tar. The biosynthesis of long-chain hydrocarbons is revealed by the metabolic pathways for the biosynthesis of iso and anteiso hydrocarbons. A close relationship is believed to exist between fatty acid and hydrocarbon metabolism. Carotenoids follow the general pattern of terpenoid biosynthesis in that they are derived from the biological isoprene precursor, isopentyl pyrophosphate, which originates from acetyl coenzyme-A (Goodwin, 1968). Aromatic compounds are commonly synthesized as quinones in microorganisms and larger plants (Thomson, 1965).

Immediate Precursors. Hydrocarbons may be formed from specific immediate precursors such as fatty acids and alcohols, and from nonspecific sources such as kerogens (q.v.) and coals (q.v.). Geochemical conversions are thought to occur both thermally and catalytically.

GORDON W. HODGSON

References

Bray, E. E., and Evans, E. D., 1961. Distribution of n-paraffins as a clue to recognition of source beds, *Geochim. Cosmochim. Acta,* 22, 2-15.

Erdman, J. G., 1965. The molecular complex comprising heavy petroleum fractions, *Hydrocarbon Analysis, Am. Soc. Testing Mater., Spec. Tech. Publ. 389,* 259-300.

Goodwin, T. W., 1968. Recent developments in the biosynthesis of carotenoids, *J. Sci. Indus. Research,* 27(3), 103-105.

Hedberg, H. D., 1968. Significance of high-wax oils with respect to genesis of petroleum, *Bull. Am. Assoc. Petrol. Geologists,* 52, 736-750.

Hills, I. R.; Smith, C. W.; and Whitehead, E. V., 1968. Hydrocarbons from fossil fuels and their relationship with living organisms, *Am. Chem. Soc. Preprints,* 13(4), B5-B20.

Hodgson, G. W.; Hitchon, B.; Taguchi, K.; Baker, B. L.; and Peake, E., 1968. Geochemistry of porphyrins, chlorins and polycyclic aromatics in soils, sediments and sedimentary rocks, *Geochim. Cosmochim. Acta,* 32, 737-772.

Hunt, J. M., 1968. How gas and oil form and migrate, *World Oil,* 167, 140-150.

Oro, J.; Nooner, D. W.; and Wikstrom, S. A., 1965, Paraffinic hydrocarbons in pasture plants, *Science,* 147, 870-873.

Smith, H. M., 1966, Crude oil: Qualitative and quantitative aspects, *U.S. Bur. Mines Inf. Circ. 8286,* 41p.

Thomson, R. H., 1965. Quinones: Nature, distribution and biosynthesis, in T. W. Goodwin, ed., *Chemistry and Biochemistry of Plant Pigments.* London: Academic Press, 309-332.

Whitehead, W. C., and Breger, I. A., 1963. Geochemistry of petroleum, in I. A. Breger, ed., *Organic Geochemistry.* New York: MacMillan, 248-332.

Cross-references: *Bitumen, Bituminous Sediments; Coal; Crude-Oil Composition and Migration; Humic Matter in Sediments; Kerogen; Organic Sediments; Petroleum—Origin and Evolution.* Vol. IVA: *Hydrocarbons.*

IGNIMBRITES

Ignimbrites are defined as the group of pyroclastic rocks composed primarily of poorly sorted pumice and shards showing flattened structures indicative of pyroclastic flow origin. In common usage, the term *ignimbrite* is synonymous with *welded tuff,* tuff flow, pumice flow, and ash flow. The latter is usually considered as the basic unit of most ignimbrites (Smith, 1960). Use of "ignimbrite" is often restricted to welded pyroclastic flows, or it may be applied to rock bodies resulting from the eruptive phase characterized by actively moving flow, whether they are welded or not (Sparks et al., 1973). Ignimbrite deposits are mostly associated with acidic lavas, but the composition, and degree of welding as well, may vary substantially within a given sequence. Some welded ignimbrites may reach the hardness and density of real lava flows. They may also exhibit columnar jointing somewhat similar to lavas. Many ignimbrite sequences show flow units having reverse grading of large pumice clasts and normal grading of large lithic clasts, with μ-sized particles commonly coexisting with blocks approaching 1 m in size (Sparks et al., 1973).

The microscopic characteristics of welded pyroclastic rocks consist of vitroclastic to flow-banded structures and foliation. These structures are indicative of compaction, flowage, recrystallization, and fusion of the fragments due to the activities of escaping gases and fluids at elevated temperatures. The pyroclastic materials or *tephra* thus originated from semiconsolidated magma which was ejected and pulverized from the explosion caused by compressed gas, or by water vapor in the vent. Ignimbrite deposits are generally accepted to be the products of Péléan-phase volcanic eruption, which is characterized by lateral spreading of hot viscous clouds along the volcanic cone. This phenomenon, described as a *nuée ardente* (q.v.) by Lacroix in 1930, is also known in the literature as *glowing cloud* and *glowing avalanche.* Pyroclastic flows which result in ignimbrite deposits come mainly from Lacroix's "nuée ardente d'explosion dirigée," which is composed of vesiculated tephra, and the height of the advancing glowing cloud grows outward from the volcanic center. Some authors, however, have taken exception to the Péléan-type nuée ardente as the universal agent for ignimbrites, especially for those of large volume. Furthermore, whereas welding of pyroclasts occurs at temperature of 500–1000°C (Smith, 1960), nonwelded ignimbrites suggest ejected fragments at temperatures no greater than 500°C. Volcanoes producing ignimbrites in fact usually show different eruptive phases.

Sparks et al. (1973) recognized the following sequence: (1) a highly explosive, often Plinian phase (i.e., gas-blast eruption of great violence) producing low-density, highly inflated pumice-fall deposits; (2) a Péléan phase, which produces higher-density pyroclastic flow deposits; and (3) an effusive phase, producing a nonvesicular lava flow. All three phases may not occur, however; and in many eruptions, only two of the three phases may be recognized.

Thus, the main distinctive characters of ignimbrite deposits stem from their pyroclastic origin in chaotic flow or avalanche mechanisms, in contrast to fluidal lava flows, which are primarily composed of nonvesiculated material. Air-fall pyroclastic rocks show better sorting and show viscous deformation only near the vent because air-borne tephra would cool rapidly enough to prevent such deformation and welding farther than 1 km from the vent (Korringa, 1973). Many ignimbrites show comagmatic series, and the various degrees of welding are indicative of different phases in the evolution of a volcanic center.

Products of ignimbrite eruptions are probably the most abundant variety of tuff (q.v.). Their extension at the earth's surface is comparable only to that of plateau basalts. According to Smith (1960), some welded tuffs or ash-flow fields distributed from a common source may reach areal dimensions greater than 31,000 km^2 and a volume of about 2000 km^3, some multiple source fields are known to have volume greater than 8000 km^3. Holocene ignimbrites are volumetrically important in circum-Pacific areas such as New Zealand, the Indonesian islands, Japan and western North and South America.

FLORENTIN MAURRASSE

References

Cook, E. F., 1966. *Tufflavas and Ignimbrites: a Survey of Soviet Studies*. New York: American Elsevier, 212p.

Korringa, M. K., 1973. Linear vent area of the Soldier Meadow Tuff, an ash-flow sheet in Northwestern Nevada, *Geol. Soc. Am. Bull.*, **84**, 3849-3866.

Lacroix, A., 1930. Remarques sur les matériaux de projection des volcans et sur la génèse des roches pyroclastiques qu'ils constituent, *Soc. Géol. France Livre Jubilaire Centenaire 1830-1930*, **2**, 431-472.

Moore, B. N., 1934. Deposits of possible nuée ardente in the Crater Lake region, Oregon, *J. Geol.*, **42**, 358-375.

Ross, C. L., and Smith, L. R., 1961. Ash flow tuffs: Their origin, geologic relations and identification, *U.S. Geol. Surv. Prof. Pap. 366*, 81p.

Smith, R. L., 1960. Ash flows, *Geol. Soc. Am. Bull.*, **71**, 795-842.

Sparks, R. S. J.; Self, S.; and Walker, G. P. L., 1973. Product of ignimbrite eruptions, *Geology*, **1**, 115-118.

Steiner, A., 1960. Origin of ignimbrites of the North Island, New Zealand, *New Zealand Geol. Surv. Bull.*, **68**, 42p.

Cross-references: *Base-Surge Deposits; Nuée Ardente; Pyroclastic Sediment; Tuff; Volcanic Ash—Deposits; Volcaniclastic Sediments and Rocks.*

IMPACT LAW

The settling velocities of small particles held in suspension are best described by *Stokes' Law* (q.v.); as the particle size increases beyond 0.1 mm, however, the *Impact Law* is more accurate (Rubey, 1933). In the case of the smaller particles (<0.1 mm), the impact of a fluid against a settling particle is primarily the result of fluid viscosity; for larger particles, the impact involves both the viscosity and the inertia. In the transition zone (0.1-1.0 mm), experimental data agrees with an average of the two laws, indicating that neither is truly expressive of de facto conditions (see Fig. 1). The significance of the difference between Stokes' Law and the Impact Law is that larger particles settle out of a fluid independently of the viscosity of the medium,

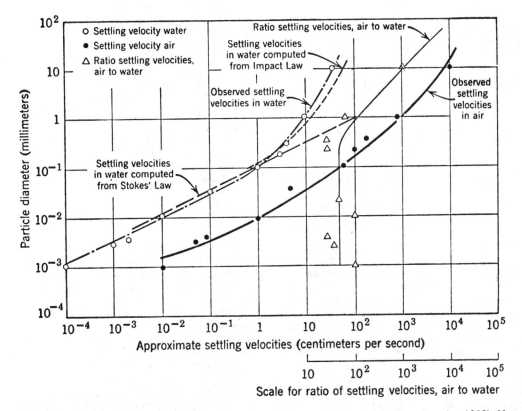

FIGURE 1. Comparison of settling velocities of particles through air and water (from Dapples, 1959). Note the similarity of the slopes of the two curves. Stokes' law holds for particles with diameters less than 0.1 mm, but for larger particles the settling velocities are controlled by the impact law. For particles up to about 0.5 mm diameter the ratio of air-water settling velocities remains uniform, but for larger sizes the ratio increases logarithmically.

whereas smaller ones remain in long-continued suspension resulting from the viscosity (internal friction due to molecular cohesion of the fluid). This difference may best be seen in a comparison of the settling velocities of various sized particles suspended in water and also in air. The ratio of the settling velocities (Fig. 1), air to water, is constant from colloidal sizes up to about 0.5 mm diameter, at which point it increases logarithmically, becoming more than a thousand times as great (Dapples, 1959).

B. CHARLOTTE SCHREIBER

References

Dapples, E. C., 1959. *Basic Geology for Science and Engineering.* New York: Wiley, 609p.
Krumbein, W. C., and Sloss, L. J., 1951. *Stratigraphy and Sedimentation.* San Francisco: Freeman, 660p.
Rubey, W. W., 1933. Settling velocities of gravel, sand, and silt particles, *Am. J. Sci.,* 224, 325-338.

Cross-references: *Hydraulic Equivalence; Settling Velocity; Stokes' Law.*

IMPACT MARKS—*See* CURRENT MARKS; TOOL MARKS

INDURATION—*See* DURICRUST

INLET FACIES—*See* TIDAL-INLET FACIES

INTERSTITIAL WATER IN SEDIMENTS

Interstitial waters are aqueous solutions that occupy the pore spaces between grains of soils, sediments, and rocks. They are sensitive indicators of particle-fluid reactions and equilibria in sediments, reflect migration pathways and origin of fluids, and supply most of the nutrient salts to terrestrial plants (via soils). They have often been suggested as important sources of reactive constituents for the formation of economically valuable deposits such as heavy-metal sulfides, phosphorites, and iron and manganese ores. Their role in the formation of oil and gas deposits is certainly significant, but still poorly known. In the larger sense, they are involved in most diagenetic (postdepositional) reactions in sedimentary and metamorphic rocks (see *Diagenesis;* also in Vol. IVA: *Fluid Inclusions; Natural Brines; Clay Membrane Phenomena;* Sayles and Manheim, 1975). Discussion below is limited largely to Holocene and subrecent deposits.

Early Studies

Perhaps the first major study of interstitial waters was published in 1866 by the French agronomist Schloesing, who displaced pore fluids from a quantity of soil and carefully analyzed the effluent. Since that time, studies of sediment-water-plant interactions have continued to be a major area of interest in soil science (Mehlich and Drake, 1958).

The first geological-oceanographic investigation of interstitial waters was published by Murray and Irvine (1895); they squeezed fluids from a "blue mud" from the Scottish coast and found that whereas the major components of sea water remained in fairly constant proportions in the mud, oxygen had been depleted, sulfate was lost, and bicarbonate alkalinity increased. The relative proportions of the constituents indicated that the (largely bacterial) oxidation of organic carbon to CO_2 had balanced most of the reduction of sulfate to form sulfide. Moreover, on realizing that there was an appreciable amount of dissolved manganese in the pore water, Murray revised his earlier concept of a largely volcanic origin for deep-sea manganese concretions. He suggested that, along with river-borne and volcanic supplies of Mn, a major source might be manganese diffusing up into sea water from terrigenous continental sediments undergoing syngenetic reduction.

Interstitial chlorinity of coastal, and especially intertidal sediments, was studied by British biologists from the late 1920s (see references in Smith, 1955). Some coastal sediments have lower interstitial chlorinities than average local bottom water because of the influence of ground-water discharge; higher salinities also arise, however, through gravitational convection and downflow of heavier solutions through permeable sediments (see Scholl, 1965). Recent interest in interstitial waters of coastal sediments centers on their role in transporting nutrients or pollutants (Lehman, 1976).

Interest in the geochemistry of pore waters revived in the middle-late 1930s partly through the writings of the Russian geochemist, V. I. Vernadsky. Soviet earth scientists, led by Vinogradov, Brujewicz, and L. A. Shchukarev studied pore waters from the Caspian and Black seas, as well as fresh and saline lakes. Following World War II, Soviet oceanographers took advantage of effective sediment squeezers developed by P. A. Kruikov and expanded studies to show enrichments of up to 1000-fold in minor ions, nutrients, organic, and absorbed constituents in sediments from the Russian northern seas, the Black, Caspian, and Baltic seas, and the Atlantic, Pacific, and Indian oceans (see Manheim, 1976).

Outside the Soviet Union, Emery and Rittenberg and coworkers made detailed studies of biogenic cycles and constituents, including gases in interstitial waters from southern California offshore basins. Ammonia, carbon dioxide, and methane were shown to dominate volatiles (see Manheim, 1976).

B. Kullenberg, the Swedish oceanographer who had developed the piston corer, utilized cores up to 10 m long to determine whether marine sediments might retain a record of the chlorinity of the water in which they were deposited. Marked decreases in chlorinity with depth in Baltic Sea sediments (approaching glacial-equivalent sediments) and parallel observations by Brujewicz in the Black Sea signalled presence of earlier, brackish-fresh stages of the water bodies in question.

Recent Studies

Beginning in the 1960s, interstitial water studies expanded to explore new aspects of sediment-water systems and sea-floor phenomena. Careful studies by Shishkina revealed few differences in composition of interstitial waters in pelagic sediments (q.v.) from overlying ocean water. Chief anomalies remaining (Si, K enrichment, Mg depletion) were later revealed to be partly due to influence of temperature on exchangeable ions in the solids, i.e., increase of temperature on raising sediments from cold (1–2°C) deep waters to shipboard temperatures, before extraction of fluid (Mangelsdorf et al., 1969). After correction for these effects, it appears that K is depleted in nearly all ocean pore waters through uptake on solids. Trace elements more soluble in reducing environments (Fe, Mn, Pb, Co, Ni, I) are known to be especially enriched in pore fluids from organic-rich, nearshore environments such as the Black, Baltic, and Russian northern seas, southern California offshore basins, estuaries, and lakes. Whereas interstitial manganese and iron may be considered the main source of these metals in lake and shallow marine ferromanganese concretions, the role of interstitial solution is clearly less pronounced for deep ocean nodules.

Decomposition of organic matter in marine sediments is signalled initially by increased interstitial alkalinity and ammonia (Eq. 1 and 2).

$$CH_2O + O_2 \rightarrow CO_2 + H_2O \rightarrow HCO_3^- + H^+ \quad (1)$$

$$CH_3NH_2CHCOOH \text{ (alanine)} + 3O_2 \quad (2)$$
$$\rightarrow 3CO_2 + NH_3 + 2H_2O$$

When sulfate is totally depleted by bacterial reduction (Eq. 3), synthesis of methane from H_2 and CO_2 may proceed (Eq. 4 and 5).

$$2CH_2O + SO_4 \rightarrow 2HCO_3 + H_2S \quad (3)$$

Methanobacillus omelianskii
$$C_2H_5OH + H_2O \rightarrow CH_3COOH + 2H_2 \quad (4)$$

$$4H_2 + CO_2 \rightarrow CH_4 + 2H_2O \quad (5)$$

H_2S is absorbed as iron sulfide in iron oxide, and exchangeable iron in silicates (Drever, 1971). Phosphate and silicate, also liberated by decomposition of phytoplankton and other detritus, are limited by solubility of a variety of components such as iron phosphate (Berner, 1973), and fluorapatite, and magnesium and more complex silicates. Maximum silica concentrations in marine sediments are less than 30 ppm. Other inorganic reactions are dolomitization (q.v.) and recrystallization of biogenic calcite to produce lower-Sr calcite which tends to enrich Sr in pore fluids.

Many reactions may have only weak reflection in pore fluids from relatively shallow sediments. "Conservative" (poorly reactive) elements such as Cl usually show little change from standard sea-water values in marine sediments, with few exceptions. Erratic chlorinity values in some studies of marine pore fluids are generally a sign of problems in sampling or analytical technique. Greater changes are observed in deeper sediments from marine drillings (below) or from special environments such as anoxic fjords (see Nissenbaum et al., 1972), sewer outfalls (Berner et al., 1970), volcanic fumaroles, and the extraordinary Red Sea hot metalliferous brine sediments (Manheim, 1974). The latter are actually hydrothermal ore deposits in the process of formation, and should not be regarded as normal sedimentary phenomena.

Recently, attention has been focused on chemical processes near the sediment/water interface and on mathematical modelling of pore-water migration and interaction (Berner, 1976). Strong chemical gradients caused by excesses or deficiencies of elements in pore fluids with respect to bottom water may cause a significant movement of components and gain or loss to (or from) the oceans. A new pore water "spear" or in-situ extraction device is able to document gradients very effectively (Sayles et al., 1973). Ions contributed in greatest quantity to the oceans are, respectively, I, Ba, F, P, Li, and Mn. In contrast, components removed from the oceans by uptake on bottom sediments are SO_4, B, K, Mg, and O_2 (Manheim, 1976). Essential to knowledge of flux calculations are diffusion coefficients of ions. In unconsolidated sediments the mobility of ions is about ½–¼ of their mobility in free solution, on average (Table 1).

Extensive references to interstitial water in sediments, including work by authors not cited

TABLE 1. Flux of Dissolved Solids to the Oceans from Sediments, Compared with River Contributions

Component	$D(4°C)$ cm^2/sec · 10^{-6}	M g/kg	C_1 g/kg	F_{C_1} g/yr	F_R g/yr	F_{C_1}/F_R
Na$^+$	7.2	10.76	<10.8?	(−14 · 10^{13})	12 · 10^{13}	1.1
K$^+$	10.7	0.387	0.38	− 2.8 · 10^{13}	5 · 10^{13}	0.56
Ca^{++}	4.3	0.413	0.415^a	+ 3 · 10^{12}	45 · 10^{13}	0.007
Mg^{++}	3.9	1.29	1.25	− 6 · 10^{13}	11 · 10^{13}	0.55
Cl$^-$	11.1	19.35	19.4	+ 8 · 10^{13}	22 · 10^{13}	0.36
SO$_4^-$	5.2	2.71	2.50	−41 · 10^{13}	42 · 10^{13}	1.0
HCO$_3^-$	6.4	0.14	0.19	+12 · 10^{13}	21 · 10^{14}	0.057
Br$^-$	11.0	0.067	0.069	+ 8 · 10^{12}	7 · 10^{11}	10.4
Sr^{++}	4.3	7600 · 10^{-6}	7800 · 10^{-6}	+ 3 · 10^{11}	18 · 10^{11}	0.27
SiO$_4^{--}$ − Si	5.4^b	3500 · 10^{-6}	9000 · 10^{-6}	+ 1.1 · 10^{13}	50 · 10^{13}	0.022
O$_2$	14.2	4500 · 10^{-6}	(100 · 10^{-6})	− 2.6 · 10^{13}	22 · 10^{13}	0.12
PO$_4^{---}$ − P^c	4.6	75 · 10^{-6}	300 · 10^{-6}	+ 3.8 · 10^{11}	7 · 10^{13}	0.54
NH$_4^+$ − N	10.8	10 · 10^{-6}	2500 · 10^{-6}	+ 1.0 · 10^{13}	21 · 10^{13}	0.048
NO$_3^-$ − N	10.4	42 · 10^{-6}	200 · 10^{-6}	+ 6.3 · 10^{11}	36 · 10^{12}	0.018
BO$_3^{---}$ ^{d}	4.6	4600 · 10^{-6}	4200 · 10^{-6}	− 6.8 · 10^{11}	70 · 10^{10}	1.0
F$^-$	7.9	1300 · 10^{-6}	2400 · 10^{-6}	+ 4.2 · 10^{12}	17 · 10^{11}	2.5
I$^-$	10.9	50 · 10^{-6}	700 · 10^{-6}	+ 2.5 · 10^{12}	7 · 10^{10}	36
Mn^{++}	3.8	1 · 10^{-6}	1000 · 10^{-6}	+ 1.4 · 10^{11}	35 · 10^{10}	0.40
Fe^{++}	3.9	3.4 · 10^{-6}	(20 · 10^{-6})	(+ 2.5 · 10^{10})	(24 · 10^{12})	<0.01 ?
Co^{++}	3.8	1 · 10^{-7}	(1 · 10^{-6})	(+ 1 · 10^9)	7 · 10^9	0.14
Cu^{++}	4.0	9 · 10^{-7}	(5 · 10^{-6})	(+ 6 · 10^9)	17 · 10^{10}	0.35
Zn^{++}	3.9	5 · 10^{-6}	(25 · 10^{-6})	(+ 3 · 10^{10})	70 · 10^{10}	0.043
Ni^{++}	3.7	4 · 10^{-6}	(5 · 10^{-6})	(+ 1.4 · 10^9)	11 · 10^9	0.13
Li^{++}	5.7	170 · 10^{-6}	(200 · 10^{-6})	(+ 6.5 · 10^{10})	12 · 10^{10}	0.54
Ba^{++}	4.6	21 · 10^{-6}	(100 · 10^{-6})	+12 · 10^{10}	36 · 10^{10}	0.37
H$^+$	63	1 · 10^{-8}	3 · 10^{-8}	+ 5 · 10^8	3 · 10^9	0.17
(UO$_2$)$^{4+}$	2.3	3 · 10^{-6}	2−60 · 10^{-6}	?	35 · 10^9	?

^{a}Sayles, et al., 1973, and Sayles, 1974 (written communication).
^{b}A directly determined value for sediment has been obtained by Fanning, 1974 (oral communication): 3.9.
^{c}Refers to H$_2$PO$_4^-$ ion.
^{d}Borate ion assumed to be similar to phosphate ion.
( ) Refers to estimated ranges for values subject to considerable uncertainty, either due to inadequate data, temperature of extraction, or analytical problems.

$$J = D_0 (\text{cm}^2/\text{sec}) \frac{\Delta c}{\Delta x} (\text{g/cm}^3)(3.1 \cdot 10^7 \text{ sec})(3.2 \cdot 10^{18} \text{ cm}^2)$$

J = flux in g/yr for total sea floor
D_0 = diffusion coefficient for individual ions in free solution (chiefly from Li and Gregory, 1974)
c = difference in concentration from mean sea water, in g/cm^3
x = migration distance in cm
M = estimated bottom-water composition
C_1 = mean composition of interstitial water at 1 m depth
F_{C_1} = flux of ions in response to the 1-m gradient. Positive values refer to gain of ion by bottom water, negative values refer to loss from bottom water to sediment. Diffusion coefficients are reduced by 2.5 (factor of) corresponding to inferred effect of sediment tortuosity.
F_R = yearly flux of salts from rivers and streams, given chiefly by Vinogradov (1967) and Turekian (1969). Sea water concentrations are chiefly as given by Culkin (1965) and Turekian (1969), in Manheim and Sayles (1974), and supplemented by Brujewicz (1966) and references cited.

with dates, above and subsequently, are given in Manheim (1976), in a Russian monograph (Shishkina, 1972), and in Gieskes (1975).

Ocean Drilling

Scientific investigation of deeper ocean sediments began with the Experimental Mohole, drilled in nearly 3600 m of water off Baja California, Mexico (Rittenberg et al., 1963; Siever et al., 1965). In JOIDES and subsequent drilling off the US Atlantic coast, fresh and brackish interstitial waters have been found as far as 120 km seaward of the coast. These are partly attributed to Pleistocene fresh water remaining from earlier, lower stands of sea level.

Increases in interstitial salt concentrations with depth are typical in areas characterized by buried evaporite deposits (especially rock salt). Such areas include much of the western and northern Gulf of Mexico, the eastern and northern Caspian Sea, the Mediterranean and Red seas, and areas in the eastern and western margins of the Atlantic Ocean. An example of pore-fluid composition above and in an evaporite body in the Red Sea is given in Table 2 (see Manheim, 1974).

The consistency with which halos of increased salt concentration are found above buried salt deposits offers a method to predict such bodies as much as 4 km below investigated strata, depending on the age of the beds (and therefore the time available for diffusional communication). This method permits distinction between salt diapirs and igneous or other intrusive bodies that have similar structure based on seismic evidence.

Special phenomena occur in marine sediments over weathering basalts or other igneous bodies. Here, large interstitial calcium enrichments up to 5 g/kg (>100 milliequivalents per kg) are partly compensated by depletions in sodium (as well as magnesium and potassium). The formation of new clay minerals in such submarine weathering also results in withdrawal of water and hence some increase in total salinity. Reactions such as these are also documented in carbon and oxygen isotope studies that reflect influence on the fluids from interaction with solids (Anderson et al., 1976).

A variety of special geochemical aspects of pore fluids including gases, and isotopic properties were investigated on Leg 15 (Caribbean Sea) and Leg 35 (SE Pacific) of the Deep Sea Drilling Project.

Geotechnical Aspects of Pore Fluids

The physical state of water-sediment configurations may change statistically under different conditions (see *Clay;* see also Vol. IVA: *Groundwater;* Vol. XIII: *Engineering Properties of Sediments*). Only one example will be given here. Clays that are poor in salt content may show

TABLE 2. Major Interstitial Constituents, Site 227, Red. Sea. Location 21°19.9'N, 38°08.0'E, depth 1,795 m. The site encountered uppermost Miocene evaporites including halite at depths below 180 m. Values are in g/kg; H_2O refers to total water content of bulk sediment, determined by loss on heating at 110–120°C.

Depth (m)	Description	Na	K	Ca	Mg	Cl	SO_4	HCO_3	Sr	Sum	H_2O	pH
Surface	Red Sea water	11.8	0.43	0.46	1.42	21.4	3.06	.15	.009	38.8	–	8.3
46	Green-gray clayey chalk	14.7	0.40	0.88	1.28	25.8	3.39	.2	.018	46.5	29	7.2
82	Clayey-silty chalk	28.7	0.22	0.98	0.84	46.1	3.57	.15	.027	80.6	19	7.0
136	Gray, silty nanno chalk	61.3	0.38	2.32	0.92	99.7	2.43	.04	.046	167.2	15	6.5
167	Gray, silty nanno chalk	80.2	0.32	2.44	0.86	129.7	1.55	<.05	.034	215.1	–	6.5
186	Gray, silty nanno chalk	93.6	0.50	1.81	0.67	149.0	1.37	<.05	–	247.0	19	–
282	Dark shale and anhydrite	71.0	1.29	4.9	12.6	157.1	1.9	<.05	.099	250.7	23	6.5
350	Dark shale between anhydrite	67.4	0.93	6.7	16.8	164.8	1.2	<.05	–	257.6	14	6.2

From Manheim et al., 1974a.

what is called "quick" behavior. For example, marine clays that are washed free of most of their salt by ground-water percolation can lose most of their strength after prolonged stress or sudden shock. Under these conditions, the sediments may suddenly slump or slide. Massive destruction along railroad embankments, hillsides, and clay-rich slopes is caused by "spontaneous liquefaction" or temporary behavior of clays as fluids (see *Fluidized Sediment Flows*). Spontaneous liquefaction has also been cited as a possible mechanism involved in submarine cable breaks.

FRANK T. MANHEIM

References

Anderson, T. F., and 7 others, 1976. Geochemistry and diagenesis of deep-sea sediments from Leg 35 of the Deep Sea Drilling Project, *Nature,* 261, 473-476.

Berner, R. A., 1973. Phosphate removal from sea water by adsorption on volcanogenic ferric oxides, *Earth Planet. Sci. Lett.,* 18, 77-86.

Berner, R. A., 1976. The benthic boundary layer from the point of view of a geochemist, in I. N. McCave, ed., *The Benthic Boundary Layer,* New York: Plenum, 33-55.

Berner, R. A.; Scott, M. R.; and Thomlinson, C., 1970. Carbonate alkalinity in the pore waters of anoxic marine sediments, *Limnol. Oceanog.,* 15, 544-549.

Brujewicz, (Bruevich), S. V., ed., 1966. Chemistry of interstitial waters in sediments of the Pacific Ocean (in Russian), *Khimiya Tikhogo Okeana, Izdat. Akad. Nauk., Moscow,* 2, 263-358.

Culkin, F., 1965. The major constituents of sea water, in J. P. Riley and G. Skirrow, eds., *Chemical Oceanography,* vol. 1. London: Academic Press, 121-161.

Drever, J. I., 1971. Magnesium-iron replacement in clay minerals in anoxic marine sediment, *Science,* 172, 1334-1336.

Gieskes, J. M., 1975. Chemistry of interstitial water of marine sediments, *Ann. Rev. Earth Planet. Sci.,* 3, 433-453.

Lehman, F. J., ed., 1976. Sediment-water interaction and its effect upon water quality: a bibliography with abstracts, *Natl. Tech. Inf. Service,* PS-76/0021, 17p.

Li, Y., and Gregory, S., 1974. Diffusion of ions in sea water and in deep-sea sediments, *Geochim. Cosmochim. Acta,* 38, 703-714.

Mangelsdorf, P. C.; Wilson, T. R. S.; and Daniel, E., 1969. Potassium enrichments in interstitial waters of recent marine sediments, *Science,* 165, 171-174.

Manheim, F. T., 1974. Red Sea geochemistry, Leg 23 (Red Sea), *Initial Reports of the Deep Sea Drilling Project,* XXIII, 975-998.

Manheim, F. T., 1976. Interstitial waters of marine sediments, in J. P. Riley and R. Chester, eds., *Treatise on Chemical Oceanography,* 6, 115-186.

Manheim, F. T., and Sayles, F. L., 1974. Composition and origin of interstitial waters of marine sediments, based on deep-sea drill cores, in E. D. Goldberg, ed., *The Sea,* vol. 5. New York: Wiley, 527-568.

Manheim, F. T.; Waterman, L. S.; Woo, C. C.; and Sayles, F. L., 1974. Interstitial water studies on small core samples, Leg 23 (Red Sea), *Initial Reports of the Deep Sea Drilling Project,* XXIII, 955-967.

Mehlich, A., and Drake, M., 1958. Soil chemistry and plant nutrition, *Am. Chem. Soc. Monogr.* 126, 286-327.

Murray, J., and Irvine, R., 1895. On the chemical changes which take place in the composition of the sea waters associated with blue muds on the floor of the ocean, *Trans. Roy. Soc. Edinburg,* 37, 481-507.

Nissenbaum, A.; Presley, B. J.; and Kaplan, I. R., 1972. Early diagenesis in reducing fjord, Saanich Inlet, British Columbia-I. Chemical and isotopic changes in major components of interstitial water, *Geochim. Cosmochim. Acta,* 36, 1007-1027.

Rittenberg, S. C., and 8 others, 1963. Biogeochemistry of sediments in Experimental Mohole, *J. Sed. Petrology,* 33, 140-172.

Sayles, F. L., and Manglesdorf, P. C., Jr., 1977. The equilibration of clay minerals with sea water, *Geochim Cosmochim. Acta,* 41, 951-960.

Sayles, F. L., and Manheim, F. T., 1975. Interstitial solutions and diagenesis in deeply buried marine sediments, *Geochim. Cosmochim. Acta,* 39, 103-128.

Sayles, F. L.; Wilson, T. R. S.; Hume, D. N.; and Mangelsdorf, P. C., Jr., 1973. In situ sampler of marine sedimentary pore waters; evidence for potassium depletion and calcium enrichment, *Science,* 181, 154-156.

Schloesing, M. T., 1866. Sur l'analyse des principes solubles de la terre végétale, *Compt. Rend. Acad. Sci.,* 63, 1007-1012.

Schmidt, G. W., 1973. Interstitial water composition and geochemistry of deep Gulf Coast shales and sandstones, *Bull. Am. Assoc. Petrol. Geologists,* 57, 321-337. *See also* discussion by M. F. Osmaston (1975), 59, 715-726.

Scholl, D. W., 1965. High interstitial water chlorinity in estuarine mangrove swamps, Florida, *Nature,* 207, 284-285.

Shishkina, O. V., 1972. *Geohkimiya morskikh i okeanicheskikh ilovykh vod.* Moscow: Izdat. Nauka, 228p.

Siever, R.; Beck, K. C.; and Berner, R. A., 1965. Composition of interstitial waters of modern sediments, *J. Geol.,* 73, 39-73.

Smith, R. I., 1955. Salinity variations in interstitial water of sand at Karnes Bay, Millport, with reference to the distribution of *Nereis diversicolor, J. Marine Biol. Assoc. U.K.,* 34, 33-46.

Turekian, K. K., 1969. The ocean, streams, and atmosphere, in K. H. Wedepohl et al., eds. *Handbook of Geochemistry.* New York: Springer-Verlag, 297-320.

Valyashko, M. G., 1963. Genesis of brines in sedimentary rocks, (in Russian), in A. P. Vinogradov, ed., *Khimiya Zimnoi Kory.* Moscow: Akad. Nauk. SSSR, 253-277.

Vinogradov, A. P., 1967. *Vvedenie v geokhimiyu okeana.* Moscow: Isdatel'stvo Nauka, 215p.

Cross-references: *Clay; Crude-Oil Composition and Migration; Diagenesis; Euxinic Facies; Evaporites; Gases in Sediments; Organic Sediments; Pelagic Sediments; Petroleum—Origin and Evolution; Water-Escape Structures in Sediments; Weathering in Sediments.*

INTERTIDAL GEOLOGY—See TIDAL-FLAT GEOLOGY

INTRACLAST

The term *intraclast* was introduced by Folk (1959) to describe "fragments of penecontemporaneous, usually weakly consolidated carbonate sediment that have been eroded from adjoining parts of the sea bottom and redeposited to form a new sediment"—the term *intraclast* signifying that they have been reworked within the area of deposition. Although generally applied to calcareous sediments, the term may reasonably be extended to muddy (clayey) sediments (Pettijohn, 1975).

Intraclasts may be torn up from sedimentary layers almost immediately after deposition or, under more severe erosion, from deeper, more consolidated layers. Intraclasts reworked from surficial carbonate mud are usually plastically deformed and commonly have vague or mashed boundaries (Folk, 1959). Grapestone has been classified as an intraclastic rock (Folk, 1959); recent work has indicated, however, that the oolite "clasts" are relict (see *Grapestone*). The most common mode of formation of intraclasts is by erosion of fragments of a widespread layer of semiconsolidated carbonate sediment, with erosion reaching to depths of a few centimeters up to a meter in the bottom sediment; these intraclasts may be formed either by submarine erosion; by mild tectonic upwarps of the sea floor; or by low-tide exposure, allowing wave attack on exposed, mudcracked flats (Folk, 1959). The term *intraclast*, then, embraces the entire spectrum of deposited, aggregated, and then reworked particles, regardless of degree of cohesion or time gap between deposition of the original layer of sediment and later reworking of parts of it (Folk, 1959).

The great majority of intraclastic rocks have a sparry calcite cement, inasmuch as currents that are strong enough to transport fairly large carbonate rock fragments are also usually capable of washing away any microcrystalline ooze matrix; thus *intrasparite* is more common than *intramicrite*, which is relatively rare (Folk, 1959). Texturally, intraclastic rocks are about equally divided between calcirudites and calcarenites (q.v.). Intraclastic rocks include the edgewise or "flat-pebble" conglomerate so common in lower Paleozoic limestones (Folk, 1959; see also *Clay-Pebble Conglomerate and Breccia*). Conglomerates and breccias consisting primarily of intraclasts are commonly called *intraformational conglomerates* (see *Breccias*).

B. CHARLOTTE SCHREIBER

References

Folk, R. L., 1959. Practical petrographic classification of limestone, *Bull. Am. Assoc. Petrol. Geologists*, **43**, 1-38.

Pettijohn, F. J., 1975. *Sedimentary Rocks,* 3rd ed. New York: Harper & Row, 628p.

Cross-references: *Allochem; Breccias, Sedimentary; Calcarenite, etc.; Clay-Pebble Conglomerate and Breccia; Grapestone; Intraformational Disturbances; Limestone Fabrics; Limestones.*

INTRAFORMATIONAL DISTURBANCES

The term *intraformational disturbances* (Fairbridge, 1947) is a collective term that has been applied to various kinds of contorted strata that are enclosed within noncontorted strata. If only one bed has been disturbed, the term *intrastratal* would be applied. On the other hand, usage generally excludes the contortion of "large" groups of strata that are considered to be formations.

The most important concept implied by the term intraformational disturbance is that both before and after the disturbance(s), strata accumulated without complications. Strictly speaking, the term *disturbances* refers to the act of disturbing and hence is more a process word than a product term. More exact usage would be "intraformationally disturbed strata."

Intraformational disturbances are a kind of penecontemporaneous deformation (q.v.) or distortion. Some common deformational processes are gravity sliding (see *Gravity-Slide Deposits*), gravity flows (q.v.), mass wasting (see *Mass-Wasting Deposits*), slumping (see *Slump Bedding*), drag created by the stranding of ice floes or by the overriding of a glacier, and differential vertical adjustments that take place on deltas or other places where sediments accumulate rapidly. In some cases, the deformation results from extreme tractional shearing applied by a current moving at high speeds, which at slower speeds deposited the sediments that were not deformed. For example, a river at peak flood may deform the sediments it deposited at lower discharges. Large-scale shearing intense enough to deform fine-grained lake sediments might be applied during an unusually large and sudden seiche, caused perhaps by a rock avalanche entering the lake or, in a proglacial lake, by the calving of a large iceberg. Intraformational disturbances were first considered to be criteria of deposition in bodies of water near active glaciers (Lahee, 1914), but other factors were later found to be involved in creating the contorted strata.

Subaqueous slump structures have attracted

much study because they are indicators of submarine slopes. Most slump axes should lie parallel to the slope contours, but the fold-axis versus slope relationship is not always readily apparent (Corbett, 1973). Where present, slide marks, formed at the base of a slumped mass and oriented downslope, may be the best slope indicator (Kuenen and Sanders, 1956).

Contorted strata resulting from glacial deformation are easily identified in Pleistocene sediments, and many examples have been recorded. In ancient bedrock, however, it may not be possible to separate the effects of glacial overriding from other disturbances.

Injection-type structures resulting from intraformational disturbances evidently involve reversed density gradients and localized zones of greater-than-normal pore pressures in the sediments (see *Volcanism, Sedimentary*).

The simultaneous effect of deposition and shearing drag from a current moving fast enough to deform the subjacent sediments creates deformational structures that are truly contemporaneous with deposition. Large-scale features involving more than one bed are the recumbent folds found in sandy deposits having large-scale cross-strata. Such folds evidently formed when the current ceased adding sand to the bed but was driving a thick traction carpet forward with such vigor that the lower boundary of the carpet deformed the subjacent cross-strata. Recumbently folded cross-strata have been created in the deposits of a laboratory flume by dragging a sand bag down the flume (McKee et al., 1962).

Where a current, carrying sand, flows across a deformable muddy substrate, the impinging of oppositely flowing separation eddies may "squeeze" the mud up into pointed wisps called *flame structures* (q.v.). Wisps that grew and were deformed while sand was accumulating around them offer important evidence bearing on the origin of convoluted laminae (Sanders, 1960, 1965). In convoluted laminae, the actively growing pointed anticlines, many of which developed by further deposition on top of ordinary current ripples, behave much like the wisps. The difference is that each lamina in the convoluted anticlines was first deposited before it was deformed. In extreme cases of deformation during convoluted laminae deposition, material flowed bodily upward into the core of the rising anticline, as in a diapiric salt plug. An alternate explanation for the origin of some convoluted laminae is that they are water-escape structures (q.v.).

JOHN E. SANDERS

References

Corbett, K. D., 1973. Open-cast slump sheets and their relationship to sandstone beds in an Upper Cambrian flysch sequence, *J. Sed. Petrology,* 43, 147-159.

Fairbridge, R. W., 1947. Possible causes of intraformational disturbances in the Carboniferous varve rocks of Australia, *J. Royal Soc. New South Wales,* 81, 99-121.

Kuenen, P. H., and Sanders, J. E., 1956. Sedimentation phenomena in Kulm and Flozleeres graywackes, Sauerland and Oberharz, Germany, *Am. J. Sci.,* 254, 649-671.

Lahee, F. H., 1914. Contemporaneous deformation: a criterion for aqueoglacial sedimentation, *J. Geol.,* 22, 786-790.

McKee, E. D.; Reynolds, M. A.; and Baker, C. H., Jr., 1962. Experiments on intraformational recumbent folds in cross-bedded sand, *U.S. Geol. Surv. Prof. Pap. 450-D*, 155-160.

Sanders, J. E., 1960. Origin of convolute laminae, *Geological Mag.,* 97, 409-421.

Sanders, J. E., 1965. Primary sedimentary structures formed by turbidity currents and related resedimentation mechanisms, *Soc. Econ. Paleont. Mineral. Spec. Publ.* 12, 192-219.

Cross-references: *Bedding Genesis; Clay-Pebble Conglomerate and Breccia; Flame Structures; Penecontemporaneous Deformation of Sediments; Pillow Structures; Secondary Sedimentary Structures; Slump Bedding; Solution Breccias; Storm Deposits; Water-Escape Structures.*

INVERSE GRADING

Inverse grading (*reverse grading,* inverted grading or negative graded bedding) is a sedimentary phenomenon in a unit bed characterized by a gradation in grain size—both the maximum and/or average grain sizes—from fine at the bottom to coarse at the top. Two types of inverse grading are recognized in clastic sediments. In one type, the maximum and/or average sizes increase upwards, but no other sedimentary structures are observed at all. This type of inverse grading tends to occur under the effect of the intergranular dispersive pressure in granular material that flows under gravity. Bagnold (1954) believed that as these grains, of mixed sizes, are sheared together, the larger grains tend to drift upward to the zone of least shear stress, and the smaller grains downward to the zone of greatest shear stress (see *Bagnold Effect*). Others believe there is a sieve effect (Middleton, 1970; see *Grain Flows*).

In the second type of inverse grading, grain sizes in each horizon are generally sorted, and the average size increases upward; thus, parallel stratification is more or less developed, particularly in lower horizons. This type of inverse grading may be primarily concerned with the difference in traction velocities of grains (Okada, 1971). Other processes involved in the formation of inverse grading are: winnowing of finer grains due to differential transportation, rotational flows of the vortex, flocculating effect of mud (Barrell, 1917; Kindle, 1917), different fall velocities between larger and lighter grains and smaller and heavier ones (e.g., pumice and crystal), and frost action (Gilluly et al., 1968, p. 43), among others. Another type of inverse grading is found in chemical sediments such as evaporites (Ogniben, 1955) and is attributed to varying growth rates of crystals.

Reverse order of grading is a typical feature of high-concentration gravity-flow deposits, such as grain-flow and debris-flow deposits (Fisher, 1971; Fisher and Mattinson, 1968). Thus, it is not rare in turbidites (q.v.; Walton, 1956; Knill and Knill, 1961; Sanders, 1965). Inverse grading is often observed in beach and stream gravels (Leopold et al., 1964, pp. 209-212); pumice-fall deposits (Lirer et al., 1973); base-surge deposits (q.v.); lahars and other pyroclastic-flow deposits (Schmincke, 1967); fluvial deposits (Costa, 1974); and others (Fisk, 1974).

HAKUYU OKADA

References

Bagnold, R. A., 1954. Experiments on a gravity-free dispersion of large solid spheres in Newtonian fluid under shear, *Roy. Soc. London Proc.*, ser. A, 225, 49-63.

Barrell, J., 1917. Rhythms and measurements of geologic time, *Geol. Soc. Am. Bull.*, 28, 745-903.

Costa, J. E., 1974. Stratigraphic, morphologic, and pedologic evidence of large floods in humic evironments, *Geology*, 2, 301-303.

Fisher, R. V., 1971. Features of coarse-grained, high-concentration fluids and their deposits, *J. Sed. Petrology*, 41, 916-927.

Fisher, R. V., and Mattinson, J. M., 1968. Wheeler Gorge turbidite-conglomerate series, California, inverse grading, *J. Sed. Petrology*, 38, 1013-1023.

Fisk, L. H., 1974. Inverse grading as stratigraphic evidence of large floods: comment, *Geology*, 2, 613-615.

Gilluly, J.; Waters, A. C.; and Woodford, A. O., 1968. *Principles of Geology*, 3d ed. San Francisco: Freeman, 687p.

Kindle, E. M., 1917. Diagnostic characteristics of marine clastics, *Geol. Soc. Am. Bull.*, 28, 905-916.

Knill, J. L., and Knill, D. C., 1961. Reversed graded beds from the Dalradian of Inishowen, Co. Donegal, *Geological Mag.*, 98, 458-463.

Leopold, L. B.; Wolman, M. G.; and Miller, J. P., 1964. *Fluvial Processes in Geomorphology*. San Francisco: Freeman, 522p.

Lirer, L.; Pescatore, T.; Booth, B.; and Walker, G. P. L., 1973. Two Plinian pumice-fall deposits from Somma-Vesuvius, Italy, *Geol. Soc. Am. Bull.*, 84, 759-772.

McCall, G. J. H., 1962. Reversed graded bedding, *New Zeal. J. Geol. Geophys.*, 5, 666.

Middleton, G. V., 1970. Experimental studies related to problems of flysch sedimentation, *Geol. Assoc. Canada Spec. Pap. 7*, 253-272.

Ogniben, L., 1955. Inverse graded bedding in primary gypsum of chemical deposition, *J. Sed. Petrology*, 25, 273-281.

Okada, H., 1971. Inverse grading in gravels and conglomerates, *J. Geol. Soc. Japan*, 74, 589-594.

Sanders, J. E., 1965. Primary sedimentary structures formed by turbidity currents and related resedimentation mechanisms, *Soc. Econ. Paleont. Mineral. Spec. Publ. 12*, 192-219.

Schmincke, H. -U., 1967. Graded lahars in the type sections of the Ellensburg Formation, south-central Washington, *J. Sed. Petrology*, 37, 438-448.

Walton, E. K., 1956. Limitations of graded bedding and alternative criteria of upward sequence in the rocks of the Southern Uplands, *Edinburgh Geol. Soc. Trans.*, 16, 262-271.

Cross-references: *Bagnold Effect; Base-Surge Deposits; Graded Bedding; Grain Flows; Gravity Flows; Turbidites.*

IRONSTONE

Ironstone is a noncherty, iron-rich sedimentary rock. It has been distinguished from *cherty iron formations* (q.v.) primarily because of its lack of chert and its age: ironstone occurs in Late Precambrian to modern sediments; iron formations are a separate, distinctive group of iron-rich sediments found mainly in Early Precambrian to Cambrian rocks (James, 1966) but also formed in later times. Other iron-rich sedimentary rocks which may be classified as ironstones include laterite (see *Ferricrete*), bog iron ore, and clay ironstone. Pettijohn (1975) reviews the nature and origin of iron-bearing sediments. Table 1 lists iron-bearing minerals found in sediments (see also *Glauconite; Phosphates in Sediments; Sulfides in Sediments*).

Types of Ironstone

Bedded Iron Sulfides. Bedded iron sulfides, primarily pyrite, are uncommon and insignificant in total volume (Pettijohn, 1975). They are, however, important environmental indicators of reducing environments (see *Euxinic Facies*) and are found in black shales and occasionally in limestones. *Bedded siderite* is found in cherty iron formations (q.v.). It may also be found in clay ironstones, consisting of argillaceous nodular siderite and commonly associated with coal measures.

Bedded Iron Oxides. The best-known iron oxide deposits, primarily hematite, are the Clinton type, one of two main categories of ironstone described by Gross (1965). These ores are typically deep red to purple, massive hematite-chamosite-siderite beds with oolitic textures; they are rich (>50%) in iron, have a large clastic component, and are demonstrably

TABLE 1. Iron-Bearing Minerals of Sediments

Sulfides:	Pyrite	FeS_2
	Marcasite	FeS_2
	Hydrotroilite	$FeS \cdot nH_2O$
Oxides:	Limonite	$FeO(OH) \cdot nH_2O$
	Goethite	$HFeO_2$
	Hematite	Fe_2O_3
	Magnetite	Fe_3O_4
Silicates:	Glauconite	$KMg(Fe, Al)(SiO_3)_6 \cdot 3H_2O$
	Chamosite	$3(Fe, Mg)O \cdot (Al, Fe)_2O_3 \cdot 2SiO_2 \cdot nH_2O$
	Stilpnomelane	$2(Fe, Mg)O \cdot (Fe, Al)_2O_3 \cdot 5SiO_2 \cdot 3H_2O$
	Minnesotaite	$(OH)_2(Fe, Mg)_3Si_4O_{10}$
	Greenalite	$FeSiO_3 \cdot nH_2O$
Carbonates:	Siderite	$FeCO_3$
	Ankerite	$Ca(Mg, Fe)(CO_3)_2$
Phosphates:	Vivianite	$Fe_3(PO_4)_2 \cdot 8H_2O$

From Pettijohn, 1975.

shallow-marine in origin (Blatt et al., 1972). *Bog iron ore* is included in the oxide category; it is an earthy, often spongy mixture of yellow to dark brown ferric hydroxides, largely limonite, deposited along the borders of lakes and bogs in recently glaciated regions. Bog and lake iron ores represent one of the few examples of modern iron-rich sediments. Lake ores are often oolitic or pisolitic, cemented into disks up to a meter across; bog ores form thin earthy or pisolitic layers at the surface or within layers of peat (Blatt et al., 1972).

Bedded Iron Silicates. Sedimentary iron silicates are very common in cherty iron formations (q.v.); glauconite (q.v.) is also treated separately in this volume. Chamosite is the common iron silicate in ironstones. It is found in the Clinton-type ores but is most typical of the Minette-type ores (Gross, 1965) occurring in several Jurassic formations in Europe. The Northampton sand ironstone of England is an example (Blatt et al., 1972); it consists of a carbonate phase of sideritic mud-, silt-, sand-, and limestone, a mixed phase of chamosite-oxide ooliths in a sideritic matrix, and an aluminosilicate phase of chamositic and kaolinitic oolites and sandstones (Taylor, 1949). Chamosite is found today in warm, shallow shelf areas (Porrenga, 1966).

Deep-Sea Ironstones. Iron oxides and hydroxides have recently been reported from deep-sea dredging and drilling samples. There are two types of occurrence: (1) iron-rich sediments associated with hydrothermal activity near active ocean ridges (Dymond et al., 1973); and (2) iron-rich crusts in hemipelagic deep-sea sediments (McGeary and Damuth, 1973). Pequegnat et al. (1972) described a latter occurrence at the sediment/water interface and attributed its formation to hydrolyzation of iron in situ by swift, oxygen-rich bottom currents; they suggested that ferruginous yellow layers in cores are genetically related to this occurrence.

Origin

Although modern occurrences of all sedimentary iron mineral classes are known, there are few modern examples of sedimentary iron accumulations, especially on the scale of ancient ironstone deposits. The origin of ironstones has been widely debated, and although it is generally accepted that iron-rich sediments are chemical precipitates, possibly with biological influence, there is strong disagreement concerning their depositional chemistry, their source of iron, and their paleogeographic significance (Pettijohn, 1975). Fig. 1 illustrates the many factors that may affect sedimentary iron accumulation.

Source of Iron. Two principal sources of iron have been suggested. The first is a terrigenous source, especially from areas of tropical weathering. Mode of transport of such iron has also been debated; transport is probably either in oxide form, adhering to the surface of clay minerals, or as colloidal suspensions in sols. The second source suggested is direct volcanic exhalation, and although some workers have been skeptical, this source has been confirmed, for at least some cases, by deep-sea research.

Environments of Deposition. Nearly all major ironstone deposits are marine in origin. The type of iron mineral precipitated reflects the prevailing physicochemical conditions at the depositional site (see descriptions above).

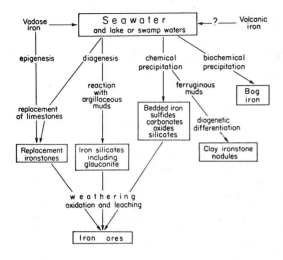

FIGURE 1. Provenance of iron-bearing sediments (from Pettijohn, 1975).

Strakhov (1967) discusses five factors which govern sedimentary iron accumulation: (1) an increased intensity of iron precipitation compared with normal sedimentation; (2) a paleogeographic situation such as a dissected shoreline, with islands, shoals, and basins, which reduces hydrodynamic disturbances and prevents dispersal of fine-grained precipitates; (3) a reduction in the supply of clastic material; (4) secondary concentration by diagenesis; and (5) reworking of the sediments during formation of the ore bed, with accumulation of oolites, or of products of diagenesis, and removal of fine clasts.

Pettijohn (1975) lists two major sedimentary iron facies: (1) cratonic, or shallow shelf, represented by the oolitic ironstones, and (2) basinal, represented by the laminated, cherty iron formations (q.v.). James (1954, 1966) and Borchert (1960, 1965) review the various sedimentary iron facies. Two investigations of shallow-marine ironstone deposits, both of which review the extensive literature, are Adeleye (1973) and Talbot (1974). Lepp (1975) has edited a collection of many of the fundamental papers on sedimentary iron deposits.

Diagenesis. Diagenetic reorganization, weathering, and metamorphism may all affect iron-bearing sediments, often enhancing their economic value. In addition, there is evidence that some sedimentary iron deposits, especially siderite and sometimes hematite, may originate by replacement of calcium carbonate.

JOANNE BOURGEOIS

References

Adeleye, D. R., 1973. Origin of ironstones, an example from the middle Niger Valley, Nigeria, *J. Sed. Petrology*, **43,** 709-727.

Blatt, H.; Middleton, G. V.; and Murray, R. C., 1972. *Origin of Sedimentary Rocks*. Englewood Cliffs, N.J.: Prentice-Hall, 634p.

Borchert, H., 1960. Genesis of marine sedimentary iron ore, *Trans. Inst. Min. Metall., 640,* **79,** 261-279.

Borchert, H., 1965. Formation of marine sedimentary iron ores, in P. Riley and G. Skirrow, *Chemical Oceanography,* vol. 2. London: Academic Press, 159-204.

Dymond, J.; Corliss, J. B.; Heath, G. R.; Field, C. W.; Dasch, E. J.; and Veeh, H. H., 1973. Origin of metalliferous sediments from the Pacific Ocean, *Geol. Soc. Am. Bull.,* **84,** 3355-3372.

Gross, G. A., 1965. Geology of iron deposits in Canada, I, general geology and evaluation of iron deposits, *Geol. Surv. Canada, Econ. Geol. Rept.* **22,** 181p.

James, H. L., 1954. Sedimentary facies of iron formation, *Econ. Geol.,* **49,** 235-293.

James, H. L., 1966. Chemistry of the iron-rich sedimentary rocks, *U.S. Geol. Surv. Prof. Pap. 440-W,* 61p.

Lepp, H., ed., 1975. *Geochemistry of Iron.* Stroudsburg, Pa.: Dowden, Hutchinson & Ross, 464p.

McGeary, D. F. R., and Damuth, J. E., 1973. Postglacial iron-rich crusts in hemipelagic deep-sea sediment, *Geol. Soc. Am. Bull.,* **84,** 1201-1212.

Pequegnat, W. E.; Bryant, W. R.; Fredericks, A. D.; McKee, T. R.; and Spalding, R., 1972. Deep-sea ironstone deposits in the Gulf of Mexico, *J. Sed. Petrology,* **42,** 700-710.

Pettijohn, F. J., 1975. *Sedimentary Rocks,* 3rd ed. New York: Harper & Row, 628p.

Porrenga, D. H., 1966. Glauconite and chamosite as depth indicators in the marine environment, *Marine Geol.,* **5,** 495-501.

Strakhov, N. M., 1967. *Principles of Lithogenesis,* vol. 1. New York: Consultants Bureau, 245p. (Original Russian edition: 1962).

Talbot, M. R., 1974. Ironstones in the Upper Oxfordian of southern England, *Sedimentology,* **21,** 433-450.

Taylor, J. H., 1949. The Mesozoic ironstones of England; petrology of the Northampton Sand Ironstone Formation, *Gr. Brit. Geol. Surv. Mem.* **6,** 111p.

Cross-references: *Cherty Iron Formation; Duricrust; Euxinic Facies; Ferricrete; Glauconite; Oolite; Phosphates in Sediments; Pisolite; Sulfides in Sediments.* Vol. IVB: *Hematite.* Vol. XI: *Soil Pans.*

K

KARA-BOGAZ GULF EVAPORITE SEDIMENTOLOGY

A 10,000 km^2 embayment off the eastern shore of the Caspian Sea, in the Turkmen Republic of the USSR, the Kara-Bogaz Gulf (Gol) is the type area for evaporite sedimentation behind a natural barrier. It was with this model that Ochsenius (1877, 1888, 1905) developed his "bar theory" to explain the celebrated Permian Stassfurt salt deposits of Germany.

The salinity of the Gulf of Kara-Bogaz has always been dependent on the precipitation/evaporation balance of its parent sea, the Caspian, whose level has oscillated repeatedly during glacial/interglacial cycles. The gulf is connected to the Caspian by a channel 10 km long, 300 m wide and mostly < 1 m deep. Because of the arid climate, the evaporation rate of the Gulf always exceeds that of the Caspian, and the channel ends in about 4 m of waterfalls over a resistant rock bar down to the level of the gulf.

During its high-level stages, the gulf was merely brackish, and the former shorelines are marked by numerous banks of *Cardium edule* shells; but the present waters are hypersaline and support only brine shrimp, certain algae, and flagellates. The incoming Caspian water averages 13‰ (rich in sulfates), but becomes concentrated to 310‰, thus making Kara-Bogaz Gulf one of the world's saltiest lakes (comparable to the Great Salt Lake, Utah; Lake Assal in Afar; and Lake Iletzk in Kazakstan). The rate of salt deposition is around 350,000 tons/day (1.3 × 10^8 tons/yr, reaching 2 × 10^8 tons/yr in 1950). The accumulated evaporite reserve is over 39×10^9 tons. Salts precipitate in the order: gypsum—mirabilite—halite; epsomite and astrakhanite also occur with the mirabilite.

The sulfate-rich evaporites distinguish the Kara-Bogaz from marine evaporites, reflecting the sulfate-dominated Caspian and its primary intake, the Volga River (which obtains much of the sulfate from leaching of preexisting anhydrite). The evaporite regime is thus distinct also from that of the Great Salt Lake of Utah, which is chloride dominated (in part of cyclic origin).

RHODES W. FAIRBRIDGE

References

Dzens-Litovsky, A. I., 1967. The problem of Kara-Bogaz-Gol, *Lithol. Miner. Resour.*, 1, 70–76.
Garbell, M. A., 1963. The sea that spills into a desert, *Sci. Am.*, 209, 94–100.
Grabau, A. W., 1920. *Principles of Salt Deposition*. New York: McGraw-Hill, 435p.
Ochsenius, C., 1877. *Die Bildung der Steinsalzlager und ihrer Mutterlaugensalze*. Halle: C. E. M. Pfeffer, 172p.
Ochsenius, C., 1888. On the formation of rock-salt beds and mother liquor salts, *Proc. Acad. Natl. Sci. Phila.*, 40, 181–187.
Ochsenius, C., 1905. Uebereinstimmung der geologischen und chemischen Bildungsverhältnisse in unsern Kalilagern, *Z. Prakt. Geol.*, 13, 167–179.
Strakhov, N. M., 1970. *Principles of Lithogenesis*, vol. 3. New York: Consultants Bureau, 577p.

Cross-references: *Black Sea Sediments; Evaporite Facies; Evaporites—Physicochemical Conditions of Origin*. Vol. III: *Kara-Bogaz Gulf*.

KEROGEN

Kerogen is generally understood to be the insoluble portion of the organic matter in sedimentary rocks. It is a complex, heterogeneous, brownish, relatively stable high polymer akin to humic matter (q.v.; Degens, 1965). Kerogen occurs in sedimentary rocks of all ages and accounts for more than 95% of all the organic matter in these rocks. Despite its quantitative importance and its considerable scientific and economic significance, kerogen has received less attention than solvent-extractable organic compounds, probably because it is more difficult to extract, analyze, and define chemically.

Analysis

Several physical and chemical methods are used to extract kerogen from rocks (Forsman, 1963), but none of them is entirely satisfactory. Physical methods of extraction from pulverized rock include flotation in liquid media in which the mineral particles sink on centrifugation, and a method based on differential wettability, whereby the kerogen is dispersed in a hydrocarbon solvent while the mineral particles are taken up by a layer of water. A widely used chemical procedure for isolating kerogen involves digestion of the rock with hydrochloric and hydrofluoric acids, which dissolve most of the mineral matrix.

Although kerogen may be analyzed by infrared spectrophotometry, mass spectrometry, and methods for determining elemental composition and functional group content, investigation of the molecular structure of kerogen has been severely limited by the extreme complexity of the substance. An indirect approach that has yielded some fragmentary information is to subject kerogen to partial degradation followed by determination of the simpler, lower molecular-weight degradation products (Forsman, 1963; Degens, 1965). Pyrolysis and partial oxidation are two of the more widely used degradative techniques for characterizing kerogen. Treatment of kerogen with ozone or other oxidizing agents yields an assortment of organic acids. Pyrolysis in a vacuum or inert atmosphere yields petroleum-like mixtures of hydrocarbons together with oxygen-, sulfur-, and nitrogen-bearing organic compounds. Degradation products can be identified by methods such as gas chromatography and mass spectrometry.

Although the exact molecular structure of kerogen is elusive, the various analytical methods listed here are useful for broad characterization of kerogens and for distinguishing between kerogens of diverse origins or postdepositional histories. The available data show that kerogens are composed chiefly of carbon, oxygen, hydrogen, and nitrogen; they consist of varying proportions of aromatic and aliphatic components together with polar groups such as hydroxyl and carbonyl groups, and polar bridges such as ether and sulfur linkages. Kerogens can be divided into at least two broad categories: (1) "coaly" kerogen, which is more highly aromatic, and (2) "noncoaly" kerogen, which is more highly aliphatic. *Coaly kerogen* has a lower hydrogen content and lower H:C ratio than *noncoaly kerogen* (Forsman, 1963; Breger and Brown, 1963), and a smaller percentage of its total carbon content is split off as volatile degradation products during pyrolysis (Gransch and Eisma, 1970). There are gradations between coaly and noncoaly character; the degree of coaly character is a function of both source material and thermal history.

Origin

Among relatively unmetamorphosed Phanerozoic kerogens, coaly character is considered evidence that the kerogen was mostly derived from lignin-rich terrestrial organic matter of vascular-plant origin, whereas noncoaly character probably means that aquatic algae were the principal starting materials (Forsman, 1963). Furthermore, terrigenous kerogen yields a greater abundance of "tar acids" on pyrolysis (Breger and Brown, 1963). These rules of thumb may not apply, however, to pre-Phanerozoic "coal" of presumably algal origin. Regardless of origin, kerogen becomes increasingly "coaly" during thermal metamorphism, and is finally converted to graphite under sufficiently severe conditions.

Kerogen is undoubtedly formed mainly by diagenetic alteration of sedimentary humic matter (Degens, 1965). This conclusion is supported by the chemical similarities between humic matter and kerogen, and by the fact that humic matter is the major component of recent sedimentary organic matter, just as kerogen is the major component of ancient sedimentary organic matter. After burial, humic matter suffers partial loss of functional groups, and becomes more highly condensed until its molecular weight is so high that it cannot be dissolved in any known solvent (Degens, 1967). At that stage it is called kerogen. Besides humic matter, substances cited as possible precursors of kerogen are polymerized unsaturated fatty acids (Abelson, 1967) and sporopollenin (Brooks and Shaw, 1971). Biochemical compounds may be complexed with, or condensed with, humic matter and thus be preserved in kerogen instead of being decomposed; formation of metal-organic complexes may play a key role in this process (E. T. Degens, pers. comm.).

Importance

Kerogen is geochemically important because it is a major sink for carbon and other elements such as nitrogen and complexed metals. Economically it is important because noncoaly kerogen is the principal source of petroleum, which is formed when aliphatic chains are split off as a result of mild metamorphism of the kerogen. Coal is a concentrated deposit of coaly kerogen. Finally, kerogen is a potential source of paleobiochemical and paleoenvironmental information, for it is ultimately derived from the remains of organisms and may bear the imprint of the biochemical peculiarities of its source material; however, little work has yet been done in this field. In paleobiochemical research, kerogen has at least one salient advantage over soluble organic matter: It is immobile and therefore almost certainly indigenous to the rock in which it is found, whereas the soluble organic matter may include contaminants which migrated into the rock from other formations. Analysis of kerogen could be especially valuable in the study of pre-Phanerozoic life (McKirdy, 1974).

TOGWELL A. JACKSON

References

Abelson, P. H., 1967. Conversion of biochemicals to kerogen and n-paraffins, in P. H. Abelson, ed., *Researches in Geochemistry,* vol. 2. New York: Wiley, 63-86.

Breger, I. A., and Brown A., 1963. Distribution and types of organic matter in a barred marine basin, *Trans. N.Y. Acad. Sci.,* ser 2., **25,** 741-755.

Brooks, J., and Shaw, G., 1971. Evidence for life in the oldest known sedimentary rocks—the Onverwacht Series chert, Swaziland System of southern Africa, *Grana,* **11,** 1-8.

Degens, E. T., 1965. *Geochemistry of Sediments.* Englewood Cliffs, N.J.: Prentice-Hall, 342p.

Degens, E. T., 1967. Diagenesis of organic matter, in G. Larsen, and G. V. Chilingar, eds., *Diagenesis in Sediments.* Amsterdam: Elsevier, 343-390.

Forsman, J. P., 1963. Geochemistry of kerogen, in I. A. Breger, ed., *Organic Geochemistry.* Oxford: Pergamon, 148-182.

Gransch, J. A., and Eisma, E., 1970. Characterization of the insoluble organic matter of sediments by pyrolysis, in G. D. Hobson and G. C. Speers, eds., *Advances in Organic Geochemistry.* Oxford: Pergamon, 407-426.

McKirdy, D. M., 1974. Organic geochemistry in Precambrian research, *Precambrian Res.,* **1,** 75-137.

Cross-references: *Bitumen, Bituminous Substances; Coal; Crude-Oil Composition and Migration; Humic Matter in Sediments; Hydrocarbons in Sediments; Organic Sediments; Petroleum—Origin and Evolution.*

KURTOSIS

Kurtosis (K) is a measure of the contrast between the sorting observed in the central part of a particle size distribution curve with that of the extremes (or "tails"). It represents the degree to which the particles are concentrated near the center of the curve (i.e., the peakedness of the curve). In a normal (Gaussian) probability curve, the phi diameter interval between the ϕ_5 and ϕ_{95} (5th and 95th percentiles in phi units) points should be exactly 2.44 times the phi diameter interval between the ϕ_{25} and ϕ_{75} points. If any particular sample plots as a straight line on probability paper, this ratio is followed and the kurtosis is normal, 1.00.

Kurtosis is variously expressed as:

$$K_{mm} = \frac{(Q_3 - Q_1)}{2}(P_{90} - P_{10})$$

$$\beta_\phi = \frac{\frac{1}{2}(\phi_{95} - \phi_5) - \sigma_\phi}{\sigma_\phi}$$

$$K_G = \frac{\phi_{95} - \phi_5}{2.44\,(\phi_{75} - \phi_{25})}$$

Where Q_1 and Q_3 are the first and third quartiles (P_{25}, P_{75}) in millimeters; P_{90} and P_{10} are the 90th and 10th percentiles in millimeters; ϕ_{95}, ϕ_{75}, ϕ_{25}, and ϕ_5 are the phi percentiles for 95, 75, 25, and 5; β_ϕ is the phi kurtosis measure; K_G is the graphic kurtosis; and σ_ϕ is the phi deviation measure.

If the central portion of a curve is better sorted that the tails, the curve is excessively peaked or *leptokurtic,* if the tails are better sorted than the central portion the term applied is *platykurtic.*

Folk (1965) suggested the following scale and descriptive terminology for K_G: < 0.67 = very platykurtic; 0.67-0.90 = platykurtic; 0.90-1.11 = mesokurtic; 1.11-1.50 = leptokurtic; 1.50-3.00 = very leptokurtic; > 3.00 = extremely leptokurtic.

B. CHARLOTTE SCHREIBER

References

Baker, R. A., 1968. Kurtosis and peakedness, *J. Sed. Petrology,* **38,** 679-680.

Cronan, D. S., 1972. Skewness and kurtosis in polymodal sediments from the Irish Sea, *J. Sed. Petrology,* **42,** 102-106.

Folk, R. L., 1965. *Petrology of Sedimentary Rocks.* Austin, Tex.: Hemphill's, 159p.

Cross-referenccs: *Grain-Size Studies; Phi Scale; Skewness; Sorting.*

L

LABILE CONSTITUENTS
—*See* MATURITY

LACUSTRINE SEDIMENTATION

A lake is any body of water surrounded by land; thus, it is a sedimentary basin. Lake basins have been variously classified by their geometry and origin; Hutchinson's (1957) classification is most widely used. A lake is a partly isolated environment and may be regarded as a microcosmos or ecosystem, with many complicated processes of biological and physicochemical nature. Two characteristic features which distinguish a lake from an open ocean are its limited dimensions and its close relationship to the relief and geology of surrounding land. Consequently, sedimentation is strongly influenced by the land and its climatic fluctuations. Multiple climatic changes during the Pleistocene, for example, left proof not only of major oscillations but also of minor ones in many lake sediments (Horie, 1974; Jopling and McDonald, 1975), whereas the sediments of the ocean only preserve major fluctuations. Another characteristic feature of a lake is its limited time of existence (Table 1). The ocean as a continuous water body has existed since early Precambrian, although, due to sea-floor spreading, the oldest ocean basins known are Jurassic. The existence of a lake, however, is always much shorter; Lake Baikal, believed to be the oldest lake in the world, has only existed since the Miocene. The Caspian Sea, although a former arm of the Tethys (ancestral Mediterranean), has only had a limnic nature since the Pliocene. Lake Biwa-ko in Japan originated in the Mio-Pliocene; Lake Tanganika and Lake Ohrid in the Balkans appeared in the Pliocene, and Lake Prespa in the Balkans and Lake Titicaca belong to the Plio-Pleistocene. Thus the oldest lake known is only about 20 m.y. old.

Sediments

Lake sediments are commonly composed of clay, silt, sand, and, occasionally, gravel. Normally, coarse-grained sediments are deposited close to shore, but subaqueous gravity flows, slumps, and slides may transport coarse and fine material together to deeper waters. Turbidity-current sedimentation (q.v.) was observed first in lakes and reservoirs. Frequently there is an

TABLE 1. Areas and Approximate Durations of Typical Lakes

Rock Unit, Lake	Location	Age	Area (in km^2)	Duration (in m. y.)
Lake Bonneville	Utah, Nevada, Idaho	Pleistocene	51,000 (max.)	1.0
Searles Lake	California	Pleistocene	1,000	0.17 (?)
Lake San Augustin	New Mexico	Pleistocene	585	0.65
Lake Uinta (Green River, lower Uinta fms.)	Utah	Eocene	20,000	13.3
Gosiute Lake (Green River Fm.)	Wyoming, Colo.	Eocene	44,000 (max.)	4.0
Flagstaff Limestone	Utah	Paleocene-Eocene	18,000 (max.)	2.75
Todilto Limestone	New Mexico, Colo., Ariz.	Late Jurassic	89,600	0.02
Popo Agie Formation	Wyo., Utah	Late Triassic	130,000 (+)	3.0 (?)
Lockatong Formation	New Jersey, Pa.	Late Triassic	5,850	5.1

Modified from Picard and High, 1972.

alternation of fairly fine material such as silt or clay with coarse material, usually sands. The mud usually contains considerable amounts of biogenous detritus produced in the lake itself.

At a river mouth, sediments frequently have the morphology of an alluvial fan, fan-like delta, or delta (see *Deltaic Sediments*). In saline waters, clay in suspension will often be deposited quickly by flocculation (q.v.); consequently, river waters develop remarkable deltas in sea water. This phenomenon is less developed in lake water, although these materials must eventually be deposited, so that at the river mouth the sediments form small deltas in which topset, foreset, and bottomset beds develop, with characteristic structures (Fig. 1). Gilbert (1885, 1891) developed his classic theory of delta formation from his studies of lake sediments. Particle analysis indicates that coarse materials generally occupy the nearshore region, and fine materials are situated in the deeper central part. Narrow lacustrine shelves surround deeper zones in the form of irregular, homocentric circles.

In glacial districts, summer meltwater transports huge loads of relatively coarse sediment to some lakes. During the winter, the load is relatively small, and fine materials settle out from suspension onto the lake bottom. This alternation of coarse and fine material produces an annual rhythm in the stratigraphic sequence (see *Varves and Varved Clays*).

Another phenomenon related to seasonal changes is the deposition of calcium carbonate in some lakes. During the summer, high temperatures cause high photosynthetic activity by phytoplankton and aquatic vascular plants; thus removal of carbon dioxide from lake water and dissociation of bicarbonate occurs. As the hydroxyl group increases the pH becomes more alkaline. This combination of high temperature and high alkalinity accelerate the deposition of calcium carbonate in the epilimnion (upper lake water), although the particles may dissolve in the cold water of the hypolimnion (deep lake water). Therefore, the amount of calcium carbonate in lake sediments may be a high-temperature indicator.

A characteristic of lakes is the variability of their chemistry; the salinity may range from <1% to >25% and reflects the chemistry of source areas (Picard and High, 1972). Lakes are sensitive to climate changes and may even completely evaporate at times. Thus evaporites, including many unusual minerals, are common in lake sediments (see *Evaporite Facies; Oil Shale*); these deposits are often cyclic (see *Cyclic Sedimentation; Varves*). Unusual authigenic minerals may also be produced, as in the Green River Formation (Milton, 1971). It has also been suggested that the ancient cherty iron formations (q.v.) and certain ironstones (q.v.) are lacustrine in origin.

Organic material, both allochthonous and autochthonous, is often an important component of lacustrine sediments (see *Oil Shales*). Autochthonous deposits (gyttja, q.v.: sapropel, q.v.) are composed of diatoms and other phytoplankton, animal microfossils, and pollen grains. Organic-rich lake sediments also some-

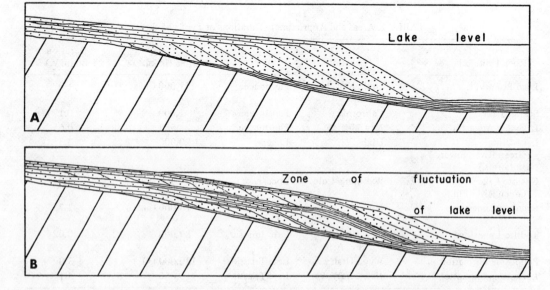

FIGURE 1. Lake deltas. (A) stable level; (B) fluctuating level (from Dunbar and Rodgers, 1957).

times show an annual rhythm because the bloom of phytoplankton occurs regularly every summer.

Lake Processes

Lakes situated in the cold regions of the earth are closed by ice most of the time and are thus termed *amictic* lakes (no mixing). In temperate zones, the lake water circulates twice a year and is thus called *dimictic*. The underlying factor in the mixing of lakes is that the maximum density of (fresh) water is reached at 4°C (at 1 atm). In spring, the surface waters warm up to 4°C and then sink to the bottom, i.e., the lake "turns over." In the autumn, the surface waters cool back down to 4°C, and again the lake turns over. After this circulation, the water having a temperature of 4°C occupies the lower part of the water body, and the upper layer is either warmer (summer) or colder (winter). Because the high-density water is then below the low-density water, the stability is good. Thus the summer and winter are stagnation periods (Fig. 2). During a stagnation period, the surface temperature and wind-induced circulation separate the uppermost layer from the lower. This upper layer is called the *epilimnion,* which is underlain by a heterogeneous water body called the *metalimnion,* below which is the *hypolimnion* (Fig. 3). In these three parts, the water temperature is relatively constant. In addition to dimictic lakes, there is an intermediate type called the *monomictic* lake that is transitional between the amictic and dimictic lakes, and between the dimictic lake and the oligomictic lake of tropical zones (see below).

Biological production is most active in the epilimnion, where the temperature is high and maximum sunshine is received. The deeper the water, the less the production and the greater the decomposition. The depth where these two processes balance each other is called the *compensation depth,* dividing the upper *trophogenic* layer from the *tropholytic* layer below (Fig. 3).

These phenomena control autochthonous lacustrine sedimentation. Organisms decompose after death; the nutrients from their bodies dissolve into the water and are then utilized for the production of the next generation. Such decomposition is most remarkable in the metalimnion, in which the temperature is comparatively high, thus accelerating the activity of bacteria. The detritus of dead organisms, however, float for a long time in the water. In dimictic lakes, as the temperature passes 4°C at the end of the summer, the metalimnion is destroyed, and most of the detritus in the epilimnion and former metalimnion drop into the hypolimnion, covering over the former sediments. They are still soft and are easily moved by disturbances, but they gradually settle to form a permanent layer. During the early stage of this stagnation period, hypolimnic water usually contains dissolved oxygen supplied from the atmosphere during the turnover. The top surface of the new sediments, therefore, has a brown color indicating the existence of Fe^{+3}. As the oxygen is consumed by the oxidation, reduction begins; needless to say, the deeper parts of the sediment are in a reducing state.

The geochemical nature of the interstitial water (q.v.), and of the bottom water at the mud/water interface, is important in the trophic stage of the lake. At a state of Eh = 0.2 volts, considerable amounts of Fe^{++}, Mn, NH_4 and P are dissolved into the lake water. Even in such a deep lake bottom as Lake Biwa-ko, the amounts of nutrients in the bottom water are comparable with those in eutrophic lakes (Fig. 3). If conditions are favorable, e.g., if the lake is relatively shallow and such nutrients are carried upward after stagnation periods, they accelerate the next generation of organisms; then the amounts of detritus increase, and deposition increases. In *meromictic* lakes, however, conditions are somewhat different; here the lowest part of the water body stagnates permanently because of the presence of more saline waters in the lake bottom. The *oligomictic* lakes of the tropical zone show a similar meromixis, but its cause is a permanent temperature stratification. Inasmuch as such water does not mix with the surface water containing dissolved oxygen, reduction occurs in the sediments, often producing FeS_2 (pyrite), giving the sediments a black color; in some cases sapropel (q.v.) develops.

Long-term Effects. Lacustrine sedimentation gradually "buries" the lake basin and works to decrease the depth, except in downwarping regions of the world, where the oldest lakes are found. For instance, Lake Baikal contains 2000

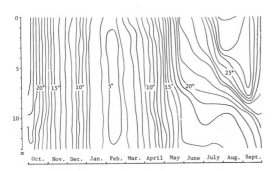

FIGURE 2. Fluctuations of water temperature at the deepest point of Lake Yogo-ko, 1961–1962 (after Horie, 1964).

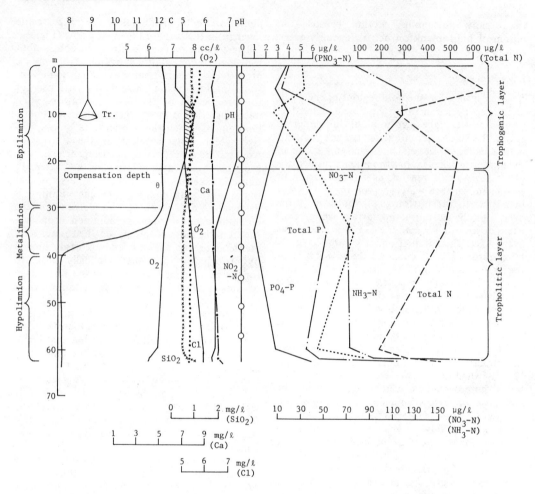

FIGURE 3. Physicochemical features at the center of Lake Biwa-ko (Dec. 10, 1963; after Horie, 1964).

m of sediments (an average accumulation rate of 0.1 mm/yr), and Lake Suwa-ko and Lake Biwa-ko in Japan are deeper than 400 m with >1000 m of sediments in each; all of these are still active lakes, located in tectonic regions. In stable regions, however, deep lakes gradually become shallow lakes as a result of sedimentation. Theinemann (1928) has given a diagrammatic scheme for such a transition (Fig. 4). In type A, dead organisms fall into the deeper part and are decomposed. The depth of the tropholytic layer is small, however, in comparison to amounts of organisms; dissolved oxygen is thus rapidly consumed, the water frequently showing no oxygen state. Consequently, redox potential drops, and large amounts of nutrients begin to dissolve into the lake water. Type A, where the volumetric ratio between the trophogenic layer and the tropholytic layer is small, is called a *eutrophic* lake type. In type B, called an *oligotrophic* lake, equal amounts of organisms are decomposed slowly in a thick tropholytic layer. Nutrients are not found in bottom water; even if they appeared at the end of the stagnation period, they could not be carried to the upper parts through a stable metalimnion. Consequently, the organic production is less than in type A. As the result of sedimentation, there is an eventual transition from the oligotrophic to the eutrophic lake type. The pattern is often more complicated in nature, however; paleolimnological studies of lake sediments frequently suggest an alternation of lake types. Some examples in northern Europe indicate a fluctuation of lake level which accompanied an alternation of climate; and a 200-m core from Lake Biwako, Japan, provides similar evidence for the last 500,000 years (Koyama et al., 1973).

The measurement of varying sedimentation rates in lakes presents a difficult problem. The sedimentation rate in Lake Biwa-ko, measured directly at 70 m depth, is approximately 1 mm/

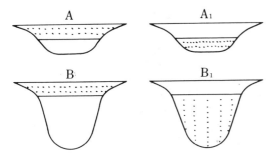

FIGURE 4. Sketch showing eutrophic (A) and oligotrophic (B) lake types (after Theinemann, 1928). Horizontal line is the compensation depth. Dots indicate organisms. (See text for further explanation.)

yr. Indirect measurement techniques include counting varves (q.v.) or dating by radiocarbon (see *Sedimentation Rates*). In oligotrophic lakes, however, sediments are mostly composed of inorganic material, and dating is not easy; consequently, estimates of annual sedimentation rates are difficult, as is any calculation of rates of diagenesis of the deposits. Geophysical surveys provide a new approach to identifying structures of buried sediments.

Pluvial Lakes. Lacustrine sedimentation goes through an important cycle in those basins that were affected by glacial-interglacial climatic oscillation, resulting in alternating fresh and evaporative conditions. Presently hypersaline Great Salt Lake in Utah was immensely larger (Lake Bonneville) during the high precipitation phase that straddled the last glacial/interglacial boundary; in Nevada, Lake Lahontan was comparable. Here, the coldest climates were marked by arid tundra conditions, and "pluviation" marked the transition to the postglacial. In the case of Lake Chad, the cool phases were marked by desert dunes, the warmest interglacial stages by a flood of alluvial material, and intermediate stages by blooms of freshwater diatoms, producing immense diatomite deposits. A 2-m bed of diatomite is found beneath Lake Maracaibo in South America; the "lake" is at present a marine gulf, but the diatomite probably represents a recent period when the depression was a true lake, isolated by lowered sea level. Small diatomite-filled lake beds were very widespread in northern Europe and elsewhere during the postglacial warm phase; most of them are now desiccated and constitute valuable local deposits.

Ancient Lake Deposits

Lakes are rarely long lived (see Table 1) and are limited in extent (usually $<25,000$ km^2); it is not surprising, therefore, that there are few extensive lake deposits recognized in the stratigraphic record, especially in the pre-Cenozoic (Feth, 1964). Picard and High (1972) suggest that there may be many unrecognized lake deposits, but it is granted that they make up at most only a few percent of the stratigraphic record. Studies of ancient lake deposits have, however, produced significant contributions to our knowledge of paleoclimatology, especially of the Pleistocene; deltaic sedimentation (q.v.); paleontology of freshwater and terrestrial forms; and geochemistry, especially of evaporite and authigenic minerals.

Picard and High (1972) review the criteria for recognizing lacustrine rocks and conclude that no criterion is definitive and that only a combination of criteria may be diagnostic. It is usually not difficult to distinguish lacustrine from fluvial rocks, but epicontinental seas and lakes have many of the same characteristics. Paleontologic, mineralogic, geochemical, and regional stratigraphic information is most useful for the identification of lacustrine rocks (Picard and High, 1972).

Beyond the Quaternary, the best-known ancient lake deposit is the Eocene Green River Formation and related deposits of Wyoming, Utah, and Colorado (see *Oil Shales*, Fig. 1). Deposits in the central part of the basin are dominantly oil shale, organic-rich calcareous and dolomitic shale, and sandy or argillaceous limestone and dolomite; shore phases are characterized by green and red claystone, calcareous argillaceous siltstone and sandstone, oosparite, algal limestones, and intraformational and chert pebble conglomerates (Picard and High, 1972). Marginal facies and perhaps even whole phases of sedimentation were shallow and evaporative in nature (Surdam and Wolfbauer, 1975). The minerals of the Green River rocks are unusual, and many are unique (Milton, 1971; Picard and High, 1972); fossils are locally abundant and diverse. The deposits are often cyclic, including both sapropelic varves and major cycles (Bradley, 1929; Picard and High, 1968).

Upper Triassic (-Jurassic?) lake deposits are well known from both western (High and Picard, 1965) and eastern (Van Houten, 1964; Sanders, 1968) North America. The Lockatong Formation of New Jersey and Pennsylvania consists primarily of dark gray, lacustrine argillite. Detrital (grain size) and chemical (carbonate) cycles are well defined in this formation (Van Houten, 1964). Van Houten compared the Lockatong with several other ancient lacustrine deposits. Lacustrine rocks in a similar setting, i.e., a fault trough basin marked by volcanism, have been identified in Connecticut and Nova Scotia. The Triassic Popo Agie Formation in Wyoming, which consists of carbonates and analcime-rich mudstones, is one of the largest recognized lacustrine units (Picard and High, 1972). Picard and High note its nonmarine fauna as evidence of a freshwater environment, and its carbonate and analcime abundance as evidence of lacustrine rather than fluvial origin.

Paleozoic lake deposits include the Devonian Caithness Flagstone series of Scotland (Rayner, 1963), the Mississippian Albert Shale of New Brunswick (Greiner, 1974), and the Carboniferous Oil Shales of Scotland (Greensmith, 1962).

Economic Deposits

Besides normal terrigenous deposits, diatomites, and evaporites, lacustrine sediments include a variety of rather specialized deposits, often economically valuable. The bulk of "Gondwanaland" coal deposits are nonmarine and strictly lacustrine or paludal in origin; they constitute major reserves in Australia, South Africa, Rhodesia, and Zambia (Duff et al., 1967, 38-47). Remarkably pure kaolinite is concentrated in some lake beds, for example, the celebrated Oligocene lake deposits of Bovey Tracey in Devonshire, England; the kaolinite is believed to have been concentrated by fluvial wash from the adjacent, deeply weathered granites of Dartmoor. Comparable, economically important kaolinites are relatively scarce, but are known from the US (e.g., Georgia), in SE Australia, and elsewhere. Some unusual examples of bauxite are also found in lake beds. Most Cenozoic bauxites appear to be related to intertidal mangrove swamp conditions, but a Carboniferous bauxitic lake deposit has been described by Wilson (1922) from Scotland, apparently formed from material eroded from a nearby lateritic surface. Other economically valuable deposits mentioned in previous sections include some evaporites, oil shales, and possibly iron deposits.

SHOJI HORIE

References

Bradley, W. H., 1929. The varves and climates of the Green River Epoch., *U.S. Geol. Surv. Prof. Pap. 158-E*, 88-95.
Bradley, W. H., 1948. Limnology and the Eocene lakes of the Rocky Mountains region, *Geol. Soc. Am. Bull.*, 59, 635-648.
Davis, W. M., 1882. On the classification of lake basins, *Proc. Boston Soc. Nat. Hist.*, 21, 315-381.
Duff, P. McL. D.; Hallam, A.; and Walton, E. K., 1967. *Cyclic Sedimentation*. Amsterdam: Elsevier, 280p.
Dunbar, C. O., and Rodgers, J., 1957. *Principles of Stratigraphy*. New York: Wiley, 356p.
Feth, H., 1964. Review and annotated bibliography of ancient lake deposits (Precambrian to Pleistocene) in the western states, *U.S. Geol. Surv. Bull.*, 1080, 119p.
Gilbert, G. K., 1885. The topographic features of lake shores, *U.S. Geol. Surv. Ann. Rept. 5*, 69-123.
Gilbert, G. K., 1891. Lake Bonneville, *U.S. Geol. Surv. Mon. 1*, 438p.
Greensmith, J. T., 1962. Rhythmic deposition in the Carboniferous Oil Shale Group of Scotland, *J. Geol.*, 70, 355-364.
Greiner, H., 1974. The Albert Formation of New Brunswick: a Paleozoic lacustrine model, *Geol. Rundschau*, 63, 1102-1113.
High, L. R., Jr., and Picard, M. D., 1965. Sedimentary petrology and origin of analcime-rich Popo Agie Member, Chugwater (Triassic) Formation, west-central Wyoming, *J. Sed. Petrology*, 35, 49-70.
Horie, S., 1964. *Nihon no mizuumi (Regional Limnology in Japan)*. Tokyo: Nihon Keizai Shinbunsha, 226p. (In Japanese).
Horie, S., ed., 1974. *Paleolimnology of Lake Biwa and the Japanese Pleistocene*, 2, 3, 4, 288p., 577p., 836p.
Hutchinson, G. E., 1957. *A Treatise on Limnology. Vol. 1–Geography, Physics, and Chemistry*. New York: Wiley, 1015p.
Jopling, A. V., and McDonald, B. C., eds., 1975. Glaciofluvial and glaciolacustrine sedimentation, *Soc. Econ. Paleont. Mineral. Spec. Publ. 23*, 320p.
Kemp, A. L. W.; Anderson, T. W.; Thomas, R. L.; and Mudrochova, A., 1974. Sedimentation rates and recent sedimentary history of Lakes Ontario, Erie and Huron, *J. Sed. Petrology*, 44, 207-218.
Koyama, T., and 10 others, 1973. Chemical studies of a 200 meter core sample from Lake Biwa-ko, *Jap. J. Limnol.*, 34, 75-88. (In Japanese with English summary).
Milton, C., 1971. Authigenic minerals of the Green River Formation, *Univ. Wyom., Laramie, Contrib. Geol.*, 10, 57-64.
Mullens, M. C., 1973. Bibliography of the geology of the Green River Formation, Colorado, Utah, and Wyoming, to March 1, 1973, *U.S. Geol. Surv. Circ. 675*, 20p.
Picard, M. D., and High, L. R., 1968. Sedimentary cycles in the Green River Formation (Eocene), Uinta Basin, Utah, *J. Sed. Petrology*, 38, 378-383.
Picard, M.D., and High, L. R., 1972. Criteria for recognizing lacustrine rocks, *Soc. Econ. Paleont. Mineral. Spec. Publ. 16*, 108-145.
Rayner, D. H., 1963. The Achanarras Limestone of the Middle Old Red Sandstone, Caithness, Scotland, *Geol. Soc. Yorkshire Proc.*, 34, 117-138.
Reeves, C. C., 1968. *Introduction to Palaeolimnology*. Amsterdam: Elsevier, 228p.
Sanders, J. E., 1968. Stratigraphy and primary sedimentary structures of fine grained, well-bedded strata, inferred lake deposits, Upper Triassic, central and southern Connecticut, *Geol. Soc. Am. Spec. Pap. 106*, 265-305.
Sly, P. G., and Thomas, R. L., 1974. Review of geological research as it relates to an understanding of Great Lakes limnology, *J. Fisheries Research Board Canada*, 31, 795-825.
Surdam, R. C., and Wolfbauer, C. A., 1975. Green River Formation, Wyoming: a playa-lake complex, *Geol. Soc. Am. Bull.*, 86, 335-345.

Theinemann, A., 1928. Der Sauerstoff im eutrophen und oligotrophen Seen, *Die Binnengewässer,* **4,** Stuttgart: Schweizerbartsche, 175p.

Toyoda, Y.; Horie, S.; and Saijo, Y., 1968. Studies on the sedimentation of Lake Biwa from the viewpoint of lake metabolism, *Mitt. Intern. Verein, Limnol.,* **14,** 243-255.

Van Houten, F. B., 1964. Cyclic lacustrine sedimentation, Upper Triassic Lockatong Formation, central New Jersey and adjacent Pennsylvania, *Kansas Geol. Surv. Bull.,* **169,** 497-531.

Wilson, G. V., 1922. The Ayrshire bauxitic clay, *Mem. Geol. Surv. Scotland,* 28p.

Cross-references: *Authigenesis; Clay Sedimentation Facies; Continental Sedimentation; Delta Sedimentation; Diatomite; Dolomite, Dolomitization; Euxinic Facies; Evaporite Facies; Fluvial Sedimentation; Gyttja, Dy; Ironstone; Lagoonal Sedimentation; Marl; Oil Shale; Oolite; Peat-Bog Deposits; Sapropel; Turbidity-Current Sedimentation; Varves and Varved Clays.* Vol. III: *Glacial Lakes; Lakes.* Vol. IVA: *Lake Geochemistry; Limnology; Paleolimnology.* Vol. IVB: *Green River Mineralogy.*

LAGOONAL SEDIMENTATION

Postglacial rise in sea level has resulted in a marine transgression that has created barrier islands parallel to the coast in many parts of the world. The elongate body of water between the barrier islands and the mainland, connected by one or more inlets to the sea, is a lagoon (q.v., Vol. III). The major source of the sediments in lagoons is usually the continental shelf. Sediment from the shelf forms the barrier islands. Most of the sediment in the lagoons has either been washed over the barrier islands from the sea, blown into the lagoon from the barrier islands, or swept by tidal currents through the inlets into the lagoon. A secondary source of lagoonal sediment is the mainland. Sediment is introduced into the lagoon by rivers or estuaries, or directly eroded from the mainland shore.

For a discussion of *tropical lagoonal sedimentation* involving largely carbonate sediments, see the entry under that title.

Sedimentary Processes

The Delaware coast (Kraft et al., 1973) provides a good example of the sedimentary processes that shape the modern barrier island-lagoon complex (see also *Barrier Island Facies*). Fig. 1 shows the leading edge of Holocene transgression and the deposits left as various lagoonal environments overlap one another toward the mainland. The source of sediments is mostly the barrier island and offshore, but there is some input from the mainland.

Lagoons are, in geologic time, ephemeral features and are soon filled in with the available sediment. Shepard (1953) found an average shoaling in Texas lagoons of 38 cm per century. A lagoon with depths of 2 or 3 m does not take long to fill.

Lagoonal Sediments. Sediments in US east-coast lagoons are usually muddy, organic-rich, fine to very fine sands on the mainland side, well-sorted clayey silts or very fine sands in the deeper centers or river mouths, and clean fine

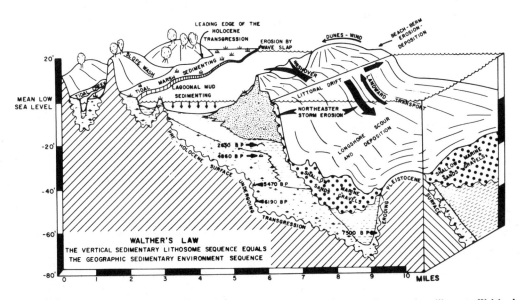

FIGURE 1. Schematic illustration showing lagoonal sedimentary processes (here used to illustrate Walther's Law; from Kraft et al., 1973). Vertical scale is greatly exaggerated.

sands on the barrier island side. Broken shells, predominantly oyster, are an important constituent of the sediment. Fig. 2 shows generalized sediment size distribution in Pamlico Sound, N.C.

In the more tropical parts of the Gulf of Mexico, evaporites and oolites may be found in shallow landward portions of lagoons such as Laguna Madre, Texas (Rusnak, 1960). Otherwise, the coarse-fraction content is similar to sediments on the Atlantic coast.

Sedimentary Structures. Evidence of reworking by organisms is characteristic of lagoonal sediments. Pamlico Sound (Pickett and Ingram, 1969) can be divided into three environments: one of textural mottling of sediments, one of homogeneous sediments, and one of texturally banded or layered sediments. The mottling is found in muddy, organic-rich sediments, principally near the mainland, and is evidence of activity by burrowing organisms. The homogeneous sediments are the deep areas of fine-grained sediments such as the middle of the lagoon or in river mouths entering the lagoon. The banded or layered sediments are found in sandy areas mostly near the barrier island, represent alternate times of low and high current action, and have not been extensively burrowed. Shepard and Moore (1955) also describe a lack of stratification in the deep central parts of bays (lagoons) on the central Texas coast.

Oyster bars, usually linear and projecting perpendicular from the mainland shore, are structures found in lagoonal sediments in the Gulf of Mexico and on the Atlantic coast.

Environments of Lagoonal Deposition

Gulf Coast. The lagoonal sediments of the central Texas coast were studied in detail by Shepard and Moore (1955), Rusnak (1960), Lankford and Rogers (1969) and others. Coarse-fraction analyses of lagoonal sediments from the Rockport area of the central Texas coast (Shepard and Moore, 1955) indicate a division into eight lagoonal sedimentary environments. The sediment distribution in these environments is influenced by salinity, waves and currents, depth of water, and source material. The bays near river mouths are characterized by a high content of plant fibers, ferruginous aggregates, $CaCO_3$, and ostracods. Montmorillonite is the dominant clay mineral. The central and deeper parts of the lagoon are noted for an abundance of foraminifera and oyster shells, and have less abundant $CaCO_3$. The lower bays (lagoons) near the inlets are characterized by high sand content (over 50%) and a higher content of clay than silt. The faunas show admixtures of open Gulf and lagoonal types. Inlets have numerous echinoid plates and spines (Shepard and Moore, 1955).

Atlantic Coast. Environments of lagoonal deposition have been delineated in Pamlico Sound, N.C. (Pickett and Ingram, 1969), the largest and best-developed lagoon on the Atlantic coast of North America. This is an area undergoing marine transgression in which the present-day barrier islands were built perhaps about 5000 years ago. The newly created lagoons proceeded to fill by washover and blowover from the barrier island and by the transportation of sediments into the lagoon through inlets, with some contribution from the mainland.

Once the sediment is in the lagoon it is distributed into environments of deposition. This process is controlled largely by wave-generated currents, water depth, and amount of available sediment. Mapping of the sedimentary facies distribution (Fig. 2) suggests a division into eleven sedimentary environments: barrier island,

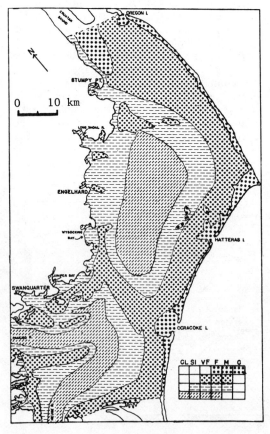

FIGURE 2. Generalized distribution of sediment sizes in Pamlico Sound, N.C., clay to coarse sand (from Pickett and Ingram, 1969).

inlet, lagoonal beach, cross-lagoon shoal, marginal lagoon, deep central basin, lagoon near narrows, finger shoal, lagoon near river mouth, protected mainland embayment, and mainland marsh. The average composition of the coarse fraction of sediment from these environments is shown in Fig. 3. The distribution of coarse-fraction constituents in Pamlico Sound is comparable to that of the Rockport area in Texas.

The sediment distribution map of Pamlico Sound (Fig. 2) shows a basin of muddy very fine to fine sands (deep central basin and marginal lagoon), with sandy silts and clays (lagoon near river mouth) projecting into it, and elongate bars of sand (finger shoals, cross lagoon shoal, lagoon near narrows, inlet, lagoonal beach). Sediment in Pamlico Sound is generally well-sorted, due to the strong wave and current action in shallow water.

Other lagoons on the Atlantic coast, such as Rehoboth Bay, Delaware (Kraft, 1971) are comparable with the model of lagoonal depositional environments shown in Pamlico Sound.

For descriptions of lagoonal sedimentation in NW Europe, see *Salt-Marsh Sedimentology* and *Tidal-Flat Geology*. Bird (1967) describes lagoonal environments in Australia.

THOMAS E. PICKETT

References

Bird, E. C. F., 1967. Coastal lagoons of southeastern Australia, in J. N. Jennings and J. A. Mabbutt, eds., *Landform Studies from Australia and New Guinea*. Canberra: Austral. Nat. Univ. Press, 365-385.

Bressau, S., 1957. Abrasion, transport and sedimentation in the Beltsee, *Die Küste*, 1, 64-102.

Emery, K. O., and Stevenson, R. E., 1957. Estuaries and lagoons, *Geol. Soc. Am., Mem.*, 67(1), 673-750.

Gierloff-Emden, H. G., 1961. Nehrungen and Lagunen; Gesetzmässigkeiten ihrer Formenbildung und Verbreitung, *Petermanns Geol. Mitt. Jg.*, 105, 81-92, 161-176.

King, C. A. M., 1972. *Beaches and Coasts*. New York: St. Martin's, 570p.

Kraft, J. C., 1971. Sedimentary facies patterns and geologic history of a Holocene marine transgression, *Geol. Soc. Am. Bull.*, 82, 2131-2158.

Kraft, J. C.; Biggs, R. B.; and Halsey, S. D., 1973. Morphology and vertical sedimentary sequence models in Holocene transgressive barrier systems, in D. R. Coates, ed., *Coastal Geomorphology*. Binghamton: St. Univ. New York, 321-354.

Lankford, R. R., and Rogers, J. J. W., 1969. *Holocene Geology of the Galveston Bay Area*. Texas: Houston Geol. Soc., 141p.

Lentacker, F., 1972. La sédimentation polderienne dans la plaine maritime flamande en Belgique et aux Pays-Bas, *Cah. Géogr. Phys. de Lille*, 1, 3-28.

Pickett, T. E., and Ingram, R. L., 1969. The modern sediments of Pamlico Sound, North Carolina, *Southeastern Geol.*, 11(2), 53-83.

Rusnak, G. A., 1960. Sediments of Laguna Madre, Texas, in F. P. Shepard, F. B. Phleger, and T. H. Van Andel, eds., *Recent Sediments, Northwest Gulf of Mexico*. Tulsa: Am. Assoc. Petrol. Geologists, 153-196.

Shepard, F. P., 1953. Sedimentation rates in Texas estuaries and lagoons, *Bull. Am. Assoc. Petrol. Geologists*, 37, 1919-1934.

Shepard, F. P., 1973. *Submarine Geology*, 3rd ed. New York: Harper & Row, 517p.

Shepard, F. P., and Moore, D. G., 1955. Central Texas coast sedimentation: Characteristics of sedimentary environments, Recent history and diagenesis, *Bull. Am. Assoc. Petrol. Geologists*, 39, 1463-1593.

Van Loon, A. J., and Wiggers, A. J., 1975. Holocene lagoonal silts (formerly called "sloef") from the Zuiderzee, *Sed. Geol.*, 13, 47-56.

Zenkovitch, V. P., 1959. On the genesis of cuspate spits along lagoon shores, *J. Geol.*, 67, 169-177.

Zenkovitch, V. P., 1967. *Processes of Coastal Development*. New York: Interscience, 738p.

Cross-references: *Barrier Island Facies; Estuarine Sedimentation; Littoral Sedimentation; Salt-Marsh Sedimentology; Tidal-Current Deposits; Tidal-Flat Geology; Tidal-Inlet and Tidal-Delta Facies; Tropical Lagoonal Sedimentation; Wadden Sea Sedimenta-*

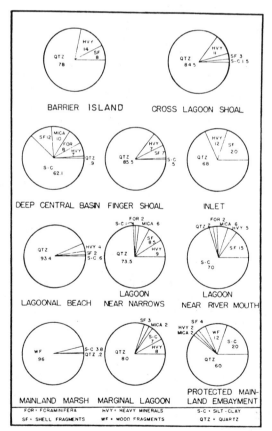

FIGURE 3. Average composition of the coarse fraction of sediments from Pamlico Sound, N.C. sedimentary environments (from Pickett and Ingram, 1969).

tion. Vol. III: *Coastal Lagoon Dynamics; Lagoon; Mangrove Swamps–Geology and Sedimentology;* Vol. IVA: *Lagoon Geochemistry.*

LAMINA, LAMINASET, LAMINITE

A lamina is the smallest megascopic layer of particles in a sedimentary sequence, made up of sediment which is the product of deposition during a single fluctuation in the competency of the transporting agent or in the environment of deposition (Apfel, 1938; Campbell, 1967). Campbell (1967) specifies the use of "megascopic," at the suggestion of L. L. Sloss, to exclude layers that can only be distinguished by the use of microscopes, x-rays, or trace element and isotopic analyses. A *lamina,* then, is a sedimentary layer, usually <1 cm thick (McKee and Weir, 1953; see *Bedding Genesis*), although it may be as thick as 2.5 cm. Thus, a bed without recognizable internal lamination might be construed as a lamina, but a study of a bed in relation to adjacent beds and in terms of thickness will decide the useful ranking of the layer under consideration.

Campbell (1967) characterizes a lamina as having (1) a relatively uniform texture and composition; (2) no internal layering (megascopically); (3) a smaller areal extent than the enclosing bed, except where laminae are parallel to bedding surfaces; (4) a period of formation of shorter duration than the encompassing bed; and (5) boundary surfaces (above and below) termed laminar surfaces. A group of conformable laminae that form a distinctive structure within a bed is termed a *laminaset* (Campbell, 1967), used as McKee and Weir (1953) used *bed* and *bedset.*

The term *laminite* was proposed by Sander (1936) and further defined by Lombard (1960, 1963) as any finely laminated detrital rock. Sander preferred this nongenetic definition for simple descriptive purposes. More specific terms include *rhythmite,* proposed by Bramlette (1946) for any sediment with a distinctive couplet arrangement in a sequence of rhythmic bedding, and *turbidite* (q.v.) and *contourite* (q.v.), both of which imply a specific genesis and are therefore subjective.

Lombard (1963) suggested that there might be two dimensional categories of laminites: (1) *first-order laminites,* thick stratal units alternating with fine-grained layers, as in a flysch sequence, where the thick bed is interpreted as a turbidite (emplaced by a single turbidity flow), and the fine-grained interval, commonly argillaceous, is interpreted as a deep-sea pelagic deposit that accumulated over a long period of time and was terminated by the next turbidite

event; and (2) *second-order laminites,* thin-bedded laminae continuously repeating throughout any bedding unit. These laminites, Lombard believed, characterize the distal parts of basins beyond the limit of coarse turbidite flows or the remote bottomset components of large deltas or submarine fans. They are also found in algal limestones.

RHODES W. FAIRBRIDGE

References

Apfel, E. T., 1938. Phase sampling of sediments, *J. Sed. Petrology,* 8, 67-68.
Bramlette, M. N., 1946. The Monterey Formation of California and the origin of siliceous rocks, *U.S. Geol. Surv. Prof. Pap. 212,* 57p.
Campbell, C. V., 1967. Lamina, laminaset, bed and bedset, *Sedimentology,* 8, 27-33.
Lombard, A., 1960. Les laminites et la stratification du flysch, *Arch. Sci. Soc. Phys. Hist. Nat. Genève,* 13(4), 567-570.
Lombard, A., 1963. Laminites: a stratum of flysch-type sediments, *J. Sed. Petrology,* 33, 14-22.
McKee, E. D., and Weir, G. W., 1953. Terminology for stratification and cross stratification in sedimentary rocks, *Geol. Soc. Am. Bull.,* 64, 381-390.
Otto, G. H., 1938. The sedimentation unit and its use in field sampling, *J. Geol.,* 46, 569-582.
Sander, B., 1936. Beiträge zur Kenntis der Anlagerungsge füge (Rhythmische Kalke and Dolomite aus der Trias), *Mineral. Pet. Mitt.,* 48, 27-209. English translation by E. B. Knopf, 1951. Tulsa: Am. Assoc. Petrol. Geologists.

Cross-references: *Bedding Genesis; Contourites; Cyclic Sedimentation; Stromatolites; Turbidites; Varves and Varved Clays.*

LATERAL AND VERTICAL ACCRETION

Geologists have long been aware of the concepts of lateral and vertical accretion. The terms were originally applied to fluvial sedimentation (q.v.; Fenneman, 1906). *Lateral accretion* specifically referred to deposition on a migrating point bar in a meandering stream, whereas *vertical accretion* referred to the accumulation of suspended material in overbank deposits that conformed to the depositional surface. Although the basic connotation of the terms has remained relatively unchanged, in present usage they are applied generally to the nature of accumulation in all depositional environments. As discussed herein, lateral accretion is a process in which sediment accumulates on a slope without changing the elevation of any topographic feature of the slope (i.e., the top or base), causing the surface to shift laterally in the downslope direction. Vertical accretion, in contrast, is an accumulation process in which topographic elements of the surface do not shift laterally but rather are extended directly upward.

The nature of accretion is readily apparent on a horizontal depositional surface that is extended upward by vertical accretion; but deposition commonly occurs on sloping surfaces where the processes of lateral and vertical accretion are more difficult to separate. For example, two different types of vertical accretion on a slope can be identified. In one, deposition occurs by overlapping horizontal depositional surfaces (a, Fig. 1); the filling of abyssal basins by turbidites (q.v.) is an example of this type of vertical accretion. The other type occurs where a uniform thickness of sediment is draped across a slope (b, Fig. 1). Note that in this example there still is no lateral accretion—the sloping depositional surface is extended directly upward and not transposed laterally. A blanket of pelagic sediment (q.v.) deposited on an irregular sea-floor surface is an example of this type of accretion.

In many deposits (for example, a slowly subsiding, prograding delta system), vertical and lateral accretion will both occur simultaneously (c, Fig. 1). If, however, subsidence were nil and sea level remained constant, the deposit would accrete laterally only (d, Fig. 1). Lateral accretion may also result solely from lateral transfer of sediment without a net increase in the volume of the deposit (as with migrating sand waves or point bars).

Lateral migration of a sequence of depth-dependent depositional environments results in a

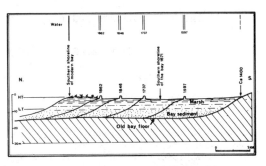

FIGURE 2. Schematic profile and section along a N-S line S of Authie Bay, France, showing historically dated time lines cutting diagonally through horizontal strata (from LeFournier and Friedman, 1974). HT = high tide; LT = low tide.

vertical sequence which has been deposited by lateral accretion (Fig. 2). This idea was introduced in Gilbert's (1890) study of deltas in Lake Bonneville, Utah. More recently, such sequences have been noted in tidal flats (see *Tidal-Flat Geology;* Van Straaten, 1952; Klein, 1971), rivers (Sundborg, 1956; Allen, 1965); deltas (see *Deltaic Sediments;* Scruton, 1960); and open coasts (LeFournier and Friedman, 1974; Sanders and Kumar, 1975). LeFournier and Friedman (1974) point out that lateral accretion rates may be as much as 10^6 times faster than vertical accretion rates (see *Sedimentation Rates*).

Obviously the process of sedimentation partly controls the mode of accretion. Lateral accretion implies lateral transport of sediment. But vertical accretion, as defined herein, need not result only from downward settling of suspended particles. An accumulation of horizontally laminated sand implies vertical accretion, even though the individual grains may have been deposited from laterally moving currents under upper flow regime conditions.

Distinction between vertically and laterally accreted sediment may be relatively easy if geometric relations between beds can be clearly defined (i.e., in excellent exposures or in acoustical profiling records). In other instances, the nature of the vertical sequence within the deposit may provide a clue. In vertical accretion, the vertical sequence is controlled primarily by depositional events and thus may show no systematic upward variation. In lateral accretion, the vertical sequence is controlled by depositional facies, wherein topographically higher deposits migrate laterally over adjacent topographically lower deposits (Fig. 2). Lateral accretion, therefore, is characterized by an internal upward trend toward shallower water (or topographically higher) deposits that is gen-

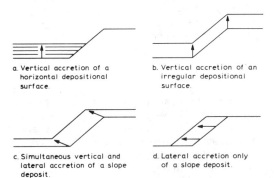

FIGURE 1. Basic types of accretion: (a) vertical accretion of a horizontal depositional surface over an irregular base; (b) vertical accretion of an irregular depositional surface (note that topographic elements of the slope have not shifted laterally); (c) simultaneous vertical and lateral accretion of a slope deposit (note lateral and upward shift of topographic elements of the slope); (d) lateral accretion only of a slope deposit (topographic elements of the slope have shifted laterally but not upward).

erally absent from sediments that result from vertical accretion.

H. EDWARD CLIFTON

References

Allen, J. R. L., 1965. A review of the origin and characteristics of Recent alluvial sediments, *Sedimentology,* 5, 89-191.
Fenneman, N. M., 1906. Flood plains produced without floods, *Am. Geog. Soc. Bull.,* 38, 89-91.
Gilbert, G. K., 1890. Lake Bonneville, *U.S. Geol. Survey Mon. 1,* 438p.
Klein, G. de V., 1971. A sedimentary model for determining paleotidal range, *Geol. Soc. Am. Bull.,* 82, 2585-2592.
LeFournier, J., and Friedman, G. M., 1974. Rate of lateral migration of adjoining sea-marginal sedimentary environments shown by historical records, Authie Bay, France, *Geology,* 2, 497-498.
Sanders, J. E., and Kumar, N., 1975. Evidence of shoreface retreat and in-place "drowning" during Holocene submergence of barriers, shelf off Fire Island, New York, *Geol. Soc. Am. Bull.,* 86, 65-76.
Scruton, P. C., 1960. Delta building and the deltaic sequence, in F. P. Shepard et al., eds., *Recent Sediments, Northwest Gulf of Mexico.* Tulsa: Am. Assoc. Petrol. Geologists, 82-102.
Sundborg, Å., 1956. The River Klarälven, a study of fluvial processes, *Geog. Annaler,* 38, 127-316.
Van Straaten, L. M. J. U., 1952. Sedimentation in tidal flat areas, *Alberta Soc. Pet. Geol. J.,* 9, 203-226.

Cross-references: *Accretion, Accretion Topography; Bedding Genesis; Delta Sedimentation; Fluvial Sedimentation; Pelagic Sedimentation, Pelagic Sediments; Sedimentation Rates.*

LIMESTONE FABRICS

Limestones are solid or indurated rocks composed primarily of calcium carbonate. In rocks containing less than about 85% $CaCO_3$, petrology and physical properties (e.g., porosity, permeability) are readily modified. This discussion will, therefore, be confined to the fabric of pure (>85%) calcium carbonate assemblages. Textural characteristics such as grain size have been purposely omitted. Granulometry is not well suited to limestones, unless they are completely allochthonous; its truly useful aspect is the distinction between calcarenites (q.v.) and calcirudites for reworked carbonate rocks. Only in such a case may limestones be treated as arenites.

Original Sediment

The characteristics displayed by the original soft sediment fall into five main categories, according to the amount of microcrystalline matrix, relative to the enclosed calcareous particles and to the voids. Examples of these five assemblages may be seen in Fig. 1: (1) A dense, uniformly fine-grained microcrystalline *ooze* can be a chemical aragonite precipitate or, in contrast, an abraded algal "flour." (2) When the number of discrete particles increases, but still remain mud-supported by the predominant matrix, the sediment may be called a *lime mud.* (3) The particles may become so numerous as to be in contact, forming a *grain-supported sediment,* still with a microcrystalline matrix. (4) Such grain- or mud-supported sediments may have *cavities,* usually formed by rapid decay of ephemeral algae or Spongiostromids; some of the particles are bound by algal mucus. (5) Persistent currents may completely wash away the matrix, creating a *clean carbonate sand;* such an assemblage may also be produced by accumulation of organisms in situ, as crinoidal piles without interstitial matrix.

Particles. The discrete calcareous particles which play a variable quantitative role in the described assemblages may be grouped in three categories: (1) True *oolites* (q.v.) display numerous envelopes or cortices, with or without nucleus, while "proto-oolites" display a single cortex; both are the result of turbulence, chemical precipitation, and algal coating. (2) *Lumps* include widely different particles and indicate an environment of reworking, algal activity or diagenetic aggregation; they include limestone lumps, grapestones (q.v.), bahamites, pseudo-oolites, intraclasts (q.v.), mud-coated grains, bothrolites, etc. (3) *Bios* includes the wide biological realm encountered in limestones. Of particular interest are algae; crinoids; pelecypods; stromatoporoids; brachiopods; sponges; and, to a lesser extent, polyps, foraminifera, bryozoans, and ostracodes.

Diagenesis

This outline of the original fabric of the sediment allows appraisal of its diagenetic processes (Fig. 1). Such mud may undergo little alteration after sedimentation. For instance, mechanically destroyed calcitic shells (*coccolithophorids*) or calcitic algal ooze are rarely subjected to diagenesis; therefore the actual rocks are still noninduarated and form loose chalk or calcareous sand; they are not limestones in the strictest sense.

Diagenesis of aragonite-calcite mud, however, usually leads to a distinctive fabric (Fig. 1; see also *Carbonate Sediments—Diagenesis*). Depending on nucleation, accretion speed, and partial pressure of CO_2, three main pathways may be followed. First, by *phase transition,* the original mud forms an interlocking mosaic of minute crystals (*micrite* of μ size) of the same

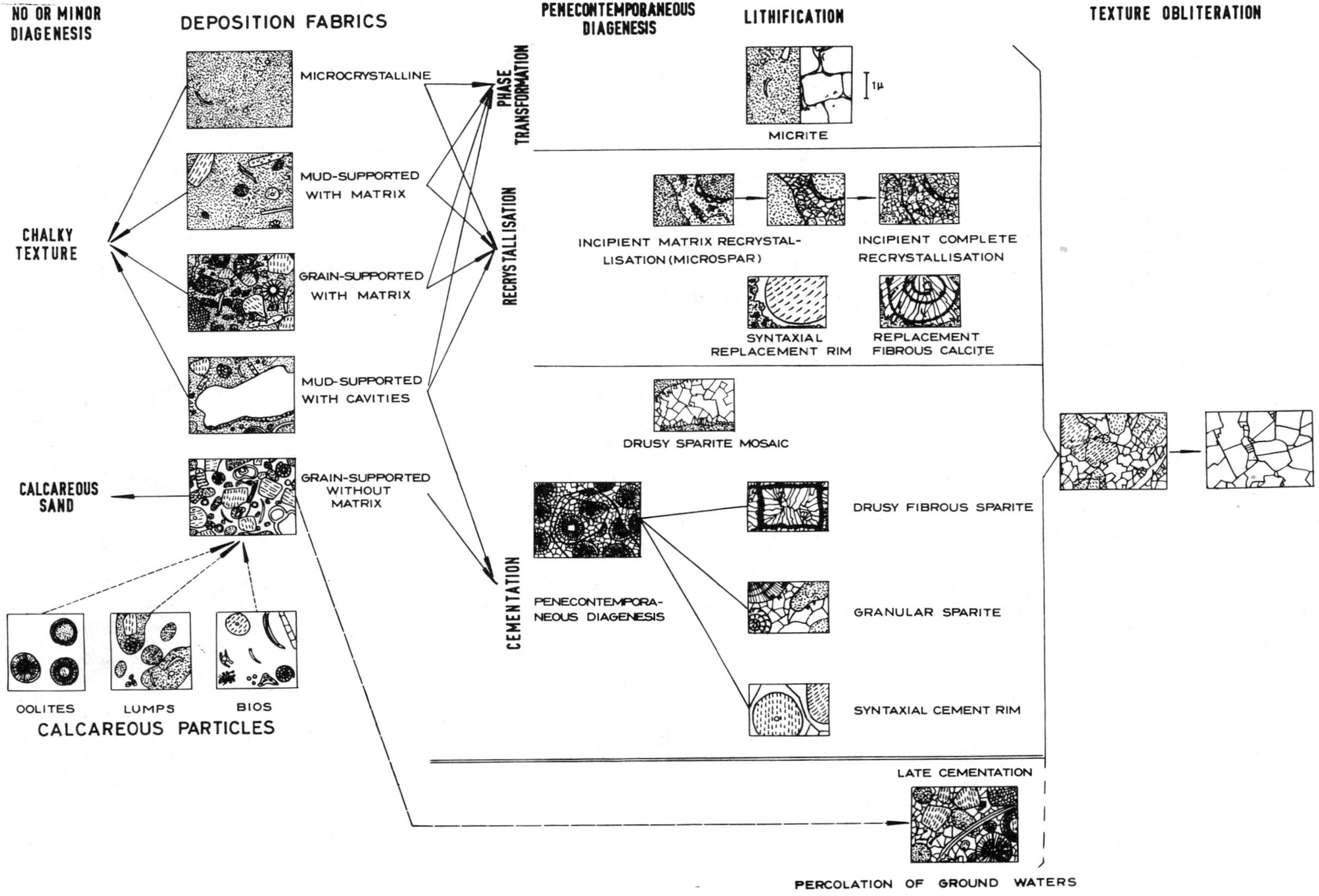

FIGURE 1. Development of limestone fabrics.

size range as the original material. The number of nucleation centers does not increase appreciably, and new grain boundaries rarely intersect original fabric; the electron microscope reveals a fine mosaic of xenomorphic crystals with little cementation.

Second, true *recrystallization* is a destructive process; new crystals are independent of the primitive mud nuclei, and the normal trend is an enlargement of crystal size. This increase is widely erratic, and the grain boundaries are often curved. Such recrystallization may occur immediately after deposition. In this case it is restricted to the matrix, which is transformed into microspar while the enclosed discrete particles remain unaffected. The process may continue until complete transformation of the rock has occurred; original structures, however, are still clearly recognizable as *ghosts*. Crystals are very irregular in outline and do not grow away from any surface. Two particular kinds of such fabrics are: (1) the "syntaxial replacement rim," where calcite is in optical continuity with a host (e.g., crinoid); it displays jagged and irregular gradation to the surrounding microspar, and micrite ghosts are conspicuous; and (2) the "replacement fibrous calcite," a late and uncommon diagenetic process which cuts into all preexisting fabrics.

The third path of induration is *cementation*. One usually restricts "drusy sparite mosaic" to void filling of cavities, while "granular sparite" or "granular cement" is restricted to cementation of grain-supported limestones. There is little difference between the two, however, because in both cases, crystals grow away from a wall, usually displaying a normal increase in size. "Syntaxial cement rim" is an optically oriented sparite on a host (e.g., crinoid) replacing a former void. "Drusy fibrous calcite" is the result of an uncommon kind of infilling, where the prismatic crystal shows obvious elongation, perpendicular to the wall. As in granular sparite, crystals are clear and have sharp, plane outlines. Such cementation may occur during sedimentation if nucleation is low and the partial pressure of CO_2 high. Typical examples are found in big *intraclasts* showing truncation of sparite cement, but in most cases, it occurs later during compaction. Moreover, calcarenites without matrix may be cemented very late by percolation of ground solutions with no relation to the conditions that occurred during sedimentation. Whereas the physical process is identical, distinction among these three cases is more structural than textural.

The preceding textures may be further obliterated and transformed by recrystallization into more or less recognizable *ghosts* or *relics;* such a process is related to important external modifications of pressure and temperature as encountered in deep burial, stress, or changes in chemical composition of percolating ground waters. Such complete obliteration can also occur during diagenesis, when as little as 2% MgO is present in the lattice. This metasomatism can drastically modify the proposed scheme, whose sequence is valid only for non-dolomitic rocks.

BERNARD L. MAMET

References

Bathurst, R. G., 1971. *Carbonate Sediments and their Diagenesis.* Amsterdam: Elsevier, 620p.

Bricker, O. P., ed., 1971. *Carbonate Cements.* Baltimore: Johns Hopkins Univ. Press, 376p.

Carozzi, A. V.; Bouroullec, J; Deloffre, R.; and Rumeau, J. L., 1972. *Introduction to the Petrography, Geochemistry and Environmental Interpretation of Carbonate Rocks.* Pau: Centre de Recherche, 300p.

Ham, W. E., ed., 1962. Classification of carbonate rocks: *A symposium, Am. Assoc. Petrol. Geologists, Mem. 1,* 279p.

Mamet, B. L., 1964. Sédimentologie des faciès "marbres noirs" du paléozoïque franco-belge, *Mem. Inst. Roy. Sci. Nat. Belgique,* **151,** 131p.

Orme, G. R., and Brown, W. W., 1963. Diagenetic fabrics in the Avonian limestone of Derbyshire and North Wales, *Proc. Yorkshire Geol. Soc.,* 34, 51-66.

Pray, L. C., and Murray, R. C., eds., 1965. Dolomitization and limestone diagenesis, *Soc. Econ. Paleont. Mineral. Spec. Publ. 13,* 180p.

Cross references: *Allochem; Bahama Banks Sedimentology; Calcarenite, etc.; Carbonate Sediments–Diagenesis; Carbonate Sediments–Lithification; Cementation; Diagenesis; Diagenetic Fabrics; Grapestone; Intraclastic Rocks; Limestones; Micrite; Oolite; Pisolite; Sparite.*

LIMESTONES

Limestones are rocks composed primarily of calcareous sediments, i.e., sediments consisting of calcium carbonate ($CaCO_3$). These sediments may be produced by biochemical and physicochemical sedimentary and diagenetic processes (see also *Carbonate Sediments–Diagenesis*). Limestones range from recently formed, slightly lithified deposits rich in both aragonite (orthorhombic $CaCO_3$) and calcite (hexagonal $CaCO_3$) to ancient, well-indurated accumulations composed solely of calcite; the mineral aragonite is unstable and changes readily to calcite with time. Among the sedimentary carbonates (Fig. 1), the calcite rocks or limestones and the dolomites (q.v.) are overwhelmingly the most widespread. Staining techniques (q.v.) permit a quick discrimination among these carbonate minerals. Marbles and carbonatites, also com-

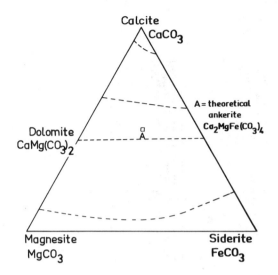

FIGURE 1. Range of chemical composition of major sedimentary carbonate minerals (after Dunbar and Rodgers, 1957). The dashed lines indicate approximate natural limits.

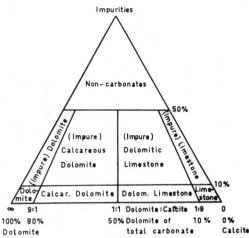

FIGURE 3. Compositional terminology for carbonate rocks (after Leighton and Pendexter, in Ham, 1962).

posed of carbonate minerals, are the products of metamorphism and magmatism, respectively, and will not be considered here.

Limestones frequently contain noncalcareous components, of which the most common are terrigenous material (e.g., clay, quartz sand) and dolomite. With an increase of these constituents limestones grade into shale and sandstone (Fig. 2) and dolomite (Fig. 3), respectively. Most dolomite (q.v.) in the sedimentary column appears to have been formed by replacement of limestones. The classification of this important carbonate rock is included in the tables and figures here, but details on its origin are found in a separate section.

Limestones range in age from the Precambrian, approximately a billion years ago, to the Holocene. Of all the sedimentary rocks, which cover 75% of the earth surface, the limestones make up 10-20% (Pettijohn, 1975). More than 50% of the world's presently known oil and gas reserves occur in limestone and dolomite reservoir rocks, and many ore deposits are associated with sedimentary carbonates.

Geographic Distribution

Both physicochemical and biochemical precipitation of calcium carbonate is enhanced by $CaCO_3$ saturation of the water medium. The degree of saturation is highest in shallow marine waters of tropical to warm temperate regions. Thus, extensive recent reefs and shallow-water limestones occur between latitudes 30°N and 30°S, of which the best known are the Great Barrier Reef and the Bahama Banks (q.v.). Some biohermal limestones exist at great depth in very cold water (Teichert, 1958), but they are quantitatively insignificant. Also, deep-sea calcareous ooze spreads approximately from latitude 70°N to 55°S (Le Blanc and Breeding, 1957) and is not strictly bound to tropical areas (see *Pelagic Sediments, Pelagic Sedimentation*).

The shifting of climatic zones during the geologic past accounts for limestone formations in regions that are at present unfavorable for carbonate deposition. For example, the Permian climate of Greenland was sufficiently warm for the development of reefs. Also, the distribution of oceans and seas relative to continents and islands varied from geologic period to period (Dott and Batten, 1976). Therefore, limestones are found at the present time in the interior of continents at localities now remote from the sea, e.g., in the Rocky Mountains and in the Alps.

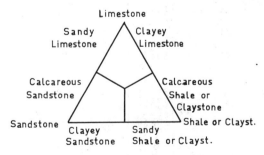

FIGURE 2. Classification of common sediments (after Pettijohn, 1957; modified after Pirsson and Schubert).

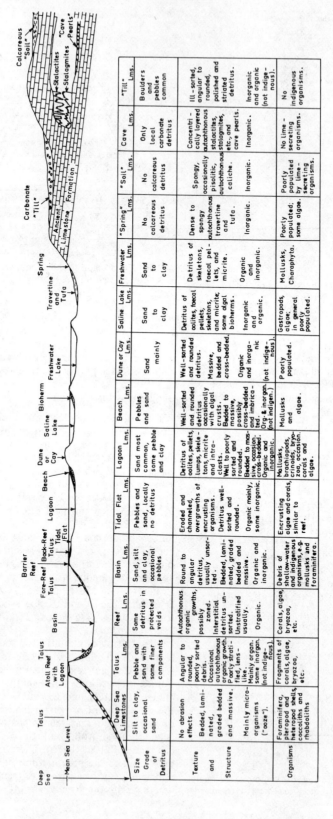

FIGURE 4. Geomorphologic distribution of limestones with some characteristics of the sediments (very diagrammatic and generalized).

Environments of Limestone Formation

Calcareous sediments form in a wide range of settings (Fig. 4). Noncalcareous components, e.g., evaporites (q.v.) and terrigenous detritus, are usually associated with the limestones. In general, more information is available on marine carbonates (Milliman, 1974); relatively little work has been done on limestones formed in lakes. Investigation by Müller et al. (1972) has led to a model of lacustrine carbonate genesis (Fig. 5), which has in turn been used to compare marine versus nonmarine carbonate sediments (Fig. 6).

The geographic variations of the different types of carbonate sediments (and associated clastics and evaporites) are related to a number of factors, such as wave energy and water circulation across a wide marine shelf (Fig. 7). The corresponding results of various limestone facies are modeled in Figs. 8 and 9, examples of environmental reconstructions used in the sedimentology of limestones. It should be noted that various processes allow the formation of carbonates in one area (e.g., shelf), and subsequent transportation of the particles may result in a wide distribution of limy sediments from the shallow-water shelf down the continental slope into the deep-water basin environments, as illustrated in one example described by Thomson and Thomasson (in Friedman, 1969; Fig. 10).

The physical, chemical, and biological factors in a particular geographic setting change through geologic time as a result of transgressions and regressions, so that the sediments making up the stratigraphic column change accordingly. The environmental changes are frequently systematic, constituting a "cycle," resulting in cyclic sediments (see *Cyclic Sedimentation*). Although the understanding of these cycles is of fundamental importance, cyclicity in carbonate sections has not been studied in detail, in contrast to the voluminous information on terrigenous rocks. One carbonate cycle recently worked out is the sabkha cycle (Wood and Wolfe, 1969), composed of the limestone lithologies formed under specific tidal milieux (Fig. 11).

Petrography and Petrogenesis

Sedimentology has as one of its principal aims the reconstruction of physical, chemical, and geographical environments. Therefore, the morphologic and genetic aspects of the individual carbonate components are of primary significance (Bathurst, 1975). Most of the features described below may be indistinct in hand specimens and are clearly discernible only in polished and thin sections under a binocular or petrographic microscope.

Two major groups of limestones exist: (1) the autochthonous deposits, which form in place by primarily biogenic processes (Fig. 12); and (2) the allochthonous or detrital limestones, composed predominantly of transported and abraded calcareous fragments (Figs. 12, 13). A third group, the so-called *calclithites,* are formed by erosion of older limestones and are considered as a type of terrigenous sediment (Fig. 13).

Autochthonites. Most autochthonous lime-

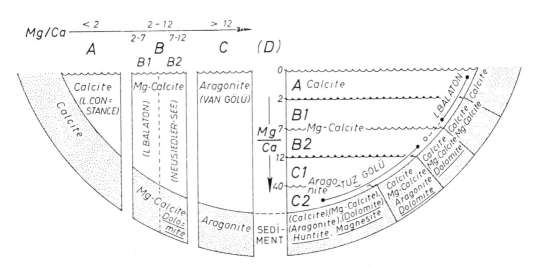

FIGURE 5. Model of lacustrine carbonate formation in "stationary" (left) and "dynamic" lakes (right) (after Müller et al., 1972). Stage D comprises the formation of primary hydrous Mg-carbonates at very high Mg/Ca concentrations.

LIMESTONES

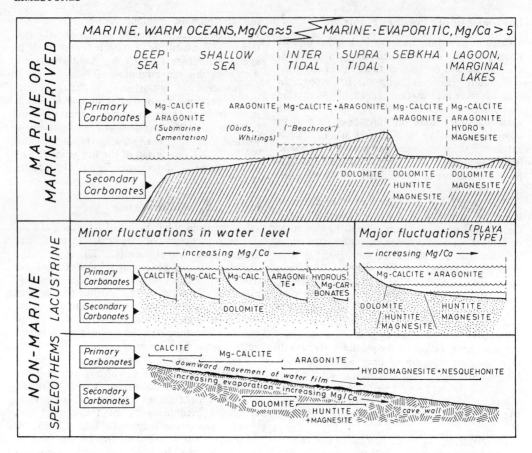

FIGURE 6. Comparison of carbonate mineralogy from different environments (after Müller et al., 1972).

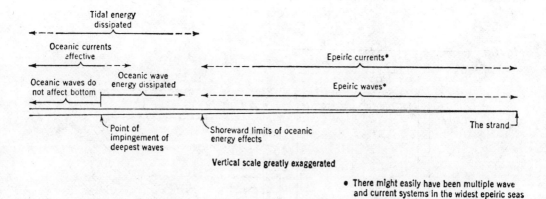

FIGURE 7. Theoretical water energy distribution and water circulation across a wide, shallow, epicontinental sea with low depositional slope (from *Time in Stratigraphy* by A. B. Shaw; ©1964 by McGraw-Hill Book Company, used with permission).

stones are composed of sessile benthonic organisms, e.g., calcareous algae, corals, and stromatoporoids, which remained in place of growth. These limestone bodies are of considerable practical and theoretical importance, and various terms are in use to describe their shape and origin (see *Reef Complex*). A *bioherm* is a body with a relatively large height:width ratio, i.e., shaped like a mound. In contrast, a *biostrome* has a low ratio and is more elongate and

438

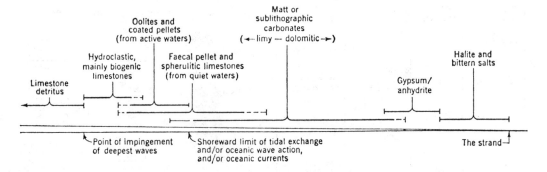

FIGURE 8. Theoretical sedimentary facies distribution related to water energy distribution and water circulation within shallow, clear water, epicontinental sea (from *Time in Stratigraphy* by A. B. Shaw; © 1964 by McGraw-Hill Book Company, used with permission).

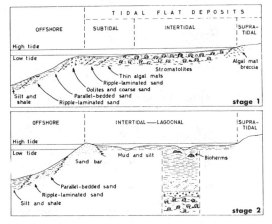

FIGURE 9. A model for sedimentation on a tidal flat which is applicable to the transition beds (from Visser and Grobler, 1972).

the distinction is important in paleoenvironmental reconstructions.

The recognition of faunal structures is relatively easy unless diagenetic obliteration has occurred. The deposits formed by calcareous algae, however, are controversial and often quite difficult to recognize, especially because most algal colonies consist primarily of finely crystalline calcite and algal-bound detritus without preserved organic structures (Wolf, 1965a, b). There has been increasing recognition, however, of the important role of calcareous algae in limestone formation (Johnson, 1961; 1967; Ginsburg et al., 1971).

Less widespread than biohermal and biostromal autochthonous limestones are travertine (q.v.), tufa, and caliche (q.v.). These limestones form under a variety of physico- and biochemical conditions. Dense, laminated *travertine* and the more spongy rock known as calcareous tufa are deposited from both cold and hot spring waters that lose CO_2 upon contact with the air (Graf, 1960; see *Spring Deposits; Speleal Sediments*). The familiar stalactites and stalagmites in caves are a form of traver-

bed-like. If these bodies are wave resistant and influence the sedimentary processes, they are biohermal reefs and biostromal *reefs*. If the bioherms and biostromes are less rigidly built and are not wave resistant, they are called biohermal and biostromal *banks*, respectively (see Nelson et al., in Ham, 1962, for new concepts and original references). The framework of these limestones can contain a high proportion of detritus in voids. These four terms are used for large stratigraphic limestone bodies and are, therefore, inapplicable in the study of hand specimens. For autochthonous growths on a small scale, Folk (1959) proposed the useful term *biolithite* (Dunham's "boundstone"). All bioherms, biostromes, reefs, and banks consist basically of biolithites. Biolithites are also found in predominantly detrital limestones, however;

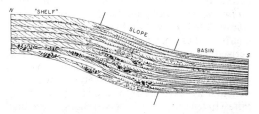

FIGURE 10. Diagrammatic cross section showing the inferred interrelations of shelf, slope, and basin facies for a short interval of the Dimple Limestone; thickness ≈ 30 m, length ≈ 25 km (from Thomson and Thomasson, in Friedman, 1969; modified from a drawing by J. L. Wilson).

439

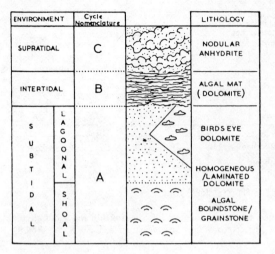

FIGURE 11. An "ideal" sabkha cycle (from Wood and Wolfe, 1969).

tine; crusts of travertine have also been reported from marine limestones, and it has been suggested that some may be of algal origin. *Caliche* is a lime-rich deposit formed in the vadose zone, especially in semi-arid regions. Capillary action may draw the lime-bearing waters to the surface where, by evaporation, the lime-rich caliche is formed. Alternately, surface waters may leach $CaCO_3$ and, when percolating downward, precipitate the carbonate above the water table.

Allochthonites. The allochthonous limestones (A. Grabau used the term *calcarenite*, q.v., here) are composed of one or more types of fragments or grains. The morphologic grain types commonly recognized are limeclasts, skeletons, oolites and pisolites, lumps, and pellets (Figs. 12, 13). Very fine material, called micrite (q.v.), is usually referred to as matrix (Table 1). Each one of the grain types and micrite can be formed by a number of very different processes. Therefore, the names given above are purely descriptive, and only detailed petrographic studies allow a genetic connotation.

Limeclasts are fragments of preexisting semi-consolidated to well-indurated limestone fragments either of intraformational origin or derived from an older carbonate rock outside the basin. They are genetically called *intraclasts* (q.v.) and *extraclasts*, respectively. If most components are extraclasts, the deposit is a *calcilithite* (Fig. 13). In many occurrences, in particular if much algal debris is present, the two may only be distinguished if the extraclasts have undergone diagenesis, e.g., cementation, recrystallization, reworking by organisms, or dolomitization, *before* transportation.

Oolites (q.v.) and *pisolites* (q.v.) are spherical

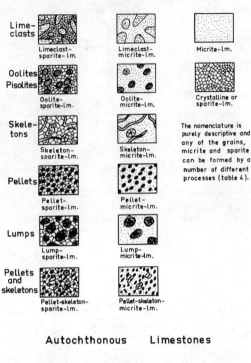

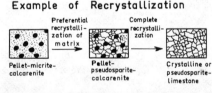

FIGURE 12. Diagrammatic scheme of allochthonous, autochthonous, and recrystallized limestones (modified after Folk, 1959).

to irregularly shaped bodies which have in cross-section two or more concentric layers and/or radially arranged calcite or aragonite mosaics. The latter are known as *spherulites*. Physico-chemical, algal, bacterial, and weathering processes may form oolites and pisolites. The recent oolitic sands of the Bahama Banks (q.v.) form in a marine environment subject to tidal currents in areas where the cool oceanic waters invading the shallow banks are warmed, becoming supersaturated with calcium carbonate. Oolite genesis can also take place in fresh and brackish waters (Carozzi, 1960). Calcareous algae are capable of producing thin to relatively thick layers around nuclei to form so-called

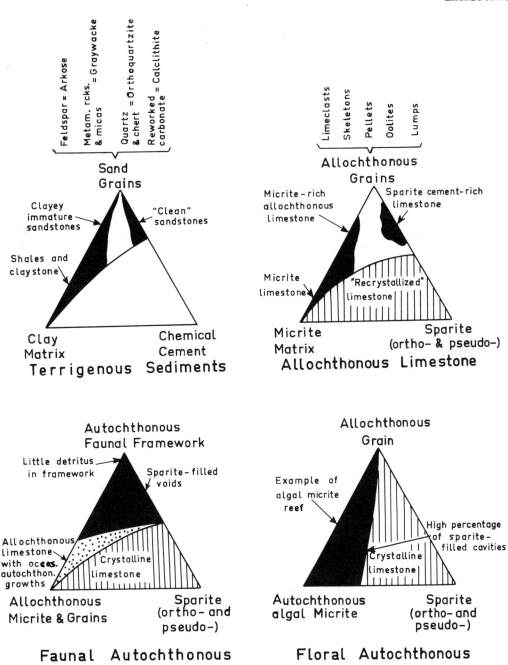

FIGURE 13. Comparison of terrigenous sediments, allochthonous limestones, and faunal and floral autochthonous limestones (after Folk, 1959, and Wolf, unpubl.).

algal *circumcrusts* (Wolf, 1965a), pisolites, and, less commonly, oolites which belong to the oncolite group. It has been suggested that bacteria can produce spherulites or oolites, but the precise processes are not well understood as yet (see *Carbonate Sediments—Bacterial Precipitation*). A fluctuating water table in warm climates may produce oolitic and pisolitic

LIMESTONES

TABLE 1. Classification of Siliciclastic and Carbonate Sediments According to Grain or Crystal Size

Size (mm)	Siliciclastic Sediments	Allochthonous Carbonate Sediments	Authigenic and Recrystallization Sparite	Authigenic or Replacement and Allochthonous Dolomite
2	conglomerate	calcirudite	macrosparite	macrodolosparite
1/16	sandstone	calcarenite	sparite	dolosparite
1/256	siltstone	calcisiltite	microsparite	microdolosparite
	shale or claystone	calcilutite (= micrite)	calcilutite (= micrite)	dolomicrite

residual soils known as *duricrust* (q.v.). Pisolitic tufa and travertine, mentioned above, can originate under similar weathering conditions.

Skeletons are the faunal and floral debris of sessile or autochthonous organisms and the wholly preserved skeletons and/or fragments of unattached forms (see Majewske, 1969; Horowitz and Potter, 1971; Johnson, 1971). Paleontologic and paleoecologic studies are concerned with the investigation of the taxonomy, evolution, and environmental controls on the occurrence of the organisms. *Biostratinomy* (q.v.) is the study of skeletons as sedimentary particles. The carbonate-secreting plants and animals are one of the most important contributors to limestones, and their occurrence ranges from the Precambrian to the Holocene (Fig. 14). As a result of evolution the many genera and species in each phylum can be used to date and correlate limestones.

Pellets are spherical to less-regularly shaped particles composed of dense to grumous, very

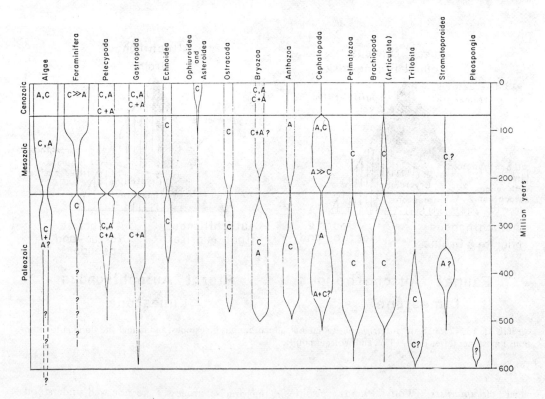

FIGURE 14. Time-stratigraphic distribution of important carbonate-secreting organisms. Width of bars indicates relative importance as contributors to sediment (from Lowenstam, 1963; copyright 1963 by the University of Chicago). C = calcite; A = aragonite.

finely crystalline calcite or aragonite (= micrite). These grains can originate in three ways and are named accordingly bahamite, fecal pellets, and algal pellets. Accretionary processes, possibly due to differences in electric charge, form small *bahamite pellets* from aragonite needles on the Bahama Banks (q.v.), for example. These pellets become increasingly more coherent by cementation, probably due to minute organic specks causing chemical changes within the pellets' microenvironment (Purdy, 1963) and by crystallographic welding (Carozzi, 1960). *Fecal pellets* are rod-like, ovoid, or spherical particles, composed primarily of finely crystalline calcite or micrite; they are the product of digestive and excretionary processes of mud-eating organisms such as gastropods (Cloud, 1962; Purdy, 1963). *Algal pellets* are formed by the disintegration and abrasion of micritic algal material such as algal bioherms, stem and leaf encrustations, and by algal corrosive destruction of calcareous fragments (Wolf, 1965a).

Lumps are formed by the agglutination or accretion of several pellets or oolites, or fragments thereof, occasionally with a high percentage of other carbonate-grain types. Accretion may take place in several ways: (1) by physicochemical agglutination that resembles the formation of bahamite pellets; (2) by physicochemical precipitation of concentric layers around oolites, etc., in contact while at rest on the substratum (Usdowski, 1962); (3) by algal precipitation of circumcrusts (Wolf, 1965a) and concentric laminae; and (4) by weathering processes.

Micrite (q.v.; often synonymous with calcilutite, lime mud, etc.) is another very important constituent of limestones comparable in some degree to the clay matrix in terrigenous sediments (Fig. 13). Micrite may form organically in situ and would be, therefore, an autochthonous sediment (= automicrite); on the other hand it may be detrital, i.e, allochthonous, in origin (= allomicrite). Four general theories of origin have been proposed for this finely crystalline calcite and aragonite: bacterial, physicochemical, algal, and "derived" (Purdy, 1963). The latter type is produced by attrition of older limestones or of intraformational calcareous accumulations. Bacterial activity has a great influence on carbonate precipitation (see *Carbonate Sediments—Bacterial Precipitation*). It has been possible, however, to demonstrate that purely physicochemical processes such as evaporation, temperature increase, and change of partial pressure can result in carbonate deposition (Bathurst, 1975). *Calcilutite* (= micrite) deposits of algal origin are of three distinct types: (1) algal-bound debris (see *Stromatolites*); (2) biogenic algal calcite formed in situ; and (3) detrital calcite or aragonite needles formed by the disintegration of algal colonies. The third type is a result of postmortem disintegration (Carozzi, 1960) of calcareous algal molds and colonies into aragonite needles which are visually identical to physicochemical or bacterially precipitated aragonite (Bathurst, 1975).

The percentage of micrite versus coarse-grained material in limestones is important in environmental studies because the two differ in origin and hydrological behavior during transportation and accumulation. It is also important in this respect to distinguish between detrital and autochthonous micrite. Fig. 12 shows that some limestones are composed of grains only, others have a fair percentage of micrite matrix, and still others are composed of micrite only.

Cement. In addition to autochthonous and allochthonous components, the third important limestone constituent is authigenic aragonite and calcite precipitated *after* accumulation of a rock framework (see *Carbonate Sediments—Lithification*). This type of carbonate is usually medium to coarsely crystalline, sparry, and colorless or slightly tinted in thin section. It is named *open-space sparite* (= orthosparite) because precipitation occurs in voids within the sediment framework. This orthosparite is distinctly different from the pseudosparite resulting from recrystallization of any calcium carbonate material. The orthosparite occurs as fibrous, drusy, and granular morphologic types, and the pseudosparite mainly as granular calcite. The latter is distinct in numerous ways from that formed in open spaces (see Bathurst, 1975; see also *Carbonate Sediments—Diagenesis*). In thin sections, therefore, it is relatively easy to discriminate between ortho- and pseudosparites. Orthosparites may reflect different environmental conditions: fibrous, and possible drusy, sparite forms penecontemporaneously in shallow water as well as subaerially, whereas granular, open-space calcite may be confined to late-diagenetic and epigenetic processes (Wolf, 1965b); exceptions do exist, however. Sparite deposition may cement a loose carbonate accumulation into a consolidated rock, or it may merely fill voids in an autochthonous limestone without actually increasing its coherence.

Classification of Sedimentary Carbonates

For the systematic examination of carbonate rocks a classification scheme that includes most of the components described above is required. It is, however, impossible to propose a scheme that would satisfy all geologists, and the reader is referred to a symposium edited by Ham (1962) for various approaches. The simplest manner of classifying limestone is based on grain size, analogous with classification of clastic sediments (q.v.; Table 1). Folk (1959, 1962; Fig. 15) made the most significant proposal, and a simplified carbonate classification based on his concepts with some modifications is given here (Table 2). It can be used both descriptively and genetically. Dunham's (1962) classification (Table 3) has also been added, as it is being widely used as an alternative scheme.

As Figs. 12 and 13 illustrate, there is a complete transition from an allochthonous and autochthonous framework, on the one hand, to rocks merely composed of micrite, on the other. In most cases, a distinct inverse relationship exists between the amounts of sparite and micrite: the more micrite matrix, the less void space is present for subsequent sparite cement deposition. In detrital accumulations, the void space between grains is approximately 30–40% (Pettijohn, 1975); the greatest amount of sparite possible would therefore be <50%. Preferential to complete recrystallization of the calcareous components, however, may result in any proportion of sparite. It is convenient, therefore, to establish a table with two parameters: (1) framework components and (2) micrite plus sparite (irrespective of genesis of the latter, which includes both ortho- and pseudosparites), as presented in Table 2. Crystalline limestones (Table 3), with or without "ghosts" of primary material, are the product of recrystallization (Figs. 12 and 13). As mentioned previously, if the majority of the organic limestone constituents can be shown to have been preserved where originally produced, the rock is a biolithite. In doubtful cases, the noncommittal term *limestone* is preferable. In using this pigeonhole classification table, all components that occur in quantities >10% are considered, naming them in order of decreasing significance. For example, a limestone composed of 15% fossils, 70% pellets, and 15% cement is a "pellet-skeleton-sparite-calcarenite." Dunham's approach to the classification of limestones has been widely accepted also. Those limestones with recognizable textures (i.e., unrecrystallized, unmetamorphosed) are subdivided into "boundstones," which formed in situ (Folk's biolithite, for example), and limestones of detrital origin. The second type is classified by those that contain lime mud and those that are made up of granular, limy particles (e.g., oolites, skeletons). Depending on the percentages of these constituents present, there are then four limestone types, namely mud-, wacke-, pack-, and grainstones. Occasionally it is possible and useful to combine the nomenclature of Folk and Dunham.

Two main types of dolomites formed by the replacement of limestones are recognizable (see *Dolomite*): (1) with a partial or complete preservation of the textures; (2) with destruction of all previous limestone features. A third type is primary dolomite, which is usually finely crystalline (= *dolomicrite*). The possible modifications of calcareous rocks by both dolomite and terrigenous constituents (Figs. 2 and 3) have been considered in Table 2.

Petrographic Classification. During less precise field and laboratory studies the nomenclature in Table 2 is descriptive only. Detailed petrographic thin-section investigations will

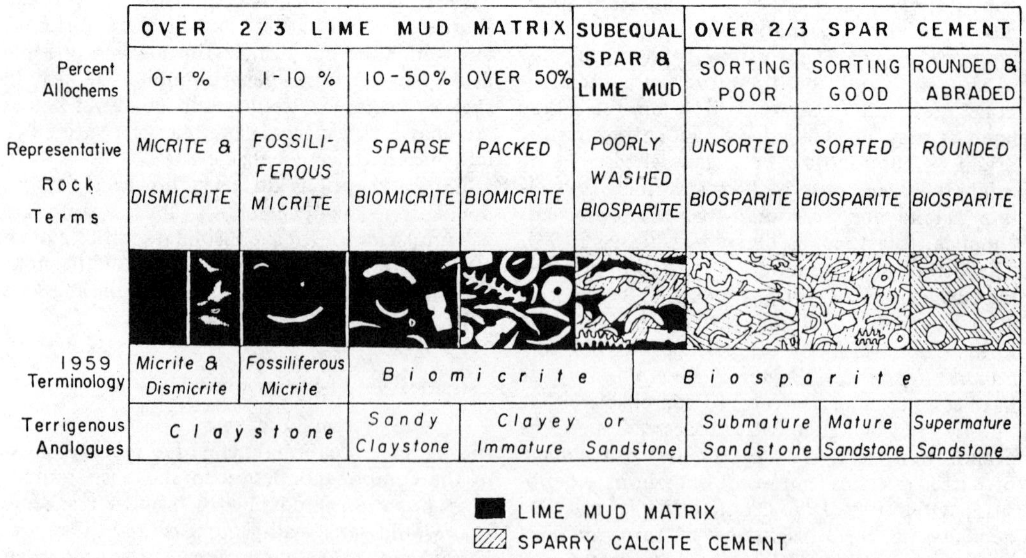

Percent Allochems	OVER 2/3 LIME MUD MATRIX				SUBEQUAL SPAR & LIME MUD	OVER 2/3 SPAR CEMENT		
	0–1%	1–10%	10–50%	OVER 50%		SORTING POOR	SORTING GOOD	ROUNDED & ABRADED
Representative Rock Terms	MICRITE & DISMICRITE	FOSSILIFEROUS MICRITE	SPARSE BIOMICRITE	PACKED BIOMICRITE	POORLY WASHED BIOSPARITE	UNSORTED BIOSPARITE	SORTED BIOSPARITE	ROUNDED BIOSPARITE
1959 Terminology	Micrite & Dismicrite	Fossiliferous Micrite	Biomicrite			Biosparite		
Terrigenous Analogues	Claystone	Sandy Claystone	Clayey or Immature Sandstone			Submature Sandstone	Mature Sandstone	Supermature Sandstone

■ LIME MUD MATRIX
▨ SPARRY CALCITE CEMENT

FIGURE 15. Folk's classification of carbonate rocks and carbonate textural spectrum (from Folk, 1962).

TABLE 2. Carbonate Rock Classification Scheme Based on Framework, Micrite Matrix, Sparite Cement and Diagenetic Sparite (Modified after Folk, 1959)

Micrite and/or sparite %	LIMESTONES						DOLOMITIZED LIMESTONES and DOLOMITES				Impurities[2] present (>10 <50%)
							PRIMARY COMPONENTS PRESERVED[1]			DESTROYED	
	Lime-clasts[3]	Skeletons	Pellets	Oolites Pisolites	Lumps	Autoch-thonous organic growths	Partially replaced by dolomite (>10 <50%)	Extensively replaced by dolomite (>50 <90%)	Completely replaced by dolomite (>90%)	Completely replaced by dolomite (>90%)	
10	Limeclast- limest.	Skeleton- limest.	Pellet- limest.	Oolite- limest.	Lump- limest.	Coral- biolithite	dolomitic Limeclast- limestone, etc.	calcareous Limeclast- dolomite, etc.	Limeclast- dolomite, etc.		pebbly, sandy, silty or clayey Limeclast- limestone, etc.
	Limeclast- micrite- limest. or Limeclast- sparite- limest.	Skeleton- micrite- limest. or Skeleton- sparite- limest.	Pellet- micrite- limest. or Pellet- sparite- limest.	Oolite- micrite- limest. or Oolite- sparite- limest.	Lump- micrite- limest. or Lump- sparite- limest.	Coral- micrite- biolithite or Coral- sparite- biolithite	dolomitic Limeclast- micrite- limestone or dolomitic Limeclast- sparite- limestone	calcareous Limeclast- micrite- dolomite or calcareous Limeclast- sparite- dolomite	Limeclast- micrite- dolomite or Limeclast- sparite- dolomite	Dolosparite	pebbly, etc. Limeclast- micrite- limestone or pebbly, etc. Limeclast- sparite- limestone
50	Micrite- limeclast- limest. or Sparite- limeclast- limest.	Micrite- skeleton- limest. or Sparite- skeleton- limest.	Micrite- pellet- limest. or Sparite- pellet- limest.	Micrite- oolite- limest. or Sparite- oolite- limest.	Micrite- lump- limest. or Sparite- lump- limest.	Micrite- coral- biolithite or Sparite- coral- biolithite	dolomitic Micrite- limeclast- limestone or dolomitic Sparite- limeclast- limestone	calcareous Micrite- limeclast- dolomite or calcareous Sparite- limeclast- dolomite	Micrite- limeclast- dolomite or Sparite- limeclast- dolomite	Dolo- micro- sparite	pebbly, etc. Micrite- limeclast- limestone or pebbly, etc. Sparite- limeclast- limestone
90	Micrite-limestone or Crystalline- or Sparite- limestone					Micrite- biolithite	dolomitic Micrite- or dolomitic Sparite-limest.	calcareous Dolomicrite or calcareous Dolosparite	Dolomicrite		pebbly, etc. Micrite- limestone

1. These columns are only examples. The nomenclature is applicable to all other types of carbonate rocks.
2. The composition of impurities should be stated, e.g., quartzose. 3. Substitute size nomenclature of table 1.

TABLE 3. Dunham's Classification According to Depositional Texture

Depositional Texture Recognizable				Depositional Texture Not Recognizable
Original components not bound together during deposition			Original components were bound together during deposition . . . as shown by intergrown skeletal matter, lamination contrary to gravity, or sediment-floored cavities that are roofed over by organic or questionably organic matter and are too large to be interstices.	Subdivide according to classifications designed to bear on physical texture or diagenesis.
Contains mud (Particles of clay and fine silt size)		Lacks mud		
Mud supported		Grain supported		
Less than 10% grains	More than 10% grains			
Mudstone	Wackestone	Packstone	Grainstone	Boundstone	Crystalline carbonate

After Dunham, 1962.

permit a genetic interpretation of the framework, matrix and sparite cement, and diagenetic products. Four convenient steps may be taken in the study of carbonate rocks: (1) hand-lens investigation; (2) binocular microscope investigation; (3) petrographic microscope investigations; and finally (4) total petrographic summary including depositional structures, stratigraphy, paleontology, etc.

The nomenclature and classification schemes presented are complicated and even cumbersome. However, they serve as a crutch for our memory, induce us to consider all likely interpretations, and are vital tools for communication among all interested scientists. The schemes are in wide use already, and the ever-increasing studies of minute physical and chemical variations of the carbonate components necessitate an unambiguous terminology.

KARL H. WOLF

References

Bathurst, R. G. C., 1975 *Carbonate Sediments and*

Their Diagenesis, 2nd Ed. Amsterdam: Elsevier, 658p.

Carozzi, A. V., 1960. *Microscopic Sedimentary Petrography.* New York: Wiley, 486p.

Cloud, P. E., 1962. Environment of calcium carbonate deposition west of Andros Island Bahamas, *U.S. Geol. Surv. Prof. Pap. 350,* 138p.

Dott, R. H., Jr., and Batten, R. L., 1976. *Evolution of the Earth.* New York: McGraw-Hill, 504p.

Dunbar, C. O., and Rodgers, J., 1957. *Principles of Stratigraphy.* New York: Wiley, 356p.

Dunham, R. J., 1962. Classification of carbonate rocks according to depositional texture, *Am. Assoc. Petrol. Geologists Mem. 1,* 108-121.

Folk, R. L., 1959. Practical petrographic classification of limestones, *Bull. Am. Assoc. Petrol. Geologists, 43,* 1-38.

Folk, R. L., 1962. Spectral subdivision of limestone types, *Am. Assoc. Petrol. Geologists Mem. 1,* 62-84.

Friedman, G. M., 1969. Depositional environments in carbonate rocks–A symposium, *Soc. Econ. Paleont. Mineral. Spec. Publ. 14,* 209p.

Ginsburg, R. N.; Rezak, R.; and Wray, J. L., 1971. Geology of calcareous algae, Notes for a short course, *Sedimenta I.* Univ. of Miami Comparative Sedimentology Laboratory, 62p.

Graf, D. L., 1960. Geochemistry of carbonate sediments and sedimentary carbonate rocks (5 parts) *Ill. Geol. Surv. Circ., 297, 298, 301, 308, 309,* 250p (total).

Hadding, A., 1958. Origin of lithographic limestones, *Kungl. Fysiogr. Saellskapets Lund Forhandl., 28,* 21-32.

Ham, W. E., ed., 1962. Classification of carbonate rocks: A symposium, *Am. Assoc. Petrol. Geologists Mem. 1,* 279p.

Heckel, P. H., 1973. Nature, origin, and significance of the Tully Limestone, *Geol. Soc. Am. Spec. Pap. 138,* 244p.

Horowitz, A. S., and Potter, P. E., 1971. *Introductory Petrography of Fossils.* New York: Springer-Verlag, 300p, 100 plates.

Johnson, J. H., 1961. *Limestone-building Algae and Algal Limestones.* Golden: Colorado School of Mines, 267p.

Johnson, J. H., 1967. Bibliography of fossil algae, algal limestones, on the geologic work of algae, *Colo. Sch. Mines Quart., 62,* 148p.

Johnson, J. H., 1971. An introduction to the study of organic limestones, revised ed., *Colo. Sch. Mines Quart., 66*(2), 185p.

LeBlanc, R. J., and Breeding, J. G., 1957. Regional aspects of carbonate deposition–symposium, *Soc. Econ. Paleont. Mineral. Spec. Publ. 5,* 178p.

Lowenstam, H. A., 1963. Biologic problems relating to the composition and diagenesis of sediments, in T. W. Donnelly, ed., *The Earth Sciences—Problems and Progress in Current Research.* Chicago: Univ. Chicago Press, 137-195.

Majewske, O. P., 1969. *Recognition of Invertebrate Fossil Fragments in Rocks and Thin Sections.* Leiden: E. J. Brill, 101p, 106 plates.

Milliman, J. D., 1974. *Marine Carbonates.* New York: Springer-Verlag, 375p.

Müller, G.; Irion, G.; and Förstner, U., 1972. Formation and diagenesis of inorganic Ca-Mg carbonates in the lacustrine environment, *Naturwissenschaften, 59,* 158-164.

Pettijohn, F. J., 1957, *Sedimentary Rocks,* 2nd ed. New York: Harper & Row, 718p.

Pettijohn, F. J., 1975. *Sedimentary Rocks,* 3rd ed. New York: Harper & Row, 628p.

Potter, P. E., 1968. A selective, annotated bibliography on carbonate rocks, *Bull. Canadian Petrol. Geol., 16,* 87-103.

Potter, P. E., and Purser, B. H., 1975. Bibliography of French language studies of carbonate rocks 1964-1974, *Bull. Canadian Petrol. Geol., 23,* 187-200.

Purdy, E. G., 1963. Recent calcium carbonate facies of the Great Bahama Bank, pts. 1 and 2, *J. Geol., 71,* 334-355, 472-497.

Shaw, A. B., 1964. *Time in Stratigraphy.* New York: McGraw-Hill, 365p.

Swineford, A.; Leonard, A. B.; and Frye, J. C., 1958. Petrology of the Pliocene pisolitic limestone in the Great Plains, *State Geol. Surv. Kansas Bull. 130,* 97-116.

Teichert, C., 1958. Cold- and deep-water coral banks, *Bull. Am. Assoc. Petrol. Geologists, 42,* 1064-1082.

Usdowski, H.-E., 1962. Die Enstehung der kalkoölithischen Fazies des norddeutschen Unteren Buntsandsteins, *Beitr. Mineral. Petrog., 8,* 141-179.

Visser, J. W. J., and Grobler, N. J., 1972. The transition beds at the base of the Dolomite Series in the Northern Cape Province, *Trans. Geol. Soc. S. Africa, 75,* 265-274.

Wilson, J. L., 1975. *Carbonate Facies in Geologic History.* New York: Springer-Verlag, 471p.

Wolf, K. H., 1965a. Petrogenesis and paleoenvironment of the Devonian algal limestones of New South Wales, *Sedimentology, 4,* 113-178.

Wolf, K. H., 1965b. Littoral environment indicated by open-space structures in algal reefs, *Palaeogeogr., Palaeoclimat., Palaeoecol., 1,* 183-223.

Wood, G. F., and Wolfe, M. J., 1969. Sabkha cycles in the Arab/Darb Formation off the Trucial coast of Arabia, *Sedimentology, 12,* 165-192.

Cross-references: *Algal-Reef Sedimentology; Allochem; Bahama Banks Sedimentology; Beachrock; Birdseye Limestone; Calcarenite, etc.; Caliche, Calcrete; Carbonate Sediments—Bacterial Precipitation; Carbonate Sediments—Diagenesis; Carbonate Sediments—Lithification; Cave Pearls; Cementation; Cementation, Submarine; Chalk; Coral Reef Sedimentology; Dedolomitzation; Diagenesis; Dolomite, Dolomitization; Evaporite Facies; Grapestone; Gypsum in Sediments; Intraclastic Rocks; Limestone Fabrics; Marine Sediments; Marl; Micrite; Oolite; Persian Gulf Sedimentology; Pisolite; Reef Complex; Sabkha Sedimentology; Solution Breccias; Sparite; Speleal Sediments; Sponges in Sediments; Spring Deposits; Staining Techniques; Stromatolites; Travertine; Tropical Lagoonal Sedimentation.* Vol. I: *Florida Bay: a Modern Site of Limestone Formation.* Vol. IVA: *Calcium Carbonate: Geochemistry.* Vol. IVB: *Calcite.* Vol. VII: *Biomineralization.* Vol. XIII: *Limestone and Dolomite.*

LIMNIC SEDIMENT—See
LACUSTRINE SEDIMENTATION

LIQUEFACTION—*See* **FLUIDIZED AND LIQUEFIED SEDIMENT FLOWS**

LITHARENITE—*See* **SUBGRAYWACKE**

LITHIFICATION—*See* **DIAGENESIS**

LITHOLOGY, LITHOFACIES, LITHOTOPE, LITHIFICATION, LITHOGENESIS

Terms based on the Greek root *lithos* (= stone) may refer to rocks in general, e.g., the *lithosphere* is the solid outer layer of the earth; most of these terms, however, are applied especially to sedimentary rocks and processes. *Lithology*, the elementary descriptive science of rocks, is most commonly applied to sedimentary clastics, but may also apply to igneous and other sedimentary rocks. The terms *petrology* and *petrography* are customarily applied to a rock's more sophisticated description or genesis. To a field geologist, the term *lithology* usually refers to the gross description of a hand specimen or outcrop, not including geometry (structure), paleontology (fossils), or other attributes.

Lithofacies describes the lithic (rock) aspect of a given stratigraphic unit which is thus recognizably mappable because of common lithic attributes, e.g., a sandstone that is quartzose, silica-cemented, cross-bedded, yellow in fresh outcrop, etc., may be said to possess distinctive characteristics, or if mapped as a continuous unit, may be properly identified by a stratigraphic formation name (see *Facies*). A distinctive lithofacies need not be restricted to a single lithology; for example, a rhythmic alternation of sandstone and shale may still maintain a constant appearance and character for a great thickness. If that original sandy lithofacies passes up, down, or laterally into a distinctively different lithofacies, however, it is proper to separate them by a formation (and facies) boundary. That boundary is frequently *diachronous*, i.e., it is a plane which runs obliquely to the time planes. A lithofacies differs from a biofacies inasmuch as the latter describes the organic (fossil) content of the rock, or part of it, and may or may not in any way coincide with the geometry of the lithofacies.

Lithotope (Gr. *topos:* place) defines an area or surface of uniform sediment or sedimentation, an area of uniform sedimentary environment (q.v.), or a place distinguished by relative uniformity of the principal environmental conditions of rock deposition, much as a *biotope* describes the ecologic environment of a given biofacies (Wells, 1947; Moore, 1948). As pointed out by Dunbar and Rodgers (1957), these terms have subsequently been used with varied connotations which could be confusing. The totality of descriptions of lithotopes and biotopes for a given timespan is expressed respectively by the paleogeography and paleoecology of that time.

Lithification (also lithifaction) is that process which converts a newly deposited sediment from its loose, unconsolidated state into a coherent or solid rock; "lapidification" is an archaic synonym. Lithification may take place at the same time as deposition, as with rocks of chemical origin such as travertine (q.v.), or much later. Sands are lithified into sandstones; clays are lithified into claystones, etc. A complex of processes are involved, including dehydration, compaction (q.v.), cementation (q.v.), and induration. All of these processes fall in the realm of diagenesis (q.v.); when the burial environment reaches sufficient temperatures and pressures, metamorphism takes place.

Lithogenesis embraces the origin and formation of rocks, especially of sedimentary rocks; the term *petrogenesis* may be considered to be synonymous but usually refers to the origin of igneous rocks. The term *lithogenesis* is commonly used in Germany and the Soviet Union (Strakhov, 1967, 1969, 1970); Vassoevich (1957) includes within it the processes of diagenesis, epigenesis, and metamorphism. Crook (1974) includes in addition *provenance* (q.v.).

RHODES W. FAIRBRIDGE

References

Crook, K. A. W., 1974. Lithogenesis and geotectonics: the significance of compositional variation in flysch arenites (graywackes), *Soc. Econ. Paleont. Mineral. Spec. Publ. 19,* 304-310.

Dunbar, C. O., and Rodgers, J., 1957. *Principles of Stratigraphy.* New York: Wiley, 356p.

Moore, R. C., 1948. Stratigraphical paleontology, *Geol. Soc. Am. Bull.,* 59, 301-325.

Strakhov, N. M., 1967, 1969, 1970. *Principles of Lithogenesis,* 3 Vols. (translated from the 1962 Russian edition). New York: Consultants Bureau; Edinburgh: Oliver and Boyd, 245p, 609p, and 577p.

Vassoevich, N. B., 1957. Terminology used for designating stages and steps of lithogenesis (in Russian), in *Geology and Geochemistry.* Leningrad: Gostoptekhizdat, 1-7.

Wells, J. W., 1947. Provisional paleoecological analysis of the Devonian rocks of the Columbus region, *Ohio J. Sci.,* 47, 119-126.

Cross-references: *Carbonate Sediments—Lithification. Cementation; Clastic Sediments—Lithification and Diagenesis; Compaction in Sediments; Diagenesis; Facies; Provenance; Sedimentary Environments.*

LITTORAL PROCESSES—See Vols. III, XV

LITTORAL SEDIMENTATION

Nowhere in the marine environment do the processes of erosion, deposition, and transportation of sediment operate more effectively than in the littoral environment. The littoral zone extends from the bottom of the surf base (Dietz, 1963; 1964)—the depth, approximately 10 m, at which wave-induced movement of sediment under normal conditions stops—to the highest level of the swash during storms.

Movement of Sediment

The direction of sediment movement is based on the neutral-line (null-point) concept proposed by Cornaglia (1891). The theory states that there are two opposing forces acting on sediment: the onshore wave drift and the offshore gravitational force due to the bottom slope. For each grain of given density, size, and shape there exists a line of equilibrium between these forces of on- and offshore motion. The larger or denser the particle, the closer to shore is its null point (Fig. 1). Shoreward of this point the grain will move onshore; seaward of it the particle will move offshore. Ippen and Eagleson (1955) give the relations determining the position as $(H/h^2)(L/H)(C/w) = 11.6$, where H = wave height, h = water depth, L = wave length, C = wave celerity, and w = the terminal fall velocity of the sediment particle in fresh water.

The quantity of sediment transported varies with the different wave parameters and the slope angle. The steepness of the foreshore and offshore slope produced by waves of constant proportions increases with the size of sand but decreases as wave size increases. A change in the wave period will influence the amount of sand transported but will leave the slope virtually unchanged. The wave steepness (H/L) is perhaps the most important variable in determining the submarine slope. Most authors fol-

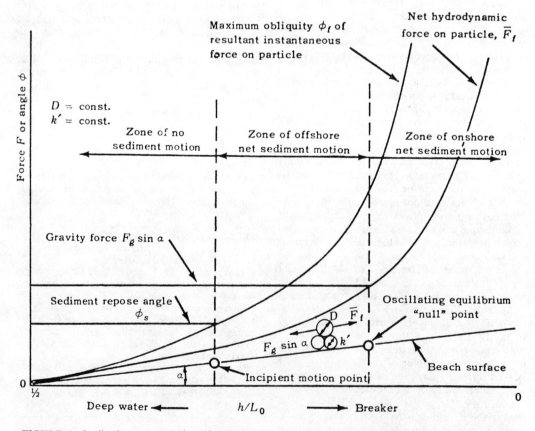

FIGURE 1. Qualitative representation of onshore-offshore bed-load mechanics (from Johnson and Eagleson, 1966, in Ippen, *Estuary and Coastline Hydrodynamics*; © 1966 by McGraw-Hill Book Company, used with permission).

low Saville (1950), who divides beach profiles into two categories: storm profiles, which are produced by waves with a steepness >0.03, and summer profiles formed by waves <0.025 steep, with a transition in between. When the waves grow steeper, during storms and generally in the winter months, sand is moved from the beach to form an offshore bar producing a storm profile. When the waves decay after the passage of the storm or under the long swell prevalent during the summer, the bar gradually moves onshore, and the sand is spread out over the beach (Sonu and van Beek, 1971). Of significance is the time difference between erosion and accretion of the beach (Fig. 2). Erosion may take place in a matter of hours, whereas accretion of the beach to its former shape and volume may take several days to several months. A profile of equilibrium (Tanner, 1958) will form if no changes in sea state occur. This *equilibrium profile* is characterized by a reduced transport rate and a smaller exchange of sediment between the various zones in the littoral zone.

Some beaches also show a seasonal trend in their sediment distribution, with the finest sediment found during the late summer and the coarsest in late winter. At any one time, finer sands occur on the upper foreshore, with coarser sediment on the inshore and the berm crest (Fig. 2).

Tidal Changes. For any section of the beach face, three phases of changes can be recognized within each tidal cycle (Fig. 3). As the swash first acts upon a section of the beachface, there is an initial phase of deposition. This phase is followed by a period of scour as the tide rises and that section of beach is under the transition zone. Finally, near high tide, there is deposition again as the area in question is under the surf regime. The ebb tide reverses the process. Deposition continues in the surf zone, erosion in the transition, and deposition in the beginning of the swash. Just before the swash zone leaves an area there is a small amount of erosion. Actual depths of scour and amount deposited depend, in part, on the breaker height. King (1951) found erosion depths to be 3–4% of the breaker height; Williams (1971) reports 20–40% or more. Brenninkmeyer (1973) found a maximum deposition of $\approx 30\%$ for a 1 m breaker.

Refraction

The submarine relief is probably the most important factor, in conjunction with the direction and length of dominant waves, in determining the plan of the coastline because it is these factors that determine the wave refraction patterns (q.v., Vol. I). As a wave begins to "feel" bottom, all its characteristics except its period begin to be modified. In overcoming friction with the bottom, the wavelength is reduced. A submarine ridge slows the waves, and the wave front is bent toward the ridge (Fig. 4); if a valley is present the waves are bent away from it. This bending of the wave front redistributes the wave energy along the shore, concentrating it on headlands and decreasing it in bays. Refraction is also responsible for aligning the breaker zone so that it tends to be parallel to shore no matter from which direction the deep-water waves arrive.

Dynamic Zones

There are five dynamic zones present in the littoral environment: the offshore, breaker, surf, transition, and swash zones (Fig. 5). In each of these belts, wave conditions are unique so that sediment transport differs, and diverse sedimentary structures are encountered. The width of the surf and transition zones is primarily a function of beach slope and tidal phase. Beaches with gentle foreshore and inshore slopes usually have wide surf and transition zones during the whole tidal cycle; whereas on steeper beaches these zones may be absent at high tide, and on very steep beaches they may never be present.

The Offshore Zone. In the offshore zone, the volume of sediment transported becomes greater as the water becomes shallower, reaching a maximum just seaward of the break point. The flatter the slope of the offshore, the larger the amount of sand that can be transported under given wave conditions. In this zone, sand movement is predominantly shoreward and is intensified with greater wave height and period. When the wave steepness increases, however, the shoreward movement may decrease and even reverse (Rector, 1954). In this zone, landward-facing, flat-crested, lunate megaripples are common in flat wave conditions. These ripples migrate slowly landward and, therefore, exhibit cross-stratification with foresets dipping steeply landward (Clifton et al., 1971). Farther offshore, asymmetric ripples predominate; their asymmetry is either landward or seaward, dependent on the predominant drift that crosses them. Seaward bottom drift is associated with onshore winds and short-period waves, while longer-period swell causes shoreward pulses to dominate. Under the latter conditions, internal structure usually consists of shoreward inclined ripple cross-lamination. Cook and Gorsline (1972) found that the ripple size is directly related to sand size and inversely to current strength. The sediment is segregated,

LITTORAL SEDIMENTATION

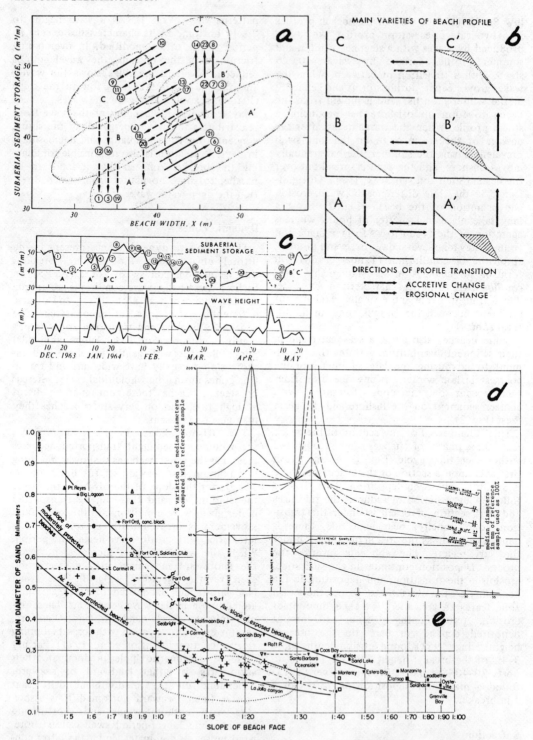

FIGURE 2. (a) Time series of subaerial sediment storage and wave height, Nags Head, North Carolina. (b) Six major types of beach profiles. (c) Profile data as a function of sediment storage, beach width, and configuration, Nags Head, North Carolina (Sonu, 1969). (d) Size distribution across a beach (Bascom, 1951 ©copyright American Geophysical Union). (e) Size slope relationship of beaches at the reference point midway between high and low tide (from Wiegel, ©1964; reprinted by permission Prentice-Hall, Inc.).

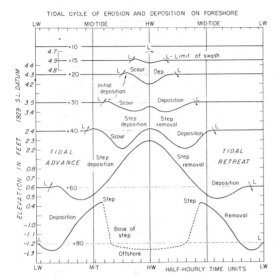

FIGURE 3. Synthesis of elevation changes in a tidal cycle (from Strahler, 1966; © copyright 1966 by the University of Chicago).

with extremes in hydraulic sizes concentrated at ripple crests and troughs.

Breaker Zone. Sand movement in the littoral environment is greatest within the breaker zone, with scour of 5-10 cm per wave not uncommon in high-energy conditions. The size of the sediment can be very coarse, especially on the bar; finer sizes occur in the trough. The breaker zone seems to be a boundary difficult for the coarser grains to cross. Sediment transport is largely bedload, with frequencies coherent with both swell and sea periods. During the breaking of each wave, the sediment moves in paths which are horizontal to flat elliptical, onshore-offshore, with small alongshore components. Both seaward and shoreward, sediment motion is toward the breaker zone (Ingle, 1966; Miller and Zeigler, 1958). Acceleration of shoreward water velocities takes place immediately offshore, and seaward acceleration occurs in the large trough developed in the water shoreward of the breaking wave; this action may produce a bar which accretes from both sides.

The predominantly back-and-forth movement of the bedload can also be shown by the suspended sediment, which may equal 5% of the total sediment moved. In the breaker zone, there can be a sharp decrease of suspended sediment with elevation above the bottom (Homma et al., 1960). The higher the waves, the less the suspended concentration changes with elevation (Fairchild, 1973). The predominant direction of suspended sediment transport is dependent on the beach slope; Nagata (1964) reports greater concentrations in the plunge point in the seaward direction if the beach is steep, whereas the opposite is true if the beach is gentle. This diffusion of sediment through the breaker zone is described by Inman et al. (1971).

Bars are very common off gently sloping beaches. Shepard (1950) found that if the foreshore slope is $>4°$, bars may be insignificant or absent. Bars form slightly seaward of the plunge point. As wave size increases, the bar is cut away, and a new bar develops seaward of the old one. The crest of the bar never grows above water level; the height of the bar above the

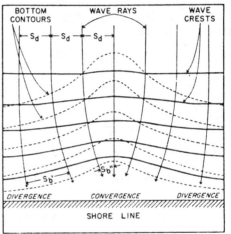
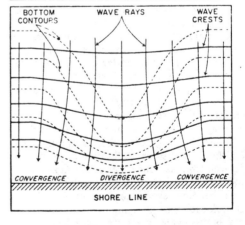

FIGURE 4. Wave-refraction diagrams showing wave convergence and divergence produced by a submarine ridge (A) and a submarine valley head (B) (from Shepard, 1963). The relative change in wave height is given by the square root (airy-wave theory) or cube root (solitary-wave theory) of the ratio of the distance between deep-water and shallow-water wave rays (s_d/s_b). Note that the ratio $s_d/s_b{'}$ (divergence) is less than one, while $s_d/s_b{''}$ (convergence) is greater than one.

451

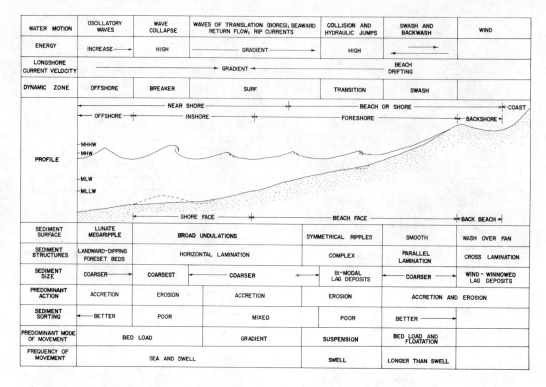

FIGURE 5. The dynamic zones of the littoral environment and the water and sediment motion and sedimentary characteristics associated with each zone (modified after Brenninkmeyer, 1973; Clifton et al., 1971; US Army Corps of Engineers CERC, 1966; Emery, 1945; Ingle, 1966; Komar, 1971).

original profile is ½ the depth of water over the bar (King, 1972). A typical bar will slope gently toward the sea, have a rounded crest, and slope steeply downward into the trough. Bars may also occur transverse to the shore, but they are generally transient in nature and location.

If no bar is formed, as under spilling breakers, broad undulations may be present, or small flat-bottomed ripples may form. These ripples are destroyed during the passage of each wave and are reformed afterwards. In this case, the sediment may be finer and more densely packed than in neighboring zones. The structures left under these conditions are horizontal or nearly horizontal laminae (Clifton et al., 1971).

Surf Zone. Landward of the breakpoint, conditions become even more complex. Under ideal conditions, in plunging breakers, the wave energy is dissipated in <1 wavelength; for spilling breakers it is in several wavelengths. In this process up to 90% of the energy may be lost (Chandler and Sorensen, 1973). Therefore, after the breaking of the wave, sediment movement in the outer surf zone is small.

Measurements by Longinov (1964) inside the surf zone show that vertical water velocities 2.5 cm above the bottom may reach 50% of the horizontal velocity. The maximum vertical velocities occur either when the horizontal velocity is zero, changing from shoreward to seaward, or just after the shoreward maximum. This action may explain the onset of suspension that has been observed by Palmer (1970) when the horizontal velocity decreases.

In the inner surf zone, suspended sediment transport increases in frequency and elevation due to the breaking of the bore. The bore front is saturated with air bubbles and foam which, because the water is shallow, can penetrate the bottom. The ascent of these bubbles has a suctional effect sufficient to draw up sand (Aibulatov, 1966).

Observations in the field and in the laboratory as well as theoretical considerations (Longuet Higgens, 1972) all show that *alongshore currents* have a symmetrical distribution with a maximum in the surf zone, decreasing rapidly toward the breaker zone, and more slowly landward. If the waves have a directional component parallel to shore, and because the fluctuating water-particle motion in waves causes an excess momentum flux, an alongshore current may be generated. The waves provide the stress that moves the sediment back and forth and into suspension; the alongshore current imparts a net transport direction. Inman and Bagnold (1963) estimate that ten times as much sediment is moved back and forth perpendicular to the beach than is moved alongshore.

The amount of sediment transported alongshore can be obtained by at least twelve different formulas based on energy or momentum balance or continuity considerations (Sonu et al., 1967). One commonly used formula is based on the US Army Corps of Engineers CERC (1966) design criteria, from which Mogel et al. (1971) derived an equation relating the amount of sand transported along shore to the angle between the line perpendicular to the wave crest and bottom topography; the breaker height; the wave period; the shoaling coefficient at breaking point; and the fraction of a month that waves of a given height, period, and direction occur in deep water. The quantity of sand transported by alongshore currents may reach a million m^3/yr (Johnson, 1956). The angle of wave approach most apt to move sand along a beach is 30-50° (Vollbrecht, 1966).

Nearshore circulation patterns are also affected by variations in the breaker height along shore. Incident waves can excite transverse waves (edge waves), surface waves trapped by refraction to the shore and having a maximum amplitude at the shoreline. The wave elevation may, however, have several smaller maxima and minima (Bowen, 1973). The interaction between these edge waves and incident waves, or the motion of the edge waves themselves, may set up a circulation pattern consisting of an onshore flow toward the breakers, an alongshore current, and an offshore flow in a strong, narrow *rip current*. These rip currents are located at the antinodes of the edge waves (Bowen and Inman, 1969). The rip currents may be stationary, especially near groins and headlands, but they can also migrate along the shore. McKenzie (1958) observed that they are best developed on gently sloping beaches exposed to large waves.

Beyond the breaker zone, the currents spread out to form a broad head. In the surf zone, they are narrow and can scour out a channel 1-3 m deeper than the adjacent bottom (Cook, 1970). Rip currents carry sediment through the breaker zone and deposit it on the offshore. Some of this sediment is returned to the beach by wave drift, but some may be permanently lost to the beach. Under normal conditions, rips transport the lighter grains, and the coarser and denser particles are left behind. This process may explain some heavy-mineral concentrations. The channel scoured by the rip current extends through the bar, if present. It is floored by a rippled lag deposit of coarse sediment and shell fragments, with large-scale foreset cross-bedding sloping seaward (Fig. 6).

Standing edge waves may also be responsible for the *crescentic bars* found in the surf zone (Bowen and Inman, 1971), or these bars may be the result of surf zone topographic undulations (Sonu, 1972). Crescentic bars may be so well developed that they completely control the nearshore flow of water, and therefore the sediment movement in the surf zone.

Transition Zone. In the transition zone, where the backwash meets the incoming bore, sediment is moved predominantly by suspension but also by bedload, reflected by the bimodal size distribution of the sediment found here (Schiffman, 1965). Most of the sediment movement in the littoral zone during long-period swell takes place in the transition zone. Here the water motion may consist of a hydraulic jump or roll wave (Grant, 1943) or of the overtopping of the bore by the backwash (Miller and Zeigler, 1964). In the breaking of this "standing" wave there is a brief suspension of much sediment as it scours out the step—the nearly horizontal section of the beach at the landward edge of the transition zone. The deposits left behind may well form the steep-sided, symmetrical ripples on steep beaches and the ridges and runnels common at the level of slack tide on flat beaches. These surface projections may cause more suspension by creating a barrier to the backwash flow, again setting up a hydraulic jump.

Most of the suspension on the foreshore takes place within half a meter of the still-water level. These clouds of suspended sediment, called sand fountains by Zenkovich (1967), are short in duration (2-12 sec) and may lift 500 g/liter of sediment up to the top of the water column. The heavier suspended concentrations can be found, not necessarily at the bottom, but 0.3-1 m above it. Over the ripples, suspension increases dramatically not only under the wave crest but also up to three other times during the

FIGURE 6. Sketch of a rip-scoured channel observed in the offshore zone at Zuma Beach, California, by SCUBA-equipped divers (from Ingle, 1966). The channel extended perpendicular to shore and was filled with sand waves or large ripples averaging 0.5 m in height and ≥1 m in wave length.

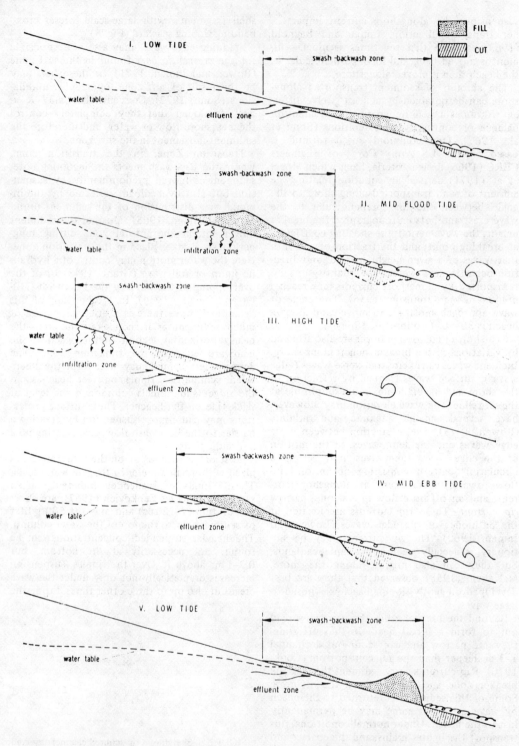

FIGURE 7. Generalized profiles of sediment distribution in the swash-backwash zone at various stages of a semidiurnal cycle from low tide to low tide (from Duncan, 1964). The successive stages shown do not indicate composite sediment distribution over the cycle; the initial distribution of sediment for each stage is indicated by a straight line.

wave period (Homma et al., 1965) coincident with the energy peaks of the bores.

The bedding in the symmetrical ripples is complex and may dip in either direction. Usually the coarser sediment is found in the first ripple outside the step and gets progressively finer seaward; the ripple height also decreases seaward. In the runnels, the seaward side contains the coarsest sand. The runnels migrate slowly seaward and so consist of seaward-dipping trough cross-laminae (Clifton et al., 1971).

Swash Zone. In the swash zone, transport of sediment reverts back to bedload. At low tide, because of the gentle inshore slope and lack of infiltration, the swash period is considerably longer than the wave period. The coarser grains settle out first; the finer grains are deposited as a blanket before the swash reaches its fullest extent. After the backwash gains momentum, it picks up these fine grains and transports them back to the transition zone. This winnowing action is aided by elutriation due to the escape of ground water, which lags 1–3 hr behind the tide. (Fig. 7).

Most sediment is transported by the sheetflow of the backwash. The amount depends on the breaker height, vertical distance between the water table and the still-water level, the beach slope, and angle of wave approach (Harrison, 1969). The direction in which sand is transported depends on the angle of wave approach. The sediment in the swash zone will move obliquely up the beach parallel to the wave crest; the backwash, driven by gravity, transports sand normal to the shoreline. Thus, beach drift has a sawtooth pattern (see Vol. III: *Beach*). Zenkovich (1967) calculated the horizontal displacement as $(h + m)\cos\alpha/1 + \sin\alpha$, where h = perpendicular distance moved by the backwash, m = slant distance moved by the swash, and α = angle of wave approach. An increase in the slope reduces the horizontal displacement; waves of medium height are more effective because they are less refracted.

As the tide advances above the ground-water level onto the steeper part of the beach, the swash zone becomes narrower, and the swash period decreases due to increase in slope and infiltration. Here some of the coarser sand is removed by flotation to the farthest reaches of the swash to form a *swash mark*. The thin wedge of water moving in front of the foam carries little sediment in traction or suspension, but is capable of picking up dry grains, floating them farther up the beach (Emery, 1945). Surface tension and wettability suggest that the size and shape and especially dryness are important in this mode of sediment movement.

Regular, nearly parallel lamination is characteristic of the swash zone (Thompson, 1937). If the sediment is not well sorted, the deposit left by one wave usually is graded upward from fine to coarse. Where different minerals are present, the basal layer commonly is of the heaviest but smallest accessory minerals, which grade upward into larger grains of lower density, and then lighter quartzose sand (Clifton, 1969).

Cusps modify the regular sorting parallel to the beach (Fig. 7). They can form in all types of sediment from fine sand to gravel, but form most readily when there is a vertical stratification of material. In bays and sheltered shores, cusps may be close together (3–20 m); on beaches facing an open sea, cusps are higher and farther apart (>30m). Cusps can appear shortly after the wave energy decreases and may originate by the interference pattern of edge waves. They are first noticeable as patches, a few centimeters thick, of gravel, shells, and other coarse material. The backwash, instead of returning in dispersed form, moves away from the patches and returns in a channel. This process is repeated and intensified until the backwash flow attains such momentum that the next swash cannot proceed against it and is projected onto the apices. There the coarsest sediment is deposited; the finer grains stay entrained as the water swings into the adjacent bays without stopping. Cusp deposits are not uniform in thickness but have a shape similar to a Gilbert delta, thinning in a seaward wedge composed of steeply dipping foreset beds. Gravel and heavy-mineral concentrations are usually exposed at the surface and pinch out below the surface closer to the coast.

BENNO M. BRENNINKMEYER, S.J.

References

Aibulatov, N. A., 1966. *Investigations of Longshore Migration of Sand in the Sea.* Moscow: Nauka Publ. Co., 159p. (In Russian).

Bagnold, R. A., 1963. Beach and nearshore processes: part 1, mechanics of marine sedimentation, in M. N. Hill, ed., *The Sea.* New York: Interscience, vol. 3, 507-528.

Bascom, W. N., 1951. The relationship between sand size and beachface slope, *Am. Geophys. Union Trans.*, 32, 866-874.

Bowen, A. J., 1973. Edge waves and the littoral environment, *Proc. 13th Conf. Coastal Eng., ASCE*, 1313-1320.

Bowen, A. J., and Inman, D. L., 1969. Rip currents, 2. Laboratory and field observations, *J. Geophys. Research*, 74, 5479-5490.

Bowen, A. J., and Inman, D. L., 1971. Edge waves and crescentic bars, *J. Geophys. Research*, 76, 8662-8671.

Brenninkmeyer, B. M., 1973. Synoptic surf zone sedimentation patterns, Univ. Southern California dissertation, 274p.

Chandler, P. L., and Sorensen, R., 1973. Transformation of waves passing a submerged bar, *Proc. 13th Conf. Coastal Eng., ASCE*, 385-404.

Clifton, H. E., 1969. Beach lamination–nature and origin. *Marine Geol.*, 7, 553-559.

Clifton, H. E.; Hunter, R. E.; and Phillips, R. L., 1971. Depositional structures and processes in the non-barred high-energy nearshore, *J. Sed. Petrology*, 41, 651-670.

Cook, D. O., 1970. The occurrence and geologic work of rip currents off southern California, *Marine Geol.*, 9, 173-186.

Cook, D. O., and Gorsline, D. S., 1972. Field observations of sand transport by shoaling waves, *Marine Geol.*, 13, 31-55.

Cornaglia, P., 1891. *Sul regime delle spiagge e sulla regolazione dei porti.* Turin.

Davidson-Arnott, R. G. D., and Greenwood, B., 1974. Bedforms and structures associated with bar topography in the shallow-water wave environment, Kouchibouguac Bay, New Brunswick, Canada, *J. Sed. Petrology*, 44, 698-704.

Dietz, R. S., 1963. Wave base, marine profile of equilibrium and wave built terraces; a critical appraisal, *Geol. Soc. Am. Bull.*, 74, 971-990.

Dietz, R. S., 1964. Wave base, marine profile of equilibrium and wave built terraces: reply, *Geol. Soc. Am. Bull.*, 75, 1275-1281.

Duncan, J. R., 1964. The effects of water table and tidal cycle on swash-backwash sediment distribution and beach profile development, *Marine Geol.*, 2, 186-197.

Emery, K. O., 1945. Transportation of marine beach sand by flotation, *J. Sed. Petrology*, 15, 84-87.

Fairchild, J. C., 1973. Longshore transport of suspended sediment, *Proc. 13th Conf. Coastal Eng., ASCE*, 1069-1088.

Grant, U. S., 1943. Waves as a sand-transporting agent, *Am J. Sci.*, 241, 117-123.

Hails, J. R., 1974. A review of some current trends in nearshore research, *Earth-Sci. Rev.*, 10, 171-202.

Hails, J. R., ed., 1975. Submarine geology, sediment distribution and Quaternary history of Start Bay, Devon, *J. Geol. Soc.*, 131(1), 101p.

Harrison, W., 1969. Empirical equations for foreshore changes over a tidal cycle, *Marine Geol.*, 7, 529-552.

Hayes, M. O., 1972. *Central Processes and Sedimentation in the New England Coast.* Amherst: Mass. Univ. Coastal Res. Center, 149p.

Homma, M.; Horikawa, K.; and Kajima, R., 1965. A study of suspended sediment due to wave action, *Coastal Eng. Japan*, 8, 85-103.

Homma, H.; Horikawa, K.; and Sonu, C., 1960. A study of beach erosion at the sheltered beaches of Katase and Kamakura, Japan, *Coastal Eng. Japan*, 3, 101-122.

Ingle, J. C., 1966. *The Movement of Beach Sand.* Amsterdam: Elsevier, 221p.

Inman, D. L., and Bagnold, R. A., 1963. Beach and nearshore processes, 2. Littoral processes, in M. N. Hill, ed., *The Sea.* New York: Interscience, vol. 3, 529-553.

Inman, D. L.; Tait, F. J.; and Nordstrom, C. E., 1971. Mixing in the surf zone, *J. Geophys. Research*, 76, 3493-3514.

Ippen, A. T., and Eagleson, P. S., 1955. A study of sediment sorting by wave shoaling on a plane beach, *U.S. Army Corps Eng. Beach Erosion Board, Tech. Mem. 63*, 83p.

Johnson, J. W., 1956. Dynamics of nearshore sediment movement, *Bull. Am. Assoc. Petrol. Geologists*, 40, 2211-2232.

Johnson, J. W., and Eagleson, P. S., 1966. Coastal processes, in A. T. Ippen, ed., *Estuary and Coastline Hydrodynamics.* New York: McGraw-Hill, 404-492.

King, C. A. M., 1951. Depth of disturbance of sand on sea beaches by waves, *J. Sed. Petrology*, 21, 131-140.

King, C. A. M., 1972. *Beaches and Coasts.* 2nd ed. New York: St. Martin's Press, 570p.

Komar, P. D., 1971. The mechanics of sand transport on beaches, *J. Geophys. Research*, 76, 713-721.

Komar, P. D., 1976. *Beach Processes and Sedimentation.* Englewood Cliffs, N.J.: Prentice-Hall, 429p.

Krumbein, W. C., and James, W. R., 1974. Spatial and temporal variations in geometric and material properties of a natural beach, *U.S. Army Coastal Eng. Research Center, Tech. Mem. 44*, 7-79.

Longinov, V. V., 1964. Some aspects of wave action on a gently sloping sandy beach, *Internat. Geol. Rev.*, 6, 212-227.

Longuet Higgins, M. S., 1972. Recent progress in the study of longshore currents, in R. E. Meyer, ed., *Waves on Beaches.* New York: Academic Press, 203-248.

McKenzie, P., 1958. Ripcurrent systems, *J. Geol.*, 66, 103-113.

Miller, R. L., and Zeigler, J. M., 1958. A model relating dynamics and sediment pattern in equilibrium in the region of shoaling waves, breaker zone and foreshore, *J. Geol.*, 66, 417-441.

Miller, R. L., and Zeigler, J. M., 1964. A study of sediment distribution in the zone of shoaling waves over complicated bottom topography, in R. L. Miller, ed., *Papers in Marine Geology.* New York: McMillan, 133-153.

Mogel, T. R.; Street, R. L.; and Perry, D., 1971. Computation of alongshore energy and littoral transport, *Proc. 12th Conf. Coastal Eng., ASCE*, 899-918.

Nagata, Y., 1964. Deformation of temporal pattern of orbital wave velocity and sediment transport in shoaling water, breaker zone and on foreshore, *Oceanogr. Soc. Japan. J.*, 20, 57-70.

Palmer, H. D., 1970. Wave induced scour around natural and artificial objects. Univ. Southern California dissertation, 172p.

Ramming, H.-G., 1973. Reproduktion physikalischer Progesse in Küstengebieten, *Die Küste*, 24, 46-59.

Rector, R. L., 1954. Laboratory study of equilibrium profiles of beaches, *U.S. Army Corps Eng., Beach Erosion Board, Tech. Mem. 41*, 38p.

Saville, T., 1950. Model studies of sand transport along an infinitely straight beach. *Am Geophys. Union Trans.*, 31, 555-556.

Schiffman, A., 1965. Energy measurements of the swash-surf system, *Limnology and Oceanography*, 10, 255-260.

Shepard, F. P., 1950. Beach cycles in southern California, *U.S. Army Corps Eng., Beach Erosion Board, Tech. Mem. 20*, 26p.

Shepard, F. P., 1963. *Submarine Geology*, 2nd ed. New York: Harper & Row, 557p.

Silvester, R., 1974. *Coastal Engineering Volume 1:*

Sonu, C. J., 1969. Collective movement of sediment in littoral environment, *Proc. 11th Conf. Coastal Eng., ASCE,* 373-400.

Sonu, C. J., 1972. Field observations of nearshore circulation and meandering currents, *J. Geophys. Research,* 77, 3232-3247.

Sonu, C. J.; McCloy, J. M.; and McArthur, D. S., 1967. Longshore currents and nearshore topographies, *Proc. 10th Conf. Coastal Eng., ASCE,* 525-549.

Sonu, C. J., and van Beek, J. L., 1971. Systematic beach changes on the Outer Banks, North Carolina, *J. Geol.,* 79, 416-425.

Strahler, A. N., 1966. Tidal cycle of changes in an equilibrium beach, Sandy Hook, New Jersey, *J. Geol.,* 74, 247-268.

Tanner, W. F., 1958. The equilibrium beach, *Am. Geophys. Union Trans.,* 39, 889-891.

Tanner, W. F., 1973. Advances in near-shore physical sedimentology: A selective review, *Shore and Beach,* 41, 22-27.

Tanner, W. F., ed., 1974. *Sediment Transport in the Nearshore Zone.* Tallahassee: Florida St. Univ., 147p.

Thompson, W. O., 1937. Original structures of beaches, bars, and dunes, *Geol. Soc. Am. Bull.,* 48, 728-752.

U.S. Army Corps of Engineers, Coastal Engineering Research Center, 1966. Shore protection, planning and design, *Tech Rept. 4,* 3d ed., 401p.

Vollbrecht, K., 1966. The relationship between wind records, energy of longshore drift and energy balance off the coast of restricted water body as applied to the Baltic, *Marine Geol.,* 4, 119-148.

Wiegel, R. L., 1964. *Oceanographical Engineering.* Englewood Cliffs, N. J.: Prentice-Hall, 532p.

Williams, A. T., 1971. An analysis of some factors involved in the depth of disturbance of beach sand by waves, *Marine Geol.,* 11, 145-158.

Zenkovich, V. P., 1967. *Processes of Coastal Development.* New York: Interscience, 738p.

Cross-references: *Attrition; Barrier Island Facies; Beach Gravels; Beach Sands; Black Sands; Continental-Shelf Sedimentation; Heavy Minerals; Lagoonal Sedimentation; Lateral and Vertical Accretion; Neritic Sedimentary Facies; Salcrete; Sands and Sandstones; Storm Deposits; Tidal-Inlet and Tidal-Delta Facies; Tracer Techniques in Sediment Transport.*

LOESS

Loess (German: *löss,* or *lösch,* Swiss German dialect; dialect; *lösch,* adj., loose, akin to Old High German *lōs,* loose) is relatively homogeneous, seemingly nonstratified, unconsolidated deposit consisting predominantly of silt-size grains, but with minor amounts of very fine sand and clay. It is commonly buff in color, but locally may be gray, red, yellow, or brown. Smalley (1975) has edited a collection of older papers on the lithology and genesis of loess.

Composition. Loess consists characteristically of angular quartz grains and generally minor quantities of feldspar, and is commonly coated with very fine ($\approx 1\mu$) particles. Scanning electron microscope studies have revealed a distinct proportion of rounded particles (Cegla et al., 1971). There also are trace amounts of other minerals of silt size. The clay minerals vary regionally, but smectite is particularly common. There may be very slight cementation with calcium carbonate, and calcium carbonate concretions or nodules (*Loess Pupen,* or *Kindchen,* or *Männchen*) are widespread, except in the upper parts of soil profiles. Many loess deposits contain shells of terrestrial mollusks and bones and teeth of mammals.

Structure. Loess occurs primarily as a thin blanket, generally <30 m thick but thicker than 100 m in parts of China, overlying irregular topography, having a massive structure and generally forming vertical slopes or cliffs. The vertical structure may be produced by shrinkage and compaction of the material after its deposition. On the other hand, it may be a function of the presence of elongate, vertical, calcium carbonate-lined holes produced by plant roots.

Distribution

Nearly all loess deposits are Quaternary in age. In North America, the largest area of loess is in the central part from the Rocky Mountains as far east as western Pennsylvania (Thorp et al., 1952). Deposits are particularly thick along the E sides of the Mississippi and Missouri river valleys, and also along the Ohio, Wabash, and Illinois river valleys. Loess also is widespread in eastern Washington, western and southern Idaho, and northeastern Oregon. Extensive deposits also occur in Alaska. In Europe, loess is well developed along the Danube, Rhine, and Rhone valleys and, in southwestern USSR, along the Bug, Dnepr, Don, and Volga rivers.

Loess is particularly thick in northern China and in the steppe region of Siberia. There are local deposits in northern Africa, but loess is not abundant anywhere in Africa. Loess is found in South Island, New Zealand, and in Argentina.

Ancient Loess. Although the term *loessite* was introduced many years ago by J. B. Woodworth, and there is positive evidence for several ancient glacial epochs, few loesses have been identified in the geologic record. M. B. Edwards (ms.) has described late Precambrian loessites from Norway and Svalbard, and it is probable that more loessites will be recognized in the future.

Genesis

Any hypothesis of the origin of loess must take into account both its uniformly fine, well-sorted texture and its distribution. Most of the early published hypotheses are summarized by Russell (1944; see also Elias, 1945). Russell (1944) proposed that lower Mississippi valley loess is formed by weathering and downslope creep of backswamp river terrace silts. This hypothesis, however, is not generally accepted as an explanation of normal loess deposits.

The eolian origin of loess is almost universally accepted, based on the fine particle size (similar to that of material deposited from dust storms) and the geographic distribution on the downwind sides of valleys of streams that carried outwash material from large glaciers, and at the borders of certain large deserts. The European, American, and New Zealand loess have long been accepted as being of glacial origin. Since the influential report of Richthofen (1882), however, the deposits of north China and Siberia were thought to have blown out from deserts. Although the origin of Chinese loess deposits is still argued (see Warnke, 1971), the recognition that Siberia and China were widely glaciated and the results of recent scanning electron microscope studies (Smalley and others) have suggested that this loess is also derived from glacial outwash.

ADA SWINEFORD

References

Cegla, J.; Buckley, T.; and Smalley, I. J., 1971. Microtextures of particles from some European loess deposits, *Sedimentology*, 17, 129-134.
Elias, M. K., ed., 1945. Symposium on loess, *Am. J. Sci.*, 243, 227-303.
Lugn, A. L., 1962. The origin and sources of loess, *Nebraska Univ. Studies*, 26, 105p.
Richthofen, F. von, 1882. On the mode of origin of the loess, *Geological Mag.*, 29, 293-305.
Russell, R. J., 1944. Lower Mississippi valley loess, *Geol. Soc. Am. Bull.*, 55, 1-40.
Schultz, C. B., and Frye, J. C., eds., 1968. Loess and related eolian deposits of the world, *7th Cong. Int. Assoc. Quat. Research Proc.*, 12, 369p.
Smalley, I. J., ed., 1975. *Loess Lithology and Genesis*. Stroudsburg, Pa.: Dowden, Hutchinson & Ross, 429p.
Smalley, I. J., and Cabrera, J. G., 1970. The shape and surface texture of loess particles, *Geol. Soc. Am. Bull.*, 81, 1591-1596.
Smalley, I. J.; Krinsley, D. H.; and Vita-Finzi, C., 1973. Observations on the Kaiserstuhl loess, *Geological Mag.*, 110, 29-36.
Thorp, J., et al., 1952. Map: "Pleistocene eolian deposits of the United States, Alaska, and parts of Canada," 2 sheets (east and west), scale 1:2,500,000. Boulder, Colo.: Geol. Soc. Am.
Warnke, D. A., 1971. The shape and surface texture of loess particles: discussion, *Geol. Soc. Am. Bull.*, 82, 2357-2360.
Zolotun, V. P., 1974. Origin of loess in the southern part of the Ukraine, *Soviet Soil Sci.*, 6, 1-12.

Cross-references: *Desert Sedimentary Environments; Eolian Sedimentation; Glacigene Sediments; Niveo-Eolian Deposits.* Vol. III: *Loess.*

LUNAR SEDIMENTATION

The development of any sediment or sedimentary rock requires an energy source. On the moon, this condition appears to be satisfied almost entirely by the small but consistent meteorite flux at the lunar surface. The energy derived from meteorite impact is partitioned such that the largest proportion is absorbed in comminuting, heating, and ejecting the impacted substrate. In the processes, two types of sedimentary deposit are formed, lunar soil and clastic rocks (Lindsay, 1976).

Lunar Soil

By far the largest proportion of the kinetic energy expended at the lunar surface is derived from small impact events which penetrate <5-10 m beneath the lunar surface. These small-scale events are largely responsible for the development of the lunar soil and the soil breccias contained therein (Fig. 1).

The lunar soil consists essentially of three components: rock, mineral, and glass fragments mixed in varying proportions (Lindsay, 1972a). The detrital particles are the product, not so much of single impact events, but of prolonged reworking of the lunar surface by large numbers of impacts with a large range of kinetic energies (King et al., 1971). Each impact event has several effects on the lunar soil: (1) a small proportion of the soil is vitrified; (2) some of the soil is lithified by sintering (McKay et al., 1971) to form soil breccias; (3) individual particles are subjected to comminution; (4) fresh bedrock material may be excavated and added to the soil; (5) the soil is spread laterally as a thin ejecta blanket in accordance with local topography.

The two most important dynamic processes resulting from meteorite impact are *vitrification* and *comminution* of the soil. The effects of the two processes are mutually opposed, such that as the glass content of the soil increases over an extended period of time, the statistical parameters of the mature soil tend to stabilize. Comminution probably plays a dominant role early in the development of the soil by reducing the mean grain size and producing a log-normal grain-size distribution. Later, the combined effects of vitrification and agglutination (Fig. 2) of the soil particles produces a bimodal distribution of glass particles such that the mean grain size of the soil stabilizes at about 3.5 ϕ and the standard deviation of the distribu-

FIGURE 1. The lunar soil close up. Notice the clods of soil, the breccia particles, and the large glass-coated particle. The glass coating is the result of a molten spray from a nearby meteorite impact. The glass-coated grain is 3.3 cm in diameter.

tion is increased to about 2ϕ. At the Apollo 15 site, the grain-size distributions of soil samples from depths >2 m (Fig. 3) are coarser grained, more poorly sorted, and more negatively skewed than more mature samples nearer the surface, thus providing clear evidence of an evolutionary sequence (Lindsay, 1973).

Clastic Rocks

Soil Breccias. Dark-colored, highly porous breccia fragments which vary widely in their degree of consolidation are abundant at all of the sites sampled by the Apollo missions. These fine-grained, poorly sorted *breccias* are characterized by an abundant glassy matrix and glass particles (including glass spheres and ropy glass fragments), with lesser amounts of angular plagioclase and pyroxene grains and lithic clasts. Accretionary *lapilli* occur in soil breccias from all of the Apollo landing sites and appear to be an intrinsic characteristic of the lithology (McKay et al., 1970; Lindsay, 1972b, 1972c).

The lapilli, which are generally poorly defined, consist of a mineral or glass fragment surrounded by concentric zones of smaller mineral fragments and glassy matrix.

Crystalline Breccias. The crystalline breccias are in general fine-grained, poorly sorted rocks that consist largely of plagioclase and pyroxene grains and varying proportions of lithic fragments in a fine-grained matrix. Most of the mineral and lithic fragments are rounded, and some, notably the plagioclase grains, are well rounded (Fig. 4; Lindsay, 1972b, 1972c). All of the detrital particles, but particularly the plagioclase grains, show evidence of shock modification. The matrix, which is largely mineralic but may contain some glass, is generally recrystallized by the impact-generated heat. The degree of crystallization is controlled by the depth of burial in the original ejecta blanket (Wilshire and Jackson, 1972). In the more highly recrystallized lithologies, small diagenetic overgrowths are present on the surface of plagioclase grains. In situations where the matrix contains small amounts of

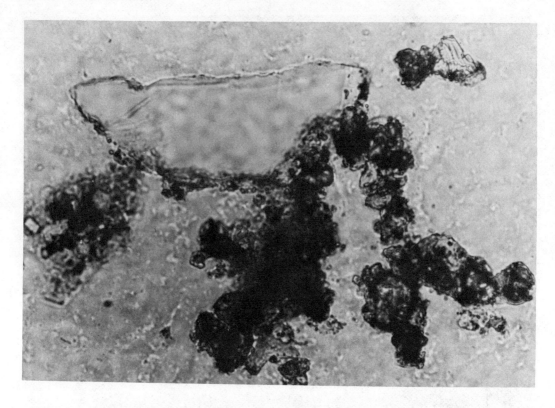

FIGURE 2. A fragile agglutinate with dendritic projections of brown glass attached to a larger feldspar grain. The brown glass contains numerous small detrital soil particles. The maximum dimension of the particle is 200μ.

glass, coarse, poorly developed accretionary lapilli are present.

Origin of Breccias. Despite the textural differences between soil breccias and crystalline breccias, they appear to be products of a single depositional mechanism: impact-generated base surge (Lindsay, 1972b; see *Base-Surge Deposits*). The differences between the two lithologies result from the difference in the scale of the events producing the rocks and a resultant difference in the available detrital materials. The soil breccias appear to be produced by the welding or sintering of surficial soils by impact-generated base surges of limited extent. The base surges probably contained a relatively low-volume concentration of particulate material, with the result that particle morphology was little modified during breccia formation. The impact events were relatively small and did not, for the most part, excavate new bedrock materials. Consequently, the texture and mineralogy of the soil breccias were determined largely by the major soil-forming processes.

The textural features of the crystalline breccias suggest that they were produced by large-scale impact-generated base surges, possibly similar in magnitude to the event that formed Mare Imbrium and the surrounding Fra Mauro Formation. In contrast to the soil breccias, the crystalline breccias were probably formed in a base surge with a high-volume concentration of solids. Consequently, particle morphology was determined largely by abrasion due to particle interaction in the base surge.

JOHN F. LINDSAY

References

Carr, M. H., and Meyer, C. E., 1974. The regolith at Apollo 15 site and its stratigraphic implications, *Geochim. Cosmochim. Acta,* 38, 1183–1198.

Fleischer, R. L., and Hart, H. R., 1974. Surface history of some Apollo 17 lunar soils, *Geochim. Cosmochim. Acta,* 38, 1615–1624.

Graf, W. H., and Warlick, D. G., 1972. Some "fluvio"-morphological features on the moon, in H. W. Shen, ed., *Sedimentation.* Fort Collins: Colo. St. Univ., Dept. Civil Eng., 25-1-25-22.

Heiken, G., and Lofgren, G., 1971. Terrestrial glass spheres, *Geol. Soc. Am. Bull.,* 82, 1045–1050.

Heiken, G. H.; McKay, D. S.; and Brown, R. W., 1974. Lunar deposits of possible pyroclastic origin, *Geochim. Cosmochim. Acta,* 38, 1703–1718.

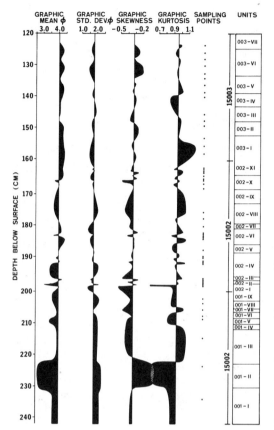

FIGURE 3. Grain-size parameters from the Apollo 15 deep drill plotted as a function of depth below the lunar surface. Black areas indicate deviations about the mean value. Note the gradual upward change, particularly of the mean and standard deviation.

FIGURE 4. A crystalline breccia from the Fra Mauro Formation (Apollo 14 site). Note the well-rounded but broken plagioclase grain which suggests a highly energetic, abrasive environment. The rounded grain has a maximum dimension of 450μ.

King, E. A., Butler, J. C. and Carman, M. F., 1971. The lunar regolith as sampled by Apollo 11 and Apollo 12: Grain size analyses, modal analyses, and origins of particles, *Geochim. Cosmochim. Acta,* suppl. 2, **1**, 737-746.

Lindsay, J. F., 1972a. Development of soil on the lunar surface, *J. Sed. Petrology,* **42**, 876-888.

Lindsay, J. F., 1972b. Sedimentology of clastic rocks from the Fra Mauro region of the moon, *J. Sed. Petrology,* **42**, 19-32.

Lindsay, J. F., 1972c. Sedimentology of clastic rocks returned from the moon by Apollo 15, *Geol. Soc. Am. Bull.,* **83**, 2957-2970.

Lindsay, J. F., 1973. Evolution of lunar soil grain-size and shape parameters, *Proc. 4th Lunar Sci. Conf.,* suppl. 4, **1**, 215-224.

Lindsay, J. F., 1974. Transportation of detrital materials on the lunar surface, *Sedimentology,* **21**, 323-328.

Lindsay, J. F., 1976. *Lunar Stratigraphy and Sedimentology.* Amsterdam: Elsevier, 302p.

McKay, D. S.; Greenwood, W. R.; and Morrison, D. A., 1970. Origin of small lunar particles and breccia from the Apollo 11 site, *Geochim. Cosmochim. Acta,* suppl. 1, **1**, 673-693.

McKay, D. S.; Morrison, D. A.; Clanton, U. S.; Ladle, G. H.; and Lindsay, J. F., 1971. Apollo 12 soil and breccia, *Geochim. Cosmochim. Acta,* Suppl. 2, **1**, 755-773.

Wilshire, H. G., and Jackson, E. D., 1972. Petrology and stratigraphy of the Fra Mauro formation at the Apollo 14 site, *U.S. Geol. Surv. Prof. Pap.* 785, 26p.

Cross-references: *Base-Surge Deposits; Breccias, Sedimentary; Extraterrestrial Material in Sediments; Martian Sedimentation; Volcaniclastic Sediments and Rocks.* Vol. II: *Moon–Lunar Geology; Moon–Lunar Soil.*

LYSOCLINE

About one half of the deep-ocean floor is covered by calcareous ooze, the other, deeper half, by sediment with only a few percent of calcium carbonate. The boundary between the two facies regimes roughly follows depth contours over large areas, owing to increased carbonate dissolution rates at depth (Murray and Renard, 1891, pp. 277-279). The boundary has been called the *calcium carbonate compensation depth* (CCD), because rate of supply and rate of dissolution are approximately compensated at this level (Bramlette, 1961; see *Pelagic Sedimentation, Pelagic Sediments*).

The rate at which the particles within the calcareous ooze dissolve varies greatly. Thick-shelled, resistant foraminifera and the more solid coccoliths are enriched within the sediment as the CCD is approached (Schott, 1935; Ruddiman and Heezen, 1967; McIntyre and McIntyre, 1971). The zone of enrichment, the R-facies (see *Paleobathymetric Analysis*), has an

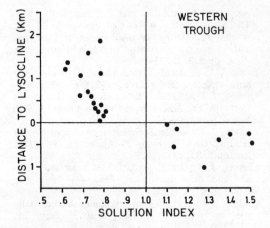

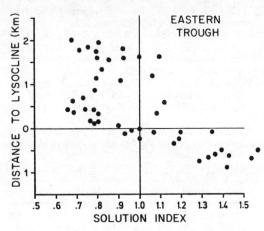

FIGURE 1. Definition of the (sedimentary) lysocline in the central Atlantic based on assessment of dissolution facies by solution indices. Position of the lysocline level coincides with the top of the Antarctic Bottom Water (taken as distance = 0). Note the sharpness of the level in the western trough where bottom-water flow is very active.

upper boundary which may be diffuse or sharp. This zone (or level, where sharp) is the *lysocline,* which is defined as the facies boundary between well-preserved and poorly preserved calcareous assemblages on the deep-sea floor (Berger, 1968).

The lysocline is well defined in the Central Atlantic, where it was first recognized on the basis of analysis of foraminiferal assemblages with respect to preservation (Fig. 1). Indices describing the state of preservation of an assemblage were formed by ranking the foraminifera with respect to their susceptibility to dissolution and calculating the weighted average rank for each sample (Berger, 1970, 1973). Simplified indices, such as percent spined foraminifera, or percent resistant forms, can also be used to determine preservation patterns. In the central and southern Atlantic, the lysocline level closely follows the boundary between Antarctic Bottom Water and North Atlantic Deep Water. In the Pacific, there is also evidence that the upper boundary of the Bottom Water is associated with a level of rapidly increasing dissolution (Fig. 2), although the change in water properties across this boundary is rather small.

Sedimentary and Hydrographic Lysocline

The term *lysocline* has been used in contexts other than sedimentary facies. Thus, from the results of the field experiments of Peterson (1966) and Berger (1967) in the Pacific, showing a drastic increase of dissolution rates near 4000 m, and from the correlation of the lysocline with the top of the Antarctic Bottom Water in the Atlantic, it appears very probable that over large areas, the sedimentary lysocline on the sea floor marks a level of increase in the propensity of the surrounding water to dissolve calcite. This level in the water column has been called *hydrographic lysocline,* to distinguish it from the sedimentologically defined *foraminiferal lysocline,* or *coccolith lysocline.* The distinction between hydrographic and sedimentary lysoclines is important both conceptually and operationally, because in fertile regions, the lysocline shallows and becomes diffuse (Fig. 1, eastern part; note high solution indices at shallow depths due to increased dissolution off Africa). In the fertile coastal areas, the high supply of organic matter leads to development of relatively CO_2-rich interstitial waters. Pore-water pH will thus be lowered with respect to the surrounding ocean water.

The reasons why the lysocline is where it is and the relationships of the lysocline model of calcite dissolution to the distribution of carbonates in the deep sea are under discussion (Heath and Culberson, 1970; Berger, 1971; Berner, 1974; Broecker and Broecker, 1974). In addition to chemical studies in the field and in the laboratory, the mapping of dissolution facies (Atlantic, see Fig. 3; Pacific, see Fig. 2 in *Paleobathymetric Analysis*) should provide the information necessary to attack these problems. Fluctuations of the lysocline through geologic time (Berger, 1972, 1973) should also shed light on its origin.

WOLFGANG H. BERGER

References

Berger, W. H., 1967. Foraminiferal ooze: solution at depth, *Science,* **156,** 383–385.
Berger, W. H., 1968. Planktonic foraminifera: selective solution and paleoclimatic interpretation, *Deep-Sea Research,* **15,** 32–43.
Berger, W. H., 1970. Planktonic foraminifera: selec-

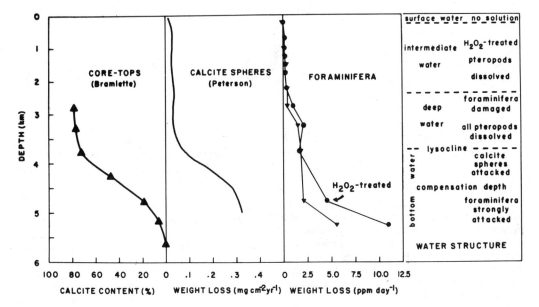

FIGURE 2. Carbonate percentages (Bramlette, 1961), weight loss of calcite spheres (Peterson, 1966), and weight loss of foraminifera (Berger, 1967) on a moored buoy in the central Pacific, compared with the regional water structure, indicating the presence of a (hydrographic) lysocline and its effect on the sediment (from Berger, 1970).

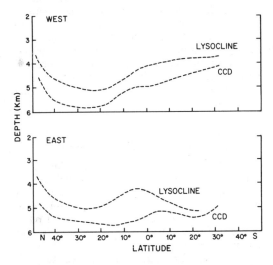

FIGURE 3. Comparison of open-ocean lysocline depth and CCD in western and eastern Atlantic. Only lower boundary of lysocline zone is drawn, corresponding to roughly 80% loss of foram assemblage.

tive solution and the lysocline, *Marine Geol.,* 8, 111-138.

Berger, W. H., 1971. Sedimentation of planktonic foraminifera, *Marine Geol.,* 11, 325-358.

Berger, W. H., 1972. Deep-sea carbonates: Dissolution facies and age-depth constancy, *Nature,* 236, 392-395.

Berger, W. H., 1973. Deep-sea carbonates: Pleistocene dissolution cycles, *J. Foram. Research,* 3(4), 187-195.

Berner, R. A., 1974. Physical chemistry of carbonates in the oceans, *Soc. Econ. Paleont. Mineral. Spec. Publ.* 20, 37-43.

Bramlette, M. N., 1961. Pelagic sediments, *Am. Assoc. Adv. Sci. Publ.* 67, 345-366.

Broecker, W. S., and Broecker, S., 1974. Carbonate dissolution on the western flank of the East Pacific Rise, *Soc. Econ. Paleont. Mineral. Spec. Publ.* 20, 44-57.

Heath, G. R., and Culberson, C., 1970. Calcite: degree of saturation, rate of dissolution, and the compensation depth in the deep oceans, *Geol. Soc. Am. Bull.,* 81, 3157-3160.

McIntyre, A., and McIntyre, R., 1971. Coccolith concentrations and differential solution in oceanic sediment, in B. M. Funnel, and W. R. Riedel, eds., *The Micropalaeontology of Oceans.* Cambridge: Cambridge Univ. Press, 253-261.

Murray, J., and Renard, A. F., 1891. *Report on Deep-Sea Deposits. H.M.S. Challenger, 1873-1876.* Reprinted, 1965. London: Johnson Reprint, 525p.

Peterson, M. N. A., 1966. Calcite: rates of dissolution in a vertical profile in the central Pacific, *Science,* 154, 1542-1544.

Ruddiman, W. F., and Heezen, B. C., 1967. Differential solution of planktonic foraminifera, *Deep-Sea Research,* 14, 801-808.

Schott, W., 1935. Die Foraminiferen in dem äquatorialen Teil des Atlantischen Ozeans, *Wiss. Ergebn. D. deutsch. Atl. Exped. "Meteor," 1925-1927,* III(3), sec. B, 43-134.

Cross-references: *Abyssal Sedimentation; Marine Sediments; Paleobathymetric Analysis; Pelagic Sedimentation, Pelagic Sediments.* Vol. I: *Calcium Carbonate Compensation Depth.*

M

MARINE EVAPORITES—DIAGENESIS AND METAMORPHISM

The terms *diagenesis* (q.v.) and *metamorphism* refer merely to the concept of change in some physical or chemical way in response to either passage of time alone or to some combination of changing physical and/or chemical environments. Quite arbitrarily, diagenesis in sedimentary rocks is usually considered to cease at 200°C and metamorphism to proceed at temperatures above 200°C. The retrograde, destructive, low-temperature weathering changes that affect rocks of all types are usually exempted from the term diagenesis. The high solubility and reactivity of most evaporite minerals and the frequent occurrence of metastable primary phases, however, commonly permits extensive alterations to occur at a very early stage in their postdepositional history. Similarly, the temperature, pressure, and chemical changes needed to alter an evaporite almost beyond recognition are much less severe than those changes needed to significantly alter most other sedimentary rock types. Consequently, "diagenesis" and "metamorphism," as used in evaporite geology, differ somewhat from more usual usage.

Many *penecontemporaneous* changes in evaporites are arbitrarily included in the discussion of primary mineral associations (see *Evaporites—Physicochemical Conditions of Origin*). Included in this category are (1) transformation of earlier gypsum to anhydrite in response to rising salinity; (2) elimination of primary kainite (or metastable primary sylvite + epsomite) by the reaction with carnallite saturated brines; and (3) the disintegration of early-formed carnallite rocks to sylvinite (sylvite + halite) owing to influx of undersaturated brines or of seawater.

Diagenetic changes in evaporites are defined here as those changes that take place without significant changes of the relevant primary intensive parameters of the environment. Diagenesis comprises (1) postdepositional transformation of primary minerals—metastable primary minerals may be replaced without physical or chemical changes having occurred since initial precipitation, or initially metastable or stable minerals may be replaced by other phases as a response to changed physical and/or chemical conditions; and (2) postdepositional isochemical and isophase recrystallization of primary chemical sediments—leading to consolidation, compaction, and minimization of internal strain energies.

Metamorphic changes in evaporites are defined here as those changes that occur in response to significant changes in the relevant primary intensive parameters of the environment. Owing to the high reactivity of salt rocks to changing temperatures, salt metamorphism may start at relatively low temperatures. Metamorphism comprises subsequent isochemical or allochemical transformations of the original rocks outside the stability fields of the primary mineral associations. *Thermometamorphism,* originating from elevated temperatures, may lead to the development of nearly isochemical (neglecting any solutions produced) but allophase associations. *Solution metamorphism* originates from invading secondary brines which differ significantly from the original parent brine. *Dynamometamorphism* originates from isophase recrystallization following plastic deformation.

Structural and Textural Characteristics

The pronounced ability of saline minerals to react and to recrystallize very often deletes primary depositional textures completely and produces typical crystalloblastic textures. One of the most prominent primary textural relics is bedding itself. Diagenetic *recrystallization* in many cases should lead to a considerable volume decrease, but unambiguous textural characteristics of shrinkage of individual beds have rarely been convincingly demonstrated.

Evidence of recrystallization textures is common, particularly in chloride rocks. The grains show a platy habit parallel to the bedding planes, with interlocking contacts and with more or less equal size distribution due to the elimination of smaller primary grains. Occasionally, however, very delicate primary structures are reported to have survived as, for instance, hopper crystals of halite originally formed at the brine surface (Dellwig, 1955). Zonation of fluid inclusions attributable to rhythmic precipitation at the brine surface has also been observed.

Evidences of replacement structures are widespread pseudomorphs, reaction rims, and rare relics of earlier minerals. Well-known examples are pseudomorphs of anhydrite after swallowtail twins of gypsum; reaction rims of polyhalite around earlier anhydrite; relics of carnallite in secondary sylvite; pseudomorphs of sylvite plus kieserite after earlier langbeinite, and many others. Numerous examples have been described by Schaller and Henderson (1932); Borchert and Baier (1953); A. Baar, and E. Bessert (see summary and references in Stewart, 1963). Complete recrystallization of anhydrite rocks, in some examples, is indicated by spherulitic aggregates. Folk and Pittman (1971) have found that length-slow chalcedony (quartzine and lutecite) are indicative of silica replacement of evaporites, confirmed by their literature review and also by subsequent studies, e.g., Siedlecka (1972) and West (1973).

The identification of the different types of metamorphism and of diagenetic recrystallization on the basis of textures is often difficult or ambiguous. A thermometamorphic fabric originating from an earlier kainite rock perhaps is represented by the so-called "Flockensalz," consisting of rounded kieserite grains in a matrix of sylvite and halite (details and references in Braitsch, 1971). Large-scale characteristics are more reliable; for example, thermometamorphism acts uniformly over relatively large areas. Solution metamorphism acts on a more local scale, however, producing a sharp zonal distribution of mineral assemblages; for example, in potash deposits, barren halite rocks typically occur in the cores of the alteration zones.

Mineral Reactions

Thermometamorphism. The stabilities of individual mineral associations, shown in Fig. 2 of *Evaporites–Physicochemical Conditions of Origin,* provide the key for the derivation of metamorphic assemblages. At the present time few or no known potash-bearing evaporite deposits have been buried at subsurface temperatures much in excess of about 100°C (4 km or so, under average surface temperature and geothermal gradient conditions). In such deposits, we can thus ignore the thermal decomposition reactions of bischofite + halite and carnallite + halite which occur at 116°C and 167°C respectively, and consider only reactions involving the less highly soluble phases of marine evaporites such as gypsum or kainite. Of the early workers in this area, including S. Arrhenius and F. Rinne, Jänecke (1923) gave a particularly detailed treatment of thermometamorphism (see references in Braitsch, 1971).

If one assumes that a "dry" evaporite sequence (without residual brines) is buried below gas-tight younger sediments with halite in excess, only stable-phase associations need consideration; and the pressure dependency of the univariant points to a first approximation may be disregarded. Thermal disintegration occurs at the upper stability limits of individual phase assemblages. The thermal decomposition of kainite may serve as an example.

The upper stability limit of the coexistence of kainite + carnallite is reached at 72°C. The reac-

$$2\,\text{KMgClSO}_4(\text{H}_2\text{O})_3 \text{ (2 kainite)} + \text{KMgCl}_3(\text{H}_2\text{O})_6 \text{ (carnallite)} \underset{72°C}{\rightleftharpoons} 2\,\text{MgSO}_4\,\text{H}_2\text{O} \text{ (2 kieserite)} + 3\,\text{KCl} \text{ (3 sylvite)} + 10\,\text{H}_2\text{O} \text{ (solution)} \quad (1)$$

$$4\,\text{KMgClSO}_4(\text{H}_2\text{O})_3 \text{ (4 kainite)} \underset{83°C}{\rightleftharpoons} \text{K}_2\text{Mg}_2(\text{SO}_4)_3 \text{ (langbeinite)} + \text{MgSO}_4\,\text{H}_2\text{O} \text{ (kieserite)} + 2\,\text{KCl} \text{ (2 sylvite)} + \text{MgCl}_2 + \text{H}_2\text{O} \text{ (solution)} \quad (2)$$

$$\text{K}_2\text{Mg}_2(\text{SO}_4)_3 \text{ (langbeinite)} + \text{KMgCl}_3(\text{H}_2\text{O})_6 \text{ (carnallite)} \rightleftharpoons 3\,\text{MgSO}_4\,\text{H}_2\text{O} \text{ (3 kieserite)} + 3\,\text{KCl} \text{ (3 sylvite)} + 3\,\text{H}_2\text{O} \text{ (solution)} \quad (3)$$

$$\text{KMgCl}_2(\text{H}_2\text{O})_6 \text{ (carnallite)} + 4\,\text{H}_2\text{O} \text{ (water)} \rightleftharpoons \text{KCl} \text{ (sylvite)} + \text{MgCl}_2 + 10\,\text{H}_2\text{O} \text{ (solution)} \quad (4)$$

$$87.4\,c + 2.6\,ks \text{ (original rock)} + 0.473\,\text{sol}_1 \text{ (invading soln.)} \underset{83°C}{\rightleftharpoons} 67.6\,sy + 48.1\,n \text{ (metamorphic rock)} + 1\,\text{sol}_2 \text{ (final soln.)} \quad (5)$$

$$\text{K}_2\text{Mg}(\text{SO}_4)_3 \text{ (langbeinite)} + \text{MgCl}_2 + 3\,\text{H}_2\text{O} \text{ (solution)} \rightleftharpoons 3\,\text{MgSO}_4\,\text{H}_2\text{O} \text{ (3 kieserite)} + 2\,\text{KCl} \text{ (2 sylvite)} \quad (6)$$

FIGURE 1. Mineral reactions of marine evaporites.

tion at this temperature may be approximated as shown in (1) of Fig. 1.

At 83°C the upper stability limit of kainite is reached, as shown in (2) of Fig. 1. (The mineral equations given for this reaction in some standard petrological texts are in error inasmuch as the phase rule is not obeyed.) In any case, the solution formed will be saturated in all the phases present, including halite; therefore the above equations only hold qualitatively. The quantitative relations may be found by considering the true composition of the coexisting brine (Braitsch, 1971). In the second example, an interesting variation is possible if carnallite is also present: if the brine originating by reaction (2) dissolves carnallite, langbeinite cannot remain stable in this brine but will react [see (3) in Fig. 1] so that the result as a whole is similar to reaction (1). This process is one possible way to explain the important association sylvite + kieserite. It requires (if calculated quantitatively) a weight proportion of 1 sylvite to 1.5 kieserite. If, in the original kainite rock, kieserite was already present, then the final kieserite content will be larger. A possible example of this reaction is found in the "Flockensalz" of the potash seam "Hessen" of the German Zechstein (Braitsch, 1971).

A second example of a fairly common thermometamorphic reaction is the burial dehydration of gypsum. If primary gypsum is buried, thermal decomposition will occur and anhydrite will replace the original gypsum. The reaction is a function of temperature and the activity of water of associated pore-space brines. It is also slightly pressure dependent and is very sensitive to brine salinity; the reaction nearly always takes place before burial exceeds 1 km, although rare occurrences of gypsum have been reported from as deep as 1400 m. A general discussion and the possible importance of this dehydration reaction to the maintenance of anomalous subsurface fluid pressures is presented by Hanshaw and Bredehoeft (1968).

Thermometamorphism is isochemical, if the solution formed remains in the rock. Lowering the temperature can reverse the reactions. In nature, however, the solution usually migrates away, so that the high-temperature mineral associations are preserved at low temperature, being stable under dry conditions.

Solution Metamorphism (Metasomatism) in "Diluted" Brines. This process is by far the most important type of allophase metamorphism in evaporitic rocks (Borchert, 1965). After early qualitative discussions of M. Naumann and M. Rosza, it was quantitatively formulated by J. D'Ans and R. Kuhn and in more detail by A. Baar and by U. Storck (see references in Stewart, 1963; Braitsch, 1971). The result depends on whether the invading brine corresponds to a "higher" or a "lower" concentration stage than the invaded rock. Because many evaporite minerals show incongruent solubility, new phases will form. In the case of a diluted invading brine, this process may be illustrated qualitatively for the carnallite distintegration [see (4) in Fig. 1].

In a quantitative equation approaching natural conditions, one must consider the true composition of the invading solution, which is normally saturated in halite, and of the final solution saturated in carnallite. This is represented in (5) of Fig. 1, where concentrations are in mol; c = carnallite; ks = kieserite; sy = sylvite; n = halite; sol_1 = invading solution saturated in NaCl = 117.8 mol NaCl/1000 mol H_2O; sol_2 = final solution saturated in carnallite = 87.4 mol $MgCl_2$ + 2.6 mol $MgSO_4$ + 19.8 mol KCl + 7.6 mol NaCl + 1000 mol H_2O (see tables of D'Ans, 1933; also Braitsch, 1971).

Two points are significant: (a) a great part of the sodium chloride content of the invading brine is precipitated as halite ("salting-out-effect"); and (b) the main part of the potassium present in the original rock is preserved in the metamorphic rock as sylvite, whereas the magnesium chloride content goes into solution. If the kieserite content of the original rock exceeds the proportion indicated in the equation, this excess will be preserved in the metamorphic rock as well. Therefore metamorphic sylvite rocks may or may not contain kieserite. Both types are abundant, the one typified by the kieserite-sylvite-halite rocks in the German Zechstein, the other by some sylvite-halite rocks (namely the "variegated" sylvites of the pre-Ural basin and many other localities). In other examples of this type, anhydrite may be present in variable amounts for example in the anhydritic "Hartsalz" (= sylvite-halite rock) of the South Harz area, Germany, and this anhydrite partly originates by "salting out" ("Reaktionsanhydrite" of Borchert, 1965).

Sylvite rocks, however, are not the final stage of the solution metamorphism because sylvite itself is soluble in solutions saturated in halite. But sylvite is dissolved congruently, and some of the sodium chloride content of the invading brine is salted out as halite, often as blue halite. In nature, the migration of the brines is slow, the quantity of invading brines is often limited, and complete dissolution of all the sylvite is not achieved. Therefore, one normally observes different zones of metasomatic reactions: the original carnallite rock surrounded by a sylvitehalite rock which is itself surrounded by a halite rock.

In the case of original (anhydrite-bearing)

kieserite-carnallite rocks, additional intermediate zones occur in the following succession: kieserite-halite-carnallite rock, kieserite-sylvite-halite rock (kieserite-polyhalite-halite rock), langbeinite-halite rock, loeweite-halite rock, and vanthoffite-halite rock. Examples of such metamorphic sequences are well represented in the classic Stassfort area (see references and additional examples in Braitsch, 1971).

Rinneite, $K_3 NaFeCl_6$, is also formed by solution metamorphism of carnallite rocks by invading iron-rich brines saturated in halite.

Solution Metamorphism in "Concentrated" Brines. If a solution saturated in carnallite is pressed off and reaches a preexisting langbeinite rock, then langbeinite may react with the solution to give kieserite and sylvite. The reaction, similar to reaction (3), may be approximated as shown in (6) of Fig. 1. Numerous examples of sylvite + kieserite pseudomorphs after langbeinite at Stassfort are probably explicable by this mechanism.

Where more concentrated brines react with minerals corresponding to an earlier evaporation stage (for instance with loeweite or vanthoffite instead of langbeinite), additional intermediate steps are possible. Reactions with more concentrated invading brines should produce a sequence of metasomatic rocks which qualitatively corresponds to the progressive order of primary evaporites. Examples of such sequences, however, are not known to occur in nature, due to the scarcity of loeweite and vanthoffite rocks.

Solution Metamorphism in Cap Rocks. At temperatures below 83°C, kainite is one of the most important products of solution metamorphism of preexisting kieserite-bearing sylvite and carnallite rocks. In the neighborhood of the earlier potash ore, irregular relics of the initial rock survive. In the kainite rock, kieserite or sylvite, depending on the original mineral proportions, may be preserved. More intensive retrograde metamorphism may produce leonite or aphthitalite (glaserite), and from originally kieserite-rich source rocks loeweite or bloedite may form. Kainite itself may be replaced in outer zones by schoenite (picromerite) and this is followed by residual gypsum.

Well-known examples containing additional minerals (some rare borates) are represented by the cap rocks of potash-bearing salt domes and by some interesting lenses and pods in alpine salt deposits (Mayrhofer, 1955; see references in Braitsch, 1971).

As can be seen, the solution metamorphism products resulting from the invasion of diluted brines correspond with the mineral succession of normal evaporites without reaction at the reaction points, but in the reversed order. The relative amounts of different minerals formed depends on the mineralogy of the source rocks, on the composition of the invading brine, and on the temperature.

In conclusion it may be said that in evaporite metamorphism, equilibrium normally is reached. Thermometamorphism acts on a large scale, whereas solution metamorphism is restricted to dimensions from cm to a few hundred m and is characterized by a sharp mineral zoning indicative of local equilibrium.

Economic Importance

The metamorphic mineral associations kieserite-sylvite-halite (= Kieseritic "Hartsalz") and anhydrite-sylvite-halite (= Anhydritic "Hartsalz") form the basis of the German potash industry. The grade of ore varies between about 10 and 20% K_2O-equivalent. The commercial value is improved if kieserite is present because kieserite is essential in the production of sulfatic fertilizers. In the last century, kainite cap rocks were also mined on a large scale.

At the present time, production from diagenetically recrystallized, primary sylvite rocks of modified marine evaporites strongly predominates outside Germany.

OTTO BRAITSCH*
DAVID J. J. KINSMAN

*Deceased.

References

Borchert, H., 1965. Principles of oceanic salt deposition and metamorphism, in J. P. Riley, and G. Skirrow, eds., *Chemical Oceanography,* vol. 2. London: Academic Press, 205–276.

Borchert, H., and Baier, E., 1953. Zur Metamorphose ozeaner Gipsablagerungen, *N. Jahrb. Mineral. Abhandl.,* **86**, 103–154.

Borchert, H., and Muir, R. O., 1964. *Salt Deposits: The Origin, Metamorphism and Deformation of Evaporites.* London: Van Nostrand, 338p.

Braitsch, O., 1971. *Salt Deposits: Their Origin and Composition.* Berlin: Springer-Verlag, 297p (original German ed., 1962).

D'Ans, J., 1933. *Die Lösungsgleichgewichte der Systeme der Salze ozeanischer Salzablagerungen.* Berlin: Verl. Ges. F. Ackerbau, 254p.

Dellwig, L. F., 1955. Origin of the Salina salt of Michigan, *J. Sed. Petrology,* **25**, 83–110.

Folk, R. L., and Pittman, J. S., 1971. Length-slow chalcedony: A new testament of vanished evaporites, *J. Sed. Petrology,* **41**, 1045–1058.

Hanshaw, B. B., and Bredehoeft, J. D., 1968. On the maintenance of anomalous fluid pressures. II Source layer at depth, *Geol. Soc. Am. Bull.,* **79**, 1107–1122.

Jänecke, E., 1923. *Die Entstehung der deutschen Kalisalzlager.* Braunschweig: Fr. Vieweg, 111p.

Kirkland, D. W., and Evans, R., 1973. *Marine Evaporites: Origin, Diagenesis, and Geochemistry.* Stroudsburg, Pa.: Dowden, Hutchinson & Ross, 426p.

Mayrhofer, H., 1955. Uber ein Langbeinit-und Kainit-Vorkommen im Ischler Salzgebirg, *Karinithin,* 30, 94-98.
Schaller, W. T., and Henderson, E. P., 1932. Mineralogy of drill cores from the potash field of New Mexico and Texas, *U.S. Geol. Surv. Bull.,* 833, 124p.
Siedlecka, A., 1972. Length-slow chalcedony and relicts of sulphates—evidences of evaporitic environments in the Upper Carboniferous and Permian beds of Bear Island, Svalbard, *J. Sed. Petrology,* 42, 812-816.
Stewart, F. H., 1963. Marine Evaporites: Data of geochemistry, 6th ed., chapt. Y, *U.S. Geol. Surv. Prof. Pap. 440-Y,* 52p.
West, I., 1973. Vanished evaporites—significance of strontium minerals, *J. Sed. Petrology,* 43, 278-279.

Cross-references: *Carbonate Sediments—Diagenesis; Diagenesis; Evaporite Facies; Evaporites—Physicochemical Conditions of Origin.*

MARINE SEDIMENTS

Marine sediments are combinations of several components, most of which can be grouped as (1) *terrigenous,* the particulate detritus eroded from the continents (allochthonous), and (2) *oceanic,* the biochemical or physicochemical precipitates from sea water (autochthonous). Gravels, sands, silts, and clays carried into the sea by runoff, wind, and ice and subsequently distributed by waves and currents exemplify allochthonous materials. Autochthonous precipitates include shells and tests of marine organisms and the authigenic minerals such as barite, glauconite, phosphorite, and manganese dioxide. Strictly speaking, the planktonic biogenic contribution is transported some distance from the surface waters to the deep ocean, but these distances are small as compared to the transport of terrigenous sediments. Detailed geochemical studies have shown the possibility that some of the fine, clay-sized sediments may also be autochthonous oceanic precipitates (see *Clay—Genesis).* Large contributions of fine clay, ash, and zeolitic sediments are derived from oceanic volcanic eruptions, and one could argue that this material should be classed separately from the terrigenous contribution (see *Volcanism—Submarine Products).* Organic compounds are contributed by marine plants and animals and from land-plant debris; natural hydrocarbon seeps contribute bitumens (q.v.).

The aspect of a given marine sediment type is governed by the rates of supply of the major components, the kinetic energy of the environment, the level of biologic activity, the availability of oxygen, the hydrogen ion concentration of the water, and the carbon dioxide content. These factors, in turn, are all functions of the depth of water, distance from land, bottom topography, circulation patterns within the water column, water chemistry, and productivity of the overlying surface waters.

Volume and Distribution

Several scientists have calculated the volume of marine sediments produced during geological time by weathering and erosion of the continents. The calculated volumes for deep-sea sediments, however, greatly exceed the amounts actually found on the sea floor. Several factors may account for this discrepancy: (1) the assumptions upon which the calculations are based may be in error; (2) the unconsolidated sediments in the sea often cover denser, consolidated sedimentary rocks; (3) the continental margins are often effective sediment traps; and (4) the deep marine sediments appear to be incorporated progressively into the continental blocks by movement and subduction of oceanic plates. Increasing amounts of data on all of these effects are being accumulated by the Deep Sea Drilling Project. In any event, the amount of unconsolidated sediment on the deep-sea floor distant from land is usually small, and large areas of the mid-ocean ridge are apparently sediment free or only thinly veneered by detritus.

Early observations of the bottom sediments of the world oceans by the *Challenger* Expedition outlined a broad pattern of zones of terrigenous detritals, pelagic biogenic oozes, and red clays. Glacial materials are important in the northern and southern polar areas. With further study, these broad patterns have been modified in detail, but the major belts are essentially as originally mapped (Fig. 1).

Inner Continental Margins

Paralic Facies. Terrigenous materials compose much of the shoreline and estuarine sediments bordering land areas (see *Estuarine Sedimentation; Littoral Sedimentation).* These peripheral sediments exhibit wide textural variations, ranging from the clays and silts of coastal marshes to the gravels and cobbles comprising shingle beaches. Until estuaries have filled, much terrigenous sediment is trapped, and little passes out to the shelves and deep-sea floor. Deltas (q.v.) are typical silt-clay environments, with sands in distributary channels.

Continental Shelf. Continental-shelf sediments (see *Continental-Shelf Sedimentation)* encompass a wide spectrum of textures, but are dominantly sands and silts. Variability of these textures reflects many sedimentation episodes matching the relatively rapid shifting of sea level during the Pleistocene. This complexity of modern shelf deposits may be anomalous and

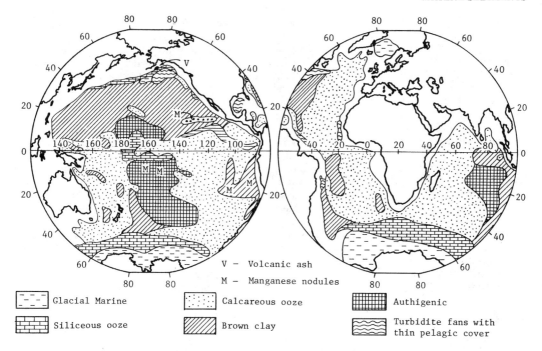

FIGURE 1. Distribution of deep-sea sediments (after Shepard, 1973, compiled with information and help from Arrhenius, Nayudu, Ericson, Menard, and Riedel). It is likely that areas with abundant manganese nodules are more widespread, and that there are many more areas with turbidite layers interbedded with pelagic deposits.

atypical of continental shelves in the geological record which were produced under relatively constant sea levels. Much of the material exposed on the present continental shelf is thus relict, from previous erosion and deposition cycles. These sediments do undergo reworking and sorting by currents and long storm waves, however, and are by no means inactive.

Because the sediments covering inner continental margins reflect the local and regional conditions of adjacent portions of the continent, the shelf sedimentary deposits can be broadly divided into geographic provinces, for example, in North America.

The Pacific coast N of Point Conception is typified by relatively narrow shelves covered by terrigenous detritus. Off southern California, the depressions in the continental borderland contain fine silts and clays typical of deeper oceanic deposits. Sands and silts cover the coastal shelf; relict and residual deposits with larger admixtures of carbonate fragments and authigenic mineral coatings and nodules form a veneer on the bank tops and shoal areas.

The western parts of the Gulf of Mexico are dominated by prodeltaic silts and clays spread over a broad shelf; the finer deltaic terrigenous deposits give way to carbonate sediments on the broad limestone plateau of the Florida Peninsula.

The central Atlantic coast S of Cape Hatteras is bordered by a broad shelf covered by coarse quartz sand with negligible fine-grained content; these sands are probably relict, as evidenced by the presence of large sand waves on the broad shelf surface. Calcareous content increases seaward; detrital phosphorite is also common. North of Cape Hatteras, the shelf widens again and is covered by terrigenous detritus of all sizes, much of which was the product of Pleistocene glacial or periglacial deposition.

It has been shown that a broad latitudinal pattern exists on all shelves, with glacial sediments in high latitudes, terrigenous sands and silts in mid-latitudes, and carbonates in tropical areas. In most nearshore waters, the dilution of carbonate by the very high rates of terrigenous deposition produces sediments with low carbonate contents. Authigenic precipitates are typically found on margins in zones of oceanic upwelling. This general pattern is usually complicated on present shelves by the effects of Pleistocene sea-level changes. These combinations illustrate the quite different products that may form in response to varying combinations of sedimentological factors working on a limited number of major components.

Shallow Marine Carbonates. Calcareous deposits may form in shallow areas marginal to

469

continents or on isolated platforms in the oceanic areas where detritals are excluded either by lack of supply or because of the presence of barriers to dispersal. These sediments are quite different in appearance from their deep-water counterparts because the contributing organisms usually differ in character. Examples would be mollusk fragment gravels and sands of the outer portions of broad continental shelves, coral reefs (see *Coral-Reef Sedimentology*), and atolls of the mid-Pacific, and carbonate platforms such as the Bahama Banks (see *Bahama Banks Sedimentology*). Organic content frequently is large in these shallow-water carbonates due to rapid burial, limited exposure to destructive processes in the water column, and larger supply than in the deep sea. Atolls are a further exception in that the reef framework is extensively reworked by later phases of biologic growth in the evolution of the reef mass.

Outer Continental Margins

Submarine slope sediments are typically gray-green silts and clays draped over underlying structures (see *Submarine-Slope Sedimentation*). They are generally thin and often show effects of mass movement. Thickness is strongly influenced by local current patterns, and suspended sediment transport may bypass some areas, leaving older sediments and rock exposed.

All outer margins are cut by submarine canyons (q.v.), which at one time or another serve as the conduits for the movement of coarse sediments from inner shelf to deep sea. They also affect the shelf-water circulation, and thus much of the fine suspended load is drawn to these channels. Evidence for this effect is the typical presence at the mouths of canyons of submarine fans, in which channel deposits, natural levees, interdistributary silt layers and distal fine sand sheets document the large flow of sediment from the canyons (see *Submarine-Fan Sedimentation*). Further documentation comes from the fact that the largest canyons and fans are associated with the world's largest rivers. The fans merge with basin-floor sediments in closed depressions or with abyssal plains on open margins.

Where continental margins merge smoothly with the deep-sea floor, as in the Atlantic, the largest fans merge to form the continental rise (see *Continental-Rise Sediments*). This feature together with the associated abyssal plains is the locus of the largest deposition of terrigenous materials on earth. Mass wasting (see *Mass-Wasting Deposits*), turbidity currents (see *Turbidity-Current Sedimentation*), and low-density turbid flows (see *Nepheloid Sediments*) contribute the sediment. Pelagic sedimentation (q.v.) is important only during periods of diminished terrigenous contribution as, for example, during interglacial epochs when the flooded shelves and coasts trap much terrigenous sediment. At these same times, the effects of near-bottom oceanic currents become important. "Contour currents" (see *Contourites*) or deep counter-current flow under the strong western boundary currents of the oceans (e.g., the Gulf Stream) may rework and redistribute the uppermost few meters of sediment on the rises.

During glacial, low-sea-level epochs, terrigenous detritals pour directly into the deep sea, and the rises and abyssal plains grow and thicken. In the rise/abyssal-plain sequence, turbidites and slump deposits interlayer with fine hemipelagic (mixed terrigenous and pelagic) deposits, probably of nepheloid and contour-current origin.

Where margins are active tectonic provinces of subduction or shear, the rises and abyssal plains are absent because the trenches (see *Hadal Sedimentation*) and borderland types of topography effectively trap terrigenous sediment. In these areas, deep oceanic pelagic sediment types may be found just beyond the marginal trenches.

Deep Sea

In the deep sea, the particulate terrigenous materials from the diffuse and low-concentration suspended load accumulate at very slow rates due to distance from continental sources, or because of barriers formed by water circulation, bottom topography, or the filtering activity of planktonic life. The productivity of the pelagic organisms is then the dominant effect, and the resulting sediments will be biogenic oozes (see *Pelagic Sedimentation, Pelagic Sediments*). Typically the organic content of such sediments is low because of the action of bacteria, scavengers, and oxidation during its slow fall through the deep-water column. If the rates of accumulation are slow, much additional reworking will take place in the sediments as benthic organisms burrow and ingest the materials many times (see *Bioturbation*).

Surface water productivity, the primary factor in the production of the biogenic oozes, is governed by the availability of nutrients, in which oceanic circulation is a basic factor. Recent work has demonstrated that nutrients are present in higher concentration in Pacific deep water as compared to the Atlantic, and that the productivity in areas of upwelling in the former ocean is therefore higher. Increased productivity produces large quantities of silica and carbonate, but it also produces a large amount of organic matter which, upon oxidation in deeper water layers, generates carbonate dioxide and lower pH. Thus, solution rates also increase. This effect, plus the position of the carbonate and siliceous oozes, and changes in oceanic circulation are the basic factors that govern the areas of carbonate ooze accumulation. The deep zone of carbonate solution has been termed the *lysocline* (q.v.), which has shifted from time to time to produce the stratigraphic zonations evident in most deep-sea cores (see *Pelagic Sedimentation, Pelagic Sediments*). Another important factor is the changing position of oceanic crustal plates with respect to latitude. Plates moving below zones of high productivity pick up oozes, and plates subsiding below the carbonate compensation depth accumulate red clays (q.v.). This factor also contributes to the stratigraphy of any given oceanic location.

The biogenic oozes can be subdivided into two major classes: (1) the calcareous oozes of the equatorial and mid-latitude sea floors, and (2) the siliceous oozes of the great Antarctic Convergence. Comparison of the patterns illustrated in Fig. 1 with a generalized bathymetric map of the world oceans reveals the added influence of bottom topography. Sediments forming in depths >4500 m are typically composed of clays or zeolites with little or no carbonate. These sediments, as noted earlier, may be of terrigenous origin (*chthonic*) or from oceanic volcanism or chemical precipitation (*halmeic*).

General Patterns

Several generalizations may be applied to the general pattern of sediment characteristics. Marine deposits typically are thickest and deposited fastest adjacent to the continents and particularly in the zone of the continental rise. Detritals are normally coarsest near the shore when the products of a single sedimentological episode are considered. The apparent heterogeneity of the present shelf sediments is the result of overlap and interlayering of several cycles of deposition and erosion, as noted earlier. Carbonate content usually increases seaward and then decreases again in the deepest parts of the ocean. Organic contents in sediments usually are large in fine-grained deposits at intermediate depths, where sedimentation rates are also intermediate. Authigenic concentrations typically occur on surfaces receiving little or no clastic sediment and in zones where the precipitated components are at or above saturation in the overlying water. This effect may be related to volcanic exhalations, temperature changes, biological activity, or current flow.

Sedimentation Rates. Generally, the rates of sedimentation are fastest for terrigenous sediments, slower for carbonate biogenic clastics, and slowest for the authigenic minerals. The range of rates can be from cm per yr to fractions of mm per thousand yr (see *Sedimentation Rates*). Red clays (q.v.), found in the deep sea, are considered to be either residues formed by solution of biogenic particles or in part to be authigenic precipitates. Some workers have noted that much of the "red clay" is authigenic zeolite. Much must have originally come from the land, however, as indicated by latitudinal variations in clay mineralogy which match the major suites formed by weathering in the major climatic belts (see *Clay as a Sediment*). Recent studies have indicated that some mineral matter is suspended in all ocean water. Radioactive element distribution studies off the Oregon and Washington coasts have shown that the planktonic and pelagic organisms are surprisingly efficient screens for suspended particulate matter. The finer material is aggregated into sand-sized fecal pellets which rapidly fall onto the sea floor. In addition, nepheloid layers (see *Nepheloid Sediments*) show patterns related to land sources.

High Latitudes. Increasing contributions of poorly sorted and relatively unaltered *glacial debris* dilute oozes within the zone of the Antarctic Convergence as the Antarctic continent is approached. In addition, high productivity plus carbonate solution generates large accumulations of diatom tests in the sediments. In the N polar areas, the icebergs calved from piedmont and shelf glaciers transport materials to the adjacent seas, but probably in much less quantity than in the S polar regions, due to the relatively landlocked condition of the Arctic Ocean. The sediments of the Arctic Ocean contain much glacial marine sediment (q.v.).

Atmospheric Contributions. Wind-borne fine particulate quartz spreads out over the deep sea where the prevailing winds are offshore and where climate permits the exposure of the land surface to the wind's action (see *Eolian Dust in Marine Sediments*). Volcanic debris is also distributed in response to the prevailing winds, witnessed by the ash layers in deep-sea sediments downwind of explosive volcanic activity. In the deep-sea areas most distant from land, rates of sedimentation may be sufficiently slow so that extraterrestrial contributions become discernable (see *Extraterrestrial Material in Sediments*). Micrometeorites and cosmic dust are commonly found in the coarse fractions of red clays; tektites are found in belts related to fall patterns that may be of extraterrestrial origin.

Authigenic Sediments. Authigenic minerals

including phosphorite, glauconite, and manganese nodules accumulate slowly around nuclei such as foraminiferal tests or sharks' teeth (see *Authigenesis*). The internal structures of the authigenic sediments give evidence of periods of reworking and recementation. Often they contain large quantities of relatively rare trace metals as a consequence of the slow rates of accretion and the chemically active nature of the precipitates. The long exposure to the water of these chemically active surfaces scavenge the rarer elements much as absorption resins sweep ions out of solution. These authigenic minerals also yield information concerning the relative depths of their depositional environments. Glauconite (q.v.) and phosphorite (see *Phosphates in Sediments*) are apparently formed in continental-shelf depths, whereas manganese nodules are typically found on surfaces at abyssal depths (see *Pelagic Sedimentation*). It has been postulated that these precipitates may be biochemically, rather than physicochemically, generated.

Investigations

Methods. The great depth of water above most of the sea floor makes it relatively inaccessible to study. The advent of deep-diving vehicles, however, offers a means of making extensive direct observations of the bottom at all depths. Traditionally, the materials of the deep-sea floor have been sampled by lowering dredges, grabs, or corers from specially equipped research vessels. The expense and time necessary for sampling has limited the rate of accumulation of information, and the data from each sample are often extrapolated over hundreds and even thousands of km^2. The development of high-energy sound systems which can be operated while the ship is underway (obtaining reflections from layers below the sediment/water interface in continuous profiles) has dramatically multiplied the volume of information on the distribution of sediment in the ocean. This technique has demonstrated the very thin cover present in much of the sea outside of the abyssal plains and continental aprons. In areas of relatively thick accumulation, new insight has been gained concerning the mode of deposition of these materials and their structural relationships. Penetration of a km or more has been routinely obtained. Photographs of the sea floor, taken by bottom-activated or automatic cameras, are now being collected at an ever-increasing rate. Underwater television systems have also been developed. These photographs show evidence of relatively strong currents in all depths of the ocean. Ripple marks and scour are noted in many areas.

Direct sampling devices are limited in the size of the sample obtained. Dredges and grabs collect only the upper few cm of sediment or rock and frequently are composite samples and thus of limited use. Gravity coring devices of many types have penetrated to a maximum of about 30 m. The best deep-sea samples are now being obtained by the Deep Sea Drilling Project (DSDP) of JOIDES (Joint Oceanographic Institutions for Deep Earth Sampling), administered by the National Science Foundation. This coring has reached depths of several hundred m below the surface and has sampled complete sections of the deep-sea sediments.

Deep-sea drilling is producing increasingly more complete sections, although disturbance of samples may still be a problem. Conventional *gravity coring* is subject to a number of mechanical problems including bypassing of material, distortion of layers and lamination, and the foreshortening of structures. Some of these effects can be eliminated or diminished by the use of *piston coring* devices, but care should always be taken in evaluating cores, and particularly old and dry samples. Piston coring always eliminates the top few tens of centimeters of sediment, which may be sampled by adjacent *trigger cores*. *Box cores* efficiently recover surface material and have the advantage of large cross-sections for study of sedimentary structures.

Analysis. After collection, the cores are split and photographed. Both visible light and x-ray photography are used to preserve a record of original structures in the cores. The cores are carefully logged, and such data as color, structures, and gross composition are recorded versus depth in the core. For chemical studies, the samples often require special treatment or storage to preserve original moisture content. Analysis of moisture content with depth and also the chemistry of the interstitial waters (see *Interstitial Water in Sediments*) are a necessity in investigating diagenetic changes in the sediments. After the core has been logged, photographed, and critical measurements obtained, representative sections of the sediments are cut from the core either at regular intervals or at changes in color or lithology. These samples are analyzed for textural properties, general composition including bulk chemical composition, clay mineralogy, and gross mineralogy. Trace element contents and isotopic analyses are determined. Biogenic components such as tests, shells, or other hard parts are separated and identified.

All of the foregoing factors are usually plotted versus depth in the core. Variation in hydrogen ion content and state of oxidation and reduction are sometimes observed and compared with the variations in other factors. Core-to-core comparisons are attempted, and the relationship to bottom topography examined. Such factors as chemistry and circulation patterns of the overlying water are examined, as are the biological contents of the water. From these data are synthesized the history of the sedimentary episodes represented in the sections; the rates of accumulation, the alteration of the deposits after burial, and evidence of reworking or of several discrete periods of sedimentation are also sought. Ideally, the end product should include a comparison of the sections with analogs from the stratigraphic record.

Areas of Study. The geochronology of deep marine deposits is a major area of interest. Biostratigraphic position of benthic and planktonic foraminifera and their ecologic associations are extensively used. New techniques employ such nannoplankton as the extinct discoasters and the modern coccoliths; radiolarians and diatoms are also useful indicators. Radiocarbon dates are usually limited in their applicability to deep-sea chronologies because long cores may include the record of deposition through much of Tertiary time. Thorium and uranium have been used, as have such well-known methods as potassium-argon and strontium-rubidium dating (see *Sedimentation Rates; Radioactivity in Sediments*). Fission-track dating has also been applied in studies of volcanic ash.

At the present time, the major problems of marine sedimentation studies are questions concerning rates of accumulation, diagenesis, and transporting processes. A major goal of these studies is the reconstruction of the history of the oceans and of their paleo-oceanography. In the continental margins, the sedimentary environments may have equivalents in the great geosynclinal accumulations that make up a large proportion of the continental geologic record. The origins of the margins and the manner in which they evolve is recorded in the geometry of the continental-margin sedimentary masses and in the physical and chemical characteristics of these materials.

Marine sediments can yield evidence of worldwide climatic changes, changes in oceanic circulation, age of features of relief, evolutionary development of the planktonic organisms, and many other elements of the natural history of the earth.

DONN S. GORSLINE

References

Andrée, K., 1920. *Geologie des Meeresbödens*. Leipzig: Gebrüder Borntraeger, 689p.
Belderson, R. H.; Kenyon, N. H.; Stride, A. H.; and Stubbs, A. R., 1972. *Sonographs of the Sea Floor*. Amsterdam: Elsevier, 185p.
Broecker, W. S., 1974. *Chemical Oceanography*. New York: Harcourt Brace Jovanovich, 214p.
Burk, C. A., and Drake, C. L., eds., 1974. *The Geology of Continental Margins*. New York: Springer-Verlag, 1009p.
Bushnell, V. C., ed., 1973. Marine sediment of the southern oceans, *Am. Geog. Soc. Antarctic Map Folio Ser.*, 17, 18p and 9 maps.
Deacon, M., 1971. *Scientists and the Sea 1650-1900*. New York: Academic, 445p.
Dott, R. H. Jr., and Shaver, R. H., eds., 1974. Modern and ancient geosynclinal sedimentation, *Soc. Econ. Paleont. Mineral. Spec. Publ. 19*, 380p.
Emery, K. O., 1960. *The Sea off Southern California*. New York: Wiley, 366p.
Emery, K. O.; Tracy, J.; and Ladd, H. S., 1954. Geology of Bikini and nearby Atolls; Part I, Geology, *U.S. Geol. Surv. Prof. Pap. 260A*, 265p.
Emery, K. O., and Uchupi, E., 1972. Western North Atlantic Ocean: Topography, rocks, structure, water life and sediments, *Am. Assoc. Petrol. Geologists Mem. 17*, 532p.
Ericson, D. B., and Wollin, G., 1964. *The Deep and the Past*. New York: Knopf, 292p.
Fairbridge, R. W., ed., 1966. *The Encyclopedia of Oceanography*. New York: Reinhold, 1021p.
Goldberg, E. D., ed., 1973. *North Sea Science*. Cambridge, Mass.: MIT Univ. Press, 500p.
Goldberg E. D., ed., 1974. *The Sea*, vol. 5, Marine Chemistry. New York: Wiley Interscience, 895p.
Goldberg, E. D.; McCave, I. N.; O'Brien, J. J.; and Steele, J. H., eds., 1977. *The Sea*, vol. 6, Marine Modeling. New York: Wiley, 1048p.
Gross, M. G., 1972. *Oceanography a View of the Earth*. Englewood Cliffs, N.J.: Prentice-Hall, 581p.
Hampton, L., ed., 1974. *Physics of Sound in Sediments*. New York: Plenum, 567p.
Hay, W. W., ed., 1974. Studies in paleo-oceanography, *Soc. Econ. Paleont. Mineral. Spec. Publ. 20*, 218p.
Hedgpeth, J. W., and Ladd, H. S., eds. 1957. Treatise on marine ecology and paleoecology (2 vols.), *Geol. Soc. Am. Mem. 67*, 1296p and 1077p.
Heezen, B. C., ed., 1967. *The Quaternary History of the Ocean Basins*. New York: Pergamon, 344p.
Heezen, B. C., and Hollister, C. D., 1971. *The Face of the Deep*. New York: Oxford Univ. Press, 659p.
Hill, M. N., ed., 1962, 1963. *The Sea*, Vols. I (1962), II, III (1963). New York: Wiley Interscience, 864p, 554p, 963p.
Hood, Donald W., 1971. *Impingement of Man on the Oceans*. New York: Wiley Interscience, 738p.
Hsü, K. J., and Jenkyns, H. C., eds., 1974. Pelagic sediments: on land and under the sea, *Internt. Assoc. Sedimentologists, Spec. Publ. 1*, 447p.
Inderbitzen, A. L., ed., 1974. *Deep-Sea Sediments: Physical and Mechanical Properties*. New York: Plenum, 1974.
JOIDES, 1969-present. *Initial Reports of the Deep Sea Drilling Project*, Vol. 1-. Washington: U.S. Govt. Printing Office.
Keen, M. J., 1968. *An Introduction to Marine Geology*. Oxford: Pergamon, 218p.
Kuenen, P. H., 1950. *Marine Geology*. New York: Wiley, 568p.

Lisitsyn, A. P., 1972. Sedimentation in the world ocean, *Soc. Econ. Paleont. Mineral. Spec. Publ. 17,* 218p.

Maxwell, A. E., ed., 1971. *The Sea,* vol. 4 (2 parts). New York: Wiley Interscience, 791p and 664p.

Menard, H. W., 1964. *Marine Geology of the Pacific.* New York: McGraw-Hill, 271p.

Miller, R. L., ed., 1964. *Papers in Marine Geology.* New York: Macmillan, 531p.

Murray, J., and Renard, A. F., 1891. Report on deep-sea deposits based on the specimens collected during the voyage of H. M. S. *Challenger* in the years 1872-1876, in *Rept. Voyage Challenger,* London: Longmans, 525p (reprinted 1965, Johnson Reprint Co., N.Y.).

Nairn, A. E. M., and Stehli, F. G., eds., 1973-. *The Ocean Basins and Margins* (7 volumes). New York: Plenum, Volumes 5-7 in prep.

Pytkowicz, R. M., 1975. Some trends in marine chemistry and geochemistry, *Earth Sci. Rev.,* **11,** 1-46.

Schäfer, W., 1972. *Ecology and Palaeoecology of Marine Environments.* Edinburgh: Oliver & Boyd, 568p.

Shepard, F. P., 1973. *Submarine Geology,* 3rd ed. New York: Harper & Row, 517p.

Shepard, F. P.; Phleger, F. B.; and Van Andel, T. H., eds., 1960. *Recent Sediments, Northwest Gulf of Mexico.* Tulsa, Okla.: Am. Assoc. Petrol. Geologists, 394p.

Stanley, D. J., ed., 1973. *The Mediterranean Sea, a Natural Sedimentation Laboratory.* Stroudsburg, Pa.: Dowden, Hutchinson & Ross, 765p.

Trask, P. D., ed., 1939. *Recent Marine Sediments.* London: Murby, 736p.

Valentine, J. W., 1973. *Evolutionary Paleoecology of the Marine Biosphere.* Englewood Cliffs, N.J.: Prentice-Hall, 511p.

Van Andel, T. H., and Shor, G. G. Jr., eds., 1964. Marine geology of the Gulf of California, *Am. Assoc. Petrol. Geologists Mem. 3,* 408p.

Walther, J., 1893/1894. *Einleitung in die Geologie als historiche Wissenschaft.* Jena: Gustav Fischer, 1055p.

Weyl, P. K., 1970. *Oceanography an Introduction to the Marine Environment.* New York: Wiley, 535p.

Cross-references: *Abyssal Sedimentary Environments; Abyssal Sedimentation; Bahama Banks Sedimentology; Barrier Island Facies; Bay of Fundy Sedimentology; Chalk; Clay as a Sediment; Clay—Genesis; Clays, Deep-Sea; Continental-Rise Sediments; Continental-Shelf Sedimentation; Continental-Shelf Sediments in High Latitudes; Contourites; Coral-Reef Sedimentology; Deltaic Sediments; Eolian Dust in Marine Sediments; Estuarine Sedimentation; Extraterrestrial Material in Sediments; Glacial Marine Sediments; Glauconite; Gravity Flows; Hadal Sedimentation; Hardground Diagenesis; Interstitial Water in Sediments; Lagoonal Sedimentation; Littoral Sedimentation; Lysocline; Nepheloid Sediments and Nephelometry; Neritic Sedimentary Facies; Outer Continental-Shelf Sediments; Paleobathymetric Analysis; Pelagic Sedimentation; Pelagic Sediments; Persian Gulf Sedimentology; Radioactivity in Sediments; Red Clay; Reef Complex; Relict Sediments; Sedimentation—Paleoecologic Aspects; Sedimentation Rates, Deep-Sea; Submarine (Bathyal) Slope Sedimentation; Submarine Canyons and Fan Valleys, Ancient; Submarine Fan (Cone) Sedimentation; Tidal-Current Deposits; Tropical Lagoonal Sedimentation; Tsunami Sedimentation; Turbidity-Current Sedimentation; Volcanism—Submarine Products; Wadden Sea Sedimentology.* Vol. I: *Benthonic Zonation; Marine Sediments; pH and Eh of Marine Sediments.* Vol III: *Sediment Transport: Long Term Net Movement; Submarine Geomorphology.* Vol. IX: *Deep-Sea Stratigraphy.* Vol. XIII: *Marine Sediments, Compaction; Marine Sediments, Engineering Properties; Ocean Engineering; Submarine Geology.*

MARL

Marl is a sediment or sedimentary rock (marlstone) composed of a friable mixture of silt- and clay-sized grains of carbonate and clay minerals. Some authors limit the definition to sediments of lacustrine origin (see Terlecky, 1974), but the term has been extended, in many studies, to marine sediments. Its color is variable, commonly buff to white but often darker. The most frequent carbonate is calcite; dolomite and aragonite are rare, and siderite rarer. The chemical composition varies widely, the carbonate content ranging from 30-70% by weight. Among clay minerals, the most common is illite; smectite and kaolinite are less common. Among accessory components are quartz, glauconite, feldspars, minerals of the apatite group, pyrite, marcasite, and iron oxides and hydroxides, as well as organic substances. Megascopic stratification is usually poorly visible; and microscopically, in thin section, layers are also difficult to recognize—only by using radiographic methods it is possible to differentiate thin layering caused by small variations in sedimentation. Microscopic examination also reveals that most of the carbonate grains are of irregular habit, characteristic of detrital origin. Sand fractions, as well as silts, as a rule exhibit traces of mechanical transport. Some carbonate grains, however, exhibit distinct crystallographic features indicative of microorganic precipitation. Foraminiferal tests and coccolith fragments are often visible in thin section; fragments of macroorganisms are comparatively rare.

The most widespread marine marls contain variable amounts of macro- and microfaunal remains, often pointing to their autochthonous origin, without traces of transportation. The complete lack, or at most the presence of only small amounts, of coarse detrital material suggests that marls were deposited in calm marine basins, without intensive ocean current activity and isolated from the mouths of great rivers transporting coarse detrital material. Marls are, however, often associated with deltaic deposits (Sarnthein and Walger, 1973). The presence of macro- and microfaunal remains, as well as biogenic carbonate connected with the process of photosynthesis, indicate that the marls were formed in marine basins of moderate depths.

The classification of marls can be based on various criteria:

1. Qualitative composition of carbonates; dolomitic and sideritic marls may be differentiated from the most common calcite marls.
2. Accompanying minerals: when the carbonates and clay minerals are accompanied by significant amounts of glauconite, collophane or quartz grains, one can distinguish glauconite, phosphorite or sandy marls.
3. Environmental conditions of sedimentation: most common are marls of marine origin, lake marls being rarer and of limited area extent, where marine marls may cover hundreds to thousands of km^2.

Lake Marls

Lake marls, sometimes called meadow marls, are known especially from Tertiary and Pleistocene sediments. They are composed of calcium carbonate occurring as calcite or aragonite, the latter connected with the presence of lamellibranchs, gastropods and other skeletal remains. Lake marls are also known from Holocene sediments. Sedimentation takes place in a shallow-water environment rich in calcium carbonate, its precipitation mainly due to plant or bacterial action. Certain plants, notably the stonewort (*Chara*), are able to obtain carbon dioxide for photosynthesis from CO_2 in solution (Pettijohn, 1975). The calcium carbonate may be precipitated on plant tissue, gradually replacing its components or sloughing off and accumulating as a pseudobrecciated freshwater limestone. Excess calcium carbonate is precipitated as lime mud, mixing with the clay and shells.

Siderite Marls

Siderite marls, often called clay siderites, may form continuous beds or occur as accumulations of spherosiderite concretions, their chemical composition distinguished by a large amount of ferrous iron which joins with the carbon dioxide contained in the sediment. Siderite marls can be formed in fresh-water lakes and marshes or shallow seas, in any reducing environment connected with the disintegration of plant remains. Many siderite marls contain marine fauna and have an oolitic structure. It is probable that diagenetic processes are involved in the genesis of these rocks. The lack of coarse detrital fragments in this type of sediment speaks for a slow, calm sedimentation. Rocks of this type are known from the Precambrian sediments near Lake Superior in North America. Marine siderite marls and spherosiderites are known from flysch assemblages. Fresh-water sideritic marls are characteristic of carbon-bearing sediments.

Marls may also be rich in silica, pyrite, marcasite, or phosphorite concretions.

KAZIMIERZ ŁYDKA

References

Bolewski, A., and Turnau-Morawska, M., 1963. *Petrografia*. Warszawa: Wydawnictwa Geol., 811p.
Houbolt, J. J. H. C., 1957. *Surface Sediments of the Persian Gulf Near the Qatar Peninsula*. The Hague: Mouton, 113p.
Pettijohn, F. J., 1975. *Sedimentary Rocks*. New York: Harper & Row, 628p.
Sarnthein, M., and Walger, E., 1973. Classification of modern marl sediments in the Persian Gulf by factor analysis, in B. H. Purser, ed., *The Persian Gulf*. New York: Springer-Verlag, 81–97.
Terlecky, P. M., Jr., 1974. The origin of a Late Pleistocene and Holocene marl deposit, *J. Sed. Petrology*, 44, 456–465.

Cross-references: *Coal; Euxinic Facies; Glauconite; Lacustrine Sedimentation; Limestones; Peat-Bog Deposits; Phosphates in Sediments.*

MARTIAN SEDIMENTATION

Erosion in one place means sedimentation in another, and Mariner 9 (1971–1972) photos showed that parts of the martian surface have been extensively eroded. However, erosional features are more easily identified by the remote sensing techniques applied to Mars than are the resulting sediments (Mutch et al., 1976). Conclusions concerning the latter are necessarily based on interpretation bordering in many instances on outright speculation.

Most sedimentation, apart from processes involving condensed volatiles from the atmosphere, requires prior disintegration or decomposition of lithospheric materials and their subsequent transport. Weathering and transportation are limited under the present rigorous climatic environment of Mars, featuring a mean annual planetary temperature in the neighborhood of $-80°C$, a predominantly CO_2 atmosphere at a mean pressure of about 6 mb (compared to 1000 mb on Earth), and no significant liquid water at the surface. As long ago as 3-4 billion yr, Mars may have had a denser atmosphere, residual from the initial heavy meteoroidal bombardment, which possibly permitted liquid water on the martian surface. Many of the erosional and depositional features of Mars may thus have a considerable antiquity.

Processes of Fragmentation

Under current conditions, fragmentation of martian lithospheric materials is primarily by meteoroidal impact, catastrophic mass movements, and volcanism, possibly by tectonism, and perhaps by thermal fracturing, although the latter is suspect (Ryan, 1962; Sharp, 1968). Frost shattering, although favored by the thermal regime, is believed to be ineffective owing to the lack of liquid water, and the same may be true of disintegration by salts (Malin, 1974). Alteration and break up of surface materials may occur to some degree under the influence of adsorbed water vapor and the intense solar radiation flux through photostimulated oxidation (Huguenin, 1973). Under earlier conditions, these and more common processes of weathering may have occurred. Volcanism is widespread on Mars (Carr, 1973) and may have generated significant amounts of pyroclastic materials.

Transportation

Wind is the principal agent of transport thought to be effective under current conditions. Mass movements, especially slides, have occurred in the past and may take place currently. Dry avalanching is suggested by U-shaped chutes on steep faces; and creep, possibly caused by thermal fluctuations (Sharp, 1968), may be a pervasive process on the martian surface.

Large scoured channels locally scarring the martian surface suggest the the action of some fluid in times past (Milton, 1973). Concrete evidence of glaciation is not widely recognized, but some investigators speculate that it has occurred. Atmospheric transfer of the volatiles composing condensed particles within sedimentary accumulations is to be expected.

Examples of Martian Sedimentary Deposits

Layered Deposits of the Polar Regions. The most striking and confidently identified sedimentary accumulations on Mars are the sheets of well-stratified materials (Fig. 1) peripheral to and presumably underlying the perennial ice caps at both poles. These layers have been the subject of much study and debate. They are exposed over roughly 1.5×10^6 km² in the S and 1.1×10^6 km² in the N polar regions. Total thickness at maximum is 2-4 km. Individual layers are remarkably uniform over extensive exposures, and they differ in thickness, one from the other, only by a factor of 2 or 3, the average being perhaps 30 m. This uniformity of layering has been speculatively attributed to

FIGURE 1. Eroded layered deposits of S polar region. Width of photo is about 60 km.

variations in the obliquity of the spin axis of Mars (Murray et al., 1973).

Constitution of the layers is a matter of debate, with water ice, solid CO_2, CO_2-clathrate, dust, and volcanic ash all considered possible. The difference in albedo between the layers is presumed to reflect different admixtures of some of the above materials. All these substances are presumed to have been carried to the polar region in the atmosphere, so they are eolian deposits.

Other Polar Deposits. Beneath the well-layered deposits in both polar regions, but of somewhat greater areal extent, are sheets of massive homogeneous materials of possibly comparable thickness. These underlying sheets may be of volcanic origin, but there is as yet no firm proof that they are not sedimentary.

Both the perennial polar ice caps and the annual polar frosts are sedimentary deposits of atmospheric origin. Arguments have been advanced to the effect that the perennial caps must be predominantly water ice except possibly for one part of the N polar cap which may be solid CO_2 (Murray and Malin, 1973). The annual polar frosts are believed to be solid CO_2, inevitably accompanied by a little water ice and some CO_2-clathrate (Miller and Smythe, 1970).

Subpolar Mantled Regions. Both the layered

and homogeneous deposits of the polar regions have been extensively eroded, presumably by eolian processes (Cutts, 1973; Sharp, 1973a). Soderblom et al. (1973) suggest that the remobilized material composes a mantle of eolian debris covering broad belts in both the northern and southern hemispheres poleward of 30–40° latitude. These mantles are presumably sheets of dust (loess) thinning outward toward lower latitudes but locally thick enough to bury craters up to a few km in diameter.

Huge storms aperiodically obscuring Mars suggest that much of its surface may be dusty, but this cover must be so thin and so easily remobilized that it does not significantly obscure topographic features in mid- and equatorial latitudes. Movements of such dust, and possibly also of eolian sand, are probably responsible for temporal changes in albedo markings on the martian surface (Sagan et al., 1973).

Dunes. The degree of eolian activity on Mars creates an expectation that dunes should be abundant. They may be, but confident identification is difficult on Mariner 9 photos unless the size of individual martian dunes is on a much larger scale than in most terrestrial dune fields. An exception would be the groups of huge longitudinal dunes of some terrestrial deserts which are large enough to be seen but which have not yet been recognized on martian photos.

There are a few instances of features believed to be martian dunes of huge size. One is a probable dune field (Fig. 2) 30 by 60 km in area on the floor of a large crater at $47.5°S, 331°W$ (Cutts and Smith, 1973). The individual forms within this area look like typical transverse terrestrial dune ridges, but they are 0.5–2 km apart, which is much greater than the spacing of ridges within most terrestrial transverse dune complexes. Similar forms of corresponding size are also seen in parts of the wind-scoured areas of the S polar layered deposits and perhaps locally elsewhere on Mars (Cutts and Smith, 1973).

Mars may have many other dune fields with individual forms on a scale comparable to terrestrial analogs. Such areas would simply appear as light (or possibly dark) spots on Mariner 9 photographs (Smith, 1972). Light areas on crater floors (Cutts et al., 1971) and the light and dark streaks seen extending outward from craters and other topographic obstructions on the martian surface following the great storm of 1971 probably represent, in many instances, lee-side accumulations of eolian material (Sagan et al., 1973), at least some of which is probably sand and may be in the form of dunes.

Mass Movement Deposits. The scars of large slides are abundant on the steep walls of some craters, calderas, and particularly the huge equatorial troughs (Valles Marineris). At the base of these features are irregularly jumbled masses of slide debris, and part of the floors of the Valles Marineris troughs are mantled with such material. Similar deposits have been formed by widespread slumping and sliding in areas of chaotic terrain (Sharp, 1973b). An escarpment, possibly 6 km high, encircles the base of Olympus Mons, and its face displays much evidence of slide, slump, and creep deposits. In one spot, a lobate mass of jumbled debris extends 35 km out onto the gentle slope at the scarp base. This mass may represent a rock-fall slide similar to but much larger than known terrestrial analogs. Many steep martian faces are based by steep but smooth looking slopes which may be talus aprons.

Alluvium. Some martian channels display large-scale braided patterns (Fig. 3), possibly analogous to features created by the Spokane and Bonneville floods of western US (Baker and Milton, 1974). If this interpretation is correct, then the patterned material should consist of alluvial deposits. Broad, relatively featureless plains in the debouchment areas of other large channels possibly created by great outflow floods may be covered by alluvium. An example may be seen around $12°N, 28°W$. Within the equatorial trough, Ganges ($7.5°S, 49°W$), is a dissected tableland composed of well-layered materials (Fig. 4), most likely of sedimentary origin but possibly composed of layered volcanics. This accumulation is at least 2 km thick, and materials of this type may be widespread on the floors of these equatorial troughs. They could be primarily alluvial in nature.

That part of the northern hemisphere of Mars

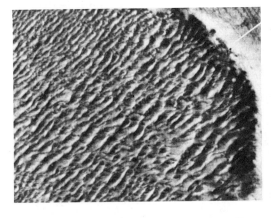

FIGURE 2. Large field of transverse dunes(?) on floor of martian crater. Width of dune field shown is 40 km.

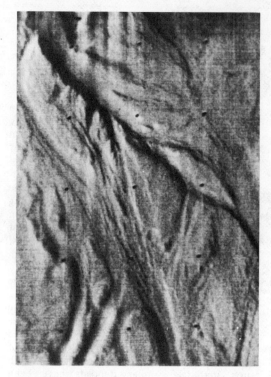

FIGURE 3. Braided pattern in possible alluvial materials in Mangala channel. Width of photo is about 25 km.

between latitudes 45° and 70° is known to be abnormally low topographically. It is expressed as a broad circum-planetary belt of nearly featureless topography, and the inference is logical that this belt is more deeply mantled by secondary deposits than most other parts of Mars. Some of these deposits are probably eolian (Soderblom et al., 1973), but this great depressed area may also be filled in part with volcanic debris and possibly alluvium.

ROBERT P. SHARP

References

Baker, V. R., and Milton, D. J., 1974. Erosion by catastrophic floods on Mars and Earth, *Icarus,* **23**, 27–41.

Carr, M. H., 1973. Volcanism on Mars, *J. Geophys. Research,* **78**, 4049–4062.

Cutts, J. A. 1973. Wind erosion in the Martian polar regions, *J. Geophys. Research,* **78**, 4211–4221.

Cutts, J. A., and Smith, R. S. U., 1973. Eolian deposits and dunes on Mars, *J. Geophys. Research,* **78**, 4139–4154.

Cutts, J. A.; Soderblom, L. A.; Sharp, R. P.; Smith, B. A.; and Murray, B. C., 1971. The surface of Mars: 3. Light and dark markings, *J. Geophys. Research,* **76**, 343–356.

Huguenin, R. L., 1973. Photostimulated oxidation of the Martian surface, *J. Geophys. Research,* **78**, 8481–8493.

Malin, M. C., 1974. Salt weathering on Mars. *J. Geophys. Research,* **79**, 3888–3894.

Miller, S. L., and Smythe, W. D., 1970. Carbon dioxide clathrate in the martian ice cap, *Science,* **170**, 531–533.

Milton, D. J., 1973. Water and processes of degradation in the martian landscape, *J. Geophys. Research,* **78**, 4037–4047.

Murray, B. C., and Malin, M. C., 1973. Polar volatiles on Mars—theory vs. observations, *Science,* **183**, 437–443.

Murray, B. C., Ward, W. R. and Young, S. C., 1973. Periodic insulation variations on Mars, *Science,* **180**, 638–640.

Mutch, T. A.; Arvidson, R. E.; Head, J. W. III; Jones, K. L.; and Saunders, R. S., 1976. *Martian Sedimentation.* New Jersey: Princeton Univ. Press, 400p.

Ryan, J. A., 1962. The case against thermal fracturing at the lunar surface, *J. Geophys. Research,* **67**, 2549–2558.

Sagan, C., and 10 others, 1973. Variable features on Mars, 2. Mariner 9 global results, *J. Geophys. Research,* **78**, 4163–4196.

Sharp, R. P., 1968. Surface processes modifying martian craters, *Icarus,* **8**, 472–480.

Sharp, R. P., 1973a. Mars: Fretted and chaotic terrains, *J. Geophys. Research,* **78**, 4073–4083.

Sharp, R. P., 1973b. Mars: South polar pits and etched terrain, *J. Geophys. Research,* **78**, 4222–4230.

Smith, H. T. U., 1972. Aeolian deposition in Martian craters, *Nature Phys. Sci.,* **238**, 72–74.

Soderblom, L. A.; Kreïdler, T. J.; and Masursky, H., 1973. Latitudinal distribution of a debris mantle on the Martian surface, *J. Geophys. Research,* **78**, 4117–4122.

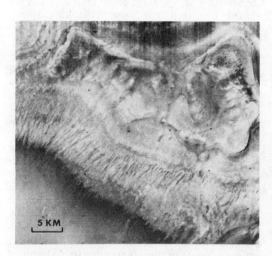

FIGURE 4. Layered materials, possibly alluvial in nature, on floor of Ganges trough. Vertical relief perhaps 2 km; bar gives horizontal scale.

Cross-references: *Braided-Stream Deposits; Eolian Sedimentation; Extraterrestrial Material in Sediments; Flood Deposits; Lunar Sedimentation; Mass-*

Wasting Deposits; Volcaniclastic Sediments and Rocks. Vol II: *Mars.*

MASS-WASTING DEPOSITS

Mass-wasting deposits accumulate by the transfer of rock debris under gravity, either subaerially or subaqueously, without the influence of water as a transporting medium. Quantitatively, they are probably the most important group of deposits after marine sediments. The type of material is the principal factor determining not only the type of mass wasting failure, but also the expression of structural features within each group, whether this be deposits resulting from falling, sliding, flowing, creeping, or subsidence. Thus, in a slide movement, a crystalline or arenaceous rock may behave as a rigid elastic solid and display features such as brittle fracture along shear zones, whereas a clay rock behaving as a plastic or quasi-plastic mass will show deformation features, and realignment of minerals in the shear matrix, without evidence of fracturing (see *Gravity-Slide Deposits*). An increase in moisture content beyond certain critical boundary values will transform the mode of movement (see *Mudflow Deposits*).

Sedimentary features within the main groups of mass wasting mentioned above can be examined in terms of their disorganization or disaggregation—i.e., the alteration of original depositional features, such as grain packing and orientation, or bedding characteristics—and of their reorganization—i.e., their sorting or progressive loss of random features. Using these two parameters and the type of material involved, the main groups are deposits formed by falls, slides, mass flows, creep, and subsidence.

Deposits Formed by Falls

This group includes the vertical or near vertical falls or slumps of rock or soil, and the rolling, bouncing, and discontinuous slides of rock debris down inclined surfaces, where the horizontal component of movement is equal to or greater than the vertical. The latter process is normally termed *avalanching,* although this term is better reserved for movements involving slush or snow (see *Avalanche Deposits*). Disorganization can be slight, as in many rockfalls, or complete where cohesionless or nearly cohesionless material such as sand is involved. In vertical falls, there is generally no reorganization, but avalanching is a process of reorganization of debris. Good sorting is only achieved by postdepositional processes such as creep on talus or scree cones. Bedded screes have been termed *grès litées* or *éboulis ordonées* by European geologists.

Slide Deposits

The presence of shear planes distinguishes this group from others. Disorganization ranges from reorientation of large blocks in rotational slips to the complete repacking of grains in many flow slides. Sorting is virtually absent in all types, although orientation of minerals of platey habit, in and around shear zones of plastic slides, may occur (see *Gravity-Slide Deposits*).

Mass-Flow Deposits

There are no obvious discrete shear surfaces in mass-flow deposits—disorganization is complete, apart from the survival of clasts retaining many features of the original deposit. Hard rock clasts may survive flow for many kilometers, but lutite clasts are rapidly broken down. Clasts display a typical dipping-girdle fabric (see *Mudflow Deposits*). Reorganization may be evident but is in general much poorer than in waterlaid deposits. Difficulty arises when comparing debris "avalanches" and debris flows, as the deposits can display many similar features.

Deposits Accumulating from Creep Movements

Creep movements involving the slow transfer of rock, rock debris, or soil have been divided into three groups: seasonal, continuous, and progressive. Only *seasonal creep* produces a distinctive deposit. Seasonal creep, involving the upper part of the soil mantle and the larger material within this zone, is often clearly seen where a stratified sequence outcrops on a slope, and horizons are attenuated downslope, ultimately passing into trails of debris or stone lines within the mantle. Movement in soil creep is through distributed shear and is essentially one of reorganization. *Talus* or *rock creep* may clearly show this reorganization when fissile debris is realigned parallel to the slope. Frost action can play an important part in such movements, and solifluction deposits accumulate in part as a result of frost creep.

Subsidence

There are many causes of subsidence (see Vol. III: *Mass Movement*), but they all involve the downward movement of rock, soil, or debris with no free side. The degree of disorganization depends to a large extent on the depth of sinking and the type of material involved, but generally is greater around the margin of movements. Any reorganization that occurs results from the concentration of water flow and seepage through or around the subsiding mass, and the movement of material in solution or by winnowing. Subsidence deposits are relatively unimportant, except in limestone areas (see *Breccias, Sedimentary*). Deposits that have sunk into the surrounding bedrock, however, may be preserved long after the strata from which they were derived have been removed; these remnants often provide evidence of the former existence of such strata and enable the depositional and denudational history of the area to be reconstructed.

Many subdivisions of the above groups have been made, e.g., Varnes (1958) and C. N. Savage (see Vol. III: *Mass Wasting*); but these subtypes are rarely discernible in sedimentological terms. Many deposits show evidence of the operation of more than one mass-wasting process and can be considered transitional in character.

PETER GASCOYNE

References

Hsü, K. J., 1975. Catastrophic debris streams (Sturzstroms) generated by rockfalls, *Geol. Soc. Am. Bull.,* **86,** 129–140.

Sharpe, C. F. S., 1938. *Landslides and Related Phenomena.* New York: Columbia University Press, 136p.

Varnes, D. J., 1950. Relation of landslides to sedimentary features, in P. D. Trask, ed., *Applied Sedimentation.* New York: Wiley, 229–246.

Varnes, D. J., 1958. Landslide types and processes, in E. B. Eckel, ed., Landslides and engineering practice, *Highway Research Bd. Spec. Rept. 29,* 20–47.

See also references under various cross-referenced articles.

Cross-references: *Avalanche Deposits; Breccias, Sedimentary; Clastic Sediments and Rocks; Conglomerates; Gravity Flows; Gravity-Slide Deposits; Mudflow, Debris-Flow Deposits; Pebbly Mudstones; Penecontemporaneous Deformation of Sediments; Slump Bedding; Submarine (Bathyal) Slope Deposits; Submarine Fan (Cone) Deposits; Talus; Turbidity-Current Sedimentation.* Vol. III: *Avalanche; Debris Flow; Earthflows; Landslides; Mass Movement; Mass Wasting; Mudflow; Soil Creep; Talus Fan or Cone.*

MATURITY

The concept of maturity in sediments was first published by Plumley (1948) in his study of the Black Hills Terrace gravels of South Dakota. These concepts were similar to the ideas expressed by Pettijohn (1949), with whom Plumley worked at the University of Chicago.

Pettijohn (1949) defined maturity of a clastic sediment as "the extent to which it approaches the ultimate end product to which it is driven by the formative processes that operate upon it." Plumley (1948) defined the three fundamental indices of maturity as *roundness, sphericity,* and *lithology.* He described a sediment at the youthful end as characterized by extreme angularity, low sphericity, and a high percentage of *labile* (chemically and/or mechanically unstable) *constituents.* On the other hand, he described a very mature clastic sediment as one consisting primarily of highly rounded, spherical particles and lacking any appreciable percentage of labile constituents.

Pettijohn (1949) expressed the maturity of a sandstone in mineralogical terms simply as a ratio of one mineral to another, e.g., quartz:feldspar, or (quartz+chert):feldspar, or (quartz+chert):(feldspar+rock fragments). He listed the maturity indices of some of the common sandstones, ranging, e.g., for a quartz:feldspar ratio, from about 1.1 for arkose, to >10 for an orthoquartzite.

Pettijohn (1949) further pointed out that one of the most important aspects of maturity is that it is a measure of *time.* Obviously if a clastic sediment has had only a brief history it should be immature. The intensity of the processes that have operated on the sediment is important as well. Hence a beach is a zone where maturity will be reached within a relatively short span of time because of the intensity of wave action in the surf zone.

Following Krynine's (1948) concept that sediments are characterized by two fundamental properties, mineral composition and texture, Folk (1951) suggested that *textural maturity* of a sediment can be expressed in terms of *clay content, sorting,* and *roundness.* He described four distinct stages of textural maturity: (1) immature stage, (2) submature stage, (3) mature stage, and (4) supermature stage. Three basic processes—removal of clay, increase in sorting, and degree of rounding—are used to define these stages. They progress from a poorly sorted, immature sediment with angular grains and high in clay content, to a supermature sediment with extremely well-sorted and rounded grains and no clay. Folk (1956; Fig. 1) showed these four stages of textural maturity and mineral composition plotted for a number of different sandstones, embracing most common sandstone types. Fig. 1 illustrates that all sandstones, no matter what their mineralogical compositions, can occur in most or all stages of textural maturity.

Interpretation

The concept of *mineralogical stability* has a great bearing on a sediment's relative maturity.

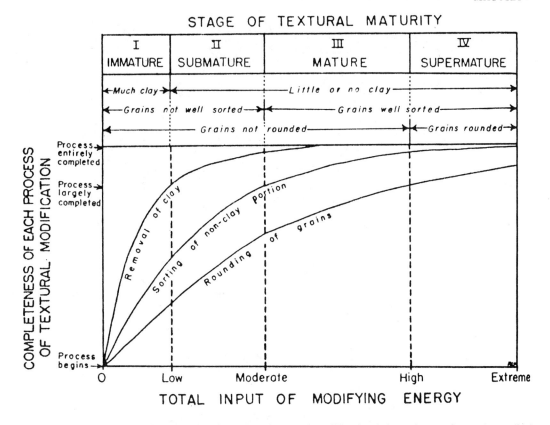

FIGURE 1. Relative completeness of each process of textural modification (winnowing, sorting, and rounding) as a function of the amount of energy expended. When a given process is largely completed, the rock passes from one stage into the next one, as shown by the broken vertical lines (from Folk, 1951).

Many recent studies of modern sediments have examined the complete problem of mineral and rock-fragment stability, revealing the numerous problems inherent in such studies, e.g., the continuous recycling of detrital grains. Some of these problems were reviewed by Blatt (1967), who concluded that judgments made on the origin of detrital grains were mostly subjective. This problem is particularly intense for ancient sedimentary rocks, yet even in modern sediments, changes in mineralogical composition do not necessarily reveal changes in maturity. For instance, Cleary and Conolly (1971), in their study of quartz in the modern sediments of piedmont and coastal plain environments, described a sand consisting of almost 100% quartz grains, many of which were well rounded. This sand is by definition an orthoquartzite and could be considered mature; these properties, however, could well be inherited from the rocks and soils from which they have just been derived, as is the case in many areas of the coastal plain of South Carolina where present-day streams cut through older coastal-plain sands. The presence of a small percentage of extremely angular grains in these sediments gives an important clue to the textural immaturity.

Conversely, some modern deep-sea sediments are composed of extremely angular fragments and contain an abundance of so-called mineralogically unstable minerals and rock fragments, e.g., the sands of the Hatteras deep-sea fan of the NW Atlantic (Cleary and Conolly, 1974). These sands have probably undergone extensive transportation and reworking, for they now rest thousands of km from their source area. They have indeed suffered a long transportation history and hence might be regarded as mature; yet, using Folk's (1951) classification, they are obviously immature.

These examples illustrate the problem inherent in the concept of maturity. Provenance (q.v.) studies and close textural examination are necessary for a reliable investigation; as shown above, however, the data are still open to subjective interpretation.

JOHN R. CONOLLY

References

Blatt, H., 1967. Provenance determinations and recycling of sediments, *J. Sed. Petrology,* 37, 1031–1044.
Cleary, W. J., and Conolly, J. R. 1971. Distribution and genesis of quartz in piedmont and coastal plain environments, *Geol. Soc. Am. Bull.,* 82, 2755–2766.
Cleary, W. J., and Conolly, J. R., 1974. Hatteras deep-sea fan, *J. Sed. Petrology,* 44, 1140–1154.
Krynine, P. D., 1948. The megascopic study and field classification of sedimentary rocks, *J. Geol.,* 56, 130–165.
Folk, R. L., 1951. Stages of textural maturity in sedimentary rocks, *J. Sed. Petrology,* 21, 127–130.
Folk, R. L. 1956. Discussion. Role of texture and composition in sandstone classification, *J. Sed. Petrology,* 26, 166–171.
Pettijohn, F. J., 1949. *Sedimentary Rocks.* New York: Harper & Row, 526p.
Plumley, W. J., 1948. Black Hills Terrace gravels: A study in sediment transport, *J. Geol.,* 56, 526–577.

Cross-references: *Arkose, Subarkose; Graywacke; Heavy Minerals; Provenance; Roundness and Sphericity; Sands and Sandstones; Sorting; Weathering in Sediments.*

MICRITE

Micrite is a term, first proposed and used by Folk (1959), to describe aphanitic limestones consisting almost entirely of a mosaic of interlocking calcite crystals 1–4μ in diameter. It is found either as an unconsolidated calcite or aragonite ooze or mud, or in a consolidated form, and may be of either chemical or mechanical origin. Sediments of this composition and grain size are equivalent to the often misused term *lithographic limestone,* a fine-grained limestone used in lithography, of which Jurassic Solenhofen is the classic example. Folk classified limestones of similar texture but with grains 5–15μ in diameter as *microspar,* which he believed was the product of secondary growth of slightly larger crystals at the expense of micrite.

If there is any appreciable fossil content (1–10%), the sediment may be termed a *fossiliferous micrite.* Increasing fossil content grades continuously up into a *biomicrite.* Similarly, some inclusions of pellets (1–10%) would be termed *pelletiferous micrite,* while above this percentage there is a transition to a *pelmicrite.* Inclusions of oolites (*oomicrite*) and intraclasts (*intramicrite*) are relatively rare, although they are sometimes seen in mixtures of various allochemical components within a micrite matrix.

Folk and others consider micrite to be the result of a rapid precipitation of microcrystalline ooze in an area lacking strong currents. Many studies have suggested that algal activity is involved (see *Limestones*).

Field identification of very fine-grained limestones and sediments should be made using the general term *calcilutite* until they may be properly examined and classified under the microscope (Folk, 1959).

B. CHARLOTTE SCHREIBER

References

Alexandersson, T. 1972. Micritization of carbonate particles: Processes of precipitation and dissolution in modern shallow-marine sediments, *Bull. Geol. Inst. Univ. Uppsala,* 3, 201–236.
Bockelie, T. G., 1973. A method of displaying sedimentary structures in micritic limestones, *J. Sed. Petrology,* 43, 537–539.
Folk, R. L., 1959. Classification of limestones. *Bull. Am. Assoc. Petrol. Geologists,* 43, 1–38.
Kobluk, D. R., and Risk, M. J., 1974. Devonian boring algae or fungi associated with micrite tubules, *Canadian J. Earth Sci.,* 11, 1606–1610.
Walther, J., 1904. Die solnhofener Plattenkalke bionomisch betrachtet, in *Festschrift zum 70ten Geburtstage von Ernst Haeckel.* Jena: Gustav Fischer, 604p.
Wolf, K. H., 1965. "Grain diminutions" of algal colonies to micrite, *J. Sed. Petrology,* 35, 420–427.

Cross-references: *Calcarenite, Calcilutite, etc.; Carbonate Sediments—Bacterial Precipitation; Limestones; Microbiology in Sedimentology.*

MICROBIOLOGY IN SEDIMENTOLOGY

The occurrence and activities of microorganisms, sometimes called *Protists,* in sedimentologic processes, have been widely documented. Few articles have appeared however, that treat the total effects of microorganisms in the processes of sedimentation and the diagenesis that takes place after deposition. The activities and types of microorganisms are as varied as the sediment composition and site of deposition, whether it be in shallow or deep water, the tropics or polar regions. A common feature, however, is that wherever there are sediments there is or has been microbial activity.

The microorganisms found in sediments are the microscopic yeasts, fungi, algae, protozoans, and bacteria. They enter the sediment adsorbed to particulate materials. They can move through interstitial pore spaces, and are redistributed by burrowing activities of higher organisms, resorting of sediments by currents or storms, and human activities such as dredging. In sediments of large porosity, microorganisms are relatively free to move about, but they tend to become entrapped in clays.

Most of the microbial activities occur early after the depositional processes. The sediment depth to which microorganisms are still active is difficult to determine. In most samples collected by conventional oceanographic coring techniques using gravity equipment (up to 80 m of core), microorganisms have been shown to be present. In such procedures, however, contamination is always possible. In deep or shallow drilling, it is relatively impossible to collect a clearly uncontaminated sample.

The diagenetic activities of microorganisms are directly related to available energy and sediment porosity. Energy is available in the form of organic matter or light in aerated, shallow-water sediments and as reduced inorganic ions in anaerobic environments, although this latter energy source is usually tied to former presence of organic matter. Generally, sediments become anaerobic whenever sufficient organic matter is present and environmental conditions permit the microbial utilization of oxygen to result in anoxic conditions. There are very few environments, such as those associated with volcanic activity, that are anaerobic through purely chemical processes.

The effects of microbial activities in sediments are complicated by the wide range of types and numbers of microorganisms present, the variable chemical and physical nature of the sediment, the type of energy and carbon source, water depth, temperature, sediment depth, etc. The following microbial activities in sediments have been observed in situ and in laboratory experiments: (1) regulation or change of pH and Eh; (2) alteration of organic and inorganic carbon, nitrogen, and phosphorous; (3) solubilization and precipitation of inorganic compounds; (4) thixotropic changes and surface potential of sediment particles; (5) production of gases; (6) concentration of ions from seawater/sediment interfaces; (7) alteration of isotope ratios; and (8) production of sediments.

The abundance and types of microorganisms are related to sediment porosity (*Lebensraum*), the depth of the sediment, and age after deposition. Larger microorganisms are generally related to the coarser sediments. Numbers may vary from a few per gram of wet sediment to hundreds of millions per gram in shallow-water sediments rich in nutrients. If the microorganisms are motile, they can migrate; and some are chemo- and phototactic. For example, in intertidal mud flats, blue-green algae and/or diatoms may change the color of the surface sediments as they move in response to light intensity.

Colloidal Phenomena

The physical forces involved in sedimentation processes are important to microorganisms. Surface changes due to Zeta potential or electrokinetic forces become significant to microorganisms with high surface-to-mass ratio such as the bacteria. There, electrokinetic changes result in a phenomenon known as *thigmotaxis,* where the charged surfaces of bacteria are attracted to sediments, organic molecules, and detritus.

This surface attraction may play a role in the movement of microorganisms through sediments by regulation of the attachment and displacement of types of microorganisms. There is also a possibility that charged microorganisms moving through the sediment may set up microspheres of electrical potential in the sediment pore spaces. This phenomenon is known as *streaming potential* and may account for the results shown by Perfeliev and Gabe (1969), where microorganisms were found associated in bands in the sediment.

Color Production

The activities of microorganisms may alter the color of sediments (q.v.) through the production of photosynthetic pigments, creation of sulfide layers, the precipitation of iron sulfide, and the diagenesis of organic matter on sediment-particle surfaces. The author has observed beach sands covered with green photosynthetic microorganisms and red photosynthetic bacteria to a depth of two centimeters. Color effects resulting from microbial activity are best demonstrated in a Winogradsky column. Sediment from shallow-water environments may be collected in a plastic or glass tube 2 in. in diameter and 12 in. long. If light conditions and organic matter in the sediment are regulated, the following series of events may be observed to take place in a few weeks; these demonstrate several activities of the microorganism in natural shallow-water environments. In the presence of subdued light, the surface of the sediment along the side of the glass may turn green or red due to the growth of various algae or photosynthetic bacteria. In such a closed system, algae do not produce enough oxygen to maintain an aerobic environment, except for small microcosms. As heterotrophic microorganisms oxidize the organic matter in the sediment, they decrease the oxygen content. When the oxygen is completely gone, the sulfate-reducing bacteria will produce sulfide, which in turn combines with iron and other cations, turning the sediment black.

The degree of color change will be related to the amount and type of organic matter present as energy. The sediment will remain black until oxygen is reintroduced, at which time it will turn the reddish brown of iron oxide. The presence of sulfide may also provide energy for the photosynthetic bacteria, and a green or red color may be associated with the black zone, depending on the species present. If the sediment is placed in direct light, diatoms or algae will grow along the glass, producing gas bubbles in the sediment and a brownish green color of the carotinoid pigments, etc., occurs. Changes in pH, etc., will also take place; but these are not apparent to the eye during the Winogradsky column experiment.

Microbial Activities

pH and Eh Changes in Sediments. One of the primary modifying forces acting on the pH and Eh of sediments are living organisms (see Vol. I: *pH and Eh of Marine Sediments*). Burrowing activities, even in the deep seas, will effectively plow up the surface, producing changes both physically and chemically. In such environments, the sediments that eventually become consolidated are the results of much physical movement. The greatest biological change in pH and Eh is produced during the activities of the living organisms, (1) as they consume and produce oxygen by photosynthesis and respiration; (2) by altering the carbonate buffer system by both consuming and producing carbon dioxide; and (3) during the production and consumption of acids and bases, and reduced and oxidized molecules or ions.

Alteration of Carbon, Nitrogen, and Phosphorous. As microorganisms metabolize, carbon is utilized and nitrogen and phosphorus ratios in sediments will change. Phosphorous released in the ion form during organic-matter diagenesis may be precipitated as Hydroxyapatite and may be resolubilized under anaerobic conditions. Nitrogen may be released as ammonia under aerobic conditions with an appropriate pH increase. Under anaerobic conditions, nitrogen gas is released. Residual carbon compounds such as humus and hydrocarbons are preserved under anaerobic conditions. Fossil hydrocarbons are thought to be the result of microbial diagenesis of protoplasm, plus time, temperature, and pressure. Some hydrocarbons (q.v.), such as the carotenoids, low molecular weight paraffins, and complex waxes are always found in recent sediment organic matter.

In highly alkaline environments, the carbon dioxide released by metabolism may be precipitated as carbonate. When large amounts of CO_2 are released, carbonate molds such as found in fossil fishes may be produced. If the metabolic oxygen consumption exceeds oxygen diffusion into the sediment, anaerobic conditions are produced which in turn cause several effects noted below. Rates of decomposition will vary with the environment following biologically dependent effects of temperature, pressure, Lebensraum, particle size, etc.

Solubilization and Precipitation of Inorganic Compounds. During metabolic activity, microorganisms produce and use up acids, bases, surfactants, and chelating substances and may, in some instances where growth is rapid and numbers are large, increase temperatures. While most microorganisms will produce weak organic acids, strong acids (up to 0.5 N sulfuric acid) may be produced by the sulfur bacteria. During photosynthesis and respiration, CO_2 uptake and production will alter the carbonate equilibrium and cause pH changes. Changes in redox potential will cause both precipitation and solubilization of minerals in sediments. Clays are altered through sorption of bacterially produced organic materials that may change the acid-base relationships by the introduction of ions or small organic molecules to the lattice, and may in some instances substitute ions to change the clay mineralogy. Some evidence has shown that clay minerals may be in part formed or destroyed through the activities of microorganisms during weathering processes.

The production of surface-active materials that collect at the water-air surface may act as flotation phenomenon to transport fine particles and later release the particles to take part in sediment formation. The same materials may affect the rate of compaction of hydrated sediments or produce change in thixotropy due to the presence of organic materials on the surface of sediment particles. Thus, microorganisms may produce or destroy colloidal micelles that act to hydrate or change the sapropel at sediment/water interfaces.

Thixotropic Changes Affecting Sedimentation. Sediments are commonly in a colloidal state. Fine sediment particles can be coated with organic materials that are attracted by surface effects. Thus, thixotropic sediments are formed and altered as the microorganisms produce and consume organic materials. Consolidation rate in areas of high organic content may be affected by microbial processes. This is especially true for clay-rich sediments in shallow-water environments such as estuaries.

Production of Gases. The metabolic byproducts of microorganisms include oxygen, carbon dioxide, nitrogen, methane, and hydrogen sulfide (see also *Gases in Sediments*). These gases may in some instances be entrapped in the sediment, producing gas bubbles. This is especially true in shallow intertidal mud flats covered by blue-green algal mats. The mat traps the gases in sediment, and often the sediment for 10 cm below the surface looks like bread dough that has risen by the action of yeast. In some sediment along the Texas coast, this phenomenon acts as a sediment heat sink, and the gaseous sediment may have a temperature

10°C higher than adjacent solid sediment. Entrapped gas in deep-sea sediments may physically change the appearance of core samples by a degassing effect accompanied by the decrease of pressure and an increase in temperature. This author found that microorganisms, through the production of gases, may produce pressures up to 500 psi in experimental chambers.

Concentration of Ions from Sea-Water/Sediment Interfaces. In general, the greatest microbial activity is found at the sediment/water interface where higher concentrations of organic matter are available, e.g., from 0.5 to 5% carbon by dry weight of sediment. During microbial activities, ions can be concentrated by cell uptake and through environmental changes such as micronucleation or change in pH. If anaerobic areas are produced, such as around a decaying fish or detritus, there may be a scavenging effect by the sulfide ions produced by sulfate-reducing bacteria. Analyses of cations in the area of anaerobic surface sediments indicates such concentration effects. Alternately, ions may be released as the organic matter is solubilized or the sediment particles weathered. It has been postulated that the manganese nodules on the sea floor are partly caused by microbial activity. On land most bog iron deposits are formed through the action of microorganisms. Thus the microbe causes complex changes at sediment water interfaces that are subject to the amount of organic matter, type of sediment, area of sea floor, etc.

Alteration of Isotope Ratios. As elements or ions pass through biological membranes, lighter isotopes are enriched in the cell, leaving the environment enriched in the heavy isotopes. Elements thus affected are hydrogen, carbon, oxygen, nitrogen, and sulfur. The total reaction is complex, however, because the above elements are also nutrients and can be recycled. Therefore, diagenesis prediction using isotope ratios must be made with caution.

Production of Sediments. Sediments may be formed through direct and indirect activities of microorganisms. Sediments are formed from the skeletons of the microscopic diatoms and silicoflagellates. Iron, manganese, carbonates, and clays may be precipitated or produced by weathering effects. Rafting of sediments on surface slicks may distribute sediments. Lithification, especially of beachrock (q.v.), may be caused by the precipitation of carbonates in situ or the production of opaline silica in pH transition microenvironments. In tropical and temperate environments, weathering of coral reefs by microorganisms produces sediments. Oolites (q.v.) have been shown to be produced by bacteria, and photosynthetic-metabolic pH changes alter carbonate sediments. Solution precipitation due to microorganisms may result in the formation of stromatolitic crusts. Some algae, e.g. *Lithothamnion,* precipitate carbonate.

These examples are but a few of the many processes that microorganisms introduce or alter in the processes of sedimentation and in the sediment.

CARL H. OPPENHEIMER

References

Alexander, M., 1961. *Soil Microbiology.* New York: Wiley, 472p.
Alexander, M., 1971. *Microbial Ecology.* New York: Wiley, 511p.
Davis, J. B., 1967. *Petroleum Microbiology.* Amsterdam: Elsevier, 604p.
Kuznetzov, S. I., 1970. *The Microflora of Lakes and Its Geochemical Activity.* Leningrad: Nauka Publishing House, 402p.
Kuznetzov, S. I.; Ivanov, M. V.; and Lyalikova, N. N., 1962. *Introduction to Geological Microbiology.* New York: McGraw-Hill, 252p.
Oppenheimer, C. H., ed., 1968. *Marine Biology. Marine Microbiology.* New York: N. Y. Acad. Sci., 485p.
Perfeliev, B. V., and Gabe, D. R., 1969. *Capillary Methods of Investigating Micro-Organisms.* Toronto: Univ. Toronto Press, 627p.
ZoBell, C. E., 1946. *Marine Microbiology.* Waltham, Mass.: Chronica Botanica Co., 240p.

Cross-references: *Beachrock; Carbonate Sediments–Bacterial Precipitation; Carbonate Sediments–Diagenesis; Cherty Iron Formation; Clay–Diagenesis; Color of Sediments; Coral-Reef Sedimentology; Diagenesis; Diatomite; Flocculation; Gases in Sediments; Hydrocarbons in Sediments; Ironstone; Lacustrine Sedimentation; Limestones; Oolite; Organic Sediments; Pelagic Sedimentation. Pelagic Sediments; Pisolite; Stromatolites.* Vol. XII: *Soil Biology, Microbiology and Biochemistry.*

MIDDEN

A midden, or kitchen-midden, is literally a garbage pile. The term is used for archeological sedimentary accumulations (anthropogenic deposits); it is sometimes used for the dump of soil left by a worm next to its burrow. The word appears to have originally been introduced as the Danish *kjøkkenmødding,* a pile of kitchen scrapings, but is mainly applied to piles of mollusk shells, together with charcoal, broken tools, and utensils.

Mesolithic (in places, also Neolithic) populations in coastal areas ate large quantities of shellfish, and vast shell middens, or *"shell mounds,"* mainly dating from 7000–3000 B.P., are known from the coasts of North and South

America, Europe, Africa, Asia, and Australia. South American aboriginal people living in the Straits of Magellan are still building such middens. Some of those measured by the writer in Brazil exceed 20 m in height, 100 m in length and contain up to 2½ billion shells. Radiocarbon dating of middens is helpful in working out Holocene littoral sedimentation rates, inasmuch as they can be used to establish beach-ridge dates on rapidly prograding shores. In the past, the middens have often been confused with raised beach deposits and glacial "shelly drift" (Charlesworth, 1957, p. 630).

A Danish midden at Erteb∅lle on the Limfjord (Jylland) measured 140 m long, 20 m wide, and 2 m high, containing some 9000 archeological items of flint and bone. Very large examples are found elsewhere in Denmark and in southern Sweden (in Bohuslän).

RHODES W. FAIRBRIDGE

References

Brinton, D. G., 1866. Artificial shell deposits of the United States, *Smithsonian Inst., Ann. Rept.,* 356–358.

Brothwell, D., and Brothwell, P., 1969. *Food in Antiquity.* London: Thomas & Hudson, 248p.

Charlesworth, J. K., 1957. *The Quaternary Era,* 2 vols. London: Edward Arnold, 1700p.

Jessen, K., 1920. Stenalderhavets Udbredelse i det nordlige Jylland, *Danmarks Geol. Unders.,* ser 2, 35, 1–112.

Meggers, B. J., 1972. *Prehistoric America.* Chicago: Aldine, 200p.

Meighan, C. W.; Pendergast, D. M.; Swartz, B. K. Jr.; and Wissler, M. D., 1958. Ecological interpretation in archeology, *Am. Antiquity,* 24, 1–23, 131–150.

Schumacher, P. 1873. Remarks on the kjökkenmöddings on the northwest coast of America, *Smithsonian Inst., Ann. Rept.,* 354–362.

Shetelig, H., and Falk, H., 1937. *Scandinavian Archaeology.* Oxford: Clarendon Press, 458p.

Weber, H. A., 1918. Über spät- und postglaziale lakustrine und fluviatile Ablagerungen bei Lobstädt und Borna und die Chronologie der Postglazialeit Mitteleuropas, *Abhandl. Naturwiss. Ver. Bremen,* 29, 189–267.

Wyman, J., 1875. Fresh water shell mounds of the St. John's River, Florida, *Mem. Peabody Acad. Sci.,* 1, 94p.

Cross-references: *Biogenic Sedimentary Structures.*

MUD BELT, MUD ZONE, MUD LINE

Areas on the continental shelves of the world which have an appreciable amount of silt- and clay-sized sediment, as opposed to the more usual sand- and gravel-sized particles, are sometimes called *mud belts* or *mud zones.* The most commonly referred to areas of this kind are the outer portions of the continental shelves, which may sometimes be at least temporary depositional areas for finer than sand-sized particles. Locally, mud deposits are often found seaward of the offshore portions of beaches, and on the inner and central portions of continental shelves. These areas are also sometimes referred to as mud belts or mud zones, especially in the European literature.

Mud line. The mud line is the boundary on the continental shelf between sand and gravel facies and facies containing appreciable but unspecified amounts of silt- and clay-sized particles, usually in addition to some coarser sediments.

LARRY J. DOYLE

Cross-references: *Continental-Shelf Sediments; Littoral Sedimentation; Neritic Sedimentary Facies; Outer Continental-Shelf Sediments.*

MUD CRACKS (CONTRACTION POLYGONS)

Mud crack (Fig. 1) is the general term used to describe a fissure that forms in fine-grained soils and sediments as volume is reduced during dewatering. Synonymous terms found in the literature are *desiccation crack* (or *fissure*), *shrinkage crack,* and *sun crack.* Mud cracks belong to a larger category of *contraction cracks* that are ubiquitous features in nature and may be seen in perennially frozen ground (permafrost), limestone, lava, cement, dried paint, and ceramic ware. The cracks in all of these media intersect to form *contraction-crack polygons.* The term *syneresis crack* is sometimes used interchangeably with mud crack, but is misleading (see *Syneresis*).

Occurrence

Mud cracks form in clay and silt, but not in pure sand because the latter undergoes little or no volume reduction upon drying. Admixtures of sand and clay do shrink upon drying and therein may display mud cracks. Mud cracks are commonly seen in modern sediments but are less frequently preserved in the geologic record because of their rapid obliteration. Where seen in the latter, they are often preserved as molds (subsequent infill in the cracks) and may be termed *fossil cracks.* Molds composed of chemical precipitates (limestone, chert) may be indicative of syneresis cracking (White, 1964). Fossil mud cracks are most often seen in shales, argillaceous limestone, and silty sandstone, all of which shrink upon drying. The downward taper of a mud crack has been oft-cited as a

FIGURE 1. Newly formed mud cracks in silty clay. Cracks are angular and intersect to form irregular, random orthogonal polygons.

top-and-bottom criterion (q.v.), but Shrock (1948) urges that caution be exercised in these interpretations.

Mud cracks have been suggested by some authors to indicate the continental origin of argillaceous deposits because subaerial drying is the most common cause. Mud cracks do begin forming under water (which can be saline); therefore, they should not be accepted as prima facie evidence of continental origin. Because of their susceptibility to rapid obliteration and inherent lack of strength in fine sediment, however, a subaqueous environment is not apt to favor their growth and perpetuation.

Dimensions

Mud cracks range in width from millimeters to about 1 m (Fig. 2), the latter being termed "giant" to distinguish them from the common surface forms that are usually 1 cm or so wide and spaced about 25 cm apart (Neal et al., 1968). In homogeneous material, crack walls taper inward as they extend to depth. The depth of mud cracks will range from a few cm in small surface forms to as much as 15 m in the giant variety.

Individual cracks intersect to form *mud-crack polygons*. Spacing between cracks is influenced by the thickness of the contracting layer; a rule-of-thumb relationship suggests that crack spacing is about ten times crack depth. Baldwin (1974) found that gastropod or other animal trails may influence the polygonal pattern. The preponderance of small mud-crack polygons in most environments can be explained by the generally shallow depth of surficial wetting and drying.

Mud-crack polygons often display curvature in both concave and convex upward directions; the difference may be caused by vertical particle-size gradation in the contrasting layer, but in other cases the reasons are obscure. Minter (1970) suggested that the rarer concave-downward form may evolve where drainage is more rapid than evaporation. In extreme situations, thin layers may actually form rolls with overlapping edges; hence the term *mud-curls* is used.

Development and Form

Lachenbruch (1962, 1963) suggested the following classification of polygon systems for ice-wedge polygons; it is also applicable to mud-crack polygons:

Polygon Systems
I. Orthogonal
  A. Random
    1. Regular random
    2. Irregular random
  B. Oriented
II. Nonorthogonal

The orthogonal system is characterized by a predominance of $90°$ fissure intersections. Lachenbruch (1963) stated that new cracks in a previously uncracked area will follow randomly distributed zones of weakness; thus, a sinuous course may develop. Where the crack curves, the tangential stress component is greater on

FIGURE 2. Orthogonal mud-crack polygons in calcareous siltstone, central Iran.

the convex side. The horizontal tension, therefore, will be anisotropic (least in the direction perpendicular to the crack) within the zone of stress relief. Secondary cracks will form perpendicular to the greatest tension, and tend to intersect the primary crack at right angles; that is, the intersections are *orthogonal*.

In *random* systems, the initial and subsequent secondary cracks are not directionally oriented. *Oriented* orthogonal shapes develop where the stress field is not generally isotropic. This type of crack form is found in ice-wedge polygons near bodies of water where temperature gradients exist. Mud cracks might display oriented polygons where moisture gradation occurs. Random orthogonal shapes are further distinguished as being either *regular* or *irregular*. For both mud-crack and ice-wedge polygons, first-forming cracks tend to produce irregular polygons. With subdivision, polygons tend to become regular as their size approaches the zone of stress relief for individual cracks.

In dried mud, "irregular random orthogonal polygons" are predominant (see Fig. 1), but they frequently approach the "regular random orthogonal pattern." Despite the ubiquitous occurrence of these orthogonal polygons, some previous authors (e.g., Hewes, 1948) have stated that mud-crack patterns should be more nearly hexagonal, basing the argument on the law of least energy. Nonorthogonal polygons (hexagons, circles, and other shapes) are seen only infrequently in dried mud. Published photographs of well-formed hexagons (such as by Longwell, 1928) are atypical. Orthogonal polygons in mud cracks are the rule, hexagons the exception.

JAMES T. NEAL

References

Baldwin, C. T., 1974. The control of mud crack patterns by small gastropod trails, *J. Sed. Petrology,* **44,** 695-697.

Hewes, L. I., 1948. A theory of surface cracks in mud and lava and resulting geometrical relations, *Am. J. Sci.,* **246,** 138-149.

Lachenbruch, A. H., 1962. Mechanics of thermal contraction cracks and ice-wedge polygons in permafrost, *Geol. Soc. Am. Spec. Pap. 70,* 69p.

Lachenbruch, A. H., 1963. Contraction theory of ice-wedge polygons, a qualitative discussion, *Natl. Acad. Sci., Natl. Research Council Publ. 1287,* 63-71.

Longwell, C. R., 1928. Three common types of desert mud cracks, *Am. J. Sci.,* 5th ser., **15,** 136-145.

Minter, W. E. L., 1970. Origin of mud polygons that are concave downward, *J. Sed. Petrology,* **40,** 755-764.

Neal, J. T.; Langer, A. M.; and Kerr, P. F., 1968. Giant desiccation polygons of Great Basin playas, *Geol. Soc. Am. Bull.,* **79,** 69-90.

Shrock, R. R., 1948. *Sequence in Layered Rocks.* New York: McGraw-Hill, 188-210.

White, W. A., 1964. Origin of fissure fillings in a Pennsylvanian Shale in Vermillion County, Illinois, *Trans. Ill. St. Acad. Sci.,* **57** (4), 208-215.

Cross-references: *Breccias, Sedimentary; Casts and Molds; Clay-Pebble Conglomerate and Breccia; Desert Sedimentary Environments; Fluvial Sediments; Shale; Syneresis; Top and Bottom Criteria.*

MUDFLOW, DEBRIS-FLOW DEPOSITS

The place of mudflows in the classification of mass movements has been disputed since early description of such flows (Endlich 1876; McGee, 1897; Blackwelder, 1928). The term has been used to cover a wide range of forms, processes, and deposits. Mudflows are now generally considered as a type of *debris flow,* and because of the similarity of deposits and type of movement, plus the necessity of a fine-grained component in all such movements, the two terms are here considered synonymous. It is important to note, however, that sandy submarine debris flows can be low in clay content (see Middleton and Hampton, 1973); hence the term mudflow may be misleading. A *debris flow* (mudflow) may be defined as any mass movement involving sand-grade material or finer, which may either be the total moved mass, or, more commonly, a medium (matrix) for the movement of larger clasts.

Mechanics

Movement appears to be the result of distributed shear in a saturated debris mass, unlike *mudslides* (see *Gravity-Slide Deposits*). At different stages in downslope progress, the mass may behave as a plastic solid or as a viscous fluid, and within one sequence of mudflow activity there is often striking evidence of depletion of the debris "reservoir" and a transition to a turbid fluid flow (Blackwelder, 1928). Mudflow viscosities can lie in the range 2×10^3 to 6×10^3 poises (Sharp and Nobles, 1953).

Many terms have been given to the types of movement within the plastic-solid flow—viscous-fluid flow spectrum (see *Mass-Wasting Deposits; Gravity Flows*), but there is general agreement that both subaerial and subaqueous debris flows are transitional in character between slides and turbid-water flows. The degree of turbulence in mudflow movement is determined by considering variations in flow viscosities. Thus when Kuenen (1956) speaks of "a kind of slow boiling agitation" he may be describing only part of a movement that also includes laminar flow in its later stages. Lindsay (1968) believes that this laminar flow imparts the typical dipping-girdle clast fabric.

Johnson (1970) developed a rheological model that describes the flow behavior of debris, as well as sediment transport by debris flow. Equations derived from his model predict the velocity profile within a debris flow and the critical conditions necessary for maintaining the flow (see Middleton and Hampton, 1973). Hampton (1975) applied this equation to hypothetical submarine debris flows and demonstrated that the amount of clay, relative to solid grains, necessary to support sand-sized material is about 10% or even less (1.5-4% for fine sand, 19% or less for coarse sand). Competence of a true debris flow is controlled by the strength and density of the clay-water fluid (Rodine and Johnson, 1976). Grains are supported by strength and buoyancy, rather than by turbulence, upward escape of fluid, or dispersive pressure, although dispersive pressure (see *Grain Flows*) is probably significant in many debris flows (Middleton and Hampton, 1973). The largest grain suspended in a debris flow has a weight that just equals the support of fluid strength and buoyancy; larger grains overcome the strength and sink downward out of the flow (Middleton and Hampton, 1973).

Mudflow movement is generally rapid (mean velocities of the order of 5 m/sec to 15 m/sec commonly, and up to 100 m/sec in volcanic mudflows), but even so the erosive power of the flow is limited, e.g., flows commonly override grass. Although mudflows are normally initiated on steep slopes, in excess of 30°, the flow can extend over slopes of only a few degrees for many km, especially where lateral spread of the flow is restricted, e.g., down a river valley or submarine gulley. One suggested mechanism that would allow movement on gentle slopes is undrained loading (Hutchinson and Bhandari, 1971). Rodine and Johnson (1976) suggest that the poor sorting of coarse clastic debris flows allows them to have a high density yet have essentially no interlocking of clasts; the high density reduces effective normal stresses between clasts, thereby reducing apparent friction of the mixture.

General Characteristics

Lists of suggested diagnostic features for mass movements of this kind have been given, but Dott (1963) points out that structures are often polygenetic in origin, and their extent of development depends almost exclusively on texture. Almost invariably the sediments themselves rather than the associated structures must be used to decide a debris-flow origin.

Various terms have been used to denote mudflow deposits; diamictite (q.v.) and diamicton, pebbly mudstone (q.v.), tilloid, and fluxoturbidite are but a few. All emphasize an essential characteristic of most mudflow deposits: their poor sorting (though in general not so poor as tills) and wide particle size range. There are

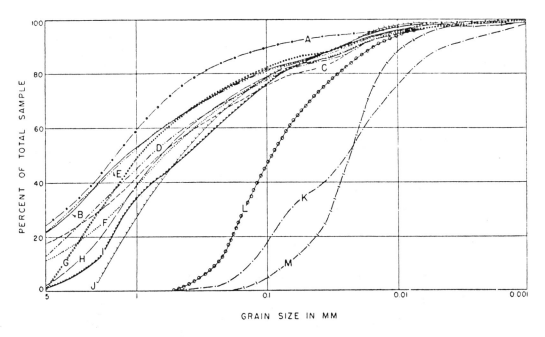

FIGURE 1. Cumulative curves for samples from the 1941 Wrightwood Mudflow (from Sharp and Nobles, 1953). Samples A to J are mudflow materials; K to M are water-laid detritus.

normally two components of mudflow deposits, *clasts* and *matrix;* exceptionally, flows such as thin silt flows may have no clast component. Bull (1963) gives the following sorting data for mudflows: Trask sorting coefficient (*So*) 5.0–25 (average, 9.7); phi standard deviation 4.1–6.2 (average, 4.7); and phi quartile deviation 2.3–4.7 (average, 3.1). Associated waterlaid deposits had *So* averages respectively of 1.8; 1.4; and 0.79. These figures compare with an *So* range of 2.67–5.03 for the Wrightwood mudflow, California (Sharp and Nobles, 1953; Fig. 1), which would seem to have been more fluid, and an *So* of 11.4 for an alpine mudflow in the Yukon (Broscoe and Thomson, 1969). Some differences between waterlaid gravels and mudflow gravels are tabulated in Table 1.

Solids usually make up 70–85% of the total volume of the moving mudflow, and of this figure clay-sized material is often <10%. Only a small amount of clay seems necessary to reduce cohesion and lubricate flow.

Stratification is normally absent, although a sequence of mudflows can be well bedded, with each flow forming an individual bed. The thickness of such beds ranges from a few cm to several m, depending on the flow viscosity and also on lateral constraints to spreading. In some mudflows, thickness can remain fairly constant over a wide area (e.g., in large volcanic mudflows); but in others there is a regular decrease in thickness away from the origin (Bull, 1963). The fronts and sides of flows can be somewhat thicker than the main mass, however, especially in channelized flows. The maximum depth of the surge body in the 1941 Wrightwood flow was about 1.3 m, yet the bouldery fronts of the surges reached heights of 3.1–4.6 m, largely consisting of boulders 0.66–1.0 m in diameter. Bagnold (1968) believes that this concentration of larger clasts at the front of mudflow deposits could be due to an inertial stage of flow, analogous to fluid turbulence. In this state, the dispersive stress increases as the square of the size and, in conditions of shear, thus promotes a drift of larger material to regions of least shear rate, i.e., the surface of the flow. Because the upper part of the mass moves faster than the lower—as a result of lower resistance to flow—coarser debris piles up at the front. Misleading impressions of clast concentration at the front and on the surface of a mudflow may be produced by surface wash and removal of fines. Thick, viscous flows can carry large clasts in suspension, but these boulders are normally moved by rolling, dragging, and pushing. Subaerial debris flows have been observed to have surprising competence, however, with low clay content (Middleton and Hampton, 1973). Curry (1966) reported boulders nearly 1 m in diameter carried in a flow with only 1.1% clay-sized material in the matrix.

If a mudflow is viscous and halts on gently sloping ground, a steep-fronted, lobate or snoutlike nose can be formed. Rapid loss of excess pore-water pressure halts the moving mass; this loss can result from settling of the solids under gravity or sieving of the more liquid part of the flow through the coarse snout; and it leaves the surge front as a blunt, steep scarp.

Types of Mudflow Deposits

Position in the landscape determines many features of mudflows and their deposits, and they may be divided into physiographic groups.

Alluvial Fan (Cone) Mudflows. Mudflows can make up almost any proportion of alluvial fans (q.v.) or talus cones. Fans largely composed of

TABLE 1. Characteristics of Waterlaid and Mudflow Deposits

Parameter	Waterlaid Sediments	Mudflow Deposits
Average sorting coefficient (So)	1.8[a]	9.7[a]
Median grain size	steady decrease from source	irregular decrease from source
Maximum grain size	gradual, steady decline from source	rapid, irregular decline from source
Particle shape (largely depends on debris type)	well rounded	subangular to subrounded
Cumulative frequency curve (see Fig. 1)	S shaped	much flatter, with bimodal peaks

[a]Figures from Bull (1963).

mudflow deposits generally have steeper surfaces than fans where fluvial deposition dominates. Small flows tend to follow previous channel networks on the fan surface, but thicker mudflows can overtop the banks and spread out as thin sheets or produce levees by forcing up the sides of the channel. Some levees in more open channel networks appear to be formed by the pushing aside of the coarser material at the snout of a surge after the critical mix for flow has been lost by seepage. Reestablishment of this solid/fluid balance will either restart the mass or push it aside. This stop-start movement, apparently the result of blockage of the flow channel or the variable production of debris in the source area, produces a "caterpillar" movement where body segments are sequentially moved toward the head before movement is restarted.

Interbedding of fluvial deposits with mudflow deposits on a fan is normal. Cut-and-fill structures, where a mudflow fills a streamcut channel and causes wandering of the stream, produces a lensoid pattern of fluvial gravels laterally within the fan. Bull (1962) used a CM graph to plot the various depositional environments within fan sequences and suggested that mudflows and fluvial deposits can be distinguished by sorting characteristics (Fig. 2).

Directed and Nondirected Mudflow Deposits. Directed mudflows are guided by the thalweg of a valley, in which case, movement can occur down relatively gentle slopes. This type is usually very long compared with its thickness or width, e.g., the Wrightwood Flow was 6.2–40.6 m wide and extended 24 km. Downstream decrease in thickness of deposits is irregular because of variations in valley width. In places, mudflows can follow a braided or anastomosing drainage pattern. Levee features are generally not well developed (Sharp and Nobles, 1953).

Valley-side flows occur on steep slopes and are undirected by any channel. Because of the small area of debris supply they tend to be short but fairly wide and thick; the thickness allows movement of large material. Preservation of deposits is more likely in this group than in valley-bottom deposits, where often all that remains is a sinuous lag-gravel trail, although this may perhaps be identified by its very poor sorting even when the fines have been removed.

Volcanic Mudflow (Lahar) Deposits. Lahars

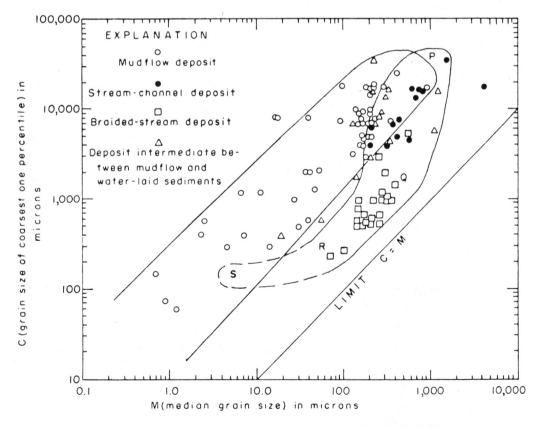

FIGURE 2. CM patterns of surficial alluvial-fan deposits (from Bull, 1962).

frequently make up a larger part of volcanic cones than true lava flows; they are sometimes as much as 50 or 60 km long.

One commonly recognized feature of volcanic mudflows is reverse or inverse grading (q.v.); Schmincke (1967) gives examples of such features (Fig. 3). The lowest thin layer of medium-grade sand is overlain abruptly by a massive, coarse, poorly sorted central portion, which may contain blocks up to 3 m in diameter; this zone grades into a cross-bedded or bedded top unit of medium-grained sand and pebbles. The sharp base of the flow may have basal protrusions similar to load casts or flute molds. Large boulders are fairly rare, and the main part of the deposits comprises coarse, subrounded or subangular sand, with angular to subrounded granules and pebbles, generally < 20 cm in diameter. Normal grading may sometimes be present in the central part. Schmincke believes that the basal layer represents water slurries which lubricated the main mass flow, and the inverse grading in the main body is the result of an inertial type of flow.

Wide, sheetlike lahars, 2-5 m in thickness, may fill in the irregularities in the landscape without any apparent changes in the upper surface level, suggesting a relatively fluid movement or overriding of the upper part on the lower. A hummocky surface, however, appears to be typical of most volcanic mudflows.

Subaqueous Mudflow Deposits. Subaqueous mudflow deposits are more abundant in the geological record than subaerial ones (see *Gravity Flows*). This predominance may be a result of better preservation or the dominance of the marine environment over the subaerial; it does not necessarily imply a greater susceptibility to this type of mass movement under water. Dott (1963) notes that many if not all groups of deformational features found in subaqueous deposits may be produced by either gravity or current drag. Because of the similarity of many structures and the present poor understanding of many subaqueous processes, many theories of underwater debris flows are based on a study of subaerial movements. Walker (1975) discusses the characteristics of conglomerates (q.v.) produced by subaqueous debris flow and other mechanisms (see also *Diamictite; Pebbly Mudstones*).

Dott (1963) explains subaqueous gravitational deposits in terms of two dynamic boundaries: the yield limit and the liquid limit. Plastic flow begins where the yield limit of cohesive sediments is exceeded; stratification tends to be distorted rather than destroyed. When the liquid limit is exceeded, viscous fluid flow destroys stratification and forms a suspension. Debris flows span these two limits.

Turbulence is an important element of debris flows, and Lindsay (1968) believes that turbulence depends on the quantity of sand in the fine component. If the matrix of the flow contains much clay, flow tends to be laminar, and debris blocks are left intact. If the matrix is sandy, pebbles, cobbles, and boulders are all that remain of the blocks, testifying to the low viscosity and high velocity of the turbulent flow. Debris-flow deposits can show laminar orientation of fines and intact clasts or a nonlaminated matrix and disaggregated clasts. Lindsay believes the pebble fabric in both types is a result of laminar flow. This fabric shows particles with their thick ends pointing downstream and their axes parallel to the direction of flow. Subaqueous debris flows do not always contain clasts, and sometimes show graded sequences in the upper part, which are formed by settling of material thrown into suspension by movement of the flow mass.

PETER GASCOYNE

References

Bagnold, R. A., 1968. Deposition in the process of hydraulic transport, *Sedimentology*, 10, 45-56.

Blackwelder, E., 1928. Mudflow as a geologic agent in semi-arid mountains, *Geol. Soc. Am. Bull.*, 39, 465-484.

Broscoe, A. J., and Thomson, S., 1969. Observations on an alpine mudflow, Steele Creek, Yukon, *Canadian J. Earth Sci.*, 6, 219-229.

Bull, W. B., 1962. Relation of textural (CM) patterns

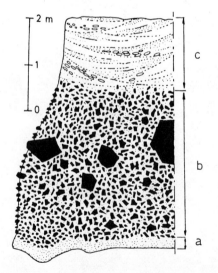

FIGURE 3. Schematic cross section through a lahar, showing the fine-grained base (a), the coarse-grained poorly sorted central section (b), and the upper part of cross-bedded tuffs with pumice pebbles (c) (from Schmincke, 1967).

to depositional environment of alluvial-fan deposits, *J. Sed. Petrology,* **32,** 211-216.

Bull, W. B., 1963. Alluvial fan deposits in W. Fresno County, California, *J. Geol.,* **71,** 243-251.

Curry, R. R., 1966. Observations of alpine mudflows in the Tenmile Range, Central Colorado, *Geol. Soc. Am. Bull.,* **77,** 771-776.

Dott, R. H., Jr., 1963. Dynamics of subaqueous gravity depositional processes, *Bull. Am. Assoc. Petrol. Geologists,* **47,** 104-128.

Endlich, F. M., 1876. Report of progress of exploration for the year 1874, *U.S. Geol. Geogr. Surv. Terr., 8th Ann. Rept.,* 515p.

Hampton, M. A., 1975. Competence of fine-grained debris flows, *J. Sed. Petrology,* **45,** 834-844.

Hutchinson, J. N., and Bhandari, R. K., 1971. Undrained loading, a fundamental mechanism of mudflows and other mass movements, *Geotechnique,* **21,** 353-358.

Johnson, A. M., 1970. Flow of ice, lava and debris, ch. 12-15 in *Physical Processes in Geology.* San Francisco: Freeman, Cooper, 432-571.

Kuenen, P. H., 1956. The difference between sliding and turbidity flow, *Deep-Sea Research,* **3,** 134-139.

Lindsay, J. F., 1968. The development of clast fabric in mudflows, *J. Sed. Petrology,* **38,** 1242-1253.

McGee, W. J., 1897. Sheetflood erosion, *Geol. Soc. Am. Bull.,* **8,** 87-112.

Middleton, G. V., and Hampton, M. A., 1973. Mechanics of flow and deposition, *Soc. Econ. Paleont. Mineral. Pacif. Sec., Short Course* (May 12, 1973), 1-38.

Rodine, J. D., and Johnson, A. M., 1976. The ability of debris, heavily freighted with coarse clastic material, to flow on gentle slopes, *Sedimentology,* **23,** 213-234.

Schmincke, H. -U., 1967. Graded lahars in the type section of the Ellensburg Formation, south-central Washington, *J. Sed. Petrology,* **37,** 438-448.

Sharp, R. P., and Nobles, L. H., 1953. Mudflow of 1941 at Wrightwood, Southern California, *Geol. Soc. Am. Bull.,* **64,** 547-560.

Walker, R. G., 1975. Conglomerate: Sedimentary structures and facies models, *Soc. Econ. Paleont. Mineral. Short Course 2,* (April, 1975), 133-161.

Winder, C. G., 1965. Alluvial cone construction by alpine mudflow in a humid temperate region, *Canadian J. Earth Sci.,* **2,** 270-277.

Cross-references: *Alluvial-Fan Deposits; Conglomerates; Diamictite; Graded Bedding; Grain Flows; Gravity Flows; Gravity-Slide Deposits; Inverse Grading; Mass-Wasting Deposits; Pebbly Mudstones; Penecontemporaneous Deformation of Sediments; Submarine Fan (Cone) Deposits; Till and Tillite; Turbidites; Volcaniclastic Sediments and Rocks;* Vol. III: *Alluvial Fan, Cone; Debris Flow; Landslides; Mass Movement; Mass Wasting; Mudflow (and Lahar); Submarine Geomorphology.*

MUDFLAT DEPOSITS—See TIDAL-FLAT GEOLOGY

MUDSTONE AND CLAYSTONE

Mudstone and *claystone* are very fine-textured argillaceous rocks which lack the thin lamination and fissility that characterize shale (q.v.). An *argillite* is a more compact or more indurated claystone or mudstone. Mudstone and claystone normally occur in beds of uniform lithology greater than approximately 1 cm thick, reflecting a more constant environment of deposition than for shales, whose laminae betoken numerous interruptions or changes of sedimentation rate and composition. It has been observed, however, that massive, moist muds, when subjected to compactive pressures, such as those associated with deep burial, can develop shaly fissility due to expulsion of water and attendant reorientation and recrystallization of clay mineral particles (Shepard and Moore, 1955).

Claystone (indurated clay) is composed primarily ($>85\%$) of particles $<4\mu$ in diameter, as prescribed by the Wentworth Scale (q.v.). Many of the particles in this size range are composed of the clay minerals. Claystones also contain various proportions of organic matter, iron and aluminum oxides, and other mineral powders, e.g, quartz. Trace quantities of water are also common in most claystones; and some claystones, especially those of dark gray color, are known to contain significant minor amounts of hydrocarbons.

A *tonstein* (literally, in German, claystone) is a kaolinite-rich claystone often associated with coal seams. Tonsteins were first described in detail by Schmitz-Dumont (1894), who believed that they were weathered volcanic ash. Not all tonsteins originate in this manner (Williamson, 1970), however, and there is some disagreement concerning origin and nomenclature (Duff and Loughnan, 1972).

Mudstone (indurated mud) resembles claystone in most respects except that silt-sized (4-62μ) particles make up a substantial proportion (15-50%) of the rock. Some "mudstones" and "claystones" are actually breccias and conglomerates composed of fragments of mudstone, claystone, or shale in a matrix of mud or clay. Owing to the common presence of finely divided organic matter, the lithification of mud is characterized by proportionally much greater water loss, and hence, much greater volume reduction than clay, in which organic components are typically minor.

Lutite (Latin: *lutum,* mud) is a general term applicable to all rocks of clay- and silt-sized particles regardless of their composition; prefixes may be added, e.g., *calcilutite,* to signify composition. *Pelite* is an equivalent Greek-derived term used mainly by Europeans; this term is also used, however, in a more specific sense to denote aluminum-rich sediments, and *pelitic* describes rocks metamorphosed from Al-rich sediments, as in *pelitic schist*.

ALISTAIR W. McCRONE

References

Boswell, P. G. H., 1961. *Muddy Sediments.* Cambridge: W. Heffer and Sons, 140p.

Duff, P. M. D., and Loughnan, F. C., 1972. Wongawilli claystones and European tonsteins, *J. Geol. Soc. Aust.,* **19**(2), 281-283.

Schmitz-Dumont, A., 1894. Die Saarbrucker Tonstein, *Tonindustriezeitung,* **18**, 714.

Shepard, F. A., and Moore, D. A., 1955. Central Texas coast sedimentation, *Bull. Am. Assoc. Petrol. Geologists,* **39**, 1463-1593.

Spears, D. A., and Rice, C. M., 1973. An Upper Carboniferous tonstein of volcanic origin, *Sedimentology,* **20**, 281-294.

Williamson, I. A., 1970. Tonsteins—their nature, origins and uses, *Mining Mag.,* **122**, 119-125, 203-211.

Cross-references: *Agrillaceous Rocks; Clay; Clay as a Sediment; Coal; Compaction in Sediments; Oil Shales; Shale.*

N

NATURAL GAS—*See* GASES IN SEDIMENTS

NEOMORPHISM—*See* CARBONATE SEDIMENTS—DIAGENESIS

NEPHELOID SEDIMENTS AND NEPHELOMETRY

A *nepheloid layer,* or near-bottom cloudy suspension of particulate matter, occurs in all deep-sea areas where soundings exceed 2500 fathoms, such as in the North American, Argentine, and North Pacific basins, the Aleutian Trench, and the Gulf of Alaska. This zone of suspended sediment is also found in shallower regions near land where strong bottom currents exist, such as along the western sides of the Argentine, Brazil, and North American basins. There are wide variations in the thickness of the layer, its vertical gradient, and the concentration of suspended particles, so that generalizations for the oceans as a whole cannot be made. In the North Pacific the thickness ranges from 55–820 m above the bottom of the deep basins and continental margin (Ewing and Connary, 1970). A system of counterclockwise gyres transport suspended sediment homogeneously. In the North American Basin of the North Atlantic a nepheloid layer with a thickness ranging from 300–2400 m covers the continental rise, abyssal plains and Bermuda Rise (Eittreim et al., 1969). The greatest thickness occurs S of the Blake Plateau and Bahama Islands. There the nepheloid layer appears to be related to the Western Boundary Undercurrent and the Antarctic Bottom Current.

The nepheloid layer has a significant role in sediment transport to the deep sea and accounts for aspects of deep-sea sediment stratification (recorded on seismic profiles) which could not be otherwise explained (Jacobs and Ewing, 1969). It has been observed that where thick accumulations of pelagic sediment are found, a nepheloid layer occurs extending several hundred meters upward from the sea bottom. A balance between gravitational settling and turbulent diffusion of particulate matter characterizes the nepheloid layer. Its load of suspended particles is replenished by sediment injections from turbidity currents and erosion of clay-sized material half-settled and half-suspended at the sediment/water interface. The clay-sized particles are activated by the movement of bottom water into a flow, a fine-grained analog of a turbidity current. The nepheloid layer follows the bottom topography and deposits blankets of fine-grained sediment which smooth out basement irregularities, as portrayed by seismic profiles. Mineralogical analysis of particles suspended in the nepheloid layer provides an approach to the larger problem of defining source and mode of transport of deep-sea sediment (Jacobs and Ewing, 1965).

Observations

Observations of the nepheloid layer have been made by light-scattering studies. Briefly described, light passes through a medium undeflected as long as the refractive index of the medium is uniform. When light encounters particles having indices of refraction different from that of the medium, sizes within a few orders of magnitude of the wavelength of light, and varying concentrations, the particles cause a portion of the light to be scattered in all directions. The index of refraction of the sea-water medium is approximately 1.33; the dispersed particles exhibit considerable variability, with mineral particles ranging from 1.40 for opal to 3.22 for hematite, a range which includes values for fibers, dusts, pollen, industrial dusts, and volcanic products. When the light source from the nephelometer camera system emits a beam of light into sea water, the light is scattered by fine particles suspended in the water and is recorded on photographic film. Thus *nephelometry* is the measurement of the cloudiness of a medium, utilizing the analysis of light scattered by particles in suspension to determine their concentration.

Nephelometers have been designed for making in-situ measurements of light scattering continuously at all depths from the surface to the bottom of the water column (Thorndike and Ewing, 1967). The nephelometer consists of a light source, an attenuator, and a camera. Light scattered by particulate matter in the water is

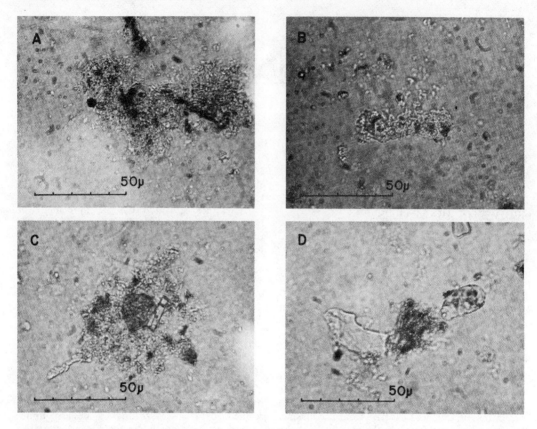

FIGURE 1. Aggregates of suspended particulate matter from nepheloid water (from Jacobs et al, 1973). A, B: Filter V26-141; fine particles coalesced into aggregates, longest dimensions are 100μ and 50μ, respectively. Dark particles are particulate iron. C: Filter V26-145; aggregate formed around large grain with 20μ diameter. D: Filter V26-143; iron oxide aggregate adhering to mineral grain.

recorded on film on either side of the direct beam for the range 8–30° from the direct beam. Time and depth are recorded on the film for the descent and ascent of the nephelometer. The nephelometer film is measured with a photodensitometer, and graphs are constructed which show the intensity of both scattered and direct light versus depth for the lowering and raising of the system. Sensitometer patches from sources of known relative intensity placed on the film before lowering are also photometered. The depth to the top of the nepheloid layer and the intensity of scattering are two basic quantities measured at each station. At a typical station, sunlight and large quantities of comminuted organic particulate matter create high light scattering in the neritic zone. This scattering decreases with depth to a minimum, which is regarded as a reference level. Below it, light scattering increases with descent to the ocean floor, where the nepheloid layer occurs, the zone of most intense light scattering being nearest the bottom. The intensity of light scattering is designated by R values, the ratio of the light scattered at the bottom to the light in the clearest water of that same station. A station with no nepheloid layer has an R of 1.0; R values increase with increasing intensity of the layer.

Particles

The nature of the particles composing the nepheloid layer have been studied microscopically and compared with those in suspension in clear water (Jacobs et al., 1973). Because light scattering has been used as an indicator of the amount of material in suspension, the cause for the observed differences should be understood. The total count of suspended particles is a basic difference between nepheloid and clearer water (Jacobs et al., 1973); i.e., the nepheloid layer of the North American Basin has four times, and the Brazil Basin three times the particle density of clearer water over the mid-ocean ridge. The concentration of material suspended in nepheloid water ranges from 0.44–0.56 mg/liter com-

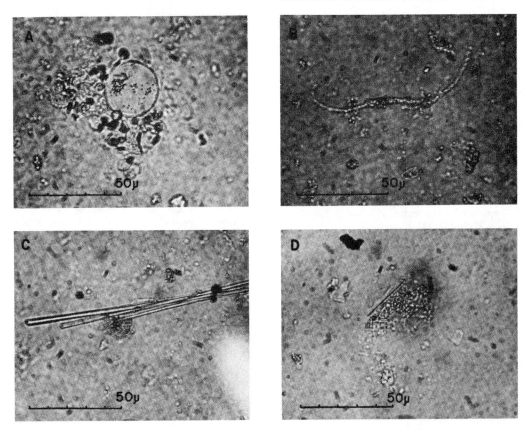

FIGURE 2. Fine particles adhering onto larger settling objects (from Jacobs et al., 1973). A: Filter V26-147; aggregate clustering around diatom of 30μ diameter. B: Filter V26-147; fiber-coated with rust-colored particles. C: Filter V26-145; aggregate adhering to silica rods, long dimension of aggregate about 20μ. D: Filter V26-16; aggregate adhering to silica rod, long dimension of aggregate about 40μ.

pared to 0.03–0.24 mg/liter for clear water. A greater percentage of particles of $<2\mu$ size prevails in nepheloid water, ranging from 85–96% as compared to a range of 76–87% in the clear water. Aggregates of mineral grains, exoskeletons, and plant debris occur in both nepheloid and clear water, although in the nepheloid water their concentration is diluted by vast quantities of $<2\mu$ particles (Jacobs et al., 1973).

Pelagic sedimentation (q.v.) from the nepheloid layer involves the settling of individual particles, the coalescence of fine particles, and their deposition by gravity (Fig. 1). Another sedimentation process involves the adhering of minute particles onto larger objects settling through the water column, such as mineral grains, silica rods, fibers, and exoskeletons (Fig. 2). Grain-by-grain deposition appears to be by far the most common phenomenon of pelagic sedimentation, although particle interaction also takes place.

In summary, the increased light scattering observed in the near-bottom nepheloid layer is produced by abundant nonopaque mineral grains of clay size. The relationship observed between the near-bottom nepheloid layer and underlying stratified pelagic deep-sea clays observed on seismic profiles suggests that the nepheloid layer has an active role in oceanic sedimentation.

MARIAN B. JACOBS

References

Betzer, P. R.; Richardson, P. L.; and Zimmerman, H. B., 1974. Bottom currents, nepheloid layers and sedimentary features under the Gulf Stream near Cape Hatteras, *Marine Geol.*, 16, 21–29.

Biscaye, P. E., and Eittreim, S. L., 1974. Variations in benthic boundary layer phenomena: Nepheloid layer in the North American Basin, in R. J. Gibbs, ed., *Suspended Solids in Water*. New York: Plenum, 227–260.

Eittreim, S.; Ewing, M.; and Thorndike, E., 1969. Suspended matter along the continental margin of

the North American Basin, *Deep-Sea Research,* **16**, 613-624.
Ewing, M., and Connary, S., 1970. Nepheloid layer in the North Pacific, *Geol. Soc. Am. Mem. 126,* 41-82.
Jacobs, M. B., and Ewing, M., 1965. Mineralogy of particulate matter suspended in sea water, *Science,* **149**, 179-180.
Jacobs, M. B., and Ewing, M., 1969. Suspended particulate matter: Concentration in the major oceans, *Science,* **163**, 380-383.
Jacobs, M. B.; Thorndike, E.; and Ewing, M., 1973. A comparison of suspended particulate matter from nepheloid and clear water, *Marine Geol.,* **14**, 117-128.
Thorndike, E., and Ewing, M., 1967. Photographic nephelometers for the deep sea, in J. B. Hersey, ed., *Deep Sea Photography.* Baltimore: Johns Hopkins Univ. Press, 113-116.

Cross-references: *Abyssal Sedimentary Environments; Clays, Deep-Sea; Hadal Sedimentation; Pelagic Sedimentation, Pelagic Sediments; Turbidity-Current Sedimentation.*

NERITIC SEDIMENTARY FACIES

Neritic sediments are produced by sedimentation between the low-tide line and the outer edge of the continental shelf, conventionally and arbitrarily placed at about the 200 m isobath. Included within the broad scope of the definition are the standard sedimentation parameters—provenance, transportation, deposition, and diagenesis—applied to the world's continental shelves (see *Continental-Shelf Sedimentation*). In marine ecology, the term *neritic* refers to the upper 100 fathoms of ocean water, even in the deep sea.

In the neritic zone, provenance may be terrestrial or marine. Climate in the areas through which sediments are transported, as well as climate in source areas, is a major factor in determining the rate and constancy of terrigenous supply. Water, both as rivers and as waves and currents, is by far the most important medium of transport of neritic sediments, although wind and ice may be locally important (see, e.g., *Continental-Shelf Sediments in High Latitudes*). Ice rafting was more important in transporting sediments during Pleistocene glacial periods than it is at present. Biologic transportation and rafting are less important agents, although they also may be locally important over geologic time (Emery, 1960).

Neritic zones of the world may have been much more extensive in the past than they are today. During the Paleozoic and Mesozic eras, vast shallow neritic seas called *epeiric* or *epicontinental* seas covered large portions of the continents of the world. Rates of neritic sedimentation vary from 1000 cm/1000 yr for large deltas to 10-20 cm/1000 yr for the outer edge of the shelf.

Neritic Sedimentary Processes

Rates and constancy of terrigenous supply and of marine dispersal and reworking are principal and strongly interacting factors in determining types of neritic sediment facies. When the supply is far larger than the rate of reworking, large deltas form. With increasing efficiency of waves and currents or decreasing rate of supply, first smaller deltas with much lateral marine dispersion of sediments (e.g., Rhone Delta), then barrier and lagoonal complexes, then beach plains form. When terrigenous sediment supply is eliminated by isolation (Bahamas, Bermuda), underground drainage (Yucatan Platform), very great shelf width (Sahul Shelf), or by lack of local supply, then neritic calcareous, authigenic, and residual sedimentary deposits become dominant.

The surf zone filters sand-sized sediment from the terrigenous supply brought to the sea by rivers (see *Littoral Sedimentation*). Efficiency of the filter increases with increasing wave energy and decreasing sediment supply. The sand fraction is transported and deposited in the nearshore zone in depths normally < 20 m. At least part of the clay and silt fraction is carried offshore in suspension in response to local current systems. Some of this material may be deposited in the neritic zone, but most bypasses the shelf completely. Under ideal condition, then, most of the world's neritic sediments should be graded from coarsest materials, gravels and sands, lying nearest the coast to progressively finer and finer sediments toward the shelf break. Exploration has shown, however, that this ideal graded case is *rare* on present-day continental shelves.

Effects of Sea-Level Change. Ideal distribution of neritic sediments is rare because changes in sea level strongly affect sediment types and distribution patterns (Van Andel and Curray, 1960). Sediments on most of the world's continental shelves reflect sedimentation regimes that existed during Pleistocene glacial stages, when sea level fluctuated at least as much as 100 m. It is the rate, not the sense, of sea level change that is most important. With a rapid shift of shoreline, either up or down, even large rivers cannot maintain deltas; and only thin, discontinuous sand bodies are deposited. At intermediate rates, deltas may form in regions of high sediment supply; in areas of lesser supply open lagoons, broken barriers, or thin beach complexes are built. Only a relatively stable sea level will allow production of the extensive deltaic-neritic deposits like those of the Gulf of Mexico and the coast of NE South America. Lateral migrations of the shoreline are reflected in the spreading of gravels and sands and winnowing of silts and clays over the neritic zone.

Many of the world's continental shelves, then, are covered with sand and gravel sheets which were derived from beach and nearshore deposits laid down, thinned, and spread as the result of sea-level fluctuations caused by glacial advances and retreats during Pleistocene time. The last major worldwide eustatic sea level change, a rapid rise, ended about 5000 yr B.P. Many river mouths were drowned, forming estuaries. At present these rivers are delivering less sediment to the world's shelves than they have in the past, and much of what is being delivered, especially the coarser fractions, is being trapped within the estuaries. Therefore, many of the sedimentary processes that affect the neritic environment today involve the reworking of sediments deposited in the Pleistocene and the dispersal and reworking of fine-grained material that escapes from the estuaries. Primary agents of reworking and dispersion are currents and waves. Tidal currents, wind-generated currents, and even oceanic currents which intercept the outer parts of continental shelves from time to time, and wind-generated and internal waves all play their part.

Silts and Clays. Deposition of fine sediments is controlled by concentration of particles near the bottom, current velocity, and magnitude and frequency of wave activity (McCave, 1971). Fine sediments may be deposited and resuspended many times as they move about the shelf until they finally settle, are lost to deeper oceanic regimes, or are carried back into the estuaries. According to McCave, only a few hours of compaction may dramatically raise the value of the limiting sheer stress necessary to resuspend silts and clays; thus, permanence is added to a deposit. Strengths of and interaction among these agents determine what pathways fine sediments will follow. Where currents and waves have enough energy, silts and clays in suspension may be carried completely across the continental shelf to be deposited on the continental slope or deep-sea floor. Under lower or sporadic energy conditions fine sediments may be deposited in mud belts on the inner, middle, or outer portions of a shelf. Thus, the neritic zone is often a transit zone or a source area where winnowing from ancient deposits provides fine-grained sediments which may be deposited in the estuaries or on the slopes (Doyle et al., 1968; Meade and Hathaway, 1974).

Sands and Gravels. Unlike silts and clays, which are moved in suspension, sands and gravels are primarily transported as bedload, rolling and saltating along the bottom. While the agents of sedimentation are the same, energy levels must generally be greater to affect the sands and gravels than to affect the finer sediments. Higher energy levels are provided by storm waves, fast tidal currents, and perhaps by breaking internal waves on the outer portions of the shelves. Southard and Cacchione (1972) have performed experiments with breaking internal gravity waves in a wave tank. In the marine environment, these waves are commonly found associated with density discontinuities such as the thermocline. Their experiments have shown that these waves break on gentle slopes like those of the outer portions of continental shelves. The process is similar to wind-generated waves breaking on beaches. The waves break and rush up the slight incline, transporting sediment at least partly in suspension. As the wave retreats back down the slope it carries sediment with it as bedload. In the wave tank, breaker retreat transport exceeded breaker advance transport, indicating that the process may tend to carry some of the shelf sediments in a net seaward direction.

Wind-generated waves and currents as well as tidal currents may rework sands in the neritic zone. Gorsline and Grant (1972) have shown that waves 2 m high and about 620 m long can erode fine sand at depths up to 100 m. Sand waves and other ripple marks as well as migrating bars also attest to the efficacy of sand transport in the neritic zone. Storms may be major factors in sedimentation on the world's shelves (see *Storm Deposits*); one large storm may have more effect than several years of normal oceanographic conditions. According to Pilkey and Field (1972), on at least some continental margins, a significant component of sediment transport may be onshore. Mineralogy, texture, and carbonate content of beach and estuarine sands along the SE Atlantic coast show affinities with corresponding parameters of shelf sediments. Ooids on central Atlantic Florida beaches and phosphorite grains on North Carolina beaches can only come from reworked shelf deposits which may lie as much as 20 km offshore. Presence of soft carbonate ooids in quartz sands are especially significant because they indicate that onshore transport must be occurring now; otherwise the ooids would long ago have been abraded away.

Neritic Sediments

Detrital. Most neritic sediments are detritally dominated; that is, the minerals that make up the deposits were derived from preexisting igneous, metamorphic, and sedimentary rocks, most of which are exposed on the continents. Shelves that have large sediment contributions from major rivers like the Nile, Mississippi, and Amazon may be forming large neritic deltaic deposits, especially if wave heights are less than about 1.5 m. Where detrital supply is less and neritic surficial sediments were brought to the shelves during Pleistocene lower sea-level stands, reworked sand and gravel sheets dominate. Some sediments may be contributed to this type of neritic system by submarine weathering from older deposits which outcrop on the shelves. Northeastern Atlantic neritic sedimentation is of this type. In high latitudes, continental shelves are often dominated by glacial debris and ice-rafted sediments, which may range from fine glacial flour to large blocks of rock weighing several tons.

On the outer portions of some continental shelves whose inner and central sectors are detritally dominated, hemipelagic sediment is deposited. The detrital component is much reduced in these zones, which are far from source areas and below the influence of most waves; thus, planktonic foraminiferal tests settling through the water column become an important component of the sediment.

Authigenic. In areas of the neritic zone with reduced detrital component, authigenesis (q.v.) may be important. Glauconite (q.v.), is an important sediment component near the edge of many of the world's neritic zones including that of the SE United States. Phosphorite is also found in some neritic zones with upwelling currents.

Biogenic. Biogenic components are a part of almost all neritic sediments. Most common is calcium carbonate secreted by many algae, protozoans, and invertebrates. In areas where there is a large supply of detrital sediment, biogenic components are masked and are often quantitatively insignificant; as detrital supply decreases, biogenic sediments become quantitatively more important. Where detrital supply is nearly eliminated, biogenic sediments dominate, as in the Caribbean and off north and NE Australia. In conjunction with biogenic deposits, direct precipitation of calcium carbonate from sea water may occur, but it is probably organically controlled (see *Bahama Banks Sedimentology*).

Methods of Study

Until recently, sedimentation in the world's neritic zones has been studied by inference from surface sediment samples of continental shelf deposits. Careful examination can allow the sedimentologist to make conclusions concerning provenance, transportation, deposition, and diagenesis of the sediments in question. Such a method of study has obvious limitations, however. Recently, investigators have studied neritic sedimentation as it occurs. For instance, fine-grained sediment in suspension may be studied while in the water column, actually involved in transportation and deposition. Sampling is done by use of large-volume water samplers or by continuous pumping and centrifugation. Nephelometers and transmissometers are also used to delineate turbid layers and to determine, at least semi-quantitatively, the amount of particulate suspensate. Current measurements and bottom photography have provided further insight into shelf sedimentation. Vibratory coring devices have been developed and are being used to core in sands and gravels. Thus, sedimentation through time can be studied and, in addition, some diagenetic changes such as cementation and recrystallization can be measured and investigated.

LARRY J. DOYLE

References

Doyle, L. J.; Cleary, W. J.; and Pilkey, O. H., 1968. Mica: Its use in determining shelf-depositional regimes, *Marine Geol.,* 6, 381-389.

Emery, K. O., 1960. *The Sea Off Southern California—A Modern Habitat of Petroleum.* New York: Wiley, 366p.

Gorsline, D. S., and Grant, D. J., 1972. Sediment textural patterns on San Pedro Shelf, California (1951-1971): Reworking and transport by waves and currents, in Swift et al., 1972, 575-600.

McCave, I. N., 1971. Wave effectiveness at the seabed and its relationship to bed-forms and deposition of mud, *J. Sed. Petrology,* 41, 89-96.

Meade, R. H., and Hathaway, J. C., 1974. Continental shelves as sources of estuarine sediments and estuaries as sources of shelf sediments, *Proc. Intern. Symp. Relations Sed. Estuar. Plateaux Contin., Inst. Geol. Bass. Aquit., Bordeaux,* 61.

Pilkey, O. H., and Field, M. E., 1972. Onshore transportation of continental shelf sediment: Atlantic southeastern United States, in Swift et al., 1972, 429-446.

Southard, J. B., and Cacchione, D. A., 1972. Experiments on bottom sediment movement by breaking internal waves, in Swift et al., 1972, 83-98.

Swift, D. J. P.; Duane, D. B.; and Pilkey, O. H., eds., 1972. *Shelf Sediment Transport Process and Pattern.* Stroudsburg, Pa.: Dowden, Hutchinson & Ross, 656p.

Van Andel, T. H., and Curray, J. R., 1960. Regional aspects of modern sedimentation in northern Gulf of Mexico and similar basins, in F. P. Shepard, et al., eds., *Recent Sediments, Northwestern Gulf of Mexico.* Tulsa, Okla.: Am. Assoc. Petrol. Geologists, 345-364.

Cross-references: *Barrier Island Facies; Continental-Shelf Sedimentation; Continental-Shelf Sediments; Continental-Shelf Sediments in High Latitudes; Delta Sedimentation; Estuarine Sedimentation; Lagoonal Sedimentation; Littoral Sedimentation; Paralic Sedimentary Facies; Phosphate in Sediments.* Vol. I: *Neritic Zone.*

NIVEO-EOLIAN DEPOSITS

Niveo-eolian, or niveolian, deposits (abbreviated as NIV), are mixed deposits of wind-driven snow and sand, silt, vegetal debris, or other detritus. Though mentioned as early as 1751 by Linnaeus, NIV began to retain attention only in 1932, when C. E. Wegmann, having observed them in East Greenland, interpreted some Quaternary sand deposits of the Netherlands as NIV (see Edelman and Crommelin, 1939). Such sands were designated as niveo-eolian by V. Van Straelen (1946; verbally at a field conference) and were subsequently reported from the Quaternary of many parts of Europe and in modern environments in Antarctica, Canada, and the United States. The geomorphology of NIV is typical and easy to observe and thus better known than the sedimentological characteristics.

Occurrence

Niveo-eolian deposits usually form in cold, rather dry climates. Some are *recent,* the snow being present at least during winter; others are *ancient,* all the snow having melted away. Some are *natural;* others *anthropogenic,* as in Poland where the sand and silt originating from naked cultivated fields are wind-deposited in nearby snow-covered areas. Some NIV are *annual,* all the snow dissipating during summer, as at Poste-de-la-Baleine (northern Quebec); others are *perennial,* as in Victoria Dry Valley (Antarctica).

NIV sands usually form patches or blankets, not well-developed dunes. In the dunes of Victoria Valley, however, one layer of perennial snow has been found beneath the sand; at Poste-de-la-Baleine typical parabolic dunes are partly enriched, during winter and spring, by annual NIV sand. In both cases, the deposit is partly eolian, partly NIV.

Characteristics

In Victoria Valley, NIV blankets appear from the air as long, flat banks, about 0.5-2 km in length. They differ from the neighboring dunes by their gentler slopes (0-10°, with a median of 2°, as compared with 5-37° for the dunes), by the lack of any crest, by their smaller thickness (0.5 to 2 or 3 m), and by the occasional presence of patches of fine gravel (3-5 mm diameter). Furthermore, they are marked by more or less vertical contraction cracks, dividing the surface into 8-20-m-wide polygons, and further by crevices and subsiding blocks along the sides due to sapping by melting water streamlets. Like the dunes, the niveo-eolian sand banks show small ripples (7-10 cm wavelengths), but they also have large asymmetric ripples (0.8-1 m long). In transverse section, sand and snow layers alternate, their thickness varying from 0.2-60 cm (Fig. 1). In the summer, the top layer always consists of pure sand about 20-30 cm thick; it appears that all the snow that might have been originally mixed with the sand in the top layer has melted or sublimated, this layer then acting as an insulator for the underlying layers. The niveo-eolian banks may occupy an area five times more extensive than the dunes and may pass laterally into them.

At Poste-de-la-Baleine, the annual NIV forms patches and, along the coast, on the ice-foot, a 1-3-m-high NIV rampart. Mixed ripples of sand and snow form on fresh snow. Denivation forms (due either to melting or to sublimating) are 2-5-mm-thick sand pellets and sand rolls, 10-30-cm-high sharp cones, or rounded hillocks over which the sand is typically cracked (Fig. 2). If preserved, these cracks, rolls, and pellets would be good criteria for NIV in ancient sediments.

In Victoria Valley, the thickness of any one layer of sand varies from place to place, so that when the snow melts one would expect to find a kind of wavy or undulating stratification. This characteristic has indeed been observed in Quaternary deposits interpreted as NIV sands, which in many parts of Europe form blankets very similar to the modern ones of Antarctica. Variations in grain-size from one layer to another and the sporadic presence of small (usually <3-5 cm) wind-driven pebbles occur in both cases. Slick-ice ("verglas") on the surface of the ground may have enhanced the sliding of these pebbles. Such sliding surfaces have been observed in Antarctica, on sea-ice and glacier-ice, and in the Netherlands on slick-ice, and have been produced experimentally in a wind tunnel by Grove and Sparks (1952).

In Quaternary NIV deposits, sand and silt layers sometimes alternate. The interpretation of some irregularly stratified silts as NIV is, however, an open question. The NIV deposits in NW Europe are sometimes related to loess (q.v.), and the two may interfinger (and be included in deposits given the field designation ("limon jaune").

FIGURE 1. Denivation in the bottom of a deflation basin. Knife is for scale. (Photo: A. Jahn.)

FIGURE 2. Surficial niveo-eolian forms: rounded hillocks with characteristic cracks. The snow beneath is not visible on this photograph.

Because NIV sand layers often vary in texture, the grain size analysis of a grab sample which includes parts of several layers shows a smaller degree of sorting than do the neighboring sand dunes (Table 1); in other cases (Poste-de-la-Baleine) there is little difference. The shapes of 0.5–2 mm sand grains are about the same as in the nearby dunes, i.e., mostly unworn at Poste-de-la Baleine, rounded and frosted by the wind in Victoria Valley (Table 1).

According to the proportion of sand and snow, and the amount of meltwater, the albedo

TABLE 1. Niveo-Eolian Sands (mean of 7 samples) Compared with 12 Fluvioglacial and 2 Dune Sands, All from Victoria Valley, McMurdo, Antarctica

	Dune	Niveo-eolian	Fluvio-glacial
Grain Size (as *Shapes of Grains*)			
Q_1 mm	0.36	0.60	0.92
Q_2 = Md mm	0.27	0.32	0.42
Q_3 mm	0.20	0.21	0.21
So (Trask)	1.31	1.56	1.92
$Qd\phi$ (Krumbein)	0.39	0.63	0.90
He (Cailleux)	0.38	0.57	0.74
Shapes of grains (quartz and feldspar)			
Angular	8%	14%	31%
Smoothed and frosted by wind	35	31	29
Rounded and frosted by wind	43	46	35
Others (reworked, etc.)	14	9	5

After Cailleux, 1962, and Michel, 1964.

of NIV may have every value between that of pure wet sand (0.15 at Poste-de-la-Baleine) and that of pure snow (0.80-0.90).

On the planet Mars, near the poles, Mariner 9 photographs show horizontally stratified plateaus darker than the polar caps but much lighter than other kinds of rock surfaces. It is known heavy dust storms occur on Mars, and it has been suggested that this circumpolar stratified terrain might be NIV (carbonic, hydric, or carbohydric) which has been compacted by its own weight into "glacio-eolian" deposits.

ANDRE CAILLEUX

References

Cailleux, A., 1962. Etudes de géologie au detroit de McMurdo (Antarctique), *CNFRA*, **1**, 41p. English translation, 1968; Periglacial of McMurdo Strait (Antarctica), *Biul. Peryglacjalny (Lodz)*, **17**, 57-90.

Cailleux, A., 1972. Les formes et dépots nivéo-éoliens actuels en Antarctique et au Nouveau-Québec, *Cah. Geogr. Québ.*, **16**(39), 377-409.

Cailleux, A., 1974. Formes précoces et albédos du nivéo-éolien, *Z. Geomorph. N. F.*, **18**, 437-459.

Edelman, C. H., and Crommelin, R. D., 1939. Over de periglaciale natuur van het jong-pleistoceen in Nederland, *Tijds. Nederl. Aardrijksk Gen.*, ser. 2, **56**, 502-513.

Grove, A., and Sparks, P. W., 1952. Le déplacement des galets par le vent sur la glace, *Rev. Géomorph. Dyn.*, **3**(1), 37-39.

Jahn, A., 1972. Niveo-eolian processes in the Sudetes Mountains, *Geographia Polonica*, **23**, 93-110.

Michel, J. -P., 1964. Contribution à l'étude sédimentologique de l'Antartique, *Com. Nat. Fr. Rech. Antarctiques*, **5**, 1-91.

Teller, J. T., 1972. Aeolian deposits of clay sand, *J. Sed. Petrology*, **42**, 684-686.

Cross-references: *Eolian Sands; Eolian Sedimentation; Loess; Martian Sedimentation.*

NODULES IN SEDIMENTS

A *nodule* is a local concentration of chemical compounds in sediments or soils, which has formed within the sediment or as a result of sediment-forming processes. Individual nodules are discrete bodies which differ from the enclosing sediment or soil matrix in hardness, color, fabric, and composition.

Nodules vary in size from <1 mm to >1 m in diameter. Although they are mostly rounded, they vary from perfectly spherical to a variety of irregular shapes (Fig. 1). According to composition, they can be grouped as siliceous—chert (q.v.), flint; oxides or hydroxides of iron, aluminum, manganese, or titanium; carbonates and sulfates of alkaline earths; and clay minerals.

FIGURE 1. Chert nodules in chalk (Israel), showing very irregular shapes. (Photo: D. H. Yaalon.)

Concretions (q.v.) in the strict sense are nodules that have developed by concentric accretion around a nucleus. *Rhizoconcretions* have formed around roots. *Oolite* (q.v.) and *pisolite* (q.v.) are accumulations of, respectively, small and large concentric concretions. *Glaebule* is a term introduced by Brewer and Sleeman (1964) as a collective designation for nodules, concretions, and related pedological features.

Various modes of origin may produce similar kinds of nodules both in terrestrial and marine environments, either during sediment accumulation or at various stages of diagenesis (q.v.). Pore filling in voids and/or authigenic precipitation with expansion of the sedimentary framework is a common process; postdepositional displacement or metasomatic replacement reactions without significant volume change are equally common. The concentric layering in concretions indicates rhythmic processes of precipitation in a variable environment. Certain nodules are good environmental indicators, e.g., of reducing conditions (Weber et al., 1964). Only a detailed study, however, will enable an interpretation of the conditions of formation.

Chert (q.v.) and *flint*, which are the common forms of siliceous nodules, are usually built up of cryptocrystalline silica and small quartz crystallites (Micheelsen, 1966). Calcareous nodules have a variety of internal structures or fabrics and form mostly both by direct precipitation and by mobilization of the $CaCO_3$ from the surrounding sediment (Sass and Kolodny, 1972; Wieder and Yaalon, 1974). Iron and other oxide nodules form preferentially in tropical environments (Ojanuga and Lee, 1973). Nodules formed by direct precipitation on the ocean floor, in deep-sea or coastal environments, include manganese nodules (q.v., Vol. IVA; see also *Pelagic Sediments*) and *oolite* (q.v.).

DAN H. YAALON

References

Brewer, R., and Sleeman, J. R., 1964. Glaebules: Their definition, classification and interpretation, *J. Soil. Sci.,* **15,** 66-78.

Bryan, W. H., 1952. Soil nodules and their significance, *Sir Douglas Mawson Anniversary Volume.* Adelaide: Univ. Adelaide, 43-53.

Johnson, D. B., and Swett, K., 1974. Origin and diagenesis of calcitic and hematitic nodules in the Jordan Sandstone of northeast Iowa, *J. Sed. Petrology,* **44,** 790-794.

Micheelsen, H., 1966. The structure of dark flint from Stevns, Denmark, *Med. Dansk. Geol. Fören.,* **16,** 285-368.

Ojanuga, A. G., and Lee, G. B., 1973. Characteristics, distribution and genesis of nodules and concretions in soils of Nigeria, *Soil Sci.,* **116,** 282-291.

Sass, E., and Kolodny, Y., 1972. Stable isotopes, chemistry and petrology of carbonate concretions (Mishash Formation, Israel), *Chem. Geol.,* **10,** 261-286.

Weber, J. N.; Williams, E. G.; and Keith, M. L., 1964. Paleoenvironmental significance of siderite nodules in some shales of Pennsylvanian age, *J. Sed. Petrology,* **34,** 814-818.

Wieder, M., and Yaalon, D. H., 1974. Effect of matrix composition on carbonate nodule crystallization, *Geoderma,* **11,** 95-121.

Zaritskii, P. V., 1974. Chemical and mineral composition of concretions and enclosing rocks, *Lithology Min. Resources,* **8,** 203-210.

Cross-references: *Authigenesis; Cementation; Chert and Flint; Concretions; Diagenesis; Geodes; Ironstone; Oolite; Pelagic Sedimentation, Pelagic Sediments; Pisolite; Weathering in Sediments.* Vol. IVA: *Manganese Nodules.* Vol. XI: *Concretions in Soils.*

NOVACULITE

Novaculite is a variety of chert composed almost entirely of granular microcrystalline quartz; in some regions the term is restricted to white chert of this character. According to Griswold (1892, pp. 2, 84), the Englishman Richard Kirwan in 1784 coined the anglicized word *novaculite* from the Latin *novacula,* a term used by Cicero, Petronius, Celius, and others, first for a sharp knife and then for a razor; then by Linnaeus to denote a fine quality of whetstone. Colloquial use of novaculite extended its meaning from a whetstone to the rock from which whetstones were made. The term *novaculite* was applied first in the Arkansas region by Schoolcraft (1819, p. 183), who referred to the mineral novaculite at Hot Spring, Arkansas; then by Featherstonhaugh (1835, p. 69), who applied the term to whetstone raw mineral. Griswold (1892) subsequently urged that novaculite be used in the petrographic sense for a variety of chert. The term was adopted later as the rock part of the binomial formation name "Arkansas Novaculite." Petrographic use of novaculite as applied to this formation, however, has not been consistent. Various uses of the term include (1) all chert in the formation, (2) whetstone-grade chert only, and (3) light-colored chert that weathers to very fine sugary granules. Use of novaculite as a petrographic term was extended to the Marathon region of Texas by Baker (in Udden et al., 1916), who commonly applied novaculite to white chert only. This usage was extended by King (1930, 1937), who established informal members of novaculite (white chert) as distinct from members of dark-colored chert.

Although novaculite is applied by some workers to white nodular chert that formed by replacement of limestone, the term is most commonly applied to thick sequences of (white) bedded chert whose origin is controversial. The Arkansas Novaculite (late Devonian to early Mississippian) of the Ouachita Mountains and Caballos Novaculite [Silurian(?) to Early Mississippian(?)] of the Marathon region of Texas are the most extensive novaculite-bearing formations of North America.

Megascopically, novaculite of the Marathon region is milk-white, opaque, nonporous chert that has a duller luster than glazed porcelain (porous white chert is commonly called *tripolitic* chert). It has poor conchoidal fracture and is locally stained by hematite along fractures. Novaculite owes its lack of color to the absence of detrital impurities and chemical pigments,

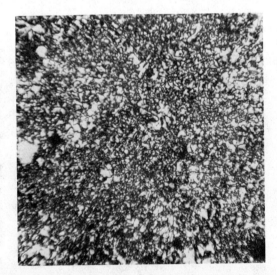

FIGURE 1. Crossed polarizer view of novaculite showing mosaic of microquartz grains. White grain at left is detrital quartz; dark speck at lower right is the core of a Radiolarian capsule. Caballos Novaculite. 150X.

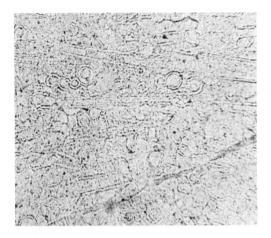

FIGURE 2. Novaculite showing relict spicules in plane polarized light. Caballos Novaculite. 150X.

and its milkiness to the dispersion of light by minute water inclusions (Folk, 1965) and reflecting faces of microgranular quartz crystals. Chemical analyses show that novaculite contains 98.6-99.6% SiO_2; x-ray diffraction and thin sections reveal that quartz is the only silica phase present.

Novaculite is composed chiefly of interlocking microquartz grains 5-25μ in diameter (Fig. 1). Grain boundaries are diffuse in most rocks, but are sharp in coarser-grained samples that have undergone incipient metamorphic recrystallization. Chalcedony is absent to rare. Coarser quartz grains occur along healed tectonic fractures. Trace amounts of detrital quartz, illite, authigenic carbonate (manganiferous calcite and rhodochrosite), and hematite are present locally. Most thin sections of novaculite show no relict texture of the original rock, but a few show clear relics of monaxon sponge spicules (Fig. 2), questionable Radiolaria, and ovoid patches of finer-than-average microquartz that resembles fecal pellets.

Explanations (Goldstein, 1959; Park and Croneis, 1969; McBride and Thomson, 1970) for the origin of bedded novaculite include: replacement of limestone, cementation of quartz silt or sand by silica, diagenetic alteration of volcanic ash or organic siliceous ooze, chemical precipitation from sea water, and hydrothermal alteration of siliceous sediment. Relict textures suggest that diagenetic alteration of siliceous, spicule-rich ooze is the principle origin; the role of volcanism cannot be assessed directly. Controversy about the Caballos Novaculite is concerned with whether the ooze accumulated very slowly in a deep marine basin that was starved of terrigenous detritus (McBride and Thomson, 1970), or whether deposition was relatively fast in shallow-marine to supratidal environments (Folk, 1973).

EARLE F. McBRIDE

References

Featherstonhaugh, G. W., 1835. Geological report of an examination made in 1834 of the elevated country between the Mississippi and Red rivers, *U.S. 23rd Congress, House of Rep. Doc. 151*, 97p.

Folk, R. L., 1965. *Petrology of Sedimentary Rocks* Austin, Texas: Hemphill's, 159p.

Folk, R. L., 1973. Evidence for peritidal deposition of Devonian Caballos Novaculite, Marathon Basin, Texas, *Bull. Am. Assoc. Petrol. Geologists,* **57,** 702-725.

Goldstein, A., Jr., 1959. Cherts and novaculites of the Ouachita facies, *Soc. Econ. Paleont. Mineral. Spec. Publ. 7,* 135-149.

Griswold, L. S., 1892. Whetstones and the novaculites of Arkansas, *Arkansas Geol. Surv., Ann. Rept, 1890,* 3, 443p.

King, P. B., 1930. The geology of the Glass Mountains, Texas, Part 1: Descriptive geology, *Univ. Texas Bull.,* **3038,** 167p.

King, P. B., 1937. Geology of the Marathon Region, Texas, *U.S. Geol. Surv. Prof. Pap. 187,* 148p.

McBride, E. F., and Thomson, A., 1970. The Caballos Novaculite, Marathon region, Texas, *Geol. Soc. Am. Spec. Pap. 122,* 129p.

Park, D. E., Jr., and Croneis, C., 1969. Origin of the Caballos and Arkansas Novaculite formations, *Bull. Am. Assoc. Petrol. Geologists,* **53,** 417-428.

Schoolcraft, H. R., 1819. *A Review of the Lead Mines of Missouri, Including Some Observations on the Mineralogy, Geology, Geography, Antiquities, Soil, Climate, Population and Production of Missouri, Arkansas, and Other Sections of the Western Country.* New York: Wilson, 299p.

Udden, J. A.; Baker, C. L; and Böse, E., 1916. Review of the geology of Texas, *Univ. Texas Bull.,* **1644,** 178p.

Cross-references: *Chert and Flint; Porcelanite; Silica in Sediments; Silicification.*

NUÉE ARDENTE

One of the most unusual agencies of sedimentation is the "glowing cloud" or *nuée ardente,* a cloud of very hot gas and ash produced by certain acidic volcanic eruptions. The first systematic description of a nuée ardente was from the eruption of Mount Pelée on Martinique in 1902 and 1903 (Lacroix, 1903, 1904). There, on several occasions, masses of incandescent, ash-laden gas burst from fissures on the flanks of the volcano and cascaded toward the sea. The velocity of the nuée ardente which destroyed the town of St. Pierre during this episode was estimated at over 150 m/sec, and its temperature was near 800°C.

Although the glowing cloud is the most visible and spectacular part of the eruption, Lacroix (1903) noted that there was a dense basal component (translated in Ross and Smith, 1960):

At the base [of a nuée ardente] is found a zone at very high temperature, in which the solid materials predominate (blocks of all dimension, very small fragments, fine cinders); each of these pieces, or the solid particles of which it is formed, radiate heat, and must be surrounded by an atmosphere of gas and vapors, extremely compressed at the beginning, but expanding rapidly; it is this atmosphere which prevents the solid particles from touching one another, maintaining the mass in a state of mobility which allows it to flow over the slope almost in the manner of a liquid.

This basal flow has sometimes been referred to as a *glowing avalanche* or *ash flow* (Ross and Smith, 1960). Ross and Smith point out that usage has not always differentiated between the clouds themselves and the dense ash- or block-and-ash-transporting basal part. This basal part constitutes the noncloud portion of a glowing cloud, which may not even be glowing; in general the cloud glows only by reflection of the underlying ash flow.

The basal flow is a type of fluidized sediment flow (q.v.), where the ash, lapilli, and occasional blocks are kept in suspension by escaping gases. As early as 1903, Anderson and Flett suggested that the nuée ardente was a turbulent mixture of expanding gases, solid material, and molten droplets which continued to give off gas during flowage. Perret (1935) also believed that the mobility of nuées ardentes was the result of gas emission during flowage, and this hypothesis has been accepted by most workers (McTaggart, 1960, 1962; see also Fisher, 1966). Sparks (1976) has pointed out that only the fine-grained components can be fluidized by gas and that the flow behaves as a dispersion of larger clasts in a medium of fluidized fines, which acts as a lubricant similar to a mud-water slurry in a mudflow (q.v.). Some nuées ardentes move out explosively from the base of a volcanic eruption column and may, strictly speaking, be classified as base surges (see *Base-Surge Deposits*).

The deposits produced by nuées ardentes (including the basal part) are termed *ash-flow tuffs, welded tuffs,* or *ignimbrites* (q.v.; see also *Tuff*). They usually have a rhyolitic composition and, because the individual particles are often still molten when they come to rest, they often are difficult to distinguish from a rhyolitic lava. Recognition criteria include thick (30-100 m or more), poorly sorted, poorly bedded units; lack of flow structures; a "layering" effect caused by variation in the degree of welding; uniformity of tuff sheets over large areas and distances; uneven bases and nearly level tops; columnar jointing; welded and deformed pumice; and vapor-phase minerals (Ross and Smith, 1960).

FREDERICK F. WRIGHT

References

Anderson, T., and Flett, J. S., 1903. Report on the eruptions of the Soufrière in St. Vincent in 1902, and on a visit to Montagne Pelée, in Martinique, Part I, *Roy. Soc. London, Phil. Trans.,* ser. A, **200**, 353-533.

Fisher, R. V., 1966. Mechanism of deposition from pyroclastic flows, *Am. J. Sci.,* **264**, 350-363.

Lacroix, A., 1903. L'éruption de la montagne Pelée en janvier 1903, *Compt. Rend., Acad. Sci. Paris,* **136**, 442-445.

Lacroix, A., 1904. *La Montagne Pelée et ses Éruptions.* Paris: Masson, 662p.

McTaggart, K. C., 1960. The mobility of *nuées ardentes, Am. J. Sci.,* **258**, 369-382.

McTaggart, K. C., 1962. *Nuées ardentes* and fluidization—a reply, *Am. J. Sci.,* **260**, 470-476.

Perret, F. A., 1935. The eruption of Mt. Pelée 1929-1932, *Carnegie Inst. Washington Publ. 458,* 126p.

Ross, C. S., and Smith, R. L., 1960. Ash-flow tuffs: Their origin, geologic relations and identification, *U.S. Geol. Surv. Prof. Pap. 366,* 81p.

Smith, R. L., 1960. Ash flows, *Geol. Soc. Am. Bull.,* **71**, 795-842.

Sparks, R. S. J., 1976. Grain size variations in ignimbrites and implications for the transport of pyroclastic flows, *Sedimentology,* **23**, 147-188.

Cross-references: *Base-Surge Deposits; Fluidized Sediment Flows; Gravity Flows; Ignimbrites; Pyroclastic Sediment; Tuff; Volcaniclastic Sediments and Rocks.*

O

OFFSHORE SPRINGS—*See* Vol. III

OIL SHALE

An oil shale is a fairly dark, dense, fine-grained, and well-bedded to thinly laminated sediment containing a significant amount (>10%) of an organic component called *kerogen* (q.v.; see also *Bitumen, Bituminous Substances*). In industry, an oil shale is any rock from which substantial quantities of oil can be extracted. Some oil shales are true shale (q.v.), e.g., the Lower Carboniferous oil shales of Scotland (Greensmith, 1962); others are marlstones and dolomitic limestones, e.g., the Eocene Green River Formation (Fig. 1). The latter were called organic marlstones prior to 1920. Retorting of oil shales at high temperatures breaks down the kerogen content into petroleum hydrocarbons.

Oil shales may be of marine or of lacustrine origin, are of widely variable and complex composition, and often grade into organic-poor sediments. Their kerogen content is a very complex organic material thought to have been derived from wind-blown plant spores and pollens and minute aquatic plants such as algae which were deposited along with the water-borne silts. In order for this material to be preserved, a stagnant, anaerobic environment is necessary (see *Euxinic Facies*). The presence of perfectly preserved fish skeletons and the absence of benthos in the Green River Formation, for example, are further proof that the center of the basin was euxinic.

Although some oil shales may be marine in origin, it appears that most known examples, notably the Green River Formation, originated in lacustrine environments. Picard and High (1972) review some of the criteria for distinguishing lacustrine and marine hydrocarbons. Hunt et al. (1954) found a correlation between certain solid hydrocarbons of the Uinta Basin and extracts from Tertiary beds believed to be their source; as the depositional environment of the lake changed, the hydrocarbons ozocerite, albertite, gilsonite, and wurtzilite formed successively. The high content of saturated hydrocarbons and low asphaltenes and tars in the Precambrian Thompson Slate and Rove Formation graywacke suggested to Swain et al. (1958) that deposition had taken place in oligotrophic lakes; in contrast, bitumens in the Rove Formation argillite and the Biwabik Formation resembled hydrocarbons in modern marine sediments in the Gulf of Mexico.

Oil shales are distinguishable from bituminous and carbonaceous shales (see *Euxinic Facies*) in that the latter tend to be odorless, less dense, more easily weathered, and brittle; and the organic content is visibly recognizable as such. Freshly broken surfaces of oil shales may sometimes have a slight petroleum-like odor, and cut surfaces and shavings of portions rich in kerogen have a pronounced slipperiness and waxy ap-

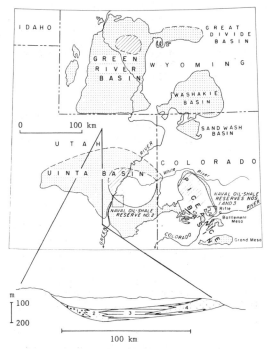

FIGURE 1. *Top:* Map showing location and grades of oil shales of the Green River Formation (permission IPAA Monthly); dots = low grade; diagonal lines = >3 m thick of 100 liters oil per ton shale. *Bottom:* Cross section through Uinta Basin (after Selley, 1971 after Osmond). (1) fluvial sand facies; (2) delta and nearshore sand facies; (3) oil shale facies; (4) mudflat deltaic deposits.

pearance. Rich bituminous and carbonaceous shales will burn on the face of an outcrop, whereas only very thin splinters and shavings of rich oil shales can be ignited with a match. Weathering is very slow and inversely proportional to the kerogen content.

A typical oil shale from the kerogen-rich zone of beds in Utah and Colorado consists of approximately 86% inorganic and 14% organic materials. The inorganic components by wt % are approximately: carbonates, 50%; feldspars, 19%; clays (mostly illites), 15%; quartz, 10%; analcite, 5%; pyrite, 1%. The kerogen appears to be the cementing material in the richer shales. It is generally composed of approximately 5–15% by weight of soluble components, chiefly paraffins and cycloparaffins, tetralin derivatives, steroids, carboxylic acids, carotenes, and tryptophanes. Overall composition of the organic content by weight is approximately 80% carbon and 10% hydrogen, with the balance consisting of nitrogen, sulfur, and oxygen. Exact analysis of the insoluble portion is difficult.

Oil shales are widely distributed throughout the stratigraphic record. Two-thirds of the estimated recoverable shale oil in the world is in the US, and one-fifth is in Brazil. The largest deposits of rich oil shales in the world occur in the Eocene Green River Formation (Fig. 1). The thickness and richness of these oil-shale beds are quite well known from measured outcrops and hundreds of core-hole and drill-cutting samples. Continuous oil shale sections 5 m to 700 m in thickness and having an average recoverable oil content of 60 liters (½ barrel) per ton underlie 3600 km^2 of NW Colorado in the Green River Basin. Retortable oil content is estimated at >1 trillion barrels. Included in the totals are some richer sections which average 100 liters/ton and contain 400 billion barrels of retortable oil. The richest section, called the "Mahogany Zone," varies from 1–35 m in thickness; thin, persistent beds within the zone average from 200–300 liters/ton.

RICHARD V. HUGHES

References

Bass, N. W., 1964. Relationship of crude oils to depositional environment of source rocks in the Uinta Basin, in E. F. Sabatka, ed., *Guidebook to the Geology and Mineral Resources of the Uinta Basin.* Intermountain Assoc. Geol., 201-206.

Bradley, W. H., 1964. Geology of Green River Formation and associated Eocene rocks in southwestern Wyoming and adjacent parts of Colorado and Utah, *U.S. Geol. Surv. Prof. Pap. 496-A,* 86p.

Bradley, W. H., 1973. Oil shale formed in desert environment: Green River Formation, Wyoming, *Geol. Soc. Am. Bull.,* 84, 1121-1124.

Brobst, D. A., and Tucker, J. D., 1973. X-ray mineralogy of the Parachute Creek Member, Green River Formation, in the northern Piceance Creek Basin, Colorado, *U.S. Geol. Surv. Prof. Pap. 803,* 53p.

Cashion, W. B., 1967. Geology and fuel resources of the Green River Formation, southeastern Uinta Basin, Utah and Colorado, *U.S. Geol. Surv. Prof. Pap. 548,* 48p.

Duncan, D. C., and Swanson, V. E., 1965. Organic rich shale of the United States and the world land areas, *U.S. Geol. Surv. Circ. 523,* 30p.

Greensmith, J. T., 1962. Rhythmic deposition in the Carboniferous oil-shale group of Scotland, *J. Geol.,* 70, 355-364.

Hunt, J. M.; Stewart, F.; and Dickey, P. A., 1954. Origin of hydrocarbons of Uinta Basin, Utah, *Bull. Am. Assoc. Petrol. Geologists,* 38, 1671-1698.

Picard, M. D., and High, L. R., 1972. Criteria for recognizing lacustrine rocks, *Soc. Econ. Paleont. Mineral. Spec. Publ. 16,* 108-145.

Sears, J. D., and Bradley, W. H., 1925. Relations of the Wasatch and Green River Formations in northwestern Colorado and southwestern Wyoming with notes on oil shale in the Green River Formation, *U.S. Geol. Surv. Prof. Pap. 132-F,* 93-108.

Selley, R. C., 1971. *Ancient Sedimentary Environments.* Ithaca, N.Y.: Cornell Univ. Press, 237p.

Smith, J. W., Trudell, L. G., and Stanfield, K. E., 1969. Characteristics of Green River Formation oil shales at Bureau of Mines Wyoming Corehole No. 1, *U.S. Bur. Mines Rept. Invest. 7172,* 92p.

Swain, F. M.; Blumentals, A.; and Prokopovich, N., 1958. Bitumens and other organic substances in Precambrian of Minnesota, *Bull. Am. Assoc. Petrol. Geologists,* 42, 173-189.

Winchester, D. E., 1923. Oil shale of the Rocky Mountain region, *U.S. Geol. Surv. Bull.,* 729, 204p.

Cross-references: *Bitumen, Bituminous Sediments; Crude-Oil Composition and Migration; Euxinic Facies; Kerogen; Lacustrine Sedimentation; Organic Sediments.*

OLISTOSTROME, OLISTOLITE

The terms *olistostrome* and *olistolite* (olistolith) originated in the Central Sicilian Basin, where they were proposed for the first time by geologists of the Gulf Italia Company, particularly by Flores (1955) during the Fourth World Petroleum Congress, held in Rome in June 1955. His summary is given in the paper by Beneo (1955).

An *olistostrome* is a mappable stratigraphic layer representing a horizon of slumped material. It comes from the Greek words *olistomai*—to slide, and *stroma*—layer. By analogy with biostrome (an organic accumulation such as shell bank), olistostrome indicates an accumulation due to sliding. An *olistolite* (olistolith) is a mass included within the olistostrome. Olistolites are usually layered; their size has not been defined, but Abbate et al. (1970) suggest a length of 4 m to 1-2 km (200-300 m maximum thickness).

The following definitions are quoted from Flores (1955):

By olistostromes we define those sedimentary deposits occurring within normal geologic sequences that are sufficiently continuous to be mappable, and that are characterized by lithologically and/or petrographically heterogeneous materials, more or less intimately admixed, that were accumulated as a semifluid body. They show no true bedding, except for possible large inclusions of previously bedded materials. In any olistostrome we distinguish a "binder" or "matrix" represented by prevalently pelitic, heterogeneous material containing dispersed bodies of harder rocks. These latter may range in size from pebble to boulder and up to several cubic km. There is no constant ratio between the total volume to the inclusions and that of the binding mass. . . .

The name "olistolith" from the Greek words *olistomai*—to slide, and *lithos*—rock, is applied to the masses included as individual elements within the binder. These masses were previously referred to as "exotics" . . .

In some olistostromes one or more types of accumulation due to flowage can be recognized ranging from the chaotic deposition of coarser elements which were bodily detached from their original position, down to "graded bedding" due to turbidity currents.

Since 1955, numerous authors have used the terms proposed by Flores, but some have occasionally deviated greatly from their original meaning (see Abbate et al., 1970). Olistostrome has sometimes been incorrectly used to describe a "gliding nappe." In fact, an olistostrome can be emplaced by synsedimentary slumping off the front of a nappe, but it is not a gliding nappe itself.

Certain authors, among them Jacobacci (1965), tend to restrict the concept. This author states: "an olistostrome is composed of a deposit or an accumulation of material originating from a landslide ('frane'). Although due to various causes this is a sedimentary phenomenon, and should be found intercalated, at least in some cases, within an environment of normal sediments or turbidites. . . . It should show, at least in part, the chaotic aspect of a landslide. A landslide should not show stratification, nor should it be limited by subparallel surfaces."

It has to be noted that in his definition, Flores would even regard turbidites as olistostromes. In general, turbidity currents are distinct phenomena from fluidized submarine slides, and although there may be transitions, the two concepts are distinct (see *Gravity Flows; Gravity-Slide Deposits; Turbidity-Current Sedimentation*).

It appears desirable that the terms olistostrome and olistolite should have very precise connotations. Their genetic characteristics have been reviewed by Jacobacci (1965), and involve synsedimentary sliding, which is frequent in sedimentary sequences of unstable and subsiding zones. Here it is possible to distinguish between *allolistostrome* and *endolistostrome*, in the sense of Elter and Raggi (1965). In a flysch sequence, the allolistostromes are generally called "wildflysch," i.e., their components are exotic, in contrast to the endolistostrome, in which the components are local.

By definition, no important tectonic influences are directly involved in the emplacement of olistostromes, so that it would be logical to assume that they can only form beds or lenses of minor thickness and area. One can conceive that landslides or mudflows of minor thickness (on the order of one to several tens of meters) could be emplaced on the sea floor, or even on dry land, without the necessity of tectonic transport; they must be emplaced by simple sliding (see *Mass-Wasting Deposits; Mudflow Deposits*). In the field it is often found that a transition exists between tectonic and sedimentary phenomena so that a conventional limit must sometimes be drawn. The same restriction also applies to olistolites, which can be nothing else than small blocks embedded in the olistostromes.

For the above reasons, and contrary to certain tendencies, it is urged that the term *olistostrome* be used only in limited and well-defined cases (Broquet, 1968, 1970a,b). On the other hand, there should be widespread usage of the term *sedimentary klippe*. Only a few klippes conform to the usual textbook definition (tectonic outliers, remnants of nappes isolated by erosion). The vast majority were connected with nappe brows but were emplaced by sedimentary sliding.

Melange

The term *melange* (French: mixture) was introduced by Greenly (1919) in the phrase "autoclastic melange," which he used to describe tectonic mixtures in Precambrian rocks on Anglesey Island, Wales (see Hsü, 1968, for original quote). Hsü defined melanges as "mappable bodies of deformed rocks characterized by the inclusion of native and exotic blocks, which range up to several miles long, in a pervasively sheared, commonly pelitic matrix. Fragmentation and mixing of melanges result from tectonic deformation under an overburden pressure." Two examples he cites are the *argille scagliose* of the Apennines and the Franciscan melange of the western US.

Since the acceptance of plate tectonics, the use of terms such as melange, olistostrome, wildflysch, and tectonite has burgeoned, especially in the study of subduction zone complexes (see

Raymond, 1975, for a long but partial list). As mentioned above, it is often difficult to distinguish tectonic and sedimentary processes in such environments. Indeed, Wood (1974) has shown that the complex studied by Greenly contains olistostromes which were later sheared, and Abbate et al. (1970) have made similar observations on the *argille scagliose*.

Although a definition of melange has not been universally accepted, the present trend favors the use of *melange* alone to describe any body of chaotically mixed rocks. If the mixture is known to be tectonically produced, it is called *tectonic melange;* if the mixture originated by slumping or sliding, *sedimentary melange* (= olistostrome) is used.

PAUL BROQUET

References

Abbate, E.; Bortolotti, V.; and Passerini, P., 1970. Olistostromes and olistoliths, *Sed. Geol.,* 4, 521-557.

Beneo, E., 1955. The results of the studies on petroleum exploration in Sicily, IV World Petrol. Congr. Roma, sect. I/A/2, *Boll. Serv. Geol. Ital.,* 78, 27-50.

Broquet, P., 1968. Etude géologique de la région des Madonies (Sicile), *Thèse Fac. Sci. Lille,* 796p.

Broquet, P., 1970a. La notion d'olistostrome et d'olistolite–Historique et étude critique, *Ann. Soc. Géol. Nord.,* 90, 77-86.

Broquet, P., 1970b. Observations on gravitational sliding: The concept of olistostrome and olistolite, in W. Alvarez and K. Gohrbandt, *Geology and History of Sicily.* Tripoli: Petrol. Explor. Soc. Libya, 255-259.

Broquet, P., 1973. Olistostrome, olistolite et klippe sédimentaire, *Ann. Sci. Univ. Besançon,* 3(20), 45-53.

DeJong, K. A., 1974. Melange (olistostrome) near Lago Titicaca, Peru, *Bull. Am. Assoc. Petrol. Geologists,* 58, 729-741.

Dimitrijevic, M. D., and Dimitrijevic, M. N., 1973. Olistostrome melange in the Yugoslavian Dinarides and Late Mesozoic plate tectonics, *J. Geol.,* 81, 328-340.

Elter, P., and Raggi, G., 1965. Contributo alla conoscenza dell'Appennino ligure: 1) osservazioni preliminari sulla posizione delle ofioliti nella zona di zignago (La Spezia); 2) considerazioni sul problema degli olistostromi, *Boll. Soc. Geol. Ital.,* 84, 303-322.

Flores, G., 1955. Discussion, in Beneo (1955), p. 47.

Flores, G., 1959. Evidence of slump phenomena (olistostromes) in areas of hydrocarbon exploration in Sicily, *Proc. 5th Petrol. Geol. Congr., Sec.* 1, 259-275.

Greenly, E., 1919. The geology of Anglesey, *Geol. Surv. Gr. Brit., Mem.* 2, 980p.

Hsü, K. J., 1968. Principles of melanges and their bearing on the Franciscan-Knoxville paradox, *Geol. Soc. Am. Bull.,* 79, 1063-1074.

Hsü, K. J., 1974. Melanges and olistostromes, *Soc. Econ. Paleont. Mineral. Spec. Publ. 19,* 321-333.

Jacobacci, A., 1965. Frane sottomarine nelle formazioni geologiche. Interpretazione dei fenomeni olistostromici e degli olistolite nell'Appennino e in Sicilia, *Boll. Serv. Geol. Ital.,* 86, 65-85.

Kay, M., 1974. Geosynclines, flysch, and melanges, *Soc. Econ. Paleont. Mineral. Spec. Publ. 19,* 377-380.

Kleist, J. R., 1974. Deformation by soft-sediment extension in the coastal belt, Franciscan complex, *Geology,* 2, 501-504.

Lamare, P., 1946. Les formations détritiques crétacées du massif de Mendibelza, *Bull. Soc. Geol. France, ser. 5,* 16, 265-312.

Nagle, F., 1972. Chaotic sedimentation in north-central Dominican Republic, *Geol. Soc. Am. Mem.,* 132, 415-428.

Raymond, L. A., 1975. Tectonite and melange–a distinction, *Geology,* 3, 7-9.

Wood, D. S., 1974. Ophiolites, melanges, blueschists, and ignimbrites: early Caledonian subduction in Wales? *Soc. Econ. Paleont. Mineral. Spec. Publ. 19,* 334-344.

Cross-references: *Gravity Flows; Gravity-Slide Deposits; Mass-Wasting Deposits; Mudflow and Debris-Flow Deposits; Slump Bedding.*

OOLITE

The term *oolite*, also spelled oölite, is generally used to describe a sedimentary rock, usually limestone, composed of spherical to ovoid, accretionary grains of sand size that are cemented together (Fig. 1A). The grains themselves are called *ooids,* ooliths, or oolites, and were so named for their close resemblance to fish roe or ova (Greek: $\omega o\nu$ = oon, egg; pronounced ō'ŏ). Oolite has also been applied by some earth scientists to present-day deposits of carbonate sand made up of ooids. The term oolitic has also been used to describe the texture peculiar to these rocks and sediments (Fig. 1B), regardless of mineralogy or chemical composition. Ooids are comparable in shape, internal structure, and composition to pisolite grains (pisoids, pisoliths; see *Pisolite*) but by convention are distinguished by their smaller size (<2 mm in diameter). Although awareness of oolitic rocks and textures dates back to the Roman Empire, the term oolite seems to have been used first by F. E. Bruckmann in 1727 and ooid by E. Kalkowsky in 1908 (Cayeux, 1935, p. 211). Detailed information about the geology and origin of oolitic sediments and rocks can be found in books by Cayeux (1935), Carozzi (1960), and Bathurst (1975). Further details of definition and terminology are discussed by Teichert (1970).

Grain Characteristics

Like pisolite grains, oolite grains have distinc-

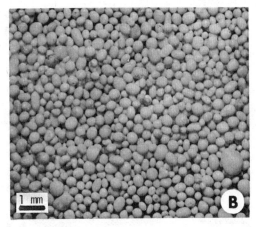

FIGURE 1. (A) Cemented oolite, freshly broken rock surface. Ste. Genevieve Limestone (Mississippian), Illinois. (B) Oolitic sand composed of polished ooids. Recent, near Browns Cay, Bahamas. Both photographs in reflected light.

tive internal structure. In cross section, the typical ooid is seen to be made up of a *nucleus* surrounded by a *cortex* composed of as many as 90–100 concentric laminae. A variety of particles may serve as nuclei depending on the supply available; grains of detrital sand, shell and other fossil fragments, and fecal pellets are among the more common nuclei (Fig. 2A, 2C). Although nuclei may be angular and of varied shapes, the accretion of laminae tends to fill in depressions in the surfaces, producing ooids that are well rounded and dominantly subspherical. Most ooids are 0.1–2 mm in diameter. Many have a high surface polish (Fig. 1B), similar in appearance and probably in origin to the polish imparted by grain-on-grain abrasion, as in a lapidary tumbler.

Most ooids and oolitic sediments and rocks consist of calcium carbonate, either calcite or aragonite. Seen with a petrographic microscope in cross-polarized light, unaltered $CaCO_3$ ooids usually have a distinctive "cross structure" (Fig. 2B) caused by the arrangement of individual rod-like crystals statistically either tangent or radial to the laminae of the cortex (Eardley, 1938; Illing, 1954).

Recent research on oolitic deposits has focused on establishing in detail the environments where marine ooids occur, the ultrastructure of ooids as seen with an electron microscope (e.g., Fabricius and Klingele, 1970; Loreau, 1973), and the organic chemistry and microbiology of ooids (e.g., Mitterer, 1968, 1971; Suess and Fütterer, 1972). Recent marine ooids have laminae made up of minute rods of aragonite (Fig. 3). The packing and orientation of these rods vary in different depositional settings, from more or less random at some sites to strongly tangential or radial at many others. Recent ooids that have been dissolved gently in very dilute acid have been reduced to a complex mass of mucilaginous material mainly composed of filamentous blue-green algae (Nesteroff, 1956; Newell et al., 1960; Shearman and Skipwith, 1965). This material is proteinaceous, made up largely of amino acids like those in biologically calcified tissues (Mitterer, 1968, 1971). Other organic matter, also proteinaceous, occurs in concentric laminae alternating with aragonite, or in radial pillar-like configurations.

Ooids and oolitic rocks composed of minerals other than $CaCO_3$ minerals are common though not as widespread. Phosphatic ooids are composed of apatite or collophane, and iron-rich ooids consist of goethite, hematite, limonite, chamosite, or more rarely siderite and ankerite. Some who have studied these mineralogical types regard them as primary, and others as replacements of $CaCO_3$ (reviews in Cayeux, 1935, 1970 translation; Carozzi, 1960). $CaCO_3$ ooids may also be replaced by silica minerals, dolomite, low-temperature feldspars, glauconite, and a few other minerals. Oolitic sand composed of halite ooids has recently been reported from constructed coastal salt pans in the Dead Sea (Weiler et al., 1974).

Modern Oolite Deposits

Calcium carbonate ooids, by far the most abundant compositional type, today are forming in a variety of conditions. The largest deposits known are accumulating in carbonate sand bodies in warm, shallow sea water and in a number of hypersaline lakes, notably Great Salt Lake (Eardley, 1938; Carozzi, 1957; Kahle, 1974; Sandberg, 1975). Many of these occur-

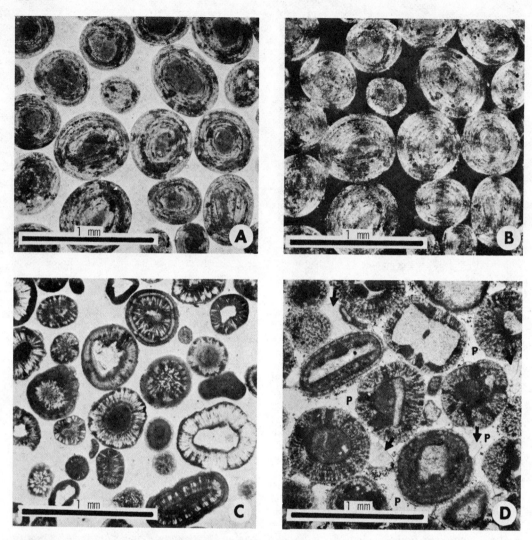

FIGURE 2. (A) Oolitic sand showing concentric laminae of cortex in ooids. (B) Same oolitic sand seen in cross-polarized light showing cross-structure apparently caused by orientation of aragonite crystals in laminae statistically tangent to laminae. Nuclei are pellets possibly of fecal origin. Both A and B are Holocene sand, near Browns Cay, Bahamas. (C) Ooids with radial structure in cortex. Nuclei are varied and include quartz grains (white) and pellets (dark, microcrystalline). Great Salt Lake, Utah. (D) Oolite showing both concentric and radial structure in cortex ooids. Rock is partially cemented by calcite (arrows show some). Much of porosity (P) of original oolitic sand has survived. Ste. Genevieve Limestone (Mississippian), Illinois. All photographs in transmitted light.

rences are discussed in monographs by Cayeux (1935) and Bathurst (1975).

Marine oolitic sand deposits have been reported on in the most detail from the Bahama Banks (q.v.) and along the Trucial Coast of the Persian Gulf (q.v.). These and other occurrences in the northern and southern hemispheres suggest that favored sites of oolitic sand accumulation are warm, shallow shelf areas of moderate to high wave and current activity, and associated protected lagoons at latitudes of <30°. Lees (1975) predicts that oolith/aggregate associations should be found in areas of relatively high salinity and temperature. Oolitic sands are characteristically most abundant in waters <10 m (often <2 m) deep, in complexes of carbonate sand bars along the edges of shallow platforms and continental shelves, and in tidal deltas, beaches, and bars associated with coastal barrier islands. Oolitic sands occur in submarine sand waves or megaripples whose surface bed forms (ripples and small dunes) and internal cross-

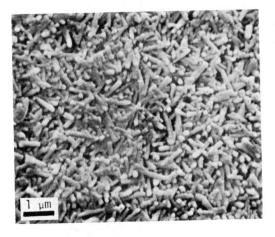

FIGURE 3. Rod-like crystals of aragonite on the surface of a recent ooid from the Persian Gulf (Trucial Coast). The aragonite crystals are statistically oriented tangent to the surface. (Photo: J.-P. Loreau.)

and repack the aragonite crystals in tightly tangential configuration in their cortex laminae, as illustrated by Fig. 3, decrease their microporosity, and impart a polish. Ooids formed in these settings are commonly blown inland onto tidal flats and dunes, or offshore into deeper water. Fig. 4 illustrates the settings in which ooids are known to occur along the Trucial Coast.

Other sites of much smaller Quaternary deposits of $CaCO_3$ ooids include many limestone caverns, many hot springs and possibly streams (McGannon, 1975), freshwater lakes, and artificial sites such as boilers and furnace coils. Cave pearls (q.v.) ranging from ooids to pisolite grains in size form in splash pools, ponds, and streams in caverns such as Carlsbad. Ooids and pisolites also are common in caliche (q.v.). In such settings, they range from symmetrical to strongly asymmetrical, and presumably form with little or no grain motion, in "soil" profiles in semiarid regions. Dunham (1972) reviews and illustrates some properties of caliche and cave ooids.

bedding suggest active transport by tides, waves, and wind-driven currents. Ooids may actually form, however, in depressions between these features and along sides of tidal deltas, as well as in nearby lagoons, after which they are swept into more turbulent environments (Loreau, 1973; Loreau and Purser, 1973). More intense agitation and jostling of ooids may then reorient

Other in-situ deposits which are not composed of $CaCO_3$ are chiefly the laterite and bauxite deposits and "shot soils" of tropical and subtropical regions. These deposits contain high proportions of hydrous iron and/or aluminum oxides as well as clays, and are generally thought to be residual products of weathering.

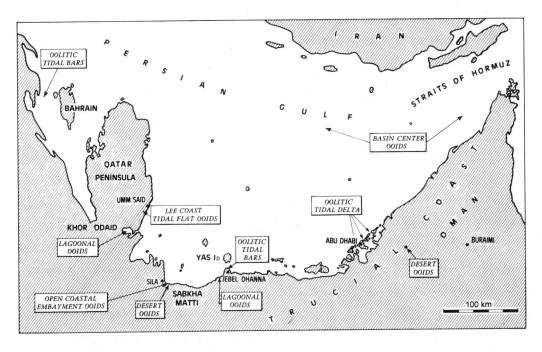

FIGURE 4. Map of the southern Persian Gulf showing environments in which recent oolitic sand occurs (from Loreau and Purser, 1973).

Origin of Ooids

Ideas about the origin of ooids can be grouped into three main categories: (1) accretion of $CaCO_3$ mud particles, "snow-ball" style (Sorby, 1879); (2) biochemical precipitation caused by bacteria (Monaghan and Lytle, 1956; Lalou, 1957), amino acids (Mitterer, 1968), or other organic matter; and (3) purely inorganic precipitation (e.g., Cayeux, 1935; Newell et al., 1960). Inorganic precipitation is at present the most widely favored mechanism. The presence of organic matter in concentric laminae alternating with $CaCO_3$, and experiments in which amino and humic acids influenced the form, internal structure, and crystal form of aragonite in artificially precipitated ooids (Mitterer, 1971; Suess and Fütterer, 1972), suggest that organic matter may be important to the formation of ooids. Further study of the ultrastructure and organic microchemistry may help resolve the present uncertainties.

Ancient Oolitic Sedimentary Rocks

Oolitic limestone and other compositional varieties of oolite are known throughout the geologic column from Precambrian into Cenozoic and from every continent. In North America, oolite is especially widespread in the Mississippian (eastern interior and Rocky Mountain region) and the Jurassic and Cretaceous (Gulf Coast region). In western Europe, oolitic limestone is widespread in the Mississippian, Middle Triassic, Jurassic, and Lower Cretaceous (Cayeux, 1935). Sandberg (1975) discusses the nonuniformitarian aspect of ancient and modern carbonate ooid precipitation.

Oolitic iron ore deposits of Precambrian age in the Lake Superior district, Silurian age in the southeastern US and in France, and of Jurassic age in England and France have been exploited for many decades (see *Ironstone*). Oolitic and pisolitic phosphorite of marine origin is widespread in the Permian Phosphoria Formation of Utah, Idaho, and western Wyoming and Montana, as well as in Miocene rocks of S Florida (see *Phosphates in Sediments*). Residual oolitic phosphate deposits such as those of Florida have commercial value for fertilizer, drugs, and other uses. The soft, easily quarried oolitic limestones of Jurassic age in France and England have been used to build great edifices in W Europe, as has building stone from the Mississippian Salem Limestone in the US. Oolitic and pisolitic limestones contain important oil and gas reserves in many petroleum provinces of the world.

PHILIP W. CHOQUETTE

References

Bathurst, R. G. C., 1975. *Carbonate Sediments and Their Diagenesis,* 2nd ed. Amsterdam: Elsevier, 658p.

Black, M., 1933. The precipitation of calcium carbonate in the Great Bahama Bank, *Geological Mag.,* 70, 455-466.

Carozzi, A. V., 1957. Contribution à l'étude des propriétés géométriques des oolithes–l'exemple du Grand Lac Salé, Utah, *Bull. Inst. Nat. Genev.,* 58, 3-52.

Carozzi, A. V., 1960. *Microscopic Sedimentary Petrography:* New York: Wiley, 485p.

Cayeux, L., 1935. *Les Roches Sédimentaires de France: Roches Carbonatées.* Annotated translation by A. V. Carozzi, 1970: *Carbonate Rocks.* Darien, Conn.: Hafner, 506p.

Dunham, R. J., 1972. Capitan reef, New Mexico and Texas: Facts and questions to aid interpretation and group discussion, *Soc. Econ. Paleont. Mineral., Permian Basin Sect., Midland, Texas, Publ. 72-14,* 180p.

Eardley, A. J., 1938. Sediments of Great Salt Lake, *Bull. Am. Assoc. Petrol. Geologists,* 22, 1305-1411.

Fabricius, F. H., and Klingele, H., 1970. Ultrastrukturen von oöiden und oölithen: zur Genese und Diagenese quartäre Flachwasserkarbonate des Mittelmeeres, *Verh. Geol., B. -A.,* 4, 594-617.

Freeman, T., 1962. Quiet water oolites from Laguna Madre, Texas, *J. Sed. Petrology,* 32, 475-483.

Illing, L. V., 1954. Bahaman calcareous sands, *Bull. Am. Assoc. Petrol. Geologists,* 38, 1-95.

Kahle, C. F., 1974. Ooids from Great Salt Lake, Utah, as an analogue for the genesis and diagenesis of ooids in marine limestone, *J. Sed. Petrology,* 44, 30-39.

Kirchmayer, M., 1962. Zur Untersuchung rezenter Ooide, *Neues Jahrb. Geol. Palaeont. Abhandl.,* 114, 245-272.

Lalou, C., 1957. Studies on bacterial precipitation of carbonates in sea-water, *J. Sed. Petrology,* 27, 190-195.

Lees, A., 1975. Possible influence of salinity and temperature of modern shelf carbonate sedimentation, *Marine Geol.,* 19, 159-198.

Loreau, J. -P., 1973. Nouvelles observations sur la génèse et la signification des oolithes, *Sci. de la terre,* 18(3), 213-244.

Loreau, J. -P., and Purser, B. H., 1973. Distribution and ultrastructure of Holocene ooids in the Persian Gulf, in B. H. Purser, ed., *The Persian Gulf.* New York: Springer-Verlag, 279-328.

McGannon, D. E., Jr., 1975. Primary fluvial oolites, *J. Sed. Petrology,* 45, 719-727.

Mitterer, R. M., 1968. Amino acid composition of organic matrix in calcareous oolites, *Science,* 162, 1498-1499.

Mitterer, R. M., 1971. Comparative amino acid composition of calcified and non-calcified polychaete worm tubes, *Comp. Biochem. Physiol.,* 38B, 405-409.

Monaghan, P. H., and Lytle, M. A., 1956. The origin of calcareous ooliths, *J. Sed. Petrology,* 26, 111-118.

Nesteroff, W. D., 1956. De l'origin des oolithes, *Compt. Rend. Acad. Sci.,* 242, 1047-1049.

Newell, N. D.; Purdy, E. G.; and Imbrie, J., 1960. Bahamian oolitic sand, *J. Geol.,* 68, 481-491.

Rusnak, G. A., 1960. Some observations of Recent oolites, *J. Sed. Petrology,* **30,** 471-480.

Sandberg, P. A., 1975. New interpretations of Great Salt Lake oöids and of ancient non-skeletal carbonate mineralogy, *Sedimentology,* **22,** 497-537.

Shearman, D. J., and Skipwith, P. D. d'E., 1965. Organic matter in recent and ancient limestones and its role in their diagenesis, *Nature,* **208,** 1310-1311.

Shearman, D. J.; Twyman, J.; and Zand Karimi, M., 1970. The genesis and diagenesis of oolites. *Proc. Geol. Assoc.,* **81,** 561-575.

Sorby, H. C., 1879. The structure and origin of limestones, *Proc. Geol. Soc. London,* **35,** 56-95.

Suess, E., and Fütterer, D., 1972. Aragonitic ooids: Experimental precipitation from seawater in the presence of humic acid, *Sedimentology,* **19,** 129-139.

Teichert, C., 1970. Oolite, oolith, ooid: discussion, *Bull. Am. Assoc. Petrol. Geologists,* **54,** 1748-1749.

Weiler, Y.; Sass, E.; and Zak, I., 1974. Halite oolites and ripples in the Dead Sea, Israel, *Sedimentology,* **21,** 623-632.

Cross-references: *Bahama Banks Sedimentology; Caliche, Calcrete; Carbonate Sediments—Bacterial Precipitation; Cave Pearls; Ironstone; Limestone Fabrics; Limestones; Persian Gulf Sedimentology; Pisolite; Spring Deposits.*

OOZES, DEEP-SEA—*See* PELAGIC SEDIMENTS

ORES IN SEDIMENTS—*See* Vol. IVA: **ALUMINUM ORE DEPOSITS; IRON—ECONOMIC DEPOSITS.** *See also* ENCYCLOPEDIA OF ORE DEPOSITS

ORGANIC SEDIMENTS

Organic sediments contain >50% (by weight) original organic tissues or their derivatives. They are further subdivided into two major groups: (1) *coal* (q.v.), which is derived primarily from the lignin fraction of land plants and is characterized by relatively high oxygen content; and (2) *bitumen* (q.v.), which is derived primarily from the lipid fraction of marine organisms, with some contribution of terrestrial material, and is characterized by low oxygen content. In addition to these two major groups, there are minor amounts of substances such as the fossil resins, of which amber is an example. Also, sedimentary rocks contain vast amounts of organic matter in a highly dispersed form. Although this material cannot be classified as organic sediment, it does represent the bulk of the organic matter in the crust of the earth.

Table 1 is a classification of these major groups of organic materials. Coal is derived from two types of organisms. The more common coal is formed from massive land plants, which become converted to an aromatic, high molecular weight (HMW) type of humus. Eventually this material forms peat, brown coal (lignite), bituminous coal, and anthracite. Each substance in this series represents an increasing molecular weight and higher rank of metamorphism (see *Coal—Diagenesis and Metamorphism*). The second and less common type of coal is formed from specialized accumulations of plant materials such as spores, resins, and organic matter derived from algae and plankton. In a swamp environment or in stagnant parts of lakes, this organic matter tends to be concentrated as a watery ooze called *sapropel* (q.v.) or *gyttja* (q.v.). Algal bogs represent this type of deposit. Sapropel is considered to form in waters low in or free of oxygen; gyttja in its broader sense may be used to refer to an organic-rich sediment deposited in open waters. Eventually, through diagenesis and metamorphism, sapropel and some forms of gyttja are converted into various ranks of both *cannel* and *torbanite coals*.

The dispersed organic matter in Table 1 represents organic debris that is deposited with sediments to form limestones, sandstones, and shales. The terms *bituminous limestone* and *oil shale* (q.v.) are used to describe rock with relatively high organic contents (>10%). The organic matter dispersed within sediments tends to be the more aliphatic, low molecular weight (LMW) *humus*. Eventually this material is converted by microbial alteration and diagenesis into *kerogen* (q.v.), which is a general term for the insoluble organic material of sedimentary rocks.

Source Materials

All organic sediments were originally derived from the remains of plants and animals which, when alive, constituted the complex aggregate of living matter called the *biosphere*. The biosphere receives its carbon from the carbon dioxide of the atmosphere and in turn deposits part of the carbon in the lithosphere as organic sediments and organic constituents of sedimentary rock (see Vol. IVA: *Carbon Cycle*). The basic substances of all plants and animals are the proteins, carbohydrates, lipids, and lignin. The *proteins* are complex polymers of amino acids. Most of the nitrogen, sulfur and, to some extent, the phosphorus in living things is incorporated with the proteins. They are the chief source of nitrogen in organic sediments. The proteins are concentrated in animal life as shown in Table 2.

ORGANIC SEDIMENTS

TABLE 1. Origin of Organic Materials

Coals		Dispersed Organic Matter	Bitumens
massive land plants	algae, plankton spores, resins	land and marine organic matter deposited with inorganic sediments	selective fraction (<3% of organic matter
↓	↓	↓	↓
concentrated as aromatic HMW humus	concentrated as jelly-like slime-sapropel (gyttja) in algal bogs	adsorbed on mineral matter as aliphatic LMW humus	dispersed lipids carried by migrating waters to reservoirs
↓	↓	↓	↓
Peat Brown coal (lignite) Bituminous coal Anthracite	cannels and torbanites (bog heads)	kerogen	petroleum waxes asphalts asphaltites pyrobitumens

HMW = High molecular weight
LMW = Low molecular weight
From Hunt, 1968.

TABLE 2. Composition of Living Matter

		Wt. % of major constitutents[a]		
Substance	Proteins	Carbohydrates	Lignin	Lipids
Plants				
Spruce Wood	1	66	29	4
Oak leaves	6	52	37	5
Scots-Pine Needles	8	47	17	28
Phytoplankton	23	66	0	11
Diatoms	29	63	0	8
Lycopodium Spores	8	42	0	50
Animals				
Zooplankton	60	22	0	18
Copepods	65	25	0	10
Higher Invertebrates	70	20	0	10

[a]Ash-free basis.

Carbohydrates make up the bulk of plant material, both terrigenous and marine. The simple carbohydrates, the sugars, are soluble in water and are important as primary products of photosynthesis. *Cellulose,* which is a HMW polymer of the sugars, is the fundamental constituent of cell walls; wood contains about 60% cellulose. *Lignin* is a tough, polymeric substance insoluble in acid and most organic solvents; it remains after hydrolyzing the cellulose of plants. Lignin comprises about 30% of the wood of trees. Small amounts of lignin residues occur in mosses and ferns but none has been found in unicellular plants such as diatoms. Decomposition products of lignin, proteins, and carbohydrates are the forerunners of coal and also contribute to the large amount of humic substances found in soils.

Chitin, which forms the fingernails of animals and the hard shell of crustaceans and insects, is a very tough material that is chemically similar to cellulose except that its building block is a sugar-amine rather than a pure sugar.

Lipids are oil-like constituents of living things, which are only slightly soluble in water but readily soluble in fat solvents such as ether, chloroform, and carbon tetrachloride. Lipids comprise waxes, fats, essential oils, and pigments. Waxes such as bee's wax, apple wax, and spermaceti are esters of complex sterols and aliphatic alcohols containing chains of 16-36 carbon atoms. *Fats,* which are a major constituent of spores and some animals, are esters of fatty acids with glycerol. Plants also contain small amounts of essential oils, such as wintergreen, rose oil, and turpentine. Many of these oils are hydrocarbons (q.v.). The pigments or colored materials of plants, such as chlorophylls, xanthophylls, and carotenes, also contain mostly carbon and hydrogen in their structure.

Transformation of Dead Material to Organic Sediments

Dead organic matter deposited on the continents and exposed to atmospheric influences is eaten by scavengers or undergoes decomposition through bacterial decay and atmospheric weathering. Usually nothing remains of animals except a few hard parts such as bones and teeth. Most plant material is destroyed, except lignin and decayed humus in the form of peat. In order to *preserve* a large quantity of organic substance, it is necessary to eliminate oxygen and suppress microbial activity. Both the microbial activity and available oxygen are greatly reduced by increasing the depth of burial; it only requires a few cm of a clay or carbonate mud or silt to effectively preserve organic matter. Also, stagnant waters in the stratified bottom layers of silled basins and lakes are very effective in preserving organic matter (see *Euxinic Facies*).

The *fossilization* of organic matter follows the general outline given in Table 3. The first step in the decomposition of organic matter is the *microbial degradation* of the biopolymers to simple biomonomers. The enzymes of microbes convert carbohydrates such as cellulose and starch to simple sugars; proteins become peptides and amino acids, fats are changed to fatty acids and glycerol, and lignin to phenols, quinones, and aromatic acids. In most areas of organic deposition the first few cm of sediment contains large quantities of bacteria that have the enzymes necessary for catalyzing these decomposition reactions. Many of the decomposition products have been identified in both lacustrine and marine sediments. Gases that form, depending on the conditions, are carbon dioxide, ammonia, methane, hydrogen, and hydrogen sulfide (see *Gases in Sediments*). Methane is a common product of organic decay in marshes, swamps, and stagnant areas of lakes. Proteins decompose most rapidly, with the carbohydrates next, and the lipids and lignin the most resistant to decay.

Many of the biomonomers listed in Table 3

TABLE 3. Fossilization of Organic Matter

Molecular Type	Examples
BIOPOLYMERS	carbohydrates, proteins, lipids, lignin
Microbial degradation, hydrolysis	
BIOMONOMERS	sugars, amino acids, fatty acids, alcohols, phenols, quinones, aromatic acids, purines, pyrimidines, hydrocarbons
Condensation, polymerization, "browning" reaction	
GEOPOLYMERS	nitrogenous and humus complexes
Diagenesis—Reduction, decarboxylation, deamination, cyclization, thermal alteration	
GEOPOLYMERS GEOMONOMERS	Coal, asphalt petroleum hydrocarbons

are highly reactive and condense with each other to form nitrogenous and humus complexes, which then undergo slow diagenetic changes with time and increasing depth of burial. Depending on the environment of deposition, the organic matrix of the sediment consists of varying quantities of the biopolymers, biomonomers, and geopolymers, which then undergo the diagenetic reactions shown in Table 3.

Coal formed in the peat and brown-coal series is considered to be autochthonous, i.e., formed in place (see *Peat Bog Deposits*). Many brown coals show an abundance of large plant remains, particularly along bedding planes, in the form of wood, stems, leaves, bark, and roots. These remains emphasize the importance of woody material as a progenitor of coal. This material, along with the humus formed from leaves, twigs, and other organic debris, is eventually converted to peat wherever there are fresh or brackish water swamps. *Peat* deposition takes place extensively where there is rapid growth and reproduction of the plants which form humic material that is relatively difficult to decompose due to the environmental conditions. Peat accumulation may be tens of meters thick and cover many km^2. The most extensive period of peat formation was during the Carboniferous when uninterrupted swamps spread over vast areas of the world. These swamps apparently went through a process of subsidence that was balanced by sedimentation to form thick sequences containing large amounts of organic matter.

If the subsiding swamp area is near the sea, it may be invaded by sea water and covered with a deposit of sand or calcareous mud. If the subsiding swamp is inland, the area may be flooded by fresh water and covered with terrestrial soil or clay. If the invasion of mineral matter is more rapid than the formation of organic matter, the peat swamp ceases to exist. Many coal measures consist of bands of coal and of sedimentary rocks arranged in a repetitive sequence known as a *cyclothem* (see Cyclic Sedimentation).

The *bitumens,* such as petroleum, asphalts, asphaltites, etc., are believed in most cases to be allochthonous, i.e., they have migrated from their fine-grained source rocks into porous, coarse-grained sediments, the reservoir rocks. In the case of petroleum, this migration can occur over hundreds of meters, whereas many asphalts occur in the fractures and fissures of the shales and carbonate rocks in which they originated. The bitumens represent very little of the original organic matter in fine-grained rocks, in contrast to coal beds, which contain most of the organic matter that was originally deposited as an organic sediment. When a shale or dense carbonate is formed in a sedimentary basin, it usually contains 0–10% organic matter. The average value for shale is about 1% and for carbonates 0.3% (Hunt, 1972). In a typical marine sediment, part of this organic debris is carried in by rivers, the rest is deposited from marine organisms. The organic matter is largely a complex HMW insoluble material that cannot be removed from the sediment except by destructive pyrolysis, or by dissolving the mineral matter. However, a small portion of this organic matter, generally <3%, comprises bitumen formed by diagenesis and thermal alteration of the organic matter. The migrating waters of a compacting basin gradually remove part of this bitumen from the source bed to the porous sections of the rock, where it accumulates. The exact process of migration of bitumens from a source bed to a reservoir is unknown, but it is believed to occur both in solution and as fine particles. The whole process of the migration of petroleum and asphalts from source beds, and their accumulation in reservoirs is very inefficient; on the average about 99% of these bitumens remain in the source rock (see Crude-Oil Composition and Migration).

Metamorphism. Metamorphism is the chemical alteration of an organic sediment due to increasing temperatures and to some extent to increasing pressures. Because organic sediments undergo drastic changes at much lower temper-

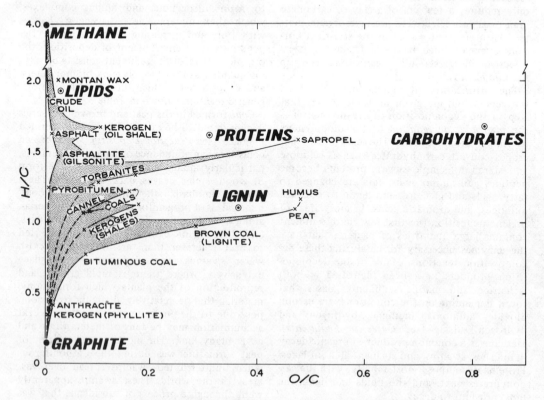

FIGURE 1. Metamorphism of organic sediments (see text).

atures (and pressures) than other sediments, the term *diagenesis* (q.v.) is usually not applicable. As organic sediments are buried more deeply, their temperature increases 2–5°C/100 m depth. The organic matter undergoes thermocatalytic cracking at an increasing rate; normally an increase in temperature of 10°C doubles the reaction rate. In addition, some organic compounds melt to liquids and others vaporize to gases, thereby increasing their chance of migration from the host rock.

Metamorphic changes are due primarily to temperature. Time alone is of minor importance—there are brown coals in the Lower Carboniferous that have never been exposed to very high temperatures or pressures. Pressure alone is not a major factor—there are coal deposits of relatively low rank found in strongly folded areas. Shear forces, however, are said to be effective in the formation of some bituminous coals.

The metamorphism of *coals* proceeds through a series of well-defined coal types (see *Coal–Diagenesis and Metamorphism*). Fig. 1 illustrates the interrelationships of the main elemental constituents of organic sediments (carbon, hydrogen, and oxygen) during metamorphism. The ratio of hydrogen to carbon atoms in the substance is plotted on the ordinate, and the ratio of oxygen atoms to carbon atoms on the abscissa. The starting materials from living organisms (lipids, proteins, carbohydrates, and lignin) are shown as circles with a central dot. The end products of all metamorphism (methane and graphite) are shown as solid circles. The metamorphism of organic matter in its late stage is really a hydrogen transfer process in which small hydrogen-rich molecules break off, leaving large hydrogen-deficient molecules which polymerize to even bigger residues. Eventually, most of the hydrogen is combined with single carbon atoms as methane gas, leaving almost pure carbon, graphite. The shaded area in Fig. 1 shows that all organic sediments are eventually converted to graphite and methane. All of the starting materials contribute in varying amounts to the formation of the organic sediments shown. The metamorphism of coal is represented by the line extending from humus through peat, brown coal, bituminous coal, and anthracite to graphite (see *Coal–Diagenesis and Metamorphism*). The curve shown does not apply to all the constituents of coal but only to the vitrinite fraction, which is the major part of coal originating from wood. The resin and wax fractions of coal start with very little oxygen and move toward graphite primarily through the loss of hydrogen as methane. The exinite fractions of coal are lower in oxygen and higher in hydrogen than the vitrinite fraction, so they tend to follow a line more comparable to the sapropelic coals. Metamorphism of the sapropelic and cannel coals follows the same slope as the humic coals.

Petroleum (q.v.) accumulated in a reservoir also undergoes metamorphism (or maturation) as the temperature of the reservoir rock increases. The principal LMW hydrogen-rich products of petroleum maturation are light oils, condensates, and gases. Heavy asphalts and pyrobitumens represent the hydrogen-deficient products. Continued maturation eventually leads to methane and graphite. Most of the petroleum produced at depths >3000 m are light petroleums or condensates due to this maturation effect.

Asphalts are also believed to go through stages of metamorphism caused by increasing temperature and possibly by shear pressures. Thus, the change from oil to asphalt to asphaltite to pyrobitumen, has been observed in such places as Peru, in going from the plains to the highly folded areas of the Andes Mountains. Not all

TABLE 4. Organic Matter in Earth's Sedimentary Rocks of Continents, Shelf and Slope

	Mass in 10^{18} Grams
Organic sediments	
Petroleum	1
Asphalts	0.5
Coal, lignite, peat	15
Organic matter finely dispersed in sedimentary rocks	
Hydrocarbons (petroleum)	200
Extractable Nonhydrocarbons (asphalt)	275
Nonextractable organic matter (Kerogen)	12,000

chemical alterations in organic sediments are caused by metamorphism. If tectonic movements bring an oil accumulation near the surface, it may lose its lighter components through volatilization. Circulating meteroric waters may degrade the oil by carrying off light hydrocarbons or by oxidizing the oil from contact with sulfate ions. Oil leaking to the surface may form an asphalt accumulation through weathering.

Most natural asphalts found in the veins and fissures of sedimentary rocks probably moved into these openings as heavy liquid bitumens, later solidifying due to polymerization, oxidation, and the loss of volatile fractions. In the Uinta Basin of Utah, the asphaltite gilsonite has been found to occur in veins as thick as 6 m. Although the average gilsonite sample contains <1% oxygen, the more weathered fractions of it may contain as much as 6%. Organic sediments, then, are very much like inorganic sediments in that they will undergo changes due to weathering and metamorphism, which will bring about partial or complete alteration in their chemical and physical characteristics. To some extent, organic sediments are much more sensitive to the environment than inorganic sediments and can be used to recognize incipient changes long before there is any obvious change in mineralogy. In cases where it is difficult to recognize disconformities in sediments due to little or no change in the bedding planes, analysis of the organic constituents can frequently show that there was an interruption in the sedimentary cycle.

Quantities of Organic Sediments in the Earth's Crust

Carbon constitutes about 0.3% by weight of the crust of the earth (Hunt, 1972). Although small in concentration, it is an indispensable part of life as well as a major source of energy in our society. Carbon in the form of wood, coal, gas, oil, and asphalt has been the prime source of fuel for centuries.

Of the total organic matter in sedimentary rocks, only a small fraction is in the form of organic sediments such as petroleum, asphalt, and coal; and most of this is coal, as shown in Table 4. The combined petroleum and asphalt reservoir accumulations in the world represent about 1/10 of the total quantity of coal, lignite, and peat. This ratio probably represents the fact that the migration and accumulation process for petroleum and asphalt is much more selective and inefficient than the accumulation process for coal. In addition, it is probable that large amounts of petroleum reserves have been destroyed by tectonic movements throughout geologic time. Although petroleum occurs in sediments of all ages from Cambrian to Pleistocene, about 80% of it is reservoired in the Mesozoic-Tertiary of the Gulf of Mexico, Caribbean, and Mesopotamia–Persian Gulf areas. About 35% of all the oil is found in the upper half of the Tertiary. In contrast, the Silurian and Triassic, which cover twice the geologic time of the upper Tertiary, contain <1% of the world's oil reserves. Asphalt accumulations also appear to be concentrated primarily in Mesozoic sediments. For example, the Cretaceous formations of the western Canada Basin contain unusually large deposits of asphalt. The Athabasca sands alone are believed to hold some 300 billion barrels of asphaltic oil of Cretaceous origin.

The oldest (common) coal beds are Devonian, and some brown coals (lignite) are found in peat deposits in Quaternary sediments of New Zealand. The thickest and most extensive deposits of coal occurred during the Carboniferous, when coal formed on all major land massses in the world. The Cretaceous Period was also very favorable for coal formation on a world-wide basis. In contrast, there is very little coal in sediments of Permian, Triassic, and Jurassic age.

In recent years, sedimentary rocks from all over the world have been analyzed in detail by oil-company research laboratories, and all but quartz sands and highly oxidized red beds are found to contain trace amounts of hydrocarbons and nonhydrocarbons having the same characteristics as the petroleum and asphalt that is found accumulated in reservoirs. Disseminated coaly particles are also common in some sedimentary rocks along with the finely dispersed organic matter. As seen in Table 4, the quantity of dispersed organic matter far exceeds all forms of organic sediments. In such a highly dispersed form, however, it is of no commerical value, although it may be very useful in paleoenvironmental studies.

JOHN M. HUNT

References

Abraham, H., 1960. *Asphalts and Allied Substances.* New York: van Nostrand, 370p.

Baker, D. R., 1970. Organic geochemistry and geological interpretations, *J. Geol. Educ.,* **20,** 221-234.

Eglinton, D., and Murphy, M. T. J., 1969. *Organic Geochemistry.* New York: Springer-Verlag, 828p.

Hunt, J. M., 1968. How gas and oil form and migrate, *World Oil,* **167,** 140, 145, 148-150.

Hunt, J. M., 1972. Distribution of carbon in crust of earth, *Bull. Am. Assoc. Petrol. Geologists,* **56,** 2273-2277.

McKirdy, D. M., 1974. Organic geochemistry in Precambrian research, *Precambr. Research,* **1,** 75-138.

Menzel, D. W., 1974. Primary productivity, dissolved and particulate organic matter and the sites of oxidation of organic matter, in E. D. Goldberg, ed., *The Sea,* vol. 5. New York: Wiley, 659-678.

Murchison, D., and Westoll, T. S., 1968. *Coal and Coal Bearing Strata.* New York: American Elsevier, 418p.

Weeks, L. G., 1958. *Habitat of Oil.* Tulsa: Am. Assoc. Petrol. Geologists, 1384p.

Cross-references: *Bitumen, Bituminous Sediments; Coal; Crude-Oil Composition and Migration; Euxinic Facies; Humic Matter in Sediments; Hydrocarbons in Sediments; Kerogen; Microbiology in Sedimentology; Oil Shale; Peat-Bog Deposits; Petroleum–Origin and Evolution; Sapropel; Shale; Tar Sands.*

ORTHOQUARTZITE—See QUARTZOSE SANDSTONE

OUTER CONTINENTAL-SHELF SEDIMENTS

The outer continental shelf is the undefined zone between the shallow marine environment of the inner shelf and shoreline and the deep marine environment of the continental slope. Generally it is taken to extend from the mid-depth region of the shelf out to the shelf edge. As a transition zone, the outer shelf is subject to the sedimentary processes typical of each adjacent environment (see *Continental-Shelf Sedimentation; Submarine Slope Sedimentation*). Bottom disturbance and sediment transport by storm waves and bottom currents, processes common in shallower water, occasionally occur out to the shelf edge. During periods of quiescence, vertical settling of suspended material duplicates depositional processes occurring on the continental slope.

Shallow-marine sediments are typically a complex mixture of particles, originating from many sources, that were deposited in different environments, at different times, and by different processes. To compound the complexity they are continually being modified by addition or removal, or simply by reorganization of the sediment. Present classification schemes reflect the ambiguity inherent in trying to categorize outer continental-shelf deposits; and unrelated terms are often used in direct comparison (age terms vs. lithology terms vs. environment terms). Emery's (1952) classification, which has been used most extensively and successfully, classifies five groups of shelf sediments *authigenic* (precipitated in place), *residual* (weathered in place), *biogenic* (derived from plants and animals), *detrital* (modern sediments derived from the continent), and *relict* (remnant from a different, previous environment of deposition). Recognizing that many shelf deposits have attributes of more than one of these groups, Swift et al. (1971) proposed the term *palimpsest* to describe sediments originally deposited in another environment but which are redeposited so that size and sorting are in equilibrium with the present marine environment.

Sediment Distribution

Regardless of whether or not present-day outer shelf sediments are in textural equilibrium with the shelf hydraulic regime, most owe their origin to an earlier, different environment (Kelling et al., 1975). About 65% of all shelf sediments are "relict" according to Emery (1968); and although much of this material has been reorganized (see *Continental-Shelf Sedimentation*), modern sedimentation on the shelf is probably quite different than at many times in the geologic past. Quaternary glaciations alternately extended the ranges of both the polar and biogenic belts of sedimentation (Emery, 1968), leaving great quantities of coarse clastics and shell material across the shelves. Modern deposits, even off tectonically active coasts, are fine by comparison and in most areas only veneer the older deposits. On coastal plain shelves, the Holocene flooding of the shelf and drowning of rivers sharply decreased supply of sediment to the shelf. Along such sediment-starved shelves, surficial sediment is generated almost solely by erosion of the retrograding shoreline and of exposed residual sediments on the shelf (Pilkey and Field, 1972). On most shelves, however, it is the cumulative effect of a host of factors rather than any single one that determines the final character of the shelf sediments. Figs. 1, 2, and 3 illustrate three types of outer continental shelves.

Factors Controlling Sediment Type

The factors that control the nature of outer continental-shelf sediment may be grouped as: (1) source and transport factors; (2) nature of the original depositional environment; and (3) modern shelf processes. On actively accreting shelves, groups 2 and 3 are the same.

Source and Transport Factors. The significant controls on sediments prior to their emplacement on the shelf are the climate, relief, and rock type of the source region. Climate determines the type (chemical vs. mechanical) and extent of weathering of source materials, and, along with rock type, controls the petrology of sediments reaching the shelf. Climate and relief both influence the mode (glacial vs. fluvial) and the rate of transport of sediments from source regions to the shelf. Hence shelf sediments off humid coastal plans are more quartzitic than those of arid borderland shelves, and shelf sediments off rocky, glaciated areas are

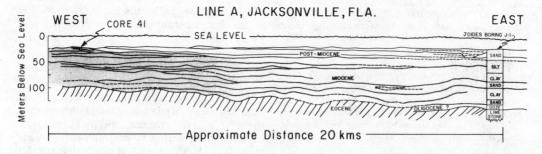

FIGURE 1. Interpreted geophysical record from the US Atlantic OCS off Jacksonville Florida, showing correlation of acoustic reflectors with lithologic information from vibratory cores and JOIDES Hole J-1 (from Meisburger and Field, 1975). Note Quaternary sediments thin in landward direction and that Miocene-Pliocene deposits are exposed on the W edge of the profile. The 20-m erosional channel on E edge of profile shows influence of fluvial agents on outer shelf during lower sea level.

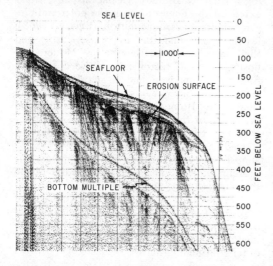

FIGURE 2. High-resolution seismic reflection profile across the southern California continental shelf (off Laguna Beach) (from Field, unpublished data). Note the Pleistocene erosional surface cut into older folded and faulted rocks and buried by thick sequences of marine progradational deposits.

coarser and less mature then those of low-lying temperate and tropical areas.

Nature of the Environment of Deposition. The single most important environmental factor influencing the nature and distribution of outer shelf sediments in the role of Pleistocene eustatic sea level fluctuations. Multiple eustatic transgressions and regressions of the sea across the continental shelves brought about by the waxing and waning of continental ice sheets during the Pleistocene led to "stacking" of various continental and shallow-marine depositional environments on the shelf (Curray, 1965). Most shelf sediments still bear evidence of their origin in these environments (Milliman et al., 1972).

Off glaciated regions, e.g., outer shelf sediments occur in till, ice-contact, and outwash deposits. In nonglaciated regions, shelf sediments were deposited in advancing and retreating coastal facies: broad low-relief shelves are covered by estuarine, lagoonal, and barrier-island sediments; steep, high-relief shelves are covered by fluvial and marine terrace materials (Field and Duane, 1976).

Modern Shelf Processes. Modern shelf processes modify and in some cases completely obscure the original character of sediment deposits on the shelf. Two principal measures that both influence and reflect the numerous shelf processes are rate of sedimentation and degree of sediment mixing. A low sedimentation rate results in an increased role in accumulation for authigenic (e.g., glauconite), biogenic (e.g., shell), and suspended fine material and also in increased sediment mixing by biological and physical processes. High sedimentation rates result in complete burial of relict deposits and restrict shelf sediment character to the sole nature of the input source (e.g., marine deltaic or carbonate reef facies). The degree of sediment mixing is dependent on the influence of tides, storm waves, and the fair-weather shelf hydraulic field, which in turn are influenced by regional shelf morphology (exposure, slope, depth, etc.). Often as significant in the redistribution of outer shelf sediments, but perhaps not as well understood, is the role of organisms which continually filter, sort, and transport individual grains in such quantities that the original details of bedding and sorting are sometimes destroyed.

Economic Potential

The outer continental shelf is a well-proven and rapidly developing resource area for hydrocarbons. Only recently, however, have solid

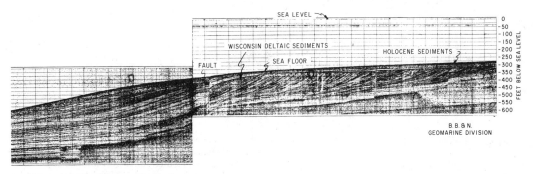

FIGURE 3. High-resolution seismic reflection profile across the outer continental shelf of the Gulf of Mexico. The profile shows the progradation of the shelf edge during the Pleistocene by deltaic deposition on the outer shelf. Faulting is common in this region, possibly related to overloading of sediments on the shelf during times of lower sea level. (Courtesy of Geomarine Division of Bolt, Beranek, and Newman, Inc., Cambridge, Massachusetts and Houston, Texas.)

minerals in surface and shallow subsurface deposits of the outer shelf received recognition as a potential resource. The voluminous quantities of sand and gravel deposited in the high-energy glacial, fluvial, and coastal environments migrating across the shelf during the Holocene transgression represent a viable resource for beach restoration and industrial uses. Residual phosphorite (W coasts of Africa, North America) as well as sulfur and other minerals associated with salt dome intrusions are also exploitable. Detrital minerals in buried stream channels and submerged beach deposits in certain areas are sources of rutile, ilmenite, gold, and other economic minerals. At present, exploitation is done on a limited or experimental basis, but advancing dredging technology coupled with dwindling supplies and increasing costs in land operations will probably lead to more emphasis on the solid-mineral potential of the outer continental shelf.

MICHAEL E. FIELD

References

Curray, J. R., 1965. Late Quaternary history, continental shelves of the world, in H. E. Wright, Jr., and D. G. Frey, eds., *The Quaternary of the United States.* Princeton: Princeton Univ. Press, 723-735.

Emery, K. O., 1952. Continental shelf sediments of southern California, *Geol. Soc. Am. Bull.,* **63,** 1105-1108.

Emery, K. O., 1968. Relict sediments on continental shelves of the world, *Bull. Am. Assoc. Petrol. Geologists,* **52,** 445-464.

Field, M. E., and Duane, D. B., 1976. Post-Pleistocene history of the United States inner continental shelf: Significance to origin of barrier islands, *Geol. Soc. Am. Bull.,* **87,** 691-702.

Kelling, G.; Sheng, H.; and Stanley, D. J., 1975. Mineralogic composition of sand-sized sediment on the outer margin off the mid-Atlantic states: Assessment of the influence of the ancestral Hudson and other fluvial systems, *Geol. Soc. Am. Bull.,* **86,** 853-862.

Meisburger, E. P., and Field, M. E., 1975. Geomorphology, shallow structure, and sediments off the Florida inner continental shelf, Cape Canaveral to Georgia, *U.S. Army Coast. Engin. Research Center, Tech. Mem. 54,* 119p.

Milliman, J. D.; Pilkey, O. H.; and Ross, D. A., 1972. Sediments of the continental margin off the eastern United States, *Geol. Soc. Am. Bull.,* **83,** 1315-1334.

Pilkey, O. H., and Field, M. E., 1972. Onshore transportation of continental shelf sediment: Atlantic southeastern United States, in D. J. P. Swift, et al., eds., *Shelf Sediment Transport: Process and Pattern.* Stroudsburg, Pa.: Dowden, Hutchinson & Ross, 429-446.

Swift, D. J. P.; Stanley, D. J.; and Curray, J. R., 1971. Relict sediments, a reconsideration, *J. Geol.,* **79,** 322-346.

Woodbury, H. O.; Murray, I. B.; Pickford, P. J., Jr.; and Akers, W. H., 1973. Pliocene and Pleistocene depocenters, outer continental shelf, Louisiana and Texas, *Bull. Am. Assoc. Petrol. Geologists,* **57,** 2428-2439.

Cross-references: *Authigenesis; Continental-Rise Sediments; Continental-Shelf Sedimentation; Continental-Shelf Sediments; Glauconite; Neritic Sedimentary Facies; Relict Sediments; Submarine (Bathyal) Slope Sedimentation; Submarine-Fan (Cone) Sedimentation.*

PACKING—See FABRIC

PALEOBATHYMETRIC ANALYSIS

According to Huxley (1873), A. Risso laid the foundation for bathymetric distribution of marine organisms, while J. V. Audouin and H. Milne Edwards were the first to recognize the implications for geology, in the early 19th century. Edward Forbes subsequently developed a nearshore depth zonation based on shelled organisms, providing a tool for reconstructing the depth of deposition of fossiliferous marine sediment (see Ladd and Gunter, in Hedgpeth and Ladd, 1957, vol. 2). In short order, sediment properties other than individual fossils were added to the list of depth indicators. A striking example of such a property is the percentage of carbonate in pelagic sediments, which decreases from maximum values near 90% on ocean rises to only traces at abyssal depths, as first documented by the *Challenger* Expedition. Of the two main calcium carbonate phases, aragonite is the more soluble and disappears first at a level called "aragonite compensation depth" (ACD). Calcite tends to disappear at much greater depths, usually 4–5 km, at a level called "calcite compensation depth" (CCD). These levels have been used in the bathymetric interpretation of Alpine Mesozoic strata (Garrison and Fischer, in Friedman, 1969).

Refinements of Paleontologic Criteria

Paleodepth analysis from marine fossils proceeds from four propositions:

1. Certain kinds of organisms are limited to shallow water because they depend on the presence of light, warm water, wave action, intermittent exposure, or similar shallow-water phenomena. Light, of course, is crucial for the photosynthesizing algae as well as organisms living in symbiosis with algae (see Wells and Yonge, in Hedgpeth and Ladd, 1957, vol. 1; Wells, in Hallam, 1967).

2. Modern species of benthic organisms show depth preferences within more or less narrow limits, and one may suppose that their fossil relatives had similar distributions. The classic work in this field has been done on benthic foraminifera (Natland, in Hedgpeth and Ladd, 1957, vol. 2; Phleger, 1960; Walton, in Imbrie and Newell, 1964; Funnell, in Hallam, 1967). The depth zonation of these forms (Fig. 1) provided the chief evidence for the displacement origin of deep-sea sands and contributed much to the reconstruction of sea-level variations during the Pleistocene, especially with regard to documentation of the Holocene transgression. Paleobathymetric analysis of benthic foraminifera has been extended as far back as early Tertiary (Frerichs, 1970) and Cretaceous time (Sliter and Baker, 1972) based on correlations of species distributions with sedimentary parameters, and in the latter case with associated megafossil assemblages.

3. Modern forms in certain groups show convergences in morphology or behavior which can be correlated with depth habitat. For example, sturdy shells, resistant to destruction, are common in nearshore assemblages exposed to wave action (see, e.g., R. H. Parker, in van Andel and Shor, 1964). Ornamentation, size, and shell structure of certain groups of benthic foraminifera appear correlated with depth (Bandy, in Imbrie and Newell, 1964), although the reasons are not clear. Eye-reduction in trilobites was suggested as an example of depth control on functional morphology (Clarkson, in Hallam, 1967). Trace fossil communities, reflecting feeding habits, are largely depth controlled, perhaps as a result of a bathymetric gradient in the amount and kind of food supplied to the sea floor (Seilacher, in Hallam, 1967).

4. Diversity, ratios between planktonic and benthic representatives of forams (or mollusks), and other statistical properties of modern faunal assemblages are correlatable with depth and give clues for the reconstruction of ancient environments (see Eicher, 1969).

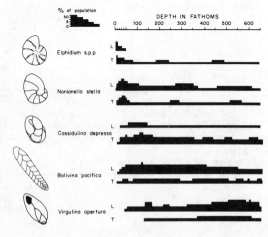

FIGURE 1. Depth distribution of several dominant benthic foraminifera of San Diego, California (adapted from Uchio, 1960). L = living, T = total fauna.

Generally, then, any paleontologic parameter that has a depth distribution definable by a range and, in cases, a most probable depth of occurrence, is useful for paleodepth analysis. The equation for the depth estimate of a given fossil assemblage is of the form: $D_{paleo} = \Sigma P_i D_i / \Sigma P_i$, where P_i is the weighting factor (e.g., the percentage of a species), and D_i is the depth at which the parameter (e.g., the species) is found in greatest abundance. The estimate can be improved by scaling the weighting factor (e.g., by dividing it through the depth range of the ith species) or by using groups of parameters for which a D_i can be specified (e.g., the fauna of a depth zone or an assemblage determined by factor analysis). The paleodepth estimate thus derived will be associated with an error that may be expressed in terms of the standard deviation of the estimate. A large error points toward mixed assemblages, for which the shallowest occurrence of the indicator for the greatest depth will give a more reliable depth estimate than a total fauna analysis.

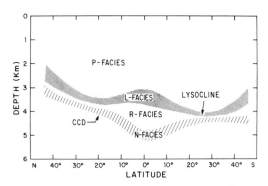

FIGURE 2. Model of bathymetric zonation of planktonic foraminifera in a N-S section through the central Pacific. P—plenty of delicate, spined foraminifera; L—lowered abundance of delicate shells, lysocline (q.v.) zone, delicate and resistant species about equally abundant; R—delicate forms rare, resistant shells greatly enriched; N—no delicate forms, nannofossils greatly enriched (zonation in the North Pacific inferred from rather scant information).

Refinements of Sedimentologic Criteria

Clues for bathymetric reconstruction for sediment properties other than fossils may be conveniently separated into geometric and compositional criteria.

Geometry. Paleoslope estimates proceed from directional structures, intraformational slumping and other gravitationally controlled indications of initial sedimentary dip, as well as the overall geometry of synchronous deposits (see *Paleocurrent Analysis*). Combined with a fixed elevation, such slope estimates can yield paleobathymetric maxima and minima (Eicher, 1969).

Composition. The bathymetric significance of compositional properties is rather analogous to the information gained from fossils. Thus, some components of rocks are limited to a shallow-water environment because of physical requirements necessary for their production, e.g., oolites (q.v.). Other components show depth preferences that are poorly understood, e.g., glauconite (q.v.), phosphorite (q.v.), and various kinds of ferromanganese concretions (see Price, in Hallam, 1967).

In the deep sea, the decreasing carbonate content at abyssal depths provides a bathymetric criterion. The usefulness of this measure has been greatly increased by detailed studies on differential preservation of calcareous shells, which led to the mapping of preservation facies (Fig. 2; see also *Lysocline*).

Conceptually, paleodepth reconstruction from sedimentologic information proceeds analogously to the paleobathymetric equation for paleontologic data.

A brief summary of criteria for distinguishing different depth zones brings out the fact that there is an abundance of indicators for very shallow depths but that zonations below, say, 100 m depth are ill defined. Hallam (1967) distinguishes the following main categories:

1. *Intertidal-supratidal*—algal mats and stromatolitic limestones, certain dolomites and evaporites, desiccation cracks.
2. *Shallow neritic*—Limestones with ooliths and/or algal micrite; chamosite; phosphorite at deeper levels. Flat- and cross-bedded sands tend to be dominant over muds in normal shelf regimes. Abundant benthic organisms; hermatypic corals; calcareous algae. Low diversity of pelagic organisms.
3. *Deep neritic, shallow bathyal*—Glauconite and phosphorite in greatest relative abundance. Ferromanganese nodules enriched in iron and poor in trace elements. Muds usually dominate over sands.
4. *Deep bathyal, abyssal*—Fine-grained sediments, low in organic matter and carbonate-poor at great depth. Sparse benthonic organisms typically surface sediment grazers. Ferromanganese nodules relatively enriched in manganese and in trace elements.

Reliability of Criteria

Some of the difficulties in estimating paleodepth using present-day bathymetric distribution of organisms and sediments are obvious from a cursory study of Fig. 1. Living *Elphidium* are restricted to shallow depths, but their empty shells occur over a wide depth range, attesting to redeposition processes. Living *N. stella* correspond well with the empty-shell distributions, but there are secondary modes of abundance, showing that factors other than depth control the patterns and suggesting that the sampling space is inhomogeneous. The range of empty shells of *V. apertura* is restricted relative to the range of living specimens, perhaps due to partial preservation. Redeposition, geographic variation, and partial preservation introduce uncertainties at both ends of paleobathymetric analysis—the inventory of modern depth correlation, as well as the interpretation of ancient sediments.

Additional difficulties are summarized by Hallam (1967), and may be paraphrased as follows: The sequence of facies produced by increasing depth in the sea is similar to the sequence due to increasing "oceanicity," such as increasing distance from river influx or other terrigenous sources [or decrease in fertility]; there seems to be a striking lack of absolute depth markers, whereas the establishment of relative depth in a given sequence may present no problem; and present geographical configurations (i.e., depth distributions of assemblages or sediments) may change drastically through time. Thus, reliance on correlations instead of an understanding of basic processes is unjustified in the long run.

In principle, the chief difficulty in paleobathymetric analysis arises from the fact that only one environmental parameter, the ambient pressure, is directly correlated with water depth. All other parameters, many of which are ecologically much more important, show nonlinear and variable correlations with depth, e.g., irradiation, food supply, temperature, turbulence, substrate texture, and chemistry of the sediment/water interface.

Paleobathymetry and Plate Tectonics

Age-Depth Constancy. The concept of sea-floor spreading (Fig. 3) implies that new oceanic crust generated at the crest of the mid-ocean ridge subsides as it moves away from its origin and slowly cools. The rate of subsidence has been investigated by Menard (1969) and by Sclater et al. (1971), who plotted sea-floor age against amount of subsidence (vertical distance to ridge crest) for various areas of the ocean. Sclater et al. found very similar plots from one area to the other and concluded that the fit of age versus depth of (basaltic) sea floor is good enough in many cases to date a given area of actively spreading sea floor to within 2 m.y., from a knowledge of bathymetry alone, for ocean floor younger than 40 m.y. The great similarity between Pacific, Atlantic, and Indian Ocean age-depth relationships suggests that the relationship is based on a fundamental law and was valid throughout Tertiary time and possibly earlier.

The hypothesis of age-depth constancy is

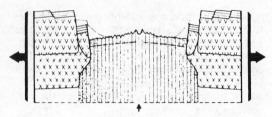

FIGURE 3. Consequences of sea-floor spreading for paleobathymetry (from Heezen, 1969). (1) As the ocean opens, the ocean floor migrates from the central ridge to deeper levels, where it becomes partially covered with hemipelagic sediments. (2) There is a tendency for the continental margins to subside and collect thick deposits of terrigenous mud.

based on this argument. The hypothesis states that at any point in geological time the age of the sea floor determines how far it moved downward from the original ridge crest. This postulate provides a simple means of reconstructing the approximate depth of deposition of ancient deep-sea sediments, independent of spreading rates of the underlying basement, by a "back-tracking" procedure (Fig. 4). The site to be backtracked is located according to its present depth and basement age (at C, in Fig. 4). The preliminary paleodepth estimate (at D) corresponds to the age of the sediment whose depth of deposition is to be found. To compensate for isostatic loading, one adds one-half the depth-in-hole of the sediment. As the sea floor subsided from P to C, sedimentation also built it upward, making it somewhat shallower than it would have been without sediments. Simultaneously, the increasing sediment load depressed the crust, presumably by an amount corresponding to about one-half its own thickness, hence the correction.

Age-depth constancy has been used for paleobathymetric construction in the Atlantic (Berger and von Rad, 1972; Sclater and McKenzie, 1973) and in the central Pacific (Winterer, 1973; Van Andel and Heath, 1973; Berger, 1973). An example of such depth reconstruction (Fig. 5) demonstrates the crucial role played by the general geographic reconstruction.

Error sources in the backtracking procedure are of two kinds: (1) Assuming that the general age-depth curve is valid in describing subsidence relative to the ridge crest, the quality of paleobathymetric estimates depends largely on the degree of uncertainty associated with basement age assignments. Sediment age and isostatic adjustment also contain some error (see Berger, 1973). (2) The general age-depth curve (Fig. 4) may not apply in cases because of additional sources of "basement," besides the ridge crest,

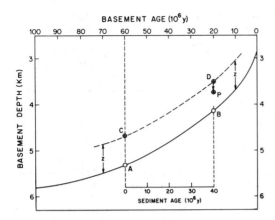

FIGURE 4. Paleodepth determination on spreading deep-sea floor, by vertical backtracking parallel to an average empirical subsidence curve, idealized (from Berger and Winterer, 1974). A and B: Present depth and paleodepth of 40 m.y. ago on idealized curve. C and D: Same on parallel curve. P: Final paleodepth after correction for isostatic loading.

or because of vertical motions associated with mantle plumes and sinks (see Menard, 1973).

Subsidence of Continental Margins. The opening of the Atlantic created continental margins at either side of the new ocean and provided the opportunity for these margins to subside as the adjacent ocean floor deepened and to collect large amounts of sediment, commonly several km thick (Fig. 3). Typically, a succession from shallow-water sediments to deep-water deposits developed on these margins. A well-known example is the area off the SE US coast where shallow-water sediments of Mesozoic age have been recovered both by conventional coring, dredging, and by drilling (see Emery et al., 1970; Rona, 1970).

Estimates of subsidence rates are obtained as the ratio between the present depth of ancient nearshore sediments and their age. Continental margins bordering the Atlantic Ocean commonly show subsidence rates near several tens of meters per million years (e.g., Blake Plateau, North Sea region, Bay of Biscay, Demerara Plateau). The subsidence rates found in this fashion, of course, are averages and may have varied greatly within the time spans considered (see, for example, the subsidence history inferred for Orphan Knoll and for Rockall Plateau in *Initial Repts., DSDP,* **12,** 78, 454–455). The depth of deposition at each time is the amount of subsidence with regard to sea level up to that time minus the thickness of sediment corrected for compaction.

Variations of Sea Level. It has been suggested (Hallam, 1963) that the vast transgression in the Late Cretaceous (Fig. 5) and the following

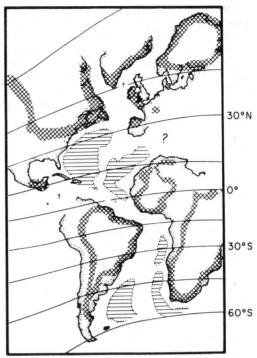

FIGURE 5. Sketch of geography (continent position after Francheteau) and paleobathymetry of the Atlantic Ocean during late Cretaceous time (from Berger and von Rad, 1972). Ancient coastline is cross-hachured. Areas deeper than 4000 m are ruled.

major transgression-regression oscillations in the Tertiary were produced by changes in the hypsographic distributions on the surface of the earth, due to differential vertical motions of continents and oceanic crust. Applying concepts derived from sea-floor spreading to this problem, Hallam proposed that the growth of a Cretaceous Mid-Atlantic Ridge between the separating continents, perhaps in conjunction with growth of submarine mountains and ridges elsewhere, displaced enough ocean water to produce the Late Cretaceous transgression. Menard (1964) expressed similar ideas in connection with a postulated ancient mid-ocean ridge, the "Darwin Rise." Because the rate of subsidence of spreading sea floor is rather independent of the rate of spreading (Fig. 4, idealized curve through A and B), slow-spreading ridges are narrow and fast-spreading ones are wide. A worldwide change of spreading rates, therefore, would change the volume of the mid-ocean rise, concomitantly causing transgression (increase in spreading rate) or regression (decrease in rate). Quantification of this process (Hays and Pitman, 1973) is of great potential use in paleobathymetric analysis. Such quantification depends crucially on a reliable absolute time scale and accurate dating of the ocean crust. The data for both these requirements need substantial improvement before depths of ancient shelf seas can be calculated from sea-floor spreading data with any degree of confidence (see Berger and Winterer, 1974).

WOLFGANG H. BERGER

References

Berger, W. H., 1973. Cenozoic sedimentation in the eastern tropical Pacific, *Geol. Soc. Am. Bull.,* **84**, 1941-1954.

Berger, W. H., and von Rad, U., 1972. Cretaceous and Cenozoic sediments from the Atlantic Ocean, *Init. Repts. Deep Sea Drilling Project,* **14**, 787-954.

Berger, W. H., and Winterer, E. L., 1974. Plate stratigraphy and the fluctuating carbonate line, in K. J. Hsü and H. C. Jenkyns, eds., *Pelagic Sediments: On Land and Under the Sea.* Oxford: Blackwell, 11-48.

Eicher, D. L., 1969. Paleobathymetry of Cretaceous Greenhorn Sea in eastern Colorado, *Bull. Am. Assoc. Petrol. Geologists,* **53**, 1075-1090.

Emery, K. O.; Uchupi, E.; Phillips, J. D.; Bowin, C. O.; Bunce, E. T.; and Knott, S. T., 1970. Continental rise off eastern North America, *Bull. Am. Assoc. Petrol. Geologists,* **54**, 44-108.

Frerichs, W. E., 1970. Paleobathymetry, paleotemperature, and tectonism, *Geol. Soc. Am. Bull.,* **81**, 3445-3452.

Friedman, G. M., ed., 1969. Depositional environments in carbonate rocks—a symposium, *Soc. Econ. Paleont. Mineral. Spec. Publ. 14,* 209p.

Hallam, A., 1963. Major epeirogenic and eustatic changes since the Cretaceous, and their possible relationship to crustal structure, *Am. J. Sci.,* **261**, 397-423.

Hallam, A., ed., 1967. Depth indicators in marine sedimentary environments, *Marine Geol.,* **5**, 329-555.

Hays, J. D., and Pitman, W. C., 1973. Lithospheric plate motion, sea level changes and climatic and ecological consequences, *Nature,* **246**, 18-22.

Hedgpeth, J. W., and Ladd, H. S., eds., 1957. Treatise on marine ecology and paleoecology, 2 vols., *Geol. Soc. Am. Mem. 67,* 1296p. and 1077p.

Heezen, B. C., 1969. The world rift system: An introduction to the symposium, *Tectonophysics,* **8**, 269-279.

Hsü, K. J., and Jenkyns, H. C., eds., 1974. Pelagic sediments on land and in the ocean, *Internat. Assoc. Sed. Spec. Publ.,* **1**, 447p.

Huxley, T. H., 1873. The problems of the deep sea, *Discourses–Biological and Geological Essays.* New York: Appleton, 37-68.

Imbrie, J., and Newell, N., 1964. *Approaches to Paleoecology.* New York: Wiley, 432p.

Larson, R. L., and Pitman, W. C., 1972. Worldwide correlation of Mesozoic magnetic anomalies and its implications, *Geol. Soc. Am. Bull.,* **83**, 3645-3662.

Menard, H. W., 1964. *Marine Geology of the Pacific.* New York: McGraw-Hill, 271p.

Menard, H. W., 1969. Elevation and subsidence of oceanic crust, *Earth Planetary Sci. Lett.,* **6**, 275-284.

Menard, H. W., 1973. Depth anomalies and the bobbing motion of drifting islands, *J. Geophys. Research,* **78**, 5128-5137.

Phleger, F. B., 1960. *Ecology and Distribution of Recent Foraminifera.* Baltimore: Johns Hopkins, 297p.

Rona, P. A., 1970. Comparison of continental margins of eastern North America at Cape Hatteras and northwestern Africa at Cap Blanc, *Bull. Am. Assoc. Petrol. Geologists,* **54**, 129-157.

Rona, P. A., 1973. Relations between rates of sediment accumulation on continental shelves, sea-floor spreading, and eustasy inferred from the central North Atlantic, *Geol. Soc. Am. Bull.,* **84**, 2851-2872.

Sclater, J. G., and McKenzie, D. P., 1973. Paleobathymetry of the South Atlantic, *Geol. Soc. Am. Bull.,* **84**, 3203-3216.

Sclater, J. G.; Anderson, R. N.; and Bell, M. L., 1971. Elevation of ridges and evolution of the central eastern Pacific, *J. Geophys. Research,* **76**, 7888-7900.

Sliter, W. V., and Baker, R. A., 1972. Cretaceous bathymetric distribution of benthic foraminifers, *J. Foram. Research,* **2**, 167-183.

Uchio, T., 1960. Ecology of living benthonic foraminifera from the San Diego, California, area, *Cushman Found. Foram. Research, Spec. Publ.* **5**, 72p.

Van Andel, T. H., and Heath, G. R., 1973. Geological results of Leg 16: the central equatorial Pacific west of the East Pacific Rise, *Init. Repts. Deep Sea Drilling Project,* **16**, 937-949.

Van Andel, T. H., and Shor, G. G., 1964. Marine geology of the Gulf of California, *Am. Assoc. Petrol. Geologists Mem. 3,* 408p.

Winterer, E. L., 1973. Sedimentary facies and plate tectonics of equatorial Pacific, *Bull. Am. Assoc. Petrol. Geologists,* **57**, 265-282.

Cross-references: *Abyssal Sedimentary Environments; Biogenic Sedimentary Structures; Biostratinomy; Bioturbation; Glauconite; Lysocline; Marine Sediments; Neritic Sedimentary Facies; Paleocurrent Analysis; Pelagic Sedimentation, Pelagic Sediments; Sedimentation–Paleoecologic Aspects.* Vol. I: *Benthonic Zonation.*

PALEOCURRENT ANALYSIS

The current systems responsible for transportation of sediments often leave record of their movement direction within the body of the sediment. This information can be used to decipher the paleocurrent. The sediment properties used for paleocurrent analysis are of two types, directional and scalar. The directional elements (also called vectorial elements) are indicators of the paleocurrent direction at the point of observation. The scalar properties show gradual change in the direction of transportation, and the paleocurrent direction is known by systematically mapping this change over an area.

Directional Elements

The directional elements useful for paleocurrent analysis are mainly of two types, planar and linear, but more complex structures are also known. Brief descriptions of some of the planar and linear structures commonly used in the study of paleocurrents are given below. Detailed accounts of these features will be found in Pettijohn (1962) and Potter and Pettijohn (1977).

Planar Structures. Cross-bedding (q.v.) of various sizes, produced either by migrating sand waves or by deposition of sloping beds in standing bodies of water, are the commonest tool for studying paleocurrents. The direction of inclination of a cross-bedding foreset (the azimuth of cross-bedding) indicates the direction of sediment transport. High and Picard (1974) review the reliability of various types of cross-strata as paleocurrent indicators.

Linear Structures. Ripple marks (q.v.) are one of the common linear structures used in paleocurrent work. Crest lines of asymmetric ripples are generally normal to the direction of sediment transport. Grooves and other types of striae produced by objects dragged on the sand bed lie parallel to the current direction (see *Current Marks*). Molds of these grooves, flute marks, skip marks, and many other similar features are preserved on the undersurface of sandstone beds as *sole marks*. Orientations of these sole marks and other current-oriented linear features such as parting lineation (q.v.), fossilized shell lineations, etc., can be used for interpreting the paleocurrent.

Orientation and imbrication of long axes of sand grains can indicate paleocurrent direction, but the technique is rarely used because measurement of direction of sand-sized grains is a very tedious job (see *Fabric*). More commonly used for paleocurrent study are the orientation and imbrication of pebbles and boulders. The long axes of pebbles generally lie in the direction of the current and they dip upstream, but perpendicular arrangements of the long axes of pebbles with respect to the flowing water are not unknown (see *Conglomerates*). In a cross-stratified sequence, orientation and plunge of the long axes of pebbles are different in different portions of the cross-strata. The elongate pebbles tend to be transverse to the current direction on the topset and bottomset units, and longitudinal on the foresets. "Current crescents" formed on sand beds by flow of water or air around obstacles, can also be used as paleocurrent indicators (Johansson, 1976).

Collection and Analysis of Directional Data

Directional elements can be treated as vectors because they have direction as well as magnitude. In the case of planar elements like cross-bedding, direction is given in three dimensions by the azimuth and inclination of the foreset. Direction of a linear structure is given by the attitude or orientation of the property concerned. The magnitude of the vector is determined arbitrarily, by assigning unit weight to each observation.

Certain procedural problems regarding collection, summation, and interpretation of data are faced in the study of paleocurrents. Special statistical techniques of analysis are needed for handling these problems because the conventional techniques are not applicable to the directional data that are spread radially on a compass dial. When the data are spread on either side of the origin, the usual method of computation of arithmetic mean gives absurd results (for example, the arithmetic mean of $340°$ and $20°$ is $180°$), and the usual standard deviation of measurements cannot also be used as a measure of dispersion. Studies of the problems concerning development of adequate mathematical techniques for the analysis of directional data have been reviewed by Rao and Sengupta (1972). The mathematical procedures discussed below are based on these studies and take into account the special needs of the "circularly distributed" data.

Tilt Correction. When the area of study is structurally tilted or folded, a correction has to be applied to restore the bed containing the directional elements to its original position. In the case of a cross-bedded unit, it is done by rotating the plane of the cross-bed around the strike of the true bedding through the angle of dip of the bedding. Tilts in the linear structures can be similarly treated. Potter and Pettijohn (1977) have discussed these techniques in detail. Dott (1974) provides a field example. The tilt correction is negligible when the amount of tilt of the bed is small ($<25°$), or when the direction of the foreset dip is nearly the same as the direction of the structural dip. The influence of structural tilt is large when the direction of foreset dip makes a large angle with the direction of the structural dip.

Sampling. When the geological formation under study contains a large number of directional features of a particular type, it is necessary to know the minimum number of observations that would give the mean direction for the whole formation with a specified precision. This is essentially a sampling problem. The optimum hierarchical sampling procedure developed by Rao and Sengupta (1972), who review earlier practices, takes into account the special requirements of the radially distributed data. Using a pilot survey in the area of study as a guide, the following problems can be solved by this sampling technique: (1) the minimum sample size required for estimating, with a desired precision, the mean direction of a formation, and (2) the optimum allocation of samples between and within the outcrops that would allow efficient sampling at minimum cost. A summary of this sampling procedure is given in Rao and Sengupta (1972).

Summation. Arithmetic averaging of the directional data might lead to erroneous results, but a meaningful average of a population of directional data can be obtained by computing the vector resultant as follows:

Each directional observation is treated as a unit vector with components $\cos \alpha_i$, and $\sin \alpha_i$. The sums of sines and cosines for the sample $\alpha_1 \ldots \alpha_n$ are computed as follows:

$$V = \sum_1^n \cos \alpha_i, \quad W = \sum_1^n \sin \alpha_i$$

Then the sample mean direction is given by $\gamma = \tan^{-1}(W/V)$. Grouping of observations should be avoided as far as possible. Where grouping of data cannot be avoided, the mean direction is computed as follows:

$$V = \sum_{i=1}^n n_i \cos x_i, \quad W = \sum_{i=1}^n n_i \sin x_i$$
$$\hat{\gamma} = \tan^{-1}(W/V)$$

where x_i is the midpoint azimuth of the ith class interval, n_i is the number of observations in the ith class, and $\hat{\gamma}$ is the azimuth of the resultant vector. The quadrant in which this $\hat{\gamma}$ lies is determined by the signs of V and W.

There is no point in computing the mean direction of a population when there is no significantly preferred direction. The uniformity (or lack of preferred direction) of a set of directional data in a sample of moderate size can be tested by a simple test developed by Rao (see Rao and Sengupta, 1972).

Computation of Dispersion. Variability of the directions within a sector is represented by the magnitude or length of the resultant vector R, where $R = \sqrt{(W^2 + V^2)}$. A useful measure of the concentration of azimuths is the consistency ratio, expressed as R/n, where n = the number of observations. This can also be expressed in terms of percent ($L = R/n \times 100$). L has been named "vector magnitude" (Curray, 1956) and "vector strength" (Pincus, 1956). When the distribution is unimodal, the quantity $(n - R)$ provides a good measure of dispersion of the sample directions (Rao and Sengupta, 1972).

Presentation. The observed directions within an area can be graphically represented in the form of a rose diagram (a circular histogram). For convenience of representation, the area of study is sometimes arbitrarily divided into a number of small sectors of equal size, and the resultant direction for each sector is represented by an arrow at the center. In each case, the length of the arrow is drawn proportional to the vector magnitude (Fig. 1).

Regional Paleocurrent Trends. While analyzing the paleocurrent system in a large area, it is desirable to emphasize the major trends of sediment transport by smoothing the local variations. In the simple "moving average" method used by Pelletier (1958) and others, the area of study is arbitrarily divided into a number of sectors of equal size; the resultant direction of all the data within the area is then moved systematically over the map with the help of

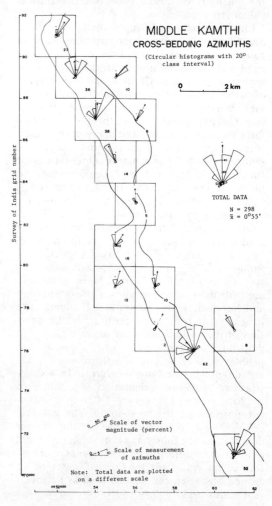

FIGURE 1. Graphic representation of cross-bedding directions, vector resultants, and vector magnitudes (from Sengupta, 1970).

a grid; and the averages are recomputed and plotted at the grid coordinates. Fig. 2 illustrates the technique of interpreting regional paleocurrent trends with the help of the moving-average method. The results obtained by this simple method are known to compare favorably (Pettijohn, 1962) with the more sophisticated, computer-based techniques of isolating the local from the regional trends. In the technique of trend surface analysis, different trends, linear, quadratic, or higher, are fitted to the observed data. The "residual" values are worked out from the difference between the computed and the observed values at each point (see Krumbein, 1959; Miller, 1956; and Fox, 1967, for detailed discussions on trend surface technique).

Tests for Homogeneity. Interesting geological conclusions can sometimes be reached by comparing the paleocurrent directions from two or more geological horizons. For this purpose, visual comparison of data might not always be enough. Statistical tests are needed to find out whether the shift in the paleocurrent direction is significant or not. A standard test for comparing the mean direction of several circular normal populations with the same concentration parameter is the F test (Watson, 1966). Computer programs for such tests and also for other analyses of directional data are now readily available (Schuenemeyer et al., 1972). When the sample size is large, the equality of resultant directions and the equality of dispersion can also be tested by the tests of homogeneity (H tests) proposed by Rao (see Rao and Sengupta, 1972).

Scalar Elements

Properties such as grain size, grain roundness, sediment composition, and bed thickness show gradual variation in the direction of sediment transport. Systematic mapping of these changes can provide clues to the direction of paleocurrent. Measurements at several points spread over an area are required for mapping the areal variations; and, unlike the directional properties, a single observation of a scalar property does not provide any clue to the paleocurrent. In case of subsurface mapping, however, the scalar properties have an advantage over the vectorial ones. Paleocurrent maps from scalar elements can be prepared on the basis of unoriented cores, whereas for interpretation of a vectorial element, an oriented core is an essential prerequisite.

Size Variation. The commonly observed phenomenon of progressive grain-size decrease downstream has been used as a clue to the paleocurrent by many workers (Schlee, 1957; Pelletier, 1958; Sengupta, 1970; and others). Determination of the mean size of a detrital population being a difficult problem, Pettijohn (1962) suggested the simpler technique of measurement of the maximum pebble size for paleocurrent determination.

In practice, the ten largest pebbles are collected in each locality, and the average values of the ten measurements are either represented graphically or plotted on a map and contoured. The trend of paleocurrent is given by the direction normal to the trend of contours. This procedure, which assumes that the maximum pebble size is a function of the mean size of the detrital population, has yielded results consistent with the cross-bedding information in many cases (e.g., Fig. 3; see also *Sternberg's Law*).

Roundness Variation. Systematic measurement of grain roundness (q.v.) has helped in paleocurrent work in many cases, but the technique is of limited use. In some cases, the stream sands remain angular or become progressively so with transportation. When roundness increases with transportation, the change is rapid at the initial stage only. Thereafter, the roundness becomes asymptotic to some particular

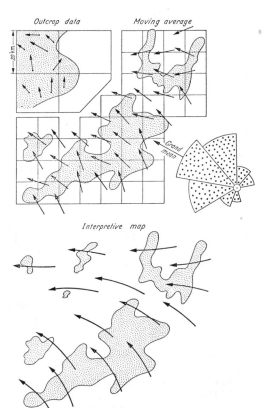

FIGURE 2. The technique of interpretation of regional paleocurrent trends from directional data (from Potter and Pettijohn, 1963).

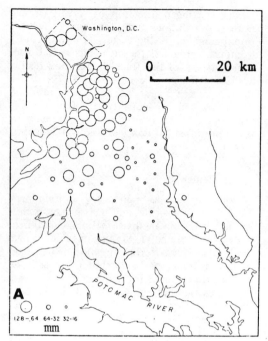 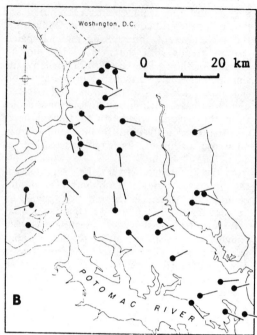

FIGURE 3. Maps showing the relation between the scalar and vectorial elements in the same area (from Pettijohn, 1962, after Schlee, 1957). A: Shows variations in maximum pebble sizes. B: Shows cross-bedding directions.

value and shows little change with further transportation (Pettijohn, 1962). In many cases, moreover, pebble roundness varies only as a function of pebble size and is not indicative of paleocurrent direction.

Other Scalar Properties. Areal variation in thickness of individual beds is another property which, under favorable conditions, can indicate the paleocurrent direction. Mineral composition of sand may show a downcurrent change due to selective abrasion or sorting. Pollen concentration in sediments is known to decrease downcurrent. These properties have been utilized by some, but the techniques have not been extensively used. A gradual decrease of cross-bedding scale and an increase in the dispersion of cross-bedding azimuths downslope have been noted by several workers, but these techniques yield useful results only when measurements are made on a regional scale.

Environmental and Paleogeographic Interpretation

Paleocurrents provide clues to the dip and strike of the beds at the time of their deposition. The depositional dip and strike in turn define the paleoslope or the slope of the ancient surface over which the sediments were transported and deposited (Hoque, 1975). A precise correlation between the patterns of paleocurrent dispersion and the depositional environments is difficult to achieve, but some broad relations between the two have been suggested (see Potter and Pettijohn, 1977). In fluvial, deltaic, and some turbidite sands the paleocurrents are generally unimodal pointing downslope. Contour currents produce alongslope indicators. Eolian paleocurrents are unrelated to paleoslope and may have any kind of dispersion. In the tidal zones, bimodal paleocurrents are common, but other patterns, all unrelated to paleoslope, are also known. The shoreline current patterns may be of different types, depending on whether the currents are of marine or continental (fluvial) origin (Selley, 1968).

Important information regarding configuration of the basin margin, trend of the shoreline, direction of the sediment source, and the pattern of sediment dispersal within a basin can be obtained from regional paleocurrent studies. This information, together with a knowledge of paleoslope and the environment of deposition can be of invaluable help in paleogeographic reconstruction (Bigarella, 1974).

SUPRIYA SENGUPTA

References

Bigarella, J. J., 1974. Continental drift and paleocurrent analysis, *Bol. Paran. Geosci.*, **30**, 73-97.

Curray, J. R., 1956. The analysis of two-dimensional orientation data, *J. Geol.,* **64,** 117-131.

Dott, R. H., Jr., 1974. Paleocurrent analysis of severely deformed flysch-type strata—a case study from South Georgia Island, *J. Sed. Petrology,* **44,** 1166-1173.

Fox, W. T., 1967. Fortran IV program for vector trend analysis of directional data, *Kansas Geol. Surv. Comp. Contr. 11,* 36p.

High, L. R., Jr., and Picard, M. D., 1974. Reliability of cross-stratification types as paleocurrent indicators, *J. Sed. Petrology,* **44,** 158-168.

Hoque, M., 1975. Paleocurrent and paleoslope—a case study, *Palaeogeogr. Palaeoclimat., Palaeoecol.,* **17,** 77-85.

Iriondo, M. H., 1973. Volume factor in paleocurrent analysis, *Bull. Am. Assoc. Petrol. Geologists,* **57,** 1341-1342.

Johansson, C. E., 1976. Structural studies of frictional sediments, *Geografiska Annaler,* **58A,** 201-300.

Krumbein, W. C., 1959. Trend surface analysis of contour-type maps with irregular control-point spacing, *J. Geophys. Research,* **64,** 823-834.

Miall, A. D., 1974. Paleocurrent analysis of alluvial sediments: A discussion of directional variance and vector magnitude, *J. Sed. Petrology,* **44,** 1174-1185.

Miller, R. L., 1956. Trend surfaces: Their application to analysis and description of environments of sedimentation, *J. Geol.,* **64,** 425-446.

Pelletier, B. R., 1958. Pocono paleocurrents in Pennsylvania and Maryland, *Geol. Soc. Am. Bull.,* **69,** 1033-1064.

Pettijohn, F. J., 1962. Paleocurrents and paleogeography, *Bull. Am. Assoc. Petrol. Geologists,* **46,** 1468-1493.

Pincus, H. J., 1956. Some vector and arithmetic operations of two-dimensional orientation variates, with applications to geologic data, *J. Geol.,* **64,** 533-557.

Potter, P. E., and Pettijohn, F. J., 1977. *Paleocurrents and Basin Analysis,* 2nd ed. New York: Springer-Verlag, 460p.

Rao, J. S., and Sengupta, S., 1972. Mathematical techniques for paleocurrent analysis: Treatment of directional data, *Math. Geol.,* **4,** 235-248.

Sanderson, D. J., 1973. Some inference problems in paleocurrent studies, *J. Sed. Petrology,* **43,** 1096-1100.

Schlee, J., 1957. Upland gravels of southern Maryland, *Geol. Soc. Am. Bull.,* **68,** 1371-1410.

Schuenemeyer, J. H.; Koch, G. S.; and Link, R. F., 1972. Computer program to analyze directional data based on the methods of Fisher and Watson, *Math. Geol.,* **4,** 177-202.

Selley, R. C., 1968. A classification of paleocurrent models, *J. Geol.,* **76,** 99-110.

Sengupta, S., 1970. Gondwana sedimentation around Bheemaram (Bhimaram), Pranhita-Godavari valley, India, *J. Sed. Petrology,* **40,** 140-170.

Shelton, J. W., and Mack, D. E., 1970. Grain orientation in determination of paleocurrents and sandstone trends, *Bull. Am. Assoc. Petrol. Geologists,* **54,** 1108-1119.

Watson, G. S., 1966. The statistics of orientation data, *J. Geol.,* **74,** 786-797.

Cross-references: *Anisotropy in Sediments; Conglomerates; Cross-Bedding; Current Marks; Fabric, Sedimentary; Paleobathymetric Analysis; Ripple Marks; Roundness and Sphericity; Sedimentary Structures; Top and Bottom Criteria.*

PALEOMAGNETISM—See REMANENT MAGNETISM IN SEDIMENTS

PALUDAL SEDIMENTS

Paludal sediments are the deposits of marshes and swamps, which are common in regions of low, irregular topography, along the banks of lakes, and on river flood plains; they may also form on deltas and elsewhere along coastlines. Under conditions of high humidity or perched water tables, they may form wherever drainage is inadequate to carry away excess water. Peat bogs are particularly abundant in recently glaciated areas where drainage is typically poor; they are considered as separate from marshes and swamps (see *Peat-Bog Deposits*).

Paralic marshes and swamps (lagoons, salt marshes) derive some of their sediment from the sea, although much is derived directly from the land (see *Lagoonal Sedimentation; Salt-Marsh Sedimentology*). It is also probable that true marine sediments may be interbedded with the paludal deposits when intermittent submergence of the coastline occurs or where mangroves are involved.

The deposits of marshes and swamps are principally organic-rich muds; iron oxides (see *Ironstone*); carbonates; and, locally, sand and marl (q.v.). Sand-filled channels in marsh and swamp deposits may indicate the course of a former stream. Paludal sediments may pass laterally into either marine (littoral or neritic) deposits or into lacustrine sediment (see *Lacustrine Sedimentation*).

Ancient paludal sediments are best represented by the coal seams occurring in Pennsylvanian, Jurassic, Cretaceous, and Tertiary sediments (see *Coal; Deltaic Sediments; Fluvial Sediments*).

DEREK W. HUMPHRIES

Cross-references: *Coal; Deltaic Sediments; Fluvial Sediments; Ironstone; Lacustrine Sedimentation; Lagoonal Sedimentation; Mudstone and Claystone; Organic Sediments; Peat-Bog Deposits; Salt-Marsh Sedimentology.*

PARAGENESIS OF SEDIMENTARY ROCKS

The mineral parageneses—the natural mineral associations—of sedimentary rocks can be divided into two main types: detrital (clastic) sedimentary rocks, the bulk of whose constituents have been deposited mechanically, and chemical sedimentary rocks, the bulk of whose constituents have been chemically precipitated.

Detrital Sedimentary Rocks

Most clearly in the detrital rocks, the natural associations of minerals are the end products of a sequence of processes—weathering (q.v.), erosion, transportation (q.v.), deposition (q.v.), and diagenesis (q.v.)—that have acted on some parent material. Variation in the character or intensity of any one of these processes or in the composition of the parent material may be reflected in the final mineral paragenesis of a detrital rock.

Most detrital rocks contain three types of constituents; following Krynine (1948), we can designate these as grains, matrix, and cement (see *Clastic Sediments and Rocks*). The grains represent relic mineral particles that have *survived* the processes of chemical weathering and so reflect directly the mineralogy of the parent material. They are the coarsest of the detrital (i.e., mechanically deposited) material in the rock, and they form what has been characterized as the *framework* of the rock. The *matrix* also is composed of detrital material, but it is composed largely or entirely of particles of minerals that are the *products* of chemical weathering of the parent material and so reflect only imperfectly its mineralogy. The matrix particles invariably are of very fine (silt or clay) size. Finally, the *cement* consists of minerals deposited chemically, from solution or colloidal suspension, between the grains or the matrix particles. It also is, at least ultimately, a product of chemical weathering of some parent material, though that parent material and weathering may be totally unrelated in time and space to the source of the detrital material.

Grains. The mineralogy of the grains present in a detrital rock depends mainly on the mineralogy of the parent material and the intensity and duration of chemical weathering. Thus, if the parent material were a biotite granite, possible grain minerals would be biotite, quartz, and alkali feldspar; pyroxene or olivine would be most improbable. However, all of the parent rock's minerals will be represented as grains only if no chemical weathering has taken place. The parent rock usually has been subject to at least some chemical weathering, and the minerals that will show up as grains will depend on how resistant the various minerals are to weathering (Fig. 1) and on how intensive and extensive weathering has been. The intensity will be controlled mainly by the climate in the source area: the rate of chemical weathering will be less in arid than in humid regions and in tem-

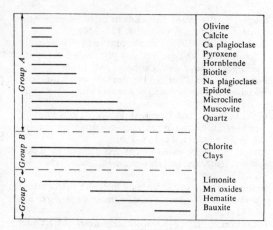

Increasing intensity of chemical weathering ⟶

FIGURE 1. Mineral paragenesis in residual deposits. Minerals in Group A are all survivors of chemical weathering; those in Group C are products of chemical weathering; those in Group B may be either.

perate than in tropical regions. The duration of chemical weathering will depend largely on the rapidity with which the minerals are removed once chemical weathering has begun, and this in turn will depend mainly on the relief in the source area: in areas of low relief, mineral particles generally will remain in situ longer than in areas of high relief and so will be subject to a more extended period of chemical weathering.

To a much smaller degree, the minerals represented as grains will depend on the other processes in the sequence mentioned earlier. The process of erosion may affect the grain mineralogy little if at all. Its most important influence is that where erosion is rapid, it may "rescue" grains of mineral matter that are unstable in a certain weathering regime (e.g., augite, wollastonite, basaltic glass) and add them to the detritus. Transportation may have an effect, for some abrasion of the grains does occur, and the minerals that compose these grains are not all equally resistant to abrasion (Fig. 2). This resistance will depend mainly on the strength of the mineral and the perfection of its cleavage. For example, quartz is more brittle than garnet and is, therefore, more rapidly abraded. Feldspar and mica, two tenacious minerals, differ in their resistance to abrasion as a result of the more perfect cleavage of the latter. These differences in ease of abrasion show up only indirectly in the grain mineralogy of the final sediment, and then only because of the sorting of detritus by grain size during deposition: where deposition results in good sorting, the coarser sediments will be enriched in the strong and the difficult-to-cleave minerals.

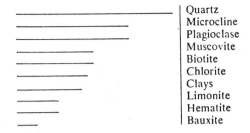

FIGURE 2. Ability of the more resistant minerals of residual deposits to withstand mechanical transportation. The distance transported increases from left to right.

Finally, diagenesis also may affect the grain mineralogy in several ways. If the detritus contains relatively large amounts of grains of minerals that normally would not have survived the process of chemical weathering in the source area, once redeposited these minerals may react with one another or with interstitial solutions to yield *pseudomorphs* composed of more stable minerals such as clay minerals or zeolites. It has also been suggested that diagenetic solution of mineral grains occurs and that as a result minerals such as kyanite, topaz, and sphene become less common in the older detrital rocks.

Matrix. The mineralogy of the matrix in a detrital sediment is also primarily a function of the processes of weathering in the source area. On the other hand, the proportion of the sediment that consists of these weathering products is mainly a function of the later processes of erosion and deposition.

The minerals produced by chemical weathering will depend on the composition of the source material, on the type of chemical weathering that takes place, and on the degree to which the process goes to completion. The two commonest types of chemical weathering are *podzolization* and *lateritization,* the former characteristic of humid temperate climates and the latter of humid tropical ones. The principal insoluble products of podzolization are the clay minerals, chlorite, and limonite; the process also results in the extensive leaching of the alkalies and alkaline earths. The principal insoluble products of lateritization are bauxite and hematite; it results in extensive leaching not only of the alkalies and alkaline earths but also of silica (see *Duricrust*).

Because the insoluble products of weathering are all stable under surface conditions, none of the other processes in the sedimentary cycle, except diagenesis, will have much effect on their composition. Diagenesis will be important where chemical weathering has not been allowed to go to completion and the original matrix minerals represent an intermediate stage in the process. In practice, this applies most obviously to the weathering products of the ferromagnesian minerals and of basic to intermediate volcanic glasses, which alter first to smectite clay minerals, which subsequently are degraded to illite or kaolinite (see *Clay–Diagenesis*). Continuous and intensive erosion will result in a matrix rich in smectite, the most reactive of the clay minerals, which during diagenesis, will take up Na, K, and Ca from the interstitial solutions, and convert into zeolites.

The degree to which these minerals will be represented in the final sediment depends heavily

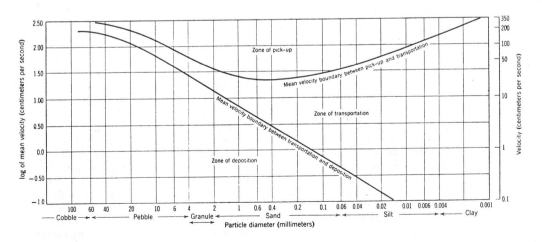

FIGURE 3. Hjulström's diagram, somewhat simplified, showing the relationship between the average stream velocity and the movement of particles of specified size (from Dapples, 1959). Note that particles of medium sand size are caused to move at lowest velocities and that velocities needed to pick up clay (from a clay layer) are as great as those for pebble size.

on the other processes in the sedimentary cycle; this dependence stems mainly from the contrast in grain size and shape of the mineral particles produced by weathering (the matrix particles) and the mineral particles that survived this process (the grains). In any residual deposit, the weathering products are very fine grained and have distinctly flaky shapes; the relic grains are usually of sand size or larger and have generally equant shapes. In other words, the matrix particles are easy to keep in suspension but difficult to pick up, whereas the opposite is true of the grains.

A stream of air or water moving over a residual deposit will first pick up small particles of equant shape, in most cases the sand- and coarse silt-sized relic grains. Only at a somewhat higher velocity will the fine-grained flaky products of chemical weathering be carried off (Fig. 3). Thus the relic minerals may be removed selectively from the weathered debris, and if deposition were to occur soon after, the detrital sediment might consist of grains accompanied by little or no matrix.

More commonly encountered is grain-size separation during deposition. As a stream slows down, the coarsest material will be deposited first, and the finest detritus may not be deposited until much later (and much farther downstream). The result will be a series of well-sorted deposits of detrital sediment, well sorted in size *and* in mineralogy.

Cement. The cement in a detrital rock is a product of diagenesis and so reflects essentially the conditions in the basin of deposition. Simply because this basin is usually marine, the cement is commonly calcite, the constituent in which sea water is most generally saturated. In other situations, the cement will be silica, which for the most part appears to have been derived from the originally deposited, finest grains of quartz produced by abrasion during transportation.

Chemical Sedimentary Rocks

The mineral parageneses of the chemical sediments more directly reflect the environment of their deposition rather than the petrography, relief, and climate of their source area. Thus, it is useful to discuss these parageneses in terms of depositional environments, distinguishing between open-marine, restricted-marine, arid-lacustrine, and humid-lacustrine parageneses. The identity and sequence of formation of the minerals in each environment will depend on the degree of saturation of the water in these minerals (Fig. 4).

Open Seas. In open marine parageneses, formed where sea waters have unrestricted ac-

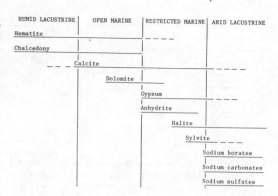

FIGURE 4. Paragenesis of minerals in chemical sediments. The total salinity of the waters in general increases from left to right.

cess to the area of deposition, the mineral paragenesis is relatively constant, for in all cases these waters are most strongly saturated (even oversaturated) in calcium carbonate, silica, and ferric oxide.

Much of the iron and silicon is brought down to the sea by streams in colloidal suspension and is flocculated when meeting the saline waters of the sea. Thus, all large-scale chemical deposits of these minerals are confined to nearshore regions, where most of the marine detrital sediments are also deposited. The SiO_2 and Fe_2O_3 may be dispersed in detritus as a cement, therefore, and no "integral" siliceous or ferruginous deposits are formed. Independent deposits of these minerals occur only under relatively quiet tectonic conditions (see *Ironstone*). The silica is initially deposited as a silica gel that hardens to opal, which may in turn be "devitrified" to form microcrystalline quartz (see *Chert and Flint*). The ferric oxide is probably deposited initially as a $Fe(OH)_3$ gel that soon dehydrates to cryptocrystalline goethite, $FeO(OH)$, or limonite, which may subsequently dehydrate to hematite, Fe_2O_3.

The solutes in which sea water is most strongly saturated, calcium carbonate and silica, may precipitate at any distance from shore. Part of this precipitation is purely physicochemical, but most is probably due to the activity of organisms. Thus radiolaria, diatoms, and some sponges extract silica from sea water; a much greater variety of plants (algae) and animals (sponges, corals, bryozoa, brachiopods, mollusks, echinoderms) extract calcium carbonate. Thus, chemical sediments composed of calcite (limestone) and of amorphous to microcrystalline quartz (chert) are characteristic of all reasonably shallow marine basins of deposition.

The limestones in particular are subject to diagenetic changes, the most common being their replacement by dolomite (q.v.) or by chert, particularly in the nearer-shore portions of the basin.

Evaporites. Under arid to semi-arid conditions, lake basins and marine basins whose access to the open sea is restricted are characterized by rates of evaporation that equal or exceed the rate of inflow of water from the sea or from streams. Thus, the concentration of dissolved constituents in the basin waters will increase until these waters become saturated in some constituent and it begins to precipitate. Such chemical sediments are called evaporite deposits.

In restricted marine basins the first-deposited mineral invariably is $CaCO_3$, and so limestones are normally the first-formed members of these evaporite sequences. The limestones are followed in relatively regular sequence by calcium sulfates (gypsum, anhydrite), by halite, and finally by salts of potassium and magnesium (e.g., sylvite, kieserite). This sequence is relatively constant from basin to basin because of the relatively constant composition of the dissolved constituents in sea water (see *Evaporites–Physicochemical Conditions of Origin*). The minerals that now are present in these evaporite deposits, however, particularly those of Mesozoic and Paleozoic age, also reflect the effects of diagenesis (see *Marine Evaporites–Diagenesis and Metamorphism*). As a result, a variety of additional minerals, such as polyhalite, carnallite, and kainite may also be present.

In desert-lake evaporites, the sequence in which the minerals are precipitated and even the minerals that form may vary much more widely because the composition of the dissolved constituents reflects local conditions of chemical weathering and of source material. In a number of areas, the latter includes the waters of hot springs that may be relatively rich in borate and sulfate ions. These ions are combined with alkalies, which are the only constituents normally removed from the parent rock during chemical weathering in arid areas, producing varied precipitates of sodium chloride (halite); sodium carbonates (natron, trona); sodium sulfates (mirabilite, thenardite); sodium borates (borax, ulexite); and, to a much lesser extent, potassium salts (mainly sylvite and hanksite). Buried desert-lake evaporites also undergo diagenesis; the process has not been so well studied as in marine evaporites, however, and almost the only information available applies to Tertiary borate deposits, where kernite and colemanite appear to be the result of diagenesis.

Humid Lakes. Lake basins in humid climates show little effect of evaporation, and the precipitation of chemical sediments here is a result almost entirely of the activity of organisms. The waters of most such lakes are very undersaturated in all dissolved constituents, however, and chemical sedimentation occurs almost solely in humid-region lakes that rest on glacial deposits or on recent lava flows. Only this type of substrate will contain large amounts of readily weathered minerals that will yield large amounts of soluble constituents, essentially amorphous silica and ferrous iron. The lacustrine chemical sediments thus consist mainly of opal in the form of diatomaceous or infusorial earth and of limonite in the form of bog iron ores.

Transitional between these humid-lake deposits and the evaporites of the desert lakes are occasional deposits of freshwater limestone, which appear to form in lakes in semiarid climates (e.g., Pyramid Lake, Nevada) where the leachate from the surrounding area is rich in calcium, and evaporation of the lake waters raises their calcium content significantly. Under these conditions, freshwater reefs and mounds of calcareous algae flourish, and limestone deposits are produced.

CHARLES P. THORNTON

References

Bathurst, R. G. C., 1971. *Carbonate Sediments and Their Diagenesis.* Amsterdam: Elsevier, 620p.

Berner, R. A., 1971. *Principles of Chemical Sedimentology.* New York: McGraw-Hill, 240p.

Blatt, H.; Middleton, G.; and Murray, R., 1972. *Origin of Sedimentary Rocks.* Englewood Cliffs, N.J.: Prentice-Hall, 634p.

Chauvel, J. -J., 1974. Les minerais de fer de l'Ordovicien inférieur du bassin de Bretagne-Anjou, France, *Sedimentology,* 21, 127-148.

Chilingar, G. V.; Bissell, H. J.; and Fairbridge, R. W., eds., 1967. *Carbonate Rocks.* Amsterdam: Elsevier, 471p.

Dapples, E. C., 1959. *Basic Geology for Science and Engineering.* New York: Wiley, 609p.

Degens, E. T., 1965. *Geochemistry of Sediments.* Englewood Cliffs, N.J.: Prentice-Hall, 342p.

Garrels, R. M., and MacKenzie, F. T., 1971. *Evolution of Sedimentary Rocks.* New York: Norton, 397p.

Krynine, P. D., 1948. The megascopic study and field classification of sedimentary rocks, *J. Geol.,* 56, 130-165.

Larsen, G., and Chilingar, G. V., 1967. *Diagenesis in Sediments.* Amsterdam: Elsevier, 551p.

Millot, G., 1970. *Geology of Clays.* New York: Springer-Verlag, 429p.

Pettijohn, F. J., 1975. *Sedimentary Rocks.* New York: Harper & Row, 628p.

Pettijohn, F. J.; Potter, P. E.; and Siever, R., 1972. *Sand and Sandstone.* New York: Springer-Verlag, 618p.

Schmalz, R. F., 1969. Deep-water evaporite deposition: A genetic model, *Bull. Am. Assoc. Petrol. Geologists,* 53, 798-823.

Cross-references: *Abrasion pH; Cementation; Chert and Flint; Clastic Sediments and Rocks; Diagenesis;*

Evaporites—Physicochemical Conditions of Origin; Graywacke; Ironstone; Marine Evaporites—Diagenesis and Metamorphism; Provenance; Transportation; Weathering in Sediments. Vol. IVA: *Paragenesis.* Vol. IVB: *Sedimentary Minerals.*

PARALIC SEDIMENTARY FACIES

The term *paralic* was first applied by Naumann (1854) to the environments of deposition of coal deposits. Unique geographic, environmental, and tectonic characteristics operate in paralic regions, which occur marginal to continents, or marginal to inland seas. They include those areas adjacent to barrier islands which divide open ocean water from landward bays, lagoons, marshes, and estuaries in which fresh and marine waters mix. Paralic regions can be either small or extensive; but in all cases, sedimentation occurs in shallow water. Excessively thick sequences of sediments are common and indicate a delicate balance between tectonic downwarp and sedimentation rate.

Tercier (1939) recognizes "paralic" as one of four fundamental types of marine sedimentation, characterized by generally thick, intracontinental or pericontinental, intensely terrigenous, partly marine, partly estuarine and continental sediments. The marine facies are exclusively neritic (see *Neritic Sedimentary Facies*) and are characterized by a predominance of detrital rocks, with organogenic rocks being subordinate. Relatively steady subsidence often results in very monotonous strata often several hundred meters thick.

Tercier's other three marine environments are (1) continental shelves with epicontinental sedimentation (which lack the terrigenous component of paralic); (2) geosynclines (marked by accentuated mountainous relief, varied bathymetric conditions, and as a result, very variable sedimentation); and (3) oceanic sediments, which he regards as exclusively marine, accumulating in abyssal depths.

Pettijohn (1957) assigns his subgraywacke suite to paralic environments, distinguishing between (1) reducing environments common to paludal-fluvial areas, containing coals and micaceous shales, and (2) subgraywacke-protoquartzites, from oxidizing environments common to fluvial areas, with red muds, shale, subarkose, and subgraywacke.

Present examples of paralic sedimentary facies include the Sunda shelf, Mississippi delta, and typical suites being deposited at and near the mouths of large rivers: the Amazon, Orinoco, La Plata, Ganges, Indus, Huang Ho, Yangtse, Rhine, Loire, and Garonne. Stratigraphic equivalents include marginal deposits of the great mountain ranges; the Alpine Molasse; the Neogene of Sumatra, Borneo, and Java; and the Tertiary of the Gulf Coast of North America.

Present Environments

Paralic sediments are deposited within and marginal to deltas, lagoons, and swampy coastal plains. Rivers commonly deposit most of their load shortly before or immediately after reaching the zone of tidal flux; sedimentation rates in this transitional zone are high, and the resulting rock formations thick, as compared with marine deposits of the same time intervals. Thick sequences of paralic rocks form most of the world's coal reserves and often important oil reservoirs. Deposits formed in paralic environments are thus more important quantitatively, interpretatively, and economically than the restricted area of present-day alluvial coasts would suggest. Principal paralic environments include estuaries, lagoons and barriers, marshes, tidal flats, and deltas.

Estuaries. Most present-day estuaries are river valleys drowned by the postglacial rise of sea level, recording < 6000 yr of sedimentation. Hydrographic conditions are usually complex (see *Estuarine Sedimentation*), and estuary deposits are variable (see *Estuarine Sediments*). In some, sand predominates; but in others, fine-grained, organic-rich muds may be characteristic. The sand may be relict or river-induced within the estuary, or ocean-induced by longshore drift and tides near the mouth. The indigenous fauna of estuaries is dominantly a reduced-marine one; crustacea and mollusks are the dominant macrofaunal groups at the present day. The salinity gradient is often recognizable in the fauna. Transported microfaunal remains and drifted fragments of land plants are common. Estuaries are probably short-lived features in geological time. Deposits in them have poor chance of survival, being liable to erosion if sea level falls, or to reworking by waves if it rises, the former estuary becoming part of an open coast.

Lagoons. Lagoons frequently occur along coasts that are both subsiding and heavily alluviated. Their deposits are, therefore, likely to be preserved. Well-known lagoons occur on the Texas Coast, described by Shepard et al. (1960). They are typical examples of their kind, passing landward into river mouths and having minor deltas built into them. Other examples occur on the Rhone delta and in the former Zuider Zee in Holland. Brackish water conditions are typical of such lagoons, favored by both heavy runoff from the rivers and a wet climate in the coastal region. If the lagoons are large, there may be a salinity gradient from almost fresh to fully saline, but the water is usually shallow (<3 m in the Texas lagoons) and tides are negligible, so that any one part of the system may have a fairly stable intermediate salinity, with minor seasonal variations, for a period of years. Conditions for life are thus less rigorous than in an estuary, and the fauna, though restricted in numbers of species, may be abundant. Catastrophic salinity changes, caused especially by fresh-water floods, cause mass mortalities in the invertebrate and fish faunas.

On the lagoon floor, sediments are mostly silts and clays, but sands occur adjacent to the seaward bounding barriers and near inlets (see *Lagoonal Sedimentation*). Varying rates of deposition, and also wave action, produce laminae in the sediment, but burrowing animals tend to destroy them. In general, the lower the salinity the sparser the fauna, and the better preserved the lamination. An important effect of wave action is to winnow out lamellibranch valves from the mud and to concentrate them into layers of convex-up valves, or eventually into shell beds, which are potential limestones.

Lagoons with dominantly carbonate sediments and hypersaline lagoons are not part of the typical paralic environment. Nevertheless, the brackish Texas lagoons are almost continuous with the hypersaline Laguna Madre, and similar close associations have occurred in the past, as in the Purbeck Beds (Upper Jurassic) of England.

Tidal Flats. The tidal flats and associated salt marshes of the Dutch Wadden Sea have been studied extensively (see *Salt-Marsh Sedimentology; Tidal-Flat Geology; Wadden Sea Sedimentology*). Similar flats are widespread along lowland coasts in temperate climates. In tropical regions, the situation is modified by the extensive growth of mangroves. The Dutch Waddens are expressed as a series of zoned salt marshes inundated only at spring tides. The tidal flats proper occur between tide marks, and the tidal channels below low-tide mark. The channels migrate laterally and erode the previously deposited tidal-flat sediments on one side and redeposit their coarse fraction on the other. As a result, their deposits contain an overrepresentation of channel sediments. Because the area is subsiding, sections in subrecent deposits usually show a characteristic sequence of thin marsh deposits, thin tidal-flat deposits, and thicker channel deposits.

To prove intertidal deposition, whole associations of structures must be found, preferably in a recognizable sequence. In the Devonian "Psammites du Condroz," Van Straaten (1954) concluded that although some intertidal deposits were present, the greater part of the formation was lagoonal. Allen and Tarlo (1963) recognize intertidal deposits as an important component of a complex series of environments in the Lower Old Red Sandstone (Devonian) of England. A sequence from high-flat to subtidal channel deposits is closely comparable to that recognized on the North Sea coast; it is repeated many times. The deposits recognized as intertidal occur in their logical intermediate position in a major facies change from shallow marine to fluviatile.

Deltas. The deltas of great rivers are important sites of paralic accumulation. The delta is an area where deposition is dominant, proximal to and in a body of standing water, at the place where a river system bifurcates and deposits its detritus. The original subdivision of deltaic deposits was proposed by Gilbert (1885) for lake deltas, and that subdivision may be applied with modification to marine deltas also (Scruton, 1960). The concept of foreset beds deposited on a slope extending from water level to the floor of the basin, bottomset beds extending basinward in front of the foreset, and topset beds extending at water level back from the delta front over the former alluvial plain has been extensively amplified and modified by many authors (see *Deltaic Sediments; Delta Sedimentation*).

Modifications take into account processes operating in the alluvial, distributary, interdistributary, and delta front areas coupled with amplifications of structures, sedimentology, and paleogeography as affected by size, changes in sediment supply, changes in physiography with time under stable sea level, and changes with time under fluctuating sea levels. Lithologies associated with delta sedimentation include prodelta muds, peats, natural-levee and barfinger sands and sandy silts, marine-shelf sands and limestones, and alluvial deposits. Many studies of ancient deltaic complexes have been reported (see *Delta Sedimentation*).

DONALD J. COLQUHOUN

References

Allen, G. P., and 13 others, 1974. Environments et processus sédimentaires sur le littoral Nord-Aquitain, *Bull. Inst. Géol. du Bassin d'Aquitaine,* **15,** 183p.

Allen, J. R. L., and Tarlo, L. B., 1963. The Downtonian and Dittonian facies of the Welsh borderland, *Geological Mag.,* **100,** 129-155.

Asquith, D. O., 1974. Sedimentary models, cycles, and deltas, Upper Cretaceous, Wyoming, *Bull. Am. Assoc. Petrol. Geologists,* **58,** 2274-2283.

Barrell, J., 1913. Criteria for the recognition of ancient delta deposits, *Geol. Soc. Am. Bull.,* **23,** 377-446.

Brown, L. F., and Wermund, E. G., 1969. *A Guidebook to the Late Pennsylvanian Shelf Sediments North-Central Texas.* Dallas, Tex.: Dallas Geol. Soc., 69p.

Gilbert, G. K., 1885. The topographic features of lake shores, *U.S. Geol. Surv., Ann. Rept.*, **5**, 69-123.

Naumann, C. F., 1854. *Lehrbuch der Geognosie*, vol. 2. Leipzig: Englemann, 571-579.

Pettijohn, F. J., 1957. *Sedimentary Rocks*, 2nd ed., New York: Harper & Row, 718p.

Reineck, H.-E., 1974. Recent and ancient shallow marine deposits, *Bull. Centre Rech. de Pau*, **8**, 295-304.

Scruton, P. C., 1960. Delta building and the deltaic sequence, in F. P. Shepard et al., eds., *Recent Sediments Northwest Gulf of Mexico*. Tulsa: Am. Assoc. Petrol. Geologists, 82-102.

Shepard, F. P.; Phleger, F. B.; and Van Andel, T. H., eds., 1960. *Recent Marine Sediments, Northwest Gulf of Mexico*. Tulsa: Am. Assoc. Petrol. Geologists, 394p.

Sheridan, R. E.; Dill, C. E., Jr.; and Kraft, J. C., 1974. Holocene sedimentary environment of the Atlantic inner shelf off Delaware, *Geol. Soc. Am. Bull.*, **85**, 1319-1328.

Tercier, J., 1939. Dépôts marins actuels et séries géologiques, *Eclogae Geol. Helv.*, **32**, 47-100.

Van Straaten, L. M. J. U., 1954. Sedimentology of recent tidal flat deposits and the Psammites du Condroz (Devonian), *Geol. Mijnbouw*, **16**, 25-47.

Cross-references: *Alluvial-Fan Sediments; Barrier Island Facies; Coal; Continental-Shelf Sediments; Deltaic Sediments; Estuarine Sediments; Fluvial Sediments; Lagoonal Sedimentation; Neritic Sedimentary Facies; Peat-Bog Deposits; Prodelta Sediments; Salt-Marsh Sedimentology; Tidal-Flat Geology; Wadden Sea Sedimentology*.

PARTING LINEATION

Parting lineation refers to linear, low-relief irregularities visible on surfaces of splitting (parting) in laminated or cross-laminated sandstone. One form of the structure was first described by Sorby (1859). Cloos (1938) referred to the feature in German and Stokes (1947) in English as "primary current lineation." The term *parting lineation* was introduced by Crowell (1955) and is preferable because it is not genetic.

Parting lineation includes two features of different origin and scale of relief. Lineation of one type appears as linear, parallel shallow grooves and ridges of low relief (<1 mm) on a lamina surface (Fig. 1). This lineation occurs on individual bedding planes and is actually "windrows" of sand shaped by current flow parallel with the lineation (Allen, 1964) and exhumed when the subsequently lithified bed is split. This lineation, called *parting-plane lineation*, is best observed where parting is perfect, following a single lamina over a wide area of the bedding surface, and when viewed in strong oblique light. *Parting-step lineation* is a larger-scale feature, with relief of 1-3 mm. Linear steps form

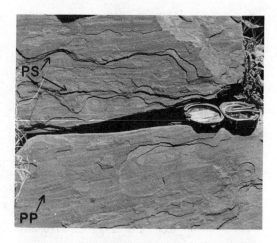

FIGURE 1. Split slab of laminated sandstone showing both types of parting lineation. Parting-step lineation (PS) is visible as prominent shadows because of its greater relief. Parting-plane lineation (PP) is the faint lineation on individual lamina. Compass is 15 cm long.

where the surface of parting follows an individual bedding plane for several mm or cm and then cuts abruptly across a lamina to a subjacent bedding plane. Steps generally have a strong preferred orientation on major parting surfaces (Fig. 1). Microscopic fabric studies (McBride and Yeakel, 1973; Allen, 1964) show that the orientation of the steps is controlled by and is parallel with the preferred orientation of the long axes of sand grains. Whereas parting-plane lineation is an exhumed primary depositional feature, parting-step lineation is secondary and forms when the rock splits. Both types of parting lineation have coincident orientations on individual parting surfaces, and both are excellent paleocurrent indicators that tell the orientation but not the sense of the depositing current (Bertsbergh, 1940).

EARLE F. McBRIDE

References

Allen, J. R. L., 1964. Primary current lineation in the Lower Old Red Sandstone (Devonian), Anglo-Welsh Basin, *Sedimentology*, **3**, 89-108.

Bertsbergh, J. W. B. von, 1940. Richtungen der Sedimenten in der rheinischen Geosynkline, *Geol. Rundschau*, **31**, 328-364.

Cloos, H., 1938. Primäre Richtungen in Sedimenten der rheinischen Geosynkline, *Geol. Rundschau*, **29**, 357-367.

Crowell, J. C., 1955. Directional-current structures from the Prealpine Flysch, Switzerland, *Geol. Soc. Am. Bull.*, **66**, 1351-1384.

McBride, E. F., and Yeakel, L. S., 1963. Relationship between parting lineation and rock fabric, *J. Sed. Petrology*, **33**, 779-782.

Sorby, H. C., 1859. On the structures produced by the

currents present during the deposition of stratified rocks, *Geologist,* **2,** 137-147.

Stokes, W. L., 1947. Primary lineation in fluvial sandstones: a criterion of current direction, *J. Geol.,* **55,** 52-54.

Cross-references: *Bedding Genesis; Fabric, Sedimentary; Fissility; Paleocurrent Analysis; Sedimentary Structures.*

PEAT-BOG DEPOSITS

Origin

Peat bogs are generally associated with enclosed basins of deposition, such as former lake basins, or raised surfaces (flat, sloping, or irregular) where plant growth and succession results in a compact accumulation of plant debris, often of considerable thickness. The controlling factor for fresh-water bogs seems to be abundant precipitation favoring the growth of peat-producing moss (generally *Sphagnum*).

Lake Basins. Almost any lake, owing its origin to a variety of depositional or erosional agents, may eventually become a peat bog through natural processes of deposition and plant growth. Many lake basins may have complicated histories, such as those originating as maritime shoreline features and finally evolving into fresh-water bodies, or those originating as bedrock basins, such as tarns or potholes. Most types have developed bogs and even dry meadows during postglacial time, under favorable climatic conditions. A number of peat bogs that have been studied by botanists and geologists occur in former kettle lakes or as polludification bogs in small damp depressions, both located on glacial deposits of Pleistocene and Holocene age. These bogs contain a more or less uniform sequence of deposition and a stratigraphic record of climatic and vegetational changes (Fig. 1). The type of sediment may not correlate in time with the pollen zones, climate, or strata in other bogs, however, because the rate of deposition and local plant succession may vary.

Raised Bogs. Where conditions permit, i.e., moist climate; high water table; rapid plant growth, particularly of *Sphagnum;* and compaction and some alteration of organic detritus, bogs may develop on (1) flat, gently sloping, or irregular surfaces, such as the topogenous bogs (ground-water control); (2) hillsides, such as the German Hochmoor, the domed or ombrogenous (surface rainfall control) bogs, and soligenous (surface-water or runoff control) bogs; (3) larger areas of undrained or poorly drained surface, such as the sub-Arctic muskeg. Because

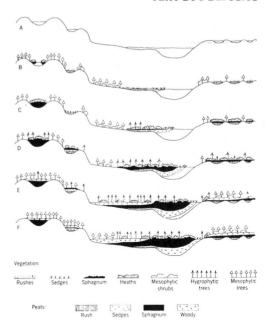

FIGURE 1. Autogenic bog succession typical of the Laurentian shielf area of Canada. Six stages of development are shown for a variety of habitats (from Strahler, 1965, after Dansereau and Segadas-Vianna).

muskeg can be regionally extensive, it may be more appropriately treated as a separate entity or type of swamp apart from the generally restricted nature of peat bogs in lake basins or other single topographic features.

Raised bogs frequently lack the sedimentary record of the lake basin bogs, because many modern examples originated during the moist climate of the sub-Atlantic climate episode, which began approximately 2500 yr ago. Some raised bogs may not be differentiated from other sites of peat growth, such as marshes and swamps.

Autogenic Bog Formation. The progressive filling of fresh-water lakes or topographic basins follows a sequence of varying chemical and vegetational characteristics. Typically, a lake changes from eutrophic, with organic-rich bottom sediments, to dystrophic, with extensive growth of sedge and moss, particularly *Sphagnum,* resulting in the buildup of peat. The vegetational succession listed below incorporates a variety of plant types depending on climate (Fig. 1): (1) open water (as in oligotrophic lakes)—algae, diatoms; (2) submerged—pondweeds; (3) floating—pondweeds, lilies; (4) emergent—sedges, rushes, reeds, cattail; (5) mat, meadow, and saturated substrates—moss, sedges, rushes, grasses, heaths, and trees favoring a wet or waterlogged soil; (6) land—shrubs and trees.

PEAT-BOG DEPOSITS

Types of Peat

As an organic deposit or sediment, peat may be *limnic* (deposited under water), *telmatic* (deposited in the zone between low and high water), or *terrestric* (or terrestrial, formed above high water). Filling of a basin follows the sequence:

1. *Fine clastic sediments* of low organic content, including clay, silt, marl, volcanic ash, authigenic minerals such as bog iron (limonite), and of numerous planktonic forms such as diatoms.
2. *Limnic peat,* including *dy*, a blackish-brown, acidic, gel-mud formed as a colloidal and amorphous precipitate of plant and animal residues in lakes poor in plant nutrients, such as dystrophic lakes; and *gyttja* (q.v.) or nekton mud, a deposit of microscopic and submicroscopic organic remains. Both dy̅ and gyttja vary in the amount of incorporated material, such as clay, silt, sand, algae, diatoms, calcium carbonate, and iron oxide. These impurities lead to gradational types, such as clay gyttja and detrital dy, among others.
3. *Telmatic peat* is represented by the fine-textured *Sphagnum* peat and coarser peats formed from hummocks of emergent vegetation such as sedge and rush interwoven with the moss. Generally, telmatic-peat mat forms above standing water, while limnic peat accumulates below (an unripe bog), until these layers meet (ripe bog).
4. *Terrestrial peat* is generally coarser in texture than limnic and telmatic varieties. It is composed of fragments of the plants dominant at this stage of the hydrosere, such as moss, alder, birch, grass, and sedge. These peats can be differentiated on the basis of texture—fibrous or woody—and composition of dominant plant remains, e.g., alder peat.

Special Characteristics. Layering or banding is indicative of depositional environment, seasonal deposition, and climatic change, as in regenerative peat or recurrence surfaces. Post-depositional chemical changes may take place, such as lignitization, due to loss of living substance and the concentration of carbon, and humidification, the slow oxidation of vegetation forming humic acid and humus:

$$2C_6H_{10}O_5 \rightleftarrows C_8H_{10}O_5 + 2CO_2 + CH_4 + H_2O$$
cellulose   humified   marsh
             residue   gas

While a lowland peat may be as much as 90% H_2O, a typical peat sample dried at 100% yields carbon, 60%; hydrogen, 6%; oxygen, 30% nitrogen, 1%; and ash, 3% (varies from 2-15%). Organic content may be determined by hydrogen peroxide oxidation (Fig. 2).

Fossils. Fossils found in peat bog sediments are varied, although not all are useful as guide fossils or paleoecologic indicators. The following categories include some readily identifiable remains: (1) macroscopic plant fossils—cones, seeds, fruits, leaves and needles, stems, twigs,

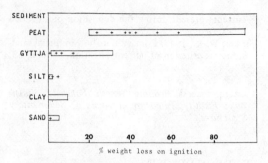

FIGURE 2. Organic content of typical bog sediments. + = % organic content determined by hydrogen peroxide leaching.

roots, bark, and tree trunks in place or as fallen logs. (2) Microscopic plant fossils—cone scales, diatoms, algae, dinoflagellates, spores and pollen—the latter are best suited for regional climatic and biogeographic studies. (3) Macroscopic animal fossils—shells of fresh-water mollusks, and mummified corpses; and fecal pellets of animals, most notably frogs, birds, mice, and mammals. (4) Microscopic animal fossils—protozoa, particularly testaceous rhizopods; crustacea, including ostracods, cladocera, insect larvae; and bryozoa.

World-wide Climate and Distribution of Peat Bogs

Climate. Humid areas (>50 cm/yr) provide the necessary conditions for bog development, such as a high water table and soil capillarity, ground-water seepage, and standing water in surface depressions. These conditions prevail in both warm and cold subhumid (50-100 cm/yr) and humid (>100 cm/yr) regions in the subtropical, middle-latitude, and subarctic climate belts, particularly in the glaciated portions of North America, Europe, northern Asia, New Zealand, and southern South America. In tropical regions, oxidation and bacterial decay of dead vegetation results in low peat production in bogs, although some do exist containing a tropical vegetation. Peat production in the tropics is more extensive in tropical rain-forest and mangrove swamps (Cohen, 1974). As shown in Fig. 3, destruction of organic material exceeds production above the mean annual temperature of 25°C.

Distribution. Peat bogs are distributed commonly through the following natural vegetation regions: (1) middle-latitude deciduous forest in eastern North America, northwestern Europe, eastern Asia; (2) subarctic needleleaf forest (taiga) in northern North America and northern Asia; (3) Arctic and Alpine tundra of high

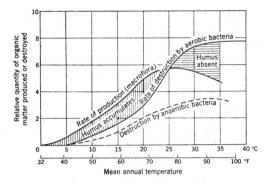

FIGURE 3. Graph showing the relative rates of production and destruction of organic matter as related to mean annual temperature (from Strahler, 1965, after Senstius).

mountain regions; (4) temperate rain-forests of New Zealand and Chile.

Pre-Holocene Peat Bogs. These bogs exist today as peat bodies interbedded in Pleistocene deposits or as lignite deposits, for example the Brandon (Vt.) lignite (Miocene?), and some pre-Cenozoic coal deposits known to be of freshwater origin (see *Coal*).

LESLIE SIRKIN

References

Cohen, A. D., 1974. Petrography and paleoecology of Holocene peats from the Okefenokee swamp-marsh complex of Georgia, *J. Sed. Petrology,* **44,** 716-726.

Cohen, A. D., and Spackman, W., 1972. Methods in peat petrology and their application to reconstruction of paleoenvironments, *Geol. Soc. Am. Bull.,* **83,** 129-142.

Connally, G. G., and Sirkin, L. A., 1973. Wisconsinan history of the Hudson-Champlain Lobe, *Geol. Soc. Am. Mem. 136,* 47-69.

Dansereau, P., 1957. *Biogeography: An Ecological Perspective.* New York: Ronald, 394p.

Ertdman, G., 1943. *An Introduction to Pollen Analysis.* New York: Ronald, 239p.

Faegri, K., and Iversen, J., 1964. *Textbook of Pollen Analysis.* New York: Hafner, 237p.

Greensmith, J. T., and Tucker, E. V., 1973. Peat balls in Late-Holocene sediments of Essex, England, *J. Sed. Petrology,* **43,** 894-897.

Koch, J., 1966. *Petrologische Untersuchungen an jungpleistozanen Schieferkohlen aus dem Alpenvorland der Schweiz und Deutschlands mit Vergleichsuntersuchungen an Holozanen Torfen.* Bonn: Druck Krupinski Mondorf/Rh., 186p.

Heusser, C. J., 1960. *Late Pleistocene Environments of North Pacific North America.* New York: Am. Geographical Soc., 308p.

Puffe, D., and Grosse-Brauckmann, G., 1963. Mikromorphologische Untersuchungen an Torfen, *Z. Kulturtechnik u. Flurbereinigung,* **4**(3), 159-188.

Strahler, A. N., 1965. *Introduction to Physical Geography.* New York: Wiley, 455p.

Teichmuller, M., 1958. Rekonstruktionenverschiedener Moortypen des Hauptfloezes der Niederrheinischen Braunkohle, *Fortschr. Geol. Rheinland u. Westfanlen,* **2,** 599-612.

Cross-references: *Coal; Gyttja, Dy; Lacustrine Sedimentation; Organic Sediments; Paludal Sediments.* Vol. XI: *Bog Soils.*

PEBBLY MUDSTONES

Mudstones containing dispersed pebbles, cobbles, and boulders are frequently encountered in the geologic record (Fig. 1). Although the mixture requires that both mud and dispersed stones were laid down at essentially the same time, deposition could not have been the result of a simple slackening of current. Interest in these rocks in recent years has been focused on criteria to reconstruct the processes of origin, including glacial deposition, turbidity currents, subaqueous slumping, and mudflows.

Among the varieties of pebbly mudstones are *tillites*, lithified till (q.v.), the unsorted accumulations of debris left by glaciers. Sound criteria are needed to recognize tillites in order to document glacial climates in the remote past (Crowell, 1964; Schermerhorn, 1974). If the mixture of mud and rock flour with stones of many sizes and shapes is discovered in ancient strata, special scrutiny is necessary to ascertain that the deposit is a true tillite, and not formed

FIGURE 1. Pebbles and cobbles intermixed and dispersed with sandy siltstone, Upper Cretaceous Pigeon Point Formation, Pigeon Point, San Mateo County, California. Scale ≈ 15 cm.

by nonglacial process. Striated pavements on distinctly older rocks lying below the accumulation are particularly helpful, but are rare, as are striated and faceted stones and till fabrics. Evidence of rafting is particularly useful in support of ancient glaciation, especially in Precambrian rocks. Most glacial sequences consist of thinly bedded layers, perhaps varved, into which stones and unsorted debris have apparently been dropped from above. Rafting of some kind is therefore implied, and if the rocks are so old that tree roots, kelp holdfasts, and large animals can be eliminated as transporting agents, a glacial origin may be acceptable. Vigorous turbidity currents, however, may move across a mud substrate and leave behind isolated, large clasts as the current wanes. Although such clasts are scattered upon a bedding plane, graded lenses of debris at the same level may attest to the turbidity-current process.

Pebbly mudstones also result from deposition by gravity flows (q.v.). A vigorous turbidity current, carrying mud, sand, gravel, and larger stones, may travel across a mud substrate and begin to deposit the material as a layer of gravel. The bulk density of the gravel, however, is significantly greater than that of the mud, with the result that deposition of the gravel on top of mud is unstable. Downslope or downcurrent mixing accordingly ensues so that megaclasts are churned into finer matrix. Stilled stages in this process are preserved in many flysch sequences (Crowell, 1957; Winterer, 1964). Subaqueous slumps (Corbett, 1973) and debris flows may produce similar deposits (Stanley, 1974).

Pebbly mudstones originate from still other processes. Normal conglomerates may weather selectively so that resistant large clasts are preserved in a fine matrix derived from the chemical alteration of interstitial sand and mud. In terrestrial environments, mudflows carry jumbled masses of debris with sparse matrix from high ground out into depositional basins, where they are preserved as intercalated layers of sedimentary breccia in finer deposits. In addition, certain pebbly shales of tectonic origin resemble pebbly mudstones, except for marked fissility, such as exposures of wildflysch and the Apennine *argille scagliose* (see *Olistostrome*).

Pebbly mudstone is a descriptive term and is partly equivalent to *diamictite* (q.v.); the latter applies to any size of large stone embedded in muddy matrix, and not only to pebbles. Schermerhorn (1966) and Harland et al. (1966) advocate the term *mixtite* for such rocks, pointing out that the word suggests the mixed characteristics of mud with dispersed clasts of all sizes stirred together. Despite the advantages of mixtite, the principle of priority of usage applies to diamictite, and many geologists now employ the latter term.

JOHN C. CROWELL

References

Corbett, K. D., 1973. Open-cast slump sheets and their relationship to sandstone beds in an Upper Cambrian flysch sequence, Tasmania, *J. Sed. Petrology*, **43**, 147-159.

Crowell, J. C., 1957. Origin of pebbly mudstones, *Geol. Soc. Am. Bull.*, **68**, 993-1010.

Crowell, J. C., 1964. Climatic significance of sedimentary deposits containing dispersed megaclasts, in A. E. M. Nairn, ed., *Problems in Palaeoclimatology*. New York: Interscience, 86-98.

Harland, W. B.; Herod, K. N.; and Krinsley, D. H., 1966. The definition and identification of tills and tillites, *Earth-Sci. Rev.*, **2**, 225-256.

Schermerhorn, L. J. G., 1966. Terminology of mixed coarse-fine sediments, *J. Sed. Petrology*, **36**, 831-835.

Schermerhorn, L. J. G., 1974. Late Precambrian mixtites: Glacial and/or nonglacial? *Am. J. Sci.*, **274**, 673-824.

Stanley, D. J., 1974. Pebbly mud transport in the head of Wilmington Canyon, *Marine Geol.*, **16**, M1-M8.

Winterer, E. L., 1964. Late Precambrian pebbly mudstone in Normandy, France: Tillite or tilloid? in A. E. M. Nairn, ed., *Problems in Palaeoclimatology*. New York: Interscience, 159-177.

Cross-references: *Conglomerates; Diamictite; Gravity Flows; Gravity-Slide Deposits; Olistostrome, Olistolite; Submarine Fan (Cone) Deposits; Till and Tillite; Turbidity-Current Sedimentation.*

PEDOLOGY—*See* Vols. XI and XII

PEEL TECHNOLOGY—*See* SEDIMENTOLOGICAL METHODS

PELAGIC SEDIMENTATION, PELAGIC SEDIMENTS

The *Challenger* Expedition (1872-1875) discovered two types of deep-sea deposits that were unfamiliar to geologists at the time: (1) biogenous oozes essentially without admixture of components obviously derived from land or from continental shelves, and (2) extremely fine-grained sediments with only a small proportion of coarse material, mainly authigenic minerals, volcanogenic clastics, and biogenous particles. These kinds of deposits, or their typical components, are generally referred to as *pelagic sediments* (Bramlette, 1961; Arrhenius,

1963; Berger, 1974, 1976). Characteristically, they have arrived at their site of deposition by settling through the water, and they have accumulated at slow rates (see *Sedimentation Rates*).

Pelagic or *eupelagic* sediments are distinguished from *hemipelagic* sediments, which possess a substantial proportion of land- or shelf-derived material and are deposited at relatively high rates, typically on continental slopes and rises, as well as on abyssal plains in places, usually involving redeposition processes such as turbidity currents or strong geostrophic currents. Murray and Renard (1891) recognized this dichotomy by separating the true pelagic deposits (red clay, q.v.; radiolarian ooze, diatom ooze, *Globigerina* ooze, pteropod ooze) from *terrigenous deposits* (blue mud, red mud, green mud, volcanic mud, coral mud). This mixed descriptive-genetic classification is still in use today, modified to accommodate first Pleistocene and then pre-Pleistocene deposits, as well as turbidites (see Olausson, 1960; Beall and Fischer, 1969; Berger and von Rad, 1972). The main constituents of these deposits are of biogenic, terrigenic, authigenic, volcanogenic, or cosmogenic origin.

Distribution

As Murray pointed out, the distribution of pelagic sediments (Fig. 1, Table 1) shows characteristic patterns that can be related to depth, latitude, temperature, fertility of surface waters, ocean currents, distance from land, amount of organic matter in sediments, and associated CO_2 production and oxidation state, as well as activity of benthic organisms. Some

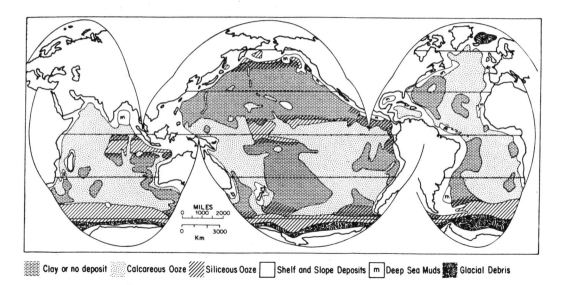

FIGURE 1. Distribution of pelagic sediments in the world ocean (from various sources; see Berger, 1974).

TABLE 1. Areas Covered by Deep-Sea Deposits (millions km²)

	Atlantic Ocean		Pacific Ocean		Indian Ocean		Total	
	Area	%	Area	%	Area	%	Area	%
Calcareous ooze[a]	39.9	54.6	61.8	36.8	39.0	58.0	140.7	45.6
Siliceous ooze	6.0	8.2	13.6	8.1	14.0	20.8	33.6	10.9
Siliceous clay	1.5	2.1	14.9	8.8	2.3	3.4	18.7	6.1
Red clay	15.5	21.2	71.5	42.5	8.3	12.4	95.3	30.9
Glacial mud	6.4	8.8	4.2	2.5	2.3	3.4	12.9	4.2
Turbidite mud	3.7	5.1	2.1	1.3	1.3	2.0	7.1	2.3
	73.0	100.0	168.1	100.0	67.2	100.0	308.3	100.0

[a]Calcareous ooze includes highly calcareous clay (about 10% of total).

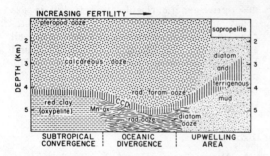

FIGURE 2. Pelagic sediments in a depth-fertility frame, eastern central Pacific (from Berger, 1974).

of these relationships can best be represented in a depth-fertility diagram (Fig. 2). The contrast between sediments below the barren subtropical gyres and those of the fertile upwelling regions (see Vol I: *Calcareous Oozes*) is akin to Murray's distinction between "mid-ocean" and "near-land" sediments, each characterized by their own faunal and floral aspects, in addition to more obvious differences in components and color. Ernst Haeckel, referring to the biological aspects alone, introduced the terms *oceanic* and *neritic* for this contrast. Bramlette (1961) stresses the difference between "highly oxidized, (eu)pelagic" versus "not highly oxidized, organic-rich, hemipelagic" deposits. All these concepts are not identical, but are sufficiently congruent so that the diagram can be read accordingly. The contrast between relatively shallow sediments—those covering the flanks of the mid-ocean ridges, plateaus, and the slopes of seamounts—and relatively deep sediments of the ocean-basin provinces, is one of carbonate content. The major facies boundary on the deep-sea floor is the *calcite compensation depth* (CCD). It separates the shallow, carbonate-rich deposits from the deep, carbonate-poor ones, which accumulate in regions where carbonate dissolution balances the incoming supply.

A bathymetric contour map of the CCD (Fig. 3) shows that the average depth of this oceanic facies boundary is near 4.5 km and that its depth range varies considerably, from a maximum (>5.5 km) in the North Atlantic to a minimum (<3.5 km) on the continental margin of the North Pacific. The equatorial CCD anomaly in the Pacific is very obvious, and has been ascribed to high productivity and concomitant high supply of calcareous shells (Arrhenius, 1963). The various factors controlling CCD distributions are discussed by Berger and Winterer (*in* Hsü and Jenkyns, 1974).

Controls on siliceous deposits are in part analogous to the controls on carbonate distributions, especially with regard to supply and to the evolution of skeletons. Oceanic fractionation favors the basin with the old, nutrient-rich deep waters as a site of silica accumulation. The facies boundary between high-fertility and low-fertility deposits ("silica line") is much less well defined than the CCD (or carbonate line), mainly because the scale of variations is different: the carbonate line varies through some 2 km of depth, the silica line along fertility gradients of several thousand km length.

Red clay (q.v.) is a residual deposit remaining after other particles are dissolved, and its ex-

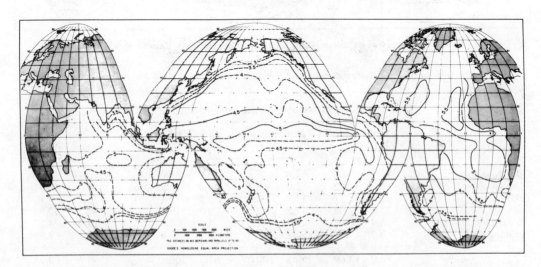

FIGURE 3. Depth (in km) at which the supply of carbonate is compensated by its rate of dissolution—Calcite Compensation Depth, or CCD (from Berger and Winterer, in Hsü and Jenkyns, 1974). Solid line, >20 control samples per 10° square; dashed line, <20 control samples per 10° square.

ceedingly low rate of sedimentation allows the formation of authigenic phases, and presumably, equilibration of certain clay minerals with sea water.

The main patterns of pelagic sediment distributions, then, are a function of fertility and associated supply of biogenic particles, of dissolution processes in the water and on the sea floor, and of the overall supply of clay-sized particles from land and from volcanism. Secondary patterns are apparent within the major sediment types, and changes through time (Pleistocene, pre-Pleistocene) can considerably alter these distributions.

Calcareous Ooze and Carbonate Dissolution

Calcareous oozes consist of coccoliths and foraminifera, and in places, of "pteropods," as well as noncarbonate matter (usually <40%). Coccoliths, tiny ($<10\mu$) skeletal plates—parts of spheroidal calcareous nannoplankton (see Vol. VII: *Coccoliths*)—constitute 30-70% of deep-sea carbonates. Tertiary calcareous oozes have distinctly higher proportions than Pleistocene ones, and the oldest known coccoliths are early Jurassic in age. Planktonic foraminifera, of coarse silt and sand size, are the most conspicuous contributors to biogenous deep-sea sediments. Benthic forams, usually <2% of an assemblage, contain calcitic, agglutinated, tectinous, and aragonitic forms; only the first two are seen in calcareous ooze (see Vol. VII: *Foraminifera, Benthonic*). Planktonic forams, which are calcitic, are essentially restricted to normal marine waters, whereas benthic forams also occur in hypersaline and brackish waters. The oldest pelagic foraminifera are reported from the Jurassic. The rapid evolution of planktonic forams during Cretaceous and their abundance in the Cenozoic suggested to Kuenen (1950) the reason for a shift of carbonate deposition from the shelves to the deep sea. This theory, however, neglects the great importance of coccolithophores in premodern calcareous oozes (Bramlette, 1958; Heath, 1969).

Classification of modern planktonic foraminifera generally follows Parker (1962), who essentially divides these forms into two groups, those having spines in life and those lacking them (see Vol. VII: *Foraminifera, Planktonic*). Roughly, this two-fold grouping also splits the foraminifera into "surface" water (spined) and "subsurface" water (nonspined) forms and into dissolution-susceptible (spined) and dissolution-resistant (nonspined) forms.

Modern coccoliths common in pelagic sediments are illustrated by McIntyre and Bé (1967). Less than 10% of the existing species of calcareous nannoplankton are found in the sediment, unlike foraminifera, where all species found in the water also are represented by their empty shells on the ocean floor. Virtually all calcareous shell production, except of benthic forams, takes place in the upper 200 m of the water column. The exact mode of shell production and the transfer process to the sea floor are poorly understood at the present time.

Rivers are essentially dilute solutions of calcium bicarbonate and silica, providing to the oceans an average of 0.34 g $CaCO_3$ per cm^2 per 1000 yr, under steady-state conditions, corresponding to a sedimentation rate of about 0.5 $cm/10^3$ yr. Because almost 50% of the deep-sea floor is below the CCD (Fig. 3), however, the other half has to accumulate carbonate at an average rate of 1 $cm/10^3$ yr. Within the realm of carbonate sedimentation, variation in this rate is controlled by depth, fertility patterns, circulation, and redeposition processes, with the highest rates in relatively shallow areas, below oceanic divergences, in basins with deep water outflow, and in regions protected from strong bottom currents.

Murray first recognized the general decrease of % $CaCO_3$ (Fig. 4) with depth (Murray and Renard, 1891), ascribing it to dissolution effects, although he also recognized the other two important factors controlling carbonate percentages: rate of carbonate supply and dilution by detrital matter. Ever since, the question of the relationship between % $CaCO_3$ and depth has maintained high visibility in deep-sea sedimentology. Except in the vicinity of CCD, however, where carbonate percentages decrease drastically (Bramlette, 1961), there is no striking correlation between carbonate content with depth of water, nor would one expect such a correlation. When carbonate is measured against an insoluble residue, the loss L through solution is given by: $L = 1 - R_o/R$. R_o is the initial percentage of the undissolved portion and R is the final one. Because R_o is generally quite small in the calcareous oozes (5-10%) a variety of dissolution-rate profiles can reproduce the

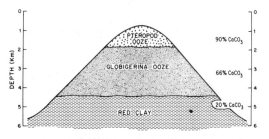

FIGURE 4. Depth control on facies distributions, according to Murray (from Berger, 1974).

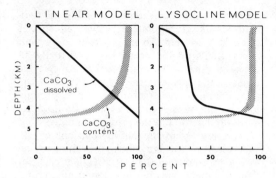

FIGURE 5. Carbonate percentages in core-top samples resulting from assumed shapes of increase of dissolution with depth (from Berger, 1971). Note that carbonate percentage distributions are not very closely related to which particular dissolution profile is adopted.

commonly encountered carbonate-depth profiles (Fig. 5). Thus the % $CaCO_3$ profiles based on core-top samples are poor indicators for dissolution-rate profiles, especially in view of other factors such as dilution and redeposition. Dilution of carbonate is mainly by clay, so that winnowing can introduce percentage changes unrelated to dissolution. The intensity of winnowing and the redeposition of the fine material is related to local topography and water movements, and is not statistically independent of depth. Great caution is required, therefore, in applying this equation to estimate losses from carbonate dissolution, especially because small changes in R_o have large effects on the final estimate.

Differential dissolution of the calcareous fossils appears to offer rather more reliable evidence for the determination of the state of preservation of calcareous ooze. Murray had already observed that dissolution acts selectively, firstly between major groups such as corals, mollusks, forams, and calcareous algae, and secondly between various species of foraminifera. By mapping the proportions of solution-resistant species, it is possible to define a boundary between well-preserved and poorly preserved calcitic assemblages. This boundary, the *lysocline* (q.v.), can be rather sharp or somewhat diffuse. The mapping of faunal and floral preservation states allows refinement of Murray's original dissolution facies scheme, by introducing three additional preservation zones: well-preserved "Globigerina ooze" above the lysocline, intermediate preservation within the lysocline, and a sublysoclinal zone where resistant species are dominant. In addition, there are examples of a facies that contains no whole foraminifera, only fine carbonate particles and nannofossils, usually close to the CCD. Solution loss of fossils may be calculated using the equation $L = 1 - R_o/R$ by entering initial and final % resistant species in R_o and R. These estimates have much less interference from redeposition processes than those based on % clay. However, the basis of assessment, the resistant species, dissolve also so that R is always too small to yield anything but a minimum loss.

The proportions of species of foraminifera

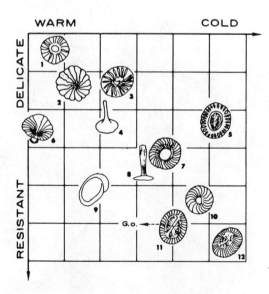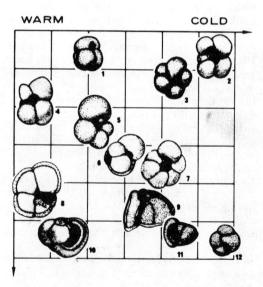

FIGURE 6. Modification of temperature aspects by differential dissolution, as indicated by the position of various species of coccoliths and foraminifera in a Temperature-Resistance diagram (coccoliths drawn from photos by McIntyre and Bé, 1967; foraminifera drawn by Parker, 1962).

and of coccoliths in calcareous ooze initially are controlled by the supply to the sea floor; differential dissolution merely modifies these original assemblages (Fig. 6). What kinds of shells are supplied depends greatly on the temperature and fertility of the overlying waters, a correspondence that has been widely used for paleoclimatic reconstruction. Pteropod ooze, with its substantial proportion of aragonitic shells from heteropod and euthecosome pelagic gastropods, can be expected to contain the best examples of initial composition of foraminiferal assemblages (the same may not be true for coccoliths, because of their high susceptibility to winnowing). In such assemblages, spined forms and delicate thin-walled forms from surface waters are dominant. Of about 250 core samples in the South Pacific studied by Parker and Berger (1971), the three that contained pteropods had 92% (580 m), 80% (2190 m), and 72% (1443 m) spined species, with the balance consisting mainly of *G. glutinata* and a remaining 3% divided between the resistant species. The resistant forms together make up a similarly small fraction in the shallower 13 of the 26 samples containing the extremely solution-susceptible form, *Hastigerina pelagica,* but lacking pteropods. The other "well-preserved" samples have about 10% resistant species.

Much less is known about redeposition processes in deep-sea carbonates than about dissolution effects. It appears that coccolith:foram ratios are very susceptible to differential redeposition processes. Winnowing of sediments—producing lag deposits—and wafting of fine material to other (lower) places—producing "chaff" deposits—have been proposed to explain observed distributions (Bramlette and Bradley, 1942; Berger and von Rad, 1972; Moore et al., 1973). Vertical mixing of sediments extends from the surface to several cm in the sediment (Arrhenius, 1963), although single burrows can span a greater length (see *Bioturbation*). Calcareous turbidites are quite common, involving shallow-water material or deep-sea sediments or both (Van Andel and Komar, 1969; Beall and Fischer, 1969; Berger and von Rad, 1972).

Siliceous Ooze and Oceanic Fertility

Siliceous ooze, by definition, is a pelagic sediment containing >30% siliceous fossils, but <30% carbonate. Depending on the predominant form, such sediments are referred to as diatom or radiolarian ooze. Both diatom and radiolarian skeletons consist of opal, a hydrated form of silica; they range in size from fine silt to fine sand. Oldest known diatoms apparently are of Jurassic age; earliest well-preserved assemblages are found in Late Cretaceous rocks (see *Diatomite*). The oldest radiolarian occurrences reportedly are of Cambrian age.

Of the radiolarian organisms in sea water, only the polycystins leave a record in the sedi-

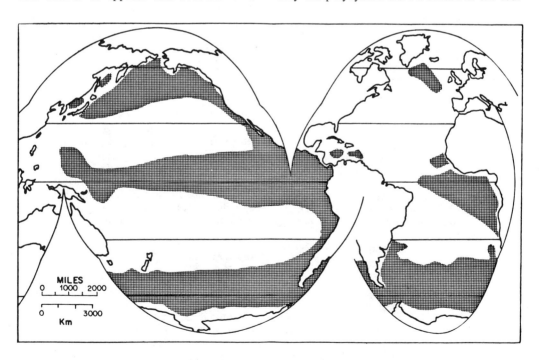

FIGURE 7. Areas where siliceous fossils are abundant (from various sources, see Berger, 1974).

ments; the acantharians and phaeodarians both dissolve easily (Riedel, 1963; see Vol. VII: *Radiolaria*). The term *radiolarian*, then, refers to *polycystins* when used by sedimentologists. Several workers have pointed out that the radiolarians in surface waters tend to be more delicate and slender than those from abyssal waters and than those found on the sea floor. Sponge spicules are another opaline component of radiolarian ooze.

The distribution of siliceous sediments (Fig. 7) is highly correlated with fertility patterns (see *Silica in Sediments*). Diatom oozes are typical for high latitudes, diatom muds for pericontinental regions, and radiolarian oozes for equatorial areas. This overall correspondence between fertility and silica in sediments can be considerably modified by redeposition processes within individual regions (Lisitzin, 1966; Moore et al., 1973). The silica frustules are light and easily transported, and the activity of benthonic animals, which tends to resuspend fine sediment, is especially pronounced in fertile areas. Thus, aided by bottom currents, siliceous frustules tend to accumulate in depressions.

Analogous to other kinds of deposits, the concentration of siliceous fossils in the sediment is a function of three variables. The first variable, production of siliceous shells, attains its maximum in coastal regions, where productivity can exceed by a factor of ten the values in the subtropical gyres. In coastal areas, upwelling currents bring up nutrients to shallow, sunlit waters, leading to the formation of a "silica ring" around the ocean basins. *Silica belts* are latitudinal, produced by oceanic divergences, which are a result of atmospheric circulation. To obtain an estimate of the amount of silica precipitated in the upper waters, one might multiply the measured amount of organic production with the ratio of solid silica to organic matter found in suspension. Most of this silica is in the shells of diatoms, which also presumably make an important contribution to the production measured (see Lisitzin, 1972). A typical fixation rate of about 200 g SiO_2/m^2/yr is suggested, with a range from <100 g (central gyres) to >500 g (Antarctic). Of this fixation, only 1 g/m^2/yr, or 0.5%, can be incorporated into sediments if the river input given by Livingstone (1963) is the only source of silica.

The second factor is the degree of dilution of the siliceous material. Whereas sedimentation rates of terrigenous material are highest around continents where silica supply rates are also high, the dilution effect in pericontinental areas cannot mask the "silica rings" around the oceans. Besides terrigenous diluents, there are the calcareous ones in oceanic regions. There is a distinct negative correlation between silica and calcite, which has been ascribed to opposing chemical requirements for preservation. A similarly opposing trend is indicated for depth relationships, with silica corrosion being greatest in the highly undersaturated upper waters. In addition, silica tends to accumulate by redeposition in deep basins, where carbonate dissolves, minimizing the importance of carbonate as a diluent.

The third factor controlling the abundance of siliceous fossils is the extent of dissolution. Maps showing both abundance and preservation patterns (Goll and Bjørklund, 1971) indicate that the preservation of siliceous shells is rather closely correlated with their relative abundance. A positive correlation between abundance and preservation may be ascribed to an increased supply of easily dissolved diatoms in fertile areas, which will "buffer" interstitial waters for the more robust skeletons, and to an otherwise favorable chemical environment in organic-rich sediments. In general, silicoflagellates and diatoms tend to dissolve well before radiolarians and sponge spicules, and the following dissolution sequence can be established (from least to most resistant): (1) silicoflagellates, (2) diatoms, (3) delicate polycystins, (4) robust polycystins (such as orosphaerids), (5) sponge spicules. Observations strongly indicate that dissolution in situ also plays an all-important role in restricting high abundances of siliceous fossils to the fertile regions. Over large areas of the ocean floor, radiolarians occur only in the uppermost few cm in Quaternary sediments (Riedel and Funnell, 1964), and concentration gradients in interstitial waters (q.v.) and differences in concentration between these waters and the overlying sea water suggest that silica is diffusing out of the sediment (Fanning and Schink, 1969), with siliceous fossils providing the source of the flux (see Fig. 8). Opaline silica that escapes this dissolution process and is buried in the sediments may participate in forming authigenic minerals (palygorskite, sepiolite, zeolite) or chert (q.v.). Pelagic cherts tend to be nodular in carbonates and bedded in clay-rich sediments, while hemipelagic cherts tend to be bedded (see Hsü and Jenkyns, 1974).

"Red Clay" and Clay in General

"Red clay" (q.v.; Murray and Renard, 1891), also called deep-sea clay, pelagic clay, or oxypelite, is uniquely characteristic of the deep-sea environment in that it accumulates only at abyssal depths, far from land, in low-fertility oceanic regions (Fig. 2; for elemental abundances in pelagic clay, see Wedepohl, 1960). The clay mineralogy of red clay is similar to

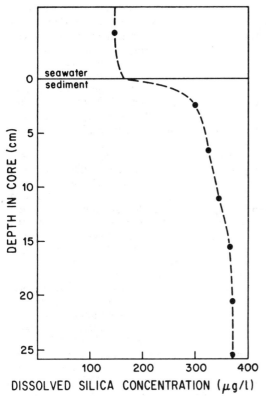
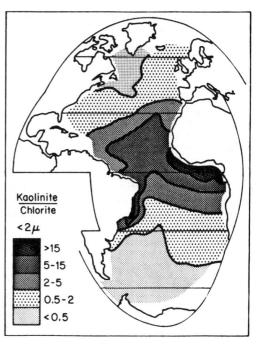

FIGURE 8. Profile of dissolved silica in a box core from the eastern Clipperton Fracture Zone area. (Courtesy T. C. Johnson, S.I.O.)

FIGURE 9. Kaolinite/chlorite ratios in the Atlantic Ocean, demonstrating climatic zonation and hence derivation from continents (adapted from Biscaye, 1965).

that of the other deposits (see *Clays, Deep-Sea*). Other clay-sized material includes ferromanganese hydroxides, quartz, feldspar, pyroxene, palagonite, and minor components.

Possible sources of these particles are (1) continental, transported by either wind or river transport; (2) in-situ decomposition of volcanic material; or (3) reconstitution of degraded clay minerals and authigenic growth. The latter possibility has stimulated much research and speculation, but unequivocal evidence for such reactions is still lacking (see *Clay–Genesis*). Evidence for a continental derivation of most of the clay-sized particles in deep-sea deposits includes distributional patterns of quartz and of associated clay minerals other than smectite (Fig. 9), as well as the isotopic composition and K-Ar age of certain of these minerals. The clay minerals most abundant in red clay are smectites and mica-illite; and this is generally true for the finest fraction of all deep-sea deposits (see *Clay as a Sediment*). The dominance of continental sources in deep-sea clay is a geologically recent phenomenon, as is evident from the study of outcropping Tertiary sediments in the central Pacific (Heath, 1969) and of Cenozoic and Cretaceous sediments recovered by JOIDES.

Ferromanganese Concretions and Trace Elements

Manganese Nodules. Areas such as red clay facies domains, where little or no sediment accumulates, characteristically are covered by *manganese nodules* (ferromanganese concretions) several cm in diameter. Micronodules are also abundant, ranging in size from microscopic specks to about 1 mm. Slabs and crusts occur on outcrops of solid rock. Ferromanganese concretions are laminated with alternating clay-rich and metal-rich layers. They consist of poorly crystallized, hydrated manganese and iron oxides with about equal amounts of Mn and Fe ($\approx 16\%$ each) on the average. Mn percentages and Mn/Fe ratios tend to be higher in the Pacific than in the Atlantic, however, and there is considerable variation within ocean basins. Ferromanganese concretions commonly have high proportions (1‰ to 1%) of Co, Ni, and Cu, as well as considerable amounts of other metals (Horn et al., 1973). Several of these (Ni,

Cu, Zn) tend to co-vary with manganese; others tend to co-vary with iron (Zr, Ti), as shown in Fig. 10.

Growth rates of ferromanganese concretions are generally agreed to be extremely slow, perhaps several cm per m.y., although exact values are under dispute. One unresolved question is why manganese nodules tend to be more abundant on and near the sea floor than deeper within the sediment, and how they keep from being covered in view of their very slow accretion rates.

Origin. Questions about the "origin" of ferromanganese concretions commonly address themselves to any number of the following aspects of ferromanganese distributions: (1) the ultimate or not-so-ultimate sources, e.g., weathering of continental or oceanic rocks or sediments, exhalations from volcanic or hydrothermal vents, or diagenetic mobilization of ferromanganese within oceanic sediments with concomitant diffusion to the interface; (2) the mode of extraction from the source, e.g., differential weathering or hydrothermal reactions; (3) the mode of transport into the ocean, whether delivery is as solute, mineral, or coating on mineral; (4) the question of transport and migration processes within the ocean, i.e., how ferromanganese gets to its place of deposition—as "colloid" or other suspensate, in solution, or locked up in organic matter and plankton shells; (5) the physicochemical and biochemical methods of precipitation and their variation through time; and (6) the processes of redeposition, physical and chemical, including movement of nodules, and scouring of paved sea floor, as well as dissolution and reprecipitation of ferromanganese, which also is commonly counted as a "source." These various aspects are summarized in Fig. 11. The problem of ultimate sources was originally raised by Murray and Renard (1891) and has dominated much of subsequent research.

The case for a *volcanic origin* of manganese nodules rests largely on two arguments: One is an intimate association of Fe-Mn deposits with volcanogenic material, such as zeolitic and palagonitic ash, which commonly forms the nucleus of nodules in large parts of the sea floor. Other materials, however, also form nuclei (whalebone, ice-rafted pebbles, fish teeth, fragments of older nodules, etc.). The second is the regional variations of trace-element composition and of morphologic characteristics, implying that sedimentary processes are unable to provide such variation—an implication that is demonstrably incorrect for concretions in shallow marine areas and in lakes, but is difficult to refute for the deep sea because of lack of information. The case for control of ferromanganese chemistry by sedimentary and diagenetic processes, rather than imprint from a proximate source, rests essentially on the parallelism of distributional and compositional patterns of manganese nodules with the patterns of sediment facies and sediment rates (Price and Calvert, 1970).

Metalliferous sediments on ridge crests, especially the East Pacific Rise, commonly are ascribed to activity such as hydrothermal exhalations. REE patterns and Sr isotopes suggest equilibrium with sea water, while Pb isotopes indicate basaltic derivation (Dymond et al., 1973). Thus, part of the metal enrichment, especially the very high iron content, would seem to derive from hydrothermal activity, while the rest is "scavenged" (Goldberg, 1954) by the freshly formed hydroxides.

Variations in trace-element content of ferromanganese deposits have been cited as evidence both for volcanic and diagenetic-sedimentary processes, following studies by Arrhenius and associates (see Arrhenius and Bonatti, 1965). These authors show that nodules in the continent near high-fertility areas off Peru, California, and Japan have relatively low cobalt values and suggest fractionation of cobalt from manganese by differential mobilization near the sediment surface to account for this observation. Similar arguments apply in part to the fractionation of manganese from iron, and to trace-element distribution in general (see Price

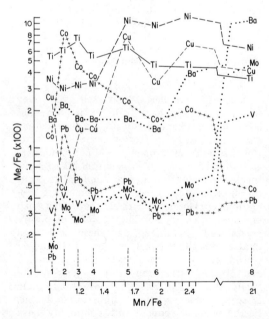

FIGURE 10. Covariance patterns of trace elements in nodules from the central Pacific (data from Cronan in Horn, 1972).

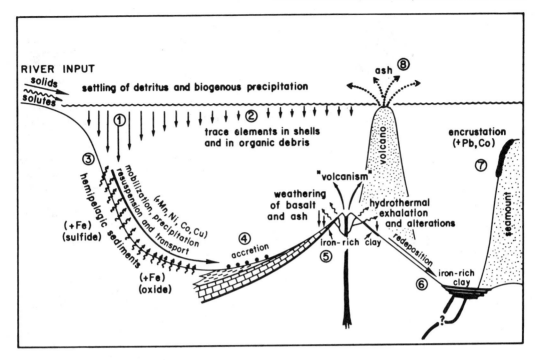

FIGURE 11. Summary of geochemical pathways proposed for manganese and associated trace elements (from Berger, 1974). (1) settling of Mn-coated particles; (2) organic precipitation; (3) partial remobilization, reprecipitation, resuspension, and transport to abyss; (4) accretion by scavenging, bacterial action, or other processes; (5) supply from ridge crest processes; (6) enrichment by redeposition below CCD; (7) encrustation of solid rock exposures; (8) supply of volcanic products containing Mn and associated elements.

and Calvert, 1970). Co-variances between elemental concentrations in manganese nodules (Fig. 10) reflect tendencies of diagenetic mobility, involvement in biological cycles, extent of detrital transport, as well as unidentified factors. The overall absolute abundances of the trace metals (chiefly the transition elements, as well as Cu and Zn) more or less reflect their availability in the crust of the planet, although there appears to be an excess of manganese, molybdenum, and lead over the amount expected from the ratio of elements in continental rocks. Either volcanic input or errors in the abundance data or in the assumptions of the steady-state calculations may account for these discrepancies.

The Deep-Sea Record

Pleistocene Sediments. Systematic investigation of pre-Holocene pelagic sediments began with the *Meteor* Expedition (in Trask, 1939) and subsequently expanded greatly due to improved coring techniques (Bramlette and Bradley, 1942; Arrhenius, 1952; Phleger et al., 1953; Emiliani 1955; Olausson, 1960). Down-core changes in paleontological and chemical aspects of pelagic sediments are interpreted in terms of glacial-interglacial fluctuations in oceanographic conditions. These fluctuations involve changes in (1) surface circulation patterns, with regional changes in the position of oceanic water mass boundaries; (2) bottom water circulation; and (3) sedimentation rates. The frequencies and amplitudes of the fluctuations also are a subject of active research (in Turekian, 1971; Hays and Moore, 1973).

The water mass and temperature association of species of shelled plankton and the distribution of their remains on the sea floor form the basis for paleoclimatic interpretations. E. Haeckel's observation on radiolarians that "the richest development of forms and the greatest number of species occurs between the tropics, whilst the frigid zones (both Arctic and Antarctic) exhibit great masses of individuals but relatively few genera and species" applies equally to the other shelled plankton. Coccoliths, however, do not extend beyond the polar front. This fact, among others, was used by Ruddiman and McIntyre (1973) to support their hypothesis of polar-front migration during deglaciation (Fig. 12).

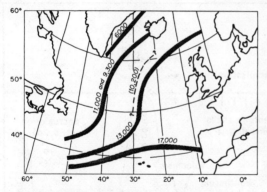

FIGURE 12. Climatic transgression on the sea floor by migration of the polar front in the North Atlantic, during postglacial time (adapted from Ruddiman and McIntyre, 1973).

With respect to the glacial-interglacial temperature amplitude, it has been shown that the differences were near zero in the central tropical parts of the North Atlantic but were significant off parts of Africa, on the basis of foraminiferal assemblages. Paleotemperature estimates are made by noting the proportion of species for which the present temperature preferences are known, and extracting statistically the most likely temperature at which a given mixture of the species was produced. The level of sophistication in these estimates varies (see Imbrie and Kipp, in Turekian, 1971). The limit of resolution is mainly a function of interference by variables unrelated to temperature, such as dissolution, redeposition, and adaptation.

The problem of changing bottom-water circulation has been discussed mainly in connection with the dissolution of deep-sea carbonates, following the demonstration by the *Meteor* scientists that Antarctic Bottom Water prevents accumulation of carbonate in the western South Atlantic. Arrhenius (1952) discovered alternating high and low carbonate stages in sediments from the eastern equatorial Pacific and discussed the possible influence of variation in fertility and of dissolution on the sea floor. It appears that the chief agent in producing the alterations is intensified dissolution during interglacials. High carbonate stages in the equatorial Pacific correspond to low carbonate stages in the Atlantic, where increased dilution of carbonate by terrigenous material during glacial appears to be the overriding factor (Broecker, in Turekian, 1971; Ruddiman, 1971).

Sedimentation rate determinations and studies of frequency spectra of the Pleistocene variations depend on a reliable time scale. The first workable absolute age scale was provided by Emiliani (1955), together with a generalized oxygen isotope curve for the tropical Atlantic. This time scale was stretched by Broecker and Van Donk (1970) to make it congruent with high sea-level stands based on coral dates. Shackleton and Opdyke (1973) accepted this stretched scale on the basis of paleomagnetic stratigraphy and oxygen-isotope stratigraphy in a long core from the western equatorial Pacific (Fig. 13).

Pre-Pleistocene Pelagic Sediments and Plate Stratigraphy. In the last few years, drilling into the deep ocean floor has provided a record of pelagic sedimentation for Cretaceous and Tertiary time. Previously, glimpses of conditions during these periods were obtained from scattered outcrops on the sea floor only. Results of the drilling venture, which is funded by the National Science Foundation and executed by the vessel *Glomar Challenger,* are published by the US Government Printing Office (Washington, D.C.) in the *Initial Reports of the Deep Sea Drilling Project.* The volumes are a rich storehouse of information and provide a ready entry into biostratigraphy of pelagic sediments. Syntheses of sediment patterns are generally provided within the volumes. Some major trends have been summarized by Moore (1972),

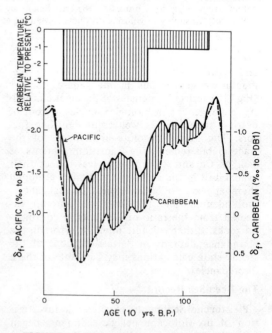

FIGURE 13. Oxygen isotope variations in planktonic foraminifera from the western tropical Pacific and from the Caribbean (after Shackleton and Opdyke, 1973). If the Pacific curve is interpreted as representing the "ice effect," the difference in the Caribbean curve could be due to temperature changes in the Caribbean.

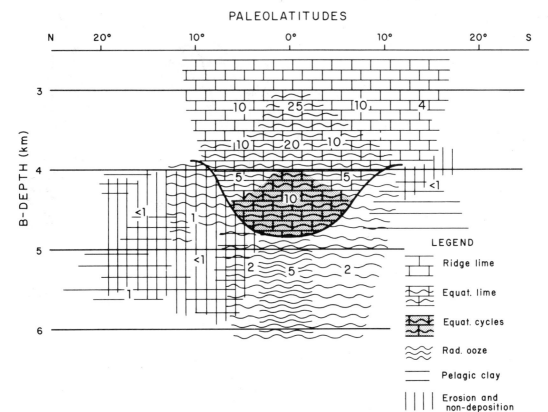

FIGURE 14. Generalized facies distributions in the eastern tropical Pacific, W of the East Pacific Rise, during post-Eocene time (from Berger, 1973).

Berger and von Rad (1972), Winterer (1973), Van Andel and Moore (1974), and Berger and Winterer (in Hsü and Jenkyns, 1974). Bernoulli (1972) has compared Mesozoic pelagic sediments from the North Atlantic and the Mediterranean.

Winterer (1973) has constructed a model of sedimentation in the Equatorial Pacific which integrates sedimentation patterns and plate motions. His model identifies the important factors that must be considered in plate stratigraphy: geographic variation in productivity and dissolution, as well as subsidence and horizontal motion of the plate receiving the deposits. The model predicts sediment thickness distributions that are in remarkable agreement with the results of Ewing et al. (1968), based on seismic profiling.

By reconstructing the depths and geographic positions of sediments recovered by drilling (see *Paleobathymetric Analysis*), it is possible to relate facies distribution to a paleo-oceanographic setting (Fig. 14). In the eastern equatorial Pacific, distributions were similar throughout post-Eocene time, indicating that supply-dissolution feed-back mechanisms tend to stabilize biogenous facies patterns along the equator. Variations do exist, however, and express themselves in fluctuations of CCD levels through time, as well as in considerable changes of sedimentation rates.

Rates during the earliest Tertiary tend to be especially low in the central Pacific, and chert-rich deposits tend to be present from the mid-Eocene and earlier (Fig. 15). Rates during the Late Cretaceous tend to be quite high. As Winterer emphasizes (in Winterer, Ewing, et al., 1973), the explanation of these major features must probably be sought in global phenomena rather than in regional subsidence or horizontal plate motions.

The global nature of major trends is apparent from a comparison of Atlantic and Pacific carbonate deposition patterns with paleotemperature information from the oxygen isotope composition of foraminiferal shells, as suggested by Moore (1972; see Fig. 16). Fluctuations of CCD levels in the extra-equatorial Pacific and in the Atlantic are parallel with paleotemperature variations which are unlikely to be coincidental. Warm, equable climates of

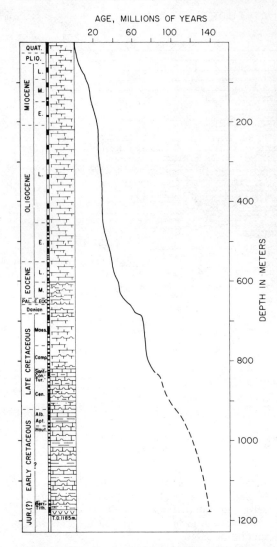

FIGURE 15. Facies succession and sediment accumulation rates in Site 167, Deep Sea Drilling Project, Central Pacific, Magellan Rise (from Winterer, Ewing, et al., 1973).

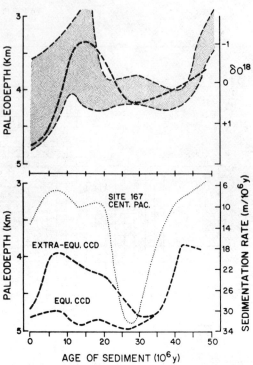

FIGURE 16. Comparison of CCD fluctuation in the Atlantic and Pacific, with oxygen isotopic composition of planktonic foraminifera from the Pacific, and with sedimentation rate variation in Site 167 (from Berger, 1974).

the past appear to have been unfavorable for carbonate deposition in the subtropical and tropical regions of the oceans, with the possible exception of the equatorial area proper. An attractive explanation for this relationship is that warm climates and widespread transgressions occur together and that transgressions lock up much carbonate on the shelves, making it unavailable for the deep sea. Such global interactions between shelf and deep ocean environments were envisaged by Kuenen (1950) and must be clarified before the trends of deep-sea deposition through geologic time can be explained.

WOLFGANG H. BERGER

References

Arrhenius, G., 1952. Sediment cores from the East Pacific, *Repts Swedish Deep Sea Exped.*, 5, 228p.

Arrhenius, G., 1963. Pelagic sediments, in M. N. Hill, ed., *The Sea*, vol. 3. New York: Interscience, 655-727.

Arrhenius, G., and Bonatti, E., 1965. Neptunism and vulcanism in the ocean, *Progress in Oceanography*, 3, 7-22.

Beall, A. O., and Fischer, A. G., 1969. Sedimentology, *Init. Repts. Deep Sea Drilling Project*, 1, 521-593.

Berger, W. H., 1971. Sedimentation of planktonic foraminifera, *Marine Geol.*, 11, 325-358.

Berger, W. H., 1973. Cenozoic sedimentation in the eastern tropical Pacific, *Geol. Soc. Am. Bull.*, 84, 1941-1954.

Berger, W. H., 1974. Deep-sea sedimentation, in C. A. Burke and C. L. Drake, eds., *The Geology of Continental Margins*. New York: Springer, 213-241.

Berger, W. H., 1976. Biogenous deep-sea sediments: production, preservation and interpretation, in J. P. Riley and R. Chester, eds., *Treatise on Chemical Oceanography*, vol. 5. London: Academic Press, 265-388.

Berger, W. H., and von Rad, U., 1972. Cretaceous and Cenozoic sediments from the Atlantic Ocean, *Init. Repts. Deep Sea Drilling Project*, 14, 787-954.

Bernoulli, D., 1972. North Atlantic and Mediterranean

Mesozoic facies: A comparison, *Init. Repts. Deep Sea Drilling Project,* **11**, 801-871.

Biscaye, P. E., 1965. Mineralogy and sedimentation of recent deep-sea clay in the Atlantic Ocean and adjacent seas and oceans, *Geol. Soc. Am. Bull.,* **76**, 803-832.

Bramlette, M. N., 1958. Significance of coccolithophorids in calcium carbonate deposition, *Geol. Soc. Am. Bull.,* **69**, 121-126.

Bramlette, M. N., 1961. Pelagic sediments, *Am Assoc. Adv. Sci. Publ. 67,* 345-366.

Bramlette, M. N., and Bradley, W. H., 1942. Geology and biology of North Atlantic deep-sea cores. Part 1. Lithology and geologic interpretation, *U.S. Geol. Surv. Prof. Pap. 196-A,* 1-34.

Broecker, W. S., and Van Donk, J., 1970. Insolation changes, ice volumes and the O^{18} record in deep-sea cores, *Rev. Geophys. Space Phys.,* **8**, 169-198.

Crerar, D. A., and Barnes, H. L., 1974. Deposition of deep-sea manganese nodules, *Geochim. Cosmochim. Acta,* **38**, 279-300.

Dymond, J.; Corliss, J. B.; Heath, G. R.; Field, C. W.; Dasch, E. J.; and Veeh, H. H., 1973. Origin of metalliferous sediments from the Pacific Ocean, *Geol. Soc. Am. Bull.,* **84**, 3355-3372.

Emiliani, C., 1955. Pleistocene temperatures, *J. Geol.,* **63**, 538-578.

Ewing, J.; Ewing, M.; Aitken, T.; and Ludwig, W. J., 1968. North Pacific sediment layers measured by seismic profiling, *Geophys. Monogr.,* **12**, 147-173.

Fanning, K. A., and Schink, D. R., 1969. Interaction of marine sediments with dissolved silica, *Limnol. Oceanogr.,* **14**, 59-68.

Goldberg, E. D., 1954. Marine geochemistry, 1. Chemical scavengers of the sea, *J. Geol.,* **62**, 249-265.

Goll, R. M., and Bjørklund, K. R., 1971. Radiolaria in surface sediments of the North Atlantic Ocean, *Micropaleontology,* **17**, 434-454.

Hays, J. D., and Moore, T. C., eds., 1973. CLIMAP program, *Quat. Research,* **3**, 154p.

Heath, G. R., 1969. Mineralogy of Cenozoic deep-sea sediments from the Equatorial Pacific Ocean, *Geol. Soc. Am Bull.,* **80**, 1997-2018.

Horn, D. R., ed., 1972. *Papers from a Conference on Ferromanganese Deposits on the Ocean Floor, New York, January, 1972.* Washington, D.C.: Natl. Sci. Found. Off. Internt. Decade Ocean Expl., 293p.

Horn, D. R.; Delach, M. N. and Horn, B. M., 1973. Metal content of ferromanganese deposits of the oceans, *NSF GX 33616, Intern. Decade Ocean Explor., Tech. Rept. 3,* 51p.

Hsü, K. J., and Jenkyns, H. C., eds., 1974. Pelagic sediments on land and under the sea, *Intern. Assoc. Sed. Spec. Publ. 1,* 447p.

Kennett, J. P., and Watkins, N. D., 1975. Deep-sea erosion and manganese nodule development in the Southeast Indian Ocean, *Science,* **188**, 1011-1013.

Kuenen, P. H., 1950. *Marine Geology.* New York: Wiley, 568p.

Lamont-Doherty Geological Observatory, Palisades, N.Y., 1973. Inter-university program of research on ferromanganese deposits of the ocean floor, *Rept. Phase 1,* 363p.

Lisitzin, A., 1966. Basic relationships in distribution of modern siliceous sediments and their connection with climatic zonation (in Russian), in *Geokhimia Kremnezema.* Moscow: Nauka Press, 90-191. Transl., 1967, in *Internat. Geol. Rev.,* 9(5), 631-652; (6), 842-865; (7), 980-1004; (8), 1114-1130.

Lisitzin, A., 1972. Sedimentation in the world ocean, *Soc. Econ. Paleont. Mineral. Spec. Publ. 17,* 218p.

Livingstone, D. A. 1963. Chemical composition of rivers and lakes, *U.S. Geol. Surv. Prof. Pap. 440-G,* 64p.

McIntyre, A., and Bé, A. W. H., 1967. Modern coccolithophoridae of the Atlantic Ocean—I. Placoliths and cyrtoliths, *Deep-Sea Research,* **14**, 561-597.

Moore, T. C., 1972. DSDP: Success, failures, proposals, *Geotimes,* **17**, 27-31.

Moore, T. C., Jr.; Heath, G. R.; and Kowsmann, R. O., 1973. Biogenic sediments of the Panama Basin, *J. Geol.,* **81**, 458-472.

Murray, J., and Renard, A. F., 1891. Report on deep-sea deposits based on the specimens collected during the voyage of H.M.S. Challenger in the years 1872 to 1876, in *Rep. Voyage "Challenger."* London: Longmans, 525p. (Reprinted in 1965 by Johnson Reprint Co., New York).

Olausson, E., 1960. Sediment cores from the West Pacific; Sediment cores from the North Atlantic Ocean; Sediment cores from the West Pacific, *Repts. Swedish Deep Sea Exped.,* **6**, 161-214; **7**, 227-286; **9**, 53-88.

Parker, F. L., 1962. Planktonic foraminiferal species in Pacific sediments, *Micropaleontology,* **8**, 219-254.

Parker, F. L., and Berger, W. H., 1971. Faunal and solution patterns of planktonic foraminifera in surface sediments of the South Pacific, *Deep-Sea Research,* **18**, 73-107.

Phleger, F. B.; Parker, F. L.; and Peirson, J. F., 1953. North Atlantic foraminifera, *Repts. Swedish Deep Sea Exped.,* **7**, 1-122.

Price, N. B., and Calvert, S. E., 1970. Compositional variation in Pacific Ocean ferromanganese nodules and its relationship to sediment accumulation rates, *Marine Geol.,* **9**, 145-171.

Riedel, W. R., 1963. The preserved record: Paleontology of pelagic sediments, in M. N. Hill, ed., *The Sea,* vol. 3. New York: Interscience, 866-887.

Riedel, W. R., and Funnell, B. M., 1964. Tertiary sediment cores and microfossils from the Pacific Ocean floor, *Quart. J. Geol. Soc. Lond.,* **120**, 305-368.

Ruddiman, W. F., 1971. Pleistocene sedimentation in the Equatorial Atlantic: Stratigraphy and faunal paleoclimatology, *Geol. Soc. Am. Bull.,* **82**, 238-302.

Ruddiman, W. F., and McIntyre, A., 1973. Time-transgressive deglacial retreat of polar waters from the North Atlantic, *Quaternary Research,* **3**, 117-130.

Shackleton, N. J., and Opdyke, N. D., 1973. Oxygen isotope and paleomagnetic stratigraphy of equatorial Pacific core V28-238; oxygen isotope temperatures and ice volumes on a 10^5 year and 10^6 year scale, *Quaternary Research,* **3**, 39-55.

Trask, P. D., ed., 1939. *Recent Marine Sediments, A Symposium.* Tulsa: Am. Assoc. Petrol. Geologists, 736p.

Turekian, K. K., ed., 1971. *The Late Cenozoic Glacial Ages.* New Haven: Yale Univ. Press, 606p.

Van Andel, T. H., and Komar, P. D., 1969. Ponded sediments of the Mid-Atlantic Ridge between $22°$ and $23°$ North latitude, *Geol. Soc. Am. Bull.,* **80**, 1163-1190.

Van Andel, T. H., and Moore, T. C., Jr., 1974. Cenozoic calcium carbonate distribution and calcite compensation depth in the central Equatorial Pacific Ocean, *Geology,* **2,** 87-92.

Wedepohl, K. H., 1960. Spurenanalytische Untersuchungen an Tiefseetonen aus dem Atlantik, *Geochim. Cosmochim. Acta,* **18,** 200-231.

Winterer, E. L., 1973. Sedimentary facies and plate tectonics of Equatorial Pacific, *Bull. Am. Assoc. Petrol. Geologists,* **57,** 265-282.

Winterer, E. L.; Ewing, J. I.; and 7 others, 1973. *Init. Repts. Deep Sea Drilling Project,* **17,** 930p.

Cross-references: *Abyssal Sedimentary Environments; Abyssal Sedimentation; Authigenesis; Bioturbation; Clay as a Sediment; Clays, Deep-Sea; Continental-Rise Sediments; Diatomite; Eolian Dust in Marine Sediments; Glacial Marine Sediments; Hadal Sedimentation; Interstitial Water in Sediments; Lysocline; Marine Sediments; Nepheloid Sediments and Nephelometry; Paleobathymetric Analysis; Red Clay; Sedimentation Rates, Deep-Sea; Submarine Fan (Cone) Deposits; Turbidity-Current Sedimentation.* Vol. I: *Calcium Carbonate Compensation Depth; Manganese Nodules; Pelagic Distribution.*

PENECONTEMPORANEOUS DEFORMATION OF SEDIMENTS

Penecontemporaneous deformation of sediments includes small- or large-scale modifications of one bed or a few successive layers shortly after the deposition of these beds. These disturbances may occur in the surficial layer or layers of the deposit (*open-cast movement*); or they may be formed somewhat later as internal deformation after deposition of subsequent strata (*closed-cast movement*), but always before lithification. The whole sequence may, of course, be affected later by tectonic folding.

Most commonly, penecontemporaneous deformation appears as *intraformational folds, load casts,* and *pseudo-nodules,* but many other less important forms are known. Intraformational folds are more or less folded beds bounded above and below by undeformed strata (see *Intraformational Disturbances*). In penecontemporaneous folds, the deformation is generally the result of lateral forces. Where lateral forces are predominant, such as in subaqueous slides, or where dragging by ice or by sliding coherent beds occurs, continued movement may bring about complete destruction of the beds to produce sedimentary breccias (q.v.).

Vertical forces may produce folds but more often cause deformation in the form of *load casts,* which are concave-upward disturbances resulting from differential gravitational settling, and *pseudo-nodules,* which are partial or complete ball-like forms at least initiated by the same force (see *Pillow Structures*).

At first considered to be of exceptional occurrence, penecontemporaneous deformations have since been found to be rather common. Their study is useful because some are continuous enough to use for stratigraphic correlation, many are good top and bottom criteria (q.v.), and most provide useful information concerning the conditions of sedimentation, such as distance from shore, local emergence, and paleoslope (Potter and Pettijohn, 1977). Penecontemporaneous deformations such as pseudo-nodules and folds cut by a local unconformity are commonly useful as geopetal criteria in folded formations (Shrock, 1948).

Intraformational Folds

Not all intraformational disturbances (q.v.) are the result of penecontemporaneous deformation, but this origin is apparent when the folds appear among subhorizontal and nonindurated beds. In such cases, and in similar examples in ancient beds, tectonically folded or not, several explanations have been given, the main ones being: (1) lateral push or drag of ice, (2) submarine sliding, and (3) drag forces from above by heavily loaded currents.

Ice-Shove Deformation. Ice may act in several ways to deform the beds beneath or beside it (Moran, 1971). Flows or icebergs drifted ashore may locally deform bottom sediments; this explanation was one of the first suggested for local deformation in ancient beds. Other local disturbances may be caused by winter ice which forms at the surface of a lake, and whose expansion pushes laterally, producing *ice ramparts* (Buckley, 1900).

More efficient and less localized, as a rule, is the work of glaciers traveling over a valley-fill or nonindurated sediments. In the Netherlands, such conditions brought about folding and faulting on a lateral scale of about 100 m, creating what is called *push moraine.* In the case of glacial readvance, the ice may also push and deform its own end moraine, and its peripheral glaciofluviatile and glaciolacustrine beds.

Deformation related to glacial action may also occur during glacial melting. Material may be scraped up and carried forward and upward by the glacier, then contorted and faulted as a result of flow movements of the ice. Sediments laid down by meltwater on the glacier may be disturbed when collapsing because of unequal melting of the ice. More frequent, however, is the deformation of kame deposits, formed in marginal meltwater lakes, and esker sediments, deposited in glacial streams. These ice-contact deposits may slide or slump when the ice walls holding them melt away.

Gravitational Sliding or Slumping. Gravity slides (q.v.) and slumps may occur on land, as illustrated by the above examples, but subaqueous slides and slumps are more likely to be preserved in the geologic record. Early studies of subaqueous slides were done by Heim (1908) in the Swiss lakes, and sliding was inferred in studies by Arkangelsky (1930) in the Black Sea. They were shown to occur on slopes as low as 2-3° and to slide either along a single basal plane or along a series of several beds, each moving over the next one, the deformations thus increasing upward. They are often cut by erosion before deposition of the next bed, producing a discontinuity. An early discussion of the mechanics of subaqueous slumps was presented by Moore (1961). Fig. 1 is a typical example of intraformational folds in ancient beds, attributed to subaqueous sliding; in addition to the discontinuity, the strong upward thinning of the sandstone beds in the two synclines to the left are proof that they were soft, i.e., sands, at the time of folding. Criteria of this kind will serve, in ancient formations, to distinguish intraformational folds from tectonic disturbances. It is much more difficult to distinguish between the various types of intraformational folds. For instance, dragging by ice is also likely to produce folds that increase upward. When ice action can be discarded by paleoenvironmental studies, however, subaqueous sliding is generally the best alternative.

Current Drag. A slight increase in the velocity of a current may disturb material just deposited. This phenomenon may explain cross-bedding steeper than the angle of repose (Fig. 2) and overturned cross-bedding (Allen and Banks, 1972).

A similar dragging effect has been suggested to explain a special kind of intraformational folding called *convolute bedding* (Ten Haaf,

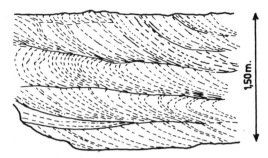

FIGURE 2. Laminae of cross-bedding overturned by current drag. Lower Tertiary of Belgium (after Macar, 1958).

1956; Fig. 3). Commonly occurring at the top of a series of thinly laminated, fine-grained sediments, the thickness of the convolutions is rarely more than about 30 cm. Convolute laminae are generally associated with turbidity currents (q.v.; Sanders, 1960), wherein the heavy load of such currents would cause the deformation. Convolute bedding is not restricted to turbidites, however (Einsele, 1963). A grain size which allows the sediments to change easily from a cohesionless state to a more cohesive one (coarse silt–very fine sand) and some form of rippling which would initiate the folds, are believed to be important factors in the formation of these folds by currents (Sanders, 1960). Recent studies have suggested that some convolute laminae may be water escape structures (q.v.; Lowe, 1975).

Sliding Under Load. In contrast to the above examples, other types of intraformational folds may be produced after more sediment has covered the layers to be folded. Movement under load of superincumbent beds which have generated excess pore-water pressure is the most significant process. There is no discontinuity above the folds; and the latter, as a rule, all pitch in the same direction, a result of the drag-

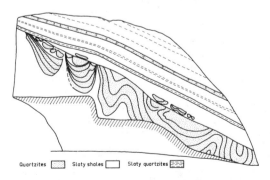

FIGURE 1. Intraformational folds due to subaquatic sliding. Heiderscheidergrund, Grand Duchy of Luxembourg. Length: about 15 m (after Macar and Antun, 1950).

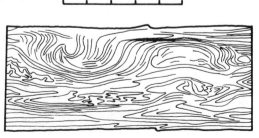

FIGURE 3. Convolute bedding. Martinsburg shale, Stroudsburg, Pennsylvania (after Miller, 1922).

FIGURE 4. Probable sliding under load; Pleistocene, Dead Sea area (Israel).

FIGURE 6. Pseudo-nodule with plane upper face; Upper Famennian, Belgium (after Macar, 1948).

ging effects of the upper beds (Fig. 4). In ancient formations affected by regional folding, intraformational folds may be mistaken for tectonic features such as drag folds or disharmonic folds, which may be distinguished from the penecontemporaneous features by slickensides, deformed fossils or nodules, and schistosity. Also, the tectonic folds should occur within the less competent beds of a sequence.

A rather special case of intraformational folding occurs when evaporite deposits undergo changes of volume by hydration or recrystallization, producing so-called *enterolithic structure*, showing very intricate folds and often disposed en echelon from one bed to the next.

Load Casts and Pseudo-Nodules

The primary movement in the formation of these penecontemporaneous deformational structures is downward and differential, occurring when water-saturated (subaqueous) sand masses sink as rounded masses into underlying, more clayey material. If the sinking is moderate, the result may be a succession of smooth curves with upward concavity on the base of a sandstone bed (resting on shale). More rarely (Kuenen, 1948), greater sinking produces rounded pockets. In both these examples of *load casts*, the bedding inside the pocket, when preserved, is parallel to the pocket walls (Fig. 5).

When rounded pockets are formed, they often sink until they are cut off from the sand body, forming isolated sand masses called *pseudo-nodules* (Macar, 1948), or *pillow structures* (q.v.). These bodies often stick to the bottom of the parent bed and present a rounded lower face, to which the internal bedding is parallel, and a plane upper face (Fig. 6). Other pseudo-nodules are entirely rounded and are generally located farther down in the underlying bed. They again show internal bedding parallel to the mass border, except at one place where they abut sharply against it, and sometimes near the center, where they may be intensely folded due to lack of space (Fig. 7). Pseudo-nodules were once thought to form mostly during storms that removed small masses from sand beds located nearshore and rolled these masses onto offshore argillaceous mud into which the nodules sank; this "storm roller" theory was largely discounted when turbidity currents were recognized as a mechanism for transporting coarse-grained material to deeper waters.

Other less common cases of penecontemporaneous deformation are found in calcareous or argillaceous oozes flanking coral reefs, where local beds dipping away from the reef are produced in the oozes as a result of differential compaction. Other examples are formed in periglacial climates, where repeated freeze-and-thaw action causes irregular, small-scale deformation called *cryoturbation;* a variety with more or less regular folds is termed *involution*. Still other forms may be caused by stones falling from icebergs or cliffs or ejected from volcanoes which may fall and penetrate beds of unconsolidated material, disturbing them locally and, if these stones are numerous enough, possibly even producing small folds. Many other, mostly small-scale deformations could be mentioned, such as small mounds caused by the rise of gas bubbles, traces of rain drops or hailstones, and collapsing of beds by underground solution, to cite a few.

PAUL J. F. MACAR

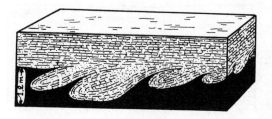

FIGURE 5. Sandstone pockets in a coal seam adhering to the roof; Ruhr area (after Kuckuk, 1936).

References

Allen, J. R. L., and Banks, N., 1972. An interpretation and analysis of recumbent-folded deformed cross-bedding, *Sedimentology,* **19,** 257–283.

FIGURE 7. Pseudo-nodule, entirely rounded; Lower Devonian, Grand Duchy of Luxembourg (after Macar and Antun, 1950).

Arkhangelsky, A. D., 1930. Slides of sediments on the Black Sea bottom and the importance of this phenomenon for geology, *Bull. Soc. Naturalists, Moscow (Sc. Géol.),* **38**, 38-82.

Beets, C., 1946. Miocene marine disturbances of strata in North Italy, *J. Geol.,* **54**, 229-245.

Buckley, E. R., 1900. Ice ramparts, *Trans. Wisc. Acad. Sci. Arts Lett.,* **13**(1), 141-161.

Cope, F. W., 1946. Intraformational contorted rocks in the Upper Carboniferous of the Southern Pennines, *Quart. J. Geol. Soc. London,* **101**, 139-176.

Einsele, G., 1963. "Convolute bedding," und änliche Sedimentstrukturen im rheinischen Oberdevon und anderen Abhagerungen, *Neues Jahrb. Geol. Paläont., Abhandl.* **116**, 162-198.

Gulinck, M., 1948. Sur des phénomènes de glissement sous-aquatique et quelques structures particulières dans les sables landeniens, *Bull. Soc. Belge Géol.,* **57**, 12-30.

Hahn, F. F., 1913. Untermeerische Gleitung bei Trenton Falls (nord-America) und ihr Verhältnis zuähnlichen Störungbildern, *Neues Jahrb. Min. Geol. Paläont.,* **36**, 1-41.

Heim, A., 1908. Ueber rezente und fossile subaquatische Rutschungen und deren lithologische Bedeutung, *Neues Jahrb. Min. Stuttgart,* **1908**, 137-157.

Kindle, E. M., 1917. Deformation of unconsolidated beds in Nova Scotia and Southern Idaho, *Geol. Soc. Am. Bull.,* **28**, 323-334

Ksiazkiewicz, M., 1950. Uwarstwienie Splywowe we Fliszu Karpackim (Slip-bedding in the Carpathian Flysch), *Ann. Soc. Géol. Pologne,* **19**, 493-504 (English summary).

Kuckuk, P., 1936. Flöszumregelmässigkeiten nichttektonischer Art im Ruhrbezirk und ihre Bedeutung fur den Betrieb unter Tage, *Glückauf,* **72**, 1021-1029.

Kuenen, P. H., 1948. Slumping in the Carboniferous rocks of Pembrokeshire, *Quart. J. Geol. Soc. London,* **104**, 365-385.

Lamont, A., 1938. Contemporaneous slumping and other problems at Bray Series, Ordovician and Lower Carboniferous horizons, in County Dublin, *Proc. Roy. Irish Acad.,* **45**(b), 1-32.

Lowe, D. R., 1975. Water escape structures in coarse-grained sediments, *Sedimentology,* **22**, 157-204.

Macar, P., 1948. Les pseudo-nodules du Famennien et leur origine, *Ann. Soc. Géol. Belgique,* **72**, 47-74.

Macar, P., 1958. Les déformations penecontemporaines de la sédimentation, *Rév. Quest. Sci.,* sér. 5, **19**, 5-33.

Macar, P., 1973. Structures sédimentaires diverses dans le caticule du bord sud du massif de Stavelot, *Ann. Soc. Géol. Belgique,* **96**, 133-155.

Macar, P., and Antun, P., 1950. Pseudo-nodules et glissement sous-aquatique dans l'Emsien inférieur de l'Oesling (Grande-Duché de Luxembourg), *Ann. Soc. Géol. Belgique,* **73**, B121-150.

Miller, W. J., 1922. Intraformational corrugated rocks, *J. Geol.,* **30**, 587-610.

Moore, D. G., 1961. Submarine slumps, *J. Sed. Petrology,* **31**, 343-357.

Moran, S. R., 1971. Glaciotectonic structures in drift, in Goldthwait, R. P., ed., *Till: A Symposium.* Columbus: Ohio State Univ. Press, 127-148.

Potter, P. E., and Pettijohn, F. J., 1977. *Paleocurrents and Basin Analysis,* 2nd. ed. New York: Springer-Verlag, 460p.

Sanders, J. E., 1960. Origin of convoluted laminae, *Geological Mag.,* **97**, 409-421.

Shrock, R. R., 1948. *Sequence in Layered Rocks.* New York: McGraw-Hill, 507p.

Ten Haaf, E., 1956. Significance of convolute lamination, *Geol. Mijnbouw,* **18**, 188-194.

Thomson, A., 1973. Soft-sediment faults in the Tesnus Formation and their relationship to paleoslope, *J. Sed. Petrology,* **43**, 525-528.

Cross-references: *Breccias, Sedimentary; Compaction in Sediments; Gases in Sediments; Gravity Flows;*

Gravity-Slide Deposits; Intraformational Disturbances; Pillow Structures; Raindrop Imprint; Slump Bedding; Till and Tillite; Top and Bottom Criteria; Turbidites; Water-Escape Structures.

PERSIAN GULF SEDIMENTOLOGY

The general geology and the Holocene sedimentary environments of the Persian Gulf (Fig. 1) have been studied in sufficient detail for the area to be fairly frequently referred to for a number of interrelated reasons: (1) the very obvious structural control of the morphological features such as shorelines, bathymetry within the Gulf, and character of the hinterlands; (2) the intense aridity of the region and its effect on marine faunal distributions, carbonate sedimentation, evaporite sabkha diagenesis, and eolian sediment distribution; (3) the well-described relict nature of many of the deeper water sediments deposited during the late Pleistocene/Holocene transgression; (4) the shelf and littoral carbonate sediment province of the Arabian shore; and, most important of all, (5) the coastal and continental sabkha environments and associated eolian sediments of the Arabian shoreline and hinterland.

The first significant study of the sediments of the Persian Gulf was published by Emery in 1956. A general description of the Persian Gulf, its oceanographic characteristics and generalized sedimentary facies has been published by Evans (1966). Purser (1973) edited a collection of papers on the Persian Gulf.

Controls and Climate

Structural Controls on Morphology and Holocene Sedimentation. The Persian Gulf is a tectonic basin of mainly Pliocene and Pleistocene age (Kassler in Purser, 1973). Most of the topographic features within the basin itself and of its hinterlands are directly controlled by these rather youthful structural elements, only slightly modified by erosion or sedimentation. The Gulf is roughly 1000 km in length and 100–300 km in width; water depths are everywhere <100 m, except locally at the Strait of Hormuz (Fig. 1). The deep-water axis lies close against the Iranian (Persian) shore, the basin in cross section being markedly asymmetrical. This asymmetry reflects the very different structural characteris-

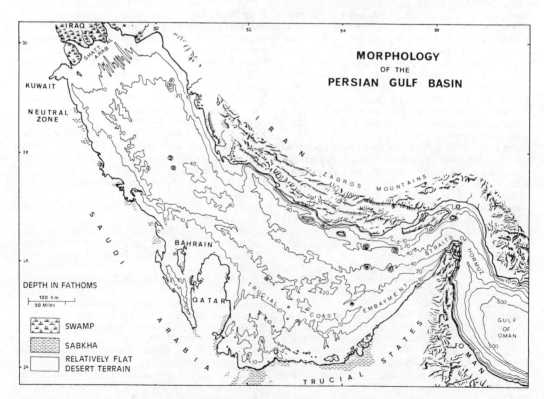

FIGURE 1. Map of the Persian Gulf and surrounding hinterland areas, showing contrasting morphologies of Iranian and Arabian shores and basin bathymetry (from Purser and Seibold in Purser, 1973). Majority of islands and circular shoals are diapiric in origin (1 fathom equals approximately 2 m).

tics of the Iranian and Arabian hinterlands of the basin. To the NE are the intensely folded ranges of the Zagros Mountains, folded on NW-SE or E-W trends. To the SW of the basin lie the low, extremely gently folded rocks of the Arabian craton. The Arabian fold trends are NE-SW and N-S and some, in spite of being very gentle features, have exerted considerable control on sedimentation patterns since Mesozoic times. Many diapiric salt structures are present throughout the region, both onshore and offshore. These structural highs have exerted a considerable control over sedimentary facies distributions (Purser, 1973).

The longitudinal axis of the Persian Gulf is a southeasterly extension of the Tigris-Euphrates valley, the whole being a Zagros feature of Plio-Pleistocene age (Kassler in Purser, 1973). The structural and temporal setting of the Persian Gulf basin would define it as a marine molasse basin; to the NE, within the Zagros fold belt, are various closed and nearly closed intermontane basins which are in reality nonmarine or continental molasse basins.

Regional Climate and Its Sedimentary Expression. The Persian Gulf is arid by any definition, and, being largely landlocked, suffers a continental climate with marked seasonality. Regional winds are westerly at the head of the Gulf, swinging to NNW in the southeastern Gulf. Gales but not hurricanes are experienced by the area, and thus storm deposits are poorly developed in comparison, for example, to Caribbean areas. The constancy of the regional wind and wave direction is responsible for the formation of extensive sand bodies to the SE of islands and shoals throughout much of the Gulf (Purser, 1973). Orientation of lagoon barriers and many other sedimentary features along the Arabian shore are related to the wave approach direction (Purser and Evans in Purser, 1973). Sand dune orientation, source areas for sand, and sand movement directions are also controlled by the regional wind direction.

Rainfall is very slight and infrequent. Little vegetative stabilization of land surfaces can occur, and thus wind activity dominates the hinterlands. The occasional flash floods may also transport vast amounts of unstabilized surface materials. Relative humidity in the Gulf and along shoreline areas is generally rather high (>60%) but may fall inland to very low values. Nighttime dew is often heavy and is the main source of water for many desert plants. Air temperatures reach 45–50°C in summer and fall to 10°C or even lower in winter; sea-water temperatures are less extreme, 20–32°C in open waters, but in nearshore areas ranging from 15°C or even less to values in excess of 35°C. The low winter temperatures are probably ecologically limiting for reef corals in many places.

The combination of fairly intense wind movement, lack of precipitation, and high temperatures leads to the development of hypersaline conditions through the Gulf, but especially along the Arabian shore. This condition is reflected, for example, in a decrease in diversity of reef coral faunas (Kinsman, 1964) and in various other faunal elements (see Purser, 1973). Browsing gastropods and other marine metazoans are unable to tolerate the high salinities which develop in many intertidal areas, and these exposed sediment surfaces are thus colonized by blue-green algal mats (Kendall and Skipwith, 1968). The evaporite environment of the sabkha is also a response to the climatic milieu of the area.

Sedimentation

Relict Transgressive Sediments. During each of the major glacial maxima of the Pleistocene, the floor of the Persian Gulf would have been mostly or entirely exposed; the Gulf would have been a valley through which the Shatt al Arab flowed en route to the sea in the Gulf of Oman. At various times during the postglacial marine transgression, a suite of shallow-water marine sediments, such as oolite sands, coral reefs, coquinas, strandline dunes, and aragonite muds, was deposited on the floor of the Gulf, but these sediments were later drowned as sea level continued to rise (Sarnthein, 1972; further papers and references in Purser, 1973). Many of these relict sediments have been subsequently intermixed with later, deeper water sediments. Along the Iranian shore, significant deposits of truly Holocene sediments are apparently limited to a very narrow nearshore belt. Transgressive sediments are also recorded from several places along the Arabian shore (Shinn in Purser, 1973) and even include thin transgressive stromatolite units.

Holocene Sedimentation Along the Arabian Shore. Studies of Holocene carbonate sedimentation along the Arabian coast of the Persian Gulf have established it as a classical example of an attached carbonate province, i.e., one attached to a continental hinterland, in contrast to an oceanic platform such as the Bahama Banks (q.v.). The shelf carbonate sediments of the Trucial Coast Embayment (Fig. 1) may pass to a high-energy beach facies and inland to a continental eolian facies, as along the northeastern Trucial Coast or parts of the Qatar and Saudi Arabian shorelines. More typically, however, lagoons and lagoon barriers fringe the shore (Fig. 2), and these are best known along the Trucial Coast. Lagoon barriers of skeletal

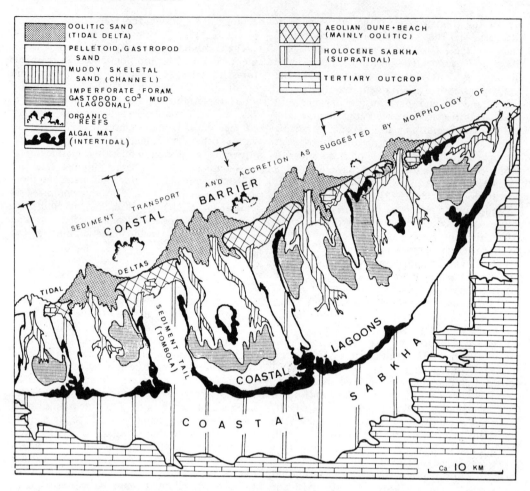

FIGURE 2. Schematic distribution of barrier and lagoonal facies, Abu Dhabi area of Trucial Coast (from Purser and Evans in Purser, 1973). Note the oolite, reef and dune sand facies of the barrier in contrast to the skeletal, pelletoid sands and aragonite muds of the lagoons. Intertidal stromatolite facies is well developed. Sedimentary offlap has developed extensive supratidal surfaces which now comprise the evaporitic sabkha environment.

sands, coral reefs or oolite sand have been developed. The oolite deltas are spectacular and are typically associated with the mouths of tidal channels, between barrier islands. Reefs are normally developed immediately seaward of the barrier island face, but some reefs also occur on the lagoon side of the barrier. In the lagoon, gastropod-rich, pelleted aragonite sands and muds are the typical sediments, with imperforate foraminifera abundant in some areas. The aragonite muds and pelletal sands are the products of nonskeletal, possibly "inorganic" precipitation of aragonite from sea water (Kinsman, 1969). Fringing the lagoon along sheltered shorelines are widespread algal mats which form a final stromatolitic infilling facies to the lagoons (Kendall and Skipwith, 1968). Continued sedimentation over the past 4000–5000 yr has drastically reduced the size of many of the lagoons, and the resulting supratidal surfaces, underlain by marine sediments, are in some places 10 km in width.

In some areas along the Saudi Arabian and Qatar coasts, eolian sand dunes are migrating into the marine environment forming thick nearshore wedges of quartz sand (Shinn in Purser 1973).

Sabkha Environments. The geological significance of the sabkha environment was first realized from studies in the Persian Gulf, and sabkhas have since been widely recognized in other areas of Holocene sedimentation, both in carbonate and terrigenous sediment provinces. Sabkha facies rocks are now widely reported from many ancient sequences (e.g., the Permian reef complex of Texas and New Mexico; Scholle and Kinsman, 1974). The sabkha concept also revolutionized much of our thought on evaporite formation and on the process of dolomitization (for more detail see *Sabkha Sedimentology*).

DAVID J. J. KINSMAN

References

Diester-Haass, L., 1973. Holocene climate in the Persian Gulf as deduced from grain-size and pteropod distribution, *Marine Geol.,* **14,** 207-223.

Emery, K. O., 1956. Sediments and water of the Persian Gulf, *Bull. Am. Assoc. Petrol. Geologists,* **40,** 2354-2383.

Evans, G., 1966. The recent sedimentary facies of the Persian Gulf region, *Roy. Soc. London, Phil. Trans.,* ser. A., **259,** 291-298.

Kendall, C. G. C., and Skipwith, P. A. E., 1968. Recent algal mats of the Persian Gulf lagoon, *J. Sed. Petrology,* **38,** 1040-1058.

Kinsman, D. J. J., 1964. Reef coral tolerance of high temperature and salinities, *Nature,* **202,** 1280-1282.

Kinsman, D. J. J., 1969. Modes of formation, sedimentary associations, and diagnostic features of shallow-water and supratidal evaporites, *Bull. Am. Assoc. Petrol. Geologists,* **53,** 830-840.

Purser, B. H., ed., 1973. *The Persian Gulf.* Berlin: Springer-Verlag, 471p.

Sarnthein, M., 1972. Sediments and history of the post-glacial transgression in the Persian Gulf and Gulf of Oman, *Marine Geol.,* **12,** 245-266.

Scholle, P. A., and Kinsman, D. J. J., 1974. Aragonitic and high-Mg calcite caliche from the Persian Gulf—a modern analog for the Permian of Texas and New Mexico, *J. Sed. Petrology,* **44,** 904-916.

Cross-references: *Bahama Banks Sedimentology; Evaporite Facies; Limestones; Oolite; Pisolite; Reef Complex; Sabkha Sedimentology; Storm Deposits; Stromatolites; Tropical Lagoonal Sedimentation.* Vol. I: *Persian Gulf.*

PETROLEUM—ORIGIN AND EVOLUTION

Petroleum is a naturally occurring, complex mixture of hydrocarbons and small quantities of oxygen, nitrogen, and sulfur-containing organic compounds as well as other minor components, some of which are complexed with metal ions. A wealth of data has been gathered concerning the typical components of petroleum, especially in the last two decades with the application of more sophisticated physicochemical techniques for the separation of complex mixtures (gas chromatography) and the identification of isolated components (mass spectrometry, etc.).

Distillation of a petroleum into broad boiling fractions is often the first step in its laboratory separation into classes of compounds. For example, the American Petroleum Institute Research Project (Mair, 1967) had isolated 295 broad distillation cuts from its representative petroleum: (1) a *gas* fraction boiling below 40°C (at 1 atmosphere pressure); (2) a *gasoline* fraction boiling from 40°-180°C; (3) a *kerosene* fraction from 180°-230°C; (4) *light gas oil,* 230°-305°C; (5) *heavy gas oil* and light lubricating distillate, 305°-405°C; (6) *lubricants,* 405°-515°C; (7) *asphalts,* distillation residue. Sometimes a partial separation of this kind takes place in nature, and thus, petroleum occurs in the form of natural gas, crude oil and asphalts.

These broad boiling fractions can be fractioned further, and eventually individual compounds can be isolated and identified. As of 1967, the American Petroleum Institute Research Project (Mair, 1967) had isolated 295 individual compounds from its reference petroleum. Saturated hydrocarbons make up a substantial part of petroleum. Cycloparaffins and aromatic hydrocarbons also constitute a significant portion. Straight-chain saturated hydrocarbons (*n*-paraffins) usually account for about 15% of the average crude oil. The distribution of *n*-paraffins in Darius Crude Oil (Fig. 1) shows the typical smooth distribution, i.e., there is no predominance of odd- over even-numbered *n*-paraffins (see *Crude Oil Composition and Migration*).

A great deal is known about the geological aspects of petroleum occurrence. Even though questions concerning petroleum generation, evolution, migration, and accumulation have attracted much research for several decades, much remains to be done.

Evidence for Biological Origin

There is little doubt that petroleum is derived predominantly from living systems. The overwhelming evidence for the biological origin of petroleum includes: (1) the general structural

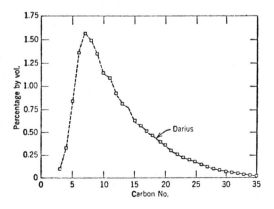

FIGURE 1. Distribution of *n*-paraffins in Darius Crude Oil (from Martin et al., 1963b).

similarity of certain petroleum hydrocarbons to the organic components synthesized by living organisms; (2) the optical activity of various petroleum fractions; (3) the fact that petroleum generally contains porphyrins which are probably derived from chlorophylls of living systems; (4) the fact that $C^{13}:C^{12}$ ratios in petroleum are more similar to those of living organic matter than to those of atmospheric or carbonate carbon; (5) the common presence in petroleum of nitrogenous compounds that are characteristic of living organisms; (6) the predominant association of petroleum with sedimentary rocks; and (7) the widespread presence of identifiable plant and animal remains in petroleum. Abiological hydrocarbons probably do exist, as for example in some *meteorites* (q.v., Vol. II), but they do not make an important quantitative contribution to petroleum.

Geologic Environment. Assuming the biological origin of petroleum, the following questions must be answered: (1) What kinds of organisms and which of the many possible components are preserved or altered to form petroleum? (2) What was the environment, and what were the conditions for the generation of petroleum? and (3) What chemical and physical processes were involved? A frame of reference can be constructed by the geologic boundary conditions of time, temperature and pressure.

Petroleum is found in rocks of all geologic periods from the Precambrian through the Pleistocene inclusive, although certain periods have been more productive than others. The temperatures to which petroleum has been exposed range from below $0°C$ to about $200°C$. The depth of burial of petroleums varies from about 1,000–10,000 m. A marine environment does not appear to be absolutely necessary for the formation of petroleum, but ideal conditions probably existed more often in the marine environment than the nonmarine. In discussing the best setting for petroleum genesis, Hedberg (1964) emphasizes factors such as abundant production of organic matter of the right kind, and conditions favorable for preservation of source material and petroleum; these conditions include a reducing (micro) environment and the absence of destructive organisms. The remains of marine and terrestrial plants and animals have probably made a contribution to the source of petroleum. At least in the case of the Precambrian petroleums, biologically simple marine organisms such as algae must have been sufficiently abundant to account for the petroleum formed. When the organisms died, their remains settled to the bottom of the body of water. Reducing conditions at the bottom or rapid burial would be an important requirement because exposure of the organic matter to oxygen and aerobic bacteria would quickly destroy it. Quiet, deep waters where organic sediments can accumulate would seem to offer the best setting for the preservation of petroleum source material.

Origin of Petroleum

There is considerable speculation about what happens to the matter after it is deposited, there being two broad schools of thought: *direct accumulation* or *transformation*.

Direct Accumulation. The first theory assumes that petroleum originates from the selective accumulation of naturally occurring hydrocarbons. It is argued that the organic matter (carbon-containing compounds) changes very little after it is deposited, and that the ubiquitous hydrocarbons found in all kinds of plants, animals, and sediments could migrate together to form petroleum. Supporting this idea of direct selective accumulation are the studies by Smith (1954) of the organic matter deposited in various types of Holocene sediments, in which he found free hydrocarbons ranging from 9000 to almost 12,000 ppm, including *n*-paraffins, aromatics, and cycloparaffins. He proved the recent origin of these hydrocarbons by radiocarbon dating. Meinschein (1959) concurred with this idea and pointed to the close resemblance of hydrocarbons in sediments and crude oils to plant and animal products. Striking examples would be the indications of the presence of steranes and triterpanes in some crude oils. Apparently, the hydrocarbons making up crude oils are obtained directly from the remains of living organisms or through only minor alterations of their remains, accomplished in part by bacterial action. Waxes, fats, oils, sterols, and isoprenoids, i.e., plant lipids, appear to be the most likely source materials. Finally, to explain how all the widely dispersed hydrocarbons can come together in petroleum formations and to account for the distributional differences between sedimentary organic matter and petroleum, Meinschein invokes the accumulation process suggested by Baker (1960). This process involves dissolution in water of the hydrocarbons in sedimentary organic matter, with later discharge at some other point, perhaps by a drop in temperature or change in the salt content of the water.

Transformation. It has been realized that there are inadequacies in the direct accumulation hypothesis for the origin of petroleum because crude oils contain hydrocarbons which are not found in living organisms and modern sediments. In particular, two differences stand out. First, while light hydrocarbons in the C_3-C_{14} range make up about half of the typical crude oil (q.v.), they are virtually absent in modern sediments and living organisms. One study bearing on this point was done by Dunton and Hunt (1962), who searched for the C_4-C_8 hydrocarbons in 21 modern sediment samples and found them lacking at the 10-ppb level. These light hydrocarbons, however, were extracted from older sediments (Miocene to Precambrian) in quantities ranging from 0.6 to 860 ppm. Furthermore, a rough correlation was established linking the amount of organic matter in the rock to the yield of C_4-C_8 hydrocarbons. Some of the rocks studied also showed that there was an increased yield of these light hydrocarbons with age.

A second difference between crude oils and modern sediments concerns odd/even carbon-number abundance. A preference for odd-numbered *n*-paraffins is observed in modern sediments, but not in the typical crude oil and ancient sediment, which exhibit smoother distributions (Fig. 2). The strong odd-numbered *n*-paraffin predominance in modern sediments is also significant in view of the fact that odd *n*-paraffins predominate in living systems along with even-numbered fatty acids. True petroleum does not yet exist in modern sediments, nor are all the requisite components present therein. Petroleum must, therefore, be generated with geologic time via the chemical change of source material(s) other than hydrocarbons.

This conclusion was confirmed by Philippi's (1965) comprehensive field study of the subsurface petroleum-generation process. He compared Holocene through Upper Miocene (15 m.y.) sediments to crude oils from the Los Angeles and Ventura basins, paying particular attention to the high-boiling normal paraffins and cycloparaffins. Philippi found that hydrocarbon content increases strongly with depth of burial and age of sediments, and that as hydrocarbons begin to form, odd-carbon number predominance among normal paraffins is substantially reduced. It was not until the Upper Miocene that sediment content began to resemble that of oil source rocks. The data also showed that the bulk of oil generation took place above $115°C$—below ≈ 2500 m in the L.A. Basin and below ≈ 3700 m in the Ventura Basin. Philippi concludes that at these temperatures the environment is sterile, and petroleum must be formed by physical (thermal) and chemical processes. Without the intervention of living systems, a relatively (but not completely) random rupture of carbon-carbon bonds would occur. Thus, the generation of *n*-paraffins would result in the observed lack of odd-carbon-number predominance.

Philippi pointed out, however, that there is no fixed depth or duration for the oil generation process. The amount of petroleum formed would be influenced by: (1) the amount and nature of the organic source material; (2) the temperature history of the rock; and (3) the effect of catalysts, if present. The amount of oil formed would vary linearly with geologic time, but exponentially with the absolute temperature.

Philippi's conclusions provide support for the transformation hypothesis, i.e., that the source materials for petroleum are subject to chemical and physical change following deposition, although they may differ as to the time and place of the changes, the source material, and the precursors involved.

Petroleum Source Material

Many types of plants and animals have contributed to the source material of petroleum, and only the broad classes of the chemical compounds involved, i.e., carbohydrates, are usually *not* considered to be significant contributors to oil formation because a great deal of chemical reduction would be required to convert them to hydrocarbons; in addition, the intermediates for this process have not been noted. Carbohydrates may, however, be involved in the formation of coal-like material, as are lignins (see *Coal*), which are not abundant in marine organisms. *Proteins* have been suggested as likely precursors for short-chain hydrocarbons via decarboxylation and reductive deamination,

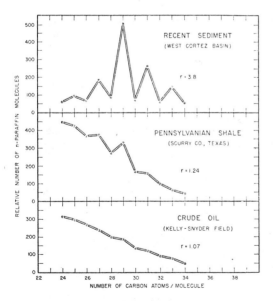

FIGURE 2. *n*-Paraffin distributions from a Holocene (Recent) sediment, a marine shale, and a crude oil (from Bray and Evans, 1961).

but the evidence is inconclusive. The *lipids*, which are major constituents of living organisms, remain as the most important source of petroleum hydrocarbons. This broad class includes fatty acids and esters, fatty alcohols, sterols, aliphatic isoprenoids, and carotenoid pigments, and, of course, hydrocarbons of all kinds.

Reactions have been written, with some experimental work, to show how lipids might be converted to petroleum components under sedimentary conditions. Carotenoids, for example, can form aromatic hydrocarbons as well as cycloparaffins under mild thermal conditions ($<200°C$) with cyclization, carbon-carbon bond cleavages, and perhaps reduction involved (Erdman, 1967). Unsaturated fatty acids may also undergo thermal cyclization and may subsequently yield aromatics or cycloparaffins. The Diels-Alder reaction may account for formation of some cyclic compounds in petroleum. Reaction involving unsaturated source compounds would have to occur relatively early in a sediment's history before the unsaturates suffer reduction.

Conversion of sterols to their corresponding steranes has been proposed by many researchers to explain some of the optical activity of high-boiling petroleum fractions. *Optical activity* (the ability of organic molecules bearing asymmetric centers to rotate plane-polarized light) is a characteristic of compounds synthesized by living systems. Sterols and steroids are widely distributed in nature and are present in all the various types of algae. Further degradation of sterols can contribute other components to petroleum. Dehydrogenation with sulfur or selenium as catalysts is known to result in aromatic compounds, especially the phenanthrenes. One of the laboratory products of cholesterol's dehydrogenation with selenium is cyclopentanophenanthrene, which has been isolated from petroleum. Confirmatory evidence for sterol precursors is found in the observations that the phenanthrenes have been found in much greater abundances in petroleums than the nonsterol related isomers, the anthracenes.

Fatty Acids. Several detailed schemes to account for the chemical transformations of lipids have recently been proposed to explain hydrocarbon genesis, with particular attention to the formation of light hydrocarbons and/or to the smooth distribution among petroleum *n*-paraffins. Cooper and Bray (1963) devised a scheme whereby the petroleum *n*-paraffins would arise from sedimentary fatty acids. The straight-chain fatty acids—major constituents of living organisms—usually show a strong predominance of even-carbon number; straight-chain hydrocarbons show some odd-carbon number preference. Both of these preferences, however, decrease significantly in ancient sediments, petroleum, and petroleum reservoir waters. Thus the distributions of both fatty acids and *n*-paraffins tend to become smoother. Also significant is the fact that fatty acid concentrations decrease with the increasing age of the sediment, whereas hydrocarbon concentrations increase. Cooper and Bray (1963) felt that these effects are related perhaps by a certain series of reactions, which could take place under geologic conditions, where, if the scheme were repeated sufficiently, little or no carbon-number preference would remain among the acids or paraffins (Fig. 3). Unfortunately, observations by other researchers indicate that the relationship between sedimentary fatty acids and hydrocarbons cannot be explained by this scheme alone. In somewhat related experiments, Jurg and Eisma (1964) heated behenic acid (n-$C_{21}H_{43}COOH$) for varying periods of time with bentonite clay (with and without water). Alkanes and fatty acids of both *longer* and shorter chain lengths were generated, but the $C_{21}H_{44}$ alkane resulting from direct decarboxylation was predominant (see Fig. 4). Olefins and branched alkanes were also generated. There is good evidence that the reactions occur via a radical mechanism initiated by the forma-

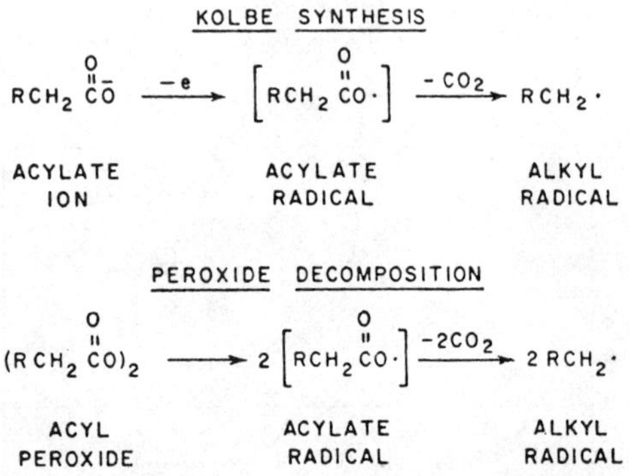

KOLBE SYNTHESIS

$$RCH_2\overset{O}{\overset{\|}{C}}O^- \xrightarrow{-e} \left[RCH_2\overset{O}{\overset{\|}{C}}O\cdot\right] \xrightarrow{-CO_2} RCH_2\cdot$$

ACYLATE ION ACYLATE RADICAL ALKYL RADICAL

PEROXIDE DECOMPOSITION

$$(RCH_2\overset{O}{\overset{\|}{C}}O)_2 \longrightarrow 2\left[RCH_2\overset{O}{\overset{\|}{C}}O\cdot\right] \xrightarrow{-2CO_2} 2RCH_2\cdot$$

ACYL PEROXIDE ACYLATE RADICAL ALKYL RADICAL

Low temperature decarboxylation of aliphatic acids

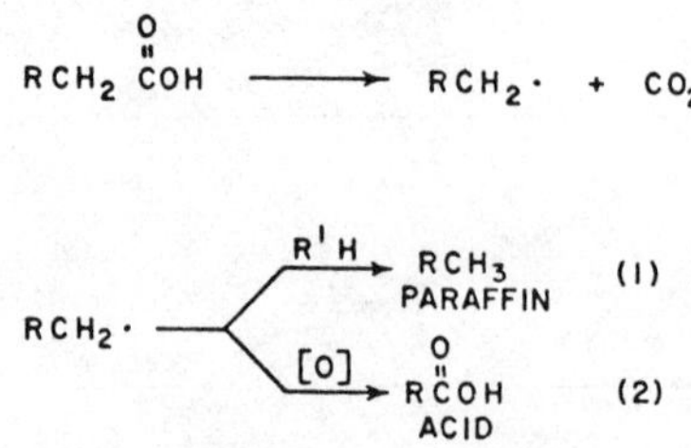

$$RCH_2\overset{O}{\overset{\|}{C}}OH \longrightarrow RCH_2\cdot + CO_2 + H\cdot$$

$$RCH_2\cdot \begin{cases} \xrightarrow{R'H} RCH_3 \text{ PARAFFIN} \quad (1) \\ \xrightarrow{[O]} R\overset{O}{\overset{\|}{C}}OH \text{ ACID} \quad (2) \end{cases}$$

Reaction paths for alkyl radicals

FIGURE 3. Fatty acid reaction proposed by Cooper and Bray (1963).

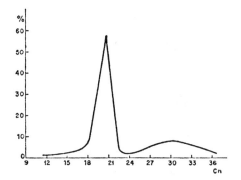

FIGURE 4. Distribution of the n-paraffins with a carbon number greater than C^{14} obtained by heating behenic acid with bentonite clay at $200°C$ (from Jurg and Eisma, 1964; Copyright 1964 by the American Association for the Advancement of Science).

tion of an alkyl radical ($C_{21}H_{44}$) upon decarboxylation of the behenic acid. In similar experiments, Henderson et al. (1968) heated an n-paraffin, $C_{28}H_{58}$, and obtained homologous series of lower weight n-alkane and alkenes as well as aromatics and branched-cyclic alkanes. The alkane and alkene distributions had little carbon-number preference. Higher concentrations of branched-cyclic alkanes were obtained through catalytic cracking (with bentonite) than by thermal cracking alone.

Kerogen. Perhaps the most comprehensive scheme for petroleum genesis has been proposed by Abelson (1967). He cites the fact that most of the organic matter in sediments, even in modern sediments, is the form of *kerogen* (q.v.), a solvent-insoluble organic polymer of high molecular weight. It is likely, then, that kerogen should play an important role in the evolution of organic matter found in petroleum deposits. Abelson traces this evolution from the formation of kerogen to its conversion to the n-paraffins of petroleum. He starts by explaining that phytoplankton, the major source of organic matter in the sea and lakes, contain as major components the C_{14}, C_{16}, C_{18}, and C_{20} fatty acids, most of which are unsaturated. These unsaturated fatty acids, however, disappear rapidly with depth in aerobic waters. Furthermore, recent sediments contain almost no unsaturated fatty acids, although reduction probably does not occur immediately after deposition. Abelson accounts for these facts by proposing that when the phytoplankton die and their cell membranes lose their integrity, the unsaturated fatty acids react with the oxygen in the water. Reactive peroxides and then the insoluble, cross-linked kerogen polymer, are formed. Abelson feels that kerogen is likely to escape destruction by microorganisms in the anaerobic sedimentary environment.

Some knowledge about what happens to the kerogen with time and increasing temperature (due to increasing depth of burial) was gained in the laboratory. Abelson and Hoering (see Hoering, 1967) have heated Green River Formation oil shale (q.v.) for varying periods of time and increasing temperatures ($25-400°C$). At lower temperatures, the predominant products released were isobutane and isopentane thought to have been formed prior to the heating. At higher temperatures, the content of C_4-C_{12} hydrocarbons was studied, and the major components were found to be straight-chain, saturated hydrocarbons of both even- and odd-carbon number. Simple branched hydrocarbons and cycloparaffins were also generated. In other words, the mixtures were like petroleum. In related work Mitterer and Hoering (1968) were able to isolate the isoprenoid hydrocarbons pristane and phytane, the C_{14}-C_{28} fatty acids, and porphyrins when they heated the kerogen of a Holocene sediment from the San Nicholas Basin. The n-paraffins obtained (C_{15}-C_{30}) had no odd-carbon number preference and a lower mean molecular weight than the n-paraffins obtained from the unheated sediment (Fig. 5). The generation of all of these compounds shows that hydrogenation has taken place. For the saturated hydrocarbons to be formed, alkyl chains are probably broken off and reduced. At the same time, the hydrogen donor, i.e., the remaining kerogen, is thought to condense to an aromatic or hydroaromatic ring.

"Protopetroleum." Another segment of the transformation school also maintains that petroleum is formed by thermal degradation, but that the precursor substance is "protopetroleum," which is thought to be a mixture of complex, high-weight organic compounds mainly of lipid character. Petroleum is pictured as a dynamic mixture of organic substances that can change throughout postdepositional history. Silverman (1964) has obtained carbon isotope data which relate to the problem. Gas phase hydrocarbons separated from petroleums had markedly lower C^{13}:C^{12} ratios than the liquids remaining behind. Methane has the lowest C^{13} content of all petroleum components, and then there is an increase in the amount of C^{13} with increasing carbon number until a plateau is reached (Fig. 6). These observations could be accounted for by the decomposition of higher molecular weight compounds to yield the lower weight hydrocarbons. In fact, it is known that C^{12}-C^{12} bonds are broken 8% more often than C^{13}-C^{12} bonds by thermal degradation of organic compounds. Hence, the product low-

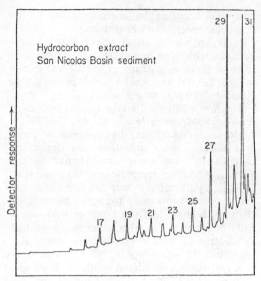

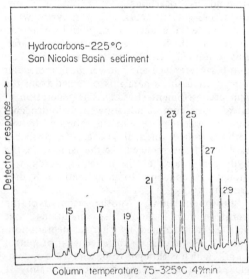

FIGURE 5. *Top:* Gas chromatogram of total hydrocarbons extracted with benzene-methanol from the unheated San Nicolas Basin sediment. *Bottom:* Gas chromatogram of normal hydrocarbons extracted with benzene-methanol from San Nicolas Basin sediment heated to 225°C. (From Mitterer and Hoering, 1968.)

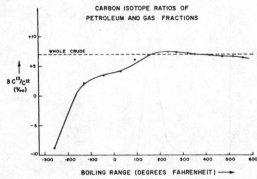

FIGURE 6. Carbon-13 content of petroleum fractions (from Silverman, 1964).

tenes (which are thought by some to be high-molecular-weight hydrocarbons of predominantly aromatic character). To be consistent with this hypothesis, aromatics and asphaltenes from petroleums should have the highest C^{13} contents of all fractions because their formation would have involved the largest number of broken C^{12}-C^{12} bonds. Indeed this effect has been observed by Silverman.

Catalysis. Besides thermal alteration, several other factors may be important in the generation and evolution of petroleum: catalysts, radioactivity, and microorganisms. Several experiments employing catalysts were mentioned above, i.e., behenic acid was heated with bentonite clay, and *n*-octacosane was heated with the same clay in another experiment. A comparison was made in the latter experiment between heating the *n*-octacosane without clay (thermally) and with bentonite clay (catalytically) at 200°C. The clay did in fact increase the yield of products ten-fold in much less time. In the first-named experiment, weight hydrocarbons should have been less C^{13} than the precursor molecules (Fig. 7). The process is pictured as a series of dehydrogenations, and as the low-molecular-weight hydrocarbons are released, double bonds are introduced into the precursor molecule. Subsequently, Silverman feels the polyenes thus formed could (1) undergo ring closure to aromatics and (2) polymerize with possible formation of asphal-

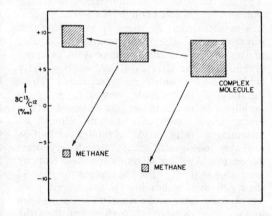

FIGURE 7. Petroleum maturation reactions (from Silverman, 1964).

no hydrocarbons at all were isolated from heating behenic acid at 200°C without the bentonite. Thus, it appears natural catalysts, including clay minerals, can play a part in petroleum generation, especially because the temperature is below 200°C in the sediments. Some geochemists have thought that perhaps the presence of water in sediments might destroy catalytic activity by blocking the active sites on the catalyst's surface. When water was added to the mixtures of behenic acid and bentonite and n-octacosane and bentonite, the yield from heating was found to be lower. Further, the reaction products were different. The percentage of branched and cyclic products is higher with the clay alone, without water. Apparently, carbonium ions are the intermediates without water, and they cause isomerization. Radicals (organic species having an unpaired electron) are probably the intermediates when water is used. These and other experiments give some insight into the influence of catalysts on petroleum, but the extent of the influence is still unknown.

Radioactivity. Very little is known about the influence of natural sedimentary radioactivity on petroleum. It has been shown, for example, that bombardment of fatty acids with alpha particles from radon will produce small amounts of low-weight hydrocarbons plus the hydrocarbon resulting from decarboxylation of the parent fatty acid: $C_{15}H_{31}COOH \rightarrow C_{15}H_{32} + CO_2$. It is doubtful, however, that radiation-induced reactions make a quantitatively significant contribution to petroleum, except perhaps in the Precambrian period.

Microorganisms. Microorganisms are known to play a large part in the initial processing of organic matter; from the time that a marine plant or animal dies, as it settles and is deposited on the sea bottom, and during the early stages of diagenesis, the organic matter is being altered by both aerobes and anaerobes. Some of these changes are shown in Fig. 8. Aerobic activity is confined to a few cm near the soil or water surface, but anaerobic activity begins below this surface and continues down usually to a depth of 5-10 m below the sea bottom. The bacteria are said to release oxygen, nitrogen, phosphorus, and sulfur from the organic material and thus start it on its way to petroleum. Anaerobic bacteria apparently are also responsible for creating and maintaining a reducing atmosphere, which helps to preserve the organic matter both before and after its conversion to petroleum. The reducing conditions can be promoted by sulfate-reducing bacteria which produce hydrogen sulfide. It has not been shown, however, that bacteria can produce significant quantities of hydrocarbons, except for methane, either from organic source material or by contributions from their own cells or metabolic processes. Furthermore, it is not known to what depth bacterial activity exists. Nevertheless, it appears unlikely that this activity could extend thousands of m to where petroleum genesis is thought to occur.

Evolution of Petroleum

It is difficult to separate the process of petroleum evolution from the process of petroleum generation, both in concept and in actual laboratory studies. Observed differences in composition between individual petroleums can be attributed to differences in sedimentary or depositional environment or source materials, as well as to different stages in an evolutionary process. Conclusive evidence that petroleum does evolve is not available. Furthermore, understanding of petroleum evolution is hampered because the generation process is still not completely understood.

As already pointed out above, some of the transformationists feel that petroleum can change at any time throughout its postdepositional history. The hypothetical evolution of a particular petroleum could presumably occur where the petroleum was generated, as well as during its migration and accumulation in a reservoir. Thermal alteration, perhaps with the influence of a catalyst, would be the most likely agent if maturation were to occur. Evolution is thought to proceed in the direction of a lighter, more paraffinic petroleum.

There is also the probability that the source rock evolves and that a series of several different kinds of petroleum are produced as conditions change, e.g., as temperature increases with increasing overburden pressure.

Migration and Accumulation

Ideally, after oil is generated—usually in a fine-grained sediment—a small portion of it migrates and finally either escapes or accumulates in a porous and permeable reservoir. This process is termed primary migration and is essential for the development of major petroleum pools. If no suitable reservoir is available, petroleum expelled by compaction of the source sediment will be otherwise dissipated and could rise to the surface and suffer oxidation. "Secondary migration" refers to movement of petroleum within the reservoir rock.

The exact mode of primary migration is still unknown, and it is probable that oil migrates in several different ways, even from the same source rock. It is also likely that the quantity of oil migrating from a particular source rock varies considerably with time (see *Crude Oil Composition and Migration*).

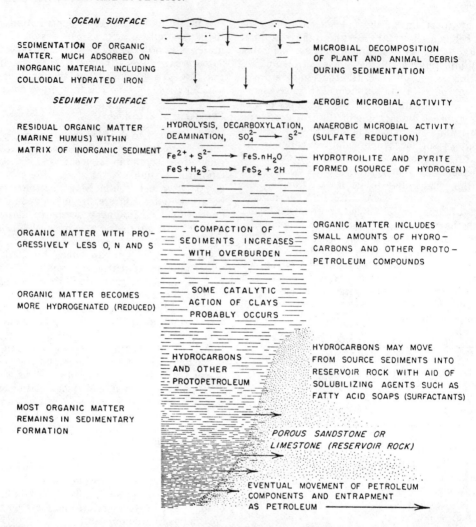

FIGURE 8. General events involved in the conversion of organic matter to petroleum (from Davis, 1967).

Compaction causes primary migration. The fine-grained source sediment initially has high porosity, but with age and increasing depth, porosity decreases. As compaction proceeds, water and other fluids in the pore spaces are pressed out. If sufficient petroleum has been generated, it too can be released from the source rock. The generated petroleum is probably trapped, adsorbed, or dissolved in the sedimentary organic matter, including the kerogen. *Adsorption* of the oil onto clay minerals is also possible. Sufficient amounts of oil, perhaps in amounts in excess of the "sorption capacity" of the source rock, are necessary for migration to occur. It has been estimated that primary migration due to compaction can continue to a depth of about 3000–6000 m when the source rock porosity is in the range of 5% to 0.

There are several divergent views concerning the form in which oil that has been squeezed out of the source rock travels to a reservoir rock. Some of the chief proposals are: (1) migration in aqueous solution; (2) migration in the gaseous state—liquid hydrocarbons dissolve in gaseous hydrocarbons and migrate in the gaseous state to a zone where pressure is lower; and (3) migration of liquid oil globules.

In a detailed scheme for the migration of hydrocarbons in water, Baker (1960) found a correspondence between the relative abundance of hydrocarbons in petroleum and the relative solubility of the same hydrocarbons in micelles formed by colloidal electrolytes. This correlation suggested that petroleum consisted of hydrocarbons which were once selectively dissolved in aqueous solution. When molecules such as organic acid salts, e.g., sodium laurate, dissolve in water, they form clusters of individual molecules known as *micelles,* which are of colloidal particle size. This size of particles is intermediate between that of those in true solution and those in coarse suspensions. The micelles increase the solubility of hydrocarbons in water because they provide hydrocarbon-like regions to which the hydrocarbons are attracted.

The micelle hypothesis has had little support in recent years, for two reasons: (1) studies of pore waters show they do not contain the 50 times more organic acid salts than hydrocarbons needed to form micelles, and (2) the micelles will pass through sands and clean silts, but not the typical clay source rock with a permeability less then 10^{-5} millidarcies.

A hypothesis gaining favor recently (Hunt, 1973; Dickey, 1975) is the concept that as compaction proceeds, the ratio of free pore water to pore water fixed to mineral surfaces decreases until the oil particles are physically expelled. The process is analogous to the freezing of ice and other crystallization processes in which foreign molecules are excluded. As yet, however, there is no migration hypothesis that has been generally accepted by petroleum geochemists.

P. C. WSZOLEK
A. L. BURLINGAME

References

Abelson, P. H., 1967. Conversion of biochemicals to kerogen and *n*-paraffins, in P. H. Abelson, ed., *Researches in Geochemistry,* vol. 2. New York: Wiley, 63-86.

Baker, E. G., 1960. A hypothesis concerning the accumulation of sediment hydrocarbons to form crude oil, *Geochim. Cosmochim. Acta,* 19, 309-317.

Baker, E. G., 1967. A geochemical evolution of petroleum migration and accumulation, in Nagy and Colombo, 1967, 299-329.

Bray, E. E., and Evans, E. D., 1961. Distribution of *n*-paraffins as a clue to recognition of source beds, *Geochim. Cosmochim. Acta,* 22, 2-15.

Colombo, U., 1967. Origin and evolution of petroleums, in Nagy and Colombo, 1967, 331-369.

Cooper, J. E., and Bray, E. E., 1963. A postulated role of fatty acids in petroleum formation, *Geochim. Cosmochim. Acta,* 27, 1113-1127.

Cordell, R. J., 1972. Depths of oil origin and primary migration: A review and critique, *Bull. Am. Assoc. Petrol. Geologists,* 56, 2029-2067.

Craig, H.; Miller, S. L.; and Wasserburg, G. J., eds., 1964. *Isotopic and Cosmic Chemistry.* Amsterdam: North Holland Publ., 553p.

Davis, J. B., 1967. *Petroleum Microbiology.* Amsterdam: Elsevier, 604p.

Dickey, P. A., 1975. Possible primary migration of oil from source rock in the oil phase, *Bull. Am. Assoc. Petrol. Geologists,* 59, 337-345.

Dunton, M. L., and Hunt, J. M., 1962. Distribution of low molecular-weight hydrocarbons in recent and ancient sediments, *Bull. Am. Assoc. Petrol. Geologists,* 46, 2246-2248.

Erdman, J. G., 1967. Geochemical origins of the low molecular weight hydrocarbon constituents of petroleum and natural gases, *7th World Petrol. Cong. Proc., Mexico,* 2, 13-24.

Fischer, A. G., and Judson, S., eds., 1975. *Petroleum and Global Tectonics.* New Jersey: Princeton Univ. Press, 322p.

Haun, J. D., ed., 1974. Origin of Petroleum II, *Bull. Am. Assoc. Petrol. Geologists, Reprint Ser. 9,* 212p.

Hedberg, H. D., 1964. Geologic aspects of the origin of petroleum, *Bull. Am. Assoc. Petrol. Geologists,* 48, 1755-1803.

Henderson, W.; Eglinton, G.; Simmonds, P.; and Lovelock, J. E., 1968. Thermal alteration as a contributory process to the genesis of petroleum, *Nature,* 219, 1012-1016.

Hoering, T. C., 1967. The organic geochemistry of Precambrian rocks, in Abelson, 1967, 87-111.

Hunt, J. M., 1973. An examination of petroleum migration processes, *Proc. 7th Intern. Geochem. Conf., OGIL, Budapest,* 1, 219-229.

Jurg, J. W., and Eisma, E., 1964. Petroleum hydrocarbons, generation from fatty acid, *Science,* 144, 1451-1452.

Lasaga, A. C.; Holland, H. D.; and Dwyer, M. J., 1971. Primordial oil slick, *Science,* 174, 53-55.

Mair, B. J., Dir., 1967. Annual report for the year ending June 30, 1967, *Am. Petrol. Inst. Research Proj.,* 6, Pittsburgh, Pa.: Carnegie Institute Techn.

Martin, R. L.; Winters, J. C.; and Williams, J. A., 1963a. Composition of crude oils by gas chromatography—Geological significance of hydrocarbon distribution, *Proc. 6th World Petrol. Cong., Frankfurt, Sec. V,* paper 13.

Martin, R. L.; Winters, J. C.; and Williams, J. A., 1963b. Distributions of *n*-paraffins in crude oils and their implications to origin of petroleum, *Nature,* 199, 110-113.

Meinschein, W. G., 1959. Origin of petroleum, *Bull. Am. Assoc. Petrol. Geologists,* 43, 925-943.

Mitterer, R. M., and Hoering, T. C., 1968. Production of hydrocarbons from the organic matter in a recent sediment, *Carnegie Inst. Yearbook,* 66, 510-514.

Nagy, B., and Colombo, U., eds., 1967. *Fundamental Aspects of Petroleum Geochemistry.* Amsterdam: Elsevier, 388p.

Palacas, J. G.; Love A. H.; and Gerrild, P. M., 1972. Hydrocarbons in estuarine sediments of Choctawhatchee Bay, Florida, and their implications for genesis of petroleum, *Bull. Am. Assoc. Petrol. Geologists,* 56, 1402-1418.

Philippi, G. T., 1965. On the depth, time and mechanism of petroleum generation, *Geochim. Cosmochim. Acta,* 29, 1021-1049.

Rossini, F. D., 1960. Hydrocarbons in petroleum, *J. Chem. Educ.,* 37, 554-561.

Silverman, S. R., 1964. Investigations of petroleum origin and evolution mechanisms by carbon isotope studies, in Craig et al., 1964, 92-102.

Smith, P. V., 1954. Studies on origin of petroleum: occurrence of hydrocarbons in recent sediments, *Bull. Am. Assoc. Petrol. Geologists,* 38, 377-404.

Cross-references: *Coal; Compaction in Sediments; Crude-Oil Composition and Migration; Diagenesis;*

Gases in Sediments; Humic Matter in Sediments; Kerogen; Oil Shale; Organic Sediments.

pH and Eh OF MARINE SEDIMENTS
—See Vol. I

PHI SCALE

The phi scale (Krumbein, 1934) is a logarithmic transformation of the Wentworth grade scale (q.v.) based on the negative logarithm to the base 2 of the particle diameter: $\phi = -\log_2 x$, where x = grain diameter on the Wentworth scale. As a result of this conversion, the higher the phi number, the finer grained are the particles. A table of conversions is available (Page, 1955; see *Grain-Size Studies*).

B. CHARLOTTE SCHREIBER

References

Krumbein, W. C., 1934. Size frequency distributions of sediments, *J. Sed. Petrology,* 4, 65-77.
Page, H. G., 1955. Phi-millimeter conversion table, *J. Sed. Petrology,* 25, 285-292.
Tanner, W. F., 1964. Modification of sediment size distributions, *J. Sed. Petrology,* 34, 156-164.

Cross-references: *Grain-Size Studies; Wentworth Scale.*

PHOSPHATE IN SEDIMENTS

Geochemical Distribution and Behavior

The earth's crust contains about 0.2% P_2O_5; and common igneous progenitors of sediments, though ranging from 0.05-0.50%, also average about 0.25-0.27% P_2O_5. Most of this phosphate occurs as fluorapatite $[Ca_{10}(PO_4)_6F_2]$; some occurs as $(PO_4)^{-3}$ substituting for $(SiO_4)^{-4}$ in silicate minerals, and a very small part occurs as monazite $(CePO_4)$ and xenotime (YPO_4).

Most of the phosphate in sedimentary rocks of the continental crust appears to be between 0.1-0.15% P_2O_5, depending on the relative volumes of sandstone, shale, and limestone assumed. By these estimates, sediments of the continents and shelves are notably deficient in P_2O_5 compared to their igneous-rock parents. The distribution of phosphate among different classes of sediment has been tabulated by McKelvey (1973).

The major sedimentary-rock categories—sandstone, shale, and carbonate rocks—have average P_2O_5 contents of 0.10%, 0.17%, and 0.07% respectively. Compilations of sandstone analyses by Pettijohn (1963) show that orthoquartzites contain < 0.2% P_2O_5, arkoses and lithic arenites (subgraywackes) average 0.1% P_2O_5, and graywackes average 0.2% P_2O_5. Thus, with increasing content of rock fragments, sediments approach igneous rocks in P_2O_5 content. This relationship is true also of tillites, reported to average 0.21% P_2O_5. The calcareous oozes of modern oceanic sediments contain an average of 0.2% P_2O_5, and siliceous oozes average 0.3%. Chert (q.v.) generally contains < 0.1% P_2O_5, but those rich in skeletal remains contain 0.2-0.8% or more (Cressman, 1962). Bituminous black shale generally has more phosphate than ordinary shale (q.v.), and black shale, chert, and carbonate rocks associated with phosphorites may contain apatite granules and pellets as major constituents.

Evaporites, whether of lacustrine or marine origin, generally contain less phosphate than average sediments. In the Eocene Green River Formation, however, thin layers of phosphatic carbonate siltstone are found associated with salt and trona in the saline facies, and the minerals bradleyite and francolite have been found in these rocks.

Sedimentary iron deposits (ironstone, q.v.) of Phanerozoic age are unusually rich in phosphate. Averages of limonite, hematite, magnetite, chamosite, and siderite ores range from 0.91-1.78% P_2O_5 (James, 1966). Analyses of 35 New Jersey green sands (glauconite, q.v.) listed by Mansfield (1940) average about 1.0% P_2O_5. Precambrian cherty iron formations (q.v.) are significantly less phosphatic. James (1966) gives 41 analyses with an average of 0.21% P_2O_5. The phosphate is in the form of apatite in many chamosite, siderite, and glauconite deposits. In bog iron ores, it is typically vivianite $[Fe_3(PO_4)_2 \cdot 8H_2O]$.

Deep-sea clays have higher phosphate contents (0.2-0.3%) than do sediments of the continents and shelves. Large volumes of hemipelagic and pelagic sediments tie up a major part of phosphate liberated by continental weathering, accounting for the deficiency of P_2O_5 in continental sediments relative to igneous rocks.

With the exception of the transport of detrital igneous fluorapatite to first-cycle coarse-grained rocks and its partial retention in recycled detrital rocks, the phosphate mineral that accounts for most sedimentary phosphate is francolite, or carbonate-fluorapatite. This fine-grained microcrystalline variety, which precipitates or replaces carbonates in the marine realm, is also called *collophane*. It composes the nodules, spherulites, and concretions of phosphatic shales and limestones, as well as the coprolites, conodonts, fossil bones, and phosphatized shell of biogenic phosphate in sediments; and it is the essential mineral of the pellets and oolites that characterize phosphorites. In contrast to simple fluorapatite [$Ca_{10}(PO_4)_6F_2$], francolite [$Ca_{10}(PO_4)_{6-x}(CO_3F)_x(F,OH)_2$] is typically deficient in P_2O_5 by about 3-6 wt %; it contains small amounts of structurally emplaced carbonate and extra fluorine or hydroxyl, thought to be due to the substitution of $(CO_3F)^{-3}$ for $(PO_4)^{-3}$. In some phosphorites, sulfate is also an important substituent for phosphate, a replacement apparently coupled with that of sodium for calcium.

Apatites as a class sustain many structural substitutions. Among them uranium (generally 0.005-0.2%), yttrium, and the lanthanides are notable. In many common rocks, the phosphate mineral, though present in minor quantity, contributes a major part of the fluorine (typically 3.5-4% in francolite), the uranium, the radioactivity, and the rare earths.

Apatite (McConnell, 1973) is slowly soluble in neutral or alkaline waters, and its solubility increases with increasing acidity and decreasing hardness. Apatite may survive rock weathering for a time to form a phosphate-rich residuum, but it breaks down under prolonged exposure. Some of it may be redeposited locally. In high-calcium soils and rocks, apatite reforms. In igneous terrane, aluminum or iron phosphates, such as variscite or strengite, are the secondary minerals. In highly weathered rocks, rich in clays and hydroxides, aluminum phosphates, such as crandallite and wavellite are formed.

Most phosphorus is carried to the sea as phosphate minerals or adsorbed on iron or aluminum hydroxides or clay, but some is carried in particulate or dissolved organic compounds, and about 17% of the total is carried in solution. The PO_4 content of most river and lake waters ranges from about 0.01-0.5 ppm. Soft acid waters, such as those in parts of Florida, may contain a few ppm, however, and the highly saline alkaline lake waters (such as Searles Lake, Calif.) contain 200-900 ppm.

The ocean as a whole is essentially saturated with phosphate, and an amount equivalent to that brought by rivers is precipitated more or less continuously. The phosphate content of the ocean is by no means uniform, however; deep, cold waters contain nearly 0.3 ppm, but warm surface waters contain only 0.01 ppm or less.

Oceanographic Controls of Phosphate Deposition. Oceanic circulation deep-plows the sea and brings phosphate-rich waters to the surface in several environments. Phosphate in such upwelling water is precipitated either inorganically or biochemically as the pH and temperature increases near the surface.

In an idealized ocean, the main elements of the current system consist of a large circulating gyre in each hemisphere (Fig. 1). In this sys-

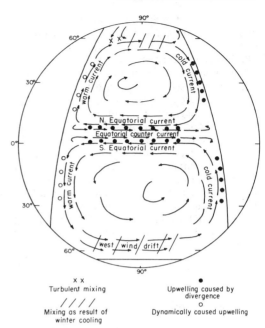

FIGURE 1. Surface currents in an idealized ocean, showing areas of ascending, nutrient-rich water (see Fleming, 1957; from McKelvey, 1967).

tem, cold, nutrient-rich waters are brought to the surface in four situations: (1) where a current diverges from a coast, or where two currents diverge from each other; the effects of cold currents moving toward the equator in bringing cold, phosphate-rich water along the coast are abetted by the seaward movement of coastal surface water that results from the combined effects of prevailing wind and Coriolis effect, for as the surface water moves seaward, cold water wells up to replace it; (2) where two currents meet to produce turbulence; (3) along the W edge of poleward-moving density currents; and (4) in high latitudes where highly saline water from the tropics tends to sink as a result of winter cooling.

Pronounced climatic, biologic, and geologic effects accompany this deep-plowing of the sea, especially in the areas of upwelling produced by divergence. The presence of cold waters along coasts produces coastal fogs and humid-air deserts, as in Peru, northern Chile, and SW Africa. The mineral-rich waters that lie along these deserts are the lushest gardens of the sea, for the upwelling cold waters there support tremendous quantities of organisms, with diatoms and other phytoplankton at one end of the food chain, and fish, whales, and fish-eating sea fowl at the other. Blooms of dinoflagellates (red tides) and diatoms are characteristic phenomena, as are the mass mortalities of fish that accompany red tides and may be a consequence of them.

Biological Concentration of Phosphate. Phosphorus is an essential component of every living cell. It is essential for the synthesis of starch, is an integral part of nucleic acids, may be essential to photosynthesis, is the principal constituent of bone and teeth, and is important in shell formation in many organisms. The availability of phosphate in soils and natural waters is a fundamental control in plant and animal growth. In turn, organic processes play a major role in concentrating phosphate, not only by forming phosphate-rich substances, such as skeletons and excrement, but also in putting phosphate in organic compounds, such as nuclear proteins, through which it can be brought to the burial environment in a relatively soluble form.

Phosphate Deposits

Concentrations of phosphate are formed at all stages of the phosphate cycle, through the processes just described. Ordinarily they merely concentrate phosphorus a few times over its crustal average, but when they operate under unusually favorable conditions, they may yield concentrations 100-125 times those of the crustal average. Particularly important in forming rich deposits is the upgrading effect of two or more of these concentrating processes acting in sequence. For example, the oceanographic processes that result in upwelling bring phosphate-rich waters to the zone of photosynthesis, where biologic or chemical processes may precipitate phosphate and form concentrations of a few percent P_2O_5. Biologic, chemical, and mechanical processes may further enrich these deposits, forming pellets or nodules through diagenetic addition or reorganization of phosphatic material within the sediment, winnowing away lighter and smaller particles such as carbonaceous matter and clay, or oxidative weathering of organic matter. The phosphorites that result from these successive processes may contain as much as 25-30% P_2O_5. The bulk of the world's production of phosphate, particularly of rock containing >30% P_2O_5, comes from deposits that have been enriched by weathering.

Sedimentary Phosphorite. Most of the world's phosphate production comes from marine phosphorites. The richest and largest deposits form at low latitudes in areas of upwelling associated with divergence, chiefly along the west coasts of the continents, or, in the case of large mediterranean seas, along the equatorial side of the basin. Lesser but significant concentrations form along the west edges of poleward-moving density currents, chiefly the warm currents along the eastern coasts of continents (Figs. 1 and 2).

The rock assemblage in the cold-current environment is the product of deposition on a shoaling bottom over which shoreward-moving waters are progressively warmed. The typical lateral sequence of rocks, in a shoreward direction, consists of dark carbonaceous shale, phosphatic shale, phosphorite, and dolomite; chert or diatomite; several facies of carbonate rock; and saline deposits and red or light-colored sandstone or shale. These rocks were deposited synchronously, and they grade laterally into each other or intertongue. Because the environments in which these rocks are deposited shift laterally with epeirogenic movements of the sea bottom, the rocks are found in vertical sequence in somewhat this same or reverse order. The thickest accumulations of phosphorite form in areas of geosynclinal subsidence, where phosphorite is generally associated with carbonaceous shale and chert. The phosphorite is typically carbonaceous and pelletal, but nodules as well as skeletal matter and phosphatic shells may be present in lesser quantities. Individual beds may be 1 m or more thick, contain up to 30% or more P_2O_5, and extend over hundreds of km^2. Phosphorites formed on adjacent platforms or stable areas are generally associated with cherty carbonate rock and light-colored sandstone or shale. They are not appreciably carbonaceous and are likely to be nodular and to contain phosphatized shells as well as naturally phosphatic organic remains. Individual beds are generally only about 1 m thick and contain <30% P_2O_5; thicker and richer deposits may occur locally but are highly lenticular.

Examples of phosphorites deposited from upwelling associated with divergence include those of the Permian Phosphoria Formation in Idaho and adjacent states, the Miocene Monterey Formation of California, the newly discovered Miocene deposits in the Sechura Desert of Peru, and extensive Cretaceous and Eocene deposits in western and northern Africa and the Middle East. Phosphate deposits are abundant on the present sea bottom in most areas of upwelling; some of those off the California coast are being considered for mining by dredging.

Phosphate deposits formed as the result of dynamic upwelling related to warm currents along the eastern coasts of the continents generally consist of phosphatic limestone or sandstone. Chert, black shale, and salines are not associated with these deposits, and they are less extensive and of lower grade than the deposits formed from cold currents. Examples include

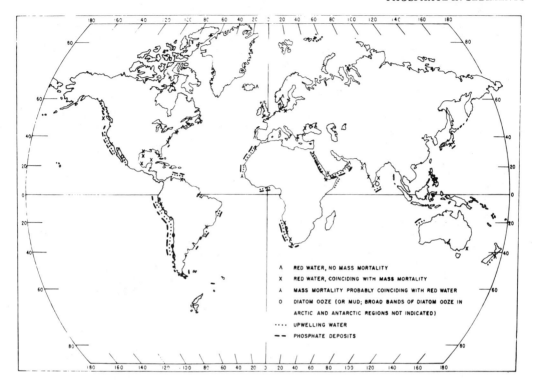

FIGURE 2. Distribution of upwelling water and related phenomena in modern oceans (from McKelvey, 1967; modified from Brongersma-Sanders, 1957, to include data on the distribution of phosphate deposits).

deposits in the Pliocene Bone Valley Formation of Florida, the newly discovered deposits in the Miocene Pungo River Formation of North Carolina and the Upper Cretaceous Gramame Formation near Recife, Brazil.

Alteration and Secondary Phosphorite. Secondary processes have played a prominent role in concentrating the richest of the marine deposits. For example, the highest grade phosphate beds in the Phosphoria Formation appear to have been extensively washed by marine currents. In addition, leaching of carbonates and sulfides, together with oxidation of carbonaceous matter, has raised their P_2O_5 content from 27-30% or so at depths of about 100 m to 32-35% near the surface. Similarly, the Bone Valley deposits originated through submarine reworking of phosphatic residuum developed during deep weathering of phosphatic marls in the Miocene Hawthorne Formation. Leaching related to recent weathering has upgraded some of the deposits still further. The common occurrence of phosphorite at unconformities also reflects the combined effects of weathering and submarine reworking.

Quaternary weathering accounts for the formation of important residual and replacement deposits from phosphatic deposits not otherwise minable. The Tennessee *brown rock phosphates* consist of present-day residuum developed through decomposition of phosphatic limestones of Ordovician age. The *river pebble deposits,* mined earlier in Florida and South Carolina, are essentially placers of phosphatic pebbles eroded from the phosphatic formations of the adjacent terrain.

The Tennessee *white rock* and Florida *hard rock* deposits are of supergene origin, and consist of redeposited as well as metasomatic apatite produced by the weathering of the overlying primary phosphorites. Supergene processes also account for the widespread formation of calcium-aluminum-phosphate and aluminum-phosphate in laterized argillaceous phosphorites. The "leached zone" of the Bone Valley field and the deeply weathered Cretaceous and Eocene deposits of west Africa are examples. These deposits are essentially replacements of clays by millisite [$(Na,K)CaAl_6(PO_4)_4(OH)_9 \cdot 3H_2O$] and crandallite [$CaAl_3(PO_4)_2(OH)_5 \cdot H_2O$]. With continued weathering, the minerals wavellite [$Al_3(OH)_3(PO_4)_2 \cdot 5H_2O$] and augelite [$Al_2(PO_4)(OH)_3$] develop.

Guano and Related Deposits. Guano consists of natural accumulations of animal excrement, altered to various degrees through dehydration, oxidation, and leaching. Most of the large accumulations are produced by sea fowl. Smaller, cave deposits are formed by bats and, subordinately, by other mammals and birds. The bat guanos are most abundant in the cave districts of temperate and tropical regions. Although many deposits were formerly mined, most of them contained only hundreds or thousands of tons, and only sporadic production is obtained from them now. Many sea-fowl guano deposits were of the order of several hundred thousand tons in size, and although most of the fossil accumulations have been mined out, production from the current crop is continuing.

The phosphate-rich upwelling waters that generated the marine phosphorites are also the source of the sea-fowl guano deposits; in addition, they are responsible for the extremely dry climate necessary for the preservation of guano. Most of the deposits are found along the W coasts of Lower California, South America, and Africa, and on islands near the equatorial currents.

Fresh sea-fowl droppings contain about 22% N and 4% P_2O_5. Decomposition proceeds rapidly, and the phosphate content increases as the nitrogen (and total organic matter) decreases. Modern guano contains 10–12% P_2O_5, but leached guano contains 20–32%. The mineralogy of guano is complex. Slightly decomposed deposits contain soluble ammonium and alkali oxalates, sulfates, and nitrates, and a variety of magnesium and ammonium-magnesium phosphates. Largely decomposed guano consists essentially of calcium phosphates [e.g., brushite ($CaHPO_4 \cdot 2H_2O$), monetite ($CaHPO_4$), or whitlockite ($Ca_3(PO_4)_2$].

Under conditions of even slight rainfall, the soluble phosphates of guano are carried to underlying rocks, where they may be deposited as cavity fillings or replacements. Through this process, phosphate from guano has accumulated over long periods of geologic time and formed relatively large deposits. For example, reserves on Nauru, an island in the equatorial region of the western Pacific, were originally about 90 million tons of rock averaging about 39% P_2O_5.

The mineralogy of guano-altered rocks depends on the composition of the host rock. Where it is limestone, as on many of the coral atolls, the terminal phosphate mineral is an apatite similar to francolite in composition but containing less fluorine. Where the underlying rock is a silicate, as is the case with islands of volcanic origin, the secondary phosphate minerals are generally members of the variscite ($AlPO_4 \cdot H_2O$) strengite ($FePO_4 \cdot 2H_2O$) series or its monoclinic dimorphs. The intermediate phase, barrandite [$(AlFe)PO_4 \cdot 2H_2O$], is very common.

VINCENT E. McKELVEY

References

Altschuler, Z. S., 1973. The weathering of phosphate deposits—geochemical and environmental aspects, in Griffith et al., 1973, 33-96.

Ames, L. L., Jr., 1959. The genesis of carbonate apatites, *Econ. Geol.*, 54, 829-840.

Baturin, G. N., and Dubinchuk, V. T., 1974. Microstructures of Agulhas Bank phosphorites, *Marine Geol.*, 16, M63-M70.

British Sulphur Corporation, 1964. *A World Survey of Phosphate Deposits*. London: Woodalls, 206p.

Bromley, R. G., 1967. Marine phosphorites as depth indicators, *Marine Geol.*, 5, 503-509.

Brongersma-Sanders, M., 1957. Mass mortality in the sea, *Geol. Soc. Am. Mem. 67*, 941-1010.

Brown, G. E., 1974. Phosphatic zone in the lower part of the Maquoketa Shale in northeastern Iowa, *J. Research U.S. Geol. Surv.*, 2, 219-232.

Cathcart, J. B., 1963. Economic geology of the Plant City quadrangle, Florida, *U.S. Geol. Surv. Bull. 1142-D*, 56p.

Cressman, E. R., 1962. Nondetrital siliceous sediments, *U.S. Geol. Surv. Prof. Pap. 440-T*, 23p.

Cressman, E. R., and Swanson, R. W., 1964. Stratigraphy and petrology of the Permian rocks of southwestern Montana, *U.S. Geol. Surv. Prof. Pap. 313-C*, 275-569.

D'Anglejan, B. F., 1967. Origin of marine phosphorites off Baja, California, Mexico, *Marine Geol.*, 5, 15-44.

Debrabant, P., and Paquet, J., 1975. L'association glauconites-phosphates-carbonates (Albien de la Sierra d'Esperuña, Espagne meridionale) *Chem. Geol.*, 15, 61-75.

Fleming, R. H., 1957. General features of the oceans, *Geol. Soc. Am. Mem. 67*, 87-108.

Gibson, T. G., 1967. Stratigraphy and paleoenvironment of the phosphatic Miocene strata of North Carolina, *Geol. Soc. Am. Bull.*, 78, 631-650.

Griffith, E. J., et al., eds., 1973. *Environmental Phosphorus Handbook*. New York Wiley, 718p.

Gulbrandsen, R. A., 1960. Petrology of the Meade Peak Phosphatic Shale member of the Phosphoria Formation at Coal Canyon, Wyoming, *U.S. Geol. Surv. Bull. 1111-C*, 74-146.

Hutchinson, G. E., 1952. The biogeochemistry of phosphorus, in L. F. Wolterink, ed., *The Biology of Phosphorus*. East Lansing: Michigan St. Univ. Press, 1-35.

James, H. L., 1966. Chemistry of the iron-rich sedimentary rocks, *U.S. Geol. Surv. Prof. Pap. 440-W*, 60p.

Lotze, E. G., 1973. *Phosphate Retention by Lake Sediments*. Orono: Maine Univ. Dept. Plant & Soil Sci., 66p.

McConnell, D., 1973. *Apatite—Its Crystal Chemistry, Mineralogy, Utilization, and Geologic and Biologic Occurrences*. New York: Springer, 111p.

McKelvey, V. E., 1967. Phosphate deposits, *U.S. Geol. Surv. Bull. 1252-D*, 21 p.

*McKelvey, V. E., 1973. Abundance and distribution of phosphorus in the lithosphere, in Griffith et al., 1973, 13-31.
Manheim, F.; Rowe, G. T.; and Jipa, D., 1975. Marine phosphorite formation off Peru, *J. Sed. Petrology,* 45, 243-251.
Mansfield, G. R., 1940. The role of fluorine in phosphate deposition, *Am. J. Sci.,* 238, 863-889.
Marlowe, J. I., 1971. Dolomite, phosphorite and carbonate diagenesis of a Caribbean seamount, *J. Sed. Petrology,* 41, 809-827.
Pettijohn, F. J., 1963. Chemical composition of sandstones—excluding carbonate and volcanic sands, *U.S. Geol. Surv. Prof. Pap. 440-S,* 21p.
Pevear, D. R., 1966. The estuarine formation of the United States Atlantic Coastal Plain phosphate, *Econ. Geol.,* 61, 251-255.
Powell, T. G.; Cook, P. J.; and McKirdy, D. M., 1975. Organic geochemistry of phosphorites: Relevance to petroleum genesis, *Bull. Am. Assoc. Petrol. Geologists,* 59, 618-632.
Sheldon, R. P., 1963. Physical stratigraphy and mineral resources of Permian rocks in western Wyoming, *U.S. Geol. Surv. Prof. Pap. 313-B,* 49-273.
Sheldon, R. P., 1964. Paleolatitudinal and paleogeographic distribution of phosphorite, *U.S. Geol. Surv. Prof. Pap. 501-C,* 106-113.

*Additional bibliographic references mentioned in the text may be found in this work.

Cross-references: *Authigenesis; Chert and Flint; Diagenesis; Diatomite; Evaporite Facies; Marine Sediments.* Vol. IVA: *Phosphatization; Phosphorus Cycle.* Vol. IVB: *Apatite; Fluorapatite.*

PILLOW STRUCTURES

Pillow structures (or the more cumbersome term *ball-and-pillow structures*) have been noted in unconsolidated sand-mud sequences, and from sandstones of all ages and various environments. The terminology applied to these ellipsoidal or irregular masses of sand or sandstone (or clastic limestone) has been varied. The oldest name is that noted above, first used by Smith (1916). The structures have since been referred to as "balled-up" structures (Jones, 1937), pseudonodules (Macar, 1948), storm rollers (Chadwick, 1931), flow rolls (Pepper et al., 1954; Sorauf, 1965), slump balls (Kuenen, 1949), and snowball structures (Hadding, 1931). In a series of works (Potter and Pettijohn, 1963; Pettijohn and Potter, 1964; Pettijohn et al., 1973), first the name ball-and-pillow structure was used, and then pillow structure, because of their nongenetic descriptive nature. *Sand pillow* is also a useful term.

Pillows are subelliptical or subrounded masses of sandstone (or siltstone, clastic limestone) more or less surrounded by finer clastic material. They occur either as isolated masses within a shaley sequence, or as more or less locally continuous layers, generally concentrated near the base of a sandstone bed overlying shale. The internal structure of the sand pillows is a basin shape, usually opening upward, and often reflecting the bottom topography of the pillow. From above, isolated pillows display an elliptical outline, which may be modified if many pillows occur together in one layer.

Sand pillows result from the penecontemporaneous deformation of sediments (q.v.). The two main factors operative in their formation appear to be (1) foundering of denser coarser clastics into underlying finer-grained muds, often with thixotropic character; and (2) downslope movement of the foundering masses or the whole mass of mud containing sand pillows. The phenomenon does not appear to be connected to sea-floor slump phenomena, as thought by some earlier workers, nor to wave action, as proposed by others.

Occurrence

Sand pillows may occur either singly or in beds. Single pillows generally have the appearance of being isolated masses of coarser clastic material "floating" in finer sediments. All types of internal structure are found in these pillows, ranging from very simple basin structure to complexly folded interiors. Occurrence of pillows in beds is frequent, and is invariably linked to apparent flowage of more shaley material up through and around overlying coarser material, sand or silt (see *Penecontemporaneous Deformation of Sediments*). Table 1 lists some stratigraphic and geographic occurrences of pillow structures; it is not certain that all these structures are of like origin.

External Features. The bottoms of all sand pillows are convex downward, but the upper surface may also be convex (upward) or may be more or less planar. They generally have a long axis, but whether the long axes have a preferred orientation in each locality or region has not yet been satisfactorily demonstrated. Measurements of long-axis orientation may be misleading unless measurements are restricted to instances where the individual pillows can be isolated from surrounding rock matrix.

Bases of sand pillows show two types of external markings; most frequent are longitudinal lines expressed as low ridges or steps that are more or less parallel to the long axis of the pillows, but modified by the shape of the individual (Fig. 1). The ridging is the result of intersection of internal laminae with the exterior

PILLOW STRUCTURES

TABLE 1. Some Occurrences of Pillow Structures

Age	Location	Description	References
Middle Silurian	North Wales	"balled-up" structure	Jones (1937)
Upper Silurian	Ohio	pillow structures	Pettijohn and Potter (1964)
L. Devonian	Luxembourg	pseudonodule	Macar and Antun (1950)
U. Devonian	Belgium	pseudonodule	Macar (1948)
U. Devonian	New York	storm roller	Chadwick (1931)
U. Devonian	New York and Pennsylvania	flow roll	Dunbar and Rodgers (1957)
Mississippian	Ohio	flow structure	Cooper (1943)
Mississippian	Ohio	flow roll	Pepper et al. (1954)
U. Pennsylvanian	Illinois	ball-and-pillow structure	Potter and Pettijohn (1963)
U. Carboniferous	SW Wales	slump ball	Kuenen (1949)
U. Triassic	Sweden	snowball structure	Hadding (1931)
U. Cretaceous	Utah	pillow structures	Howard and Lohrengel (1969)
Holocene	Zuider Zee, Holland	pseudonodule	Macar (1951)

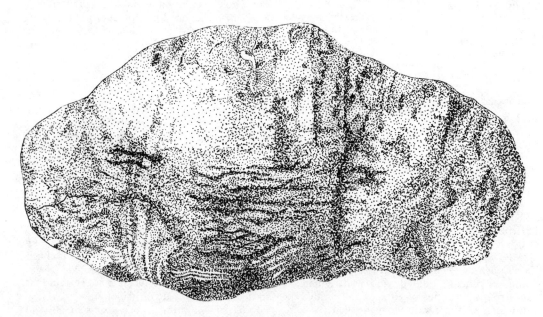

FIGURE 1. Under surface of a pillow from Upper Devonian (Frasnian) of New York (specimen approx. 35 cm long).

surface of the nodule. Ridging perpendicular to the outline of the pillow is well developed in some specimens. As shown in Fig. 1, this feature occurs as ridges and intervening grooves. These ridges and grooves are believed to result from directional pressure exerted by flowing mud in an unconsolidated or semiconsolidated state and are best developed in discrete pillows isolated in shale.

Internal Features. Macar (1948) divided "pseudonodules" in Belgium into two general types depending on their internal structure, noting that each appears to be a stage of the same phenomenon, and that numerous intermediate forms have been recorded: (1) sharply flattened, asymmetrical pillows possessing a lower surface that is well rounded, and an upper surface that is either planar or almost so (Fig. 2). The internal lamination conforms very well to the lower surface of the pillows and abuts against the upper, planar surface; (2) more globular although frequently flattened pillows with a convex upper surface; the internal laminae have a somewhat concentric aspect (Fig. 3).

The significance of different types of internal structures is not completely clear but may be related to the consistency of the materials in-

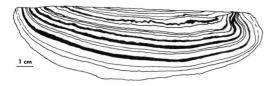

FIGURE 2. Transverse section of a pillow structure from Chemung strata (Frasnian) of New York.

volved and the rapidity with which deformation was accomplished. It should also be noted that evidence of erosional truncation of some laminae occurs commonly within pillows, which may be inherited, or erosion could have taken place during sand foundering.

Origin

Jones (1937) related "balled-up" structures to submarine slumps. Hadding (1931) had previously stated his opinion that "snowball structures" in Rhaetic sandstones of Sweden had originated as a result of sea-floor slumps. Cooper (1943) described what he called "flow structures" in the Berea Sandstone of Ohio, and stated that they were the result of contemporaneous flowage of subaqueous sediments. He regarded the surface slope as the fundamental cause of the Berea flows and thus inferred that they are the result of downslope flowage.

Kuenen (1949) noted similar structures in Carboniferous beds of SE Wales. His figured "slump balls" appear quite similar in internal structure to some pillows of the Devonian of New York and Belgium, but do differ in their attitude. Some structures suggest overturning (rotational movement) during horizontal displacement. It is likely that these ball-like masses were formed by crumpling during slumping, and although similar in appearance to usual pillow structures, did not have an identical origin.

Macar (1948) thought that "pseudonodules" in the Devonian of Belgium formed by deformation of soft, highly plastic sediments with a certain amount of tenacity either by slow vertical descent of a sandy mass into a shaley mass or by horizontal displacement of the sediments. Macar also suggested that a pocket-like load cast may be detached from the bottom of a sandstone bed by a slight movement of the overlying bed.

Emery (1950) attributed similar structures in the Pleistocene of southern California to the simple vertical descent of sand or silt into underlying finer-grained and more mobile clastics. This descent is activated by the higher specific gravity of a mixture of coarse clastics and water overlying less dense finer clastics and water. Macar (1951) also described lenticular masses of sand that had foundered, with resultant warping of internal laminae and flowage of surrounding mud without any horizontal movement. Kuenen (1958) experimentally tested the foundering hypothesis of Macar by loading sand on mud in an aquarium. He found that, with simulated seismic vibrations, sand loading and mud flowage resulted, forming kidney-shaped or nodular masses similar in internal structure to the pseudonodules described by Macar. He also showed that the "wrapped-up" shapes of the nodules can result directly from mud flowage up and around the mass of sand.

It would appear, then, that pillow structures have been subjected to one or more of three processes: (1) foundering of an isolated clastic mass in underlying finer clastics; (2) load deformation, resulting from a sheet of coarser clastics being deposited on still mobile finer clastics; and (3) horizontal displacement, submarine gliding or sliding of a nonrotational type, of coarser clastics on a glide plane of finer mud.

The pillows in the Devonian sandstones of Belgium have a preferred long axis orientation (Macar, 1948), and those examined in southern New York appear to have a preferred orientation also, at least at each outcrop (Sorauf, 1965). This orientation would indicate that: (1) there is lineation due to elongation of the nodules by the motive force connected with horizontal displacement as suggested by Macar (1948); or (2) the lineation is due to the origin of the sand bodies as originally possessing a linear shape with directional orientation (Sorauf, 1965).

Kelling and Williams (1966) believe that pillow structures in the Lower Carboniferous of Wales were oriented by later downslope movement of sediment containing already formed pillows. Thus, it is probable that in many cases both foundering and downslope movement (perhaps even slumping) have played a role in the formation of pillow structures.

JAMES E. SORAUF

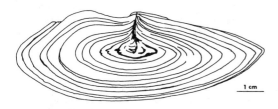

FIGURE 3. Transverse section of a pillow from the Fammenian (Upper Devonian) strata of Belgium. (Specimen courtesy of C. Ek, University of Liege.)

References

Chadwick, G. H., 1931. Storm rollers, *Geol. Soc. Am. Bull.*, **42**, 242.

Cooper, J. R., 1943. Flow structures in the Berea Sandstone and Bedford Shale of Central Ohio, *J. Geol.*, **51**, 190-203.

Dunbar, C. O., and Rogers, J., 1957. *Principles of Stratigraphy*. New York: Wiley, 356p.

Emery, K. O., 1950. Contorted Pleistocene strata at Newport Beach, California, *J. Sed. Petrology*, **20**, 111-115.

Hadding, A., 1931. On subaqueous slides, *Geol. Foren.*, **53**, 377-393.

Howard, J. D., and Lohrengel, C. F., II, 1969. Large non-tectonic deformational structures from Upper Cretaceous rocks of Utah, *J. Sed. Petrology*, **39**, 1032-1039.

Jones, O. T., 1937. On the sliding or slumping of submarine sediments in Denbigshire, North Wales, during the Ludlow Period, *Quart. J. Geol. Soc. London*, **93**, 241-283.

Kaye, C. A., and Power, W. R., Jr., 1954. A flow cast of very recent date from northeastern Washington, *Am. J. Sci.*, **252**, 309-310.

Kelling, G., and Williams, B. P. J., 1966. Deformation structures of sedimentary origin in the Lower Limestone Shales (basal Carboniferous) of Pembrokeshire, Wales, *J. Sed. Petrology*, **36**, 927-939.

Kuenen, P. H., 1949. Slumping in the Carboniferous rocks of Pembrokeshire, *Quart. J. Geol. Soc. London*, **104**, 365-385.

Kuenen, P. H., 1958. Experiments in geology, *Geol. Soc. Glasgow Trans.*, **23**, 1-28.

Macar, P., 1948. Les pseudo-nodules du Fammenien et leur origin, *Ann. Soc. Géol. Belg.*, **72**, 47-74.

Macar, P., 1951. Pseudo-nodules en terrains meubles, *Ann. Soc. Géol. Belgique*, **75**, 111-115.

Macar, P., and Antun, P., 1950. Pseudo-nodules et glissement sous-aquatique dans l'Emsien inférieur de l'Oesling (Grand-duché de Luxembourg), *Ann. Soc. Géol. Belgique*, **73**, 121-149.

Pepper, J. F.; DeWitt, W. Jr.; and Demarest, D. F., 1954. Geology of the Bedford Shale and Berea Sandstone in the Appalachian Basin, *U.S. Geol. Surv. Prof. Pap. 259*, 109p.

Pettijohn, F. J., and Potter, P. E., 1964. *Atlas and Glossary of Primary Sedimentary Structures*. New York: Springer-Verlag, 370p.

Pettijohn, F. J.; Potter, P. E.; and Siever, R., 1973. *Sand and Sandstone*. New York: Springer-Verlag, 618p.

Potter, P. E., and Pettijohn, F. J., 1963. *Paleocurrents and Basin Analysis*. New York: Academic Press, 296p.

Smith, B., 1916. Ball- or pillow-form structures in sandstones, *Geological Mag.*, 6th ser., **3**, 146-156.

Sorauf, J. E., 1965. Flow rolls of Upper Devonian rocks of south-central New York, *J. Sed. Petrology*, **35**, 553-563.

Cross-references: *Gravity Flows; Intraformational Disturbances; Penecontemporaneous Deformation of Sediments; Slump Bedding; Water-Escape Structures.*

PIPETTE ANALYSIS—See GRANULOMETRIC ANALYSIS

PISOLITE

The term *pisolite* is used to describe certain rather distinctive grain types found in sediments and sedimentary rocks. Pisolite grains (pisoids, pisoliths) are comparable to *oolite* (q.v.) grains but distinguished by their pea size (generally 2-10 mm diameter). In cross section, pisolite grains are seen to be made up of concentric laminae; in some cases, a nuclear grain serves as an initial site for the accretion of the laminae (Fig. 1a). The composition varies, but commonly consists of calcium carbonate (calcite or aragonite); occurrences of bauxite, gibbsite, hematite, limonite, silica and phosphate pisolites are also known (Carozzi, 1960).

Origin

Three modes of origin have been suggested for pisolite grains: (1) chemical precipitation from agitated solution; (2) chemical or biochemical precipitation from nonagitated or quiescent solutions; and (3) during soil formation, especially in humid tropical or subtropical climates, e.g., in calcareous hardpan (see *Caliche*) and lateritic crusts (see *Duricrust*).

Calcareous pisolites which are *precipitated from agitated solution* are relatively common in caves and mines (see *Cave Pearls*) and are also found in geyser and hot spring deposits. In the former, drops of water falling from the ceilings of caves and mines constantly agitate and move the growing grains. Thus, the envelope of concentric laminae grows by chemical precipitation and is abraded by contact with neighboring grains, resulting in a nearly spherical shape with continued growth of concentric laminae (Donahue, 1969). In hot springs, loss of CO_2 from hot, mineralized waters causes rapid precipita-

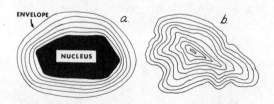

FIGURE 1. A: Cross section of a pisolite grain formed under agitated conditions; original irregularities in the nucleus are rounded by the accretion of concentric laminae. B: Cross section of a pisolite grain formed under quiescent conditions; irregularities are increased by the addition of new concentric laminae.

tion of calcium carbonate, e.g., at the Sprudel of Karlsbad (Karlovy Vary, Czechoslovakia); there is a similar pisolitic sinter at Vichy in France. Water currents and boiling within the hot springs and geysers keep the accreting grains in motion and cause abrasion (see *Spring Deposits*).

Pisolite grains *formed in nonagitated solutions* have distinctly different characteristics, mainly due to a lack of abrasion. In cross section, they tend to have an irregular shape (Fig. 1b). Original irregularities tend to be increased with the addition of new laminae; the presence of a nucleus is not essential. Calcareous pisolites of this sort form either chemically or biochemically (see *Oolite;* Bathurst, 1971). Phosphatic, ferruginous, and siliceous compositions are also known; in most cases, they are due to replacement of an originally calcareous grain. The ferruginous pisolites (and oolites) are problematic. In Britain, the Penrhyn Ironstone (Ordovician), composed of hematite and limonite, is believed to have been altered during early diagenesis from chamosite. The Pisolitic Ironstones of North Wales (Cambrian/Ordovician) are black due to finely divided magnetite and pyrite, probably the product of slight metamorphism of chamosite (Pulfrey, 1933).

In humid tropical and subtropical climates, pisolitic *gravels* often develop in the evolution of ferruginous duricrust from a laterite soil. These soils are rich in the oxides of iron and aluminum, which after a complex evolution of climatic change, form limonite, hematite, gibbsite, and bauxite. Quite often during arid-phase dehydration, the grains are broken into small fragments that provide further nuclear centers. Additional oxide is deposited on these fragments as concentric layers (Dunham, 1969). Thus, thick sections of the soil profile develop a pisolitic texture. Inasmuch as the grains develop without movement, they closely resemble calcareous grains formed in quiescent solutions. Scholle and Kinsman (1974) believe that the mineralogy of pisolitic caliche described by Dunham could be produced in a hypersaline vadose zone, as is found in a sabkha (q.v.) environment (also see *Caliche*). Assereto and Kendall (1977) have reviewed tepee—pisolite occurrences in Silurian through Holocene carbonates. A *subaqueous* origin for Dunham's "vadose pisolite" and associated tepees, as exemplified by the shelf-crest facies of the Permian Reef Complex, Guadalupe Mts., has been proposed by Esteban and Pray (1977).

A somewhat similar process produces pisolitic textures in clay minerals such as kaolinite and calcareous caliche (q.v.) soils in desert and semiarid regions. An ancient example of a pisolitic limonite is found in the Etruria Marl (Upper Carboniferous) of Staffordshire in England, which Hallimond (1925) has interpreted as an ancient laterite. The pisolites have quartz grains as nuclei. Possibly related are the sphaerosiderites found in some underclays of the Carboniferous in England, interpreted as products of stagnant swamp soils (Hatch et al., 1971). One recent report on bauxite pisolites associates them with diagenesis of intertidal tropical mangrove swamps (Valenton et al., 1973).

JACK DONAHUE

References

Assereto, R. L. A. M., and Kendall, C. G. St. C., 1977. Nature, origin and classification of peritidal tepee structures and related breccias, *Sedimentology,* 24, 153-210.

Bathurst, R. G., 1971. *Carbonate Sediments and Their Diagenesis.* Amsterdam: Elsevier, 620p.

Carozzi, A. V., 1960. *Microscopic Sedimentary Petrography.* New York: Wiley, 485p.

Donahue, J., 1969. Genesis of oolite and pisolite grains: An energy index, *J. Sed. Petrology,* 39, 1399-1411.

Dunham, R. J., 1969. Vadose pisolite in the Capitan Reef (Permian), New Mexico and Texas, *Soc. Econ. Paleont. Mineral. Spec. Publ. 14,* 182-191.

Esteban, M., and Pray, L. C., 1977. Origin of the pisolite facies of the shelf crest, *Soc. Econ. Paleont. Mineral. Permian Basin Section 1977 Field Conf. Guidebook (Publ. 77-16),* 479-483.

Hallimond, A. F., 1925. Iron ores: Bedded ores of England and Wales, *Gr. Brit. Geol. Surv. Spec. Repts. Mineral Resources,* 29, 139p.

Hatch, F. H., and Rastall, R. H., rev. by Greensmith, J. T., 1971. *Petrology of the Sedimentary Rocks,* 5th ed. London: Unwin Brothers, 502p.

Pulfrey, W., 1933. The iron-ore oolites and pisolites of North Wales, *Quart. J. Geol. Soc. London,* 89, 401-430.

Read, J. F., 1974. Calcrete deposits of Quaternary sediments, Edel Province, Shark Bay, Western Australia, *Am. Assoc. Petrol. Geologists Mem. 22,* 250-282.

Scholle, P. A., and Kinsman, D. J. J., 1974. Aragonitic and high-Mg calcite caliche from the Persian Gulf—a modern analog for the Permian of Texas and New Mexico, *J. Sed. Petrology,* 44, 904-916.

Valenton, I.; Jürgens, U.; and Khoo, F., 1973. Prebauxite red sediments and sedimentary relicts in Surinam bauxites, *Geol. Mijnbouw,* 52, 317-334.

Cross-references: *Caliche, Calcrete; Carbonate Sediments—Diagenesis; Cave Pearls; Duricrust; Ironstone; Limestones; Oolite; Sabkha Sedimentology; Speleal Sediments; Spring Deposits.*

PLAYA—See DESERT SEDIMENTARY ENVIRONMENTS; LACUSTRINE SEDIMENTATION; SALAR, SALAR STRUCTURES

PORCELANITE

The term *porcelanite* (porcellanite) was introduced into sedimentary petrology by Taliaferro (1934). After some refinements by Bramlette (1946), Ernst and Calvert (1969), and Calvert (1971), it is generally used today to describe a marine siliceous rock in which the major constituent is α-cristobalite with various degrees of stacking disorders (see *Chert*). The term is somewhat unfortunate because it describes different rocks in other branches of geology: (1) in metamorphic petrology, "porcelanite" applies to a light-colored porcelaneous rock from contact metamorphism of marl (q.v.); and (2) in coal geology, it describes fused shales and clay that occur in the roof and floor of burned coal seams. Even in sedimentology, "porcelanite" has been given more than one meaning, some authors applying it merely to a transitory rock type between siliceous shale and chert. Porcelanites that have been recovered by the DSDP were often termed in their initial reports as "cherts," "lussatite cherts" and "porcelanitic cherts." To avoid ambiguity, the definition quoted above should be retained. In the Russian literature the terms *opoka* and *trepel* (*tripoli*), which are apparently equivalent to porcelanite, are widely used (see G. I. Bushinskii in Rukhin, 1958).

Mineralogy-petrography

The color of porcelanite varies between white and gray; as the name implies, it has the outward appearance of unglazed porcelain. Porcelanite is brittle and extremely porous (up to 50% by volume, bulk density 0.9-1.8), but unpermeable.

In recent years, the understanding of the nature of the siliceous phase in porcelanites has vastly increased. What has before been referred to as "opal" has been shown to consist of several different phases (see *Silica in Sediments*). Opal as it occurs in porcelanites exhibits characteristic x-ray diffraction patterns and infrared absorption spectra. Both have been interpreted as representing α-cristobalite with various degrees of stacking disorder producing tridymite maxima. The peak width indicates crystallite size of 50-130Å.

Opal as it occurs in porcelanite has been termed by Jones and Segnit (1971) *opal-CT*. It differs distinctly from the highly disordered, nearly amorphous opal of siliceous organisms, diatomites, and geyserites (opal-A). The term *lussatite* has been used in European literature as describing a phase identical with opal-CT. Weaver and Wise (1972) showed by scanning electron microphotography that α-cristobalite in porcelanites from the deep sea occurs as fine cristobalite blades which grow in spherules 3-10μ in diameter. Porcelanite contains 50-90% opal-CT. Nonopaline constituents are clays, calcite, apatite and dolomite. In transmitted polarized light, thin sections of porcelanites are mostly isotropic. The refractive index of the opal varies from 1.458-1.462. The major rock type that occurs in paragenesis with porcelanite is chert (q.v.), in which granular microcrystalline quartz is the predominant mineral.

Occurrence

Porcelanites occur in marine formations of Upper Mesozoic and Cenozoic age, throughout the globe, generally intimately associated with chert. Porcelanites are most conspicuous in Cretaceous and Tertiary formations; none has been reported from pre-Mesozoic rocks. The on-land formations in which porcelanites were extensively studied are the Miocene Monterey Formation of California (Bramlette, 1946) and the Cretaceous Mishash Formation in Israel (Kolodny et al., 1965).

Extensive layers of deep-sea porcelanite and chert have been encountered in many of the Deep Sea Drilling Project (DSDP) drill-holes. Porcelanites were recovered from both the Atlantic and Pacific oceans; in several locations, they were identified with reflector Horizon A. The oldest reported porcelanite is Lower Cretaceous (Keene, 1975).

Genesis

The problem of genesis of porcelanite, as the more general problem of formation of marine siliceous rock, can be subdivided into three major questions: the source of silica (see *Silica in Sediments*); the mode of silica emplacement; and the relationships between amorphous silica, porcelanite, and chert.

A polygenetic origin applies to the mode of porcelanite formation as well. Whereas there are some porcelanites that are definitely of replacement origin, there are others that seem to be direct precipitates (Calvert, 1971). The formation of porcelanite in either mode is the result of mobilization of highly soluble amorphous silica (at room temperature, ≈ 120 ppm) and precipitation of much-less soluble opal-CT (10-15 ppm).

The transformation of porcelanite to chert (q.v.) has been achieved experimentally by Ernst and Calvert (1969), who found it to be a zero-order reaction. Heath and Moberly (1971) also present evidence for a solid-solid transition in this case. Among the major observations which support the concept of "maturation" of porcelanite into chert is the reported predominance of porcelanite in young rocks, and of chert in pre-Cretaceous formations. Lancelot (1973) questions this evidence and claims that the major factor determining the mineralogy of the predominant silica phase is the lithology of the host sediments (clays causing porcelanite formation, whereas chert forms in carbonates). Kolodny et al. (1965) proposed a penecontemporaneous origin of porcelanite and chert, both early diagenetically replacing a carbonate mud. The crucial factor may be, as suggested by Lancelot, the presence of foreign cations during the crystallization of silica.

YEHOSHUA KOLODNY

References

Bramlette, M. N., 1946. The Monterey Formation of California and the origin of its siliceous rocks, *U.S. Geol. Surv. Prof. Pap. 212*, 57p.

Calvert, S. E., 1971. Composition and origin of North Atlantic deep sea cherts, *Contr. Mineral. Petrology*, 33, 273-288.

Ernst, W. B., and Calvert, S. E., 1969. An experimental study of the recrystallization of porcelanite and its bearing on the origin of some bedded cherts, *Am. J. Sci.*, 267-A, 114-133.

Heath, G. R., and Moberly, R., 1971. Cherts from the Western Pacific Leg 7, Deep Sea Drilling Project, *Init. Repts. Deep Sea Drilling Proj.*, 7, 991-1007.

Jones, J. B., and Segnit, E. R., 1971. The nature of opal. I: Nomenclature and constituent phases, *J. Geol. Soc. Australia*, 18, 57-68.

Keene, J. B., 1975. Cherts and porcellanites from the North Pacific, DSDP, Leg 32, *Init. Repts. Deep Sea Drilling Proj.*, 32, 429-509.

Kolodny, Y.; Nathan, Y.; and Sass, E., 1965. Porcelanite in the Mishash Formation, Negev, southern Israel, *J. Sed. Petrology*, 35, 454-463.

Lancelot, Y., 1973. Chert and silica diagenesis in sediments from the Central Pacific, *Init. Repts. Deep Sea Drilling Proj.*, 17, 577-405.

Rukhin, L. B., ed., 1958. *Spravochnoe Rukovodstvo po Petrografii Osadochnykh Porod*, vol. 2. (Handbook of petrography of sedimentary rocks, in Russian). Leningrad: Gosudar. Nauk.-Teckh. Izd. Neft. Gorno-Topliv. Lit., 520p.

Taliaferro, N. L., 1934. Contraction phenomena in cherts, *Geol. Soc. Am. Bull.*, 45, 189-231.

Weaver, F. M., and Wise, S. W., 1972. Ultramorphology of deep sea cristobalitic chert, *Nature, Phys. Sci.*, 237, 56-57.

Cross-references: *Chert and Flint; Diatomite; Novaculite; Silica in Sediments; Silicification.*

PRESSURE SOLUTION AND RELATED PHENOMENA

Pressure solution (solution transfer or *lösungsumatz*) involves solution around points of contact between mineral grains in response to pressure, usually the weight of the overburden. Because stressed solids are more soluble than unstressed ones, the increased local pressure at grain contacts increases the solubility of the minerals. The principle of increased solubility with nonuniform pressure is known as the *Riecke Principle* (E. Riecke, 1894, in Trurnit, 1968). As a sediment is buried to considerable depths, appreciable excess pressure is generated at horizontal grain contacts and pressure solution becomes increasingly important. Certain minerals are more susceptible to pressure solution than others, e.g., the development of impressions in limestone pebbles requires burial depths of only 30-40 m; in sandstones, smooth pressure-solution contacts appear at 800-1050 m, and differentiated ones at 1000,1250, and 1600 m (Trurnit, 1968).

These observations have led to the formulation of a series of minerals showing declining relative pressure solubilities (see Trurnit, 1968): (1) halite and potassium salts; (2) calcite; (3) dolomite; (4) anhydrite; (5) gypsum; (6) amphibole; (7) chert; (8) quartzite; (9) quartz, glauconite, rutile, hematite; (10) feldspars and cassiterite; (11) mica and clay minerals; (12) arsenopyrite; (13) tourmaline and sphene; (14) pyrite; (15) zircon; and (16) chromite. Thus, *macrostylolites* usually develop in carbonates and *microstylolites* in sandstones; the former begin to form early in diagenesis, and the latter require greater pressures and temperatures, which usually result from deeper burial but can result from tectonic stress and/or intrusions (Habermehl, 1970).

In both carbonates and sandstones, pressure solution may act as an important process for supplying carbonate (Bathurst, 1971) and silica (Berner, 1971) for cement; the carbonate or silica in solution may be precipitated in pore spaces adjacent to pressure solution locations or they may migrate away (see *Cementation*).

Pressure-Solution Phenomena

Stylolites. Stylolites (Greek: *stylos*, pillar; *lithos*, stone) are thin contacts of discontinuity; in cross section they have undulating to zigzagged braces (Fig. 1); in plan they consist of conical or columnar projections with intervening depressions such that the opposing sides fit together in a complementary manner. They occur most commonly in homogeneous rocks, especially carbonates, cherts, sandstones, and more rarely in certain igneous rocks (Park and Schot, 1968). They vary from microscopic sutured contacts (microstylolites), as between adjacent quartz grains, up to 15 m or more in length (Park and Schot, 1968); amplitudes up to 10 m have been recorded in ice (Plessman, 1972).

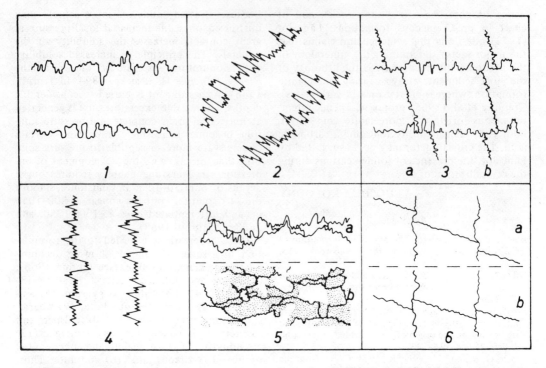

FIGURE 1. Classification of stylolites in relation to the bedding plane (from Park and Schot, 1968). (1) horizontal stylolites, (2) inclined stylolites, (3) horizontal-inclined (vertical)-cross-cutting stylolites, (4) vertical stylolites, (5) interconnecting network stylolites, (6) vertical-inclined (horizontal)-cross-cutting stylolites.

Park and Schot (1968) define two types of stylolites: (1) aggregated stylolites with amplitudes greater than the grain diameter of the rock; and (2) intergranular stylolites with amplitudes smaller than the grain size of the host rock, which may be microscopic, as between two sand grains (microstylolites) or macroscopic, as between pebbles. They classify stylolites on a geometric basis, first in cross section and second by their relationship to the bedding plane of the host rock (Fig. 1). Trurnit (1968) has published a detailed classification of intergranular pressure solution contacts (Fig. 2) and has produced a set of formulae to account for the types encountered, based on their geometry and genesis.

Stylolites commonly run parallel to the bedding but may be oblique or perpendicular. Because the projections lie parallel to the direction of principal stress (Plessman, 1972) and normal to the stylolitic surface, the parallel stylolite type occurs where the principal stress was caused by the overburden pressure and the oblique type occurs in tectonic areas.

The apex of each of the projections is characterized by the presence of a grain, or group of grains, less susceptible to pressure solution than the grains on the opposite side of the stylolite seam; the configuration of the differentiated surface develops in direct response to the relative pressure solubility contrast of the vertically adjacent grains that come into contact across the stylolite (Mossop, 1972). In contrast, in intergranular stylolites the greatest differentiations are seen in rocks where the two sides have similar solubilities (Trurnit, 1968).

The contact is lined with insoluble material derived from the host rock; commonly this is clay with organic or carbonaceous material, silica, and iron minerals. All minerals that have a lower pressure solubility than calcite may become part of the solution residue of a pressure-solution surface in limestone, and those with a pressure solubility lower than quartz as the residue in a sandstone (Trurnit, 1968). Authigenic minerals may also be formed along the seams, e.g., chlorite. The residual material may be thicker at the crests and troughs; it may occur either as one dark seam, or as many semiparallel sets of very thin seams indicating periodic intervals of dissolution.

Laterally, many stylolite seams either die out in a bundle of small stylolites, or pass into clay partings, several cm thick. In many cases, evidence for an origin from a stylolite precursor comes from remnant stylolitic projections protruding from those residual seams. It appears that the insoluble material has inhibited

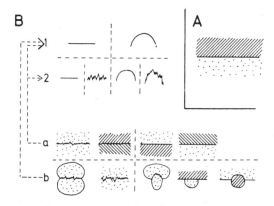

FIGURE 2. A: Pressure-solution phenomena are composed of two partners and a contact surface (pressure-solution plane). B: The contact surface may be (1) plain or curved, (2) smooth or differentiated (sutured, stylolitic); the partners may have (a) equivalent or different relative pressure solubility in the direction of stress, (b) equivalent or different radii of curvature at the contact. The geometry of the contact surface depends on the properties of the partners in contact. While point (1) is influenced by (a) and (b), point (2) is determined only by (a). (From Trurnit, 1968.)

the continued formation of the stylolites. Stylolitic surfaces may truncate fossils, ooliths, mineral grains, veins, and other stylolites. This truncation, together with the presence of the relatively insoluble residues derived from the host rock, has supported an origin by solution during compaction of the sediment. In many cases the limestone undersurface of a clay parting shows slickolitic slickensides; the stylolites may also be associated with microfaulting displacing vertical diagenetic calcite veinlets (Park and Schot, 1968). Thus, stylolites may act as glide planes to relieve stress along the surface (Wanless, 1973). Stylolites tend to be barriers to fluids migrating perpendicular to them, but are channelways for fluids moving laterally along the stylolite surface, including petroleum (see Dunnington, 1967).

Cone-in-Cone Structures. Cone-in-cone structures consist of stacks of inverted right circular cones, the axes of which are normal to the bedding plane (Fig. 3). They generally occur in the peripheral few cm of thin beds and lenses of fine-grained sedimentary rocks, especially impure carbonates, sandstones, mudstones, clay ironstones, and occasionally coals and sulfates. These structures commonly occur along the upper surface only, where the concentric arrangement of the cones can be seen in transverse section on the bedding plane, and occasionally also along the lower surface. The altitude is typically 10–100 mm, exceptionally 200 mm. The slope of the sides typically varies from $30°-60°$; the diameter of the cone varies from about 1/3 to near equality with the altitude of the cone; in some cases, the cones have flaring bases (Pettijohn, 1957). The sides of the cones may be warped and marked with grooves parallel to the slant height of the cone, and may be lined with argillaceous material. Many cones are marked by annular depressions which are strongest near the base (Pettijohn, 1957). Most commonly, the cones are formed

FIGURE 3. Cone-in-cone structure surrounding a bedding-plane concretion (magnification approx. 2X). (Photo: courtesy B. M. Shaub.)

from fibrous calcite, the fibers tending to be parallel to the axis of the cone. MacKenzie (1972) described fibrous calcite beds containing cone-in-cone structures up to 7 cm in height occurring at a constant stratigraphic position over 24,000 km^2.

Several theories have been proposed to account for the origin of cone-in-cone, the general consensus being that they are diagenetic. MacKenzie (1972) suggested that the fibrous calcite beds develop in partly consolidated muds, close to the sediment/water interface, in response to changes in the physicochemical environment; organic remains concentrated along bedding planes probably acted as catalysts to initiate the growth of calcite fibers. He concluded that cone-in-cone structures in fibrous calcite are a secondary phenomenon formed in response to stresses in the rock, induced following lithification. The fibrous structure antedates the formation of the cones because the cone structure transgresses the fibers and normally extends across the fibrous seam; but the fibrous character is not essential to the structure because cone-in-cone structures occur in coals (Pettijohn, 1957). Indeed there is abundant evidence that these conical structures are a response to pressure from the interior of the bed, directed outward, usually upward; but they may in some cases be analogous structures to percussion cones. For a discussion of the differences between cone-in-cone structures, which are of diagenetic origin, and shatter cones, which form by impact, see French and Short (1968). The pressures built up by the growth of the fibrous layers is responsible for the conical shearing.

Usdowski (1963) showed that trace elements were expelled during recrystallization of the carbonate. Because the cone-in-cone structures are usually calcareous, solution occurs along the conical shear surfaces in response to pressure and is responsible for the annular corrugations and clay films between the cones; such features are related in origin to stylolites.

Associated Features. Zones of intense stylolite development are commonly associated with cementation and the lining or filling of voids with sparry calcite or quartz (Mossop, 1972; Park and Schot, 1968), suggesting that all these structures owe their formation to pressure solution during early diagenesis (Park and Schot, 1968). This is supported by the fact that voids and cavities are thought to result from chemical erosion by waters squeezed out of the sediment during compaction. The process of pressure solution could well account for the redistribution and differentiation of large quantities of material dissolved during stylolitization, by their use in the reduction of pore space and the compaction of the sediment.

Mode of Origin of Pressure-Solution Phenomena

The reality of the pressure solution process is supported by both physical and chemical considerations (Blatt et al., 1972); the main points of controversy are the factors that influence and promote pressure solution and the mechanism by which it occurs. There is still some controversy, however (see, e.g., Deelman, 1975).

When a sediment is buried, the overburden pressure and the consequent compaction provide the pore solution migration necessary for pressure solution to occur. Because solubility increases with increasing pressure, a decreasing concentration gradient is built up between the grain contact zone and the adjacent lower-pressure pore space; consequently, outward diffusion occurs, resulting in the dissolution of the particles in the contact zone (Berner, 1971). The rate of diffusion is controlled by the grain size, the effective normal stress between the grains, the diffusion constant in the diffusion film, the film thickness, clay concentration, and the stress coefficient of the sol (Blatt et al., 1972).

Weyl (1959) suggested that, in sandstones, a film of water a few molecules in thickness held by the attraction of the quartz grain surfaces serves as a migration path for the diffusion of the dissolved silica. Stylolitization may be enhanced by a high content of insoluble material in the original rock (Flügel, 1968; Mossop, 1972; Wanless, 1973). The correlation between illite occurrences and pressure solution in many sandstones led Thomson (1959) to suggest that K^+ is leached from the illite and replaced by H^+, resulting in increased alkalinity in the interstitial solution, pH decreasing away from the clay. Because quartz is more soluble above pH 9, dissolution will occur at the clay/quartz contact. Weyl (1959) considered that clay increased the diffusion rate, because a 10μ clay film will contain about 5000 water films compared with 1 or 2 between clean grains. Apparently the lower coefficients of diffusion in the adsorbed and interlayer water are compensated by the greater cross section of diffusion (Berner, 1971). To promote pressure solution, the requirements are preferably a porous material saturated with water (not necessarily clay) of a small size compared with the grain size of the rock, and it must not itself be subject to rapid pressure solution (Weyl, 1959).

Pressure solution is effective in the bulk

reduction of carbonate rocks. For example, Mossop (1972) studied a reef complex and showed that stylolitization accounted for a total reduction of 13% (55 m) on the rim and 24% (100 m) in the lagoon; he attributed the difference to the higher initial clay content of the lagoonal carbonates. Appreciable amounts of bulk volume reduction must accompany pressure solution if material for cementation is to be released (Manus and Coogan, 1974). Important parameters in pressure solution are the shape and size of the grains, the surface tension, temperature, and pore fluid composition (Trurnit, 1968).

The episodic development of stylolites may be accounted for as follows: the lack of processes for causing the large-scale movement of fluids results in difficulties in changing the composition of the interstitial water (Blatt et al., 1972); once solutions have become saturated, stylolitization will cease until either these solutions are replaced by unsaturated ones, or until increased pressures and temperatures increase the capacity of the original fluids. The need for constant flushing explains why pressure solution and cementation usually occur in sandstones and limestones rather than in muds. Pressure solution, including stylolitization, is a large, drawn-out process which operates throughout the diagenetic history of its host rock (Park and Schot, 1968).

Although present knowledge is not comprehensive enough to dictate a consistent set of textural and/or compositional parameters to determine which sediments are preferentially susceptible to pressure solution, the important fact is that the original sediments control the extent to which pressure solution will proceed given similar conditions of temperature and pressure.

DIANA R. THURSTON

References

Bathurst, R. G. C., 1971. *Carbonate Sediments and Their Diagenesis.* Amsterdam: Elsevier, 620p.

Berner, R. A., 1971. *Principles of Chemical Sedimentology.* New York: McGraw-Hill, 240p.

Blatt, H; Middleton, G.; and Murray, R., 1972. *Origin of Sedimentary Rocks.* Englewood Cliffs, N.J.: Prentice-Hall, 634p.

Deelman, J. C., 1975. "Pressure solution" or indentation? *Geology,* 3, 23-24. (See also discussion—and reply—by Atkinson and Rutter, 477-478.)

Dunnington, H. V., 1967. Aspects of diagenesis and shape change in stylolitic limestone reservoirs, *7th World Petrol. Congress,* 2, 39-352.

Flügel, H. W., 1968. Some notes on isoluble residues in limestones, in G. Muller and G. M. Friedman, eds., *Carbonate Sedimentology in Central Europe.* New York: Springer-Verlag, 46-54.

French, B. M., and Short, N. M., eds., 1968. *Shock Metamorphism of Natural Materials.* (Proceedings of the first Conference, Goddard Space Flight Center, 1966.) Baltimore: Mono Book Corp., 644p.

Habermehl, M. A., 1970. Depositional history and diagenesis quartz-sand bars and lime-mud environments in the Devonian Basibe Formation (Central Pyrenees, Spain), *Leidse Geol. Meded.,* 46, 1-55.

MacKenzie, W. S., 1972. Fibrous calcite, a Middle Devonian geologic marker with stratigraphic significance, District of MacKenzie, Northwest Territories, *Canadian J. Earth Sci.,* 9, 1431-1440.

Manus, R. W., and Coogan, A. H., 1974. Bulk volume reduction and pressure solution derived cement, *J. Sed. Petrology,* 44, 466-471.

Mossop, G. D., 1972. Origin of the peripheral rim, Redwater Reef, Alberta, *Bull. Canadian Petrol. Geol.,* 20, 238-280.

Park, W. C., and Schot, E. H., 1968. Stylolites: Their nature and origin, *J. Sed. Petrology,* 38, 175-191.

Pettijohn, F. J., 1957. *Sedimentary Rocks.* New York: Harper & Row, 718p.

Plessman, W., 1972. Horizontal-Stylolithen im französischschweizerischen Tafekind Faltenjura und ihre Einpassung in den regionalen Rahmen, *Geol. Rundschau.,* 61, 332-347.

Thomson, A., 1959. Pressure solution and porosity, *Soc. Econ. Paleont. Mineral., Spec. Publ. 7,* 92-110.

Trurnit, P., 1968. Pressure solution phenomena in detrital rocks, *Sed. Geol.,* 2, 89-114.

Usdowski, H. E., 1963. Die Genese der Tutenmergel oder Nagelkalke (cone-in-cone), *Beitr. Mineral. Petrog.,* 9, 95-110.

Van Tassel, R., 1971. Le constituent carbonate des concrétions cone-in-cone du Houiller belge, *Bull. Soc. Belge Géol., Paléont. Hydrol.,* 80, 21-29.

Wanless, H. R., 1973. Microstylolites, bedding and dolomitization, *Bull. Am. Assoc. Petrol. Geologists,* 57, 811.

Weyl, P. K., 1959. Pressure solution and the force of crystallization—a phenomenological theory, *J. Geophys. Research,* 64, 2001-2025.

Cross-references: *Carbonate Sediments—Diagenesis; Cementation; Clastic Sediments—Lithification and Diagenesis; Compaction in Sediments; Diagenesis; Limestone Fabrics; Silicification.*

PRIMARY SEDIMENTARY STRUCTURES

Structures that develop in sediments during the process of deposition are called primary sedimentary structures (Table 1). These depositional structures may be of physical or of biologic origin (see *Biogenic Sedimentary Structures*). Primary sedimentary structures of physical origin depend mainly on current speed and the rate of sedimentation for their formation. Sedimentation is determined by the rate of sediment supply. It is customary to include in this category only those primary sedimentary

TABLE 1. Classification of Common Primary Sedimentary Structures

Physical Structures	Biologic Structures

A. Bedding: geometry
   1. Laminae (q.v.)
   2. Wavy bedding
B. Bedding: internal structures
   1. Cross-bedding (q.v.)
   2. Ripple marks (q.v.)
   3. Graded bedding (q.v.)
   4. Growth bedding
C. Bedding: plane marks
   1. Scour marks (q.v.)
   2. Tool marks (q.v.)
   3. Parting lineation (q.v.)
   4. Wave and swash marks
   5. Pits and prints
      e.g., raindrop imprints (q.v.)
D. Deformed bedding
   1. Load and founder structures
      e.g., pillow structures (q.v.)
   2. Synsedimentary folds and breccias (q.v.)
   3. Clastic dikes (q.v.) and sills
	A. Petrifactions
	
	B. Layering
	   e.g., stromatolites (q.v.)
	
	C. Miscellaneous
	   1. Tracks and trails
	   2. Casts and molds (q.v.)
	   3. Borings
	   4. Coprolites and fecal pellets

After Pettijohn, 1975.

structures that are large enough to be studied in outcrop rather than in hand specimens or under the microscope.

Primary sedimentary structures of the *planar bedding* type are useful in paleocurrent analysis (q.v.); in flow regime (q.v.) studies; and, in some cases, in determining the nature and magnitude of sediment supply just before deposition. *Linear* and *planar* structures are indicative of the current dynamics at the sediment/fluid interface. The linear structures may be parallel to the current flow (striae, etc.), or perpendicular to it (ripple marks, etc.). Some structures and marks, such as spatulate depressions, striae, and grooves, are preserved only as molds on the underside of the overlying stratum.

Deformed, disrupted structures may result from disturbances prior to the deposition of the overlying bed or may be penecontemporaneous with deposition (see *Penecontemporaneous Deformation of Sediments*). Deformation or disruption under these circumstances may show characteristics of mobilization, folding, faulting, and subsequent injection of mobile materials into the newly created voids.

Primary sedimentary structures of biologic origin include fossils preserved in various degrees of alteration. Biogenic structures may be minor features, or, as in the case of limestones, the dominant structures present (see *Stromatolites*). Structures produced by organisms, e.g., tracks, trails, borings, and other similar marks are also included here.

IMRE V. BAUMGAERTNER

References

Conybeare, C. E. B., and Crook, K. A. W., 1968. Manual of sedimentary structures, *Australia Dept. Nat. Dev. Brit. Min. Resources, Geology Geophys. Bull.*, 102, 327p.

Pettijohn, F. J., 1975. *Sedimentary Rocks,* 3rd ed. New York: Harper & Row, 628p.

Pettijohn, F. J., and Potter, P. E., 1964. *Atlas and Glossary of Primary Sedimentary Structures.* New York: Springer-Verlag, 370p.

Shrock, R. R., 1948. *Sequence in Layered Rocks.* New York: McGraw-Hill, 507p.

Cross-references: *Backset Bedding; Bedding Genesis; Biogenic Sedimentary Structures; Casts and Molds; Clastic Dikes; Cross-Bedding; Current Marks; Fabric, Sedimentary; Flame Structures; Graded Bedding; Intraformational Disturbances; Inverse Grading; Lamina, etc.; Mud Cracks; Parting Lineation; Penecontemporaneous Deformation of Sediments; Pillow Structures; Ripple Marks; Scour Marks; Secondary Sedimentary Structures; Sedimentary Structures; Stromatolites; Tool Marks; Varves and Varved Clays.*

PRODELTA SEDIMENTS

The *prodelta* is that part of the deltaic environment that is seaward of the delta front, the open-water marine (or lacustrine) basin into which the delta is advancing. Prodelta deposits are the first terrigenous sediments introduced into a depositional area by an advancing delta. The vast majority of these materials are deposited in a broad fan about the delta front. In large deltas, the deposits consist of fine-grained, highly plastic clays at distance from the active delta front and silty or sandy clays just seaward of the delta front.

The term *prodelta clay* is somewhat correlative with Gilbert's *bottomset beds* (see *Deltaic Sediments*). Both apply to the basal beds of a deltaic sequence deposited ahead of the advancing delta. Distinctive *topset* and *foreset* beds are rare in large delta masses, however, and a more useful terminology involving *prodelta, intradelta, interdistributary* environments has developed. Prodelta clays typically grade upward and sometimes laterally into delta deposits consisting of: (1) interdistributary clays which settle out in the shallow brackish to freshwater areas between active delta distributaries; (2) intradelta complex deposits which include a host of deltaic environments consisting of the coarsest deltaic sediments associated with an advancing distributary mouth; and (3) organic marsh deposits which form as the delta front builds up to or above sea level. Fig. 1 shows the relationship of these depositional environments to one another in the Mississippi River Delta.

Although environmental types overlying the prodelta often differ from delta to delta, a prodelta facies is almost invariably found at the base of a deltaic mass. The distribution and nature of the prodelta environment are particularly well known in the Mississippi Delta. Data presented below are summarized from work on Mississippi prodelta clays by Kolb and Van Lopik (1958) and Kolb and Kaufman (1967).

Each day, the Mississippi carries an average of 1.5 million tons of sediment to the Gulf. Of this amount, 50% consists of material $< 5\mu$ in diameter. The vast majority of this clay is deposited in a broad fan that thins seaward. Studies by Scruton and Moore (1953) conclude

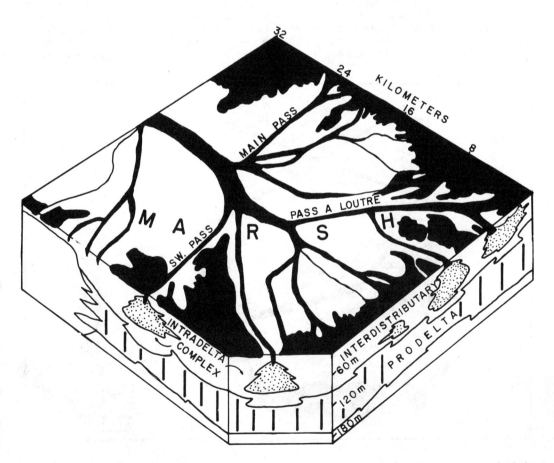

FIGURE 1. Sedimentary framework of the modern Mississippi Delta, showing position of the prodelta facies (after Fisk et al., 1954).

that most prodelta deposits settle within 8 km of the mouths of the Mississippi's major distributaries in a direction generally paralleling current flow and that the amount of clay settling from river-derived waters >16 km offshore is insignificant. The very fine clays settle slowly at great distances from the delta front, either as clay-sized particles or as larger-sized flocculants (see *Flocculation*). The coarser, siltier fraction is deposited on the fine-clay facies and settles within a few km of the distributary mouth. It, in turn, is rapidly buried beneath the advancing deposits associated with the delta front.

The Mississippi Deltaic Plain consists of the present active delta and several abandoned deltas (see Vol. III: *Deltaic Evolution*, Fig. 2); several waves of prodelta deposition, each associated with a specific delta advance, have been superposed above one another in SE Louisiana. The result has been the formation of a massive, persistent clay stratum that thickens seaward (Fig. 2). The stratum, during the past 3000 yr, has reached over 100 m in thickness and an aggregate estimated volume 400 km^3 in SE Louisiana (Kolb and Kaufman, 1967).

Prodelta clays of the Mississippi Delta are gray to dark gray, homogeneous and highly plastic. A fairly common and distinctive constituent is vivianite, hydrous ferrous phosphate, which changes from drab blue-green to vivid blue-green on exposure. Diagnostic microforms, particularly foraminifera, are plentiful and help to identify the prodelta environment in soil samples. The faunal assemblage varies with the depth of water in which the prodelta clays are deposited. Phleger (1955) has made a comprehensive study of the foraminifera on the Gulf bottom just seaward of the Mississippi Delta.

To the unaided eye, samples of prodelta clay appear to be entirely devoid of structure. X-radiographs of the deposits, however, disclose a wide variety of laminae, fractures, contorted bedding, and complex displacements. These features were formed at the time of deposition or as water was squeezed from the pores of the material as it consolidated beneath the load of the advancing delta. Fracturing and distortion of the layers appear to depend chiefly upon the rapidity of deposition. The slowly deposited basal clays consolidate as they are loaded, and distortion and fracturing are at a minimum. Gross distortion is characteristic of the upper, siltier, and more rapidly deposited portion of each prodelta unit. Fracturing undoubtedly occurs concurrently with distortion; fracturing and plastic flow, however, probably continue to

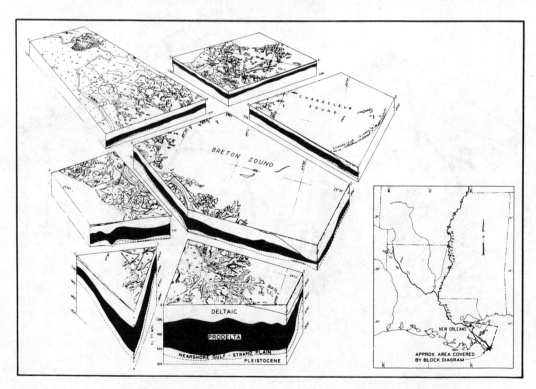

FIGURE 2. Distribution and relative thickness of prodelta clays in extreme southeastern Louisiana (from Kolb and Kaufman, 1967).

be active for some time after the original distortion and warping of bedding takes place. Deposition of a given prodelta unit begins at a relatively slow rate, <30cm/100yr, and increases twenty-fold before advancing intradelta-interdistributary deposits rapidly bury the sequence.

CHARLES R. KOLB

References

Bernard, H. A., and LeBlanc, R. J., 1965. Resume of the Quaternary geology of the northwestern Gulf of Mexico province, in H. E. Wright, Jr., and D. G. Frey, eds., *The Quaternary of the United States.* Princeton: Princeton Univ. Press, 137-185.

Coleman, J. M., and Gagliano, S. M., 1964. Cyclic sedimentation in the Mississippi River deltaic plain, *Gulf Coast Assoc. Geol. Soc. Trans.,* 14, 67-80.

Fisk, H. N.; McFarlan, E., Jr.; Kolb, C. R.; and Wilbert, L. J., Jr., 1954. Sedimentary framework of the modern Mississippi Delta, *J. Sed. Petrology,* 24, 76-99.

Kolb, C. R., and Kaufman, R. I., 1967. Prodelta clays of southeast Louisiana, in F. Richards, ed., *Marine Geotechnique.* Urbana: Univ. Illinois Press, 3-21.

Kolb, C. R., and Van Lopik, J. R., 1958. Geology of the Mississippi River deltaic plain, southeastern Louisiana, *U.S. Army Eng. Waterways Exper. Sta.* [Vicksburg, Miss.], *TR 3-483* (2 vols.), 120p. A condensed version (1966) appears in M. L. Shirley, ed., *Deltas in Their Geologic Framework.* Houston: Houston Geol. Soc., 17-62.

Phleger, F. B., 1955. Ecology of foraminifera in southeastern Mississippi Delta area, *Bull. Am. Assoc. Petrol. Geologists,* 39, 712-752.

Scruton, P. C., 1960. Delta building and the deltaic sequence, in F. P. Shepard et al., eds., *Recent Sediments, Northwest Gulf of Mexico.* Tulsa, Okla.: Am. Assoc. Petrol. Geologists, 82-102.

Scruton, P. C., and Moore, D. G., 1953. Distribution and surface turbidity off Mississippi delta, *Bull. Am. Assoc. Petrol. Geologists,* 37, 1067-1074.

Cross-references: *Deltaic Sediments; Deltaic Sedimentation; Flocculation; Paralic Sedimentary Facies.*

PROVENANCE

The term *provenance* (provenience) derived from the French *provenir,* meaning to originate or to come forth, has been used to encompass all the factors relating to the production or "birth" of a sediment (Pettijohn et al., 1972). The word *provenance* is defined as "a place of origin; specifically the area from which the constituent materials of a sedimentary rock or facies are derived" (AGI *Glossary,* 1972). Its synonyms are provenience, source area, and sourceland.

Concepts

The word "provenance" appears to have been introduced into the English language in 1861 (OED) in the sense of "derivation" or "the fact of coming from some particular source or quarter." Its introduction into geological terminology seems to have occurred in the early 1920s (see *Sedimentary Petrography*). Twenhofel (1950) says: "'Provenance' refers to the terranes or parent rocks from which any association of sediments was derived." Pettijohn (1957) includes also the nature of the climate and the relief of the source area in his conception of provenance. He uses provenance in the sense of "origin" to describe the mode of formation of sedimentary rocks, including chemical deposits formed by precipitation from sea water (e.g., cherts and ironstones).

While sedimentologists agree that "provenance" means "source," there has been wide variation in the interpretation of the significance of "source." For example, Ross (1955) uses the term in the sense of the place of occurrence (i.e., the geographical location) of the material described. Brammall (1928) uses the term in the sense of a particular granite being the source of specific minerals found in a variety of sediments. Krynine (1946) has discussed the sources (provenances) of tourmaline in a wide variety of rock types (igneous, metamorphic, pegmatitic, sedimentary authigenic, and reworked sedimentary). The implication of the term "provenance" and hence "source" depends to some extent on whether a single mineral grain, a rock fragment, or a pebble is being considered or whether the concern is with the total rock mass itself. For a single mineral grain, rock fragment, or pebble, provenance may imply (1) the rock from which the grain, fragment, or pebble was taken, independent of stratigraphic horizon or geographical location; (2) the immediate source rock from which the grain, fragment, or pebble was derived; (3) the ultimate (primary) rock type from which the grain was ultimately derived; or (4) the *specific* primary rock from which the grain was derived. For rocks, provenance may be (1) the place of occurrence of the rock *at the present time;* (2) the general region or the direction of the region from which the grains composing the rock were derived; (3) the general nature of the terrane from which the detritus was derived; or (4) the specific rocks present in the region from which the detritus was derived.

Distributive Province. In order to avoid some of the ambiguities that arise from the use of the term *provenance,* the term *distributive province* was proposed by Brammall (in Milner, 1922) and subsequently redefined by Twenhofel

(1950) as "the environment embracing all rocks that contribute to the formation of a contemporaneous sedimentary deposit and includes the agent responsible for their distribution." Hutton (1950) coined the phrase "distributive provenances" to describe those areas from which heavy minerals in (modern) rivers were derived.

The distributive province may be entirely sedimentary, and hence its denudation leads to the formation of sediments whose minerals are predominantly "second hand" (second cycle or polycyclic). In contrast, igneous rocks in a distributive province will, on disintegration, give rise to primary or first-cycle sediments. It is, of course, possible that the distributive province will comprise mainly metamorphic rocks. The minerals derived from such a source will be dominantly primary (of metamorphic origin), although some highly refractory minerals, in particular zircon, may have survived from earlier sedimentary cycles, unchanged by the metamorphic episode.

Petrologic Province. The erosion of the rocks of a distributive province and the transportation of the debris to the site of deposition leads to the concept of a "sedimentary province" or *sedimentary petrologic province,* which Edelmann (1933) defined as a group of sediments which constitute a natural unity by age, origin, and distribution. Essentially, it is the counterpart of provenance. The petrologic province defines the area of eventual sedimentation, its salinity, currents, water depth, paleogeography, and other synsedimentary environmental factors. Whereas much of the information deduced from studies of the sedimentary material in defining the petrologic province is rather direct, the deductions required for establishing the provenance are frequently very indirect; the erosional and transport processes tend to deface original textures, and complex paleogeographic reconstructions tend to be rather subjective and oversimplified (Suttner, 1974). Systematic sampling of a sedimentary region is carried out on a map constructed to bring out the associations of related mineral grains. The first regional petrology of the North Sea was constructed by Baak (1936).

The mere fact of the formation of a sediment implies that the source must have been partially or completely destroyed, and the precise determination of the provenance of a sediment is dependent upon the chance that some part of the source rock escaped destruction. Where destruction is complete, the determination of the provenance of a sediment must be largely hypothetical. Indeed, the determination of the provenance of sediments is an intrinsic part of the reconstruction of the paleogeography of an area of which little or nothing has been preserved.

Indicators to Provenance

Early studies of sediments were concerned mainly with their characteristics as seen in thin sections. The apparent uniformity of detrital sediments led to the attitude that little or nothing could be learned from their microscopic examinations. The development in the 1880s of techniques for the concentration of accessory minerals led to a revival of interest in the microscopy of sediments. The accessory minerals, more familiarly known as heavy minerals (q.v.), were shown to display a considerable variety in composition, form, color, and other properties. By the end of the 19th century, many papers had appeared describing the features of these minerals. Few workers were, however, concerned with the source of the minerals (see *Sedimentary Petrography*).

Illing (1916) demonstrated the feasibility of using heavy mineral assemblages for correlative purposes, and with increasing interest in this aspect of heavy minerals, the problems of provenance had become of secondary interest by the early 1930s. Since the 1940s, interest has revived, although studies have been more concerned with the direction of derivation and the general character of the terrane than with the specific rock types acting as the source. In other words, provenance is becoming a tool for determining regional paleogeology for any particular epoch.

In addition to the evidence furnished by the constituent grains of the clastic sediments and rocks, we may use other evidence gained from stratigraphy and facies studies in our analysis of provenance or source areas. A better understanding of the broad geologic framework of a region will provide clues to provenance. The source areas are those of positive relief and tectonic activity. Paleocurrent analysis (q.v.) may also contribute in a significant way to provenance studies. Therefore, we definitely need to know a great deal more than just the mineralogy of a clastic sediment or rock to determine its provenance. The analysis of provenance is undoubtedly one of the most difficult problems that the sedimentary petrologist tries to solve.

Detrital Quartz. The ubiquity of quartz in sediments has led many investigators to study the possible use of quartz grains as an indicator to provenance. As early as 1877, Sorby had observed that quartz grains derived from granites were more or less equiaxial, whereas those from schists were more or less flattened. He also observed that quartz grains from granites were simple (monogranular), whereas those from schists were complex (polygranular). Mackie (1899) claimed that "acicular and irregular inclusions pre-eminently abound in the quartz of granites," and hence the granitic origin of a quartz grain could be deduced. Other authors have supported this view, but Krynine (1940) believed that acicular inclusions were characteristic of grains from metamorphic rocks. Thus inclusions in quartz grains are not reliable indicators.

Bokman (1952) reintroduced Sorby's view (although he credited it to Krynine) that (1) the grain shape and (2) simple vs. composite grains were the only valid provenance criteria for quartz grains. It has sometimes been suggested that the shape of a quartz grain can be markedly modified by abrasion and is thus an indicator of environment rather than provenance. Experimental studies by Ingerson and Ramisch (1942), however, support the conclusion that the elongation of quartz grains is due to original shape.

The presence of "strain shadows" (undulatory extinction) in quartz grains has been held to be indicative of a metamorphic origin. A historical review of this aspect of provenance has been given by Blatt and Christie (1963) and Blatt (1967), who show that "strained" quartz is also relatively common in plutonic igneous rocks, although the quartz of extrusive igneous rocks is almost entirely strain free. They also show that strained quartz grains are less stable than unstrained ones; hence an excess of unstrained quartz may be indicative of several episodes of recycling, rather than provenance (see *Maturity*). They also note that large quantities of polycrystalline quartz grains are derived from both plutonic and metamorphic rocks and that the presence or absence of undulatory extinction and polycrystallinity in clastic quartz is of very limited usefulness in provenance studies.

Detrital Feldspars. Feldspars in sedimentary rocks are generally accepted as indicators to a coarse-grained plutonic source, such as a granite or granite gneiss. To a lesser extent, feldspar is derived from volcanic sources. Pittman (1970) has shown that oscillatory zoning in plagioclase is indicative of either a volcanic or a hypabyssal rock. Pittman found that progressive zoning had no value in distinguishing between a volcanic-hypabyssal and a plutonic provenance. The feldspar of acid (or rhyolitic) volcanics is most likely to be sanidine, however, whereas either orthoclase or microcline is indicative of acid plutonic rock origin. The presence of zoning may be taken as an indication of an igneous rather than a metamorphic source. Rimsaite (1967), on the basis of study of the optical heterogeneity of feldspars in potential source rocks, defined nine classes of feldspar grains.

The occurrence of feldspar in a sediment has been used to infer the climatic conditions under which the source rock was eroded. The relative instability of feldspar led to the belief that chemical weathering must be inhibited in any area where feldspar was being released in large quantity. Accordingly, very arid or cold climates were postulated. This supposition must be regarded as inadequate in view of the observation by Krynine (1935) of the accumulation of detrital feldspar under humid tropical conditions with average temperatures of 25°C and annual rainfall of 300 cm. The significant features of the accumulation of feldspar appear to be rapid mechanical erosion, limited transportation, and rapid burial. In general, it seems that relief is more important than climate in the production of highly feldspathic sediments.

Rock fragments. Rock fragments are perhaps the best criterion to be used in interpreting provenance. Conglomerates and immature sandstones (graywackes and subgraywackes) commonly contain considerable amounts of rock fragments in addition to usual quartz and feldspar. Large, easily identified rock fragments in conglomerates are the most useful guides to provenance.

Rock fragments in immature sandstones are generally divided into four groups: (1) the argillaceous group, including shale, siltstone, slate, phyllite, and schist; (2) volcanic rocks; (3) polycrystalline quartz (various sources) and chert; and (4) carbonate rocks. These fragments are identifiable by careful study of thin sections of both ancient and modern sands. Volcanic rock fragments are particularly common in graywackes (q.v.). If a sandstone contains a considerable amount of chert and carbonate rock fragments, it generally suggests that a carbonate terrain dominates the source area.

Heavy Minerals. The accessory or "heavy" minerals have long been used as valuable indices to provenance or source areas. Certain detrital heavy mineral species (or suites) are well known to the sedimentary petrologists as indicative of a major class of source rocks (Krumbein and Pettijohn, 1938; Feo-Codecido, 1956; Baker, 1962; Milner, 1962; Hatch and Rastall, 1971).

Among the nonopaque heavies, very few are restricted to igneous rocks. Nearly all heavies reported from intrusives are also reported frequently from schists and gneisses. Many diagnostic metamorphic minerals, however, e.g., kyanite, staurolite, sillimanite, and wollastonite, do not occur in magmatic rocks.

Heavy minerals have proved to be very useful in determining the provenance of many Tertiary sedimentary basins in the world. The usefulness of heavy minerals is greatly reduced, however, in provenance studies of Mesozoic and older sedimentary rocks because many of them are most likely to be derived from older sediments.

Provenance Studies by Scalar and Vector Properties

The scalar properties of a sediment (i.e., size, sorting, shape, roundness, and composition) depend on a number of factors, among which the nature of the parent rock varies in importance according to the maturity (q.v.) of the sediment. As maturity increases, the character of the source rock becomes progressively obscured but is not completely obliterated, as the discussion of the shape of quartz grains has shown. The composition of the sediment also changes as the less stable minerals (including many heavy minerals) are destroyed. The size of particles may indicate the proximity of the source, especially in the case of large fragments (see *Sternberg's Law*). Fine-grained sediments usually indicate remoteness from source, but a fine-grained, poorly consolidated rock could give rise to fine-grained debris close to the source.

The size of particles also gives some indication of the climatic conditions and physiography of the terrane being denuded. Coarse, angular, polymictic sediments would in general be regarded as being much closer to their source than those in which the particles were small, well rounded, and consisting dominantly of quartz. Unfortunately, scalar properties give little indication of the absolute distance of a source, except in rare instances of talus. Nor do they give a clear indication of the direction of the source unless a marked trend can be demonstrated over a considerable area.

Vector properties indicate the specific direction of movement of sediment and hence can be used to infer the direction of the source of supply (see *Paleocurrent Analysis*). In this case, the source is to be regarded as the immediate source and not the ultimate, or parent source. An early example of the application of current bedding features to the determination of the direction of the source area was Gilligan's (1919) study of the Millstone Grit of Yorkshire. Potter and Olson (1954) demonstrated, however, that whereas the current bedding directions of the basal Pennsylvanian sands of Indiana and Illinois indicate the Canadian Shield as the source area, heavy minerals suites pointed to an Appalachian source. It must be remembered that indicators of direction of transport are only locally significant. If directional data are available over a large area, however, the general direction of the source (or sources) can probably be inferred with reasonable accuracy. As with scalar properties, the extrapolation of vector data over considerable distances must be approached with extreme caution.

In summary, the problem of provenance of a clastic sediment or clastic sedimentary rock unit can best be solved by an integrated study of the detrital mineral constituents, stratigraphy, regional lithofacies pattern, and paleocurrents.

DEREK W. HUMPHRIES

References

Baak, J. A., 1936. *Regional Petrology of the Southern North Sea*. Wageningen, Netherlands: Veenman and Zonen, 127p.

Baker, G., 1962. Detrital heavy minerals in natural accumulates, *Austral. Inst. Min. Metal., Monogr. 1*, 146p.

Blatt, H., 1967. Provenance determinations and recycling of sediments, *J. Sed. Petrology*, 37, 1031-1044.

Blatt, H., and Christie, J. M., 1963. Undulatory extinction in quartz of igneous and metamorphic rocks and its significance in provenance studies of sedimentary rocks, *J. Sed. Petrology*, 33, 559-579.

Bokman, J., 1952. Clastic quartz particles as indices of provenance, *J. Sed. Petrology*, 22, 17-24.

Brammall, A., 1928. Dartmoor detritals: A study in provenance, *Proc. Geol. Assoc.*, 39, 27-48.

Edelmann, C. H., 1933. *Petrologische Provincies in het Nederlands Kwartair*. Amsterdam: Centen Publ. Co., 104p.

Feo-Codecido, G., 1956. Heavy-mineral techniques and their application to Venezuelan stratigraphy, *Bull. Am. Assoc. Petrol. Geologists*, 40, 984-1000.

Gilligan, A., 1919. The petrography of the Millstone Grit of Yorkshire, *Quart. J. Geol. Soc. London*, 75, 251-294.

Hatch, F. H., and Rastall, R. H., 1971. *Petrology of the Sedimentary Rocks* (5th ed., revised by J. T. Greensmith). New York: Hafner, 502p.

Hutton, C. O., 1950. Studies of heavy detrital minerals, *Geol. Soc. Am. Bull.*, 61, 635-716.

Illing, V. C., 1916. The oilfields of Trinidad, *Proc. Geol. Assoc.*, 27, 115.

Ingerson, E., and Ramisch, J. L., 1942. Origin and shapes of quartz grains, *Am. Mineralogist*, 27, 595-606.

Krumbein, W. C., and Pettijohn, F. J., 1938. *Manual of Sedimentary Petrography*. New York: Appleton-Century, 549p.

Krynine, P. D., 1935. Arkose deposits in the humid tropics. A study in sedimentation in southern Mexico, *Am. J. Sci.*, 29, 353-363.

Krynine, P. D., 1940. Petrology and genesis of the Third Bradford Sand, *Penn. St. Coll. Min. Indust. Sta. Bull.*, 29, 134p.

Krynine, P. D., 1946. The tourmaline group of sediments, *J. Geol.*, 54, 65-87.

Mackie, W., 1899. Sands and sandstones of Eastern Moray, Scotland, *Trans. Edinburgh Geol. Soc.*, 7, 148-172.

Milner, H. B., 1922. The nature and origin of the Pliocene deposits of the county of Cornwall and their bearing on the Pliocene geography of the southwest of England, *Quart. J. Geol. Soc. London*, 78, 348-377.

Milner, H. B. 1962. *Sedimentary Petrography*, vol. 2. London: Allen and Unwin, 425-456.

Pettijohn, F., 1957. *Sedimentary Rocks*. New York: Harper & Row, 718p.

Pettijohn, F. J.; Potter, P. E.; and Siever, R., 1972.

Sand and Sandstone. New York: Springer-Verlag, 618p.
Pittman, E. D., 1970. Plagioclase feldspar as an indicator of provenance in sedimentary rocks, *J. Sed. Petrology,* **40,** 591–598.
Potter, P. E., and Olson, J. S., 1954. Variance components of cross-bedding direction in some basal Pennsylvanian sandstones of the Eastern Interior Basin, geological application, *J. Geol.,* **62,** 50–73.
Rimsaite, J., 1967. Optical heterogeneity of feldspars observed in diverse Canadian rocks, *Schweiz, Min. Pet. Mitt.,* **47,** 61–76.
Ross, C. S., 1955. Provenience of pyroclastic materials, *Geol. Soc. Am. Bull.,* **66,** 427–434.
Sorby, H. C., 1877. Anniversary address to the Royal Microscopical Society, *Monthly Microscop. J.,* **7,** 113–136.
Suttner, L. J., 1974. Sedimentary petrographic provinces: An evaluation, *Soc. Econ. Paleont. Mineral. Spec. Publ. 21,* 75–84.
Twenhofel, W. H., 1950. *Principles of Sedimentation.* New York: McGraw-Hill, 673p.

Cross-references: *Clastic Sediments and Rocks; Heavy Minerals; Maturity; Paleocurrent Analysis; Roundness and Sphericity; Sedimentary Petrography; Sedimentary Petrology.*

PSEUDOMORPHS

The term *pseudomorph* is commonly used in geology to indicate various tridimensional bodies consisting of a mineral or mineral aggregates which bear the crystal habit of another mineral or the external shape proper to another object. Pseudomorph derives from the Greek words *pseudes* meaning false, and *morphos* meaning form. The term is indeed used to designate the "false form" or the natural external reproduction of an original object. It implies that the natural replica is an exact duplicate of the surface, the size, as well as the shape of the original object. It should be emphasized that a pseudomorph does not preserve any of the preexisting internal structure. It is merely a solid occupying the space once bounded by the external surface of the original object, and should as much as possible not be confused with a cast (see *Casts and Molds*). It is not uncommon indeed that the two terms are taken as synonymous, and pseudomorphs are then referred to as natural casts.

The origin and genesis of pseudomorphs are directly related to the conditions of deposition and subsequent diagenetic processes within a rock or sediment. Thus, pseudomorphs depend essentially on the nature of the embedding rock and in-situ changes that may have taken place while the sediment was being deposited or anytime subsequent to deposition. Although the subsequent changes that lead to the genesis of pseudomorphs may be directly related to certain physicochemical properties of the embedding rocks, the resulting mineral fillings are often completely unrelated to the original constituents of the replaced object. Silica, calcite, calcium phosphate, dolomite, pyrite, hematite, limonite, and glauconite are commonly involved in pseudomorphism. Other minerals such as barite, sulfur, and even silver are also known to produce perfect pseudomorphs of embedded fossils. Pseudomorphism is also quite common among minerals where one may faithfully replace the other as is often the case when calcite is replaced by quartz that retains the crystal shape of the calcite. Some rather unusual agate pseudomorphs of plants occur in lava flows in the Pacific Northwest, where trees engulfed by hot lava were burned away but left an exact mold in the hardened lava, which was subsequently filled by variegated chalcedony and opal.

Many trace fossils such as burrow fillings and coprolites are also often found as pseudomorphs. Interpretation of many such features as pseudomorphs remains precarious, however, because diagenetic processes may occasionally produce concretions which can be confused with similar animal activities. Also excluded from pseudomorphs are the so-called pseudofossils, which also show superficial resemblance to fossils. They are merely mineral aggregates which have no genetic relationship whatsoever with any preexisting object. Some of the most common such occurrences are the dendrites of pyrolusite (MnO_2) frequently observed on the surface of fractures and coating pebbles. "Moss agate," for instance, is a variety of agate in which the moss-like manganiferous dendrites occur, but this is only a fortuitous appearance of the mineral aggregates and not a pseudomorph.

Thus, pseudomorphs are essentially formed by the natural filling of cavities once occupied by an object with which they share only the common external shape. Pseudomorphs are important in sedimentology not only because they faithfully reproduce the shape of the original object, but also the mineral content of the filling materials allows one to understand the diagenetic processes that took place within the rock.

FLORENTIN MAURRASSE

References

Cloud, P., 1973. Pseudofossils: A plea for caution, *Geology,* **1**(3), 123–127.
Moore, R.; Lalicker, C. G.; and Fischer, A. G., 1952. *Invertebrate Fossils.* New York: McGraw-Hill, 766p.

Shrock, R. R., 1948. *Sequence in Layered Rocks.* New York: McGraw-Hill, 507p.

Shrock, R. R., and Twenhofel, W. H., 1953. *Principles of Invertebrate Paleontology.* New York: McGraw-Hill, 816p.

Cross-references: *Casts and Molds; Concretions; Diagenesis; Silicification.* Vol. IVB: *Pseudomorphs.* Vol. VII: *Pseudofossils.*

PYROCLASTIC SEDIMENT

Pyroclastic sediment consists chiefly of volcanic particles eroded from unconsolidated pyroclastic deposits and redeposited elsewhere by wind, running water, or gravity (for discussion of classification, see *Volcaniclastic Sediments and Rocks*). The volcanic particles contained in pyroclastic sediment vary widely in structure and composition, but virtually all the material is originally fragmented as a direct result of volcanic activity. Fragments of volcanic rock and whole or broken crystals are present in most deposits, and fragile shards of pumice and smaller bits of volcanic glass are abundant in many. Most of this debris is mechanically and chemically unstable and is preserved as pyroclastic sediment chiefly because of the speed and ease with which it is eroded and redeposited.

Nonvolcanic material is commonly mixed with pyroclastic debris as it is reworked and transported. With increasing dilution by the nonvolcanic material, pyroclastic sediment grades into ashy sediment containing moderate amounts of pyroclastic debris, and finally into normal sediment containing only sparse amounts of volcanic debris.

Mode of Deposition

Primary pyroclastic debris can be erupted onto land or directly into water, and therefore the pyroclastic sediment derived from these loose materials can also be deposited on land and water. In addition, large amounts of pyroclastic debris are eroded from land and carried into water, adding to the waterlaid deposits of pyroclastic sediment.

Pyroclastic Sediment Deposited on Land. Pyroclastic sediment deposited on land is familiar to geologists because in many places, the processes of initial deposition, erosion, and redeposition can be observed. Pyroclastic debris erupted into the air and showered upon the ground is highly susceptible to erosion, especially if the material falls upon a hilly terrain (Segerstrom, 1950). Most of the vegetation that normally tends to retard erosion is smothered as thick deposits of airborne pyroclastic debris accumulate, causing the rate of erosion to increase many times. Runoff from heavy rains can strip this loose material from wide areas and wash it into streams and rivers which then redeposit it as alluvial or flood plain material at lower altitudes. During these periods of heavy runoff, many streams become choked with volcanic ash (q.v.) and lapilli and are transformed into thick slurries that carry huge amounts of pyroclastic debris.

Large amounts of pyroclastic debris are also transported and deposited on land by thick, fluid mudflows (Anderson, 1933). Many mudflows (q.v.) are produced as a direct result of pyroclastic eruptions, and these are properly considered to be pyroclastic flows. Many are not directly related to eruptions of pyroclastic debris, however, but are important transporters of volcanic sediment. Mudflows not related to volcanic eruptions are triggered in many ways, although the most common origins fall into two main catagories: failure of temporary dams, and collapse of water-saturated pyroclastic deposits.

Temporary dams are commonly formed in volcanic terrains when lava flows, pyroclastic flows, or large landslides come to rest on valley floors and block the local stream's drainage. In time, many of these dams fail, and if the failure is sufficiently sudden, the impounded water surges downstream, sweeping up parts of the dam and other debris to form destructive mudflows. Similarly, the walls enclosing crater lakes can collapse, releasing large volumes of water and volcanic sediment that produce the same type of mudflows.

Thick deposits of pyroclastic debris saturated by heavy rains or by meltwater from snow or ice are potential sources of mudflows, especially if they are partly undermined by flooded rivers or jarred by earthquake shocks. When a large volume of this water-saturated debris collapses, the abundant pore water enables the entire mass to mobilize and surge downslope as a mudflow, often progressively diluted downslope with nonvolcanic debris. When mudflows finally come to rest, they are highly susceptible to erosion by running water and wind, and the reworked material can be carried even farther downslope.

Wind is also able to erode the fine debris from all types of unconsolidated pyroclastic deposits and redeposit it elsewhere as pyroclastic sediment. The wind is most effective when the pyroclastic debris is thoroughly dry; for example, during the dry season near Paricutin volcano, Mexico, visibility was at times reduced to <100 m by dense clouds of windblown ash (Segerstrom, 1950). The total volume of debris transported by wind is comparatively small, however, and in most areas is considerably less than that carried by running water.

Pyroclastic Sediment Deposited Under Water. Water-laid pyroclastic sediment can also originate in many different ways. Much pyroclastic debris is eroded from land by running water and is carried into lakes or seas, where it is interbedded with nonvolcanic sediment. Pyroclastic flows that travel from land into water and clouds of airborne debris that fall into water also supply pyroclastic debris that is easily reworked and incorporated into the sedimentary section. But perhaps the greatest volume of waterlaid volcanic sediment is supplied directly to basins of deposition by submarine volcanoes (see *Volcanism—Submarine Products*). Pyroclastic debris heaped around submarine volcanic vents is readily eroded and redeposited by ocean currents. In addition, large amounts of debris erupted by submarine volcanoes are repeatedly shaken free by earthquake shocks and carried into deeper water by turbidity currents and other gravity flows (q.v.). Very fine ash and floating fragments of pumice are dispersed over wide areas by surface currents, but this material is usually so diluted by nonvolcanic debris when it finally settles to the sea floor that it seldom forms deposits of pyroclastic sediment.

Physical and Chemical Characteristics

Physical Characteristics. Deposits of pyroclastic sediment vary widely in texture, thickness, lateral extent, and internal structure, depending chiefly upon the environment in which they were deposited. Pyroclastic sediment deposited on dry land by rivers and streams has many of the same features as nonvolcanic material deposited in the same environment (Hay, 1956). Individual deposits range in thickness from ≈ 1 cm to 1 m or more and in overall geometry from stubby channel fillings to more extensive lenses and sheets that pinch out and interfinger with other deposits in a complex manner. In areas close to the source volcanoes, many deposits of pyroclastic sediment consist chiefly of gravel or coarse breccia. Farther away, the deposits generally become finer grained, and sand- and silt-sized material predominates. Cut-and-fill structures and cross-bedding are common, reflecting the relatively high current velocities of the streams and rivers that transport and deposit the debris.

Mudflows carrying volcanic sediment form extensive tabular deposits ranging in thickness from one to several meters (see *Mudflow Deposits*).

Pyroclastic debris erupted into shallow water is sorted and dispersed by currents of water, and the resulting deposits of pyroclastic sediment have the same general geometry and internal current structures as deposits of nonvolcanic sediment deposited in the same environment. Pyroclastic debris dislodged into deeper water can form turbidity currents which, in turn, deposit extensive beds that exhibit graded bedding and other features common to nonvolcanic turbidities.

Chemistry. Detailed studies of the chemistry of pyroclastic sediment have not been made, although it is generally assumed that the pyroclastic sediment deposited in a given area has about the same composition as the source deposits of pyroclastic debris from which it was derived. This assumption may be valid if all of the source deposits of pyroclastic debris have nearly the same composition and if this debris is transported and redeposited with little or no winnowing or sorting. The individual particles in the primary pyroclastic deposits, however, vary widely in size and density (e.g., dense lithic fragments vs. lightweight pumice), and this debris is readily sorted and winnowed according to grain size and grain density as it is transported and redeposited. Consequently, some beds of pyroclastic sediment may be unusually rich in pumice lapilli or in tiny fragments of glass, whereas others may be rich in crystals or lithic fragments. The bulk chemical composition of many of these individual beds will therefore be significantly different than the composition of the source deposits of pyroclastic debris.

An additional complication is introduced if all of the source deposits are not of the same composition, as is generally so. This heterogeneous debris is readily mixed in varying proportions as it is transported and redeposited, also causing the deposits of pyroclastic sediment to have bulk compositions somewhat different than the individual pyroclastic deposits from which they were derived.

Admixtures of nonvolcanic sediment can greatly modify the bulk chemical composition of pyroclastic sediment. The hybrid nature of such deposits must therefore be recognized before their chemistry can be properly interpreted.

Recognition of Pyroclastic Sediment. Many deposits of pyroclastic sediment superficially resemble deposits of nonvolcanic sediment in overall geometry and in internal structure, and attention must therefore be focused upon the composition and texture of the debris itself if pyroclastic sediment is to be recognized. The ease with which reworked pyroclastic debris can be recognized depends chiefly upon the amount of winnowing and abrasion the material has suffered during transport and redeposition. Material that has been transported only a short distance is, in general, the least modified and is the easiest to recognize as pyroclastic. Fragments of volcanic rock and characteristic volcanic minerals such as biotite, hornblende, and pyroxene are generally conspicuous. If there has been little abrasion, some of the original crystal faces will be preserved on many of the mineral grains, and narrow rims of volcanic glass might be found partly jacketing others. Tiny glass shards and larger fragments of glass bubbles are readily broken by impact and abrasion, but small bits and pieces of this glassy debris can often be recognized. Even with little transport, however, the jagged edges of frothy pumice fragments are smoothed, and the fragments are quickly worn into rounded pebbles and cobbles.

With continued reworking and abrasion, especially by running water, the fragments of pumice and glass are worn down to small bits and shards. Volcanic minerals are readily broken along cleavage surfaces, and the original crystal faces of all mineral grains are chipped and scarred as the grains are reduced in size. Angular fragments of volcanic rock are broken into smaller pieces and rounded into pebbles and cobbles.

As transport and abrasion is prolonged, the less stable minerals, fragments of pumice, and smaller bits of glass are destroyed; then scattered minerals of volcanic origin and rounded fragments of volcanic rock may be the only clues suggesting a pyroclastic origin. Moreover, as the material is mixed with debris of nonvolcanic origin, its true identity becomes even less certain.

RICHARD S. FISKE*

*Publication authorized by the Director, U.S. Geological Survey.

References

Anderson, C. A., 1933. The Tuscan Formation of northern California, *Calif. Univ. Publ. Geol. Sci.,* **23**, 215-276.

Fisher, R. V., 1961. Proposed classification of volcaniclastic sediments and rocks. *Geol. Soc. Am. Bull.,* **72**, 1409-1414.

Fisher, R. V., 1966. Rocks composed of volcanic fragments and their classification, *Earth Sci. Rev.,* **1**, 287-298.

Hay, R. L., 1956. Pitchfork Formation, detrital facies of early basic breccia, Absaroka Range, Wyoming, *Bull. Am. Assoc. Petrol. Geologists,* **40**, 1863-1898.

Segerstrom, K., 1950 Erosional studies at Paricutin, state of Michoacan, Mexico, *U.S. Geol Surv. Bull. 965-A,* 164p.

Wentworth, C. K., and Williams, H., 1932. The classification and terminology of pyroclastic rocks, *Natl. Research Council Bull.,* **89**, 19-53.

Cross-references: *Base-Surge Deposits; Clastic Sediments and Rocks; Ignimbrites; Mudflow, Debris-Flow Deposits; Nuée Ardente; Sands and Sandstones; Tuff; Volcanic Ash Deposits; Volcaniclastic Sediments and Rocks; Volcanism−Submarine Products.*

Q

QUARTZOSE SANDSTONE

Quartz-rich sandstone is variously described as quartzite, orthoquartzite, quartzose sandstone, quartzitic sandstone, quartz arenite, and so on. These terms are used synonymously in many instances, although *quartzite* is generally used in a metamorphic context. The term *quartzose sandstone* in sandstone nomenclature has been provided with various definitions, as shown in Table 1.

It is recommended here that *quartzose sandstone* be defined as a sandstone containing >75% detrital quartz. Sandstone with >95% quartz is more properly called *quartz sandstone*. In this definition, chert is not incorporated in the quartz content. Quartzose sandstone is further subdivided into arenite and wacke types, according to clay matrix content; *quartzose* or *quartz arenite* is characterized by <15% matrix, *quartzose* or *quartz wacke* by >15% matrix. Thus, quartz arenite (or quartzarenite) is synonymous with *orthoquartzite*. As a recent trend in the literature, the term quartz arenite is replacing the term orthoquartzite.

Quartzose sandstone is cemented commonly by silica and/or carbonate, less commonly by iron oxide, and rarely by anhydrite, barite, celestite, etc. Krynine (1940) restricted "quartzose sandstone" to a sandstone not cemented by silica, and instead used "quartzitic sandstone" for a silica-cemented one. Pressure-solution phenomena (q.v.) are common in silica-cemented quartzose sandstone.

Quartzose sandstone is well known in stable shelf, paralic, and subaerial environments, but is not uncommon in geosynclinal or orogenic belts. Particularly, quartz arenite is regarded as an end product in sandstone sedimentation. It is characterized by its supermature petrological nature: the best sorting, the highest roundness, the highest concentration of quartz, and the most stable heavy mineral suite (see *Maturity*).

Chert sandstone often found in eugeosynclinal belts cannot be regarded as a variety of quartzose sandstone. It is instead a lithic sandstone (see *Sand and Sandstone*).

HAKUYU OKADA

TABLE 1. Various Definitions of Quartzose Sandstone by Quartz Content

Quartz Content	Authors
>99%	Shrock (1948)
>95%	Krynine (1940); Pettijohn et al. (1972)
>90%	Dapples et al. (1953) Dunbar and Rodgers (1957); van Andel (1958); Crook (1960); Valin (1967); Chen (1968);
>75%	Mazanov (1965); Okada (1971)
>50%	Baisert et al. (1967)
50-70%	Hubert (1960)

References

Baisert, D.; Illers, K.; Kaemmel, T.; and Stelnike, K., 1967. Zu den Prinzipien der Bezeichnung sedimentärer Gestein, *Z. angew. Geologie,* 13, 329-334.

Chen, P. -Y., 1968. A modification of sandstone classification, *J. Sed. Petrology,* 38, 54-60.

Crook, K. A. W., 1960. Classification of arenites, *Am. J. Sci.,* 258, 419-428.

Dapples, E. C.; Krumbein, W. C.; and Sloss, L. L., 1953. Petrographic and lithologic attributes of sandstones, *J. Geol.,* 61, 291-317.

Dunbar, C. O., and Rodgers, J., 1957. *Principles of Stratigraphy.* New York: Wiley, 356p.

Hubert, J. F., 1960. Petrography of the Fountain and Lyons formations, Front Range, Colorado, *Colo. Sch. Mines Quart.,* 55, 1-242.

Krynine, P. D., 1940. Petrology and genesis of the Third Bradford Sand, *Penn. St. Coll. Min. Indus. Expt. Sta. Bull.,* 29, 134p.

Mazanov, D. D., 1965. To the problem of a classification of arenaceous rocks (in Russian), *Doklady Nauk. Azerbaidjanskoi, SSR,* 21, 18-22.

Okada, H., 1971. Classification of sandstone: Analysis and proposal, *J. Geol.,* 79, 509-525.

Pettijohn, F. J.; Potter, P. E.; and Siever, R., 1972. *Sand and Sandstone.* Berlin: Springer-Verlag, 618p.

Shrock, R. R., 1948. A classification of sedimentary rocks, *J. Geol.,* 56, 118-129.

Valin, F., 1967. Navrh nove klasifikace piskovcu, *Vestnik Ustredniho Ustavu Geologickeho,* 42, 293-295.

Van Andel, T. H., 1958. Origin and classification of Cretaceous, Paleocene and Eocene sandstones of western Venezuela, *Bull. Am. Assoc. Petrol. Geologists,* 42, 734-763.

Cross-references: *Arenite; Maturity; Pressure Solution and Related Phenomena; Sands and Sandstones.*

R

RADIOACTIVITY IN SEDIMENTS

The intensity of nuclear radiation from sediment ranges widely from nearly undetectable amounts in common anhydrite to the very high radioactivity of uranium ore deposits in sandstone. Most of the radioactivity in sediments is caused by uranium and thorium, with their host of radioactive daughter products, and by potassium; rubidium, carbon, and beryllium may also produce radioactivity, but generally in very small amounts.

Radioisotopes

The parent isotopes and the daughter products in the three major naturally occurring radioactive decay series, and the half-life and the sequential position during decay for each radioisotope, are indicated in Fig. 1. Radioactive equilibrium, i.e., when the rate of formation of any daughter product is equal to its rate of decay, is not commonly established in sediments because geochemical processes may distribute some of the daughter products in different phases or sites than those in which the parent isotopes occur. The distribution in sediments of the most important radioisotopes are noted below.

Uranium-238 and *Uranium-235* occur in nearly constant proportions to each other; at the present time the radioactivity of U^{238} is 22 times that of U^{235}.

In many sediments, *Uranium-234* is not in radioactive equilibrium with its parent U^{238} because of isotope fractionation and redistribution during the disintegration sequence of U^{238} to U^{234}.

The daughter products of *Thorium-232* have short half-lives, and they generally are present in relatively constant proportion to Th^{232} in sediments.

As a daughter product of U^{234}, *Thorium-230* is not always in equilibrium with U^{234} in sediments because of the difference in geochemical behavior of the two.

Protactinium-231 is not always in equilibrium with its parent U^{235}. The daughter products of Pa^{231} are short-lived, and thus their contribution is proportional to Pa^{231}. The geochemical behavior of protactinium is more nearly like that of thorium than uranium.

Because of its relatively short half-life, the abundance and radioactivity of *Radium-226* and its daughter products are primarily controlled by the distribution of its parent, Th^{230}. Secondary redistribution of Ra^{226} occurs in some sediments where geochemical processes have been effective in recent time.

Exclusive of mineralized rock, the most abundant radioactive isotope in sediments is *Rubidium-87;* but because it has a very low specific activity its contribution to the total radioactivity of a sediment is less than that of other less-abundant radioisotopes. Rb^{87} has a half-life of 47 billion years, and decays by beta particle emission to Sr^{87}, which is a stable isotope.

In many sediments, *Potassium-40* contributes more radioactivity than other individual radioisotopes. It decays by beta particle emission and electron capture, accompanied by relatively energetic gamma emission, and has a 1.25-billion year half-life. Its stable decay products are Ca^{40} and Ar^{40}.

Carbon-14 and *Beryllium-10*, produced in the atmosphere by the bombardment of nitrogen with cosmic-ray-produced particles, are the only such radioisotopes of significant abundance in sediments (Lal and Peters, 1967). C^{14}, 5730-year half-life, decays by beta emission to N^{14}. Be^{10}, 2.5-million year half-life, decays to B^{10} by beta emission.

Modern Sediments

Sediments now being deposited and Holocene sediments can be discussed conveniently in terms of their environment of deposition, either on land or at the bottom of oceans.

Marine Sediments. The ocean bottoms are largely covered by pelagic sediments (q.v.), which consist of several components, mineral and biotic (see *Marine Sediments*). Along ocean margins, terrigenous sediments dominate, composed of debris transported from the continents. The major components of some ocean sediments are halmeic, produced by authigenesis (q.v.). Most measurements of radioisotopes have been made on (1) red clay (q.v.), which includes sediment of the halmeic and terrigenous types;

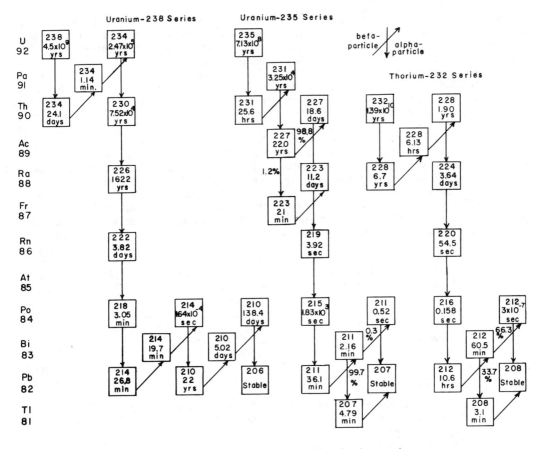

FIGURE 1. Three naturally occurring radioactive decay series.

and (2) *Globigerina ooze,* which includes sediment of the calcareous biotic and terrigenous types. The radioisotopes Th^{230}, Pa^{231}, and their numerous daughter products account for most of the radioactivity emitted from the upper layers of these sediments. Other sources of radioactivity in ocean sediments are K^{40}, the members of the Th^{232} series, the uranium isotopes, Rb^{87}, C^{14}, and Be^{10}.

In the pelagic environment, the radiation intensity is greatest in the uppermost layers of red clay, which has a very slow rate of accumulation (as low as 0.3 mm/1000 yr; Goldberg and Koide, 1962). Compared to the other radioisotopes, K^{40} has the most uniform distribution, although the other radioisotopes with long half-lives are only slightly less uniformly distributed. The concentrations and radioactivities of these other radioisotopes are compared to those of K^{40} in Table 1. The average concentration of K^{40} in pelagic red clay is about 2.5 ppm, which yields radioactive emissions of about 40/min/g of sediment. In contrast to the long-lived isotopes, Th^{230} and Pa^{231} and their radioactive daughters (see Fig. 1) are nonuniformly distributed with depth in the top m of deep-sea

TABLE 1. Distribution of Long-Lived Radioisotopes in Sediments Compared to K^{40} Distribution
(radioactivity in terms of disintegrations per unit time)

Radioisotopes	Approximate Concentrations	Radioactivity
Rb^{87}	10 times K^{40}	1/10 of K^{40}
Th^{232} and daughters	2 times K^{40}	1/3 of K^{40}
U isotopes	1/2 of K^{40}	1/10 of K^{40}

sediments. The greatest concentrations occur in the uppermost layer of red clay. The radioactivity of Pa^{231} in this sediment is about one-tenth than that of Th^{230}; their combined radioactivity, about 40 times that of K^{40}, does not persist with depth into the sediment because of the short half-lives of these isotopes. Thus the radioactivity from Th^{230} and Pa^{231} progressively decreases with depth, and at a depth of about 50 cm, it is only about one-fourth of the K^{40} radioactivity. *Globigerina* ooze, which has a more rapid rate of accumulation, has less radioactivity than does red clay; the radioactivity in the uppermost layer of a typical *Globigerina* ooze is lower by a factor of about 25, and it decreases with depth in the sediment until it is about one-fourth of K^{40} radioactivity at a depth of about 500 cm.

The sources of the high concentrations of Th^{230} and Pa^{231} in the surface layers of deep-sea sediments are the soluble parent uranium isotopes present in the overlying column of sea water, with an average concentration of 3 µg/liter. Presumably the two isotopes produced are adsorbed on particles settling to the bottom.

Some oceanic sediments contain measurable amounts of C^{14} and Be^{10}, both introduced to the ocean through the atmosphere. Calcareous marine organisms that have settled to the ocean bottom after assimilating CO_2 from sea water provide the mechanism of transfer of C^{14} to calcareous biotic components of sediments. B^{10} presumably is adsorbed on particles settling to the bottom. Radioactivity from both of these sources decreases with depth in the sediment because of their relatively rapid radioactive decay.

Significant radioactivity from isotopes in the uranium decay series occurs in some special types of oceanic sediments; and in some cases, the radioactivity emitted is about the same order of magnitude as that emitted by Th^{230} and its daughter products in uppermost layers of red clay. Concentrations as high as 130 ppm uranium in organic-rich black clay deposited in inland seas have been reported (Koczy and Rosholt, 1962); an average of 75 ppm uranium was reported for phosphorite dredged from the bottom of the Pacific Ocean.

Continental Sediments. (See also *Continental Sediments.*) Eolian sediments commonly emit the least amount of radioactivity of any type of sediment. The radioactivity of dunes composed of gypsum is nearly undetectable, and dunes composed of quartz sand are only very slightly radioactive. Dunes of calcareous sand contain as much as 3 ppm uranium, which was incorporated in the $CaCO_3$ at the time of marine deposition, prior to its accumulation in coastal dunes. The radioactivity of these wind-laid sediments increases with the proportion of fine-grained clay and other fine-grained minerals in the sediment.

In general, fluvial and lacustrine sediments reflect the radioactivity of the rock from which the sediments were derived. Distribution of the total radioactivity is similar to that of average crustal material, with U series, Th series, and K^{40} each contributing approximately equal amounts of radioactivity; Rb^{87} contributes <10% of the total. Radioactivity from the U and Th series is usually increased in the clay components relative to K^{40} and decreased in the sand components relative to K^{40}.

Immediately after deposition, the organic components in bog-type sediments have a uranium content slightly greater than the average uranium content of crustal rocks. As would be expected, a greater proportion of the total initial radioactivity is contributed by C^{14} because the bogs contain a larger amount of carbon-bearing plant remains. The C^{14} radioactivity decreases with time and depth of burial because of its short-lived radioactive decay, whereas the U series radioactivity increases with time and depth of burial because of gradual assimilation of uranium and the progressive increase of its radioactive daughter products.

Radioactive Disequilibrium of Continental vs. Oceanic Deposits. During the erosion cycle, the radioactive substances containing Th^{230}, Ra^{226}, and their daughter products rise and sometimes reach the surface of the ground to form gossan and hot spring deposits. Thus, the overall patterns of radioactive disequilibrium at many places near the surface of continents are similar to those at the surface of sediments on the ocean floor. The direction of change of radioactivity is the opposite, however, as radioactivity decrease-by-decay occurs with emergence on the continents, whereas it occurs with burial in the oceanic sediments. The reason for this difference is the opposing direction of the geochemical processes affecting the distribution of the daughter products produced by the radioactive disintegration of uranium. Prior to accumulation in oceanic sediments, particles adsorbing the newly generated daughter products pass through sea water containing dissolved uranium; on the continents, migratory uranium—intermittently, in the form of dissolved species in surface water and ground water—passes through the sediments, leaving behind the remnant daughter product radioactivity.

Ancient Sediments

Ancient sediments can be divided into: (1) marine sediments, which include common shales, sandstones, and limestones and the less common phosphatic shales, phosphatic limestones, and black shales; (2) nonmarine sediments, which include fluvial sandstones, lacustrine shales, and coal; and (3) littoral and beach sediments. Common concentrations of the major sources of radioactivity in the major lithologic types of sedimentary rocks are shown in Table 2.

Generally marine sediments are slightly more radioactive than fluvial sediments. Amounts of radioactivity contributed by the U series, the Th series, K^{40}, and Rb^{87} are difficult to calculate because of the irregular distribution of these substances in rocks from the same geologic formation; therefore, common concentrations of radioactive sources in unmineralized sedimentary rocks are given. These values are for rock types of large areal extent, and radioactive equilibrium is usually attained in units of this size. Higher radioactivity occurs in local areas in these sediments, such as in uranium deposits in fluvial sandstones, and in placer sands containing primarily thorium. The highly radioactive rocks are discussed in the section on mineralized deposits.

Coal ranks as a sediment with intermediate radioactivity, usually greater than for common sedimentary rocks but less than for radioactive mineralized deposits. Uranium and its daughter products contribute most of the radioactivity in the more radioactive coal deposits. Presumably the uranium associated with coaly carbonaceous rocks is assimilated from ground water in a manner similar to that in which peat assimilates uranium. Among the various types of coaly carbonaceous rocks, impure lignite, and lignitic shale generally contain the most uranium.

Mineral Deposits of Radioactive Elements

Abnormally high radioactivity in sediments is a consequence of secondarily introduced material, concentrated by accumulation of detrital material or concentrated by chemical precipitation from ground water or by evaporation. The radioactive materials so concentrated are limited to K^{40} in potash deposits, and the U and Th decay series in coarse-grained sediments such as conglomerate and sandstone (Finch et al., 1973).

Placer Deposits. Some placer sands have relatively intense radioactivity from the contained Th^{232} series in the sediment. These radioactive placer deposits, either alluvial or littoral, are among the most widespread of radioactive mineral deposits. The common radioactive resistate minerals are monazite, zircon, euxenite, samarskite, thorite, thorianite, and betafite (see *Heavy Minerals*). Sands bearing these minerals generally contain more thorium than uranium.

Radioactive Conglomerates. Precambrian conglomerates in Africa, Canada, and Brazil contain significant concentrations of uranium and lesser concentrations of thorium. The common uranium content of the conglomerates is $\approx 0.05\%$, thorium is $\approx 0.02\%$, both of which are produced as a by-product of gold mining and production. Pyrite is abundant in these gold-bearing conglomerates, and the common radioactive minerals are uraninite, monazite, and thorium-uranium-bearing hydrocarbons. It is uncertain whether these deposits are consolidated and metamorphosed equivalents of a placer sand, whether the radioactive minerals are secondarily introduced, or whether the complex is a mixture of materials of both origins (Davidson, 1957).

Uranium-Sandstone Deposits. The most intensely radioactive sediments are sandstones that contain secondarily introduced uranium. These uranium deposits are classified as low-temperature disseminated deposits in clastic rocks (Heinrich, 1958); they do not contain significantly more Th^{232}, K^{40}, or Rb^{87} than do common clastic rocks. In sandstone-type ore deposits, uranium is primarily associated with sediments of fluvial origin. The largest uranium deposits of this type are in the semiarid western US, mainly in lenticular sandstone and conglomerate. The sandstone-type deposits occur in many stratigraphic units ranging from Pennsylvanian to Holocene in age.

In general, rocks saturated with water or otherwise protected from oxidation give the appearance of unoxidized ore deposits. The primary

TABLE 2. Common Content of Major Radioactive Isotopes, in Unmineralized Sedimentary Rocks (ppm)

Rock Type	U^{238}	Th^{232}	Rb^{87}	K^{40}
Limestone	2	1.7	<5	0.1
Sandstone	0.5	1.7	17	1
Shale	3.7	12	40	2
Black shale	50	12	–	2
Phosphate rock	100	4	<10	0.2

or "black" ore has a high proportion of low-valent mineral assemblages, but slight weathering of the primary ore has produced a host of different high-valent uranium and vanadium minerals. The primary ores were deposited in a reducing environment and probably remained stable until exposed to oxidizing conditions, which were usually related to the lowering of the water table. During transition from reducing to oxidizing environment, significant quantities of uranium isotopes and some daughter products migrated in the host sandstone and often produced extreme radioactive disequilibrium. In the completely oxidized zones that lack vanadium minerals, uranium was leached out, leaving behind radioactive Th^{230}, Pa^{231}, and their daughter products. Before complete oxidation was accomplished, however, significant quantities of U^{234} often were preferentially leached from the parent U^{238}. In rocks above the water table and partly protected from air oxidation, deficiencies of U^{234} of 40–60% are not uncommon in ore deposits. Deficiencies of U^{234} of as much as 22% (Rosholt et al., 1964) have been observed in less-oxidized deposits below the water table.

Use of Radioactivity in Sediments

Radioactive elements in sediments have two basic uses: (1) economic and (2) academic—in studies of geologic, geochemical, and archeologic problems.

Radioactive Ores. Most of the thorium that has been allocated for nuclear energy has been recovered from placer deposits in beach sands. A secondary source has been from ancient conglomerates as a by-product with recovery of gold and uranium from South Africa and Canada. A preponderance of the uranium allocated for nuclear energy has been obtained from sedimentary rocks. In the earlier stages of the nuclear age, uranium was recovered mostly from vein-type deposits. In the last ten years, the proportion of uranium recovered from sandstone-type deposits has greatly increased; and the majority of reserves in the US is in this type of deposit (Finch et al., 1973).

Dating of Sediments. The application of radiocarbon (C^{14}) dating to archeological studies has been the most useful. In Pleistocene studies that require estimating the age of materials, particularly carbonate sediments, between 30,000 and 200,000 years old, Th^{230}/U^{234} and Pa^{231}/U^{235} dating methods are being developed, and Pa^{231}/Th^{230} methods have been used for deep-sea sediments.

The ratio of radiogenic Ar^{40} to radioactive K^{40} in several minerals such as biotite, hornblende, sanidine, and glauconite, and in volcanic ash, has been used to date sedimentary horizons. The ratio of radiogenic Sr^{87} to radioactive Rb^{87} in biotite and sanidine in bentonitic shales has also been used to date ancient sediments.

Studies of Geochemical Processes Using Natural Tracers. The radioactive daughter products of the U^{238} and U^{235} decay series have been used to study the Pleistocene geochemistry of uranium in sandstones. In these studies, the primary natural tracers that have been investigated are U^{234}, Th^{230}, and Pa^{231}. Osmond et al. (1968) have investigated the usefulness of U^{234}-U^{238} disequilibrium to evaluate mixing proportions of waters from differing sources that contribute to ground water contained in sedimentary formations.

JOHN N. ROSHOLT

References

Davidson, C. F., 1957. On the occurrence of uranium in ancient conglomerates, *Econ. Geol.*, 52, 668–693.

Finch, W. I.; Bulter, A. P., Jr.; Armstrong, F. C.; Weissenborn, A. E.; Staatz, M. H.; and Olson, J. C., 1973. Nuclear fuels, *U.S. Geol. Surv. Prof. Pap. 820*, 455–476.

Goldberg, E. D., and Koide, M., 1962. Geochronological studies of deep sea sediments by the ionium/thorium method, *Geochim. Cosmochim. Acta*, 26, 417–435.

Heinrich, E. W., 1958. *Mineralogy and Geology of Radioactive Raw Materials*. New York: McGraw-Hill, 654p.

Koczy, F. F., and Rosholt, J. N., 1962. Radioactivity in oceanography, in H. Israel and A. Krebs, eds., *Nuclear Radiation in Geophysics*. Berlin: Springer-Verlag, 18–46.

Lal, D., and Peters, B., 1967. Cosmic ray produced radioactivity on the earth, in *Handbuch der Physik*. Berlin: Springer-Verlag, 551–612.

Osmond, J. K.; Rydell, H. S.; and Kaufman, M. I., 1968. Uranium disequilibrium in groundwater: An isotope dilution approach in hydrologic investigations, *Science*, 162, 997–999.

Rosholt, J. N.; Harshman, E. N.; Shields, W. R.; and Garner, E. L., 1964. Isotopic fractionation of uranium related to roll features in sandstone, Shirley Basin, Wyoming, *Econ. Geol.*, 59, 570–585.

Cross-references: *Fluvial Sediments; Heavy Minerals; Organic Sediments; Pelagic Sedimentation, Pelagic Sediments; Sedimentation Rates; Spring Deposits; Tracer Techniques in Sediment Transport.* Vol. I: *Radionuclides in Oceans and Sediments.* Vol. IVB: *Carnotite.*

RAINDROP IMPRINT

A raindrop imprint is a small, circular or roughly elliptical crater-like depression (12 mm maximum diameter, 3 mm maximum depth) which forms most readily from the impact of a falling raindrop of soft, fine-grained, semiwet sediment, and which, under certain conditions (see below), may be preserved as a *primary sedimentary structure* (q.v.). The original surface feature and the rock-equivalent cast or mold (see *Casts and Molds*) are both referred to as raindrop imprint, although the latter may be referred to as *rain cast* or *rain mold* (Fig. 1). Raindrops may grow as large as 5 mm in diameter and approach a free-fall velocity of 7.6 m/sec in still air and 27 m/sec in turbulent air; the impact of a raindrop on a soft sediment surface (i.e., mud or sand) is thus certainly sufficient to form a crater. Vertically falling raindrops produce circular imprints, whereas the introduction of a horizontal component to the raindrop trajectory is sufficient to produce an asymmetrical, ellipse-like imprint (Fig. 2).

Desor (1850), Lyell (1851), Twenhofel (1921), Grabau (1924), Bucher (1938), Shrock (1948), and Pettijohn and Potter (1964) are prominent among those who have described the occurrence of raindrop imprints in the geologic column. Desor, Lyell, and Twenhofel pointed out long ago, however, that not all raindrop-like depressions are actual raindrop imprints; other mechanisms can account for the presence of some of these features, notably the escape of gas bubbles from sediments (Moussa, 1974; see *Gases in Sediments*).

Raindrop imprints preserved in a sedimentary rock unit indicate that the environment at the time of deposition of that unit was one of subaerial exposure and subject to rainfall. The relative frequency of these structures, as well as their orientation and associated features (e.g., mud cracks, q.v.), tells us something about prevailing wind and weather conditions at the time.

FIGURE 2. Laboratory-produced elliptical raindrop impressions in pure St. Peter sand. Artificial rain impacted at a high oblique angle from the right, thus producing elongate shape and displacement toward the left (see arrows). Impressions were poorly preserved upon complete desiccation.

Laboratory and field experiments have shown that raindrop impressions are best formed in semiwet sediment and are preserved only if the sediment retains cohesion upon drying. Approximately 15% (by wt.) of clay is necessary in sandy sediment to reach the cohesion threshold necessary for imprint preservation. If desiccation cracks are present on the same surface, the raindrop imprint must have formed first, that is, shortly after subaerial exposure but before appreciable desiccation.

THOMAS E. EASTLER

References

Bucher, W. H., 1938. Key to papers published by an institute for the study of modern sediments in shallow seas, *J. Geol.*, 46, 726-755.

Desor, E., 1850. On fossil raindrops, *Edinburgh New Phil. J.*, 49(98), 246-248.

Grabau, A. W., 1924. *Principles of Stratigraphy*. New York: Seiler, 1185p.

Lyell, C., 1851. On fossil rain-marks of the Recent, Triassic, and Carboniferous periods, *Quart. J. Geol. Soc. London*, 7, 238-247.

Moussa, M. T., 1974. Raindrop impressions? *J. Sed. Petrology*, 44, 1118-1121.

Pettijohn, F. J., and Potter, P. E., 1964. *Atlas and Glossary of Primary Sedimentary Structures*. New York: Springer-Verlag, 370p.

Shrock, R. R., 1948. *Sequence in Layered Rocks*. New York: McGraw-Hill, 507p.

Twenhofel, W. H., 1921. Impressions made by bubbles, rain-drops, and other agencies, *Geol. Soc. Am. Bull.*, 32, 359-371.

FIGURE 1. Rain molds (counterparts of raindrop imprints) on bottom of medium-grained sandstone unit on North End, Change Islands, Newfoundland, Canada. Rain prints are in association with nearby "mud cracks" and "groove molds."

Cross-references: *Casts and Molds; Gases in Sediments; Mud Cracks; Primary Sedimentary Structures; Top and Bottom Criteria.*

RAPID SEDIMENT ANALYZER

The term *rapid sediment analyzer* may be applied to any automated method of particulate analysis suitable for use on natural materials, which gives the results of analysis relatively quickly as compared to conventional methods. Many modern methods of analysis make use of a direct plotting technique that produces a record as the analysis proceeds. Over a period of years, however, the term *rapid sediment analyzer* has come to have a more specific meaning, being applied to the technique where the density differential (caused by introduction of a sample) between a sedimentation column and a reference column is measured. As the particles fall out of the system, the density imbalance is gradually restored; plotting this density differential against time gives the distribution of particles with respect to fall velocity. The method and a technique was first described by Appel (1953) and subsequently improved by Zeigler et al. (1960) and Schlee (1966), both of Woods Hole Oceanographic Institution. The Woods Hole design has been used to produce a commercially available unit called Rapid Sediment Analyzer Type 341, produced by Benthos. This name is unfortunate because it implies that the instrument is capable of analysis of all sediment sizes; it is capable only of analysis within a small range, namely sand-sized particles. A more appropriate name would be Rapid Sand Analyzer.

Outline of Operation

The sediment column (Fig. 1) is filled with de-aired water and installed in a location protected from draught and vibration, which may affect the transducer (Lawrence, 1971). A dry, disaggregated sediment sample of appropriate weight (0.5–10 g), depending on the instrument and the scale sensitivity set on the X-Y plotter (Fig. 2), is placed on the gauze tray (Fig. 3) of the sample holder and then wetted down so that the sand grains are held to the gauze by capillary action. The sample introduction mechanism is then activated, inverting the sample tray and bringing it in contact with the water surface at the top of the sedimentation column. At this time, the capillary holding action is broken, and the particles begin to fall within the column. At the instant the sedimentation starts the X-Y plotter is automatically started. The pressure transducer (Fig. 4) located between the sedimentation and the reference columns now senses a pressure change because the effective density of the sedimentation column is greater than the reference column to which no sediment has been added. The electrical signal generated in the transducer is fed to the elec-

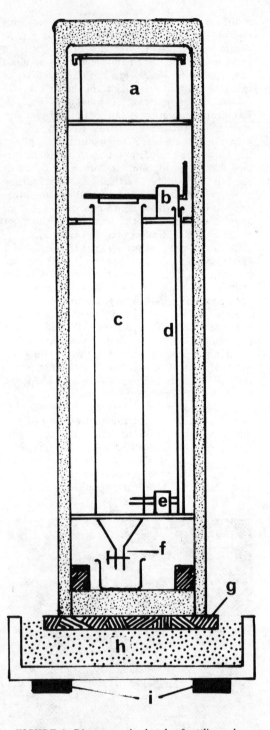

FIGURE 1. Diagrammatic sketch of settling-column module (not to scale). (a) water reservoir, (b) sample introduction mechanism, (c) settling column, (d) reference column, (e) transducer, (f) sample recovery mechanism, (g) 2-inch (5 cm) concrete levelling platform, (h) 5 inches (12.5 cm) of uniform sand, (i) isomode antivibration pads.

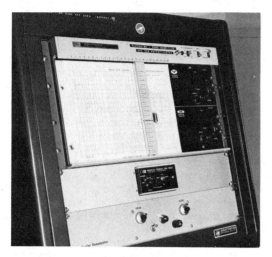

FIGURE 2. Rapid Sediment Analyzer recording module. (Photo: GSC No. 2016219-A.)

FIGURE 4. Lower part of Analyzer showing the sedimentation and reference columns and the pressure transducer. Accumulated particles from previous runs may be seen in the bottom of the sedimentation column. (Photo: GSC 201630-A.)

tronics console where it is amplified and fed to the recorder where the pressure differential (weight of sample in the sedimentation column) is plotted against time. The particles with the greatest fall velocities pass the transducer and no longer contribute to the effective weight of the column. There is a reduction of the differential pressure between the columns, and thus the curve being generated on the recorder decays with time as the successively finer particles (lower fall velocities) fall past the pressure sensor. If all the particles in the sample were of the same density and shape, the curve generated would be a grain-size distribution curve. The above-mentioned factors, however, especially shape and density, have a strong influence on the rate of fall of the particles. Thus if there is a wide range of particle densities and shapes the curve will deviate from a size-distribution curve. This curve of fall velocity is, however, more useful to many investigations than is the size distribution, because the process of sedimentation has been duplicated; thus interpretation of the conditions of sedimentation can be inferred, and ancient environment can be postulated with a higher degree of certainty.

When all of the sample has fallen past the pressure sensor, it is effectively out of the system, and there is no need to remove the sample from the column. Before introduction of the next sample, however, sufficient time must elapse for all fine particles from previous samples to settle. To speed this procedure, the sample may be sieved through a screen of lower limiting size, usually 50μ. The rate of fall of this size of particle is approximately known, and under average conditions all particles would be out of the system in seven minutes.

Theory of Operation. The basic theory of operation of the rapid sediment analyzer derives from theories related to the rate of fall of particles through a liquid medium. The rate of fall of a particle is dependent on several variables and may be expressed by the equations of Stokes' Law (q.v.) and the Impact Law (q.v.). These equations may be combined to give Rubey's (1933) general formula, which may be used for all particles (see *Settling Velocity*). The equations assume that the particles are spherical. Shape, although not a factor considered in Stokes' Law, is a significant element in the sedimentation of larger particles. One of the main benefits of measuring fall velocity, rather than measuring mean diameter (as in sieving), is that

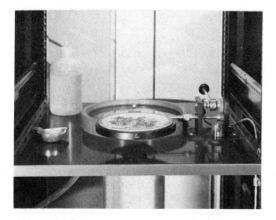

FIGURE 3. Sample introduction mechanism showing sample on gauze holder. (Photo: GSC No. 201629-C.)

the natural process under which the grains were originally deposited is duplicated; fall velocities, therefore, give direct information as to the depositional history of the sample. Schlee (1966) compares the settling times determined in the modified Woods Hole Rapid Sediment Analyzer to theoretical fall velocities, with similar but slightly displaced curves (Fig. 5).

History of Development

Rapid sediment analysis was first used on a production basis by Scripps Institute of Oceanography when sieve analysis proved impractical for use aboard ship, as well as too slow for volume production (Emery, 1938). Emery's device and technique was based on earlier work by Owens (1911). The sand particles falling through a water column approximately 150 cm in length are accumulated in a visual accumulation tube 15 cm long and 7 mm in diameter. The rate of accumulation was observed and recorded by Emery, who observed discrepancies due to changes in temperature, density currents within the tube, and also due to the varying concentration of the sample. The time required for analysis by this method is approximately 5 min compared with approximately 1 hr for a sieve analysis. Developments of the basic method have been principally with the sample introduction mechanism and in automation of the recording of the rate of sample accumulation. The visual-accumulation-tube method is a standard method of size analysis used by many agencies and is suggested by the US Federal Interagency Sedimentation Project as a standard method of analysis.

The development of the modern rapid sediment analyzer, involving the measurement of pressure differential between a sedimentation column and a reference column, is first mentioned by Appel (1953). The method was further developed in the late 1950s at Woods Hole Oceanographic Institution (Zeigler et al., 1960). The size of the tube used by Woods Hole is larger than for the visual accumulation tube, thus allowing a greater separation and less interference between particles as they fall. The original sample introduction mechanism was by means of a knife-edged gate. The great advantage of the method was that the sample size could vary between wide limits (approx. 2-30 g), and the sample need not be weighed. The pressure differential sensed by the transducer is amplified and recorded on a chart recorder. The pressure (weight of sample within the system) is plotted against time, and a curve of fall velocity is generated.

The modifications by Schlee (1966) include: (1) an increase in the size of the settling tube from 2 to 2½ inches (5.08 to 6.35 cm), allowing the grains a larger volume through which to settle; (2) the sample introduction mechanism was changed to a porous disk which holds the sample by capillary action until the disk touches the surface of the water and is released; and (3) a commercial pressure-sensing device is used. The pressure transducer must be sufficiently sensitive and also have the ability to exclude extraneous vibration (see Schlee, 1966); the recorder must be compatible with the transducer and should have a fast response time. These modifications have served to make the rapid sediment analyzer a faster and more efficient instrument. Modifications have also been made by Benthos, Inc., who now produce a commercially available unit; it uses a fall distance of 1 m in a 5-inch (12.7-cm) inside diameter acrylic tube and should be useful in the size range 50μ to 4 mm. It uses a X-Y plotter rather than a continuous-chart recorder.

D. E. LAWRENCE

References

Appel, D. W., 1953. An instrument for rapid size frequency analysis of sediment, Ph.D. Thesis, St. Univ. Iowa.
Buckley, D. E., 1964. Mechanical analysis of microscopic dispersal systems by means of a settling tube, *Bedford Inst. Oceanogr. Rept. 64-2*, 43p.
Carver, R. E., 1971. Heavy-mineral separation, in R. E. Carver, ed., *Procedures in Sedimentary Petrology*. New York: Wiley, 427-452.
Emery, K. O., 1938. Rapid method of mechanical analysis of sands, *J. Sed. Petrology*, 8, 105-111.
Lawrence, D. E., 1971. Modification to rapid sediment analyser, *Geol. Surv. Canadian Pap. 71-1(8)*, 120-123.
Owens, J. S., 1911. Experiments on the settlement of solids in water, *Geogr. J.*, 37, 59-77.
Rubey, W. W., 1933. Settling velocities of gravel, sand and silt particles, *Am. J. Sci.*, 25, 325-338.

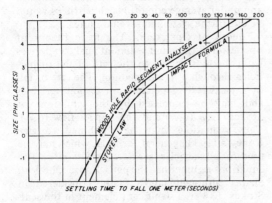

FIGURE 5. Comparison of settling curves: Woods Hole Rapid Sediment Analyzer and the Impact Formula-Stokes' Law (from Schlee, 1966).

Schlee, J., 1966. A modified Woods Hole Rapid Sediment Analyzer, *J. Sed. Petrology,* 36, 403–413.

Zeigler, J. M.; Whitney, G. G., Jr.; and Hayes, C. R., 1960. Woods Hole Rapid Sediment Analyzer, *J. Sed. Petrology,* 30, 490–495.

Cross-references: *Grain-Size Studies; Granulometric Analysis; Impact Law; Sedimentary Petrology; Settling Velocity; Stokes' Law.*

RAUHWACKE

Rauhwackes (rauchwackes) are thoroughly weathered breccias with a preponderantly calcitic to dolomitic composition. They show a yellowish brown to gray color and are characterized by their finely porous to coarsely cellular structure. Rauhwackes may be subdivided into two main varieties: (1) *monomict* rauhwackes, showing breccia structures formed by fragmentation of only one type of rock (usually dolostone); and (2) *polymict* rauhwackes, formed by fragmentation and mixing of two or more types of rock (Leine, 1968). Both varieties usually occur in zones parallel to the general stratification and are in many cases associated with gypsum or anhydrite deposits. Rauhwackes form a characteristic part of Triassic formations in the Alpine mountain chains of Europe. They have also been reported in Permian and Triassic formations of Europe outside the Alpine mountain ranges, and from N African mountain chains.

The literal translation of the German term *Rauhwacke* (*Rauchwacke*) is "rough stone" or "rough rock." These terms would appear to have originated in Thuringia (Central Germany) as a special name for cavernous dolomitic rocks, corresponding to the present monomict rauhwackes. The French word for rauhwacke is *cargneule* or *cornieule;* the Spanish word is *carniola.*

Origin

A number of theories have been propounded regarding the genesis of rauhwackes. Many geologists assume that these rocks represent sedimentary breccias. Others are of the opinion that the brecciated structure of rauhwackes was formed by collapse or by volumetric increase as a result of the conversion of anhydrite into gypsum. Recent investigation has suggested that rauhwackes represent true tectonic breccias which have acquired their porous to cavernous appearance as result of later calcitization, recrystallization, and selective weathering (Leine, 1968).

Monomict rauhwackes are carbonate rocks showing an erratic cavernous to *boxwork* structure (see *Boxwork*). In the vast majority of cases, they were formed from brecciated dolostones; in rare cases they originated from fragmented dolomitic marls. Fissures between the rock fragments become filled with calcite, and the matrix of crush breccias is dedolomitized. In many cases, the fragments themselves have also undergone dedolomitization (q.v.); thus all stages may occur—from fragments replaced only along their borders to fully dedolomitized fragments. As a consequence of selective weathering, the calcitic framework of these breccias stands out prominently, thus forming a cellular structure which represents the most striking feature of monomict rauhwackes (see *Boxwork*).

The cavities of monomict rauhwackes may contain pulverulent dolomite or dolomitic sand (German *Dolomitasche*), consisting of a loose fabric of dolomite crystals with a fragile cohesion. This material is formed by partial weathering of dolostone fragments. Monomict rauhwackes may be separated from associated carbonate rocks by a more or less sharp contact, but they usually merge gradually into veined dolostones or dolomitic limestones. These latter rocks originated from cracked dolostones by means of intense dedolomitization.

Polymict rauhwackes vary from weathered, porous crush breccias to true mylonites. They possess a yellow to brown, primarily calcareous matrix. Besides carbonate fragments the matrix often contains fragments of pelitic and psammitic rocks and occasionally gypsum. The fragments in polymict rauhwackes may be angular to well rounded. Polymict rauhwackes may range from microbreccias to megabreccias, containing rock bodies up to tens of meters in length. Polymict rauhwackes sometimes contain exotic fragments derived from underlying or overlying nappes, picked up during sliding movements along the contacts of the nappes.

Polymict rauhwackes form discontinuous bands and lenticular masses usually arranged parallel or subparallel to the stratification. These rauhwackes represent interformational crush breccias which were formed by tectonic sliding parallel to the general bedding in zones containing specific types of carbonate rocks with a slight intergranular cohesion, i.e., rocks which possess the property of easily becoming deformed as a result of differential penetrative movements, a consequence of sliding. A large portion of the carbonate matrix of the polymict rauhwackes was formed by pulverization of the above-mentioned rocks.

Polymict rauhwackes found as intercalations in regionally metamorphosed rocks such as mica schists, calcareous schists, micaceous marbles, and gneisses, often show remarkably nonmetamorphic appearance because the brecciated structure of these rauhwackes originated essentially as result of tectonic movements which took place after the metamorphism. Subsequent to this brecciation, the rauwackes have undergone recrystallization, producing a random fabric. The matrix of polymict rauhwackes is usually extensively recrystallized, a process which is attended with dedolomitization and sometimes with dolomitization.

The carbonate rocks with comparatively slight intergranular cohesion, to which the vast majority of rauhwackes owe their origin, were probably confined to a saline environment of deposition. This would explain the fact that rauhwackes are often associated with evaporites and also why they are mostly restricted to specific stratigraphic horizons.

LOUIS LEINE

References

Brücker, W., 1941. Uber die Entstehung der Rauhwacken und Zellendolomite, *Eclogae Geol. Helv.,* 34, 117-134.

Leine, L., 1968. Rauhwackes in the Betic Cordilleras, Spain—Nomenclature, description and genesis of weathered carbonate breccias of tectonic origin, Amsterdam. (Printed Thesis, Univ. Amsterdam), Culemborg: N.V. Princo, 112p.

Leine, L., and Egeler, C. G., 1962. Preliminary note on the origin of the so-called "konglomeratische Mergel" and associated "rauhwackes," in the region of Menas de Seron, Sierra de los Filabres (SE Spain), *Geol. Mijnbouw,* 41, 305-314.

Riedmüller, G., 1976. Genese und charakteristik der Rauhwacken im Pittental (Niederosterreich), *Geol. Rundschau,* 65, 290-332.

Speed, R. C., 1975. Carbonate breccia (rauhwacke) nappes of the Carson Sink region, Nevada, *Geol. Soc. Am. Bull.,* 86, 473-486.

Weidmann, M., 1971. Cargneule ou cornieule? *Eclogae Geol. Helv.,* 64, 47-51.

Cross-references: *Boxwork; Breccias, Sedimentary; Carbonate Sediments—Diagenesis; Dedolomitization; Dolomite, Dolomitization; Gypsum in Sediments; Intraformational Disturbances; Limestones; Olistostrome, Olistolite.*

RED CLAY

Red clay is a reddish-brown deep-sea deposit formed from sediment of the finest grain size transported by ocean currents and slowly accumulated on basin floors at depths generally >3500 m. The color of this pelagic sediment is due to the presence of ferric oxide and ranges from brick red in the North Atlantic to chocolate brown in the South Pacific, where the color is darkened because of the presence of finely divided manganese dioxide. The median grain size of deep-sea red clays is 1.0μ, with about 83% of constituent grains of clay size $<2.0\mu$ and 17% of silt size.

The word *clay* (q.v.) should not be construed as a mineralogical term; it connotes a loose, extremely fine-grained sediment composed of clay-sized or colloidal particles, characterized by high plasticity and containing a large proportion of clay minerals. The finely crystalline clay minerals which commonly constitute red clay are illite, smectite, kaolinite, and chlorite; amorphous hydrous aluminum silicates such as allophane and colloidal complexes may also be present. The silt-sized mineral particles are either detrital (quartz, feldspars, micas, amphiboles, pyroxenes); authigenic (zeolites, barite, manganites); or volcanogenic. Numerous heterogeneous constituents are found in the fine-grained red clay matrix: volcanic ash and pumice, shark teeth, whale bones, ice-rafted rock debris, nodules of manganese and phosphorus, particles of cosmic origin, and biogenous skeletal fragments of silica and calcium carbonate. Red clay is the residual material derived from the accumulation and dissolution of continental, atmospheric, volcanogenic, and biogenic particles that have settled slowly through the water column. It is found in areas where biogenic sedimentation is minimal (see *Pelagic Sedimentation, Pelagic Sediments*). It covers about 30% of the total area of the deep sea and is more prevalent in the Pacific than in the Atlantic and Indian oceans.

Red mud is a reddish-brown terrigenous marine deposit of unconsolidated sediment consisting of sand, silt, and lesser amounts of clay, mixed together with water. Red mud is found on parts of the ocean bottom adjacent to the mouths of tropical rivers, at depths of 900–1800 m. Its color is due to the presence of ferric oxides and hydroxides and indicates an oxidizing sedimentary environment. Two regions where a large extent of red mud occurs are along the Brazilian coast at the mouth of the Amazon River and in the Yellow Sea at the mouth of the Yangtze River.

MARIAN B. JACOBS

References

Goldberg, E. D., and Arrhenius, G., 1958. Chemistry of Pacific pelagic sediments, *Geochim. Cosmochim. Acta,* 13, 153-212.

Griffin, J. J., and Goldberg, E. D., 1963. Clay mineral distributions in the Pacific Ocean, in M. N. Hill, ed., *The Sea,* vol. 3. New York: Interscience, 728-741.

Sverdrup, H. U.; Johnson, M. W.; and Fleming, R. H., 1963. *The Oceans.* Englewood Cliffs, N.J.: Prentice-Hall, 946-1049.

Twenhofel, W. H., 1961. *Treatise on Sedimentation,* vol. 1. New York: Dover, 250-254.

Cross-references: *Clay; Clay as a Sediment; Clays, Deep-Sea; Deltaic Sediments; Pelagic Sedimentation, Pelagic Sediments.*

REDEPOSITION, RESEDIMENTATION

Redeposition implies that a sediment is deposited at some place and time, is eroded and transported from that place, then deposited at some other point. The basic meaning is always clear, but there is great variation in the time scale over which redeposition occurs. For the upper few cm of the intertidal zone of a moderate energy beach, redeposition is a twice-daily occurrence; redeposition of floodplain sediment may occur at intervals of one to several thousand years; and redeposition of cratonic, marine sands may occur only after a lapse of several hundred million years.

In general, *redeposition* applies to sediments, rather than rocks, and involves time intervals that are geologically short. *Reworking* indicates the erosion, transport, and finally deposition of preexisting rock fragments, fossils, or authigenic grains in a sediment younger than the parent material. *Recycling* applies to the erosion, transport, and deposition of single grains of quartz, or other detrital minerals, derived from a preexisting sedimentary rock.

In the most common usage, resedimentation is synonymous with redeposition. The term is most frequently encountered in the recent literature on turbidity current and other subaqueous current and slide deposits. Some turbidites and related deposits appear to be the end product of several episodes of retransportation and redeposition (see, e.g., *Submarine Fan (Cone) Deposits*). Unfortunately, the term *resedimentation* is also used to indicate sedimentation in pore spaces of preexisting sedimentary rocks or to indicate either reworking or recycling of sedimentary rocks or sedimentary materials. The meaning embodying the concept of erosion and redeposition of unconsolidated sediment, without major change in the overall character of the sediment, is preferred.

The term *reworked sediment* normally describes sediment that has been eroded and redeposited, most commonly sediment that has been redeposited several times. Some degree of winnowing may be suggested by the term, but not properly. As in the case of "reworked fossil," or "reworked glauconite," "reworked" indicates the erosion and transport of some entity, intact. Reworked sediment is, therefore, sediment that has been redeposited without major alteration of its texture or composition by winnowing.

ROBERT E. CARVER

Cross-references: *Deposition; Maturity; Provenance; Scour-and-Fill; Sedimentation.*

REDUCTION, REDUCTION NUMBER

The reduction number (Trask and Hammar, 1936) is defined as the number of cm^3 of 0.4 normal chromic acid that are reduced by 100 mg of sediment under certain standard conditions. It is actually a titration number, similar to the saponification or iodine number and is employed to give an estimate of the organic content of a sediment. The procedure is based on the Schollenberger (1928) method of determining soil organic matter, with various adaptations which depend on the different conditions encountered (see Trask and Patnode, 1942).

The reduction number is influenced by several factors: (1) the quantity of organic matter in the sediments; (2) the state of oxidation of this organic matter; (3) the presence of oxidizable inorganic substances; (4) the extent to which the type of organic matter present can be oxidized by chromic acid under the standard conditions of analysis. The standard conditions (Trask and Patnode, 1942) include provisions for time, temperature, and quantity controls, to insure uniformity of results.

B. CHARLOTTE SCHREIBER

References

Schollenberger, C. J., 1928. A rapid method for determining soil organic matter, *Soil Sci.,* **24,** 65-68.

Trask, P. D., and Hammar, H. E., 1936. Degrees of reduction and volatility as indices of source beds, *Drilling Production Practice, 1935,* 250-266.

Trask, P. D., and Patnode, H. W., 1942. *Source Beds of Petroleum.* Tulsa: Am. Assoc. Petrol. Geologists, 566p.

Cross-references: *Color of Sediments; Organic Sediments.*

REEF COMPLEX

Various compounds with "reef" (*reef breccia, reef conglomerate, reef limestone,* etc.) are applied to the wide variety of lithic types that may be collectively referred to in paleoecologic terms as a "reef association" or in structural terms as a "reef complex" or *biohermal complex.* A reef is commonly defined as a wave-resistant framework and may be dominated by several different framework builders—corals, algae or, on a smaller scale, serpulids, oysters, rudistids, and others (see *Algal-Reef Sedimentology; Coral-Reef Sedimentology*). The term *bioherm* (Cummings, 1932) is sometimes preferred inasmuch as it implies no wave-resistant properties, merely a structural dome or framework, which may lie *below* wave base (see *Limestones*).

Reef complex implies the totality of structures that may include fringing, barrier, and atoll forms, together with smaller units—patch, hummock, platform reefs, down to single colonies such as individual pinnacles or coral heads (Fairbridge, 1950). Coral heads range from a few cm up to about 5 m in diameter; modern barrier reefs range up to 4000 km in length. In large reef complexes, there may be many sea-level reefs surrounded by deeper water, which is often a site of noncarbonate terrigenous sedimentation. Each reef usually comprises a polarized assemblage: (1) roughwater (high wave energy) species and sedimentary features facing the prevailing wind (in trade wind belts), and (2) calm-water (low energy) species and features on the lee side. Circular or ovoid reefs mark regions of alternating seasons (roughly speaking, latitudes $15°N-15°S$). The trade-wind types are most characteristic of $15°-30°N$ and S.

Reef limestone is a general term applicable to any phase of the reef complex. *Rag, coral rag,* or *reef rock* (or *coral reef rock,* as appropriate) is usually applied to reef limestones where the coral, algae, or other colonies for the most part retain their original positions of growth. The term *biolithite* is a useful collective term for this material (see *Limestone*). *Reef breccia* (or *rubble* if uncemented) refers to an angular calcirudite of bioclastic origin; it develops on the windward or deep-water side of the reef. The off-reef basinward structure is called the *reef talus* (a geomorphic term). Reef fragments thrown onto the *reef flat* or *reef platform* become more or less rounded by wave action and after cementation become a *reef conglomerate*. This accumulation is commonly in the form of an intertidal *rampart* so that cementation is also often intertidal, and the conglomerate so formed frequently passes laterally into a cemented *beachrock* (q.v.) that represents a former sand bar or sand cay. In the case of fringing reefs abutting continental coasts, the reef flat and beach deposits may contain high proportions of noncarbonates, e.g., in Queensland (Fairbridge, 1950), in the Comoro Islands (Guilcher, 1965), and on the Red Sea shores (Friedman, 1968). It should be stressed that the "clean," almost pure carbonate sands and lutites of mid-ocean atolls (e.g., Emery et al., 1954; Ginsburg et al., 1963) are rarely encountered on continental coasts.

Oceanic atolls (with foundations of volcanic piles on oceanic crust) have not been found in ancient rocks, but *shelf atolls* (Fairbridge, 1950) are not infrequent. The best known are the Horseshoe Atoll and Scurry-Snider reefs in the Pennsylvanian of W Texas (Leversen, 1967) and the Cretaceous El Abra reef in Mexico (Guzman, 1967). In these instances, there are generally much purer carbonate sediments than in nearshore reefs. A quite exceptional reef complex is the Devonian Hamat Lagdad reef in the Tafilalet basin of Morocco (Massa, 1965), where the buildup has occurred on a freshly erupted volcanic foundation (but *not* over oceanic crust).

To the interior, calm-water, shelf side of the reef there is a rapid gradation from a zone of massive reef rock, through a belt of scattered coral heads, separated by *calcarenites (skeletal sands)*. In modern reefs, the latter grade shelfward (or inward in an atoll) into *calcilutites* (McKee et al., 1959) or *lagoon limestone* (see *Tropical Lagoonal Sedimentation*). In lagoon areas there may be traces of minor eustatic oscillations marked by terraces and rows of intermediate reefs, and by inactive zones of bioclastics. In ancient reef environments, the shelf facies are very varied, ranging from terrigenous red beds to evaporites, stromatolitic limestones, or sabkha facies. These nonreef carbonates and shelf facies are not as a rule classified with the reef complex, although they often owe their existence to the barrier provided by the carbonate buildup (Selley, 1970).

Ancient Reef Complexes

Proterozoic-Cambrian. Very early carbonate "catchers" were the stromatolitic algae, their role being limited to stabilizing carbonate muds (see *Stromatolites*). Massive stromatolitic horizons constitute biostromes, but massive frame-building bioherms are not known. Some Cambrian *Archaeocyathus* patches have been described as reef-formers (in South Australia), but the term seems ill-advised.

Ordovician. This period saw the appearance of algal bioherms; albeit small by Quaternary standards, some were large enough to modify the paleogeography in a minor way.

Silurian-Devonian-Carboniferous-Permian. True coral frame-builders evolved, generating a wide range of reef patches ("reef knolls"). Some of the earliest recognized were the Silurian reefs in Gotland in Sweden (Hadding, 1950; Manten, 1971), while some of the best known are in the Niagaran of the Midwest and along the southern edge of the Canadian Shield, forming discontinuous complexes in trends that mark shelf margins and other boundaries between shallow and deeper basins (Cummings, 1932; Lowenstam, 1950; Briggs and Briggs,

1974). Silurian reefs also reflect structural trends in an extensive belt along the Tasman Geosyncline of eastern Australia.

Also well studied and economically important are the Devonian reefs of Alberta; they tend to mark shelf margins and structural "highs" (Davies, 1975). A similar situation is occupied in the Devonian of the Fitzroy Valley, W Kimberley of western Australia (Playford and Lowry, 1966). The Devonian examples probably represent the greatest reef development of the geological past. Subsequent buildups were fewer, but some complexes were spectacular.

Some of the best-known Carboniferous reef complexes are the structurally controlled reef knolls of Clitheroe, Cracoe and Craven, in the north of England, often crowning a submarine fault-scarp (Parkinson, 1944; Bond, 1950). The Pennsylvanian shelf atolls of W Texas, noted earlier, reflect mid-basin "highs."

Permian reefs were much less widespread than those of earlier periods, but nevertheless include one of the most remarkable carbonate buildups ever seen, the Capitan reef complex of the Guadalupe Mountains of New Mexico and W Texas (Newell et al., 1953; Silver and Todd, 1969; Tyrell, 1969). Very favorable field relations in the structurally almost undisturbed area make it possible to trace lagoonal facies through every phase of the main reef buildup through to the fore-reef facies and to the deeper basin deposits. The reef rock is particularly interesting inasmuch as its principal components are algae. Organic boundstones and submarine cement are believed to be important components in the massive reef facies (see articles in Hileman and Mazzullo, 1977).

Triassic-Jurassic-Cretaceous. The Mesozoic was marked by the evolution of more vigorous frame-builders among the corals, whereas algal reef-builders remained generally similar to the Paleozoic families. The most famous are the celebrated Triassic reefs of the Dolomite Alps in the Italian Tyrol; both coral and algae contribute. The outcrops are spectacular, but the complex Alpine tectonics have until recently tended to render the sedimentologic facies difficult to correlate. Following their recognition as giant reefs by Mojsisovičs (1879), the critical mapping work here was done initially by the M. M. Ogilvie-Gordons and more recently by Leonardi (1967; see Bosellini and Rossi, 1974).

Reef facies in the Jurassic and Cretaceous were curiously small and unimportant, bearing in mind the world-wide mild climates and extensive carbonate facies of those periods. Small clusters of rudistid mollusks built minor bioherms in the Cretaceous of Mexico and Texas and in the Tethyan (Mediterranean) belt. Coral reefs started building up during the early stages of orogeny in the late Cretaceous of the Zagros Mountains of Iraq and Iran (Henson, 1950).

Tertiary reefs are quite extensive and economically important, especially along the southern edge of the Mediterranean and the Mesopotamian-Persian Gulf Basin. Those of Iraq and Iran (Henson, 1950) are mainly located near the crests of anticlines, which were tectonically active throughout reef growth.

RHODES W. FAIRBRIDGE

References

Bond, G., 1950. The Lower Carboniferous reef limestones of northern England, *J. Geol.*, **58**, 430-487.

Bosellini, A., and Rossi, D., 1974. Triassic carbonate buildups of the Dolomites, northern Italy. *Soc. Econ. Paleont. Mineral. Spec. Publ. 18*, 209-233.

Briggs, L. I., and Briggs, D., 1974. *Silurian Reef–Evaporite Relationships.* Lansing: Mich. Basin Geol. Soc., 111p.

Cummings, E. R., 1932. Reefs or bioherms? *Geol. Soc. Am. Bull.*, **43**, 331-352.

Davies, G. R., ed., 1975. Devonian reef complexes of Canada, I (Rainbow, Swan Hills), *Canadian Soc. Petrol. Geol., Reprint Ser.*, **1**, 229p.

Emery, K. O.; Tracey, J. I.; and Ladd, H. S., 1954. Geology of Bikini and nearby atolls, Part I, Geology, *U.S. Geol. Surv. Prof. Pap. 260-A*, 265.

Fairbridge, R. W., 1950. Recent and Pleistocene coral reefs of Australia, *J. Geol.*, **58**, 336-401.

Friedman, G. M., 1968. Geology and geochemistry of reefs, carbonate sediments and waters, Gulf of Akaba (Elat) Red Sea, *J. Sed. Petrology*, **38**, 895-919.

Ginsburg, R. N.; Lloyd, R. M.; Stockman, K. W.; and McCallum, J. S., 1963. Shallow water carbonate sediments, in M. N. Hill, ed., *The Sea*, vol. 3. New York: Wiley, 544-582.

Guilcher, A., 1965. Coral reefs and lagoons of Mayotte Island, Comoro Archipelago, Indian Ocean, and of New Caledonia, Pacific Ocean, in W. F. Whittard and R. Bradshaw, eds., *Submarine Geology and Geophysics.* London: Butterworths, 21-45.

Guzman, E. J., 1967. Reef type stratigraphic traps in Mexico, *Proc. 7th Petrol. Congr.*, **2**, 461-470.

Hadding, A., 1950. Silurian reefs of Gotland, *J. Geol.*, **58**, 402-409.

Henson, F. R. S., 1950. Cretaceous and Tertiary reef formations and associated sediments in the Middle East, *Bull. Am. Assoc. Petrol. Geologists*, **34**, 215-238.

Hileman, M. E., and Mazzullo, S. J., eds., 1977. Upper Guadalupian facies, Permian Reef Complex, Guadalupe Mountains, New Mexico and West Texas, *Permian Basin Section, Soc. Econ. Paleont. Mineral., 1977 Field Conf. Guidebook (Publ. 77-16)*, 508p.

Jones, O. A., and Endean, R., 1973. *Biology and Geology of Coral Reefs,* vol. 1, Geology 1. New York: Academic Press, 410p.

Laporte, L. F., ed., 1974. Reefs in time and space—selected examples from the recent and ancient, *Soc. Econ. Paleont. Mineral. Spec. Publ. 18*, 256p.

Leonardi, P., 1967. *Le Dolomiti. Geologia dei monti tra Isarco e Piave.* Rome: Natl. Research Council, 2 vols., 1019p.

Leverson, A. I., 1967. *Geology of Petroleum.* San Francisco: Freeman, 724p.

Lowenstam, H. A., 1950. Niagaran reefs of the Great Lakes area, *J. Geol.,* **58**, 430–487.

McKee, E. D.; Chronic, J.; and Leopold, E. B., 1959. Sedimentary belts in lagoon of Kapingamarangi Atoll, *Bull. Am. Assoc. Petrol. Geologists,* **43**, 501-562.

Manten, A. A., 1971. *Silurian Reefs of Gotland.* Amsterdam: Elsevier, 539p.

Massa, D., 1965. Observations sur les séries Siluro-Devoniennes des confins Algéro-Marocains du Sud, *Compagnie Fr. Pétroles,* **8**, 187p.

Mojsisovičs, E. M., 1879. *Die Dolomit-Riffe von Sudtirol und Venetien.* Wien: Hölder, 551p.

Newell, N. D.; Rigby, J. K.; Fischer, A. G.; Whiteman, A. J.; Hickox, J. E.; and Bradbury, J. S., 1953. *The Permian Reef Complex of the Guadalupe Mountains Region, Texas and New Mexico.* San Francisco: Freeman, 236p.

Parkinson, D., 1944. The origin and structure of the Lower Visean reef knolls of the Clitheroe District, Lancashire, *Quart. J. Geol. Soc. London,* **99**, 155–168.

Playford, P. E., and Lowry, D. C., 1966. Devonian reef complexes of the Canning Basin, Western Australia, *Geol. Surv. W. Australia Bull.,* **118**, 150p.

Selley, R. C., 1970. *Ancient Sedimentary Environments.* Ithaca, N.Y.: Cornell U. Press, 237p.

Silver, B. A., and Todd, R. G., 1969. Permian cyclic strata, Northern Midland and Delaware basins, West Texas and Southeastern New Mexico, *Bull. Am. Assoc. Petrol. Geologists,* **53**, 2223-2251.

Tyrell, W. W., 1969. Criteria useful in interpreting environments of unlike but time-equivalent carbonate units (Tansill-Capitan-Lamar), Capitan Reef Complex, West Texas and New Mexico, *Soc. Econ. Paleont. Mineral. Spec. Publ. 14,* 80–97.

Cross-references: *Bahama Banks Sedimentology; Calcarenite, Calcilutite, Calcirudite, Calcisiltite; Carbonate Sediments–Diagenesis; Coral-Reef Sedimentology; Evaporite Facies; Limestones; Sabkha Sedimentology; Stromatolites; Talus; Tropical Lagoonal Sedimentation.* Vol. III: *Atolls; Coral Reefs.* Vol. VII: *Reefs and Other Carbonate Buildups.*

RELICT SEDIMENTS

Unconsolidated sediments on the sea floor that do *not* reflect the contemporary sedimentary regime are classified as *relict sediments.* Most commonly, such materials are relatively coarse (sand and gravel) nearshore deposits which have been drowned by the Holocene rise in sea level and then protected from reworking or from further deposition by the action of currents (Emery, 1968). Relict sediments may grade into what are termed *palimpsest sediments,* similar material which has been significantly reworked and altered by contemporary agencies (see *Continental Shelf Sediments*). A particularly obvious example of relict sediments would be the gravels which constitute many of the sills punctuating high-latitude fjords. These sills usually represent marine moraines, places where the retreat of a tidal glacier margin stopped for a time, and the coarsest material could accumulate. These sills often are much shallower than the surrounding basins and thus are kept clear of later sedimentation by tidal currents.

Relict sediments are not restricted to the continental shelves of glaciated regions. Off the southern California coast there are large accumulations of manganese crusts and nodules which appear to be related to earlier Cenozoic conditions; in the deep sea beyond the present northern limit of sea ice and bergs in the Antarctic, there are large areas of glacial marine sediment (q.v.) reflecting the much greater former extent of icebergs.

Recognition of relict sediments can be difficult. Perhaps the most useful criterion is simply the discovery of incongruous sediments, such as coarse material in deep, calm waters where fine debris would be expected. Physiographic anomalies like sills or submerged beach ridges and dunes may indicate the likelihood of a relict history. Sometimes geochemical or biological effects such as an iron oxide film on sea floor gravels, or extensive encrustation by animals on boulders and cobbles, both suggesting present quiescence, may be useful criteria.

Some economic mineral occurrences are associated with the relict sediments of continental shelves, including cassiterite (tin ore) in Indonesia and diamonds off SW Africa, both in drowned river deposits, and gold in submerged beach ridges near Nome, Alaska.

FREDERICK F. WRIGHT

References

Emery, K. O., 1968. Relict sediments on continental shelves of world, *Bull. Am. Assoc. Petrol. Geologists,* **52**, 445-464.

Milliman, J. D., and Barretto, H. T., 1975. Relict magnesian calcite oolite and subsidence of the Amazon shelf, *Sedimentology,* **22**, 137–146.

Siesser, W. G., 1972. Relict algal nodules (rhodolites) from the South African continental shelf, *J. Geol.,* **80**, 611–616.

Siesser, W. G., 1974. Relict and recent beachrock from Southern Africa, *Geol. Soc. Am. Bull.,* **85**, 1849–1854.

Slatt, R. M., 1974. Formation of palimpsest sediments, Conception Bay, southeastern Newfoundland, *Geol. Soc. Am. Bull.,* **85**, 821–826.

Cross-references: *Continental-Shelf Sedimentation; Continental-Shelf Sediments; Neritic Sedimentary Facies; Outer Continental-Shelf Sediments.*

REMANENT MAGNETISM IN SEDIMENTS

The natural remanent magnetism of sediments that is due to the preferred alignment of detrital, magnetic grains by a magnetic field is called *detrital remanent magnetism* (DRM). The DRM of sediment is typically of low intensity, usually less than 10^{-4} Gauss, and is commonly carried by small particles of magnetite that constitute a small fraction of the sediment.

Two types of DRM have been distinguished: depositional and postdepositional. In depositional DRM, the magnetic particle alignment attained in the water column is preserved to some extent when these particles are deposited; postdepositional DRM refers to magnetic grain alignment that occurs after deposition but prior to consolidation of the sediment. The relative importance of each type in contributing to the remanent magnetism of natural sediment is still open to debate and may depend on the characteristics of both the sediment and the depositional environment.

Depositional DRM

The acquisition of depositional DRM by sediment has been investigated under controlled laboratory conditions by Johnson et al (1948), King (1955), and Griffiths et al. (1960). Generally, it was found that the depositional DRM of sediment redeposited in still water faithfully records the azimuth, but the inclination is consistently shallower than the dip of the ambient magnetic field. The difference is referred to as the "inclination error" and amounts to as much as $32°$ (at ambient field dips of about $58°$). Redeposition experiments performed in running water and on sloped surfaces show other systematic deflections of the depositional DRM from the direction of the ambient field. These experiments point to the apparent importance of gravitational and hydrodynamic forces, in competition with the force of the magnetic field, in controlling the net alignment of magnetic particles during sedimentation (see *Anisotropy in Sediments*). The properties of laboratory-produced depositional DRM, however, may not be characteristic of the process that occurs in natural sediment. For example, an "inclination error" has not been demonstrated conclusively to occur in natural sediment deposits. Theoretical models that can account for the observed characteristics of depositional DRM are reviewed in King and Rees (1966) and Collinson (1965).

Postdepositional DRM

One of the first demonstrations of post depositional DRM was reported in Clegg et al. (1954). Some fine red sandstone was powdered and mixed with water until a slurry containing about 75% by weight of water was obtained. By applying a magnetic field at various stages of the drying process it was found that a remanence of comparable intensity to that of the parent rock could be produced so long as the water content was $>50\%$. Subsequent experiments by Irving and Major (1964) showed that the direction of laboratory-produced postdepositional DRM agreed with that of the ambient magnetic field over a large range of field inclinations, in contrast to the observations of "inclination error" in depositional DRM experiments. Other experimental work has shown that a postdepositional DRM can modify an existing depositional DRM (Løvlie, 1974) and that a postdepositional DRM that accurately records the ambient magnetic field may be acquired even in sediment that has been affected by bioturbation (Kent, 1973).

Application

Paleomagnetic studies on sediment can provide information on the past behavior of the earth's magnetic field and also can be used for stratigraphy and correlation purposes. In the context of the latter, the paleomagnetic reversal stratigraphy has become a standard and widely used method of correlation and age dating of Neogene strata, particularly in deep-sea sediments (see reviews by Opdyke, 1972; Harrison, 1974) but also in terrestrial deposits (e.g., Johnson et al., 1975) and elsewhere. Because a reversal of polarity of the earth's magnetic field would be experienced essentially simultaneously over the entire earth, the applicability of paleomagnetic reversal stratigraphy is ostensibly independent of geographic location and of sedimentary facies. This characteristic has enabled the correlation of different Neogene biostratigraphic zonations and their placement into an absolute time framework (Ryan et al., 1974; Berggren and Van Couvering, 1974). Other applications of the remanent magnetism of sediment are discussed in Irving (1964) and McElhinny (1973).

DENNIS V. KENT

References

Berggren, W. A., and Van Couvering, J. A., 1974. The Late Neogene: biostratigraphy, geochronology and paleoclimatology of the last 15 million years in marine and continental sequences, *Palaeogeogr. Palaeoclimat. Palaeoecol.*, 16, 1-216.

Clegg, J. A.; Almond, M.; and Stubbs, P. S., 1954. The remanent magnetism of some sedimentary rocks in Britain, *Phil. Mag.*, 45, 583-598.

Collinson, D. W., 1965. Depositional remanent magne-

tization in sediments, *J. Geophys. Research,* **70,** 4663-4668.
Griffiths, D. H.; King, R. F.; Rees, A. I.; and Wright, A. E., 1960. The remanent magnetism of some recent varved sediments, *Proc. Roy. Soc., London,* **A256,** 359-383.
Harrison, C. G. A., 1974. The paleomagnetic record from deep-sea sediment cores, *Earth Sci. Rev.,* **10,** 1-36.
Irving, E., 1964. *Paleomagnetism and Its Application to Geological and Geophysical Problems.* New York: Wiley, 399p.
Irving, E., and Major, A., 1964. Post-depositional remanent magnetization in a synthetic sediment, *Sedimentology,* **3,** 135-143.
Johnson, E. A.; Murphy, T.; and Torreson, O. W., 1948. Pre-history of the earth's magnetic field, *Terr. Magn. Atmos. Elect.,* **53,** 349-372.
Johnson, N. M.; Opdyke, N. D.; and Lindsay, E. H., 1975. Magnetic polarity stratigraphy of Pliocene-Pleistocene terrestrial deposits and vertebrate faunas, San Pedro Valley, Arizona, *Geol. Soc. Am. Bull.,* **86,** 5-12.
Kent, D. V., 1973. Post-depositional remanent magnetisation in deep-sea sediment, *Nature,* **246,** 32-34.
King, R. F., 1955. The remanent magnetism of artifically deposited sediment, *Roy. Astron. Soc. Mo. Not., Geophys. Suppl.,* **7,** 115-134.
King, R. F., and Rees, A. I., 1966. Detrital magnetism in sediments: An examination of some theoretical models, *J. Geophys. Research,* **71,** 561-571.
Løvlie, R., 1974. Post-depositional remanent magnetization in a redeposited deep-sea sediment, *Earth Planetary Sci. Lett.,* **21,** 315-320.
McElhinny, M. W., 1973. *Paleomagnetism and Plate Tectonics.* London: Cambridge Univ. Press, 358p.
Opdyke, N. D., 1972. Paleomagnetism of deep-sea cores, *Rev. Geophys. Space Phys.,* **10,** 213-249.
Ryan, W. B. F.; Cita, M. B.; Dreyfus Rawson, M.; Burckle, L. H.; and Saito, T., 1974. A paleomagnetic assignment of Neogene stage boundaries and the development of isochronous datum planes between the Mediterranean, the Pacific and Indian oceans in order to investigate the response of the world ocean to the Mediterranean "salinity crisis," *Riv. Ital. Paleont.,* **80,** 631-688.
Turner, P., 1975. Depositional magnetization of Carboniferous limestones from the Craven Basin of northern England, *Sedimentology,* **22,** 563-581.

Cross-references: *Anisotropy in Sediments; Fabric, Sedimentary; Pelagic Sedimentation, Pelagic Sediments.*

RESIN AND AMBER

Natural resins, secreted by trees or shrubs upon injury, appear either as balsam (a resin solution in esters) or as an oleoresin (a resin solution in essential oils). Resins are volatile and nonvolatile terpinoids and may occur with alcohols, aldehydes, esters, resenes, and other minor nonterpenoid substances. Amber and other fossil or subfossil resins of geologic interest apparently are derived chiefly from nonvolatile terpinoids through oxidation and polymerization. The details are poorly known, but the process apparently requires a relatively long lapse of time.

Kauri Gum and Copal

Kauri gum and copal are solid resins from trees but may be found in soils and alluvial deposits. Kauri gum, which also is known as *kauri copal,* is produced by the coniferous araucarian *Agathis australis,* a species confined to New Zealand. *Agathis alba,* widespread in Indonesia and the Philippines, yields a closely related resin. Both are characterized by containing agathic acid, melt at 150-210°C, range from almost black to clear white, and have a specific gravity of 1.07-1.08. Copals that are geographically distinguished may be chemically or botanically distinct. They range from colorless and transparent to yellow-brown or black, are resinous, and have conchoidal fracture; their specific gravity is about 1.07, and they melt at 180-245°C.

The plant source of Kauri gum and most copals is known, and most of the physical and chemical intermediates between fresh resin and subfossil or fossil resins have been found in nature. Subfossil and fossil Kauri gum and copal occur in lumps and irregular masses from several cm to 1 m or more below the surface. Both may be found in areas where there is no other trace of the source tree and where the plants have not been known during recorded history. It is generally assumed that these resins are at least several thousand years old. It should be noted, however, that at one time commercially significant quantities of Sierra Leone copal were gathered from streambeds and beaches as "pebble gum," suggesting that deposits not related to forests may be alluvial rather than relict.

Resinite

Coal petrographers refer to fossil resins and waxes as "resinite." Much of this material occurs as cylindrical hollow or solid rods called resin rodlets. These rodlets presumably were secreted in canals or ducts of the coal forming plant. Less commonly, coal beds as much as 1 m thick may be more than half composed of resin bodies and are referred to as "resinite" coal. Chemical composition of resinites ranges widely, depending upon the original plant source, the degree and mode of decay before coalification, the age of the coal and the maximum temperature to which the coal has been subjected (see *Coal—Diagenesis and Metamorphism*). In general, Lower Cretaceous coals have a higher proportion of coniferous material and a higher ratio of resin to wax than do Upper Cretaceous and Tertiary coals, which contain more angiosperm material. Older resins also tend to be less soluble in organic solvents, reflecting a higher degree of polymerization and oxidation.

Amber

Resins are widespread in unmetamorphosed sedimentary rocks containing woody debris and also occur in almost all coals ranking below medium bituminous. Amber has long been noted for fossil inclusions (Bachofen-Echt, 1949; Carpenter et al., 1937; Hurd et al., 1962). Resins associated with coal or carbonaceous sediments probably were deposited at or near their point of origin. Much amber, however, occurs in sandstone, claystone or the so-called "blue earth," and is associated with marine fossils. This material most probably has been transported either as fresh or subfossil resin. The Baltic amber has yielded several specimens with coral inclusions, and one specimen with a partially embedded oyster shell is known from Chiapas, suggesting transport as unsolidified resin or secretion by plants growing in or over salt water. Isolated amber fragments in marine sedimentary rocks, however, may be reworked, either from subfossil accumulations or from older rocks; the amber in the Pleistocene deposits of northern Europe and Arctic Alaska definitely is redeposited.

Any "lump resin" in coal or sedimentary rock may be referred to as "amber" but it should be noted that the name is restricted by many authorities to *succinite*—the well-known Baltic amber. The almost infinite range of chemically and botanically different ambers has been divided among a succinite group characterized by succinic acid, a retinite group lacking succinic acid, and a tasmanite series of sulfur-bearing resins. In addition, many resin "minerals" have been distinguished by chemical, physical, botanical, geographic, or temporal criteria. Most mineralogists, however, no longer include resins amongst minerals because they are amorphous mixtures of organic compounds.

Paclt (1953) treats fossil resins as organic rocks or *caustolites*. In his classification, all fossil and subfossil resins are classified into (1) a coniferous group, including most named and described ambers, Kauri gum, and Manila copal; and (2) a nonconiferous, angiosperm group, including sieburgite, the copals, jaffaite, and subfossil para-damarite. The highly oxygenated and polymerized ambers contain organic acids, the most important of which are succinic, formic, butyric, and cinnamonic acid. The specific composition of individual specimens from the same or different localities may differ greatly, however, because of the wide range of possible botanic origin and subsequent geologic history for any specimen. Amber generally ranges from 1–3 on Moh's scale of hardness, with most specimens falling approximately at 2. Specific gravity ranges from 1–1.25, generally between 1.05 and 1.08. The most prized amber is clear and yellow, with conchoidal fracture and resinous luster. Most specimens are pale to golden yellow, but red to orange-red amber is well known, and yellow-green, deep-red, brown, and black ambers also are known. In addition, there are all gradations from transparent resinous material to opaque, dull or earthy substances. Fossil resins are brittle, but most varieties soften at around 150°C and melt at 250–300°C. Most amber occurs as irregularly rounded lumps with brown to black opaque crusts. Although lumps as large as 8 kg have been reported, most pieces are small, and a fist-sized lump is considered large. Many pieces have ropy, swirled surface patterns, and frozen runnels and droplets are widespread.

Although the shape, surface texture, color, luster, and other physical properties, along with chemical analyses, easily establish amber as a fossilized resin, this information has been inadequate to definitely identify the plant source for most amber. Physical properties and conventional chemical analyses range widely in fossil resins from a given locality as well as from different localities and are not specifically comparable with data from modern resins. Thus, in the past, associated or included fossil plant remains have been the basis for inferring a botanical origin for ambers. Infrared spectrophotometry has been extensively applied to fossil and recent resins by Langenheim (1969), who finds that the infrared spectra of ambers in many cases closely match those of resins derived from living trees. These results allow direct comparison of fossil and living resins and has served to relate numerous ambers, chiefly Cretaceous but including the early Tertiary Baltic amber, with resin of modern *Auracaria* and/or *Agathis*. Other, chiefly Tertiary, ambers have been related to a wide variety of tropical trees, including *Liquidambar, Shorea, Copaifera,* and *Hymenaea*. In addition x-ray diffraction analysis has indicated an angiospermous origin for the fossil resins, Highgate copalite, glessite, and guayaquillite.

Baltic amber has been an article of trade in Europe almost since the dawn of history. Specimens have been recovered from Egyptian tombs, and the ancient Greeks and Romans both wrote of Baltic amber. In the new world, Chiapanecan amber has been found at Mayan archeological sites in Mexico and Central America. Burmese amber has long been mined and exported to China for use in jewelry and medicine. During the nineteenth and early twentieth centuries, Baltic amber was widely used for jewelry and other ornamental objects.

RALPH L. LANGENHEIM, JR.

References

Bachofen-Echt, A., 1949. *Der Bernstein und Seine Einschlüsse*. Vienna: Springer, 204p.

Carpenter, F. M.; Folsom, J. W.; Essig, E. O.; Kinsey, A. C.; Brues, C. T.; Boesel, M. W.; and Ewing, H. E., 1937. Insects and arachnids from Canadian Amber, *Univ. Toronto Studies, Geol. Ser.*, 40, 7-62.

Frondel, J. W., 1967. X-ray diffraction study of some fossil and modern resins, *Science*, 155, 1411-1413.

Hollick, C. A., 1905. The occurrence and origin of amber in the eastern United States, *Am. Naturalist*, 39(459), 137-145.

Howes, F. N., 1949. *Vegetable Gums and Resins*. Waltham, Mass.: Chronica Botanica, 188p.

Hurd, P. D.; Smith, F. R.; and Durham, J. W., 1962. The fossiliferous amber of Chiapas, Mexico, *Ciencia*, 21, 107-118.

Langenheim, J. H., 1969. Amber: A botanical inquiry, *Science*, 163, 1157-1169.

Langenheim, R. L., Jr.; Smiley, C. J.; and Gray, J., 1960. Cretaceous amber from the Arctic Coastal Plain of Alaska, *Geol. Soc. Am. Bull.*, 71, 1345-1356.

Paclt, J., 1953. A system of caustolites, *Tschermaks Mineral. Petrog. Mitt.*, 3(4), 332-347.

White, D., 1914. Resins in Paleozoic plants and in coals of high rank, *U.S. Geol. Surv. Prof. Pap. 85*, 65-96.

Williamson, G. C., 1932. *The Book of Amber*. London: Ernest Benn, 268p.

Cross-references: *Coal; Coal-Diagenesis and Metamorphism; Organic Sediments*. Vol. IVB: *Amber*.

REYNOLDS AND FROUDE NUMBERS

Dimensionless numbers are pure numbers, i.e., they have no units of measure attached to them. When scientific observations are made, units of measure are nearly always present, but it is commonly desirable to eliminate such units, either to incorporate more information in the number which is being used or to permit comparisons despite important changes of scale. For Reynolds and Froude numbers, the three basic dimensions necessary for a mechanical scheme are mass, length, and time. Derived entities are made up out of basic dimensions: velocity is composed of length/time; acceleration is length/time-time (or length/time^2); and density is mass/volume (or mass/length^3). Dimensionless numbers are numbers that have been constructed in such a way that the dimensions cancel. Any ratio is a dimensionless number; the expression tv/a (where t = time, v = velocity, and a = acceleration) is dimensionless, as can be seen by inspection.

Of the large variety of dimensionless numbers theoretically available, only a few have proved useful in geology. The most widely used one is the *Reynolds number*, named after Osborne Reynolds, who in 1883 published the results of his work applying dimensional principles to the flow of fluid in pipes. He showed that the character of the motion of fluids bounded by solid surfaces does not depend on any single parameter, but rather on the cluster of measurements that now carries his name. In more recent years, scientists have extended this early work to flow not confined to pipes and to characteristic behavior best described by other dimensionless numbers as well (see also *Bernoulli's Theorem*). Fluid flow in a small brook may be hydrodynamically identical with that in a much larger stream, even though sizes are quite different. Conversely, two creeks of identical width may have quite unlike regimes. It is necessary, them, to be able to describe fluid flow with measures that are independent of size. Experience has further shown that such measures will be most useful if they are completely free of all units of measure, i.e., if they are pure numbers, or dimensionless ratios. The two dimensionless numbers that have wide applicability in studies of moving fluids are the *Reynolds number* and the *Froude number*.

Reynolds Number

The Reynolds number can be defined by the expression $R = vl\rho/\psi$, where the symbols on the right represent velocity, length, mass density, and viscosity (mass/length/time).

To maintain the dimensionless character of R, dynamic viscosity must be expressed in poises (where viscosity values are reported in centipoises, the viscosity must be multiplied by 10^{-2} to convert it to poises). In the cm-g-sec (cgs) system, the pertinent dimensions are: velocity in cm/sec; length in cm; mass density in g/cm^3; and dynamic viscosity in poises (dyne-sec/cm^2). If measures of velocity, length, etc., are obtained in the m-kg-sec (mks) system (or the International System), any Reynolds number computed from them would still be dimensionless.

For water at ordinary temperatures, mass density is close to unity, and dynamic viscosity is close to 1.0 cp (or 0.01 poise). Hence the dimensionless Reynolds number can be reduced, *in practice*, and where *crude* results are acceptable, in *water*, using centipoises, to R_W = 100 vl, or using poises, to $R_W = vl$. This is still a dimensionless number, inasmuch as the dimensions of density and viscosity are contained in it, but it is much simpler to compute than the longer version. For work with *air*, $R_a = 6.6\, vl$, where the factor 6.6 incorporates the differences in density and viscosity between water and air. For precise results, and where temperature differences may be important, the full expression must be used.

The Reynolds number is widely used as a criterion to separate the turbulent flow regime (high R) from the laminar one (low R). It has been common practice for a century or so to place the dividing line in the neighborhood of $R = 4000$, but this value was based on the flow of water through pipes. A natural stream of moving water is not directly comparable to water in a pipe, inasmuch as the latter exhibits incompressible flow, whereas a creek has an air-covered upper surface and therefore behaves in certain ways as if it were a compressible system. There is also a question concerning the choice of a length to be used in the Reynolds formulation. In pipes, the length that represents the geometry of the system is obviously the diameter of the flow. In natural streams, diameter has been replaced by various terms (such as width, depth, and hydraulic radius), but none of them is universally acceptable. The wisest procedure is to adopt a specified R in each case, using that length which seems most pertinent: perhaps a grain diameter (for sediment transport studies), a spacing (for ripple mark studies), a depth, or others. The Reynolds numbers specified by this choice should be properly designated: R_r, R_s, R_h.

Furthermore, the criterion is actually a transition zone rather than a single number or even a narrow band. In general, R less than unity defines laminar flow, at ordinary transport velocities, and R between one and perhaps 1500 or 2000 defines the transition zone, in which local turbulence will depend on the geometry and nature of channel wall roughness (Fig. 1). Higher values of R specify general turbulence.

Froude Number

The Froude ("frood") number can be defined by the expression $F = v^2/gl$, in which the symbols on the right represent velocity, acceleration of gravity, and length. The Froude number can be written alternatively as $F = v/\sqrt{gl}$, inasmuch as the Froude criterion is roughly unity, a value that is not affected by squaring or taking square roots. The second formulation is perhaps more common than the first, and is the one used here. The Reynolds number presents, uniquely, information about density and viscosity effects; the Froude number is a measure of

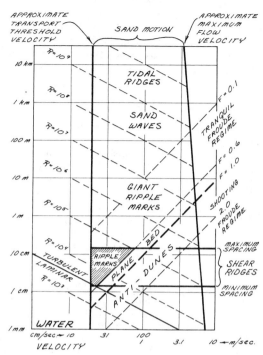

FIGURE 1. Reynolds and Froude numbers in water.

gravity effects. The velocity and length measures, which are common to both, allow us to specify the conditions under which fluid flow is being investigated.

The Froude criterion separates tranquil flow (low F) from shooting flow (high F). The transition zone occupies the region between about $F = .07$ and $F = 1.0$. It is commonly not possible to distinguish between tranquil and shooting flow by simple observation of a current moving across a fixed bed. If a movable bed is used, such as ordinary sand, certain bed effects characterize the two regimes. The transition zone can be assumed when planar motion of bed materials occurs: a thin sheet of sand moves uniformly along the bottom, without visible surface markings. Shear ridges and ripple marks typically appear between $F = 0.3$ and $F = 0.7$. Froude numbers higher than unity may produce antidunes (having steep faces and direction of migration upstream).

The Froude number is particularly important because it distinguishes between two regimes, having different velocities but identical bed slopes and depths (see *Flow Regimes*). The lower velocity is associated with ripple marks, a high friction factor (much boundary layer turbulence), slow sediment motion, and small sediment discharge (F commonly between 0.3 and 0.7). The higher velocity is associated with either flat beds or antidunes, little or no boundary layer turbulence, rapid sediment motion, and great sediment discharge (F above 0.7, and probably above 1.0). The computed shear, at the bed, may appear to be identical in the two cases.

Application

Important differences in various mobile bed features can be studied best by the use of the R and F criteria, which may be plotted conveniently in terms of length and velocity, provided a single fluid and temperature are specified (see Fig. 1).

Concrete and other fixed-bed channels exhibit the two Froude regimes in a different way. On a fixed bed, the change from high-Froude to low-Froude flow (i.e., from shooting to tranquil flow) is accompanied by a change in depth; this change is referred to as the hydraulic jump. In appearance, it looks like a local turbulent stretch of water between two more or less laminar reaches. The local water surface may slope quite steeply, unlike the slope in the adjacent reaches.

The Froude number becomes particularly important at high velocities or at shallow depths. Waves on beaches, waves and currents on sand bars which break (or almost reach) the water surface, and flow in shallow streams commonly produce interesting Froude effects. It is possible for water flowing on a sandy bed to sink into that bed (and/or evaporate) in such a fashion that high-Froude effects, such as the production of a planar sand surface (sheet flow), continues until all sand motion is stopped; i.e., the bare sand surface is smooth and featureless (no ripple marks, and no other current marks). Such a surface can be flooded, later, and buried under other sediment, thus producing a smooth bedding plane (see *Flow Regimes*). Froude effects may also appear where there is an abrupt change in water depth, relative to total water depth, i.e., where one depth is a rather small fraction of the other. These two depths may be associated with the two Froude regimes. High-velocity Froude effects are observed most commonly in narrow swift channels, such as certain tidal inlets, and in very small streams, down to a depth of about 1 cm. The supercritical flow ($F > 1.0$) is characterized by waves which break up-current, forming the so-called "rooster tail" of spray and foam, and second-order waves which are generated by their collapse, thus producing a wave pattern having no wind origin. These larger waves may be associated with antidunes on the bed, or with other sediment transport phenomena, some of which appear to be preserved in lithified sandstones.

The Reynolds number is important in considering turbulence on the grand scale (such as in large rivers), local boundary layer separation and the production of gyres or eddies as a result, and other behavior in the vicinity of roughness elements of various kinds and at various scales. The latter includes effects as small as the eddy system (three-dimensional von Karman trail, or street) which is set up by the fall of a quartz grain of medium-sand size in ordinary water, as well as eddy production adjacent to ripple marks. Roughness elements cannot be evaluated by use of the classical Reynolds number, which excludes considerations of geometry other than the one "representative" length, and therefore must be computed and interpreted in terms of local Reynolds numbers, as stated in a previous paragraph. Grain entrainment, for transport, is much easier to understand in terms of local Reynolds numbers than in terms of ordinary fluid velocity measures.

Both Reynolds and Froude effects appear to be important in studies of erosion, such as soil erosion, where the small water depth provides for high Froude values, and the various roughness elements provide for large local Reynolds numbers.

WILLIAM F. TANNER

References

Duncan, W. J., 1953. *Physical Similarity and Dimensional Analysis.* London: Edward Arnold, 156p.

Leliavsky, S., 1959. *An Introduction to Fluvial Hydraulics.* London: Constable, 256p.

Middleton, G. V., and Southard, J. B., 1977. Mechanics of sediment movement, *Soc. Econ. Paleont. Mineral. Short Course,* **3**, 246p.

Pao, R. H. F., 1961. *Fluid Mechanics.* New York: Wiley, 502p.

Rouse, H., ed., 1959. *Advanced Mechanics of Fluids.* New York: Wiley, 444p.

Simons, D. B.; Richardson, E. V.; and Albertson, M. L., 1961. Flume studies using medium sand (0.45 mm), *U.S. Geol. Surv. Water Supply Pap. 1498-A,* 76p.

Sundborg, Å., 1956. The River Klarälven: a study of fluvial processes, *Geograf. Ann.,* **38**, 127-316.

Cross-references: *Bedding Genesis; Bed Forms in Alluvial Channels; Bernoulli's Theorem; Experimental Sedimentology; Flow Regimes; Fluvial Sediment Transport; Ripple Marks.* Vol. II: *Reynolds, Froude and Other Dimensionless Numbers.*

RIPPLE MARKS

Ripple marks are regular, ridge-like structures, transverse to the current, which arise and are maintained at the interface between a moving, viscous fluid (water, air) and a moveable, noncohesive sediment (usually sand) by interaction between fluid and transported sediment (Allen, 1968).

The effect of the moving fluid is to shear the sediment bed, and the shear may be applied steadily, as in steady unidirectional currents, or unsteadily and with reversal in direction, as in the more complex case of reversing flows induced at the bottoms of water bodies by the overhead passage of surface gravity waves. Given certain boundary conditions, ripple marks as self-maintaining features exist in relation to regular, downstream perturbations in fluid pressure, velocity, and sediment transport rate (Kennedy, 1964). The perturbations are most obviously revealed in the eddy cells engaged with the downstream faces of ripple marks (Raudkivi, 1964). Bottom irregularities other than already existing ripples can initiate these perturbations; and, depending on the character of the fluid flow, the scale and pattern of the ripples may be determined either by the grain size of the sediment or by the geometry of the whole fluid flow (Figs. 1, 2; Sundborg, 1956; Yalin, 1964; Southard and Dingler, 1971; Harms, 1975). In the case of ripples formed under conditions of steady shear in unidirectional currents, the eddy is engaged with the steep or lee-side of each ripple and moves downstream with the advancing ripple. When the bottom tractive currents are wave-induced, the eddy becomes engaged with first one and then the other face of the ripple, as the bottom flow is reversed in direction (Inman and Bowen, 1963). Different patterns and states of fluid flow give rise to ripple marks differing in size and geometry (Fig. 2). Because of the eddies mentioned, ripple marks offer a form of resistance to the current in addition to the grain resistance generated by the grains composing the ripples.

Geometry

Ripple marks are not solitary structures but occur in squads or trains of ridges similar in form and size (Kindle, 1917; Bucher, 1919; Allen, 1963, 1968). In vertical profile transverse to the current and to the ridges, the ripples may be *asymmetrical* or *symmetrical*, with sharp to well-rounded *crests* and *troughs*. When the profile is asymmetrical, the short, steeply sloping face of the ripple is referred to as the *leeside,* inclined downcurrent (Fig. 3). The long, gently sloping face is the *stoss-side,* which faces and is eroded by the oncoming flow. The degree of symmetry of such profiles is given by the *symmetry index,* the ratio of the horizontal stoss-side length to the leeside length. The *ripple length* is the horizontal distance,

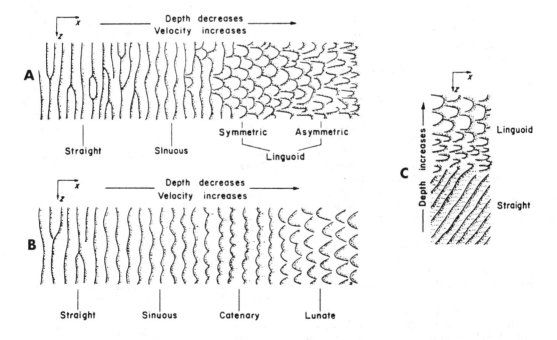

FIGURE 1. Schematic sequences of ripple types with changing flow velocity and/or depth (from Allen, 1968). A, small-scale ripples; B, large-scale ripples; C, small-scale ripples transverse to flow. Leesides are stippled.

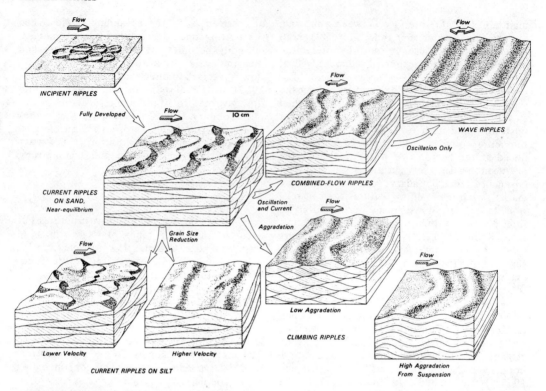

FIGURE 2. Ripples and ripple stratification related to a range of flow conditions, grain size, and aggradation rate (after Harms, 1975).

transverse to the ripple crest, between the foot of the ripple leeside and stoss-side; the *ripple height* is the greatest vertical distance between the ripple crest and trough (Fig. 3). Relief of the ripple is described by the *vertical form index,* the ratio of the ripple length to ripple height.

Ripple marks vary in plan (Fig. 4). Those with long and essentially straight crests can be described as *straight,* and flow over them is two-dimensional. Other ripples have short, curved crests, which close downcurrent, described as *linguoid,* or upcurrent, termed *lunate.* Flow over ripples with curved crests is three-dimensional. A fourth crestline pattern, closing downcurrent, resembles the linguoid type in closure, but is sharply pointed and is termed *rhomboid.* Intermediate between straight and curved forms are *sinuous* and *catenary* crests. Ripple form in plan is described by the *horizontal form index,* the ratio of the length of the ripple transverse to the current, to the ripple length.

Ripple trains may be described as simple, compound or superposed. A *simple* train consists of ripples similar in shape, size, and orientation. *Compound* trains are formed of two or more trains of ripples of similar vertical profile but differing in size and/or orientation; the trains formed more or less simultaneously. In *superposed* ripple trains, the component ripple systems did not form simultaneously and generally differ in vertical profile.

Among simple, asymmetrical ripples two scales of development can be recognized, there being a range of size poorly represented by actual ripple marks. A somewhat arbitrary height of 4 cm divides small-scale from large-scale ripples, the latter also being known as dunes or sand waves (see *Bed Forms in Alluvial Channels*).

Classification

Present geological needs are best served by classifying ripple marks according to geometrical characteristics, with ripple genesis, about which much still has to be learned, taking a minor role. The classification given in Table 1 is based on earlier proposals (Kindle, 1917; Bucher, 1919; Allen, 1963) and takes account of ripple symmetry, ripple plan, and the nature of the train. It is not comprehensive but covers the commoner occurrences of ripple marks. Although eolian ripple marks and dunes are included in the discussion below, it must be em-

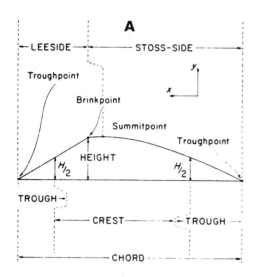 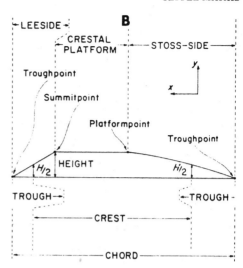

FIGURE 3. Ripple marks in profile (from Allen, 1968). A—characteristic ripple; B—ripple formed in shallow flows, when the upper part of the mound becomes planed down to a flat surface.

phasized that because of differences in fluid density and viscosity, the processes of ripple formation are somewhat different in the eolian (vs. aqueous) environment (Bagnold, 1941).

Asymmetrical Ripple Marks. Small-scale, *straight ripples* formed in water give horizontal form indexes generally of 10–20, although exceptionally reaching 100, and vertical form indexes of about 10. They are widespread phenomena, recorded from bioclastic as well as quartzose sands, and from environments ranging from the deep sea to beaches and shallow streams (Kindle, 1917; Heezen and Hollister, 1964). Steady unidirectional flows may create asymmetrical ripples, as may alternating, wave-induced bottom currents characterized by strong net fluid displacement. The ripples of the same general size formed in air have vertical form indexes much higher than those formed in water.

Sand waves (q.v.), giant ripples, dunes and megaripples have been used as alternative names for large-scale, straight ripples formed in water. They yield vertical form indexes of 10–100, and are relatively less elevated than the otherwise similar small ripples. The horizontal form index ranges up to about 25. Again, the structures are widespread (Jordan, 1963; Allen, 1968), being recorded from many streams and rivers, estuaries, tidal flats, partly enclosed tidal seas, and open tide-swept coastal shelves. Ripple size is often considerable, but depends on water depth, heights ranging up to 25 m and lengths to 500 m. The transverse dunes with slightly sinuous crests of deserts are an eolian form of straight, large-scale ripple.

Small-scale, *linguoid ripples* seem restricted in production to shallow and rather turbulent environments, particularly stream and tidal channels (Kindle, 1917; Bucher, 1919). In vertical form index, they agree with small, straight ripples but give horizontal form indexes generally < 3. Ripple trains show two contrasted patterns of ripple arrangement. The more common shows ripples that are offset, in rows across the current. Less common is a "nested" arrangement of the ripples, when the term *cuspate* is sometimes applied. Small linguoid ripples are not known from eolian environments.

Streams and tidal channels are the only aqueous environments to have yielded examples of

TABLE 1. Classification of Ripple Marks

I. SIMPLE TRAINS
    1. Asymmetrical ripples
        a. straight (large and small scale)
        b. linguoid (large and small scale)
        c. lunate (large scale)
        d. rhomboid
    2. Symmetrical ripples

II. COMPOUND TRAINS
    1. Symmetrical oblique
    2. Symmetrical parallel
    3. Asymmetrical parallel

III. SUPERPOSED TRAINS
    1. Asymmetrical-symmetrical oblique
    2. Symmetrical-symmetrical oblique
    3. Asymmetrical-asymmetrical oblique

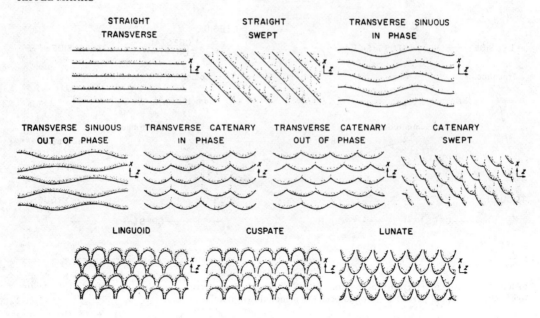

FIGURE 4. Idealized ripple trains (from Allen, 1968). Leesides and spurs are stippled.

large-scale, linguoid ripples (Allen, 1963). The ripples compare in size with large-scale, straight ripples but yield horizontal form indexes of 0.35–2.2. The conditions required for their generation are poorly known but probably involve a highly turbulent and shallow fluid in which three-dimensional flow is possible. Superficially resembling them are the plant-anchored parabolic dunes of coastal sand dune and desert areas.

Regular trains of natural ripples *lunate* in plan are known only on the large scale, from beaches and estuarine tidal flats (Allen, 1963). Trains of somewhat isolated ripples of this type, resembling certain eolian dunes in their sparse distribution on hard pavements, are reported from shallow undersea benches in the Bahamas. Conditions of formation are unknown, but flow over the ripples would seem to be three-dimensional. Large-scale lunate ripples have long been known from deserts as barchan dunes (Bagnold, 1941), in some areas reaching heights of 10 m or more. Like the underwater lunate ripples of the Bahamas, many desert barchans are isolated mounds of sand migrating over pavements of rock or coarse stones.

Rhomboid ripples are low, scale-like structures ranging in length from a few cm to meters, and in height from < 1 cm to 2–3 cm. They are commonly developed on beaches and sandy tidal flats where thin wave backwash and other shallow flows operate. Generating currents are < 10 cm deep and fast moving, commonly > 1 m/sec, in the upper flow regime. The ripples form in conjunction with a cross-laced pattern of low waves on the water surface.

Symmetrical ripples. Simple trains of symmetrical ripples ("oscillation ripples") are known from beaches, tidal flats, sheltered bays, and lakes, and are developed extensively to depths up to 200 m on sandy beds lying off exposed sea or lake shores. In length, these ripples vary from 2–5 cm up to 1–2 m. Heights vary up to 15–25 cm. Crests are long and generally very slightly sinuous. Vertical profiles are of two types. Ripples of one type have broad, flat troughs; the other type have rounded troughs. Ripple size tends to increase with increasing coarseness of the sand. Wave depth and intensity of wave action also influence ripple size (Komar, 1974). Ripples in deeper water off exposed coasts tend to be larger than those formed in shallow or protected areas.

Compound Trains. *Interference ripples* are compound ripple trains formed of two simultaneously produced systems of symmetrical ripples, the crests making a steep angle. Ripples of this type arise most often when a system of water waves is reflected from the side of a sand bar or bank. The ripples seldom reach a large size and are sometimes referred to as *tadpole nests*. Small- and large-scale asymmetrical ripples agreeing in crest orientation are sometimes formed together under a single flow system, giving a further type of compound train (Sundborg, 1956).

Superposed Trains. A form often encountered, especially from beaches and tidal flats, comprises a series of small, asymmetrical ripples, in the troughs of which small, symmetrical ripples have been developed at a steep angle. Often both trains are the products of wave action, the asymmetrical ripples representing the larger and more powerful waves associated with the flood tide or an early stage in the ebb, and the symmetrical ripples small waves arising by refraction or reflection as the tide ebbed off the shoal-strewn beach or flat. A not uncommon form well known from the shores of the North Sea comprises a system of large-scale ripples across which small, asymmetrical or symmetrical ripples trend obliquely.

J. R. L. ALLEN

References

Aario, R., 1972. Exposed bed forms and inferred three-dimensional flow geometry in an esker delta, Finland, *24th Intern. Geol. Congr., Sec.* 12, 149-158.

Allen, J. R. L., 1963. Asymmetrical ripple marks and the origin of water-laid cosets of cross-strata, *Liverpool Manchester, Geol. J.*, **3**, 187-236.

Allen, J. R. L., 1968. *Current Ripples*. Amsterdam: North-Holland, 447p.

Allen, J. R. L., 1973a. Features of cross-stratified units due to random and other changes in bed forms, *Sedimentology*, **20**, 189-202.

Allen, J. R. L., 1973b. A classification of climbing-ripple cross-lamination, *J. Geol. Soc.*, **129**, 537-542.

Bagnold, R. A., 1941. *The Physics of Blown Sand and Desert Dunes*. London: Methuen, 265p.

Baker, R. A., 1970. Oscillation ripplemarks and orbital diameter, *Geol. Soc. Am. Bull.*, **81**, 1589-1590.

Banks, N. L., and Collinson, J. D., 1975. The shape and size of small-scale current ripples: An experimental study using medium sand, *Sedimentology*, **22**, 583-599.

Bucher, W., 1919. On ripples and related sedimentary surface forms and their palaeogeographic interpretation, *Am. J. Sci., Ser. 4*, **42**, 149-209, 241-269.

Harms, J. C., 1969. Hydraulic significance of some sand ripples, *Geol. Soc. Am. Bull.*, **80**, 363-396.

Harms, J. C., 1975. Stratification produced by migrating bed forms, *Soc. Econ. Paleont. Mineral. Short Course*, **2**, 45-61.

Heezen, B. C., and Hollister, C., 1964. Deep-sea current evidence from abyssal sediments, *Marine Geol.*, **1**, 141-174.

Inman, D. L., and Bowen, A. J., 1963. Flume experiments on sand transport by waves and currents, *Proc. 8th Conf. Coastal Engin. Council Wave Research*, 137-150.

Jopling, A. V., and Walker, R. G., 1968. Morphology and origin of ripple-drift cross lamination, with examples from the Pleistocene of Massachusetts, *J. Sed. Petrology*, **38**, 971-984.

Jordan, G. F., 1963. Large submarine sand waves, *Science*, **136**, 839-848.

Kachel, N. B., and Sternberg, R. W., 1971. Transport of bedload as ripples during an ebb current, *Marine Geol.*, **10**, 229-244.

Kennedy, J. F., 1964. The formation of sediment ripples in closed rectangular conduits and in the desert, *J. Geophys. Research*, **69**, 1517-1524.

Kindle, E. M., 1917. Recent and fossil ripple-mark, *Mus. Bull., Geol. Surv. Canada*, **25**, 121p.

Komar, P. D., 1974. Oscillatory ripple marks and the evaluation of ancient wave conditions and environments, *J. Sed. Petrology*, **44**, 169-180.

McKee, E. D., 1965. Experiments on ripple lamination, *Soc. Econ. Paleont. Mineral. Spec. Publ.* 12, 66-83.

Mercer, A. G., Haque, M., 1973. Ripple profiles modeled mathematically, *Proc. ASCE, J. Hydraul. Div.*, **99(HY3)** *Pap. 9618*, 441-459.

Middleton, G. V., ed., 1965. Primary sedimentary structures and their hydrodynamic interpretation, *Soc. Econ. Paleont. Mineral. Spec. Publ.* 12, 265p.

Moss, A. J., 1975. Initiation and inhibition of the formation of asymmetrical sand ripples, *J. Geol. Soc. Australia*, **22**, 79-90.

Raudkivi, A. J., 1964. Study of sediment ripple formation, *Proc. ASCE, J. Hydraul. Div.*, **86(HY6)**, 15-33.

Southard, J. B., and Dingler, J. R., 1971. Flume study of ripple propagation behind mounds on flat sand beds, *Sedimentology*, **16**, 251-263.

Stanley, K. O., 1974. Morphology and hydraulic significance of climbing ripples with superimposed micro-ripple-drift cross-lamination in Lower Quaternary lake silts, Nebraska, *J. Sed. Petrology*, **44**, 472-483.

Sundborg, Å., 1956. The river Klarälven: A study of fluvial processes, *Geograf. Ann.*, **38**, 127-316.

Trenhaile, A. S., 1973. Near-shore ripples: Some hydraulic relationships, *J. Sed. Petrology*, **43**, 558-568.

Yalin, M. S., 1964. Geometrical properties of sand waves, *Proc. ASCE, J. Hydraul. Div.*, **90(HY5)**, 105-119.

Cross-references: *Bedding Genesis; Bed Forms in Alluvial Channels; Cross-Bedding; Current Marks; Flow Regimes; Fluvial Sediment Transport: Primary Sedimentary Structures; Sand Waves; Tidal-Current Deposits.*

RIVER GRAVELS

Gravels are characteristic deposits of high-velocity streams. Typical occurrences of fluvial gravels include actively eroding channels in areas of high relief, proximal braided streams, glacial outwash plains, and alluvial fans. They have been occasionally observed in channel and point-bar deposits of meandering streams as well.

River gravels consist of a framework of mutually touching gravel-sized clasts (> 2 mm) and interstitial voids usually filled with a matrix of sand and finer detritus. Matrix is usually incorporated into the gravel framework during reduced flows when the gravel is stationary on the stream bed. During rapid deposition, however, some gravels become buried before deposition of fines, and their voids remain unfilled.

Such *openwork* gravels are relatively uncommon in fluvial deposits, however, and are usually confined to a few thin discontinuous beds within the dominantly matrix-filled sediments.

Coarse river gravels tend to have bimodal grain size distributions, with the principal mode in the gravel fraction and a secondary mode of sand (Pettijohn, 1957). The depression between the two modes generally falls in the 1-6 mm interval, commonly, 1-2 mm. Two interpretations have been most often forwarded to explain this deficiency: (1) The deficiency results from polymodal populations, naturally produced by source rock weathering; e.g., weathering of granite is considered to yield modes of coarse gravel, sand, and clay with deficiencies of very coarse sand granules and silt. (2) The deficiency is a product of hydrodynamic sorting involving preferential entrainment and transport of 1-6 mm particles out of coarse gravel sediments. A third interpretation involving selective fragmentation of this size fraction during transport has been proposed by Moss (1972).

Composition of river gravels depends on several factors, including source bedrock lithology, rates and styles of weathering and erosion, and transport variables. Chert, quartz, and quartzite are both mechanically and chemically stable relative to other common pebble types, hence tend to concentrate in regions of strong chemical weathering and with increasing distance from source. Relatively unstable lithologies such as limestones and most igneous and metamorphic rocks may characterize fluvial deposits in areas of high relief, rapid erosion, and principally physical weathering. Dal Cin (1968) shows that percentage and roundness of quartz pebbles in gravelly deposits are strongly influenced by climate.

Gravel particle shape is primarily determined by the initial shape of the fragment inherited from the source, which in turn is controlled by such factors as cleavage, jointing, bedding planes, and weathering intensity. Thus, flat pebbles are derived from such rocks as shales and thin-bedded limestones, whereas blocky or massive lithologies such as chert, granite, and quartzite tend to yield more equant shapes. Shape is modified during transport by breaking, abrasion, and chemical weathering. Field and experimental studies have both shown that rates of mechanical attrition depend on such variables as durability, size, and initial roundness of the clasts, transport velocity, and the substrate over which the clasts are moving. In general, gravels become rounder and finer grained with increased transport distance. High roundness values are usually attained within a few km of transport for the coarser clasts, and considerable rounding can occur while clasts remain virtually stationary on the stream bed (Schumm and Stevens, 1973).

Several workers have investigated the effects of fluvial transport upon particle shape sorting, but the results appear to be conflicting and inconclusive (see Bradley et al., 1972, for a review of earlier works). For example, some studies have concluded that spherical pebbles are preferentially transported to other shapes, whereas other studies indicate the same conclusion for discoidal or elongate shapes. Most of the interpretations have been drawn from plots of pebble-shape factors against distance or time of transport. The diverse results obtained so far are probably in part due to the failure to consider that other factors such as mode of transport, flow variability, size-shape relations, and substrate characteristics (size, shape, packing) are also important in shape sorting.

Most fluvial gravels show preferred clast orientations that represent their most stable positions relative to flow direction. Discoidal clasts tend to be oriented with flat sides facing the current and dipping upstream. Elongate clasts have preferred orientations with long axes either parallel or normal to flow, in some cases both. Rust (1972) observed that small size and high pebble concentrations tend to reduce preferred orientations of elongate clasts, whereas imbrication of discoidal clasts is enhanced by high concentrations. Preferred clast orientations are useful for interpreting paleoflow directions in ancient river gravel deposits.

Available data suggest that most fluvial gravels are transported slowly and sporadically during infrequent high discharges (e.g., Tricart and Vogt, 1967). Deposits occur mostly in channels and channel bars, yielding internal stratification that is usually massive to crudely horizontal within irregular or lenticular beds. Some stream gravels, particularly finer grained fractions, show high and low angle cross-stratification resulting from deposition at slip-face and riffle margins of channel bars and on the sides of channels.

NORMAN D. SMITH

References

Bradley, W. C.; Fahnestock, R. K.; and Rowekamp, E. T., 1972. Coarse sediment transport by flood flows on Knik River, Alaska, *Geol. Soc. Am. Bull.,* **83,** 1261-1284.

Dal Cin, R., 1968. Climatic significance of roundness and percentage of quartz in conglomerates, *J. Sed. Petrology,* **38,** 1094-1099.

Moss, A. J., 1972. Initial fluviatile fragmentation of granitic quartz, *J. Sed. Petrology,* **42,** 905-916.

Pettijohn, F. J., 1957. *Sedimentary Rocks.* New York: Harper & Row, 718p.

Rust, B. R., 1972. Pebble orientation in fluvial sediments, *J. Sed. Petrology,* **42,** 384-388.

Schumm, S. A., and Stevens, M. A., 1973. Abrasion in place: A mechanism for rounding and size reduction of coarse sediments in rivers, *Geology,* **1,** 37-40.

Tricart, J., and Vogt, E., 1967. Quelques aspects du transport des alluvions grossières et du façonnement des lits fluviaux, *Geogr. Ann.* **49(A),** 351-366.

Cross-references: *Alluvial-Fan Sediments; Alluvium; Beach Gravels; Braided-Stream Deposits; Conglomerates; Fabric, Sedimentary; Fluvial Sediments; Glacial Gravels; Glacigene Sediments.*

RIVER SANDS

Rivers are hydraulically complicated and can vary greatly in source materials. Therefore, nearly all river sands show a broad spectrum of textures within the same deposit, whereas many other environments are much more homogeneous in this respect. This wide textural variation within the deposit is the most diagnostic criterion for river sands.

In general, river deposits are composed of different grain populations: (1) A *coarsest* population, derived from the underlying bedrock—The competency of the river can be insufficient to transport all the components of the bedrock material and the coarsest particles can remain in situ as an admixture of the channel-floor deposit. This population reveals lag characteristics. (2) A *coarse* population, derived as bottom load from sources higher upstream—Due to the decreasing downstream competency, this material is deposited in the channel-floor and lower point-bar sediments. (3) A *finer* population, which, due to capacity changes, drops out from the saltation and suspension loads. These particles often form the mode, especially in the upper point-bar sequence. During transport, this population is vertically segregated by size, shape, and density. Accordingly, material from this population varies from the lower point bar toward the top of the point bar. Of a same grain size, spherical and dense grains contribute to the lower point-bar deposition; lighter and less spherical grains are more abundant in the upper point bar, in the levees, and especially in floodplain deposits. (4) A *still finer* population, which was transported as suspended load. It may filter into the coarse sediments of the channel floor and the lower point bar, but it is more abundant in the upper point bar, where it can form lamina of its own. It forms the bulk of floodplain and oxbow-lake deposits. The maximum grain size of this population varies from about 35–100μ, depending on the average turbulence in the river.

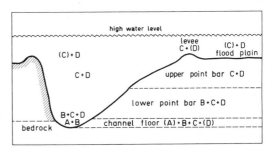

FIGURE 1. Schematic river section, showing the different grain populations in the water (left) and in the deposits (right). A—residual bedrock population; B—bottom (bed) load population; C—saltating population; D—uniformly dispersed population.

Textural Characteristics

In between broad limits, each type of river deposit is characterized by its specific mixture of populations (Fig. 1). Udden (1914) and Doeglas (1946) were among the first to try to relate the textural properties of river sands to the different transport populations.

There are several ways to distinguish these populations. In histograms, river sands nearly always show a broad spectrum of grain sizes, indicating the presence of different populations. This distribution is better revealed on a cumulative curve on log-probability paper. Sediments, deposited from one population, produce a grain size distribution which plots approximately as a straight line on this paper; truncation points in the distribution curves mean

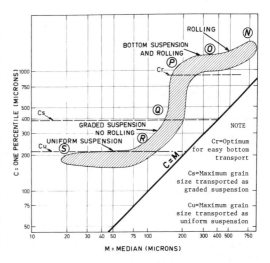

FIGURE 2. C/M pattern of a river deposit (after Passega, 1964). The typical shape is obtained due to the different ratios of populations in the different samples.

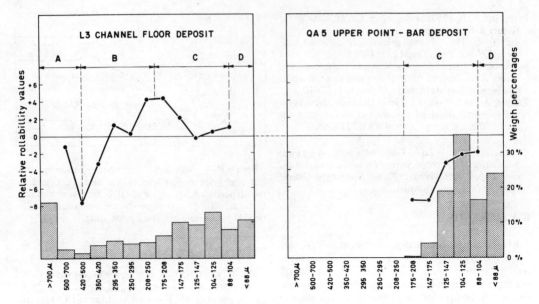

FIGURE 3. Shape distribution curves with histograms of a lower Oligocene river sand from Belgium (after Winkelmolen, 1972). The letters A–D indicate the different populations as in Fig. 1. Note the complementary character of population C in the channel floor and the upper point bar.

different populations (see *Grain-Size Studies;* Visher, 1969). By plotting a C/M pattern of many small samples of the same river deposit (Passega, 1964), a specific pattern is obtained (Fig. 2) due to the different mixtures of populations typical for river deposits. Populations are not only defined by the size of their grains, but as well by their shape and density. When, for a small sample, the average grain shape in each size fraction is plotted against these size fractions, a *shape distribution curve* (SDC) is obtained (Winkelmolen, 1972). Such an SDC also shows inflection points, separating different grain populations with different shape-sorting characteristics (Fig. 3). The SDC also reveals the depositional characteristics of each population. When the SDC slopes down toward finer sizes, it indicates lag conditions. An up-sloping curve indicates accumulating conditions.

Rivers are a special case of confined, unidirectional traction systems. Therefore, river sands show strong textural resemblance to deposits from similar systems. For example, flood-plain and tidal-flat deposits, or river and tidal channel sands can be very much alike in texture due to their analogous mode of origin (see *Tidal-Current Deposits*).

Some environments imprint a special surface texture on sand grains. Krinsley and Donahue (1968) studied many river samples for surface textures. They concluded that river abrasion of sand-sized material in general is insufficient to produce distinctive new textural surface patterns. Consequently, observed surface textures on grains of river sands are largely inherited from former cycles and could at most be used to relate the grains to their source materials.

A. M. WINKELMOLEN

References

Binda, P. L., and Lerbekmo, J. F., 1973. Grain-size distribution and depositional environment of Whitemud sandstones, Edmonton Formation (Upper Cretaceous) Alberta, Canada, *Bull. Canadian Petrol. Geol.,* 21, 52-80.

Doeglas, D. J., 1946. Interpretation of the results of mechanical analyses, *J. Sed. Petrology,* 16, 19-40.

Krinsley, D. H., and Donahue, J., 1968. Environmental interpretation of sand grain surface textures by electron microscopy, *Geol. Soc. Am. Bull.,* 79, 743-748.

Passega, R., 1964. Grain size representation by CM patterns as a geological tool, *J. Sed. Petrology,* 34, 830-847.

Udden, J. A., 1914. Mechanical composition of clastic sediments, *Geol. Soc. Am. Bull.,* 25, 655-744.

Visher, G. S., 1969. Grain size distributions and depositional processes, *J. Sed. Petrology,* 39, 1074-1106.

Winkelmolen, A. M., 1972. Shape sorting in Lower Oligocene, Northern Belgium, *Sed. Geol.,* 7, 183-227.

Cross-references: *Alluvium; Beach Sands; Eolian Sands; Fluvial Sediments; Glacial Sands; Grain-Size Studies; Sands and Sandstones; Sand Surface Textures.*

ROUNDNESS AND SPHERICITY

Roundness and sphericity are two measures of particle shape. Morphometric, or particle shape, analyses are important for two reasons: (1) the shape of a particle will have a definite effect on the reaction of the particle to any method of granulometric analysis (q.v.), and (2) the shape will affect the hydraulic properties of a particle in a moving fluid. An analysis that is made for either of these reasons can be based on two- or three-dimensional measurements of the particles. Particle shape is usually expressed as a "shape factor," i.e., some ratio of axial measurements of the particle. The shape factor permits the two-dimensional particle sizes obtained by microscopic methods, for example, to be compared with the three-dimensional sizes that are obtained by sieving (see, e.g., Friedman, 1958, 1962). A different shape factor would permit comparison between microscopic and sedimentation methods. Fourier shape analysis may be used for stratigraphic analysis and correlation (Mrakovich et al., 1976).

The shape of particles has sometimes been separated into various components such as *sphericity*—the approach toward a spherical form; *roundness*—the approach toward rounded edges; and other components. The particles may be compared to illustrations of a series of standard particles portraying certain degrees of sphericity, roundness, etc., in order to classify the shapes of the particles. Two or three axes of the particle, or some other distance or area, may be measured for calculation of a shape factor by use of one of several equations.

Roundness

Roundness describes the degree of abrasion of a clastic fragment as shown by the sharpness of its edges and corners, independent of shape. Properly defined (Wadell, 1932),

$$\text{roundness} = \frac{\text{av. radius of corners and edges}}{\text{radius of max. inscribed circle}}$$

Thus spherical particles are perfectly rounded, but well-rounded objects need not be spherical (Fig. 1). Tabular, bladed, and prolate fragments may show beautifully rounded corners and edges, but by no means are they spherical.

Roundness may provide evidence of time or distance of transport. The grain is usually angular when freshly introduced (unless recycled) and becomes progressively more rounded as it moves downstream. Experiments on particle rounding, begun by Daubrée (1879), and continued by Wentworth (1919), Krumbein (1941), Kuenen (1956), and Bradley (1970), have a certain unreality owing to the synthetic

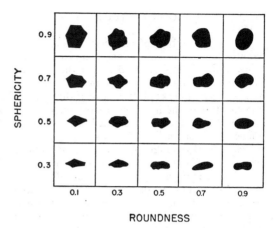

FIGURE 1. Chart for visual estimation of roundness and sphericity of sand grains (from *Stratigraphy and Sedimentation*, Second Edition, by W. C. Krumbein and L. L. Sloss. W. H. Freeman and Company. Copyright © 1963).

nature of ball mills, tumbling barrels, flumes, and so on. Pearce (1971) showed that cobbles in an actual stream can become rounded in one-fifth to one-tenth the distance predicted by the experiments. Closer studies by Schumm and Stevens (1973) showed that with turbulent flow the particles tend to move up and down, vibrating in place, becoming rounded in situ with time, until they eventually break loose and move downstream by saltation until a new semistable slot is occupied. In contrast, in some ephemeral fluvial regimes, e.g., the arid Lake Eyre region of Australia, the flow is so infrequent that the time-distance factor is greatly extended (Williams, 1971).

Sphericity

The sphericity of any fragment is an expression of the degree to which its shape approaches the form of a sphere. The original concept was defined by Wadell (1932):

$$\text{True sphericity} = \frac{\text{surface area of the particle}}{\text{surface area of a sphere of the same volume}}$$

Whereas the measurement of the surface area of an irregular particle is not feasible, however, Wadell (1933) proposed a practical definition:

$$\text{Operational sphericity} = \sqrt[3]{\frac{\text{volume of a particle}}{\text{volume of a circumscribing sphere}}}$$

From the second equation, Wadell developed a simple method applicable to large particles, in which the volume of the particle is measured by water displacement (Krumbein and Sloss, 1963).

For large numbers of particles, it is easier to group them into convenient shape classifications. Zingg (1935) showed that if the ratio of the intermediate to the maximum intercept (b/a) of a particle is plotted against the ratio of the shortest to the intermediate intercept (c/b), the particle may be classified according to shape. The four shape classes (Fig. 2) provide an easy visual grouping of particle classification. This shape classification is related to the operational sphericity of a particle in terms of particle intercepts, assuming the particle shape as atri-axial-ellipsoid. Lines of equal operational sphericity, termed *intercept sphericity*, may be plotted on a grid (Fig. 2).

Wadell's formula does not always indicate accurately how a particle behaves while settling in a fluid medium, e.g., a rod settles faster than a dish of equal volume. Sneed and Folk (1958), therefore, defined a new expression, *maximum projection sphericity,* as the cube root of the ratio c^2/ab (the ratio of the square of the shortest intercept to the product of the long and intermediate intercepts). This number shows a higher linear statistical correlation with observed settling velocity than does Wadell's operational sphericity.

Microscopic methods for obtaining the intercepts of sand grains have been developed, so that the same concepts may be applied to sand grains (Rittenhouse, 1943; Curray and Griffiths, 1955; Wright, 1957).

Hydraulic Equivalence

Measurement of a hydraulic shape factor (Fig. 3) appears to offer the most significant approach for morphometric analysis beyond the stage of equilibrating granulometric methods (see *Hydraulic Equivalence*). In the hydraulic method of morphometric analysis, shape is not subdivided into arbitrary components, but is measured as a single effect on hydraulic properties of settling particles of a given size (Briggs et al., 1962). Another morphometric method that may have genetic significance is based on the descent velocity of a particle rolling down an inclined surface or placed on a vibrating surface. This method is said to measure the "pivotability" of the particle. As with the hydraulic shape factor, the pivotability measurement treats shape as a single, indivisible factor.

B. CHARLOTTE SCHREIBER

References

Bradley, W. C., 1970. Effect of weathering on abrasion of granitic rocks, Colorado River, Texas, *Geol. Soc. Am. Bull.,* 81, 61-80.

Briggs, L. I.; McCulloch, D. S.; and Moser, F., 1962. The hydraulic shape of sand particles, *J. Sed. Petrology,* 32, 645-656.

Brock, E. J., 1974. Coarse sediment morphometry: A comparative study, *J. Sed. Petrology,* 44, 663-672.

Cestre, G., and Cailleux, A., 1970. Indice d'émoussé du ler au 4e ordre, *Rev. Géomorph. Dyn.,* 19, 15-16.

Curray, J. R., and Griffiths, J. C., 1955. Sphericity and roundness of quartz grains in sediments, *Geol. Soc. Am. Bull.,* 66, 1075-1096.

Daubrée, A., 1879. *Etudes Synthétiques de Géologie Expérimentale,* 2 vols. Paris: Dunod, 828p.

Friedman, G. M., 1958. Determination of sieve-size distribution from thin-section data, *J. Geol.,* 66, 394-416.

Friedman, G. M., 1962. Comparison of moment measures for sieving and thin-section data, *J. Sed. Petrology,* 32, 15-25.

Glezen, W. H., and Ludwick, J. C., 1963. An automated grain-shape classifier, *J. Sed. Petrology,* 33, 23-40.

Krumbein, W. C., 1941. The effects of abrasion on the size, shape and roundness of rock fragments, *J. Geol.,* 49, 482-520.

Krumbein, W. C., and Sloss, L. L. 1963. *Stratigraphy and Sedimentation,* 2nd ed. San Francisco: Freeman, 660p.

Kuenen, P. H., 1956. Experimental abrasion of pebbles. 2. Rolling by current, *J. Geol.,* 64, 336-368.

Mrakovich, J. V.; Ehrlich, R.; and Weinberg, B., 1976. New techniques for stratigraphic analysis and correlation—Fourier grain shape analysis Louisiana offshore Pliocene, *J. Sed. Petrology,* 46, 226-233.

Patro, B. C., and Sahee, B. K., 1974. Factor analysis of sphericity and roundness data of clastic quartz grains: Environmental significance, *Sed. Geol.,* 11, 59-78.

Pearce, T. H., 1971. Short distance fluvial rounding of volcanic detritus, *J. Sed. Petrology,* 41, 1069-1072.

Perez-Rosales, C., 1972. A new method for determining sphericity to Zingg's *J. Sed. Petrology,* 42, 975-977.

Pryor, W. A., 1971. Grain shape, in R. E. Carver, ed., *Procedures in Sedimentary Petrology.* New York: Wiley, 131-150.

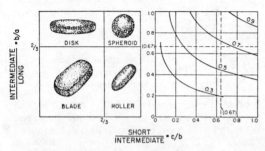

FIGURE 2. *Left:* Zingg's classification of pebble shapes, based on ratios of intercepts. *Right:* Relation of intercept sphericity to Zingg's classification of pebble shapes. (From *Stratigraphy and Sedimentation,* Second Edition, by W. C. Krumbein and L. L. Sloss. W. H. Freeman and Company. Copyright © 1963.)

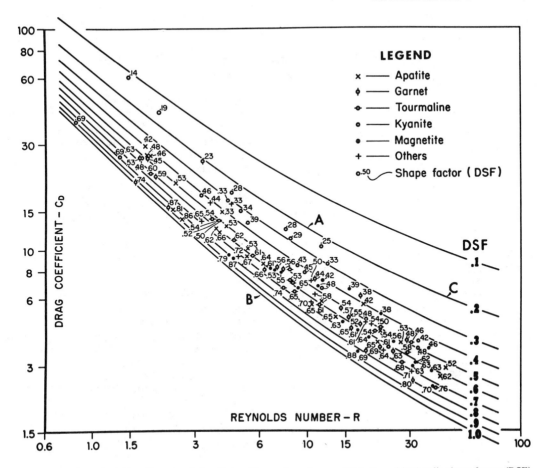

FIGURE 3. A nicuradze diagram relating Reynolds number, drag coefficient, and hydraulic shape factor (DSF) obtained on heavy minerals (from Briggs et al., 1962). Spheres are represented by a shape factor of 1.0.

Rittenhouse, G., 1943. The transportation and deposition of heavy minerals, *Geol. Soc. Am. Bull.,* **54,** 1725-1780.

Schumm, S. A., and Stevens, M. A., 1973. Abrasion in place: A mechanism for rounding and size reduction of coarse sediments in rivers, *Geology,* **1,** 37-40.

Sneed, E. D., and Folk, R. L. 1958. Pebbles in the lower Colorado River, Texas, a study in particle morphogenesis, *J. Geol.,* **66,** 114-150.

Spalletti, L. A., 1976. The axial ratio C/B as an indicator of shape selective transportation, *J. Sed. Petrology,* **46,** 243-248.

Swan, B., 1974. Measures of particle roundness: A note, *J. Sed. Petrology,* **44,** 572-577.

Thébault, J. -V., 1969. Contribution à l'étude des formes des galets, *Bull. Bur. Rech. Géol. Min.,* ser. 2, sect. IV, **2,** 1-105.

Wadell, H., 1932. Volume, shape and roundness of rock particles, *J. Geol.,* **40,** 443-451.

Wadell, H., 1933. Sphericity and roundness of quartz particles, *J. Geol.,* **41,** 310-331.

Wentworth, C. K., 1919. A laboratory and field study of cobble abrasion, *J. Geol.,* **27,** 507-522.

Whalley, W. B., 1972. The description and measurement of sedimentary particles and the concept of form, *J. Sed. Petrology,* **42,** 961-965.

Williams, G. E., 1971. Flood deposits of the sand-bed ephemeral streams of central Australia, *Sedimentology,* **17,** 1-40.

Wright, A. E., 1957. Three-dimensional shape analysis of fine-grained sediments, *J. Sed. Petrology,* **27,** 306-312.

Zingg, T., 1935. Beiträge zur Schotteranalyse, *Schweiz. Min. Petrog. Mitt.,* **15,** 39-140.

Cross-references: *Faceted Pebbles and Boulders; Grain-Size Studies; Granulometric Analysis; Hydraulic Equivalence; Maturity; Sediment Parameters; Settling Velocity.*

SABKHA SEDIMENTOLOGY

Sabkha is a phonetic rendering of an Arabic word connotating salt flat; many variants of this spelling are to be found, for example, *sebkha* in French-dominated Arabic areas, *sebjet* in Spanish-dominated Arabic areas. The term *sabkha* is applied to both coastal and interior, or continental, salt flats (Kinsman, 1969). In Spanish-speaking regions, the term *salina* is synonymous with *coastal sabkha*. The *solonchak* salt flats of the Caspian Sea region are similarly coastal sabkhas. In North America, the term *playa* is synonymous with *continental sabkha;* in South America, *salar* is used (see *Salar, Salar Structures*). Other equivalents for continental sabkha are the *shotts* of North Africa, the *kevir* of Iran and Transcaspian USSR, the *salt pans* and *vleis* of South Africa, and most of the "*lakes*" of Australia.

This group of geomorphic forms is generally characterized by very low surface slope (<1/2000). The ground-water table lies 1-2 m below the surface. Occasionally, sheets of flood water inundate the surface, forming ephemeral lakes which then desiccate and leave behind a layer of salts. Continued evaporation from the surface draws water upward through the capillary zone above the ground-water table, and near-surface salt minerals are also precipitated by this mechanism.

Sabkhas may be described as geomorphic surfaces that are attemping to reach or maintain deflational equilibrium. The surface elevation is controlled dominantly by the local ground-water level. Sabkhas are developed where eolian sand supply is limited and where bedrock does not crop out. Coastal sabkhas are supratidal surfaces developed by normal processes of depositional offlap of nearshore marine sediments under semiarid climatic conditions; offlap rates of 1-2 m/yr are recorded from the Trucial Coast of the Persian Gulf (Kinsman, 1969). The subtidal and intertidal marine sediments of coastal sabkhas are capped by a supratidal unit, the non-evaporitic part of which accumulates through storm accretion and adhesion of windblown sediment grains.

Coastal sabkhas are the arid zone analogue of the supratidal fresh- and brackish-water marsh of climatically wetter regions.

Persian Gulf Sabkhas

Sabkhas are well developed along the Arabian margin of the Persian Gulf, and those most intensely studied are located along the Trucial Coast (Evans et al., 1964; Illing et al., 1965; Kinsman, 1966, 1969; Butler, 1969; see also *Persian Gulf Sedimentology*). The coastal sabkhas have developed mainly behind island barriers by the infilling of lagoons with marine sediments over the past 5000 years. Stromatolitic sediments are typically the final infilling facies (see *Stromatolites*). There is evidence in this area of a relative sea-level fall of 120 ± 10 cm over the past 4000-5000 yr, which is a contributing factor to the rate of offlap.

Of great historical interest in sedimentology were the reports by Wells (1962) of dolomite in the coastal sabkhas of Qatar and by Curtis et al. (1964) of dolomite and anhydrite in the coastal sabkhas of the Trucial Coast, near Abu Dhabi. These were some of the earliest reports of modern dolomite and the first record of anhydrite actually forming in an evaporite environment at the earth's surface. Subsequent studies have demonstrated the widespread development of early diagenetic dolomite, gypsum, anhydrite, and magnesite, with minor amounts of celestite, huntite and halite (see *Dolomite, Dolomitization*).

The hydrological framework of a coastal sabkha is critical to an understanding of processes of interstitial evaporite mineral precipitation and dolomitization (Kinsman, 1966, 1969). Evaporation from the sabkha surface leads to interstitial brine development. In a seaward zone 2-3 km wide, there is storm flooding by sea water (Butler, 1969), which recharges the subsurface brine reservoir and balances losses by evaporation and seaward subsurface flow or *reflux* of dense brines. Intrusion of continental ground waters occurs in landward parts of the coastal sabkha; brines are generated from these waters once they enter the sabkha. As progradation continues, the zone of sea-water flooding moves seaward, and the intrusion of continental ground water also moves seaward. Associated with brine development is the interstitial precipitation of gypsum crystals and the later development of anhydrite (see *Gypsum in Sediments*). Some anhydrite is a primary precipitate, and some replaces earlier gypsum. Aragonite muds may be intensively dolomitized by reaction with the sea-water-derived sabkha brines. The framework wedge of sediments, the subaerial upper surface of which forms the sabkha, is obviously diachronous. Similarly, the belts of differing brines and the belts of evaporite mineral precipitation and dolomitization sweep diachronously across the sabkha, giving a complex end product.

Continental sabkhas are extensively developed in Arabia and are also sites of evaporite mineral formation. They represent a largely neglected sedimentary environment and have been unrecognized in ancient sedimentary rock sequences. For a discussion of continental sabkhas, see Kinsman (1969; see also *Salar, Salar Structures*).

North African Sabkhas

The sabkhas of the North African coast are apparently dominated by continentally derived ground waters. Subsidence has deepened many of the basins since initial infilling began. The *shotts* (Chott Djerid, etc.) of Tunisia and Algeria are extreme examples; the surface there today is >50 m below sea level. It has been postulated that these shotts are deflation hollows, fed by ground water from the Atlas Mountains (Gautier, 1908).

Deep borings in the sabkhas of southernmost Tunisia disclose a complex history during the Quaternary. For example, in the Sabkha el Melah, a mid-Quaternary basin was delimited by faulting and differential movement and contains two dolomitized horizons related to low stands of sea level (Perthuisot, 1974). The second dolomitization phase was followed by a gypsum and later a halite evaporite sequence. Marginal gypsum is in places completely replaced by polyhalite in a manner analogous to that described by Holser (1966) from Baja California. The replacement probably occurred during the latest halite phase, when K^+ and Mg^{++} were enriched in the lagoonal barriers. The Sabkha el Melah is obviously quite distinctive from the coastal sabkhas of the Persian Gulf and includes in addition a lagoonal evaporite.

Other Contemporary Sabkhas

In the arid subtropical climatic belt, coastal sabkhas are to be widely found today, where gently shelving offshore areas and lagoons are progressively infilled with sediments to form supratidal surfaces. The sediments constructing the supratidal surfaces may range from almost pure marine carbonates to almost pure noncarbonate terrigenous clays, silts, and sands. Few sabkhas have been studied in great detail, but almost every one possesses certain distinctive features. Some guidance to the literature may be helpful: Laguna Madre, Texas (Masson, 1955); Andros Island, Bahamas (Shinn et al., 1965; see *Bahama Banks Sedimentology*); Shark Bay, Western Australia (see *Shark Bay Sedimentology*) Coorong, South Australia (Alderman, 1965); Baja California, Mexico (Holser, 1966; Kinsman, 1969; Phleger, 1969).

Ancient Sabkhas

Studies in the Persian Gulf showed that extensive dolomitization of lagoonal and stromatolitic carbonates and widespread development of gypsum and nodular anhydrite could occur in the subaerial coastal sabkha environment. This environmental interpretation had not even been hinted at in earlier studies of ancient dolomicrites and nodular sulfate deposits. The environments of dolomitization and evaporite formation had always been assumed to be lagoonal or brine pan and to have required the presence of a standing surface body of brine. The sabkha concept demonstrated that the brine body was accommodated within the pore spaces of the previously deposited marine sediments and that the development of the brines was related to their confinement and their lack of dilution by adjacent marine waters.

Although supratidal surfaces are developed in basin margin positions and are therefore likely to be very frequently destroyed, it has become obvious that a vast amount of ancient supratidal coastal sabkha of all ages is preserved. It might be conceptually difficult to understand how, if a stack of sabkha cycles can be superposed, the evaporites at the top of one cycle are not dissolved when the surface is inundated by the sea once again and the marine sediments of the succeeding cycle accumulate. However, ancient sequences tell us that this has frequently been achieved: the evidence suggests slowly submerged basin margins, modulated by eustatic oscillation, preservation being aided by rapid diagenesis, clay "seals," and arid climates.

The supratidal and intertidal facies may be identified by the association of microdolomite, discoidal gypsum crystals, nodular anhydrite with felted-lath texture, clay pebble conglomerates, etc. (Lucia, 1972). Littoral fauna and associated trace fossils are sometimes present. The uppermost, nonevaporitic, supratidal sediments are commonly eolian or fluvial silts and are often represented as red beds.

Ancient sabkha environments are most commonly recognized in the dolomite-sulfate facies, but noncarbonate analogues are also known. Extensive nodular anhydrite units of sabkha origin are now known in several areas to be oilfield reservoir seals; for example, the famous Upper Jurassic reservoirs of Arabia are sealed by the overlying Hith Anhydrite, which is an extensive sabkha anhydrite unit.

DAVID J. J. KINSMAN
RHODES W. FAIRBRIDGE

References

Alderman, A. R., 1965. Dolomitic sediments and their environment in the south-east of South Australia, *Geochim. Cosmochim. Acta*, 29, 1355-1365.
Butler, G. P., 1969. Modern evaporite deposition and geochemistry of coexisting brines, the sabkha, Trucial Coast, Arabian Gulf, *J. Sed. Petrology*, 39, 70-89.
Curtis, R.; Evans, G.; Kinsman, D. J. J.; and Shearman, D. J., 1964. Association of dolomite and anhydrite in the Recent sediments of the Persian Gulf, *Nature*, 197, 679-680.
Evans, G.; Kendall, C. G. St. C.; and Skipwith, P., 1964. Origin of the coastal flats, the sabkha, of the Trucial Coast, Persian Gulf, *Nature*, 202, 759-761.
Gautier, E. -F., 1908. *Le Sahara Algérien* (vol. 1 of *Missions au Sahara*, Gautier, and Chudeau, R., eds.). Paris: Colin, 371p.
Gavish, E., 1974. Geochemistry and mineralogy of a recent sabkha along the coast of Sinai, Gulf of Suez, *Sedimentology*, 21, 397-414.
Holser, W. T., 1966. Diagenetic polyhalite in recent salt from Baja California, *Am. Mineralogist*, 51, 99-109.
Illing, W. V.; Wells, A. J.; and Taylor, J. C. M., 1965. Penecontemporary dolomite in the Persian Gulf, *Soc. Econ. Paleont. Mineral. Spec. Publ. 13*, 89-111.
Kinsman, D. J. J., 1966. Gypsum and anhydrite of Recent age, Trucial Coast, Persian Gulf, *Proc. 2nd Salt Symp.*, 1, 302-326.
Kinsman, D. J. J., 1969. Modes of formation, sedimentary associations and diagnostic features of shallow-water and supratidal evaporites, *Bull. Am. Assoc. Petrol. Geologists*, 53, 830-840.
Kogan, V. D., 1972. Sabkha analogs in the Permian in the south of the Russian platform, *Lithology and Mineral Resources* 7(6), 779-787 (transl. from Russian).
Lucia, F. J., 1972. Recognition of evaporite-carbonate shoreline sedimentation, *Soc. Econ. Paleont. Mineral. Spec. Publ. 16*, 160-191.
Masson, P. H., 1955. An occurrence of gypsum in southwest Texas, *J. Sed. Petrology*, 25, 72-77.
Perthuisot, J. P., 1974. Les dépots salins de la sebkha el Melah de Zarzis; conditions et modalités de la sédimentation évaporitique, *Rév. Géogr. Phys. Géol. Dyn.*, 16, 177-188.
Phleger, F. S., 1969. A modern evaporite deposit in Mexico, *Bull. Am. Assoc. Petrol. Geologists*, 53, 824-829.
Renfro, A. R., 1974. Genesis of evaporite-associated stratiform metalliferous deposits—a sabkha process, *Econ. Geol.*, 69, 33-45.
Shinn, E. A.; Ginsburg, R. N.; and Lloyd, R. M., 1965. Recent supratidal dolomite from Andros Island, Bahamas, *Soc. Econ. Paleont. Mineral. Spec. Publ. 13*, 112-123.
Wells, A. J., 1962. Recent dolomite in the Persian Gulf, *Nature*, 194, 274-275.

Cross-references: *Carbonate Sediments—Diagenesis; Desert Sedimentary Environments; Dolomite, Dolomitization; Evaporite Facies; Gypsum in Sediments; Limestones; Oolite; Persian Gulf Sedimentology; Pisolite; Salar, Salar Structures; Shark Bay Sedimentology; Stromatolites.* Vol. III: *Sabkha or Sebkha.*

SALAR, SALAR STRUCTURES

The Spanish term *salar* (to salt) is used widely in South America as a name for salt-encrusted flats or playas (Fig. 1) and salt-filled basins that are particularly abundant in northern Chile and bordering regions of Argentina and Bolivia (see Stoertz and Ericksen, 1974). Similar features in the southwestern US and northern Mexico are generally termed *salt flats, salinas, salt pans,* or *dry lakes.* A salar may be considered a type of continental sabkha (see *Sabkha Sedimentology*). Typical salars occupy the lowest parts of basins of interior drainage, areas that formerly were sites of perennial or ephemeral lakes; and many salars are underlain by unconsolidated lake or playa sediments. The salar represents salt deposition during a phase of increasing aridity in a given climatic cycle. Salars similar to those of North and South America are found in semiarid to arid regions of other continents, being particularly noteworthy in N Africa, southern USSR (e.g., Kazakhstan), western China (e.g., Sinkiang and Tibet), Australia, and S Asia (India, Pakistan, Afghanistan, and Iran).

Salt crusts on salars generally range in thickness from a few cm to a few m; in some salt-filled basins, crusts may attain thicknesses of a few tens of m. Borings have shown that Salar Grande, Chile, is underlain by rock salt up to 150 m thick. On the basis of mode of formation, three types of salars may be recognized: (1) residual evaporites consisting of saline

Salar Structures

Salar crusts have a variety of textures, forms, and microrelief features. Primary structures form during initial crystallization of the saline minerals from brines and desiccation of the newly formed crust; secondary structures form through modification of the crust by rain, surface-water flooding, wind, temperature changes, and tectonic deformation. Perennial structures are preserved mainly on hard salar crusts, chiefly those consisting of halite; few, if any, can be observed on soft crusts, where winds and seasonal rains tend to obliterate the perennial structures, leaving only ephemeral features that survive a season or two.

Mineral grains in the typical salar crust tend to be equigranular and <5 mm in diameter; a sugary texture is characteristic of many salar crusts. Some salars, however, locally contain well-formed crystals several cm to several tens of cm long. Salar surfaces range from smooth, like those of the Bonneville Salt Flats in Utah, to extraordinarily rugged, like those of the Devils Golf Course in Death Valley, California (see Hunt et al., 1966). A salar surface that has been modified by rain, fog, or flooding may be slabby, nodular or knobby, pinnacled, hummocky, and pitted to cellular; it may consist of loose nodules, slabs, or irregular clinker-like fragments.

Salar structures may be grouped into five categories on the basis of processes that formed them: (1) depositional and desiccation structures; (2) structures resulting from fresh-water flooding; (3) structures resulting from rain and fog; (4) structures resulting from diurnal temperature changes; and (5) other structures.

Depositional and Desiccation Structures. The most widespread structures of hard salar crusts are those formed during deposition and desiccation. Salars typically show depositional layering and desiccation polygons (Fig. 2). Successive salt layers of similar mineralogical composition, generally ranging from a few mm to several cm in thickness, result from seasonal flooding and desiccation. Layering may also result from capillary evaporation of saline ground water; such layering tends to form within or at the base of the salar crust and tends to be local and lenticular, individual layers having distinctly different chemical and mineralogical composition at some localities. Polygons, outlined by tensional cracks that form as the saline crust desiccates, commonly are a few tens of cm to 1.5 m in diameter. Where the crust is thin and rests on brine-saturated mud, capillary evaporation results in deposition of new saline material, causing the polygons to expand and thus

FIGURE 1. Small salar (Andes, Northern Argentina). White crust in center of salar is chiefly halite; light-gray marginal zone is covered with a thin layer of sulfate minerals and halite. Stratified sediments at far side of salar are tectonically deformed Quaternary playa and lacustrine deposits.

minerals that crystallized from shallow saline ponds or lakes; (2) efflorescences formed by evaporation at or near a moist to wet playa surface; and (3) saline-cemented soil formed by capillary evaporation above near-surface saline ground water—such crusts may contain 50% or more saline cement. Salar surfaces range from soft to hard, moist to dry, and smooth to rugged.

Most salars seen today are of continental origin and are supplied by leaching of salt from soil and rock within the closed basin in which the salar occurs; locally, this source is augmented by saline spring waters that originate outside the basin or by thermal springs associated with volcanism within the basin. Other salars are found along the margins of large inland seas (Caspian Sea) or saline lakes (Great Salt Lake).

The dominant saline minerals found in varying proportions in all salars are halite, gypsum, and thenardite (Na_2SO_4). Other minerals that are widespread and abundant in some salars include trona [$Na_3H(CO_3)_2 \cdot 2H_2O$], mirabilite ($Na_2SO_4 \cdot 10H_2O$), and ulexite ($NaCaB_5O_9 \cdot 8H_2O$). Many other chlorides, sulfates, and carbonates of calcium and sodium have been found in salars, as well as a few noteworthy borates, phosphates, and nitrates, including some rare and new minerals; in only a few localities, however, are these less common minerals in sufficient concentrations to be economically recovered.

FIGURE 2. Hard halite crust (Salar de Pocitos, Argentina) showing *polygons* bordered by extruded salt and reticulate *salt sheets*. Within polygons are many *salt veins*. Salt cone beside geologic hammer is over an open *salt tube* extending to water level about 30 cm below surface.

FIGURE 3. Pressure ridges in thin layer of newly deposited salt; Salar del Carmen, Chile.

become *thrust polygons*. In some places, a thin salt layer may act as a coherent sheet in which expansion due to crystallization is transmitted for distances of several or tens of m, forming irregular pressure ridges (Fig. 3) or rounded to elongate blister-like crusts.

Structures Resulting from Fresh-Water Flooding. Unusual structures formed by the leaching of hard halite crusts by fresh water from springs or by seasonal floods have been observed at several Andean salars. Salt pools form by leaching of the crust by newly formed springs in sediments below the crust. The size of the pool is determined by the amount of fresh water supplied, and the pool becomes stable when inflow becomes equal to loss by evaporation; leaching ceases when the ground water becomes a saturated brine. Sinkholes (see Krinsley, 1970) may form in a soft crust or saline-cemented soil as the result of solution of saline material during an exceptional seasonal rise of the ground-water level. Crenulate and cuspate margins (Fig. 4A) occur at flood boundaries; they result from wave action and leaching along the margin of older salt crusts. Cusps may be as much as several m across. Elongate channels (Fig. 4B) may form in a salt crust on the leeward side of fresh-water springs where windblown sheets of water leach narrow channels. Windblown water also may cause extensive leaching of the crust, leaving low mounds of older salt in the form of barchan

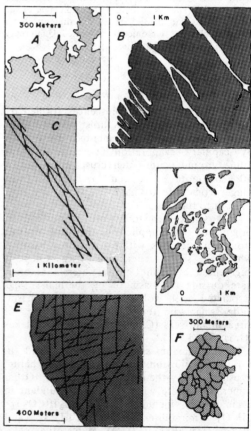

FIGURE 4. Large structures in salars having hard halite crusts, northern Chile, as seen in aerial photographs. Old salt crust is shaded; new or leached crust is white. A, crenulate margin; B, salt channels; C, polygons along fault zone; D, salt pseudobarchans; E, oriented nonorthogonal giant polygons; F, random orthogonal giant polygons.

or irregular multiple barchan dunes (Fig. 4D) that have been called *pseudobarchans* (Stoertz and Ericksen, 1974).

Structures Resulting from Rain or Fog. In regions of sparse intermittent rainfall or of seasonal heavy fogs (such as the coastal region of Chile), several structures of solutional or solutional-depositional origin may form. *Salt saucers* are basin-shaped polygons (Fig. 5) that form by accumulation of extruded salt on the margin of the polygon (Fig. 2) and/or by solution in its center (Fig. 5). *Salt nodules* are rounded masses of white granular salt that formed from fragments of the salar crust as a result of frequent wetting by fog or dew. *Salt stalactites*, which form in cavities in the salar crusts, consist of salt carried downward from the surface by sporadic rainfall.

Structures Resulting from Diurnal Temperature Changes. Structures resulting primarily from diurnal temperature changes include *salt veins*, *salt tubes*, *salt cones* and *cups*, and *salt sheets*. Veins of granular salt form in massive salt crusts along cracks that have opened because of thermal expansion and contraction. These processes cause granulation of salt along the crack; capillary rise of near-surface brine may result in formation of a welt or nodule of moist granular salt on the vein surface (Fig. 2). A *salt cone* may form from one of these nodules where the rising brine dissolves a vertical open tube in the vein to connect the nodule to the brine level. Diurnal pumping of moist air then causes the nodule to expand slowly into a hollow cone.

FIGURE 6. Gypseous rampart on lee side of saline pond; Salar de San Martin, Chile.

FIGURE 5. *Salt saucers* penetrated by *salt tubes* (T) that were formerly capped by *salt cones* or *cups*. Upraised rims of saucers consist of recemented extruded salt (see Figure 2), and centers have been leached by rainwater that drains through the salt tubes.

Expansion continues until an opening forms at some spot on the cone, generally at the top. Ultimately, the opening widens, and the thin salt rim bends outward to form a cup-like structure. Cones and cups are widespread in Andean salars where diurnal temperature changes are great and where the ground-water level is near the surface. The largest salt cones seen there are 30-40 cm in diameter and in height. *Salt sheets* (Fig. 2) are formed by extrusion of newly crystallized moist salt at the margins of polygons in a hard rock-salt crust. Sheets are as thin as 1 mm, and some show delicate, open reticulate structure.

Other Structures. Other noteworthy structures have diverse origins. *Gypseous ramparts* (Fig. 6) are low nodular ridges that form on the lee sides of saline ponds, where saline material, chiefly gypsum, accumulates by evaporation of windblown spray. The ramparts have an irregular nodular surface, a maximum height of about 1 m, and a width of about 5 m. *Giant polygons* (Fig. 4C,E,F) have been observed in aerial photographs of many salars; they commonly are not readily visible on the ground. Some may originate by deep desiccation of sediments (see Neal, 1965) under the salar crusts, whereas others appear to be associated with fault movement (Fig. 4C). Some have the randomly oriented outlines of the desiccation polygons that are common to all salars (Fig. 4F), whereas others, particularly those related to recent faulting, are bounded by straight intersecting lines that form an oriented nonorthogonal network (Fig. 4E). *Scarps* and *sag basins* occur along recent faults that displace salar crusts; good examples of these features are found at Salar Grande, northern Chile.

GEORGE E. ERICKSEN
GEORGE E. STOERTZ

References

Hunt, C. B.; Robinson, T. W.; Bowles, W. A.; and Washburn, A. L., 1966. General geology of Death Valley, California–Hydrologic basin, *U.S. Geol. Surv. Prof. Pap. 494-B*, 138p.

Krinsley, D. B., 1970. *A Geomorphological and Paleoclimatological Study of the Playas of Iran*, 2 vols. Washington, D. C.: U.S. Govt. Print. Office, 486p.

Neal, J. T., ed., 1965. Geology, mineralogy, and hydrology of U.S. playas, *U.S. Air Force Cambridge Research Labs., Environ. Research Pap. 96*, 176p.

Neal, J. T., ed., 1975. *Playas and Dried Lakes*. Stroudsburg, Pa.: Dowden, Hutchinson & Ross, 411p.

Reeves, C. C., ed., 1972. *Playa Lake Symposium 1970*. Lubbock, Texas: Texas Tech. Univ. Intern. Center for Arid and Semi-Arid Land Studies, 334p.

Stoertz, G. E., and Ericksen, G. E., 1974. Geology of salars of northern Chile, *U.S. Geol. Survey Prof. Pap. 811*, 65p.

Cross-References: *Desert Sedimentary Environments; Evaporite Facies; Lacustrine Sedimentation; Mud Cracks; Sabkha Sedimentology.*

SALCRETE

As originally proposed by Yasso (1966), *salcrete* is a white to light gray surface crust of marine beach sand cemented by sodium chloride, and lesser amounts of other marine salts, concentrated by evaporation of swash or of spray blown onshore by breaking waves. Salcretes form parallel to the shoreline. They are thickest (up to several mm) in the upper foreshore and berm areas. Thickness and landward extent depend primarily on breaker characteristics and onshore wind velocity and duration.

Salt-cemented sediments (salcretes) may also form in sabkhas (q.v.). Gavish (1974) reports the presence of interstitial sodium chloride crystals dispersed in the upper 20 cm of medium to coarse sand in a young sabkha of the El Belayim region in the western Sinai. Highly saline water from the Gulf of Suez migrates 200 m or more laterally to supply ground water at shallow depth beneath the sabkha surface. Evaporation in this arid, subtropical environment causes concentration of sodium chloride to a depth of about 20 cm throughout the A and B soil horizons. Gypsum is concentrated in the thinner C horizon about 10 cm above ground-water table.

Amiel and Friedman (1971) describe a partially soft, encrusted area of silty and sandy clay in the central barren zone of the Yotrata Sebkha. Chlorinity increases from the groundwater table through the capillary zone to the sabkha surface. In one of the four exploration holes chlorinity reached 120.6 m/kg for the upper 15 cm of sediment.

Surface crusts of sodium chloride (halite) and resultant salcretes are reported as supratidal evaporation deposits of coastal sabkhas of Baja California and the Persian Gulf (Kinsman, 1969). Here the marine salt is supplied by coastal flooding.

WARREN E. YASSO

References

Amiel, A. V., and Friedman, G. M., 1971. Continental sabkha in Arava Valley between the Dead Sea and the Red Sea: Significance for origin of evaporites, *Bull. Am. Assoc. Petrol. Geologists*, 55, 581-592.

Gavish, E., 1974. Geochemistry and mineralogy of a Recent sabkha along the coast of Sinai, Gulf of Suez, *Sedimentology*, 21, 397-414.

Kinsman, D. J. J., 1969. Modes of formulation, sedimentary associations, and diagnostic features of shallow water and supratidal evaporites, *Bull. Am. Assoc. Petrol. Geologists*, 53, 830-840.

Yasso, W. E., 1966. Heavy mineral concentration and sastrugi-like deflation furrows in a beach salcrete at Rockaway Point, N.Y., *J. Sed. Petrology*, 36, 836-838.

Cross-references: *Cementation; Duricrust; Sabkha Sedimentology.*

SALTATION

Bagnold (1941) used the term saltation to denote the prevalent motion of sand in air; he observed that when a moving grain strikes a hard surface (another grain or a pebble), it bounces off it with almost perfect resilience (Fig. 1) and may rise several hundred or even a thousand grain diameters. When the risen grain is acted on by the wind, it moves in a curved, flattened path. It strikes the ground at a flat angle of 10–16°, depending on the size of the grain, its height of rise, and wind speed (Fig. 2; Bagnold, 1941). There is still considerable uncertainty about the physics of saltation (Middleton and Southard, 1977). Observed trajectories (Bagnold, 1941; Chepil, 1945; Zingg, 1952) support the idea that a grain is lifted from the bed as a result of a short-lived impact, and that its subsequent path results from the action of gravity and fluid drag (Middleton and Southard, 1977).

Gilbert (1914) first used the term saltation to describe the similar motion of sand in water,

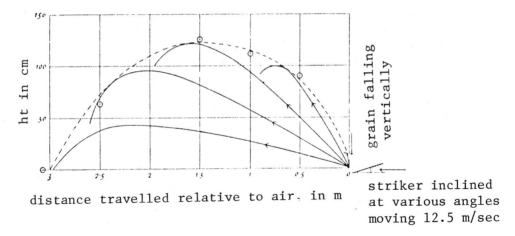

FIGURE 1. Comparison of calculated and observed limiting paths of 0.3 mm diameter grains on rebound after impact with ground pebbles (still air and moving ground); the circles denote the limiting possibility of actual grains found experimentally (after Bagnold, 1941).

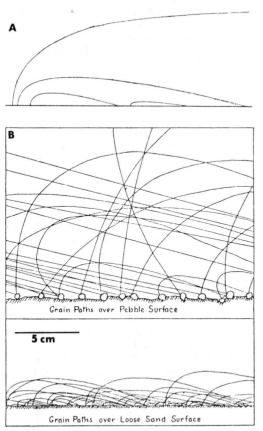

FIGURE 2. A: Typical saltation grain paths. B: Difference between the saltation over sand and over pebbles. (After Bagnold, 1941.)

but Bagnold (1941) and others have noted that the mechanism is much better developed in air, where sand grains leap about 800 times higher than sand grains in water, a result of fluid drag (Kalinske, 1943). There is a large disparity in the density ratio for quartz in air (2000:1) and quartz in water (1.65:1). Bagnold noted that the momentum of a sand grain moving at the same speed as the transporting fluid is 2000 times as great as that of its own volume of air, but is only 1.65 times that of its own volume of water. For air to impart to one stationary grain a velocity equal to its own, it must lose momentum equivalent to 2000 grain volumes; for water, only 1.65 grain volumes are necessary. In the case of air, then, this reduction in velocity close to the surface is so great that the residual drag due to the underlying stationary surface can be neglected, and the velocity of air near the surface is controlled entirely by the intensity of saltation of sand above the surface. In water, the reverse is true: the direct effect of saltation being negligible, the velocity of the water is largely controlled by unevenness of the stationary surface (Bagnold, 1941).

Bagnold listed three possible types of grain motion: suspension, saltation, and surface creep. Sand grains are rarely suspended in air, and grains in surface creep have little effect on the wind; only saltating grains offer significant resistance to the wind (see above). The great bulk of sand grains moves by saltation, and it is this process which is the basis for most eolian sedimentation (q.v.).

Bagnold (1941) also related the origin of wind ripples to saltation; because the angle

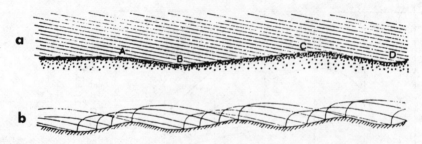

FIGURE 3. a: Differential intensity of bombardment on windward and lee slopes (after Bagnold, 1941). On lee side AB there are few impacts, but on the windward side (BC) where they are much closer, bombardment is more intense. The grains that are excavated from the hollow will accumulate at C, because grains are not removed as quickly as they accumulate, and lee slope CD develops; thus the process perpetuates itself. b: Coincidence of ripple wavelength and range of characteristic path of grain (Bagnold, 1941).

of incidence of saltating grains is constant, even very small irregularities in the underlying surface will affect the number of impacts per unit area; small hollows will thus be accentuated by local sand removal (Fig. 3a), producing incipient ripples. The random irregularities grow bigger and multiply until the two processes—saltation and local sand removal—fall in phase (Fig. 3b). Thus the ripple wavelength depends on the length of the "characteristic path" of the grain, which, in turn, is dependent on wind strength (Bagnold, 1941).

JOANNE BOURGEOIS

References

Bagnold, R. A., 1941. *The Physics of Blown Sand and Desert Dunes.* London: Methuen, 265 p.

Chepil, W. S., 1945. Dynamics of wind erosion, I, Nature of movement of soil by wind, *Soil. Sci., 60*, 305–320.

Gilbert, G. K., 1914. The transportation of debris by running water, *U.S. Geol. Surv. Prof. Pap. 86*, 263p.

Kalinske, A. A., 1943. Turbulence and the transport of sand and silt by wind, *Annals N.Y. Acad. Sci., 44*, 41–54.

Middleton, G. V., and Southard, J. B., 1977. Mechanics of sediment movement, *Soc. Econ. Paleont. Mineral. Short Course 3*, 246p.

Zingg, A. W., 1952. Wind tunnel studies of the movement of sedimentary material, *St. Univ. Iowa Studies in Engineering Bull. 34*, 111–136.

Cross-references: *Eolian Sedimentation; Fluvial Sediment Transport; Sediment Transport—Initiation and Energetics.*

SALT-MARSH SEDIMENTOLOGY

The term (coastal) *salt marsh* is generally restricted to intertidal areas where the bottom (consisting of loose sedimentary material) is more or less closely covered by a low, treeless vegetation of salt-tolerant or salt-requiring, predominantly phanerogamic plants. In many cases, the intertidal halophyte flora starts directly at the low-tide line, mostly with *Zostera* plants, which may cover large surfaces of tidal flats. Although botanists normally include these *Zostera* flats into the salt marsh environments, most geologists prefer to think of salt marshes as beginning at a higher level, where the vegetation starts to have a more pronounced effect on the morphology of the sediment surface. On their landward side, the salt (or brackish) marshes often pass very gradually into fresh-water marshes or swamps. The highest parts of salt marshes may be covered by water only a few times each year, during exceptionally strong tides or when wind action has raised the water level along the coast to abnormal heights. Marshes in deltas or estuaries may, moreover, be inundated (by fresh water) during river floods.

Salt marshes are only formed on parts of coasts that are sheltered from the direct influence of high waves, i.e., along bays, lagoons, tidal-flat areas, etc., and, in some cases (e.g., Louisiana), also along the open sea, where excessive muddiness of the water or shallowness of the sea floor damps the waves.

The bulk of the marsh sediment (sand, silt, and clay) is normally supplied by the sea or river waters flooding the area. Less commonly, winds contribute sand and dust from neighboring beaches and dunes. In some instances, where the supply of inorganic material is very small, the marsh bottom may locally consist of peat, produced by the marsh plants themselves, e.g., *Spartina patens* in the New England marshes.

In the initial stages of salt-marsh formation (in the restricted sense) isolated groups of Spartina or other "marsh pioneers" appear at places where the surface of an intertidal area has been raised by sedimentation to close below mean high-tide level (Fig. 1). Within these tufts of plants, the water is relatively quiet, and the deposition of mud and sand suspended in the water takes place at a higher rate than on the bare intervening surfaces. Low elevations are thus formed, which by the action of waves and currents soon acquire more or less rounded outlines. Between these elevations, which tend to enlarge both horizontally and vertically, the flow of water is concentrated in strongly winding courses, which form the beginnings of marsh creeks (Fig. 2).

In these early stages, many changes still take place in the pattern of the creeks, and in their depths and the size of their curves. As the creek banks grow up by further sedimentation, however, and the bank sediments are held together by an increasingly tight network of plant roots, the creeks become gradually more stabilized. Creek meanders in mature marshes show as a rule only a small tendency to migrate.

In many marsh areas, little systematic orientation can be detected in the creek patterns. In other cases the major creeks have comparatively straight, parallel courses. The latter are inherited from channels or other elongate depressions originally present in the area before it became covered with plants.

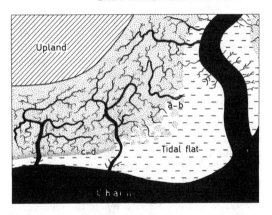

FIGURE 2. Idealized sketch map of salt marsh area with stages of horizontal accretion (a,b) and of formation of outer marsh ridge (c,d).

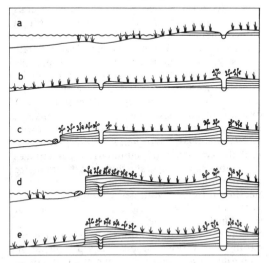

FIGURE 1. Possible successive stages of salt-marsh development. (a),(b) Horizontal accretion (towards the left) and upward growth of marsh; formation of natural levees along creeks. (c),(d) Continued upward growth of marsh and wave erosion of outer edge (at left, combined with formation of marsh ridge and closing of minor creeks). (e) Continued upgrowth and renewal of horizontal accretion.

Most of the sediment that is carried to salt marshes is trapped by the vegetation as the water passes over the outer edge or over the banks of the creeks. In consequence, these places tend to be raised above their surroundings. Natural levees are formed along the creeks, which eventually may lie 30 cm or more above the adjoining parts of the marsh surface. Where owing to a gradual silting up of the whole area, the marsh grows out over the tidal flats, a transition zone of generally low relief is found between the two. At places where the marsh edge remains stationary, or where it is cut back by wave erosion (usually marked by a miniature cliff), a high ridge may be produced, blocking the way out for all but the largest creeks.

While the sediment of the creek levees often contains appreciable amounts of sand or silt (in the outer parts of the marshes), the material deposited in interlevee basins is mostly very clayey. The bottom of these basins, therefore, has a very low permeability, and the water left over after (marine) floodings in the deepest parts may disappear mainly by evaporation, notably in regions with warm climates. Then hypersaline conditions arise, and there may even be formation of salt crusts. Under such conditions all higher plants are killed off, and bare surfaces, *salt pans,* are the result. Where the water is drained off by underground percolation (toward the creeks) before the salinities in the basins become too high, vegetation may be able to maintain itself. This motion of the ground water, however, is often so slow that decomposition of the vegetable remains results in high CO_2 concentrations whereby the bottom sediments may become completely decalcified.

Marine or brackish bottom animals are very scarce in the salt-marsh environment. They are usually restricted to the creek bottoms and to interlevee basins that are frequently inundated by salt water. On the higher parts of the marshes, which are flooded only occasionally, and in parts where the bottom (surface) is well drained, e.g., on creek banks and in relatively sandy or silty intercreek areas, they are usually altogether absent. Burrowing land animals are also lacking, because of the salt content of the bottom. At most places, accordingly, the depositional structures of the marsh sediments are disturbed only by the penetration of plant roots. Where the layers are composed of different grain-size classes, the original lamination is often still clearly visible in, e.g., eroded creek banks or wave-cut marsh edges. At each place in the marsh, the duration of the inundations and the intensity of the water movements decrease as the marsh grows up, the average thickness of the laminae in such sections is usually seen to diminish upward, and the relative clay content increases in the same direction.

L. M. J. U. VAN STRAATEN

References

Chapman, V. J., 1960. *Salt Marshes and Salt Deserts of the World.* New York: Wiley, 392p.

Proceedings, Salt Marsh Conference, Sapelo Island, Georgia, 1958. Athens, Ga.: Marine Institute, Univ. Georgia, 133p.

Steers, J. A., 1953. *The Sea Coast.* London: Collins, 276p.

Stevenson, R. E., and Emery, K. O., 1958. Marshlands at Newport Bay, Calif., *Allan Hancock Found. Publ., Occ. Pap. 20,* 109p.

Treadwell, R. C., 1955. *Sedimentology and Ecology of Southeast Coastal Louisiana.* Baton Rouge: Louisiana St. Unv., 78p.

Van Lopik, J. R., 1955. *Recent Geology and Geomorphic History of Central Coastal Louisiana.* Baton Rouge: Louisiana St. Univ., 89p.

Van Straaten, L. M. J. U., 1954. Composition and structure of Recent marine sediments in the Netherlands, *Leidse Geol. Med.*, 19, 1-110.

Cross-references: *Fluvial Sediments; Lagoonal Sedimentation; Paralic Sedimentary Facies; Tidal-Flat Geology; Wadden Sea Sedimentology.*

SANDS AND SANDSTONES

Sand is an accumulation of grains ranging in size from 1/16 to 2 mm in diameter; its lithified equivalent is *sandstone.* Most sands contain accessory quantities of gravel, silt, and clay. The properties of sand include grain size, grain roundness, grain shape, and mineral composition. Large-scale accumulations of sand are described in terms of both primary structures and geometry.

Hydraulically, sand-sized particles are the most mobile of particle sizes (Hjulström, 1939; Sundborg, 1956) because they occur in a transition stage between cohesionless and cohesive grain diameters (Sanders, 1960). Accordingly, lower entrainment velocities and bed-shear are required to transport sand, regardless of fluid viscosity.

Sands

Grain-Size Distribution. Sand is deposited and associated with a mixture of several grain sizes characterized by a variable range of sorting. Folk (1954) classified sand-gravel-silt-clay mixtures; in his scheme, mixtures of sand and gravel are considered to be sand if >70% by weight of the sediment is within sand size. In mixtures of sand and clay, 50% of the sediment must be sand to qualify as a "sand." Qualifying terms are added to indicate volumetrically significant amounts of gravel, silt, or clay (Fig. 1, Table 1).

Sand is eroded, transported, and deposited by wind, water, and ice. The difference in viscosity of each of these agents controls the size distribution of the deposited sediment, assuming the availability of a heterogeneous size distribution. Wind, because of its low viscosity, transports fine and medium-grained sands. In contrast, ice, being the most viscous of the transporting agents, will deposit a heterogeneous mixture of sand and other particle sizes. Water, having intermediate viscosity, deposits a broad range of particle sizes. Attempts have been made to subdivide the distributions of sand-sized sediment deposited by rivers, waves, and currents, but overlap often exists because of provenance, i.e., sizes available for transport (see *Grain-Size Parameters–Environmental Interpretation*). Probability plots of cumulative percentages of grain sizes suggest that most grain-size distributions are composite populations reflecting different volumes of sediment transport by rolling or saltation bed load, or suspended load (see *Grain-Size Frequency Studies*). These populations differ when comparing the various modes of sand dispersal, including wave action, open-channel flow, longshore currents, tidal currents, longshore drift, ocean currents, and turbidity currents.

Mineralogy. The mineralogy of sands is highly variable. Any conceivable combination and permutation of minerals are present, depending primarily on source-area composition and secondarily on maturity (q.v.), sorting

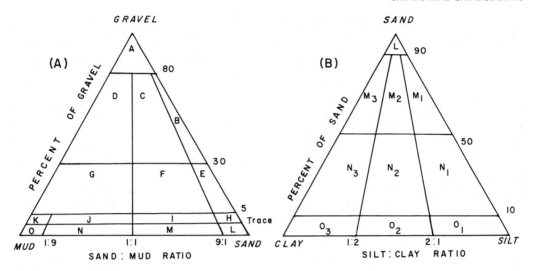

FIGURE 1. Grain-size nomenclature according to Folk (1954, 1968). Letters in ternary diagram correspond to letters in Table 1.

TABLE 1. Terminology of Mixtures of Gravel, Sand, Silt and Clay[a]

	Sediment	Lithified Equivalent
A.	Gravel	conglomerate
B.	Sandy Gravel	sandy conglomerate
C.	Muddy sandy gravel	muddy sandy conglomerate
D.	Muddy gravel	muddy conglomerate
E.	Gravelly sand	conglomeratic sandstone
F.	Gravelly muddy sand	conglomeratic muddy sandstone
G.	Gravelly mud	conglomeratic mudstone
H.	Slightly gravelly sand	slightly conglomeratic sandstone
I.	Slightly gravelly muddy sand	slightly conglomeratic muddy sandstone
J.	Slightly gravelly sandy mud	slightly conglomeratic sandy mudstone
K.	Slightly gravelly mud	slightly conglomeratic mudstone
L.	Sand	sandstone
M_1.	Silty sand	silty sandstone
M_2.	Muddy sand	muddy sandstone
M_3.	Clayey sand	clayey sandstone
N_1.	Sandy silt	sandy siltstone
N_2.	Sandy mud	sandy mudstone
N_3.	Sandy clay	sandy claystone
O_1.	Silt	siltstone
O_2.	Mud	mudstone
O_3.	Clay	claystone

[a]Folk, 1954
Note: This table is summarized in Fig. 1.

(q.v.), and diagenesis. Experiments (Kuenen, 1959) and field studies (Pollack, 1961) show that many of these minerals are not differentiated by abrasion over distances comparable to the length of a major stream (averaging 1000 km). Mineral differentiation appears to be controlled by recycling of sediments (Klein, 1963), by weathering, or by extreme long distances of grain transport within localized areas (Balazs and Klein, 1972).

The mineral variability of sand appears to be controlled in part by the specific gravity of individual grains. Most of the darker, ferromagnesian minerals, which are characterized by a density greater than quartz, are concentrated in finer particle sizes than are quartz grains. Such differences in density give rise to hydraulically equivalent particle sizes (see *Hydraulic Equivalence*). Thus, the particle density may in part control the mineral distribution in different size ranges.

Occurrence. Sand occurs in many depositional environments (see cross-references). It accumulates as dunes along coastal areas or in deserts (see *Eolian Sands*). It is plentiful in rivers (see *River Sands*) lakes, shallow marine areas (see *Beach Sands; Deltaic Sediments; Lagoonal Sedimentation; Tidal-Current Deposits*), continental shelves (see *Continental-Shelf Sediments*), and the deep-sea floor. In lakes and the deeper portions of oceans, turbidity currents or grain flows are the most probable mechanism for deposition and transport of sands (see *Gravity Flows; Turbidity-Current Sedimentation*), whereas wave and tidal currents predominate along shores (see *Littoral Sedimentation*). Channel flow regimes are operative in rivers and in the channel systems occurring in estuaries and on tidal flats (see *Fluvial Sediment Transport*).

Sands show a variety of geometric configurations. Lenses, pods, sheets, linear shoestrings, and channels are the most common geometric forms and are usually controlled by depositional processes and environment (Peterson and Osmond, 1961; Potter, 1967; Reineck, 1970; LeBlanc, 1972; Pettijohn et al., 1972; Klein, 1975). Associated primary structures usually are characterized by a distinct directional aspect from which dispersal patterns of sedimentary rocks can be determined; these patterns are environmentally controlled (Klein, 1967).

Sandstones

Sandstones are lithified equivalents of sands and therefore also consist of fragments whose particle diameters range in size from 1/16 mm to 2 mm. Sandstones may also be mixed with gravel- and clay-sized particles, so it is customary to classify such mixtures according to their lithified equivalents (Folk, 1954, 1968; see also Fig. 1 and Table 1).

Previous remarks concerning dynamics, descriptive parameters, and occurrence of sand are just as pertinent to sandstones. Because postdepositional changes in many of these parameters obscure some of the comparative relations between sands and sandstones, direct comparison, especially of grain size, leads to conflicting conclusions. Mineral replacement of particles and other diagenetic phenomena obscure textural comparison (see *Clastic Sediments–Lithification and Diagenesis*). Because of diagenetic changes in sandstone properties, considerable emphasis has been placed on sandstone mineralogy to interpret sandstone genesis. Many mineralogically oriented classifications have been proposed. Between 1940 and 1960, seventeen sandstone classifications were proposed in the North American geological literature (Klein, 1963). Huckenholz (1963) reported four additional quantitative sandstone classifications appearing in the European literature since 1958. McBride (1963), Dott (1964), Folk (1968), Okada (1971), and Pettijohn et al. (1972) have since contributed to sandstone classification.

The authors of sandstone classifications have proposed six factors on which to base their classification: mineral composition (as an indicator of provenance), diastrophism, mineralogical maturity, textural maturity (index of sorting and rounding), fluidity factor (measure of fluid density and viscosity), and primary structures. No more than a combination of four of these factors have been used in a single classification.

The criteria of provenance and mineralogical maturity (or increasing mineralogical stability) seem to be the only ones that can define mineralogical groups of sandstones. Texture can be included in the descriptive classification of sandstones as a separate, modifying term, as proposed by Folk (1954). It has

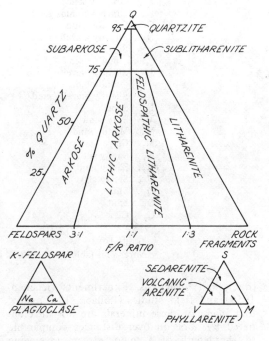

FIGURE 2. Sandstone classification according to Folk (1968). Q = quartz; F = feldspar; R = rock fragments; S = sedimentary rock fragments; M = igneous rock fragments; V = volcanic rock fragments.

furthermore been shown by many that all textural properties function independently of mineralogy. Thus Folk's (1968; Fig. 2) approach to sandstone classification should be followed. Another classification that has gained some acceptance (Pettijohn et al., 1972) was proposed by Dott (1964).

Three major classes of sandstones, together with three transitional classes, have been recognized by all proponents of sandstone classification. The major classes are *quartzose sandstone* (quartz arenite, quartzite), *arkose*, and *litharenite*. Transitional classes have received a variety of names (see Klein, 1963), but here, Folk's (1968) classes (Fig. 2) are used. Folk's transitional classes are subarkose and sublitharenite.

Arkose (q.v.) is a sandstone containing >25% feldspar and whose feldspar:rock fragments ratio is >1. Arkoses are derived from granitic, granodioritic, syenitic, and volcanic terrains and from metamorphic rocks of the pyroxene-granulite facies. Volcanic-derived arkoses are rare on land exposures, but perhaps more common to oceanic sediments. Arkoses are deposited in any sedimentary basin and any sedimentary environment whose area is enriched in feldspar. The type "graywacke" is itself an arkose (Huckenholz, 1963). Many geosynclinal sandstones are arkose.

The term *graywacke* has been variously defined, and its environmental and tectonostratigraphic significance argued (see *Graywacke*, also *Subgraywacke*). No one petrographic definition can be assigned to it; by most definitions it would be a special kind of arkose or subarkose.

Litharenites are sandstones that consist of >25% recognizable rock fragments and whose feldspar:rock fragment ratio is <1. In Folk's (1968) classification, litharenites are further subdivided according to provenance; nomenclature is based on the predominance of sedimentary, metamorphic, or igneous rock fragments (Fig. 2).

The *quartz arenites* (or *orthoquartzites*) contain >95% quartz and detrital chert. They represent the most stable chemical residue after prolonged weathering and recycling of preexisting sedimentary rocks, during which time the relatively unstable feldspars and recognizable rock fragments are progressively removed. Several workers have proposed that progressive recycling and weathering of graywacke, litharenites, and arkose generates the intermediate sublitharenites and subarkoses before generating a quartz arenite. Perhaps prolonged abrasion over long distances of transport in a confined locality produces quartz arenites (Balazs and Klein, 1972; Swett et al., 1971). The quartzose sandstones, as well as the intermediate sandstone classes, occur under a variety of depositional conditions which flank deeply weathered or sedimentary terranes.

The lithological association of sandstones is variable, thus contributing to a flexible understanding of sandstone associations in space and time. Sandstones are interbedded with mudstones and with limestones in rocks ranging in age from Precambrian to Holocene in a variety of nonmarine and marine environments. The variable occurrence of sandstone mineralogical groups makes generalizations about tectonic and environmental associations and sandstone classes premature. Our present knowledge confirms that provenance and mineralogical maturity (or mineral differentiation) control sandstone composition.

GEORGE DE VRIES KLEIN

References

Balazs, R. J., and Klein, G. deV., 1972. Roundness-mineralogical relations of some intertidal sands, *J. Sed. Petrology*, 42, 425-433.

Davies, D. K., and Ethridge, F. G., 1975. Sandstone composition and depositional environment, *Bull Am. Assoc. Petrol. Geologists*, 59, 239-264.

Dott, R. H., Jr., 1964. Wacke, graywacke and matrix – What approach to immature sandstone classification? *J. Sed. Petrology*, 34, 625-632.

Folk, R. L., 1954. The distinction between grain size and mineral composition in sedimentary rock nomenclature, *J. Geol.*, 62, 344-359.

Folk, R. L., 1968. *Petrology of Sedimentary Rocks*. Austin: Hemphills, 170p.

Forgotson, J. M., Sr., 1974. Calcitic sandstone, *Bull. Am. Assoc. Petrol. Geologists*, 58, 1838.

Hjulström, F., 1939. Transportation of detritus by moving water, in P. D. Trask, ed., *Recent Marine Sediments*. Tulsa: Soc. Econ. Paleont. Mineral., 736p.

Huckenholz, H. G., 1963. A contribution to the classification of sandstones, *Geol. Fören. Förh.*, 85, 156-172.

Klein, G. deV., 1963. Analysis and review of sandstone classification in the North American geological literature, 1940-1960, *Geol. Soc. Am. Bull.*, 74, 555-576.

Klein, G. deV., 1967. Paleocurrent analysis in relation to modern marine sediment dispersal patterns, *Bull. Am. Assoc. Petrol. Geologists*, 51, 366-382.

Klein, G. deV., 1975. *Sandstone Depositional Models for Exploration for Fossil Fuels*. Champaign, Ill.: Continuing Educ. Publ., 109p.

Kuenen, P. H., 1959. Experimental abrasion. 3. Fluviatile action on sand, *Am. J. Sci.*, 257, 172-190.

LeBlanc, R. J., 1972. Geometry of sandstone reservoir bodies, *Am. Assoc. Petrol. Geol. Mem. 18*, 133-189.

McBride, E. F., 1963. A classification of common sandstones, *J. Sed. Petrology*, 33, 664-670.

Mainguet, M., 1972. *Le Modelé des Grès. Problèmes Generaux*. Paris: Inst. Géogr. National, 657p.

Okada, H., 1971. Classification of sandstone: Analysis and proposal, *J. Geol.,* 79, 509-525.
Perriaux, J., 1974. Généralités sur les grès, *Bull. Centre Rech. de Pau,* 8, 161-185.
Peterson, J. A., and Osmond, J. C., 1961. *Geometry of Sandstone Bodies.* Tulsa, Okla.: Am. Assoc. Petrol. Geologists, 240p.
Pettijohn, F. J.; Potter, P. E.; and Siever, R., 1972. *Sand and Sandstone.* New York: Springer-Verlag, 618p.
Pollack, J. M., 1961. Significance of compositional and textural properties of South Canadian River channel sands, New Mexico, Texas and Oklahoma, *J. Sed. Petrology,* 31, 15-37.
Potter, P. E., 1967. Sand bodies and sedimentary environments: a review, *Bull. Am. Assoc. Petrol. Geologists,* 51, 337-367.
Reineck, H.-E., 1970. Marine Sandkörper, rezent and fossil, *Geol. Rundschau,* 60, 302-321.
Sanders, J. E., 1960. Origin of convolute laminae, *Geological Mag.,* 97, 409-421.
Shelton, J. W., 1973. Models of sand and sandstone deposits: A methodology for determining sand genesis and trend, *Okla. Geol. Surv. Bull.,* 118, 122p.
Smalley, I. J., 1966. Formation of quartz sand, *Nature,* 211, 476-479.
Stone, W. J., and Erickson, J. M., 1970. A FORTRAN program for Folk's sandstone classification, *Compass,* 47(3), 163-168.
Sundborg, Å., 1956. The River Klaralven, *Geograf. Ann.,* 38, 127-316.
Swett, K.; Klein, G. deV.; and Smit, D. E., 1971. A Cambrian tidal sand body—Eriboll Sandstone of northwest Scotland: An ancient-Recent analog, *J. Geol.,* 79, 400-415.

Cross-references: *Alluvium; Arenite; Arkose, Subarkose; Barrier Island Facies; Bay of Fundy Sedimentology; Beach Sands; Bed Forms in Alluvial Channels; Black Sands; Blanket Sand; Calcarenite, etc.; Channel Sands; Clastic Sediments and Rocks; Clastic Sediments—Lithification and Diagenesis; Continental-Shelf Sediments; Cross-Bedding; Deltaic Sediments; Eolian Sands; Flow Regimes; Fluvial Sediments; Fluvial Sediment Transport; Glacial Sands; Glauconite; Grain Flows; Grain-Size Studies; Graywacke; Hydraulic Equivalence; Littoral Sedimentation; Marine Sediments; Maturity; Paleocurrent Analysis; Provenance; Quartzose Sandstone; Relict Sediments; Ripple Marks; River Sands; Roundness and Sphericity; Sand Surface Textures; Sedimentary Petrography; Sedimentary Rocks; Silica in Sediments; Subgraywacke; Submarine-Fan (Cone) Sedimentation; Tar Sands; Tidal-Current Deposits; Tidal-Inlet and Tidal-Delta Facies; Traction; Turbidites; Volcaniclastic Sediments and Rocks.*

SAND SURFACE TEXTURES

Particles set free as a result of the decomposition of rock may be weathered in place, transported and abraded, deposited and weathered again, and finally subjected to diagenesis. These processes give rise to differing textures on the surfaces of the quartz grains that frequently make up the bulk of the sediment so formed. A number of workers, among them Heald (1956) and Hull (1957), have examined quartz sand-grain surfaces with the light microscope to see if texture could be related to the environment of transportation and deposition; little definitive information has been obtained to date.

Greater success in eliciting information as to environment has been obtained in recent years with the introduction of electron microscopy to the study of sand surface textures. The transmission electron microscope (TEM) had been used to examine sand grain surface textures up to 1968; since that time, most work of this type has been accomplished with the scanning electron microscope (SEM). Attempts have been made to relate surface features to specific environments of transportation and deposition, with varying degrees of success (Krinsley and Doornkamp, 1973). Initially, these studies were done without an understanding of the physical and chemical mechanisms involved. More recently, physical and chemical textures have been explained in terms of fracture, cleavage, and solution chemistry (Krinsley and Smalley, 1973; Margolis and Krinsley, 1974).

It is important to realize that many grains have been exposed to more than one environment and may, therefore, exhibit complex surface textures.

Environments of Transport

Glacial Textures. Characteristic surface textures on Pleistocene glacial grains have been studied in very general terms (see *Glacial Sands*); recently, sand grains from a number of modern glacial subenvironments (supraglacial, englacial, subglacial, fluvioglacial) in Switzerland have been described (Whalley and Krinsley, 1974). The features fall into two groups: inherited characters and features subsequently impressed on sand grains during a particular glacial stage. Eight mechanical features have been described: conchoidal textures include arc-shaped steps, parallel and subparallel steps, arc-shaped grooves, and microsteps; cleavage forms consist of upturned plates, cleavage flakes, micro-blocks, and facets. Three solution-precipitation features can be described: precipitation platelets (Krinsley and Doornkamp, 1973), precipitation rounding, and carapaces (see Whalley and Krinsley, 1974, for detailed descriptions and photographs).

Generally, no surface textures were observed that could characterize a particular glacial

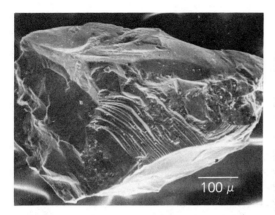

FIGURE 1. Modern glacial grain from Norwegian fjord. Note angular corners, conchoidal breakage patterns, and flat cleavage surfaces. Also note lack of rounding, which indicates that hardly any solution-precipitation has taken place.

subenvironment. A combination of the mechanical features mentioned above does, however, suggest a general glacial environment, and it is possible that statistical work on the various textures may provide subenvironmental information. Figs. 1 and 2 indicate the great variation in single samples of quartz grains. Pleistocene glacial grains have been described by Krinsley and Newman (1965) and Krinsley et al. (1973).

Eolian Textures. Wind abrasion on large (400-500μ) modern sand grains produces a number of characteristic textures (Fig. 3); in addition most of the grains are well rounded (see also *Eolian Sands*). Single or multiple

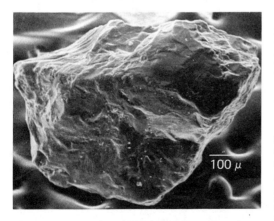

FIGURE 2. Grain is from the same sample as Fig. 1, but shows solution-precipitation effects in addition to mechanical action. The edges are slightly rounded, and the surface somewhat smoothed; but the upper left surface contains conchoidal breakage patterns and the facing upper right surface is flat.

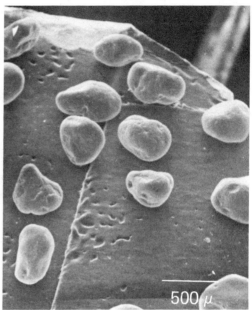

FIGURE 3. Windblown sand from Libyan desert. Note extreme roundness and smoothness, typical of mature dessert sand >400-500μ in diameter.

rounded, dish-shaped concavities may cover as much as 1/6 of grain surfaces and are probably formed during periods of violent abrasion related to strong wind storms. They represent conchoidal chips removed by a single mechanical event, perhaps during saltation (q.v.). Upturned plates tend to cover most of the rounded surfaces of larger grains and are produced by quartz cleavage during mechanical abrasion. Upturned plates and dish-shaped concavities are modified by solution and precipitation phenomena in hot deserts; the upturned plates tend to be subdued, producing a rolling topography. The dish-shaped features acquire a smooth patina that is probably initially amorphous silica (for details see Krinsley and Doornkamp, 1973; Margolis and Krinsley, 1974).

The smaller grains in hot deserts (<400-500 μ) are generally rather flat and elongated, as compared to larger grains, and contain irregular edges indicating cleavage (Fig. 4). These grains are carried in suspension and thus are not abraded; they eventually acquire a precipitated layer of silica which may crack and craze, depending upon the amount of water available. This material never includes the dish-shaped concavities because momentum is never sufficient to produce breakage. An unconsolidated New Red Sandstone locality (Permo-Triassic) has been found in southern England with hot

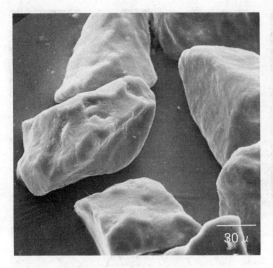

FIGURE 4. Quartz grains from same sample as Fig. 3 with diameters <400–500μ. Note the difference in shape and surface texture. These angular and somewhat flattened grains cannot be mistaken for those in Fig. 3.

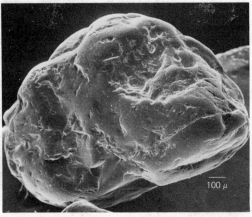

FIGURE 5. Large quartz sand grain from the beach at Sandy Hook, New Jersey. Grain shows edges rounded by subaqueous abrasion, probably in the littoral zone. Note large, irregular depression with precipitated silica on the upper right part of the grain; also note that even though the corners are rounded, the grain is still elongated.

desert features almost exactly duplicating those described above (Krinsley et al., 1976); diagenesis seems almost to have been inoperative here.

Subaqueous Textures. Irregularly oriented, V-shaped depressions that have somewhat irregular sides are present on subaqueous quartz grains where appreciable energy is available; rather vigorous grain-to-grain contacts occur in moving water under these circumstances (Fig. 5, 6, 7). These features appear to reach a saturation level of about 7 V's/μ^2 (Margolis and Kennett, 1971), based on laboratory experiments; beyond this point, as many V's are removed as are created. V's are simply notches cut into the tops of upturned cleavage plates, which suggests that perhaps more than one mechanical breakage episode may be necessary to produce them. The smaller V's are generally 5μ in diameter; little is known about their distribution. V's are commonly produced in the littoral zone and perhaps on the continental shelf, most likely above wave base, but are not usual in river deposits. Very little is known about the density of V-shaped patterns in various subaqueous environments, although Margolis and Kennett (1971) have indicated that the lowest percentage of V's/μ^2 appears on river deposits, with frequency increasing on low-turbulence beach, moderate- to high-turbulence beach, and shelf, in that order. This particular problem should be studied in great detail in terms of environmental definition (see also *Beach Sands; River Sands*).

Straight or slightly curved grooves or scratches usually are present on subaqueous grains subject to high abrasion levels and only rarely on grains in environments with little energy input. Grooves are a very good indicator of subaqueous abrasion as they are never observed on grains from other environments. The grooves are usually <25μ in length, average about 1–5μ in width, and are generally observed on material >400–500μ in diameter, perhaps

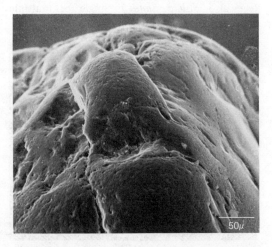

FIGURE 6. Closeup of the extreme-right middle side of Fig. 5. The depression immediately below the most prominent topographic high in the middle of the photograph contains irregular, precipitated silica, as does the depression to its left. The high, on the other hand, has been abraded and rounded and contains typical mechanical V-shaped patterns.

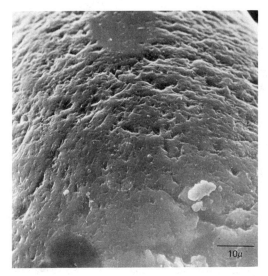

FIGURE 7. Closeup of Fig. 6. Overall roundness is still apparent, but a series of parallel ridges are visible, cut by mechanical, V-shaped notches. The apparent height of these cleavage plates changes with the viewing angle; they are most obvious at a viewing angle of 45° and are difficult to distinguish at the top and bottom of the micrograph. Only the center of the picture is in perfect focus because of the extreme depth of the field. The flat area at the top of the photograph suggests solution postdating the mechanical features.

because smaller grains lack the momentum to cause surface penetration (Margolis and Krinsley, 1974). These mechanical features are almost always found on grain edges and very rarely in depressions. They are more common on well-rounded than on angular edges; the former can also be used as a criterion for vigorous subaqueous action (Krinsley and Doornkamp, 1973).

The above features are found in ancient rocks; the oldest example of V's and grooves known to the authors are sands from the Upper Cretaceous of New Jersey.

Diagenesis

Numerous chemical effects are produced by diagenesis (Figs. 8, 9); see Margolis and Krinsley (1974) for details. Features resulting from chemical weathering and diagenesis consist of various etch features, overgrowths, and pitting of various types. The main kinds of etching are oriented etch pits, which apparently form slowly, and nonoriented pits, related to very rapid removal of silica. Overgrowths may occur in crystallographic continuity with existing quartz crystals (Pittman,

FIGURE 8. Sand grain from granite regolith, South Dakota. Note sharp corners and block textures, probably produced during release of the grain from its parent rock. Also note lower right side, which has broken away from main part of the grain, a common occurrence after release. Debris on surface has been cemented by solution-precipitation of silica in the weathering zone.

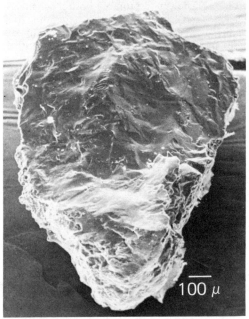

FIGURE 9. Sand grain from granite regolith, Santa Barbara, California. This grain is quite angular and shows little if any solution precipitation. The amount of chemical action here is probably a function of the type of climate during weathering and the length of time exposed.

1972) or may consist of disordered amorphous silica (see Krinsley and Doornkamp, 1973).

Some chemical action is always observed on quartz grains, even on those that have just been released from parent rock. Chemical action may modify or obliterate previous textures, either mechanical or chemical, resulting in a smooth featureless plane or an irregular pitted and etched surface. In some cases, if the layer of precipitated silica is thin, the original textures may be visible under the surface crust.

Silica precipitation and mechanical abrasion may occur simultaneously, as in the case of grains rolled about in the littoral zone. Here, abrasion may occur on the protruding edges, rounding them, while deposition can occur in hollows (Le Ribault, 1971). It is difficult if not impossible to distinguish between prediagenetic and postdiagenetic solution-precipitation, but perhaps detailed studies of the type that Le Ribault has been pursuing may give partial answers.

If grains have been ground mechanically, either wet or dry, the disrupted lattice layer produced on grain surfaces will be more reactive to chemical action than the original quartz. Additionally, this structure is more susceptible to mechanical abrasion than ordinary silica, so that abrasion may promote both mechanical breakage and chemical action, depending on conditions. Very little is known about any of these processes at present, and the entire subject is a fertile field for research.

It is thus possible to investigate the sedimentary history of sand deposits by studying surface textures with the electron microscope. The method is, unfortunately, limited by the progressive diagenetic alteration of grain surfaces with time, so that only a few deposits of pre-Mesozoic age still retain original mechanical textures. Detailed studies of diagenesis, however, could be undertaken with the new techniques now available.

DAVID KRINSLEY
TARO TAKAHASHI

References

Barbaroux, L.; Bousquet, B.; Brosse, R.; Nobrega-Coutinho, P.; and Jovic, P., 1972. Examen au microscope électronique à balayage de grains de sable de diverses origines. Essai de typologie, signification environmentale, *Bull. Bur. Rech. Geol. Min.*, sec. 4, **4**, 3-31.

Heald, M. T., 1956. Cementation of the Simpson and St. Peter sandstone in parts of Oklahoma, Arkansas and Missouri, *J. Geol.*, **64**, 16-30.

Hull, J. P. D., 1957. Petrogenesis of Permian Delaware Mountain Sandstone, Texas and New Mexico, *Bull. Am. Assoc. Petrol. Geologists*, **41**, 278-307.

Krinsley, D., and Doornkamp, J., 1973. *Atlas of Quartz Sand Grain Surface Textures*. New York: Cambridge Univ. Press, 91p.

Krinsley, D., and Newman, W., 1965. Pleistocene glaciation: a criterion for recognition of its onset, *Science*, **149**, 442-443.

Krinsley, D., and Smalley, I., 1973. The shape and nature of small sedimentary quartz particles, *Science*, **180**, 1277-1279.

Krinsley, D.; Biscaye, P.; and Turekian, K., 1973. Argentine basin sediment sources as indicated by quartz surface textures, *J. Sed. Petrology*, **43**, 251-257.

Krinsley, D. H.; Friend, P. F.; and Klimentidis, R., 1976. Eolian transport textures on the surfaces of sand grains of Early Triassic age, *Geol. Soc. Am. Bull.*, **87**, 130-132.

Le Ribault, L., 1971. Comporément de la pellicule de silice amorphe sur les cristaux de quartz en fonction des différents millieux evolutifs, *Compt. Rend. Acad. Sci.*, **272**, 2649-2652.

Lin, I. J.; Rohrlich, V.; and Slatkine, A., 1974. Surface microtextures of heavy minerals from the Mediterranean coast of Israel, *J. Sed. Petrology*, **44**, 1281-1295.

Margolis, S., and Kennett, J., 1971. Cenozoic paleoglacial history of Antarctica recorded in subantarctic deep sea cores, *Am. J. Sci.*, **271**, 1-36.

Margolis, S., and Krinsley, D., 1974. Processes of formation and environmental occurrence of microfeatures on detrital quartz grains, *Am. J. Sci.*, **274**, 449-464.

Pittman, E., 1972. Diagenesis of quartz as revealed by scanning electron microscopy, *J. Sed. Petrology*, **42**, 507-519.

Whalley, W. B., and Krinsley, D., 1974. A scanning electron microscope study of surface textures of quartz grains from a glacial environment, *Sedimentology*, **21**, 87-105.

Cross-references: *Beach Sands; Eolian Sands; Glacial Sands; Grain-Size Studies; River Sands; Roundness and Sphericity; Sands and Sandstones; Sediment Parameters.*

SAND WAVES

The term *sand waves* has been applied to sand ripples, bars, and dunes of all types and sizes but in particular has been used in connection with very large subaqueous structures (Lane and Eden, 1940; Twenhofel, 1950; see *Bed Forms in Alluvial Channels*). Kennedy (1963) noted that in true sand waves, transport and sedimentation occur across the entire wave, resulting in slopes less than the angle of repose for sand. Many authors prefer to use the terms *dune, antidune, sand ripple*, etc., but Visher (1965) again suggested that one clear term be used, especially to avoid confusion with eolian dune forms.

More recently, it has been suggested by

TABLE 1. Most Important Features of Dunes and Sand Waves in Intertidal Environments[a]

	Dunes	Sand Waves
Spacing	≈1-5 m	≈5-100 m
Height:spacing ratio	relatively large	relatively small
Geometry	sinuous to highly three-dimensional; prominent scour pits in troughs	straight to sinuous; uniform scour in troughs
Characteristic flow velocity	high (>70-80 cm/sec, <100-150 cm/sec)	moderate (>30-40 cm/sec, <70-80 cm/sec)
Velocity asymmetry	negligible to substantial	usually substantial

[a]Southard, 1975

Boothroyd and Hubbard (1971) from field observations and by Costello and Southard (1974) from laboratory experiments that a third lower-flow regime bed form, called *sand waves* (or bars), should be distinguished from dunes (see *Flow Regimes*). Southard (1975) compared the characteristics of sand waves with dunes in intertidal environments, where sand waves are most common (see Table 1). Sand waves tend to be larger in deeper water, although Boothroyd and Hubbard (1974) described trains of sand waves migrating up tidal channels into shallower water with no significant changes in either size or morphology. Southard (1975) noted that unlike dunes, sand waves tend not to reverse direction with the tides but remain oriented in the direction of dominant flow. Hawkins and Sebbage (1972), however, found evidence for the reversal of sand waves in the Bristol Channel (see also *Tidal-Current Deposits*).

B. CHARLOTTE SCHREIBER

References

Boothroyd, J. C., and Hubbard, D. K., 1971. Genesis of estuarine bedforms and crossbedding, *Geol. Soc. Am. Progr. Abstr.*, 3(7), 509-510.

Boothroyd, J. C., and Hubbard, D. K., 1974. Bedform development and distribution pattern, Parker and Essex estuaries, Massachusetts, *U.S. Army Corps Engin. Coastal. Engin. Research Cen. Misc. Paper 1-74*, 39p.

Costello, W. R., and Southard, J. B., 1974. Development of sand bed configurations in coarse sands, *Ann. Meet. Am. Assoc. Petrol. Geologists, Abstr.*, 1, 20-21.

Hawkins, A. B., and Sebbage, M. J., 1972. The reversal of sand waves in the Bristol Channel, *Marine Geol.*, 12, M7-M9.

Kennedy, J. F., 1963. The mechanics of dunes and antidunes in erodible bed channels, *J. Fluid Mech.*, 16, 521-544.

Lane, E. W., and Eden, E. W., 1940. Sand waves in the lower Mississippi River, *J. West. Soc. Engineers*, 45, 281-291.

Southard, J. B., 1971. Representation of bed configurations in depth-velocity-size diagrams, *J. Sed. Petrology*, 41, 903-915.

Southard, J. B., 1975. Bed configurations, *Soc. Econ. Paleont. Mineral. Short Course*, 2, 5-44.

Twenhofel, W. H., 1950. *Principles of Sedimentology*, 2nd ed. New York: McGraw-Hill, 673p.

Visher, G. S., 1965. Fluvial processes as interpreted from ancient and recent fluvial deposits, *Soc. Econ. Paleont. Mineral Spec. Publ. 12*, 116-132.

Cross-references: *Bed Forms in Alluvial Channels; Flow Regimes; Tidal-Current Deposits.* Vol. I: *Subaqueous Sand Dunes.*

SAPROPEL

Sapropel is a black, organic-rich sediment formed in lakes and marine basins by decomposition of organic detritus exclusively through the activity of anaerobic bacteria. Sapropels are rich in chlorophyll derivatives and other unstable organic compounds. The higher-than-normal organic carbon content is generally explained by enhanced preservation of organic matter under anoxic conditions. Richards (1970) suggests that the higher contents may represent faster rates of sedimentation and accumulation rather than slower rates of decomposition. When the oxygen/sulfide interface is found in the free water above the bottom, benthonic invertebrates are absent, and thus the detritus is mainly planktonic. Most sapropels show a microlamination, the sedimentary pattern not being disturbed by burrowing or mud-eating benthos.

Certain trace metals are enriched in fossil sapropels; V, Ni, and Mo seem to be typical (Krejci-Graf, 1966). Molybdenum is enriched in recent sapropels of Saanich Inlet (British Columbia), presumably due to biological activity or precipitation by hydrogen sulfide (see Richards, 1965); molybdenum also plays an important role in nitrogen fixation and in nitrate reduction, which takes place on a large scale in anoxic and nearly anoxic waters. When free oxygen is exhausted by decomposition of organic matter, denitrification takes place, with reduction of nitrate and nitrite ions to free nitrogen. Once these ions are consumed, sulfate reduction ensues; there may be an intermediate buildup of nitrite ions, as in Saanich Inlet and the Pacific off Mexico and Peru.

Anoxic waters develop where the biochemical oxygen demand exceeds the circulatory replenishment of oxygen (see *Euxinic Facies*). The replenishment is unusually low in meromictic lakes, and in marine basins of humid climates where the vertical circulation is limited by strong haloclines (Black Sea, q.v., bays and fjords of Norway and British Columbia). The oxygen demand is high in subsurface layers of lakes and seas with highly productive surface waters.

In the oceans, the areas of highest productivity occur in the belt of trade winds and monsoons. The trades cause an eastward up-slope of the thermocline and its disappearance in areas of upwelling. Both a shallow position of the thermocline and the process of upwelling lead to high fertility of surface waters; in adjacent basins the same wind systems may cause an estuarine-like circulation, enhancing the fertility. High productivity is the chief cause of oxygen deficiency in the eastern tropical and subtropical Pacific and Atlantic oceans. Huge bodies of nearly anoxic water extend off the coasts of Peru, Mexico, and SW Africa; the thickness off Peru is > 800 m, off Mexico > 1200 m, the upper boundary coming to within 50 m of the sea surface. The oxygen minimum occurs at 300–500 m depth. This minimum probably develops in shallow water; its position is determined thereafter by mixing and diffusion (Menzel and Ryther, 1970). When the oxygen minimum layer enters into neighboring basins where oxygen replenishment is restricted by geomorphology, and additional organic matter is settling from fertile surface waters, the situation deteriorates, and complete anoxia may be reached.

Extreme fertility coincides with extreme aridity (Peru, SW Africa, to a lesser extent off Venezuela). The eastern boundaries of the oceans in low latitudes bring about a circulation in the sea which is the mirror image of the circulation in the air (for maps, see *Phosphates in Sediments*). They result in coastal divergences in water and air, which implies upwelling in the sea and subsidence in the air; moreover, they are the primary cause of the eastward up-slope of the thermocline in the sea, and of the eastward down-slope of the temperature inversion in the air. Upwelling and the high position of the thermocline lead to extreme fertility and anoxic conditions. The low position of the inversion and maximum subsidence result in a rainless climate which favors deposition of evaporites. In this way a contiguous occurrence of sapropelites and evaporites becomes comprehensible (Brongersma-Sanders, 1970).

MARGARETHA BRONGERSMA-SANDERS

References

Brongersma-Sanders, M., 1970. Origin of major cyclicity of evaporites and bituminous rocks: an actualistic model, *Marine Geol.*, **11**, 123–144.

Krejci-Graf, K., 1966. Geochemische Faziesdiagnostik, *Freiberger Forschungsh.*, **C224**, 1–80.

Menzel, A. W., and Ryther, J. H., 1970. Distribution and cycling of organic matter in the oceans, *Occ. Publ. Inst. Mar. Sci. Alaska*, **1**, 31–54.

Richards, F. A., 1965. Anoxic basins and fjords, in J. P. Riley and G. Skirrow, eds., *Chemical Oceanography*, vol. 1. New York: Academic Press, 611–645.

Richards, F. A., 1970. The enhanced preservation of organic matter in anoxic marine environments, *Occ. Publ. Inst. Mar. Sci. Alaska*, **1**, 399–411.

Cross-references: *Bitumen, Bituminous Sediments; Black Sea Sediments; Euxinic Facies; Evaporite Facies; Gyttja, Dy; Humic Matter in Sediments; Organic Sediments; Phosphate in Sediments.*

SCOUR-AND-FILL

Scour-and-fill is the name for a sedimentary process (and the resultant structure) that consists of a scour or channel-forming erosional phase, followed directly or later by a filling or depositional phase. This feature is known under various names. Most authors consider *cut-and-fill* a true synonym; terms such as *washout* and *channel* are often used for identical features of larger scale. Smaller features are called *scour marks* (q.v.).

Special prefixes can be added for certain environments, such as glacial (scour-and-) fill, tidal (scour-and-) fill or tidal channel. Channel fill is usually restricted to river action.

River and glacial valleys, as well as submarine canyons (q.v.), normally are too large to fit the scour-and-fill terminology.

The processes of scouring and filling can be subaerial or submarine. The erosional as well as the depositional phase can alternate several times. They can result from massive transport, e.g., by ice, lava flow, or slump, or by sediment-laden water/air, e.g., a stream, current, tide, or grain flow.

The fill may consist of material similar or foreign to the rock or sediment in which the scouring action took place. The fill can form layers that are more or less parallel to the bottom, as is often the case when motion ceases and the material is fine grained. Small deltaic and oxbow lakes may reveal such type of filling. The fill bedding is usually horizontal when low transport velocities are involved. In most instances, a combination of these processes produces layers that become thinner to the sides and often reveal a decrease in grain size in the same direction.

A third type of fill consists of inclined beds (German: *Schrägschichtung*) resulting from higher velocities in laterally migrating channels such as meandering rivers and tidal channels and gullies in estuaries and mudflats. A lag deposit at the base is the most conclusive feature for separating this type of channel fill from large-scale foreset bedding (see *Bedding Genesis*).

Combinations and successions of these three types of fill are normally encountered (Bouma, 1962), often together with an upward fining of the sediment. Scour-and-fill structures may often be used as top and bottom criteria (q.v.; Shrock, 1948). When an outcrop is small compared to the size of the scour-and-fill structure, however, an inaccurate interpretation can easily result.

ARNOLD H. BOUMA

References

Bouma, A. H., 1962. *Sedimentology of Some Flysch Deposits.* Amsterdam: Elsevier, 168p.

Selley, R. C., 1970. *Ancient Sedimentary Environments.* Ithica, N. Y.: Cornell Univ. Press, 237p.

Shrock, R. R., 1948. *Sequence in Layered Rocks.* New York: McGraw-Hill, 507p.

Cross-references: *Alluvium; Bedding Genesis; Casts and Molds; Current Marks; Fluvial Sedimentation; Scour Marks; Sedimentary Structures; Submarine Canyons and Fan Valleys, Ancient; Tidal-Inlet and Tidal-Delta Facies; Tidal-Current Deposits; Tidal-Flat Geology.*

SCOUR MARKS

Scour marks are sedimentary structures produced as a result of erosion of a cohesive sediment surface, i.e., of a soft, muddy surface (see *Current Marks*). They are often best studied by their counterparts (*molds*) found in the lower surfaces of the overlying sandstone bed. Such molds in sedimentary rocks are often referred to as *sole marks* (Kuenen, 1957). Molds have sometimes incorrectly been called casts, e.g., flute cast (see *Casts and Molds*).

Important contributions to the understanding of scour marks have been made by Rücklin (1938), Vassoevich (1953), Ten Haaf (1959, 1964), Kuenen (1957), Dzulynski and Walton (1965), Sanders (1965), and Allen (1971, 1973, 1975).

Scour marks originate in turbulent flows at large Reynolds numbers (q.v.). The erosion of the muddy surface occurs due to formation of vortices in a turbulent flow related to the process of flow separation, and is initiated on a passive bed or at the points of defects in the bed. Erosion takes place by the abrasive action of sediment grains suspended in the eddies or through the direct action of fluid stresses.

Types of Scour Marks

Scour-mark nomenclature is based on the shape of the associated molds and their orientation with respect to the direction of flow. Usually, similar marks occur together in an assemblage. They can be oriented transverse, oblique, or parallel to the flow direction.

Flute Marks. Molds of flute marks look

FIGURE 1. Molds of flute marks; direction of flow is away from observer; Vindhyan sandstone (shallow marine), India.

like discontinuous bulbous bodies, elongated in shape (Fig. 1). Each flute may show varied form: linguiform, conical, twisted (corkscrew), generally elongated parallel to current flow. The upstream end of a flute mark is usually deeper, the corresponding mold more bulbous. Flutes usually occur in groups arranged in en-echelon fashion. In the genesis of flute marks, the muddy surface is eroded mainly by horizontal eddies produced behind some chance scour.

Transverse Scour Marks. Transverse scour marks are series of parallel, regularly spaced hollows running normal to the current direction. They are produced as a result of erosion, coupled with a shearing process, by the current flowing over a soft muddy bottom.

Flute Rill Marks. Flute rill marks are made up of continuous, narrow, slightly meandering scours, elongated parallel to the flow direction. They are produced when scouring vortices move in filaments, coalescing and dividing during the flow.

Longitudinal Furrows and Ridges. Structures made up of closely spaced, continuous ridges alternating with furrows, arranged parallel to the current direction are shown in Fig. 2; these ridges may sometimes branch. These structures are believed to be produced by longitudinal stringers in the flow. In each stringer, fluid moves in the form of two helical spirals possessing an opposite sense of motion (*Langmuir circulation*). Stringers erode the material and pile it up in the form of longitudinal ridges.

Whitaker (1973) proposed the term *gutter cast* (properly gutter mold) for a number of scour-and-fill sole structures previously called large groove molds, elongated flute molds, parallel scour structures, priels, erosional channels, runnel molds, and *Rinnen*. He included any down-bulge on the soles of sedimentary strata, of great length (>1 m) compared with their width and depth (a few to tens of cm). In cross-section, they have the form of small channels (Fig. 3).

Pillow-like Scour Marks. Small bulbous bodies with indistinct orientation are thought to be produced as a result of the action of cell vortices formed due to current interference. (See also *Pillow Structures*, which are load structures.)

Other. A final type of scour-mark structure is that of the very flat triangular marks with ends pointing down-current. In shape, these resemble very flat triangular flute marks, but with opposite orientation.

Distribution in Environments

Various types of scour marks are abundantly found as molds on the undersurface of sandy layers overlying muddy layers. Rarely, they are also present on the upper surface of the mud layers. Scour marks have been often considered as a characteristic feature of turbidites (q.v.). They are, however, also commonly found in the sediments of other environments, e.g., fluvial, tidal flat, lagoon, deposited under the influence of ordinary currents (Reineck and Singh, 1973).

INDRA BIR SINGH

FIGURE 2. Longitudinal furrows and ridges (molds); Simla slates, India.

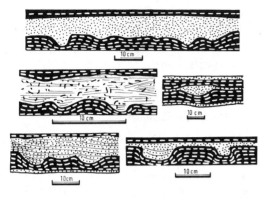

FIGURE 3. "Gutter cast" (properly, gutter mold) from the Lower Silurian of southern Norway (Whitaker, 1973, reproduced with permission of the Norwegian Geological society).

References

Allen, J. R. L., 1971. Transverse erosional marks of mud and rock: Their physical basis and geological significance, *Sed. Geol.,* **5,** 167-385.

Allen, J. R. L., 1973. Development of flute-mark assemblages, 1. Evolution of pairs of defects, *Sed. Geol.,* **10,** 157-178.

Allen, J. R. L., 1975. Development of flute-mark assemblages, 2. Evolution of trios of defects, *Sed. Geol.,* **13,** 1-26.

Dzulynski, S., and Walton, E. K., 1965. *Sedimentary Features of Flysch and Greywackes.* Amsterdam: Elsevier, 274p.

Karcz, I., 1974. Reflections on the origin of source of small-scale longitudinal streambed scours, in M. Morisawa, ed., *Fluvial Geomorphology.* Binghamton: St. Univ. N. Y., 149-173.

Kuenen, P. H., 1957. Sole markings of graded greywacke beds, *J. Geol.,* **65,** 231-258.

Reineck, H.-E., and Singh, I. B., 1973. *Depositional Sedimentary Environments.* Berlin: Springer-Verlag, 439p.

Rücklin, H., 1938. Strömungs-Marken im Unteren Muschelkalk des Saarlandes, *Senckenberg. Leth.,* **20,** 94-114.

Sanders, J. E., 1965. Primary sedimentary structures formed by turbidity currents and related resedimentation mechanisms, *Soc. Econ. Paleont. Mineral. Spec. Publ. 12,* 192-219.

Ten Haaf, E., 1959. Graded beds of the Northern Apennines, Thesis, Univ. of Groningen, 102p.

Ten Haaf, E., 1964. Flysch formations of the northern Apennines, in A. H., Bouma and A. Brouwer, eds., *Turbidites.* Amsterdam: Elsevier, 127-136.

Vassoevich, N. B., 1953. O nekotorykh flishevykh teksturakh (znakakh), *L'vov Geol. Obshch. Geol. Sbornik,* geol. ser. 3, 17-85.

Whitaker, J. H. McD., 1973. "Gutter casts," a new name for scour-and-fill structures, with examples from the Llandoverian of Ringerike and Malmöya, southern Norway, *Norsk Geol. Tidsskr.,* **53,** 403-417.

Cross-references: *Current Marks; Scour-and-Fill; Tool Marks; Top and Bottom Criteria; Turbidites.*

SECONDARY SEDIMENTARY STRUCTURES

Secondary or epigenetic sedimentary structures refer to internal and external features that were generated after deposition of the sediment (see *Sedimentary Structures*). In contrast to features that were produced by processes active within the environment at the time of deposition (see *Primary Sedimentary Structures*), secondary sedimentary structures result primarily from processes related either to physical stresses or to chemical changes that occur after deposition, e.g., compaction (q.v.) and diagenesis (q.v.). These processes may occur from hundreds to tens of millions of years after deposition. Thus, these structures give useful information only about disturbances and internal changes relating to the postdepositional history of the rock body, whether the features occurred at the original site of deposition or not.

Structures originating from physical stresses include features such as folds, faults, and cone-in-cone structure and other pressure-solution phenomena (q.v.). Caution must be exercised in interpreting some folded structures because convolute laminae can be confused with drag folds (see *Penecontemporaneous Deformation of Sediments*). Whereas the former are contemporaneous or penecontemporaneous with sedimentation, the latter are essentially postdepositional. Drag folds frequently develop within less competent layers of a large-scale fold. They are usually asymmetrical toward the fold hinge; and, because they are formed from the drag produced by relative movements of layers of different competence, they show confinement to particular strata. These strata can be very thin, and the drag structures display planar upper and lower surfaces, criteria usually used in differentiating convolute laminae. Perhaps the best criterion to differentiate convolute laminae is that they bear no relation to the larger structures or the tectonic pattern of the region.

Similar complex fold structures occur in evaporitic rocks, and are also referred to as "enterolithic structures" (Grabau, 1913), from the resemblance of the distorted layers to the convolutions of an intestine. These structures are also secondary and are produced by forces that acted from within the mass during diagenesis (see *Marine Evaporites-Diagenesis and Metamorphism*). Recrystallization and hydration are the common causes, owing to substantial increase in volume that may ensue; e.g., anhydrite changing into gypsum involves 63% increase in volume. The essential distinctive features attributable to secondary structures related to internal stresses would be that the deformation is in all directions, not in certain ones, as would have been the case in tectonic or in gliding deformations (Hahn, 1912).

Secondary sedimentary structures originating from pressure solution and/or remobilization of minerals include features such as stylolites, geodes (q.v.), vugs, septaria, loess-Kindchen, marlekor or imatra stone, chert (q.v.), concretions (q.v.), pseudomorphs (q.v.), boxwork (q.v.), corrosion surfaces (q.v.), and color banding. Color banding and rings also known as "liesegang bands" and "liesegang rings" occur randomly in a sedimentary body. Their extension is limited and confined around or at the proximity of the causing agent. They result from postdepositional diffusion or weathering processes, and are usually better developed in fine-grained and porous materials.

FLORENTIN MAURRASSE

References

Blatt, H.; Middleton, G.; and Murray, R., 1972. *Origin of Sedimentary Rocks.* Englewood Cliffs, N. J.: Prentice-Hall, 634p.
Grabau, A. W., 1913. *Principles of Stratigraphy.* New York: Seiler, 1185p.
Hahn, F. F., 1912. The form of salt deposits, *Econ. Geol.,* 7, 120-135.
Krumbein, W. C., and Sloss, L. L., 1963. *Stratigraphy and Sedimentation.* San Francisco: Freeman, 660p.
Pettijohn, F. J., 1975. *Sedimentary Rocks,* 3rd ed. New York: Harper & Row, 628p.

Cross-references: *Boxwork; Carbonate Sediments–Diagenesis; Chert and Flint; Color of Sediments; Concretions; Corrosion Surface; Diagenesis; Diagenetic Fabrics; Geodes; Loess; Nodules in Sediments; Penecontemporaneous Deformation of Sediments; Pressure Solution and Related Phenomena; Primary Sedimentary Structures; Pseudomorphs; Sedimentary Structures; Slump Bedding; Water-Escape Structures; Weathering in Sediments.*

SEDIMENTARY ENVIRONMENTS

A *sedimentary environment* is a part of the planet's surface which is physically, chemically, and biologically distinct from adjacent areas. Examples of sedimentary environments are deserts, deltas, and abyssal plains. The three defining parameters listed above include the fauna and flora of the environment, its geology, geomorphology, climate, weather, and, if subaqueous, the depth, temperature, chemistry, and current system of the water.

Modern sedimentary environments are sites of erosion, equilibrium (i.e., they are nonerosional and nondepositional), and/or deposition. As a broad generalization, subaerial environments tend to be sites of erosion, and subaqueous environments are largely areas of deposition.

Depositional Environments

A *depositional environment* is a particular type of sedimentary environment, one in which net sedimentation occurs over a span of time. The product of a depositional environment is a sedimentary *facies* (q.v.). The relationship between environments and facies are shown in Fig. 1. The recognition of depositional environments from sedimentary rocks is an important field of study which has been extensively documented (e.g., Laporte, 1968; Selley, 1970; Rigby and Hamblin, 1972). Environmental analysis of ancient sediments is important, not only as an academic pursuit. It also has diverse economic applications in the fields of water supply and in the search for fossil fuels and certain types of mineral deposits.

The interpretation of ancient depositional environments is based on the application of the principle of uniformitarianism. The essential argument used is that an ancient sedimentary facies can be defined by a particular set of variables: geometry, lithology, sedimentary structures, paleocurrent pattern, and fossils. In most cases, the parameters of an ancient sedimentary facies can be closely matched with a modern environment. Inevitably such comparisons are never perfect, but neither are the sedimentary products of two modern reefs, or two modern deltas, identical. The concept of the sedimentary model (see *Sedimentology–Models*) is implicit in environmental analysis.

The basic approach to the interpretation of depositional environments is summarized in Fig. 2.

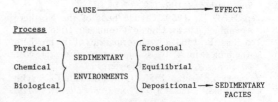

FIGURE 1. The relationship between environments and facies. A sedimentary facies is generated in a depositional environment, a particular type of sedimentary environment.

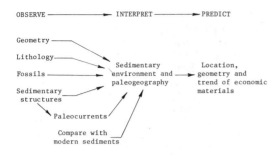

FIGURE 2. Illustrative of the basic approach to finding out how a sediment was deposited.

Classification of Depositional Environments

Numerous classifications of sedimentary environments have been proposed (e.g., Crosby, 1972; Blatt et al., 1972, p. 195). The classification of depositional environments shown in Fig. 3 is comparable to most classifications of sedimentary environments but is more restricted because it lists only those environments that have been extensively recognized in ancient sedimentary rocks. These environments have generated clearly recognizable facies. Descriptions of these environments and their sedimentary products will be found in appropriate places in this volume (see cross-references).

The following outline (after Selley, 1970) summarizes the diagnostic parameters of the products of the major depositional environments that can be recognized in sedimentary rocks.

Fluvial 1. (Alluvium of braided rivers). *Geometry:* The facies as a whole may be prismatic, fan-shaped, or blanket-shaped. Internally it will consist of a down-slope trending complex of channel shoestrings of various grades of sediment. *Lithology:* Predominantly coarse-grained clastics with conglomerates and coarse sandstones, often red-colored; rare finer sands and silts. *Sedimentary structures:* Large channels of cross-bedded and plane-bedded sandstones with occasional quicksand structures. Rare conglomerate-floored abandoned channels infilled by laminated silts with thin rippled sand units and desiccation cracks. *Paleocurrents:* Very low scatter of data at individual sample stations; regionally may describe fan-shaped arcs. *Fossils:* Very rare abraded disarticulated bones of terrestrial vertebrates and plant debris. See *Alluvial Fan Sediments; Braided-Stream Deposits*.

Fluvial 2. (Alluvium of meandering rivers). *Geometry:* Sheet; the facies may be composed of alternating sand and shale beds in units of about 1-3 m. Sand units may be composed of coalesced channel complexes or occur as discrete shoestrings enclosed in shale. *Lithology:* Coarse to fine-grained sands and shales present in about equal proportions or sand subordinate; some conglomerates and coals, occasionally predominantly red-colored. *Sedimentary structures:* Tendency for beds to be arranged in fining-upward sequences of grain size with the following structures—scoured and channelled erosion surface overlain by massive or rudely bedded conglomerate; overlain by cross-bedded, flat-bedded, and rippled sands which grade up into siltstones. These are laminated and contain desiccation cracks and thin laminated and rippled fine-sand units. *Paleocurrents:* Unimodal at a point with a wide scatter of readings, regionally arranged parallel to the paleoslope. *Fossils:* Channel sand may contain plant debris, rolled disarticulated bones of terrestrial animals and fresh-water aquatic vertebrates and shellfish. Overbank floodplain silts may contain plant debris, rootlet horizons, and coal beds. See *Alluvial Sediments; Fluvial Sediments*.

Eolian. *Geometry:* Irregular sheet. *Lithology:* Typically clean, well-sorted medium to fine sandstones; occasionally gypsum, shell-sand, or silt (loess). *Sedimentary structures:* Large-scale cross-bedding with set heights of up to 30 m. Plane bedding and rare penecontemporaneous deformation may be present. Widely spaced low-amplitude ripples. *Paleocurrents:* Dip directions of cross-bedding may be very variable at outcrop due to complex dune morphology. Regional trend of paleocurrents may be present but are not slope controlled. *Fossils:* Generally absent. Rare footprints and plant debris. Coastal dunes may contain, and be exclusively composed of, abraded marine fossils. See *Desert Sedimentary Environments; Eolian Sedimentation; Loess*.

Lacustrine. *Geometry:* Irregular sheet. *Li-*

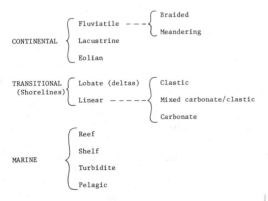

FIGURE 3. A classification of depositional environments (after Selley, 1970).

thology: Predominantly fine grained, argillaceous, calcareous, or evaporitic; coarse marginal clastics may be present. *Sedimentary structures:* Lamination, varves, turbidites, ripples, and desiccation cracks. Cross-bedding and channelling in marginal facies. *Paleocurrents:* Mainly unidirectional and centripetal to the deepest part of the lake. Along the shore, bipolar paleocurrents may indicate onshore flow. Wave-generated currents may move in any direction regardless of bottom slope. *Fossils:* Fresh-water often diverse, and very well preserved. Land plants and animals may be washed in. Fine-grained marginal facies may contain coals. See *Lacustrine Sedimentation.*

Lobate (Deltaic) Shorelines. *Geometry:* Prismatic or fan shaped. *Lithology:* Clay, silt, and sand present in varying proportions, with a tendency to be arranged in upward-coarsening sequences. *Sedimentary structures:* Shales in the lower part of a sequence (prodelta) are laminated, rarely rippled. Central part of the sequence may have fans or channels filled with turbidite sands. Slumps and slides may be present. Overlying delta platform consists of a radiating complex of cross-bedded channel sands with intervening interlaminated, laminated, rippled, and bioturbated clay, silt, and very fine sand. *Paleocurrents:* Unimodal, may be regionally radiating both in the slope turbidites and channel sands. *Fossils:* Marine fauna in lower fine-grained part of sequence; brackish and fresh-water fossils present in upper part together with plant debris, rootlet beds, and coals. See *Deltaic Sediments; Prodelta Sediments.*

Clastic Linear Shorelines. Barrier-island and shoal complexes consist of four vertically or laterally juxtaposed facies: alluvial, lagoonal and tidal flat, barrier island, and open marine. The first of these has been described already, the last is described in the next section. The two intermediate facies will now be described separately. See *Barrier Island Facies; Continental-Shelf Sedimentation.*

Lagoon and Tidal-Flat Complex. *Geometry:* Sheets or shoestrings parallel to the paleostrike. *Lithology:* Predominantly clays, silts, and fine sands. *Sedimentary structures:* Laminated, rippled, and bioturbated. Rarely with desiccation cracks and conglomerate-lined channels infilled with obliquely inclined shale due to tidal gullies. *Paleocurrents:* Extremely variable, little to measure. *Fossils:* Vertebrates and invertebrates ranging from marine, through brackish to fresh-water. Shell reefs (especially of oysters) are particularly characteristic. See *Lagoonal Sedimentation; Tidal-Flat Geology.*

Barrier Island Complex. *Geometry:* Sheets or shoestrings parallel to the paleostrike. *Lithology:* Coarse, medium, fine, and very fine sands. Tendency to be matrix free, and well sorted. Either terrigenous (predominantly quartzose) or calcarenitic. *Sedimentary structures:* For a regressive barrier, the following sequence of structures and grain size may be present—transitional base with laminated argillaceous open-marine (sub-wave base) facies, grading up into interlaminated, rippled, and burrowed argillaceous silt and very fine sand. This in turn passes up into fine and medium sands with regular parallel stratification, horizontal or gently seaward dipping. Rare troughs and tabular planar cross-beds. Occasional channels (scoured by tidal currents surging in and out through gaps in the barrier). Abrupt upper contact with lagoonal facies. For a transgressive barrier, the sequence previously described is reversed. *Paleocurrents:* Cross-beds dip predominantly onshore, sometimes bipolar. Bipolar pattern typical of tidal channels. *Fossils:* Bioturbation well developed in transitional zone between barrier sand and open marine facies. Sand itself contains, and may be largely composed of, shallow marine benthonic fossils. See *Barrier Island Facies; Tidal-Current Deposits; Tidal-Inlet and Tidal-Delta Facies.*

Carbonate Shorelines and Shelves. *X Zone—open-marine sub-wave base:* Shales, argillaceous, and siliceous limestones, thinly bedded and laminated. Intraformational conglomerate-covered penecontemporaneous erosion surfaces may be present (hardgrounds, see *Hardground Diagenesis*). Well-preserved fauna of pelagic and, rarer, benthonic organisms. Algae absent. *Y Zone—Barrier Island Complex,* see above. *Z Zone—Lagoon and sabkha:* Fecal pellet limestones, occasionally laminated, desiccation cracked, and with intraformational conglomerates. Often burrowed, with sparse well-preserved, marine fauna. Interbedded with, and pass shoreward into, microcrystalline dolomite with laminae and nodules of anhydrite (may be replaced by gypsum or calcite and other evaporite minerals). See *Bahama Banks Sedimentology; Sabkha Sedimentology; Tropical Lagoonal Sedimentation.*

Reef. *Geometry:* In plan linear, subcircular or, rarely, atoll-shaped. *Lithology:* Three lithofacies generally recognizable—(1) calcilutites, calcarenites, and pellet limestones (back-reef lagoon); (2) biolithite, i.e., in-situ skeletons of calcium carbonate-secreting organisms that may be completely recrystallized, often dolomitized, obliterating original fabric (*reef core*); (3) skeletal calcarenites and calcirudites with micrite matrices (*reef talus*). Back-reef facies may grade into sabkha-type evaporites. Reef front may be laterally equivalent to "basinal" evaporites. *Sedimentary structures:* Back-reef facies: laminated, rarely bioturbated and with desiccation cracks. Reef core—massive. Reef talus—poorly developed bedding dipping off reef front; slides and slumps. May grade basinward into carbonate turbidites. *Paleocurrents:* Seldom anything to measure. Slumps, slides, and turbidites may indicate paleoslope often of only local (reef-generated) significance. *Fossils:* Reef core (if not extensively recrystallized) characterized by an abundant fauna with large populations of many species. Reef framework formed of calcareous algae, stromatoporoids, bryozoa, and corals, associated with many other sessile and mobile organisms. See *Coral Reef Sedimentology; Reef Complex.*

Flysch. (*Marine? turbidite deposits.*) *Geometry:* Sheets, fans, or channels. *Lithology:* Alternations of sand and shale; sands very variable in composition ranging from skeletal shell sands, through protoquartzites to graywackes, which are probably most typical, especially in older (pre-Tertiary) examples. *Sedimentary structures:* Typical bedding aspect shows monotonous alternations of sands and shales. Sand units seldom > 3 m, generally < 0.5 m. Bases of sands show diverse suite of erosional structures; graded internally often with a set sequence of structures from base to top (see *Bouma Sequence*). Slides, slumps, and channels may also be present. *Fossils:* Shales may contain pelagic fossils, possibly of deep water aspect. Sands contain abraded shallow-water benthonic fossils and, sometimes, plant debris. See *Graywacke; Submarine Fan (Cone) Deposits; Turbidites.*

Pelagic. *Geometry:* Often hard to determine because this facies is typical of tectonically disturbed mountain chains. Typically occurs in thin sequences above shelf carbonates and below flysch facies. *Lithology:* Calcilutites, sometimes red and nodular; red clays, highly ferruginous and manganese rich, radiolarian cherts. *Sedimentary structures:* Thinly bedded and laminated. Occasionally with ripples and "hardground" horizons. *Paleocurrents:* Seldom anything to measure. *Fossils:* Characterized by an absence of benthos, though some burrowing may be present. Typically with a sparse but well-preserved pelagic fauna of radiolaria, foraminifera, thin-shelled mollusks, and marine vertebrates. Tendency for fossils to show penecontemporaneous corrosion and preservation of shells of original calcitic composition, over those originally aragonitic. See *Pelagic Sedimentation, Pelagic Sediments.*

RICHARD C. SELLEY

References

Anstey, R. L., and Chase, T. L., 1974. *Environments Through Time.* Minneapolis: Burgess, 136p.
Blatt, H.; Middleton, G.; and Murray, R., 1972. *Origin of Sedimentary Rocks.* Englewood Cliffs, N.J.: Prentice-Hall, 634p.
Crosby, E. J., 1972. Classification of sedimentary environments, *Soc. Econ. Paleont. Mineral. Spec. Publ. 16,* 4-11.
Laporte, L. F., 1968. *Ancient Environments.* Englewood Cliffs, N. J.: Prentice-Hall, 116p.
Reineck, H. E., and Singh, I. B., 1973. *Depositional Sedimentary Environments.* New York: Springer Verlag, 439p.
Rigby, J. K., and Hamblin, W. K., eds., 1972. Recognition of ancient sedimentary environments, *Soc. Econ. Paleont. Mineral. Spec. Publ. 16,* 340p.
Selley, R. C., 1970. *Ancient Sedimentary Environments.* Ithaca, N. Y.: Cornell Univ. Press, 237p.

Cross-references: *Abyssal Sedimentary Environments; Alluvial-Fan Sediments; Bahama Banks Sedimentology; Bay of Fundy Sedimentology; Black Sea Sediments; Clay Sedimentation Facies; Continental Sedimentation; Desert Sedimentary Environments; Euxinic Facies; Evaporite Facies; Facies; Fluvial Sediments; Geosynclinal Sedimentation; Glacigene Sediments; Hadal Sedimentation; Kara-Bogaz Gulf Evaporite Sedimentology; Limestones; Littoral Sedimentation; Marine Sediments; Neritic Sedimentary Facies; Pelagic Sedimentation, Pelagic Sediments; Persian Gulf Sedimentology; Reef Complex; Sands and Sandstones; Sedimentation—Paleoecologic Aspects; Sedimentology—Models; Shark Bay Sedimentology; Volcaniclastic Sediments and Rocks.* See also cross-references in last section.

SEDIMENTARY FACIES AND PLATE TECTONICS

Prior to the establishment of the theory of plate tectonics in the late 1960s, many geologists had related sedimentary facies and associations of facies, in particular those of geosynclines, to tectonic environments. The American viewpoints are well summarized by Krumbein and Sloss (1963) and Pettijohn (1957), and the traditional European approach is exemplified by Aubouin (1965). In most of these studies, geosynclinal sedimentation was considered within the confines of ancient "geosynclines" and "orogenic belts" with little reference to either modern sediments or to situations in which similar rocks accumulate today. For example, words like *preflysch, flysch,* and *molasse* were coined to describe the sedimentary facies of ancient geosynclines with only rare attempts to discover their modern equivalents (see *Geosynclinal Sedimentation*).

Those who attempted to use a uniformitarian or actualistic approach included Dietz (1963), who suggested that the present Atlantic margin of North America could be used as a model for ancient geosynclinal sedimentation, and many Dutch authors who followed Haug in advocating that the deep-sea trenches associated with island arcs were modern geosynclines (Kuenen, 1950).

The advent of plate tectonics gave a tremendous impetus to this actualistic approach. Mitchell and Reading (1969) and Dewey and Bird (1970) showed how many sedimentary and tectonic features of ancient geosynclinal and orogenic belts could be explained by comparison with modern plate and continental margins, in particular indicating where sediments similar to *preflysch, flysch,* and *molasse* might be accumulating today. The symposium edited by Dott and Shaver (1974) summarized and expanded on this approach (see introduction by Dott for a discussion of actualism).

Plate Tectonics

The lithosphere consists of a number of rigid plates, separated from each other by junctions which are of three types: (1) *Divergent,* as at mid-oceanic ridges, where two lithospheric plates are moving apart. Both plate margins are growing with the addition of new lithosphere by igneous activity. (2) *Convergent,* as at oceanic trenches, where one lithospheric plate descends beneath another along a Benioff (subduction) zone. Here the descending plate is being destroyed as the overriding plate is growing by a combination of igneous activity and the tectonic accumulation of material from the descending plate. (3) *Strike-slip,* where two plates move laterally without substantial divergence or convergence. Both plate margins are conserved as lithosphere is being neither created nor destroyed.

Some continental margins, for example the western margin of South America, coincide with plate margins and are active. Others, such as those of the present Atlantic, are not linked to plate margins, and are passive.

Consideration of sedimentary facies in relation to plate tectonic theory demonstrates how the accumulation of various facies depends on their position relative to plate or continental margins or their position within plates (Fig. 1). In addition, some facies are more likely than others to be preserved.

Divergent Plate Junction Sedimentary Environments

Continental (Red Sea-Type) Environment. An early stage of a divergent plate junction is exemplified by the Red Sea (Fig. 1A). Here, a sequence of arching, compound rifting, and separation of continental plates has occurred. The early stages of arching and rifting were accompanied by basaltic igneous activity. Drainage was directed away from the rift by the elevation of its margins, and sedimentation was limited to fault-bounded troughs in which lakes subsequently became evaporitic basins (Fig.1A). Thus the sedimentary facies associated with an initial continental breakup in an arid tropical zone are lacustrine, piedmont fan, and thick evaporitic successions, deposited in fault-bounded basins between blocks of continental basement associated with basaltic volcanism.

Ancient examples are the Carboniferous to Triassic horst and graben structures and associated continental red beds and basalts that occur around the North Atlantic. They are well seen in the Canadian Maritimes and eastern Appalachians.

Oceanic Environment. At the crest of oceanic ridges, new crust is being generated by the formation of igneous and metamorphic rocks, mostly basalt, gabbro, peridotite, serpentinite, and greenschists. These rocks become blanketed by pelagic sediments (q.v.) composed either of abyssal clay, derived from fine-grained terrigenous detritus and pyroclastic material, or of biogenic ooze.

The organisms are either calcareous or siliceous. At the compensation depths, first calcareous and then siliceous organisms are dissolved so that in deeper water no calcareous or siliceous organisms are found on the sea floor (see *Pelagic Sedimentation*). At the relatively shallow depth of some oceanic ridges, carbonate oozes are forming. As the crust moves away from the ridge it subsides, and successive layers of siliceous ooze and abyssal clay accumulate. If the crust moves across the zone of equatorial upwelling, where high productivity raises the compensation depths, the sequence may reverse, provided the depth is no greater than about 5000 m.

Where there is a supply of clastic material on the ridge, turbidites become the dominant sediment type. For example, on the mid-Atlantic ridge, very fine-grained turbidites collect in small ponded depressions; and in the adjoining abyssal plains, coarse, graded sandstone turbidites accumulate (Fig. 1B). In the Bay of Bengal, a vast pile of turbidites forming the Bengal Delta Fan extends over oceanic crust some 2000 km from its source.

Oceanic sediments are rare as sedimentary rocks because most are lost by subduction. Occasionally, small portions have been uplifted and appear within mountain belts,

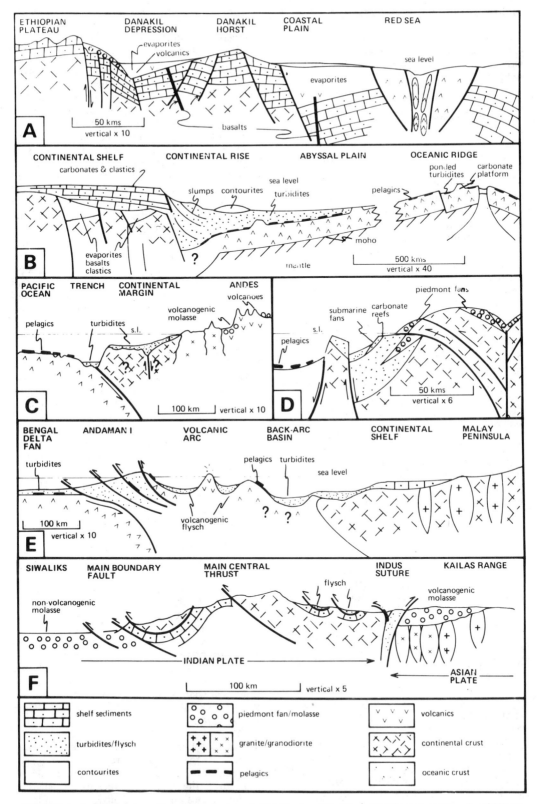

FIGURE 1. A: Divergent continental environment; Red Sea (modified from Hutchinson and Engels, 1972). B: Oceanic and passive continental margin environment; Atlantic (modified from Dewey and Bird, 1970). C: Convergent continent-ocean environment; Peruvian Andes (modified from Scholl et al., 1970). D: Strike-slip environment. E: Island arc and back-arc environment; Andaman Sea (modified from Rodolfo, 1969). F: Collision environment; Himalayas (modified from Gansser, 1964.)

e.g., the Alps, as *preflysch* pelagic shales, carbonates, and cherts associated with the basic and ultra-basic igneous rocks known as ophiolites. Thick turbidite accumulations may also be scraped off and emplaced above the subduction zone to form *flysch* belts.

Passive Continental Margin (Atlantic-Type). As a rift zone of Red Sea-type develops, the continents become separated by a widening ocean, and a passive continental margin forms. On the continental rise, which, to a limited extent, is building out across ocean floor, one of the most important depositional agents is thermohaline ocean bottom currents that flow along the rise parallel to the shoreline. These currents deposit fine sands (see *Continental-Rise Sediments*), known as contourites (q.v.). They not only form a large part of the continental rise but may also build a huge body of sediment, away from the continental margin, e.g., the "outer ridge" off Florida (Fig. 1B).

On the continental slope, sediments are rather unstable; and, although normally continuous sedimentary draping of the slope may take place, at intervals large masses of sediment slump downslope to give a confused pattern of rotational slump faults. In addition, there is some deposition by turbidity currents associated especially with submarine canyons [see *Submarine (Bathyal) Slope Sedimentation; Submarine Fan (Cone) Deposits*].

On the continental shelf (q.v.), evaporites, shallow-water carbonates, and clastic sediments form. Substantial thicknesses, up to 15,000 m, indicate great subsidence of the continental margin. Off South America large deltaic accumulations also occur.

If the continental margin were to continue to build out, a vertical stratigraphical succession would consist of (from bottom to top): oceanic crust; pelagics and fine turbidites, coarse turbidites and pelagics; contourites, slumps and turbidites; shelf sediments. Little is known about the transitional zone between continental and oceanic crust, however, and at some point, as the continental margin is approached, Red Sea-type facies occur beneath younger sediments.

An Atlantic margin and its associated sedimentary facies can be uplifted and exposed at the surface only when it changes to a convergent or strike-slip plate junction. Thus, ancient sedimentary facies of Atlantic type occur in some early stages of geosynclines, such as in the Cambro-Ordovician of the Appalachians, where an orthoquartzite/carbonate belt passes eastward into a *flysch* belt of mass-flow and turbidite deposits.

Convergent Plate Junction Sedimentary Environments

Continent/Ocean (Andean-Type) Environment. Where an oceanic plate underrides a plate with continental crust, a trench such as the Peru-Chile trench and adjacent mountain belt develops. In this environment, sedimentation is taking place in three main belts—the mountain range, the coastal plain and submarine slope, and the submarine trench.

In the Peruvian Andes, block faulting and uplift result in rapid erosion of Mesozoic and older metamorphic rocks, Mesozoic and Early Tertiary andesitic and tonalitic igneous rocks, and later andesitic and silicic volcanoes. Coarse detritus forms conglomeratic piedmont fans, and rock-slide and fluvial deposits. Some of this material accumulates in local grabens, and some is transported beyond the mountain range; in Peru and Chile much of it is deposited in a longitudinal valley or intermontane plateau. This thick succession of continental deposits has been termed *molasse* (q.v.), but it differs from the comparable Alpine molasse by its volcanogenic nature (Fig. 1C).

The coastal plain of Peru and Chile, consisting of alluvial deposits mantling metamorphic bedrock, passes westward into a nearshore sedimentary shelf. Beyond the shelf, a belt up to 100 km wide includes sediment-draped slopes and turbidite-filled, fault-controlled troughs, which trap sediment and prevent it from reaching the trench.

The inner trench slope is usually steep and largely bare of sediment. Within the trench, the sedimentary succession, consisting of pelagic sediments and turbidites, varies in thickness from a few hundred m to more than a km. These deposits are mostly carried under the continental margin by the underthrusting lithosphere.

At other Andean-type junctions, where considerable thicknesses of terrigenous sediment reach the ocean floor and are carried into the trench, underthrust slices may be tectonically emplaced above and on the landward side of the trench, forming thick belts of deformed *flysch* and ocean floor *preflysch* pelagic facies. Examples are the Tertiary Mentawai Islands west of Sumatra and the Late Mesozoic Kodiak Shelf of Alaska (see *Hadal Sedimentation*).

The presence of these elevated *flysch* belts results in sedimentation in the zone between them and the volcanic arc. Examples occur between the Mentawai Islands and Sumatra, and form the Cenozoic fluvial and shallow marine facies in the Central Lowlands of Burma. Thus an entire arc-trench gap succession in an Andean-type setting would include forearc-basin deposits beside the arc and (seaward) an accretion wedge of trench-slope basin and trench deposits, part or all of the succession underlain by subduction complex (melange).

Ocean/Ocean (Island Arc-Type) Environment. Three main sedimentary environments can be recognized in island arcs—the flanks of the volcanic arc, the zone between the trench and volcanic arc, and the submarine trench. Sedimentation is most rapid in and around the active volcanic arc. Steep volcanic slopes and frequent earthquakes result in volcanic mudflows and rock slides; part of this material accumulates on the lower slope of stratovolcanoes, but much of it is carried into deep water either directly as submarine mass flows, or subsequently as a result of river and wave erosion. At low latitudes, carbonate reefs supply talus to the adjacent ocean floor. These processes result in thick sedimentary aprons of volcanogenic turbidites and other mass-flow deposits interbedded with pyroclastics, mudstones, and limestones. In the more complex arcs, for example Japan, these sediments include plutonic and metamorphic detritus derived from continental fragments within arcs. Examples of ancient volcanic arcs are known from the British Lake District and Newfoundland.

Sedimentation in the zone between the trench and volcanic arc broadly resembles that in the similar zone in Andean-type environments. Although in some arcs the sedimentary belt is very narrow, in others, e.g., the Aleutian, Mariana and Tonga arcs, the zone is occupied by an irregular seaward-dipping slope with local troughs, blanketed in turbidite and pelagic sediments.

Trench sediments, like those in the trench bordering the Andes, are mostly subducted. However, where large volumes of sediment, mostly continent-derived, reach the trench or adjacent ocean floor they may be preserved as deformed *flysch* belts, forming, e.g., the Andaman and Nicobar islands in the Bay of Bengal (Fig. 1E) and the island of Barbados in the Lesser Antilles.

Back-Arc Basin Environment. Behind most island arcs there is a marine basin which separates them either from another arc or from a continental margin. Some of these basins have resulted from subsidence; many have probably formed largely by the outward movement of the island arc and the creation of crust similar to oceanic crust. Depending on the location of the arc two main types of sedimentary basin are found: interarc basins and arc-continent basins.

Interarc basins, for example the Lau Basin between the Tonga and Lau arcs, and the Mariana Trough, are characteristically bordered by steep scarps and have a highly irregular floor with only a thin sedimentary cover.

Arc-continent basins (Fig. 1E) are present around the western margin of the Pacific. Near the island arc, volcanic-arc apron facies form a thick sedimentary wedge, and in the center of the basin a relatively thin sedimentary layer consists largely of pelagic material. Adjacent to the continental margin there is a thick succession commonly similar to that found on passive continental margins with a prism of *flysch* passing shoreward into shallow marine deposits. Because the basins are isolated from oceanic currents, the sediments on the continental margin undergo less redistribution than on Atlantic-type margins, and thick accumulations may build out, as in the Yellow Sea.

Strike-Slip Plate Junction Sedimentary Environments

Associated with major strike-slip junctions are numerous blocks that move vertically relative to each other and lead to a complex pattern of rapidly sinking troughs and uplifted massifs (Fig. 1D). Examples of this type of junction are the San Andreas fault zone—with its associated mountains and valleys and the Channel Island area of small basins off southern California—and E-W trending margins of the Caribbean plate. Differential vertical movement may have been as much as 10 km since the Early Pliocene.

At these junctions, there is an absence of igneous activity, and metamorphism is negligible. Sedimentation, on the other hand, is extremely rapid and variable over short vertical and lateral distances. The result is very thick successions of coarse conglomeratic piedmont fans, which pass seaward, through a narrow shelf zone where shallow-marine sandstones and limestones may accumulate, to the mass-flow and turbiditic deposits of deep-sea fans. Some basins, such as the distal Californian ones, may be starved of clastic material so that only pelagic sediments accumulate. Because of rapid changes in vertical movement, breaks in sedimentation are common, but these are not laterally persistent.

One example of the sedimentary facies associated with a strike-slip junction is that of the Upper Carboniferous of the Cantabrian Mountains of northern Spain, where a total thickness of nearly 10 km of very variable sediments ranging from turbidites and slide blocks to coastal and fluvial facies is developed as a series of sequences separated from each other by angular unconformities.

Continental Collision-Type Junction Sedimentary Environments

Vertical movements resulting from collision between two continents lead to accumulation of thick conglomeratic piedmont fans and fluvial deposits generally known as *molasse*. *Molasse* in this sense may be of two main types—volcanogenic and nonvolcanogenic. Nonvolcanogenic continental deposits form a thick sedimentary prism on the Swiss Plain, the site of the classical *molasse*. The succession, of Late Cenozoic age, occupies a "foredeep" on the northern or underthrusting continent immediately in front of folded rocks and nappe structures. Similar Late Cenozoic *molasse* facies comprise the Siwalik succession S of the Main Boundary Fault in the Himalayas (Fig. 1F). In both the Alps and the Himalayas the nonvolcanogenic *molasse* is a late synorogenic and postorogenic facies.

Volcanogenic continental deposits in collision belts occur as a synorogenic facies and are also known as *molasse*. The facies is poorly developed in the Alps, but forms a major stratigraphic unit on the overriding continental plate in the Tibetan Himalayas. Thick conglomerates and boulder beds of Early Tertiary age consist entirely of acidic to intermediate volcanic and plutonic detritus derived from a magmatic arc. The tectonic setting in which the Himalayan volcanogenic *molasse* accumulated was thus analogous to that of the Andean *molasse*.

An example of ancient volcanogenic *molasse* is the Old Red Sandstone of the Midland Valley of Scotland, derived from volcanic rocks related to northward subduction. In Wales, the Old Red Sandstone forms a nonvolcanogenic *molasse* facies which accumulated on continental crust of the subducting plate.

Facies and Tectonic Environments

The use of present-day sedimentary models indicates that facies of ancient mountain belts, known as *preflysch*, *flysch*, and *molasse*, may have formed in several environments. *Preflysch* has mainly formed either within an oceanic environment or within a back-arc basin. A similar but not identical facies, because it lacks volcanics, may have formed in the distal basins of a strike-slip junction.

Flysch may have formed on oceanic crust, e.g., the Bengal Fan, on an Andean margin, around an island arc, or near a strike-slip junction. *Flysch* may also have formed as the clastic wedge of a passive continental margin, although this is sometimes referred to as a *flyschoid facies*, because it is not associated with orogeny.

Molasse is a continental facies related to convergent plate margins; it may be either volcanogenic, forming on the overriding plate, or nonvolcanogenic, deposited on the underriding plate following collision. A sedimentologically similar, nonvolcanogenic *molasse* may also form close to a strike-slip junction.

Although all these facies may have formed in widely separated environments they may be tectonically brought together as the result of both subduction and continental collision.

A. H. G. MITCHELL
HAROLD G. READING

References

Aubouin, J., 1965. *Geosynclines*. Amsterdam: Elsevier, 355p.

Baker, B. H.; Mohr, P. A.; and Williams, L. A. J., 1972. Geology of the eastern rift system of Africa, *Geol. Soc. Am. Spec. Pap. 136*, 67p.

Dewey, J. F., and Bird, J. M., 1970. Mountain belts and the new global tectonics, *J. Geophys. Research*, 75, 2625-2647.

Dietz, R. S., 1963. An actualistic concept of geosynclines and mountain building, *J. Geol.*, 71, 314-343.

Dott, R. H., Jr., and Shaver, R. H., eds., 1974. Modern and ancient geosynclinal sedimentation, *Soc. Econ. Paleont. Mineral. Spec. Publ. 19*, 380p.

Fischer, A. G., and Judson, S., 1975. *Petroleum and Global Tectonics*. N. J.: Princeton Univ. Press, 322p.

Gansser, A., 1964. *Geology of the Himalayas*. London: Interscience, 289p.

Hutchinson, R. W., and Engels, G. G., 1972. Tectonic evolution in the southern Red Sea and its possible significance to older rifted continental margins, *Geol. Soc. Am. Bull.*, 83, 2989-3002.

Kasbeer, T., 1973. Bibliography of continental drift and plate tectonics, *Geol. Soc. Am. Spec. Pap, 142*, 96p.

Kellogg, H. E., 1975. Tertiary stratigraphy and tectonism in Svalbard and continental drift, *Bull. Am. Assoc. Petrol. Geologists*, 59, 465-485.

Krumbein, W. C., and Sloss, L. L., 1963. *Stratigraphy and Sedimentation*. San Francisco: Freeman, 660p.

Kuenen, P. H., 1950. *Marine Geology*. New York: Wiley, 568p.

Mitchell, A. H., and Reading, H. G., 1969. Continental margins, geosynclines and ocean floor spreading, *J. Geol.*, 77, 629-646.

Pettijohn, F. J., 1957. *Sedimentary Rocks*, 2nd ed. New York: Harper & Row, 718p.

Rodolfo, K. S., 1969. Bathymetry and marine geology of the Andaman Basin, and tectonic implications for Southeast Asia, *Geol. Soc. Am. Bull.*, 80, 1203-1230.

Rona, P. A., 1973. Relations between rates of sedimentation on continental shelves, sea-floor spreading, and eustasy inferred from the central North Atlantic, *Geol. Soc. Am. Bull.*, 84, 2851-2872.

Schneider, E. D., 1972. Sedimentary evolution of rifted continental margins, *Geol. Soc. Am. Mem. 132*, 109-118.

Scholl, D. W.; Christenson, M. N.; Von Huene, R.; and Marlow, M. S., 1970. Peru-Chile trench sediments and sea-floor spreading, *Geol. Soc. Am. Bull., 81*, 1339-1360.

Schwab, F. L., 1974. Ancient geosynclinal sedimentation, paleogeography, and provinciality: A plate tectonics perspective for British Caledonides and Newfoundland Appalachians, *Soc. Econ. Paleont. Mineral. Spec. Publ. 21*, 54-74.

Cross-references: *Geosynclinal Sedimentation; Sedimentary Environments, Sedimentation-Tectonic Controls.*

SEDIMENTARY MINERALS—
See Vol. IVB

SEDIMENTARY PETROGRAPHY

Sedimentary petrography is the systematic description of sediments, implying both their lithology and mineralogy and, by interpretation of these functions in a wider sense, their genesis. This science is essentially the study of these rocks *in the laboratory,* embracing all consolidated and incoherent sediments composing the earth's crust, irrespective of their mode of origin. Research into the natural history of sediments has become the province of *sedimentology* (q.v.). Chronological phases of geology, apart from paleontological evidences throughout the record, are enlightened and enriched by contemporary sedimentological investigations of worldwide pursuit and significance; and increasingly important contributions made by sedimentary petrography to this relatively new discipline are today universally recognized. It was, however, not always thus, as a review of the history of the subject reveals (Milner, 1962).

General History

In 1815, appeared the first general geological map of England and Wales, by William Smith. His identification and segregation of sedimentary rocks of varying geological age led to the concept of correlation and delineation of the strata cropping out across the country. Georges Cuvier, working in the Paris Basin, also contributed to the development of stratigraphy. Before the close of the 19th century, the general stratigraphy of Britain, Europe, North America, parts of Asia, and elsewhere, was well established. Curiously enough, despite the detailed work and prolific publication of geological maps and memoirs, little attention was paid to mineralogical and physical constituents of sediments for their own sake. Fossils were of paramount, host rocks of secondary, importance. A simple, descriptive nomenclature was developed, e.g., clay, coal, limestone, sandstone, etc., preceded by stratigraphic formation group names that were assigned in the field to help establish position on the geological time scale and as labels for the geographical distribution of outcrops.

It was not until 1879 that an intensive petrographic study of sedimentary rocks first appeared. In that year, Sorby wrote about the structure and origin of limestones and then, in 1880, of noncalcareous stratified rocks. These two papers pioneered the techniques of sedimentary petrography (see Folk, 1973). By contrast, petrology of igneous rocks was by this time a well-established branch of geology. Even with Sorby's work as a pointer, it was not in fact until some twenty years later that sedimentary petrography really came into its own (Bonney, 1900). Bonney's work was the start of intensive study of detrital sediments, followed by important contributions by Thomas (1902, 1909). These two papers firmly established the role of the heavy mineral in petrographic studies of sediments, a cult destined to attract many geologists, not only in Britain, but equally in the United States and W Europe.

This new technique focused attention on a far-reaching potential for studies of the wider aspects of geologic history, e.g., provenance (q.v.) and paleogeography. It also revitalized the whole concept of sedimentary petrography and gave to this comparatively new science a methodology certainly needed at the time to speed its evolution.

The first systematic account in the English language of the detrital heavy minerals in sediments appears to have been by T. Crook, in an appendix to a volume by Hatch and Rastall (1913) on the petrology of sedimentary rocks, a unique textbook at the time. In Harker's (1895) petrology text, which was devoted mainly to igneous rocks, there was only minor consideration given for sedimentary types, and detrital heavy minerals received only a casual mention.

Another revolutionary change in sedimentary petrographic studies was in the making, produced in Britain by exigencies of World War I (1914-1918) and its serious economic consequences. Shortages of essential raw materials, e.g., glass, foundry, moulding, and refractory sands previously imported from Europe, had to be met by exploitation of indigenous resources. Boswell (1916; for the Ministry of

Munitions) played an important role in revealing the domestic resources of sands, etc., that had previously been neglected, in a work on the application of petrologic quantitative methods to stratigraphy; he prophetically summed up the academic possibilities of sedimentary petrography at that time.

At this exciting stage in history, the search for oil, often in remote territories, also attracted geologists trained in sedimentary petrography. The geology of petroleum, both its surface exploration and the elucidation of subsurface conditions, was a relatively well-established science in North America by World War I. Those rather rough-and-ready field techniques were not so effective or adaptable, however, in the investigation of the complex and deformed fossil-poor sequences in oil-bearing regions such as Trinidad, Venezuela, Rumania, the Middle East, and the East Indies. In 1916, Illing used detailed heavy-mineral analysis to unravel the stratigraphy and structure of highly disturbed oil-bearing sediments in southern Trinidad; his successful recognition of key horizons characterized by indicative heavy minerals was an original approach to solving problems of correlation and differentiation of detrital deposits hitherto defying solution. Boswell's contemporary work on industrial sands, e.g., in the London Basin (1915), also showed that the identification of detrital sediments, especially in younger rocks, was facilitated by distinctive heavy mineral assemblages. These petrographic correlation methods were at first confined to unconsolidated deposits but later embraced lithified sediments. By 1920, they had become a vogue, although not without raising questions of validity. The sedimentary petrographic technique for stratigraphic correlation, *cautiously applied*, has nevertheless stood worldwide tests over the years, and is today universally acknowledged.

The first textbook devoted especially to the use of sedimentary petrography in stratigraphic correlation was by Milner (1922a). In that year also he confirmed the value of these methods in correlating scattered Pliocene outliers in the West of England (Milner, 1922b). In two decades, 1920-1940, the intensity of research on sedimentary rocks rose in a remarkable way, and from world-wide sources; the science became truly international, but mainly concentrated in Britain, Europe, and the United States (full bibliography in Milner, 1962).

In Britain, original work by Boswell continued at an accelerating rate, covering mineralogical aspects with field and laboratory studies of a high order (see Boswell, 1933). Brammall and Harwood (1923) initiated a fundamental petrologic program on the Dartmoor Granite (SW England), profoundly influencing concepts of provenance (q.v.) and significance of sedimentary rock-minerals derived therefrom; their work was followed by a masterly paper by Groves (1931). Double's papers (1924-1927) on petrography of various ancient formations are noteworthy for their detail of sedimentary mineral constitution. One of the revered names in Scottish sedimentary petrography is that of Mackie (1897), who from 1896 onward wrote a succession of papers on heavy minerals in various Scottish rocks; each paper was a perfect little study in itself, going beyond mere description to theories of their wider implication, e.g., as climatic indicators at time of formation. Smithson's first studies were of sedimentary petrography; he did pioneer work on the application of statistical methods (1939). Also in this period, Woolridge (1924, 1927), a versatile and original thinker, did regional petrographic studies that are models of their kind.

In continental Europe during this important 1920-1940 period, there was comparable activity to which a voluminous literature in many different languages bears testimony. Reference to this international work of the highest quality can only be very briefly noted here (detailed references in Milner, 1962). Arkhangel'skii (1927) studied the Black Sea sediments in detail and discussed their importance for the study of sedimentary rocks. Artini, well known for his work on detrital mineralogy, wrote much on Italian sediments (e.g., 1926) following pioneering work by Chelussi (1924). Clerici (1924) wrote a useful bibliography of Italian studies in sedimentary petrology.

In France, Lucien Cayeux, one of the greatest pioneers in sedimentology, produced a series of masterpieces; his published work on sedimentary rocks dates back to 1891, but he is best known for his general textbooks on sedimentary petrography and his series on the sedimentary rocks of France. Deverin (1924) contributed a general study on the lithology of sedimentary rocks, and Doyen (1926) dealt especially with the heavy minerals.

A series of particularly stimulating papers were those of Edelman (e.g., 1938) describing his research on sediments in the Netherlands. Baak (1936) produced a regional petrology of the southern North Sea, a trail blazer for this important continental shelf region. In Germany, Müller (1931) worked on quantitative mineralogy of Tertiary sediments in the Hamburg region; his fundamental paper on sedimentary petrography was published in 1933.

In the United States, the two decades 1920-1940 represented an era of unprecedented activity by sedimentary petrologists. Their impetus to advancement of knowledge of this science at this time was internationally acclaimed; even the briefest survey of the literature makes impressive reading, if only to illustrate the wide range of topics covered (see bibliography in Milner, 1962). Foremost in the early days was Goldman (1923), well known for his works on the Catahoula Sandstone of Texas (1915). Bramlette made notable studies on heavy minerals (e.g., 1934); Dryden's (1935) successful application of statistical methods to heavy mineral correlation was a timely directive.

Krumbein wrote a remarkable series of original papers during a period of many years (1932 on), notably on the physical characteristics of sedimentary particles on a statistical basis. Martens is recalled for his extensive and illuminating petrographic investigations of beach sands (e.g., 1935). Tester's (1931) work on particle and pebble shapes is also noteworthy. Tickell's (1924) analysis of the correlative value of heavy minerals was followed by his work on the examination of fragmental rocks (1939).

A key role in the United States at this period was played by the work of the Committee on Sedimentation (1930-1934) of the National Research Council in Washington (see notes in *Sedimentology–History*). Pettijohn's detailed work on sediments during the 1930s, much of it in connection with the Committee, was marked by Krumbein's and his well-known and very useful *Manual of Sedimentary Petrography* (1938). Wadell (1932, 1935) lucidly discussed shape of rock particles, sedimentation formulas, etc. (see *Roundness and Sphericity*). In similar context, Wentworth's systematic work on mechanical analysis of sediments, also on the controversial issues of terminology (e.g., 1922), were always original approaches to these and other problems. This brief summary would be incomplete without mention of the first edition of *Data of Geochemistry* by Clarke (1924), a brilliant compilation to which geochemists, mineralogists, and petrographers alike still turn for information and references.

After World War II (1939-1945), up to the present, a phase of *consolidation* can be discerned in the classical science of sedimentary petrography. Exceptions are the revolutionary developments in the rather special field of clay mineralogy and its applications to soil studies. Early work on clay constitution and minerals was done, e.g., by Gruner (1935) who worked on crystal structures of kaolinite, dickite, nacrite, vermiculites, nontronites, and smectites. Marshall (1935) studied clays in general, as did Ross (1928). The greatest advances have occurred since World War II, however.

Clay mineralogy and sedimentary petrography is truly an international field. It was undoubtedly pioneered by Bragg's (1937) work on x-ray diffraction analysis. The early work (1940s), particularly in the United States, of such investigators as W. F. Bradley, C. H. Edelman, F. A. Van Baren, J. C. L. Favejee and G. W. Brindley, is listed in Carroll (1962). Carroll herself should be noted for early work on sediments, including soils, first in Australia and later in the United States. Grim is well-known for his text *Clay Mineralogy*, first published in 1953. Hendricks (1938, 1942) did early work on lattice structure, properties of clay minerals, and base exchange of silicates, etc. P. F. Kerr worked on clay mineral standards and differential thermal analysis (see references in Carroll, 1962), and Mackenzie (1957) on differential thermal investigation of clays. Marshall's (1949) text on colloid chemistry and Ross' (1943) study of soils in relation to geological processes are also noteworthy. Much common ground exists between the two disciplines of clay and soil mineralogy, evidenced by writings of some well-known authors in both fields; Carroll (1962) summarizes these studies.

Thus, through the course of over 80 yr, the history of sedimentary petrography is portrayed by the research of past and contemporary international petrologists (see also *Sedimentology–History*). The literature supplied by the chronicles of their work make impressive reading (Milner, 1962). These are permanent milestones in the evolution of a natural science once characterized by neglect. As the future unfolds and research in sedimentology expands, as it most certainly will, it is not unreasonable to anticipate commensurate advancement of practical knowledge of sediments of all types, from geologically ancient to the most recent, the province of this petrography (see *Sedimentology–Yesterday, Today, and Tomorrow*). There is now also the wider vista afforded by investigation of ocean-bottom deposits, which modern exploratory methods of location and perfection in depth sampling have brought within our reach. There are in store rare and exciting opportunities for new research projects in this largely unknown and hitherto mostly inaccessible submarine terri-

Methods in Sedimentary Petrography

Petrographic methods constitute an essential part of this science. They have developed both in kind and scope with its progress to a point today when practically all weapons of the border sciences to geology are brought to bear in solving the simple to the most complex sedimentary problems. Many of them constitute, in fact, a science within a science. The traditional petrological microscope with built-in polarized light; the conventional thin section of the rock; later the technique of isolating and examining "heavy" minerals, are still with us, and likely to remain for a long time to come. But the higher degree of accuracy in diagnosis and measurement of sedimentary particles so long desired has only been made possible by appeal to other techniques and instruments, notably those available in the varied disciplines of applied physics. Analysis of sedimentary rocks may still, for routine purposes, follow the traditional codes of practice; but intensive studies, particularly of the finer argillaceous materials, consolidated and coherent alike, today demand methods of a high order of precision wherein the personal equation is reduced to a minimum.

Almost all the long-cherished methods of mineralogical and mechanical analysis of sediments are still recognized, with but a few obsolete exceptions. The sort of "petrographic methods and calculations" so ably discussed by Holmes (1921) in his book under that title are generally as valid now as ever; but we have gone a very long way since then. Both technique and interpretations tend to assume more and more complicated designs and expressions geared to a more fundamental understanding of sedimentary processes and the products arising therefrom. This is only natural when it is realized that the petrologist and co-workers on sediments have at their disposal the powerful tools of enquiry offered by: chemical analysis (short-cut methods), including ion-exchange capacity determinations; differential thermal analysis, especially as applied to clays; electrochemical methods; electron microscopy; fluorescence analysis; infra-red absorption (spectrophotometer) studies; microchemical analysis, including the aid of paper chromotography; nuclear methods, e.g., atomic absorption; radiometry; spectrographic analysis; and the mathematical approach, implied in a deeper development of statistical assessments of the quantitative elements concerned in this subject. In short, *inexactness,* proverbially the criticism of geology, is fast becoming outmoded. If we can successfully date rocks in time, as we are now doing, then rationalization of the methods employed in their study must follow; and this applies with full force to sediments.

HENRY B. MILNER*

*deceased

References

Arkhangel'skii, A. D., 1927. On the Black Sea sediments and their importance for the study of sedimentary rocks, *Mosk. Ovo. Lspyt. Prir., Byull., Otd. Geol.,* 5, (3-4) (n. s. 35), 199-298. (Russian with English summary, 264-289).

Artini, E., 1926. Sulla composizione mineralogica di quattro campioni di Sabbia Provenienti dalle dune dei dintorni di Chisimaia nell' Oltregiuba, *Agric. Colon.,* 20, 101-102.

Baak, J. A., 1936. *Regional Petrology of the Southern North Sea.* Holland: Veenman, Wageningen, 127p.

Berthois, L., 1975. *Étude Sédimentologique des Roches Meubles: Techniques et Méthodes.* Paris: Doin, 280p.

Bonney, T. G., 1900. The Bunter pebble beds of the Midlands and the source of their materials, *Quart. J. Geol. Soc. London,* 56, 287-306.

Boswell, P. G. H., 1915. Deposits of the north-eastern part of the London Basin, *Quart. J. Geol. Soc. London,* 71, 536-591.

Boswell, P. G. H., 1916. The application of petrological and quantitative methods of stratigraphy, *Geological Mag.,* 53, 105-111, 163-169.

Boswell, P. G. H., 1933. *On the Mineralogy of Sedimentary Rocks.* London: Murby, 393p.

Bragg, W. L., 1937. *Atomic Structure of Minerals.* Ithaca, N. Y.: Cornell Univ. Press, 292 p.

Bramlette, M. N., 1934. Heavy mineral studies on correlation of sands at Kettleman Hill, California, *Bull. Am. Assoc. Petrol. Geologists,* 18, 1559-1576.

Brammall, A., and Harwood, H., F., 1923. The occurrence of rutile, brookite and anatase on Dartmoor, *Mineralog. Mag.,* 20, 20-26.

Carroll, D., 1962. The clay minerals, in H. B. Milner, ed., *Sedimentary Petrography,* vol. 2. London: Allen and Unwin, 288-371.

Chelussi, I., 1924. Psammografia di due pozzi trivellati di Novara e di Carisio, *Bull. Soc. Geol. Ital.,* 43, 161-169.

Clarke, F. W., 1924. Data of Geochemistry, *U. S. Geol. Surv. Bull. 770,* 841p.

Clerici, E., 1924. Saggio bibliografico di psammografia italiana, *Bull. Soc. Geol. Ital.,* 43, 21-31.

Deverin, L., 1924. L'Etude lithologique des roches sédimentaires, *Bull. Suisse Min. Petrog.,* 4, 29-50.

Double, I. S., 1924. The petrography of Late Tertiary deposits of the East of England, *Proc. Geol. Assoc.,* 35, 332-358.

Double, I. S., 1926. The petrography of the Triassic rocks of the Vale of Clwyd, *Proc. Liverpool Geol. Soc.,* 14, 249-262.

Double, I. S., 1927. The microscopic characters of certain horizons of the upper chalk, *J. Roy. Micros. Soc.,* 47, 226-231.

Doyen, A., 1926. Sur la distribution des éléments lourds dans quelques sédiments anciens, *Ann. Soc. Géol. Belge,* 49, B48-B49.

Dryden, L., 1935. A statistical method for comparison of heavy mineral suites, *Am. J. Sci.,* 393-408.

Edelman, C. H., 1938. Petrology of recent sands of the Rhine and the Meuse in the Netherlands, *J. Sed. Petrology,* 8, 59-66.

Folk, R. L., 1973. Carbonate petrography in the post-Sorbian Age, in R. N., Ginsburg, ed., *Evolving Concepts in Sedimentology.* Baltimore: Johns Hopkins Univ. Press, 118-158.

Goldman, M. I., 1915. Petrographic evidence on the origin of the Catahoula sandstone of Texas, *Am. J. Sci.,* 189, 261-287.

Goldman, M. I., 1923. Review of *An Introduction to Sedimentary Petrography, Bull. Am. Assoc. Petrol. Geologists,* 7, 194.

Grim, R. E., 1953. *Clay Mineralogy.* New York: McGraw-Hill, 384p.

Groves, A. W., 1931. The unroofing of the Dartmoor granite and distribution of its detritus in southern England, *Quart. J. Geol. Soc. London.* 87, 62-96.

Gruner, J. W., 1935. The structural relationship of glauconite and mica, *Am. Mineralogist,* 20, 699-714.

Harker, A., 1895. *Petrology for Students: An Introduction to the Study of Rocks Under the Microscope.* Cambridge: University Press, 305p.

Hatch, F. H., and Rastall, R. H., 1913. *Textbook of Petrology, Vol. 2: Petrology of the Sedimentary Rocks* (appendix by T. Crook on the systematic examination of loose detrital sediments, 339-414), 425p.

Hendricks, S. B., 1938. On the crystal structure of the clay minerals: Dickite, halloysite and hydrated halloysite, *Am. Mineralogist,* 23, 295-301.

Hendricks, S. B., 1942. Lattice structure of clay minerals and some properties of clays, *J. Geol.,* 50, 276-291.

Holmes, A., 1921. *Petrographic Methods and Calculations.* London: Murby, 512p.

Illing, V. C., 1916. The oilfields of Trinidad (correlation by means of "heavy" residues), *Proc. Geol. Assoc.,* 27, 115.

Krumbein, W. C., 1932. A history of the principles and methods of mechanical analysis, *J. Sed. Petrology,* 2, 89-124.

Krumbein, W. C., and Pettijohn, F. J., 1938. *Manual of Sedimentary Petrography.* New York: Appleton-Century Crofts, 549p.

Mackenzie, R. C., 1957. *The Differential Thermal Investigation of Clays.* London: Miner. Soc., 456p.

Mackie, W., 1897. On the laws that govern the rounding of particles of sand, *Trans. Edin. Geol. Soc.,* 7, 298-311.

Marshall, C. E., 1935. Mineralogical methods for the study of clays, *Z. Krist.,* 90, 8-34.

Marshall, C. E., 1949. *The Colloid Chemistry of the Silicate Minerals.* New York: Academic Press, 195p.

Martens, J. H. C., 1935. Beach sands between Charleston, South Carolina, and Miami, Florida, *Geol. Soc. Am. Bull.,* 46, 1563-1596.

Milner, H. B., 1922a. *An Introduction to Sedimentary Petrography.* London: Murby, 125p.

Milner, H. B., 1922b. Pliocene deposits of Cornwall (Petrography and correlation), *Quart. J. Geol. Soc. London,* 78, 348-377.

Milner, H. B., 1962. *Sedimentary Petrography,* 2 vols. London: Allen and Unwin, 643 and 715 p.

Müller, H., 1931. Über die quantitative mineralische Zusammensetzung tertiärer Sande im Untergrunde von Hamburg und Umgebung, *Centr. Mineral. Geol.,* A, 278-296.

Müller, H., 1933. Sedimentpetrographie und Geologie, *Z. Deut. Geol. Ges.,* 85, 719-720.

Ross, C. S., 1928. The mineralogy of clays, *Proc. 1st Int. Congr. Soil Sci.,* 4, 555-561.

Ross, C. S., 1943. Clays and soils in relation to geologic processes, *J. Washington. Acad. Sci.,* 33, 225-235.

Smithson, F., 1939. Statistical methods in sedimentary petrology, *Geological Mag.,* 76, 297-309, 348-360, 417-427.

Sorby, H. C., 1879. The structure and origin of limestones, *Quart. J. Geol. Soc. London,* 35, 56-95.

Sorby, H. C., 1880. The structure and origin of non-calcareous stratified rocks, *Quart. J. Geol. Soc. London,* 36, 46-92.

Tester, A. C., 1931. The measurement of the shapes of rock particles, *J. Sed. Petrology,* 1, 3-11.

Thomas, H. H., 1902. The mineralogical constitution of the finer material of the Bunter pebble beds in the west of England, *Quart. J. Geol. Soc. London,* 58, 620-632.

Thomas, H. H., 1909. A contribution to the petrography of the new red sandstone in the west of England, *Quart. J. Geol. Soc. London,* 65, 229-245.

Tickell, R. G., 1924. Correlative value of the heavy minerals, *Bull. Am. Assoc. Petrol. Geologists,* 8, 158-168.

Tickell, R. G., 1939. *The Examination of Fragmental Rocks.* Stanford: Stanford Univ. Press, 154p.

Wadell, H., 1932. Volume, shape and roundness of rock particles, *J. Geol.,* 40, 443-451.

Wadell, H., 1935. Volume, shape and roundness of quartz particles, *J. Geol.,* 43, 250-280.

Wentworth, C. K., 1922. A scale of grade and class terms for clastic sediments, *J. Geol.* 30, 377-392.

Woolridge, S. W., 1924. Bagshot beds of Essex, *Proc. Geol. Assoc.* 35, 359-383.

Woolridge, S. W., 1927. The Pliocene history of the London Basin, *Proc. Geol. Assoc.,* 38, 49-133.

Cross-references: *Clastic Sediments and Rocks; Grain-Size Studies; Limestones; Provenance; Sedimentary Petrology; Sedimentary Rocks; Sedimentology; Sedimentology—History; Sedimentology—Yesterday, Today, and Tomorrow; Sediment Parameters.*

SEDIMENTARY PETROLOGY

Sedimentary petrology is the scientific study of sedimentary rocks. It differs from sedimentary petrography (q.v.) in that it involves not only description but interpretation of every aspect of the rock or unconsolidated sediment.

The direct objective of sedimentary petrology is to determine, in detail, the history of the sediment: its source (provenance, q.v.), its transport, its mode of deposition, the character of the environment of deposition (see *Sedimentary Environments*), and the nature of its postdepositional alterations (diagenesis, q.v.). The ultimate objective is to reveal some portion of the history of the earth.

Sedimentary rocks can be divided into three major groups; the *clastic* (or *detrital*) *rocks,* the *carbonates,* and the *chemical rocks.* Clastic rocks consist dominantly of grains of minerals weathered from preexisting igneous, metamorphic, or sedimentary rocks. Further subdivision of the clastic rocks depends most importantly on grain size and mineralogy (see *Clastic Sediments and Rocks*). The carbonates consist, in large part, of biochemically produced calcium carbonate and its alteration products. Classification of the carbonates depends on the proportion of shell fragments (allochems), fecal pellets, oolites, and carbonate mud; on the arrangement of these constituents; and on the subsequent history of replacement and alteration of the original carbonate (see *Limestones*). Dolomites (q.v.), apparently usually resulting from the replacement of original calcite or aragonite, are volumetrically important carbonates. The chemical rocks: chert (q.v.), evaporites (q.v.), ironstone (q.v.), phosphorites (see *Phosphate in Sediments*), and others are classified on the basis of chemical composition and mode of origin, although the origins of many chemical rocks are less than perfectly understood.

Methods of study of sedimentary rocks vary with rock type. In the case of clastic rocks, sediment source and mode of transport of the sediment are primary questions. On the other hand, the source of the chemical evaporite halite is usually sea water confined in an evaporative basin; thus the questions of sediment source and transport are essentially answered and are replaced by equally critical and difficult questions about the origin of the evaporitive basin, character of the basin waters at the time of deposition, and many others (see *Evaporites–Physicochemical Conditions of Origin*). The techniques required to answer questions of source and mode of transport for the clastic rocks are entirely different from, and no less complex than, the techniques required to answer critical questions about the origin of halite.

There is a different set of critical questions and techniques of study for each of six broadly defined groups of sedimentary rocks: the coarse and fine clastics; the carbonates; and the evaporite, precipitate, and resistate subdivisions of the chemical rocks. Because the broadest, but not necessarily the most complex, spectrum of critical questions and methods of study is associated with the clastics, most of the following material relates directly to clastics; but problems that are particularly important to some other group of rocks are discussed in appropriate sections.

Source Area: Location and Character

The source of clastic materials is of primary importance in paleogeographic reconstruction. The object is to determine the distribution, or at least the relative abundance, of igneous, metamorphic, and sedimentary rocks in the source area. It is also desirable to know the approximate composition of the igneous rocks, the grade of the metamorphic rocks, and the types of sedimentary rocks involved.

The amount of information about *provenance* (q.v.) that can be obtained from mineralogical analysis of a clastic sediment, through optical microscopy or other means, is largely dependent on the history of weathering and transport. Rock fragments and characteristic minerals (e.g., calcic plagioclase) are commonly destroyed, either in the process of initial weathering, by weathering during transport, or as a result of intrastratal solution. The question is, To what extent have the less stable minerals been removed from the sedimentary rock? Although the generality of intrastratal solution is subject to question, it is probable that the mineralogy of most sedimentary rocks reflects neither the mineralogy of the source area, nor the mineralogy of the sediment as originally deposited, in any quantitative sense. Nonetheless, *coarse fraction analysis* is among the most powerful tools available to the sedimentary petrologists.

The presence of a single recognizable rock ("lithic") fragment is more valuable than a volume of speculation about rock types based on the presence of single mineral fragments in the finer sizes. Rock fragments composed of fine-grained micaceous minerals are particularly important. Fragments of slate and phyllite derived from low-grade metamorphic rocks are abundant constituents of sediments associated with tectonically active continental margins, but they may not be recognized as such because the shape and internal texture tend to become considerably distorted during compaction of the sediment.

To the extent that feldspars survive the weathering and transport processes, they are strongly indicative of general sialic or mafic character of the source rocks. However, plagioclase feldspars weather so much faster than potassium feldspars that sediment feldspar contents rarely reflect the true proportion of plagioclase and potassium feldspars in the source area. Because feldspars are much more abundant than quartz in igneous and most metamorphic source rocks, the feldspar content of a sediment is considered to be a good index of the intensity of weathering during derivation, and weathering and abrasion during transport of the sediment. Sands consisting almost entirely of quartz (orthoquartzites) are said to be "mineralogically mature" because the less stable (and heavy) minerals have been successively leached or sorted out. Mineralogically mature sediments are considered to be the products of either intense weathering and long transport, or multiple recycling of clastic materials (see *Maturity*). Sediments with high feldspar content are thought to have been carried rapidly to the site of final deposition along streams with high gradients such as in tectonically active areas.

Heavy minerals are among the best indicators of provenance. In relatively immature sediments many heavy minerals indicative of provenance (e.g., pyroxenes, kyanite) may survive; but in more thoroughly weathered sediments, the less stable minerals will have been destroyed, leaving a heavy-mineral suite that is only partially representative of the source area or of the sediment as it was originally deposited. For many sedimentary rocks, however, it appears that the effects of intrastratal solution are negligible. In this case, the sum of percentages of the ultrastable minerals zircon, tourmaline, and rutile appears to be an excellent index of mineralogical maturity of the sediment.

In summary, the location and character of the source area for a given sediment are, to some extent, indicated by the petrologic and mineralogic character of the sediment. Unfortunately, weathering of the parent rock and weathering and size reduction during transport tend to destroy much information about the character of rocks in the source area, and the fact that weathering may occur either in situ or during transport obscures much information about the climate of the source area.

Transport

The materials of clastic rocks are carried from their point of origin to the local stream network commonly by mass-weathering processes, typically soil creep, and eventually become part of the stream load. Very fine fragments move quickly along the network as suspended load, but the downstream progress of larger fragments is usually very slow. Thus, the weathering process does not end in the source area but continues to operate during the long process of stream transport.

Stream-channel and floodplain deposits are geologically temporary phenomena, most of which are destroyed by channel shifts or general erosion of the area of deposition. Those deposits that remain as long-term records of processes occurring in the past represent the last few events in the history of some ancient stream. Ancient fluvial deposits are, however, surprisingly common in the rock record and are ordinarily identified by their complex suite of sedimentary structures (especially trough cross-bedding), by analysis of the channel and flood-plain facies relationships, and by characteristics of the sediment size distribution.

If the sediment reaches the sea, it is subject to a different and very complex transport system. In the coastal zone, wave action may carry sediment offshore, onshore, or along the beach by longshore drift (see *Littoral Sedimentation*). As in desert environments, the wind is an important agent of transport, and large dune fields are characteristic of many coasts.

Currents capable of moving sand exist at surprisingly great depths, and downslope and contour currents distribute clastic sediments far into major ocean basins. Perhaps the most remarkable of all sediment transport mechanisms are turbidity currents, the downslope flow of relatively dense mixtures of water and suspended sediment that is characteristic of subsea areas of high relief and rapid sedimentation. Turbidity-current deposits are recognized by a combination of graded bedding and sole marks, sedimentary structures at the base of each unit representing a single turbidity-current event (see *Bouma Sequence*).

Identification of the specific transport mechanism that carried a sediment to its final resting place has been among the most difficult problems in sedimentary petrology, but quite powerful methods of interpretation are now available. These include detailed analysis of sedimentary structures (q.v.; see also *Flow Regimes*), and their orientation (see *Paleocurrent Analysis*); grain size analysis of the sediment as a whole, or of specific mineral constituents (see *Grain-Size Studies*); and analysis of regional trends in mineralogy or other characters of the sediment. It is impossible to separate entirely studies of source, transport, and environment of deposition. A regional pattern of sediment transport suggested by orientation of sedimentary structures often identifies the location of the source area, and specific identification of the environment of deposition may provide much information on the possible and most probable modes of transport.

Sedimentary Environment

There is little question that fossils, where they are abundantly present and benthonic, offer more information about sedimentary environments (q.v.) than any other character of the sediment or sedimentary rock. For that reason, we probably understand more about the environments of deposition of ancient limestones than any other rock type. A few of the assumptions involved might not be entirely accurate (e.g., hermatypic corals are not always restricted to shallow, warm water of "normal" salinity), but most correlative data tend to support the concept that benthic faunas, regardless of age, are excellent indicators of environment.

Where a significant fossil fauna is absent, or has been destroyed during diagenesis, other indicators of sedimentary environment must be used. The suite of sedimentary structures and the presence of environmentally significant authigenic minerals may be completely indicative of the environment of deposition. Facies analysis alone may reveal a modern analog that is in all important respects identical to the ancient environment.

In most cases, however, no single criterion, or set of criteria, fully defines the sedimentary environment. Every scrap of evidence relative to the environment must be utilized to build a consistent, convincing, and realistic picture of the conditions that existed at the instant of sedimentation of a particular fragment or thin bed of rock. Many such analyses must be combined to obtain an accurate view of the paleogeography of an area through a short segment of geologic time (see *Sedimentary Environments*).

Lithification

Few ancient sediments accurately reflect, in their final physical character, the condition in which they existed at the time of deposition. All have suffered some alteration through processes of postdepositional change that are known, collectively, as lithification, or diagenesis (q.v.). Compaction (q.v.) begins shortly after deposition and may continue, or at least remain incomplete, for long periods of time. A very large number of mineralogical changes may also occur. Aragonite commonly dissolves and may, like high-Mg calcite, be replaced by low-Mg calcite (see *Carbonate Sediments—Diagenesis*). With increasing depth of burial, and time, smectite tends to alter to illite (see *Clay—Diagenesis*). Organic constituents are altered; often with the formation of H_2S and metal sulfides (see *Coal—Diagenesis and Metamorphism*). Solution and redeposition of silica and calcium carbonate leads to the formation of chert (q.v.) and to cementation of sandstones (see *Clastic Sediments—Lithification and Diagenesis*).

The list of diagenetic processes can be expanded to great length, and a major part of sedimentary petrology consists of identifying the nature and sequence of diagenetic changes. Again, every available tool from field observation of sedimentary structures through microscope examination of thin sections, and chemical analysis of the rock, must be employed to determine the nature of the changes and the processes involved.

Summary

Sedimentary petrology is an infinitely diverse and complex subject involving all of sedimentology and many areas of mineralogy, stratigraphy, paleontology, paleoecology, geochemistry, and other subdisciplines of geology. The necessarily brief discussion above touches on only a few aspects of a field of study that becomes increasingly complex as new methods of analysis are devised.

Because of the great breadth of the subject, no specific references to concepts and definitions have been included in the above material. The following references include recent books or long publications that will provide a guide to the literature on the subject. For historical references see *Sedimentary Petrography*. The cross-reference list, necessarily abbreviated, includes the most directly applicable titles of this volume.

ROBERT E. CARVER

References

Allen, J. R. L., 1969. *Physical Processes of Sedimentation.* New York: American Elsevier, 248p.
*Bathurst, R. G. C., 1975. *Carbonate Sediments and their Diagenesis.* Amsterdam: Elsevier, 658p.
*Blatt, H.; Middleton, G.; and Murray, R., 1972. *Origin of Sedimentary Rocks.* Englewood Cliffs., N.J.: Prentice-Hall, 634p.
Bouma, A. H., 1969. *Methods for the Study of Sedimentary Structures.* New York: Wiley, 458p.
Carozzi, A., 1960. *Microscopic Sedimentary Petrography.* New York: Wiley, 485p.
Carver, R. E., ed., 1971. *Procedures in Sedimentary Petrology.* New York: Wiley, 653p.
*Chilingar, G. V.; Bissel, H. J.; and Fairbridge, R. W., eds., 1967. *Carbonate Rocks: Origin, Occurrence and Classification.* Amsterdam: Elsevier, 471p.
*Chilingar, G. V.; Bissel, H. J.; and Fairbridge, R. W., eds., 1967. *Carbonate Rocks: Physical and Chemical Aspects.* Amsterdam: Elsevier, 413p.
*Degens, E. T., 1965. *Geochemistry of Sediments.* Englewood Cliffs, N.J.: Prentice-Hall, 342p.
*Friedman, G. M., and Sanders, J. E., 1978. *Principles of Sedimentology.* New York: Wiley, 700p.
Füchtbauer, H., 1974. *Sediments and Sedimentary Rocks, 1.* New York: Halsted, 464p.
Garrels, R. M., and Mackenzie, F. T., 1971. *Evolution of Sedimentary Rocks.* New York: Norton, 397p.

Griffiths, J. C., 1967. *Scientific Method in Analysis of Sediments.* New York: McGraw-Hill, 508p.

Grubić, A., and Obradović, J., 1975. *Sedimentologija.* Belgrade: Gradevinska Knjiga, 331p.

*Ham, W. E., ed., 1962. Classification of carbonate rocks, *Am. Assoc. Petrol. Geologists Mem. 1*, 279p.

*Hatch, F. H., and Rastall, R. H., rev. by J. T. Greensmith, 1971. *Petrology of the Sedimentary Rocks*, 5th ed. New York: Hafner, 502p.

Kukal, Z., 1971. *Geology of Recent Sediments.* Prague: Academia, 490p. (Distributed by Academic Press, London.)

Larsen, G., and Chilingar, G. V., 1967. *Diagenesis in Sediments.* Amsterdam: Elsevier, 551p.

Loewinson-Lessing, F. V., 1954. *A Historical Survey of Petrology* (transl. from Russian by S. I. Tomkieff). Edinburgh: Oliver and Boyd, 112p.

Lucas, G.; Cros, P.; and Lang, J., 1976. *Etude Microscopique des Roches Meubles et Consolidées. Série: Les Roches Sédimentaires.* Paris: Doin, 506p.

Majewske, O. P., 1969. *Recognition of Invertebrate Fossil Fragments in Rocks and Thin Sections.* Leiden: Brill, 101p.

Middleton, G. V., ed., 1965. Primary sedimentary structures and their hydrodynamic interpretation, *Soc. Econ. Paleont. Mineral. Spec. Publ. 12*, 265p.

*Muller, G., 1964. *Methoden der Sediment-Untersuchung.* Stuttgart: E. Schweizerbart'sch Verlagbuchhandlung, 303p.

Muller, G., 1967. *Methods in Sedimentary Petrology* (transl. by H. -U. Schmincke). New York: Hafner, 283p.

*Pettijohn, F. J., 1975. *Sedimentary Rocks*, 3d ed. New York: Harper & Row, 628p.

*Pettijohn, F. J.; Potter, P. E.; and Siever, R., 1972. *Sand and Sandstone.* Berlin: Springer-Verlag, 618p.

*Reineck, H. -E., and Singh, I. B., 1973. *Depositional Sedimentary Environments.* Berlin: Springer-Verlag, 439p.

Rodrigo, L. A., and Coumes, F., 1973. *Manual of Sedimentology* (in Spanish). La Paz, Bolivia: Univ. Major de San Andrés, 150p.

Scolari, G., and Lille, R., with D. Giot, 1973. Nomenclature et classification des roches sédimentaires (Roches détritiqués terrigenes et roches carbonatées). *Bull. Bur. Rech. Géol. Min.*, sec. 4, **2**, 58-127.

Selley, R. C., 1970. *Ancient Sedimentary Environments.* Ithaca, N.Y.: Cornell Univ. Press, 237p.

Till, R., 1974. *Statistical Methods for the Earth Scientist: an Introduction.* London: Macmillan, 154p.

*Additional bibliographic references may be found in this work. For older references, see *Sedimentary Petrography.*

Cross-references: *Argillaceous Rocks; Authigenesis; Bedding Genesis; Clastic Sediments and Rocks; Clay as a Sediment; Coal; Compaction in Sediments; Deposition; Diagenesis; Evaporite Facies; Facies; Flow Regimes; Grain-Size Studies; Limestones; Marine Sediments; Maturity; Organic Sediments; Paleobathymetric Analysis; Paleocurrent Analysis; Paragenesis of Sedimentary Rocks; Provenance; Sedimentary Environments; Sedimentary Petrography; Sedimentary Rocks; Sedimentary Structures; Sedimentation; Sedimentological Methods; Sedimentology; Sediment Parameters; Transportation; Volcaniclastic Sediments and Rocks; Weathering in Sediments.*

SEDIMENTARY ROCKS

Sedimentary rocks are the products of the accumulation of previously weathered and eroded rocks carried to a place of deposition either as discrete particles or in solution, to be deposited, usually in layers, by physical, chemical, and biological agencies which may act independently or in concert. Such accumulations may undergo some chemical reorganization immediately after deposition (diagenesis, q.v.); further modification can occur during lithification, which results from compaction (q.v.) and/or cementation (q.v.) by chemical precipitation. It is generally held that the processes that lead to sedimentation today are the same as those that prevailed in the geological past. It is possible to postulate the nature of past events and conditions by studying modern sedimentary processes and the sedimentary features that result from definable circumstances.

Sedimentary rocks cover 75% of Earth's land surface, but comprise only an infinitesimally small part of the volume of our planet, or, indeed, even of its lithosphere ($\approx 5\%$), for sediments are the product surface processes. Sedimentary rocks are, however, of great interest for four reasons: (1) Because of their wide distribution they literally form the physical basis for many human activities, from agriculture to civil engineering. (2) Certain sedimentary processes have led to deposits of great economic value: coal, lignite, oil, shales, evaporite deposits, ironstones, phosphates, sand and gravels, clays, building stones, limestones, and placer deposits of valuable minerals. (3) The pore spaces and voids that exist in sedimentary rocks, together with the impermeable condition of other rocks, allow the accumulation of oil, natural gas, tar, and water. (4) The compositions, textures, structures, and included fossils of sedimentary rocks permit the reconstruction of Earth history. It is hardly surprising that sedimentary rocks are the subject of study, directly or indirectly, of the majority of geologists.

In order to understand sedimentary rocks, it is necessary to refer to the process of weathering (q.v.), the disintegration and decomposition at normal temperatures and pressures of the rocks at the earth's surface, for it is the weathering of rocks which provides the raw material of sedimentary rocks. Disintegration, a mechanical action, and decomposition, a chemical action, usually act concurrently, each action facilitating the other. Mechanical actions include the effects of natural heating and cooling of constituent minerals with differing coefficients of expansion, the volume change of minerals on chemical weathering, the expansion of freezing water, and the action of rootlets. Chemical weathering is caused principally by meteoric water, containing dissolved CO_2 and other chemical components, creeping through cracks caused by mechanical forces and natural passages and dissolving the minerals of which the rock is composed. Most rocks, particularly igneous rocks, are composed of minerals that are only metastable at normal temperatures; indeed only quartz can be described as a stable common mineral, and even this mineral can be slowly dissolved by water. Higher-temperature minerals dissolve more readily than lower-temperature minerals, and this relative susceptibility to weathering permits the concept of mineralogical *maturity* (q.v.) in sedimentary rocks. The aluminosilicate minerals are broken down into clay minerals, with part of the original mineral going into solution. Carbonate rocks are particularly soluble in waters rich in CO_2, and weather rapidly.

Thus, chemical weathering can break rocks down into (1) resistant particles, which give rise to the *sandstones;* (2) aqueous solutions, which are later precipitated by chemical and biochemical means, with *limestones* as the chief representative; and (3) the clay minerals, which are produced by alteration, and are the major constituents of a broad group of sedimentary rocks called *shales.* These three major types comprise 99% of the total volume of sedimentary rocks. The relative proportions of sandstones, limestones, and shales have been calculated from published descriptions of lithologic types in accounts of sedimentary successions, although various studies have produced differing results. Geochemists have attempted, by various processes but principally by calculations based on the composition of the average igneous rock, to determine the proportions of these three major types, also with varying results. Calculations of the compositions of the average sandstone, the average shale, and the average limestone are not easy to make, for sedimentary rocks are themselves mixtures. Garrels and Mackenzie (1971) have discussed this topic, quoting the works of the principal investigators. It is also clear that the various types of sandstones, shales, and limestones have not always been present in the same relative abundance through geological time.

A simple but fundamental classification which permits a discussion of depositional processes divides sedimentary rocks into two groups on the basis of the origin of their components: (1) *clastic particles;* and (2) chemically, including biochemically, *precipitated crystals.* Most sedimentary rocks are mixtures of the two components, however, and complicating factors are introduced; e.g., a clastic rock may have a chemical cement, and biochemically produced limestones may be subsequently eroded into clastic fragments and thus produce a clastic limestone. These two major groups may be further subdivided—the clastic group by texture (grain size, grain shape, grain roundness, and sorting) and mineral composition, the chemical group either by composition or by the biological agencies responsible for deposition. The subject of classification is complex, and further mention of it is made elsewhere in the volume.

Clastic Rocks

Clastic rocks are subdivided primarily on the basis of size (see *Clastic Sediments and Rocks*). *Rudaceous rocks* are composed mainly of particles >2 mm in diameter; *arenaceous rocks,* i.e., the *sandstones,* have particle diameters between 2 and 1/16 mm; the *argillaceous rocks* consist of particles <1/16 mm in diameter. The boundary sizes given are taken from the Wentworth-Udden scale of particle sizes (see *Grain-Size Studies*). The grain sizes of sediments may be determined by various methods of granulometric analysis (q.v.), but it is more difficult to deal with indurated samples; direct measurement from thin sections of rock, with appropriate mathematical corrections, is one method used.

Grain shape and roundness (q.v.) can also be used in the classification of clastic sediments. Both relative and absolute methods of determining these parameters have been derived; generally speaking, relative methods are the more widely adopted (see *Roundness and Sphericity*).

The textural features are modified by energy input and can reflect, in relative terms, the duration and/or intensity and/or variability of the application of energy. It is thus possible to talk of the *textural maturity* of sediments and sedimentary rocks, the more mature sediments being better sorted and their particles more rounded, and to compare this with mineralogical maturity (see *Maturity*). The texture of sedimentary rocks can reflect the agents of erosion and deposition: an extreme case is the contrast between the well-sorted, well-rounded grains of an eolian deposit with the poorly sorted, often only partially rounded constituents of a glacial deposit. Additionally, the surface texture of sedimentary particles, particularly when observed by using the scanning electron microscope, can reveal evidence of conditions of weathering, erosion, transport, and diagenesis (see *Sand Surface Textures*).

Rudaceous Rocks. Rudaceous rocks (*rudites*) consist of breccias (q.v.) and conglomerates (q.v.). Breccias are texturally and mineralogically immature, and the affinity between a source rock and a breccia is usually obvious. This link is less clear in conglomerates, which are more mature. Oligomictic conglomerates consist of one principal (resistant) pebble type, usually quartz or chert; polymictic conglomerates are composed of mixtures of pebble types. Diamictites (q.v.) consist of a mixture of pebbles and clay-sized material.

Arenaceous Rocks. The sandstones, as has been mentioned, are one of the three major sedimentary rock types (see *Sands and Sandstones*). Although sandstone can be used as a general name, these rocks exhibit a wide range of compositions. Quartz is the common constituent, but feldspar and calcite can be present and even predominant. Sands are usually lithified by the precipitation of cement from trapped solutions; calcareous and siliceous cements predominate, and argillaceous matrix bonding is not uncommon (see *Clastic Sediments—Lithification and Diagenesis*).

Four major groups of sandstones are commonly recognized by sandstone petrologists, although not all users agree on the precise definition of the terms. *Graywacke* (q.v.) is a widely used term and the subject of much controversy, even confusion. First coined before the use of the petrological microscope to describe gray, poorly sorted sandstones of a characteristic field appearance, such sediments are mineralogically and texturally immature; although the origin of the fine matrix of graywackes is a matter of debate, with some workers preferring a diagenetic origin (see *Graywacke*). *Orthoquartzite* is more precisely defined as a sandstone composed mainly (75% or more) of quartz or chert grains; the particles of this type of sandstone may be cemented by silica or carbonate (see also *Quartzose Sandstone*). *Arkose* (q.v.) is an old term to describe sandstones with an obvious (>25% of the recognizable grains) component of feldspar grains. *Lithic sandstone* (*litharenite*) is a more recent term used to describe sandstones in which rock fragments are more abundant than feldspar grains, although both may be minor to quartz grains. Classification problems have given rise to much discussion (see *Sands and Sandstones*).

A fifth group of sandstones exists—the *calcarenites* (q.v.), which are best described as limestones; these limestones are composed of detrital particles of calcium carbonate of sand size and thus are clastic rocks. This term is allied to the names for other detrital limestones: calcilutites, calcirudites, and calcisiltites.

Argillaceous Rocks. This group of sedimentary rocks includes the siltstones, marls, mudstones, mudrocks, shales, clays, and lutites (see *Argillaceous Rocks*). Two grades of particle sizes are involved: the silt grade from 1/16 to 1/256 mm diameter and the clay grade of particles, <1/256 mm diameter. Mudrock, mudstone (q.v.), lutite, and shale (q.v.) are, in fact, mixtures of the two grades; marl (q.v.) is a mixture of calcium carbonate and insoluble silt and clay. Composed of the finest of eroded rock particles and the wide variety of clay minerals—the complex hydro-alumino-silicates produced by weathering—argillaceous rocks have widely varying characteristics. It is unfortunate that the term *clay* (q.v.) is used for both a grade size and a range of special minerals. The composition of the latter, best studied by x-ray and electron microscopy, are usually related to parent rock, although some sedimentologists believe that diagenetic changes in the place of deposition are important (see *Clay as a Sediment; Clay—Genesis; Clay—Diagenesis; Clay Sedimentation Facies*).

Nonclastic Rocks

Nonclastic sediments, having been deposited by chemical or biochemical action from solution, can only provide evidence of depositional conditions, in contrast with the clastic sediments, which can provide additional information about source rocks, conditions of weathering, and the agencies of erosion and transport. Deposits of nonclastic sediments, which give rise to rocks such as limestones, dolomites, gypsum, salt, phosphates, and others, only become significant when clastic sedimentation is absent or at low level. It is not always easy to distinguish between chemical and biochemical agencies in these rocks.

Limestones. The most abundant nonclastic rocks are the carbonates, particularly the calcium carbonate rocks—the limestones (q.v.). Some limestones are clearly the product of biological activity; the remains of the organisms can be clearly seen, and such names as *foraminiferal limestone* or *coral limestone* may be used when appropriate. Other limestones may be the result of direct chemical precipitation, for example *oolitic limestone* (see *Oolite*). In detailed study, most limestones are shown to be mixtures of both biochemically and chemically produced carbonates. Also, as has been noted, the calcarenites are composed of clastic particles of calcium carbonate.

Carbonate petrology is a broad subject, but a practical limestone classification is based on the relative proportions of *allochems* (q.v.), which include all types of transported carbonate particles—clasts, shells and oolites; *microcrystalline ooze*, composed of minute particles caused by rapid chemical precipitation and possibly organic precipitation due to bacteria and algae (see *Micrite*); and *sparry calcite cement*, clear, pore-filling calcite crystals (see *Sparite*).

Dolomite. Most dolomites are formed by *dolomitization*, the postdepositional alteration of limestones to dolomite (q.v.). Some modern environments of dolomite precipitation have also been found. Various mechanisms are proposed for the necessarily increased Mg:Ca ratio—alteration by circulating sea water, direct precipitation, and diffusion from Mg-rich organic remains (see *Dolomite, Dolomitization*).

Evaporites. Other nonclastic sediments include a wide range of *evaporites* (q.v.), i.e., rocks produced by evaporation. The concentration of sea water necessary for such direct precipitation is only achieved in isolated seas, lagoons, saline lakes, or in the vadose zone of sediments partially brine-saturated and partially air-saturated. Some brines are created in nonmarine conditions by rainwater which dissolves soluble rocks; evaporation may then occur in landlocked basins. As evaporation concentrates the brine, direct precipitation takes place, and higher concentrations lead to rarer evaporites in the sequence: iron oxide, calcium carbonate, magnesium chloride, sodium bromide, and potassium chloride, succeeded by rare bittern salts. Conditions that would cause the total evaporation of a body of water would not lead to significant quantities of some of these compounds; rather it seems to be the case of certain balances being achieved between rates of evaporation and brine influx. Cycles of precipitation are not uncommon in large accumulations of evaporites. Much discussion has centered on the hydrated calcium sulfates, anhydrite, and gypsum, in particular on which of the two is the original mineral. See *Evaporite Facies; Evaporites—Physicochemical Conditions of Origin; Marine Evaporites—Diagenesis and Metamorphism*.

Iron-Rich Rocks. Iron is present in most sedimentary rocks, and some sedimentary rocks are iron rich. The composition and origin of the iron are varied, and clearly a wide variety of conditions can permit the accumulation of iron minerals. There are two basic types of iron-rich sedimentary rocks; the *cherty iron formations* (q.v.) and *ironstone* (q.v.).

Silica. Silica in the form of *chert and flint* (q.v.) occurs in many sedimentary rocks, particularly in limestones, and as bedded deposits. Both direct precipitation and biochemical action have been invoked and, certainly in many rocks, a fair degree of reprecipitation and even replacement has occurred (see *Silica in Sediments*).

Phosphate-Rich Rocks. Phosphate occurs in sedimentary rocks in many forms, including detrital phosphates. Ancient phosphorites occur in rocks of all geological periods, including the Precambrian; and phosphatic minerals occur in a wide variety of circumstances, including terrestrial conditions (see *Phosphates in Sediments*).

Coal and Petroleum. Coal (q.v.) and petroleum (q.v.) are special sedimentary deposits. Both are organic in origin, and each reflects special conditions of deposition and diagenesis (see *Organic Sediments*). Coals and lignites (brown coals) are accumulations of plant matter. Deposition takes place at or near the site where the plants grew. The ratio of volatiles and moisture to fixed carbon controls the rank of coal, with anthracite, composed of nearly 100% carbon, having the highest range (see *Coal—Diagenesis and Metamorphism*). Coal-bearing sequences of sedimentary rocks are usually characterized by cyclothems, i.e., repeated cycles of lithologies possibly related to sea-level changes (see *Cyclic Sedimentation*). The origin and accumulation of petroleum is more complex and still the subject of much discussion (see *Crude-Oil Composition and Migration; Petroleum—Origin and Evolution*).

Petroleum, like gas and water, is almost wholly recovered from sedimentary rocks. Except where they have been closely fractured, igneous and metamorphic rocks do not possess spaces capable of holding or permitting the passage of liquid and gases. Sedimentary rocks, on the other hand, may naturally have 30–40% pore space. When deposited, clastic sediments, including clastic carbonates, may contain 30–90% by volume of trapped water. As compaction (q.v.) takes place, the water is squeezed out until grains are in contact and capable of supporting the superincumbent load. In the absence of later cementing precipitates, sandstones and shales have an average porosity of about 30%. The actual linked porosity may finally be about 15%. The original porosity of clastics may be increased, and carbonate porosity developed, by solution by moving water or by postdepositional changes such as the dolomitization of limestones (see *Carbonate Sediments—Diagenesis*). The measure of the ease of transmission of fluids is called *permeability*. Movement can be retarded by friction against grains, and consequently fine-grained rocks, with minute linked pore spaces, have the greatest resistance, which in clays may effectively prevent any movement of fluids or even gas. The alternation of permeable and impermeable layers, in the absence of fractured rocks, can effectively prevent the easy migration of oil, gas, or water; these conditions, in conjunction with suitable geological structures, produce valuable reservoirs.

Sedimentary Structures

In addition to their significance in the migration and accumulation of oil (see above), sedimentary structures (q.v.) provide evidence of transportational and depositional conditions. They are usually subdivided into *primary sedimentary structures* (q.v.), formed during deposition and *secondary sedimentary structures* (q.v.), formed after the deposition of the sediment and when some chemical reorganization has been initiated (see *Sedimentary Structures*).

Primary Sedimentary Structures. Primary structures range, by the definitions of some workers, from the size and shape of the sedimentary body to structures associated with the simplest readily distinguishable features in exposures of sedimentary rocks, the *bed*. Indeed the shape and size of the bed itself and the sedimentary structures within the bed offer vital clues to the conditions at the time of deposition (see *Bedding Genesis*). *Laminae*, thin layers within the bed, may be parallel, inclined, rippled, or convoluted; inclined laminae are called current or *cross-bedding* (q.v.). *Graded bedding* (q.v.) is the upward decrease of particle size within a bed. This structure and the primary grain orientation (see *Fabric*) offer clues about the condition of the transporting medium. The upper and lower surfaces of beds—the bedding planes—also exhibit a wide variety of structures. Examples on the upper surface include mud cracks (q.v.), raindrop imprints (q.v.), pseudomorphs (q.v.), swash and rill marks, and ripple marks (q.v). *Sole marks,* found on the underside of beds, represent the counterparts (molds) of current marks (q.v.). Sole marks include the counterparts of scour marks (q.v.), produced by current scours, and tool marks (q.v.), made by particles being dragged along or impacting onto the depositional interface. A wide variety of other structures is caused by sediment loading and the movement of trapped water and unconsolidated sediment between freshly deposited beds (see *Penecontemporaneous Deformation of Sediments; Water-Escape Structures*).

In addition to the above mechanical structures, structures caused by browsing or burrowing creatures are preserved on bedding planes and within beds. These structures, which do not include the skeletal remains of the organism itself, have been called *trace fossils.* Such tracks, trails, and burrows can give information about water depth, sedimentation rates, and water conditions (see *Biogenic Sedimentary Structures*).

All primary structures, both mechanical and biogenic, give a guide, often very precise, to the conditions that prevailed at the time of sedimentation. Many structures provide evidence of current directions and most clearly point to the original attitude and hence to the correct sequence of beds, thus aiding the unravelling of tectonically complicated areas (see *Paleocurrent Analysis; Top and Bottom Criteria*). Certain structures occur together in suites, being the product of essentially similar conditions.

Secondary Sedimentary Structures. Some larger mechanical structures are formed soon after deposition (penecontemporaneous) and often affect several beds. Postdepositional structures include slump bedding (q.v.), intraformational disturbances (q.v.), and clastic dikes (q.v.). All of these structures are due to movements in response to gravity and to excess pore pressure.

Secondary structures of a chemical nature include concretions (q.v.), nodules (q.v.), geodes (q.v.), corrosion surfaces (q.v.), and pressure-solution phenomena (q.v.). All of these structures are created in response to chemical changes following deposition (*diagenesis,* q.v.), often through the interaction between connate, i.e., trapped, water and the sediments, or due to changes in solubility and precipitation caused by very local chemical environments within the rock.

Geometry. Large-scale structures include the shape and size of the sedimentary formation. Some are thin in relation to their lateral extent and thus called *sheet* or *blanket* deposits; others, though long, have only a limited width in relation to their thickness and are often called *shoestring* deposits. These terms apply especially to sandstones. Other bodies, thicker and wedge shaped, are described as *prismatic* deposits. All these forms reflect controls and constraints at the time of deposition of clastic sediments.

Biogenic sediments sometimes create a mound of skeletal material composed of skeletons, organic debris, and associated chemically deposited materials: such a structure is called *bioherm* or *reef* (q.v.). Subsequent layered sediments on and around this structure may, on compaction, exhibit quaquaversal dips. A contrasting sheet-like mass of organic material is called a *biostrome* (see *Limestones*).

Facies

The term *facies* (q.v.) refers to the sum total of features present in a sedimentary rock which characterize it as having been deposited in a particular environment. Geologists can recognize what a certain assemblage of textures, structures, lithologies, sequences of beds, and formation geometry represent in terms of sedimentary environments (q.v.). It is known that several different facies can be formed concurrently in adjacent locations, the facies can cross time planes (i.e., can be diachronous), and that a particular facies can be repeated in units of different ages or in different parts of the world at the same time.

Studies of facies of modern sedimentary environments and ancient sedimentary rocks have led to the realization that given situations are repeated in Earth history in response to certain controls—hence the recognition of *facies models* (see *Sedimentary Environments*). Particularly detailed work has been done on fluvial, deltaic, and turbidite deposits (see *Fluvial Sediments; Deltaic Sediments; Turbidites*). In making such studies, sedimentary geologists seek to construct a picture of the paleogeography, i.e., the relative positions of land and water at the time of sedimentation, by studying (1) the distribution of facies over one short time span; (2) the shape of the formation; (3) the pattern of current directions from sedimentary structures; and (4) the regional variations in grain size sorting, rounding, and so on—all being used to ascertain the shape of the basin and the nature and size of the source areas (see *Paleocurrent Analysis*). The remains and traces of organisms all provide extra information of conditions (see *Sedimentation—Paleoecologic Aspects*).

In all studies of past and present sedimentation, it is recognized that several major controls are interacting to produce the final sedimentary rocks. Tectonic controls are obviously important, for without tectonic changes there would be no source areas above a universal sea and no basins of definite geometry (see *Sedimentary Facies and Plate Tectonics; Sedimentation—Tectonic Controls*). The concept of broad associations of sedimentary rock types, sequences, and structures is useful in this light (Fairbridge, 1958). Broad and useful distinctions have been drawn, for example, between the "arkosic suite," "graywacke suite," and the "orthoquartzite-limestone suite." Such a scheme is an oversimplification, but it offers a useful starting point for a broad, unifying approach to the study of sedimentary rocks. Many geologists have been concerned with the concept of geosynclines and the sediments they contain (see *Geosynclinal Sedimentation*). Flysch and molasse are terms covering affinities of sedimentary rocks associated with the evolution of geosynclines; these terms are useful but have caused much confusion.

Climatic controls of sedimentation can be recognized by, among other things, evaporites, red beds, and tillites. Climatic controls, along with sea-level changes, can also be inferred as controls for some sedimentary cycles (see *Cyclic Sedimentation*). Changes in sea level can be clearly recognized through the nature of sediments, sedimentary structures, and faunal assemblages.

By synthesizing many sedimentary data, a geologist can begin to interpret rocks in a very broad framework. Blatt et al. (1972), in their textbook on sedimentary rocks, clearly state the desirable approach: "The sedimentary geologist's first task, after careful observation and description, should be to interpret his observations in terms of mechanisms. A number of such interpretations of related strata suggest larger scale interpretations in terms of processes, and a final synthesis of all the observations and interpretations and the influence of major variables such as tectonics and climate."

ALEC J. SMITH

References

Blatt, H.; Middleton, G.; and Murray, R., 1972. *Origin of Sedimentary Rocks.* Englewood Cliffs, N.J.: Prentice-Hall, 634p.
Fairbridge, R. W., 1958. What is a consanguineous association? *J. Geol.,* 66, 319-324.
Folk, R. L., 1974. *Petrology of Sedimentary Rocks.* Austin, Texas: Hemphills, 184p.
Garrels, R. M., and Mackenzie, F. T., 1971. *Evolution of Sedimentary Rocks.* New York: Norton, 397p.
Krynine, P. D., 1948. The megascopic study and field classification of sedimentary rocks, *J. Geol.,* 56, 130-165.
Pettijohn, F. J., 1975. *Sedimentary Rocks,* 3rd ed. New York: Harper & Row, 628p.
Selley, R. C., 1976. *An Introduction to Sedimentology.* London: Academic Press, 408p.
Shrock, R. R., 1948. *Sequence in Layered Rocks.* New York: McGraw-Hill, 507p.

Cross-references: *Argillaceous Rocks; Bedding Genesis; Cementation; Chert and Flint; Clastic Sediments and Rocks; Diagenesis; Evaporite Facies; Facies; Ironstone; Limestones; Organic Sediments; Paragenesis of Sedimentary Rocks; Phosphate in Sediments; Sands and Sandstones; Sedimentary Petrology; Sedimentary Structures; Volcaniclastic Sediments and Rocks.* See also numerous cross-references in text.

SEDIMENTARY STRUCTURES

Sedimentary structures are features that originate within layers of sediment or along the sediment/fluid interface in response to various processes that are active prior to lithification. Sedimentary structures produced during deposition and determined by the conditions of deposition are *primary sedimentary structures* (q.v.), e.g., bedding (q.v.), cross-bedding (q.v.), graded bedding (q.v.), and current marks (q.v.). Structures produced by penecontemporaneous deformation of sediments (q.v.) are also considered to be primary. Postdepositional sedimentary structures, produced by such processes as compaction (q.v.) and diagenesis (q.v.), are *secondary sedimentary structures* (q.v.).

The sizes of sedimentary structures range from millimeter to kilometer dimensions. Many sedimentary structures are readily apparent in field inspection of exposures; others can be seen only by applying special laboratory techniques such as etching, making peels, applying x-rays, and preparing thin sections (see *Sedimentological Methods*).

Study of sedimentary structures began early in the 19th century with the development of modern geology. These structures provided important evidence in support of the ideas that many strata of ancient bedrock had originated by the cementation of ordinary sediments (Lyell, 1830). This fundamental linkage between modern sediments and ancient sedimentary bedrock formed the cornerstone of the widespread geologic motto that "the present is the key to the past," an idea later incorporated into the doctrine of uniformitarianism.

Sedimentary structures next were applied to the problems of structural geology. In the 1840s and 1850s, diverse viewpoints were expressed about the origin and thickness of the Triassic-Jurassic (Newark) strata in NE North America. According to one view, these inclined strata had been deposited with an initial dip equal to their present attitudes and hence (1) were not very thick and (2) had never been subjected to tectonic activity or extensive postdepositional erosion (Rogers, 1858; Whelpley, 1845; Whitney, 1860). Study of sedimentary structures such as footprints, ripple marks, and raindrop impressions (Redfield, 1843; Hitchcock, 1858; Russell, 1878; Cook, 1889) demonstrated convincingly that the Newark strata had been deposited as horizontal sediments and afterwards had been deformed by tectonic activity.

Another application of sedimentary structures in structural geology was for determining the original top directions of deformed strata. Sedimentary structures used for this purpose first included features found on the bedding surfaces of the Alpine Flysch (Haidinger, 1848; Fuchs, 1895; Paul, 1898). Widespread application of sedimentary structures for determining the correct order of superposition of deformed strata began again independently in North America with geologists mapping intensely folded Precambrian strata (e.g., Van Hise, 1896), but was not confined to studies of Precambrian rocks (Cox and Dake, 1916). Renewed interest among European geologists appeared in the 1930s and 1940s (Tanton, 1930; Sander, 1936; Bailey, 1934; Fiege, 1942). Awareness of sedimentary structures as integral adjuncts to structural studies became practically universal as a result of Shrock's monograph, *Sequence in Layered Rocks* (1948).

Sedimentary structures contain much information of value in studies of paleogeography and in the analysis of depositional processes. Of particular importance in such studies are the features created in cohesionless sediments, especially in sands, by currents of water and of air. In the latter part of the 19th century, H. C. Sorby published a series of brilliant studies on structures generated in sands by water currents, from which he inferred the nature and direction of current flow, coastal configuration, and wave tide interaction (see Summerson, 1976). Sorby's work had little influence upon his contemporaries, however, who either failed to recognize sedimentary structures or else tended to regard them as curiosities. Sedimentary structures were employed in paleogeographic studies only sporadically during the first half of the 20th century but were generally ignored, or relegated to a poor second place after fossils. Systematic paleogeographic applications of sedimentary structures were not commonly used until after 1950 and the brilliant new insights of J. L. Rich, P. H. Kuenen, and C. I. Migliorini. Detailed paleogeographic applications of sedimentary structures in both modern sediments and ancient sedimentary rocks are now being carried out in all parts of the world by numerous investigators.

Detailed studies of the interactions between organisms and sediments were inaugurated at the Senckenberg Museum and its North Sea research station during the 1930s, under the direction of R. Richter (see *Biostratinomy; Bioturbation*). Numerous researchers are continuing these studies, in Germany and elsewhere.

Modern studies of sedimentary structures are diverse in both method and application (see cross-references). Middleton (1965) summarizes the reasons for the study of sedimentary structures:

1. A careful description of structures in sedimentary rocks, in recent sediments and in experimental studies is a necessary prerequisite to other studies; especially important is the description of structures in three dimensions and their recognition in two-dimensional outcrop.
2. Structures may then be used as top and bottom criteria (q.v.) or may be mapped and used statistically for paleocurrent (q.v.) and paleotectonic analysis.
3. Surveys of sedimentary structures found in modern sedimentary environments may aid in paleoenvironmental studies.
4. The study of the formation of sedimentary structures, including a detailed analysis of the physical,

chemical, and biological process involved, may be an end in itself.

JOHN E. SANDERS

References

Bailey, E. B., 1934. West Highland tectonics, *Quart. J. Geol. Soc. London*, 90, 462-525.
Cook, G. H., 1889. On the Triassic or Red Sandstone rocks, *N.J. State Geologist, Ann. Rept.*, 1888, 11-44.
Conybeare, C. E. B., and Crook, K. A. W., 1968. Manual of sedimentary structures, *Australian Dept. Nat. Dev., Bur. Min. Resources, Geol. and Geophys., Bull.*, 102, 327p.
Cox, G. H., and Dake, C. L., 1916. Geologic criteria for determining the structural position of sedimentary beds, *Missouri Univ. School of Mines Bull.*, 2(4), 1-59.
Fiege, K., 1942. Hilfsmittel zur Erkennung normaler und inverser Lagerung in tektonisch stark gestörten Gebieten, *Senckenbergiana*, 25, 292-325.
Fuchs, T., 1895. Studien über Fucoiden und Hieroglyphen, *K. Akad. Wiss., Wien*, 62, 369-448.
Haidinger, W., 1848. Theirfährten aus dem Wiener- oder Karpathensandsteine, *Freunden Naturwiss. Wien, Berichte Mitt.*, 3, 284-288.
Hitchcock, E., 1858. *Ichnology of New England*. A report on the sandstone of the Connecticut Valley, especially its fossil footmarks. Boston: William White, 220p.
Lyell, C., 1830. *Principles of Geology*, vol. 1. London: John Murray, 511p.
Middleton, G. V., 1965. Introduction, in Primary sedimentary structures and their hydrodynamic interpretation, *Soc. Econ. Paleont. Mineral. Spec. Publ. 12*, 1-4.
Paul, C. M., 1898. Der Wienerwald. Ein Beiträg zur Kenntis der nordalpinen Flyschbildungen, *Kon. K. Geol. Reichanstalt, Jahrb.*, 48, 53-176.
Pettijohn, F. J., and Potter, P. E., 1964. *Atlas and Glossary of Sedimentary Structures*. New York: Springer, 370p.
Redfield, W. C., 1843. Notice of newly discovered fish beds and a fossil foot mark in the Red Sandstone Formation of New Jersey, *Am. J. Sci.*, 43, 134-136.
Rogers, H. D., 1858. *The Geology of Pennsylvania*. A government survey. . . , vol. 2, pt. 2. Philadelphia: Lippincott, 677-1045.
Russell, I. C., 1878. The physical history of the Triassic formation of New Jersey and the Connecticut Valley, *Ann. N.Y. Acad. Sci.*, 1, 220-254.
Sander, B., 1936. Beiträge zur Kenntis der Anlagerungsgefüge (Rhythmische Kalk und Dolomit aus der Trias), *Min. Petrog. Mitt.*, 48, 27-208. Translated by E. B. Knopf, 1951. Tulsa, Okla.: Am. Assoc. Petrol. Geologists, 207p.
Shrock, R. R., 1948. *Sequence in Layered Rocks*. New York: McGraw-Hill, 507p.
Summerson, C. H., ed., 1976. *Sorby on Sedimentology*, a collection of papers from 1851 to 1908 by Henry Clifton Sorby. Miami: U. Miami Comparative Sedimentology Lab., 225p.
Tanton, T. L., 1930. Determination of age relations in folded rocks, *Geological Mag.*, 67, 73-76.
Van Hise, C. R., 1896. Principles of North American Pre-Cambrian geology, *U.S. Geol. Survey, Ann. Rept.* 16(1), 571-843.
Whelpley, J. D., 1845. Trap and sandstone of the Connecticut Valley; theory of their relation, *Proc. Am. Assoc. Geol. Natur.*, 6th Mtg., 61-64.
Whitney, J. D., 1860. On the stratigraphic position of the sandstone of the Connecticut Valley, *Ann. Sci. Discovery*, 1860, 322.

Cross-references: *Backset Bedding; Bedding Genesis; Bed Forms in Alluvial Channels; Biogenic Sedimentary Structures; Biostratinomy; Bioturbation; Birdseye Limestone; Bouma Sequence; Casts and Molds; Compaction in Sediments; Concretions; Cross-Bedding; Current Marks; Diagenesis; Diagenetic Fabrics; Experimental Sedimentology; Fabric, Sedimentary; Fissility; Flame Structures; Fucoid; Graded Bedding; Intraformational Disturbances; Lamina, Laminaset, Laminite; Limestone Fabrics; Mud Cracks; Nodules in Sediments; Oolite; Paleocurrent Analysis; Parting Lineation; Penecontemporaneous Deformation of Sediments; Pillow Structures; Pisolite; Pressure Solution and Related Phenomena; Primary Sedimentary Structures; Pseudomorphs; Raindrop Imprint; Ripple Marks; Salar, Salar Structures; Sand Waves; Scour-and-Fill; Scour Marks; Secondary Sedimentary Structures; Sedimentary Environments; Sedimentary Petrology; Sedimentological Methods; Slump Bedding; Stromatolites; Tool Marks; Top and Bottom Criteria; Varves and Varved Clays; Water-Escape Structures.*

SEDIMENTATION

Sedimentation is the act or process of forming or accumulating material in layers from mobile media at surface temperatures and atmospheric pressures. Indeed, it encompasses the entire complex of interrelated geological, physical, chemical, and biological processes that bring about the accumulation of sediments in almost every environmental niche on the face of the earth. Sedimentation involves the liberation of established rock and mineral constituents of the earth's crust (see *Weathering in Sediments*), their transportation (q.v.) and recombination as sediments, and their alteration, or diagenesis (q.v.). Fresh sediments are composed of chemically and physically altered fragments of previously existing mineral and organic matter, mixed with contemporary organic matter and new chemical precipitates. Despite the fact that many sediments and sedimentary rocks are mixtures of fragmental and precipitated components, it is useful to view chemically precipitated sediments simply as aggregates of fragmental material of molecular dimensions.

Under the omnipresent pull of gravity, and given sufficient time, all sediments tend to be shifted toward, and to accumulate at the bases of slopes. Consequently, most sediments accumulate layer upon layer in sedimentary basins of all sizes and shapes. The tendency to spread sediments in broad layers is aided by all agents such as winds and wind-induced water currents, which function with components of force acting roughly parallel to the earth's surface. The ultimate basin of accumulation, toward which all sediments tend to gravitate, is the deepest part of the ocean; but this tendency is frequently interrupted by "temporary" deposition in any of a number of sedimentary environments (q.v.).

Some sediments come to rest very near the source of their constituents, but others are transported great distances before coming to rest. Some sediments have chemical and mineral composition very similar to their source rocks; others are much more complex than any known source rock, because components from many sources often become mixed in new sediments. Conversely, by processes of selective erosion, transportation, and deposition (i.e., sorting), material from complex source rocks may yield relatively simple sediments.

Few places in the world are sterile, so that sedimentation virtually always occurs in the presence of microorganisms such as bacteria, fungi, and algae, which have a profound effect upon all chemical reactions that occur in sedimentary environments (see *Microbiology in Sedimentology*). Likewise, few places in the world are perfectly dry, so that water almost always plays a role in sedimentation. In desert sedimentary environments (q.v.), where air is the dominant sedimentary medium, water is present in quantities scarcely sufficient to sustain simple microorganic life; but elsewhere, as in the oceans, water is an overwhelmingly forceful chemical and physical agent of erosion, transportation, and deposition of sediments.

In sedimentation, time is the great unifying factor that magnifies the effects of minute reactions and quantities. Thus, in the billions of years since life began on earth the sedimentary effects of microorganic enzymes and thin films of water loom large in the chemical alteration and recombination of sedimentary matter.

It is difficult to distinguish sharply between geochemical processes that affect sediments during their transportation and processes that occur after their deposition. It is also extremely difficult to assess the complex of chemical, physical, and biological factors that interact to convert loose sediment to sedimentary rock. *Diagenesis* (q.v.) is the term that refers to all changes (at normal temperatures and pressures) that occur in a sediment after the moment of its deposition, and which often tend to convert it to rock. Excluded from diagenesis are the indefinitely higher temperatures and pressures that would be assigned to *metamorphism*. Many diagenetic processes have their chemical, physical, and biological origin *before* deposition of a sediment, commonly during transportation; thus, it is impossible, in terms of environmental factors, to set diagenesis totally apart from sedimentary weathering, erosion, transportation, and deposition. Consequently, as a broad scientific discipline, the study of sedimentation must deal with all of these special topics plus *lithification,* which is the conversion of loose sediments into sedimentary rocks.

Clearly, any detailed inquiry into sedimentation requires study of its various geological, chemical, biological, and physical participants and their interrelationships.

Understanding of the geochemical aspects of sedimentation requires assessment of the chemical properties of, and reactions between, both the mobile medium and the sedimentary particles contained therein. Chemical reactions that occur before, during, and after deposition all must be taken into account. The chemical framework of the sedimentary environment is also shaped by the biochemical activity of the organisms that dwell in the mobile medium and in the pore spaces between sedimentary particles. Conversely, the biota of the sedimentary environment is regulated by available chemical nutrients and by geochemical controls of metabolic functions such as osmoregulation and photosynthesis.

The physical aspects of sedimentary environments both govern and depend upon their chemical and biological properties. For example, the temperature of the mobile medium greatly affects rates of sedimentary chemical reaction. Also, by its relations to thermal convection currents, temperature influences the mixing and interaction of physico- and biochemical reagents. Hydraulic physical properties (viscosity, density, shear, etc.) of the mobile medium are similarly significant, because, through the influence of variable turbulence and current velocity, sediments are selectively eroded, transported, and allowed to settle. Such hydraulic selection of sediments (according to composition, size, and shape) greatly affects the nature, speed, and continuity of chemical and biological activities in accumulated sediments. Likewise, the simple burrowing of worms through a sediment alters it both physically and chemically by changing permeability and by adding catalytic enzymes. These changes control, to a large degree, the chemical diffusion in both the solid and liquid phases of sediments.

All measureable aggregate properties of sediments are the combined result of extremely complex interactions of many variable factors. Consequently, measurements of properties such as hydrogen ion concentration (pH) or oxidation-reduction potential (Eh) of sedimentary aggregates or fluids, are, at best, approximations of the quasi-equilibria established in sediments at the moment of measurement. With the expulsion of sedimentary fluids during lithification and diagenesis, great changes in aggregate physical and chemical (and hence, biological) properties occur in sediments. The usual geological result of these changes is *sedimentary rocks* (q.v.), whose properties bear the stamp of the many interdependent factors that figure in their history. At any instant of study, the properties of a sedimentary rock may appear very permanent, but it must be remembered that in the span of geologic time, even the most durable rocks experience change. Thus, the major differences between early and late diagenetic changes in sediments is the rate at which they progress—rapid at first, then progressively slower.

In broad perspective, it is natural to notice the philosophic parallel between the evolution of organisms and the evolution of continents, which are largely mantled by sedimentary rocks. Both kinds of evolutionary processes are characterized by slow, continuous, irreversible activities, which embody changes of form and composition of materials that are used over and over again. Sediments that are used again and again in the formation of new ones are said to be *reworked*. Such changes in rock and mineral material, in constant exchange with elements of the atmosphere and lithosphere, is obviously a manifestation of the Law of Conservation of matter and energy; nothing is created or destroyed, it is merely changed from one form to another and mixed in various proportions.

ALISTAIR W. McCRONE

Cross-references: *Aggradation; Authigenesis; Bioturbation; Cementation; Compaction in Sediments; Continental Sedimentation; Cyclic Sedimentation; Deposition; Diagenesis; Dispersal; Lateral and Vertical Accretion; Marine Sediments; Redeposition; Resedimentation; Sedimentary Environments; Sedimentary Petrology; Sedimentary Rocks; Sedimentation-Paleoecologic Aspects; Sedimentation Rates; Sedimentation-Tectonic Controls; Sedimentology; Transportation; Weathering in Sediments.*

SEDIMENTATION—PALEOECOLOGIC ASPECTS

The role played by organisms in sedimentation may be considered under two headings: (1) their role as one of the major controlling processes in sedimentation, and (2) the more specific effects of organisms in various sedimentary systems where they may be a major or minor component.

Organisms as a Sedimentation Control

Organisms exert a control on sedimentation in eight ways. First, organisms are sediment generators of shell and other skeletal materials from animals and plants. Such materials may accumulate where the organisms actually lived and grew autochthonously (e.g., a coal seam integrated with a root horizon, or an ecologically complex coral reef). Closely related are pelagic sediments, such as chalk (q.v.), formed largely by the accumulation from settling of the microscopic skeletons of coccoliths. Calcareous skeletons did not evolve until the Phanerozoic, so that in pre-Phanerozoic sediments there are only pseudoskeletal structures (stromatolites and thrombolites) formed by the trapping of sediment by filamentous and coccoid algae (see *Stromatolites*).

Second, organisms are liable to undergo physical transport, abrasion, and sorting during life and, especially, after death (see *Biostratinomy*). Such physical processes may lead to the formation of calcareous breccias peripheral to a coral reef, or shelly sandstones where the skeletal debris is concentrated and mixed with terrigenous sediment. These deposits are allochthonous.

Third, the chemical regime external to that under which the skeleton was formed is liable to change, either during the life of the organism or, especially, after death. This change often affects the skeleton, which may be partly or wholly chemically modified; differential solution of aragonite and calcite is particularly frequent. It can also affect the local environment, leading to precipitation or solution in the surrounding sediment. The formation of concretions and nodules including flint, chert, and pyrite are typical examples.

Fourth, organisms are important agents of destruction and erosion. On the macroscopic scale, certain mollusks, sponges, and polychaetes and, to a lesser extent, other groups of organisms (Warme, in Frey, 1975) corrode and bore, by chemical and/or mechanical means, into most rock types, but particularly into calcareous sediment. Microscopic boring fungi and algae attack rock (Schneider, 1976), and in terrestrial environments humic acid and root penetration aid rock weathering. The principal means of recognizing fossil *hardgrounds* (q.v.) is by their encrusting and boring fauna. Excretion by predators and grazers of calcareous organisms and sediment is also responsible for much carbonate mud.

Fifth, organisms modify unconsolidated sediment by burrowing, mainly in search of food and shelter, thereby destroying the primary sedimentary structures. This process of *bioturbation* (q.v.) may lead to the complete destratification of the sediment, obliterating sediment lamination and fabric. In interbedded sands and muds, a mottled and poorly sorted sediment is the typical end product. Particle size and sorting analyses of such sediment can have little hydraulic significance. Where bioturbation has been less intense, the secondary structures produced (trace fossils) may be informative of the depositional environment (see *Biogenic Sedimentary Structures*). A number of sedimentary facies have been recognized on the basis of trace fossil assemblages (see Frey, 1975).

Sixth, organisms influence the physical processes of sedimentation by baffling, and then by trapping and binding the sediment. Grasses, sea grasses, many types of algae, mangroves, and, indeed, all sessile organisms can "locally" modify the hydraulic regime to induce sedimentation. Alternatively, massive organic stands may obstruct and divert flow and thus indirectly change the sedimentary pattern. This effect is particularly evident in present-day fluvial and deltaic environments where increased runoff is associated with deforestation and overgrazing. Allen (1965) has contrasted the tropical delta with its mangrove-influenced distributaries and intertidal system with the temperate, sparsely vegetated delta and tidal flats. The retention of sediment is perhaps most important at its source, and the increase in continental vegetative cover, particularly following the evolution of rhizomatous plants (especially grasses) in the late Mesozoic and Tertiary must have considerably decreased surface runoff in otherwise comparable conditions.

Seventh, the evolution of organisms through time has led to the formation of different environments and sediments, and often to entirely novel ones, such as reefs and vegetated swamps (inland and peripheral), and indirectly, to lagoons.

Eighth, the indirect biochemical role of organisms on otherwise stable chemical equilibria is probably the most important control, although geologically slow in its action. The present proportions of oxygen, carbon dioxide, and calcium carbonate have been organically determined, and indeed, all elements and compounds that are concentrated, isolated, or generated by living organisms must ultimately affect the chemical equilibrium of the earth, insofar as complete recycling does not occur.

Role of Organisms in Various Environments

The particular role of organisms in various sedimentary environments is not easy to synthesize. The only environments in which organisms play no significant role are those in which the limits of organism tolerance to temperature, oxygen, and toxins (e.g., heavy metals, detergents, acids) are exceeded: hot and cold deserts and anaerobic environments, particularly anaerobic substrates, and volcanic vents. In other environments, conditions may be so inhospitable that life is unable to be supported locally, for instance, as tends to occur in the surf zone, due to high turbulence and substrate instability. Only derived organic particles will survive in such environments; they may be locally abundant, as in bone or shell beds.

Oxygen and Turbulence. Availability of oxygen is the most fundamental factor determining the distribution of organisms, although other factors may be the limiting ones. Schäfer has proposed five environmental models applicable to aqueous environments with turbulence as the principal environmental factor (Fig. 1). Turbulence nearly always has a positive correlation with oxygenation and, in degree, is generally reflected in sedimentation by the form of the sedimentary structures (cross-stratification, ripple bedding, discontinuities).

Physiological and Related Aspects. In order to understand the role organisms play in sedimentation, it is important to fully appreciate how organisms develop, build their skeletons, move and behave, reproduce and die (Schäfer, 1972). Following death, two aspects are particularly important in sedimentation: how skeletons disarticulate and how shells, bones, teeth, etc. are oriented by currents and wave action (see *Biostratinomy*). The latter aspect is to some extent predictable, inasmuch as the shape of organisms often approximates a model of known hydraulic properties. However, disarticulation and fragmentation must mostly be determined experimentally. Their proper understanding is particularly important in locating and extracting large fossil vertebrates.

Substrate Significance. The nature of the

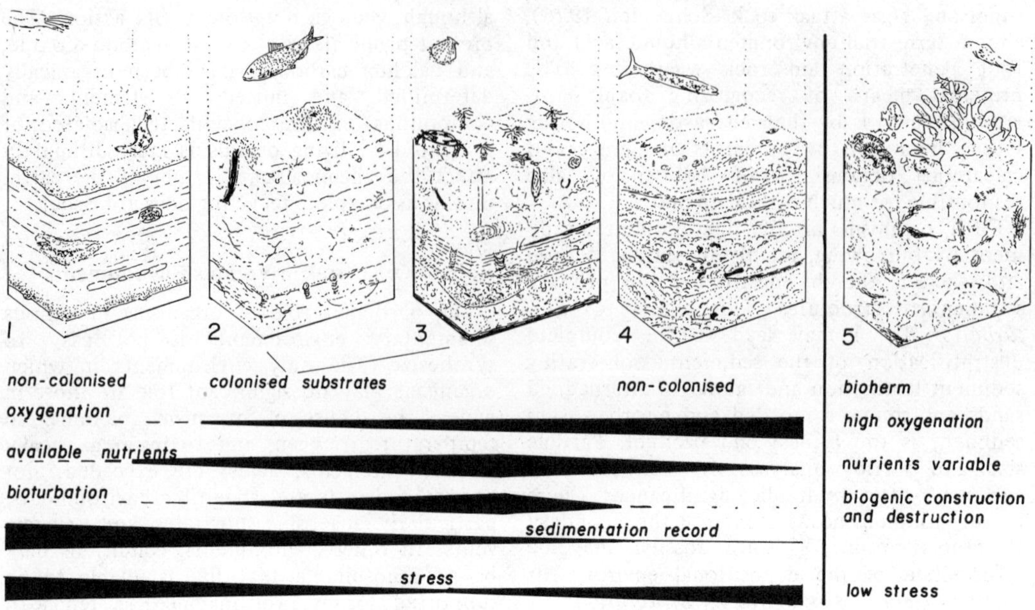

FIGURE 1. Generalized organism-sediment relationships in marine environments (based on Schäfer, 1972) and applicable to nonmarine environments with modification. Oxygenation increases left to right. Nutrients available to organisms are at a maximum in environment 3. The sedimentary record is most complete in environment 1. Bioturbation is absent in 1 and 4 and most typical of 3. Stratification is essentially absent in environment 5. In general, organisms live under less stress in environments 3 and 5 than in environments 1, 2, and 4. (1) In environment 1 there is no benthos. The complete sedimentary record (mud and occasional graded silts) shows an entombed cephalopod, and a concretion has enclosed a fish. Another fish decays at the sediment/water interface (Black Sea type, see *Euxinic Facies*). (2) In environment 2 there is sufficient oxygen for a deposit-feeding holothurian and surrounding suspension-feeding polychaetes. The sedimentary record is fairly complete (offshore bay, e.g., Cape Cod Bay). (3) Environment 3 is well oxygenated. The muddy sand carries sufficient nutrients for a rich and diverse epifauna and endofauna of several trophic types. Occasionally sedimentation may be so rapid that the infauna has to escape upward. Turbulence may also occasionally be sufficient to produce local discontinuities (inner shelf, e.g., Gulf of Mexico). (4) Environment 4 is too unstable to support much life, and the repeated discontinuities record high turbulence. Although the sedimentation rate is high the record is very incomplete. The sediment is relatively coarse (high-energy river mouths, bars, oolite shoals). (5) The almost wholly organic environment 5 is "reefal" and highly oxygenated. It supports a diverse assemblage with many specialized ecological niches. At depth, diagenetic changes tend to mask the growth forms.

substrate is, after oxygenation, an important factor. A fairly close correlation exists between particle size and trophic type, with suspension feeders typical of sandier substrates (associated with greater water movement) and deposit feeders typical of muddier substrates (Rhoads, 1974). In one particular study, e.g., Rhoads and Young (1970) describe how deposit-feeding nuculid bivalves produce a fluid-rich surface which tends to clog structures of filter feeders, buries or discourages settling of suspension-feeding larvae, and prevents sessile epifauna from settling. A short distance above the amensal substrate, conditions may be suitable for suspension feeders. However, deposit-feeding holothurians have elsewhere created an irregular substrate where the more elevated parts (fecal mounds) are densely populated by suspension-feeding polychaetes, forming a consolidating mat attractive to other suspension feeders.

One can identify fossil assemblages in Mesozoic and Tertiary shelf sediments with trophic type and sediment composition associations similar to Holocene assemblages. In Paleozoic shelf environments it has been claimed that the distribution of brachiopods was not, in general, related to sediment composition.

The settlement of epizoan larvae seems to be determined by the attractiveness of the substrate as well as by previous settlement. The condition of the substrate will determine whether the larvae are able to attach and resist abrasion. The possibilities of burrowing are also determined by grain size and sorting, and there is a relationship between the sediment type and method of penetration employed (e.g., digging by arthropod claws, or protrusion of foot or proboscis and utilizing dilatency effects). Sediment cohesiveness determines the need for the organism to line the burrow (e.g., *Callianassa major* and the trace fossil *Ophiomorpha*). Particle-feeding crustaceans can cope with only a limited range of grain size, and there is often good correlation between species frequency and grain size (Warren and Sheldon, 1968).

Evolutionary Aspects. Whereas the physical and chemical evolution of the earth was essentially stabilized in the Precambrian, organic evolution has been continuous. Hence, all environments in which organisms are a factor must be experiencing some evolutionary changes. Evolution of corals and calcareous algae (Newell, 1972; Wilson, 1975) provides an example of organism control on facies evolution and diversification. Certain facies may be characteristic of particular periods (e.g., Waulsortian bioherms in the Ordovician to Permian, or chalk, which is dominantly Cretaceous). There is constant pressure by organisms to extend and modify their habitats, and humans are no exception; indeed we are responsible for the main (continental and coastal) sedimentation changes today.

ROLAND GOLDRING

References

Allen, J. R. L., 1965. Coastal geomorphology of eastern Nigeria: Beach-ridge barrier islands and vegetated tidal flats, *Geol. Mijnbouw.*, **44**, 1-21.

Frey, R. W., ed., 1975. *The Study of Trace Fossils.* New York: Springer, 562p.

Newell, N. D., 1972. The evolution of reefs, *Sci. Am.*, **226**(6), 54-65.

Rhoads, D. C., 1974. Organism-sediment relations and the muddy sea floor, *Oceanogr. Mar. Biol. Ann. Rev.*, **12**, 263-300.

Rhoads, D. C., and Young, D. K., 1970. The influence of deposit-feeding organisms on sediment stability and community trophic structure, *J. Marine Research*, **28**, 150-178.

Schäfer, W., 1972. *Ecology and Palaeoecology of Marine Environments.* Edinburgh: Oliver and Boyd, 568p.

Schneider, J., 1976. Biological and inorganic factors in the destruction of limestone coasts. *Contributions to Sedimentology 6.* Stuttgart: Schweizerbart, 112p.

Warren, P. J., and Sheldon, R. W., 1968. Association between *Pandalus borealis* and fine-grained sediments off Northumberland, *Nature*, **217**, 579-580.

Wilson, J. L., 1975. *Carbonate Facies in Geologic History.* New York: Springer, 471p.

Cross-references: *Biogenic Sedimentary Structures; Biostratinomy; Bioturbation; Chalk; Coal; Euxinic Facies; Facies; Limestones; Organic Sediments; Reef Complex; Sedimentary Environments.*

SEDIMENTATION RATES

Sedimentation rates generally cannot be expressed in absolute data because periods of rapid sedimentation alternate with periods of slower deposition, nonsedimentation, or erosion. Nevertheless, it is important to gain some understanding of the *average* values of *net* sedimentation in various depositional environments in order to better comprehend the geological and chemical processes that take place on the surface of the earth. An understanding of net sedimentation rates has become increasingly valuable with the onset of intensive water pollution studies, because sedimentation is one of the most important processes in the removal of pollutants from natural waters. Also, accurate data on the rate of sedimentation provide estimates of the material exchange between water and sediment and are necessary for the correct interpretation of the entire sedimentation process.

Combinations of the biological, chemical, geological, and geographical factors that influence sedimentation rates are almost infinite, are different for each depositional environment, and have continuously fluctuated throughout the past. Consequently, the sedimentation rates listed in Table 1 should only be considered as average values of *net* sedimentation.

Methods to Determine Sedimentation Rates

Direct Observations. Early measures of net sedimentation rates were estimated from direct observation involving a systematic comparison of old and new soundings and geologic surveys (Shepard, 1953; Arnal. 1961). Direct observation methods are best suited to river deltas, lagoons, bays, and estuaries, where sedimentation rates are high; but the results are complicated by several factors, such as tectonic movement and sediment compaction.

Theoretical Calculations. Detrital sedimentation rates may be obtained by subtracting the amount of sediment leaving from the amount of sediment entering a basin over a specific time interval. Rates thus obtained assume deposition in a closed or steady-state system.

Radiometric Dating. Carbon-14 has a 5730-yr half-life and is continuously produced in the atmosphere; it is rapidly oxidized to CO_2 and becomes distributed in the carbon of all living organisms. If the organic material in a sediment was deposited at the same time as the organism's death, the residual carbon of the fossil material can be used to estimate sediment ages as old as 50,000 years or more. Other radionuclides produced in the outer atmosphere and used for radiometric dating are Be^{10}, Al^{26}, and Si^{32} (Broecker, 1965; see *Radioactivity in Sediments*).

Long-lived uranium and thorium isotopes (U^{238}, U^{235}, and Th^{232}) decay through a series of intermediate daughter products to stable isotopes of lead. The daughter products most suitable for radiometric dating are U^{234}, Th^{230}, Pa^{231}, and Pb^{210}, with half-lives of 250,000, 75,000, 35,000, and 22.3 yr respectively (Broecker, 1965; Koide et al., 1972). Sedimentation rates are measured by using the decay constant of the radionuclides and determining their decrease in concentration, to secular equilibrium values, with depth in the sediment. The long half-lives of U^{234}, Th^{230}, and Pa^{231} make these nuclides most suitable for dating oceanic cores where sedimentation rates are relatively low (Broecker, 1965; Goldberg and Koide, 1962; see also *Sedimentation Rates, Deep-Sea*). The shorter-lived Pb^{210} is used to date sedimentary processes with time periods on the order of a century, such as lake sedimentation (Krishnaswami et al., 1971) and nearshore shelf sedimentation (Koide et al., 1972.)

Many of the common rock-forming minerals (muscovite, biotite, K-feldspar) contain potassium-40 and may thus be used for dating by the K-Ar method. The K-Ar method is not directly applicable to obtaining sedimentation rates, however, because it is used to date minerals at the time of their formation, rather than at the time of their deposition as sediments. Nevertheless, several deep-sea sedimentation rates have been determined by K-Ar dating of individual volcanic ash layers which arrive at the depositional site soon after their formation during volcanic episodes (Dymond, 1966).

Paleontological Methods. The last glacial period ended about 11,000 yr ago, and this abrupt change in climate is reflected in deep-sea sediments by relative abundances of warm-versus cold-water species of foraminifera (see *Sedimentation Rates, Deep-Sea*) and in lake sediments by different pollen types.

Amino Acid Racemization. Various marine sediments have been found to contain small quantities of amino acids commonly associated with the proteins of organisms. These amino acids undergo slow racemization (a process in which amino acids of the L-configuration convert to a D-configuration). Sedimentation rates based on the rate of racemization of particular amino acids may be calculated by determining the degree of racemization with increasing depth in a core (Bada et al., 1970). As in all chemical reactions, the rate of racemization is temperature dependent and thus is best suited for dating sediments that have maintained a relatively constant temperature.

Varves. In glacial lakes and in certain other environments, the rate of sedimentation may be estimated directly from the number and thickness of varves (q.v.).

Pollutants. The distribution of pollutants in sediments has recently been used to determine sedimentation rates in rivers, estuaries, and bays. The rate is determined by measuring the thickness of the sediment layer containing a pollutant which has been introduced into the depositional environment at a known time. Pollutants most often used are: DDT (dichlorodiphenyltrichloroethane), introduced into the environment with modern insecticides; ABS (alkylbenzenesulphonates), widely used in synthetic detergents (Ambe, 1972); Fe^{55} and Cs^{137}, radionuclides produced by nuclear testing (Ritchie et al., 1973); chromium (Capuzzo and Anderson, 1973) and other trace metals introduced by industry; and coal, introduced to estuarine and river systems with the onset of barge transportation (Moore, 1971).

Recent Sedimentation Rates

Numerous investigators have determined sedimentation rates in various depositional environments; some of these are compiled in Table 1. Most of the sedimentation rates reported in Table 1 are based on vertical accumulation. It has been pointed out, however, that lateral sedimentation rates can be orders of magnitude faster than vertical rates (LeFournier and Friedman, 1974; see *Lateral and Vertical Accretion*). Representative values of lateral sedimentation rates are listed at the end of Table 1.

Fluvial. The most extensive vertical deposition of sediment by rivers occurs during floods (see *Fluvial Sedimentation*). Coleman (1969) investigated channel deposition and erosion patterns of the braided Brahmaputra River in E Pakistan during flooding and found that as the current velocity decreased, rapid sedimentation occurred, and as much as 3 m of sediment was deposited along the channel bottom. When a meandering river floods its banks, its velocity is rapidly checked, and sediment deposition occurs adjacent to the banks. The rate of floodplain deposition usually ranges from several mm to several cm/yr (Kukal, 1971).

Lacustrine. Significant differences exist in lacustrine sedimentation rates due to differences in the supply of terrigenous material, in the amount of biological productivity, and in the intensity of chemical sedimentation. Also, rates may vary within the same lake; the

TABLE 1. Sedimentation Rates in Various Depositional Environments

Depositional Environment	Sedimentation Rate (cm/1000 yr)	Reference
Fluviatile		
Nile River Flood Plain	900	Kukal, 1971
Ohio River Flood Plain (US)	450	Moore, 1971
Lacustrine		
Lake Leman (glacial, Switzerland)	120	Krishnaswami et al., 1971
Lake Pavin (crater lake, France)	130	Krishnaswami et al., 1971
Lake Michigan (US)	300	Kukal, 1971
Lake Tahoe (US)	100	Koide et al., 1972
Trout Lake (US)	400	Koide et al., 1972
Mule Creek Reservoir (artificial, US)	10,000	Ritchie et al., 1973
Salton Sea (saline, US)	500–1,400	Arnal, 1961
Eolian		
Loess (glacial)	20–100	Kukal, 1971
Deltaic		
Alamo River Delta (US)	5,000	Arnal, 1961
Fraser River Delta (Canada)	5,000–30,000	Kukal, 1971
Mississippi River Delta (US)	6,000–45,000	Kukal, 1971
Orinoco River Delta (Venezuela)	1,000	Kukal, 1971
Tama River Delta (Japan)	3,000–7,000	Ambe, 1972
Estuarine		
Great Bay Estuary (US)	160–780	Capuzzo and Anderson, 1973
Hamton Estuary (US)	100–230	Keene, 1970
James River Estuary (US)	150–300	Nichols, 1972
Bays, Gulfs, Lagoons and Sounds		
Corpus Cristi Bay (US)	≈380	Shepard, 1953
Florida Bay (calcareous, US)	46	Scholl, 1966
Kiel Bay (Germany)	150–200	Kukal, 1971
San Francisco Bay (US)	30–130	Story et al., 1966
Mobile Bay (US)	560	Ryan and Goodell, 1972
Delaware Bay (US)	150	Oostdam and Jordan, 1972
Gulf of California (US)	60–100	Kukal, 1971
Gulf of Paria (Venezuela)	0–100	Kukal, 1971
Long Island Sound (US)	600	Thomson and Turekian, 1973
Tidal Marshes and Peat Bogs		
Tidal flats (The Wash, England)	1,000–2,000	Evans, 1965
Tidal flats (Netherlands)	1,000–2,000	Kukal, 1971
Salt marsh (Skallingen Penninsula, Denmark)	360	Schou, 1967
Salt marsh (US)	200–650	Harrison and Bloom, 1974
Peat bogs (North America)	550	Kukal, 1971
Swabian high moors (Germany)	150–180	Kukal, 1971
Inland Seas		
Adriatic Sea	1	Kukal, 1971
Baltic Sea	20–200	Alhonen, 1966
Black Sea	5–40	Ross et al., 1970
Caspian Sea	10–18	Kukal, 1971
Mediterranean Sea	10–20	Kukal, 1971
Marine		
Siliceous sediments	0.5–2	Kuznetsov, 1969
Calcareous sediments	1–4	Ku et al., 1968
Red clays	0.1–0.4	Kuznetsov, 1969
Eolian sediments	0.06	Delany et al., 1967
Detrital sediments		
Shelf (North America)	0–400	Kukal, 1971
Rise (North America)	6.8	Emery et al., 1970
Abyssal plain (North America)	2	Emery et al., 1970

TABLE 1 continued

Depositional Environment	Sedimentation Rate (cm/1000 yr)	Reference
Authigenic		
Mn nodules and crusts	0.0003	Bender et al., 1970
Postdepositional CaCO$_3$ concretions	0.4-2	Berner, 1968
Evaporites (Kara-Bogaz Gulf)	500	Fairbridge, 1968
Organic and Biogenic		
Organic phosphates	2,000-4,000	Kukal, 1971
Coral reefs	33-4,000	Kukal, 1971
Algal reefs	200-700	Kukal, 1971
Stromatolites (maximum growth rate)	1.1 mm/day	Gebelein, 1969
Lunar		
Lunar regolith	0.0001	Lindsay, 1975
Lateral Sedimentation		
Long Island barrier	65 m/yr	Kumar, 1973
Sandy Hook spit (N.J.)	12 m/yr	Sanders, 1970
Picardy coast spit (France)	10 m/yr	LeFournier and Friedman, 1974
Fluvial	30-80 m/yr	Wolman and Leopold, 1975

rate is generally maximum where the supply of detrital material is greatest, i.e., near river mouths and along the lake periphery. The average rate of deposition in oligotrophic, fresh-water lakes is on the order of 300 cm/1000 yr; but this rate is highly affected by changes in climate, thermal condition, salinity, productivity, and topography (Twenhofel and McKelvey, 1941; see *Lacustrine Sedimentation*).

Eolian. Delany et al. (1967), made collections of wind-borne dust on the island of Barbados and estimated its contribution to the rate of sedimentation in the western Tropical Atlantic as 0.06 cm/1000 yr. On the continents, wind storms can deposit several mm to cm of dust during a few hours, but such storms are generally separated by long periods of nonsedimentation and erosion. During the Pleistocene glacial stages, wind-blown dust (loess, q.v.) was deposited at a rate of 20-100 cm/1000 yr (Kukal, 1971).

Deltaic. Deltas of major rivers are sites of long-term deposition and subsidence and have some of the highest rates of sedimentation, averaging 3000-10,000 cm/1000 yr. The rate of sedimentation varies in different areas of a delta. The lowest rates occur near the seaward (or lakeward) end of the delta. Depositional rates increase as the mouth of the river is approached and are maximum near the top of the foreset slope (Scruton, 1960).

Estuarine. Estuaries are generally sites of rapid sedimentation due to several factors: (1) an abundant supply of detritus from the river drainage system, (2) variations in the current velocity associated with alternating tides, (3) flocculation effects, and (4) estuarine dynamics (see *Estuarine Sedimentation*). Sedimentation rates in modern estuaries average 150-600 cm/1000 yr (Rusnak, 1967).

Bays, Lagoons, Gulfs, and Sounds. Bays, lagoons, gulfs, and sounds, like estuaries, act as settling basins for river-borne sediments. Sedimentation rates are generally lower than those recorded in estuaries, however, because bays, lagoons, gulfs, and sounds are more exposed to the sea. Rates of deposition average 100-500 cm/1000 yr and may vary in different areas of the same water body, due to the amount of sea exposure and the proximity to land.

Tidal Marshes, Mud Flats, Swamps, and Peat Bogs. Average rates of deposition for mud and fine-grained suspended material on tidal marshes range from 1000-2000 cm/1000 yr (Kukal, 1971; see *Tidal-Flat Geology*). Kukal estimates the average rate of peat-bog accumulation at 2 mm/yr (200 cm/1000 yr).

Inland Seas. The rate of deposition in inland seas is generally indirectly proportional to their dimensions (Kukal, 1971). The larger the basin, the slower the rate of sedimentation because the supply of terrigenous material is spread over a larger area. Average rates of deposition for inland seas are 5-30 cm/1000 yr.

Marine. The rate of sedimentation in the marine environment varies with climatic, circum-continental and bathymetric zonalities, organic productivity, and the type of sediment deposited (see Table 1 and *Sedimentation Rates, Deep-Sea*). The highest rates are associated with detrital sedimentation near the continents, and the lowest rates are associated with red-clay sedimentation and manganese nodule formation in the deep, S central Pacific.

Authigenic. Authigenic sediments (q.v.) form by direct precipitation from water and occur most often in areas with extremely low detrital and biogenic sedimentation rates or also in highly saline depositional environments. *Manganese nodules* and crusts may grow at rates on the order of 3 mm/10^6 yr (Bender et al., 1970); postdepositional concretions precipitate from slowly flowing ground water, supersaturated in $CaCO_3$ by 10 ppm, at rates of 0.4-2 cm/1000 yr (Berner, 1968). *Evaporites* precipitate from the Gulf of Karabogaz at a rate of 1.3×10^8 tons/yr (Fairbridge, 1968); assuming an average density of 2.5 gm/cm^3 for evaporites and given the area of Karabogaz (10,500 km^2), this rate of salt deposition corresponds to 500 cm/1000 yr.

Organic and Biogenic. Gebelein (1969) has studied the accretion rate of subtidal algal *stromatolites* on the Bermuda Islands. He found that one complete lamina (composed of a thick—1 mm—sediment-rich layer capped by a thin—0.1 mm—algae-rich layer) forms every 24 hours. Bird and bat *guano* may be a major contributor to sedimentation rates on various islands and in some caves, and the rate of deposition of organic *phosphates* may be as high as several cm/yr (Kukal, 1971). The average growth rate of *coral reefs* in the upper parts of atolls is about 14 mm/yr; whereas on the submerged outer reef it ranges from 0.33-0.91 mm/yr, and in the lagoonal environment attains 3.8 mm/yr (Kukal, 1971).

Lunar. The lunar surface is covered by a weakly coherent fragmental material, termed lunar regolith, which is probably a product of occasional meteorite impacts. At the Apollo 12 site this loose material is approximately 3.7 m thick and overlies a more coherent substrate which has been dated at 3.26 billion yr (Lindsay, 1975). This suggests that the net average rate of regolith accumulation is on the order of 1 mm/10^6 yr.

CURTIS R. OLSEN

References

Alhonen, P., 1966. Baltic Sea, in Fairbridge, 1966, 87-91.
Ambe, Y., 1972. ABS as a geological tracer, *Nature (London), Phys. Sci.,* 239, 24-25.
Arnal, R. E., 1961. Limnology, sedimentation, and microorganisms of the Salton Sea, California, *Geol. Soc. Am. Bull.,* 72, 427-478.
Bada, J. L.; Luyendyk, B. P.; and Maynard, J. B., 1970. Marine sediments: dating by racemization of amino acids, *Science,* 170, 730-732.
Bender, M. L.; Ku, T.-L.; and Broecker, W. S., 1970. Accumulation rates of manganese in pelagic sediments and nodules, *Earth Planetary Sci. Lett.,* 8, 143-149.
Berner, R. A., 1968. Rate of concretion growth, *Geochim. Cosmochim. Acta,* 82, 477-483.
Broecker, W. S., 1965. Isotope geochemistry and the Pleistocene climatic record, in H. E. Wright, Jr. and D. G. Frey. *The Quaternary of the United States.* Princeton, N. J.: Princeton Univ. Press, 737-753.
Capuzzo, J. M., and Anderson, F. E., 1973. The use of modern chromium accumulation to determine estuarine sedimentation rates, *Marine Geol.,* 14, 225-235.
Coleman, J. M., 1969. Brahmaputra River: Channel processes and sedimentation, *Sed. Geol.,* 3, 129-239.
Delany, A. C.; Parkin, D. W.; Griffin, J. J.; Goldberg, E. D.; and Reimann, B. E. F., 1967. Airborne dust collected at Barbados, *Geochim. Cosmochim. Acta,* 31, 885-909.
Dymond, J. R., 1966. Potassium-argon geochronology of deep-sea sediments, *Science,* 152, 1239-1241.
Emery, K. O.; Uchupi, E.; Phillips, J. D.; Bowin, C. O.; Bunce, B. T.; and Knott, S. T.; 1970. Continental rise off eastern North America, *Bull. Am. Assoc. Petrol. Geologists,* 54, 44-103.
Evans, G., 1965. Intertidal flat sediments and their environments of deposition in the Wash, *Quart. J. Geol. Soc. London,* 121, 208-245.
Fairbridge, R. W., ed, 1966. *The Encyclopedia of Oceanography.* Stroudsburg, Pa.: Dowden, Hutchinson & Ross, 1021p.
Fairbridge, R. W., ed., 1968a. *The Encyclopedia of Geomorphology.* Stroudsburg, Pa.: Dowden, Hutchinson & Ross, 1295p.
Fairbridge, R. W., 1968b. Karabogaz Gulf, in Fairbridge, 1968a, 579-581.
Gebelein, C. D., 1969. Distribution, morphology and accretion rate of recent subtidal algal stromatolites, Bermuda, *J. Sed. Petrology,* 39, 49-69.
Goldberg, E. D., and Koide, H., 1962. Geochronological studies of deep-sea sediments by the ionium-thorium method, *Geochim. Cosmochim. Acta,* 26, 417-450.
Harrison, E. Z., and Bloom, A. L., 1974. The response of Connecticut salt marshes to the recent rise in sea level, *Geol. Soc. Am. Abstr.,* 6(1), 35-36.
Keene, H. W., 1970. Salt marsh evolution and postglacial submergence in New Hampshire, Thesis, Univ. New Hampshire, Durham, N. H., 87p.
Klenova, M. V., 1968. Caspian Sea, in Fairbridge, 1968a, 109-116.
Koide, M.; Soutar, A.; and Goldberg, E. D., 1972. Marine geochronology with ^{210}Pb, *Earth Planetary Sci. Lett.,* 14, 442-446.
Krishnaswami, S.; Lal, D.; Martin, J.; and Meybeck, M., 1971. Geochronology of lake sediments, *Earth Planetary Sci. Lett.,* 11, 407-414.
Ku, T.-L.; Broecker, W. S.; and Opdyke, N., 1968. Comparison of sedimentation rates measured by paleomagnetic and ionium methods of age determination, *Earth Planetary Sci. Lett.,* 4, 1-16.
Kukal, Z., 1971. *Geology of Recent Sediments.* New York: Academic Press (in Czechoslovakia: Prague, Czechoslovak Academy Sci.), 490p.
Kumar, N., 1973. Modern and ancient barrier sedi-

ments: New interpretation based on stratal sequence in inlet-filling sands and on recognition of nearshore storm deposits, *N.Y. Acad. Sci. Ann.*, **220**, 245-340.

Kuznetsov, Y. V., 1969. Rates of recent sedimentation in the oceans, *Geokhimiya*, **3**, 251-260.

LeFournier, J., and Friedman, G. M., 1974. Rate of lateral migration of adjoining sea-marginal sedimentary environments shown by historical records, Authie Bay, France, *Geology*, **2**, 497-498.

Lindsay, J. F., 1975. A steady-state model for the lunar soil, *Geol. Soc. Am. Bull.*, **86**, 1661-1670.

Moore, B. R., 1971. The distribution of Pennsylvanian-age coal particles in recent river sediments, Ohio River, Kentucky, as age and sediment rate indicators, *Sedimentology*, **17**, 129-134.

Nichols, M. M., 1972. Sediments of the James River estuary, Virginia, *Geol. Soc. Am. Mem. 133*, 169-212.

Oostdam, B. L., and Jordan, R. R., 1972. Suspended sediment transport in Delaware Bay, *Geol. Soc. Am. Mem. 133*, 143-149.

Ritchie, J. C.; McHenry, R. J.; and Gill, A. C., 1973. Dating recent reservoir sediments, *Limnol. Oceanogr.*, **18**, 254-263.

Ross, D. A.; Degens, E. T.; and MacIlvaine, J., 1970. Black Sea: Recent sedimentary history, *Science*, **170**, 163-165.

Rusnak, G. A., 1967. Rates of sediment accumulation in modern estuaries, *Am. Assoc. Advanc. Sci. Publ. 83*, 180-184.

Ryan, J. J., and Goodell, H. G., 1972. Marine geology and estuarine history of Mobile, Alabama, I. Contemporary sediments, *Geol. Soc. Am. Mem. 133*, 517-554.

Sanders, J. E., 1970. Coastal-zone geology and its relationship to water pollution problems, in A. A. Johnson, ed., *Water Pollution in the Greater New York Area*. New York: Gordon and Breach, 23-35.

Scholl, D. W., 1966. Florida Bay: A modern site of limestone formation, in Fairbridge, 1966, 282-287.

Schou, A., 1967. Estuarine research in the Danish Moraine Archipelago, *Am. Assoc. Advanc. Sci. Publ. 83*, 129-145.

Scruton, P. C., 1960. Delta building and the deltaic sequence, in F. P. Shepard et al., eds., *Recent Sediments Northwest Gulf of Mexico*. Tulsa: Am. Assoc. Petrol. Geologists, 82-102.

Shepard, F. P., 1953. Sedimentation rates in Texas estuaries and lagoons, *Bull. Am. Assoc. Petrol. Geologists*, **37**, 1919-1934.

Story, J. A.; Wessels, V. E.; and Wolfe, J. A., 1966. Radiocarbon dating of recent sediments in San Francisco Bay, *Calif. Div. Mines Geol. Mineral Inf. Service*, **19**(3), 47-50.

Thomson, J., and Turekian, K. K., 1973. Sediment accumulation rates in Long Island Sound by Pb-210 dating and an estimate of the Ra-228 flux, *EOS Trans.* (abstr.), **54**(4), 337.

Twenhofel, W. H., and McKelvey, V. E., 1941. Sediments of fresh water lakes, *Bull. Am. Assoc. Petrol. Geologists*, **25**, 826-849.

Wolman, M. G., and Leopold, L. B., 1957. River flood plains: Some observations on their formation, *U.S. Geol. Surv. Prof. Pap. 282-C*, C87-C107.

Cross-references: *Authigenesis; Continental Sedimentation; Deposition; Lateral and Vertical Accretion; Marine Sediments; Radioactivity in Sediments; Sedimentation-Paleoecologic Aspects; Sedimentation Rates, Deep-Sea; Sedimentation-Tectonic Controls; Varves and Varved Clays.*

SEDIMENTATION RATES, DEEP-SEA

One of the first reliable calculations of deep-sea sedimentation rates was made by Schott (1955), who identified stratigraphic layers in cores based on variations in abundance of warm- and cold-water species of pelagic foraminifera. On this basis, he calculated the rate of accumulation of blue mud, globigerina ooze, diatom ooze, and red clay from the tropical Atlantic and southwestern Indian oceans. Although Schott did not account for compaction during sampling and assumed that the length of time since the last glaciation is about twice that a recent estimates, his calculated sedimentation rates are in fair agreement with more recent estimates. Other earlier estimations based on, for example, sediment cover on transoceanic telegraph cables, abundance of cosmic spherules in deep-sea sediments, and coccolith production, generally have resulted in unreliable estimates of sedimentation rates (Kuenen, 1950).

Development of radiometric methods of dating have made possible more accurate determinations of sedimentary rates (see *Sedimentation Rates; Radioactivity in Sediments*). Radiocarbon and the uranium and thorium decay series are used most often. The K-Ar method, although not directly applicable to dating deep-sea sediments, can be used to data ash layers in the sediment.

A relatively new method of correlating and dating deep-sea sediments relies on the fact that the earth's magnetic field has changed polarity numerous times throughout the earth's history. Reversals of the earth's magnetic field have been recorded in deep-sea deposits, thus providing a method of worldwide correlation of sediment cores (Opdyke et al., 1966; Opdyke, 1972). Correlation of the reversal sequence (paleomagnetic stratigraphy) in a core with the geomagnetic reversal stratigraphy worked out on lava sequences on land makes it possible to determine the average rates of deposition between reversals. One normal or reversed epoch is difficult to distinguish from another, however, so that paleontological control is helpful. This method gives results similar to several radioactive methods (Ku et al., 1968; Dymond, 1969).

Based on the above methods it appears that the sedimentation rate for red clay (q.v.) varies from <1 to ≈4 mm/10^3 yr; globigerina ooze, 1–3 cm/10^3 yr; siliceous ooze, 1–10 mm/10^3 yr; and clays and silts (hemipelagic sediments) of the continental margins up to 60 cm/10^3 yr or higher (see *Pelagic Sedimentation, Pelagic Sediments*).

Distribution

Lisitsin (1971) constructed a map showing worldwide sedimentation rates for recent deep-sea sediments (Fig. 1), calculated primarily from paleomagnetic stratigraphies and by the ionium (excess Th^{230}) dating methods. Fossil and C^{14} data were also used. According to Lisitsin, the generalized distribution of sedimentation rates is determined primarily by a combination of climatic, circumcontinental, and vertical sedimentation zonalities. Superimposed on the overall pattern is an uneven distribution of bottom sediment, particularly in the regions of topographic highs where areas of zero or negative sedimentation (erosion) alternate with pockets having very high sedimentation rates. The lowest sedimentation rates may be <1 mm/10^3 yr. The regions of most rapid sedimentation form three latitudinal belts: the equatorial zone and the northern and southern humid zones. Here, sedimentation rates are usually >10 mm/10^3 yr and locally 30 or even 100 mm/10^3 yr. The high sedimentation rates in the humid zones are due to a high biogenic fraction and also to a large supply of terrigenous material. In addition, there are minor areas of slightly higher sedimentation rates in areas of upwelling off the western shores of the continents (e.g., SW Africa, SW South America).

Within the climatic zones, common to all the oceans, there are appreciable differences in sedimentation rates due to uneven supply of sediment to each ocean. The Atlantic Ocean is known to receive, on the average, the most sediment per unit area, followed by the Indian and Pacific oceans. Areas with rates as low as 1 mm/10^3 yr are small and very rare in the Atlantic and reach their maximum development in the Pacific, particularly in areas of red clay (q.v.).

Lisitsin (1971) notes that, all other conditions being equal, circumcontinental zonality is manifested by an increase in sedimentation rates with proximity to the continents. It appears, however, that the highest sedimentation rates do not generally occur on the continental shelves but rather at the base of the continental slope, i.e., the continental rise (see *Continental Rise Sediments*), where sedimentation rates can be as high as 60 cm/10^3 yr or higher. Near the mouths of large rivers, sedimentation rates are much higher. Abyssal plains also have accumulation rates higher than the surrounding ocean floor due to bottom-transported sediments (turbidites, q.v.).

The third major pattern of variation in

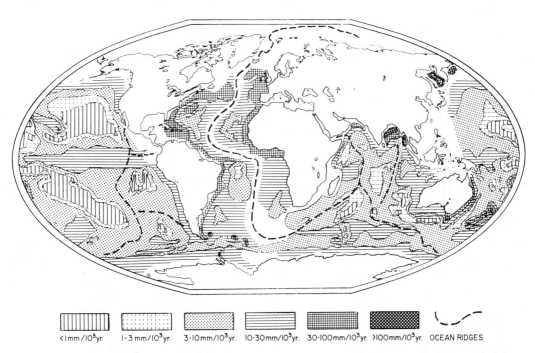

FIGURE 1. Sedimentation rates in the world's oceans (modified after Lisitzin, 1972).

sedimentation rates is associated with vertical or bathymetric zonality, primarily to a marked increase in calcium carbonate solution below a certain depth known as the critical or compensation depth (see *Pelagic Sedimentation, Pelagic Sediments*). The carbonate-rich sediments in many cases have sedimentation rates 10–20 times higher than adjacent red clays in waters below the calcium carbonate compensation depth. Extremely low sedimentation rates (particularly in the south-central Pacific) allow the production of predominantly authigenic deposits.

Variation with Time

Heezen and Hollister (1971) point out that in 10 m.y., close to the crest of the mid-ocean ridge, the crust receives as much sediment as it does in 100 m.y., deep on the ridge flank. Because the crust appears to be continuously moving into deeper water, higher rates of ooze accumulation on the new crust are later compensated by the lower rate of clay accumulation on the subsiding flanks, resulting in a sediment layer which only slightly increases in thickness toward the continental margins.

Radiometric dating techniques have shown that clay accumulation was highest during Pleistocene glacial periods and lowest during the nonglacial periods (Broecker et al., 1958). Reliable sedimentation rates for Tertiary and older sediments have been obtained only recently from cores drilled by JOIDES. Analyses of the results indicate that there is a dearth of Paleocene sediments and a peak in accumulation rates corresponding to highly calcareous Oligocene sediments (Moore, 1972). The late Campanian to early Maestrichtian (late Cretaceous) also seems to have been a time of high accumulation rates. Low accumulation rates, usually associated with hiatuses and reworked older microfossils, occurred in the late Miocene to Pliocene, the Paleocene, and around the Eocene-Oligocene boundary. Recent evidence suggests that there may be a relationship between deep-sea changes in sedimentation rates, fluctuations in the calcium carbonate compensation depth, and the rate of sea-floor spreading.

BILLY P. GLASS
L. J. ROSEN

References

Broecker, W. S., 1965. Isotope geochemistry and the Pleistocene climatic record, in H. E. Wright, Jr., and D. G. Frey, eds., *The Quaternary of the United States*. Princeton, N.J.: Princeton Univ. Press, 737–753.

Broecker, W. S.; Turekian, K. K.; and Heezen, B. C., 1958. The relation of deep-sea sedimentation rates to variations in climate, *Am. J. Sci.*, 256, 503–517.

Dymond, J., 1969. Age determinations of deep-sea sediments: A comparison of three methods, *Earth Planetary Sci. Lett.*, 6, 9–14.

Heezen, B. C., and Hollister, C. D., 1971. *The Face of the Deep.* New York: Oxford Univ. Press, 659p.

Ku, T.; Broecker, W.; and Opdyke, N. D., 1968. Comparison of sedimentation rates measured by paleomagnetic and ionium methods of age determination, *Earth Planetary Sci. Lett.*, 4, 1–16.

Kuenen, P. H., 1950. *Marine Geology.* New York: Wiley, 568p.

Lisitsin, A. P., 1971. The rate of sedimentation in the oceans, *Oceanology*, 11, 790–802.

Lisitsin, A. P., 1972. Sedimentation in the world ocean, *Soc. Econ. Paleont. Mineral. Spec. Publ. 17*, 218p.

Moore, T. C., Jr. 1972. DSDP: Successes, failures, proposals, *Geotimes*, 17(7), 27–31.

Opdyke, N. D., 1972. Paleomagnetism of deep-sea cores, *Rev. Geophys. Space Phys.*, 10, 213–294.

Opdyke, N. D.; Glass, B.; Hays, J. D.; and Foster, J., 1966. Paleomagnetic study of Antarctic deep-sea cores, *Science*, 154, 349–357.

Schott, W., 1955. Rate of sedimentation of recent deep-sea sediments, *Soc. Econ. Paleont. Mineral. Spec. Publ. 4*, 409–415.

Turekian, K. K., 1968. *Oceans.* Englewood Cliffs, N.J.: Prentice-Hall, 119p.

Cross-references: *Authigenesis; Continental-Rise Sediments; Pelagic Sedimentation, Pelagic Sediments; Radioactivity in Sediments; Red Clay; Remanent Magnetism in Sediments; Sedimentation Rates; Submarine Fan (Cone) Sedimentation; Turbidity-Current Sedimentation.*

SEDIMENTATION–TECTONIC CONTROLS

The close relationship between deformation of the earth's crust and sedimentary processes has only been fully appreciated within the last three decades. Almost all early structural geologists and stratigraphers assumed that the deposition of sedimentary rocks, and their subsequent deformation, were separate and discrete events. The concept was rarely stated explicitly; more frequently it was implied in such statements as "but the record . . . has been constantly interrupted; now by upheaval, now by volcanic outbursts, now by protracted and extensive denudation. These interruptions serve as natural divisions of the chronicle. . ." (Geikie, 1882). During this same period, however, James Hall and James Dana propounded the famous dictum that "the line of greatest accumulation is the line of the mountain chain," thus clearly connecting sedimentation with subsequent orogeny (see *Geosynclinal Sedimentation*). Nevertheless, for the next fifty years, although the geosynclinal concept became firmly established and greatly elaborated, the possibility that orogeny could proceed simultaneously with sedimentation appears not to have been envisaged. The role of earth movement was regarded simply as a determinant of depth of water, and as a provider of an elevated source of eroded material.

The recognition by Jones (1937) that large-scale sediment movement had immediately followed deposition in the Silurian rocks of N Wales stimulated a reappraisal of many geological theories, and slump bedding (q.v.) began to be recognized in many parts of the world. In 1948, Italian geologists, led by Migliorini, began to propound their revolutionary theory of the structure of the Apennines, in which they envisaged subaqueous sliding on a gigantic scale ("orogenic landslips") developed as a result of differential uplift of a succession of tectonic ridges (Migliorini, 1952; Fig. 1). Thus, a thick breccia sheet, formerly regarded as a tectonic nappe, was reinterpreted as a sedimentary deposit, driven forward by uplift of the basement as it accumulated (see *Olistostrome, Olistolite*).

Also at this time, Kuenen and his associates postulated the existence of large-scale turbidity currents in the modern oceans, and attributed many ancient graded sandstones to their activity (see *Turbidity-Current Sedimentation*). As early as 1936, Bailey had connected such sandstones with earthquake, and, therefore tectonic, activity; but it was not until the 1950s that the full implications were realized. It is now generally accepted that the flysch facies of the Alps and similar thick sandstones in other orogenic belts are of turbidity-current origin; that these turbidity currents were initiated by the seismicity of the orogenic period; and that they were forming during the orogenic process. This essentially kinematic view of sedimentation has been used to portray a coherent picture of mountain building in many parts of the world.

Controlling Factors

Although the closest tectonic control is observed in mountain ranges, it is also a major factor in foreland areas perhaps far removed from major orogeny. In the following, an attempt is made to classify genetically the results of such controls.

Depth of Water. On a small scale, local tectonic activity, e.g., the differential movement of fault blocks, is believed to be responsible for minor facies changes. Thus, in the Upper Devonian of the Rhenish Schiefergebirge, Rabien (1956) has described the stratigraphy in terms of two facies, a Cypridenschiefer or ostracode/shale facies developed in deep basins, and a Cephalopodenkalk facies characterizing the uplifted block between. In northern England, the facies changes occurring across the important Craven Fault belt have long been interpreted as reflecting the influence of underlying structure, the stable block to the N inducing the slow deposition of massive bioclastic carbonates, the subsiding trough to the S accumulating a great thickness of shales. In the Tertiary rocks of NW Europe, the accumulation of littoral sands and lacustrine clays has been pictured as occurring within downfaulted basins, as in SW England, S Belgium, and the Auvergne of France.

Tectonic Uplift as a Source of Detritus. The classical view of the origin of many major clastic formations is that they were derived from the erosion of a recently elevated mountain chain, e.g., the Old Red Sandstone of N Europe is supposedly derived from the breaking down of the Caledonide mountain ranges, and the Tertiary Alpine Molasse bears a similar relationship to the Alps. It appears that phases of mountain uplift were gradual and continually renewed, and that in consequence supply of clastic material was intermittent and varied. Many sequences show an alternation of coarse and fine sediment, referred to as "cyclic" or "cyclothemic," which has been ascribed to this fundamental cause

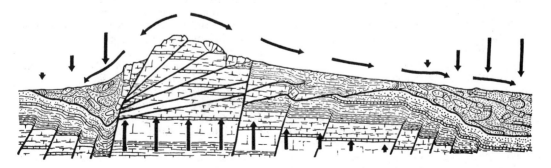

FIGURE 1. Diagram to show development of composite wedges in the basement, and "orogenic landslides" in the cover; an interpretation of the structure of the Apennines (Migliorini, 1952).

(e.g., Weller, 1956; see *Cyclic Sedimentation*). The influence of differential compaction, deltaic advance (progradation), and sea-level changes, however, may also have a significant effect.

A more immediate effect of tectonics is seen in the development of gravity-slide deposits (q.v.). Landslides and avalanches (see *Avalanche Deposits*) are included here but are not often preserved in the geologic record. Submarine landslides produced by the breaking down of submarine fault-scarps during their formation, however, are more commonly preserved. Such rock masses are usually chaotic and unsorted, the blocks they contain being angular and often of gigantic size (see *Olistostrome, Olistolite*). A particularly convincing example of successive developments of this kind has been described by Williams (1962) from the Ordovician of SW Scotland.

Tectonic Movement as an Activator of Gravity Flows. Submarine landslides generated by faulting may well extend their influence far from their origin, giving rise to slumps. Such movement could also generate instability in a pile of sediment, which would then flow down any available submarine slope as a gravity flow (see *Gravity Flows*), e.g., a turbidity current, thus redepositing the material at a great distance and in great depth of water. The common association of turbidites, resedimented conglomerates, and similar deposits, with orogenic belts supports this theory; these rocks are often themselves strongly folded, suggesting that sedimentation of this kind occurred during the orogeny. Paleocurrent analysis typically indicates depositional currents moving *along* the axis of the trough, with only infrequent movements from the side. The question, then, of the origin of the clastic material has given rise to the suggestion that such troughs are normally filled from the end, and that the currents are nourished by the uprising of "internal cordilleras" within the trough (Graham et al., 1975).

Orogeny and Sedimentation

On the basis of an understanding of the importance of tectonic control, it has been possible to discern several *consanguineous associations* of rock types, each of which has been related to a particular phase in the history of an orogeny. These associations have been described as preorogenic, synorogenic, and postorogenic, related to the classical geosynclinal theory (see *Geosynclinal Sedimentation*). Orogenies, however, are slow processes having neither clear beginnings nor well-defined ends, and it is better to use entirely noncommittal names as follows.

First phase (sometimes known as preorogenic or eugeosynclinal): Subsidence creates a deep trough unaccompanied by mountain uplift, so that the supply of clastic material is sparse. Thus, fine-grained rocks with a large organic content, mostly derived from planktonic or pelagic organisms, are characteristic. The black graptolitic shales of the Paleozoic in all parts of the world; the radiolarian cherts of the Variscan geosyncline in Europe, and of the Cordilleran area in the North American Mesozoic; and the *Schistes Lustres* of the Alps are good examples. The association with extrusive igneous rocks of the ophiolitic suite was first noted by G. Steinmann in the Apennines and the Alps, and is now known to be very common.

Second phase (synorogenic): The Alpine stratigraphic term *flysch* has been widely used for this rock association. It owes its special characters to the combination of a subsiding trough, an actively rising mountain range, and abundant seismicity. Under these circumstances a maximum development of turbidites is to be expected. Dzulynski (1963) and others have attempted to define these rocks.

Thick masses of this type are found in almost every geosyncline: in the Lower Paleozoic of Britain and Scandinavia; in Upper Paleozoic of Central Europe and eastern North America; in the Mesozoic of the Carpathians; and in the Tertiary of the Alps, western US, and Australasia. Smaller accumulations of turbidites in basins in cratonic areas (e.g., in the Devonian and Carboniferous of W Europe) should not be described as flysch.

Third phase (postorogenic). The Alpine *Molasse* is the standard reference here. Accumulated in the large piedmont basin below the newly erected mountain range, it displays the effects of rapid erosion and accumulation of clastics in a shallow, often subaerial zone. Its chief features are described in Bersier (1950) and others. Some, but not all, of the characteristics of this facies can be recognized in the Old Red Sandstone of Northern Europe, and in the Upper Carboniferous of the Northern Hemisphere.

The advent of the plate tectonic theory of earth structures and sedimentological investigation of the deep-sea floor imposes some modifications on the concept of the geosyncline but does not substantially change the relationships described above (see *Sedimentary Facies and Plate Tectonics*). For example, the ophiolite-chert association may be associated with the subduction-zone type of plate margin; pelagic sedimentation not associated with volcanic activity could be the product of a tectonically quiet ocean floor. Turbidite deposition, associated with submarine fans (q.v.) and abyssal plains, is linked with a widening ocean or with laterally moving plate margins.

Full elucidation of these relationships is yet to be realized; but it is clear they emphasize rather than reduce the measure of the dependence of sedimentation upon tectonic control (Dickinson, 1974).

J. E. PRENTICE

References

Bailey, E. B., 1936. Sedimentation in relation to tectonics, *Geol. Soc. Am. Bull.,* **47,** 1723-1726.

Bersier, A., 1950. Les sédimentations rhythmique synorogeniques dans l'avant fossé molassique Alpine, *18th Internat. Geol. Congr.,* **4,** 83-93.

Dickinson, W. R., ed., 1974. Tectonics and sedimentation, *Soc. Econ. Paleont. Mineral. Spec. Publ. 22,* 204p.

Dzulynski, S., 1963. Directional structures in flysch, *Studia Geol. Polon.,* **12,** 136p.

Geikie, A., 1882. *Textbook of Geology.* London: Macmillan, 971p.

Graham, S. A.; Dickinson, W. R.; and Ingersoll, R. V., 1975. Himalayan-Bengal model for flysch dispersal in the Appalachian-Ouachita system, *Geol. Soc. Am. Bull.,* **86,** 273-286.

Jones, O. T., 1937. On the sliding or slumping of submarine sediments in Denbighshire, *Quart. J. Geol. Soc. London,* **93,** 241-283.

Ksiazkewicz, M., 1960. Pre-orogenic sedimentation in the Carpathian geosyncline, *Geol. Rundschau,* **50,** 8-31.

Kuenen, P. H., 1951. Properties of turbidity currents of high density, *Soc. Econ. Paleont. Mineral. Spec. Publ. 2,* 14-33.

Lonsdale, P., 1975. Sedimentation and tectonic modification of Samoan Archipelagic Apron, *Bull. Am. Assoc. Petrol. Geologists,* **59,** 780-798.

Migliorini, C. I., 1952. Composite wedges and orogenic landslips in the Apennines, *18th Internat. Geol. Congr.,* **13,** 186-198.

Rabien, A., 1956. Zur Stratigraphie und Facies des Oberdevons in der Waldecker Hauptmulde, *Abhandl. hess. L.-Amt Bodenforsch.,* **16,** 1-83.

Weller, J. M., 1956. Argument for diastrophic control of Late Paleozoic cyclothems, *Bull. Am. Assoc. Petrol. Geologists,* **40,** 17-50.

Williams, A., 1962. The Barr and Lower Ardmillan Series (Caradoc) of the Girvan district, southern Ayrshire, with descriptions of the brachiopoda, *Geol. Soc. London Mem. 3,* 267p.

Cross-references: *Arkose, Subarkose; Breccias, Sedimentary; Clastic Dikes; Conglomerates; Cyclic Sedimentation; Geosynclinal Sedimentation; Gravity Flows; Gravity-Slide Deposits; Graywacke; Mass-Wasting Deposits; Olistostrome, Olistolite; Penecontemporaneous Deformation of Sediments; Rauhwacke; Sedimentary Facies and Plate Tectonics; Slump Bedding; Submarine Fan (Cone) Sedimentation; Turbidites.*

SEDIMENTOLOGICAL METHODS

Different techniques must be applied in order to study various sediments and sediment properties. Detailed comparisons between consolidated and unconsolidated deposits, however, is difficult when not approached in the same manner. For example, if the original grain arrangement is to be maintained, unconsolidated sediments must be lithified prior to making thin or thick sections and polished surfaces. Sedimentary rocks and artificially lithified sediments may be studied by many of the same techniques. Most techniques are very simple to apply, but practice is needed before good results can be expected.

By spraying a certain binding product onto a fresh surface of unconsolidated sandy material, a sedimentary peel can be made whose relief provides information about grain size and sedimentary structures. A polished surface of a rock may not always reveal what is expected or needed, and methods such as acetate peels can be applied to obtain additional information.

There are deposits that give the impression of being homogeneous, but with the application of x-ray radiography reveal important inhomogeneities. Radiography can be applied to any type of sediment, being hard or soft, wet or dry, coarse or fine grained, impregnated or embedded.

Demonstration collections of samples of hard rock can be stored easily without much risk of damage, but soft and wet materials tend to break or crack due to drying out. Impregnation or embedding can prevent such destructive processes.

Methods used in sedimentology to preserve and/or enhance sedimentary structures fall into some major groups. Each group will be discussed separately and the most common techniques briefly described. Special equipment is needed to collect undisturbed samples. This will not be discussed here except a few simple ones. For deep-sea coring techniques the reader is referred to Vol. 1: *Sediment Coring.* For other techniques, see *Sedimentary Petrology.* Extensive detailed information on methods and sampling is given by Bouma (1969).

None of the techniques described can be applied to all kinds of investigations a scientist may want to carry out. Each method has its own merit, and often more than one should be applied to obtain all results desired (see Bouma, 1969; Carver, 1971). For example, radiographs are very useful for scanning purposes and for studying the exact shape of structures or for revealing certain enclosures, but not for petrological data. Each research worker has to decide which technique(s) will serve best.

Sedimentary Peels

A film of material can be removed from the surface of an unconsolidated deposit without disturbing sedimentary structures by spraying or pouring a liquid adhesive onto the surface to bind the grains in their original fabric setting (Klein, 1971). Moisture may prevent the chemicals from penetrating the surface and/or hardening, and for many techniques wet sediments must be sampled and then dried in the laboratory. Nitrocellulose lacquer is normally used for noncohesive deposits, and lacquer-plastic combinations are recommended for sandy unconsolidated deposits that contain clay. Araldite peels with high relief can be collected by box core. When dealing with hard rock, acetate peels are advised.

Lacquer Peels of Dry Material. This method was introduced by Voigt (1936). Necessary equipment: spade, trowel or machete, nails, wire, thin rope, spray can, colorless nitrocellulose lacquer with thinner, flat paint brush, mixing can, stirring rod, bandage or cheese cloth, compass and marker. The lacquer should produce a flexible film without too much shrinking.

Part of the wall of an outcrop in unconsolidated gravelly, sandy, or silty material is removed with a spade and smoothed with a trowel or machete to produce a flat surface that is a little larger than the desired peel size; gravelly parts or shells often protrude. Normally it is impossible to make a surface absolutely vertical, and a dip of about 70° is recommended. Completely dry sediments should be moistened slightly by spraying water to avoid running of sediment particles. If the section is too wet, thinner or acetone should be sprayed on the surface and then ignited; two or three times usually is sufficient. Then a coat of thinned lacquer is sprayed on the surface; the viscosity should be such that the lacquer is just sprayable. Once dry, the spraying can be repeated one to three times.

Four nails are pressed into the future corners of the peel, and a thin rope is fixed between these nails to indicate the size of the peel and to prevent loosening of the corners and sides. Pieces of U-bent wire are pressed around the rope at regular intervals and at places where the rope does not touch the sediment. A mixture of lacquer and thinner, just thin enough to be brushed, is now applied by pressing a brush gently against the sediment, thereby being careful that no dripping occurs. Each next spot should not touch the foregoing ones. Once the lacquer spots are more or less dry, the areas in between can be handled similarly until the whole surface, including the rope, is covered. When this coat is dry, the whole surface should be brushed with a slightly thicker mixture. After this film is dry, strips of cheesecloth or bandage are painted against the peel to give it strength and to prevent breaking. The strips should overlap each other and also cover the rope.

It is necessary to wait until the lacquer is absolutely dry and all thinner has evaporated before the peel can be removed. After locality, orientation, etc., are indicated, the peel can be collected by taking one or two upper corners and pulling them from the wall. The peel should be transported carefully and allowed to dry for at least one day. With a trimming knife the peel has to be cut to size, after which it should be mounted on a board with a flexible glue (rubber base). Excess particles can be removed with a soft brush or with compressed air, which makes the structures clearly visible (Fig. 1). After an extra day of drying, the peel should get a protective finish with a crystal-clear acrylic spray coating.

Peels of Wet Unconsolidated Sands. This technique is described by McMullen and Allen (1964) using quick-setting polyester resins that are heavier than water. Some epoxy techniques may also be used on wet sediments (see below).

A sample is collected with a special sampling box (see under epoxy peels). The two parts of the box can be turned such that a vertical plane of the sample is exposed. This plane is levelled

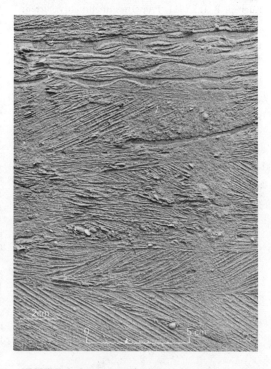

FIGURE 1. Lacquer peel of estuarine deposits. The sandy parts show relief based on permeability differences between laminae. Of the clayey parts only a very thin clay coat hangs onto the bandage. Excavation in the Haringvliet Delta Project, The Netherlands.

to permit uniform distribution of the resin over the surface. A piece of bandage or cheesecloth, slightly larger than the box, is placed over this surface. The resin mixture is poured carefully and uniformly over this bandage. The mix is chosen such that a gelling time of 20 min is obtained, allowing the liquid to penetrate slightly. The peel can be removed about half an hour after the pouring.

Lacquer-Resin Peels. Maarse and Terwindt (1964) succeeded in making sedimentary peels from clay-sand estuarine deposits by using polyester resin and nitrocellulose lacquer. The mixture can be used for peels of sections or cores with clayey sediments. Two types of resin mixture (A and B) are used depending on the type of sediment. Each mixture contains polyester resin, styrene monomer, methyl ethyl ketone peroxyde as catalyst, and cobalt octoate with 1% Co as accelerator. The percentages per volume for A are: 60-40-2-1 and for B: 85-15-4-2 (for mixing, see under impregnation).

Type A is used for profiles containing layers of sand and clay. Mixture B is used for sandy profiles. When clay is present, a thick pretreatment coat of nitrocellulose lacquer should be painted onto the sandy strata. Four or five hours later the resin mixture A can be applied over a piece of gauze. After five days drying time, an additional coating of thinned nitrocellulose lacquer should be applied to strengthen the gauze. The obtained peel is hard and stiff, and care must be taken during removal; the clay layers can be cut loose with a long knife. The peel can then be immersed in water for 24 hr, after which nonhardened sediment can easily be washed off. Mounting and finishing are similar to normal sedimentary peels.

Epoxy Peels. Epoxy peels are useful for unconsolidated sandy and gravelly sediments, wet or dry, in the lab or in the field (Burger et al., 1969). To collect a sample, various types of box cores, e.g., the Senckenberg box (Fig. 2), can be used. Further descriptions and references are given in Bouma (1969). Split cores and trench walls (Barr et al., 1970) may also be sampled.

The exposed surface is scraped a little, leaving small rims along the sides. The following peeling technique was introduced by Reineck (1962). Different products can be used (for more information see Bouma, 1969; Klein, 1971); only the application of Araldite Casting Resin 6010 from CIBA Company with hardener HT 951 is given here. First the sample is dried for 24 hr in an oven at a temperature of 100-110°C. After this the resin is mixed at a ratio of 100 resin to 13 hardener by weight. The sample is removed from the oven and the resin poured evenly on it. For sandy sediments, 35 ml of resin mixture

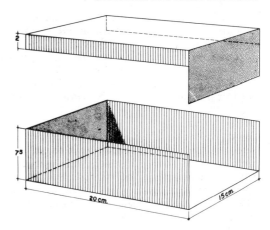

FIGURE 2. The two parts of the Senckenberg sample box.

per 100 cm² sample surface are used; for clayey sediments, 50 ml/100 cm² are used. The sample is then returned to the oven and cured for 24 hr at a temperature of 40°C, followed by 2-8 hr at 100°C. When cooled, the peel is first loosened from the box sides with a knife. The tape is removed and the box parts pulled apart. The peel is then removed and washed under running water. If thin parts are seen and the

FIGURE 3. Araldite peel of modern beach deposits within a bay; Mission Bay, San Diego, California, USA.

risk of breaking is present, the peel should be mounted on masonite. Like a lacquer peel, it should be sprayed with a acrylic resin. The result is a peel with high relief (Fig. 3), presenting a somewhat three dimensional picture.

Instead of this epoxy resin, paraffin can be used in the same quantities. No curing is necessary. When the sample is completely cooled, additional paraffin must be poured on the back for strengthening. This method only works for sand, and the peels are fragile. When the sand is well sorted and no differences exist between adjacent laminae, no differential penetration can be expected of the paraffin. Burger et al. (1969) have described an epoxy mixture that can be applied to wet, sandy sediments in the field (see also Klein, 1971).

Acetate Peels of Sedimentary Rocks. Thin transparent acetate peels can be made from consolidated sediments. Nonporous sandstones and shales have to be cut and the obtained surface polished with carborundum powder up to a fineness of 600–800 mesh. The final polish should be done on a glass plate. The sample is then cleaned carefully in water and allowed to dry. The polished surface is placed absolutely horizontal, moistened with acetone and a piece of Kodatrace is pressed onto it immediately, taking care that no air bubbles are trapped. After 1 hr the peel can be removed.

For porous sandstones and shales a cellulose-acetate mixture can be used. Twenty grams cellulose-acetate powder is dissolved overnight in 130 ml acetone. Just prior to application, this solution is mixed with an equal quantity of tetrachlorethane. (This mixture should be handled carefully and under a fume hood because tetrachlorethane is toxic.) The polished surface of the sample is placed absolutely horizontal, moistened with acetone and then painted with a thin coat of cellulose-acetone-tetrachlorethane mixture. Care should be taken that no solution flows over the edge of the sample. The sample is left to dry overnight. For very porous samples, a second coat must be applied. When dry, the edges are loosened with a razor blade and the peel can be removed. The peels can be placed between two sheets of glass for protection. They can be used directly for study, projection, or as a negative for making prints.

The same techniques can be used for limestones, which have to be etched lightly after the polishing and washing, either in a 5% solution of formic acid or a 3% solution of acetic acid for 3–15 sec, depending on the type of limestone. The sample is then washed carefully in a basin with slow running water. After the sample has dried, the same procedure is followed as described above. Staining after etching makes it possible to identify the minerals (see *Staining Techniques*).

Impregnation

Unconsolidated and badly consolidated sediments should be impregnated in order to obtain samples than can be treated as hard rock. A number of methods are known, but only two will be described. The first can be applied to dry samples, the second to wet material. In both cases, care should be taken not to change specified quantities or curing times without making some experiments.

Many types of products and combinations, such as Canada balsam, shellac, Vernicolor, and lakeside, have been used as impregnants. Methods variations and extensive references are given by Jongerius and Heitzberger (1963); Altemüller (1962); Reineck (1963); Bouma (1969); and Stanley (1971). Insufficient impregnation or disruption of microstructures during the preparation of thin sections were no exceptions.

Impregnation of Dry Sediments. Many types of unsaturated polyester resin can be used for impregnation. When dealing with fine-grained sediments, the resin should be diluted with styrene monomer to lower its viscosity. Polymerization of a plastic mixture is not a spontaneous reaction, and catalysts such as methyl ethyl ketone peroxide (MEK), benzoyl peroxide, or cyclohexanone peroxide are necessary to initiate this process. An accelerator (cobalt octoate or naphthenate, dimethyl aniline has to be added to split the peroxides before catalysis will begin.

There are special combinations of catalysts and accelerators. The description below will deal only with MEK and cobalt octoate (1 or 6% Co), as this combination is easy to handle. *Catalyst and accelerator should never be mixed at the same time as this may produce a violent explosion.*

The hardening process is an exothermic reaction. Curves made from temperature versus time vary for the different types of resins, amount of styrene monomer, and types and concentrations of accelerator and catalyst. The slower the curve rises, the easier it is to carry off the heat, and the larger the total volume to be handled at once can be (Bouma, 1969); this property is especially important for embedding (see below). The addition of fillers, in this case the sample, usually slows down the process of polymerization.

The sample should be placed in a container, e.g., aluminum foil, so that it does not fall apart. After it is air-dry, its surface can be prehardened with a 1:1 mixture of cellulose varnish and acetone. A number of containers may be placed in a larger (e.g., cardboard) box lined with plastic or aluminum foil; it is advisable to give this liner a covering of a mold release agent.

Jongerius and Heintzberger (1963) use for four soil samples, each one 15 × 8 × 5 cm, the following mixture: 1500 ml resin (Vestopal-H; in US, Plaskon 951 from Allied Chemical); 1000 ml styrene monomer; 4 ml catalyst (MEK); and 2 ml accelerator (cobalt naphthenate with 1% Co).

For smaller samples, or more porous ones, Altemüller (1962) uses a mixture of 300 ml resin, 250 ml monomer, 1.65 ml catalyst (cyclohexanone peroxide) and 0.8 ml accelerator (with 1% Co). Also, less vacuum is applied, and thus a vacuum desiccator will work well.

The large box is placed in an impregnation cylinder (e.g., Fig. 4). For half an hour, a rather high vacuum (60 cm Hg) is applied to the cylinder and to the funnel, which is filled with the resin mixture, to extract most of the air. The tap of the funnel is next opened to permit a flow of about 0.5 liter/hr, maintaining the vacuum. When the samples are well covered with mixture, the tap is closed, but the vacuum is kept for another hour.

The container is then placed in a running fume hood to allow the styrene monomer to evaporate. A new mixture (100 ml resin, 0.3 ml catalyst, 0.3 ml 1% Co accelerator) must be added regularly to prevent the samples from becoming dry and trapping air. It take about 6 months for the total sample to harden, after which it can be cut. If tacky, it can be postcured in an oven. (The mixture and quantities used by Altemüller only require a few weeks to harden.) For mounting the impregnated object to an object glass, and also for placing the cover glass, a mixture of 10 ml resin, 0.2 ml catalyst, and 0.2 ml accelerator (1% Co) can be used.

Impregnation of Wet Sediments. A square (5 × 5 cm and 1 cm high) or rectangular form of nonrusting mosquito netting is pressed into a sediment core (see Reineck, 1963; Bouma, 1969). Both open sides are smoothed, leaving the sample just within the netting. The total sample is then embedded in plaster of Paris. A thin cover of 0.4 cm is enough, as it only serves to support the sample. The four corners of the wire frame should be indicated by holes. These later serve also as sawing marks.

When the plaster is hard the sample is placed in a solution of Arigal C (CIBA Company) for 5 days (25 g Arigal C in 100 ml of water; dissolving temperature 80°C). After the 5 days, an Arigal catalyst is added (10% in volume of the grams Arigal C dissolved in water; e.g., 37.5 g Arigal C in 150 ml water needs 3.7 ml catalyst). It should be stirred well to obtain a good distribution throughout the solution. After about 2 days, a grayish slimy precipitate becomes visible, indicating the end of the process. The embedded sample is then placed, in wet condition, in a polyvinyl bag, sealed, and placed in an oven for 48 hr (65°C). After the 2 days, the oven is switched off. When room temperature is reached, the sample can be removed from the bag and laid down to dry for another 48 hr.

Silty-sandy samples are now placed in a container and further hardened in Araldite Casting Resin and hardener (see under epoxy peels). After this cure, the sample can be cut along the four holes. It is recommended, however, that

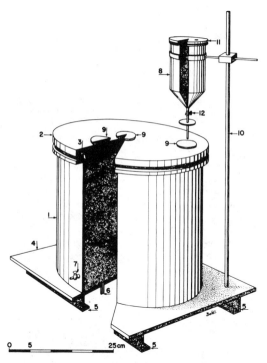

FIGURE 4. Impregnation cylinder (1) made of steel with a plexiglas top (2). An O ring (3) in between prevents leaking of air. The cylinder is welded to base plate (4) with two slides (5) underneath. Tube (6) must be connected with a vacuum pump. Tap (7) serves for release of vacuum. In the top plate, three holes are located to allow more possibilities for placing the mixture funnel (8). The holes can be closed with mushroom-shaped stops (9) with an O ring underneath. The funnel (8) can be moved in all directions along the standard (10). It contains a plexiglas top (11) with one hole to which a vacuum tube should be connected. An O ring is put in between funnel and cover. The vacuum tap (12) is made out of steel to allow the operator to apply more force than can be applied to a glass one.

the sample be given another 2 days prior to additional cutting.

Silty-clayey samples should be impregnated at a temperature of 40°C under 30-40 cm Hg vacuum for about 1 hr. The sample is then placed in an oven at 40-50°C for 16 days and then at 120°C for 2 days; after which, it can be cut. About 1-2 additional weeks should elapse before thin sections are made. The samples can be cut and treated as hard rock.

Embedding

Fragile consolidated samples, as well as badly consolidated ones, can be embedded completely or partly in a clear medium for preservation. Unconsolidated wet sediments should be embedded completely if they are to be kept in their original condition.

It is easiest to use unsaturated polyester resins (for other possibilities see Bouma, 1969). The first step is to make a mold of proper size, at least 2 cm wider than the sediment slice and 4-5 cm longer. The mold can be made of plate glass strips mounted with tape. A saturated watery solution of polyvinyl alcohol or other mold release agent may be used.

First, a bottom layer, not thicker than 5-8 mm, is poured in the mold; any type of clear plastic is suitable. If no data concerning the mixing amounts are known one can use 100 ml plastic, 2 ml catalyst, and 1 ml (1% Co) accelerator (see under impregnation). It is essential that all air bubbles be removed (with or without vacuum application) before the mixture is poured. The humidity of the laboratory should be as low as possible because moisture makes the free surface of the cast sticky; the surface becomes hard overnight in a 50°C oven.

When the base is hard, a 1-mm layer of new mixture (same composition) is poured on top, and the sample slice is placed, taking care that no air bubbles are trapped. (For wet, unconsolidated samples, see Bouma, 1969). The surface of a moist sediment should be sprayed several times with an acrylic lacquer (Krylon) to prevent drying out and to prevent intermixing of moisture and plastic mixture, which results in a milky appearance.

After some hours, the thin second layer should be hard enough to permit pouring of another layer about 4-6 mm thick covering the whole sample. When this layer is absolutely hard, the cast can be removed from the mold. The upper surface normally is tacky and irregular due to shrinkage. If the cast is bent, it can be placed in an oven at 50°C for a few hours, and then placed on a flat surface with a flat weight on top. If necessary, the sides can be straightened by cutting the slab on a diamond saw. The upper surface can be polished with water-proof abrasive paper (#300 and 600) and finished with copper polish. Excellent results can be obtained in this way.

X-Ray Radiography

An x-ray radiograph can be made of hard rocks, unconsolidated dry or wet sediments, and impregnated or embedded samples. It is important to use a plane parallel slice which is as thin as possible. The radiograph is a shadow image from differences in radiation that passes the specimen, a result of the different absorption properties of various parts of the sediment (Fig. 5). Normally it will reveal information on

FIGURE 5. Print of an x-ray radiograph made of a vertical slide of a clayey marsh sample (after Bouma, 1969). Thickness 5 mm; 40 kV, 5mA, 45 sec., Kodak Industrial film AA, tube distance 1 m. Mission Bay, San Diego, California, U.S.A. 1. Fragment of a *Spartina* stem. 2. Roots of Salicornia. 3. Lamination. 4. Lens form. 5. Concretions of iron oxide.

sedimentary structures that are not seen in visual observation (Hamblin, 1971). Thick slices project too many data, and a sterographic exposure should be made; nonparallel samples should be embedded in fine sand (see Bouma, 1969).

As x-ray apparatus, the technical unit used should have a low kilovoltage (30-150 kV range). A higher kV gives more penetration, but also produces more scattering. The tube current need not be over 4 or 5 mA. The distance between tube and object should be about 1 m; the radiation cone should cover the whole object. A technical not a medical film is recommended, as the latter gives less distinct results on nonorganic material.

The film is placed in an x-ray exposure holder or a cassette; the sample should be placed directly on or close to the cassette. Lead letters can be used for identification. The amount of radiation necessary depends on the distance between the tube and specimen, type and thickness of the sample, type of x-ray unit and film (Bouma, 1969; Hamblin, 1971). The film should be developed in special tanks and in special x-ray chemicals. Each manufacturer gives directions for its products.

This method is fast and easy to carry out, and compared to the time required for other techniques, is not expensive. Attention must be given however, to protection against radiation.

ARNOLD H. BOUMA

References

Altemüller, H. J., 1962. Verbesserung der Einbettungs— und Schleiftechnik bei der Herstellung von Bodendünnschliffen mit Vestopal, *Z. Pflanzenernaehr., Düng., Bodenkd.,* 99, 164-177.

Barr, J. L.; Dinkelman, M. G.; and Sandusky, C. L., 1970. Large epoxy peels, *J. Sed. Petrology,* 40, 445-449.

Bouma, A. H., 1969. *Methods for the Study of Sedimentary Structures.* New York: Wiley, 458p.

Burger, J. A.; Klein, G. deV.; and Sanders, J. E., 1969. A field technique for making epoxy relief-peels in sand sediments saturated with salt water, *J. Sed. Petrology,* 39, 338-341.

Carver, R. E., ed., 1971. *Procedures in Sedimentary Petrology.* New York: Wiley, 653p.

Hamblin, W. K., 1971. X-ray photography, in Carver, 1971, 251-284.

Jongerius, A., and Heitzberger, G., 1963. The preparation of mammoth-sized thin sections, *Netherland Soil Surv. Pap.,* 1963, 37p.

Klein, G. deV., 1971. Peels and impressions, in Carver, 1971, 217-250.

Maarse, H., and Terwindt, J. H. J., 1964. A new method of making lacquer peels, *Marine Geol.,* 1, 98-103.

McMullen, R. M., and Allen, J. R. L., 1964. Preservation of sedimentary structures in wet unconsolidated sands using polyester resin, *Marine Geol.,* 1, 88-97.

Price, I., 1975. Acetate peel techniques applied to cherts, *J. Sed. Petrology,* 45, 215-216.

Reineck, H. E., 1962. Reliefsguesse ungestörter Sandproben, *Z. Pflanzenernaehr., Düng. Bodenk.,* 99, 151-153.

Reineck, H. E., 1963. Naszhartung von ungestörter Bodenproben im Format 5 x 5 cm für projizierbare Dickschliffe, *Senckenbergiana Lethaea,* 44, 357-363.

Stanley, D. J., 1971. Sample impregnation, in Carver, 1971, 183-216.

Voigt, E., 1936. Die Lackfilmmethode, ihre Bedeutung und Anwendung in der Palaeontologie, Sedimentpetrographie und Bodenkunde, *Deutsche Geol. Ges. Z.,* 88, 272-292.

Cross-references: *Grain-Size Studies; Sedimentary Petrology; Staining Techniques.* Vol. I: *Sediment Coring—Unconsolidated Materials.*

SEDIMENTOLOGY

Sedimentology is the study of sedimentary deposits and their genesis. It is applied to various types of formations: ancient and recent, marine and terrestrial, including their faunas, floras, minerals, textures, diagenesis, and evolution in time and space. Sedimentology is based on observation of the numerous and intricate features of soft and hard rocks in natural sequence, with the goal of reconstructing their original environment in both a stratigraphic and a tectonic frame (Fig. 1). It involves the use of many other sciences (Fig. 2). Also called "sedimentary geology," sedimentology is a part of stratigraphy, the study of Earth history through the investigation of stratified rocks (Krumbein and Sloss, 1963; Fig. 3). Specifically, sedimentology examines the geometry and dynamics of these stratified formations.

There are no precise subdivisions of sedimentology, although various fields (e.g., carbonate sedimentology) have very well-defined areas of research.

Philosophy and Methods

Laws and Hypotheses. Sedimentology is strongly influenced by the law of uniformitarianism (actualism). Consequently, studies of oceanography, geophysics, and geochemistry are frequently applied to interpret ancient formations and their original environment. Experiments and mathematical expression are now the tools of modern methods. As a consequence, the former status of sedimentology and lithostratigraphy, which was essentially qualitative, has become increasingly quantitative.

The previous and classical study of sedimentary sequences produced a static picture of successive stages of deposition, a series of "stra-

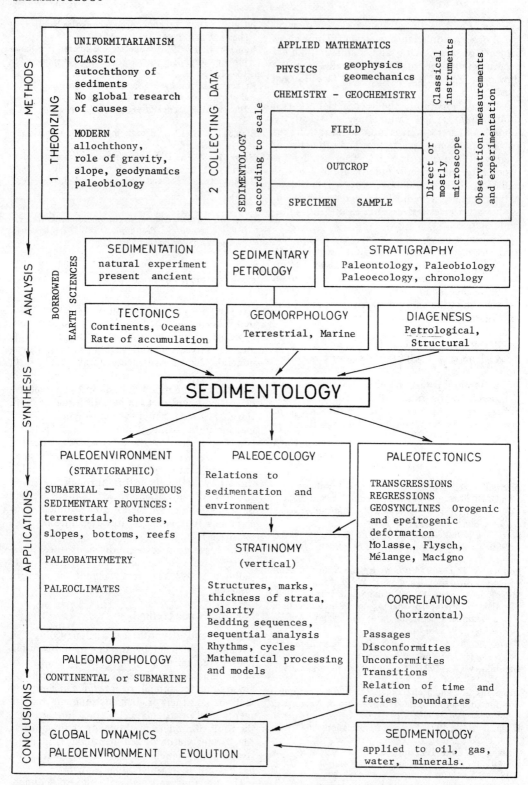

FIGURE 1. Sedimentology among the sciences and its interconnection with the earth sciences.

Biology		Mathematics and Statistics
Paleontology and Paleoecology	Oceanography	Hydraulics and Fluid Mechanics
	SEDIMENTOLOGY	
Stratigraphy		Geomorphology
Geochemistry	Petrography	Tectonics and Geophysics
Chemistry		Physics

FIGURE 2. An "egocentric" view of sedimentology (after Potter, 1974).

tigraphic landscapes" (Gignoux, 1950). Modern trends of interpretation present a more dynamic picture of the environment, with an evolution from one deposit to the next, and introducing such concepts as gravity flow of sediments, mass transport and sliding, resedimentation, levels of energy, penecontemporaneous deformation, reciprocal influences of sediments on faunas and floras, etc. Diagenesis is considered a revealing factor of conditions during and after deposition.

Methods. The methods of sedimentology rely largely on the scale of observations. The *megascopic* scale concerns paleosedimentary provinces extending to continents, marine basins, or geosynclines. Data are provided directly from the field or indirectly by subsurface borings or photogeology, more recently from satellite pictures. Without sharp limits, one passes into *macroscopic* studies, applied to the outcrop, the specimen, or to cores. Careful observation and simple physical and chemical tests are used to study the textures, sequences, rhythms, and fossils, leading to bio- and stratigraphical zonation, and to paleogeographical syntheses of a higher scale. The *microscopic* scale, based on the use of the microscope and other very sensitive equipment, reveals the elementary particles of the sediments: minerals, matrix, cement, and organisms—their relations, dimensions, quantity, degree of sorting, and diagenesis. Microscopic analysis of particles is a quick and widespread method that can often replace delicate and time-consuming chemical operations. *Chemical analyses* are useful supplements to either qualitative or quantitative studies of consolidated and unconsolidated sediments. Chemical studies are especially important in the study of evaporites, carbonates, organic sediments, and diagenesis. *Physical methods* are numerous and of wide application: e.g., electronic and scanning microscope, x-rays; gamma rays; magnetic separation; soil mechanics; sieving; hydrosorting; and, on a larger scale, gravity, seismic, magnetic, and thermal properties of sedimentary provinces. Such information provides data for mathematical models.

Sedimentology Among the Earth Sciences

Sedimentology includes data and methods borrowed from several branches of the earth sciences (Figs. 1 and 4).

Sedimentation. The study of sedimentation (q.v.) investigates the mechanisms and processes of the deposition of present or past deposits.

MACROGEOMETRY
thickness, shape and orientation of sedimentary bodies

STRATIGRAPHY

SEDIMENTOLOGY

(field)   (laboratory)

MICRO-PETROGRAPHY
VECTOR PROPERTIES
(structures: cross-bedding, etc.)

MACRO-PETROGRAPHY
SCALAR PROPERTIES
(composition and textures)

FIGURE 3. Relations of sedimentology to stratigraphy and sedimentary petrology (after Pettijohn, 1975).

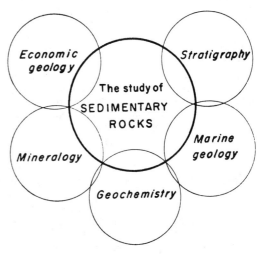

FIGURE 4. Relation of sedimentology to the other geological sciences (from Pettijohn, 1975).

The most stimulating field is provided by researches in ongoing process: eolian, fluvial, colluvial, littoral, deltaic, shallow and deep marine. Comparisons between modern and fossil sediments have provided valuable clues to paleotectonic interpretations (e.g., Bernoulli and Laubscher, 1972).

Sedimentary Petrology. Sedimentary petrology (q.v.) is the study of sedimentation through the examination of sediments and sedimentary rocks. For example, the polarizing microscope is used to examine sedimentary particles and thin sections cut in consolidated sediments. Identification may indicate the petrology of the source rocks (see *Provenance; Sedimentary Petrography*). Measurements can be made of the size, degree of rounding, and shape of grains (see *Sediment Parameters*). Several classifications of sedimentary rocks are based on the relative proportion of matrix and debris and upon the sorting of the fragments of organisms and minerals. Sedimentary petrology also includes the study of diagenetic minerals.

Stratinomy. Stratinomy deals with the accumulation and the succession of layered sequences. The thickness of different strata is measured, e.g., to obtain the sand:shale ratio. Rhythms and cycles are evinced by the study of ratios and thicknesses, giving clues to the origin of the sequence. Sedimentary structures of the strata, particularly in detrital rocks, are related to the dynamics of the depositional environment. Examples include cross-bedding, graded bedding, slumping, and laminae; current marks on the bedding plane also belong to this group.

Stratigraphy. Stratigraphy, as well as being the parent science of sedimentology, is of significant value in sedimentological research. For example, correct chronology is necessary to determine whether limits of formations are diachronous or synchronous.

Paleoecology. Macro- and microfossils are not as important in sedimentology for their chronological value, as for their significance in the reconstruction of past environments in terms of energy, temperature, depth of water, and/or salinity (see *Sedimentation–Paleoecologic Aspects*). Associations of forms are analyzed in assemblages and often treated by statistical methods.

Tectonics. Tectonics is also an important field of sciences which is integrated into sedimentology. The composition and the stratinomy of sedimentary sequences depends on crustal movements that occur during the phase of deposition and sometimes before. Many formations are marked by a tectonic style: epeirogenic, orogenic, marginal, intracratonic, or geosynclinal (see *Sedimentary Facies and Plate Tectonics; Sedimentation–Tectonic Controls*). Depending on the relative speed of subsidence and the rate of sedimentation, a sedimentary basin may be filled up or become deeper.

Morphology. There are no sharp limits between tectonic deformation of the crust and the resulting configuration (morphology) of continents and oceanic depressions. It is well known that the coarseness of detrital sediments is usually due to high relief, with a torrential drainage and a short distance of transportation. On the other hand, reefs represent organic buildups on a flat, stable, and shallow platform. The geometry of the sedimentary bodies is a function of their geotectonic postition and of the form of their bottom. We know various shapes and styles of such sedimentary bodies: shore deposits, wedges, shoestrings, sheets, internal basins, deltas, slope and oceanic basin formations, volcanic arcs and platforms.

Climate. Continental and marine climates may give a certain imprint to a sedimentary sequence, or they may even be the cause of a certain deposit. In the first case, red beds and platform carbonates may serve as examples. To the second category belong evaporites, bauxites, laterites, and tillites.

Synthesis in Sedimentology: Applications

The foregoing are the prevalent disciplines of the earth sciences upon which the methods and activities of sedimentology are based. In sedimentology, these other fields are used to produce syntheses of information, which could be considered as "offspring" studies.

Paleogeography. This science has acquired a renewal of scope. Instead of providing merely static pictures of past environmental provinces, it now includes information such as source of the sediments, direction of transportation, paleocurrent or ecological associations, etc. The succession of formations is considered to have evolved through time and changing tectonic conditions. Instead of depicting a succession as a series of still photographs, paleogeography now gives a motion picture of their history. An excellent review of the subject has been published by Gubler (1972), and recent studies review the relations between vertical and horizontal suites of formations.

Paleotectonics. Sediments ultimately repose in the structural depressions of the earth's surface or in marine basins. Their lithofacies, thicknesses, and structures give a record of basinal deformation and its consequences: paleocurrents, paleodirections of mass sliding, river channels of braided rivers, subsidence of deltas as shown in the cyclothemic formations of

Illinois (Wanless, 1969) or in the Molasse deposits on the front of the western Alps (Bersier, 1958).

Correlation of Formations. The sequential analysis of sedimentary sections is a relatively new development of sedimentology. It has given a great impulse to their analysis and correlation (Lombard, 1956; Gubler, 1972). Sequences are suites of several lithologies (as in cyclothems) occurring at various scales and repeated in a rhythmic style (see *Cyclic Sedimentation*). Megasequences may be followed for kilometers and may be used as correlation marks. Such analyses are descriptive or genetic. Many causes are at the origin of sequences: glacial (varves), chemical (evaporites), eustatic (cyclothems), or variations in the ratio of subsidence to sedimentation. Sequential analysis has led to an intense study of cyclic and rhythmic series in various environments and scales of magnitude.

Applied Sedimentology. These fundamental researches in sedimentology have greatly improved our skill in the quest for oil, gas, coal, sedimentary ores, placer deposits, and water.

Conclusion

The complementary role of sedimentology in stratigraphy is great and has increased steadily for over 30 years (see *Sedimentology–History*). It has produced dynamic and evolutionary models at various scales. The methods of analysis in the field or in the laboratory are now better grouped and coordinated, thus giving a clearer insight into an interpretation of the results (see *Sedimentology–Yesterday, Today, and Tomorrow*). Historical geology is no longer the elaboration of a suite of static stages of environment but has instead become a kinetic view of sedimentary evolution.

AUGUSTIN LOMBARD

References

Bernoulli, D., and Laubscher, H., 1972. The palinspastic problem of the Hellenides, *Eclogae Geol. Helv.,* **65,** 107-118.

Bersier, A., 1958. Séquences détritiques et divagations fluviales, *Eclogae Geol. Helv.,* **51,** 854-893.

Bur. Rech. Géol. Min., 1972. Colloque sur les méthodes et tendances de la stratigraphie, *Bur. Rech. Géol. Min. Mém.,* **77**(2), 523-1010.

Gignoux, M., 1950. *Géologie Stratigraphique.* Paris: Masson, 709p.

Ginsburg, R. N., 1973. *Evolving Concepts in Sedimentology.* Baltimore: Johns Hopkins Univ. Press, 191p.

Gubler, Y., 1972. Stratigraphie et sédimentologie. Introduction et rapport de synthèse, *Bur. Rech. Géol. Min. Mém.,* **77**(2), 523-534.

Krumbein, W. C., and Sloss, L. L., 1963. *Stratigraphy and Sedimentation,* 2nd ed. San Francisco: Freeman, 660p.

Lombard, A. E., 1956. *Géologie Sédimentaire–Séries Marines.* Paris: Masson, 722p.

Pettijohn, F. J., 1975. *Sedimentary Rocks.* New York: Harper & Row, 628p.

Picard, M. D., 1975. *Grit and Clay.* Amsterdam: Elsevier, 258p.

Potter, P. E., 1974. Sedimentology: Past, present and future, *Naturwissenschaften,* **61,** 461-467.

Wadell, H. A., 1932. Sedimentation and sedimentology, *Science,* **75,** 20.

Wanless, H. R., 1969. Marine and non-marine facies of the Upper Carboniferous of North America, *C.R. 6e Congr. Int. Strat. Geol. Carbonif., Sheffield,* **1,** 293-336.

Cross-references: See text.

SEDIMENTOLOGY–HISTORY

Sedimentology (q.v.) may be defined as the study of natural sediments, both lithified (sedimentary rocks, q.v.) and unlithified, and of the processes by which they were formed. The term *sedimentology* was first used by A. C. Trowbridge in 1925 (Wadell, 1933), but an interest in sedimentation and the origin of sedimentary rocks dates back to the very beginnings of geology.

In the following discussion, the history of the subject has been divided, somewhat arbitrarily, into five main periods separated by the dates, 1830, 1894, 1931, 1950. The criteria used to separate the periods are not uniform. The first two periods are defined conceptually, but the third period is defined mainly in terms of a social phenomenon: the definition of "sedimentology" as a subdiscipline of geology, with its own professional specialists (professionalization).

The major developments marking the end of each period were: (1) general acceptance of actualism as a basis for geology (Lyell's *Principles,* 1830); (2) the theoretical elaboration of actualism as the basis of a specific science of sedimentary rocks ("comparative lithology" of Walther's *Einleitung,* 1893-1894) and the practical demonstration of actualistic methods by such major investigations as the Funafuti boring and the *Challenger* expedition; (3) growth in professionalization culminating in the publication of the first sedimentological journal (*Journal of Sedimentary Petrology,* 1931); (4) beginning of large-scale studies of modern sediments and introduction of many new techniques, particularly in oceanography. These developments coincided with major conceptual advances in geochemistry and in the interpretation of carbonate rocks, sandstones, and sedimentary structures.

First Period

Rise of Actualism in Geology. The end of the first period (1830) is marked by the date when uniformitarianism, or the actualistic method, became generally accepted as the basis of geology. As with most changes in science, this was a gradual process. Uniformitarianism had been proposed before James Hutton and John Playfair, and its general acceptance was not achieved by them. It was not even achieved at first by Lyell (Cannon, 1960), but the date of the appearance of the first volume of Lyell's *Principles*—1830—is a convenient one with which to limit a history of sedimentology. At some time not long after 1830, there was widespread acceptance of the general actualistic paradigm: "that, during the ages contemplated by geology, there has never been any interruption to the agency of the same, uniform laws of change. The same assemblage of general causes . . . may have been sufficient to produce by their various manifestations, the endless diversity of effects of which the shell of the earth has preserved the memorials. . . ." It was also at this time that geology itself became professionalized.

Although actualism is the central paradigm for the science of sedimentology, the science itself can scarcely be said to have begun with Lyell. Lyell only made the science possible, he did not develop its principles. Like many geologists of the time, Lyell was more concerned with establishing the basic working method of historical geology as a whole, than he was with the investigation in detail of the origin of sedimentary rocks. Lyell developed stratigraphy by the use of the powerful paradigm furnished by William Smith's law of "strata identified by fossils," and greatly advanced understanding of the evolution of landscape and of the origin, in general, of the main classes of rocks. Lyell's *Principles* contained many observations that we would now regard as within the scope of sedimentology; for example, his discussion (borrowed in part from C. E. A. Von Hoff) of rivers, deltas, and coastal processes. But these phenomena were presented more to prove uniformitarianism than to form the basis for an interpretation of sedimentary rocks. Lyell's own systematic discussion of sedimentary rocks (Vol. 3 of the *Principles*, which later became the *Elements*) tended to follow Smith's stratigraphic paradigm rather than his own actualistic one. Most of the *Elements* could equally well have been written by a catastrophist such as William Smith himself, Georges Cuvier, or Adam Sedgwick.

For many years, the great successes of stratigraphy, which determined the sequence of sedimentary rocks by Smith's two "laws," diverted the attention of most geologists, away from the investigation of sedimentary rocks themselves, to the study of their correlation and enclosed fossils.

Second Period (1830-1894)

Founding of an Actualistic Science of Sedimentary Rocks. Almost as soon as it was realized that fossils characterized sedimentary rocks of a given age, it was also acknowledged that different types of sediments, deposited at the same time, give rise to sedimentary sequences that differ in aspect (facies) and that contain different types of fossils. This idea (the concept of facies) was developed particularly by the Swiss geologist, Amand Gressly (1938; see *Facies;* Wegmann, 1963).

Perhaps the first major success of the actualistic method was, however, the glacial theory developed by Agassiz (1840) from observations of modern Swiss glaciers and their sediments. It was strongly supported by William Buckland and, at first, by Lyell. Early investigation of modern sediments was limited to accessible areas, but was extended by investigations of coral reefs (Darwin, 1837, 1842), deserts (Walther, 1890), modern evaporites (Ochsenius, 1877), and deep-sea sediments (Murray and Renard, 1891). The origin of coal was a subject of intense investigation (the allochthonous vs. autochthonous debate).

The earliest microscopical investigations, begun a little earlier with investigations of coal, were extended to other types of sedimentary rocks with spectacular results. Ehrenberg (1854) showed the organic origin of diatomites, radiolarian cherts, and chalk. In 1850, Sorby began the use of the petrographic microscope, and by the end of the period he had, almost single-handedly, established the science of sedimentary petrology (Folk, 1965).

The first work to draw together the many scattered observations on modern sediments and to document the actualistic method as the basis for a specific science of sedimentary rocks ("comparative lithology") was Johannes Walther's *Introduction to Geology as Historical Science* (1893-1894; see Middleton, 1973). The method was demonstrated by major investigations, including Walther's own studies of deserts (1891, 1900); the Report of the Royal Society of London's expedition to

Funafuti (Bonney, 1904); and, most important of all, publication by Murray and Renard (1891) of the sediment studies made by the *Challenger* expedition to the deep seas (1872-1876).

These works, with pioneer geomorphological studies such as Gilbert's work in the Henry Mountains (1877) and Lake Bonneville (1890) and the first papers of Davis (1884, 1889), completely changed the character of sedimentological studies and set them on a firm foundation of observation and theory (Loewinson-Lessing, 1954; Cross, 1902; Teall, 1902; Chorley et al., 1964). Thus, 1894, the date of publication of the third volume of Walther's book, a volume prophetically titled *Modern Lithogenesis,* seems a suitable date to choose as the beginning of a systematic science of sedimentology.

Third Period (1894-1931)

Consolidation and Professionalization of the New Science of Sedimentology. This period was marked mainly by the further development of lines of investigation established by pioneers during the previous period. Investigations of modern sediments continued, and were summarized in the *Treatise on Sedimentation* (edited by Twenhofel, 1926, revised, 1932), and in *Recent Marine Sediments* (edited by Trask, 1939). Both volumes were produced under the sponsorship of the U.S. National Research Council Committee on Sedimentation, which had been established in 1920. There were important applications to sedimentary rocks by such men as A. W. Grabau, J. Barrell and C. Schuchert.

The major new techniques developed were mainly in textural studies: size analysis (J. A. Udden, C. K. Wentworth, A. Atterberg; see *Grain-Size Studies*) and shape analysis (Wentworth, Wadell, T. Zingg; see *Roundness and Sphericity*). The petrographic microscope was applied mainly to the study of heavy minerals, though Lucien Cayeux in France continued the study of sedimentary rocks in thin section. The techniques of heavy mineral separation were pioneered by Thoulet (1881) and others, and the use of heavy minerals for the determination of provenance was convincingly demonstrated, first by Thomas (1902) and then by many others (see *Sedimentary Petrography*).

A German tradition of chemical studies, already established toward the end of the previous period by K. G. Bischof, J. Roth, and others was continued by the monumental studies of J. H. Van't Hoff on the evaporation of sea water (Eugster, in Ginsburg, 1973; see *Evaporites—Physicochemical Conditions of Origin*).

Physical experimentation also began during the preceding period with the work of Daubrée (1879; see *Experimental Sedimentology*). By the end of the 19th century there was a considerable body of quantitative empirical knowledge on the hydraulics of rivers. Gilbert's interest in the physiographic activity of rivers, combined with a practical problem (sediment pollution produced by hydraulic mining of gold placers in California), led him to carry out his extensive flume experiments (1914).

Perhaps the main characteristic of the period, however, is not to be found in the investigations themselves, so much as in a trend toward defining sedimentology as a distinct discipline within geology. This professionalization of sedimentology, which can be traced through the appearance of textbooks, journals, societies, and research institutions, took place in America, Russia, and, to varying degrees, in England and Germany at about the same time, during the 1920s and 1930s. A convenient date is 1931, marking the appearance of the first journal devoted exclusively to sedimentology, the *Journal of Sedimentary Petrology,* published by the Society of Economic Paleontologists and Mineralogists, founded in 1927 as an offshoot of the American Association of Petroleum Geologists (see Russell, 1970).

In Russia, the influence of the *Challenger* reports and Walther's work was strong. Strakhov (1970) dates the "independence of lithology as a scientific discipline" from 1923, the date when Y. V. Semoilov clearly expressed the "endeavour to systematically develop lithology as a science. . . ." The main period of professionalization, however, seems to have come somewhat later, during the late 1920s and 1930s (Anonymous, 1967). The first general textbooks on sedimentology in Russian were produced by D. V. Nalivkin in 1932 and M. S. Shvetsov in 1934. National lithological conferences, however, were not held until 1952, and it was not until 1963 that there appeared *Lithology and Mineral Resources,* a journal devoted exclusively to sediments and sedimentary rocks.

In Germany, there were many early sedimentological studies (reviewed by Brinkmann, 1948). Marine institutes existed at Naples, Helgoland, and Kiel, but most of their work was biological. In 1929 Rudolf Richter established the Senckenberg am Meer as an institute for the study of modern shallow-water sediments to aid in the study of ancient rocks (Spencer, 1957). In 1937, Erich Wasmund established a marine geological station at Kiel. In 1918, Richter had begun publication of a new journal *Senckenbergiana,* in which most of the sedimentological research carried out at the marine institute was published. But there is still no national sedimentological journal or society of sedimentologists in Germany.

The European countries were active in the founding of the International Association of Sedimentologists, and the first Congress was held in Belgium in 1946. The IAS journal, *Sedimentology*, appeared for the first time in 1962 (see *Sedimentology—Organizations and Associations*).

An active group of sedimentologists existed in Great Britain at the beginning of the century, and the first English textbook on sedimentary petrology, by Hatch and Rastall, appeared in 1913. Following this date, however, British sedimentology became dominated by the study of heavy minerals, so that this field of specialization increasingly became identified with the whole discipline of sedimentary petrography. Textbooks that dealt mainly with heavy minerals appeared in the 1920s and early 1930s (see *Sedimentary Petrography*), but professionalization of sedimentology proceeded no further until the IAS was founded.

In summary, the professionalization of sedimentology took place first in the US in the 1920s and 1930s. The first results, particularly the new journal and the two NRC-sponsored symposia, exerted a profound influence on the development of sedimentology outside the US. Some degree of professionalization was achieved at about the same time in Russia and Europe, but only recently have active national and international groups been set up.

Fourth Period (1931-1950)

The Early Professional Period. This period was characterized by the revival of studies pioneered earlier (petrography, diagenesis) and the development of new techniques (x-ray methods for clay mineralogy, statistical methods). Techniques were standardized for routine sedimentological analysis, particularly of modern sediments. These techniques and the dissemination of sedimentological observations and ideas through the publication of the many books that appeared at the close of the period (and particularly the work by Pettijohn, 1949) set the stage for the rapid progress in and growth of sedimentology in the modern period.

In Germany, the first systematic basin studies, making use of paleocurrents and facies analysis were begun by R. Brinkmann and E. Cloos (Pettijohn, 1962). In England and America, attention had earlier been focused mainly on paleogeography (by A. J. Jukes-Browne and Schuchert; see Schuchert, 1910) but shifted under Twenhofel's influence to the recognition of sedimentary environments. Facies (q.v.), long neglected by English-speaking geologists, were rediscovered in America (and popularized at a meeting of the Geological Society of America held in 1948), and geosynclinal theory was elaborated first in Europe and then in America (see *Geosynclinal Sedimentation*).

Tectonic theories provided a further impetus to petrographic studies; and there was a revival of such studies, which had been continued during the previous period mainly by the lone efforts of Lucien Cayeux (Taylor, 1950). Shvetsov and others in Russia developed theories of the control of sedimentary mineralogy by tectonics, and these theories were further developed and popularized in America by P. D. Krynine (Folk and Ferm, 1966), E. C. Dapples, W. C. Krumbein, L. L. Sloss and others. Heavy mineral and insoluble residue studies continued but began to wane in popularity, as their use for correlation purposes in the petroleum industry was replaced by the use of geophysical logging techniques.

Statistical techniques were developed for the processing of petrographic (and particularly grain size) data. Clay mineralogical studies, using x-ray techniques, revolutionized the study of fine-grained sediments; and investigations began on organic matter in sediments and the origin of petroleum.

Studies of physical processes of sedimentation were carried out by hydraulic engineers (e.g., Shields, 1936; Vanoni, 1946), but most of these studies were not well known to geologists. An exception is the work of R. A. Bagnold on wind-blown sands.

Fifth Period (1950-present)

Modern Studies. A number of books appeared at the close of the fourth period (Shrock, 1948; Shepard, 1948; Pettijohn, 1949; Twenhofel, 1950; Trask, 1950; Kuenen, 1950; Krumbein and Sloss, 1951), and a rereading of these books is a simple way to assess the "state of the art" just before the beginning of the contemporary era.

There are several notable features which distinguish the present from the previous period. The first is the great expansion in the scale of investigations of modern sediments. For example, the API project on the Gulf Coast and most of H. N. Fisk's work on the Mississippi Delta postdate 1950. L. V. Illing's work on the Bahama Banks (q.v.) was the first of many investigations of modern carbonate sediments in the Caribbean and, more recently, the Persian Gulf (q.v.). These investigations had a revolutionary effect on the study of ancient limestones, dolomites, and evaporites. Even more recently, the scope of deep-sea sediment investigations has been dramatically enlarged by the JOIDES program of drilling.

Secondly, there have been major advances in both theory and techniques of sedimentary geochemistry. Krumbein and Garrels' 1952 paper set the stage for many subsequent experimental and field studies of chemical sedimentation and diagenesis under low temperature, aqueous conditions. Modern isotopic techniques (C^{14} and carbon, oxygen, and sulfur stable isotope studies) were also developed and applied to problems of sedimentation and diagenesis only after 1950.

Third, the concept of high-density turbidity currents proposed by Kuenen and Migliorini in 1950 (the first entirely new hypothesis of sediment transport and deposition since the glacial hypothesis) revolutionized the study of *flysch* sediments (see *Turbidity-Current Sedimentation*). Application of the turbidity-current hypothesis led to a great increase in studies of sedimentary structures (analysis, classification, and mapping) and to the examination of other subaqueous sedimentary processes.

Fourth, experimental work on sediment transport and bed forms (q.v.), carried out by engineers since the 1930s, culminated in large-scale studies in the 1960s that gained the attention of geologists because of the light they threw on the origin of cross-bedding and other sedimentary structures (see *Bedding Genesis*). This work, and the example of workers such as Kuenen, led to an increase in physical experimentation by geological sedimentologists themselves.

It is probably too early to gain any true historical perspective on the accomplishments of the modern era. In retrospect, perhaps the development of plate tectonics in the mid-1960s will be seen to mark the beginning of a new period for sedimentology, as it does, almost certainly, for a revitalized stratigraphy and tectonics (see *Sedimentology—Yesterday, Today, and Tomorrow*).

GERARD V. MIDDLETON

References

Agassiz, L., 1840. *Etudes sur les Glaciers.* Neuchatel: Petitpierre, 346p. (Reprinted, 1966. London: Dawson of Pall Mall.) Transl. and ed. by A. V. Carozzi. New York: Hafner, 1967, 213p.

Anonymous, 1967. The development of lithology in the USSR and its immediate objectives, *Lith. Mineral Res.,* 1967(5), 527-544.

Bonney, T. G., ed., 1904. *The Atoll of Funafuti:* Borings into a coral reef and the results being the report of the Coral Reef Committee of the Royal Society. London: Royal Society London, 428p.

Brinkmann, R., 1948. Die allgemeine Geologie (insbes. die exogene dynamik) in letzten Jahrhundert, *Z. deut. geol. Gesell.,* 100, 25-49.

Cannon, W. F., 1960. The uniformitarian-catastrophist debate, *Isis,* 51, 38-55.

Chorley, R. J.; Dunn, A. J.; and Beckinsale, R. P., 1964. *The History of the Study of Landforms or the Development of Geomorphology.* Vol. I, *Geomorphology Before Davis.* London: Methuen, 678p.

Cross, W., 1902. The development of systematic petrography in the nineteenth century, *J. Geol.,* 10, 331-376, 451-499.

Darwin, C., 1837. On certain areas of elevation and subsidence in the Pacific and Indian oceans, as deduced from the study of coral formations, *Geol. Soc. London Proc.,* 2, 552-554.

Darwin, C., 1842. *Structure and Distribution of Coral Reefs.* London: Smith, Elder, 214p. Reprinted 1962, Univ. California Press (from 1851 ed.), 214p.

Daubrée, A., 1879. *Études Synthetiques de Géologie Experimentale.* Paris: Dunod, 828p.

Davis, W. M., 1884. Geographic classification, illustrated by a study of plains, plateaus and their derivatives, *Proc. Am. Assoc. Advanc. Sci.,* 33, 428-432.

Davis, W. M., 1889. Geographic methods in geologic investigation, *Natl. Geogr. Mag.,* 1, 11-26.

Dunoyer de Segonzac, G., 1968. The birth and development of the concept of diagenesis (1886-1966), *Earth Sci. Rev.,* 4, 153-201.

Ehrenberg, C. G., 1854. *Mikrogeologie.* Leipzig: Voss, 374p.

Folk, R. L., 1965. Henry Clifton Sorby (1826-1908), the founder of petrography, *J. Geol. Educ.,* 13, 43-47.

Folk, R. L., and Ferm, J. C., 1966. A portrait of Paul D. Krynine, *J. Sed. Petrology,* 36, 851-863.

Gilbert, G. K., 1877. *Report on the Geology of the Henry Mountains.* Washington, D. C.: U.S. Geog. Geol. Survey Rocky Mt. Reg., 160p.

Gilbert, G. K., 1890. Lake Bonneville, *U.S. Geol. Surv. Monogr. 1,* 438p.

Gilbert, G. K., 1914. Transportation of debris by running water, *U.S. Geol. Surv. Prof. Pap. 86,* 263p.

Ginsburg, R. N., ed., 1973. *Evolving Concepts in Sedimentology.* Baltimore: Johns Hopkins Univ. Press, 191p.

Gressly, A., 1838. Observations géologiques sur le Jura Soleurois, *Nouveaux Mem. Soc. Helv. Sci. Natur.,* 2, 349p.

Hatch, F. H., and Rastall, R. H., 1913. *The Petrology of the Sedimentary Rocks.* London: George Allen, 425p.

Krumbein, W. C., and Garrels, R. M., 1952. Origin and classification of chemical sediments in terms of pH and oxidation-reduction potentials, *J. Geol.,* 60, 1-34.

Krumbein, W. C., and Sloss, L. L., 1951. *Stratigraphy and Sedimentation.* San Francisco: Freeman, 497p. 2nd ed., 1963, 660p.

Kuenen, P. H., 1950. *Marine Geology.* New York: Wiley, 568p.
Kuenen, P. H., and Migliorini, C. I., 1950. Turbidity currents as a cause of graded bedding, *J. Geol.,* 58, 91-127.
Loewinson-Lessing, F. Y., 1954. *A Historical Survey of Petrology.* (Transl. from the Russian by S. I. Tomkeieff.) Edinburgh: Oliver and Boyd, 112p.
Lyell, C., 1830-1833. *Principles of Geology, Being an Attempt to Explain the Former Changes of the Earth's Surface by Reference to Causes Now in Operation.* London: Murray, 3 vols.
Middleton, G. V., 1973. Johannes Walther's Law of the correlation of facies, *Geol. Soc. Am. Bull.,* 84, 979-988.
Murray, J., and Renard, A., 1891. Deep-sea deposits, *Report on the Scientific Results of the Exploring Voyage of H. M. S. Challenger, 1872-1876.* London: Longmans, 525p.
Ochsenius, C., 1877. *Die Bildung der Steinsalzlager und ihrer Mutterlaugensalze.* Halle: Pfeffer, 172p.
Pettijohn, F. J., 1949. *Sedimentary Rocks.* New York: Harper & Row, 526p. 2nd ed., 1957, 718p. 3rd ed., 1975, 628p.
Pettijohn, F. J., 1962. Paleocurrents and paleogeography, *Bull. Am. Assoc. Petrol. Geologists,* 46, 1468-1493.
Russell, D., 1970. History of the S. E. P. M., *J. Sed. Petrology,* 40, 7-28.
Schuchert, C., 1910. Paleogeography of North America, *Geol. Soc. Am. Bull.,* 20, 427-606.
Shepard, F. P., 1948. *Submarine Geology.* New York: Harper & Row, 348p. 2nd ed., 1963, 557p. 3rd ed., 1973.
Shields, A., 1936. Anwendung der Ahnlichkeitsmechanik und der Turbulenzforschung auf die Geschiebebewegung, *Mitteilungen der Preuss. Versuch anst. f. Wasserbau. u. Schiffbau, Berlin, Heft 26,* 26p. See also translation by W. P. Ott and J. C. van Uchelen, U.S. Dept. Agriculture, Soil Conservation Service Coop Lab., Calif. Inst. Tech.
Shrock, R. R., 1948. *Sequence in Layered Rocks. A Study of Features and Structures Useful for Determining Top and Bottom or Order of Succession in Bedded and Tabular Rock Bodies.* New York: McGraw-Hill, 507p.
Spencer, L. J., 1957. Obituary of Rudolph Richter, *Proc. Geol. Soc. London,* 1554, 137-188.
Sorby, H. C., 1851. On the microscopial structure of the calcareous grit of the Yorkshire Coast, *Quart. J. Geol. Soc. London,* 7, 1-6.
Strakhov, N. M., 1970. Evolution of concepts of lithogenesis in Russian geology (1870-1970), *Lith. Mineral Res.,* 1970(2), 157-177.
Taylor, J. H., 1950. The contribution of petrology to the study of sedimentation, *Sci. Progress,* 38, 652-667.
Teall, J. J. H., 1902. The evolution of petrological ideas, *Proc. Geol. Soc. London,* 58, lxiii-lxxviii.
Thomas, H. H., 1902. The mineralogical constitution of the finer material of the Bunter Pebble Beds in the west of England, *Quart. J. Geol. Soc. London,* 58, 620-661, discussion, 631-632.
Thoulet, J., 1881. Etude mineralogique d'un sable du Sahara, *Bull. Soc. Min. France,* 4, 262-268.
Trask, P. D., ed., 1939. *Recent Marine Sediments,* A symposium, prepared under the direction of a subcommittee of the Committee on Sedimentation of the Division of Geology and Geography of the National Research Council, Washington, D.C. Tulsa, Okla.: Am. Assoc. Petro. Geol., 736p. Second printing, 1955, contains preface and statement of progress in studies of recent marine sediments, 1939-1954 by P. D. Trask, ix-lvi.
Trask, P. D., ed., 1950. *Applied Sedimentation.* New York: Wiley, 707p.
Twenhofel, W. H., ed., 1926, 1932. *Treatise on Sedimentation.* Baltimore: Williams & Wilkins, 661p. (2nd ed., 921p.) Reprinted by Dover.
Twenhofel, W. H., 1939, 1950. *Principles of Sedimentation.* New York: McGraw-Hill, 610p. 2nd ed., 1950, 673p.
Vanoni, V. A., 1946. Transportation of suspended sediment by water, *Am. Soc. Civ. Engin. Trans.,* 3, 67-133.
Wadell, H., 1933. Sedimentation and sedimentology, *Science,* 77, 536-537.
Walther, J., 1891. Die Denudation in der Wüste und ihre geologische Bedeutung. Untersuchungen die Bildung der Sedimente in den ägyptischen Wüsten, *Abhandl. Königlich-sächsichen Ges. Wiss. Wiss., mat.-phys. Classe,* 16, 347-369.
Walther, J., 1893-1894. *Einleitung in die Geologie als historische Wissenschaft.* Jena: Gustav Fischer, 3 vols., 1055p.
Walther, J., 1900. *Das Gesetz der Wüstenbildung in Gegenwart und Vorzeit.* Berlin: Reimer (Vohsen), 175p.
Wegmann, E., 1963. L'exposé original de la notion de faciès par A. Gressly (1814-1865), *Sci. de la Terre,* 9, 83-119.

Cross-references: *Geosynclinal Sedimentation; Sedimentary Petrography; Sedimentology-Yesterday, Today, and Tomorrow.*

SEDIMENTOLOGY-MODELS

The scientific method forms the basis for all geological enquiries, both in the field and in the laboratory. Although our instrumental, mathematical, and interpretive techniques are becoming increasingly refined, the philosophy underlying the scientific approach changes very little. Crucial to the method are various types of models, discussed by Potter and Pettijohn (1963), Krumbein and Graybill (1965), Griffiths (1967), Chorley (1964), and Wolf (1973). But what is a model? Krumbein and Graybill (1965) point out that a model "provides a framework for organizing or 'structuring' a study." Such techniques are a necessity because "the geologist studies multivariate problems by selecting and integrating from innumerable details those elements that appear to control a given geological phenomenon. The internal consistency of the data and the comparison of several lines of evidence simultaneously commonly provide a basis for selecting some single set of conditions that most satisfactorily accounts for the phenomenon." The model assists in this selection of the single set of conditions. A sedimentary model is referred to by Potter and Pettijohn (1963) as "an intellectual construct . . . based on a prototype." They discussed, for example, glacial, molasse, turbidite, delta, and reef models, and stated that the model concept relates all "geological elements to one another to provide an integrated description of a recurring pattern of sedimentation." They use only five elements, but many more factors and parameters can be considered, depending upon the problem to be analyzed (Wolf, 1973).

Potter and Pettijohn made it clear that a particular model, applied to a specific location, may prevail for a certain length of time; with a change of conditions, a new pattern may be established which is represented by a new model. Some models may succeed in an orderly, predictable fashion, as in the geosynclinal cycle, where molasse facies follows flysch (see *Geosynclinal Sedimentation*). On the other hand, several models, e.g., fluvial, deltaic, and turbidite, may all be applied at approximately the same time to a particular basin.

Chorley (1964) has said that the need for models, as well as for classification schemes in general, comes from the "high degree of ambiguity" presented by the various subject matters of the different scientific disciplines, and

the attendantly large "elbow room" which the researcher has for the manner in which this material may be organized and interpreted. This characteristic is a necessary result of the relatively small amount of available information which has been extracted in a very partial manner from a large and multivariate reality and leads not only to radically conflicting "explanations" . . . but to differing opinions regarding the significant aspects of . . . reality. . . . Where such ambiguity exists, scholars commonly handle the associated information either by means of classifications or models. . . . Model building, which sometimes may even precede the collection of a great deal of data, involves the association of supposedly

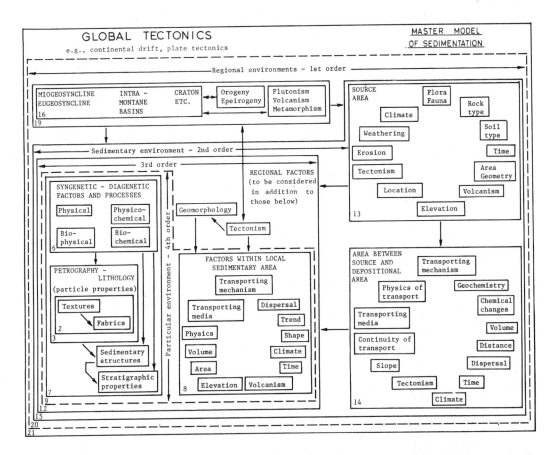

FIGURE 1. Hierarchy of factors and processes that singly or in combination determine the properties of sediments (after Wolf, 1973).

significant aspects of reality into a system which seems to possess some special properties of intellectual stimulation.

According to Kendall and Buckland (1960), a model is "a formalized expression of a theory or the causal situation which is regarded as having generated observed data. In statistical analysis the model is generally expressed . . . in a mathematical form, but diagrammatic models are also found." The diagrammatic or conceptual models are mental images, pictorial presentations of phenomena, situations, etc.; and the "initial model is commonly simplified to retain the essential features of the phenomenon, with extraneous details omitted" (Chorley, 1964). The models are based on observations, expressing some segment of the real world in idealized form. These initial models are often qualitative, but quantitative models are increasingly being developed in the various fields of geology (e.g., Schwarzacher, 1975). Nevertheless, qualitative models will hardly ever be replaced, as many of the quantitative ones are derived from them. "Conceptual models in diagrammatic form are largely qualitative, but they point the way toward a choice of observations and measurements that can be used to implement the model, either to test its validity in terms of real-world phenomena or to use the model as a predicting device" (Krumbein and Graybill, 1965).

Two parameters in earth-science models deserve special consideration, namely, the "time" and "scale" elements. The *time factor* is related to sequences of events, either projected back into the geologic past or into the future; these sequences may make up a succession of phenomena each of which is a process-response pair (e.g., Whitten, 1964; Wolf and Ellison, 1971). The nature of the *scale factor* can be exemplified in sedimentological-environmental studies that begin with the investigation of small components in sandstones, then gradually expand to include textures, fabrics, structures, outcrops, local stratigraphy, regional aspects, finally including the whole geosynclinal or cratonic environment, and may even go beyond that stage in the study of continental drift where the information from all branches of geology are employed as complemental data. Fig. 1 represents the scales by "boxes" or "nests" that are grouped together into different "orders" (see also Vol. III: *Geomorphology: Expanded Theory; Land Mass and Major Landform Classification*).

No comprehensive classification scheme seems to have been offered that would combine the various types of models, but the following will present a cross section of those widely used in sedimentology and related fields. There are (1) models in the form of *flow charts* which may be linear (Fig. 2), circular (Fig. 3), or hierarchical (e.g., Fig. 1); (2) *verbal models*, either tabular or nontabular, classificational or nonclassificational, some of which may be comparative (e.g., Allen, 1965, Tables 2 and 3); (3) *pictorial models* (e.g., Allen, 1965, Figs. 35 and 36; Wolf and Ellison, 1971, Fig. 16); and (4) *mathematical models*, some of which may be in graphical form, others purely statistical in the form of equations (see Krumbein and Graybill, 1965; Griffiths, 1967; Schwarzacher, 1975).

Models have numerous uses: (1) to demonstrate interrelationships between two or more factors, e.g., processes and settings; (2) as a "mental crutch" to assist in memorizing complex phenomena; (3) to assist in realizing new problems by formulating correct, meaningful questions (Griffiths, 1967); (4) to improve the critical appraisal and testing of ideas in research and exploration studies, to improve problem-solving procedures, to find workable

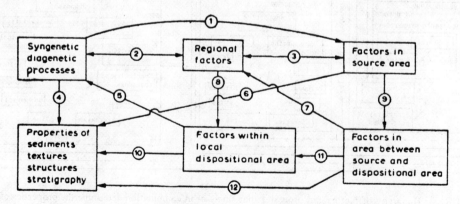

FIGURE 2. General model of factors that determine properties of sediments (after Wolf, 1973).

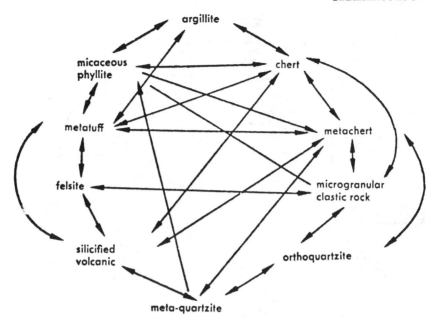

FIGURE 3. Circular diagram to show ten lithologies between which there are possible textural and compositional transitions that make identification and discrimination in sand grains and pebbles often difficult or impossible (after Wolf, 1971).

solutions, and to form prognostications (Griffiths, 1967); (5) to compel oneself to consider all possible parameters and various possible interpretations; (6) to improve the "visualization of phenomena, as models may give a bird's eye view" and summarize a complex subject; and (7) to prepare flow diagrams in computer geology.

Figs. 1-4 present some models, most of which are of the conceptual variety. Fig. 2, for example, has six "boxes" listing factors that must be considered in the interpretation of sediments and in the reconstruction of their environments of deposition. The twelve arrows indicate genetic relationships, e.g., arrow 6 points to the fact that the climatic factor operating in the source-rock locality will determine the degree of weathering and the type of sand and clay particles that are transported to the site of sedimentation. Arrow 5 indicates that all chemical, biological, and physical processes in the depositional environment will determine the type of syngenetic and diagenetic processes affecting the sediments as indicated by arrow 4. On the other hand, the hydrodynamic factors will determine the textures and structures of the sedimentary accumulations (arrow 10). Fig. 2 is only a very simplified view of the more complex possible interrelationships that number in the hundreds. For example, a separate conceptual model illustrates 40 of the complex genetic relationships during formation of a soil in a source area (Fig. 4). The same diagram can be used for provenance (q.v.) studies because the erosion of soil and the transportation to a site of sedimentation leads to the origin of various pebble, sand, silt, and clay deposits. Under ideal conditions, this type of study may enable the sedimentologists to reconstruct the conditions that prevailed in the source area.

Fig. 1 is in effect an expansion of Fig. 2, presenting as many factors as possible that control the properties of sediments. The "starting point" in reading the hierarchical model lies near the lower left-hand corner, i.e., at *particle properties*. These properties determine the textural characteristics, which in turn determine the fabric. The latter two are also closely related and are, therefore, combined to form box 2. The study of particle properties, textures, and fabrics comes under the heading of petrography and lithology and are grouped together to form box 3. It is clear that the properties of the parameters in box 3 determine *sedimentary structures*, and so on (Wolf, 1973).

The complex interrelationships between the numerous parameters in individual boxes are quite complex (as in Fig. 4). Many relationships also exist between parameters from different

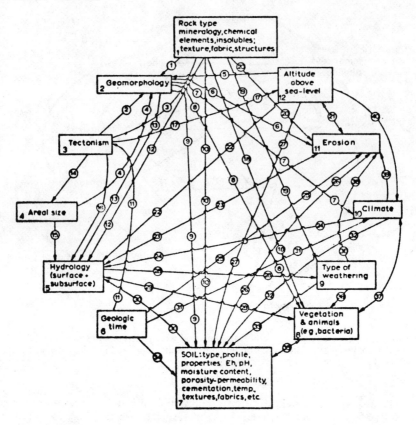

FIGURE 4. Model relating all parameters important in the genesis of soils (also useful in provenance studies; from Wolf, 1973).

boxes, however, and it is up to the investigator to select parameters that will help to solve the problem chosen and to prepare more detailed models. One could isolate two or more of the factors and processes, such as those in box 6, and draw up a new conceptual model demonstrating the interdependencies of the physical, biophysical, physicochemical, and biochemical processes listed. Another example would be to isolate boxes 13 and 3 and establish diagrammatically the numerous relationships between the processes and factors in the source area and the mineralogic composition of the particles in a sedimentary rock. The cross-correlations between the major boxes as in Fig. 4 are so numerous that they are not shown in the master scheme (see Wolf, 1973).

In studying the textures of rock fragments that are frequently present in conglomerates and coarser sandstones, problems of identification arise during microscopic examinations of thin sections because (1) the areas of the grains are small; (2) the textures in the grains are microscopic in size; and (3) there are numerous mineralogical and textural transitions from one rock type to another. To simplify the problem of identification, a circular, conceptual model (Fig. 3) was prepared. The arrows indicate rock types that have various transitional varieties which make correct genetic recognition very difficult.

An overview of all major sedimentary environments is useful as a model in attempts to reconstruct ancient paleogeographic conditions; Wolf (1973) has prepared such a model. Detailed conceptual models have been prepared for glacial environments (Harland et al., 1966), fluvial environments (Allen, 1965), placer gold mineralization in Precambrian conglomerates (Pretorius, 1966), formation of pyroclastic sediments (Wolf and Ellison, 1971), cyclothems in sedimentary basins (Duff et al., 1967; Asquith, 1974), and on the application of computers in model studies (Harbaugh and Merriam, 1968).

KARL H. WOLF

SEDIMENTOLOGY—ORGANIZATIONS AND ASSOCIATIONS

National groups in various countries active in sedimentology include the Japanese Association of Sedimentologists, the Association of Southern African Sedimentologists, the Sedimentology Specialists Group in Australia, the Grupo Español de Sedimentologia, The British Sedimentologists Research Group, and the Association des Sédimentologistes Française. These groups hold meetings and issue newsletters. In addition, informal groups meet in various countries; an example is the Carbonate Research Group in Israel, which conducts field trips and research and which is centered around the Oil Division of the Geological Survey of Israel.

Two organizations, both international in scope, set the tone in sedimentology: (1) the Society of Economic Paleontologists and Mineralogists (SEPM), a division of the American Association of Petroleum Geologists (AAPG), and (2) the International Association of Sedimentologists (IAS). The SEPM, headquartered in North America, holds one annual convention jointly with the AAPG; attendance at recent meetings has been estimated at 4500 persons. The IAS holds its congress (the International Sedimentological Congress) in Europe once every four years; attendance at recent meetings has been approximately 300.

Society of Economic Paleontologists and Mineralogists

The SEPM, founded in 1926, publishes the *Journal of Sedimentary Petrology* and the *Journal of Paleontology*. The founders of SEPM were paleontologists, some of whom worked as micropaleontologists in the oil business. In 1932, SEPM became a technical division of the AAPG. The total membership of the society is over 4000.

The business of the SEPM is conducted through a council which is elected annually. The SEPM has six regional sections throughout the US; regional presidents are eligible for nomination to the national presidency of SEPM. These six regional sections conduct field trips, hold symposia, and publish guidebooks. Research groups in the various specialties in sedimentology and paleontology, presently nine, meet at the society's annual convention. These research groups offer the members a forum in which they can express new ideas through a mutual interchange of information. Research groups include (1) evaporites, (2) coastal sedimentation, (3) environmental geology, (4) fine-grained sediments, (5) fusulinids, (6) geochemistry of

References

Allen, J. R. L., 1965. A review of the origin and characteristics of Recent alluvial sediments, *Sedimentology*, 5, 89-191.

Allen, J. R. L., 1974. Reaction, relaxation and lag in natural sedimentary systems: general principles, examples and lessons, *Earth Sci. Rev.*, 10, 263-342.

Asquith, D. O., 1974. Sedimentary models, cycles and deltas, Upper Cretaceous, Wyoming, *Bull. Am. Assoc. Petrol. Geologists*, 58, 2274-2283.

Chorley, R. J., 1964. Geography and analogue theory, *Ann. Assoc. Am. Geographers*, 54, 127-137.

Duff, P. M.; Hallam, A.; and Walton, E. K., 1967. *Cyclic Sedimentation*. Amsterdam: Elsevier, 280p.

Griffiths, J. C., 1967. *Scientific Method in Analysis of Sediments*. New York: McGraw-Hill, 508p.

Harbaugh, J. W., and Merriam, D. F., 1968. *Computer Applications in Stratigraphic Analysis*. New York, Wiley, 282p.

Harland, W. B.; Herod, K. N.; and Krinsley, D. H., 1966. The definition and identification of tills and tillites, *Earth Sci. Rev.*, 2, 225-256.

Kendall, M. G., and Buckland, W. R., 1960. *A Dictionary of Statistical Terms*. London: Oliver and Boyd, 590p.

Krumbein, W. C., and Graybill, F. A., 1965. *An Introduction to Statistical Models in Geology*. New York: McGraw-Hill, 475p.

Kuenen, P. H., 1966. Geosynclinal sedimentation, *Geol. Rundschau*, 56, 1-19.

Mizutani, S., and Hattori, I., 1972. Stochastic analysis of bed-thickness distribution in sediments, *Math. Geol.*, 4, 123-146.

Potter, P. E., and Pettijohn, F. J., 1963. *Paleocurrent and Basin Analysis*. New York: Springer-Verlag, 296p.

Pretorius, D. A., 1966. Conceptual geological models in the exploration of gold mineralization in the Witwatersrand basin, *Univ. Witwatersrand, Johannesburg, Econ. Geol. Research Unit., Inf. Circ. 33*, 39p.

Schwarzacher, W., 1975. *Sedimentation Models and Quantitative Stratigraphy*. Amsterdam: Elsevier, 396p.

Whitten, E. H. T., 1964. Process-response models in geology, *Geol. Soc. Am. Bull.* 75, 455-464.

Wolf, K. H., 1971. Textural and compositional transitional stages between various lithic grain types, *J. Sed. Petrology*, 41, 328-332.

Wolf, K, H., 1973. Conceptual models. 1. Examples in sedimentary petrology, environmental and stratigraphic reconstruction, and soil, reef, chemical and placer sedimentary ore deposits. 2. Fluvial-alluvial, glacial, lacustrine, desert, and shorezone (bar-beach-dune-chenier) sedimentary environments, *Sed. Geol.,* 9, 153-193, 261-281.

Wolf, K. H., and Ellison, B., 1971. Sedimentary geology of the zeolitic volcanic lacustrine Pliocene Rome Beds, Oregon, Part 1, *Sed. Geol.,* 6, 271-302.

Cross-references: *Experimental Sedimentology*. Vol. III: *Geomorphology: Expanded Theory; Land Mass and Major Landform Classification.*

sediments, (7) oceanic plankton, (8) trace fossils, and (9) turbidites and deep-sea sedimentation. In addition, a carbonate research group, formed originally as a subcommittee of the Research Committee of the AAPG, holds informal meetings with the SEPM. In 1952, the SEPM inaugurated a series of special publications; over 25 have been published, about 2/3 of which are sedimentology oriented, the others paleontology oriented.

SEPM has 5 standing committees: (1) Research, (2) Publication, (3) Finance, (4) Convention, and (5) Headquarters Advisory. Four named awards recognize achievement: (1) the Outstanding-Paper Award in each of the society's journals, (2) honorary membership, (3) the Francis P. Shepard Medal for excellence in marine geology, and (4) the William H. Twenhofel Medal, the highest recognition of the society, for excellence in sedimentary geology. Russell (1970) provides an excellent and very personable history of the society's first 22 years (1926-1948).

International Association of Sedimentologists

The IAS, first meeting in 1946, exists to promote the study of sedimentology and the interchange of research, particularly where international cooperation is desirable (Doeglas, 1976). Members of the association receive the journal *Sedimentology,* published since 1962, and news circulars which cover association news, new books and journals, and activities in sedimentology throughout the world. National correspondents from each country in which the field of sedimentology is active report news for inclusion in the newsletter. Recently, the association has initiated the publication of a series of special publications, the first of which appeared in 1974. IAS is affiliated with the International Union of Geological Societies. The association has two standing committees: (1) Membership and (2) Publications Policy. Business of the association is conducted through its elected council. Elections for council are held every four years at a meeting of the General Assembly. The General Assembly consists of members that are present at the business session of the International Sedimentological Congress. The association sponsors regional meetings or field trips, which are generally held in conjunction with the meetings of other societies. The total membership of the association is approximately 1000.

GERALD M. FRIEDMAN

References

Doeglas, D. J., 1976. The first fifteen years of the International Association of Sedimentologists 1952-1967, *Sedimentology,* 23, 5-16.

Russell, R. D., 1970. SEPM history, *J. Sed. Petrology,* 40, 7-28.

Cross-references: S*edimentary Petrography; Sedimentology—History; Sedimentology—Yesterday, Today, and Tomorrow.*

SEDIMENTOLOGY—YESTERDAY, TODAY, AND TOMORROW

Throughout most of the 19th century, the study of sediments was largely based on field work—the definition, age, and correlation of lithologic units—and analysis was virtually limited to megascopic field description, so that the term *stratigraphy* adequately described the study of sediments. The megascopic description of sediments in the field remains important, but other disciplines have been added so that now the word *sedimentology* (q.v.) best describes the study and origin of sediments (see also *Sedimentology—History*).

There were various stages in the development of stratigraphy into sedimentology: first, the use of chemical analyses, followed by the petrographic microscope, the x-ray, optical and mass spectrographs, the electron microscope and its derivate the scanning electron microscope (SEM). The laboratory study of the magnetic properties of sediments made it possible to estimate their paleolatitude from their natural remanent magnetism (q.v.). In the field, boots, backpack, compass, notebook, and hammer are now supplemented by remote sensing devices designed to measure geophysical properties of sediments, to which we give a sedimentological interpretation, e.g., a wide variety of wire-line logs obtained from boreholes—electric, sonic, gamma-ray, neutron, dip, and caliper logs. Without these data our perception of the subsurface correlation of lithologic units and their three-dimensional geometry would be vastly restricted.

Other significant contributions to modern field geology are made by the reflection seismograph (*seismic stratigraphy*), particularly along continental margins and in the deep sea, and the ability to obtain long cores from the deep sea. Indirect observation by aerial photography, or side-looking radar for subaerial photography and side-looking sonar for submarine landscapes, is now fairly commonplace. Even more exciting is the contribution of space research (see *Lunar Sedimentation; Martian Sedimentation*).

TABLE 1. Some Previous Reviews of Sedimentology

Author and Year	Remarks
Boswell, 1916	Primarily a report on how heavy minerals can help stratigraphy.
Milner, 1931	Brief comments on how sedimentary petrology, in spite of Sorby's emphasis upon thin sections, evolved from the study of heavy minerals to its more modern (1931) state, i.e., emphasis on *particles*.
Twenhofel, 1934	Key idea: paleoecology, diagenesis, petrology, and sediments as long chains of complex processes.
Pettijohn, 1935	Reviews of 35 papers published during 1932–34 (some abstracts reproduced). Notes appearance of first isopleth map of sedimentary mineral composition.
Trowbridge, 1935	Report on activities of the Committee on Sedimentation, plus a list of 17 future problems.
Trask, 1939	Summaries of seven topics of special interest, including an annotated bibliography of Russian papers by P. D. Krynine and German papers by Correns.
Halbouty et al., 1940	This short note gives an outline of all the things that could be studied in the Tertiary of the Gulf Coast: provenance, transgressions and regressions, old shorelines, pinchouts, finding stratigraphic traps, use of isopach maps, and laboratory techniques. Also a good source of references to reports of the Committee on Sedimentation, 1928–1938.
Levorson, 1940	A world-famous petroleum geologist notes how techniques of oil-finding change with time. He suggests that (in 1940) more intense study of stratigraphy, sedimentation, reservoir rocks, and environments of deposition is in vogue.
Knopf, 1941	A short history of sedimentary petrology and brief comments on some major activities: geochemistry, petrofabrics, and carbonate rocks. Several interesting historical papers cited.
Trask, 1941	Current problems of interest included origin of black shales, rates of sedimentation, shelf sedimentation, alluvial hydraulics, x-ray identification of minerals, organic geochemistry, precipitation of $CaCO_3$.
Twenhofel, 1941	Chiefly considers provenance of sands and to a lesser extent authigenic minerals and depositional environments.
Pettijohn, 1942	Quantification in sedimentology and history thereof, largely based on the study of sands and sandstones. Suggests all sedimentologists should study calculus and hydrodynamics and gives sedimentology curriculum at University of Chicago, still very relevant today. Many references.
Krumbein, 1945	Quantitative aspects of sedimentary environments, primarily using shape and roundness as examples. Suggests three major factors are involved: boundary conditions, materials, and energy of the depositional system. Stresses need for additional types of maps. 64 references.
Lowman, 1947	How to organize research in sedimentology and petroleum exploration; proposes specific topics for study.
Pettijohn, 1956	Discusses the relative importance of laboratory and field work to sedimentology and in so doing reviews current advances in sedimentology: turbidites, paleocurrents, the stratigraphic framework, and reefs.
Russell, 1957	Nine major categories of research, most of which are sedimentary. Detailed, extended discussion.
Guilcher, 1964	Topics in recent marine sediments include: coral reefs, shelf sedimentation, undersea processes, deep-sea physiography, and deep-sea sediments. Over 120 references.
Gubler, 1972	An introduction and overview by Gubler, followed by 14 papers which give a good insight into some of sedimentology's many diverse aspects.
Füchtbauer, 1973	Compact treatment of terrigenous sediments (9 topics) and carbonate sediments (5 topics) plus cherts (65 references and 5 figures).

Shortened from Potter, 1974.

Past Appraisals

To appraise the present and future in a field as diverse and as large as sedimentology is a most difficult task; nonetheless, many have made the attempt. Table 1 lists some papers that review the potentialities and problems of all or some major part of sedimentology, papers that discuss new "frontiers" and "modern points of view." Many are reports by committee chairmen, especially of the Committee on Sedimentation of the National Research Council (Washington, 1919-1940), whereas a few, e.g., Pettijohn (1956), are explicit statements of a personal philosophy. It is sobering to read some of these appraisals and see how much we are still working on problems that were well identified 30 years ago and more.

As was noted by Levorsen (1940), oil exploration techniques change with time, any one technique attaining a maximum and then declining. The same is true for the different subdivisions of sedimentology. The following is a review of the more significant present and likely future developments in sedimentology, fields that should offer, per unit of effort expended, promise of rising rather than declining rewards.

Promising Fields

Megasedimentology. The study of the sedimentology of basins and larger areas, e.g., ancient or modern continental margins, or of entire ancient fold belts, e.g., the Alpine or Andean mountain chains, has evolved, in large part, from the basin analysis of Potter and Pettijohn (1963), where prime interest was focused on the basin itself: how variation in basin shape, paleocurrents, basin fill, and lithologic arrangement of that fill produced some of the major basin types. Much remains to be done, for many of the world's basins have not been described in these terms, and a classification of basin types has not been constructed. For example, How many types of coal basins are there and what are their basic sedimentologic parameters? If the latter could be properly identified, it should be possible to represent or describe all the world's coal basins in terms of a few essential parameters. Such an analysis would be a useful activity for sedimentologists during the current energy crisis.

Today, thanks to the discovery of sea-floor spreading and the development of theories of plate tectonics, megasedimentology has the additional possibility of "explaining" some of the major types of basins. There is much activity concerning both sedimentation and basin development along modern continental margins, as well as attempts to reinterpret ancient basins in terms of plate-tectonic theories. It seems clear that this activity will yield high rewards in coming years, primarily to sedimentologists who excel at deductive "big thinking" (see, e.g., *Sedimentary Facies and Plate Tectonics*).

Provenance. The determination of the source, or provenance (q.v.), of detrital terrigenous debris will always be one of sedimentologists' major activities. Typical questions are: From what kind of terrain? In what geographic location? and, In how many steps or cycles? The paleoclimate of the source region is commonly also included in provenance studies. Evidence of continental drift due to sea-floor spreading has renewed interest in provenance studies, which help define former continental configurations. Systematic studies of petrology, paleocurrents, and facies are required; a good example is the comparison of paleocurrents between Africa and South America (Bigarella, 1970).

Twenty years ago, a major step forward occurred when the systematic petrology and paleocurrents of a single formation were studied over an entire basin (see *Paleocurrent Analysis*). Today and tomorrow, the corresponding "forward leap" will take in groups of related basins or even an entire geosyncline or continental margin. These are, of course, large and expensive undertakings; yet if marine geologists can fund full-time research vessels, should not land-based sedimentologists think in comparable terms, perhaps of several provenance teams equipped with helicopters?

New provenance techniques should include very sophisticated studies, e.g., the determination of temperature of formation of alkali feldspars (Wright, 1968), which could be very informative for the discrimination of some source regions. As with any other mineral, however, many observations must be collected over a wide area before this or other new techniques can be significant.

Depositional Environments. The identification of primary depositional environments continues to occupy much of the attention of sedimentologists (see *Sedimentary Environments*). Inferring ancient depositional environments is a very old activity, one that attracted the attention of Leonardo da Vinci. After a period of rapid advance, especially for carbonates, environmental recognition appears to have reached a plateau, and future research will probably be directed toward (1) identification of depositional environments from wire-line geophysical logs and (2) detailing subenvironments. The identification of depositional environments directly from geophysical logs is a promising area of research, especially for interbedded sandstones and shales. Currently, what is required is, first, the careful use of cuttings and cores to establish the depositional environment of a particular unit and, second, correlation of these results with several key geophysical logs, which serve as "local standards." Universal generalization still lies in the future, although some studies already suggest it (Pirson, 1970).

Very little has been done with environmental recognition of carbonates from wire-line logs, although the possibility of recognizing "shoaling" upward and "fining" upward cycles does exist. Future years will also see more effort at environmental interpretation directly from seismic sections.

Campbell and Oaks' (1973) study of fluvial and tidal environments in a Cretaceous channel fill in Wyoming, primarily based on field observations, is a good example of the progressive discrimination of closely related environments that is now possible for many outcropping terrigenous sediments. An ancient environment that remains difficult to identify is the eolian one; for example, many eolian skeletal limestones in ancient platform carbonates have probably been overlooked. An outstanding example of a study of modern carbonate environments and of the vast amount of information that such study can yield is the collection of 23 papers on the Persian Gulf (Purser, 1973).

Another major problem in environmental recognition is the controversy over the shallow- versus deep-water origin of evaporites (see Dean, 1975). The evaporites recently discovered in the Mediterranean by the JOIDES drilling program may well resolve this controversy. If all evaporites are not of shallow-water origin, a prime problem for sedimentologists is to find some lithologic or chemical evidence that will distinguish between the two.

Coming years will also see more use made of paleolatitude (determined from paleomagnetics) in studies of ancient depositional environments, e.g., Dott (1974); not to use it would be to discard a valuable insight into the ancient climate of the depositional site.

Electronic Data Processing (EDP). EDP includes tabulation and analysis of the quantitative data that is the earmark of most modern sedimentology. Stated differently, EDP includes the storage, retrieval, and processing of geologic information. The analysis of such data is carried out mostly by statistical methods; there are many such methods available, and it is simply a matter of more sedimentologists learning to use them. EDP may make a large contribution once more sedimentologists place their data on magnetic tape in a data bank and then use it to generate some of geology's most fundamental tools, i.e., graphic displays of contour maps, cross sections, and fence and perspective diagrams. EDP will not make its full impact until we utilize the punch card (first introduced into sedimentology by Parker, 1946) to obtain instantaneous maps and cross sections of a sedimentary body, e.g., a large basin or a small area containing many different coal beds. The technology is at hand, and some use has been made of it (Kaas, 1969), but its full potential is still to be realized.

Sedimentary Geochemistry and Diagenesis. Recent years have seen much progress in the application of equilibrium thermodynamics to the common sedimentary minerals at low temperatures and pressures, as well as some progress in the use of chemical kinetics. The activity diagrams of Helgeson and Mackenzie (1970) are good examples of the use of equilibrium thermodynamics in sedimentology (Fig. 1). Such diagrams set forth possible equilibrium mineral assemblages with fluids of variable composition. This particular pair of diagrams shows which major detrital minerals are in equilibrium with warm (25°C) and cold (0°C) sea water at 1 atm. Because they do not appear in the water diagram, plagioclase, muscovite, and analcime are not in equilibrium with sea water (slow reaction rates explain their persistence in sea water). Short residence time in the ocean, and a mineral composition generally compatible with sea water explain why detrital minerals predominate among alumina silicates. It has been calculated that only about 3% of the sediments of the ocean are authigenic (Berner, 1971).

Calculated stability fields of different minerals will be incorrect, of course, if the choice of simplifying assumptions is wrong or the thermodynamic data are wrong. After an equilibrium study has demonstrated what reactions are possible, chemical kinetics can be used to predict whether these reactions will indeed occur. Berner (1971) cites a wide range of phys-

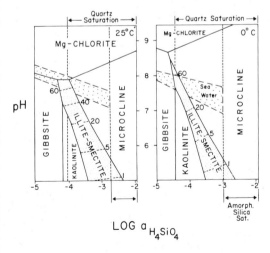

FIGURE 1. Activity diagram showing equilibrium relations of aluminosilicates and sea water in an idealized, nine-component model (from Potter, 1974, after Helgeson and MacKenzie). Heavier stipple represents the compositional range of sea water at 25°C and 0°C. Notice effect of temperature on boundaries.

ical chemical principles, including the application of kinetics to the study of sediments and their associated fluids. A good example of the integration of laboratory kinetic data and field observation is Morse and Berner's (1972) study of calcium carbonate dissolution in the ocean.

Porosity and permeability have long been considered to be the best pair of variables to produce an overall assessment of diagenesis. For both the essential questions are: How much? What types? and, How are various types related to original depositional environment and/or later burial history? These questions have too long remained outside the scope of academic sedimentologists, most of the work having been done by those in industry. Answers to these questions seem likely to be a major preoccupation of sedimentologists for many years to come. Because carbonates are very reactive, we have learned a great deal about their original porosity from the study of modern environments. A most useful appraisal of porosity in ancient carbonates, and one that shows the complexity of the problem, is that of Choquette and Pray (1970). For both carbonates and sandstones the fundamental processes are *cementation* (q.v.) and *diagenesis* (q.v.). In future years, the *systematic mapping* of the areal distribution of kinds and amounts of both cement and porosity will perhaps give rise to a *cement-porosity stratigraphy* that will rely heavily on the luminescence petrography pioneered by Sippel (1968), the SEM probably being an important supplement. An excellent example of the systematic study of cementing agents and porosity in a sandstone over a large part of a basin is that of Levandowski et al. (1973). Many more studies are needed. It is always essential to discriminate between the effects of primary depositional environment and later burial history; one has to write a history of the deposit including, if possible, some information on the composition of its pore waters, past and present, and such additional variables as depth of burial, detrital mineralogy, grain size, internal surface area, and flow rate of the pore fluid (Fig. 2).

An unexpected development in shale diagenesis is the idea, that pressured shale–shale with abnormally high pore-water pressures–can cause significant faulting in thick, marginal basins (Bruce, 1972). The mineralogical transformation of smectite to illite with depth releases water which, if not conducted away by permeable interbedded sandstones, remains in the shale to cause differential compaction and hence faulting. Instead of tectonic deformation causing diagenesis, as is usually the case, a diagenetic process produces large-scale tectonic features—a conclusion that not long ago prob-

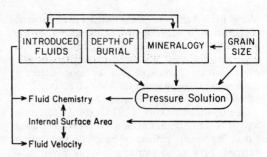

FIGURE 2. Probable interrelationships between some of the major variables that affect cementation and porosity in granular sediments (from Potter, 1974).

ably would have provoked laughter. (See also *Authigenesis; Compaction in Sediments; Diagenesis;* and Vol IVA, *Encyclopedia of Geochemistry and Environmental Sciences*).

Air-Sea-Particle Interactions. This broad heading includes most marine sedimentology, the interaction of air and sea with both inorganic particles and organisms; the subject and its literature are vast. Sedimentation along a shoreline has always been most important for sedimentology (see *Littoral Sedimentation; Paralic Sedimentary Facies*), but it has always been difficult to quantify, particularly at the level of coastal landforms, which are the modern equivalents of ancient lithologic units or sedimentary facies. Two questions are relevant: How does one estimate the relative importance of wave energy and sediment input? and How does one relate their ratio to shoreline morphology? A major contribution to sedimentology was made by Wright and Coleman (1973), who attempted to answer these questions concerning deltas. Although their results involve approximations and are, in the final analysis, only semiquantitative, they effectively relate this ratio to delta shape and morphology. Imbrie and Kipp (1971) have quite a different approach, using equations in matrix notation to relate variations in modern fossil plankton to the physical parameters of their water masses. Pleistocene plankton from a deep-sea core were used to infer a paleooceanography that determines surface temperature and salinity for the winter and summer of the Pleistocene Atlantic Ocean. With appropriate modifications, this method has considerable significance for paleoecology in general and perhaps for much of sedimentology as well.

Sedimentology and Society

Beyond its importance in the exploration and development of petroleum, ground water, and mineral resources, sedimentology is playing an increasing role in studies of humans and their environment. A paper by Guy and Jones (1972) reviews some of these applications. Other examples of socially useful studies are: (1) silting of reservoirs, harbors, rivers, and canals, one of sedimentology's first "socially significant" activities; (2) dispersion and effect of chemical pollutants on organisms in lakes, rivers, and coastal waters, a subject which involves modern ecology, water circulation, and water chemistry; (3) development of a systems analysis for deltas that includes river discharge, biota, climate, ground water, and geomorphology, so enabling artificial alterations to deltas to be planned more effectively; (4) research on the physical and chemical characteristics of sediment dredged from channels and canals, particularly their ecological impact on the area where they are placed and their long-term stability (the Office of Dredged Material Research, U.S. Corps of Engineers, Vicksburg, Mississippi, USA, was recently opened to study this problem); and (5) soil erosion, gulleying and sediment yield, for which Heinemann and Piest (1975) have prepared an excellent review.

An interesting aspect of such applications is how sedimentologic concepts, many of which might be considered totally academic, unexpectedly turn up. Debris accumulation in cities, both modern and ancient, is a good example because it involves the local rate of sedimentation (Fig. 3). To compare rates of sedimentation between modern cities and ancient ones is certainly a useful way to put our present problems of waste accumulation into a better perspective. For example, Gunnerson (1973) reports that Bronze-Age Troy and Manhattan, New York, both have about the same rate of solid waste accumulation: 140 cm per century—*plus les choses changent, plus elles restent les mêmes?*

Conclusion

Twenty to forty years ago some sedimentologists were writing of the need for quantification, mostly relative to the description of particles. Today, more sophisticated instrumentation facilitates such description of particles and sediments, and now the problem is to *quantify the sedimentological process*. Quantification of sedimentary processes, be they chemical, physical, or biological, will probably become the dominant activity of sedimentologists in years to come.

As quantification increases, it may be easy to forget that we study sedimentary processes to explain sediments and that the latter always fill a three-dimensional volume. Hence *mapping* will always be a major activity for sedimentologists. Advances in remote sensing and computer graphics by EDP should play a significant role in helping sedimentologists to make better use of maps—geology's oldest and most fundamental device.

PAUL EDWIN POTTER

Note: This entry is largely abstracted from Potter (1974), with the kind permission of Springer-Verlag.

References

Berner, R. A., 1971. *Principles of Chemical Sedimentology*. New York: McGraw-Hill, 240p.

Bigarella, J. J., 1970. Continental drift and paleocurrent analysis, in *Second Gondwana Symposium, Proceedings and Papers* (Intern. Union Geol. Sci., Comm. on Stratigraphy), 73-98.

Boswell, P. G. H., 1916. The application of petrological and quantitative methods to stratigraphy, *Geological Mag.*, 53, 105-111, 163-169.

Bruce, C. H., 1973. Pressured shale and related sediment deformation: Mechanism for development of regional contemporaneous faults, *Bull. Am. Assoc. Petrol. Geologists*, 57, 878-886.

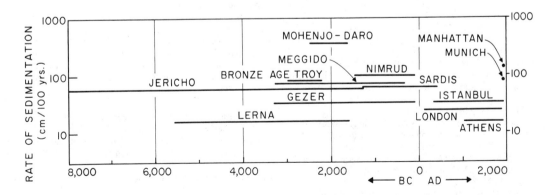

FIGURE 3. Rate of "debris sedimentation" for some cities during the last 10,000 years (from Potter, 1974, after Gunnerson).

Campbell, C. V., and Oaks, R. Q., Jr., 1973. Estuarine sandstone filling tidal scours, Lower Cretaceous Fall River Formation, Wyoming, *J. Sed. Petrology,* **43,** 765-778.

Choquette, P. W., and Pray, L. C., 1970. Geologic nomenclature and classification of porosity in sedimentary carbonates, *Bull. Am. Assoc. Petrol. Geologists,* **54,** 207-250.

Dean, W. E., 1975. Shallow-water versus deep-water evaporites: Discussion, *Bull. Am. Assoc. Petrol. Geologists,* **59,** 534-542.

Dott, R. H., Jr., 1974. Cambrian tropical storm waves in Wisconsin, *Geology,* **2,** 243-246.

Füchtbauer, H., 1973. Neuere Entwicklungen und Ergebnisse aus dem Gebiet der Sedimentpetrographie, *Forschr. Mineral.,* **50,** 188-204.

Gubler, Y., 1972. Stratigraphie et sédimentologie. Introduction et rapport de synthèse, *Bur. Rech. Géol. Min. Mem.,* 77(2), 521-529.

Guilcher, A., 1964. Present-time trends in the study of recent marine sediments and in marine physiography, *Marine Geol.,* **1,** 4-15.

Gunnerson, C. G., 1973. Debris accumulation in ancient and modern cities, *J. Environm. Engin., Div., ASCE 99, Proc. Pap. 9774,* 229-243.

Guy, H. P., and Jones, D. E., Jr., 1972. Urban sedimentation in perspective, *J. Hydr. Div. ASCE 98, Proc. Pap. 9420,* 2099-2116.

Halbouty, M. T., and 8 others, 1940. Sedimentation—Report of Houston Geological Society group, *Bull. Am. Assoc. Petrol. Geologists,* **24,** 374-376.

Heinemann, H. G., and Piest, R. F., 1975. Soil erosion—sediment yield research in progress, *Eng. and Sci.,* **56,** 149-159.

Helgeson, H. C., and Mackenzie, F. T., 1970. Silicate-sea water equilibria in the ocean system, *Deep-Sea Research,* **17,** 877-892.

Imbrie, J., and Kipp, N. G., 1971. A new micropaleontological method for quantitative paleoclimatology; application to a late Pleistocene Caribbean core, in K. K. Turekian, ed., *The Late Cenozoic Glacial Ages.* New Haven: Yale Univ. Press, 71-147.

Kaas, L. M., 1969. Computer graphics: A new tool for exploration and mining, in A. Weiss, ed., *A Decade of Digital Computing in the Mineral Industry.* New York: Am. Inst. Min. Metall. Petrol. Eng., 9-22.

Knopf, A., 1941. Progress of systematic petrography, in *Geology, 1888-1938. Geol. Soc. Am. 50th Annivers. Vol.,* 335-359.

Krumbein, W. C., 1945. Recent sedimentation and the search for petroleum, *Bull. Am. Assoc. Petrol. Geologists,* **29,** 1233-1261.

Levandowski, D. W.; Kaley, M. E.; Silverman, S. R.; and Smalley, R. G., 1973. Cementation in Lyons Sandstone and its role in oil accumulation, Denver Basin, Colorado, *Bull. Am. Assoc. Petrol. Geologists,* **57,** 2217-2244.

Levorsen, A. I., 1940. Petroleum geology—symposium on new ideas in petroleum exploration, *Bull. Am. Assoc. Petrol. Geologists,* **24,** 1355-1360.

Lowman, S. W., 1947. Fundamental research in sedimentology, *Bull. Am. Assoc. Petrol. Geologists,* **31,** 501-512.

Milner, H. B., 1931. Sedimentary petrology in retrospect and prospect, *J. Sed. Petrology,* **1,** 66-72.

Morse, J. W., and Berner, R. A., 1972. Dissolution kinetics of calcium carbonate in sea water: II. A kinetic origin for the lysocline, *Am. J. Sci.,* **272,** 840-851.

Parker, M. A., 1946. Use of International Business Machine technique in tabulating drilling data, *Ill. St. Acad. Sci. Trans.,* **39,** 92-95.

Pettijohn, F. J., 1935. The mineralogy of the sedimentary rocks, *Natl. Research Council Bull.,* **98,** 162-171.

Pettijohn, F. J., 1942. Quantitative and analytical sedimentation, *Natl. Research Council Comm. Sed. Rept., 1940/1941,* 43-61.

Pettijohn, F. J., 1956. In defense of outdoor geology, *Bull. Am. Assoc. Petrol. Geologists,* **40,** 1455-1461.

Pirson, S. J., 1970. *Geologic Well Log Analysis.* Houston: Gulf Publ., 370p.

Potter, P. E., 1974. Sedimentology: Past, present and future, *Naturwissenschaften,* **61,** 461-467.

Potter, P. E., and Pettijohn, F. J., 1963. *Paleocurrents and Basin Analysis.* New York: Springer-Verlag, 296p, 2nd ed., 1977, 460p.

Purser, B. H., 1973. *The Persian Gulf.* Berlin: Springer-Verlag, 471p.

Reading, H. G., 1974. Essay review: What have we learned in sedimentology in the past seventy years, *Geological Mag.,* **111,** 455-461.

Russell, R. D., 1957. Research needs in petroleum geology, *Bull. Am. Assoc. Petrol. Geologists,* **41,** 1854-1876.

Sippel, R. F., 1968. Sandstone petrology, evidence from luminescence petrography, *J. Sed. Petrology,* **38,** 530-554.

Trask, P. D., 1939. Report of the committee on sedimentation, 1938-1939, *Natl. Research Counc., Div. Geol. Geog. Ann. Rept.,* **1938-1939.**

Trask, P. D., 1941. Sedimentation. *Geol. Soc. Am., 50th Annivers. Vol.,* 223-239.

Trowbridge, A. C., 1935. Introduction to report of the committee on sedimentation, 1932-1934, *Natl. Research Counc. Bull.,* **98,** 5-15.

Twenhofel, W. H., 1934. Sedimentation and stratigraphy from modern points of view, *J. Paleont.,* **8,** 456-468.

Twenhofel, W. H., 1941. The frontiers of sedimentary mineralogy and petrology, *J. Sed. Petrology,* **11,** 53-63.

Wright, L. D., and Coleman, J. M., 1973. Variations in morphology of major river deltas as functions of ocean wave and river discharge regimes, *Bull. Am. Assoc. Petrol. Geologists,* **57,** 370-398.

Wright, T. L., 1968. X-ray and optical study of alkali feldspar: II. An x-ray method for determining the composition and structural state from measurement of 2θ values in three reflections, *Am. Mineralogist,* **53,** 88-104.

Cross-references: *Sedimentary Facies and Plate Tectonics; Sedimentary Petrography; Sedimentology; Sedimentology–History; Sedimentology–Models;* See also numerous cross-references in text.

SEDIMENT PARAMETERS

An object (or a population or aggregate) possesses certain characteristic properties, and the values associated with these properties are the *parameters* of an object; for example, a rock specimen has a weight, which we assume is a fixed constant (i.e., a true value = θ_w). By performing an experiment, e.g., weighing on an analytical balance, we obtain a value $\hat{\theta}_w$, an estimate of this true value θ_w. This *estimator* $\hat{\theta}_w$ is a statistic which estimates the parametric value θ_w.

The estimator should possess certain characteristics if it is to be an appropriate estimator (Fisher, 1948); the estimator (statistic) should be *consistent*, preferably *unbiased, efficient,* and *sufficient*. A consistent estimator converges (stochastically) on a fixed value (θ'); if the estimator is also unbiased, the fixed value is the true value, i.e., $\theta' = \theta$. Of all consistent estimators, the one with the lowest variance is an *efficient estimator;* efficient estimators are usually the "best" estimators of population parameters. An efficient estimator that contains all the information on the parameter it estimates is a *sufficient estimator*. Sufficient estimators, when they exist, may be found by the *method of maximum likelihood* (Brownlee, 1960).

An object may possess many parameters, and it is necessary to decide upon some criterion to select those parameters that are to be estimated. To devise a criterion, it is necessary to decide upon the objective of the analysis. Suppose it is desired to find an unique index (P) which will characterize the object of interest; then P will be some function of the properties (i.e., parameters) of the object or, formally,

$$P = f(\theta_i) \quad i = 1 \ldots r$$

In practice, the values will be estimators not parameters, so the equation becomes

$$\hat{P} = f(\hat{\theta}_i) \quad i = 1 \ldots r$$

Any object (population) is an aggregate of elements; e.g., a rock is an aggregate of minerals. Then the object (population, or rock) is characterized by certain parameters associated with the properties of the elements. An individual element (e.g., a grain or crystal) may be identified as belonging to a specific kind, such as quartz, feldspar, mica, etc., by a set of diagnostic criteria; and, when all the kinds have been enumerated, it is necessary to estimate their mutual proportions; suppose there are m kinds of minerals (elements), then there will be m estimators for the property subsumed under the name *mineral composition*. The index P may now be written in terms of parameters as $P = f(\theta_1, \theta_2, \theta_3, \ldots, \theta_m, \ldots)$.

Suppose we select a single kind of mineral (element), then measure the properties of size and shape; shape may be considered as length in different directions through the grain, usually defined by three axes (a, b, c; see *Roundness and Sphericity*). The identity, size, and shape of a single grain or element is then written in parametric terms as $P = f(\theta_1, \theta_2, \theta_3, \ldots, \theta_m, \theta_a, \theta_b, \theta_c, \ldots)$; and this may be extended to many grains of the same kind by using expected values for the θ_i and to grains of different kinds by simple extension.

After the element (grain) is identified and its size and shape specified, we have, in effect, determined its mass and volume; it then becomes necessary to specify its position in space. In theory, it is possible to achieve this by measuring the angles between fixed (orthogonal) planes and the three grain axes, and this is frequently accomplished for elements that are large (> 5 mm in length). For smaller elements, it is difficult to perform the measurements and simultaneously to maintain the appropriate position of the element; in this case, the measurements are performed in thin section as the angle between a fixed reference direction and the apparent long, a, axis. Two parameters are estimated, the (mean) direction of the long axis of many grains of the same kind, e.g., quartz, and the spread or dispersion (standard deviation) of these inclinations. These two measures specify the individual position of a single kind of element (see *Anisotropy in Sediments; Fabric*).

The fact that the rock is an aggregate leads to a second aspect of position, the mutual positions of several neighboring elements (grains) or *packing*. This property may also be specified by two parameters, one for the packing and one for the spread in packing.

In summary then, the overall unique index is a function of r parameters which embrace the five properties, composition, size, shape, orientation, and packing of the elements; and this may be symbolized (see Griffiths, 1961) as $P = f(m,s,sh,o,p)$. In parametric terms, this may be written as

$$P = f(\theta_1, \theta_2, \ldots, \theta_m, \theta_a, \theta_b, \theta_c, \theta_o, \theta_{\sigma_o}, \theta_p, \theta_{\sigma_p}, \ldots)$$

All other properties of the aggregate (rock) are *derived* properties in the sense that they require a knowledge of the *fundamental* properties of the elements, above described, for their understanding; or, alternately, variations in the derived properties may be represented as dependent upon variation in the fundamental properties. In this case, the dependent property (Y_i) may be substituted for the index P, and the relationships between fundamental and derived properties may be evaluated.

Composition

The mineral composition of sedimentary rocks is determined by an areal sampling technique, *point counting* (Chayes, 1956; Griffiths, 1960), which leads to estimates of the mutual proportions of the various constituents. The essential mineral elements in clastic (detrital) sediments (q.v.) are quartz, feldspar, and lithic fragments; the various kinds of mica, chlorite, etc. are varietal minerals. Two other kinds of constituents are generally recognized: matrix and cement, which are often conceptually defined in terms of their genesis. The character used to identify them, however, is their morphology in thin section; for example, the external morphology (bounding surfaces) of matrix and cement is irregular, i.e., they may be both concave and convex, and they are usually discontinuous, in contrast to the bounding surfaces of constituent minerals, which are, at least approximately, continuous and everywhere convex. The difference between cement and matrix is largely expressed in their internal morphology, the former being continuous and usually monomineralic, whereas the latter is discontinuous and multiphase.

Using the four detrital rock groups named by Krynine (1948), the composition of some representative specimens determined by point counting in thin section is listed in Table 1. It is obvious that a wide range in proportions exists; perhaps the most important subdivision is between the felspathic detritals (arkose and high-rank graywacke) and the feldspar-poor sediments (low-rank graywacke and quartzite). The break-point is at or below 5% feldspar (likely about 2%), where the figure represents a mean value based on 20 or more specimens. Quartzites, as a whole, are characterized by >95% silica minerals, where the silica morphology varies from quartz through rock fragments of quartzite and chert to silica cement (usually overgrowths on quartz).

The feldspathic rock types, arkose and high-rank graywacke, are supposedly differentiated on the basis of the kind of feldspar; the potash feldspars exceed the plagioclase in arkoses, and high-rank graywackes should possess the opposite relationship. In practice, however, the Roslyn arkose possesses more plagioclase than K-feldspar, and the Franciscan high-rank graywacke yields a granodiorite norm (Taliaferro, 1943).

It seems likely that the most consistent basis for differentiating all four rock types (Krynine, 1948) is the proportions of different kinds of rock fragments. The quartzite group contains fragments of preexisting quartzites and chert; the low-rank graywacke contains, in addition, rock fragments of shale, slate, and phyllite and other low-rank metamorphic rocks. The arkoses include fragments of granitic rocks and related schistose types; whereas the high-rank graywacke is typified, as its name implies, by high-rank metamorphic rock types. Volcanic rock fragments are most abundant in high-rank graywackes but are quite common in arkoses and are not by any means rare in low-rank graywackes (see *Graywacke*). There are no other obviously consistent differences in proportions of the constituents in different rock types.

Size and Shape

In detrital sediments, the size distributions of the detritus are considered to be related to the characteristics of the transporting and depositing media, such as gravity, air, water, and ice. The statistical estimator, mean, and standard deviation of quartz grains, as measured in thin section, for a selected set of detrital sediments is presented in Table 2. The effects of different sampling arrangements on the estimators are reflected in the values of asymmetry ($\sqrt{b_1}$) and peakedness (b_2); unbiased estimators are very difficult to obtain from stratified populations in which the strata are ill defined (see *Grain-Size Studies*). The estimators are related to the (log) normal distribution and are therefore sufficient estimators when appropriately used.

The shapes of quartz grains from a representative set of detrital sediments are described by Curray and Griffiths (1955), and again the estimators tend to be biased unless the sampling design is appropriately adjusted to the patterned variation in the different detrital sedimentary rock groups. The population frequency distribution of Krumbein sphericity (Krumbein, 1941) is likely to be normal in detrital sediments, with the mean ($\bar{X} = \hat{\theta}_s \to \theta_s$) equal to 0.750 and the standard deviation ($\hat{\sigma} = \hat{\theta}_\sigma \to \theta_{\sigma_s}$) equal to 0.083; 99.7% of the quartz grains in detrital sediments are likely to possess three-dimensional sphericity of 0.500–1.000 φ units (Curray and Griffiths, 1955). Sedimentary processes are not random selectors; thus, the biases associated with subpopulations corresponding to the four rock types are likely to represent characteristics of the selective sorting process and are, therefore, useful in identifying environments of deposition.

In thin section, the relationships between the apparent long, a, axis and the apparent short, b, axis tends to become constant when means of small sets of quartz grains are used as estimators. A representative set expressed in terms of the four statistical estimators is presented in Table 3; again the estimators are biased, and in this case the most common bias appears in the

TABLE 1. Mineral Composition of Some Selected Detrital Sediments

	Arkose						Quartzite						Low Rank Graywacke						High Rank Graywacke	
	Roslyn[a]		Fountain[a]		Salt Wash[b]		Oriskany[a]		Homewood[a]		Tuscarora[a]		Cow Run[c]		Berea[a] Graywacke		Pocono[a] Graywacke		Welsh Grits[a]	
Element	$\bar{X}$	$\hat{\sigma}$	$\bar{X}$[d]	$\hat{\sigma}$	$\bar{X}$	$\hat{\sigma}$	$\bar{X}$	$\hat{\sigma}$	$\bar{X}$	$\hat{\sigma}$	$\bar{X}$	$\hat{\sigma}$	$\bar{X}$	$\hat{\sigma}$	$\bar{X}$	$\hat{\sigma}$	$\bar{X}$	$\hat{\sigma}$	$\bar{X}$	$\hat{\sigma}$
Quartz	30.7	6.6	27.4		60.9	9.9	77.7		70.1	9.6	77.9	3.3	39.7	7.0	63.4	8.6	61.7	10.9	27.6	
Feldspar	23.7[e]	8.7	29.3		13.1	9.0					0.9	0.4	1.0	0.9	0.4		0.5		7.1	
Rock fragments	18.2	12.1	20.2				1.4		15.9	10.0			40.1	8.9	2.1		7.7		28.9	
Micas	7.3	6.2	4.4										0.7	0.9	1.3		1.6		<1.	
Matrix	13.7	16.9	14.3		14.5	12.2			9.5	8.3	5.7	1.0	9.8	6.3	16.2		25.9		33.3	
Silica			0.1		11.9	8.1	9.5		3.7	1.8	14.5	3.4	8.4	3.7	<0.1		<0.1			
Carbonate	5.2[f]	6.7	2.6		7.6	8.1	3.8						4.0	7.0	11.6		3.1		1.3	
Others	1.2	1.1	1.6				7.1[g]								5.1				1.8[h]	
Total	100.0		99.9		108.0		99.5		99.2		99.0		103.7		100.2		101.6		100.0	
Specimens	40		48		25		80		20		20		39		16		16		9	
Travs/spec	2		6		6		20		80		80		6		5		5		5	
Points/travs	100		100		100		25		100		100		200		100		100		100	

[a]From unpublished theses, Pennsylvania State University.
[b]From Griffiths et al., 1956.
[c]From Griffiths, 1958.
[d]weighted means, no standard deviations recorded.
[e]12.2% plagioclase, 9.5% K-feldspar, others undifferentiated.
[f]carbonate and sulfate.
[g]includes iron hydroxide and phosphate.
[h]mostly chloritic matter.

TABLE 2. Long a Axis of Quartz Grains in Sandstones; Measured in Thin Section

	$\bar{X}a_\phi$	σa_ϕ	$\sqrt{b_1}$	b_2	% Qtz.	Specimens	Grains	Total	Locality	Rock Type
Tuscarora[a]	1.823	0.966	−0.279**	2.564NS	—	46	25	1175	N.Y. to Ky.	quartzite
Oriskany[a]	1.659	1.040	−0.075**	3.650**	87.0	87	various	9743	Ottawa, Ill.	" "
St. Peter[a]	1.833	0.906	0.291**	2.717	—	13	40	520	W. Va.	" "
Cow Run sand[b]	1.240	0.70	0.196**	3.494**	39.7	38	40	1520	Ky.	low-rank graywacke
Weir sand[a]	3.650	0.37	0.301**	2.920**	51.5	19	40	760	N.Y.–Pa.	" " "
Chipmunk sand[a]	2.760	0.660	−0.076**	3.007	53.5	71	35	2485	Colo. Plat.	" " "
Salt Wash[c]	2.758	0.662	−0.075**	3.007	61.0	50	300	15000	Colo.	arkosic quartzite
Fountain[a]	0.810	1.093	—	—	40.5	16	36	576	Colo.	arkose
Waynesburg[a]	1.452	1.024	1.228	2.333	49.8	75	20	1500	Penna.	arkose
Paleozoic[a] "grits"	1.888	1.111	0.223**	3.201**	—	7	40	280	U.K.	high-rank graywacke

**Significant at the 1% level
[a]unpublished theses, Pennsylvania State University
[b]Griffiths, 1958.
[c]Griffiths et al., 1956.

TABLE 3. Axial Ratio Shape of Quartz Grains for a Selected Set of Detrital Sediments

Sediment	Samples	Grains	Mean	Standard Deviation	$\sqrt{b_1}$	b_2	Rock Type
Morrison[1]	NR	784	.688	.157	−.102	2.574**	Arkosic quartzite
Dakota[1]	NR	694	.729	.140	−.318**	2.879	Quartzite
Oriskany[2]	87	9981	.677	.142	−.068*	2.648**	Quartzite
Aux Vases[1]	NR	750	.679	.148	.0003	2.658*	Quartzite
Bradford Sand[3]	1	1250	.651	.150	NR	NR	Low-rank graywacke
Pocono[4]	16	800	.661	.174	NR	NR	Low-rank graywacke
Cow Run Sand[5]	38	1520	.671	.168	−.121*	2.529**	Low-rank graywacke
Rec. Holly Beach, La.[1]	NR	801	.700	.159	−0.345**	3.014	?
Rec. Topsail Beach, N.C.[1]	NR	277	.608	.160	−.107	2.007**	?
Rec. Mississippi River[1]	NR	802	.678	.161	−.152*	2.562**	?

* Significant at 5% level
** Significant at 1% level

Sources:
1. Bokman, 1957; 2. Rosenfeld, 1953, unpublished; 3. Griffiths and Rosenfeld, 1953; 4. Emery, 1954, unpublished; 5. Griffiths, 1958.

fourth-moment statistic, the peakedness. A large part of this bias arises from scale restriction in that sphericity expressed as a ratio [either b/a or $\sqrt[3]{(bc/a^2)}$] cannot exceed unity; reversing the ratio to a/b does not remove this source of bias. It is some advantage to avoid the use of ratios and, instead, to use a linear compound of the axes in phi units; a linear compound of normally distributed variates is itself normally distributed, and bias due to restrictions of scale does not then arise. Relationships between the axes are closely (log) linear, and, where means of small sets of grains or of thin sections are used as estimators, the coefficient of determination (the square of the coefficient of correlation = r^2) rapidly approaches values of 0.95 or more. This implies that the axial ratio shape of quartz grains in detrital sediments shows a very limited range of variation; to all intent, the shape is constant over quite wide ranges of size.

Fabric and Packing

The population frequency distribution of orientation, as measured in thin section, displays a range from rectangular (no preferred orientation) to normal (see *Fabric*). If the population is homogeneous, then, using Chayes' minimum variance (Chayes, 1954; Griffiths et al., 1956) the preferred orientation varies from $\sigma = 52°$ (no preferred orientation) to $\sigma = 30°$ (normal frequency distribution). Homogeneous orientation is likely to be a very local property of small sets of grains; the values for most sediments will vary over this entire range. No preferred orientation is typically displayed by sediments with disturbed texture (e.g., slumped sediments), and such sediments are typically described as "massive." Most sediments in which the internal texture is primary in origin will show some degree of preferred orientation; mean inclination ($\bar{X} = \hat{\theta}_o$) is characteristic of local events associated with the last phases of deposition or early diagenesis, and degree of preferred orientation ($\hat{\sigma} = \hat{\theta}_\sigma o$) is likely to be characteristic of the depositional process (Cailleux, 1938).

Measurement of packing in detrital sediments has been evaluated in some detail by Kahn (1956; Miller and Kahn, 1962). The two measures defined by Kahn as *packing density* and *packing proximity* are suitable estimators of the parametric value of packing; values for representative specimens of three of Krynine's four main rock types are listed in Table 4. It seems clear that packing may vary widely within one rock type; it is a local characteristic much like orientation and, in the main, reflects the effects of processes operating during the late stages of deposition and early diagenesis. Subsequent diagenetic effects, e.g., solution, reprecipitation, and replacement, modify primary textures; and orientation and packing may be changed by these late events.

Parametric Analysis

In order for parametric analysis to be meaningful, it is usually necessary to obtain unbiased estimators which also fulfill the requirements of consistency, efficiency, and, where possible, sufficiency. To achieve these desirable characteristics, the random variables underlying the estimators should have known frequency distributions. Often, random samples from homogeneous populations will adequately fulfill these requirements; e.g., size and shape, measured either directly on individual elements or as apparent axes in thin section, will approximate log normality and the estimators will then be sufficient. Similarly, inclination measured as recommended will range in frequency distribution from rectangular to normal; and, under the central-limit theorem (Miller and Kahn, 1962), the conventional statistical estimators will fulfill the required criteria for, at least, efficient estimation of population parameters.

The properties of mineral composition and packing differ from the properties of size, shape, and inclination in that they are counts rather than measurements; i.e., the variates are discrete rather than continuous. Again, under appropriate sampling conditions, the counts will converge on binomial, Poisson, etc. frequency distributions and, once more invoking the central-limit law, the estimators will achieve efficiency and in some cases will be sufficient estimators. It is of considerable advantage to adopt measurement procedures and sampling arrangements that will lead to estimators with these desirable characteristics because testing for bias is simplified; and, when unbiased efficient estimators are achieved, more sophisticated statistical analysis may be undertaken without violating the more rigorous requirements that accompany the use of such analytical tools.

When the appropriate set of r estimators are obtained on a set of n specimens, the first step in parametric analysis is to determine whether redundancy is present in the $(n \times r)$ matrix of observations. Factor analysis using principal components in the R mode is a suitable algebraic tool for this purpose (Griffiths, 1961, 1963), and it will generally be found that considerable redundancy exists, many of the statistics estimating factors that are common to more than one variate. Ideally, each variate should reflect the effects of a single factor such as source area, erosion, transportation, deposition, and dia-

TABLE 4. Statistical Estimators of Packing in Some
Representative Detrital Sediments

Formation	Packing Density	Packing Proximity	Rock Type
Tuscarora	87.45	63.28	
Oriskany	60.93	27.93	
Cuche	55.73	36.85	
			Quartzite
Rock-type mean	68.00	43.00	
Rock-type standard deviations	15.75	18.50	
Pocono	72.59	39.85	
Oswego	82.47	56.98	
Bradford	70.77	33.61	
			Low-rank Graywacke
Rock-type mean	74.00	43.00	
Rock-type standard deviations	8.72	12.77	
Portland	76.64	46.86	
Roslyn	90.54	73.52	
Stockton	78.52	58.54	
			Arkose
Rock-type mean	82.00	60.33	
Rock-type standard deviations	9.00	14.32	

After Kahn, 1956.

genesis; in practice, all the estimators are affected by the selective sorting process, which dominates the formation of detrital sediments. Upon removing the effects of this process, the remaining variation may reflect differences in source area, and/or diagenesis, etc. (see, e.g., Griffiths, 1963). Similarly many other aspects of multivariate analysis including multiple regression (Griffiths, 1958), factor analysis in the Q mode (Imbrie and Purdy, 1962), trend surface fitting (Krumbein, 1959), etc., depend essentially on the achievement of efficient estimators as the initial starting point in the data matrix. To this end, the sedimentary parameters described above and their corresponding statistical estimators play a very important role in the analysis of sediments. This approach may be generalized, by redefining the population and its elements, to include parametric analysis of most populations; it will be found that some r estimates of the five fundamental properties (composition, size, shape, orientation, packing) will generally be necessary, and for most problems they also tend to be sufficient.

JOHN C. GRIFFITHS

References

Bokman, J., 1957. Comparison of two and three dimensional sphericity of sand grains, *Geol. Soc. Am. Bull.*, 68, 1689-1692.

Brownlee, K. A., 1960. *Statistical Theory and Methodology in Science and Engineering.* New York: Wiley, 570p.

Cailleux, A., 1938. La disposition individuelle des galets dans les formations détritiques, *Rev. Géogr. Phys. Géol. Dyn.*, 11, 171-198.

Chayes, F., 1954. Effect of change of origin on mean and variance of two dimensional fabrics, *Am. J. Sci.*, 252, 567-570.

Chayes, F., 1956. *Petrographic Modal Analysis.* New York: Wiley, 113p.

Curray, J. R., and Griffiths, J. C., 1955. Sphericity and roundness of quartz grains in sediments, *Geol. Soc. Am. Bull.*, 66, 1075-1096.

Fisher, R. A., 1948. *Statistical Methods for Research Workers*, 10th ed. Edinburgh: Oliver and Boyd, 354p.

Griffiths, J. C., 1958. Petrography and porosity of the Cow Run Sandstone, St. Marys, W. Va., *J. Sed. Petrology*, 28, 13-30.

Griffiths, J. C., 1960. Modal analysis of sediments, *Rev. Géogr. Phys. Géol. Dyn.*, 3, 29-48.

Griffiths, J. C., 1961. Measurement of the properties of sediments, *J. Geol.*, 69, 487-498.

Griffiths, J. C., 1962. Translating sedimentary petrography into petrology (Abstr.), *Geol. Soc. Am. Spec. Pap. 68*, 186.

Griffiths, J. C., 1963. Statistical approach to the study of potential oil reservoir sandstones, *Comp. Min. Industr., Proc. 3d Ann. Conf.*, 9(2), 637-668.

Griffiths, J. C., 1967. *Scientific Method in the Analysis of Sediments.* New York: McGraw-Hill, 508p.

Griffiths, J. C., and Rosenfeld, M. A., 1953. A further test of dimensional orientation of quartz grains in

Bradford Sand, *Am. J. Sci.,* **251,** 192-214.

Griffiths, J. C., and Smith, C. M., 1964. Relationships between volume and axes of some quartzite pebbles from Olean Conglomerate, Rock City, New York, *Am. J. Sci.,* **262,** 497-512.

Griffiths, J. C.; Cochran, J. A.; Hutta, J. J.; and Steinmetz, R., 1956. Petrographical investigations of the Salt Wash Sediments, *U.S. Atomic Energy Comm., R.M.E.* 3122 (2 parts), 66p and 84p.

Imbrie, J., and Purdy, E. G., 1962. Classification of modern Bahamian carbonate sediments, *Am. Assoc. Petrol. Geologists Mem.* **1,** 253-272.

Kahn, J. S., 1956. The analysis and distribution of the properties of packing in sand-size sediments. 1. On the measurement of packing in sandstones. 2. The distribution of the packing measurements and an example of packing analysis, *J. Geol.,* **64,** 385-395, 578-606.

Krumbein, W. C., 1941. Measurement and geological significance of shape and roundness of sedimentary particles, *J. Sed. Petrology,* **11,** 64-72.

Krumbein, W. C., 1959. Trend surface analysis of contour type maps with irregular control point spacing, *J. Geophys. Research,* **64,** 823-834.

Krynine, P. D., 1948. The megascopic study and field classification of sedimentary rocks, *J. Geol.,* **56,** 130-165.

Miller, R. L., and Kahn, J. S., 1962. *Statistical Analysis in the Geological Sciences.* New York: Wiley, 483p.

Taliaferro, N. L., 1943. Franciscan-Knoxville problem, *Bull. Am. Assoc. Petrol. Geologists,* **27,** 109-219.

Cross-references: *Anisotropy in Sediments; Clastic Sediments and Rocks; Fabric, Sedimentary; Grain-Size Studies; Roundness and Sphericity; Sedimentary Petrology; Sedimentary Rocks; Sedimentological Methods.*

SEDIMENT TRANSPORT, FLUVIAL AND MARINE—*See* Vol. I

SEDIMENT TRANSPORT—INITIATION AND ENERGETICS

The processes of initiation of transport and maintenance of transport are quite different, as illustrated by the case of clay, which once in suspension, is easily moved by any current; in a cohesive deposit, however, clay is extremely resistant to erosion. Sand and fine gravel are also commonly placed into transport by a mechanism that differs from the one that continues the transport. Initiation of transport generally requires a greater energy input than does maintenance of transport, as illustrated by the Hjulström diagram (Fig. 1); but the processes are also different.

Initiation

Transport may be started by either local turbulence, including hydraulic lift, or local bed shear (drag). Sand moved initially by the wind is usually moved by a bed shear effect, and the eolian ripple marks produced in this fashion have an amplitude which is hydrodynamically negligible, so that no significant boundary layer separation takes place and no local turbulence develops, unless the wind strength increases well beyond that necessary to start the motion. At higher wind velocities, saltation (q.v.) may replace sliding or rolling on the sand surface (Bagnold, 1941), but this is not initiation of transport.

Under water, on a flat smooth bed, sand is moved initially by bed shear (drag). This early movement produces a series of parallel low ridges (shear ridges), which have sufficient amplitude to provide for boundary layer separation and the generation of lee turbulence (Tanner, 1963, 1967). If the water flow is unidirectional, the lee eddies have relatively long lives and hence decay into short-axis cones. The cascading of water, in the bed layer, from one string of cones to the next, provides for transport maintenance and in this process shapes the irregular, short-crested, nonparallel, low-amplitude ripple marks that are characteristic of strong water currents.

If, on the other hand, the flow is bidirectional (as under simple wave action), the lee eddies have relatively short lives and do not decay, but rather reverse the sense of motion while they are still parallel and linearly continuous. This reversing is done in parallel sets of two, so that the water motion, as seen in cross section, marks the form of a figure eight. Various figure-eights drift over the ripple-mark field, treating all ripple marks alike. The symmetry or smooth asymmetry, continuity, parallelism, crest height, and great crest length are the results of sediment transport by the mature but undecayed eddies that make up the bottom eddy layer.

In the swash zone, under ordinary conditions, a shear effect provides for both initiation and maintenance because the sheet of moving water is too thin to develop turbulence. Unlike the shear ridges under wind or unidirectional water flow, however, the shear ridges which appear in the swash and backwash are bidirectional and hence symmetrical. In some instances, these ridges may produce high-Froude results, which can be observed on the beach as tiny, long, parallel hydraulic jumps; again, this is not initiation. Something very close to swash-zone conditions, but with unidirectional flow, can be seen in the "rain waves" which commonly form in sheet runoff on sloping paved surfaces immediately after a rain (Tanner, 1969).

The layer that underlies the actual surges is taken as a stripped wall layer (that is, without a deep channel full of water above it), but it is not the same as the bottom-eddy layer, which

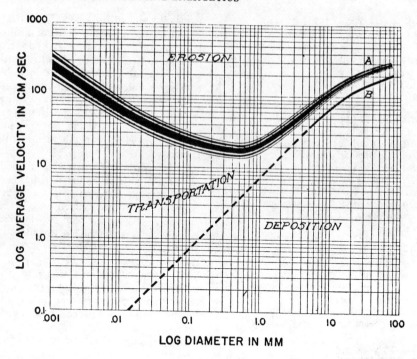

FIGURE 1. Hjulström's diagram of the relations among erosion, transportation and deposition of sedimentary particles (after Hjulström, 1939).

contains eddies, nor does it extend upward into the boundary layer. In a river, the boundary layer is commonly taken as reaching to, or nearly to, the water surface. The bottom-eddy layer is only two or three times as thick as the roughness element heights. The wall layer is even thinner, lying completely below the bottom eddies. Flow in the boundary layer drives the bottom-eddy layer, which in turn exerts a local bed shear through the very thin wall layer.

In some settings, initiation results from local turbulence. Sand and fine gravel are put into motion in the surf zone, for example, by the collapse of the wave orbit at that point where water volume (visually evident as water depth) is insufficient to fill the core of the orbit and hence to support the water in the uppermost part of the circle. In running water, flow past obstacles (such as corners of boulders or rock ledges) creates local turbulence. Under these conditions, the local bed shear is a third-order effect in the following chain: obstacle, local eddy, local bed shear. Similar effects obtain in the air, especially around rocks and clumps of vegetation.

The deformation of streamlines passing over a particle (Fig. 2), causing hydrodynamic (hydraulic) lift, is one of the primary forces in initiating the rise of particles; Blatt et al. (1972) discuss this phenomenon. Where streamlines are crowded together, the velocity must increase because fluid particles do not cross streamlines. Bernoulli's theorem (q.v.) states that along a streamline, the sum of the energy components due to pressure head and velocity must be constant; therefore, where streamlines are close together and velocity is high, the pressure must be smaller than in the region where streamlines are farther apart (Fig. 2). Head may be neglected because of the small change in height (Blatt et al., 1972). From symmetry it is clear that this pressure must act upward, i.e., it is a lift force; and it has been calculated, and demonstrated experimentally, that lift force is almost as large as drag force and is indeed capable of lifting a particle up from a bed (Raudkivi, 1967). Once the particle is lifted, the streamline patterns become symmetrical, and there is no further lift force (Blatt et al., 1972).

Bed shear is commonly computed for flowing water as $\tau_o = \gamma\rho i h$, where γ is the acceleration of gravity, ρ is the mass density of the fluid, i is the energy gradient (commonly the water surface slope) in radians, and h is the water depth. Although a convenient index for various kinds of work, the bed shear cannot be applied directly and realistically in studies of sand transport by flowing water because velocities within the bottom-eddy layer are diminished and augmented in ways that cannot be projected from considerations of the overall flow, in some

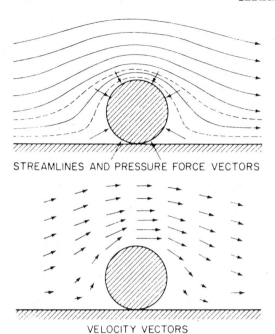

FIGURE 2. Flow pattern computed for an ideal (inviscid) fluid moving past a cylinder lying on the bottom with its axis at right angles to the direction of flow (from Blatt et al., © 1972; reprinted by permission of Prentice-Hall, Inc.). *Top:* streamlines and the relative magnitude of pressures acting on the surface of the cylinder. *Bottom:* the direction and relative magnitude of velocity vectors.

instances being momentarily up to an order of magnitude greater than would be deduced from gross measurements of velocity, slope, or depth.

Studies of the microstructure of the water velocity field within the bottom-eddy layer show accelerations quite capable of initiating grain transport under conditions where even the surface velocity does not appear to be great enough. For this reason, measures of water velocity, whether taken as an average or at some selected depth in the main flow, can be related only in very loose fashion to sediment transport.

The shear velocity is an arbitrary measure, having the dimensions of velocity, which is obtained by taking the square root of the ratio of bed shear to water density. Computationally, the shear velocity occurs at $<1/2$ cm above the sediment surface. The bottom-eddy layer typically has a thickness two or three times the heights of the ripple marks or other local roughness elements, however; the computed shear velocity would, therefore, fall within this layer with its highly developed mature or decayed eddies. The shear velocity, then, can only be used realistically at the moment of shear initiation, prior to the earliest appearance of the bottom-eddy layer.

Energetics

Sediment is transported by several important agencies: water currents, wind, glacial ice, water-wave action, and the direct influence of gravity (including mudflows, landslides, creep, and rock falls). There are also other agencies, such as various organisms, including humans.

The motive power for transport is of two basic types: (1) *gravity*—streams, ice, mudflows, landslides, tidal currents, density currents; and (2) *wind*—eolian transport, water waves, some water currents. The second category is, more basically, due to other causes, such as rotation and solar heating of the earth, but is here attributed to the intermediate agency, the wind.

Gravity transport, of whatever kind, is due directly—or indirectly, through the mediation of flowing water—to gravitational attraction. Wind transport is not a gravity phenomenon; in fact, wind may move sediment such as sand uphill. Sand on a lake or sea floor may also be moved up a gentle slope by either wave action or by certain water currents.

Very fine sediment such as clay and silt is easily moved by currents of all kinds once it is put into suspension. In lake or sea water, wave action may stir fine sediments, or rivers or other agencies may introduce fine sediments, which are then transported by currents otherwise too feeble to have put the sediment into motion. In stream flow, such fine sediment is stirred by turbulence, and then carried in suspension: it can be designated as the *suspension*, or *wash*, *load* (see *Fluvial Sediment Transport*).

Coarse sediment such as sand and gravel is not commonly put into suspension for any significant period of time, but is moved by rolling, sliding, and bouncing on or very close to the bed: it can be designated as *bed*, or *traction*, *load*. With very high energy levels, such as in river floods or storm surf, sand or very fine gravel may be put into suspension, becoming for a while wash load (see *Flood Deposits; Storm Deposits*).

Very coarse gravel may be present as erratics (whether of glacial or some other origin), not necessarily in transport. The large boulder, or block that slides down a cliff and comes to rest in stream or breaker zone is not moving in the stream or the surf, but it may be put into transport at some later date if energy levels are increased sufficiently, or if the boulder is worn down in size. Such material may be completely *relict*, if the present potential transporting agency has recently invaded the site where the boulder was deposited by some other method of transport; a good example of this is *moraine gravel,* not now moving, on the floor of a small lake (see *Relict Sediments*).

The concept of sediment transport does not require that each grain be in motion at each moment, but merely that each grain move intermittently, perhaps at intervals as long as 10^2 or 10^3 yr. Sand grains stored in a river bank, a beach, an island, or a shoal may move at this rate, or at more frequent intervals.

The movement of suspended (wash) load under gravity conditions makes additional potential energy available to the system; this energy may be utilized in good part in creating turbulence, to be dissipated as heat (see *Turbidity-Current Sedimentation*). The movement of bed load, on the other hand, requires an energy input. Under gravity conditions, this energy is supplied by the water (and wash load) which flows downhill. It is not an easy matter to partition the total energy available into that part that transports bed load, and that part that creates turbulence (Raudkivi, 1967); but a few preliminary studies have been made. The Arkansas River, between Tulsa, Okla., and its mouth, apparently uses $10^{-3} - 10^{-2}$ of its available energy for transporting bed load (0.1–1.0%); the rest is translated through turbulence into heat.

Example: Littoral Energetics. Under waterwave action, the total energy available in a current-free sea comes from the wind. The partitioning of this energy has been studied in detail (Gulf of Mexico, near Cape San Blas, Florida; Stapor, 1971). With waves having an average breaker height of 20 cm and an average period of 5 sec, one specific mass of sand has been transported 3.6 km in 68 yr. The contribution of water currents, although very small, is not known. The mass of sand measures 1.36×10^{12} cm^3. The energy used in net unidirectional transport was 10^{19} ergs (not including excursions of the sand in any other direction, or any other energy sinks). The work done, under these conditions, to move 1 cm^3 of sand 1 cm along its unidirectional path, was 20.4 ergs—approximately 0.1% of the total wave energy available in the nearshore belt. That is, about 99% of the available wave energy was used in driving water eddies near the bottom and in breaker-zone turbulence. Of the remaining 1%, 0.9% was used for sand shuffling, and only 0.1% for net unidirectional sand transport, an efficiency, for net transport along one path, of about 10^{-3}. The uncertainty is less than one order of magnitude. A totally different method for computing efficiency of moving the sand mass produced a result of 10^{-4}, again with an uncertainty of less than one order of magnitude. The correct answer probably lies between these two results.

This work, for a specific stretch of coast, covering an interval of 68 yr, cannot be extrapolated to other coasts having different wave climates, sediment characteristics, and bathymetries. Nevertheless, it furnishes an example of how energy partitioning can be approximated, and how long-term efficiencies can be estimated. The study area was taken to have an average breaker height of 20 cm, which is a compromise between two data sets collected in two different ways. A "zero-energy" coast would have, in contrast, a much smaller efficiency, i.e., essentially all of the available wave energy would go into bottom-eddy production, with almost no energy in the surf zone (the average breaker is typically about 1 cm high) and virtually nothing going into any kind of *net* sediment transport. A high-energy coast, on the other hand, should have much higher efficiencies, especially for short-term measurements, reaching values of 0.1 or perhaps more, and an appreciable proportion of total available wave energy should be transferred through the bottom-eddy system into sediment transport; the extent to which net transport would be realized would depend on the wave approach direction and its variability. No similar study is known for wind transport.

The transporting work of organisms, other than humans, is interesting in many instances, but must be exceptionally small, although local efficiencies might be large.

WILLIAM F. TANNER

References

Bagnold, R. A., 1941. *The Physics of Blown Sand and Desert Dunes.* London: Methuen, 265p.

Blatt, H.; Middleton, G.; and Murray, R., 1972. *Origin of Sedimentary Rocks.* Englewood Cliffs, N.J.: Prentice-Hall, 634p.

Hjulström, F., 1939. Transportation of detritus by moving water, in P. D. Trask, ed., *Recent Marine Sediments.* Tulsa, Okla.: Am. Assoc. Petrol. Geologists, 5-31.

LeFournier, J., and Friedman, G. M., 1974. Rate of lateral migration of adjoining sea-marginal sedimentary environments shown by historical records, Authie Bay, France, *Geology,* **2,** 497-498.

May, J. P., and Tanner, W. F., 1973. The littoral power gradient and shoreline changes, in D. Coates, ed., *Coastal Geomorphology.* Binghamton: St. Univ. New York, 43-60.

May, J. P., and Tanner, W. F., 1975. Estimates of net wave work along coasts, *Z. Geomorph. Supplbd.,* **22,** 1-7.

Raudkivi, A. J., 1967. *Loose Boundary Hydraulics.* Oxford: Pergamon, 331p.

Stapor, F. W., 1971. Sediment budgets on a compartmented low-to-moderate energy coast in northwest Florida, *Marine Geol.,* **10,** M1-M7.

Sundborg, Å., 1956. The River Klarälven—A study of fluvial processes, *Geograf. Ann.,* **38,** 125-316.

Tanner, W. F., 1963. Origin and maintenance of ripple marks, *Sedimentology*, **2**, 307-311.
Tanner, W. F., 1967. The ripple mark analog: Preliminary results, *Southeastern Geol.*, **8**, 67-72.
Tanner, W. F., 1969. Surge flow: A model of the wall layer, *Southeastern Geol.*, **10**, 93-110.
Tanner, W. F., 1973. Advances in near-shore physical sedimentology: A selective review, *Shore and Beach*, **41**(1) (April), 22-27.
Trask, P. D., ed., 1955. Recent marine sediments, *Soc. Econ. Paleont. Mineral. Spec. Publ. 4*, 736p.

Cross-references: *Bernoulli's Theorem; Dispersal; Flood Deposits; Fluvial Sediment Transport; Littoral Sedimentation; Relict Sediments; Reynolds and Froude Numbers; Ripple Marks; Saltation; Stokes' Law; Storm Deposits; Transportation; Turbidity-Current Sedimentation.*

SEDIMENT TRANSPORT—LONG-TERM NET MOVEMENT—*See* Vol. III

SETTLING VELOCITY

Settling velocity, or more correctly *settling speed*, is defined as the constant terminal speed at which a particle falls through a fluid medium which is essentially at rest. The settling speed of a particle depends on the viscosity and specific gravity of the medium and upon the size, shape, and specific gravity of the particle. According to the classical equation of Stokes (1851), the settling speed (w) of a particle with spherical shape is given by

$$w = \frac{2}{9} g \frac{\rho_1 - \rho_2}{\mu} r^2$$

where ρ_1 and ρ_2 are the densities of the sphere and the liquid respectively, g is the acceleration of gravity, r is the radius of the particle, and μ is the dynamic viscosity of the liquid (see *Stokes' Law*).

If the settling speed is great, i.e., if the viscosity is small, the radius of the particle or the density difference is large because of the turbulence set up by the particle itself, and the above relationship is not valid. Various corrections and empirical formulas have been developed to correct the above formula for larger spheres (Waddel, 1934, 1936; Krumbein and Pettijohn, 1938; Gibbs et al., 1971).

Stokes' equation may also not be applied to very fine-grained material entering salty waters. Clay and colloidal particles tend to flocculate into units whose settling speed are equivalent to those of quartz spheres 5-15μ in diameter. This flocculation (q.v.) is related to the salt content of the water and the composition of the clay minerals.

TABLE 1. Settling Speeds for Quartz Spheres in Distilled Water at 20°C

Size (mm)	Settling Speed (cm/sec) Waddel	Stokes
2		
Very coarse sand	16	89.2
1		
Coarse sand	7.7	22.3
1/2		
Medium sand	3.4	5.58
1/4		
Fine sand	1.2	1.39
1/8		
Very fine sand		301 m/day
1/16		
Silt		75.2-12 m/day
1/256		
Clay		<1.2 m/day

Some characteristic settling speeds for coarse and finer material are given in Table 1. Studies have shown that settling speed may be a good measure of depositional environment (Reed et al., 1975; see also *Hydraulic Equivalence*).

IMRE V. BAUMGAERTNER

References

Allen, H. S., 1900. The motion of a sphere in a viscous fluid, *Philos. Mag.*, **50**, 323-338, 519-534.
Gibbs, D. J.; Matthews, M. D.; and Link, D. A., 1971. The relationship between sphere size and settling velocity, *J. Sed. Petrology*, **41**, 7-18.
Krumbein, W. C., and Pettijohn, F. J., 1938. *Manual of Sedimentary Petrography.* New York: Appleton-Century, 549p.
Reed, W. E.; LeFever, R.; and Moir, G. J., 1975. Depositional environment interpretation from settling-velocity (Psi) distributions, *Geol. Soc. Am. Bull.*, **86**, 1321-1328.
Romanovsky, V. V., 1972. Experimental investigation of the fall velocity of sediments (in Russian), *Trudy Gosud. Gidrol. Inst.*, **191**, 111-136.
Stokes, G. G., 1851. On the effect of internal friction on the motion of pendulums, *Cambridge Philos. Trans.*, **9**(2), 8-106.
Waddel, H., 1934. Some new sedimentation formulas, *Physics*, **5**, 381-391.
Waddel, H., 1936. Some practical sedimentation formulas, *Geol. Fören. Stockholm Förh.*, **58**, 397-408.
Warg, J. B., 1973. An analysis of methods for calculating constant terminal-settling velocities of spheres in liquids, *J. Internat. Assoc. Math. Geol.*, **5**, 59-72.

Cross-references: *Flocculation; Grain-Size Studies; Granulometric Analysis; Hydraulic Equivalence; Impact Law; Stokes' Law.*

SHALE

The term *shale* (*argile* in French, *Tonschiefer* in German) is applied as a generic term to the fine-grained group, of the three major types of sedimentary rocks; lutite and pelite are Latin- and Greek-derived names for fine-grained rocks (see *Sedimentary Rocks*). Some 40–60% of the measured volume of sedimentary rocks is classified as shale. Various methods of geochemical calculation indicate that their total abundance should be ≈ 80% of crustal weathering products. They are ubiquitously distributed, but most characteristic in subsiding basins, where shale facies several hundred m thick have been measured. They are usually composed of a mixture of detrital, authigenic, and altered minerals, and contain at least 30–40% and up to almost 100% of the *clay minerals*, which impart to the shale their characteristic properties (see *Clay*). Illite is the most abundant mineral, especially in ancient shales; smectite, kaolinite, chlorite, and the other clay minerals are present in smaller and more variable quantities (Weaver, 1959). Quartz is the common and most abundant of the nonclay minerals, especially in the silt fraction; feldspars and carbonates are common accessories (Table 1). *Color* is a conspicuous attribute of shales and is often indicative of environmental conditions (see *Color in Sediments*). Black shales are usually rich in organic material. Green and red colors are indicative of the state of reduction or oxidation of the iron-containing minerals.

A more specific definition restricts the term *shale* to the fissile clayey sediments with laminae parallel to the bedding and without any obvious postdepositional changes except those due to diagenesis and/or compaction. *Claystone* or *mudstone* is then used for the massive nonfissile rock, *clay* or *mud* for the unconsolidated sediment, and *argillite* for rocks showing incipient metamorphism (see *Mudstone and Claystone*).

Compaction and Sedimentary Fabric

The lamination of shales is primarily due to the parallel or preferred orientation of the constituent clay minerals. *Preferred orientation*, which is usually better developed in the larger kaolinite or illite minerals than in smectite, forms during the dewatering and compaction of the clayey sediment by overburden pressures. It seems to be favored by small amounts of organic matter and by the presence of packets or domains of particles oriented already at the onset of the compaction (Fig. 1).

The compaction and burial of clayey sediment is accompanied by a considerable reduction of its porosity, from 50–80% in muds to about 10% in indurated shales (Fig. 2; see *Compaction in Sediments*). The gradual removal of the interstitial water and, subsequently, of the envelope of adsorbed water on the clay surface increases the number of contacts between the particles. The plastic sediment gradually consolidates. When porosity decreases to about 35% (approximately 300 m depth), the number of continuous face-to-face contacts between the platy particles has increased to such an extent that consolidation and lithification result. *Permeability* of indurated shales of 10–20% porosity is extremely low, 10^{-6} or $10^{-7}-10^{-10}$ millidarcy, and they often serve, therefore, as an aquiclude or as petroleum caprock.

Electrochemical Structure. The original electrolyte concentration of the pore solution undergoes changes during compaction. The electrochemical properties of the cation-exchanging clays—smectite, chlorite, illite, etc.—repel electrolytes, which result in a diminish-

TABLE 1. Mineral Composition of the Average Shale (wt %)

Clay minerals	59
Quartz and chert	20
Feldspars	8
Carbonates	7
Fe-oxides	3
Other minerals	2
Organic matter	1

After Yaalon, 1962.

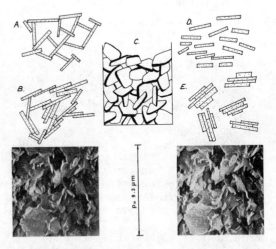

FIGURE 1. *Top:* Schematic arrangements of clay mineral particles in shales—A, edge-to-face flocculated clay; B, face-to-face flocculated clay; C, spatial model of edge-to-edge fabric; D, preferentially oriented clay; E, aggregates with preferred orientation (from Meade, 1964). *Bottom:* Scanning electron micrograph (stereopair) of a montmorillonitic clay from Lod, Israel (from Tovey, 1970).

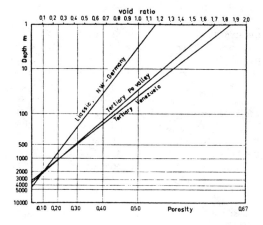

FIGURE 2. Dependence of porosity and void ratio of several shale formations on depth of burial (from Engelhardt and Gaida, 1963). Porosity (P) is volume of pores in total volume; void ratio (E) relates volume of pores to volume of solids, $E = P/(1-P)$. The function of its dependence on depth of burial (m) is $E_m = E_O - \log bm$. The coefficient (b) is dependent on the initial porosity and hence is related to the nature of the sediment.

ing electrolyte concentration of the interstitial solution and in the formation of saline brines in the porous beds surrounding the compacted clays.

The negatively charged shale acts as a *semipermeable membrane,* which restricts the passage of the anions, but slowly transmits the water molecules in response to the hydraulic pressure head in the adjoining beds. The electrochemical properties of the shales may be responsible for the formation of highly saline brines in deep basins. They are also evident in the shape of the resistivity and self-potential curves used in well-log interpretation (Wyllie, 1955).

Geochemistry

The average chemical composition of shales (Table 2) compared to that of igneous rocks shows differences due to additions of water, carbonates, and organic matter. There are decreases in the Na:K, Ca:Mg, and Si:Al ratios due to losses of sodium to the ocean, calcium to limestones, and silica to sandstones. There is also an increase in the oxidation state of the iron oxide component, reflecting the influence of atmospheric weathering on clay-mineral formation.

Depending on the nature and quantity of the clay minerals, shales possess a cation exchange capacity ranging from about 4 to 50 meq/100 g, with Na, Ca, Mg, and H as the main exchangeable ions. The pH range is 3-10, usually reflecting the diagenetic environment.

TABLE 2. Average Chemical Composition of Major Constituents in Shales (%)

	Average Shale[a]	Shales of the Russian Platform[b]	Pierre Shale Cretaceous[c]
SiO_2	58.10	52.21 ± 6.37	59.68 ± 3.49
TiO_2	0.65	0.73 ± 0.14	0.60 ± 0.05
Al_2O_3	15.40	14.64 ± 2.57	15.40 ± 1.52
Fe_2O_3	4.02		4.56 ± 1.09
		6.19 ± 0.87	
FeO	2.45		0.96 ± 0.64
MgO	2.44	3.21 ± 1.48	2.11 ± 0.47
CaO	3.11	7.22 ± 4.47	1.52 ± 1.12
Na_2O	1.30	0.87 ± 0.31	1.09 ± 0.42
K_2O	3.24	3.39 ± 0.96	2.49 ± 0.26
CO_2	2.63	5.95 ± 4.10	0.87 ± 1.35
C	0.80		-
H_2O^-		5.40 ± 1.54	3.73 ± 1.09
	5.00		
H_2O^+			4.77 ± 0.83

After Tourtelot, 1962.
[a] Avg. of 2 composites of 78 individual samples.
[b] Avg. based on 249 composites from 6769 individual samples.
[c] Avg. of 17 samples, US Western Interior.

Several *trace elements* are often concentrated in shales as compared to igneous rock averages, mainly as a result of sorption by the clay colloidal particles and by organic matter in the environment of deposition (Table 3). They are the most promising *environmental discriminants* and indicators (Wedepohl, 1970). B, V, and Cr are generally more abundant in marine shales than in continental clays or even nearshore marine shales. The black shale or stagnant-water environment usually brings about significant enrichments of V, As, Mo, Pb, Cu, Ni, Zn, and U (see *Euxinic Facies*). The overwhelming part of fossil organic matter is contained in shales (see *Oil Shales*).

Chemical changes taking place in the *pore solution* during compaction involve the reduction of sulfate, the elimination of carbonate by precipitation, and fixation of magnesium and potassium by dolomite and clay minerals, respectively. During initial stages of diagenesis, cation exchange reactions take place, and degraded illite and chlorite are reconstituted by adsorption of K or Mg between the layers. As the overburden pressure increases, the expanded clay minerals contract, and certain more penetrating structural alterations may take place with time, forming first mixed-layer minerals and then illite, which is the most common mineral in ancient shales. Transformation of smectite and mixed-layer minerals to illite may release silica to the overlying beds. The clay mineral suite of any particular shale thus reflects the combined effects of the source material, of the depositional environment, and of the diagenetic history (Fig. 3).

Classification

As yet, no uniform classification of shales is in general use. Some classifications use subdivisions according to the major clay mineral component, grouping them into micaceous, chloritic, and kaolinitic shales. Siliceous, cherty,

TABLE 3. Average Chemical Composition of Minor Elements in Shales

Element	Atomic Number	Symbol	Abundance[a]	
			ppm by Weight	Atoms per 10^6 Atoms Oxygen
Arsenic	33	As	13	5
Barium	56	Ba	580	130
Boron	5	B	100	290
Chlorine	17	Cl	180	160
Chromium	24	Cr	90	53
Cobalt	27	Co	19	10
Copper	29	Cu	45	22
Fluorine	9	F	740	1200
Gallium	31	Ga	19	8
Lead	82	Pb	20	3
Lithium	3	Li	66	290
Manganese	25	Mn	850	480
Nickel	28	Ni	68	36
Niobium	41	Nb	11	4
Phosphorus	15	P	700	700
Rare earths	57–71		210	45
Rubidium	37	Rb	140	50
Scandium	21	Sc	13	9
Strontium	38	Sr	300	110
Sulfur	16	S	2400	2300
Thorium	90	Th	12	2
Vanadium	23	V	130	80
Yttrium	39	Y	26	9
Zinc	30	Zn	95	45
Zirconium	40	Zr	160	54

[a]Weight abundance data from Turekian and Wedepohl (1961). Only elements >10 ppm in abundance are included. Relative atomic abundance is calculated on the basis of average content of 51.8% oxygen in shales.

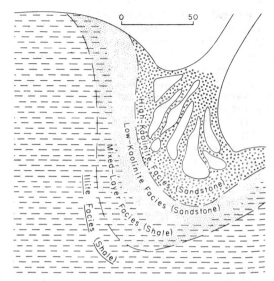

FIGURE 3. Schematic chart of the relation between clay facies and environment of deposition; Illinois Basin (after Smoot, 1960). A mixed sediment has been supplied from the same general source area and deposited in a marine environment with a decreasing rate of sedimentation with distance from the shore. Degraded clay minerals, which settled slowly far from the shore, were reconstituted or altered during and soon after deposition by adsorption of K and Mg. In the rapidly deposited sandstone facies the mixed and kaolinitic clay mineral suites are attributed to postdepositional degradation and changes by circulating formation fluids. The slowly permeable shales have been subjected to less postdepositional structural alteration, and the illitic and mixed-layer clay-mineral suites are attributed to inheritance and reconstitution.

or diatomaceous shales contain up to 70% of unbound, mainly opaline silica.

Other groupings are based on the characteristic accessory components. Calcareous or marly shales contain some 15-40% $CaCO_3$, usually fine-grained or as microfossils (see *Marl*). Black shales are rich in organic matter (1-15% C) and are also noted for their unusual concentration of certain trace elements. Phosphatic, glauconitic, and ferruginous shales represent other widespread types and reflect special environmental conditions under which the respective compounds were incorporated into the sediment.

A uniform descriptive terminology based on texture, mineral composition, and other attributes has been proposed by Picard (1971).

DAN H. YAALON

References

Boswell, P. G. H., 1961. *Muddy Sediments.* Cambridge: Heffer, 140p.
Engelhardt, W. von, and Gaida, K. H., 1963. Concentration changes of pore solutions during the compaction of clay sediments, *J. Sed. Petrology,* **33**, 919-930.
Englund, J. -O., and Jorgensen, P., 1973. A chemical classification system for argillaceous sediments and factors affecting their composition, *Geol. Fören. Stockholm Förh.,* **95**, 87-98.
Meade, R. H., 1964. Removal of water and rearrangement of particles during the compaction of clayey sediments—review, *U.S. Geol. Surv. Prof. Pap. 497-B,* 23p.
Mielenz, R. C., and King, M. E., 1955. Physical-chemical properties and engineering performance of clays, *Calif. Div. Mines Bull.,* **169**, 196-254.
Millot, G., 1970. *Geology of Clays.* London: Chapman and Hall, 429p.
Müller, G., 1967. Diagenesis in argillaceous sediments, in G. Larsen and G. V. Chillingar, eds., *Diagenesis in Sediments.* Amsterdam: Elsevier, 127-177.
Picard, M. D., 1971. Classification of fine-grained sedimentary rocks, *J. Sed. Petrology,* **41**, 179-195.
Skempton, A. W., 1970. The consolidation of clays by gravitational compaction, *Quart. J. Geol. Soc. London,* **125**, 373-412.
Smoot, T. W., 1960. Clay mineralogy of pre-Pennsylvanian sandstones and shales of the Illinois Basin. Part III. Clay minerals of various facies of some Chester formations, *Ill. St. Geol. Surv., Circ. 293,* 19p.
Tourtelot, H. A., 1960. Origin and use of the word shale, *Am. J. Sci.,* **258A**, 335-343.
Tourtelot, H. A., 1962. Preliminary investigation of the geologic setting and chemical composition of the Pierre shale, Great Plains region, *U.S. Geol. Surv. Prof. Pap. 390,* 74p.
Tovey, N. K., 1970. Electron microscopy of clays, Ph.D. Thesis, Cambridge Univ.
Turekian, K. K., and Wedepohl, K. H., 1961. Distribution of the elements in some major units of the earth's crust, *Geol. Soc. Am. Bull.,* **72**, 175-191.
Van Moort, J. C., 1973. The magnesium and calcium contents of sediments, especially pelites, as a function of age and degree of metamorphism, *Chem. Geol.,* **12**, 1-38.
Weaver, C. E., 1959. The clay petrology of sediments, *Clays Clay Minerals,* **6**, 154-187.
Wedepohl, K. H., 1970. Environmental influences on the chemical composition of shales and clays, *Phys. Chem. Earth,* **8**, 305-333.
Wyllie, M. R. J., 1955. Role of clay in well-log interpretation, *Calif. Div. Mines Bull.,* **169**, 282-305.
Yaalon, D. H., 1962. Mineral composition of the average shale, *Clay Minerals Bull.,* **5**, 31-36.

Cross-references: *Argillaceous Rocks; Clastic Sediments and Rocks; Clay; Clay Sedimentation Facies; Compaction in Sediments; Euxinic Facies; Fissility; Flocculation; Marl; Mudstone and Claystone; Oil Shale; Sedimentary Rocks; Varves and Varved Clays.*

SHAPE—See ROUNDNESS AND SPHERICITY

SHARK BAY SEDIMENTOLOGY

Shark Bay, Western Australia, offers exceptional opportunities for research in carbonate sedimentation and paleoecology, as first suggested by Curt Teichert in 1946. The bay has a wide range of environments, biotic communities, and sediment types available for study. Sedimentation is controlled by hydrologic factors; waves; currents; and organisms, which play a major role by contribution of skeletal parts and by intervention in sedimentation (Logan, in Logan et al., 1970). Studies of Shark Bay were initiated by workers from the University of Western Australia; two memoirs have been published on this research (Logan et al., 1970; Logan et al., 1974).

Environment

Shark Bay is a large embayment at latitude 25°S on the W coast of Australia (Fig. 1). In dimension, Shark Bay approaches a small epicontinental sea, approximately 13,000 km² in area with a 10-m average depth. The water mass is cut off from the Indian Ocean by a barrier ridge of calcareous eolianite and is subdivided internally into numerous inlets, gulfs, and basins by dune ridges and submerged banks (sills). Oceanic water influx is only through openings in the northern part of the outer barrier.

The embayment is adjacent to a low-relief, arid to semiarid hinterland. Runoff influx is negligible, and annual evaporation (≈ 230 cm) greatly exceeds precipitation (≈ 23 cm). These factors, combined with hydrologic structure and restriction imposed by banks and sills, result in increasing salinity gradients into the closed southern embayment; the salinity varies from 36‰ to as high as 90‰. The water mass has a layered structure with nearly vertical attitude. Major clines subdivide the water body into three major types (Fig. 2): oceanic (36–40‰), metahaline (40–56‰), and hypersaline (56–72‰).

The water types are limiting on the biota (Logan and Cebulski, in Logan et al., 1970). There are three biotic zones, the distribution of which is essentially similar to the distribution of water types. In the broad environmental zones, numerous local environments are limited by such factors as wave action; tides and tidal currents; and depth, which is of major importance because most other parameters are linked to it.

Quaternary History

The geomorphology of the embayment reflects the geometry and lithology of underlying rock units. Western and central areas are underlain by dune rocks of Pleistocene age, and numerous inlets and gulfs are interdune depressions which were flooded by the postglacial transgression. The eastern shore is an alluvial coastal plain in the N and a terrain of flat-lying Cretaceous limestone and dolomite in the S.

There were two dune-building phases in the early Pleistocene. The Peron Sandstone, which was deposited during the first phase, underlies the central area and is composed of red quartz sandstone. The second dune phase completed the basic architecture of the embayment, building large ridges of calcareous eolianite (Tamala Eolianite) along the western perimeter.

The ancestral dune landscape was flooded in three separate marine transgressions during the Quaternary. The Dampier and Bibra transgressions occurred during the Pleistocene; they are represented by marine carbonate strata which crop out around the embayment margins and also lie beneath postglacial sediments in offshore areas. The Pleistocene rocks are intertidal and

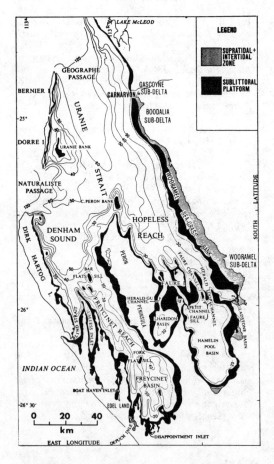

FIGURE 1. Bathymetric map of Shark Bay (from Logan et al., 1970). Depth contours in feet.

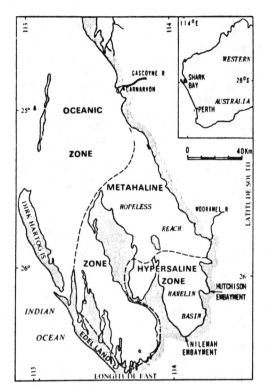

FIGURE 2. Principal biotic zones of Shark Bay (from Logan et al., 1974).

shallow subtidal facies. The embayment during both Dampier and Bibra phases was essentially similar to the present embayment, caused by the Holocene transgression. Transgressive deposits are separated by erosional unconformities marked by weathering phenomena, including soils and calcrete.

Holocene Sedimentation. Holocene sediments are chiefly calcareous biogenic-skeletal fragment grainstone, packstone, and wackestone; but coquina, ooid grainstone, and lithoskel grainstone also are abundant. In hypersaline zones, there is widespread lithification of sediments by precipitation of aragonitic cements. Organisms play a major role in sedimentation not only by their postmortem contribution of skeletal detritus, but also through their intervention in depositional and diagenetic processes. Notable products of organic intervention are carbonate banks and platforms, and cryptalgal (stromatolite, q.v.) structures (Logan, 1961; Logan et al., 1974).

Sea-grass communities that live in oceanic and metahaline biotic zones have had a major part in the evolution of sedimentary environments throughout the Holocene interval. Grass communities have been responsible for construc-

tion of carbonate banks—patch, fringing, and barrier types. Barrier banks partition the embayment to form closed basins, while prograding fringing banks (e.g., Davies, in Logan et al. 1970) diminish the embayment cross section. This growth has gradually restricted tidal influx and exchange and has led to hypersaline conditions in southern areas.

BRIAN W. LOGAN

References

Logan, B. W., 1961. Cryptozoon and associate stromatolites from the Recent, Shark Bay, Western Australia, *J. Geol.,* 69, 517-533.

Logan, B. W.; Davies, G. R.; Read, J. F.; and Cebulski, D. E., 1970. Carbonate sedimentation and environments, Shark Bay, Western Australia, *Am. Assoc. Petrol. Geologists Mem. 13,* 223p.

Logan, B. W.; Read, J. F.; Hagan, G. M.; Hoffman, P.; Brown, R. G.; Woods, P. J.; and Gebelein, C. D., 1974. Evolution and diagenesis of Quaternary carbonate sequences, Shark Bay, Western Australia, *Am. Assoc. Petrol. Geologists Mem. 22,* 358p.

Cross-references: *Bahama Banks Sedimentology; Beachrock; Caliche, Calcrete; Carbonate Sediments–Diagenesis; Eolianite; Limestones; Persian Gulf Sedimentology; Stromatolites.*

SHEET SAND—*See* BLANKET SAND

SHOESTRING SANDS—*See* BARRIER ISLAND FACIES; CHANNEL SANDS; FLUVIAL SEDIMENTS

SILCRETE

Silcrete, like chalcedony, flint, etc., is a highly siliceous rock of secondary origin (see *Chert and Flint*). It is derived by solution from the soil and the deeper weathering mantle, followed by precipitation after movement in drainage waters. Its matrix is composed largely of fine-grained crystalline quartz, but frequently large grains are present, and occasionally small crystals growing in voids are visible with a lens or to the naked eye. Sands and gravels are commonly included within this matrix, so that the rock often displays some of the characteristics of quartzites and conglomerates. Sometimes the included material is either water-worn or brecciated silcrete, thus indicating repeated stages in its formation. Like other secondary siliceous materials, silcrete varies in color from pale grey and buff to dark red and brown, but it is predominantly grey. Its hardness is about 7, and it exhibits a marked conchoidal fracture. The less clastic forms, where unweathered, often show botryoidal or mammillated surfaces. Internal concretionary structure occurs but is visible infrequently.

The name *silcrete* was first used by Lamplugh (see *Duricrust*), and this word is today increasing in usage. In South Africa the term *surface quartzite,* and in Australia *billy, grey billy,* or *grey wether,* and *duricrust* (q.v.)—which has a wider connotation—have been used extensively in the past. The type locality, described by Lamplugh (1907), is in the vicinity of the Victoria Falls in Zambia.

Occurrence

Silcrete is most common in Africa (Frankel and Kent, 1937; Millot et al., 1959) and in Australia (Williamson, 1957; Stephens, 1971), where it occurs mainly as a capping on ancient peneplain and terrace remnants. It has also been recorded in Europe and doubtfully in North America (Kerr, 1955). Occurrences may be extensive and regional or small and local. Regionally it lies mainly in arid or strongly seasonal zones, there largely as elevated surface remnants being consumed by headward erosion, which leaves a pavement of abraded and polished silcrete boulders and pebbles on the pediment below. It also occurs within laterite profiles, and more local and scattered examples are found on river terraces, in stream beds of low gradient (as depression deposits), and beneath rocks such as basalt. In-situ silcrete also occurs at various depths, particularly in laterite profiles in which the secondary silica hardens the relevant horizon (Mabbutt, 1965). Normally there is only one silcrete horizon, but up to three have been observed in the same section.

Genesis

The varied geomorphic setting of silcrete, and the frequent association of opal with it, point to its derivation during Tertiary and Quaternary times by deposition of silica from both surface and ground waters, with progressive replacement, cementation, and recrystallization within the associated sedimentary or residual horizon. Where found in the laterite profile, silcrete appears to be either the result of relative accumulation of silica released by further weathering therein, or of absolute accumulation following invasion by silica-bearing waters. Its occurrence as a surface formation, however, especially where it incorporates water-worn gravels and sands, has been interpreted in arid Australia by Stephens (1971) as absolute accumulation following lateral drainage from adjacent and extensive, but geographically separate, lateritic regions. Many small occurrences also appear to be absolute accumulations following other forms of weathering and soil formation, and associated release of silica, in nearby catchment areas. The occurrences under basalt are attributed by some to hydrothermal action, but there seems to be a much greater likelihood of burial of earlier silcrete formations by later basalt flows, or the filling of sub-basaltic cavities by silica released by weathering of the basalt itself.

Silcrete has been quarried for refractory brick manufacture in both South Africa and Australia but otherwise seems to have little modern use. In earlier times, it was used extensively by primitive people for the manufacture of stone weapons and tools, for which its sharp conchoidal fracture made it particularly suitable.

C. G. STEPHENS

References

Frankel, J. J., and Kent, L. E., 1937. Grahamstown surface quartzites (silcretes), *Trans. Geol. Soc. S. Africa,* **40,** 1-42.

Kerr, M. H., 1955. On the occurrence of silcretes in southern England, *Proc. Leeds Phil. Soc.,* **6,** 328-337.

Lamplugh, G. W., 1907. The geology of the Zambezi Basin around the Batoka Gorge (Rhodesia), *Quart. J. Geol. Soc., London,* **63,** 162-216.

Mabbutt, J. A., 1965. Stone distribution in a stony tableland soil, *Austral. J. Soil Research,* **3,** 131-142.

Millot, G.; Radier, H.; Muller-Feuga, R.; Defossez, M.; and Wey, R., 1959. Sur la géochimie de la silice et les silicifications sahariennes, *Bull. Serv. Carte Géol. Alsace Lorraine,* **12,** 3-14.

Smale, D., 1973. Silcretes and associated silica diagenesis in southern Africa and Australia, *J. Sed. Petrology,* **43,** 1077-1089.

Stephens, C. G., 1971. Laterite and silcrete in Australia, *Geoderma,* **5,** 1-52.

Williamson, W. O., 1957. Silicified sedimentary rocks in Australia, *Am. J. Sci.,* **255,** 23-42.

Cross-references: *Chert and Flint; Duricrust; Silica in Sediments; Silicification; Weathering in Sediments.*

SILICA IN SEDIMENTS

Silica minerals make up 12% of the earth's crust; only feldspars are more abundant. Silica exists in a number of different polymorphs, of which α quartz is by far the most abundant. Tridymite and cristobalite are much less common; melanophlogite, coesite, and stishovite are rare. Keatite, a synthetic form of SiO_2, is not known in nature. Alpha quartz and possibly melanophlogite are stable at ordinary temperatures and pressures, and tridymite, cristobalite, coesite, and stishovite can exist in a metastable state almost indefinitely under these conditions.

Except for stishovite, the silicon of all silicates is surrounded by four oxygens at the corners of a nearly regular tetrahedron. The geometry of tetrahedral linkage varies in the silica polymorphs. In stishovite, silicon is octahedrally coordinated with oxygen. See Vol. IVB: *Chalcedony; Opal; Quartz*.

Varieties

The most common varieties of silica in sediments are quartz, chalcedony, and opal. For other varieties of silica, see Frondel (1962).

Quartz. Quartz has a very small range of variation in composition. Impurities are either mechanically intermixed or in solid solution, in particular with Li, Na, Al, and Ti (Frondel, 1962). Quartz is found in all types of sediments ranging from clay size to pebbles. Its abundance varies from traces to >95% in quartz arenites and cherts. The great majority of quartz grains are of detrital origin. Authigenic quartz as overgrowth, cement, and replacement is widespread. The commonly observed undulatory extinction in quartz is due either to mechanical deformation or to intergrowth of individuals in subparallel position.

Chalcedony. A cryptocrystalline variety of quartz, with submicroscopic pores, chalcedony exhibits a fibrous structure under the microscope. It may contain more impurities than quartz, and has a lower density and lower indices of refraction. Chalcedony is an important component of chert (q.v.) and is found as cement and in silicified fossils and wood.

Opal. Opal is a general term for amorphous silica and submicrocrystalline aggregates of disordered low cristobalite, usually containing 3-9% nonessential water. Opal is an important variety of silica in sediments. Based on x-ray diffraction patterns, there are three distinct opals (Jones and Segnit, 1971). *Opal C* is well-ordered α-cristobalite. X-ray patterns show only a small amount of line broadening and some evidence of tridymite stacking. Found in opals associated with lava flows, it has been observed once in deep-sea-drilled cherts. *Opal CT* is disordered α-cristobalite α-tridymite. It shows varying degrees of one-dimensional stacking disorder of cristobalite and tridymite domains which affect the relative intensities of reflections, cause line broadening, and result in additional reflections which belong to tridymite (e.g., lussatite; Flörke, 1955; Jones and Segnit, 1971). Scanning electron micrographs have shown that much opal CT is composed of 2-30μ spherules which consist of blade-shaped, radiating crystals, of tens to a few thousand Å units (Wise et al., 1972). Most porcelanites (see *Porcelanite*) of Mesozoic and Tertiary age are opal CT, which is also found as cement. *Opal A* is highly disordered and x-ray amorphous. It shows a prominent diffuse x-ray pattern centered at ≈4.1 Å. Siliceous skeletal remains, precious opal, hyalite, and geyserites are opal A.

Silica in Natural Water

Solubility. At ordinary temperatures and at pH values <9, silica is mostly in true solution as monosilicic acid, $Si(OH)_4$. Above pH 9, $Si(OH)_4$ dissociates to $SiO(OH)_3^-$ and $SiO_2(OH)_2^{2-}$ (Iler, 1955). In addition to pH, the solubility of silica is a function of the crystalline state of the solid phase, and of temperature and pressure. The solubility of amorphous silica at various temperatures is given in Table 1. At 2°C, the solubility in sea water increases from 56 ppm at 1 atm to 70 ppm at 1000 atm (Jones and Pytkowicz, 1973). At 25°C, the solubility of quartz is about 5 ppm and increases with increasing temperature. Solubility curves for quartz and amorphous silica appear to converge at higher temperatures, as shown in Fig. 1 (Siever, 1962). The solubility of other silica varieties ranges between that of quartz and amorphous silica (Fournier, 1973; Stöber, 1967). Solution can be retarded by organic-matter films or metallic-surface complexes (Iler, 1955); organo-silicon complexing, however, might increase removal of silica from solution (Siever, 1962).

Most natural waters are undersaturated with respect to amorphous silica, and much of the ocean is also undersaturated with respect to quartz. The world average for dissolved silica in river water is 13.1 ppm SiO_2 (Livingstone, 1963). Dissolved silica in oceans is highly variable (Kido and Nishimura, 1975). Surface waters are extremely depleted due to rapid biological fixation, mainly by diatoms and radiolarians, of any influx of silica into the oceans, but most of the biogenic silica is redissolved and recycled in the water column; hence silica concentration increases with depth (Fig. 2). Only about 4% of siliceous skeletal remains survive to be buried, and about 2% redissolves in interstitial waters and is used up partly by authigenic silicates. As a result, a maximum 2%

TABLE 1. Solubility of Amorphous Silica as a Function of Temperature in Water at pH Values <9[a]

Temperature °C	$m_{H_4SiO_4} \times 10^3$	ppm SiO_2
0	1.0-1.3	60-80
25	1.7-2.3	100-140
90	5.0-6.3	300-380

[a]Krauskopf, 1956.

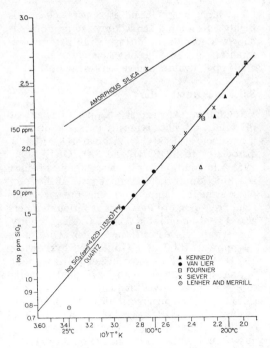

FIGURE 1. The solubility of quartz and amorphous silica as a function of temperature (from Siever, 1962).

enters the geological record (Hurd, 1973; Heath, 1974; Wollast, 1974). Oceanic circulation readily explains the great difference in silica concentration of intermediate and deep waters in the Atlantic and Pacific Oceans (Fig. 2). The deep water of the North Pacific is the oldest water mass. In general, the Atlantic can be described as "lagoonal"—i.e., it loses deep water and gains silica-poor surface water—and the Pacific as "estuarine"—i.e., it gains deep water and loses silica-poor surface water (Berger, 1970).

The Silica Budget. The major silica sources are river influx; low-temperature silicate reactions at the sediment/water interface; diffusive flux from interstitial waters of deep-sea sediments; submarine volcanism and hydrothermal activity. An increase in silica concentration in interstitial water (q.v.) in the upper few cm of deep-sea sediments (see Schink et al., 1974) is causing a diffusive flux into sea water. Median dissolved silica values range between 10 ppm in sediments devoid of siliceous organisms to 60–70 ppm in radiolarian oozes. The silica flux problem is still very controversial.

The major mechanism of silica removal is biological fixation of silica. MacKenzie et al. (1967) suggested that clay minerals partially control the silica budget of sea water, but recent experiments by Siever and Woodford (1973) indicate that adsorption or release of silica by clays is a doubtful mechanism for partial control of silica of sea water, although it could be an important mechanism in silica-rich interstitial waters. Table 2 summarizes the approximate dissolved silica budget in the ocean.

Sodium carbonate lakes have high values of dissolved silica (up to 2700 ppm) at pH values >10 (Jones et al., 1967).

Siliceous Sediments

Siliceous sediments consist largely or almost entirely of SiO_2 varieties. The most important representatives are sandstones (see *Sands and Sandstones*), in particular quartz arenites, cherts (q.v., including siliceous ooze, q.v., and porcelanites, q.v.), siliceous shale, volcanic ash beds (see *Volcanic Ash Deposits*), and silcrete (q.v.). The source of silica is detrital, biological, or diagenetic from sediment-water interactions. Combinations of the three sources in all proportions are possible. Since the Paleozoic, the major mechanism of silica precipitation at ordinary temperatures and pressures has been biochemi-

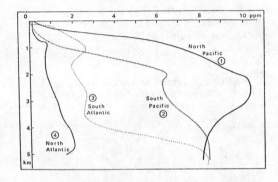

FIGURE 2. Vertical profiles of dissolved silica in the Pacific and Atlantic oceans (after Heath, 1974).

TABLE 2. Geochemical Mass Balance of Silica in the Ocean[a]

	10^{14} g SiO_2 per yr
Supply	
Rivers	4.27
Submarine weathering	0.03
Submarine volcanism	0.0003
	4.30
Removal	
Siliceous sediments	3.60
Inorganic uptake	0.43
	4.03

[a]Wollast, 1974.

cal (for a review, see Ramsay, 1973). Other mechanisms are adsorption, organo-silicon complexing, evaporation or cooling of silica-rich waters, and neutralization of strongly alkaline solutions.

Sandstone. Quartz grains are the product of mechanical weathering of igneous, sedimentary, and metamorphic rocks. Due to the physical anc chemical properties of quartz, in particular hardness, lack of cleavage, and extremely low solubility, it predominates in sandstones (see *Sands and Sandstones*). An average sedimentary rock contains 59.7% SiO_2 (Garrels and Mackenzie, 1971) and an average sandstone 78.0% (Pettijohn, 1957). Secondary quartz overgrowth, cementation, and lithification by quartz or chalcedony, mainly the result of chemical precipitation or pressure solution, are widespread in Tertiary and older sandstones but rare in Holocene ones. Reaction of quartz precipitation below 150°C is very slow, but Mackenzie and Gees (1971) successfully crystallized quartz directly from sea water at 20°C after three years.

Siliceous Shale. An average shale contains 58% SiO_2; siliceous shales (see *Shale*) contain as much as 85% (Pettijohn, 1957). The abnormally high SiO_2 is not detrital, but is derived either from remobilization and redistribution of biochemical silica, as in the Monterey Formation siliceous shales (Bramlette, 1946), or from volcanic ash (q.v.).

Siliceous Ooze, Porcelanite, and Chert. The order of decreasing importance of organisms that contribute large quantities of siliceous skeletal remains to pelagic sediments is: diatoms, radiolarians, sponge spicules, and silicoflagellates. They constitute up to about 70% of some Antarctic pelagic sediments in DSDP cores. Diatoms are most abundant in high latitudes and in regions of upwelling (see *Diatomite*); radiolarians are more abundant in warmer waters (see *Siliceous Ooze*). Diatoms are also abundant in some fresh-water lakes. Most of these skeletal remains dissolve in the open ocean, or, in nearshore sediments, are masked by terrigenous material and carbonate oozes. Therefore, the highest concentration of siliceous oozes are found below regions where nutrient-rich deep waters upwell and in deep waters below the Carbonate Compensation Depth (see *Pelagic Sedimentation, Pelagic Sediments*). Hence, oceanographic rather than geochemical controls are responsible for the deposition of biogenic opal. In Holocene pelagic sediments, their distribution reflects surface production rates, and it can be assumed that similar controls have been effective since the early Paleozoic, and possibly even in the later Precambrian (Siever, 1957). There is an accumulation of 10^{13} g of biogenic opal each year in the central Gulf of California. Calvert (1966) showed that sufficient silica is available in sea water, and volcanic sources of silica are not necessary, even for such extensive accumulations.

Evidence from deep-sea sediments supports the following diagenetic sequence from siliceous oozes to quartzose chert: Siliceous ooze (opal A) → porcelanite (opal CT) → chert (chalcedony, cryptocrystalline quartz) → chert (megaquartz). The opal A to opal CT transformation proceeds by a solution-precipitation mechanism (see Carr and Fyfe, 1958; Kastner et al., 1977). Based on experimental work, Ernst and Calvert (1969) concluded that the opal CT to quartz transformation is a solid-solid-zero order reaction. According to Carr and Fyfe (1958) and Stein and Kirkpatrick (1976), however, it proceeds by dissolution-precipitation.

A description of the geological distribution of porcelanites and cherts is needed for discussion of the physicochemical controls of the above maturation process. *Porcelanite* (q.v.) may be defined as a marine siliceous rock of which the major silica phase is opal CT. *Chert* is a dense, hard, brittle, and fine-grained rock consisting mainly of SiO_2 in the form of chalcedony and cryptocrystalline quartz. Chert occurs in two principal modes: (1) nodular and (2) bedded (see *Chert and Flint*).

Porcelanites and Cherts in Deep-Sea Sediments. Horizon A, a prominent and widespread reflector of early to middle Eocene found in the western North Atlantic, was the first porcelanite layer to be encountered by the Deep Sea Drilling Project. A large number of porcelanite and chert layers have been penetrated, mostly concentrated in the Upper Jurassic, Upper Cretaceous, Middle Eocene, and Middle Miocene. The youngest porcelanite is of Pliocene age, and the youngest chert of Middle Eocene associated with carbonates. The oldest porcelanite has been found in the Lower Cretaceous of a Leg 32 hole (Keene, 1975).

Origin of Cherts. There is no simple way of relating either the crystalline state of silica or the texture of porcelanite and chert horizons to the age of surrounding sediments, although in general, quartz becomes increasingly abundant in the older cherts. The composition of the surrounding sediments has a major influence on the conversion rate of opal A; Lancelot (1973) showed that porcelanite composed of disordered cristobalite is found in clay-rich sediments, whereas quartzose nodules are restricted to carbonates of the same age. Conversion of opal CT to quartz is more rapid in carbonates than in noncarbonates. Foreign cations, in particular alkali metals, and aluminum inhibit the transformation from opal CT to chalcedony or quartz. The permeability of clay-rich sediments is rather low, and, therefore, the silica to metallic cations ratio remains low, and cristobalite precipitates; in carbonates, the higher permeability causes an increase in this ratio, and quartz can form directly (Lancelot, 1973). Chemical analyses of opal CT by Millot (1964) indicate that they do indeed contain appreciable amounts of alkali metals.

Nodular Chert. There is overwhelming evidence that chert nodules (see *Chert and Flint*) are diagenetic products of replacement, largely deriving the silica from biogenic opal deposited with the carbonate matrix, although in some instances the density of chert nodules implies introduction of silica into the bed by circulation of silica-rich waters. The reason for localized silica precipitation is unclear. Siever (1962) suggested that an uneven distribution of organic matter and its subsequent decay will cause lateral pH variations and thus localized silica precipitation. Lack of compression of the biogenic remains within the nodules and compaction of the carbonate beds around the nodules suggests that many nodules form prior to major compaction. Deep-sea diagenesis resulting in chert nodule formation has been studied by Weaver and Wise (1973). They found that nodules from Tertiary chalk are cristobalite-rich except where metastable silica has gone to quartz. They describe several transition zones separating the chert nodules from the unsilicified host rock: a weakly silicified chalk zone, several cm wide, in which the interstices are partly filled with 10μ cristobalite microspherulites (lepispheres). The displaced calcite may be reprecipitated as ultrafine euhedral calcite grains adjacent to nodules. Wise and Kelts (1972) suggest that cristobalite spherules represent the initial stage of silicification of carbonate rock in the deep-sea environment, silica being derived from the dissolution of siliceous microfossils and volcanic glass. In contrast, Lancelot (1973) found cristobalite only at the periphery of quartzose nodules; the boundary of quartzose chert to cristobalite is sharp and is outlined by a concentration of impurities, whereas the cristobalite/carbonate boundary is diffuse. Lancelot attributes the formation of a nodule not to transformation of cristobalite to quartz, but to direct precipitation of quartz in the carbonate matrix where clay minerals are rare and dispersed; as the nodule develops outward by accretion, all the clay minerals and cations that cannot be accommodated in the quartz structures will be excluded and move along a "quartzification front," achieved because of high permeability of the sediment and allowing free circulation of the interstitial waters.

Bedded Cherts. Bedded cherts are widely distributed from the Precambrian to the Tertiary. Biogenous opal is also the principal source of silica for Mesozoic and Tertiary bedded cherts, and many of them are associated with unaltered siliceous skeletal remains. Notably extensive recrystallization has erased the clues to the source of silica of many bedded cherts, especially of Precambrian and Paleozoic cherts. It may, however, be assumed that the origin of Paleozoic cherts is similar to that of the younger ones (Ramsay, 1973). Silica may also be produced by the alteration of volcanic detritus. The association of cristobalite-smectite-clinoptilolite is often used as evidence for a volcanic origin, but no direct evidence of volcanic cherts has been found in deep-sea sediments. Smectite and zeolites, not chert, are the principal alteration products of volcanic glass in sea water (see *Volcanic Ash—Diagenesis*). Bramlette (1946) described volcanic chert layers in the Monterey Formation which could have been silicified by silica migration from the surrounding diatomaceous ooze. Volcanic cherts have also been described from alkaline lakes, as in the Barstow Formation, California (Sheppard and Gude, 1969). An additional possible silica source is the alteration of smectite to hydrous mica or chlorite (Keene and Kastner, 1974); a similar mechanism of subaerial weathering of ferric-rich smectite to chert kaolinite and goethite was described by Altschuler et al. (1963) from central Florida. Textural and structural evidence, e.g., slump structures and intraformational breccias, points to the conclusion that bedded cherts are early diagenetic, prelithification phenomena. The variation with time of the chert $^{18}O/^{16}O$ values may be explained in terms of past climatic temperature fluctuations (Knauth and Epstein, 1976).

Recent and Precambrian Cherts. Inorganic precipitation of silica from volcanic weathering in alkali Magadi and Natron Lakes, E Africa has been demonstrated by Eugster (1967), Eugster and Jones (1968), and Hay (1968). Sodium hydrous silicates, magadiite, and kenyaite, are the precursors from which percolating waters remove the sodium, to form the chert. Recent inorganic precipitation of silica from dissolution of detrital silicates during high pH (9.5–10.2) intervals in alkaline lakes associated with the Coorong Lagoon, S Australia have been described by Peterson and von der Borch (1965).

In the early Precambrian, silica-secreting organisms had not yet evolved. Therefore, the bedded cherts interbedded with iron-rich bands, e.g., the taconites of the Lake Superior region, North America, probably formed inorganically (see *Cherty Iron Formations*). Eugster (1967) suggested that similar to recent Magadi Lake cherts, silica precipitated during the lower pH or dilute intervals of the Precambrian lake, and iron precipitated during the dry intervals.

MIRIAM KASTNER

References

Altschuler, Z. S.; Dwornik, E. J.; and Kramer, H., 1963. Transformation of montmorillonite to kaolinite during weathering, *Science,* 141, 148-152.

Berger, W. H., 1970. Biogenous deep-sea sediments; fractionation by deep-sea circulation, *Geol. Soc. Am. Bull.,* 81, 1385-1401.

Berner, R. A., 1971. *Principles of Chemical Sedimentology.* New York: McGraw-Hill, 240p.

Bramlette, M. N., 1946. The Monterey Formation of California and the origin of siliceous rocks, *U.S. Geol. Surv. Prof. Pap. 212,* 57p.

Calvert, S. E., 1966. Accumulation of diatomaceous silica in the sediments of the Gulf of California, *Geol. Soc. Am. Bull.,* 77, 569-596.

Calvert, S. E., 1974. Deposition and diagenesis of silica in marine sediments, *Internat. Assoc. Sed. Spec. Publ. 1,* 273-300.

Carr, R. M., and Fyfe, W. S., 1958. Some observations on the crystallization of amorphous silica, *Am. Mineralogist,* 43, 908-916.

Degens, E. T., and Epstein, S., 1962. Relationship between O^{18}/O^{16} ratios in coexisting carbonates, cherts, and diatomites, *Bull. Am. Assoc. Petrol. Geologists,* 46, 534-542.

Ernst, W. B., and Calvert, S. E., 1969. An experimental study of the recrystallization of porcelanite and its bearing on the origin of some bedded cherts, *Am. J. Sci.,* 267-A, 114-133.

Eugster, H. P., 1967. Hydrous sodium silicates from Lake Magadi, Kenya: Precursors of bedded chert, *Science,* 157, 1177-1180.

Eugster, H. P., and Jones, B. F., 1968. Gels composed of sodium-aluminum silicate, Lake Magadi, Kenya, *Science,* 161, 160-163.

Flörke, O. W. von, 1955. Zur Frage des "Hoch"– Cristobalit in Opalen, Bentoniten, and Gläsern, *N. Jahrb. Mineral. Monatsch.,* 10, 217-223.

Fournier, R. O., 1973. Silica in thermal waters: laboratory and field investigations, *Internatl. Symp. Hydrochem. Biochem., Tokyo, 1970, Proc. 1,* 122-139.

Frondel, C., 1962. *Dana's System of Mineralogy. Vol. III. Silica Minerals,* 7th ed. New York: Wiley, 334p.

Garrels, R. M., and Mackenzie, F. T., 1971. *Evolution of Sedimentary Rocks.* New York: Norton, 397p.

Hay, R. L., 1968. Chert and its sodium-silicate precursors in sodium-carbonate lakes of East Africa, *Contr. Mineralogy Petrology,* 17, 255-274.

Heath, G. R., 1974. Dissolved silica and deep-sea sediments, *Soc. Econ. Paleont. Mineral. Spec. Publ. 20,* 77-93.

Hurd, D. C., 1973. Interaction of biogenic opal, sediment and seawater in the central Equatorial Pacific, *Geochim. Cosmochim. Acta,* 37, 2257-2282.

Iler, R. K., 1955. *The Colloid Chemistry of Silica and Silicates.* Ithaca: Cornell Univ. Press, 324p.

Jones, B. F.; Rettig, S. L.; and Eugster, H. P., 1967. Silica in alkaline brines, *Science,* 158, 1310-1314.

Jones, J. B., and Segnit, E. R., 1971. The nature of opal, I. Nomenclature and constituent phases, *J. Geol. Soc. Australia,* 18, 57-68.

Jones, M. M., and Pytkowicz, R. M., 1973. Solubility of silica in seawater at high pressures, *Bull. Soc. Roy. Sci. Liege,* 42, 118-120.

Kastner, M.; Keene, J. B.; and Gieskes, J. M., 1977. Diagenesis of siliceous oozes–I. Chemical controls on the rate of opal-A to opal-CT transformation–an experimental study, *Geochim. Cosmochim. Acta,* 41, 1041-1059.

Keene, J. B., 1975. Cherts and porcelanites from the North Pacific, DSDP, Leg 32, *Init. Repts. Deep-Sea Drilling Proj.,* 32, 429-507.

Keene, J. B., and Kastner, M., 1974. Clays and formation of deep sea cherts, *Nature, Phys. Sci.,* 249, 754-755.

Kido, K., and Nishimura, M., 1975. Silica in the sea– Its forms and dissolution rates, *Deep-Sea Research,* 22, 323-338.

Knauth, L. P., and Epstein, S., 1976. Hydrogen and oxygen isotope ratios in nodular and bedded cherts, *Geochim. Cosmochim. Acta,* 40, 1095-1108.

Krauskopf, K. B., 1956. Dissolution and precipitation of silica at low temperatures, *Geochim. Cosmochim. Acta,* 10, 1-26.

Lancelot, Y., 1973. Chert and silica diagenesis in sediments from the Central Pacific, *Init. Repts. Deep Sea Drilling Proj.,* 17, 377-405.

Livingstone, D. A., 1963. Chemical composition of rivers and lakes, *Data of Geochemistry, U.S. Geol. Surv. Prof. Pap. 440-G,* 6th ed., 61p.

Mackenzie, F. T., and Gees, R., 1971. Quartz: Synthesis at Earth-surface conditions, *Science,* 173, 533-534.

Mackenzie, F. T.; Garrels, R. M.; Bricker, O. P.; and Bickley, F., 1967. Silica in sea water: control by silica minerals, *Science,* 155, 1404-1405.

Millot, G., 1964. *Géologie des Argiles.* Paris: Masson, 499p.

Peterson, M. N. A., and von der Borch, C. C., 1965. Chert: Modern inorganic deposition in carbonate precipitating locality, *Science,* 149, 1501-1503.

Pettijohn, F. J., 1957. *Sedimentary Rocks.* New York: Harper & Row, 718p.

Ramsay, A. T. S., 1973. A history of organic siliceous sediments in oceans, in Organisms and Continents Through Time, *Spec. Pap. Palaeontol. 12,* 199-234.

Schink, D. R.; Fanning, K. A.; and Pilson, M. E. Q., 1974. Dissolved silica in the upper pore waters of the Atlantic Ocean floor, *J. Geophys. Res.,* 79, 2243-2250.

Sheppard, R. A., and Gude, A. J., III, 1969. Diagenesis of tuffs in the Barstow Formation, Mud Hills, San Bernardino County, California, *U.S. Geol. Surv. Prof. Pap. 634,* 34p.

Siever, R., 1957. Pennsylvanian sandstones of the eastern interior coal basin, *J. Sed. Petrology,* 27, 227-250.

Siever, R., 1962. Silica solubility, 0-200°C, and the diagenesis of siliceous sediments, *J. Geol.,* 70, 127-150.

Siever, R., and Woodford, N., 1973. Sorption of silica by clay minerals, *Geochim. Cosmochim. Acta,* 37, 1851-1880.

Stein, C. L., and Kirkpatrick, R. J., 1976. Experimental porcelanite recrystallization kinetics: a nucleation and growth model, *J. Sed. Petrology,* **46,** 430–435.

Stöber, W., 1967. Formation of silicic acid in aqueous suspensions of different silica modifications, in W. Stumm, ed., *Equilibrium Concepts in Natural Water Systems.* Washington, D.C.: Am. Chem. Soc., 161–182.

Weaver, F. W., and Wise, S. W., 1973. Ultramorphology of carbonate and silicate phases associated with deep-sea chert, *Bull. Am. Assoc. Petrol. Geologists,* **57,** 811.

Wise, S. W., Jr.; Buie, B. F.; and Weaver, F. M., 1972. Chemically precipitated sedimentary cristobalite and the origin of chert, *Eclogae Geol. Helv.,* **65,** 157–163.

Wise, S. W., and Kelts, K. R., 1972. Inferred diagenetic history of weakly silicified deep sea chalk, *Bull. Am. Assoc. Petrol. Geologists,* **56,** 1906.

Wollast, R., 1974. The silica problem, in E. D. Goldberg, ed., *The Sea,* vol. 5. New York: Wiley, 359–392.

Cross-references: *Cementation; Chalk; Chert and Flint; Cherty Iron Formation; Diagenesis; Diatomite; Interstitial Water in Sediments; Novaculite; Pelagic Sedimentation, Pelagic Sediments; Porcelanite; Sands and Sandstones; Shale; Silcrete; Siliceous Ooze; Sponges in Sediments; Spring Deposits; Tripoli, Tripolite; Volcanic Ash Deposits; Volcanic Ash–Diagenesis.* Vol. IVA: *Geochemistry of Sedimentary Silica.* Vol. IVB: *Chalcedony; Opal; Quartz.*

SILICEOUS OOZE

A siliceous ooze is a soft, porous, water-saturated sediment whose particulate components (>30% of wet weight) consist either entirely or partially of organically derived silica from one or a combination of diatoms, radiolarians, and sponge spicules; in addition to these three main components, silicoflagellates common in cold-water areas may occasionally be dominant in some older diatom oozes. Thus, siliceous oozes are generated essentially by the accumulation of opaline silica shells of microorganisms which extract dissolved silica from the water (see *Silica in Sediments*). The most extensive siliceous oozes of the present world ocean are primarily diatomaceous and radiolarian oozes which lie beneath the main areas of upwelling; i.e., the Antarctic polar front, the Equatorial Pacific and Indian oceans, and the NW Pacific Ocean (see *Pelagic Sedimentation, Pelagic Sediments*). Some less extensive siliceous oozes composed of sponge spicules are particularly well developed in the North Atlantic, Rockall Bank, off Ireland, and also the SE Atlantic and SW Indian oceans. Siliceous oozes may contain all kinds of terrigenous and/or biogenic calcareous impurities in varying proportions. In water below the carbonate compensation depth, calcareous components are dissolved, and siliceous oozes may become nearly pure, with an opaline silica content >90% of the dry weight.

FLORENTIN MAURRASSE

References

Murray, J., and Renard, A. F., 1901. *Report on Deep-Sea Deposits Based on Specimens Collected During the Voyage of H.M.S. Challenger in the Years 1872-1876.* London: Government Printing Office, 525p.

Revelle, R. R.; Bramlette, M.; Arrhenius, G.; and Goldberg, E. D., 1955. Pelagic sediments of the Pacific, *Geol. Soc. Am. Spec. Pap.* **62,** 221–263.

Stadum, C. J., and Burkle, L. H., 1973. A silicoflagellate ooze from the east Falkland Plateau, *Micropaleontology,* **19,** 104–109.

Cross-references: *Chert and Flint; Diatomite; Pelagic Sedimentation, Pelagic Sediments; Silica in Sediments; Sponges in Sediments.*

SILICIFICATION

Silicification is the addition of secondary silica to a rock or sediment by direct precipitation in voids, or by replacement of some preexisting material, either organic or inorganic; commonly it is a combination of the two processes. The secondary silica occurs in the form of normal, low-temperature quartz, or chalcedony, or opal (see *Silica in Sediments*).

A simple example of silicification is precipitation of silica cement in sandstones (see *Cementation*), where the silica is added in optical continuity with the detrital quartz grains. Probably the best-known example of silicification is petrified wood; the organic material in the woody tissues is replaced molecule for molecule by silica in the form of opal or chalcedony, so that even the minutest details of the organic tissues are preserved.

By far the most common example of silicified rock is chert (see *Chert and Flint*), a dense rock composed of chalcedonic and/or microcrystalline quartz; young (Tertiary) cherts may also contain metastable opal. Chert is a common product of replacement in carbonate rocks. The replacement is volume for volume; silica is precipitated simultaneously with solution of carbonate minerals in the host sediments, without intervening development of open spaces.

The source of secondary silica in silicified rocks usually is not obvious; consequently, geologists differ in their interpretation of its origin. In light of present knowledge of geochemistry of silica, however, it is likely that most of it is derived from silica-bearing minerals and rock fragments that occur either within silicified formations themselves, or in associated formations (see *Silica in Sediments*).

Formation waters which are charged with silica are potential agents for silicification wherever the waters migrate. Whether or not silicification actually occurs depends on factors which influence the solubility of silica, including the pH of interstitial water (see *Cementation*), temperature (see *Silica in Sediments*), and pressure (see *Pressure Solution Phenomena*). The redistribution of silica is complicated because opal, chalcedony, and quartz have different solubilities.

THEODORE R. WALKER

References

Berner, R. A., 1971. *Principles of Chemical Sedimentology*. New York: McGraw-Hill, 240p.

Heald, M. T., 1956. Cementation of Simpson and St. Peter Sandstone in parts of Oklahoma, Arkansas, and Missouri, *J. Geol.*, **64**, 16-30.

Iler, R. K., 1955. *Colloid Chemistry of Silica and Silicates*. Ithaca, N. Y.: Cornell Univ. Press, 324p.

Krauskopf, K. B., 1959. The geochemistry of silica in sedimentary environments, *Soc. Econ. Paleont. Mineral. Spec. Publ.* **7**, 4-19.

Siever, R., 1957. The silica budget in the sedimentary cycle, *Am. Mineralogist*, **42**, 821-841.

Walker, T. R., 1960. Carbonate replacement of detrital crystalline silicate minerals as a source of authigenic silica in sedimentary rocks, *Geol. Soc. Am. Bull.*, **71**, 145-152.

Walker, T. R., 1962. Reversible nature of chert-carbonate replacement in sedimentary rocks, *Geol. Soc. Am. Bull.*, **73**, 237-242.

See also numerous references in cross-referenced entries.

Cross-references: *Authigenesis; Carbonate Sediments–Diagenesis; Cementation; Chert and Flint; Clastic Sediments–Lithification and Diagenesis; Interstitial Water in Sediments; Novaculite; Porcelanite; Pressure Solution and Related Phenomena; Silcrete; Silica in Sediments; Siliceous Ooze; Spongolite; Volcanic Ash Deposits.*

SILT

Silt is defined as an unconsolidated clastic sediment, most of the particles falling within the range of 1/16 mm (62.5μ) to 1/256 mm (3.9μ) in diameter. *Siltstone* is the consolidated clastic sediment composed essentially of silt-sized particles. *Loam* is a mixture of approximately equal parts of sand, silt, and clay; the term *loam* is used almost exclusively in soil survey and is defined as soil containing 7-27% clay, 28-50% silt, and < 52% sand.

The precise size of particles making up silt varies slightly according to usage of engineers, geologists, and pedologists; some scales increase geometrically, others are expressed decimally and cyclically (see *Grain Size Studies*). Most geologists accept the Wentworth Scale limits noted above. Intermediate geometric subdivisions, in increasing size of 1/128, 1/64, and 1/32 mm, are qualitatively described as very fine silt, fine silt, medium silt, and coarse silt, respectively. Silt on the phi scale (q.v.) ranges from 4-8. As used by the US Department of Agriculture, the diameter range of silt is 0.05-0.002 mm.

Most silts are composed of clastic quartz together with a small proportion (5-10%) of feldspar and mica. As the mean grain size decreases, the proportion of quartz decreases, with a concomitant increase in the clay content. In mature silts, rock fragments are rare, but in immature silts, rock dust produced by crushing and abrasion may be present in considerable quantity. Silt-sized material of volcanic origin may occur, but the fine grain size inevitably leads to rapid diagenetic alteration of such materials. Silts composed entirely of limestone particles are also known.

The chemical composition of silts generally shows a high Si:Al ratio, indicating a high quartz content. With decreasing grain size, the Al, K, and Fe content rises. The color of silts depends on the content and state of oxidation of the iron and on the free-carbon content. The mineralogical and chemical composition in siltstones may be modified by the presence of cementing materials such as iron oxides and carbonates.

Silts may result from the prolonged weathering of fine-grained parent rocks in which only the more stable minerals survive, or they may result from the crushing and abrasion of coarse particles during transportation. In either case the particles will tend to be angular. The smallness of silt particles enables them to be carried in currents of low competence, and hence they are likely to be deposited in relatively quiet waters. Silts may show sedimentary features such as lamination and cross-bedding, although more commonly they are structureless.

A silt of wide distribution is geologically designated *loess* (q.v.). Generally unconsolidated, loess is characterized by a lack of stratification and ability to stand in vertical slopes. Most loess deposits are well sorted and occur almost exclusively within the particle size range of silt. These silts are primarily eolian in origin (see *Loess*).

True silts and siltstones are comparatively rare; the coarse layers of glacial varves (q.v.), however, appear to be true silts. Most rocks classified as silts or siltstones, are, according to the definitions given here, either fine sands or sandstones, clays, or shales.

ROY J. SHLEMON

References

Lane, E. W. and 8 others, 1947. Report of the subcommittee on sediment terminology, *Am. Geophys. Un. Trans.,* **28,** 936-938.

Pettijohn, F. J., 1975. *Sedimentary Rocks,* 3rd ed. New York: Harper & Row, 628p.

Shepard, F. P., 1954. Nomenclature based on sand-silt-clay ratios, *J. Sed. Petrology,* **24,** 151-158.

Soil Surv. Staff, 1951. Soil survey manual, *U.S. Dept. Agriculture Handbook* **18,** 503p.

Cross-references: *Argillaceous Rocks; Clastic Sediments and Rocks; Grain-Size Studies; Loess; Mudstone and Claystone; Shale.*

SIZE ANALYSIS—See GRAIN-SIZE STUDIES

SKEWNESS

Skewness is a term applied to the symmetry (or asymmetry) of the particle distribution of a given sediment. It may be measured by various methods in both the phi (ϕ) and millimeter scales (see *Grain-Size Studies*). The Trask skewness coefficient (1932) is expressed as

$$Sk = \frac{Q_1 Q_3}{Md^2}$$

and the phi quartile skewness (Krumbein, 1936) as

$$Sk_{q\phi} = \tfrac{1}{2}(Q_{1\phi} + Q_{3\phi}) - Md_\phi$$

where $Q_{1\phi}$ is the 25th percentile, $Q_{3\phi}$ is the 75th percentile, and Md_ϕ is the median, all in phi units. A positive (+) value indicates that the sediment has an excess of fines (tail to the right), and a negative (−) value indicates an excess of coarse sediment (tail to the left).

Graphic skewness, suggested by Inman (1952), is expressed as

$$Sk_G = \frac{\phi_{16} + \phi_{84} - 2\phi_{50}}{\phi_{84} - \phi_{16}}$$

or may be written as $\alpha_\phi = (M_\phi - Md_\phi)/\sigma_\phi$, where $M_\phi = \tfrac{1}{2}(\phi_{16} + \phi_{84})$, $Md_\phi = \phi_{50}$, and $\sigma_\phi = \tfrac{1}{2}(\phi_{84} - \phi_{16})$.

Inclusive graphic skewness (Folk and Ward, 1957) covers 90% of the curve, whereas the previous approaches cover only 68%. It is given as

$$Sk_I = \frac{\phi_{16} + \phi_{84} - 2\phi_{50}}{2(\phi_{84} - \phi_{16})} + \frac{\phi_5 + \phi_{95} - 2\phi_{50}}{2(\phi_{95} - \phi_5)}$$

and was developed by averaging Inman's (1952) measures of skewness α_ϕ and $\alpha_{2\phi}$.

Symmetrical curves have an Sk_I of 0.00; those with an excess of fines have a positive (tail to the right) skewness, and those with an excess of coarse material have a negative skewness. Folk (1965) suggests that Sk_I values from +1.00 to +0.30 are strongly fine-skewed; +0.30 to +0.10, fine-skewed; +0.10 to −0.10, near symmetrical; −0.10 to −0.30, coarse-skewed; and −0.30 to −1.00, strongly coarse-skewed.

B. CHARLOTTE SCHREIBER

References

Folk, R. L., 1962. Of skewnesses and sands, *J. Sed. Petrology,* **32,** 145-146.

Folk, R. L., 1965. *Petrology of Sedimentary Rocks.* Austin: Hemphill's 159p.

Folk, R. L., and Ward, W. C., 1957. Brazos River bar: A study in the significance of grain size parameters, *J. Sed. Petrology,* **27,** 3-26.

Inman, D. L., 1952. Measures for describing the size distribution of sediments, *J. Sed. Petrology,* **22,** 125-145.

Krumbein, W. C., 1936. The use of quartile measures in describing and comparing sediments, *Am. J. Sci.,* **32,** 98-111.

Krumbein, W. C., and Pettijohn, F. J., 1938. *Manual of Sedimentary Petrography.* New York: Appleton-Century, 549p.

Trask, P. D., 1932. *Origin and Environment of Source Sediments of Petroleum.* Houston: Gulf Publishing Co., 323p.

Warren, G., 1974. Simplified form of the Folk-Ward skewness parameter, *J. Sed. Petrology,* **44,** 259.

Cross-references: *Grain-Size Studies; Kurtosis; Phi Scale; Sorting.*

SLUMP BEDDING

Slump bedding is a genetic term that indicates a penecontemporaneous deformation or disturbance of bedding due to subaqueous horizontal movement of sediment, without loss of cohesion. Many investigators, however, apply the term more loosely in a descriptive sense, only indicating that the deformation is mechanical and not biological or chemical. As a result, many more or less synonymous expressions are used in the literature, such as slip bedding, slurry bedding, curly bedding, glide bedding, hassock structure, distorted bedding, and deformed bedding. The latter two are very general and often cover anything that involves secondary displacement.

Slump bedding is produced by subaqueous or

subaerial mass slippage on a slope of semiconsolidated or unconsolidated material under gravity action. Many authors recognize two terms for the motion, namely slumping and sliding. Sliding is considered to involve a large lateral displacement, in contrast to slumping, which is a local phenomenon (Dott, 1963). Once the instability of a slope is reached, material will move downslope. Only a restricted number of sliding or motion planes are involved; and often original parts of the sedimentary deposit can be distinguished, even when plastic motion is involved. The restricted number of motion planes, indicating that not each particle is involved individually, makes the movement different from grain flow, debris flow, and other gravity flows (q.v.; see also *Gravity-Slide Deposits*).

The factors that determine the degree of instability necessary to result in downslope motion can be numerous and are often a combination of effects. Blatt et al. (1972) list several causes, e.g., denser layers overlapping less-dense layers; chemically induced plasticity of clays underlying limestone; rapid deposition preventing sufficient dewatering; earthquakes; shear stress exerted on recently deposited sediments by a flow moving over it; loss of support due to erosion or motion of deeper material, e.g. submarine canyon side deposits moving toward and onto the canyon axis; and ground-water rise or excessive rain.

The large variety of reasons for motion results in an excessive number of sizes, shapes, and intensities of structures (see *Intraformational Disturbances; Penecontemporaneous Deformation of Sediments*). The most common slump bedding is asymmetrical and overturned in the direction of lateral movement. Other processes may become involved, making the result even more complex. The slump bedding may be restricted to one layer or even to part of it, to the material between two successive layers, to a few layers, or to a considerable portion of a sediment package. The motion may have been parallel to the bedding planes, leaving little or no evidence of motion. Slump bedding is bounded by a plane of décollement below and a depositional contact above. Distinction between slump folds and small tectonic structures is often difficult (Kuenen, 1953); the only definitive criterion is the truncation of folds by an overlying depositional contact (Jones, 1937). Folding and faulting or a complete conglomeratic texture may be observed locally.

Slump bedding is a very common phenomenon, and examples and/or descriptions can be found in many books and articles (see references).

ARNOLD H. BOUMA

References

Blatt, H.; Middleton, G. V.; and Murray, R. C., 1972. *Origin of Sedimentary Rocks*. Englewood Cliffs, N.J.: Prentice-Hall, 634p.

Dott, R. H., Jr., 1963. Dynamics of subaqueous gravity depositional processes, *Bull. Am. Assoc. Petrol. Geologists*, **47**, 104-128.

Dzulynski, S., and Walton, E. K., 1965. *Sedimentary Features of Flysch and Greywackes*. Amsterdam: Elsevier, 274p.

Helwig, J., 1970. Slump folds and early structures, northeastern Newfoundland Appalachians, *J. Geol.*, **78**, 172-187.

Jones, O. T., 1937. On the sliding or slumping of submarine sediments in Denbighshire, North Wales, during the Ludlow Period, *Quart. J. Geol. Soc. London*, **93**, 241-283.

Kennedy, W. J., and Juignet, P., 1974. Carbonate banks and slump beds in the Upper Cretaceous (Upper Turonian-Santonian) of Haute Normandie, France, *Sedimentology*, **21**, 1-42.

Kuenen, P. H., 1953. Graded bedding, with observations on Lower Paleozoic rocks of Britain, *Verh. Koninkl. Ned. Akad. Wet.*, sec. 1, **22**, 1-47.

Lajoie, J., 1972. Slump fold axis orientations: An indication of paleoslope? *J. Sed. Petrology*, **42**, 584-586.

Laury, R., 1971. Stream bank failure and rotational slumping: Preservation and significance in the geologic record, *Geol. Soc. Am. Bull.*, **82**, 1251-1266.

Lewis, K. B., 1971. Slumping on a continental slope inclined $1°-4°$, *Sedimentology*, **16**, 97-110.

McKee, E. D. 1974. Small-scale structures in Coconino Sandstone of northern Arizona, *J. Geol.*, **53**, 313-325.

Sturm, E., 1971. Subaqueous slump structures, *Geol. Soc. Am. Bull.*, **82**, 481-484.

Walker, J. R., and Massingill, J. V., 1970. Slump features on the Mississippi Fan, northeastern Gulf of Mexico, *Geol. Soc. Am. Bull.*, **81**, 3101-3108.

Cross-references: *Gravity Flows; Gravity-Slide Deposits; Intraformational Disturbances; Mass-Wasting Deposits; Penecontemporaneous Deformation of Sediments; Sedimentary Structures; Water-Escape Structures.*

SOILS—*See* Vols. XI and XII

SOLUTION BRECCIAS

Solution breccias (collapse breccias, founder breccias) are sedimentary rocks formed in the subsurface by the gravitational fragmentation and downward movement of sedimentary strata into voids created by the solution of underlying strata. Most breccias result from the solution of the evaporites gypsum, anhydrite, and halite (evaporite-solution breccias); the remainder, from the solution of limestone (limestone- or karst-solution breccias).

SOLUTION BRECCIAS

Process of Formation

Limestone-solution breccia forms within caverns by the collapse of wall- and roof-rock onto the cavern floor. The geometry of the cavern, and thus of the breccia, is controlled by jointing, which determines the pattern of solution, and by the strength, bed thickness, and jointing in the rock, which determines how the rock will collapse into the cavern (Renault, 1967, 1968). The formation of evaporite-solution breccia is generally different in that flowage of the evaporite and subsidence of the brecciating interbedded and overlying strata take place concurrently with solution, so that caverns are probably seldom formed. This conclusion is confirmed by flowage structures within evaporites that have been only partially dissolved (Stanton, 1966) and by broken and undulating but continuous bedding present in many breccias. Hundt (1950) and Montoriol Pous (1954) discuss the process in detail.

Solution brecciation is limited to the area in which soluble rocks—evaporite or limestone—occur, and, within that, to the area in which there is an adequate flow of subsurface water to dissolve the rock. Limestone-solution breccias are thus forming in modern karst regions and are common in the geologic column beneath ancient erosion surfaces. An example is the widespread limestone-solution breccia that occurs in Montana and Wyoming beneath a Mississippian-Pennsylvanian erosional surface (Roberts, 1966).

Evaporite-solution breccias are forming at present in areas where evaporite strata crop out and where adjacent strata are sufficiently permeable for surface water to flow into the subsurface and dissolve the evaporite. As a result, the breccias exposed at and near the surface pass into evaporite strata in the subsurface. This relation has been well documented for breccias derived from Pennsylvanian anhydrite flanking the Black Hills (Bowles and Braddock, 1963), Mississippian anhydrite flanking the mountain ranges of central and western Montana (Roberts, 1966), and Silurian salt in the Michigan basin (Landes, 1945). These modern conditions must have been satisfied frequently in the past, and evaporite-solution breccias below ancient erosion surfaces have been reported in Mississippian strata of Nova Scotia (Clifton, 1967), in Ordovician strata of Tennessee (Hill and Wedow, 1971), in Mississippian strata of Montana and Wyoming (Sando, 1974), and in Lower Mesozoic strata in Nevada (Speed, 1975).

Characteristics

An evaporite solution breccia is generally a laterally extensive, tabular unit at the position of the preexisting evaporite. The basal contact is smooth and coincides with the basal surface of the evaporite. The upper contact is commonly gradational upward through a sequence of breccia, then breccia in which vestiges of original stratification are present, highly fractured and then less fractured rock, and finally to intact strata which only reflect the presence of the underlying breccia by undulations in the bedding. The contact is irregular, and breccia-filled pipes may extend upward above the breccia for >100 m. A limestone-solution breccia, in contrast, is more localized and may be highly irregular, reflecting the geometry of the cavern in which it formed.

The thickness of an evaporite-solution breccia and of the overlying fractured strata can be related qualitatively to such factors as evaporite thickness, lithology of the strata overlying the evaporite, depth of burial of the evaporite being dissolved, and the amount of strata interbedded with the evaporite. Because of the concurrent effects of these several factors, it is not possible to estimate with any accuracy the original thickness of the evaporite from the thickness of a solution breccia, or vice versa. Brecciated ore formed by block caving typically has a porosity of about 30% (Bucky, 1956); as a first approximation, then, removal by solution of an evaporite bed 30 m thick should result in a breccia 100 m thick. Breccia thickness, however, is generally not much more than evaporite thickness because subsidence of the overlying rock extends all the way to the surface.

Breccia clasts are typically very angular because of lack of weathering and the limited, predominantly downward, transportation. Sorting is typically poor for the same reasons and because (1) large amounts of fine insoluble residue may be derived from the dissolved rock and (2) the brecciation process generates a large amount of fine material (Middleton, 1961). Bedding, sedimentary structures, and better sorting observed in some breccias indicate that lateral transportation by water flowing within the solution cavern does rarely occur.

The criteria by which solution breccias can be distinguished from breccias of other origins have been tabulated by Blount and Moore (1969).

Economic Importance

Solution brecciation is of economic importance because it results in both porosity and permeability and in deformation of overlying strata. Pore volume created is potentially equal to the volume of evaporite or limestone dissolved but is generally much less because (1) subsidence may extend to the surface, (2) subsequent cementation may destroy much of the porosity, and (3) pore space may be filled by fine sediment, which is either insoluble residue from the dissolved rock or is introduced from the surface. Nevertheless, solution breccia and the overlying fractured strata may provide a reservoir for petroleum (Stanton, 1966; Beales and Oldershaw, 1969). Deformation of overlying strata due to differential solution and subsidence may create structural traps for petroleum (Anderson and Hunt, 1964). The permeability created by brecciation may channel the flow of hydrothermal fluids and thus control the location of ore bodies. As an example, the lead-zinc deposits of eastern Tennessee are within solution breccias in Ordovician strata below a buried Silurian erosion surface (Harris, 1969).

ROBERT J. STANTON, JR.

References

Anderson, S. B., and Hunt, J. B., 1964. Devonian salt solution in north central North Dakota, *3rd Internat. Williston Basin Symp. Proc.*, 93-104.

Beales, F. W., and Oldershaw, A. E., 1969. Evaporite-solution brecciation and Devonian carbonate reservoir porosity in western Canada, *Bull. Am. Assoc. Petrol. Geologists*, **53**, 503-512.

Blount, D. N., and Moore, C. H., Jr., 1969. Depositional and non-depositional carbonate breccias. Chiantla Quadrangle, Guatemala, *Geol. Soc. Am. Bull.*, **80**, 429-442.

Bowles, C. G., and Braddock, W. A., 1963. Solution breccias of the Minnelusa Formation in the Black Hills, South Dakota and Wyoming, *U.S. Geol. Surv. Prof. Pap. 475-C*, C91-C95.

Bucky, P., 1956. Mining by block caving, *Colo. Sch. Min., Quarterly*, **51**, 129-143.

Clifton, H. E., 1967. Solution-collapse and cavity filling in the Windsor Group, Nova Scotia, Canada, *Geol. Soc. Am. Bull.*, **78**, 819-832.

Harris, L. D., 1969. Kingsport Formation and Mascot Dolomite (Lower Ordovician) of east Tennessee, *Tenn. Div. Geol. Rept. Invest.*, **23**, 1-39.

Hill. W. T., and Wedow, H., Jr., 1971. An early Middle Ordovician age for collapse breccias in the east Tennessee zinc districts as indicated by compaction and porosity features, *Econ. Geol.*, **66**, 725-734.

Hundt, R., 1950. *Erdfalltektonik*. Halle: Willhelm Knappe, 145p.

Landes, K. K., 1945. The Mackinac Breccia, *Mich. Geol. Surv. Publ. 44*, 121-154.

Middleton, G. V., 1961. Evaporite solution breccias from the Mississippian of southwest Montana, *J. Sed. Petrology*, **31**, 189-195.

Montoriol Pous, J., 1954. Resultado de nuevas observaciones sobre los procesos clasticos hipogeos, *Rassegna Speleologica Italiana*, **6**, 103-114.

Renault, P., 1967, 1968. Contribution à l'étude des actions mécaniques et sédimentologiques dans la spéléogenèse, *Annl. Speleol.*, pt. 1, **22**, 209-267; pt. 2, **33**, 260-307.

Roberts, A. E., 1966. Stratigraphy of Madison Group near Livingston, Montana, and discussion of karst and solution breccia features, *U.S. Geol. Surv. Prof. Pap. 526-B*, B1-B23.

Sando, W. J., 1974. Ancient solution phenomena in the Madison Limestone (Mississippian) of north-central Wyoming, *J. Research U.S. Geol. Surv.*, **2**, 133-141.

Speed, R. C., 1975. Carbonate breccia (rauhwacke) nappes of the Carson Sink region, Nevada, *Geol. Soc. Am. Bull.*, **86**, 473-486.

Stanton, R. J., Jr., 1966. The solution brecciation process, *Geol. Soc. Am. Bull.*, **77**, 843-848.

Cross-references: *Breccias, Sedimentary; Limestones; Rauhwacke.*

SORTING

Sorting is a measure of the amount of uniformity of the grain sizes in a sample. It is defined by the dispersion of the grain sizes on either side of the average (see also *Grain-Size Studies; Kurtosis; Skewness*). A well-sorted sample has a small range of grain sizes. The various factors that control sorting of sediments include nature of the source material, degree of turbulence of the transporting agency, distance of transportation, amount of abrasion, and the time of transportation (Inman, 1949). These transportational and depositional processes are reflected in the patterns of size distributions, and the tails of the distributions are thought to be particularly sensitive to the environment of deposition (see *Grain-Size Frequency Studies*). Of the different measures of sorting that have been developed, therefore, those that take into account the tail values (e.g., the 5% and 95% values) are more popular. One such measure, developed by Folk and Ward (1957), is as follows:

$$\sigma_1 = \frac{\phi_{84} - \phi_{16}}{4} + \frac{\phi_{95} - \phi_5}{6.6}$$

where ϕ_{84}, e. g., represents the 84% size value expressed in the phi scale (q.v.). Another measure popular among sedimentologists is the one proposed by Inman (1952):

$$\sigma_1 = \frac{\phi_{84} - \phi_{16}}{2}$$

Folk (1966) has reviewed the other available measures of sorting and has also listed the sorting nomenclature proposed by him and by Friedman (1962).

The term *deviation index* is applied to any one of several possible related factors which indicate the statistical spread (deviation) of given sets of data around their respective mean points, as applied to size-frequency measurements of sediments. Included are Trask's *sorting coefficient; So* = $\sqrt{(Q_3/Q_1)}$, and its analog in ϕ scale, $QD\phi$, Krumbein's (1936) *phi quartile deviation;* $QD\phi = (Q_3\phi - Q_1\phi)/2$.

Sorting is best when the same material is repeatedly available for reworking, as in the surf zone, and poorest when the grains, once deposited, are no longer available for transportation, e.g., in glacial and turbidity-current deposits. A positive correlation between sorting and sediment sizes (expressed in phi scale) is a commonly observed phenomenon, i.e., the finer sediments are better sorted and vice versa. Inman (1949) attempted to explain this phenomenon in terms of the fluid dynamics of transportation.

SUPRIYA SENGUPTA

References

Folk, R. L., 1966. A review of grain-size parameters, *Sedimentology,* **6,** 73-93.

Folk, R. L., and Ward, W. C., 1957. Brazos river bar, a study in the significance of grain-size parameters, *J. Sed. Petrology,* **27,** 3-27.

Friedman, G. M., 1962. On sorting, sorting coefficients, and the lognormality of the grain-size distribution of sandstones, *J. Geol.,* **70,** 737-756.

Inman, D. L., 1949. Sorting of sediments in the light of fluid mechanics, *J. Sed. Petrology,* **19,** 51-70.

Inman, D. L., 1952. Measures for describing the size distribution of sediments, *J. Sed. Petrology,* **22,** 125-145.

Krumbein, W. C., 1936. The use of quartile measures in describing and comparing sediments, *Am. J. Sci.,* **32,** 98-111.

Krumbein, W. C., 1939. Graphic presentation and statistical analysis of sedimentary data, in P. D. Trask, ed., *Recent Marine Sediments.* Tulsa: Am. Assoc. Petrol. Geologists; London: Murby, 558-591.

Pomerol, C., 1967. Comparative study of deviation index, Hqa, Qd_ϕ and So, *Sedimentology,* **8,** 153-157.

Sengupta, S., 1975. Size-sorting during suspension transportation—lognormality and other characteristics, *Sedimentology,* **22,** 257-274.

Trask, P. D., 1932. *Origin and Environment of Source Sediments of Petroleum.* Houston, Texas: Am. Petrol. Inst. (Gulf Publ. Co.), 323p.

Cross-references: *Grain-Size Frequency Studies; Grain-Size Studies; Kurtosis; Skewness.*

SPARITE

The term *sparite* has been used by Folk (1959) to encompass the group of rocks he terms "sparry allochemical limestones" (see *Allochem; Limestones*). The matrix of such limestone is a sparry calcite, in which there are any one or more of four allochem components: intraclasts, oolites, fossils, and pellets. This group of sediments represent, for the most part, sediments of textural maturity, produced, e.g., along beaches, in areas of shallow neritic sedimentation, along raised submarine banks, in active tidal channels, and on barrier bars.

Folk also applies a second series of terms to cover the grain-size variation of the included allochems. A biosparite with fossils of a grain size >1 mm would be termed biosparrudite. The presence of mixed allochems is provided for by additional terms such as "fossiliferous oosparite," or a specialized composition may be called, e.g., a "gastropod biosparite." *Microspar* is a term Folk has applied to a recrystallized micrite (q.v.), which has a similar grain size to a sparite.

The admixture of terrigenous components, in quantities ranging from 10-50%, is described with a composite terminology, in the following manner: a fossiliferous pellet spar with 10-50% clay is a "clayey biopelsparite"; an intraclastic dolomitized spar with 10-50% sand is a "sandy dolomitized intrasparite."

B. CHARLOTTE SCHREIBER

References

Folk, R. L., 1959. Classification of limestones, *Bull. Am. Assoc. Petrol. Geologists,* **43,** 1-38.

Cross-references: *Allochem; Carbonate Sediments-Lithification; Limestone Fabrics; Limestones.*

SPELEAL SEDIMENTS

Caves form a unique microenvironment and can act as the site of deposition for a wide variety of sediments. Chemical deposition produces mainly carbonate minerals, and its mechanisms and chemistry have been extensively investigated and are well understood (Curl, 1962; Holland et al., 1964). The clastic sediments of caves are surprisingly complex and are not well understood. The stratigraphy and petrology of clastic sediments of a few European caves have been studied in some detail (Schmid, 1958; Kukla and Lozek, 1958). Much less work has been done in the US, and most of that is in the form of unpublished theses, of which Wolfe's (1973) work on the Central Appalachians is most extensive.

Constituents of Cave Sediments

Classification. There is no generally accepted classification of cave sediments. The classification in Table 1 is based on White (1963), with some but not all of the modifications proposed

TABLE 1. Classification of Speleal Sediments

Clastic Sediments	Chemical Sediments

Autochthonous Deposits
  Breakdown
  Weathering detritus
  Organic debris
Allochthonous Deposits
  Infiltrates
  Fluvial sediments
  Eolian Sediments
	Carbonates
	Evaporites
	Phosphates, Nitrates
	Oxides, Hydrates
	Ice

by Wolfe (1973). Ford (1975) has presented yet another classification. Cave sediments have not yet been arranged by "rock type." End-member constituents are identifiable, however, and are listed in Table 1. The chemical sediments usually appear in consolidated form, whereas the clastic sediments are typically unconsolidated.

Breakdown. A number of processes in caves cause weakening of the roof with resultant failure and dropping of blocks (White and White, 1968). These gross chunks of raw bedrock material dropped into the cavern opening from the roof and walls are termed "breakdown." Breakdown varies in size from chips and splinters a few mm across to beams measuring tens of m on an edge. Breakdown occurs at all stages of cavern development and thus is found at all levels in the sedimentary sequence. The final stages of cavern truncation and destruction involve processes that produce only breakdown, and it is common in many caves for an almost pure breakdown layer to make up the topmost bed of the sedimentary column (see *Solution Breccias*).

Weathering Detritus. All insoluble residues left behind by the dissolution of the limestone are classified as weathering detritus. Their exact nature depends entirely on the composition of the bedrock. The very coarse fraction is dominantly chert nodules that occur profusely in many limestones. In some stream passages where all finer material has been winnowed out, the stream bed may consist entirely of chert. The intermediate-size fraction is usually quartz sand, although dolomite sand occurs in some caves, e.g., in the Black Hills of South Dakota. The finer silt and clay-sized weathering detritus is extremely varied. Fine-grained quartz, sericite, and clay minerals are probably the most common. Analysis of red, unctuous clays from three Missouri caves showed that they were composed of quartz, illite, and kaolinite with traces of hematite; almost exactly the same composition was found for the insoluble fraction of the limestone.

Some clay minerals may be authigenic, having formed in the cavern environment either directly or by the alteration of other clay minerals. Large amounts of endellite that occurs in the New Mexico Room of Carlsbad Caverns are of authigenic origin (Davies and Moore, 1957).

Organic Debris. Caves form convenient dwelling places for bats and certain species of birds. In many caves, particularly in the tropics and in SW United States, the droppings from roosting bats and birds have built layers of guano many m thick. These stratified layers contribute a significant part of the sedimentary deposits in some caves. Decomposition of the guano releases nitre, ammonia, and urea, which react with the wall rock of the cave and produce a suite of calcium phosphate minerals that eventually make up a significant portion of the deposit.

Infiltrates. Insoluble materials can be transported into a cave depositional site both horizontally and vertically. Clastic materials are transported vertically under the direct influence of gravity, with or without the aid of flowing water. Soils in karst regions are washed, piped, or slumped into open crevices or sinkholes. They are eventually discharged into the cavern system without much chemical modification. Solution crevices, chimneys, and vertical shafts may reach through the cavernous bedrock into overlying strata, fragments of which fall down the shafts to become part of the infiltrate sediment.

Fluvial Sediments. Stream-borne deposits, which may be transported through filled pipes as well as in open channels, make up the greatest part of most cavern sediments and are derived from many sources. Some are reworked autochthonous cave deposits, some have been derived from rocks higher in the stratigraphic column, and some have been transported from nonkarstic borderlands by sinking streams. The material usually consists of stream-rounded rock fragments, sand, silt, and clay. The finer-grained material cannot usually be distinguished from weathering detritus or the infiltrates without extensive analyses. The composition and grain size merely reflect the character of the surrounding bedrock and the load-carrying capacity of the cavern streams. In the Appalachian caves, sandstone, cobble, and boulder fills sometimes occur in thicknesses of tens of m. In the Mammoth Cave region of Kentucky, sand and silt with quartz pebbles are predominant on the upper levels, while clay and silt predominate on the lower levels. In Missouri, thick clay fills are common, but the classic red, unctuous clays are exceptional (Reams, 1968).

Eolian Sediments. Loess has not been identified in the caves of the United States but occurs fairly frequently in the alpine caves of central Europe (Schmid, 1958). It occurs mainly near the cave entrances and was apparently blown into the cave during the Pleistocene glacial maxima. Reworked loess is a constituent of fluvial sediments in some caves of the American Midwest.

Carbonates. Under this heading are included all forms of freshwater carbonates deposited in the cavern environment, e.g., stalactites, stalagmites (dripstone), flowstone sheets, and a variety of minor forms. Calcite is the most common mineral, but aragonite also occurs frequently. Dolomite is extremely rare. The growth of dripstone deposits is discussed in some detail by Moore (1962). Dripstone deposits exhibit growth rings, grain size, purity, and grain morphology changing from ring to ring. Alternating bands of calcite and aragonite occur, for example, in some central Missouri caves. Grain size varies from that of an entire stalactite consisting of a single crystal down to submicroscopic. At the lower limit of grain size is a slippery, doughy mass known as "moonmilk" or "montmilch." This peculiar material often consists of calcite in cold or alpine environments and hydromagnesite in temperate environments. Magnesite, aragonite, huntite, and nesquehonite moonmilks also have been reported.

Holland et al. (1964) divided carbonate transport into four stages: equilibration of percolating ground water with a CO_2-rich atmosphere in the soil horizon; solution of calcite from the bedrock at the base of the soil; percolation of these supersaturated solutions downward into the cave; and, finally, the deposition of calcite by CO_2 loss into the cave atmosphere as the solutions readjust their equilibrium to the lower CO_2 pressure. Holland's mechanism explains why the most active travertine deposition takes place in sealed caves at high relative humidities and why travertine deposits are sparse in caves in the arctic and at high altitudes, become more abundant in temperate climates, and are very massive in caves in the tropics. The metastable deposition of aragonite from freshwater solution occurs because of kinetic factors which allow the necessary supersaturation to build up. Many of the proposed mechanisms are summarized by Curl (1962; see also *Travertine*).

Evaporites. The evaporite minerals that occur in caves as distinct sediments are mainly sulfates, of which gypsum is predominant, and rarely halides. Epsomite and mirabilite frequently occur in minor amounts. The sulfate minerals occur in caves as dripstone, as wall and floor crusts, as clusters of crystals (gypsum flowers or oulopholites), and as crystals dispersed in the clastic fills. In most caves they form only a minor part of the sedimentary sequence. At least three mechanisms of sulfate deposition have been distinguished. In many caves of the US, the sulfate minerals are derived from overlying evaporite beds and are transported into the cave by percolating water and deposited by evaporation. In central Kentucky, the sulfates appear to originate from the weathering of pyrite in the overlying clastic rocks and to be deposited by the reaction of sulfate-rich solutions with the carbonate wall rock. In central Tennessee and in many Appalachian caves, the sulfates seem to originate in the cave fills, and their primary origin is unknown.

Phosphates and Nitrates. These minerals usually arise by secondary reactions in the cave soils and, although they are important minor constituents, do not in general form distinct sedimentary masses by themselves. Phosphate deposits in some tropical caves are exceptions. Extensive phosphate deposits are found in the caves of Mona Island off the Coast of Puerto Rico (Kaye, 1959). In these caves the guano deposits are very old, and alteration to phosphorite is essentially complete. The phosphate minerals include hydroxyapatite, crandallite, brushite, and monetite (see *Phosphates in Sediments*).

Oxides and Hydrates. Brown and black crusts of hydrated iron and manganese oxides are common in many caves. The manganese crusts are usually extremely thin, often only fractions of a mm. They usually form in flowing, free-surface streams, and their occurrence in the sedimentary sequence is evidence for the former existence of free-surface streams. More rarely, thick deposits of these oxides occur; manganese oxides on the order of 1 m thick occur in Jewel Cave, South Dakota, and thick iron oxide deposits occur in some Iowa caves. Of the iron oxides, goethite appears to be the most common mineral. The manganese oxides are so fine-grained that identification by x-ray diffraction is difficult. Psilomelane, todorokite, and perhaps hollandite are known to occur.

Ice. Perennial ice occurs in a number of high alpine caves, particularly in Central Europe, and must be regarded as a sediment. In the Eisriesenwelt Cave in Austria, for example, the ice deposit is thick and dates at least from the Pleistocene.

Transport and Deposition of Clastic Sediments

Depositional Environments. Cave sedimentary sequences are composed of the materials previously described. Most clastics appear to be channel deposits, intermixed with weathering detritus and infiltrates. Deposition occurs along protected reaches of cave passages by fluvial processes (equivalent to the alluvium of surface streams). When passages cross under vertical openings (fissures, chimneys, vertical shafts), large contributions of gravitationally deposited infiltrates (equivalent to colluvium) occur. These are found as massive mounds in some caves, nearly blocking the passage. A third and important environment occurs where cave passages intersect the land surface. The entrance-area environment is characterized by talus fans of colluvial material from hill slopes, breakdown activated by the more active weathering near the surface, and other inwash materials. So different is the entrance talus from the other cave sediments that Kukla and Lozek (1958) divide all cave sediments into an entrance facies and an interior facies. The entrance talus migrates inward as the hill slopes retreat, removing old deposits and generating new ones with concurrent shortening of the cave. There need be no opening penetrable by humans in the entrance environment, and indeed most surface intersections are completely blocked by entrance facies sediments.

The character of cavern sedimentation also changes with the stage of the cave's development. Early phreatic stages are characterized by fine-grained clays and silt and weathering detritus from the limestone. Flood-water stages may result in extensive breakdown and coarser clastic materials. Late vadose stages are characterized by carbonates, evaporites, organic debris, and more breakdown.

Depositional Features. Fluvial sediments in caves exhibit a complex stratigraphy of interbedded sands, gravels, cobbles, and clays sometimes showing a very delicate layering. These have been examined in a number of caves in Kentucky (Davies and Chao, 1959), Missouri (Reams, 1968; Helwig, 1964), and West Virginia (Wolfe, 1973). Sand pockets, old channel fills, and many other features are observed, and attempts to laterally map the individual layers have not been successful. The details of the stratigraphy vary greatly over distances of a few m. Such rapid facies change is best interpreted as evidence for deposition by streams accompanied by much reworking by later processes. Wolfe (1973) shows that provenance type and marker materials are much more useful for cross-correlating cave sediments.

Most cave passages, as observed today, are floored with clastic sediments. The surface of the sediments exhibits most of the features of surface channel deposits. There is often a meander channel incised below a flat floor section analogous to a flood plain. Point bars occur, and there may be alternating reaches of pools and riffles (or their record impressed in now-dry passages). Active stream passages are frequently armored with a layer of cobble near the surface, with finer-grained material underneath.

Transport Mechanisms. Some cave systems are highly interconnected and are (or have been) linked with a large recharge area to permit water flow of sufficient velocity to move a sediment load laterally. The less interconnected, or low-velocity, systems must of necessity depend on gravitational deposition for most of their sediments. Infiltrates may indeed completely fill cave systems at shallow depth when the caves are overlain by a thick regolith.

Deposition by lateral transport requires water, either in pipe-full or open channel regimes, with sufficient velocity to move the observed clastic debris. Application of usual engineering criteria (White and White, 1968) requires flow velocities in the range of 0.1–1 m/sec to move the sediments as bed load. Transport by suspended load also requires velocities at least in the turbulent range. Flood flows appear to be extremely important in accounting for most of the fluvial sediments found in cave passages. It has been observed that a single flood can flush most of the material from an active cave passage or can refill a previously open passage. Random flood events also provide an explanation for the chaotic stratigraphy observed in most fill sequences.

Relation of Sedimentation to Cave Passage Development. If the cave system is or has been an integral part of an underground drainage system, it can be argued that the underground drainage must carry a sediment load and that all insoluble weathering material both from the dissolving limestone and from the eroding clastic rocks is deposited elsewhere in the catchment. In many karst drainage systems, all discharge is by way of underground routes, and there is no alternative way of removing the products of erosion (White and White, 1968). Wolfe (1973), however, finds evidence that many of the fluvial sediments in the caves of West Virginia postdate cavern development and appear to have been deposited by high discharge events, perhaps related to periglacial climates of the Pleistocene.

The intrusion of clastic sediments into a carbonate aquifer is of importance in modifying the flow behavior. Flow paths will remain open only along those routes through which water velocities are high enough for sediment transport. Deeper fractures and openings will be silted up; and among the lateral passages, there will be a balance between siltation and erosion in equilibrium with the available discharge.

Age of Cave Sediments

Methods of dating cavern sediments include stratigraphic correlations, vertebrate fossils, pollen analysis, isotope dating, and paleomagnetics. From the cave-sediment dating done thus far, one interesting conclusion stands out. Holocene and Wisconsin (and Würm) age sediments are very common; relatively few sediments have been definitely proved to be of Sangamon age; and Illinoian (Riss) or pre-Illinoian sediments seem to be extremely rare.

Stratigraphy. The stratigraphic section of cave fills records not only changes in the local environment but changes in surface climate. Layers of travertine mark times of higher rainfall on the surface. Rock temperatures during the glacial maxima fell below the freezing point of water in many regions of Europe, and frost-pry of cave walls and ceilings has resulted in a peculiarly angular breakdown that can be easily recognized and forms marked beds by which the Würm maximum may be recognized (Schmid, 1958).

Vertebrate Fossils. The larger animals, particularly the cave bear, have been studied for a long time in Europe, and cave sediments have dated back to the Würm-Riss interglacial. Important recent work in Pennsylvania by Guilday et al. (1964), utilizing smaller animals which are much more sensitive to climatic change, has provided a fairly detailed picture of the climate back to about the close of the Illinoian glaciation.

Pollen Analysis. Pollen are best preserved in the sediments of the entrance talus where they are directly deposited by wind transport. Analyses from several European caves gives results consistent with other methods. Pollen are sparse in the deeper cave sediments, and attempts to identify them have not met with much success.

Carbon-14 Dating. C^{14} dating of travertine deposits has had some success in the American West (Broecker and Orr, 1958), where the method was used to date wave-cut caves above Lake Lahontan and Lake Bonneville. Chemical reactions take place mainly in the soil horizon, however, so that there is always danger of isotope exchange with resulting ages much younger than they should be. A few dates have been obtained on guano deposits. Many cave travertines are too old for radiocarbon dating.

Uranium/Thorium Dating. A newer method for the dating of cave travertines makes use of the tiny (1-10 ppm) concentration of uranium found in natural calcites (Thompson et al., 1974). Th^{230}/U^{234} isotope ratios can be used to calculate age of deposition, while O^{18} enrichment gives a measure of paleotemperature. Dates have been obtained to nearly 300,000 B.P. and indicate that much travertine deposition takes place (at least in northern temperate-climate caves) during the warmer and wetter interglacial periods.

Paleomagnetic Dating. Cave sediments of both entrance and interior facies have been found to preserve good magnetic remanence, making them amenable to paleomagnetic dating and to interpretation by several magnetic parameters (Kopper, 1975). In practice, it has been found that secular changes in the magnetic field recorded in cave sediments can often be compared over the entire Mediterranean region to provide dating of archaeological strata for the past 30,000 yr and accurate to within 150-1400 yr. Concurrently, the imprint of magnetic field behavior recovered from cave sediments promises to resolve disputed reversal events during at least the latter part of the Brunhes epoch (c. 690,000 B.P. to present).

WILLIAM B. WHITE

References

Broecker, W. S., and Orr, P. C., 1958. Radiocarbon chronology of Lake Lahontan and Lake Bonneville, *Geol. Soc. Am. Bull.,* 69, 1009-1032.

Curl, R. L., 1962. The aragonite-calcite problem, *Bull. Natl. Speleol. Soc.,* 24, 57-73.

Davies, W. E., and Chao, E. C. T., 1959. Report on sediments in Mammoth Cave, Kentucky, *Admin. Rept. U.S. Geol. Surv. to Natl. Park Serv.,* 117p.

Davies, W. E., and Moore, G. W., 1957. Endellite and hydromagnesite from Carlsbad Caverns, *Bull. Natl. Speleol. Soc.,* 19, 24-27.

Ford, T. D., 1975. Sediments in caves, *Trans. Brit. Cave Res. Assoc.,* 2, 41-46.

Guilday, J. E.; Martin, P. E.; and McCrady, A. D., 1964. New Paris No. 4: A late Pleistocene cave deposit in Bedford County, Pennsylvania, *Bull. Natl. Speleol. Soc.,* 26, 121-194.

Helwig, J., 1964. Stratigraphy of detrital fills of Carroll Cave, Camden Country, Missouri, *Missouri Speleol.,* 6, 1-15.

Holland, H. D.; Kirsipu, T. V.; Huebner, J. S.; and Oxburgh, U. M., 1964. On some aspects of the chemical evolution of cave waters, *J. Geol.,* 72, 36-67.

Jones, W. K., 1971. Characteristics of the underground floodplain, *Bull. Natl. Speleol. Soc.,* 33, 105-114.

Kaye, C. A., 1959. Geology of Isla Mona, Puerto Rico, and notes on the age of Mona Passage, *U.S. Geol. Surv. Prof. Pap. 317-C,* 141-178.

Kopper, J. S., 1975. Dating and interpretation of archaeological cave deposits by the paleomagnetic method, Ph.D. Thesis, Columbia University, New York, N.Y., 175p.

Kukla, J., and Lozek, V., 1958. Problems of investigations of cave deposits, *Československý Kras,* 11, 19-83.

Miskovsky, J.-C., 1966. Les principaux types de dépots des grottes et les problèmes que pose leur étude, *Rev. Géomorph. Dyn.,* 16(1), 1-11.

Moore, G. W., 1962. The growth of stalactites, *Bull. Natl. Speleol. Soc.,* 24, 95-106.

Reams, M. W., 1968. Cave sediments and the geomorphic history of the Ozarks, Ph.D. Thesis, Washington Univ., St. Louis, Missouri, 167p.

Schmid, E., 1958. Höhlenforschung und Sedimentanalyse, *Schriften des Institutes für Ur- und Frühgeschichte der Schweiz* (Basel), 13, 186p.

Thompson, P.; Schwarcz, H. P.; and Ford, D. C., 1974. Continental Pleistocene climatic variations from speleothem age and isotopic data, *Science,* **184,** 893-895.

White, W. B., 1963. Sedimentation in caves: A review, *Natl. Speleol. Soc. News,* **21,** 152-153.

White, E. L., and White, W. B., 1968. Dynamics of sediment transport in limestone caves, *Bull. Natl. Speleol. Soc.,* **30,** 115-129.

Wolfe, T. E., 1973. Sedimentation in karst drainage basins along the Allegheny Escarpment in southeastern West Virginia, USA, Ph.D. Thesis, McMaster Univ., Hamilton, Ontario, 455p.

Cross-references: *Cave Pearls; Limestones; Solution Breccias; Spring Deposits; Travertine.* Vol. III: *Speleology.* Vol. IVB: *Cave Minerals.*

SPHERICITY—See ROUNDNESS AND SPHERICITY

SPONGES IN SEDIMENTS

Sponges with spicular skeletons can contribute significantly to sediments. The roles played by them fall into four categories.

Dissociated Spicules

Upon decay of sponges with nonrigid skeletons, the spicules fall apart unless densely interwoven; if dispersed by current action, they may be included in sediments as dissociated units. Many forms with rigid frameworks have additional loose spicules which are subject to similar dispersal.

Rocks with significant dissociated spicules are mainly limestones or sandstones, and silicification or occurrence of chert nodules are often associated features. The spicules are almost always of siliceous sponges and are usually megascleres, though microscleres also occur in a Tertiary example (New Zealand). Spicular content ranges from subordinate to predominant, and some examples are formed almost wholly of spicules (see *Spongolite*). Demosponge spicules are generally commonest, but Hexactinelids or Heteractinids may occur. Spicules of different types of sponges are often mixed together, but sorting may lead to prevalence of one type (usually monaxon) and size grade (see *Biostratinomy*). Examples are parts of the Lower Carboniferous limestone of Britain and Ireland; Permian limestones of the Delaware Basin, New Mexico and west Texas; Oxfordian gaize of N France; and some Aptian-Albian greensands of N France and S England.

Spicules may also be foci for emplacement of secondary minerals, which may fill axial canals or cavities left by solution of spicules, or replace spicular calcite or silica directly.

Skeletal Frameworks

Rigid structures, which do not collapse after death, contribute to sediments in other ways.

Calcarea. Thalamida ("sycon sponges") and pharetronids are important bioherm framebuilders in many Permian reefs of the Delaware Basin, N.M., and intergrown specimens also form traps for loose sediment (Newell et al., 1953). Both are sometimes minor associates of Mesozoic reef-corals. The Faringdon Sponge-Gravel (Aptian-Albian) of Berkshire, S England, is a frameless biostromal deposit formed largely from drifted examples.

Demospongia and Hexactinellida. Lithistids and dictyonines, mostly calcified, crowd upper Jurassic sponge-limestone of the Franco-Swabian region, Europe (Fig. 1). Some of the limestones are bedded, others are massive stromatolites (Massenkalk) forming banks and mound-like features known as "sponge-reefs" (e.g., Gwinner, 1958; 1971). Piled skeletal frameworks, mainly lithistid, in siliceous preservation form the Senonian sponge bank of St. Cyr, S France. Gwinner (1971) illustrates how these unique algal-sponge reefs, forming stillwater bioherms and biostromes, interfinger with normal but more compacted shelf facies. The latter are mainly marls (micritic with bioclastic particles), and the reefs are more carbonate-rich, consisting mainly of algal crusts growing between and over the sponges, associated also with brachiopods, mollusks, serpulids, and benthonic foraminifera. Some horizons are almost entirely stromatolites. Presence of amino acids confirms the organic nature of the crusts (Hiller, 1968; see also *Algal-Reef Sedimentology*).

Sponge Phosphorites. Dictyonines predomi-

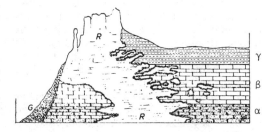

FIGURE 1. A classical sketch of one of the celebrated algal-sponge reefs in the Swabian Alb of SW Germany; Upper Jurassic age (from an old sketch by E. Fraas). The immediate off-reef sediments should appear *draped,* with steep dips. α = shelf marls; β = well-bedded Oxfordian limestones; γ = laminated Kimmeridgian limestones; R = massive reef limestone; G = gravel and solifluction debris (Quaternary).

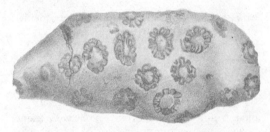

FIGURE 2. Hexactinellid sponge *Polyblastidium racemosum* (T. Smith) in a flint nodule; Upper Cretaceous (Senonian), Upper Chalk, Kent, England.

nate in beds of phosphatic sponge-nodules seen in greensands basal to the Chalk in SE England (Glauconitic Marl, L. Cenomanian), N Ireland (Hibernian Greensand "Spongarian Zone," L. Senonian). Phosphatization may affect only matrix, filling skeletal meshes and canals, or additional adherent or enclosed external matrix in reworked examples. Spicular silica is usually lost, and calcite is the commonest replacement.

Secondary Silica

Siliceous sponge remains are widely regarded as a source of silica responsible for diagenetic silicification, including chert formation (e.g., Hinde, 1885; Cayeux, 1929; Newell et al., 1953), although this is disputed in the case of chalk flints (e.g., Nestler, 1961; see *Silica in Sediments*). Skeletal frameworks have also sometimes been foci for silicification, so that chert now fills skeletal meshes and canals or encloses the sponge within a nodule (Fig. 2).

Boring Sponges

Monaxonid demosponges of the family Clionidae contribute indirectly to sediments by boring in limestone and calcareous skeletal structures of macroscopic size, e.g., mollusk shells, and coral (Fütterer, 1974). Excavated lime is converted to fine sediment by removal in small pellets (e.g., 40μ diameter), which the sponge expels through its osculum. Objects attacked may also be weakened to the point of collapse to form coarse debris; study of a modern Jamaican coral-reef (Goreau and Hartman, 1963) has shown collapse of clionid-bored coral as a major source of talus in the deeper fore-reef region (>30 m), beyond the range of algal cementation.

ROBIN REID

References

Cayeux, L., 1929. *Les Roches Sédimentaires de France. Roches Siliceuses*. Paris: Impr. Nationale, 774p.

Finks, R. M., 1960. Late Paleozoic sponge faunas of the Texas region—the siliceous sponges, *Am. Mus. Natl. Hist. Bull.*, 120, 1-160.

Fütterer, D. K., 1974. Significance of the boring sponge Cliona for the origin of fine grained material of carbonate sediments, *J. Sed. Petrology*, 44, 79-84.

Goreau, T. F., and Hartman, W. D., 1963. Boring sponges as controlling factors in the formation and maintenance of coral reefs, *Am. Assoc. Adv. Sci. Publ. 75*, 22-54.

Gwinner, M. P., 1958. Schwammbänke, Riffe and submarines Relief im oberen weissen Jura der Schwäbischen Alb (Württemberg), *Paläont. Rundschau*, 47(1), 402-418.

Gwinner, M. P., 1971. Carbonate rocks of the Upper Jurassic in SW-Germany, in G. Müller, ed., *Sedimentology of Parts of Central Europe*. Frankfurt: Kramer, 193-207.

Hiller, K., 1968. Proof and significance of amino-acids in Upper Jurassic algal-sponge-reefs of the Swabian Alb (SW-Germany), in G. Müller and G. M. Friedman, eds., *Recent Developments in Carbonate Sedimentology*. New York: Springer-Verlag, 136-137.

Hinde, G. J., 1885. On beds of sponge-remains in the Lower and Upper Greensands of the South of England, *Roy. Soc. London, Phil. Trans.*, 1885, 403-453.

Nestler, H., 1961. Spongien aus der weissem Schreibkreide der Insel Rügen, *Paläont. Abhandl.* 1(1), 1-70.

Newell, N. D., et al., 1953. *The Permian Reef Complex of the Guadalupe Mountain Region, Texas and New Mexico—A Study of Paleoecology*. San Francisco: Freeman, 236p.

Cross-references: *Algal-Reef Sedimentology; Chalk; Chert and Flint; Limestones; Silica in Sediments; Siliceous Ooze; Spongolite*. Vol. VII: *Porifera*.

SPONGOLITE

A spongolite (French: spongolithe; German: spongilit) is a sedimentary deposit composed almost exclusively of sponge spicules. The individual spicules are dissociated, and usually no complete specimens of the original sponges are present. The interstices between the spicules are filled with a matrix either of the shale, marl, or glauconitic sediment typical of the sequence in which they occur, but which has usually been subjected to diagenetic alteration, or of secondarily deposited silica. The remains of other types of fossils are virtually absent.

The term *spongolithe* was introduced by Cayeux (1897) to distinguish rocks "*qui sont pour ainsi dire exclusivement formées de spicules des Spongiaires*" from those containing a smaller proportion of the remains of various siliceous organisms (= "gaizes"). The type material was taken from the Belgian Meule de Bracquegnies of Upper Cretaceous age. In his major work, Cayeux (1929) described spongolite specimens from various geological systems. In all cases, the rock consisted of a stratified felt of sponge spicules and a matrix which was clearly developed though always subordinate to the spicules. The term as defined by Cayeux carried no implications as to the conditions of deposition or subsequent diagenesis of the rock, nor did it limit the composition of the groundmass or of the spicules themselves. In most cases, however, spongolites are either primarily or secondarily siliceous.

Synonyms

Other terms have been improperly used to describe highly spicular deposits. The term *gaize* may still be applied to spicular rocks, but following Cayeux's distinction, this term should denote only deposits containing abundant remains of other organisms, although these may pass laterally into true spongolites.

Some carboniferous spongolites from Belgium described by Cayeux (1929) had been previously described as "phthanites" ("phtanite," Bellière, 1922). This word was originally created by Haüy (1822) to denote bedded, siliceous shaly rocks "*quartz compacte, argileux, kieselschiefer,*" and the term as such is still widely accepted by European geologists. Other authors, however, have used phthanite to denote siliceous nodules of secondary origin such as are found in the Carboniferous Limestone, whether they contain sponge spicules or not. Confusion has inevitably resulted. Although Haüy's term has priority, spongolite is preferable, as it is descriptive of the constitution of the rocks. Many spongolites may be phthanites, but few phthanites are spongolites.

The terms *spiculite* and *spiculitic chert* have also been used to denote rocks composed of sponge spicules (e.g., Carozzi, 1960). The use of these terms is pointless because Cayeux's term is more descriptive of the rock, and inadmissable because "spiculite" was used as early as 1920 as a petrological term for a spindle-shaped crystallite. The only advantage in the use of spiculite would be that the term could still be used if there were doubt as to the zoological affinities of the original spicule-bearing animal.

Petrography

Spongolite deposits cover a range of facies from fresh-water beds of Upper Carboniferous, through the nerito-bathyal deposits of the Holocene, Upper Cretaceous, and Permian. In some cases, it is possible that the parent sponges were able to tolerate euxinic conditions, as in some Upper Carboniferous deposits. Spongolites typically form bedded cherts of a light or dark, semitranslucent brown color or an opaque black. Rarely, a calcareous matrix is known. They can form a local bioherm-like deposit or

TABLE 1. Distribution of Spongolites

Holocene:	Brazil, Mid-Atlantic Ridge (30°N), Western Australia, USSR (Sea of Okhotsk)
Tertiary:	Caucasus
Cretaceous:	Senonian: SW Paris Basin Turonian: Belgium Albian: Belgium, Germany, S England
Jurassic:	Oxfordian: Belgium
Permian:	Spitzbergen US (Delaware Basin, W Texas and N.M.)
Carboniferous:	Pennsylvanian: Appalachian Plateau Stephanian: France Mississippian–Namurian: Belgium, N England, Ireland Visean: N Wales
Devonian	US (Virginia)

be interbedded with other rocks. They are typically much harder than the "normal" rocks of the sequences in which they occur, but may become porous and friable if solution of the spicules has occurred. They are then light in color, as in the type material from the Cretaceous of Belgium. Chert nodules, if composed mainly of sponge spicules, may also be referred to as "spongolite nodules."

The sponge spicules themselves may be of any type and may occur in association. A simple monaxon type is the most common, however, which is to be expected because monaxonid sponges readily disintegrate at death to form a pile of loose spicules (see *Sponges in Sediments*). They may accumulate in situ or may be dispersed and sorted by currents. They are often comminuted, but a high percentage of complete spicules may be present.

The individual spicules are usually siliceous, in the form of the original opal or altered to fibrous chalcedony or microcrystalline silica. All stages in this continuous process have been recognized (Carozzi, 1960). Occasionally spicules may be replaced by calcite and eventually dolomite. The axial canals may remain hollow or be infilled by secondary silica, calcite, pyrite, limonite, glauconite, undifferentiated carbonaceous material (?derived from the organic material of the parent sponges), or sapropel, according to the depositional environment of the sediment and subsequent diagenesis. The canals are often enlarged by solution or molecular replacement and infilled by secondary material or matrix. Occasionally the spicule has been completely replaced by secondary material, leaving only an outer rim and the central canal clear. Solution after burial produces hollow spicule molds containing solid canal casts if infilled previously. Molds may be destroyed by compaction, leaving only canal casts, or filled by calcite or glauconite to give solid casts. Sometimes matrix material fills the spaces left by solution of the spicules, producing "ghost" spicules.

The matrix may be of undifferentiated opal, chalcedony, secondary calcite, amorphous carbonaceous matter, or sapropel. Argillaceous material and occasional grains of quartz, glauconite, or collophane may be present.

Distribution

Sponges are found throughout the geological column from the Paleozoic to Holocene, but spongolite deposits are rarer. Examples of their occurrence are listed in Table 1.

GILLIAN C. LEWARNE-SHEEHAN

References

Bellière, M., 1922. Contribution à l'étude lithologique de l'Assise de Chokier, *C.R. Cong. Géol. Int., 13th Sess.*, 1202-1229.
Carozzi, A. L., 1960. *Microscopic Sedimentary Petrography.* New York: Wiley, 485p.
Cavaroc, V. V., Jr., and Ferm, J. C., 1968. Siliceous spiculites as shoreline indicators in deltaic sequences, *Geol. Soc. Am. Bull.,* 79, 263-272.
Cayeux, L., 1897. Contribution à l'étude micrographique des terrains sédimentaires, *Mém. Soc. Géol. Nord.,* 4, 1-206.
Cayeux, L., 1929. *Les Roches Sédimentaires de France. Roches Silicieuses.* Paris: Impr. Nationale, 696p.
Haüy, R. J., 1822. *Traité de Mineralogie,* 2nd ed. Paris: Conseil des Mines.
Lewarne, G. C., 1963. Spongolites from the Arnsbergian of County Limerick, Ireland, *Geological Mag.,* 100, 289-298.

Cross-references: *Silica in Sediments; Siliceous Ooze; Sponges in Sediments.*

SPRING DEPOSITS

Spring deposits are formed by precipitation of mineral matter from solution in hot or cold springs that emerge from permeable rocks such as sandstone and gravel, or limestone containing solution channels, or from faults, fractures, or fissures in the ground. Rare hot springs erupt discontinuously as geysers. Among the factors causing precipitation are escape of dissolved gases (perhaps by relief of pressure); cooling; evaporation; chemical reactions; and, especially, activity of organisms, notably algae. The most common deposits are calcium carbonate and hydrated silica (or opal).

Travertine

Travertine (q.v.; French, from Italian travertino, from Latine tiburtine, from Tibur, region in ancient Latium corresponding to Tivoli) is the most common spring deposit. It consists of calcium carbonate, $CaCO_3$, generally of light color; it may be concretionary, compact, porous, spongy, cellular, banded, layered, or earthy. Sometimes the term is used interchangeably with *tufa, calc-sinter* or *calcareous sinter.* Travertine is a particularly common spring deposit because calcium bicarbonate is a common constituent of ground water; and spring water, on relief of pressure, loses carbon dioxide, thus precipitating calcium carbonate. Other factors causing precipitation are evaporation, agitation, and activity of algae. Well-known examples of travertine spring deposits include Mammoth Hot Springs, Yellowstone (Weed, 1889); Carlsbad, Czechoslovakia; Hierapolis, W Asia Minor; Hammon Meschoutin, Algeria; Hot Springs, Arkansas; Auvergne, France; and near Rome, Italy.

Tufa (from Italian tufo, from Latin tophus, tofus, probably from Osco-Umbrian), according to many authors, is a spongy, porous travertine, as opposed to a more dense, banded deposit (Pettijohn, 1975). Weed (1889) restricts the term *tufa* to deposits of calcium carbonate formed by evaporation alone, without the presence of algae. Some authors also include siliceous deposits in the term tufa.

Flowstone, dripstone, and calcium carbonate formed in certain rivers and lakes are also known as travertine and tufa. A well-known example is the Tivola River (Italy) which deposits travertine as a result of agitation and loss of carbon dioxide.

Onyx marble, or *Mexican onyx,* is dense, banded calcite resembling onyx. It forms as spring deposits, or dripstone in caves, or vein filling, and is used as an ornamental stone.

Sinter

The term *sinter* (German, for Old High German sintar, Old Slav sedra) generally refers to *siliceous sinter,* a deposit from hot spring water and geysers. The term is also used for travertine (calc-sinter, calcareous sinter). Sinter is a variety of opal, or hydrated silica, occurring as a grayish white to brownish incrustation about hot springs. It may be soft and friable to dense and flinty. Measured rates of deposition include 0.0025 cm/yr and 10 cm/yr per year (Weed, 1889). Some of the best known deposits are in Iceland, New Zealand (Fig. 1), and Yellowstone (Wyoming). Algal activity is an important factor in the rapid growth of sinter.

Geyserite (French, from geyser, from Icelandic Geysir) is siliceous sinter that is deposited

FIGURE 1. White sinter terraces of Rotomahana, New Zealand. These terraces were destroyed by a volcanic eruption in 1886. (Photo: C. Spencer; courtesy Alexander Turnbull Library, Wellington, New Zealand.)

around geysers. Spring deposits of sinter commonly are also called geyserite, however, and the two terms are essentially interchangeable. Weed (1889) uses the term *geyserite* for dense sinter formed by evaporating water rather than by algal activity.

Fiorite (from Santa Fiora, Tuscany, its locality) is another name for siliceous sinter formed on evaporation of siliceous waters of hot springs and geysers. The incrustations are fibrous and pearly.

Miscellaneous Spring Deposits

Other spring deposits, predominantly from hot springs, include sulfur, limonite, sulfates (particularly gypsum), and arsenates (Clarke, 1924, 181-217). Barite, fluorite, and the manganese oxides psilomelane and wad have been found in association with travertine. Gold, silver, and sulfides of mercury (cinnabar), antimony (stibnite, metastibnite), and iron (pyrite, marcasite) have been found in various siliceous sinter deposits from hot springs. Also reported are realgar (arsenic sulfide), aragonite (calcium carbonate), hokutolite (a mixture of barium and lead sulfates), phosphate, jarosite (hydrated potassium iron sulfate), and various zeolites (chabazite, mesolite, analcime, stilbite). Apophyllite, chabazite, opal, chalcedony, tridymite, fluorite, and calcite have been deposited in Roman bricks from hot spring waters (White, 1955).

Sources of the thermal waters are believed to be in part connate and in part meteoric.

ADA SWINEFORD

References

Clarke, F. W., 1924. The data of geochemistry, *U.S. Geol. Surv. Bull.,* 770, 841p.

Pettijohn, F. J., 1975. *Sedimentary Rocks,* 3rd ed. New York: Harper & Row, 628p.

Weed, W. H., 1889. Formation of travertine and siliceous sinter by vegetation of hot springs, *U.S. Geol. Surv. 9th Ann. Rept. 1887-1888,* 613-676.

White, D. E., 1955. Thermal springs and epithermal ore deposits, *Economic Geology, Fiftieth Anniversary Vol., 1905-1955,* 1, 99-154.

Cross-references: *Cave Pearls; Limestones; Oolite; Pisolite; Silica in Sediments; Speleal Sediments; Travertine.*

STAINING TECHNIQUES

Staining is among the most useful techniques in the analysis of sedimentary rocks, particularly in carbonate analysis. It is indispensable in identifying the minerals that make up limestones and dolostones, and hence serves as a useful tool in fabric analysis. It can be used in modern carbonate sediments and in partially or wholly consolidated Pleistocene rocks to identify the metastable carbonate minerals aragonite and magnesium calcite. It is equally useful in the differentiation of calcite, dolomite, gypsum, and anhydrite in modern and ancient limestones, dolostones, and evaporites. Identification of carbonate minerals is effective in hand specimen, polished surface, and thin section; in addition, stained peels of carbonate rocks can be prepared simply and effectively.

Staining methods have been used in carbonate analysis since Lemberg published his classical paper in 1887. The literature on staining techniques in the study of carbonate rocks has been reviewed by Friedman (1959, 1971).

Etching

Hand samples, cores, or drill cutting should be etched with dilute hydrochloric acid and washed in running water before staining. The acid solution is made up of 8-10 parts by volume of concentrated hydrochloric acid diluted with water to 100 parts (Lamar, 1950; Ives, 1955). Exposure to acid should vary depending on fabric and mineralogy. An etching period of 2-3 min is usually effective.

Staining of Carbonates

Staining schemes have been proposed for routine carbonate staining analysis using a combination of alizarine red S and Feigl's solution (Fig. 1). The staining procedure (see Friedman, 1959) can be used to differentiate between dolomite, calcite, aragonite, magnesium calcite, gypsum, and anhydrite. Warne (1962) extended this scheme to include witherite, rhodochrosite, smithsonite, strontianite, cerussite, and siderite.

Organic Stains Specific for Calcite. Alizarine red S is the most effective stain for calcite. It is prepared (see Friedman, 1959) by dissolving 0.1 g in 100 ml 0.2% hydrochloric acid. Calcite is stained deep red within 2-3 min, whereas dolomite is not stained, except on excessive exposure. Staining time should be considered flexible, however, depending on the texture of the rock studied. This stain is suited particularly for thin sections.

Organic Stains Specific for Dolomite. Alizarine red S, alizarine cyanine green, and titan yellow are the recommended organic stains, although 16 other organic stains have been used with equal success (see Friedman, 1959). Each is prepared by dissolving 0.2 g of dye in 25 ml methanol, if necessary by heating; methanol lost by evaporation should be replenished. Then 15 ml of 30% NaOH (70 ml of water + 30 gm of sodium hydroxide) is added to the solution and this is brought to a boil. The sample is immersed in this boiling solution for about 5 min (occasionally it may take even more time). Dolomite is stained purple in alizarine red, deep green in alizarine cyanine green, and deep orange-red in titan yellow alkaline solution. Inadequate staining imparts a yellow to yellow-orange color with titan yellow.

Inorganic Stain Specific for Calcite and Dolomite Containing Ferrous Iron. This procedure (see Heeger, 1913; Friedman, 1959; Evamy, 1963; Katz and Friedman, 1965) is a routine analytical test for iron. A staining solution is prepared by dissolving 5 g of potassium ferricyanide in distilled water containing 2 ml of concentrated hydrochloric acid, followed by dilution to 1 liter with distilled water. A black color will be imparted to the specimen, the deepness of color being proportional to the Fe^{++} concentration. This test is described as specific for dolomite containing ferrous iron (see Friedman, 1959); it is, however, a test for iron, and both ferroan calcite and ferroan dolomite are stained, as pointed out by Evamy (1963). The stain is suitable for use on thin sections.

Organic-Inorganic Stains Specific for Calcite, Ferroan Calcite, and Ferroan Dolomite. A solution consisting of alizarine red S and potassium ferricyanide will stain calcite as well as ferroan calcite and ferroan dolomite (see Evamy, 1963). The reactions of alizarine red S and potassium ferricyanide in the combined reagent are the same as those in the individual stain solutions. According to Evamy (1963), this solution consist of 0.2% hydrochloric acid, 0.2% alizarine red S, and 0.5-1.0% potassium ferricyanide. Katz and Friedman (1965) recommend that the solution be made up as follows: dissolve 1 g of alizarine red S with 5 g of potassium ferricyanide in distilled water containing 2 ml concentrated hydrochloric acid and bring the solution to 1 liter with distilled water. The following colors are obtained: iron-free calcite, red; iron-poor calcite, mauve; iron-rich calcite, purple; iron-free dolomite (dolomite *sensu stricto*), not stained; ferroan dolomite, light blue; ankerite, dark blue.

Stains Specific for Aragonite. Of several stains specific for aragonite (see Friedman, 1959), the most sensitive was developed by Feigl (1937) and named *Feigl's solution* by Friedman (1959). To prepare this solution, add 1 g of solid (commercial grade) Ag_2SO_4 to a solution of 11.8 g $MnSO_4 \cdot 7H_2O$ in 100 ml of water and boil. After cooling, the suspension is filtered, and one or two drops of diluted sodium hydroxide solution is added. The precipitate is filtered off after 1-2 hr (Feigl, 1958). It is important that only distilled water be used; tap water leaves a white precipitate of silver chloride (Katz and Friedman, 1965).

Stains Specific for Magnesium Calcite. Magnesium calcite stains with alizarine red S in acid solution like any other calcite, i.e., the stain imparts a deep red color to the mineral (see Friedman, 1959). In alkaline solution, however, as described above under "Organic Stains Specific for Dolomite," magnesium calcite reacts in the same manner as dolomite. Alizarine red S in alkaline solution imparts a purple color to magnesium calcite; the intensity of staining reflects the amount of magnesium present in the mineral.

Staining of Sulfates

Stains Specific for Gypsum. Alizarine red S (0.1-0.2 g) or of one of several other organic stains (see Friedman, 1959) is dissolved in 25 ml of methanol and 50 ml of 5% sodium hydroxide are added. The specimen is immersed in the cold solution. Staining imparts a deep color to gypsum within a few minutes and a very faint tint of the same color to dolomite. Alizarine red will stain gypsum purple. Heating the solution increases the intensity of the stain. Anhydrite and calcite are not stained.

Stains Specific for Anhydrite. No effective stains have been developed for anhydrite, but schemes have been worked out (see Friedman, 1959) whereby the presence of anhydrite can be determined by the process of elimination, as indicated in Fig. 1.

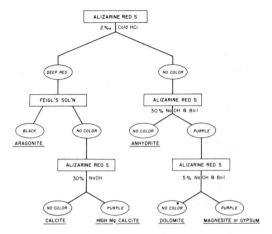

FIGURE 1. Recommended staining procedure (from Friedman, 1959). Asterisk (*) = no color or faint stain.

Staining of Feldspars

In sandstone petrography, staining of feldspars is a common practice, using the standard technique (Bailey and Stevens, 1960). The rock or thin section is etched in 52% hydrofluoric acid, then placed in sodium cobaltnitrate for 15 sec. After rinsing, it is dipped in 5% $BaCl_2$ shortly, rinsed, and then covered with potassium acid rhodizonate until plagioclase is the desired pink. K-feldspar will be stained yellow.

GERALD M. FRIEDMAN

References

Bailey, E. H., and Stevens, R. E., 1960. Selective staining of K-feldspar and plagioclase on rock slabs and thin sections, *Am. Mineralogist*, 45, 1020-1025.

Evamy, B. D., 1963. The application of a chemical staining technique to a study of dedolomitization, *Sedimentology*, 2, 164-170.

Feigl, F., 1937. *Qualitative Analysis by Spot Tests*. New York: Nordemann, 400p.

Feigl, F., 1958. *Spot Tests in Inorganic Analysis*. Amsterdam: Elsevier, 600p.

Friedman, G. M., 1959. Identification of carbonate minerals by staining methods, *J. Sed. Petrology*, 29, 87-97.

Friedman, G. M., 1971. Staining, in R. E. Carver, ed., *Procedures in Sedimentary Petrology*. New York: Wiley, 511-530.

Heeger, J. E., 1913. Ueber die mikrochemische Untersuchung fein verteilter Carbonate im Gesteinsschliff, *Centralbl. Mineral.*, 1913, 44-51.

Ives, W., Jr., 1955. Evaluation of acid etching of limestone, *Kansas Geol. Surv. Bull.*, 114(1), 1-48.

Katz, A., and Friedman, G. M., 1965. The preparation of stained acetate peels for the study of carbonate rocks, *J. Sed. Petrology*, 35, 248-249.

Lamar, J. E., 1950. Acid etching in the study of limestones and dolomites, *Ill. Geol. Surv., Circ.* 156, 47p.

Lemberg, J., 1887. Zur microchemischen Untersuchung von Calcit, Dolomit und Predazzit, *Z. Deutsch. Geol. Gesell.*, 39, 489-492.

Warne, S., 1962. A quick field or laboratory staining scheme for the differentiation of the major carbonate minerals, *J. Sed. Petrology*, 32, 29-38.

Whitlach, R. B., and Johnson, R. G., 1974. Methods for staining organic matter in marine sediments, *J. Sed. Petrology*, 44, 1310-1312.

Cross-references: *Carbonate Sediments-Diagenesis; Carbonate Sediments-Lithification; Dedolomitization; Dolomite, Dolomitization; Gypsum in Sediments; Limestone Fabrics; Limestones; Sedimentary Petrology; Sedimentological Methods.*

STERNBERG'S LAW

The downcurrent decrease in the size of the clastic elements carried by a stream was first studied by Sternberg (1875). Sternberg measured the maximum and average size of the pebbles he found along a 250 km section of the Rhine and noted a decline in size in the downcurrent direction; he concluded that the decline was proportional to the weight of the pebble in water and to the distance travelled. This relationship has been expressed mathematically (Barrell, 1925) as: $W = W_o e^{-as}$, where W is weight at any distance s; W_o is the initial weight of the pebble; and a is a coefficient of size reduction. This equation may also be used if the pebbles are measured in terms of diameter rather than weight. Pettijohn (1975) notes that it applies to alluviating gravels but not to those in the channels of downcutting streams. Such declines in size have been mapped in ancient deposits to deduce paleocurrents (q.v.; Bluck, 1965) and to estimate basin size (Pelletier, 1958; Yeakel, 1962).

Sternberg also noted that the change in gradient (downstream) of many streams also follows an exponential curve in the same manner as the pebble size decreases (see also Unrug, 1957; Bradley et al., 1972). Plumley (1948) studied streams that did not show an exponential decrease in gradient, and interestingly the reduction in related grain size varied in the same manner, suggesting some sort of direct relationship between the two factors. Hack (1957) made similar observations, and rapid declines in boulder size, followed by less rapid declines, have been noted on alluvial fans and in other areas subject to flash flooding (see Pettijohn, 1975).

Systematic decrease of grain size has also been reported on shorelines subject to littoral drift (see *Attrition*).

B. CHARLOTTE SCHREIBER

References

Barrell, J., 1925. Marine and terrestrial conglomerates, *Geol. Soc. Am. Bull.*, 36, 279-342.

Bluck, B. J., 1965. The sedimentary history of some Triassic conglomerates in the Vale of Glamorgan, South Wales, *Sedimentology*, 4, 225-245.

Bradley, W. C.; Fahnestock, R. K.; and Rowehamp, E. T., 1972. Coarse sediment transport by flood flows on Knik River, Alaska, *Geol. Soc. Am. Bull.*, 83, 1261-1284.

Hack, J. T., 1957. Studies of longitudinal stream profiles in Virginia and Maryland, *U.S. Geol. Surv. Prof. Pap. 294-B*, 45-97.

Pelletier, B. R., 1958. Pocono paleocurrents in Pennsylvania and Maryland, *Geol. Soc. Am. Bull.*, 69, 1033-1064.

Pettijohn, F. J., 1975. *Sedimentary Rocks*, 3d ed. New York: Harper & Row, 628p.

Plumley, W. J., 1948. Black Hills terrace gravels: A study in sediment transport, *J. Geol.*, 56, 526-577.

Sternberg, H., 1875. Untersuchungen über langen- und Querprofil geschiebeführende Flusse, *Z. Bauwesen*, 25, 483-506.

Unrug, R., 1957. Recent transport and sedimentation of gravels in the Dunajec valley (western Carpathians), *Acta Geol. Polonica*, 7, 217-257 (Polish with English summary).

Yeakel, L. S., Jr., 1962. Tuscarora, Juniata, and Bald Eagle paleocurrents and paleogeography in the Central Appalachians, *Geol. Soc. Am. Bull.*, 73, 1515-1540.

Cross-references: *Alluvial Fan Sediments; Attrition; Conglomerates; Flood Deposits; Fluvial Sediments; Paleocurrent Analysis; Roundness and Sphericity; Sediment Parameters.*

STOKES' LAW

One of the laws controlling settling velocities of particles moving in a fluid medium was formulated by Stokes (1851) and has consequently been called *Stokes' law*, which states that if a large number of very small spherical particles all of common density are permitted to sink through a motionless fluid, the velocity with which any individual particle sinks is directly proportional to the square of its radius (see also *Settling Velocity*).

Stokes demonstrated this relationship in the following manner (Dapples, 1959): The resistance that a fluid exerts against the movement of a suspended sphere can be expressed as $R = 6\pi rnv$, where R = resistance in g-cm/sec^2; r = radius of the sphere in cm; n = viscosity of the fluid in dynes-cm/sec^2; and v = velocity of the sphere in cm/sec.

A small sphere sinking through the medium is acted upon by the force of gravity acting downward, $(4/3)\pi r^3 d_1 g$, and the buoyant force of the liquid acting upward $(4/3)\pi r^3 d_2 g$, which leaves a resultant force acting downward of $(4/3)\pi r^3 (d_1 - d_2)g$, where d_1 = density of the solid; d_2 = density of the liquid; and g = acceleration due to gravity = 980 cm/sec^2.

At the instant that the fluid resistance is equal to the net downward force, the velocity of the particles is constant, and

$$R = 6\pi rnv = \frac{4}{3}\pi r^3 (d_1 - d_2)g$$

Solving the equation for v yields the following expression

$$v = \frac{2\pi g(d_1 - d_2)r^2}{9n}$$

which is the equation of *Stokes' law*.

If the temperature of the water is held constant, and the spheres have uniform density, Stokes' law is simplified to $v = Cr^2$, in which C is a constant equal to $2\pi g(d_1 - d_2)/9n$. C has

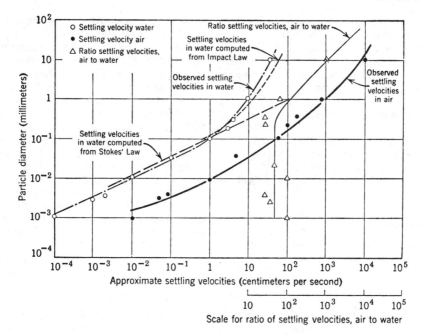

FIGURE 1. Comparison of settling velocities of particles through air and water (from Dapples, 1959). Note the similarity of the slopes of the two curves. Stokes' law holds for particles with diameters <0.1 mm; but for larger particles, the settling velocities are controlled by the impact law (q.v.). For particles up to about 0.5 mm diameter, the ratio of air-water settling velocities remains uniform; but for larger sizes, the ratio increases logarithmically.

a value of 3.57×10^4 for water at 20°C and a specific gravity of 2.65 for the solid spheres (Dapples, 1959).

Stokes' law holds well for small particles (0.001-0.1 mm), but for diameters >0.1 mm, settling velocities are substantially slower than predicted by Stokes' calculation (Fig. 1). For grain sizes >0.1 mm, a more accurate approximation is obtained by the application of the impact law (q.v.).

Similarly, certain very small particles of fine clay sizes also exhibit retarded settling rates, due to intergranular repulsion resulting from the accumulation of a similar surface charge on the surfaces of the grains themselves. The effects of this condition may sometimes be reduced by the addition of appropriate chemicals to the fluid medium, facilitating loss of this surface charge or preventing its formation.

B. CHARLOTTE SCHREIBER

References

Dapples, E. C., 1959. *Basic Geology for Science and Engineering.* New York: Wiley, 609p.

Krumbein, W. C., and Sloss, L. L., 1951. *Stratigraphy and Sedimentation,* 2nd ed. San Francisco: Freeman, 660p.

Stokes, G. G., 1851. On the effect of the internal friction of fluids on the motion of pendulums, *Cambridge Philos. Soc.,* 9(2), 8-106.

Cross-references: *Grain-Size Studies; Granulometric Analysis; Impact Law; Settling Velocity.*

STORM DEPOSITS

The activities of storms in coastal areas have long been subject of geologic concern and study. For many years, however, the chief emphasis of these studies has been on the enormous destructive and erosive effects of the storms and on the resulting great changes in coastal morphology and, to some extent, on the ability of a damaged coast to recover from the effects of a storm (see Kumar and Sanders, 1976).

In more recent times, attention has shifted to the depositional aspects of hurricanes, typhoons, and other severe storms. Such studies began in tropical regions where carbonate sediments were affected (McKee, 1959; Ball et al., 1967; High, 1969; Ball, 1971). Subsequently, the effects of severe storms on a few coastal areas containing only noncarbonate terrigenous sediments have been examined (Swift, 1969; Walton, 1970;

Reineck and Singh, 1972; Kumar and Sanders, 1976; see also *Barrier Island Facies*).

What is more, in recent years the geologic philosophy of *uniformitarianist catastrophism* has continued to gain adherents and respectability (Ager, 1973; see also *Flood Deposits; Tsunami Sedimentation*). The geologic record of ancient nearshore sediments has recently been interpreted in terms of the effects of storms by various workers in several areas (Hobday and Reading, 1972; Goldring and Bridges, 1973; Brenner and Davies, 1973; Dott, 1974; Ager, 1974; Scott et al., 1975; Kelling and Mullin, 1975).

Characteristics

Distinctive nearshore sediments off Long Island, New York, have been interpreted by the authors as characteristic storm deposits (Kumar and Sanders, 1976). The distinctive sediments consist of a sequence of three parts from bottom to top: (a) Basal lag, consisting of pebbles (or possibly even of larger sizes) or coarse shell debris; generally 50 cm or so thick. (b) Laminated sand, of variable thickness, composed of sediment dropped rapidly from suspension as the storm waned and while bottom shear was intense enough to exclude rhythmic bed forms (but possibly including a few cross-stratified or ripple-laminated beds if rhythmic bed forms were present); generally 1–2 m thick and finer grained than unit (a). (c) A fair-weather capping, of variable thickness, consisting either of wave-ripple laminae (if landward of fair-weather wave base) or of burrow-mottled sediment (if seaward of fair-weather wave base); generally 30 cm or so thick. Cores of modern sediments from the inner shelf off Long Island, New York display units (a) and (b) in some cores and units (c) and (b) in others. Fig. 1 shows CERC cores taken off Long Island in water depths ranging from 9–20.7 m. These cores consist largely of fine-grained, plane-parallel laminated fine sand, unit (b). Core C65 is composed entirely of unit (b), whereas C72 includes unit (b) from 0–1.0 m (measuring down from the top), and C74 contains a few pebbles as large as 2 cm in diameter, unit (a), scattered within the laminated sand. The bottom part (1.0–1.6 m) of C72 contains only very coarse gravel, unit (a).

Although the well-developed plane laminae of the cores displayed in Fig. 1 are distinctive and are conspicuous features of these cores, the absence of certain other features found in sediments from other barrier environments is equally noteworthy. Not present in these cores are the following: (1) segregated layers of heavy minerals that characterize certain beach sediments and may also display conspicuous plane-parallel laminae; (2) ripple cross-laminae, common in many nearshore sediments; (3) large-scale, high-angle cross-strata, common in many barrier sediments; (4) bioturbation structures; and (5) skeletal debris from marine organisms.

Fig. 2 illustrates epoxy relief-peels from cores GMF-13 and GMF-22, taken from off Long Island in a 17-m water depth. These cores contain burrow-mottled sediment about 30 cm thick, unit (c), overlying plane-parallel laminated sand, unit (b). The origin of these distinctive units (a) and (b) cannot be explained through fair-weather processes operating in the nearshore zone. Fair-weather conditions in the nearshore zone favor either growth of ripples (between breaker zones and wave base) or burrowing (seaward of wave base). Alternatively, units (a) and (b) can be explained in the light of storm processes as we know them. Any conclusion based

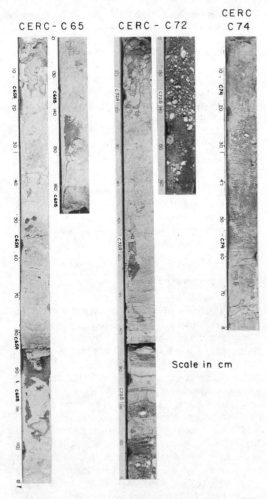

FIGURE 1. Epoxy relief-peels of CERC cores C65, C72, and C74 from the inner shelf off Long Island, New York displaying units (a) and (b) of the storm sequence (see text).

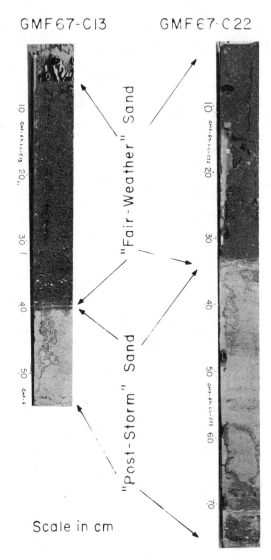

FIGURE 2. Epoxy relief-peels of GMF cores 13 and 22, from the inner shelf off Long Island, New York displaying units (b) and (c) of the storm sequence (from Sanders and Kumar, 1975).

on the present state of knowledge about storm processes, however, can at best be termed tentative.

Interpretation

The basal gravel, unit (a), is interpreted as a lag deposit, which, during the height of a severe storm, may have formed a continuous blanket at the water-sediment interface. The nearshore sand is probably in suspension when the storm is at its peak. The laminated unit (b) is inferred to be the product of deposition from suspension when the storm begins to wane and the sand begins to fall out of suspension. Unit (b) is deposited under conditions of intense bottom shear, which may be somewhat comparable to conditions within the transition between the lower flow regime and the upper flow regime (see *Flow Regimes*). Under such conditions of bottom shear, the stable bed morphology is a plane surface. Sediments deposited at such times show plane-parallel laminae throughout. If the conditions of bottom shear become slightly less intense and reach something comparable with the lower flow regime of fluvial channels, then sand waves and/or various ripples would form, and unit (b) would contain cross-laminae. The sediment of unit (c), which is coarser grained than unit (b), is deposited as a normal result of sorting by fair-weather wave action during which medium sand is transported landward in ripples and very fine sand is transported seaward (see *Littoral Sedimentation*). Unit (c) would be cross-laminated if the sediment contained in it lies landward of fair-weather wave base and would be burrow-mottled if it lies seaward of the fair-weather wave base.

The modern sediments on the shelf of Long Island containing this characteristic sequence always lie within a depth range of 5-21 m. This depth range coincides with the shoreface at the S shore of Long Island. Hence, the sediments presumably deposited by storms may be referred to as "shoreface storm deposits." The depth range in which these sediments occur is so characteristic that even if their origin is contested, these sediments can be considered as significant indicators of depth of deposition.

Ancient Storm Deposits

Examples of sequences identical to the "storm sequence" described above are known from rocks and sediments ranging in age from Holocene to Cambrian. This sequence has been recognized in the relict Holocene sediments on the shelf off Long Island, New York, and in the Pleistocene sediments underneath the modern Fire Island barrier off the S shore of Long Island (Kumar and Sanders, 1976), and also in the Norfolk Formation (Pleistocene) at Benns Church, SE Virginia (Fig. 3) and in the Eocene sandstones and conglomerates in the Tehachapi and San Emigdio Mountains, southern California (Sanders and Kumar, 1976). Some examples of sequences similar to the one described here have been identified as storm deposits by other authors and include recent sediments, southern North Sea (Reineck and Singh, 1972), offshore Virginia (Swift, 1969); Upper Sundance Formation (Jurassic), Wyoming and Montana (Brenner and Davies, 1973); Upper Jurassic clastic carbon-

FIGURE 3. Gravel and sand interbeds seen in a pit at Benns Church, SE Virginia. Compare these interbeds with sequences seen in cores in Figs. 1 and 2. The gravel dips gently eastward, in the former offshore direction. Each sequence of units (a) and (b) is interpreted as the consequence of a single storm. Parts of unit (b), however, deposited during one storm, may be eroded by a subsequent storm, and a new layer of gravel, unit (a), may be deposited either on unit (a) or unit (b) of the previous storm. Unit (c) is only sparsely represented in the sequence (from Kumar and Sanders, 1976).

ates in Morocco (Ager, 1974); Upper Cambrian, Baraboo district, Wisconsin (Dott, 1974); and various "sublittoral sheet sandstones" (Goldring and Bridges, 1973).

If the interpretation of this characteristic sequence is correct, and if the suggestion that the storms were responsible for creating the characteristic sequence recognized in all of the above mentioned examples is valid, then it is apparent that storm deposits are far more common in the geologic record than has been heretofore recognized. It is quite possible that once the criteria for recognizing storm-deposited sediment are well understood, the geologic record of the nearshore zone will be found to consist largely of storm deposits, and the products of the longer-lasting, fair-weather conditions will be found to compose only a minor proportion.

NARESH KUMAR
JOHN E. SANDERS

References

Ager, D. V., 1973. *The Nature of the Stratigraphical Record.* New York: McMillan, 114p.

Ager, D. V., 1974. Storm deposits in the Jurassic of the Moroccan High Atlas, *Palaeogeogr., Palaeoclimat., Palaeoecol.,* 15, 83-93.

Ball, M. M., 1971. The Westphalia Limestone of the northern mid-continent: a possible ancient storm deposit, *J. Sed. Petrology,* 41, 217-232.

Ball, M. M.; Shinn, E. A.; and Stockman, K. W., 1967. The effects of hurricane Donna in South Florida, *J. Geol.,* 75, 583-597.

Brenner, R. L., and Davies, D. K., 1973. Storm-generated coquinoid sandstone; genesis of high-energy marine sediments from the upper Jurassic of Wyoming and Montana, *Geol. Soc. Am. Bull.,* 84, 1685-1698. See also discussion by R. P. Wright (1974), 85, 837.

Dott, R. H., Jr., 1974. Cambrian tropical storm waves in Wisconsin, *Geology,* 2, 243-246.

Goldring, R., and Bridges, P., 1973. Sublittoral sheet sandstones, *J. Sed. Petrology,* 43, 736-747.

High, L. R., Jr., 1969. Storms and sedimentary processes along the northern British Honduras coast, *J. Sed. Petrology,* 39, 235-245.

Hobday, D. K., and Reading, H. G., 1972. Fair weather versus storm processes in shallow marine sand bar sequences in the Late Precambrian of Finnmark, North Norway, *J. Sed. Petrology,* 42, 318-324.

Kelling, G., and Mullin, P. R., 1975. Graded limestones and limestone-quartzite couplets: Possible storm-deposits from the Moroccan Carboniferous, *Sed. Geol.,* 13, 161-190.

Kumar, N., and Sanders, J. E., 1976. Characteristics of shoreface storm deposits: Modern and ancient examples, *J. Sed. Petrology,* 46, 145-162.

McKee, E. D., 1959. Storm sediments on a Pacific atoll, *J. Sed. Petrology,* 29, 354-364.

Reineck, H.-E., and Singh, I. B., 1972. Genesis of laminated sand and graded rhythmites in storm-sand layers of shelf mud, *Sedimentology,* 18, 123-128.

Sanders, J. E. and Kumar, N., 1975. Evidence of shoreface retreat and in-place drowning during Holocene submergence of barriers, shelf off Fire Island, New York, *Geol. Soc. Am. Bull.,* 86, 65-76.

Scott, R. W.; Laali, H.; and Fee, D. W., 1975. Density current strata in Lower Cretaceous Washita group, north-central Texas, *J. Sed. Petrology,* 45, 562-575.

Swift, D. J. P., 1969. Inner shelf sedimentation: Processes and products; Outer shelf sedimentation: Processes and products; and Evolution of the shelf surface, and the relevance of the modern shelf studies to the rock record, in D. J. Stanley, ed., *The NEW Concepts of Continental Margin Sedimentation.* Washington, D.C.: Am. Geological Inst., various paging.

Walton, W. R., 1970. Modern and ancient hurricane deposits, *2nd Ann. Offshore Technol. Conf., Proc.,* 1, 37-39.

Cross-references: *Barrier Island Facies; Flood Deposits; Littoral Sedimentation; Tsunami Sedimentation.* Vol. I: *Storm, Storminess; Storm Surge.*

STRENGTH OF SEDIMENTS
—*See* Vol. XIII

STROMATOLITES

Stromatolites have the distinction of being by far the oldest indicators of organized life on Earth, ranging back over 3 billion yr. They occur on all continents in rocks from middle Precambrian to Holocene age. *Stromatolites* (term proposed by Kalkowsky, 1908) are laminated limestone structures of simple to complex form commonly attributed to debris-binding and biochemical processes of benthonic blue-green, green, and possibly, red algae. Some other organisms (e.g., bacteria; Doemel and Brock, 1974) and inorganic processes may also create laminated structures that resemble algal stromatolites. Aitken (1967) introduced the term *cryptalgal* to describe carbonates in which the work of noncalcareous algae is largely inferred. The individual laminae of a stromatolite are usually a fraction of a mm thick and are composed of algal-bound mud and silt, minute pellets, some inorganic debris, and sometimes algal-secreted calcite. Thicker, algally precipitated crusts are also known. Although most stromatolites are composed of calcium carbonate (unless diagenetically replaced, e.g., by dolomite), Walter (1972) and Walter et al. (1972) have described modern, siliceous algal and bacterial stromatolites in hot springs and geyser effluents of Yellowstone National Park, US, and have compared the "geyserite" precipitations with some Precambrian occurrences.

World-wide investigation of various modern stromatolite-forming environments is required to understand ancient stromatolites and to reconstruct their paleogeographic and sedimentologic settings (see Hoffman, 1973, for historical review). Hommeril and Rioult (1965) have examined stromatolites composed of trapped calcareous silt and argillaceous muds by association of filamentous algae; Friedman et al. (1973) discuss the mechanisms involved in the origin of modern algal mats in hypersaline pools in the Red Sea; the Florida algal deposits have been studied for some time by numerous researchers, e.g., Ginsburg et al. (1954; see Monty, 1972, for older references). Two other well-known modern sites of stromatolite formation are the Persian Gulf (q.v.) and Shark Bay (q.v.), W Australia.

The stromatolite-forming algae can withstand a wide range of physical and chemical conditions; they are found in fresh water (Bradley, 1928), marine, and hypersaline environments, where they often form extensive biohermal and biostromal reefs (Fig. 1). Most stromatolites originate in shallow water (i.e., littoral and supralittoral), but under conducive geochemical and geological conditions, "deep-water" stromatolites can grow down to 150 m below water level (Playford and Cockbein, 1969; Hoffman, 1974).

Construction

Stromatolites are constructed where algal filaments and cells form a slimy mat to which small, foreign particles easily adhere and become part of the sediments when the algae reestablish themselves above the sedimentary accumulation (Fig. 2). In addition to the trapping of foreign particles, calcium carbonate is directly precipitated by some algae as molds around filaments and cells. As a result of various changes in the environment, e.g., fluctuations in the rate of sedimentation, the algae produce laminae which often have a tendency to arch upwards as the algal mats expand laterally (Fig. 3). The initially high porosity of these calcareous masses is often responsible for the destruc-

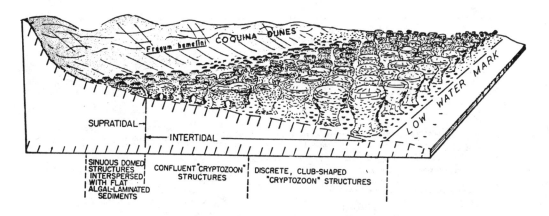

FIGURE 1. Idealized form zonation in stromatolite reef at Flint Cliff, Hamilton, western Australia (from Logan, 1961).

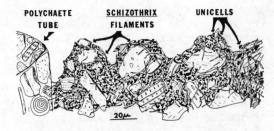

FIGURE 2. Microstructure of *Schizothrix* mat (from Neumann et al., 1970). Abundance of fine filaments and mucilage create a rough-surfaced mat lacking rigid fibrous structure but within which grains are completely enmeshed.

tion of the minute molds by solution and incipient recrystallization together with the possible upward movement of the algae through the fine-grained sediments. Algal cellular and filamentous features may also be destroyed by various diagenetic processes. However, the major outlines of the individual laminae and the colony remain unaffected. In some rare cases, however, filaments and cells are preserved. Grégoire and Monty (1962) used an electron microscope in their investigation of the carbonate textures of ancient stromatolites, and Carozzi and Davis (1964) performed a very detailed petrographic study of the lamina-by-lamina evolution of a small stromatolite reef and its surrounding sediments.

Form and Classification

The external shapes of the stromatolitic growths are frequently remarkably constant in both space and time in the stratigraphic record, and attempts have been made to use them in local and regional correlation (Cloud and Semikhatov, 1969; Donaldson, 1963; Hofmann, 1973). The stromatolites range from simple, horizontal laminae to wavy, bulbous, columnar, nodular, and spherical forms, varying from a cm to over 1 m in thickness or diameter. It is thought that the environment is the primary control of the shape of stromatolites, but little precise information is available (Chafetz, 1973; Ahr, 1971).

The generic names that have been given to particular forms (Hofmann, 1969) are useful for description, but are taxonomically invalid because it has been found that algal mats composed of more than one species or genera can produce stromatolites (Neumann et al., 1970; Hofmann, 1973). Hence, changes in the nomenclature have been suggested. Johnson (1961) employed the term *Stromatolithi* (often loosely called stromatolites) as a collective taxonomic name of a subdivision of Spongiostromata. Stromatolithi comprise those laminated or massive algal accumulations which form in situ, in contrast to the *Oncolithi* (or oncolites), which are unattached at the time of formation. To make a clear distinction between laminated and massive Stromatolithi, it has been suggested that the former be called orthostromatolites and the latter, pseudostromatolites (Wolf, 1965).

Logan et al. (1964) offered a classification scheme for algal stromatolites that is now being widely employed, although other approaches have been suggested (e.g., Aitken, 1967). Hofmann (1969, 1973) has provided and compared detailed information on various classification schemes on growth patterns of stromatolites, and on configuration and structural variations of laminae. The study by Logan et al. (1964), together with earlier investigations, indicates that algal stromatolites are organosedimentary structures rather than fossil organisms, and that stromatolites and oncolites are composed of

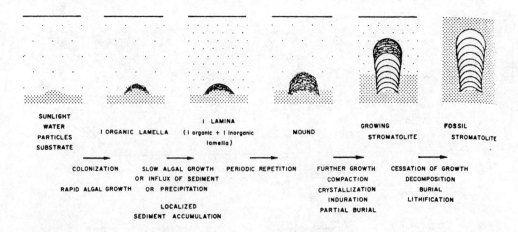

FIGURE 3. Schematic representation of development of an algal stromatolite (from Hofmann, 1969).

DESCRIPTION	VERTICAL SECTION OF STROMATOLITE STRUCTURES
Space-linked hemispheroids with close-linked hemispheroids as a microstructure in the constituent laminae.	
Discrete, vertically stacked hemispheroids composed of close-linked hemispheroidal laminae on a microscale.	
Initial space-linked hemispheroids passing into discrete, vertically stacked hemispheroids with upward growth of structures and continued differentiation of domes and interareas.	
Initial discrete, vertically stacked hemispheroids passing into close-linked hemispheroids by upward growth expansion.	
Alternation of discrete, vertically stacked hemispheroids and spaced-linked hemispheroids due to periodic sediment infilling of interstructure spaces.	
Initial space-linked hemispheroids passing into discrete, vertically stacked hemispheroids; both with laminae of close-linked hemispheroids.	
Initial discrete, vertically stacked hemispheroids passing into close-linked hemispheroids; both with laminae of close-linked hemispheroids.	
Concentrically stacked spheroids with laminae composed of close-linked hemispheroids.	

FIGURE 4. Compound stromatolite structures (after Logan et al., 1964).

basic geometric units (hemispheroids and spheroids) arranged in various structural configurations (Fig. 4).

Precambrian Stromatolites

Investigations carried out on Precambrian stromatolites include the work by Cloud and Semikhatov (1969), who reviewed the nomenclature, methods of study, and stratigraphic zonation of Proterozoic stromatolites (see also Gebelein, 1974). Truswell and Eriksson (1973) and Krüger (1969) investigated the Precambrian dolomitized stromatolites of southern Africa; Donaldson (1963) examined those of the Labrador Geosyncline in Canada and was able to establish six distinct stromatolite forms useful for correlation over a distance of >3 km. Hoffman (1974) conducted a basin analysis by identifying shallow- and deep-water stromatolites.

KARL H. WOLF

References

Ahr, W. M., 1971. Paleoenvironment, algal structures, and fossil algae in the Upper Cambrian of Central Texas, *J. Sed. Petrology*, 41, 205-216.

Aitken, J. D., 1967. Classification and environmental significance of cryptalgal limestones and dolomites, with illustrations from the Cambrian and Ordovician of southwestern Alberta, *J. Sed. Petrology*, 37, 1163-1178.

Bradley, W. H., 1928. Algae reefs and oolites of the Green River Formation, *U.S. Geol. Surv. Prof. Pap. 154*, 203-223.

Carozzi, A. V., and Davis, R. A., Jr., 1964. Pétrographie et paléoécologie d'une série de dolomies à stromatolithes de l'Ordovician Inférieur du Wisconsin, U.S.A., *Archives Sci.*, 17, 47-63.

Chafetz, H. S., 1973. Morphological evolution of Cambrian algal mounds in response to a change in depositional environment, *J. Sed. Petrology*, 43, 435-446.

Cloud, P. E., Jr., and Semikhatov, M. A., 1969. Proterozoic stromatolite zonation, *Am. J. Sci.*, 267, 1017-1061.

Doemel, W. N., and Brock, T. D., 1974. Bacterial stromatolites: Origin of laminations, *Science*, 184, 1083-1085.

Donaldson, J. A., 1963. Stromatolites in the Denault Formation, Marion Lake, coast of Labrador, Newfoundland, *Geol. Surv. Canada Bull.*, 102, 1-33.

Friedman, G. M.; Amiel, A. J.; Braun, M.; and Miller, D. S., 1973. Generation of carbonate particles and laminites in algal mats—example from sea-marginal hypersaline pool, Gulf of Aqaba, Red Sea, *Bull. Am. Assoc. Petrol. Geologists*, 57, 541-557.

Gebelein, C. D., 1969. Distribution, morphology, and accretion rate of Recent subtidal algal stromatolites, Bermuda, *J. Sed. Petrology*, 39, 49-69. See also comment by J. H. Sharp and reply (1970), 40, 521-522.

Gebelein, C. D., 1974. Biologic control of stromatolite microstructure: Implication for Precambrian time stratigraphy, *Am. J. Sci.*, 274, 575-598.

Ginsburg, R. N.; Isham, L. B.; Bein, S. J.; and Kuperberg, J., 1954. Laminated algal sediments of south Florida, and their recognition in the fossil record, *Univ. Miami Marine Lab. Rept.*, 54.21 (unpubl.), 33p.

Grégoire, C., and Monty, C. L. V., 1962. Observations au microscope électronique sur le calcaire à pate fine entrant dans la constitution de structures stromatolithiques du Viséen Moyen de la Belgique, *Ann. Soc. Géol. Belg.*, 85, 389-397.

Hoffman, P., 1973. Recent and ancient algal stromatolites: Seventy years of pedagogic cross-pollination, in R. N. Ginsburg, ed., *Evolving Concepts in Sedimentology*. Baltimore: Johns Hopkins, 178-191.

Hoffman, P., 1974. Shallow and deepwater stromatolites in lower Proterozoic platform-to-basin facies change, Great Slave Lake, Canada, *Bull. Am. Assoc. Petrol. Geologists*, 58, 856-867.

Hofmann, H. J., 1969. Attributes of stromatolites, *Geol. Surv. Can., Pap. 69-39*, 1-58.

Hofmann, H. J., 1973. Stromatolites: Characteristics and utility, *Earth-Sci. Rev.*, 9, 339-373.

Hommeril, P., and Rioult, M., 1965. Etude de la fixation des sédiments meubles par deux algues marines: *Rhodothamniella Florula* (Dillwyn), J. Feldm. et *Microcoleus Chtonoplastes* Thur., *Marine Geol.*, 3, 131-155.

Johnson, J. H., 1961. *Limestone-Building Algae and Algal Limestones*. Golden: Colo. Sch. Mines, 297p.

Kalkowsky, E., 1908. Oolith und Stromatolith in norddeutschen Buntsandstein, *Z. Deutsch. Geol. Gesell.,* **60,** 68-125.

Krüger, L., 1969. Stromatolites and oncolites in the Otavi Series, South West Africa, *J. Sed. Petrology,* **39,** 1046-1056.

Logan, B. W., 1961. *Cryptozoon* and associated stromatolites from the Recent, Shark Bay, Western Australia, *J. Geol.,* **69,** 517-533.

Logan, B. W.; Rezak, R.; and Ginsburg, R. N., 1964. Classification and environmental significance of algal stromatolites, *J. Geol.,* **72,** 68-83.

Maslov, V. P., 1960. Stromatolites and facies, *Doklady Acad. Sci. USSR,* **125,** 1-6.

Monty, C. L. V., 1972. Recent algal stromatolitic deposits, Andros Island, Bahamas, preliminary report, *Geol. Rundschau,* **61,** 742-783.

Monty, C. L. V., 1973. Precambrian background and Phanerozoic history of stromatolitic communities, *Ann. Soc. Géol. Belg.,* **96,** 585-624.

Neumann, A. C.; Gebelein, C. D.; and Scoffin, T. P., 1970. The composition, structure, and erodability of subtidal mats, Bahamas, *J. Sed. Petrology,* **40,** 274-297.

Playford, P. E., and Cockbein, A. E., 1969. Algal stromatolites: Deepwater forms in the Devonian of Western Australia, *Science,* **165,** 1008-1010.

Rezak, R., 1957. Stromatolites of the Belt Series in Glacier National Park and vicinity, Montana, *U.S. Geol. Surv. Prof. Pap. 294-D,* 127-154.

Schwartz, H. -U.; Einsele, G.; and Herm, D., 1975. Quartz-sandy, grazing-contoured stromatolites from coastal embayments of Mauritania, West Africa, *Sedimentology,* **22,** 539-561.

Serebryakov, S. N., and Semikhatov, M. A., 1974. Riphean and Recent stromatolites: A comparison, *Am. J. Sci.,* **274,** 556-574.

Truswell, J. F., and Eriksson, K. A., 1973. Stromatolitic associations and their paleo-environmental significance: A re-appraisal of a Lower Proterozoic locality from the northern Cape Province, South Africa, *Sed. Geol.,* **10,** 1-23.

Walter, M. R., 1972. A hot spring analog for the depositional environment of Precambrian iron formations of the Lake Superior region, *Econ. Geol.,* **67,** 965-980.

Walter, M. R., ed., 1976. *Stromatolites.* Amsterdam: Elsevier, 790p.

Walter, M. R.; Bauld, J.; and Brock, T. D., 1972. Siliceous algal and bacterial stromatolites in hot spring and geyser effluents of Yellowstone National Park, *Science,* **178,** 402-405.

Wolf, K. H., 1965. Gradational sedimentary products of calcareous algae, *Sedimentology,* **5,** 1-38.

Cross-references: *Dolomite, Dolomitization; Limestones; Persian Gulf Sedimentology; Shark Bay Sedimentology.*

STRUCTURES—See SEDIMENTARY STRUCTURES

SUBGRAYWACKE

The term *subgraywacke* was first proposed by Pettijohn (1949) to designate sandstones that are intermediate in composition between graywacke and orthoquartzite; subgraywacke would contain 0-10% feldspar, 15-85% quartz, and 20-75% detrital matrix (mica and chlorite), plus rock fragments. The term was redefined by Pettijohn (1954, 1957) himself, on the basis of textural and compositional characteristics, as a sandstone containing <15% matrix and >25% unstable sand grains (feldspar and rock fragments), in which rock fragments exceed feldspar. Thus, the sandstone has little or no clay matrix but has empty voids which may be filled with carbonate or silica cement or both.

The term *subgraywacke* has been defined in somewhat different ways by others, and in recent literature it is being replaced by the term *lithic arenite,* or *litharenite,* or *sublitharenite* (see *Sands and Sandstones*).

HAKUYU OKADA

References

Okada, H., 1971. Classification of sandstone: analysis and proposal, *J. Geol.,* **79,** 509-525.

Pettijohn, F. J., 1949. *Sedimentary Rocks.* New York: Harper & Row, 526p.

Pettijohn, F. J., 1954. Classification of sandstones, *J. Geol.,* **62,** 360-365.

Pettijohn, F. J., 1957. *Sedimentary Rocks,* 2nd ed. New York: Harper & Row, 718p.

Pettijohn, F. J.; Potter, P. E.; and Siever, R., 1972. *Sand and Sandstone.* Berlin: Springer-Verlag, 618p.

Cross-references: *Graywacke; Sands and Sandstones.*

SUBMARINE (BATHYAL) SLOPE SEDIMENTATION

The submarine slope, also often called the *continental slope,* occupies the outer continental margin and generally floors the bathyal zone. The slope lies between the outer continental shelf (q.v.) and the continental rise (q.v.). Its upper limit is defined by the shelf break, which may occur from 20 to 500 m below sea level, but most frequently is found at about 130 m (70 fathoms). Its lower limit is loosely defined by the 1000-3000 m bathymetric contours; in major ocean basins, the deeper figure is generally used.

The bathyal province lies below wave base and is largely aphotic, although the upper part may sustain some photosynthetic activity in tropical waters where high sun angles prevail. Bottom currents affecting sediment distribu-

SUBMARINE (BATHYAL) SLOPE SEDIMENTATION

SEDIMENTATION PATTERNS ON THE SLOPE PROPER	FACTORS INFLUENCING SLOPE DEPOSITION						
	TECTONIC ACTIVITY	EUSTATIC SEA LEVEL CHANGES	GRADIENT	GEOGRAPHIC SETTING OF SLOPE IN RELATION TO SHELF AND COAST	RATE AND AMOUNT OF SEDIMENTATION	SEDIMENT TRANSPORT PROCESSES	BIOTURBATION
PROGRADATION AND PERMANENT ACCUMULATION OF DEPOSITS ON SLOPE.	Stable tectonic setting; low level of seismicity	High sea level stands; outer shelf submerged	Relatively low	Shelf-break in deep water; little or no coastal influence	Low sedimentation rates; or large supply accumulation with low pore water pressure	Pelagic and hemipelagic sedimentation dominant	Low rate of sediment turnover
SLOPE SERVES AS TEMPORARY SEDIMENTATION SURFACE: EROSIONAL PROCESSES DOMINANT	Unstable tectonic setting; frequent earthquakes, faulting, etc.	Lower sea level-stands; outer shelf subaerially exposed, or at shallow depth	Relatively high	Shelf-break shallow and close to coast; influence of seasonal floods and wave and tidal currents	High sedimentation rates; or sediment accumulating with high pore-water pressure	Sediment failure by mass-gravity processes dominant	Rapid and thorough reworking of bottom sediment

FIGURE 1. Interpretive diagram showing interplay of factors influencing slope sedimentation: deposition vs. erosion (Stanley, 1969).

tion are largely of two types: (1) turbidity and other density currents which flow down the continental slope, generally within depressions such as gullies and submarine canyons; and (2) geostrophic currents that tend to flow parallel or oblique to the base of the continental slope in response to sea-water density difference on a rotating earth (see *Contourites*). Additional, locally important factors include density flows generated by higher-than-normal salinity brines, either in high polar latitudes where sea ice is forming, or in low latitudes where tropical lagoons promote rapid evaporation. Another regional factor, apparently related to tidal forces and internal gravity waves (Southard and Cacchione, 1973), results in rapid current reversal, often producing strong upcanyon flow; this phenomenon has actually been observed from deep submersible vessels (Shepard et al., 1974). Tectonics, sea-level changes, slope gradient, and other factors also influence slope sedimentation, as outlined in Fig. 1.

Sediments

Subsequent to the *Challenger* expedition (1872–1876), the report on sediments by Murray and Renard (1891) recognized that sediments from the bathyal slope province are typified by *gray-green muds*—a mixture of terrigenous organic clays, silts, and pelagic ooze (see Fig. 2). It has been suggested that these *hemipelagites* (see *Pelagic Sediments*) may be deposited nepheloid sediments (q.v.) and fine-grained mud turbidites (Stanley, 1969). Remains of bathyal benthonic organisms are uncommon in slope sediments; Murray and Renard did note the presence of shark teeth and fish scales. Subsequent oceanographic work, especially coring,

dredging, and bottom photography, has disclosed regions of rich bottom life, including an "in-fauna" of thin-shelled bivalves, brachiopods and various echinoderms. Furthermore, close studies of submarine canyons and trench walls have shown that there are frequently large sectors of nondeposition or erosion. Here, bottom sampling has disclosed outcrops ranging from granitic basement to a wide range of sedimentary formations dating from Cretaceous to Quaternary (Heezen et al., 1959; Emery et al., 1970). The processes and sediments on a modern submarine slope are illustrated in Fig. 3.

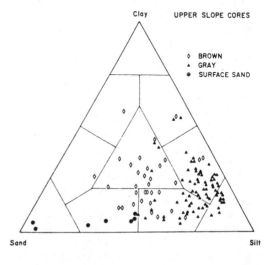

FIGURE 2. Textural analysis of some Nova Scotian continental-slope sediments plotted on a triangular diagram (Stanley, 1969). The clay-rich matrix of some silty and sandy sediments (e.g., those plotted in the central triangle) may be likened to modern graywackes.

SUBMARINE (BATHYAL) SLOPE SEDIMENTATION

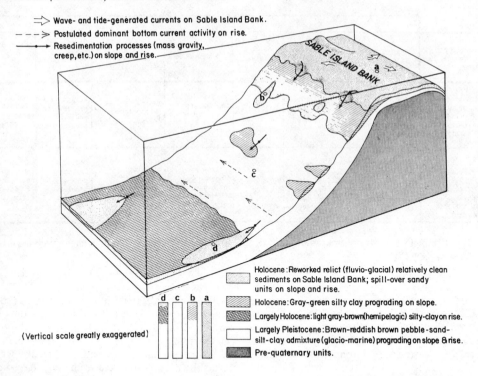

FIGURE 3. Schematic diagram showing progradation of Quaternary facies and modern dispersal patterns on the continental slope and rise off Sable Island Bank (Stanley, 1969).

Recognition

In ancient sediments, the identification of former bathyal slope conditions is often a matter of questionable interpretation. On a large scale, numbers of deep cut-and-fill structures and even full-scale "fossil" submarine canyons have been recognized (see *Submarine Canyons and Fan Valleys, Ancient*). Channel deposits, the coarse fill of submarine canyons and valleys (*fluxoturbidites*), are important here (Stanley and Unrug, 1972).

With the possible exception of in-situ benthic forms with restricted depth ranges and associated organic tracks and markings (*Lebensspuren*), however, there are few facies truly unique to submarine slopes (Stanley, 1969, 1970). An attempt to reevaluate the criteria used to define submarine slopes is presented in Fig. 4. The dominant slope assemblage comprises fine-grained pelagic deposits, hemipelagic materials influenced by bottom current activity (and often bioturbated), turbidites (q.v.), contorted slumped units, and large allochthonous slices (Stanley and Unrug, 1972). Stanley (1969) has found that wedges of pebbly mudstone (see *Pebbly Mudstones*) in association with submarine channels and slump deposits, are a diagnostic base-of-slope indicator. Stanley and Unrug (1972) describe examples of ancient submarine slope deposits from the French Alps (Tertiary Annot Sandstone) and the Polish Carpathians (Tertiary Flysch) and compare them with modern slope facies.

RHODES W. FAIRBRIDGE
JOANNE BOURGEOIS

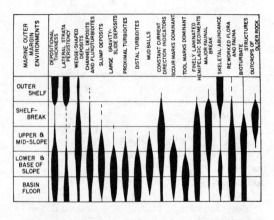

FIGURE 4. Diagnostic lithologic assemblages in submarine outer-margin environments (Stanley, 1969).

References

Dzulynski, S.; Ksiazkiewicz, M.; and Kuenen, P. H., 1959. Turbidites in flysch of the Polish Carpathian mountains, *Geol. Soc. Am. Bull.*, 70, 1089-1118.

Emery, K. O.; Uchupi, E.; Phillips, J. D.; Bowin, C. O.; Bunce, E. T.; and Knott, S. T., 1970. Continental rise off eastern North America, *Bull. Am. Assoc. Petrol. Geologists*, 54, 48-108.

Gonthier, E., and Klingebiel, A., 1973. Facies et processus sédimentaires dans le canyon sous-marin Gascoyne 1, *Bull. Inst. Geol. Bassin d'Aquitaine*, 13, 163-262.

Heezen, B. C.; Tharp, M.; and Ewing, M., 1959. The floors of the oceans, 1. The North Atlantic, *Geol. Soc. Am. Spec. Pap.* 65, 122p.

Lewis, K. B., and Kohn, B. P., 1973. Ashes, turbidites, and rates of sedimentation on the continental slope off Hawkes Bay, *New Zealand J. Geol. Geophys.*, 16, 439-454.

Lowe, D. R., 1972. Implications of three submarine mass-movement deposits, Cretaceous, Sacramento Valley, California, *J. Sed. Petrology*, 42, 89-101.

Minter, L. L.; Keller, G. H.; and Pyle, T. E., 1975. Morphology and sedimentary processes in and around Tortugas and Agassiz Sea valleys, southern Straits of Florida, *Marine Geol.*, 18, 47-69.

Murray, J., and Renard, A. F., 1891. Report on deep-sea deposits based on specimens collected during the voyage of the H.M.S. *Challenger* in the years 1872-1876, *Report on the Voyage of H.M.S. Challenger*. London: H.M.S.O., 525p.

Rona, P. A., 1969. Middle Atlantic continental slope of the United States: deposition and erosion, *Bull. Am. Assoc. Petrol. Geologists*, 53, 1453-1465.

Schmitz, W. J., 1974. Observations of low frequency fluctuations on the continental slope and rise near site D, *J. Marine Research*, 32, 233-251.

Shepard, F. P.; Marshall, N. F.; and McLoughlin, P. A., 1974. Currents in submarine canyons, *Deep-Sea Research*, 21, 691-706.

Southard, J. B., and Cacchione, D. A., 1973. Experiments on bottom sediment movement by breaking internal waves, in D. J. P. Swift, et al., eds., *Shelf Sediment Transport—Process and Pattern*. Stroudsburg, Pa.: Dowden, Hutchinson & Ross, 83-97.

Stanley, D. J., 1969. Sedimentation in slope and base-of-slope environments. *The NEW Concepts of Continental Margin Sedimentation*. Washington, D.C.: American Geol. Inst., lecture 8.

Stanley, D. J., 1970. Flyschoid sedimentation on the outer Atlantic margin off northeast North America, *Geol. Assoc. Canada, Spec. Pap.* 7, 179-210.

Stanley, D. J., and Unrug, R., 1972. Submarine channel deposits, fluxoturbidites and other indicators of slope and base-of-slope environments in modern and ancient marine basins, *Soc. Econ. Paleont. Mineral. Spec. Publ.* 16, 287-340.

Cross-references: *Continental-Rise Sediments; Contourites; Olistostrome, Olistolite; Outer Continental-Shelf Sediments; Slump Bedding; Submarine Canyons and Fan Valleys, Ancient; Submarine Fan (Cone) Sedimentation; Turbidity-Current Sedimentation.* Vol. I: *Bathyal Zone.*

SUBMARINE CANYONS AND FAN VALLEYS, ANCIENT

The rapidly accumulating data on the sedimentology of modern submarine canyons and fan valleys were summarized by Shepard and Dill (1966), and many papers have appeared since that date (see *Submarine (Bathyal) Slope Sedimentation; Submarine Fan (Cone) Sedimentation*). Ancient (pre-Pleistocene) canyons and fan valleys are much less well known, and thus far examples have been described from fewer than 40 areas. Even this limited number includes some that are poorly substantiated owing to lack of data.

Stanley (1967) compares the sedimentology of some modern and ancient canyons, Stanley and Unrug (1972) deal with submarine channel deposits as slope and base-of-slope indicators in modern and ancient basins, and Whitaker (1974) lists all known ancient canyons and fan valleys, analyzing their similarities and differences and stating criteria for recognizing ancient examples. For full bibliographies, these papers should be consulted. The ages of old canyons range from Tertiary (when many were formed) to Precambrian. See also Whitaker (1976).

Although few ancient submarine canyons and fan valleys are known, their morphology and sedimentary fills compare closely with the various parts of some modern canyons and fan channels. There are canyon heads cut in shallow water, long and wide sinuous canyons leading to deeper water, and fan valleys which traverse the deep-sea submarine fans built out from the lower ends of many ancient canyons.

Canyon Heads

Cut-and-filled canyon heads are well exposed near Leintwardine, Welsh Borderland (Whitaker, 1962). In late Silurian time, six parallel canyon heads, fault controlled, deepened rapidly from shelf to basin, with axial gradients up to 10°. Walls had inclinations up to 35°, but some were vertical where lithified limestones cropped out. After rapid cutting, the channels began to fill with finely laminated calcareous siltstones showing concave-up bedding due to axial parts compacting more than the flanks (see Fig. 1). Slumps developed downflank or downchannel, and axial currents produced ripple marks trending across the canyon heads, produced scour marks, and oriented some fossils. Fig. 1 shows these features and lists an unusual indigenous fauna concentrated in the channels. Shelf faunas (such as *Dayia navicula*) were swept in from the shelf at times of greater current activity, adding a derived fauna to the indigenous fauna of the fill; occasionally, limestone boulders up to 2 m

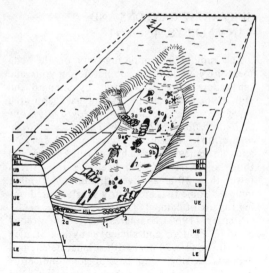

FIGURE 1. Diagrammatic reconstruction of a submarine canyon head of Ludlovian age, Welsh Borderland (from Whitaker, 1962, reproduced with the permission of the Council of the Geological Society of London). Data combined from six channels. Not to scale. 1—concave-up bedding of channel fill; 2—slump structures (a, down-flank; b, down-channel); 3—boulder beds (a, imbricate down-flank; b, imbricate down-channel); 4—ripple marks at right angles to axis; 5—grooves filled with broken fossils, parallel with axis; 6—skip casts parallel with axis; 7—prod casts parallel with axis; 8—oriented fossils a, *Saetograptus leintwardinensis* parallel with axis; b, *Sphaerirhynchia wilsoni* and *Camarotoechia nucula,* umboes upchannel); 9—unusual fauna concentrated in channel (a, Eurypterida; b. Phyllocarida; c, Asterozoa; d, Echinoidea; e, Annelida; f, Xiphosura).

long and fossil-bearing mud lumps fell from the canyon walls, depressed the underlying siltstone laminae, and contributed an older derived fauna to the fill. While the channels were filling, occasional flushing produced minor unconformities, and derived fossils became rederived.

Although these Leintwardine canyon heads are known in detail, the largest of the six channels is only 4 km long, 1 km wide, and 185 m deep; and they do not provide good examples of downcanyon morphology and sedimentation.

Canyons

Well-studied examples occur in the uppermost Eocene and Lower Oligocene Annot and Contes channels, SE France (Stanley, 1967, 1975). The latter is 8 km long, 2 km wide, >350 m deep. The fillings of these channels are coarser than in the previous example, consisting of massive single or composite sand layers, some >20 m thick, with pebbles showing some downslope imbrication and decreasing in size away from the canyon heads. Large sandstone spheroids, some with mudstone nuclei, and slumps >8 m thick occur, directed both downflank and downchannel. There are interbedded shales. Also present are cut-and-fill structures, ripple marks at right angles to the axis, scour marks behind irregularities, and cross-laminated silt and sand as discontinuous sheets; the coarsest material occupies isolated channel tongues in the axes of the main and tributary canyons. Stanley designates these deposits as fluxoturbidites or proximal turbidites and notes that graded bedding improves downslope toward the canyon fans.

Several large canyons are found in the US in the thick Tertiary sequences of California, Texas, and Louisiana. They are almost entirely subsurface and have been located and mapped by seismic methods and deep drilling. Some contain oil or gas, either in the canyon fill itself, if sandy, or trapped beneath the fill, if shaly and resting on sand layers truncated by the fill. All trend down the slope from shelf toward the basin environments. The Middle Wilcox or Yoakum Middle Eocene canyon in Texas is one of the largest known. It is sinuous, 80 km long; 16 km wide, increasing downdip; and about 915 m deeper than the shelf into which it was cut. The fill is almost uniformly of silty shale, but there is more sand updip: there has been considerable compaction of this fill material (Hoyt, 1959). Below the Great Valley of California lies the late Miocene Rosedale canyon (Martin, 1963). It is 10 km long, up to 2.5 km wide, and at least 370 m deep; its axis slopes at 2½°, its walls 11–22°. It is filled mainly with sandstones, some graded, but there are also conglomerates and interbedded siltstones. The apparent increase in shale content downcanyon makes subsurface recognition of the fill difficult. Shallow mega- and microfossils are mixed with indigenous foraminifera, which indicate a water depth of at least 400 m at the time of filling of the canyon.

At least three canyons are known below the N Sacramento Valley. One is a 64-km long, narrow, sinuous canyon cut into Upper Cretaceous and filled with marine Eocene over 615 m thick (Frick et al., 1959). Another, the Markley Gorge, cuts into late Cretaceous and is filled with Oligocene and possibly Miocene shales, sandstones, and conglomerates 770 m thick (Almgren and Schlax, 1957). The lower part of the fill has Oligocene foraminifera with rare upper Eocene forms derived from the channel walls. The third canyon, described in detail by Dickas and Payne (1967), is the large Meganos Channel, its characteristics determined from both outcrop and subsurface data. It is 80 km long, 9.6 km wide, 620 m thick, cutting late Cretaceous and filled with late Paleocene sediments

SUBMARINE CANYONS AND FAN VALLEYS, ANCIENT

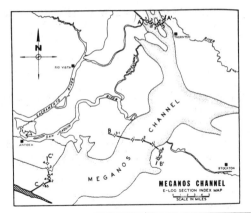

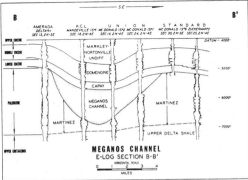

FIGURE 2. *Top:* Location of electric-log sections and wells used, Meganos channel, Sacramento Valley, California. *Bottom:* Cross section B–B' showing differential compaction of channel shale beds. (After Dickas and Payne, 1967.)

(Fig. 2). Its axis slopes about 2°, and its walls slope 5–15°. The fill consists almost entirely of silty shales, but sandstones are more abundant upcanyon. As with the Middle Wilcox channel, compaction is considerable, estimated at 35–60%. The silty shale fill rests mainly on massive sands: four oil and gas fields occur along its length. The canyon was apparently cut in relatively short time by extensive slumping and turbidity currents in water depths from neritic to upper bathyal (Dickas and Payne, 1967).

Tertiary canyons are being discovered on the continental shelf by seismic methods. For example, in the East Bass Strait, off SE Australia, an early Oligocene canyon fill is overlain by fill of early to middle Miocene age (Conolly, 1968).

Fan Valleys

Beyond the lower end of an ancient canyon is found a leveed fan valley, perhaps branching into less-incised distributaries in a downfan direction (see *Submarine Fan (Cone) Deposits*).

The Upper Carboniferous Grindslow Shales of Derbyshire, England, are interpreted as the fan deposits debouching from canyons (Walker, 1966).

A combination of ancient canyons and fan valleys is seen in the Permian of the Delaware Basin, described by Jacka et al. (1968). The canyons have fills of limestone blocks derived from reef masses, and the deep-sea fans have recognizable inner (proximal), middle (intermediate), and outer (distal) parts, each part displaying characteristic fan-valley, levee, and overbank deposits. The authors compare the Permian canyons and fans with modern counterparts off southern California.

A well-documented deep-sea fan, the Eocene Butano Fan in California, has been described in detail by Nilsen and Simoni (1973). Study of clast orientation, sole marks, and internal structures has given paleocurrent directions, and measurements from contorted strata have given paleoslopes; northward transport of finer grained sediments outward and away from channels by overspilling is deduced (Fig. 3). Nelson and Nilsen (1974) compare the Butano Fan with the modern Astoria Fan.

A spectacular cross section of the Miocene Doheny Channel seen in the coastal cliffs at Dana Point, southern California, was interpreted by Bartow (1966) as a distributary channel, emerging from a submarine canyon, crossing a deep-sea fan. Foraminifera indicate that the depth of formation was 615 m. Normark and Piper (1969) studied this fan valley in greater detail, compared it with modern equivalents, and examined it again (Piper and Normark, 1971) after the cliff exposure had been cut back artificially.

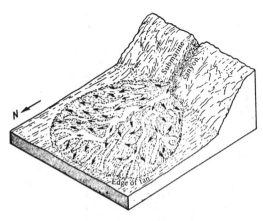

FIGURE 3. Sediment-dispersal model for the Eocene Butano Sandstone, a deep-sea fan deposit (from Nilsen and Simoni, 1973). Arrows represent direction of sediment transport; shaded areas represent major channels.

779

Other submarine canyons and fan valleys from the geological record could have been selected from Italy, the Pyrenees, the Carpathians, or Japan (see Whitaker, 1974, 1976), but the examples given are representative of the wide range of types so far described. One advantage of studying ancient rather than modern valleys is that they often show evidence of their complete evolutionary history: a rapid downcutting episode followed by a period of canyon filling by a variety of sediments and showing many different primary structures—filling which may be delayed by flushing-out processes. Further studies of ancient submarine channels—their morphology, relationship with contemporaneous faulting, types of sediment fillings, indigenous and derived faunas, and evolutionary histories—should throw light on the origins and probable histories of some of the modern canyons and fan valleys.

J. H. McD. WHITAKER

References

Almgren, A. A., and Schlax, W. N., 1957. Post-Eocene age of "Markley Gorge" fill, Sacramento Valley, California, *Bull. Am. Assoc. Petrol. Geologists,* 41, 326-330.

Bartow, J. A., 1966. Deep submarine channel in Upper Miocene, Orange County, California, *J. Sed. Petrology,* 36, 700-705.

Conolly, J. R., 1968. Submarine canyons of the continental margin, east Bass Strait (Australia), *Marine Geol.,* 6, 449-461.

Dickas, A. B., and Payne, J. L., 1967. Upper Paleocene buried channel in Sacramento Valley, California, *Bull. Am. Assoc. Petrol. Geologists,* 51, 873-882.

Frick, J. D.; Harding, T. P.; and Marianos, A. W., 1959. Eocene gorge in northern Sacramento Valley (abstr.), *Bull. Am. Assoc. Petrol. Geologists,* 43, 255.

Hoyt, W. V., 1959. Erosional channel in the Middle Wilcox near Yoakum, Lavaca County, Texas, *Gulf Coast Assoc. Geol. Soc. Trans.,* 9, 41-50.

Jacka, A. D.; Beck, R. H.; St. Germain, L. C.; and Harrison, S. C., 1968. Permian deep-sea fans of the Delaware Mountain Group (Guadalupian), Delaware Basin, *Soc. Econ. Paleont. Mineral. Perm. Basin Sec. Publ. 68-11,* 49-90.

Martin, B. D., 1963. Rosedale Channel—evidence for Late Miocene submarine erosion in Great Valley of California, *Bull. Am. Assoc. Petrol. Geologists,* 47, 441-456.

Nelson, C. H., and Nilsen, T. H., 1974. Depositional trends of modern and ancient deep-sea fans, *Soc. Econ. Paleont. Mineral. Spec. Publ. 19,* 66-91.

Nilsen, T. H., and Simoni, T. R., Jr., 1973. Deep-sea fan paleocurrent patterns of the Eocene Butano Sandstone, Santa Cruz Mountains, California, *J. Research U.S. Geol. Surv.,* 1, 439-452.

Normark, W. R., and Piper, D. J. W., 1969. Deep-sea fan-valleys, past and present, *Geol. Soc. Am. Bull.,* 80, 1859-1866.

Piper, D. J. W., and Normark, W. R., 1971. Re-examination of a Miocene deep-sea fan and fan-valley, southern California, *Geol. Soc. Am. Bull.,* 82, 1823-1830.

Shepard, F. P., and Dill, R. F., 1966. *Submarine Canyons and Other Sea Valleys.* Chicago: Rand McNally, 381p.

Stanley, D. J., 1967. Comparing patterns of sedimentation in some modern and ancient submarine canyons, *Earth Planetary Sci., Lett.,* 3, 371-380.

Stanley, D. J., 1975. Submarine canyon and slope sedimentation (Grès d'Annot) in the French Maritime Alps, *Guidebook 9th Int. Cong. Sed. (Nice),* 130p.

Stanley, D. J., and Unrug, R., 1972. Submarine channel deposits, fluxoturbidites and other indicators of slope and base-of-slope environments in modern and ancient marine basins, *Soc. Econ. Paleont. Mineral. Spec. Publ. 16,* 287-340.

Walker, R. G., 1966. Shale Grit and Grindslow Shales: transition from turbidite to shallow water sediments in the Upper Carboniferous of Northern England, *J. Sed. Petrology,* 36, 90-114.

Whitaker, J. H. McD., 1962. The geology of the area around Leintwardine, Herefordshire, *Quart. J. Geol. Soc. London,* 118, 319-351.

Whitaker, J. H. McD., 1974. Ancient submarine canyons and fan valleys, *Soc. Econ. Paleont. Mineral. Spec. Publ. 19,* 106-125.

Whitaker, J. H. McD., ed., 1976. *Submarine Canyons and Deep-Sea Fans, Modern and Ancient.* Stroudsburg, Pa.: Dowden, Hutchinson & Ross, 460p.

Cross-references: *Continental-Rise Sediments; Gravity Flows; Outer Continental-Shelf Sediments; Slump Bedding; Submarine (Bathyal) Slope Sedimentation; Submarine Fan (Cone) Sedimentation; Turbidites.*

SUBMARINE FAN (CONE) SEDIMENTATION

Cone-shaped mounds called submarine fans are commonly encountered along the outer continental margins of the world oceans and in smaller seas, and are best developed seaward of major rivers. Submarine fans most frequently occupy base-of-slope environments on continental terraces and continental borderlands and also form on slopes in island arc settings, enclosed seas and in deep-sea trenches. Similar features also occur in some lakes (Forel, 1885; Nelson, 1967). Other terms applied to these mounds include *subsea* or *deep-sea fans* or *cones, submarine deltas,* or *abyssal cones* or *fans.*

Fans, in most instances, are defined as depositional structures resulting from the vertical and seaward progradation of largely terrigenous material. Their radial symmetry is attributed to confined flow of sediments downslope through a submarine canyon and migrating channel feeder system and their subsequent deposition

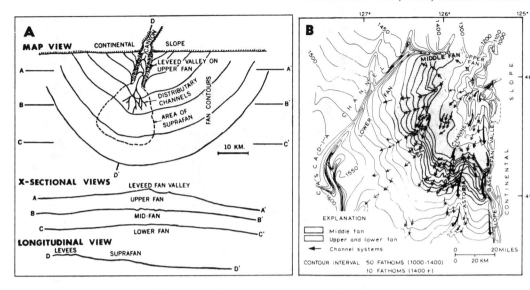

FIGURE 1. A: Physiographic division of fan. Note the convex-up bulge and distributary channel system on the suprafan (after Normark, 1970). B: Radial sediment dispersal pattern of the Astoria Fan off northwestern states (from Nelson and Nilsen, 1974).

from a radiating distributary fan-valley complex in a zone of decreasing gradient at the end of the canyon (Fig. 1). Sediment accumulation results in cone-shaped build-ups generally 0.5–3 km thick (medium sized fans). The Ganges Bengal Fan in the Indian Ocean (Fig. 2), the world's largest fan, is 3000 km long, 1000 km wide, and as much as 12 km thick (Curray and Moore, 1971). In a few instances, submarine fans have been identified as essentially structural features, e.g., the Balearic Rise, a large, cone-shaped foundered block lying S of the Balearic Platform in the western Mediterranean (Stanley et al., 1974), or as depositional–tectonic bodies, e.g., the La Jolla Fan, a thin, sediment-covered basement block off southern California (Shepard et al., 1969).

Morphology

The size, thickness, and general morphology of fans are most closely related to the volume of sediment transported by the canyon-valley system, but the tectonic and geographic framework of the site of deposition also influences the ultimate shape of the deposit.

Downslope and cross-slope echosounder and seismic transects across submarine fans allow the definition of four subenvironments downslope away from the fan apex (Fig. 1): (1) The *upper fan* zone is defined by a concave-up profile and a steep gradient approximating 1:100; the surface is incised by one main, deep, levee-bounded channel (Fig. 2). (2) The *middle fan*, or *suprafan* (Normark, 1970), zone is distinguished by a lower gradient (1:500) and convex-up profile; it is on this major fan bulge that the major channel splits into numerous meandering distributaries with less well-developed levees. (3) The *lower fan* zone, with a concave-up profile of low gradient (1:1000), is characterized by a smooth surface of low relief crossed by a broad, shallow, braided channel system. (4) The distal *fan fringe* extends between the fan lobes and outer fan and the nearly flat abyssal (or basin) plain surface beyond; its smooth surface is crossed by gentle channels only a few m wide and deep. Descriptions of modern and ancient fan morphology are provided by Shepard et al. (1969), Stanley (1969), Nelson et al. (1970), Normark (1970, 1974), Curray and Moore (1971), Normark and Piper (1972), Nelson and Kulm (1973), Nelson and Nilsen (1974), Cleary and Conolly (1974), Stanley and Kelling (1978).

Transport Mechanisms and the Fan Sedimentation Model

Downslope transport mechanisms channelize fine to coarse sediment types in, and close to, submarine valleys, while finer-grained units (mostly turbidites and hemipelagic deposits) are deposited downslope and in interchannel areas away from fan valleys as a result of overbank flow and suspension mechanisms (Fig. 3). The predominant transport agent is turbidity current flow (Menard, 1955), which in the case

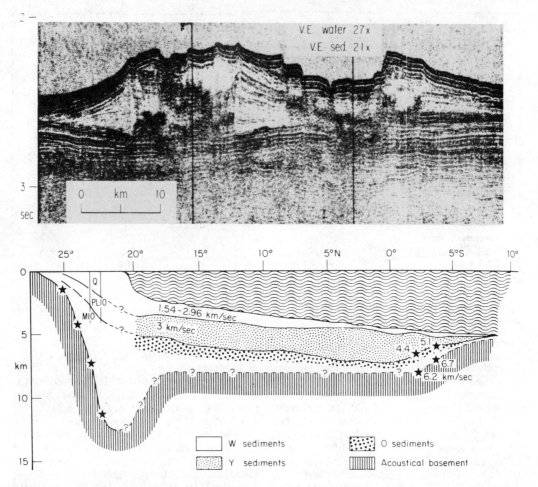

FIGURE 2. *Top:* Reflection profile across the northern apex of the Bengal Fan showing the major channel, terraces, the more acoustically transparent levee sediments, and the old surface of the fan depressed beneath the older fan system. *Bottom:* North to south longitudinal section from the Ganges-Brahmaputra delta seaward across the Bengal Fan showing interpretation of the Quaternary, Pliocene, and late Miocene, and pre-late Miocene sediments (W, Y, and O, respectively) lying above the acoustical basement. (From Curray and Moore, 1971.)

of large open ocean basins is deflected by the Coriolis effect and by flow of deep-water masses along the bottom. In some instances these bottom currents move parallel and subparallel to contours (see *Contourites*). These factors influence the rate of channel migration across the inclined fan surface and, thus, the rates of interchannel deposition and, ultimately, the overall shape of the fan.

In a comparative analysis of the Quaternary Astoria Fan off the Oregon-Washington margin and the Eocene Butano Fan, Nelson and Nilsen (1974) identify five major facies: (1) chaotic, massive-bedded, and very poorly sorted sand and gravel units at the canyon mouth and in upper fan channels are transported by *debris flow* (see *Mudflow*) from the adjacent slope and canyon walls (Stanley, 1974). Massive-bedded, better-sorted, low-matrix sands and gravels primarily in the upper fan channels, are deposited by *liquefied* and/or *grain flow* processes. (3) Basal *a-b* turbidite sequences (see *Bouma Sequence*) in middle and lower fan channels are deposited by *turbidity currents* flowing downslope under upper flow-regime conditions. (4) Well-developed turbidite *c* and *d* sequences (see *Bouma Sequence*) occur in and outside fan valleys, including natural levees (Hall and Stanley, 1973); they are deposited from turbidity currents flowing under a lower flow regime than (3). (5) Hemipelagic sequences consisting of a predominant clay and silt fraction of terrigenous origin and a sand-sized biogenic fraction are transported by various

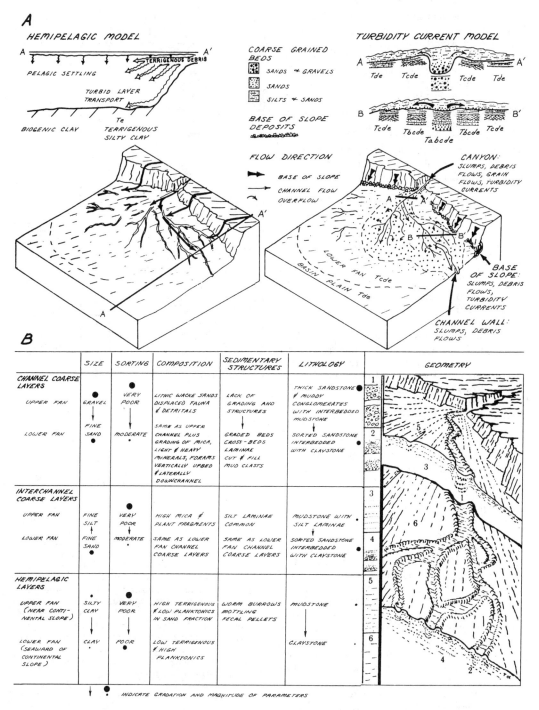

FIGURE 3. A: Sedimentation models showing hemipelagic and turbidity current processes on a fan surface. Cross-sections A-A' and B-B' show the distribution of sediment type and sequence of structures in a deposit laid down in a single flow. Other mechanisms are also illustrated in the feeder canyon and adjacent slope. B: Sedimentary facies and depositional trends in the different fan sectors. (After Nelson and Kulm, 1973.)

suspension mechanisms including *turbid layer flows* (Moore, 1969), low-density turbidity-current flows (*mud turbidites,* see Piper, 1972; Rupke and Stanley, 1974), nepheloid layers (q.v.), and suspension layers transported by contour-following bottom currents (see *Contourites*). The latter facies predominates in interchannel and distal (lower fan and fan fringe) environments.

Distinguishing Fan Deposits. The five major lithologic facies discussed above, and modifications thereof, are also recognized in ancient fan deposits (Sullwold, 1960; Walker, 1966; Mutti and Ricci Lucchi, 1972; Stanley and Unrug, 1972; Mutti, 1974; see also *Submarine Canyons and Fan Valleys, Ancient*). In typical fan sections, lenticular, coarser-grained debris define fan-valley deposits; laterally more extensive and thinner sheets of finer-grained units are interpreted as overflow facies that were deposited between channels (Figs. 2, 3). Deep-sea cores, high-resolution subbottom profiles, and detailed mapping of land exposures serve to distinguish mid- and lower-fan facies from most other types of deposits: most characteristic is the vertical and lateral alternation of channel and interchannel deposits that result from the meandering and lateral migration of the channels across a low gradient slope (Fig. 4).

The presence of channels migrating across the middle and lower fan surface precludes a bed-by-bed correlation; there is an abrupt facies change between channel and interchannel deposits. Downslope channelization of sediment in fan valleys can result in the by-passing of coarser-grade material so that proximal facies (1), (2), and (3) are encountered within more distal, fine-grade turbidite and hemipelagite sequences. In general, lateral gradation from channel to interchannel facies becomes less abrupt in lower fan settings because the importance of overbank flows increases. Also characteristic of fans is the *downchannel* and downfan decrease in the thickness of the basal part of turbidite layers (see *Turbidites*). A consistent decrease downslope also occurs in maximum clast size and percent of sand coincident with improved sorting (Walker and Mutti, 1973), less positive skewness, and more mica and plant fragments away from the fan apex (Fig. 3).

The influence of bottom-water movement, including contour-following currents, can modify to varying degrees the slope and surface sediment of a fan. An example is the smoothing of the upper surface of the large (>100 km wide) coalescing *megafans* that form the continental rise off the mid-Atlantic states of North America (see *Continental-Rise Sediments*). Southwest flowing water masses

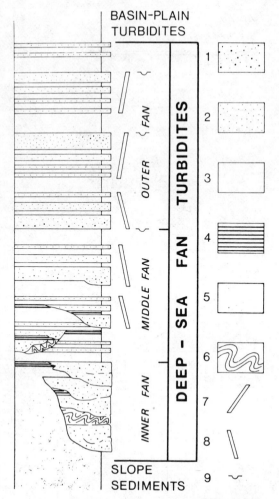

FIGURE 4. Hypothetical model of a fan progradational sequence based on facies analyses of ancient fan deposits (from Mutti, 1974): (1) thick-bedded graded and/or crudely laminated coarse-grained sandstone and conglomerate; (2) thick-bedded graded sandstone and varying amounts of finer-grained current-laminated deposits, including mudstone; (3) mudstone and parallel bedded, thoroughly current-laminated sandstone and siltstone; (4) thin, lenticular, and discontinuous beds of fine- to coarse-grained sandstone displaying grading and poorly developed current laminae; (5) hemipelagic mudstone; (6) chaotic sediment; (7) thickening-upward cycle; (8) thinning-upward cycle; (9) shallow (<2 m) channels.

(Western Boundary Undercurrent) are capable of reworking silt and sand units as well as of depositing clay-silt fractions on the fan surface.

The vertical growth of fan bulges is continuously changing with time, so that the youngest deposits in any fan can be found in the upper, middle, or lower fan region, depending on such

Importance of Fans in the Geologic Record

Fan development in the world oceans was considerably accelerated in the Pleistocene as a result of the marked climatic and sea-level oscillations related to the waxing and waning of glaciers. During the low stands of the sea, considerably larger volumes of terrigenous material were transferred from the subaerially exposed shelves directly into canyon heads than at present. The growth of the large submarine deltas of the Rhone (Bellaiche, 1975), Nile (Maldonado and Stanley, 1976), Amazon (Damuth and Kumar, 1975), and other major rivers owe much of their present configuration to these Quaternary events.

Tectonic activity such as the uplift of mountain chains or hinterland also results in increased rates of denudation and the more rapid build-up of fans offshore. For instance, the close relation between the orogenic phases in the Himalayas and the growth of the Bengal Fan has been demonstrated (Curray and Moore, 1971). Graham et al. (1975) have used the Bengal Fan as a model for Carboniferous flysch sedimentation in the Ouachitas.

Fans in deep-sea trenches may display a long, linear channel facies oriented parallel to the trench axis, and smaller fan lobes commonly build out perpendicularly to the steep trench walls (Piper et al., 1973). Fans that develop in open ocean basins tend to be well-developed, large and thick; those in smaller and more restricted basins are often smaller, thinner (with the exception of those off large rivers, e.g., the Nile and Mississippi cones), and irregular in shape due to the limiting modification of adjacent topographic or tectonic barriers.

Individual submarine fans and coalescing fan complexes forming rise wedges at the base of the continental slope account for the thickest and most extensive accumulations of terrigenous sediment in the world oceans. Recent work shows that fan facies consisting of turbidite sands and muds, hemipelagites, and lenticular sequences (debris flow, grain/liquefied flow, and turbidity-current flow) are probably also volumetrically the most extensive type of deep-marine sediment deposits preserved in the ancient rock record (see *Gravity Flows*). Investigations of exposures in the circum-Mediterranean, western North and South American, and other mobile belts indicate that submarine fan deposits represent the most important sediment facies in flysch and related deep-water geosynclinal formations.

DANIEL JEAN STANLEY

References

Bellaiche, G., 1975. Sur l'origine et l'age des levées sédimentaires profondes: cas du delta sous-marin du Rhône (Mediterranée nord-occidentale, *Marine Geol.*, **19**, M1-M6.

Cleary, W. J., and Conolly, J. R., 1974. Hatteras deep-sea fan, *J. Sed. Petrology*, **44**, 1140-1154.

Curray, J. R., and Moore, D. G., 1971. Growth of the Bengal deep-sea fan and denudation in the Himalayas, *Geol. Soc. Am. Bull.*, **82**, 563-572.

Damuth, J. E., and Kumar, N., 1975. Amazon Cone: Morphology, sediments, age, and growth pattern, *Geol. Soc. Am. Bull.*, **86**, 863-878.

Forel, F., 1885. Les ravins sous-lacustres des fleuves glaciaires, *Compt. Rend. Acad. Sci.*, **101**, 725-728.

Graham, S. A.; Dickinson, W. R.; and Ingersoll, R. V., 1975. Himalayan-Bengal model for flysch dispersal in the Appalachian-Ouachita System, *Geol. Soc. Am. Bull.*, **86**, 273-286.

Hall, B. A., and Stanley, D. J., 1973. Levee-bounded submarine base-of-slope channels in the Lower Devonian Seboomook Formation, northern Maine, *Geol. Soc. Am. Bull.*, **84**, 2101-2110.

Kruit, C.; Brouwer, J.; and Ealey, P., 1972. A deep-water sand fan in the Eocene Bay of Biscay, *Nature*, **240**, 59-61.

Maldonado, A., and Stanley, D. J., 1976. The Nile Cone: Submarine fan development by cyclic sedimentation, *Marine Geol.*, **20**, 27-40.

Menard, H. W., 1955. Deep-sea channels, topography and sedimentation, *Bull. Am. Assoc. Petrol. Geologists*, **39**, 236-255.

Moore, D. G., 1969. Reflection profiling studies of the California continental borderland: structure and Quaternary turbidite basins, *Geol. Soc. Am. Spec. Pap. 107*, 142p.

Mutti, E., 1974. Examples of ancient and deep-sea fan deposits from circum-Mediterranean geosynclines, *Soc. Econ. Paleont. Mineral. Spec. Publ. 19*, 92-105.

Mutti, E., and Ricci Lucchi, F., 1972. Le torbiditi dell'Appennino settentrionale: Introduzione all'analisi di facies, *Soc. Geol. Ital. Mem.* **11**, 161-199.

Nelson, C. H., 1967. Sediments of Crater Lake, Oregon, *Geol. Soc. Am. Bull.*, **78**, 833-848.

Nelson, C. H., and Kulm, L. D., 1973. Submarine fans and deep-sea channels, in G. V. Middleton and A. H. Bouma, eds., *Turbidites and Deep-Water Sedimentation*. Anaheim: SEPM, Pacific Sec., 39-70.

Nelson, C. H., and Nilsen, T. H., 1974. Depositional trends of modern and ancient deep-sea fans, *Soc. Econ. Paleont. Mineral. Spec. Publ. 19*, 69-91.

Nelson, C. H.; Carlson, P. R.; Byrne, J. V.; and Alpha, T. R., 1970. Development of the Astoria canyon-fan physiography and comparison with similar systems, *Marine Geol.*, **8**, 259-291.

Normark, W. R., 1970. Growth patterns of deep-sea fans, *Bull. Am. Assoc. Petrol. Geologists*, **54**, 2170-2195.

Normark, W. R., 1974. Submarine canyons and fan valleys: Factors affecting growth patterns of deep-sea fans, *Soc. Econ. Paleont. Mineral. Spec. Publ. 19*, 56-68.

Normark, W. R., and Piper, D. J., 1972. Sediments and growth pattern of Navy deep-sea fan, San Clemente Basin, California borderland, *J. Geol.,* **80,** 198-223.

Piper, D. J. W., 1972. Turbidite origin of some laminated mudstones, *Geological Mag.,* **109,** 115-126.

Piper, D. J. W.; von Huene, R.; and Duncan, J. R., 1973. Late Quaternary sedimentation in the eastern Aleutian Trench, *Geology,* **1,** 19-22.

Rupke, N. A., and Stanley, D. J., 1974. Distinctive properties of turbiditic and hemipelagic mud layers in the Algéro-Balearic Basin, western Mediterranean Sea, *Smithsonian Contr. Earth Sci.,* **13,** 40p.

Shepard, F. P.; Dill, R. F.; and von Rad, U., 1969. Physiography and sedimentary processes of La Jolla submarine fan and fan-valley, California, *Bull. Am. Assoc. Petrol. Geologists,* **53,** 390-420.

Stanley, D. J., 1974. Pebbly mud transport in the head of Wilmington Canyon, *Marine Geol.,* **16,** M1-M8.

Stanley, D. J., ed., 1969. *The NEW Concepts of Continental Margin Sedimentation.* Washington, D.C.: Am. Geol. Inst., 400p.

Stanley, D. J., and Unrug, R., 1972. Submarine channel deposits, fluxoturbidites, and other indicators of slope and base-of-slope environments in modern and ancient basins, *Soc. Econ. Paleont. Mineral Spec. Publ. 16,* 287-340.

Stanley, D. J.; Got, H.; Leenhardt, O.; and Weiler, Y., 1974. Subsidence of the western Mediterranean basin in Pliocene-Quaternary time: Further evidence, *Geology,* **2,** 345-350.

Stanley, D. J., and Kelling, G., eds., 1978. *Sedimentation in Submarine Canyons, Fans, and Trenches.* Stroudsburg, Pa.: Dowden, Hutchinson & Ross, 416p.

Sullwold, H. H., 1960. Tarzana Fan, deep submarine fan of late Miocene age, Los Angeles County, California, *Bull. Am. Assoc. Petrol. Geologists,* **44,** 433-457.

Walker, R. G., 1966. Shale grit and Grindslow shales: Transition from turbidite to shallow-water sediments in the Upper Carboniferous of northern England, *J. Sed. Petrology,* **36,** 90-114.

Walker, R. G., and Mutti, E., 1973. Turbidite facies and facies associations, *Soc. Econ. Paleont. Mineral. Pacific Sec., Anaheim, Short Course Lecture Notes,* 119-157.

Cross-references: Bouma Sequence; Contourites; Fluidized Sediment Flows; Geosynclinal Sedimentation; Grain Flows; Gravity Flows; Nepheloid Sediments and Nephelometry; Pebbly Mudstones; Submarine (Bathyal) Slope Sedimentation; Submarine Canyons and Fan Valleys, Ancient; Turbidity-Current Sedimentation.

SULFIDES IN SEDIMENTS

Sediments and sedimentary rocks ranging in age from Precambrian to Holocene contain sulfide minerals that are of theoretical and practical interest inasmuch as the more highly concentrated sulfide units constitute some of the world's best-known ore deposits. The host rock in which the sulfides are formed range widely in origin and composition; they may be claystones or shales, siltstones, sandstones, and conglomerates, as well as carbonates. Also important are pyroclastic rocks and associated volcanic flows. The sulfides, together with their host rocks, can originate in both marine and nonmarine milieux and under geosynclinal and cratonic tectonic conditions. Although widely occurring, thick units of sulfides (i.e., of economic ore grade) are usually recognized to have been formed under particular tectonic requirements, especially where volcanism is pronounced. On the other hand, minor, uneconomic sulfide concentrations are relatively independent of large-scale environmental, tectonic, and volcanic factors; the main controls instead appear to be more local geochemical and biochemical variables.

Sulfides can originate more or less at the same time as their host sediments, i.e., as *syngenetic* and/or *diagenetic* minerals. Much later, the metallic minerals may be precipitated from intraformational solutions after the host rock has been partly lithified; the sulfide is then termed *epigenetic*. Old sedimentary deposits together with their sulfide minerals, especially in Precambrian Shield areas of Canada, Africa, Australia, and South America, for example, have undergone high-temperature and high-pressure metamorphic alterations in both mineralogy and textures, with minor remobilization of the metallic components and reprecipitation into fractures to form veins. It is important to realize that the primary concentration of the metals was syngenetic-diagenetic; only the last-stage precipitation was epigenetic in nature.

If sulfide mineralization occurs at the same time as the host rock is formed, the ore deposits are parallel to the host rock and are often termed *stratabound ore*. Stratabound ores include some of the world's largest and best-known ore districts.

Mineralogy

Some of the most common sulfide minerals are listed in Table 1, and all are of economic interest if certain requirements are fulfilled. Most ores consist of two or more minerals, so that several metals may be commercially extracted. In considering mineralogy, it must be made clear that terrigenous sediments, and their associated volcanic rocks, are composed of silicate minerals, whereas chemical sediments may be made up of carbonates, oxides,

TABLE 1. Some Common Sedimentary Sulfides and Their Mode of Origin

Mineral	Formula	Hy	OC	Pl	We	Pr
Realgar	AsS	o	x			
Orpiment	As_2S_3	o	x			
Galena	PbS	x				x
Argentite	Ag_2S	x	x			
Chalcocite	Cu_2S	o	x			
Covellite	CuS	x	x			
Chalcopyrite	$CuFeS_2$	x	x			x
Bornite	Cu_5FeS_4	x	x			x
Cinnabar	HgS	x	x	o		
Sphalerite	ZnS	x	x			x
Pyrite	FeS_2	x				x
Millerite	NiS	x			x	

After Schneiderhöhn, 1962.
Hy = Hydrothermal deposits
OC = Oxidation and cementation zones.
Pl = Placer.
We = Weathering products.
Pr = Precipitated from fresh or salt water.
x = main occurrences.
o = less important occurrences.

and many others. The sulfide minerals are, therefore, not basic components of sediments but are the results of special geological and geochemical conditions. It is to be realized that most minerals formed under sedimentary conditions can also originate in either plutonic or volcanic rocks; some minerals are also stable in metamorphic milieux.

Sulfides are, of course, only one group of minerals that constitute metallic concentrations. Under other geological conditions, the metallic ions may combine with CO_3 or O_2 to form carbonates and oxides, respectively. It is also quite common for originally formed sulfides to be subsequently exposed to new geochemical environments (e.g., from reducing to oxidizing Eh) and to be altered to oxides and carbonates. This process is especially prevalent near the earth's surface, so that chalcopyrite—$CuFeS_3$ is changed to malachite—$CuCO_3Cu(OH)_2$.

The textures, fabrics, and structures of the sulfide concentrations range widely. The grain size varies from microscopic particles to a few mm in diameter; the latter are present in those rocks that have undergone metamorphic recrystallization. The sulfide minerals are anhedral, subhedral, and/or euhedral in crystal or grain shape, depending on the condition related to their growth. The fabrics may be framboidal, colloform, laminated and bedded, or structureless internally (massive). Depending on the mode of origin, the sulfide either fills partially to wholly open spaces, or replaces previously existing material, e.g., carbonate minerals. Sulfide minerals may also change composition by an atom-by-atom replacement, e.g., the Fe in pyrite can be replaced by Cu, so that the secondary sulfide mineral is of economic value, whereas this was not the case of the primary pyrite.

Mode of Occurrence

As to large-scale structures and style of distribution of the sulfides, the following are the more common modes of occurrence: disseminations, clots, nodules, concretions, irregular masses, laminae or thin lenses, and beds. Depending on the proportions of sulfide minerals, the ratio of metallic versus host-rock and gangue minerals, the sulfide may range from uneconomic microconcentrations to large ore bodies.

A number of geochemical and geological variables determine the localization of the sulfide mineralization: (1) karst and other cavities; (2) porosity and permeability distribution pattern in both host (=reservoir) and channel (=conduit) rocks; (3) stratigraphic and facies variations; (4) structural controls; and (5) physicochemical and biochemical conditions. In particular, gradational or transitional variations in factors (2), (3), and (5) can result in local to regional (up to several km) zonation along one horizon and/or in a vertical stratigraphic succession.

A number of sedimentary processes can give rise to sulfide mineral concentrations. In the zone of cementation where oxidizing chemical conditions usually prevail, a change to reducing conditions can cause several sulfide minerals to form below the earth's surface at depths of several mm to hundreds or thousands of m. Under certain geochemical conditions, direct syngenetic-diagenetic precipitation from natural surface fluids can concentrate galena, chalcopyrite, bornite, pyrite, and sphalerite. Surface weathering can form millerite, whereas mechanical erosion may transport and accumulate the most resistant minerals, such as cinnabar.

Although the hydrothermal processes have in the past been listed as igneous or magmatic mechanisms, more recently it has been recognized that many of the so-called hydrothermal sulfide ore deposits may have been precipitated by several subtypes of solutions, or mixtures thereof: (1) Surface rain water may pass down to hot igneous rocks, become heated, and then forced through the subsurface rock units as a result of the formation of steam accompanied by increasing pressure. Solutions and gases expelled from a magma or an igneous rock may mix with the heated connate fluids. If the superheated solutions moving through the rocks reach the surface, either on land or in the sea, they are considered to be of a *volcanic-exhalative* nature. (2) Connate fluids in sedimentary basins, which represent trapped original sea water, become gradually heated to form the second variety of hydrothermal water. The fluid is heated as a result of burial and increase of the geothermal gradient. (3) Compaction fluids are related to (2); but there are numerous differences to warrant separate listing, and Wolf (1976a) has given a comprehensive summary of the relationship between compaction and ore-forming processes. During subsidence of a sedimentary sequence, fluids are expelled which may also have undergone an increase in temperature during burial. Many of the stratabound sulfide ore deposits in both sedimentary and volcanic rocks are believed to have formed from either one or a combination of the three hydrothermal processes.

Origin

There are four possible sources for the metallic ions required for the sulfide minerals: (1) transport in solution in ionic or complexed form from the land and thus derived from an older rock, such as a plutonic rock, by subaerial chemical weathering; (2) from the sea water or concentrated saturated surface waters (lake or lagoon); (3) supply by volcanic-exhalative processes and derived from a frequently unknown and/or surmised, deep-seated magmatic source; or (4) from source beds within the sedimentary basin, such as clay-rich beds and volcanic sediments that have metallic ions adsorbed on the fine constituents (e.g., clay minerals per se and organic matter), or the metals are part of the glassy volcanic fragments. On release during diagenesis into subsurface interstitial fluids, the solutions may migrate (possibly as a result of compaction of the sedimentary-volcanic pile) to localities where the physicochemical conditions are conducive to the formation of sulfide minerals (Wolf, 1976a).

The sulfur may have been derived from the following sources: (1) from the mantle of the earth or a deep-seated igneous magma; (2) from volcanic sources, which may be related to the mantle source, but more commonly the sulfur from the so-called volcanic processes is of mixed origin; (3) from sea-water sulfate and related connate intrastratal fluids; (4) from sedimentary evaporite deposits; and (5) from hydrocarbon sources within the sedimentary unit. In many instances, bacterial activity has resulted in the transformation of S-containing compounds to pure sulfur, H_2S, or some other product that was used in the formation of sulfide minerals.

As to the precise mechanisms of formation of the sulfide deposits, much information has been obtained from laboratory and field geochemical (isotope, trace element, phase equilibria, secondary changes during metamorphism, etc.), bacteriological, sedimentological, and stratigraphical studies of modern and ancient rocks as well as from volcanological, tectonic (e.g., plate tectonic and geosynclinal evolution), and comparative studies (e.g., oceanographic versus Paleozoic rock data). Although the factors and conditions that control the origin of sulfide minerals in sediments can be subdivided according to the scale on which they operate and vary in a corresponding manner from case to case, some basic requirements must be fulfilled. The pH and Eh of the ore-forming solutions are most fundamental. In particular, Eh determines whether, all factors equal, chemical elements will combine to form an oxide or a sulfide mineral, inasmuch as a negative Eh (absence of oxygen and presence of sulfur) is required for the precipitation of sulfides. Although purely physicochemical precipitation is possible, bacterial processes have been proven to be very important.

As mentioned above, upon formation of sulfide minerals in a sedimentary environment, the mineralogy, texture, and structure of these deposits usually exhibit characteristics that identify them as low-temperature and low-pressure stratabound ores. On the other hand, secondary changes during diagenesis, burial epigenesis, and metamorphism progressively obliterate and finally destroy many of these features.

Specific combinations of sulfide minerals and host rocks that appear to have originated under particular geologic and geochemical conditions occur on nearly all continents in the world so that it is possible to establish a number of ore types. These types have certain basic common features that allow them to be grouped together. Despite many years of investigations, however, many disagreements persist as to the details of the various ore types. Very detailed petrographic and geochemical examinations linked to regional studies are relatively recent, and this part of the field of ore petrology is still in a developmental state.

Ore Types. Investigations of sedimentary sulfide deposits are of economic importance, for many of them have supplied the world with large volumes of metals. Inasmuch as there is a genetic relationship between the host rocks and the sulfides, it has become customary to classify the ores accordingly and assign a type locality to each of the deposits. Although this system of classification varies among researchers and is in a state of flux as a result of increasing information in ore petrology, the following gives examples of the better known types: (1) sulfides in shales or slates: Mount Isa, Australia; Rammelsberg-Meggen, Germany; Kupferschiefer, Germany; (2) sulfides in carbonates: Mississippi Valley-type lead-zinc-barite-fluorite ore deposits of the Tri-State district, US, and many other parts of the world; (3) sulfides in sandstones: Rhodesian-Zambian Copper Belt, South Africa; Colorado Plateau-Wyoming (=Western State) type of uranium-vanadium deposits, US; White Pine copper deposits, Michigan, US; (4) sulfides in conglomerates: uranium-gold deposits of the Witwatersrand South Africa; uranium deposits of Blind River, Ontario, Canada.

KARL H. WOLF

References

Amstutz, G. C., and Bernard, A. J., eds., 1973. *Ores in Sediments.* Berlin: Springer-Verlag, 350p.

Berner, R. A., 1971. *Principles of Chemical Sedimentology.* New York: McGraw-Hill, 240p.

Filipov, V. A., 1974. Position of stratified deposits of ores in sedimentation cycle, *Akad. Nauk SSSR Doklady,* 216, 655-657.

Garlick, W. G., 1972. Sedimentary environment of Zambian copper deposition, *Geol. Mijnbouw,* 51, 277-298.

Gwosdz, W.; Krüger, H.; Paul, D.; and Baumann, A., 1974. Die Liegendschichten der devonischen Pyrit- u. Schwerspat-Lager von Eisen (Saarland), Meggen und des Rammelsberges, *Geol. Rundschau,* 63, 23-40.

Renfro, A. R., 1974. Genesis of evaporite-associated stratiform metalliferous deposits—a sabkha process, *Econ. Geol.,* 69, 34-45.

Rickard, D. T., 1973. Limiting conditions for synsedimentary sulfide ore formation, *Econ. Geol.,* 68, 605-617.

Sawkins, F.-E. J., 1972. Sulfide ore deposits in relation to plate tectonics, *J. Geol.,* 80, 377-397.

Saxby, J. D., 1973. Diagenesis of metal-organic complexes in sediments: formation of metal sulfides from cystine complexes, *Chem. Geol.,* 12, 241-248.

Schneiderhöhn, H., 1962. *Erzlagerstätten.* Stuttgart: Gustav Fischer, 371p.

Stanton, R. L., 1972. *Ore Petrology.* New York: McGraw-Hill, 713p.

Thornber, M. R., 1975. Supergene alteration of sulphides, I. A chemical model based on massive nickel sulphide deposits at Kambulda, Western Australia, *Chem. Geol.,* 15, 1-14.

Van Eden, J. G., 1974. Depositional and diagenetic environment related to sulfide mineralization, Mulfulira, Zambia, *Econ. Geol.,* 69, 59-79.

Wolf, K. H., 1976a. Ore genesis influenced by compaction, in G. V. Chilingar, and K. H. Wolf, eds., *Compaction of Coarse-grained Sediments,* Vol. 2. Amsterdam: Elsevier, 475-675.

Wolf, K. H., ed., 1976b. *Handbook of Strata-bound and Strataform Ore Deposits,* vol. 1, 2, 3, and 4. Amsterdam: Elsevier, 338p., 363p., 353p., 325p.

Cross-references: *Diagenesis; Euxinic Facies.*

SYNERESIS

Syneresis refers to the spontaneous liberation of liquid and concomitant shrinkage of a gel during aging. The liquid that is extruded is a dilute colloidal sol derived from the original concentrated colloid system. It is caused by the contraction of the internal units of the gel and probably results from the formation of additional bonds between different parts of the colloidal structure (Dean, 1948). Syneresis is probably the most characteristic property of a gel that is aging and is exhibited to a greater or lesser degree by all gels. That syneresis occurs upon aging is an indication that gels are thermodynamically unstable. The absence of syneresis does not prove the converse to be true, however, because marked time lags occur in gel systems. In general, syneresis decreases with increasing concentration of the gelling agent. It is an essentially irreversible process.

Syneresis is not entirely synonymous with dehydration. *Dehydration* and associated shrinkage may occur through evaporation without the release of liquid from the gel (see *Mud*

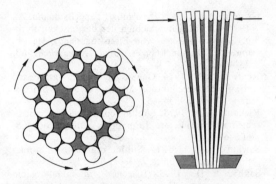

FIGURE 1. The forces that cause silica gel to shrink during drying resemble those that cause wet plates of glass to draw together. Reprinted from Ralph K. Iler: *The Colloid Chemistry of Silica and Silicates*. Copyright 1955 by Cornell University. Used by permission of Cornell University Press.

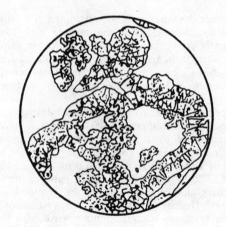

FIGURE 3. Colloform pitchblende in a gangue of barite; Joahimsthal, Bohemia, 35X (after Schneiderhöhn and Ramdohr, 1931). Radial shrinkage cracks are well developed.

Cracks). Silica gel that is dried in air shrinks because the tension of the water menisci in the capillary pores exceeds the tensile strength of the gel structure (Fig. 1). A characteristic feature of contracting gels is the formation of cracks and fissures.

Syneresis is observed in both natural and artificial gels, such as silica gel, and in clays, colloidal ore deposits, and many organic substances. It is an important process in the natural *diagenesis* (q.v.) of aging colloidal gels, leading to the end result of cryptocrystalline aggregates such as chert. Similar effects occur in the hardening of Portland cement. Gel replacement of rock-forming minerals and sulfides is an important aspect of metasomatism in solid rocks. Microstructures in rocks that are indicative of gel origin are colloform structures expressed by curved, reniform, or spheroidal form (Lindgren, 1925; Taliaferro, 1934; Rust, 1935).

The term *syneresis cracks* appears in the geologic literature (e.g., Bastin, 1950) to describe shrinkage cracks that form in contracting gels. Although these gels may exhibit syneresis, the cracking is not necessarily a direct result of the liberation of liquid that occurs during syneresis. That is, cracks can form in shrinking gels where syneresis is absent. Nonetheless, the term does indicate the shrinking of a gel, as opposed to surface mud cracks that form in silt, etc.

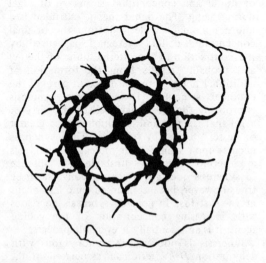

FIGURE 2. Septarian structure. Length of specimen, about 12 cm (from Pettijohn, 1957).

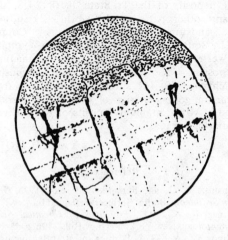

FIGURE 4. Shrinkage cracks developed in successive bands of chalcopyrite deposited on the walls of a fracture in dolomite; Cornwall mine, Missouri, 10X (after Rust, 1935). Cracks in lower band were developed before upper band was deposited.

perpendicular to boundaries, or to crystallographic orientation; and (6) random, irregular cracks (Fig. 5).

As a rule, cracks attributed to shrinkage of natural gels are smaller than mud cracks, but some meter-long cracks in argillaceous limestone have been called syneresis cracks. The recognition of "syneresis" shrinkage cracks and associated colloform structure is evidence of gel deposition in a colloidal state. The form alone is not restricted to colloids, as mud cracks are often similar in appearance, so that caution must be exercised in their interpretation.

JAMES T. NEAL

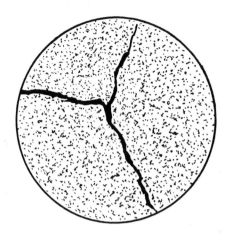

FIGURE 5. Bifurcating shrinkage crack developed in globular area of sulfides that has replaced a dolomite nodule in sandstone; Cornwall mine, Missouri, 23X (after Rust, 1935). The sulfides are aggregates of tiny chalcopyrite spherules, often with pyrite cores, in a bornite matrix.

The geometry of syneresis cracks often resembles that of *mud cracks* (q.v.) because of similar mechanical properties that lead to fissuring. A major factor controlling the geometry is that mud cracks generally form on essentially flat or slightly inclined surfaces exposed to the air, whereas syneresis cracks in natural gels may form in completely confined spaces, at depth, and without reference to the earth's gravity field. Consequently, syneresis cracks, as a rule, are much less regular in form.

Some of the more common forms of shrinkage cracks that occur in gels are: concentric and radial cracks, associated with spheroidal-shaped colloidal structures; (2) septarian structures (Fig. 2), often formed in nodules of aluminous gel in shales; (3) "cracked porcelain" textures, and radial shrinkage forms, resembling mud crack patterns (Fig. 3); (4) tapered cracks, often perpendicular to the surface of deposition (Fig. 4); (5) lenticular cracks, often aligned

References

Bastin, E. S., 1950. Interpretation of ore textures, *Geol. Soc. Am. Mem. 45*, 101p.

Berkson, J. M., and Clay, C. S., 1973. Possible syneresis origin of valleys on the floor of Lake Superior, *Nature,* **245,** 89-91.

Dean, R. B., 1948. *Modern Colloids.* New York: Van Nostrand, 303p.

Donovan, R. N., and Foster, R. J., 1972. Subaqueous shrinkage cracks from the Caithness Flagstone series (Middle Devonian) of Northeast Scotland, *J. Sed. Petrology,* **42,** 309-317.

Iler, R. K., 1955. *The Colloid Chemistry of Silica and Silicates.* Ithaca, N.Y.: Cornell Univ. Press, 324p.

Lindgren, W., 1925. Gel replacement, a new aspect of metasomatism, *Proc. Natl. Acad. Sci.,* **11,** 5-11.

Pettijohn, F. J., 1957. *Sedimentary Rocks,* 2nd ed. New York: Harper & Row, 718p.

Rust, G. W., 1935. Colloidal primary copper ores at Cornwall Mines, southeastern Missouri, *J. Geol.,* **43,** 398-426.

Schneiderhöhn, H., and Ramdohr, P., 1931. *Lehrbuch der Erzmikroskopie,* vol. 2. Berlin: Borntraeger, 714p.

Taliaferro, N. L., 1934. Contraction phenomena in cherts, *Geol. Soc. Am. Bull.,* **45,** 189-232.

Cross-references: *Chert and Flint; Concretions; Diagenesis; Mud Cracks; Nodules in Sediments; Silica in Sediments; Silicification.*

T

TALUS

The fragments of rock that accumulate by the direct influence of gravity at the base of a mountain slope unprotected by vegetation are known as *talus* in American usage or *scree* in Britain (see *Mass-Wasting Deposits*). Submarine accumulations of reef debris at the base of a reef are called *reef talus*. The size and shape of the individual bits of rock debris are extremely variable, depending upon the characteristics of the underlying bedrock. Fracture tendencies of the rock, varying susceptibility to weathering of the component minerals, and the geomorphic agencies of the environment all influence the appearance of a given talus. In general, the coarser and more angular the fragments, the steeper the slope of the talus. Often these accumulations will approach or even exceed the theoretical angle of repose of the materials; hence they may be somewhat unstable and likely to slide if disturbed. Slides on talus slopes, however, are usually not of very great magnitude, for the loose mass of debris is inherently close to an equilibrium position.

Great masses of talus are particularly characteristic of mountainous areas where either aridity or other factors inhibit the development of a solid vegetative cover. They are produced primarily by mechanical processes, especially the thermal shocks of alternate freeze-thaw cycles; and once the inevitable cover of shrubs and trees develops, the talus slopes merge into the landscape. In many recently deglaciated mountain areas, talus is very prominent and may form accumulations ("rock glaciers") which move downslope in valleys far from their origin very much like true glaciers. Lithified talus would be a type of sedimentary breccia (q.v.).

FREDERICK F. WRIGHT

Cross-references: *Breccias, Sedimentary; Mass-Wasting Deposits; Reef Complex*. Vol. III: *Talus*.

TANGUE

Tangue is a calcareous silt found as a sediment in relatively shallow water along the Atlantic coast of France, especially in Brittany. Highly permeable, tangue has a high calcareous content (35-75% $CaCO_3$) and is very coherent. Its origin is organic. Although tangue has been considered a recent sediment, older deposits have been discovered near Mont-Saint-Michel, and French farmers have used tangue as a fertilizer for generations.

Tangue deposits are of a silvery to pearly gray color, quite different from sands, silts, and muds; they occur in a succession of thin beds (± 2 mm) reminiscent of varves (q.v.). Each layer corresponds to a sedimentary stage; separating the layers is a colloidal-coated surface corresponding to an emergence stage. If the emergence lasts long enough, eolian material is added, and desiccation of the top layer may even take place. Deposits have been observed only in littoral and deltaic environments.

The calcareous component of tangue originates from the grinding of coquina (*Ostrea*) facilitated by the destructive action—on the shells—of some mushrooms (*Litophythium gangliiformis; Ostracobable implexa*) and the agglutination of spicules of sponges. The largest concentration of calcium observed is 80%. The noncalcareous fraction is composed of fine sand, diatoms, and quartzose sand.

Tides bring the tangue to its depositional site, at the limit of the beach or at the foot of such obstacles as a seawall or a microcliff. Tangue particles have average diameters of 0.002-2 mm; sizes increase from the shore toward deeper water. This anomalous size distribution results because the oyster banks, source of the material, are offshore, and incoming tides do the sorting. The smallest diameters of tangue particles are observed in estuaries and at the top of the deposit (colloidal); size increases toward the bottom of the deposit. Although the ferrohumic colloidal role of the tangue is of little importance, grain size influences the color: fine tangue is much darker than coarse tangue.

The tangue-carrying tidal currents erode at depth, carrying material landward; deposits occasionally outcrop at low tide. Deposits may attain considerable thickness, exceeding that of the beach sand (> 15 m) due to their coherence, which provides considerable resistance to all types of erosion.

The varved nature of tangue permits dating of the deposits; old tangue contains less calcareous material than contemporary tangue, and more spicules of calcareous and siliceous sponges.

ROGER H. CHARLIER

References

Bourcart, J.; Jacquet, J.; and Francis-Boeuf, C., 1944. Sur la nature du sédiment marin appelé tangue, *Compt. Rend. Acad. Sci.*, 218, 469-470.

Bourcart, J., and Charlier R. H., 1959. The tangue, a "nonconforming" sediment, *Geol. Soc. Am. Bull.*, 70, 565-568.

Francis-Boeuf, C., 1974. Recherches sur le milieu fluvio-marin et les dépôts d'estuaires, *Ann. Inst. Océanogr.*, 23, 149-344.

Cross-references: *Estuarine Sedimentation; Littoral Sedimentation; Marl.*

TAR SANDS

Tar sands is an expression commonly used in the petroleum industry to describe sandstone reservoirs impregnated with a very heavy, viscous crude oil (see *Crude Oil Composition and Migration*). Two other terms, *bituminous sands* and *oil sands*, are rapidly gaining favor and may ultimately replace the older term as the true nature of these deposits is more widely recognized. The heavy, viscous petroleum substances impregnating the "tar sands" are called asphaltic oils. Other names used to describe these oils are *maltha, brea,* and *chapapote.* Asphaltic petroleums are most commonly confused with but are not related to the asphaltites, the asphaltic pyrobitumens, native mineral wax, or the pyrogenous distillates of bituminous substances (see *Bitumen, Bituminous Sediments*).

Nature of the Deposits

Tar sands or petroleum reservoirs impregnated with heavy asphaltic oils are commonly found at or near the surface of the ground in stratigraphic traps at the updip margins of large sedimentary basins. The reservoir sand is commonly very porous and permeable, and even in the oldest rocks it commonly lacks a mineral cement. The associated formation waters are low in chloride and high in carbonate ions. The oil varies in consistency but is mostly soft and viscous with rheological properties intermediate in nature between a flowing crude oil and semisolid asphalt; it is probably best described as maltha. Although the oil is heavier than pure water, it always overlies the water at the edge of a reservoir. "Tar" seals are reported to be present at the edgewater of many oil fields.

The mineral fraction of tar sands consists mainly of quartzose material of gravel, sand, or silt size, which in most reservoirs has been well sorted before being deposited in a fluvial, lacustrine, or deltaic environment. The porosity of such deposits commonly reaches 40%; thus the amount of oil in tar sands rarely exceeds 36% volume or 18% by weight. The heavy asphaltic oils of tar sands commonly contain a higher proportion of sulfur, nitrogen, oxygen, and trace metals such as vanadium, nickel, iron, and uranium than do lighter oils and are also richer in porphyrin pigments and their metallic complexes.

Because tar sands contain the largest accumulations of liquid hydrocarbon in the earth's crust their origin and relationship to the crude oils of conventional oil fields is of considerable interest to geologists and geochemists. It has been established that a continuous series of petroleum hydrocarbons exists with physical and chemical properties intermediate between the heaviest asphalts and the lightest crude oils; thus the oil of the tar sands probably has the same origin as the crude oil produced from wells (see *Petroleum–Origin and Evolution*). Several hypotheses have been advanced to explain the origin of the heavy oils. One theory postulates that they are not far removed from the hypothetical protopetroleum, the light crude oils being derived from them by physical and chemical changes induced by deep burial and tectonic forces. Another hypothesis postulates that they are the most mobile hydrocarbons, which have been polymerized and immobilized (inspissated) by low-temperature chemical changes—aided by mineral catalysts or by biochemical activities—as they migrated laterally toward the margins of the sedimentary basins. Whatever the origin of the heavy asphaltic oil in these deposits, once formed it is very resistant to attack by mineral acids and alkalies and to normal weathering processes. It readily forms emulsions with water and migrates freely, filling newly formed reservoirs or forming asphalt lakes.

Occurrence

The most extensive deposits of tar sands in the world are located in Canada in the northern part of the province of Alberta (#1 in Fig. 1). The largest and best known of these is the Athabasca deposit, which outcrops in the valley of the lower Athabasca River for a distance of 160 km. Heavy oil impregnates the lower Cretaceous McMurray Formation under an area of $> 26,000$ km^2, mainly to the W and S of the outcrops. Other large reservoirs of similar oil have been discovered in the subsurface of N

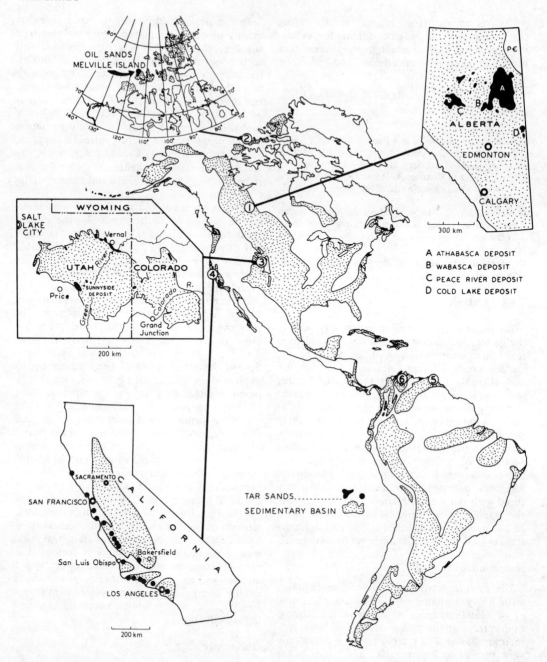

FIGURE 1. Map showing the location of the major "tar sand" deposits of the western hemisphere. Base map from Pratt and Good (1950.) *Source of Information:* (1) Anonymous, 1965; (2) Carrigy, 1963; (3) Wells, in Weeks, 1958; (4) Adams and Beatty, 1962; (5) Weeks, in Weeks, 1958; (6) Miller et al., in Weeks, 1958.

Alberta in Lower Cretaceous strata down to depths of 760 m. These are known as the Peace River, Wabasca, and Cold Lake deposits. The total amount of oil in all of these deposits exceeds 900 billion barrels of oil. Tar sands are also known to occur in the Bjorne Formation of Triassic age on Melville Island in the Canadian Arctic Archipelago near the southern margin of the Sverdrup sedimentary basin (#2 in Fig. 1).

In the US, many small tar-sand deposits are known, and a high proportion of these were quarried in the early part of this century and sold under the name *asphalt rock* as paving material for road construction. The most extensive areas of tar sands in the US are located in Utah in the Uinta sedimentary basin (#3 in Fig. 1). Next in importance are the tar-sand deposits in the Tertiary sedimentary basins of California (#4 in Fig. 1). Other minor deposits are found in Alabama, Arkansas, Colorado, Kansas, Kentucky, Louisiana, Missouri, Montana, New Mexico, New York, Oklahoma, Texas, and Wyoming.

In South America, tar-sand deposits known to contain at least 700 billion barrels of heavy oil are present in a "tar-belt" 40 km wide S of the Oficina-producing trend in the eastern Venezuelan sedimentary basin (#5 in Fig. 1; Galavis and Velarde, 1968). Also numerous large seepages of asphaltic oil form pitch lakes on the island of Trinidad and on the mainland at Guanoco and in the Orinoco delta NE of Pedernales (see *Volcanism, Sedimentary*). Another large reservoir of heavy oil in Venezuela is present in the NE part of the Maracaibo sedimentary basin (#6 in Fig. 1). Other deposits of heavy oil in South America are reported to be present in sedimentary basins in Brazil, Argentina, Colombia, Ecuador, and Peru.

Outcroppings of tar sand or asphaltic rock have been extensively quarried in Europe and Asia for many centuries. However, only two deposits have been exploited successfully as a source of crude oil. One is located at Pechelbronn in Alsace and the other at Wietze in Hanover. In the early part of this century, large quantities of a semiliquid asphalt were produced from the tar sands located at Tataros, in beds of Pliocene age, in W Romania.

In the Middle East, an extensive "tar belt" containing large reservoirs of heavy sulfurous oil is believed to exist in the Mesopotamian-Persian Gulf sedimentary basin parallel to the Euphrates River. Heavy oils have been found in Miocene and Upper Cretaceous limestone reservoirs at Qaiyarah and in seepages at Hit on the Euphrates and at many other localities. Bitumen from these deposits was extensively used in ancient times for building, waterproofing, medicine, and magic.

Lesser known tar sands are present in Nigeria, on the W side of the island of Madagascar, on the island of Leyte in the Philippines, in Japan on Boeton Island (Indonesia), and on the island of Sakhalin, and many other localities throughout the world.

MAURICE A. CARRIGY

References

Abraham, H., 1960. *Asphalts and Allied Substances*, vol. 1. *Historical Review and Natural Raw Materials.* New York: Van Nostrand, 370p.

Adams, E. W., and Beatty, W. B., 1962. Bituminous rocks in California, *Calif. Div. Mines Geol. Min. Inf. Serv.*, 15(4), 1-9.
Anonymous, 1965. Alberta "tar sands," *Oilweek*, 16(11), 27.
Carrigy, M., ed., 1963. *The K. A. Clark Volume; a collection of papers on the Athabasca oil sands.* Edmonton: Res. Council of Alberta, 241p.
Carrigy, M., 1967a. The physical and chemical nature of a typical tar sand—Bulk properties and behaviour, *7th World Petrol. Congr.*, 3, 573-581.
Carrigy, M., 1967b. Some sedimentary features of the Athabasca oil sands, *Sed. Geol.*, 1, 327-352.
Carrigy, M. A., 1976. Fuel from tar sands, in D. M. Considine, ed., *Energy Technology Handbook.* New York: McGraw-Hill, 3-159-3-163.
Demaison, G. J., 1977. Tar sands and supergiant oil fields, *Bull. Am. Assoc. Petrol. Geologists*, 61, 1950-1961.
Fanning, L. M., 1945. *Our Oil Resources.* New York: McGraw-Hill, 331p.
Galavis, J., and Velarde, H. M., 1968. Geological study and preliminary evaluation of potential reserves of heavy oil of the Orinoco tar belt Eastern Venezuelan Basin, *7th World Petrol. Congr.*, 3, 229-234.
Pratt, W. E., and Good, D., eds., 1950. World geography of petroleum, *Am. Geog. Soc. Spec. Publ. 31*, 464p.
United States Bureau of Mines, 1965. Surface and shallow oil-impregnated rocks and shallow oil fields in the United States, *U.S. Bur. Mines Mon. 12*, 375p.
Weeks, L. G., ed., 1958. *Habitat of Oil*, vol. 1. Tulsa: Am. Assoc. Petrol. Geologists, 384p.

Cross-references: *Bitumen, Bituminous Sediments; Crude-Oil Composition and Migration; Oil Shales; Organic Sediments; Petroleum—Origin and Evolution; Volcanism, Sedimentary.*

TECTONICS—See SEDIMENTARY FACIES AND PLATE TECTONICS; SEDIMENTATION—TECTONIC CONTROLS

TIDAL-CURRENT DEPOSITS

Tidal-current deposits include all sedimentary accumulations whose form and placement result from sediment transportation primarily induced by tidal water flow. This entry emphasizes the major types of tidal-current deposits found in sandy environments, which may be classified into two groups: (1) estuarine (subdivided by tidal range) and (2) shelf.

Estuarine Tidal Deposits

Microtidal and Mesotidal Estuarine Deposits. Characteristic tidal-current deposits are found at inlets and estuary mouths in microtidal (tidal range up to 2 m) and mesotidal (2-4 m) environments (Davies, 1964). Prominent among these

TABLE 1. Examples of Tidal-Current Deposits

Micro- Mesotidal Estuaries (tidal range 0–4 m)
  E coast of US
  Dutch coast
  Gulf of Mexico

Macrotidal Estuaries (tidal range >4 m)
  Bay of Bengal
  N end of Persian Gulf
  Gulf of Cambay, India
  Bay of Fundy, Nova Scotia
  Bristol Bay, Alaska
  Amazon River Estuary
  Rio Guayas, Ecuador

Shelf Deposits (not associated with large river deltas)
  Gulf of Korea
  Tongue of the Ocean, Bahamas
  Lily Bank, Bahamas
  E coast of Borneo
  North Sea
  S coast of Africa
  Georges Bank
  Prince of Wales Strait, N coast of Australia

are the sedimentary accumulations grouped under the terms *ebb* and *flood tidal deltas*. Examples are numerous from around the world (see Table 1); particularly fine examples are found in New England, on the New Jersey coast, and in the Carolinas and Georgia (see *Tidal-Inlet and Tidal-Delta Facies*).

The ebb tidal delta is composed of sediment outside the tidal entrance whose overall morphology is due mainly to ebb tidal flow and wave action. It is sometimes referred to as the *outer shoal* or *sea shoal*. Major components (Fig. 1) include: (1) *channel-margin linear bars* flanking either side of the *main ebb channel*, within which ebb tidal flow is concentrated; (2)

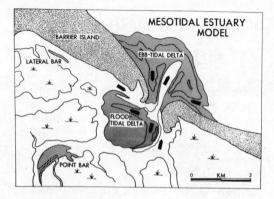

FIGURE 1. A typical mesotidal inlet entrance with arrows showing areas of flood- and ebb-tidal dominance. Such extensive areas of *Spartina alterniflora* low marsh are typical along the southeastern US coast.

marginal flood channels, located between the linear bars and the adjacent shore; early flood flow is restricted to these channels as a result of concentrated, residual ebb in the main channel as the tide turns; (3) the *terminal lobe,* forming the outer margin of the main ebb channel. Swash bars, usually parallel to shore, occur adjacent to the channel-margin linear bars, but these are largely wave-formed features (Hayes et al., 1973). An alternate terminology has been developed by Oertel (1972) during work on Georgia estuary entrances (see *Tidal-Inlet and Tidal-Delta Facies*).

The flood tidal delta, sometimes termed *bay shoal* or *inner shoal,* owes its form mainly to flood tidal flow and includes the following components: (1) a *flood ramp,* or gradually shoaling, bed-form-covered channel, leading to the topographically higher inner parts of the delta; (2) an *ebb shield* or high inner margin, built up by flood-current sediment transport and so named because it protects the flood delta from the later stages of ebb flow; (3) a trailing *ebb spit,* located at the margin of the feature, formed by ebb flow but not obliterated during the flood cycle; and (4) a *spillover lobe,* a finger-like lobe of sediment with a central channel formed where ebb flow has broken through the ebb shield or other topographically higher part of the flood tidal delta (see *Tidal-Inlet and Tidal-Delta Facies*).

Bed forms themselves comprise an important type of tidal-current deposit and occur on both intertidal and subtidal portions of the tidal deltas, as well as on the tidal-creek and estuary bottom. Orientation may change with each half tidal cycle or remain substantially the same, depending upon the relative flow strength of the ebb current and the flood current. One classification of tidal bed forms has the following divisions, based on spacing (Boothroyd, 1969): ripples, up to 60 cm; megaripples, 60 cm to 6 m; sand waves (q.v.), over 6 m.

Point bars are found attached to the convex banks of tidal-creek or estuary meanders (Land and Hoyt, 1966). The writers have observed several examples in South Carolina which show a flood-current-dominated inner area with a trailing, ebb-oriented spit on the outer flank of the bar. *Lateral bars* occur within tidal-creek or estuary reaches that are relatively straight. They may or may not occur in conjunction with a point bar. Lateral bars result from a reworking of other sand deposits, such as point bars, ebb spits, flood tidal deltas, and beach ridges. Channel *lag deposits* result from tidal-current winnowing of the finer sands and mud from a sedimentary accumulation. The result may be a gravel or shell pavement on the channel bottom; the latter is common in the SE United

States where abundant oyster shell (*Crassostrea virginica*) is found.

Macrotidal Estuarine Deposits. Macrotidal estuaries (tidal range > 4 m) are coastal embayments commonly having large, active river deltas located at the embayment apex. The funneling effect of the embayment creates these large tidal ranges, which generate strong tidal currents (up to 250–300 cm/sec).

The sediments discharged from the river are reworked by the tidal currents so that the river delta is essentially an estuarine-tidal delta complex, consisting of *intertidal and subtidal linear sand shoals, intertidal deltaic sand flats,* and *fringing mud tidal flats* (Fig. 2). Large *point bars* and *lateral bars* commonly occur within the upper estuary. An example of such a tidal-delta complex is the Nushagak and Kvichak Bay complex of Bristol Bay, Alaska.

The primary sedimentary deposits of macrotidal estuaries are the large linear (can also be curved or S-shaped) sand bodies. They divide the distal portion of the estuary into a series of ebb- and flood-current-dominated channels. Once established, the linear sand shoals are maintained by the tidal flow, remain relatively stable, and migrate very little laterally. Sand is transported along the length of the sand body in and out of the estuary by bed-load transport in the form of sand waves.

On the lower energy margins of the estuary are large, wide, mud tidal flats. These features can be tens of km wide and are largely exposed at low water. Suspended sediments are carried into these shallows during flood, fall out of suspension, and remain on the flats. These zones are natural traps for fine-grained sediments, and they grow rapidly both vertically and laterally (see *Tidal-Flat Geology*).

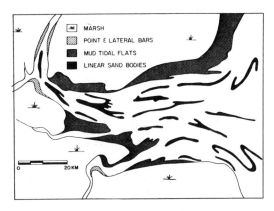

FIGURE 2. Model for macrotidal estuary. Dominant tidal current deposits are linear sand bodies and fringing mud tidal flats. Point bars and lateral bars are common in upper portions of the estuary.

Tidal-Current-Generated Shelf Deposits

Tidal-current-generated deposits on the continental shelves and shallow shelf-like submarine platforms are primarily linear sand bodies whose orientations are either transverse or parallel to the ebb- and flood-tidal flow. *Transverse linear sand bodies* are generally located along open, unrestricted, lower-energy shelf margins and are probably influenced to a significant degree by wave-generated currents. Sand bodies oriented parallel with tidal flow are located on more restricted, higher-energy shelves where continental land masses, islands, or submarine topographic features channelize tidal flow and consequently increase flow strength. Generally, these bodies occur in hydrodynamic regimes where tidal currents exceed 100 cm/sec for > 50% of the tidal cycle.

Tidal-current-generated shelf sand bodies should not be confused with deeper, relict, open-shelf, linear sand bodies (*ridge and swale topography*) that exist on the continental shelf off the E coast of the US (see *Continental-Shelf Sedimentation*). The sand bodies are formed by wave-generated currents in the nearshore zone, and their orientation is governed by the approach direction of dominant waves. During the Holocene transgression, these features formed along the retreating shoreline and eventually became inactive but remained intact as shelf depth increased due to sea-level rise.

The best examples of transverse, tidal-current-dominated sand bodies are the oolitic *marine sand belts* (Ball, 1967) that lie parallel to the platform margins of the Bahama Banks (q.v.). The dominant bed-form features of these linear sand bodies are large (spacing 10–50 m), asymmetrically shaped sand waves which commonly have megaripples superimposed on them. This combination of bed forms indicates active sand transport responding to tidal currents. Net sand transport appears to be toward the shelf in most areas on this type of sand body. Undoubtedly, these open-shelf, transverse marine sand belts are significantly affected by the local wave regime and certainly by major storms. However, quantitative studies have not been made to determine relative importance of these processes.

Restricted-shelf linear sand bodies (*tidal bar belts,* Ball, 1967) (*tidal current ridges,* Off, 1963) are much more common than open-shelf sand bodies and occur most commonly in large coastal embayments (off the W coast of Korea), submarine shelf embayments (surrounding the cul de sac of the Tongue of the Ocean, Bahamas), between land masses (English Channel and North Sea), or locally between small islands (salt diapirs in Persian Gulf) where tidal currents are relatively strong. The linear sand bodies associated with macrotidal estuaries can be considered in this group but have been discussed separately because they occur primarily in embayments containing large, active river deltas.

Restricted-shelf linear sand bodies not associated with macrotidal estuaries have been formed from sediments deposited on the shelves during lower stands of sea level (fluvial or glacial) or formed from sediments being generated in situ (ooids) as in the carbonate environment. They are commonly several km in length (up to 70 km), 5-10 m in height (up to 35 m), and spaced 1-2 km apart (up to 10 km). The sand bodies are occasionally asymmetrical in cross section, indicating some lateral movement, and are frequently covered with sand waves and megaripples, indicating active sand transport.

Although in-depth field process studies have never been done, the more widely published ideas concerning how these sand bodies form and maintain themselves invoke either refracted tidal-current flow (Caston, 1972) or a system of converging helical flow cells (Houbolt, 1968).

Ancient Tidal-Current Deposits

Better definition of modern tidal-current deposits has led to the identification of ancient examples (see also *Tidal-Inlet and Tidal-Delta Facies; Tidal-Flat Geology*). One of the best-studied examples is found in the Late Precambrian-Cambrian in Europe (Banks, 1973; Goodwin and Anderson, 1974; Johnson, 1975). Other examples include the Cretaceous in Colorado (MacKenzie, 1972) and the Eocene near the Pyrenees (Piudefabregas, 1974). Klein (1971, 1972) has reviewed ancient tidal sand occurrences and proposed a tidal cycle and a means of determining paleotidal range.

ALBERT C. HINE III
ROBERT J. FINLEY
MILES O. HAYES

References

Ball, M. M., 1967. Carbonate sand bodies of Florida and the Bahamas, *J. Sed. Petrology*, 37, 556-597.

Banks, N. L., 1973. Tide-dominated offshore sedimentation, Lower Cambrian, North Norway, *Sedimentology*, 20, 213-228.

Boothroyd, J. C., 1969. Hydraulic conditions controlling the formation of estuarine bedforms, *Coastal Res. Group, Dept. Geol., Univ. Mass.*, 1, 417-429.

Caston, V. N. D., 1972. Linear sand banks in the southern North Sea, *Sedimentology*, 18, 63-78.

Davies, J. L., 1964. A morphogenic approach to world shorelines, *Z. Geomorph.*, 8, Mortensen Sonderheft, 127-142.

Ford, J., 1975. The earth, the moon and tidal sediments, *Mercian Geol.*, 5, 205-239.

Goodwin, P. W., and Anderson, E. J., 1974. Associated physical and biogenic structures in environmental subdivision of a Cambrian tidal sand body, *J. Geol.*, 82, 779-794.

Gordon, C. M., 1975. Sediment entrainment and suspension in a turbulent tidal flow, *Marine Geol.*, 18, M57-M64.

Hayes, M. O.; Owens, E. H.; Hubbard, D. K.; and Abele, R. W., 1973. The investigation of form and process in the coastal zone, in D. R. Coates, ed., *Coastal Geomorphology*. Binghamton: St. Univ. New York, 11-41.

Houbolt, J. J. H. C., 1968. Recent sediments in the southern bight of the North Sea, *Geol. Mijnbouw*, 47, 245-273.

Johnson, H. D., 1975. Tide and wave-dominated inshore and shoreline sequences from the Late Precambrian, Finmark, North Norway, *Sedimentology*, 22, 45-74.

Klein, G. DeV., 1971. A sedimentary model for determining paleotidal range, *Geol. Soc. Am. Bull.*, 82, 2585-2592.

Klein, G. DeV., 1972. Sedimentary model for determining paleotidal range: Reply, *Geol. Soc. Am. Bull.*, 83, 539-546.

Land, L. S., and Hoyt, J. H., 1966. Sedimentation in a meandering estuary, *Sedimentology*, 6, 191-207.

Ludwick, J. C., 1974. Tidal currents and zig-zag sand shoals in a wide estuary entrance, *Geol. Soc. Am. Bull.*, 85, 717-726.

Ludwick, J. C., 1974. Variations in the boundary-drag coefficient in the tidal entrance to Chesapeake Bay, Virginia, *Marine Geol.*, 19, 19-28.

MacKenzie, D. B., 1972. Tidal sand flat deposits in Lower Cretaceous Dakota Group near Denver, Colorado, *Mn. Geol.*, 9, 296-277.

Oertel, G. F., 1972. Sediment transport of estuary entrance shoals and the formation of swash platforms, *J. Sed. Petrology*, 42, 858-863.

Off, T., 1963. Rhythmic linear sand bodies caused by tidal currents, *Bull. Am. Assoc. Petrol. Geologists*, 47, 324-341.

Piudefabregas, C., 1974. Les sédiments de Marée du bassin éocène sudpyrénéen, *Bull. Centre Rech. de Pau*, 8, 305-325.

Swift, D. J. P., 1975. Tidal sand ridges and shoal-retreat massifs, *Marine Geol.*, 18, 105-134.

Ulrich, J., and Pasenau, H., 1973. Morphologische Untersuchungen zum Problem der tidebedingten Sandbewegung im Lister Tief, *Die Küste*, 24, 95-112.

Cross-references: *Bay of Fundy Sedimentology; Continental-Shelf Sedimentation; Estuarine Sedimentation; Lagoonal Sedimentation; Sand Waves; Tidal-Flat Geology; Tidal-Inlet and Tidal-Delta Facies; Wadden Sea Sedimentology.*

TIDAL-FLAT GEOLOGY

Tidal flats are sandy to muddy or marshy flats emerging during low tide and submerging during high tide (= *intertidal zone*). The sediment consists of either siliciclastics or carbonates. The area and the sediments below the low-water line is the *subtidal zone*. The area and the sediments above the high-water line, but sometimes under water during extremely high tides, is the *supratidal zone*. In many cases, there are no clear limits between the intertidal and the supratidal zone.

Siliciclastic Tidal Flats

The best-known siliciclastic tidal flats are those of the North Sea (Evans, 1965; Reineck, 1972), including the Wadden Sea (q.v.); the Gulf of California (Thompson, 1968); the Bay of Fundy (q.v.) and Saint-Michel, France (Dolet et al., 1965).

Morphology. The surface is flat and gently inclined, but slightly irregular, toward low-water line (Fig. 1). Tidal currents have cut meandering gullies and channels into the tidal flats. Cut sides and point bars with longitudinal cross-bedding are developed.

Supratidal Zone. In a moderate climate, a salt marsh, characterized by clayey sand and clayey silt layers with nodular appearance (see *Salt-Marsh Sedimentology*) develops in the supratidal zone. In an arid climate, chaotic layers develop due to crystal growths; there are abundant mud pebbles, mud chips, and desiccation cracks, and no plant roots (Thompson, 1968).

Intertidal Zone. The intertidal zone lies between marsh and low-water line (Fig. 1). Here are mudflats, the significant features of which include mud layers with thin, flat sand lenses and thin interlamination of mud and sand, with moderate to strong bioturbation. Mixed flats are characterized by flaser and lenticular bedding, and shells in living position; bioturbation is moderate, sometimes strong in surface layers, and weak in point-bar deposits. Sand flats consist of plane laminated sand and small-scale cross-bedding, sometimes with herringbone structure, with mostly weak bioturbation. Surface structures of the intertidal zone include rain and hail imprints; desiccation cracks; numerous minor and large-scale erosional features; and traces of insects, birds, and other animals.

The most common grain size in the intertidal zone is clayey silt to fine sand, with many fecal pellets; in areas with extreme tidal range and currents the sand may be coarser. Thinly interlayered sand and mud, and lenticular and flaser bedding are the result of rhythmically changing tidal currents and slack water. Changes in bedding type are due to changing wave height and wave direction (Reineck and Wunderlich 1969).

Subtidal Zone. Gullies in the subtidal zone are marked by longitudinal cross-bedding and channel fill, with intercalations of mud and sand in many variations with weak bioturbation; channel lag deposits are found at the base.

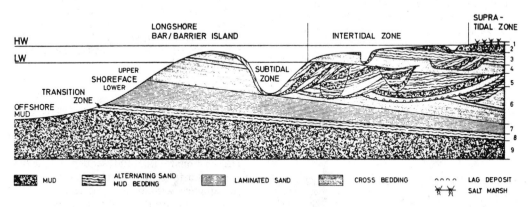

FIGURE 1. Schematic cross section across the tidal flats of the North Sea in a situation on a hypothetical prograding shoreline (from Reineck and Singh, 1973, based partly on data from Van Straaten). The North Sea tidal flats, in reality, rest on Quaternary deposits. 1. *Salt marsh (supratidal zone):* Very fine sand and mud, interbedded shell layers, plant roots, irregular wavy bedding. 2. *Mud flat (intertidal zone):* Mud, occasional very fine sand layers, lenticular bedding with flat lenses, strong bioturbation. 3. *Mixed flat (intertidal zone):* Sandy mud, thinly interlayered sand/mud bedding, lenticular bedding, flaser bedding, shell layers, bioturbation strong to weak. 4. *Sand flat (intertidal zone):* Very fine sand, small-ripple bedding, sometimes herringbone structure, flaser bedding, laminated sand, occasional strong bioturbation. 5. *Subtidal zone:* Medium to coarse sand, mud pebbles, megaripple bedding, small-ripple bedding, laminated sand, weak bioturbation. 6. *Upper shoreface:* Beach bar- and ripple cross-bedding, laminated sand, weak bioturbation. 7. *Lower shoreface:* Laminated sand, bioturbation stronger than in the upper shoreface. 8. *Transition zone:* Alternating sand/mud bedding, i.e., flaser and lenticular bedding, thinly and thickly interlayered sand/mud bedding, moderate bioturbation. 9. *Shelf mud:* Mud with storm silt layers, moderate bioturbation.

Channel walls are characterized by huge, soft, angled longitudinal cross-bedding of intercalations of sand and mud, e.g., flaser bedding and small cross-bedding. Channel lag deposits (mud pebbles and shells), with megaripples and giant ripples, are found at the channel bottom; there is weak bioturbation in sandy areas, and no shells in living position, but in certain parts, escape traces of shells and other bottom living animals often occur.

Carbonate Tidal Flats

Carbonate tidal flats are related to warm climates. The best known carbonate flats are those of Bahama Banks (q.v.), Shark Bay (q.v.), and the Persian Gulf (q.v.).

Andros Island, Bahamas. The supratidal zone on NW Andros Island (exposed > 75% of a year; see Shinn et al., 1969, and Ginsburg, 1975), is characterized by a beach ridge on the seaside of the tidal-flat belt (Fig. 2), with festoon cross-bedded gastropod sand and laminated pellets with many birdseye structures. Levee crests along channels contain mm laminae in pelleted lime muds. The high algal marsh area has crinkled mm laminae. The inland algal marsh consists of light-colored layers of pelleted lime mud and dark layers of algal tufa; birdseye vugs are abundant; root structures of marsh grass and of black and red mangroves are common. In the intertidal zone (exposed 54-75% of a year), the dominating structures are bioturbated pelleted mud with roots of red mangrove (Fig. 2). In the subtidal zone, the shallow marine sediments are built up of soft elliptical pellets of aragonitic muds. The channel sediments vary from pelleted mud to sediments of gastropod shells, foram tests, and mud lumps. Pond sediments are very muddy and soft. In all these three subenvironments, primary sedimentary structures are completely or nearly completely destroyed by bioturbation (see *Bahama Banks Sedimentology*).

Shark Bay, Western Australia. In the Nilemah and Hutchinson embayment around Shark Bay (Brown and Woods, 1974; Hagan and Logan, 1974), the supratidal zone is marked by a beach ridge of *Fragum* coquinas (Fig. 3) brought in during high-energy episodes from the subtidal zone. Major features are intraclasts from broken indurated crusts, ranging from the intertidal zone of flat-pebble breccias to progressively finer sand. These intraclasts are either fragments of aragonitic pelmicrite crust or finely cemented, skeletal pellet grainstone. In the intertidal zone, desiccation cracks, gypsiferous sediments, and laminoid-fenestral fabric produced beneath pustular mats are characteristic of the upper intertidal zone; dominant grain types are aragonitic pellets. Distinctive criteria of the middle and lower intertidal zone are well laminated, algal-bound sediments. The subtidal zone consists of planar to cross-bedded, unconsolidated sands of aragonite ooids and pellets with autochthonous forams and coarse shells (*Fragum* valves), with cryptalgal structures (see *Shark Bay Sedimentology*).

Trucial Coast, Persian Gulf. Along the Trucial Coast (see Evans, 1970; Shinn, 1973), the supratidal zone consists of nodules of anhydrite with fine quartzose carbonate sand and pellet-rich mud; structures include supratidal lamination and birdseye vugs, stromatolites, and mud cracks. Intertidal zone algal mud flats contain algal mats and stromatolites. Tidal flats consist of skeletal, pelletal carbonate sand and mud; gypsum; birdseye vugs; and high bioturbation. The subtidal zone is marked by soft, pelleted mud and silt-sized carbonate with varying amounts of skeletal grains (see *Persian Gulf Sedimentology*).

The Vertical Sequence

The siliciclastic tidal-flat sequence is a marine fining-upward sequence: sand–mud–peat, with flaser bedding, intraclasts, and channel-fill structures.

The carbonate tidal-flat sequence is a limestone-evaporite sequence or a limestone-algal mat sequence with plant roots and with birdseye vugs, stromatolites, and intraclasts.

Ginsburg (1975) has compiled studies of recent and ancient tidal flats, and Reineck (1973) contains an extensive bibliography.

HANS-ERICH REINECK

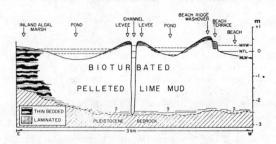

FIGURE 2. Cross section of tidal deposits, NW Andros Island, Bahamas. Generalized distribution of sedimentary structures (from R. N. Ginsburg and L. A. Hardie in Ginsburg, 1975).

References

Brown, R. G., and Woods, P. J., 1974. Sedimentation and tidal-flat development, Nilemah Embayment, Shark Bay, Western Australia, *Am. Assoc. Petrol. Geologists, Mem.* 22, 316-340.

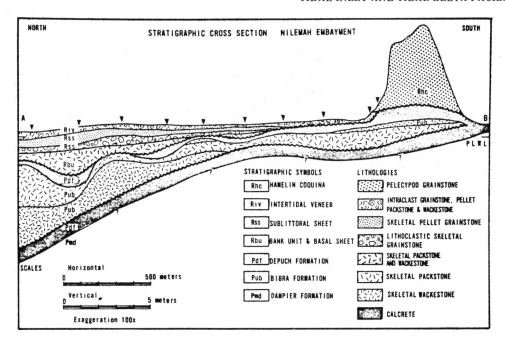

FIGURE 3. Stratigraphic cross section of Nilemah Embayment (from Brown and Woods, 1974).

Dolet, M.; Giresse, P.; and Larsonneur, C., 1965. Sédiments et sédimentation dans la baie du Mont Saint-Michel, *Soc. Linn. Norm. Bull.,* 6, 51-65.

Evans, G., 1965. Intertidal flat sediments and their environment of deposition in the Wash, *Quart. J. Geol. Soc. London,* 121, 209-245.

Evans, G., 1970. Coastal and nearshore sedimentation; a comparison of clastic and carbonate deposition, *Proc. Geol. Assoc.,* 81, 493-508.

Ginsburg, R. N., ed., 1975. *Tidal Deposits,* a source book of recent examples and fossil counterparts. New York: Springer, 428p.

Hagan, G. M., and Logan, B. W., 1974. History of Hutchinson Embayment tidal flat, Shark Bay, Western Australia, *Am. Assoc. Petrol. Geol. Mem.* 22, 283-316.

Reineck, H.-E., 1972. Tidal flats, *Soc. Econ. Paleont. Mineral. Spec. Publ.* 16, 146-159.

Reineck, H.-E., 1973. Bibliographie geologischer Arbeiten über rezente und fossile Kalk- und Silikatwatten, *Cour. Forsch. -Inst. Senck.,* 6, 1-57.

Reineck, H.-E., 1975. Tidal flats, German North Sea, in Ginsburg, 1975, 5-12.

Reineck, H.-E., and Wunderlich, F., 1969. Die Entstehung von Schichten und Schichtenbänken im Watt, *Senck. marit.,* 50, 85-106.

Reineck, H.-E., and Singh, I. B., 1973. *Depositional Sedimentary Environments.* New York: Springer, 439p.

Shinn, E. A., 1973. Recent intertidal and nearshore carbonate sedimentation around Rock Highs, E Qatar, Persian Gulf, in B. H. Purser, ed., *The Persian Gulf.* New York: Springer, 193-209.

Shinn, E. A.; Lloyd, R. M.; and Ginsburg, R. N., 1969. Anatomy of a modern carbonate tidal flat, Andros Island, Bahamas, *J. Sed. Petrology,* 39, 1202-1228.

Thompson, R. W., 1968. Tidal flat sedimentation on the Colorado River Delta, northwestern Gulf of California, *Geol. Soc. Am. Mem.* 107, 133p.

Cross-references: *Bahama Banks Sedimentology; Bay of Fundy Sedimentology; Bedding Genesis; Persian Gulf Sedimentology; Salt-Marsh Sedimentology; Shark Bay Sedimentology; Tidal-Current Deposits; Tidal-Inlet and Tidal-Delta Facies; Wadden Sea Sedimentology.*

TIDAL-INLET AND TIDAL-DELTA FACIES

Tidal inlets are breaks in an otherwise continuous chain of barrier islands. A tidal inlet may be defined as a major passage from an open ocean into a lagoon and having a spit on one of its banks. Tidal inlets differ from tidal channels in that the latter are relatively small features compared to tidal inlets and occur in the tidal-flat and barrier-flat (backbarrier) environments. Inlets typically occur on submerging coasts. Depending upon the direction from which the waves approach the coast, the tidal inlets may migrate systematically in one direction. Alongshore currents, powered by waves, deposit sediment on one bank of the

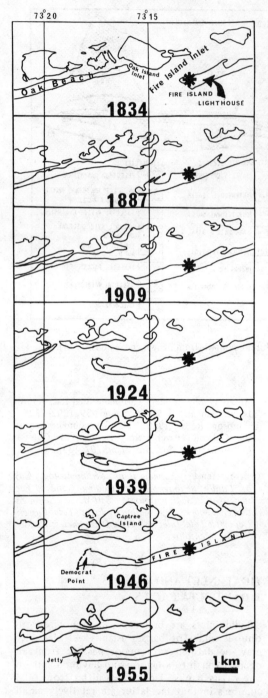

inlet, causing the tidal currents to erode the other side. By this mechanism, tidal inlets may migrate a considerable lateral distance during a relatively short period. For example, the Fire Island Inlet, one of seven major inlets on the S shore of Long Island, New York, migrated 8 km westward during the 115-yr period of 1825–1940 (Fig. 1).

Inlet Sequence

Sediments deposited in tidal inlets are characteristic of their environment and are preserved in the rock record as an "inlet sequence" (Kumar and Sanders, 1974; Fig. 2). Because the hydraulic conditions vary in the several parts of an inlet, from the bottom of the channel up to the spit, the sedimentary structures and textures differ in sediments deposited at different depths in the inlet. During the lateral migration of an inlet, therefore, a sequence of sedimentary structures and textures, i.e., an inlet

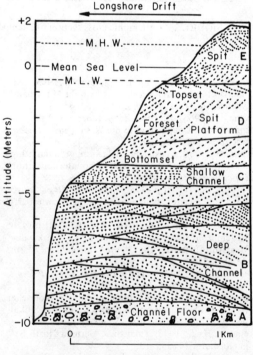

FIGURE 1. Map showing the successive westward migration of Fire Island, Long Island, New York, during the period 1834 to 1955 (from Kumar and Sanders, 1970). A jetty constructed in 1939 stopped further migration of Fire Island for some time, but by 1950, the area east of the jetty had overflowed with sand, and the spit had started to grow farther W again. The position of inlet is now maintained by dredging.

FIGURE 2. Profile and section through E bank of Fire Island Inlet showing sequence of strata deposited as inlet shifts westward (from Kumar and Sanders, 1974). View is N along axis of inlet. Letters A through E designate units in inlet sequence. Cross-strata, not drawn to scale, are shown in views revealing their distinctive aspects. Most cross-strata in all units except spit-platform foresets show maximum dips in a N-S line, parallel to the axis of the inlet and at right angles to the plane of this figure.

sequence, is deposited under those parts of the barrier through which the inlet has migrated (Kumar, 1973). This sequence has a high potential for preservation in the geologic record. Thus, inlet sediments should be recognized in the rock record not by a single type of sediment but rather by a characteristic sequence.

The criteria for recognizing the various units in an inlet sequence have been established from the sediments in Fire Island Inlet, and this sequence has been recognized in bore holes drilled on those parts of Fire Island through which the inlet is known to have migrated (Fig. 1). The inlet sequence is an important example of lateral accretion (see *Lateral and Vertical Accretion;* Van Straaten, 1954); this sequence is analogous to the "point-bar sequence" formed in meandering streams (Bernard et al., 1970; see *Fluvial Sediments*).

Even though criteria to distinguish "inlet sediments" from sediments in other environments have been mentioned by Shepard (1960) and Shepard and Moore (1955), the concept of an inlet sequence or "channel sequence" is a relatively new one (Hoyt and Henry, 1965, 1967; Bernard et al., 1970; LeBlanc, 1972). Kumar and Sanders (see references) have carried out studies describing the details of the inlet sequence and its geologic significance.

The significance of the inlet sequence lies in the fact that this sequence is probably far more common among the strandline deposits than hitherto has been suspected. Inlet sediments have not been recognized in the geologic record because definite criteria to distinguish such sediments from sediments deposited in other environments did not exist until recently. Calculations show that 2-4% of all coastlines may be underlain by the inlet sequence (Kumar and Sanders, 1974). Presumably, this frequency of occurrence also applies to the ancient geologic record. Inlet sediments may also be used to decipher the history of a transgression. The geometry of "basal sands" (which may consist chiefly of inlet-filling sands, Kumar and Sanders, 1970) can be used to determine the history of barrier behavior during a transgression (see *Barrier Island Facies*). Unfortunately, the geologic significance of inlet-filling sediments has not been sufficiently emphasized even in relatively recent publications on barrier sediments.

The inlet sequence established for Fire Island Inlet contains five major units (Fig. 2). In ascending order the units are: (A) *channel floor*, characterized by a lag gravel composed of large shells, pebbles, and other coarse particles; (B) *deep channel* (-10.0 to -4.5 m), characterized by lenticular sets of ebb-oriented "reactivation" surfaces; (C) *shallow channel* (-4.5 to -3.75 m), characterized by plane, parallel laminae; (D) *spit platform* (-3.75 to -0.6 m), characterized by steep and gentle seaward-dipping laminae and steep and gentle landward-dipping laminae in its various subenvironments; and (E) spit. Fig. 2 is a schematic representation of sedimentary structures in various units of the inlet sequence in Fire Island Inlet.

Swift currents in the channel bottom produce the concentrate that is characteristic of the *channel floor*. Large sand waves 60-100 m in length and 0.5-2.0 m in height form and migrate within the depth range of the *deep channel*. These sand waves do not form and migrate in waters shallower than 4.5 m. The thickness of the deep-channel unit is, therefore, a function of the depth of the channel; the upper limit is always at a depth of -4.5 m. The internal laminae in these sand waves are characteristic of periodically reversing flow (tidal cycle; Klein, 1970; Boersma, 1969). The sediments in the *shallow channel* are deposited in the upper flow regime and hence contain planeparallel laminae.

The *spit platform* is a shallow submerged platform built by waves before a spit can form and grow subaerially (Meistrell, 1972). The spit platform resembles a Gilbert-type delta (see *Deltaic Sediments*) with characteristic topset, foreset, and bottomset surfaces; each surface has characteristic sedimentary structures. A *spit* forms and grows over the topset surface of the spit platform. The processes operating on the spit are very similar to the processes operating on the foreshore and berm of the main barrier beach except that, because of refraction, waves approach various parts of the spit from several directions. Consequently, on a spit the "landward" and "seaward" cross-strata may span an arc of up to 90° or more.

Barwis and Makurath (1978) have described a sedimentary sequence from the Upper Silurian Keyser Limestone as an ancient analogue of the inlet sequence described here. More ancient examples will certainly be recognized once the criteria for recognizing inlet-filling sediments become widely known.

Tidal Deltas

As tidal currents leave the tidal inlet, they experience a sudden drop in velocity. Consequently, tidal deltas may form at either end of a tidal inlet; the delta at the landward end of the inlet is termed the "flood tidal delta," the delta at its seaward end the "ebb tidal delta." The ebb tidal delta is seldom developed because the longshore currents generally prevent the growth of a tidal delta at the seaward end of an inlet. Shoals and sand sheets, however, occur at the seaward end of many inlets (see *Tidal-Current*

Deposits; also see Hayes and Kana, 1976, for discussion and bibliography on modern barriers, tidal deltas, and tidal inlets).

Flood Tidal Delta. Caldwell (1971) has studied a section of the tidal delta associated with the Moriches Inlet, Long Island, New York. He has recognized the following depositional environments in the Moriches tidal delta: channels, mussel beds, active sand lobes, tidal flats, and marshes. Cross-strata are produced in the channel and active sand-lobe environments where the energy level and the rate of sediment supply are moderate to high. Active sand lobes are characterized by brown medium sand (15–20 cm thick) overlying (with a gradational color boundary) gray sand of the same average size. Channels are characterized by brown, medium sand overlying gray, fine silty sand with a sharp color and size boundary. Horizontal strata, which form when either the energy level or the rate of sediment supply is low, are found in all environments except the mussel beds. Burrow structures predominate in the tidal-flat environment because of high level of organic activity and low rates of sediment supply.

Hayes (1969) with others investigated the tidal deltas of New England coastal inlets and were able to distinguish the zones of flood-current dominance in a flood tidal delta from the zones of ebb-current dominance by distinct differences in the scale and type of cross-bedding formed (see *Tidal-Current Deposits*).

Ebb Tidal Delta. Oertel (1972, 1973, 1975) has studied the sediment transport at the seaward end of tidal inlets in Georgia. He uses the term *ramp-to-the-sea* for the landward-sloping surface that extends from the entrance of the inlet to the deepest part of the channel. Sand shoals called *ramp-margin shoals* develop on the borders of the ramp-to-the-sea. Linear bed forms, with planar upper surfaces, called *swash bars* (which generally extend parallel to the main beach) and *swash platforms* (which generally extend perpendicular to the main beach) develop on the ramp-margin shoals (Fig. 3). Sedimentation on a swash platform is influenced by the patterns of interfering waves, whereas sedimentation on a swash bar is influenced by straight-crested waves. High-angle foresets associated with swash bars generally have a preferred dip orientation toward the beach and away from wave approach. Low-angle laminae and simple and planar cross-stratification of swash platforms do not have a preferred orientation relative to the shoreline. Hayes et al. (1969) found the seaward ends of inlets in the New England area are also characterized by swash bars and submerged ebb-dominated sand sheets. Wave-generated flow over the intertidal

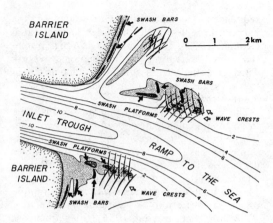

FIGURE 3. Generalized sketch map illustrating the topographic features at tidal inlets along the Georgia Coast (from Oertel, 1972). Sand shoals bordering the ramp to the sea are ramp-margin shoals. Refracted wave crests generally interfere along the axis of swash platforms. Depth is in m below mean low water during spring tides.

bars creates an abundance of large-scale (up to 6 m thick) planar cross-beds oriented landward.

Preservation. Ideally, the features associated with an inlet at its seaward and landward ends should migrate with the inlet. It is questionable, however, whether the features associated with the seaward end of an inlet would be preserved in the nearshore region after the inlet has migrated laterally in the downdrift direction of longshore drift. Probably the sand accumulated in the ramp-margin shoals of the earlier site of an inlet would be redistributed to form new ramp-margin shoals in equilibrium with the new site of the inlet. Thus, whereas the inlet sequence would underlie the barrier for the entire distance of inlet migration, sedimentary structures associated with ramp-margin shoals may be seen near the downdrift end only of a preserved inlet sequence. In contrast, the flood tidal delta may migrate laterally with the tidal inlet. After the tidal inlet has migrated from a site, the tidal delta associated with that site of the inlet is rendered "inactive" and hence may later be overlain by backbarrier marsh and peat. Thus, the tidal-delta facies may be present in the backbarrier environment of a barrier along the entire distance of inlet migration.

The chances for preservation of flood and ebb tidal-delta facies are excellent during a regression or a transgression by in-place drowning of barriers. If the transgression takes place through shoreface retreat of barriers, however, the delta facies may be destroyed during the barrier retreat, and the only record of inlets preserved may be the inlet sequence (see *Barrier Island Facies*).

NARESH KUMAR

References

Barwis, J. H., and Makurath, J. H., 1978. Recognition of ancient tidal inlet sequences: An example from the Upper Silurian Keyser Limestone in Virginia, *Sedimentology,* 25, in press.

Bernard, H. A.; Major, C. F., Jr.; and Parrott, B. S., 1970. Recent sediments of southwest Texas; a field guide to the Brazos alluvial and deltaic plains and the Galveston barrier island complex, *Univ. Texas Bur. Econ. Geol. Guidebook,* 11, 16p.

Boersma, J. R., 1969. Internal structure of some tidal megaripples on a shoal in the Westerschelde Estuary, the Netherlands, *Geol. Mijnbouw.,* 48, 409-414.

Caldwell, D. M., 1971. *A Sedimentological Study of an Active Part of a Modern Tidal Delta, Moriches Inlet, Long Island, New York.* Master's Thesis, Columbia Univ., New York, 70p.

Hayes, M. O., ed., 1969. Coastal-environments, NE Massachusetts and New Hampshire, *Coastal Res. Group, Univ. Mass Contrib. 1, (Eastern Section SEPM Guidebook, May 9-11),* 462p.

Hayes, M. O., and Kana, T. W., eds., 1976. Terrigenous clastic depositional environments (A.A.P.G. Short Course), *Univ. S. Carolina Dept. Geology,* 315p.

Hobday, D. K., and Tankard, A. J., 1977. Tide-dominated back-barrier sedimentation, early Ordovician Cape Basin, Cape Peninsula, South Africa, *Sed. Geol.,* 18, 155-159.

Hoyt, J. H., and Henry, V. J., Jr., 1965. Significance of inlet sedimentation in the recognition of ancient barrier islands, *Wyoming Geol. Assoc. Guidebook 19th Field Conf.,* 190-194.

Hoyt, J. H., and Henry, V. J., Jr., 1967. Influence of island migration on barrier-island sedimentation, *Geol. Soc. Am. Bull.,* 78, 77-86.

Johnson, J. W., 1974. Bolinas Lagoon inlet, California, *U.S. Army Coastal Eng. Res. Center, Misc. Pap. 3-74,* 5-46.

Klein, G. deV., 1970. Depositional and dispersal dynamics of intertidal sand bars, *J. Sed. Petrology,* 40, 1095-1127.

Kumar, N., 1973. Modern and ancient barrier sediments: New interpretations based on stratal sequence in inlet-filling sands and on recognition of nearshore storm deposits, *N.Y. Acad. Sci. Ann.,* 220, 245-340.

Kumar, N., and Sanders, J. E., 1970. Are basal transgressive sands chiefly inlet-filling sands, *Maritime Sed.,* 6, 12-14.

Kumar, N., and Sanders, J. E., 1974. Inlet sequence: A vertical succession of sedimentary structures and textures created by the lateral migration of tidal inlets, *Sedimentology,* 21, 491-532.

LeBlanc, R. J., 1972. Geometry of sandstone reservoir bodies, *Am. Assoc. Petrol. Geol. Mem. 18,* 133-190.

Meistrell, F. J., 1972. The spit-platform concept: Laboratory observation of spit development, in M. L. Schwartz, ed., *Spits and Bars.* Stroudsburg, Pa.: Dowden, Hutchinson & Ross, 225-283.

Morton, R. A., and Donaldson, A. C., 1973. Sediment distribution and evolution of tidal deltas along a tide-dominated shoreline, Wachapreague, Virginia, *Sed. Geol.,* 10, 285-300.

Oertel, G. F., 1972. Sediment transport of estuary entrance shoals and the formation of swash platforms, *J. Sed. Petrology,* 42, 858-863.

Oertel, G. F., 1973. Examination of textures and structures of mud in layered sediments at the entrance of a Georgia tidal inlet, *J. Sed. Petrology,* 43, 33-41.

Oertel, G. F., 1975. Post Pleistocene island and inlet adjustment along the Georgia coast, *J. Sed. Petrology,* 45, 150-159.

Sanders, J. E., and Kumar, N., 1975. Evidence of shoreface retreat and in-place "drowning" during Holocene submergence of barriers, shelf off Fire Island, New York, *Geol. Soc. Am. Bull.,* 86, 65-76.

Shepard, F. P., 1960. Gulf coast barriers, in F. P. Shepard et al., eds., *Recent Sediments, Northwest Gulf of Mexico.* Tulsa, Okla.: Am. Assoc. Petrol. Geologists, 338-344.

Shepard, F. P., and Moore, D. G., 1955. Central Texas coast sedimentation: characteristics of sedimentary environment, recent history, and diagenesis, *Bull. Am. Assoc. Petrol. Geologists,* 39, 1463-1593.

Van Straaten, L. M. J. U., 1954. Composition and structure of Recent marine sediments in the Netherlands, *Leid. Geol. Meded.,* 19, 1-110.

Winkelmolen, A. M., and Veinstra, H. J., 1974. Size and shape sorting in a Dutch tidal inlet, *Sedimentology,* 21, 107-126.

Cross-references: *Barrier Island Facies; Lagoonal Sedimentation; Lateral and Vertical Accretion; Littoral Sedimentation; Paralic Sedimentary Facies; Tidal-Current Deposits; Tidal-Flat Geology.* Vol. III: *Tidal Delta; Tidal Inlet.*

TILL AND TILLITE

The term *till* was originally applied in Scotland to a stiff, hard clay subsoil, generally impervious and unstratified, often containing gravel and boulders (AGI *Glossary,* 1972). In present usage, the word *till* means a clastic glacial deposit, usually poorly sorted and nonstratified, and derived from glacial drift; it consists of a heterogeneous mixture of rock and mineral fragments of varied lithologic composition, size, and shape. Though nonweathered rock and mineral fragments dominate, weathered material and even organic particles may be present. Local rocks prevail among clasts, but till matrix may contain more distantly derived material; some rock and mineral fragments may have been transported >1000 km by ice sheets. Particle sizes in till may range from clay size to large-scale blocks. Structurally, till is either massive or displays dynamic-metamorphic or flow structures; it is often jointed and fissile, and a more or less distinct fabric is common in most tills (see also Vol. III: *Glacial Geology;* Flint, 1971; Goldthwait, 1972; Francis, 1975; Dreimanis, 1976; Lavrushin, 1976).

If till is indurated by cementation or by metamorphism, it is called *tillite* (a term introduced by Penck, 1906); a more specific term for metamorphosed tillite is *metatillite.* Though some authors, when defining till or tillite, stress absence of meltwater during its deposition, and also lack of sorting and stratification, some types of till deviate from such restrictions (Flint, 1971).

The British literature often uses *boulder clay* as a synonym for till; its German equivalent is *Geschiebemergel* or *Geschiebelehm;* these terms are less comprehensive than till, as many tills contain very little clay (Fig. 2) or hardly any boulders. In various languages of continental Europe, most commonly used terms are *moraine, ground moraine,* or *morainic deposit* (see Vol. III: *Moraines*), e.g., the German *Moräne, Grundmoräne,* the Russian *morena, donnaia morena.* These terms are more appropriate for landforms of glacial origin. *Hardpan* is a well driller's term for resistant basal till.

Numerous nongenetic terms may include till. Various geologic processes, such as gravity flows (q.v.), gravity slides (q.v.), and others may produce till-like materials (see Table 1 in *Diamictite;* Flint, 1971; Harland et al., 1966; Jago, 1974). If it is not certain whether the sediment or rock in question is true till or tillite, some nongenetic terms are used, such as diamictite (q.v.), mixtite, paraconglomerate (see *Conglomerate*), muddy conglomerate, mudstone conglomerate, shale-matrix conglomerate, shale-matrix breccia conglomerate, polymictic conglomerate, mudstone, cobbly or pebbly mudstone (q.v.), conglomerate mudstone, conglomeratic sandy mudstone, sandy clay containing boulders, pebble loam, stony loam, and tilloid.

Incorporation of Drift Material in or on a Glacier

Glacial drift or debris (see *Glacigene Sediments*), from which till is derived after melting of glacial ice, may become incorporated into a glacier at its base as basal debris, or on its surface as superglacial or supraglacial debris (Fig.1).

Superglacial Drift. Superglacial drift is either extraglacial (talus, rock slides, etc.) or englacial in origin. When carried on the surface of glaciers, drift may be carried englacially for part of the distance, but prior to deposition as till, it reemerges by the process of ablation (Fig. 1). On the glacial surface, superglacial drift is subjected to repeated frost activity, and also to washing by water. Because of frost shattering, rock fragments are mostly angular; and due to washing, most superglacial debris are coarser textured than the englacially and basally transported debris (Fig. 2).

Basal Drift. The glacial drift that becomes incorporated or entrained into glacial ice at its base consists either of loose material picked up by the glacier or, more commonly, of products of glacial erosion: abrasion, quarrying, and even entrainment of large-scale blocks of bedrock or nonconsolidated sediments, some of them measuring as much as 2 by 4 km horizontally and up to 120 m vertically (Woldstedt, 1954; Moran, 1972). The large inclusions are called *floes,* or

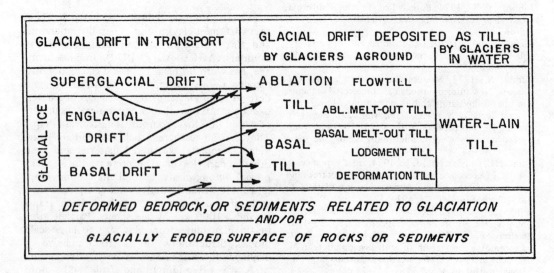

FIGURE 1. Genetic classification of tills, and their relationship to glacial drift in transport.

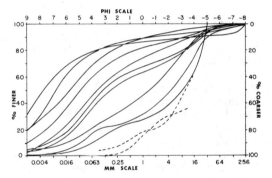

FIGURE 2. Cumulative granulometric curves of selected tills of various lithologic composition and origin. *Solid lines:* basal late-Wisconsin tills from Ontario (Vagners, 1966 and 1969; see Dreimanis and Vagners, 1971 for references). *Dashed lines:* ablation tills derived from superglacial drift in an end moraine of a modern glacier (from preprint of Elson, 1961). *Note:* Boulders, and in some analyses also cobbles, are excluded from the analyses.

rafts in English *Schollen* in German, *krupniye ottorzhentsi* in Russian.

The thickness of the dark or dirty basal drift layer usually ranges from a few centimeters to several meters. It either consists of numerous dark dirt bands of varied thickness separated by bands of relatively clean ice, or looks like a dynamically metamorphosed rock (see Goldthwait, 1972; Lavrushin, 1976). The dark bands and mylonitic structures have probably resulted from differential movement of glacial ice.

Differential movement is also responsible for comminution of drift particles by crushing and abrasion. Instead of producing a gradual progression from coarse to fine particles, comminution is most rapid at the transition from small rock fragments to their constituent minerals or to splinters of well-cleaved and fractured minerals. Because of this process, each lithic component of glacially transported drift attains bimodal distribution: one mode is in the clast grade, while another is in the terminal mineral grade, which is characteristic for each mineral and represents the final size of its glacial comminution (Dreimanis and Vagners, 1972); for instance igneous quartz, feldspars, and amphiboles are concentrated in the 31–250μ range, carbonate minerals in the 2–62μ and clay minerals in the 1–2μ range. With increasing distance of glacial transport, or with increasing degrees of comminution (more vigorous in basal than in englacial drift), the terminal modes, which make up the matrix of glacial drift, increase in volume because of comminution of clasts. Because glacial drift and the resulting till consist of many lithic components which have traveled various distances in the glacier, the final particle size distribution is usually multimodal; or, if the components are well distributed, the resulting granulometric cumulative curve may become a nearly straight line (Fig. 2).

Many constituents of the matrix of glacial drift, or the resulting till, are silt sized; tills, therefore, usually contain at least 25% silt, whereas the clay-size fraction may be as low as 1%, and sand-size as low as 3% (Fig. 3).

Mineral fragments in the terminal grades are usually angular, but clasts tend to become first subangular, then subrounded, polished, and striated; many of them show impact marks, and some have faceted surfaces. Most clasts, and also sand grains, become oriented with their long axes either parallel or transverse to the direction of glacial movement, and the *a-b* planes are parallel to the foliation in ice (Boulton, 1972).

Englacial Drift. In areas of compressive glacial flow, the dirt bands of basal drift become sheared up into glacial ice—from the basal into an englacial position—up to about 100 m from the base (Goldthwait, 1972). Here they are separated by wide bands of ice which reduce the chance of collision of clasts and cause less crushing; friable and soft rocks and sediments, even clay lumps, may thereby survive glacial transport for hundreds of km while in the englacial position (Dreimanis, 1976). Differential movement still takes place, resulting mainly in abrasion and rounding of clasts. Both parallel and transverse fabrics have been found in englacial drift, depending upon the dominant stress system: transverse fabric in compressive

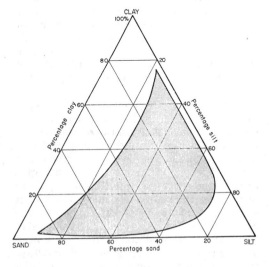

FIGURE 3. Granulometric composition of till matrix (< 2 mm) in ternary diagram, based upon > 1000 analyses from North America (from various published and unpublished sources).

stress zones, parallel in tensile zones (Boulton, 1972). Although upshearing is responsible for taking basal drift into englacial position, the further englacial transport may also be parallel to the base of the glacier or even downward, returning the englacial drift into basal ice, particularly in areas of deposition at the glacier base.

Downmelting of the glacial surface in the area of ablation will cause the upper portion of englacial drift to become superglacial. Such superglacial drift shows evidence of englacial transport.

Deposition of Till

Till may become deposited (1) at the base of a glacier as basal till; (2) on or from its surface as ablation till; or (3) beneath ice shelves, at the contact of glacier and water, as "waterlain" till (see *Glacigene Sediments;* see also Vol. III: *Glaciation,* Table 1).

Basal Till. Basal till, often called *lodgment till* (see Flint, 1971) may be deposited in at least three different ways: (1) as basal melt-out till, (2) as lodgment till *sensu stricto,* and (3) as deformation till.

Basal Melt-Out Till. The general term *melt-out till* was introduced by Boulton in 1970 (see Boulton, 1972). This till may become deposited at the base of actively moving or stagnant temperate ice by slow melting, due to the ability of geothermal heat to melt up to 1 cm of ice per yr. It is called *subglacial ablation till* by Elson (1961). Because of the load of overlying ice, the basal melt-out till is denser than the ablation melt-out till, but still not as dense as lodgment till. Its fabric and structure derives from glacial ice and therefore reflects the basal or the englacial drift; as the volume of the melted-out ice exceeds the volume of drift, however, the fabric and structures tend to become flattened parallel to the plane of deposition, and the resulting till looks foliated.

Lodgment Till. Plastering or lodgment under an actively moving glacier is responsible for formation of lodgment till (*sensu stricto*). The process is aided by pressure melting, heat from shear friction and crushing of rocks, and geothermal heat, providing just enough meltwater to lodge and pack the various grain sizes, particularly in till matrix, for maximum density and least porosity. Such dense till, consisting mainly of crushed rocks and minerals, is called *comminution till* by Elson (1961). The process of lodgment or plastering may create fissile structure, which in some cases becomes visible after subsequent frost activity. Various inclusions, e.g., lenses of nonconsolidated sediments, and platy clasts, are imbricated, dipping upglacier, because of compressive flow present along the surfaces of lodgment.

Deformation Till. Elson (1961) suggested the term *deformation till* for till derived from blocks and slices of bedrock and underlying unconsolidated sediments which had been deformed by glacial thrust in situ, or transported for short distances. Such till has not gone through prolonged glacial comminution and packing, as has lodgment till, and contains small amounts of distantly derived drift material. Primary structures of the original material are usually recognizable, but they are deformed to various degrees by shearing, faulting, folding, overthrusting, injections, etc. Deformation till is less dense than lodgment till, either because the original material had higher porosity, or faulting and crushing had increased voids, for instance in local tills composed of friable rocks. Lavrushin's (1976) dynamic facies of scaly moraines, or erratic masses, and those glaciotectonic structures of various authors, which had been incorporated in the base of glaciers, may belong to this type of till. Other glaciotectonic structures without any drift material added by the deforming glacier should be considered merely as deformed bedrock or deformed nonconsolidated sediments.

Ablation Till. Ablation till is deposited by ablation from superglacial or englacial drift either on or adjacent to a stagnant glacier, or in the terminal zone of an active, though shrinking glacier. The resulting till is loose and less compact than basal till, and its texture and composition and shapes of clasts reflect the superglacial or englacial transport (see above). Clayey and silty ablation tills may acquire compactness by later desiccation. Angularity of clasts is due not only to superglacial transport, but also to repeated freezing and thawing during and after deposition. Ablation till may form (1) a melt-out till in situ on the surface of glacier or (2) a flowtill.

Ablation Melt-Out Till. Usually described as *ablation till* or *ablation moraine,* ablation melt-out till may be derived from superglacial drift only (in the lateral and medial moraines of mountain glaciers); in which case, it is particularly coarse textured (Fig. 2) and loose. Its clasts are usually angular, and their fabric unrelated to glacial movement. If it is derived from englacial drift by slow ablation on the glacial surface, however, it resembles basal melt-out till in its composition and fabric (see above), but it is less compact, unless it has undergone desiccation at some later time.

Flowtill. The term *flowtill* was proposed by Hartshorn (1958). Flowtill is usually derived from ablation melt-out till which had moved down the glacier surface. Depending upon its derivation, flowtill resembles in its composition either superglacial, or, more often, englacial drift. The nonglacial flow produces its own fabrics, both parallel and transverse, depending upon local conditions, but unrelated to glacial movement. In very mobile and liquid flows, clasts tend to sink toward their base, and also stratification develops, while semiplastic flows produce little stratification. If the movement of flowtill is merely a downslope creep, then downslope shear planes and fissility may develop, and also a parallel fabric; such flowtill resembles basal till, and it may even be derived from the surface portion of basal till on a slope.

Waterlain Till. Dreimanis proposed the term *waterlaid till* in 1967 (see Dreimanis and Vagners, 1972). Waterlain till corresponds in part to the para-till of Harland et al. (1966), and to some of the glaciomarine sediments and the so-called lacustro-tills and lacustrine ablation materials (see also Jago, 1974). It is deposited along the edge of a glacier terminating in water or beneath floating ice shelves by sedimentation of drift material from the base of a wet-base glacier, with admixture of glaciolacustrine, glaciomarine, or glaciofluvial sediments. It may grade, laterally or vertically, into basal, particularly deformation, till, wherever the glacier overrode it, or into dominantly glaciolacustrine or glaciomarine sediments containing dropstones (see *Glacial Marine Sediments*). The term *waterlain till* should be used only in cases where till-like material dominates over lacustrine or marine. Waterlain till shows disrupted and irregular bedding, and flow structures; some parts may have originated as subaqueous flows; it is less dense and more porous than other tills. Either the fabric is random or flow fabric has developed. The abundance of clay pebbles suggests rapid sedimentation, and the dominance of till-like material over lacustrine or marine sediments indicates that the distance between the place of deposition and the edge of the base of the glacier was small.

Tills Deposited in Arid Polar Environments. Shaw (1977), from observations in Antarctica, has concluded that in arid polar environments till is deposited passively by sublimation, rather than by melting of ice. The result is preservation of primary structures related to the transport of glacial drift in ice. This model helps to understand better the complex glaciodynamic classification of basal till or ground moraine as proposed by Lavrushin (1976 and in his earlier publications).

Ancient Tills

Some of the most distinctive features of tills, e.g., striated pavements and striated and faceted clasts, are difficult to recognize in lithified rocks. It is also likely that there were situations in the past where continental glaciers reached the sea and may have contributed not only to marine tills but also to unstable sediment piles, leading to subaqueous slumps, slides, and flows. For these reasons, the glacial origin of some diamictites has been refuted (e.g., Dott, 1961; Schermerhorn, 1974). Nevertheless, several ancient glaciations have been recognized and widely accepted (see Harland and Herod, 1975).

The oldest evidence for widespread glaciation is in Middle Precambrian (≈ 2150-2500 m.y.; Pettijohn, 1975) of the Canadian Shield. The best-known tillite from this time is the Gowganda Formation, which extends over an area of many thousands of km^2 (Lindsey, 1969; Young, 1973).

Glacial deposits of the Eocambrian or Late Precambrian (≈ 800-600 m.y. ago) seem to be worldwide, although the synchroneity and glacial origin of these deposits has been reviewed and questioned by Schermerhorn (1974). Spencer (1975) reviewed critical evidence for this glaciation; the Varangian Tillite Formation (660-680 m.y.) of the North Atlantic region is the best known and most conclusive.

Evidence has been found of an Ordovician glaciation in central Sahara (Beuf et al., 1971), which may have extended also over other areas.

The most famous ancient glacial deposits are Permo-Carboniferous in age, found in the Southern Hemisphere (then presumably polar Gondwanaland). Permo-Carboniferous tillites and related deposits are known from South Africa, South America, Australia, Tasmania, the Falkland Islands, and Antarctica. Frakes and Crowell have reviewed these occurrences in a series of papers (see Crowell and Frakes, 1972, and Frakes et al., 1975, for older references).

ALEKSIS DREIMANIS

References

Beuf, S., and 5 others, 1971. Les grès du Paléozoïque inférieur au Sahara, *Publ. Inst., Franc. Pétrol 18*, Paris, 464p.

Boulton, G. S., 1972. Till genesis and fabric in Svalbard, Spitsbergen, in Goldthwait, 1972, 41-72.

Crowell, J. C., and Frakes, L. A., 1972. Late Paleozoic glaciation. Part V, Karroo Basin, South Africa, *Geol. Soc. Am. Bull., 83*, 2887-2917.

Dott, R. H., Jr., 1961. Squantum "tillite," Massachusetts—evidence of glaciation or subaqueous movements? *Geol. Soc. Am. Bull., 72*, 1289-1306.

Dreimanis, A., 1976. Tills: Their origin and properties, *Roy. Soc. Canada Spec. Publ. 12*, 11-49.

Dreimanis, A., and Vagners, U. J., 1972. Bimodal distribution of rock and mineral fragments in basal tills, in Goldthwait, 1972, 237-250.

Elson, J. A., 1961. The geology of tills, *Proc. 14th*

Canadian Soil Mech. Conf., Natl. Res. Counc. Canad. Com. Soil Snow Mech., Techn. Mem. 69, 5-36.
Flint, R. F., 1971. *Glacial and Quaternary Geology.* New York: Wiley, 892p.
Frakes, L. A.; Kemp, E. M.; and Crowell, J. C., 1975. Late Paleozoic glaciation: Part VI, Asia, *Geol. Soc. Am. Bull.,* 86, 454-464.
Francis, E. A., 1975. Glacial sediments: A selective review, *Geol. J. Spec. Iss. 6,* 43-68.
Goldthwait, R. P., ed., 1972. *Till: A Symposium.* Columbus: Ohio State Univ. Press, 402p. See also introduction by editor, 3-26.
Harland, W. B.; Herod, K. N.; and Krinsley, D. H., 1966. The definition and identification of tills and tillites, *Earth-Sci. Rev.,* 2, 225-256.
Harland, W. B., and Herod, K. N., 1975. Glaciations through time, *Geol. J. Spec. Issue 6,* 189-216.
Hartshorn, J. H., 1958. Flowtill in southeastern Massachusetts, *Geol. Soc. Am. Bull.,* 69, 477-482.
Jago, J. B., 1974. The terminology and stratigraphic nomenclature of proven and possible glaciogenic sediments, *J. Geol. Soc. Austral.,* 21, 471-474. See also discussion by M. Schwarzbach (1975), 22, 255-256, and reply.
Lavrushin, Y. A., 1976. *Structure and Development of Ground Moraines of Continental Glaciations* (in Russian). Moscow: Nauka, 237 p.
Lindsey, D. A., 1969. Glacial sedimentology of the Precambrian Gowganda Formation, Ontario, Canada, *Geol. Soc. Am. Bull.,* 80, 1685-1702.
Moran, S. R., 1972. Glaciotectonic structures in drift, in Goldthwait, 1972, 127-148.
Penck, A., 1906. Süd-Afrika und Sambesifälle, *Geogr. Zeitschr.,* 12, 600-611.
Pettijohn, F. J., 1975. *Sedimentary Rocks,* 3rd ed. New York: Harper & Row, 628p.
Rukhina, E. V., 1973. *Litologiia Lednikovikh Otozhenii,* Leningrad: Nedra, 176p.
Schermerhorn, L. J. G., 1974. Late Precambrian mixtites: Glacial and/or nonglacial? *Am. J. Sci.,* 274, 673-824.
Shaw, J., 1977. Till deposited in arid polar environments, *Canad. J. Earth Sci.,* 14, 1239-1245.
Spencer, A. M., 1975. Late Precambrian glaciation in the North Atlantic region, *Geol. J. Spec Iss. 6,* 217-236.
Woldstedt, P., 1954. *Das Eiszeitalter,* 2nd ed., vol. 1. Stuttgart: Ferdinana Enke Verlag, 374p.
Young, G. M., 1973. Tillites and aluminous quartzites as possible time markers for middle Precambrian (Aphebian) rocks of North America, *Geol, Assoc. Canada Spec. Publ. 12,* 92-127.

Cross-references: *Conglomerates; Diamictite; Fabric, Sedimentary; Faceted Pebbles and Boulders; Glacial Gravels; Glacial Marine Sediments; Glacial Sands; Glacigene Sediments; Gravity Flows; Pebbly Mudstones; Sand Surface Textures.* Vol. III: *Glacial Deposits.*

TILLOIDS—See DIAMICTITE; PEBBLY MUDSTONES

TOOL MARKS

Tool marks are the structures produced on soft sediment surfaces by moving or stationary objects (tools) during deposition (see *Current Marks*). Tool marks can be divided into three kinds: stationary tool marks, obstacle marks, and moving tool marks (Reineck and Singh, 1973).

Kinds of Tool Marks

Stationary Tool Marks. Impressions on a soft sediment surface made by temporarily resting passive objects are termed stationary tool marks. Various objects, e.g., pebbles, wood, shells, and ice blocks, resting on a soft sediment surface produce marks below them. When these objects are carried away, they leave behind the impressions on the sediment surface.

Obstacle Marks. Objects lying in the path of a current produce turbulence in the flow, resulting in erosion and/or deposition behind them. Such marks are termed obstacle marks (see also *Current Marks*). Best known of the obstacle marks are *current crescents,* formed as a result of accumulation of sand behind objects like shells, pebbles, etc. Sometimes, contemporaneous erosion produces a small depression behind the obstacle; termed a *longitudinal obstacle scour,* this is commonly preserved as a mold on the undersurface of sandstone. In the wind regime, sand is deposited behind small irregularities in the form of *sand tails.*

Moving Tool Marks. Various types of marks are made by objects being carried by a current along a soft sediment surface (see also *Scour Marks*). Important studies on moving tool marks have been made by Ten Haaf (see *Scour Marks*), Dzulynski and Sanders (1962), and Dzulynski and Walton (1965). Moving tool marks develop when flow conditions are such that an object is lifted and moved along the soft cohesive sediment surface.

The morphology of a tool mark is controlled by the shape and size of the tool, the angle at which the tool approaches the sediment surface, and the consistency of the sediment bottom. Depending upon the shape, several kinds of moving tool marks are recognized (see *Current Marks,* Fig. 3).

Groove marks are long, straight, gutter-like troughs produced by tools rolling and dragging continuously on a soft sediment surface.

Chevron marks are made up of continuous, open V marks arranged to form a straight ridge, the V forms closing downcurrent; they are produced by a tool moving just above the cohesive sediment surface (not touching the surface).

Prod marks are asymmetrical, elongated depressions with a shallow, pointed upcurrent part and a broad, deeper downcurrent part; they are formed when a tool reaches the sediment surface at a high angle and is lifted away after striking.

Bounce marks are almost symmetrical depressions, tapering in both directions, produced when a tool approaches the sediment surface at a rather low angle and bounces back into the current.

Brush marks are elongated shallow depressions with a small rounded ridge of mud at the downcurrent end; they are produced when a tool with its axis inclined in the upcurrent direction reaches the sediment surface at a very low angle and is lifted away.

Skip and roll marks are produced when a tool makes impact at regular intervals (saltation), thus producing a set of markings arranged in a row. If a tool rolls on a soft sediment surface, a rather continuous track is produced, known as a roll mark.

Distribution in Environments

Tool marks are conveniently produced on soft muddy bottoms and are often preserved as molds on the under surface of sandy layers (sole marks). Moving tool marks are abundantly found in flysch facies as sole marks. Tool marks are also abundant in shallow-water environments, e.g., intertidal flats, and natural levees, but there they are not often preserved because the sedimentation surfaces undergo active reworking. In deep water however, tool marks are well preserved because intermittent reworking of the sediment surface is rare.

INDRA BIR SINGH

References

Dzulynski, S., and Sanders, J. E., 1962. Current marks on firm mud bottoms, *Conn. Acad. Arts Sci. Trans.,* 42, 57-96.

Dzulynski, S., and Walton, E. K., 1965. *Sedimentary Features of Flysch and Greywackes.* Amsterdam: Elsevier, 274p.

Reineck, H.-E., and Singh, I. B., 1973. *Depositional Sedimentary Environments.* Berlin: Springer-Verlag, 439p.

Cross-references: *Casts and Molds; Current Marks; Scour Marks; Sedimentary Structures; Top and Bottom Criteria.*

TOP AND BOTTOM CRITERIA

Inasmuch as sedimentary rocks are commonly very diverse in structure and composition, a competent geologist generally has no trouble determining their order of superposition. In heavily folded, overturned, partly metamorphosed, and poorly exposed sections, precise observations are necessary. Shrock's (1948) description of the top and bottom criteria used to determine this order of superposition still remains the most complete compendium of these criteria. Most of these criteria are sedimentary structures that reflect the primary influence of the earth's gravitational field at the time of their formation.

Geopetality

From the Greek *geos* (earth) plus Latin *petere* (to seek), the term *geopetal* refers to features or structures that reflect the direction of gravitational acceleration at the time of their formation. Proposed by Sander (1936), it applies to any feature that enables the determination of the relation of top to bottom at the time when the feature was formed.

The *Law of Superposition,* probably the oldest and most basic law of geology, going back to the time of Steno in the 17th century (see Dunbar and Rodgers, 1957), requires that in any undisturbed sedimentary sequence the layer on top is younger than the layer beneath. In contorted, overturned, and overthrust rocks, it is not intuitively obvious what was formerly on top and what was beneath. Hence there is a need for geopetality studies in the establishment of former superposition.

There are four major categories of geopetal criteria (Shrock, 1948; Hills, 1963): Biologic, igneous, tectonic, and sedimentologic.

Biologic Criteria. Fossils and paleobiologic evidences disclose many features of geopetality—in-situ organisms tend to grow upward, sediment fills empty shells from the bottom up, and so on (see *Biogenic Sedimentary Structures; Biostratinomy*).

Igneous Criteria. The disposition of many igneous features, which may be intercalated in a sedimentary complex, often displays excellent geopetal criteria. A basic sill, when injected into a flat-lying sequence, will commonly develop a density segregation or gradation, the heaviest minerals settling to the base of the cooling body. Volcanic flows display vesicles (former gas bubbles) and surface flow structures (scoria, aa, and pahoehoe lavas) at the top (Butler and Burbank, 1929). Pillow lavas tend to be round on top, pointed below (Wilson, 1942).

Tectonic Criteria. Drag folds ideally (in open folds) have axial planes that dip more steeply than the stratification but less steeply on overturned limbs (Hills, 1963). Slaty cleavage tends to parallel axial planes and may be used to indicate this relationship, provided the strike of the strata and of the axial planes do not differ by >45°. Dunbar and Rodgers (1957) illustrate complex cases caused by a second deformation and note that the sedimentologic criteria are given priority. In disharmonic folds, the structures die out downwards (on the slip planes).

Sedimentologic Top and Bottom Criteria

Some diverse examples of sedimentologic geopetal criteria are: ripple marks (q.v.), which generally point up; graded bedding (q.v.); cross-bedding (q.v.), which attenuates and flattens downdip and is truncated above; numerous minor diapiric features (see, e.g., *Pillow Structures; Water-Escape Structures*) are geopetal (Ten Haaf, 1956; Prentice and Kuenen, 1957); slumps (q.v.) are truncated above; unconformities commonly truncate underlying features and are often capped by conglomerates or breccias of the underlying members; cyclic bedding such as a cyclothem is frequently in a distinctive sequence (see *Cyclic Sedimentation*); paleosols may contain geochemical gradients, reflecting former soil profiles; duricrusts (q.v.) attenuate downward; raindrop imprints (q.v.) are rare but distinctive; mud cracks (q.v.) wedge down, and the edges curl up.

Top and bottom criteria can be simply classified into microscopic, macroscopic, and megascopic categories.

Microscopic. Microscopic criteria include all those inhomogeneities that can generally only be recognized with a hand lens or a microscope. Included in this group are: (1) details of bedding relationships in shales and siltstones such as micro-cross-lamination, varved bedding, and very small graded beds; (2) fillings of small cavities or cracks, such as those that frequently occur in limestones; original microcavities are commonly partially filled with sediment during deposition, and the open space at the top of the cavity is generally later filled with a cement or some different material.

Macroscopic. Macroscopic criteria include all those criteria that can be easily recognized within a single bed or small outcrop and comprise the bulk of top and bottom criteria. These criteria can be subdivided into: (1) features on the upper or under surfaces of beds, i.e., criteria based on inhomogeneities between beds; and (2) criteria based on inhomogeneities within beds.

Features formed on the upper or under surface of beds (group 1) are the most widely used criteria and include: scour-and-fill (q.v.); ripple marks (q.v.); swash and rill marks; pit and mound structure and gas pits (see *Gases in Sediments*); hailstone and raindrop imprints (q.v.); imprints, casts, and pseudomorphs of crystals such as salt and ice; flow, flute, and groove molds; tracks, trails, and footprints, mud cracks (q.v.) and other superficial cracks; bored or burrowed surfaces; load casted structures; and rolling and tumbling marks, and bounce, brush, or prod marks (see *Current Marks*). Some of these criteria are illustrated in Fig. 1.

Included in the second group are: cross-stratification (see *Cross-Bedding*); imbricated-pebble structure in conglomerates (q.v.); intraformational breccias or conglomerates, sometimes referred to as basal conglomerates (see *Intraformational Disturbances*); graded bedding (q.v.); contorted, slumped, or convoluted bedding; and organic structures, such as stromatolite heads, trees buried in situ, and fossils preserved in situ.

Megascopic. The ease with which one can determine the top of sedimentary strata in folded and contorted sediments generally depends upon one's understanding of the mode and environment of deposition of the sediments. Most sedimentary sequences are characterized by some kind of pattern, rhythm or cycle which is often repeated. The recognition of part of this pattern generally makes it possible to determine the original depositional attitude of the rocks. Such criteria are generally recognized in fairly large outcrops or a series of outcrops and can be referred to as megascopic. Included in this group are:

1. Alluvial sedimentary cycles deposited by meandering streams: the coarsest material is deposited in the channel, point bar, or levee and is generally topped by the successively finer backswamp or flood-plain silts or clays (see *Fluvial Sediments*).

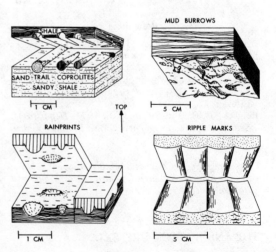

FIGURE 1. Examples of top and bottom criteria observed at the junction between two sediment types (redrawn from Shrock, 1948, and Ten Haaf, 1959).

2. Details of deltaic sedimentary patterns: once again, the coarsest material is deposited within or at the mouth of the distributary channel or close to the distributary levees; facies changes occur away from these areas with much interfingering of the coastal plain or bay sediments with the silts and clays deposited in the immediate vicinity of the active distributaries (see *Deltaic Sediments*).
3. Coal sequences: coal is generally deposited in cyclic repetitions within one of the two sedimentary sequences mentioned above and is commonly underlain by a "seat-earth" or underclay (see *Coal; Cyclic Sedimentation*).
4. Salinastone (evaporite) sequences: when large bodies of water are evaporated, the salt deposits generally consist of particular cycles of sediment; e.g., anhydrite or gypsum may merge upward into halite or downward to limestones. Repetition of two of these members is common, but three is not so common (see *Evaporite Facies*).
5. Turbidite and graded bed sequences: beds deposited rapidly by slumps, turbid flow, and by diminishing currents are characterized by coarser-grained sediments at the base making a sharp contact with the underlying unit, which is commonly a finer-grained sediment (see *Graded Bedding*). The upper part of the graded bed is commonly burrowed and may be topped with layers of sediment containing pelagic organisms. Displaced shallow-water organisms are commonly found in the lower and coarsest portion of the graded bed.

JOHN R. CONOLLY

References

Butler, B. S., and Burbank, W. S., 1929. The copper deposits of Michigan, *U.S. Geol. Surv. Prof. Pap. 144*, 238p.
Dunbar, C. O., and Rodgers, J., 1957. *Principles of Stratigraphy.* New York: Wiley, 356p.
Friedman, G. M., and Sanders, J. E., 1974. Positive-relief bedforms on modern tidal flat that resemble molds of flutes and grooves; implications for geopetal criteria and for origin and classification of bedforms, *J. Sed. Petrology,* **44**, 181-189.
Hills, E. S., 1963. *Elements of Structural Geology.* New York: Wiley, 483p.
Prentice, J. E., and Kuenen, P. H., 1957. Flow-markings and load casts, *Geological Mag.,* **94**, 173-174.
Sander, B., 1936. Beiträge zur Kenntnis der Anlagerungsgelfüge (Rhythmische Kalke und Dolomite aus der Trias), *Mineralog. Petrog. Mitt.,* **48**, 27-209. (English Transl.: E. B. Knopf, 1951, Tulsa: Am. Assoc. Petrol. Geologists.)
Shrock, R. R., 1948. *Sequence in Layered Rocks.* New York: McGraw-Hill, 507p.
Ten Haaf, E., 1959. *Graded Beds of the Northern Apennines.* Rijksuniversiteit te Groningen, 102p.
Wilson, M. E., 1942. Structural features of the Keewatin volcanic rocks of Western Quebec, *Geol. Soc. Am. Bull.,* **53**, 54-69.

Cross-references: *Bedding Genesis; Biogenic Sedimentary Structures; Bouma Sequence; Casts and Molds; Cross-Bedding; Current Marks; Cyclic Sedimentation; Fabric, Sedimentary; Flame Structures; Graded Bedding; Inverse Grading; Mud Cracks; Paleocurrent Analysis; Penecontemporaneous Deformation of Sediments; Pillow Structures; Primary Sedimentary Structures; Raindrop Imprint; Ripple Marks; Scour-and-Fill; Scour Marks; Secondary Sedimentary Structures; Slump Bedding; Tool Marks; Turbidites; Water-Escape Structures.*

TRACE FOSSILS—See Vol. VII; Also see BIOGENIC SEDIMENTARY STRUCTURES

TRACE METALS IN SEDIMENTS— See Vol. IVA

TRACER TECHNIQUES IN SEDIMENT TRANSPORT

Many kinds of natural and artificial materials have been used as tracer particles in the study of sediment transport in rivers and on beaches. Water-soluble dyes, such as potassium permanganate, rhodamine-B, and sodium fluorescein, have also been used to estimate water flow characteristics and, in some cases, for comparison of water flow and movement of tracer particles. Suitability of fluorescent and radioactive tracers for surf mixing studies is reviewed by Verwey and McMurray (1964). A summary of dye studies of water flow in streams is presented by Boning (1974). The present report, however, is limited to a description of artificial and natural sediment tracer particles that have actually found wide use in laboratory and field investigations.

Tracer particles fall into three basic categories: (1) *natural tracers,* which differ in some observable characteristic from the indigenous sediment of beaches and streams being investigated; (2) *fluorescent tracers,* made by introducing fluorescent pigments into the formulation for artificial tracer particles or by applying a fluorescent coating to natural materials; natural sediment having fluorescent properties is also included in this category; and (3) *radioactive tracers,* made by irradiating natural particles in an atomic pile or by applying radioactive liquids for impregnating or coating natural sediment; natural sediment containing radioelements is also included in this category.

In a few studies, exotic materials such as ferromagnetic coatings (Griessier, 1964) or combination of two tracer categories (Rathbun et al., 1971) provided an appropriate tracer material. Simulating important properties (size, shape, sorting, density, etc.) of the indigenous sediment or turning the indigenous sediment into a tracer without serious modification of its physical properties is a common desire. In the reality of field studies, however, there is often a trade-off between such factors as cost, need for large volumes of tracer, length of the study, and the type of tracer chosen. Increasing sophistication in tracer technology has been accompanied by better knowledge of variables and improved theory.

Natural Tracers

Color, mineralogy, and specific gravity are the three characteristics most often used to distinguish tracer particles from their indigenous counterparts in flume, wave-tank, and field investigations. Using optical identification of heavy minerals in the $63-125\mu$ size fraction, Judge (1970) determined that the ratio of less stable augite and hornblende to the more stable garnet and zircon in coastal sediment could be used to establish direction of littoral drift on the California coast. Decreasing percentage of heavy minerals in the beach sands of Long Island, N.Y., westward from Montauk Point was interpreted in Colony (1932) as evidence for predominantly westward littoral drift.

Carr (1971) used beach pebbles and cobbles from Scotland and Wales as tracers of different color and mineralogy from natural sediment on Chesil Beach, England. He also used basalt as a tracer of higher density than flint or chert pebbles and cobbles found on Chesil Beach. Jolliffe (1961) summarized early attempts to use coal, manganese ore, artificially colored sands, and magnetic sands as natural tracers.

Fluorescent Tracers

Beginning in 1958, Russian investigators used fluorescent tracers in the study of beach sand movement (Zenkovich 1960, 1967). In this technique, a colloidal film containing organic, fluorescent dyes, such as anthracene or lumogene, is applied to natural sand grains. Wright (1962) used anthracene itself as a coating for sand grains in a study of littoral transport at Sandy Hook, N.J. Jolliffe (1961) reports on the preparation and use of fluorescent tracers in experiments that began in 1959 at the Hydraulic Research Station, Wallingford. Two types of fluorescent tracers were produced. One type was a concrete, containing granulated plastic which incorporated organic dyes such as anthracene and rhodamine-B. A second type of fluorescent tracer used natural sand coated with a resin, containing any one of a variety of fluorescent pigments (uvitex, kiton yellow, eosine, or rhodamine-B) to allow color coding of the tracer. The resulting cakes of sand and hardened urea resin were then granulated and sieved (Jolliffe, 1963). Further development of this resin-fluorescent organic dye technique is reported by Teleki (1966).

Experiments begun in 1961 with different fluorescent pigments and coating vehicles resulted in tracers that avoided the daylight invisibility, relatively long setting time, and granulation procedures encountered in most of the earlier, resin-based techniques (Yasso, 1962, 1966). These tracers used Day-Glo Acrylic Laquer, available in 10 colors, in combination with two essentially nonclumping resin vehicles to form a rapidly drying, thin coating on sand and coarser sediment. Sampling and evaluation procedures were simplified because the highly colored coating was visible both in daylight and in long-wave ultraviolet illumination. This Day-Glo fluorescent technique has been used in studies of the size-velocity relationship for beach sand movement (Yasso 1965; Boon 1969); onshore-offshore movement of different sand sizes (Boon 1968); sediment movement patterns within enclosed bays (Schwartz, 1966); cusp formation (Williams, 1973); and, with minor modification, for tracing movement of stream sediment (Rathbun et al., 1971).

Ingle (1966) used two colors of fluorescent tracers, coated by an unrevealed proprietary method, to trace sediment movement in the surf zone at five S California beaches.

Radioactive Tracers

Sediment containing natural radioelements has been used in a study of suspended matter in the Gironde estuary (Martin et al., 1970). In a study of littoral transport along part of the California coast, Kamel (1962) measured natural radiation from thorium contained in heavy minerals. Other minerals, containing thorium or uranium, which could be used as natural radioactive tracers are listed by Martin (1970).

Naturally radioactive elements in sediment have the disadvantage that their long half lives generally prevent repeated experiments along river stretches or littoral environments. Reactor activation of radioelements can provide short-lived tracers for sediment transport studies. One of the first tracer experiments using reactor activation of natural sediment was performed by Goldberg and Inman (1955). Slow-neutron activation of small quantities of the St. Peter sandstone yielded a phosphorus-32 tracer with half-life of 15 days.

Early experiments with tracers made by irradiation of glass or plastic, incorporation of radioelements into glass or plastic, and coating sediment with solution of radioelements are discussed by Arlman et al. (1958). In later experiments, a special glass melt was made to equal the density of quartz. Granulated and sieved particles were then irradiated to produce a sodium-24 tracer. Comparison of the transport behavior of glass tracer with equivalent size irradiated quartz particles was made in a movable-bed flume. Even though the glass tracer was of more angular shape the transport results were within the range of experimental error.

Large volumes of beach sand were labeled by applying a coating of radioactive ammonium phosphate (phosphorus 32) in a casein solution (Gibert and Cordeiro, 1964). Fluorescent tracers provided a check on the movement of this coated sediment along the Portuguese coast.

In 1966, a cooperative project between the U.S. Army Coastal Engineering Research Center and the U.S. Atomic Energy Commission was begun to find and evaluate a method of producing radioactive tracers for large-scale, littoral transport studies. It was desired to create a short-lived tracer using indigenous sediment, without need of a coating procedure, and without altering the hydraulic characteristic of the sediment. As described by Duane and Judge (1969), the radioelement chosen was xenon 133 with a half-life of 5.27 days. In this process, carbonate-free sand, heated to 900°C in a furnace having a xenon-133 atmosphere, absorbs the tracer radioelement. Liquid nitrogen is used as a cooling agent. This continuing project is called RIST, an acronym for *Radioisotopic Sand Tracer Study*.

Later field tests with xenon-133-tagged sand and a new technique for surface labeling of sand with gold-198 and gold-199 isotopes (2.7 day half-life) are discussed by Duane (1970). Carbonate-free sand is washed with a solvent and dried. Then it is treated with a solution of gold chloride containing gold 198 and 199. After decanting the excess solution, the sand is cured by heating to 1000°C.

Judge (1975) provides a summary of RIST field-trial designs, data-acquisition procedures, and preliminary interpretation of littoral transport regime for experiments at Pt. Mugu and Oceanside, California.

Procedures for introducing tracer sediment into the hydraulic system to be studied, sampling procedures, and evaluative models for interpreting experimental data are well described in the papers already cited.

WARREN E. YASSO

References

Arlman, J. J.; Santema, P.; and Svasek, J. N., 1958. Movement of bottom sediment in coastal waters by currents and waves; measurements with the aid of radioactive tracers in the Netherlands, *U.S. Army,*

Beach Erosion Board, Tech. Memo., 105, 56p.

Boning, C. W., 1974. Generalization of stream travel rates and dispersion characteristics from time-of-travel measurements, *J. Research U.S. Geol. Surv.,* **2,** 495-499.

Boon, J. D., III., 1968. Trend surface analysis of sand tracer distributions on a carbonate beach, Bimini, B.W.I., *J. Geol.,* **76,** 71-87.

Boon, J. D., III., 1969. Quantitative analysis of beach sand movement, Virginia Beach, Virginia, *Sedimentology,* **13,** 85-103.

Carr, A. P., 1971. Experiments on longshore transport and sorting of pebbles: Chesil Beach, England, *J. Sed. Petrology,* **41,** 1084-1104.

Colony, R. J., 1932. Source of the sands on the south shore of Long Island and the coast of New Jersey, *J. Sed. Petrology,* **2,** 150-159.

Duane, D. B., 1970. Tracing sand movement in the littoral zone: Progress in the Radioisotopic Sand Tracer (RIST) study, *U.S. Army, Coastal Engin. Research Center, Misc. Pap. 4-70,* 46p.

Duane, D. B., and Judge, C. W., 1969. Radioisotope sand tracer study, Point Conception, California, *U.S. Army, Coastal Engin. Research Center, Misc. Pap. 2-69,* 194p.

Gibert, A., and Cordeiro, S., 1962. A general method for sand labelling with radioactive nuclides, *Internat. J. Applied Radiation and Isotopes,* **13,** 41-45.

Goldberg, E. D., and Inman, D. L., 1955. Neutron-irradiated quartz as a tracer of sand movements, *Geol. Soc. Am. Bull.,* **66,** 611-613.

Griessier, H., 1964. Über den Linzotz von Magnetischen Werkstoffen blim Studium der Littoralen Materialbewegungen, *Monatsber. Deutsch. Akad. Wiss. Berlin,* **6,** 171-173.

Ingle, J. C., Jr., 1966. *The Movement of Beach Sand.* Amsterdam: Elsevier, 221p.

Jolliffe, I. P., 1961. The use of tracers to study beach movements; and the measurement of littoral drift by a fluorescent technique, *Rev. Géomorphologie Dynam.,* **2,** 81-98.

Jolliffe, I. P., 1963. A study of sand movements on the Lowestoft Sandbank using fluorescent tracers, *Geogr. J.,* **129,** 480-493.

Judge, C. W., 1970. Heavy minerals in beach and stream sediments as indicators of shore processes between Monterey and Los Angeles, California, *U.S. Army Coastal Engin. Research Center, Tech. Memo. 33,* 44p.

Judge, C. W., 1975. Use of the Radioisotope Sand Tracer (RIST) system, *U.S. Army Coastal Engin. Res. Center, Tech. Memo. 53,* 75p.

Kamel, A. M., 1962. Littoral studies near San Francisco using tracer techniques, *U.S. Army Beach Erosion Board, Tech. Memo. 131,* 86p.

Krishnamurthy, K., and Rao, S. M., 1973. Comparison of radioactive glasses of scandium and iridium as tracers in sediment transport studies, *J. Hydrol.,* **19,** 189-204.

Martin, J. M., 1970. Utilization de radioéléments naturels en sédimentologie, *Univ. Paris, Centre Rech. Géodynam.,* **14,** 24p.

Martin, J. M.; Meybeck, M.; and Heuzel, M., 1970. A study of the dynamics of suspended matter by means of natural radioactive tracers: An application to the Gironde Estuary, *Sedimentology,* **14,** 27-37.

Rathbun, R. E.; Kennedy, V. C.; and Culbertson, J. K., 1971. Transport and dispersion of fluorescent tracer particles for the flat-bed condition, Rio Grande conveyance channel, near Bernardo, New Mexico, *U.S. Geol. Surv. Prof. Pap. 56 Z-I*, 156p.

Schwartz, M. L., 1966. Fluorescent tracer: Transport in distance and depth in beach sands, *Science,* **151**, 701-702.

Teleki, P. G., 1966. Fluorescent sand tracers, *J. Sed. Petrology,* **36**, 468-485.

Verwey, C. J., and McMurray, W. R., 1964. Tracers for the study of mixing in the surf, *CSIR (South Africa) Research Rept. 222*, 45-49.

Williams, A. T., 1973. The problem of beach cusp development, *J. Sed. Petrology,* **43**, 857-866.

Wright, F. F., 1962. The development and application of a fluorescent marking technique for tracing sand movement on beaches, *U.S. Navy, Off. Naval Research, Geogr. Branch, Proj. NR 338-057, Tech. Rept. 2*, 29p.

Yasso, W. E., 1962. Fluorescent coatings on coarse sediments: an integrated system, *U.S. Navy, Office Nav. Research Geogr. Branch, Project NR 338-057, Tech. Rept. 1*, 55p.

Yasso, W. E., 1965. Fluorescent tracer particle determination of the size-velocity relation for foreshore sediment transport, Sandy Hook, New Jersey, *J. Sed. Petrology,* **35**, 989-993.

Yasso, W. E., 1966. Formulation and use of fluorescent tracer coatings in sediment transport studies, *Sedimentology,* **6**, 287-301.

Zenkovich, V. P., 1960. Fluorescent substances as tracers for studying the movement of sand on the sea bed; experiments conducted in the U.S.S.R., *Dock & Harbour Auth.,* **40**.

Zenkovich, V. P., 1967. *Processes of Coastal Development*. Edinburgh: Oliver and Boyd, 738p.

Cross-references: *Bed Forms in Alluvial Channels; Experimental Sedimentology; Littoral Sedimentation; Sediment Transport–Initiation and Energetics; Transportation.*

TRACTION

The term *traction* was introduced by Gilbert (1914) for hydraulic transportation (sweep and drag) of sediment along a bed. It includes rolling, sliding, dragging, pushing, and saltation (q.v.). It applies to both water and air transport (see *Fluvial Sediment Transport; Saltation*). The tractive force is the drag or shear that is developed on the wetted area of a stream channel and develops in the direction of the flow. This force, per unit area, is called the unit tractive force (τ_0) and can be expressed as $\tau_0 = \gamma RS$, where γ is the specific weight of water, R is the hydraulic radius, and S is the slope of the channel bed.

Sediment traction in moving fluids results in distinct depositional forms depending on the water volume, current velocity, flow resistance, etc. (see *Bed Forms in Alluvial Channels*).

B. CHARLOTTE SCHREIBER

References

Allen, J. R. L., 1970. *Physical Processes of Sedimentation*. London: Allen and Unwin, 248p.

Gilbert, G. K., 1914. The transportation of debris by running water, *U.S. Geol. Surv. Prof. Pap. 86*, 263p.

Cross-references: *Bed Forms in Alluvial Channels; Flow Regimes; Fluvial Sediment Transport; Saltation; Transportation.*

TRANSPORTATION

Transportation is the movement of rock material from one place to another on or near the surface of the earth. The products of weathering, in the form of particles or in solution, may be transported by one or more agents such as water, atmosphere, ice, gravity, and organisms. Transportation may be effected directly from the source to the ultimate site of deposition in a brief span of time, or in stages covering an enormously long geologic interval. Chemical and physical changes may occur in the material during transport, and detrital sediments are generally sorted in proportion to the distance traveled.

Transportation by Water

Detrital sediments are transported in water by traction or suspension and, depending on particle size, may ultimately form lutites, arenites, or rudites (see *Clastic Sediments and Rocks*). Colloids are a special case of minute particles carried in suspension, which may settle out by flocculation (q.v.) to form clay deposits. Material in solution may be transported by water until a change in conditions causes the formation of precipitates or evaporites.

Rain Wash. Surface sheet flow of rain water occurs in temperate regions where the ground is moist, or in semiarid areas with cemented surface crusts. Fine particles are carried along by the water, while larger particles remain as lag. Rill marks and miniature cliffs and terraces are formed on the slopes as the silty sediments are moved downgrade. In truly arid locales, the surface flow mixes with the sediment to form a slurry which then moves downgrade as a viscous mudflow (q.v.).

Ground Water. Removal and transportation by ground water is almost exclusively in the form of solution. The process by which compounds are leached from the upper soil zones and concentrated in the lower ones is *illuviation;* compounds removed from the lower soil zones and concentrated in the upper layers by capillarity and evaporation result in the formation of a duricrust (q.v.). In regions underlain by limestone, removal and transportation by ground water may result in a karst topography. Where none of the foregoing occur, material carried in solution by ground water may be transported to streams, and by them to the sea, causing a depletion of the soil known as *podzolization.*

Streams and Rivers. Sediment is transported in a linear direction by water flowing between two banks either as bed load or as suspended load (see *Fluvial Sediment Transport*). Terms used in connection with stream transport are: *competency,* the ability of a stream to transport in terms of particle size; *capacity,* the maximum load a stream can carry; and *load,* the actual quantity of material carried at any one time. Although many other relationships may be considered, these three decrease in magnitude as the velocity of the water diminishes.

Bed-load transportation progresses in stages according to current velocities. Individual sediment particles, depending on their size, shape, and density, slide or roll along on an almost smooth bottom at low current velocities. As the velocity increases, bed forms form and move downstream as particles eroded on the gently sloping upstream face are deposited on the steep downstream face (see *Bed Forms in Alluvial Channels*). With a greater increase in velocity, antidunes appear to move upstream due to erosion on the downstream face and deposition on the upstream face (see *Flow Regimes*).

In the early stages, some of the rolling and sliding particles may occasionally bounce along in a motion called *saltation* (q.v.). Caught up by the current and turbulence, these particles settle at rates dependent mainly on their size, shape, and density. While supported by the motion of the current they are in suspension. Through impact, abrasion, and grinding, the angularity of individual particles is reduced and an increase in roundness (q.v.) follows. In general, flatness with rounded corners predominates in those larger particles shaped along cleavage, bedding, or parting surfaces; particles of more homogeneous structure approach sphericity (see *Roundness and Sphericity*). Because both bed load and settling rates depend on particle size, shape, and density, transported debris tends to be sorted according to these parameters whenever a change occurs in the transporting efficency of the stream (see *Sorting*).

Solutions, and colloids in suspension, are transported by streams until a chemical or physical change takes place in the regime to cause deposition. In the case of solutions, current turbulence and waterfalls may deposit travertine from soluble carbonate in a stream (due to expulsion of CO^2); evaporation in a saline, stream-fed bay may result in the deposition of halite; biochemical reactions may induce the precipitation of limestone where a river enters the sea; and springs and streams fed by hydrothermal sources, on cooling, often precipitate siliceous sinter or geyserite. Colloids, on the other hand, may settle out slowly upon reaching the stillness of a pond or lake, or become flocculated by the electrolytes met with in the oceans (see *Flocculation*).

Marine Currents. Currents may be generated in standing bodies of water by wind, tide, and density differences. The friction of wind on the water surfaces and the motion of tides often induce, separately and in combination, strong currents at bottom depths. These currents transport sediments and solutions in much the same manner as streams. Investigations on continental shelves have revealed the transport of tremendous quantities of current-driven sand (see *Continental-Shelf Sedimentation*). Density currents may be caused by a difference in temperature, salinity, or turbidity between two adjacent bodies of water. The denser water will tend to flow downgrade below the less dense. Most noteworthy of the three types as an agent of transportation is the turbidity current. Heavily laden currents are developed by the discharge of rivers with large loads, or the collapse and slumping of sediments on a subaqueous slope. Great quantities of sediment may be transported in this manner, with sorting less prominent than that found in streams and rivers. Greywackes and some conglomerates are typical of turbidity-current transport and deposition (see *Turbidity-Current Sedimentation; Turbidites*).

Waves. The effect of waves is, essentially, repeated turbulence within a general area rather than linear motion in one direction. The results are evidenced in various ways. Wave turbulence is first felt on the offshore bottom as a slightly circular motion, when the water depth is equal to one-half the wave length. This motion develops pointed and symmetrical oscillation ripples. Considerable sorting and rounding takes place in the zone of breaking waves. A form of transport called beach drift develops as particles travel laterally in a scalloped pattern along the swash zone with repeated swash and backwash (see *Littoral Sedimentation*). When waves approach at an angle to the shoreline, longshore currents are generated parallel to the beach. Littoral drift, the net effect of beach drift and longshore current transport, is the force that builds beaches, spits, tombolos, and bay-mouth bars.

Transportation in the Atmosphere

In arid regions, wind plays a major role in the transportation of sediments. Sand is moved by traction, while dust is carried in suspension. The dynamics are somewhat similar to those found in water transport. Individual sand grains may be caught up by wind turbulence and carried forward in the bouncing motion of saltation (q.v.), the energy of their impact propelling other grains into the same action. Most of these particles rise only about 1 m above the ground. Other grains are rolled along under the force of the wind and saltation impact, in a motion termed *surface creep*. Impact and abrasion in the atmosphere, without the cushioning effect present in water, cause considerable rounding and sphericity of small particles. Multiple minute impact craters give these particles a frosted texture (see *Sand Surface Textures*). Sorting is developed to a high degree, as wind is more selective in its transport capacity than water.

Traction. Dunes of various shapes develop, depending on the sand supply, topography, and wind regime. As sand collects on the protected lee side of an object, the accumulation itself forms a trap for more sand. With scarce sand and moderate winds in a constant direction, over smooth terrain, crescentric barchan dunes form. Sand is eroded from their gently sloping windward side and deposited on the steep lee side, where it lies at the angle of repose (30-34°). More plentiful sand and higher winds generate transverse dunes normal to the wind direction, with an asymmetrical profile as in the barchans. Abundant sand and strong winds result in longitudinal or seif dunes parallel to the wind, which attain great size in desert regions. Small wind ripples may develop on the surfaces of all the larger dune forms. Cross-stratification exists at diverse angles within the dunes, formed by the moving sands and changing winds.

Suspension. Dust carried into the upper atmosphere by turbulent air currents may be transported great distances until it settles or is washed out by precipitation. In the lee of desert regions, or of extensive glacial outwash deposits, dust may settle to great thicknesses in a homogeneous (nonstratified) accumulation called loess (q.v.). In the same manner, while large fragments ejected into the atmosphere by volcanic explosions fall to earth in close proximity to the cone, pyroclastic dust and ash may drift for long periods and encircle the globe before settling. A spectacular demonstration of this phenomenon was seen in the 1883 eruption of Krakatoa and in the 1963 eruption of Mt. Agung, Bali (see *Eolian Dust in Marine Sediments; Volcanic Ash Deposits*).

It can be seen from the foregoing that transport in the atmosphere is one of the few means by which material may be moved to a higher altitude than its source.

Transportation by Ice

Continental ice sheets and mountain glaciers transport immense quantities of material that are scoured or plucked from the surfaces over which they travel, or become lodged in the ice through landslides from adjoining higher landforms. Once incorporated, particles of all sizes are subject to the internal or bottom grinding and abrasion of the plastic ice flow. Thus, flattened surfaces, striae, and broken, angular fragments are developed in contrast to the rounding of water and atmospheric transportation. Very little chemical decomposition occurs while the sediment is frozen in the ice. Deposition directly from the melting ice results in poor sorting (see *Till and Tillite*). In retrospect, country rock striae, chatter marks, depositional orientation, and the fabric of deposits indicate the direction of transport.

Meltwater and local precipitation may cause glaciofluvial transportation of products of abrasion and grinding of fine, scoured sediments, with subsequent rounding and sorting typical of water transport. Debris-laden ice floes or bergs, formed by the calving of ice sheets or glaciers where they meet the sea, transport sediment great distances. Thus the ocean floors some distance from the poles may be covered with unsorted particles of all sizes comprising a marine till, and even huge boulders called erratics (see *Glacial Marine Sediments*). Similar ice rafting may also occur in streams, rivers, and lakes.

Transportation Due to Gravity

The force of gravity, sometimes assisted by changes in temperature or in pore-water pressure, acts continually to move particles downgrade once they are freed from their parent formation. Rate of transport varies with the consistency of the material and steepness of the slope (see *Gravity Flows; Gravity-Slide Deposits; Mass-Wasting Deposits*).

Temperature Change. As particles alternately expand and contract with changes in temperature, any movement involved is downgrade in the direction of least resistance. This action, coupled with disturbances and the pull of gravity, generates surficial creep on slopes. In cold regions, the expansion and contraction of debris and interstitial ice develops rock glaciers, which resemble regular mountain glaciers.

Earth Flow. The movement of moist sediments on steep slopes results in various types of earth flowage. Localized surficial thawing of frozen soil and regolith may cause a viscous, lobate, downgrade flow known as *solifluction*. A slumping earth flow in the form of a tongue-like mass develops a depression bordered by curved scarps at its uphill end, and a bulging "toe" of material at its downhill end. Should the process occur beneath a rigid surface unit, a slump block may develop, rotating backward on a curved slip plane. With increased saturation, a true mudflow develops, capable of transporting huge boulders as it follows stream channels in its downgrade plunge.

Landslides. Landslides occur as large surface units start to slide on weak structural planes and break into fragmental debris which piles up at the bottom of the slope. Closely akin to this are the rock slides of loosely consolidated fragments on oversteep or undercut slopes.

Rock Fall. In all of the foregoing forms of gravity transport, particle sorting is markedly absent although limited polishing, scratching, and rounding of fragments may occur. When a continual rock fall of fragments is funneled on a cliff face, however, a talus cone is formed at the foot, with the larger particles rolling to the base.

Tectonic Sliding. Low-angle gravitational sliding may take place on a regional scale, involving units measured in km. This form of transport produces the complex nappes and klippes found in the European Alps.

Subaqueous Slumping. Subaqueous slumps are the marine or lacustrine counterpart of terrestrial gravity collapse. They differ from turbidity currents in that the transported strata retains its internal bedded relationships when it comes to rest in highly contorted folds at the bottom of the slope (see *Gravity-Slide Deposits*).

Transportation by Organisms

Transportation of sedimentary particles by organisms occurs by various means. The burrowing of terrestial or marine organisms and the growth of plant roots displaces considerable sediment and often makes it more easily available for other agents of transportation. Rock fragments attached to the roots or holdfasts of floating plants, or carried in the stomachs of animals as gastroliths, may be transported great distances. These erratic particles are then deposited in sediments of entirely different character (Emery, 1963).

MAURICE SCHWARTZ

References

Allen, J. R. L., 1970. *Physical Processes of Sedimentation*. London: Allen and Unwin, 248p.
Bagnold, R. A., 1963. Mechanics of marine sedimentation, in M. N. Hill, ed., *The Sea*. New York: Interscience, 507-528.
Blatt, H., Middleton, G. and Murray, R., 1972. *Origin of Sedimentary Rocks*. Englewood Cliffs, N.J.: Prentice-Hall, 634p.
Emery, K. O., 1963. Organic transportation of marine sediments, in M. N. Hill, ed., *The Sea*, New York: Interscience, 776-793.

Cross-references: *Bedding Genesis; Bed Forms in Alluvial Channels; Clastic Sediments and Rocks; Continental Sedimentation; Current Marks; Deposition; Dispersal; Eolian Sedimentation; Flow Regimes; Fluvial Sediment Transport; Gravity Flows; Littoral Sedimentation; Marine Sediments; Mass-Wasting Deposits; Sedimentation; Sediment Transport—Initiation and Energetics;* see also numerous cross-references in text. Vol III: *Sediment Transport—Fluvial and Marine; Sediment Transport—Long-Term Net Movement*.

TRAVERTINE

Travertines (German: *sinter*) are fresh-water carbonate deposits found in caves, near spring mouths, as hot spring deposits, as coatings on cliffs, and as fracture fillings in carbonate rocks. Massive dense deposits in caves are known as *cave onyx*. More impure and porous varieties are known as *calcareous tufa*. The principal mineral is calcite, although aragonite and opal sometimes occur along with a variety of included clays and quartz (Mills, 1965). Travertine deposits are layered with the layers parallel to the growth surfaces. Growth bands are distinguished by variations in color and in grain size. Colors vary from pure white through shades of yellow, tan, and brown. A few cave travertines show unusual blood red, black, lemon yellow, or green colorations from dissolved transition metal ions or from organic stains (White, 1976). The much more common browns and tans arise from included iron oxides and hydrates.

Travertines deposited by springs and rivers can build up into mounds of massive size. Spring mouths in the Grand Canyon spill from the sides of cliffs over mounds of travertine 5-15 m in height. Masses of travertine in caves are known as *flowstone*. Travertine also makes up stalactites, stalagmites, and other minor depositional forms (Hill, 1976). Travertines tend to grow in the form of dams, and thus travertine-depositing streams and springs are frequently a series of pools separated by rimstone dams (see *Spring Deposits*, Fig. 1).

Most travertines are composed of low-Mg

TABLE 1. Some Analyses of Travertines
(wt %)

	SiO$_2$	Al$_2$O$_3$	Fe$_2$O$_3$	MgO	SrO
Calcareous tufa, Mammoth Hot Springs, Yellowstone National Park[a]	0.05	—0.06—		0.26	n.a.
Calcareous tufa, Redding Spring, Great Salt Lake Desert[a]	8.40	1.31		3.54	n.a.
Calcite stalactite, New River Cave, Virginia[b]	0.57	—0.36—		1.68	n.a.
Calcite stalactite, Wind Cave, South Dakota[c]	n.a.	0.13	0.23	0.38	n.a.
Calcite flowstone, Timpanogos Cave, Utah[d]	0.11	0.007	0.007	1.61	0.023

[a] Analyses 34 and 36 in Graf, 1960.
[b] Murray, 1954.
[c] White and Deike, 1962.
[d] White and Van Gundy, 1974.
n.a. = no analysis reported.

calcite and would be classified as an exotic type of fresh-water limestone. Few chemical analyses and no complete ones have been reported. Table 1 presents summary analyses of a few specimens.

Travertines are deposited from solutions supersaturated with respect to carbonate minerals. The mechanism by which spring and cave waters reach this level of supersaturation as developed by Holland et al. (1964), Thrailkill (1971), and others begins with rain-water striking the earth in equilibrium with atmospheric CO_2 pressure of $10^{-3.5}$ atm. That fraction of the water that infiltrates through the soil absorbs CO_2 in the biologically active zone, and CO_2 pressures may reach 10^{-1} atmospheres. Most of the uptake of calcium carbonate occurs at the base of the soil where the highly acidulated soil water first makes contact with the underlying carbonate rock. Depending on many details of infiltration rate, availability of CO_2, temperature, bedrock, and soil composition, the waters may reach saturation with respect to calcite at a concentration level of 400 ppm. Carbonate-laden waters percolate through the limestone aquifer with little change in chemistry. When such waters emerge into air-filled cave passages they must adjust to a new environment with a CO_2 partial pressure between $10^{-3.5}$ and $10^{-2.5}$ atm by deposition of the $CaCO_3$ that becomes the cave travertine deposits. If the waters make their way entirely through the drainage system to emerge at big springs, the waters must again adjust to atmospheric CO_2 levels, and in addition evaporation may play an important role. All these effects help account for the observation that travertine deposits in tropical springs, rivers, and caves are much more massive than those in temperate climates where cave deposits are less massive and spring and river deposits nearly nonexistent.

WILLIAM B. WHITE

References

Graf, D. L., 1960. Geochemistry of carbonate sediments and sedimentary carbonate rocks, IVA. Isotopic composition—chemical analyses, *Ill. St. Geol. Surv. Circ.* 308. 1-42.

Hill, C. A., 1976. *Cave Minerals*. Hunstville, Ala.: Natl. Speleol. Soc., 137p.

Holland, H. D.; Kirsipu, T. V.; Huebner, J. S.; and Oxburgh, U. M., 1964. On some aspects of the chemical evolution of cave waters, *J. Geol.*, 72, 36-67.

Mills, J. P., 1965. Petrography of selected speleothems of carbonate caverns, M.S. Thesis, Univ. of Kansas, 47p.

Murray, J. W., 1954. The deposition of calcite and aragonite in caves, *J. Geol.*, 62, 481-492.

Thrailkill, J., 1971. Carbonate deposition in Carlsbad Caverns, *J. Geol.*, 79, 683-695.

White, W. B., 1976. Cave minerals and speleothems, in T. D. Ford and C. H. D. Cullingford, eds., *The Science of Speleology*. London: Academic Press, 265-327.

White, W. B., and Deike, G. H. III, 1962. Secondary mineralization in Wind Cave, South Dakota, *Bull. Natl. Speleol. Soc.*, 24, 74-87.

White, W. B., and Van Gundy, J. J., 1974. Reconaissance geology of Timpanogos Cave, Wasatch County, Utah, *Bull. Natl. Speleol. Soc.*, 36, 5-17.

Cross-references: *Limestones; Speleal Sediments; Spring Deposits.*

TRENCH SEDIMENTATION
—See **HADAL SEDIMENTATION**

TRIPOLI, TRIPOLITE

Tripoli designates a number of more or less similar types of porous, friable siliceous rocks of sedimentary origin, which are the end product of the weathering of siliceous, cherty, or argillaceous limestones. It appears that chemical weathering removes the calcium carbonate, leaving the finely divided residue. There is question, however, as to whether a chert must form first and then be converted into tripoli or whether tripoli is derived by direct decay from siliceous impurities, during the initial weathering of limestone.

Considerable confusion exists as to what constitutes "tripoli," and a type of diatomaceous earth, *tripolite*, found in Tripoli, North Africa, has also been called tripoli (see *Diatomite*). Tripoli, though, usually contains few or no diatoms. Good examples of tripoli are found in the Missouri-Oklahoma region, U.S.A., from the so-called *rottenstone* at Hull, England, and in the Yoredale rocks of Derbyshire. Tripoli is also a formation name (Messinian stage).

Tripoli is variously used as a buffing compound producing a high luster finish for copper, brass, aluminum, and zinc; as an absorbant for fats, grease, and stains in dry cleaning; and in scouring powder and soaps.

B. CHARLOTTE SCHREIBER

References

Heinz, C. E., 1937. Tripolite, in *Directory of Minerals and Rocks.* New York: Am. Inst. Mining Met. Eng., 911-922.
Metcalf, R. W., 1946. Tripoli, *Dept. Interior, Bur. Mines, Inf. Circ. 7371,* 1-25.
Tarr, W. A., 1938. Terminology of the chemical siliceous sediments, *Rept. Comm. Sed. 1937-1938, Natl. Research Council Ann. Rept. 8-27.*

Cross-references: *Chert and Flint; Diatomite; Silica in Sediments; Siliceous Ooze.*

TROPICAL LAGOONAL SEDIMENTATION

Tropical lagoonal sedimentation is marked by the deposition of predominantly calcium carbonate sediments from warm-water calcareous marine organisms that, upon dying, contribute their skeletal remains to a great variety of environments called tropical lagoons.

A combination of factors dictates the type of sedimentation in tropical lagoons. If there is a significant influx of terrigenous clastic material into tropical lagoons, carbonate sedimentation will not occur, and clastic sediments will be deposited (see *Lagoonal Sedimentation*). Carbonate deposition is favored by warm marine waters, a lack of rivers supplying terrigenous material to the marine environment, and low relief in adjacent land masses. The following discussion treats solely modern carbonate sedimentation in lagoons.

The term *lagoon* lacks clear definition, and its usage in geological literature is hardly consistent. So many factors are involved in defining lagoons, that it is difficult to find a general definition of the term that has few exceptions. The following definition would be agreed upon by most workers: *Lagoon*–a shallow body of marine water that may be brackish, normal, or hypersaline, that is protected from high marine energies by any of several types of barriers, that has somewhat restricted water circulation, and that collects muddier and/or finer-grained sediment than adjacent higher-energy depositional environments.

Generally, five major types of modern tropical lagoons are recognized: coastal lagoons, backreef lagoons, interreef lagoons, shelf lagoons, and atoll lagoons. Criteria for defining these classes are the nature of the barrier(s) that provide protection and the way that these barriers are related to the mainland or the landward side of the lagoon. Table 1 lists the general characteristics of the five lagoonal types. Certainly, there are combination-type lagoons that are difficult to place in a single category. Table 2 (A-D) summarizes Holocene lagoonal studies on a global basis and documents the natural variability of tropical lagoons.

Setting

Barriers on lagoon margins restrict water circulation patterns and, as in Florida Bay, may significantly alter lagoon salinities. In humid climates, lagoon barriers permit less mixing of fresh and marine waters by confining freshwater runoff from swamps. In arid climates, the presence of extensive barriers permits deposition of evaporites, e.g. gypsum, by concentrating and finally precipitating certain minerals that are in a dissolved state in sea water of normal salinities.

The presence of barriers on at least some sides of a lagoon is an essential feature. Open oceanic currents, swell, and even wind-driven waves

TABLE 1. General Characteristics of Recent Carbonate Lagoons

Lagoon Type	Barrier Types		Relationship Between Barrier and Major Landmass[a]	Shape	Marine Energy	Sediments
	Major Element	Optional Accessories				
Coastal lagoons	mainland peninsulas	islands, reefs	barriers are either connected to the landmass or are very close[b] to it	circular to triangular	low	lime muds with minor skeletal fragments
Backreef lagoons	*barrier reefs*	islands, shoals	barrier is close[b] and subparallel to landmass	linear band	moderate to high	*skeletal sands* with little or no lime mud
Interreef lagoons	reef complexes (esp. patch reefs)	shoals	barriers occur on broad *shelf* or *platform* and may not have a landmass nearby	variable	moderate	skeletal sands with lime mud
Shelf lagoons	islands	barrier reefs, *mudbanks*	barrier is distant from the landmass (if it exists); results in a wide lagoon across inner and middle portions of the shelf	oval to linear bands	low	pure lime muds and those with skeletal fragments
Atoll lagoons	barrier reefs	islands	no major landmass nearby	circular	moderate to low	ranges from muddy skeletal sands to pure lime muds

[a]"Major Landmass" can include continental mainland, large islands, or island chains.
[b]Typically, barriers of this type are less than 20 km offshore from landmass.

are considerably dampened by lagoonal barriers. Thus, the energy regime of lagoons is usually low or at least lower than adjacent high-energy environments. Good examples of this latter case are the backreef lagoons of Table 2C and 2D, where moderate to high wave and current energies occur behind a high-energy reef tract with pounding surf and large ocean swells. Thus, backreef lagoons are the highest energy form of lagoons and should dispel the idea that all lagoons are completely quiet-water environments.

Lagoonal Plants and Animals

Sediments which are found in tropical lagoons are derived primarily from the skeletons of plants and animals living in or near the lagoon. The living lagoonal flora and fauna are directly controlled by lagoonal salinities (and their variation), type of substrate on the lagoon floor, and the energy conditions in the lagoon. Major plants are *Halimeda, Penicillus* (both algae), and *Thalassia,* a grass; mollusks and foraminifera are the major animal groups. A distinction must be made between plants and animals actually living in the lagoon (indigenous) and skeletons found in lagoonal sediments (indigenous and transported). A direct correspondence between living and dead assemblages usually occurs in low-energy lagoons, whereas transportation of skeletal fragments from the adjacent environment into a lagoon requires high marine energies (e.g., backreef lagoons or the shallow fringes of atoll lagoons).

Corals and gorgonians (sea fans and sea whips) are generally associated with reef faunas, but may exist in more open lagoons either on patch reefs or on patches of hard, rocky substrate; they usually are not considered important in lagoonal faunas. A notable exception, however, is the delicately branching coral *Oculina,* where extending branches support the organism on a muddy substrate (see Table 2A, Harrington Sound). Echinoids (sea urchins), as well as corals, are present to a minor degree in some lagoons and are positive indications of normal-salinity waters.

TABLE 2A. CHARACTERISTICS OF RECENT TROPICAL LAGOONS

JORDAN, 1975

LAGOON TYPE	COASTAL LAGOONS				
LAGOON	[1] ABU DHABI LAGOON AND [2] KHOR AL BAZAM	[1] LAGARTOS LAGOON	[2] YALAHAU LAGOON [3] BOCA IGLESIA LAG. [4] BLANCA LAGOON [5] NICHUPTE LAGOON	NORTH SOUND, GRAND CAYMAN	HARRINGTON SOUND
LOCATION	PERSIAN GULF	YUCATAN, MEXICO		BRITISH WEST INDIES	BERMUDA
MAPS — LEGEND: LAND ■ WATER □ MUDBANKS ▨ REEFS & REEF FLATS ▧	(map: Khor Al Bazam, Persian Gulf, Abu Dhabi Lagoon)	(map: Gulf of Mexico, Yucatan Mainland)		(map: North Sound, Caribbean Sea)	(map: Bermuda, Harrington Sound)
SIZE (SQ. KM.)	[1] 450 [2] 1850	[1] 80	[2] 270 [4] 70 [3] 28 [5] 31	88	3.1
BARRIER	REEFS & ISLANDS	ISLAND	ISLANDS, SHOALS, AND MANGROVE SWAMPS	REEFS AND MAINLAND PENINSULA	PENINSULAS OF THE MAIN BERMUDA ISLAND
MAXIMUM DEPTH (Meters)	11–12	[1] 1	[2] 3.0 [4] 3.5 [3] 1.5 [5] 4.0	4.5	25
CLIMATE	HOT, ARID	HUMID, TROPICAL		HUMID, TROPICAL	HUMID, TROPICAL
SALINITY 0/00 NORMAL MARINE SALINITY IS 35–36 0/00	HYPERSALINE 42–66	[1] HYPERSALINE 118–200	[2] 36–42 [4] 47 [3] NO DATA [5] 28	PROBABLY NORMAL MARINE TO SLIGHTLY HYPERSALINE	NORMAL MARINE
SIGNIFICANT LIVING LAGOONAL PLANTS AND ANIMALS (T) = TRACE (R) = RARE (C) = COMMON (A) = ABUNDANT	ALGAL MATS AND CRABS (INTERTIDAL); GASTROPODS COMMON IN LAGOON	MOLLUSCS FORAMINIFERA SPONGES VARIOUS ALGAE AND BACTERIA (WHICH COLOR THE WATER REDDISH–BROWN)	THALASSIA (A) DASYCLADACEAN AND CODIACEAN ALGAE HALIMEDA (R–C) FORAMINIFERA MOLLUSCS CORALS (T)	THALASSIA HALIMEDA MOLLUSCS FORAMINIFERA	THALASSIA GREEN AND BLUE–GREEN ALGAE MOLLUSCS (A) HALIMEDA OCULINA CORALS (C) ECHINOIDS
RECENT UNCONSOLIDATED LAGOONAL SEDIMENTS	ALGAL–STABILIZED PELLETED LIME SANDS; SOME LIME MUD WITH SHELLS	LIME MUD WITH SOME SKELETAL FRAGMENTS THIN SURFACE LAMINAE OF GYPSUM–ALGAL MAT–LIME MUD	ORGANIC–RICH PELLETED LIME MUD, COMMONLY CONTAINING SKELETAL FRAGMENTS	MUDDY BIOCLASTIC SANDS & LIME MUDS WITH SKELETAL FRAGMENTS	MUDDY SKELETAL SANDS AND LIME MUDS WITH SHELL FRAGMENTS
LAGOONAL DEPOSITS THEORETICAL ROCK EQUIVALENTS: TERMS BY DUNHAM, 1962	MAINLY PELLETAL PACKSTONE, SOME MUDSTONE & MOLLUSCAN WACKESTONE	FORAM–MOLLUSCAN WACKESTONE, MUDSTONE AND THIN EVAPORITES	PELLETAL MUDSTONE AND WACKESTONE, MOLLUSC–FORAM WACKESTONE & SOME MOLLUSC–HALIMEDA WACKESTONE	ALGAL PLATE–FORAM WACKESTONE	MOLLUSCAN WACKESTONE AND PACKSTONE
LAGOON MARGINS	MANGROVES ALGAL FLATS	ALGAL MATS, SOME BLACK MANGROVES	RED MANGROVES, SOME SAWGRASS MARSHES	THIN SEDIMENT COVER ON ROCKY SHORE ZONE	PLEISTOCENE LIMESTONE CLIFFS
REMARKS	AT LOW TIDE, WINDS EXPOSE MUCH OF LAGOON FLOOR; EXTREMELY WARM WATER; TIDAL RANGE = 0.7 METERS.	NUMEROUS SHOALS & ISLANDS IN LAGOON; SEPARATE WET AND DRY SEASONS; MAXIMUM THICKNESS OF RECENT SEDIMENTS IS 2.3 METERS.	SLUGGISH LAGOONAL CIRCULATION; MUDBANKS PRESENT IN [2], [4], AND [5]; CALCAREOUS GREEN ALGAE ACCOUNT FOR MOST OF THE LIME MUD.	EASTERN PART OF LAGOON IS MORE RESTRICTED; LAGOONAL LIME MUDS ARE PRIMARILY PRODUCTS OF IN SITU DEGRADATION OF COARSE MATERIAL.	SMALL SINGLE INLET IN SW CORNER OF SOUND; DEFINITE THERMOCLINE AT 17 M; 3 DEPTH–DEPENDENT ECOLOGIC ZONES RECOGNIZED IN LAGOON, INCLUDING AN OCULINA ZONE.
REFERENCES	EVANS & BUSH, 1967 KENDALL & SKIPWITH, 1969	BRADY, 1971		ROBERTS, 1971	NEUMANN, 1965

TABLE 2B. CHARACTERISTICS OF RECENT TROPICAL LAGOONS

JORDAN, 1975

LAGOON TYPE	COASTAL LAGOONS Continued	ATOLL LAGOONS			
LAGOON	WHITEWATER BAY	BIKINI	KAPINGAMARANGI	MIDWAY	KURE
LOCATION	SOUTHERN FLORIDA	MARSHALL ISLANDS	CAROLINE ISLANDS	HAWAIIAN ISLANDS	HAWAIIAN ISLANDS
MAPS LEGEND: LAND ■ WATER □ MUDBANKS ▦ REEFS & REEF FLATS ▨	FLORIDA, WH BAY, CAPE SABLE, GULF OF MEXICO, 10 KM.	PACIFIC OCEAN, BIKINI LAGOON, BIKINI ISLAND, 10 KM.	PACIFIC OCEAN, KAPINGAMARANGI LAGOON, 1 KM.	PACIFIC OCEAN, MIDWAY LAGOON, 1 KM.	PACIFIC OCEAN, KURE LAGOON, 1 KM.
SIZE (SQ. KM.)	130	629	40	54	46
BARRIER	MANGROVE ISLANDS	REEFS AND ISLANDS	REEFS AND ISLANDS	REEFS AND ISLANDS	REEFS AND ISLANDS
MAXIMUM DEPTH (Meters)	2	58	73	21	14
CLIMATE	HUMID, SUBTROPICAL	HUMID, TROPICAL	HUMID, TROPICAL	HUMID, SUBTROPICAL	HUMID, SUBTROPICAL
SALINITY 0/00 NORMAL MARINE SALINITY IS 35-36 0/00	VARIABLE; FRESH TO NORMAL MARINE	NORMAL MARINE	NORMAL MARINE	NORMAL MARINE	NORMAL MARINE
SIGNIFICANT LIVING LAGOONAL PLANTS AND ANIMALS (T) = TRACE (R) = RARE (C) = COMMON (A) = ABUNDANT	MOLLUSCS	HALIMEDA MOLLUSCS FORAMINIFERA CORALS	HALIMEDA CODIACEAN ALGAE FORAMINIFERA CORALS MOLLUSCS	HALIMEDA MOLLUSCS FORAMINIFERA	HALIMEDA MOLLUSCS FORAMINIFERA
RECENT UNCONSOLIDATED LAGOONAL SEDIMENTS	LIME MUD WITH SHELL FRAGMENTS AND PURE LIME MUD	MUDDY HALIMEDA SANDS	SILTY BIOCLASTIC SANDS IN SHALLOW WATER; SANDY LIME MUDS IN WATER DEEPER THAN 45 METERS.	MUDDY SKELETAL SANDS	MUDDY SKELETAL SANDS
LAGOONAL DEPOSITS THEORETICAL ROCK EQUIVALENTS: TERMS BY DUNHAM, 1962	MOLLUSCAN WACKESTONE AND MUDSTONE	ALGAL PLATE PACKSTONE	SKELETAL PACKSTONE (SHALLOW), SKELETAL WACKESTONE AND MUDSTONE (DEEP)	RED ALGAL– FORAMINIFERAL PACKSTONE	RED ALGAL– FORAMINIFERAL PACKSTONE
LAGOON MARGINS	MANGROVES AND SAWGRASS MARSHES	ENCIRCLING REEF FLATS AND SMALL ISLANDS	EXTENSIVE ENCIRCLING REEF FLATS AND SMALL ISLANDS	LAGOON CENTER IS SURROUNDED BY AN EXTENSIVE LAGOON TERRACE WHICH IS RIMMED WITH ISLANDS AND REEF FLATS.	LAGOON CENTER IS SURROUNDED BY AN EXTENSIVE LAGOON TERRACE WHICH IS RIMMED WITH ISLANDS AND REEF FLATS.
REMARKS	NUMEROUS MANGROVE ISLANDS IN THE LAGOON	PATCH REEFS COMMON IN THE LAGOON	PATCH REEFS COMMON IN LAGOON; CORAL THICKETS WITH CORAL DEBRIS AT DEPTHS OF 9-32 METERS; SIX DEPTH–DEPENDENT LAGOONAL FACIES RECOGNIZED.	PATCH REEFS COMMON; LAGOONAL SEDIMENTS ARE MAINLY REEF–DERIVED AND REFLECT A SMALL SHALLOW LAGOON	PATCH REEFS COMMON; LAGOONAL SEDIMENTS ARE MAINLY REEF–DERIVED AND REFLECT A SMALL SHALLOW LAGOON
REFERENCES	TAFT & HARBAUGH, 1964	EMERY ET AL., 1954	MC KEE ET AL., 1959	GROSS ET AL., 1969	GROSS ET AL., 1969

TABLE 2C. CHARACTERISTICS OF RECENT TROPICAL LAGOONS

JORDAN, 1975

LAGOON TYPE	SHELF LAGOONS				BACKREEF LAGOONS
LAGOON	FLORIDA BAY	GULF OF BATABANO	BRITISH HONDURAS SHELF	LEEWARD LAGOON, WEST OF ANDROS ISLAND	BACKREEF LAGOON, EAST OF ANDROS ISLAND
LOCATION	SOUTHERN FLORIDA	CUBA	BRITISH HONDURAS	BAHAMA BANK	BAHAMA ISLAND
MAPS — LEGEND: LAND (black), WATER (white), MUDBANKS (dotted), REEFS & REEF FLATS (hatched)	FLORIDA BAY (10 KM.)	ISLA DE PINOS, GULF OF BATABANO, CUBA (10 KM.)	BRITISH HONDURAS, BELIZE, AMBERGRIS CAY, CARIBBEAN SEA (10 KM.)	PREVAILING WIND; OOLITE BARS; GREAT BAHAMA BANK; LEEWARD LAGOON; ANDROS (10 KM.)	BACKREEF LAGOON; ANDROS IS.; TONGUE OF THE OCEAN
SIZE (SQ. KM.)	1750	20,725 APPROX.	7000 TOTAL, 20–30 KM. WIDE	10,000	1–6 KM. WIDE 175 KM. LONG
BARRIER	FLORIDA REEF TRACT, FLORIDA KEYS, AND NUMEROUS MUDBANKS IN FLORIDA BAY	REEFS, ISLA DE PINOS, AND SMALL ISLANDS	BARRIER REEFS AND ISLANDS	LAGOON LIES IN THE WIND SHADOW OF ANDROS ISLAND	DISCONTINUOUS REEFS, SHOALS, AND ISLANDS
MAXIMUM DEPTH (Meters)	4	13	64	8	5
CLIMATE	HUMID, SUBTROPICAL	HUMID, TROPICAL	HUMID, TROPICAL	HUMID, TROPICAL	HUMID, TROPICAL
SALINITY 0/00 NORMAL MARINE SALINITY IS 35–36 0/00	VARIABLE 10–40	27–39 (EXCEPT NEAR RIVER MOUTH TO THE NE)	18–34 HYPOSALINE	38–45 HYPERSALINE	NORMAL MARINE (BRACKISH LOCALLY)
SIGNIFICANT LIVING LAGOONAL PLANTS AND ANIMALS (T) = TRACE (R) = RARE (C) = COMMON (A) = ABUNDANT	THALASSIA FORAMINIFERA PENICILLUS MOLLUSCS	THALASSIA HALIMEDA PENICILLUS MOLLUSCS FORAMINIFERA	THALASSIA HALIMEDA FORAMINIFERA MOLLUSCS	THALASSIA HALIMEDA MOLLUSCS FORAMINIFERA VARIOUS GREEN ALGAE	THALASSIA HALIMEDA VARIOUS RED AND GREEN ALGAE CORALS ECHINODERMS MOLLUSCS FORAMINIFERA (A)
RECENT UNCONSOLIDATED LAGOONAL SEDIMENTS	PURE LIME MUDS, LIME MUDS WITH SKELETAL FRAGMENTS AND MUDDY SKELETAL SANDS	MUDDY PELLETAL SANDS, SHELLY PELLETED LIME MUD—UP TO 6 METERS THICK	SANDY LIME MUDS AND PURE LIME MUDS	LIME MUD, COMMONLY PELLETED—UP TO 3 METERS THICK	FINE SKELETAL SANDS
LAGOONAL DEPOSITS THEORETICAL ROCK EQUIVALENTS: TERMS BY DUNHAM, 1962	MUDSTONE AND FORAM-MOLLUSCAN WACKESTONE AND PACKSTONE	MUDSTONE, PELLETAL AND MOLLUSCAN WACKESTONE, AND PELLETAL PACKSTONE	FORAM-MOLLUSCAN MUDSTONE AND WACKESTONE	MUDSTONE, PELLETAL WACKESTONE AND PACKSTONE	CORAL–RED ALGAL PACKSTONE AND GRAINSTONE
LAGOON MARGINS	MANGROVES AND COASTAL SWAMPS	MANGROVES	FINE TERRIGENOUS SEDIMENTS ARE BEING DEPOSITED ALONG LAGOON'S WEST SHORE	TIDAL FLATS OF ANDROS ISLAND; GRAPESTONE AND OOLITIC FACIES OF BAHAMA BANK	SCATTERED PLEISTOCENE OOLITE OUTCROPS AND NUMEROUS TIDAL CREEKS ON EAST COAST OF ANDROS ISLAND
REMARKS	MUDBANKS ARE VERY COMMON AND HELP RESTRICT BAY WATER CIRCULATION; MOST LIME MUD ORIGINATES FROM PENICILLUS.	REEF BELT IS NOT AS CONTINUOUS AS SHOWN ON MAP; FIVE LAGOONAL FACIES RECOGNIZED.	LIME MUD ORIGINATES FROM SKELETAL DESTRUCTION; LAGOON MAY ALSO BE CONSIDERED A VERY WIDE BACKREEF LAGOON.	WEAK CIRCULATION OF BANK WATER IS CAUSED BY THE WIND SHADOW OF ANDROS ISLAND AND THE GREAT EXPANSE OF THE SHALLOW BAHAMA BANKS.	MODERATE TO HIGH WAVE ENERGY; PATCH REEFS COMMON IN LAGOON; THIN SEDIMENTS IN LAGOON, BARE ROCK COMMON; LAGOON IS FILLING MAINLY WITH FINE REEF DEBRIS.
REFERENCES	GINSBURG, 1956 STOCKMAN ET AL., 1967	DAETWYLER & KIDWELL, 1959 HOSKINS, 1964 BANDY, 1964	MATTHEWS, 1966 EBANKS, 1967	CLOUD, 1962 PURDY, 1963	NEWELL ET AL., 1951 BATHURST, 1971

TABLE 2D. CHARACTERISTICS OF RECENT TROPICAL LAGOONS

JORDAN, 1975

LAGOON TYPE	BACKREEF LAGOONS Continued		INTERREEF LAGOONS		
LAGOON	BACKREEF LAGOON AT AMBERGRIS CAY	BACKREEF LAGOON OF FLORIDA REEF TRACT	BLUE LAGOON OFF HERON ISLAND	[1] MURRAY'S ANCHORAGE AND [2] SEVERAL INTERREEF LAGOONS ON PLATFORM	[1] CENTRAL LAGOON AND [2] INTERREEF LAGOONS
LOCATION	BRITISH HONDURAS	SOUTHERN FLORIDA	GREAT BARRIER REEF AUSTRALIA	BERMUDA PLATFORM	ALACRAN REEF COMPLEX, CAMPECHE BANK MEXICO
MAPS	(map)	(map)	(map)	(map)	(map)
SIZE (SQ. KM.)	0.5 KM. WIDE 70 KM. LONG	6–12 KM. WIDE 250 KM. LONG	5.4	[1] 33 [2] VARIABLE, 0.2–2.5	[1] 41 [2] VARIABLE, PROBABLY 0.5–2.0
BARRIER	BARRIER REEF	REEFS	REEF FLATS	[1] PATCH REEF COMPLEXES [2] PATCH REEFS	[1] CELLULAR REEF COMPLEX AND BARRIER REEF [2] CELLULAR REEFS
MAXIMUM DEPTH (Meters)	5	12	6	[1] 17 [2] 18	[1] 8–10 [2] 12
CLIMATE	HUMID, TROPICAL	HUMID, SUBTROPICAL	HUMID, TROPICAL	HUMID, SUBTROPICAL	HUMID, TROPICAL
SALINITY 0/00 NORMAL MARINE SALINITY IS 35–36 0/00	NORMAL MARINE	NORMAL MARINE 32–38	NORMAL MARINE	NORMAL MARINE	NORMAL MARINE
SIGNIFICANT LIVING LAGOONAL PLANTS AND ANIMALS (T) = TRACE (R) = RARE (C) = COMMON (A) = ABUNDANT	HALIMEDA (A) THALASSIA MOLLUSCS PENICILLUS	HALIMEDA (A) PENICILLUS CORAL (R-C) MOLLUSCS RED ALGAE (R-C) FORAMINIFERA	HALIMEDA MOLLUSCS FORAMINIFERA	HALIMEDA PENICILLUS MOLLUSCS FORAMINIFERA	THALASSIA HALIMEDA MOLLUSCS FORAMINIFERA
RECENT UNCONSOLIDATED LAGOONAL SEDIMENTS	SKELETAL SANDS	MUDDY SKELETAL SANDS	MUDDY CORAL–RED ALGAL SANDS	MUDDY SKELETAL SANDS	PELLETED, SANDY LIME MUDS
LAGOONAL DEPOSITS THEORETICAL ROCK EQUIVALENTS: TERMS BY DUNHAM, 1962	ALGAL PLATE–MOLLUSCAN GRAINSTONE	ALGAL PLATE–MOLLUSCAN PACKSTONE	SKELETAL WACKESTONE AND PACKSTONE	[1] ALGAL PLATE–MOLLUSCAN PACKSTONE [2] ALGAL PLATE PACKSTONE	ALGAL PLATE–PELLETAL WACKESTONE
LAGOON MARGINS	LAGOONAL SEDIMENTS GRADE EASTWARD INTO AN APRON OF REEF–DERIVED SEDIMENTS	FLORIDA KEYS AND DISCONTINUOUS REEFS OF THE REEF TRACT	LAGOON TERRACE BORDERS BLUE LAGOON TO THE SOUTHWEST; REEF FLATS ENCIRCLE TERRACE & LAGOON	FLANK SEDIMENTS OF PATCH REEFS	FLANK SEDIMENTS OF CELLULAR PATCH REEFS AND BARRIER REEF
REMARKS	MODERATE TO HIGH ENERGY LAGOON; SOME SEDIMENT TRANSPORTED FROM REEF INTO LAGOON.	CORAL & RED ALGAL PARTICLES TRANSPORTED FROM REEFS INTO LAGOON; PATCH REEFS COMMON; SOME BARE ROCK ON LAGOON FLOOR.	SMALL PATCH REEFS VERY ABUNDANT; MUCH SEDIMENT TRANSPORT FROM REEFS INTO LAGOON; WATER ON LAGOON TERRACE IS ONLY 1 METER DEEP.	LARGE REEF AREAS ARE MAINLY PATCH REEF COMPLEXES AND CONTAIN MANY INTERREEF LAGOONS TOO SMALL TO BE SHOWN ON MAP.	INTERREEF LAGOONS AND CELLULAR REEFS SHOWN DIAGRAMMATICALLY ON MAP.
REFERENCES	EBANKS, 1967	GINSBURG, 1956	MAXWELL ET AL., 1964	UPCHURCH, 1970 JORDAN, 1973	KORNICKER & BOYD, 1962 HOSKIN, 1966

Sediments

Deposits in tropical lagoons are, in most cases, muddy skeletal or pelletal deposits. They display a continuous range from pure lime muds to muddy carbonate sands. A major difference between carbonate and clastic lagoonal sedimentation is that coarse sediment is produced locally within carbonate lagoons by plants and animals, without the necessity of having strong currents to transport coarse material. A combination of locally derived and transported sediment is found in moderate- to high-energy backreef lagoons, where much sediment originates from reef tracts and is washed into the lagoon.

Fine carbonate mud, common in most tropical lagoons, has been the subject of several studies to determine the origin of the mud. Several possible origins probably occur at various places and even act simultaneously in single lagoons. Stockman et al. (1967) and Brady (1971) have found that postmortem disintegration of certain calcified green algae (especially *Penicillus*) yields micron-sized needles of aragonite. They have direct evidence of such needles in the lime muds of Florida Bay and of some Yucatan lagoons. Matthews (1966) demonstrated that lime muds on the British Honduras shelf were derived mainly from decomposition (abrasion, weakening by borers) of mollusks and foraminifera. The complex nature of the origin of lime mud is far from resolved, but researchers are retracting ideas suggesting direct chemical precipitation of fine $CaCO_3$ from sea water.

Ancient Lagoons

Recent lagoons and their sediments have been studied primarily to gain further insight into the problems of recognizing and interpreting ancient, lithified lagoonal deposits in the geologic record. The problems of petroleum geologists and stratigraphic researchers boil down to the basic questions of (1) What depositional environment does this rock outcrop or drilled subsurface section represent? (2) What does a given environmental interpretation indicate for extrapolations into distant areas where there are no outcrops or drilled wells?

The significance of barriers to lagoons has been emphasized, but these barriers are usually small in comparison with the areal extent of the lagoon proper. As a result, they are difficult to observe directly in ancient deposits. More commonly, a stratigrapher is faced with the problem of interpreting broad expanses of muddy skeletal or pelletal sands or fossiliferous lime mudstones. Usually, in such a case, a hypothetical barrier is presumed to exist (because there may be no direct evidence), and the deposits are classified as "inner or middle shelf deposits." "Shelf" is another loosely defined term, but certainly many carbonate shelf deposits were probably once tropical lagoons.

CLIFTON F. JORDAN

References

Bandy, O. L., 1964. Foraminiferal biofacies in sediments of Gulf of Batabano, Cuba, and their geologic significance, *Bull. Am. Assoc. Petrol. Geologists*, 48, 1666-1679.

Bathurst, R. G. C., 1971. *Carbonate Sediments and Their Diagenesis*. New York: Elsevier, 620p.

Brady, M. J., 1971. Sedimentology and diagenesis of carbonate muds in coastal lagoons of NE Yucatan, Ph.D. Thesis, Rice Univ., Houston, Texas, 288p.

Castanares, A. A., and Phleger, F. B., eds., 1967. *Coastal Lagoons, A Symposium*, Memoir of the International Symposium on Coastal Lagoons. UNAM-UNESCO, Universidad Autonoma de Mexico, 686p.

Cloud, P. E., 1962. Environment of calcium carbonate deposition west of Andros Island, Bahamas, *U.S. Geol. Surv. Prof. Pap. 350*, 138p.

Daetwyler, C. C., and Kidwell, A. L., 1959. The Gulf of Batabano, a modern carbonate basin, *5th World Petrol. Congr. Proc.*, 1, 1-21.

Dunham, R. J., 1962. Classification of carbonate rocks according to depositional texture, *Am. Assoc. Petrol. Geologists Mem. 1*, 108-121.

Ebanks, J. W., 1967. Recent carbonate sedimentation and diagenesis, Ambergris Cay, British Honduras, Ph.D. Thesis, Rice Univ. Houston, Texas, 189p.

Emery, K. O.; Tracey, J. I.; and Ladd, H. S., 1954. Geology of Bikini and nearby atolls, Part 1, geology, *U.S. Geol. Surv. Prof. Pap. 260-A*, 265p.

Evans, G., and Bush, P., 1967. Some oceanographical and sedimentological observations on a Persian Gulf lagoon, in Castanares and Phleger, 1967, 155-170.

Ginsburg, R. N., 1956. Environmental relationships of grain size and constituent particles in some south Florida carbonate sediments, *Bull. Am. Assoc. Petrol. Geologists*, 40, 2384-2427.

Gross, M. G.; Milliman, J. D.; Tracey, J. I.; and Ladd, H. S., 1969. Marine geology of Kure and Midway Atolls, Hawaii: A preliminary report, *Pacific Sci.* 23, 17-25.

Hoskin, C. M., 1966. Coral pinnacle sedimentation, Alacran Reef lagoon, Mexico, *J. Sed. Petrology*, 36, 1058-1074.

Hoskins, C. W., 1964. Molluscan biofacies in calcareous sediments, Gulf of Batabano, Cuba, *Bull. Am. Assoc. Petrol. Geologists*, 48, 1680-1704.

Jordan, C. F., 1973. Carbonate facies and sedimentation of patch reefs off Bermuda, *Bull. Am. Assoc. Petrol. Geologists*, 57, 42-54.

Kendall, C. G., and Skipwith, P. A., 1969. Holocene shallow-water carbonate and evaporite sediments of Khor al Bazam, Abu Dhabi, southwest Persian Gulf, *Bull. Am. Assoc. Petrol. Geologists*, 53, 841-869.

Kornicker, L. S., and Boyd, D. W., 1962. Shallow-water geology and environments of Alacran reef complex, Campeche Bank, Mexico, *Bull. Am. Assoc. Petrol. Geologists*, 46, 640-673.

McKee, E. D.; Chronic, J.; and Leopold, E. B., 1959. Sedimentary belts in lagoon of Kapingamarangi Atoll, *Bull. Am. Assoc. Petrol. Geologists,* 43, 501-562.

Matthews, R. K., 1966. Genesis of Recent lime mud in southern British Honduras, *J. Sed. Petrology,* 36, 428-454.

Maxwell, W. G.; Jell, J. S; and McKellar, R. G., 1964. Differentiation of carbonate sediments in the Heron Island reef, *J. Sed. Petrology,* 34, 294-308.

Neumann, A. C., 1965. Processes of Recent carbonate sedimentation in Harrington Sound, Bermuda, *Bull. Marine Sci. Gulf Caribbean,* 15, 987-1035.

Newell, N. D.; Rigby, J. K.; Whiteman, A. J.; and Bradley, J. S., 1951. Shoal-water geology and environments, eastern Andros Island, Bahamas, *Bull. Am. Mus. Natl. Hist.* 97, 30p.

Purdy, E. G., 1963. Recent calcium carbonate facies of the Great Bahama Bank, 1. Petrography and reaction groups, 2. Sedimentary facies, *J. Geol.,* 71, 334-335, 472-497.

Roberts, H. H., 1971. Mineralogical variation in lagoonal carbonates from North Sound, Grand Cayman Island (British West Indies), *Sed. Geol.,* 6, 201-213.

Stockman, K. W.; Ginsburg, R. N.; and Shinn, E. A., 1967. The production of lime mud by algae in South Florida, *J. Sed. Petrology,* 37, 633-648.

Taft, W. H., and Harbaugh, J. W., 1964. Modern carbonate sediments of southern Florida, Bahamas, and Espiritu Santo Island, Baja California: A comparison of their mineralogy and chemistry, *Stanford Univ. Publ. Geol. Sci.,* 8, 133p.

Upchurch, S. B., 1970. Sedimentation on the Bermuda platform, *Army Corps Engin., U.S. Lake Surv., Research Rept.* 2-2, 172p.

Cross-references: *Bahama Banks Sedimentology; Evaporite Facies; Lagoonal Sedimentation; Limestones; Persian Gulf Sedimentology; Reef Complex.* Vol I: *Florida Bay: A Site of Modern Limestone Formation.* Vol. III: *Mangrove Swamps–Geology and Sedimentology.*

TSUNAMI SEDIMENTATION

Tsunamis are neglected but important agents of sedimentation. They are complex "long waves," many features of which are as yet poorly understood, differing markedly in their behavior from other large sea waves, such as those induced by great storms. A tsunami is actually a wave system, formed as a result of an enormous but brief disturbance of the sea surface and involving great volumes of water (see Vol I: *Tsunami;* Van Dorn, 1965; Cox, 1963). Once formed, as Van Dorn puts it, "the wave system resembles nothing so much as that produced by tossing a stone into the middle of a large shallow pond."

In the open sea, a tsunami wave series may pass unnoticed, for the wave amplitude is extremely small compared to the wave length (1 m or so compared with several hundred km), and it travels at great speeds, as much as 800 km/hr. The topography of the sea floor affects the initial symmetry of the system. In particular, as it approaches a susceptible coast and shoaling occurs, the wave amplitude increases markedly; and a complex interaction takes place between the incoming heightening wave series and the coastal shelf and coastline. In effect, there follows a small and variable number of oscillations of sea level along the coast, each resulting in an invasion of and retreat from the land. Such invading "waves" may be tens of m high and extend inland several km, especially within bays and up river valleys. There is tremendous disturbance of nearshore waters. Currents exceeding 10 knots may persist for hours, and destructive sloshing or seiche movements take place in semienclosed bays and harbors (Shepard et al., 1950).

Most large tsunamis, of the kind that are propagated across whole ocean basins, are thought to be brought about by profound seismic disturbance, sometimes involving the sudden vertical shift of large areas of sea floor as a result of faulting. Explosive volcanic eruption may also produce large tsunamis, for example, that which followed the explosion of Krakatoa in 1883. Subaerial land slumps, submarine slumping of masses of sediment, and small-scale fault movements along coastlines may also result in large waves. Such "small" tsunamis may produce effects over localized areas more than comparable with those of larger oceanic tsunamis.

Bailey and his colleagues pioneered the notion that seismic events, and hence tsunamis, were connected with the formation of so-called *chaotic sediments* (Bailey and Weir, 1932; Harrison, 1935; Lamont, 1936). Bucher (1940) suggested that tsunamis caused oscillatory currents which carved or sawed out submarine canyons over coastal shelves. He also implied that tsunamis were capable of shifting vast quantities of sediment.

These ideas were generally neglected, and no unified hypothesis on the constructive geological work of tsunamis had been proposed until recently (Coleman, 1968). It has three main parts: (1) tsunami activity may build up sequences of chaotic sediments in shallow water; (2) it assists, at least, in the formation and maintenance of submarine canyons; and (3) it primes and fires powerful turbidity currents, which carry coarse and fine sediments out on to the fan areas of canyons. Under this hypothesis, tsunamis would be most effective along the coasts of tectonically active orogenic regions providing large supplies of sediment, the abrasive medium. Tsunami action offers solutions to many sedimentary problems, particularly those involving the source and depositional environment of chaotic sediments and turbidites.

The agents that produce tsunamis must have existed throughout the great part of geological time. Tsunamis will have occurred with great frequency, roughly 100,000/m.y., to extrapolate from historical records. We may assume that there have been even larger ones than are on record, which is important, for an extremely large tsunami would accomplish much more work than a number of smaller ones. To concede the general incidence of tsunamis also means that, with fluctuations of sea level, they have operated over a swathe of land and nearshore surfaces. This effect has particular relevance to the Quaternary with its large eustatic oscillations, and hence to modern marine-sediment studies and such problems as the origin of submarine canyons.

Tsunami attack is inherently erratic. Apart from the varying nature of these waves, coasts are not equally susceptible. Wide shelves, for example, effectively soak up their energy and inhibit their action. Susceptible coasts have narrow shelves, receive large quantities of sediment, and front open ocean. Such coasts are a feature of tectonically active orogenic regions, as exemplified by island arcs today and geosynclinal areas of the past. These regions also have a high incidence of local tsunamis, which will accomplish the same work as those traveling from afar.

Tsunamis and Sediments

The effects of the interaction between the coast and the incoming wave series, including the flooding of coastal areas, are described here as the *on-surge,* which plays primarily a softening-up role, affecting already deposited shallow-water sediment. It is conceivable that it may actually shift such sediment, but just what happens at the sea floor during the on-surge is not known. The *run-up* (or *set-up*) is the accepted term for the maximum height above sea level reached by an invading wave. The *off-surge* is the seaward movement of water that follows the on-surge; more descriptively, it is the rush downslope of the piled-up water from the coast back into the trough which heralds the next wave of the series.

It is the off-surge and associated currents that are thought to be the main mechanism by which sediments are shifted seaward. Such sediments may be extremely varied, originating from tidal flats, piedmont, flood plains, deltas and lagoons, beaches, sand bars and fringing reefs. Coastal vegetation, especially of swamps and deltas, may be stripped and removed seaward in a single episode.

The off-surge rushes out from the coast as an extensive, turbulent bottom layer loaded with sediment. If the offshore area is shallow and smooth and *not* cut by a well-developed submarine gully and canyon system, then this mixture loses energy comparatively rapidly; the sediment load is then slopped forth and dumped within km of the shore, just how close depending on the size of the tsunami and the configuration of the coastal shelf. The work of a particular off-surge need not be arrested by the on-surge of the next wave. Each wave of the series is commonly separated by a lapse of many minutes.

Although prone to reworking by later off-surges and tsunamis (Von Koenigswald, 1971), suitable subsiding environments may allow the accumulation into sequences of these chaotically deposited sediments. Such shallow-water sinks are common enough in tectonically active geosynclinal areas and the island arcs of today. Similar sequences, not so intensely reworked, may accumulate on deeper-water shelves off actively faulted coasts.

These tsunami-produced, nearshore sediments would have at least some of the aspects of "wildflysch" and "tilloidal" and "chaotic" sequences, reflecting the convulsive character of the off-surge and its sedimentary jumble. Such features would include crudely graded rudaceous beds with basal clasts of pebble to boulder size, layers of paraconglomerates with boulders set in fine-grained matrix; layers of unfossiliferous, noncalcareous, land-derived sediment alternating with calcareous fossiliferous layers; beds or "mats" of concentrated vegetation with both fine- and coarse-grained sandy interbeds; evidence of current directions conflicting strongly with the direction of slope; and the presence of undoubted shallow-water indicators such as mud cracks, scour channels, and so on.

There may also be a lesser development of turbiditic features—graded beds, flute casts, "pelagic" layers (see *Turbidites*)—which reflect both the less-violent closing stages of the depositional event, when grading and sorting can take place, and interdepositional periods. Such sequences may be but short distances (in a lateral, nonstratigraphic sense) from shell beds and fringing reef masses.

Tsunamis, Submarine Canyons and Turbidity Currents

Tsunami action can also be considered to assist in the explanation for the shifting of miscellaneous sediments, of the kind described, into deeper, even abyssal, water areas far from shore. At this point, the suggested relationship between tsunami action, powerful turbidity currents and submarine canyons must be considered.

A stretch of coast that is susceptible to tsunami attack, exciting extremely large oscillations, and is strongly digitate will receive enormous volumes of water at the peak of the on-surge. The consequent off-surge will have a scouring and probing or selective action on the adjacent seabed. The scouring follows because of its speed and load of sediment and the probing because relatively greater volumes of water have been piled up in inlets and up river valleys: the action will be strongest opposite the larger, narrow inlets and valleys because they have accumulated the most water. As well, nearshore marine sediments will have been unsettled by the on-surge and associated currents and so are more easily shifted and eroded by the off-surge.

It is suspected that, over a long period of tsunami visitations, this selective, probing off-surge eventually carves out a submarine-channel "drainage" system in a way analogous to the *wadi* or *arroyo* system set up by intermittent flash floods in arid regions. The ultimate development would be a submarine canyon and its commonly associated features: its tributary system; the usual relation of river valleys and canyon head; the frequent presence of large exotic blocks; the hummocky-surfaced deep-sea sediment fan at the canyon exit; the concave profile, steepest near the head (where the off-surge still has great energy); and its steep transverse profile. Even resistant bedrock such as granite may be so carved, joints and/or faults assisting the process. The varying distance of the head area from shore, or even extension of the canyon inland (Starke and Howard, 1968), would be partly explained by the operation of tsunamis during the major fluctuations of Pleistocene sea level, or by localized tectonic movements, which bring about vertical movements of particular stretches of coast.

This process of canyon formation may proceed at various points along a coast. Associations of submarine canyons are becoming a recognized feature of today's offshore shelves.

Some of the features observed in today's canyons may be irrelevant to the notion that tsunamis are a formative agent. Tsunami attack is certainly severe but nevertheless sporadic. There may be tens or hundreds of years between attacks, and during this time the canyon is left to its own quiet evolution. Once the canyon system is set up, probably off a young embayed coast receiving much sediment, it will be maintained by its own internal agents (Shepard and Dill, 1966). It may receive sediment from other nearshore processes such as storm waves, currents and gravity slumps. This sediment may fill the canyon to varying depths and will tend to flow or creep down the canyon floor. Eventually, however, a big tsunami will come and the canyon will be at once flushed and further eroded. If there be a change in geologic and tectonic regime so that the canyon is spared further tsunami attack, then it may fill and be buried, becoming fossilized (see *Submarine Canyons and Fan Valleys, Ancient*).

The notion of flushing of the canyon suggests the transport of coastal and shallow water sediment out to deep water. In the description of the accumulation by tsunamis of shallow-water chaotic sediments, the smoothness of the offshore area was stressed. If this area is stable long enough, however, a gully-to-canyon system will be set up, and different conditions will prevail. The suggestion is that a large part of the sediment-laden off-surge is spared energy loss by being captured by the nearshore gullies and channels, which are of steeper slope than the intervening sea floor. These contribute to the larger tributaries. The tributary system, therefore, captures and diverts a large part of the off-surge and funnels it and its sediment load into a single sluice, the main canyon. The sediment, given this initial impetus, pours downslope to the exit as a powerful turbidity current and there discharges onto the deep-sea fan.

The larger the tsunami and supply of coastal and nearshore sediment, the more drastic will be the action proposed. A small tsunami will have less total energy, so that the off-surge and its resultant turbidity current will also be small and fail to reach the lower reaches of the canyon, dropping its load within the canyon or expelling only the finer material out over the fan. Intermediate examples may lead to the build-up of levees and the reshaping of the canyon profile and fan.

The all-important distinction between the turbidity current described and the one normally envisaged is that the former has a higher initial impetus: the off-surge and its load is concentrated into a single sluice and there becomes translated into a turbidity current. The off-surge, as it were, "primes and fires" its offspring.

Selected Applications

Tsunami action may help to elucidate some long-standing sedimentary problems, a few of which can be cited.

The absence of delta-front sediments in geologically ancient deposits is at least partly explained, for deltas are peculiarly susceptible to tsunami attack. Those enigmatic sediments, paraconglomerates and edgewise conglomerates, may in part be tsunami products. The former have already been described as chaotic sediments slopped forth by the off-surge; they are deposited extremely rapidly from a chaotic mass. The flaggy or tablet clasts of edgewise conglomerate result from the shattering and redeposition of a surface crust of semilithified sediment (Fischer and Garrison, 1967). The pseudoturbidite features, especially graded bedding, of the carbonate sands of offshore swells and banks can be accounted for by the disturbance occasioned by the passage of a tsunami. The sediments on such shoals, it is suggested, may be churned to depths within the sediment beyond the capacity of either burrowing benthos or hurricane surge (see *Storm Deposits*). In reef environments, the scattering over extensive areas of fore-reef talus and its crude sorting (Mountjoy et al., 1972), may be due to tsunamis as is also, at least in part, the dumping of massive quantities of reef breccia and exotic blocks in lagoonal areas and on fringing reef platforms.

Evidence of energetic reworking of deep-water turbidite sequences, together with the presence of exotics, conglomerates, and massive quantities of nearshore sediment types in these sequences can be explained by the action of forceful, tsunami-produced turbidity currents. The distinction between shallow-water tsunami deposits and canyon-fan turbidites may not always be clear, but the criteria cited, plus a realization of the roles of tsunamis as depositors in their own right and as instigators of turbidity currents, should aid in field studies of turbidite, turbiditic, tilloidal and wildflysch sequences.

Sudden changes in nearshore bathymetry along coasts visited by tsunamis are well known and amount to scouring or shoaling, of amounts to 20 m or more, over areas of 1 km^2 or thereabouts (Houtz, 1963; Jordan, 1963; Menard, 1964). The proposal here is that such changes are a normal expression of the reworking and deposition of sediment in shallow water by tsunamis.

Tsunami action may possibly perform another work. A tsunami wave series as it approaches a coast may produce violent deep-water currents at the point where shoaling begins to bring about increase in wave height, i.e., near the edge of the coastal or even continental shelf (see Wong et al., 1963). The action envisaged is opposite but yet parallel to the "weir effect" of a string of islands lying across the passage of tides, another type of "long wave." In this action may lie the explanation of the deep-water effects observed to follow the passage of a tsunami (Bernstein, 1954). By implication, the deeper parts of submarine canyons and accumulations of sediment in deep water would be affected.

The hypothesis presented here does not claim an exclusive role for tsunamis in the production of turbidites, chaotic sediments, and submarine canyons. Rather, tsunamis are presented as a complementary agent, acting separately and in conjunction with others.

PATRICK J. COLEMAN

References

Bailey, E. B., and Weir, J., 1932. Submarine faulting in Kimmeridgian time: East Sutherland, *Trans. Roy. Soc. Edinburgh*, 47, 429-454.

Bernstein, J., 1954. Tsunamis, *Sci. Am.*, 191, 60-64.

Bucher, W. H., 1940. Submarine valleys and related geologic problems of the North Atlantic, *Geol. Soc. Am. Bull.*, 51, 489-512.

Coleman, P. J., 1968. Tsunamis as geological agents, *J. Geol. Soc. Austral.*, 15, 267-273.

Cox, D. C., ed., 1963. Proceedings of tsunamic meetings, Tenth Pacific Science Congress, Honolulu, 1961, *Internat. Un. Geodesy. Geophys. Monogr. 24*, 265p.

Fischer, A. G., and Garrison, R. E., 1967. Carbonate lithification on the sea floor, *J. Geol.*, 75, 488-496.

Harrison, S. M. K., 1935. Ordovician submarine disturbances in the Girvan district, *Trans. Roy. Soc. Edinburgh*, 58, 487-507.

Houtz, R. E., 1963. The 1953 Suva earthquake and tsunami, *Internat. Un. Geodesy. Geophys. Monogr. 24*, 73-76.

Jordan, G. F., 1963. Major redistribution of sediments in certain bays and inlets of Alaska (abstr.) *Pac. Sci. Cong.* (Honolulu, 1961), 374.

Lamont, A., 1936. Palaeozoic seismicity, *Nature (London)*, 138, 243-244.

Menard, H. W., 1964. *Marine Geology of the Pacific.* New York: McGraw-Hill, 271p.

Mountjoy, E. W.; Cook, H. E.; Pray, L. C.; and McDaniel, P. N., 1972. Allochthonous carbonate debris flows—worldwide indicators of reef complexes, banks or shelf margins, *Internat. Geol. Congr. 24th (Montreal)*, 6, 172-189.

Shepard, F. P., and Dill, R. F., 1966. *Submarine Canyons and Other Sea Valleys.* Chicago: Rand McNally, 381p.

Shepard, F. P., Macdonald, G. A.; and Cox, D. C., 1950. The tsunami of April 1, 1946, *Scripps Inst. Oceanogr. Bull.*, 5, 391-528.

Starke, G. W., and Howard, A. D., 1968. Polygenetic origin of Monterey Submarine Canyon, *Geol. Soc. Am. Bull.*, 79, 813-826.

Van Dorn, W. G., 1965. Tsunamis, *Adv. Hydroscience*, 2, 1-48.

Von Koenigswald, G. H. R., 1971. Tsunami disturbed Tertiary sediments on the northern coast of Crete, *1st Internat. Sci. Congr. on Volcano of Thera, Sept. 1969, Acta*, 283-287.

Wong, K. K., Ippen, A. T. and Harleman, D. R. F., 1963. Interaction of tsunamis with oceanic islands and submarine topographies, *Mass. Inst. Tech., Hydrodyn. Lab. Rept.*, 62, 86p.

Cross-references: *Diamictite; Flood Deposits; Gravity Flows; Storm Deposits; Submarine Canyons and Fan Valleys, Ancient; Submarine Fan (Cone) Sedimentation; Turbidites; Turbidity-Current Sedimentation.* Vol. I: *Tsunami.*

TUFA— See SPRING DEPOSITS

TUFF

Tuff is an indurated volcanic ash deposit, essentially composed of aerially transported fragments formed during volcanic eruptions Much of this ejected, pyroclastic material is formed either by the explosive fragmentation of crystals and rocks or from expelled particles of liquid lava which have rapidly cooled to form volcanic glass. Welded tuff is deposited from a glowing ash flow, or nuée ardente (q.v.; see *Ignimbrite;* Ross and Smith, 1961).

The main factors controlling the deposition of the ash are prevailing winds and gravity sorting by both size and density. The coarser or heavier fragments fall more rapidly, are rather poorly sorted, and normally constitute the basal part of the deposit. The finer, lighter materials are carried farther, are the last to be deposited, and form a well-sorted upper layer in the deposit (see *Volcanic Ash Deposits*).

Tuffs belong to the general family of volcaniclastic sediments and rocks (q.v.), which may be classified either by the size or composition of the constituent fragments. Size classifications for the nonconsolidated materials and the corresponding indurated rocks are shown in Figs. 1 and 2 in *Volcaniclastic Sediments and Rocks.* In compositional classification, the term *essential* is used for the fragments formed from the erupting lava. If they are derived from volcanic rocks previously expelled from the same cone, they are termed *accessory;* and if they are composed of other rocks, they are termed *accidental.*

The ash- and dust-sized fragments may consist of small particles of crystals, rocks, or volcanic glasses. The resulting tuffs are therefore termed *crystal, lithic,* or *vitric* tuffs respectively. Tuffs which contain more than one of these components have been classified by O'Brien (1963), as shown in Fig. 1.

Tuffs may also be subdivided according to their magmatic derivation, e.g., rhyolitic, trachytic, andesitic, and basaltic tuffs. In a similar manner they have also been named after the dominant component mineral or mineral type, e.g., feldspathic and palagonite tuffs.

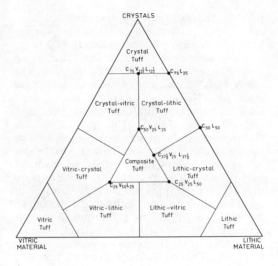

FIGURE 1. Compositional classification of tuffs containing crystals (C), vitric material (V), and lithic material (L). Percentage limits for crystal-lithic and composite tuffs are shown.

Types of Tuffs

Vitric Tuffs. The fragments of volcanic glass that typify vitric tuffs usually represent broken pieces of vesicular lava which have solidified during flight. These fragments include curved sections of isolated vesicle walls, known as *shards,* doubly concave particles which form between two adjacent vesicles, and other particles which represent the material formed between three or more vesicles.

Where molten particles are stretched apart during eruption, the vesicles become elongated into tube-like structures. In the case of more fluid lavas, this process frequently leads to the formation of many fine glass threads, known as Pelé's hair. Microscopic studies of these threads sometimes reveal small, elongated vesicles.

Viscous, usually silicic lavas which are highly charged with volatile components may generate such high pressures that they burst through the top or flank of the volcano. The release of pressure generates clouds of incandescent, gas-charged material known as *nuées ardentes* (q.v.), which rapidly descend the sides of the vent until loss of volatiles allows the suspended material to settle, generally as thick, lava-like deposits. The individual glass shards and other fragments in these deposits are typically fused together by the intense heat, and the rock is known as a *welded tuff* or *ignimbrite* (q.v.).

Vitric tuffs may be subdivided into acid, intermediate, and basic types on the basis of their chemical composition. Acid and intermediate vitric tuffs are more common than basic types and are derived from magmas which are cooler, more viscous, and have a higher volatile content. While chemical analyses may be required to determine the magmatic derivation of a tuff, the glass of acid and intermediate tuffs is colorless and has a refractive index near 1.5. Basic vitric tuffs are relatively rare; the glasses are dark-colored and have a refractive index nearer 1.6 (Carozzi, 1960).

Crystal Tuffs. The crystalline material of crystal tuffs may be either essential, accessory, or accidental, and individual tuffs frequently contain all three types. As with vitric tuffs, acid or intermediate magmas are more common sources than basic magmas for crystal tuffs.

Essential crystalline material is derived from magmas that contain a high proportion of crystals at the time of the eruption. The crystals are frequently fractured, but all gradations between perfectly formed and extremely fractured varieties may be found. They frequently exhibit signs of resorption, and the corroded areas may be filled with glass.

Both accessory and accidental crystals are usually minor with respect to those derived directly from the magma. The matrix of crystal tuffs consists of fine vitric material and debris resulting from the fracturing of the crystals.

Lithic Tuffs. Lithic tuffs are formed by the explosive shattering of previously existing rocks, which may range from essential lava to accessory lava, to accidental igneous, sedimentary, or metamorphic rocks which have become involved in the volcanic eruption. Essential fragments are more common than the other two types.

As with vitric and crystal components, angularity and evidence of brittle fracture are marked features of the lithic fragments. They are also usually somewhat larger than the crystals of crystal tuffs, which themselves are usually larger than the rather fragile glass fragments in vitric tuffs. The matrix of lithic tuffs may consist of fine vitric, crystal, and lithic components.

Some difficulty may be encountered in proving a pyroclastic origin for crystal or lithic tuffs, because, in contrast to vitric particles, crystals and lithic fragments may be found in other rock types. The recognition of vitroclastic textures in the matrix areas of fine-grained pyroclastic rocks may give the first indication of the origin of these rocks.

Alteration. The volcanic glasses in tuffs and ash deposits are particularly prone to alteration (see *Volcanic Ash—Diagenesis*). The glass may be present as either vitric or lithic fragments. Nonglassy crystal or lithic constituents are less susceptible to alteration and, when so affected, respond in a manner comparable to that of similar constituents of igneous rocks. These alterations may completely change the character of the rock and conceal its origin.

Admixture of Ash with Other Sediments. While the coarser pyroclastic ejecta settle near the vent, the finer materials, particularly vitric fragments, may travel great distances and may settle in water, where they become mixed with varying amounts of other sedimentary detritus before final consolidation and lithification. Such rocks correspond to *tuffites,* as defined by Mügge (1893). Hay (1952) proposed that when the sediments contain < 50% pyroclastic material, they should be qualified by the term *tuffaceous.*

REGINALD T. O'BRIEN

References

Blyth, F. G. H., 1940. The nomenclature of pyroclastic deposits, *Bull. Volcan.,* ser. 2, **6,** 145-156.

Carozzi, A. V., 1960. *Microscopic Sedimentary Petrography.* New York: Wiley, 485p.

Hay, R. L., 1952. The terminology of fine-grained detrital volcanic rocks, *J. Sed. Petrology,* **22,** 119-120.

Heiken, G., 1972. Morphology and petrography of volcanic ashes, *Geol. Soc. Am. Bull.,* **83,** 1961-1988.

Lorenz, V., 1974. Vesiculated tuffs and associated features, *Sedimentology,* **21,** 273-291.

Mügge, O., 1893. Untersuchungen über die "Lenneporphyre" in Westfalen und den angrenzenden Gebieten, *Neues Jahrb. Mineral Geol. Palaeontol.,* BB, **8,** 535-716.

O'Brien, R. T., 1963. Classification of tuffs, *J. Sed. Petrology,* **33,** 234-235.

Pirsson, L. V., 1915. The microscopical character of volcanic tuffs—a study for students, *Am. J. Sci.,* ser. 4, **40,** 191-211.

Ross, C. S., and Smith, R. L., 1961. Ash-flow tuffs—their origin, geologic relations, and identification, *U.S. Geol. Surv. Prof. Pap. 366,* 81p.

Cross-references: *Ignimbrites; Nuée Ardente; Pyroclastic Sediment; Volcanic Ash Deposits; Volcanic Ash—Diagenesis; Volcaniclastic Sediments and Rocks; Volcanism—Submarine Products.*

TURBIDITES

Turbidites are sedimentary rocks that have been deposited by turbidity currents. Characteristically they occur as beds of sandstone, siltstone, or clastic limestone alternating with mudstone, each bed sitting with a sharp lower contact upon the underlying mudstone. They are also common in volcaniclastics, and have even been found in evaporites (Mutti and Ricchi-Lucchi, 1972). Turbidity currents, driven by gravity due to excess density caused by a suspended load of sediment, flow down any available slope (see *Turbidity-Current Sedimentation*). They have been known to geologists since 1936, when they provided R. A. Daly with a process by which submarine canyons could be excavated. It was not until 1950, however, when Kuenen and Migliorini attributed the Oligocene graywackes of the northern Apennines to the action of turbidity currents that their importance as agents of transport and deposition of ancient sediments was appreciated.

During the following decade, the action of turbidity currents was invoked to explain the features of many similar formations such as the Plio-Pleistocene succession in the Ventura Basin of southern California, the Silurian Aberystwyth Grits of Wales, and the Cretaceous and Paleogene successions of the Polish Carpathians. For most of this time, these sequences were called graywacke or flysch because they had the texture of graywacke (q.v.) or were found in flysch basins within orogenic belts, and much confusion arose through the indiscriminate use of the words *graywacke* (a petrographic term) and *flysch* (a tectonic facies). Meanwhile, Kuenen (1957) suggested the word *turbidite* for the deposits of turbidity currents.

Characteristics

The recognition of a turbidite is difficult and its diagnosis a matter of opinion. Certain features, however, are characteristic:

1. Turbidites are part of a succession made up of a marked alternation of fine sediments such as shales, marls, mudstones, and siltstones and coarse sediments such as sandstones or detrital limestones; each bed has a considerable lateral extent.
2. They are often moderately or poorly sorted and contain a notable amount of clay-grade material.
3. They show sharply defined bottom surfaces frequently covered by a profusion of sole marks of both inorganic and organic origin. Inorganic sole marks consist either of scour marks (q.v.) or of tool marks (q.v.). Organic sole marks are either grazing patterns and burrows formed on the muddy sea floor and preserved beneath the succeeding turbidity current or burrows formed by animals at the base of the turbidite after it was deposited.
4. The top surfaces are usually indistinct, and there is a transition from sandstone into shale.
5. They often show graded bedding (q.v.), and a sequence of sedimentary structures indicating deposition by a waning current.
6. Apart from burrows, fossils are rare. Shallow-water benthonic fossils are usually absent or rare in non-calcareous turbidite formations. Some displaced or redeposited fossils occur within the turbidites, and pelagic or relatively deep-water benthonic organisms are found in the intervening shales, where displaced benthonics are normally absent. In the carbonates, displaced and pelagic fossils can be very abundant (e.g., graded bedding of Nummulites).
7. There is an absence of evidence for emergence or shallow water in the associated beds. Rootlet beds, desiccation cracks, and wave ripple marks are characteristically lacking. Transported plant remains are often incorporated.

The essential characteristic is that, within a background of mud or silt deposited from suspension in very quite water, there is a sudden incursion of coarse material deposited from a powerful bottom current which was initially fast and erosive and rapidly died out.

In 1962, Bouma recognized a well-defined sequence of textures and sedimentary structures in each turbidite bed. Although in most cases the lower part or the upper part is absent, there is an "ideal" turbidite sequence now known as the *Bouma sequence* (q.v). Walker (1965), using data from experimental studies, interpreted the sequence in the terms of flow regimes (q.v.). He showed that the Bouma sequence is due to a waning turbidity current passing from an upper flow regime to lower flow regimes (see *Bouma Sequence*).

Turbidite successions are of different kinds, depending to some extent on their closeness to the source of material and initiation point of the turbidity current. With increasing distance from the source, individual beds become thinner, grain size decreases, fewer beds have sharp tops, and there are fewer signs of bottom erosion marks. Bedding may become more regular; there is a higher proportion of well-graded beds and of beds beginning with Bouma divisions C and sometimes B, and there is an increase in the number of tool marks relative to scour marks.

Interpretation

The distinction between "proximal" and "distal" turbidites has led to the concept of proximality in turbidites, a concept that has proved valuable in interpreting the paleogeography of some turbidite basins. It must be applied with care, however, because a particular turbidite facies depends as much on contemporaneous tectonics, the sediment available, and paleoslope as upon distance from source. In addition, it is probable that most ancient turbidites were deposited on submarine fans (q.v.) where depositional environments are complex, rather than upon monotonous abyssal plains. Present research is concentrating on the recognition of sequences and facies associations in turbidites. Sequences where the sandstone beds become thicker upward are thought to have formed in the lower part of submarine fans, while thinning-upward sequences form on the upper part, with the thicker, more "proximal" turbidites forming in the fan channels and the thinner, more "distal" turbidites forming between the channels as interchannel deposits (Walker and Mutti, 1973).

There have been two major geological revelations from the recognition of turbidites. One has been the discovery that sandstones, hitherto thought to be indicators of shallow-water environments, could have been transported into deeper water, and thus most geosynclinal sedimentation (q.v.) took place below wave base. The other has been the measurement of current structures, in particular the sole marks formed during the erosive phase of the turbidity current and preserved as molds on the base of the turbidite. These indicate the final direction of flow of the currents. Paleocurrent analyses (q.v.) must be applied with care, as the measurements indicate only the direction of current flow at the point of deposition; but when tied to other data such as lateral facies changes and compositional petrography, they help in the building of a picture of the regional environment.

In the early years of the turbidite hypothesis, opposition to it came largely from die-hards who seemed incapable of imagining a new process. Since about 1964, however, a more informed opposition has arisen because of the realization that there are important normal currents running along or close to the bottom of present oceans and that the current directions measured in ancient turbidites are frequently rather variable, even opposed to each other, and sometimes suggest that the currents flowed across the paleoslope rather than down it. It has, therefore, been argued that the features generally accepted as indicators of turbidity currents could be produced by fluctuating normal oceanic currents (see *Contourites*). Contour currents are important today in the carriage and deposition of silt and fine sand on the continental slopes and rises of the deep seas (see *Continental-Rise Sediments*); presumably, they were important in the past in the formation of thick siltstone sequences and of the sometimes rippled and laminated, silty mudstones between the coarser sandstones. They are, however, incapable of producing the coarser, sharp-based sandstones interbedded with fine-grained mudstones, the typical turbidite sequence, and it is generally possible to distinguish between them (Bouma and Hollister, 1973).

Another problem is the mechanism of formation of thick, massive, poorly graded sandstone beds with sharp tops, as well as bottoms, sometimes showing "dish" structure and lacking the typical features both of turbidites and of traction currents. They are frequently associated with slump and slide deposits and have been called fluxoturbidites by Dzulynski et al. (1959), who thought they were "watery slides" formed by "a mechanism related to both turbidity flow and to sliding." They may be partly produced either by fine debris flows, or by liquefied sediment flows. It is difficult to separate the mechanism by which a particular bed was formed, and because flows change from one type to another during flow, it is often a matter of semantics whether these massive beds are called turbidites or not (see *Gravity Flows*).

Recent turbidites are found mainly near continental margins in submarine fans (q.v.) or in the neighboring abyssal plains. They also occur in lakes. Ancient turbidites have been recognized mainly in marine flysch facies. They also occur in association with intracratonic deltaic and lacustrine successions (Raaf, 1968). In some cases, beds with turbidite features have been deposited in shallow water by flooding rivers entering either the sea or a lake or breaking into an interfluvial area (see *Flood Deposits*). In other cases storms have generated a catastrophic current which, as it wanes, produces beds indistinguishable from turbidites (see *Storm Deposits*). These deposits, however, are not true turbidites because the motive force of the current is the river or the storm, not the turbidity of the current. On the other hand, if, as in present Lake Geneva, the river flood carries on downslope due to gravity acting on the suspension, a river-generated turbidite results.

HAROLD G. READING

References

Bouma, A. H., 1962. *Sedimentology of Some Flysch Deposits. A Graphic Approach to Facies Interpretation.* Amsterdam: Elsevier, 168p.

Bouma, A. H., and Brouwer, A., eds., 1964. *Turbidites.* Amsterdam: Elsevier, 264p.

Bouma, A. H., and Hollister, C. D., 1973. Deep ocean basin sedimentation, in Middleton and Bouma, 1973, 79-118.

Dzulynski, S., and Walton, E. K., 1965. *Sedimentary Features of Flysch and Greywackes.* Amsterdam: Elsevier, 274p.

Dzulynski, S.; Ksiazkiewicz, M.; and Kuenen, P. H., 1959. Turbidites in flysch of the Polish Carpathian Mountains, *Geol. Soc. Am. Bull.,* 70, 1089-1118.

Hesse, R., 1975. Turbiditic and non-turbiditic mudstone of Cretaceous flysch sections of the East Alps and other basins, *Sedimentology,* 22, 387-416.

Kuenen, P. H., 1957. Sole markings of graded graywacke beds, *J. Geol.,* 65, 231-258.

Kuenen, P. H., 1964. Deep sea sands and ancient turbidites, in A. H. Bouma, and A. Brouwer, eds., *Turbidites.* Amsterdam: Elsevier, 3-33.

Kuenen, P. H., and Migliorini, C. I., 1950. Turbidity currents as a cause of graded bedding, *J. Geol.*, **58**, 91-127.

Middleton, G. V., and Bouma, A. H., eds., 1973. Turbidites and deep water sedimentation, *Soc. Econ. Paleont. Mineral., Pacif. Sect., Short Course, Anaheim*, 157p.

Mutti, E., and Ricchi-Lucchi, F., 1972. Le torbiditi dell'Apennino settentrionale: Introduzione all' analisi di facies, *Mem. Soc. Geol. Italiana*, **11**, 161-199. Trans., 1978 *Int. Geol. Rev.*, **20**, 125-166.

Parkash, B., 1970. Downcurrent changes in sedimentary structures in Ordovician turbidite greywackes, *J. Sed. Petrology*, **40**, 572-590.

Raaf, J. F. M. de, 1968. Turbidites et associations sédimentaires apparentées, *Koninkl. Ned. Akad. Wetenschap., Proc.*, ser. B., **71**, 1-23.

Ricchi-Lucchi, F., 1975. Depositional cycles in two turbidite formations of northern Apennines (Italy), *J. Sed. Petrology*, **45**, 3-43.

Walker, R. G., 1965. The origin and significance of the internal sedimentary structures of turbidites, *Proc. Yorkshire Geol. Soc.*, **35**, 1-32.

Walker, R. G., and Mutti, E., 1973. Turbidite facies and facies associations, in Middleton and Bouma, 1973, 119-158.

Cross-references: *Bouma Sequence; Continental-Rise Sediments; Contourites; Current Marks; Flow Regimes; Geosynclinal Sedimentation; Graded Bedding; Gravity Flows; Graywacke; Scour Marks; Sedimentary Facies and Plate Tectonics; Submarine Canyons and Fan Valleys, Ancient; Submarine Fan (Cone) Sedimentation; Tool Marks; Turbidity-Current Sedimentation.*

TURBIDITY-CURRENT SEDIMENTATION

The term *turbidity current* was introduced by Johnson (1939a) to designate density currents whose excess density is the product of turbidly suspended solids, to differentiate these currents from other *density currents* caused either by excess salinity (dissolved load) or by cooling. Johnson (1939a, b) cited earlier observations in Europe, notably in Swiss lakes (see Walker, 1973). Daly (1936) was probably the first geologist to focus attention on the potential for such currents, although he used the term *density current* to describe the process. His work was based on observations, principally in lakes, of the behavior of turbid flood waters entering deep-water bodies; it was supported by the contemporary and later work of civil engineers and engineering geologists on the transport of fine sediments in artificial lakes produced by the damming of rivers, the major example being Lake Mead behind Hoover Dam (Grover and Howard, 1938). Later, detailed studies by the U.S. Geological Survey of the deposits and processes operating in this large lake contributed an important body of data on the ability of density and turbidity currents to move fine sediment over distances of many km (Gould, 1953).

Although this earlier work is not often cited (see Walker, 1973), and the major impetus to study turbidity-current processes and products came primarily from the classic paper by Kuenen and Migliorini (1950), it is interesting to note that Kuenen's experimental studies were inspired by Daly's paper. Kuenen's extensive contributions (see references) to the literature on the characteristics of turbidity-current processes and on the resulting deposits (see *Turbidites*) are the basis for much of the major emphasis given them by sedimentologists, marine geologists, stratigraphers, and engineers since the 1950 paper.

Most definitions would restrict turbidity currents to the marine and lacustrine environments, but currents consisting of turbidly-suspended material may be generated in dry-snow avalanches (q.v.), base surges (q.v.), and floods (q.v.). Only flows that are gravity induced can be strictly classified as turbidity currents, a definition which excludes wave- and storm-generated turbid suspensions.

Although no generally accepted upper limit density has been agreed upon, fluid-sediment mixtures with densities of >1.1 gm/cm^3 have excess density sufficient to move the turbid mass to all depths of the ocean and contain 3% solids by volume, which is an appreciable load. Several workers using sediment-water mixtures in laboratory experiments (e.g., Kuenen, 1950; Natland, 1963) have used denser mixtures, with 10-20% solids by volume and densities of 1.3 gm/cm^3 and higher.

Related Processes

Recent studies have shown that the turbidity current is part of a spectrum of sediment-moving processes that range from dominantly fluid impact transport, as in rivers, to mass movements such as slides and slumps (see *Gravity Flows; Gravity-Slide Deposits*). The unifying characteristic is the changing proportion of fluid to solid: as the proportion of solids increases, the driving force changes from the gravity-induced flow of a fluid to the gravity-induced motion of the solids. Hampton (1972) suggested that turbidity flow evolves from slumping processes and formalized the ideas of many earlier workers who noted the need for a process to allow the assemblage of diverse particles, which could then be released as a turbid, high-density mass that would rapidly be fluidized and then flow as a true turbidity current. Field work has shown the striking association of the resulting turbidites (q.v.) with such features as submarine canyons, slump scars, and slope gullies. Several mechanisms for the generation of the mass release have been offered, including discharge of heavily sediment-laden river water into a lake or ocean, and slumps from slope or canyon, triggered by storm-wave pressures, tsunamis (q.v.), or earthquake shocks. Modern observations of the character of sedimentary deposits in submarine canyons has shown that the canyon is usually the scene of mass movement, either as grain flows (q.v.) or slumps. Heezen et al. (1964) noted the association of cable breaks on the sea floor near the mouths of canyons with times of exceptional river discharge. Conversely, the submarine fan (q.v.) typically contains deposits of the turbidite type in which the grading, lamination, cross-lamination, and sorting indicate a well-mixed, turbulent-flow process.

Turbidity Flow

Actual observation of turbidity currents has been limited to lakes (Gould, 1953; Lambert et al., 1976) and laboratory experiments (Kuenen, 1950; Natland, 1963; Middleton, 1966, 1967); this limitation is a matter of our time and place. It is evident that the process occurs infrequently on a human time scale because it requires the accumulation of large volumes of material in areas adjacent to relatively steep slopes, or of exceptional river discharge. Of equal effect, the present time of high sea level and broad shelves is one of few turbidity currents because most sediment is trapped on the inner shelf, in estuaries and in deltas. Most submarine canyons appear to be relatively inactive. Thus, lakes are probably the best locale for field observation of the process, and recent work with current meters and turbidity meters emplaced in active lake-floor channels has recorded some interesting data (Lambert et al., 1976).

Most of our impressions of the form of the flow come from experimental work. These experiments show that once a well-mixed sediment-water slurry is allowed to move down a slope the flow can be visualized as three zones which probably merge imperceptibly from one to the next. These zones have been defined by Middleton and Hampton (1973) as the *head, body,* and *tail.* The head is usually thicker than the rest of the flow; a zone of nearly uniform thickness (the body) follows the head, and the tail is a zone of thin and dilute flow. Mixing, perhaps primarily from the upper part of the head, forms an entrained layer above the main turbidity flow; this dilute layer is dragged along by the lower unit. It is suggested that the entrained layer may continue to flow after the main current has passed by and may rework the newly deposited fine upper portions of the turbidite layer left by the passing current.

The head of the flow moves more slowly than the body due to loss of volume by eddies at its top, which form the entrained layer (Middleton and Hampton, 1973). The head is recharged with dense material while releasing fines to the entrained layer and back to the body and tail, thus concentrating the coarser sediment in the head. The head may also be the main scouring agent, producing the flutes and grooves typical of bottoms of turbidites and which are buried by deposition from the following body (Middleton and Hampton, 1973). Allen (1971) has noted that the head overhangs the bed due to frictional resistance to the flow at its base; by this means, water may be incorporated into the flow by overriding and entrainment. This head is a zone of highly turbulent flow, in any event, and this turbulence is the main force supporting the moving sediment suspension.

Once a turbidity flow has been started, the gradient in density maintained by the turbulent flow is probably the main force maintaining motion. Bagnold (1962) defined this process as *autosuspension,* where the density is maintained by turbulent motion of the fluid, which is generated by its flow, which in turn is the product of the density gradient produced by the suspended sediments. The main loss in momentum is caused by frictional interaction with the bottom, and if this loss is compensated by a sufficient gravity gradient (slope), the flow will continue indefinitely. This "feedback loop" (Middleton and Hampton, 1973) explains the long distances over which the flows must move to lay down the turbidites that are a common feature of deep and nearly flat abyssal plains. Deposition of the contained sediment also reduces the density in the suspension and slows the process, but the effect is probably sufficient to maintain the flows over an essentially flat surface for appreciable distances (1000 km on some abyssal plains). The relatively low frequency of the deposition of these beds on basin floors and abyssal plains also suggests that only the largest turbidity currents reach these distances.

In a model of "rebounding" turbidity currents within ponds on the Mid-Atlantic Ridge, Van Andel and Komar (1969) called for a hydraulic jump at the base of the initial slope which would convert what was essentially a slump or slide into a thick, low-density turbidity current and would provide the energy for dispersal of the material. Komar (1971) examined the possible effect of hydraulic jumps in flows emerging from submarine canyons onto basin floors and concluded that a high-density slump would not thereby be converted to a low-density turbidity current.

Deposition

Rates of turbidity-current deposition are probably highly variable and depend on the size and speed of the current, the amount and size distribution of particles in suspension, whether the flow is channelized or moving as sheet flow, and the rate of supply. It would seem likely that the same criteria also determine whether the current will erode or deposit sediment. There is strong evidence for erosion in the submarine canyons (q.v.) and upper submarine fans (q.v.; Nelson and Kulm, 1973), and equally compelling evidence for deposition with little or no erosion in the central and lower fan and basin floor. Preservation of trails, fine sediment layers, and microrelief under some turbidites suggests an absence of erosion in channels or surfaces of low gradient. Incision of channels, truncation of older deposits, and adjustments of tectonically warped channels suggest erosive power is high when slope is increased.

The lower portion of turbidity currents may approach grain flow (q.v.) transport, suggested by such phenomena as reverse grading, massive deposition of coarse material, and poor sorting in the basal portions of turbidites. Field evidence shows these features to be common in areas where a large and relatively coarse supply of sediment is available (Chipping, 1972).

Submarine fan (q.v.) morphology provides boundary dimensional data for turbidity currents from such evidence as channel form and associated natural levee dimensions, channel gradient, range of thickness and grain-size distribution in the turbidites (q.v.), amount of matrix, and the dimensions of bed structures produced by dewatering of the low bulk density deposits immediately following rapid deposition (see *Water-Escape Structures*). In rare euxinic basins, the varved sediments also provide a means of determining the frequency of flows, and with close coring intervals the approximate volumes of individual flows can be determined. These data provide a means for marine geologists to test and use the theoretical equations developed by experimentalists.

Turbidity currents are most effective where the relief is high; their occurrence in the marine environment thus implies an initial slope and deep ocean-floor depths. A relatively shallow basin could provide sufficient slope for the generation of these currents, as in lakes, provided that the margins are relatively steep close to shore; Pleistocene low stands of sea level near the shelf edge were, therefore, more conducive to turbidity-current sedimentation than the present.

DONN S. GORSLINE
JOANNE BOURGEOIS

References

Allen, J. R. L., 1971. Mixing at turbidity current heads and its geological implications, *J. Sed. Petrology*, **41**, 97-113.
Bagnold, R. A., 1962. Autosuspension of transported sediments: Turbidity currents, *Proc. Roy. Soc. London* ser. A, **265**, 315-319.
Chipping, D. H., 1972. Sedimentary structure and environment of some thick sandstone beds of turbidite type, *J. Sed. Petrology*, **42**, 587-595.
Daly, R. A., 1936. Origin of submarine canyons, *Am. J. Sci.*, ser. 5, **31**, 401-420.
Gould, H. R., 1953. Lake Mead sedimentation, Ph.D. Thesis, Univ. Southern California, 333p.
Grover, N. C., and Howard, C. S., 1938. The passage of turbid water through Lake Mead, *Trans. Am. Soc. Civ. Eng.*, **103**, 720-790.
Hampton, M. A., 1972. The role of subaqueous debris flow in generating turbidity currents, *J. Sed. Petrology*, **42**, 775-793.
Heezen, B. C.; Menzies, R. J.; Schneider, E. D.; Ewing, M.; and Granelli, N. C. L., 1964. Congo Submarine Canyon, *Bull. Am. Assoc. Petrol. Geologists*, **48**, 1126-1149.
Hurley, R. J., 1964. Analysis of flow in Cascadia deep-sea channel, in R. L. Miller, ed., *Papers in Marine Geology*. New York: Macmillan, 117-132.
Johnson, D., 1939a. The origin of submarine canyons (part 6), *J. Geomorph.*, **2**, 213-236.
Johnson, D., 1939b. *The Origin of Submarine Canyons*. New York: Columbia Univ. Press, 126p.
Komar, P. D., 1969. The channelized flow of turbidity currents with application to Monterey deep-sea fan channel, *J. Geophys. Research*, **74**, 4544-4558.
Komar, P. D., 1970. The competence of turbidity current flow, *Geol. Soc. Am. Bull.*, **81**, 1555-1562.
Komar, P. D., 1971. Hydraulic jumps in turbidity currents, *Geol. Soc. Am. Bull.*, **82**, 1477-1488.
Komar, P. D., 1972. Significance of head and body spill from a channelized turbidity current, *Geol. Soc. Am. Bull.*, **83**, 1151-1156.
Kuenen, P. H., 1950. Turbidity currents of high density, *18th Internat. Geol. Cong., London*, **8**, 44-52.
Kuenen, P. H., 1951. Properties of turbidity currents of high density, *Soc. Econ. Paleont. Mineral. Spec. Publ.*, **2**, 14-33.
Kuenen, P. H., 1952. Estimated size of the Grand Banks turbidity current, *Am. J. Sci.*, **250**, 874-884.
Kuenen, P. H., 1966. Matrix of turbidites: Experimental approach, *Sedimentology*, **7**, 267-297.
Kuenen, P. H., 1967. Emplacement of flysch-type sand beds, *Sedimentology*, **9**, 203-243.
Kuenen, P. H., and Migliorini, C. I., 1950. Turbidity currents as a cause of graded bedding, *J. Geol.*, **58**, 91-127.
Lambert, A. M.; Kelts, K. R.; and Marshall, N. F., 1976. Measurements of density underflows from Walensee, Switzerland, *Sedimentology*, **23**, 87-105.

Middleton, G. V., 1966. Experiments on density and turbidity currents, I: Motion of the head, *Canadian J. Earth Sci.*, **3**, 523-546.

Middleton, G. V., 1967. Experiments on density and turbidity currents, III: Deposition of sediments, *Canadian J. Earth Sci.*, **4**, 475-505.

Middleton, G. V., and Hampton, M. A., 1973. Mechanics of flow and deposition, in G. V. Middleton and A. H. Bouma, eds., *Turbidites and Deep Water Sedimentation*. Anaheim: SEPM Pacific Sec. Short Course Notes, 1-38.

Natland, M. L., 1963. Paleoecology and turbidites, *J. Paleont.*, **37**, 225-230.

Nelson, C. H., and Kulm, L. D., 1973. Submarine fans and channels, in G. V. Middleton and A. H. Bouma, eds., *Turbidites and Deep Water Sedimentation*. Anaheim: SEPM Pacific Sec. Short Course Notes, 39-78.

Normark, W. R., and Dickson, F. H., 1976. Man-made turbidity currents in Lake Superior, *Sedimentology*, **23**, 815-831.

Romanovskiy, S. I., 1973. Role of turbidity current in sedimentary processes, *Litol. Poleznyye Iskop.*, **3**, 29-37.

Van Andel, T. H., and Komar, P. D., 1969. Ponded sediments of the Mid-Atlantic Ridge between 22° and 23° North latitude, *Geol. Soc. Am. Bull.*, **80**, 1163-1190.

Walker, R. G., 1973. Mopping up the turbidite mess, in R. N. Ginsburg, ed., *Evolving Concepts in Sedimentology*. Baltimore: Johns Hopkins Univ. Press, 1-37.

Cross-references: *Abyssal Sedimentation; Base-Surge Deposits; Bouma Sequence; Continental-Rise Sediments; Contourites; Current Marks; Fluidized Sediment Flows; Graded Bedding; Grain Flows; Gravity Flows; Mass-Wasting Deposits; Nepheloid Sediments and Nephelometry; Redeposition, Resedimentation; Submarine (Bathyal) Slope Sedimentation; Submarine Canyons and Fan Valleys, Ancient; Submarine Fan (Cone) Sedimentation; Tsunami Sedimentation; Turbidites.*

UDDEN SCALE

In 1898, J. A. Udden introduced to geologists the concept of the geometrically graded size scale for fragmental materials. The original scale, which appears in a long footnote (Udden, 1898), was based on a maximum diameter of 16 mm, with grade limits decreasing by a factor of ½ (Table 1). Udden acknowledged that the grade scale was adapted from one devised by soil scientist Milton Whitney. Whitney may also have been responsible for Udden's appreciation of the importance of the mode (medium grade, chief ingredient, or maximum in Udden's usage) and the "coarse and fine admixtures."

By 1914, Udden had greatly expanded his scale (Table 1), run 371 sieve analyses of sands and gravels, and formulated some general rules relating to grain-size distributions and their significance. A number of other classifications were proposed between 1898 and 1914, and one had been proposed as early as 1875 (see Wentworth, 1922), but all lacked the essential geometric regularity of the Udden Scale universally adopted by geologists in later years (see *Grain-Size Studies*).

ROBERT E. CARVER

References

Udden, J. A., 1898. The mechanical composition of wind deposits, *Augustana Libr. Publ. 1*, 5-69.

Udden, J. A., 1914. Mechanical composition of clastic sediments, *Geol. Soc. Am. Bull.*, **25**, 655-744.

Wentworth, C. K., 1922. A scale of grade and class terms for clastic sediments, *J. Geol.*, **30**, 377-392.

Cross-references: *Alling Scale; Atterberg Scale; Grain-Size Studies; Granulometric Analysis; Phi Scale; Sediment Parameters; Wentworth Scale.*

TABLE 1. The Udden Grade Scale, 1898 and 1914

Grade Limits (mm)	Grade Term (1898)	Grade Term (1914)
256		
128		large boulders
64		medium boulders
32		small boulders
16		very small boulders
8		very coarse gravel
4	coarse gravel	coarse gravel
2	gravel	medium gravel
1	fine gravel	fine gravel
1/2	coarse sand	coarse sand
1/4	medium sand	medium sand
1/8	fine sand	fine sand
1/16	very fine sand	very fine sand
1/32	coarse dust	coarse silt or dust
1/64	medium dust	medium silt or dust
1/128	fine dust	fine silt or dust
1/256	very fine dust	very fine silt or dust
1/512		coarse clay
1/1024		medium clay
1/2048		fine clay

V

VARVES AND VARVED CLAYS

A *varve* (Swedish: *varv*, layer) is any sedimentary bed or lamination deposited within the period of one year, or any pair of contrasting laminae representing seasonal sedimentation (as summer and winter) within the period of 1 yr. In distinction to varves (*sensu stricto*), alternating beds and laminae that do not represent a 1-yr period are known as *rhythmites* (see *Cyclic Sedimentation*). Besides clastic sedimentary varves, there are chemical varves (precipitates and evaporites such as carbonate concretions, stalagmites, some salt and gypsum beds) and other varves, such as tree rings, zoological growth beds, and layers in ice cores.

Varved clay is any pair of contrasting laminae representing seasonal sedimentation within one year, such as coarser and lighter summer units and finer and darker winter units (Fig. 1). Besides varved clay proper, there are varved silt, varved sand, and perhaps even varved gravel. The deposition of varved sediments is either caused by seasonal variations in the glacial meltwater discharge (glacial varves) or by seasonal variations in the river discharge (postglacial varves). The freezing over of the basin during winters, creating calm conditions that allow suspended particles to settle, is another important factor in the formation of varves. Varved clay is a common Quaternary deposit in glaciated regions, but is also found associated with earlier glaciations, such as the Carboniferous, Ordovician, and Precambrian glaciations.

The study of varved clay goes back to about 1880, when De Geer (Fig. 1), working in the Stockholm area in Sweden, "was struck by the marked cyclical banding of the varved clay or so-called *hvarfig lera* (old spelling), so denominated from its alternating, tiny layers of fine sand and clay. From the obvious similarity with the regular, annual rings of the trees I got at once the impression that both ought to be annual deposits" (De Geer, 1940, p. 13). The first varve measurements were done by De Geer in 1884, and the first correlation between varve diagrams from different localities dates from 1904 (see De Geer, 1940).

FIGURE 1. Varved clay: a sequence of normal distal varves with light summer units of sandy silt and dark winter units of clay being measured by De Geer in 1920 (as in 1891), Essex Junction, USA.

Glacial Varves

Glacial varves (clay, silt, and sand) are formed as a function of the seasonal variations in meltwater outflow from continental or alpine glaciers and the freezing over of the water body during winters. Where a meltwater stream debouches into a body of water, stratified drift is deposited, grading from coarse glaciofluvial material in the proximal zone to clay in the very

841

distal zone. These zones migrate with the recession and advance of the ice margin and with the seasonal environmental changes, e.g., velocity, material supply, freezing over. The summer units of varves deposited in the vicinity of a meltwater outflow are, therefore, thick and coarse grained; the first summer unit may be several m thick. The varve thickness decreases parabolically in distal directions. According to De Geer's classification, one may speak of *proximal varves*, thick varves formed close to the ice margin; *distal varves*, thickness measured in cm; *microdistal varves* or *microvarves*, thickness measured in mm; and *ultradistal varves*, too thin to be measured.

According to Sauramo (1923), there are four characteristics of a varve: the grain size, the color, the plasticity, and the chemical composition. The dominating rhythmic arrangement of a varve is alternation between coarser and lighter summer units and finer and darker winter units (Fig. 1). These granulometrically graded varves are termed *diatactic* varves. There are also nongraded varves, with the seasonal rhythm expressed only in the color changes (darker winters units). These varves are known as *symmict* varves and are common in brackish water environment.

In fresh water, free from electrolytes, deposition results from the settling of each particle separately, the rate of sinking depending upon the size, form and density of the particle and all factors influencing the rate of settling. In salt or brackish water, on the other hand, suspended particles group together into larger aggregates which settle more rapidly (see *Flocculation*). The salinity of the water also influences the vertical spreading of the meltwater, which rises to the surface in salt water and sinks to the bottom in fresh water (Kuenen, 1951). Glacial clays deposited in fresh water are, therefore, generally diatactic, and they are symmict or unvarved in brackish and marine environments. Despite this general rule, marine organisms such as *Portlandia arctica*, are found in *varved clay* in, for example, Sweden.

Postglacial Varves. In northern Sweden more than 8500 postglacial varves have been measured (Liden, 1938). These varves are formed by seasonal variations in river discharge, the heavy meltwater discharge during springs creating coarser and thicker spring units. The seasonal rainfall variations may create similar varving, especially in regions with large contrast between dry and wet seasons. Postglacial varved sediments have been reported from various deltas. Varves have also been found in postglacial lacustrine gyttja beds, e.g., Faulenseemoos in Switzerland and in Lake of the Clouds in Minnesota, USA.

Drainage Varves. In glacial varved sequences, there often occur thick varves called drainage varves because they can usually be directly correlated with the drainage of a glacially ponded lake. The term *secondary drainage varve* has been applied to thick varves within the basin that drained, in contrast to primary drainage varves formed within the basin into which the ice lake drained. The drainage varves are important markers in varve chronology, e.g., De Geer's Zero-varve and varve-1073, marking the drainage of the Baltic Ice Lake.

Glacial Varve Chronology. In deglaciated areas, the varves are arranged as tiles, where the end of each varve marks the position of the ice margin during a winter halt. As demonstrated by De Geer (1940), these winter halts often correspond with terminal moraines and esker centers. Lines combining varve localities with the same bottom varve are termed *equicesses*.

In the early 20th century, De Geer started to count and correlate varve sequences in Sweden, thus reconstructing the recession of the last ice sheet in detail and establishing an absolute time scale, known as the Swedish Time Scale or the Swedish Varve Chronology (De Geer, 1912, 1940); De Geer subdivided his time scale into four periods. Various opinions have been expressed about the extension and reliability of the Swedish Time Scale. Today, we may conclude that it is reliably brought down to the Stockholm region, or 10,000 B.P., with a recent extension to 10,500 B.P.; it may possibly be extended down to Lake Bolmen or 11,800 B.P., with uncertain portions, but it is not yet reliably extended down to southernmost Sweden, or 13,000 B.P., as sometimes stated.

In Finland, Sauramo (1923) built up a similar Finnish varve chronology. It is not fixed with respect to the present, however, so that absolute dates have to be inferred via the Swedish Time Scale, and this correlation has been very much debated.

In eastern North America, Antevs (1928) built up a varve chronology based on 4000 varves in Connecticut Valley and 1500 varves in Ontario-Quebec. This chronology is not related to the present, however, and includes a gap between the two chronological sequences.

Minor varve chronological sequences have been established in other parts of northern Europe and North America, in New Zealand, in Patagonia (South America), in Kenya (Africa), and in the Himalayas. "Floating" varve sequences have become useful in combination with other dating methods such as radiocarbon.

Besides direct dating, varve chronology has become important for the establishment of seasonal and annual changes (chemical, climatic, paleomagnetic), various cycles, climatic fluctuation, and deviation in other dating methods (e.g., radiocarbon, dendrochronology).

Nonglacial Varves

Varved sediments can also be formed in lacustrine and marine environments due to nonglacial seasonal changes in biological productivity (e.g., mass blooming of diatoms); in chemical-physical environment (temperature, precipitation, pH/Eh, evaporation); and in vertical and horizontal water motions. Lacustrine nonglacial varves have been described by many authors (see Bradley, 1937; *Lacustrine Sedimentation*). Similar but lithified varves have been described from different geological periods, e.g., the Eocene Green River beds (Bradley, 1929) and the Jurassic Todilto Formation (Anderson and Kirkland, 1960). Marine nonglacial varves have been described from, e.g., the Saanich Inlet, British Columbia (Gross et al., 1953), where seasonal variations are caused by mass diatom blooms in late spring/early summer (see also *Sapropel*).

NILS-AXEL MÖRNER

References

Anderson, R. Y., and Kirkland, D. W., 1960. Origin, varves and cycles of Jurassic Todilto formation, New Mexico, *Bull. Am. Assoc. Petrol. Geologists*, 44, 37-52.
Antevs, E., 1928. The last glaciation, *Am. Geogr. Soc. Research Ser.*, vol 17, 292p.
Ashley, G. M., 1975. Rhythmic sedimentation in Glacial Lake Hitchcock, Massachusetts-Connecticut, *Soc. Econ. Paleont. Mineral. Spec. Publ. 23*, 304-320.
Bradley, W. H., 1929. The varves and climate of the Green River epoch, *U.S. Geol. Surv. Prof. Pap. 159-E*, 87-110.
Bradley, W. H., 1937. Non-glacial varves with selected bibliography, *Natl. Res. Council, Ann. Rept., App. A., Comm. Geologic Time*, 32-42.
De Geer, G., 1912. A geochronology of the last 12,000 years, *C.R. 9th Congr. Geol. Intern., Stockholm, 1910*, 1, 241-258.
De Geer, G., 1940. Geochronologia Suecica Principles, *Kungl. Sven. Vetenskapsakad. Handl.*, ser. 3, 18, 367p.
Gross, M. G.; Gucluer, S. M.; Creager, J. S.; and Dawson, W. A., 1963. Varved marine sediments in a stagnant fjord, *Science*, 141, 918-919.
Jopling, A. V., and McDonald, B. C., eds., 1975. Glaciofluvial and glaciolacustrine sedimentation, *Soc. Econ. Paleont. Mineral. Spec. Publ. 23*, 320p.
Kuenen, P. K., 1951. Mechanics of varve formation and the action of turbidity currents, *Geol. Fören. Förhandl.*, 73, 69-84.
Liden, R., 1938. Den senkvartära strandförskjutningens förlopp och kronologi i Angermanland, *Geol. Fören. Förhandl.*, 60, 397-404.
Mörner, N.-A., 1973. Postglacial–A term with three meanings, *J. Glaciol.*, 12(64), 139-140.
Mörner, N.-A., ed., 1978. *Varve Chronology*. Stroudsburg, Pa.: Dowden, Hutchinson & Ross, in press.
Rayner, D. H., 1963. The Achanarras limestone of the Middle Old Red Sandstone, Caithness, Scotland, *Proc. Yorkshire Geol. Soc.*, 34, 117-138.
Richter-Bernburg, G., 1957. Isochrone Warven im Anhydrit des Zechstein, 2, *Geol. Jahrb.*, 74, 601-610.
Sauramo, M., 1923. Studies of the Quaternary varve sediments in southern Finland, *Bull. Comm. Geol. Finlande*, 60, 164p.
Stewart, F. H., 1963. The petrology of the Permian Lower Evaporites of Fordon in Yorkshire, *Proc. Yorkshire, Geol. Soc.*, 34, 1-44.

Cross-references: *Cyclic Sedimentation; Evaporite Facies; Glacigene Sediments; Lacustrine Sedimentation; Oil Shale; Sapropel.*

VERTICAL SEDIMENTATION
—See LATERAL AND VERTICAL ACCRETION

VOLCANIC ASH DEPOSITS

Volcanic ash is the fine-grained product of an explosive volcanic eruption in which fragmental rock debris (*pyroclastic debris*) is expelled from a vent or fissure with considerable energy. The effluent material either sweeps downslope as an *ash flow* or *pyroclastic* flow (a gas-rich, avalanche-like body of molten or glassy ash, with high mobility) or rises into the air as an *ash cloud* or *column* (a billowing mixture of gas and finely comminuted, partially or completely solidified, particulate matter).

An ash flow, under the influence of gravity, may spread over a vast region. Many flows retain their heat long enough to effect annealing and flattening of the constituent glassy particles. The flow's product is a compact, consolidated mass known as *welded tuff* or *ignimbrite* (q.v.; see also *Nuée Ardente* for references).

An ash cloud, on the other hand, is subject to transport by wind currents in the atmosphere, and its constituent particles cool before settling to the surface under the influence of gravity. An extensive deposit of loose ash may accumulate by eolian sedimentation from the passage of a dense ash plume, and it is this type of deposit that is of interest here. Lithified ash deposits are called *tuff* (q.v.). The great bulk of the ash accumulates close to the vent, but very fine particles are carried farther. Some have been known to stay aloft for months, or even years, encircling the globe following gigantic eruptions, but these do not form appreciable deposits.

Geologists have long been interested in the

study of volcanic ash-fall deposits. They form useful marker units for basin-wide stratigraphic correlation; they provide minerals for radioactive age dating of sedimentary sections; and their composition reveals information about the processes of magmatic evolution (Smith and Bailey, 1966).

The term *ash,* as used here, is restricted to pyroclastic fragments < 4 mm in diameter. Coarser fragments, known as *lapilli, blocks,* and *bombs,* are sometimes found in ash-fall deposits close to their source (see *Volcaniclastic Sediments and Rocks* for description and classification).

Composition

The constituent particles of a volcanic ash-fall deposit may consist of glass (*vitric ash*), rock (*lithic ash*) or crystals (*crystal ash*). Ash containing a mixture of these materials is referred to as *composite ash.* If the lithic fragments are derived from the walls of the vent from which the explosion cloud issues, rather than from liquid lava, they are described as *accessory.* Fragments torn from the volcano's foundation rocks and caught up in the outrush of material are known as *accidental* debris. They may or may not be of volcanic origin.

The range of chemical composition of ash-producing lavas is similar to that of lava *flows,* but the relative frequency of occurrence of ash of a particular composition differs from that of flows of the same composition. Fluid basalt, the most abundant of the flow-forming lavas, yields ash much less often then do lavas of rhyolitic or dacitic composition.

Origin

Most volcanic ash particles are produced by the fragmentation of molten or solidified lava in the throat of the volcano. Fragmentation of *liquid* lava takes place as the result of nucleation, growth, and coalescence of vapor bubbles evolving from the lava. It generally gives rise to vitric ash. As vapor bubbles coalesce and burst, the cooling lava is shattered into splinters or *shards* of glass, with concave edges. Sometimes these glass fragments contain large numbers of bubbles that have failed to burst. The resulting material is known as *pumice.*

Verhoogen reviewed the mechanics of vitric ash formation and summarized the process as follows:

> If a few bubbles form and rise swiftly by buoyancy to the surface, the lava will "boil" as it does in lava lakes. If, on the contrary, a large number of bubbles expand more rapidly than they rise, a time may come when neighboring bubbles begin to coalesce. When such coalescence becomes general the lava will be fragmented into small shreds of liquid or glass between adjacent bubbles, and may lose cohesion. If, in addition, the [vapor] pressure in the bubbles is still large at this time, the fragments will be blown apart, producing an explosion the intensity of which will depend essentially on the magnitude of this residual pressure. If it is small, we might observe a mild motion of a slowly expanding [ash] cloud . . . if the pressure is large a violent outburst . . . might occur. (Verhoogen, 1951, p. 730)

Sheridan (1973) has suggested that most of the primary and secondary nonballistic features of observed pyroclastic eruptions can be explained by fluidization in both the vent and the eruptive column. Lava that has begun to crystallize before a volcanic explosion produces ash containing crystals as well as glass fragments. The crystals may have glass adhering to them, and some may be partly or wholly mantled with glass. The adherence of the glass is generally regarded as identifying primary crystals in the deposit, a point of some importance insofar as radiometric dating is concerned.

Explosive fragmentation of *solidified* lava results from the accumulation of gas evolving from molten lava at depth or from steam generated by contact between ground water and molten, or very hot, lava. Explosions of the last type are termed *phreatic* explosions. Fragmentation occurs when the difference between atmospheric or confining pressure and the pressure of the accumulating gases exceeds the strength of the enclosing rocks. Ash produced in this manner is generally lithic, although ultracomminution may result in the generation of crystal debris.

Molten lava flowing into the sea sometimes shatters explosively as a result of rapid quenching (see *Volcanism–Submarine Products*). Accumulations of pyroclastic debris originating in this manner are known as *littoral cones.*

Intensity of Ash-Producing Explosions. The intensity of a volcanic explosion is best judged from its energy, which, in turn, may be estimated from the summed kinetic energy of ejected fragments or from the seismic disturbance created by the explosion. The explosive eruption of Asama volcano, Japan, on June 7, 1938, had an energy of 1.7×10^{20} ergs, equal, in terms of violence, to the detonation of a 10 kiloton-equivalent nuclear device (Minakami, 1950). On March 29, 1946, Hekla volcano in Iceland began an eruptive episode of notable proportions with an explosion of magnitude 4×10^{22} ergs; and on March 30, 1956, Bezymianny volcano on the Kamchatkan Peninsula exploded with an energy of 4×10^{23} ergs (Gorshkov, 1959). Vast quantities of ash were ejected during two of these eruptions, and the plumes of ash that were carried away by the wind blanketed regions to the lee of the volcano with a choking cover of ash.

The well-known eruptions of Mount Katmai, Alaska, in 1912, and of the Indonesian volcanoes of Tamboro in 1915, and Krakatau in 1883, were more violent yet. Ash from these eruptions encircled the globe, and ash accumulated in the region near the volcanoes in deposits many m thick. Ash from Tamboro and Katmai was reported to be 60 cm thick at distances as great as 150 km.

Transportation and Deposition of Ash

The horizontal distance that ash is transported in the atmosphere and, therefore, the extent of the surface area over which it is deposited, depends on four factors: (1) the maximum height of the ash column; (2) the direction and velocity of the wind from the surface of the ground to the top of the ash column; (3) the distribution of ash particles in the column and the settling velocities of these particles; and (4) the presence or absence of precipitation in the path of the plume.

The height to which a column of ash will rise depends on its exit velocity and temperature. In general, phreatic explosions produce cooler debris clouds than those of liquid, gas-rich lava. They rise more slowly into the atmosphere and generally cease rising at lower elevations (Fig. 1). As an ash column rises, it cools by adiabatic expansion and entrainment of air from the atmosphere. The incorporation of this air by turbulent mixing results in radial growth of the column with increasing elevation. The buoyant component of rise continues to act until the temperature of the cloud approximately equals ambient air temperature. Heights of rise of 5-15 km above the vent are not uncommon. The ash-laden cloud from Krakatau in 1883 rose nearly 32 km.

Most of the particulate matter starts to fall back to the surface from an ash cloud as soon as the cloud stops rising. In eruptions where a stream of gas and ash issues from the vent continuously, ash falls with steady regularity from the plume that trails downwind. The trajectories assumed by the falling particles are influenced by the wind fields through which they fall. If winds at various levels in the atmosphere are blowing in different directions, the path of a given particle may be quite irregular.

The rate at which the particles fall depends on their size, shape, and density. Large, smooth, dense particles fall relatively rapidly, whereas those of irregular shape, low density, and small size, fall slowly (see *Settling Velocity*). Particles of disparate size or density may occur together in a deposit because they have equivalent terminal fall velocities. Turbulence in the rising eruption cloud tends to keep particles of a wide variety of sizes, shapes, and densities mixed. As a result, unlike particles may start to fall from the same point in the cloud or plume, but they will fall at different rates and land at widely separated points on the ground. Particles with relatively high settling velocities will land close to the vent and those with lower settling velocities, farther away. This sometimes produces a regular gradation in particle size, shape, and density away from the vent toward the distal end of the deposit. A similar gradation is often observed from the base of the deposit upward.

Physical Characteristics

Most deposits of volcanic ash are elongate in plan in the region near their source, the longitudinal axis of a deposit trending away from its parent vent in the direction of the vector resultant wind (Eaton, 1963). If the winds above the volcano are similar in direction at all elevations, the resulting deposit will be narrow and relatively long. If the wind directions vary with altitude, or with time, the configuration of the deposit will be broad or irregular. Some deposits accumulating under these circumstances display several depositional lobes. Beyond the area of thickest accumulation there may be a thin blanket of very fine-grained ash covering tens of thousands of km². The areal distribution and thickness of this blanket are dependent on winds over the entire region of accumulation and, in continental mountainous areas, may be quite irregular. In addition to such a blanket, the extended region beyond it may experience

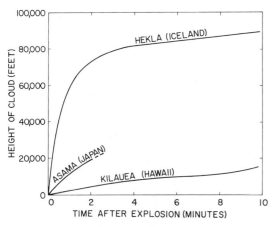

FIGURE 1. Curves comparing rates of rise of three volcanic explosion clouds. The explosion at Kilauea was phreatic, and the rate of rise and maximum elevation of the relatively cool debris cloud were less than those of the Asama and Hekla clouds. (The uppermost part of the Hekla cloud is believed to have been ash-free water vapor.)

additional accumulation of ash from the same eruption, but as particles finely dispersed throughout a section of sediments from other sources. If one includes this in the total area of accumulation, it may be as large as 2–4 million km^2 (Huang et al., 1973).

Several of the physical characteristics of a typical *near-source* ash-fall deposit are illustrated by an accumulation of pumiceous ash near Crater Lake National Park, Oregon. The culminating eruption of Mount Mazama, the collapsed top of which is occupied by Crater Lake, resulted in the outpouring of vast quantities of fragmental pumice (Williams, 1942). This pumice accumulated in a bilobed deposit which is elongated both to the NNE and to the SE (Fig. 2, left).

There is a systematic decrease in thickness away from the source along both depositional lobes. Mathematical analysis of longitudinal profiles of many such deposits reveals that they are geometric or exponential in form. The determining factor is primarily the wind structure.

Typical examples of geographic variations in thickness were illustrated by Eaton (1964). For major deposits, >40 km long, the *average* downwind distance to the point where the accumulation is 10^{-2}–10^{-3} m thick is about 300–400 km. Ninkovich et al. (1964) and Lamb (1968) both suggest a limit 1000 km for the maximum downwind distance to which ash can be transported by wind and still accumulate as discrete and observable strata; but observations by G. A. Izett and R. E. Wilcox (pers. comm.) indicate that this limit is conservative. They have studied and identified discrete layers of windborne ash 1250–1750 km from their source. Huang et al. (1973) have developed a method for separating and counting finely dispersed volcanic glass particles in deep-sea sediments sampled by piston coring. They have tracked pulses of ash in this manner as far as 3800 km downwind from their source.

Samples collected from the Mazama deposit were subjected to mechanical analysis by Moore (1937), and the results, as supplemented and analyzed by Fisher (1964), are plotted in Fig. 2 (middle and right). Particle size decreases systematically away from the source, as does the Inman sorting coefficient; these variations are both systematic and continuous.

Ash deposits displaying variations in physical properties of this type sometimes though not

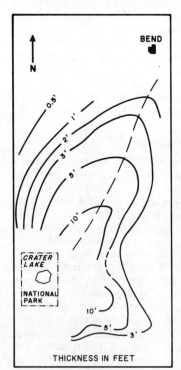 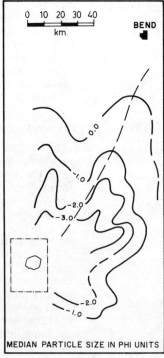 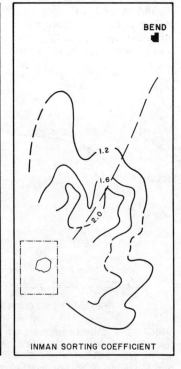

FIGURE 2. Variations in thickness, grain size, and sorting of a pumice deposit near Crater Lake National Park, Oregon (data from Williams 1942, and Fisher, 1964). The interested reader is referred also to Kittleman (1973) for further details of the mineralogy and size-frequency distributions of particles in this deposit.

always reveal systematic variations in composition, as well. Thus, the proportion of heavy to light minerals, or of minerals to glass, may be found to decrease in a downwind direction. On the other hand, Kittleman's (1973) study of Mazama ash in S central Oregon indicates that it is mineralogically homogeneous. Variations are generally regarded as a consequence of gravity sorting, which gives rise to a fractionation of unlike particles and was termed *eolian differentiation* by Larsson (1935).

Many ash deposits reveal *vertical* variations of a similar nature. *Graded bedding* (q.v.), in which coarser, more crystalline debris near the base passes upward into finer, more glass-rich debris toward the top, is characteristic of some ash-fall deposits. Thick deposits of ash frequently display a succession of graded beds, each representing a separate pulse of deposition.

Pumiceous ash that has accumulated in subaqueous environments may display *inverse graded bedding*. Pumice, because of its cellular structure, is very light and, unless saturated, floats in water. The smallest particles of floating pumice become water logged before the larger ones and hence sink to the bottom first. As a result, this type of deposit is usually fine grained at the base and coarse at the top.

The passage of rain through an ash plume may result in the scavenging and premature deposition of ash particles. If the precipitation is restricted locally to one part of the plume, thickness and particle-size variations in the resulting deposit will be irregular. A further result may be the formation of ovoid, accretionary pellets of ash. Some deposits yield large numbers of these *accretionary lapilli*.

Marine Ash Deposits

Many early investigations of marine sediments (see *Pelagic Sedimentation, Pelagic Sediments*) noted that volcanic ash is a common constituent of sediments on the sea floor in some areas. A map showing the distribution of several deposits of ash in Indonesia is shown in Figure 3. Numerous modern oceanographic investigations have employed these ash occurrences in their studies.

Ash falling in the sea may undergo transportation and sorting by oceanic currents in addition to that experienced in the atmosphere; the patterns of areal distribution of the ash are, therefore, sometimes different from those of continental deposits, though not always. Bramlette and Bradley (1942), Neeb (1943),

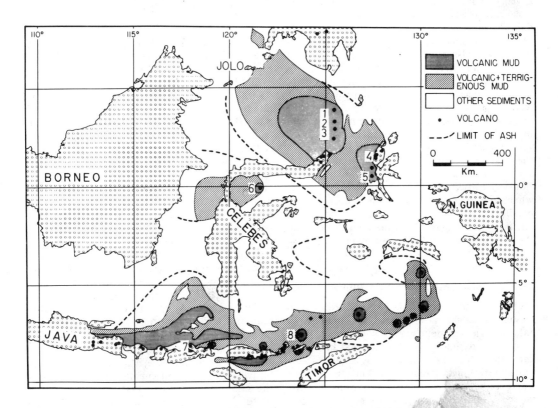

FIGURE 3. Deposits of volcanic ash on the sea floor in Indonesia (after Neeb, 1943).

Ewing et al. (1959), Nayudu (1964), Ninkovich and Heezen (1965), and others have cited examples of deep-sea ash that lack the systematic variations in thickness or grain size which, on land, are often a function of direction and distance from source. In part, this absence may be the result of transport by surface or bottom currents, turbidity currents, or even an effect of bottom topography; but in several instances it is probably because the volcanic source is so distant that effective measurable variations in particle size are absent. The layers at these distances are thin enough that reworking, contamination, and mixing have also affected their apparent thickness. The presence of mineral suites derived from adjacent river drainage basins and the inclusion of shallow-water benthonic foraminifera and plant fragments led Nelson et al. (1968) to conclude that deep-sea Mazama ash had slumped with debris from the continental slope and was redistributed by turbidity currents.

Quasi-systematic variations in particle size analogous to those observed on land have been recorded in marine ash in areas within distances of 800 km or less of the source. Ninkovich et al. (1964), Ryan et al. (1965), and others have cited examples where eolian transport is believed to have been the dominant agent responsible for observed distribution or variation in marine ash. Variations in the accumulation rate of very fine ash dispersed in epiclastic sediments have also been noted to be systematic in a downwind direction (Huang et al., 1973). Transport by marine currents must dominate in other areas, however, particularly where the water column is appreciably thick, for the settling rate of ash is much slower in water than in air. Marine currents will disperse widely the very fine-grained material, but the great bulk of the coarser material will be deposited near the source and display systematic downwind variations in physical properties. In those areas where the prevailing atmospheric currents are parallel to oceanic currents (e.g., the North Pacific), it is impossible to distinguish between the effects imparted to the ash by each of these agents.

Use of Volcanic Ash in Geologic Studies

Volcanic ash deposits play an important role in some geologic investigations, and it is fortunate that ancient deposits are widespread, both in space and time (Fig. 4). Ash that has blanketed a sedimentary basin may constitute a valuable time-stratigraphic marker. In a basin where there are pronounced variations in sedimentary facies, such beds often provide the only means of stratigraphic correlation. Whit-

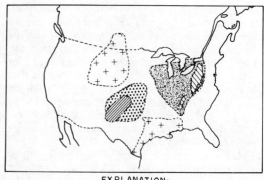

FIGURE 4. Distribution of volcanic ash and bentonite in ancient rocks of the continental US. Each of the areas outlined includes a succession of beds of volcanic ash-fall material (data from Hass, 1948; Flowers, 1952; Ross, 1955; Marsh, 1960; and Reeside and Cobban, 1960).

comb (1932), Rosenkrans (1934), Knechtel and Patterson (1956), Healy et al. (1964), and Slaughter and Earley (1965) have all presented examples of stratigraphic correlation based on discrete beds of ash or bentonite, the principal alteration product of vitric ash (see *Volcanic Ash–Diagenesis*). In addition, because the deposition of ash is essentially contemporaneous with its generation at the volcano, and because it commonly contains one or more minerals amenable to radioactive age dating, it also provides a means of determining the absolute age of an ash-bearing stratigraphic section. Only the primary minerals in the deposit can be used for such a purpose. As noted earlier, these are identified by adhering glass.

According to Wilcox and Izett (1973), reliable correlation of separated parts of a given ash-fall deposit in a region with many such deposits requires compatibility of their stratigraphic, paleontologic, paleomagnetic, and/or radiometric age relations. It also requires the identification of characteristics by which the deposit may be distinguished from the other ash falls. Such characteristics as the habit, internal structures, and chemical compositions of the shards and primary crystals are particularly useful. In their studies of marine ash, Ninkovich and others have used shard color, index of refraction, magnetic properties, chemical composition, and mineralogy for correlation purposes.

The petrologist interested in the composition of magmas and lavas, and particularly in their evolution during the process of crystallization, also gains valuable information from studies of volcanic ash. The shards of glass in such a deposit represent chilled samples of the melt remaining after early crystallization of phenocrysts.

Suggestions that volcanic ash deposits might be used in the construction of planetary circulation maps for the geologic past and for determining continental plate rotations relative to the pole have also been made (Eaton, 1963, 1964). Present-day deposits faithfully reflect tropospheric and lower stratospheric circulation patterns, and preliminary studies of Cretaceous bentonites support the view that the upper-air circulation pattern during the Cretaceous period was essentially like that of the present (Slaughter and Earley, 1965).

Cultural Effects

There are, during the period of deposition following an ash-producing volcanic eruption, a number of effects resulting from the accumulation of ash that are of importance to humans, plants, and animals in the area. These effects have been described in detail by Wilcox (1959).

GORDON P. EATON

References

Bennett, F. D., 1974. On volcanic ash formation, *Am. J. Sc.,* **274,** 648-661.

Bramlette, M. N., and Bradley, W. H., 1942. Geology and biology of North Atlantic deep-sea cores between Newfoundland and Ireland, Part 1, Lithology and geologic interpretation, *U.S. Geol. Surv. Prof. Pap. 196-A,* 1-55.

Eaton, G. P., 1963. Volcanic ash deposits as a guide to atmospheric circulation in the geologic past, *J. Geophys. Research,* **68,** 521-528.

Eaton, G. P., 1964. Windborne volcanic ash: A possible index to polar wandering, *J. Geol.,* **72,** 1-35.

Ewing, W. M.; Heezen, B. C.; and Ericson, D. B., 1959. Significance of the Worzel deep-sea ash, *Natl. Acad. Sci. Proc.,* **45,** 355-361.

Fisher, R. V., 1964. Maximum size, median diameter, and sorting of tephra, *J. Geophys. Research,* **69,** 341-355.

Flowers, R. R., 1952. Lower Middle Devonian metabentonite in West Virginia, *Bull. Am. Assoc. Petrol. Geologists,* **36,** 2036-2038.

Gorshkov, G. S., 1959. Gigantic eruption of Volcano Bezymianny, *Bull. Volcan.,* ser. 2, **20,** 77-109.

Hass, W. H., 1948. Upper Devonian bentonite in Tennessee, *Bull. Am. Assoc. Petrol. Geologists,* **32,** 816-819.

Healy, J.; Vucetich, C. G.; and Pullar, W. A., 1964. Stratigraphy and chronology of late Quaternary volcanic ash in Taupo, Rotorua, and Gisborne districts, *New Zealand Geol. Surv., Bull.,* **73,** 88p.

Huang, T. C.; Watkins, N. D.; Shaw, D. M.; and Kennett, J. P., 1973. Atmospherically transported volcanic dust in South Pacific deep sea sedimentary cores at distances over 3,000 km from the eruptive source, *Earth Planet. Sci. Lett.,* **20,** 119-124.

Kittleman, L. R., 1973. Mineralogy, correlation, and grain size distributions of Mazama tephra and other postglacial pyroclastic layers, Pacific Northwest, *Geol. Soc. Am. Bull.,* **84,** 2957-2980.

Knechtel, M. M., and Patterson, S. H., 1956. Bentonite deposits in marine Cretaceous formations of the Hardin district, Montana and Wyoming, *U.S. Geol. Surv. Bull.,* **1023,** 116p.

Lamb, H. H., 1968. Volcanic dust, melting of ice caps, and sea levels, *Palaeogeogr., Palaeoclimat., Palaeoecol.,* **4,** 219-222.

Larsson, W., 1935. Vulkanische asche vom Ausbruck des chilenischen Vulkans Quizapu (1932) in Argentina gesammelt, *Bull. Geol. Inst. Univ. Uppsala,* **26,** 27-52.

Marsh, O. T., 1960. Geology of the Orchard Peak area, California, *Calif. Dept. Nat. Res., Div. Mines, Spec. Rept. 62,* 42p.

Minakami, T., 1950. On explosive activities of andesite volcanoes and their forerunning phenomena, *Bull. Volcan.,* ser 2, **10,** 59-87.

Moore, B. N., 1937. Non-metallic resources of eastern Oregon, *U.S. Geol. Surv. Bull.,* 875, 180p.

Nayudu, Y. R., 1964. Volcanic ash deposits in the Gulf of Alaska and problems of correlation of deep-sea ash deposits, *Marine Geol.,* **1,** 194-212.

Neeb, G. A., 1943. The composition and distribution of the samples, *The Snellius Expedition,* 5(3), Leyden: Brill, 265p.

Nelson, C. H.; Kulm, L. D.; Carlson, P. R.; and Duncan, J. R., 1968. Mazama ash in the northeastern Pacific, *Science,* **161,** 47-49.

Ninkovich, D., and Heezen, B. C., 1965. Santorini tephra, *Proc. 17th Sympos. Colston Research Soc.,* 413-452.

Ninkovich, D.; Heezen, B. C.; Conolly, J. R.; and Burckle, L. H., 1964. South Sandwich tephra in deep-sea deposits, *Deep-Sea Research,* **11,** 605-619.

Ninkovich, D.; Opdyke, N.; Heezen, B. C.; and Foster, J. H., 1966. Paleomagnetic stratigraphy, rates of deposition and tephrachronology in North Pacific deep-sea sediments, *Earth Planet. Sci. Lett.,* **1,** 476-492.

Reeside, J. B., Jr., and Cobban. W. A., 1960. Studies of the Mowry Shale (Cretaceous) and contemporary formations in the United States and Canada, *U.S. Geol. Surv. Prof. Pap. 355,* 126p.

Rosenkrans, R. R., 1934. Correlation studies of the central and south central Pennsylvania bentonite occurrences, *Am. J. Sci.,* ser 5, **27,** 113-134.

Ross, C. S., 1955. Provenience of pyroclastic materials, *Geol. Soc. Am. Bull.,* **66,** 427-434.

Ryan, W. B. F.; Workum, F., Jr.; and Hersey, J. B., 1965. Sediments on the Tyrrhenian Plain, *Geol. Soc. Am. Bull.,* **76,** 1261-1282.

Sheridan, M. F., 1973. A fluidization model for pyroclastic eruptions, *Geol. Soc. Am. Abstr. Progr.,* **5,** 806.

Slaughter, M., and Earley, J. W., 1965. Mineralogy

and geological significance of the Mowry bentonites, Wyoming, *Geol. Soc. Am. Spec. Pap. 83*, 95p.
Smith, R. L., and Bailey, R. A., 1966. The Bandelier tuff: A study of ash-flow eruption cycles from zoned magma chambers, *Bull. Volcan.*, ser. 2, **29**, 83-103.
Verhoogen, J., 1951. Mechanics of ash formation, *Am. J. Sci.*, **249**, 729-739.
Whitcomb, L., 1932. Correlation by Ordovician bentonite, *J. Geol.*, **40**, 522-534.
Wilcox, R. E., 1959. Some effects of recent volcanic ash falls, with especial reference to Alaska, *U.S. Geol. Surv. Bull.*, **1028-N**, 409-476.
Wilcox, R. E., and Izett, G. A., 1973. Criteria for the use of volcanic ash beds as time-stratigraphic markers, *Geol. Soc. Am. Abstr. Progr.*, **5**, 863.
Williams, H., 1942. The geology of Crater Lake National Park, Oregon, *Carnegie Inst. Washington Publ. 540*, 162p.

Cross-references: *Bentonite; Eolian Dust in Marine Sediments; Ignimbrites; Nuée Ardente; Pyroclastic Sediment; Tuff; Volcanic Ash—Diagenesis; Volcaniclastic Sediments and Rocks; Volcanism—Submarine Products.*

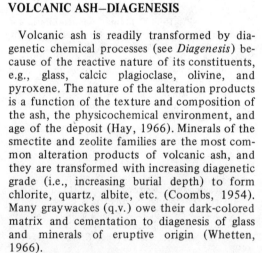

FIGURE 1. Vitric particle diagenetically altered to clinoptilolite in the John Day Formation (Miocene) of Oregon (Hay, 1963). Crystals of clinoptilolite (Cl) form margin enclosing unfilled central cavity (Cv). Ash particle is in matrix of vitric material replaced by clinoptilolite and montmorillonite. Length of bar is 0.5 mm.

VOLCANIC ASH—DIAGENESIS

Volcanic ash is readily transformed by diagenetic chemical processes (see *Diagenesis*) because of the reactive nature of its constituents, e.g., glass, calcic plagioclase, olivine, and pyroxene. The nature of the alteration products is a function of the texture and composition of the ash, the physicochemical environment, and age of the deposit (Hay, 1966). Minerals of the smectite and zeolite families are the most common alteration products of volcanic ash, and they are transformed with increasing diagenetic grade (i.e., increasing burial depth) to form chlorite, quartz, albite, etc. (Coombs, 1954). Many graywackes (q.v.) owe their dark-colored matrix and cementation to diagenesis of glass and minerals of eruptive origin (Whetten, 1966).

Diagenetic alteration of glass, incorrectly termed devitrification, is a result of chemical reaction with a fluid, and it proceeds by a solution-precipitation process, which commonly results in hollow pseudomorphs from which the glass has been wholly dissolved (Fig. 1). Mafic (e.g., basaltic) and silicic (e.g., rhyolitic) glasses generally yield different alteration products, particularly in the early stages of diagenesis. Mafic glasses alter to a resinous-appearing substance of brown or golden color termed *palagonite*. It varies in chemical composition and is generally noncrystalline (x-ray amorphous). Alteration of palagonite involves a large gain of water and significant losses of alkali and alkaline-earth ions, silica, and, not uncommonly, alumina. In an appropriate chemical environment, constituents lost from mafic glass in palagonitization are precipitated as zeolites and calcite to form palagonite tuff, as on Oahu, Hawaii (Hay and Iijima, 1968). Palagonite is readily transformed to a smectite, which with increasing burial depth may be altered to chlorite. Similarly, the first-formed zeolites (e.g., phillipsite, chabazite) are commonly altered to analcime, which may yield laumontite, wairakite, or albite at higher grades of diagenesis (Coombs, 1954; Hay, 1966; Mumpton, 1977).

The alteration of silicic glasses is highly variable and depends primarily on the pH of the interstitial fluid. In solutions with a pH ≥ 9.5, as in many lakes of arid regions, vitric ash is altered in 10,000 yr or less to alkali-rich zeolites (e.g., phillipsite, clinoptilolite), which may involve an intermediate gel stage (Mariner and Surdam, 1970). If salinity is high, the first-formed zeolites are characteristically transformed either to analcime ($\pm$ quartz) or to K-feldspar ($\pm$ quartz) (Hay, 1966). Analcime can be replaced by K-feldspar, albite, quartz, etc. Examples are found in tuffs of the Miocene Barstow Formation of California (Sheppard and Gude, 1969) and the Eocene Green River Formation of Wyoming (Hay, 1966; Surdam and Parker, 1972). Alkali feldspars and quartz are the thermodynamically stable phases in the saline, alkaline environment; glass alters initially to zeolites for kinetic reasons (Hay, 1966). At a lower pH, as in fresh-water or marine sediments, silicic glass alters chiefly to a phyllosilicate, usually a smectite, which may be accompanied by opal and/or zeolites. Relatively pure deposits of a smectite formed from vitric ash are termed *bentonites* (q.v.). At a pH of 7-8, the alteration of vitric ash to bentonite usually requires at least several million yr. With increasing age and/or burial depth, the smectite, opal, and zeolites are generally transformed to quartz, mixed-layer phyllosilicates, alkali feldspars, etc.

Thick sequences of silicic tuffaceous sediments, either fresh-water or marine, exhibit a vertical zonation of alteration products somewhat analogous to the zoning of saline-lake deposits in response to a salinity gradient. Phyllosilicates and opal are associated with fresh glass in the topmost zone, which is as much as 1 km thick. Below, glass is altered to alkali-rich zeolites, usually accompanied by opal and a phyllosilicate. Analcime, heulandite, quartz, and K-feldspar are characteristic of the next lower zone; and the lowermost zone of diagenesis may contain laumontite, wairakite, prehnite, and albite. The zones represent a sequence of mineralogic reactions in response to changes in water chemistry and/or increasing temperature with depth. Examples are found in the Green Tuff regions of Japan (Iijima and Utada, 1966).

RICHARD L. HAY

References

Boles, J. R., and Coombs, D. S., 1975. Mineral reactions in zeolitic Triassic tuff, Hokonui Hills, New Zealand, *Geol. Soc. Am. Bull., 86*, 163-173.

Coombs, D. S., 1954. The nature and alteration of some Triassic sediments from Southland, New Zealand, *Roy. Soc. New Zealand Trans., 82*(1), 65-109.

Hay, R. L., 1963. Stratigraphy and zeolitic diagenesis of the John Day Formation of Oregon, *Univ. Calif. Publ. Geol., Sci., 42*, 199-262.

Hay, R. L., 1966. Zeolites and zeolitic reactions in sedimentary rocks, *Geol. Soc. Am. Spec. Pap. 85*, 130p.

Hay, R. L., and Iijima, A., 1968. Nature and origin of palagonite tuffs of the Honolulu Group on Oahu, Hawaii, *Geol. Soc. Am. Mem. 116*, 331-376.

Iijima, A., and Utada, M., 1966. Zeolites in sedimentary rocks, with reference to the depositional environment and zonal distribution, *Sedimentology, 7*, 327-357.

Mariner, R. H., and Surdam, R. C., 1970. Alkalinity and formation of zeolites in saline, alkaline lakes, *Science, 170*, 977-980.

Mumpton, F. A., ed., 1977. Mineralogy and geology of natural zeolites, *Mineral. Soc. Am. Short Course Notes, 4*, 233p.

Scafe, D. W., 1973. Bentonite characteristics from deposits near Rosalind, Alberta, *Clays Clay Minerals, 21*, 437-450.

Sheppard, R. A., and Gude, A. J., 1969. Diagenesis of tuffs in the Barstow Formation, Mud Hills, San Bernardino County, California, *U.S. Geol. Surv. Prof. Pap. 634*, 35p.

Sheppard, R. A., and Gude, A. J., 1973. Zeolites and associated authigenic silicate minerals in tuffaceous rocks in the Big Sandy Formation, Mohave County, Arizona, *U.S. Geol. Surv. Prof. Pap. 830*, 36p.

Surdam, R. C., and Parker, R. D., 1972. Authigenic aluminosilicate minerals in the tuffaceous rocks of the Green River Formation, Wyoming, *Geol. Soc. Am. Bull., 83*, 689-700.

Walton, A. W., 1975. Zeolitic diagenesis in Oligocene volcanic sediments, Trans-Pecos Texas, *Geol. Soc. Am. Bull., 86*, 615-624.

Whetten, J. T., 1966. Sediments from the lower Columbia River and the origin of graywacke, *Science, 152*, 1057-1058.

Cross-references: *Bentonite; Diagenesis; Volcanic Ash Deposits.*

VOLCANICLASTIC SEDIMENTS AND ROCKS

Volcaniclastic rocks encompass the entire spectrum of fragmental volcanic rocks containing particles formed by any mechanism of fragmentation, emplaced by any means of dispersal in any physiographic environment, or mixed in any significant portion with nonvolcanic fragment types (Fisher, 1961, 1966). Thus *epiclastic* volcanic fragments, produced by the weathering and erosion of volcanic terranes, as well as *pyroclastic* fragments, which are explosively produced and projected from volcanoes or volcanic vents either vertically or along the ground as flows, are considered to be volcaniclastic materials. Other processes of fragmentation include: *autoclastic fragmentation,* involving self-breakage by mechanical friction or gaseous explosion during movement of lava or crumbling of spines and domes; and *alloclastic fragmentation,* which refers to the disruption of nonvolcanic rocks by volcanic processes beneath the earth's surface, with or without intrusion of fresh magma. *Hyaloclastic* fragments form by granulation caused by steam explosions or mechanical breaking of quickly chilled magma or lava that comes in contact with water or water-saturated sediments (see *Volcanism—Submarine Products*). Granulation by tectonic process is another means by which volcanic rocks may become fragmented.

The need to use similar limiting grade sizes, in order to include all volcaniclastic rocks within a single scheme, necessitates a slight change in the grade-size limits originally applied by Went-

VOLCANICLASTIC SEDIMENTS AND ROCKS

Grade size (mm)	Epiclastic fragments	Pyroclastic fragments	
— 256 —	Boulders (and "blocks")	Coarse	Blocks and bombs
— 64 —	Cobble	Fine	
	Pebble	Lapilli	
— 2 —			
— 1/16 —	Sand	Coarse	Ash
— 1/256 —	Silt	Fine	
	Clay		

FIGURE 1. Grade size limits for epiclastic and pyroclastic fragments.

worth and Williams (1932) to pyroclastic fragments. This change is advisable because it is not always easy to determine whether particles within a particular volcaniclastic deposit are pyroclastic. Corresponding grade-size limits for epiclastic and pyroclastic fragments are given in Fig. 1.

Fig. 2 gives a classification of pyroclastic and epiclastic volcanic rocks based upon the sediment grade size limits shown in Fig. 1. Also included is a nongenetic category based only upon particle size and upon the presence of particles of volcanic composition, whatever their origin. Rock terms for size mixtures of various pyroclastic fragments are given in Fig. 3. Parsons (1968) presents a more detailed classification of volcanic breccias. Further classification of tuff (q.v.) may be based upon the type of transporting agent (fluvial, eolian), upon environment of deposition (lacustrine, marine), upon mode of origin of pyroclastic

Predominant grain size (mm)	Pyroclastic * Primary or reworked	Epiclastic * +	Equivalent nongenetic terms * +
— 256 —	Pyroclastic breccia	Epiclastic volcanic breccia	Volcanic breccia
— 64 —	Agglomerate	Epiclastic volcanic conglomerate	Volcanic conglomerate
— 2 —	Lapillistone		
— 1/16 —	Coarse	Epiclastic volcanic sandstone	Volcanic sandstone
— 1/256 —	Tuff	Epiclastic volcanic siltstone	Volcanic siltstone
	Fine	Epiclastic volcanic claystone	Volcanic claystone

* May be mixed with nonvolcanic clastic material
\+ Add adjective "tuffaceous" to rocks containing pyroclastic material < 2mm in size

FIGURE 2. Pyroclastic, epiclastic, and nongenetic volcaniclastic rock terms using grade size limits of Fig. 1.

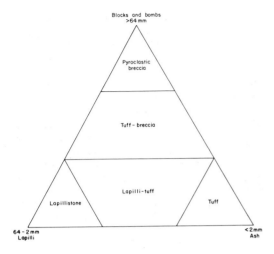

FIGURE 3. Rock terms for size mixtures of pyroclastic fragments.

fragments (juvenile, cognate, accidental), upon composition (crystal, lithic, vitric), etc.

Pyroclastic material deposited in water or in areas where it is easily reworked may become mixed with epiclastic material. The epiclastic material may itself be composed of debris which is nonvolcanic or volcanic, including tuff fragments weathered from lithified tuff deposits. In Anglo-American usage, epiclastic rocks containing pyroclastic mixtures are termed tuffaceous (Fig. 2). Russian and German geologists generally use terms for pyroclastic-epiclastic mixtures as follows (Blokhina, et al., 1959): Pyroclastic rocks, 100% pyroclastic fragments; tuffites, > 50% pyroclastic, < 50% epiclastic; and tuffogenic rocks, < 50% pyroclastic, > 50% epiclastic.

Clastic products of volcanic action have contributed vast quantities of sediment to the stratigraphic record throughout geologic time. Accordingly, the chemical, mineralogical, and physical characteristics of volcaniclastic sediments and rocks are useful tools for solving many kinds of stratigraphic and structural problems. Detailed sedimentological studies may provide important insights into volcanic mechanisms and emplacement processes, or may provide a basis for determining the crystallization history of parent magmas from close analyses of layered sequences. Moreover, volcaniclastic analysis may give important clues of past volcanic and tectonic events involving such broad topics as the growth and development of island arcs, or may be used in solving specific stratigraphic problems such as determining ancient wind patterns or the location and type of ancient volcanism which in itself may have tectonic or stratigraphic significance. Unlike most sedimentary deposits, topography forms no barrier to transportation, and ash may fall within many widely separated depositional basins at the same time (see *Volcanic Ash Deposits*). The fact that an ancient ash fall may have been distributed over a wide area and have spanned several climatic zones makes it an important tool in paleoclimatic and paleoecological problems (Wilcox, 1965).

In ancient geosynclinal sequences, volcaniclastic strata may be abundant. The sequences can be treated just as other sedimentary units, but important genetic conclusions may be gained by determining their volcanic petrology, especially if related to their original source. These stratigraphic units, together with their source rocks, are collectively termed *volcanogenic* units and include lavas and cogenetic intrusives, as well as volcaniclastic rocks of pyroclastic, epiclastic, or other origin (Dickinson, 1968). Some authors have tended to equate "volcanogenic" with "volcaniclastic," but the terms should be clearly distinguished.

RICHARD V. FISHER

References

Blokhina, L. I.; Kopter-Dvornikov, V. S.; Lomize, M. G.; Petrova, M. A.; Tikhomirova, I. E.; Frolova, I. I.; and Yakovleva, E. B., 1959. Principles of classification and nomenclature of the ancient volcanic clastic rocks, *Internat. Geol. Rev.*, 1, 56-61.

Dickinson, W. R., 1968. Sedimentation of volcaniclastic strata of the Pliocene Koroimavua Group in northwest Viti Levu, Fiji, *Am. J. Sci.*, 266, 440-453.

Fisher, R. V., 1961. Proposed classification of volcaniclastic sediments and rocks, *Geol. Soc. Am. Bull.*, 72, 1409-1414.

Fisher, R. V., 1966. Rocks composed of volcanic fragments and their classification, *Earth-Sci. Rev.*, 1, 287-298.

Hay, R. L., 1952. The terminology of fine-grained detrital volcanic rocks, *J. Sed. Petrology*, 22, 119-120.

Parsons, W. H., 1968. Criteria for the recognition of volcanic breccias: review, *Geol. Soc. Am. Mem. 115*, 263-304.

Pettijohn, F. J., 1975. *Sedimentary Rocks*, 3rd ed. New York: Harper & Row, 628p.

Vlodavets, V. I., ed., 1963. A classification of pyroclastic rocks: Draft of the proposal submitted to the Commission of the Classification of Pyroclastic Rocks elected at the First All-Union Volcanology Conference, *Intern. Geol. Rev.*, 5, 516-524.

Wentworth, C. K., and Williams, H., 1932. The classification and terminology of the pyroclastic rocks, *Rept. Comm. Sed., Natl. Research Council Bull.*, 89, 19-53.

Wilcox, R. E., 1965. Volcanic-ash chronology, in H. E. Wright, Jr., and D. G. Frey, eds., *The Quaternary of the United States*. Princeton, N.J.: Princeton Univ. Press, 807-816.

Cross-references: *Agglomerate, Agglutinate; Base-Surge Deposits; Bentonite; Breccias, Sedimentary; Clastic Sediments and Rocks; Eolian Dust in Marine Sediments; Graywacke; Ignimbrites; Mudflow, Debris-Flow Deposits; Nuée Ardente; Pyroclastic Sediment; Sedimentary Facies and Plate Tectonics; Tuff; Volcanic Ash Deposits; Volcanic Ash–Diagenesis; Volcanism–Submarine Products.*

VOLCANISM, SEDIMENTARY

Buried rocks saturated with gas, oil, or water under pressure carry a latent energy capable of producing geological phenomena similar to those caused by magmatic intrusions and eruptions. The term *sedimentary volcanism* has been applied to these phenomena (Kugler, 1933). They are best known from folded Tertiary sediments and have been particularly well studied in the West Indian island of Trinidad.

Causes

Like all complex geological processes, sedimentary volcanism owes its existence to a variety of causes. Some geologists (e.g., Goubkin, 1934) stress the relationship between diapiric structures and mud volcanoes, but it is as wrong to relate all mud volcanoes to piercement structures as it is to assume they are all the result of escaping gas. The essential prerequisite in all cases, however, appears to be the presence of hydrocarbons trapped under supranormal pressures. For an explanation of all types of sedimentary volcanism we must consider both exogenic and endogenic causes. The latter includes (1) orogenic compression; (2) postorogenic isostatic movements and tension faults; (3) differences in density and thickness of sediments; and a combination of 1, 2, and 3.

Exogenic Forces. Artificial tapping of a high-pressure reservoir, particularly in the presence of fractured formations, can lead to the formation of mud volcanoes, craters (e.g., Caddo Lake, northern Louisiana), and even to gas and oil eruptions > 100 m from the borehole, whereby sediments are squeezed into every joint and fissure. Desiccation cracks above shallow gas deposits, or erosion of their capping clay, can cause the formation of mud volcanoes, if water is present. Seasonal removal of sediments by marine currents can induce violent outbreaks of methane stored in decaying masses of vegetable matter (see *Gases in Sediments*). Mudlumps, found in the Mississippi delta by Morgan (1961), are upwellings of plastic clay arising from the static pressure of accumulating sediments, in places accompanied by vents discharging mud, methane, and salt water (see Hedberg, 1974).

Endogenic Forces. Diagenetic processes such as lithification, or normal compaction of pelites, may cause fissures along which gas, fluids, and sediments can escape (see *Compaction in Sediments; Water-Escape Structures*). Diastrophism, however, is the main source of all the spectacular mud volcanoes and basins known from Romania, the southern USSR (Jakubov et al., 1971), Iran, Burma, Indonesia, Mexico, Ecuador, Colombia, Venezuela and Trinidad (Higgins and Saunders, 1974). They are usually located on anticlines with cores of plastic sediments rising under high pressure. This pressure, in places exceeding the geostatic pressure, is transmitted to any superposed reservoir with gas or fluids. The argillaceous sediments of the mobile core can carry large blocks of rocks older than the matrix. Previously, it was thought conceivable that diastrophically liberated gas and water alone would be able to widen tension fissures of competent sediments above the diapiroid core, and by the interaction of gas, oil, and water would form a puddle of gritty and pebbly shales. Gas was considered the main driving force responsible for the ascent of the mud, which, on its way up, would break, entrain, and lift large blocks of country rock (Kugler, 1933); in Trinidad, older formations are often thrust on Miocene sediments, and the presence in the mudflows of fragments and often huge blocks of Older Tertiary and Cretaceous rocks fit into this conception. Subsequent studies of these pebbly and gritty shales have, however, established that they are in fact poorly sorted hydroclastic sediments which were formed on submarine slopes or in basinal depressions; they often envelop large submarine slip masses. Under orogenic pressure, these abnormally thick accumulations of incompetent sediments bulged upwards and produced cores of a specific gravity lighter than that of the graywackes, silts, and shales syn- or postorogenically deposited on their flanks. The bulk density of the calcareous mudflow is about 2.0 or less, of the younger flanking silts, about 2.4 or more. The source of the mudflows is, therefore, a *density inversion layer,* the prerequisite for a mobile belt. Detailed gravity surveys have found pronounced minima above the Trinidad Pitch Lake and above mud volcanoes (Nettleton, 1949). The former is readily explicable by the density difference between core and country rock, but some of the latter are too great to be accounted for in this way, unless one assumes the presence underground of huge masses of gas-cut mud, which does not seem compatible with the narrowness of the feeding channels (Gansser, 1960).

The foredeep depressions into which turbidites and slip masses come to rest are often environments of hydrocarbon generation. Gas, oil, and salt water help to lubricate the upward movement of the core, which already in its initial stage receives little or no sediments in its axial region. An excellent example for vertical migration is the core of Miocene rubble and Cruse clays of the Forest Oil Field in Trinidad (Bower, 1951). The shape and size of the 600,000,000 m^3 "laccolith" is reasonably well known; slip masses (olistoliths) of Middle Eocene marl are > 30 m thick. The connection with its root zone, the deep-seated Lengua rubble bed, has not yet been located, but the neck of the denuded mud volcano of the axial region was mapped, and with it also the sheets of mudflow, up to 65 m thick, interbedded in the Upper Miocene Morne L'Enfer Formation (Fig. 1). It appears that an initially thick, lenticular mass of pelitic rubble accumulated during Lower Miocene (Lengua) time and was then mobilized during the deposition of the Cruse Formation. A submarine mud volcano formed on it during the sedimentation of the Lower Forest Clay, as is documented by the mudflow material interbedded with layers of clay carrying large marine mollusks. New eruptions occurred during Upper Morne l'Enfer times, concomitant with late orogenic movements.

Related to such mobilized cores are the *clay dikes* and *sills* that fill the numerous tension fissures in the well-bedded Moruga Formation of Guayaguayare (Kugler, 1933). These dikes vary in thickness from 1–30 cm and more and consist of prismatically fracturing, tough clay with scattered small pebbles of silt. The mud-filled, parallel joints are in places so numerous as to imitate bedding. The joints must have been caused by the pressure of the mobile core of the anticline, and the mobilized mud must have migrated over considerable distances to fill the fissures that were formed.

When the gas-saturated core reaches a stage where the geostatic pressure is less than the gas pressure, eruptions occur. Thousands of tons of pelites and entrained blocks are hurled into the air, or form new islands and mud banks. The

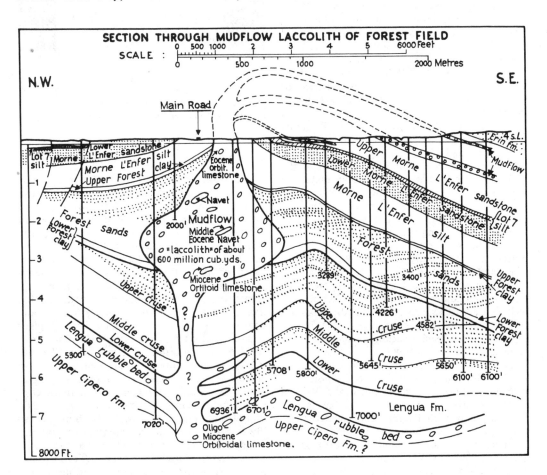

FIGURE 1. Cross section through the anticline of the Forest Reserve Oil Field, Trinidad, West Indies.

quantity of mud ejected in 20 minutes in November 1930, at Tabaquite, Trinidad, was at least 500,000 m^3. If the escaping gas catches fire, the similarity to pyrogenic volcanism is even more impressive; in March 1924, the flames of the mud volcano Touragai in South Kabristan could be seen from a distance of hundreds of km (Kugler, 1939). Submarine indications of sedimentary volcanism have been observed since the dawn of history as "boiling water," "burning seas," or suddenly appearing islands, known from Apsheron Peninsula (USSR), Indonesia, and Trinidad. In November, 1911, a new island appeared off Chatham Bay, on S coast of Trinidad, accompanied by an exceptionally violent eruption. The common presence of chert and pyrite among the ejecta was probably responsible for the ignition of the gas by sparking. The flames rose to a height of 30 m. A further eruption took place at the same locality in 1964 (see Higgins and Saunders, 1974).

The vents of mud volcanoes migrate back and forth along the axis of the mobile uplift. *Fossil mudflow* is mainly unbedded and can be recognized by its porphyry-like texture of pebbly marl, silt, rounded, and angular polygenic components in a dark clay matrix (Fig. 2; see also Suter, 1951, Plate 2). The texture of a mudflow deposit (q.v.) is reminiscent of tillites, for which it could easily be mistaken, particularly in the presence of scratched pebbles as found in Trinidad, and also reported from Romania (Krejci-Graf, 1927). Typical also are the small, highly polished chert and quartz pebbles of brilliant luster, the result of movement of the hydroplastic clay mass. Many pebbly clays and breccias of this nature have been wrongly interpreted as "tectonites" squeezed out at the front of overthrusts—although the evidence clearly points to their being the product of solifluction and sliding on marine slopes—orogenically mobilized, and eventually expelled with the help of gas and water.

Along the S coast of Trinidad, large blocks of Eocene and Miocene orbitoidal limestones, as well as other rocks, rest on steeply dipping Miocene sandstones and siltstones. These blocks represent the back-weathered remnants of former, extensive mudflows.

Mud basins are formed by subsidence induced by the underground loss of material upon mud-volcano eruptions. Lagon Bouffe and Moruga Bouffe in Trinidad are typical mud basins of a subsiding area with its anastomosing river system and swamps. Photos of these and other Trinidad mud volcanoes and given by Higgens and Saunders (1974). Gansser (1960) has photographed the caldera-like, collapsed craters of large mud volcanoes in NE Iran.

FIGURE 2. Clay pebbles in a 10 cm core from 1400 m (4612 ft), Antilles Merimac No. 14, Trinidad. Matrix: marine Miocene. Pebbles: Paleocene and Eocene foraminiferal marl, silt, and shale.

Sand Dikes, Hydrocarbon Dikes and Sills

Most sandstone dikes are sedimentary phenomena associated with compaction and extrusion of water (see *Clastic Dikes; Water-Escape Structures*). Sand of a high-pressure reservoir can be forced into a fissure, either slowly, or in an explosive manner. Jenkins (1930) suggested the term *intruclast* instead of sand dikes, because of the latter's connotation of igneous origin. He reported sandstone dikes from California up to 1 km long and 6 m thick. In Trinidad a 2.5-m oilsand dike carries angular and edge-worn pieces of well-bedded silt, with plant remains derived from a horizon 30–60 m below their present position. Silt and clay dikes are known from Burma; they are believed to have acted as feeding channels for mud volcanoes. Mud dikes are reported from Huronian shales in a gas field near Cleveland, Ohio, and clastic dikes from metamorphic rocks. Both may possibly be proof of former sedimentary volcanism.

Inspissated oil is commonly present in clastic dikes. Every stage of polymerization, from a highly viscous oil to grahamite and impsonite, can occur in fissures many km long, like the famous gilsonite dikes of Utah. Ozokerite dikes are known from Romania. The term *hydrocarbon dike* (Kugler, 1933) has been suggested for this type of occurrence of bitumen, which in its ultimate state may be metamorphosed to graphite.

A special case of migration of hydrocarbons is the *Asphalt Lake* of Trinidad. Previously it was assumed to represent an infill of a mud basin in which oil and salt water were mechanically mixed by the churning action of gas. The resulting lake-asphalt (parianite) consists of about 40% bitumen, 30% colloidal clay, and 30% salt water. It is a plastic mass kept in slow motion by convection currents and is characterized by abundant vugs (cheese pitch) filled with sulfurated gas and salt water. The paucity of trapped animal remains, compared to Rancho la Brea in California, threw the first doubts on the hypothesis of basinal origin. The foraminiferal content and mineral residue of the asphalt, as well as of the asphalt residue surrounding the "lake," are the result of a fossil submarine mixture of heavy sulfurous oil with clay and silt of the Miocene Nariva Formation, suggesting that the oil, originally in a dome of Cretaceous shale, seeped onto the Miocene sea floor. The original oil had a density close to that of sea water and, after becoming mixed with sediment, was able to remain on the sea bottom. Agitated by gas, the mixture of inspissated oil, sediments, and water formed a large lenticular body. The low density (1.23) of the impure asphalt, surrounded by accumulating masses of Nariva beds (density 2.08), was sufficient to initiate a gradual rising of the bitumen, akin to a salt plug. Finally, the mass of more than 10 million tons came to rest at its present position about 1200 m above the place of origin (Kugler, 1956).

In the Pitch Lake area and elsewhere, the highly viscous oil leaving the Cretaceous reservoirs during the latest orogenic stresses filled fissures of the overlying sediments. They are the grahamite (manjak) dikes, formerly mined in Trinidad and Barbados. At San Fernando, Trinidad, there is a vein of manjak 10 m thick, and the bitumen was sufficiently solid to carry entrapped blocks of country rock.

The phenomena described above are classified in Table 1.

Fossil Sedimentary Volcanism

Mud-volcano activity during Tertiary time is proven. Some of the tillites and similar rocks described from various countries may turn out to be such mudflows (see *Diamictite*). Buschback and Ryan (1963) illustrate rocks from an Ordovician explosion structure in Illinois that can be matched with mudflow rocks from Trinidad. Some of the structures described as "cryptovolcanic" may belong to this class. Freeman (1968) described a convincing example of a Middle Tertiary mud-volcano complex in Texas.

A. Gansser (pers. comm.) described the Decaturville structure, about 80 km W of Rolla, Missouri. This complexly built, circular uplift of almost 2.5 km diameter is surrounded by slightly domed Ordovician limestone. The center of the uplift is formed by Cambrian quartzite, with a dike of tourmaline-bearing pegmatite, and a lenticular body—over 200 m long and 30 m wide—which appeared to Gansser to be a lithified, sulfide-bearing mudflow with a texture akin to those he had observed in Trinidad.

HANS G. KUGLER

TABLE 1. Classification of the Phenomena Produced by Sedimentary Volcanism

Incompetent masses of mobilized pelites with block rubble (olistostromes)	Diapiroid structures with sills and dikes
Density inversion layer of mobilized pelites plus hydrocarbons and water	a. Laccoliths, sills, and clay dikes b. Asphalt lake of Trinidad c. Large, block-carrying mud extrusions and mud volanoes, "sedimentary lava" d. Mud lumps
Gas and water (or oil) accumulations under pressure	a. Mud volcanoes with clay silt, sand, and subordinate small blocks b. Oil-sand dikes c. Hydrocarbon dikes

References

Bower, T. H., 1951. Mudflow occurrence in Trinidad, *Bull. Am. Assoc. Petrol. Geologists,* 35, 908-912.

Buschbach, T. C., and Ryan, R., 1963. Ordovician explosion structure at Glasford, Illinois, *Bull. Am. Assoc. Petrol. Geologists,* 47, 2015-2022.

Dionne, J.-C., 1973. Monroes: A type of so-called mud volcanoes in tidal flats, *J. Sed. Petrology,* 43, 848-856.

Freeman, P. S., 1968. Exposed Middle Tertiary mud diapirs and related features in South Texas, *Am. Assoc. Petrol. Geologists Mem. 8,* 162-182.

Gansser, A., 1960. Über Schlammvulkane und Salzdome, *Vierteljahresschr. Naturf. Gesell. Zürich,* 105, 1-46.

Goubkin, I. M., 1934. Tectonics of south-western Caucasus and its relation to the production oilfields, *Bull. Am. Assoc. Petrol Geologists,* 18, 603-671.

Hedberg, H. D., 1974. Relation of methane generation to undercompacted shales, shale diapirs and mud volcanoes, *Bull. Am. Assoc. Petrol. Geologists,* 58, 661-673.

Higgins, G. E., and Saunders, J. B., 1974. Mud volcanoes—their nature and origin, *Verh. Naturf. Gesell. Basel,* 84, 101-152.

Jakubov, A. A.; Ali-Zadi, A. A.; and Zeinalov, M. M., 1971. *Mud Volcanoes of the Azerbaijan SSR.* Bakus Acad. Sciences, 257p (photo atlas, 15p English summary).

Jenkins, O. P. 1930. Sandstone dykes as conduits for oil migration through shales, *Bull. Am. Assoc. Petrol. Geologists,* 14, 411-422.

Krejci-Graf, K., 1927. Zur Kritik von Vereisungs-Anzeichen, *Senckenbergiana,* 5, 157-169.

Kugler, H. G., 1933. Contribution to the knowledge of sedimentary volcanism, *J. Inst. Petrol. Technologists,* 19, 743-772.

Kugler, H. G., 1939. Visit to Russian oil districts, *J. Inst. Petrol. Technologists,* 25, 68-88.

Kugler, H. G., 1956. Trinidad, *Geol. Soc. Am. Mem. 65,* 355-365.

Levorsen, A. I., 1954. *Geology of Petroleum.* San Francisco: Freeman, 703p.

Lovell, J. P. B., 1974. Sand volcanoes in the Silurian Rocks of Kirkcudbrightshire, *Scottish J. Geol.,* 10, 161-162.

Morgan, J. P., 1961. Mudlumps at the mouths of the Mississippi River, *Louisiana Geol. Surv. Bull.,* 35, 116p.

Nettleton, L. L. 1949. Geophysics, geology and oil finding, *Am. Bull. Assoc. Petrol. Geologists,* 33, 1154-1160.

Suter, H. H., 1951, 1952. The general and economic geology of Trinidad, B.W.I., *Colon. Geol. Mineral Resour.* 2 (3-4); 3(1), 134p.

Tamrazyan, G. P., 1972. Peculiarities in the manifestation of gaseous-mud volcanoes, *Nature,* 240, 406-408.

Cross-references: *Clastic Dikes; Compaction in Sediments; Diagenesis; Gases in Sediments; Mudflow, Debris-Flow Deposits; Olistostrome, Olistolite; Petroleum—Origin and Evolution; Water-Escape Structures.*

VOLCANISM—SUBMARINE PRODUCTS

Submarine volcaniclastic sedimentation is a widespread phenomenon in island arc settings (Mitchell, 1970; Dickinson, 1974). Because island arcs are especially characteristic of colliding plate margins, their products may be preserved on the accreting margin of a continent; the dynamic forces involved in plate collision and continental accretion, however, may make later recognition of these deposits difficult. Not only are island arcs prolific producers of volcaniclastic material, but they are also set in tectonically active zones and are bordered or rimmed by relatively steep and unstable slopes. This setting is favorable for the production of gravity slides (q.v.), debris flows, turbidity flows, and other types of gravity flows (q.v.).

Submarine volcaniclastics are also found along the mid-ocean ridges and around volcanic island chains such as Hawaii; these sediments are less likely to be preserved in the geologic record. Islands such as Hawaii have been the site of actual observation of submarine volcanism and deposition, and the FAMOUS expedition observed mid-ocean-ridge products (Ballard and Moore, 1977). Subaqueous volcanism or the flow of lava into water may also occur in lakes (Sanders, 1968). The intrusion of magma or lava into water-soaked sediments, both subaerial and subaqueous, may produce phenomena similar to subaqueous volcanic eruptive products, or, on the other hand, may be difficult to distinguish from lahars.

Submarine Eruption

Submarine eruption may be quiet or explosive (Fig. 1). Lava may flow quietly out onto the sea floor, protected from the cold water by a glassy crust and producing either pillow lavas or lava closely resembling pahoehoe (Jones, 1968). The primary products of explosive eruption are hyaloclastites—shattered volcanic glass fragments, composed primarily of sideromelane and palagonite. Bonatti (1968) attributed the difference in eruptive process to the viscosity of the lava: very-hot fluid melts erupt quietly, whereas more viscous lavas produce hyaloclastites.

Pillow Lava. Pillow lava is probably the most abundant form of volcanic rock on earth (Moore, 1975), although most of it is beneath the oceans and is often mantled by sediments. It is generally accepted that pillow lava forms when fluid lava chills upon contact with water. Only recently has this process been directly observed, however, by divers in Hawaii and on the Mid-Atlantic Ridge (Moore, 1975). The term

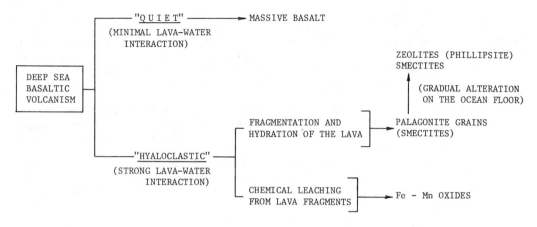

FIGURE 1. Scheme of mechanisms of deep-sea volcanism, its products and alteration products (from Bonatti, 1968).

pillow lava originates from the shape of the masses produced, which in cross section look like independent, sack-like or ellipsoidal bodies. Direct observation, however, as well as the discovery of good three-dimensional outcrops of ancient pillow lavas, indicates that many, if not most, "pillows" are actually tubes—cylindrical, interconnected flow lobes (Moore, 1975). Despite the confusing terminology, Moore suggested that the term *pillow* be retained to refer to part of or an entire flow lobe, with *pillow bud* and *pillow lobe* describing smaller units.

McBirney (1963) found it likely that magma intruding wet sediment would most likely produce sills and dikes, and extrusions would be rare. Intrusion into wet sediments does not, however, preclude the formation of pillow-type lava, although this material would be less freely formed than the pillows formed in water. The vesicularity of pillows has been related to water content of the magma (McBirney, 1963), depth of eruption (Moore, 1965), and site of initial eruption—subaerial or subaqueous (Jones, 1969).

Pillow Breccia. Carlisle (1963) described pillow breccias as rocks consisting of whole or disaggregated volcanic pillows in an abundant matrix of cogenetic basic tuff. This *aquagene tuff* consisted of broken shards of basaltic glass. In a Triassic eugeosynclinal sequence in British Columbia, Carlisle described successions grading from pillow lava at the base, through deformed, isolated pillows (*isolated pillow breccia*), up into broken pillows (*broken pillow breccia*), the matrix increasing to as much as 80% of the total volume. He concluded that the pillow lava formed in clear water and the isolated pillows in turbid, vapor-charged water, and that the broken pillow breccia was deposited by penecontemporaneous slumping. In addition, autobrecciation of pillows may occur by explosion, implosion (Moore, 1975), or collapse of hollow lava tubes. It may be difficult to distinguish broken pillow breccias from autobrecciated pillow breccias, and from redeposited pillow fragments.

Hyaloclastites. Hyaloclastites are tuffs composed principally of shattered basaltic glass (sideromelane) and hydrated basaltic glass (palagonite) erupted under water or ice. *Palagonite tuff* is not a synonym because not all hyaloclastites are palagonite-rich; moreover, in ancient, altered examples, it is difficult to identify the original glass. The term *hyaloclastite (ialoclasti)* was introduced by Rittmann (1958), and Cucuzza-Silvestri (1963) expanded the definition to include shattered or broken pillows. The term is usually restricted to tuffs, although it appears that in some cases, pillows and hyaloclastites may be produced contemporaneously (Carlisle, 1963).

It has been apparent from direct observation that lavas reaching the sea or erupting subaqueously may explode violently, producing *littoral cones* on shorelines and *seamounts* (Bonatti, 1968) under water. The cause of the explosion, however, and even that explosion is necessary to produce glass fragments, has been questioned. Studies by Rittmann (1962) and McBirney (1963) assumed that aquagene tuffs are formed by the explosive release of volatiles from lava, as is subaerial volcanic ash (q.v.). If volatiles are the controlling factor, then at depths where the hydrostatic pressure exceeds the critical pressure of water, volatiles would not be released explosively. McBirney (1963) estimated that for lavas of average wt% water, both basaltic and rhyolitic, the limiting depth would be 500 m; lavas with unusually high water content might explode at depths up to 2000 m. He noted, however, that many cores from areas deeper than 2000 m (with no nearby shallow sources) contained fragmental basaltic glass. Finding no evidence of explosive eruption—vesicles or concave surfaces—he suggested an essentially nonexplosive origin: the chilling effect of cold water on silicate melts, known to cause rapid and intense shattering. Carlisle (1963) described this quenching effect in some detail, and it is this process that suggested to Bonatti (1968) that "explosive" eruptions, i.e., eruptions producing hyaloclastites, were produced from more viscous melts than those of quiet eruptions.

The shape of hyaloclastic particles is variable. Carlisle (1963) described globules—deformed droplets; granules—similar to globules but bounded in part by conchoidal fractures; and shards—irregularly broken fragments. Fiske (1963) did not find globules or granules in the Tertiary Ohanopecosh Formation and attributed their absence to a more viscous lava (andesitic dacitic vs. Carlisle's basaltic), which he concluded would be more explosive, with some quench-spalling. There is some question, however, whether the Ohanopecosh tuffs were erupted subaqueously or not. Piccoli (1966) described hyaloclastic particles as very regular polyhedrons of nearly uniform size.

Palagonite. Whether palagonite is distinctive of subaqueous eruption has been debated. Palagonite consists essentially of hydrated glass together with its alteration products, usually smectitic clays. Whereas normal basaltic glass contains 2–3% water, palagonite may contain 20% or even more. Palagonite may be found as the matrix of a basaltic tuff; it may be found on the crusts of pillows, or it may be found as small grains making up hyaloclastites. It has been thought that any subaqueous ash, whether subaerially or subaqueously erupted, could be palagonitized upon contact with water. Palagonite has been found in several localities on land, but always associated with eruptions believed to have occurred under water or ice. Marshall (1961) showed that hydration of volcanic glasses in sea water is a very slow process, even on a geologic time scale; Bonatti (1965, 1968) used this information to argue that almost all palagonite forms at the time of subaqueous lava eruption. One other possibility is that sideromelane could be altered by later hydrothermal action, certainly a possibility in subaqueous volcanic settings. Bonatti points out that sideromelane and palagonite may be cogenetic, and Carlisle (1963) found subaqueously erupted glass which was almost entirely sideromelane.

Alteration. Hyaloclastites composed of palagonite are easily devitrified. Alteration of palagonite and other volcanic glasses in a submarine environment may produce smectites and zeolites, notably phillipsite (see *Volcanic Ash—Diagenesis*). In a few instances, but not diagnostically, lava-water reactions can affect feldspar grains, e.g., labradorite may be altered to calcite and zeolites (Bonatti, 1968). Many hyaloclastites, and other volcaniclastics, are cemented by calcite; although this cementation is common in submarine environments, it may readily occur in terrestrial settings and is not diagnostic.

Bonatti (1968) suggested that concentrations of Fe and Mn oxides may be diagnostically associated with marine eruptions. The leaching of Mn and Fe from lavas, which is enhanced at high temperatures, is favored by the presence of acidic magmatic gases dissolved in sea water during the eruption, and should also be dependent on the extent of comminution (Bonatti, 1968), i.e., hyaloclastites will contain more Fe and Mn than will pillow-lava formations (Fig. 1). The convection currents set up by submarine eruption will also help to supply enough oxygen to precipitate the Mn and Fe oxides.

Subaqueous Volcaniclastic Deposition

The material supplied to a subaqueous volcaniclastic deposit needn't be erupted subaqueously; possible sources include: (1) pyroclastic flows reaching water; (2) air-fall of subaerially erupted debris; (3) sloughing of subaerially erupted debris along shorelines; (4) subaqueous phreatomagmatic eruptions; (5) subaqueous hyaloclastic eruptions or fragmentation of lava reaching the sea; (6) collapse of lava domes, spines, etc.; and (7) erosion of older volcanic material. Subaqueously deposited tuffs may originate in several ways: (1) direct gravity settling; (2) submarine glowing avalanches (Williams, 1957); (3) debris flows; and (4) turbidity currents; other gravity flow (q.v.) mechanisms may also play a role.

Direct settling of material should produce a graded deposit; reverse grading may be produced if pumice is present. Fiske (1969) cited the presence of pumice at the top of thick tuff beds as evidence of marine deposition. Fiske and Matsuda (1964) pointed out that settling of volcanic material in water should produce more distinct grading than material settled in air because of the differences in density of the medium; and they describe "doubly graded tuffs," which they believe are characteristic of subaqueous eruption and deposition.

Subaqueous volcaniclastic breccias (and tuff breccias) may be produced by autobrecciation; by debris flows carrying coarse material (rubble flows of Jones, 1967); by the collapse of volcanic domes, spines, or lava flows; and by slumps and slides. The distinction of pyroclastic from epiclastic deposits may be extremely difficult in the case of breccias.

JOANNE BOURGEOIS

References

Ballard, R. D., and Moore, J. G., 1977. *Photographic Atlas of the Mid-Atlantic Ridge Rift Valley*. New York: Springer-Verlag, 114p.

Bonatti, E., 1965. Palagonite, hyaloclastites, and the alteration of volcanic glass in the ocean, *Bull. Volcan.,* 28, 257-269.

Bonatti, E., 1968. Mechanisms of deep sea volcanism in the South Pacific, in P. H. Abelson, ed., *Researches in Geochemistry*. New York: Wiley, 453-491.

Carlisle, D., 1963. Pillow breccias and their aquagene tuffs, Quadra Island, British Columbia, *J. Geol.,* 71, 48-71.

Cucuzza-Silvestri, S. C., 1963. Proposal for a genetic classification of hyaloclastites, *Bull. Volcan.,* 25, 315-321.

Dickinson, W. R., 1974. Sedimentation within and beside ancient and modern magmatic arcs, *Soc. Econ. Paleont. Mineral. Spec. Publ. 19,* 230-239.

Fiske, R. S., 1963. Subaqueous pyroclastic flows in the Ohanopecosh Formation, Washington, *Geol. Soc. Am. Bull.,* 74, 391-406.

Fiske, R. S., 1969. Recognition and significance of pumice in marine pyroclastic rocks, *Geol. Soc. Am. Bull.,* 80, 1-8.

Fiske, R. S., and Matsuda, T., 1964. Submarine equivalents of ash flows in the Miocene Tokiwa Formation, *Am. J. Sci.,* 262, 76-106.

Jones, J. G., 1967. Clastic rocks of Espiritu Santo Island, *Geol. Soc. Am. Bull.,* 78, 1281-1288.

Jones, J. G., 1968. Pillow lava and pahoehoe, *J. Geol.,* 76, 485-488.

Jones, J. G., 1969. Pillow lavas as depth indicators, *Am. J. Sci.,* 267, 181-195.

McBirney, A. R., 1963. Factors governing the nature of submarine volcanism, *Bull. Volcan.,* 26, 455-469.

Marshall, B. R., 1961. Devitrification of natural glass, *Geol. Soc. Am. Bull.,* 72, 1493-1520.

Mitchell, A. H. G., 1970. Facies of an early Miocene volcanic arc, Malekula Island, New Hebrides, *Sedimentology,* 14, 201-244.

Moore, J. G., 1965. Petrology of deep-sea basalt near Hawaii, *Am. J. Sci.,* 263, 40-52.

Moore, J. G., 1975. Mechanism of formation of pillow lava, *Am. Sci.,* 63, 269-277.

Piccoli, G., 1966. Subaqueous and subaerial basic volcanic eruptions in the Paleogene of the Lessinian Alps, *Bull. Volcan.,* 29, 253-270.

Rittmann, A., 1958. Il mechanismo di formazione delle lave a pillows e dei cosidetti tufi palagonitici, *Boll. Accad. Sci. Nat.,* ser. 4, 4, 311-318.

Rittmann, A., 1962. *Volcanoes and Their Activity* (translated from 1960 German edition). New York: Wiley, 305p.

Sanders, J. E., 1968. Stratigraphy and primary sedimentary structures of fine-grained, well-bedded strata, inferred lake deposits, Upper Triassic, central and southern Connecticut, *Geol. Soc. Am. Spec. Pap. 106,* 265-305.

Williams, H., 1957. Glowing avalanches in the Sudbury basin, *Ontario Dept. Mines, Ann. Rept.,* 65(3), 57-89.

Cross-references: *Breccias, Sedimentary; Gravity Flows; Gravity-Slide Deposits; Ignimbrites; Mudflow, Debris-Flow Deposits; Nuée Ardente; Pyroclastic Sediment; Tuff; Turbidites; Volcanic Ash Deposits; Volcanic Ash–Diagenesis; Volcaniclastic Sediments and Rocks.*

WADDEN SEA SEDIMENTOLOGY

The Wadden Sea is a belt of tidal flats and channels along the North Sea coast of the Netherlands, Germany, and Denmark. It is the largest tidal flat area in the world, its length being about 500 km, its width varying from 3 to 35 km. The range of the tides increases from the W end (average at Den Helder 1.3 m) to the inner corner of the Bay of Heligoland (>3 m) and then decreases again northward (1.3 m in the Graadyb, Denmark).

Along most of its length, the Wadden Sea is separated from the North Sea by a series of wave-built barrier islands. The tidal currents have scoured out deep inlets between these islands. Owing to the combined effects of ebb currents coming out of the Wadden Sea and of wave action in the North Sea, arcuate sand bars have been formed on the seaward side of the inlets. Parts of them emerge at low tide. In the inner corner of the Bay of Heligoland, where the tides are strongest and the large tidal channels converge toward the North Sea (owing to the curvature of the Wadden Sea belt) only a few isolated, small islets are present.

The tidal channels in the Wadden Sea form branching systems (Fig. 1), starting in the inlets (or on the seaward edge of the tidal flat area). Adjoining channel systems are separated from each other by "tidal divides," low unchanneled areas which in most cases connect the barrier islands with the mainland shore.

The position of the individual channels varies continuously, owing to lateral migration. Moreover, new channels are formed while other ones are filled up. The strength of the tidal currents, and, concomitantly, the depth and width of the channels diminishes in general from the inlet toward the interior. The maximum depth, >40 m, in the whole Wadden Sea is found in the inlet near Den Helder. The maximum current velocity in this inlet reaches, for average tides, about 190 cm/sec at the surface. With strong gales the maximum velocity may be twice as large.

In the Wadden Sea, three main environments may be distinguished (Fig. 2): (1) the channel floors, below mean low tide level; (2) the tidal flats, between the lines of mean low and mean high water; and (3) the salt marshes, above mean high tide level. The tidal flats occupy the greatest surface. They normally lack vegetation, except for scattered algae and abundant

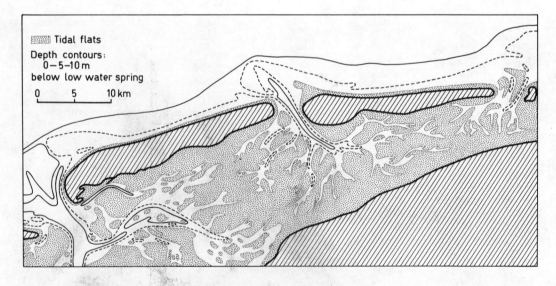

FIGURE 1. Simplified map of tidal-flat area with barrier islands Terschelling and Ameland.

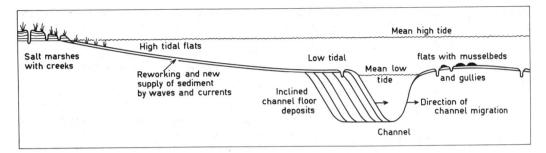

FIGURE 2. Diagram of main Wadden Sea environments.

benthonic diatoms. The lower part of the flats, bordering the tidal channels, may be rich in small, meandering "gullies," especially in muddy regions, and in mussel beds (see *Tidal-Flat Geology*). Most of the higher parts of the flats are extremely flat and monotonous. Where the bottom rises to a little below mean high-tide level, the first halophytes are seen, which in the salt marsh environment mainly determine the character of the sedimentation (see *Salt-Marsh Sedimentology*).

Sediments

The bulk of the sediment in the Wadden Sea consists of sand, which is brought in by the flood currents from the outer shores of the barrier islands and from the shallow parts of the adjoining North Sea floor. A subordinate part of the sand is derived from Pleistocene deposits, eroded in the Wadden Sea itself.

The sand is chiefly deposited against the advancing sides of migrating channels, as inclined layers, that extend from the channel banks to the deepest parts. In this way, the shifting of the channel leads to the gradual replacement of the older sediments by new Wadden Sea deposits. Older sediments that have not (yet) been eroded by the channels are covered, at most places, by thin layers of mud or sand, spread out from the channel banks under the influence of waves and currents. In consequence of these processses, the thickness of the "Wadden Sea formation" varies strongly from one place to another. The depths of the channels increase on the average toward the North Sea side, and the intensity of migration of the channel augments in the same direction; the general shape of the Wadden formation is, therefore, an irregular wedge, thinning landward, with many elongate, downward bulges cutting into the underlying formations.

The second important constituent of the Wadden sediments is mud, which is also mostly brought in from the North Sea. Much of it comes originally from large rivers draining into the North Sea. Another part is supplied by coastal erosion. The deposition of mud in the Wadden Sea is enhanced by various coagulating processes, including flocculation (q.v.) and the production of fecal pellets by mud-eating organisms, as well as by conditions of rapid sedimentation, whereby part of the mud gets covered under new sand before it can be resuspended. The hydrological properties of the Wadden Sea environment cause the muddy material to be concentrated along the inner shores and on the tidal divides (Van Straaten and Kuenen, 1957).

The third main constituent of the Wadden Sea sediments is mollusk shells, mostly produced in the area itself; a few are washed in by flood currents from the North Sea. Only few mollusk species live in the Wadden Sea, among them the cockle, *Cardium edule*, and the mussel, *Mytilus edulis*, but these species are represented by enormous numbers of specimens. The same is true for the other fauna: few organisms can stand the strong variations in temperature, salinity, and turbidity, and the tidal emergence of the bottom, but species that can survive profit from the relative scarcity of predators and from the wealth of food.

Development

Favorable circumstances for the origin of the Wadden Sea tidal flat belt were:

1. The intensity of wave action on and along the coast.
2. The relative rise of the sea.
3. The low elevation of the hinterland.
4. The availability of much loose sediment.
5. The strength of the tides.
6. The occasional occurrence of storm surges.

Owing to factor 1, a dune-covered coastal barrier system was produced, separating a low area (factors 2 and 3) from the open sea. Under influence of factors 1, 4, and 5, enough sediment could be carried into this area to keep its surface for the greater part above (the rising) low-tide level. At certain times during the late

Holocene, considerable areas were even silted up to marsh level, whereby the inlets in the barrier system deteriorated. Increased storm surge activity then reopened or widened the inlets; new channels were scoured out; and new sediments were deposited on the old marsh surface in between.

L. M. J. U. VAN STRAATEN

References

Gierloff-Emden, H. G., 1961. Luftbild und Küstengeographie am Beispiel der deutschen Nordseeküste, *Landeskundliche Luftbildauswertung im Mitteleuropäischen Raum*, Bad Godesberg, vol. 4, 117p.

Klein, G. DeV., and Sanders, J. E., 1964. Comparison of sediments from Bay of Fundy and Dutch Wadden Sea tidal flats, *J. Sed. Petrology*, **34**, 18-24.

Linke, O., 1939. Die Biota des Jadebusenwattes. *Helgoländer Wiss. Meeresunters.*, **1**(3), 201-348.

Postma, H., 1961. Transport and accumulation of suspended matter in the Dutch Wadden Sea, *Netherlands J. Sea Research*, **1**, 148-190.

Reineck, H.-E., ed., 1970. *Das Watt*. Frankfurt: Kramer, 142p.

Van Straaten, L. M. J. U., 1954. Composition and structure of recent marine sediments in the Netherlands, *Leidse Geol. Meded.*, **19**, 1-110.

Van Straaten, L. M. J. U., and Kuenen, P. H., 1957. Accumulation of fine grained sediments in the Dutch Wadden Sea, *Geol. Mijnbouw*, **19**, 329-354.

Van Voorthuysen, J. H., ed., 1960. Das Ems-Estuarium (Nordsee), *Verhand. K. Neder. Geol. Mijnbouwkd. Genoot.*, **19**, 1-300.

Vosjan, J. H., 1974. Sulphate in water and sediment of the Dutch Wadden Sea, *Netherlands J. Sea Research*, **8**, 208-213.

Cross-references: *Barrier Island Facies; Bay of Fundy Sedimentology; Lagoonal Sedimentation; Paralic Sedimentary Facies; Salt-Marsh Sedimentology; Tidal-Current Deposits; Tidal-Flat Geology.*

WATER-ESCAPE STRUCTURES

Water-escape structures are postdepositional sedimentary structures formed in loose, unconsolidated deposits as a result of pore-water escape. They may represent modified primary or previously formed postdepositional structures or entirely new structures formed during the rearrangement of sediment grains by escaping water. Deformation may result directly from pore-water movement or from current or gravity forces acting on a bed whose strength has been significantly reduced by pore-water movement. The general result of water escape is sediment consolidation.

Consolidation Processes

Sediment consolidation includes three processes of water-particle interaction. Most beds dewater by *seepage*, i.e., escaping water percolates through existing pore space without disturbing the sediment grains. In some cases, however, consolidation may involve the temporary breakdown of the sediment's grain-supported framework, accompanied by a rapid rise in pore-fluid pressure. *Liquefaction* occurs when a loosely packed sediment collapses, the grains losing contact with each other and settling briefly through their own pore water. The particles fall a short distance, fluid is displaced upward, and a more tightly packed, grain-supported framework is established. *Fluidization* occurs when a fluid forced vertically through loose sediment exerts an upward drag force on the grains equal to their immersed weight (Davidson and Harrison, 1963; Kunii and Levenspiel, 1969). The range of fluidization extends from a superficial fluid-escape velocity equal to the settling velocity of the grain aggregate to that equal to the fall velocity of a single sediment grain within the pore fluid (see *Fluidized and Liquefied Sediment Flows*).

Classification

Water-escape structures represent one type of soft-sediment deformation structure, a broad group of structures which form by the movement of masses of unlithified sediment and which include water escape, slump, depositional loading, and some tectonically induced structures (see *Penecontemporaneous Deformation of Sediments*). Four general geometric varieties of water-escape structures are recognized (Fig. 1; Lowe, 1975): (1) soft-sediment mixing bodies; (2) soft-sediment intrusions; (3) soft-sediment folds; and (4) consolidation or diagenetic laminae.

Soft-sediment mixing bodies represent layers or lenses of sediment that have been internally restructured during water escape but have not moved significantly relative to surrounding sediment. Based on the nature of solid-liquid interaction during deformation, hydroplastic mixing bodies, liquefaction layers and pockets (Fig. 1a), and fluidization layers and channels (Fig. 1b, d) can be defined. Hydroplastic mixing bodies, including layers of convolute laminae, generally appear as zones of soft-sediment folds. Liquefaction layers and pockets may show deformed primary structures, such as oversteepened cross-bedding, but are more commonly structureless or characterized by dish and pillar structures. Fluidization channels, representing well-defined local channels of fluid escape, are also termed *pillar structures* because of their appearance in vertical cross-section (Figs. 1b, d). As noted by Laird (1970), in some cases these "channels"

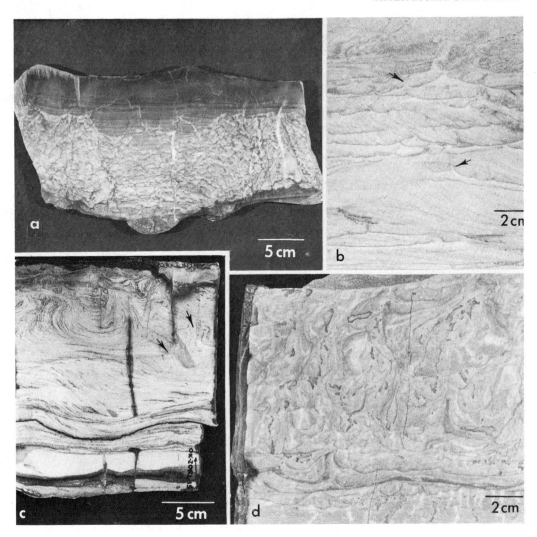

FIGURE 1. Water-escape structures. a: *Liquefaction layer and liquefaction pockets.* A single turbidite composed of graded, fine- to medium-grained, flat-laminated quartzite shows the disruptive effects of postdepositional liquefaction. Liquefaction was not uniform but marked by the development of small liquefaction pockets. Within each pocket, differential settling of heavy mineral grains formed a dark consolidation lamina along the base of the pocket overlain by clean, white structureless sand. Uncompahgre Group, Precambrian, Colorado. b: *Dish and pillar structures.* Top of a cross-laminated turbidite showing dark cross-cutting dish structures, marking secondary concentrations of fluidized clay and organic particles, and pillar structures (arrows), representing vertical water-escape channels. Jackfork Group, Pennsylvanian, Oklahoma. c. *Soft-sediment folds.* Convolute lamination and oversteepened cross-lamination (arrows) in a turbidite. Jackfork Group, Pennsylvania, Oklahoma. d. *Soft-sediment intrusions.* Load structures formed by the foundering of slightly denser sediment masses during liquefaction. Loaded masses are defined along their sides and lower boundaries by consolidation laminae representing secondary concentrations of fluidized clay and organic grains. White streaks near bottom of photograph are fluidization channels. Jackfork Group, Pennsylvania, Oklahoma.

are elongated in the third dimension; he therefore called them "sheet structures."

Soft-sediment intrusions form where hydroplastic, liquefied, or fluidized sediment is mobilized and intruded into adjacent strata. Geometrically, soft-sediment intrusions are commonly termed *clastic dikes* (q.v.) or *sills*. Hydroplastic intrusions tend to be more-or-less concordant with structures, generally deformed primary laminae, in the intruded sediment; fluidized intrusions are typically discordant; and liquefied intrusions may be concordant or discordant or both.

Intrusion is driven in most cases by gravity

forces resulting from density instabilities; dense sediment is present above low-strength, less-dense unconsolidated layers. Unequal gravity loading may cause local sediment foundering and associated upward movement (Fig. 1d); or sudden pressure release on confined sediment, as through joint formation, may result in rapid fluid escape, sediment fluidization, and the injection of large-scale clastic dikes (see *Clastic Dikes*). Where escaping pore fluids and fluidized sediment reach the surface, sand or mud volcanoes may develop (see *Volcanism, Sedimentary*).

Although many soft-sediment intrusions form during sediment consolidation and water escape, the formation of many others results from depositional density instabilities unrelated to fluid escape. Most such structures are termed *load structures*.

Soft-sediment folds represent deformed primary or diagenetic laminae. Most are geometrically irregular and complex, and form in association with differential loading, soft-sediment intrusion, or downslope movement. A few are relatively consistent in geometry and genesis. Convolute laminae are related to simultaneous water escape from liquefied sediment and foundering of supra jacent hydroplastic layers (Fig. 1c). Oversteepened and recumbently folded cross-bedding has been shown to form as a result of current drag on beds of cohesionless liquefied sand (Allen and Banks, 1972; Fig. 1c).

Consolidation or *diagenetic laminae* are entirely new laminae formed as a result of particle rearrangement by escaping pore water. In a liquefied layer, larger particles and dense heavy mineral grains may resediment through lighter and finer debris and accumulate as a compositionally or texturally distinct layer at the base of the liquefied zone (Fig. 1a). Also, very fine grains and easily moved mica and organic particles may be fluidized within the pore space between resedimenting larger liquefied grains and carried upward or laterally by escaping pore water. Such mobile grains commonly accumulate along the bases of dense, unliquefied sediment masses (Fig. 1b,d), particularly in systems within which liquefaction and/or resedimentation is nonuniform. Dish structures are laminae formed in this way (Lowe and LoPiccolo, 1974); they are typically thin, darker-colored than surrounding sediment, and concave upward as a result of subsidence of the unliquefied sediment pockets (Fig. 1b). Flat consolidation laminae may develop in poorly liquefied layers such as the B-subdivision of turbidites.

Consolidation and Water Escape

Causes. Numerous external disturbances can trigger the collapse of loosely packed cohesionless sediment. Cyclic earthquake vibrations have been cited as a major cause of sediment liquefaction and water escape (Seed, 1968; Seed and Lee, 1966), but most water-escape structures appear to form during the late stages of or very closely following the deposition of individual sedimentation units. Apparently, stresses related to current drag on the bed surface, sediment deposition, bed-form migration, waves, or other current activities can be equally important in initiating disruptive consolidation. Also, in deposits consisting of alternating sand and mud layers, such as turbidite and deltaic sequences, the abrupt deposition of a sand layer may induce rapid consolidation of subjacent clay and mud, a process termed loading consolidation (Lowe, 1975). Water squeezed upward through the sand may in turn trigger liquefaction of the sand itself.

Controls. There are at least four basic controls on the kinematics and dynamics of water escape and associated sediment deformation (Fig. 2): (1) grain size, (2) packing, (3) permeability and strength, and (4) hydrodynamic instability.

Grain size exerts a direct influence on the liquefaction and fluidization potential of sediments, as well as on other sediment properties such as strength and permeability. Coarse-grained sand and gravel are difficult to fluidize and resediment quickly if liquefied. The most common water-escape structure in such coarse deposits is oversteepened cross-bedding (Fig. 2). Very fine-grained silt and clay are usually cohesive and not readily liquefied or fluidized. Hydroplastic intrusions and soft-sediment folds may form, but are usually not related to water escape. The most easily liquefied and fluidized material is fine- to medium-grained sand, and such deposits typically display the best-developed water-escape structures.

Sediment packing controls the susceptibility of sediment to liquefaction and fluidization. Tightly packed sediment, with porosity below 40%, will generally offer much greater resistance to liquefaction than one more loosely packed. Fluidization requires a porosity of about 50%. Sediment packing is determined largely during deposition; cross-bedded and other rapidly deposited sands tend to be loosely packed, whereas flat-laminated sand deposited by high velocity currents is usually more tightly packed and difficult to liquefy (Allen, 1972).

Strength and permeability also play fundamental roles in controlling consolidation. In general, sediments with a high strength, either cohesive or frictional, resist liquefaction and fluidization and dewater instead by seepage. Nonuniform permeability within most sediments will localize water escape to the most

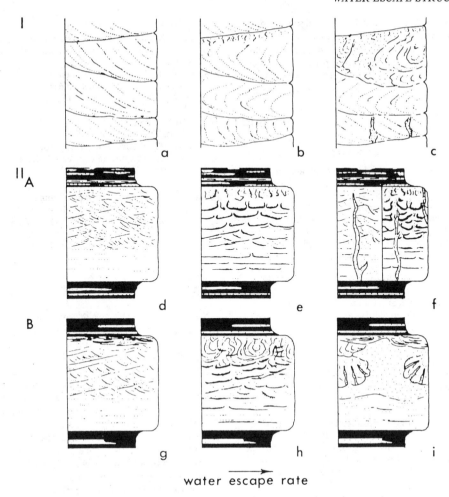

FIGURE 2. Common relationships among sediment grain size, strength, primary sedimentary structures, water-escape rates, and consolidation structures in natural deposits (from Lowe, 1975). I: Sequence of medium- to coarse-grained cross-stratified sands as occur in many alluvial and shallow marine deposits—(a) unmodified primary structures; (b) repeated liquefaction coupled with current drag acts to produce oversteepened and recumbent-folded deformed cross-bedding; (c) at higher water-discharge rates, local liquefied intrusives and small pillars evolve. II: Turbidite-type beds showing Bouma B, C, D, and E subdivisions. (A) Bed with noncohesive C and D subdivisions—(d) unmodified primary structures; (e) at discharges just above those characterizing pure seepage, elutriation, and redistribution of mobile grains in B subdivision form flat consolidation laminae and small pillars. Collapse of loosely packed sand in C subdivision results in partial liquefaction and wide development of pillars and strongly curved dishes. Bed surface is fully fluidized forming layer of free-surface pillars; (f) at highest discharge rates, structures formed depend on rate of discharge increase. If discharge rises abruptly, large pillars may develop across unmodified primary structures (left); if discharge increases gradually, large pillars can form, cross-cutting lower-discharge consolidation structures (right). (B) Bed with cohesive upper C and D subdivisions—(g) unmodified primary structures; (h) similar structures evolve as in (e) at low dewatering rates in noncohesive layers. Cohesive upper layers resist fluidization and deform hydroplastically into convolute laminae; (i) at highest water-escape rates, bed is fully liquefied; large liquefied intrusives and coupled subsidence lobes showing stress pillars develop.

permeable zones, which may be liquefied or fluidized, while surrounding less permeable sediment remains intact.

Hydrodynamic instability, in the form of Rayleigh-Taylor instability, plays a fundamental role in the formation and geometry of most water-escape structures (Anketell et al., 1970; Anketell and Dzulynski, 1968). Density or gravitational loading represents the primary driving mechanism of water escape; and the scale and geometric form of most soft-sediment intrusions, soft-sediment folds, pillar structures, and dish structures can be related to instabilities within moving, fluid-like masses of sediment. The main bulk sediment properties regulating the geometric and dynamic properties of Rayleigh-Taylor instabilities are viscosity and density respectively (Chandrasekhar, 1961; Ramberg, 1968a, b).

DONALD R. LOWE

References

Allen, J. R. L., 1972. Intensity of deposition from avalanches and the loose packing of avalanche deposits, *Sedimentology,* **18,** 105-111.

Allen, J. R. L., and Banks, N. L., 1972. An interpretation and analysis of recumbent-folded deformed cross-bedding, *Sedimentology,* **19,** 257-283.

Anketell, J. M., and Dzulynski, S., 1968. Patterns of density controlled convolutions involving statistically homogeneous and heterogeneous layers, *Rocznik Polskie. Towarz. Geol.,* **38,** 401-410.

Anketell, J. M; Cegla, J.; and Dzulynski, S., 1970. On the deformational structures in systems with reversed density gradients, *Rocznik Polskie. Towarz. Geol.,* **40,** 3-30.

Burne, R. V., 1970. The origin and significance of sand volcanoes in the Bude Formation (Cornwall), *Sedimentology,* **15,** 211-228.

Chandrasekhar, S., 1961. *Hydrodynamic and Hydromagnetic Stability.* London: Cambridge Univ. Press, 654p.

Davidson, J. F., and Harrison, D., 1963. *Fluidizied Particles.* London: Cambridge Univ. Press, 155p.

Florin, V. A., and Ivanov, P. L., 1961. Liquefaction of saturated sandy soils, *Proc. 5th Internat. Conf. Soil Mechan. Found. Engineers,* **1,** 107-111.

Kunii, D., and Levenspiel, O., 1969. *Fluidization Engineering.* New York: Wiley, 534p.

Laird, M. G., 1970. Vertical sheet structures; a new indicator of sedimentary fabric, *J. Sed. Petrology,* **40,** 428-434.

Lee, K. L., and Seed, H. B., 1967. Cyclic stress conditions causing liquefaction of sand, *Am. Soc. Civ. Engineers Proc., J. Soil Mechan. Found. Div.,* **93** (SMI), 47-70.

Lowe, D. R., 1975. Water escape structures in coarse-grained sediments, *Sedimentology,* **22,** 157-204.

Lowe, D. R., and LoPiccolo, R. D., 1974. The characteristics and origins of dish and pillar structures, *J. Sed. Petrology,* **44,** 484-501.

Ramberg, H., 1968a. Instability of layered systems in the field of gravity, I, *Phys. Earth Planetary Interiors,* **1,** 427-447.

Ramberg, H., 1968b. Instability of layered systems in the field of gravity, II, *Phys. Earth Planetary Interiors,* **1,** 448-474.

Seed, H. B., 1968. Landslides during earthquakes due to liquefaction, *Am. Soc. Civ. Engineers Proc., J. Soil Mechan. Found. Div.,* 94(SM5), 1053-1122.

Seed, H. B., and Lee, K. L., 1966. Liquefaction of saturated sands during cyclic loading, *Am. Soc. Civ. Engineers Proc., J. Soil Mechan. Found. Div.,* 92(SM6), 105-134.

Selley, R. C., 1969. Torridonian alluvium and quicksands, *Scottish J. Geol.,* **5,** 328-346.

Selley, R. C.; Shearman, D. J.; Sutton, J.; and Watson, J., 1963. Some underwater disturbances in the Torridonian of Skye and Raasay, *Geological Mag.,* **100,** 224-243.

Stanley, D. J., 1974. Structures en coupelle et coulées de sable dans les anciennes vallées sous-marines des Alpes maritime francaises, *Bull. Centre Rech. de Pau,* **8,** 351-371.

Wallis, G. B., 1969. *One-Dimensional Two-Phase Flow.* New York: McGraw-Hill, 408p.

Wentworth, C. M., Jr., 1967. Dish structure, a primary sedimentary structure in coarse turbidites, *Bull. Am. Assoc. Petrol. Geologists,* **51,** 485.

Cross-references: *Clastic Dikes; Compaction in Sediments; Fluidized Sediment Flows; Grain Flows; Gravity Flows; Penecontemporaneous Deformation of Sediments; Pillow Structures; Turbidity-Current Sedimentation; Volcanism, Sedimentary.*

WEATHERING IN SEDIMENTS

Weathering is the breakdown and alteration of materials near the earth's surface to products that are more in equilibrium with newly imposed physicochemical conditions (Ollier, 1969). This definition is easy to follow for a plutonic rock; the rock is formed under conditions of high temperature and high pressure in the absence of air and water, and when exposed at the earth's surface the rock is subjected to reduced pressure and temperature and the presence of air and water. This process also affects sediments, in a more limited way, for although temperature and pressure may not change much, air and water content and the course of chemical reactions can undergo considerable changes.

Chemical alteration in situ can be accepted as part of weathering but must be distinguished from diagenesis (q.v.), where chemical changes build up new minerals, largely as a result of burial pressure and the expulsion of air and water. Weathering should also be distinguished from soil formation, which is the production of certain layers or horizons within weathered material.

Weathering processes are conventionally classified as physical or chemical. Biological weathering is sometimes separated, but acts through physical or chemical methods. Physical weathering has little effect on sediments, for the sediment is free to expand or shrink, and volume changes in individual grains can be accommodated in voids. The only process of significance is "soil ripening" (Pons and Zonneveld, 1965; Ollier, 1969), a name given to the irreversible processes that occur when a water-laid sediment is exposed to the air for the first time. Physical ripening consists of the loss of water by compaction, evaporation, and transpiration.

Factors

If abrasion of mineral grains is counted as a weathering phenomenon, it is worth noting that minerals with good cleavage are most prone to weathering. This factor is important in windblown materials, where feldspars, amphiboles, and especially mica are rare. When eolian dunes do contain mica, e.g., those of California described by Reed (1931), it is a sure indication that they are immature and have not been transported far. Abrasion by glaciers is very great, affecting a wide range of minerals, and this process is probably the source of much fine-grained quartz. In running water, abrasion is much reduced.

Salt weathering may be important in coastal areas where gravels of quartz and other rocks may be splintered into angular fragments. Disintegration is caused by wedging of crystal growth, accompanied by some chemical weathering.

Chemical processes are much more important than physical ones in the weathering of sediments, with various processes operating in different hydrological zones. There is an aerated zone where air fills pore space almost permanently; a saturated zone below the water table where pores are permanently filled with water; and in a situation with variable water table level, there may be a zone with alternating air- or water-filled pores. The aerated zone is an oxidation zone; the saturated zone is a reducing zone. The aerated zone is also a leaching zone, from which material is lost by the downward passage of solution; the saturated zone may be an accumulation zone for some of the materials derived from the leaching zone.

The course of weathering of a sediment will depend on its mineral composition. There is no absolute scale of weatherability of minerals, but minerals may be ranked into a *weathering series*, according to ease of weathering. The order may change in different conditions; but for common minerals in leaching conditions a series might be, from most to least resistant: zircon, tourmaline, garnet, apatite, ilmenite, quartz, staurolite, kyanite, epidote, muscovite, hornblende, augite, biotite, feldspar, and olivine.

A rock with a high content of easily weathered minerals is more weatherable than one with a high proportion of resistant minerals and is said to have a greater *weathering potential*. A quartz sand has a very low weathering potential, a volcanic ash a high weathering potential.

Weathering of sediments is very much controlled by porosity and permeability, which govern the ease with which water can enter and weathering products be removed. Bedded sediments may have marked anisotropy of water movement, caused by the flat-lying platy minerals, which, coupled with variations in grain size and mineral composition, can lead to quite complex weathering profiles.

Weathering of Clay

Clay (q.v.) consists of the clay minerals together with varying amounts of water and minor quantities of other minerals. A water-laid deposit of clay is laminated, with the flaky clay minerals taking up a horizontal position. If the deposit is undisturbed, the clay slowly compacts and water is expelled, and the minerals may be affected by a certain amount of ion exchange and migration of salts if conditions permit (see *Compaction in Sediments*). If a clay dries out, there will be shrinkage, and the formation of desiccation cracks allows further penetration of air and perhaps some oxidation. Most changes will be reversed upon wetting. In waterlogged clays, bacterial activity is common and may lead to segregation of certain elements, the formation of secondary minerals, and consequent alteration of clay minerals. Waterlogged clays tend to be bluish-gray because iron is present in the ferrous state, while well-aerated clays are brown. In transition zones, there is usually color mottling (see *Color of Sediments*).

Impeded drainage promotes the formation of gley (or glei) soils, in which the three typical zones of ground water are reflected in soil and weathering horizons: (1) an aerobic zone, gray or brown in color; (2) a zone of ground-water fluctuation, gray, with yellow, red, or brown mottles or ferruginous concretions or iron pan; secondary gypsum, manganese dioxide and carbonates may also occur; and (3) a permanently anaerobic zone colored blue gray and containing iron sulfide. The three horizons tend to be thick in porous materials, thin in impermeable materials, and therein less distinct. Tropical gley soils (vlei, mbuga) have a black topsoil, but the typical mottling and gray clay is present in the lower part of the profile.

In soils, the clay minerals are not generally oriented horizontally but are arranged in far more complex ways, making up crumbs, blocks, and other soil structure units called peds. The development of soil structure allows more rapid leaching and therefore more rapid weathering than a simple, laminated clay deposit. Biotic activity is also increased in the soil zone, again enhancing weathering due to organic stirring and mixing, and chemical effects brought about by respiration, ion exchange, chelation and bacterial activity.

Clay aggregates can also be formed in dry salt lakes (playas). The crumbs are susceptible to wind transport and may be blown into clay dunes on the downwind shore of the lake bed. There the clay is reweathered by leaching out of salt, and oxidation.

In some soil and weathering profiles, there may be mechanical transport of clay particles down the profile (*eluviation*) and accumulation of clay in a lower situation (*illuviation*). This process was traditionally believed to cause the formation of *podzols*: the upper sandy part of the soil was thought to be an eluvial horizon from which clay had been removed, and the clay-rich subsoil was thought to be an illuvial horizon where clay had accumulated. Several detailed studies of podzol genesis have shown that in reality the clay in the subsoil is largely derived by weathering in place, and that clay produced by weathering in the topsoil has been completely removed, not transported to a lower zone. In studying soil and weathering profiles in sediments, it is often very difficult to determine which layers are original sedimentary layers, which ones are modified by simple weathering processes, and which are produced by pedological processes of eluviation and illuviation. Probably the best tool available for study of such problems is thin-section study of soils, *micropedology*. Illuviated clay that lines cavities (void argillans) is easily recognized, and the percentage of illuviated clay can be quantified (Brewer, 1972).

Clay minerals can themselves be weathered, that is one clay mineral can change into another which is more stable in the changed environmental conditions (see also *Clay—Genesis; Clay—Diagenesis*). The common clay minerals are built of layers of silica and alumina. The 2:1 clay minerals have a structure of one alumina layer sandwiched between two silica layers; montmorillonite and sometimes illite are examples. The 1:1 clay minerals consist of a stack of repeated layers of one alumina and one silica layer; kaolinite and halloysite are examples. If one silica layer is removed from a 2:1 sheet, a 1:1 sheet results, so that in a situation where silica can be removed, smectite minerals will be weathered into kaolinite minerals. In weathered sediments, smectite minerals are commonest in conditions of waterlogging, high pH, and presence of large amounts of cations. Many silicate minerals weather to smectites as a first stage of weathering. Illite may be derived clay mineral, but much illite is formed in marine sediments. In leaching, acidic conditions, kaolinite minerals will be produced by weathering, the mineral species kaolinite being the dominant material.

The formation of kaolin is favored if Na, K, Ca, Mg, and Fe ions are completely leached away and H ions introduced. It is especially important to remove the divalent ions Ca, Mg, and Fe. In conditions of extreme leaching, the kaolinite minerals can be further desilicified to form bauxites.

In many geomorphic situations, quite different processes affect various parts of a slope. This variance is best exemplified in the humid tropics, where hillslopes are usually thoroughly leached to form products rich in kaolinite and iron oxides. The weathering products are washed into swamps or alluvial flats, where accumulation of cations and silica produces montmorillonite or other smectites.

Weathering of Other Sediments

Sand. Sands are very porous and easily leached, so weatherable minerals are readily exposed, and weathering products removed. Well-drained sand thus tends to get ever cleaner with weathering, that is, minerals other than quartz are steadily removed. A pure quartz sand is virtually immune to weathering.

In waterlogged sites, the removal of weathering products is hindered even though movement of solutions through sand is still fairly rapid. In impure sands and sandy clays, weathering will be mainly concerned with the nonquartz fraction.

Alluvium. Sheets of detritus laid down by rivers, although in any one spot usually well sorted, can exhibit overall a wide range of grain sizes. Clays and sands behave as described above. Waterlogged soils are common on flood plains; but old alluvium of river terraces is usually well drained, oxidized, and leached. Ground water moves down the valley through alluvium just like the surface water, but much more slowly. Moving ground water is conducive to more rapid weathering because weathering products are carried away in solution.

An important factor in weathering of alluvium is that it is bedded, and successive beds may be of different texture and mineral composition. Great care is required in studying alluvial weathering profiles to distinguish layers of depositional origin from layers produced by weathering. For example, a layer of nonmicaceous silt over a layer of micaceous silt might be due to weathering of mica from the upper layer, or due to deposition of an upper layer that never contained mica.

Weathering products tend to be precipitated at junctions between layers of very different permeability. Quite frequently, iron oxide accumulation binds the basal gravels into a ferruginous conglomerate, sometimes firmly cemented to the bedrock.

A mixture of sand and clay does not behave like either individual. Sand prevents "panning" of clay and enhances leaching, clay alteration, and eluviation. In periglacial climates the process of frost heaving is best displayed in mixing deposits, and although this achieves little weathering itself, it affects other weathering processes.

Changes in mineral content of alluvium in a downstream direction are mostly brought about by attrition or sorting, although weathering phenomena such as solution may also be operative.

Loess. Two features of loess (q.v.) weathering must be discussed: (1) what happens when loess weathers, and (2) to what extent loess itself is formed by weathering. The weathering of loess is the simpler of these problems. Loess is very porous, so there is strong leaching and eluviation. Decalcification occurs in the leaching zone, and there is sometimes precipitation of a carbonate layer at depth. Soils are formed very rapidly on loess and have been used extensively in stratigraphy (Schultz and Frye, 1968).

Regarding the second problem, although it is generally accepted that loess deposits form by eolian action, the concept that they can be produced by a kind of weathering or "loessification" has at times been supported, notably by Berg (1964) and Russell (1944). These theories of in-situ formation of loess have been effectively refuted by Smalley (1971), who concluded that the fine quartz in loess could only be produced by glacial grinding and that loess deposits were produced by eolian sorting and deposition of this material. The universal presence of carbonate in loess is not so easily explained, and Smalley only says that it is deposited on quartz grains in a later stage.

Till. Till (q.v.) is poorly sorted but commonly contains a large amount of clay, which often makes till impermeable and so affects the course of weathering; hydromorphic soil types such as surface water gleys are common. Where the original till is calcareous, decalcification is an important weathering process. Weathering tills are commonly found in the podzol zone, and various kinds of podzol and podzolic soil occur. Iron pans are common. In North America, the unctuous, sticky soils that result from weathering of till are known as *gumbotil*. The depth of weathering of gumbotil is believed to be dependent upon time. Many detailed studies of the weathering of till are available, e.g., Willman et al. (1966).

Coastal Sediments. Coastal sand dunes behave much as other sand, except that they may be more calcareous, and so carbonate leaching assumes greater importance. The same applies to most beach sands.

Coastal sediments may be initially salt rich, but leaching rapidly removes salt when weathering starts. The salt may, however, affect the course of clay weathering and eluviation, leading to solodic soils as a result of the sodium-induced deflocculation of clay. It is possible that windborne salts may affect the weathering of sediments within several km of the coast.

Uplifted marine sediments will be weathered in much the same way, with variations depending on original composition and texture. Included salts will be leached very rapidly, followed by decalcification and, eventually, weathering of clays and mineral grains. Illite is commonly produced in the marine environment and is likely to be converted into other clay minerals during the course of weathering.

Volcaniclastic Deposits. Sediments containing a large amount of volcanic ash undergo very rapid weathering, especially if the ash is basic in composition. Volcanic glass is very rapidly weathered, but the base-rich silicate minerals are not far behind. The rapidity of weathering is such that a distinct type of soil may be formed, an *andosol,* which is remarkable in having a larger clay content in the topsoil (where weathering is strongest).

Duricrusts. Duricrusts (q.v.) are hard layers within weathering profiles, and are classified as ferricrete, calcrete, silicrete, or alcrete depending on the dominant chemical composition (Goudie, 1973). Various weathering processes of relative or absolute concentration can produce duricrusts in a wide range of sediments, as well as on bedrock. In some instances, a sediment consisting largely of duricrust fragments may be recemented to make a secondary duricrust.

CLIFFORD D. OLLIER

References

Berg, N. D., 1964. Loess as a product of weathering and soil formation (trans. from Russian). Jerusalem: Israel Program for Scientific Translations, 207p.

Brewer, R., 1972. Use of macro- and micromorphological data in soil stratigraphy to elucidate surficial geology and soil genesis, *J. Geol. Soc. Australia,* **19,** 331-334.

Goudie, A., 1973. *Duricrusts in Tropical and Subtropical Landscapes.* Oxford: Clarendon Press, 174p.

Hester, N. C., 1974. Post-depositional subaerial weathering effects on the mineralogy of an Upper Cretaceous sand in southeastern United States, *J. Sed. Petrology,* **44,** 363-373.

Mann, W. R., and Cavaroc, V. V., 1973. Composition of sand released from three source areas under humid, low relief weathering in the North Carolina piedmont, *J. Sed. Petrology,* **43,** 870-88.

Mousinho De Meis, M. R., and Amador, E. S., 1974. Note on weathered arkosic beds, *J. Sed. Petrology,* **44,** 727-737.

Ollier, C. D., 1969. *Weathering.* Edinburgh: Oliver and Boyd, 304p.

Pons, L. J., and Zonneveld, I. S., 1965. Soil ripening and soil classification, *Internat. Inst. Land Reclam. Improvement, Netherlands, Publ. 13*, 128p.

Reed, R. D., 1930. Recent sands of California, *J. Geol.*, 38, 223-245.

Russell, R. J., 1944. Lower Mississippi Valley loess, *Geol. Soc. Am. Bull.*, 55, 1-40.

Schultz, C. B., and Frye, J. C., eds., 1968. Loess and related eolian deposits of the world, *Proc. 7th INQUA Congress*, 12, 369p.

Smalley, I. J., 1971. "In situ" theories of loess formation and the significance of the calcium-carbonate content of loess, *Earth-Sci. Rev.*, 7, 67-85.

Willman, H. B.; Glass, H. D.; and Frye, J. C., 1966. Mineralogy of glacial tills and their weathering profiles in Illinois, *Ill. St. Geol. Surv. Circ. 400*, 76p.

Cross-references: *Carbonate Sediments—Diagenesis; Clastic Sediments—Lithification and Diagenesis; Clay—Diagenesis; Clay—Genesis; Color of Sediments; Compaction in Sediments; Diagenesis; Duricrust; Loess; Maturity; Volcanic Ash—Diagenesis;* Vol. III: *Regolith and Saprolite; Weathering.* Vol. XI: *Weathering and Soil Formation.*

WELDED TUFF—*See* IGNIMBRITES

WENTWORTH SCALE

The Wentworth grade scale of sediment sizes is a modification of the Udden Scale (q.v.) and is often called the Udden-Wentworth Scale (see also *Grain-Size Studies*). The scale was proposed by Wentworth (1922) as a means of bringing order to the terminology for coarse-grained sediments and rocks. Wentworth added the terms granule, pebble, and cobble to Udden's scale and clearly defined separate terms for grains, granular aggregates, and rocks.

The importance of maintaining a constant geometric ratio between grades and the desirability of using 2 as the ratio were recognized and discussed at length by Wentworth. He considered the starting point of the scale to be 1 mm and suggested that the scale could be further subdivided by ratios of the square root, or fourth root of 2. In order to preserve the

TABLE 1. The Wentworth Scale

The Pieces	The Aggregate	The Indurated Rock
Boulder[a]	boulder gravel	boulder conglomerate
256 mm		
Cobble	cobble gravel	cobble conglomerate
64 mm		
Pebble	pebble gravel	pebble conglomerate
4 mm		
Granule	granule gravel	granule conglomerate
2 mm		
Very coarse sand grain	very coarse sand	very coarse sandstone
1 mm		
Coarse sand grain	coarse sand	coarse sandstone
1/2 mm		
Medium sand grain	medium sand	medium sandstone
1/4 mm		
Fine sand grain	fine sand	fine sandstone
1/8 mm		
Very fine sand grain	very fine sand	very fine sandstone
1/16 mm		
Silt particle	silt	siltstone
1/256 mm		
Clay particle	clay	claystone

[a]Wentworth (1922) used the spelling "bowlder" throughout.

rational basis of the scale, Wentworth set the size limits of the granule through boulder classes at the nearest power of 2 that conformed to general usage of the time. The Wentworth Scale, with the associated grain, aggregate, and rock terms are presented in Table 1.

ROBERT E. CARVER

Reference

Wentworth, C. K., 1922. A scale of grade and class terms for clastic sediments, *J. Geol.*, **30,** 377-392.

Cross-references: *Alling Scale; Atterberg Scale; Grain-Size Studies; Granulometric Analysis; Phi Scale; Udden Scale.*

Z

ZEBRA DOLOMITE

This term has been applied to a banded dolomite, found in the Leadville district of Colorado, which was believed to have been formed by hydrothermal alteration of the Mississippian Leadville Limestone (Emmons) et al., 1927). Other mechanisms have since been proposed that seem more accurate (Beales and Onasick, 1970; see also *Sulfides in Sediments*). The zebra rock is made up of an inequigranular dolomite in which the striped appearance is caused by subparallel veinlets of light gray to white, coarsely crystalline dolomite in a darker, finer-grained, dense matrix. The coarse-grained light phase also contains numerous vugs, filled with white dolomite and quartz crystals. These stripes tend to be parallel to the bedding, although they may diverge by as much as 10–30°.

The gray color of the fine-grained dolomite may range to almost black, and Engel et al. (1958) have found that it is caused almost entirely by the presence of carbonaceous matter. The only other difference noted in the composition of the two types of layers is that of oxygen isotope concentrations. The origin of this unusual textural alteration begins with the formation of incipient fractures along or subparallel to the original bedding of the dark-colored limestone (relict bedding). These fractures then served as loci for later crystallization. Some solution of dolomite took place during and after recrystallization, followed by the growth of crystals of quartz, dolomite, and some pyrite in the cavities.

B. CHARLOTTE SCHREIBER

References

Banks, N. G., 1968. Geology and geochemistry of the Leadville limestone (Mississippian, Colorado) and its diagenetic supergene, hydrothermal and metamorphic derivations, Dissertation abst., Section B, *Sci. and Eng.,* 28(9), 3747B–3748B.

Beales, F. W., and Onasick, E. P., 1970. Stratigraphic habitat of Mississippi Valley type ore bodies, *Inst. Min. Metal. Trans.,* 79, B145–B154.

Emmons, S. F.; Irving, J. D.; and Loughlin, G. F., 1927. Geology and ore deposits of the Leadville mining district, Colorado, *U.S. Geol. Surv. Prof. Pap. 148,* 368p.

Engel, A. E. J.; Clayton, R. N.; and Epstein, S., 1958. Variations in isotopic composition of oxygen and carbon in Leadville limestone (Mississippian, Colorado) and in its hydrothermal and metamorphic phases, *J. Geol.,* 66, 374–393.

Cross-references: *Carbonate Sediments–Diagenesis; Dolomite, Dolomitization; Limestones; Sulfides in Sediments.*

ZEOLITES—See VOLCANIC ASH–DIAGENESIS

ZINGG SHAPE CLASSES—See ROUNDNESS AND SPHERICITY

Index

Ablation till, 362, 808
Abrasion, 869
  in attrition process, 18
  of detrital sedimentary rock, 534
  eolian, 323
Abrasion pH, 1
Abyssal plains, 202, 395, 396. *See also* Deep sea
  and plate tectonics, 663
  sedimentation, 1-4, 693
Abyssal zone, 1, 246
Accretion, 4
  of alluvial sediments, 12
  of fluvial sedimentation, 336
  lateral and vertical, 430
Accretionary lapilli, 847
Accretion topography, 4
Acetate peels, 700
Acoustic anisotropy, 321
Actualism, 708
Adhesian ripples, 52
Aeolian. *See* Eolian
Agglomerates, 5
Agglutinates, 5
Aggradation, 5, 154, 155. *See also* Accretion
Aggraded valley plains, 5
Aggrading neomorphism, 101, 102
Air-sea-particle interactions, 722
Alacran Reef Complex, 826
Algae
  of carbonate sediments, 90
  endolithic, 7
  sedimentary role of, 31-32, 403, 685
Algal banks, 6
Algal bench, 6
Algal crusts, 7
Algal mats, 54
  blue-green, 32
  of coastal desert sediments, 250
  stromatolites produced by, 772
Algal reefs, 5-8
Algal-sponge reefs, 759
Algal tufa, 5
Alling scale, 8-9, 377
Allochems, 9, 677
Allochthonous shelf sedimentation, 195-196
Alloclastic fragmentation, 851
Allogene, 20
Allothigene, 20
Alluvial breccias, 86

Alluvial channels
  deposits, 118
  flow regimes in, 331
  fluvial sediment transport in, 339-343
Alluvial-fans
  deposits, 337
  mudflows, 490-491
  sediments, 10-11
Alluvial islands, 336
Alluvial plains, 227
Alluvial sediments, 11-13, 82, 335, 447, 870
Aluminosilicate, 137
Aluminum, in clastic sediments, 129
Amber, 618-620
American Association of Petroleum Geologists (AAPG), 717
Amino acid racemization, 688
Ammonia, in authigenesis, 21
Anadiagenesis, 21, 252
Analog models, 311, 313
Anamorphism, 252
Andean-type junctions, plate tectonics of, 664
Andosol, 871
Andros Island, 30, 31, 32, 825
Anhydrite, 113, 765
Anisotropy, 13-14
Ankerite, 267
Anoxic systems, 297
Antarctica
  chlorite in, 145
  glacial marine sediments, 355, 356
Antarctic Bottom Water, 554
Antarctic Convergence, 471
Antarctic-shelf sediments, 200
Anthracite stages, 170-171
Anthropogenic wastes, 295
Antidune bedform, in Bouma sequence, 81
Antidunes, 28, 39, 49, 56, 58
Anomalous magnetic fabrics, 14
Apatites, 575
API gravity, 213
Appalachian sedimentary sequence, 351
Applied sedimentology, 718
Appositional fabric, 320
Aragonite, 112
  and bacterial precipitation, 90
  on Bahama Banks, 32
  in beachrock, 45
  and bedding genesis, 53
  in carbonate sediments, 105, 106
  micrite-sparite range of, 99

in microbial cultures, 91
morphology of, 96
stains specific for, 764
Aragonite compensation depth (ACD), 524
Aragonite needles, 31, 91, 93, 443
Aragonite skeletons, 105
Archeolithophyllum, 8
Arctic Ocean
chlorite in, 145
glacial marine sediments of, 355, 356
Arctic-shelf sediments, 200
Arenaceous rocks, 15, 677
Arenite group, 130
Argillaceous rocks, 15–17, 677
Argillaceous sediments, 854
Argillites, 130, 493, 736
Arguate delta, 236
Arkansas Novaculite, 504
Arkose, 17, 160, 352, 647, 727
Armored mud balls, 10
Arnenite, 15
Aromatics, in crude oil, 215
Arroyo system, 830
Ash, 844. *See also* Volcanism
Ash flows, 38, 843
Asphalt rock, 794, 795
Asphalts, 519, 520
Atlantic coast
barrier systems of, 35
lagoonal sedimentation in, 428–429
Atlantic Ocean
algal reefs in, 6
atmospheric dusts for, 277–278
geosyncline of, 352
glacial marine sediments of, 355
kaoline content of, 144, 145
Atmosphere, transport processes in, 818
Attapulgite, 60, 163
Atterberg scale, 18, 377
Attrition, 18–19
Authigenesis, 20–24, 254
of abyssal sedimentation, 2
of clay minerals, 152, 161
and diagenetic facies, 22
of smectite, 146
Authigenic environment, sedimentation rate in, 691
Authigenic minerals, 91, 116, 198, 471
Autochthonous sedimentation, 193–196, 199
Autoclastic breccias, 84
Autoclastic fragmentation, 851
Autosuspension, 837
Avalanche deposits, 24–27
Avalanching, 479

Backbarrier, 34
Back-barrier deposits, 197
Backbarrier facies, 35–36

Back filling, 70
Backset bedding, 28–29, 49
Back-swamp sediments, 12
Bacterial processes
aerobic, 21
anaerobic, 21, 75
and bedding genesis, 54
in carbonate cements, 112
in carbonate sediments, 441
and crude oil deposition, 214
denitrifying calcium, 92
in evaporite facies, 301
in interstitial waters, 409
as microenvironmental regulators, 92
in organic sediments, 517
sedimentary role of, 31–32
and submarine cementation, 115
Bagnold effect, 29
Bahama Banks, 30–33, 92, 101, 270, 280, 384, 435, 500, 512, 513, 797, 800, 825
Bahamite, 384, 443
Bajada breccias, 85
Ball clays, 137, 138
Ball-and-pillow structures, 579. *See also* Pillow structures
Baltic amber, 619–620
Banded iron formation (BIF), 124, 225
Banding, of beach sands, 37
Barchan dune, 286
Bar-finger sands, 238
Barnacles, 7
Barrier environments, 34–35
Barrier flats, 35–36
Barrier island facies, 33–37, 428, 564, 862
Barrier-lagoon complex, 34
Barrier rim, of Bahama Banks, 30
Barrier toe, 34
Bars
and bedding genesis, 49
in bed forms, 56
in braided-stream deposits, 82
littoral, 451
Basal breccias, 86
Basal drift, 806
Basal till, 808
Base-surge deposits, 38–40
Base surges, 506
Basin analysis, 352
Basins, lake, 421
Bassanite, 392
Bathyal zone, 1, 246
Bathymetry, of Persian Gulf, 562
Bauxite, 426
Bay of Fundy, 41–43, 799
Bays
clay sedimentation facies in, 161–162
sedimentation rates in, 689, 670
Beach deposition, 245–246
Beach environments, 34, 36

Beach facies, 35-36
Beach gravels, 43-44, 323
Beach profiles, 450
Beachrock, 44-45, 87, 90, 93, 112, 281
Beach sands, 45-47, 74, 375, 376
Beach sediments, 73
Bedded cherts, 54, 122, 746
Bedded iron oxides, 415
Bedded iron silicates, 416
Bedded iron sulfides, 415
Bedding, 47-55
Bedding planes, 48
Bed forms, 56-59, 220, 332
Bed interface, 58
Bed load, 339, 340, 448
Bed-load equations, 342
Beds, 679
Beekite, 120
Bengal Delta Fan, 662, 663, 781, 782, 785
Benthic organisms, 470, 524
Bentho-abyssal scheme of zonation, 1
Bentonite, 59-60, 137, 848, 851
Benzene, 405
Bering Sea, 356
Berm crests, 46
Bermudas
    carbonate sediments of, 101
    contourites, 202
    cup reef, 8
    eolianite, 280
    submarine cementation in, 114
Bermuda Platform, tropical lagoons of, 826
Bernoulli's theorem, 60-61, 620, 732
Bikini atoll, 6, 205, 206, 824
Bioconstructional lip, 6
Bioerosion, 684, 760, 792
    algal reef, 7
    coral reef, 7
Biofacies, 324, 447
Biogenic debris, in estuarine sediments, 295
Biogenic oozes, 2, 470
Biogenic precipitation, 536
Biogenic structures, 61-64, 590
Biohermal complex, 438, 613. *See also* Reef complex
Biolithite, 439
Biological systems, Eh-pH characteristics of, 21
Biologic criteria, 811
Biophases, 223
Biosphere, 515
Biostratification structures, 68
Biostratinomy, 65-67, 442
Biostrome, 438
Biotite, 600
Bioturbation, 14, 61, 62, 68-71, 254, 685, 869
    on Bahama Banks, 31
    in Bouma sequence, 80
    of continental rise sediments, 188
    coral reefs, 205

of deltaic sediments, 238
and estuarine sediments, 294
in euxinic facies, 299
in outer continental shelf sediments, 522
and sedimentary transport, 65
of tidal flats, 800
Birdfoot delta, 236, 240
Bischofite, in evaporite sequences, 308, 309
Bitumen, 72
    in crude oil composition, 218
    in organic sediments, 515
    origin of, 516
Bituminous sediments, 72, 170-171, 224. *See also* Coal
Bivalves, in Black Sea sediments, 76. *See also* Mollusks
Biwa-ko, Lake, 424
Black sands, 72-75, 173
Black Sea, 300, 668
    anoxic conditions in, 297
    bathymetric chart, 76
    benthic invertebrate groups in, 299
    geological history of, 76
    marine muds from, 179
Black shales, 61, 72, 296
Blake-Bahama Outer Ridge, 202, 348
Blake Plateau, 114
Blow-outs, 286
Blue clay, 138
Bog iron ore, 416
Bogs, raised, 541
Boilers, 6
Boiling water, 856
Bonaire, 270, 272
Bone beds, 67
Bone breccias, 84
Boring organisms
    algae, 7, 257
    sponge, 8, 760
Borings, 93, 590
Bosses, 6
Bottom currents, and continental rise sediments, 186, 187. *See also* Currents
Bottomset beds, 591
Bottom shear stresses, 190
Boulder clay, 361
Boulders
    in avalanche deposits, 25
    in mudflows, 490
Bouma sequence, 51, 78-81, 834
Bounce marks, 811
Bounce molds, 108
Boundstones, 444
Box cores, 472
Boxwork, 81-82
Brackish water shales, 16
Brahmaputra river system, 12, 82
Braided streams, 329
    and aggradation, 5

in alluvial-fan sediments, 11
deposits, 82-83, 363
Braunkohle stage, of coal diagenesis, 168-170
Breaker zone, 6, 56, 451
Breccias, 156-157, 189. *See also* Rauhwackes
alluvial, 86
collapse, 234, 264-265
fault, 264-265
lunar, 460
pillow, 859
sedimentary, 84-86
soil, 459
solution, 751-752
volcanic, 264-265, 852
Brick earth, 138
British Honduras, 209, 825
Bromine, in evaporite sequences, 310
Brownian movement, 18
Brown rock deposits, 577
Brush marks, 811
Burning seas, 856
Burrowing organism, 61, 484
in biogenic sediments, 62
in graded bedding, 367
in lagoonal sediments, 428

Cahcedony, 119, 133
Calcarea, 759
Calcarenites, 9, 15, 87-88, 279, 440, 677
Calcareous aggregates, in barrier flats, 36
Calcareous algae, 6, 7
Calcareous oozes, 2, 54, 364, 435, 471, 547
Calcareous sand, radioactivity in, 604
Calcareous tufa, 819
Calcilutite, 8, 87-88, 443
Calcirudite, 9, 87-88
Calcisiltite, 87-88
Calcite, 112
on Bahama Banks, 32
in beachrock, 45
and bedding genesis, 53
in carbonate sediments, 111
Dolomite replacement of, 267
micrite-sparite range of, 99
morphology of, 96
stains specific for, 764
Calcite compensation depth (CCD), 2, 524, 546
Calcithites, 437, 440
Calcitization, 233, 234
Calcium carbonate, 3, 96
Calcium carbonate compensation depth (CCCD), 461
Calcium sulfate, 305
Calcrete, 88-89, 325
Caliche, 84, 88-89, 440, 583
Canonical plotting, 381
Canyon heads, 777-778
Canyons, submarine, 368, 775, 777-780, 829-830

Carbohydrates, in organic sediments, 516
Carbon, in sediments, 484, 520
Carbon-14, 602, 758
Carbonaceous compounds, coloring of, 174
Carbonate cements, 96, 111
Carbonate dissolution, 548
Carbonate grains, of Bahama Banks, 31
Carbonate oozes, of pelagic sedimentation, 3
Carbonates, 435, 438
Bermuda holocene, 106
chemical composition of, 435
and clay genesis, 155
compressibility, 178
in concretions, 182
grain-size scale, 87
in marine deposits, 471
in petrology, 672
Pleistocene, 106
sedimentary processes, 32
shallow marine, 469
in speleal sediments, 756
staining of, 764
sulfides in, 789
Carbonate sedimentation, 821-828
from Bahama Banks, 30, 660
bacterial precipitation, 90
diagenesis, 94-104
lithification of, 105-107
porosity types in, 98
Carbonate shelf deposits, 827
Carbonate shorelines, 660
Carbonate tidal flats, 800
Carbon dioxide, 21, 215
Carbonification, 168, 171
Caribbean Sea, 6, 146, 271
Carnallite, in evaporite sequences, 307-309, 310
Caspian Sea, 300, 411
Casts, 107-109
Catalysis, in petroleum genesis, 570
Catskill sequence, 338
Caustolites, 619
Cave onyx, 819
Cave pearls, 109-110
Caves, 754, 757, 758
Cave sediments. *See* Speleal sediments
Cay rock, 30, 279
Cellulose, in organic sediments, 516
Cementation, 8, 110-114, 675, 722
in algal reefs, 7
in authigenesis, 21
of carbonate sediments, 100
in clastic sediments, 133
in coral-reef sediments, 204-205
and diagenesis, 254, 259
of grapestone, 384
of hardgrounds, 399
in intertidal zone, 44
of limestone fabric, 434
rim, 260

shallow-marine, 93
and stylolites, 588
submarine, 114–115
Cements
of breccias, 84
carbonate, 111
in limestone, 443
for lithification, 105
in optical continuity, 203
silica, 112–113
types, 101
Cenuglomerate, 85
Ceramics, 138
Chalcedony, 112, 743
Chalk, 115–117
flint nodules, 119
hardgrounds, 398
skeletal components of, 117
Challenger expedition, 4, 316, 468, 524, 544, 554, 707, 709
Channel fill, 11, 337
Channels, 654
deposits, 12, 776
rip-scoured, 453
Channel sands, 118
Chaotic sediments, 828
Chelating agents, humic matter as, 402
Chemistry, of anoxic systems, 297
Cherts, 54, 119–123
in carbonate sediments, 103
of chalk sediments, 117
origin of, 745
sandstone, 601
tripolite, 504
Cherty iron formation (BIF), 124. *See also* Ironstone
Chesapeake Bay, 195
Chevron marks, 810
Chitin, in organic sediments, 516
Chlorite, 16
in bentonite, 60
in clay diagenesis, 149, 150
in clay sedimentation facies, 160
in deep-sea clays, 158
in world ocean, 145
Chlorite/kaolinite ratios, 158
Chondrites, in biogenic sediments, 61
Chute-and-pool phase, of base-surge deposits, 39
Chute-and-pool structures, 28, 39
Chutes, 49, 57
Circumcrusts, algal, 441
Clastic-calcareous cycles, 228
Clastic dikes, 125, 865
Clastic limestones, 78
Clastic linear shorelines, 660
Clastic particles, 676
Clastic ratio, 131
Clastic rocks, 459–460, 676
Clastic sediments, 176, 756

classification of, 126–131
lithification and diagenesis, 132–136
in petrology, 672
Clasts, definition, 126
Clathrates, 347
Clay diagenesis, 144, 217
Clayel materials, 180
Clay-marl-lime cycle, 228
Clay minerals
in ancient argillaceous rocks, 16
authigenesis, 143
in chalk sedimentation, 116
fabric types in, 135
Clay-pebble conglomerates, 85, 156–157
Clay-quartz contact, 588
Clays, 136–139
clay-mineral suites in, 16
commercial uses of, 138
compressibility, 178
deep-sea, 157–159
deposition of sediments, 499
diagenesis of, 149–151
genesis of, 152–155
grade scales for, 127
in marine sediments, 471
prodelta, 591, 592
properties, 136
red, 612–613
as sediment, 139–148
shales, 137, 139
varved, 841–843
weathering of, 869
Clay sedimentation facies, 159–165
Claystones, 15, 122, 493–494, 736
Climate, 706
and clay sedimentation facies, 164
and deltaic sedimentation, 241
and genesis of clays, 154
and outer continental shelf sediments, 521
and peat bogs, 542
and sedimentation, 30
Climbing ripples, 80
Closed-cast movement, 558
Coacervates, 328
Coal, 165–167
birdseye, 71
bituminous, 72
classification of, 166–167
in delta sedimentation, 242, 243
diagenesis of, 167–172
in earth's crust, 520
and humic matter, 402
and kerogen, 419
in lacustrine sedimentation, 426
metamorphism of, 519
in organic sediments, 515, 517
origin of, 516
radioactivity in, 605
in sedimentary rock, 678
torbanite, 515

Coal analysis, 167
Coal beds, 54
Coalification, 166, 168, 169, 170, 172
Coal-measure facies, 338
Coal seams, 533
Coal sequences, 227, 813
Coaly kerogen, 419
Coarse fraction analysis, 672
Coarsening upward sequences, 224, 226, 242
Coastal deposits, mineral concentration in, 401
Coastal sediments, 249, 538, 871
Coasts
  and lateral accretions, 431
  submerging of, 35
Cobalt, in iron spherules, 317
Coccolith ooze, in Black Sea sediments, 75
Coccoliths, 3, 116, 553
Colloform structures, 790
Colloidal phenomena, 483
Collophane, 207, 208, 574
Color
  of euxinic sediments, 297
  of sediments, 173-175, 483
Color charts, 175
Comminution, of soil, 458
Compaction, 675
  of argillaceous sediments, 149
  in authigenesis, 21
  of carbonate, 101
  differential, 176
  and interstitial water chemistry, 179
  and petroleum migration, 572
  in sediments, 176-182
  in submarine canyons, 777
Compaction fluids, 788
Compensation depth, 423
Compressibility, 177
Concentrate, 72. *See also* Black sands
Conceptual models, 714
Concretions, 4, 64, 182-183
  in authigenesis, 21
  calcium carbonate, 92
  cave pearls, 109
  in cyclic sedimentation, 227
Cone-in-cone structures, 587, 657
Conglomerates, 84, 183-186
  edgewise, 85, 157
  in submarine canyons, 778
  sulfides in, 789
Consolidation, 176
Contact load, in fluvial sediment transport, 339, 340
Continental clays, 16
Continental environment, and plate tectonics, 662, 663
Continental margins
  plate tectonics of, 663, 664
  subsidence of, 527
Continental rise, 3, 186-189, 693, 784

Continental sedimentation, 189, 604
Continental shelf
  and lagoonal sedimentation, 427
  outer, 521
  petrology of, 668
  and plate tectonics, 663
  sand and gravel sheets on, 499
  sedimentation, 190-196
  sediments, 196-201, 468, 693
  stratigraphic column on, 197
  tidal deposits on, 797
Continental slope, 693, 774-777, 785
Continental/ocean environment, plate tectonics of, 664
Contorted bedding, 53
Contour currents, 201, 835
Contourites, 3, 187, 201-202, 430
Contraction cracks, 486
Convergent plates, 662
Convoluted bedding, 79, 559
Coorong Lagoon, 268, 270, 271
Copal, 618
Copper, in black shale, 298, 299
Coprogenic fossitexture, 70
Coquina, 54, 203
Coralline algae, 6, 8
Coral-reef sedimentology, 204-207, 691
Corals, 7, 31, 822
Coring techniques, 483
Coriolis effect, 187, 782
Corniche, 6
Cornieule, 234
Cornstone, 89
Correlation methods, petrographic, 668
Corrensite, 150, 155
Corrosion surface, 207-209
Cosmic spherules, 316
Cracks
  desiccation, 486
  syneresis, 790, 791
Crandallite, 575
Creep movements, deposits accummulating from, 479
Crescentic bars, 453
Cretaceous system, 115
Crevasse-splay deposits, 337
Crinoidal sedimentation, 66
Criquina, 203
Cristobalite, 122
Cross-bedding, 209-212, 679, 812
  in alluvial sediments, 13
  in deltaic sediments, 237
  eolian, 212
  of fluvial sediments, 338
  graphic representation of, 530
  in littoral sedimentation, 453
  marine, 212
  of oolite sand, 31
  overturned, 559
  trough, 279

Cross-lamination, 50, 209
Cross-stratification, 209, 210
    of braided-stream deposits, 83
    tabular, 211
    trough, 211
Crude bedding, 54
Crude oils
    classes of, 213, 214
    composition of, 212-216
    hydrocarbons in, 405
    interstitial water chemistry, 179
    migration, 216-220
    in tar sands, 793
Crustose, 6, 8
Cryoturbation, 560
Cryoturbation structures, 68
Cryptalgal, 771
Cryptalgalaminite, 55
Cryptobioturbation, 70. See also Bioturbation
Cryptopores, 119
Crystal euhedra, 22
Crystalline breccias, 459
Crystalloids, 116
Crystals
    gypsum, 392
    precipitated, 676
    in salar structures, 637
Crystal tuffs, 833
Cuba, tropical lagoons of, 825
Cup reefs, 6
Current crescents, 529
Current drag, 559
Current marks, 220-222
Current meters, 4
Currents
    alongshore, 452-453
    and bedding genesis, 49
    bottom, 186, 187

Dachbank cycle, 224
Dead Sea, evaporite facies of, 301
Debris avalanche, 24-25
Debris cones, 25
Debris flows, 10, 385, 488-493, 782, 835
Debris slides, 387
Dedolomitization, 233, 611
Deep sea
    ash in, 848
    chert in, 121
    clays in, 144, 147, 157-159
    ironstones, 416
    pelagic sediments, 553-556
    silica sources in, 746
    submarine cementation in, 114
    trenches of, 395
    volcanism of, 859
Deep-Sea Drilling Project, 147, 272, 411, 468, 472, 554, 745

Deep-sea floor, paleodepth determination on, 527
Deep-sea sediments, 55, 276-278, 469, 470, 604
Deformational structures, 414
Deformation till, 808
Degrading neomorphism, 102
Dehydration, 21, 789
Deltaic deposition, 245, 363
Deltaic sediments, 235-240
Deltas
    of alluvium, 13
    and bedding genesis, 49
    classification of, 236, 292
    and clay sedimentation facies, 162
    and coal deposits, 165
    and current marks, 222
    cuspate, 236, 240
    ebb tidal, 796, 804
    formation of, 291
    Gilbert, 455
    oolite, 249
    paludal sediments of, 533
    paralic sediments in, 538, 539
    proglacial, 238
    sedimentation, 240-244, 689, 690
    tidal, 803
Demospongia, 759
Density currents, 2
Density inversion layer, 854
Deposit feeders, 63, 686
Deposition, 34, 244-246, 372. See also Sedimentation
    calcium sulfate, 305
    evaporite, 300-301
    modes of, 371
    of sand, 286
    subaqueous volcaniclastic, 860
    of till, 808-809
    of turbidity-currents, 838
Depositional environments, 720, 756
    defined, 658-661
    sedimentation rates in, 689
Deposits. See also Sediments
    flood, 329
    gravity-slide, 386-389
    mass-wasting, 479-480
    modern oolite, 511
    peat-bog, 541-543
    spring, 762-763
Depth-fertility diagram, 546
Depth zones, 525-526
Deserts. See also Dunes
    fulgurites in, 344
    role of organisms, 685
    sedimentary environments, 247-251
    sediments, 323
    stony, 284
Desiccation breccia, 85
Desiccation conglomerates, 156

Desiccation crack, 486
Desilicification, 142
Detrital feldspars, 595
Detrital limestones, 437
Detrital quartz, 594
Detrital remanent magnetism (DRM), 617
Detrital sediments, 727
Detritus
  airborne, 278
  tectonic uplift as source of, 695
  weathering, 755
Detritus feeders, 2
Deviation index, 753
Diagenesis, 20, 23, 61, 111, 252–256, 464, 683, 772
  assessment of, 721–722
  carbonate, 94
  cavity-filling fabrics of, 259
  of chert, 119–120, 121
  in clays, 149
  coal, 167–172
  of concretions, 183
  destructive, 101
  in evaporite deposits, 304
  graywacke in, 390
  and gypsum, 391, 393
  hardground, 397–400
  and iron deposition, 417
  limestone, 432–434
  and organic sediments, 519
  porosity development through, 98, 100
  postlithification, 399
  preburial, 399
  prelithification, 398–399
  and pseudomorphs, 597
  and sand surface textures, 651
  subaerial, 97
  syneresis in, 790
  in volcanic ash, 850–851
Diagenetic evolution, 20, 22
Diagenetic fabrics, 22, 256–261
Diagenetic facies, 255
Diagenetic grade, 255
Diagenetic zone, 152
Diamictite, 262, 544
Diamonds, 400
Diastem, 207, 398
Diatactic varves, 842
Diatomite, 263–266
Diatom ooze, 549
Diatoms, 31, 122
Diatom tests, 471
Dielectric anisotropy, 321
Diels-Alder reaction, 568
Dikes, 126
  clay, 855
  hydrocarbon, 856–857
  sand, 856–857
Diluvial theory, 329
Dimensional orientation, measurement of, 320

Dimictic lakes, 423
Diplocraterion, 62, 64
Disconformity, 207
Dicontinuity surfaces, 207, 208
Dish structures, 79, 866
Disilicate, 136
Dismicrite, 71
Dispersion systems, 266, 327, 530. *See also* Transport processes
Dissolution-rate profiles, 547
Dissolution sequence, 550
Distributive province, 593
Divergent plates, 662
Dolomite, 54, 82, 112, 266–273, 678
  on Bahama Banks, 32
  in carbonate sediments, 111
  stains specific for, 764
  zebra, 874
Dolomitic chalk, 116
Dolomitization, 84, 97, 100, 233, 266, 273, 634
  of beachrock, 45
  and diagenesis, 254
Dolostone, 266–273
Dopplerite, 394
Drag, 413
Drewite, 91
Dripstone, 109, 762
Dropstones, 184–185, 356
Drumlins, 362
Drusy vug, 348
Dune facies, 35–36
Dunes, 36, 286. *See also* Sands
  backset bedding on, 28
  and barrier islands, 34
  of base-surge deposits, 38, 39
  bed forms, 49, 56
  characteristics of, 287
  in intertidal environments, 653
  in laboratory flume, 57
  on Mars, 477
  radioactivity in, 604
  sediments from, 36
Durability index, 19
Duricrust, 84, 88, 113, 245, 274, 442, 812, 871
Dust, 276–278, 285
Dust storms, 458
Dy, 394
Dynamometamorphism, 464
Dysaerobic range, 298

Earth flow, 819
Earthquake shocks, 598
Earth sciences, 704, 705–706
Echinoids, 822
Éclatement, 84
Edgewise conglomerate, 85, 157
Einstein's bed-load function, 342

Electronic data processing, 721
Electron microscopy, 261, 383, 457, 648
Elutriation, 275
Eluviation, 140
Embedding, 700, 702
End-member concept, 129
Englacial drift, 807
Enhydrites, 349
Enterolithic structure, 560
Environmental analysis, 658, 723
Eolian deposition, 245
Eolian deposits, 287, 476, 478
Eolian dust, 276-278
Eolian environment, 659, 690
Eolianite, 87, 279-283
Eolian sands, 283-284
Eolian sedimentation, 52-53, 284, 641
Eolian sediments, 248-249, 604, 755-756
Eolian textures, 649
Eolian transport, 848. *See also* Transport processes
Epeiric seas, 32
Epeirogenic facies, 351
Epiclastic rocks, 853
Epiclastic volcanic fragments, 851
Epicontinental sea, 438, 439
Epidiagenesis, 22, 252
Epigenesis, 22, 98
Epilimnion, 422, 423
Epoxy peels, 699, 768, 769
Equicesses, 842
Equilibrium configuration, for bed forms, 58
Equilibrium profile, 449
Erosion. *See also* Bioerosion
　desert, 247
　shoreface, 193, 195
　shoreline, 41
Erratics, 353
Escape structures, 68
Eskers, 362
Estuaries, 288-299
　classification of, 292
　clay sediments in, 143
　and coal deposits, 165
　continental-shelf sediments in, 196
　homogeneous, 290
　hydrology of, 289
　and lagoonal sediments, 427
　paralic sediments in, 538
　sedimentation, 288-293, 689, 690
　tidal-current deposits in, 795
Estuarine delta, 235, 236
Estuarine sediments, 144, 293-296
Etching, 764
Eupelagic sediments, 545
Entrophic lakes, 423, 424
Euxinic environment, 246
Euxinic facies, 75, 76, 296-300
Evaporite basins, 302, 303
Evaporite facies, 300-303

Evaporites, 246, 537, 678
　and bedding genesis, 54
　of Black Sea sediments, 76
　in Bouma sequence, 78
　in cyclic sedimentation, 230
　of desert sediments, 249
　diagenetic processes in, 252
　gypsum, 393
　and intraformational folding, 560
　in lake sediments, 422
　marine, 308-309, 464-467
　origin of, 303-311
　phosphate in, 574
　in speleal sediments, 756
　of sulfate lakes, 233
　in tropical lagoons, 821
Evaporite sequences, 225, 813
Exotic breccias, 85
Experimental sedimentology, 311-316, 331
Extraclasts, 440

Fabric
　of glacial gravels, 353
　of glacial sands, 359
　limestone, 432-434
　measurement of, 725
　primary magnetic, 13
　sedimentary, 320-322
　statistical parameters of, 729
　of sulfide sediments, 787
　till, 807
Faceted pebbles and boulders, 322-323
Facies, 323-325
　defined, 658
　distribution of, 32
　geosynclinal cycles, 351
　neritic sedimentary, 498-500
　paralic sedimentary, 538-540
　of sedimentary rock, 679
　and tectonic environments, 666
Factor analysis, 729, 730
Fall diameter, 404
Falls, deposits formed by, 479
Fan deposits, 784
Fans, 780
　of abyssal-sedimentary environments, 2
　alluvial, 10-11, 337, 490-491
　submarine, 470
Fanglomerates, 11, 85
Fan valleys, 777-780
Fat clay, 138
Fatty acids, in hydrocarbon genesis, 568
Fatty soil, 394
Faults, 657
Fauna
　in Black Sea sediments, 75
　in euxinic facies, 298
　lagoonal, 822
Fecal pellets, 31, 863

Feldspar:quartz ratio, 358
Feldspars
  in clastic sediments, 128
  detrital, 595
  staining of, 765
Ferralite, 325
Ferricrete, 325
Ferromanganese concretions, 551–553
Ferromanganese nodules, 409
Ferrous iron, in glauconite grains, 365
Fine-material load, in fluvial sediment transport, 339
Fining-upward sequence, 224, 225, 226, 227, 336, 338, 339, 800
Fiorite, 763
Fireclays, 137, 138
Fire Island barrier, 34
Fissility, 17, 325–326, 361
Fjords, 297
Flame structures, 326–327, 414
Flandrian transgression, 77, 288
Flaser bedding, 52
Flat bed, 49
Flint, 117, 119, 120
Flint clay, 137, 138, 143
Flocculation, 162, 327–329
Flood deposits, 237, 329, 337
Flood plains
  of alluvial sediments, 12
  clay sedimentation facies in, 161–162
  and current marks, 222
  delta, 237, 239
  paludal sediments of, 533
Flood ramp, 796
Floods
  and deltaic sedimentation, 241
  flash, 248, 830
  and salar structures, 637
Flood tidal delta, 804
Florida, 209, 469
  calcium carbonate muds in, 116
  carbonate sediments of, 101
  straits of, 207
  tropical lagoons of, 824, 825, 826
Flotzgebirge, 389
Flow casts, 109
Flow regimes, 330–333, 590
  backset bedding in, 28
  in Bouma sequence, 81
Flow rolls, 579
Flow slides, 386
Flowstone, 109, 762, 819
Flowtill, 362, 808
Flow tubes, 221
Flow, turbidity, 837. *See also* Turbidity currents
Fluidization, 25, 333–334, 864
Fluid mechanics, of sediment transport, 731
Flume experiments, 48, 79, 312, 314, 315, 332, 709

Flumes
  backset bedding in, 28–29
  in bed forms, 59
Fluorapatite, 574
Fluorescent tracers, 814
Flute casts, 107
Flute marks, 655
Flute molds, 107, 108
Fluvial deposits, 142, 244
  backset bedding in, 28
  petrology of, 673
Fluvial sedimentation, 315, 335–338
Fluvial sediments, 338, 755
  of alluvium, 12
  of Bay of Fundy, 42
  black sands in, 73
  in deserts, 248
Flysch, 188, 661, 666, 680, 711
Fog, and salar structures, 639
Folds, 657
Footprints, 681
Foraminifer, 7. *See also specific organisms*
  and lysocline, 462
  of pelagic sedimentation, 3
  and sedimentary facies, 32
Foreset beds, 591
Foreshore, beach sands on, 46
Foreshore berm, 36
Formations, correlation of, 707
Fossil cracks, 486
Fossiliferous cherts, 120
Fossilization, 65, 517
Fossils, 9, 590
  in fluvial environment, 659
  fragmentation of, 66
  internal sedimentation of, 66–67
  of lobate shorelines, 660
  of loess, 457
  paleodepth analysis from, 524
  in peat-bog sediments, 542
  in salt deposits, 301
  sedimentary transport of, 65–66
  siliceous, 549
  silicified, 103
  sorting of, 66
  in speleal sediments, 758
  in submarine cements, 115
  trace, 343
Fossitexture, 68, 69
Foundering hypothesis, 581
Froude numbers, 620–622
Fucoid, 343–344
Fulgurites, 344–345
Funafuti expedition, 709
Fundy, Bay of, 41–43, 799
Fungi, and diagenetic fabrics, 257

Ganges river system, 12
Garnet, 400, 401

INDEX

Gases. *See also* Natural gas
　in carbonate rocks, 103
　chromatogram, 570
　of ferromanganese concretions, 201
　generation of, 172, 484
　interstitial waters in, 408
　in sediments, 346–348, 514
Gaussian assumption, 380
Geochemical cycle, 151, 152
Geochemical gelfication, 170
Geochemistry, sedimentary, 721, 737–738
Geochronology, of deep-marine deposits, 473
Geodes, 348–350
Geography, and sedimentation, 30. *See also* Paleogeography
Geologic record, 189, 785
Geology, 312, 566
Geometry, of sedimentary structures, 7, 8, 34, 102, 679
Geopetality, 7, 8, 102, 811
Geostrophic flow, 190
Geosynclinal formations, 785
Geosynclinal theory, 710
Geosyncline, 350, 351, 680
Geröllton, 262
Geyserite, 120, 762, 763
Geysers, 762
Ghosts, 434
Giant polygons, 639
Gibber plain, 86
Gibbsite, 144
Gilbert-type delta, 236, 238
Glacial clasts, 323
Glacial deposition, 245
Glacial deposits, 28
　classification of, 361
　clastic, 805
　quaternary, 142
Glacial drift, 806
Glacial gravels, 353–355
Glacial lake-burst, 329
Glacial-marine drift, 264–265
Glacial-marine sediments, 355–357, 363
Glacial outwash, 82
Glacial sands, 357–360
Glacial sediments, 558
Glacial theory, 708
Glacial varves, 841–843
Glaciers, dry vs. wet-base, 355
Glacigene sediments, 360–364
Glaciofluvial sediments, 362–363
Glaciolacustrine sediments, 363
Glaebule, 503
Glassy spherules, 317–318
Glauconite, 22, 364–366
　in authigenesis, 21, 22
　in calcareous cycles, 229
　and clay genesis, 155
Glauconitic beds, 54
Globigerina ooze, 40, 116, 603–604

Glowing avalanche, 406, 506
Glowing cloud, 406, 505–506
Gold, 400
Gondwanaland, 357
Gossans, 81
Gouy-Chapman theory, 1
Graded bed, 366
Graded bedding, 225, 679, 812, 845, 847
　in Bouma sequence, 78, 79
　and geosynclinal sedimentation, 350
　in glacigene sediments, 362
　negative, 414
Graded-bed sequences, 813
Graded varves, 842
Grade scales, 127
Grading
　of conglomerates, 184
　of continental rise sediments, 187
　inverse, 414–415
Grain-flow deposition, 369
Grain flows, 368–370, 385
Grain motion, 641
Grain orientation, 320–321
Grains. *See also* Sands
　measurement of, 377
　origin of, 31
Grain-size analysis, 275, 376–382
　frequency studies, 370–374
　measurement of, 8
　parameters of, 374–376, 383
　of sands, 644–645
　scales for, 378–379
Grainstones, 444
Grain-supported sediment, porosity development of, 98
Grand Cayman, tropical lagoons of, 823
Granulometric analysis, 382–384
Granulometry, 432
Grapestone, 31, 102, 384–385, 413
Gravels, 83
　beach, 43–44
　glacial, 353–355
　grade scales for, 127
　river, 627–629
　in storm deposits, 769
Gravitational compaction, 176. *See also* Compaction
Gravitational theory, 342
Gravity, and sediment transport, 733, 818–819
Gravity coring, 472
Gravity flows, 385–386, 413, 696
Gravity-slide deposits, 386–389
Gravity sliding, 413
Gravity waves, in bed forms, 59
Gray-green muds, 775
Graywackes, 17, 353, 389–391
　in clay sedimentation facies, 160
　coloring of, 174
　and geosynclinal sedimentation, 350
　mineral composition of, 727

Great Bahama Bank, 32
Great Barrier Reef, 435
  submarine cementation in, 114
  tropical lagoons of, 826
Great Salt Lake, 425
  eolianite of, 280
  oolite deposits in, 384, 511
Green River Formation, 425
Grès dunaire, 279
Grinding, in attrition process, 18
Groove casts, 107
Groove marks, 810
Groove and spur systems, 205
Ground moraines, 361. See also Moraines
Ground water
  dolomitization by, 272
  resource development for, 722
  transportation by, 816
Guano, 577–578, 691, 755
Gulf of Alaska, glacial marine sediments of, 356
Gulf of Batabano, tropical lagoons of, 825
Gulf of California, 392
Gulf of Mexico
  barriers in, 34
  evaporite facies of, 301
  lagoonal sedimentation in, 428
  ocean drilling in, 411
Gulfs, sedimentation rates in, 689, 690
Gutter casts, 656
Gypseous ramparts, 639
Gypsum, 113, 640
  of caliche, 89
  in sediments, 391–394
  in tropical lagoons, 821
Gyttja, 394

Hadal sedimentation, 395
Hadal zone, 1
Halimeda, 93, 204, 205, 206
Halite, in evaporite sequences, 308, 309, 310
Halite crusts, 638
Halmyrolysis, 20, 253
Hamada, 284
Hardgrounds, 117, 207, 229, 397–400
Hardpan, 806
Hard rock deposits, 577
Harrington Sound, 101, 823
Heavy minerals, 400–402, 595
  in littoral sedimentation, 453
  in petrology, 673
Helmholtz instability, of bed forms, 58
Hematite, 415
Hemipelagic sediments, 80, 187, 188
Hemipelagic sequences, 782
Hemipelagites, 775
Hexactinellida, 759
Hilt's rule, 171
Histograms, 377–380

History, of sedimentology, 707–712
Hjulström's diagram, 535, 731, 732
Holland, barrier sediments in, 34
Holocene sediments
  of Bahama Banks, 30
  of Persian Gulf, 563
  Shark Bay, 741
Homogeneity, tests for, 531
Homogenization, of sediments, 69
Hornblende, 400, 600
Hot springs, 582, 762, 763
Huascarán avalanche deposit, 25
Humic matter, in sediments, 166, 170, 402–403
Humin, 402
Hyaloclastic fragments, 851
Hyaloclastites, 859
Hydrates, in speleal sediments, 756
Hydration, 657
Hydraulic equivalence, 403–404
Hydraulics, 330
Hydrocarbons
  alkane, 218
  in bituminous sediments, 72
  classification of, 565
  in crude oils, 215, 217
  in mudstones, 493
  in oil shales, 508
  in organic sediments, 519
  origin of, 405
  in outer continental shelf sediments, 522
  petroleum, 507. See also Petroleum
  in sediments, 346, 404–405
Hydrodynamic instability, 867
Hydrodynamic laws, 127
Hydrodynamic processes, 382
Hydrodynamic sorting, 49. See also Sorting
Hydrogen, in authigenesis, 20, 21
Hydrography, and sedimentation, 30
Hydrothermal processes, 552, 788
Hydrotroilite, 174
Hypolimnion, 422, 423

Ice
  in speleal sediments, 756
  transportation by, 818
Ice caps, Martian, 476
Ice-rafting, 200, 355, 356, 359, 544
Ice-Shove deformation, 558
Ichnocoenoces, 61
Ichnofossils, 343
Igneous breccias, 84
Igneous criteria, 811
Ignimbrites, 406–407, 506, 832
Illite, 16, 137, 328
  in bentonite, 60
  in clay diagenesis, 149, 150
  in clay sedimentation facies, 164
  crystallinity of, 172
  in deep-sea clays, 159

degraded, 143
distribution of, 145
in world ocean, 146
Illite-smectite, 16, 60
Illitic clays, 137
Illuviation, 140, 245
Ilmenite, 74, 400
Imbrication, in alluvial sediments, 13
Impact, in attrition process, 18
Impact law, 407–408, 609, 767
Impregnation cylinder, 701
Impregnation of wet sediments, 701
Inclusions, in authigenesis, 22
Indian Ocean
  atmospheric dusts for, 277–278
  kaolinite content of, 144
  smectite in, 146, 147
Indus river system, 12
Injection-type structures, 414
Inland seas, sedimentation rates in, 690
Inlets, and barrier islands, 34
Inlet sequences, 35, 802
Inman sorting coefficient, 846
Insolation, 284
Intermontane basins, fining-upward cycles for, 227
International Association of Sedimentologists (IAS), 717, 718
Interstitial fluids
  during compaction, 178
  of diagenetic fabrics, 261
  gases in, 347
  in sediments, 408–412
Intertidal zone, 798
Intraclasts, 9, 384, 413, 434, 440
Intraformational breccia, 85
Intraformational conglomerates, 156
Intraformational disturbances, 413
Intraformational folds, 558–560
Intrasparite, 413
Intrazonal soils, 140
Inverse grading, 367, 369, 414–415. *See also* Grading
Inversion, 100, 102
Involution, 560
Ionic potential, of authigenesis, 20
Iron chlorites, in authigenesis, 22
Iron compounds, color of, 173
Iron deposits, 54
  in lacustrine sedimentation, 426
  phosphate in, 574
  source of, 416
Iron hydroxides, in authigenesis, 22
Iron-manganese deposits, in authigenesis, 21
Iron-ore facies, in carbonate diagenesis, 95
Iron-oxide cements, 113
Iron-rich rocks, 678
Iron spherules, 316
Ironstones, 124, 229, 415–417
Island arcs, in hadal sequences, 397

Islands
  alluvial, 336
  barrier, 33
  sand, 206
Isohalines, in evaporite sequences, 307
Itabarite, 120

Jamaica, 114, 209
Jasper, 120
Joint Oceanographic Institutions for Deep Earth Sampling (JOIDES), 472, 721
Journal of Paleontology, 717
Journal of Sedimentary Petrology, 717
Juan Basin, coal deposits in, 165

Kalkkruste, 88
Kames, 362
Kandites, in clay diagenesis, 150
Kaolinite, 6, 60, 137, 138, 139, 141
  in clay sedimentation facies, 162–163
  in deep-sea clays, 158
  in Indian Ocean, 278
  in lacustrine sedimentation, 426
  in world ocean, 145
Kaolinite/chlorite ratios, 551
Kaolinization, 141
Kaolins, 137
Kara-Bogaz Gulf, 301, 418
Karst breccias, 84
Karst development, eolianite, 281–282
Katamorphism, 252
Kauri gum, 618
Kerogen, 72, 418–420, 507, 515
  and humic matter, 402
  in petroleum genesis, 569
Kieselguhr, 263
Kieserite, in evaporite sequences, 309
Klippe, sedimentary, 509
Kunkur, 88
Kurkar, 279
Kurtosis, 46, 420

Lacquer peels, 698
Lacquer-resin peels, 699
Lacustrine clays, 142
Lacustrine environments, 421–427, 437, 507, 841. *See also* Lakes
Lag deposits, 52, 796, 799
Lag gravel, 184
Lagoon complexes, 33, 301. *See also* Barrier Island facies
Lagoons
  atoll, 824
  backreef, 825, 826
  carbonate, 822
  clay sediments in, 143, 161–162
  and coal deposits, 165

coastal, 245
continental-shelf sediments in, 196
interreef, 826
ooids in, 513
paralic sediments in, 538
reef-bordered, 206
sedimentation, 427–429, 689, 690
shelf, 825
tropical, 821–828
Laguna Madre, 393, 428
Lahar deposits, 492
Lake basins, 421, 541
Lake Chad, 425
Lake-floor deposits, 363
Lake marls, 475
Lakes
aggressive, 143
alkaline, 122, 143
amictic, 423
circulation of, 423
clay sediments of, 142–143, 161–162
and coal deposits, 165
and current marks, 222
depositional environments of, 244, 659
desert, 248
glacial, 363
humid, 537
pitch, 795
saline, 309
sedimentation rates for, 688–689
typical, 421
Laminae, 40, 48, 430, 679, 866
Laminaset, 430
Laminated sediments, 299
Lamination, 48, 79, 80
Laminites, 52, 430
Landslides, 25, 85, 819
Landweber method, 191
Langmuir cells, 221
Lapilli, of base-surge deposits, 39
Lateral accretion, 4, 430–432, 803
Lateritization, 535
Lava, pillow, 858
Lava flow, 406. *See also* Volcanism
Leaching, 141
Lean clay, 138
Lechatelierite, 344
Lee eddies, 56, 287
Lensoid crystals, 392
Lenticular cracks, 791
Levees, natural, 12, 643
Liesegang-type banding, 175
Lignin, in organic sediments, 516
Lignite
bituminous, 72
in earth's crust, 520
radioactivity in, 605
Lime gyttja, 394
Lime muds, 31, 87, 94, 102, 432

Limestones, 115, 434–446, 676, 677
allochthonous, 441
birdseye, 70
bituminous, 515
classification of, 10, 443–445
coastal, 279
crinoidal, 203
cyclic, 225
dikes, 126
distribution of, 436
epoxy peels of, 700
freshwater, 537
lithographic, 87, 482
ore precipitation in, 103
as petroleum reservoir, 105
stromatolites, 771–774
Limestone fabrics, 432–434
Limnic peat, 542
Linear structures, 590
Lineation, parting, 540–541
Lipids, in organic sediments, 516
Liquefaction, 333–334, 864
Liquified sediment flows, 385
Litharenites, 647
Lithic arenite, 17
Lithic tuffs, 833
Lithification, 447, 675
in carbonate diagenesis, 95
of carbonate sediments, 98, 105–107
of clastic sediments, 132–136
petrology of, 674
and pH, 105
in sedimentation, 683
Lithofacies, 324, 447
Lithogenesis, 447
Lithographic limestone, 87
Lithology, 324, 480
Lithophase, 223
Lithotope, 447
Littoral deposits, 73
Littoral energetics, 734
Littoral environment, 449, 452
Littoral sedimentation, 401, 448–457, 498
Load casts, 81, 109, 326, 560
Load structures, 866
Loam, 479
Lobate, depositional environment of, 660
Lobate delta, 236, 240
Lodgment till, 361, 808
Loess, 457–458, 749
in eolian sediments, 248, 284–287
weathering of, 871
Loessite, 457
Log-probability plots, 371
Loire Estuary, 291
Longitudinal dune, 286
Lower Cretaceous carbonates, 30
Lunar environment, sedimentation rate in, 691
Lunar sedimentation, 458–461

Lunate deltas, 240
Lucite, 119
Lutite, 493
Lydite, 120
Lysocline, 461–463, 471, 548

Maar volcanoes, 39
Magnacycles, 226
Magnesium, in carbonate sediments, 111
Magnesium calcite, 6, 91, 93, 105, 106
Magnetic spherules, 316–317
Magnetic susceptivity anisotropy, 13
Magnetite, 400
Manganese, in iron spherules, 317
Manganese nodules, 316, 485
 of abyssal-sedimentary environments, 2
 in authigenic mineral formation, 3
 in pelagic sedimentation, 551–553
Manganese oxide, 175
Mangrove trees, 92, 93, 239, 241, 279, 426, 539, 583, 685
Mapping, 723
Marine ash deposits, 847
Marine brines, evaporites from, 303, 304
Marine deposition, 246
Marine environments
 organism-sediment relationships in, 686
 sedimentation rates in, 690
Marine sediments, 276–278, 468–474
Marine shales, 16
Markov processes, 55
Marls, 474–475, 542, 611
Mars
 glacio-eolian deposits on, 503
 sedimentation, 475–479
Marshes
 of Bay of Fundy, 42
 and coal deposits, 165
 deposition, 245
 paludal sediments of, 533
Mass-flow deposits, 479
Massive bedding, 52
Mass wasting, 413
Mass-wasting deposits, 479–480
Mathematical models, 714
Maturity, concept of, 480–482, 676
Mecropedology, 870
Mediterranean
 algal reefs in, 6
 eolianite of, 280
 geosyncline of, 352
 hardgrounds of, 398
 ocean drilling in, 411
 submarine cementation in, 114
Megabreccias, 25, 85
Megacycles, 226
Megafans, 784
Megaripples, 51

Megasedimentology, 720
Melange, 85, 509–510
Melikaria, 82
Melt-out till, 362
Meniscus fill, 70
Meromictic lakes, 423
Messinian stage evaporites, 76
Metalimnion, 423
Metamorphic zone, 152
Metamorphism, 20, 464
 of coals, 168
 and diagenesis, 252
 of organic sediment, 518
 in sedimentation, 683
Metamorphization, 304
Metasomatism, 20, 254, 466
Meteor Expedition, 553
Meteorites, 316, 317
Methane, 295, 346, 347, 404
Metharmosis, 253
Methods
 in petrography, 670
 of sedimentology, 697–703, 705
Methylcyclohexane, in crude oil, 216
Micelles, 573
Michigan Basin, 302
Micrite, 87, 99, 111, 443, 444, 482
Micritic envelope, 105
Microatolls, 6
Microbiology, 347, 482
Microcrystalline ooze, 677
Microorganisms. *See also* Foraminifera
 and diagenetic fabrics, 257
 in petroleum genesis, 571
Microscope, petrological, 670
Microspar, 482, 754
Microsparite, 99
Microstylolites, 257
Microtektites, 317, 318
Middle Atlantic, morphologic elements of, 193. *See also* Atlantic Ocean
Miliolite, 279
Mineral composition, measurement of, 725, 726
Mineral deposits, in authigenesis, 21
Mineralization, epigenic, 20
Mineralogy, of sands, 644
Minerals, 400. *See also* Heavy minerals: *specific minerals*
 authigenic, 23
 in beach sands, 46
 in evaporite facies, 302, 304
 resource development for, 722
 saline, 306–307
Minette-type iron ores, 124
Minus-three-power law, 59
Mississippi Delta, 238, 591. *See also* Deltas
Mississippi Valley type, ore deposits of, 103
Mohole, 411
Molasse, 666, 680

Molasse deposition, 352
Moldic porosity, 105
Molds, 105, 107–109
Mollusks, 398, 863
Molybdenum, in black shales, 298, 299
Monazite, 74, 400, 574
Monohydrocalcite, in microbial cultures, 91
Monomictic lakes, 423
Montmorillonite, 137, 163, 178
Moon, avalanche deposits on, 27
Moraine breccia, 86
Moraines, 806
Moss agate, 597
Mottled limestone, 70
Mud basins, 856
Mud belt, 486, 499
Mud cracks, 157, 486–488, 790, 791, 812
Mud crust, 157
Mud-curls, 487
Mud dunes, 202
Muddy sediments, 413
Mudflows, 386, 854, 855
  in alluvial-fan sediments, 11
  of avalanche deposits, 25
  breccias, 85
  deposits, 184–185, 488–493
Mud lime, 486
Muds
  deep-sea, 16
  red, 612
Mudstones, 15, 262, 444, 493–494, 543–544, 736
Mud zone, 486
Muer pebble deposits, 577
Multiple regression analysis, 730
Multivariate analysis, 730
Munsell system, 175
Muscovite, in clay sedimentation facies, 164
Muskeg, 541

Nappe, gliding, 509
Natural gas. *See also* Gases
  and coal deposits, 171
  and diagenesis, 253
Near-bottom currents, 3
Nearshore currents, 235
Necrolysis, 65
Nepheloid layer, 496–497
Nepheloid sediments, 495–498
Nephelometry, 495–498
Neomorphic fabrics, 259, 260, 261
Neomorphism, 100, 102, 254
Neritic deposition, 246
Neritic sedimentary facies, 498–500
Neutral-line concept, 448
New Zealand, White sinter terraces of Rotomahana, 763
Nickel, in iron spherules, 317. *See also* Minerals

Nicuradze diagram, 633
Niger Delta, 239
Nitrates, in speleal sediments, 756
Nitrogen, in sediments, 484
Niveo-eolian deposits, 501–503
Noachian Flood, 77
Nodules. *See also* Manganese nodules
  in authigenesis, 21
  in sedimentation, 503–504
Nomenclature, 445, 667. *See also* Terminology
Nonclastic rocks, 677
Nonsequential beds, 208
North Africa, sabkhas of, 635
North Pacific, kaoline content of, 145. *See also* Pacific Ocean
North Sea
  black sands in, 75
  petrology of, 668
Novaculite, 119, 120, 504–505
n-paraffins, 569
Nuee Ardente, "glowing cloud," 505–506
Null-line concept, 51

Ocean basins, 421
Ocean environment, plate tectonics of, 662, 664, 665
Ocean floor, age-depth relationships for, 526
Oceans
  pelagic sediments in, 545
  phosphate deposits in, 575–576, 577
  sedimentation rates for, 693
  sediments, 410, 468
  silica in, 744
Offshore bars, 51
Offshore zone, 449
Off-surge, 829
Oil. *See also* Crude oil; Petroleum
  in carbonate rocks, 103
  dating of, 134
  and diagenesis, 253
  of ferromanganese concretions, 201
  generation of, 172
  and interstitial waters in, 408
  in oolitic sediments, 514
  search for, 668
Oil exploration, 720
Oil shales, 72, 224, 298, 426, 507–508, 515
Old Red Sandstone, 338
Oligomictic conglomerates, 183
Oligomictic lakes, 423
Olistolites, 508–510
Olistostromes, 85, 508–510
Olivine, 400
Omission surfaces, 398
Oncolites, 772
On-surge, 829
Onyx marble, 762
Ooids, 4, 268
Oolites, 9, 102, 109, 432, 503, 510–515

INDEX

Oolitic shoal, of Bahama Banks, 30, 31
Ooze, in limestone fabric, 432. *See also specific oozes*
Opal, 112, 119, 133, 536, 584, 743
Open-cast movement, 558
Open channel flow, 60
Open-space sparite, 443
Open-space structures, in carbonate sediments, 103
Optical orientation, measurement of, 320
Ore deposits, 22, 97, 103. *See also specific ores*
Organic matter
   carbonate-secreting, 442
   in lake sediments, 422–423
   in sediments, 515–521
   in speleal sediments, 755
Organisms. *See also* Boring organisms; Burrowing organisms
   as agents of sedimentation, 204–205, 684–685
   transportation by, 819
Orogenic facies, 351, 661
Orogeny, and sedimentation, 696
Orthochemical constituents, 9
Orthoconglomerates, 183
Orthogonal system, 487
Orthoquartzites, 78, 161
Outer continental shelf sediments, 521–523
Outwash gravels, 42
Outwash plains, 363
Overbank deposits, 13, 118, 430, 779
Overburden pressure, 177, 178
Oxbows, 336
Oxides
   in authigenesis, 20, 21
   in speleal sediments, 756
Oxygen isotopes, 269, 272
Oxykertschenite, 22
Oyster bars, 428

Pacific Ocean
   atmospheric dusts for, 277, 278
   clay sediments in, 147
   geosyncline of, 352
   glacial marine sediments of, 355
   kaolinite content of, 144
Pack ice, 363
Packing, 321–322, 333
   measurement of, 725
   statistical parameters of, 729
Packstones, 444
Palagonite, 850, 860
Paleobathymetric analysis, 66, 524–528
Paleoclimates, evidence of, 17, 158, 274
Paleocurrent analysis, 528–533, 590, 594, 696, 835
   of cross-bedding, 210
   current marks used in, 221, 222
   indicators, 321, 540
   and sedimentary structures, 681

Paleocurrents
   in fluvial environment, 659
   of lobate shorelines, 660
   and magnetic susceptibility anisotropy, 14
Paleoecology, 684–687, 706
Paleogeography, 594, 667, 706
   of polar shelves, 200
   of volcanic ash deposits, 849
Paleomagnetic dating, 758
Paleontology
   of conglomerates, 185
   of contourites, 202
   of eolian sedimentation, 287
   of euxinic facies, 299
   of evaporite facies, 302
   of lagoons, 827
   of lake deposits, 425–426
   of oolitic limestone, 514
   of storm deposits, 769
   of tidal current deposits, 798
   of tills, 809
Paleo-oceanography, 473, 555
Paleoslope, 66, 212, 532, 558, 779
Paleotectonics, 706. *See also* Plate tectonics
Palimpsest sediments, 199, 616
Paludal sediments, 533
Parabolic dunes, 286
Paraconglomerates, 183, 262
Paracontinuity, 207
Parafacies, 255
Paraffins, in crude oil composition, 215, 218
Paragenesis, 22, 533–538
Paragonite, in clay diagenesis, 150
Paralic deposition, 245
Paralic facies, 468, 539–540
Parametric analysis, 729
Paramoudras, 117
Paris Basin, 116
Particle shape, sphericity, 631–633. *See also* Sphericity
Parting lineation, 540–541
Peat, 166
   in backbarrier sediments, 36
   and coastal eolianites, 279
   in earth's crust, 520
   in organic sediments, 517
   radioactivity in, 605
Peat-bog deposits, 541–543
Peat stage, of coal diagenesis, 168
Pedalfers, 139
Pedocals, 89, 139
Pedogenesis, 88, 139
Peels, sedimentary, 697–698
Pelagic deposition, 430, 661
Pelagic organisms, 7
Pelagic sedimentation, 3, 246, 544–558, 612
   of abyssal sediments, 2
   in nepheloid layer, 497
Pelagic sediments, 187, 409, 544–558, 602
Pelecypodichnus, 62

Pele's hair, 832
Pelite, 493
Pellets, 9
  fecal, 31, 863
  limestone, 442–443
Pelsparites, 71
Penecontemporaneous breccias, 84
Peri-azoic zone, 298
Permeability, 678
  of brecciated ore, 752
  of carbonate rocks, 97
  of clastic sediments, 128
  of shales, 736
Persian Gulf, 711, 800
  dolomites, 270, 271
  evaporite facies of, 301
  hardgrounds of, 398
  lagoons of, 564
  oolitic sand deposits of, 512, 513
  sedimentology of, 562–565
  stromatolites of, 771
  submarine cementation in, 114
  tropical lagoons of, 823
Peru-Chile trench, 664
Petrocalcic zones, in caliche, 89
Petrogenesis, of limestones, 437
Petrography, 447
  of continental-shelf sediments, 198–199
  of limestones, 437
  of oolite, 511, 512
  of porcelanite, 584
  sedimentary, 351, 667–671
  of spongolite deposits, 761–762
Petroleum. *See also* Oil
  accumulation of, 571–573
  from barrier islands, 33
  biological origin of, 565–566
  from carbonate rocks, 97
  classification of, 565
  and coal deposits, 171
  in delta sedimentation, 243
  in earth's crust, 520
  evolution of, 570, 571
  gases associated with, 347
  and humic matter, 402
  hydrocarbons in, 404–405
  and interstitial fluid chemistry, 181
  and kerogen, 419
  migration of, 216–220, 518, 571–573, 793
  in oolite sediments, 514
  in organic sediments, 519
  origins of, 348, 566
  resource development for, 722
  in sedimentary rock, 678
  source material for, 567–571
  in tropical lagoons, 827
Petroleum analysis, 565
Petrologic province, 594
Petrology, sedimentary, 447, 671–675, 705

Phenanthrenes, 568
Phengites, in clay diagenesis, 150
Phi scale, 127, 374, 377, 574
Phosphate-rich rocks, 678
Phosphates
  and clay genesis, 155
  economic deposits of, 576
  in sediments, 574–579
  in speleal sediments, 756
Phosphatic nodules, 116
Phosphatic remains, 67
Phosphorites, 21, 759
Phosphorous, in sediments, 484
Photic zone, of carbonate sediments, 90
Phreatic explosions, 844
Phreatic zones, 96
Physiography, and sedimentation, 30
Phytogenic structures, of biogenic sediments, 61
Phytoturbation structures, 68
Pictorial models, 714
Pillow breccia, 859
Pillow lava, 858
Pillow structures, 560, 579–582
Pisolites, 109, 503, 510, 582–583
Pisoliths, 4
Pisolitic gravels, 583
Piston coring, 4, 472
Pitchblende ore, 82
Placer deposits, radioactivity in, 605–606
Planar structures, 590
Plankton
  bathymetric zonation of, 525
  of deltaic sediments, 238
Plate tectonics, 471, 711
  and continental rise sediments, 187
  paleobathymetry and, 526
  pelagic sediments and, 554
  and sedimentary facies, 661–667
  and sedimentary structures, 681
Platinum, 400
Pleistocene, pelagic sediments of, 553
Pleistocene karst, 32
Pluvial lakes, 425
Podzolization, 535
Podzols, 870
Point-bar deposits, 12, 237
Point bars, 12, 50, 430, 431, 757, 796
Point-bar sequences, 336
Point counting, 726
Poisons, environmental, 402
Poisson's ratio, 177
Polar deposits, Martian, 476
Polar environments, 809
Pollen analyses, 77, 758
Polludification bogs, 541
Pollutants, and sedimentation rates, 688
Polycyclic compounds, in crude oil, 215
Polycystins, 550

Polyhalite, significance of, 309
Polymictic breccias, 84, 85
Polymictic conglomerates, 183
Polymorphic transformation, 254
Porcelanite, 120, 122, 584-585, 745
Porcelanitic chert, 119
Pore geometry, 128
Pore pressure, 177
Pore waters, 722
Porosity
  of algal reefs, 7
  of avalanche deposits, 25
  of brecciated ore, 752
  in carbonate rocks, 97, 105
  in clastic sediments, 128, 132, 133
  and diagenesis, 254
  minus-cement, 133
  of natural sediments, 176
  of shales, 736
Postglacial varves, 842
Potash, in evaporite sequences, 308
Potash industry, 467
Potassium-40, 602
Potholes, 220
Precambrian, stromatolites of, 773
Precipitation. *See also* Transport processes
  of inorganic compounds, 484
  travertine mode of, 820
Prefossilization, 67
Pressure-solution phenomena, 113, 585-589, 657
  for clastic sediments, 133
  of diagenesis, 259
Probability plots, 380
Process-response relationships, 312
Prodeltas, 239, 243, 590-593
Prodelta sequences, 202. *See also* Deltas
Prod marks, 810
Progradation, 235, 241, 242, 784
Propane, in crude oil, 216
Protactinium-231, 602
Proteins
  in hydrocarbon genesis, 567-568
  in organic sediments, 515
Protodolomite, 30, 267, 269
Protopetroleum, 569, 793
Provenance, 593-597, 667, 672, 715, 720
Pseudobedding, 53
Pseudomorphs, 22, 597-598
Pseudonodules, 560
Pteropods, of pelagic sedimentation, 3
Pumice, 600, 844, 846, 860
Pumice-fall deposits, 406
Pumiceous ash, 847
Push moraines, 558
Pyrite, 22, 113
Pyrobitumens, 72
Pyroclastic accumulation, 5
Pyroclastic flow, 843

Pyroclastic fragments, 851, 853
Pyroclastic rocks, 406
Pyroclastic sediment, 589-600
Pyroxene, 600

Qualitative experiments, 315
Qualitative models, 714
Quantitative models, 714
Quartz, 112, 743
  in clastic sediments, 128
  cryptocrystalline, 119
  precipitation of, 113
  silcrete in, 741
  transport of, 401
Quartz arenites, 78, 647
Quartz grains, 133, 728
Quartzites, 351, 601, 727
Quartzose sandstone, 601
Quartz spheres, settling speeds for, 735
Quaternary
  alluvial breccias of, 86
  Black Sea sediments during, 76
  clay sediments of, 147
  coastal eolianite of, 279-280
  deserts of, 247
  occurrences of dolomite, 269-271
  Shark Bay during, 740-741
Quick behavior, 412

Radioactive decay series, 603
Radioactive ores, 606
Radioactive tracers, 191, 814
Radioactivity
  in petroleum genesis, 571
  in sediments, 602-606
Radiocarbon (C-14), 606
Radioisotopes, 602-603
Radiolaria, 3, 505
Radiolarian cherts, 122
Radiolarian ooze, 549, 550
Radiometric dating, 687
Radium-226, 602
Rafting, 807
  biologic, 3, 356, 819
  ice, 3, 200, 355, 356, 359, 544
Rain, and salar structures, 639
Raindrop imprints, 606-607, 681
Rain wash, 816
Ramleh, 279
Ramp-to-the-sea, 804
Randanite, 263
Rann of Cutch, 300
Rapid sediment analyzer, 608-610
Raroia atoll, 6
Rauhwackes, 82, 234, 611-612
Rayleigh-Taylor instability, 867

Recrystallization, 21, 100, 101, 102, 254, 434, 657
Red beds, 226
Red clay, 612–613
  of abyssal-sedimentary environments, 2
  coloring of, 174
  deep-sea, 16
  in pelagic sediments, 3, 546, 550–551
Redeposition, 613
Red mud, 612
Red Sea
  carbonate sediments of, 101
  evaporite facies of, 301
  hardgrounds of, 398
  ocean drilling in, 411
  plate tectonics of, 662, 663
  salt pans along, 300
  stromatolites of, 771
  submarine cementation in, 114
Red tides, 575
Reduction number, 613
Reef breccias, 85
Reef complexes, 204, 439, 613–615
  and bacterial precipitation, 90
  of Bahama Banks, 30, 31
  and coastal eolianites, 279
  depositional environment of, 660–661
  stromatolite, 771
Reef crests, 205
Reef limestone, 614
Refraction studies, 449
Reg, 86
Regressive facies, 242
Regressive sequences, 37, 197
Relics, 434
Relict deposits, 200
Relict seas, 301
Relict sediments, 42, 198–199, 563, 616
Remanent magnetis, detrital (DRM), 617–618
Replacement fabrics, 259
Resedimentation, 613
Residual breccias, 86
Resin, fossil, 618–620
Resinite, 618–619
Resting traces, 69
Retortable oil content, 508
Retrograde diagenetic zone, 153–154
Reynolds numbers, 620–622, 633
Rhizoconcretions, 279, 281, 503
Rhizocorallium, 62
Rhodoids, 6
Rhodolites, 6
Rhodoliths, 6
Rhodophyta, 6
Rhythmite, 430, 841
Riecke Principle, 585
Rill marks, 656
Rip currents, 453
Ripple lamination, 79

Ripple marks, 622–627, 681, 812
  classification of, 624–625
  flow conditions, 623
  geometry of, 623–624
  occurrence of, 625–627
Ripples
  backset bedding on, 28
  and Bagnold effect, 29
  and bedding genesis, 49
  in bed forms, 56
  cross-bedding produced by, 209
  eolian, 249
  in laboratory flume, 57
Ripple trains, 624
RIST (Radioisotope Sand Tracer Study), 815
River gravels, 627–629
Rivers
  clay sediments of, 142–143
  depositional environment for, 659
  and lateral accretions, 431
  sedimentation rates for, 688–689
  sediment flux to, 410
  transportation by, 817
River sands, 375, 376, 629–630
Rock fall, 819
Rocks, *See also specific rocks*
  beachrock, 44–45
  carbonate, 444
  clastic, 459–460
  dune, 279
  elastic, 126–131
  faceted, 323
  sedimentary, 675–680
Rock slides, 387
Rootlet structures, 61
Rose diagram, 211
Rotational slides, 387
Roundness, 480, 631–633
Roundstone, 84
Rubey scale, 377
Rubidium-87, 602
Rudaceous rocks, 677
Rudites, 129
Run-up, 829
Rutile, 74, 400

Sabkha cycle, 437, 440
Sabkhas
  coastal, 250
  inland, 248
  in Persian Gulf, 562, 564, 634–635
  salcretes in, 640
  sedimentology of, 634–636
Sable Island Bank, 776
Saint Peter sandstone, 78
Salar structures, 636–640
Salcrete, 113, 640

INDEX

Salinas, 300, 301, 303
Saline minerals, stability relations of, 306
Salinity
  of Black Sea, 77
  and carbonate precipitate, 97
  and lithification, 105
Saltation, 285, 339, 640–642
Salt basins, 300
Salt crusts, 643
Salt flats, 636–640
Salting-out effect, 466
Salt-marsh sedimentology, 642–644
Salt metamorphism, 464
Salt pans, 643
Salts, of evaporite facies, 300
Salt weathering, 869
Salt-wedge estuary, 289, 290
Sampling, for paleocurrent analysis, 529–530
Sandar, 363
Sand avalanching, 368
Sand bars, 43
Sand blasting, 19
Sand dikes, 856–857
Sand dunes, 28. *See also* Dunes
Sand fulgurites, 344
Sand injection, 125
Sand pillows, 579–582
Sand ridges, 198, 237
Sands, 285, 644–648
  beach, 45–47
  blanket, 35, 36, 37, 78
  brown over gray, 36
  channel, 118
  compressibility, 178
  coral, 206
  dune, 251, 284, 376
  eolian, 249, 283–284
  glacial, 357–360
  inlet-filling, 37
  oolitic, 522, 512
  river, 629–630
  shoestring, 35, 37, 243
  sorting classification for, 375
  in submarine canyons, 778
  tar, 793
  transport, 732
  turbidite vs. contourite, 188
  weathering of, 870
Sand shows, 286
Sand sheet, and continental shelf sediments, 196
Sand-silt-clay mixtures, 129
Sandstone limestone mixtures, 129
Sandstones, 126, 130, 644–648, 676
  bituminous, 72
  compaction of, 132–134
  compressibilities, 177
  epoxy peels of, 700
  porosity of, 132
  quartzose, 601
  silica in, 745
  "sublittoral sheet," 770
  sulfides in, 789
Sand streams, 194
Sand surface textures, 648–652
Sand units, storm-deposited, 62
Sand waves, 652–653, 769, 797, 803
  backset bedding on, 28
  and bedding genesis, 49
  cross-bedding produced by, 209
Sapropel, 72, 515, 653–654
Sapropelic mud, in Black Sea sediments, 76
Sapropelic series, 166
Scanning electron microscope, 648. *See also* Electron microscopy
Schizothrix mat, microstructure of, 772
Schulze-Hardy Rule, 327
Scour-and-fill, 50, 55, 654–655, 812
Scour-and-fill structures, 81, 83, 237
Scour marks, 107, 221, 367, 655–657
Scour molds, 108
Scree breccias, 85
Scripps Institute of Oceanography, 610
Sea floor, photographs of, 472
Sea level, variations of, 498–499, 527
Seamounts, 859
Sea spray, 640
Sea water, progressive evaporation of, 309
Sea-water/sediment interfaces, 485
Sediment analysis, 312
Sedimentary environments, 658–661
Sedimentary facies, 32, 224. *See also specific species*
Sedimentary klippes, 85
Sedimentary petrology, 706
Sedimentary prisms, 351
Sedimentary rocks, 675–680
  chemical, 534, 536–537
  detrital (clastic), 533, 534–536
Sedimentary sequences, 812
Sedimentary structures, 657–658, 679, 680–682
Sedimentation, 682–684
  continental shelf, 190–196
  cyclic, 223–232
  delta, 240–244
  estuarine, 288–293
  evaporite, 418
  fluvial, 335–338
  geosynclinal, 350–353
  hadal, 395
  lacustrine, 421–427
  lagoonal, 427–429, 821–828
  littoral, 448–457
  lunar, 458–461
  Martian, 475–479
  nodules in, 503–504
  paleoecologic aspects of, 684–687

pelagic, 544–558
in radioactivity, 602–606
submarine fan, 780–786
submarine slope, 774–777
tectonic controls on, 694–697
turbidity current, 836–839
urban, 723
Sedimentation models, 783
Sedimentation rates, 63, 159, 687–692
for continental rise sediments, 188
deep sea, 692–694
and deep-sea clays, 159
determination of, 687–688
pelagic, 3
thixtropic changes affecting, 484
Sedimentation zone, 152
Sediment flows, 333
Sediment/fluid interface, 590
Sedimentology, 718
Sedimentology, 718
applied, 707
comparative, 324
defined, 667
experimental, 311–316
flocculation in, 328
history of, 707–712
models for, 712–717
philosophy of, 703–705
reviews of, 719
sabkha, 634–636
salt marsh, 642–644
Shark Bay, 740–741
Sediments
carbonate, 442
color of, 173–175
continental shelf, 196–199
detrital remanent magnetism (DRM) in, 617–618
estuarine, 293–296
euxinic, 297–299
estraterrestrial material in, 316–319
fluvial, 338, 755
gases in, 346–348
glacial marine, 355–357
glacigene, 360–364
gypsum in, 391–394
humic matter in, 402–403
hydrocarbons in, 404–405
immature, 357
interstitial water in, 408–412
marine, 468–474
mechanical analysis of, 669
microbial production of, 485
organic, 515–521
outer continental shelf, 521–523
pelagic, 544–558
penecontemporaneous deformation of, 558
primary structures of, 589–590
prodelta, 590–593
pyroclastic, 598–600

relict, 616
remanent magnetism in, 617–618
reworked, 684
silica in, 742–748
siliceous, 744
siliciclastic, 442
speleal, 754–759
submarine slope, 470
of tropical lagoons, 827
weathering in, 868–872
"Seepage refluxion," 271, 272
Seiche, 413
Seif dune, 286
Seismic reflection profiling, 4, 395–396, 472
Seismic sections, 721
Sepiolite, 163
Settling method, of granulometric analysis, 383
Settling-time measurements, 377
Settling velocity, 735, 766, 767
Setulfs, 108, 222
Shales, 15, 676, 736–739
bituminous, 72
chemical composition of, 16, 160
classification of, 738–739
clay minerals in, 16, 159
color of, 17, 161
compaction of, 134
compressibility, 178
continental vs. marine, 15
diagenetic sequence in, 135
epoxy peels of, 700
euxinic, 296
mineralogical composition, 15
minor elements in, 738
oil, 72, 224, 298, 426, 507–508, 515
pebbly, 854
porosity of, 132
pre-Silurian, 135
radioactivity in, 605
siliceous, 745
sulfides in, 789
Shark Bay, 740–741, 771, 800
Sharpstone, 84
Shear instability, of bed forms, 58
Sheet deposits, 118
Sheetflood deposits, 11
Sheetfloods, 335, 336
Sheet sand, 78
Shelf valley complexes, 194
Shingle, 43. *See also* Gravels, beach
Shoal breccias, 85
Shoal-retreat massifs, 194
Shoals, 796, 804
Shoestring sands, 33, 118
Shoreface storm deposits, 769
Shoreline, 799. *See also* Coasts
Shrinkage crack, 486
Siderites, 22, 82, 113
Sieve deposits, 11

Sieve effect, 414
Sieving method, of granulometric analysis, 383
Sieving techniques, 376
Silcrete, 113, 325, 741-742
Silica, 112, 678
  in clastic sediments, 129
  inorganic precipitation of, 122
  secondary, 760
  in sediments, 742-748
Silica belts, 550
Silica budget, 744
Silica cements, 112-113
Silica line, 546
Silicates, 155
Siliceous oozes, 2, 3, 54, 471, 549, 745, 748
Siliceous organisms, 122
Siliceous shales, 161
Silicification, 748-749
Sills, 855, 856-857
Silt, 749-750. *See also* Soils
  in beach sands, 36
  deposition of fine sediments, 499
  grade scales for, 127
  turbidite vs. contourite, 188
Siltstones, 132, 749, 778
Silty clay, 138
Sinkholes, 638
Sinter, 120, 762
Skeletal materials
  of carbonate sediments, 442
  in sedimentation, 684
  from sponges, 759
Skeletal sands, 31, 107
Skewness, of beach sands, 46
Skip marks, 223
Skip and roll marks, 811
Skolithos, 62
Slab slides, 387
Slates, 15, 789
Slide deposits, 479
Slides
  features of, 387
  on Mars, 477
Sliding, gravitational, 559
Sliding under load, 559-560
Slope failures, 25
Slump balls, 579
Slump bedding, 750-751
Slumping, 413
  and sedimentary transport, 65
  subaqueous, 819
Slump structures, breccias associated with, 85
Smectites, 16, 137
  in bentonite, 60
  in clay diagenesis, 149, 150
  in clay sedimentation facies, 163
  in deep-sea clays, 158-159
  in world ocean, 146
Smithsonite, boxwork, 82
Snowball structure, 579

Society, sedimentology and, 722-723
Society of Economic Paleontologists and
    Mineralogists (SEPM), 717
Soft-sediment intrusions, 865
Soft-sediment mixing bodies, 864
Sohlbank cycle, 224
Soils, 139-142
  black earth, 142
  carbonate-rich, 141
  chest, 142
  of continental sedimentation, 189
  lunar, 459-461
  model for, 716
  podsolic, 141
  red, 142
  silcrete from, 741
  tropical, 141
  tundra, 142
  world distribution of, 140
Sole marks, 529, 655
Solifluction, 819
Solifluction breccias, 85
Solubility constant, of calcium carbonate, 111
Solubilization, of inorganic compounds, 484
Solution breccias, 751-752
Solution metamorphism, 464, 467
Solution pipes, 281
Sorting, 753-754, 807, 816
  in bedding genesis, 50
  of clasts, 25
  of conglomerates, 183
Sorting coefficient, 381
Sorting process, 730
South Pacific
  clay sedimentation facies in, 163
  smectite in, 146
Sparite, 99, 444, 754
Specific gravity, for crude oil, 213
Spectrophotometry, 403
Speleal sediments, 754-759
Sphagnum peat, 542
Sphericity, 480, 631-633
  of beach gravels, 44
  of clastic sediment particles, 128
  statistical parameters for, 726
Spherulites, 22, 440
Spicules, sponge, 759, 760
Spilite, 390
Spillover lobe, 796
Spit, ebb, 796
Spit platform, 803
Spokane flood, 329
Sponge-reefs, 759
Sponges, 759-760
Sponge spicules, 505
Spongolite, 760-762
Spring deposits, 762-763
Spur and groove systems, 205
Stability diagrams, 153
Staining techniques, 764-765

Stalactites, 439
Stalagmite, 109, 439
Standing waves, 28. *See also* Waves
Statistical analysis, 714, 724–731
  for grain-size studies, 380–382
  of grain orientation, 321
  mathematical models for, 311
Steinmann association, 122
Sternberg's Law, 531, 765–766
Stokes' Law, 275, 383, 404, 407, 609, 766–767
Stony spherules, 316
Storm deposits, 767–770
Storm rollers, 560, 579
Storms
  in Bahamas, 30
  and continental shelf sediments, 192, 194, 196
  and coral reefs, 206
Strandline deposits, 33, 363
Stratabound ore, 786, 789
Stratification, 48, 184
Stratigraphy, 705, 706, 718
Stratinomy, 706
Stream-channel deposits, 11
Streamfloods, 335, 336
Streaming potential, 483
Streams, and transport processes, 817
Strengite, 575
Strike-slip plates, 662, 665
Stromatolites, 5, 691, 759, 771–774
"Strudel scout," 200
Struvite, 91
Stylolites, 257, 585, 586
Subaqueous mudflow deposits, 492
Subaqueous textures, 650
Subarkose, 17
Subbottom reflection profiles, 200
Subglacial deposits, 361–362
Submarine channels, 2
Submarine delta, 242
Submarine fan sedimentation, 780–786
Submarine lithification, 64
Submarine slope sedimentation, 774–777
Submarine slumping, 75, 84
Submarine terrace, of Bahama Banks, 30
Submergent coast, barrier island on, 33
Submersibles, and abyssal sedimentation, 4
Subsidence, 479–480, 526
Subsurface studies, 210
Subtidal zone, 799
Succinite, 619
Sulfate cements, 113
Sulfates
  in authigenesis, 21
  and clay genesis, 155
  staining of, 765
Sulfide cements, 113
Sulfides, 82, 786–789

Sulfur content, in crude oil composition, 219
Sun crack, 486
Superposition, law of, 811
Suprafan, 781
Supraglacial deposits, 362
Supratidal zone, 271, 799
Surface creep, 285, 286
Surf zone, 452
Suspended load, 339, 340
Suspended-load equation, 342
Suspended particulate matter, from nepheloid water, 496
Suspension, 285, 818
Suspension feeders, 63, 686
Swale-fill deposits, 337
Swamps. *See also* Mangroves
  clay sediments of, 142–143
  and coal deposits, 165
  paludal sediments of, 533
Swamp sediments, 12
Swash-backwash zone, 454
Swash mark, 455
Swash platforms, 804
Swash zone, 449, 455
Sylvite, in evaporite sequences, 307, 308, 309, 310
Symmetrical cycle, 224
Symmictite, 262
Symmict varves, 842
Syndiagenesis, 252
Syneresis, 789–791
Syneresis crack, 486
Syntaxial cement, 260

Taal Volcano, 38
Taconite, 120
Talus, 792
Talus breccia, 85
Talus cones, 245
Tangue, 792–793
Taphonomy, 65
Tar acids, 419
Tar sands, 793
Tectonic criteria, 811
Tectonic melange, 510
Tectonics, 706, 711, 751. *See also* Plate tectonics
Tectonic sliding, 819
Tectonic theory, 710
Tectonites, 856
Teepee structures, of caliche, 89
Tektites, 317, 318
Telmatic peat, 542
Temperature-resistance diagram, 548
Tephra, 406
Terminal grades, 358
Terminology, layer, 48. *See also* Nomenclature
Terrestrial deposition, 244–245

Terrestrial peat, 542
Terrigenous sediments, 2, 468, 786
Textural analysis, 374
Texture, 320
    carbonate, 444, 445
    and color, 174
    crystalloblastic, 464
    diagenetic, 256. See also Diagenesis
    modification of, 481
    of niveo-eolian sands, 502–503
    sand surface, 648–652
Thalassinoides, 62, 64
Thermocline, 499
Thermodynamic diagrams, 152
Thermometamorphism, 464, 465
Thigmotaxis, 483
Thixotropic sediments, 484
Thorium, 602
Thrust polygons, 638
Tidal channels
    sand waves in, 653
    in Wadden Sea, 862
Tidal currents
    of Bay of Fundy, 43
    deposits, 795–798
    and lagoonal sediments, 427
Tidal cycle, and littoral sedimentation, 449
Tidal delta facies, 801–805
Tidal-flat deposits, 367
Tidal flats, 798–801
    clay sedimentation facies in, 161–162
    dolomites in, 271
    and lateral accretions, 431
    and salt marshes, 643
    sedimentation on, 439
Tidal inlets, 801–805
Tidal marshes, sedimentation rates in, 689, 690
Tidal swamps, 239. See also Swamps
Tides
    of Bay of Fundy, 41
    and bedding genesis, 50
    and continental shelf, 190
    and continental shelf sediments, 192, 194, 196
    and deltaic sedimentation, 241
    and estuarine sedimentation, 288–289, 291
    and estuarine sediments, 294
    and ice foot, 356
    and oolite deposits, 513
Tillites, 86, 262, 264–265, 361, 543, 806
Tills, 264–265, 355, 360, 805–806, 811. See also Glacial sediments
Tilt correction, 529
Time-stratigraphic unit, 48
Tin, 400
Titaniferous magnetite, 317
Toluene, 405
Tonstein, 493
Tool marks, 810–811

Top and bottom criteria, 811
Topography
    and clay genesis, 155
    ridge and swale, 194
Topset beds, 591
Trace elements
    in black sands, 75
    in black shales, 298
    in clay sedimentation facies, 161, 162
    in evaporite sequences, 310
    in interstitial waters, 409
    in pelagic sedimentation, 551–553
    in sapropel, 654
    in sedimentary carbonates, 96
    in shales, 738
Trace fossils, 61, 62, 63, 70, 343, 597, 776, 834. See also Fossils
Traction, 818
    and Bagnold effect, 29
    in fluvial sediment transport, 339
Tractional currents, 65–66
Traction load, and sediment transport, 733
Transformation, process of, 152, 154, 155
Transformation hypothesis, for petroleum, 566, 567
Transgressions
    of barrier islands, 35, 37
    for continental shelf sedimentation, 197
    Flandrian, 77, 288
    pitchy soil formed by, 394
    during Quaternary, 740
Transgressive facies, 242
Transition zone, 453
Translational slides, 387
Transport processes, 816–819
    biological rafting, 356. See also Rafting
    in cave systems, 757
    for continental shelf sediments, 191
    and deep-sea clays, 159
    dispersal, 266
    eolian, 285–286
    for fluvial sediments, 335, 339–343
    and gases in sediments, 346
    glacial drift, 806
    and grain-size frequency studies, 371
    for iron, 416
    for Martian sedimentation, 476
    model for, 313
    and petrology, 673
    and sand grains, 644
    for sediment, 50, 731–735
    settling speed, 735
    tracer techniques in, 813–816
Transverse dunes, 286
Transverse scour marks, 656
Trask sorting coefficient, 490
Travertine, 762
Trenches, deep-sea, 395
Trench walls, 775

Trend surface analysis, 531
Trigger cores, 472
Trinidad
  Asphalt Lake in, 857
  Pitch Lake, 854
Tripolite, 263, 821
Trophogenic layer, 423
Tropholytic layer, 423
Tropical lagoonal sedimentation, 821–828
Trottoir, 6
Truncation points, 370–371, 372
Tsunami sedimentation, 828–832
Tuff, 832–833, 843
  aquagene, 859
  classification of, 852
  doubly graded, 860
  palagonite, 859
Tuff beds, of base-surge deposits, 39
Tuffites, 833, 853
Tuff matrix, and agglomerates, 5
Turbidite beds, backset bedding in, 28
Turbidites, 51, 52, 225, 368, 430, 833–836
  and abyssal sedimentation, 3
  calcareous, 549
  geometry of, 81
  as olistostromes, 509
  scour marks of, 656
Turbidite sequences, flame structures in, 326
Turbidity currents, 385, 828, 829–830
  of abyssal-sedimentary environments, 2, 3
  and bedding genesis, 51–52
  and clay sedimentation facies, 162
  and continental rise sediments, 186, 187
  and current marks, 222
  sedimentation of, 836–839
Turbid layer flows, 784
Turbulence
  of debris flows, 492
  and grain size, 373
  and oxygen, 685
  in sediment transport, 734

Udden scale, 127, 377, 840
Underclays, 138, 143
Uniformitarianism, principle of, 658, 708
Unit bars, 83
Uranium, 298, 299, 602
Uranium/thorium dating, 758

Vadose environment, cementation in, 399
Valley fill, and aggradation, 5
Valley-side flows, 491
Valley trains, 363
Variscite, 575
Varves, 49, 224, 363, 841–843
  drainage, 842
  and sedimentation rates, 688
Vaterite, 96

Vector analysis, of cross-bedding, 211
Vegetational succession, for peat bogs, 541
Velocity profile, for debris flow, 489
Ventifacts, 248, 322
Vermetid gastropods, 7, 8
Vertical accretion, 4, 430–432
Vertical mixing, 549
Viscosity, sabolt, 219
Vitric tuffs, 832
Vitrification, of soil, 458
Vitrinite, 170, 171
Vivianite, 592
Vodose zone, 96
Void ratio, and pressure, 180
Volcanic ash
  in bentonite, 59
  deposits, 843–850
  and diagenesis, 850–851
  in hadal sedimentation, 396
Volcanic debris, 598
  in marine sediments, 471
  on Mars, 478
Volcanic environments, role of organisms in, 685
Volcanic-inorganic chert formation, 122
Volcaniclastic deposits, 871
Volcaniclastics, 78
Volcanism
  base surges produced by, 38
  fossil sedimentary, 857
  and manganese nodules, 552
  marine sediments derived from, 468
  and microbiology, 483
  mud volcanoes, 856
  mudflows from, 490, 491–492
  nueé ardente, 505–506
  and salar structures, 637
  sedimentary, 854–858
  submarine products, 858
  tuff, 832
Volcanogenic units, 853
Von Karman-Prandtl velocity profile equation, 191
von Karman vortex trails, 377

Wackestones, 444
Wadden Sea, 42, 799
  fauna, 863
  sedimentology of, 862–864
Wadis, 248
Walther's Law, 37, 325
Wash load, 339
Washout, 654
Washover fans, 36
Waste accumulation, 723
Waste discharge, and estuarine sediments, 294
Water
  and desert sediments, 250

INDEX

and diagenesis, 253
transportation by, 816
Water depth
  and grain size, 373
  and tectonic movement, 695
Water-escape structures, 414, 864–868
Waterlaid deposits, 490
Waterlain till, 809
Water-stones, 349
Water washing, and crude oil deposition, 214
Wave activity
  of Bay of Fundy, 41
  and beach sands, 46
  and continental shelf, 190, 192
  flow regimes, 331
Wave energy, and beach gravels, 44
Wavellite, 575
Wave refraction, 451
Waves
  and bedding genesis, 50
  and deltaic sedimentation, 241
  and estuarine sediments, 294
  and oolite deposits, 513
  sand, 652–653. *See also* Sand waves
  tsunami, 828
Waxes, hydrocarbon, 405
Weathering, 22, 816
  and continental shelf, 190
  and pebbly mudstones, 544
  of sedimentary rocks, 676
  in sediments, 683, 868–872
Weathering profiles, 869
Weathering zone, 152, 154
Weat herring, 175
Weir effect, 831
Welded chert, 120
Wentworth scale, 127, 377, 574, 872–873
West Indies, calcium carbonate muds in, 116
Whistling sand, 47
White rock deposits, 577
Whitewater Bay, tropical lagoons of, 824
Wildflysch, 85, 509
Wind action. *See also* Transport processes
  and bedding genesis, 52
  and desert sediments, 250
  and oolite deposits, 513
  and sediment transport, 733
Wind abrasion, 649
Windrift currents, and continental shelf sedimentation, 195
Windrows, 540
Wind tunnels, 315
Winnowing, 275, 549
Winogradsky column, 483
Woods Hole Rapid Sediment Analyzer, 610
Wrightwood Mudflow, 489
Wüstite, 317

X Cap rocks, 467
Xenotime, 574
X-ray diffraction analysis, 403, 669
X-ray techniques, 261, 702–703

Yellow River system, 12, 83
Yucatan
  carbonate sediments of, 101
  cementation of, 282
  submarine cementation in, 114
  tropical lagoons of, 823

Zebra dolomite, 874
Zellendolomit, 234
Zeolite facies, 252
Zeolites, 54, 255, 851
  of abyssal-sedimentary environments, 2
  in authigenesis, 3, 21
  in clay diagenesis, 150
Ziolitic clays, 147
Zingg's classification, of pebble shapes, 632
Zircon, 74, 400
Zoogenic structures, of biogenic sediments, 61
Zoophycos, 62
Zooturbation structures, 68